Fully Revised **Edition**

AF889368

Understanding Physics

JEE Main & Advanced

Electricity and Magnetism

Fully Revised **Edition**

Understanding Physics

JEE Main & Advanced

Electricity and Magnetism

DC Pandey
[B.Tech, M.Tech, Pantnagar, ID 15722]

arihant
SUCCESS AND NOTHING LESS

ARIHANT PRAKASHAN (Series), MEERUT

卐 **Administrative & Production Offices**

Regd. Office
'Ramchhaya' 4577/15, Agarwal Road, Darya Ganj, New Delhi -110002
Tele: 011- 47630600, 43518550

Head Office
Kalindi, TP Nagar, Meerut (UP) - 250002
Tel: 0121-7156203, 7156204

卐 **Sales & Support Offices**
Agra, Ahmedabad, Bengaluru, Bareilly, Chennai, Delhi, Guwahati, Hyderabad, Jaipur, Jhansi, Kolkata, Lucknow, Nagpur & Pune.

卐 **PRICE** ₹650.00

卐 **PO No :** TXT-XX-XXXXXXX-X-XX

Published by Arihant Publications (India) Ltd.

For further information about the books published by Arihant, log on to www.arihantbooks.com or e-mail at info@arihantbooks.com

Follow us on

Preface

"The overwhelming response to the previous editions of this book gives me an immense feeling of satisfaction, and I take this as an opportunity to introduce you to the new version of the book, "Electricity and Magnetism."

"Electricity and Magnetism" has undergone meticulous revision to ensure accuracy and relevance. We take pride in affirming that this book serves as a standalone resource, encompassing all essential topics without any extraneous content.

This book is designed with a singular focus on providing concise yet complete theory, coupled with an ample number of practice questions covering all pertinent concepts essential for success in JEE Mains & Advanced.

With each chapter, or as we call them, units, subdivided into modular formats, "Electricity and Magnetism" ensures an organized and structured approach to learning. Each module within a unit is self-contained, comprising theory, illustrative examples, and exercises aimed at reinforcing understanding.

At the end of every unit, readers will find unit exercises presented in two levels, offering ample opportunities for practice and self-assessment. Furthermore, to aid in problem-solving, most exercises are accompanied by helpful hints.

Our commitment to excellence extends to the inclusion of logical problems sourced from reputable resources like **Irodov** and **previous years' question papers.** The level of questions spans from foundational to 1.25 times the highest difficulty level of the JEE Advanced examination, ensuring comprehensive preparation.

In conclusion, I am confident that "Electricity and Magnetism" will prove to be an indispensable companion for aspirants aiming for success in JEE. It is my belief that with dedication and diligent practice using this book, students will be well-equipped to excel in both JEE Mains and Advanced examinations.

I take this opportunity to express my gratitude to **Prof. Lalit Kumar,** who gave me valuable advice and was kind enough to find time for fruitful discussions. I would also like to thank my student Dheeraj Yadav for managing this project as per my expectations. I am also thankful to Mr. Anoop Dhyani and Mr. Ashish, due to their tireless efforts this book could be transformed into the present form.

To err is human, hence despite such tireless efforts, there may still be scope for improvement in this book. So, comments and criticism from readers will be highly appreciated and incorporated in subsequent editions." I also request the readers to drop me a mail at **arihantcorrections@gmail.com,** if you find any corrections in the book. Your feedback is greatly appreciated and will help to improve the quality of the book.

Thankyou for your support and contribution in making this book the **Best it can be.**

DC Pandey

[B.Tech, M.Tech, Pantnagar, ID 15722]

Contents

UNIT 1	**Current Electricity**	**1 - 107**
	Module 1. Electric Current	2-4
	Module 2. Electric Currents in Conductors	5-8
	Module 3. Resistance of a Wire	9-11
	Module 4. Temperature Dependence of Resistance	12-14
	Module 5. Ohm's Law and Kirchhoff's Laws	15-28
	Module 6. Heating Effects of Current	29-36
	Module 7. Grouping of Cells	37-43
	Module 8. Electrical Measuring Instruments	44-57
	Module 9. Colour Codes for Resistors (For other Examination)	58-59
	Cumulative Exercise	60-73
	Hints & Solutions	74-107
UNIT 2	**Electrostatics**	**109-240**
	Module 1. Electric Charge	110-112
	Module 2. Coulomb's Law	113-118
	Module 3. Electric Field	119-127
	Module 4. Electric Potential Energy	128-132
	Module 5. Electric Potential	133-142
	Module 6. Relation Between Electric Field and Potential	143-148
	Module 7. Equipotential Surfaces	149-150
	Module 8. Electric Dipole	150-156
	Module 9. Gauss's Law	157-164
	Module 10. Properties of a Conductor	165-166
	Module 11. Charged Spheres	167-177
	Cumulative Exercise	178-193
	Hints & Solutions	194-240
UNIT 3	**Capacitors**	**241-349**
	Module 1. Capacitance	242-243
	Module 2. Energy Stored in a Charged Capacitor	244-247
	Module 3. Capacitors	248-260
	Module 4. Mechanical Force on a Charged Conductor	261
	Module 5. Capacitors in Series and Parallel	262-267
	Module 6. Two Laws in Capacitors	268-272
	Module 7. *C-R* Circuits	273-283
	Module 8. Methods of Finding Equivalent Resistance and Capacitance	284-293
	Cumulative Exercise	294-309
	Hints & Solutions	310-349

UNIT 4 **Magnetics** 351-465

Module 1. Magnetic Force on a Moving Charge 352-355
Module 2. Path of a Charged Particle in Uniform Magnetic Field 356-370
Module 3. Magnetic Force on a Current Carrying Conductor 371-374
Module 4. Magnetic Dipole 375-380
Module 5. Biot Savart Law 381-389
Module 6. Ampere's Circuital Law 390-395
Module 7. Force Between Parallel Current Carrying Wires 396-402
Module 8. Magnetic Poles and Bar Magnets 403-405
Module 9. Earth's Magnetism 406-410
Module 10. Magnetic Materials 411-417
Module 11. Moving Coil Galvanometer 418-419
Cumulative Exercise 420-433
Hints & Solutions 434-465

UNIT 5 **Electromagnetic Induction** 467-578

Module 1. Faraday and Lenz Law 468-476
Module 2. Motional Electromotive Force 477-489
Module 3. Self Inductance and Inductor 490-499
Module 4. Mutual Inductance 500-503
Module 5. Growth and Decay of Current in an *L-R* Circuit 504-512
Module 6. Oscillations in *L-C* Circuit 513-517
Module 7. Induced Electric Field 518-521
Cumulative Exercise 522-541
Hints & Solutions 542-578

UNIT 6 **Alternating Current** 579-630

Module 1. Alternating Currents and Phasors 580-583
Module 2. Current and Potential Relations 584-592
Module 3. Power in an AC Circuit 593-602
Cumulative Exercise 603-613
Hints & Solutions 614-630

This book

is dedicated to

my Honourable Grandfather

(Late) Sh. Pitamber Pandey

a Kumaoni poet and a resident of Village

Dhaura (Almora), Uttarakhand

UNIT–1

CURRENT ELECTRICITY

Learning Modules

1. Electric Current
2. Electric Currents in Conductors
3. Resistance of a Wire
4. Temperature Dependence of Resistance
5. Ohm's Law and Kirchhoff's Laws
6. Heating Effects of Current
7. Grouping of Cells
8. Electrical Measuring Instruments
9. Colour Codes for Resistors (For other Examination)

MODULE 1

Electric Current

Note *Some of the terms introduced in the present unit will be explained in subsequent units.*

Introduction

An electrical circuit consists of some active and passive elements. The active elements such as a battery or a cell, supply electric energy to the circuit. On the contrary, passive elements consume or store the electric energy. The basic passive elements are resistor, capacitor and inductor.

A resistor opposes the flow of current through it and if some current is passed by maintaining a potential difference across it, some energy is dissipated in the form of heat. A capacitor is a device which stores energy in the form of electric potential energy. It opposes the variations in voltage. An inductor opposes the variations in current. It does not oppose the steady current through it.

Fundamentally, electric circuits are a means for conveying energy from one place to another. As charged particles move within a circuit, electric potential energy is transferred from a source (such as a battery or a cell) to a device in which that energy is either stored or converted to another form, like sound in a stereo system or heat and light in a toaster or light bulb. Electric circuits are useful because they allow energy to be transported without any moving parts (other than the moving charged particles themselves).

Electric Current

Flow of charge is called electric current. The direction of electric current is in the direction of flow of positive charge or in the opposite direction of flow of negative charge.

Current is defined quantitatively in terms of the rate at which net charge passes through a cross-section area of the conductor.

Thus, $$I = \frac{dq}{dt} \quad \text{or} \quad i = \frac{dq}{dt}$$

We can have the following two concepts of current, **instantaneous current** and **average current.**

Instantaneous current i or $I = \frac{dq}{dt}$ = current at any point of time and Average current i_{av} or $I_{av} = \frac{q}{t}$

Hence-forth unless otherwise referred to, current would signify instantaneous current. By convention, the direction of the current is assumed to be that in which positive charge moves. In the SI system, the unit of current is ampere (A).

$$1\text{ A} = 1\text{ C/s}$$

Household currents are of the order of few amperes.

Flow of Charge

If a current is passing through a wire then it implies that a charge is flowing through that wire. Further,

$$i = \frac{dq}{dt} \Rightarrow dq = i\,dt \qquad \text{...(i)}$$

Here, are three possible cases :

Case 1 If given current is constant, then from Eq. (i) we can see that flow of charge can be obtained directly by multiplying that constant current with the given time interval. Or,

$$\Delta q = i \times \Delta t$$

Case 2 If given current is a function of time, then charge flow can be obtained by integration. Or,

$$\Delta q = \int_{t_i}^{t_f} i\,dt$$

Case 3 If current *versus* time graph is given, then flow of charge can be obtained by the area under the graph.

$$\Delta q = \text{area under } i\text{-}t \text{ graph}$$

Important POINTS

- The current is the same for all cross-sections of a conductor of non-uniform cross-section. Similar to the water flow, charge flows faster where the conductor is smaller in cross-section and slower where the conductor is larger in cross-section, so that charge rate remains unchanged.
- Electric current is very similar to water current, consider a water tank kept at some height and a pipe is connected to the water tank. The rate of flow of water through the pipe depends on the height of the tank. As the level of water in the tank falls, the rate of flow of water through the pipe also gets reduced. Just as the flow of water depends on the height of the tank or the level of water in the tank, the flow of current through a wire depends on the potential difference between the end points of the wire. As the potential difference is changed, the current will change.

  For example, during the discharging of a capacitor potential difference and hence, the current in the circuit decreases with time. To maintain a constant current in a circuit a constant potential difference will have to be maintained and for this a battery is used which maintains a constant potential difference in a circuit.
- Though conventionally a direction is associated with current (opposite to the motion of electrons), it is not a vector as the direction merely represents the sense of charge flow and not a true direction. Further, current does not obey the law of parallelogram of vectors, i.e. if two currents i_1 and i_2 reach a point we always have $i = i_1 + i_2$ whatever be the angle between i_1 and i_2. It is independent of θ, the angle between two currents.

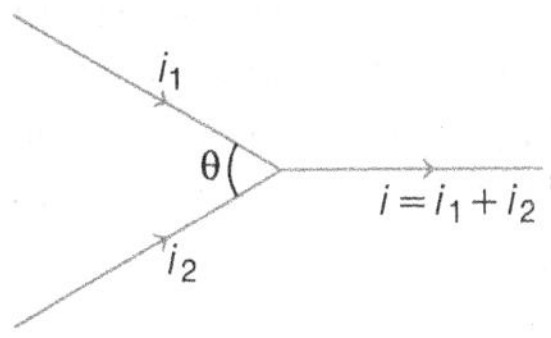

- According to its magnitude and direction, current is usually divided into two types :
  (i) **Direct current (DC)** If the magnitude and direction of current does not vary with time, it is said to be direct current (DC). Cell, battery or DC dynamo are its sources.
  (ii) **Alternating current (AC)** If a current is periodic (with constant amplitude) and has half cycle positive and half negative, it is said to be alternating current (AC). AC dynamo is the source of it.
- If a charge q revolves in a circle with frequency f, the equivalent current,

$$\boxed{i = qf}$$

- In a conductor, normally current flow or charge flow is due to flow of free electrons.
- Charge is quantised. The quantum of charge is e. The charge on any body will be some integral multiple of e, i.e.

$$\boxed{q = \pm ne}$$

where, $e = 1.6 \times 10^{-19}$ C

and $\quad n = 1, 2, 3 \ldots$

Example 1 In a given time of 10 s, 40 electrons pass from right to left. In the same interval of time 40 protons also pass from left to right. Is the average current zero? If not, then find the value of average current.

Sol. No, the average current is not zero. Direction of current is the direction of motion of positive charge or in the opposite direction of motion of negative charge. So, both currents are from left to right and both currents will be added.

$$\therefore \quad I_{av} = I_{electron} + I_{proton}$$
$$= \frac{q_1}{t_1} + \frac{q_2}{t_2}$$
$$= \frac{40\,e}{10} + \frac{40\,e}{10} \qquad (\because q = ne)$$
$$= 8\,e$$
$$= 8 \times 1.6 \times 10^{-19}\ \text{A}$$
$$= 1.28 \times 10^{-18}\ \text{A}$$ **Ans.**

Example 2 A constant current of 4A passes through a wire for 8 s. Find total charge flowing through that wire in the given time interval.

Sol. Since, i = constant

$$\therefore \quad \Delta q = i \times \Delta t$$
$$= 4 \times 8 = 32\ \text{C}$$

Example 3 A wire carries a current of 2.0 A. What is the charge that has flown through its cross-section in 1.0s? How many electrons does this correspond to?

Sol. $\therefore \ i = \frac{q}{t}$

$$\Rightarrow \quad q = it = (2.0\ \text{A})\,(1.0\ \text{s})$$
$$= 2.0\ \text{C}$$ **Ans.**
$$\therefore \quad q = ne$$
$$\Rightarrow \quad n = \frac{q}{e} = \frac{2.0}{1.6 \times 10^{-19}}$$
$$= 1.25 \times 10^{19}$$ **Ans.**

Example 4 The current in a wire varies with time according to the relation

$$i = (3.0\text{ A}) + (2.0\text{ A/s})\,t$$

(a) How many coulombs of charge pass a cross-section of the wire in the time interval between $t = 0$ and $t = 4.0$ s?

(b) What constant current would transport the same charge in the same time interval?

Sol. (a) $i = \dfrac{dq}{dt}$

$\therefore \quad \int_0^q dq = \int_0^4 i\,dt$

$\therefore \quad q = \int_0^4 (3 + 2t)\,dt$

$= [3t + t^2]_0^4 = [12 + 16]$

$= 28$ C **Ans.**

(b) $i = \dfrac{q}{t} = \dfrac{28}{4} = 7$ A **Ans.**

Example 5 Current passing through a wire decreases linearly from 10 A to 0 in 4 s. Find total charge flowing through the wire in the given time interval.

Sol. Current *versus* time graph is as shown in figure.

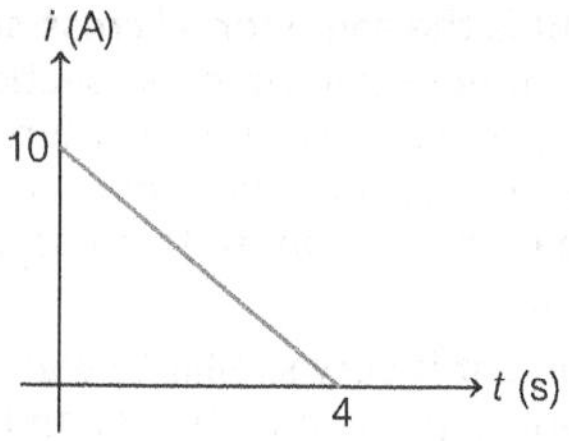

Area under this graph will give us net charge flow.

Hence, $\Delta q = \text{area} = \dfrac{1}{2} \times \text{base} \times \text{height}$

$= \dfrac{1}{2} \times 4 \times 10$

$= 20$ C **Ans.**

MODULE 1 Exercise

Single Correct Option

1. A current through a wire depends on time as $i = \alpha_0 t + \beta t^2$, where $\alpha_0 = 20$ A/s and $\beta = 8$ As^{-2}. Find the charge crossed through a section of the wire in 15 s. **(2021 Main)**

(a) 2250 C (b) 11250 C (c) 2100 C (d) 260 C

Subjective Questions

2. How many electrons per second pass through a section of wire carrying a current of 0.7 A?

3. A current of 3.6 A flows through an automobile headlight. How many coulombs of charge flow through the headlight in 3.0 h?

4. A current of 7.5 A is maintained in wire for 45 s. In this time,

(a) how much charge and

(b) how many electrons flow through the wire?

5. In the Bohr model, the electron of a hydrogen atom moves in a circular orbit of radius 5.3×10^{-11} m with a speed of 2.2×10^6 m/s. Determine its frequency f and the current I in the orbit.

6. The current through a wire depends on time as, $i = (10 + 4t)$.

Here, i is in ampere and t in seconds. Find the charge crossed through a section in time interval between $t = 0$ to $t = 10$ s.

7. In an electrolyte, the positive ions move from left to right and the negative ions from right to left. Is there a net current? If yes, in what direction?

Answers

1. (b) 2. 4.375×10^{18} 3. 38880 C 4. (a) 337.5 C (b) 2.1×10^{21} 5. 6.6×10^{15} Hz, 1.06 mA
6. 300 C 7. Yes, from left to right

See hints on page 74.

MODULE 2

Electric Currents in Conductors

Conductors are those materials which can conduct electricity. Conductors can be broadly classified into two groups :

(i) Solid conductors

(ii) Electrolyte conductors

Normally in atoms and molecules, the negatively charged electrons and the positively charged nuclei are bound to each other and are thus not free to move. In solid conductors (notably metals), some of the electrons (called free electrons) are free to move within the bulk materials. In these conductors, current flow takes place due to these free electrons. Positive ions in these conductors are almost fixed. They do not move. So, they do not contribute in the current. In electrolyte solutions however, both positive and negative ions can move.

In our following discussions, we will focus only on solid conductors so that the current is carried by the negatively charged free electrons in the background of fixed positive ions.

Theory of Current Flow through Solid Conductors

At room temperature, the free electrons in a conductor move randomly with speeds of the order of 10^5 m/s. Since, the motion of the electrons is random, there is no net charge flow in any direction. For any imaginary plane passing through the conductor, the number of electrons crossing the plane in one direction is equal to the number crossing it in the other direction. Therefore, net current is zero from any section.

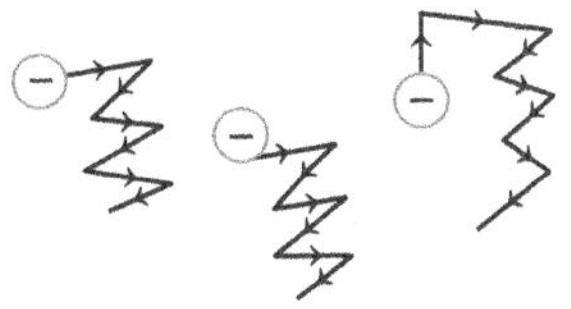

When a constant potential difference V is applied between the ends of the conductor as shown in figure below, an electric field **E** is produced inside the conductor. The conduction electrons within the conductor are then subjected to a force $-e\mathbf{E}$ and move overall in the direction of increasing potential.

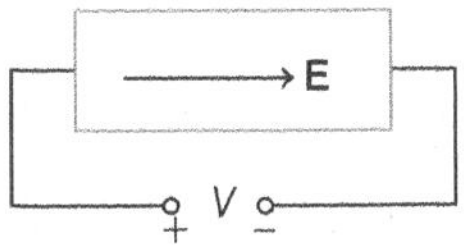

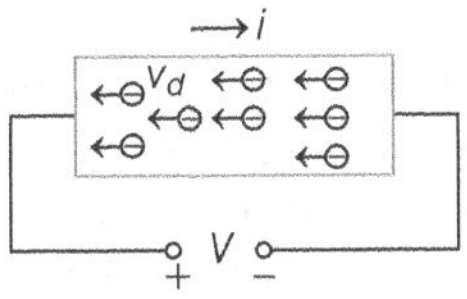

However, this force does not cause the electrons to move faster and faster. Instead, a conduction electron accelerates through a very small distance (about 5×10^{-8} m) and then collides with fixed ions or atoms of the conductor. Each collision transfers some of the electron's kinetic energy to the ions (or atoms). Because of the collision, electron moves slowly along the conductor or we can say that it acquires a **constant drift velocity v_d** in the direction opposite to **E** (in addition to its random motion).

The drift motion of free electrons produce an electric current in the opposite direction of this motion or in the direction of electric field (from higher potential to lower potential). It is interesting to note that the magnitude of the drift velocity is of the order of 10^{-4} m/s or about 10^9 times smaller than the average speed of the electrons of their random (or thermal) motion. The above discussion can be summarized as follows :

1. Free electrons inside a solid conductor can have two motions:
   (i) random or thermal motion (speed of the order of 10^5 m/s)
   (ii) drift motion (speed of the order of 10^{-4} m/s)
2. Net current due to random (or thermal motion) is zero from any section, whereas net current due to drift motion is non-zero.
3. In the absence of any electric field (or a potential difference across the conductor) free electrons have only random motion. Hence, net current from any section is zero.
4. In the presence of an electric field (or a potential difference across the conductor) free electrons have both motions (random and drift). Therefore, current is non-zero due to drift motion.
5. Drift motion of free electrons is opposite to the electric field. Therefore, direction of current is in the direction of electric field from higher potential to lower potential.

Example 1 Electric field inside a conductor is always zero. Is this statement true or false?

Sol. False. Under electrostatic conditions when there is no charge flow (or no current) in the conductor, electric field is zero. If current is non-zero, then electric field is also non-zero. Because the drift motion (of free electrons) which produces a net current starts only due to electric force on them.

Drift Velocity and Relaxation Time

As discussed before, in the presence of electric field, the free electrons experience an electric force of magnitude.

$$F = qE \text{ or } eE \qquad (\text{as } q = e)$$

This will produce an acceleration of magnitude,

$$a = \frac{F}{m} = \frac{eE}{m} \qquad (m = \text{mass of electron})$$

Direction of force (and acceleration) is opposite to the direction of electric field.

After accelerating to some distance an electron will suffer collisions with the heavy fixed ions. The collisions of the electrons do not occur at regular intervals but at random times.

Relaxation time τ is the average time between two successive collisions. Its value is of the order of 10^{-14} second.

After every collision, let us assume that drift motion velocity of electron becomes zero. Then, it accelerates for a time interval τ, then again it collides and its drift motion velocity becomes zero and so on.

If v_d is the average constant velocity (called drift velocity) in the direction of drift motion, then relation between v_d and τ is given by

$$\boxed{v_d = \frac{eE\tau}{m}}$$

Key Point

- In some standard books of Indian authors, the relation is given as

  $$v_d = \frac{eE\tau}{2m}$$

  Initially, I was also convinced with this expression. But later on after consulting many more literatures in this I found that $v_d = \frac{eE\tau}{m}$ is correct. But at this stage it is very difficult for me to give its correct proof. Because the correct proof requires a knowledge of high level of physics which is difficult to understand for a class XII student.

Current and Drift Velocity

Consider a cylindrical conductor of cross-sectional area A in which an electric field E exists. Drift velocity of free electrons is v_d and n is number of free electrons per unit volume (called free electron density).

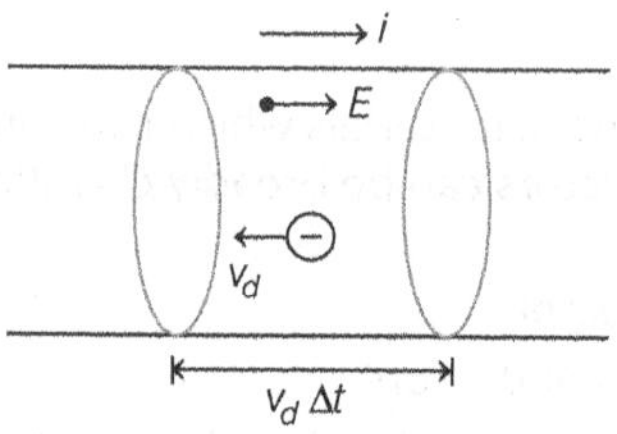

Consider a length $v_d \Delta t$ of the conductor.

The volume of this portion is $Av_d \Delta t$.

Number of free electrons in this volume = (free electron density) × (volume)

$$= (n)(Av_d \Delta t)$$
$$= nAv_d \Delta t$$

All these electrons cross the area A in time Δt.

Thus, the charge crossing this area in time Δt is

$$\therefore \quad \Delta q = (nAv_d \Delta t)(e)$$

or $$i = \frac{\Delta q}{\Delta t} = neAv_d$$

or $$\boxed{i = neAv_d}$$

Thus, this is the relation between current and drift velocity.

Current Density

Current per unit area (taken normal to the current), i/A is called current density and is denoted by j. The SI units of current density are A/m^2. Current density is a vector quantity **j** directed along **E**.

$$\boxed{j = \frac{i}{A}}$$

But $i = neAv_d$, therefore

$$\boxed{j = nev_d}$$

Important POINTS

- Drift velocity of electrons in a conductor is of the order of 10^{-4} m/s, then question arises in everybody's mind that why a bulb glows instantly when switched on?
  Reason is : when we close the circuit, electric field is set up in the entire closed circuit instantly (with the speed of light).
  Due to this electric field, the free electrons instantly get drift velocity in the entire circuit and a current is established in the circuit instantly. The current so set up does not wait for the electrons to flow from one end of the conductor to the other end.
- If a current i is flowing through a wire of non-uniform cross-section, then current will remain constant at all cross-sections. But drift velocity and current density are inversely proportional to the area of cross-section. This is because

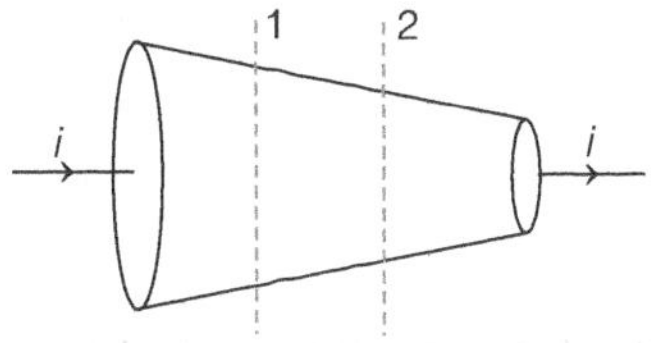

$$i = neAv_d$$

or $$v_d = \frac{i}{neA}$$

or $$v_d \propto \frac{1}{A}$$

Further, $$j = \frac{i}{A}$$

or $$j \propto \frac{1}{A}$$

So, in the figure $i_1 = i_2 = i$ but, $(v_d)_2 > (v_d)_1$ and $j_2 > j_1$ because $A_2 < A_1$

Note *Later on we can also prove that electric field at 2 is also more than electric field at 1.*

- **Mobility** The physical significance of mobility means how mobile the charge carriers are for the current flow. If mobility of charge carriers is more than we can say that current flow will be more.

In metals, the mobile charge carriers are electrons. In an ionised gas they are electrons and positive charged ions. In an electrolyte, these can be positive and negative ions. In semiconductors, charge carriers are electrons and holes. Later, we will see that mobility of electrons (in semiconductor) is more than the mobility of holes.

Mobility (μ) is defined as the magnitude of drift velocity per unit electric field.

Thus, $$\boxed{\mu = \frac{v_d}{E}}$$

But, $$v_d = \frac{eE\tau}{m}$$

$\therefore$ $$\boxed{\mu = \frac{e\tau}{m}}$$

The SI units of mobility are $m^2/V\text{-}s$. Practical units of mobility is $cm^2/V\text{-}s$. Mobility of any charge carrier (whether it is electron, ion or hole) is always positive.

- Conductivity $(\sigma) = \dfrac{1}{\text{Resistivity }(\rho)}$
- $\boxed{E = J\rho}$

Example 2 An electron beam has an aperture of 1.0 mm^2. A total of 6.0×10^{16} electrons go through any perpendicular cross-section per second. Find (a) the current and (b) the current density in the beam.

Sol. (a) The current is given by

$$i = \frac{q}{t} = \frac{ne}{t}$$

Substituting the values, we have

$$i = \frac{(6.0 \times 10^{16})(1.6 \times 10^{-19})}{1}$$

$$= 9.6 \times 10^{-3}\ A$$ **Ans.**

(b) The current density is

$$j = \frac{i}{A} = \frac{9.6 \times 10^{-3}}{10^{-6}}$$

$$= 9.6 \times 10^{3}\ A/m^2$$ **Ans.**

Example 3 Calculate the drift speed of the electrons when 1A of current exists in a copper wire of cross-section 2 mm^2. The number of free electrons in 1 cm^3 of copper is 8.5×10^{22}.

Sol. n = free electron density

$$= 8.5 \times 10^{22} \text{ per cm}^3 \quad \text{(Given)}$$

$$= (8.5 \times 10^{22})(10^6) \text{ per m}^3$$

$$= 8.5 \times 10^{28} \text{ per m}^3$$

From $i = neAv_d$, we get

$$v_d = \frac{i}{neA}$$

Substituting the values in SI units, we have

$$v_d = \frac{1}{(8.5 \times 10^{28})(1.6 \times 10^{-19})(2 \times 10^{-6})}$$

$$= 3.6 \times 10^{-5}\ m/s$$ **Ans.**

MODULE 2 Exercise

True/False

1. All points of a conductor are always at same potential. Is this statement true or false?

Single Correct Option

2. Through an electrolyte an electrical current is due to drift of
   (a) free electrons
   (b) positive and negative ions
   (c) free electrons and holes
   (d) protons

3. The electric field E, current density J and conductivity σ of a conductor are related as
   (a) $\sigma = E/J$ (b) $\sigma = J/E$
   (c) $\sigma = JE$ (d) None of these

4. A current of 10 A exists in a wire of cross-sectional area of 5 mm^2 with a drift velocity of 2×10^{-3} ms^{-1}. The number of free electrons in each cubic metre of the wire is **(2021 Main)**
   (a) 2×10^{25} (b) 2×10^{23}
   (c) 625×10^{25} (d) 2×10^{6}

5. A current of 5 A passes through a copper conductor (resistivity $= 1.7 \times 10^{-8}$ Ω-m) of radius of cross-section 5 mm. Find the mobility of the charges, if their drift velocity is 1.1×10^{-3} m/s. **(2019 Main)**
   (a) 1.5 m^2/V-s
   (b) 1.3 m^2/V-s
   (c) 1.0 m^2/V-s
   (d) 1.8 m^2/V-s

6. Drift speed of electrons, when 1.5 A of current flows in a copper wire of cross-section 5 mm^2 is v. If the electron density in copper is 9×10^{28} / m^3, the value of v in mm/s is close to (Take, charge of electron to be $= 1.6 \times 10^{-19}$ C) **(2019 Main)**
   (a) 0.02 (b) 0.2
   (c) 2 (d) 3

Subjective Questions

7. When a wire carries a current of 1.20 A, the drift velocity is 1.20×10^{-4} m/s. What is the drift velocity when the current is 6.00 A?

8. Find the velocity of charge leading to 1 A current which flows in a copper conductor of cross-section 1cm^2 and length 10 km. Free electron density of copper is $8.5 \times 10^{28}/m^3$. How long will it take the electric charge to travel from one end of the conductor to the other?

9. Copper has one conduction electron per atom. Its density is 8.89 g/cm^3 and its atomic mass is 63.54 g/mol. If a copper wire of diameter 1.0 mm carries a current of 2.0 A, what is the drift speed of the electrons in the wire?

10. Find the total momentum of electrons in a straight wire of length $l = 1000$ m carrying a current $I = 70$ A. **(Irodov 3.207)**

Integer Type Question

11. The number density of free electrons in copper is nearly 8×10^{28} m^{-3}. A copper wire has its area of cross-section $= 2 \times 10^{-6}$ m^2 and is carrying a current of 3.2 A. The drift speed of the electrons is .... $\times 10^{-6}$ ms^{-1}. **(2023 Main)**

Answers

1. False **2.** (b) **3.** (b) **4.** (c) **5.** (c) **6.** (a) **7.** 6.0×10^{-4} m/s
8. 0.735 μm/s, 431.4 yr **9.** 1.9×10^{-4} m/s **10.** 0.4 kg - ms^{-1} **11.** 125

See hints on pages 74-75.

MODULE 3

Resistance of a Wire

The resistance of a wire is typically specified or measured between two defined points or surfaces (say P and Q).

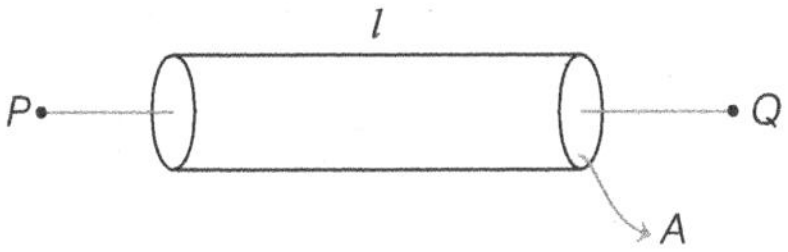

Here, l = length of wire, A = area of cross-section.

Now, $$R \propto l \qquad \text{..(i)}$$

and $$R \propto \frac{1}{A} \qquad \text{..(ii)}$$

Combining Eqs. (i) and (ii), we get

$$\boxed{R = \rho \frac{l}{A}} \qquad \text{..(iii)}$$

Here, ρ is called the resistivity of the material of the wire. This depends upon number of free electrons present in the material. With increase in number of free electrons, the value of ρ decreases.

Note *(i)* $\rho = \frac{1}{\sigma}$, *where* σ = *conductivity.*

(ii) SI units of resistivity are Ω*-m (ohm-metre).*

(iii) SI units of conductivity are $(\Omega\text{-m})^{-1}$.

(iv) In Eq. (iii), l is that dimension of conductor which is parallel to P and Q and A is that cross-sectional area, which is perpendicular to P and Q.

(v) From P to Q if A is constant then direct formula can be applied. Otherwise we will have to take the help of integration.

Example 1 Two copper wires of the same length have got different diameters,

(a) which wire has greater resistance?

(b) greater specific resistance?

Sol. (a) For a given wire, $R = \rho \frac{l}{A}$, i.e. $R \propto \frac{1}{A}$

So, the thinner wire will have greater resistance.

(b) Specific resistance (ρ) is a material property. It does not depend upon l or A.

So, both the wires will have same specific resistance.

Example 2 A wire has a resistance R. What will be its resistance if it is stretched to double its length?

Sol. Let V be the volume of wire, then

$$V = Al \quad \therefore \quad A = \frac{V}{l}$$

Substituting this in $R = \rho \frac{l}{A}$, we have

$$R = \rho \frac{l^2}{V}$$

So, for given volume and material (i.e. V and ρ are constants)

$$R \propto l^2$$

When l is doubled, resistance will become four times, or the new resistance will be $4R$.

Example 3 The dimensions of a conductor of specific resistance ρ are shown below. Find the resistance of the conductor across AB, CD and EF.

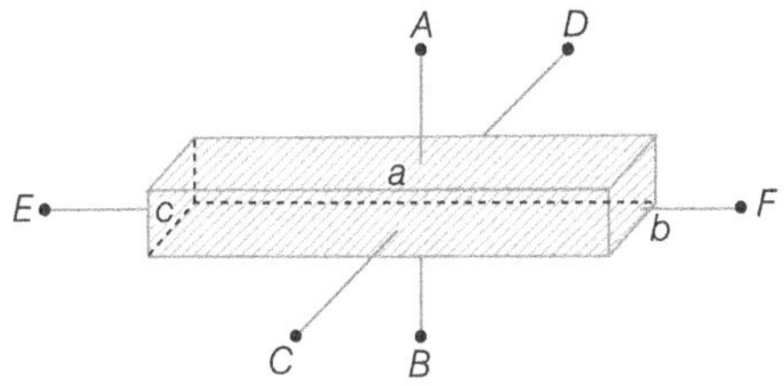

Sol. $R = \rho \frac{l}{A}$

Resistance across AB, CD and EF in tabular form is shown below.

Table 1.

	l	A	R
AB	c	$a \times b$	$\rho \frac{c}{ab}$
CD	b	$a \times c$	$\rho \frac{b}{ac}$
EF	a	$b \times c$	$\rho \frac{a}{bc}$

Example 4 A copper wire is stretched to make it 0.1% longer. What is the percentage change in its resistance? **(1978 JEE)**

Sol. $$R = \rho \frac{l}{A} = \frac{\rho l}{V/l} \qquad (V = \text{volume of wire})$$

$$= \frac{\rho l^2}{V}$$

$$\therefore \quad R \propto l^2 \qquad (\rho \text{ and } V = \text{constant})$$

For small percentage change,

% change in $R = 2$ (% change in l) $= 2\,(0.1\%) = 0.2\%$

Since $R \propto l^2$, with increase in the value of l, resistance will also increase.

Example 5 A homogeneous poorly conducting medium of resistivity ρ fills up the space between two thin coaxial ideally conducting cylinders. The radii of the cylinders are equal to a and b, with $a < b$, the length of each cylinder is l. Neglecting the edge effects, find the resistance of the medium between the cylinders. **(Irodov 3.154)**

Sol. Using $R = \frac{\rho l}{A}$

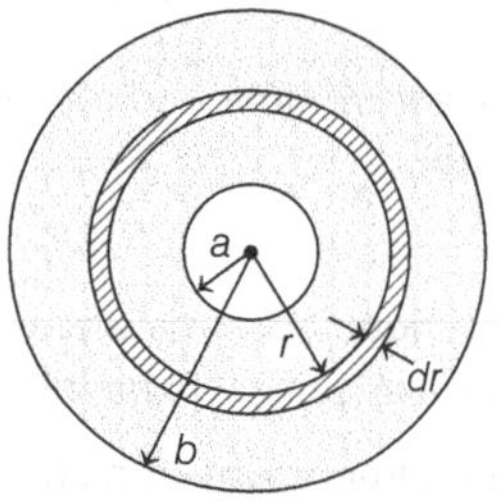

Cylindrical cross-section

Small resistance of thin layer of thickness dr,

$$dR = \frac{\rho\,(dr)}{2\pi r l}$$

Integrating dR from $r = a$ to $r = b$, we get the total resistance which comes out to be,

$$R = \frac{\rho}{2\pi l}\ln\left(\frac{b}{a}\right)$$

Example 6 A metal ball of radius a is surrounded by a thin concentric metal shell of radius b. The space between these electrodes is filled up with a poorly conducting homogeneous medium of resistivity ρ. Find the resistance of the inter electrode gap. Analyse the obtained solution at $b \to \infty$. **(Irodov 3.155)**

Sol. Using $R = \rho\frac{l}{A}$, small resistance of thin layer of thickness dr shown in figure is,

$$dR = \frac{\rho(dr)}{4\pi r^2}$$

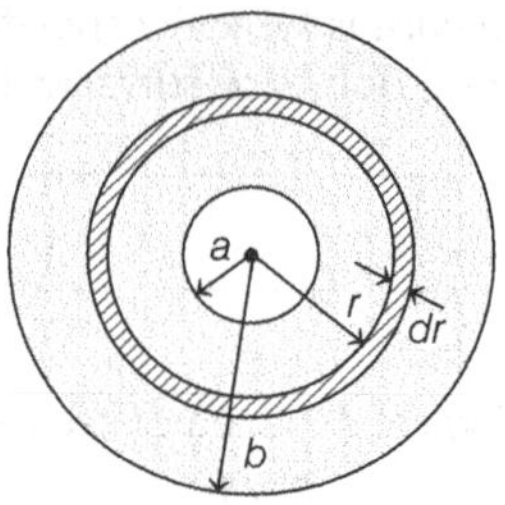

Spherical cross-section

Integrating dR from $r = a$ to $r = b$, we get the total resistance which comes out to be,

$$R = \frac{\rho(b-a)}{4\pi ab}$$

For $b \to \infty$

R can also be written as,

$$R = \frac{\rho}{4\pi}\left(\frac{1}{a} - \frac{1}{b}\right)$$

As $b \to \infty, \frac{1}{b} \to 0.$

So, $R = \frac{\rho}{4\pi a}$

MODULE 3 Exercise

True/False

1. The product of resistivity and conductivity of a conductor is constant. Is this statement true or false?

Single Correct Option

2. Consider a thin square sheet of side L and thickness t, made of a material of resistivity ρ. The resistance between two opposite faces, shown by the shaded areas in the figure is **(2010 JEE)**

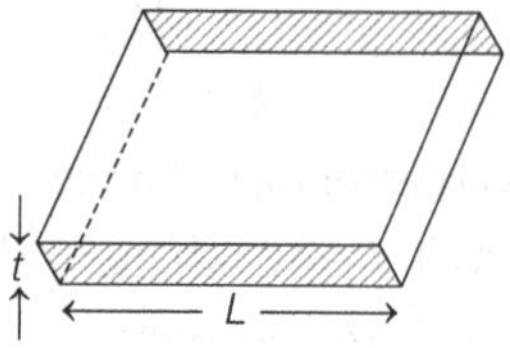

(a) directly proportional to L
(b) directly proportional to t
(c) independent of L
(d) independent of t

3. Conductivity increases in the order of
(a) Al, Ag, Cu (b) Al, Cu, Ag
(c) Cu, Al, Ag (d) Ag, Cu, Al

4. The resistance of a wire is $10\,\Omega$. Its length is increased by 10% by stretching. The new resistance will now be
(a) $12\,\Omega$ (b) $1.2\,\Omega$ (c) $13\,\Omega$ (d) $11\,\Omega$

5. The specific resistance of a wire is ρ, its volume is 3 m^3 and its resistance is $3\,\Omega$, then its length will be
(a) $\sqrt{\frac{1}{\rho}}$ (b) $\frac{3}{\sqrt{\rho}}$ (c) $\frac{1}{\rho}\sqrt{3}$ (d) $\rho\sqrt{\frac{1}{3}}$

6. In order to increase the resistance of a given wire of uniform cross-section to four times, a fraction of its length is stretched uniformly till the full length of the wire becomes $\frac{3}{2}$ times the original length. What is the value of this fraction?
(a) $\frac{1}{4}$ (b) $\frac{1}{8}$ (c) $\frac{1}{16}$ (d) $\frac{1}{6}$

7. The resistance of a wire is 5 Ω. It's new resistance in ohm, if stretched to 5 times of it's original length will be **(2023 Main)**
(a) 5 (b) 25 (c) 125 (d) 625

8. In an aluminium (Al) bar of square cross section, a square hole is drilled and is filled with iron (Fe) as shown in the figure. The electrical resistivities of Al and Fe are 2.7×10^{-8} Ω-m and 1.0×10^{-7} Ω-m, respectively. The electrical resistance between the two faces P and Q of the composite bar is **(2015 Adv.)**

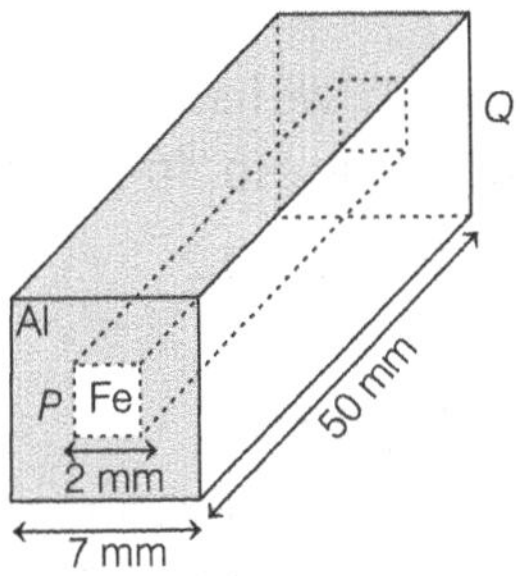

(a) $\frac{2475}{64}$ μΩ (b) $\frac{1875}{64}$ μΩ (c) $\frac{1875}{49}$ μΩ (d) $\frac{2475}{132}$ μΩ

9. In a conductor, if the number of conduction electrons per unit volume is 8.5×10^{28} m^{-3} and mean free time is 25 fs (femto second), it's approximate resistivity is (Take, $m_e = 9.1 \times 10^{-31}$ kg) **(2019 Main)**
(a) 10^{-7} Ω-m (b) 10^{-5} Ω-m
(c) 10^{-6} Ω-m (d) 10^{-8} Ω-m

Subjective Questions

10. In household wiring, copper wire 2.05 mm in diameter is often used. Find the resistance of a 35.0 m long wire. Specific resistance of copper is 1.72×10^{-8} Ω-m.

11. You need to produce a set of cylindrical copper wires 3.50 m long that will have a resistance of 0.125 Ω each. What will be the mass of each of these wires? Specific resistance of copper $= 1.72 \times 10^{-8}$ Ω-m, density of copper $= 8.9 \times 10^3$ kg/m^3.

12. A conductor of length l has a non-uniform cross-section. The radius of cross-section varies linearly from a to b. The resistivity of the material is ρ. Find the resistance of the conductor across its ends.

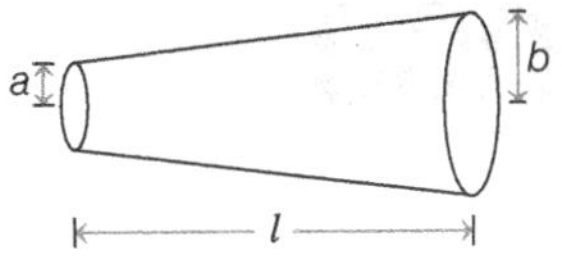

13. A rectangular block of metal of resistivity ρ has dimensions $d \times 2d \times 3d$. A potential difference V is applied between two opposite faces of the block.
(a) To which two faces of the block should the potential difference V be applied to give the maximum current density? What is the maximum current density?
(b) To which two faces of the block should the potential difference V be applied to give the maximum current? What is this maximum current?

14. An electrical conductor designed to carry large currents has a circular cross-section 2.50 mm in diameter and is 14.0 m long. The resistance between its ends is 0.104 Ω.
(a) What is the resistivity of the material?
(b) If the electric field in the conductor is 1.28 V/m, what is the total current?
(c) If the material has 8.5×10^{28} free electrons per cubic metre, find the average drift speed under the conditions of part (b).

Integer Type Questions

15. The length of a metallic wire is increased by 20% and its area of cross-section is reduced by 4%. The percentage change in resistance of the metallic wire is .... **(2023 Main)**

16. A rectangular parallelopiped is measured as 1 cm × 1 cm × 100 cm. If its specific resistance is 3×10^{-7} Ω-m, then the resistance between its two opposite rectangular faces will be .... $\times 10^{-7}$ Ω. **(2023 Main)**

17. A hollow cylindrical conductor has length of 3.14 m, while its inner and outer diameters are 4 mm and 8 mm respectively. The resistance of the conductor is $n \times 10^{-3}$ Ω. If the resistivity of the material is 2.4×10^{-8} Ω-m. The value of n is .... **(2023 Main)**

Answers

1. True **2.** (c) **3.** (b) **4.** (a) **5.** (b) **6.** (b) **7.** (c) **8.** (b) **9.** (d) **10.** 0.18 Ω
11. 15 g **12.** $\frac{\rho l}{\pi ab}$ **13.** (a) $2d \times 3d$, $\frac{V}{\rho d}$ (b) $2d \times 3d$, $\frac{6Vd}{\rho}$ **14.** (a) 3.65×10^{-8} Ω-m (b) 172.3 A (c) 2.58×10^{-3} m/s
15. (25) **16.** (3) **17.** (2)

See hints on pages 75-77.

MODULE 4

Temperature Dependence of Resistance

If we increase the temperature of any material, the following two effects can be observed :

(i) Numbers of free electrons increase. Due to this effect conductivity of the material increases. So, resistivity or resistance decreases.

(ii) The ions of the material vibrate with greater amplitude and the collision between electrons and ions become more frequent. Due to this effect resistivity or resistance of the material increases.

In Conductors

There are already a large number of free electrons. So, with increase in temperature effect-(i) is not so dominant as effect-(ii). Hence, resistivity or resistance of conductors increase with increase in temperature.

Over a small temperature range (upto 100°C), the resistivity of a metal (or conductors) can be represented approximately by the equation,

$$\rho(T) = \rho_0 [1 + \alpha (T - T_0)] \quad \ldots(i)$$

where, ρ_0 is the resistivity at a reference temperature T_0 (often taken as 0°C or 20°C) and $\rho(T)$ is the resistivity at temperature T, which may be higher or lower than T_0. The factor α is called the **temperature coefficient of resistivity.**

The resistance of a given conductor depends on its length and area of cross-section besides the resistivity. As temperature changes, the length and area also change. But these changes are quite small and the factor l/A may be treated as constant.

Then, $R \propto \rho$ and hence,

$$R(T) = R_0 [1 + \alpha (T - T_0)] \quad \ldots(ii)$$

In this equation, $R(T)$ is the resistance at temperature T and R_0 is the resistance at reference temperature T_0. The temperature coefficient of resistance α is the same constant that appears in Eq. (i), if the dimensions l and A in equation $R = \rho \frac{l}{A}$ do not change with temperature.

In Semiconductors

At room temperature, numbers of free electrons in semiconductors (like silicon, germanium etc.) are very less. So, with increase in temperature, effect-(i) is very dominant. Hence, resistivity or resistance of semiconductors decreases with increase in temperature or we can say that temperature coefficient of resistivity α for semiconductors is negative.

Example 1 The resistance of a thin silver wire is 1.0 Ω at 20°C. The wire is placed in a liquid bath and its resistance rises to 1.2 Ω. What is the temperature of the bath? α for silver is 3.8×10^{-3}/°C.

Sol. $R(T) = R_0[1 + \alpha(T - T_0)]$

Here, $R(T) = 1.2\,\Omega, R_0 = 1.0\,\Omega, \alpha = 3.8 \times 10^{-3}/°C$

and $T_0 = 20°C$

Substituting the values, we have

$$1.2 = 1.0\,[1 + 3.8 \times 10^{-3}(T - 20)]$$

or $$3.8 \times 10^{-3}(T - 20) = 0.2$$

Solving this, we get

$$T = 72.6°C$$ **Ans.**

Example 2 Read the following statements carefully **(1993 JEE)**

Y : The resistivity of semiconductor decreases with increase of temperature.

Z : In a conducting solid, the rate of collisions between free electrons and ions increases with increase of temperature.

Select the correct statement (s) from the following

(a) Y is true but Z is false

(b) Y is false but Z is true

(c) Both Y and Z are true

(d) Y is true and Z is the correct reason for Y

Sol. Resistivity of conductors increases with increase in temperature because rate of collisions between free electrons and ions increase with increase of temperature. However, the resistivity of semiconductors decreases with increase in temperature, because more and more covalent bonds are broken at higher temperatures and free electrons increase with increase in temperature. Therefore, the correct option is (c).

Example 3 An electric toaster uses nichrome for its heating element. When a negligibly small current passes through it. Its resistance at room temperature 27.0°C is found to be 75.3 Ω. When the toaster is connected to a 230 V supply, the current settles, after a few seconds, to a steady value of 2.68 A. What is the steady temperature of the nichrome element? The temperature coefficient of resistance of nichrome averaged over the temperature range involved, is $1.70 \times 10^{-4}\,°C^{-1}$.

Sol. Given, $T_0 = 27°C$ and $R_0 = 75.3\,\Omega$

At temperature T, $R_T = \dfrac{V_T}{i_T}$ $\left(\text{as } R = \dfrac{V}{i}\right)$

$$= \frac{230}{2.68} = 85.82\,\Omega$$

Using the equation $R_T = R_0\,[1 + \alpha\,(T - T_0)]$, we have

$$85.82 = 75.3\,[1 + (1.70 \times 10^{-4})(T - 27)]$$

Solving this equation, we get $T \approx 850\,°C$ **Ans.**

Thus, the steady temperature of the nichrome element is 850°C.

Example 4 Two resistors with temperature coefficients of resistance α_1 and α_2 have resistances R_{01} and R_{02} at 0°C. Find the temperature coefficient of the combined resistance consisting of the two resistors connected

(a) in series and

(b) in parallel

Sol. (a) **In Series**

At 0°C R_{01} R_{02} $R_0 = R_{01} + R_{02}$

At t°C $R_{01}\,(1 + \alpha_1 t)$ $R_{02}\,(1 + \alpha_2 t)$ $R_0\,(1 + \alpha t)$

Further, $R = R_1 + R_2$

$$\therefore\; R_{01}\,(1 + \alpha_1 t) + R_{02}\,(1 + \alpha_2 t) = R_0\,(1 + \alpha t)$$

or $R_{01}\,(1 + \alpha_1 t) + R_{02}\,(1 + \alpha_2 t) = (R_{01} + R_{02})\,(1 + \alpha t)$

$$\therefore\; R_{01} + R_{01}\alpha_1 t + R_{02} + R_{02}\alpha_2 t = R_{01} + R_{02} + (R_{01} + R_{02})\alpha t$$

or $$\alpha = \frac{R_{01}\alpha_1 + R_{02}\alpha_2}{R_{01} + R_{02}}$$ **Ans.**

(b) **In Parallel**

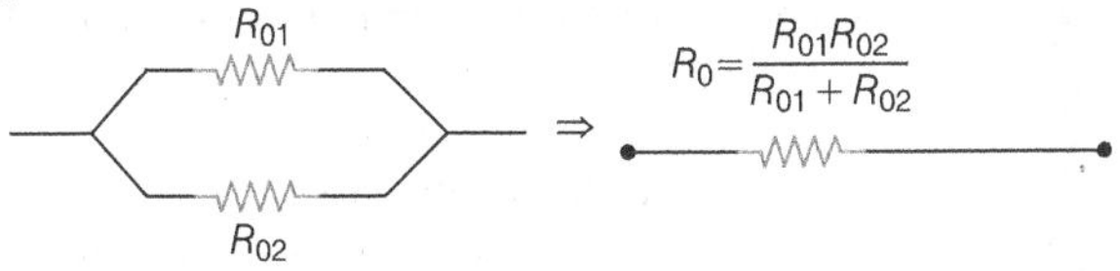

At t°C, $$\frac{1}{R_0\,(1 + \alpha t)} = \frac{1}{R_{01}\,(1 + \alpha_1 t)} + \frac{1}{R_{02}\,(1 + \alpha_2 t)}$$

or $$\frac{R_{01} + R_{02}}{R_{01}R_{02}\,(1 + \alpha t)} = \frac{1}{R_{01}\,(1 + \alpha_1 t)} + \frac{1}{R_{02}\,(1 + \alpha_2 t)}$$

Using the Binomial expansion, we have

$$\frac{1}{R_{02}}(1 - \alpha t) + \frac{1}{R_{01}}(1 - \alpha t) = \frac{1}{R_{01}}(1 - \alpha_1 t) + \frac{1}{R_{02}}(1 - \alpha_2 t)$$

i.e. $$\alpha t\left(\frac{1}{R_{01}} + \frac{1}{R_{02}}\right) = \frac{\alpha_1}{R_{01}}t + \frac{\alpha_2}{R_{02}}t$$

or $$\alpha = \frac{\alpha_1 R_{02} + \alpha_2 R_{01}}{R_{01} + R_{02}}$$ **Ans.**

Important POINTS

- **Thermistor** The temperature coefficient of resistivity is negative for semiconductors.

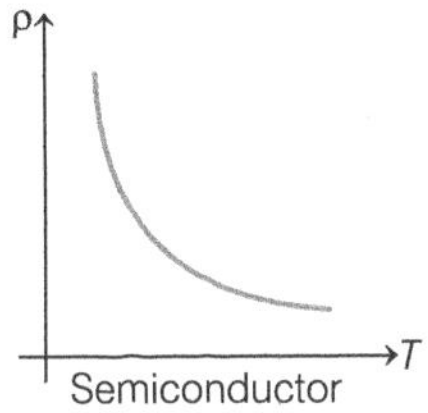

Semiconductor

This means that the resistivity decreases as we raise the temperature of such a material. The magnitude of the temperature coefficient of resistivity is often quite large for a semiconducting material.

This fact is used to construct thermometers to detect small changes in temperatures. Such a device is called a thermistor. The variation of resistivity of a semiconductor with temperature is shown in figure. A typical thermistor can easily measure a change in temperature of the order of 10^{-3} °C.

- **Superconductors** Superconductivity was first discovered in 1911 by the Dutch physicist Heike Kamerlingh Onnes. There are certain materials, including several metallic alloys and oxides for which as the temperature decreases, the resistivity first decreases smoothly, like that of any metal. But then at a certain critical temperature T_c a phase transition occurs, and the resistivity suddenly drops to zero as shown in figure.

Once, a current has been established in a superconducting ring, it continues indefinitely without the presence of any driving field. Possible applications of superconductors are ultrafast computer switches and transmission of electric power through superconducting power lines.

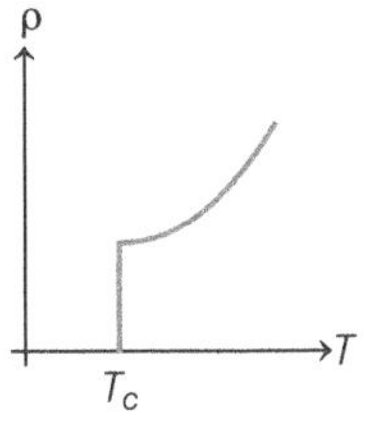

Superconductor

However, the requirement of low temperature is posing difficulty. For instance the critical temperature for mercury is 4.2 K. Scientists are putting great effort to construct compounds and alloys which would be superconducting at room temperature (300 K). Superconductivity at around 125 K has already been achieved.

MODULE 4 Exercise

Single Correct Option

1. A piece of copper and another of germanium are cooled from room temperature to 80 K. The resistance of **(1988 JEE)**
   (a) each of them increases
   (b) each of them decreases
   (c) copper increases and germanium decreases
   (d) copper decreases and germanium increases

2. Resistance of the wire is measured as 2 Ω and 3 Ω at 10°C and 30°C respectively. Temperature coefficient of resistance of the material of the wire is **(2022 Main)**
   (a) $0.033°C^{-1}$ (b) $-0.033°C^{-1}$
   (d) $0.011°C^{-1}$ (d) $0.055°C^{-1}$

Subjective Questions

3. The resistance of a copper wire and an iron wire at 20°C are 4.1 Ω and 3.9 Ω, respectively. Neglecting any thermal expansion, find the temperature at which resistances of both are equal.
$$\alpha_{Cu} = 4.0 \times 10^{-3}\ K^{-1}$$
and $$\alpha_{Fe} = 5.0 \times 10^{-3}\ K^{-1}.$$

4. An aluminium wire 7.5 m long is connected in parallel with a copper wire 6 m long. When a current of 5 A is passed through the combination, it is found that the current in the aluminium wire is 3 A. The diameter of the aluminium wire is 1 mm. Determine the diameter of the copper wire. Resistivity of copper is 0.017 μΩ-m and that of the aluminium is 0.028 μΩ-m.

Hint : *In parallel current distributes in inverse ratio of resistance.*

5. It is desired to make a 20.0 Ω coil of wire which has a zero thermal coefficient of resistance. To do this, a carbon resistor of resistance R_1 is placed in series with an iron resistor of resistance R_2. The proportions of iron and carbon are so chosen that $R_1 + R_2 = 20.00\ \Omega$ for all temperatures near 20°C. How large are R_1 and R_2? Given, $\alpha_C = -0.5 \times 10^{-3}\ K^{-1}$ and $\alpha_{Fe} = 5.0 \times 10^{-3}\ K^{-1}$.

Integer Type Question

6. The current flowing through a conductor connected across a source is 2A and 1.2 A at 0°C and 100°C respectively. The current flowing through the conductor at 50°C will be .... $\times 10^2$ mA. **(2023 Main)**

Answers

1. (d) **2.** (a) **3.** 85°C **4.** 0.569 mm **5.** $R_1 = 18.18\Omega$, $R_2 = 1.82\Omega$ **6.** (15)

See hints on page 77.

MODULE 5

Ohm's and Kirchhoff's Law

Ohm's Law

The equation, $V = iR$ is not Ohm's law. It is a mathematical relation between current passing through a resistance, value of resistance R and the potential difference V across it.

According to Ohm's law, there are some of the materials (like metals or conductors) or some circuits for which, current passing through them is proportional to the potential difference applied across them. Or,

$$i \propto V \quad \text{or} \quad V \propto i$$

$$\Rightarrow \quad V = iR \quad \text{or} \quad \frac{V}{i} = R = \text{constant}$$

or V-i graph for such materials and circuits is a straight line passing through origin. Slope of this graph is called its resistance. The materials or circuits which follow this law are called **ohmic.**

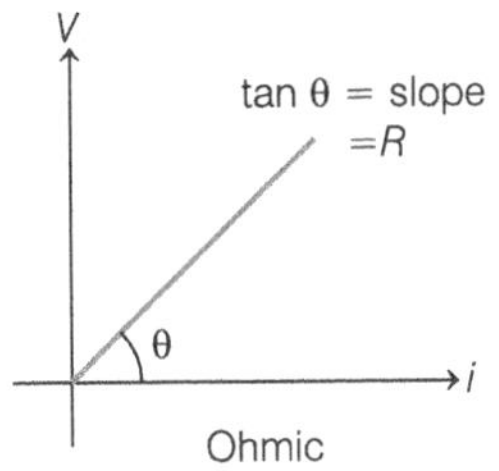

Ohmic

The materials or circuits which do not follow this law are called **non-ohmic**. V-i graph for non-ohmic circuits is not a straight line passing through origin. $\frac{V}{i}$ or R is not constant and i is not proportional to V.

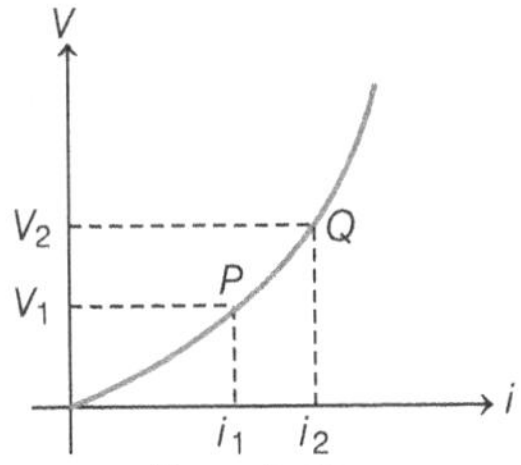

Non-ohmic

Note *The equation $V = iR$ holds true not only for ohmic circuits but also for non-ohmic circuits.*

For example, $\frac{V_1}{i_1} = R_1$ = *resistance at P*

$\frac{V_2}{i_2} = R_2$ = *resistance at Q, but*

$R_1 \neq R_2$

Example 1 The current-voltage graphs for a given metallic wire at two different temperatures T_1 and T_2 are shown in the figure. The temperature T_2 is greater than T_1. Is this statement true or false? **(1985 JEE)**

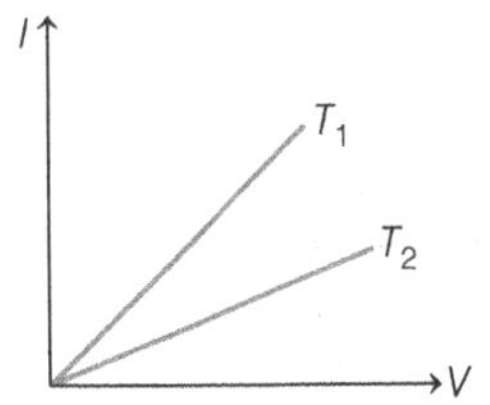

Sol. $\frac{I}{V}$ = Slope of given graph $= \frac{1}{R}$

or $R = \frac{1}{\text{Slope}} \Rightarrow (\text{Slope})_{T_2} < (\text{Slope})_{T_1}$

$$\Rightarrow \quad \frac{1}{(\text{Slope})_{T_2}} > \frac{1}{(\text{Slope})_{T_1}}$$

Resistance of a metallic wire increases with increase in temperature.

or $\quad R_{T_2} > R_{T_1}$ or $T_2 > T_1$

Therefore, the statement is true.

Example 2 The variation of applied potential and current flowing through a given wire is shown in figure. The length of wire is 31.4 cm. The diameter of wire is measured as 2.4 cm. The resistivity of the given wire is measured as $x \times 10^{-3}$ Ω-cm. What is the value of x ?

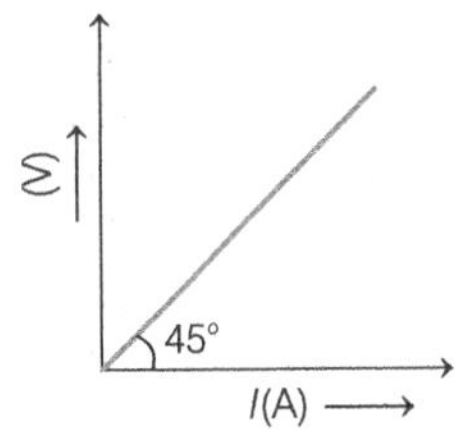

(2022 Main)

Sol. According to the given graph,

R = slope of curve $= \tan 45° = 1\,\Omega$

As, $R = \frac{\rho l}{A}$

$$\Rightarrow \quad \rho = \frac{RA}{l} = \frac{1 \times \pi \times (1.2)^2}{31.4}$$

$$= 144 \times 10^{-3}\ \Omega\text{-cm}$$

Hence, $x = 144$

Important POINTS

- **Deduction of Ohm's law** We know that

$$i = neAv_d$$

where, $v_d = \frac{e\tau E}{m}$

$$\therefore \quad i = \frac{ne^2\tau AE}{m}$$

If V is the potential difference across the conductor and l is its length, then

$$E = \frac{V}{l}$$

$$\therefore \quad i = \frac{ne^2\tau AV}{ml} \quad \text{or} \quad V = \left(\frac{m}{ne^2\tau}\right)\frac{l}{A}i$$

Here, $\frac{m}{ne^2\tau} = \frac{1}{\sigma}$ or ρ

$$\therefore \quad V = \left(\frac{\rho l}{A}\right)\cdot i \quad \text{or} \quad V = Ri$$ **Hence proved.**

where, $R = \rho\frac{l}{A}$ is the resistance of the conductor.

The Battery and the Electromotive Force

Before studying the electromotive force (emf) of a cell let us take an example of a pump which is more easy to understand. Suppose we want to recycle water between a overhead tank and a ground water tank. Water flows from overhead tank to ground water tank by itself (by gravity).

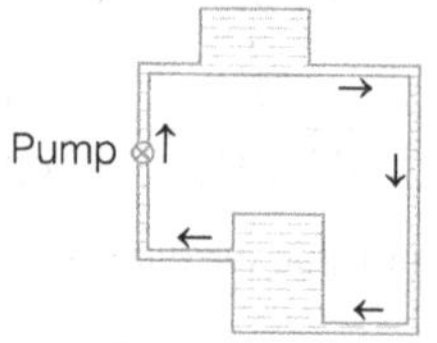

No external agent is required for this purpose. But to raise the water from ground water tank to overhead tank a pump is required or some external work has to be done.

In an electric circuit, a battery or a cell plays the same role as the pump played in the above example.

Suppose a resistance (R) is connected across the terminals of a battery. A potential difference is developed across its ends. Current (or positive charge) flows from higher potential to lower potential across the resistance by itself. But inside the battery, work has to be done to bring the positive charge from lower potential to higher potential. The influence that makes current flow from lower to higher potential (inside the battery) is called **electromotive force** (abbreviated **emf**).

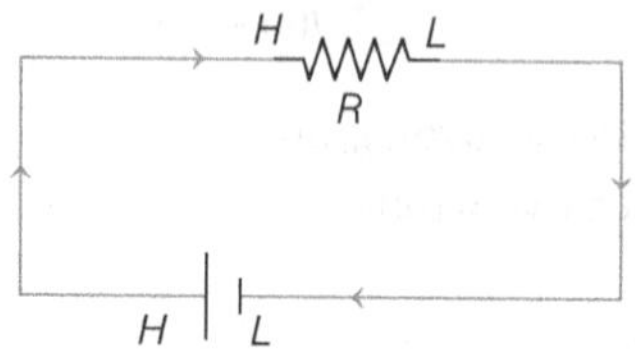

If W work is done by the battery in taking a charge q from negative terminal to positive terminal, then work done by the battery per unit charge is called emf (E) of the battery.

Thus, $$E = \frac{W}{q}$$

The name electromotive force is misleading in the sense that emf is not a force it is work done per unit charge. The SI unit of emf is J/C or V ($1\text{ V} = 1\text{ J/C}$).

Note *Emf is also sometimes denoted by V.*

Simple Circuits

Single current in a simple circuit (single loop) can be found by the relation,

$$i = \frac{\text{net emf}}{\text{net resistance}} = \frac{E_{net}}{R_{net}}$$

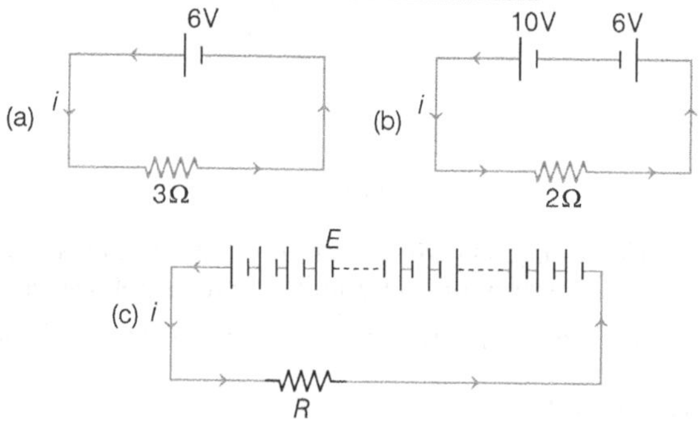

For example :

In Fig. (a) Net emf is 6 V and net resistance is 3 Ω. Therefore,

$$i = \frac{6}{3} = 2\text{ A}$$

In Fig. (b) Net emf = (10 − 6) V = 4 V

and Net resistance = 2 Ω

Therefore, $$i = \frac{4}{2} = 2\text{ A}$$

In Fig. (c) We have n cells each of emf E. Of these polarity of m cells (where $n > 2m$) is reversed. Then, net emf in the circuit is $(n-2m)E$ and resistance of the circuit is R.

Therefore,

$$i = \frac{(n-2m)E}{R}$$

Resistors in Series and in Parallel

In series :

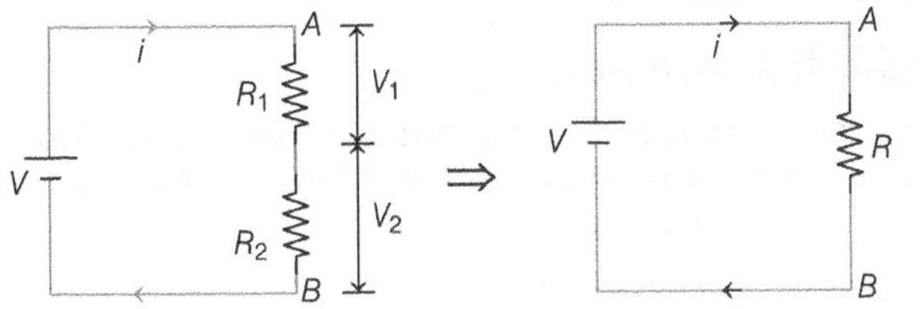

Figure represents a circuit consisting of a source of emf and two resistors connected in series. We are interested in finding the net resistance R of the network lying between A and B.

Because there is only one path for electric current to follow, i must have the same value everywhere in the circuit. The potential difference between A and B is V. This potential difference must somehow be divided into two parts V_1 and V_2 as shown.

$\therefore \quad V = V_1 + V_2 = iR_1 + iR_2$

or $\quad V = i(R_1 + R_2)$ ..(i)

Let R be the equivalent resistance between A and B, then

$$V = iR \quad ..(ii)$$

From Eqs. (i) and (ii),

$$R = R_1 + R_2 \text{ for resistors in series}$$

This result can be readily extended to a network consisting of n resistors in series.

$\therefore \quad R = R_1 + R_2 + \ldots\ldots + R_n$

In Parallel

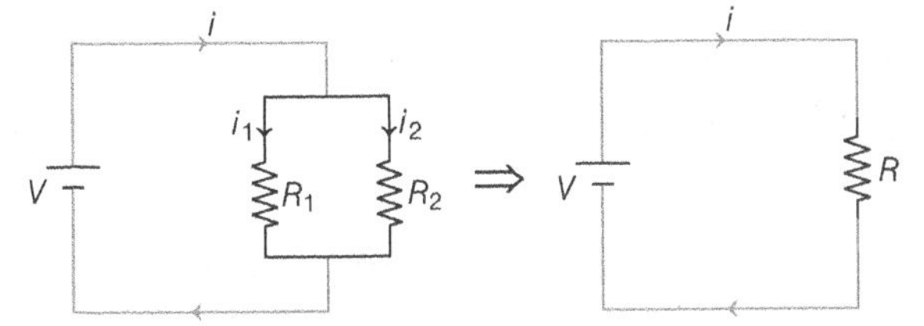

In above figure, the two resistors are connected in parallel. The voltage drop across each resistor is equal to the source voltage V. The current i, however, divides into two branches, which carry currents i_1 and i_2.

$$i = i_1 + i_2 \quad ...(iii)$$

If R be the equivalent resistance, then

$$i = \frac{V}{R}, \quad i_1 = \frac{V}{R_1} \text{ and } i_2 = \frac{V}{R_2}$$

Substituting in Eq. (iii), we get

$$\frac{1}{R} = \frac{1}{R_1} + \frac{1}{R_2} \text{ for resistors in parallel}$$

or $$R = \frac{R_1 R_2}{R_1 + R_2}$$

This result can also be extended to a network consisting of n resistors in parallel. The result is

$$\frac{1}{R} = \frac{1}{R_1} + \frac{1}{R_2} + \ldots\ldots + \frac{1}{R_n}$$

Note *(i) In parallel, net resistance is less than all individual resistances.*

(ii) If n resistors each of resistance R are connected in parallel, then net resistance is $\frac{R}{n}$.

(iii) Following combination of resistances are frequently used in different problems.

$R_1 (\Omega)$	$R_2 (\Omega)$	$R = (R_1 R_2)/(R_1 + R_2) \Omega$
6	3	2
12	4	3
20	5	4

Example 3 Compute the equivalent resistance of the network shown in figure and find the current i drawn from the battery.

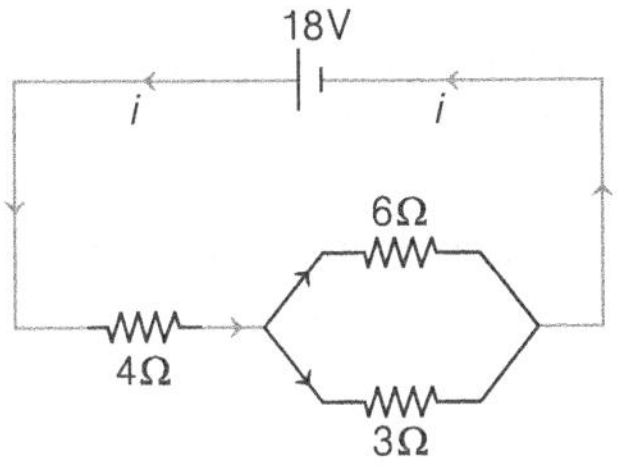

Sol. The 6 Ω and 3 Ω resistances are in parallel. Their equivalent resistance is

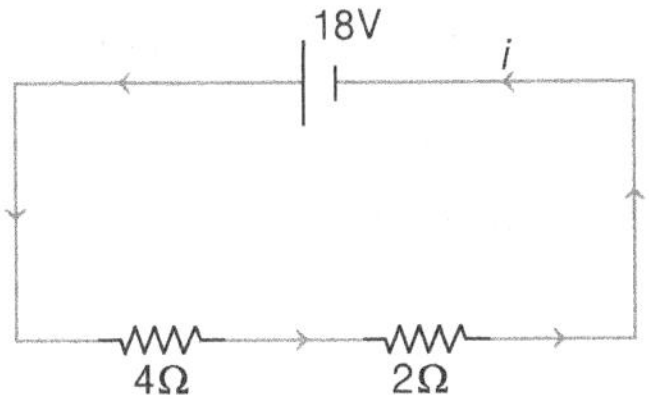

$$\frac{1}{R} = \frac{1}{6} + \frac{1}{3} \quad \text{or} \quad R = 2\,\Omega$$

Now, this 2 Ω and 4 Ω resistances are in series and their equivalent resistance is $4 + 2 = 6\,\Omega$.

Therefore, equivalent resistance of the network $= 6\,\Omega$.

Ans.

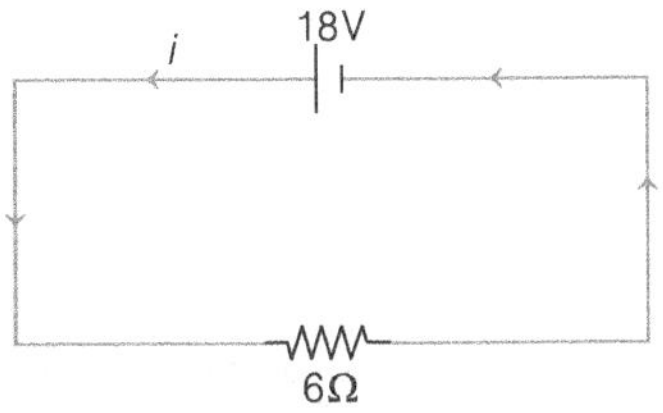

Current drawn from the battery is

$$i = \frac{\text{net emf}}{\text{net resistance}} = \frac{18}{6} = 3 \text{ A}$$ **Ans.**

Complex Circuits : Kirchhoff's Laws

Many electric circuits cannot be reduced to simple series-parallel combinations. For example, two circuits that cannot be so broken down are shown in figure.

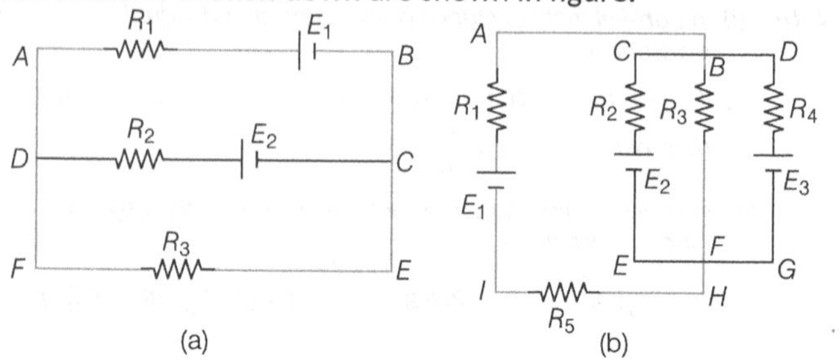

However, it is always possible to analyze such circuits by applying two rules, devised by Kirchhoff in 1845 and 1846 when he was still a student.

First there are two terms that we will use often.

Junction

A junction in a circuit is a point where three or more conductors (or wires) meet. Junctions are also called nodes or branch points.

For example, in Fig. (a) points D and C are junctions. Similarly, in Fig. (b) points B and F are junctions.

Loop

A loop is any closed conducting path. For example, in Fig. (a) *ABCDA*, *DCEFD* and *ABEFA* are loops. Similarly, in Fig. (b), *CBFEC*, *BDGFB* are loops.

Kirchhoff's rules consist of the following two statements :

Kirchhoff's Junction Rule

The algebraic sum of the currents into any junction is zero.

That is, $\sum_{\text{junction}} i = 0$

This law can also be written as, "the sum of all the currents directed towards a point in a circuit is equal to the sum of all the currents directed away from that point."

Thus, in figure $i_1 + i_2 = i_3 + i_4$

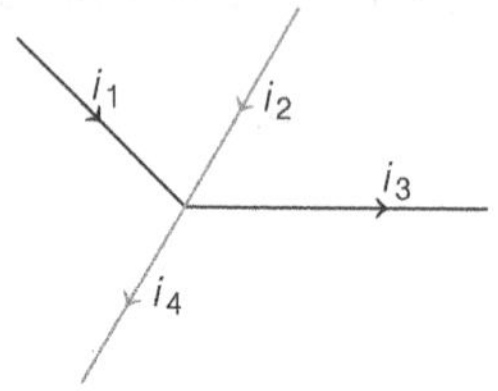

The junction rule is based on **conservation of electric charge.** No charge can accumulate at a junction, so the total charge entering the junction per unit time must equal to charge leaving per unit time. Charge per unit time is current, so if we consider the currents entering to be positive and those leaving to be negative, the algebraic sum of currents into a junction must be zero.

Kirchhoff's Loop Rule

The algebraic sum of the potential differences in any loop including those associated emf's and those of resistive elements, must be zero.

That is, $\sum_{\text{closed loop}} \Delta V = 0$

Kirchhoff's second rule is based on the principle of **conservation of energy.**

Method of Applying the Loop Law

It is advised to write H (for higher potential) and L (for lower potential) across all the batteries and resistances of the loop under consideration while using the loop law. Then write – while moving from H to L and + for L to H. Across a battery write H on positive terminal and L on negative terminal. Across a resistance keep in mind the fact that current always flows from higher potential (H) to lower potential (L). For example, in the loop shown in figure we have marked H and L across all batteries and resistances. Now let us apply the second law in the loop *ADCBA*.

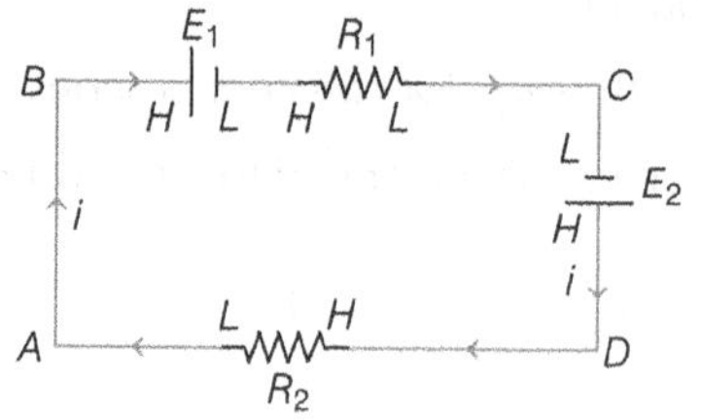

The equation will be $+iR_2 - E_2 + iR_1 + E_1 = 0$

Example 4 Find currents in different branches of the electric circuit shown in figure.

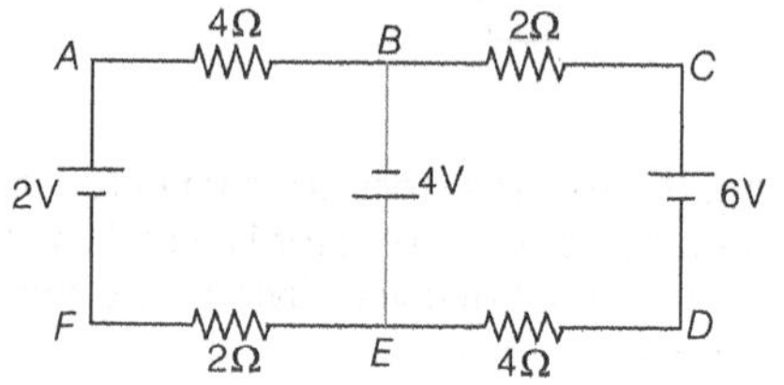

HOW TO PROCEED In this problem there are three wires *EFAB*, *BE* and *BCDE*. Therefore, we have three unknown currents i_1, i_2 and i_3. So, we require three equations. One equation will be obtained by applying Kirchhoff's junction law (either at B or at E) and the remaining two equations, we get from the second law (loop law). We can make three loops *ABEFA*, *ACDFA* and *BCDEB*. But we have to choose any two of them.

Initially, we can choose any arbitrary directions of i_1, i_2 and i_3.

Sol. Applying Kirchhoff's first law (junction law) at junction B,

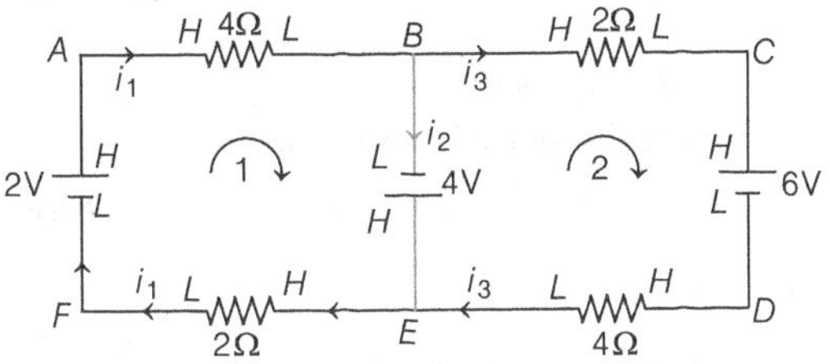

$$i_1 = i_2 + i_3 \quad \text{...(i)}$$

Applying Kirchhoff's second law in loop 1 ($ABEFA$),

$$-4i_1 + 4 - 2i_1 + 2 = 0 \quad \text{...(ii)}$$

Applying Kirchhoff's second law in loop 2 ($BCDEB$),

$$-2i_3 - 6 - 4i_3 - 4 = 0 \quad \text{...(iii)}$$

Solving Eqs. (i), (ii) and (iii), we get

$$i_1 = 1\text{ A}$$

$$\Rightarrow \quad i_2 = \frac{8}{3}\text{ A}$$

$$i_3 = -\frac{5}{3}\text{ A} \quad \textbf{Ans.}$$

Here, negative sign of i_3 implies that current i_3 is in opposite direction of what we have assumed.

Example 5 In example 4, find the potential difference between points F and C.

HOW TO PROCEED To find the potential difference between any two points of a circuit you have to reach from one point to the other *via* any path of the circuit. It is advisable to choose a path in which we come across the least number of resistors, preferably a path which has no resistance.

Sol. Let us reach from F to C *via* A and B,

$$V_F + 2 - 4i_1 - 2i_3 = V_C$$

$$\therefore \quad V_F - V_C = 4i_1 + 2i_3 - 2$$

Substituting, $i_1 = 1$ A and $i_3 = -\frac{5}{3}$ A, we get

$$V_F - V_C = -\frac{4}{3}\text{ volt} \quad \textbf{Ans.}$$

Here, negative sign implies that $V_F < V_C$.

Internal Resistance (r) and Potential Difference (V) across the Terminals of a Battery

The potential difference across a real source in a circuit is not equal to the emf of the cell. The reason is that charge moving through the electrolyte of the cell encounters resistance. We call this the **internal resistance** of the source, denoted by r.

As the current moves through r, it experiences an associated drop in potential equal to ir.

Thus, when a current is drawn through a source, the potential difference between the terminals of the source is

$$\boxed{V = E - ir}$$

This can also be shown as below :

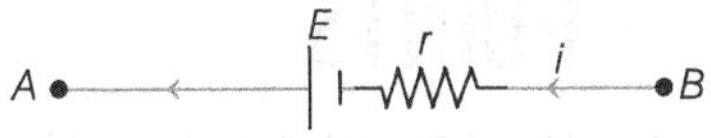

$$V_A - E + ir = V_B$$

or $$V_A - V_B = E - ir$$

Here are three possible cases :

(i) If the current flows in opposite direction (as in case of charging of a battery), then $V = E + ir$

(ii) $V = E$, if the current through the cell is zero.

(iii) $V = 0$, if the cell is short circuited. Short circuited means terminals of the battery are directly connected by a conducting wire.

$$R_{ext} = 0 \text{ but } R_{in} = r \neq 0$$

This is because current in the circuit

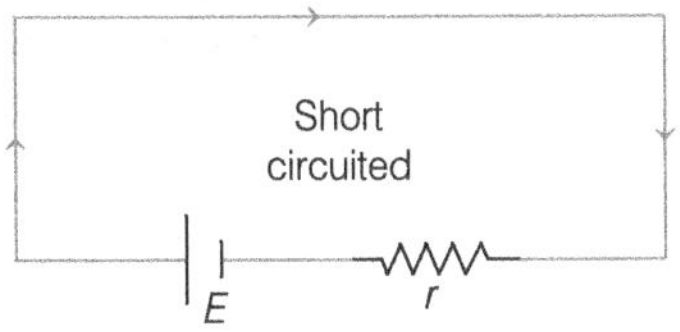

$$i = \frac{E}{r} \quad \text{or} \quad E = ir$$

$$\therefore \quad E - ir = 0 \quad \text{or} \quad V = 0$$

Thus, we can summarise the results as follows :

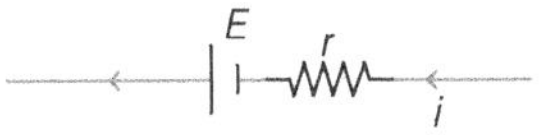

$V = E - ir$ or $V < E$

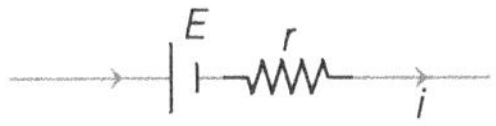

$V = E + ir$ or $V > E$

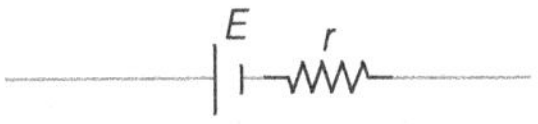

$V = E$ if $i = 0$

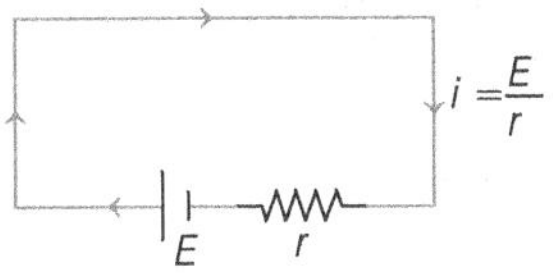

$V = 0$ if short circuited

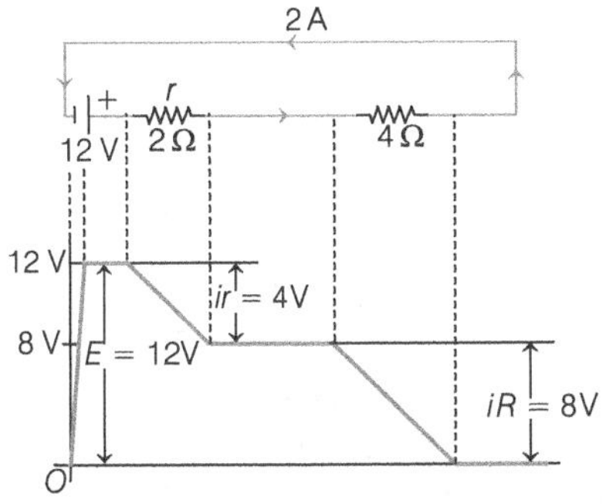

Potential rise and fall in a circuit

Important POINTS

- **In figure (a) :** There are eight wires and hence, will have eight currents or eight unknowns. The eight wires are $AB, BC, CE, EA, AD, BD, CD$ and ED.

(a)

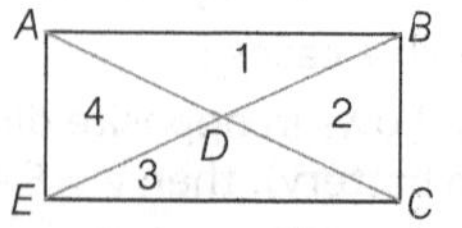

Number of independent loops are four.Therefore, from the second law we can make only four equations. Total number of junctions are five (A, B, C, D and E). But by using the first law, we can make only four equations (one less). So, the total number of equations are eight.

In figure (b) : Number of wires are six (AB, BC, CDA, BE, AE and CE). Number of independent loops are three so, three equations can be obtained from the second law.

(b)

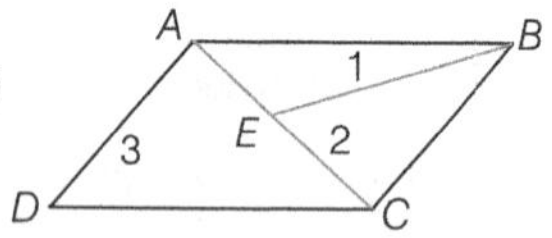

Number of junctions are four (A, B, C and E) so, we can make only three (one less) equations from the first law. But total number of equations are again six.

- **Short circuiting :** Two points in an electric circuit directly connected by a conducting wire are called short circuited. Under such condition both points are at same potential.

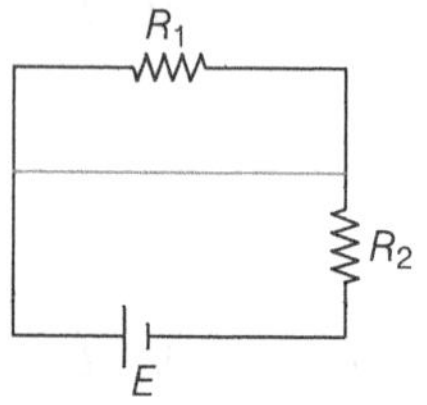

For example, resistance R_1 in the given circuit is short circuited, i.e. potential difference across it is zero. Hence, no current will flow through R_1 and the current through R_2 is therefore, E/R_2.

- **Earthing :** If some point of a circuit is earthed, then its potential is taken to be zero.

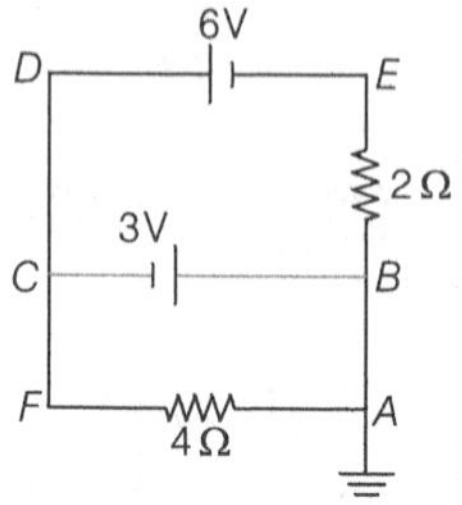

For example, in the figure,

$$V_A = V_B = 0$$
$$V_F = V_C = V_D = -3\text{ V}$$
$$V_E = -9\text{ V}$$

$\therefore \quad V_B - V_E = 9\text{ V}$

or current through $2\,\Omega$ resistance is

$$\frac{V_B - V_E}{2} \text{ or } \frac{9}{2}\text{ A} \qquad \text{(from } B \text{ to } E\text{)}$$

Similarly, $V_A - V_F = 3\text{ V}$

and the current through $4\,\Omega$ resistance is $\dfrac{V_A - V_F}{4}$

or $\dfrac{3}{4}$ A (from A to F)

- For a current flow through a resistance there must be a potential difference across it but between any two points of a circuit the potential difference may be zero.

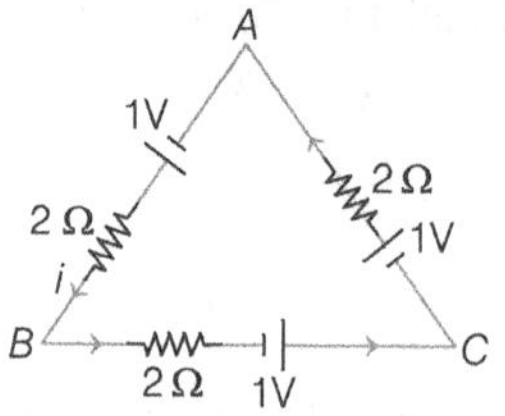

For example, in the circuit,

net emf $= 3$ V

and net resistance $= 6\,\Omega$

$\therefore$ current in the circuit, $i = \dfrac{3}{6} = \dfrac{1}{2}$ A

$\boldsymbol{V_A - V_B} \qquad V_A + 1 - 2 \times \dfrac{1}{2} = V_B$

or $\qquad V_A - V_B = 0$

or by symmetry, we can say that

$$V_A = V_B = V_C$$

So, the potential difference across any two vertices of the triangle is zero, while the current in the circuit is non-zero.

- **Distribution of current in parallel connections**

When more than one resistances are connected in parallel, the potential difference across them is equal and the current is distributed among them in inverse ratio of their resistance as

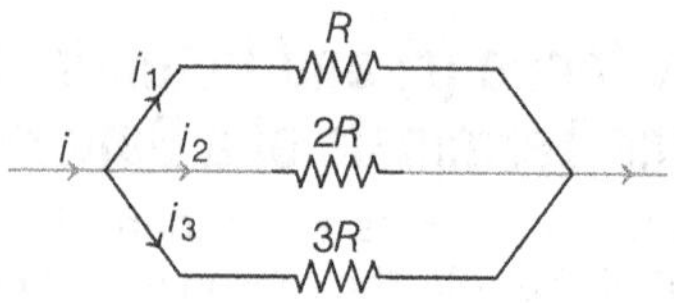

$$i = \frac{V}{R}$$

or $\qquad i \propto \dfrac{1}{R}$ for same value of V

e.g. in the figure,

$$i_1 : i_2 : i_3 = \frac{1}{R} : \frac{1}{2R} : \frac{1}{3R} = 6 : 3 : 2$$

$$\therefore \quad i_1 = \left(\frac{6}{6+3+2}\right) i = \frac{6}{11} i$$

$$i_2 = \left(\frac{3}{6+3+2}\right) i = \frac{3}{11} i$$

and $$i_3 = \left(\frac{2}{6+3+2}\right) i = \frac{2}{11} i$$

Note *In case of only two resistances,* $\frac{i_1}{i_2} = \frac{R_2}{R_1}$

- **Distribution of potential in series connections**
  When more than one resistances are connected in series, the current through them is same and the potential difference is distributed in the direct ratio of their resistance as

$$V = iR \quad \text{or} \quad V \propto R \quad \text{for same value of } i.$$

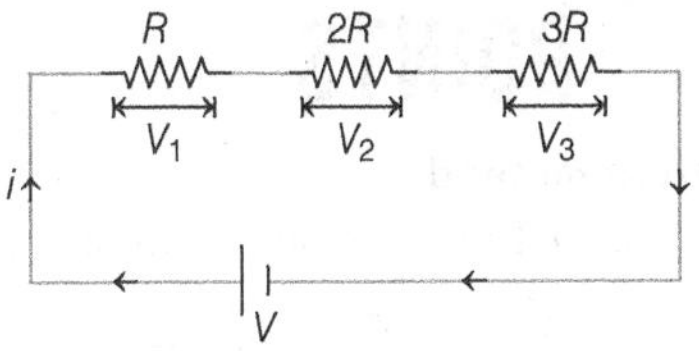

For example in the figure,

$$V_1 : V_2 : V_3 = R : 2R : 3R = 1 : 2 : 3$$

$$\therefore \quad V_1 = \left(\frac{1}{1+2+3}\right) V = \frac{V}{6}$$

$$V_2 = \left(\frac{2}{1+2+3}\right) V = \frac{V}{3}$$

and $$V_3 = \left(\frac{3}{1+2+3}\right) V = \frac{V}{2}$$

Example 6 In the circuit shown in figure,

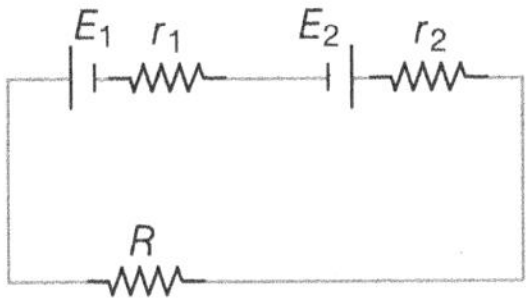

$E_1 = 10$ V, $E_2 = 4$ V, $r_1 = r_2 = 1\,\Omega$ and $R = 2\,\Omega$.
Find the potential difference across battery 1 and battery 2.

Sol. Net emf of the circuit $= E_1 - E_2 = 6$ V

Total resistance of the circuit $= R + r_1 + r_2 = 4\,\Omega$

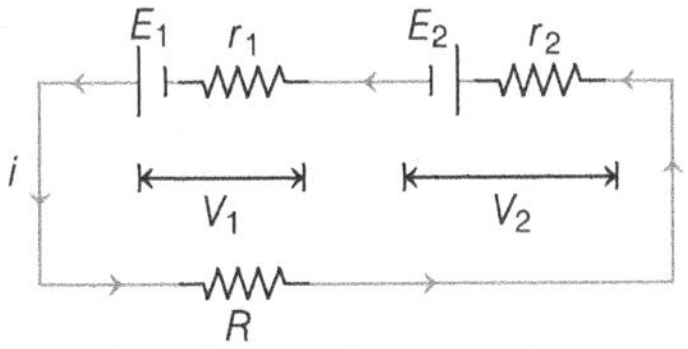

$\therefore$ Current in the circuit, $i = \dfrac{\text{net emf}}{\text{total resistance}} = \dfrac{6}{4} = 1.5$ A

Now, $V_1 = E_1 - ir_1 = 10 - (1.5)(1)$
$= 8.5$ V **Ans.**

and $V_2 = E_2 + ir_2 = 4 + (1.5)(1)$
$= 5.5$ V **Ans.**

INTRODUCTORY Exercise

1. Find the current through $2\,\Omega$ and $4\,\Omega$ resistance.

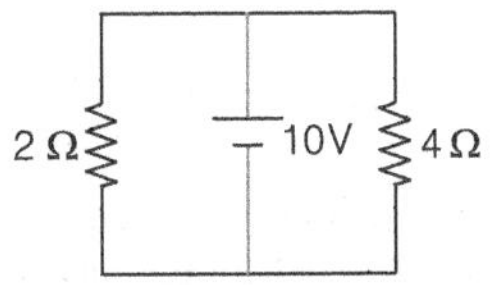

2. In the circuit shown in figure, find the potentials of A, B, C and D and the current through $1\,\Omega$ and $2\,\Omega$ resistance.

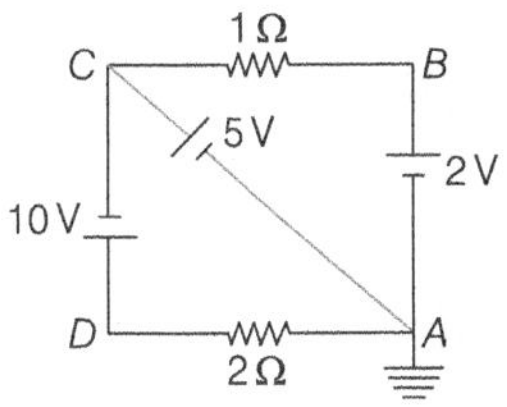

3. For what value of E the potential of A is equal to the potential of B?

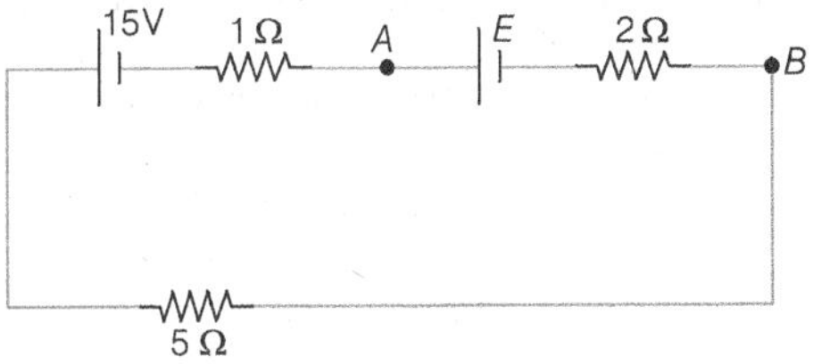

4. Ten cells each of emf 1 V and internal resistance $1\,\Omega$ are connected in series. In this arrangement, polarity of two cells is reversed and the system is connected to an external resistance of $2\,\Omega$. Find the current in the circuit.
5. In the circuit shown in figure, $R_1 = R_2 = R_3 = 10\,\Omega$. Find the currents through R_1 and R_2.

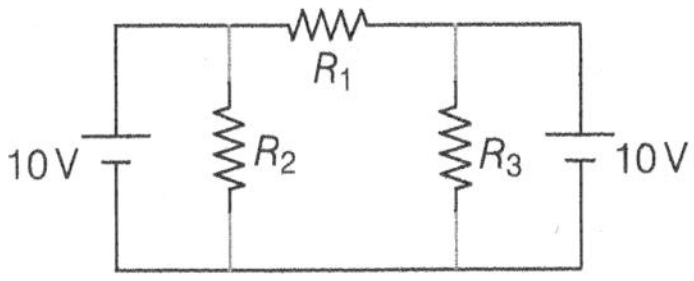

Answers

1. 5 A, 2.5 A
2. 0, 2 V, 5 V, 15 V, 3 A from C to B, 7.5 A from D to A.
3. 5 V **4.** 0.5 A
5. Zero, 1 A

See hints on page 78.

Important POINTS

- **Loop current method**

  Currents in each branches can be obtained as below :

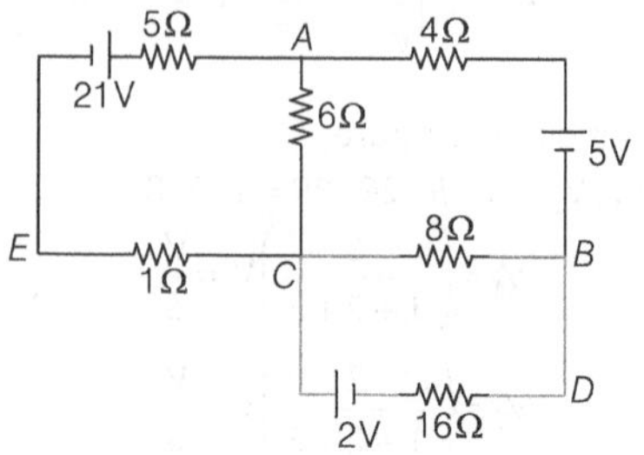

It is possible to use Kirchhoff's laws in a slightly different form, which may simplify the solution of certain problems. This method of applying Kirchhoff's laws is called the **loop current method.**

In this method, we assign a current to every closed loop in a network.

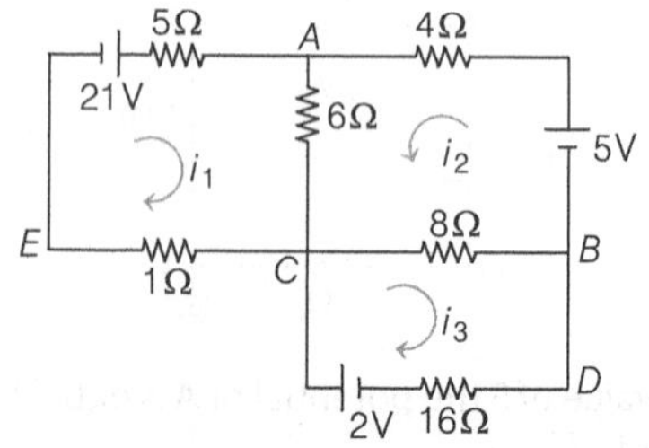

Suppose currents i_1, i_2 and i_3 are flowing in the three loops. The clockwise or anti-clockwise sense given to these currents is arbitrary. Applying Kirchhoff's second law to the three loops, we get

$$21 - 5i_1 - 6(i_1 + i_2) - i_1 = 0 \quad \text{...(i)}$$

$$5 - 4i_2 - 6(i_1 + i_2) - 8(i_2 + i_3) = 0 \quad \text{...(ii)}$$

and $$2 - 8(i_2 + i_3) - 16i_3 = 0 \quad \text{...(iii)}$$

Solving these three equations, we get

$$i_1 = 2\text{ A},\ i_2 = -\frac{1}{2}\text{ A} \text{ and } i_3 = \frac{1}{4}\text{ A}$$

Therefore, current in different branches are as shown in figure given below.

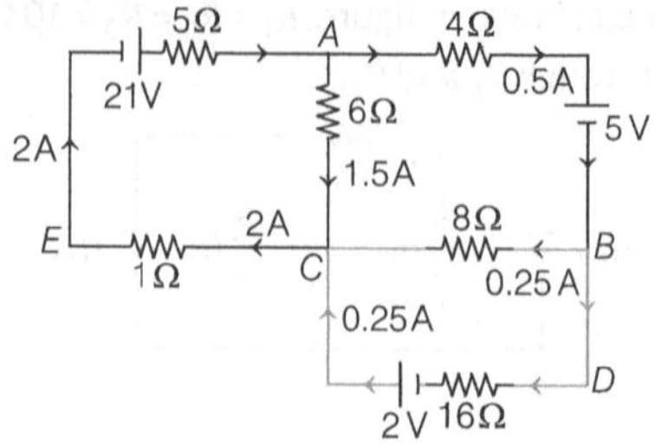

Note *In wire AC, current is $i_1 + i_2$ and in CB it is $i_2 + i_3$.*

- **The principle of superposition**

  Figure shows a network with two loops. The currents in various branches can be calculated using Kirchhoff's laws.

We can get the same solution by considering only one battery at a time and then superposing the two solutions. If a battery has an internal resistance, it must be left in place when the emf of the battery is removed. Figure shows how the superposition principle can be applied to the present problem

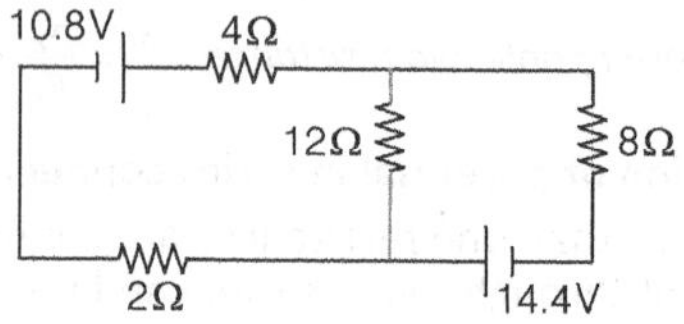

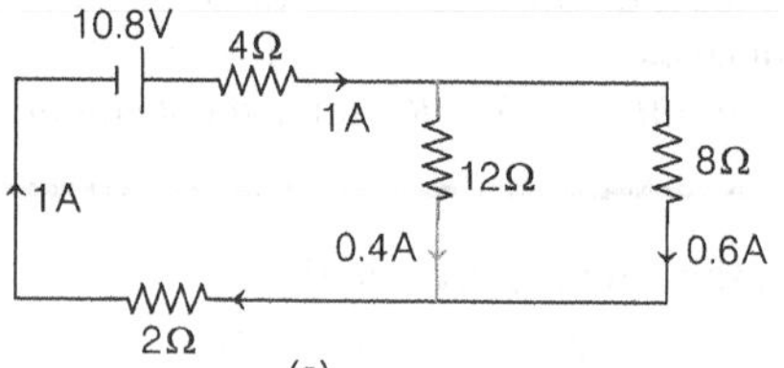

(a)

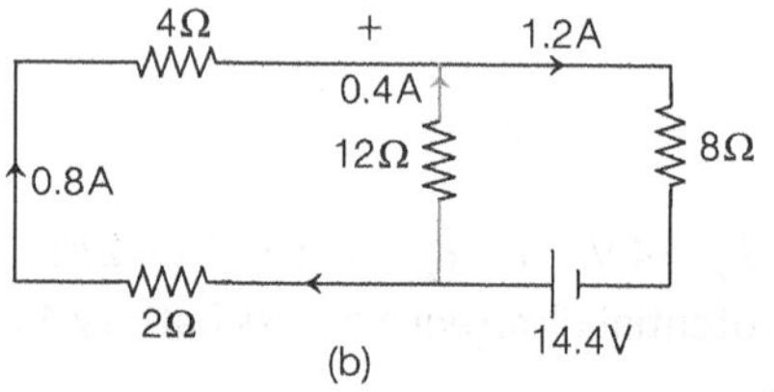

(b)

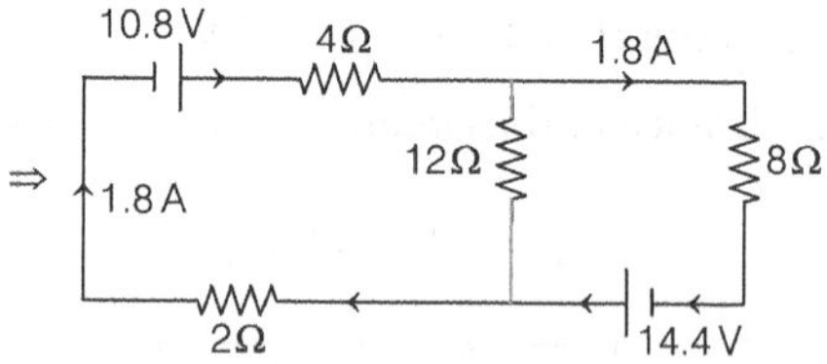

(c)

- The current values in Fig. (a) and (b) are easily verified. For example when the 10.8 V battery alone is acting, the total resistance in the circuit is

$$4 + \frac{12 \times 8}{12 + 8} + 2 = 10.8\,\Omega$$

This makes the total current $\dfrac{10.8\text{ V}}{10.8\,\Omega} = 1\text{ A}.$

This current splits between $8\,\Omega$ and $12\,\Omega$ in the ratio 12 : 8 or 3 : 2. Similarly, the total resistance when only the 14.4 V battery is acting is

$$8 + \frac{12 \times 6}{12 + 6} = 12\,\Omega$$

Therefore, the total current is $\dfrac{14.4\text{ V}}{12\,\Omega} = 1.2\text{ A}.$

The superposition principle shows that there is no current in the $12\,\Omega$ resistance. Only a current of 1.8 A flows through the outer loop. All these conclusions can be verified by analyzing the circuit using Kirchhoff's laws.

Typed Problems

Type 1. *Based on potential difference across the terminals of a battery.*

Concept

Potential difference V across the terminals of a battery is given by

$V = E$ if $i = 0$

$V = 0$ if battery is short-circuited

$V = E - ir$ if normal current ($\leftarrow\!\dashv\vdash\!\overset{i}{\leftarrow}$) flows through the battery and

$V = E + ir$ if current flows in the opposite direction ($\rightarrow\!\dashv\vdash\!\overset{i}{\rightarrow}$)

Ex. 1 Find the potential difference across each of the four batteries B_1, B_2, B_3 and B_4 as shown in the figure.

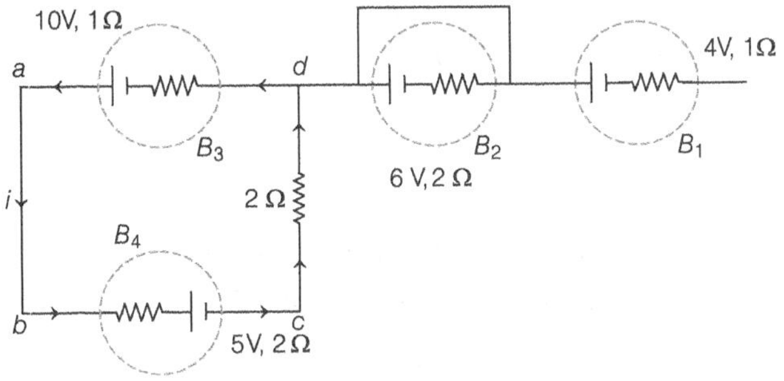

Sol. Across B_1 This battery does not make any closed circuit.

$$\therefore \qquad i = 0$$

or $$V = E = 4\text{ volt} \qquad \textbf{Ans.}$$

Across B_2 This battery is short-circuited. Therefore,

$$V = 0 \qquad \textbf{Ans.}$$

Across B_3 and B_4 A current in anti-clockwise direction flows in the closed loop *abcda*. This current is

$$i = \frac{\text{net emf}}{\text{net resistance}} = \frac{10 - 5}{1 + 2 + 2}$$

$$= 1\text{A}$$

Now, current flows through B_3 in normal direction. Hence,

$$V = E - ir = 10 - 1 \times 1$$

$$= 9 \text{ volt} \qquad \textbf{Ans.}$$

From B_4, current flows in opposite direction. Hence,

$$V = E + ir = 5 + 1 \times 2 = 7 \text{ volt} \qquad \textbf{Ans.}$$

Ex. 2 Draw (a) current *versus* load and (b) current *versus* potential difference (across its two terminals) graph for a cell.

Sol. (a) $i = \dfrac{E}{R + r}$

$$i \to 0 \text{ as } R \to \infty$$

$$i = \frac{E}{r} \text{ at } R = 0$$

i versus R graph is shown in figure given below.

(a)

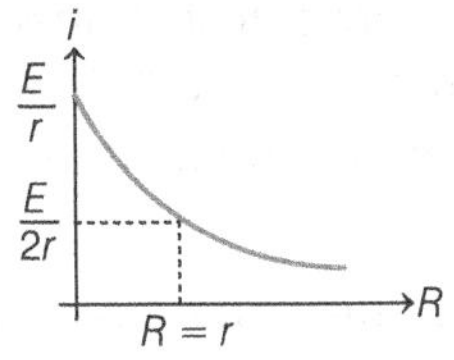

(b) $V = E - ir$

$$V = E \text{ at } i = 0$$

$$V = 0 \text{ at } i = \frac{E}{r} \text{ or short circuit.}$$

Further, *V versus i* graph is straight line.

(b)

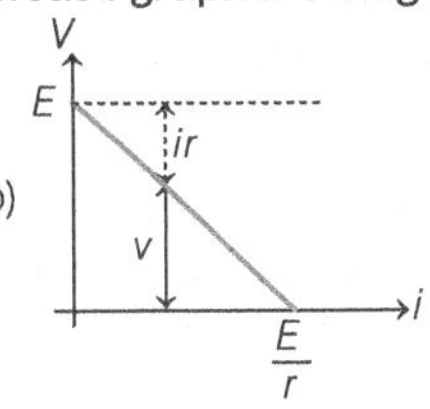

V versus i graph is shown in figure given above.

Type 2. *To find values of V and i across all resistors of a complex circuit if values across one resistance are known.*

Concept

(i) In series, current remains same. But the potential difference distributes in the direct ratio of resistance.

(ii) In parallel, potential difference is same. But the current distributes in the inverse ratio of resistance.

Ex. 3 In the circuit shown in figure potential difference across 6Ω resistance is 4 volt. Find V and i values across each resistance. Also find emf E of the applied battery.

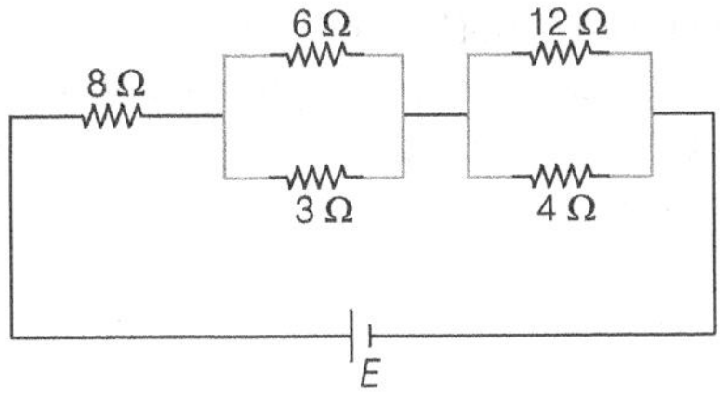

Sol.

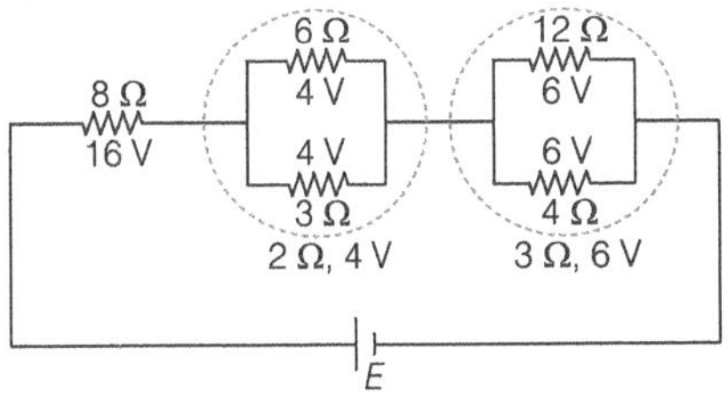

8Ω, 2Ω (resultant of 6Ω and 3Ω) and 3Ω (resultant of 12Ω and 4Ω) are in series. Therefore, potential drop across them should be in direct ratio of resistance. So, using this concept we can find the potential difference across other resistors. For example, potential across 6Ω was 4 V. Therefore potential difference across 3Ω or 2Ω (resultant of 6Ω and 3Ω) will also be 4V. So, potential difference across 8Ω (which is four times of 2Ω) should be 16 V. Similarly, potential difference across 3Ω (which is 1.5 times of 2Ω) should be 1.5 times or 6 V.

Once V and R are known, we can find i across that resistance. For example,

$$i_{12\Omega} = \frac{6}{12} = \frac{1}{2}\text{ A} \qquad \left(\text{as } i = \frac{V}{R}\right)$$

$$i_{8\Omega} = \frac{16}{8} = 2\text{ A etc.}$$

Miscellaneous Examples

Ex. 4 Two sources of current of equal emf are connected in series and have different internal resistances r_1 and r_2 $(r_2 > r_1)$. Find the external resistance R at which the potential difference across the terminals of one of the sources becomes equal to zero.

Sol. $V = E - ir$

E and i for both the sources are equal. Therefore, potential difference (V) will be zero for a source having greater internal resistance, i.e. r_2.

$\therefore \quad 0 = E - ir_2$

or $\quad E = ir_2 = \left(\dfrac{2E}{R + r_1 + r_2}\right) \cdot r_2$

$\therefore \quad 2r_2 = R + r_1 + r_2$

or $\quad R = r_2 - r_1$ **Ans.**

Ex. 5

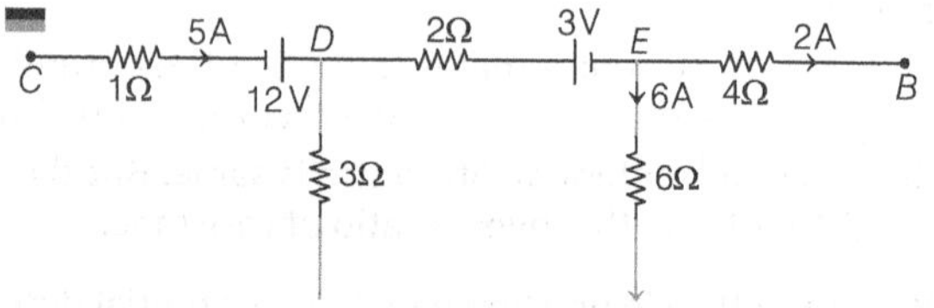

Figure shows the part of a circuit. What is the current through 3Ω resistance and potential difference $V_C - V_B$?

Sol. (i) Applying Kirchhoff's junction law at E current in wire DE is 8A from D to E. Now, further applying junction law at D, the current in $3\,\Omega$ resistance will be 3A towards D.

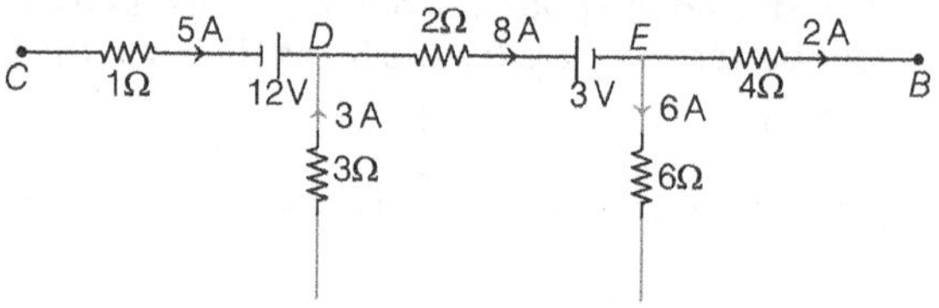

(ii) $V_C - V_B$

$$V_C - 5 \times 1 + 12 - 8 \times 2 - 3 - 4 \times 2 = V_B$$

$\therefore \quad V_C - V_B = 5 - 12 + 16 + 3 + 8$

or $\quad V_C - V_B = 20\text{ V}$ **Ans.**

Ex. 6 The emf of a storage battery is 90 V before charging and 100 V after charging. When charging began the current was 10 A. What is the current at the end of charging if the internal resistance of the storage battery during the whole process of charging may be taken as constant and equal to $2\,\Omega$?

Sol. The voltage supplied by the charging plant is here constant which is equal to,

$$V = E_i + (i_i)(r) = (90) + (10)(2)$$
$$= 110\text{ V}$$

Let i_f be the current at the end of charging.

Then, $\quad V = E_f + i_f r \quad$ or $\quad i_f = \dfrac{V - E_f}{r}$

$$= \frac{110 - 100}{2} = 5\text{ A}$$

MODULE 5 Exercise

Assertion and Reason

Directions : Choose the correct option

*(a) If both **Assertion** and **Reason** are true and the **Reason** is correct explanation of the **Assertion**.*

*(b) If both **Assertion** and **Reason** are true but **Reason** is not the correct explanation of **Assertion**.*

*(c) If **Assertion** is true, but the **Reason** is false.*

*(d) If **Assertion** is false but the **Reason** is true.*

*(e) If both **Assertion** and **Reason** are false.*

1. **Assertion :** In the part of a circuit shown in figure, given that $V_a > V_b$. The current should flow from a to b.

Reason : Direction of current inside a battery is always from negative terminal to positive terminal.

Single Correct Option

2. Given below are two statements.

Statement I : The equivalent resistance of resistors in a series combination is smaller than least resistance used in the combination.

Statement II : The resistivity of the material is independent of temperature. **(2023 Main)**

In the light of above statements, choose the most appropriate answer from the options given below.

(a) Statement I is false but Statement II is true
(b) Statement I is true but Statement II is false
(c) Both Statement I and Statement II are true
(d) Both Statement I and Statement II are false

3. Three resistances are joined together to form a letter Y, as shown in figure. If the potentials of the terminals A, B and C are 6 V, 3 V and 2 V respectively, then the potential of the point O will be

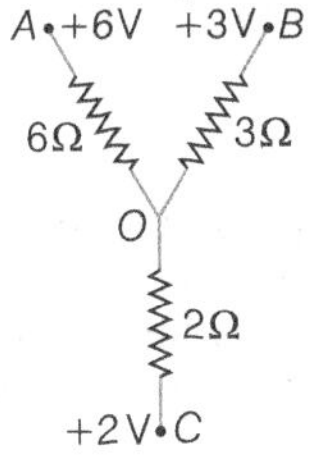

(a) 4 V (b) 3 V (c) 2.5 V (d) 0 V

4. A uniform wire of resistance 4 Ω is bent into a circle of radius r. A specimen of the same wire is connected along the diameter of the circle. What is the equivalent resistance across the ends of this wire?

(a) $\frac{4}{(4+\pi)}\,\Omega$ (b) $\frac{3}{(3+\pi)}\,\Omega$ (c) $\frac{2}{(2+\pi)}\,\Omega$ (d) $\frac{1}{(1+\pi)}\,\Omega$

5. Two batteries one of the emf 3 V, internal resistance 1 Ω and the other of emf 15V, internal resistance 2 Ω are connected in series with a resistance R as shown. If the potential difference between points a and b is zero, the resistance R in Ω is

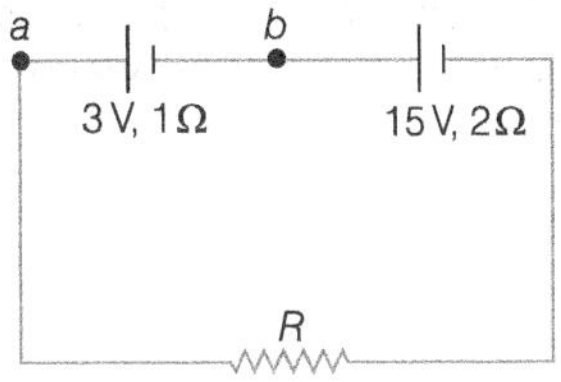

(a) 5 (b) 7
(c) 3 (d) 1

6. The potential difference between points A and B of the following figure is

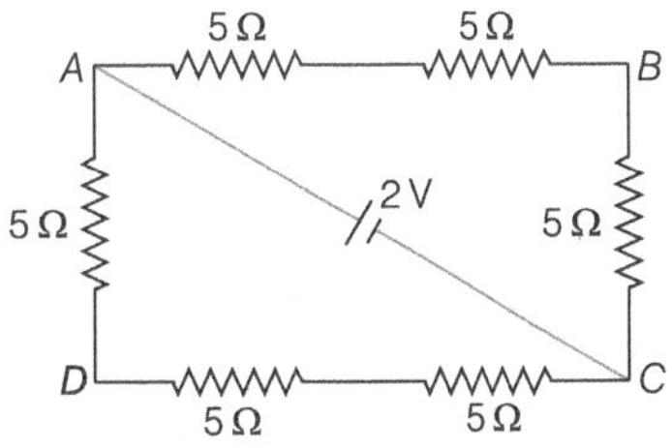

(a) $\frac{2}{3}$ V (b) $\frac{8}{9}$ V
(c) $\frac{4}{3}$ V (d) 2 V

7. In the figure, current through the 3 Ω resistor is 0.8 A, then potential drop through 4 Ω resistor is

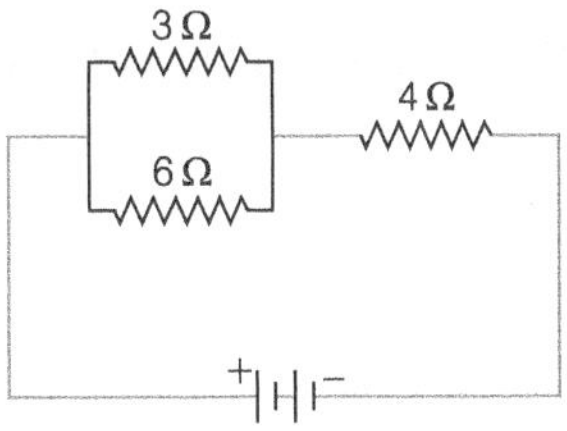

(a) 9.6 V
(b) 2.6 V
(c) 4.8 V
(d) 1.2 V

8. Consider an electric circuit given below. What is the potential difference across the resistor R_4? (Assume that internal resistance of battery is zero.)

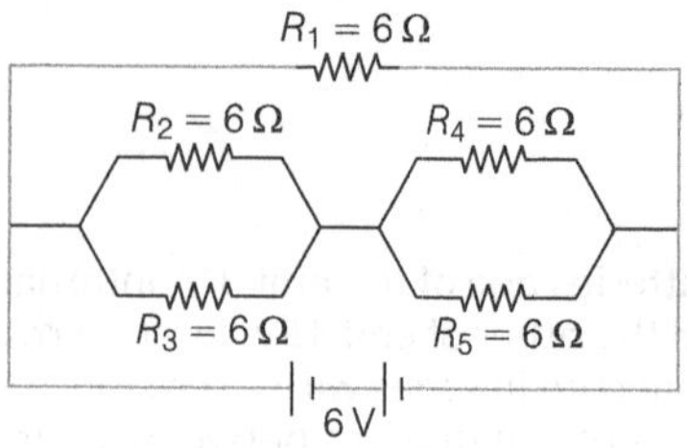

(a) 6 V (b) 4.5 V (c) 3 V (d) 1.5 V

9. In the circuit shown, each resistance is $R = 2\,\Omega$. If a DC source of 10 V is connected between A and B, then the current I is

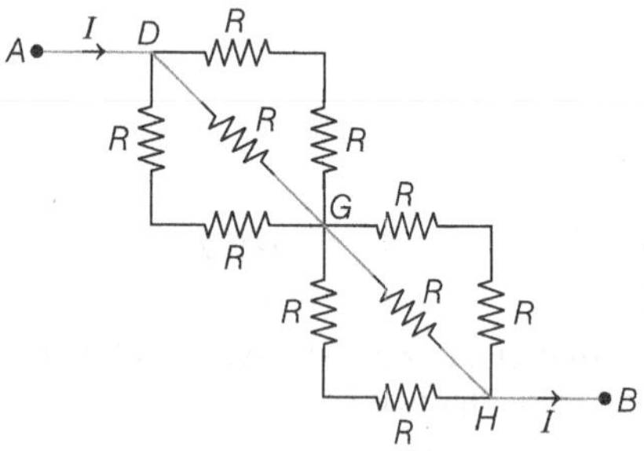

(a) 2.5 A (b) 10 A
(c) 1.25 A (d) 5 A

10. Two batteries are joined as shown. The internal resistance of 6 V battery is $2\,\Omega$ and 4 V battery is $8\,\Omega$. The potential difference between the points X and Y is

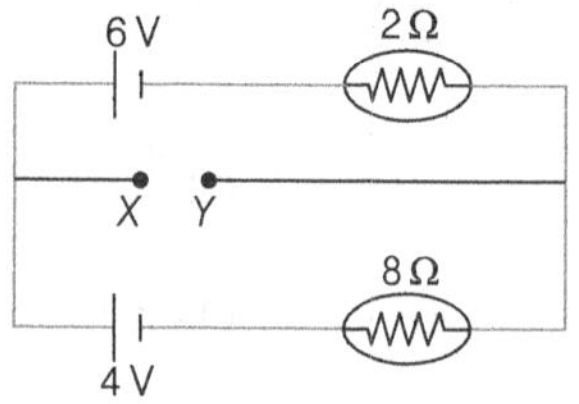

(a) 2 V (b) zero
(c) 5.6 V (d) 3 V

11. The potential difference across $4\,\Omega$ will be

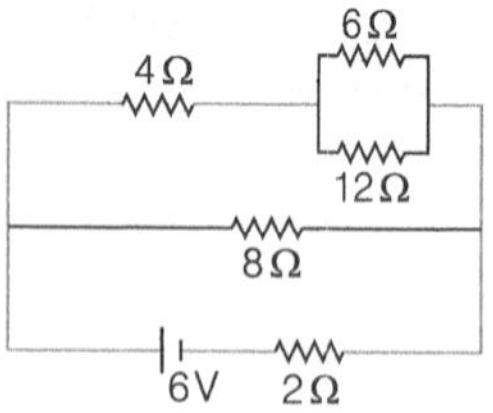

(a) 2 V (b) 4 V
(c) 3 V (d) 6 V

12. A uniform wire of resistance R is shaped into a regular n-sided polygon (n is even). The equivalent resistance between any two corners can have

(i) the maximum value $\dfrac{nR}{(n-1)}$

(ii) the minimum value $\dfrac{R}{n}$

(iii) the minimum value $R\left(\dfrac{n-1}{n^2}\right)$

(iv) the maximum value $\dfrac{R}{n}$

(a) (i) and (ii) (b) (iii) and (iv)
(c) (i) only (d) (i) and (iv)

13. A wire of resistance 9Ω is broken in two parts. The length ratio being 1 : 2. The two pieces are connected in parallel. The net resistance will be

(a) 2Ω (b) 3Ω (c) 4Ω (d) 6Ω

14. The network shown in figure is an arrangement of nine identical resistors. The resistance of the network between points A and B is $1.5\,\Omega$. The resistance r is

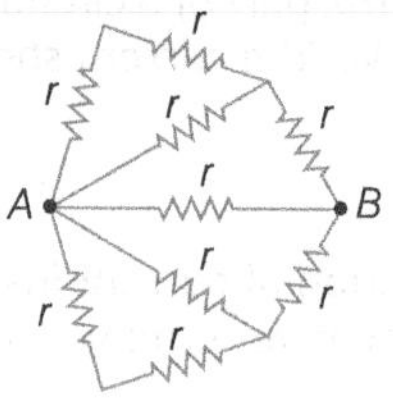

(a) 1.1Ω (b) 3.3Ω (c) 1.8Ω (d) 1.6Ω

15. A uniform wire of resistance $18\,\Omega$ is bent in the form of a circle. The effective resistance across the points a and b is

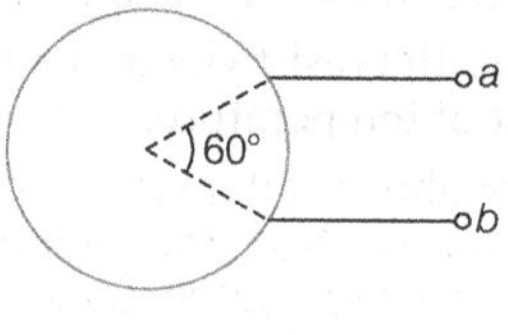

(a) 3Ω (b) 2Ω (c) 2.5Ω (d) 6Ω

16. Two wires A and B made of same material and having their lengths in the ratio 6 : 1 are connected in series. The potential difference across the wires are 3 V and 2 V respectively. If r_A and r_B are the radii of A and B respectively, then $\dfrac{r_B}{r_A}$ is

(a) $\dfrac{1}{4}$ (b) $\dfrac{1}{2}$ (c) 1 (d) 2

17. For a cell, the graph between the potential difference (V) across the terminals of the cell and the current (I) drawn from the cell is shown in the figure. The emf and the internal resistance of the cell are

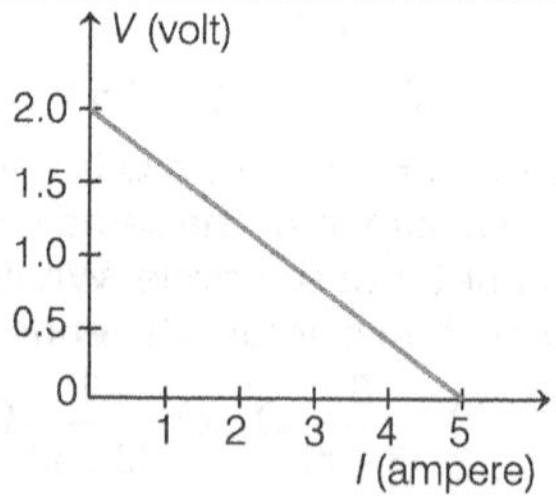

(a) 2 V, $0.5\,\Omega$ (b) 2 V, $0.4\,\Omega$
(c) > 2 V, $0.5\,\Omega$ (d) > 2 V, $0.4\,\Omega$

18. A wire of resistance R and length L is cut into 5 equal parts. If these parts are joined parallelly, then resultant resistance will be **(2024 Main)**

(a) $25R$ (b) $\frac{1}{25}R$ (c) $5R$ (d) $\frac{1}{5}R$

19. What will be the most suitable combination of three resistors $A = 2\Omega$, $B = 4\Omega$, $C = 6\Omega$, so that $\left(\frac{22}{3}\right)\Omega$ is equivalent resistance of combination? **(2022 Main)**

(a) Parallel combination of A and C connected in series with B.
(b) Parallel combination of A and B connected in series with C.
(c) Series combination of A and C connected in parallel with B.
(d) Series combination of B and C connected in series with A.

20. Six equal resistances are connected between points P, Q and R as shown in the figure. Then, the net resistance will be maximum between **(2004 JEE)**

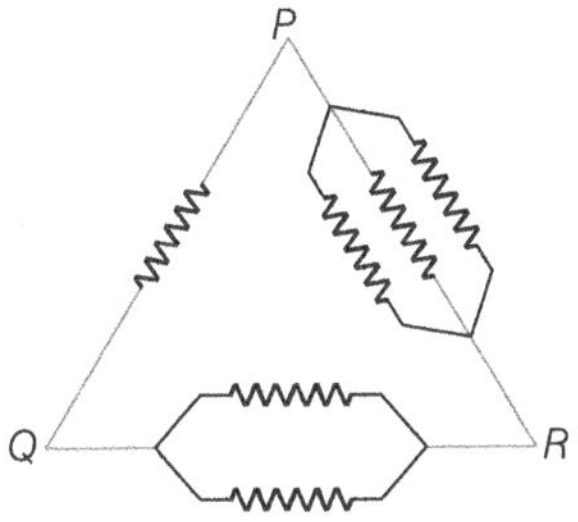

(a) P and Q (b) Q and R
(c) P and R (d) any two points

21. Figure shows a part of an electric circuit. The potentials at points a, b and c are 30 V, 12 V and 2 V respectively. The current through the 20 Ω resistor will be **(2023 Main)**

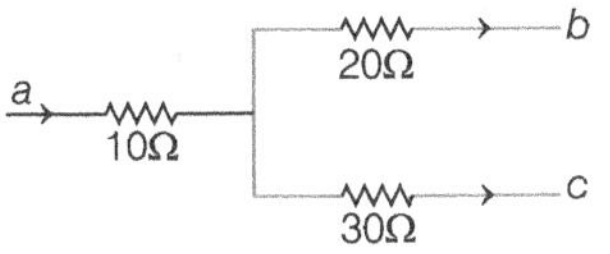

(a) 1.0 A (b) 0.4 A (c) 0.2 A (d) 0.6 A

22. When the switch S in the circuit shown is closed, then the value of current i will be **(2019 Main)**

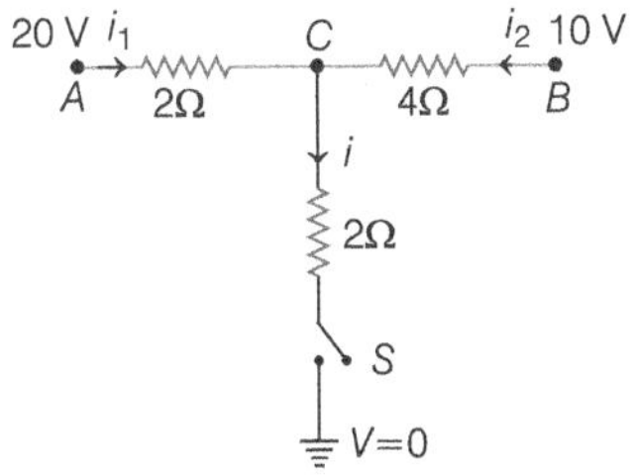

(a) 4A (b) 3A (c) 2A (d) 5A

23. The current i in the circuit (see figure) is **(1983 JEE)**

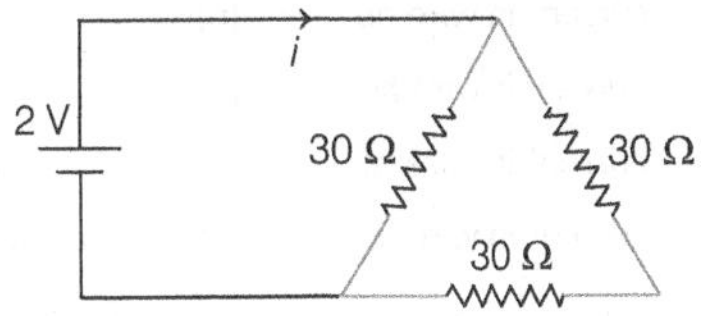

(a) $\frac{1}{45}$ A (b) $\frac{1}{15}$ A (c) $\frac{1}{10}$ A (d) $\frac{1}{5}$ A

Multiple Correct Options

24. Two cells of emf $E_1 = 6$ V and $E_2 = 5$ V are joined in parallel with same polarity on same side, without any external load. If their internal resistances are $r_1 = 2\Omega$ and $r_2 = 3\,\Omega$ respectively, then

(a) terminal potential difference across any cell is less than 5 V
(b) terminal potential difference across any cell is 5.6 V
(c) current through the cells is 0.2 A
(d) current through the cells is zero if $E_1 = E_2$

25. In the figure shown,

(a) current will flow from A to B
(b) current may flow A to B
(c) current may flow from B to A
(d) the direction of current will depend on E

26. In the circuit shown in figure it is given that $V_b - V_a = 2$ volt. Choose the correct options.

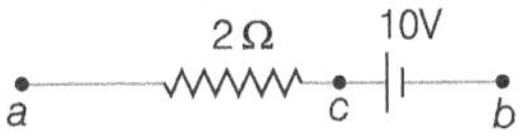

(a) Current in the wire is 6 A
(b) Direction of current is from a to b
(c) $V_a - V_c = 12$ volt
(d) $V_c - V_a = 12$ volt

Match the Columns

27. For the circuit shown in figure, match the two columns.

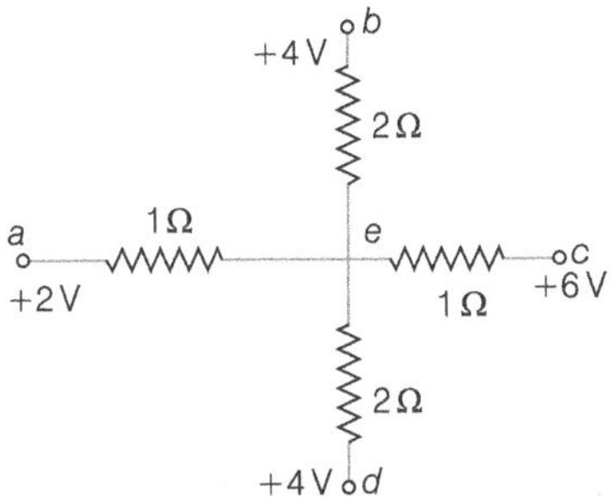

	Column I		Column II
(a)	current in wire *ae*	(p)	1 A
(b)	current is wire *be*	(q)	2 A
(c)	current in wire *ce*	(r)	0.5 A
(d)	current in wire *de*	(s)	None of these

28. In the circuit shown in figure, after closing the switch *S*, match the following two columns.

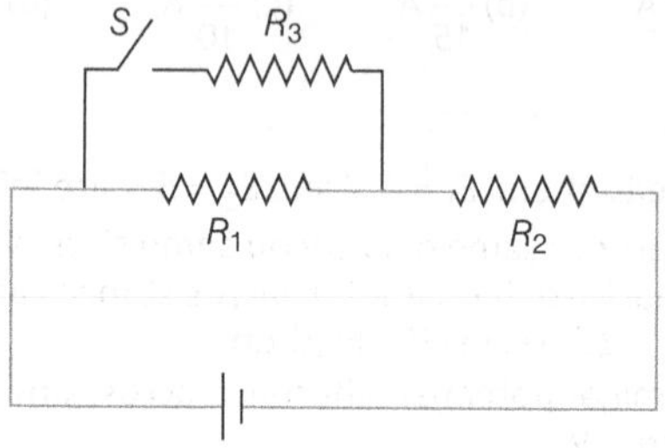

	Column I		Column II
(a)	current through R_1	(p)	will increase
(b)	current through R_2	(q)	will decrease
(c)	potential difference across R_1	(r)	will remain same
(d)	potential difference across R_2	(s)	data insufficient

Subjective Question

29. The potential difference across the terminals of a battery is 8.4 V when there is a current of 1.50 A in the battery from the negative to the positive terminal. When the current is 3.50 A in the reverse direction, the potential difference becomes 9.4 V.
(a) What is the internal resistance of the battery?
(b) What is the emf of the battery?

Integer Type Questions

30. A set of *n* equal resistors of resistance *R* each are connected in series to a battery of emf *E* and internal resistance *R*. A current *I* is observed to flow in the circuit. Then, *n*-resistors are connected in parallel to the same battery. It is observed that current is increased 10 times. What is the value of *n*?

31. A network of four resistances is connected to 9 V battery, as shown in figure. The magnitude of voltage difference between the points *A* and *B* is .... V. **(2023 Main)**

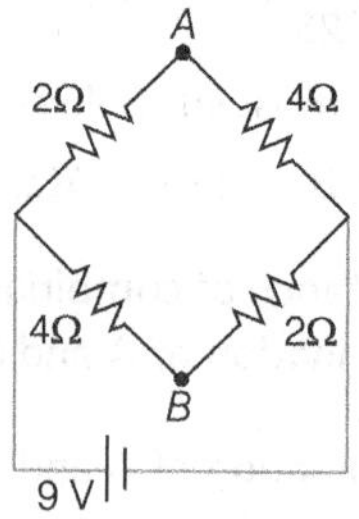

32. For the network shown below, the value $V_B - V_A$ is .... V. **(2022 Main)**

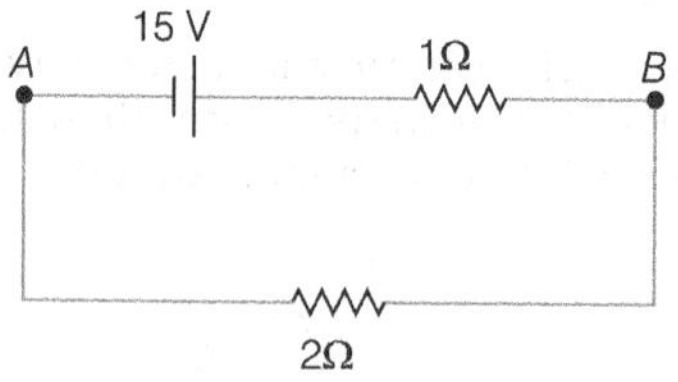

Answers

1. (c) **2.** (d) **3.** (b) **4.** (a) **5.** (c)
6. (c) **7.** (c) **8.** (c) **9.** (d) **10.** (c)
11. (a) **12.** (b) **13.** (a) **14.** (b) **15.** (c)
16. (b) **17.** (b) **18.** (b) **19.** (b) **20.** (a)
21. (b) **22.** (d) **23.** (c) **24.** (b,c,d) **25.** (b,c,d)
26. (a,d)
27. (a) → q, (b) → s, (c) → q, (d) → s
28. (a) → q, (b) → p, (c) → q, (d) → p
29. (a) 0.20 Ω (b) 8.7 V **30.** (10)
31. (3) **32.** (10)

See hints on pages 78-81.

MODULE 6

Heating Effects of Current

An electric current through a resistor increases its thermal energy. Also, there are other situations in which an electric current can produce or absorb thermal energy.

Power Supplied or Power Absorbed by a Battery

When charges are transported across a source of emf, their potential energy changes. If a net charge Δq moves through a potential difference E in a time Δt, the change in electric potential energy of the charge is $E\,\Delta q$. Thus, the source of emf does work,

$$\Delta W = E\,\Delta q$$

Dividing both sides by Δt, then taking the limit as $\Delta t \to 0$, we find

$$\frac{dW}{dt} = E\frac{dq}{dt}$$

By definition, $\dfrac{dq}{dt} = i$, the current through the battery and $\dfrac{dW}{dt} = P$, the power output of (or input to) the battery. Hence,

$$\boxed{P = Ei}$$

The quantity P represents the rate at which energy is transferred from a discharging battery or to a charging battery.

In the figure shown, energy is transferred (or supplied) from the source at a rate Ei

In the figure shown, energy is absorbed by the source at a rate Ei

Power Dissipated Across a Resistance

Now, let's consider the power dissipated in a conducting element. Suppose it has a resistance R and the potential difference between its ends is V. In moving from higher to lower potential, a positive charge Δq loses energy $\Delta U = V\,\Delta q$. This electric energy is absorbed by the conductor through collisions between its atomic lattice and the charge carriers, causing its temperature to rise. This effect is commonly called **Joule heating.** Since, power is the rate at which energy is transferred, we have,

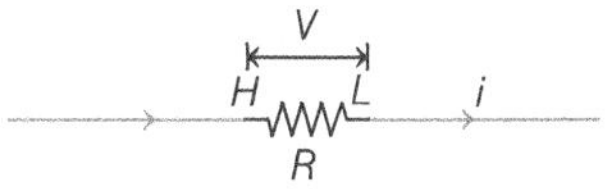

$$P = \frac{\Delta U}{\Delta t} = V\cdot\frac{\Delta q}{\Delta t} = Vi$$

$\therefore$ $$\boxed{P = Vi}$$

which with the help of equation $V = iR$ can also be written in the forms,

$$\boxed{P = i^2R} \quad \text{or} \quad \boxed{P = \frac{V^2}{R}}$$

Power is always dissipated in a resistance. With this rate, the heat produced in the resistor in time t is

$$H = Pt \quad \text{or} \quad \boxed{H = Vit = i^2Rt = \frac{V^2}{R}t}$$

Joule heating occurs whenever a current passes through an element that has resistance. To prevent the overheating of delicate electronic components, many electric devices like video cassette recorders, televisions and computer monitors have fans in their chassis to allow some of the heat produced to escape.

Important POINTS

- We have seen above that power may be supplied or consumed by a battery. It depends upon the direction of current.

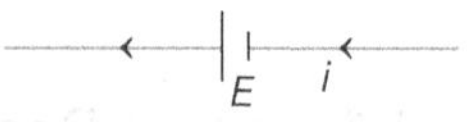

In the above direction of current, power is supplied by the battery (=Ei)

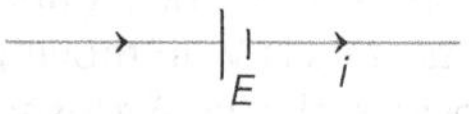

In the opposite direction of current as shown in figure, power is consumed by the battery. This normally happens during charging of a battery.

- A resistance always consumes power. It does not depend on the direction of current.

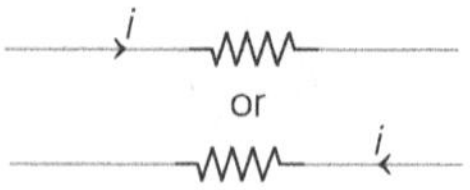

In both cases shown in figure, power is only consumed and this power consumed is given by the formula.

$$P = i^2R = \frac{V^2}{R} = Vi$$

In the above equations V and i are the values across a resistance in which we wish to find the power consumed.

- In any electrical circuit, law of conservation of energy is followed.

Net power supplied by all batteries of the circuit = net power consumed by all resistors in the circuit.

- In series current i is same. So, from the relation

$$P = i^2R$$

We have,

$$p \propto R$$

$\therefore$ $$P_1 : P_2 : P_3 = R_1 : R_2 : R_3$$

- In parallel potential difference V is same. So, from the relation,

$$P = \frac{V^2}{R}$$

We have, $P \propto \frac{1}{R}$

$\therefore$ $$\frac{P_1}{P_2} = \frac{R_2}{R_1}$$

or, $$p_1 : P_2 : P_3 = \frac{1}{R_1} : \frac{1}{R_2} : \frac{1}{R_3}$$

Example 1 In the circuit shown in figure, find

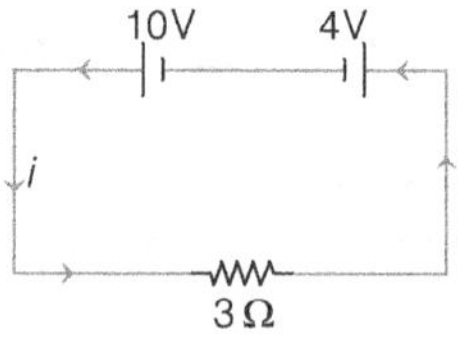

(a) the power supplied by 10 V battery
(b) the power consumed by 4 V battery and
(c) the power dissipated in 3 Ω resistance.

Sol. Net emf of the circuit = (10 − 4) V = 6 V

$\therefore$ Current in the circuit

$$i = \frac{\text{net emf}}{\text{total resistance}} = \frac{6}{3} = 2\text{ A}$$

(a) Power supplied by 10 V battery = Ei = (10) (2) = 20 W **Ans.**

(b) Power consumed by 4 V battery = Ei = (4) (2) = 8 W **Ans.**

(c) Power consumed by (or dissipated in) 3 Ω resistance

$$= i^2R = (2)^2(3)$$

$$= 12\text{ W}$$ **Ans.**

Note *Here, we can see that total power supplied by 10 V battery (i.e. 20 W) = power consumed by 4 V battery and 3 Ω resistance. Which proves that conservation of energy holds good in electric circuits also.*

Example 2 In the circuit shown in figure, find the heat developed across each resistance in 2 s.

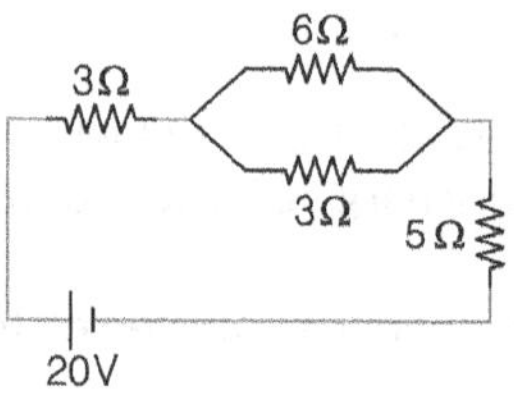

Sol. The 6 Ω and 3 Ω resistances are in parallel. So, their combined resistance is

$$\frac{1}{R} = \frac{1}{6} + \frac{1}{3} = \frac{1}{2} \quad \text{or} \quad R = 2\,\Omega$$

The equivalent simple circuit can be drawn as shown below.

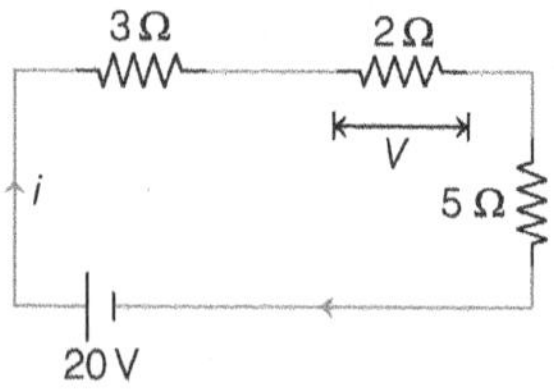

Current in the circuit,

$$i = \frac{\text{net emf}}{\text{total resistance}} = \frac{20}{3+2+5} = 2\text{ A}$$

$$V = iR = (2)(2) = 4\text{ V}$$

i.e. Potential difference across $6\,\Omega$ and $3\,\Omega$ resistances are 4 V.

Now, $H_{3\Omega}$ (which is connected in series)

$$= i^2Rt = (2)^2\,(3)\,(2) = 24\ \text{J}$$

$$H_{6\Omega} = \frac{V^2}{R}t = \frac{(4)^2}{6}(2) = \frac{16}{3}\ \text{J}$$

$H_{3\Omega}$ (which is connected in parallel)

$$= \frac{V^2}{R}t = \frac{(4)^2\,(2)}{3} = \frac{32}{3}\ \text{J}$$

and $\quad H_{5\Omega} = i^2Rt = (2)^2(5)\,(2)$

$$= 40\ \text{J}$$ **Ans.**

INTRODUCTORY Exercise

1. In the circuit shown in figure, a 12 V battery with unknown internal resistance r is connected to another battery with unknown emf E and internal resistance $1\,\Omega$ and to a resistance of $3\,\Omega$ carrying a current of 2 A. The current through the rechargeable battery is 1 A in the direction shown. Find the unknown current i, internal resistance r and the emf E.

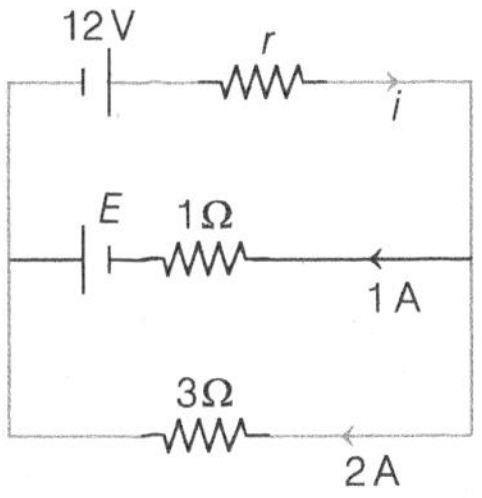

2. In the above example, find the power delivered by the 12 V battery and the power dissipated in $3\,\Omega$ resistor.

3. Ratio of thermal energy released in two resistors R and $3R$ connected in parallel in an electric circuit is **(2023 Main)**
   (a) 1 : 27 (b) 3 : 1
   (c) 1 : 1 (d) 1 : 3

4. Different combination of 3 resistors of equal resistance R are shown in the figures. The increasing order for power dissipation is **(2023 Main)**

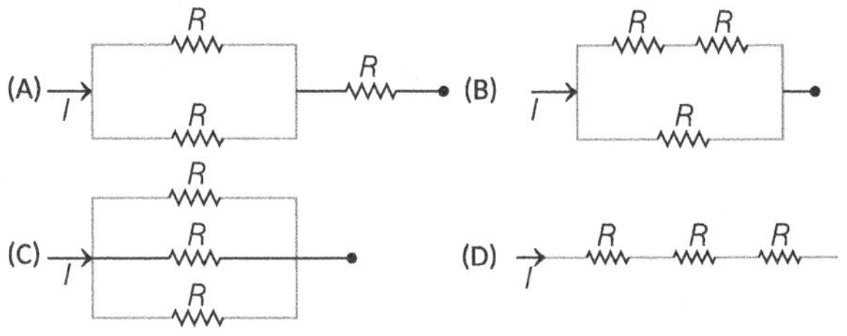

   (a) $P_A < P_B < P_C < P_D$ (b) $P_C < P_D < P_A < P_B$
   (c) $P_C < P_B < P_A < P_D$ (d) $P_B < P_C < P_D < P_A$

5. A rigid container with thermally insulated walls contains a coil of resistance $100\,\Omega$, carrying current 1 A. Change in internal energy after 5 min will be **(2005 JEE)**
   (a) zero (b) 10 kJ
   (c) 20 kJ (d) 30 kJ

Answers

1. 3A, $2\,\Omega$, – 5V 2. 36 W, 12 W
3. (b) 4. (c)
5. (d)

See hints on page 81.

Typed Problems

Type 1. *Theory of bulbs or heater etc.*

Concept

(i) From the rated (written) values of power (P) and potential difference (V) we can determine resistance of filament of bulb.

$$P = \frac{V^2}{R} \Rightarrow \boxed{R = \frac{V^2}{P}} \qquad \ldots\text{(i)}$$

For example, if rated values on a bulb are 220 V and 60 W, it means this bulb will consume 60 W of power (or 60 J in 1 s) if a potential difference of 220 V is applied across it. Resistance of this bulb will be

$$R = \frac{(220)^2}{60} = 806.67\,\Omega$$

(ii) Normally, rated value of V remains same in different bulbs. In India, it is 220 volt. Therefore, from Eq. (i)

$$R \propto \frac{1}{P} \quad \text{or} \quad R_{60\,\text{watt}} > R_{100\,\text{watt}}$$

(iii) Actual value of potential difference may be different from the rated value. Therefore, actual power consumption may also be different.

(iv) After finding resistance of the bulb using Eq. (i), we can apply normal Kirchhoff's laws for finding current passing through the bulb or actual power consumed by the bulb.

Ex. 1 Prove that 60W bulb glows more brightly than 100W bulb if by mistake they are connected in series.

Sol. In series, we can use the formula

$$P = i^2R \quad \text{for the power consumption}$$

$$\Rightarrow \quad P \propto R \quad (\text{as } i \text{ is same in series})$$

we have seen above that,

$$R_{60\,\text{W}} > R_{100\,\text{W}}$$

$$\therefore \quad P_{60\,\text{W}} > P_{100\,\text{W}}$$ **Hence Proved.**

Note *In parallel 100 W bulb glows more brightly than 60 W bulb. Think why?*

Ex. 2 The rated values of two bulbs are (P_1, V) and (P_2, V). Find actual power consumed by both of them if they are connected in

(a) series (b) parallel

and V potential difference is applied across both of them.

Sol. (a) $\because \; R_1 = \dfrac{V^2}{P_1}$ and $R_2 = \dfrac{V^2}{P_2}$

In series, $$P = \frac{V^2}{R_{net}} = \frac{V^2}{R_1 + R_2} = \frac{V^2}{\frac{V^2}{P_1} + \frac{V^2}{P_2}} = \frac{P_1 P_2}{P_1 + P_2}$$

or $$\frac{1}{P} = \frac{1}{P_1} + \frac{1}{P_2}$$ **Ans.**

(b) In parallel,

$$P = \frac{V^2}{R_{net}} = V^2\left(\frac{1}{R_{net}}\right) = V^2\left(\frac{1}{R_1} + \frac{1}{R_2}\right)$$

or $$P = V^2\left(\frac{P_1}{V^2} + \frac{P_2}{V^2}\right)$$

or $$P = P_1 + P_2$$ **Ans.**

Ex. 3 Heater-1, takes 3 minutes to boil a given amount of water. Heater-2 takes 6-minutes. Find the time taken if,

(a) they are connected in series

(b) they are connected in parallel.

Potential difference V in all cases is same.

Sol. Here, the heat required (say H) to boil water is same. Let P_1 and P_2 are the powers of the heaters. Then,

$$H = P_1 t_1 = P_2 t_2$$

$$\therefore \quad P_1 = \frac{H}{t_1} = \frac{H}{3}$$

$$P_2 = \frac{H}{t_2} = \frac{H}{6}$$

(a) **In series,**

$$P = \frac{P_1 P_2}{P_1 + P_2} = \frac{(H/3)(H/6)}{(H/3) + (H/6)} \quad \text{(Refer Example 2)}$$

$$= \frac{H}{9}$$

Now, $$t = \frac{H}{P} = \frac{H}{(H/9)} = 9 \text{ min}$$ **Ans.**

(b) **In parallel,** $$P = P_1 + P_2 = \frac{H}{3} + \frac{H}{6} = \frac{H}{2}$$

$$\therefore \quad t = \frac{H}{P} = \frac{H}{H/2} = 2 \text{ min}$$ **Ans.**

Ex. 4 A 100 W bulb B_1, and two 60 W bulbs B_2 and B_3, are connected to a 250 V source as shown in the figure. Now W_1, W_2 and W_3 are the output powers of the bulbs B_1, B_2 and B_3 respectively. Then, **(2002 JEE)**

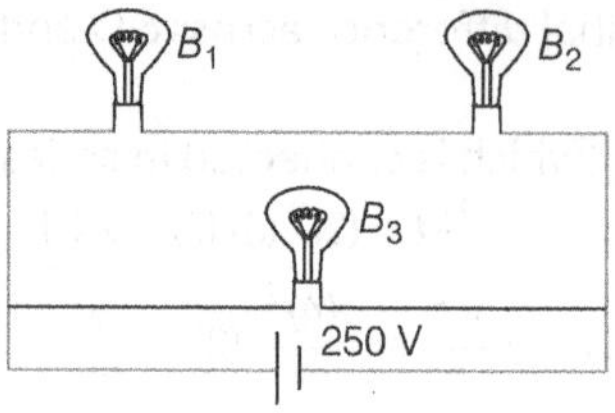

(a) $W_1 > W_2 = W_3$ (b) $W_1 > W_2 > W_3$

(c) $W_1 < W_2 = W_3$ (d) $W_1 < W_2 < W_3$

Sol. B_1 and B_3 are in series.

$$\therefore \quad P_{60\,W} > P_{100\,W}$$

or $$W_2 > W_1$$

B_2 and B_3 both are 60 W bulbs. B_3 is getting 250 V and B_2 is getting less than 250 V. Therefore, $W_3 > W_2$.

or, $W_1 < W_2 < W_3$.

$\therefore$ The correct option is (d).

Ex. 5 An electric bulb rated for 500 W at 100 V is used in a circuit having a 200 V supply. The resistance R that must be put in series with the bulb, so that the bulb delivers 500 W is .....Ω. **(1987 JEE)**

Sol. Resistance of the given bulb

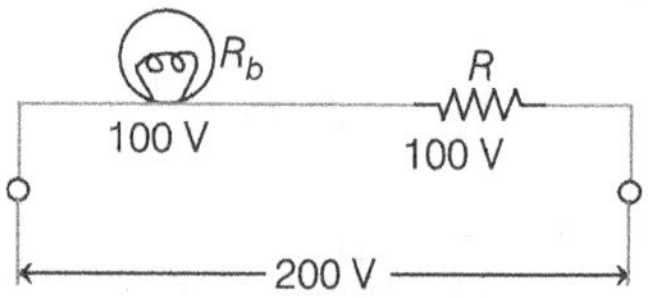

$$R_b = \frac{V^2}{P} = \frac{(100)^2}{500} = 20\Omega$$

For 500 W power we need 100 V across the bulb.

$\therefore$ To get 100 V out of 200 V across the bulb,

$$R = R_b = 20\Omega$$ **Ans.**

Ex. 6 A heater is designed to operate with a power of 1000 W in a 100 V line. It is connected in combination with a resistance of 10 Ω and a resistance R, to a 100 V mains as shown in the figure. What will be the value of R so that the heater operates with a power of 62.5 W? **(1978 JEE)**

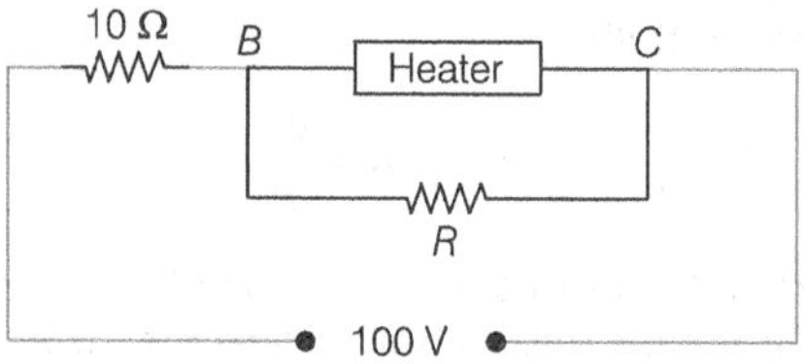

Sol. From $P = \dfrac{V^2}{R}$,

Resistance of heater, $$R = \frac{V^2}{P} = \frac{(100)^2}{1000} = 10\,\Omega$$

From $$P = i^2 R$$

Current required across heater for power of 62.5 W,

$$i = \sqrt{\frac{P}{R}} = \sqrt{\frac{62.5}{10}} = 2.5 \text{ A}$$

Main current in the circuit, $I = \dfrac{100}{10 + \dfrac{10R}{10 + R}}$

$$= \frac{100\,(10+R)}{100 + 20R} = \frac{10\,(10+R)}{10+2R}$$

This current will distribute in inverse ratio of resistance between heater and R.

$\therefore \quad i = \left(\dfrac{R}{10+R}\right) I$

or $\quad 2.5 = \left(\dfrac{R}{10+R}\right)\left[\dfrac{10\,(10+R)}{10+2R}\right] = \dfrac{10R}{10+2R}$

Solving this equation, we get

$R = 5\,\Omega$ **Ans.**

Miscellaneous Examples

Ex. 7 In which branch of the circuit shown in figure a 11 V battery be inserted so that it dissipates minimum power. What will be the current through the $2\,\Omega$ resistance for this position of the battery?

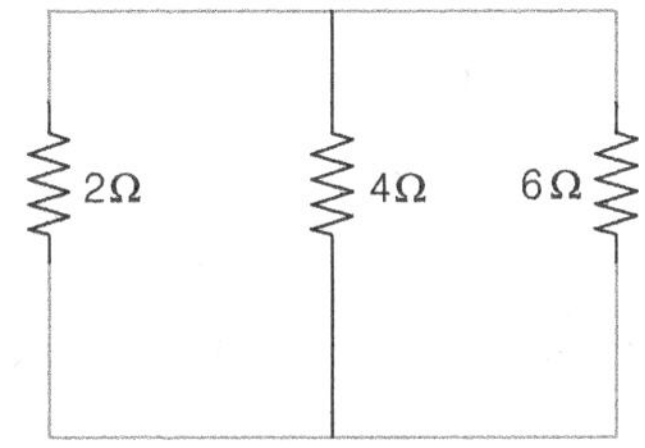

Sol. Suppose, we insert the battery with $2\,\Omega$ resistance. Then, we can take $2\,\Omega$ as the internal resistance (r) of the battery and combined resistance of the other two as the external resistance (R). The circuit in that case is shown in figure,

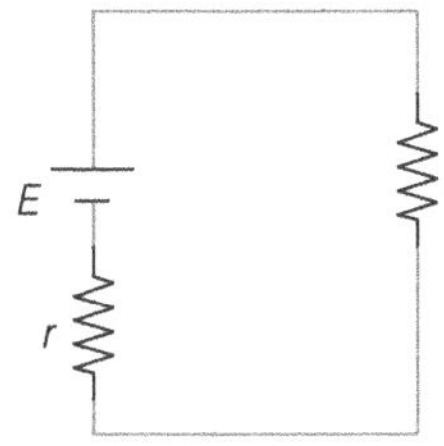

Now power, $P = Ei = E\left(\dfrac{E}{R+r}\right) = \dfrac{E^2}{R+r}$

This power will be minimum where $R + r$ is maximum and we can see that $(R + r)$ will be maximum when the battery is inserted with $6\,\Omega$ resistance as shown in figure.

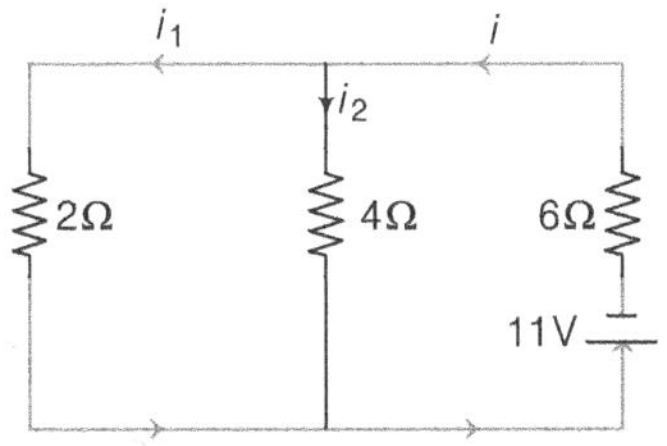

Net resistance in this case is

$$6 + \frac{2 \times 4}{2+4} = \frac{22}{3}\,\Omega$$

$\therefore \quad i = \dfrac{11}{22/3} = 1.5 \text{ A}$

This current will be distributed in $2\,\Omega$ and $4\,\Omega$ in the inverse ratio of their resistances.

$\therefore \quad \dfrac{i_1}{i_2} = \dfrac{4}{2} = 2$

$\therefore \quad i_1 = \left(\dfrac{2}{2+1}\right)(1.5) = 1.0 \text{ A}$ **Ans.**

Ex. 8 What amount of heat will be generated in a coil of resistance R due to a charge q passing through it if the current in the coil **(Irodov 3.189)**

(a) decreases down to zero uniformly during a time interval t_0?

(b) decreases down to zero exponentially halving its value every t_0 seconds?

HOW TO PROCEED Heat generated in a resistance is given by

$$H = i^2Rt$$

We can directly use this formula if i is constant. Here, i is varying. So, first we will calculate i at any time t, then find a small heat dH in a short interval of time dt. Then by integrating it with proper limits we can obtain the total heat produced.

Sol. (a) The corresponding i-t graph will be a straight line with i decreasing from a peak value (say i_0) to zero in time t_0.

i-t equation will be as

$$i = i_0 - \left(\frac{i_0}{t_0}\right)t \qquad (y = -mx + c) \quad \text{...(i)}$$

Here, i_0 is unknown, which can be obtained by using the fact that area under i-t graph gives the flow of charge. Hence,

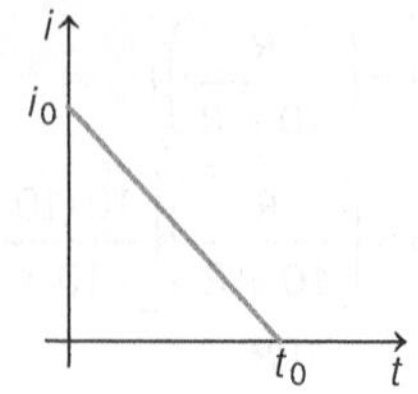

$$q = \frac{1}{2}(t_0)(i_0)$$

$$\therefore \quad i_0 = \frac{2q}{t_0}$$

Substituting in Eq. (i), we get

$$i = \frac{2q}{t_0}\left(1 - \frac{t}{t_0}\right)$$

or

$$i = \left(\frac{2q}{t_0} - \frac{2qt}{t_0^2}\right)$$

Now, at time t, heat produced in a short interval dt is

$$dH = i^2 R\,dt$$

$$= \left(\frac{2q}{t_0} - \frac{2qt}{t_0^2}\right)^2 R dt$$

$\therefore$ Total heat produced $= \int_0^{t_0} dH$

or

$$H = \int_0^{t_0}\left(\frac{2q}{t_0} - \frac{2qt}{t_0^2}\right)^2 R\,dt$$

$$= \frac{4}{3}\frac{q^2 R}{t_0}$$ **Ans.**

(b) Here, current decreases from some peak value (say i_0) to zero exponentially with half life t_0.

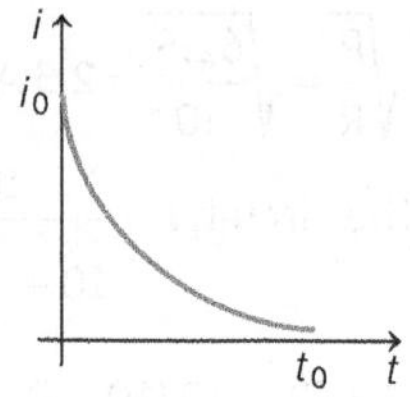

i-t equation in this case will be

$$i = i_0 e^{-\lambda t}$$

Here, $\lambda = \frac{\ln(2)}{t_0}$

Now, $q = \int_0^{\infty} i\,dt = \int_0^{\infty} i_0\,e^{-\lambda t}\,dt$

$$= \left(\frac{i_0}{\lambda}\right)$$

$$\therefore \quad i_0 = \lambda q$$

$$\therefore \quad i = (\lambda q)\,e^{-\lambda t}$$

$$\therefore \quad dH = i^2 R\,dt = \lambda^2 q^2 e^{-2\lambda t} R\,dt$$

or $H = \int_0^{\infty} dH = \lambda^2 q^2 R \int_0^{\infty} e^{-2\lambda t}\,dt$

$$= \frac{q^2 \lambda R}{2}$$

Substituting $\lambda = \frac{\ln(2)}{t_0}$, we have

$$H = \frac{q^2 R \ln(2)}{2t_0}$$ **Ans.**

Note *In radioactivity, half-life is given by*

$$t_{1/2} = \frac{\ln 2}{\lambda} \quad \therefore \quad \lambda = \frac{\ln 2}{t_{\frac{1}{2}}}$$

MODULE 6 Exercise

Assertion and Reason

See directions on page 25.

1. **Assertion :** In the part of the circuit shown in figure, maximum power is produced across R.

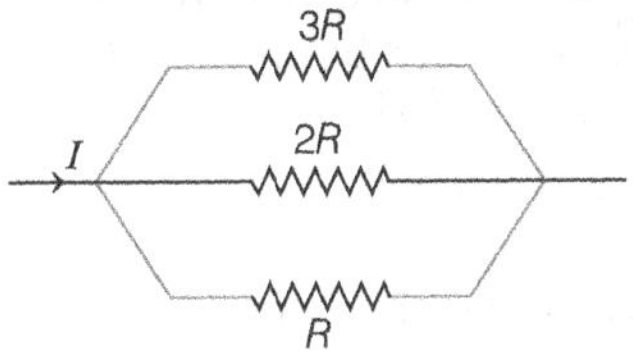

Reason : In parallel circuit power is dissipated in inverse ratio of resistance.

2. **Assertion :** If variation in resistance due to temperature is taken into consideration, then current in the circuit I and power produced across the resistance P both will decrease with time.

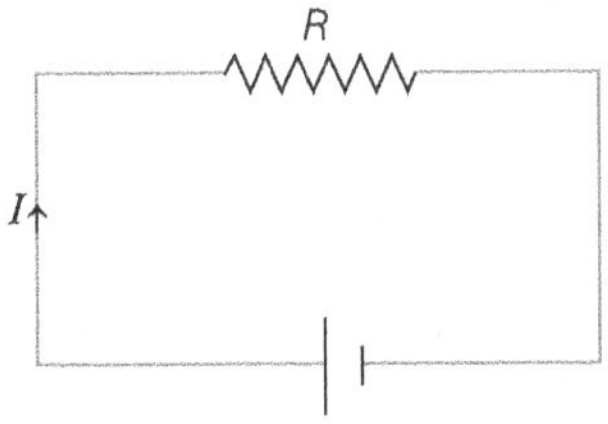

Reason : $V = IR$ is Ohm's law.

3. **Assertion :** In the circuit shown in figure, battery is ideal. If a resistance R_0 is connected in parallel with R, power across R will increase.

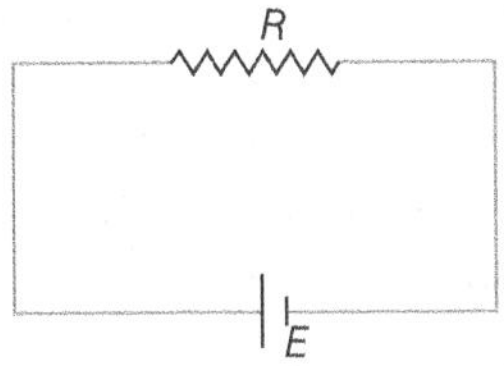

Reason : Current drawn from the battery will increase.

4. **Assertion :** Two resistance wires shown in figure are of same material. They have equal length. More heat is generated in wire A.

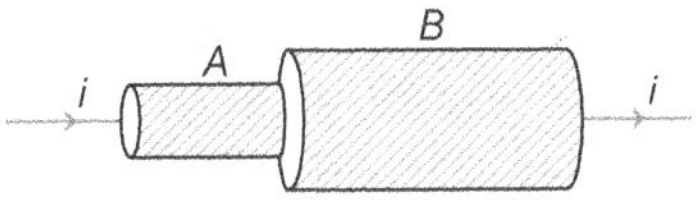

Reason : In series $H \propto R$ and resistance of wire A is more.

5. **Assertion :** Kilovolt-ampere (kVA) and kilowatt-hour have the same dimensions.

Reason : Both are the units of energy.

Single Correct Option

6. If the length of the filament of a heater is reduced by 10%, the power of the heater will
(a) increase by about 9% (b) increase by about 11%
(c) increase by about 19% (d) decrease by about 10%

7. A resistor R has power of dissipation P with cell voltage E. The resistor is cut in n equal parts and all parts are connected in parallel with same cell. The new power dissipation is
(a) nP (b) nP^2 (c) n^2P (d) n/P

8. Some electric bulbs are connected in series across a 220 V supply in a room. If one bulb is fused then remaining bulbs are connected again in series (after removing the fused bulb) across the same supply. The illumination in the room will
(a) increase (b) decrease
(c) remain the same (d) not continuous

9. For what ratio of R_1, R_2 and R_3 power developed across each resistor is equal?

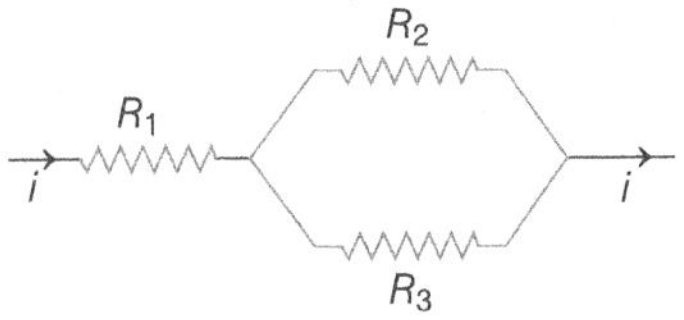

(a) 1 : 1 : 1 (b) 4 : 4 : 1
(c) 4 : 1 : 1 (d) 1 : 4 : 4

10. Ratio of power developed across R, $2R$ and $6R$ will be

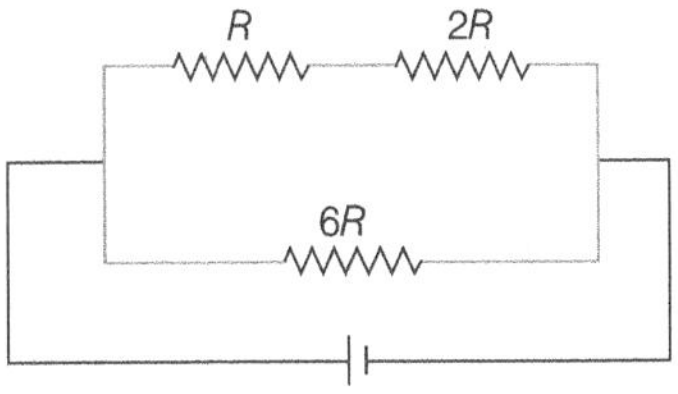

(a) 2 : 4 : 3 (b) 1 : 2 : 6
(c) 2 : 4 : 6 (d) 1 : 2 : 3

11. In the above question if resistance R is removed from the circuit. Then,
(a) power consumed by $2R$ will increase
(b) power consumed by $6R$ will decrease
(c) Both (a) and (b) are correct
(d) Both (a) and (b) are wrong

Multiple Correct Options

12. Two heaters designed for the same voltage V have different power ratings. When connected individually across a source of voltage V, they produce H amount of heat each in time t_1 and t_2 respectively. When used together across the same source, they produce H amount of heat in time t

(a) If they are in series, $t = t_1 + t_2$

(b) If they are in series, $t = 2(t_1 + t_2)$

(c) If they are in parallel, $t = \frac{t_1 t_2}{(t_1 + t_2)}$

(d) If they are in parallel, $t = \frac{t_1 t_2}{2(t_1 + t_2)}$

Match the Columns

13. Three bulbs A, B and C having rated power P_A, P_B and P_C, respectively. Each bulb is designed to operate at rated voltage of V volts.

It is given that $P_A > P_B > P_C$. In Column I, different configurations of bulbs are given while in Column II, intensities are mentioned. Match entries of Column I with entries of Column II, neglecting the change of resistances of bulbs due to temperature change.

	Column I		Column II
(a)	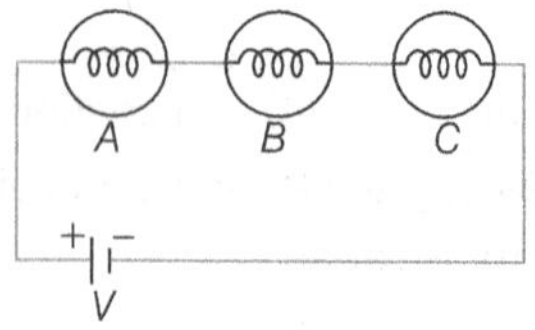	(p)	A glows with minimum intensity
(b)	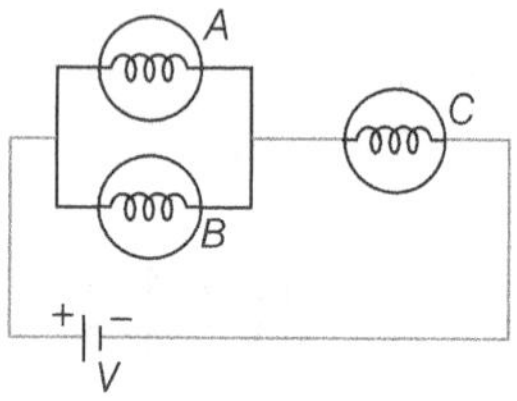	(q)	C is glowing with maximum intensity
(c)	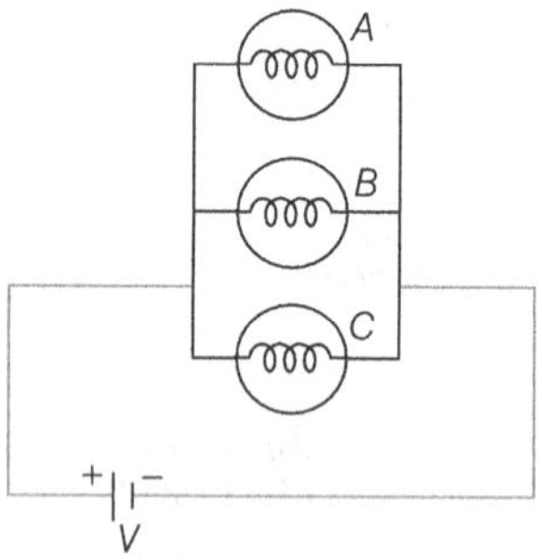	(r)	B is glowing with minimum intensity
		(s)	A is glowing with maximum intensity

Subjective Questions

14. A 120 V house circuit has the following light bulbs switched on : 40 W, 60 W and 75 W. Find the equivalent resistance of these bulbs.

15. Assume that the batteries in figure have negligible internal resistance. Find

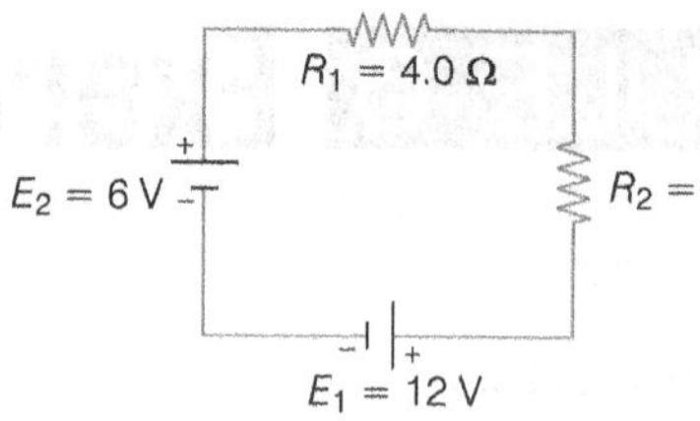

(a) the current in the circuit,

(b) the power dissipated in each resistor and

(c) the power of each battery, stating whether energy is supplied by or absorbed by it.

16. In figure, $E_1 = 12$ V and $E_2 = 8$ V.

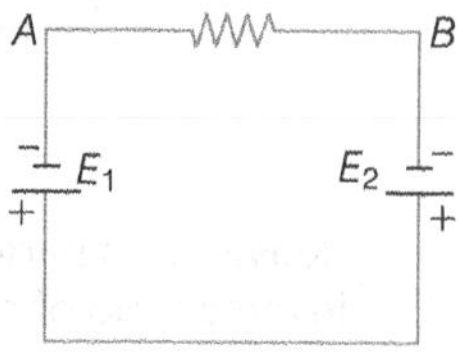

(a) What is the direction of the current in the resistor?

(b) Which battery is doing positive work?

(c) Which point, A or B, is at the higher potential?

17. In figure, circuit section AB absorbs energy at the rate of 5.0 W when a current $i = 1.0$ A passes through it in the indicated direction.

(a) What is the potential difference between points A and B?

(b) Emf device X does not have internal resistance. What is its emf?

(c) What is its polarity (the orientation of its positive and negative terminals)?

Integer Type Questions

18. A resistor develops 400 J of thermal energy in 10 s when a current of 2A is passed through it. When the current is increased to 4A, the energy dissipated in 20 s as heat is found to be 400 xJ. Find the value of x.

19. One kilowatt electric heater is to be used with 220 V DC supply. It converts $\frac{N \times 10}{3}$ g of water at 100°C into steam at 100°C in one minute. Find the value of N.

Answers

1. (a) **2.** (c) **3.** (d) **4.** (a) **5.** (e)
6. (b) **7.** (c) **8.** (a) **9.** (d) **10.** (a)
11. (a) **12.** (a,c) **13.** (a) → p, q ; (b) → q, r ; (c) → s
14. 82Ω **15.** (a) 0.5 A (b) 1W, 2W (c) 6W (supplied), 3W (absorbed)
16. (a) Anti-clockwise (b) E_1 (c) Point B **17.** (a) 5V (b) 3V (c) Positive terminal on left side **18.** (8) **19.** (8)

See hints on pages 81-83.

MODULE 7

Grouping of Cells

Cells are typically grouped into three main categories :

Series Grouping

Suppose n cells each of emf E and internal resistance r are connected in series as shown in figure.

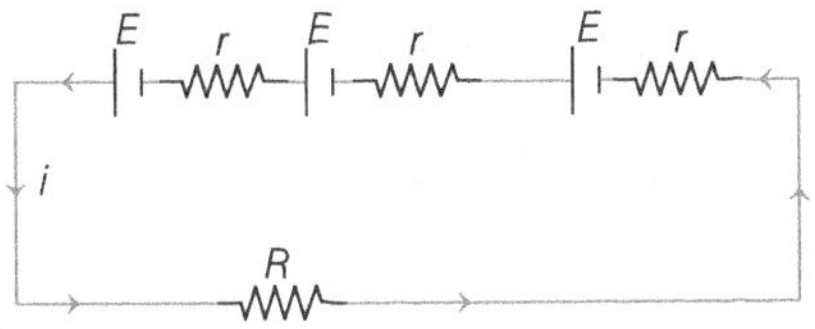

Then, Net emf $= nE$

Total resistance $= nr + R$

$\therefore$ Current in the circuit, $i = \dfrac{\text{net emf}}{\text{total resistance}}$

or
$$\boxed{i = \frac{nE}{nr + R}}$$

Note *If polarity of m cells is reversed, then equivalent emf $= (n - 2m)\,E$, while total resistance is still $nr + R$*

$$\therefore \quad i = \frac{(n - 2m)\,E}{nr + R}$$

Parallel Grouping

Here are three possible cases.

Case 1 When E and r of each cell has same value and positive terminals of all cells are connected at one junction while negative at the other.

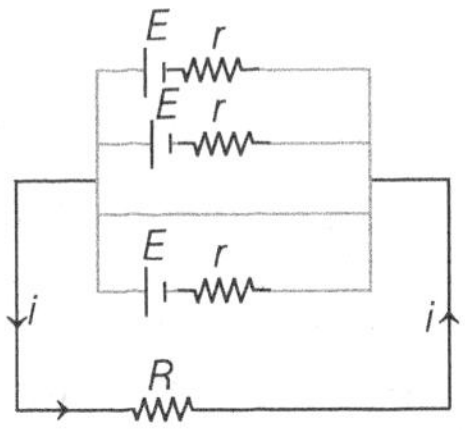

In this situation, the net emf is E. The net internal resistance is $\dfrac{r}{n}$ as n resistances each of r are in parallel.

Net external resistance is R. Therefore, total resistance is $\left(\dfrac{r}{n} + R\right)$ and so the current in the circuit will be,

$$i = \frac{\text{Net emf}}{\text{Total resistance}} \quad \text{or} \quad \boxed{i = \frac{E}{R + r/n}}$$

Note *A comparison of series and parallel grouping reveals that to get maximum current, cells must be connected in series if effective internal resistance is lesser than external and in parallel if effective internal resistance is greater than external.*

Case 2 If E and r of each cell are different but still the positive terminals of all cells are connected at one junction while negative at the other.

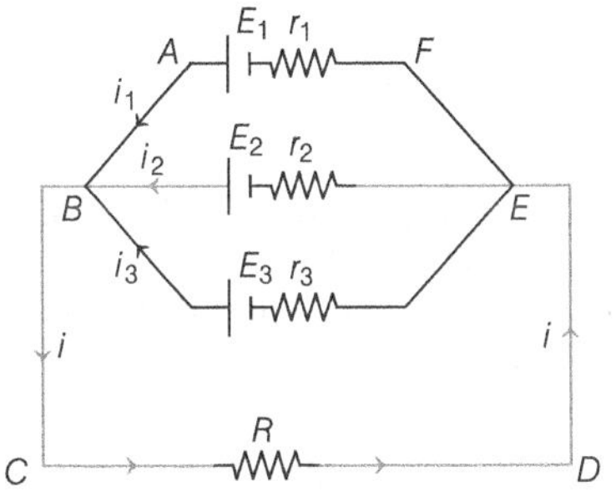

Applying Kirchhoff's second law in closed loop $ABCDEFA$,

$$E_1 - iR - i_1 r_1 = 0 \quad \text{or} \quad i_1 = -i\frac{R}{r_1} + \frac{E_1}{r_1} \qquad \text{...(i)}$$

Similarly, we can write

$$i_2 = -i\frac{R}{r_2} + \frac{E_2}{r_2} \qquad \text{...(ii)}$$

$$\dots \quad \dots \quad \dots \quad \dots$$

Adding all above equations, we have

$$(i_1 + i_2 + \dots + i_n) = -iR\,\Sigma\left(\frac{1}{r}\right) + \Sigma\left(\frac{E}{r}\right)$$

But $\quad i_1 + i_2 + \dots + i_n = i$

$$\therefore \quad i = -iR\,\Sigma\left(\frac{1}{r}\right) + \Sigma\left(\frac{E}{r}\right)$$

$$\therefore \quad i = \frac{\Sigma\,(E/r)}{1 + R\,\Sigma\,(1/r)}$$

$$= \frac{\Sigma(E/r)/\Sigma(1/r)}{\{1/\Sigma(1/r)\} + R} = \frac{E_{eq}}{R_{eq}}$$

where, $\quad E_{eq} = \dfrac{\Sigma\,(E/r)}{\Sigma\,(1/r)}$

and $\quad R_{eq} = R + \dfrac{1}{\Sigma\,(1/r)} = R_{ext} + R_{in}$

From the above expression, we can see that $i = \dfrac{E}{R + r/n}$ if n cells of same emf E and internal resistance r are connected in parallel. This is because,

$$\Sigma\,(E/r) = nE/r$$

and $\qquad \Sigma (1/r) = n/r$

$\therefore \qquad i = \dfrac{nE/r}{1 + nR/r}$

Multiplying the numerator and denominator by r/n, we have

$$i = \frac{E}{R + r/n}$$

$\therefore \qquad E_{eq} = E$

$\qquad R_{ext} = R$

and $\qquad R_{in} = \dfrac{r}{n}$

Case 3 This is the most general case of parallel grouping in which E and r of different cells are different and the positive terminals of few cells are connected to the negative terminals of the others as shown figure.

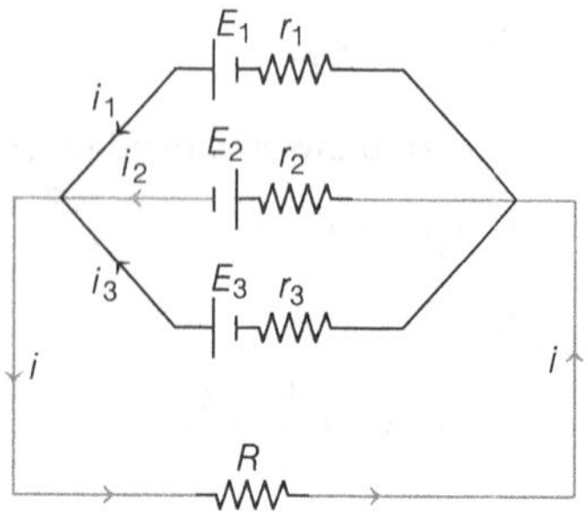

Kirchhoff's second law in different loops gives the following equations :

$$E_1 - iR - i_1 r_1 = 0$$

or $\qquad i_1 = \dfrac{E_1}{r_1} - \dfrac{iR}{r_1} \qquad$ ...(i)

$$-E_2 - iR - i_2 r_2 = 0$$

or $\qquad i_2 = -\dfrac{E_2}{r_2} - \dfrac{iR}{r_2} \qquad$ ...(ii)

Similarly, $\qquad i_3 = \dfrac{E_3}{r_3} - \dfrac{iR}{r_3} \qquad$ ...(iii)

Adding Eqs. (i), (ii) and (iii), we get

$$i_1 + i_2 + i_3 = (E_1/r_1) - (E_2/r_2) + (E_3/r_3) - iR\,(1/r_1 + 1/r_2 + 1/r_3)$$

or $\quad i\,[1 + R\,(1/r_1 + 1/r_2 + 1/r_3)] = (E_1/r_1) - (E_2/r_2) + (E_3/r_3)$

$$i = \frac{(E_1/r_1) - (E_2/r_2) + (E_3/r_3)}{1 + R\,(1/r_1 + 1/r_2 + 1/r_3)}$$

or $\quad i = \dfrac{(E_1/r_1 - E_2/r_2 + E_3/r_3)/(1/r_1 + 1/r_2 + 1/r_3)}{R + 1/(1/r_1 + 1/r_2 + 1/r_3)}$

$\therefore$ In this case, $\qquad E_{eq} = \dfrac{\dfrac{E_1}{r_1} - \dfrac{E_2}{r_2} + \dfrac{E_3}{r_3}}{\dfrac{1}{r_1} + \dfrac{1}{r_2} + \dfrac{1}{r_3}}$

If E_{eq} comes out to be negative then negative terminal of equivalent battery will be on left hand side.

Mixed Grouping

The situation is shown in figure.

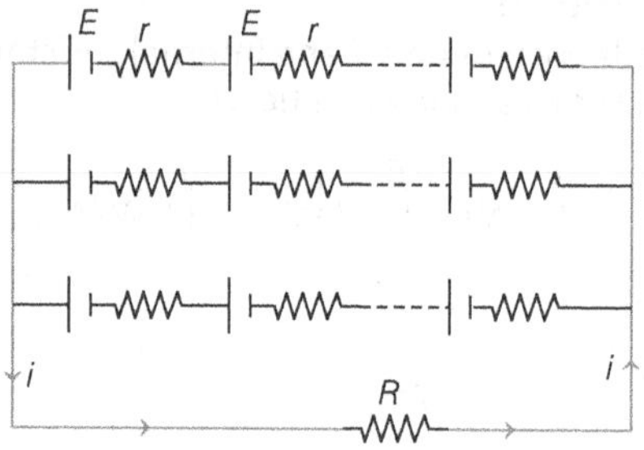

There are n identical cells in a row and number of rows are m. Emf of each cell is E and internal resistance is r. Treating each row as a single cell of emf nE and internal resistance nr, we have

$$\text{Net emf} = nE$$

$$\text{Total internal resistance} = \frac{nr}{m}$$

$$\text{Total external resistance} = R$$

$\therefore$ Current through the external resistance R is

$$i = \frac{nE}{R + \dfrac{nr}{m}}$$

This expression after some rearrangements can also be written as

$$i = \frac{mnE}{(\sqrt{mR} - \sqrt{nr})^2 + 2\sqrt{mnrR}}$$

If total number of cells are given then mn is fixed. E, r and R are also given. Suppose we have liberty to arrange the given number of cells in different rows. Then in the above expression the numerator mnE and in the denominator $2\sqrt{mnrR}$ all are fixed. Only the square term in the denominator is variable. Therefore, i is maximum when,

$$\sqrt{mR} = \sqrt{nr} \quad \text{or} \quad R = \frac{nr}{m}$$

or $\quad$ total external resistance = total internal resistance

Thus, we can say that the current and hence power transferred to the load is maximum when load resistance is equal to the internal resistance. This is known as **maximum power transfer theorem.**

Important POINTS

- Regarding maximum current in the circuit or maximum power consumed by the external resistance R, there are three special cases. One we have discussed above, where we have to arrange the cells in such a manner that current and power in the circuit should be maximum.

  And this happens when we arrange the cells in such a manner that total internal resistance comes out to be equal to total external resistance. Rest two cases are discussed below.

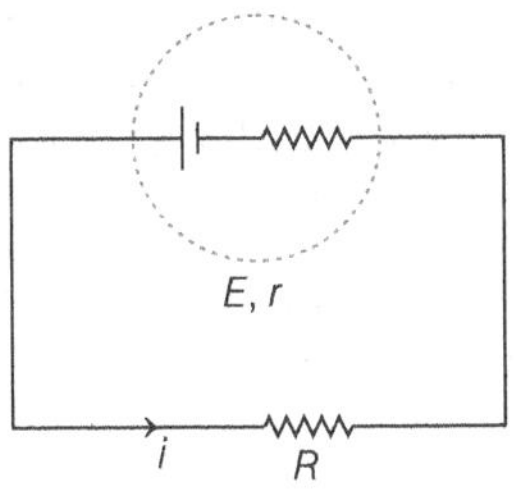

In the figure shown, $i = \dfrac{E}{R+r} \Rightarrow P_R = i^2R = \left(\dfrac{E}{R+r}\right)^2 R$

(i) If r is variable. E and R are fixed, then i and P_R both are maximum when $r = 0$.

(ii) If R is variable. E and r are fixed. Then, current in the circuit is maximum when $R = 0$.
But P_R will be maximum, when $R = r$
or external resistance = internal resistance

- The equivalent emf of a cell can also be found by the following method.

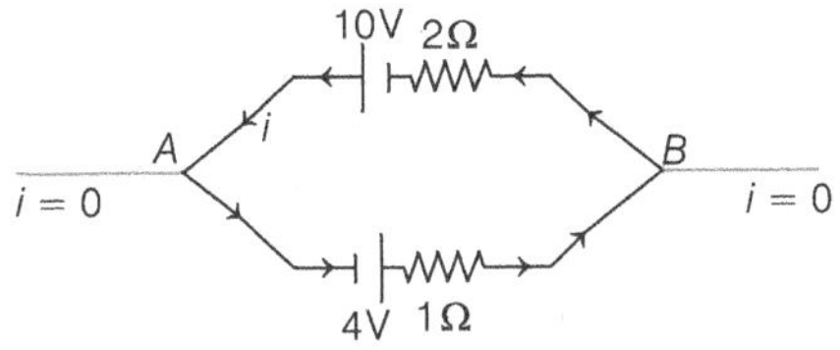

$$\Downarrow$$

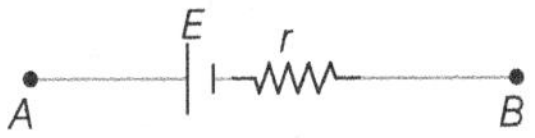

Suppose we wish to find the equivalent emf of the above circuit. We apply the fact that

$$E = V$$

When no current is drawn from the cell. But current in the internal circuit may be non-zero. This current is,

$$i = \frac{10+4}{2+1} = \frac{14}{3}\text{ A}$$

Now, $V_A + 4 - 1 \times \dfrac{14}{3} = V_B$

$$\therefore \quad V_A - V_B = \frac{14}{3} - 4 = \frac{2}{3}\text{ V}$$

$$\therefore \quad E = V_A - V_B = V = \frac{2}{3}\text{ V}$$

$$\therefore \quad E = \frac{2}{3}\text{ V}$$

Further, $V_A - V_B$ is positive, i.e. $V_A > V_B$ or A is connected to the positive terminal of the battery and B to the negative.

Internal resistance of the equivalent battery is found by the normal procedure. For example, $2\,\Omega$ and $1\,\Omega$ resistances are in parallel. Hence, their combined resistance is

$$\frac{1}{r} = \frac{1}{1} + \frac{1}{2} = \frac{3}{2}$$

or $\quad r = \dfrac{2}{3}\,\Omega$

Example 1 Find the emf and internal resistance of a single battery which is equivalent to a combination of three batteries as shown in figure.

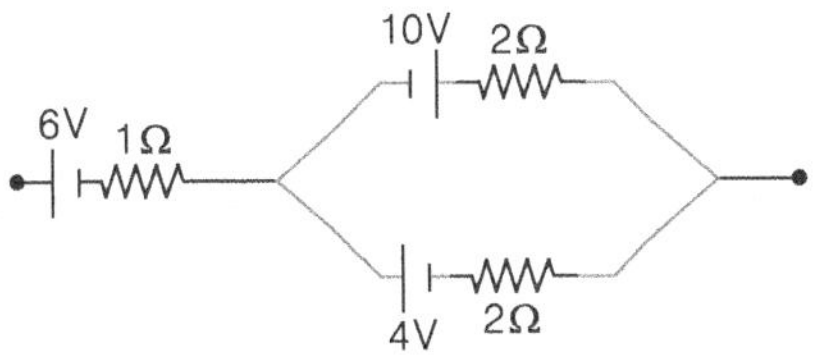

Sol. The given combination consists of two batteries in parallel and resultant of these two in series with the third one.

For parallel combination we can apply,

$$E_{eq} = \frac{\dfrac{E_1}{r_1} - \dfrac{E_2}{r_2}}{\dfrac{1}{r_1} + \dfrac{1}{r_2}} = \frac{\dfrac{10}{2} - \dfrac{4}{2}}{\dfrac{1}{2} + \dfrac{1}{2}} = 3\text{ V}$$

Further, $\dfrac{1}{r_{eq}} = \dfrac{1}{r_1} + \dfrac{1}{r_2} = \dfrac{1}{2} + \dfrac{1}{2} = 1$

$$\therefore \quad r_{eq} = 1\,\Omega$$

Now this is in series with the third one, i.e.

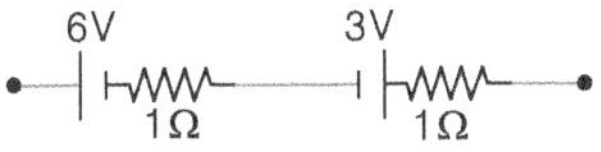

The equivalent emf of these two is (6 – 3) V or 3 V and the internal resistance will be (1 + 1) or 2 Ω

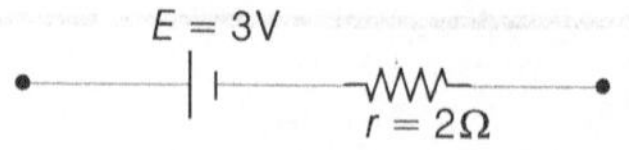

Example 2 In the circuit shown in figure $E_1 = 3\text{V}, E_2 = 2\text{V}$, $E_3 = 1\text{ V}$ and $R = r_1 = r_2 = r_3 = 1\Omega$. **(1981 JEE)**

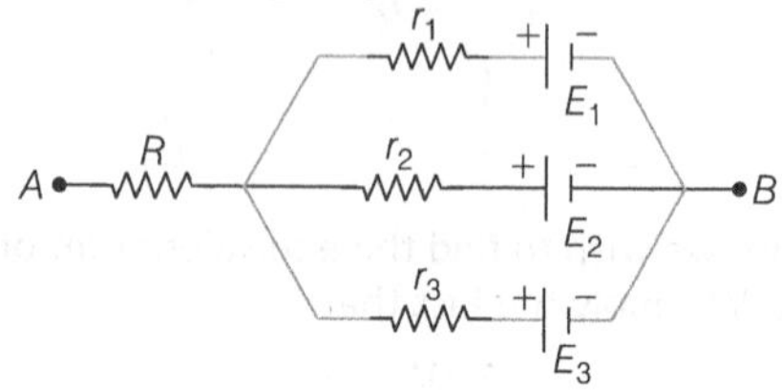

(a) Find the potential difference between the points A and B and the currents through each branch.

(b) If r_2 is short-circuited and the point A is connected to point B, find the currents through E_1, E_2, E_3 and the resistor R.

Sol. (a) Equivalent emf of three batteries in parallel would be

$$E_{eq} = \frac{\Sigma(E/r)}{\Sigma(1/r)} = \frac{(3/1 + 2/1 + 1/1)}{(1/1 + 1/1 + 1/1)} = 2\text{V}$$

Further r_1, r_2 and r_3 each are of 1Ω. Therefore, internal resistance of the equivalent battery will be $\frac{1}{3}\Omega$ as all three are in parallel.

The equivalent circuit is therefore shown in the figure.

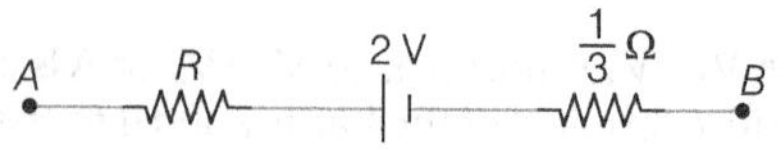

Since, no current is taken from the battery.

$$V_{AB} = E = 2\text{V} \qquad \text{(From } V = E - ir\text{)}$$

Further, $V_A - V_B = E_1 - i_1 r_1$

$$\therefore \qquad i_1 = \frac{V_B - V_A + E_1}{r_1} = \frac{-2+3}{1} = 1\text{ A}$$

Similarly, $$i_2 = \frac{V_B - V_A + E_2}{r_2} = \frac{-2+2}{1} = 0$$

and $$i_3 = \frac{V_B - V_A + E_3}{r_3} = \frac{-2+1}{1} = -1\text{ A}$$

(b) r_2 is short circuited means resistance of this branch becomes zero. Making a closed circuit with a battery and resistance R. Applying Kirchhoff's second law in three loops so formed.

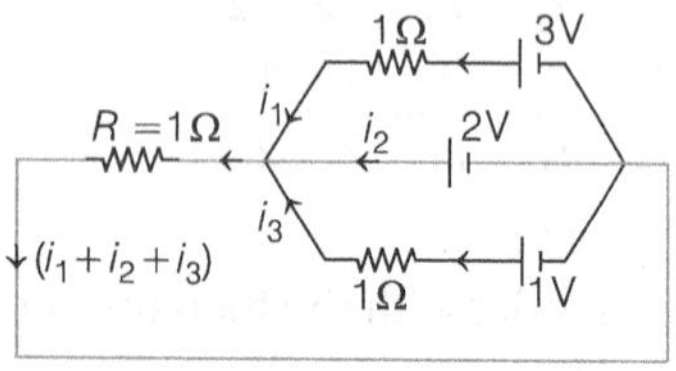

$$3 - i_1 - (i_1 + i_2 + i_3) = 0 \qquad \text{...(i)}$$

$$2 - (i_1 + i_2 + i_3) = 0 \qquad \text{...(ii)}$$

$$1 - i_3 - (i_1 + i_2 + i_3) = 0 \qquad \text{...(iii)}$$

From Eq. (ii), $\quad i_1 + i_2 + i_3 = 2\text{A}$

Substituting in Eq. (i), we get

$$i_1 = 1\text{A}$$

Substituting in Eq. (iii), we get

$$i_3 = -1\text{A}$$

$$\therefore \qquad i_2 = 2\text{ A}$$

Example 3 A battery has an open circuit potential difference of 6 V between its terminals. When a load resistance of 60 Ω is connected across the battery, the total power supplied by the battery is 0.4 W. What should be the load resistance R, so that maximum power will be dissipated in R. Calculate this power. What is the total power supplied by the battery when such a load is connected?

Sol. When the circuit is open, $V = E$

$$\therefore \qquad E = 6\text{ V}$$

Let r be the internal resistance of the battery.

Power supplied by the battery in this case is

$$P = Ei = E\left(\frac{E}{R+r}\right) = \frac{E^2}{R+r}$$

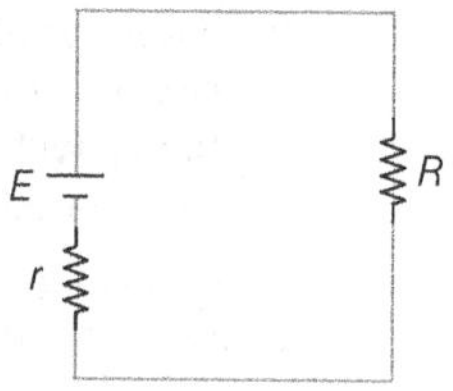

Substituting the values, we have

$$0.4 = \frac{(6)^2}{60 + r}$$

Solving this, we get

$$r = 30\,\Omega$$

Maximum power is dissipated in the circuit when net external resistance is equal to net internal resistance or

$$R = r$$

$$\therefore \qquad R = 30\,\Omega \qquad \textbf{Ans.}$$

Further, total power supplied by the battery under this condition is

$$P_{\text{Total}} = \frac{E^2}{R+r}$$

$$= \frac{(6)^2}{30 + 30}$$

$$= 0.6\text{ W} \qquad \textbf{Ans.}$$

Of this 0.6 W half of the power is dissipated in R and half in r. Therefore, maximum power dissipated in R would be

$$\frac{0.6}{2} = 0.3\text{ W} \qquad \textbf{Ans.}$$

INTRODUCTORY Exercise

1. Find the emf (V) and internal resistance (r) of a single battery which is equivalent to a parallel combination of two batteries of emfs V_1 and V_2 and internal resistances r_1 and r_2 respectively, with polarities as shown in figure **(1997 JEE)**

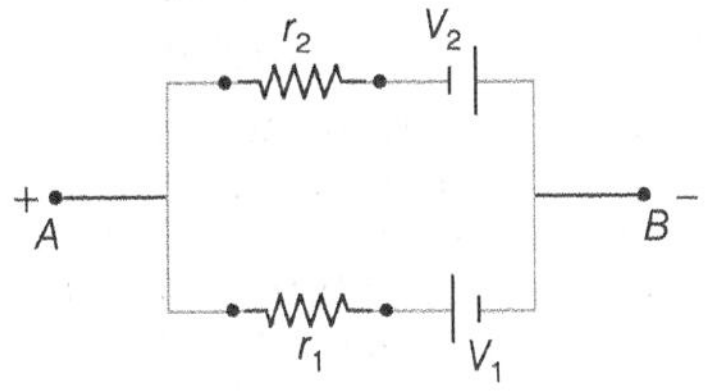

2. Find the net emf of the three batteries shown in figure.

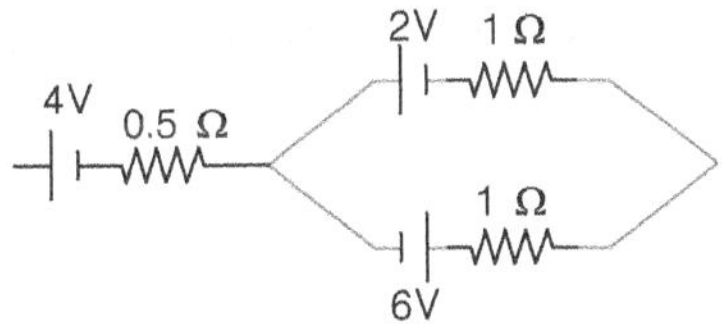

3. Find the equivalent emf and internal resistance of the arrangement shown in figure.

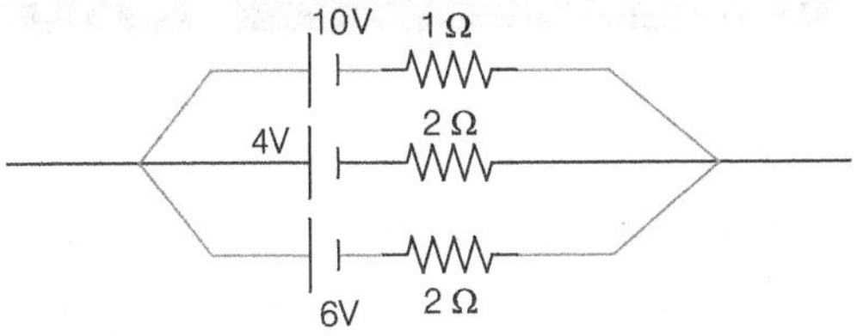

Answers

1. $V = \dfrac{V_1 r_2 - V_2 r_1}{r_1 + r_2}, r = \dfrac{r_1 r_2}{r_1 + r_2}$
2. 2 V
3. 7.5 V, 0.5 Ω

See hints on page 83.

Typed Problem

Type 1. ***To find current through a single external resistance in a complex circuit of batteries using the concepts of equivalent value of emf of a single battery.***

Concept

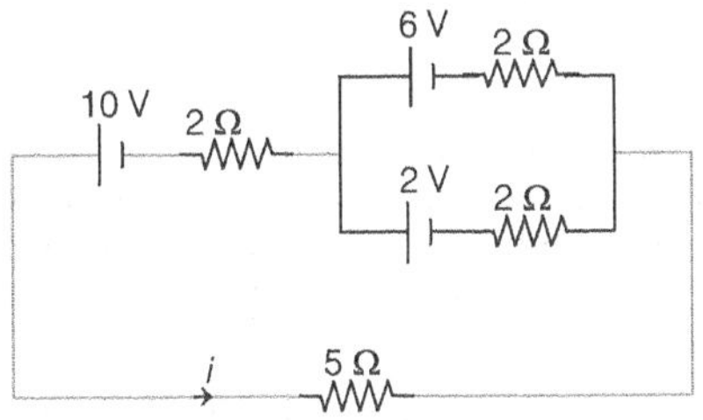

In the above circuit, if we have to find only i then two parallel batteries may be converted into a single battery. This battery is in series with the third battery of emf 10 volt. Now, we can find current i from the equation,

$$i = \frac{\text{Net emf}}{\text{Total resistance}}$$

Ex. 1 Find the value of i in the circuit shown above.

Sol. Equivalent emf of the parallel combination is

$$E = \frac{E_1/r_1 + E_2/r_2}{1/r_1 + 1/r_2}$$

$$= \frac{6/2 + 2/2}{1/2 + 1/2} = 4 \text{ volt}$$

Equivalent internal resistance of the parallel combination is

$$r = \frac{r_1 r_2}{r_1 + r_2} = \frac{(2)(2)}{2+2} = 1\,\Omega$$

Now, the equivalent simple circuit is as shown below

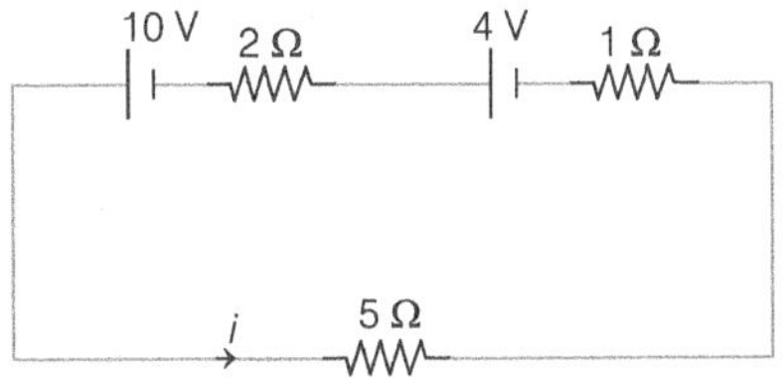

$$i = \frac{\text{Net emf}}{\text{Total resistance}}$$

$$= \frac{10 + 4}{2 + 1 + 5}$$

$$= 1.75 \text{ A}$$ **Ans.**

Note *By this concept, we can find only i. To find other currents (across 6V battery or 2V battery) we will have to apply Kirchhoff's laws.*

MODULE 7 Exercise

Assertion and Reason

See directions on page 25.

1. **Assertion :** Two non-ideal batteries are connected in parallel with same polarities on same side. The equivalent emf is smaller than either of the two emfs.

   Reason : Two non-ideal batteries are connected in parallel, the equivalent internal resistance is smaller than either of the two internal resistances.

Single Correct Option

2. Two identical batteries, each of emf 2 V and internal resistance $r = 1\,\Omega$ are connected as shown. The maximum power that can be developed across R using these batteries is

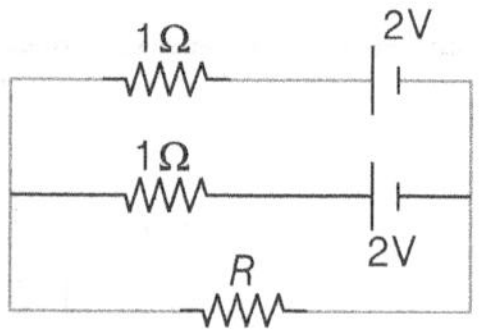

   (a) 3.2 W
   (b) 8.2 W
   (c) 2 W
   (d) 4 W

3. The current in the resistance R will be zero, if

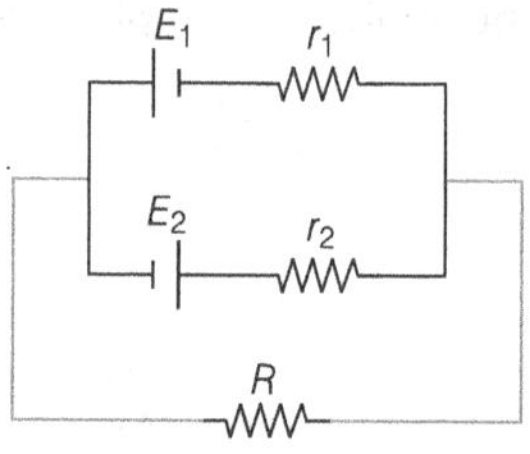

   (a) $E_1 r_1 = E_2 r_2$ (b) $\frac{E_1}{r_1} = \frac{E_2}{r_2}$
   (c) $(E_1 + E_2) r_1 = E_1 r_2$ (d) $(E_1 - E_2) r_1 = E_2 r_1$

4. The n rows each containing m cells in series are joined in parallel. Maximum current is taken from this combination across an external resistance of 3Ω resistance. If the total number of cells used are 24 and internal resistance of each cell is $0.5\,\Omega$, then

   (a) $m = 8, n = 3$ (b) $m = 6, n = 4$
   (c) $m = 12, n = 2$ (d) $m = 2, n = 12$

5. The combination of two identical cells, whether connected in series or parallel combination provides the same current through an external resistance of $2\,\Omega$. The value of internal resistance of each cell is **(2022 Main)**

   (a) 2Ω (b) 4Ω
   (d) 6Ω (d) 8Ω

6. Twelve cells each having the same emf are connected in series and are kept in a closed box. Some of the cells are wrongly connected. This battery of cells is connected in series with an ammeter and two cells identical with the others. The current is 3 A when the cells and the battery support each other and is 2 A when the cells and the battery oppose each other. How many cells in the battery are wrongly connected?

   (a) 1
   (b) 6
   (c) 4
   (d) 8

7. Two batteries with emf 12 V and 13 V are connected in parallel across a load resistor of $10\,\Omega$. The internal resistances of the two batteries are $1\,\Omega$ and $2\,\Omega$, respectively. The voltage across the load lies between **(2016 Adv.)**

   (a) 11.7 V and 11.8 V
   (b) 11.6 V and 11.7 V
   (c) 11.5 V and 11.6 V
   (d) 11.4 V and 11.5 V

Subjective Questions

8. If a battery of emf E and internal resistance r is connected across a load of resistance R. Show that the rate at which energy is dissipated in R is maximum when $R = r$ and this maximum power is $P = E^2/4r$.

9. Two identical batteries each of emf $E = 2$ volt and internal resistance $r = 1$ ohm are available to produce heat in an external resistance by passing a current through it. What is the maximum power that can be developed across an external resistance R using these batteries?

10. In a circuit shown in figure if the internal resistances of the sources are negligible then at what value of resistance R will the thermal power generated in it will be the maximum. What is the value of maximum power?

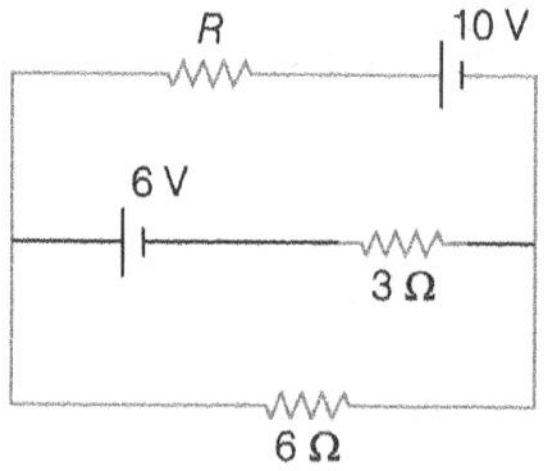

11. In the circuit shown in figure, the emfs of batteries are E_1 and E_2 which have internal resistances R_1 and R_2. At what value of the resistance R will the thermal power generated in it be the highest? What it is? **(Irodov 3.197)**

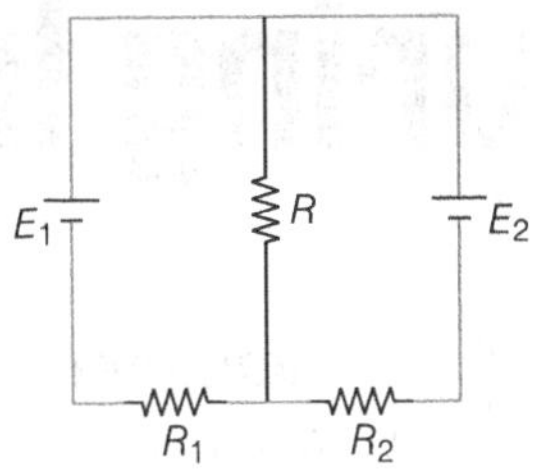

12. A circuit shown in figure has resistances $R_1 = 20\ \Omega$ and $R_2 = 30\ \Omega$. At what value of the resistance R_x will the thermal power generated in it be practically independent of small variations of that resistance? The voltage between the points A and B is supposed to be constant in this case. **(Irodov 3.196)**

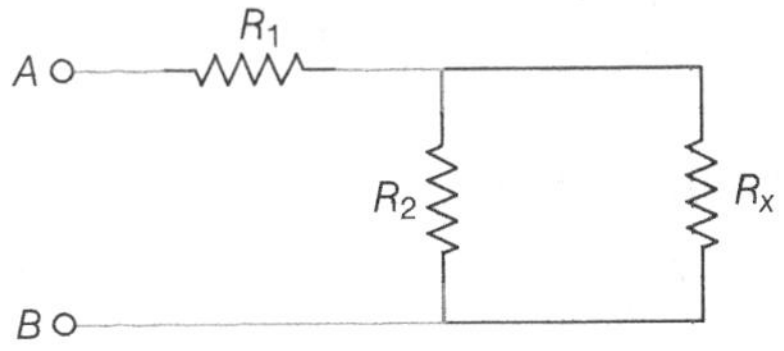

Answers

1. (d) **2.** (c) **3.** (b) **4.** (c) **5.** (a) **6.** (a) **7.** (c) **9.** 2W **10.** 2 Ω, 4.5W

11. $R = \dfrac{R_1 R_2}{R_1 + R_2}$, $P_{max} = \dfrac{(E_1 R_2 + E_2 R_1)^2}{4R_1 R_2 (R_1 + R_2)}$ **12.** 12 Ω

See hints on pages 83-85.

MODULE 8

Electrical Measuring Instruments

So far we have studied about current, resistance, potential difference and emf. Now, in this module we will study how these are measured. The basic measuring instrument is galvanometer, whose pointer shows a deflection when current passes through it.

A galvanometer can easily be converted into an ammeter for measuring current, into a voltmeter for measuring potential difference. For accurate measurement of potential difference or emf, a potentiometer is more preferred.

Resistances are accurately measured by using post office box or meter bridge which are based on the principle of "Wheatstone bridge". All these are discussed here one by one in brief.

Galvanometer

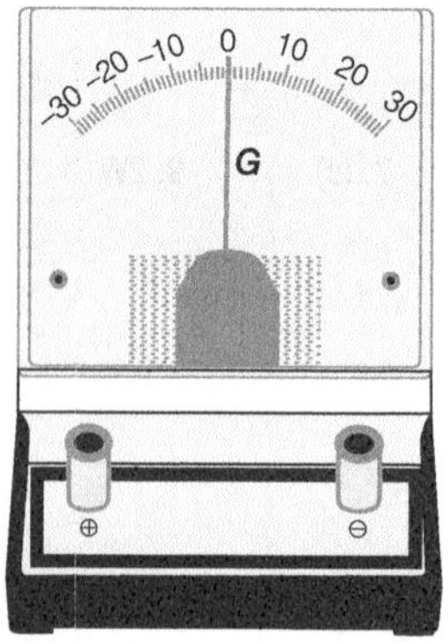

Many common devices including car instrument panels, battery chargers measure potential difference, current or resistance using d'Arsonval Galvanometer. It consists of a pivoted coil placed in the magnetic field of a permanent magnet.

Attached to the coil is a spring. In the equilibrium position, with no current in the coil, the pointer is at zero and spring is relaxed. When there is a current in the coil, the magnetic field exerts a torque on the coil that is proportional to current.

As the coil turns, the spring exerts a restoring torque that is proportional to the angular displacement. Thus, the angular deflection of the coil and pointer is directly proportional to the coil current and the device can be calibrated to measure current.

The maximum deflection, typically 90° to 120° is called full scale deflection. The essential electrical characteristics of the galvanometer are the current i_g required for full scale deflection (of the order of 10 μA to 10 mA) and the resistance G of the coil (of the order of 10 to 1000 Ω).

The galvanometer deflection is proportional to the current in the coil. If the coil obeys Ohm's law, the current is proportional to potential difference. The corresponding potential difference for full scale deflection is

$$\boxed{V = i_g G}$$

Ammeter

A current measuring instrument is called an ammeter. A galvanometer can be converted into an ammeter by connecting a small resistance S (called shunt) in parallel with it.

Suppose we want to convert a galvanometer with full scale current i_g and coil resistance G into an ammeter with full scale reading i.

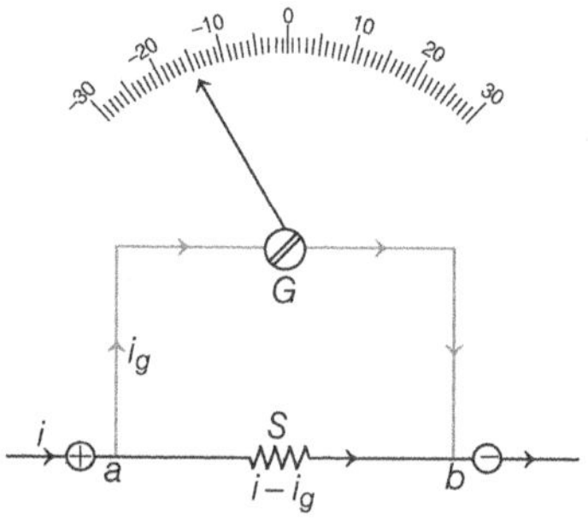

To determine the shunt resistance S needed, note that, at full scale deflection the total current through the parallel combination is i, the current through the galvanometer is i_g and the current through the shunt is $i - i_g$.

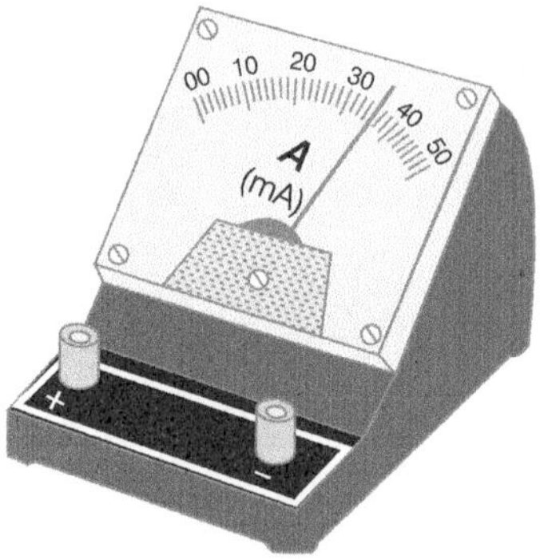

The potential difference V_{ab} $(= V_a - V_b)$ is the same for both paths, so

$$i_g G = (i - i_g) S$$

$$\therefore \quad \boxed{S = \left(\frac{i_g}{i - i_g}\right) G}$$

Voltmeter

A voltage measuring device is called a voltmeter. It measures the potential difference between two points.

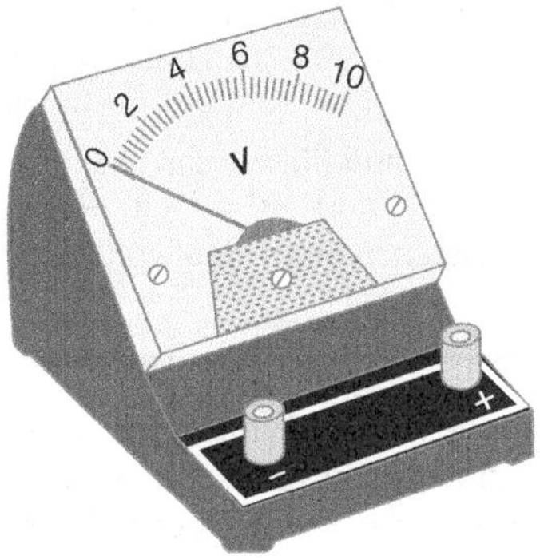

A galvanometer can be converted into a voltmeter by connecting a high resistance (R) in series with it. The whole assembly called the voltmeter is connected in parallel between the points where potential difference has to be measured.

For a voltmeter with full scale reading V, we need a series resistor R such that

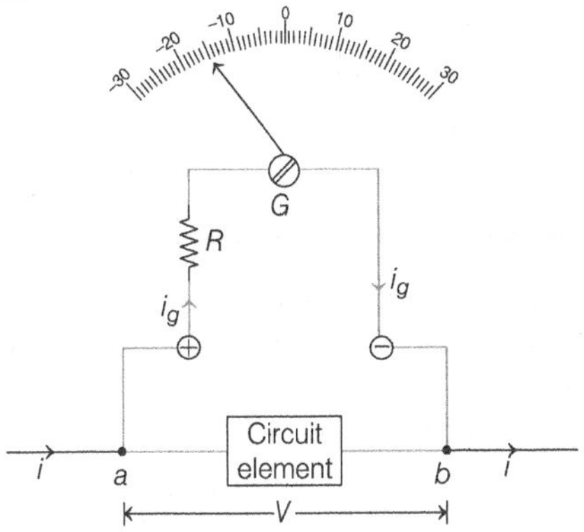

$$V = i_g\,(G + R) \quad \text{or} \quad \boxed{R = \frac{V}{i_g} - G}$$

Important POINTS

- **Conversion of galvanometer into an ammeter.**
  (i) A galvanometer is converted into an ammeter by connecting a low resistance (called shunt) in parallel with galvanometer. This assembly (called ammeter) is connected in series in the wire in which current is to be found. Resistance of an ideal ammeter should be zero.

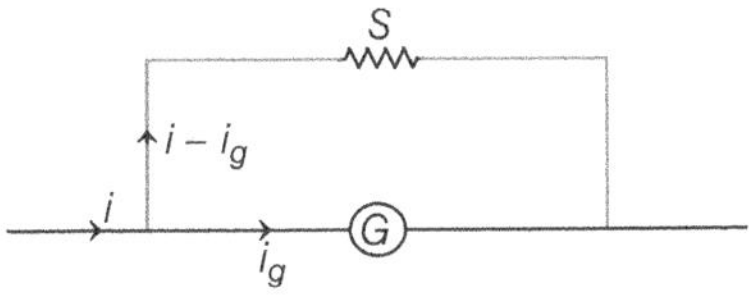

In parallel, current distributes in the inverse ratio of resistance. Therefore,

$$\frac{S}{G} = \frac{i_g}{i - i_g}$$

$$\therefore \qquad S = \text{shunt} = \left(\frac{i_g}{i - i_g}\right) G$$

(ii) Resistance of an ammeter is given by

$$\frac{1}{A} = \frac{1}{G} + \frac{1}{S} \Rightarrow A = \frac{GS}{G + S}$$

(iii) The reading of an ammeter is always less than actual current in the circuit.

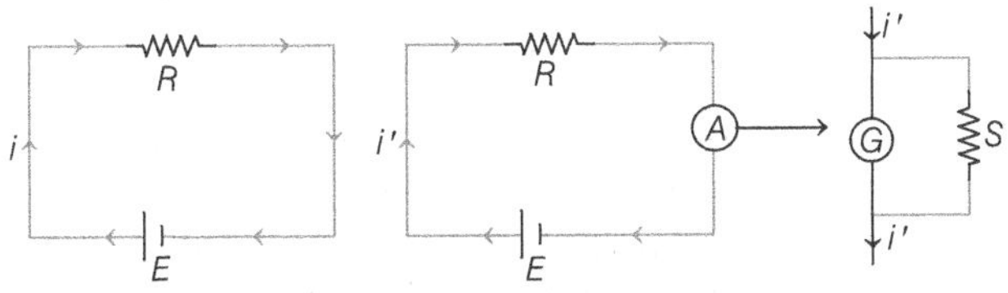

(a) (b)

For example, in Fig. (a), actual current through R is

$$i = \frac{E}{R} \qquad \text{...(i)}$$

while the current after connecting an ammeter of resistance $A\left(= \frac{GS}{G + S}\right)$ in series with R is

$$i' = \frac{E}{R + A} \qquad \text{...(ii)}$$

From Eqs. (i) and (ii), we see that

$$i' < i \text{ and } i' = i \text{ when } A = 0$$

i.e. resistance of an ideal ammeter should be zero.

- **Conversion of a galvanometer into a voltmeter**
  (i) A galvanometer is converted into a voltmeter by connecting a high resistance in series with galvanometer. The whole assembly called voltmeter is connected in parallel across the two points between which potential difference is to be found.

Resistance of an ideal voltmeter should be infinite.

$$V = I_g (G + R)$$

$$\therefore \quad R = \text{high resistance required in series} = \frac{V}{i_g} - G$$

(ii) Resistance of a voltmeter is $R_V = R + G$

(iii) The reading of a voltmeter is always less than the true value.

For example, if a current i is passing through a resistance r, the actual value is

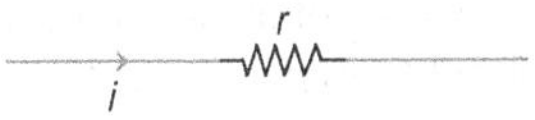

$$V = ir \qquad \text{...(i)}$$

Now, if a voltmeter of resistance R_V $(=G+R)$ is connected across the resistance r, the new value will be

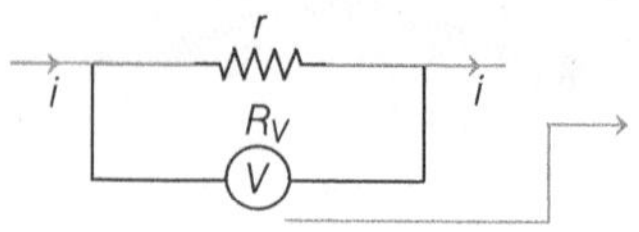

$$V' = \frac{i \times (rR_V)}{r + R_V} \quad \text{or} \quad V' = \frac{ir}{1 + \frac{r}{R_V}} \qquad \text{...(ii)}$$

From Eqs. (i) and (ii), we can see that, $V' < V$ and $V' = V$ if $R_V = \infty$

Thus, resistance of an ideal voltmeter should be infinite.

Example 1 Find the value of shunt resistance required to make the 1.00 mA, 20 Ω galvanometer into an ammeter with a range of 0 to 50.0 mA?

Sol. Here, $i_g = 1.00\text{ mA} = 10^{-3}\text{ A}$, $G = 20\,\Omega$, $i = 50.0 \times 10^{-3}\text{ A}$

$$S = \left(\frac{i_g}{i - i_g}\right) G = \frac{(10^{-3})(20)}{(50.0 \times 10^{-3}) - (10^{-3})}$$

$$= 0.408\,\Omega \qquad \textbf{Ans.}$$

Note *The resistance of ammeter is given by*

$$\frac{1}{A} = \frac{1}{G} + \frac{1}{S} = \frac{1}{20} + \frac{1}{0.408} \quad \text{or} \quad A = 0.4\,\Omega$$

The shunt resistance is so small in comparison to the galvanometer resistance that the ammeter resistance is very nearly equal to the shunt resistance. This shunt resistance gives us a low resistance ammeter with the desired range of 0 to 50.0 mA. At full scale deflection $i = 50.0$ mA, the current through the galvanometer is 1.0 mA while the current through the shunt is 49.0 mA. If the current i is less than 50.0 mA, the coil current and the deflection are proportionally less, but the ammeter resistance is still 0.4 Ω.

Example 2 How can we make a galvanometer with $G = 20\,\Omega$ and $i_g = 1.0$ mA into a voltmeter with a maximum range of 10 V?

Sol. Using $R = \frac{V}{i_g} - G$, we have

$$R = \frac{10}{10^{-3}} - 20 = 9980\,\Omega \qquad \textbf{Ans.}$$

Thus, a resistance of 9980 Ω is to be connected in series with the galvanometer to convert it into a voltmeter of desired range.

Note *At full scale deflection current through the galvanometer, the voltage drop across the galvanometer*

$$V_g = i_g G = 20 \times 10^{-3} \text{ volt} = 0.02 \text{ volt}$$

and the voltage drop across the series resistance R is

$$V = i_g R = 9980 \times 10^{-3} \text{ volt}$$
$$= 9.98 \text{ volt}$$

or we can say that most of the voltage appears across the series resistance R.

Example 3 Resistance of a milliammeter is R_1 of an ammeter is R_2 of a voltmeter is R_3 and of a kilovoltmeter is R_4. Find the correct order of R_1, R_2, R_3 and R_4.

Sol. To increase the range of an ammeter a low resistance has to be connected in parallel with galvanometer. Therefore, net resistance decreases.

To increase the range of voltmeter, a high resistance has to be connected in series. So, net resistance further increases. Therefore, the correct order is

$$R_4 > R_3 > R_1 > R_2$$

Example 4 A galvanometer gives full scale deflection with 0.006 A current. By connecting it to a 4990 Ω resistance, it can be converted into a voltmeter of range 0-30 V. If connected to a $\frac{2n}{249}$ Ω resistance, it becomes an ammeter of range 0-1.5 A. What is the value of n ? **(2014 JEE)**

Sol.

$$i_g(G + 4990) = V$$

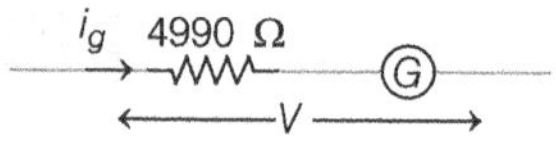

$$\Rightarrow \frac{6}{1000}(G + 4990) = 30$$

$$\Rightarrow G + 4990 = \frac{30000}{6} = 5000$$

$$\Rightarrow G = 10\,\Omega$$

$$V_{ab} = V_{cd}$$

$$\Rightarrow i_g G = (1.5 - i_g)\,S$$

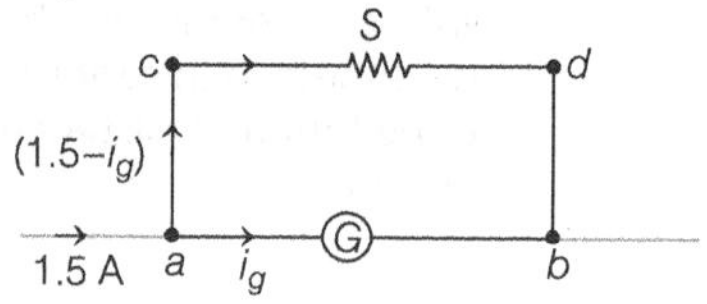

$$\Rightarrow \frac{6}{1000} \times 10 = \left(1.5 - \frac{6}{1000}\right) S$$

$$\Rightarrow S = \frac{60}{1494} = \frac{2n}{249}$$

$$\Rightarrow n = \frac{249 \times 30}{1494} = 5 \qquad \textbf{Ans.}$$

INTRODUCTORY Exercise 1

1. A microammeter has a resistance of 100 Ω and full scale range of 50 μA. It can be used as a voltmeter or as a higher range ammeter provided a resistance is added to it. Pick the correct range and resistance combination (s) **(1991 JEE)**
   (a) 50 V range with 10 kΩ resistance in series
   (b) 10 V range with 200 kΩ resistance in series
   (c) 5 mA range with 1 Ω resistance in parallel
   (d) 10 mA range with 1 Ω resistance in parallel
2. The full scale deflection current of a galvanometer of resistance 1 Ω is 5 mA. How will you convert it into a voltmeter of range 5 V?
3. A voltmeter has a resistance G and range V. Calculate the resistance to be used in series with it to extend its range to nV.

Answers

1. (b,c)
2. By connecting a resistance of 999 Ω in series with galvanometer
3. (n – 1) G

See hints on pages 85-86.

Potentiometer

The potentiometer is an instrument that can be used to measure the emf or the internal resistance of an unknown source. It also has a number of other useful applications.

Principle of Potentiometer

The principle of potentiometer is schematically shown in figure.

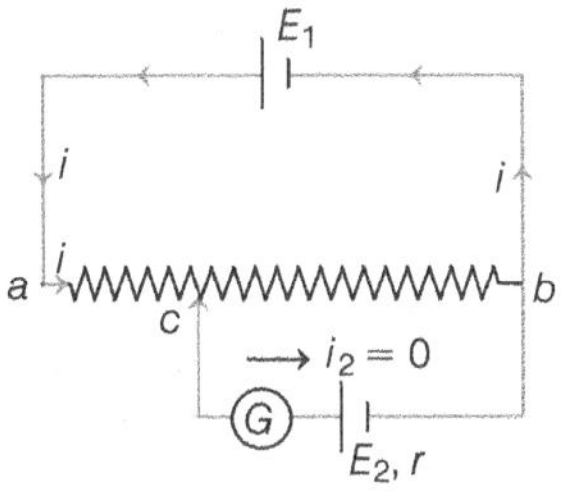

A resistance wire ab of total resistance R_{ab} is permanently connected to the terminals of a source of known emf E_1. A sliding contact c is connected through the galvanometer G to a second source whose emf E_2 is to be measured. As contact c is moved along the potentiometer wire, the resistance R_{cb} between points c and b varies. If the resistance wire is uniform R_{cb} is proportional to the length of the wire between c and b. To determine the value of E_2, contact c is moved until a position is found at which the galvanometer shows no deflection.

This corresponds to zero current passing through E_2. With $i_2 = 0$, Kirchhoff's second law gives

$$\boxed{E_2 = iR_{cb}}$$

With $i_2 = 0$, the current i produced by the emf E_1 has the same value no matter what the value of emf E_2. A potentiometer has the following applications.

To find emf of an unknown battery

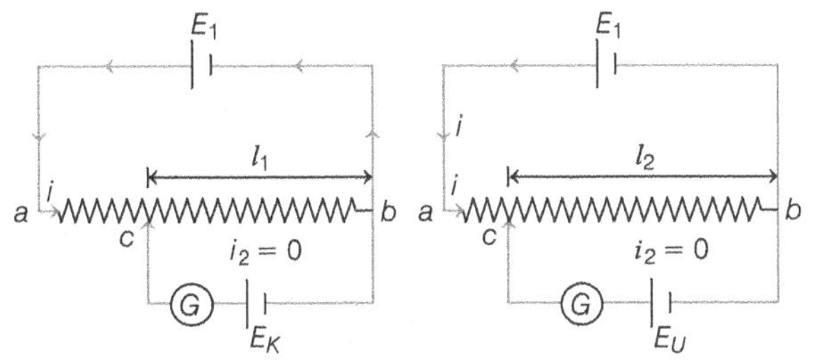

We calibrate the device by replacing E_2 by a source of known emf E_K and then by unknown emf E_U.

Let the null points (means $i_2 = 0$) are obtained at lengths l_1 and l_2. Then,

$$E_K = i(\rho l_1)$$

and $$E_U = i(\rho l_2)$$

Here, ρ = resistance of wire ab per unit length.

$$\therefore \quad \frac{E_K}{E_U} = \frac{l_1}{l_2}$$

or $$\boxed{E_U = \left(\frac{l_2}{l_1}\right) E_K}$$

So, by measuring the lengths l_1 and l_2, we can find the emf of an unknown battery.

To find the internal resistance of an unknown battery

To find the internal resistance of an unknown battery let us derive a formula.

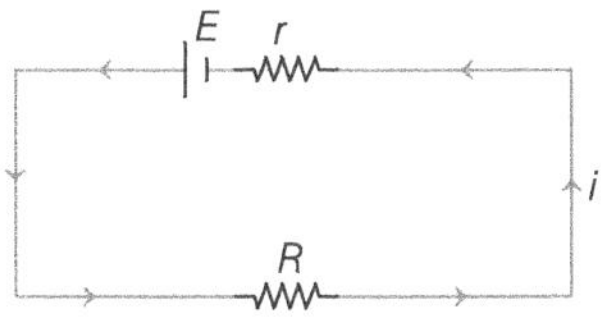

In the circuit shown in figure,

$$i = \frac{E}{R + r} \quad \text{...(i)}$$

and V = potential difference across the terminals of the battery

or $$V = E - ir = iR \quad \text{...(ii)}$$

From Eqs. (i) and (ii), we can prove that

$$\boxed{r = R\left(\frac{E}{V} - 1\right)}$$

Thus, if a battery of emf E and internal resistance r is connected across a resistance R and the potential difference across its terminals comes out to be V then the internal resistance of the battery is given by the above formula.

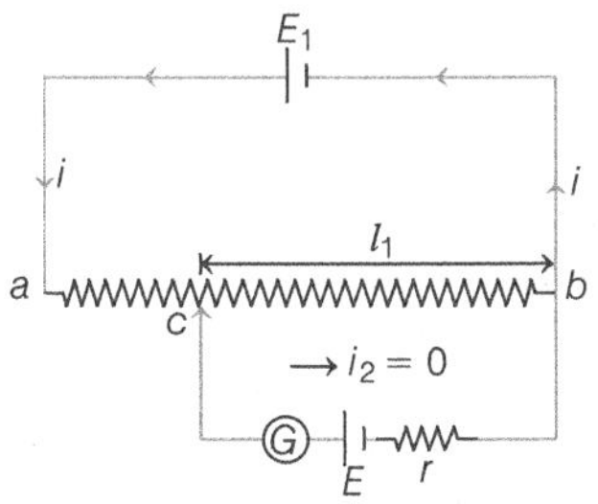

Now, let us apply it in a potentiometer for finding the internal resistance of the unknown battery. The circuit shown in figure is similar to the previous one.

Hence, $$E = i\rho l_1 \quad \text{...(i)}$$

Now, a known resistance R is connected across the terminals of the unknown battery as shown in figure.

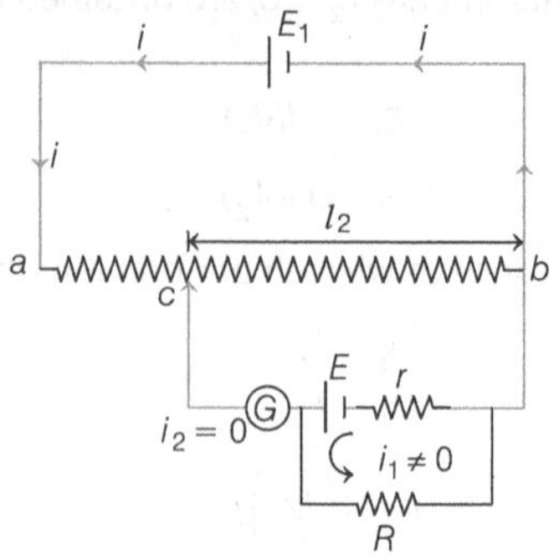

This time $V_{cb} \neq E$, but $V_{cb} = V$ because $i_1 \neq 0$

where, V = potential difference across the terminals of the unknown battery.

Hence, $V_{cb} = V = i \rho l_2$ ...(ii)

From Eqs. (i) and (ii), we get

$$\frac{E}{V} = \frac{l_1}{l_2}$$

Substituting in $r = R\left(\frac{E}{V} - 1\right)$, we get

$$r = R\left(\frac{l_1}{l_2} - 1\right)$$

So, by putting R, l_1 and l_2 we can determine the internal resistance r of unknown battery.

Important POINTS

Note *Forget about the abbreviations used above.*

- 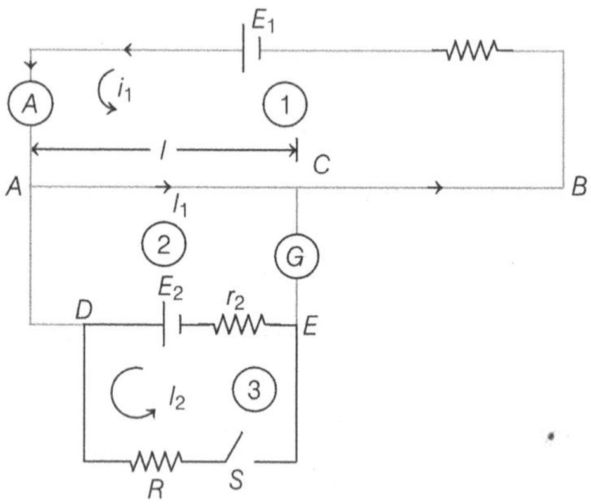

  Under balanced condition (when $I_G = 0$) loop-1 and loop-3 are independent with each other. All problems in this condition can be solved by a single equation,

  $$V_{AC} = V_{DE}$$

  or $\quad i_1 R_{AC} = E_2 - i_2 r_2$

  or $\quad \boxed{i_1 \lambda l = E_2 - i_2 r_2}$ ...(i)

  Here, λ is the resistance per unit length of potentiometer wire AB. Length l is called balance point length. Currents i_1 and i_2 are independent with each other. Current $i_2 = 0$, if switch is open.
- Under balanced condition, a part of potential difference of E_1 is balanced by the lower circuit. So, normally $E_2 < E_1$ for taking balance point length. Similarly, $V_{AC} = V_{DE}$ $\Rightarrow V_A = V_D$ and $V_C = V_E$. Therefore, positive terminals of both batteries should be on same side and negative terminals on the other side.
- From Eq. (i), we can see that null point length is

  $$l = \frac{E_2 - i_2 r_2}{i_1 \lambda}$$

  Now, suppose E_1 is increased then i_1 will also increase and null point length l will decrease. Similarly, we can make some other cases also.
- If we do not get any balanced condition ($I_G \neq 0$), then the given circuit is simply a three loops problem, which can be solved with the help of Kirchhoff's laws.
- Potential drop per unit length across the potentiometer wire is called the potential gradient.

Example 5 A potentiometer wire of length 100 cm has a resistance of 10 Ω. It is connected in series with a resistance R and a cell of emf 2 V and of negligible internal resistance. A source of emf 10 mV is balanced against a length of 40 cm of the potentiometer wire. What is the value of R?

Sol. From the theory of potentiometer, $V_{cb} = E$, if no current is drawn from the lower battery

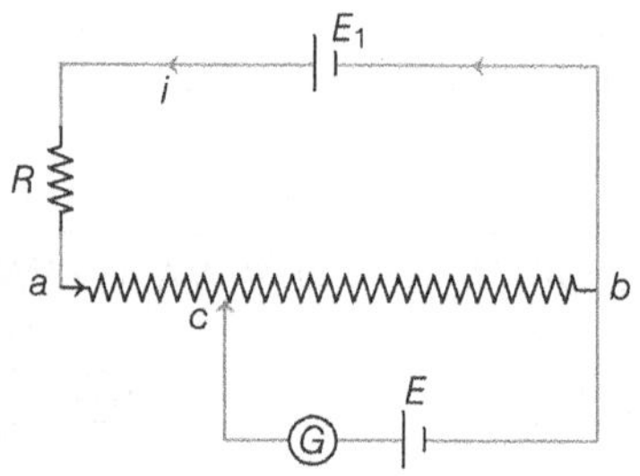

or $\left(\dfrac{E_1}{R+R_{ab}}\right)R_{cb} = E$

Here, $E_1 = 2\text{ V}, R_{ab} = 10\,\Omega,\ R_{cb} = \left(\dfrac{40}{100}\right) \times 10 = 4\,\Omega$

and $E = 10 \times 10^{-3}\text{ V}$

Substituting the values in the above equation, we get

$R = 790\,\Omega$ **Ans.**

Example 6 When the switch is open in lowermost loop of a potentiometer, the balance point length is 60 cm. When the switch is closed with a known resistance of $R = 4\Omega$, the balance point length decreases to 40 cm. Find the internal resistance of the unknown battery.

Sol. Using the result,

$$r = R\left(\frac{l_1}{l_2} - 1\right)$$

$$= 4\left(\frac{60}{40} - 1\right)$$

$$= 2\Omega$$ **Ans.**

Example 7 In the figure shown, wire AB has a length of 100 cm and resistance 8Ω. Find the balance point length l.

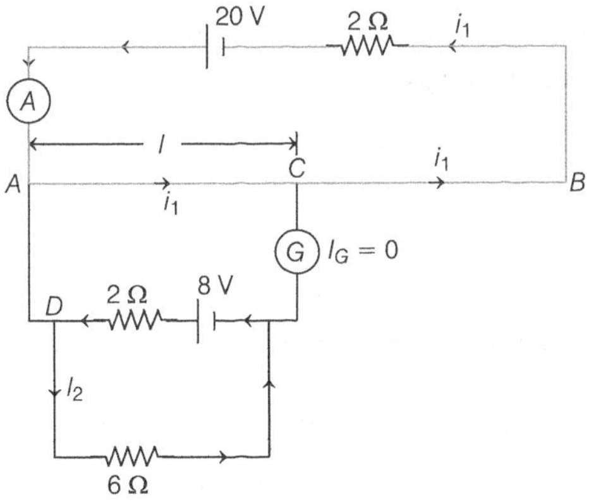

Sol. Using the equation, $i_1\lambda l = E_2 - i_2 r_2$

We have,

$$\left(\frac{20}{2+8}\right)\left(\frac{8}{100}\right)l = 8 - \left(\frac{8}{2+6}\right)(2)$$

Solving this equation, we get

$l = 37.5$ cm **Ans.**

INTRODUCTORY Exercise 2

1. In a potentiometer experiment it is found that no current passes through the galvanometer when the terminals of the cell are connected across 0.52 m of the potentiometer wire. If the cell is shunted by a resistance of $5\,\Omega$ a balance is obtained when the cell is connected across 0.4 m of the wire. Find the internal resistance of the cell.

2. The potentiometer wire AB is 600 cm long.

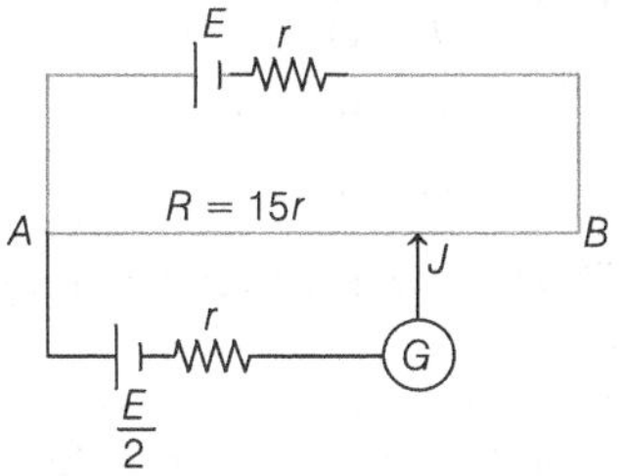

(a) At what distance from A should the jockey J touch the wire to get zero deflection in the galvanometer.

(b) If the jockey touches the wire at a distance 560 cm from A, what will be the current through the galvanometer.

3. Two identical cells each of emf 1.5 V are connected in series across a $10\,\Omega$ resistance. An ideal voltmeter connected across $10\,\Omega$ resistance reads 1.5 V. The internal resistance of each cell is ....Ω. **(2023 Main)**

Answers

1. $1.5\,\Omega$

2.(a) 320 cm (b) $\dfrac{3E}{22r}$

3. (5)

See hints on page 86.

Principle of Wheatstone's Bridge

The scientist Wheatstone designed a circuit to find unknown resistance. Such a circuit is popularly known as Wheatstone's bridge. This is an arrangement of four resistances which can be used to measure one of them in terms of the rest.

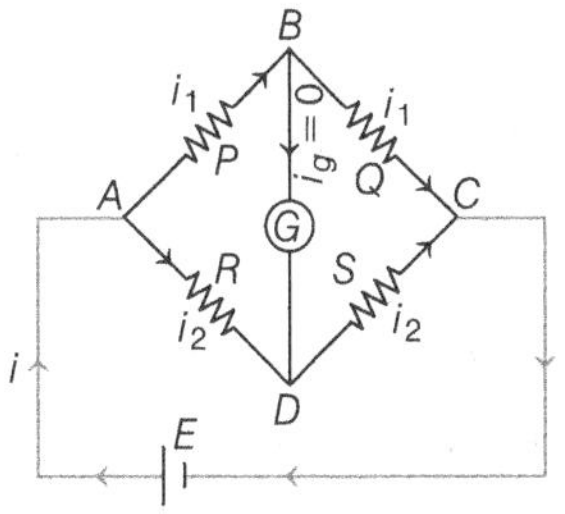

The figure shows the circuit designed by him. The bridge is said to be **balanced** when deflection in galvanometer is zero, i.e. $i_g = 0$, and hence,

$$V_B = V_D$$

Under this condition, $V_A - V_B = V_A - V_D$

or $i_1 P = i_2 R$

or $\dfrac{i_1}{i_2} = \dfrac{R}{P}$ ...(i)

Similarly, $V_B - V_C = V_D - V_C$

or $i_1 Q = i_2 S$ or $\dfrac{i_1}{i_2} = \dfrac{S}{Q}$ ...(ii)

From Eqs. (i) and (ii), $\frac{R}{P} = \frac{S}{Q}$

or $$\boxed{\frac{P}{Q} = \frac{R}{S}}$$

So, this is a condition for which a Wheatstone's bridge is balanced.

To measure the resistance of an unknown resistor, it is connected as one of the four resistors in the bridge. One of the other three should be a variable resistor. Let us suppose P is the unknown resistance and Q is the variable resistance. The value of Q is so adjusted that deflection through the galvanometer is zero. In this case, the bridge is balanced and

$$P = \left(\frac{R}{S}\right) \cdot Q$$

Knowing R, S and Q, the value of P is calculated. Following two points are important regarding a Wheatstone's bridge.

(i) In Wheatstone's bridge, cell and galvanometer arms are interchangeable.

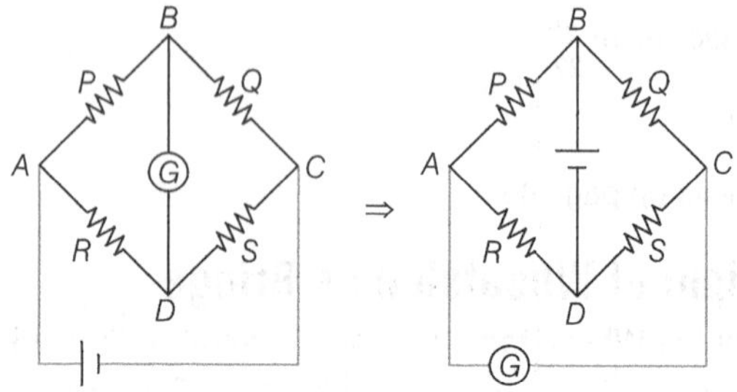

In both the cases, condition of balanced bridge is

$$\frac{P}{Q} = \frac{R}{S}$$

(ii) If bridge is not balanced, current will flow from D to B in above figure, if

$$PS > RQ$$

Exercise Prove the above two results.

Meter Bridge Experiment

Meter bridge works on Wheatstone's bridge principle and is used to find the unknown resistance (X) and its specific resistance (or resistivity).

Theory

As the meter bridge wire AC has uniform material density and area of cross-section, its resistance is proportional to its length. AB and BC are called the ratio arms and their resistances correspond to P and Q respectively.

Thus, $$\frac{\text{Resistance of } AB}{\text{Resistance of } BC} = \frac{P}{Q}$$

$$= \frac{\lambda l}{\lambda(100 - l)} = \frac{l}{100 - l}$$

Here, λ is the resistance per unit length of the bridge wire.

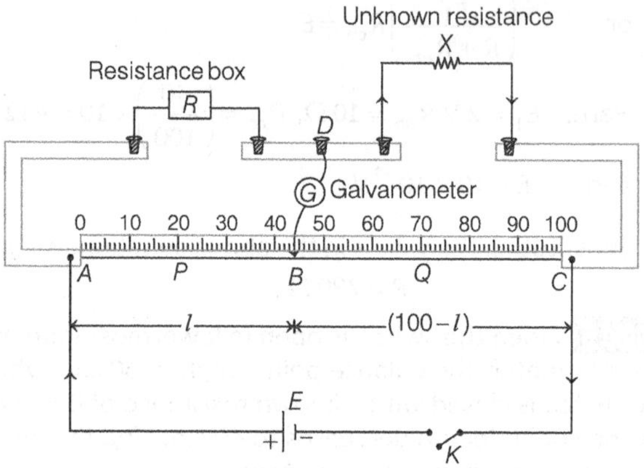

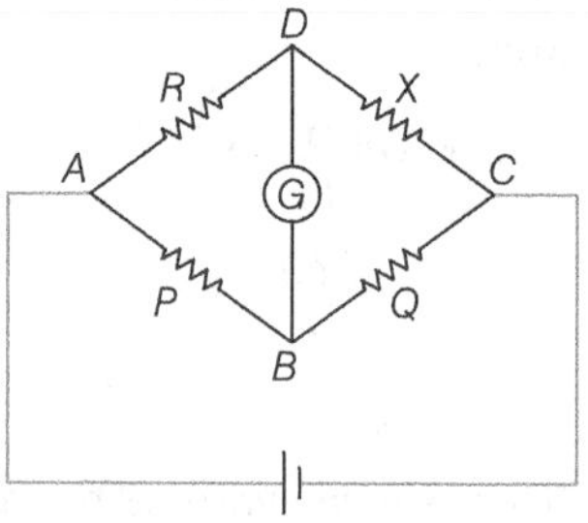

Jockey is so adjusted that the current through galvanometer = 0. Hence, according to Wheatstone's bridge principle,

When current through galvanometer is zero or bridge is balanced, then

$$\frac{P}{Q} = \frac{R}{X} \quad \text{or} \quad X = \frac{Q}{P}R$$

$$\therefore \quad X = \left(\frac{100 - l}{l}\right)R \quad \text{...(i)}$$

So, by knowing R and l unknown resistance X can be determined.

Specific Resistance From resistance formula,

$$X = \rho\frac{L}{A} \quad \text{or} \quad \rho = \frac{XA}{L}$$

For a wire of radius r or diameter $D = 2r$,

$$A = \pi r^2 = \frac{\pi D^2}{4}$$

or $$\rho = \frac{X\pi D^2}{4L} \quad \text{...(ii)}$$

By knowing X, D and L we can find specific resistance of the given wire by Eq. (ii).

Precautions

1. The connections should be clean and tight.
2. Null point should be brought between 40 cm and 60 cm.
3. At one place, diameter of wire (D) should be measured in two mutually perpendicular directions.
4. The jockey should be moved gently over the bridge wire so that it does not rub the wire.

End Corrections

In meter bridge, some extra length (under the metallic strips) comes at points A and C. Therefore, some additional length (α and β) should be included at the ends. Here, α and β are called the end corrections. Hence, in place of l we use $l+\alpha$ and in place of $100-l$ we use $100-l+\beta$.

To find α and β, use known resistors R_1 and R_2 in place of R and X and suppose we get null point length equal to l_1. Then,

$$\frac{R_1}{R_2}=\frac{l_1+\alpha}{100-l_1+\beta} \quad ...(i)$$

Now, we interchange the positions of R_1 and R_2 and suppose the new null point length is l_2. Then,

$$\frac{R_2}{R_1}=\frac{l_2+\alpha}{100-l_2+\beta} \quad ...(ii)$$

Solving Eqs. (i) and (ii), we get

$$\alpha=\frac{R_2l_1-R_1l_2}{R_1-R_2}$$

and

$$\beta=\frac{R_1l_1-R_2l_2}{R_1-R_2}-100$$

Example 8 If resistance R in resistance box is $300\,\Omega$, then the balanced length is found to be 75.0 cm from end A. The diameter of unknown wire is 1 mm and length of the unknown wire is 31.4 cm. Find the specific resistance of the unknown wire.

Sol. $\frac{R}{X}=\frac{l}{100-l}$

$$\Rightarrow \quad X=\left(\frac{100-l}{l}\right)R$$

$$=\left(\frac{100-75}{75}\right)(300)$$

$$=100\,\Omega$$

Now, $$X=\frac{\rho l}{A}=\frac{\rho L}{(\pi D^2/4)}$$

$$\therefore \quad \rho=\frac{\pi D^2X}{4L}$$

$$=\frac{(3.14)(10^{-3})^2(100)}{(4)(0.314)}$$

$$=2.5\times10^{-4}\,\Omega\text{-m}$$ **Ans.**

Example 9 In a meter bridge, null point is 20 cm. When the known resistance R is shunted by $10\,\Omega$ resistance, null point is found to be shifted by 10 cm. Find the unknown resistance X.

Sol. $\frac{R}{X}=\frac{l}{100-l}$

$$\therefore \quad X=\left(\frac{100-l}{l}\right)R$$

or $$X=\left(\frac{100-20}{20}\right)R=4R \quad ...(i)$$

When known resistance R is shunted, its net resistance will decrease. Therefore, resistance parallel to this (i.e. P) should also decrease or its new null point length should also decrease.

$$\therefore \quad \frac{R'}{X}=\frac{l'}{100-l'}$$

$$=\frac{20-10}{100-(20-10)}=\frac{1}{9}$$

or $$X=9R' \quad ...(ii)$$

From Eqs. (i) and (ii), we have

$$4R=9R'=9\left[\frac{10R}{10+R}\right]$$

Solving this equation, we get

$$R=\frac{50}{4}\,\Omega$$

Now, from Eq. (i), the unknown resistance

$$X=4R=4\left(\frac{50}{4}\right)$$

or $$X=50\,\Omega$$ **Ans.**

Note *R' is resultant of R and $10\,\Omega$ in parallel.*

$$\therefore \quad \frac{1}{R'}=\frac{1}{10}+\frac{1}{R} \quad \text{or} \quad R'=\frac{10R}{10+R}$$

Example 10 If we use $100\,\Omega$ and $200\,\Omega$ in place of R and X we get null point deflection, $l=33$ cm. If we interchange the resistors, the null point length is found to be 67 cm. Find end corrections α and β.

Sol. $\alpha=\frac{R_2l_1-R_1l_2}{R_1-R_2}$

$$=\frac{(200)(33)-(100)(67)}{100-200}=1\text{ cm}$$ **Ans.**

$$\beta=\frac{R_1l_1-R_2l_2}{R_1-R_2}-100$$

$$=\frac{(100)(33)-(200)(67)}{100-200}-100$$

$$=1\text{ cm}$$ **Ans.**

INTRODUCTORY Exercise 3

1. A resistance of 2Ω is connected across one gap of a meter bridge (the length of the wire is 100 cm) and an unknown resistance, greater than 2Ω, is connected across the other gap. When these resistances are interchanged, the balance point shifts by 20 cm. Neglecting any corrections, the unknown resistance is **(2007 JEE)**

   (a) 3Ω (b) 4Ω (c) 5Ω (d) 6Ω

2. A meter bridge is setup as shown in figure, to determine an unknown resistance X using a standard $10\,\Omega$ resistor. The galvanometer shows null point when tapping key is at 52 cm mark. The end corrections are 1 cm and 2 cm respectively for the ends A and B. The determined value of X is **(2011 JEE)**

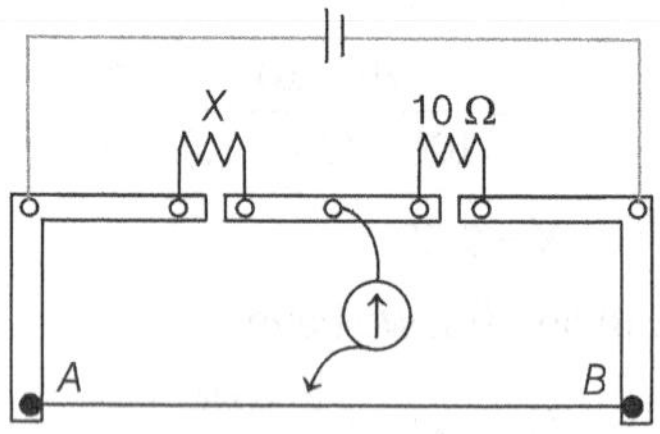

   (a) 10.2Ω (b) 10.6Ω (c) 10.8Ω (d) 11.1Ω

3. R_1, R_2, R_3 are different values of R.
   A, B and C are the null points obtained corresponding to R_1, R_2 and R_3 respectively. For which resistor, the value of X will be the most accurate and why? **(2005 JEE)**

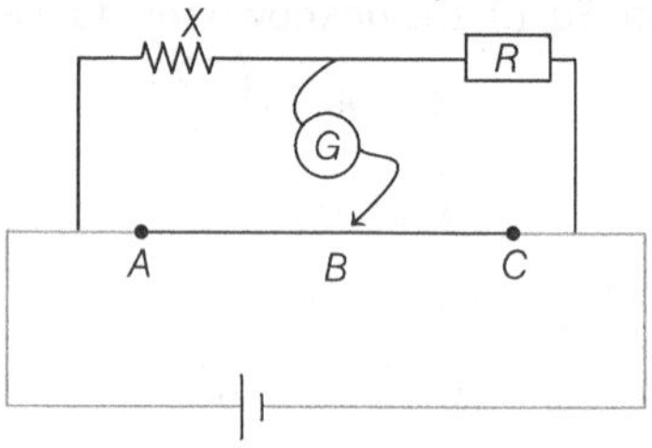

Answers

1. (a) **2.** (b) **3.** B is most accurate

See hints on page 86.

Post Office Box

Post office box also works on the principle of Wheatstone's bridge.

In a Wheatstone's bridge circuit, if $\frac{P}{Q} = \frac{R}{X}$ then the bridge is balanced. So, unknown resistance $X = \frac{Q}{P}R$.

P and Q are set in arms AB and BC where we can have, 10Ω, 100Ω or $1000\,\Omega$ resistances to set any ratio $\frac{Q}{P}$.

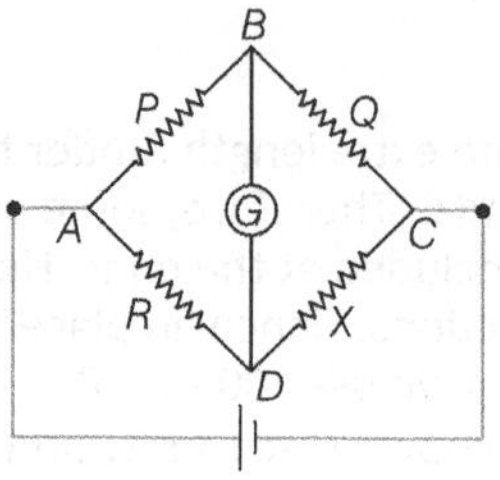

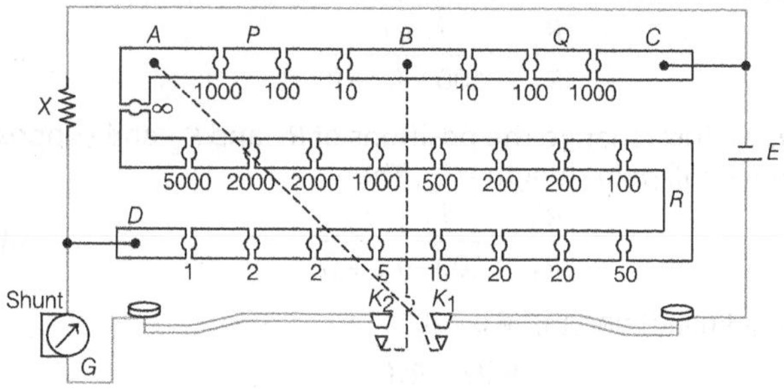

These arms are called ratio arm, initially we take $Q = 10\,\Omega$ and $P = 10\,\Omega$ to set $\frac{Q}{P} = 1$. The unknown resistance (X) is connected between C and D and battery is connected across A and C.

Now, adjust resistance in part A to D such that the bridge gets balanced. For this, keep on increasing the resistance with $1\,\Omega$ interval, check the deflection in galvanometer by first pressing key K_1 then galvanometer key K_2.

Suppose at $R = 4\Omega$, we get deflection towards left and at $R = 5\Omega$, we get deflection towards right. Then, we can say that for balanced condition, R should lie between $4\,\Omega$ to $5\,\Omega$, because :

$$X = \frac{Q}{P}R = \frac{10}{10}R$$
$$= R = 4\,\Omega \text{ to } 5\,\Omega$$

To get closer value of X, in the second observation, let us choose $\frac{Q}{P} = \frac{1}{10}$ by choosing $Q = 10\,\Omega$ and $P = 100\,\Omega$.

Suppose, now at $R = 42$ we get deflection towards left and at $R = 43$ deflection is towards right.

So R now lies between $4.2\,\Omega$ to $4.3\,\Omega$, because :

$$X = \frac{Q}{P}R = \frac{10}{100}R = \frac{1}{10}R.$$

Hence, $R \in (4.2, 4.3\,\Omega)$.

Now, to get further closer value take $\frac{Q}{P} = \frac{1}{100}$ and so on.

The observation table is shown below.

Table

S.No	Resistance in the ratio arm AB (P) (ohm)	BC (Q) (ohm)	Resistance in arm (AD (R) (ohm)	Direction of deflection	Unknown resistance $X = \frac{Q}{P} \times R$ (ohm)
1	10	10	4	Left	4 to 5
	–	–	5	Right	-
2	100	10	40	Left (large)	(4.2 to 4.3)
	–	–	50	Right (large)	-
	–	–	42	Left	-
	–	–	43	Right	-
3	1000	10	420	Left	4.25
	–	–	424	Left	-
	–	–	425	No deflection	-
	–	–	426	Right	-

So, the correct value of X is $4.25\,\Omega$.

Example 11 To locate null point, deflection battery key (K_1) is pressed before the galvanometer key (K_2). Explain why?

Sol. If galvanometer key K_2 is pressed first then just after closing the battery key K_1, current suddenly increases. So, due to self-induction, a large back emf is generated in the galvanometer, which may damage the galvanometer.

Example 12 What are the maximum and minimum values of unknown resistance X, which can be determined using the post office box.

Sol. $X = \frac{QR}{P}$

$$\therefore \quad X_{max} = \frac{Q_{max} R_{max}}{P_{min}}$$

$$= \frac{1000}{10}(11110)$$

$$= 1111\ \text{k}\Omega \qquad \textbf{Ans.}$$

$$X_{min} = \frac{Q_{min} R_{min}}{P_{max}}$$

$$= \frac{(10)(1)}{1000}$$

$$= 0.01\,\Omega \qquad \textbf{Ans.}$$

INTRODUCTORY Exercise 4

1. In post office box experiment, $\frac{Q}{P} = \frac{1}{10}$. If $R = 142\,\Omega$ is used then we get deflection towards right and if $R = 143\,\Omega$, then deflection is towards left. What is the range of unknown resistance?
2. What is the change in post office box experiment if battery is connected between B and C and galvanometer is connected across A and C?
3. For the post office box arrangement to determine the value of unknown resistance, the unknown resistance should be connected between **(2004 JEE)**

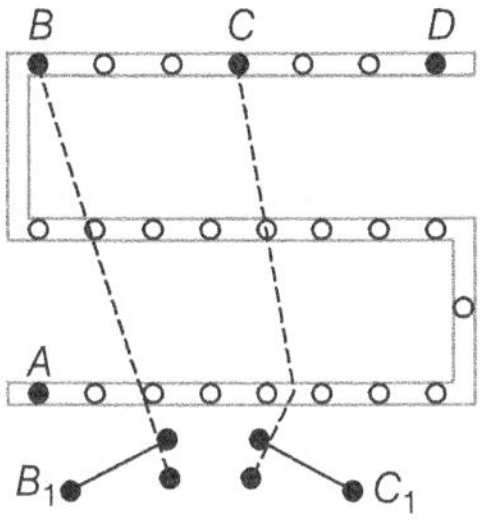

(a) B and C (b) C and D (c) A and D (d) B_1 and C_1

Answers

1. $14.2\,\Omega$ to $14.3\,\Omega$ **2.** See the hints **3.** (c)

See hints on page 87.

Miscellaneous Examples

Ex. 1 Draw the circuit for experimental verification of Ohm's law using a source of variable DC voltage, a main resistance of 100 Ω, two galvanometers and two resistances of values 10^6 Ω and 10^{-3} Ω respectively.

Clearly show the positions of the voltmeter and the ammeter. **(2004 JEE)**

Sol.

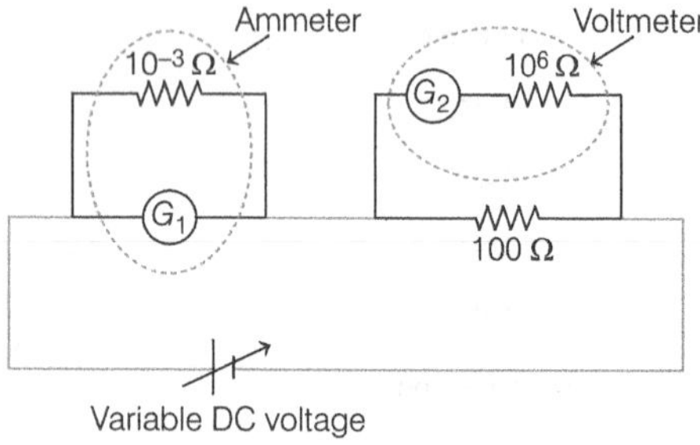

For verification of Ohm's law $\left(\frac{V}{i} = \text{constant} = R\right)$, we need an ohmic resistance, which follows this law. A voltmeter which will measure potential difference across this resistance, an ammeter which will measure current through this resistance and a variable battery which can provide a variable current in the circuit. Now, for different values of i, we have to measure different values of V and then prove that,

$$V \propto i \quad \text{or} \quad \frac{V}{i} = \text{constant}$$

and this constant is called resistance of that.

Ex. 2 An ammeter and a voltmeter are connected in series to a battery of emf $E = 6.0$ V. When a certain resistance is connected in parallel with the voltmeter, the reading of the voltmeter decreases two times, whereas the reading of the ammeter increases the same number of times. Find the voltmeter reading after the connection of the resistance. **(Irodov 3.173)**

Sol. Let R = resistance of ammeter

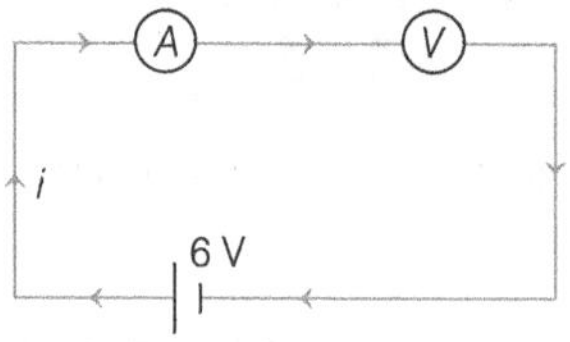

Potential difference across voltmeter = 6 − potential difference across ammeter

In first case, $V = 6 - iR$ ...(i)

In second case, $\frac{V}{2} = 6 - (2i)R$ ...(ii)

Solving these two equations, we get

$$V = 4 \text{ volt}$$

$\therefore$ $V/2 = 2$ volt **Ans.**

Ex. 3 A voltmeter of resistance R_1 and an ammeter of resistance R_2 are connected in series across a battery of negligible internal resistance. When a resistance R is connected in parallel to voltmeter, reading of ammeter increases three times while that of voltmeter reduces to one third. Find R_1 and R_2 in terms of R.

Sol. Let E be the emf of the battery.

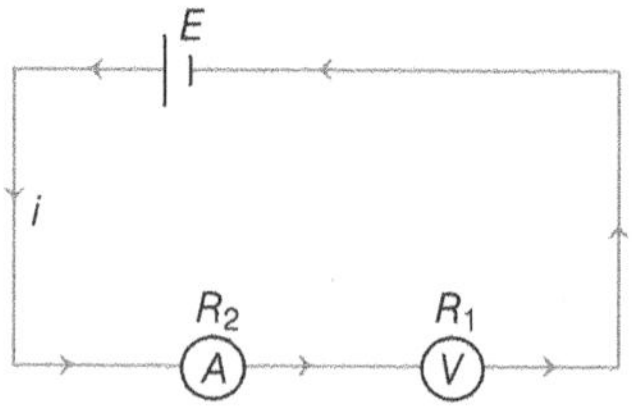

In the second case main current increases three times while current through voltmeter will reduce to $i/3$. Hence, the remaining $3i - i/3 = 8i/3$ passes through R as shown in figure.

$$V_C - V_D = \left(\frac{i}{3}\right)R_1 = \left(\frac{8i}{3}\right)R$$

or $R_1 = 8R$ **Ans.**

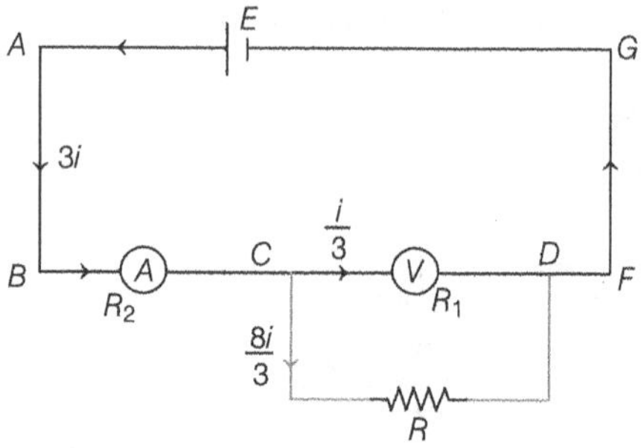

In the second case, main current becomes three times. Therefore, total resistance becomes $\frac{1}{3}$ times or

$$R_2 + \frac{RR_1}{R + R_1} = \frac{1}{3}(R_1 + R_2)$$

Substituting $R_1 = 8R$, we get

$$R_2 = \frac{8R}{3}$$ **Ans.**

MODULE 8 Exercise

Assertion and Reason

See directions on page 25.

1. **Assertion :** In the circuit shown in figure ammeter and voltmeter are non-ideal. When positions of ammeter and voltmeter are changed, reading of ammeter will increase while that of voltmeter will decrease.

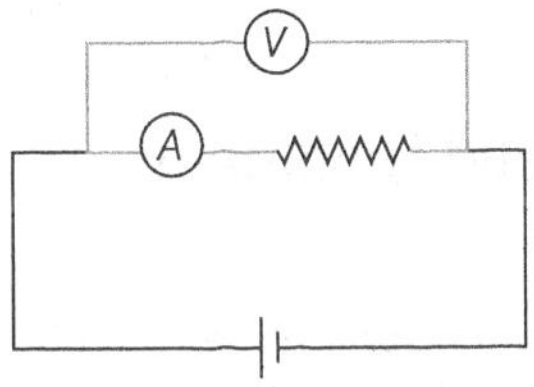

Reason : Resistance of an ideal ammeter is zero while that of an ideal voltmeter is infinite.

2. **Assertion :** In case of potentiometer experiment if emfs of both known and unknown batteries are made two times, null point length will remain unchanged.

Reason : Null point length does not depend on emf of either of the two batteries.

3. **Assertion :** In potentiometer experiment, null point cannot be obtained, if emf of unknown battery is more than the emf of known battery.

Reason : By increasing the emf of known battery, null point length decreases.

Single Correct Option

4. By mistake, a voltmeter is connected in series and an ammeter in parallel (they interchange their positions in the circuit). When the circuit is switched ON, then
   (a) both ammeter and voltmeter will be damaged
   (b) neither the ammeter nor the voltmeter will be damaged
   (c) only the ammeter will be damaged
   (d) only the voltmeter will be damaged

5. The potential difference between points A and B is

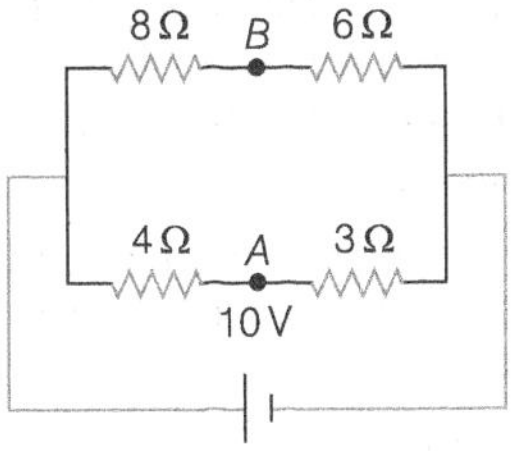

(a) $\frac{20}{7}$ V (b) $\frac{40}{7}$ V
(c) $\frac{10}{7}$ V (d) zero

6. The deflection in a galvanometer falls from 50 divisions to 20 divisions, when a 12 Ω shunt is applied. The galvanometer resistance is
   (a) 18 Ω (b) 24 Ω (c) 30 Ω (d) 36 Ω

7. If 2% of the main current is to be passed through the galvanometer of resistance G, the resistance of shunt required is
   (a) $\frac{G}{49}$ (b) $\frac{G}{50}$ (c) $49G$ (d) $50G$

8. The wire of potentiometer has resistance 4 Ω and length 1 m. It is connected to a cell of emf 2 volt and internal resistance 1 Ω. If a cell of emf 1.2 volt is balanced by it, the balancing length will be
   (a) 90 cm (b) 60 cm (c) 50 cm (d) 75 cm

9. Potentiometer wire of length 1 m is connected in series with 490 Ω resistance and 2 V battery. If 0.2 mV/cm is the potential gradient, then resistance of the potentiometer wire is approximately
   (a) 4.9 Ω (b) 7.9 Ω (c) 5.9 Ω (d) 6.9 Ω

10. A potentiometer having the potential gradient of 2 mV/cm is used to measure the difference of potential across a resistance of 10 Ω in some circuit. If a length of 50 cm of the potentiometer wire is required to get the null point, the current passing through the 10 Ω resistor is (in mA)
    (a) 1 (b) 2 (c) 5 (d) 10

11. A, B and C are voltmeters of resistances R, $1.5R$ and $3R$ respectively. When some potential difference is applied between x and y, the voltmeter readings are V_A, V_B and V_C, then

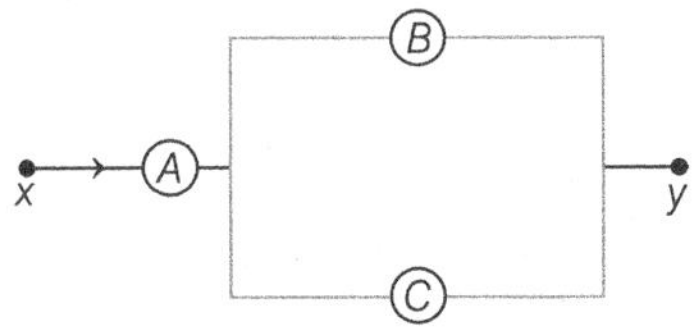

(a) $V_A = V_B = V_C$ (b) $V_A \neq V_B = V_C$
(c) $V_A = V_B \neq V_C$ (d) $V_A + V_B = V_C$

12. The figure shows a meter bridge circuit with $AB = 100$ cm, $X = 12\ \Omega$ and $R = 18\ \Omega$ and the jockey J in the position of balance. If R is now made 8 Ω, through what distance will J have to be moved to obtain balance?

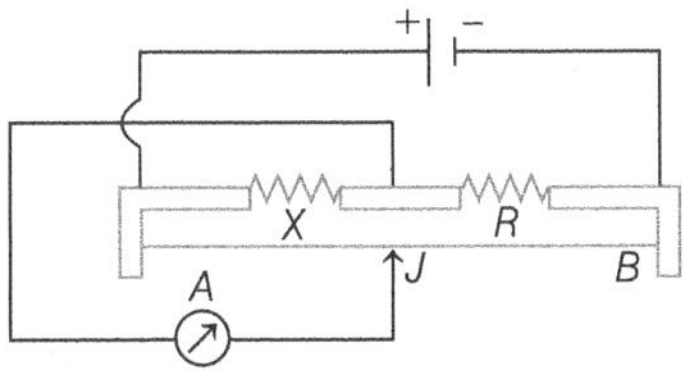

(a) 10 cm (b) 20 cm (c) 30 cm (d) 40 cm

13. A battery of emf $E_0 = 12$ V is connected across a 4 m long uniform wire AB having resistance per unit length 4 Ω/m. Two cells of small emfs $\varepsilon_1 = 2$ V and $\varepsilon_2 = 4$ V having internal resistance 2 Ω and 6 Ω respectively are connected as shown in the figure. If galvanometer shows no deflection at the point N, the distance of point N from the point A is equal to

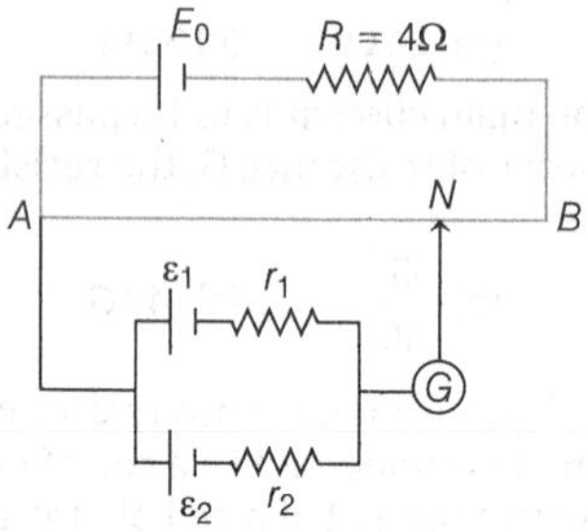

(a) $\frac{5}{3}$ m (b) $\frac{4}{3}$ m
(c) $\frac{3}{2}$ m (d) None of these

14. A galvanometer of resistance 50 Ω is connected to a battery of 3 V along with resistance of 2950 Ω in series. A full scale deflection of 30 divisions is obtained in the galvanometer. In order to reduce this deflection to 20 divisions, the above series resistance should be

(a) 4450 Ω (b) 5050 Ω (c) 5550 Ω (d) 6050 Ω

15. Figure shows a potentiometer arrangement with $R_{AB} = 10\,\Omega$ and rheostat of variable resistance x. For $x = 0$ null deflection point is found at 20 cm from A. For unknown value of x null deflection point was at 30 cm from A, then the value of x is

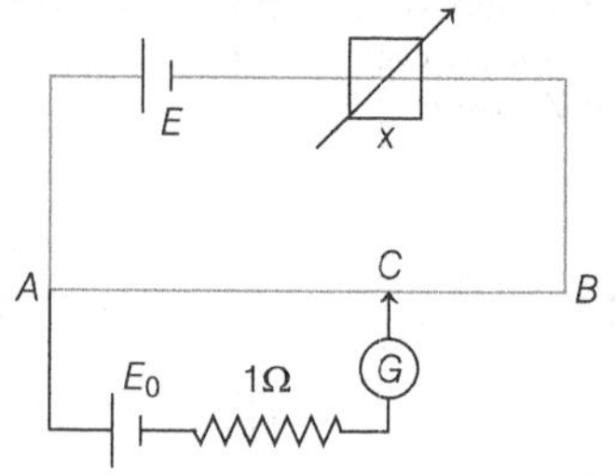

(a) 10 Ω (b) 5 Ω (c) 2 Ω (d) 1 Ω

16. In the shown arrangement of the experiment of the meter bridge if AC corresponding to null deflection of galvanometer is x, what would be its value if the radius of the wire AB is doubled?

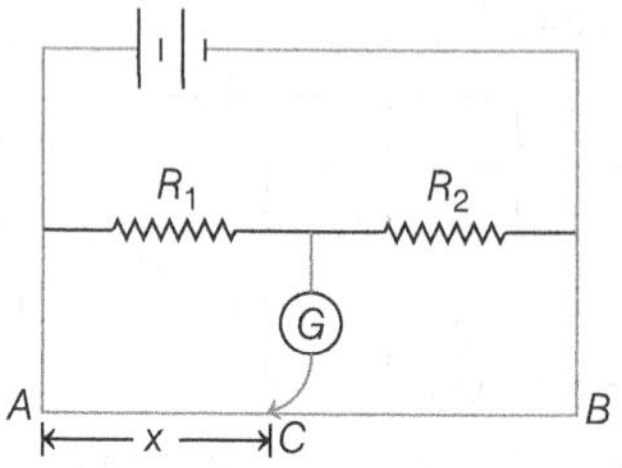

(a) x (b) $x/4$ (c) $4x$ (d) $2x$

Multiple Correct Options

17. Three voltmeters all having different resistances, are joined as shown. When some potential difference is applied across A and B, their readings are V_1, V_2 and V_3. Then,

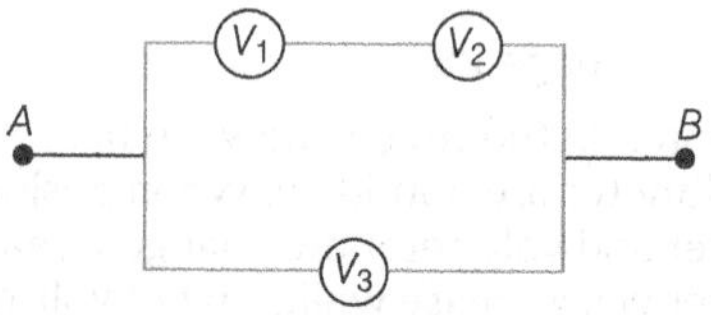

(a) $V_1 = V_2$ (b) $V_1 \neq V_2$
(c) $V_1 + V_2 = V_3$ (d) $V_1 + V_2 > V_3$

Passage Based Questions

Passage 1 (Q.18 to 19)

The length of a potentiometer wire is 600 cm and it carries a current of 40 mA. For a cell of emf 2V and internal resistance 10 Ω, the null point is found to be at 500 cm. On connecting a voltmeter across the cell, the balancing length is decreased by 10 cm.

18. The voltmeter reading will be

(a) 1.96 V (b) 1.8 V
(c) 1.64 V (d) 0.96 V

19. The resistance of the voltmeter is

(a) 500 Ω (b) 290 Ω
(c) 490 Ω (d) 20 Ω

Passage 2 (Q.20 to 23)

Consider a potentiometer arrangement having a uniform wire AB of resistance 99Ω.

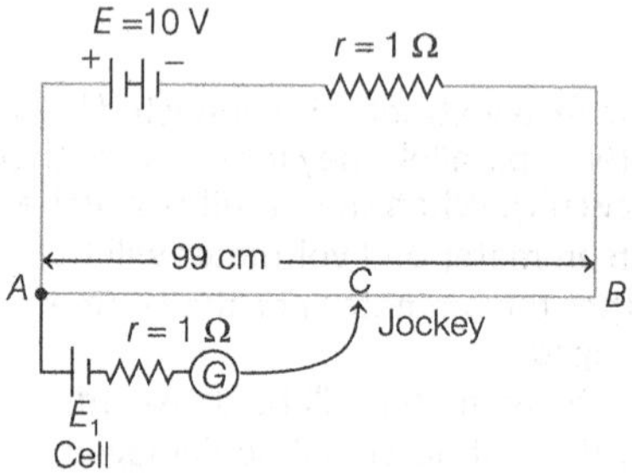

A driving battery of 10 V emf and 1 Ω internal resistance is used in the potentiometer. A cell of unknown emf E_1 and internal resistance of 1 Ω is balanced against AC length of potentiometer wire.

20. When jockey is touched with wire AB at point C such that $AC = 40$ cm, there is no deflection in galvanometer. The emf of cell is

(a) 4 V (b) 5 V (c) $\frac{5}{4}$ V (d) 6 V

21. If jockey is touched at point B, then potential difference across wire AB is

(a) 6.93 V (b) 8.93 V (c) 9.93 V (d) 0 V

22. If jockey is touched at point A, then potential difference across the cell is
(a) 3 V (b) 2.5 V
(c) 2 V (d) None of these

23. When jockey is touched at point A, then potential difference across the terminals of battery of 10 V is
(a) 9.9 V (b) 9.0 V (c) 8 V (d) 7.8 V

Subjective Questions

24. The emf E and the internal resistance r of the battery shown in figure are 4.3 V and $1.0\,\Omega$ respectively. The external resistance R is $50\,\Omega$. The resistances of the ammeter and voltmeter are $2.0\,\Omega$ and $200\,\Omega$, respectively.

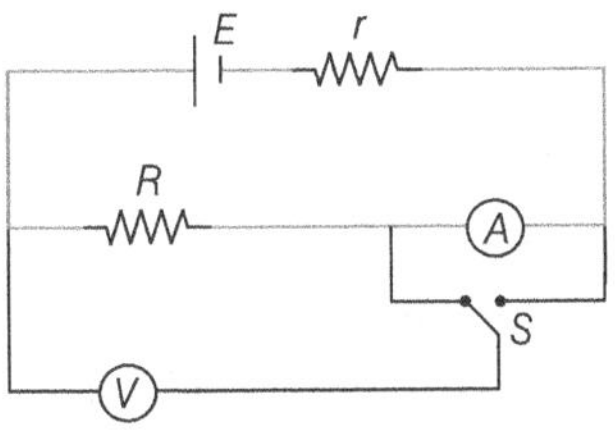

(a) Find the readings of the two meters.
(b) The switch is thrown to the other side. What will be the readings of the two meters now?

25. A resistance box, a battery and a galvanometer of resistance G ohm are connected in series. If the galvanometer is shunted by resistance of S ohm, find the change in resistance in the box required to maintain the current from the battery unchanged.

26. Resistances R_1 and R_2, each $60\,\Omega$, are connected in series. The potential difference between points A and B is 120 V. Find the reading of voltmeter connected between points C and D, if its resistance $r = 120\,\Omega$.

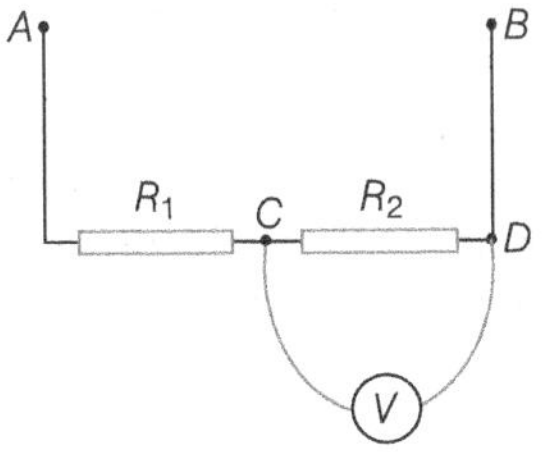

27. A moving coil galvanometer of resistance $20\,\Omega$ gives a full scale deflection when a current of 1 mA is passed through it. It is to be converted into an ammeter reading 20 A on full scale. But the shunt of $0.005\,\Omega$ only is available. What resistance should be connected in series with the galvanometer coil?

28. A cell of emf 3.4 V and internal resistance $3\,\Omega$ is connected to an ammeter having resistance $2\,\Omega$ and to an external resistance of $100\,\Omega$. When a voltmeter is connected across the $100\,\Omega$ resistance, the ammeter reading is 0.04 A. Find the voltage reading by the voltmeter and its resistance. Had the voltmeter been an ideal one what would have been its reading?

29. A 6 V battery of negligible internal resistance is connected across a uniform wire AB of length 100 cm. The positive terminal of another battery of emf 4 V and internal resistance $1\,\Omega$ is joined to the point A as shown in figure. Take the potential at B to be zero.

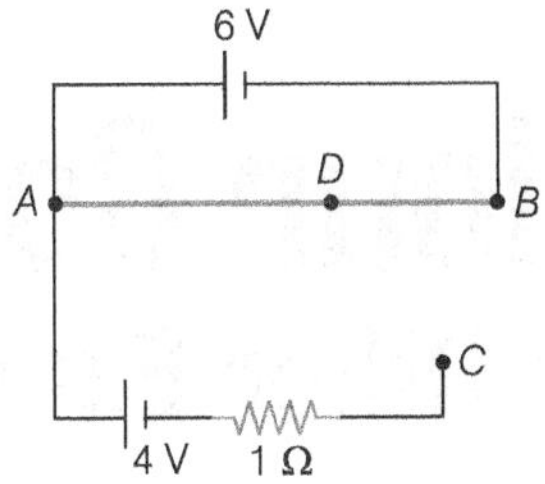

(a) What are the potentials at the points A and C?
(b) At which point D of the wire AB, the potential is equal to the potential at C?
(c) If the points C and D are connected by a wire, what will be the current through it?
(d) If the 4V battery is replaced by 7.5 V battery, what would be the answers of parts (a) and (b)?

30. A thin uniform wire AB of length 1 m, an unknown resistance X and a resistance of $12\,\Omega$ are connected by thick conducting strips, as shown in the figure. A battery and a galvanometer (with a sliding jockey connected to it) are also available. Connections are to be made to measure the unknown resistance X. Using the principle of Wheatstone bridge answer the following questions : **(2002 JEE)**

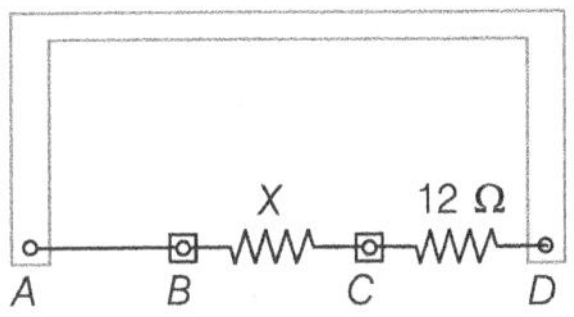

(a) Are there positive and negative terminals on the galvanometer?
(b) Copy the figure in your answer book and show the battery and the galvanometer (with jockey) connected at appropriate points.
(c) After appropriate connections are made, it is found that no deflection takes place in the galvanometer when the sliding jockey touches the wire at a distance of 60 cm from A. Obtain the value of the resistance X.

Answers

1. (b)	**2.** (c)	**3.** (b)	**4.** (b)	**5.** (d)
6. (a)	**7.** (a)	**8.** (d)	**9.** (a)	**10.** (d)
11. (a)	**12.** (b)	**13.** (d)	**14.** (a)	**15.** (b)
16. (a)	**17.** (b,c)	**18.** (a)	**19.** (c)	**20.** (a)
21. (a)	**22.** (d)	**23.** (a)		

24. (a) 0.1 A, 4.0 V (b) 0.08 A, 4.2 V **25.** $\dfrac{G^2}{G+S}$ **26.** 48 V

27. $80\,\Omega$ **28.** $400\,\Omega$, 3.2 V, 3.238 V

29. (a) 6 V, 2V (b) $AD = 66.67$cm (c) Zero (d) 6 V, -1.5 V, no such point D exists

30. (a) No (c) $8\,\Omega$

See hints on pages 87-90.

MODULE 9

Colour Codes For Resistors
(For Other Examination)

Resistors are of the following two major types :

(i) wire bound resistors and (ii) carbon resistors

First type of resistors are made by winding the wires of an alloy like nichrome, manganin or constantan etc. Materials are so chosen that their resistivities are relatively less sensitive to temperature.

In carbon resistors, carbon with a suitable binding agent is molded into a cylinder. Wire leads are attached to this cylinder and the entire resistor is encased in a ceramic or plastic jacket. The two leads connect the resistor to a circuit. Carbon resistors are compact and inexpensive. Their values are given using a colour code.

Table

Colour	1st digit and 2nd digit	Multiplier	Tolerance (%)
Black	0	1	-
Brown	1	10^1	-
Red	2	10^2	-
Orange	3	10^3	-
Yellow	4	10^4	-
Green	5	10^5	-
Blue	6	10^6	-
Violet	7	10^7	-
Gray	8	10^8	-
White	9	10^9	-
Gold	-	10^{-1}	5
Silver	-	10^{-2}	10
No colour	-		20

The resistors have a set of four (or three) co-axial coloured rings, whose significance are listed in above table.

The colours are noted from left to right.

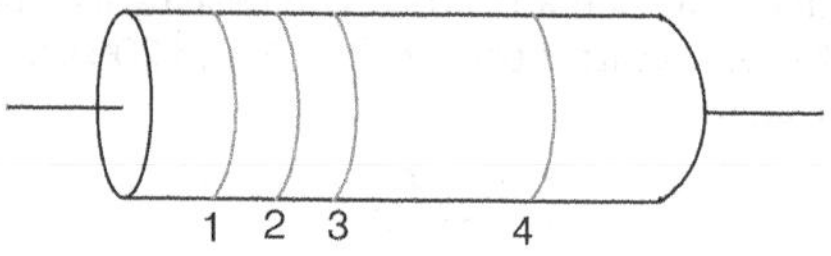

Colour 1 $\rightarrow$ First significant figure

Colour 2 $\rightarrow$ Second significant figure

Colour 3 $\rightarrow$ Decimal multiplier

Colour 4 (or no colour) $\rightarrow$ Tolerance or possible variation in percentage.

Important POINTS

- To remember the value of colour coding used for carbon resistor, the following sentences are found to be of great help (where bold letters stand for colours)
  B B ROY **G**reat **B**ritain **V**ery **G**ood **W**ife wearing **Gold Silver** necklace.
  OR
  Black **B**rown **R**ods **O**f **Y**our **G**ate **B**ecome **V**ery **G**ood **W**hen **G**iven **S**ilver colour

Example 1 The four colours on a resistor are : brown, yellow, green and gold as read from left to right. What is resistance corresponding to these colours?

Sol. From the table we can see that

Brown colour $\rightarrow 1$

Yellow colour $\rightarrow 4$

Green colour $\rightarrow 10^5$ and

Gold colour $\rightarrow 5\,\%$

$\therefore \quad R = (14 \times 10^5 \pm 5\%)\,\Omega$ **Ans.**

MODULE 9 Exercise

Single Correct Option

1. Carbon resistors used in electronic circuits are marked for their value of resistance and tolerance by a colour code. A given resistor has colour scheme brown, red, green and gold. Its value in ohm is
   (a) $5.2\times 10^6 \pm 10\%$
   (b) $2.4\times 10^5 \pm 5\%$
   (c) $1.2\times 10^4 \pm 10\%$
   (d) $1.2\times 10^6 \pm 5\%$

2. A 200 Ω resistor has a certain colour code. If one replaces the red colour by green in the code, the new resistance will be **(2019 Main)**
   (a) 100 Ω
   (b) 400 Ω
   (c) 300 Ω
   (d) 500 Ω

3. The Wheatstone bridge shown in figure here, gets balanced when the carbon resistor is used as R_1 has the colour code (orange, red, brown). The resistors R_2 and R_4 are 80 Ω and 40 Ω, respectively.
   Assuming that the colour code for the carbon resistors gives their accurate values, the colour code for the carbon resistor is used as R_3 would be **(2019 Main)**

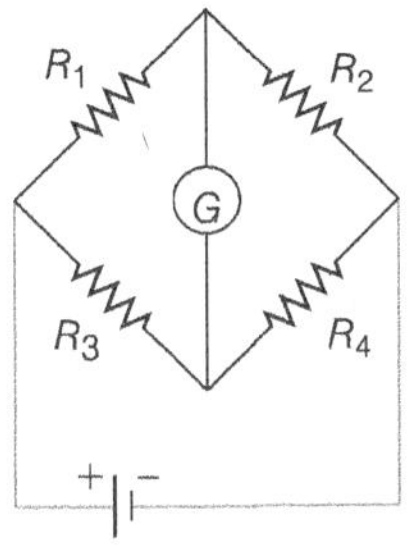

   (a) brown, blue, black
   (b) brown, blue, brown
   (c) grey, black, brown
   (d) red, green, brown

Subjective Questions

4. For the given carbon resistor, let the first strip be yellow, second strip be red, third strip be orange and fourth be gold. What is its resistance?

5. The resistance of the given carbon resistor is $(24\times 10^6 \pm 10\%)\Omega$. What is the sequence of colours on the strips provided on resistor?

Answers

1. (d) **2.** (d) **3.** (b) **4.** $(42\times 10^3 \pm 5\%)\Omega$ **5.** Red, Yellow, Blue, Silver

See hints on page 90.

Unit Exercise (Cumulative)

Note *Attempt starred (*) problems after completing Unit 3, Module 8.*

Level 1

Fill in the Blank

1. The equivalent resistance between points A and B of the circuit given below is .... **(1997 JEE)**

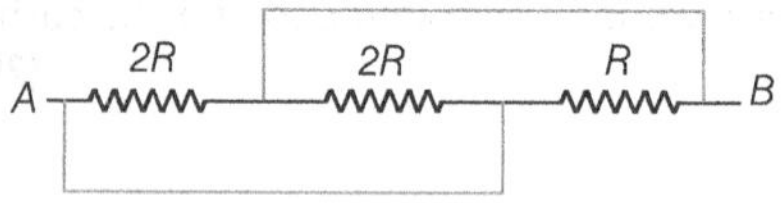

Assertion and Reason

See directions on page 25.

2. **Assertion :** In a meter bridge experiment, null point for an unknown resistance is measured. Now, the unknown resistance is put inside an enclosure maintained at a higher temperature. The null point can be obtained at the same point as before by decreasing the value of the standard resistance.
Reason : Resistance of a metal increases with increase in temperature. **(2008 JEE)**

Single Correct Option

3. N identical current sources each of emf E and internal resistance r are connected to form a closed loop as shown in figure. The potential difference between points A and B which divides the circuit into n and $(N - n)$ units is

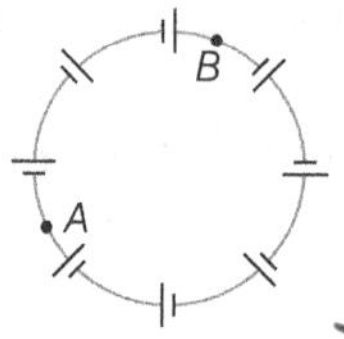

(a) NE (b) $(N - n)E$ (c) nE (d) zero

4. The drift velocity of free electrons in a conductor is v, when a current i is flowing in it. If both the radius and current are doubled, then the drift velocity will be
(a) v (b) $v/2$ (c) $v/4$ (d) $v/8$

5. In the given circuit the current flowing through the resistance $20\,\Omega$ is 0.3 A, while the ammeter reads 0.8 A. What is the value of R_1?

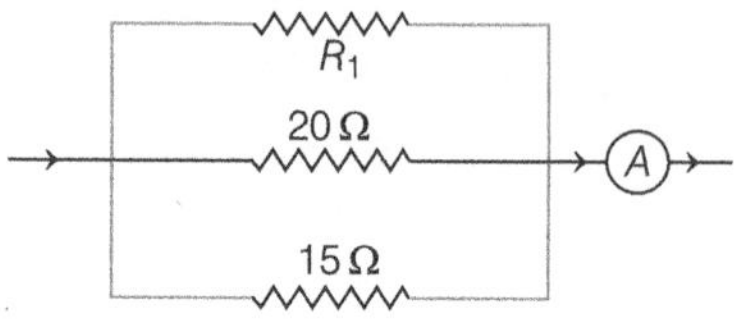

(a) $30\,\Omega$ (b) $40\,\Omega$ (c) $50\,\Omega$ (d) $60\,\Omega$

6. In the circuit diagram shown in figure, a fuse bulb can cause all other bulbs to go out. Identify the bulb

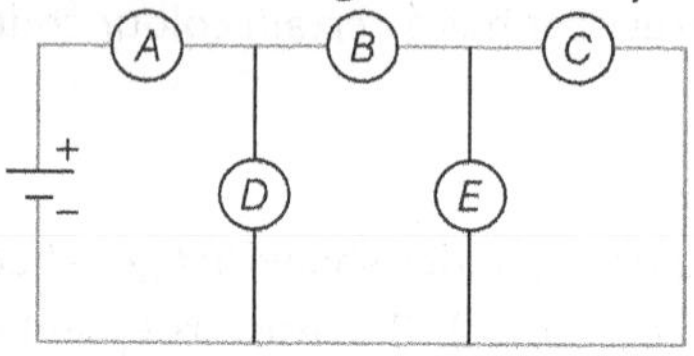

(a) B (b) C (c) A (d) D or E

7. A part of a circuit is shown in figure. Here reading of ammeter is 5 A and voltmeter is 100 V. If voltmeter resistance is 2500 ohm, then the resistance R is approximately

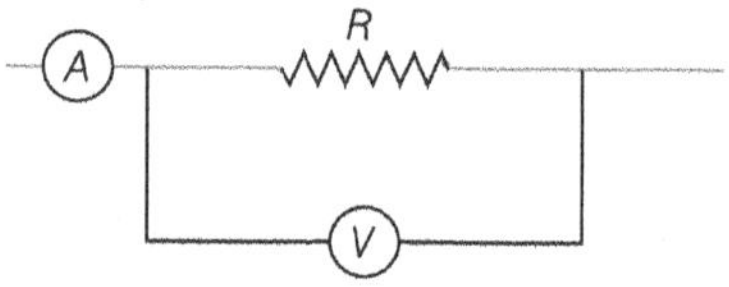

(a) $20\,\Omega$ (b) $10\,\Omega$ (c) $100\,\Omega$ (d) $200\,\Omega$

8. Two resistances are connected in two gaps of a meter bridge. The balance point is 20 cm from the zero end. A resistance of $15\,\Omega$ is connected in series with the smaller of the two. The null point shifts to 40 cm. The value of the smaller resistance in Ω is
(a) 3 (b) 6 (c) 9 (d) 12

9. In the given circuit, the voltmeter records 5 volt. The resistance of the voltmeter in Ω is

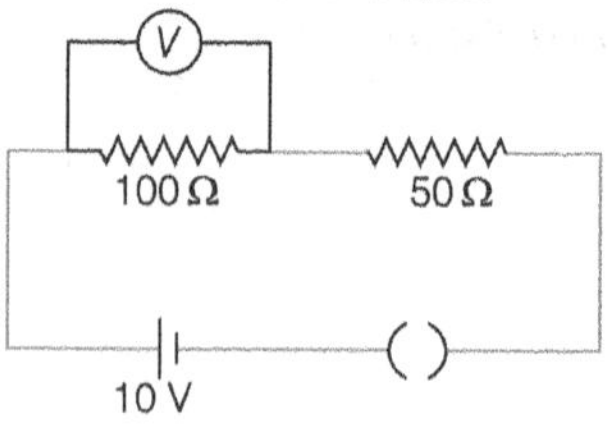

(a) 200 (b) 100 (c) 10 (d) 50

10. Find the ratio of currents as measured by ammeter in two cases when the key is open and when the key is closed

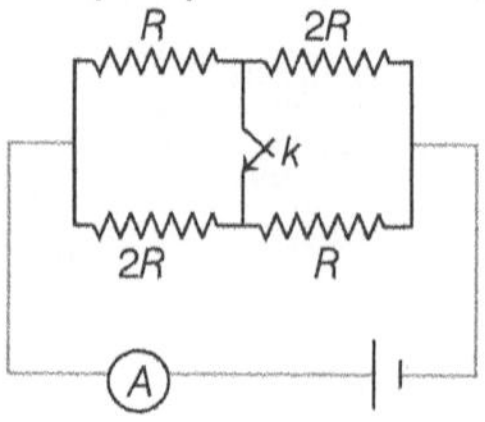

(a) 9/8 (b) 10/11
(c) 8/9 (d) None of these

11. The reading of the ammeter as per figure shown is

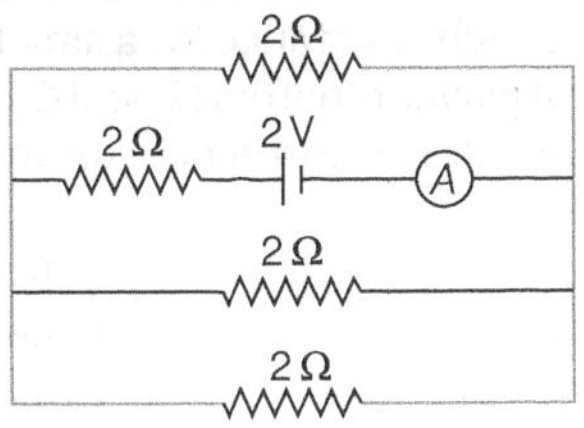

(a) $\frac{1}{8}$ A (b) $\frac{3}{4}$ A (c) $\frac{1}{2}$ A (d) 2 A

12. Five identical resistors, each of value 1100 Ω are connected to a 220 V battery as shown. The reading of the ideal ammeter A is

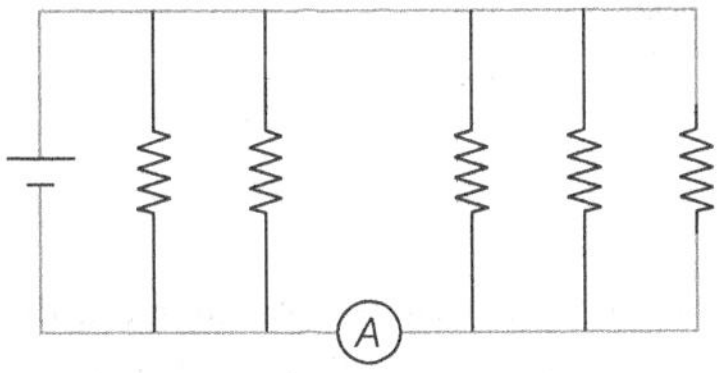

(a) 1/5 A (b) 2/5 A (c) 3/5 A (d) 4/5 A

13. A constant voltage is applied between the two ends of a uniform metallic wire. Some heat is developed in it. The heat developed is doubled, if
(a) both the length and radius of the wire are halved
(b) both the length and radius of the wire are doubled
(c) the radius of the wire is doubled
(d) the length of the wire is doubled

14. In the circuit shown here, the readings of the ammeter and voltmeter (both are ideal) are

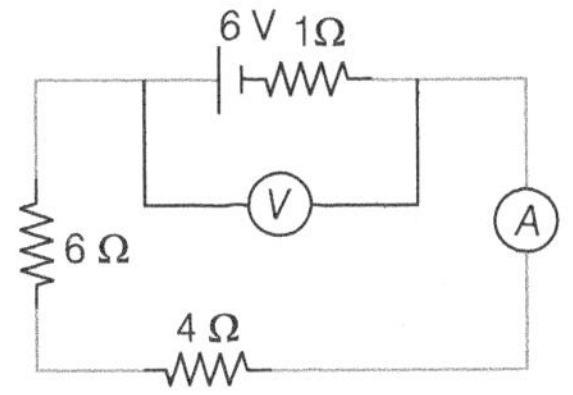

(a) 6 A, 60 V (b) 0.6 A, 6 V
(c) 6 A, 6 V (d) 6/11 A, 60/11 V

15. If a resistance R_2 is connected in parallel with a resistance R in the circuit as shown, then possible value of current through R and the possible value of R_2 will be

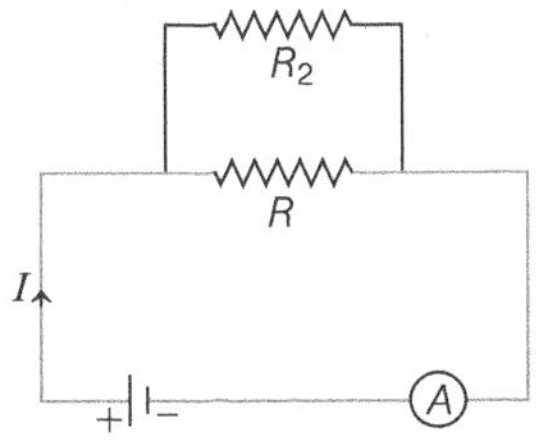

(a) $\frac{I}{3}$, R (b) $\frac{I}{2}$, $2R$ (c) $\frac{I}{3}$, $2R$ (d) $\frac{I}{2}$, R

16. An ammeter A, a voltmeter V and a resistance R are connected as shown in the figure. If the voltmeter reading is 1.6 V and the ammeter reading is 0.4 A, then R is

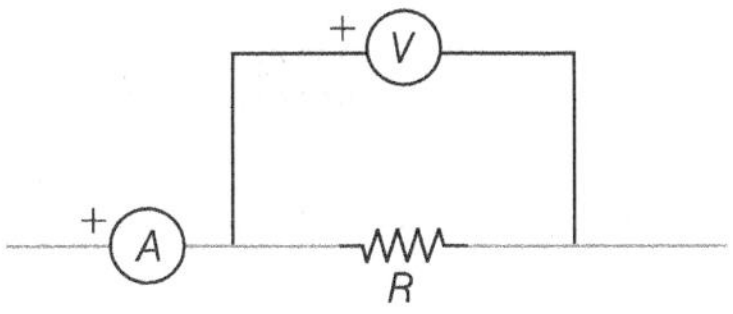

(a) equal to 4 Ω (b) greater than 4 Ω
(c) less than 4 Ω (d) between 3 Ω and 4 Ω

17. In the given circuit, the resistances are given in ohm. The current through the 10 Ω resistance is 3 A while that through the resistance X is 1 A. No current passes through the galvanometer. The values of the unknown resistances X and Y are respectively (in ohm)

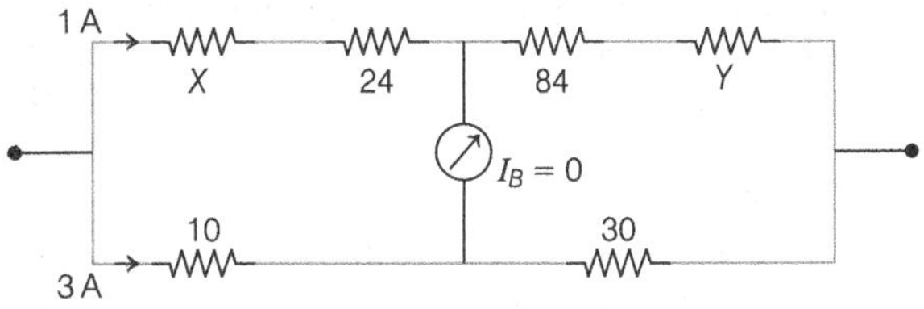

(a) 14 and 54 (b) 12 and 6
(c) 6 and 12 (d) 6 and 6

18. Two batteries, one of emf 18 volt and internal resistance 2Ω and the other of emf 12 volt and internal resistance 1Ω, are connected as shown below. The ideal voltmeter V will record a reading of

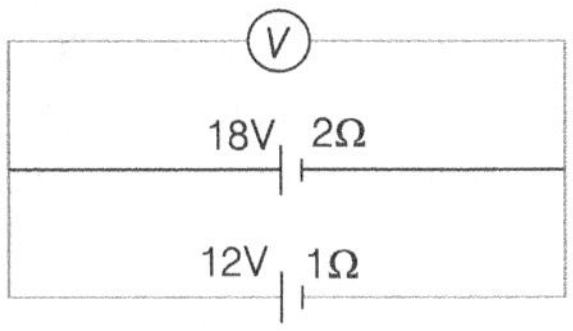

(a) 11 V (b) 8 V
(c) 14 V (d) 10 V

19. A resistance of 4 Ω and a wire of length 5 m and resistance 5Ω are joined in series and connected to a cell of emf 10 V and internal resistance 1 Ω. A parallel combination of two identical cells is balanced across 3 m of the wire. The emf E of each cell is

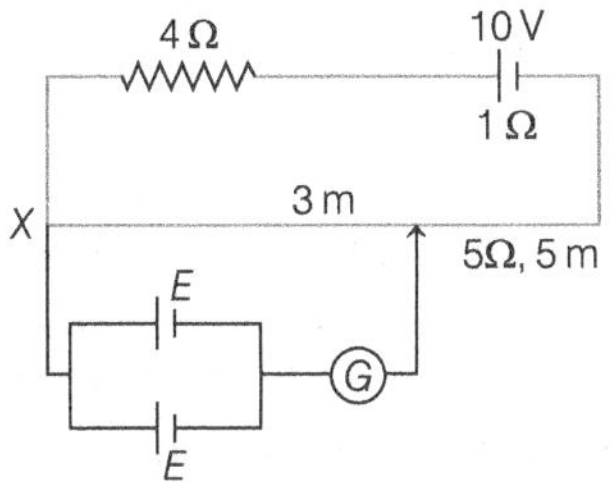

(a) 1.5 V (b) 3.0 V
(c) 0.67 V (d) 1.33 V

20. The potential difference across the $100\,\Omega$ resistance in the circuit is measured by a voltmeter of $900\,\Omega$ resistance. The percentage error made in reading the potential difference is

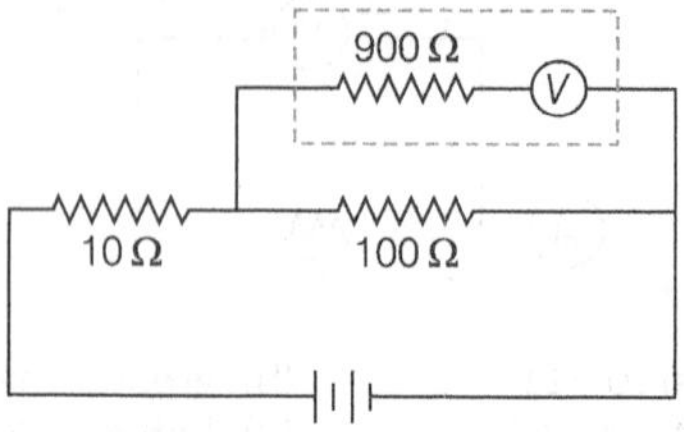

(a) $\frac{10}{9}$ (b) 0.1 (c) 1.0 (d) 10.0

21. An electric cable of copper has just one wire of radius 9 mm. Its resistance is $14\,\Omega$. If this single copper wire of the cable is replaced by seven identical well insulated copper wires each of radius 3 mm connected in parallel, then the new resistance of the combination will be **(2022 Main)**

(a) $9\,\Omega$ (b) $18\,\Omega$
(d) $28\,\Omega$ (d) $126\,\Omega$

22. A 1 m long wire is broken into two unequal parts X and Y. The X part of the wire is stretched into another wire W. Length of W is twice the length of X and the resistance of W is twice that of Y. Find the ratio of length of X and Y. **(2022 Main)**

(a) 1 : 4 (b) 1 : 2
(c) 4 : 1 (d) 2 : 1

23. A metal wire of resistance $3\,\Omega$ is elongated to make a uniform wire of double its previous length. This new wire is now bent and the ends joined to make a circle. If two points on this circle make an angle 60° at the centre, the equivalent resistance between these two points will be **(2019 Main)**

(a) $\frac{7}{2}\Omega$ (b) $\frac{5}{2}\Omega$
(c) $\frac{12}{5}\Omega$ (d) $\frac{5}{3}\Omega$

24. The drift velocity of electrons for a conductor connected in an electrical circuit is v_d. The conductor is now replaced by another conductor with same material and same length but double the area of cross-section. The applied voltage remains same. The new drift velocity of electrons will be **(2023 Main)**

(a) v_d (b) $v_d / 2$
(c) $v_d / 4$ (d) $2v_d$

25. When 5V potential difference is applied across a wire of length 0.1 m, the drift speed of electrons is $2.5 \times 10^{-4} \text{ms}^{-1}$. If the electron density in the wire is $8 \times 10^{28}\ \text{m}^{-3}$ the resistivity of the material is close to **(2015 Main)**

(a) $1.6 \times 10^{-8}\ \Omega\text{-m}$ (b) $1.6 \times 10^{-7}\ \Omega\text{-m}$
(c) $1.6 \times 10^{-5}\ \Omega\text{-m}$ (d) $1.6 \times 10^{-6}\ \Omega\text{-m}$

26. To verify Ohm's law, a student is provided with a test resistor R_T, a high resistance R_1, a small resistance R_2, two identical galvanometers G_1 and G_2, and a variable voltage source V. The correct circuit to carry out the experiment is **(2010 JEE)**

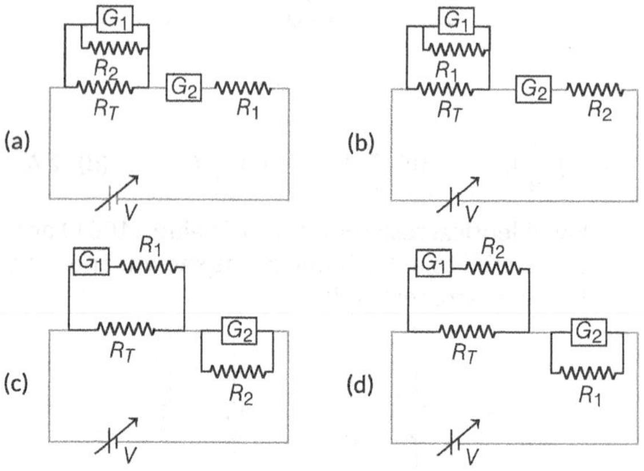

27. A steady current flows in a metallic conductor of non-uniform cross-section. The quantity/quantities constant along the length of the conductor is/are **(1997 JEE)**

(a) current, electric field and drift speed
(b) drift speed only
(c) current and drift speed
(d) current only

28. As shown in the figure, a network of resistors is connected to a battery of 24V with an internal resistance of $3\,\Omega$. The currents through the resistors R_4 and R_5 are I_4 and I_5 respectively. The values of I_4 and I_5 are **(2023 Main)**

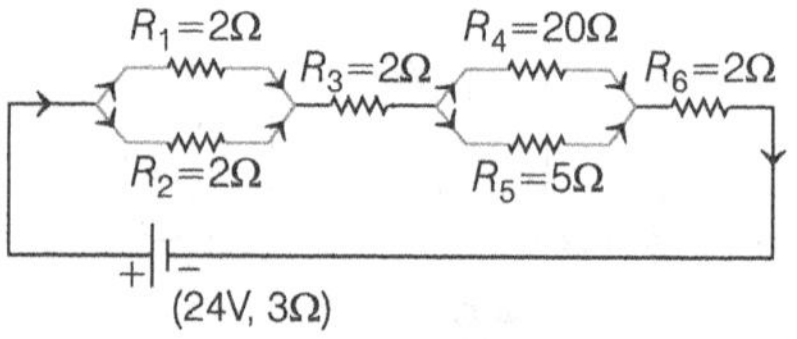

(a) $I_4 = \frac{8}{4}$ A and $I_5 = \frac{2}{5}$ A (b) $I_4 = \frac{2}{5}$ A and $I_5 = \frac{8}{5}$ A
(c) $I_4 = \frac{24}{5}$ A and $I_5 = \frac{6}{5}$ A (d) $I_4 = \frac{6}{5}$ A and $I_5 = \frac{24}{5}$ A

29. In the figure shown, what is the current (in ampere) drawn from the battery? You are given : $R_1 = 15\,\Omega, R_2 = 10\,\Omega, R_3 = 20\,\Omega, R_4 = 5\,\Omega, R_5 = 25\,\Omega, R_6 = 30\,\Omega, E = 15\,\text{V}$ **(2019 Main)**

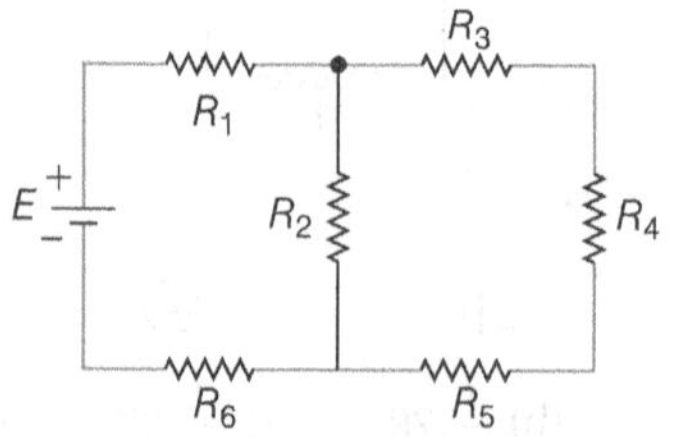

(a) 13/24 (b) 7/18 (c) 20/3 (d) 9/32

30. In the given circuit diagram, the currents $I_1 = -0.3$ A, $I_4 = 0.8$ A and $I_5 = 0.4$ A, are flowing as shown. The currents I_2, I_3 and I_6 respectively, are **(2019 Main)**

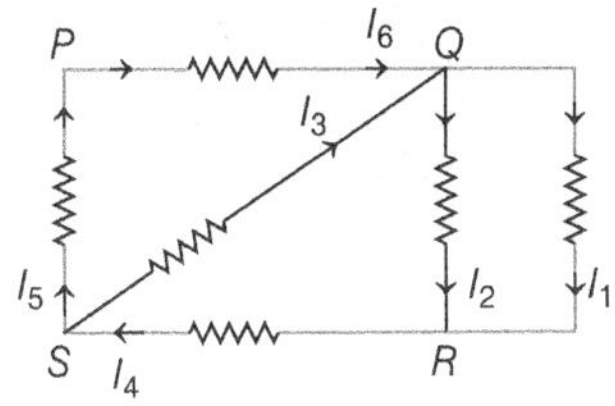

(a) 1.1 A, 0.4 A, 0.4 A (b) 1.1 A, −0.4 A, 0.4A
(c) 0.4 A, 1.1 A, 0.4 A (d) −0.4 A, 0.4 A, 1.1A

31. In the below circuit, the current in each resistance is **(2017 Main)**

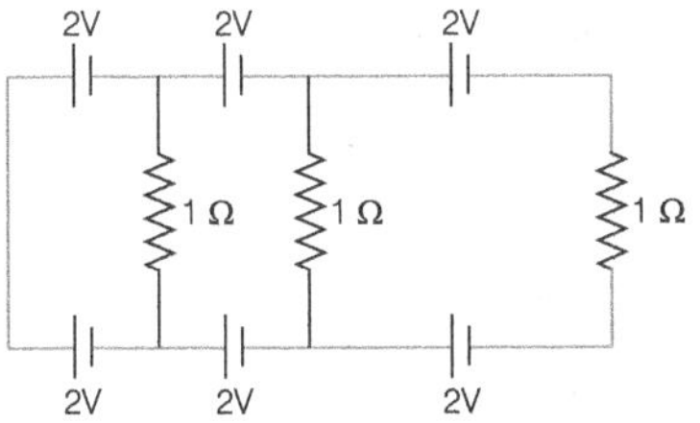

(a) 0.25 A (b) 0.5 A
(c) 0 A (d) 1 A

32. In the circuit shown below, the current in the 1Ω resistor is **(2015 Main)**

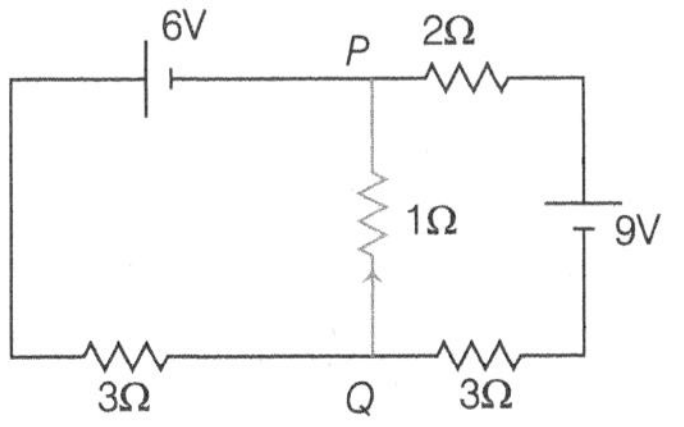

(a) 1.3 A, from P to Q (b) 0.13 A, from Q to P
(c) 0 A (d) 0.13 A, from P to Q

33. Find out the value of current through 2 Ω resistance for the given circuit. **(2005 JEE)**

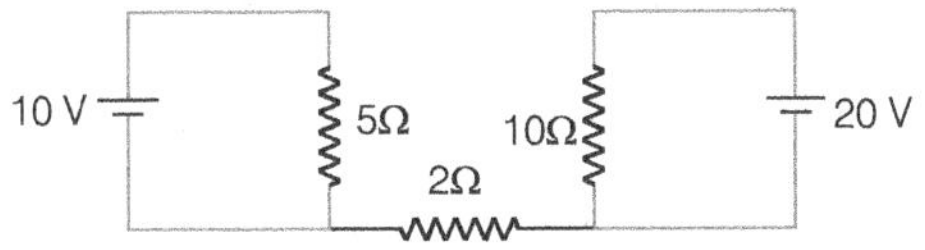

(a) 5 A (b) 2 A (c) zero (d) 4 A

34. In the circuit shown in the figure, the current through **(1998 JEE)**

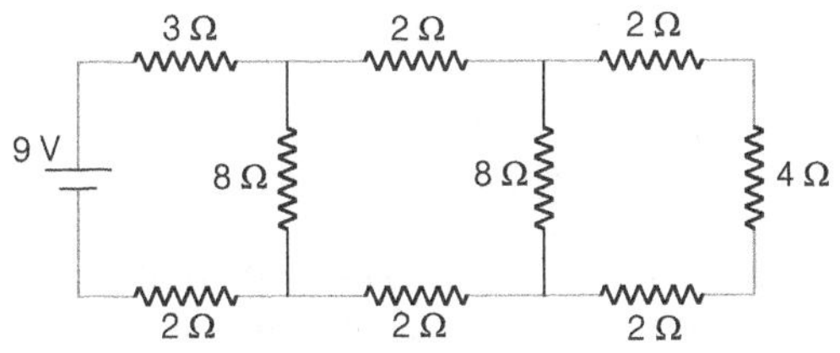

(a) the 3 Ω resistor is 0.50 A
(b) the 3 Ω resistor is 0.25 A
(c) the 4 Ω resistor is 0.50 A
(d) the 4 Ω resistor is 0.25 A

35. A wire of length 10 cm and radius $\sqrt{7} \times 10^{-4}$ m is connected across the right gap of a meter bridge. When a resistance of 4.5 Ω is connected on the left gap by using a resistance box, the balance length is found to be at 60 cm from the left end. If the resistivity of the wire is $R \times 10^{-7}$ Ωm, then value of R is **(2024 Main)**

(a) 63 (b) 70
(c) 66 (d) 35

36. A potentiometer wire AB having length L and resistance $12r$ is joined to a cell D of EMF ε and internal resistance r. A cell C having emf $\varepsilon/2$ and internal resistance $3r$ is connected. The length AJ at which the galvanometer as shown in figure shows no deflection is **(2019 Main)**

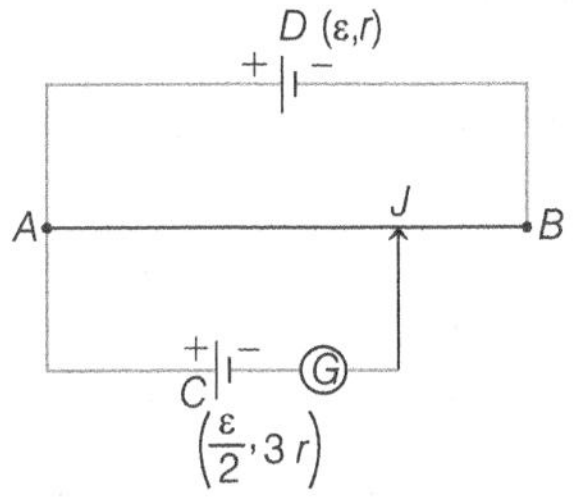

(a) $\frac{5}{12}L$ (b) $\frac{11}{12}L$
(c) $\frac{13}{24}L$ (d) $\frac{11}{24}L$

37. On interchanging the resistances, the balance point of a meter bridge shifts to the left by 10 cm. The resistance of their series combination is 1 kΩ. How much was the resistance on the left slot before interchanging the resistances? **(2018 Main)**

(a) 990 Ω (b) 910 Ω
(c) 900 Ω (d) 550 Ω

38. Which of the following statements is **false**? **(2017 Main)**

(a) In a balanced Wheatstone bridge, if the cell and the galvanometer are exchanged, the null point is disturbed
(b) A rheostat can be used as a potential divider
(c) Kirchhoff's second law represents energy conservation
(d) Wheatstone bridge is the most sensitive when all the four resistances are of the same order of magnitude

39. Two electric bulbs rated at 25 W, 220 V and 100 W, 220 V are connected in series across a 220 V voltage source. If the 25 W and 100 W bulbs draw powers P_1 and P_2 respectively, then **(2019 Main)**

(a) $P_1 = 16$ W, $P_2 = 4$ W (b) $P_1 = 4$ W, $P_2 = 16$ W
(c) $P_1 = 9$ W, $P_2 = 16$ W (d) $P_1 = 16$ W, $P_2 = 9$ W

40. All bulbs in the circuit shown in figure are identical. Which bulb glows most brightly?

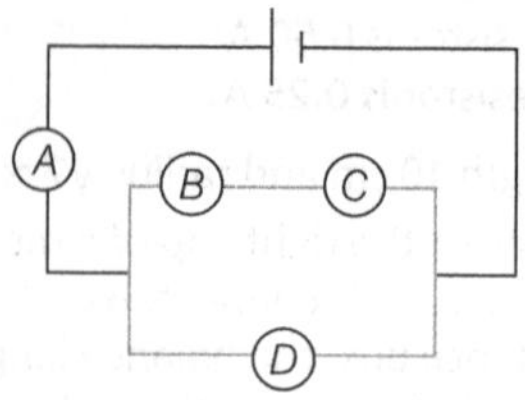

(a) B (b) A (c) D (d) C

41. A student connects an ammeter A and a voltmeter V to measure a resistance R as shown in figure. If the voltmeter reads 20 V and the ammeter reads 4 A, then R is

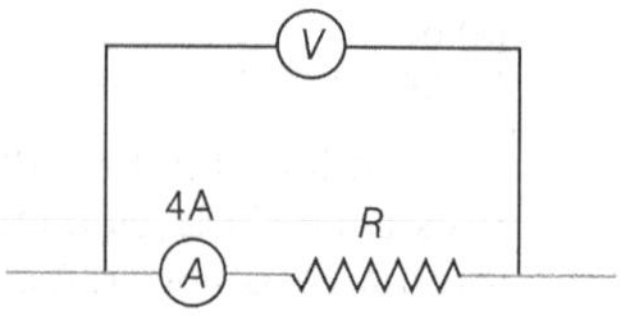

(a) equal to $5\,\Omega$
(b) greater than $5\,\Omega$
(c) less than $5\,\Omega$
(d) greater or less than $5\,\Omega$ depending upon the direction of current

42. A moving coil galvanometer is converted into an ammeter reading up to 0.03 A by connecting a shunt of resistance $\frac{r}{4}$. What is the maximum current which can be sent through this galvanometer, if no shunt is used ? (Here, r = resistance of galvanometer)
(a) 0.004 A (b) 0.005 A (c) 0.006 A (d) 0.008 A

43. In the given potentiometer arrangement, the null point

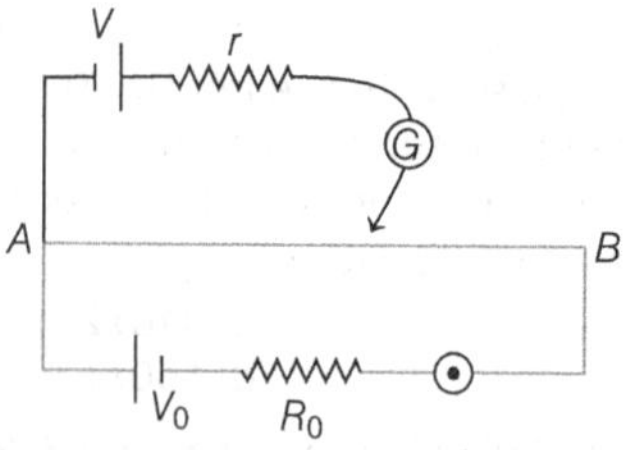

(a) can be obtained for any value of V
(b) can be obtained only if $V < V_0$
(c) can be obtained only if $V > V_0$
(d) can never be obtained

44. In the circuit shown in figure, the resistance of voltmeter is $6\,k\Omega$. The voltmeter reading will be

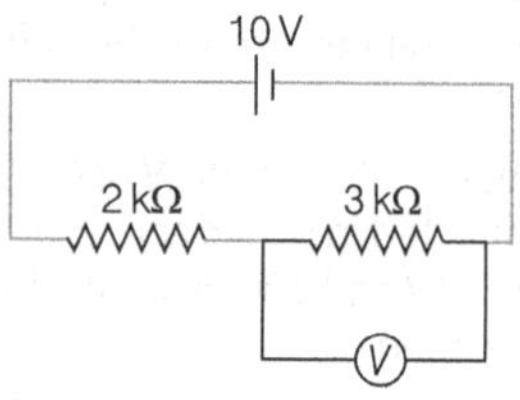

(a) 6 V (b) 5 V (c) 4 V (d) 3 V

45. In the circuit shown $P \neq R$, the reading of galvanometer is same with switch S open or closed. Then **(1999 JEE)**

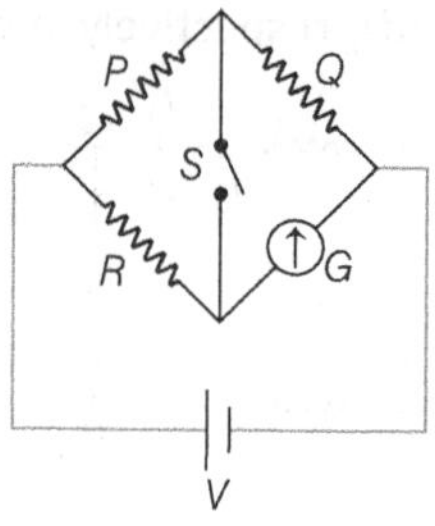

(a) $I_R = I_G$ (b) $I_P = I_G$
(c) $I_Q = I_G$ (d) $I_Q = I_R$

Match the Columns

46. Match the following two columns.

	Column I		Column II
(a)	Electrical resistance	(p)	$[MLT^{-2}A^{2}]$
(b)	Electric potential	(q)	$[ML^{2}T^{-3}A^{-2}]$
(c)	Specific resistance	(r)	$[ML^{2}T^{-3}A^{-1}]$
(d)	Specific conductance	(s)	None of these

47. In the circuit shown in figure, match the following two columns :

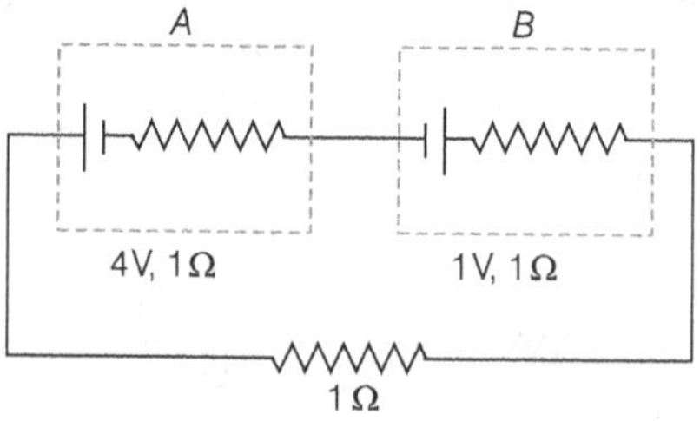

	Column I		Column II (In SI units)
(a)	potential difference across battery A	(p)	zero
(b)	potential difference across battery B	(q)	1
(c)	net power supplied/ consumed by A	(r)	2
(d)	net power supplied/ consumed by B	(s)	3

Subjective Questions

48. An aluminium wire carrying a current has diameter 0.84 mm. The electric field in the wire is 0.49 V/m. What is
(a) the current carried by the wire?
(b) the potential difference between two points in the wire 12.0 m apart?
(c) the resistance of a 12.0 m length of this wire? Specific resistance of aluminium is $2.75 \times 10^{-8}\,\Omega\text{-m}$.

49. Find the current supplied by the battery in the circuit shown in figure.

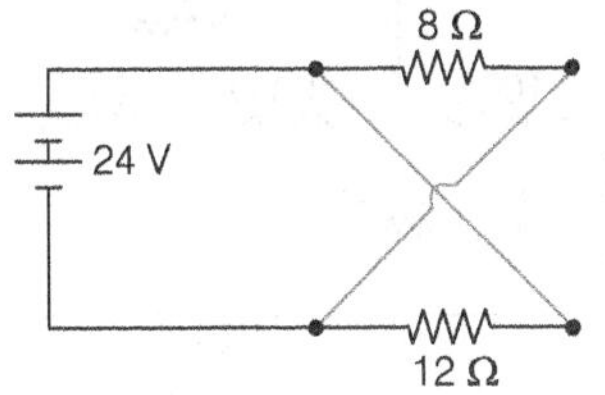

50. Calculate battery current and equivalent resistance of the network shown in figure.

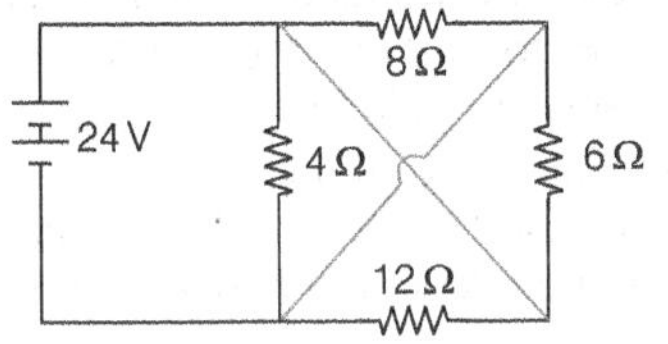

51. Compute total circuit resistance and battery current as shown in figure.

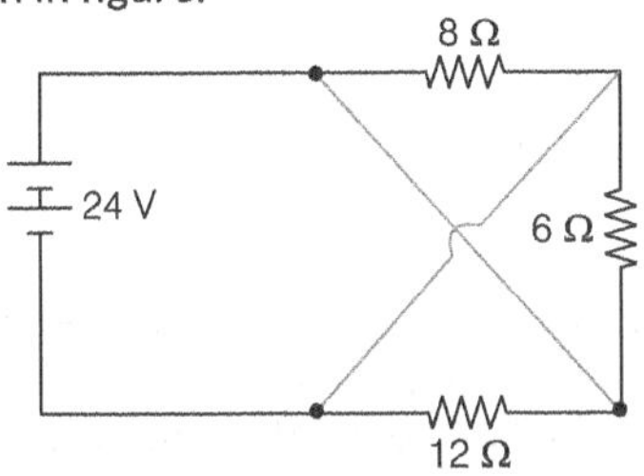

52. Compute the value of battery current i shown in figure. All resistances are in ohm.

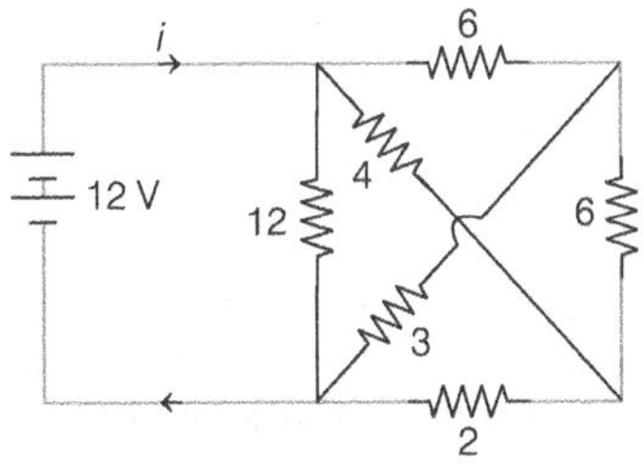

53. Calculate the potentials of points A, B, C and D as shown in Fig. (a). What would be the new potential values if connections of 6 V battery are reversed as shown in Fig. (b)? All resistances are in ohm.

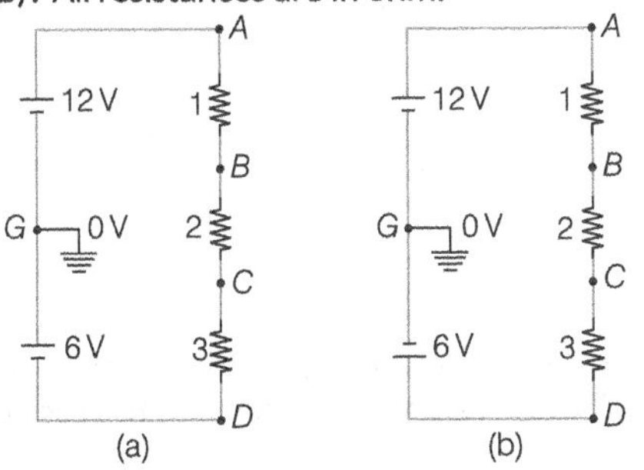

(a) (b)

54. Give the magnitude and polarity of the following voltages in the circuit of figure :

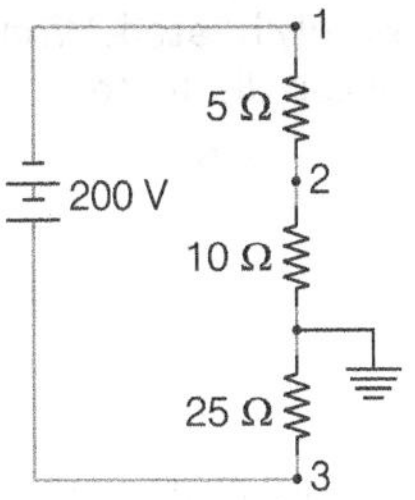

(i) V_1 (ii) V_2
(iii) V_3 (iv) V_{3-2}
(v) V_{1-2} (vi) V_{1-3}

55. In the circuit shown in figure V_1 and V_2 are two voltmeters of resistances 3000 Ω and 2000 Ω, respectively. In addition $R_1 = 2000\ \Omega$, $R_2 = 3000\ \Omega$ and $E = 200$ V, then

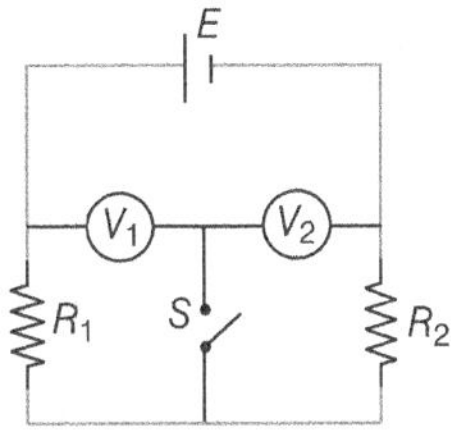

(a) Find the reading of voltmeters V_1 and V_2 when
  (i) switch S is open
  (ii) switch S is closed
(b) Current through S, when it is closed
(Disregard the resistance of battery)

56. Determine the resistance r if an ammeter shows a current of $I = 5$ A and a voltmeter 100 V. The internal resistance of the voltmeter is $R = 2{,}500\ \Omega$.

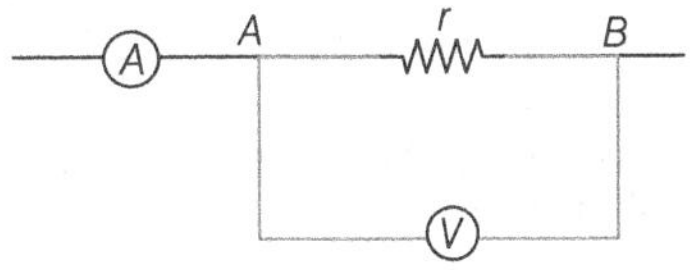

57. In the circuit, a voltmeter reads 30 V when it is connected across 400 Ω resistance. Calculate what the same voltmeter will read when it is connected across the 300Ω resistance?

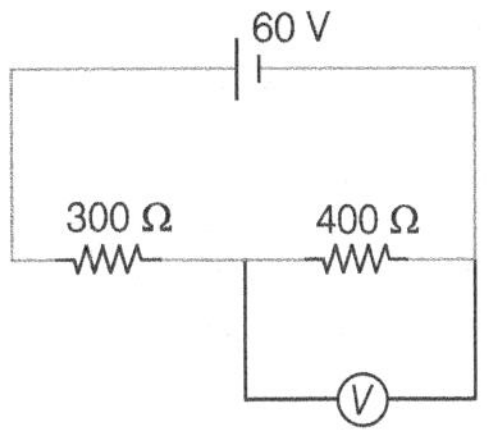

58. Each of three resistors in figure has a resistance of 2.4 Ω and can dissipate a maximum of 36 W without becoming excessively heated. What is the maximum power the circuit can dissipate?

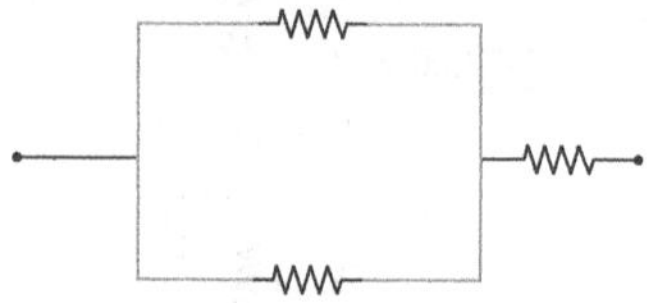

59. A storage battery with emf 2.6 V loaded with external resistance produces a current 1 A. In this case, the potential difference between the terminals of the storage battery equals 2 V. Find the thermal power generated in the battery and the net power supplied by the battery for external circuit. **(Irodov 3.192)**

60. In the circuit shown in figure $E_1 = 7$ V, $E_2 = 1$ V, $R_1 = 2\,\Omega$, $R_2 = 2\,\Omega$ and $R_3 = 3\,\Omega$ respectively. Find the power supplied by the two batteries.

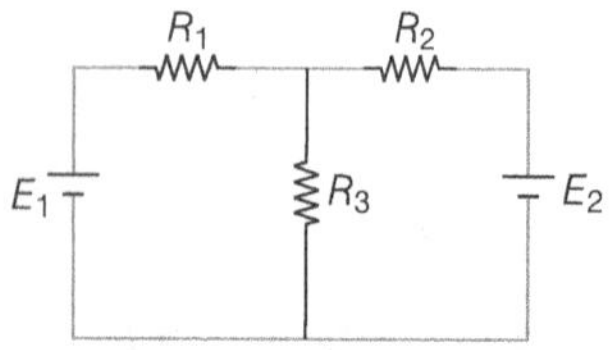

61. In the circuit shown in figure, find

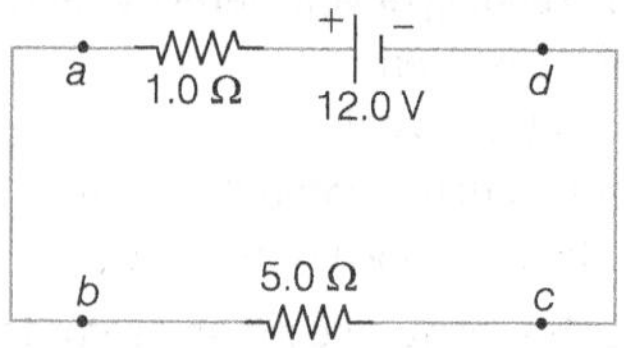

(a) the rate of conversion of internal (chemical) energy to electrical energy within the battery
(b) the rate of dissipation of electrical energy in the battery
(c) the rate of dissipation of electrical energy in the external resistor.

62. (a) The power of resistor is the maximum power the resistor can safely dissipate without too rise in temperature. The power rating of a 15 kΩ resistor is 5.0 W. What is the maximum allowable potential difference across the terminals of the resistor?
(b) A 9.0 kΩ resistor is to be connected across a 120 V potential difference. What power rating is required?

63. An electrical circuit is shown in figure. Calculate the potential difference across the resistor of 400 Ω as will be measured by the voltmeter V of resistance 400 Ω either by applying Kirchhoff's rules or otherwise. **(1996 JEE)**

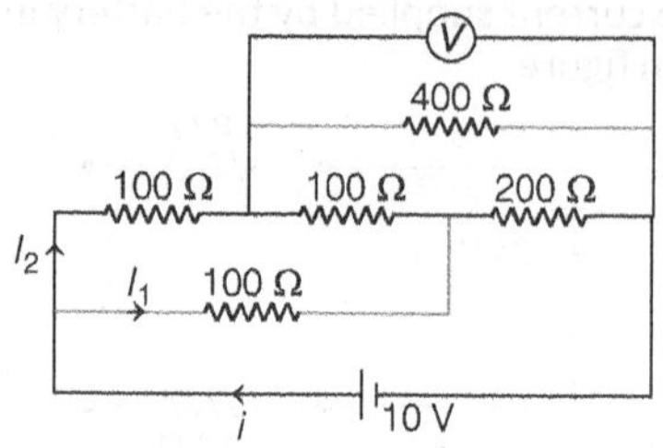

64. N sources of current with different emf's are connected as shown in figure. The emf's of the sources are proportional to their internal resistances, i.e. $E = \alpha R$, where α is an assigned constant. The lead wire resistance is negligible. Find: **(Irodov 3.176)**
(a) the current in the circuit;
(b) the potential difference between points A and B dividing the circuit in n and N − n links.

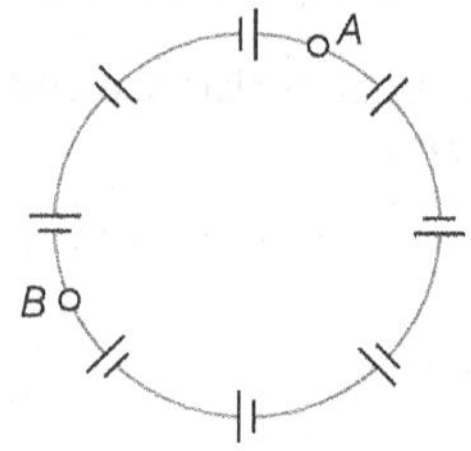

65. In the circuit shown in figure the emf of the source is equal to $E = 5.0$ V and the resistances are equal to $R_1 = 4.0\,\Omega$ and $R_2 = 6.0\,\Omega$. The internal resistance of the source equals $R = 0.10\,\Omega$. Find the currents flowing through the resistances R_1 and R_2. **(Irodov 3.178)**

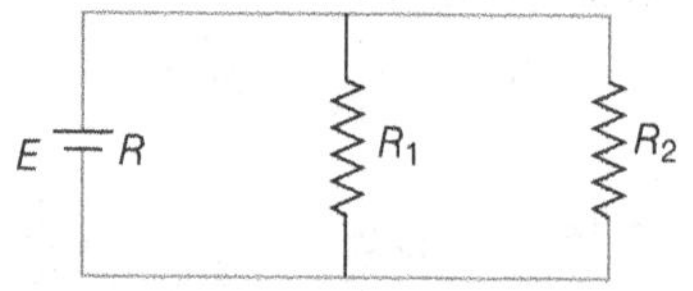

66. Find the magnitude and direction of the current flowing through the resistance R in the circuit shown in figure, if the emf's of the sources are equal to $E_1 = 1.5$ V and $E_2 = 3.7$ V and the resistances are equal to $R_1 = 10\,\Omega$, $R_2 = 20\,\Omega$, $R = 5.0\,\Omega$. The internal resistances of the sources are negligible. **(Irodov 3.181)**

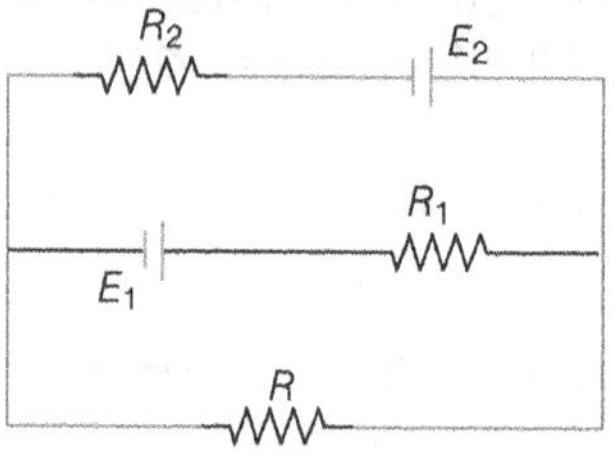

67. In the circuit shown in figure the source have emf's $E_1 = 1.5$ V, $E_2 = 2.0$ V, $E_3 = 2.5$ V and the resistances are

equal to $R_1 = 10\,\Omega$, $R_2 = 20\,\Omega$, $R_3 = 30\,\Omega$. The internal resistances of the sources are negligible.
Find:

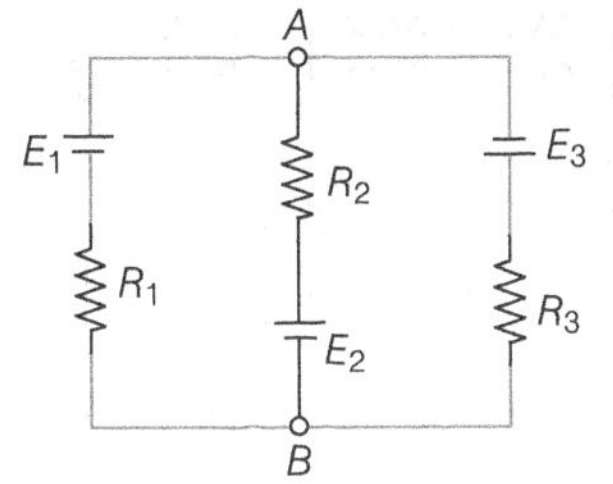

(a) the current flowing through the resistance R_1;
(b) a potential difference $V_A - V_B$ between the points A and B. **(Irodov 3.182)**

68. A constant voltage $V = 25$ V is maintained between points A and B of the circuit. Find the magnitude and direction of the current flowing through the segment CD if the resistances are equal to $R_1 = 1.0\,\Omega$, $R_2 = 2.0\,\Omega$, $R_3 = 3.0\,\Omega$, and $R_4 = 4.0\,\Omega$. **(Irodov 3.186)**

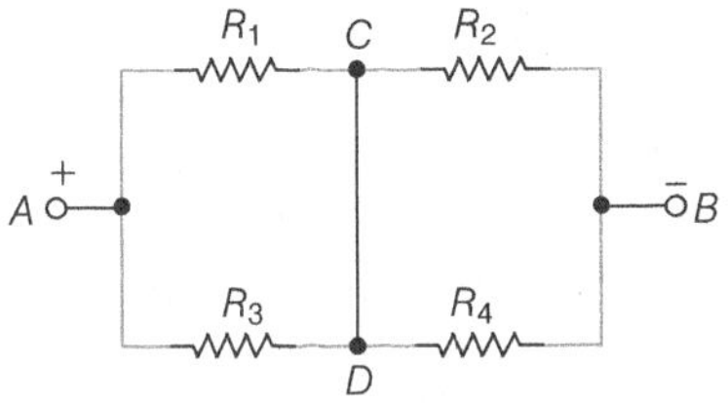

69. A series-parallel combination battery consisting of a large number $N = 300$ of identical cells, each with an internal resistance $r = 0.3\,\Omega$, is loaded with an external resistance $R = 10\,\Omega$. Find the number n of parallel groups consisting of an equal number of cells connected in series, at which the external resistance generates the highest thermal power. **(Irodov 3.198)**

Numerical Value/Integer Type Questions

70. 10 resistors each of resistance $10\,\Omega$ can be connected in such a way so as to get maximum and minimum equivalent resistance. The ratio of maximum and minimum equivalent resistance will be .... **(2023 Main)**

71. In the circuit diagram shown in figure given below, the current flowing through resistance $3\,\Omega$ is $\frac{x}{3}$ A.
The value of x is .... **(2023 Main)**

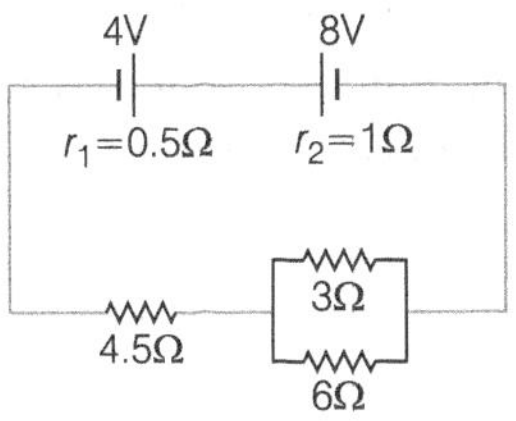

72. If the potential difference between B and D is zero, the value of x is $\frac{1}{n}\,\Omega$. The value of n is .... **(2023 Main)**

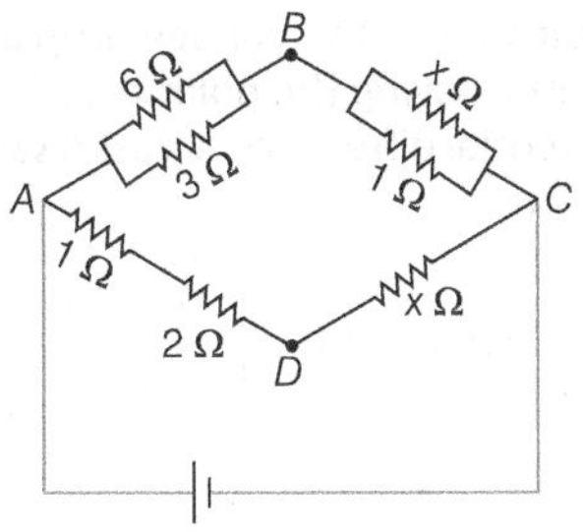

Level 2

Single Correct Option

73. Two cells A and B of emf 1.3 V and 1.5 V respectively are arranged as shown in figure. The voltmeter reads 1.45 V. The voltmeter is assumed to be ideal. Then

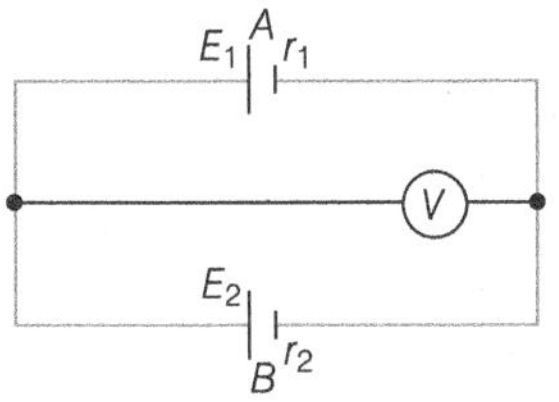

(a) $r_1 = 2r_2$
(b) $r_1 = 3r_2$
(c) $r_2 = 2r_1$
(d) $r_2 = 3r_1$

74. A voltmeter connected in series with a resistance R_1 to a circuit indicates a voltage $V_1 = 198$ V. When a series resistor $R_2 = 2R_1$ is used, the voltmeter indicates a voltage $V_2 = 180$ V. If the resistance of the voltmeter is $R_V = 900\,\Omega$, then the applied voltage across A and B is

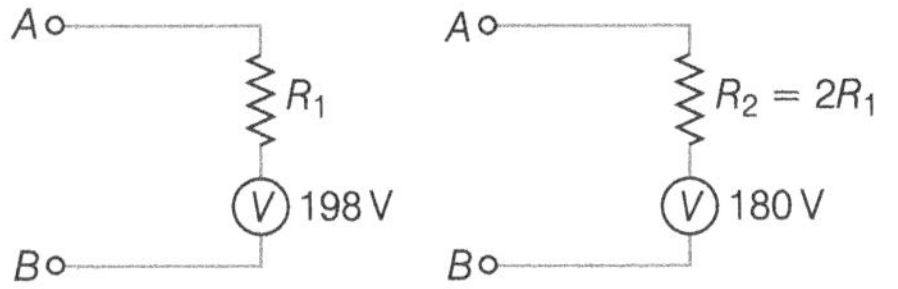

(a) 210 V
(b) 200 V
(c) 220 V
(d) 240 V

75. Switch S is closed at time $t = 0$. Which one of the following statements is correct?

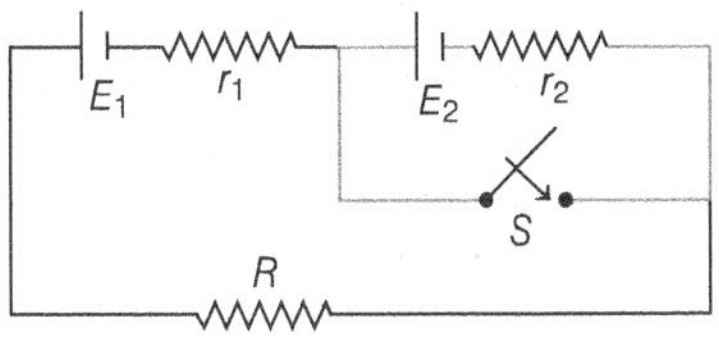

(a) Current in the resistance R increases if $E_1 r_2 < E_2(R + r_1)$
(b) Current in the resistance R increases if $E_1 r_2 > E_2(R + r_1)$
(c) Current in the resistance R decreases if $E_1 r_2 > E_2(R + r_1)$
(d) Current in the resistance R decreases if $E_1 r_2 = E_2(R + r_1)$

76. In the circuit shown, the voltage drop across the 15 Ω resistor is 30 V having the polarity as indicated. The ratio of potential difference across 5 Ω resistor and resistance R is

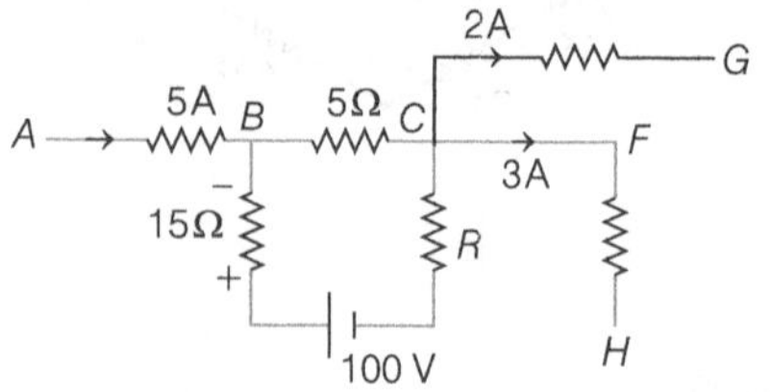

(a) 2/7 (b) 0.4 (c) 5/7 (d) 1

77. A source of emf $E = 10$ V and having negligible internal resistance is connected to a variable resistance. The resistance varies as shown in figure. The total charge that has passed through the resistor R during the time interval from t_1 to t_2 is

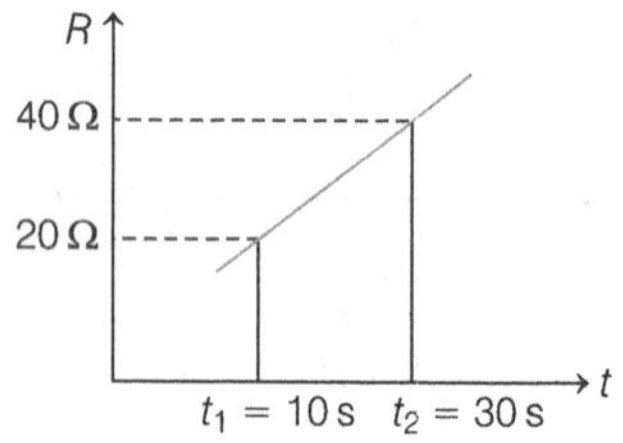

(a) $40 \log_e 4$ (b) $30 \log_e 3$
(c) $20 \log_e 2$ (d) $10 \log_e 2$

78. A milliammeter of range 10 mA and resistance 9 Ω is joined in a circuit as shown. The meter gives full scale deflection for current I when A and B are used as its terminals, i.e. current enters at A and leaves at B (C is left isolated). The value of I is

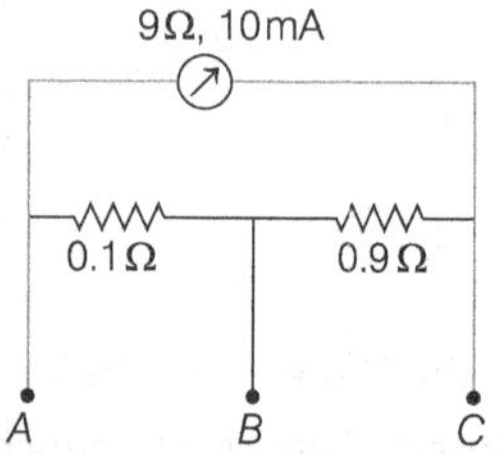

(a) 100 mA (b) 900 mA
(c) 1 A (d) 1.1 A

79. In the circuit shown, when keys K_1 and K_2 both are closed, the ammeter reads I_0. But when K_1 is open and K_2 is closed, the ammeter reads $I_0/2$. Assuming that ammeter resistance is much less than R_2, the values of r and R_1 in Ω are

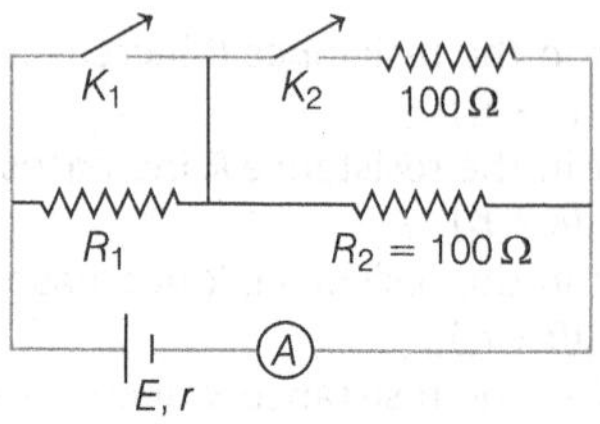

(a) 25, 50 (b) 25, 100 (c) 0, 100 (d) 0, 50

80. In the circuit shown in figure, V must be

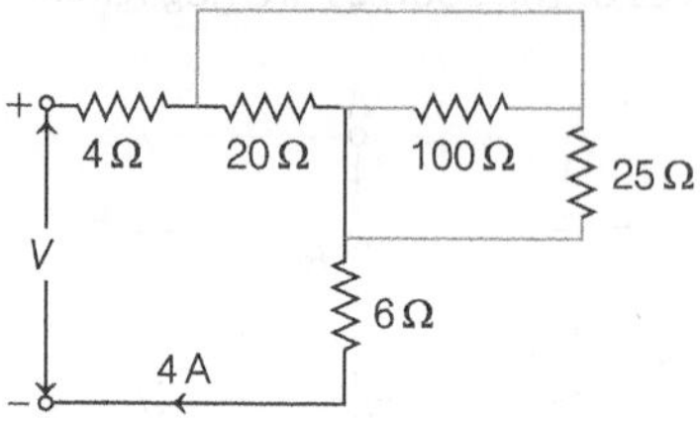

(a) 50 V (b) 80 V (c) 100 V (d) 1290 V

81. In the circuit shown in figure ammeter and voltmeter are ideal. If $E = 4$ V, $R = 9$ Ω and $r = 1$ Ω, then readings of ammeter and voltmeter are

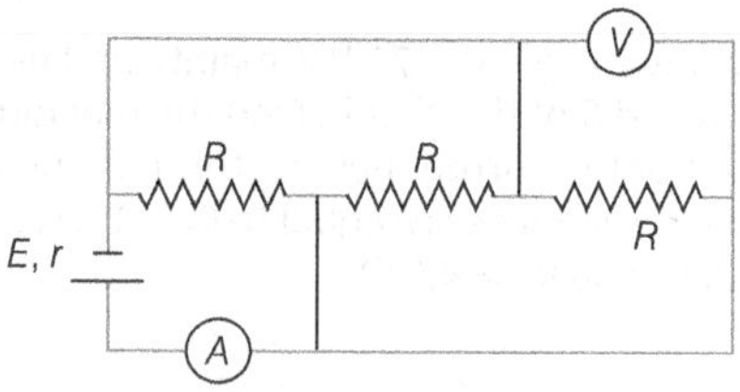

(a) 1 A, 3 V (b) 2 A, 3 V
(c) 3 A, 4 V (d) 4 A, 4 V

82. In the given figure the current through 4 Ω resistor is

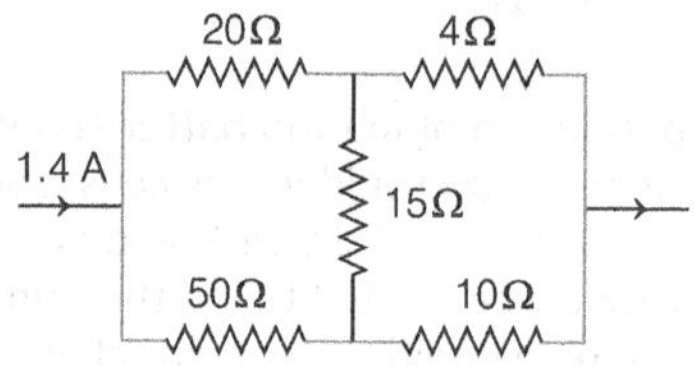

(a) 1.4 A (b) 0.4 A
(c) 1.0 A (d) 0.7 A

83. All resistances shown in circuit are 2 Ω each. The current in the resistance between D and E is

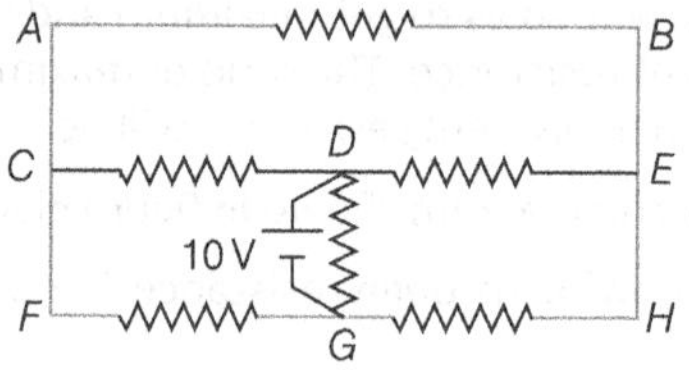

(a) 5 A (b) 2.5 A (c) 1 A (d) 7.5 A

84. When some potential difference is maintained between A and B, current I enters the network at A and leaves at B

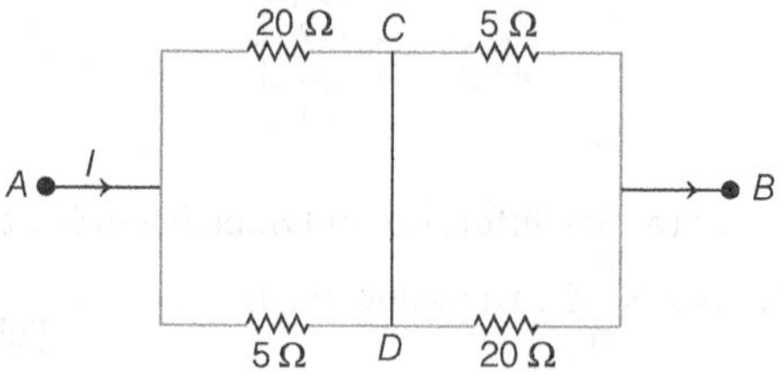

(i) The equivalent resistance between A and B is $8\ \Omega$.
(ii) C and D are at the same potential.
(iii) No current flows between C and D.
(iv) Current $3I/5$ flows from D to C.
Choose the correct option.

(a) (i), (ii), (iii) (b) (i), (iii), (iv)
(c) (i), (ii), (iv) (d) (ii), (iii), (iv)

85. For the circuit, V_1 and V_2 are two voltmeters having resistances $6000\ \Omega$ and $4000\ \Omega$, respectively. EMF of battery is 250 V, having negligible resistance.

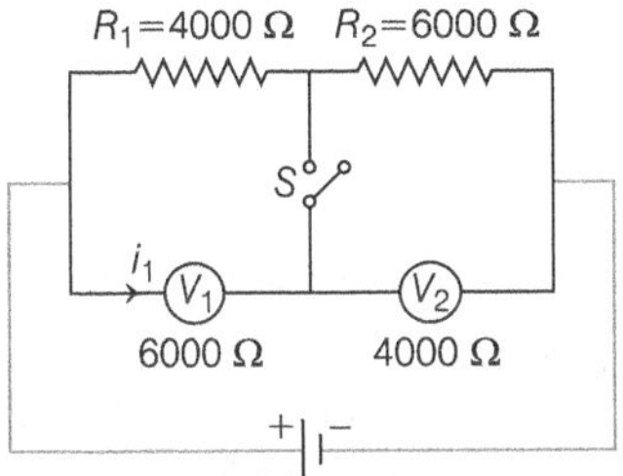

Then,

i. when switch S is open, $\frac{V_1}{V_2} = 1$

ii. when switch S is open, $\frac{V_1}{V_2} = 1.5$

iii. when switch S is closed, $\frac{V_1}{V_2} = 1$

iv. when switch S is closed, $\frac{V_1}{V_2} = 1.5$

Choose the correct option.

(a) i, ii (b) ii, iii
(c) iii, iv (d) i, iv

86. During an experiment with a meter bridge, the galvanometer shows a null point when the jockey is pressed at 40.0 cm using a standard resistance of $90\ \Omega$, as shown in the figure. Least count of the scale used in the meter bridge is 1 mm. The unknown resistance is **(2014 Adv.)**

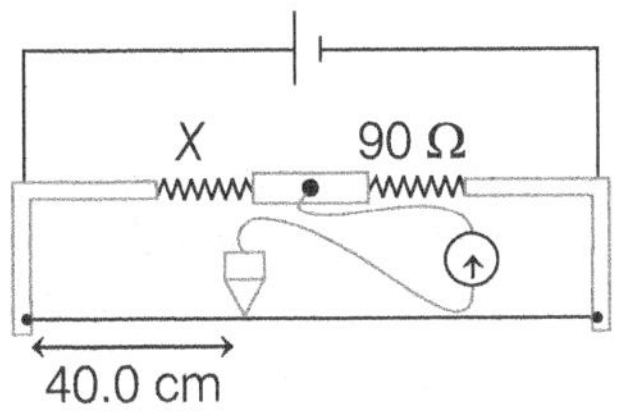

(a) $60 \pm 0.15\ \Omega$ (b) $135 \pm 0.56\ \Omega$
(c) $60 \pm 0.25\ \Omega$ (d) $135 \pm 0.23\ \Omega$

87. Figure shows three resistor configurations R_1, R_2 and R_3 connected to 3 V battery. If the power dissipated by the configuration R_1, R_2 and R_3 is P_1, P_2 and P_3, respectively, then **(2008 JEE)**

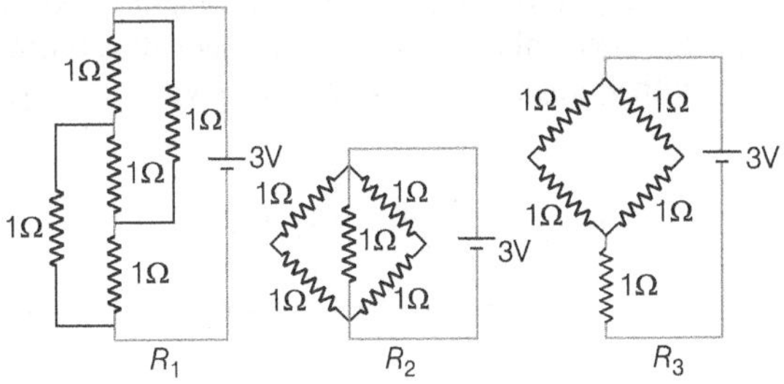

(a) $P_1 > P_2 > P_3$ (b) $P_1 > P_3 > P_2$
(c) $P_2 > P_1 > P_3$ (d) $P_3 > P_2 > P_1$

Multiple Correct Options

88. Three ammeters A, B and C of resistances R_A, R_B and R_C respectively are joined as shown. When some potential difference is applied across the terminals T_1 and T_2, their readings are I_A, I_B and I_C respectively. Then,

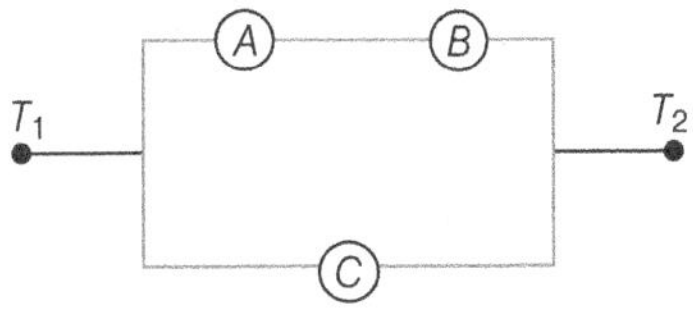

(a) $I_A = I_B$ (b) $I_A R_A + I_B R_B = I_C R_C$
(c) $\frac{I_A}{I_C} = \frac{R_C}{R_A}$ (d) $\frac{I_B}{I_C} = \frac{R_C}{R_A + R_B}$

89. Two conductors made of the same material have lengths L and $2L$ but have equal resistances. The two are connected in series in a circuit in which current is flowing. Which of the following is/are correct?

(a) The potential difference across the two conductors is the same
(b) The drift speed is larger in the conductor of length L
(c) The electric field in the first conductor is twice that in the second
(d) The electric field in the second conductor is twice that in the first

90. In the circuit shown in figure, reading of ammeter will

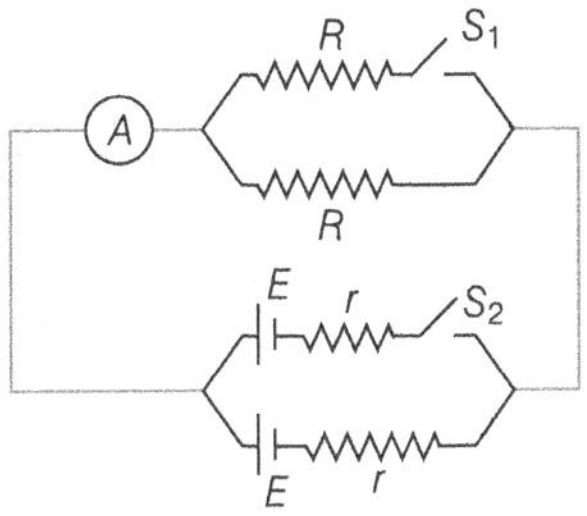

(a) increase if S_1 is closed (b) decrease if S_1 is closed
(c) increase if S_2 is closed (d) decrease if S_2 is closed

91. The figure shows a circuit having eight resistances of $1\,\Omega$ each, labelled R_1 to R_8, and two ideal batteries with voltages $\varepsilon_1 = 12$ V and $\varepsilon_2 = 6$ V. **(2022 Adv.)**

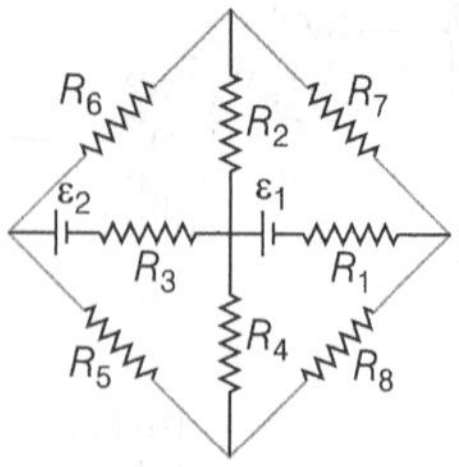

Which of the following statement(s) is(are) correct?
(a) The magnitude of current flowing through R_1 is 7.2 A.
(b) The magnitude of current flowing through R_2 is 1.2 A.
(c) The magnitude of current flowing through R_3 is 4.8 A.
(d) The magnitude of current flowing through R_5 is 2.4 A.

92. For the resistance network shown in the figure, choose the correct option(s). **(2012 JEE)**

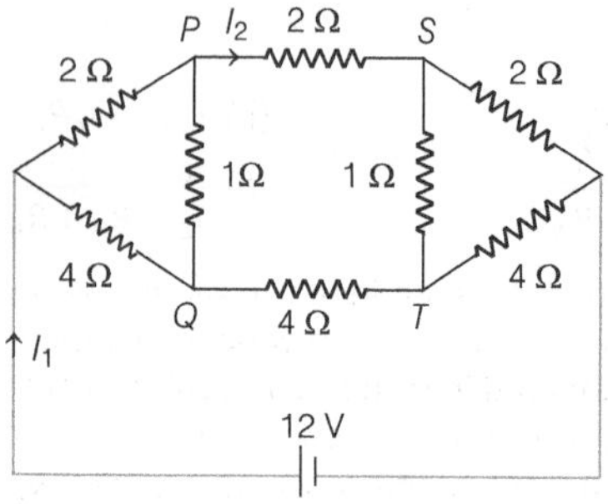

(a) The current through PQ is zero
(b) $I_1 = 3$ A
(c) The potential at S is less than that at Q
(d) $I_2 = 2$ A

93. For the circuit shown in the figure **(2009 JEE)**

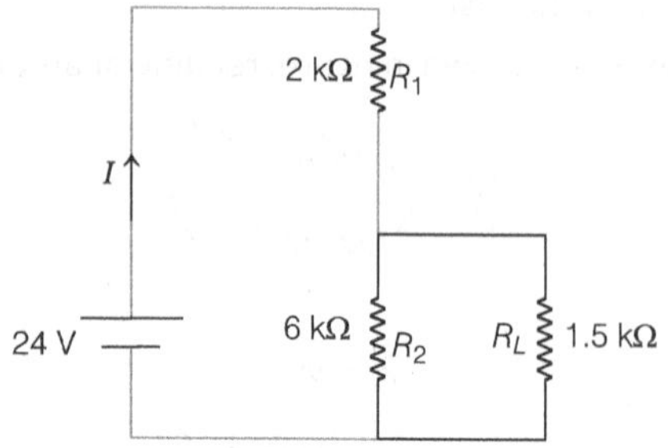

(a) the current I through the battery is 7.5 mA
(b) the potential difference across R_L is 18 V
(c) ratio of powers dissipated in R_1 and R_2 is 3
(d) if R_1 and R_2 are interchanged, magnitude of the power dissipated in R_L will decrease by a factor of 9

Subjective Questions

94. Find the equivalent resistance of the triangular bipyramid between the points.

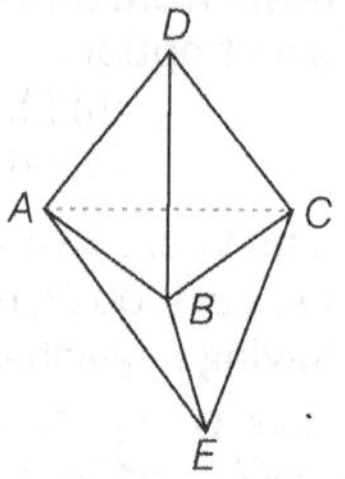

(a) A and C
(b) D and E
Assume the resistance of each branch to be R.

95. Nine wires each of resistance r are connected to make a prism as shown in figure. Find the equivalent resistance of the arrangement across

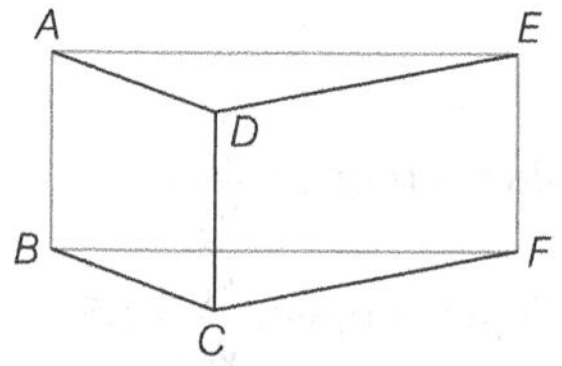

(a) AD (b) AB

96. A galvanometer (coil resistance 99 Ω) is converted into an ammeter using a shunt of 1 Ω and connected as shown in figure (a). The ammeter reads 3 A. The same galvanometer is converted into a voltmeter by connecting a resistance of 101 Ω in series. This voltmeter is connected as shown in figure (b). Its reading is found to be 4/5 of the full scale reading. Find :

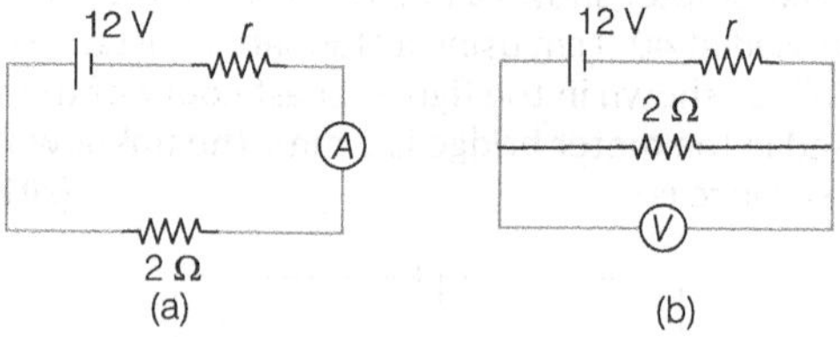

(a) internal resistance r of the cell
(b) range of the ammeter and voltmeter
(c) full scale deflection current of the galvanometer.

97. In the circuit shown in figure, find :

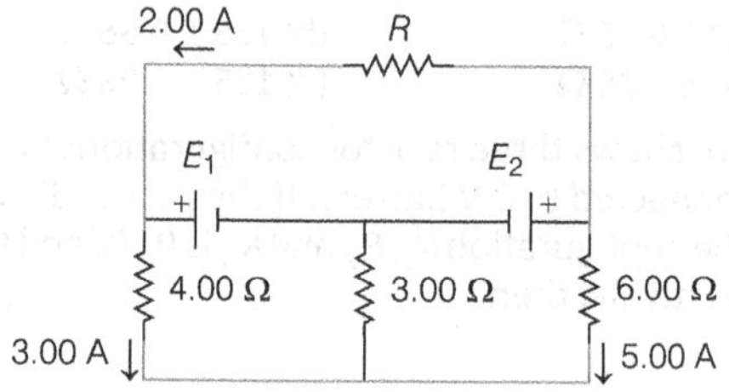

(a) the current in the 3.00 Ω resistor,
(b) the unknown emfs E_1 and E_2 and
(c) the resistance R.

98. In the circuit shown, all the ammeters are ideal.

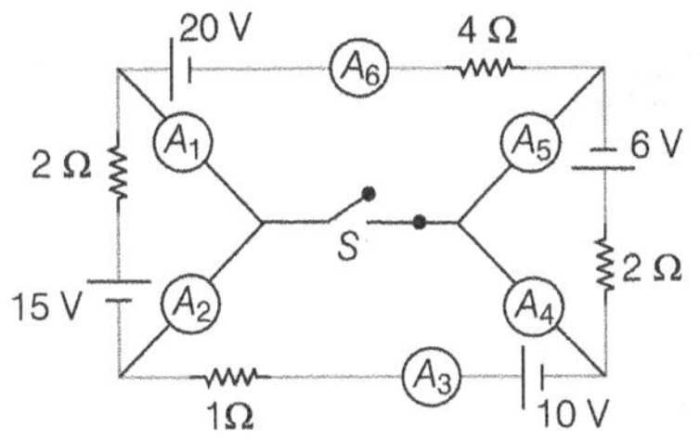

(a) If the switch S is open, find the reading of all ammeters and the potential difference across the switch.
(b) If the switch S is closed, find the current through all ammeters and the switch also.

99. An accumulator of emf 2 V and negligible internal resistance is connected across a uniform wire of length 10 m and resistance 30 Ω. The appropriate terminals of a cell of emf 1.5 V and internal resistance 1 Ω is connected to one end of the wire and the other terminal of the cell is connected through a sensitive galvanometer to a slider on the wire. What is the length of the wire that will be required to produce zero deflection of the galvanometer? How will the balancing length change?
(a) When a coil of resistance 5 Ω is placed in series with the accumulator.
(b) The cell of 1.5 V is shunted with 5 Ω resistor?

100. A long round conductor of cross-sectional area S is made of material whose resistivity depends only on a distance r from the axis of the conductor as $\rho = \alpha / r^2$, where α is a constant. Find: **(Irodov 3.169)**
(a) the resitance per unit length of such a conductor;
(b) the electric field strength in the conductor due to which a current I flows through it.

Hint: $E = \frac{V}{l}$

101. Figure illustrates a potentiometer circuit by means of which we can vary a voltage V applied to a certain device possessing a resistance R. The potentiometer has a length l and a resistance R_0 and voltage V_0 is applied to its terminals. Find the voltage V fed to the device as a function of distance x. Analyse separately the case $R \gg R_0$. **(Irodov 3.179)**

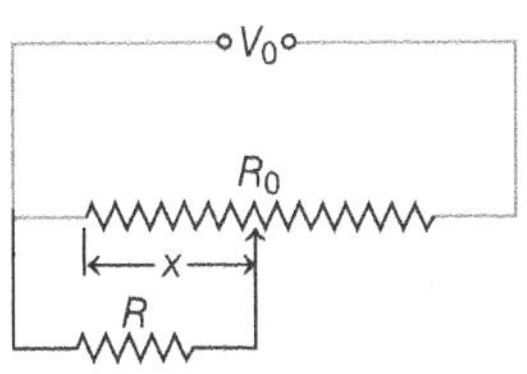

102. Find the current flowing through the resistance R in the circuit shown in figure. The internal resistances of the batteries are negligible. **(Irodov 3.183)**

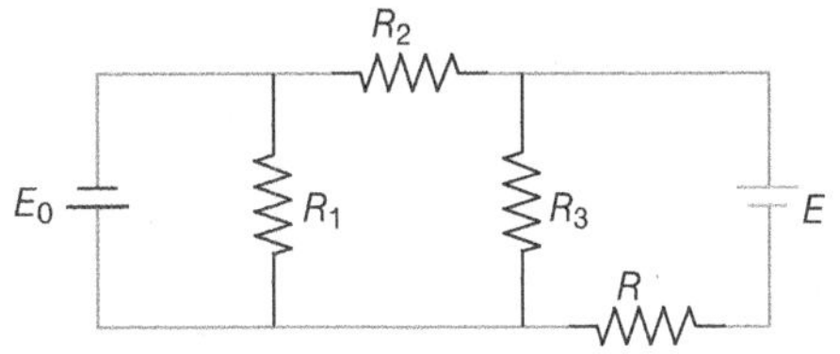

103. How much (in per cent) has a filament diameter decreased due to evaporation if an increase of voltage by 1% is required to keep the temperature constant. The amount of heat transferred from the filament into surrounding space is assumed to be proportional to the filament surface area. **(Irodov 3.194)**

104. A conductor has a temperature-independent resistance R and a total heat capacity C. At the moment $t = 0$ it is connected to a DC voltage V. Find the time dependence of the conductor's temperature T assuming the thermal power dissipated into surrounding space to vary as $q = k(T - T_0)$, where k is a constant, T_0 is the environmental temperature (equal to the conductor's temperature at the initial moment). **(Irodov 3.195)**

105. A straight copper wire of length $l = 1000$ m and cross-sectional area $S = 1.0$ mm^2 carries a current $I = 4.5$ A. Assuming that one free electron corresponds to each copper atom, find:
(a) the time it takes an electron to travel from one end of the wire to the other
(b) the sum of electric forces acting on all free electrons in the given wire. **(Irodov 3.209)**

Numerical Value/Integer Type Questions

106. In the following circuit, the magnitude of current I_1, is .... A. **(2023 Main)**

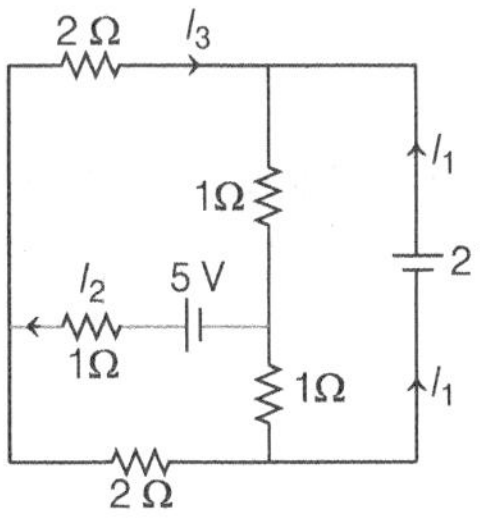

107. In the given circuit, the value of $\left| \frac{I_1 + I_3}{I_2} \right|$ is .... **(2023 Main)**

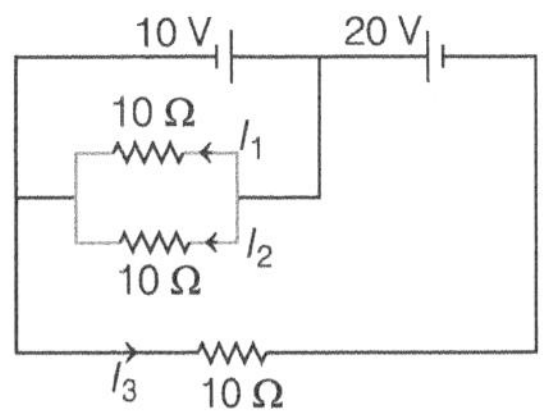

108. All resistances in figure are 1Ω each. The value of current I is $\frac{a}{5}$ A. The value of a is .... **(2022 Main)**

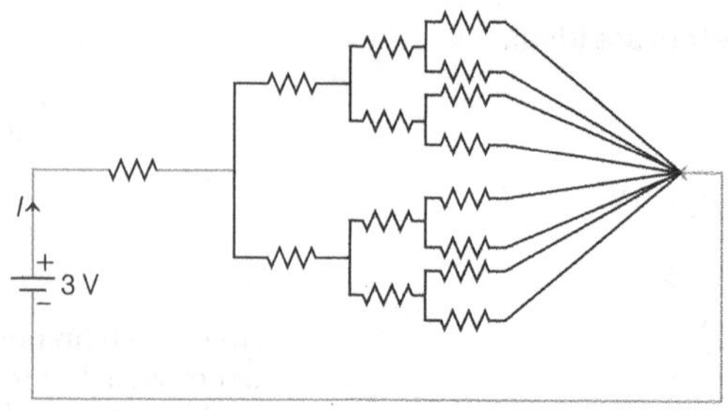

109. In the following circuit, the current through the resistor $R(=2\Omega)$ is I amperes. The value of I is **(2015 Adv.)**

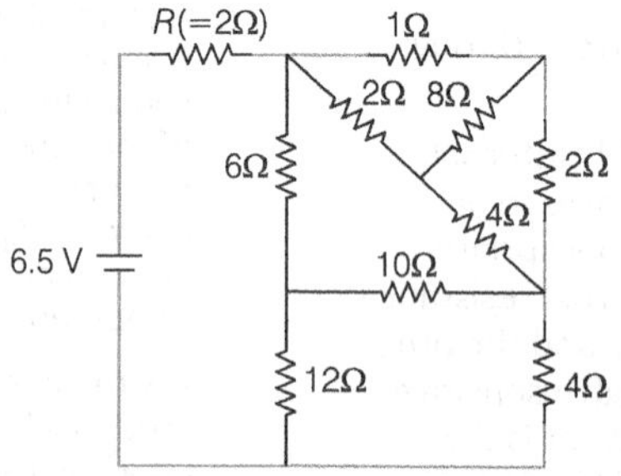

110. In order to measure the internal resistance r_1 of a cell of emf E, a meter bridge of wire resistance $R = 50\,\Omega$, a resistance $R_0/2$, another cell of emf $E/2$ (internal resistance r) and a galvanometer G are used in a circuit, as shown in the figure. If the null point is found at $l = 72$ cm, then the value of r_1 = ... Ω. **(2021 Adv.)**

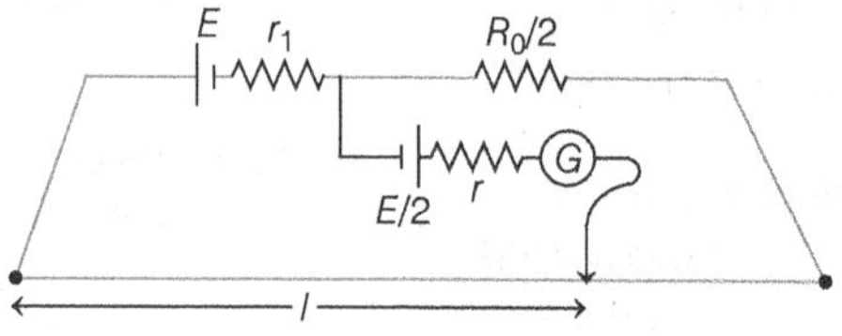

111. In the balanced condition, the values of the resistances of the four arms of a Wheatstone bridge are shown in the figure below. The resistance R_3 has temperature coefficient $0.0004°C^{-1}$. If the temperature of R_3 is increased by $100°C$, the voltage developed between S and T will be .... volt. **(2020 Adv.)**

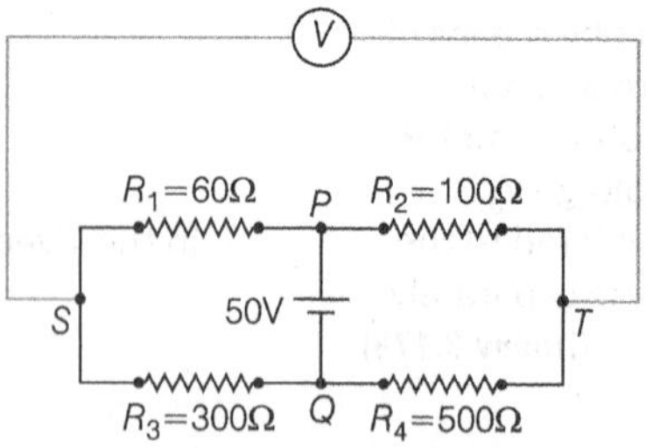

Answers

Level 1

1. ($R/2$)	**2.** (d)	**3.** (d)	**4.** (b)	**5.** (d)	**6.** (c)	**7.** (a)	**8.** (c)	**9.** (b)	**10.** (c)
11. (b)	**12.** (c)	**13.** (b)	**14.** (d)	**15.** (d)	**16.** (b)	**17.** (d)	**18.** (c)	**19.** (b)	**20.** (c)
21. (b)	**22.** (b)	**23.** (d)	**24.** (a)	**25.** (c)	**26.** (c)	**27.** (d)	**28.** (b)	**29.** (d)	**30.** (a)
31. (c)	**32.** (b)	**33.** (c)	**34.** (d)	**35.** (c)	**36.** (c)	**37.** (d)	**38.** (a)	**39.** (a)	**40.** (b)
41. (c)	**42.** (c)	**43.** (d)	**44.** (b)	**45.** (a)	**46.** (a-q; b-r; c-s; d-s)		**47.** (a-s; b-r; c-s; d-r)		

48. (a) 9.9 A (b) 5.88 V (c) 0.60 Ω **49.** 5A **50.** 15A, $\frac{8}{5}\Omega$ **51.** $\frac{8}{3}\Omega$, 9A **52.** $\frac{13}{3}$ A

53. $V_A = 12\text{V}, V_B = 9\text{V}, V_C = 3\text{V}, V_D = -6\text{V}, V_A' = 12\text{V}, V_B' = 11\text{V}, V_C' = 9\text{V}, V_D' = 6\text{V}$

54. –75V, – 50V, 125V, 175V, –25V, –200V **55.** (a) (i) 120 V, 80 V (ii) 100 V, 100 V (b) $\frac{1}{60}$ A **56.** 20.16 Ω

57. 22.5 V **58.** 54 W **59.** 0.6W, 2W **60.** +14 W, – 1W **61.** (a) 24W (b) 4W (c) 20W

62. (a) 273.8V (b) 1.6W **63.** $\frac{20}{3}$ V **64.** (a) $I = \alpha$ (b) $V_A - V_B = 0$ **65.** $I_{R_1} = 1.2$ A and $I_{R_2} = 0.8$ A

66. $I = 0.02$ A, the current is directed from the right to the left **67.** (a) $I_1 \approx 0.06$ A (b) $V_A - V_B = 0.9$ volt

68. $I_{CD} = 1$ A from C to D **69.** $n = 3$ **70.** 100 **71.** 1 **72.** 2

Level 2

73. (b)	**74.** (c)	**75.** (b)	**76.** (d)	**77.** (d)	**78.** (c)	**79.** (d)	**80.** (b)	**81.** (a)	**82.** (c)
83. (b)	**84.** (c)	**85.** (b)	**86.** (c)	**87.** (c)	**88.** (a,b,d)	**89.** (a,b,c)	**90.** (a,c)	**91.** (a,b,c,d)	

92. (a,b,c,d) **93.** (a,d) **94.** (a) $\frac{2}{5}R$ (b) $\frac{2}{3}R$ **95.** $\frac{8}{15}r$ (b) $\frac{3}{5}r$

96. (a) 1.01 Ω (b) 5A, 9.95V (c) 0.05A **97.** (a) 8A (b) 36V, 54V (c) 9 Ω

98. (a) 9.5A, 9.5A, 2A, 5A, 5A, 2A, 12V (b) 12.5A, 2.5A, 10A, 7A, 8A, 5A, 15A **99.** 7.5 m (a) 8.75 m (b) 6.25 m

100. (a) $\frac{R}{l} = \frac{2\pi\alpha}{S^2}$ (b) $E = \frac{2\pi\alpha l}{S^2}$ **101.** $V = \frac{V_0 R x}{Rl + R_0(l-x)\frac{x}{l}}$; for $R >> R_0$, $V = \frac{V_0 x}{l}$ **102.** $I_R = \frac{E(R_2 + R_3) + E_0 R_3}{R(R_2 + R_3) + R_2 R_3}$

103. 2% **104.** $T = T_0 + \frac{V^2}{kR}(1 - e^{-kt/C})$ **105.** (a) $t \approx 3$ ms (b) $F = 1$ MN **106.** 1.5 **107.** 2 **108.** 8

109. 1 **110.** 3 **111.** 0.27

See hints on pages 90-107.

Hints & SOLUTIONS

Module 1.

1. $i = 20t + 8t^2$

$\therefore \quad i = \frac{dq}{dt}$

$\Rightarrow \quad \int dq = \int i dt$

$\Rightarrow \quad q = \int_0^{15} (20t + 8t^2)\, dt$

$\Rightarrow \quad q = \left(\frac{20t^2}{2} + \frac{8t^3}{5}\right)_0^{15}$

$\Rightarrow \quad q = 10 \times (15)^2 + \frac{8(15)^3}{3}$

$\Rightarrow \quad q = 11250\text{ C}$

2. $\therefore \; i = \frac{q}{t} = \frac{ne}{t}$

$\Rightarrow \quad n = \frac{it}{e} = \frac{(0.7)(1)}{1.6 \times 10^{-19}}$

$= 4.375 \times 10^{18}$

3. $\therefore \; q = it$

$= (3.6)(3 \times 60 \times 60) = 38880\text{ C}$

4. (a) $\therefore q = it = (7.5)(45) = 337.5\text{ C}$

(b) $n = \frac{q}{e} = \frac{337.5}{1.6 \times 10^{-19}} = 2.1 \times 10^{21}$

5. $f = \frac{v}{2\pi r} = \frac{2.2 \times 10^6}{(2\pi)(5.3 \times 10^{-11})}$

$= 6.6 \times 10^{15}\text{ Hz}$

$I = qf$

$= (1.6 \times 10^{-19})(6.6 \times 10^{15})$

$= 1.06 \times 10^{-3}\text{ A}$

$= 1.06\text{ mA}$

6. $\Delta q = \int_0^{10} i dt = \int_0^{10} (10 + 4t)\, dt$

$= 300\text{ C}$

7. Current due to both is from left to right. So, the two currents are additive.

Module 2.

1. False. Only under electrostatic conditions (when $i = 0$) all points of a conductor are at same potential.

3. Resistance, $R = \rho \cdot \frac{l}{A}$

$\therefore \quad \rho = \frac{RA}{l}$

But, $\quad R = \frac{V}{i}$

$\therefore \quad \rho = \frac{VA}{il} = \frac{V/l}{i/A}$

But, $\quad \frac{V}{l} = E$ and $\frac{i}{A} = J$

$\therefore \quad \rho = \frac{E}{J}$ or $E = \rho \cdot J$ or $E = \frac{1}{\sigma} J$

where, $\quad \sigma = \frac{1}{\rho}$

Thus, $\quad \sigma = J/E$

4. We know that, $I = nev_d A$

$\Rightarrow \quad n = \frac{I}{Aev_d}$

$= \frac{10}{5 \times 10^{-6} \times 1.6 \times 10^{-19} \times 2 \times 10^{-3}}$

$= 625 \times 10^{25}$

5. Mobility of charges in a conductor is given by

$$\mu = \frac{v_d}{E} \quad \text{...(i)}$$

and resistivity is given by

$\rho = \frac{E}{J} = \frac{E}{I/A}$

or $\quad \rho = \frac{EA}{I}$

or $\quad E = \frac{\rho I}{A} \quad$ ...(ii)

From Eqs. (i) and (ii), we get

$\mu = \frac{v_d A}{\rho I}$

Substituting the given values, we get

$= \frac{1.1 \times 10^{-3} \times \pi \times (5 \times 10^{-3})^2}{1.7 \times 10^{-8} \times 5}$

$= \frac{86.9 \times 10^{-9}}{8.5 \times 10^{-8}}$

$= 10.16 \times 10^{-1}$

$\Rightarrow \quad \mu \approx 1\text{ m}^2/\text{V-s}$

6. We know that,

$v_d = \frac{I}{neA}$

Substituting the given values, we get

$$v = \frac{1.5}{9 \times 10^{28} \times 1.6 \times 10^{-19} \times 5 \times 10^{-6}}$$

or $$v = \frac{1.5 \times 10^{-3}}{72} \text{ m/s}$$

$$= 0.2 \times 10^{-4} \text{ m/s}$$

or $$v = 0.02 \text{ mm/s}$$

7. $i = neAv_d$

i.e. $v_d \propto i$

When current has increased from $i = 1.2$ A to $i = 6.0$ A, i.e five times, then drift velocity will also increase to five times.

8. From $i = neAv_d$

We have, $v_d = \frac{i}{neA}$

or $$v_d = \frac{1}{8.5 \times 10^{28} \times 1.6 \times 10^{-19} \times 10^{-4}}$$

$$= 0.735 \times 10^{-6} \text{ m/s}$$

$$= 0.735 \ \mu\text{m/s}$$

$$t = \frac{l}{v_d} = \frac{10 \times 10^3}{0.735 \times 10^{-6}} \text{ s}$$

$$= \frac{10 \times 10^3}{0.735 \times 10^{-6} \times 60 \times 60 \times 24 \times 365} \text{ yr}$$

$$= 431.4 \text{ yr}$$

9. $\rho = 8.89 \times 10^3 \text{ kg/m}^3$

Mass of $1\text{m}^3 = 8.89 \times 10^3$ kg

$$= 8.89 \times 10^6 \text{ g}$$

$\therefore$ Number of gram moles $= \frac{8.89 \times 10^6}{63.54}$

$$= 1.4 \times 10^5$$

Number of atoms $= 1.4 \times 10^5 \times 6.02 \times 10^{23}$

$$= 8.42 \times 10^{28}$$

One atom emits one conduction electron. Therefore, number of free electrons in unit volume (or 1m^3 volume)

$$n = 8.42 \times 10^{28} \text{ per m}^3$$

Now, $i = neA\,v_d$

$$\therefore \quad v_d = \frac{i}{neA} = \frac{i}{ne\,\pi r^2}$$

$$= \frac{2.0}{(8.42 \times 10^{28})(1.6 \times 10^{-19})(\pi)(0.5 \times 10^{-3})^2}$$

$$= 1.9 \times 10^{-4} \text{ m/s}$$

10. Drift velocity, $v_d = \frac{I}{neA}$

Total momentum,

$p =$ (Total number of free electrons) $(v_d)(m)$

$= (n)$ (Total volume of wire) $(v_d)(m)$

$$= (n)(Al)\left(\frac{I}{neA}\right)(m)$$

$$p = \frac{Ilm}{e}$$

Here, m is mass of electron.

Substituting the given values of I and l with

$$e = 1.6 \times 10^{-19} \text{ C}$$

and $m = 9.1 \times 10^{-31}$ kg, we get

$$p = 0.4 \text{ kg-ms}^{-1}$$

11. Using $I = nv_dAe$

We get, $v_d = \frac{I}{neA}$

$$v_d = \frac{3.2}{8 \times 10^{28} \times 1.6 \times 10^{-19} \times 2 \times 10^{-6}}$$

$$= \frac{2}{16 \times 10^3} = 125 \times 10^{-6} \text{ m/s}$$

Module 3.

1. $\rho = \frac{1}{\sigma} \Rightarrow \rho\sigma = \text{constant}$

2. $R = \frac{\rho\,(L)}{A} = \frac{\rho L}{tL} = \frac{\rho}{t}$

i.e. R is independent of L.

Hence, the correct option is (c).

3. Specific resistance of silver, copper and aluminium are $1.6 \times 10^{-8}\,\Omega\text{m}$, $1.7 \times 10^{-8}\,\Omega\text{m}$ and $2.7 \times 10^{-8}\,\Omega\text{m}$ respectively. Conductivity is opposite.

4. If there is a increase in length, then

$$R \propto l^2$$

$$\frac{R'}{R} = \left(\frac{l'}{l}\right)^2 = \left(\frac{1.1\,l}{l}\right)^2 = 1.21$$

$$\therefore \quad R' = 1.21R = 1.21 \times 10 = 12.1\,\Omega$$

5. Volume of wire $= Al$

Given, $3 = Al$

$$\therefore \quad A = \frac{3}{l}$$

Now, resistance, $R = \rho\frac{l}{A}$

$$3 = \frac{\rho l}{(3/l)} = \frac{\rho l^2}{3}$$

$$\therefore \quad l^2 = \frac{9}{\rho}$$

$$\therefore \quad l = \frac{3}{\sqrt{\rho}}$$

6. Suppose $n\,(<1)$ fraction of length is stretched to m times.

Then, $(1-n)l + (nl)\,m = 1.5l$

or $$nm - n = 0.5 \qquad \text{...(i)}$$

$$R = \frac{\rho l}{A} = \frac{\rho l}{(V/l)} \qquad (V = \text{volume})$$

$$= \frac{\rho l^2}{V} \text{ or } R \propto l^2 \qquad (\text{if } V = \text{constant})$$

Now, the second condition is

$$(1-n)R + (nR)m^3 = 4R$$

$\therefore$ $$nm^2 - n = 3 \qquad \text{...(iii)}$$

Solving these two equations, we get

$$n = \frac{1}{8}$$

7. $R \propto l^2$ (if ρ and V are constant)

$$R_2 = 5^2 R_1 \qquad (\because R_2 = n^2 R_1)$$

$$= 25 \times 5 = 125\,\Omega$$

8. $$\frac{1}{R} = \frac{1}{R_{Al}} + \frac{1}{R_{Fe}} = \left(\frac{A_{Al}}{\rho_{Al}} + \frac{A_{Fe}}{\rho_{Fe}}\right)\frac{1}{l}$$

$$= \left[\frac{(7^2 - 2^2)}{2.7} + \frac{2^2}{1.0}\right]\frac{10^{-6}}{10^{-8}} \times \frac{1}{50 \times 10^{-3}}$$

Solving, we get

$$R = \frac{1875}{64} \times 10^{-6}\,\Omega = \frac{1875}{64}\,\mu\Omega$$

9. Resistivity of a conductor is

$$\rho = \frac{m_e}{ne^2\tau} \qquad \text{... (i)}$$

Substituting values in Eq. (i), we get

$$\rho = \frac{9.1 \times 10^{-31}}{8.5 \times 10^{28} \times (1.6 \times 10^{-19})^2 \times 25 \times 10^{-15}}$$

$$= \frac{9.1 \times 10^{-6}}{8.5 \times 2.56 \times 25}$$

$$= 0.016 \times 10^{-6}$$

$$= 1.6 \times 10^{-8}\,\Omega\text{-m}$$

$$\simeq 10^{-8}\,\Omega\text{-m}$$

10. $$R = \rho\frac{l}{A} = \frac{(1.72 \times 10^{-8})(35)}{(\pi/4)(2.05 \times 10^{-3})^2}$$

$$= 0.18\,\Omega$$

11. $$R = \frac{\rho l}{A}$$

$\therefore$ $$A = \frac{\rho l}{R}$$

$m = (Vd)$, where V = volume and d = density

$\therefore$ $$m = (Ald) = \frac{\rho l^2}{R} d$$

$$= \frac{(1.72 \times 10^{-8})(3.5)^2(8.9 \times 10^3)}{0.125}$$

$$= 15 \times 10^{-3}\text{ kg}$$

$$= 15\text{ g}$$

12. Radius at distance x from end P,

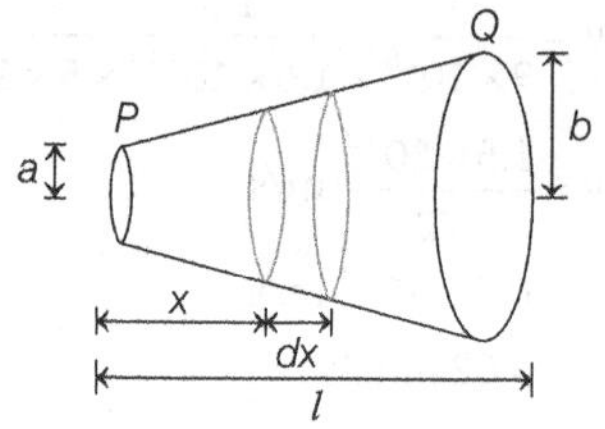

$$r = a + \left(\frac{b-a}{l}\right)x$$

Resistance of element of thickness dx is

$$dR = \frac{\rho\,(dx)}{\pi r^2} \qquad \left(\text{Using } R = \frac{\rho l}{A}\right)$$

$\therefore$ $$R = \int_{x=0}^{x=l} dR$$

13. (a) $$J = \frac{i}{A} = \frac{V}{RA} = \frac{V}{\left(\frac{\rho l}{A}\right)A}$$

$$J = \frac{V}{\rho l} \qquad \text{...(i)}$$

or $$J \propto \frac{1}{l}$$

$l_{min} = d$. So, J is maximum. Hence, potential difference should be applied across the face $(2d \times 3d)$

From Eq. (i),

$$J_{max} = \frac{V}{\rho d}$$

(b) $$i = \frac{V}{R} = \frac{V}{(\rho l/A)} = \frac{VA}{\rho l}$$

or $$i \propto \frac{A}{l}$$

Across face $(2d \times 3d)$, area of cross-section is maximum and l is minimum. Hence, current is maximum.

$$i_{max} = \frac{V\,(2d \times 3d)}{\rho\,(d)} = \frac{6Vd}{\rho}$$

14. (a) $$\rho = \frac{RA}{l} = \frac{(0.104)(\pi/4)(2.5 \times 10^{-3})^2}{14}$$

$$= 3.65 \times 10^{-8}\,\Omega\text{-m}$$

(b) $V = El = 1.28 \times 14 = 17.92$ V

$\therefore$ $$i = \frac{V}{R} = \frac{17.92}{0.104}$$

$$= 172.3\text{ A}$$

(c) $$v_d = \frac{i}{neA}$$

$$= \frac{172.3}{(8.5 \times 10^{28})(1.6 \times 10^{-19})\left(\frac{\pi}{4}\right)(2.5 \times 10^{-3})^2}$$

$$= 2.58 \times 10^{-3}\text{ m/s}$$

15. $R = \frac{\rho l}{A}$

where, $l' = 1.2l, A' = 0.96A$

$$R' = \frac{\rho \times 1.2l}{0.96A} = \frac{10}{8}R = 1.25R$$

It mean 25% increase in resistance.

16. Given, specific resistance,

$$\rho = 3 \times 10^{-7}\ \Omega\text{-m}$$

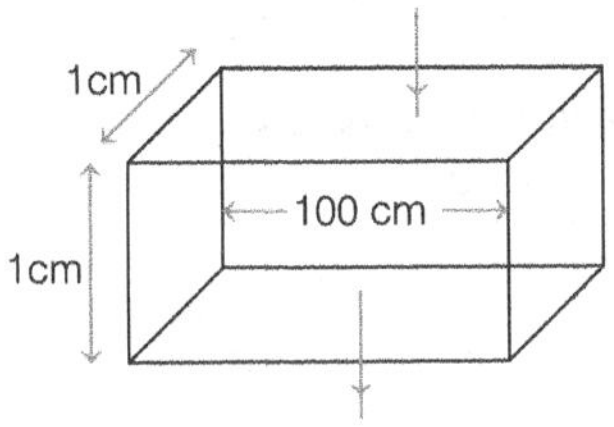

Resistance, $R = \rho\frac{l}{A}$.

Resistance through its two opposite faces will be

$$R = \frac{3 \times 10^{-7} \times (1 \times 10^{-2})}{100 \times 1 \times 10^{-4}}$$

$$= 3 \times 10^{-7}\ \Omega$$

17.

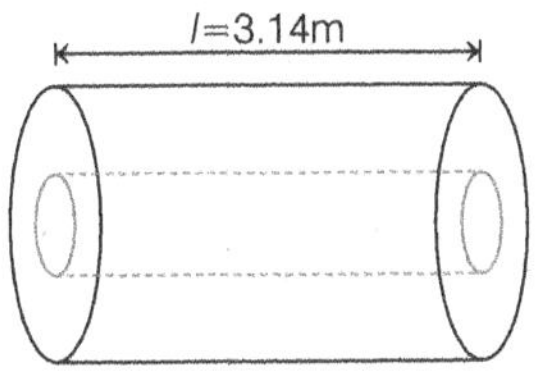

Inner diameter, $d_1 = 4\text{ mm} = 4 \times 10^{-3}$ m

Outer diameter, $d_2 = 8\text{ mm} = 8 \times 10^{-3}$ m

Resistance of conductor given is

$$R = \rho\frac{l}{A}$$

$$= \frac{2.4 \times 10^{-8} \times 3.14}{\frac{\pi}{4}[(8 \times 10^{-3})^2 - (4 \times 10^{-3})^2]}$$

$$= 2 \times 10^{-3}\ \Omega$$

Given, $R = n \times 10^{-3}\ \Omega$

So, $n = 2$

Module 4.

1. Copper is metal and germanium is semiconductor. Resistance of a metal decreases and that of a semiconductor increases with decrease in temperature.

2. We know that, $R = R_0(1 + \alpha\Delta T)$

So, $2 = R_0(1 + \alpha \times 10)$ ...(i)

$3 = R_0(1 + \alpha \times 30)$ ...(ii)

Dividing Eq. (ii) by Eq. (i), we get

$$\frac{3}{2} = \frac{1 + 30\alpha}{1 + 10\alpha}$$

$\Rightarrow$ $3 + 30\alpha = 2 + 60\alpha$

$\Rightarrow$ $1 = 30\alpha$

$\Rightarrow$ $\alpha = \frac{1}{30} = 0.033°\text{C}^{-1}$

3. $4.1[1 + 4.0 \times 10^{-3}(\theta - 20)] = 3.9[1 + 5.0 \times 10^{-3}(\theta - 20)]$

Solving we get,

$$\theta \approx 85°\text{C}$$

4. In parallel, current distributes in inverse ratio of resistance 1 $\rightarrow$ Aluminium 2 $\rightarrow$ Copper

$$\frac{R_1}{R_2} = \frac{i_2}{i_1} \Rightarrow \frac{\rho_1 l_1 / A_1}{\rho_2 l_2 / A_2} = \frac{2}{3}$$

$\therefore$ $\frac{\rho_1 l_1 d_2^2}{\rho_2 l_2 d_1^2} = \frac{2}{3}$

$\therefore$ $d_2 = \left(\sqrt{\frac{2\rho_2 l_2}{3\rho_1 l_1}}\right) d_1$

$$= \left(\sqrt{\frac{2 \times 0.017 \times 6}{3 \times 0.028 \times 7.5}}\right)(1\text{ mm})$$

$$= 0.569\text{ mm}$$

5. $R_1 + R_2 = 20$ ...(i)

$\alpha_{eq} = \frac{R_1\alpha_1 + R_2\alpha_2}{R_1 + R_2}$ (in series)

$$0 = \frac{R_1(-0.5 \times 10^{-3}) + R_2(5.0 \times 10^{-3})}{20}$$

$\therefore$ $R_1 = 10R_2$ ...(ii)

Solving Eqs. (i) and (ii), we get

$$R_2 = R_{Fe} = \frac{20}{11}\Omega$$

$$= 1.82\ \Omega$$

$$R_1 = R_{Cu} = 10\ R_2$$

$$= 18.18\ \Omega$$

6. For the same source, we have

$$i_0 R_0 = i_{100} R_{100}$$

$\Rightarrow$ $2R_0 = 1.2\,R_0(1 + 100\,\alpha)$ ... (i)

$\Rightarrow$ $1 + 100\,\alpha = \frac{5}{3}$

$\Rightarrow$ $100\,\alpha = \frac{2}{3}$

$\Rightarrow$ $50\,\alpha = \frac{1}{3}$

$\therefore$ $i_{50}R_{50} = i_0 R_0$ [Same source]

$\Rightarrow$ $i_{50} = \frac{i_0 R_0}{R_{50}}$

$$= \frac{2 \times R_0}{R_0(1 + 50\,\alpha)} = \frac{2}{1 + \frac{1}{3}}$$

$$= 1.5\text{ A}$$

$$= 15 \times 10^2\text{ mA}$$

Module 5.

Introductory Exercise

1. PD across each resistance is 10 V.

$$\therefore \quad i_{2\Omega} = \frac{10}{2} = 5\,\text{A}$$

$$i_{4\Omega} = \frac{10}{4} = 2.5\,\text{A}$$

2. $V_A = 0\,\text{V}$ (as it is earthed)

$$V_C - V_A = 5\,\text{V}$$
$$\therefore \quad V_C = 5\,\text{V}$$
$$V_B - V_A = 2\,\text{V}$$
$$\therefore \quad V_B = 2\,\text{V}$$
$$V_D - V_C = 10\,\text{V}$$
$$\therefore \quad V_D = 10 + V_C = 15\,\text{V}$$
$$i_{1\Omega} = \frac{V_C - V_B}{1} = 3\,\text{A from } C \text{ to } B \text{ as } V_C > V_B$$
$$i_{2\Omega} = \frac{V_D - V_A}{2} = 7.5\,\text{A from } D \text{ to } A \text{ as } V_D > V_A$$

3. $V_A = V_B$

$$\therefore \quad V_{AB} = 0$$
$$\text{or} \quad E - ir = 0$$
$$\therefore \quad E - \left(\frac{15 + E}{8}\right)(2) = 0$$

Solving this equation, we get

$$E = 5\,\text{V}$$

4. Net emf $= (n - 2m)E$

$$= (10 - 2 \times 2)(1) = 6\,\text{V}$$
$$i = \frac{\text{Net emf}}{\text{Net resistance}} = \frac{6}{10 + 2}$$
$$= 0.5\,\text{A}$$

5. $V_{R_1} = 0$

$$\therefore \quad i_{R_1} = 0$$
$$V_{R_2} = V_{R_3} = 10\,\text{V}$$
$$\therefore \quad i_{R_2} = i_{R_3} = \frac{10}{10} = 1\,\text{A}$$

Exercise

1. If current flows from a to b, then equation will become

$$V_a - ir - E = V_b$$
$$\text{or} \quad V_a - V_b = E + ir$$

So, $V_a - V_b$ is always positive. Hence, V_a is always greater than V_b.

2. In series combination, equivalent resistance,

$$R = R_1 + R_2 + R_3$$

Hence, equivalent resistance of resistors in series combination is more than largest resistance used in combination.

Therefore, statement I is incorrect. Resistivity of the material is dependent on temperature.

For metals, it increases with increase in temperature and for semiconductor , it decreases with increase temperature. Thus, statement II is also incorrect.

3. Let $V_0 = V$

$$\text{Now,} \quad I_{AO} + I_{BO} = I_{OC}$$
$$\therefore \quad \frac{6 - V}{6} + \frac{3 - V}{3} = \frac{V - 2}{2}$$

Solving this equation, we get

$$V = 3\,\text{V}$$

4.

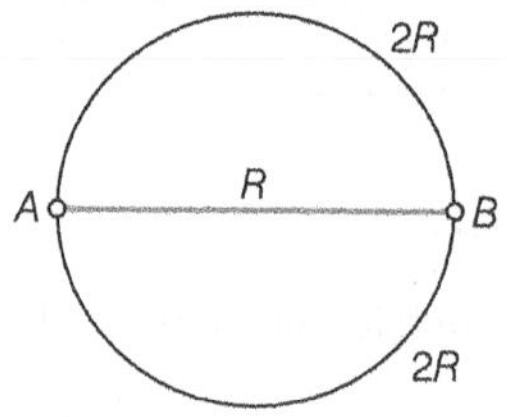

$$R = \left(\frac{4}{2\pi r}\right)(2r) = \frac{4}{\pi}$$

Now $2\Omega, 2\Omega$ and R are in parallel.

5. $E - ir = 0$

$$\therefore \quad 3 - \left(\frac{3 + 15}{1 + 2 + R}\right)(1) = 0$$

Solving this equation, we get

$$R = 3\Omega$$

6. Given circuit can be redrawn as follows

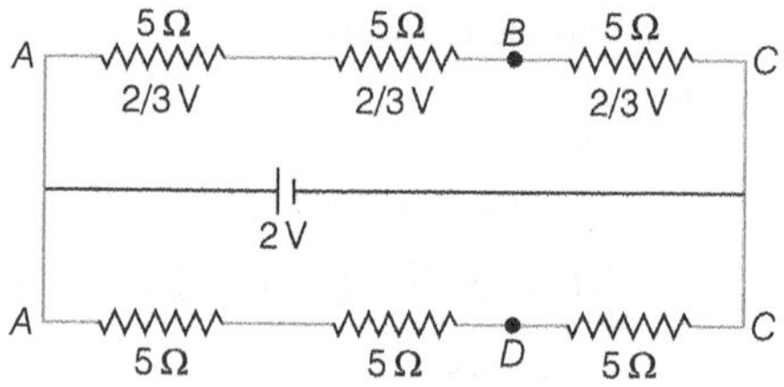

$$\therefore \quad V_A - V_B = \frac{4}{3}\,\text{V}$$

7. Current in 6Ω resistance is $\frac{0.8}{2} = 0.4\,\text{A}$

So, total current in circuit is $0.8 + 0.4 = 1.2\,\text{A}$

$\therefore$ Potential drop across 4Ω resistance

$$= 1.2 \times 4 = 4.8\,\text{V}$$

8. 6 volt potential difference will equally distribute in the two parallel combination of resistors shown in figure.

9. $R_{net} = R = 2\Omega$

$$\therefore \quad I = \frac{10}{2} = 5\,\text{A}$$

10. The given circuit can be redrawn as,

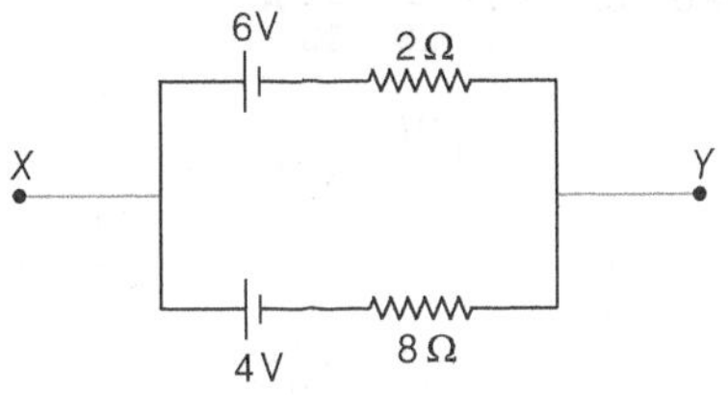

$$E_{net} = \frac{E_1/r_1 + E_2/r_2}{1/r_1 + 1/r_2} = \frac{(6/2)+(4/8)}{(1/2)+(1/8)}$$

$$= \frac{28}{5} = 5.6 \text{ volt}$$

Since, no current is taken from this battery.

So, $V_{XY} = E_{net} = 5.6$ volt $(\because V = E - ir)$

11.

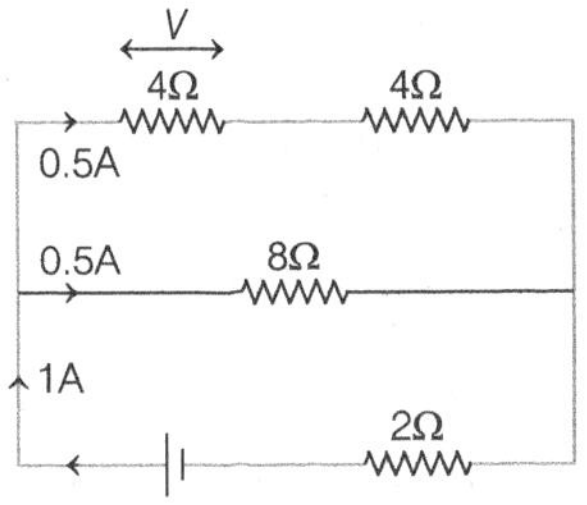

$$V = iR = (0.5)(4) = 2 \text{ volt}$$

12.

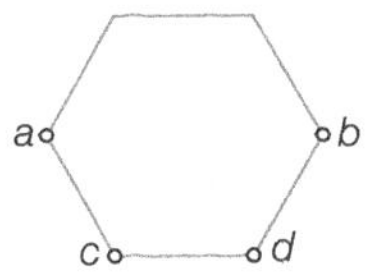

$$R_{max} = \frac{\left(\frac{R}{n/2}\right)}{2} = \frac{R}{n} = R_{ab}$$

$$R_{min} = \frac{\left(\frac{R}{n}\right)\left[(n-1)\frac{R}{n}\right]}{\frac{R}{n} + \frac{(n-1)R}{n}} = R_{cd} = \frac{(n-1)R}{n^2}$$

13. $R \propto l \Rightarrow \frac{R_1}{R_2} = \frac{l_1}{l_2} = \frac{1}{2}$

$\therefore$ $R_1 = \frac{1}{3} \times 9 = 3\Omega$

$R_2 = \frac{2}{3} \times 9 = 6\Omega$

$R_{net} = \frac{R_1 R_2}{R_1 + R_2} = 2\Omega$

14. Simple series and parallel grouping of resistors.

15. $R_{ab} = \frac{(3 \times 15)}{3 + 15} = 2.5\,\Omega$

As $R_{60^\circ} = \left(\frac{18}{360^\circ}\right)(60^\circ) = 3\Omega$

16. $V = iR$

$\therefore$ $V \propto R$ (as i = constant)

$\therefore$ $\frac{V_A}{V_B} = \left(\frac{\rho l_A}{\pi r_A^2}\right)\left(\frac{\pi r_B^2}{\rho l_B}\right)$

or $\frac{r_B}{r_A} = \sqrt{\frac{V_A}{V_B} \times \frac{l_B}{l_A}} = \sqrt{\frac{3}{2} \times \frac{1}{6}} = \frac{1}{2}$

17. $V = E - Ir$

Slope $= E = 2.0$ volt

Slope $= -r = -\frac{2}{5} = -0.4\Omega$

$\therefore$ $r = 0.4\ \Omega$

18. As $R \propto l$

$\therefore$ Resistance of each part will be $\frac{R}{5}$.

Parallel arrangement is,

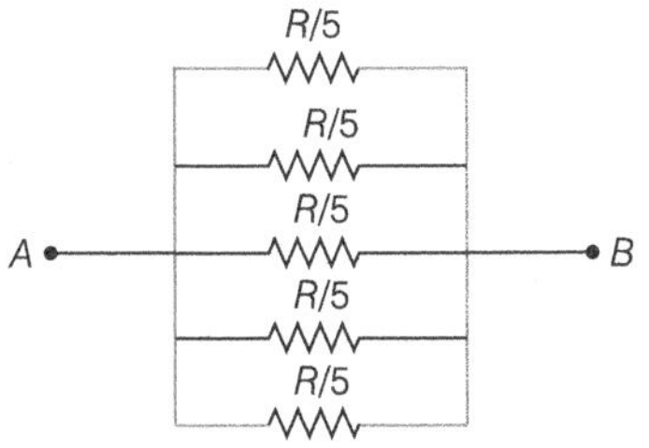

$$R_{eq} = \frac{R/5}{5} = \frac{R}{25}$$

19. In option (b),

$$R_{eq} = \frac{2 \times 4}{2+4} + 6 = \frac{8}{6} + 6$$

$$= \frac{4}{3} + 6 = \frac{22}{3}\Omega$$

20. $R_{PQ} = \frac{5}{11}r$, $R_{QR} = \frac{4}{11}r$

and $R_{PR} = \frac{3}{11}r$

$\therefore R_{PQ}$ is maximum.

21. Given,

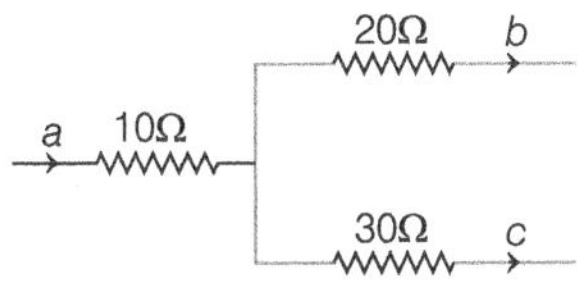

Let potential of the junction be x volts using junction law

$$i_1 + i_2 + i_3 = 0$$

or $$\frac{x-30}{10} + \frac{x-12}{20} + \frac{x-2}{30} = 0$$

$$\Rightarrow \frac{1}{60}[6x-180+3x-36+2x-4]=0$$

$$\frac{1}{60}[11x-220]=0$$

$$\Rightarrow \quad x=\frac{220}{11}=20\text{ V}$$

Current through 20Ω is $=\frac{x-12}{20}$

$$i_2=\frac{20-12}{20}=0.4\text{ A}$$

22. When the switch '*S*' is closed the circuit, hence formed is given in the figure below

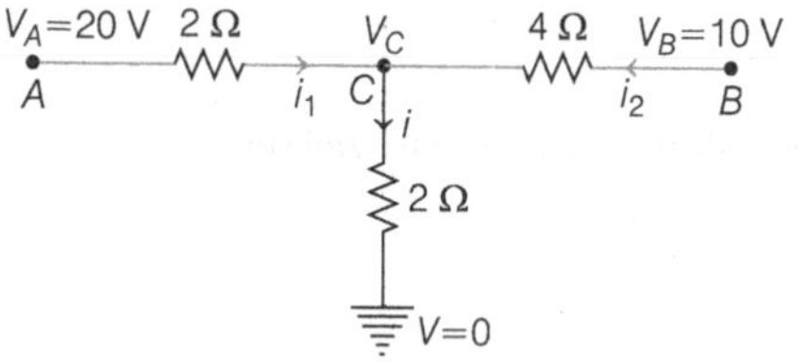

Then, according to Kirchhoff's current law,

$$i_1+i_2=i$$

$$\Rightarrow \quad \frac{V_A-V_C}{2}+\frac{V_B-V_C}{4}=\frac{V_C-V}{2}$$

(using Ohm's law, $V=iR$)

or $$\frac{20-V_C}{2}+\frac{10-V_C}{4}=\frac{V_C-0}{2}$$

$$\Rightarrow \quad 20-V_C+(10-V_C)2=V_C$$

$$40=V_C+3V_C$$

$$40=4V_C$$

or $$V_C=10\text{ V}$$

$\therefore$ The current, $i=\frac{V_C}{2}=\frac{10}{2}=5\text{A}$

23. The simplified circuit is shown in the figure.

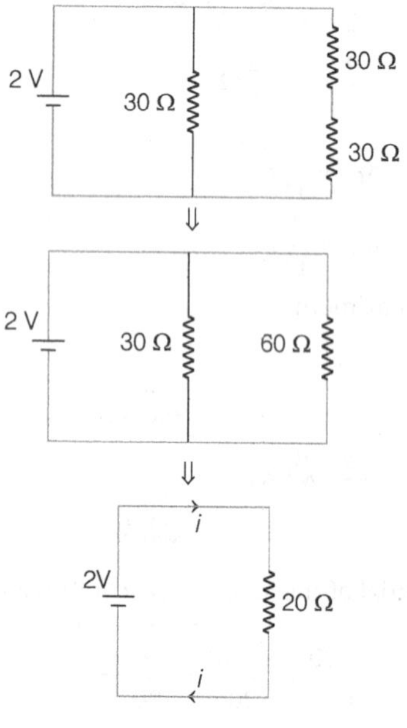

Therefore, current $i=\frac{2}{20}=\frac{1}{10}$ A

24.

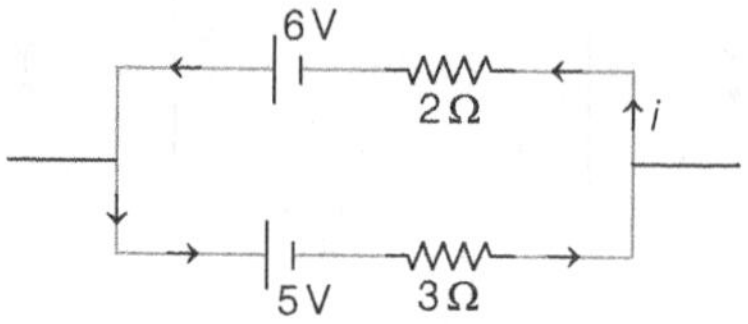

$$i=\frac{6-5}{2+3}=0.2\text{ A}$$

$$V_1=E_1-ir_1$$
$$=6-0.2\times 2$$
$$=5.6\text{ V}$$

26.

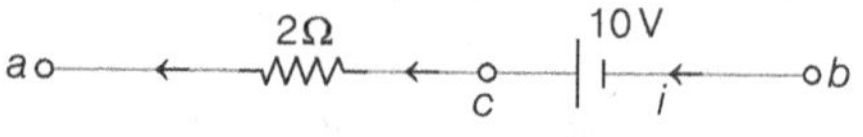

$$V_b+10-2i=V_a$$

$$V_b-V_a=2i-10=2\text{ V}$$

$\therefore$ $$i=6\text{ A}$$

Now, $$V_c-V_a=2\times 6=12\text{ V}$$

Since $V_c>V_a$, current is from *c* to *a* or from *b* to *a*.

27. Let potential of point *e* is *V* volts. Then,

$$I_{ae}+I_{be}+I_{ce}+I_{de}=0$$

$$\therefore \left(\frac{2-V}{1}\right)+\left(\frac{4-V}{2}\right)+\left(\frac{6-V}{1}\right)+\left(\frac{4-V}{2}\right)=0$$

or $$V=4\text{ V}$$

Now current through any wire can be obtained by the equation,

$$I=\frac{PD}{R}$$

28. By introducing parallel resistance R_3 in the circuit, total resistance of the circuit will decrease. Hence, main current *i* will increase.

Now, $V_{R1}=E-V_{R_2}=E-iR_2$

Since, *i* is increasing, so V_{R_2} will increase. Hence, V_{R_1} or current passing through R_1 will decrease.

29. $8.4=E-1.5\,r$ ...(i)

$$9.4=E+3.5\,r \quad \text{...(ii)}$$

Solving these two equations, we get

$$r=0.20\ \Omega$$

and $$E=8.7\text{ V}$$

30. $I=\frac{E}{R+nR}$

and $$\frac{E}{R+\frac{R}{n}}=10I \Rightarrow 10I=\frac{nE}{R+nR}$$

$$\therefore \quad n\left(\frac{E}{R+nR}\right)=10\left(\frac{E}{nR+R}\right)$$

$$\Rightarrow \quad n=10$$

31. Applying KVL in loop $ADBCA$,

$$2I_1 + 4I_1 - 2I_2 - 4I_2 = 0$$

$$\Rightarrow \quad I_1 = I_2 \quad \text{...(i)}$$

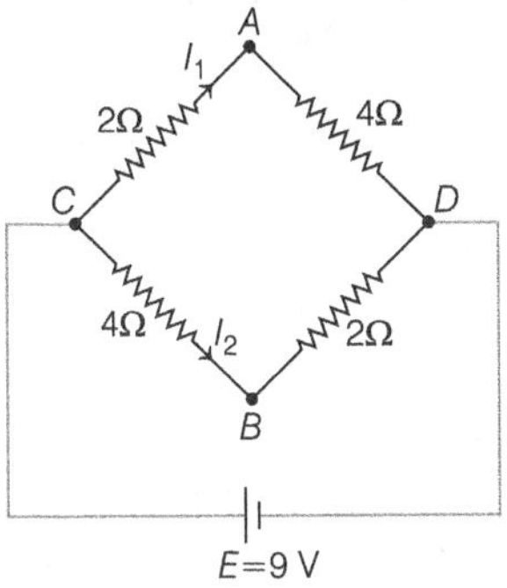

Applying KVL in loop $CBDEC$, we have

$$4I_2 + 2I_2 - 9 = 0 \Rightarrow I_2 = \frac{9}{6}$$

$$I_2 = 1.5\text{A}$$

From Eq. (i), $I_1 = 1.5$ A

$\therefore$
$$V_A = 9 - I_1 \times 2$$
$$= 9 - 1.5 \times 2 = 6 \text{ V}$$
$$V_B = 9 - I_2 \times 4$$
$$9 - 1.5 \times 4 = 3 \text{ V}$$

$\therefore$
$$V_A - V_B = 6 - 3 = 3 \text{ V}$$

32. We have, $i = \frac{E}{R_{eq}} = \frac{15}{1+2} = 5$ A

and
$$V_A - V_B = -15 + iR$$
$$= -15 + 5 \times 1 = -10 \text{ V}$$

So,
$$V_B - V_A = 10 \text{ V}$$

Module 6.

Introductory Exercise

1. Applying loop law equation in upper loop, we have

$$E + 12 - ir - 1 = 0 \quad \text{...(i)}$$

Applying loop law equation in lower loop, we have where

$$i = 1 + 2 = 3\text{A}$$

$$E + 6 - 1 = 0 \quad \text{...(ii)}$$

Solving these two equations, we get

$$E = -5 \text{ V and } r = 2\Omega$$

2. Power delivered by a battery $= Ei$

$$= 12 \times 3 = 36 \text{ W}$$

Power dissipated in resistance

$$= i^2 R = (2)^2 (3) = 12 \text{ W}$$

3.

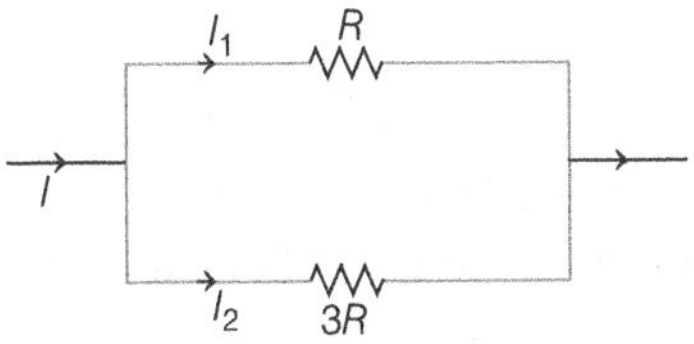

According to current division rule,

$$I_1 = \frac{3R}{(R + 3R)} I = \frac{3I}{4}$$

and
$$I_2 = \frac{RI}{(R + 3R)} = \frac{I}{4}$$

Hence, ratio of thermal energies released in resistors is

$$\frac{H_1}{H_2} = \frac{I_1^2 R}{I_2^2 \times 3R}$$

$$= \frac{\frac{9}{16}}{\frac{1}{16} \times 3} = 3:1$$

4. $P = i^2 R_{net}$

$\therefore$
$$P \propto R_{net}$$

5. $W = 0$. Therefore, from first law of thermodynamics,

$$\Delta U = \Delta Q = i^2 Rt$$
$$= (1)^2 (100) (5 \times 60) \text{ J}$$
$$= 30 \text{ kJ}$$

Exercise

1. In parallel, $V =$ constant

$\therefore$ From the equation

$$P = \frac{V^2}{R} \Rightarrow P \propto \frac{1}{R}$$

2. Resistance will increase with temperature on heating. Hence current will decrease.

Further $P = \frac{V^2}{R}$ or $P \propto \frac{1}{R}$

Resistance is increasing. Hence, power consumed across R should decrease.

$V = IR$ is just an equation between PD across a resistance current passing through it and its resistance. This is not Ohm's law.

3. In both cases potential difference across R is E.

$\therefore$
$$P = \frac{E^2}{R}$$

In second case net resistance will decrease. Therefore, main current will increase.

4. $R = \rho \frac{l}{A}$ or $R \propto \frac{1}{A}$

Area of cross-section of first wire is less. Hence, its resistance is more.

5. kVA is the unit of power. While kWh is the unit of energy.

6. $P = \frac{V^2}{R}$ or $P \propto \frac{1}{R}$

$$\frac{P_2}{P_1} = \frac{R_1}{R_2} = \frac{l_1}{l_2} \quad (\text{as } R \propto l)$$

$\therefore \quad P_2 = \left(\frac{I_1}{I_2}\right) P_1 = \left(\frac{I}{0.9\,I}\right) P_1 = 1.11\,P_1$

So, power will increase by 11%.

7. $P = \frac{V^2}{R}$ or $P \propto \frac{1}{R}$ ...(i)

Resistor is cut in n equal parts. Therefore, each resistance will become $\frac{R}{n}$. Now, these are connected in parallel. Therefore, net resistance will become $\frac{1}{n}$ times $\frac{R}{n}$ or $\frac{R}{n^2}$.

Now, from Eq. (i), power will become n^2 times.

8. $P = \frac{V^2}{R}$, R_{net} will decrease. Therefore, P will increase.

9. $R_2 = R_3$ as $P = \frac{V^2}{R}$ and in parallel V is same.

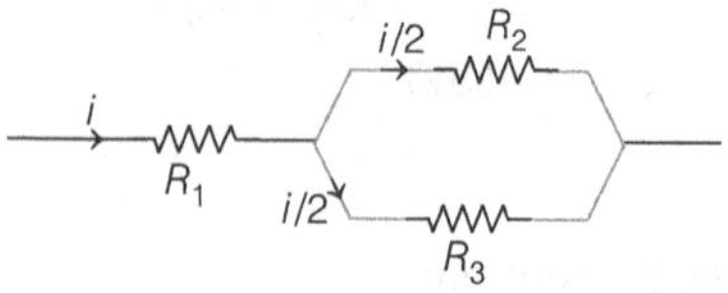

Hence, $P_{R_2} = P_{R_3}$

If $R_2 = R_3$

Now, current through R_1 is double so R_1 should be $\frac{1}{4}$th of R_2 or R_3 for same power. As $P = i^2 R$

10. $P_R = i^2 R$, $P_{2R} = i^2 (2R)$

$$P_{6R} = \left(\frac{i}{2}\right)^2 (6R)$$

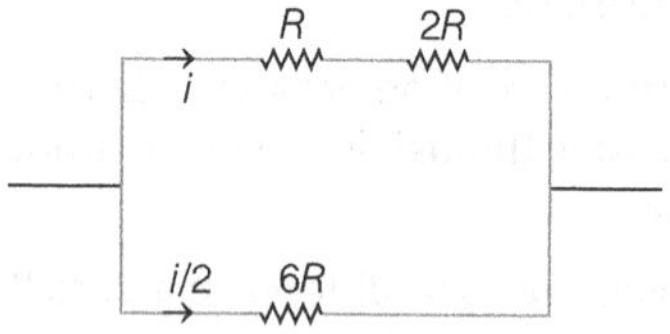

$$\therefore \quad P_R : P_{2R} : P_{6R} = 1 : 2 : 1.5$$
$$= 2 : 4 : 3$$

11. More current will flow through $2R$ while current through $6R$ will remain unchanged.

12. $H = \frac{V^2}{R_1} t_1 \Rightarrow R_1 = \frac{V^2 t_1}{H}$

Similarly, $R_2 = \frac{V^2 t_2}{H}$

In series, $H = \left(\frac{V^2}{R_1 + R_2}\right) t$

$$t = \frac{H(R_1 + R_2)}{V^2}$$

Substituting the values of R_1 and R_2, we get

$$t = t_1 + t_2$$

In parallel, $H = \frac{V^2}{R_{\text{net}}} t = V^2 t \left(\frac{1}{R_1} + \frac{1}{R_2}\right)$

$$= V^2 t \left(\frac{H}{V^2 t_1} + \frac{H}{V^2 t_2}\right)$$

Solving we get,

$$t = \frac{t_1 t_2}{t_1 + t_2}$$

13. (a) → (p,q), (b) → (r), (c) → (s)

In (a) : Current through each bulb is same.

So, power output

$$= I^2 R = I^2 \left(\frac{V^2}{P}\right)$$

∴ Power output or brightness $\propto P_{\text{out}} \propto \frac{1}{P_{\text{rated}}}$

Most bright bulbs is of least rated power. Hence, C is glowing with maximum intensity.

In (b) : If we compare with A and B, then brightness of B will be less as compared to A. Further R_C and I_C is maximum.

So, C will be glowing with maximum intensity.

In (c) : Voltage across each bulb is same. So, power output

$$= \frac{V^2}{R} = \frac{V^2}{\frac{V^2_{\text{rated}}}{P_{\text{rated}}}} = \frac{V^2 P_{\text{rated}}}{V^2_{\text{rated}}}$$

So, power output or brightness $\propto P_{\text{rated}}$

Most bright bulbs are of largest power. Hence, A is most bright.

14. $P = \frac{V^2}{R}$

$$\therefore \quad R = \frac{V^2}{P}$$

$$R_1 = \frac{(120)^2}{40} = 360\,\Omega$$

$$R_2 = \frac{(120)^2}{60} = 240\,\Omega$$

$$R_3 = \frac{(120)^2}{75} = 192\,\Omega$$

Now, all these resistors are in parallel.

15. (a) $i = \frac{12 - 6}{4 + 8} = 0.5$ A

(b) $P_{R_1} = i^2 R_1 = 1$ W

$\Rightarrow \quad P_{R_2} = i^2 R_2 = 2$ W

(c) Power supplied by $E_1 = E_1 i = 6$ W

and power consumed by $E_2 = E_2 i = 3$ W

16. (a) $E_1 > E_2$

Therefore, net current is anti-clockwise or from B to A.

(b) Current through E_1 is normal. Hence, it is doing the positive work.

(c) Current flows from B to A

$\therefore \quad V_B > V_A$

17. Power absorbed by resistor is i^2R or 2W. Therefore, remaining 3W is absorbed by the battery $(=Ei)$. Hence, E is 3 V and current of 1A enters from the position terminal as shown below.

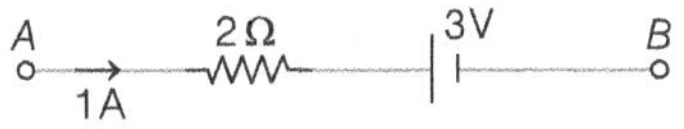

(a) $V_{AB} = E + ir$

$= E + iR$ (Here, $r = R$)

$= 3 + (1)(2) = 5$ V

(b) $E = 3$V

(c) Positive terminal on left side

18. Rate of energy developed $= I^2R$

So, energy developed in 10 s

$= I^2Rt = 400$

$$\Rightarrow \quad R = \frac{400}{2 \times 2 \times 10} = 10\,\Omega$$

When current is increased to 4 A, energy dissipated will be

$E = I^2Rt$

$= 4^2 \times 10 \times 20$

$= 3200$ joules

$= 8 \times 400$ joules

$\therefore \quad x = 8$

19. Amount of heat delivered by heater = Amount of heat required by water in phase change

$\Rightarrow$ power × time = mass × latent heat

$\Rightarrow$ 1000×60 (watt × s) $= m$ (g) $\times 540$ (cal × g)

As 1 cal = 4.2 J

$\Rightarrow \quad 60000 = m\,(\text{g}) \times 540 \times 4.2$ J/g

$$\Rightarrow \quad \frac{60000}{540 \times 4.2} = m$$

$$\Rightarrow \quad m\,(\text{in grams}) = \frac{80}{3}\,\text{g}$$

$\therefore \quad N = 8$

Module 7.

Introductory Exercise

1. (a) **Equivalent emf (V) of the battery**

PD across the terminals of the battery is equal to its emf when current drawn from the battery is zero. In the given circuit,

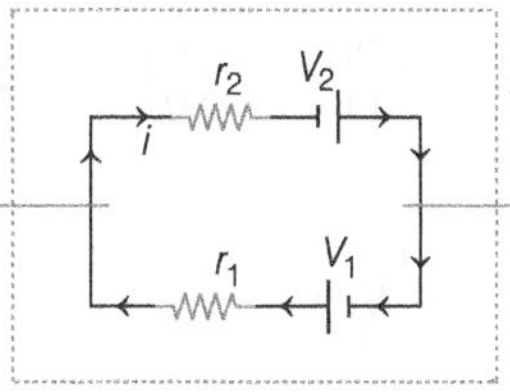

Current in the internal circuit,

$$i = \frac{\text{Net emf}}{\text{Total resistance}} = \frac{V_1 + V_2}{r_1 + r_2}$$

Therefore, potential difference between A and B would be

$$V_A - V_B = V_1 - ir_1$$

$$\therefore \quad V_A - V_B = V_1 - \left(\frac{V_1 + V_2}{r_1 + r_2}\right) r_1 = \frac{V_1 r_2 - V_2 r_1}{r_1 + r_2}$$

So, the equivalent emf of the battery is

$$V = \frac{V_1 r_2 - V_2 r_1}{r_1 + r_2}$$

Note that if $V_1 r_2 = V_2 r_1 : V = 0$

If $V_1 r_2 > V_2 r_1 : V_A - V_B =$ Positive i.e. A side of the equivalent battery will become the positive terminal and *vice-versa*.

(b) **Internal resistance (r) of the battery**

r_1 and r_2 are in parallel. Therefore, the internal resistance r will be given by

$$1/r = 1/r_1 + 1/r_2$$

or $$r = \frac{r_1 r_2}{r_1 + r_2}$$

2. 

$$E = \frac{\Sigma(E/r)}{\Sigma(1/r)} = \frac{(6/1) - (2/1)}{(1/1) + (1/1)} = 2\text{ V}$$

Now, net emf of E and 4 V is 2V as they are oppositely connected.

3. $$E_{eq} = \frac{\Sigma(E/r)}{\Sigma(1/r)} = \frac{(10/1) + (4/2) + (6/2)}{(1/1) + (1/2) + (1/2)} = 7.5\text{ V}$$

$$\frac{1}{r} = \frac{1}{1} + \frac{1}{2} + \frac{1}{2}$$

$\therefore \quad r = 0.5\,\Omega$

Exercise

1. $$E_{eq} = \frac{E_1/r_1 + E_2/r_2}{(1/r_1) + (1/r_2)} = E_1\left[\frac{(1/r_1) + (E_2/E_1)(1/r_2)}{(1/r_1) + (1/r_2)}\right]$$

So, E_{eq} may be greater than E_1 also, if $E_2/E_1 > 1$

r_1 and r_2 are in parallel.

Hence, r_{eq} is less than both r_1 and r_2 individually.

2.

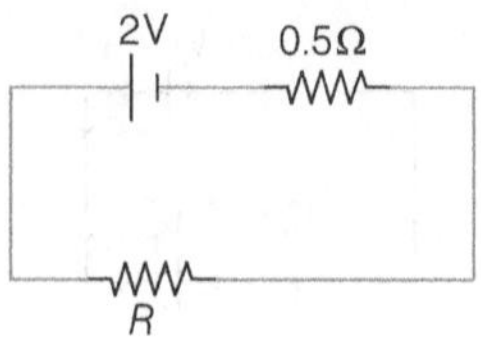

Equivalent simple circuit is given as maximum power across R is obtained

when $R = r = 0.5\,\Omega$

$$i = \frac{2}{R+r} = 2\,\text{A} \quad = i^2R = (2)^2\,(0.5)$$

$$= 2\,\text{W}$$

3. $E_{net} = \dfrac{E_1/r_1 - E_2/r_2}{1/r_1 + 1/r_2}$

Current through R will be zero, if $E_{net} = 0$

or $$\frac{E_1}{r_1} = \frac{E_2}{r_2}$$

4. There are n rows each containing m cells.

$\therefore$ Total cells $= m \times n = 24$ ...(i)

For maximum current in the circuit, external resistance should be equal to net internal resistance.

$$R = \frac{mr}{n}$$

$\therefore$ $$3 = \frac{m}{n}(0.5)$$

$\therefore$ $m = 6n$ ...(ii)

From Eqs. (i) and (ii), we get

$$m = 12, n = 2$$

5. Current flowing through the circuits,

$$i_s = \frac{2E}{2r+2}$$

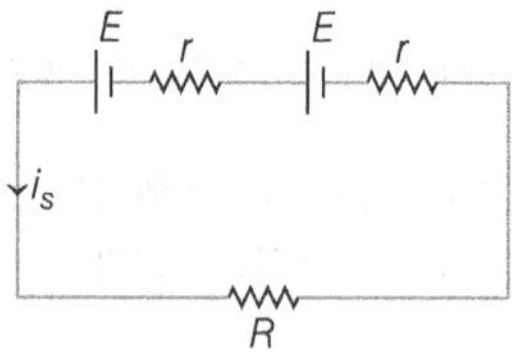

$$i_p = \frac{E}{\frac{r}{2}+2} = \frac{2E}{r+4}$$

Since, $i_s = i_p$

$\Rightarrow$ $$\frac{2E}{2r+2} = \frac{2E}{r+4} \Rightarrow r = 2\Omega$$

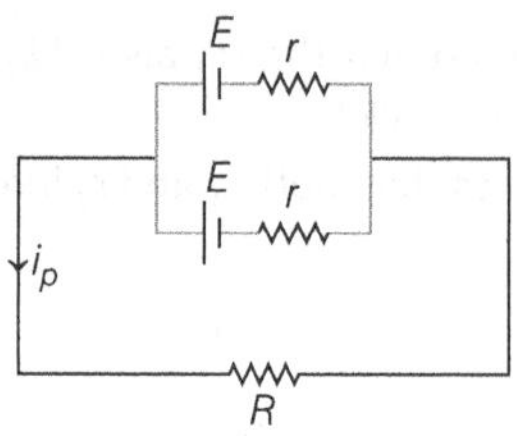

6. Let E be the emf of each cell and let n be the number of cells connected wrongly.

Then, the net emf of the battery is equal to the emf of correctly connected cells – emf of wrongly connected cells.

$$= (12 - n)E - nE$$
$$= (12 - 2n)E$$

When the two cells aid the battery, the net emf is

$$E_1 = (12 - 2n)E + 2E$$
$$= (14 - 2n)E$$

If R is the resistance in the circuit, the current is

$$I_1 = \frac{E_1}{R} = \frac{(14-2n)E}{R} \quad \text{...(i)}$$

When the two cells oppose the battery, the net emf is

$$E_2 = (12 - 2n)E - 2E$$
$$= (10 - 2n)E \quad \text{...(ii)}$$

Now, the current in the circuit is

$$I_2 = \frac{(10-2n)E}{R}$$

Dividing Eqs. (i) by Eq. (ii), we get

$$\frac{I_1}{I_2} = \frac{(14-2n)}{(10-2n)} = \frac{(7-n)}{(5-n)}$$

Given, $I_1 = 3\,\text{A}$

and $I_2 = 2\,\text{A}$

Thus, $\dfrac{3}{2} = \dfrac{(7-n)}{(5-n)}$

Which gives $n = 1$.

Hence, one cell in the battery is wrongly connected.

7.

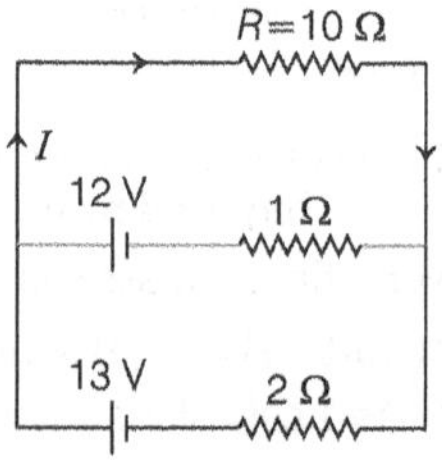

$$E_{eq} = \frac{\Sigma E/r}{\Sigma(1/r)} = \frac{\frac{12}{1}+\frac{13}{2}}{\frac{1}{1}+\frac{1}{2}} = \frac{37}{3}\,\text{V}$$

$$r_{eq} = \frac{r_1 r_2}{r_1 + r_2} = \frac{2}{3}\Omega$$

$\Rightarrow$ $$I = \frac{\frac{37}{3}}{\frac{2}{3}+10} = \frac{37}{32}\,\text{A}$$

Voltage across load

$$= IR = \left(\frac{37}{32}\right)(10)$$

$$= 11.56\,\text{V}$$

Hence, voltage across the load lies between 11.5 V and 11.6 V.

8. $i = \frac{E}{R+r} \Rightarrow P = \text{power across } R = i^2 R$

$$P = \left(\frac{E}{R+r}\right)^2 R \qquad \text{...(i)}$$

For power to be maximum,

$$\frac{dP}{dR} = 0$$

By putting $\frac{dP}{dR} = 0$, we get

$$R = r$$

Further, by putting $R = r$ in Eq. (i), we get

$$P_{max} = \frac{E^2}{4r}$$

9. As derived in the above question,

$$P_{max} = \frac{E^2}{4r}$$

Here, $E =$ net emf $= 2 + 2 = 4\text{ V}$

and $r =$ net internal resistance $= 1 + 1 = 2\Omega$

$$\therefore \quad P_{max} = \frac{(4)^2}{(4)(2)} = 2\text{ W}$$

10. Applying Kirchhoff's laws in two loops, we have

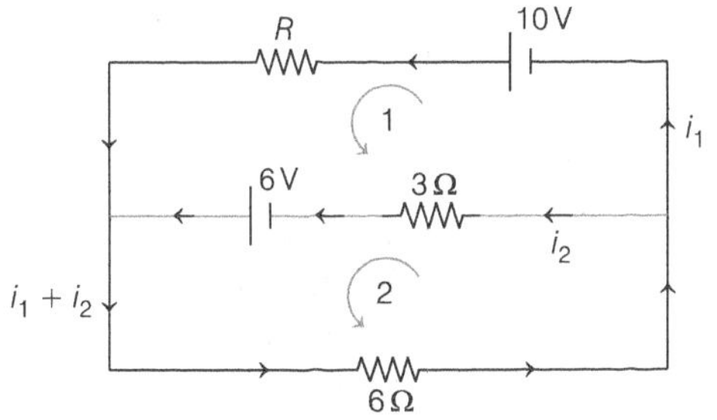

$$10 - i_1 R - 6 + 3i_2 = 0 \qquad \text{...(i)}$$

$$6 - 6(i_1 + i_2) - 3i_2 = 0 \qquad \text{...(ii)}$$

Solving these two equations, we get

$$i_1 = \frac{6}{R+2}$$

Power developed in R,

$$P = i_1^2 R = \frac{36}{(R+2)^2} R \qquad \text{...(iii)}$$

For power to be maximum,

$$\frac{dP}{dR} = 0$$

$$\therefore \quad (R+2)^2(36) - (36R)(2)(R+2) = 0$$

or $$R + 2 - 2R = 0$$

$$\therefore \quad R = 2\Omega \qquad \textbf{Ans.}$$

For maximum power from Eq. (iii), we have

$$P_{max} = 4.5\text{ W} \qquad \textbf{Ans.}$$

11. The two batteries are in parallel. Thermal power generated in R will be maximum when,

total internal resistance = total external resistance

or $$R = \frac{R_1 R_2}{R_1 + R_2}$$

$$E_{eq} = \frac{\left(\frac{E_1}{R_1} + \frac{E_2}{R_2}\right)}{\left(\frac{1}{R_1} + \frac{1}{R_2}\right)} = \left(\frac{E_1 R_2 + E_2 R_1}{R_1 + R_2}\right)$$

$$R_{net} = \frac{2R_1 R_2}{R_1 + R_2}$$

$$\therefore \quad i = \frac{E_{eq}}{R_{net}} = \frac{E_1 R_2 + E_2 R_1}{2R_1 R_2}$$

Maximum power through R

$$P_{max} = i^2 R = \frac{(E_1 R_2 + E_2 R_1)^2}{4R_1 R_2 (R_1 + R_2)}$$

12. We can make a simple circuit as shown below.

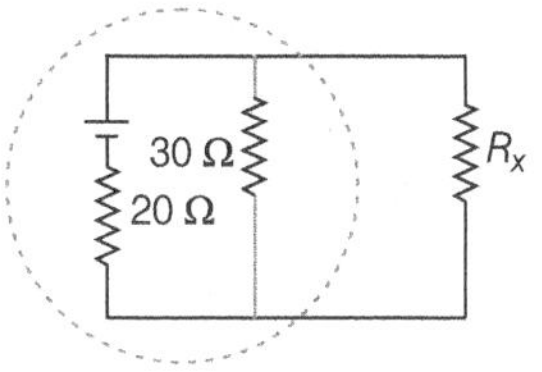

Let us assume a battery of internal resistance say R.

$$R = \frac{R_1 R_2}{R_1 + R_2} = \frac{(20)(30)}{(20)+(30)} = 12\Omega$$

We know that power across R_x is maximum or $\frac{dP}{dR_x} = 0$, when

total external resistance = total internal resistance

or $$R_x = R = 12\Omega$$

For P to be constant (of small variations of R_x) condition will be same or $\frac{dP}{dR_x} = 0$. Hence,

$$R_x = 12\Omega$$

Module 8.

Introductory Exercise 1

1. To increase the range of ammeter a parallel resistance (called shunt) is required which is given by

$$S = \left(\frac{i_g}{i - i_g}\right) G$$

For option (c), $S = \left(\frac{50 \times 10^{-6}}{5 \times 10^{-3} - 50 \times 10^{-6}}\right)(100) \approx 1\Omega$

To change it in voltmeter, a high resistance R is put in series, where R is given by $R = \frac{V}{i_g} - G$

For option (b), $R = \frac{10}{50 \times 10^{-6}} - 100 \approx 200\text{ k}\Omega$

Therefore, options (b) and (c) are correct.

2. $V = i_g (G + R)$

$\therefore\ R = \frac{V}{i_g} - G$ = series resistance connected with galvanometer

$$= \left(\frac{5}{5 \times 10^{-3}}\right) - 1 = 999\,\Omega$$

3. $V = i_g G$

$\therefore\qquad i_g = \frac{V}{G}$

Now, $\qquad nV = i_g (G + R) = \frac{V}{G}(G + R)$

$\therefore\qquad R = (n - 1)G$

Introductory Exercise 2

1. $r = R\left(\frac{l_1}{l_2} - 1\right)$

$$= 5\left(\frac{0.52}{0.4} - 1\right)$$

$$= 1.5\,\Omega$$

2. (a) $V_{AJ} = \frac{E}{2}$ or emf of lower battery

$\therefore\qquad iR_{AJ} = \frac{E}{2}$

or $\qquad \left(\frac{E}{15r + r}\right)\left(\frac{15r}{600}\right)(l) = \frac{E}{2}$

Solving this equation, we get

$$l = 320\text{ cm}$$

(b) Resistance of 560 cm $= \left(\frac{15r}{600}\right)(560)$

$$= 14r$$

Now the circuit is as under,

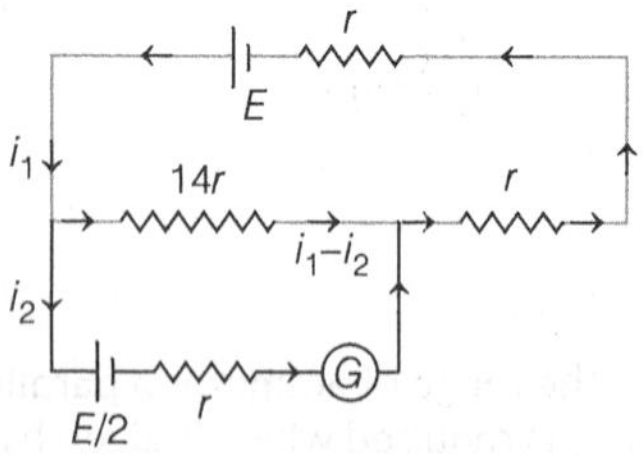

Applying loop law in upper loop

we have, $\quad E - 14r(i_1 - i_2) - i_1 r - i_1 r = 0 \qquad$ ...(i)

Applying loop law in lower law loop

we have, $\quad -\frac{E}{2} - i_2 r + (i_1 - i_2)(14r) = 0 \qquad$ ...(ii)

Solving these two equations

we get, $\qquad i_2 = \frac{3E}{22r}$

3.

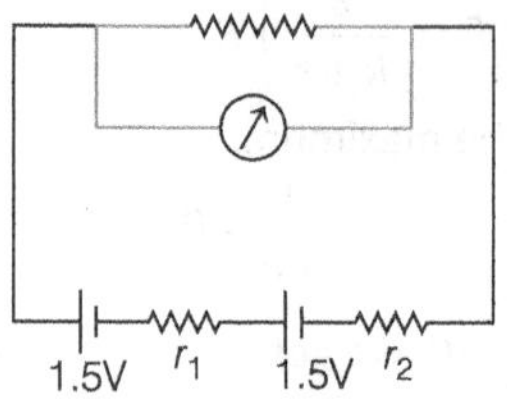

$$r = \left(\frac{E}{V} - 1\right)R$$

$$r = \left(\frac{3.0}{1.5} - 1\right) \times 10 = \left(\frac{30}{15} - 1\right) \times 10$$

$$= (2 - 1) \times 10$$

$$V_{eq} = 10$$

$$r_1 = 5\,\Omega \text{ or } r_2 = 5\,\Omega$$

Introductory Exercise 3

1. $R > 2\,\Omega \Rightarrow 100 - x > x$

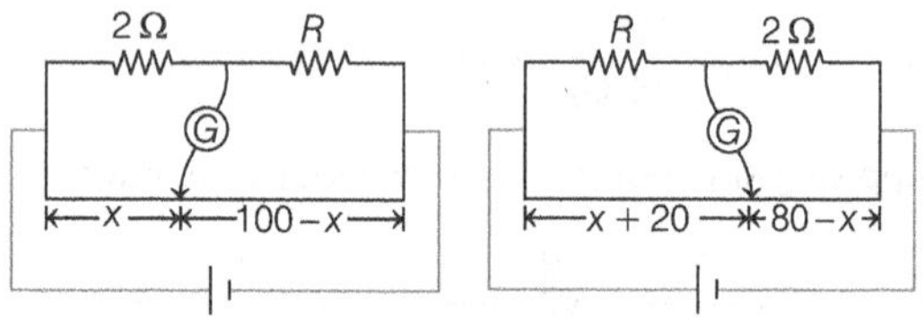

Applying $\quad \frac{P}{Q} = \frac{R}{S}$

We have $\qquad \frac{2}{R} = \frac{x}{100 - x} \qquad$ ...(i)

$$\frac{R}{2} = \frac{x + 20}{80 - x} \qquad \text{...(ii)}$$

Solving Eqs. (i) and (ii), we get

$$R = 3\,\Omega$$

$\therefore$ Correct option is (a).

2. Using the concept of balanced, Wheatstone bridge, we have,

$$\frac{P}{Q} = \frac{R}{S} \Rightarrow \frac{X}{(52 + 1)} = \frac{10}{(48 + 2)}$$

$\therefore\qquad X = \frac{10 \times 53}{50} = 10.6\,\Omega$

$\therefore$ Correct option is (b).

3. Slide wire bridge is most sensitive when the resistance of all the four arms of bridge is same.

Hence, B is the most accurate answer.

Introductory Exercise 4

1. $\frac{P}{Q}=\frac{R}{X} \Rightarrow X=\left(\frac{Q}{P}\right)R=\left(\frac{1}{10}\right)R$

R lies between 142 Ω and 143 Ω.

Therefore, the unknown resistance X lies between 14.2Ω and 14.3 Ω.

2. Experiment can be done in similar manner but now K_2 should be pressed first then K_1.

3. 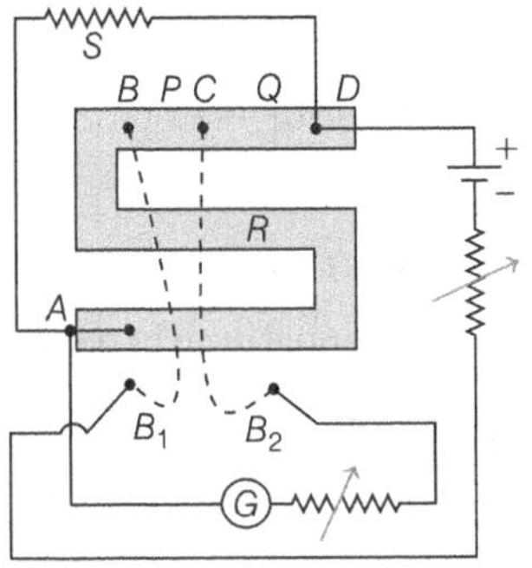

BC, CD and BA are known resistances.
The unknown resistance is connected between A and D.

Exercise

1. Even if ammeter is non-ideal, its resistance should be small and net parallel resistance is less than the smallest individual resistance.

∴ R_{net} < resistance of ammeter in the changed situation. Hence, net resistance of the circuit will decrease. So, main current will increase.

But maximum percentage of main current will pass through ammeter (in parallel combination) as its resistance is less. Hence, reading of ammeter will increase.

Initial voltmeter reading = emf of battery

Final voltmeter reading = emf of battery − potential drop across shown resistance.

Hence, voltmeter reading will decrease.

2. $i\lambda l=E_U \Rightarrow l=\frac{E_U}{i\lambda}$ (E_U = EMF of unknown battery)

E_U and i both will become two times. Therefore, l will remain unchanged.

3. A part of potential difference of known battery is balanced by emf of unknown battery. Hence,

E of unknown battery < E of known battery.

Further, E of unknown battery $=i\lambda l$

$$\Rightarrow \quad l=\frac{E}{(i\lambda)}$$

i = current from known battery

and λ = resistance per unit length of potentiometer wire.

If emf of known battery increases, i increases. So, l decreases.

4. When voltmeter is connected in series, resistance of the circuit increases. So, current drastically decreases.

5. $\frac{8\Omega}{6\Omega}=\frac{4\Omega}{3\Omega}$

$$\therefore \quad V_A=V_B$$

or $$V_{AB}=0$$

6. In parallel current distributes in increase ratio of resistance

$$\therefore \quad \frac{I_G}{I_S}=\frac{S}{G}$$

$$\therefore \quad G=\left(\frac{I_S}{I_G}\right)(S)=\left(\frac{50-20}{20}\right)(12)$$

$$=18\,\Omega$$

7. $\frac{I_G}{I_S}=\frac{S}{G}$

$$\therefore \quad S=\left(\frac{I_G}{I_S}\right)G=\left(\frac{2}{98}\right)G=\frac{G}{49}$$

8. $V_{AC}=V_{DE}$

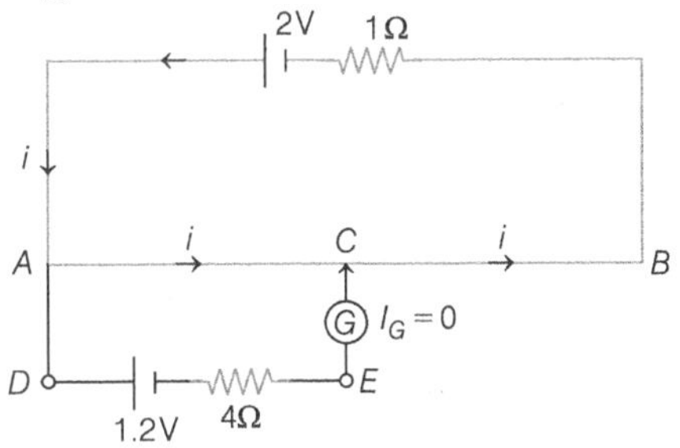

$$i(R_{AC})=E=1.2$$

$$\therefore \quad \left(\frac{2}{4+1}\right)\left(\frac{4}{100}\times l\right)=1.2$$

Solving this equation, we get

$$l=75\text{ cm}$$

9. Potential drop across potentiometer wire

$$=(0.2\times10^{-3})(100)=0.02\text{ V}$$

Now, given resistance and potentiometer wire are in series with given battery. So, potential will drop in direct ratio of resistance.

$$\therefore \quad \frac{0.02}{2-0.02}=\frac{R}{490}$$

$$\therefore \quad R=4.9\,\Omega$$

10. $(2\times10^{-3})(50)=10\times i$

$$i=10\times10^{-3}\text{ A}=10\text{ mA}$$

11. B and C are in parallel.

$$\therefore \quad V_B=V_C$$

Further $$R_A=R$$

$$R_{BC}=\frac{(1.5R)(3R)}{1.5R+3R}=R$$

or $$R_A=R_{BC}$$

$\therefore \quad V_A = V_{BC}$

Because $\quad iR_A = iR_{BC}$

12. Initially, $\dfrac{l_1}{100 - l_1} = \dfrac{X}{R} = \dfrac{2}{3}$

$\therefore \quad l_1 = \dfrac{2}{5} \times 100 = 40\text{ cm}$

Finally, $\dfrac{l_2}{100 - l_2} = \dfrac{X}{R'} = \dfrac{12}{8} = \dfrac{3}{2}$

$\therefore \quad l_2 = \dfrac{3}{5} \times 100 = 60\text{ cm}$

$\therefore$ J is displaced by

$$l_2 - l_1 = 20\text{ cm}$$

13. Equivalent emf of two batteries ε_1 and ε_2 is

$$\varepsilon = \frac{\varepsilon_1/r_1 + \varepsilon_2/r_2}{1/r_1 + 1/r_2} = \frac{(2/2) + (4/6)}{(1/2) + (1/6)}$$

$$= 2.5\text{ V}$$

Now, $\quad V_{AN} = \varepsilon$

$\therefore \quad (I_{AN})(R_{AN}) = \varepsilon$

or $\quad \left(\dfrac{12}{4 + 4 \times 4}\right)(4)(l) = 2.5$

Solving this equation, we get

$$l = \frac{25}{24}\text{ m}$$

14. Current decreases $\dfrac{20}{30}$ times or $\dfrac{2}{3}$ times. Therefore, net resistance should become $\dfrac{3}{2}$ times.

$\therefore \quad R + 50 = \dfrac{3}{2}(2950 + 50)$

Solving we get,

$$R = 4450\,\Omega$$

15. $E_0 = V_{AC} = (i)_{AC}\,(R)_{AC}$

$$= \left(\frac{E}{10}\right)\left(\frac{10}{1} \times 0.2\right) = \frac{E}{5} \quad \text{...(i)}$$

In second case,

$$E_0 = \left(\frac{E}{10 + x}\right)\left(\frac{10}{1} \times 0.3\right) \quad \text{...(ii)}$$

Solving Eqs. (i) and (ii), we get

$$x = 5\,\Omega$$

16. The ratio $\dfrac{AC}{CB}$ will remain unchanged.

17. The correct options are

(b) $V_1 \neq V_2$

(c) $V_1 + V_2 = V_3$

The potential difference across AB is given by

$$(V_{AB} = V_1 + V_2 = V_3)$$

Now, current through the upper branch is flowing through (V_1) and (V_2).

But as their resistances are different their readings won't be equal.

18-19. $r = R\left(\dfrac{l_1}{l_2} - 1\right)$

$\therefore \quad 10 = R\left(\dfrac{500}{490} - 1\right)$

Solving this equation, we get

$$R = 490\,\Omega$$

Further $\quad r = R\left(\dfrac{E}{V} - 1\right)$

or $\quad 10 = 490\left(\dfrac{2}{V} - 1\right)$

Solving, we get

$$V = 1.96\text{ V}$$

20. The emf of cell = potential drop of balance length

$$= xl_{(AC)}$$

Here, $x = \dfrac{99\,\Omega}{99\text{ cm}} = \dfrac{1\,\Omega}{\text{cm}}$

I = current in potentiometer wire

$$= \frac{V_{AB}}{R_{AB}} = \frac{10\text{ V}}{100\,\Omega} = \frac{1}{10}\text{ A}$$

$\therefore \quad E = 1\,\Omega/\text{cm} \times \dfrac{1}{10}\text{ A} \times 40\text{ cm}$

$$= 4\text{ V}$$

21. When jockey is touched at point B, the circuit will be

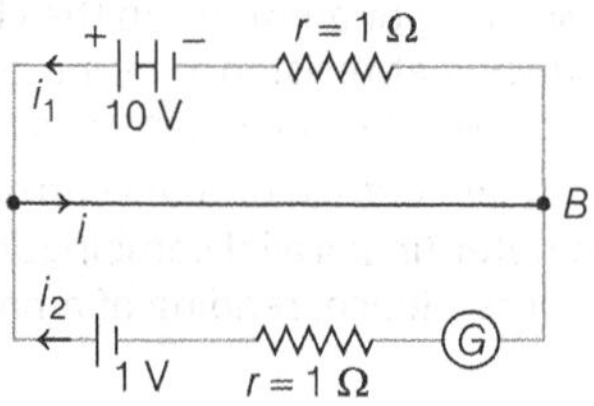

Applying Kirchhoff's law,

$$-99i - i_1 + 10 = 0 \quad \text{...(i)}$$

$$-99i - i_2 + 1 = 0 \quad \text{...(ii)}$$

and $\quad i_1 + i_2 = i \quad \text{...(iii)}$

On solving these equations, we get i and potential drop across AB length of wire

$$= iR_{AB} = V_{AB} = 6.93\text{ V}$$

22. When jockey is touched at point A, then there are two separate loops.

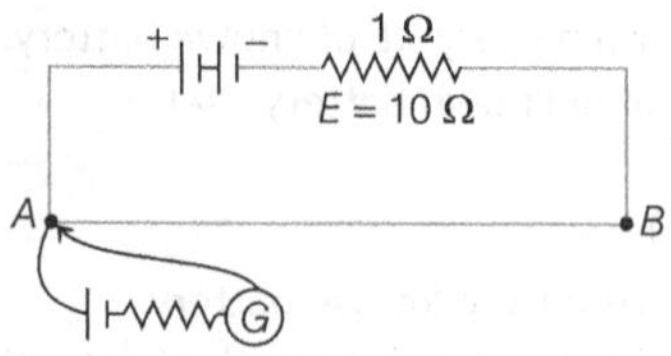

So, potential difference across the cell is zero.

23. When jockey is touched at point A, then potential difference across the terminals of driver cell is given

$$V = E - Ir = 9.9\text{ V}$$

24. (a)

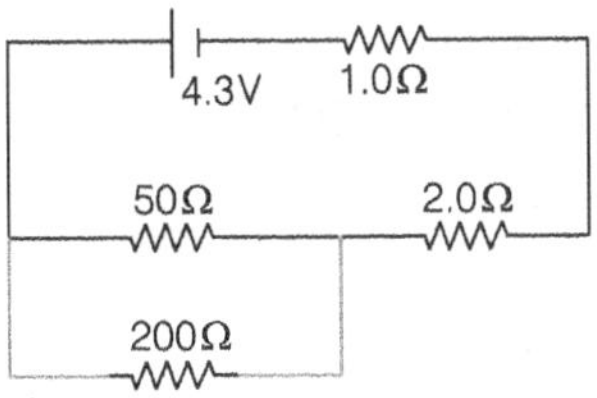

$$R_{net} = 1.0 + 2.0 + \frac{50 \times 200}{50 + 200} = 43\ \Omega$$

$\therefore$ $i = \frac{4.3}{43} = 0.1\text{ A} = \text{Readings of ammeter}$

Readings of voltmeter

$= (i) \times$ net resistance of $50\ \Omega$ and $200\ \Omega$

$$= (0.1)\left(\frac{50 \times 200}{50 + 200}\right) = 4\text{ V}$$

(b)

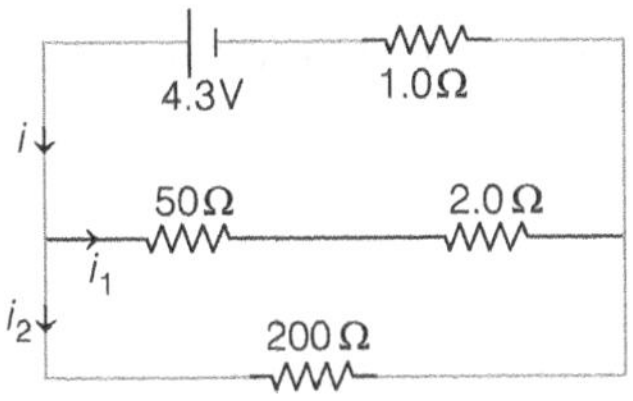

$$R_{net} = 1.0 + \frac{52 \times 200}{52 + 200} = 42.27\ \Omega$$

$\therefore$ $i = \frac{4.3}{42.27} \approx 0.1\text{ A}$

Now, $\frac{i_1}{i_2} = \frac{200}{52}$

$\therefore$ $i_1 = \left(\frac{200}{252}\right)(0.1) = 0.08\text{ A}$

$=$ Reading of ammeter

$\therefore$ Reading to voltmeter

$=$ Potential difference across $50\ \Omega$ and $2.0\ \Omega$

$= 0.08 \times 52 \approx 4.2\text{ V}$

25. Net resistance should remain unchanged.

$\therefore$ $R + G = R' + \frac{GS}{G + S}$

$\therefore$ $R' - R = G - \frac{GS}{G + S} = \frac{G^2}{G + S}$

26. $R_{net} = 60 + \frac{(60)(120)}{60 + 120} = 100\ \Omega$

$\therefore$ $i = \frac{120}{100} = 1.2\text{ A}$

Now, reading of voltmeter

$= 120 -$ potential drop across R_1

$= 120 - (60)(1.2) = 48\text{ V}$

27. In parallel current distributes in inverse ratio of resistance.

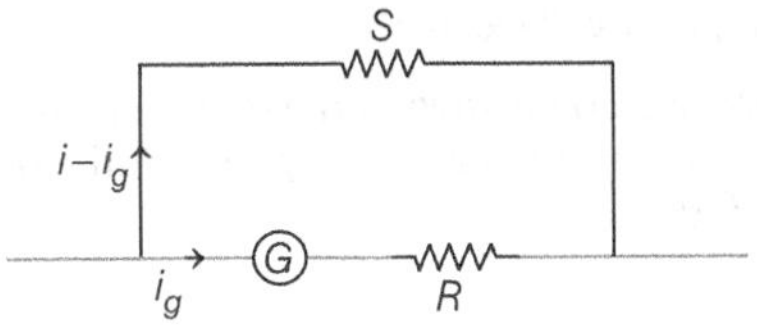

$\therefore$ $\left(\frac{i - i_g}{i_g}\right) = \frac{G + R}{S}$

$\therefore$ $R = \left(\frac{i - i_g}{i_g}\right)S - G$

$$= \left(\frac{20}{10^{-3}}\right)(0.005) - 20 = 80\ \Omega$$

Note *In calculations, we have taken* $i - i_g \approx i$.

28.

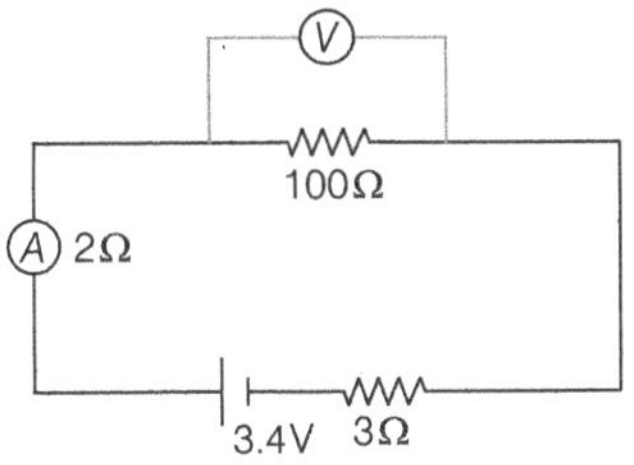

Reading of voltmeter $= 3.4 -$ Voltage drop across ammeter and 3Ω resistance

$= (3.4) - 0.04 \times 2 - 0.04 \times 3$

$= 3.2\text{ V}$

Now, $3.2\text{ V} = \frac{(0.04)(100\,R)}{(100 + R)}$

where, $R =$ resistance of voltmeter

$\therefore$ $R = 400\ \Omega$

If voltmeter is ideal, then

$$i = \frac{3.4}{2 + 100 + 3} = 0.03238\text{ A}$$

Reading of voltmeter $= 100\,i = 3.238\text{ V}$

29. (a) $V_A - V_B = 6\text{ V}$

Since, $V_B = 0$

$\therefore$ $V_A = 6\text{ V}$

$V_A - V_C = 4\text{ V}$

$\Rightarrow$ $V_C = V_A - 4 = 2\text{ V}$

(b) $V_A - V_D = V_A - V_C = 4\text{ V}$

From unitary method, we can find that,

$$AD = \left(\frac{100}{6}\right)(4) = 66.67\text{ cm}$$

(c) Since, they are at same potential, no current will flow through it.

(d) $V_A - V_B$ is still 6 V

$\therefore$ $V_A = 6\text{ V}$

Further, $V_A - V_C = 7.5\text{ V}$

$\therefore$ $V_C = -1.5\text{ V}$

Since, EMF of the battery in lower circuit is more than the EMF of the battery in upper circuit. No such point will exist.

30. (a) There are no positive and negative terminals on the galvanometer because only zero deflection is needed.

(b)

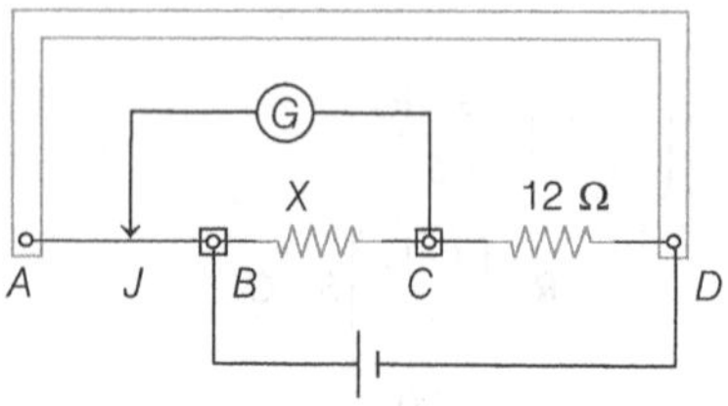

(c) $AJ = 60$ cm $\Rightarrow$ $BJ = 40$ cm

If no deflection is taking place. Then, the Wheatstone bridge is said to be balanced. Hence,

$$\frac{X}{12} = \frac{R_{BJ}}{R_{AJ}} \quad \text{or} \quad \frac{X}{12} = \frac{40}{60} = \frac{2}{3}$$

or $X = 8\,\Omega$ **Ans.**

Module 9.

1. Brown $\rightarrow$ 1, Red $\rightarrow$ 2, Green $\rightarrow 10^5$, Gold $\rightarrow$ 5 %

$\therefore$ $R = (12 \times 10^5\,\Omega) \pm 5\%$

$= (1.2 \times 10^6\,\Omega) \pm 5\%$

2. Given, resistance is $200\,\Omega$.

$= 20 \times 10^1\,\Omega$

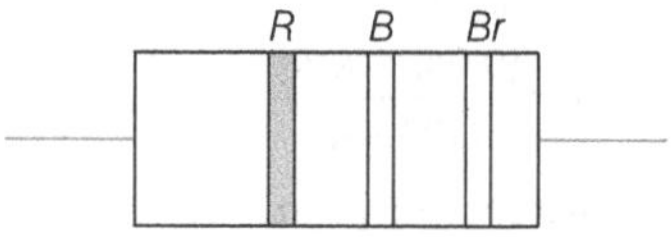

So, colour scheme will be red, black and brown.
Significant figure of red band is 2 and for green is 5. When red (2) is replaced with green (5), new resistance will be

200 ohm $\longrightarrow$ 500 ohm

3. The value of R_1 (orange, red, brown)

$= 32 \times 10 = 320\,\Omega$

Given, $R_2 = 80\,\Omega$ and $R_4 = 40\,\Omega$

In balanced Wheatstone bridge condition,

$$\frac{R_1}{R_2} = \frac{R_3}{R_4} \Rightarrow R_3 = R_4 \times \frac{R_1}{R_2}$$

$\Rightarrow$ $R_3 = \frac{40 \times 320}{80}$

or $R_3 = 160\,\Omega = 16 \times 10^1$

Comparing the value of R_3 with the colours assigned for the carbon resistor, we get

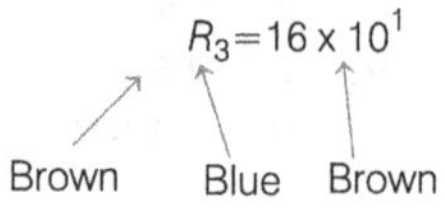

4. Yellow $\rightarrow$ 4

Red $\rightarrow$ 2

Orange $\rightarrow 10^3$

Gold $\rightarrow$ 5

$\therefore$ $R = (42 \times 10^3 \pm 5\%)\,\Omega$

5. 2 $\rightarrow$ Red

4 $\rightarrow$ Yellow

$10^6 \rightarrow$ Blue

10% $\rightarrow$ Silver

Unit Exercise (Cumulative)

Level 1

1. All the three resistances are in parallel.

Therefore, $\frac{1}{R_{AB}} = \frac{1}{2R} + \frac{1}{2R} + \frac{1}{R} = \frac{2}{R}$

$\therefore$ $R_{AB} = \frac{R}{2}$

2. With increase in temperature, the value of unknown resistance will increase.

In balanced Wheatstone bridge condition,

$$\frac{R}{X} = \frac{l_1}{l_2}$$

Here, R = value of standard resistance

X = value of unknown resistance.

To take null point at same point or $\frac{l_1}{l_2}$ to remain unchanged, $\frac{R}{X}$ should also remain unchanged.

Therefore, if X is increasing R should also increase.

3. By symmetry, $V_A = V_B$ or $V_{AB} = 0$

4. $v_d = \frac{i}{neA} = \frac{i}{ne(\pi r^2)} \Rightarrow v_d \propto \frac{i}{r^2}$

5. $V_{20\,\Omega} = V_{\text{Total}}$

$\therefore$ $(20)(0.3) = (R_{\text{Total}})(0.8)$

$\therefore$ $R_{\text{Total}} = \frac{30}{4}\,\Omega$

$\therefore$ $\frac{4}{30} = \frac{1}{R_1} + \frac{1}{20} + \frac{1}{15}$

Solving we get $R_1 = 60\,\Omega$

6. If A is fused, then complete circuit is broken.

7. $100 = (5)\left[\frac{(2500\,R)}{2500 + R}\right]$

Solving this equation, we get

$R = 20\,\Omega$

8. $\frac{P}{Q} = \frac{20}{100 - 20}$ or $Q = 4P$ ...(i)

$\therefore$ $P < Q$

Now, $\frac{P+15}{Q} = \frac{40}{100-40}$

or $\frac{P+15}{Q} = \frac{2}{3}$ ...(ii)

Solving these two equations, we get

$$P = 9\Omega$$

9. Total potential of 10V equally distributes between 50 Ω and other parallel combination of 100 Ω and voltmeter. Hence, their net resistance should be same. Or

$$\frac{100 \times R}{100 + R} = 50$$

∴ Resistance of voltmeter, $R = 100\,\Omega$

10. When K is open

$$R_{net} = 3R/2$$

$$\therefore \quad i_1 = E/(3R/2) = \frac{2E}{3R}$$

When K is closed

$$R_{net} = 2\left[\frac{R \times 2R}{R + 2R}\right] = \frac{4}{3}R$$

$$\therefore \quad i_2 = E/(4R/3) = \frac{3E}{4R}$$

$$\therefore \quad \frac{i_1}{i_2} = \frac{8}{9}$$

11. The circuit can be redrawn as follows

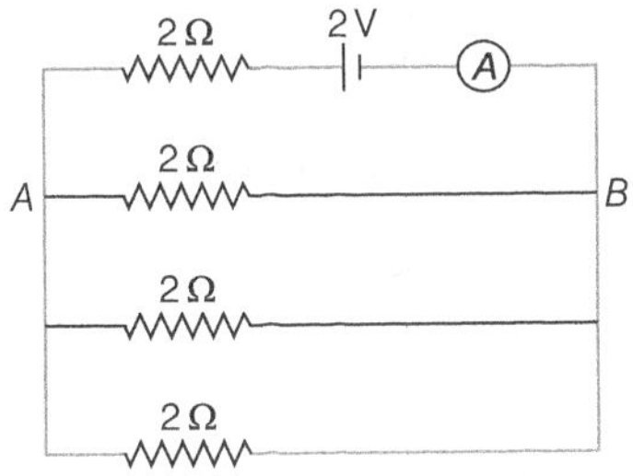

Resistance across AB,

$$R_{AB} = \frac{2}{3}\Omega$$

Total resistance of circuit,

$$R_T = 2 + \frac{2}{3} = \frac{8}{3}\Omega$$

Current through ammeter,

$$i = \frac{2}{(8/3)} = \frac{6}{8} = \frac{3}{4}\text{ A}$$

12. Reading of ammeter, $i = 3\left(\frac{220}{1100}\right) = \frac{3}{5}\text{ A}$

13. V is constant, $H = \frac{V^2}{R}t$ or $H \propto \frac{1}{R}$.

14. $i = \frac{6}{1+6+4} = \frac{6}{11}\text{ A}$

$$V = E - ir = 6 - \frac{6}{11} \times 1$$

$$= \frac{60}{11}\text{ volt}$$

15. In parallel, current distributes in inverse ratio of resistance.

16. If voltmeter is ideal, then R should be $\frac{1.6}{0.4} = 4\Omega$. If it is non-ideal R should be greater than 4Ω.

17. $\frac{24+x}{84+y} = \frac{10}{30} = \frac{1}{3}$ ...(i)

$$\frac{3}{1} = \frac{24+84+x+y}{10+30}$$

$$= \frac{108+x+y}{40}$$ ...(ii)

Solving this equation, we get

$$x = y = 6\Omega$$

18. $E_{net} = \frac{E_1/r_1 + E_2/r_2}{1/r_1 + 1/r_2}$

$$= \frac{(18/2)+(12/1)}{(1/2)+(1/1)}$$

$$= (21)\left(\frac{2}{3}\right) = 14\text{ volt}$$

Since we are not taking any current from this group of batteries. So,

$V = E_{net} = 14$ volt $\quad (V = E - ir)$

19. $E_{net} = E = i\lambda l$

$$\therefore \quad E = \left(\frac{10}{1+4+5}\right) \times \left(\frac{5}{5}\right) \times (3) = 3\text{V}$$

20.

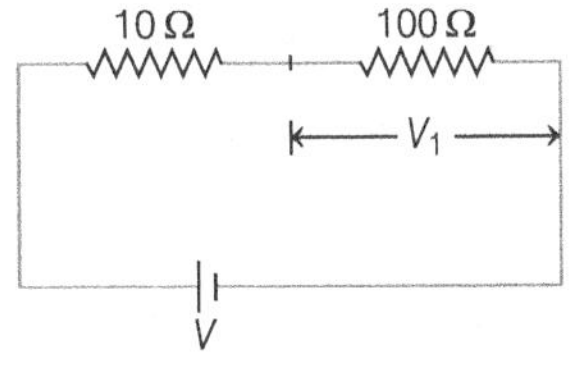

Before connecting the voltmeter let the potential difference across 100 Ω is V_1.

$$\therefore \quad V_1 = \left(\frac{100}{100+10}\right) \times V = \frac{10}{11}V$$

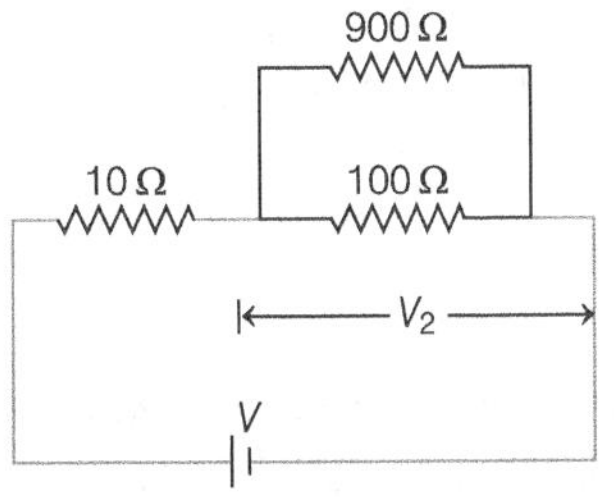

After connecting the voltmeter across 100Ω.

Equivalent resistance,

$$R_{eq} = \frac{100 \times 900}{100 + 900} = 90\Omega$$

Let this time potential difference is V_2.

$$\therefore \quad V_2 = \left(\frac{90}{90+10}\right)V = \frac{9}{10}V$$

$$\text{Magnitude of \% error} = \frac{\frac{10}{11}V - \frac{9}{10}V}{\frac{10}{11}V} \times 100$$

$$= 1.0\,\%$$

21. As we know that, $R = \frac{\rho l}{A}$

$$\Rightarrow \quad R_i = 14\Omega = \frac{\rho l_0}{\pi r_0^2}$$

and
$$R_f = \frac{\rho l_0}{\pi\left(\frac{r_0}{3}\right)^2} = 9\frac{\rho l_0}{\pi r_0^2} = 9R_i$$

$$= 9 \times 14 = 126\Omega$$

$$\therefore \quad R_{eq} = \frac{R_f}{7} = \frac{126}{7} = 18\,\Omega$$

22. The given situation is shown below

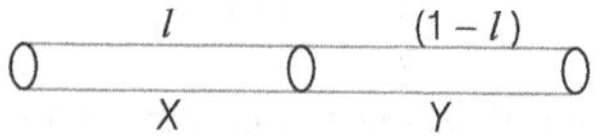

Thus,
$$\frac{R_X}{R_Y} = \frac{l_X}{l_Y} \qquad \left(\because R = \frac{\rho l}{A}\right)$$

When wire is stretched to double of its length, then resistance becomes 4 times. $(\because R' = n^2R)$

$$R_W = 4R_X = 2R_Y \qquad [\text{Given, } R_W = 2R_Y]$$

$$\frac{R_X}{R_Y} = \frac{1}{2}$$

So,
$$\frac{l_X}{l_Y} = \frac{1}{2}$$

23. Initial resistance of wire is 3Ω. Let its length is l and area is A.

Then,
$$R_{\text{initial}} = \rho\frac{l}{A} = 3\Omega \qquad \text{...(i)}$$

When wire is stretched twice its length, then its area becomes A', on equating volume, we have

$$Al = A'2l$$

$$\Rightarrow \quad A' = \frac{A}{2}$$

So, after stretching, resistance of wire will be

$$R' = R_{\text{final}} = \rho\frac{l'}{A'} = 4\rho\frac{l}{A} = 12\Omega \qquad [\text{using Eq. (i)}]$$

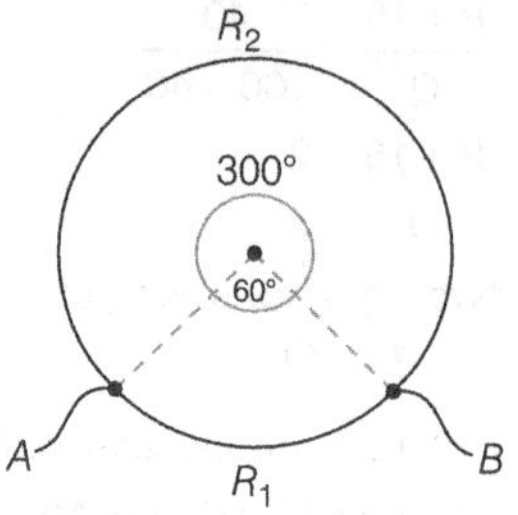

Now, this wire is made into a circle and connected across two points A and B (making 60° angle at centre) as

Now, above arrangement is a combination of two resistances in parallel,

$$R_1 = \frac{60 \times R'}{360} = \frac{1}{6} \times 12 = 2\Omega$$

and
$$R_2 = \frac{300}{360} \times R' = \frac{5}{6} \times 12 = 10\Omega$$

Since, R_1 and R_2 are connected in parallel.

So,
$$R_{AB} = \frac{R_1R_2}{R_1 + R_2} = \frac{2 \times 10}{2 + 10} = \frac{5}{3}\Omega$$

24. Drift velocity of electrons through conductor is given as,

$$v_d = \frac{I}{neA} \quad \text{or} \quad v_d \propto \frac{I}{A}$$

$\because$ For same potential difference,

$$I \propto \frac{1}{R} \propto A \qquad \left(\because R = \frac{\rho l}{A}\right)$$

$$\Rightarrow \quad v_d \propto \frac{A}{A}$$

$$\Rightarrow \quad v_d \propto 1$$

So, drift velocity will not change.

25. $i = neAv_d$ or $\frac{V}{R} = neAv_d$

$$\Rightarrow \quad \frac{V}{\left(\frac{\rho l}{A}\right)} = neAv_d \qquad \left[\because R = \frac{\rho l}{A}\right]$$

$$\therefore \quad \rho = \frac{V}{nelv_d} = \text{resistivity of wire}$$

Substituting the given values, we have

$$\rho = \frac{5}{(8 \times 10^{28})(1.6 \times 10^{-19})(0.1)(2.5 \times 10^{-4})}$$

$$\approx 1.6 \times 10^{-5}\ \Omega\text{-m}$$

26. We will require a voltmeter, an ammeter, a test resistor and a variable battery to verify Ohm's law.

Voltmeter which is made by connecting a high resistance with a galvanometer is connected in series with the test resistor. Further, an ammeter which is formed by connecting a low resistance in parallel with galvanometer is required to measure the current through test resistor.

The correct option is (c).

$$i_4 = \frac{100}{3000} = \frac{1}{30}\text{ A}$$

If we apply junction law at P, then current through switch

$$= i_2 - i_4 = \frac{1}{60}\text{ A in upward direction.}$$

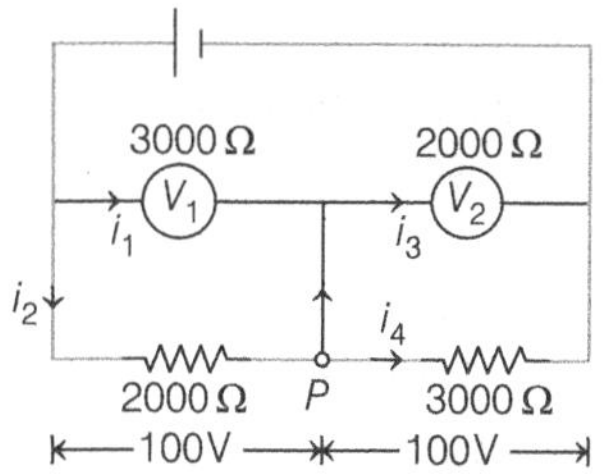

56. Current through voltmeter

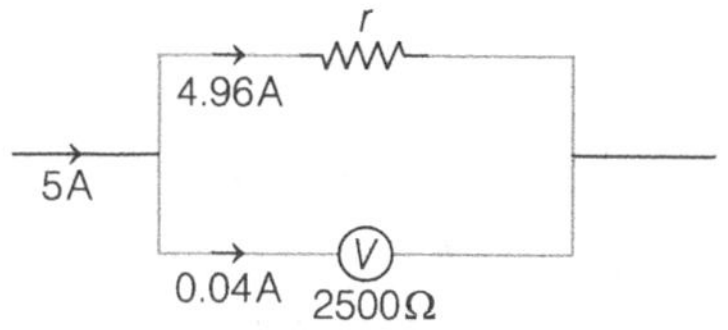

$$= \frac{V}{R} = \frac{100}{2500} = 0.04\text{ A}$$

In parallel current distribution in inverse ratio of resistors. Hence,

$$\frac{4.96}{0.04} = \frac{2500}{r}$$

$\therefore \quad r = 20.16\ \Omega$

57. Voltmeter reads 30 V, half of 60 V. Hence, resistance of 400 Ω and voltmeter is also equal to 300 Ω.

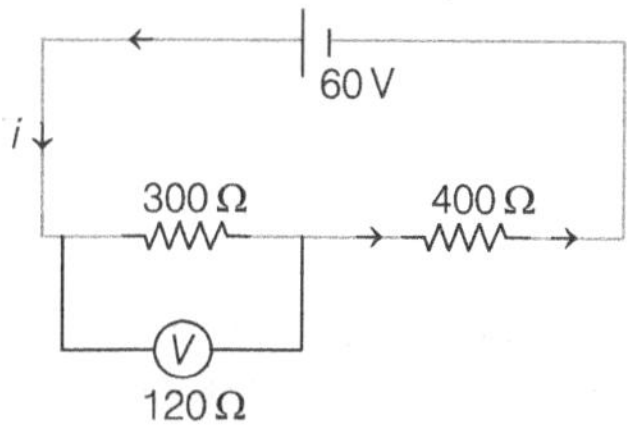

$$\therefore \quad 300 = \left(\frac{400 \times R}{400 + R}\right)$$

where, R = resistance of voltmeter.

Solving the above equation, we get

$$R = 1200\ \Omega$$

In the new situation,

$$R_{net} = 400 + \frac{(300)(1200)}{300 + 1200} = 640\ \Omega$$

$$\therefore \quad i = \frac{60}{640} = 0.09375\text{ A}$$

Now, voltage drop across

Voltmeter = 60 − potential drop across 400 Ω resistor

$$= 60 - (400)i = 60 - (400)(0.009375)$$

$$= 22.5\text{ V}$$

58. $P = i^2R$

$$\therefore \quad i_{max} = \sqrt{\frac{P_{max}}{R}} = \sqrt{\frac{36}{2.4}} = \sqrt{15}\text{ A}$$

$$\text{Total maximum power} = (i_{max})^2 \left(\frac{3R}{2}\right)$$

$$= (15)(1.5)(2.4)$$

$$= 54\text{ W}$$

59. $V = E - ir$

$$\therefore \quad r = \frac{E - V}{i} = \frac{2.6 - 2}{1} = 0.6\ \Omega$$

Now, power generated in the battery

$$P = i^2r$$

$$= (1)^2(0.6) = 0.6\text{ W}$$

Power supplied by the battery $= Ei$

$$= 2.6\text{ W}$$

$\therefore$ Net power supplied for external circuit

$$= 2.6 - 0.6$$

$$= 2.0\text{ W}$$

60.

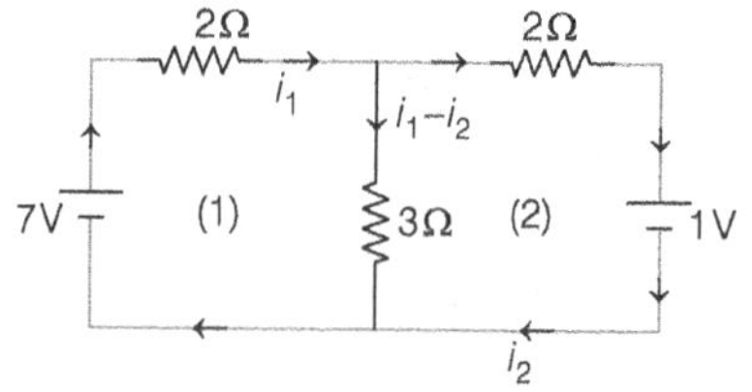

Loop equation in loop (1)

$$+7 - 2i_1 - 3(i_1 - i_2) = 0 \qquad \text{...(i)}$$

Loop equation in loop (2)

$$-1 + 3(i_1 - i_2) - 2i_2 = 0 \qquad \text{...(ii)}$$

Solving Eqs. (i) and (ii), we get

$$i_1 = 2\text{ A} \quad \text{and} \quad i_2 = 1\text{ A}$$

Power supplied by $E_1 = E_1 i_1 = 14\text{ W}$

But power consumed by $E_2 = E_2 i_2 = 1\text{W}$

61. (a) $i = \frac{12}{5 + 1} = 2\text{ A}$

$\Rightarrow \quad P_1 = Ei = 24\text{ W}$

(b) $P_2 = i^2r = (2)^2(1) = 4\text{W}$

(c) $P = P_1 - P_2 = 20\text{ W}$

62. (a) $P = \frac{V^2}{R}$

$$\therefore \quad V = \sqrt{PR} = \sqrt{5 \times 15 \times 10^3}$$

$$= 273.8\text{ V}$$

(b) $P = \frac{V^2}{R} = \frac{(120)^2}{9 \times 10^3}$

$$= 1.6\text{ W}$$

27. We know that,

$$i = neAv_d$$

Drift speed, $\quad v_d = \frac{1}{neA} \propto \frac{1}{A}$

Therefore, for non-uniform cross-section (different values of A), drift speed will be different at different sections. Only current (or rate of flow of charge) will be same.

28. Given circuit is

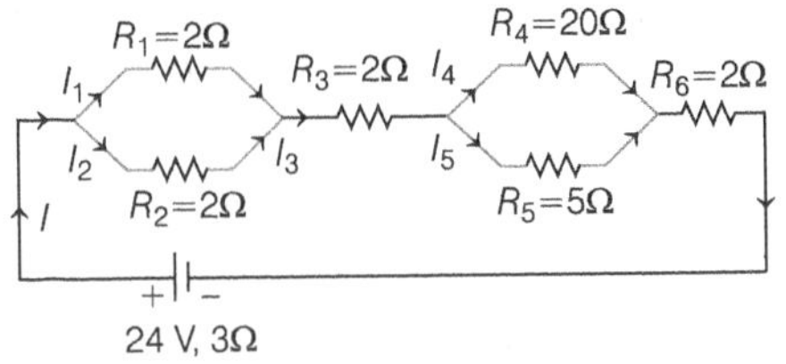

Total resistance of circuit

$$= (2\|2) + 2 + (20\|5) + 2 + 3$$

$$= \left(\frac{2 \times 2}{2 + 2}\right) + 2 + \left(\frac{20 \times 5}{20 + 5}\right) + 2 + 3$$

$$= 3 + 4 + 5 = 12\ \Omega$$

Current through battery,

$$I = \frac{24}{12} = 2\text{A}$$

So, current through resistor R_4,

$$I_4 = \left(\frac{5}{25}\right) \times 2 = \frac{2}{5}\text{ A}$$

and current through resistor R_5,

$$I_5 = \frac{20}{25} \times 2 = \frac{8}{5}\text{ A}$$

29. R_3, R_4, R_5 are in series and the combination is parallel to R_2, R_1 and R_6 are in series.

$$\therefore \quad R_{net} = \frac{160}{3}$$

$$i = \frac{V}{R} = \frac{15}{(160/3)} = \frac{9}{32}\text{ A}$$

30. Given circuit with currents as shown in the figure below

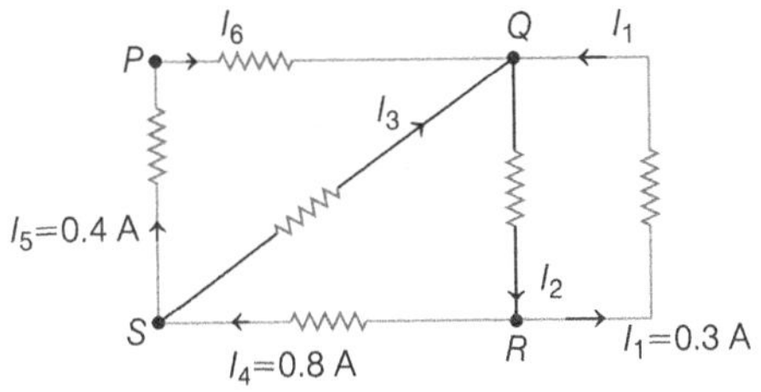

From Kirchhoff's junction rule, $\Sigma I = 0$

At junction S, $\quad I_4 = I_5 + I_3$

$\Rightarrow \quad 0.8 = 0.4 + I_3$

$\Rightarrow \quad I_3 = 0.4\text{ A}$

At junction P, $\quad I_5 = I_6$

$\Rightarrow \quad I_6 = 0.4\text{ A}$

At junction Q,

$$I_2 = I_1 + I_3 + I_6$$

$$= 0.3 + 0.4 + 0.4$$

$$= 1.1\text{ A}$$

31. Potential drop across each resistor is zero, so the current through each of resistor is zero.

32.

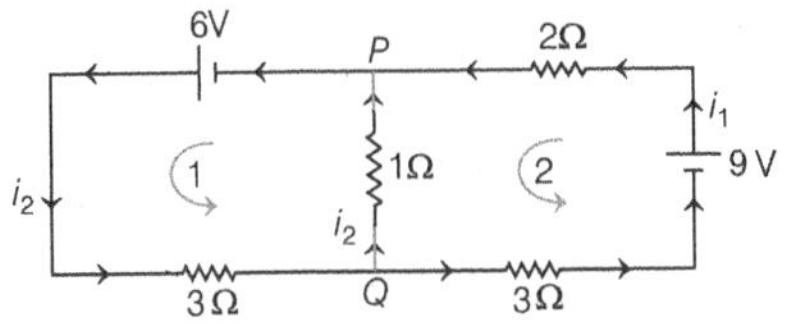

Applying Kirchhoff's loop law in loops 1 and 2 in the directions as shown in figure, we have

$$6 - 3(i_1 + i_2) - i_2 = 0 \qquad \text{...(i)}$$

$$9 - 2i_1 + i_2 - 3i_1 = 0 \qquad \text{...(ii)}$$

Solving Eqs. (i) and (ii), we get

$$i_2 = 0.13\text{ A}$$

Hence, the current in 1 Ω resister is 0.13 A from Q to P.

33. Current in the respective loop will remain confined in the loop itself.

Therefore, current through 2 Ω resistance = 0

34. Net resistance of the circuit is 9 Ω.

$\therefore$ Current drawn from the battery,

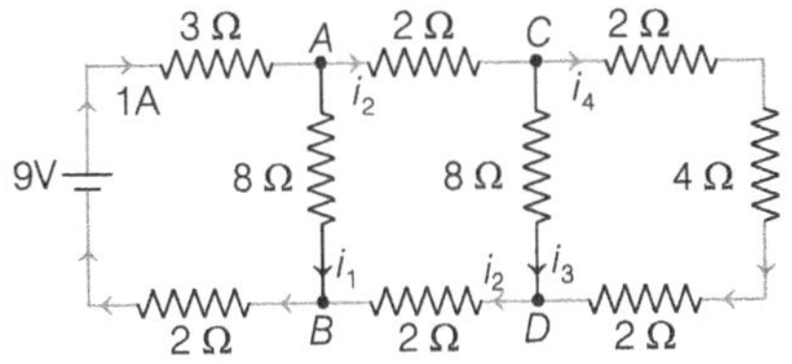

$$i = \frac{9}{9} = 1\text{ A} = \text{current through 3 Ω resistor}$$

Potential difference between A and B is

$$V_A - V_B = 9 - 1(3 + 2)$$

$$= 4\text{V} = 8i_1$$

$\therefore \quad i_1 = 0.5\text{ A}$

$\Rightarrow \quad i_2 = 1 - i_1 = 0.5\text{ A}$

Similarly, potential difference between C and D is

$$V_C - V_D = (V_A - V_B) - i_2(2 + 2)$$

$$= 4 - 4i_2$$

$$= 4 - 4(0.5)$$

$$= 2\text{V} = 8i_3$$

$\Rightarrow \quad i_3 = 0.25\text{ A}$

Therefore, $\quad i_4 = i_2 - i_3$

$$= 0.5 - 0.25$$

$$i_4 = 0.25\text{ A}$$

35.

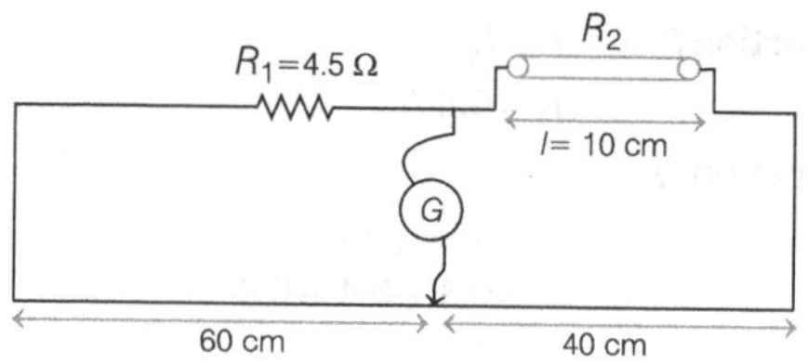

Given, $r=\sqrt{7}\times 10^{-4}$ m,

$l = 10 \text{ cm} = 10\times 10^{-2}\text{m}$

$= \dfrac{1}{10}\text{m}$

Under balance condition $\dfrac{R_1}{l_1}=\dfrac{R}{l_2}$

Here, $R_1 = 4.5\,\Omega, l_1 = 60$ cm

$l_2 = 100 - l_1 = 40$ cm

$R_2 = R$

$\therefore \quad \dfrac{l_1}{l_2}=\dfrac{R_1}{R_2} \Rightarrow \dfrac{60}{40}=\dfrac{4.5\,\Omega}{R}$

$\Rightarrow \quad R = 3\,\Omega$

$R=\rho\dfrac{l}{A} \Rightarrow 3=\rho\times\dfrac{1}{\pi r^2}$

$3=\rho\times\dfrac{1}{10\times\pi\times 7\times 10^{-8}}$

$\Rightarrow \quad \rho = 21\pi\times 10^{-7}$

$\rho = 21\pi\times 10^{-7}$

$\rho = 21\times\dfrac{22}{7}\times 10^{-7}$

$\rho = 66\times 10^{-7}$

$\Rightarrow \quad R = 66$

36. $i_{AB}=\dfrac{E}{13r}$

$i_{AJ}=\left(\dfrac{x}{L}12r\right)\times i_{AB}=\dfrac{E}{2}$

$\dfrac{x}{L}12r\times\dfrac{E}{13r}=\dfrac{E}{2}$

$\Rightarrow \quad x=\dfrac{13}{24}L$

37. $\dfrac{R_1}{R_2}=\dfrac{l}{100-l}$ ...(i)

and $\dfrac{R_2}{R_1}=\dfrac{l-10}{100-(l-10)}$ ...(ii)

From Eqs. (i) and (ii),

$l = 55$ cm

Substituting in Eq. (i), we get

$\dfrac{R_1}{R_2}=\dfrac{55}{45}$ ...(iii)

Further, $R_1 + R_2 = 100\,\Omega$ ...(iv)

Solving Eqs. (iii) and (iv),

$R_1 = 55\,0\Omega$

38. In a balanced Wheatstone bridge, there is no effect on position of null point, if we exchange the battery and galvanometer. So, option (a) is incorrect.

39. Resistance of the given two bulbs are

$R_1=\dfrac{V^2}{P_1}=\dfrac{(220)^2}{25}$

and $R_2=\dfrac{V^2}{P_2}=\dfrac{(220)^2}{100}$

$\therefore$ Current in circuit is

$i=\dfrac{V}{R_{\text{Total}}}=\dfrac{220}{\dfrac{(220)^2}{25}+\dfrac{(220)^2}{100}}=\dfrac{1}{11}$A

Power drawn by bulbs are respectively,

$P_1 = i^2R_1 = \left(\dfrac{1}{11}\right)^2\times\dfrac{220\times 220}{25}$

$= 16$ W

and $P_2 = i^2R_2 = \left(\dfrac{1}{11}\right)^2\times\dfrac{220\times 220}{100}$

$= 4$ W

40. Maximum current will pass through A.

$P = i^2R$

or $P\propto i^2$ (R is same)

41. $4(R+R_A) = 20$ V

$\therefore \quad R = 5 - R_A$

where, R_A = resistance of ammeter

42.

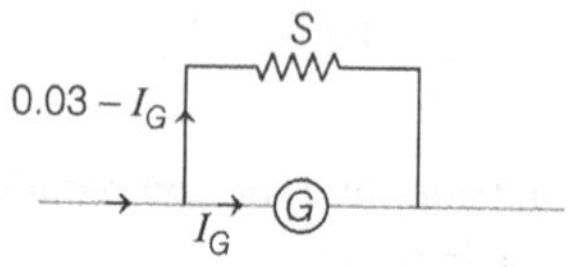

In parallel, current distributers in inverse ratio of resistance.

$\dfrac{0.03-I_G}{I_G}=\dfrac{G}{S}=\dfrac{r}{(r/4)}=4$

Solving this equation, we get

$I_G = 0.006$ A

43. V and V_0 are oppositely connected.

44. Net resistance of 3 kΩ and voltmeter is also 2 kΩ. Now, the applied 10 V is equally distributed between 2 kΩ and 2 kΩ. Hence, reading of voltmeter

$=\dfrac{10}{2}=5$ V

45. As there is no change in the reading of galvanometer with switch S open or closed. It implies that bridge is balanced. Current through S is zero and $I_R = I_G, I_P = I_Q$.

46. (a) $R=\dfrac{H}{i^2t}$

$\therefore \quad [R] = [ML^2T^{-2}]/[A^2T] = [ML^2T^{-3}A^{-2}]$

(b) $V = iR$

$\therefore \quad [V] = [A][ML^2T^{-3}A^{-2}] = [ML^2T^{-3}A^{-1}]$

(c) $\rho=\dfrac{RA}{l}$

$\therefore [\rho]=\dfrac{[ML^2T^{-3}A^{-2}][L^2]}{[L]}=[ML^3T^{-3}A^{-2}]$

(d) $[\sigma]=\left[\dfrac{1}{\rho}\right]=[M^{-1}L^{-3}T^3A^2]$

47. $i=\dfrac{4-1}{1+1+1}=1$A (anti-clockwise)

(a) $V_A = E - ir = 4 - 1\times 1 = 3$ V

(b) $V_B = E + ir = 1 + 1\times 1 = 2$ V

(c) $|P_A| = Ei - i^2r = (4\times 1) - (1)^2(1) = 3$ W

(d) $|P_B| = Ei + i^2r = (1)(1) + (1)^2(1) = 2$ W

48. (a) In 1 m, potentials difference,

$V = 0.49 \quad \therefore \quad V = iR$

$\therefore \quad i=\dfrac{0.49}{R}=\dfrac{(0.49)A}{\rho l}$

$=\dfrac{(0.49)(\pi/4)(0.84\times 10^{-3})^2}{(2.75\times 10^{-8})(1)}$

$= 9.9$ A

(b) PD between two points, 12 m apart

$= (0.49 \text{ V/m})(12\text{m}) = 5.88$ V

(c) $R=\dfrac{V}{i}=\dfrac{5.88}{9.9}=0.60\Omega$

49. 8Ω and 12Ω resistors are in parallel.

$\therefore \quad R_{\text{net}}=\dfrac{8\times 12}{8+12}=4.8\,\Omega$

$\therefore \quad i=\dfrac{24}{4.8}=5$A

50. All four resistors are in parallel.

$\therefore \quad \dfrac{1}{R}=\dfrac{1}{8}+\dfrac{1}{4}+\dfrac{1}{6}+\dfrac{1}{12}$

or $R=\dfrac{8}{5}\Omega$

$\therefore \quad i=\dfrac{24}{8/5}$

$= 15$ A

51. All these resistors are in parallel.

52. The given network is as shown below.

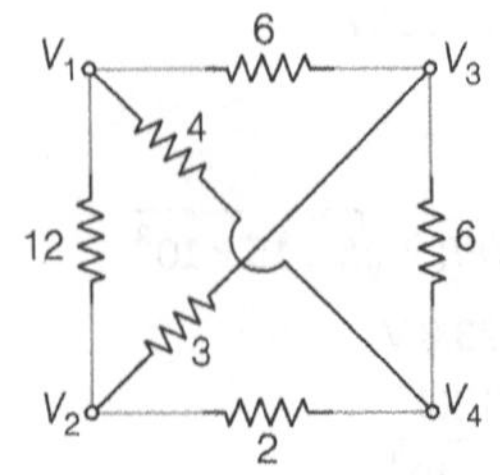

The simple circuit is as shown below.

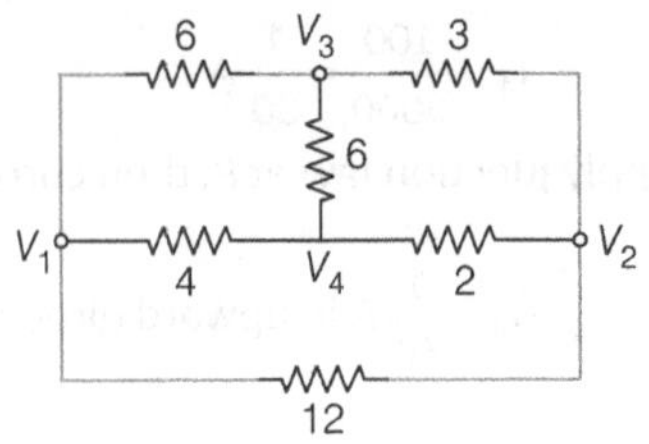

Now, this is a balanced Wheatstone bridge in parallel with 12 Ω resistance.

53. **First case**

$i=\dfrac{12+6}{1+2+3}=3$A (clockwise)

Now, $V_A - V_G = 12$ V

$\therefore \quad V_A = 12$ V, as $V_G = 0$

$V_A - V_B = 1\times 3 = 3$ V

$\therefore \quad V_B = V_A - 3 = 9$ V

$V_B - V_C = 2\times 3 = 6$ V

$\therefore \quad V_C = V_B - 6 = 3$ V

$V_G - V_D = 6$ V

$\therefore \quad V_D = -6$ V, as $V_G = 0$

In the second case,

$i=\dfrac{12-6}{1+2+3}=1$A

Rest procedure is same.

54. $i=\dfrac{200}{5+10+25}=5$A (anti-clockwise)

$V_3 - V_G = 25\times 5 = 125$

$\therefore \quad V_3 = 125$ V (as $V_G = 0$)

$V_G - V_2 = 10\times 5 = 50$

$\therefore \quad V_2 = -50$ V

$V_2 - V_1 = 5\times 5 = 25$ V

$\therefore \quad V_1 = V_2 - 25 = -75$ V

Now, $V_{3-2} = V_3 - V_2$

55. (a) (i) When switch S is open, V_1 and V_2 are in series, connected to 200 V battery. Potential will drop in direct ratio of their resistors.

$\therefore \quad V_1 : V_2 = R_{V_1} : R_{V_2} = 3000 : 2000$

$= 3 : 2$

$\therefore \quad V_1=\dfrac{3}{5}\times 200=120$ V

$V_2=\dfrac{2}{5}\times 200=80$ V

(ii) When S is closed then V_1 and R_1 are in parallel. Similarly, V_2 and R_2 are also in parallel. Now, they are in series and they come out to be equal. So, 200 V will equally distribute between them.

$\therefore \quad V_1 = V_2 = \dfrac{200}{2}=100$ V each

(b) $i_2=\dfrac{100}{2000}=\dfrac{1}{20}$A

63. The given circuit actually forms a balanced Wheatstone's bridge (including the voltmeter) as shown below

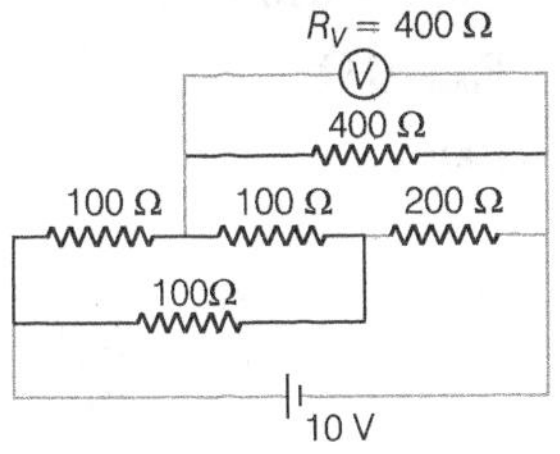

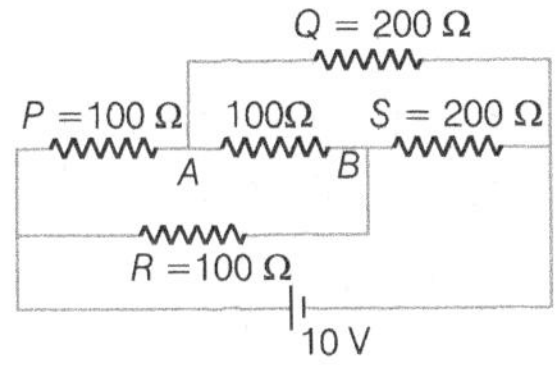

Here, we see that $\frac{P}{Q}=\frac{R}{S}$

Therefore, resistance between A and B can be ignored and equivalent simple circuit can be drawn as below.

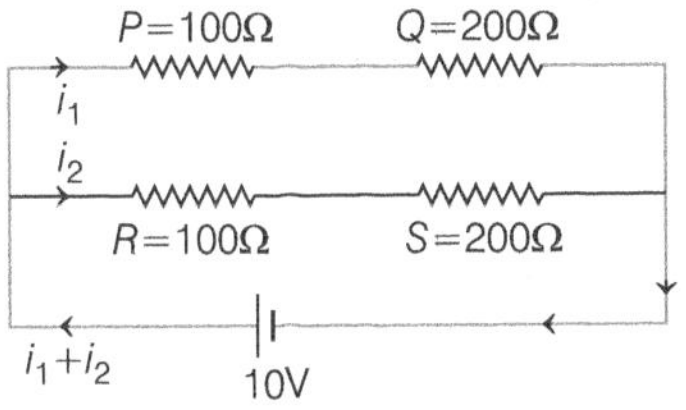

The voltmeter will read the potential difference across resistance Q.

Currents $i_1=i_2=\frac{10}{100+200}$

$$=\frac{1}{30}\text{A}$$

∴ Potential difference across voltmeter

$$=Qi_1=(200)\left(\frac{1}{30}\right)\text{V}$$

$$=\frac{20}{3}\text{V}$$

64. (a) $I=\frac{\text{Net emf}}{\text{Net resistance}}=\frac{E_1+E_2+\ldots+E_N}{R_1+R_2+\ldots+R_N}$

(R = internal resistance)

$$=\frac{\alpha R_1+\alpha R_2+\ldots+\alpha R_N}{R_1+R_2+\ldots+R_N}$$

∴ $I=\alpha$

(b) From A to B, if we move in clockwise direction, then

$$V_A+(E_1-IR_1)+(E_2-IR_2)+\ldots+(E_n-IR_n)=V_B$$

∴ $V_A-V_B=(E_1+E_2+\ldots+E_n)-I(R_1+R_2+\ldots+R_n)$

$=(\alpha R_1+\alpha R_2+\ldots+\alpha R_n)-\alpha(R_1+R_2+\ldots+R_n)$

or $V_A-V_B=0$

65. Net resistance of the circuit,

$$R_{\text{net}}=0.1+\frac{6\times 4}{6+4}=2.5\,\Omega$$

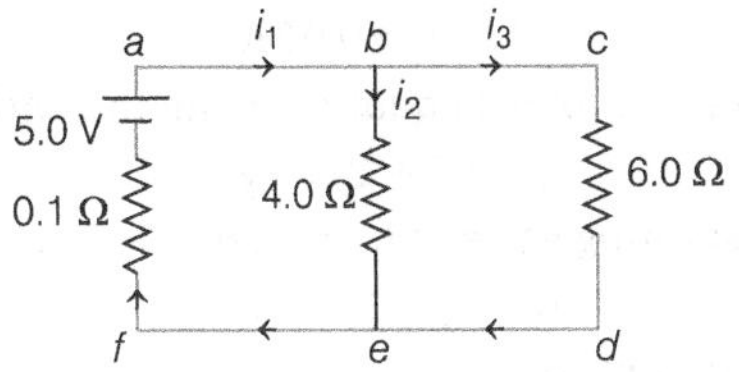

Therefore main current, $i_1=\frac{5.0}{2.5}=2\,\text{A}$

This current i_1 will distribute in 4.0 Ω and 6.0 Ω in inverse ratio of resistance. Thus,

$$i_2=\left(\frac{6}{6+4}\right)(2) \quad \text{or} \quad i_2=1.2\text{ A}$$

and $$i_3=\left(\frac{4}{6+4}\right)(2) \quad \text{or} \quad i_3=0.8\text{ A}$$

Thus, $I_{R_1}=1.2$ A and $I_{R_2}=0.8$ A

66. Using the concept of equivalent battery

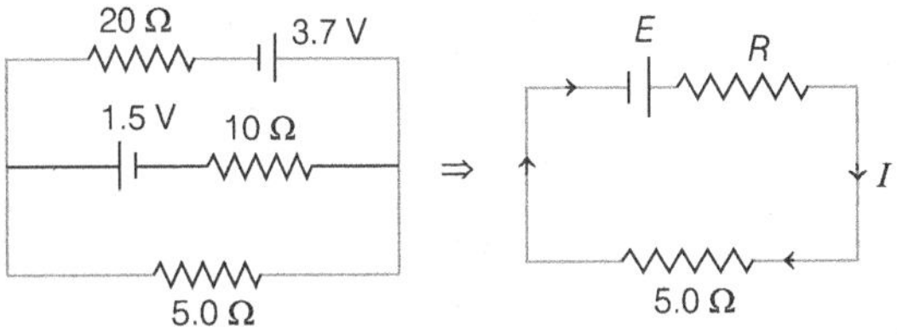

$$E=\frac{E_2R_1-E_1R_2}{R_1+R_2}$$

$$=\frac{3.7\times 10-1.5\times 20}{10+20}$$

⇒ $E=\frac{7}{30}$ A

⇒ $R=\frac{R_1R_2}{R_1+R_2}=\frac{20}{3}\,\Omega$

Now, $I=\frac{E}{R+5}$

$$=\frac{7/30}{(20/3+5)}=0.02\text{ A}$$

Thus, $I=0.02$ A

67. (a)

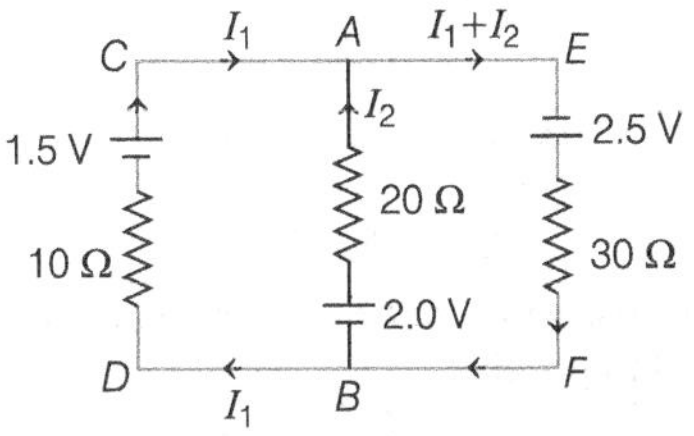

Applying Kirchhoff's loop law in $ABDCA$,

$$20I_2-2.0-10I_1+1.5=0 \quad \ldots\text{(i)}$$

Applying Kirchhoff's loop law in *AEFBA*,

$$2.5 - 30(I_1 + I_2) + 2.0 - 20I_2 = 0 \quad ...(ii)$$

Solving these two equations, we get

$$I_1 \approx 0.06 \text{ A}$$

(b) Writing potential equation along the path *ACDB*,

$$V_A - 1.5 + 10I_1 = V_B$$

Substituting value of I_1, we get

$$V_A - V_B = 0.9 \text{ volt}$$

Alternate method

Find equivalent values of *E* and *r* of all three batteries. Then $V_{AB} = E$.

Further by putting $V_{AB} = E_1 \pm I_1 r_1$ etc. we can find different currents through different batteries.

68. Resistors 1Ω and 3Ω are in parallel. Similarly 2Ω and 4Ω are also in parallel.

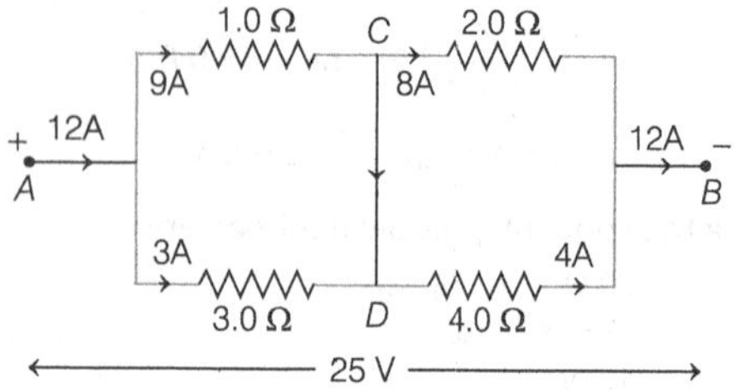

Therefore, $R_{net} = \dfrac{1 \times 3}{1+3} + \dfrac{2 \times 4}{2+4} = \dfrac{25}{12} \Omega$

$\therefore \quad I = \dfrac{25}{25/12} = 12 \text{ A}$

This current will distribute in inverse ratio of resistors. So, different currents are as shown in figure. If we apply junction law at *C*, then

$$I_{CD} = (9 - 8) \text{ A}$$

or $\quad I_{CD} = 1 \text{ A}$

69. Let us keep *m* cells in each row. Then,

$$m \times n = 300 \quad ...(i)$$

For maximum power across *R*,

total internal resistance = total external resistance

or $\quad \dfrac{mr}{n} = R$

or $\quad \dfrac{0.3\, m}{n} = 10 \quad ...(ii)$

Solving Eqs. (i) and (ii), we get

$$n = 3$$

70. Maximum resistance occurs when resistors are connected in series,

$$R_{max} = \Sigma R_i = 100 \Omega$$

Minimum resistance occurs when resistors are connected in parallel.

$\Rightarrow \quad \dfrac{1}{R_{min}} = \Sigma \dfrac{1}{R_i} = 1 \Omega^{-1}$

$$\Sigma \frac{1}{R_i} = 10\left(\frac{1}{10}\right) = 1 \Omega$$

$\Rightarrow \quad R_{min} = 1 \Omega$

Hence, $\dfrac{R_{max}}{R_{min}} = 100 : 1$

71.

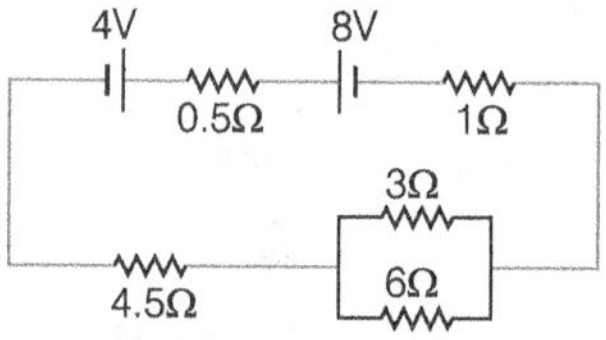

$$R_{eq} = 0.5 + 1 + 4.5 + 2$$

$$R_{eq} = 6 + 2 = 8 \Omega$$

$$I = \frac{8 - 4}{8} = \frac{1}{2} \text{ A}$$

$$= 0.5 \text{ A}$$

$\therefore \quad I_1 : I_2 = \dfrac{1}{3} : \dfrac{1}{6}$

$$I_1 : I_2 = 2 : 1$$

and $\quad I_1 + I_2 = 0.5 \text{ A}$

$$I_1 = \frac{2}{3} \times 0.5$$

$$= \frac{1}{3} \text{ A}$$

So, $\quad \dfrac{1}{3} = \dfrac{x}{3} \Rightarrow x = 1$

72. The potential difference between points *B* and *D* is zero, hence it is the balanced Wheatstone bridge circuit.

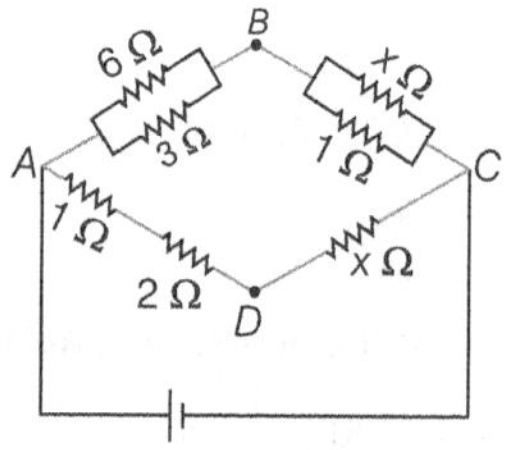

$\therefore \quad \dfrac{R_{AB}}{R_{BC}} = \dfrac{R_{AD}}{R_{DC}}$

$\Rightarrow \quad \dfrac{\dfrac{6 \times 3}{6+3}}{\dfrac{x \times 1}{x+1}} = \dfrac{1+2}{x}$

$\Rightarrow \quad \dfrac{2(x+1)}{x} = \dfrac{3}{x}$

$\Rightarrow \quad 2x^2 + 2x - 3x = 0 \Rightarrow 2x^2 - x = 0$

$\Rightarrow \quad x(2x - 1) = 0$

$\Rightarrow \quad x = 1/2 = \dfrac{1}{n}$ [Given]

$\therefore \quad n = 2$

Level 2

73. No current will flow through voltmeter. As it is ideal (infinite resistance). Current through two batteries

$$i=\frac{1.5-1.3}{r_1+r_2}=\frac{0.2}{r_1+r_2}$$

Now, $V=E_2-ir_2$

$$\therefore \quad 1.45=1.5-\left(\frac{0.2}{r_1+r_2}\right)(r_2)$$

Solving this equation, we get

$$r_1=3r_2$$

74. In series, PD distributes in direct ratio of resistance.

In first case,

$$\frac{198}{V_{AB}-198}=\frac{900}{R_1} \quad \text{...(i)}$$

In second case, $$\frac{180}{V_{AB}-180}=\frac{900}{2R_1} \quad \text{...(ii)}$$

Solving these two equations, we get

$$V_{AB}=220\text{ V}$$

75. Initial current, $i_1=\dfrac{E_1+E_2}{R+r_1+r_2}$

Final current, when second battery is short circuited is

$$i_2=\frac{E_1}{R+r_1}$$

$$i_2>i_1, \text{ if } \frac{E_1}{R+r_1}>\frac{E_1+E_2}{R+r_1+r_2}$$

or $$E_1R+E_1r_1+E_1r_2>E_1R+E_1r_1+E_2R+E_2r_1$$

or $$E_1r_2>E_2(R+r_1)$$

76.

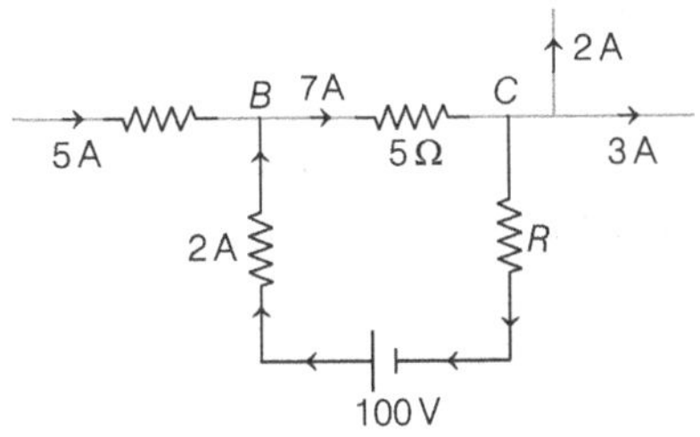

Applying loop equation in closed loop we have,

$$100-30-35-2R=0$$

$$\therefore \quad 2R=35\text{ V}=V_R$$

$$V_{5\Omega}=7\times 5=35\text{ V}$$

$$\therefore \quad \frac{V_{5\Omega}}{V_R}=1$$

77. Let $R=at+b$

At $t=10$ s, $R=20\ \Omega$

$$\therefore \quad 20=10a+b \quad \text{...(i)}$$

At $t=30a+b$ ...(ii)

Solving these two equations, we get

$$a=1.0\ \Omega/\text{s}$$

and $$b=10\ \Omega$$

$$\therefore \quad R=(t+10)$$

$$i=\frac{E}{R}=\frac{10}{t+10}$$

$$\Delta q=\int_{10}^{30} i\,dt$$

$$=\int_{10}^{30}\left(\frac{10}{t+10}\right)dt$$

$$=10\log_e(2)$$

78. In parallel, current distributes in inverse ratio of resistance.

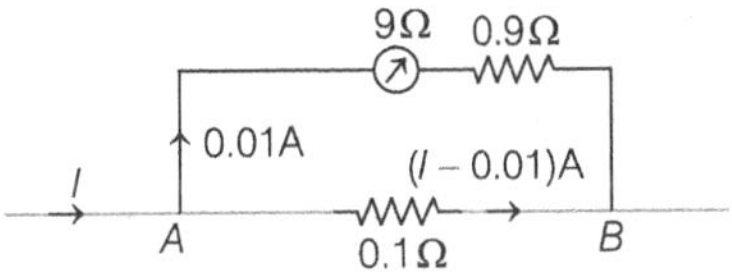

$$\therefore \quad \frac{0.01}{I-0.01}=\frac{0.1}{9+0.9}$$

Solving we get, $I=1$ A

79. When K_1 and K_2 both are closed R_1 is short-circuited,

$$R_{net}=(50+r)\ \Omega$$

When K_1 is open and K_2 is closed, current remains half. Therefore, net resistance of the circuit becomes two times.

or $$(50+r)+R_1=2(50+r)$$

Of the given options, the above equation is satisfied if

$$r=0 \quad \text{and} \quad R_1=50\ \Omega$$

80. $100\ \Omega$, $25\ \Omega$ and $20\ \Omega$ are in parallel.

Their, net resistance is $10\ \Omega$

$$\therefore \quad R_{net}=4\Omega+10\Omega+6\Omega=20\Omega$$

$$V=iR_{net}=80\text{ V}$$

81. All these resistors are in parallel.

$$\therefore \quad R_{net}=\frac{R}{3}+r=4\Omega$$

Hence, the main current

$$i=\frac{E}{R_{net}}=1\text{ A}$$

Current through either of the resistance is

$$\frac{i}{3} \text{ or } \frac{1}{3}\text{ A}$$

$$\therefore \quad V=iR=\left(\frac{1}{3}\right)(9)=3\text{ V}$$

82. Balanced Wheatstone bridge. Hence, $1.5\ \Omega$ resistance can be removed from the circuit.

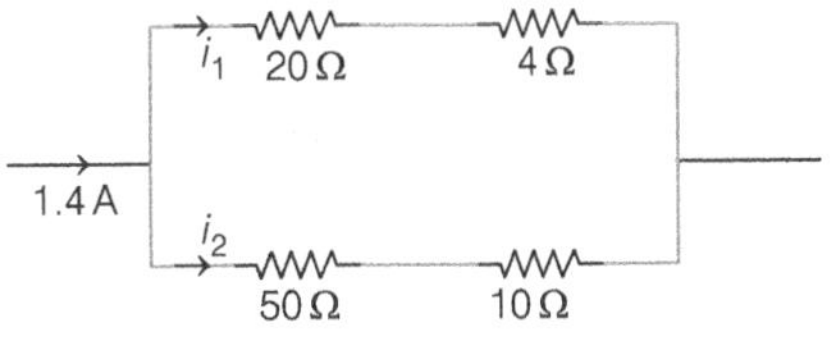

$$\frac{i_1}{i_2}=\frac{50+10}{20+4}=\frac{2.5}{1}$$

$\therefore$ $$i_1=\left(\frac{2.5}{2.5+1}\right)(1.4)=1\text{ A}$$

83. Resistance between A and B can be removed due to balanced Wheatstone bridge concept. Now, R_{DE} and R_{GH} are in series and they are connected in parallel with 10 V battery.

$\therefore$ $$I_{DE}=\frac{10}{R_{DE}+R_{HG}}=\frac{10}{2+2}=2.5\text{ A}$$

84. $$R_{net}=\frac{20\times 5}{20+5}+\frac{5\times 20}{5+20}=8\Omega$$

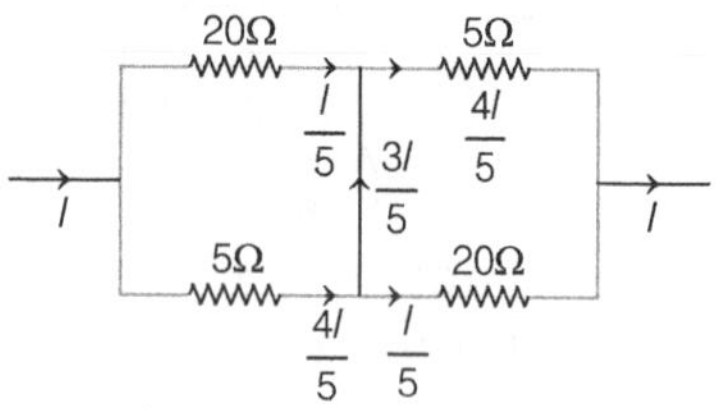

$$R_{CD}=0$$

$\therefore$ $$V_{CD}=(I_{CD})(R_{CD})=0$$

or $$V_C=V_D$$

In parallel, current distributes in inverse ratio of resistance.

$\therefore$ $$I_{20\,\Omega}=\left(\frac{5}{5+20}\right)I=\frac{I}{5}$$

85. When S is open, R_1 and R_2 are in series, total resistance

$$R_1=4000+6000$$
$$=10000\,\Omega$$

Voltmeters are also in series, then resistance of voltmeters

$$R_2=6000+4000=10000\,\Omega$$

So, $$R_{eq}=\frac{R_1R_2}{R_1+R_2}=5000\,\Omega$$

$$I=\frac{E}{R_{eq}}=\frac{250}{5000}=\frac{1}{20}\text{ A}$$

Current i_1 (in voltmeters) $=\frac{1}{2}\times\frac{1}{20}=\frac{1}{40}$ A

$\therefore$ Potential drop across, $V_1=\frac{1}{40}\times 6000=150$ V

and potential drop across, $V_2=\frac{1}{40}\times 4000=100$ V

So, $$\frac{V_1}{V_2}=\frac{150}{100}=1.5$$

and when switch is closed,

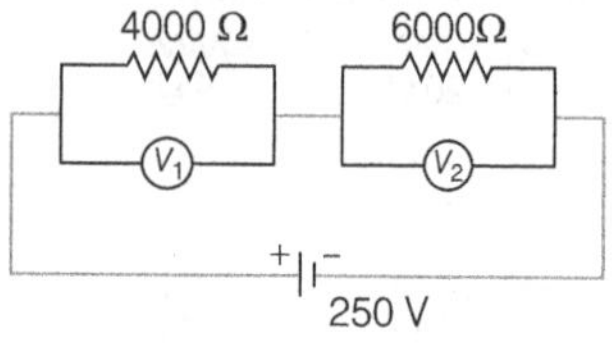

$$V_1=125\text{ V}, V_2=125\text{ V}$$

$\therefore$ $$\frac{V_1}{V_2}=1$$

86. For balanced meter bridge

$$\frac{X}{R}=\frac{l}{(100-l)}$$

$\therefore$ $$\frac{X}{90}=\frac{40}{100-40}\qquad\text{(where, } R=90\,\Omega)$$

$$X=R\frac{l}{(100-l)}$$

$\therefore$ $$X=60\,\Omega$$

$$\frac{\Delta X}{X}=\frac{\Delta l}{l}+\frac{\Delta l}{100-l}$$

$$=\frac{0.1}{40}+\frac{0.1}{60}$$

$$\Delta X=0.25$$

So, $$X=(60\pm 0.25)\,\Omega$$

87. Applying $P=\frac{V^2}{R}$, $R_1=1\Omega, R_2=0.5\Omega$

and $$R_3=2\Omega$$

$$V_1=V_2=V_3=3\text{V}$$

$\therefore$ $$P_1=\frac{(3)^2}{1}=9\text{ W}$$

$$P_2=\frac{(3)^2}{0.5}=18\text{ W}$$

and $$P_3=\frac{(3)^2}{2}=4.5\text{W}$$

$\therefore$ $$P_2>P_1>P_3$$

88. (a) In series, current is same.

$\therefore$ $$I_A=I_B$$

(b) $$V_A+V_B=V_C$$

$\therefore$ $$I_AR_A+I_BR_B=I_CR_C$$

(d) In parallel, current distributes in inverse ratio of resistance.

$\therefore$ $$\frac{I_B}{I_C}=\frac{I_A}{I_C}=\frac{R_C}{R_A+R_B}$$

89. (a) $V=iR$

In series i is same. Hence, V is also same as R is given same.

(b) $R=\frac{\rho l}{A}$

R is same. Hence, A should be smaller in first wire.

Secondly, $v_d=\frac{i}{neA}$ or $v_d\propto\frac{1}{A}$

A of first wire is less. Hence, its drift velocity should be more.

(c) $E=\frac{V}{l}$

or $$E\propto\frac{1}{l}\qquad(V\rightarrow\text{same})$$

90. By closing S_1, net external resistance will decrease. So, main current will increase.

By closing S_2, net emf will remain unchanged but net internal resistance will decrease. Hence, main current will increase.

91.

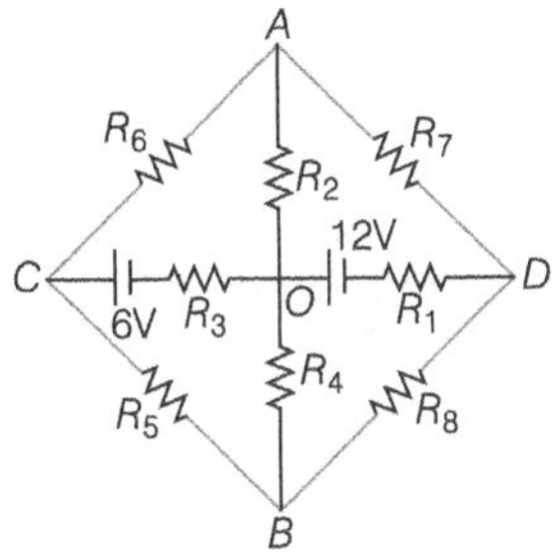

Points A and B are at same potential, so they can be merged/folded.

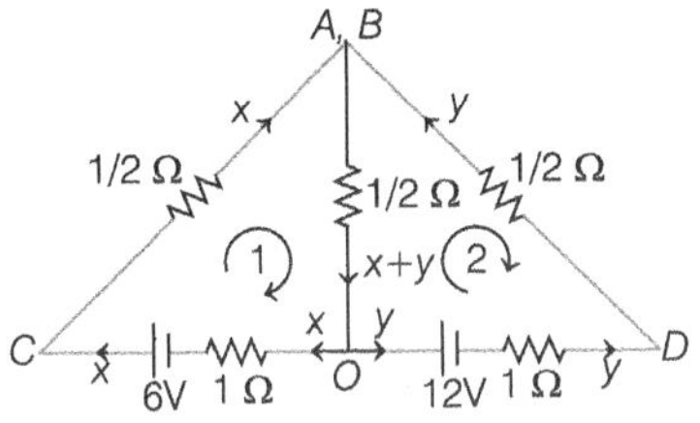

For loop-1,

$$-\frac{1}{2}x-\frac{1}{2}(x+y)-x+6=0$$

$$\Rightarrow \quad -2x-\frac{y}{2}=-6$$

$$\Rightarrow \quad -4x-y=-12$$

$$\Rightarrow \quad 4x+y=12 \quad \text{...(i)}$$

For loop-2,

$$\frac{1}{2}y+1y+12+\frac{1}{2}(x+y)=0$$

$$\Rightarrow \quad \frac{3}{2}y+\frac{y}{2}+\frac{x}{2}=-12$$

$$\Rightarrow \quad 2y+\frac{x}{2}=-12$$

$$\Rightarrow \quad 4y+x=-24 \quad \text{...(ii)}$$

From Eqs. (i) and (ii), we get

$$4y+x-16x-4y=-24-48$$

$$-15x=-72 \Rightarrow x=\frac{72}{15}$$

$$\Rightarrow \quad 4\left(\frac{72}{15}\right)+y=12 \Rightarrow y=12-\frac{288}{15}$$

$$=\frac{180-288}{15}=\frac{-108}{15}=-7.2\text{ A}$$

Current in $R_1=7.2$ A

Current in $R_2=\frac{x+y}{2}=\left(\frac{72}{15}-\frac{108}{15}\right)\frac{1}{2}$

$$=\frac{1}{2}\left(\frac{36}{15}\right)=\frac{2.4}{2}\text{ A}=1.2\text{ A}$$

Current in $R_3=x=4.8$ A

Current in $R_5=\frac{1}{2}x=2.4$ A

92. Due to symmetry on upper side and lower side, points P and Q are at same potentials. Similarly, points S and T are at same potentials. Therefore, the simple circuit can be drawn as shown in figure

$$I_2=\frac{12}{2+2+2}=2\text{ A}$$

$$I_3=\frac{12}{4+4+4}=1\text{ A}$$

$$\therefore \quad I_1=I_2+I_3=3\text{ A}$$

$$I_{PQ}=0$$

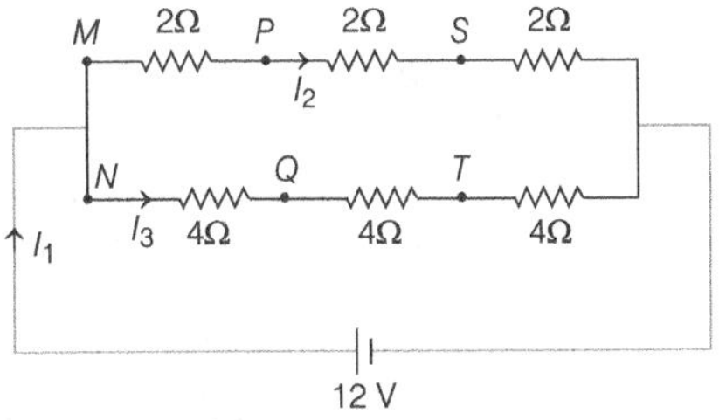

because $\quad V_P=V_Q$

$$V_{MS}=2\times 4=8\text{ V}$$

$$V_{NQ}=1\times 4=4\text{ V}$$

or $\quad V_S<V_Q$

93. $R_{\text{Total}}=2+\frac{6\times 1.5}{6+1.5}=3.2\,\text{k}\Omega$

(a) $I=\frac{24\text{ V}}{3.2\,\text{k}\Omega}=7.5\text{ mA}=I_{R_1}$

$$I_{R_2}=\left(\frac{R_L}{R_L+R_2}\right)I$$

$$\Rightarrow \quad I_{R_2}=\frac{1.5}{7.5}\times 7.5=1.5\text{ mA}$$

$$\Rightarrow \quad I_{R_L}=6\text{ mA}$$

(b) $V_{R_L}=(I_{R_L})(R_L)=9$ V

(c) $\frac{P_{R_1}}{P_{R_2}}=\frac{(I_{R_1}^2)R_1}{(I_{R_2}^2)R_2}=\frac{(7.5)^2(2)}{(1.5)^2(6)}=\frac{25}{3}$

(d) When R_1 and R_2 are interchanged, then

$$\frac{R_2R_L}{R_2+R_L}=\frac{2\times 1.5}{3.5}=\frac{6}{7}\text{k}\Omega$$

Now potential difference across R_L will be

$$V_L=24\left[\frac{6/7}{6+6/7}\right]=3\text{ V}$$

Earlier it was 9 V

Since, $\quad P=\frac{V^2}{R}$ or $P\propto V^2$

In new situation potential difference has been decreased three times. Therefore, power dissipated will decrease by a factor of 9.

94. (a) Points D and E are symmetrically located with respect to points A and C. The circuit can be redrawn as shown in figure.

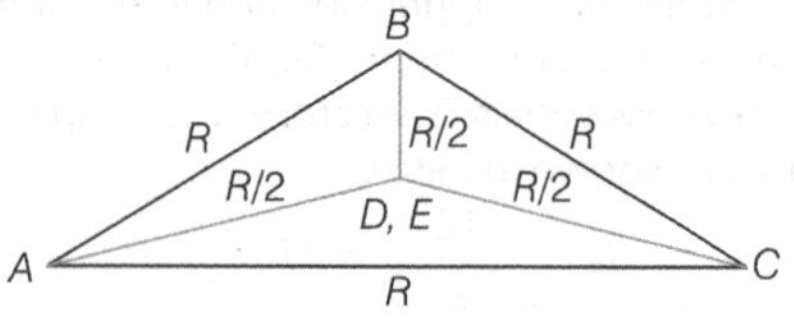

This is a combination of a balanced Wheatstone bridge in parallel with a resistance R. So, the resistance between B and D (or E) can be removed.

$$\frac{1}{R_{AC}}=\frac{1}{R}+\frac{1}{\frac{R}{2}+\frac{R}{2}}+\frac{1}{2R}$$

or $$R_{AC}=\frac{2R}{5}$$ **Ans.**

(b) With respect to D and E, points A, B and C all are symmetrically located. Hence, the simplified circuit can be drawn as shown in figure.

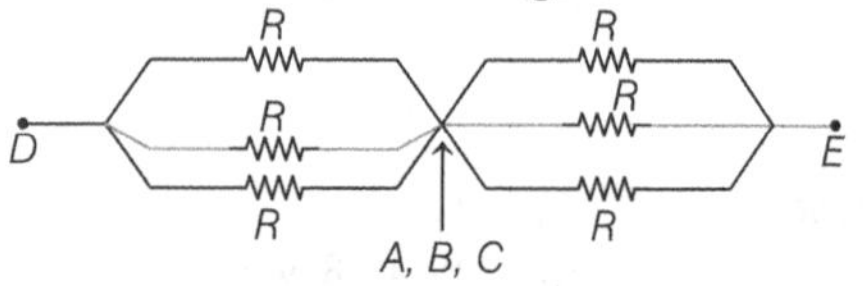

$\therefore$ $$R_{DE}=\frac{R}{3}+\frac{R}{3}=\frac{2R}{3}$$ **Ans.**

95. (a) Due to symmetry about the shaded plane, current distribution on either side of the plane will be identical and points E and F will be at same potential and no current will flow through it.

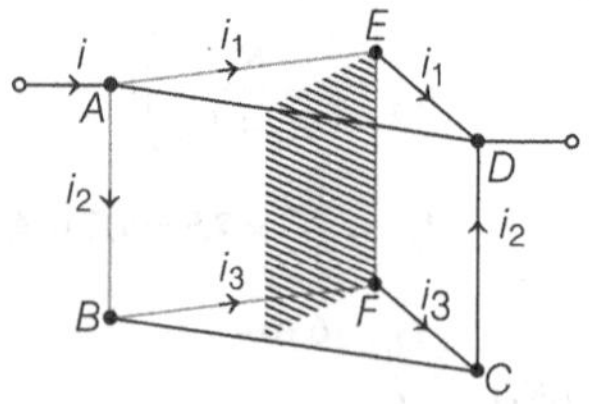

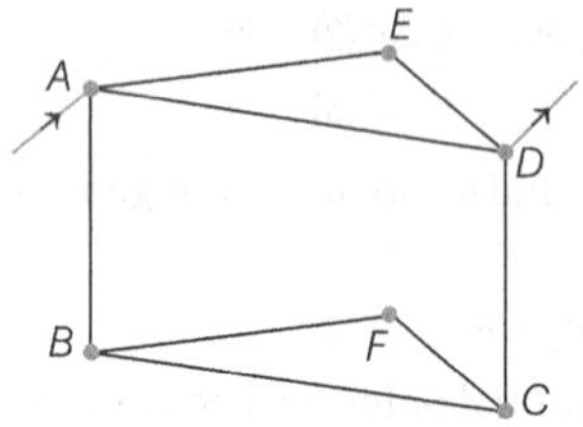

$\therefore$ $$R_{AD}=\frac{\frac{2}{3}r\times\frac{8}{3}r}{\frac{2}{3}r+\frac{8}{3}r}=\frac{8}{15}r$$ **Ans.**

(b) Redrawing the given arrangement for resistance across AB. Potentials $V_D=V_E$

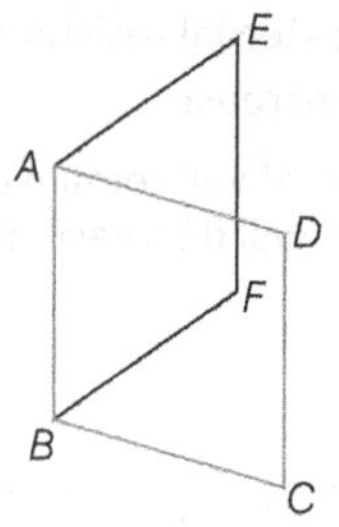

$$V_C=V_F$$

$\therefore$ No current flows through DE and CF.

$\therefore$ $$R_{AB}=\frac{\frac{3}{2}r\times r}{\frac{3}{2}r+r}=\frac{3}{5}r$$ **Ans.**

96. For ammeter $99I_g=(I-I_g)1$

or $$I=100\,I_g \quad \text{...(i)}$$

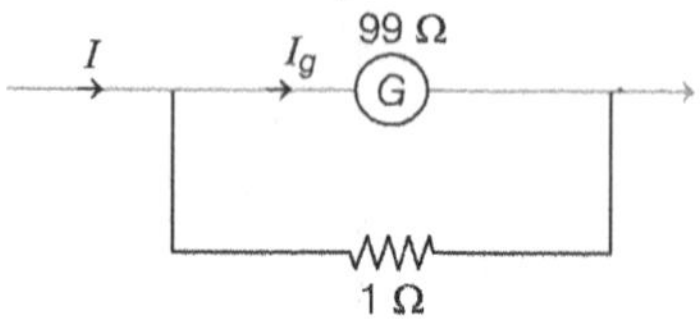

Fig. 1

I_g is the full scale deflection current of the galvanometer and I the range of ammeter.

For the circuit in Fig.1, given in the question

$$\frac{12\text{ V}}{2+r+\frac{99\times 1}{99+1}}=3\text{ A}$$

$\Rightarrow$ $$r=1.01\,\Omega$$ **Ans.**

For voltmeter, range

$$V=I_g\,(99+101)$$

$$V=200\,I_g \quad \text{...(ii)}$$

Also resistance of the voltmeter

$$=99+101=200\,\Omega$$

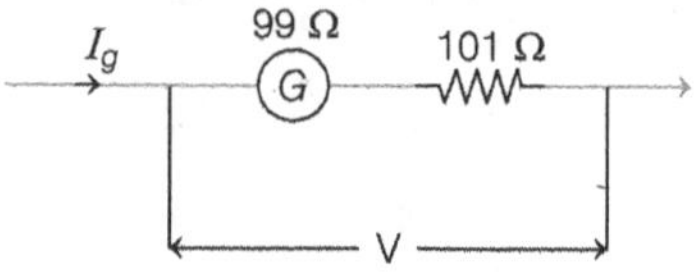

Fig. 2

In Fig. 2, resistance across the terminals of the battery

$$R_1=r+\frac{200\times 2}{200+2}=2.99\,\Omega$$

$\therefore$ Current drawn from the battery,

$$I_1=\frac{12}{2.99}=4.01\text{ A}$$

$\therefore$ Voltmeter reading

$$\frac{4}{5}V=12-I_1r=12-4.01\times 1.01$$

$$V=7.96\times\frac{5}{4}=9.95\text{ V}$$ **Ans.**

Using Eq. (ii), $I_g = \dfrac{9.95}{200} = 0.05\ \text{A}$

Using Eq.(i), range of the ammeter

$$I = 100\, I_g = 5\ \text{A}$$ **Ans.**

97. Applying loop law in loops 1, 2 and 3, we have

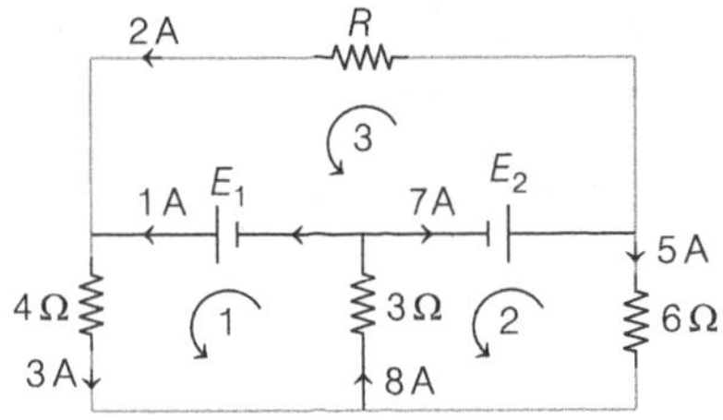

$$E_1 - 12 - 24 = 0$$

$\therefore$ $$E_1 = 36\ \text{V}$$

$$-E_2 + 24 + 30 = 0$$

or $$E_2 = 54\ \text{V}$$

$$-2R - E_1 + E_2 = 0$$

or $$R = \frac{E_2 - E_1}{2}$$

$$= 9\,\Omega$$

98. Using the loop current method,

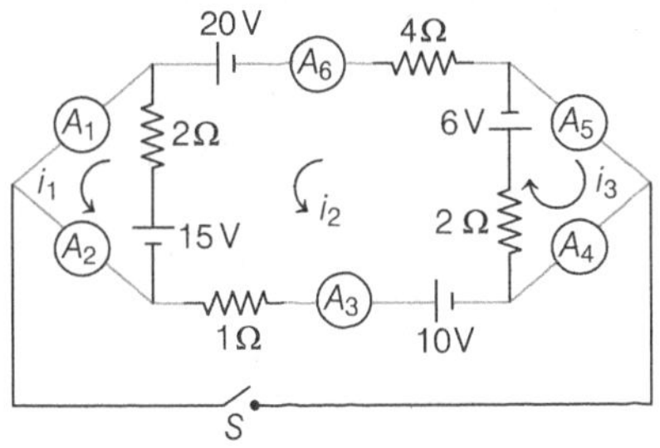

(a) $-2(i_1 - i_2) + 15 = 0$ ...(i)

$$-4i_2 + 20 - 2(i_2 - i_1) - 15 - i_2$$
$$-10 - 2(i_2 + i_3) - 6 = 0$$

or $2i_1 - 9i_2 - 2i_3 - 11 = 0$ ...(ii)

$$-2(i_2 + i_3) - 6 = 0$$

or $i_2 + i_3 + 3 = 0$ ...(iii)

Solving these equations, we get

$i_1 = 9.5\ \text{A}, i_2 = 2\ \text{A}$

and $i_3 = -5\ \text{A}$

Ammeter	A_1	A_2	A_3 and A_6	A_4 and A_5
Reading (amp)	9.5	9.5	2	5

PD across switch $= 10 + (1)\, i_2 = 10 + 2 = 12\ \text{V}$ **Ans.**

(b)

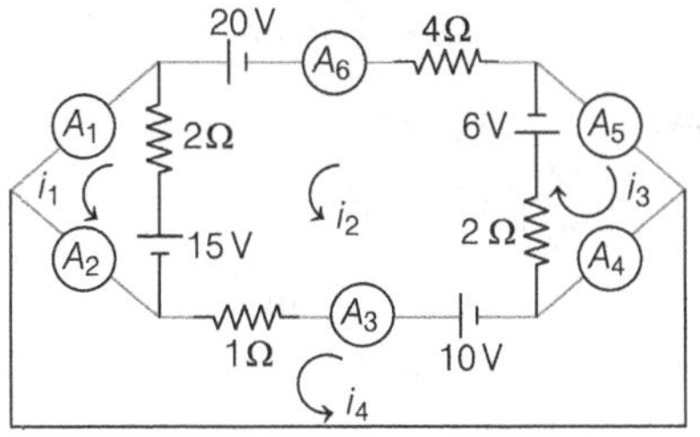

When switch is closed

$-2(i_1 - i_2) + 15 = 0$ ...(i)

$2i_1 - 9i_2 - 2i_3 - 11 + i_4 = 0$ ...(ii)

$i_2 + i_3 + 3 = 0$ ...(iii)

$10 - (i_4 - i_2) = 0$ ...(iv)

Solving these four equation, we get

$i_1 = 12.5\ \text{A}, i_2 = 5.0\ \text{A}, i_3 = -8.0\ \text{A}$

$i_4 = 15\ \text{A}$

Ammeter	A_1	A_2	A_3	A_4	A_5	A_6
Reading (amp)	12.5	2.5	10	7	8	5

And the current through switch is 15 A. **Ans.**

99.

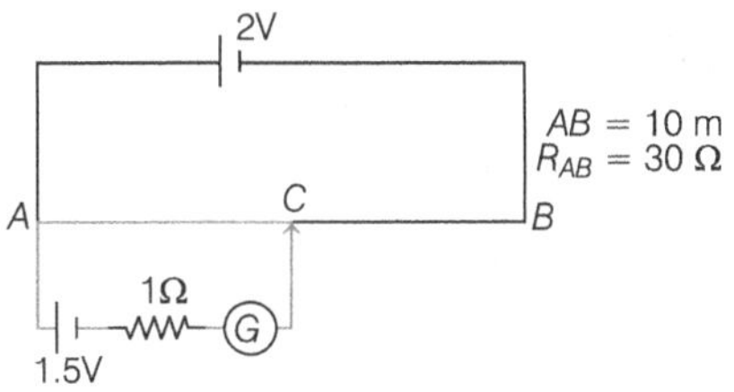

Potential gradient across wire,

$$AB = \frac{2}{10} = 0.2\ \text{V/m}$$

Now, $V_{AC} = 1.5\ \text{V}$ or $(0.2)(AC) = 1.5$

$\therefore$ $AC = 7.5\ \text{m}$ **Ans.**

(a) $V_{AB} = \left(\dfrac{R_{AB}}{R_{AB} + 5}\right) \times 2 = \left(\dfrac{30}{30 + 5}\right) \times 2 = \dfrac{12}{7}\ \text{V}$

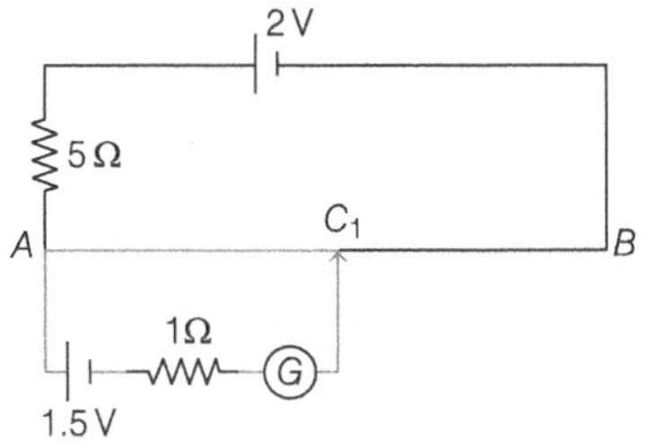

$\therefore$ Potential gradient across

$$AB = \frac{12}{70}\text{ V/m}$$

Now, $\quad V_{AC} = 1.5\text{ V}$

$\therefore \quad \left(\frac{12}{70}\right)(AC_1) = 1.5$

$\therefore \quad AC_1 = 8.75\text{ m}$ **Ans.**

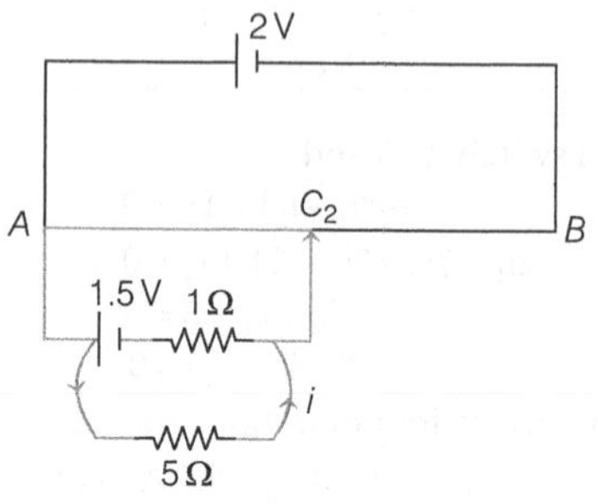

(b) $\quad V_{AC_2} = V$

or $\quad (0.2)(AC_2) = \left(\frac{5}{5+1}\right)(1.5)$

or $\quad AC_2 = 6.25\text{ m}$ **Ans.**

100. (a) Let a be the radius of cylindrical conductor and l its length.

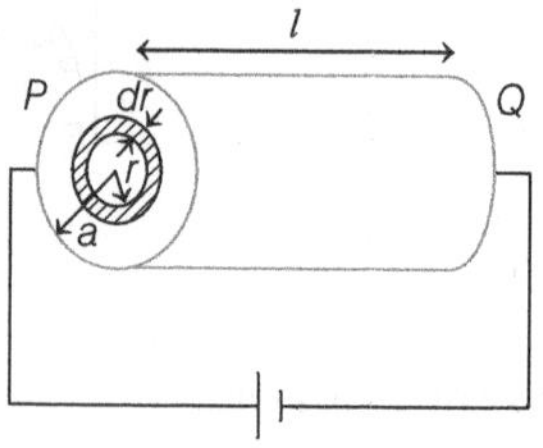

Then, $S = \pi a^2$

or $\quad a = \sqrt{\frac{S}{\pi}}$

Using $R = \frac{\rho l}{A}$, small resistance of the element shown in figure is,

$$dR = \frac{\left(\frac{\alpha}{r^2}\right) l}{(2\pi r)\, dr} = \frac{\alpha l}{2\pi r^3 dr}$$

The system may be assumed as a parallel combination of a number of such types of elementary resistances. Thus, net resistance may be given as,

$$\frac{1}{R} = \int_{r=0}^{r=a} \frac{1}{dR}$$

Completing the calculations, we get

$$R = \frac{2\pi\alpha l}{S^2}$$

So, resistance per unit length,

$$\frac{R}{l} = \frac{2\pi\alpha}{S^2}$$

(b) $E = \frac{V}{l} = I\frac{R}{l} \quad \therefore \quad E = \frac{2\pi\alpha I}{S^2}$

101. The simple circuit is as shown in figure.

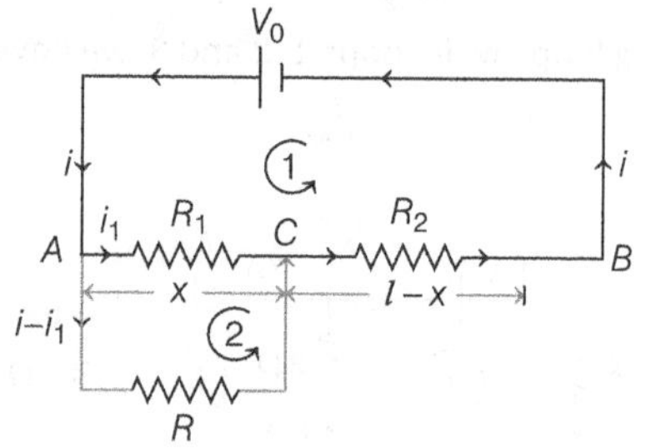

Resistance per unit length of potentiometer wire AB is,

$$\lambda = \frac{R_0}{l}$$

$\therefore \quad R_1 = \lambda x = \frac{R_0}{l}x$

and $\quad R_2 = \lambda(l - x) = \frac{R_0}{l}(l - x)$

Applying KVL in loops 1 and 2, we have

$$V_0 - i_1 R_1 - iR_2 = 0$$

or $\quad V_0 - i_1\left(\frac{R_0 x}{l}\right) - i\left(\frac{R_0}{l}\right)(l - x) = 0 \quad$ ...(i)

and $\quad (i - i_1)R - i_1 R_1 = 0$

or $\quad (i - i_1)R - i_1\left(\frac{R_0}{l}\right)x = 0 \quad$ ...(ii)

Solving these two equations, we get

$$i_1 = \frac{V_0 R l}{R_0\left[Rl + R_0(l - x)\frac{x}{l}\right]}$$

Now, $V_{AC} = i_1 R_1$

Substituting the values of i_1 and R_1, we get

$$V_{AC} \text{ or } V = \frac{V_0 R x}{Rl + R_0(l - x)\frac{x}{l}}$$

Note *We can also find net resistance along AC and then divide V_0 in direct ratio of resistance.*

For $R >> R_0$

We can ignore the term $R_0(l - x)\frac{x}{l}$ in the denominator.

So, $\quad V \approx \frac{V_0 x}{l}$

102. Potential difference across R_1 is E_0.

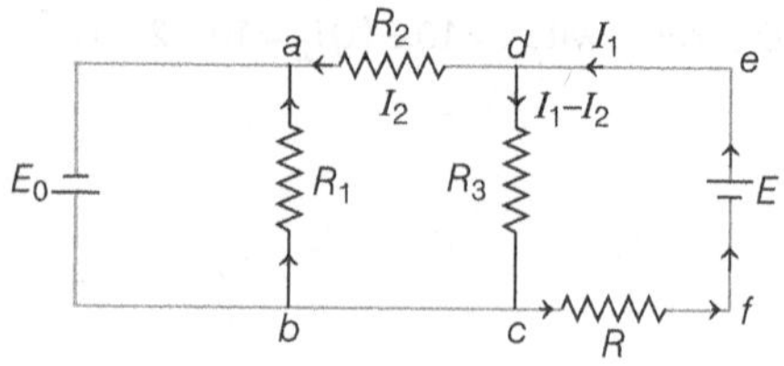

or $\quad V_b - V_a = E_0$

Applying loop law in *abcda*,

$$E_0 + (I_1 - I_2)R_3 - I_2R_2 = 0 \quad \text{...(i)}$$

Again applying loop law in closed loop *fedcf*,

$$E - (I_1 - I_2)R_3 - I_1R = 0 \quad \text{...(ii)}$$

Solving these two equations, we get

$$I_R \text{ or } I_1 = \frac{E(R_2 + R_3) + E_0R_3}{R(R_2 + R_3) + R_2R_3}$$

103. Heat or power transferred to atmosphere,

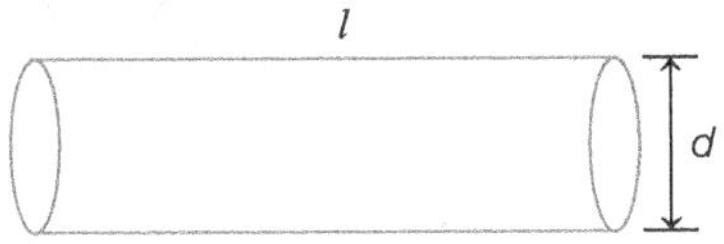

$$P \propto \text{surface area}$$

or $$P \propto \pi dl$$

or $$P = \alpha \pi dl \quad (\alpha = \text{a constant})$$

Now, $$\frac{V^2}{R} = \alpha\pi dl$$

or $$\left(\frac{V^2}{\rho l}\right)\left(\frac{\pi d^2}{4}\right) = \alpha\pi dl \quad \left(R = \frac{\rho l}{A} \text{ and } A = \frac{\pi d^2}{4}\right)$$

$\therefore$ $$d = 4\alpha l^2 \rho V^{-2}$$

or $$\ln d = \ln(4\alpha l^2\rho) - 2\ln V$$

Here d and V are variables. So, differentiating, we get

$$\frac{\Delta d}{d} = -2\frac{\Delta V}{V}$$

Given $\frac{\Delta V}{V} \times 100 = +1\%$

$\therefore$ $$\frac{\Delta d}{d} \times 100 = -2\%$$

$\therefore$ Diameter will have to be decrease by 2%.

104. $$\frac{V^2}{R} = k(T - T_0) + C\left(\frac{dT}{dt}\right)$$

or $$\frac{dT}{\frac{V^2}{R} - k(T - T_0)} = \frac{dt}{C}$$

or $$\int_{T_0}^{T} \frac{dT}{\frac{V^2}{R} - k(T - T_0)} = \int_0^t \frac{dt}{C}$$

(at $t = 0$, temperature of conductor $T = T_0$)

Solving this equation, we get

$$T = T_0 + \frac{V^2}{kR}(1 - e^{-kt/C})$$

105. (a) $$t = \frac{l}{v_d} = \frac{l}{(I/neS)} = \frac{neSl}{I} \quad \text{...(i)}$$

Here, n = number of free electrons per unit volume.

Mass of unit volume = density d

Gram moles of unit volume $= \frac{d}{M}$

Number of free electrons in unit volume or n = number of atoms in unit volume

$$= \left(\frac{d}{M}\right)N_A$$

where N_A is Avogadro number and d = density of copper.

Substituting in equation (i), we get

$$t = \frac{dN_AeSl}{MI}$$

Substituting the proper values, we get

$$t \approx 3 \text{ ms}$$

(b) Electric field $E = \frac{V}{l}$

$$= \frac{IR}{l} = \frac{I\left(\frac{\rho l}{S}\right)}{l}$$

$$E = \frac{\rho I}{S}$$

Here, ρ = resistivity of copper.

Total electrostatic force

F = (total number of free electrons) qE

or F = (number of free electrons per unit volume) (volume)(eE)

or $$F = \left(\frac{dN_A}{M}\right)(Sl)(e)\left(\frac{\rho I}{S}\right)$$

$$= \frac{\rho dN_A Iel}{M}$$

Substituting the values, we get

$$F = 1.0 \text{ MN}$$

106.

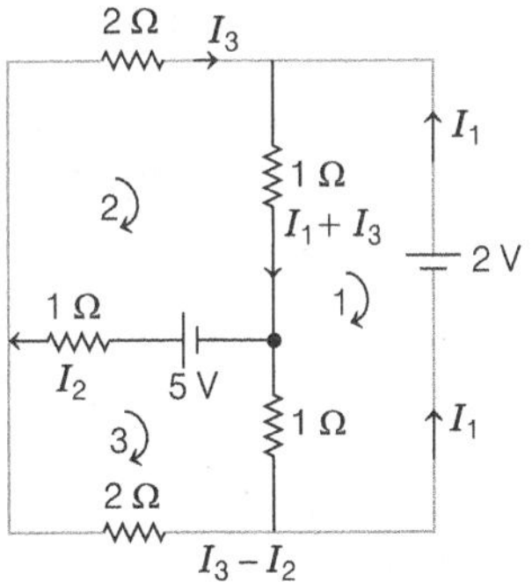

For first loop,

$$(I_1 + I_3) + (I_1 + I_3 - I_2) = 2$$

$\Rightarrow$ $$2I_1 - I_2 + 2I_3 = 2 \quad \text{... (i)}$$

For second loop,

$$2I_3 + I_1 + I_3 + I_2 = 5$$

$$I_2 + I_2 + 3I_3 = 5 \quad \text{... (ii)}$$

For third loop,

$$I_2 - 5 = 2(I_3 - I_2) + (I_1 + I_3 - I_2)$$

$\Rightarrow$ $$I_1 - 4I_2 + 3I_3 = -5 \quad \text{... (iii)}$$

On subtracting Eq. (iii) from Eq. (ii), we get

$$5I_2 = 10$$

$\Rightarrow$ $$I_2 = 2\text{A}$$

On putting in Eq. (i) and Eq. (ii), we get

$$2I_1 + 2I_3 = 4$$

$\Rightarrow \quad I_1 + I_3 = 2$

$\Rightarrow \quad I_1 = 2 - I_3 \quad \text{... (iv)}$

and $\quad I_1 + 3I_3 = 3$

$\Rightarrow \quad 2 - I_3 + 3I_3 = 3 \quad$ [From Eq. (iv)]

$\Rightarrow \quad 2 + 2I_3 = 3$

$\Rightarrow \quad I_3 = \frac{1}{2}\text{ A}$

From Eq. (iv), we get

$$I_1 = \frac{3}{2}\text{ A} = 1.5\text{ A}$$

107. The given circuit may be reduced in following form

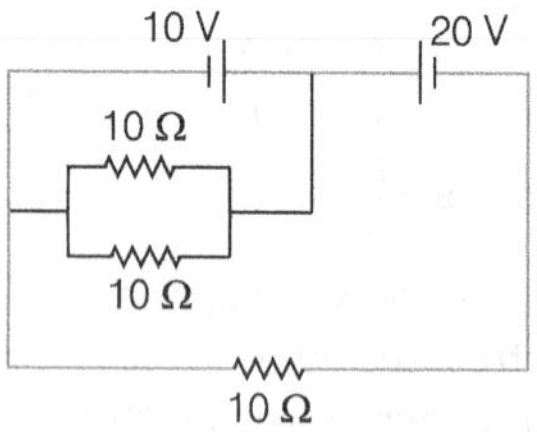

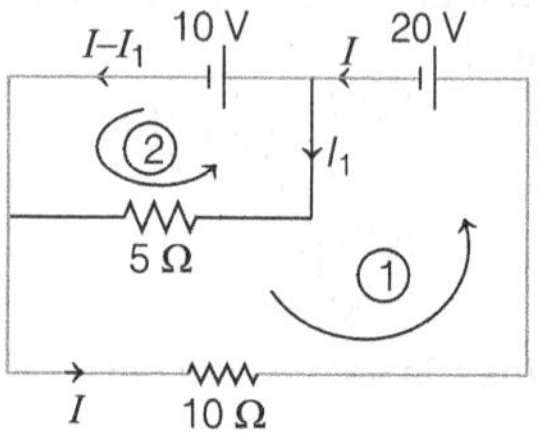

On applying KVL in loop-1 along the direction shown, we get

$$20 - 10 - 10I = 0 \Rightarrow I = 1\text{ A}$$

On applying KVL in loop-2 along the direction shown, we get

$$-10 + 5I' = 0 \Rightarrow I' = 2\text{A}$$

On comparing, we get $I_3 = 1\text{A}$

$$= I_1 = \frac{I'}{2} = 1\text{A}$$

So, the value of $\left|\frac{I_1 + I_3}{I_2}\right| = \left|\frac{1+1}{1}\right| = 2$

108.

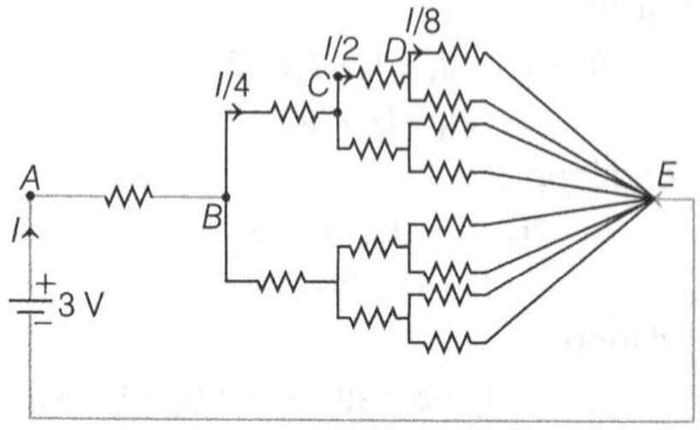

Using KVL in loop *ABCDEA*, we have

$$-I \times R - \frac{I}{2} \times R - \frac{I}{4} \times R - \frac{I}{8} \times 8 + 3 = 0$$

$\Rightarrow \quad IR\left(1 + \frac{1}{2} + \frac{1}{4} + \frac{1}{8}\right) = 3$

Here, $R = 1\Omega$, $\quad$ so $I\left(\frac{8+4+2+1}{8}\right) = 3$

$\Rightarrow \quad I \times \frac{15}{8} = 3$

$\Rightarrow \quad I = \frac{24}{15} = \frac{8}{5}\text{ A} = \frac{a}{5}\text{ A} \quad$ (given)

So, $\quad a = 8$

109.

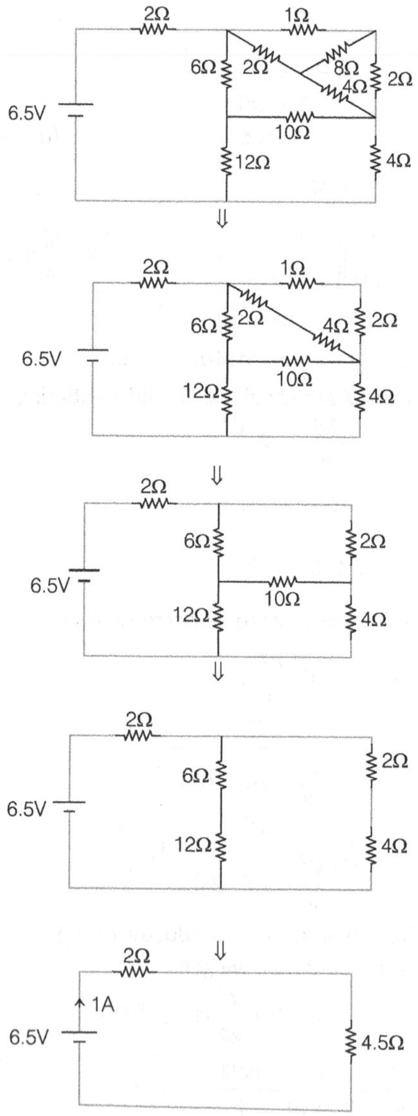

$$\therefore \quad I = \frac{V}{R_{eq}} = \frac{6.5}{4.5+2} = 1\text{ A}$$

110.

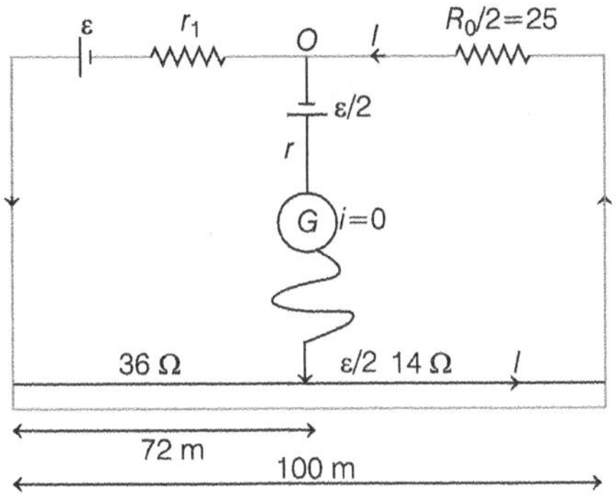

Resistance of potential wire, $R_0 = 50\Omega$

Resistance of 100 m wire $= 50\Omega$

So, resistance of 72 cm wire

$$= \frac{50}{100} \times 72 = 36\Omega$$

Current, $$I = \frac{\frac{\varepsilon}{2}}{14\Omega + 25\Omega}$$

$$= \frac{\frac{\varepsilon}{2}}{r_1 + 36\Omega}$$

$$\Rightarrow \quad r_1 = 39 - 36 = 3\Omega$$

111. $R_3' = 300(1 + \alpha\Delta T) = 312\Omega$

Now,

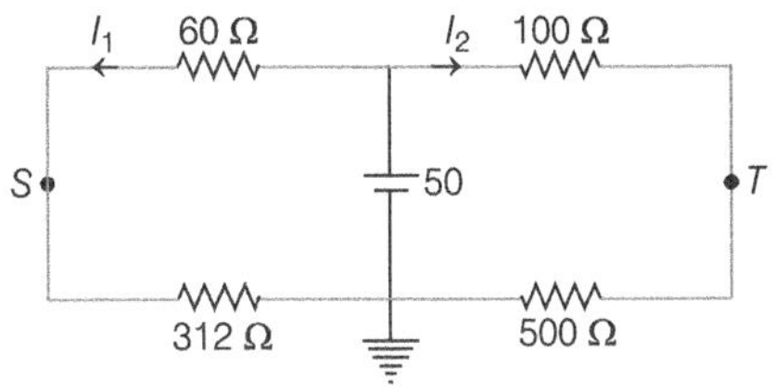

$$I_1 = \frac{50}{372} \text{ and } I_2 = \frac{50}{600}$$

$$V_S - V_T = 312I_1 - 500I_2$$

$$= 41.94 - 41.67$$

$$= 0.27 \text{ V}$$

UNIT-2

ELECTROSTATICS

Learning Modules

1. Electric Charge
2. Coulomb's Law
3. Electric Field
4. Electric Potential Energy
5. Electric Potential
6. Relation Between Electric Field and Potential
7. Equipotential Surfaces
8. Electric Dipole
9. Gauss's Law
10. Properties of a Conductor
11. Charged Spheres

MODULE 1

Electric Charge

Introduction

When we comb our hair on a dry day and bring the comb close to tiny pieces of paper, we note that they are swiftly attracted by the comb. Similar phenomena occur if we rub a glass rod or an amber rod with a cloth or with a piece of fur. Why does this happens? What really happens in an electric circuit? How do electric motors and generators work?

The answers to all these questions come from a branch of physics known as **electromagnetism**, the study of electric and magnetic interactions. These interactions involve particles that have a property called electric charge, an inherent property of matter that is as fundamental as mass.

We begin our study of electromagnetism in this chapter by the electric charge. We will see that it is quantized and obeys a conservation principle. Then we will study the interactions of electric charges that are at rest, called **electrostatic** interactions. These interactions are governed by a simple relationship known as **Coulomb's law.** This law is more conveniently described by using the concept of **electric field.**

Electric Charge

The electrical nature of matter is inherent in atomic structure. An atom consists of a small, relatively massive nucleus that contains particles called protons and neutrons. A proton has a mass 1.673×10^{-27} kg, while a neutron has a slightly greater mass 1.675×10^{-27} kg. Surrounding the nucleus is a diffuse cloud of orbiting particles called electrons. An electron has a mass of 9.11×10^{-31} kg.

Like mass, **electric charge** is an intrinsic property of protons and electrons, and only two types of charge have been discovered positive and negative. A proton has a positive charge, and an electron has a negative charge. A neutron has no net electric charge.

The magnitude of the charge on the proton exactly equals the magnitude of the charge on the electron. The proton carries a charge $+e$ and the electron carries a charge $-e$. The SI unit of charge is coulomb (C) and e has the value

$$\boxed{e = 1.6 \times 10^{-19}\ \text{C}}$$

Regarding charge the following points are worth noting:

1. Like charges repel each other and unlike charges attract each other.
2. Charge is a scalar and can be of two types positive or negative.
3. Charge is quantized. The quantum of charge is e. The charge on any body will be some integral multiple of e, i.e.

$$\boxed{q = \pm ne} \qquad \text{where,} \quad n = 1, 2, 3 \ldots$$

Charge on any body can never be $\left(\frac{1}{3}e\right)$, $1.5e$, etc.

Note (i) *Apart from charge, energy, angular momentum and mass are also quantized. The quantum of energy is $h\nu$ and that of angular momentum is $\frac{h}{2\pi}$. Quantum of mass is not known till date.*

(ii) *The protons and neutrons are combination of other entities called **quarks**, which have charges $\pm\frac{1}{3}e$ and $\pm\frac{2}{3}e$.*

However, isolated quarks have not been observed. So, quantum of charge is still e.

4. During any process, the net electric charge of an isolated system remains constant or we can say that charge is conserved. Pair production and pair annihilation are two examples of conservation of charge.
5. A charged particle at rest produces electric field. A charged particle in an unaccelerated motion produces both electric and magnetic fields and does not radiate energy. But an accelerated charged particle not only produces an electric and magnetic fields but also radiates energy in the form of electromagnetic waves.

Example 1 How many electrons are there in one coulomb of negative charge?

Sol. The negative charge is due to the presence of excess electrons, since they carry negative charge. Because an electron has a charge whose magnitude is $e = 1.6 \times 10^{-19}$ C, the number of electrons is equal to the charge q divided by the charge e on each electron. Therefore, the number n of electrons is

$$n = \frac{q}{e} = \frac{1.0}{1.6 \times 10^{-19}} = 6.25 \times 10^{18} \qquad \textbf{Ans.}$$

Conductors and Insulators

For the purpose of electrostatic theory, all substances can be divided into two main groups, conductors and insulators. In conductors, electric charges are free to move from one place to another, whereas in insulators they are tightly bound to their respective atoms. In an uncharged body, there are equal number of positive and negative charges.

The examples of conductors of electricity are the metals, human body and the earth and that of insulators are glass, hard rubber and plastics. In metals, the free charges are free electrons known as conduction electrons.

Semiconductors are a third class of materials and their electrical properties are somewhere between those of insulators and conductors. Silicon and germanium are well known examples of semiconductors.

Charging of a Body

There are primarily three methods of charging a body :

1. Charging by Rubbing

The simplest way to experience electric charges is to rub certain bodies against each other. When a glass rod is rubbed with a silk cloth, the glass rod acquires some positive charge and the silk cloth acquires negative charge by the same amount.

The explanation of appearance of electric charge on rubbing is simple. All material bodies contain large number of electrons and equal number of protons in their normal state. When rubbed against each other, some electrons from one body pass onto the other body.

The body that donates the electrons becomes positively charged while that which receives the electrons becomes negatively charged. For example, when glass rod is rubbed with silk cloth, glass rod becomes positively charged because it donates the electrons while the silk cloth becomes negatively charged because it receives electrons. Electricity so obtained by rubbing two objects is also known as **frictional electricity.**

The other places where the frictional electricity can be observed are when amber is rubbed with wool or a comb is passed through a dry hair. Clouds also become charged by friction.

2. Charging by Contact

When a negatively charged ebonite rod is rubbed on a metal object, such as a sphere, some of the excess electrons from the rod are transferred to the sphere. Once the electrons are on the metal sphere, where they can move readily, they repel one another and spread out over the sphere's surface.

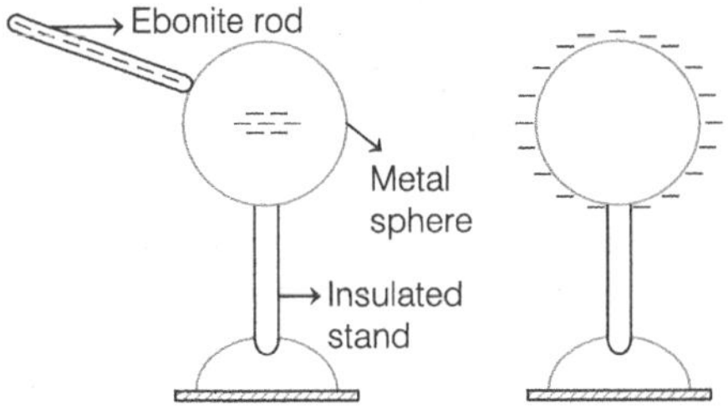

The insulated stand prevents them from flowing to the earth. When the rod is removed, the sphere is left with a negative charge distributed over its surface. In a similar manner, the sphere will be left with a positive charge after being rubbed with a positively charged rod. In this case, electrons from the sphere would be transferred to the rod.

The process of giving one object a net electric charge by placing it in contact with another object that is already charged is known as **charging by contact.**

3. Charging by Induction

It is also possible to charge a conductor in a way that does not involve contact.

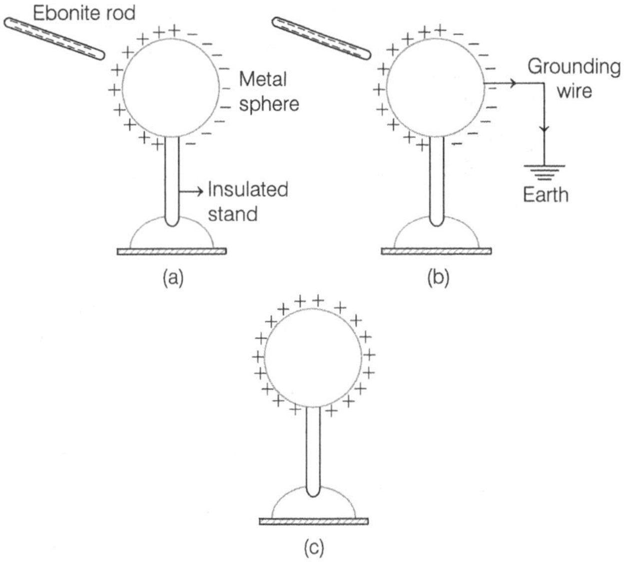

In Fig. (a), a negatively charged rod is brought close to (but does not touch) a metal sphere. In the sphere, the free electrons close to the rod move to the other side (by repulsion). As a result, the part of the sphere nearer to the rod becomes positively charged and the part farthest from the rod negatively charged. This phenomenon is called induction.

Now, if the rod is removed, the free electrons return to their original places and the charged regions disappear. Under most conditions the earth is a good electric conductor. So, when a metal wire is attached between the sphere and the ground as in figure (b) some of the free electrons leave the sphere and distribute themselves on the much larger earth. If the grounding wire is then removed, followed by the ebonite rod, the sphere is left with a net positive charge.

The process of giving one object a net electric charge without touching the object to a second charged object is called **charging by induction.** The process could also be used to give the sphere a net negative charge, if a positively charged rod were used. Then, electrons would be drawn up from the ground through the grounding wire and onto the sphere.

If the sphere were made from an insulating material like plastic, instead of metal, the method of producing a net charge by induction would not work, because very little charge would flow through the insulating material and down the grounding wire. However, the electric force of the charged rod would have some effect as shown in figure.

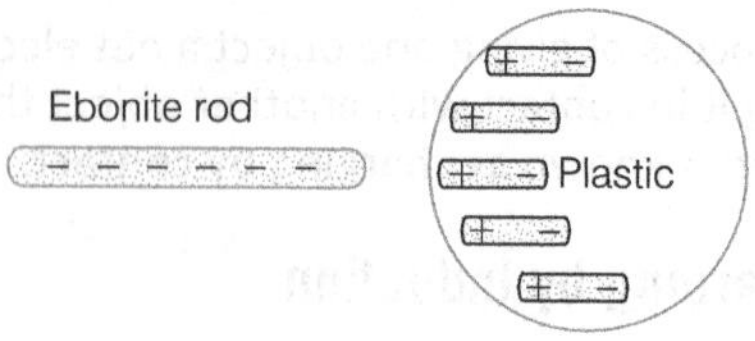

The electric force would cause the positive and negative charges in the molecules of the insulating material to separate slightly, with the negative charges being pushed away from the negative rod. The surface of the plastic sphere does acquire a slight induced positive charge, although no net charge is created.

Example 2 If we comb our hair on a dry day and bring the comb near small pieces of paper, the comb attracts the pieces, why?

Sol. This is an example of frictional electricity and induction. When we comb our hair, it gets positively charged by rubbing. When the comb is brought near the pieces of paper some of the electrons accumulate at the edge of the paper piece which is closer to the comb.

At the farther end of the piece there is deficiency of electrons and hence, positive charge appears there. Such a redistribution of charge in a material, due to presence of a nearby charged body is called **induction.**

The comb exerts larger attraction on the negative charges of the paper piece as compared to the repulsion on the positive charge. This is because the negative charges are closer to the comb. Hence, there is a net attraction between the comb and the paper piece.

Example 3 Does the attraction between the comb and the piece of papers last for longer period of time?

Sol. No, because the comb loses its net charge after some time. The excess charge of the comb transfers to earth through our body after some time.

Example 4 Can two similarly charged bodies attract each other?

Sol. Yes, when the charge on one body (q_1) is much greater than that on the other (q_2) and they are close enough to each other so that force of attraction between q_1 and induced charge on the other exceeds the force of repulsion between q_1 and q_2. However, two similar point charges can never attract each other because no induction will take place here.

Example 5 Does in charging, the mass of a body change?

Sol. Yes, as charging a body means addition or removal of electrons and electron has a mass.

Example 6 Why a third hole in a socket provided for grounding?

Sol. All electric appliances may end with some charge due to faulty connections. In such a situation charge will be accumulated on the appliance. When the user touches the appliance, he may get a shock. By providing the third hole for grounding all accumulated charge is discharged to the ground and the appliance is safe.

MODULE 1 Exercise

Single Correct Option

1. Five conducting balls numbered 1 to 5 are suspended using separate threads. Pairs (1, 2), (2, 4) and (4, 1) show electrostatic attraction, while pairs (2, 3) and (4, 5) show repulsion. Therefore, ball 1 must be
   (a) positively charged (b) negatively charged
   (c) neutral (d) data insufficient

Subjective Questions

2. Is attraction a true test of electrification?
3. Is repulsion a true test of electrification?
4. Why does a phonograph record attract dust particles just after it is cleaned?
5. What is the total charge, in coulombs, of all the electrons in three gram mole of hydrogen atom?

Answers

1. (c) **2.** No, as attraction can take place between a charged and an uncharged body too. **3.** Yes
4. Record gets charged when cleaned and then by induction, it attracts dust particles. **5.** -2.89×10^5 C

See hints on page 194.

MODULE 2

Coulomb's Law

The law that describes how charges interact with one another was discovered by Charles Augustin de Coulomb in 1785.

The electric force $\mathbf{F}_e$ exerted by one point charge on another acts along the line between the charges. It varies inversely as the square of the distance separating the charges and is proportional to the product of charges. The force is repulsive if the charges have the same sign and attractive if the charges have opposite signs.

The magnitude of the electric force exerted by a charge q_1 on another charge q_2 a distance r away is thus, given by

$$F_e = \frac{k|q_1 q_2|}{r^2} \qquad \text{...(i)}$$

The value of the proportionality constant k in Coulomb's law depends on the system of units used. In SI units the constant k is

$$k = 8.987551787 \times 10^9 \frac{\text{N-m}^2}{\text{C}^2}$$

$$\approx 8.988 \times 10^9 \frac{\text{N-m}^2}{\text{C}^2}$$

This value is closely related to the speed of light in vacuum. The numerical value of k is defined in terms of c (the speed of light in vacuum) to be precisely.

$$k = \left(10^{-7} \frac{\text{N-s}^2}{\text{C}^2}\right) c^2$$

This constant k is often written as $\frac{1}{4\pi\varepsilon_0}$, where ε_0 ("epsilon-nought") is another constant.

Thus, Eq. (i) can be written as

$$\boxed{F_e = \frac{1}{4\pi\varepsilon_0} \frac{|q_1 q_2|}{r^2}} \qquad \text{...(ii)}$$

Here, $$\frac{1}{4\pi\varepsilon_0} = \left(10^{-7} \frac{\text{N-s}^2}{\text{C}^2}\right) c^2$$

Substituting value of $c = 2.99792458 \times 10^8$ m/s, we get

$$\boxed{\frac{1}{4\pi\varepsilon_0} = 8.99 \times 10^9 \text{N-m}^2/\text{C}^2}$$

In examples and problems, we will often use the approximate value

$$\frac{1}{4\pi\varepsilon_0} = 9.0 \times 10^9 \text{ N-m}^2/\text{C}^2$$

Here, the quantity ε_0 is called the **permittivity of free space.** It has the value,

$$\boxed{\varepsilon_0 = 8.854 \times 10^{-12} \text{C}^2/\text{N-m}^2}$$

Regarding Coulomb's law, the following points are worth noting:

1. Coulomb's law stated above describes the interaction of two point charges. When two charges exert forces simultaneously on a third charge, the total force acting on that charge is the vector sum of the forces that the two charges would exert individually. This important property, called the **principle of superposition of forces,** holds for any number of charges. Thus,
$$\mathbf{F}_{\text{net}} = \mathbf{F}_1 + \mathbf{F}_2 + \ldots + \mathbf{F}_n$$
2. The electric force is an action reaction pair, i.e. the two charges exert equal and opposite forces on each other.
3. The electric force is conservative in nature.
4. Coulomb's law as we have stated above can be used for point charges in **vacuum.** If some dielectric (an electrical insulator that can be polarised by an applied electric field) is present in the space between the charges, the net force acting on each charge is altered because charges are induced in the molecules of the intervening medium. We will describe this effect later. Here at this moment it is enough to say that the force decreases K times if the medium extends till infinity.

   Here, K is a dimensionless constant which depends on the medium and called dielectric constant of the medium. Thus,

$$F_e = \frac{1}{4\pi\varepsilon_0} \cdot \frac{q_1 q_2}{r^2} \qquad \text{(in vacuum)}$$

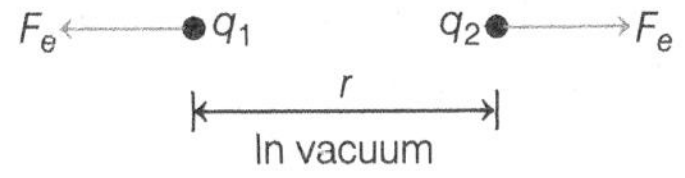

$$F_e' = \frac{F_e}{K} = \frac{1}{4\pi\varepsilon_0 K} \cdot \frac{q_1 q_2}{r^2} = \frac{1}{4\pi\varepsilon} \cdot \frac{q_1 q_2}{r^2} \qquad \text{(in medium)}$$

Here, $\varepsilon = \varepsilon_0 K$ is called permittivity of the medium.

Important POINTS

- In some of the problems of electrostatics **Lami's theorem** is very useful.

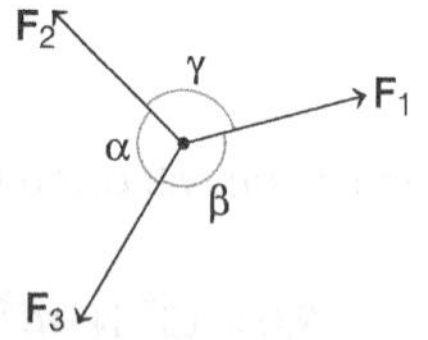

According to this theorem, "if three concurrent forces $\mathbf{F}_1, \mathbf{F}_2$ and $\mathbf{F}_3$ as shown in figure are in equilibrium or if $\mathbf{F}_1 + \mathbf{F}_2 + \mathbf{F}_3 = 0$, then

$$\frac{F_1}{\sin\alpha} = \frac{F_2}{\sin\beta} = \frac{F_3}{\sin\gamma}$$

- Suppose the position vectors of two charges q_1 and q_2 are $\mathbf{r}_1$ and $\mathbf{r}_2$, then electric force on charge q_1 due to charge q_2 is,

$$\mathbf{F}_1 = \frac{1}{4\pi\varepsilon_0} \cdot \frac{q_1 q_2}{|\mathbf{r}_1 - \mathbf{r}_2|^3} (\mathbf{r}_1 - \mathbf{r}_2)$$

Similarly, electric force on q_2 due to charge q_1 is

$$\mathbf{F}_2 = \frac{1}{4\pi\varepsilon_0} \cdot \frac{q_1 q_2}{|\mathbf{r}_2 - \mathbf{r}_1|^3} (\mathbf{r}_2 - \mathbf{r}_1)$$

Here, q_1 and q_2 are to be substituted with sign. $\mathbf{r}_1 = x_1\hat{\mathbf{i}} + y_1\hat{\mathbf{j}} + z_1\hat{\mathbf{k}}$ and $\mathbf{r}_2 = x_2\hat{\mathbf{i}} + y_2\hat{\mathbf{j}} + z_2\hat{\mathbf{k}}$ where (x_1, y_1, z_1) and (x_2, y_2, z_2) are the coordinates of charges q_1 and q_2.

Example 1 What is the smallest electric force between two charges placed at a distance of 1.0 m?

Sol. $$F_e = \frac{1}{4\pi\varepsilon_0} \cdot \frac{q_1 q_2}{r^2} \quad \text{...(i)}$$

For F_e to be minimum $q_1 q_2$ should be minimum. We know that

$$(q_1)_{min} = (q_2)_{min} = e = 1.6 \times 10^{-19} \text{ C}$$

Substituting in Eq. (i), we have

$$(F_e)_{min} = \frac{(9.0 \times 10^9)(1.6 \times 10^{-19})(1.6 \times 10^{-19})}{(1.0)^2}$$

$$= 2.304 \times 10^{-28} \text{ N}$$

Ans.

Example 2 Three charges $q_1 = 1\,\mu\text{C}$, $q_2 = -2\,\mu\text{C}$ and $q_3 = 3\,\mu\text{C}$ are placed on the vertices of an equilateral triangle of side 1.0 m. Find the net electric force acting on charge q_1.

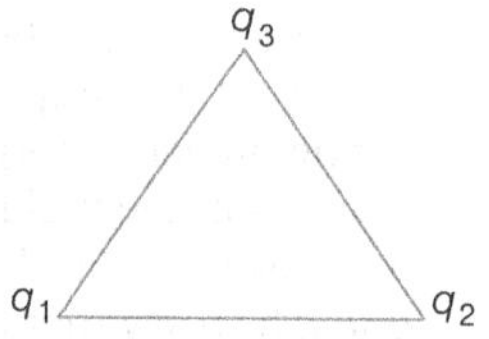

HOW TO PROCEED Charge q_2 will attract charge q_1 (along the line joining them) and charge q_3 will repel charge q_1. Therefore, two forces will act on q_1, one due to q_2 and another due to q_3. Since, the force is a vector quantity both of these forces (say $\mathbf{F}_1$ and $\mathbf{F}_2$) will be added by vector method. There are following two methods of their addition.

Sol. Method 1. In the figure,

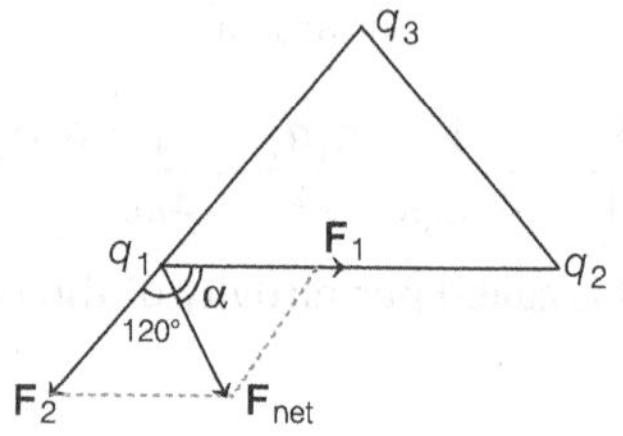

$$|\mathbf{F}_1| = F_1 = \frac{1}{4\pi\varepsilon_0} \cdot \frac{q_1 q_2}{r^2}$$

= magnitude of force between q_1 and q_2

$$= \frac{(9.0 \times 10^9)(1.0 \times 10^{-6})(2.0 \times 10^{-6})}{(1.0)^2}$$

$$= 1.8 \times 10^{-2} \text{ N}$$

Similarly, $$|\mathbf{F}_2| = F_2 = \frac{1}{4\pi\varepsilon_0} \cdot \frac{q_1 q_3}{r^2}$$

= magnitude of force between q_1 and q_3

$$= \frac{(9.0 \times 10^9)(1.0 \times 10^{-6})(3.0 \times 10^{-6})}{(1.0)^2}$$

$$= 2.7 \times 10^{-2} \text{ N}$$

Now, $$|\mathbf{F}_{net}| = \sqrt{F_1^2 + F_2^2 + 2F_1F_2 \cos 120°}$$

$$= \left(\sqrt{(1.8)^2 + (2.7)^2 + 2(1.8)(2.7)\left(-\frac{1}{2}\right)}\right) \times 10^{-2} \text{ N}$$

$$= 2.38 \times 10^{-2} \text{ N}$$

and $$\tan\alpha = \frac{F_2 \sin 120°}{F_1 + F_2 \cos 120°}$$

$$= \frac{(2.7 \times 10^{-2})(0.87)}{(1.8 \times 10^{-2}) + (2.7 \times 10^{-2})\left(-\frac{1}{2}\right)}$$

or $$\alpha = 79.2°$$

Thus, the net force on charge q_1 is 2.38×10^{-2} N at an angle $\alpha = 79.2°$ with a line joining q_1 and q_2 as shown in the figure. **Ans.**

Method 2. In this method let us assume a coordinate axes with q_1 at origin as shown in figure. The coordinates of q_1, q_2 and q_3 in this coordinate system are (0, 0, 0), (1 m, 0, 0) and (0.5 m, 0.87 m, 0) respectively. Now,

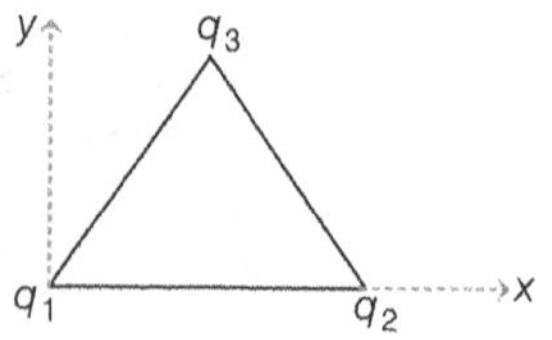

$\mathbf{F}_1$ = force on q_1 due to charge q_2

$$= \frac{1}{4\pi\varepsilon_0} \cdot \frac{q_1 q_2}{|\mathbf{r}_1 - \mathbf{r}_2|^3} (\mathbf{r}_1 - \mathbf{r}_2)$$

$$= \frac{(9.0 \times 10^9)(1.0 \times 10^{-6})(-2.0 \times 10^{-6})}{(1.0)^3}$$

$$[(0-1)\,\hat{\mathbf{i}} + (0-0)\,\hat{\mathbf{j}} + (0-0)\hat{\mathbf{k}}]$$

$$= (1.8 \times 10^{-2}\,\hat{\mathbf{i}})\text{ N}$$

and $\mathbf{F}_2$ = force on q_1 due to charge q_3

$$= \frac{1}{4\pi\varepsilon_0} \cdot \frac{q_1 q_3}{|\mathbf{r}_1 - \mathbf{r}_3|^3} (\mathbf{r}_1 - \mathbf{r}_3)$$

$$= \frac{(9.0 \times 10^9)(1.0 \times 10^{-6})(3.0 \times 10^{-6})}{(1.0)^3}$$

$$[(0-0.5)\,\hat{\mathbf{i}} + (0-0.87)\,\hat{\mathbf{j}} + (0-0)\,\hat{\mathbf{k}}]$$

$$= (-1.35\,\hat{\mathbf{i}} - 2.349\,\hat{\mathbf{j}}) \times 10^{-2}\text{ N}$$

Therefore, net force on q_1 is

$$\mathbf{F} = \mathbf{F}_1 + \mathbf{F}_2$$

$$= (0.45\,\hat{\mathbf{i}} - 2.349\,\hat{\mathbf{j}}) \times 10^{-2}\text{ N}$$ **Ans.**

Note *Once you write a vector in terms of $\hat{\mathbf{i}}$, $\hat{\mathbf{j}}$ and $\hat{\mathbf{k}}$, there is no need of writing the magnitude and direction of vector separately.*

Example 3 Two identical balls each having a density ρ are suspended from a common point by two insulating strings of equal length. Both the balls have equal mass and charge. In equilibrium each string makes an angle θ with vertical. Now, both the balls are immersed in a liquid. As a result the angle θ does not change. The density of the liquid is σ. Find the dielectric constant of the liquid.

Sol. Each ball is in equilibrium under the following three forces :

(i) tension, (ii) electric force and (iii) weight

So, Lami's theorem can be applied.

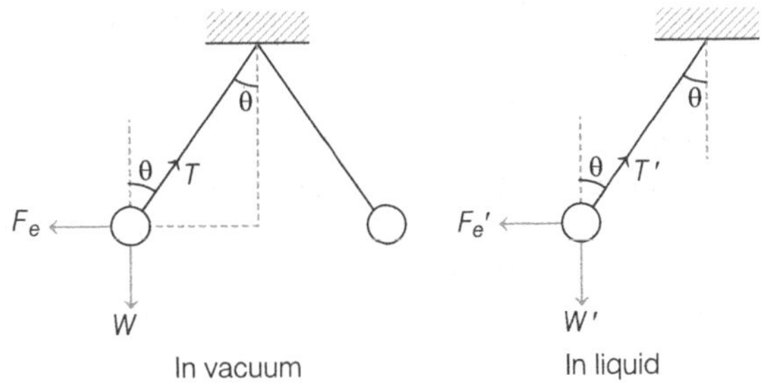

In the liquid, $F_e' = \dfrac{F_e}{K}$

where, K = dielectric constant of liquid and $W' = W -$ upthrust

Applying Lami's theorem in vacuum

$$\frac{W}{\sin(90° + \theta)} = \frac{F_e}{\sin(180° - \theta)}$$

or $$\frac{W}{\cos\theta} = \frac{F_e}{\sin\theta} \quad \text{...(i)}$$

Similarly in liquid, $$\frac{W'}{\cos\theta} = \frac{F_e'}{\sin\theta} \quad \text{...(ii)}$$

Dividing Eq. (i) by Eq. (ii), we get

$$\frac{W}{W'} = \frac{F_e}{F_e'}$$

or $$K = \frac{W}{W - \text{upthrust}} \quad \left(\text{as } \frac{F_e}{F_e'} = K\right)$$

$$= \frac{V\rho g}{V\rho g - V\sigma g} \quad (V = \text{volume of ball})$$

or $$K = \frac{\rho}{\rho - \sigma}$$ **Ans.**

Note *In the liquid F_e and W have changed. Therefore, T will also change.*

Example 4 Five point charges each of value $+q$ are placed on five vertices of a regular hexagon of side 'a' metre. What is the magnitude of the force on a point charge of value $-q$ coulomb placed at the centre of the hexagon? **(1992 JEE)**

Sol.

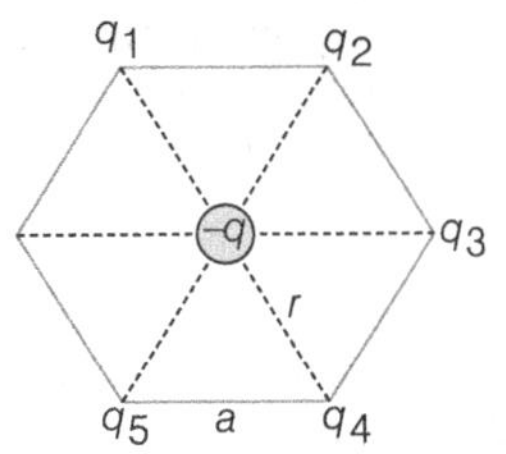

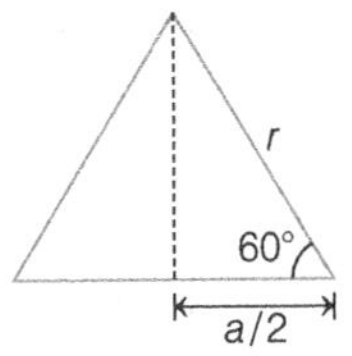

$$\frac{a/2}{r} = \cos 60° = \frac{1}{2}$$

$\therefore$ $$a = r$$

$$q_1 = q_2 = \ldots = q_5 = q$$

Net force on $-q$ is only due to q_3 because forces due to q_1 and due to q_4 are equal and opposite so cancel each other. Similarly, forces due to q_2 and q_5 also cancel each other. Hence, the net force on $-q$ is

$$F = \frac{1}{4\pi\varepsilon_0} \cdot \frac{(q)(q)}{r^2} \quad (\text{towards } q_3)$$

or $$F = \frac{1}{4\pi\varepsilon_0} \cdot \frac{q^2}{a^2}$$ **Ans.**

Important POINTS

- **Permittivity** Permittivity or absolute permittivity relates to a material's ability to resist an electric field (while unfortunately, the word "permit" suggests the inverse quantity).

  The permittivity of a medium describes how much electric field (more correctly, flux) is generated per unit charge in that medium. More electric flux (per unit charge) exists in a medium with a low permittivity. Vacuum has the lowest permittivity (therefore maximum electric flux per unit charge). Any other dielectric medium has K-times (K = dielectric constant) the permittivity of vacuum.

  This is because, due to polarization effects electric flux per unit charge deceases K-times ($K > 1$).

- **Dielectric constant (K)** Also known as relative permittivity of a given material is the ratio of permittivity of the material to the permittivity of vacuum. This is the factor by which the electric force between the two charges is decreased relative to vacuum. Similarly, in the topic of capacitors we will see that it is the ratio of capacitance of a capacitor using that material as a dielectric compared to a similar capacitor that has vacuum as its dielectric.

MODULE 2 Exercise

Note *You can write* $\frac{1}{4\pi\varepsilon_0} = k$

Fill in the Blank

1. Two small balls having equal positive charges Q (coulomb) on each are suspended by two insulating strings of equal length L (metre) from a hook fixed to a stand. The whole set up is taken in a satel1ite into space where there is no gravity (state of weightlessness). The angle between the two strings is .... and the tension in each string is .... newtons. **(1986 JEE)**

Single Correct Option

2. Two positive ions, each carrying a charge q, are separated by a distance d. If F is the force of repulsion between the ions, the number of electrons missing from each ion will be (e being the charge on an electron)

   (a) $\frac{4\pi\varepsilon_0 Fd^2}{e}$ (b) $\sqrt{\frac{4\pi\varepsilon_0 Fe^2}{d^2}}$

   (c) $\sqrt{\frac{4\pi\varepsilon_0 Fd^2}{e^2}}$ (d) $\frac{4\pi\varepsilon_0 Fd^2}{e^2}$

3. Two copper balls, each weighing 10 g, are kept in air 10 cm apart. If one electron from every 10^6 atoms is transferred from one ball to the other, the coulomb force between them is (atomic weight of copper is 63.5)

   (a) 2.0×10^{10} N (b) 2.0×10^4 N

   (c) 2.0×10^8 N (d) 2.0×10^6 N

4. Charge Q is divided into two parts which are then kept some distance apart. The force between them will be maximum if the two parts are

   (a) $\frac{Q}{2}$ and $\frac{Q}{2}$ (b) $\frac{Q}{4}$ and $\frac{3Q}{4}$ (c) $\frac{Q}{3}$ and $\frac{2Q}{3}$ (d) $\frac{Q}{5}$ and $\frac{4Q}{5}$

5. A particle A having a charge of 2.0×10^{-6} C and a mass of 100 g is fixed at the bottom of a smooth inclined plane of inclination 30°. Where should another particle B having same charge and mass, be placed on the inclined plane so that B may remain in equilibrium?

   (a) 8 cm from the bottom
   (b) 13 cm from the bottom
   (c) 21 cm from the bottom
   (d) 27 cm from the bottom

6. Two point charges $+q$ and $-q$ are placed a distance x apart. A third charge is so placed that all the three charges are in equilibrium. Then,

   (a) unknown charge is $-4q/9$
   (b) unknown charge is $-9q/4$
   (c) it should be at $(x/3)$ from smaller charge between them
   (d) None of the above

7. A charge Q is placed at each of the two opposite corners of a square. A charge q is placed at each of the other two corners. If the resultant force on each charge q is zero, then

   (a) $q = \sqrt{2}Q$ (b) $q = -\sqrt{2}Q$

   (c) $q = 2\sqrt{2}Q$ (d) $q = -2\sqrt{2}Q$

8. Two point charges q_1 and q_2 are placed at a distance of 50 m from each other in air, and interact with a certain force. The same charges are now put in oil whose relative permittivity is 5. If the interacting force between them is still the same, their separation now is

   (a) 16.6 m (b) 22.3 m

   (c) 28.4 m (d) 25.0 m

9. Two similar small spheres having $+q$ and $-q$ charge are kept at a certain distance. F force acts between the two. If in the middle of two spheres, another

similar small sphere having $+q$ charge is kept, then it will experience a force in magnitude and direction as

(a) zero
(b) $8F$ towards $+q$ charge
(c) $8F$ towards $-q$ charge
(d) $4F$ towards $+q$ charge

10. Point charges $+4q$, $-q$ and $+4q$ are kept on the x-axis at points $x = 0$, $x = a$ and $x = 2a$ respectively. Then,

(a) only $-q$ is in stable equilibrium
(b) none of the charges are in equilibrium
(c) all the charges are in unstable equilibrium
(d) all the charges are in stable equilibrium

11. Four identical charges, q each are placed at the corners of a square of side a. The charge Q that must be placed at the centre of the square such that the whole system of charges is in equilibrium is

(a) $-\frac{q}{4}$ (b) $\frac{q}{4}[2\sqrt{2}+1]$
(c) $-\frac{q}{4}[2\sqrt{2}+1]$ (d) $-\frac{q}{4}[3\sqrt{3}+1]$

12. Three charges $-q_1$, $+q_2$ and $-q_3$ are placed as shown in the figure below. The X-component of the force on $-q_1$ is proportional to

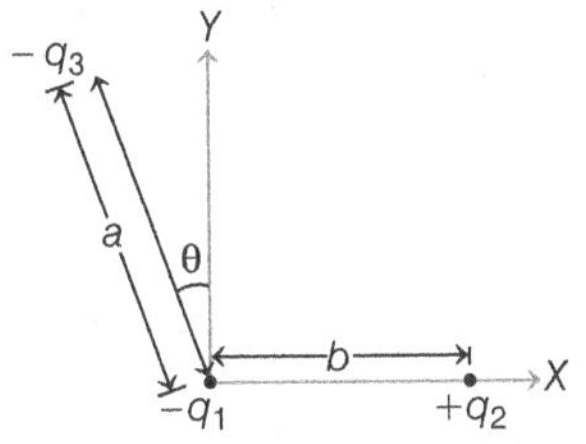

(a) $\frac{q_2}{b^2}-\frac{q_3}{a^2}\cos\theta$ (b) $\frac{q_2}{b^2}+\frac{q_3}{a^2}\sin\theta$
(c) $\frac{q_2}{b^2}+\frac{q_3}{a^2}\cos\theta$ (d) $\frac{q_2}{b^2}-\frac{q_3}{a^2}\sin\theta$

13. If two charges q_1 and q_2 are separated with distance d and placed in a medium of dielectric constant k. What will be the equivalent distance between charges in air for the same electrostatic force? **(2023 Main)**

(a) $d\sqrt{k}$ (b) $2d\sqrt{k}$ (c) $1.5d\sqrt{k}$ (d) $k\sqrt{d}$

14. Two identical positive charges Q each are fixed at a distance of $2a$ apart from each other. Another point charge q_0 with mass m is placed at midpoint between two fixed charges. For a small displacement along the line joining the fixed charges, the charge q_0 executes SHM. The time period of oscillation of charge q_0 will be **(2022 Main)**

(a) $\sqrt{\frac{4\pi^3\varepsilon_0 ma^3}{q_0Q}}$ (b) $\sqrt{\frac{q_0Q}{4\pi^3\varepsilon_0 ma^3}}$
(c) $\sqrt{\frac{2\pi^3\varepsilon_0 ma^3}{q_0Q}}$ (d) $\sqrt{\frac{8\pi^3\varepsilon_0 ma^3}{q_0Q}}$

15. Two charges, each equal to q, are kept at $x = -a$ and $x = a$ on the x-axis. A particle of mass m and charge $q_0 = \frac{q}{2}$ is placed at the origin. If charge q_0 is given a small displacement y ($y << a$) along the y-axis, the net force acting on the particle is proportional to **(2013 JEE)**

(a) y (b) $-y$ (c) $\frac{1}{y}$ (d) $-\frac{1}{y}$

16. Two equal negative charges $-q$ are fixed at points $(0, -a)$ and $(0, a)$ on y-axis. A positive charge Q is released from rest at the point $(2a, 0)$ on the x- axis. The charge Q will **(1984 JEE)**

(a) execute simple harmonic motion about the origin
(b) move to the origin and there it remains at rest
(c) move to infinity
(d) execute oscillatory but not simple harmonic motion

Subjective Questions

17. The mass of an electron is 9.11×10^{-31} kg, that of a proton is 1.67×10^{-27} kg. Find the ratio F_e/F_g of the electric force and the gravitational force exerted by the proton on the electron.

18. Find the dimensions and units of ε_0.

19. Three point charges q are placed at three vertices of an equilateral triangle of side a. Find magnitude of electric force on any charge due to the other two.

20. Three point charges each of value $+q$ are placed on three vertices of a square of side a metre. What is the magnitude of the force on a point charge of value $-q$ coulomb placed at the centre of the square?

21. Coulomb's law states that the electric force becomes weaker with increasing distance. Suppose that instead, the electric force between two charged particles were independent of distance. In this case, would a neutral insulator still be attracted towards the comb.

22. A metal sphere is suspended from a nylon thread. Initially, the metal sphere is uncharged. When a positively charged glass rod is brought close to the metal sphere, the sphere is drawn towards the rod. But if the sphere touches the rod, it suddenly flies away from the rod. Explain, why the sphere is first attracted then repelled?

23. Is there any lower limit to the electric force between two particles placed at a certain distance?

24. Does the force on a charge due to another charge depend on the charges present nearby?

25. The electric force on a charge q_1 due to q_2 is $(4\hat{i} - 3\hat{j})$ N. What is the force on q_2 due to q_1?

26. Point charges q_1 and q_2 lie on the x-axis at points $x = -a$ and $x = +a$ respectively.

(a) How must q_1 and q_2 be related for the net electrostatic force on point charge $+Q$, placed at $x = +a/2$, to be zero?

(b) With the same point charge $+Q$ now placed at $x = +3a/2$.

27. Two identical beads each have a mass m and charge q. When placed in a hemispherical bowl of radius R with frictionless, non-conducting walls, the beads move, and at equilibrium they are a distance R apart (figure). Determine the charge on each bead.

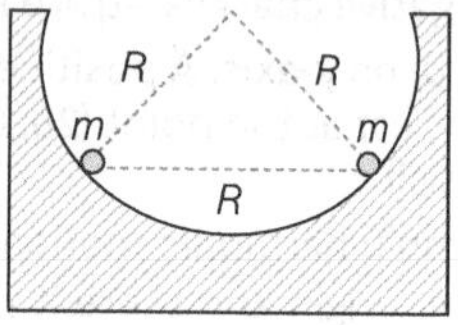

28. Three identical small balls, each of mass 0.1 g are suspended at one point on silk thread having a length of $l = 20$ cm. What charges should be imparted to the balls for each thread to form an angle of $\alpha = 30°$ with the vertical?

Numerical Value/Integer Type Questions

29. Three point charges $q, -2q$ and $2q$ are placed on x-axis at a distance $x = 0, x = 3/4R$ and $x = R$ respectively from origin. If $q = 2 \times 10^{-6}$ C and $R = 2$ cm, the magnitude of net force experienced by the charge $-2q$ is .... N. **(2023 Main)**

30. Two equal positive point charges are separated by a distance $2a$. The distance of a point from the centre of the line joining two charges on the equatorial line (perpendicular bisector) at which force experienced by a test charge q_0 becomes maximum is $\frac{a}{\sqrt{x}}$. The value of x is .... **(2023 Main)**

Answers

1. $180°, \frac{kQ^2}{4L^2}$ 2. (c) 3. (c) 4. (a) 5. (d) 6. (d) 7. (d) 8. (b) 9. (c)
10. (c) 11. (c) 12. (c) 13. (a) 14. (a) 15. (a) 16. (d) 17. 2.27×10^{39}
18. $[M^{-1}L^{-3}T^4A^2]$, $\frac{C^2}{N\text{-}m^2}$ 19. $\frac{\sqrt{3}}{4\pi\varepsilon_0}\left(\frac{q}{a}\right)^2$ 20. $\frac{1}{2\pi\varepsilon_0}\left(\frac{q}{a}\right)^2$ 21. No
22. Induction $\rightarrow$ Conduction $\rightarrow$ Repulsion 23. Yes 24. No 25. $(-4\hat{i} + 3\hat{j})$ N
26. (a) $q_1 = 9q_2$ (b) $q_1 = -25q_2$ 27. $\left(\frac{4\pi\varepsilon_0 mgR^2}{\sqrt{3}}\right)^{1/2}$ 28. 3.3×10^{-8} C 29. 5440 30. 2

See hints on pages 194-198.

MODULE 3

Electric Field

A charged particle cannot directly interact with another particle kept at a distance. A charge produces something called an electric field in the space around it and this electric field exerts a force on any other charge (except the source charge itself) placed in it.

Thus, the region surrounding a charge or distribution of charge in which its electrical effects can be observed is called the **electric field** of the charge or distribution of charge. Electric field at a point can be defined in terms of either a vector function **E** called **'electric field strength'** or a scalar function *V* called **'electric potential'**. The electric field can also be visualised graphically in terms of **'lines of force'**. Note that all these are functions of position **r** (*x*, *y*, *z*). The field propagates through space with the speed of light, *c*. Thus, if a charge is suddenly moved, the force it exerts on another charge a distance *r* away does not change until a time r/c later. In our forgoing discussion, we will see that electric field strength **E** and electric potential *V* are interrelated. It is similar to a case where the acceleration, velocity and displacement of a particle are related to each other.

Electric Field Strength (E)

Like its gravitational counterpart, the electric field strength (often called electric field) at a point in an electric field is defined as the electrostatic force $\mathbf{F}_e$ per unit positive charge. Thus, if the electrostatic force experienced by a small test charge q_0 is $\mathbf{F}_e$, then field strength at that point is defined as

$$\mathbf{E} = \lim_{q_0 \to 0} \frac{\mathbf{F}_e}{q_0}$$

The electric field is a vector quantity and its direction is the same as the direction of the force $\mathbf{F}_e$ on a positive test charge. The SI unit of electric field is N/C. Here, it should be noted that the test charge q_0 should be infinitesimally small so that it does not disturb other charges which produces **E**. With the concept of electric field, our description of electric interactions has two parts. First, a given charge distribution acts as a source of electric field. Second, the electric field exerts a force on any charge that is present in this field.

An Electric Field Leads to a Force

Suppose there is an electric field strength **E** at some point in an electric field, then the electrostatic force acting on a charge $+q$ is qE in the direction of **E**, while on the charge $-q$ it is qE in the opposite direction of **E**.

Example 1 An electric field of 10^5 N/C points due west at a certain spot. What are the magnitude and direction of the force that acts on a charge of $+2\ \mu C$ and $-5\ \mu C$ at this spot?

Sol. Force on $+2\ \mu C = qE = (2 \times 10^{-6})(10^5)$

$= 0.2$ N (due west) **Ans.**

Force on $-5\ \mu C = (5 \times 10^{-6})(10^5)$

$= 0.5$ N (due east) **Ans.**

Electric Field Due to a Point Charge

The electric field produced by a point charge *q* can be obtained in general terms from Coulomb's law. First note that the magnitude of the force exerted by the charge *q* on a test charge q_0 is

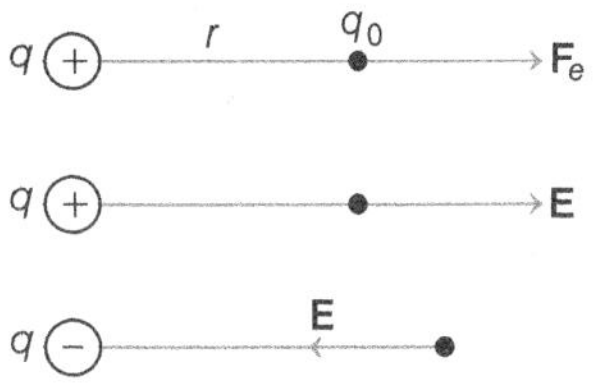

$$F_e = \frac{1}{4\pi\varepsilon_0} \cdot \frac{qq_0}{r^2}$$

then divide this value by q_0 to obtain the magnitude of the field.

$$E = \frac{1}{4\pi\varepsilon_0} \cdot \frac{q}{r^2}$$

E versus r graph is as shown in figure.

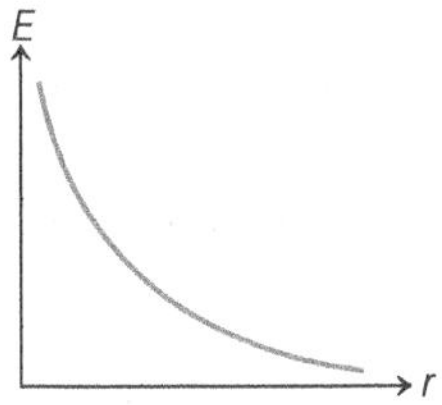

If *q* is positive, **E** is directed away from *q*. On the other hand, if *q* is negative, then **E** is directed towards *q*.

The electric field at a point is a vector quantity. Suppose $\mathbf{E}_1$ is the field at a point due to a charge q_1 and $\mathbf{E}_2$ is the field at the same point due to a charge q_2. The resultant field when both the charges are present is

$$\mathbf{E} = \mathbf{E}_1 + \mathbf{E}_2$$

If the given charge distribution is continuous, we can use the technique of integration to find the resultant electric field at some point.

Example 2 Two positive point charges $q_1 = 16\ \mu C$ and $q_2 = 4\ \mu C$, are separated in vacuum by a distance of 3.0 m. Find the point on the line between the charges where the net electric field is zero.

Sol. Between the charges the two field contributions have opposite directions, and the net electric field is zero at a point (say P) where the magnitudes of $\mathbf{E}_1$ and $\mathbf{E}_2$ are equal. However, since $q_2 < q_1$, point P must be closer to q_2, in order that the field of the smaller charge can balance the field of the larger charge.

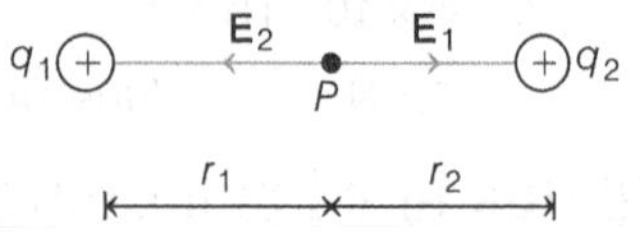

At P, $E_1 = E_2$

or $$\frac{1}{4\pi\varepsilon_0}\frac{q_1}{r_1^2} = \frac{1}{4\pi\varepsilon_0}\cdot\frac{q_2}{r_2^2}$$

$\therefore$ $$\frac{r_1}{r_2} = \sqrt{\frac{q_1}{q_2}} = \sqrt{\frac{16}{4}} = 2 \qquad \text{...(i)}$$

Also, $$r_1 + r_2 = 3.0\ \text{m} \qquad \text{...(ii)}$$

Solving these equations, we get

$$r_1 = 2\ \text{m} \quad \text{and} \quad r_2 = 1\ \text{m}$$

Thus, the point P is at a distance of 2 m from q_1 and 1 m from q_2. **Ans.**

Note *Suppose a charge q is placed at a point whose position vector is $\mathbf{r}_q$ and we want to find the electric field at a point P whose position vector is $\mathbf{r}_P$. Then, in vector form the electric field is given by*

$$\mathbf{E} = \frac{1}{4\pi\varepsilon_0}\cdot\frac{q}{|\mathbf{r}_P - \mathbf{r}_q|^3}(\mathbf{r}_P - \mathbf{r}_q)$$

Here, $\mathbf{r}_P = x_P\hat{\mathbf{i}} + y_P\hat{\mathbf{j}} + z_P\hat{\mathbf{k}}$

and $\mathbf{r}_q = x_q\hat{\mathbf{i}} + y_q\hat{\mathbf{j}} + z_q\hat{\mathbf{k}}$

In this equation, q is to be substituted with sign.

Example 3 A charge $q = 1\ \mu C$ is placed at point (1 m, 2 m, 4 m). Find the electric field at point P (0, –4 m, 3 m).

Sol. Here, $$\mathbf{r}_q = \hat{\mathbf{i}} + 2\hat{\mathbf{j}} + 4\hat{\mathbf{k}}$$

and $$\mathbf{r}_P = -4\hat{\mathbf{j}} + 3\hat{\mathbf{k}}$$

$\therefore$ $$\mathbf{r}_P - \mathbf{r}_q = -\hat{\mathbf{i}} - 6\hat{\mathbf{j}} - \hat{\mathbf{k}}$$

or $$|\mathbf{r}_P - \mathbf{r}_q| = \sqrt{(-1)^2 + (-6)^2 + (-1)^2} = \sqrt{38}\ \text{m}$$

Now, $$= \frac{1}{4\pi\varepsilon_0}\cdot\frac{q}{|\mathbf{r}_P - \mathbf{r}_q|^3}(\mathbf{r}_P - \mathbf{r}_q)$$

Substituting the values, we have

$$\mathbf{E} = \frac{(9.0\times10^9)(1.0\times10^{-6})}{(38)^{3/2}}(-\hat{\mathbf{i}} - 6\hat{\mathbf{j}} - \hat{\mathbf{k}})$$

$$= (-38.42\,\hat{\mathbf{i}} - 230.52\,\hat{\mathbf{j}} - 38.42\,\hat{\mathbf{k}})\ \text{N/C} \qquad \textbf{Ans.}$$

Electric Field of a Ring of Charge

A conducting ring of radius R has a total charge q uniformly distributed over its circumference. We are interested in finding the electric field at point P that lies on the axis of the ring at a distance x from its centre.

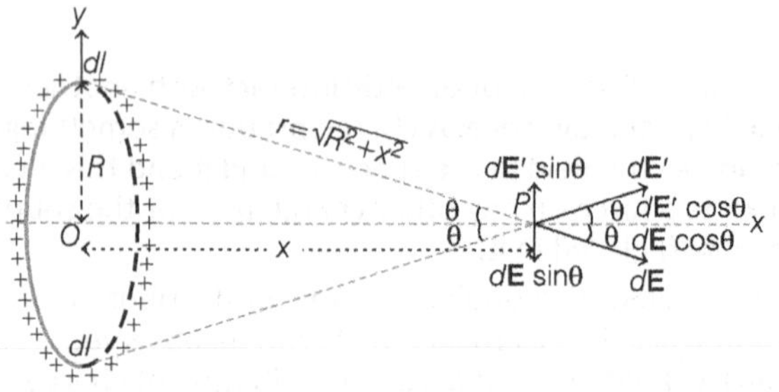

We divide the ring into infinitesimal segments of length dl. Each segment has a charge dq and acts as a point charge source of electric field.

Let $d\mathbf{E}$ be the electric field from one such segment; the net electric field at P is then the sum of all contributions $d\mathbf{E}$ from all the segments that make up the ring. If we consider two ring segments at the top and bottom of the ring, we see that the contributions $d\mathbf{E}$ to the field at P from these segments have the same x-component but opposite y-components. Hence, the total y-component of field due to this pair of segments is zero. When we add up the contributions from all such pairs of segments, the total field $\mathbf{E}$ will have only a component along the ring's symmetry axis (the x-axis) with no component perpendicular to that axis (i.e. no y or z-component). So, the field at P is described completely by its x-component E_x.

Calculation of E_x

$$dq = \left(\frac{q}{2\pi R}\right)\cdot dl$$

$$dE = \frac{1}{4\pi\varepsilon_0}\cdot\frac{dq}{r^2}$$

$\therefore$ $$dE_x = dE\cos\theta$$

$$= \left(\frac{1}{4\pi\varepsilon_0}\right)\left(\frac{dq}{x^2 + R^2}\right)\left(\frac{x}{\sqrt{x^2 + R^2}}\right)$$

$$= \frac{1}{4\pi\varepsilon_0}\cdot\frac{(dq)\,x}{(x^2 + R^2)^{3/2}}$$

$\therefore$ $$E_x = \int_{\text{whole ring}} dE_x = \frac{x}{4\pi\varepsilon_0(x^2 + R^2)^{3/2}}\int_{\text{whole ring}} dq$$

or $$\boxed{E_x = \left(\frac{1}{4\pi\varepsilon_0}\right)\frac{qx}{(x^2 + R^2)^{3/2}}}$$

From the above expression, we can see that

(i) $E_x = 0$ at $x = 0$, i.e. field is zero at the centre of the ring. We should expect this, charges on opposite sides of the ring would push in opposite directions on a test charge at the centre, and the forces would add to zero.

(ii) $E_x = \frac{1}{4\pi\varepsilon_0}\cdot\frac{q}{x^2}$ for $x \gg R$, i.e. when the point P is much farther from the ring, its field is the same as that of a point charge. To an observer far from the ring, the ring would appear like a point charge and the electric field reflects this.

(iii) E_x will be maximum where $\frac{dE_x}{dx} = 0$. Differentiating E_x w.r.t. x and putting it equal to zero we get $x = \frac{R}{\sqrt{2}}$

and E_{max} comes out to be, $\frac{2}{(\sqrt{3})^3}\left(\frac{1}{4\pi\varepsilon_0}\cdot\frac{q}{R^2}\right)$.

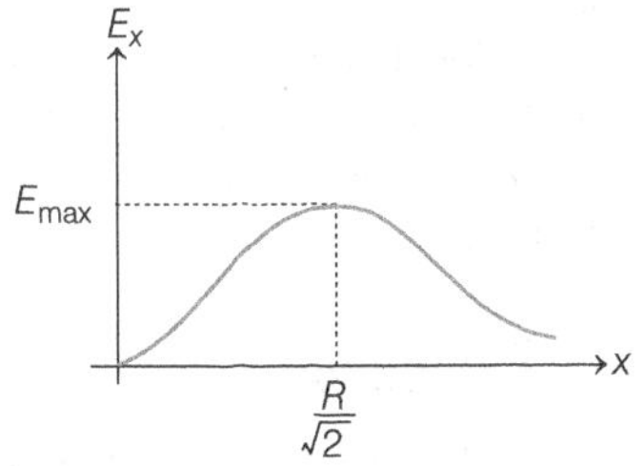

Electric Field of a Line Charge

Positive charge q is distributed uniformly along a line with length $2a$, lying along the y-axis between $y = -a$ and $y = +a$. We are here interested in finding the electric field at point P on the x-axis.

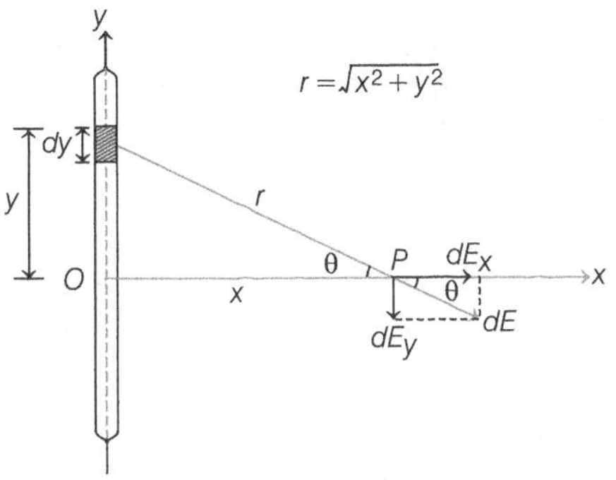

$$\lambda = \text{charge per unit length} = \frac{q}{2a}$$

$$dq = \lambda\, dy = \frac{q}{2a}\, dy$$

$$dE = \frac{1}{4\pi\varepsilon_0}\cdot\frac{dq}{r^2} = \frac{q}{4\pi\varepsilon_0}\,\frac{dy}{2a\,(x^2+y^2)}$$

$$dE_x = dE\cos\theta = \frac{q}{4\pi\varepsilon_0}\cdot\frac{x\,dy}{2a\,(x^2+y^2)^{3/2}}$$

$$dE_y = -dE\sin\theta = -\frac{q}{4\pi\varepsilon_0}\cdot\frac{y\,dy}{2a\,(x^2+y^2)^{3/2}}$$

$$\therefore\quad E_x = \frac{1}{4\pi\,\varepsilon_0}\cdot\frac{qx}{2a}\int_{-a}^{a}\frac{dy}{(x^2+y^2)^{3/2}} = \frac{q}{4\pi\varepsilon_0}\cdot\frac{1}{x\sqrt{x^2+a^2}}$$

and $$E_y = -\frac{1}{4\pi\varepsilon_0}\cdot\frac{q}{2a}\int_{-a}^{a}\frac{y\,dy}{(x^2+y^2)^{3/2}} = 0$$

Thus, electric field is along x-axis only and which has a magnitude,

$$\boxed{E_x = \frac{q}{4\pi\varepsilon_0 x\sqrt{x^2+a^2}}} \qquad \text{...(i)}$$

From the above expression, we can see that :

(i) if $x >> a$, $E_x = \frac{1}{4\pi\varepsilon_0}\cdot\frac{q}{x^2}$, i.e. if point P is very far from the line charge, the field at P is the same as that of a point charge.

(ii) if we make the line of charge longer and longer, adding charge in proportion to the total length so that λ, the charge per unit length remains constant. In this case, Eq. (i) can be written as

$$E_x = \frac{1}{2\pi\varepsilon_0}\cdot\left(\frac{q}{2a}\right)\cdot\frac{1}{x\sqrt{x^2/a^2+1}}$$

$$= \frac{\lambda}{2\pi\varepsilon_0\, x\sqrt{x^2/a^2+1}}$$

Now, $x^2/a^2 \to 0$ as $a >> x$, $E_x = \frac{\lambda}{2\pi\varepsilon_0 x}$

Thus, the magnitude of electric field depends only on the distance of point P from the line of charge, so we can say that at any point P at a perpendicular distance r from the line in any direction, the field has magnitude

$$\boxed{E = \frac{\lambda}{2\pi\varepsilon_0 r}} \qquad \text{(due to infinite line of charge)}$$

or $$E \propto \frac{1}{r}$$

Thus, E-r graph is as shown in figure.

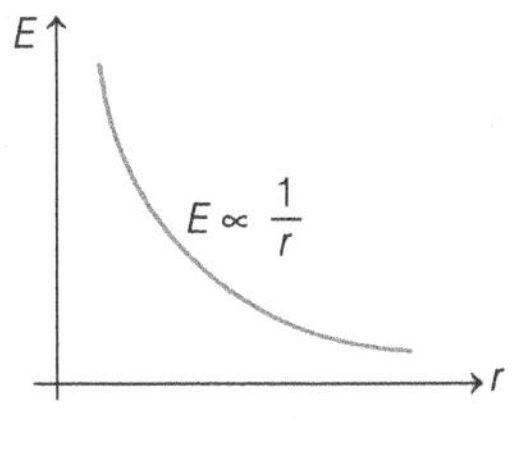

The direction of **E** is radially outward from the line.

Electric Field Lines

As we have seen, electric charges create an electric field in the space surrounding them. It is useful to have a kind of "map" that gives the direction and indicates the strength of the field at various places. **Field lines,** a concept introduced by Michael Faraday, provide us with an easy way to visualize the electric field.

"An **electric field line** is an imaginary line or curve drawn through a region of space so that its tangent at any point is in the direction of the electric field vector at that point. The relative closeness of the lines at some place give an idea about the intensity of electric field at that point."

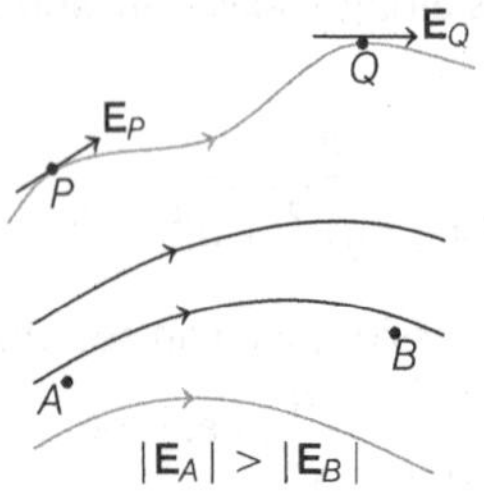

The electric field lines have the following properties :

1. The tangent to a line at any point gives the direction of **E** at that point. This is also the path on which a positive test charge at rest will tend to move if free to do so.
2. Electric field lines always begin on a positive charge and end on a negative charge and do not start or stop in mid-space.
3. The number of lines leaving a positive charge or entering a negative charge is proportional to the magnitude of the charge. This means, for example that if 100 lines are drawn leaving a $+4\ \mu C$ charge then 75 lines would have to end on a $-3\ \mu C$ charge.
4. Two lines can never intersect. If it happens then two tangents can be drawn at their point of intersection, i.e. intensity at that point will have two directions which is absurd.
5. In a uniform field, the field lines are straight parallel and uniformly spaced.

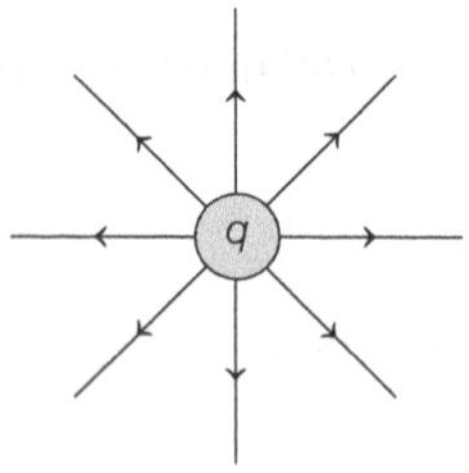

(a)

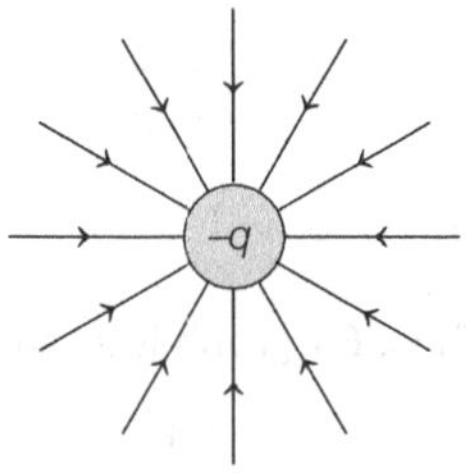

(b)

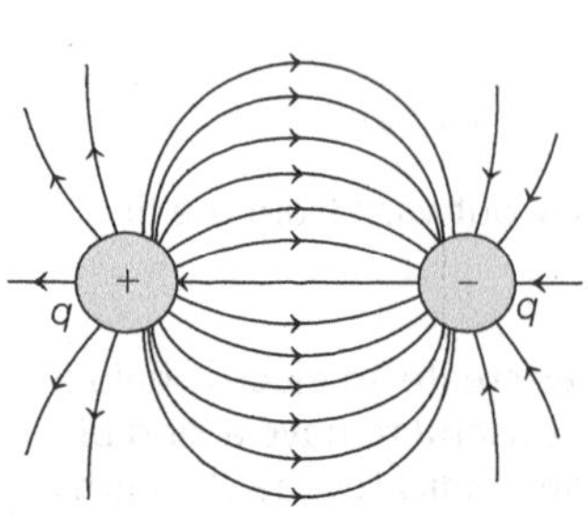

(c)

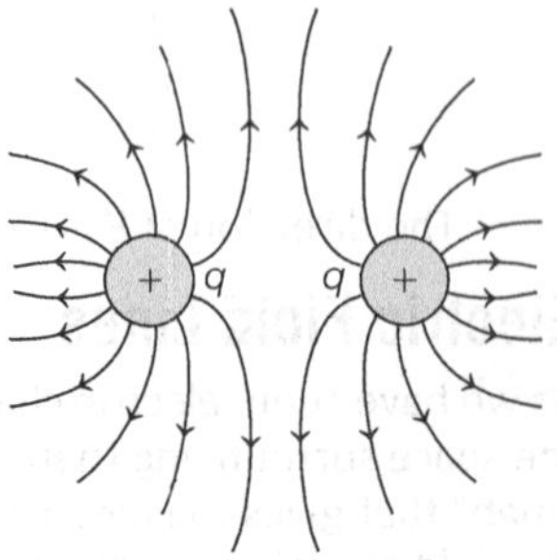

(d)

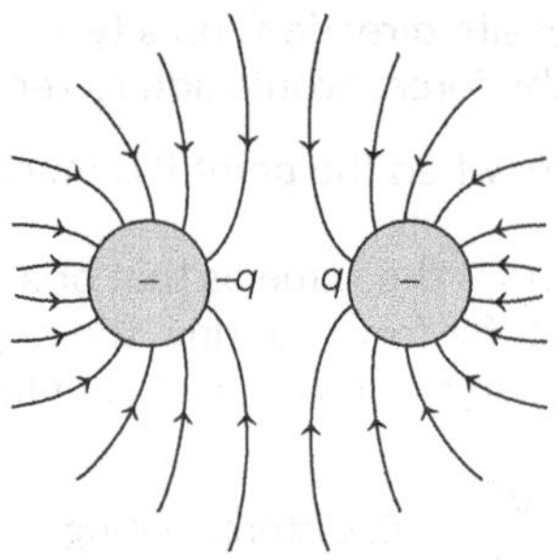

(e)

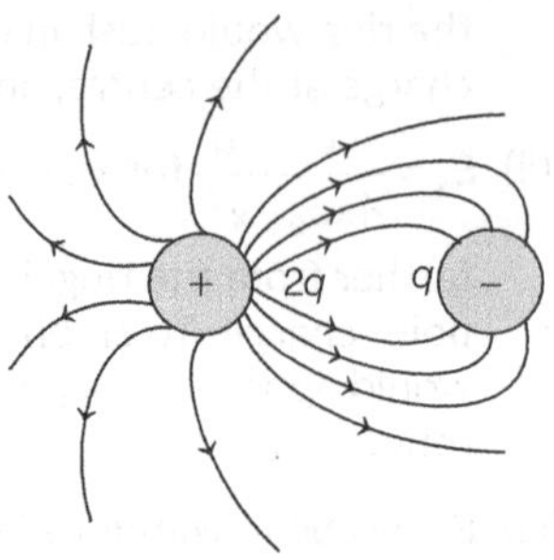

(f)

6. The electric field lines can never form closed loops as a line can never start and end on the same charge.
7. Electric field lines always flow from higher potential to lower potential.
8. In a region where there is no electric field, lines are absent. This is why inside a conductor (where electric field is zero) there, cannot be any electric field line.
9. Electric lines of force ends or starts normally from the surface of a conductor.

INTRODUCTORY Exercise

1. Electric field lines are shown in figure. State whether the electric potential is greater at *A* or *B*.

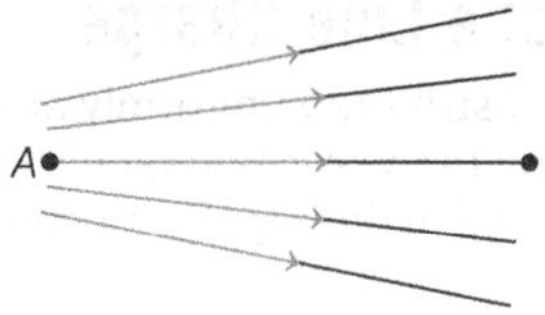

2. Figure shows some of the electric field lines due to three point charges q_1, q_2 and q_3 of equal magnitude. What are the signs of each of the three charges?

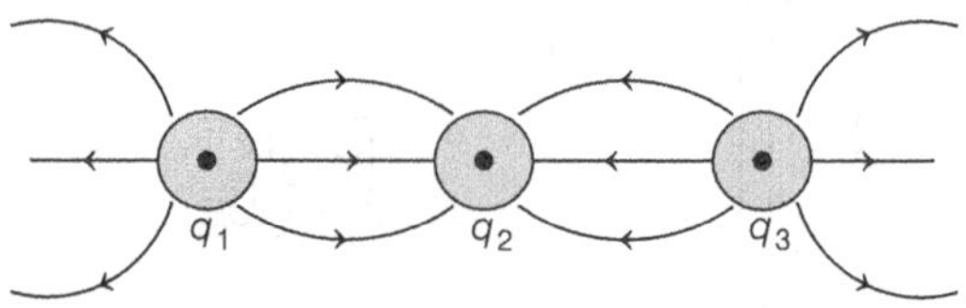

3. Four particles each having a charge q, are placed on the four vertices of a regular pentagon. The distance of each corner from the centre is a. Find the electric field at the centre of the pentagon.
4. A charge $q = -2.0\ \mu C$ is placed at origin. Find the electric field at (3 m, 4 m, 0).

Answers

1. At *A*
2. q_1 and q_3 are positive and q_2 is negative
3. $\frac{1}{4\pi\varepsilon_0} \cdot \frac{q}{a^2}$
4. $-(4.32\ \hat{i} + 5.76\ \hat{j}) \times 10^2$ N/C

See hints on page 198.

Important POINTS

Millikan's oil drop experiment

- This experiment was conducted by Millikan and Fletcher in 1909.
  It determined the value of the fundamental **elementary charge.**
- A find mist of **oil drops** is sprayed into a chamber
  Oil is used instead of water because it does not evaporate quickly.
  This means the **mass** of the drops will remain **constant**.
- As the drops pass out of the spray nozzle they are charged by friction (alternatively, they can also be **ionised** by X-rays).
  Some drops **lose** electrons and become positively charged.
  Some drops **gain** electrons and become negatively charged.
- The drops pass into a region between **two metal plates** and are viewed using a **microscope.**

Equipment set up for Millikan's oil drop experiment

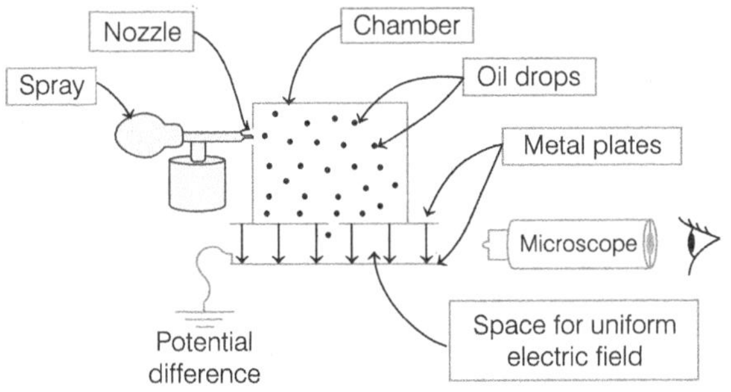

- In Millikan's oil drop experiment, oil is sprayed into a chamber before passing between metal plates where the electric and gravitational forces are compared.

Electric *versus* gravitational force

No Electric Field

- The oil drops fall **under gravity** between the metal plates.
- They reach a **terminal velocity** when the air resistance and gravitational force acting on the drop are **equal.**

With Electric Field

- A **potential difference** is applied between the metal plates which creates an electric field.
- The charged oil drops begin to **rise** when the electric field is strong enough.
- This means the upward electrical force is greater than the gravitational force.
- The equation for electric force is :

$$F = Eq$$

where,

E = electric field strength (NC^{-1})

F = electrostatic force on the charge (N)

q = charge (C)

- The distance the drops rise depends upon their **mass.**
  With the correct potential difference applied, the electric and gravitational forces can become **equal** and **opposite.**
- The equation for gravitational force, which comes from Newton's second law, is :

$$w = mg$$

where,

w = weight of drop (N)

m = mass of drop (kg)

g = gravitational field strength ($N\ kg^{-1}$)

- By equating the electric and gravitational forces of the drops, the value of fundamental charge was determined to be 1.60×10^{-19} C.
- The magnitude of the charge on any object is found to be a **multiple** of 1.60×10^{-19} C.
- Therefore, Millikan's experiment provides evidence for the **quantisation of charge**.

Typed Problems

Type 1. *Based on motion of a charged particle in uniform electric field*

Concept

(i) In uniform electric field, force on the charged particle is

$$F = qE$$

or qE force acts in the direction of electric field if q is positive and in the opposite direction of electric field if q is negative.

(ii) Acceleration of the particle is therefore,

$$a = \frac{F}{m} = \frac{qE}{m}$$

This acceleration is constant. So, path is therefore either a straight line or parabola. If initial velocity is zero or parallel to acceleration or antiparallel to acceleration, then path is straight line. Otherwise in all other cases, path is a parabola.

Ex. 1 An electron with a speed of 5×10^6 m/s enters a uniform electric field of magnitude 10^3 N/C, travelling along the field lines in the direction that retards its motion.

(a) How far will the electron travel in the field before stopping momentarily?

(b) How much time will have elapsed?

(c) If the region with the electric field is only 8.00 mm long (too short from the electron to stop with in it), what fraction of the electron's initial kinetic energy will be lost in that region?

Sol. (a) $s = \frac{u^2}{2a} = \frac{u^2}{2(qE/m)} = \frac{mu^2}{2qE}$

$$= \frac{(9.1 \times 10^{-31})(5 \times 10^6)^2}{2 \times 1.6 \times 10^{-19} \times 10^3}$$

$$= 7.1 \times 10^{-2} \text{ m}$$

$$= 7.1 \text{ cm}$$ **Ans.**

(b) $t = \frac{u}{a} = \frac{u}{qE/m} = \frac{mu}{qE}$

$$= \frac{(9.1 \times 10^{-31})(5 \times 10^6)}{(1.6 \times 10^{-19})(10^3)}$$

$$= 2.84 \times 10^{-8} \text{ s}$$ **Ans.**

(c) Loss of energy (in fraction)

$$= \frac{\frac{1}{2}mu^2 - \frac{1}{2}mv^2}{\frac{1}{2}mu^2} = 1 - \frac{v^2}{u^2}$$

$$= 1 - \frac{u^2 - 2as}{u^2} = \frac{2as}{u^2} = \frac{2qEs}{mu^2}$$

$$= \frac{2 \times 1.6 \times 10^{-19} \times 10^3 \times 8 \times 10^{-3}}{9.1 \times 10^{-31} \times (5 \times 10^6)^2}$$

$$= 0.11$$ **Ans.**

Ex. 2 A charged particle of mass $m = 1$ kg and charge $q = 2\ \mu$C is thrown from a horizontal ground at an angle $\theta = 45°$ with speed 20 m/s. In space a horizontal uniform electric field $E = 2 \times 10^7$ V/m exist. Find the range on horizontal ground of the projectile thrown.

Sol. The path of the particle will be a parabola, but along x-axis also motion of the particle will be accelerated. Time of flight of the projectile is

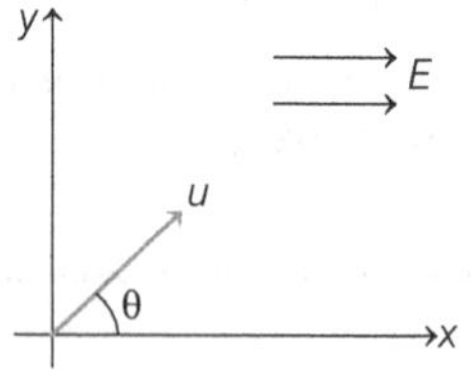

$$T = \frac{2u_y}{a_y} = \frac{2u_y}{g}$$

$$= \frac{2 \times 20 \sin 45°}{10} = 2\sqrt{2} \text{ s}$$

Horizontal range of the particle will be

$$R = u_x T + \frac{1}{2} a_x T^2$$

Here, $a_x = \frac{qE}{m}$

$$= \frac{(2 \times 10^{-6})(2 \times 10^7)}{1} = 40 \text{ m/s}^2$$

$\therefore$ $R = (20 \cos 45°)(2\sqrt{2}) + \frac{1}{2}(40)(2\sqrt{2})^2$

$$= 40 + 160 = 200 \text{ m}$$ **Ans.**

Type 2. *To find electric field by integration.*

Ex. 3 Find the electric field caused by a disc of radius a with a uniform surface charge density σ (charge per unit area), at a point along the axis of the disc a distance x from its centre.

Sol. We can assume this charge distribution as a collection of concentric rings of charge.

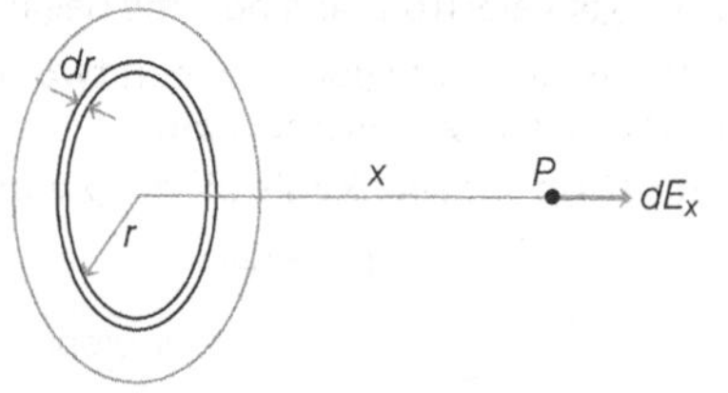

$$dA = (2\pi r)\, dr$$

$$dq = \sigma\, dA = (2\pi\sigma r)\, dr$$

$$dE_x = \frac{1}{4\pi\varepsilon_0} \cdot \frac{(dq)x}{(x^2 + r^2)^{3/2}}$$

$$= \left(\frac{1}{4\pi\varepsilon_0}\right) \frac{(2\pi\sigma r\, dr)\, x}{(x^2 + r^2)^{3/2}}$$

$\therefore$ $E_x = \int_0^a dE_x = \int_0^a \frac{(2\pi\sigma r\, dr)\, x}{4\pi\varepsilon_0 (x^2 + r^2)^{3/2}}$

$$= \frac{\sigma x}{2\varepsilon_0} \int_0^a \frac{r\, dr}{(x^2 + r^2)^{3/2}}$$

or $E_x = \frac{\sigma}{2\varepsilon_0}\left[1 - \frac{1}{\sqrt{a^2/x^2 + 1}}\right]$

For $a >> x$, the term $\frac{1}{\sqrt{a^2/x^2 + 1}}$ becomes negligibly small, and we get $E = \frac{\sigma}{2\varepsilon_0}$.

Thus, we can say that electric field produced by an infinite plane sheet of charge is independent of the distance from the sheet. Thus, the field is uniform, its direction is everywhere perpendicular to the sheet.

MODULE 3 Exercise

True/False

1. The electric field of a point charge is uniform. Is it true or false?
2. A charged particle always move in the direction of electric field. Is this statement true or false?
3. The trajectory of a charged particle is the same as a field line. Is this statement true or false?

Single Correct Option

4. A neutral pendulum oscillates in a uniform electric field as shown in figure. If a positive charge is given to the pendulum, then its time period

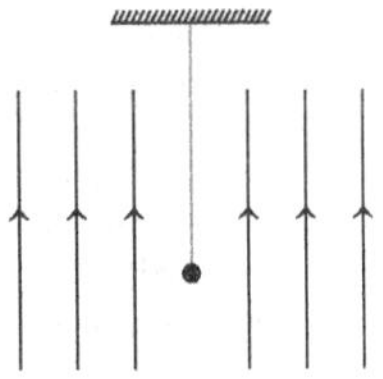

(a) will increase
(b) will decrease
(c) will remain constant
(d) will first increase then decrease

5. Identify the correct statement about the charges q_1 and q_2,

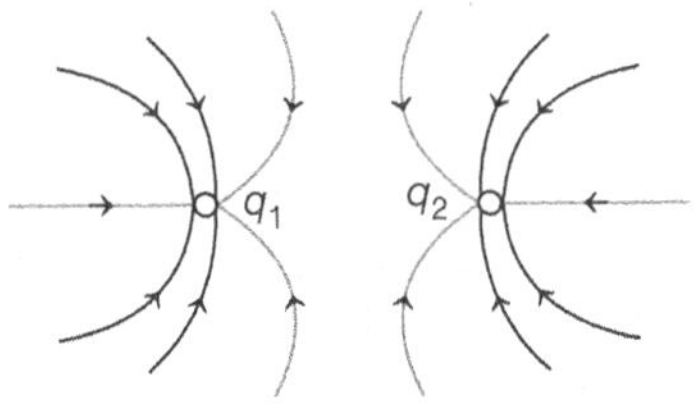

(a) q_1 and q_2 both are positive
(b) q_1 and q_2 both are negative
(c) q_1 is positive q_2 is negative
(d) q_2 is positive and q_1 is negative

6. A proton is released from rest, 10 cm from a charged sheet carrying charged density of $-2.21\times10^{-9}\ C/m^2$. It will strike the sheet after the time (approximately)

(a) $4\ \mu s$ (b) $2\ \mu s$
(c) $2\sqrt{2}\ \mu s$ (d) $4\sqrt{2}\ \mu s$

7. A pendulum bob of mass m carrying a charge q is at rest with its string making an angle θ with the vertical in a uniform horizontal electric field E. The tension in the string is

(a) $\dfrac{mg}{\sin\theta}$ (b) mg (c) $\dfrac{qE}{\sin\theta}$ (d) $\dfrac{qE}{\cos\theta}$

8. A wire is bent in the form of a regular hexagon and a total charge q is distributed uniformly on it. The electric field at the centre will be

(a) $\dfrac{q}{4\pi\varepsilon_0 a^2}$ (b) zero
(c) $\dfrac{3q}{4\pi\varepsilon_0 a^2}$ (d) $\dfrac{4q}{4\pi\varepsilon_0 a^2}$

9. A charged particle of mass m and charge q is released from rest from the position $(x_0, 0)$ in a uniform electric field $E_0\,\hat{j}$. The angular momentum of the particle about origin

(a) is zero (b) is constant
(c) increases with time (d) decreases with time

10. A small ball of mass m and charge $+q$ tied with a string of length l, rotating in a vertical circle under gravity and a uniform horizontal electric field E as shown. The tension in the string will be minimum for

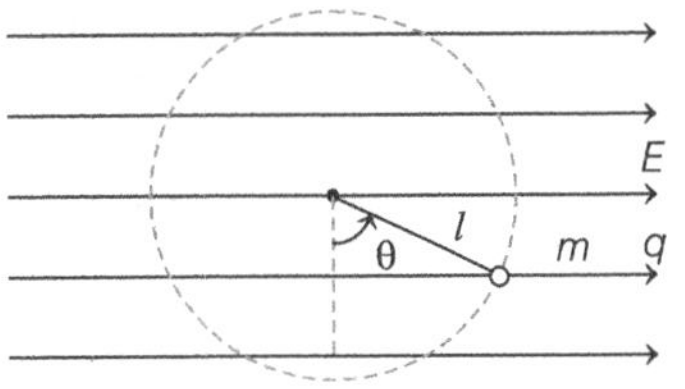

(a) $\theta=\tan^{-1}\left(\dfrac{qE}{mg}\right)$ (b) $\theta=\pi$
(c) $\theta=0°$ (d) $\theta=\pi+\tan^{-1}\left(\dfrac{qE}{mg}\right)$

11. A particle of charge $-q$ and mass m moves in a circle of radius r in gravity free space around an infinitely long line charge of linear charge density $+\lambda$. Then, time period will be

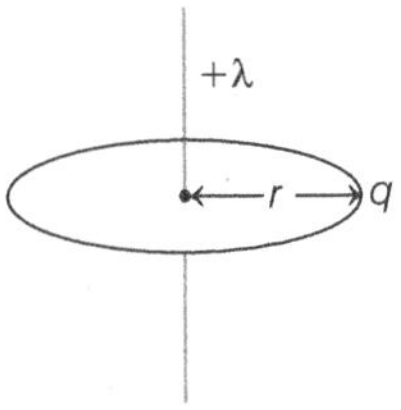

(a) $T=2\pi r\sqrt{\dfrac{m}{2k\lambda q}}$
(b) $T^2=\dfrac{4\pi^2 m}{2k\lambda q}r^3$
(c) $T=\dfrac{1}{2\pi r}\sqrt{\dfrac{2k\lambda q}{m}}$
(d) $T=\dfrac{1}{2\pi r}\sqrt{\dfrac{m}{2k\lambda q}}$

where, $k=\dfrac{1}{4\pi\varepsilon_0}$

12. Charges $2q$ and $-q$ are placed at $(a, 0)$ and $(-a, 0)$ as shown in the figure. The coordinates of the point at which electric field intensity is zero will be $(x, 0)$, where

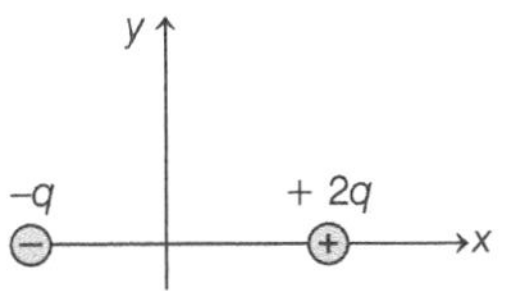

(a) $-a<x<a$ (b) $x<-a$
(c) $x>-a$ (d) $0<x<a$

13. Charge Q is given a displacement $\mathbf{r} = a\hat{i} + b\hat{j}$ in an electric field $\mathbf{E} = E_1\hat{i} + E_2\hat{j}$. The work done by electrostatic force is

(a) $Q(E_1a + E_2b)$ (b) $Q\sqrt{(E_1a)^2 + (E_2b)^2}$
(c) $Q(E_1 + E_2)\sqrt{a^2 + b^2}$ (d) $Q\sqrt{E_1^2 + E_2^2}\sqrt{a^2 + b^2}$

14. A particle having charge q and mass m is projected with velocity $\mathbf{v} = 2\hat{i} - 3\hat{j}$ in a uniform electric field $\mathbf{E} = E_0\hat{j}$. Change in momentum $|\Delta\mathbf{p}|$ during any time interval t is given by

(a) $\frac{qE_0t}{2m}$ (b) qE_0t (c) $\frac{qE_0t}{m}$ (d) zero

15. Figures below show regular hexagons, with charges at the vertices. In which of the following cases the electric field at the centre is not zero?

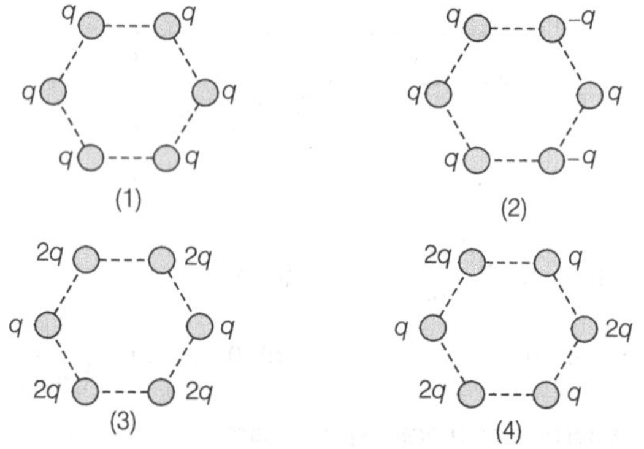

(a) 1 (b) 2 (c) 3 (d) 4

16. The figure shows four situations in which charged particles are at equal distances from the origin. If E_1, E_2, E_3 and E_4 be the magnitude of the net electric fields at the origin in four situations (i), (ii), (iii) and (iv) respectively, then

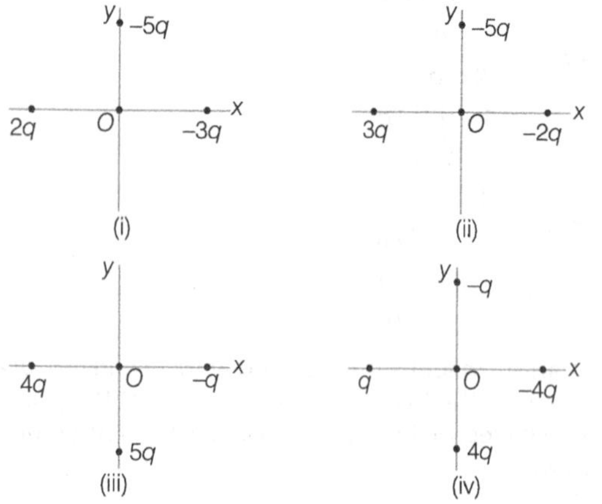

(a) $E_1 = E_2 = E_3 = E_4$ (b) $E_1 = E_2 > E_3 > E_4$
(c) $E_1 < E_2 < E_3 = E_4$ (d) $E_1 > E_2 = E_3 < E_4$

17. A point charge is located at origin. At point (a, a), electric field is $\mathbf{E}_1$. At point $(-a, a)$ electric field is $\mathbf{E}_2$ and at point $(-a, -a)$ electric field is $\mathbf{E}_3$. Choose the correct option.

(a) $\mathbf{E}_1 \cdot \mathbf{E}_2 = 0$ (b) $|\mathbf{E}_1 \times \mathbf{E}_3| = 0$
(c) Both are correct (d) Both are wrong

18. Two point charges $q_1 = 2\ \mu C$ and $q_2 = 1\ \mu C$ are placed at distances $b = 1$ cm and $a = 2$ cm from the origin of the y and x-axes as shown in figure. The electric field vector at point $P(a, b)$ will subtend an angle θ with the x-axis given by

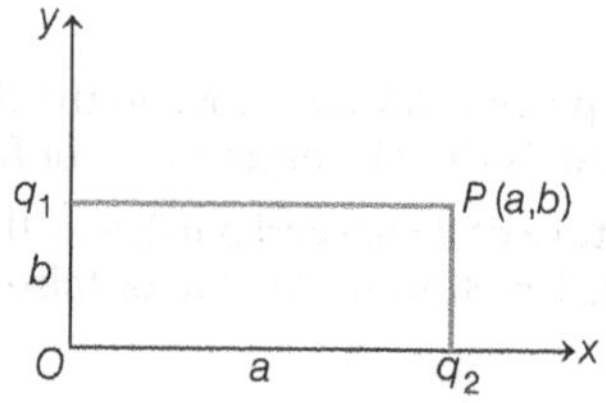

(a) $\tan\theta = 1$ (b) $\tan\theta = 2$
(c) $\tan\theta = 3$ (d) $\tan\theta = 4$

19. Six charges, three positive and three negative of equal magnitude are to be placed at the vertices of a regular hexagon such that the electric field at O is double the electric field when only one positive charge of same magnitude is placed at R.

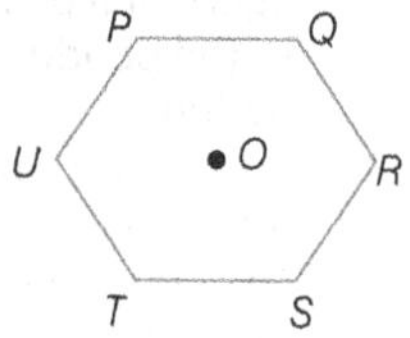

Which of the following arrangements of charge is possible for, P, Q, R, S, T and U respectively? **(2004 JEE)**

(a) +, +, +, −, −, − (b) +, −, +, −, +, −
(c) −, +, +, +, −, − (d) −, +, +, −, +, −

20. Let σ be the uniform surface charge density of two infinite thin plane sheets shown in figure. Then the electric fields in three different region E_I, E_{II} and E_{III} are **(2023 Main)**

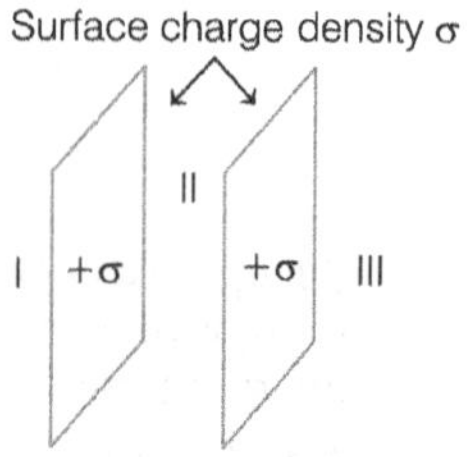

(a) $\mathbf{E}_I = -\frac{\sigma}{\varepsilon_0}\hat{\mathbf{n}}, \mathbf{E}_{II} = 0, \mathbf{E}_{III} = \frac{\sigma}{\varepsilon_0}\hat{\mathbf{n}}$

(b) $\mathbf{E}_I = \frac{\sigma}{2\varepsilon_0}\hat{\mathbf{n}}, \mathbf{E}_{II} = 0, \mathbf{E}_{III} = \frac{\sigma}{2\varepsilon_0}\hat{\mathbf{n}}$

(c) $\mathbf{E}_I = 0, \mathbf{E}_{II} = \frac{\sigma}{\varepsilon_0}\hat{\mathbf{n}}, \mathbf{E}_{III} = 0$

(d) $\mathbf{E}_I = \frac{2\sigma}{\varepsilon_0}\hat{\mathbf{n}}, \mathbf{E}_{II} = 0, \mathbf{E}_{III} = \frac{2\sigma}{\varepsilon_0}\hat{\mathbf{n}}$

21. Three infinitely long charge sheets are placed as shown in figure. The electric field at point P is **(2005 JEE)**

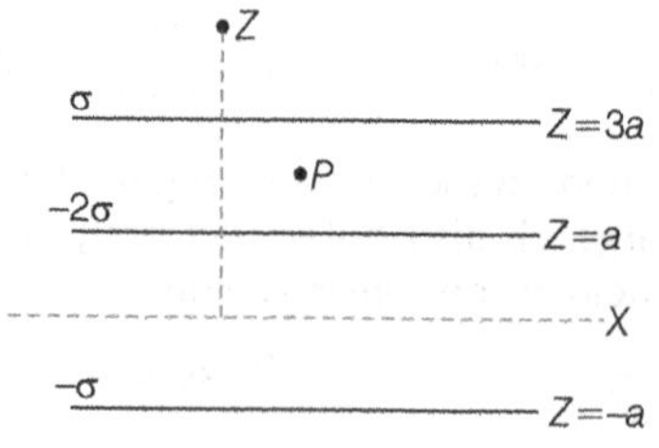

(a) $\frac{2\sigma}{\varepsilon_0}\hat{\mathbf{k}}$ (b) $\frac{4\sigma}{\varepsilon_0}\hat{\mathbf{k}}$ (c) $-\frac{2\sigma}{\varepsilon_0}\hat{\mathbf{k}}$ (d) $-\frac{4\sigma}{\varepsilon_0}\hat{\mathbf{k}}$

Match the Columns

22. Five identical charges are kept at five vertices of a regular hexagon. Match the following two columns at centre of the hexagon. If in the given situation electric field at centre is E. Then,

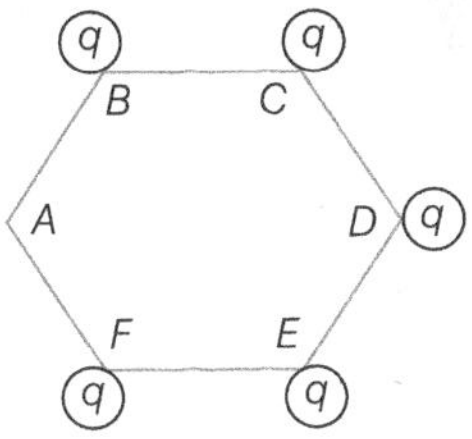

Column I		Column II
(a) If charge at B is removed, then electric field will become	(p)	$2E$
(b) If charge at C is removed, then electric field will become	(q)	E
(c) If charge at D is removed then electric field will become	(r)	zero
(d) If charges at B and C both are removed, then electric field will become	(s)	$\sqrt{3}\,E$

Note *Only magnitudes of electric field are given.*

Subjective Questions

23. A circular wire loop of radius R carries a total charge q distributed uniformly over its length. A small length x ($<< R$) of the wire is cut off. Find the electric field at the centre due to the remaining wire.

24. Three point charges q, $-2q$ and q are located along the x-axis as shown in figure. Show that the electric field at P ($y >> a$) along the y-axis is,

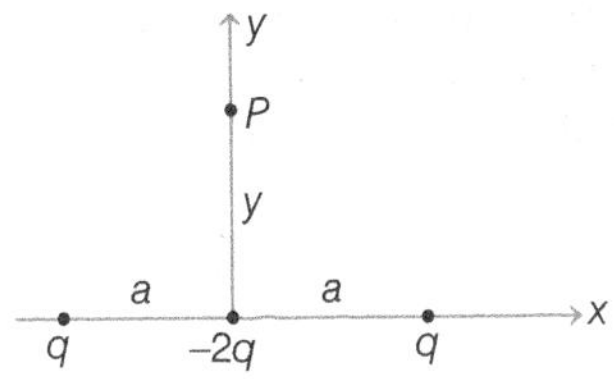

$$\mathbf{E} = -\frac{1}{4\pi\varepsilon_0}\frac{3qa^2}{y^4}\hat{\mathbf{j}}$$

Note *This charge distribution which is essentially that of two electric dipoles is called an electric quadrupole. Note that* **E** *varies as r^{-4} for a quadrupole compared with variations of r^{-3} for the dipole and r^{-2} for a monopole (a single charge).*

25. Find the electric field at the centre of a uniformly charged semicircular ring of radius R. Linear charge density is λ.

26. Find the electric field at a point P on the perpendicular bisector of a uniformly charged rod. The length of the rod is L, the charge on it is Q and the distance of P from the centre of the rod is a.

27. Find the direction of electric field at point P for the charge distribution as shown in figure.

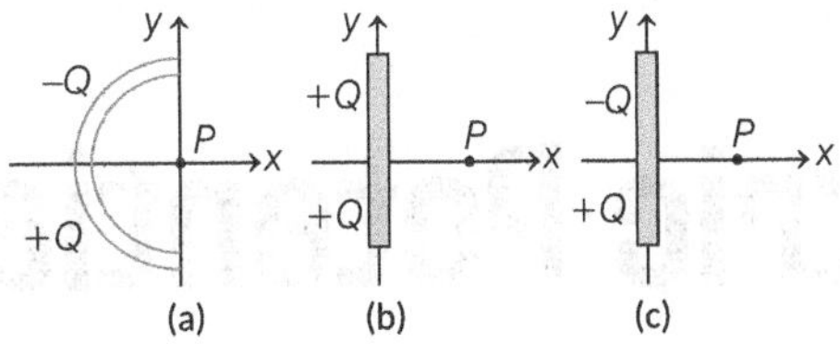

28. A clock face has charges $-q, -2q, -3q, \ldots -12q$ fixed at the position of the corresponding numerals on the dial. The clock hands do not disturb the net field due to point charges. At what time does the hour hand point in the direction of the electric field at the centre of the dial.

29. Protons are projected with an initial speed $v_i = 9.55 \times 10^3$ m/s into a region where a uniform electric field $\mathbf{E} = (-720\,\hat{\mathbf{j}})$ N/C is present, as shown in figure.

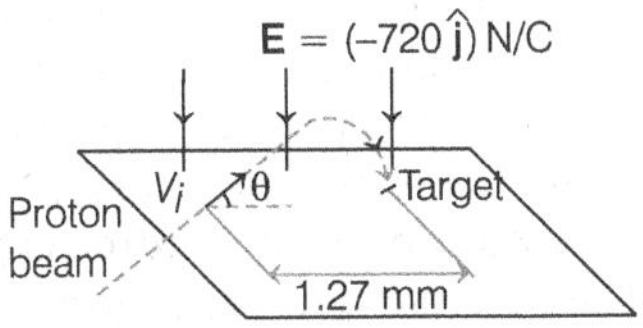

The protons are to hit a target that lies at a horizontal distance of 1.27 mm from the point where the protons are launched. Find

(a) the two projection angles θ that result in a hit and

(b) the total time of flight for each trajectory.

30. At some instant the velocity components of an electron moving between two charged parallel plates are $v_x = 1.5 \times 10^5$ m/s and $v_y = 3.0 \times 10^6$ m/s. Suppose that the electric field between the plates is given by $\mathbf{E} = (120\ \text{N/C})\,\hat{\mathbf{j}}$.

(a) What is the acceleration of the electron?

(b) What will be the velocity of the electron after its x-coordinate has changed by 2.0 cm?

Answers

1. False	**2.** False	**3.** False	**4.** (a)	**5.** (b)
6. (a)	**7.** (c)	**8.** (b)	**9.** (c)	**10.** (d)
11. (a)	**12.** (b)	**13.** (a)	**14.** (b)	**15.** (b)
16. (a)	**17.** (c)	**18.** (b)	**19.** (d)	**20.** (a)

21. (c) **22.** (a → s; b → q; c → r; d → p) **23.** $\dfrac{qx}{8\pi^2\varepsilon_0 R^3}$

25. $\dfrac{\lambda}{2\pi\varepsilon_0 R}$ **26.** $\dfrac{Q}{2\pi\varepsilon_0 a\sqrt{L^2+4a^2}}$ **27.** (a) Along positive y-axis

(b) Along positive x-axis (c) Along positive y-axis

28. 9 : 30

29. (a) 37° and 53° (b) 1.66×10^{-7}s, 2.2×10^{-7}s

30. (a) $(-2.1 \times 10^{13}\,\hat{\mathbf{j}})\,\text{m/s}^2$ (b) $(1.5\,\hat{\mathbf{i}} + 2.0\,\hat{\mathbf{j}}) \times 10^5$ m/s

See hints on pages 198-201.

MODULE 4

Electric Potential Energy

Like the gravitational force, the electric force is conservative, so there is a potential energy function U associated with it.

Since the electrostatic force **F** is conservative, the work done by **F** can always be expressed in terms of a **potential energy U**. When the particle moves from a point where the potential energy is U_a to a point where it is U_b, the change in potential energy is, $\Delta U = U_b - U_a$. This is related by the work $W_{a \to b}$ as

$$W_{a\to b} = U_a - U_b = -(U_b - U_a) = -\Delta U \qquad ...(i)$$

Here, $W_{a\to b}$ is the work done in displacing the particle from a to b by the **conservative force** (here electrostatic) not by us. Moreover we can see from Eq. (i) that if $W_{a\to b}$ is positive, ΔU is negative and the potential energy decreases. So, **whenever the work done by a conservative force is positive, the potential energy of the system decreases and *vice-versa*.** That's what happens when a particle is thrown upwards, the work done by gravity is negative, and the potential energy increases.

Example 1 A uniform electric field E_0 is directed along positive y-direction. Find the change in electric potential energy of a positive test charge q_0 when it is displaced in this field from $y_i = a$ to $y_f = 2a$ along the y-axis.

Sol. Electrostatic force on the test charge,

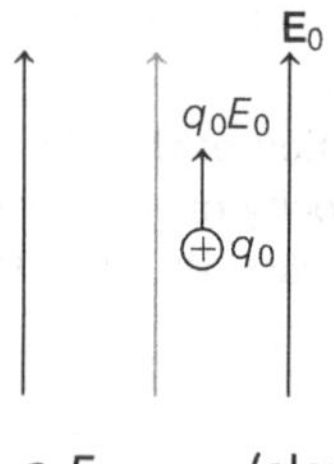

$$F_e = q_0E_0 \qquad \text{(along positive } y\text{-direction)}$$

$$\therefore \quad W_{i-f} = -\Delta U$$

or $$\Delta U = -W_{i-f} = -[q_0E_0\,(2a - a)]$$

$$= -q_0E_0a \qquad \textbf{Ans.}$$

Note *Here, work done by electrostatic force is positive. Hence, the potential energy is decreasing.*

Electric Potential Energy of Two Charges

The idea of electric potential energy is not restricted to the special case of a uniform electric field as in example 1. Let us now calculate the work done on a test charge q_0 moving in a non-uniform electric field caused by a single, stationary point charge q.

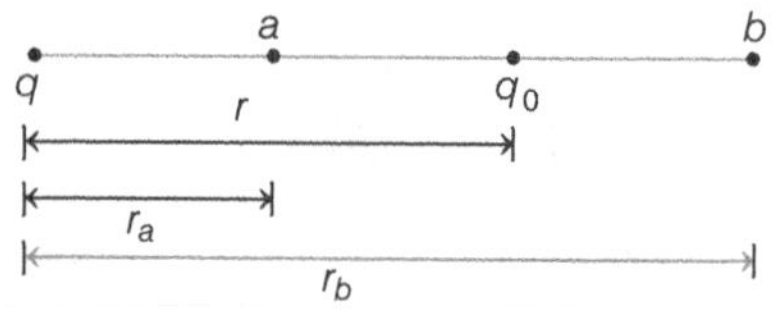

The Coulomb's force on q_0 at a distance r from a fixed charge q is

$$F = \frac{1}{4\pi\varepsilon_0}\cdot\frac{qq_0}{r^2}$$

If the two charges have same signs, the force is repulsive and if the two charges have opposite signs, the force is attractive. The force is not constant during the displacement, so we have to integrate to calculate the work $W_{a\to b}$ done on q_0 by this force as q_0 moves from a to b.

$$\therefore \quad W_{a\to b} = \int_{r_a}^{r_b} F\,dr = \int_{r_a}^{r_b} \frac{1}{4\pi\varepsilon_0}\cdot\frac{qq_0}{r^2}\,dr = \frac{qq_0}{4\pi\varepsilon_0}\left(\frac{1}{r_a} - \frac{1}{r_b}\right)$$

Being a conservative force this work is path independent. From the definition of potential energy,

$$U_b - U_a = -W_{a-b} = \frac{qq_0}{4\pi\varepsilon_0}\left(\frac{1}{r_b} - \frac{1}{r_a}\right)$$

We choose the potential energy of the two charge system to be zero when they have infinite separation. This means $U_\infty = 0$. The potential energy when the separation is r is U_r

$$\therefore \quad U_r - U_\infty = \frac{qq_0}{4\pi\varepsilon_0}\left(\frac{1}{r} - \frac{1}{\infty}\right)$$

or $$\boxed{U_r = \frac{qq_0}{4\pi\varepsilon_0}\frac{1}{r}}$$

This is the expression for electric potential energy of two point charges kept at a separation r. In this expression both the charges q and q_0 are to be substituted with sign. The potential energy is positive if the charges q and q_0 have the same sign and negative if they have opposite signs.

Electric Potential Energy of a System of Charges

The electric potential energy of a system of charges is given by

$$\boxed{U = \frac{1}{4\pi\varepsilon_0}\sum_{i<j}\frac{q_iq_j}{r_{ij}}}$$

This sum extends over all pairs of charges. We don't let $i = j$, because that would be an interaction of a charge with itself, and we include only terms with $i < j$ to make sure that we count each pair only once.

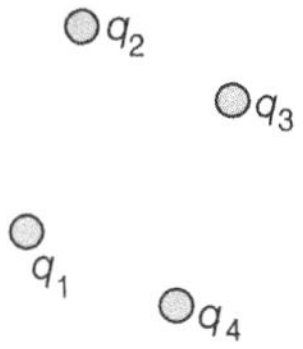

Thus, to account for the interaction between q_5 and q_4, we include a term with $i = 4$ and $j = 5$ but not a term with $i = 5$ and $j = 4$.

For example, electric potential energy of four point charges q_1, q_2, q_3 and q_4 would be given by

$$U = \frac{1}{4\pi\varepsilon_0}\left[\frac{q_4q_3}{r_{43}} + \frac{q_4q_2}{r_{42}} + \frac{q_4q_1}{r_{41}} + \frac{q_3q_2}{r_{32}} + \frac{q_3q_1}{r_{31}} + \frac{q_2q_1}{r_{21}}\right] \quad \text{...(ii)}$$

Here, all the charges are to be substituted with sign.

Note *Total number of pairs formed by n point charges are* $\frac{n(n-1)}{2}$.

Example 2 Four charges $q_1 = 1\,\mu C$, $q_2 = 2\,\mu C$, $q_3 = -3\,\mu C$ and $q_4 = 4\,\mu C$ are kept on the vertices of a square of side 1m. Find the electric potential energy of this system of charges.

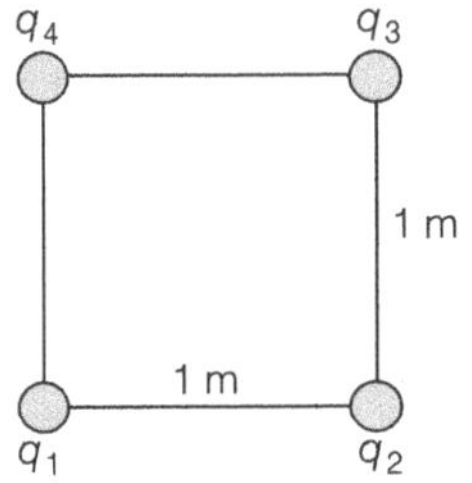

Sol. In this problem,

$$r_{41} = r_{43} = r_{32} = r_{21} = 1\text{ m}$$

and

$$r_{42} = r_{31} = \sqrt{(1)^2 + (1)^2} = \sqrt{2}\text{ m}$$

Substituting the proper values with sign in Eq. (ii), we get

$$U = (9.0 \times 10^9)(10^{-6})(10^{-6})$$

$$\left[\frac{(4)(-3)}{1} + \frac{(4)(2)}{\sqrt{2}} + \frac{(4)(1)}{1} + \frac{(-3)(2)}{1} + \frac{(-3)(1)}{\sqrt{2}} + \frac{(2)(1)}{1}\right]$$

$$= (9.0 \times 10^{-3})\left[-12 + \frac{5}{\sqrt{2}}\right]$$

$$= -7.62 \times 10^{-2}\text{ J}$$ **Ans.**

Note *Here, negative sign of U implies that positive work has been done by electrostatic forces in assembling these charges at respective distances from infinity.*

Example 3 Two point charges are located on the x-axis, $q_1 = -1\,\mu C$ at $x = 0$ and $q_2 = +1\,\mu C$ at $x = 1$ m.

(a) Find the work that must be done by an external force to bring a third point charge $q_3 = +1\,\mu C$ from infinity to $x = 2$ m.

(b) Find the total potential energy of the system of three charges.

Sol. (a) The work that must be done on q_3 by an external force is equal to the difference of potential energy U when the charge is at $x = 2$ m and the potential energy when it is at infinity.

$$\therefore \quad W = U_f - U_i$$

$$= \frac{1}{4\pi\varepsilon_0}\left[\frac{q_3q_2}{(r_{32})_f} + \frac{q_3q_1}{(r_{31})_f} + \frac{q_2q_1}{(r_{21})_f}\right] - \frac{1}{4\pi\varepsilon_0}\left[\frac{q_3q_2}{(r_{32})_i} + \frac{q_3q_1}{(r_{31})_i} + \frac{q_2q_1}{(r_{21})_i}\right]$$

Here, $(r_{21})_i = (r_{21})_f$

and $(r_{32})_i = (r_{31})_i = \infty$

$$\therefore \quad W = \frac{1}{4\pi\varepsilon_0}\left[\frac{q_3q_2}{(r_{32})_f} + \frac{q_3q_1}{(r_{31})_f}\right]$$

Substituting the values, we have

$$W = (9.0 \times 10^9)(10^{-12})\left[\frac{(1)(1)}{(1.0)} + \frac{(1)(-1)}{(2.0)}\right]$$

$$= 4.5 \times 10^{-3}\text{ J}$$ **Ans.**

(b) The total potential energy of the three charges is given by,

$$U = \frac{1}{4\pi\varepsilon_0}\left(\frac{q_3q_2}{r_{32}} + \frac{q_3q_1}{r_{31}} + \frac{q_2q_1}{r_{21}}\right)$$

$$= (9.0 \times 10^9)\left[\frac{(1)(1)}{(1.0)} + \frac{(1)(-1)}{(2.0)} + \frac{(1)(-1)}{(1.0)(1.0)}\right](10^{-12})$$

$$= -4.5 \times 10^{-3}\text{ J}$$ **Ans.**

Example 4 Two point charges $q_1 = q_2 = 2\,\mu C$ are fixed at $x_1 = +3$ m and $x_2 = -3$ m as shown in figure. A third particle of mass 1 g and charge $q_3 = -4\,\mu C$ are released from rest at $y = 4.0$ m. Find the speed of the particle as it reaches the origin.

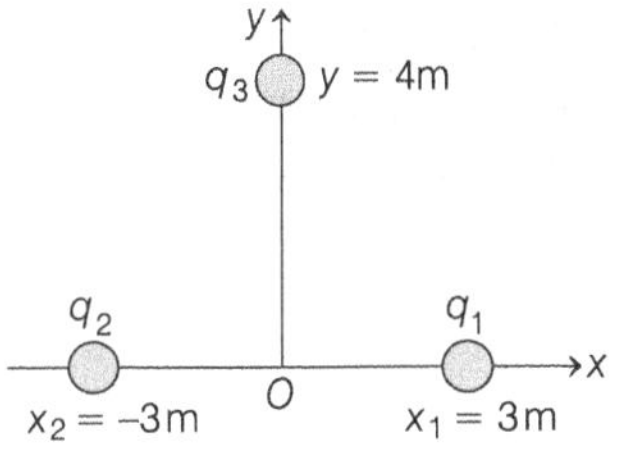

HOW TO PROCEED Here, the charge q_3 is attracted towards q_1 and q_2 both. So, the net force on q_3 is towards origin.

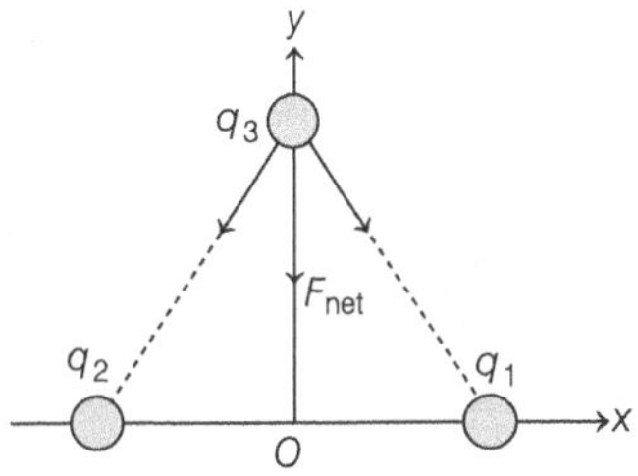

By this force, charge is accelerated towards origin, but this acceleration is not constant. So, to obtain the speed of particle at origin by kinematics we will have to first find the acceleration at some intermediate position and then will have to integrate it with proper limits. On the other hand, it is easy to use energy conservation principle, as the only forces are conservative.

Sol. Let v be the speed of particle at origin. From conservation of mechanical energy,

$$U_i + K_i = U_f + K_f$$

or $$\frac{1}{4\pi\varepsilon_0}\left[\frac{q_3q_2}{(r_{32})_i} + \frac{q_3q_1}{(r_{31})_i} + \frac{q_2q_1}{(r_{21})_i}\right] + 0$$

$$= \frac{1}{4\pi\varepsilon_0}\left[\frac{q_3q_2}{(r_{32})_f} + \frac{q_3q_1}{(r_{31})_f} + \frac{q_2q_1}{(r_{21})_f}\right] + \frac{1}{2}mv^2$$

Here, $(r_{21})_i = (r_{21})_f$

Substituting the proper values, we have

$$(9.0\times 10^9)\left[\frac{(-4)\,(2)}{(5.0)} + \frac{(-4)\,(2)}{(5.0)}\right]\times 10^{-12}$$

$$= (9.0\times 10^9)\left[\frac{(-4)\,(2)}{(3.0)} + \frac{(-4)\,(2)}{(3.0)}\right]\times 10^{-12} + \frac{1}{2}\times 10^{-3}\times v^2$$

$$\therefore\quad (9\times 10^{-3})\left(-\frac{16}{5}\right) = (9\times 10^{-3})\left(-\frac{16}{3}\right) + \frac{1}{2}\times 10^{-3}\times v^2$$

$$(9\times 10^{-3})\,(16)\left(\frac{2}{15}\right) = \frac{1}{2}\times 10^{-3}\times v^2$$

$\therefore$ $v = 6.2$ m/s **Ans.**

Example 5 A point charge $q_1 = 9.1\ \mu C$ is held fixed at origin. A second point charge $q_2 = -0.42\ \mu C$ and a mass 3.2×10^{-4} kg is placed on the x-axis, 0.96 m from the origin. The second point charge is released at rest. What is its speed when it is 0.24 m from the origin?

Sol. From conservation of mechanical energy, we have

Decrease in electrostatic potential energy = Increase in kinetic energy

or $$\frac{1}{2}mv^2 = U_i - U_f$$

$$= \frac{q_1q_2}{4\pi\,\varepsilon_0}\left(\frac{1}{r_i} - \frac{1}{r_f}\right)$$

$$= \frac{q_1q_2}{4\pi\,\varepsilon_0}\left(\frac{r_f - r_i}{r_ir_f}\right)$$

$$\therefore\quad v = \sqrt{\frac{q_1q_2}{2\pi\,\varepsilon_0 m}\left(\frac{r_f - r_i}{r_ir_f}\right)}$$

$$= \sqrt{\frac{(9.1\times 10^{-6})\,(-0.42\times 10^{-6})\times 2\times 9\times 10^9}{3.2\times 10^{-4}}\left(\frac{0.24 - 0.96}{(0.24)\,(0.96)}\right)}$$

$= 26$ m/s **Ans.**

Example 6 A point charge $q_1 = -5.8\ \mu C$ is held stationary at the origin. A second point charge $q_2 = +4.3\ \mu C$ moves from the point (0.26m, 0, 0) to (0.38m, 0, 0). How much work is done by the electric force on q_2?

Sol. Work done by the electrostatic forces $= U_i - U_f$

$$= \frac{q_1q_2}{4\pi\,\varepsilon_0}\left(\frac{1}{r_i} - \frac{1}{r_f}\right) = \frac{q_1q_2}{4\pi\,\varepsilon_0}\left(\frac{r_f - r_i}{r_i\,r_f}\right)$$

$$= \frac{(-5.8\times 10^{-6})\,(4.3\times 10^{-6})\,(9\times 10^9)\,(0.38 - 0.26)}{(0.38)\,(0.26)}$$

$= -0.272$ J **Ans.**

Example 7 An alpha particle with kinetic energy 10 MeV is heading towards a stationary tin nucleus of atomic number 50. Calculate the distance of closest approach. Initially they were far apart.

[1 eV = 1.6×10^{-19} J]

Sol. Due to repulsion by the tin nucleus, the kinetic energy of the α-particle gradually decreases at the expense of electrostatic potential energy.

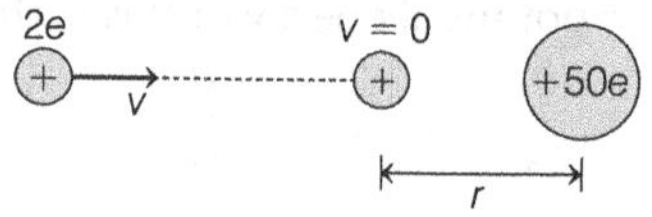

$\therefore$ Decrease in kinetic energy = increase in potential energy

or $$\frac{1}{2}mv^2 = U_f - U_i$$

or $$\frac{1}{2}mv^2 = \frac{1}{4\pi\varepsilon_0}\cdot\frac{q_1q_2}{r} - 0$$

$$\therefore\quad r = \frac{1}{4\pi\varepsilon_0}\cdot\frac{(2e)\,(50e)}{(KE)}$$

Substituting the values,

$$r = \frac{(9\times 10^9)\,(2\times 1.6\times 10^{-19})\,(1.6\times 10^{-19}\times 50)}{10\times 10^6\times 1.6\times 10^{-19}}$$

$= 14.4\times 10^{-15}$ m **Ans.**

Example 8 Three point charges of 1C, 2C and 3C are placed at the corners of an equilateral triangle of side 1m. Calculate the work required to move these charges to the corners of a smaller equilateral triangle of side 0.5 m.

Sol. Work done $= U_f - U_i$

$$= \frac{1}{4\pi\,\varepsilon_0}\left(\frac{1}{r_f} - \frac{1}{r_i}\right)[q_3q_2 + q_3q_1 + q_2q_1]$$

$$= 9\times 10^9\left(\frac{1}{0.5} - \frac{1}{1}\right)[(3)(2) + (3)(1) + (2)(1)]$$

$= 99\times 10^9$ J **Ans.**

Note *Work done by electrostatic forces is $U_i - U_f$ but work done by external forces is $U_f - U_i$. Sometimes in a simple way it is asked, find the work done. It means $U_f - U_i$.*

MODULE 4 Exercise

Assertion and Reason

See directions on page 25.

1. **Assertion :** When two unlike charges are brought nearer, their electrostatic potential energy decreases.

   Reason : All conservative forces act in the direction of decreasing potential energy.

Single Correct Option

2. Three identical charges are placed at corners of an equilateral triangle of side l. If force between any two charges is F, the work required to double the dimensions of triangle is

   (a) $-3Fl$ (b) $3Fl$ (c) $(-3/2)Fl$ (d) $(3/2)Fl$

3. Four charges $+q, -q, +q$ and $-q$ are placed in order on the four consecutive corners of a square of side a. The work done in interchanging the positions of any two neighbouring charges of the opposite sign is

   (a) $\frac{q^2}{4\pi\varepsilon_0 a}(-4+\sqrt{2})$ (b) $\frac{q^2}{4\pi\varepsilon_0 a}(4+2\sqrt{2})$

   (c) $\frac{q^2}{4\pi\varepsilon_0 a}(4-2\sqrt{2})$ (d) $\frac{q^2}{4\pi\varepsilon_0 a}(4+\sqrt{2})$

4. Four identical charges $+50\,\mu C$ each are placed, one at each corner of a square of side 2 m. How much external energy is required to bring another charge of $+50\,\mu C$ from infinity to the centre of the square?

   (a) 64 J (b) 41 J (c) 16 J (d) 10 J

5. A particle A of mass m and charge Q moves directly towards a fixed particle B, which has charge Q. The speed of A is v when it is far away from B. The minimum separation between the particles is proportional to (Choose the **incorrect** option). Neglect gravitational force.

   (a) Q^2 (b) $\frac{1}{m}$

   (c) $\frac{1}{v}$ (d) None of these

6. Four point charges A, B, C and D are placed at the four corners of a square of side a. The energy required to take the charges C and D to infinity (they are also infinitely separated from each other) is

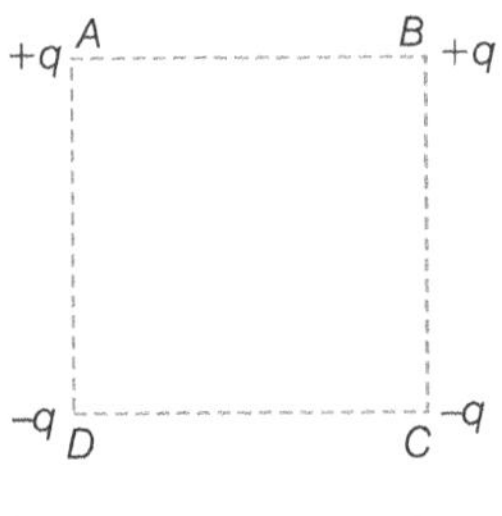

   (a) $\frac{q^2}{4\pi\varepsilon_0 a}$ (b) $\frac{2q^2}{\pi\varepsilon_0 a}$

   (c) $\frac{q^2}{4\pi\varepsilon_0 a}(\sqrt{2}+1)$ (d) $\frac{q^2}{4\pi\varepsilon_0 a}(\sqrt{2}-1)$

7. Two identical charges are placed at the two corners of an equilateral triangle. The potential energy of the system is U. The work done in bringing an identical charge from infinity to the third vertex is

   (a) U (b) $2U$ (c) $3U$ (d) $4U$

8. Two charges q_1 and q_2 are placed 30 cm apart, as shown in figure. A third charge q_3 is moved along the arc of a circle of radius 40 cm from C to D. The change in the potential energy of the system is $\frac{q_3}{4\pi\varepsilon_0}k$, where k is

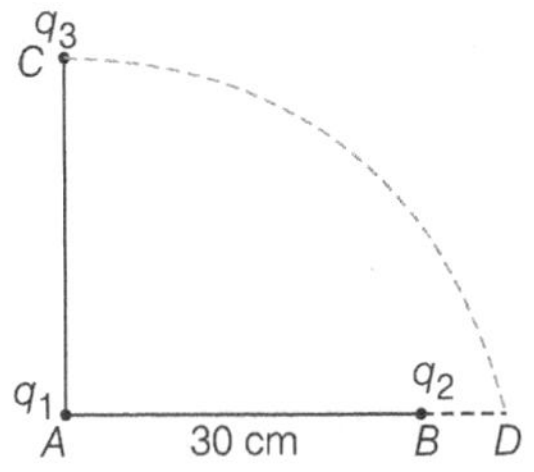

   (a) $8q_2$ (b) $8q_1$
   (c) $6q_2$ (d) $6q_1$

9. Three point charges q, $-2q$ and $-2q$ are placed at the vertices of an equilateral triangle of side a. The work done by some external force to increase their separation to $2a$ will be

   (a) $\frac{1}{4\pi\varepsilon_0}\frac{2q^2}{a}$ (b) $\frac{1}{4\pi\varepsilon_0}\frac{q^2}{2a}$

   (c) $\frac{1}{4\pi\varepsilon_0}\frac{8q}{a^2}$ (d) zero

10. Three charges Q, $+q$ and $+q$ are placed at the vertices of a right angle triangle (isosceles triangle) as shown in the figure. The net electrostatic energy of the configuration is zero, if Q is equal to

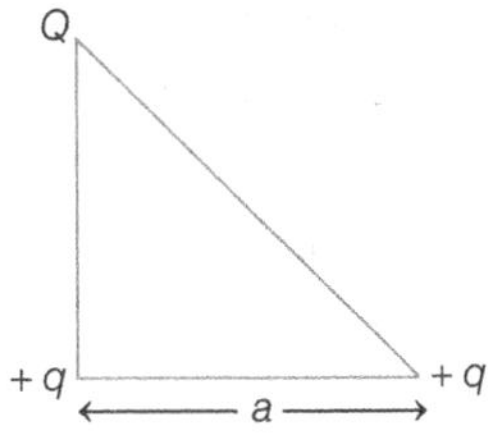

   (a) $\frac{-q}{1+\sqrt{2}}$ (b) $\frac{-2q}{2+\sqrt{2}}$ (c) $-2q$ (d) $+q$

11. Two equal point charges are fixed at $x = -a$ and $x = +a$ on the X-axis. Another point charge Q is placed at the origin. The change in the electrical potential energy of Q, when it is displaced by a small distance x along the X-axis, is approximately proportional to

(a) x (b) x^2 (c) x^3 (d) $1/x$

Multiple Correct Options

12. Two charges $+Q$ each are fixed at points C and D. Line AB is the bisector line of CD. A third charge $+q$ is moved from A to B and then from B to C.

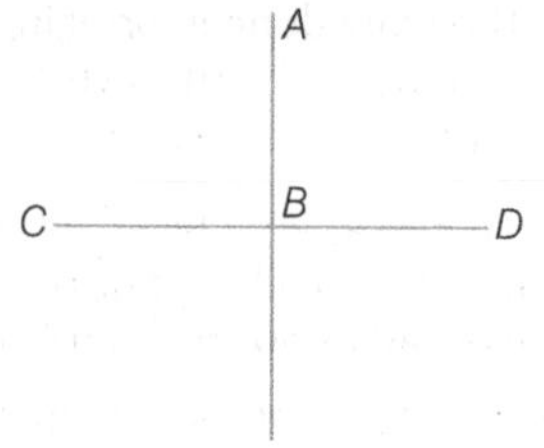

(a) From A to B electrostatic potential energy will decrease
(b) From A to B electrostatic potential energy will increase
(c) From B to C electrostatic potential energy will increase
(d) From B to C electrostatic potential energy will decrease

Passage Based Questions

Passage (Q.13 to 16)

Two fixed equal positive charges, each of magnitude $q = 5 \times 10^{-5}$ C are located at points A and B, where AB is 6 m. An equal and opposite charge moves towards line AB along the perpendicular bisector of AB.

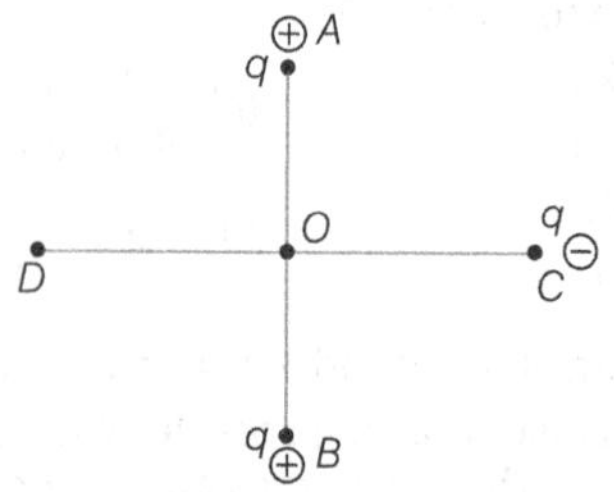

The moving charge, when reaches at point C at a distance of 4 m from O, has a kinetic energy of 4 J.

13. Total energy of charge $-q$ at C is

(a) 4 J (b) 9 J (c) −9 J (d) −5 J

14. Total energy of charge $-q$ at D is

(a) 4 J (b) 9 J (c) −9 J (d) −5 J

15. Total potential energy of charge $-q$ at D is

(a) around 4 J (b) around −9 J
(c) around −4 J (d) around −5 J

16. The distance of point D at which $-q$ charge will reach before returning towards C, is

(a) less than 9 m (b) more than 9 m
(c) less than 9.5 m (d) more than 9.5 m

Subjective Questions

17. A point charge q_1 is held stationary at the origin. A second charge q_2 is placed at a point a, and the electric potential energy of the pair of charges is -6.4×10^{-8} J. When the second charge is moved to point b, the electric force on the charge does 4.2×10^{-8} J of work. What is the electric potential energy of the pair of charges when the second charge is at point b?

18. Is it possible to have an arrangement of two point charges separated by finite distances such that the electric potential energy of the arrangement is the same as if the two charges were infinitely far apart? What if there are three charges?

19. A point charge $q_1 = 4.00$ nC is placed at the origin, and a second point charge $q_2 = -3.00$ nC is placed on the x-axis at $x = +20.0$ cm. A third point charge $q_3 = 2.00$ nC is placed on the x-axis between q_1 and q_2. (Take as zero the potential energy of the three charges when they are infinitely far apart).
(a) What is the potential energy of the system of the three charges if q_3 is placed at $x = +10.0$ cm?
(b) Where should q_3 be placed to make the potential energy of the system equal to zero?

20. Three point charges, which initially are infinitely far apart, are placed at the corners of an equilateral triangle with sides d. Two of the point charges are identical and have charge q. If zero net work is required to place the three charges at the corners of the triangles, what must the value of the third charge be?

Answers

1. (a)	**2.** (c)	**3.** (c)	**4.** (a)	**5.** (c)
6. (c)	**7.** (b)	**8.** (a)	**9.** (d)	**10.** (b)
11. (b)	**12.** (b,c)	**13.** (d)	**14.** (d)	**15.** (d)
16. (a)	**17.** -10.6×10^{-8} J		**18.** No, Yes	
19. (a) -3.6×10^{-7} J (b) $x = 0.0743$ m				**20.** $-\frac{q}{2}$

See hints on pages 201-203.

MODULE 5

Electric Potential

As we have discussed in Module 3 that an electric field at any point can be defined in two different ways:

(i) by the field strength **E**, and

(ii) by the electric potential V at the point under consideration.

Both **E** and V are functions of position and there is a fixed relationship between these two. Of these, the field strength **E** is a vector quantity while the electric potential V is a scalar quantity. In this module, we will discuss about the electric potential and in the next, the relationship between **E** and V.

"**Potential** is the potential energy per unit charge." Electric potential at any point in an electric field is defined as the potential energy per unit charge, same as the field strength is defined as the force per unit charge. Thus,

$$V = \frac{U}{q_0} \quad \text{or} \quad U = q_0 V$$

The SI unit of potential is **volt** (V) which is equal to joule per coulomb. So,

$$1\text{ V} = 1\text{ J/C}$$

The work done by the electrostatic force in displacing a test charge q_0 from a to b in an electric field is defined as the negative of change in potential energy between them, or

$$\Delta U = -W_{a-b}$$

$$\therefore \quad U_b - U_a = -W_{a-b}$$

We divide this equation by q_0

$$\frac{U_b}{q_0} - \frac{U_a}{q_0} = -\frac{W_{a-b}}{q_0}$$

or
$$V_a - V_b = \frac{W_{a-b}}{q_0}$$

as
$$V = \frac{U}{q_0}$$

Thus, the work done per unit charge by the **electric force** when a charged body moves from a to b is equal to the potential at a minus the potential at b. We sometimes abbreviate this difference as $V_{ab} = V_a - V_b$.

Another way to interpret the potential difference V_{ab} is that the potential at a minus potential at b, equals the work that must be done by some external agent to move a unit positive charge slowly from b to a against the electric force.

$$V_a - V_b = \frac{(W_{b-a})_{\text{external force}}}{q_0} = \frac{-(W_{b\to a})_{\text{electric force}}}{q_0}$$

Absolute Potential at Some Point

Suppose we take the point b at infinity and as a reference point assign the value $V_b = 0$, the above equations can be written as

$$V_a - V_b = \frac{(W_{a-b})_{\text{electric force}}}{q_0} = \frac{(W_{b-a})_{\text{external force}}}{q_0}$$

or
$$V_a = \frac{(W_{a-\infty})_{\text{electric force}}}{q_0} = \frac{(W_{\infty-a})_{\text{external force}}}{q_0}$$

Thus, the absolute electric potential at point a in an electric field can be defined as the work done in displacing a unit positive test charge from infinity to a by the external force or the work done (by electric force) per unit positive test charge in displacing it from a to infinity.

Note *The following three formulae are very useful in the problems related to work done in electric field.*

$$(W_{a-b})_{\text{electric force}} = q_0 (V_a - V_b)$$

$$(W_{a-b})_{\text{external force}} = q_0 (V_b - V_a) = -(W_{a-b})_{\text{electric force}}$$

$$(W_{\infty-a})_{\text{external force}} = q_0 V_a$$

Here, q_0, V_a and V_b are to be substituted with sign.

Example 1 The electric potential at point A is 20 V and at B is -40 V. Find the work done by an external force and electrostatic force in moving an electron slowly from B to A.

Sol. Here, the test charge is an electron, i.e.

$$q_0 = -1.6 \times 10^{-19}\text{ C}$$

$$V_A = 20\text{ V}$$

and
$$V_B = -40\text{ V}$$

Work done by external force

$$(W_{B-A})_{\text{external force}} = q_0 (V_A - V_B)$$
$$= (-1.6 \times 10^{-19})[(20) - (-40)]$$
$$= -9.6 \times 10^{-18}\text{ J} \qquad \textbf{Ans.}$$

Work done by electric force

$$(W_{B-A})_{\text{electric force}} = -(W_{B-A})_{\text{external force}}$$
$$= -(-9.6 \times 10^{-18}\text{ J})$$
$$= 9.6 \times 10^{-18}\text{ J} \qquad \textbf{Ans.}$$

Note *Here, we can see that the electron (a negative charge) moves from B (lower potential) to A (higher potential) and the work done by electric force is positive. Therefore, we may conclude that whenever a negative charge moves from a lower potential to higher potential work done by the electric force is positive or when a positive charge moves from lower potential to higher potential the work done by the electric force is negative.*

Example 2 Find the work done by some external force in moving a charge $q = 2\,\mu C$ from infinity to a point where electric potential is 10^4 V.

Sol. Using the relation,

$$(W_{\infty-a})_{\text{external force}} = qV_a$$

We have, $$(W_{\infty-a})_{\text{external force}} = (2 \times 10^{-6})(10^4)$$

$$= 2 \times 10^{-2}\ \text{J}$$ **Ans.**

Electric Potential Due to a Point Charge q

From the definition of potential, $$V = \frac{U}{q_0} = \frac{\frac{1}{4\pi\varepsilon_0}\cdot\frac{q q_0}{r}}{q_0}$$

or $$\boxed{V = \frac{1}{4\pi\varepsilon_0}\cdot\frac{q}{r}}$$ or $$\boxed{V \propto \frac{1}{r}}$$

Here, r is the distance from the point charge q to the point at which the potential is evaluated.

If q is positive, the potential that it produces is positive at all points; if q is negative, it produces a potential that is negative everywhere. In either case, V is equal to zero at $r = \infty$.

V versus r graph is as shown below.

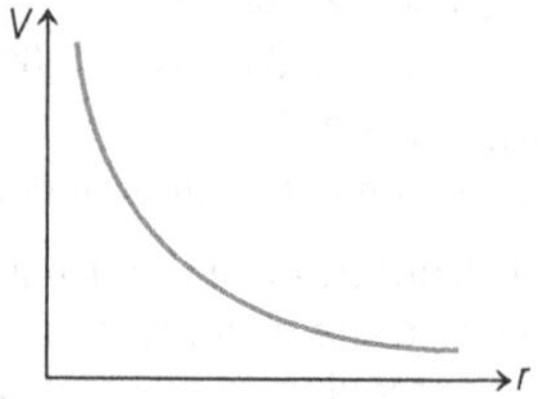

Electric Potential Due to a System of Charges

Just as the electric field due to a collection of point charges is the vector sum of the fields produced by each charge, the electric potential due to a collection of point charges is the scalar sum of the potentials due to each charge.

$$\boxed{V = \frac{1}{4\pi\varepsilon_0}\sum_i \frac{q_i}{r_i}}$$

In this expression, r_i is the distance from the ith charge, q_i, to the point at which V is evaluated. For a continuous distribution of charge along a line, over a surface or through a volume, we divide the charge into elements dq and the sum in the above equation becomes an integral,

$$\boxed{V = \frac{1}{4\pi\varepsilon_0}\int\frac{dq}{r}}$$

Note *In the equation $V = \frac{1}{4\pi\varepsilon_0}\sum_i \frac{q_i}{r_i}$ or $V = \frac{1}{4\pi\varepsilon_0}\int\frac{dq}{r}$, if the whole charge is at equal distance r_0 from the point where V is to be evaluated, then we can write,*

$$V = \frac{1}{4\pi\varepsilon_0}\cdot\frac{q_{net}}{r_0}$$

where, q_{net} is the algebraic sum of all the charges of which the system is made.

Here there are few examples :

Ex I. Four charges are placed on the vertices of a square as shown in figure. The electric potential at centre of the square is zero as all the charges are at same distance from the centre and

$$q_{net} = 4\,\mu C - 2\,\mu C + 2\,\mu C - 4\,\mu C = 0$$

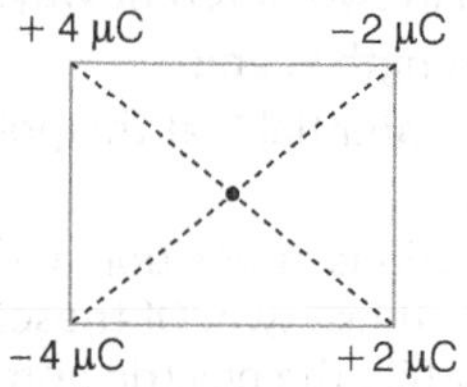

Ex II. A charge q is uniformly distributed over the circumference of a ring in Fig. (a) and is non-uniformly distributed in Fig. (b).

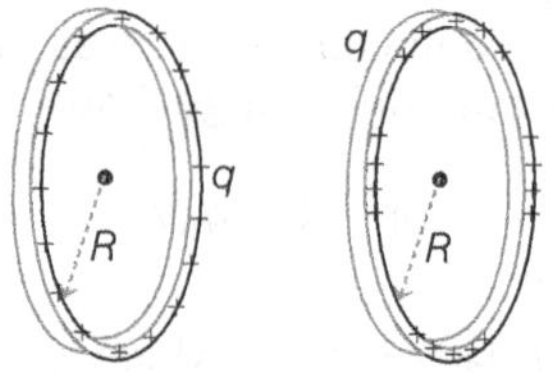

(a) (b)

The electric potential at the centre of the ring in both the cases is

$$V = \frac{1}{4\pi\varepsilon_0}\cdot\frac{q}{R} \quad \text{(where, } R = \text{radius of ring)}$$

and at a distance r from the centre of ring on its axis would be

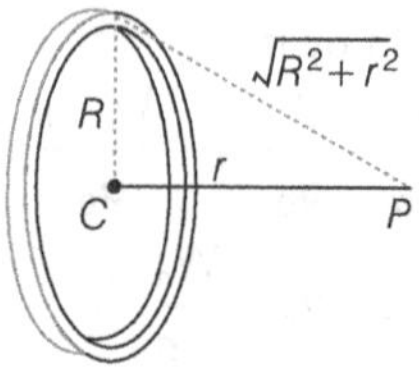

$$V = \frac{1}{4\pi\varepsilon_0}\cdot\frac{q}{\sqrt{R^2+r^2}}$$

Example 3 Three point charges $q_1 = 1\,\mu C$, $q_2 = -2\,\mu C$ and $q_3 = 3\,\mu C$ are placed at (1m, 0, 0), (0, 2m, 0) and (0, 0, 3m) respectively. Find the electric potential at origin.

Sol. The net electric potential at origin is

$$V = \frac{1}{4\pi\varepsilon_0}\left[\frac{q_1}{r_1} + \frac{q_2}{r_2} + \frac{q_3}{r_3}\right]$$

Substituting the values, we have

$$V = (9.0 \times 10^9)\left(\frac{1}{1.0} - \frac{2}{2.0} + \frac{3}{3.0}\right) \times 10^{-6}$$

$$= 9.0 \times 10^3\ \text{V}$$ **Ans.**

Example 4 A charge $q = 10\ \mu C$ is distributed uniformly over the circumference of a ring of radius 3m placed on *x-y* plane with its centre at origin. Find the electric potential at a point *P* (0, 0, 4 m).

Sol. The electric potential at point *P* would be

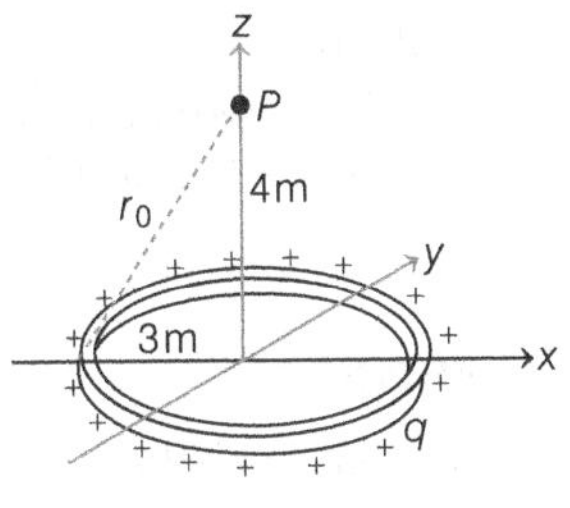

$$V = \frac{1}{4\pi\varepsilon_0} \cdot \frac{q}{r_0}$$

Here, r_0 = distance of point *P* from the circumference of ring

$$= \sqrt{(3)^2 + (4)^2} = 5\text{ m}$$

and $q = 10\ \mu C = 10^{-5}$ C

Substituting the values, we have

$$V = \frac{(9.0 \times 10^9)(10^{-5})}{(5.0)}$$

$$= 1.8 \times 10^4 \text{ V}$$ **Ans.**

Variation of Electric Potential on the Axis of a Charged Ring

We have discussed earlier that the electric potential at the centre of a charged ring (whether charged uniformly or non-uniformly) is $\frac{1}{4\pi\varepsilon_0} \cdot \frac{q}{R}$ and at a distance *r* from the centre on the axis of the ring is $\frac{1}{4\pi\varepsilon_0} \cdot \frac{q}{\sqrt{R^2 + r^2}}$. From these expressions, we can see that electric potential is maximum at the centre and decreases as we move away from the centre on the axis. Thus, potential varies with distance *r* as shown in figure.

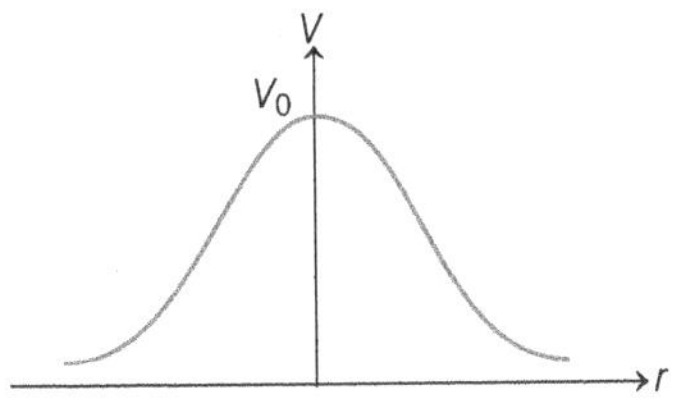

In the figure,

$$V_0 = \frac{1}{4\pi\varepsilon_0} \cdot \frac{q}{R}$$

Electric Potential on the Axis of a Uniformly Charged Disc

Let us find the electric potential at any point *P*, a distance *x* on the axis of a uniformly charged circular disc, having surface charge density σ. Let us divide the disc into a large number of thin circular strips and consider a strip of radius *r* and width *dr*. Each point of this strip can be assumed to be at equal distance $\sqrt{r^2 + x^2}$ from point *P*. Potential at *P* due to this circular strip is

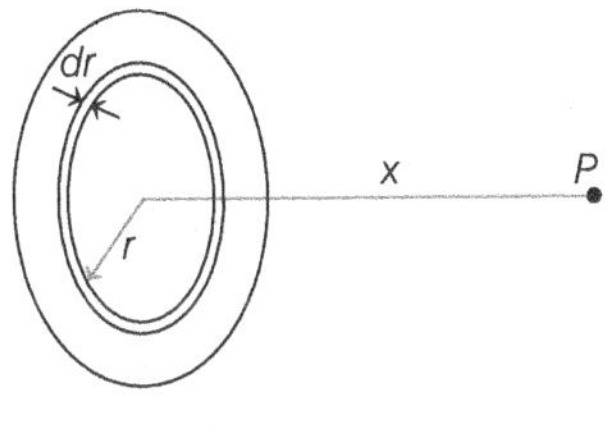

$$dV = \frac{1}{4\pi\varepsilon_0} \cdot \frac{dq}{\sqrt{r^2 + x^2}}$$

Here, $dq = \sigma$ (area of strip) or $dq = \sigma(2\pi r dr)$

$$\therefore \quad dV = \frac{1}{4\pi\varepsilon_0} \cdot \frac{\sigma(2\pi r dr)}{\sqrt{r^2 + x^2}}$$

Thus, the potential due to the whole disc is

$$V = \int_0^R dV = \frac{\sigma}{2\varepsilon_0} \int_0^R \frac{r dr}{\sqrt{r^2 + x^2}}$$

or

$$V = \frac{\sigma}{2\varepsilon_0}[\sqrt{R^2 + x^2} - x]$$

(i) At the centre of the disc, $x = 0$

$$\therefore \quad V\text{ (centre)} = \frac{\sigma R}{2\varepsilon_0} \quad \text{...(i)}$$

(ii) For $x >> R$, using the Binomial expansion for

$$\sqrt{R^2 + x^2} = x\left(1 + \frac{R^2}{x^2}\right)^{1/2} \approx x + \frac{R^2}{2x}$$

$$\therefore \quad V = \frac{\sigma}{2\varepsilon_0}\left(x + \frac{R^2}{2x} - x\right) = \frac{\sigma R^2}{4\varepsilon_0 x} = \frac{\pi R^2 \sigma}{4\pi\varepsilon_0 x}$$

or

$$V = \frac{q}{4\pi\varepsilon_0 x}$$

as $\pi R^2 \sigma = q$, the total charge on the disc.

This is the relation as obtained due to a point charge. Thus, at far away points, the distribution of charge becomes insignificant. It is difficult to calculate the potential at the points other than on the axis. However, potential on the edge of the disc can be calculated as under.

Potential on the Edge of the Disc

To calculate the potential at point *P*, let us divide the disc in large number of rings with *P* as centre. The potential due to one segment between *r* and *r* + *dr* is given as

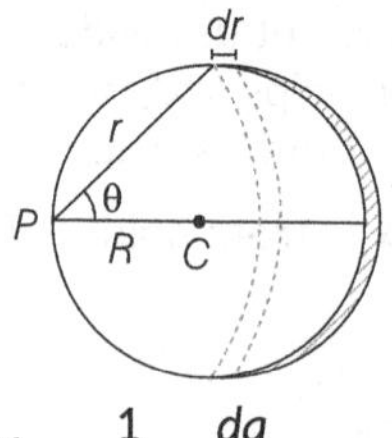

$$dV = \frac{1}{4\pi\varepsilon_0} \cdot \frac{dq}{r}$$

Here, $dq = \sigma$ (Area of ring)

$$= \sigma (2r\ \theta)\ dr$$

$\therefore$ $$dV = \frac{1}{4\pi\varepsilon_0} \cdot \frac{\sigma (2r\ \theta)\ dr}{r}$$

$$= \frac{\sigma}{2\pi\varepsilon_0} \cdot \theta\ dr$$

Further, $r = 2R \cos\theta$

$\therefore$ $$dr = -2R \sin\theta\ d\theta$$

Hence, $$dV = -\frac{\sigma}{2\pi\varepsilon_0} 2R\theta \sin\theta\ d\theta$$

$\therefore$ $$V = \int_{\pi/2}^{0} dV = \frac{\sigma R}{\pi\varepsilon_0} \int_0^{\pi/2} \theta \sin\theta\ d\theta$$

Solving, we get $$V = \frac{\sigma R}{\pi\varepsilon_0} \qquad \text{...(ii)}$$

Comparing Eqs. (i) and (ii), we see that potential at the centre of the disc is greater than the potential at the edge.

Example 5 Find out the points on the line joining two charges + *q* and – 3*q* (kept at a distance of 1.0 m) where electric potential is zero.

Sol. Let *P* be the point on the axis either to the left or to the right of charge + *q* at a distance *r* where potential is zero. Hence,

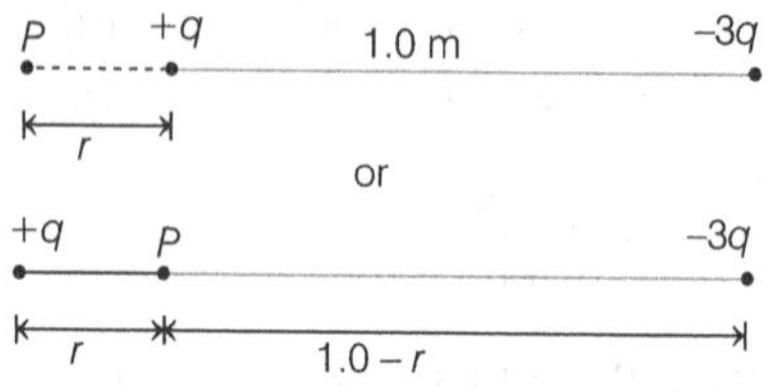

$$V_P = \frac{q}{4\pi\varepsilon_0 r} - \frac{3q}{4\pi\varepsilon_0 (1+r)} = 0$$

Solving this, we get $r = 0.5$ m

Further, $$V_P = \frac{q}{4\pi\varepsilon_0 r} - \frac{3q}{4\pi\varepsilon_0 (1-r)} = 0$$

which gives $r = 0.25$ m

Thus, the potential will be zero at point *P* on the axis which is either 0.5 m to the left or 0.25 m to the right of charge + *q*. **Ans.**

INTRODUCTORY Exercise

1. Find V_{ba} if 12 J of work has to be done against an electric field to take a charge of 10^{-2} C from *a* to *b*.
2. A rod of length *L* lies along the *x*-axis with its left end at the origin. It has a non-uniform charge density $\lambda = \alpha x$, where α is a positive constant.
   (a) What are the units of α?
   (b) Calculate the electric potential at point *A* where $x = -d$.
3. A charge *q* is uniformly distributed along an insulating straight wire of length 2*l* as shown in figure. Find an expression for the electric potential at a point located a distance *d* from the distribution along its perpendicular bisector.

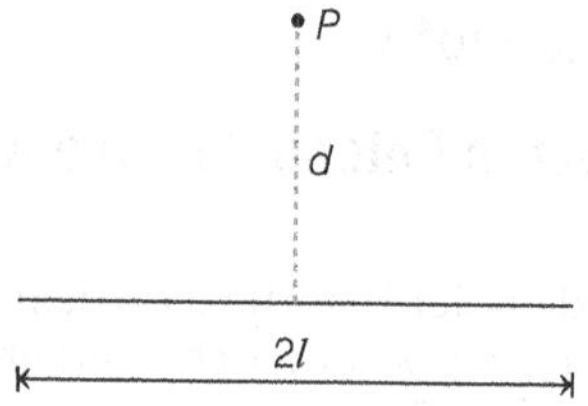

4. A cone made of insulating material has a total charge *Q* spread uniformly over its sloping surface. Calculate the work done in bringing a small test charge *q* from infinity to the apex of the cone. The cone has a slope length *L*.

Answers

1. 1.2×10^3 V 2. (a) $\frac{C}{m^2}$ (b) $\frac{\alpha}{4\pi\varepsilon_0}\left[L - d \ln\left(1 + \frac{L}{d}\right)\right]$

3. $V = \frac{1}{4\pi\varepsilon_0} \cdot \frac{q}{2l} \ln\left(\frac{\sqrt{l^2 + d^2} + l}{\sqrt{l^2 + d^2} - l}\right)$ 4. $\frac{Qq}{2\pi\varepsilon_0 L}$

See hints on pages 203-204.

Typed Problems

Type 1. *Based on E-r and V-r graphs due to two point charges.*

Concept

(i) $E = \dfrac{kq}{r^2}$ $\left(k = \dfrac{1}{4\pi\varepsilon_0}\right)$

and $V = \pm\dfrac{kq}{r}$ (due to a point charge)

(ii) As $r \to 0, E \to \infty$ and $V \to \pm\infty$

As $r \to \infty, E \to 0$ and $V \to 0$

(iii) E is a vector quantity. Due to a point charge, its direction is away from the charge and due to negative charge it is towards the charge. Along one dimension if one direction is taken as positive direction then the other direction is taken as the negative direction.

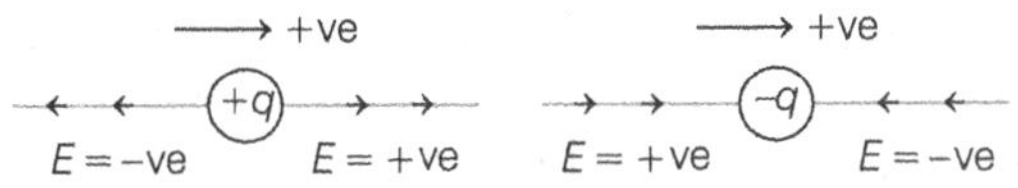

(iv) V is a scalar quantity. On both sides of a positive charge it is positive and it is negative due to negative charge.

(v) Between zero and zero value, normally we get either a maximum or minimum value.

Ex. 1 Draw E - r and V - r graphs due to two point charges $+q$ and $-2q$ kept at some distance along the line joining these two charges.

Sol. ***E* - *r* graph**

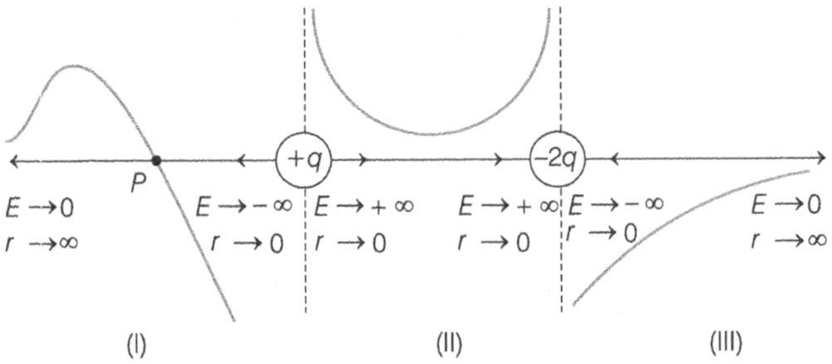

In region I E due to $+q$ is towards left (so negative) and E due to $-2q$ is towards right (so positive). Near $+q$, electric field of $+q$ will dominate. So, net value will be negative. At some point say P both positive and negative values are equal. So, $E_p = 0$. Beyond this point, electric field due to $-2q$ will dominate due to its greater magnitude. So, net value will be positive. $E_p = 0$ and E_∞ (towards left) is also zero. Between zero and zero we will get a maximum positive value.

In region II E due to $+q$ and due to $-2q$ is towards right (so positive). Between the value $+\infty$ and $+\infty$ the graph is as shown in figure.

In region III E due to $+q$ is towards right (so positive) and E due to $-2q$ is towards left (so negative). But electric field of $-2q$ will dominate due to its greater magnitude and lesser distance. Hence, net electric field is always negative.

***V* - *r* graph**

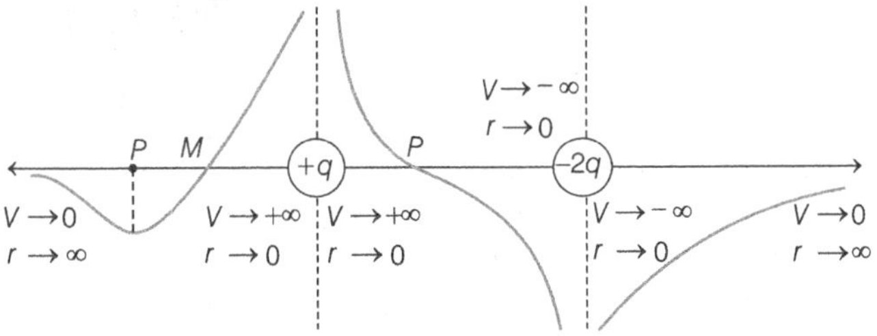

The logics developed in E - r graph can also be applied here with V - r graph. At point M, positive potential due to $+q$ is equal to negative potential due to $-2q$. Hence, $V_M = 0$, and this point is near $+q$.

At point P, electric field is zero.

Therefore slope of V - r graph is zero. This is because,

$$E = -\frac{dV}{dr}$$

Type 2. *Based on the work done (by external forces) in moving a charge from one point to another point.*

Concept

If kinetic energy of the particle is not changed, then

$$W = \Delta U = U_f - U_i = q(V_f - V_i)$$

or $q(\Delta V)$

Here, q is the charge to be displaced and V_i and V_f are the initial and final potentials.

Ex. 2 Two identical thin rings, each of radius R, are coaxially placed a distance R apart. If Q_1 and Q_2 are respectively the charges uniformly spread on the two rings, the work done in moving a charge q from the centre of one ring to that of the other is **(1992 JEE)**

(a) zero

(b) $\dfrac{q(Q_1 - Q_2)(\sqrt{2}-1)}{\sqrt{2}\,(4\pi\varepsilon_0 R)}$

(c) $\dfrac{q\sqrt{2}(Q_1 + Q_2)}{(4\pi\varepsilon_0 R)}$

(d) $q(Q_1/Q_2)(\sqrt{2}+1)\sqrt{2}\,(4\pi\varepsilon_0 R)$

Sol. $V_{C_1} = V_{Q_1} + V_{Q_2} = \dfrac{1}{4\pi\varepsilon_0}\dfrac{Q_1}{R} + \dfrac{1}{4\pi\varepsilon_0}\dfrac{Q_2}{R\sqrt{2}}$

$$= \frac{1}{4\pi\varepsilon_0 R}\left(Q_1 + \frac{Q_2}{\sqrt{2}}\right)$$

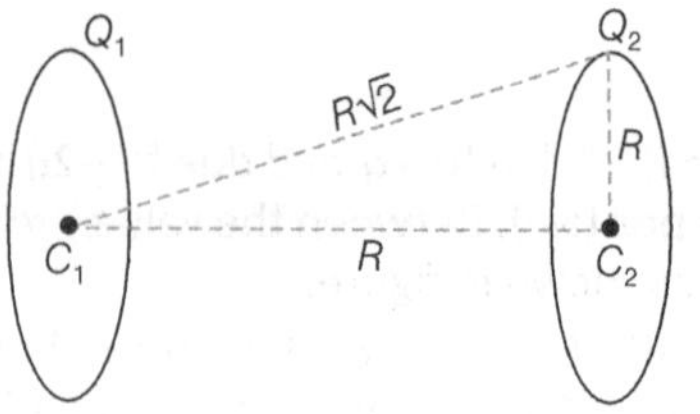

Similarly, $V_{C_2} = \frac{1}{4\pi\varepsilon_0 R}\left(Q_2 + \frac{Q_1}{\sqrt{2}}\right)$

$\therefore$ $\Delta V = V_{C_1} - V_{C_2}$

$$= \frac{1}{4\pi\varepsilon_0 R}\left[(Q_1 - Q_2) - \frac{1}{\sqrt{2}}(Q_1 - Q_2)\right]$$

$$= \frac{Q_1 - Q_2}{\sqrt{2}(4\pi\varepsilon_0 R)}(\sqrt{2} - 1)$$

$$W = q\Delta V = \frac{q(Q_1 - Q_2)(\sqrt{2} - 1)}{\sqrt{2}(4\pi\varepsilon_0 R)}$$

$\therefore$ The correct option is (b).

Miscellaneous Examples

Ex. 3 A uniformly charged thin ring has radius 10.0 cm and total charge +12.0 nC. An electron is placed on the ring's axis a distance 25.0 cm from the centre of the ring and is constrained to stay on the axis of the ring. The electron is then released from rest.

(a) describe the subsequent motion of the electron.

(b) find the speed of the electron when it reaches the centre of the ring.

Sol. (a) The electron will be attracted towards the centre C of the ring. At C net force is zero, but on reaching C, electron has some kinetic energy and due to inertia it crosses C, but on the other side it is further attracted towards C. Hence, motion of electron is oscillatory about point C.

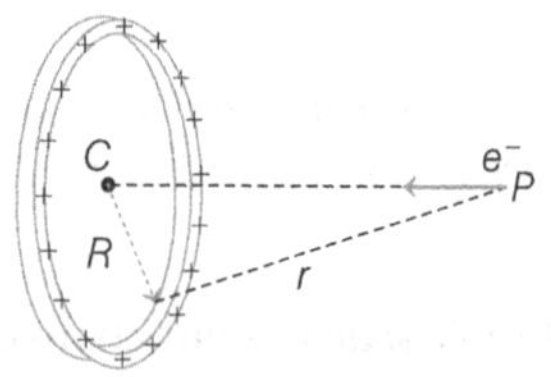

(b) As the electron approaches C, its speed (hence, kinetic energy) increases due to force of attraction towards the centre C. This increase in kinetic energy is at the cost of electrostatic potential energy. Thus,

$$\frac{1}{2}mv^2 = U_i - U_f$$

$$= U_P - U_C = (-e)[V_P - V_C] \quad \text{...(i)}$$

Here, V is the potential due to ring.

$$V_P = \frac{1}{4\pi\varepsilon_0}\cdot\frac{q}{r} \quad (q = \text{charge on ring})$$

$$= \frac{(9\times10^9)(12\times10^{-9})}{(\sqrt{(10)^2 + (25)^2})\times10^{-2}} = 401\text{ V}$$

$$V_C = \frac{1}{4\pi\varepsilon_0}\cdot\frac{q}{R}$$

$$= \frac{(9\times10^9)(12\times10^{-9})}{10\times10^{-2}} = 1080\text{ V}$$

Substituting the proper values in Eq. (i), we have

$$\frac{1}{2}\times9.1\times10^{-31}\times v^2 = (-1.6\times10^{-19})(401 - 1080)$$

$\therefore$ $v = 15.45\times10^6$ m/s **Ans.**

Ex. 4 Consider a spherical surface of radius 4 m centred at the origin. Point charges + q and $-2q$ are fixed at points A (2 m, 0, 0) and B (8 m, 0, 0) respectively. Show that every point on the spherical surface is at zero potential.

Sol. Let P (x, y, z) be any point on the sphere. From the property of the sphere,

$$x^2 + y^2 + z^2 = (4^2) = 16 \quad \text{...(i)}$$

Further, $PA = \sqrt{(x-2)^2 + y^2 + z^2}$ ...(ii)

and $PB = \sqrt{(x-8)^2 + y^2 + z^2}$ ...(iii)

$$V_P = \frac{1}{4\pi\varepsilon_0}\left[\frac{q}{PA} - \frac{2q}{PB}\right]$$

$$= \frac{1}{4\pi\varepsilon_0}\left[\frac{q}{\sqrt{(x-2)^2 + y^2 + z^2}} - \frac{2q}{\sqrt{(x-8)^2 + y^2 + z^2}}\right]$$

$$= \frac{1}{4\pi\varepsilon_0}\left[\frac{q}{\sqrt{x^2 + y^2 + z^2 + 4 - 4x}} - \frac{2q}{\sqrt{x^2 + y^2 + z^2 + 64 - 16x}}\right]$$

$$= \frac{1}{4\pi\varepsilon_0}\left[\frac{q}{\sqrt{16 + 4 - 4x}} - \frac{2q}{\sqrt{16 + 64 - 16x}}\right]$$

$$= \frac{1}{4\pi\varepsilon_0}\left[\frac{q}{\sqrt{20 - 4x}} - \frac{q}{\sqrt{20 - 4x}}\right]$$

$= 0$ **Proved.**

Ex. 5 A non-conducting disc of radius a and uniform positive surface charge density σ is placed on the ground with its axis vertical. A particle of mass m and positive charge q is dropped, along the axis of the disc from a height H with zero initial velocity. The particle has $q/m = 4\varepsilon_0\, g/\sigma$.

(a) Find the value of H if the particle just reaches the disc.

(b) Sketch the potential energy of the particle as a function of its height and find its equilibrium position.

Sol. Potential at a height H **on the axis of the disc $V(P)$**. The charge dq contained in the ring shown in figure, $dq = (2\pi r dr)\sigma$

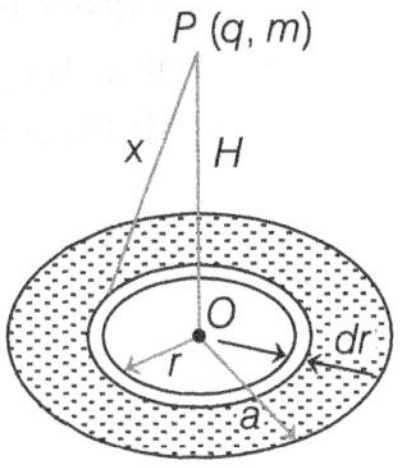

Potential of P due to this ring

$$dV = \frac{1}{4\pi\varepsilon_0} \cdot \frac{dq}{x},$$

where $\quad x = \sqrt{H^2 + r^2}$

$$dV = \frac{1}{4\pi\varepsilon_0} \cdot \frac{(2\pi r dr)\,\sigma}{\sqrt{H^2 + r^2}} = \frac{\sigma}{2\varepsilon_0} \frac{r dr}{\sqrt{H^2 + r^2}}$$

$\therefore$ Potential due to the complete disc,

$$V_P = \int_{r=0}^{r=a} dV = \frac{\sigma}{2\varepsilon_0} \int_{r=0}^{r=a} \frac{r dr}{\sqrt{H^2 + r^2}}$$

$$V_P = \frac{\sigma}{2\varepsilon_0} [\sqrt{a^2 + H^2} - H]$$

Potential at centre, O will be

$$V_O = \frac{\sigma a}{2\varepsilon_0} \qquad (H = 0)$$

(a) Particle is released from P and it just reaches point O. Therefore, from conservation of mechanical energy decrease in gravitational potential energy = increase in electrostatic potential energy

($\Delta KE = 0$ because $K_i = K_f = 0$)

$\therefore \quad mgH = q\,[V_O - V_P]$

or $\quad gH = \left(\frac{q}{m}\right)\left(\frac{\sigma}{2\varepsilon_0}\right)[a - \sqrt{a^2 + H^2} + H] \qquad$ ...(i)

$$\frac{q}{m} = \frac{4\varepsilon_0 g}{\sigma} \Rightarrow \frac{q\sigma}{2\varepsilon_0 m} = 2g$$

Substituting in Eq. (i), we get

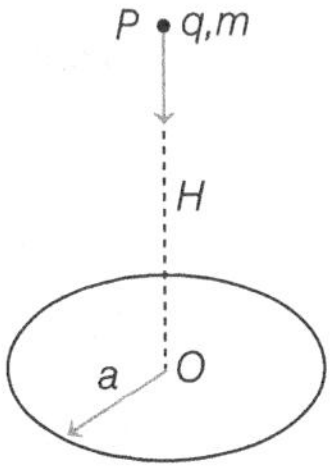

$$gH = 2g\,[a + H - \sqrt{a^2 + H^2}]$$

or $\quad \frac{H}{2} = (a + H) - \sqrt{a^2 + H^2}$

or $\quad \sqrt{a^2 + H^2} = a + \frac{H}{2}$

or $\quad a^2 + H^2 = a^2 + \frac{H^2}{4} + aH$

or $\quad \frac{3}{4}H^2 = aH$

or $\quad H = \frac{4}{3}a$ and $H = 0$

$\therefore \quad H = (4/3)a \qquad$ **Ans.**

(b) Potential energy of the particle at height H
= Electrostatic potential energy
+ gravitational potential energy

$\therefore \quad U = qV + mgH$

Here, $\quad V =$ Potential at height H

$\therefore \quad U = \frac{\sigma q}{2\varepsilon_0}[\sqrt{a^2 + H^2} - H] + mgH \qquad$ ...(ii)

At equilibrium position, $\quad F = \frac{-dU}{dH} = 0$

Differentiating Eq. (ii) w.r.t. H,

or $\quad mg + \frac{\sigma q}{2\varepsilon_0}\left[\left(\frac{1}{2}\right)(2H)\frac{1}{\sqrt{a^2 + H^2}} - 1\right] = 0$

$$\left[\text{as } \frac{\sigma q}{2\varepsilon_0} = 2mg\right]$$

$\therefore \quad mg + 2mg\left[\frac{H}{\sqrt{a^2 + H^2}} - 1\right] = 0$

or $\quad 1 + \frac{2H}{\sqrt{a^2 + H^2}} - 2 = 0$

$\Rightarrow \quad \frac{2H}{\sqrt{a^2 + H^2}} = 1$

or $\quad \frac{H^2}{a^2 + H^2} = \frac{1}{4}$ or $3H^2 = a^2$

or $\quad H = \frac{a}{\sqrt{3}} \qquad$ **Ans.**

From Eq. (ii), we can see that

$U = 2\,mga$ at $H = 0$ and

$U = U_{min} = \sqrt{3}\,mga$ at $H = \frac{a}{\sqrt{3}}$

Therefore, U-H graph will be as shown.

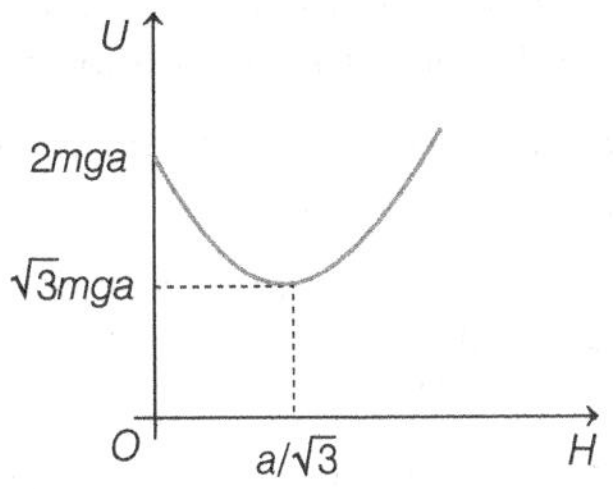

Note that at $H = \frac{a}{\sqrt{3}}$, U is minimum.

Therefore, $\quad H = \frac{a}{\sqrt{3}}$ is stable equilibrium position.

Important POINTS

- A charged particle can be accelerated or retarded by applying a potential difference V.
- Electron-volt is unit of energy.

$$1\text{ eV} = 1.6 \times 10^{-19}\text{ J}$$

- Free positive charge accelerates in moving from higher potential to lower potential and negative charge in moving from lower potential to higher potential.
- Change in kinetic energy = change in potential energy.

$\therefore \quad |\Delta K| = |\Delta U| = |qV|$

Here V is the potential difference.

- If a charge $q (= \pm ne)$ is accelerated by a potential difference V and initially the particle is at rest, then

$$KE = \frac{1}{2}mv^2 = (qV)\text{ J}$$

or $\quad KE = (nV)\text{ eV}$

and $\quad v = \sqrt{\frac{2qV}{m}}$

In the above equations V is in volt.

Ex. 6 An α-particle $q = +2e$ moves from 100V to 60V. If initially particle is at rest then find the kinetic energy in joule and electron-volt.

Sol. Positive charge is moving from higher potential to lower potential. So, it will accelerate.

Here, $\quad V = (100 - 60) = 40\text{ V} = \text{PD}$

(i) $\quad KE = (qV)\text{ J}$

$= (2 \times 1.6 \times 10^{-19})(40)$

$= 1.28 \times 10^{-17}\text{ J}$

(ii) $\quad KE = (nV)\text{ eV}$

$= (2)(40)\text{ eV}$

$= 80\text{ eV}$ **Ans.**

Exercise : An electron at rest moves from -20 V to 10V. Find kinetic energy in joule and electron-volt.

Ans. 4.8×10^{-18} J, 30 eV

MODULE 5 Exercise

Assertion and Reason

See directions on page 25.

1. **Assertion :** An independent negative charge moves itself from point A to point B. Then, potential at A should be less than potential at B.

   Reason : While moving from A to B kinetic energy of electron will increase.

2. **Assertion :** Due to two point charges electric field and electric potential can't be zero at some point, simultaneously.

   Reason : Field is a vector quantity and potential a scalar quantity.

Single Correct Option

3. Charges $2q$, $-q$ and $-q$ lie at the vertices of an equilateral triangle. The value of E and V at the centroid of the triangle will be
   (a) $E \neq 0$ and $V \neq 0$
   (b) $E = 0$ and $V = 0$
   (c) $E \neq 0$ and $V = 0$
   (d) $E = 0$ and $V \neq 0$

4. A ring of radius R is having two charges q and $2q$ distributed on its two half parts. The electric potential at a point on its axis at a distance of $2\sqrt{2}\,R$ from its centre is $\left(k = \frac{1}{4\pi\varepsilon_0}\right)$
   (a) $\frac{3kq}{R}$ (b) $\frac{kq}{3R}$ (c) $\frac{kq}{R}$ (d) $\frac{kq}{\sqrt{3}R}$

5. Two point charges $+q$ and $-q$ are held fixed at $(-a, 0)$ and $(a, 0)$ respectively of a x-y coordinate system, then
   (a) the electric field **E** at all points on the x-axis has the same direction
   (b) **E** at all points on the y-axis is along $\hat{\mathbf{i}}$
   (c) positive work is done in bringing a test charge from infinity to the origin
   (d) All of the above

6. At a certain distance from a point charge, the field intensity is 500 V/m and the potential is -3000 V. The distance to the charge and the magnitude of the charge respectively are
   (a) 6 m and 6 μC (b) 4 m and 2 μC
   (c) 6 m and 4 μC (d) 6 m and 2 μC

7. An electrical charge 2×10^{-8} C is placed at the point (1, 2, 4) m. At the point (4, 2, 0) m, the electric
(a) potential will be 36 V
(b) field will be along y-axis
(c) field will increase if the space between the points is filled with a dielectric
(d) All of the above

8. A particle A has charge $+q$ and a particle B has charge $+4q$ with each of them having the same mass m. When allowed to fall from rest through the same electric potential difference, the ratio of their speed $\frac{v_A}{v_B}$ will become
(a) 2 : 1 (b) 1 : 2 (c) 1 : 4 (d) 4 : 1

9. Four identical charges are placed at the points (1, 0, 0), (0, 1, 0), (– 1, 0, 0) and (0, –1, 0).
(i) The potential at the origin is zero
(ii) The field at the origin is zero
(iii) The potential at all points on the z-axis, other than the origin, is zero
(iv) The field at all points on the z-axis, other than the origin, acts along the z-axis
(a) (i), (ii) (b) (i), (iii) (c) (ii), (iii) (d) (ii), (iv)

10. Charges $+q$ and $-q$ are placed at points A and B respectively which are a distance $2L$ apart, C is the mid-point between A and B. The work done in moving a charge $+Q$ along the semicircle CRD is

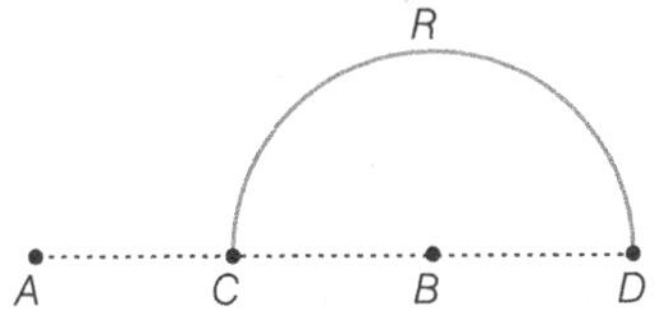

(a) $\frac{qQ}{4\pi\varepsilon_0 L}$ (b) $-\frac{qQ}{2\pi\varepsilon_0 L}$ (c) $\frac{qQ}{6\pi\varepsilon_0 L}$ (d) $-\frac{qQ}{6\pi\varepsilon_0 L}$

11. Six point charges are kept at the vertices of a regular hexagon of side L and centre O, as shown in the figure. Given that $K = \frac{1}{4\pi\varepsilon_0}\frac{q}{L^2}$, which of the following statement is **wrong**?

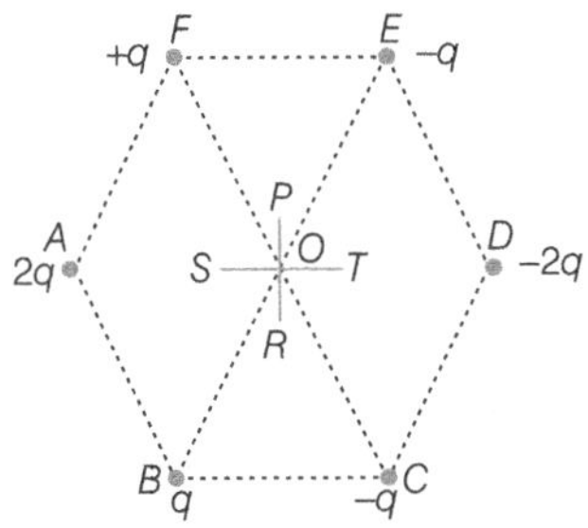

(a) The electric field at O is 6 K along OD
(b) The potential at O is zero
(c) The potential at all points on the line PR is same
(d) The potential at all points on the line ST is same

12. The curve represents the distribution of potential along the straight line joining the two charges Q_1 and Q_2 (separated by a distance r) then which of the following statements are correct?

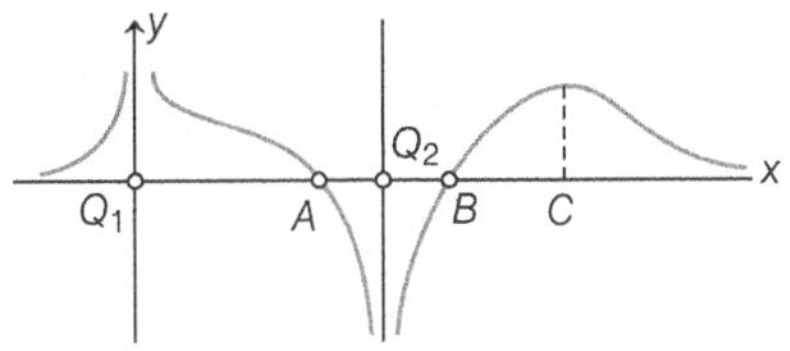

1. $|Q_1| > |Q_2|$
2. Q_1 is positive in nature
3. A and B are equilibrium points
4. C is a point of unstable equilibrium
(a) 1 and 2
(b) 1, 2 and 3
(c) 1, 2 and 4
(d) 1, 2, 3 and 4

13. A point charge $q_1 = q$ is placed at point P. Another point charge $q_2 = -q$ is placed at point Q. At some point $R (R \neq P, R \neq Q)$, electric potential due to q_1 is V_1 and electric potential due to q_2 is V_2. Which of the following is correct?
(a) Only for some points $V_1 > V_2$
(b) Only for some points $V_2 > V_1$
(c) For all points $V_1 > V_2$
(d) For all points $V_2 > V_1$

14. The variation of electric field between two charges q_1 and q_2 along the line joining the charges is plotted against distance from q_1 (taking rightward direction of electric field as positive) as shown in the figure. Then, the correct statement is

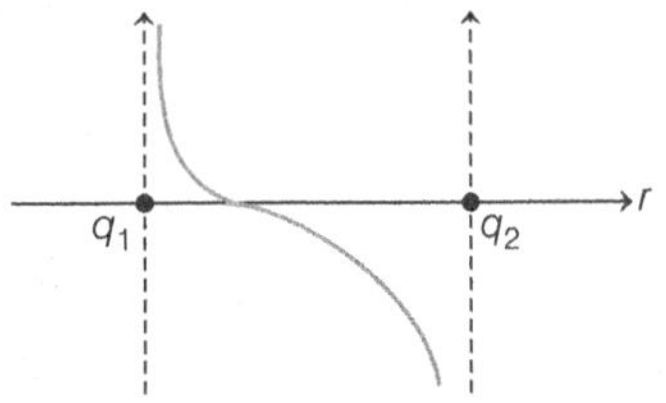

(a) q_1 and q_2 are positive and $q_1 < q_2$
(b) q_1 and q_2 are positive and $q_1 > q_2$
(c) q_1 is positive and q_2 is negative $q_1 < |q_2|$
(d) q_1 and q_2 are negative and $|q_1| < |q_2|$

15. Two identical coaxial rings each of radius R are separated by a distance of $\sqrt{3}R$. They are uniformly charged with charges $+Q$ and $-Q$ respectively. The minimum kinetic energy with which a charged particle (charge $+q$) should be projected from the centre of the negatively charged ring along the axis of the rings such that it reaches the centre of the positively charged ring is
(a) $\frac{Qq}{4\pi\varepsilon_0 R}$ (b) $\frac{Qq}{2\pi\varepsilon_0 R}$
(c) $\frac{Qq}{8\pi\varepsilon_0 R}$ (d) $\frac{3Qq}{4\pi\varepsilon_0 R}$

16. Two identical positive charges are placed at $x = -a$ and $x = a$. The correct variation of potential V along the x-axis is given by

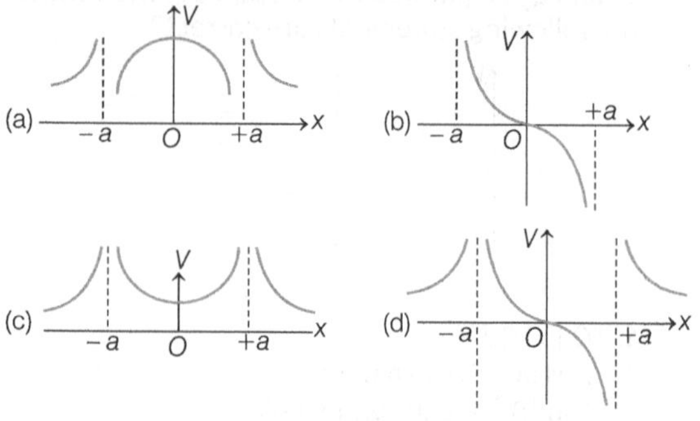

17. The electrostatic potential due to the charge configuration at point P as shown in figure for $b << a$ is

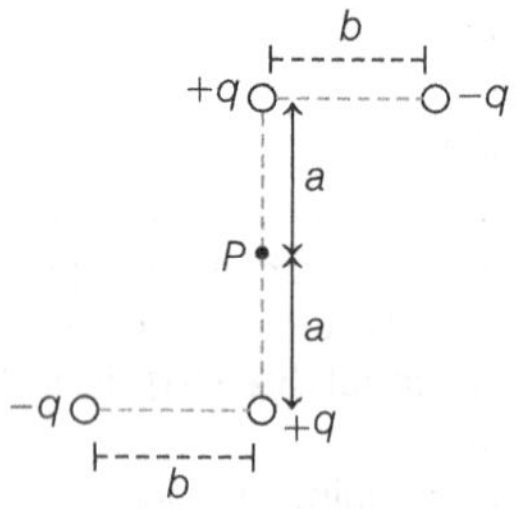

(a) $\frac{2q}{4\pi\varepsilon_0 a}$ (b) $\frac{2qb^2}{4\pi\varepsilon_0 a^3}$ (c) $\frac{qb^2}{4\pi\varepsilon_0 a^3}$ (d) zero

Multiple Correct Options

18. A free electron enters in higher potential region V_2 from lower potential region V_1, then its velocity
(a) will increase
(b) will change in direction but not in magnitude
(c) will not change in direction of field
(d) will not change in direction perpendicular to field

Subjective Questions

19. A point charge $q_1 = +2\mu C$ is placed at the origin of coordinates. A second charge, $q_2 = -3\ \mu C$, is placed on the x-axis at $x = 100$ cm. At what point (or points) on the x-axis will the absolute potential be zero?

20. A charge Q is spread uniformly in the form of a line charge density $\lambda = \frac{Q}{3a}$ on the sides of an equilateral triangle of perimeter $3a$. Calculate the potential at the centroid C of the triangle.

21. A small particle has charge $-5.00\ \mu C$ and mass 2.00×10^{-4} kg. It moves from point A, where the electric potential is $V_A = +200$ V, to point B, where the electric potential is $V_B = +800$ V. The electric force is the only force acting on the particle. The particle has speed 5.00 m/s at point A. What is its speed at point B? Is it moving faster or slower at B than at A? Explain.

22. A plastic rod has been formed into a circle of radius R. It has a positive charge $+Q$ uniformly distributed along one-quarter of its circumference and a negative charge of $-6Q$ uniformly distributed along the rest of the circumference (figure). With $V = 0$ at infinity, what is the electric potential

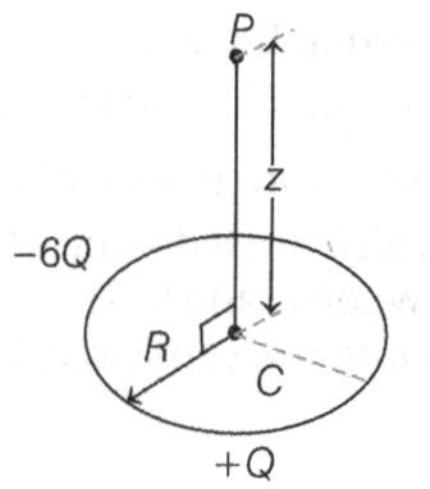

(a) at the centre C of the circle and
(b) at point P, which is on the central axis of the circle at distance z from the centre?

23. A point charge $q_1 = +2.40\ \mu C$ is held stationary at the origin. A second point charge $q_2 = -4.30\ \mu C$ moves from the point $x = 0.150$ m, $y = 0$ to the point $x = 0.250$ m, $y = 0.250$ m. How much work is done by the electric force on q_2?

Answers

1. (b) **2.** (b) **3.** (c) **4.** (c) **5.** (b) **6.** (d) **7.** (a) **8.** (b) **9.** (d) **10.** (d)
11. (d) **12.** (a) **13.** (c) **14.** (a) **15.** (a) **16.** (c) **17.** (c) **18.** (a,d)
19. At $x = 40$ cm and $x = -200$ cm **20.** $V = 2.634 \frac{Q}{4\pi\varepsilon_0 a}$ **21.** 7.42 m/s, Faster **22.** (a) $\frac{-5Q}{4\pi\varepsilon_0 R}$ (b) $\frac{-5Q}{4\pi\varepsilon_0\sqrt{R^2+z^2}}$
23. -0.356 J

See hints on pages 204-206.

MODULE 6

Relation Between Electric Field and Potential

As we have discussed above, an invisible space is produced across a charge or system of charges in which any other test charge experiences an electrical force.

The vector quantity related to this force is known as electric field. Further, a work is done by this electrostatic force when this test charge is moved from one point to another point.

The scalar quantity related to this work done is called potential. Electric field (**E**) and potential (V) are different at different positions. So, they are functions of position.

In a cartesian coordinate system, position of a particle can be represented by three variable coordinates x, y and z. Therefore, **E** and V are functions of three variables x, y and z.

In physics, we normally keep least number of variables. So, sometimes E and V are the functions of a single variable x or r. Here, x is the x-coordinate along x-axis and r is normally distance from a point charge or from the centre of a charged sphere or charged spherical shell. From the x-coordinate, we can cover only x-axis. But, from the variable r, we can cover the whole space.

Now, **E** and V functions are related to each other either by differentiation or integration. As far as differentiation is concerned, if there are more than one variables then partial differentiation is done and in case of single variable direct differentiation is required. In case of integration, some limit is required.

Limit means value of the function which we get after integration should be known to us at some position.

For example, after integrating **E**, we get V. So, value of V should be known at some given position. Without knowing some limit, an unknown in the form of constant of integration remains in the equation. One known limit of V is : potential is zero at infinity.

Conversion of *V* function into E function

This requires differentiation.

Case 1 When variables are more than one

In this case,

$$\mathbf{E} = E_x \hat{\mathbf{i}} + E_y \hat{\mathbf{j}} + E_z \hat{\mathbf{k}}$$

Here, $E_x = -\dfrac{\partial V}{\partial x} = -$ (partial derivative of V w.r.t. x)

$E_y = -\dfrac{\partial V}{\partial y} = -$ (partial derivative of V w.r.t. y)

$E_z = -\dfrac{\partial V}{\partial z} = -$ (partial derivative of V w.r.t. z)

$$\therefore \quad \mathbf{E} = -\left[\frac{\partial V}{\partial x}\hat{\mathbf{i}} + \frac{\partial V}{\partial y}\hat{\mathbf{j}} + \frac{\partial V}{\partial z}\hat{\mathbf{k}}\right]$$

This is also sometimes written as

$$\mathbf{E} = -\text{ gradient } V = -\text{ gradient } V = -\nabla V$$

Example 1 The electric potential in a region is represented as

$$V = 2x + 3y - z$$

obtain expression for electric field strength.

Sol. $\mathbf{E} = -\left[\dfrac{\partial V}{\partial x}\hat{\mathbf{i}} + \dfrac{\partial V}{\partial y}\hat{\mathbf{j}} + \dfrac{\partial V}{\partial z}\hat{\mathbf{k}}\right]$

Here, $\dfrac{\partial V}{\partial x} = \dfrac{\partial}{\partial x}(2x + 3y - z) = 2$

$\dfrac{\partial V}{\partial y} = \dfrac{\partial}{\partial y}(2x + 3y - z) = 3$

$\dfrac{\partial V}{\partial z} = \dfrac{\partial}{\partial z}(2x + 3y - z) = -1$

$\therefore \quad \mathbf{E} = -2\hat{\mathbf{i}} - 3\hat{\mathbf{j}} + \hat{\mathbf{k}}$ **Ans.**

Case 2 When variable is only one In this case, electric potential is function of only one variable (say r or x) and we can write the expression like :

$$E = -\frac{dV}{dr} = -\frac{dV}{dx}$$

or $E = -$ slope of V-r graph

$= -$slope of V-x graph

Example Electric potential due to a point charge q at distance r is given as

$$V = \frac{1}{4\pi\varepsilon_0}\cdot\frac{q}{r} \Rightarrow \frac{dV}{dr} = -\frac{1}{4\pi\varepsilon_0}\cdot\frac{q}{r^2}$$

$$\therefore \quad E = -\frac{dV}{dr} = \frac{1}{4\pi\varepsilon_0}\cdot\frac{q}{r^2}$$

and we know that this is the expression of electric field due to a point charge.

Note *E is a vector quantity. In the above method, if single variable is x and E comes out to be positive, then direction of E is towards positive x-axis. Negative value of E means direction is towards negative x-axis. If variable is r, then positive value* *of E means away from the point charge or away from the centre of charged spherical body and negative value of E means towards the charge or towards the centre of charged spherical body.*

Let us take one more example : We wish to find E-r graph corresponding to V-r graph as shown in figure.

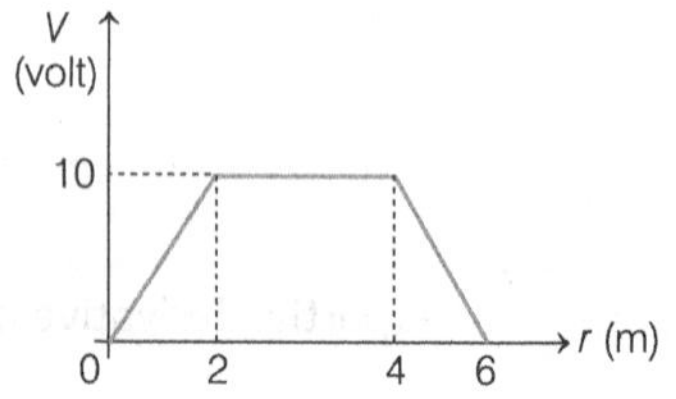

$$E=-\frac{dV}{dr}=-(\text{Slope of } V\text{-}r \text{ graph})$$

Electric field $E=-5$ V/m for $0 \le r \le 2$ m as slope of V-r graph is 5 V/m. $E=0$ for $2 \text{ m} \le r \le 4$ m as slope of V-r graph in this region is zero. Similarly, $E=5$ V/m for $4 \text{ m} \le r \le 6$ m as slope in this region is -5 V/m.

So, the corresponding E-r graph is as shown in figure given below.

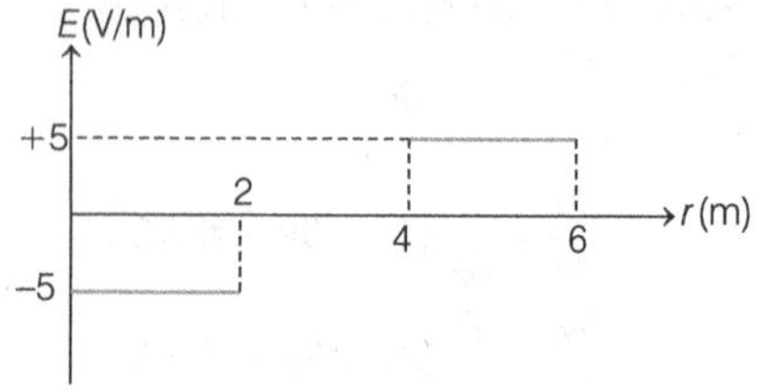

Example 2 The electric potential V at any point x, y, z (all in metre) in space is given by $V=4x^2$ volt. The electric field at the point (1m, 0, 2 m) is ....V/m. **(1992 JEE)**

Sol. $\mathbf{E}=-\left[\frac{\partial V}{\partial x}\hat{i}+\frac{\partial V}{\partial y}\hat{j}+\frac{\partial V}{\partial z}\hat{k}\right] \Rightarrow V=4x^2$

Therefore, $\frac{\partial V}{\partial x}=8x$ and $\frac{\partial V}{\partial y}=0=\frac{\partial V}{\partial z}$

$$\mathbf{E}=-8x\hat{i}$$

or $\mathbf{E}$ at (1 m, 0, 2 m) is $-8\hat{i}$ V/m.

Note *Here the variable is only x. So, we can write,*

$$E=-\frac{dV}{dx}=-8x.$$

At given point $E=-8$ V/m.

Negative sign means along negative x- direction.

Conversion of E into V

We have learnt, how to find electric field **E** from the electrostatic potential V. Let us now discuss how to calculate potential difference or absolute potential if electric field **E** is known. For this, use the relation

$$dV=-\mathbf{E}\cdot d\mathbf{r}$$

or $$\int_A^B dV=-\int_A^B \mathbf{E}\cdot d\mathbf{r}$$

or $$V_B-V_A=-\int_A^B \mathbf{E}\cdot d\mathbf{r}$$

Here, $$d\mathbf{r}=dx\,\hat{i}+dy\,\hat{j}+dz\,\hat{k}$$

When E is Uniform

Let us take this case with the help of an example.

Example 3 Find V_{ab} in an electric field $\mathbf{E}=(2\hat{i}+3\hat{j}+4\hat{k})\frac{N}{C}$,

where $\mathbf{r}_a=(\hat{i}-2\hat{j}+\hat{k})$ m and $\mathbf{r}_b=(2\hat{i}+\hat{j}-2\hat{k})$ m

Sol. Here, the given field is uniform (constant). So using,

$$dV=-\mathbf{E}\cdot d\mathbf{r}$$

or $$V_{ab}=V_a-V_b=-\int_b^a \mathbf{E}\cdot d\mathbf{r}$$

$$=-\int_{(2,1,-2)}^{(1,-2,1)}(2\hat{i}+3\hat{j}+4\hat{k})\cdot(dx\,\hat{i}+dy\,\hat{j}+dz\,\hat{k})$$

$$=-\int_{(2,1,-2)}^{(1,-2,1)}(2\,dx+3\,dy+4\,dz)$$

$$=-[2x+3y+4z]_{(2,1,-2)}^{(1,-2,1)}$$

$$=-1\text{ V}$$ **Ans.**

Note *In uniform electric field, we can also apply*

$$\boxed{V=Ed}$$

Here, V is the potential difference between any two points, E is the magnitude of uniform electric field and d is the projection of the distance between two points along the electric field.

For finding the potential difference between points A and B we will have to keep two points in mind,

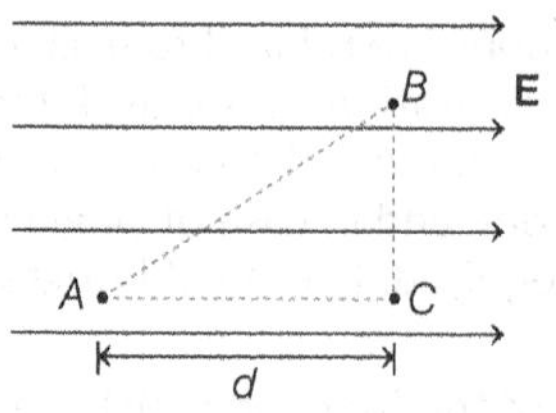

(i) $V_A > V_B$ as electric lines always flow from higher potential to lower potential.

(ii) $d \ne AB$ but $d=AC$

Hence, in the above figure,

$$V_A-V_B=Ed$$

Example 4 In uniform electric field $\mathbf{E}=10$ N/C, find

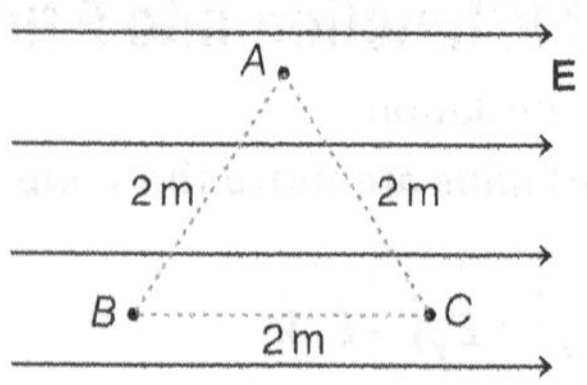

(a) V_A-V_B (b) V_B-V_C

Sol. (a) $V_B > V_A$,

so V_A-V_B will be negative.

Further $d_{AB}=2\cos 60°=1$ m

$\therefore \quad V_A - V_B = -Ed_{AB} = (-10)(1)$

$= -10$ volt **Ans.**

(b) $V_B > V_C$, so $V_B - V_C$ will be positive.

Further, $d_{BC} = 2.0$ m

$\therefore \quad V_B - V_C = (10)(2)$

$= 20$ volt **Ans.**

Example 5 A uniform electric field of 100 V/m is directed at 30° with the positive x-axis as shown in figure. Find the potential difference V_{BA} if $OA = 2$ m and $OB = 4$ m.

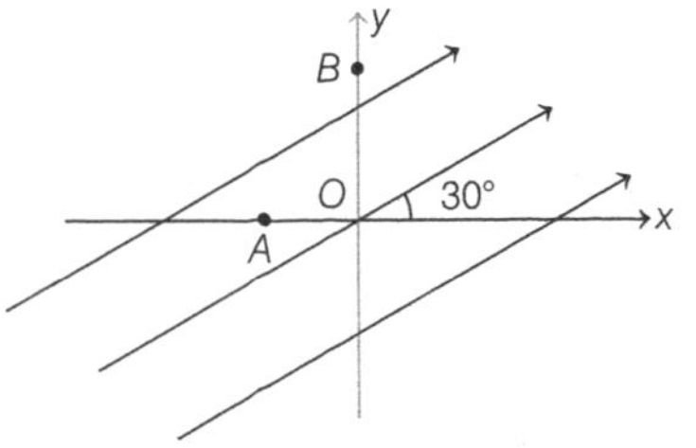

Sol. This problem can be solved by both the methods discussed above.

Method 1. Electric field in vector form can be written as

$$\mathbf{E} = (100 \cos 30° \hat{\mathbf{i}} + 100 \sin 30° \hat{\mathbf{j}}) \text{ V/m}$$

$$= (50\sqrt{3}\,\hat{\mathbf{i}} + 50\hat{\mathbf{j}}) \text{ V/m}$$

$$A = (-2\text{m}, 0, 0)$$

and $B = (0, 4\text{m}, 0)$

$$\therefore \quad V_{BA} = V_B - V_A = -\int_A^B \mathbf{E}\cdot d\mathbf{r}$$

$$= -\int_{(-2\text{m},0,0)}^{(0,4\text{m},0)} (50\sqrt{3}\,\hat{\mathbf{i}} + 50\hat{\mathbf{j}})\cdot(dx\,\hat{\mathbf{i}} + dy\,\hat{\mathbf{j}} + dz\,\hat{\mathbf{k}})$$

$$= -[50\sqrt{3}x + 50y]_{(-2\text{m},0,0)}^{(0,4\text{m},0)}$$

$$= -100(2 + \sqrt{3}) \text{ V}$$ **Ans.**

Method 2. We can also use, $V = Ed$

With the view that $V_A > V_B$ or $V_B - V_A$ will be negative.

Here, $d_{AB} = OA \cos 30° + OB \sin 30°$

$$= 2 \times \frac{\sqrt{3}}{2} + 4 \times \frac{1}{2} = (\sqrt{3} + 2)$$

$$\therefore \quad V_B - V_A = -Ed_{AB}$$

$$= -100(2 + \sqrt{3})$$ **Ans.**

Example 6 A uniform electric field pointing in positive x-direction exists in a region. Let A be the origin, B be the point on the x-axis at $x = +1$ cm and C be the point on the y-axis at $y = +1$ cm. Then, the potentials at the points A, B and C satisfy **(2001 JEE)**

(a) $V_A < V_B$

(b) $V_A > V_B$

(c) $V_A < V_C$

(d) $V_A > V_C$

Sol. Potential decreases in the direction of electric field. Dotted lines are equipotential lines (lines of same potentials).

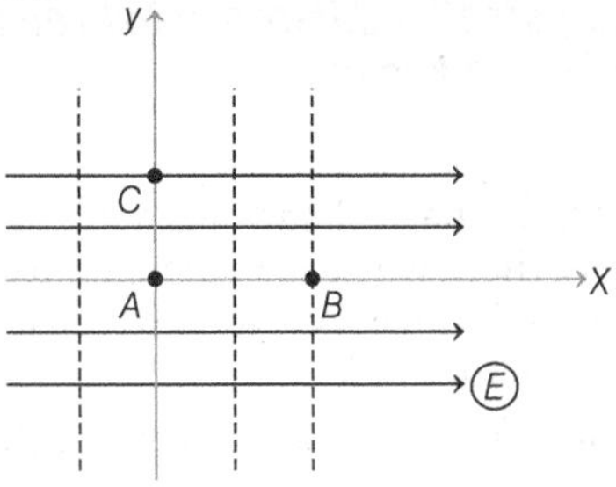

$\therefore \quad V_A = V_C$ and $V_A > V_B$

Hence, the correct option is (b).

Example 7 A non-conducting ring of radius 0.5 m carries a total charge of 1.11×10^{-10} C distributed non-uniformly on its circumference producing an electric field **E** everywhere in space. The value of the integral $\int_{l=\infty}^{l=0} -\mathbf{E}\cdot d\mathbf{l}$ ($l = 0$ being centre of the ring) in volt is **(1997 JEE)**

(a) + 2 (b) −1 (c) −2 (d) zero

Sol. $-\int_{l=\infty}^{l=0} \mathbf{E}\cdot d\mathbf{l} = \int_{l=\infty}^{l=0} dV = V(\text{centre}) - V(\text{infinity})$

but $V(\text{infinity}) = 0$

$\therefore -\int_{l=\infty}^{l=0} \mathbf{E}\cdot d\mathbf{l}$ corresponds to potential at centre of ring.

and $V(\text{centre}) = \dfrac{1}{4\pi\varepsilon_0}\cdot\dfrac{q}{R}$

$$= \frac{(9\times10^9)(1.11\times10^{-10})}{0.5} \approx 2\text{V}$$

Therefore, the correct answer is (a).

Example 8 Two points A and B are 2 cm apart and a uniform electric field E acts along the straight line AB directed from A to B with $E = 200$ N/C. A particle of charge $+10^{-6}$ C is taken from A to B along AB. Calculate

(a) the force on the charge

(b) the potential difference $V_A - V_B$ and

(c) the work done on the charge by **E**

Sol. (a) Electrostatic force on the charge,

$$F = qE = (10^{-6})(200)$$

$$= 2 \times 10^{-4} \text{ N}$$ **Ans.**

(b) In uniform electric field,

$$V = E \cdot d$$

or $V_A - V_B = 200 \times 2 \times 10^{-2}$

$= 4$ V **Ans.**

(c) $W = Fs\cos\theta = (2\times10^{-4})(2\times10^{-2})\cos 0°$

$= 4 \times 10^{-6}$ J **Ans.**

Typed Problems

Type 1. ***To find potential difference between two points when electric field is known.***

Concept

In the theory part, we have already read the relation between **E** and V. There we have taken a simple case when electric field was uniform. Here, two more cases are possible depending on the nature of **E**.

When E has a Function Like $f_1(x)\hat{i} + f_2(y)\hat{j} + f_3(z)\hat{k}$

In this case also, we will use the same approach. Let us take an example.

Ex. 1 Find the potential difference V_{AB} between A (2m, 1m, 0) and B (0, 2m, 4m) in an electric field,

$$\mathbf{E} = (x\hat{i} - 2y\hat{j} + z\hat{k})\ \text{V/m}$$

Sol. $dV = -\mathbf{E} \cdot d\mathbf{r}$

$$\int_B^A dV = -\int_B^A \mathbf{E} \cdot d\mathbf{r}$$

$$= -\int_{(0,\,2,\,4)}^{(2,\,1,\,0)} (x\hat{i} - 2y\hat{j} + z\hat{k}) \cdot (dx\hat{i} + dy\hat{j} + dz\hat{k})$$

$$\therefore \quad V_A - V_B = -\int_{(0,\,2,\,4)}^{(2,\,1,\,0)} (xdx - 2ydy + zdz)$$

or $$V_{AB} = -\left[\frac{x^2}{2} - y^2 + \frac{z^2}{2}\right]_{(0,\,2,\,4)}^{(2,\,1,\,0)}$$

$$= 3 \text{ volt}$$ **Ans.**

When E · dr becomes a Perfect Differential.

Same method is used when $\mathbf{E} \cdot d\mathbf{r}$ becomes a perfect differential. The following example will illustrate the theory.

Ex. 2 Find potential difference V_{AB} between A (0, 0, 0) and B (1m, 1m, 1m) in an electric field (a) $\mathbf{E} = y\hat{i} + x\hat{j}$ (b) $\mathbf{E} = 3x^2y\,\hat{i} + x^3\,\hat{j}$

Sol. (a) $dV = -\mathbf{E} \cdot d\mathbf{r}$

$$\therefore \quad \int_B^A dV = -\int_{(1,\,1,\,1)}^{(0,\,0,\,0)} (y\hat{i} + x\hat{j}) \cdot (dx\hat{i} + dy\hat{j} + dz\hat{k})$$

or $$V_A - V_B = -\int_{(1,\,1,\,1)}^{(0,\,0,\,0)} (y\,dx + x\,dy)$$

or $$V_{AB} = -\int_{(1,\,1,\,1)}^{(0,\,0,\,0)} d\,(xy) \qquad [\text{as } y\,dx + x\,dy = d\,(xy)]$$

$$\therefore \quad V_{AB} = -[xy]_{(1,1,1)}^{(0,0,0)} = 1\text{ V}$$ **Ans.**

(b) $dV = -\mathbf{E} \cdot d\mathbf{r}$

$$\therefore \quad \int_B^A dV = -\int_{(1,\,1,\,1)}^{(0,\,0,\,0)} (3x^2y\hat{i} + x^3\hat{j}) \cdot (dx\hat{i} + dy\hat{j} + dz\hat{k})$$

or $$V_A - V_B = -\int_{(1,\,1,\,1)}^{(0,\,0,\,0)} (3x^2\,ydx + x^3dy)$$

$$= -\int_{(1,\,1,\,1)}^{(0,\,0,\,0)} d\,(x^3y)$$

$$\therefore \quad V_{AB} = -[x^3y]_{(1,1,1)}^{(0,0,0)} = 1\text{V}$$ **Ans.**

MODULE 6 Exercise

Assertion and Reason

See directions on page 25.

1. **Assertion :** At a point electric potential is decreasing along x-axis at a rate of 10 V/m. Therefore, x-component of electric field at this point should be 10 V/m along x-axis.

   Reason : Magnitude of $E_x = \left|\frac{\partial V}{\partial x}\right|$

2. **Assertion :** In the electric field $\mathbf{E} = (4\hat{\mathbf{i}} + 4\hat{\mathbf{j}})$ N/C, electric potential at A (4m, 0) is more than the electric potential at B (0, 4m).

   Reason : Electric lines of forces always travel from higher potential to lower potential.

Single Correct Option

3. A uniform electric field exists in x-y plane. The potential of points A (2m, 2m), B (−2m, 2m) and C (2m, 4m) are 4V, 16V and 12V, respectively. The electric field is
   (a) $(4\hat{\mathbf{i}} + 5\hat{\mathbf{j}})$ V/m
   (b) $(3\hat{\mathbf{i}} + 4\hat{\mathbf{j}})$ V/m
   (c) $-(3\hat{\mathbf{i}} + 4\hat{\mathbf{j}})$ V/m
   (d) $(3\hat{\mathbf{i}} - 4\hat{\mathbf{j}})$ V/m

4. In a uniform electric field, the potential is 10 V at the origin of coordinates and 8 V at each of the points (1, 0, 0), (0, 1, 0) and (0, 0, 1). The potential at the point (1, 1, 1) will be
   (a) 0 (b) 4 V
   (c) 8 V (d) 10 V

5. The electric potential at a point (x, y) in the x-y plane is given by $V = -kxy$. The field intensity at a distance r in this plane, from the origin is proportional to
   (a) r^2
   (b) r
   (c) $1/r$
   (d) $1/r^2$

6. For two points A and B in an electric field, if ΔW is work done by electrostatic forces in taking a test charge q' from A to B, $\Delta U = U_B - U_A$ is change in potential energy of a test charge when it is moved from A to B. $\Delta V = V_B - V_A$ is change in potential due to field if we move along path AB from A to B and ds is differential displacement along path AB.

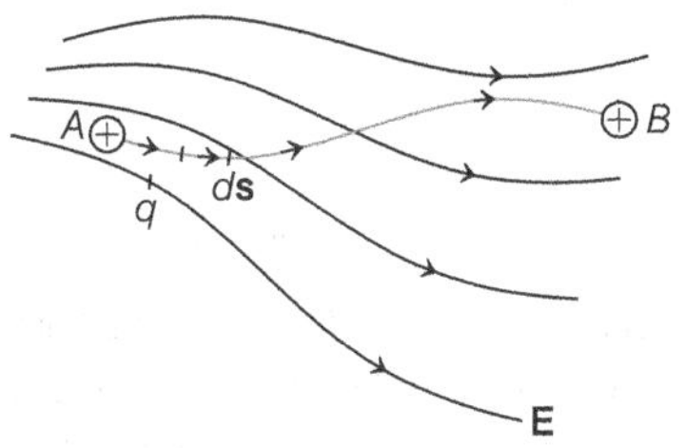

Then, which of the following options is correct?
   (a) $V_B - V_A = \int_A^B \mathbf{E} \cdot d\mathbf{s}$
   (b) $\Delta W = -q' \int_A^B \mathbf{E} \cdot d\mathbf{s}$
   (c) $\Delta V = -\frac{\Delta U}{q'}$
   (d) $\Delta V = -\frac{\Delta W}{q'} = \frac{\Delta U}{q'}$

Match the Columns

7. In an electric field $\mathbf{E} = (2\hat{\mathbf{i}} + 4\hat{\mathbf{j}})$ N/C, electric potential at origin is 0 V. Match the following two columns.

	Column I		Column II
(a)	Potential at (4m, 0)	(p)	8 V
(b)	Potential at (– 4m, 0)	(q)	– 8 V
(c)	Potential at (0, 4m)	(r)	16 V
(d)	Potential at (0, – 4 m)	(s)	– 16 V

Subjective Questions

8. Determine the electric field strength vector if the potential of this field depends on x, y coordinates as
   (a) $V = a(x^2 - y^2)$ (b) $V = axy$

   where, a is a constant.

9. The electrical potential function for an electrical field directed parallel to the x-axis is shown in the given graph.

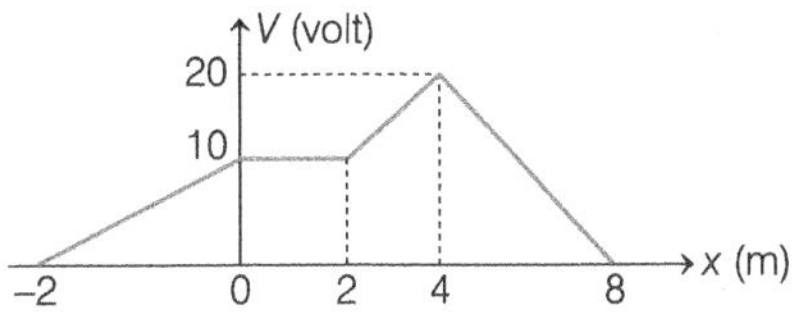

Draw the graph of electric field strength.

10. In the uniform electric field shown in figure, find :

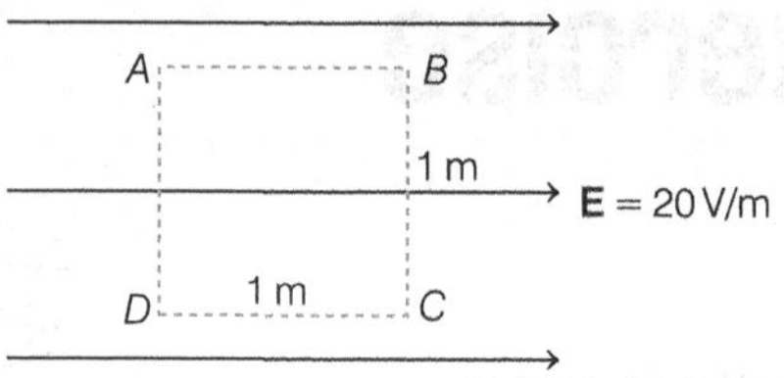

(a) $V_A - V_D$ (b) $V_A - V_C$ (c) $V_B - V_D$ (d) $V_C - V_D$

11. A uniform electric field of magnitude 250 V/m is directed in the positive x-direction. A + 12μC charge moves from the origin to the point $(x, y) = (20.0\text{ cm}, 5.0\text{ cm})$.
(a) What was the change in the potential energy of this charge?
(b) Through what potential difference did the charge move?

Answers

1. (d) 2. (d) 3. (d) 4. (b) 5. (b) 6. (d)
7. (a → q; b → p; c → s; d → r) 8. (a) $\mathbf{E} = -2a\,(x\hat{\mathbf{i}} - y\hat{\mathbf{j}})$ (b) $\mathbf{E} = -a(y\hat{\mathbf{i}} + x\hat{\mathbf{j}})$
9.

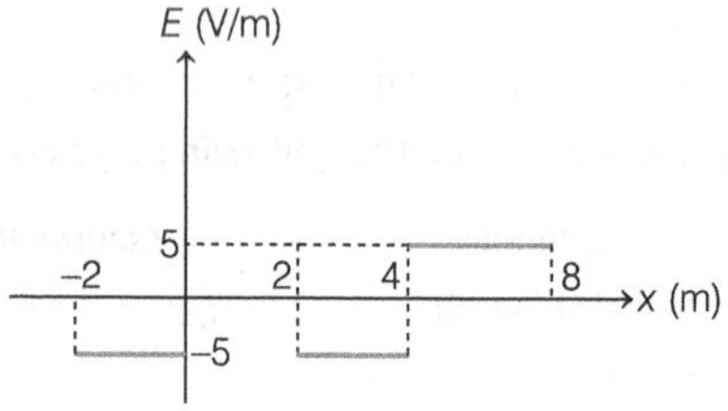

10. (a) Zero (b) 20 V (c) – 20 V (d) – 20 V 11. (a) -6×10^{-4} J (b) 50 V

See hints on page 206.

MODULE 7

Equipotential Surfaces

The equipotential surfaces in an electric field have the same basic idea as topographic maps used by civil engineers or mountain climbers. On a topographic map, contour lines are drawn passing through the points having the same elevation. The potential energy of a mass m does not change along a contour line as the elevation is same everywhere.

By analogy to contour lines on a topographic map, an **equipotential surface** is a three-dimensional surface on which the electric potential V is the same at every point on it. An equipotential surface has the following characteristics.

1. Potential difference between any two points in an equipotential surface is zero.
2. If a test charge q_0 is moved from one point to the other on such a surface, the electric potential energy q_0V remains constant.
3. No work is done by the electric force when the test charge is moved along this surface.
4. Two equipotential surfaces can never intersect each other because otherwise the point of intersection will have two potentials which is of course not possible.
5. As the work done by electric force is zero when a test charge is moved along the equipotential surface, it follows that **E** must be perpendicular to the surface at every point so that the electric force q_0**E** will always be perpendicular to the displacement of a charge moving on the surface. Thus, **field lines and equipotential surfaces are always mutually perpendicular.** Some equipotential surfaces are shown in figure given below.

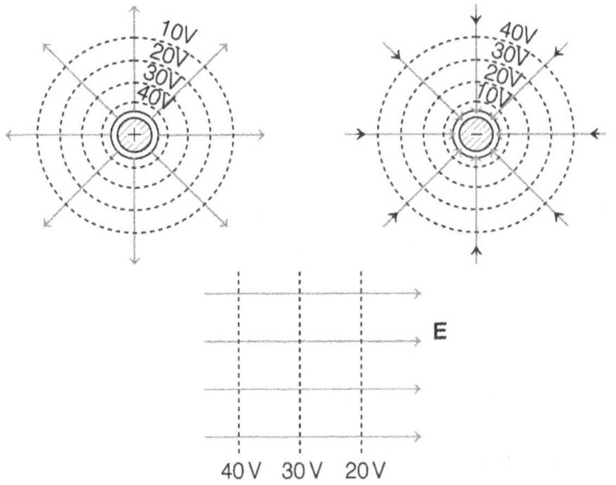

The equipotential surfaces are a family of concentric spheres for a point charge or a sphere of charge and are a family of concentric cylinders for a line of charge or cylinder of charge. For a special case of a uniform field, where the field lines are straight, parallel and equally spaced the equipotential surfaces are parallel planes perpendicular to the field lines.

Note *While drawing the equipotential surfaces, we should keep in mind the two main points.*

(i) These are perpendicular to field lines at all places.

(ii) Field lines always flow from higher potential to lower potential.

Example 1 Equipotential spheres are drawn round a point charge. As we move away from the charge, will the spacing between two spheres having a constant potential difference decrease, increase or remain constant.

Sol. $V_1 > V_2$

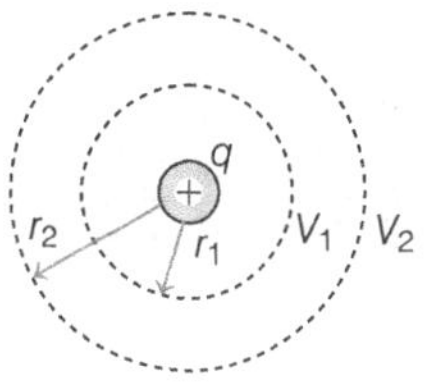

$$V_1 = \frac{1}{4\pi\varepsilon_0} \cdot \frac{q}{r_1}$$

and $$V_2 = \frac{1}{4\pi\varepsilon_0} \cdot \frac{q}{r_2}$$

Now, $$V_1 - V_2 = \frac{q}{4\pi\varepsilon_0}\left(\frac{1}{r_1} - \frac{1}{r_2}\right)$$

$$= \frac{q}{4\pi\varepsilon_0}\left(\frac{r_2 - r_1}{r_1 r_2}\right)$$

$$\therefore \quad (r_2 - r_1) = \frac{(4\pi\varepsilon_0)(V_1 - V_2)}{q}(r_1 r_2)$$

For a constant potential difference $(V_1 - V_2)$,

$$r_2 - r_1 \propto r_1 r_2$$

i.e. the spacing between two spheres $(r_2 - r_1)$ increases as we move away from the charge, because the product $r_1 r_2$ will increase.

Example 2 Some equipotential surfaces are shown in figure. The magnitude and direction of electric field is

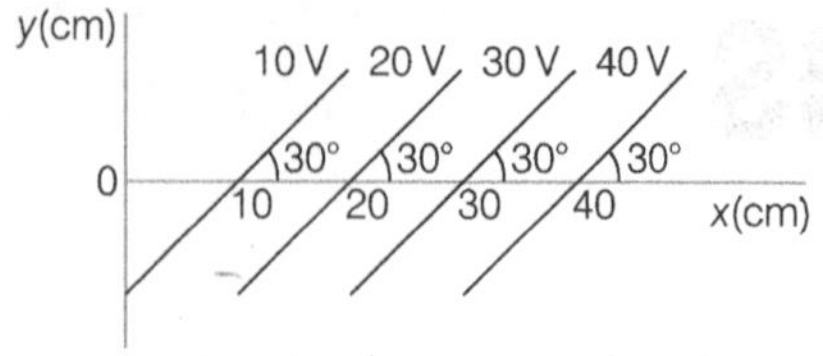

(a) 2 V/cm at angle 120° with +ve x-axis in anti-clockwise direction
(b) 2 V/cm at angle 120° with +ve x-axis in clockwise direction
(c) 1 V/cm at angle 150° with +ve x-axis in anti-clockwise direction
(d) 1 V/cm at angle 120° with +ve x-axis in clockwise direction

Sol. Electric lines of forces always flow from higher potential to lower potential.

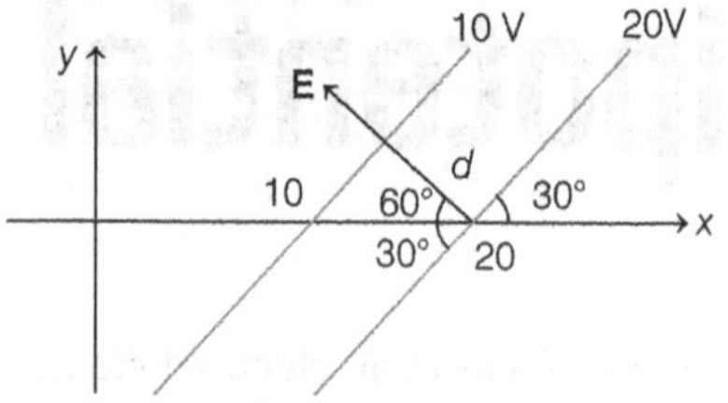

$$d = 10\cos 60° = 5 \text{ cm}$$
$$= 5 \times 10^{-2} \text{ m}$$
$$E = \frac{V}{d} = \frac{20 - 10}{5 \times 10^{-2}} = 200 \text{ V/m}$$
$$= 2 \text{ V/cm}$$

$\therefore$ The correct option is (a).

MODULE 8

Electric Dipole

A pair of equal and opposite point charges $\pm q$, that are separated by a fixed distance is known as **electric dipole.** Electric dipole occurs in nature in a variety of situations. The hydrogen fluoride molecule (HF) is typical.

Every electric dipole is characterized by its **electric dipole moment** which is a vector quantity **p** directed from the negative to the positive charge.

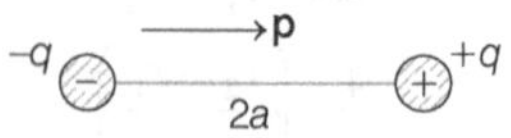

The magnitude of dipole moment is

$$p = (2a)\,q$$

Here, $2a$ is the distance between the two charges.

Electric Potential and Field Due to an Electric Dipole

Consider an electric dipole lying along positive y-direction with its centre at origin.

$$\mathbf{p} = 2aq\,\hat{\mathbf{j}}$$

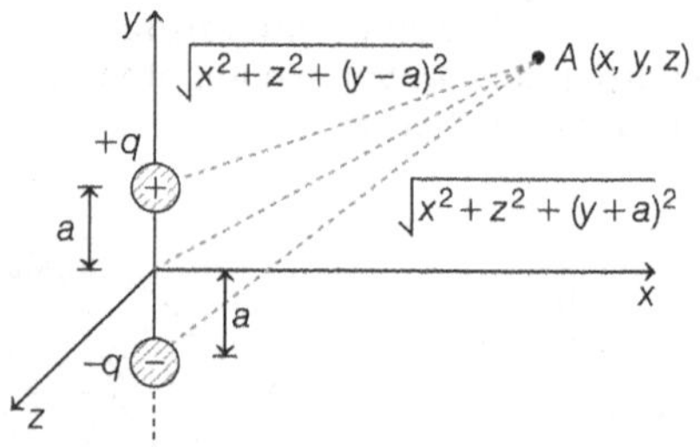

The electric potential due to this dipole at point $A\,(x, y, z)$ as shown is simply the sum of the potentials due to the two charges. Thus,

$$V = \frac{1}{4\pi\varepsilon_0}\left[\frac{q}{\sqrt{x^2 + (y-a)^2 + z^2}} - \frac{q}{\sqrt{x^2 + (y+a)^2 + z^2}}\right]$$

By differentiating this function, we obtain the electric field of the dipole.

$$E_x = -\frac{\partial V}{\partial x} = \frac{q}{4\pi\varepsilon_0}\left\{\frac{x}{[x^2+(y-a)^2+z^2]^{3/2}} - \frac{x}{[x^2+(y+a)^2+z^2]^{3/2}}\right\}$$

$$E_y = -\frac{\partial V}{\partial y} = \frac{q}{4\pi\varepsilon_0}\left\{\frac{y-a}{[x^2+(y-a)^2+z^2]^{3/2}} - \frac{y+a}{[x^2+(y+a)^2+z^2]^{3/2}}\right\}$$

$$E_z = -\frac{\partial V}{\partial z} = \frac{q}{4\pi\varepsilon_0}\left\{\frac{z}{[x^2+(y-a)^2+z^2]^{3/2}} - \frac{z}{[x^2+(y+a)^2+z^2]^{3/2}}\right\}$$

Special Cases

1. On the axis of the dipole (say, along y-axis)

$$x = 0,\ z = 0$$

$$\therefore \quad V = \frac{q}{4\pi\varepsilon_0}\left[\frac{1}{y-a} - \frac{1}{y+a}\right] = \frac{2aq}{4\pi\varepsilon_0 (y^2 - a^2)}$$

or $$V = \frac{p}{4\pi\varepsilon_0 (y^2 - a^2)} \qquad (\text{as } 2aq = p)$$

i.e. at a distance r from the centre of the dipole ($y = r$)

$$V_r = \frac{p}{4\pi\varepsilon_0 (r^2 - a^2)}$$

or $$V_{axis} \approx \frac{p}{4\pi\varepsilon_0 r^2} \qquad \text{for } r >> a)$$

V is positive when the point under consideration is towards positive charge and negative if it is towards negative charge.

Moreover the components of electric field are as under

$$E_x = 0,\ E_z = 0 \qquad (\text{as } x = 0,\ z = 0)$$

and $$E_y = \frac{q}{4\pi\varepsilon_0}\left[\frac{1}{(y-a)^2} - \frac{1}{(y+a)^2}\right] = \frac{4ayq}{4\pi\varepsilon_0 (y^2 - a^2)^2}$$

or $$E_y = \frac{1}{4\pi\varepsilon_0}\frac{2py}{(y^2 - a^2)^2}$$

Note that E_y is along positive y-direction or parallel to **p**.

Further, at a distance r from the centre of the dipole ($y = r$).

$$E_r = \frac{1}{4\pi\varepsilon_0}\frac{2pr}{(r^2 - a^2)^2}$$

or $$E_{axis} \approx \frac{1}{4\pi\varepsilon_0}\cdot\frac{2p}{r^3} \qquad (\text{for } r >> a)$$

2. On the perpendicular bisector of dipole

Say along x-axis (it may be along z-axis also).

$$y = 0,\ z = 0$$

$$\therefore \quad V = \frac{1}{4\pi\varepsilon_0}\left[\frac{q}{\sqrt{x^2+a^2}} - \frac{q}{\sqrt{x^2+a^2}}\right] = 0$$

or $$V_{\perp\ bisector} = 0$$

Moreover the components of electric field are as under,

$$E_x = 0,\ E_z = 0$$

and $$E_y = \frac{q}{4\pi\varepsilon_0}\left\{\frac{-a}{(x^2+a^2)^{3/2}} - \frac{a}{(x^2+a^2)^{3/2}}\right\} = \frac{-2aq}{4\pi\varepsilon_0 (x^2+a^2)^{3/2}}$$

or $$E_y = -\frac{1}{4\pi\varepsilon_0}\cdot\frac{p}{(x^2+a^2)^{3/2}}$$

Here, negative sign implies that the electric field is along negative y-direction or antiparallel to **p**.

Further, at a distance r from the centre of dipole ($x = r$), the magnitude of electric field is

$$E_r = \frac{1}{4\pi\varepsilon_0}\frac{p}{(r^2+a^2)^{3/2}}$$

or $$E_{\perp\ bisector} \approx \frac{1}{4\pi\varepsilon_0}\cdot\frac{p}{r^3} \qquad (\text{for } r >> a)$$

Electric Dipole in Uniform Electric Field

As we have said earlier also, uniform electric field means, at every point the direction and magnitude of electric field is constant. A uniform electric field is shown by parallel equidistant lines. The field due to a point charge or due to an electric dipole is non-uniform in nature. Uniform electric field is found between the plates of a parallel plate capacitor. Now, let us discuss the behaviour of a dipole in uniform electric field.

Force on Dipole

Suppose an electric dipole of dipole moment $|\mathbf{p}| = 2aq$ is placed in a uniform electric field **E** at an angle θ. Here, θ is the angle between **p** and **E**. A force $\mathbf{F}_1 = q\mathbf{E}$ will act on positive charge and $\mathbf{F}_2 = -q\mathbf{E}$ on negative charge. Since, $\mathbf{F}_1$ and $\mathbf{F}_2$ are equal in magnitude but opposite in direction.

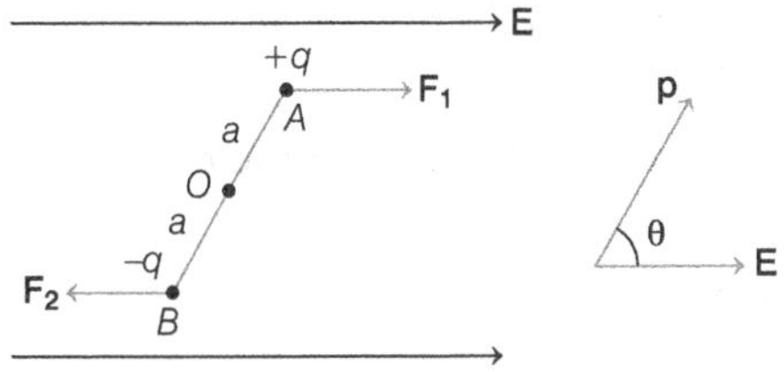

Hence, $\mathbf{F}_1 + \mathbf{F}_2 = 0$ or $\mathbf{F}_{net} = 0$

Thus, net force on a dipole in uniform electric field is zero. While in a non-uniform electric field it is not be zero.

Torque on Dipole

The torque of $\mathbf{F}_1$ about O, $\quad \tau_1 = \mathbf{OA} \times \mathbf{F}_1 = q\,(\mathbf{OA} \times \mathbf{E})$

and torque of $\mathbf{F}_2$ about O is, $\quad \tau_2 = \mathbf{OB} \times \mathbf{F}_2 = -q(\mathbf{OB} \times \mathbf{E})$

$$= q(\mathbf{BO} \times \mathbf{E})$$

The net torque acting on the dipole is

$$\tau = \tau_1 + \tau_2 = q(\mathbf{OA} \times \mathbf{E}) + q\,(\mathbf{BO} \times \mathbf{E})$$
$$= q\,(\mathbf{OA} + \mathbf{BO}) \times \mathbf{E}$$
$$= q\,(\mathbf{BA} \times \mathbf{E})$$

Here $q\,(\mathbf{BA}) = \mathbf{p}$

$\therefore \quad \boxed{\tau = \mathbf{p} \times \mathbf{E}}$

Thus, the magnitude of torque is $\tau = pE \sin\theta$. The direction of torque is perpendicular to the plane of paper inwards. Further this torque is zero at $\theta = 0°$ or $\theta = 180°$, i.e. when the dipole is parallel or antiparallel to **E** and maximum at $\theta = 90°$.

Note *If net force on a body is zero then net torque about any point has the same value. Here we have obtained net torque about point O.*

Potential Energy of Dipole

Suppose uniform electric field **E** starts from AB.

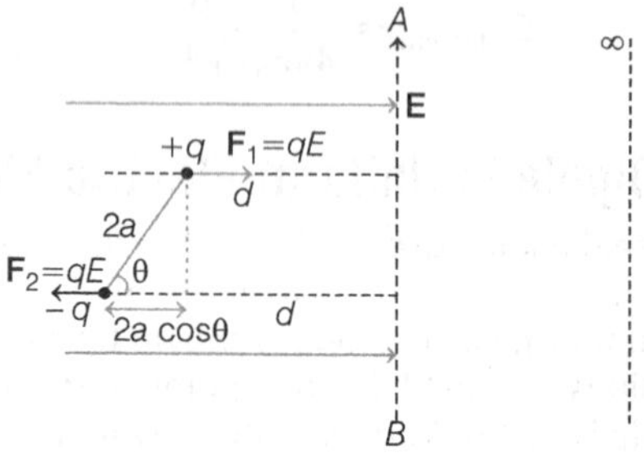

Work done by electric force in bringing the dipole from ∞ to AB is zero.

Because,

$$\mathbf{E} = 0 \;\Rightarrow\; \mathbf{F} = 0 \;\Rightarrow\; W = 0$$

From AB outwards :

$$W_{F_1} = F_1 s_1 \cos\alpha = (qE)\,(d) \cos 180° = -qEd$$
$$W_{F_2} = F_2 s_2 \cos\alpha = (qE)\,(d + 2a\cos\theta) \cos 0°$$
$$= qE\,(d + 2a\cos\theta)$$

$\therefore \quad W_{net} = q\,(2a)\,E\cos\theta = pE\cos\theta$

From the definition of potential energy,

$$U = -(\text{work done by conservative forces})$$
$$= -pE\cos\theta$$
$$= \mathbf{p} \cdot \mathbf{E}$$

Thus,

$$\boxed{U_\theta = \mathbf{p} \cdot \mathbf{E} = -pE\cos\theta}$$

If the dipole is rotated from an angle θ_1 to θ_2, then

Work done by external forces $= \Delta U = U(\theta_2) - U(\theta_1)$

or $\quad W_{\text{external forces}} = -pE\cos\theta_2 - (-pE\cos\theta_1)$

or $\quad \boxed{W_{\text{external force}} = pE\,(\cos\theta_1 - \cos\theta_2)}$

and work done by electric forces,

$$\boxed{W_{\text{electric force}} = -W_{\text{external force}} = pE\,(\cos\theta_2 - \cos\theta_1)}$$

Equilibrium of Dipole

When an electric dipole is placed in a uniform electric field net force on it is zero for any position of the dipole. But torque acting on it is zero only at $\theta = 0°$ and $180°$. Thus, we can say that at these two positions net force and torque both are zero or the dipole is in equilibrium.

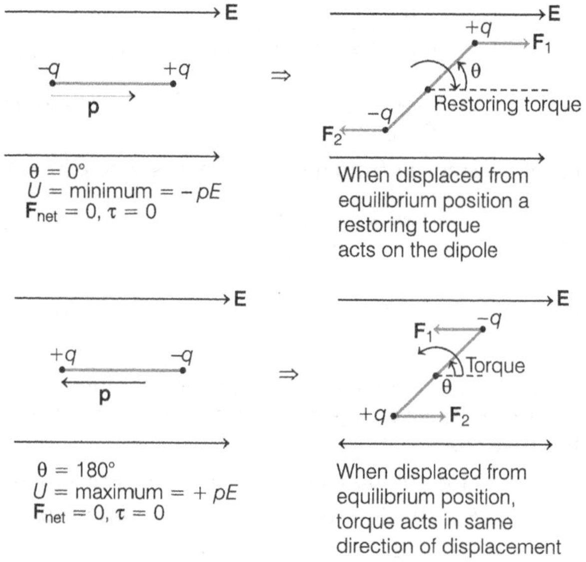

Of this, $\theta = 0°$ is the **stable equilibrium position** of the dipole because potential energy in this position is minimum ($U = -pE\cos 0° = -pE$) and when displaced from this position a torque starts acting on it which is restoring in nature and which has a tendency to bring the dipole back in its equilibrium position.

On the other hand, at $\theta = 180°$, the potential energy of the dipole is maximum ($U = -pE\cos 180° = +pE$) and when it is displaced from this position, the torque has a tendency to rotate it in other direction. This torque is not restoring in nature. So, this equilibrium is known as **unstable equilibrium position.**

List of Formulae :

1. $|\mathbf{p}| = (2a)\,q$

   Direction of $\mathbf{p}$ is from $-q$ to $+q$.

2. If a dipole is placed along y-axis with its centre at origin, then

$$V(x, y, z) = \frac{1}{4\pi\varepsilon_0}\left[\frac{q}{\sqrt{x^2 + (y-a)^2 + z^2}} - \frac{q}{\sqrt{x^2 + (y+a)^2 + z^2}}\right]$$

$$E_x = -\frac{\partial V}{\partial x},\quad E_y = -\frac{\partial V}{\partial y}$$

and $E_z = -\dfrac{\partial V}{\partial z}$

3. **On the axis of dipole** $x = 0,\ z = 0$

   (i) $V = \dfrac{1}{4\pi\varepsilon_0}\dfrac{p}{(y^2 - a^2)}$

   $= \dfrac{1}{4\pi\varepsilon_0}\cdot\dfrac{p}{r^2 - a^2}$ if $y = r$

   or $V_{axis} \approx \dfrac{1}{4\pi\varepsilon_0}\dfrac{p}{r^2}$ if $r >> a$

   (ii) $E_x = 0 = E_z$

   and $E = E_y = \dfrac{1}{4\pi\varepsilon_0}\cdot\dfrac{2py}{(y^2 - a^2)^2}$ (along $\mathbf{p}$)

   $= \dfrac{1}{4\pi\varepsilon_0}\dfrac{2pr}{(r^2 - a^2)^2}$ if $y = r$

   or $E_{axis} \approx \dfrac{1}{4\pi\varepsilon_0}\cdot\dfrac{2p}{r^3}$ for $r >> a$

4. **On the perpendicular bisector of dipole**

   Along x-axis, $y = 0,\ z = 0$

   (i) $V_{\perp\ \text{bisector}} = 0$

   (ii) $E_x = 0,\quad E_z = 0$

   and $E_y = -\dfrac{1}{4\pi\varepsilon_0}\cdot\dfrac{p}{(x^2 + a^2)^{3/2}}$

   or $E = \dfrac{1}{4\pi\varepsilon_0}\dfrac{p}{(r^2 + a^2)^{3/2}}$ (opposite to $\mathbf{p}$)

   $\approx \dfrac{1}{4\pi\varepsilon_0}\cdot\dfrac{p}{r^3}$ (for $r >> a$)

5. **Dipole in uniform electric field**

   (i) $\mathbf{F}_{net} = 0$

   (ii) $\boldsymbol{\tau} = \mathbf{p}\times\mathbf{E}$ and $|\tau| = pE\sin\theta$

   (iii) $U(\theta) = -\mathbf{p}\cdot\mathbf{E} = -pE\cos\theta$ with $U(90°) = 0$

   (iv) $(W_{\theta_1\to\theta_2})_{\text{external force}} = pE(\cos\theta_1 - \cos\theta_2)$

   (v) $(W_{\theta_1\to\theta_2})_{\text{electric force}} = pE(\cos\theta_2 - \cos\theta_1)$

   $= -(W_{\theta_1\to\theta_2})_{\text{external force}}$

   (vi) At $\theta = 0°$, $\mathbf{F}_{net} = 0$, $\tau_{net} = 0$, $U =$ minimum (stable equilibrium position)

   (vii) At $\theta = 180°$, $\mathbf{F}_{net} = 0$, $\tau_{net} = 0$, $U =$ maximum (unstable equilibrium position)

Example 1 Draw electric lines of forces due to an electric dipole.

Sol. Electric lines of forces due to an electric dipole are as shown in figure.

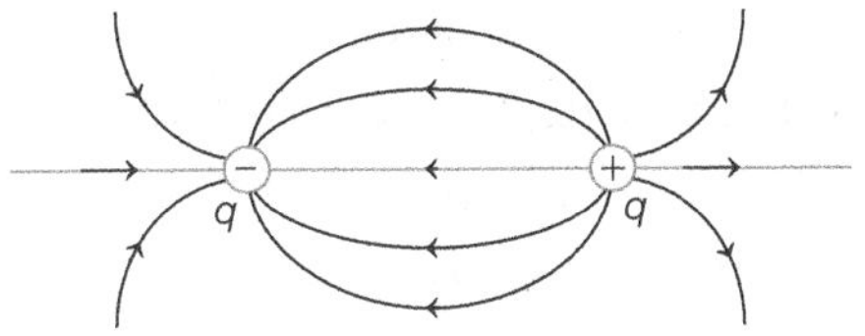

Example 2 Along the axis of a dipole, direction of electric field is always in the direction of electric dipole moment $\mathbf{p}$. Is this statement true or false?

Sol. False. In the above figure, we can see that direction of electric field is in the opposite direction of $\mathbf{p}$ between the two charges.

Example 3 At a far away distance r along the axis from an electric dipole electric field is $\mathbf{E}$. Find the electric field at distance $2r$ along the perpendicular bisector.

Sol. Along the axis of dipole,

$$E = \frac{1}{4\pi\varepsilon_0}\frac{2p}{r^3} \qquad \text{...(i)}$$

This electric field is in the direction of $\mathbf{p}$. Along the perpendicular bisector at a distance $2r$,

$$E' = \frac{1}{4\pi\varepsilon_0}\frac{p}{(2r)^3} \qquad \text{...(ii)}$$

From Eqs. (i) and (ii), we can see that

$$E' = \frac{E}{16}$$

Moreover, E' is in the opposite direction of $\mathbf{p}$. Hence,

$$\mathbf{E}' = -\frac{\mathbf{E}}{16} \qquad \textbf{Ans.}$$

Oscillations of a Dipole in Uniform Electric Field

In uniform electric field, net force on a dipole is zero at all angles. But net torque is zero for $\theta = 0°$ or $180°$. Here, $\theta = 0°$ is the stable equilibrium position and $\theta = 180°$ is unstable equilibrium position. If the dipole is released from any angle other than $0°$ or $180°$, it rotates towards $0°$. In this process electrostatic potential energy of the dipole decreases. But rotational kinetic energy increases. At two angles θ_1 and θ_2, we can apply the equation

$$U_{\theta_1} + K_{\theta_1} = U_{\theta_2} + K_{\theta_2}$$

or $$-pE\cos\theta_1 + \frac{1}{2}I\omega_1^2 = -pE\cos\theta_2 + \frac{1}{2}I\omega_2^2$$

Moreover, if the dipole is displaced an angle θ from stable equilibrium position ($\theta = 0°$), then it starts rotational oscillations between $+\theta$ and $-\theta$. For small value of θ, these oscillations are simple harmonic in nature.

Example 4 An electric dipole of dipole moment p is placed in a uniform electric field E in stable equilibrium position. Its moment of inertia about the centroidal axis is I. If it is displaced slightly from its equilibrium position, find the period of small oscillations.

Sol. When displaced at an angle θ from its equilibrium position, the magnitude of restoring torque is

$$\tau = -pE \sin\theta$$

For small angular displacement $\sin\theta \approx \theta$

$$\tau = -pE\,\theta$$

The angular acceleration is

$$\alpha = \frac{\tau}{I} = -\left(\frac{pE}{I}\right)\theta = -\omega^2\theta$$

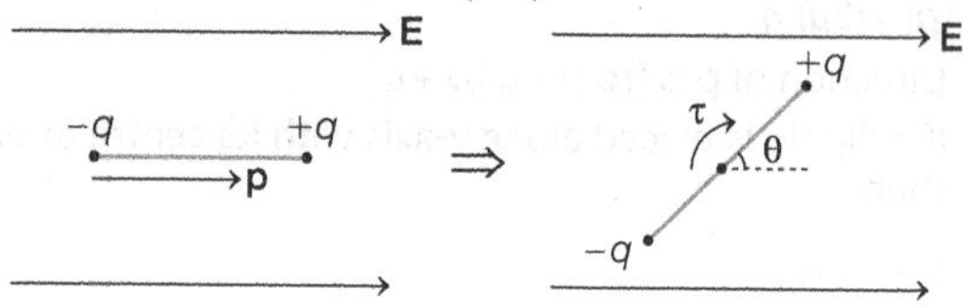

where, $\omega^2 = \dfrac{pE}{I}$

$$\Rightarrow \quad T = \frac{2\pi}{\omega} = 2\pi\sqrt{\frac{I}{pE}}$$ **Ans.**

Important POINTS

1. **Electric field and potential due to a dipole at polar coordinates (r, θ)**

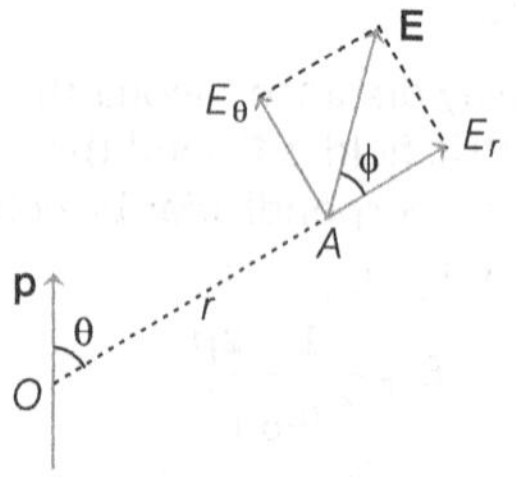

$$V = \frac{p\cos\theta}{4\pi\varepsilon_0 r^2}$$

The electric field **E** can be resolved into two components E_r and E_θ, where

or $$E_r = \frac{1}{4\pi\varepsilon_0}\cdot\frac{2p\cos\theta}{r^3}$$

and $$E_\theta = \frac{1}{4\pi\varepsilon_0}\frac{p\sin\theta}{r^3}$$

The magnitude of resultant electric field $E = \sqrt{E_r^2 + E_\theta^2}$

or $$E = \frac{p}{4\pi\varepsilon_0 r^3}\sqrt{1 + 3\cos^2\theta}$$

Its inclination ϕ to OA is given by

$$\tan\phi = \frac{E_\theta}{E_r} = \frac{p\sin\theta/4\pi\varepsilon_0 r^3}{2p\cos\theta/4\pi\varepsilon_0 r^3}$$

or $$\tan\phi = \frac{\tan\theta}{2}$$

Note *Derivation is not required.*

2. **Force between two dipoles**

The force between two dipoles varies inversely with the fourth power of the distance between their centres or

$$F \propto \frac{1}{r^4}$$

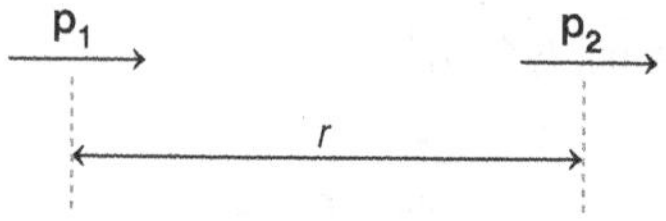

In the figure, a dipole on left with dipole moment $\mathbf{p}_1$ interacts with the dipole on the right with dipole moment $\mathbf{p}_2$. We assume that the distance between them is quite large. The electric field of the dipole on the left hand side exerts a net force on the dipole on the right hand side. Let us now calculate the net force on the dipole on right hand side.

The electric field at the centre of this dipole

$$E = \frac{1}{4\pi\varepsilon_0}\cdot\frac{2p_1}{r^3}$$

$$\therefore \quad dE = -\frac{1}{4\pi\varepsilon_0}\cdot\frac{6p_1}{r^4}dr$$

Now, the electric field at the point where $-q$ charge of the dipole lies is given by

$E_1 = E + |dE|$ and force on $-q$ is qE_1 (towards left)

Similarly, electric field at the point where $+q$ charge of the dipole lies is

$E_2 = E - |dE|$ and force on $+q$ is qE_2 (towards right)

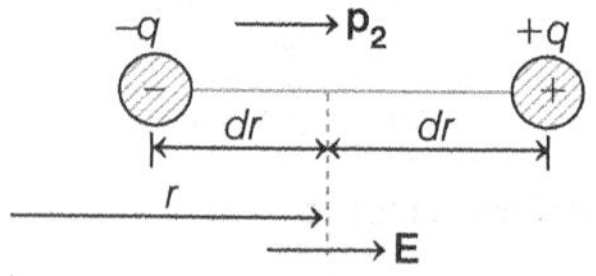

$\therefore$ Net force on the dipole is

$$F = qE_1 - qE_2 \quad \text{(towards left)}$$

$$= 2q|dE|$$

$$= \frac{6(2qdr)\,p_1}{4\pi\varepsilon_0 r^4} \quad \text{or} \quad F = \frac{6p_1p_2}{4\pi\varepsilon_0 r^4} \quad [\text{as } 2q(dr) = p_2]$$

Thus, if $\mathbf{p}_1 \parallel \mathbf{p}_2$, the two dipoles attract each other with a force given by the above relation.

MODULE 8 Exercise

Assertion and Reason

See directions on page 25.

1. **Assertion :** A small electric dipole is moved translationally from higher potential to lower potential in uniform electric field. Work done by electric field is positive.

   Reason : When a positive charge is moved from higher potential to lower potential, work done by electric field is positive.

Single Correct Option

2. The potential at a point due to an electric dipole will be maximum and minimum when the angles between the axis of the dipole and the line joining the point to the dipole are respectively
   (a) 90° and 180°
   (b) 0° and 90°
   (c) 90° and 0°
   (d) 0° and 180°

3. An electric dipole is placed perpendicular to an infinite line of charge at some distance as shown in figure. Identify the correct statement.

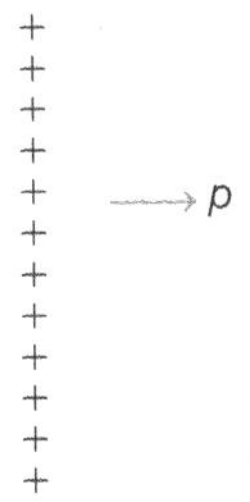

   (a) The dipole is attracted towards the line charge
   (b) The dipole is repelled away from the line charge
   (c) The dipole does not experience a force
   (d) The dipole experiences a force as well as a torque

4. Electric charges q, q, $-2q$ are placed at the corners of an equilateral triangle ABC of side l. The magnitude of electric dipole moment of the system is
   (a) ql (b) $2ql$
   (c) $\sqrt{3}ql$ (d) $4ql$

5. The electric field due to an electric dipole at a distance r from its centre in axial position is E. If the dipole is rotated through an angle of 90° about its perpendicular axis, the magnitude of electric field at the same point will be
   (a) E (b) $E/4$
   (c) $E/2$ (d) $2E$

6. An electric dipole consists of two charges $+q$ and $-q$ separated by a small distance $2a$. Its centre is on x-axis at $x = x_1$ and it points along the x-axis in positive x-direction. The dipole is in non-uniform field which is given by $\mathbf{E} = Cx\hat{\mathbf{i}}$, where $C = \mathbf{a}$ constant. Then,
   (a) net force on dipole is $Cp\hat{\mathbf{i}}$
   (b) net force on dipole is $-Cp\hat{\mathbf{i}}$
   (c) net force on dipole is $CEx_1\hat{\mathbf{i}}$
   (d) net force on dipole is $-CEx_1\hat{\mathbf{i}}$

7. Two charges each of magnitude 0.01 C and separated by a distance of 0.4 mm constitute an electric dipole. If the dipole is placed in a uniform electric field 'E' of 10 dyne/C making 30° angle with E, the magnitude of torque acting on dipole is **(2023 Main)**
   (a) 4.0×10^{-10} Nm (b) 1.0×10^{-8} Nm
   (c) 2.0×10^{-10} Nm (d) 1.5×10^{-9} Nm

8. The electric field due to a short electric dipole at a large distance (r) from centre of dipole on the equatorial plane varies with distance as **(2023 Main)**
   (a) $\frac{1}{r^2}$ (b) $\frac{1}{r^3}$ (c) $\frac{1}{r}$ (d) r

9. An electric dipole is formed by two charges $+q$ and $-q$ located in xy-plane at (0, 2) mm and (0, – 2) mm respectively, as shown in the figure. The electric potential at point P (100, 100) mm due to the dipole is V_0. The charges $+q$ and $-q$ are then moved to the points (– 1, 2) mm and (1, – 2) mm, respectively. What is the value of electric potential at P due to the new dipole? **(2023 Adv.)**

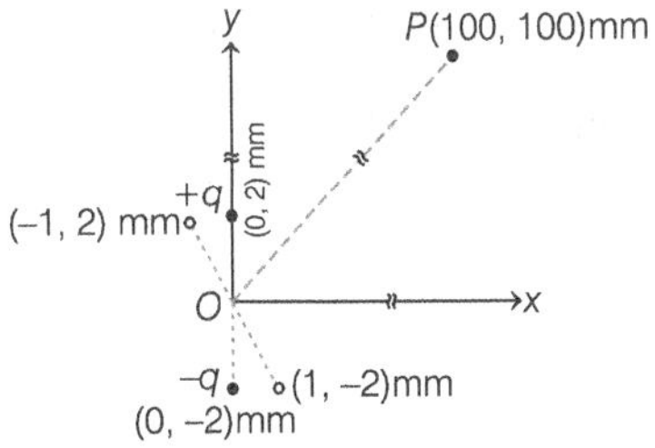

   (a) $V_0/4$ (b) $V_0/2$
   (c) $V_0/\sqrt{2}$ (d) $3V_0/4$

10. Two point charges A and B of magnitude $+8 \times 10^{-6}$ C and -8×10^{-6} C respectively are placed at a distance d apart. The electric field at the middle point O between the charges is 6.4×10^4 NC^{-1}. The distance d between the point charges A and B is **(2022 Main)**
    (a) 2.0 m (b) 3.0 m
    (c) 1.0 m (d) 4.0 m

11. Two identical electric point dipoles have dipole moments $\mathbf{p}_1 = p\hat{\mathbf{i}}$ and $\mathbf{p}_2 = -p\hat{\mathbf{i}}$ and are held on the x-axis at distance a from each other. When released, they move along x-axis with the direction of their dipole moments remaining unchanged. If the mass of each dipole is 'm', their speed when they are infinitely far apart is **(2020 Main)**

(a) $\frac{p}{a}\sqrt{\frac{1}{\pi\varepsilon_0 ma}}$ (b) $\frac{p}{a}\sqrt{\frac{1}{2\pi\varepsilon_0 ma}}$

(c) $\frac{p}{a}\sqrt{\frac{2}{\pi\varepsilon_0 ma}}$ (d) $\frac{p}{a}\sqrt{\frac{2}{2\pi\varepsilon_0 ma}}$

12. An electric dipole is formed by two equal and opposite charges q with separation d. The charges have same mass m. It is kept in a uniform electric field E. If it is slightly rotated from its equilibrium orientation, then its angular frequency ω is **(2019 Main)**

(a) $\sqrt{\frac{qE}{md}}$ (b) $\sqrt{\frac{2qE}{md}}$

(c) $2\sqrt{\frac{qE}{md}}$ (d) $\sqrt{\frac{qE}{2md}}$

13. An electric field of 1000 V/m is applied to an electric dipole at angle of 45°. The value of electric dipole moment is 10^{-29} Cm. What is the potential energy of the electric dipole? **(2019 Main)**

(a) -20×10^{-18} J (b) -7×10^{-27} J

(c) -10×10^{-29} J (d) -9×10^{-20} J

14. A point dipole $\mathbf{p} = \mathbf{p}_0\hat{\mathbf{x}}$ is kept at the origin. The potential and electric field due to this dipole on the y-axis at a distance d are respectively (Take, $V = 0$ at infinity) **(2019 Main)**

(a) $\frac{|\mathbf{p}|}{4\pi\varepsilon_0 d^2}, \frac{\mathbf{p}}{4\pi\varepsilon_0 d^3}$ (b) $0, \frac{-\mathbf{p}}{4\pi\varepsilon_0 d^3}$

(c) $0, \frac{|\mathbf{p}|}{4\pi\varepsilon_0 d^3}$ (d) $\frac{|\mathbf{p}|}{4\pi\varepsilon_0 d^2}, \frac{-\mathbf{p}}{4\pi\varepsilon_0 d^3}$

15. An electric dipole has a fixed dipole moment $\mathbf{p}$, which makes angle θ with respect to x-axis. When subjected to an electric field $\mathbf{E}_1 = E\hat{\mathbf{i}}$, it experiences a torque $\mathbf{T}_1 = \tau\hat{\mathbf{k}}$. When subjected to another electric field $\mathbf{E}_2 = \sqrt{3}E_1\hat{\mathbf{j}}$ it experiences torque $\mathbf{T}_2 = -\mathbf{T}_1$. The angle θ is **(2017 JEE)**

(a) 60° (b) 90°

(c) 30° (d) 45°

16. Two point charges $+q$ and $-q$ are held fixed at $(-d, 0)$ and $(d, 0)$ respectively of a x-y coordinate system. Then, **(1995 JEE)**

(a) the electric field E at all points on the x-axis has the same direction.

(b) work has to be done in bringing a test charge from ∞ to the origin.

(c) electric field at all point on y-axis is along x-axis.

(d) the dipole moment is $2qd$ along the x-axis.

Answers

1. (d)	2. (d)	3. (a)	4. (c)	5. (c)	6. (a)	7. (c)	8. (b)	9. (b)
10. (b)	11. (b)	12. (b)	13. (b)	14. (b)	15. (a)	16. (c)		

See hints on pages 206-208.

MODULE 9

Gauss's Law

Gauss's law is a tool of simplifying electric field calculations where there is symmetrical distribution of charge. Many physical systems have symmetry, for example a cylindrical body doesn't look any different if we rotate it around its axis.

Before studying the detailed discussion of Gauss's law let us understand electric flux.

Electric Flux (ϕ)

(i) Electric flux is a measure of the field lines crossing a surface.

(ii) It is a scalar quantity with SI units $\frac{N}{C}\text{-m}^2$ or V-m.

(iii) Electric flux passing through a small surface dS is given by

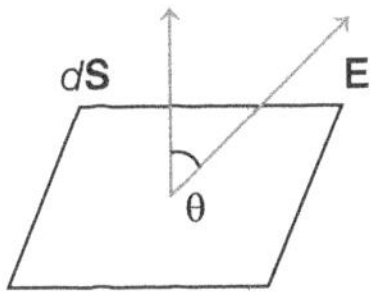

$$d\phi = \mathbf{E} \cdot d\mathbf{S} = EdS \cos \theta \qquad \ldots\text{(i)}$$

Here, $d\mathbf{S}$ is an area vector, whose magnitude is equal to dS and whose direction is perpendicular to the surface.

Note *If the surface is open, then* $d\mathbf{S}$ *can be taken in either of the two directions perpendicular to the surface, but it should not change even if we rotate the surface.*

If the surface is closed then by convention, $d\mathbf{S}$ *is normally taken in outward direction.*

(iv) From Eq. (i), we can see that maximum value of $d\phi$ is EdS, if $\theta = 0°$ or electric lines are perpendicular to the surface. Electric flux is zero, if $\theta = 90°$ or electric lines are tangential to the surface.

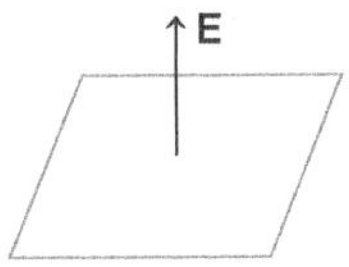

$d\phi = E\,dS$

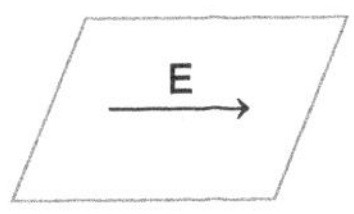

$d\phi = 0$

(v) Electric flux passing through a large surface is given by

$$\phi = \int d\phi = \int \mathbf{E} \cdot d\mathbf{S} = \int EdS \cos \theta \qquad \ldots\text{(ii)}$$

This is basically surface integral of electric flux over the given surface. But normally we do not study surface integral in detail in physics.

Here, are two special cases for calculating the electric flux passing through a surface S of finite size (whether closed or open)

Case 1 $\phi = ES$

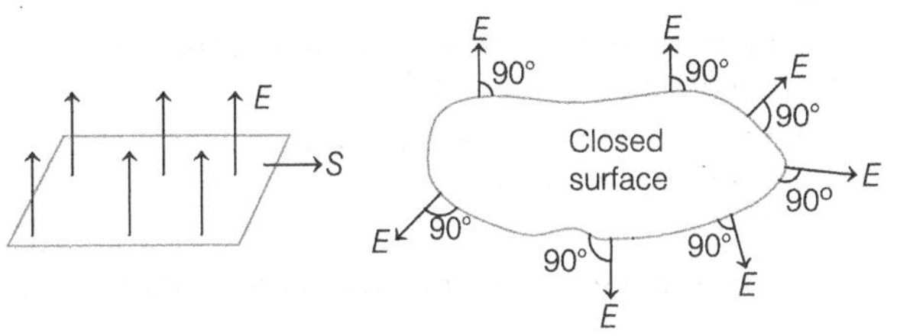

If at every point on the surface, the magnitude of electric field is constant and perpendicular (to the surface).

Case 2 $\phi = 0$

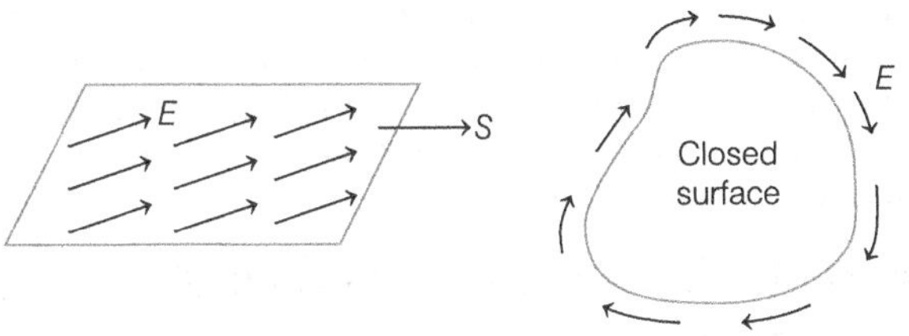

If at all points on the surface the electric field is tangential to the surface.

Gauss's Law

This law gives a relation between the net electric flux through a closed surface and the charge enclosed by the surface. According to this law,

"The net electric flux through any closed surface is equal to the net charge inside the surface divided by ε_0." In symbols, it can be written as

$$\phi_e = \oint_S \mathbf{E} \cdot d\mathbf{S} = \frac{q_{in}}{\varepsilon_0} \qquad \ldots\text{(i)}$$

where, q_{in} represents the net charge inside the closed surface and $\mathbf{E}$ represents the electric field at any point on the surface.

The above mathematical expression can also be written as :

" Surface integral of electric field over a closed surface is equal to net charge inside the surface divided by ε_0."

In principle, Gauss's law is valid for the electric field of any system of charges or continuous distribution of charge. In practice however, the technique is useful for calculating the electric field only in situations where the degree of symmetry is high. Gauss's law can be used to evaluate the electric field for charge distributions that have spherical, cylindrical or plane symmetry.

Simplified Form of Gauss's Theorem

Gauss's law in simplified form can be written as under

$$ES = \frac{q_{in}}{\varepsilon_0} \quad \text{or} \quad E = \frac{q_{in}}{S\varepsilon_0} \qquad ...(ii)$$

but this form of Gauss's law is applicable only under the following two conditions :

(i) The electric field at every point on the surface is either perpendicular or tangential.

(ii) Magnitude of electric field at every point where it is perpendicular to the surface has a constant value (say E).

Here, S is the area where electric field is perpendicular to the surface.

Applications of Gauss's Law

As Gauss's law does not provide expression for electric field but provides only for its flux through a closed surface. To calculate E we choose an imaginary closed surface (called Gaussian surface) in which Eq. (ii) can be applied easily. But before using Eq. (ii) we should have an idea about the electric field lines due to different charge distributions. Let us discuss few simple cases.

Electric Field due to a Point Charge

The electric field due to a point charge is everywhere radial. We wish to find the electric field at a distance r from the charge q. We select Gaussian surface, a sphere at distance r from the charge.

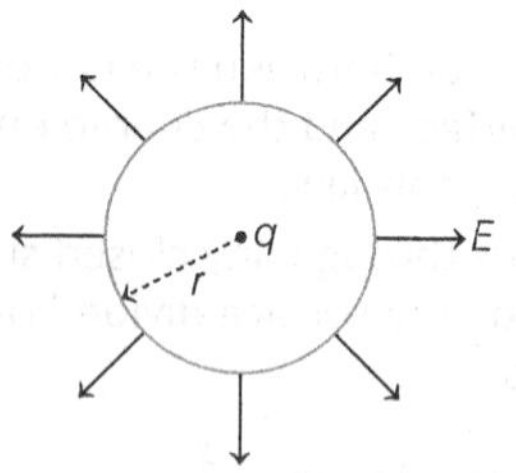

At every point of this sphere the electric field has the same magnitude E and it is perpendicular to the surface itself. Hence, we can apply the simplified form of Gauss's law,

$$ES = \frac{q_{in}}{\varepsilon_0}$$

Here, S = area of sphere $= 4\pi r^2$ and

q_{in} = net charge enclosing the Gaussian surface $= q$

$$\therefore \quad E(4\pi r^2) = \frac{q}{\varepsilon_0}$$

$$\therefore \quad E = \frac{1}{4\pi\varepsilon_0} \cdot \frac{q}{r^2}$$

It is nothing but Coulomb's law.

Electric Field due to a Linear Charge Distribution

Consider a long line charge with a linear charge density (charge per unit length) λ. We have to calculate the electric field at a point, a distance r from the line charge. We construct a Gaussian surface, a cylinder of any arbitrary length l of radius r and its axis coinciding with the axis of the line charge.

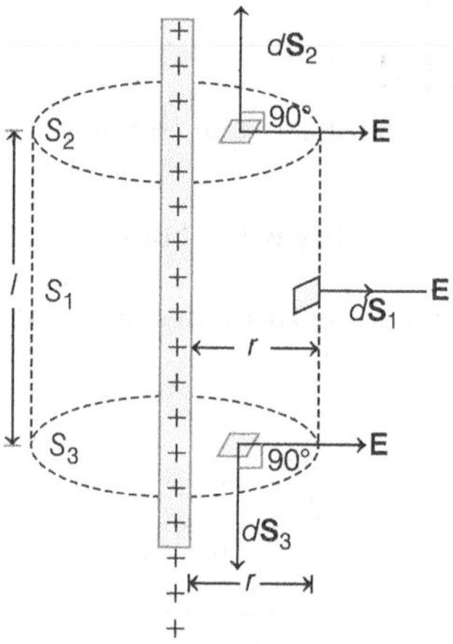

This cylinder have three surfaces. One is curved surface S_1 and the two plane parallel surfaces S_2 and S_3. Field lines at plane parallel surfaces are tangential (so flux passing through these surfaces is zero). The magnitude of electric field is having the same magnitude (say E) at curved surface and simultaneously the electric field is perpendicular at every point of this surface.

Hence, we can apply the Gauss's law as

$$ES = \frac{q_{in}}{\varepsilon_0}$$

Here, S = area of curved surface $= (2\pi rl)$

and q_{in} = net charge enclosing this cylinder $= \lambda l$

$$\therefore \quad E(2\pi rl) = \frac{\lambda l}{\varepsilon_0}$$

$$\therefore \quad E = \frac{\lambda}{2\pi\varepsilon_0 r}$$

i.e. $E \propto \dfrac{1}{r}$

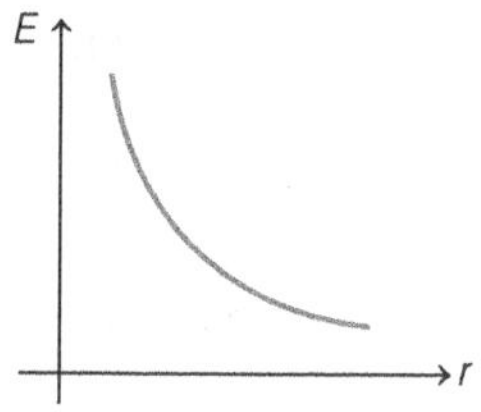

or E-r graph is a rectangular hyperbola as shown in figure.

Electric Field due to a Plane Sheet of Charge

Figure shows a portion of a flat thin sheet, infinite in size with constant surface charge density σ (charge per unit area). By symmetry, since the sheet is infinite, the field must have the same magnitude and the opposite directions at two points equidistant from the sheet on opposite sides. Let us draw a Gaussian surface (a cylinder) with one end on one side and other end on the other side and of cross-sectional area S_0. Field lines will be tangential to the curved surface, so flux passing through this surface is zero. At plane surfaces electric field has same magnitude and perpendicular to surface.

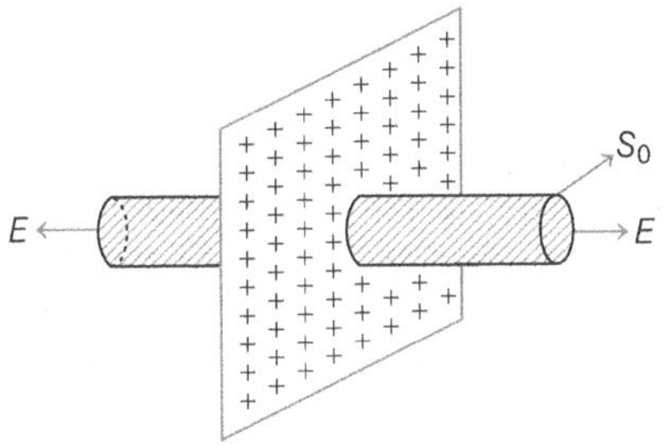

Hence, using $$ES = \frac{q_{in}}{\varepsilon_0}$$

$$\therefore \quad E(2S_0) = \frac{(\sigma)(S_0)}{\varepsilon_0}$$

$$\therefore \quad \boxed{E = \frac{\sigma}{2\varepsilon_0}}$$

Thus, we see that the magnitude of the field is independent of the distance from the sheet. Practically, an infinite sheet of charge does not exist. This result is correct for real charge sheets if points under consideration are not near the edges and the distances from the sheet are small compared to the dimensions of sheet.

Electric Field Near a Charged Conducting Surface

When a charge is given to a conducting plate, it distributes itself over the entire outer surface of the plate. The surface density σ is uniform and is the same on both surfaces if plate is of uniform thickness and of infinite size.

This is similar to the previous one the only difference is that this time charges are on both sides.

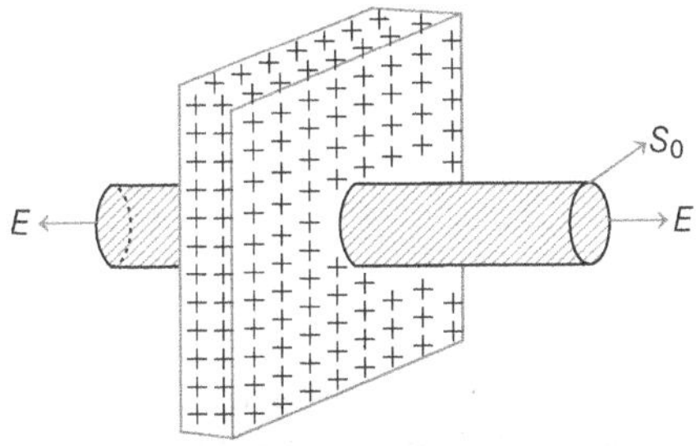

Hence, applying $$ES = \frac{q_{in}}{\varepsilon_0}$$

Here, $$S = 2S_0$$

and $$q_{in} = (\sigma)(2S_0)$$

$$\therefore \quad E(2S_0) = \frac{(\sigma)(2S_0)}{\varepsilon_0}$$

$$\therefore \quad \boxed{E = \frac{\sigma}{\varepsilon_0}}$$

Thus, field due to a charged conducting plate is twice the field due to plane sheet of charge. It also has same limitations.

Later, we will see that the electric field near a charged conducting surface of any shape is σ/ε_0 and it is normal to the surface.

Note *(i) In case of closed symmetrical body with charge q at its centre, the electric flux linked with each half will be $\frac{\phi}{2} = \frac{q}{2\varepsilon_0}$. If the symmetrical closed body has n identical faces with point charge at its centre, flux linked with each face will be $\frac{\phi}{n} = \frac{q}{n\,\varepsilon_0}$.*

(ii) Net electric flux passing through a closed surface in uniform electric field is zero.

Example 1 An electric dipole is placed at the centre of a sphere. Find the electric flux passing through the sphere.

Sol. Net charge inside the sphere $q_{in} = 0$. Therefore, according to Gauss's law net flux passing through the sphere is zero. **Ans.**

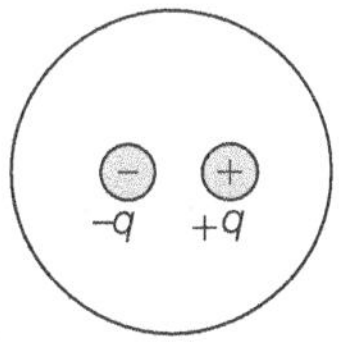

Example 2 A point charge q is placed at the centre of a cube. What is the flux linked

(a) with all the faces of the cube?

(b) with each face of the cube?

(c) if charge is not at the centre, then what will be the answers of parts (a) and (b)?

Sol. (a) According to Gauss's law,

$$\phi_{total} = \frac{q_{in}}{\varepsilon_0} = \frac{q}{\varepsilon_0} \quad \textbf{Ans.}$$

(b) The cube is a symmetrical body with 6 faces and the point charge is at its centre, so electric flux linked with each face will be

$$\phi_{each\ face} = \frac{\phi_{total}}{6} = \frac{q}{6\varepsilon_0} \quad \textbf{Ans.}$$

(c) If charge is not at the centre, the answer of part (a) will remain same while that of part (b) will change.

INTRODUCTORY Exercise

1. In figure (a), a charge q is placed just outside the centre of a closed hemisphere. In figure (b), the same charge q is placed just inside the centre of the closed hemisphere and in figure (c), the charge is placed at the centre of hemisphere open from the base. Find the electric flux passing through the hemisphere in all the three cases.

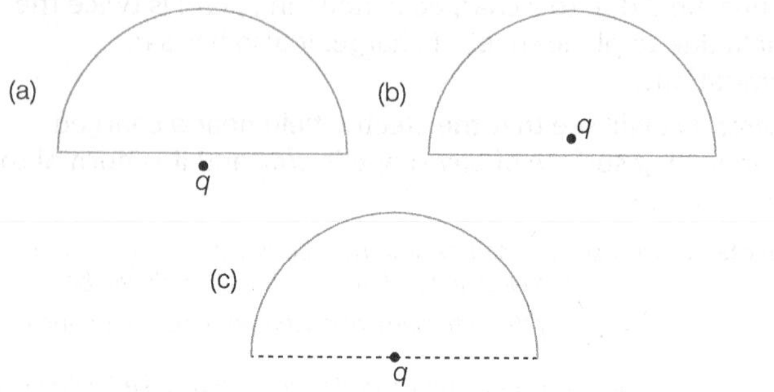

2. A hemispherical body of radius R is placed in a uniform electric field E. What is the flux linked with the curved surface, if the field is (a) parallel to the base, (b) perpendicular to the base.

3. A cube has sides of length $L = 0.2$ m. It is placed with one corner at the origin as shown in figure. The electric field is uniform and given by $\mathbf{E} = (2.5\ \text{N/C})\,\hat{\mathbf{i}} - (4.2\ \text{N/C})\,\hat{\mathbf{j}}$. Find the electric flux through the entire cube.

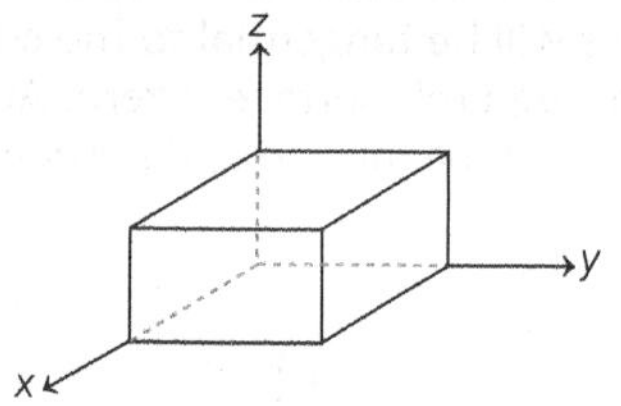

4. The electric charges are distributed in a small volume. The flux of the electric field through a spherical surface of radius 10 cm surrounding the total charge is 20 V-m. The flux over a concentric sphere of radius 20 cm will be
(a) 20 Vm (b) 10 Vm (c) 40 Vm (d) 5 Vm

Answers

1. (a) Zero (b) $\dfrac{q}{\varepsilon_0}$ (c) $\dfrac{q}{2\varepsilon_0}$
2. (a) Zero (b) $\pi R^2 E$
3. Zero
4. (a)

See hints on page 208.

Typed Problems

Type 1. *Based on calculation of electric flux.*

Concept

You will get two types of problems. In first case charge will be given and in the second case electric field will be given.

In first case point numbers (i) and (ii) discussed below can be used and in second case point numbers (iii) and (iv) are used.

(i) To find electric flux from any closed surface, direct result of Gauss's theorem can be used,

$$\phi = \frac{q_{in}}{\varepsilon_0}$$

(ii) To find electric flux from an open surface, result of Gauss's theorem and concept of symmetry can be used.

(iii) To find electric flux from a plane surface in uniform electric field,

$$\phi = \mathbf{E} \cdot \mathbf{S} \quad \text{or} \quad ES\cos\theta$$

can be used.

(iv) Net electric flux from a closed surface in uniform electric field is always zero.

Ex. 1 The electric field in a region is given by $\mathbf{E} = a\hat{\mathbf{i}} + b\hat{\mathbf{j}}$. Here, a and b are constants. Find the net flux passing through a square area of side l parallel to y-z plane.

Hint *Point (iii) can be used.*

Sol. A square area of side l parallel to y-z plane in vector form can be written as,

$$\mathbf{S} = l^2\hat{\mathbf{i}}$$

Given, $\mathbf{E} = a\hat{\mathbf{i}} + b\hat{\mathbf{j}}$

$\therefore$ Electric flux passing through the given area will be,

$$\phi = \mathbf{E} \cdot \mathbf{S}$$
$$= (a\hat{\mathbf{i}} + b\hat{\mathbf{j}}) \cdot (l^2\hat{\mathbf{i}})$$
$$= al^2$$

Ans.

Ex. 2 Figure shows an imaginary cube of side a. A uniformly charged rod of length a and charge per unit length λ moves towards right at a constant speed v. At $t = 0$, the right end of the rod just touches the left face of the cube. Plot a graph between electric flux passing through the cube *versus* time.

Hint : *Point (i) is used.*

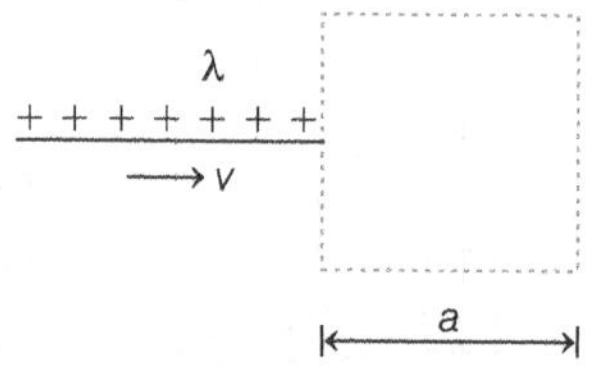

Sol. The electric flux passing through a closed surface depends on the net charge inside the surface. Net charge in this case first increases, reaches a maximum value and finally decreases to zero. The same is the case with the electric flux. The electric flux ϕ *versus* time graph is as shown in figure below.

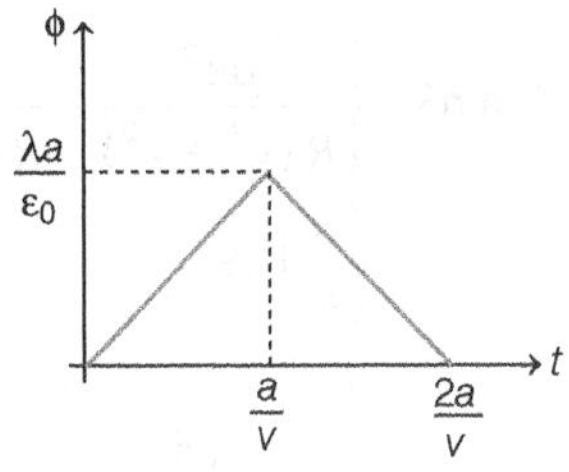

Here, $\phi_{max} = \frac{(q_{in})_{max}}{\varepsilon_0} = \frac{\lambda l_{max}}{\varepsilon_0} = \frac{\lambda a}{\varepsilon_0}$

Ex. 3 The electric field in a region is given by $\mathbf{E} = \alpha x \hat{\mathbf{i}}$. Here, α is a constant of proper dimensions. Find

(a) the total flux passing through a cube bounded by the surfaces, $x = l$, $x = 2l$, $y = 0, y = l$, $z = 0$, $z = l$.

(b) the charge contained inside the above cube.

Sol. (a) Electric field is along positive x-direction. Therefore, field lines are perpendicular to faces *ABCD* and *EFGH*. At all other four faces field lines are tangential. So, net flux passing through these four faces will be zero.

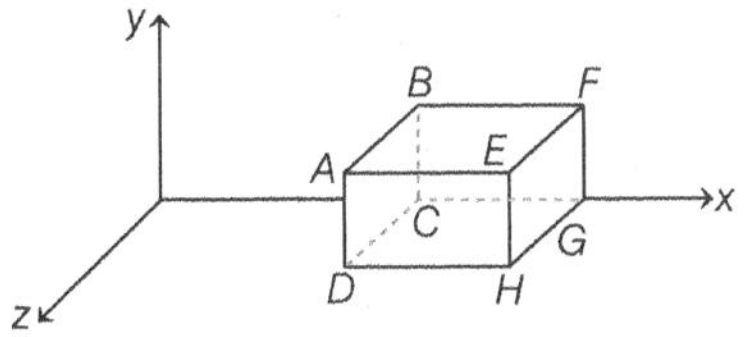

Flux entering at face *ABCD*

At this face $x = l$

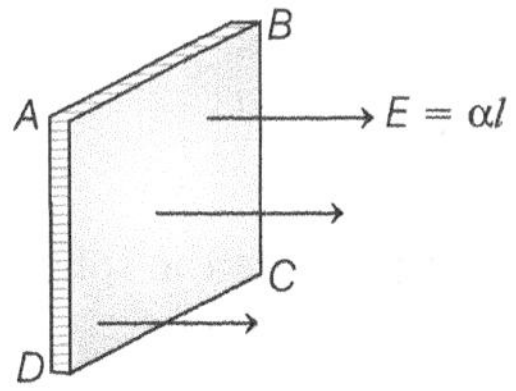

$$\mathbf{E} = \alpha l \hat{\mathbf{i}}$$

$\therefore$ Flux entering the cube from this face,

$$\phi_1 = ES = (\alpha l)(l^2) = \alpha l^3$$

Flux leaving the face *EFGH*

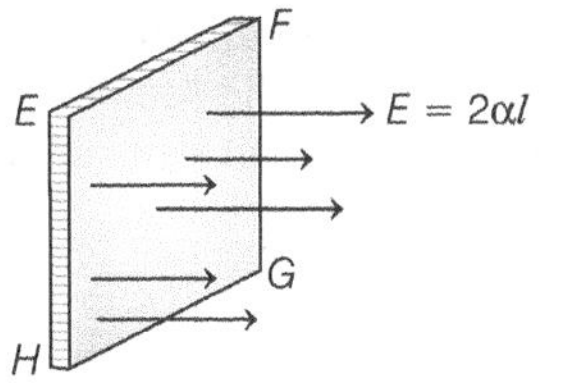

At this face $x = 2l$

$\therefore$ $\mathbf{E} = 2\alpha l \hat{\mathbf{i}}$

$\therefore$ Flux coming out of this face

$$\phi_2 = ES = (2\alpha l)(l^2)$$
$$= 2\alpha l^3$$

$\therefore$ Net flux passing through the cube,

$$\phi_{net} = \phi_2 - \phi_1 = 2\alpha l^3 - \alpha l^3$$
$$= \alpha l^3 \qquad \textbf{Ans.}$$

(b) From Gauss's law,

$$\phi_{net} \text{ from any closed surface} = \frac{q_{in}}{\varepsilon_0}$$

$$q_{in} = (\phi_{net})(\varepsilon_0)$$
$$= \alpha \varepsilon_0 l^3 \qquad \textbf{Ans.}$$

Ex. 4 Consider the charge configuration and a spherical Gaussian surface as shown in the figure. When calculating the flux of the electric field over the spherical surface, the electric field will be due to **(2004 JEE)**

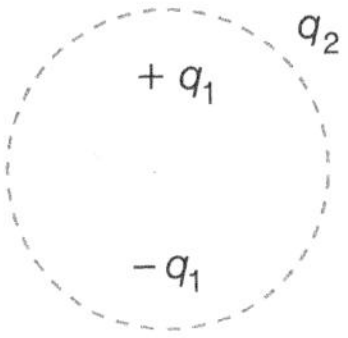

(a) q_2

(b) only the positive charges

(c) all the charges

(d) $+q_1$ and $-q_1$

Sol. At any point over the spherical Gaussian surface, net electric field is the vector sum of electric fields due to $+q_1, -q_1$ and q_2.

$\therefore$ The correct option is (c).

Note *Don't confuse with the net electric flux which is zero passing over the Gaussian surface as the net charge enclosing the surface is zero.*

Ex. 5 A point charge q is placed on the top of a cone of semi vertex angle θ. Show that the electric flux through the base of the cone is $\frac{q(1-\cos\theta)}{2\varepsilon_0}$.

Hint : *Point (ii) can be used.*

HOW TO PROCEED This problem can be solved by the method of symmetry. Consider a Gaussian surface, a sphere with its centre at the top and radius the slant length of the cone. The flux through the whole sphere is q/ε_0. Therefore, the flux through the base of the cone can be calculated by using the following formula,

$$\phi_e = \left(\frac{S}{S_0}\right) \cdot \frac{q}{\varepsilon_0}$$

Here, S_0 = area of whole sphere

and S = area of sphere below the base of the cone *ACB*.

Sol. Let R = slant length of cone = radius of Gaussian sphere

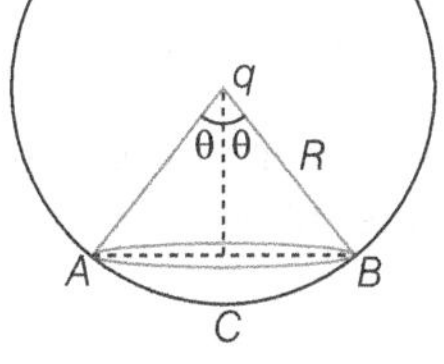

$\therefore\ S_0$ = area of whole sphere $=(4\pi R^2)$

S = area of sphere below the base of the cone

$=2\pi R^2\,(1-\cos\theta)$

$\therefore$ The desired flux is, $\phi=\left(\dfrac{S}{S_0}\right)\cdot\dfrac{q}{\varepsilon_0}$

$$=\frac{(2\pi R^2)(1-\cos\theta)}{(4\pi R^2)}\cdot\frac{q}{\varepsilon_0}$$

$$=\frac{q(1-\cos\theta)}{2\varepsilon_0}$$

Proved

Note *$S=2\pi R^2\,(1-\cos\theta)$ can be calculated by integration.*

At $\theta=0°,\quad S=2\pi R^2\,(1-\cos 0°)=0$

$\theta=90°,\quad S=2\pi R^2\,(1-\cos 90°)=2\pi R^2$

and $\theta=180°,\quad S=2\pi R^2\,(1-\cos 180°)=4\pi R^2$

Proof $dS=(2\pi r)\,R d\alpha$

$=(2\pi R\sin\alpha)\,R d\alpha$ as $r=R\sin\alpha$

$=(2\pi R^2)\sin\alpha\,d\alpha$

$\therefore\quad S=\int_0^{\theta}(2\pi R^2)\sin\alpha\,d\alpha$

$\therefore\quad S=2\pi R^2\,(1-\cos\theta)$

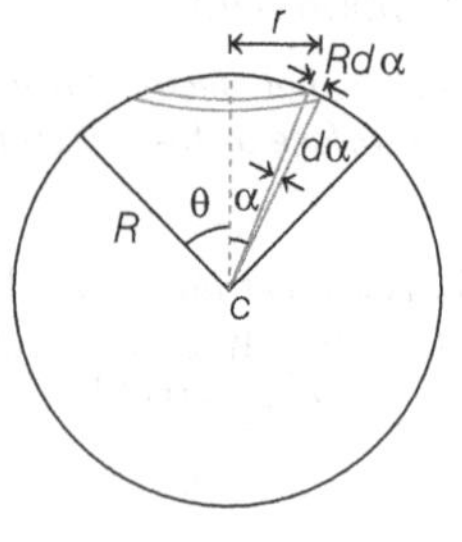

Students are advised to remember this result.

Ex. 6 The intensity of an electric field depends only on the coordinates x and y as follows

$$\mathbf{E}=\frac{a(x\hat{\mathbf{i}}+y\hat{\mathbf{j}})}{x^2+y^2}$$

where, a is a constant and $\hat{\mathbf{i}}$ and $\hat{\mathbf{j}}$ are the unit vectors of the x and y-axes. Find the charge within a sphere of radius R with the centre at the origin. **(Irodov 3.24)**

Sol. At any point $P\,(x, y, z)$ on the sphere a unit vector perpendicular to the sphere radially outwards is

$$\hat{\mathbf{n}}=\frac{x}{\sqrt{x^2+y^2+z^2}}\hat{\mathbf{i}}+\frac{y}{\sqrt{x^2+y^2+z^2}}\hat{\mathbf{j}}+\frac{z}{\sqrt{x^2+y^2+z^2}}\hat{\mathbf{k}}$$

$$=\frac{x}{R}\hat{\mathbf{i}}+\frac{y}{R}\hat{\mathbf{j}}+\frac{z}{R}\hat{\mathbf{k}}\quad\text{as}\quad x^2+y^2+z^2=R^2$$

Let us find the electric flux passing through a small area dS at point P on the sphere,

$$d\phi=\mathbf{E}\cdot\hat{\mathbf{n}}\,dS=\left\{\frac{ax^2}{R(x^2+y^2)}+\frac{ay^2}{R(x^2+y^2)}\right\}dS$$

$$=\left(\frac{a}{R}\right)dS$$

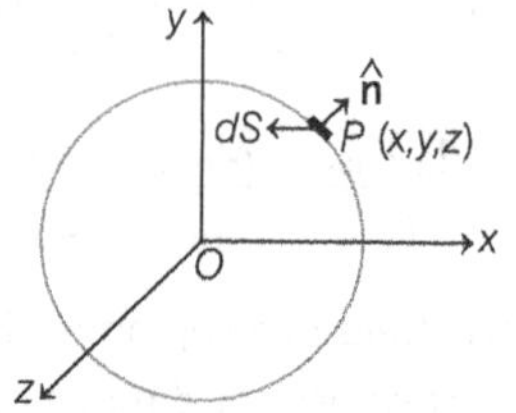

Here, we note that $d\phi$ is independent of the coordinates x, y and z. Therefore, total flux passing through the sphere

$$\phi=\int d\phi=\frac{a}{R}\int dS=\left(\frac{a}{R}\right)(4\pi R^2)$$

$$=4\pi aR$$

From Gauss's law,

$$\phi=\frac{q_{in}}{\varepsilon_0}\quad\text{or}\quad(4\pi aR)=\frac{q_{in}}{\varepsilon_0}$$

$\therefore\quad q_{in}=4\pi\varepsilon_0 aR$ **Ans.**

Type 2. *Based on finding electric field due to spherical charge distribution.*

Concept

According to Gauss's theorem, at a distance r from centre of sphere,

$$E=\frac{kq_{in}}{r^2}\qquad\left(k=\frac{1}{4\pi\varepsilon_0}\right)$$

Here, q_{in} is the net charge inside the sphere of radius r. If volume charge density (say ρ) is constant, then

$$q_{in}=(\text{volume of sphere of radius } r)(\rho)=\frac{4}{3}\pi r^3\rho$$

If ρ is variable, then q_{in} can be obtained by integration.

Passage (Ex. 7 to Ex. 9)

The nuclear charge (Ze) is non-uniformly distributed within a nucleus of radius R. The charge density $\rho(r)$ (charge per unit volume) is dependent only on the radial distance r from the centre of the nucleus as shown in figure. The electric field is only along the radial direction.

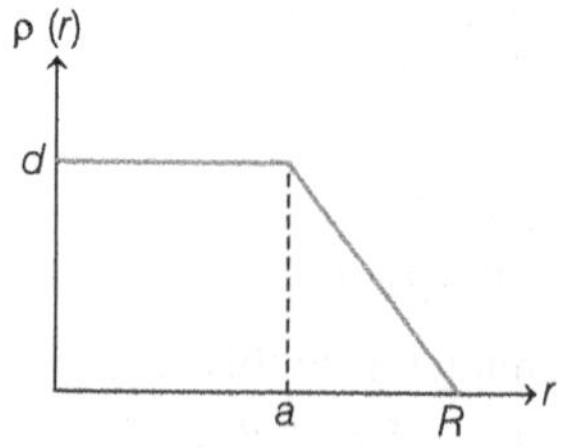

Ex. 7 The electric field at $r = R$ is **(2008 JEE)**

(a) independent of a
(b) directly proportional to a
(c) directly proportional to a^2
(d) inversely proportional to a

Sol. At $r = R$, from Gauss's law

$$E(4\pi R^2) = \frac{q_{in}}{\varepsilon_0} = \frac{Ze}{\varepsilon_0}$$

or $$E = \frac{1}{4\pi\varepsilon_0}\cdot\frac{Ze}{R^2}$$

E is independent of a.

$\therefore$ The correct option is (a).

Ex. 8 For $a = 0$, the value of d (maximum value of ρ as shown in the figure) is **(2008 JEE)**

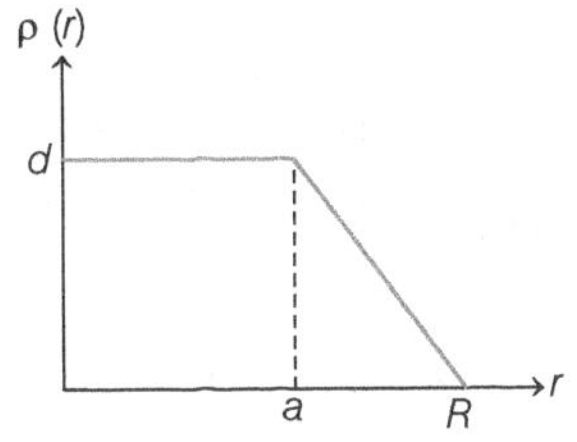

(a) $\frac{3Ze}{4\pi R^3}$ (b) $\frac{3Ze}{\pi R^3}$ (c) $\frac{4Ze}{3\pi R^3}$ (d) $\frac{Ze}{3\pi R^3}$

Sol. For $a = 0$,

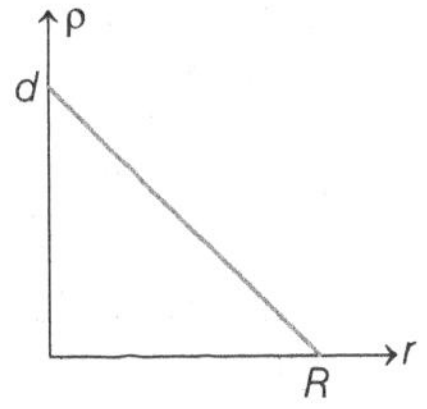

$$\rho(r) = \left(-\frac{d}{R}\cdot r + d\right)$$

Now, $$\int_0^R (4\pi r^2)\left(d - \frac{d}{R}r\right)dr = \text{net charge} = Ze$$

Solving this equation, we get

$$d = \frac{3Ze}{\pi R^3}$$

$\therefore$ The correct option is (b).

Ex. 9 The electric field within the nucleus is generally observed to be linearly dependent on r. This implies **(2008 JEE)**

(a) $a = 0$ (b) $a = \frac{R}{2}$

(c) $a = R$ (d) $a = \frac{2R}{3}$

Sol. In case of solid sphere of charge of uniform volume density

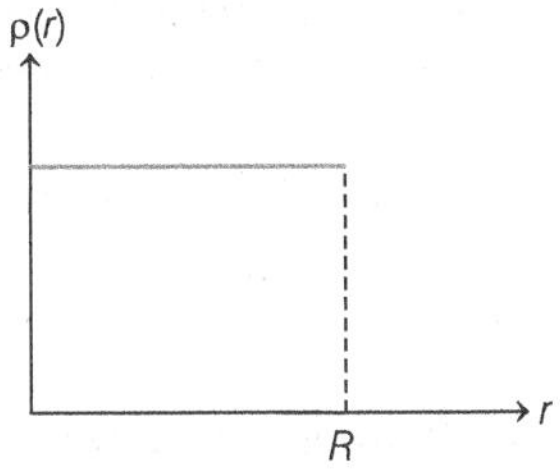

$$E = \frac{1}{4\pi\varepsilon_0}\cdot\frac{q}{R^3}\cdot r$$

or $$E \propto r$$

Thus, for E to be linearly dependent on r, volume charge density should be constant.

or $$a = R$$

$\therefore$ The correct option is (c).

MODULE 9 Exercise

True/False

1. Net charge within an imaginary cube drawn in a uniform electric field is always zero. Is this statement true or false?

Assertion and Reason

See directions on page 25.

2. **Assertion :** Gauss's theorem can be applied only for a closed surface.

   Reason : Electric flux can be obtained passing from an open surface also.

Multiple Correct Options

3. An electric dipole is placed at the centre of a sphere. Mark the correct options.
   (a) The flux of the electric field through the sphere is zero
   (b) The electric field is zero at every point of the sphere
   (c) The electric field is not zero anywhere on the sphere
   (d) The electric field is zero on a circle on the sphere
4. Mark the correct options.
   (a) Gauss's law is valid only for uniform charge distributions
   (b) Gauss's law is valid only for charges placed in vacuum
   (c) The electric field calculated by Gauss's law is the field due to all the charges lying inside or outside the Gaussian surface
   (d) The flux of the electric field through a closed surface due to all the charges is equal to the flux due to the charges enclosed by the surface
5. If the flux of the electric field through a closed surface is zero, then
   (a) the electric field must be zero everywhere on the surface
   (b) the electric field may be zero everywhere on the surface
   (c) the charge inside the surface must be zero
   (d) the charge in the vicinity of the surface must be zero

Subjective Questions

6. A charge q is placed at point D of the cube. Find the electric flux passing through the face $EFGH$ and face $AEHD$.

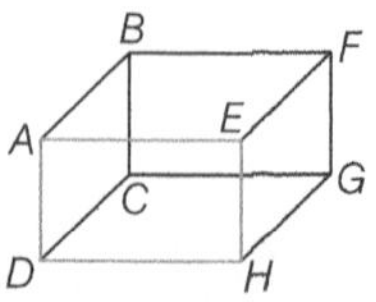

7. A sphere centered at the origin has radius 0.200 m. A $-500\ \mu C$ point charge is on the x-axis at $x = 0.300$ m. The net flux through the sphere is 360 N-m^2/C. What is the total charge inside the sphere?
8. The electric field in a region is given by $\mathbf{E} = \frac{E_0 x}{l}\hat{\mathbf{i}}$. Find the charge contained inside a cubical volume bounded by the surfaces $x = 0, x = a, y = 0, y = a, z = 0$ and $z = a$. Take $E_0 = 5 \times 10^3$ N/C, $l = 2$ cm and $a = 1$ cm.
9. A point charge Q is located on the axis of a disc of radius R at a distance b from the plane of the disc (figure). Show that if one-fourth of the electric flux from the charge passes through the disc, then $R = \sqrt{3}b$.

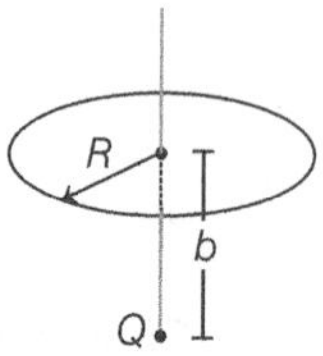

10. A cube has sides of length L. It is placed with one corner at the origin as shown in figure. The electric field is uniform and given by $\mathbf{E} = -B\hat{\mathbf{i}} + C\hat{\mathbf{j}} - D\hat{\mathbf{k}}$, where B, C and D are positive constants.

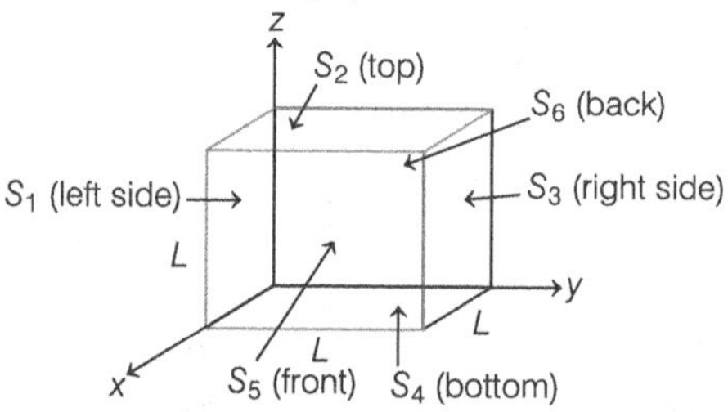

   (a) Find the electric flux through each of the six cube faces S_1, S_2, S_3, S_4, S_5 and S_6.
   (b) Find the electric flux through the entire cube.
11. Two point charges q and $-q$ are separated by a distance $2l$. Find the flux of electric field strength vector across the circle of radius R placed with its centre coinciding with the mid-point of line joining the two charges in the perpendicular plane.
12. A point charge q is placed at the origin. Calculate the electric flux through the open hemispherical surface : $(x-a)^2 + y^2 + z^2 = a^2, x \ge a$.
13. A charge q_0 is distributed uniformly on a ring of radius R. A sphere of equal radius R is constructed with its centre on the circumference of the ring. Find the electric flux through the surface of the sphere.

Answers

1. True **2.** (b) **3.** (a,c) **4.** (c,d) **5.** (b,c)

6. $\frac{q}{24\,\varepsilon_0}$, zero **7.** 3.19 nC **8.** 2.2×10^{-12} C

10. (a) $\phi_{S_1} = -CL^2$, $\phi_{S_2} = -DL^2$, $\phi_{S_3} = CL^2$, $\phi_{S_4} = DL^2$, $\phi_{S_5} = -BL^2$, $\phi_{S_6} = BL^2$ (b) Zero

11. $\frac{q}{\varepsilon_0}\left[1 - \frac{1}{\sqrt{1+\left(\frac{R}{l}\right)^2}}\right]$ **12.** $\frac{q}{2\varepsilon_0}\left(1 - \frac{1}{\sqrt{2}}\right)$ **13.** $\frac{q_0}{3\varepsilon_0}$

See hints on pages 208-210.

MODULE 10

Properties of a Conductor

Conductors (such as metals) possess free electrons. If a resultant electric field exists in the conductor these free charges will experience a force which will set a current flow. When no current flows, the resultant force and the electric field must be zero. Thus, under **electrostatic conditions** the value of **E** at all points within a conductor is zero. This idea, together with the Gauss's law can be used to prove several interesting facts regarding a conductor.

Excess Charge on a Conductor Resides on its Outer Surface

Consider a charged conductor carrying a charge q and no currents are flowing in it. Now, consider a Gaussian surface inside the conductor everywhere on which $\mathbf{E} = 0$.

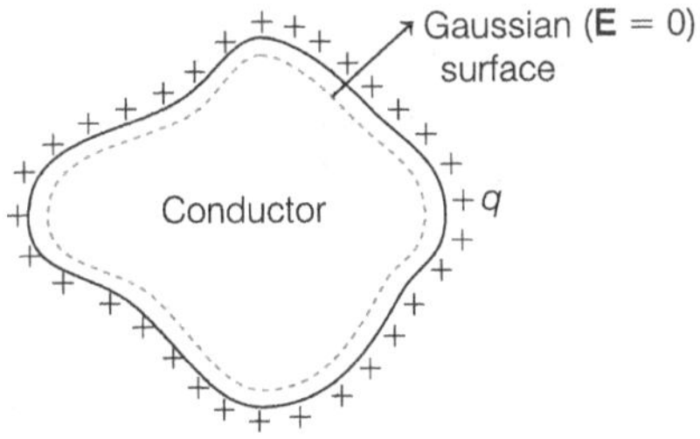

Thus, from Gauss's law, $\oint_S \mathbf{E} \cdot d\mathbf{S} = \dfrac{q_{in}}{\varepsilon_0}$

We get, $\quad q_{in} = 0, \text{ as } \mathbf{E} = 0$

Thus, the sum of all charges inside the Gaussian surface is zero. This surface can be taken just inside the surface of the conductor. Hence, any charge on the conductor must be on the surface of the conductor. In other words,

"Under electrostatic conditions, the excess charge on a conductor resides on its outer surface."

Electric Field at Any Point Close to the Charged Conductor is $\dfrac{\sigma}{\varepsilon_0}$

Consider a charged conductor of irregular shape. In general, surface charge density will vary from point to point. At a small surface ΔS, let us assume it to be a constant σ. Let us construct a Gaussian surface in the form of a cylinder of cross-section ΔS. One plane face of the cylinder is inside the conductor and other outside the conductor close to it.

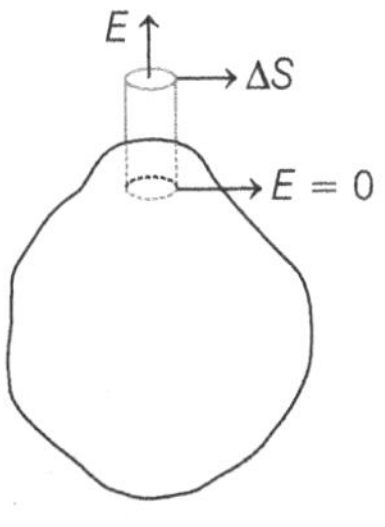

The surface inside the conductor does not contribute to the flux as **E** is zero everywhere inside the conductor. The curved surface outside the conductor also does not contribute to flux as **E** is always normal to the charged conductor and hence parallel to the curved surface. Thus, the only contribution to the flux is through the plane face outside the conductor. Thus, from Gauss's law,

$$\oint_S \mathbf{E} \cdot d\mathbf{S} = \frac{q_{in}}{\varepsilon_0}$$

or

$$E\Delta S = \frac{(\sigma)(\Delta S)}{\varepsilon_0}$$

or

$$\boxed{E = \frac{\sigma}{\varepsilon_0}}$$

Note *(i) Electric field changes discontinuously at the surface of a conductor. Just inside the conductor it is zero and just outside the conductor it is* $\dfrac{\sigma}{\varepsilon_0}$. *In fact, the field gradually decreases from* $\dfrac{\sigma}{\varepsilon_0}$ *to zero in a small thickness of about* 4 *to 5 atomic layers at the surface.*

(ii) For a non-uniform conductor the surface charge density (σ) *varies inversely as the radius of curvature* (R) *of that part of the conductor, i.e.*

$$\boxed{\sigma \propto \frac{1}{\text{Radius of curvature } (R)}}$$

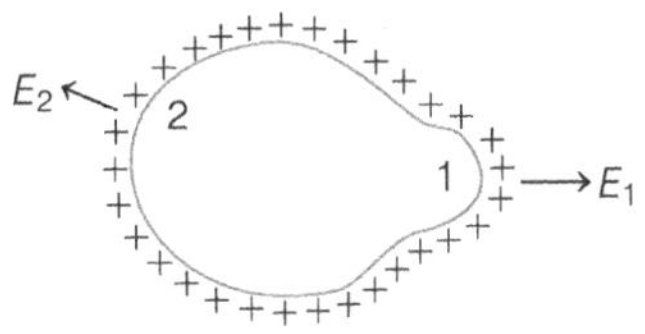

For example in the figure, $R_1 < R_2$

$\therefore \quad \sigma_1 > \sigma_2$

or $\quad E_1 > E_2$

as $\quad E = \dfrac{\sigma}{\varepsilon_0}$

Electric Field and Field Lines are Normal to the Surface of a Conductor

Net field inside a conductor is zero. It implies that no field lines enter a conductor. On the surface of a conductor, electric field and hence field lines are normal to the surface of the conductor.

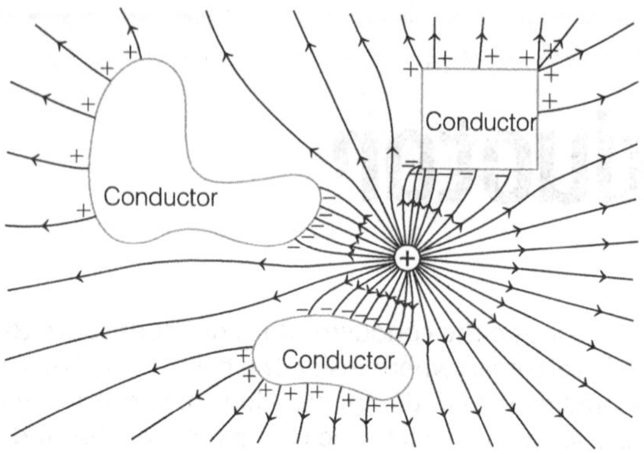

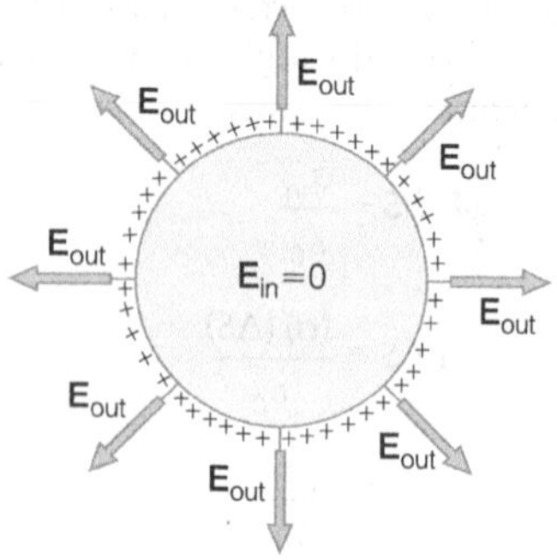

If a conductor is immersed in an electric field, the field lines near the conductor are somewhat distorted. Similarly, if a conductor is positively charged, the field lines originate from the surface and are normal at every point and if it is negatively charged the field lines terminate on the surface normally at every point.

Cavity Inside a Conductor

Consider a charge $-q$ suspended in a cavity in a conductor. Consider a Gaussian surface just outside the cavity and inside the conductor. $\mathbf{E} = 0$ on this Gaussian surface as it is inside the conductor. Hence, from Gauss's law,

$$\oint \mathbf{E} \cdot d\mathbf{S} = \frac{q_{in}}{\varepsilon_0} \quad \text{gives } q_{in} = 0$$

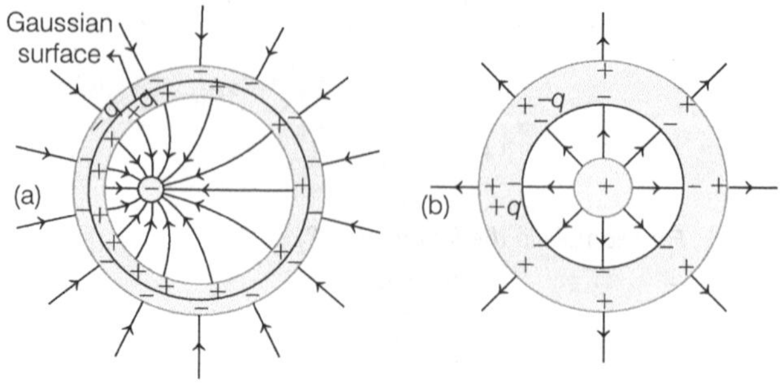

This concludes that a charge of $+q$ must reside on the metal surface of the cavity so that the sum of this induced charge $+q$ and the original charge $-q$ within the Gaussian surface is zero. In other words, a charge $-q$ suspended inside a cavity in a conductor induces an equal and opposite charge $+q$ on the surface of the cavity. Further as the conductor is electrically neutral a charge $-q$ is induced on the outer surface of the conductor. As field inside the conductor is zero, the field lines cannot penetrate into the conductor. The field lines will be as shown in Fig. (a).

In Fig. (b) $+q$ charge is placed at centre of cavity. So, induced charge distribution will be symmetrical as shown.

Electrostatic Shielding

Suppose we have a very sensitive electronic instrument that we want to protect from external electric fields that might cause wrong measurements. We surround the instrument with a conducting box or we keep the instrument inside the cavity of a conductor. By doing this charge in the conductor is so distributed that the net electric field inside the cavity becomes zero and the instrument is protected from the external fields. This is called electronic shielding.

The Potential of a Charged Conductor Throughout its Volume is Same

In any region in which $\mathbf{E} = 0$ at all points, such as the interior of a charged conductor, the line integral of $\mathbf{E}$ is zero along any path. It means that the potential between any two points in the conductor is same or the interior of a charged conductor is an equipotential region. Potential difference between any two points is zero.

Example 1 A metallic shell has a point charge q kept inside its cavity. Which one of the following diagrams correctly represents the electric lines of forces? **(2003 JEE)**

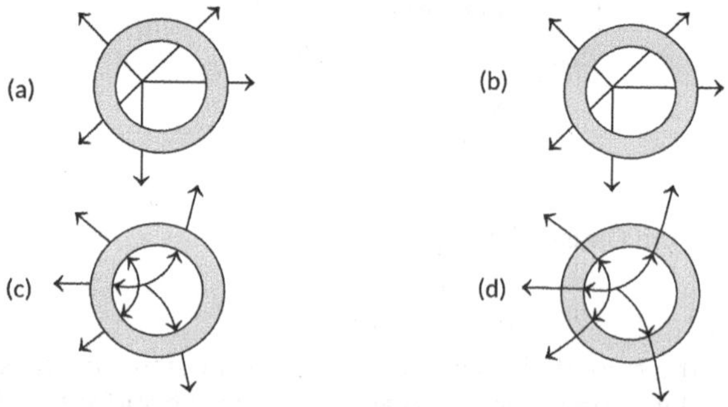

Sol. Electric field inside the metallic portion (conductor) of the shell is zero.

All electric field lines are perpendicular to the surface.

Hence, option (c) is correct.

Example 2 A metallic solid sphere is placed in a uniform electric field. The lines of force follow the path(s) shown in figure as **(1996 JEE)**

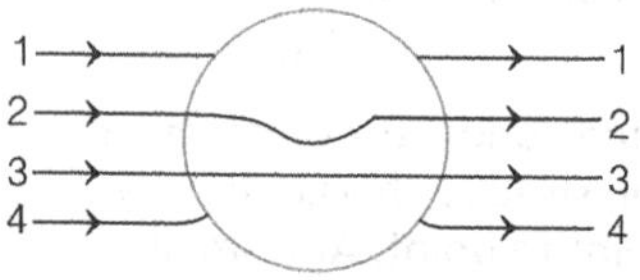

(a) 1 (b) 2 (c) 3 (d) 4

Sol. Inside a metallic sphere, $E = 0$

This, electric field lines will not enter the solid sphere. Also all electric field lines are always normal to the metallic surface. Hence, option (d) is correct.

MODULE 11

Charged Spheres

Electric Field and Potential Due to Charged Spherical Shell or Solid Conducting Sphere

Electric Field

At all points inside the charged spherical conductor or hollow spherical shell, electric field $\mathbf{E} = 0$, as there is no charge inside such a sphere. In an isolated charged spherical conductor any excess charge on it is distributed uniformly over its outer surface same as that of charged spherical shell or hollow sphere.

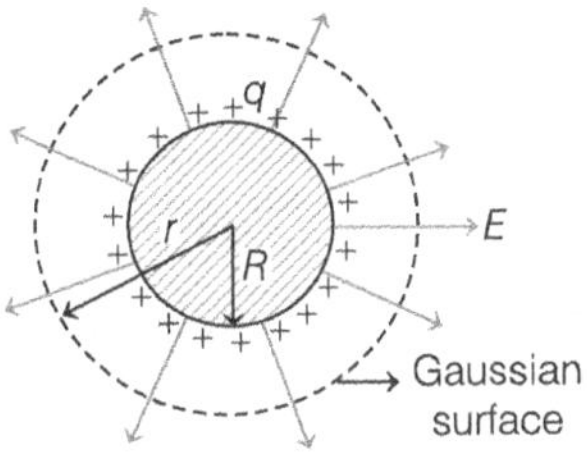

The field at external points has the same symmetry as that of a point charge. We can construct a Gaussian surface (a sphere) of radius $r > R$. At all points of this sphere the magnitude of electric field is the same and its direction is perpendicular to the surface.

Thus, we can apply

$$ES = \frac{q_{in}}{\varepsilon_0}$$

or $$E(4\pi r^2) = \frac{q}{\varepsilon_0}$$

$\therefore$ $$E = \frac{1}{4\pi\varepsilon_0} \cdot \frac{q}{r^2}$$

Hence, the electric field at any external point is the same as if the total charge is concentrated at centre.

At the surface of sphere $r = R$,

$\therefore$ $$E = \frac{1}{4\pi\varepsilon_0} \cdot \frac{q}{R^2} = \frac{\sigma}{\varepsilon_0}$$

As $\frac{q}{4\pi R^2}$ = surface charge density $= \sigma$

Thus, we can write

$$E_{inside} = 0$$

$$E_{surface} = \frac{1}{4\pi\varepsilon_0} \frac{q}{R^2} = \frac{\sigma}{\varepsilon_0}$$

$$E_{outside} = \frac{1}{4\pi\varepsilon_0} \cdot \frac{q}{r^2}$$

The variation of electric field (E) with the distance from the centre (r) is as shown in figure.

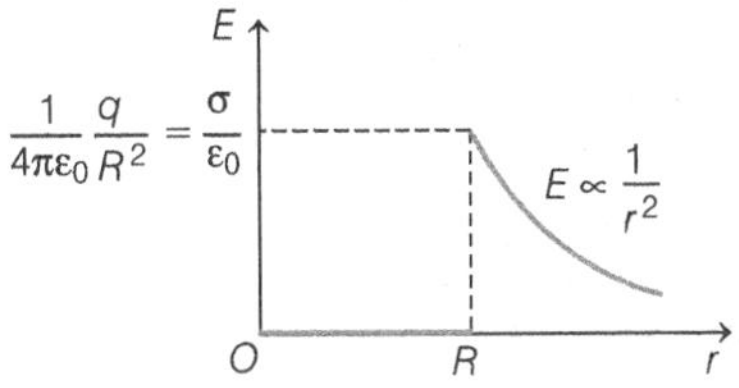

Note *(i) At the surface graph is discontinuous*

(ii) $E_{surface} = \frac{1}{4\pi\varepsilon_0} \cdot \frac{q}{R^2} = \frac{q/4\pi R^2}{\varepsilon_0} = \frac{\sigma}{\varepsilon_0}$

Potential

As we have seen, $$E_{outside} = \frac{1}{4\pi\varepsilon_0} \cdot \frac{q}{r^2}$$

$\therefore$ $$\left(\frac{-dV_{outside}}{dr}\right) = \frac{1}{4\pi\varepsilon_0} \cdot \frac{q}{r^2} \qquad \left(E = -\frac{dV}{dr}\right)$$

$\therefore$ $$\int_0^V dV_{outsidede} = \frac{-q}{4\pi\varepsilon_0} \int_\infty^r \frac{dr}{r^2} \qquad (V_\infty = 0)$$

$\therefore$ $$V = \frac{1}{4\pi\varepsilon_0} \cdot \frac{q}{r} \quad \text{or} \quad V \propto \frac{1}{r}$$

Thus, at external points, the potential at any point is the same when the whole charge is assumed to be concentrated at the centre. At the surface of the sphere, $r = R$

$\therefore$ $$V = \frac{1}{4\pi\varepsilon_0} \cdot \frac{q}{R}$$

At some internal point electric field is zero everywhere, therefore, the potential is same at all points which is equal to the potential at surface. Thus, we can write

$$V_{inside} = V_{surface} = \frac{1}{4\pi\varepsilon_0} \cdot \frac{q}{R}$$

and $$V_{outside} = \frac{1}{4\pi\varepsilon_0} \cdot \frac{q}{r}$$

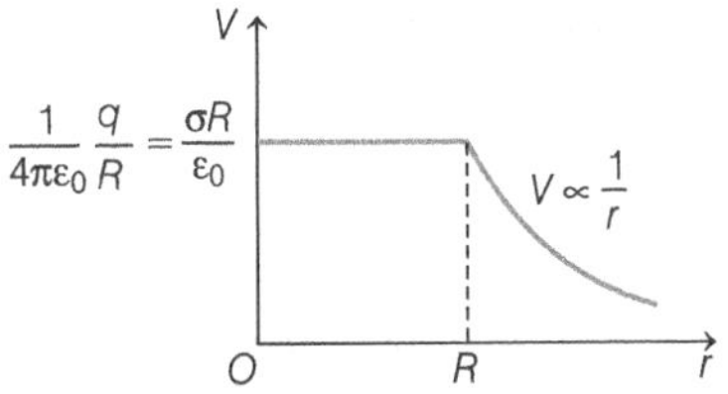

The potential (V) varies with the distance from the centre (r) as shown in figure.

Electric Field and Potential Due to a Solid Sphere of Charge

Electric Field

Positive charge q is uniformly distributed throughout the volume of a solid sphere of radius R. For finding the electric field at a distance $r\ (<R)$ from the centre let us choose as our Gaussian surface a sphere of radius r, concentric with the charge distribution.

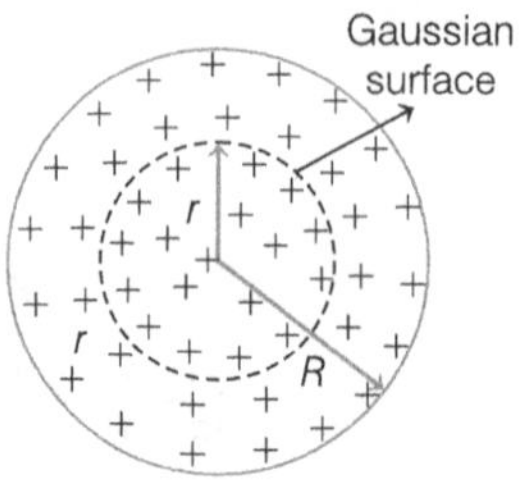

From symmetry, the magnitude E of electric field has the same value at every point on the Gaussian surface and the direction of **E** is radial at every point on the surface. So, applying Gauss's law

$$ES = \frac{q_{in}}{\varepsilon_0} \qquad \text{...(i)}$$

Here, $S = 4\pi r^2$ and $q_{in} = (\rho)\left(\frac{4}{3}\pi r^3\right)$

Here, $\rho = \text{charge per unit volume} = \dfrac{q}{\frac{4}{3}\pi R^3}$

Substituting these values in Eq. (i)

We have, $E = \dfrac{1}{4\pi\varepsilon_0}\cdot\dfrac{q}{R^3}\cdot r$ or $E \propto r$

At the centre $r = 0$, so $E = 0$

At surface $r = R$, so $E = \dfrac{1}{4\pi\varepsilon_0}\cdot\dfrac{q}{R^2}$

To find the electric field outside the charged sphere, we use a spherical Gaussian surface of radius $r\ (>R)$. This surface encloses the entire charged sphere, so $q_{in} = q$, and Gauss's law gives

$$E(4\pi r^2) = \frac{q}{\varepsilon_0} \quad \text{or} \quad E = \frac{1}{4\pi\varepsilon_0}\cdot\frac{q}{r^2} \quad \text{or} \quad E \propto \frac{1}{r^2}$$

Notice that if we set $r = R$ in either of the two expressions for E (outside and inside the sphere), we get the same result,

$$E = \frac{1}{4\pi\varepsilon_0}\cdot\frac{q}{R^2}$$

this is because E is **continuous function** of r in this case. By contrast, for the charged conducting sphere the magnitude of electric field is discontinuous at $r = R$ (it jumps from $E = 0$ to $E = \sigma/\varepsilon_0$).

Thus, for a uniformly charged solid sphere we have the following formulae for magnitude of electric field :

$$E_{inside} = \frac{1}{4\pi\varepsilon_0}\cdot\frac{q}{R^3}\cdot r$$

$$E_{surface} = \frac{1}{4\pi\varepsilon_0}\cdot\frac{q}{R^2}$$

$$E_{outside} = \frac{1}{4\pi\varepsilon_0}\cdot\frac{q}{r^2}$$

The variation of electric field (E) with the distance from the centre of the sphere (r) is shown in figure.

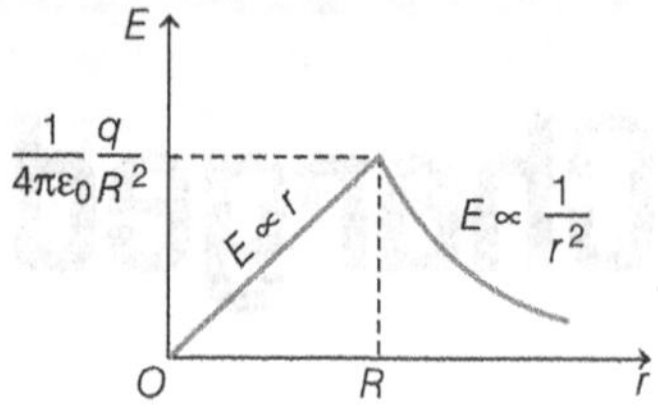

Note $E_{surface} \neq \dfrac{\sigma}{\varepsilon_0}$, *because in this case charge is distributed over a volume not on a surface.*

Potential

The field intensity outside the sphere is

$$E_{outside} = \frac{1}{4\pi\varepsilon_0}\cdot\frac{q}{r^2}$$

$$\frac{dV_{outside}}{dr} = -E_{outside}$$

$$\therefore \quad dV_{outside} = -E_{outside}\, dr$$

$$\text{or} \quad \int_\infty^V dV_{outside} = -\int_\infty^r \frac{1}{4\pi\varepsilon_0}\cdot\frac{q}{r^2}\, dr$$

$$\therefore \quad V = \frac{1}{4\pi\varepsilon_0}\cdot\frac{q}{r} \quad \text{as} \quad V_\infty = 0 \text{ or } V \propto \frac{1}{r}$$

At $r = R$, $V = \dfrac{1}{4\pi\varepsilon_0}\cdot\dfrac{q}{R}$

i.e. at the surface of the sphere potential is $V_S = \dfrac{1}{4\pi\varepsilon_0}\cdot\dfrac{q}{R}$

The electric intensity inside the sphere,

$$E_{inside} = \frac{1}{4\pi\varepsilon_0}\cdot\frac{q}{R^3}\cdot r \Rightarrow \frac{dV_{inside}}{dr} = -E_{inside}$$

$$\therefore \quad dV_{inside} = -E_{inside}\, dr$$

$$\therefore \quad \int_{V_S}^V dV_{inside} = -\frac{1}{4\pi\varepsilon_0}\cdot\frac{q}{R^3}\int_R^r r\, dr$$

$$\therefore \quad V - V_S = -\frac{1}{4\pi\varepsilon_0}\cdot\frac{q}{R^3}\left[\frac{r^2}{2}\right]_R^r$$

Substituting $V_S = \dfrac{1}{4\pi\varepsilon_0}\cdot\dfrac{q}{R}$, we get

$$V = \frac{1}{4\pi\varepsilon_0}\frac{q}{R^3}(1.5R^2 - 0.5r^2)$$

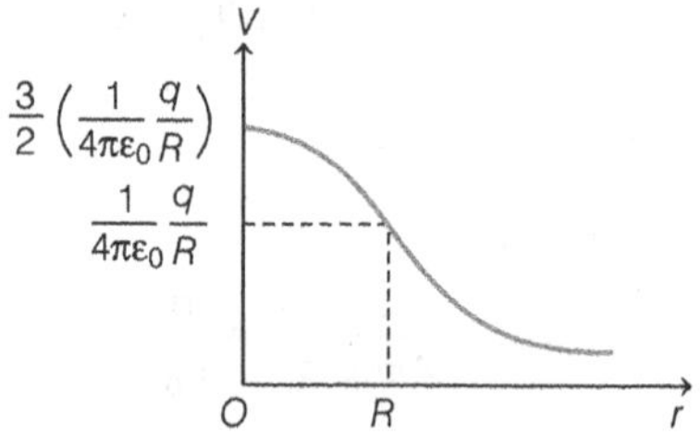

At the centre $r = 0$ and $V_C = \dfrac{3}{2}\left(\dfrac{1}{4\pi\varepsilon_0}\cdot\dfrac{q}{R}\right) = \dfrac{3}{2}V_S$, i.e. potential at the centre is 1.5 times the potential at surface.

Thus, for a uniformly charged solid sphere we have the following formulae for potential :

$$V_{\text{outside}} = \frac{1}{4\pi\varepsilon_0} \cdot \frac{q}{r}$$

$$V_{\text{surface}} = \frac{1}{4\pi\varepsilon_0} \cdot \frac{q}{R}$$

and

$$V_{\text{inside}} = \frac{1}{4\pi\varepsilon_0} \cdot \frac{q}{R}\left[\frac{3}{2} - \frac{1}{2}\frac{r^2}{R^2}\right]$$

The variation of potential (V) with distance from the centre (r) is as shown in figure. For inside points variation is parabolic.

List of formulae for field strength E and potential V $\left(k = \dfrac{1}{4\pi\varepsilon_0}\right)$

Table 1

S.No.	Charge Distribution	E: Formula	E: Graph	V: Formula	V: Graph
1	Point charge	$E = \frac{kq}{r^2}$	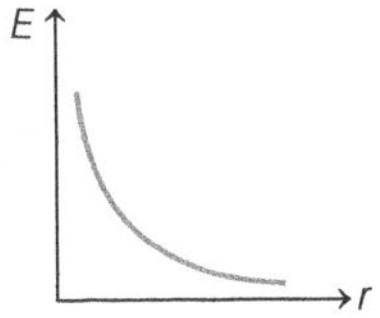	$V = \frac{kq}{r}$	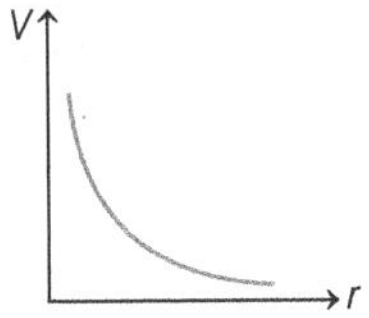
2	Uniformly charged spherical shell or Solid charged conducting sphere	$E_i = 0$ $E_s = k \cdot \frac{q}{R^2} = \frac{\sigma}{\varepsilon_0}$ $E_o = \frac{kq}{r^2}$	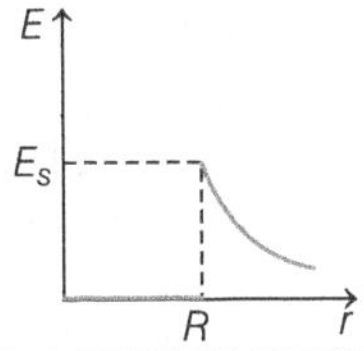	$V_i = V_s = \frac{kq}{R}$ $= \frac{\sigma R}{\varepsilon_0}$ $V_o = \frac{kq}{r}$	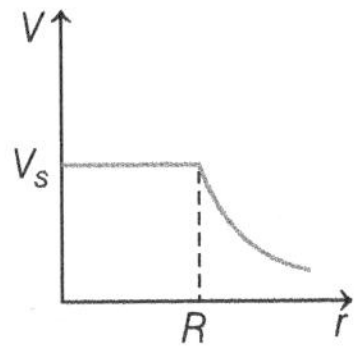
3	Solid sphere of charge	$E_i = \frac{kqr}{R^3}$ $E_s = \frac{kq}{R^2}$ $E_o = \frac{Kq}{r^2}$	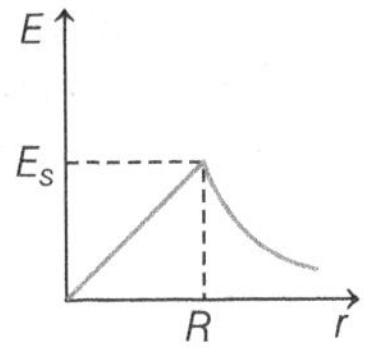	$V_i = \frac{kq}{R^3}(1.5 R^2 - 0.5 r^2)$ $V_s = \frac{kq}{R}$ $V_o = \frac{kq}{r}$	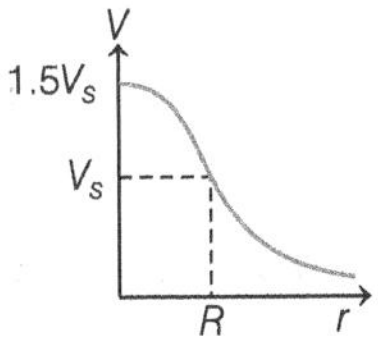
4	On the axis of uniformly charged ring	$E = \frac{kqx}{(R^2 + x^2)^{3/2}}$ At centre $x = 0$ $\therefore$ $E = 0$	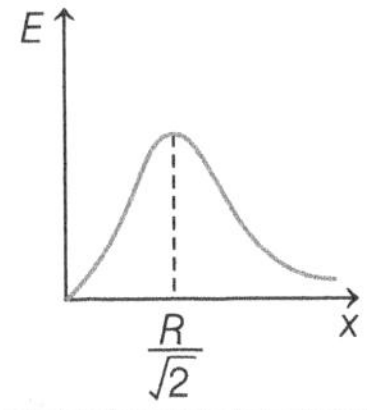	$V = \frac{kq}{\sqrt{R^2 + x^2}}$ At centre $x = 0$ $\therefore$ $V = \frac{kq}{R}$	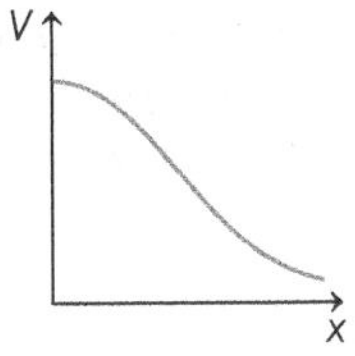
5	Infinitely long line charge	$E = \frac{\lambda}{2\pi \varepsilon_0 r}$	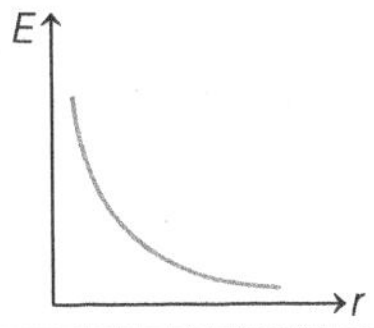	$PD = V_{r_1} - V_{r_2} = \frac{\lambda}{2\pi \varepsilon_0} \ln\left(\frac{r_2}{r_1}\right)$	Not required

Note (i) ***Earthing a conductor*** *Potential of earth is often taken to be zero. If a conductor is connected to the earth, the potential of the conductor becomes equal to that of the earth, i.e. zero. If the conductor was at some other potential, charges will flow from it to the earth or from the earth to it to bring its potential to zero.*

(ii) *In the above table magnitudes of E and V are taken on y-axis. This is because* **E** *is a vector quantity. So, it has some direction also and V is positive due to positive charge and negative due to negative charge.*

(iii) *For infinitely long line charge there is no formula for absolute potential.*

Potential difference can be obtained by using the formula,

$$dV = -E dr$$

and

$$E = \frac{\lambda}{2\pi\varepsilon_0 r}$$

Typed Problems

Type 1. *To find electric potential due to charged spherical shells.*

Concept

To find the electric potential due to a conducting sphere (or shell) we should keep in mind the following two points

(i) Electric potential on the surface and at any point inside the sphere is

$$V = \frac{1}{4\pi\varepsilon_0}\cdot\frac{q}{R} \qquad (R = \text{radius of sphere})$$

(ii) Electric potential at any point outside the sphere is

$$V = \frac{1}{4\pi\varepsilon_0}\cdot\frac{q}{r}$$

(r = distance of the point from the centre)

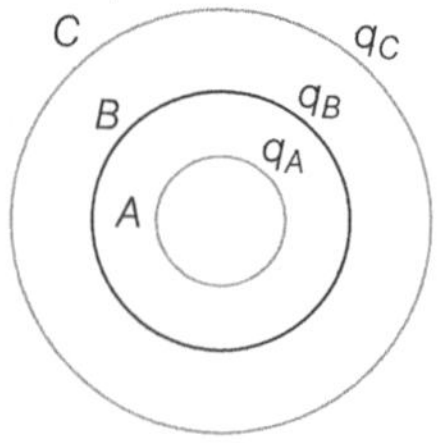

For example, in the figure shown, potential at A is

$$V_A = \frac{1}{4\pi\varepsilon_0}\left[\frac{q_A}{r_A} + \frac{q_B}{r_B} + \frac{q_C}{r_C}\right]$$

Similarly, potential at B is $V_B = \dfrac{1}{4\pi\varepsilon_0}\left[\dfrac{q_A}{r_B} + \dfrac{q_B}{r_B} + \dfrac{q_C}{r_C}\right]$

and potential at C is, $V_C = \dfrac{1}{4\pi\varepsilon_0}\left[\dfrac{q_A}{r_C} + \dfrac{q_B}{r_C} + \dfrac{q_C}{r_C}\right]$

Ex. 1 Three conducting spherical shells have charges q, $-2q$ and $3q$ as shown in figure. Find electric potential at point P as shown in figure.

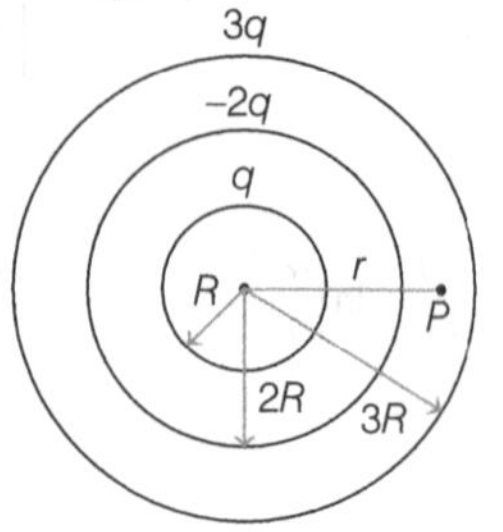

Sol. Potential at P,

$$V_P = V_q + V_{-2q} + V_{3q}$$

$$= \frac{kq}{r} - \frac{k(2q)}{r} + \frac{k(3q)}{3R}$$

$$= kq\left(\frac{1}{R} - \frac{1}{r}\right) \qquad \textbf{Ans.}$$

Here, $k = \dfrac{1}{4\pi\varepsilon_0}$

Ex. 2 Figure shows two conducting thin concentric shells of radii r and $3r$. The outer shell carries a charge q. Inner shell is neutral. Find the charge that will flow from inner shell to earth after the switch S is closed.

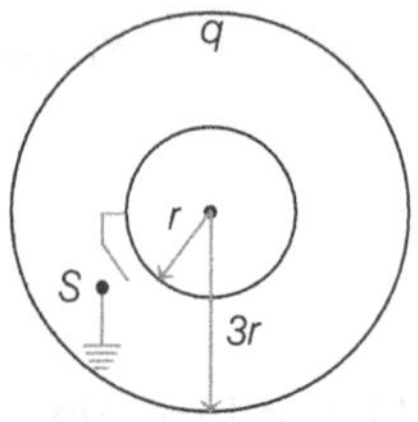

Sol. Let q' be the charge on inner shell when it is earthed.

Potential of inner shell is zero.

$$\therefore \qquad \frac{1}{4\pi\varepsilon_0}\left[\frac{q'}{r} + \frac{q}{3r}\right] = 0$$

$$\therefore \qquad q' = -\frac{q}{3}$$

i.e. $+\dfrac{q}{3}$ charge will flow from inner shell to earth. **Ans.**

Type 2. *Based on the principle of generator.*

Concept

A generator is an instrument for producing high voltages. Its design is based on the principle that if a charged conductor (say A) is brought into contact with a hollow conductor (say B), all of its charge transfers to the hollow conductor no matter how high the potential of the latter may be. This can be shown as under:

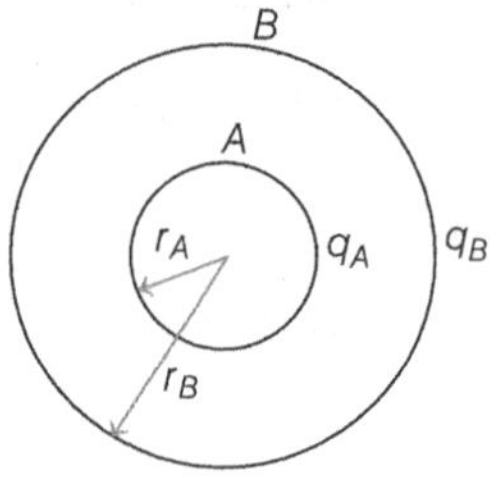

In the figure, $\quad V_A = \dfrac{1}{4\pi\varepsilon_0}\left[\dfrac{q_A}{r_A} + \dfrac{q_B}{r_B}\right]$

and $\quad V_B = \dfrac{1}{4\pi\varepsilon_0}\left[\dfrac{q_A}{r_B} + \dfrac{q_B}{r_B}\right]$

$$\therefore \qquad V_A - V_B = \frac{q_A}{4\pi\varepsilon_0}\left[\frac{1}{r_A} - \frac{1}{r_B}\right]$$

Alternate method :

From Gauss law E between A and B depends upon q_{in} or q_A only. Potential difference depends upon E. So, it depends upon q_A.

Between A and B,

$$E = \left(\frac{1}{4\pi\varepsilon_0}\right)\left(\frac{q}{r^2}\right)$$

$$dV = -E\,dr$$

or

$$\int_B^A dV = -\int_{r_B}^{r_A}\left(\frac{1}{4\pi\varepsilon_0}\right)\left(\frac{q_A}{r^2}\right)dr$$

$$\therefore \qquad V_A - V_B = \frac{q_A}{4\pi\varepsilon_0}\left[\frac{1}{r_A} - \frac{1}{r_B}\right]$$

From this expression the following conclusions can be drawn :

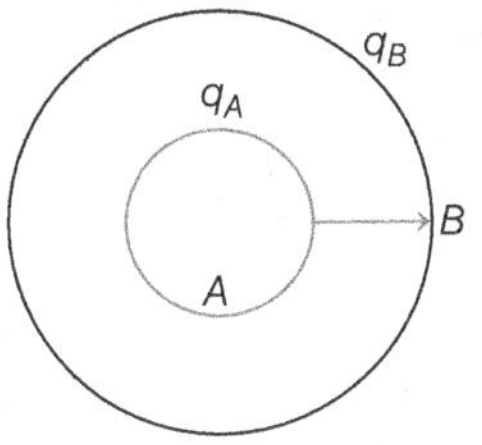

(i) The potential difference (PD) depends on q_A only. It does not depend on q_B.

(ii) If q_A is positive, then $V_A - V_B$ is positive (as $r_A < r_B$), i.e. $V_A > V_B$. So if the two spheres are connected by a conducting wire charge flows from inner sphere to outer sphere (positive charge flows from higher potential to lower potential) till $V_A = V_B$ or $V_A - V_B = 0$. But potential difference will become zero only when $q_A = 0$, i.e. all charge q_A flows from inner sphere to outer sphere.

(iii) If q_A is negative, $V_A - V_B$ is negative, i.e. $V_A < V_B$. Hence, when the two spheres are connected by a thin wire all charge q_A will flow from inner sphere to the outer sphere. Because negative charge flows from lower potential to higher potential. Thus, we see that the whole charge q_A flows from inner sphere to the outer sphere, no matter how high q_B is. Charge always flows from A to B, whether $q_A > q_B$ or $q_B > q_A$, $V_A > V_B$ or $V_B > V_A$.

(iv) Using the alternate method discussed above, we can prove that,

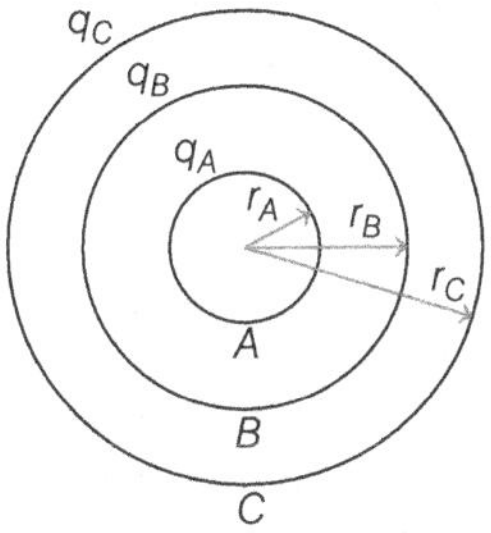

$$V_A - V_B = \frac{q_A}{4\pi\varepsilon_0}\left[\frac{1}{r_A} - \frac{1}{r_B}\right] \text{ and}$$

$$V_B - V_C = \frac{(q_A + q_B)}{4\pi\varepsilon_0}\left[\frac{1}{r_B} - \frac{1}{r_C}\right]$$

Ex. 3 Initially the spheres A and B are at potentials V_A and V_B. Find the potential of A when sphere B is earthed.

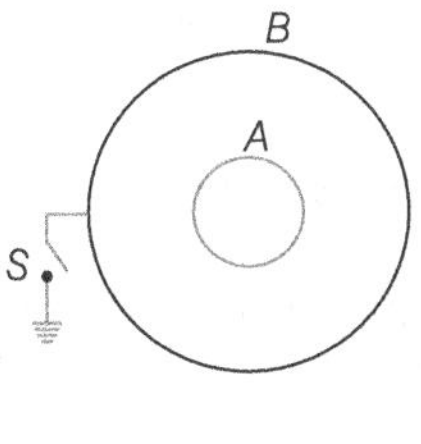

Sol. As we have studied above that the potential difference between these two spheres depends on the charge on the inner sphere only. Hence, the PD will remain unchanged, because by earthing the sphere B charge on A remains constant. Let V_A' be the new potential at A. Then,

$$V_A - V_B = V_A' - V_B'$$

but $V_B' = 0$ as it is earthed. Hence,

$$V_A' = V_A - V_B$$ **Ans.**

Type 3. *Based on the charges appearing on different surfaces of concentric spherical shells.*

Concept

Figure shows three concentric thin spherical shells A, B and C of radii a, b and c. The shells A and C are given charges Q_1 and Q_2 and the shell B is earthed. We are interested in finding the charges on inner and outer surfaces of A, B and C. To solve such type of problems we should keep the following points in mind :

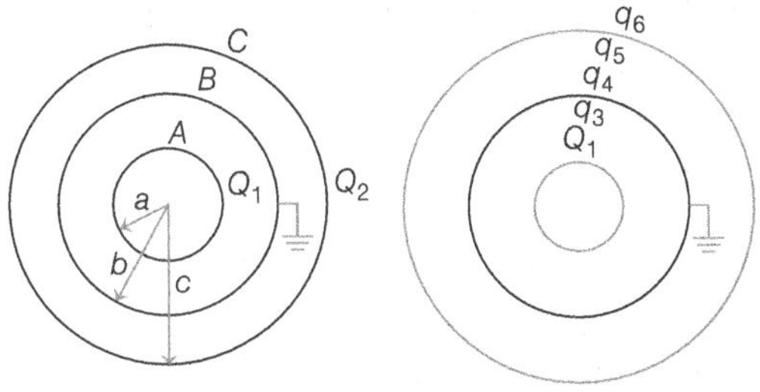

(i) The whole charge Q_1 will come on the outer surface of A unless some charge is kept inside A. To understand it let us consider a Gaussian surface (a sphere) through the material of A. As the electric field in a conducting material is zero. The flux through this Gaussian surface is zero. Using Gauss's law, the total charge enclosed must be zero. So, Q_1 should be on outer surface of A.

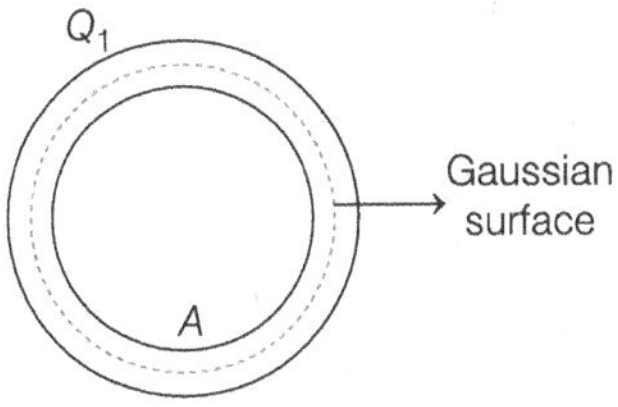

(ii) Similarly, if we draw a Gaussian surface through the material of B we can see that

$$q_3 + Q_1 = 0$$

or $$q_3 = -Q_1$$

and if we draw a Gaussian surface through the material of C, then

$$q_5 + q_4 + q_3 + Q_1 = 0$$

or $$q_5 = -q_4$$

(iii) $q_5 + q_6 = Q_2$. As Q_2 charge was given to shell C.

(iv) Potential of B should be zero, as it is earthed. Thus,

$$V_B = 0$$

or $$\frac{1}{4\pi\varepsilon_0}\left[\frac{Q_1}{b}+\frac{q_3+q_4}{b}+\frac{q_5+q_6}{c}\right]=0$$

So, using the above conditions we can find charges on different surfaces.

Following points should be remembered while solving such problems.

1. Net charge inside a closed Gaussian surface drawn in any conducting shell is zero.
2. Potential of the shell which is earthed is zero.
3. If two shells are connected, they are at the same potential.
4. In any shell,

   $q_{\text{inner surface}} + q_{\text{outer surface}} =$ charge given to the shell

   Like $q_5 + q_6 = Q_2$
5. Charge on the inner surface of the innermost shell is zero if no charge is kept inside it.

   Otherwise, $q_{\text{innermost surface}} = -(q_{\text{cavity}})$

   In all other shells charge resides on both the surfaces.
6. Equal and opposite charges appear on opposite faces.

   Like $q_4 = -q_5$ or $Q_1 = -q_3$
7. If outermost shell is earthed then $q_{\text{outermost}} = 0$.

   Ex. 4, given below illustrates this point.
8. Normally no charge is given to a shell. Because charge comes on this shell from earth to make its potential = 0.

Ex. 4 A charge q is distributed uniformly on the surface of a solid sphere of radius R. It is covered by a concentric hollow conducting sphere of radius $2R$. Find the charges on inner and outer surfaces of hollow sphere if it is earthed.

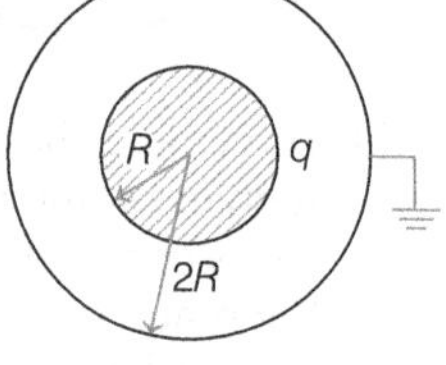

Sol. Charge q given to the solid sphere will reside on its surface. The charge on the inner surface of the hollow sphere should be $-q$, because if we draw a closed Gaussian surface through the material of the hollow sphere the total charge enclosed by this Gaussian surface should be zero. Let q' be the charge on the outer surface of the hollow sphere.

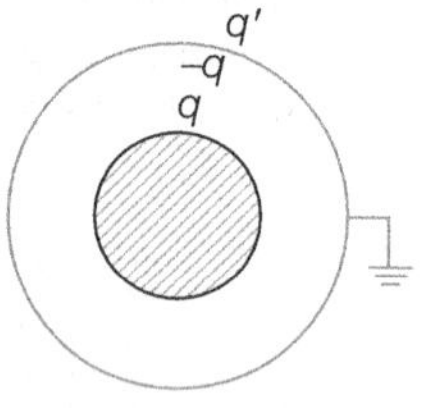

Since, the hollow sphere is earthed, its potential should be zero. The potential on it is due to the charges q, $-q$ and q', Hence,

$$V=\frac{1}{4\pi\varepsilon_0}\left[\frac{q}{2R}-\frac{q}{2R}+\frac{q'}{2R}\right]=0$$

$\therefore$ $q' = 0$ **Ans.**

Therefore, there will be no charge on the outer surface of the hollow sphere.

Ex. 5 Solve the above problem if thickness of the hollow sphere is considerable.

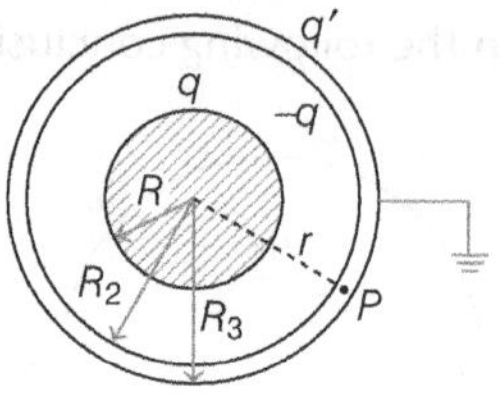

Sol. In this case, we can set $V = 0$ at any point on the hollow sphere. Let us select a point P at distance r from the centre, were $R_2 < r < R_3$. So,

$$V_P = 0$$

$$\therefore \quad \frac{1}{4\pi\varepsilon_0}\left[\frac{q}{r}-\frac{q}{r}+\frac{q'}{R_3}\right]=0$$

$\therefore$ $q' = 0$ **Ans.**

i.e. in this case also there will be no charge on the outer surface of the hollow sphere.

Ex. 6 Figure shows three concentric thin spherical shells A, B and C of radii R, $2R$ and $3R$. The shell B is earthed and A and C are given charges q and $2q$, respectively. Find the charges appearing on all the surfaces of A, B and C.

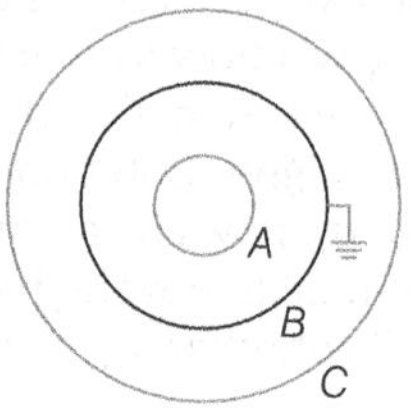

Sol. Since, there is no charge inside A. The whole charge q given to the shell A will appear on its outer surface. Charge on its inner surface will be zero. Moreover if a Gaussian surface is drawn on the material of shell B, net charge enclosed by it should be zero. Therefore, charge on its inner surface will be $-q$. Now let q' be the charge on its outer surface, then charge on the inner surface of C will be $-q'$ and on its outer surface will be, $2q-(-q') = 2q + q'$ as total charge on C is $2q$.

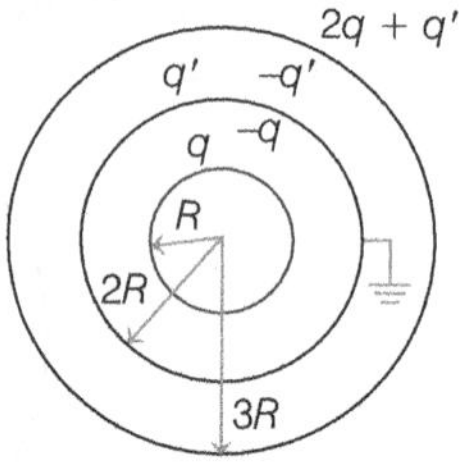

Shell B is earthed. Hence, its potential should be zero.

$$V_B = 0$$

$$\therefore \quad \frac{1}{4\pi\varepsilon_0}\left[\frac{q}{2R}-\frac{q}{2R}+\frac{q'}{2R}-\frac{q'}{3R}+\frac{2q+q'}{3R}\right]=0$$

Solving this equation, we get

$$q' = -\frac{4}{3}q$$

$$\therefore \quad 2q + q' = 2q - \frac{4}{3}q = \frac{2}{3}q$$

Therefore, charges on different surfaces in tabular form are given below :

Table 2.

	A	B	C
Inner surface	0	$-q$	$\frac{4}{3}q$
Outer surface	q	$-\frac{4}{3}q$	$\frac{2}{3}q$

Total charge on B is therefore $-q-\frac{4}{3}q=-\frac{7}{3}q$, which comes from ground.

Type 4. *E - r and V - r graphs due to charged spherical shells of negligible thickness.*

Concept

According to Gauss's theorem,

$$E=\frac{kq_{in}}{r^2} \qquad \left(k=\frac{1}{4\pi\varepsilon_0}\right)$$

So, only inside charges contribute in the electric field.

$$V=\frac{kq}{R}=\text{constant} \qquad \text{(inside the shell)}$$

$$V=\frac{kq}{r}\neq\text{constant} \qquad \text{(outside the shell)}$$

Here, q is the charge on shell.

Ex. 7 Draw E - r and V - r graphs due to two charged spherical shells as shown in figure (along the line between C and ∞).

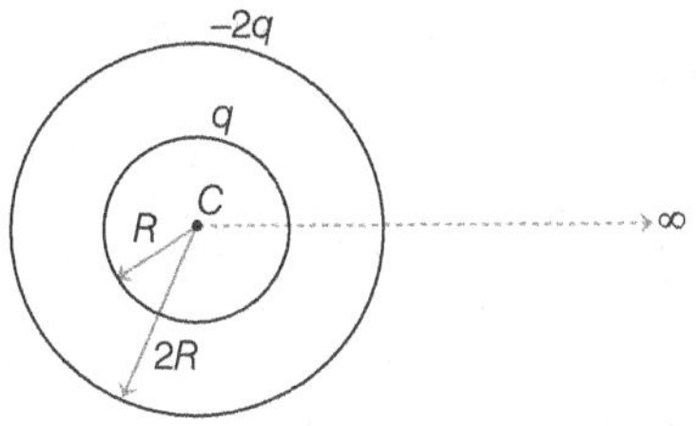

Sol. ***E - r* graph**

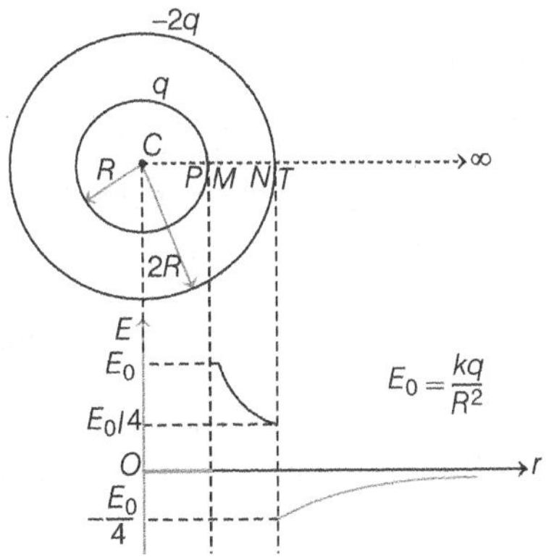

C to P $\qquad q_{in}=0$

$\Rightarrow \qquad E=0$

At M $\quad E=\frac{kq}{R^2}$ $\qquad$ (radially outwards, say positive)

$\qquad =E_0$ (say)

At N $\quad E=\frac{kq}{(2R)^2}=\frac{kq}{4R^2}=\frac{E_0}{4}$ $\qquad$ (radially outwards)

From M to N Value will decrease from E_0 to $\frac{E_0}{4}$

At T $\quad E=\frac{k(-2q+q)}{(2R)^2}$ $\qquad$ (radially inwards)

$\qquad =-\frac{E_0}{4}$

From T to ∞ Value changes from $-\frac{E_0}{4}$ to zero.

***V* - *r* graph**

From C to P Points are lying inside both the shells. Hence, potential due to both the shells is constant.

$$\therefore \qquad V=\frac{kq}{R}-\frac{k(2q)}{2R}=0$$

From M to N Potential of $-2q$ will remain constant but potential of q will decrease. So, net value comes out to be negative. At N or T

$$V=\frac{kq}{2R}-\frac{k(2q)}{2R}=-\frac{kq}{2R}=-V_0 \qquad \text{(say)}$$

From T to ∞ Value will change from $-V_0$ to zero. The correct graph is as shown below.

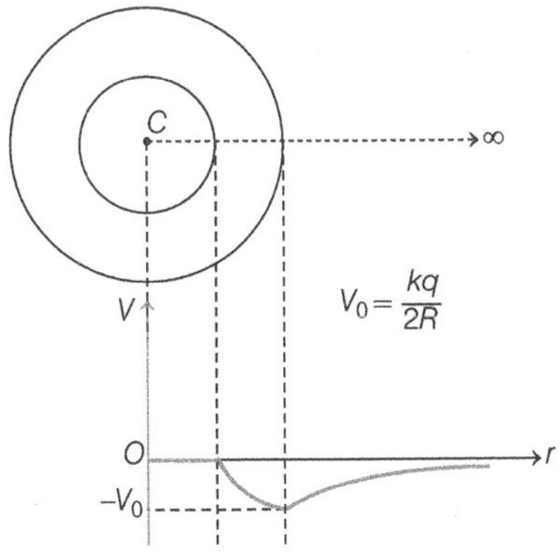

Note that on the boundary of a charged shell E - r graph is discontinuous because value suddenly changes at boundary. But V - r graph is continuous. In the above example charge on T does not contribute in E inside the outer shell but it contributes in E outside this shell. This is the reason why value of E suddenly changes at boundary of a shell.

MODULE 11 Exercise

Assertion and Reason

See directions on page 25.

1. **Assertion :** Electric potential on the surface of a charged sphere of radius R is V. Then electric field at a distance $r = \frac{R}{2}$ from centre is $\frac{V}{2R}$. Charge is distributed uniformly over the volume.

   Reason : From centre to surface, electric field varies linearly with r. Here, r is distance from centre.

2. **Assertion :** A charged sphere has a radius R. At a distance $\frac{R}{2}$ from the surface towards centre potential is V_1 and a distance $\frac{R}{2}$ from surface towards infinity potential is V_2. Then, $V_1 > V_2$.

   Reason : Inside a solid charged sphere V-r variation is parabolic. Where r is the distance from centre.

3. **Assertion :** We have a solid charged sphere. There are two points P and Q. P is a point lying between centre and surface of the sphere. Q is at finite and non-zero distance from the surface outside the sphere. There will be no situation where electric potential is equal at P and Q.

   Reason : Rate of change of electric field with distance from centre merely represents, magnitude of electric potential at that point.

4. **Assertion :** Spherical shell A shown in the figure has a charge $-q$ and shell B has charge $+q$. When these two are connected by a thin conducting wire, whole of the electrostatic potential energy stored will be lost.

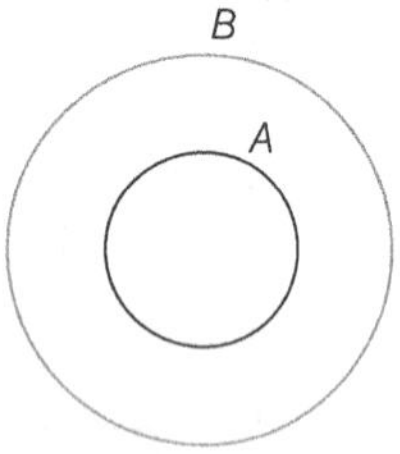

   Reason : Whole of the inner charge will transfer to outer shell B.

Single Correct Option

5. Charges Q, $2Q$ and $-Q$ are given to three concentric conducting spherical shells A, B and C respectively as shown in figure. The ratio of charges on the inner and outer surfaces of shell C will be

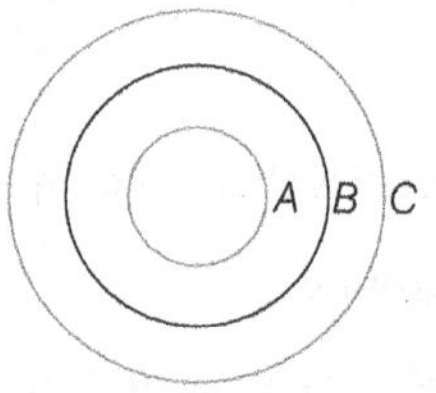

   (a) $+\frac{3}{4}$ (b) $\frac{-3}{4}$
   (c) $\frac{3}{2}$ (d) $\frac{-3}{2}$

6. A solid sphere of radius R has charge 'q' uniformly distributed over its volume. The distance from its surface outside the sphere at which the electrostatic potential is equal to half of the potential at the centre is
   (a) R (b) $2R$
   (c) $\frac{R}{3}$ (d) $\frac{R}{2}$

7. Two concentric metallic spherical shells are given equal amount of positive charges. Then,
   (a) the outer sphere is always at a higher potential
   (b) the inner sphere is always at a higher potential
   (c) both the spheres are at the same potential
   (d) no prediction can be made about their potentials unless the actual value of charges and radii are known

8. A small sphere is charged to a potential of 50 V and a big hollow sphere is charged to a potential of 100 V. Current will flow from the smaller sphere to the bigger one when
   (a) The smaller one is placed inside the bigger one and connected by a wire
   (b) Both are placed side by side and connected by a wire
   (c) Both are correct
   (d) Both are wrong

9. A solid sphere of radius R is charged uniformly. At what distance from its surface inside the sphere is the electrostatic potential half of potential at centre?
   (a) R (b) $\frac{R}{2}$
   (c) $\frac{R}{3}$ (d) None of these

10. If the electric potential of the inner shell is 10 V and that of the outer shell is 5 V, then the potential at the centre will be

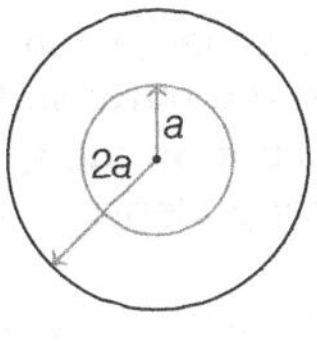

(a) 10 V (b) 5 V (c) 15 V (d) zero

11. Two concentric conducting thin spherical shells A and B having radii r_A and $r_B (r_B > r_A)$ are charged to Q_A and $-Q_B (|Q_B| > |Q_A|)$. The electrical field along a line passing through the centre is

(a)
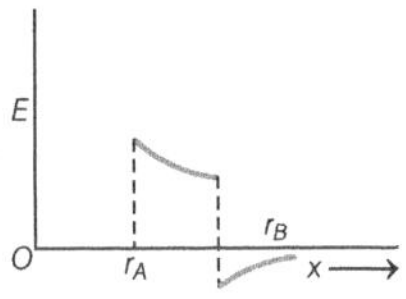

(b)
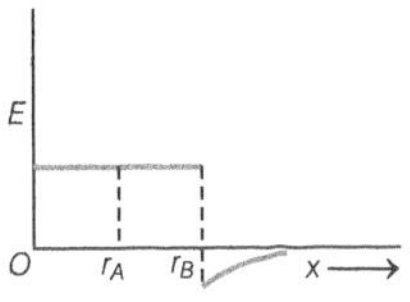

(c)
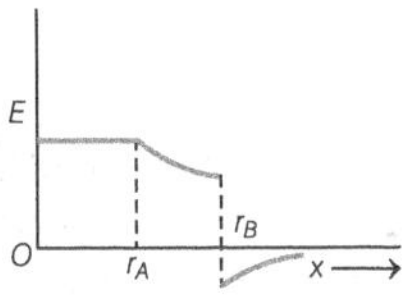

(d) None of these

12. A charge Q is distributed over two concentric hollow spheres of radii r and $R (R > r)$ such that the surface densities are equal. The potential at the common centre is $1/4\pi\varepsilon_0$ times

(a) $Q\left(\frac{r+R}{r^2+R^2}\right)$ (b) $\frac{Q}{2}\left(\frac{r+R}{r^2+R^2}\right)$

(c) $2Q\left(\frac{r+R}{r^2+R^2}\right)$ (d) zero

13. Electric potential at a point P, r distance away due to a point charge q kept at point A is V. If q is removed and twice of this charge is distributed uniformly on the surface of a hollow sphere of radius $4r$ with centre at point A, the potential at P now is

(a) V (b) $V/2$

(c) $V/4$ (d) $V/8$

14. Two concentric spheres of radii R and $2R$ are charged. The inner sphere has a charge of 1 μC and the outer sphere has a charge of 2 μC of the same sign. The potential is 9000 V at a distance $3R$ from the common centre. The value of R is

(a) 1 m (b) 2 m

(c) 3 m (d) 4 m

15. A and B are two concentric spherical shells. If A is given a charge $+q$ while B is earthed as shown in figure, then

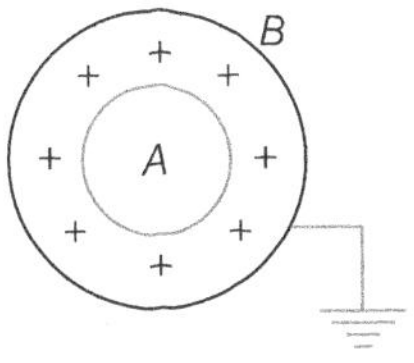

(a) charge on the outer surface of shell B is zero

(b) the charge on B is equal and opposite to that of A

(c) the field inside A and outside B is zero

(d) All of the above

16. Two isolated charged conducting spheres of radii a and b produce the same electric field near their surfaces. The ratio of electric potentials on their surfaces is

(a) $\frac{a}{b}$ (b) $\frac{b}{a}$ (c) $\frac{a^2}{b^2}$ (d) $\frac{b^2}{a^2}$

17. A conducting shell S_1 having a charge Q is surrounded by an uncharged concentric conducting spherical shell S_2. Let the potential difference between S_1 and that S_2 be V. If the shell S_2 is now given a charge $-3Q$, the new potential difference between the same two shells is

(a) V (b) $2V$ (c) $4V$ (d) $-2V$

18. If the potential at the centre of a uniformly charged hollow sphere of radius R is V, then electric field at a distance r from the centre of sphere will be $(r > R)$

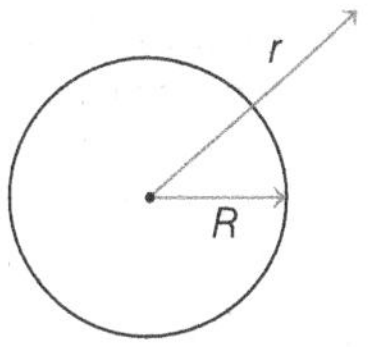

(a) $\frac{VR}{r^2}$ (b) $\frac{Vr}{R^2}$ (c) $\frac{VR}{r}$ (d) $\frac{VR}{R^2+r^2}$

19. Eight small drops, each of radius r and having same charge q are combined to form a big drop. The ratio between the potentials of the bigger drop and the smaller drop is

(a) 8 : 1 (b) 4 : 1 (c) 2 : 1 (d) 1 : 8

20. If solid sphere of charge of radius R has charge density ρ. The electric field at a distance r from the centre of sphere $(r < R)$ will be

(a) $\frac{\rho R}{3\varepsilon_0}$ (b) $\frac{\rho r}{\varepsilon_0}$ (c) $\frac{\rho r}{3\varepsilon_0}$ (d) $\frac{3\rho r}{\varepsilon_0}$

Match the Columns

21. Electric potential on the surface of a solid sphere of charge is V. Radius of the sphere is 1m. Match the following two columns.

	Column I		Column II
(a)	Electric potential at $r = \frac{R}{2}$	(p)	$\frac{V}{4}$
(b)	Electric potential at $r = 2R$	(q)	$\frac{V}{2}$
(c)	Electric field at $r = \frac{R}{2}$	(r)	$\frac{3V}{4}$
(d)	Electric field at $r = 2R$	(s)	None of these

22. Match the following two columns.

	Column I		Column II
(a)	Electric field due to charged spherical shell	(p)	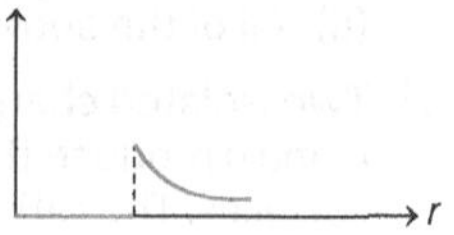
(b)	Electric potential due to charged spherical shell	(q)	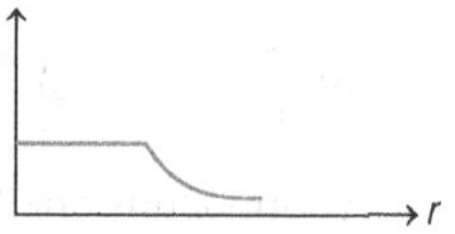
(c)	Electric field due to charged solid sphere	(r)	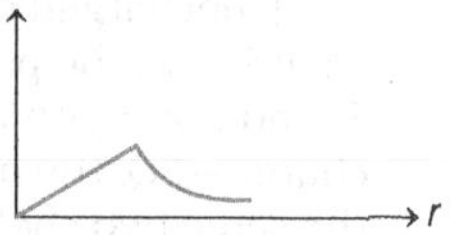
(d)	Electric potential due to charged solid sphere	(s)	None of these

Subjective Questions

23. Two concentric conducting shells A and B are of radii R and $2R$. A charge $+q$ is placed at the centre of the shells. Shell B is earthed and a charge q is given to shell A. Find the charge on outer surface of A and B.

24. Three concentric metallic shells A, B and C of radii a, b and c $(a < b < c)$ have surface charge densities, $\sigma, -\sigma$ and σ respectively.

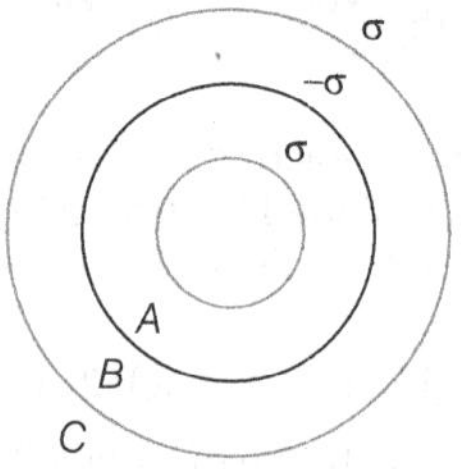

(a) Find the potentials of three shells A, B and C.
(b) It is found that no work is required to bring a charge q from shell A to shell C, then obtain the relation between the radii a, b and c.

25. A charge Q is placed at the centre of an uncharged, hollow metallic sphere of radius a,

(a) Find the surface charge density on the inner surface and on the outer surface.
(b) If a charge q is put on the sphere, what would be the surface charge densities on the inner and the outer surfaces?
(c) Find the electric field inside the sphere at a distance x from the centre in the situations (a) and (b).

26. Three spherical shells have radii R, $2R$ and $3R$ respectively. Total charge on A and C is $3q$. Find the charges on different surfaces of A, B and C. The connecting wire does not touch the shell B.

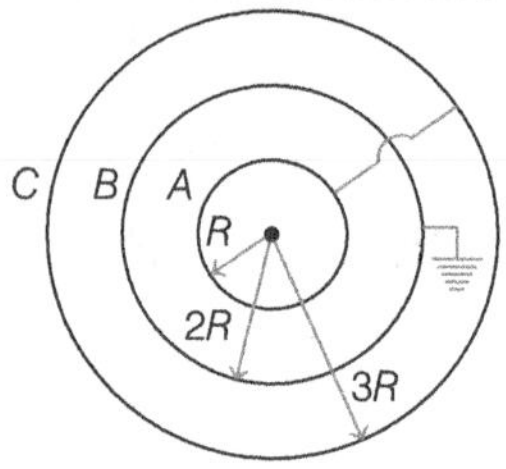

27. In the above problem, find the charges on different surfaces if a charge q is placed at the centre of the shell with all other conditions remaining the same.

28. A solid sphere of radius R has a charge $+2Q$. A hollow spherical shell of radius $3R$ placed concentric with the first sphere that has net charge $-Q$.

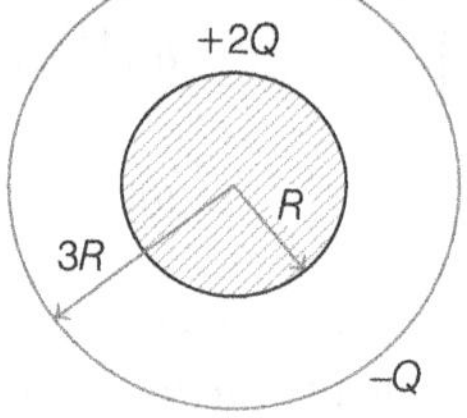

(a) Find the electric field between the spheres at a distance r from the centre of the inner sphere. $[R < r < 3R]$
(b) Calculate the potential difference between the spheres.
(c) What would be the final distribution of charges, if a conducting wire joins the spheres?
(d) Instead of (c), if the inner sphere is earthed, what is the charge on it?

Answers

1. (b) **2.** (d) **3.** (c) **4.** (a) **5.** (d) **6.** (c) **7.** (b) **8.** (a) **9.** (d) **10.** (a)
11. (a) **12.** (a) **13.** (b) **14.** (a) **15.** (d) **16.** (a) **17.** (a) **18.** (a) **19.** (b) **20.** (c)
21. (a $\rightarrow$ s; b $\rightarrow$ q; c $\rightarrow$ q; d $\rightarrow$ p) **22.** (a $\rightarrow$ p; b $\rightarrow$ q; c $\rightarrow$ r; d $\rightarrow$ s) **23.** $2q$, zero

24. (a) $V_A = \frac{\sigma}{\varepsilon_0}(a - b + c), V_B = \frac{\sigma}{\varepsilon_0}\left(\frac{a^2}{b} - b + c\right), V_C = \frac{\sigma}{\varepsilon_0}\left(\frac{a^2}{c} - \frac{b^2}{c} + c\right)$ (b) $c = a + b$

25. (a) $\frac{-Q}{4\pi a^2}, \frac{Q}{4\pi a^2}$ (b) $\frac{-Q}{4\pi a^2}, \frac{Q+q}{4\pi a^2}$ (c) $\frac{Q}{4\pi\varepsilon_0 x^2}$ in both situations

26.

	A	B	C
Inner Surface	0	$-\frac{6}{11}q$	$\frac{18}{11}q$
Outer Surface	$\frac{6}{11}q$	$-\frac{18}{11}q$	$\frac{9}{11}q$

27.

	A	B	C
Inner Surface	$-q$	$-\frac{8}{11}q$	$\frac{24}{11}q$
Outer Surface	$\frac{8}{11}q$	$-\frac{24}{11}q$	$\frac{12}{11}q$

28. (a) $\frac{Q}{2\pi\varepsilon_0 r^2}$ (b) $\frac{Q}{3\pi\varepsilon_0 R}$ (c) Zero on inner and Q on outer (d) $\frac{Q}{3}$

See hints on pages 210-212.

Unit Exercise (Cumulative)

Level 1

Fill in the Blanks

1. 512 identical drops of mercury are charged to a potential of 2 V each. The drops are joined to form a single drop. The potential of this drop is .... V. **(2021 Main)**

2. The electric potential V at any point x, y, z (all in metres) in space is given by $V = 4x^2$ volts. The electric field at the point (1m, 0,2 m) is .... V/m. **(1992 JEE)**

3. Figure shows line of constant potential in a region in which an electric field is present. The values of the potential are written in brackets. Of the points A, B and C, the magnitude of the electric field is greatest at the point .... **(1984 JEE)**

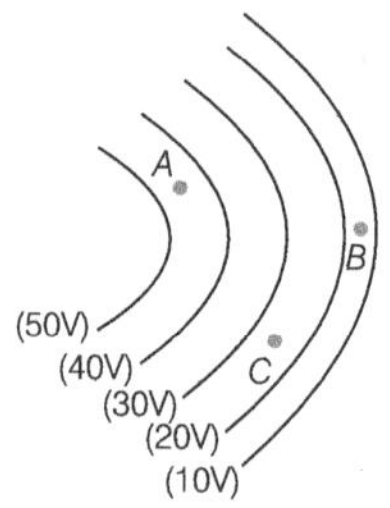

True/False

4. An electric line of forces in the x-y plane is given by the equation $x^2 + y^2 = 1$. A particle with unit positive charge, initially at rest at the point $x = 1$, $y = 0$ in the xy-plane, will move along circular line of force. **(1988 JEE)**

5. A small metal ball is suspended in a uniform electric field with the help of an insulated thread. If high energy X-ray beam falls on the ball, the ball will be deflected in the direction of the field. **(1983 JEE)**

Statements Type Question

6. **Statement I :** For practical purposes, the earth is used as a reference at zero potential in electrical circuits.

   and

   Statement II : The electrical potential of a sphere of radius R with charge Q uniformly distributed on the surface is given by $\frac{Q}{4\pi\varepsilon_0 R}$. **(2008 JEE)**

   (a) Statement I is True, Statement II is True; Statement II is a correct explanation for Statement I.
   (b) Statement I is True, Statement II is True; Statement II is not a correct explanation for Statement I
   (c) Statement I is True, Statement II is False
   (d) Statement I is False, Statement II is True

Single Correct Option

7. Two point charges a and b whose magnitudes are same, positioned at a certain distance along the positive x-axis from each other. a is at origin. Graph is drawn between electrostatic field strength and distance x from a. E is taken positive if it is along the line joining from a to b. From the graph it can be decided that

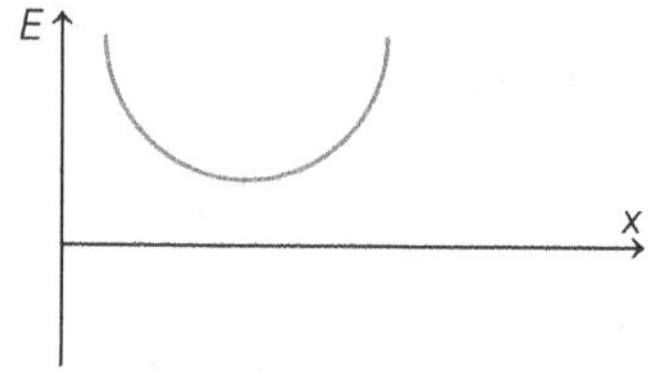

   (a) a is positive, b is negative
   (b) a and b both are positive
   (c) a and b both are negative
   (d) a is negative, b is positive

Note *Graph is drawn only between a and b.*

8. A positively charged disc is placed on a horizontal plane. A charged particle is released from a certain height on its axis. The particle just reaches the centre of the disc. Select the correct alternative.
   (a) Particle has negative charge on it
   (b) Total potential energy (gravitational + electrostatic) of the particle first increases, then decreases
   (c) Total potential energy of the particle first decreases, then increases
   (d) Total potential energy of the particle continuously decreases

9. A charge q is placed at O in the cavity in a spherical uncharged conductor. Point S is outside the conductor. If q is displaced from O towards S (still remaining within the cavity)

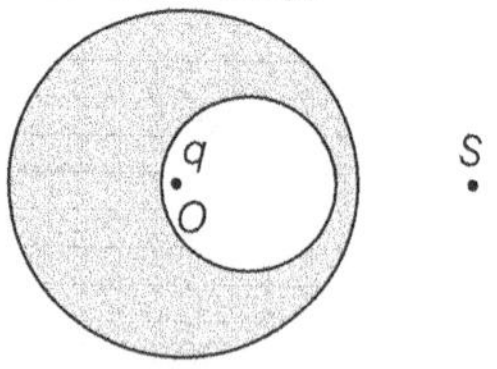

   (a) electric field at S will increase
   (b) electric field at S will decrease

(c) electric field at S will first increase and then decrease
(d) electric field at S will not change

10. A thin conducting ring of radius R is given a charge $+Q$. The electric field at the centre O of the ring due to the charge on the part AKB of the ring is E. The electric field at the centre due to the charge on the part $ACDB$ of the ring is (K is mid-point of AB)

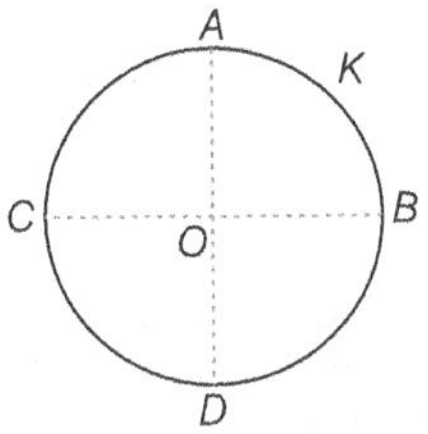

(a) E along KO (b) $3E$ along OK
(c) $3E$ along KO (d) E along OK

11. Charges $2q$ and $8q$ are placed at the end points A and B respectively of a 9 cm long straight line. A third charge q is placed at point C of AB such that the potential energy of the system is minimum. The distance of C from A is
(a) 2 cm (b) 3 cm (c) 4 cm (d) 5 cm

12. A fired ring of radius R carries a charge $+q$. A test charge $-q_0$ is released on its axis at a distance $\sqrt{3}R$ from its centre. How much kinetic energy will be acquired by the test charge when it reaches at the centre of the ring? Neglect gravity.
(a) $\frac{1}{4\pi\varepsilon_0}\left(\frac{qq_0}{R}\right)$ (b) $\frac{1}{4\pi\varepsilon_0}\left(\frac{qq_0}{2R}\right)$
(c) $\frac{1}{4\pi\varepsilon_0}\left(\frac{qq_0}{\sqrt{3}R}\right)$ (d) $\frac{1}{4\pi\varepsilon_0}\left(\frac{qq_0}{3R}\right)$

13. An oil drop having a mass 4.8×10^{-10} g and charge 24×10^{-18} C stands still between two charged horizontal plates separated by a distance of 1 cm. If now the polarity of the plates is changed, instantaneous acceleration of the drop is ($g = 10\ \text{ms}^{-2}$)
(a) $5\ \text{ms}^{-2}$ (b) $10\ \text{ms}^{-2}$
(c) $15\ \text{ms}^{-2}$ (d) $20\ \text{ms}^{-2}$

14. A proton, a deuteron and an alpha particle are accelerated through potentials of 1V, 2 V and 4 V respectively. Their velocity will bear a ratio
(a) $1:1:1$ (b) $1:\sqrt{2}:1$
(c) $\sqrt{2}:1:1$ (d) $1:1:\sqrt{2}$

15. Two identical small conducting spheres having unequal positive charges q_1 and q_2 are separated by a distance r. If they are now made to touch each other and then separated again to the same distance, the electrostatic force between them in this case will be
(a) less than before
(b) same as before
(c) more than before
(d) zero

16. There is an electric field E in x-direction. If the work done on moving a charge of 0.2 C through a distance of 2 m along a line making an angle 60° with x-axis is 4 J, then what is the value of E?
(a) $\sqrt{3}$ N/C (b) 4 N/C
(c) 5 N/C (d) 20 N/C

17. A charged ball B hangs from a silk thread S, which makes an angle θ with a large charged conducting sheet P, as shown in the figure. The surface charge density σ of the sheet is proportional to

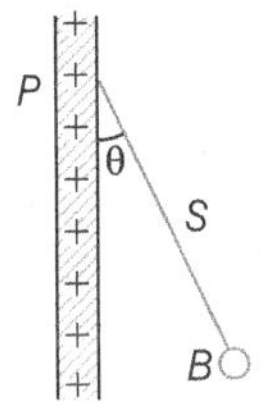

(a) $\sin\theta$ (b) $\tan\theta$
(c) $\cos\theta$ (d) $\cot\theta$

18. Figure shows a closed dotted surface which intersects a conducting uncharged sphere. If a positive charge is placed at the point P, the flux of the electric field through the closed dotted surface

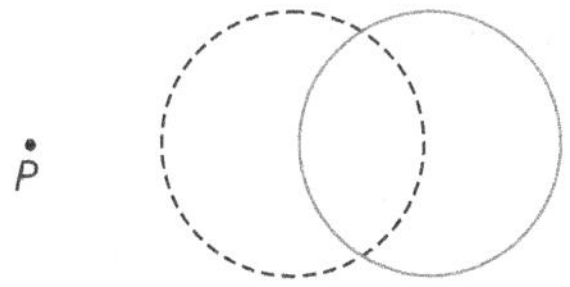

(a) will remain zero (b) will become positive
(c) will become negative (d) data insufficient

19. An infinite line of charge λ per unit length is placed along the y-axis. The work done in moving a charge q from $A(a, 0)$ to $B(2a, 0)$ is
(a) $\frac{q\lambda}{2\pi\varepsilon_0}\ln 2$ (b) $\frac{q\lambda}{2\pi\varepsilon_0}\ln\left(\frac{1}{2}\right)$
(c) $\frac{q\lambda}{4\pi\varepsilon_0}\ln\sqrt{2}$ (d) $\frac{q\lambda}{4\pi\varepsilon_0}\ln 2$

20. In Millikan's oil drop experiment, an oil drop carrying a charge Q is held stationary by a potential difference 2400 V between the plates. To keep a drop of half the radius stationary the potential difference had to be made 600 V. What is the charge on the second drop?
(a) $\frac{Q}{4}$ (b) $\frac{Q}{2}$
(c) Q (d) $\frac{3Q}{2}$

21. An infinitely long positively charged straight thread has a linear charge density λCm^{-1}. An electron revolves along a circular path having axis along the length of the wire. The graph that correctly represents the variation of the kinetic energy of electron as a function of radius of circular path from the wire is

(2024 Main)

(a)
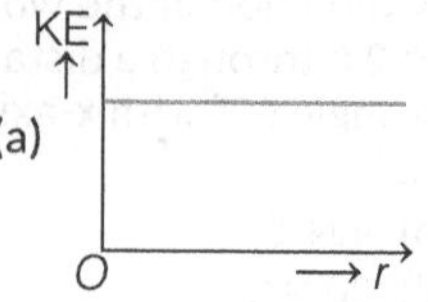

(b)
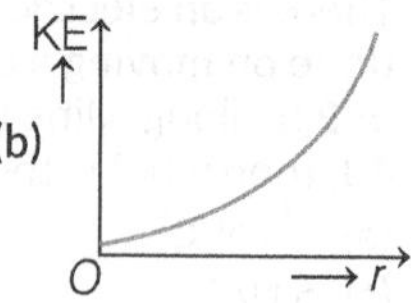

(c)
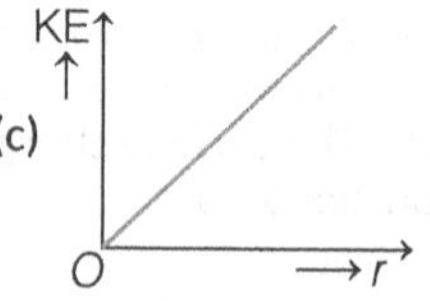

(d)
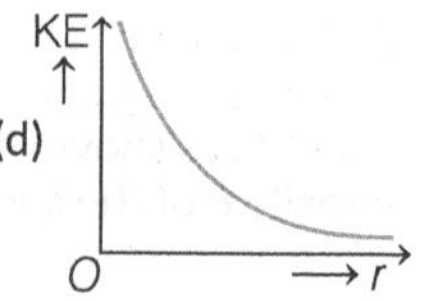

22. A point charge of 10 μC is placed at the origin. At what location on the x-axis should a point charge of 40 μC be placed so that the net electric field is zero at $x = 2$ cm on the x-axis ? **(2023 Main)**

(a) $x = 8$ cm
(b) $x = -4$ cm
(c) $x = 4$ cm
(d) $x = 6$ cm

23. As shown in the figure, a point charge Q is placed at the centre of conducting spherical shell of inner radius a and outer radius b. The electric field due to charge Q in three different regions I, II and III is given by ($\text{I}: r < a$, $\text{II}: a < r < b$, $\text{III}: r > b$) **(2023 Main)**

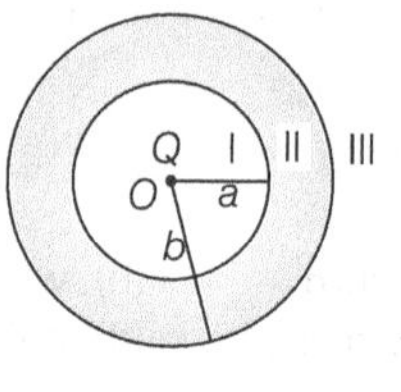

(a) $E_{\text{I}} \neq 0, E_{\text{II}} = 0, E_{\text{III}} = 0$ (b) $E_{\text{I}} = 0, E_{\text{II}} = 0, E_{\text{III}} \neq 0$
(c) $E_{\text{I}} \neq 0, E_{\text{II}} = 0, E_{\text{III}} \neq 0$ (d) $E_{\text{I}} = 0, E_{\text{II}} = 0, E_{\text{III}} = 0$

24. A cube of side a has point charges $+Q$ located at each of its vertices except at the origin, where the charge is $-Q$. The electric field at the centre of cube is **(2021 Main)**

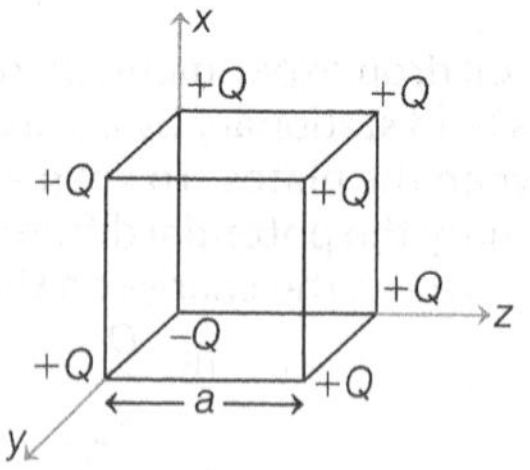

(a) $\frac{-Q}{3\sqrt{3}\pi\varepsilon_0 a^2}(\hat{x}+\hat{y}+\hat{z})$ (b) $\frac{Q}{3\sqrt{3}\pi\varepsilon_0 a^2}(\hat{x}+\hat{y}+\hat{z})$
(c) $\frac{-2Q}{3\sqrt{3}\pi\varepsilon_0 a^2}(\hat{x}+\hat{y}+\hat{z})$ (d) $\frac{2Q}{3\sqrt{3}\,\pi\varepsilon_0 a^2}(\hat{x}+\hat{y}+\hat{z})$

25. The electric potential at the centre of two concentric half-rings of radii R_1 and R_2, having same linear charge density λ is **(2023 Main)**

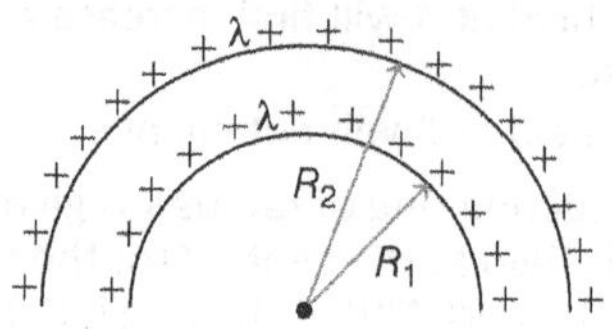

(a) $\frac{2\lambda}{\varepsilon_0}$ (b) $\frac{\lambda}{4\varepsilon_0}$
(c) $\frac{\lambda}{\varepsilon_0}$ (d) $\frac{\lambda}{2\varepsilon_0}$

26. A charge $+q$ is fixed at each of the points $x = x_0$, $x = 3x_0$, $x = 5x_0$, ... $x = \infty$ on the x-axis, and a charge $-q$ is fixed at each of the points $x = 2x_0$, $x = 4x_0$, $x = 6x_0$, ...$x = \infty$. Here x_0 is a positive constant. Take the electric potential at a point due to a charge Q at a distance r from it to be $Q/(4\pi\varepsilon_0 r)$. Then, the potential at the origin due to the above system of charges is **(1998 JEE)**

(a) 0 (b) $\frac{1}{8\pi\varepsilon_0 x_0 \ln 2}$
(c) ∞ (d) $\frac{q\ln 2}{4\pi\varepsilon_0 x_0}$

27. A hollow metal sphere of radius 5 cm is charged such that the potential on its surface is 10 volts. The potential at the centre of the sphere is **(1983 JEE)**

(a) zero
(b) 10 volts
(c) same as at a point 5 cm away from the surface
(d) same as at a point 25 cm away from the surface

Multiple Correct Options

28. Charges Q_1 and Q_2 lie inside and outside respectively of a closed surface S. Let E be the field at any point on S and ϕ be the flux of E over S.

(a) If Q_1 changes, both E and ϕ will change
(b) If Q_2 changes, E will change but ϕ will not change
(c) If $Q_1 = 0$ and $Q_2 \neq 0$, then $E \neq 0$ but $\phi = 0$
(d) If $Q_1 \neq 0$ and $Q_2 = 0$, then $E = 0$ but $\phi \neq 0$

29. Two concentric spherical shells have charges $+q$ and $-q$ as shown in figure. Choose the correct options.

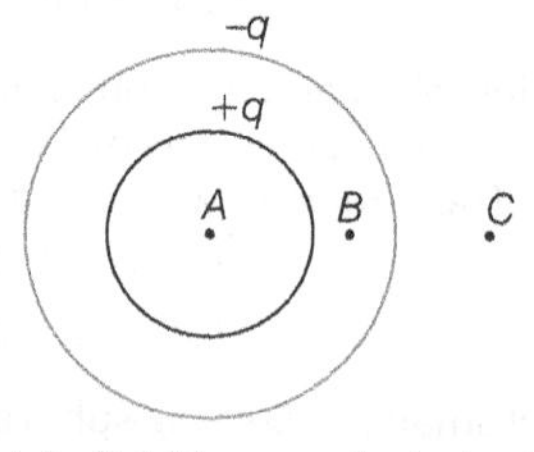

(a) At A electric field is zero, but electric potential is non-zero
(b) At B electric field and electric potential both are non-zero
(c) At C electric field is zero but electric potential is non-zero
(d) At C electric field and electric potential both are zero

30. A rod is hinged (free to rotate) at its centre O as shown in figure. Two point charges $+q$ and $+q$ are kept at its two ends. Rod is placed in uniform electric field E as shown. Space is gravity free.

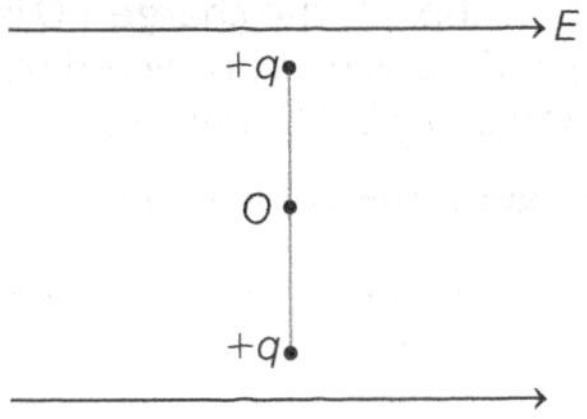

Choose the correct options.
(a) Net force from the hinge on the rod is zero
(b) Net force from the hinge on the rod is leftwards
(c) Equilibrium of rod is neutral
(d) Equilibrium of rod is stable

31. A non-conducting solid sphere of radius R is uniformly charged. The magnitude of the electric field due to the sphere at a distance r from its centre **(1998 JEE)**
(a) increases as r increases, for $r < R$
(b) decreases as r increases, for $0 < r < \infty$
(c) decreases as r increases, for $R < r < \infty$
(d) is discontinuous at $r = R$

32. A positively charged thin metal ring of radius R is fixed in the x-y plane with its centre at the origin O. A negatively charged particle P is released from rest at the point $(0, 0, z_0)$ where $z_0 > 0$. Then the motion of P is **(1998 JEE)**
(a) periodic for all values of z_0 satisfying $0 < z_0 < \infty$
(b) simple harmonic for all values of z_0 satisfying $0 < z_0 \le R$
(c) approximately simple harmonic provided $z_0 << R$
(d) such that P crosses O and continues to move along the negative z-axis towards $z = -\infty$

Passage Based Questions

Passage 1 (Q.33 to Q.35)

Two equal point charges of 8 μC are fixed at points (2 m, 0), (−2 m, 0). Another charge q with a mass of 91 mg is released at point (0, 0.1 m) at t = 0. It is observed that q oscillates simple harmonically about the origin.

At $t = 0$, force experienced by q is 9×10^{-3} N.

33. Charge q is
(a) $-8\,\mu C$ (b) $-6.5\,\mu C$
(c) $5\,\mu C$ (d) $+6.5\,\mu C$

34. Amplitude of simple harmonic motion is
(a) 10 cm (b) 20 cm
(c) 30 cm (d) 40 cm

35. Frequency of oscillation is
(a) 8 (b) 10
(c) 5 (d) 2

Passage 2 (Q.36 to Q.38)

An empty thick conducting shell of inner radius a and outer radius b is shown.

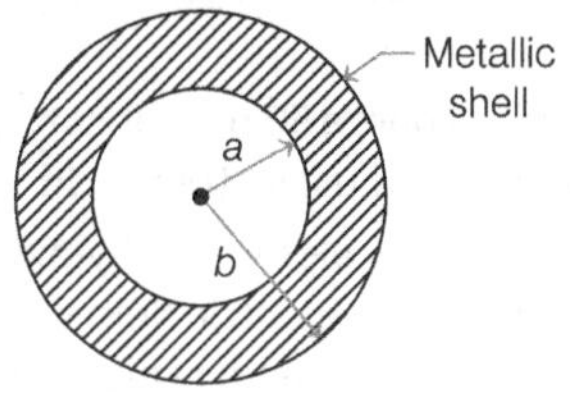

A charge q_A is placed at centre of the shell. It is observed that inner face of shell has a uniform charge density of $-\sigma$ and the outer surface has a charge density of $+\sigma$.

36. Charge q_A must be
(a) zero
(b) negative
(c) having a magnitude $4\pi\sigma(b^2 - a^2)$
(d) having a magnitude $4\pi\sigma a^2$

37. If a charge q_B is placed at a distance c, $(c > b)$, then
(a) force on charge A is $\dfrac{\sigma q_A b^2}{\varepsilon_0 c^2}$
(b) force on charge A is zero
(c) force on charge B is $\dfrac{\sigma q_A b^2}{\varepsilon_0 c^2}$
(d) force on charge B is $\dfrac{k q_A q_B}{c^2}$

38. If charge q_A is moved slowly inside the shell, then
(a) charge distribution on inner and outer surfaces of the shell changes
(b) force on charge B changes
(c) charge B also starts moving
(d) None of the above

Match the Columns

39. Match the following two columns.

	Column I		Column II
(a)	Electric potential	(p)	$[MLT^{-3}A^{-1}]$
(b)	Electric field	(q)	$[ML^3T^{-3}A^{-1}]$
(c)	Electric flux	(r)	$[ML^2T^{-3}A^{-1}]$
(d)	Permittivity of free space	(s)	None of these

Subjective Questions

40. Calculate the ratio of the electrostatic to gravitational interaction forces between two electrons, between two protons. At what value of the specific charge q/m of a particle would these forces become equal (in their absolute values) in the case of interaction of identical particles? **(Irodov 3.1)**

41. What would be the interaction force between two copper spheres, each of mass 1 g, separated by the distance 1 m, if the total electronic charge in them differed from the total charge of the nuclei by one percent? **(Irodov 3.2)**

42. A positive point charge 50 μC is located in the xy-plane at the point with radius vector $\mathbf{r}_0 = 2\hat{\mathbf{i}} + 3\hat{\mathbf{j}}$, where $\hat{\mathbf{i}}$ and $\hat{\mathbf{j}}$ are the unit vectors along x and y-axis respectively. Find the vector of the electric field strength $\mathbf{E}$ and its magnitude at the point with radius vector $\mathbf{r} = 8\hat{\mathbf{i}} - 5\hat{\mathbf{j}}$. Here $\mathbf{r}_0$ and $\mathbf{r}$ are expressed in metres. **(Irodov 3.6)**

43. Point charges q and $-q$ are located at the vertices of a square with diagonals $2l$ as shown in figure. Find the magnitude of the electric field strength at a point located symmetrically with respect to the vertices of the square at a distance x from its centre. **(Irodov 3.7)**

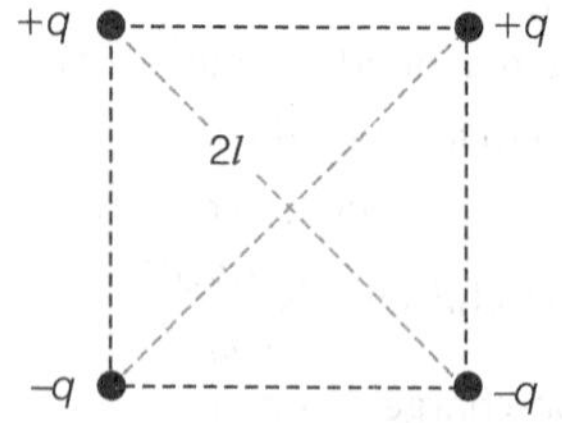

44. A system consists of a thin charged wire ring of radius R and a very long uniformly charged thread oriented along the axis of the ring, with one of its ends coinciding with the centre of the ring. The total charge of the ring is equal to q. The charge of the thread (per unit length) is equal to λ.
Find the interaction force between the ring and the thread. **(Irodov 3.11)**

45. Three concentric, thin, spherical, metallic shells have radii 1, 2, and 4 cm and they are held at potentials 10, 0 and 40 V respectively. Taking the origin at the common centre, calculate the following:
(a) Potential at $r = 1.25$ cm
(b) Potential at $r = 2.5$ cm
(c) Electric field at $r = 1.25$ cm

Integer Type Questions

46. A point charge $q_1 = 4q_0$ is placed at origin. Another point charge $q_2 = -q_0$ is placed at $x = 12$ cm. Charge of proton is q_0. The proton is placed on x-axis so that the electrostatic force on the proton is zero. In this situation, the position of the proton from the origin is .... cm. **(2023 Main)**

47. An infinite number of point charges, each carrying 1 μC charge, are placed along the y-axis at $y = 1$ m, 2 m, 4 m, 8 m ....
The total force on a 1 C point charge, placed at the origin, is $x \times 10^3$ N. **(2021 Main)**
The value of x, to the nearest integer, is ....
$\left[\text{Take, } \frac{1}{4\pi\varepsilon_0} = 9 \times 10^9 \text{ Nm}^2/\text{C}^2\right]$

Level 2

Single Correct Option

48. In the diagram shown, the charge $+Q$ is fixed. Another charge $+2q$ and mass m is projected from a distance R from the fixed charge. Minimum separation between the two charges if the velocity becomes $\frac{1}{\sqrt{3}}$ times of the projected velocity, at this moment is (Assume gravity to be absent)

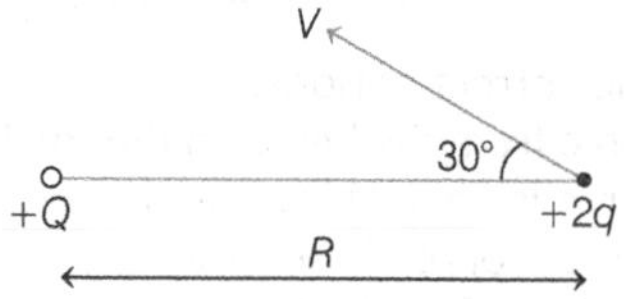

(a) $\frac{\sqrt{3}}{2}R$ (b) $\frac{1}{\sqrt{3}}R$
(c) $\frac{1}{2}R$ (d) None of these

49. A uniform electric field of strength $\mathbf{E}$ exists in a region. An electron enters a point A with velocity v as shown. It moves through the electric field and reaches at point B. Velocity of particle at B is $2v$ at 30° with x-axis. Then,

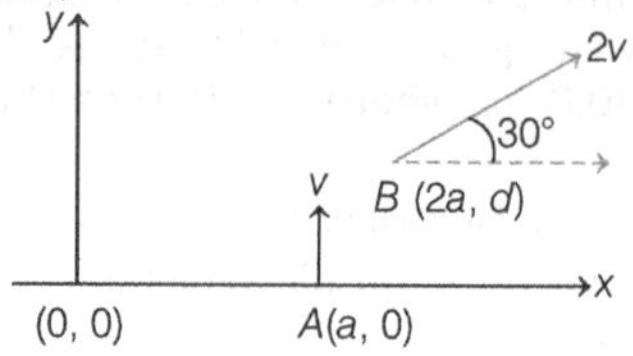

(a) electric field $\mathbf{E} = -\frac{3mv^2}{2ea}\hat{\mathbf{i}}$
(b) rate of doing work done by electric field at B is $\frac{3mv^3}{2ea}$
(c) Both (a) and (b) are correct
(d) Both (a) and (b) are wrong

50. Six charges are placed at the vertices of a regular hexagon as shown in the figure. The electric field on the line passing through point O and perpendicular to the plane of the figure as a function of distance x from point O is ($x >> a$)

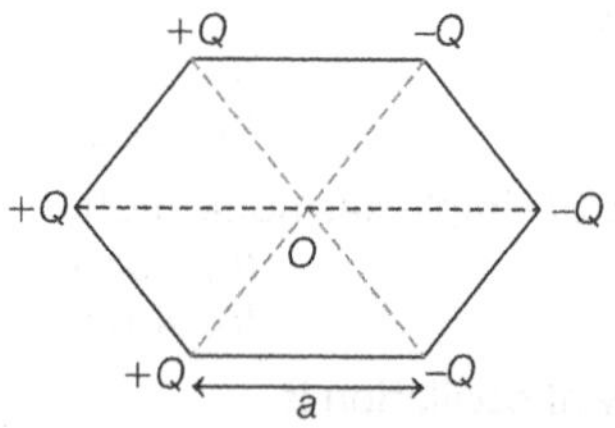

(a) 0 (b) $\frac{Qa}{\pi\varepsilon_0 x^3}$
(c) $\frac{2Qa}{\pi\varepsilon_0 x^3}$ (d) $\frac{\sqrt{3}Qa}{\pi\varepsilon_0 x^3}$

51. There are four concentric shells A, B, C and D of radii $a, 2a, 3a$ and $4a$ respectively. Shells B and D are given charges $+q$ and $-q$ respectively. Shell C is now earthed. The potential difference $V_A - V_C$ is $\left(k = \frac{1}{4\pi\varepsilon_0}\right)$

(a) $\frac{kq}{2a}$ (b) $\frac{kq}{3a}$ (c) $\frac{kq}{4a}$ (d) $\frac{kq}{6a}$

52. Potential difference between centre and surface of the sphere of radius R and uniform volume charge density ρ within it will be

(a) $\frac{\rho R^2}{6\varepsilon_0}$ (b) $\frac{\rho R^2}{4\varepsilon_0}$ (c) $\frac{\rho R^2}{3\varepsilon_0}$ (d) $\frac{\rho R^2}{2\varepsilon_0}$

53. A particle of mass m and charge q is fastened to one end of a string of length l. The other end of the string is fixed to the point O. The whole system lies on a frictionless horizontal plane. Initially, the mass is at rest at A. A uniform electric field in the direction shown is then switched on. Then,

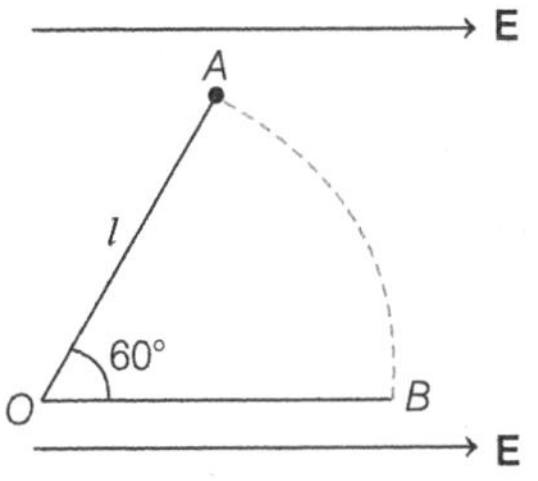

(a) the speed of the particle when it reaches B is $\sqrt{\frac{2qEl}{m}}$

(b) the speed of the particle when it reaches B is $\sqrt{\frac{qEl}{m}}$

(c) the tension in the string when the particle reaches at B is qE

(d) the tension in the string when the particle reaches at B is zero

54. The electrostatic potential inside a charged spherical ball is given by $\phi = ar^2 + b$, where r is the distance from the centre and a, b are constants. Then, the charge density inside the ball is

(a) $-6a\,\varepsilon_0 r$ (b) $-24\,\pi a\,\varepsilon_0$
(c) $-6a\,\varepsilon_0$ (d) $-24\pi a\,\varepsilon_0 r$

55. A particle of mass m and charge $-q$ is projected from the origin with a horizontal speed v into an electric field of intensity E directed downward. Choose the **wrong** statement. Neglect gravity

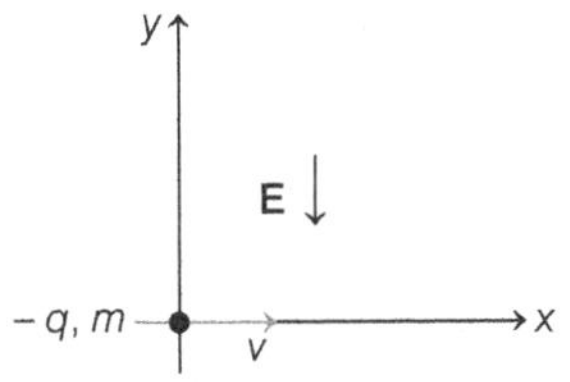

(a) The kinetic energy after a displacement y is qEy

(b) The horizontal and vertical components of acceleration are $a_x = 0, a_y = \frac{qE}{m}$

(c) The equation of trajectory is $y = \frac{1}{2}\left(\frac{qEx^2}{mv^2}\right)$

(d) The horizontal and vertical displacements x and y after a time t are $x = vt$ and $y = \frac{1}{2}a_y t^2$

56. Two conducting concentric, hollow spheres A and B have radii a and b respectively, with A inside B. Their common potential is V. A is now given some charge such that its potential becomes zero. The potential of B will now be

(a) 0 (b) $V(1-a/b)$
(c) Va/b (d) Vb/a

57. There are two uncharged identical metallic spheres 1 and 2 of radius r separated by a distance d ($d >> r$). A charged metallic sphere of same radius having charge q is touched with one of the sphere. After some time it is moved away from the system. Now, the uncharged sphere is earthed. Charge on earthed sphere is

(a) $+\frac{q}{2}$ (b) $-\frac{q}{2}$ (c) $-\frac{qr}{2d}$ (d) $-\frac{qd}{2r}$

58. A hollow charged metal sphere has radius r. If the potential difference between its surface and a point at a distance $3r$ from the centre is V, then electric field intensity at a distance $3r$ is

(a) $\frac{V}{2r}$ (b) $\frac{V}{3r}$
(c) $\frac{V}{6r}$ (d) $\frac{V}{4r}$

59. A and B are two thin concentric hollow conductors having radii a and $2a$ and charges $2Q$ and Q, respectively. If potential of outer sphere is 5 V, then potential of inner sphere is

(a) 20 V (b) 10 V
(c) $\frac{25}{3}$ V (d) $\frac{50}{3}$ V

60. A spherical conductor of radius 2 m is charged to a potential of 120 V. It is now placed inside another hollow spherical conductor of radius 6 m. Calculate the potential of bigger sphere if the smaller sphere is made to touch the bigger sphere

(a) 120 V (b) 60 V (c) 80 V (d) 40 V

61. Three identical metallic uncharged spheres A, B and C of radius a are kept on the corners of an equilateral triangle of side d ($d >> a$). A fourth sphere (radius a) which has charge Q touches A and is then removed to a position far away. B is earthed and then the earth connection is removed. C is then earthed. The charge on C is

(a) $\frac{Qa}{2d}\left(\frac{2d-a}{2d}\right)$ (b) $\frac{Qa}{2d}\left(\frac{2d-a}{d}\right)$
(c) $\frac{Qa}{2d}\left(\frac{a-d}{d}\right)$ (d) $\frac{2Qa}{d}\left(\frac{d-a}{2d}\right)$

62. A point charge q is placed at a distance r from the centre O of an uncharged spherical shell of inner radius R and outer radius $2R$. The distance $r < R$. The electric potential at the centre of the shell will be

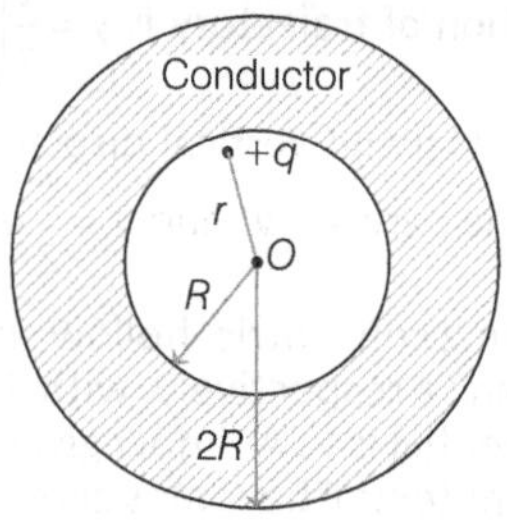

(a) $\frac{q}{4\pi\varepsilon_0}\left(\frac{1}{r}-\frac{1}{2R}\right)$ (b) $\frac{q}{4\pi\varepsilon_0 r}$
(c) $\frac{q}{4\pi\varepsilon_0}\left(\frac{1}{r}+\frac{1}{2R}\right)$ (d) $\frac{q}{4\pi\varepsilon_0}\left(\frac{1}{r}-\frac{1}{R}\right)$

63. Potential to which a soap bubble of radius R be charged, so that pressure outside and inside is same will be (take, p_0 = atmospheric pressure and T = surface tension of soap film)
(a) $V=\sqrt{\frac{Tr}{\varepsilon_0}}$ (b) $V=\sqrt{\frac{3Tr}{\varepsilon_0}}$
(c) $V=\sqrt{\frac{8Tr}{\varepsilon_0}}$ (d) $V=\sqrt{\frac{5Tr}{\varepsilon_0}}$

64. Two large vertical and parallel metal plates having a separation of 1 cm are connected to a DC voltage source of potential difference X. A proton is released at rest midway between the two plates. It is found to move at 45° to the vertical just before touching the plate having lower potential. Then, X is nearly
(a) 10^5 V (b) 10^7 V
(c) 10^9 V (d) None of these

65. A solid sphere of radius R_1 and volume charge density $\rho = \frac{\rho_0}{r}$ is enclosed by a hollow sphere of radius R_2 with negative surface charge density σ, such that the total charge in the system is zero. ρ_0 is positive constant and r is the distance from the centre of the sphere. The ratio R_2/R_1 is
(a) $\frac{\sigma}{\rho_0}$ (b) $\sqrt{\frac{2\sigma}{\rho_0}}$
(c) $\sqrt{\frac{\rho_0}{2\sigma}}$ (d) $\frac{\rho_0}{\sigma}$

66. For a solid sphere of radius R and total charge Q, let the charge distribution be $\rho = \frac{Qr}{\pi R^4}$. Then, for a point P distant r_1 from centre of sphere ($r_1 < R$) field is
(a) 0 (b) $\frac{Q}{4\pi\varepsilon_0 r_1^3}$
(c) $\frac{Qr_1^2}{3\pi\varepsilon_0 R^4}$ (d) $\frac{Qr_1^2}{4\pi\varepsilon_0 R^4}$

67. Consider an electric quadrapole of length $2l$. Then, for a point P on perpendicular bisector of quadrapole at a distance r from the centre.

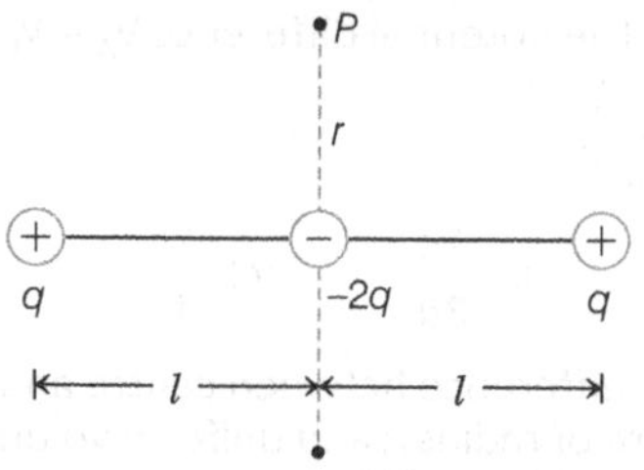

(a) Field shows $\frac{1}{r^2}$ dependence and points towards negative charge
(b) Field shows $\frac{1}{r^3}$ dependence and points away from negative charge
(c) Field shows $\frac{1}{r^4}$ dependence and points towards negative charge
(d) Field shows $\frac{1}{r^2}$ dependence and points towards negative charge

68. What will be the magnitude of electric field at point O as shown in figure? Each side of the figure is l and perpendicular to each other. **(2021 Main)**

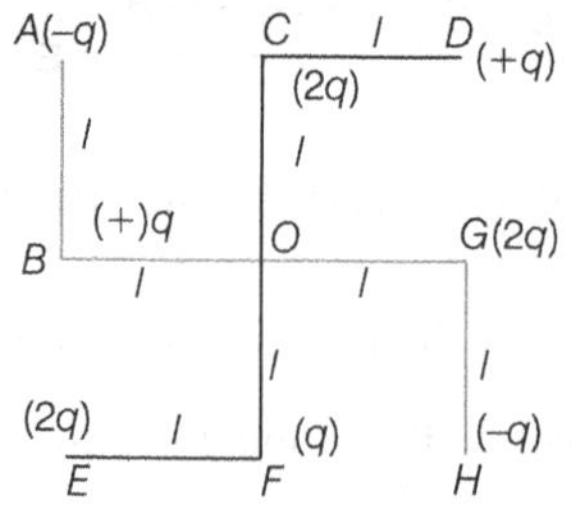

(a) $\frac{1}{4\pi\varepsilon_0}\frac{q}{l^2}$ (b) $\frac{1}{4\pi\varepsilon_0}\frac{q}{(2l^2)}(2\sqrt{2}-1)$
(c) $\frac{q}{4\pi\varepsilon_0 (2l)^2}$ (d) $\frac{1}{4\pi\varepsilon_0}\frac{q}{l^2}(\sqrt{2})$

69. Consider a uniform spherical charge distribution of radius R_1 centred at the origin O. In this distribution, a spherical cavity of radius R_2, centred at P with distance $OP = a = R_1 - R_2$ (see figure) is made. If the electric field inside the cavity at position; is $\mathbf{r}$ is $E(\mathbf{r})$, then the correct statement(s) is (are) **(2015 Adv.)**

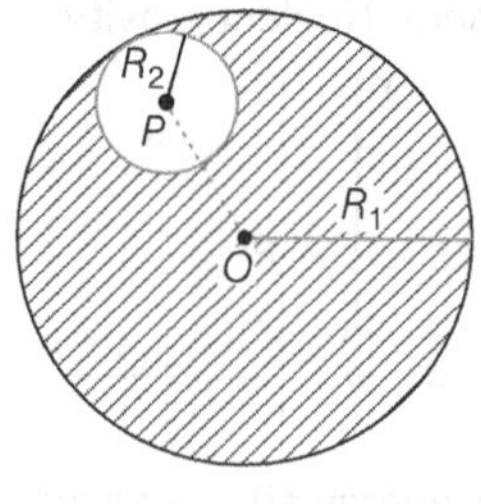

(a) **E** is uniform, its magnitude is independent of R_2 but its direction depends on **r**
(b) **E** is uniform, its magnitude depends on R_2 and its direction depends on **r**
(c) **E** is uniform, its magnitude is independent of **a** but its direction depends on **a**
(d) **E** is uniform and both its magnitude and direction depend on **a**

70. Charges Q, $2Q$ and $4Q$ are uniformly distributed in three dielectric solid spheres 1, 2 and 3 of radii $R/2$, R and $2R$ respectively, as shown in figure. If magnitude of the electric fields at point P at a distance R from the centre of sphere 1, 2 and 3 are E_1, E_2 and E_3 respectively, then **(2014 Adv.)**

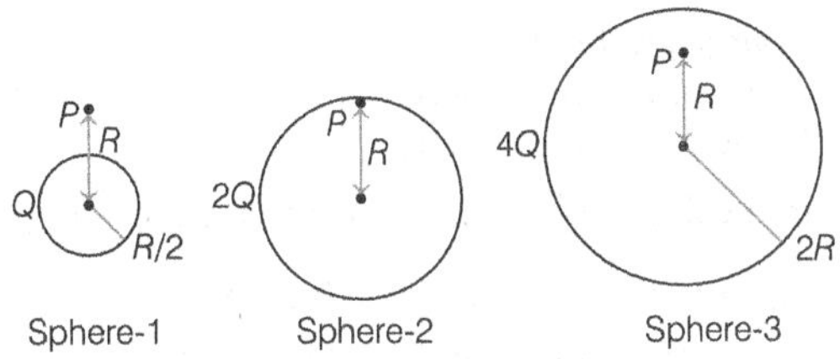

(a) $E_1 > E_2 > E_3$ (b) $E_3 > E_1 > E_2$
(c) $E_2 > E_1 > E_3$ (d) $E_3 > E_2 > E_1$

71. Let $E_1(r)$, $E_2(r)$ and $E_3(r)$ be the respective electric fields at a distance r from a point charge Q, an infinitely long wire with constant linear charge density λ, and an infinite plane with uniform surface charge density σ. If $E_1(r_0) = E_2(r_0) = E_3(r_0)$ at a given distance r_0 then **(2014 Adv.)**

(a) $Q = 4\sigma\pi r_0^2$ (b) $r_0 = \dfrac{\lambda}{2\pi\sigma}$
(c) $E_1(r_0/2) = 2E_2(r_0/2)$ (d) $E_2(r_0/2) = 4E_3(r_0/2)$

Multiple Correct Options

72. Two concentric shells have radii R and $2R$ charges q_A and q_B and potentials $2V$ and $(3/2)$ V respectively. Now, shell B is earthed and let charges on them become q_A' and q_B'. Then,

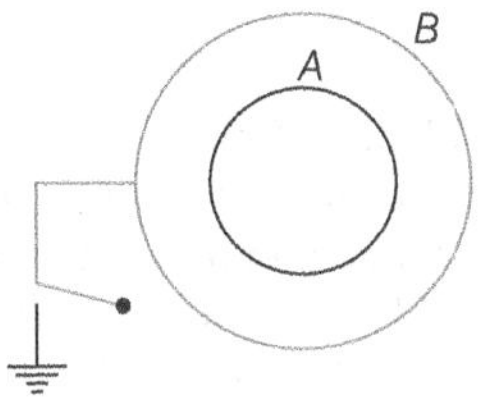

(a) $q_A/q_B = 1/2$
(b) $q_A'/q_B' = 1$
(c) potential of A after earthing becomes $(3/2)$ V
(d) potential difference between A and B after earthing becomes $V/2$

73. A uniformly charged non-conducting sphere of radius r_0 and charge density ρ C/m^3, has

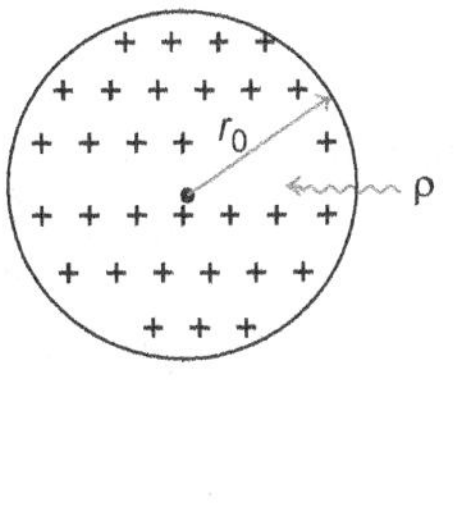

(a) a radially outward pointing field of magnitude $E = \dfrac{\rho r_0^3}{3\varepsilon_0 r^2}$, where $r > r_0$
(b) a field of magnitude $E = \dfrac{\rho r_0}{3\varepsilon_0}$ at surface
(c) a field of magnitude $E = \dfrac{33}{8}\dfrac{\rho r_0^3}{\varepsilon_0}$ at a distance $r = \dfrac{r_0}{2}$ from centre
(d) no field inside its surface

74. At a distance of 5 cm and 10 cm from surface of a uniformly charged solid sphere, the potentials are 100 V and 75 V respectively. Then,
(a) potential at its surface is 150 V
(b) the charge on the sphere is $\dfrac{50}{3} \times 10^{-10}$ C
(c) the electric field on the surface is 1500 V/m
(d) the electric potential at its centre is 25 V

75. Two small charged spheres are suspended on cords of equal length. Cords make angles θ_1 and θ_2 with the vertical.
Which of the following statements is/are correct?

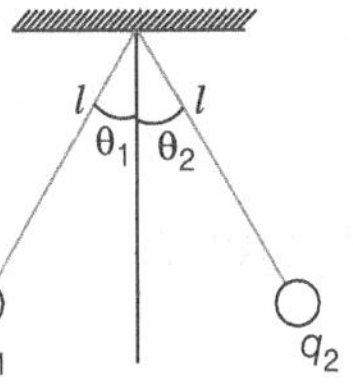

(a) If $q_1 = q$, $q_2 = 2q$ and balls have equal masses $m_1 = m_2 = m$, then $\dfrac{\theta_1}{\theta_2} = 1$
(b) If $q_1 = q$, $q_2 = 2q$ and $m_1 = m$, $m_2 = 2m$, then $\dfrac{\theta_1}{\theta_2} = 2$
(c) $\theta_1 = \theta_2$ when spheres have equal mass
(d) $\theta_1 = \theta_2$ when spheres have equal charge on them

76. Six charges are placed around a regular hexagon of side length a as shown in the figure. Five of them have charge q each and the remaining one has charge x. The perpendicular from each charge to the nearest hexagon side passes through the centre O of the hexagon and is bisected by the side.

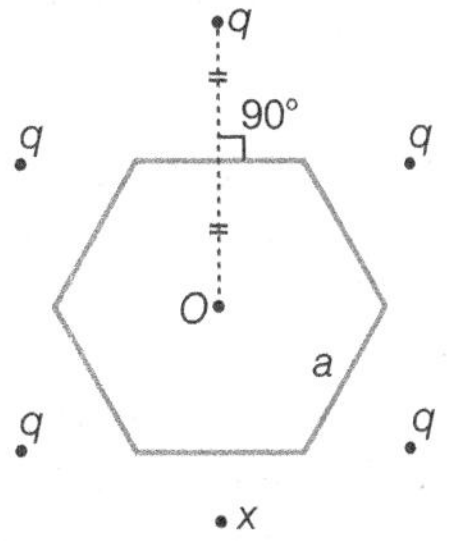

Which of the following statement(s) is(are) correct in SI unit? **(2022 Adv.)**
(a) When $x = q$, the magnitude of the electric field at O is zero.

(b) When $x = -q$, the magnitude of the electric field at O is $\frac{q}{6\pi\varepsilon_0 a^2}$.

(c) When $x = 2q$, the potential at O is $\frac{7q}{4\sqrt{3}\pi\varepsilon_0 a}$

(d) When $x = -3q$, the potential at O is $-\frac{3q}{4\sqrt{3}\pi\varepsilon_0 a}$

77. Two non-conducting spheres of radii R_1 and R_2 and carrying uniform volume charge densities $+\rho$ and $-\rho$ respectively are placed such that they partially overlap as shown in the figure. At all points in the overlapping region **(2013 Adv.)**

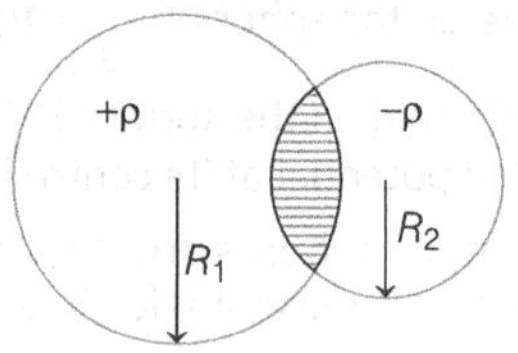

(a) the electrostatic field is zero
(b) the electrostatic potential is constant
(c) the electrostatic field is constant in magnitude
(d) the electrostatic field has same direction

Passage Based Questions

Passage 1 (Q.78 to Q.80)

A solid metallic sphere of radius a is surrounded by a conducting spherical shell of radius b ($b > a$). The solid sphere is given a charge Q. A student measures the potential at the surface of the solid sphere as V_a and the potential at the surface of spherical shell as V_b. After taking these readings, he decides to put charge of $-4Q$ on the shell. He then noted the readings of the potential of solid sphere and the shell and found that the potential difference is ΔV. He then connected the outer spherical shell to the earth by a conducting wire and found that the charge on the outer surface of the shell as q_1.

He then decides to remove the earthing connection from the shell and earthed the inner solid sphere. Connecting the inner sphere with the earth he observes the charge on the solid sphere as q_2. He then wanted to check what happens if the two are connected by the conducting wire. So he removed the earthing connection and connected a conducting wire between the solid sphere and the spherical shell. After the connections were made he found the charge on the outer shell as q_3.

Answer the following questions based on the readings taken by the student at various stages.

78. Potential difference (ΔV) measured by the student between the inner solid sphere and outer shell after putting a charge $-4Q$ is
(a) $V_a - 3V_b$ (b) $3(V_a - V_b)$
(c) V_a (d) $V_a - V_b$

79. q_2 is
(a) Q (b) $Q\left(\frac{a}{b}\right)$ (c) $-4Q$ (d) zero

80. q_3 is
(a) $\frac{Q(a+b)}{a-b}$ (b) $\frac{Qa^2}{b}$ (c) $\frac{Q(a-b)}{b}$ (d) $-\frac{Qb}{a}$

Passage 2 (Q.81 to Q.84)

A thin non-conducting ring is placed in x-y plane with its axis along z-axis and centre at origin. Ring is charged such that it acquires a charge density of $\lambda = \lambda_0 \cos\theta$. Here, λ_0 is constant and θ is angle made with positive direction of x-axis, measured anti-clockwise.

81. Charge on the ring is
(a) positive everywhere
(b) negative everywhere
(c) positive in $\frac{3}{4}$ th part and negative in $\frac{1}{4}$ th part
(d) total charge on the ring is zero

82. X-component of electric field at centre of the ring is
(a) $\frac{k\lambda_0}{R}$ (b) $\frac{2k\lambda_0}{R}$ (c) $\frac{\lambda_0}{4\varepsilon_0 R}$ (d) $\frac{\lambda_0}{2\varepsilon_0 R}$

83. Y-component of electric field at the centre of the ring is
(a) $\frac{k\lambda_0}{R}$ (b) $\frac{2k\lambda_0}{R}$ (c) $\frac{\lambda_0}{4\varepsilon_0 R}$ (d) 0

84. Electric field at a distance x from centre of ring will be
(a) $\frac{\lambda_0 R^2}{4\varepsilon_0 x^3}$ along X-axis (b) $\frac{\lambda_0 R^2}{\varepsilon_0 x^3}$ along X-axis
(c) $\frac{\lambda_0 R^2}{4\varepsilon_0 x^3}$ along Y-axis (d) $\frac{\lambda_0 R^2}{\varepsilon_0 x^3}$ along Y-axis

Passage 3 (Q.85 to Q.87)

A block of mass m and charge +Q is connected to a spring of spring constant k. The block is placed on a frictionless surface and the system is placed in the region of a uniform field E. The block is released from rest when the spring is in unstretched position at $x = 0$.

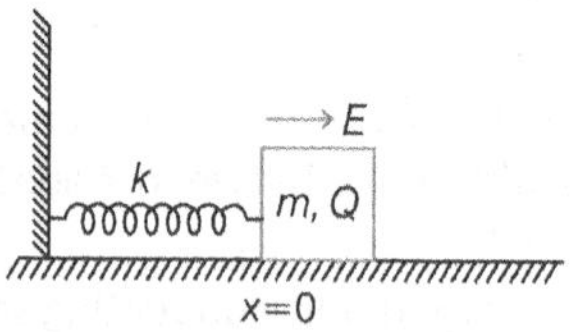

85. Maximum displacement of mass m is
(a) $\frac{QE}{k}$ (b) $\frac{2QE}{k}$ (c) $\frac{QE}{2k}$ (d) $\frac{QE}{4k}$

86. Equilibrium position of block is
(a) $x = \frac{2QE}{k}$ (b) $x = \frac{QE}{4k}$
(c) $x = \frac{QE}{k}$ (d) $x = \frac{QE}{2k}$

87. Motion of the block is

(a) simple harmonic with time period $T = 2\pi\sqrt{\frac{m}{k}}$

(b) periodic but not simple harmonic

(c) simple harmonic with time period $T = 2\pi\sqrt{\frac{2m}{k}}$

(d) periodic with time period $T = 2\pi\sqrt{\frac{m}{2k}}$

Passage 4 (Q.88 to Q.90)

Two fixed charges $-2Q$ and Q are located at the points $(-3a, 0)$ and $(+3a, 0)$ in the xoy-plane.

88. Locus of all the points with $V = 0$ in xy-plane
(a) is a straight line of slope $4a$
(b) is a circle of radius $4a$
(c) is an ellipse of semi-major axis $4a$
(d) is a parabola with focal length $4a$

89. If a particle with charge $+q$ is released at point $(5a, 0)$, then it
(a) oscillates between points $(a, 0)$ and $(9a, 0)$
(b) crosses the point $(9a, 0)$
(c) remains stationary at $(5a, 0)$
(d) rotates in a circle of radius $2a$

90. Potential function $V(x)$ due to the charges
(a) does not have any minima for $x > 9a$
(b) does not have any minima at a
(c) have a maxima at $x = a$
(d) have a minima at $x = a$

Passage 5 (Q.91 to Q.92)

Two point charges $-Q$ and $+Q/\sqrt{3}$ are placed in the xy-plane at the origin (0, 0) and a point (2, 0), respectively, as shown in the figure. This results in an equipotential circle of radius R and potential $V = 0$ in the xy-plane with its centre at (b, 0). All lengths are measured in meters. **(2020 Adv.)**

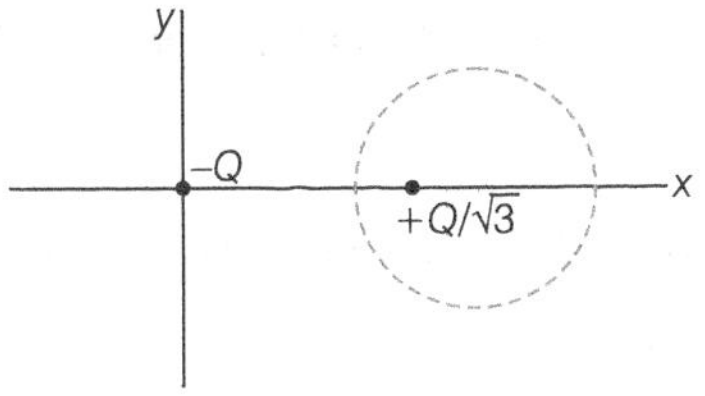

91. The value of R is .... m.

92. The value of b is .... m.

Match the Columns

93. Four charges Q_1, Q_2, Q_3 and Q_4 of same magnitude are fixed along the x-axis at $x = -2a, -a, +a$ and $+2a$ respectively. A positive charge q is placed on the positive y-axis at a distance $b > 0$. Four options of the signs of these charges are given in List-I. The direction of the forces on the charge q is given in List-II. Match List-I with List-II and select the correct answer using the codes given below the lists. **(2014 Adv.)**

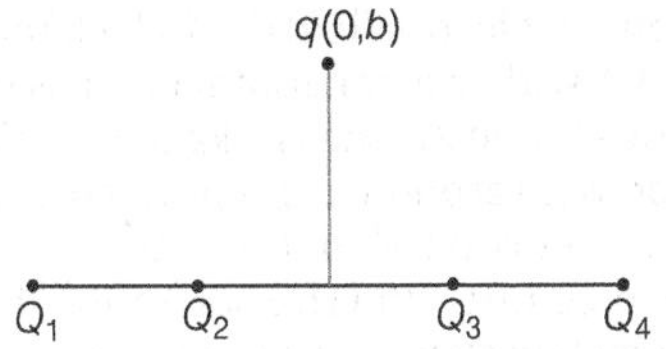

	List-I		List-II
P.	Q_1, Q_2, Q_3, Q_4 all positive	1.	$+x$
Q.	Q_1, Q_2 positive; Q_3, Q_4 negative	2.	$-x$
R.	Q_1, Q_4 positive; Q_2, Q_3 negative	3.	$+y$
S.	Q_1, Q_3 positive; Q_2, Q_4 negative	4.	$-y$

Codes
(a) P-3, Q-1, R-4, S-2
(b) P-4, Q-2, R-3, S-1
(c) P-3, Q-1, R-2, S-4
(d) P-4, Q-2, R-1, S-3

Subjective Questions

94. A 4.00 kg block carrying a charge $q = 50.0\ \mu C$ is connected to a spring for which $k = 100$ N/m. The block lies on a frictionless horizontal track, and the system is kept in a uniform electric field of magnitude $E = 5.00 \times 10^5$ V/m, directed as shown in figure. If the block is released from rest when the spring is unstretched (at $x = 0$).

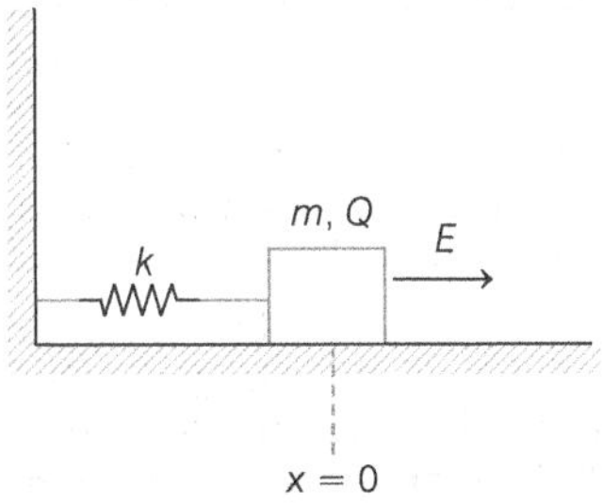

(a) By what maximum amount does the spring expand?
(b) What is the equilibrium position of the block?
(c) Show that the block's motion is simple harmonic and determine its period.
(d) Repeat part (a) if the coefficient of kinetic friction between block and surface is 0.2.

95. A particle of mass m and charge $-Q$ is constrained to move along the axis of a ring of radius a. The ring carries a uniform charge density $+\lambda$ along its length. Initially, the particle is at the centre of the ring where the force on it is zero. Show that the period of oscillation of the particle when it is displaced slightly from its equilibrium position is given by

$$T = 2\pi\sqrt{\frac{2\varepsilon_0 m a^2}{\lambda Q}}$$

96. In a region, an electric field $E = 15$ N/C making an angle of 30° with the horizontal plane is present. A ball having charge 2C, mass 3 kg and coefficient of restitution with ground 1/2 is projected at an angle of 30° with the horizontal in the direction of electric field with speed 20 m/s. Find the horizontal distance travelled by ball from first drop to the second drop.

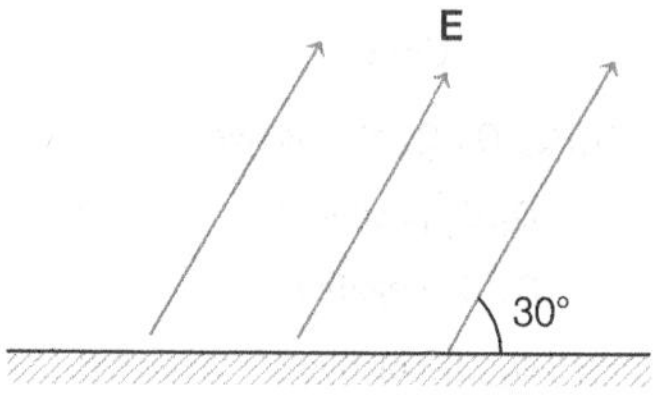

97. A long non-conducting, massless rod of length L is pivoted at its centre and balanced with a weight w at a distance x from the left end. At the left and right ends of the rod are attached small conducting spheres with positive charges q and $2q$ respectively. A distance h directly beneath each of these spheres is a fixed sphere with positive charge Q.
(a) Find the distance x where the rod is horizontal and balanced.
(b) What value should h have so that the rod exerts no vertical force on the bearing when the rod is horizontal and balanced?

Note *Ignore the force between Q (beneath q) and 2q and the force between Q (beneath 2q) and q. Also the force between Q and Q.*

98. The electric potential varies in space according to the relation $V = 3x + 4y$. A particle of mass 10 kg starts from rest from point (2, 3.2) m under the influence of this field. Find the velocity of the particle when it crosses the x-axis. The charge on the particle is $+1\ \mu$C. Assume V (x, y) are in SI units.

99. A simple pendulum with a bob of mass $m = 1$ kg, charge $q = 5\ \mu$C and string length $l = 1$m is given a horizontal velocity u in a uniform electric field $E = 2 \times 10^6$ V/m at its bottommost point A, as shown in figure. It is given that the speed u is such that the particle leaves the circle at point C. Find the speed u (Take $g = 10\ m/s^2$)

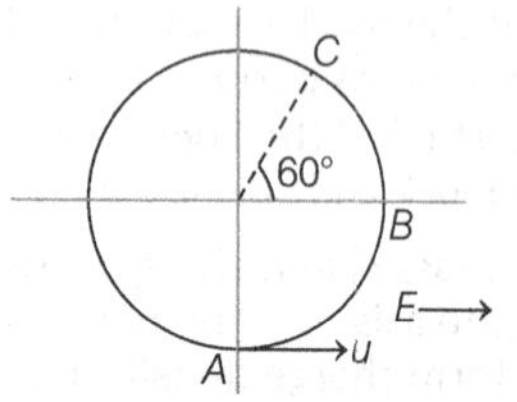

100. Eight point charges of magnitude Q are arranged to form the corners of a cube of side L. The arrangement is made in such a manner that the nearest neighbour of any charge has the opposite sign. Initially, the charges are held at rest. If the system is let free to move, what happens to the arrangement? Does the cube-shape shrink or expand? Calculate the velocity of each charge when the side-length of the cube formation changes from L to nL. Assume that the mass of each point charge is m.

101. There are two concentric spherical shells of radii r and $2r$. Initially, a charge Q is given to the inner shell. Now, switch S_1 is closed and opened then S_2 is closed and opened and the process is repeated n times for both the keys alternatively. Find the final potential difference between the shells.

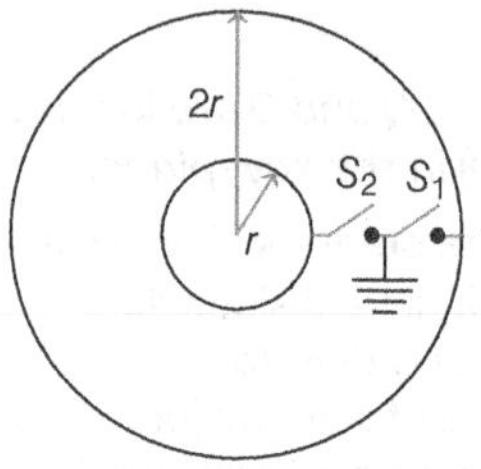

102. Two point charges Q_1 and Q_2 are positioned at points 1 and 2. The field intensity to the right of the charge Q_2 on the line that passes through the two charges varies according to a law that is represented schematically in the figure. The field intensity is assumed to be positive if its direction coincides with the positive direction on the x-axis. The distance between the charges is l.

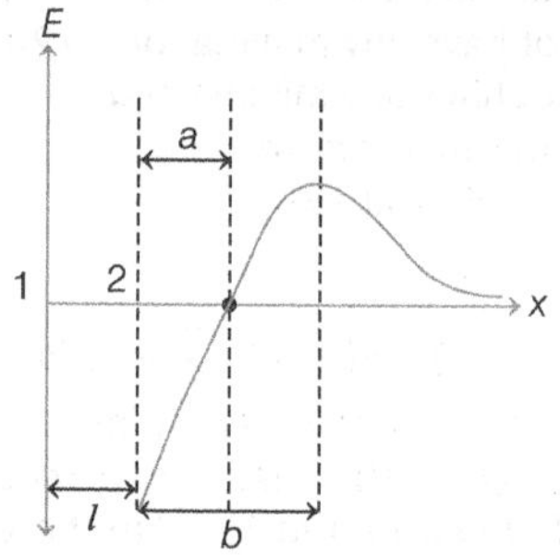

(a) Find the sign of each charge.
(b) Find the ratio of the absolute values of the charges $\left|\frac{Q_1}{Q_2}\right|$
(c) Find the value of b where the field intensity is maximum.

103. A rectangular tank of mass m_0 and charge Q over it is placed on a smooth horizontal floor. A horizontal electric field E exists in the region. Rain drops are falling vertically in the tank at the constant rate of n drops per second. Mass of each drop is m. Find velocity of tank as function of time.

104. A non-conducting ring of mass m and radius R is charged as shown. The charge density, i.e. charge per unit length is λ. It is then placed on a rough non-conducting horizontal plane. At time $t = 0$, a uniform electric field $\mathbf{E} = E_0\hat{\mathbf{i}}$ is switched on and the ring starts rolling without sliding. Determine the friction force (magnitude and direction) acting on the ring when it starts moving.

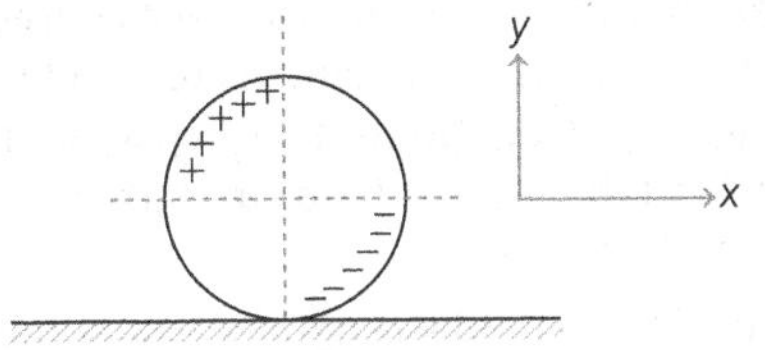

105. The region between two concentric spheres of radii a and b ($> a$) contains volume charge density $\rho(r) = \frac{C}{r}$, where C is a constant and r is the radial distance as shown in figure. A point charge q is placed at the origin, $r = 0$.
Find the value of C for which the electric field in the region between the spheres is constant (i.e. r independent).

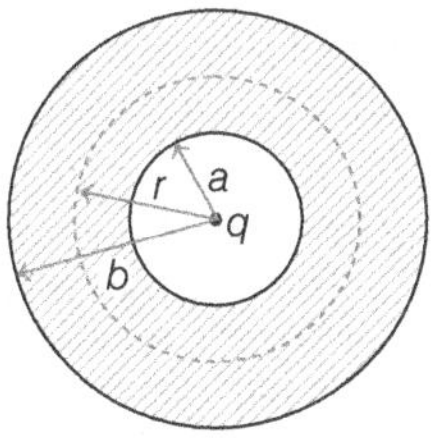

106. Positive charge Q is uniformly distributed throughout the volume of a sphere of radius R. A point mass having charge $+q$ and mass m is fired towards the centre of the sphere with velocity v from a point A at distance $r (r > R)$ from the centre of the sphere. Find the minimum velocity v so that it can penetrate $R/2$ distance of the sphere. Neglect any resistance other than electric interaction. Charge on the small mass remains constant throughout the motion.

107. Two concentric rings are placed in a gravity free region in yz-plane. One of radius R carries a charge $+Q$ and second of radius $4R$, a charge $-8Q$ distributed uniformly over it. Find the minimum velocity with which a point charge of mass m and charge $+q$ should be projected from a point at a distance $3R$ from the centre of rings on its axis so that it will reach to the centre of the rings.

108. An electric dipole is placed at a distance x from centre O on the axis of a charged ring of radius R and charge Q uniformly distributed over it.

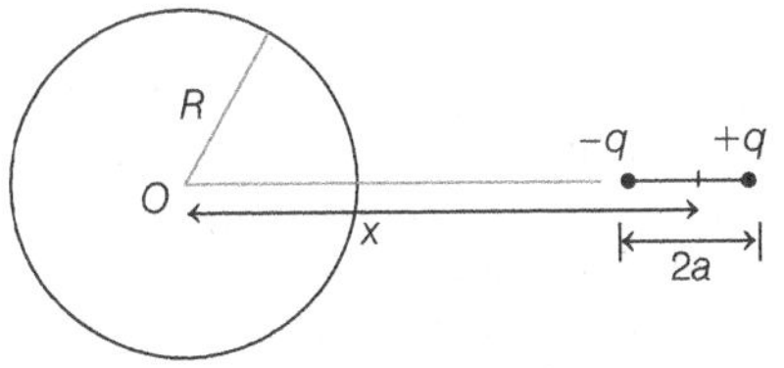

(a) Find the net force acting on the dipole.
(b) What is the work done in rotating the dipole through 180°?

109. A point charge $-q$ revolves around a fixed charge $+Q$ in elliptical orbit. The minimum and maximum distance of q from Q are r_1 and r_2, respectively. The mass of revolving particle is m. $Q > q$ and assume no gravitational effects. Find the velocity of q at positions when it is at r_1 and r_2 distance from Q.

110. A thin insulating wire is stretched along the diameter of an insulated circular hoop of radius R. A small bead of mass m and charge $-q$ is threaded onto the wire. Two small identical charges are tied to the hoop at points opposite to each other, so that the diameter passing through them is perpendicular to the thread (see figure). The bead is released at a point which is a distance x_0 from the centre of the hoop. Assume that $x_0 << R$.

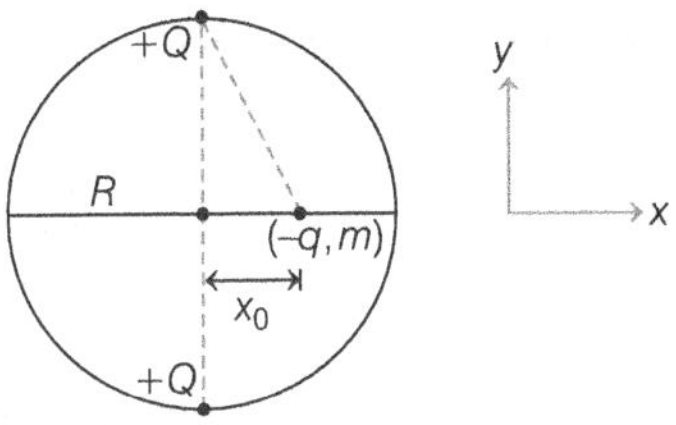

(a) What is the resultant force acting on the charged bead?
(b) Describe (qualitatively) the motion of the bead after it is released.
(c) Use the assumption that $\frac{x}{R} << 1$ to obtain an approximate equation of motion, and find the displacement and velocity of the bead as functions of time.
(d) When will the velocity of the bead will become zero for the first time?

111. Three particles, each of mass 1 g and carrying a charge q, are suspended from a common point by insulated massless strings, each 100 cm long. If the particles are in equilibrium and are located at the corners of an equilateral triangle of side length 3 cm, calculate the charge q on each particle.
(Take, $g = 10 \text{ m/s}^2$) **(1988 JEE)**

112. A thin fixed ring of radius 1 metre has a positive charge 1×10^{-5} coulomb uniformly distributed over it. A particle of mass 0.9 g and having a negative charge of 1×10^{-6} coulomb is placed on the axis at a distance of 1m from the centre of the ring. Show that the motion of the negatively charged particle is

approximately simple harmonic. Calculate the time period of oscillations. **(1982 JEE)**

113. A rigid insulated wire frame in the form of right triangle ABC is set in a vertical plane. Two beads of equal masses m each carrying charges q_1 and q_2 are connected by a chord of length l and can slide without friction on the wires. Considering the case.

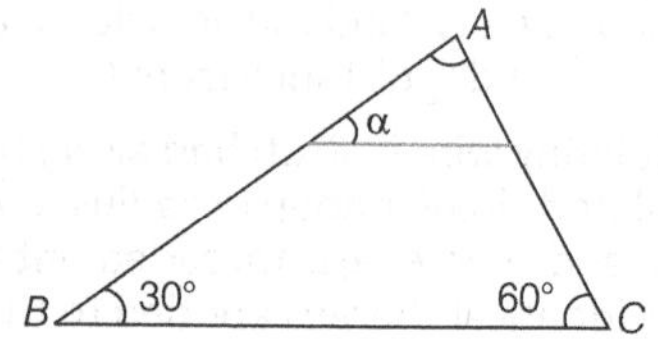

when the beads are stationary, determine : **(1978 JEE)**
(i) the angle α
(ii) the tension in the chord, and
(iii) the normal reactions on the beads.
If the chord is now cut, what are the values of the charges for which the beads continue to remain stationary?

114. A thin wire ring of radius r has an electric charge q. What will be the increment of the force stretching the wire if a point charge q_0 is placed at the ring's centre ? **(Irodov 3.5)**

115. Two positive charges q_1 and q_2 are located at the points with radius vector $\mathbf{r}_1$ and $\mathbf{r}_2$. Find a negative charge q_3 and a radius vector $\mathbf{r}_3$ of the point at which it has to be placed for the force acting on each of the three charges to be equal to zero. **(Irodov 3.4)**

116. A thin nonconducting ring of radius R has a linear charge density $\lambda = \lambda_0 \cos \phi$, where λ_0 is a constant, ϕ is the azimuthal angle. Find the magnitude of the electric field strength
(a) at the centre of the ring;
(b) on the axis of the ring as a function of the distance x from its centre. Investigate the obtained function at $x >> R$. **(Irodov 3.12)**

117. A very long straight uniformly charged thread carries a charge λ per unit length. Find the magnitude and direction of the electric field strength at a point which is at a distance y from the thread and lies on the perpendicular passing through one of the thread's ends. **(Irodov 3.14)**

118. A thread carrying a uniform charge λ per unit length has the configurations shown in Fig. (a) and Fig. (b). Assuming a curvature radius R to be considerably less than the length of the thread, find the magnitude of the electric field strength at the point O. **(Irodov 3.15)**

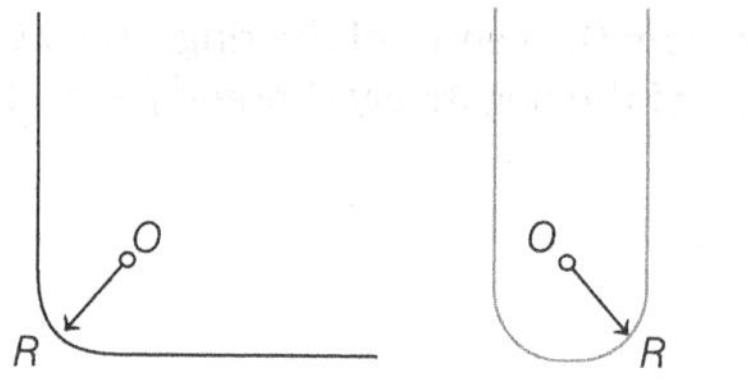

(a) (b)

119. A sphere of radius r carries a surface charge of density $\sigma = \mathbf{a} \cdot \mathbf{r}$, where $\mathbf{a}$ is a constant vector, and $\mathbf{r}$ is the radius vector of a point of the sphere relative to its centre. Find the electric field strength vector at the centre of the sphere. **(Irodov 3.16)**

120. Find the electric field strength vector at the centre of a ball of radius R with volume charge density $\rho = \mathbf{a} \cdot \mathbf{r}$, where $\mathbf{a}$ is a constant vector, and $\mathbf{r}$ is a radius vector drawn from the ball's centre. **(Irodov 3.18)**

121. A very long uniformly charged thread oriented along the axis of a circle of radius R rests on its centre with one of the ends. The charge of the thread per unit length is equal to λ. Find the flux of the vector $\mathbf{E}$ across the circle area. **(Irodov 3.19)**

122. A ball of radius R is uniformly charged with the volume density ρ. Find the flux of the electric field strength vector across the ball's section formed by the plane located at a distance $r_0 < R$ from the centre of the ball. **(Irodov 3.21)**

123. An infinitely long cylindrical surface of circular cross-section is uniformly charged lengthwise with the surface density $\sigma = \sigma_0 \cos \phi$, where ϕ is the polar angle of the cylindrical coordinate system whose Z-axis coincides with the axis of the given surface. Find the magnitude and direction of the electric field strength vector on the Z-axis. **(Irodov 3.23)**

124. A ball of radius R carries a positive charge whose volume density depends only on the distance r from the ball's centre as $\rho = \rho_0(1 - r/R)$, where ρ_0 is a constant. Assuming the permittivities of the ball and the environment to be equal to unity, find:
(a) the magnitude of the electric field strength as a function of the distance r both inside the outside the ball
(b) the maximum intensity E_{max} and the corresponding distance r_m. **(Irodov 3.25)**

125. A system consists of a ball of radius R carrying a spherically symmetric charge and the surrounding space filled with a charge of volume density $\rho = \alpha/r$, where α is a constant, r is the distance from the centre of the ball. Find the ball's charge at which the magnitude of the electric field strength vector is independent of r outside the ball. How high is this strength? The permittivities of the ball and the surrounding space are assumed to be equal to unity. **(Irodov 3.26)**

126. A space is filled up with a charge with volume density $\rho = \rho_0 e^{-\alpha r^3}$, where ρ_0 and α are positive constants, r is the distance from the centre of this system. Find the magnitude of the electric field strength vector as a function of r. Investigate the obtained expression for the small and large values of r, i.e. at $\alpha r^3 << 1$ and $\alpha r^3 >> 1$. **(Irodov 3.27)**

127. Inside a ball charged uniformly with volume density ρ, there is a spherical cavity. The centre of the cavity is displaced with respect to the centre of the ball by a

distance **a**. Find the field strength **E** inside the cavity, assuming the permittivity equal to unity. **(Irodov 3.28)**

128. Inside an infinitely long circular cylinder charged uniformly with volume density ρ, there is a circular cylindrical cavity. The distance between the axes of the cylinder and the cavity is equal to a. Find the electric field strength **E** inside the cavity. The permittivity is assumed to be equal to unity. **(Irodov 3.29)**

129. Find the electric field strength vector, if the potential of this field has the form $V = \mathbf{a} \cdot \mathbf{r}$, where **a** is a constant vector and **r** is the radius vector of a point of the field. **(Irodov 3.35)**

130. Determine the electric field strength vector if the potential of this field depends on x and y coordinates as
(a) $V = a(x^2 - y^2)$
(b) $V = axy$, **(Irodov 3.36)**
where a is a constant. Draw the approximate shape of these fields using lines of force (in the x, y plane).

131. Two thin parallel threads carry a uniform charge with linear densities λ and $-\lambda$. The distance between the threads is equal to l. Find the potential of the electric field and the magnitude of its strength vector at a distance $r \gg l$ at an angle θ to the radius vector l. **(Irodov 3.42)**

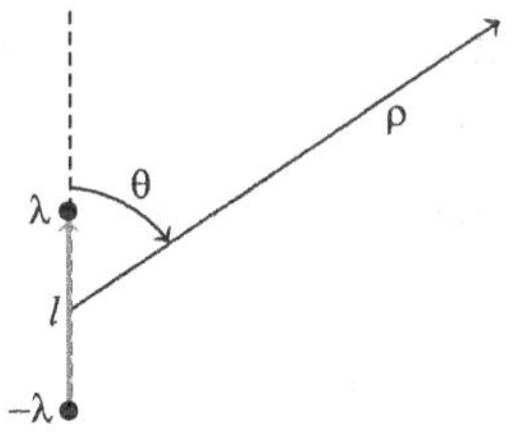

132. A dipole with an electric moment **p** is located at a distance r from a long thread charged uniformly with a linear density λ. Find the force **F** acting on the dipole if the vector **p** is oriented **(Irodov 3.46)**
(a) along the thread
(b) along the radius vector **r**
(c) at right angles to the thread and the radius vector **r**.

133. Find the interaction force between two water molecules separated by a distance $l = 10$ nm if their electric moments are oriented along the same straight line. The moment of each molecule equals $p = 0.62 \times 10^{-29}$ C-m. **(Irodov 3.47)**

134. The field potential in a certain region of space depends only on the x-coordinate as $V = -ax^3 + b$, where a and b are constants. Find the distribution of the space charge $\rho(x)$. **(Irodov 3.51)**

135. A uniformly distributed space charge fills up the space between two large parallel plates separated by a distance d. The potential difference between the plates is equal to ΔV. At what value of charge density ρ is the field strength in the vicinity of one of the plates equal to zero? What will then be the field strength near the other plate? **(Irodov 3.52)**

136. The field potential inside a charged ball depends only on the distance from its centre as $V = ar^2 + b$, where a and b are constants. Find the space charge distribution $\rho(r)$ inside the ball. **(Irodov 3.53)**

137. A system consists of two concentric conducting spheres, with the inside sphere of radius a carrying a positive charge q_1. What charge q_2 has to be deposited on the outside sphere of radius b to reduce the potential of the inside sphere to zero? How does the potential V depend in this case on a distance r from the centre of the system? Draw the approximate plot of this dependence. **(Irodov 3.65)**

138. A metal ball of radius $R = 1.5$ cm has a charge $q = 10\ \mu$C. Find the modulus of the vector of the resultant force acting on a charge located on one half of the ball. **(Irodov 3.69)**

139. When an uncharged conducting ball of radius R is placed in an external uniform electric field, a surface charge density $\sigma = \sigma_0 \cos\theta$ is induced on the ball's surface (here σ_0 is a constant and θ is the polar angle). Find the magnitude of the resultant electric force acting on an induced charge of the same sign. **(Irodov 3.70)**

140. A charge q is distributed uniformly over the volume of a ball of radius R. Assuming the permittivity to be equal to unity, find: **(Irodov 3.135)**
(a) the electrostatic self energy of the ball
(b) the ratio of the energy W_1 stored in the ball to the energy W_2 pervading the surrounding space.

141. A particle with specific charge q/m moves rectilinearly due to an electric field $E = E_0 - \alpha x$, where α is a positive constant, x is the distance from the point where the particle was initially at rest. Find: **(Irodov 3.374)**
(a) the distance covered by the particle till the moment it came to a standstill.
(b) the acceleration of the particle at that moment.

Integer Type Questions

142. A solid sphere of radius R has a charge distribution $\rho = kr^a$ with a total charge of Q coulombs. k and a are constants and r is distance from centre.
If electric field at $r = \frac{R}{2}$ is $\frac{1}{8}$ times of that at $r = R$, find the value of a.

143. An infinitely long thin wire, having a uniform charge density per unit length of 5 n C/m, is passing through a spherical shell of radius 1 m, as shown in the figure.
A 10 n C charge is distributed uniformly over the spherical shell. If the configuration of the charges remains static, the magnitude of the potential difference between points P and R, in volt, is ______.
[Given, in SI units $\frac{1}{4\pi\varepsilon_0} = 9 \times 10^9$, ln 2 = 0.7. Ignore the area pierced by the wire.] **(2024 Adv.)**

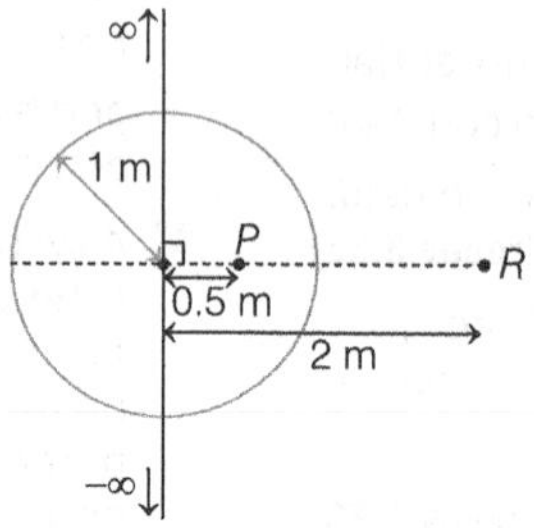

Answers

Level 1

1. 128	**2.** −8	**3.** B	**4.** False	**5.** True	**6.** (b)	**7.** (a)	**8.** (c)	**9.** (d)	**10.** (d)
11. (b)	**12.** (b)	**13.** (d)	**14.** (d)	**15.** (c)	**16.** (d)	**17.** (b)	**18.** (c)	**19.** (b)	**20.** (b)
21. (a)	**22.** (d)	**23.** (c)	**24.** (c)	**25.** (d)	**26.** (d)	**27.** (b)	**28.** (a,b,c)	**29.** (a,b,d)	**30.** (b,c)
31. (a,c)	**32.** (a,c)	**33.** (c)	**34.** (a)	**35.** (c)	**36.** (d)	**37.** (b)	**38.** (d)		

39. (a-r ; b-p; c-q; d-s) **40.** $\frac{F_e}{F_g} = \frac{4 \times 10^{42}}{1 \times 10^{26}} \cdot \frac{q}{m} = 0.86 \times 10^{-10}$ C/kg **41.** About 2×10^{15} N

42. $E = (2.7\,\hat{i} - 3.6\,\hat{j})$ kV/m, 4.5 kV/m **43.** $E_{net} = \frac{ql}{\sqrt{2}\,\pi\varepsilon_0 (x^2 + l^2)^{3/2}}$ **44.** $F = \frac{q\lambda}{4\pi\varepsilon_0 R}$

45. (a) 6V (b) 16V (c) 1280 V/m **46.** (24) **47.** (12)

Level 2

48. (a)	**49.** (a)	**50.** (b)	**51.** (d)	**52.** (a)	**53.** (b)	**54.** (c)	**55.** (a)	**56.** (b)	**57.** (c)
58. (c)	**59.** (c)	**60.** (d)	**61.** (c)	**62.** (a)	**63.** (c)	**64.** (d)	**65.** (c)	**66.** (d)	**67.** (c)
68. (b)	**69.** (d)	**70.** (c)	**71.** (c)	**72.** (a,d)	**73.** (a,b)	**74.** (a,b,c)	**75.** (a,b,c)	**76.** (a,b,c)	**77.** (c,d)
78. (d)	**79.** (b)	**80.** (c)	**81.** (d)	**82.** (c)	**83.** (c)	**84.** (a)	**85.** (b)	**86.** (c)	**87.** (a)
88. (b)	**89.** (b)	**90.** (b)	**91.** (1.73)	**92.** (3)	**93.** (a)				

94. (a) 0.5 m (b) 0.25 m (c) 1.26 s (d) 0.34 m **96.** $70\sqrt{3}$ m **97.** (a) $\frac{L}{2}\left[1 + \frac{Qq}{(4\pi\varepsilon_0)Wh^2}\right]$ (b) $\sqrt{\frac{3Qq}{(4\pi\varepsilon_0)W}}$

98. 2.0×10^{-3} m/s **99.** 6 m/s **100.** shrink, $\sqrt{\frac{Q^2(1-n)(3\sqrt{6} + \sqrt{2} - 3\sqrt{3})}{4nm\pi L\varepsilon_0\sqrt{6}}}$ **101.** $\frac{1}{2^{n+1}}\left[\frac{Q}{4\pi\varepsilon_0 r}\right]$

102. (a) Q_2 is negative and Q_1 is positive (b) $\left(\frac{l+a}{a}\right)^2$ (c) $\frac{l}{\left(\frac{l+a}{a}\right)^{2/3} - 1}$ **103.** $v = QE\left(\frac{t}{m_0 + mnt}\right)$

104. $f = \lambda R E_0$ along positive x-axis **105.** $C = \frac{q}{2\pi a^2}$ **106.** $\left[\frac{1}{2\pi\varepsilon_0}\frac{Qq}{Rm}\left(\frac{r-R}{r} + \frac{3}{8}\right)\right]^{\frac{1}{2}}$

107. $\sqrt{\frac{Qq}{2\pi\varepsilon_0 mR}\left(\frac{3\sqrt{10} - 5}{5\sqrt{10}}\right)}$ **108.** (a) $\frac{aqQ}{2\pi\varepsilon_0}\frac{R^2 - 2x^2}{(R^2 + x^2)^{5/2}}$ (b) $\frac{aqQx}{\pi\varepsilon_0 (R^2 + x^2)^{3/2}}$

109. $\sqrt{\dfrac{Qqr_2}{2\pi\varepsilon_0 mr_1(r_1+r_2)}}$, $\sqrt{\dfrac{Qqr_1}{2\pi\varepsilon_0 mr_2(r_1+r_2)}}$

110. (a) $F = -\dfrac{2kQqx}{(R^2+x^2)^{3/2}}$ (b) Periodic between $\pm x_0$ (c) $x = x_0\cos\omega t$, $v = -\omega x_0 \sin\omega t$, where $\omega = \sqrt{\dfrac{2Qqk}{mR^3}}$

(d) $t = \sqrt{\dfrac{\pi^2 mR^3}{2Qqk}}$ Here $k = \dfrac{1}{4\pi\varepsilon_0}$ **111.** 3.16×10^{-9} C **112.** 0.628 s

113. (i) 60° (ii) $\left(\dfrac{kq_1q_2}{l^2}+mg\right)$ (iii) $N_1 = \sqrt{3}mg$, $N_2 = mg$, $q_1q_2 = -\dfrac{mgl^2}{k}$

114. $\Delta T = \dfrac{qq_0}{8\pi^2\varepsilon_0 r^2}$ **115.** $\mathbf{r}_3 = \dfrac{\mathbf{r}_1\sqrt{q_2}+\mathbf{r}_2\sqrt{q_1}}{\sqrt{q_1}+\sqrt{q_2}}$, $q_3 = \dfrac{-q_1q_2}{(\sqrt{q_1}+\sqrt{q_2})^2}$

116. (a) $E = |E_x| = \dfrac{\lambda_0}{4\varepsilon_0 R}$ (b) $E = |E_z| = \dfrac{\lambda_0 R^2}{4\varepsilon_0(R^2+x^2)^{3/2}}$

117. $E = \dfrac{\lambda\sqrt{2}}{4\pi\varepsilon_0 y}$. The vector **E** is directed at an angle 45° from the thread. **118.** (a) $E = \dfrac{\sqrt{2}\lambda}{4\pi\varepsilon_0 R}$, (b) $E = 0$

119. $\mathbf{E} = -\dfrac{\mathbf{a}r}{3\varepsilon_0}$ **120.** $\mathbf{E} = -\dfrac{\mathbf{a}r}{3\varepsilon_0}$ **121.** $|\phi| = \dfrac{\lambda R}{2\varepsilon_0}$ **122.** $\phi = \dfrac{\rho\pi r_0}{3\varepsilon_0}(R^2 - r_0^2)$ **123.** $\mathbf{E} = -\dfrac{\sigma}{2\varepsilon_0}\hat{\mathbf{i}}$

124. (a) $E = \dfrac{\rho_0 r}{3\varepsilon_0}\left(1-\dfrac{3r}{4R}\right)$ for $r \le R$, $E = \dfrac{\rho_0 R^3}{12\varepsilon_0 r^2}$ for $r \ge R$ (b) $E_{max} = \dfrac{1}{9}\dfrac{\rho_0 R}{\varepsilon_0}$ for $r = \dfrac{2}{3}R$ **125.** $q = 2\pi\alpha R^2$, $E = \dfrac{\alpha}{2\varepsilon_0}$

126. $E = \dfrac{\rho_0}{3\varepsilon_0\alpha r^2}(1-e^{-\alpha r^3})$; Accordingly, $E \approx \dfrac{\rho_0 r}{3\varepsilon_0}$, $E = \dfrac{\rho_0}{3\varepsilon_0\alpha r^2}$ **127.** $\mathbf{E} = \dfrac{\rho\mathbf{a}}{3\varepsilon_0}$ **128.** $\mathbf{E} = \dfrac{\rho\mathbf{a}}{2\varepsilon_0}$

129. $\mathbf{E} = -\mathbf{a}$, i.e. the field is uniform. **130.** (a) $\mathbf{E} = -2a(x\hat{\mathbf{i}} - y\hat{\mathbf{j}})$ (b) $\mathbf{E} = -a(y\hat{\mathbf{i}} + x\hat{\mathbf{j}})$

131. $E \approx \dfrac{\lambda l\cos\theta}{2\pi\varepsilon_0 r^2}$, $V \approx \dfrac{\lambda l\cos\theta}{2\pi\varepsilon_0 r}$ **132.** (a) $\mathbf{F} = 0$, (b) $\mathbf{F} = -\dfrac{\lambda\mathbf{p}}{2\pi\varepsilon_0 r^2}$, (c) $\mathbf{F} = \dfrac{\lambda\mathbf{p}}{2\pi\varepsilon_0 r^2}$ **133.** $F \approx 2.1\times10^{-16}$ N

134. $\rho(x) = (6ax)\varepsilon_0$ **135.** $E_B = \dfrac{\rho d}{\varepsilon_0}$, $\rho = \dfrac{2\varepsilon_0\Delta V}{d^2}$ **136.** $\rho = -6\varepsilon_0 a$

137. $q_2 = -\dfrac{b}{a}q_1$, $V = \dfrac{q_1}{4\pi\varepsilon_0}\left(\dfrac{1}{r}-\dfrac{1}{a}\right)$ if $a \le r \le b$, $V = \dfrac{q_1}{4\pi\varepsilon_0 r}\left(1-\dfrac{b}{a}\right)$ if $r \ge b$ **138.** $F_{net} = 0.5$ kN

139. $F_{net} = \dfrac{\pi R^2\sigma_0^2}{4\varepsilon_0}$ **140.** (a) $W = \dfrac{3q^2}{20\pi\varepsilon_0 R}$ (b) $\dfrac{W_1}{W_2} = \dfrac{1}{5}$ **141.** (a) $x = \dfrac{2E_0}{\alpha}$ (b) $|a| = \dfrac{qE_0}{m}$

142. (2) **143.** (171)

See hints on pages 213-240.

Hints & SOLUTIONS

Module 1.

1. Repulsion is conformation of like charges. While attraction may be possible if one ball is charged and other is neutral (due to induced charges).
   (2, 3) have same type of charges.
   Similarly (4, 5) have also similar type of charges. (2, 4) are therefore charged but have unlike natures.
   Ball-1, should be therefore electrically neutral.

2. Due to induction effect, a charged body can attract a neutral body as shown below.

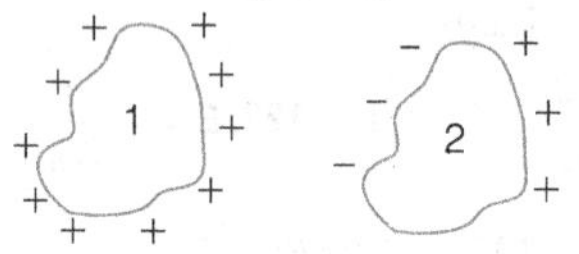

   Body-1 is positively charged and body-2 is neutral. But we can see that due to distance factor attraction is more than the repulsion.

5. Number of atoms in three gram mole of hydrogen atom = number of electrons in it

$$= 3N_0 = (3 \times 6.02 \times 10^{23})$$

   where, N_0 = Avogadro number
   $\therefore$ Total charge

$$= -(1.6 \times 10^{-19})(3 \times 6.02 \times 10^{23})$$
$$= -2.89 \times 10^5 \text{C}$$

Module 2.

1. Here, only electrostatic force of repulsion is acting which will take the two balls as far as possible. The angle between the two strings will be 180°.
   The tension in each string will be equal to the electrostatic force of repulsion.

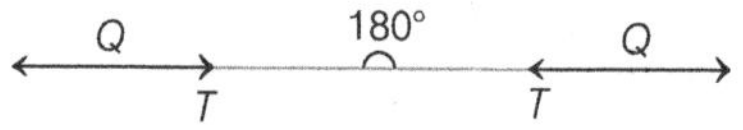

   Here, $Q_1 = Q_2 = Q$ and $R = 2L$

$$T = \frac{1}{4\pi\varepsilon_0} \times \frac{Q \times Q}{(2L)^2} = \frac{1}{4\pi\varepsilon_0} \times \frac{Q^2}{4L^2}$$

2. Two positive ions each carrying a charge q are kept at a distance d, then it is found that force of repulsion between them, is

$$F = \frac{kqq}{d^2} = \frac{1}{4\pi\varepsilon_0}\frac{q^2}{d^2}$$

   where, $q = ne$

$$\therefore \quad F = \frac{1}{4\pi\varepsilon_0}\frac{n^2e^2}{d^2} \Rightarrow n = \sqrt{\frac{4\pi\varepsilon_0 Fd^2}{e^2}}$$

3. Number of electrons,

$$n = \frac{6 \times 10^{23}}{63.5} \times 10 \times \frac{1}{10^6} = \frac{6 \times 10^{18}}{63.5}$$

$$\therefore \quad q = \frac{6 \times 10^{18} \times 1.6 \times 10^{-19}}{63.5}$$

or $\quad q = 1.5 \times 10^{-2}$ C

$$\therefore \quad F = \frac{9 \times 10^9 \times 1.5 \times 10^{-2} \times 1.5 \times 10^{-2}}{\left(\frac{10}{100}\right)}$$

$$= 2.0 \times 10^8 \text{ N}$$

4. $F = \frac{1}{4\pi\varepsilon_0}\frac{q(Q-q)}{r^2}$

   For F to be maximum, $\frac{dF}{dq} = 0.$

   By putting $\frac{dF}{dq} = 0$, we get $q = \frac{Q}{2}$

5. $\frac{1}{4\pi\varepsilon_0}\frac{q^2}{r^2} = mg \sin\theta$

$$\therefore \quad r = \sqrt{\left(\frac{1}{4\pi\varepsilon_0}\right)\left(\frac{q^2}{mg\sin\theta}\right)}$$

$$= \sqrt{\frac{(9 \times 10^9)(2.0 \times 10^{-6})^2}{(0.1)(9.8)\sin 30°}}$$

$$= 27 \times 10^{-2}\text{m}$$

   = 27 cm from the bottom

6.

   Between A and B two forces on third charge will act in same direction. So, this charge cannot remain in equilibrium.
   To the right of B or left of A forces are in opposite directions but their magnitudes are different. Because charges have equal magnitudes but distances are different.

7.

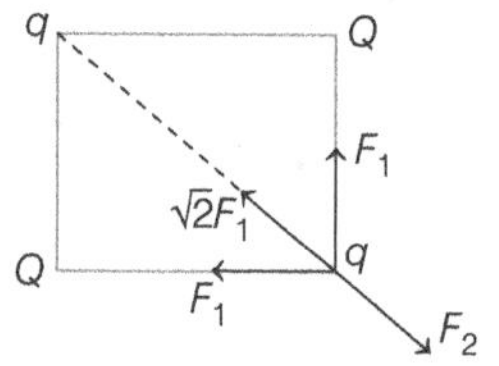

   F_2 = Force between q and q

$$= \frac{1}{4\pi\varepsilon_0}\frac{q \times q}{(\sqrt{2}\,a)^2}$$

   F_1 = Force between Q and q. For net force on q to be zero .

$$F_2 = \sqrt{2} F_1$$

$$\therefore \quad \frac{1}{4\pi\varepsilon_0} \frac{q^2}{2a^2} = \sqrt{2}\left[\frac{1}{4\pi\varepsilon_0} \cdot \frac{qQ}{a^2}\right]$$

$$\therefore \quad |q| = 2\sqrt{2}Q$$

with sign, $q = -2\sqrt{2}\,Q$

8. $$\frac{1}{4\pi\varepsilon_0} \cdot \frac{q_1 q_2}{r_1^2} = \frac{1}{4\pi\varepsilon_0 k} \cdot \frac{q_1 q_2}{r_2^2}$$

$$\therefore \quad r_2 = \frac{r_1}{\sqrt{k}} = \frac{50}{\sqrt{5}} = 22.36 \text{ m}$$

9. Situation is shown in figure.

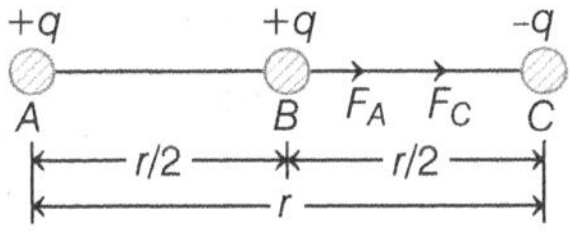

Force between A and C,

$$F = \frac{1}{4\pi\varepsilon_0} \cdot \frac{q^2}{r^2}$$

When sphere B is kept at the mid-point of line joining A and C, then net force on B is

$$F_{net} = F_A + F_C$$

$$= \frac{1}{4\pi\varepsilon_0} \cdot \frac{q^2}{(r/2)^2} + \frac{1}{4\pi\varepsilon_0} \cdot \frac{q^2}{(r/2)^2}$$

$$= 8 \cdot \frac{1}{4\pi\varepsilon_0} \cdot \frac{q^2}{r^2} = 8F$$

10. At the present position all the charges are in equilibrium. But when they are displaced slightly from their present position, they do not return back. So, they are in unstable equilibrium position.

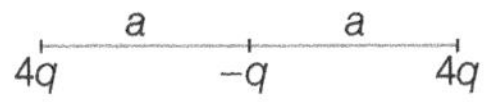

11. Q is in equilibrium for all value of q. So, we require condition of equilibrium of q.

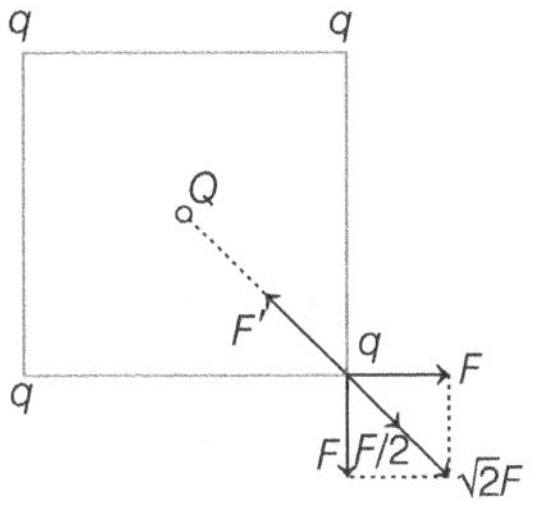

$$F = \frac{1}{4\pi\varepsilon_0} \frac{q^2}{a^2} \rightarrow \text{between } q \text{ and } q \text{ at distance } a$$

$\frac{F}{2}$ = force between q and q at distance $\sqrt{2}\,a$

F' = force between Q and q.

F' should be attractive. So, Q should be negatively charged. Magnitude wise.

$$F' = \sqrt{2}F + \frac{F}{2}$$

$$\therefore \quad \frac{1}{4\pi\varepsilon_0} \frac{Q \cdot q}{(a/\sqrt{2})^2} = \left(\sqrt{2} + \frac{1}{2}\right) \frac{1}{4\pi\varepsilon_0} \cdot \frac{q}{a^2}$$

$$|Q| = \frac{q}{4}(2\sqrt{2} + 1)$$

or $$Q = -\frac{q}{4}(2\sqrt{2} + 1)$$

12. Forces on q_1 are

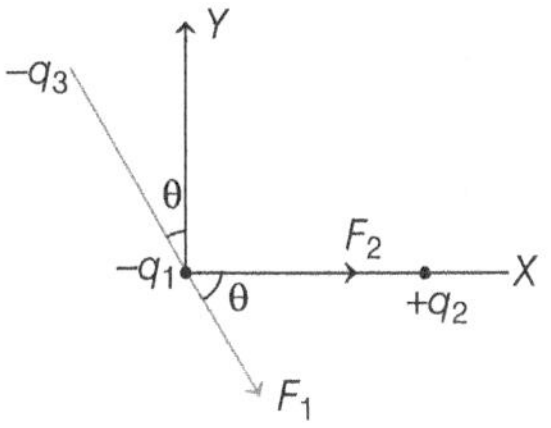

F_1 has a component along X-axis $= F_1 \cos\theta$

$$= \frac{kq_1 q_3}{a^2} \cos\theta$$

Also, F_2 is along X-axis, $F_2 = \frac{kq_1 q_2}{b^2}$

$\therefore$ Total force on q_2 along X-axis is

$$k\left(\frac{q_1 q_3}{a^2}\cos\theta + \frac{q_1 q_2}{b^2}\right) = kq_1\left(\frac{q_3}{a^2}\cos\theta + \frac{q_2}{b^2}\right)$$

13. Initially charges are placed in a medium of dielectric constant k.

$$F = \frac{q_1 q_2}{4\pi\varepsilon d^2} = \frac{q_1 q_2}{4\pi k \varepsilon_0 d^2} \quad \text{...(i)}$$

Here, $k = \frac{\varepsilon}{\varepsilon_0}$

Without dielectric, when placed in air, force between charges,

$$F' = \frac{1}{4\pi\varepsilon_0} \frac{q_1 q_2}{r^2} \quad \text{...(ii)}$$

On dividing Eq. (i) by Eq. (ii), we get

$$\frac{F}{F'} = \frac{1}{k} \times \frac{r^2}{d^2}$$

When $F = F'$, we get

$$1 = \frac{r^2}{kd^2}$$

$$r^2 = kd^2 \Rightarrow r = d\sqrt{k}$$

14. Force on charge Q

$$K = \frac{KQq_0}{(a-x)^2} - \frac{KQq_0}{(a+x)^2}$$

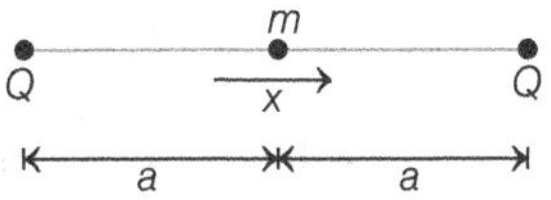

$$= KQq_0 \frac{4ax}{(a^2 - x^2)^2} \text{ (as } a << x)$$

So, $F = \dfrac{KQq_0 \times 4ax}{a^4} = \dfrac{4Kxq_0Q}{a^3}$

As, $F = m\omega^2 x$

So, $m\omega^2 = \dfrac{4KQq_0}{a^3}$

$$\Rightarrow \quad \omega = \sqrt{\frac{4KQq_0}{ma^3}}$$

$$\therefore \quad T = \sqrt{\frac{4\pi^3\varepsilon_0 ma^3}{q_0 Q}}$$

15. $F_{net} = 2F\cos\theta$

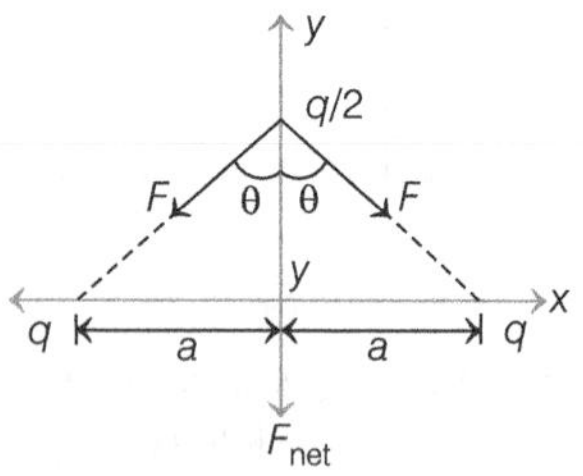

$$F_{net} = \frac{2kq\left(\dfrac{q}{2}\right)}{(\sqrt{y^2+a^2})^2}\cdot\frac{y}{\sqrt{y^2+a^2}}$$

or $$F_{net} = \frac{2kq\left(\dfrac{q}{2}\right)y}{(y^2+a^2)^{3/2}} \qquad (\because y << a)$$

$$= \frac{kq^2y}{a^3}, \text{ so } F \propto y$$

16. Let us consider the positive charge Q at a distance x from the origin. Let force between Q and $-q$ be F. On resolving this force $F\sin\theta$ cancels out. The resultant force $F_R = 2F\cos\theta$

$$= 2\times\frac{kQq}{(x^2+a^2)}\times\frac{x}{\sqrt{x^2+a^2}}$$

$$= \frac{2kQqx}{(x^2+a^2)^{3/2}}$$

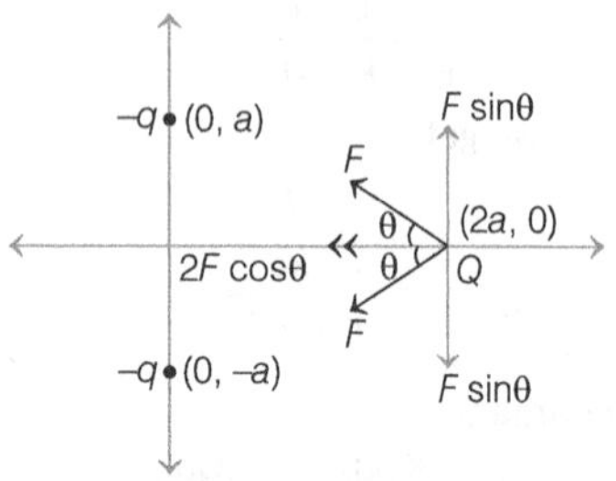

Since F_R is not proportional to x, the motion is not simple harmonic. The charge Q will accelerate till the origin and gain velocity. At the origin the net force is zero but due to momentum it will cross the origin and move towards left. As it comes on negative x-axis, the force is again towards the origin. Hence charge Q will execute oscillatory but not simple harmonic motion.

17. $F_e = \dfrac{1}{4\pi\varepsilon_0}\dfrac{q_1q_2}{r^2}$

and $F_g = G\dfrac{m_1 m_2}{r^2}$

$$\therefore \quad \frac{F_e}{F_g} = \frac{(1/4\pi\varepsilon_0)\,q_1q_2}{G\,m_1m_2}$$

$$= \frac{(9\times10^{+9})(1.6\times10^{-19})^2}{(6.67\times10^{-11})(9.11\times10^{-31})(1.67\times10^{-27})}$$

$$= 2.27\times10^{39}$$

18. $F = \dfrac{1}{4\pi\varepsilon_0}\dfrac{q_1q_2}{r^2}$

$$\therefore \quad \varepsilon_0 = \frac{q_1q_2}{4\pi F r^2}$$

Units and dimensions can be found by above equation.

19.

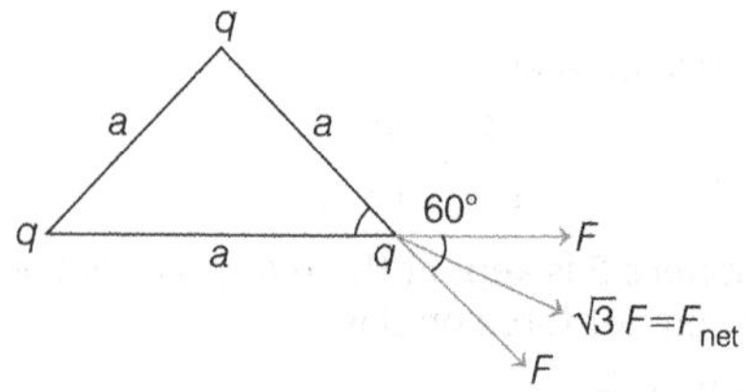

F = Force between two point charges

$$= \left(\frac{1}{4\pi\varepsilon_0}\right)\left(\frac{q\times q}{a^2}\right)$$

$$F_{net} = \sqrt{F^2+F^2+2FF\cos 60^\circ}$$

$$= \sqrt{3}F$$

$$= \left(\frac{\sqrt{3}}{4\pi\varepsilon_0}\right)\left(\frac{q}{a}\right)^2$$

20.

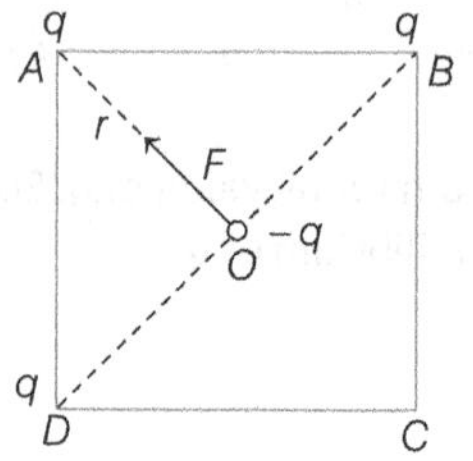

Net force on $-q$ from the charges at B and D is zero. So, net force on $-q$ is only due to the charge at A.

$$F = \left(\frac{1}{4\pi\varepsilon_0}\right)\frac{q\times q}{r^2}$$

where, $r = \dfrac{\sqrt{2}a}{2} = \dfrac{a}{\sqrt{2}}$

$$\therefore \quad F = \left(\frac{1}{4\pi\varepsilon_0}\right)\frac{q^2}{(a/\sqrt{2})^2}$$

$$= \left(\frac{1}{2\pi\varepsilon_0}\right)\left(\frac{q}{a}\right)^2$$

21. The charged body attracts the natural body because attraction (due to the distance factor) is more than the repulsion.

23. $F = \frac{1}{4\pi\varepsilon_0} \cdot \frac{q_1 q_2}{r^2}$

$$(q_1)_{min} = (q_2)_{min} = e_2$$

$$\therefore \quad F_{min} = \frac{1}{4\pi\varepsilon_0} \frac{e^2}{r^2}$$

25. Two forces are equal and opposite.

26. (a)

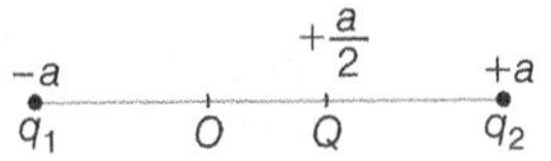

q_1 and q_2 should be of same sign.

Further, $\frac{Kq_1Q}{(3a/2)^2} = \frac{Kq_2Q}{(a/2)^2}$

$$\therefore \quad q_1 = 9q_2$$

(b)

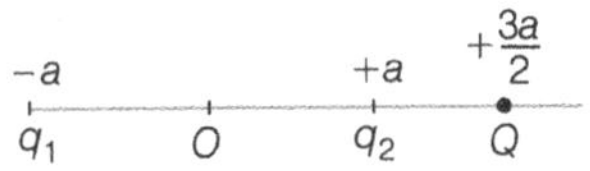

Therefore, q_1 and q_2 should be of opposite signs. Further,

$$\frac{kq_1Q}{(5a/2)^2} = \frac{kq_2Q}{(a/2)^2}$$

Or magnitude wise $q_1 = 25\, q_2$ with sign

$$q_1 = -25q_2$$

27.

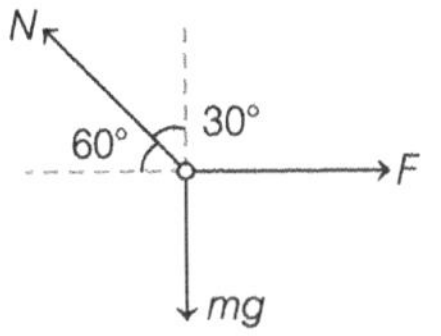

Using Lami's theorem, we have

$$\frac{mg}{\sin(90° + 30°)} = \frac{F}{\sin(90° + 60°)}$$

$$\therefore \quad F = \frac{mg}{\sqrt{3}}$$

or $\frac{1}{4\pi\varepsilon_0} \frac{(q)(q)}{R^2} = \frac{mg}{\sqrt{3}}$

$$\therefore \quad q = \sqrt{\frac{4\pi\varepsilon_0\, mg\, R^2}{\sqrt{3}}}$$

28.

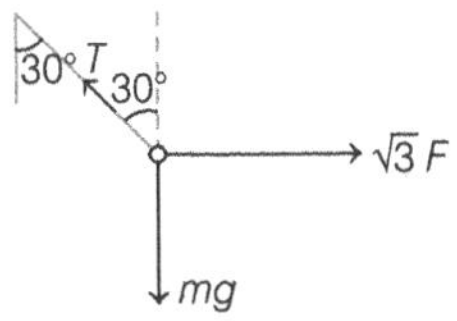

$F =$ Electrostatic force between two charged balls,
$\sqrt{3}F =$ Resultant of electrostatic force on any one ball from rest two balls.

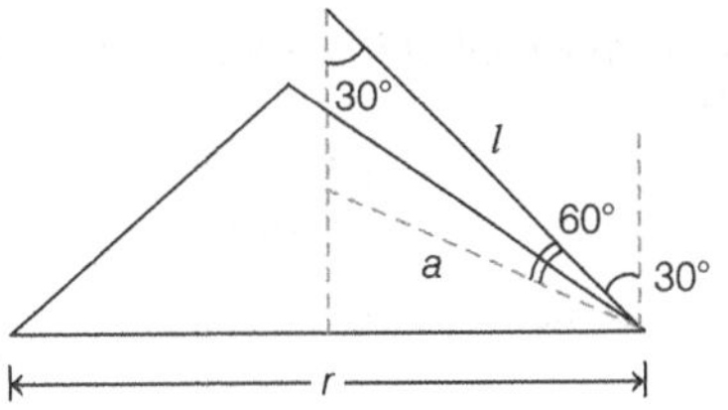

$$a = l \cos 60° = 10 \text{ cm}$$

$$r = 2\, a \cos 30°$$

$$= (2)(10)\left(\frac{\sqrt{3}}{2}\right) = 10\sqrt{3} \text{ cm}$$

Now, applying Lami's theorem for the equilibrium of ball, we have

$$\frac{mg}{\sin(90° + 30°)} = \frac{\sqrt{3}F}{\sin(90° + 60°)}$$

$$F = \frac{mg}{3} = \frac{1}{4\pi\varepsilon_0} \frac{(q)(q)}{r^2} = \frac{mg}{3}$$

$$\therefore \quad q = \sqrt{\frac{mgr^2}{3(1/4\pi\varepsilon_0)}}$$

$$= \sqrt{\frac{(0.1 \times 10^{-3})(9.8)(10\sqrt{3} \times 10^{-2})^2}{3 \times 9 \times 10^9}}$$

$$= 3.3 \times 10^{-8} \text{ C}$$

29. Given,

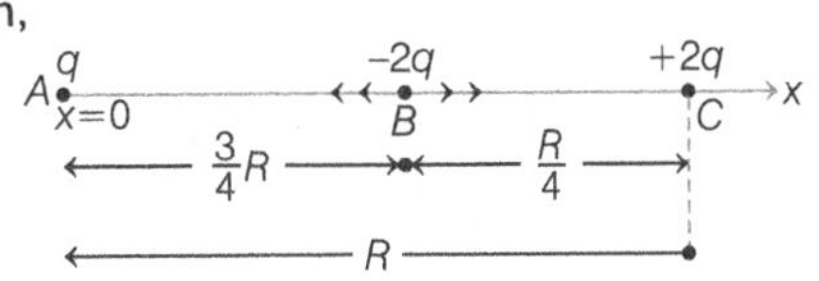

$$F_{BA} = \frac{kq \times 2q}{\left(\frac{3}{4}R\right)^2} \Rightarrow F_{BA} = \frac{16k \times 2q^2}{9R^2}$$

$$F_{BA} = \frac{32kq^2}{9R^2}$$

$$F_{BC} = \frac{k4q^2}{(R/4)^2} = \frac{64kq^2}{R^2}$$

$$F_B = F_{BC} - F_{BA} = \frac{64kq^2}{R^2} - \frac{32kq^2}{9R^2}$$

$$= \frac{kq^2}{R^2}\left(64 - \frac{32}{9}\right) = \frac{9 \times 10^9 \times (2 \times 10^{-6})^2}{(2 \times 10^{-2})^2} \times \frac{544}{9}$$

$$= 5440 \text{ N}$$

30. The situation given in question is as follows.

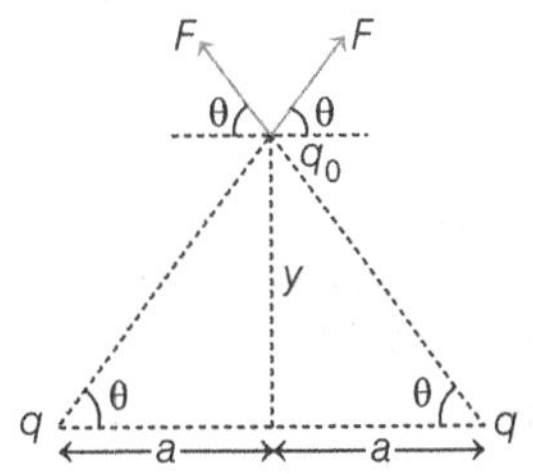

Resolving component of force acting on q_0 in x and y-direction.

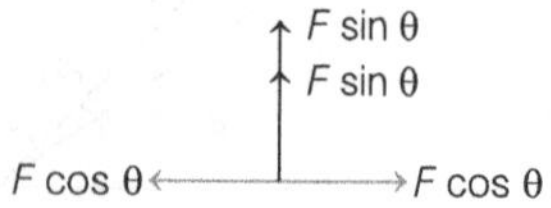

$$F_{net} = 2F\sin\theta$$
$$= \frac{2kqq_0}{(\sqrt{a^2+x^2})^2}\frac{y}{\sqrt{a^2+y^2}}$$
$$= \frac{2kqq_0 y}{(a^2+y^2)^{3/2}}$$

For maximum value of F_{net}, $\frac{dF_{net}}{dy} = 0$

$$2kqq_0\frac{d}{dy}\left(\frac{y}{(y^2+a^2)^{3/2}}\right) = 0$$
$$\Rightarrow \frac{d}{dy}\left(\frac{y}{(y^2+a^2)^{3/2}}\right) = 0$$
$$\Rightarrow (y^2+a^2)^{3/2}\frac{d}{dy}(y) - y\frac{d}{dy}(y^2+a^2)^{3/2} = 0$$
$$\Rightarrow (y^2+a^2)^{\frac{3}{2}}\cdot 1 - y\left\{\frac{3}{2}(y^2+a^2)^{\frac{1}{2}}\cdot 2y\right\} = 0$$
$$\Rightarrow (y^2+a^2)^{\frac{1}{2}}[(y^2+a^2) - 3y^2(1)] = 0$$
$$\Rightarrow (y^2+a^2)^{\frac{1}{2}}[a^2 - 2y^2] = 0$$
$$\Rightarrow a^2 - 2y^2 = 0 \text{ or } y = \frac{a}{\sqrt{2}} = \frac{a}{\sqrt{x}}$$

So, $x = 2$

Module 3.

Introductory Exercise

1. Electric lines flow higher potential to lower potential.

   $\therefore \quad V_A > V_B$

2. Electric field lines start from positive charge and terminate on negative charge.

3. In case of five charges at five vertices of regular pentagon net electric field at centre is zero. Because five vectors of equal magnitudes from a closed regular pentagon as shown in Fig. (i).

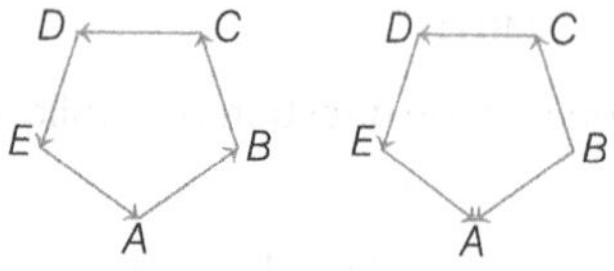

(i) (ii)

Where one charge is removed. Then, one vector (let **AB**) is deceased hence the net resultant is equal to magnitude of one vector

$$= |\mathbf{BA}| = \frac{1}{4\pi\varepsilon_0}\frac{q}{a^2}$$

4. $\mathbf{E} = \left(\frac{1}{4\pi\varepsilon_0}\right)\left(\frac{q}{r^3}\right)(\mathbf{r}_p - \mathbf{r}_q)$

   Here, $r = \sqrt{(3)^2 + (4)^2} = 5\text{ m}$

   $\therefore \quad \mathbf{E} = \frac{(9\times10^9)(-2\times10^{-6})}{(5)^3}(3\hat{i}+4\hat{j})$

   $= -(4.32\,\hat{i} + 5.76\,\hat{j})\times10^2\text{ N/C}$

Exercise

1. Electric field lines are not parallel and equidistant.

2. If charged particle is positive, and at rest. Electric field lines are straight then only it will move in the direction of electric field.

3. See the hint of above question.

4. $g_e = g - \left(\frac{qE}{m}\right)$

   or g_e will decrease. Hence, $T = 2\pi\sqrt{\frac{l}{g_e}}$ will increase.

5. Electric lines terminate on negative charge.

6. $|a| = \left|\frac{qE}{m}\right| = \left|\frac{q\sigma}{2\varepsilon_0 m}\right|$

   $t = \sqrt{\frac{2s}{a}} = \sqrt{\frac{4s\,\varepsilon_0 m}{|q||\sigma|}}$

   $= \sqrt{\frac{4\times0.1\times8.86\times10^{-12}\times1.67\times10^{-27}}{1.6\times10^{-19}\times2.21\times10^{-9}}}$

   $= 4\times10^{-6}\text{ s} = 4\,\mu\text{s}$

7. 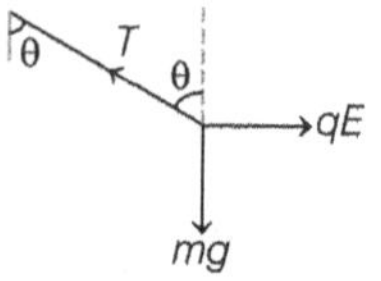

   $T\sin\theta = qE$

   or $\quad T = \frac{qE}{\sin\theta}$

   Similarly, $\quad T\cos\theta = mg$

8. Equal and opposite pair of electric fields are formed.

9. $L = mv\,r_\perp = m\,(at)\,(x_0)$

   $= m\left(\frac{qE_0}{m}\right)t\,(x_0) \quad \text{or} \quad L \propto t$

10. 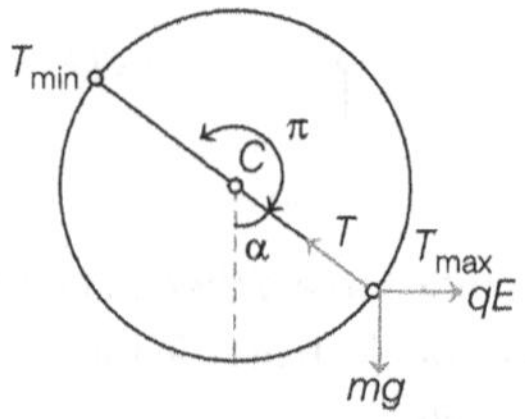

$$T \sin \alpha = qE$$
$$T \cos \alpha = mg$$
$$\therefore \qquad \alpha = \tan^{-1}\left(\frac{qE}{mg}\right)$$

Minimum tension will be obtained at $\alpha + \pi$.

11. $qE = \dfrac{mv^2}{r}$

$$q\left(\frac{\lambda}{2\pi\varepsilon_0 r}\right) = \frac{mv^2}{r}$$
$$\therefore \qquad v = \sqrt{\frac{q\lambda}{2\pi\varepsilon_0 m}} = \sqrt{\frac{2kq\lambda}{m}}$$
$$\text{Now,} \qquad T = \frac{2\pi r}{v}$$

12. Between $2q$ and $-q$, two electric fields are in same direction. So their resultant can't be zero. To the right of $2q$ left of $-q$ they are in opposite directions. So, net field will be zero nearer to charge having small magnitude.

13. $W = \mathbf{F} \cdot \mathbf{r}$

$$= (Q\mathbf{E}) \cdot \mathbf{r} = Q(\mathbf{E} \cdot \mathbf{r})$$
$$= Q(E_1 a + E_2 b)$$

14. If force is constant,

$$|\Delta \mathbf{p}| = |\mathbf{F}| t = qE_0 t$$

15. In figures (1), (3) and (4), net electric field is zero, because electric field at a point due to positive charge is away from the charge and due to negative charge it is towards the charge for Fig. (2), net electric field is not zero.

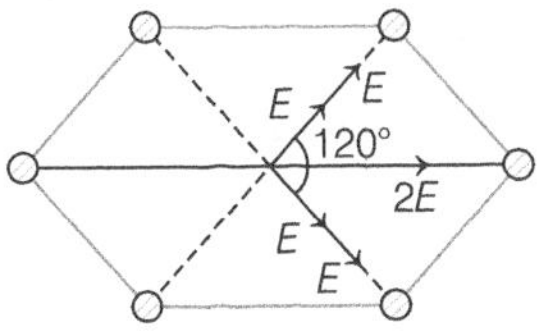

Here, net electric field in Fig. (2) is

$$= \sqrt{(2E)^2 + (2E)^2 + (2E)(2E) \cdot 2 \cos 120°} = 2E$$

16. Let E = magnitude of electric field at origin due to charge $\pm q$. Then,

(i)

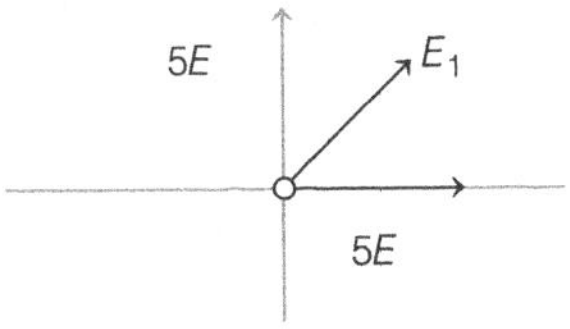

$$E_1 = \sqrt{(5E)^2 + (5E)^2} = 5\sqrt{2}E$$

(ii)

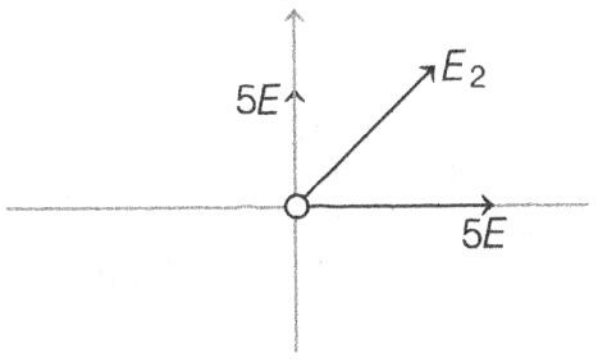

E_2 is again $5\sqrt{2}E$.

Similarly, we can find E_3 and E_4 also.

17. $\mathbf{E}_1 \perp \mathbf{E}_2 \Rightarrow \mathbf{E}_1 \cdot \mathbf{E}_2 = 0$

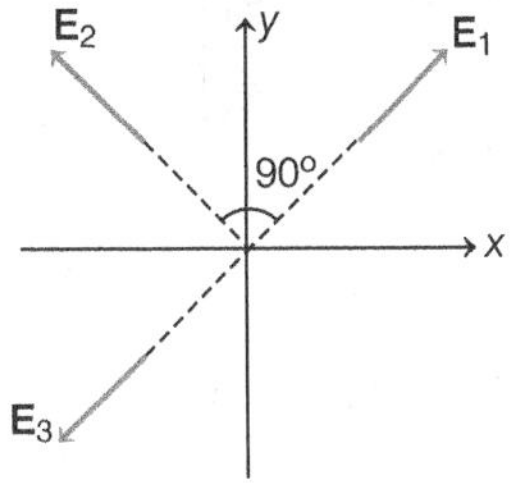

$$\therefore \qquad \mathbf{E}_1 \downarrow\uparrow \mathbf{E}_3$$
$$\Rightarrow \qquad |\mathbf{E}_1 \times \mathbf{E}_3| = 0$$

18. $E \propto \dfrac{q}{r^2}$

$$\Rightarrow \qquad \tan\theta = \frac{E_0}{E_0/2} = 2$$

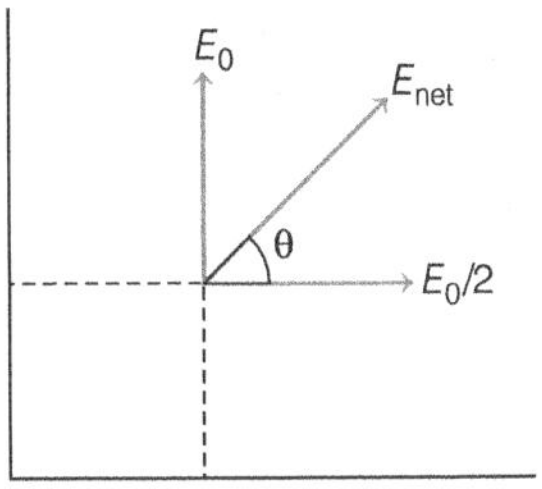

19. According to option (d) the electric field due to P and S and due to Q and T add to zero. While due to U and R will be added up. Hence, the correct option is (d).

20. Electric field intensity due to an infinite thin plane sheets of uniform surface charge density σ is given as

$$\mathbf{E} = \frac{\sigma}{2\varepsilon_0}\hat{\mathbf{n}}$$

Situation given is shown below.

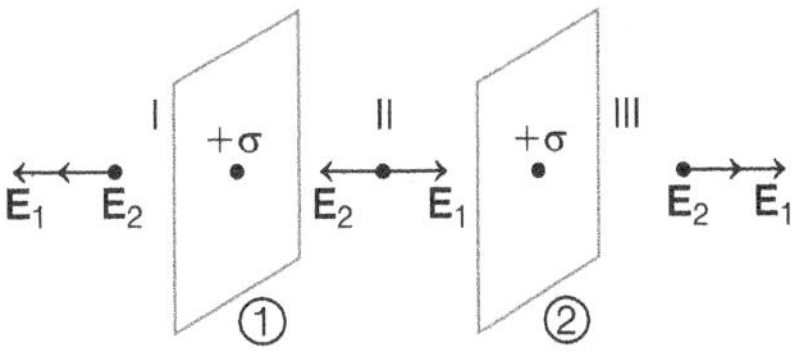

In region I, net field,

$$\mathbf{E}_\text{I} = \mathbf{E}_1 + \mathbf{E}_2$$
$$\mathbf{E}_\text{I} = \left(-\frac{\sigma}{2\varepsilon_0}\hat{\mathbf{n}}\right) + \left(-\frac{\sigma}{2\varepsilon_0}\hat{\mathbf{n}}\right) = -\frac{\sigma}{\varepsilon_0}\hat{\mathbf{n}}$$

Similarly, in region II and III,

$$\mathbf{E}_\text{II} = \left(-\frac{\sigma}{2\varepsilon_0}\hat{\mathbf{n}}\right) + \left(\frac{\sigma}{2\varepsilon_0}\hat{\mathbf{n}}\right) = 0$$
$$\mathbf{E}_\text{III} = \left(\frac{\sigma}{2\varepsilon_0}\hat{\mathbf{n}}\right) + \left(\frac{\sigma}{2\varepsilon_0}\hat{\mathbf{n}}\right) = \frac{\sigma}{\varepsilon_0}\hat{\mathbf{n}}$$

21. The direction of electric fields are according to the charge on the sheets i.e. along negative z-axis.

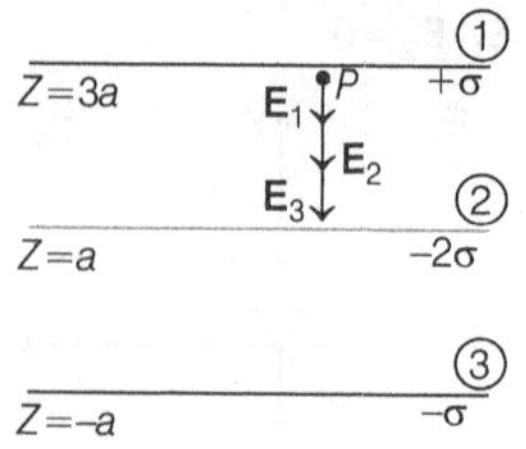

The total electric field at point P

$$\mathbf{E} = \mathbf{E}_1 + \mathbf{E}_2 + \mathbf{E}_3 = E_1(-\hat{\mathbf{k}}) + E_2(-\hat{\mathbf{k}}) + E_3(-\hat{\mathbf{k}})$$

$$= \left[\frac{\sigma}{2\varepsilon_0} + \frac{2\sigma}{2\varepsilon_0} + \frac{\sigma}{2\varepsilon_0}\right](-\hat{\mathbf{k}}) = -\frac{2\sigma}{\varepsilon_0}\hat{\mathbf{k}}$$

22. (a) $\mathbf{E_C}$ and $\mathbf{E_F}$ are cancelled. $\mathbf{E_E}$ and $\mathbf{E_D}$ at 60°

(b) $\mathbf{E_B}$ and $\mathbf{E_E}$ are cancelled. $\mathbf{E_F}$ and $\mathbf{E_D}$ at 120°.

(c) $\mathbf{E_B}$ and $\mathbf{E_E}$ are cancelled. Similar, $\mathbf{E_F}$ and $\mathbf{E_C}$ are cancelled.

(d) $\mathbf{E_F}$ and $\mathbf{E_D}$ at 120°. So, their resultant is E in the direction of $\mathbf{E_E}$. Hence, net is $2E$.

23.

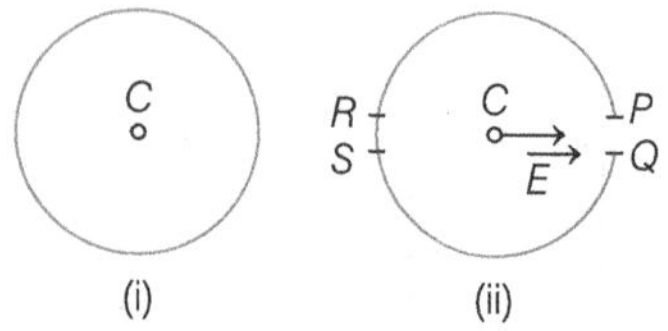

If loop is complete, then net electric field at centre C is zero. Because equal and opposite pair of electric field vectors are cancelled.

If PQ portion is removed as shown in figure, then electric field due to portion RS is not cancelled. Hence, electric field is only due to the option RS.

$$\therefore \quad E = \frac{1}{4\pi\varepsilon_0}\frac{q_{RS}}{R^2} = \left(\frac{1}{4\pi\varepsilon_0}\right)\frac{(q/2\pi R)\,x}{R^2}$$

$$= \frac{qx}{8\pi^2\varepsilon_0 R^3}$$

24.

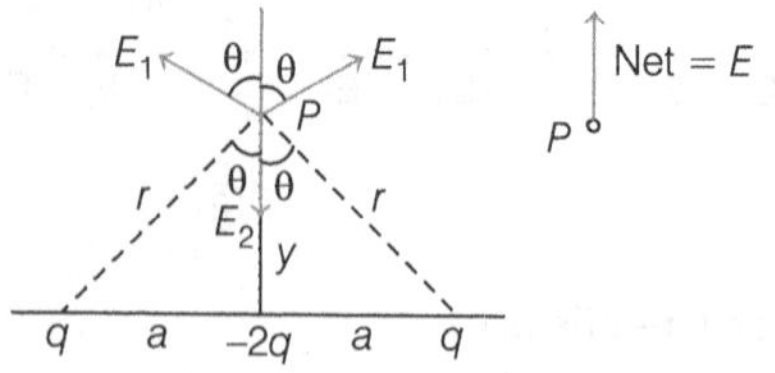

$E_1 =$ Electric field at P due to q

$$= \left(\frac{1}{4\pi\varepsilon_0}\right)\frac{q}{r^2}$$

$E_2 =$ Electric field at P due to $-2q$

$$= \left(\frac{1}{4\pi\varepsilon_0}\right)\frac{2q}{y^2}$$

$\therefore$ Net electric field at P,

$$E = 2E_1\cos\theta - E_2$$

$$= 2\left(\frac{1}{4\pi\varepsilon_0}\right)\left(\frac{q}{r^2}\right)\left(\frac{y}{r}\right) - \left(\frac{1}{4\pi\varepsilon_0}\right)\left(\frac{2q}{y^2}\right)$$

$$= \frac{2q}{4\pi\varepsilon_0}\left\{\left[\frac{y}{r^3}\right] - \left[\frac{1}{y^2}\right]\right\}$$

$$= \frac{2q}{4\pi\varepsilon_0}\left[\frac{y}{(a^2+y^2)^{3/2}} - \frac{1}{y^2}\right]$$

$$= \frac{2q}{4\pi\varepsilon_0}\left[\frac{y}{y^3}\left(1 + \frac{a^2}{y^2}\right)^{-3/2} - \frac{1}{y^2}\right]$$

Applying binomial expression for $\frac{a^2}{y^2} << 1$ we get,

$$E = \frac{2q}{4\pi\varepsilon_0}\left(-\frac{3}{2}\frac{a^2}{y^4}\right) = -\frac{3qa^2}{4\pi\varepsilon_0 y^4}$$

or $$\mathbf{E} = -\frac{3qa^2}{4\pi\varepsilon_0 y^4}\hat{\mathbf{j}}$$

25.

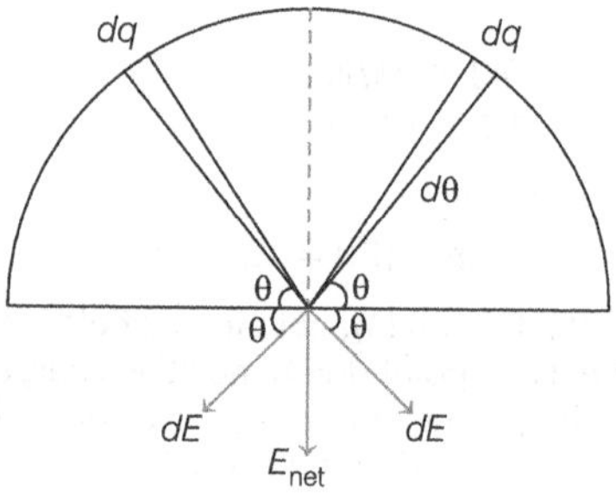

$$E_{net} = 2\int_0^{90°} dE\sin\theta$$

$$= 2\int_0^{90°}\left(\frac{1}{4\pi\varepsilon_0}\right)\left(\frac{dq}{R^2}\right)\sin\theta$$

$$= 2\int_0^{90°}\left(\frac{1}{4\pi\varepsilon_0}\right)\left(\frac{\lambda R d\theta}{R^2}\right)\sin\theta = \frac{\lambda}{2\pi\varepsilon_0 R}$$

26.

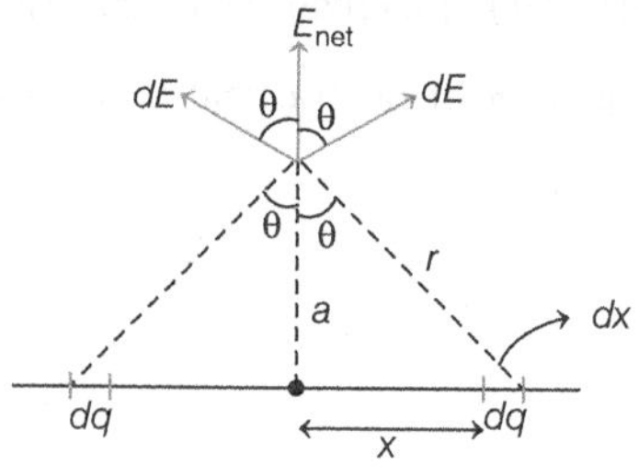

$$E_{net} = 2\int_{x=0}^{x=L/2} dE\cos\theta s$$

$$= 2\int_0^{L/2}\left(\frac{1}{4\pi\varepsilon_0}\right)\left(\frac{dq}{r^2}\right)\cdot\left(\frac{a}{r}\right)$$

$$= 2\int_0^{L/2}\left(\frac{1}{4\pi\varepsilon_0}\right)\left(\frac{\frac{Q}{L}\cdot dx}{a^2+x^2}\right)\frac{a}{\sqrt{a^2+x^2}}$$

Solving this integration, we find the result.

27. (a)

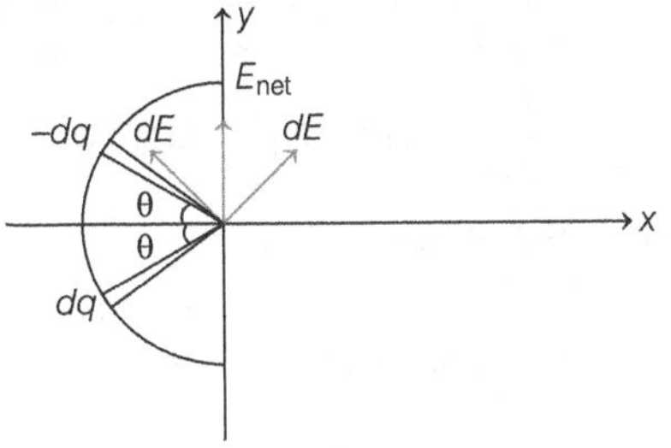

(b)

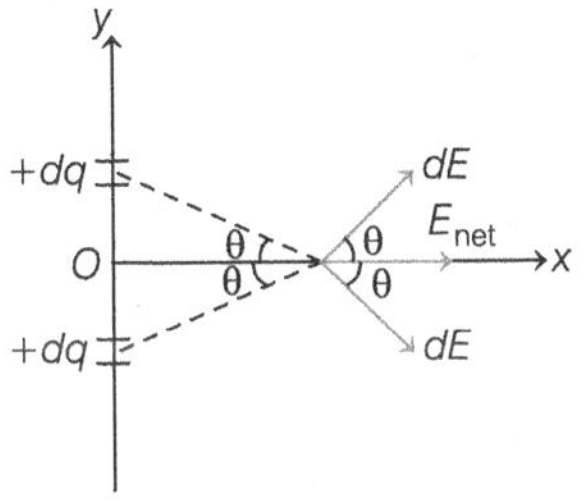

(c)

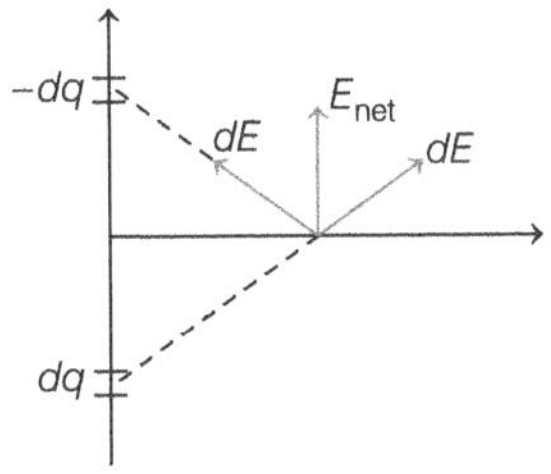

28. Six vectors of equal magnitudes are as shown in figure.

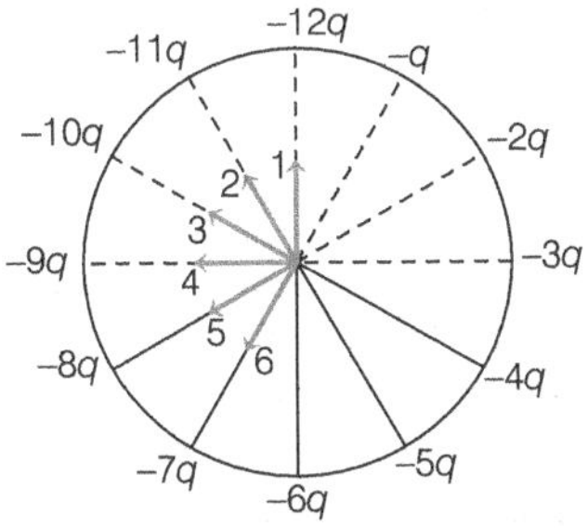

Now, resultant of two vectors of equal magnitudes ($=E$ say) at 120° is also E and passing through their bisector line.

So, resultant of 1 and 5 is also E in the direction of 3. Similarly, resultant of 2 and 6 is also E in the direction of 4.

Finally, resultant of E in the direction of 3 and E in the direction of 4 passes through the bisector line of 3 and 4 (or 9:30).

29. (a) $g_e = g + \dfrac{qE}{m}$

$$= 10 + \frac{(1.6\times10^{-19})(720)}{1.67\times10^{-27}}$$

$$\approx 6.9\times10^{100}\ \text{m/s}^2$$

Now, $$R = \frac{u^2 \sin 2\theta}{g_e}$$

$$\therefore \quad (1.27\times10^{-3}) = \frac{(9.55\times10^3)^2 \sin 2\theta}{6.9\times10^{10}}$$

Solving this equation, we get

$$\theta = 37° \text{ and } 53°$$

(b) Apply $T = \dfrac{2u\sin\theta}{g_e}$

30. (a) $\mathbf{a} = \dfrac{q\mathbf{E}}{m} = -\dfrac{(1.6\times10^{-19})(120\hat{\mathbf{j}})}{(9.1\times10^{-31})}$

$$= -(2.1\times10^{13}\hat{\mathbf{j}})\ \text{m/s}^2$$

(b) $t = \dfrac{X}{v_x} = \dfrac{2\times10^{-2}}{1.5\times10^5} = \dfrac{4}{3}\times10^{-7}\ \text{s}$

$\mathbf{v} = \mathbf{u} + \mathbf{a}t$

$$= (1.5\times10^5\hat{\mathbf{i}} + 3.0\times10^6\hat{\mathbf{j}}) - \left(2.1\times10^{13}\times\frac{4}{3}\times10^{-7}\right)\hat{\mathbf{j}}$$

$$= (1.5\hat{\mathbf{i}} + 2\hat{\mathbf{j}})\times10^5\ \text{m/s}$$

Module 4.

1. Two unlike charges come together when left freely. Conservative forces always act towards decreasing value of potential energy.

2. $W = \Delta U_i$

$$= U_f - U_i$$

$$= 3\left[\frac{1}{4\pi\varepsilon_0}\cdot\frac{qq}{2l}\right] - 3\left[\frac{1}{4\pi\varepsilon_0}\cdot\frac{qq}{l}\right]$$

$$= \left(-\frac{3}{2}\right)\left(\frac{1}{4\pi\varepsilon_0}\cdot\frac{q^2}{l^2}\right)l$$

$$= -\frac{3}{2}Fl$$

3.

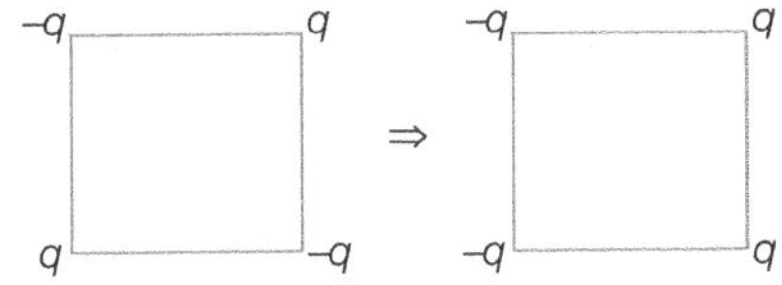

$$W = \Delta U = U_f - U_i$$

$$= \frac{1}{4\pi\varepsilon_0}\left[\left(\frac{-q^2}{a} - \frac{q^2}{\sqrt{2}\,a} + \frac{q^2}{a} + \frac{q^2}{a} - \frac{q^2}{\sqrt{2}\,a} - \frac{q^2}{a}\right) - \left(\frac{-q^2}{a} + \frac{q^2}{\sqrt{2}\,a} - \frac{q^2}{a} - \frac{q^2}{a} + \frac{q^2}{\sqrt{2}\,a} - \frac{q^2}{a}\right)\right]$$

$$= \frac{q^2}{4\pi\varepsilon_0 a}[4 - 2\sqrt{2}]$$

4. 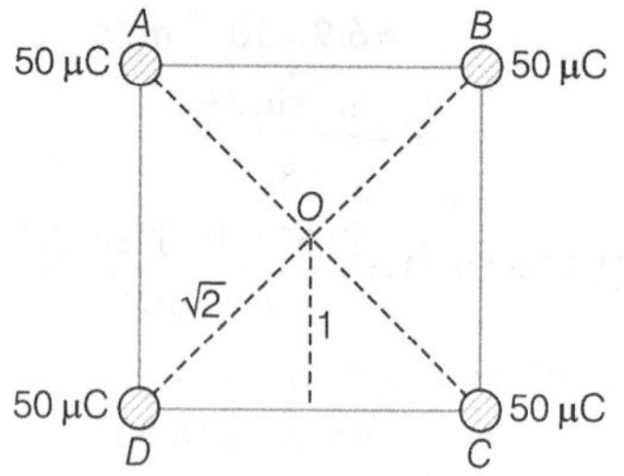

$$V_A = V_B = V_C = V_D$$
$$= \frac{1}{4\pi\varepsilon_0} \cdot \frac{(50 \times 10^{-6})}{(\sqrt{2})}$$

$\therefore$ Potential at the centre of square,

$$V_0 = 4 \times \left[\frac{9 \times 10^9 \times 50 \times 10^{-6}}{\sqrt{2}}\right]$$
$$= 90\sqrt{2} \times 10^4 \text{ V}$$

Work done in bringing a charge ($q = 50\,\mu$C) from ∞ to centre of the square is

$$W = qV_0 = 50 \times 10^{-6} \times 90\sqrt{2} \times 10^4 = 64 \text{ J}$$

5. $K_i + U_i = K_f + U_f$

$$\frac{1}{2}mv^2 + 0 = 0 + \frac{1}{4\pi\varepsilon_0}\frac{Q^2}{r_{min}}$$

$$\Rightarrow \quad r_{min} = \frac{Q^2}{2\pi\varepsilon_0 mv^2}$$

6. Energy required $= \Delta U = U_f - U_i$

$$= \frac{1}{4\pi\varepsilon_0}\left[\left(\frac{q^2}{a}\right) - \left(\frac{q^2}{a} - \frac{q^2}{\sqrt{2}a} - \frac{q^2}{a} - \frac{q^2}{a} - \frac{q^2}{\sqrt{2}a} + \frac{q^2}{a}\right)\right]$$

$$= \frac{q^2}{4\pi\varepsilon_0 a}[\sqrt{2} + 1]$$

7. $U = \frac{1}{4\pi\varepsilon_0} \cdot \frac{q^2}{a}$ (a = side of triangle)

$$W = U_f - U_i = 3\left(\frac{1}{4\pi\varepsilon_0}\frac{q^2}{a}\right) - U$$
$$= 3U - U = 2U$$

8. $\Delta U = U_f - U_i$

$$= \left[\frac{kq_1q_2}{0.3} + \frac{kq_1q_3}{0.4} + \frac{kq_2q_3}{0.1}\right] - \left[\frac{kq_1q_2}{0.3} + \frac{kq_1q_3}{0.4} + \frac{kq_2q_3}{0.5}\right]$$
$$= 8\,kq_2q_3$$

9. $W = U_f - U_i$

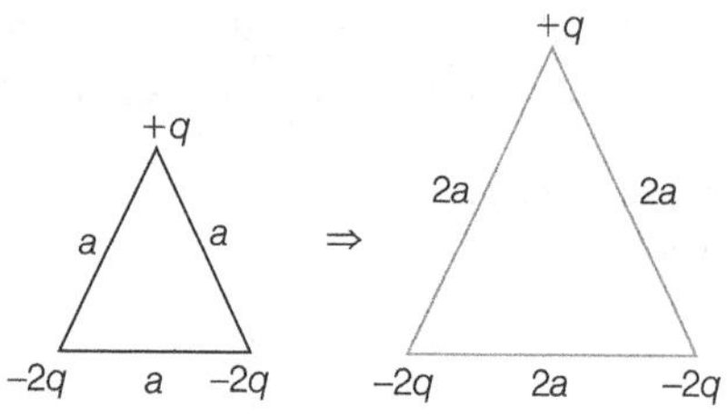

Here, $U_i = \frac{1}{4\pi\varepsilon_0}\left[\frac{q(-2q)}{a} + \frac{q(-2q)}{a} + \frac{(-2q)(-2q)}{a}\right]$

$$= \frac{1}{4\pi\varepsilon_0 a}[-2q^2 - 2q^2 + 4q^2] = 0$$

Similarly, U_f is also zero.

Hence, $W = 0$

10. Net electrostatic energy of the configuration will be

$$U = k\left[\frac{q \cdot q}{a} + \frac{Q \cdot q}{\sqrt{2}a} + \frac{Q \cdot q}{a}\right]$$

Here, $k = \frac{1}{4\pi\varepsilon_0}$

Putting $U = 0$, we get $Q = \frac{-2q}{2 + \sqrt{2}}$

11. 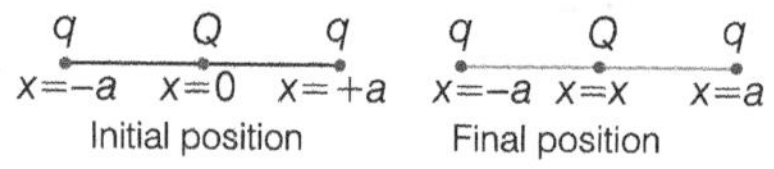

$$U_i = \frac{2kQq}{a} + \frac{k.q.q}{2a}$$

and $U_f = kQq\left[\frac{1}{a+x} + \frac{1}{a-x}\right] + \frac{k.q.q}{2a}$

Here, $k = \frac{1}{4\pi\varepsilon_0}$

$$\Delta U = U_f - U_i$$

or $|\Delta U| = \frac{2kQqx^2}{a^3}$

For $x << a$

$\therefore$ $\Delta U \propto x^2$

12. Along the line AB, charge q is at unstable equilibrium position at B (When displaced from B along AB, net force on it is away from B, whereas force at B is zero). Hence, potential energy at B is maximum.

Along CD equilibrium of q is stable. Hence, potential energy at B is minimum along CD.

13. At C, KE of charge $-q = K_C = 4$ J

and PE of charge $-q$

$$U_C = 2\left(\frac{1}{4\pi\varepsilon_0} \cdot \frac{q(-q)}{AC}\right)$$
$$= \frac{-2 \times 9 \times 10^9 \times (5 \times 10^{-5})^2}{5} = -9 \text{ J}$$

$\therefore$ Total energy at $C = K_C + U_C$

$$= 4 \text{ J} + (-9 \text{ J}) = -5 \text{ J}$$

14. Total energy of charge $-q$ is conserved.

$\therefore$ Total energy at C = total energy at D

$\therefore$ $E_D = -5$ J

15. At D, energy of charges is only potential, it is on the verge of turning back.

$\therefore$ KE = 0

So PE at $D = U_D = -5$ J

16. Also, $K_C + U_C = K_D + U_D$

$\Rightarrow$ $-5 = 0 + 2\left(\frac{1}{4\pi\varepsilon_0}\frac{q(-q)}{AD}\right)$

$\Rightarrow$ $-5 = \frac{2 \times 8.99 \times 10^9 \times (5 \times 10^{-5})^2}{AD}$

$\Rightarrow$ $-5 \approx \frac{-45}{AD}$

$\Rightarrow$ $AD \approx 9$ m

and $OD = \sqrt{AD^2 - OA^2}$

$\sqrt{9^2 - 3^2} = \sqrt{81 - 9}$

$= 8.48$ m

17. Work done by electrostatic forces

$W = -\Delta U = U_i - U_f$

$\therefore$ $U_f = U_i - W$

$= (-6.4 \times 10^{-8}) - (4.2 \times 10^{-8})$

$= -10.6 \times 10^{-8}$ J

18. $U_\infty = 0$

$U_r = \frac{1}{4\pi\varepsilon_0}\cdot\frac{q_1 q_2}{r}$ (For two charges)

$U_r \neq U_\infty$

For $r \neq \infty$

$U_r = \frac{1}{4\pi\varepsilon_0}\left[\frac{q_1 q_2}{r_{12}} + \frac{q_2 q_3}{r_{23}} + \frac{q_3 q_1}{r_{31}}\right]$

Now, U_r can be equal to U_∞ for finite value of r.

19. (a) $U = k\left(\frac{q_1 q_2}{r_{12}} + \frac{q_1 q_3}{r_{13}} + \frac{q_2 q_3}{r_{23}}\right)$ ...(i)

where, $k = \frac{1}{4\pi\varepsilon_0}$

(b) Suppose q_3 is placed at coordinate x (> 0.2 m or 20 cm), then in Eq. (i) of part (a), put

$U = 0,\ r_{12} = 0.2$ m, $r_{13} = x$

and $r_{23} = (x - 0.2)$

Now, solving Eq. (i) we get the desired value of x.

20. $U = 0$

$\therefore$ $k\left(\frac{q \times q}{d} + \frac{q \times Q}{d} + \frac{q \times Q}{d}\right) = 0$

or $Q = -\frac{q}{2}$

Module 5.

Introductory Exercise

1. $W_{ab} = \Delta U_{a-b} = U_b - U_a = q(V_b - V_a) = qV_{ba}$

$\therefore V_{ba} = \frac{W_{ab}}{q} = \frac{12}{10^{-2}} = 1.2 \times 10^3$ V

2. (a) $\alpha = \frac{\lambda}{x} = \frac{\text{C/m}}{\text{m}} = \frac{\text{C}}{\text{m}^2}$

(b)

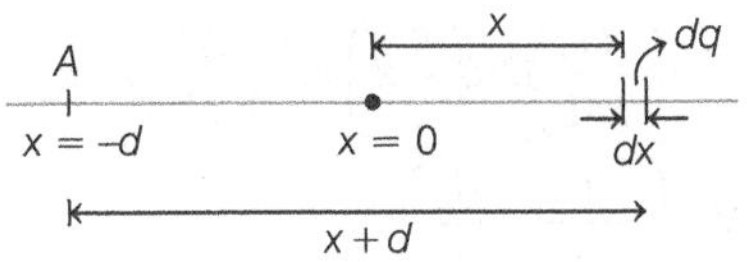

$dV = \frac{1}{4\pi\varepsilon_0}\cdot\frac{dq}{x + d}$

$= \left(\frac{1}{4\pi\varepsilon_0}\right)\left(\frac{\lambda dx}{x + d}\right)$

$= \left(\frac{1}{4\pi\varepsilon_0}\right)\left(\frac{\alpha x\, dx}{x + d}\right)$

$\therefore$ $V = \int_{x=0}^{x=L} dV = \frac{\alpha}{4\pi\varepsilon_0}\left[L - d\ln\left(1 + \frac{L}{d}\right)\right]$

3.

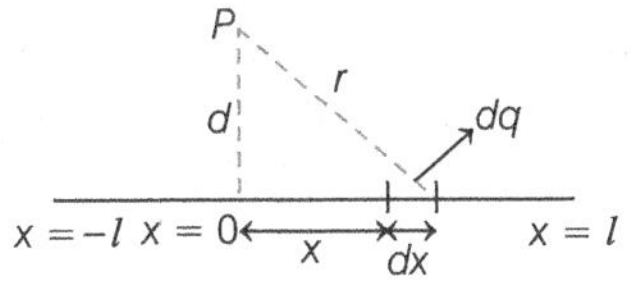

$dq = \left(\frac{q}{2l}\right)dx$

At point P,

$dV = \left(\frac{1}{4\pi\varepsilon_0}\right)\left(\frac{dq}{r}\right)$

$= \left(\frac{1}{4\pi\varepsilon_0}\right)\left(\frac{\frac{q}{2l}\cdot dx}{\sqrt{d^2 + x^2}}\right)$

$\therefore$ $V = 2\int_{x=0}^{x=l} dV$

4.

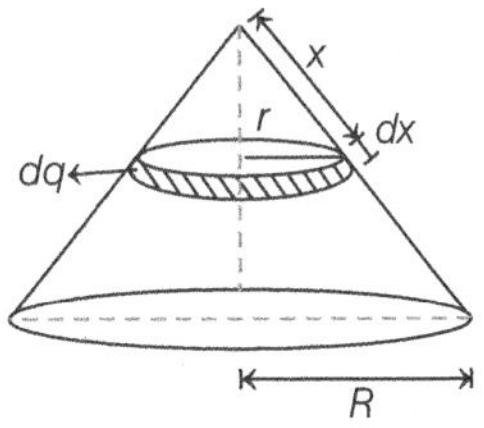

$W = -\Delta U = U_{apex} - U_\infty$

$\therefore$ $W = qV_{apex}$ ...(i)

as $U_\infty = 0$

V_{apex} $\frac{r}{x} = \frac{R}{L}$

$\therefore$ $r = \left(\frac{R}{L}\right)x$

Surface charge density,

$\sigma = \frac{Q}{\pi Rl}$

$dq = (\sigma)(dA) = \left(\frac{Q}{\pi Rl}\right)(2\pi r)\,dx$

$$=\left(\frac{Q}{\pi RL}\right)(2\pi)\left(\frac{R}{L}x\right)dx=\left(\frac{2Q}{L^2}\right)x\,dx$$

Now, $dV=\left(\frac{1}{4\pi\varepsilon_0}\right)\left(\frac{dq}{x}\right)$

$$=\left(\frac{2Q}{4\pi\varepsilon_0 L^2}\right)dx$$

$\therefore$ $V=\int_0^L dV=\frac{Q}{2\pi\varepsilon_0 L}$

Substituting in Eq. (i), we have

$$W=\frac{Qq}{2\pi\varepsilon_0 L}$$

Exercise

1. An independent negative charge moves from lower potential to higher potential. In this process, electrostatic potential energy decreases and kinetic energy increases.

2. $E\propto\frac{1}{r^2}$ and $V\propto\frac{1}{r}$

3. $V=\frac{1}{4\pi\varepsilon_0}\left[\frac{2q}{r}-\frac{q}{r}-\frac{q}{r}\right]=0$

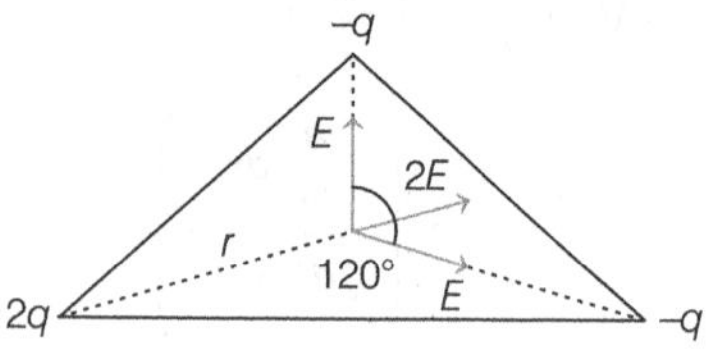

Resultant of these three vectors is 3*E* in the direction of 2*E*, where

$$E=\frac{1}{4\pi\varepsilon_0}\frac{q}{r^2}$$

4.

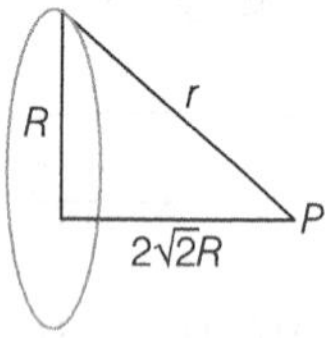

$$r=3R$$

$$V_p=\frac{kq_{net}}{r}=\frac{k(3q)}{3R}=\frac{kq}{R}$$

5.

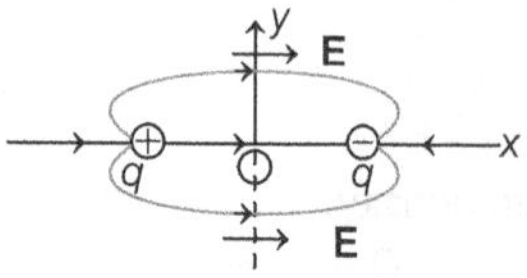

Potential is zero at infinite and at origin. Therefore, PD = 0. Hence, the work done asked in part (c) is also zero.

6. $500=\frac{k|q|}{r^2}$ ...(i)

$$-3000=\frac{k(-q)}{r}$$ ...(ii)

Solving these two equations, we get

$$r=6\text{ m}$$

$\therefore$ $|q|=\frac{500\,r^2}{k}=\frac{(500)(6)^2}{9\times10^9}$

$$=2\times10^{-6}\text{ C}$$

$$=2\ \mu\text{C}$$

7. $r=\sqrt{(4-1)^2+(2-2)^2+(0-4)^2}$

$$=5\text{ m}$$

$$V=\frac{1}{4\pi\varepsilon_0}\cdot\frac{q}{r}$$

$$=\frac{(9\times10^9)(2\times10^{-8})}{5}=36\text{ V}$$

Field is in the direction of $\mathbf{r}=\mathbf{r}_p-\mathbf{r}_q$

8. When charged particle enters in a potential field, then

$$\frac{1}{2}mv^2=qV$$

$\therefore$ $v=\sqrt{\left(\frac{2qV}{m}\right)}$

$$v\propto\sqrt{q}$$

$\therefore$ $\frac{v_A}{v_B}=\sqrt{\frac{q}{4q}}=\frac{1}{2}$

9.

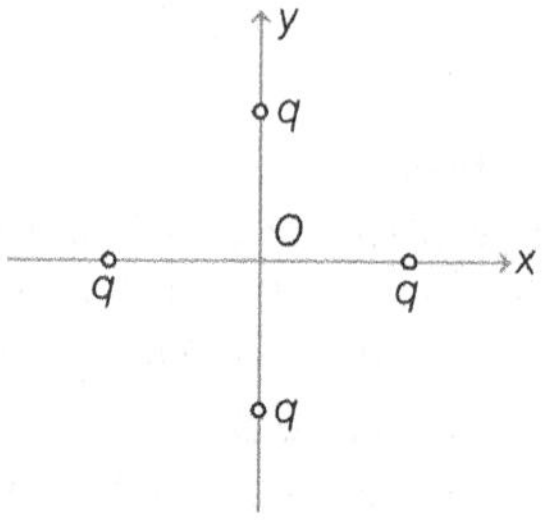

$$E_0=0,\ V_0=4\left(\frac{1}{4\pi\varepsilon_0}\frac{q}{1}\right)$$

$E_{(0,\,0,\,z)}$ is along *z*-axis.

10. $W=\Delta U$

$$=U_f-U_i$$

$$=QV_f-QV_i$$

$$=Q\left[\frac{1}{4\pi\varepsilon_0}\cdot\frac{q}{3L}-\frac{1}{4\pi\varepsilon_0}\cdot\frac{q}{L}\right]=\frac{-Qq}{6\pi\varepsilon_0 L}$$

Note *Potential due to –q at B is same at points C and D. So, we have taken change only due to +q*

11. Potential at all points on line *PR* will be zero. Because positive and equal negative charge is at same distance at every point. This is not the case with line *ST*.

12. Over Q_1, potential is $+\infty$. Hence, Q_1 is positive.

$V_A = 0$ and A point is nearer to Q_2. Therefore, Q_2 should be negative and $|Q_1| > |Q_2|$.

At A and B, potential is zero, not the force.

Equilibrium at C will depend on the nature of charge which is kept at C.

13. V_1 is positive and V_2 is negative. Hence at all points,

$$V_1 > V_2$$

14. Just to the right of q_1, electric field is $+\alpha$ or in positive direction (away from q_1). Hence, q_1 is positive. Just to the left of q_2, electric field is $-\alpha$ or towards left (or away from q_2). Hence, q_2 is also positive.

Further $E = 0$ near q_1. Hence,

$$q_1 < q_2$$

15.

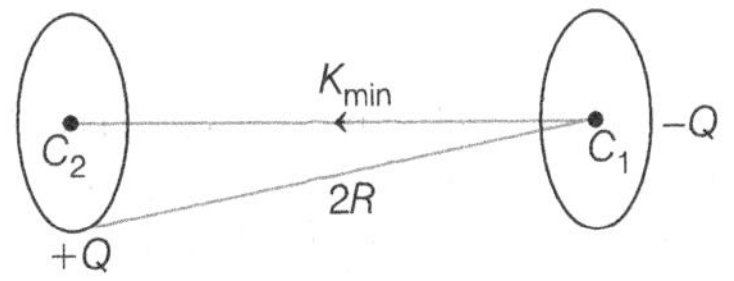

$$k_{C_1} + U_{C_1} = k_{C_2} + U_{C_2}$$

$$k_{min} + qV_{C_1} = 0 + qV_{C_2} \quad \text{...(i)}$$

$$V_{C_1} = \frac{1}{4\pi\varepsilon_0}\left(\frac{Q}{2R} - \frac{Q}{R}\right)$$

$$V_{C_2} = \frac{1}{4\pi\varepsilon_0}\left(\frac{Q}{R} - \frac{Q}{2R}\right)$$

Substituting these values in Eq. (i), we can find K_{min}.

16. On both sides of the positive charge $V = +\infty$ just over the charge.

17. $$V_p = 2\left(\frac{1}{4\pi\varepsilon_0}\frac{q}{a}\right) - 2\left[\frac{1}{4\pi\varepsilon_0}\cdot\frac{q}{\sqrt{b^2 + a^2}}\right]$$

$$= \frac{q}{4\pi\varepsilon_0 a}\left[2 - \left(2\left(1 + \frac{b^2}{a^2}\right)^{-1/2}\right)\right]$$

Since, $b << a$, we can apply binomial expansion

$\therefore$ $$V_p = \frac{q}{4\pi\varepsilon_0 a}\left[2 - \left(2\left(1 - \frac{b^2}{2a^2}\right)\right)\right]$$

$$= \frac{qb^2}{4\pi\varepsilon_0 a^3}$$

18. A negative charge when moves from higher potential to lower potential, its velocity increases.

19. $$\frac{kq_1}{r_1} = \frac{k|q_2|}{r_2}$$

$\therefore$ $$\frac{r_1}{r_2} = \frac{q_1}{|q_2|} = \frac{2}{3}$$

where, r_1 = distance from q_1 and

r_2 = distance from q_2

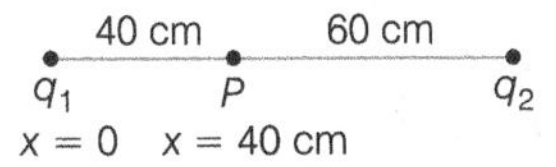

First point is at $x = 40$ cm where, $\frac{r_1}{r_2} = \frac{2}{3}$

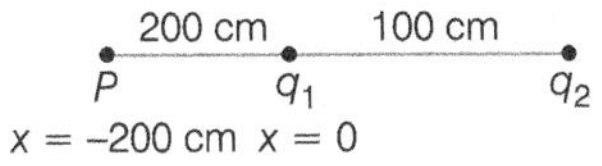

Second point is at $x = -200$ cm, where $\frac{r_1}{r_2} = \frac{2}{3}$..

20.

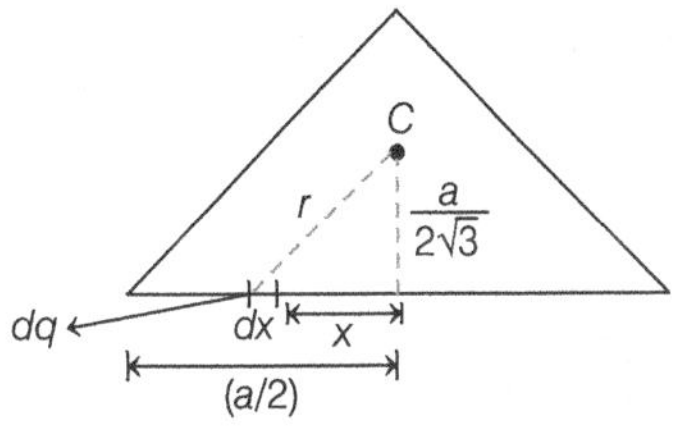

$$dq = \lambda dx$$

Potential at C, due to charge dq is

$$dV = \frac{k(dq)}{r} \qquad \left(\because k = \frac{1}{4\pi\varepsilon_0}\right)$$

$\therefore$ Total potential $= 6\int_{x=0}^{x=a/2} dV$

21. $k_A + U_A = k_B + U_B$

$\therefore$ $$\frac{1}{2}mv_A^2 + qV_A = \frac{1}{2}mv_B^2 + qV_B$$

$\therefore$ $$v_B = \sqrt{v_A^2 + \frac{2}{m}q(V_A - V_B)}$$

$$= \sqrt{(5)^2 + \frac{2\times(-5\times10^{-6})}{2\times10^{-4}}(200 - 800)}$$

$$= 7.42 \text{ m/s}$$

$v_B > v_A$ as the negative charge is moving (freely) from lower potential at A to higher potential at B. So, its electrostatic PE will decrease and kinetic energy will increase.

22. (a) $$V_C = \frac{k.q_{net}}{R} = \left(\frac{1}{4\pi\varepsilon_0}\right)\left(\frac{-5Q}{R}\right)$$

(b) $$V_p = \frac{k\,q_{net}}{r}$$

where, r = distance of P from any point on circumference

$$= \left(\frac{1}{4\pi\varepsilon_0}\right)\left(\frac{-5Q}{\sqrt{R^2 + z^2}}\right)$$

23. $W = -\Delta U = U_i - U_f = q_2 V_i - q_2 V_f$

$$= q_2\left(\frac{kq_1}{r_i}\right) - q_2\left(\frac{kq_1}{r_f}\right)$$

$$= kq_1 q_2 \left(\frac{1}{r_i} - \frac{1}{r_f}\right)$$

$$= (9 \times 10^9)(2.4 \times 10^{-6})(-4.3 \times 10^{-6})\left(\frac{1}{0.15} - \frac{1}{0.25\sqrt{2}}\right)$$

$$= -0.356 \text{ J}$$

Module 6.

1. $\mathbf{E} = -\left(\frac{\partial V}{\partial x}\hat{\mathbf{i}} + \frac{\partial V}{\partial y}\hat{\mathbf{j}} + \frac{\partial V}{\partial z}\hat{\mathbf{k}}\right)$

$\therefore \quad |\mathbf{E}| = \sqrt{\left(\frac{\partial V}{\partial x}\right)^2 + \left(\frac{\partial V}{\partial y}\right)^2 + \left(\frac{\partial V}{\partial z}\right)^2}$

$\therefore \quad |\mathbf{E}| \geq \frac{\partial V}{\partial x}$

or $\quad = 10 \text{ V/m}$

2. $V_A - V_B = -\int_B^A \mathbf{E} \cdot d\mathbf{r}$

$$= -\int_{(0,4)}^{(4,0)} (4\hat{\mathbf{i}} + 4\hat{\mathbf{j}}) \cdot (dx\hat{\mathbf{i}} + dy\hat{\mathbf{j}})$$

$$= -\int_{(0,4)}^{(4,0)} (4dx + 4dy) = 0$$

$\therefore \quad V_A = V_B$

3. $\mathbf{E} = E_x\hat{\mathbf{i}} + E_y\hat{\mathbf{j}}$

Now we can use, $\int dV = -\int \mathbf{E} \cdot d\mathbf{r}$

two times and can find values of E_x and E_y.

4. Let $\mathbf{E} = E_x\hat{\mathbf{i}} + E_y\hat{\mathbf{j}} + E_z\hat{\mathbf{k}}$

Apply $\int dV = -\int \mathbf{E} \cdot d\mathbf{r}$

three times and find values of E_x, E_y and E_z. Then, again apply the same equation for given point.

5. $\mathbf{E} = -\left[\frac{\partial V}{\partial x}\hat{\mathbf{i}} + \frac{\partial V}{\partial y}\hat{\mathbf{j}}\right]$

$$= -[(-ky)\hat{\mathbf{i}} + (-kx)\hat{\mathbf{j}}]$$

$\therefore \quad |\mathbf{E}| = \sqrt{(ky)^2 + (kx)^2}$

$$= k\sqrt{x^2 + y^2} = kr$$

$\therefore \quad |\mathbf{E}| \propto r$

6. For small displacement $d\mathbf{s}$, work done by the electric field $(\Delta W)_{\text{electric force}} = F_e \cdot d\mathbf{r}$

The change in potential ΔV is given

$$\Delta V = \frac{-[\Delta W]_{\text{electric force}}}{q'} = \frac{\Delta U}{q'}$$

7. $\int dV = -\int \mathbf{E} \cdot d\mathbf{r}$

$\therefore \quad V_A - V_B = -\int_B^A \mathbf{E} \cdot d\mathbf{r}$

8. (a) $\mathbf{E} = -\left[\frac{\partial V}{\partial x}\hat{\mathbf{i}} + \frac{\partial V}{\partial y}\hat{\mathbf{j}}\right]$

$$= -a[(2x)\hat{\mathbf{i}} - (2y)\hat{\mathbf{j}}] = -2a[x\hat{\mathbf{i}} - y\hat{\mathbf{j}}]$$

(b) Again

$$\mathbf{E} = -\left[\frac{\partial V}{\partial x}\hat{\mathbf{i}} + \frac{\partial V}{\partial y}\hat{\mathbf{j}}\right] = -a[y\hat{\mathbf{i}} + x\hat{\mathbf{j}}]$$

9. $E = -\frac{dV}{dx} = -$ Slope of V-x graph.

From $x = -2$m to $x = 0$, slope $= +5$ V/m

$\therefore \quad E = -5$ V/m

From $x = 0$ to $x = 2$m, slope $= 0$

$\therefore \quad E = 0$

From $x = 2$m to $x = 4$m, slope $= +5$ V/m

$\therefore \quad E = -5$ V/m

From $x = 4$m to $x = 8$m, slope $\equiv -5$ V/m

$\therefore \quad E = +5$ V/m

Corresponding E-x graph is as shown in answer.

10. $V_A = V_D$ and $V_B = V_C$ as the points A and D or B and C are lying on same equipotential surface ($\perp$ to electric field lines). Further, V_A or $V_D > V_B$ or V_C as electric lines always flow from higher potential to lower potential

$$V_A - V_B = V_D - V_C = Ed$$

$$= (20)(1) = 20 \text{ V}$$

11. (a) $\quad \Delta U = -q(\Delta V)$

$$= -(12 \times 10^{-6})(50)$$

$$= -6.0 \times 10^{-4} \text{ J}$$

(b) $\Delta V = Ed = E(\Delta x)$

$$= (250)(0.2) = 50 \text{ V}$$

Module 8.

1. In uniform electric field, net force on an electric dipole $= 0$

Therefore, no work is done in translational motion of the dipole.

Electric lines also flow from higher potential to lower potential. Electrostatical force on positive charge acts in the direction of electric field. Therefore, work done is positive.

2. Potential at a point due to electric dipole,

$$V = \frac{p\cos\theta}{4\pi\varepsilon_0 r^2}$$

If $\theta = 0°$, then $V =$ maximum

If $\theta = 180°$, then $V =$ minimum

3.

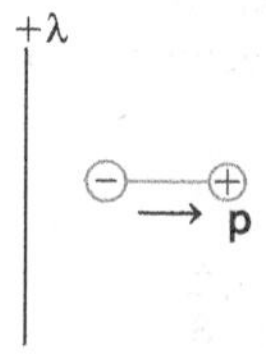

Negative charge of dipole is near to positively charge line charge. Hence, attraction is more.

4.

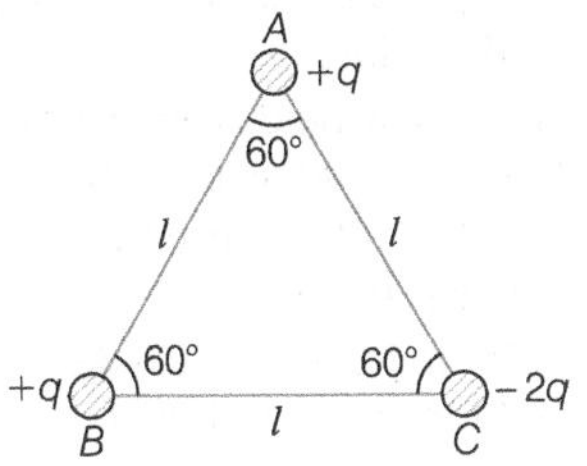

The direction of dipole moment is shown in figure.

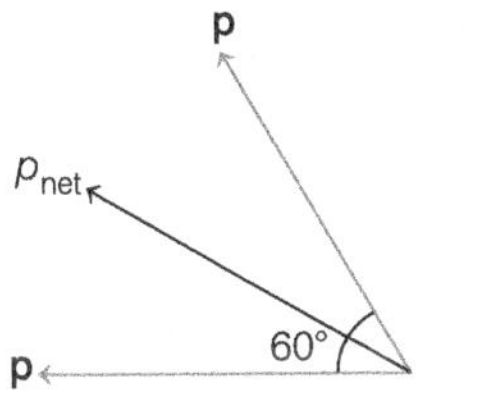

Net electric dipole moment,

$$p_{net} = \sqrt{p^2 + p^2 + 2pp\cos 60°}$$

$$= \sqrt{3}p = \sqrt{3}\,ql \qquad (\because p = ql)$$

5. If dipole is rotated through an angle of 90° about its perpendicular axis, then given point comes on equatorial line. So, field becomes half of previous value, i.e. $E/2$.

6. Given situation is

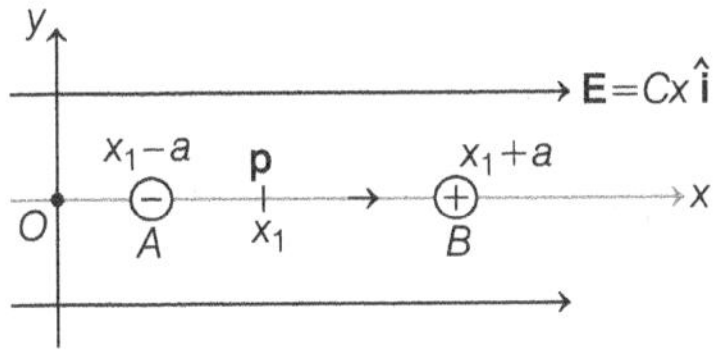

Force on negative charge

$$F_A = qE_A = qC(x_1 - a) \qquad \text{(Towards origin)}$$

and force on positive charge is

$$F_B = qC(x_1 + a) \qquad \text{(Along x-axis away from origin)}$$

$$\therefore F_{net} = F_B - F_A$$

$$= qC(x_1 + a) - qC(x_1 - a)$$

$$= qC\,2a$$

$$= 2qa \cdot C = pC\,\hat{\mathbf{i}}$$

(As net force is along positive x-axis)

Also, $\frac{d}{dx}E_x\,p\hat{\mathbf{i}} = p\hat{\mathbf{i}}\left(\frac{d}{dx}E_x\right) = p\hat{\mathbf{i}}\left(\frac{d}{dx}Cx\right)$

$$= Cp\hat{\mathbf{i}}\left(\frac{d}{dx}x\right) = Cp\,\hat{\mathbf{i}}$$

7. Electric dipole moment

$$p = q \times 2l = 0.01 \times 0.4 \times 10^{-3}$$

$$= 4 \times 10^{-6}\ \text{C-m}$$

$$E = 10 \times 10^{-5}\ \text{N/C}$$

$\therefore$ Torque $\tau = pE\sin\theta$

$$= 4 \times 10^{-6} \times 10 \times 10^{-5} \times \sin 30°$$

$$= 2 \times 10^{-10}\ \text{N-m}$$

8. Electric field due to electric dipole on equatorial plane at a distance r from its centre is given as

$$E = \frac{1}{4\pi\varepsilon_0} \cdot \frac{p}{r^3} \Rightarrow E \propto \frac{1}{r^3}$$

9. $V_1 \propto \frac{p_1 \cos\theta_1}{r_1^3}$

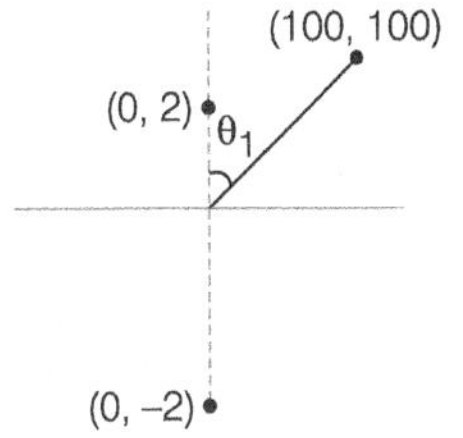

$$V_2 \propto \frac{p_2\cos\theta_2}{r_2^3}$$

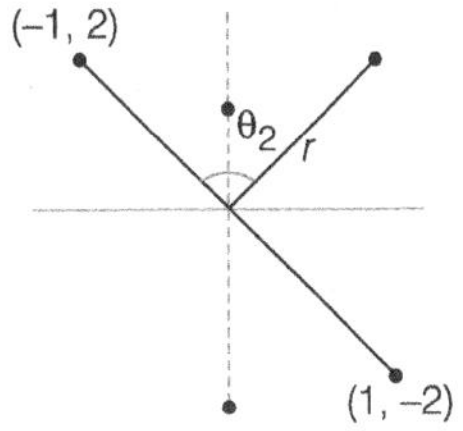

$$\therefore \quad \frac{V_2}{V_1} = \frac{p_2\cos\theta_2}{p_1\cos\theta_1}$$

$$\because \quad \frac{V_2}{V_1} = \frac{q(-2\hat{\mathbf{i}} + 4\hat{\mathbf{j}})\cdot(\hat{\mathbf{i}} + \hat{\mathbf{j}})}{q(0\hat{\mathbf{i}} + 4\hat{\mathbf{j}})\cdot(\hat{\mathbf{i}} + \hat{\mathbf{j}})} = \frac{1}{2}$$

or $\quad V_2 = \frac{V_1}{2}$ or $\frac{V_0}{2}$

10. $E_{net} = E_+ + E_- = \frac{2kq}{(d/2)^2} = \frac{8kq}{d^2}$

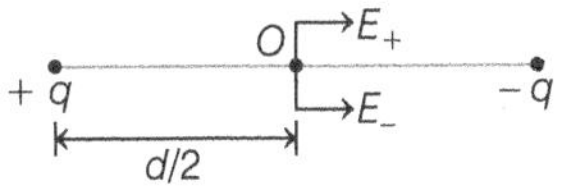

$$\Rightarrow \quad 6.4 \times 10^4 = 8 \times 9 \times 10^9 \times \frac{8 \times 10^{-6}}{d^2}$$

$$\Rightarrow \quad d^2 = 9 \times 1 \Rightarrow d = 3\ \text{m}$$

11. Let v be the speed of dipole.

Using law of conservation of energy

$$K_i + U_i = K_f + U_f$$

$$\Rightarrow \quad 0 - \frac{2k\,.\,p_1}{r^3}p_2\cos(180°) = \frac{1}{2}mv^2 \times 2 + 0$$

($\because$ Potential energy of interaction between dipoles

$$-\mathbf{p}_1 \cdot \mathbf{E}_2 = -\mathbf{p}_2 \cdot \mathbf{E}_1 = \frac{-2p_1p_2\cos\theta}{4\pi\varepsilon_0 r^3}\Bigg)$$

$$\Rightarrow \quad mv^2 = \frac{2kp_1p_2}{r^3}$$

$$\Rightarrow \quad v = \sqrt{\frac{2kp_1p_2}{mr^3}}$$

When $p_1 = p_2 = p$ and $r = a$, then

$$v = \frac{p}{a}\sqrt{\frac{1}{2\pi\varepsilon_0 ma}}$$

12. $\therefore \tau = -pE\sin\theta$

For small θ, $\sin\theta = \theta$

$$I\alpha = -pE(\theta)$$

$\Rightarrow$ $$\alpha = \frac{pE}{I}(-\theta) \quad ...(i)$$

Also, $$\alpha = -\omega^2\theta \quad ...(ii)$$

From Eqs. (i) and (ii),

$$\omega = \sqrt{\frac{pE}{I}} = \sqrt{\frac{qdE}{2m\left(\frac{d}{2}\right)^2}} = \sqrt{\frac{2qE}{md}}$$

13. Potential energy of a dipole is given by

$$U = -\mathbf{p}\cdot\mathbf{E} = -pE\cos\theta$$

Here, θ = angle between dipole and perpendicular to the electric field.

$\therefore$ $$U = -(10^{-29}(10^3)\cos 45°$$
$$= -0.707 \times 10^{-26}\ \text{J}$$
$$U = -7 \times 10^{-27}\ \text{J}$$

14. In a dipole, electric potential at the bisector is zero. The electric field is antiparallel to the dipole moment **p**.

$\therefore$ $$V = 0;\ \mathbf{E} = \frac{1}{4\pi\varepsilon_0}\left(\frac{-\mathbf{p}}{d^3}\right)$$

15. Torque on a dipole placed in an external electric field

$$\tau = \mathbf{p}\times\mathbf{E} \text{ or } \tau = p\cdot E\sin\theta$$
$$\mathbf{p} = p\cos\theta\hat{\mathbf{i}} + p\sin\theta\hat{\mathbf{j}};\ \mathbf{E}_1 = E\hat{\mathbf{i}}$$
$$\mathbf{T}_1 = \mathbf{p}\times\mathbf{E}_1 = (p\cos\theta\hat{\mathbf{i}} + p\sin\theta\hat{\mathbf{j}})\times E(\hat{\mathbf{i}})$$
$$\tau\hat{\mathbf{k}} = pE\sin\theta(-\hat{\mathbf{k}}) \quad ...(i)$$
$$\mathbf{E}_2 = \sqrt{3}\,E_1\,\hat{\mathbf{j}}$$
$$\mathbf{T}_2 = p\cos\theta\hat{\mathbf{i}} + p\sin\theta\hat{\mathbf{j}} \times \sqrt{3}E_1\hat{\mathbf{j}}$$
$$\tau\hat{\mathbf{k}} = \sqrt{3}\,pE_1\cos\theta\hat{\mathbf{k}} \quad ...(ii)$$

From Eqs. (i) and (ii),

$$pE\sin\theta = \sqrt{3}pE\cos\theta$$
$$\tan\theta = \sqrt{3}$$
$\therefore$ $$\theta = 60°$$

16. The diagrammatic representation of the given question is shown in figure.

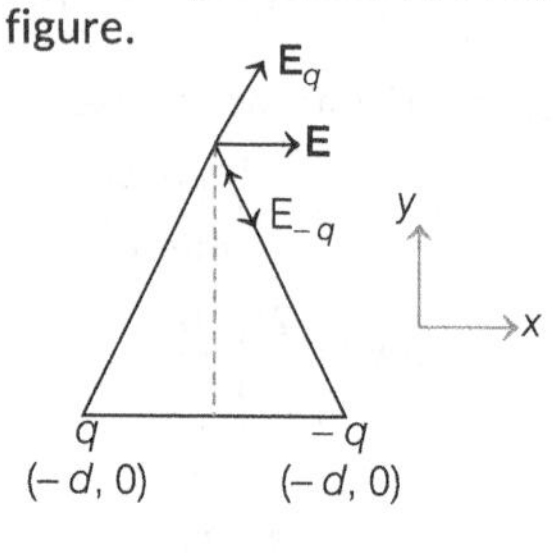

The electric field **E** at all points on the x-axis will not have the same direction.

For $-d \le x \le d$, electric field is along positive x-axis while for all other points it is along negative x-axis.

The electric field **E** at all points on the y-axis will be parallel to the x-axis (i.e. $\hat{\mathbf{i}}$).

The electric potential at the origin due to both the charges is zero, hence no work is done in bringing a test charge from infinity to the origin.

Dipole moment is directed from the $-q$ charge to the $+q$ charge (i.e. $-\hat{\mathbf{i}}$ direction).

Module 9.

Introductory Exercise

1. (a) Given surface is a closed surface. Therefore, we can directly apply the result.

$$\phi = \frac{q_{in}}{\varepsilon_0} = 0 \quad \text{as} \quad q_{in} = 0$$

(b) Again given surface is a closed surface. Hence, we can directly apply the result.

$$\phi = \frac{q_{in}}{\varepsilon_0} = \frac{q}{\varepsilon_0} \quad \text{as} \quad q_{in} = q$$

(c) Given surface is not closed surface. Hence, we cannot apply the direct result of Gauss's theorem. If we draw a complete sphere, then

$$\phi \text{ through complete sphere} = \frac{q}{\varepsilon_0}$$

$\therefore$ $$\phi \text{ through hemisphere} = \frac{1}{2}\left(\frac{q}{\varepsilon_0}\right)$$

2. (a)

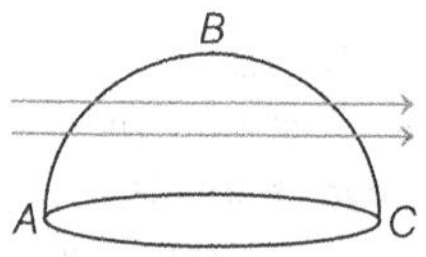

Net flux entering from AB = net flux entering from BC.

(b)

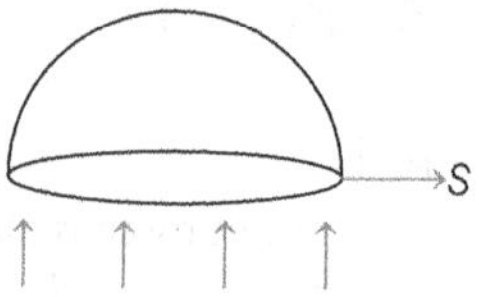

$$\phi = ES = E(\pi R^2)$$

3. Given electric field is uniform electric field. Net flux from any closed surface in uniform electric field = 0.

4. Same flux will pass through two concentric spherical shells if charge is kept at centre.

Exercise

1. Net charge from any closed surface in uniform electric field = 0

$\therefore$ Net charge inside any closed surface in uniform electric field = 0

3. Flux from any closed surface $= \dfrac{q_{in}}{\varepsilon_0}$,

$q_{in} = 0$, due to a dipole.

6. If we make a bigger cube comprising of eight small cubes of size given in the question with charge at centre (or at D).

Then, total flux through large closed cube $= \dfrac{q}{\varepsilon_0}$.

There are 24 symmetrical faces like $EFGH$ on outermost surface of this bigger cube.

Total flux from these 24 faces is $\dfrac{q}{\varepsilon_0}$. Hence, flux from

anyone force $= \dfrac{1}{24}\left(\dfrac{q}{\varepsilon_0}\right)$.

Electric lines are tangential to face $AEHD$. Hence, flux is zero.

7. Sphere is a closed surface. Therefore, Gauss's theorem can be applied directly on this.

$$\therefore \quad \phi_{total} = \frac{q_{in}}{\varepsilon_0}$$

$$\text{or} \quad q_{in} = (\phi_{total})(\varepsilon_0)$$
$$= (360)(8.86 \times 10^{-12})$$
$$= 3.19 \times 10^{-9}\ \text{C}$$
$$= 3.19\ \text{nC}$$

8. Electric flux enters from the surface parallel to y-z plane at $x = 0$. But $\mathbf{E} = 0$ at $x = 0$.

Hence, flux entering the cube $= 0$.

Flux leaves the cube from the surface parallel to y-z plane at $x = a$.

Flux leaving the cube, $\phi = ES$

$$= \left(\frac{E_0 x}{l}\right)(a^2) = \left(\frac{E_0 a}{l}\right)(a^2) \qquad (\text{at } x = a)$$

Substituting the values, we get

$$\phi = \frac{(5 \times 10^3)(10^{-2})^3}{(2 \times 10^{-2})}$$
$$= 2.5 \times 10^{-1}$$
$$= 0.25\ \text{N-m}^2/\text{C}$$

At all other four surfaces, electric lines are tangential. Hence, flux is zero.

$$\therefore \quad \phi_{net} = (0.25)\ \text{N-m}^2/\text{C} = \frac{q_{in}}{\varepsilon_0}$$

$$\therefore \quad q_{in} = (0.25)(\varepsilon_0)$$
$$= (0.25)(8.86 \times 10^{-12})$$
$$= 2.2 \times 10^{-12}\ \text{C}$$

9. We know that the flux passing through a segment which makes angle at the centre of the circle is

$$\phi = \frac{(1 - \cos\theta)Q}{2\varepsilon_0}$$

$$\text{where, } \cos\theta = \frac{b}{\sqrt{b^2 + r^2}}$$

$$\text{But,} \quad \frac{Q}{\varepsilon_0} = \phi_{total}$$

$$\therefore \quad \phi = (\phi_{total})\left(\frac{1 - \cos\theta}{2}\right) \qquad \text{...(i)}$$

Given that $\phi = \dfrac{1}{4}(\phi_{total})$

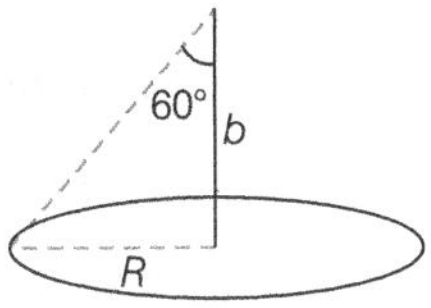

Substituting in Eq. (i), we get

$$\theta = 60°$$
$$\frac{R}{b} = \tan 60° = \sqrt{3}$$
$$\therefore \quad R = \sqrt{3}b$$

10. (a) $\mathbf{S}_1 = (L^2)(-\hat{\mathbf{j}})$

$$\therefore \quad \phi s_1 = \mathbf{E} \cdot \mathbf{S}_j = -CL^2$$

Similarly, we can find flux from other surfaces.

Note *Take area vector in outward direction of the cube.*

(b) Total flux from any closed surface in uniform electric field is zero.

11. We have $\phi = \dfrac{q(1 - \cos\theta)}{2\varepsilon_0}$

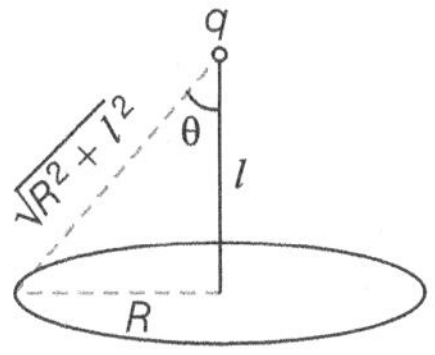

Here, flux due to $+q$ and $-q$ are in same direction.

$$\therefore \quad \phi_{total} = 2\phi = \frac{q(1 - \cos\theta)}{\varepsilon_0}$$
$$= \frac{q}{\varepsilon_0}\left[1 - \frac{l}{\sqrt{R^2 + l^2}}\right]$$

12.

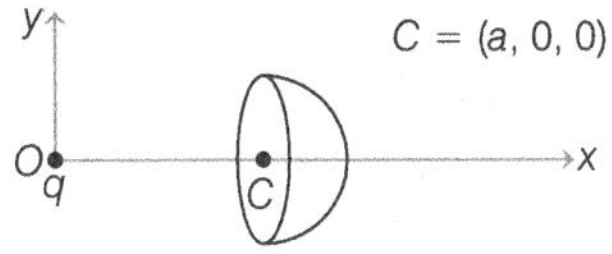

Flux passing through hemisphere = flux passing through circular surface of the hemisphere.

For finding flux through circular surface of hemisphere we can again use the concept used in above problem.

$$\phi = \frac{q(1 - \cos\theta)}{2\varepsilon_0}$$
$$= \frac{q}{2\varepsilon_0}(1 - \cos 45°)$$
$$= \frac{q}{2\varepsilon_0}\left(1 - \frac{1}{\sqrt{2}}\right)$$

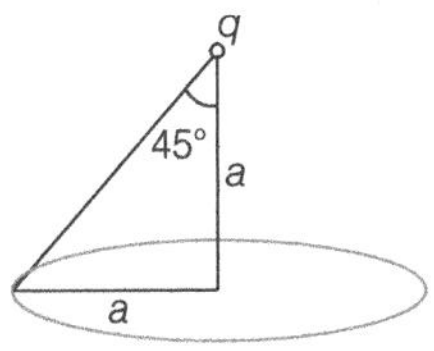

13.

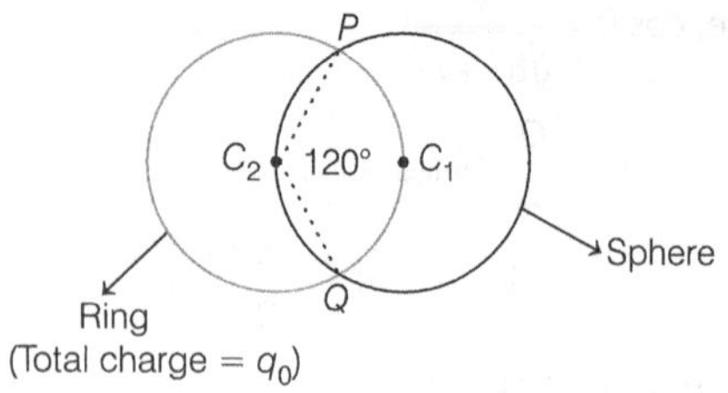

Charge on arc PQ of ring $= \frac{q_0}{3}$

This is also the charge lying inside the closed sphere.

$\therefore$ ϕ through closed sphere $= \frac{q_{in}}{\varepsilon_0}$

$$= \frac{(q_0/3)}{\varepsilon_0} = \frac{q_0}{3\varepsilon_0}$$

Module 11.

1. $V = \frac{kq}{R}$ or $kq = VR$

For inside points ($r \le R$),

$$E = \frac{kq}{R^3} r \quad \text{or} \quad E \propto r$$

At distance $r = \frac{R}{2}$,

$$E = \frac{(VR)}{R^3}\left(\frac{R}{2}\right) = \frac{V}{2R}$$

2. It depends upon the nature of charge whether it is positive or negative.

3. Rate of change of electric field with distance from centre at some point represents magnitude of electric potential at that point, that is **wrong**. Because $E = -\frac{dV}{dr}$.

Hence, the correct is $E =$ – (rate of change of electric potential at that point). Further, electric potential inside the sphere varies from $1.5\frac{Kq}{R}$ to $\frac{Kq}{R}$ inside the sphere, whereas outside the sphere it varies from $\frac{Kq}{R}$ to zero. Hence, $V_P \neq V_Q$ on surface. Here, $K = \frac{1}{4\pi\varepsilon_0}$

4. Charge on both will become zero.

$\therefore$ $U_f = 0$

Hence, U_i is completely lost.

5.

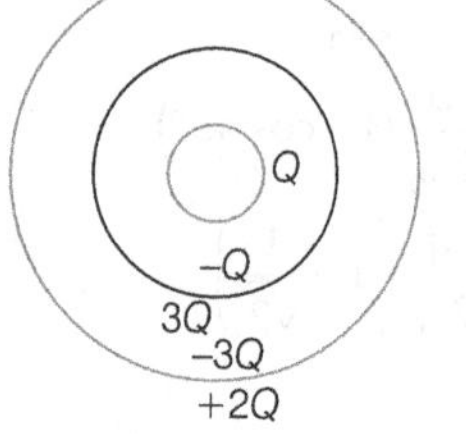

The desired ratio is $-\frac{3Q}{2Q} = -\frac{3}{2}$

6. $\frac{kq}{r} = \frac{3}{2}\left(\frac{kq}{R}\right)\Big/2$ or $r = \frac{4}{3}R$

$\therefore$ Distance from surface $= r = R$

$$= \frac{R}{3}$$

7. $PD = V_i - V_o = \frac{q_i}{4\pi\varepsilon_0}\left(\frac{1}{r_i} - \frac{1}{r_o}\right)$

If q_i is positive, $V_i - V_o =$ positive or $V_i > V_o$.

8. Principle of generator.

9. Let $V_{surface} = V_0$

Then $V_{centre} = 1.5\,V_0$

So, values varies from $1.5\,V_0$ to V_0.

$0.75\,V_0$ value is not possible.

10. From centre to the surface of inner shell, potential will remain constant = 10 V (given).

11.

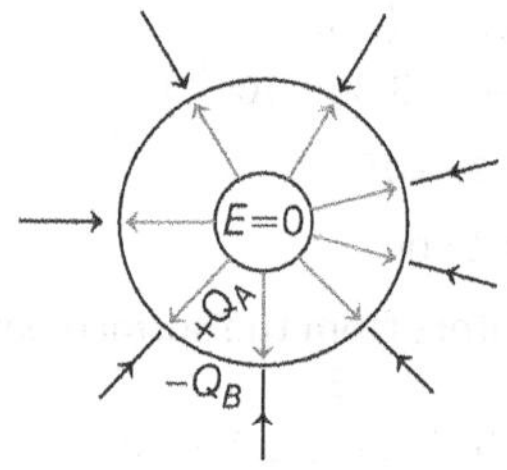

Since $|Q_B| > Q_A$, electric field outside sphere B is inwards (say negative). From A to B enclosed charge is positive. Hence, electic field is radially outwards (positive).

12. $\rho(4\pi r^2) + \rho(4\pi R^2) = Q$

$\therefore$ $\rho = \frac{Q}{4\pi(r^2 + R^2)}$

$$q_1 = (4\pi r^2)\rho = \frac{r^2}{r^2 + R^2} Q$$

= charge on sphere of radius r

$$q_2 = (4\pi R^2)\rho = \left(\frac{R^2}{r^2 + R^2}\right) Q$$

= charge on sphere of radius R

Potential at common centre $= \frac{1}{4\pi\varepsilon_0}\left[\frac{q_1}{r} + \frac{q_2}{R}\right]$

$$= \frac{1}{4\pi\varepsilon_0} Q\left[\frac{r + R}{r^2 + R^2}\right]$$

13. $V = \frac{1}{4\pi\varepsilon_0} \cdot \frac{q}{r}$

$$V' = \left(\frac{1}{4\pi\varepsilon_0}\right)\left(\frac{2q}{4r}\right) = \frac{V}{2}$$

14. $V = \frac{1}{4\pi\varepsilon_0}\cdot\frac{q_1}{3R} + \frac{1}{4\pi\varepsilon_0}\cdot\frac{q_2}{3R}$

$$= \frac{q_{net}}{(4\pi\varepsilon_0)(3R)}$$

$$\therefore \quad R = \left(\frac{1}{4\pi\varepsilon_0}\right)\left(\frac{q_{net}}{3V}\right)$$

$$= \frac{(9\times10^9)(3\times10^{-6})}{(3)(9000)}$$

$$= 1\,\text{m}$$

15. $V_B = 0$

$$\therefore \quad \frac{kq_A}{r_B} + \frac{kq_B}{r_B} = 0$$

$$\therefore \quad q_B = -q_A = -q$$

Charge distribution is as shown below.

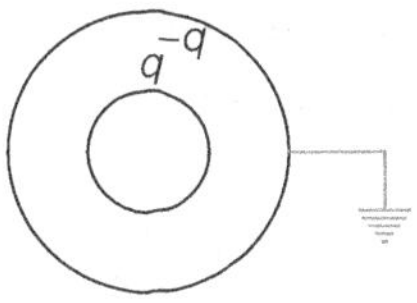

From Gauss's theorem, electric field at any point is given by

$$E = \frac{kq_{in}}{r^2}$$

q_{in} inside A and outside B is zero. Therefore, $E = 0$

16. $E = \frac{\sigma}{\varepsilon_0}$

$$E_1 = E_2$$

$$\therefore \quad \sigma_1 = \sigma_2$$

$$V = \frac{\sigma R}{\varepsilon_0}$$

or $\quad V \propto R \quad$ (as $\sigma \to$ same)

$$\therefore \quad \frac{V_1}{V_2} = \frac{R_1}{R_2} = \frac{a}{b}$$

17. According to principle of generator PD in this case only depends on the charge on inner shell.

18. $V = \frac{kq}{R}$

$$\therefore \quad kq = VR$$

$$E = \frac{Kq}{r^2} = \frac{VR}{r^2}$$

19. Let radius of big drop is R and radius of small drop is r.

$\therefore$ Volume of big drop $= 8\times$ volume of small drop

or $\quad \frac{4}{3}\pi R^3 = 8\times\frac{4}{3}\pi r^3$

or $\quad R = 2r$

$$C = 4\pi\varepsilon_0 R$$

$\Rightarrow \quad C \propto$ Radius

Potential of big drop,

$$V = \frac{Q}{C} = \frac{8q}{(8)^{1/3}C} \quad (q \text{ is charge on small drop})$$

$$\therefore \quad V = (8)^{2/3}v$$

$$\therefore \quad \frac{V_{big}}{V_{small}} = (8)^{2/3} = \frac{4}{1}$$

20. $q = \left(\frac{4}{3}\pi R^3\right)\rho \Rightarrow \frac{q}{4\pi R^3} = \frac{\rho}{3}$

$$E = \frac{1}{4\pi\varepsilon_0}\frac{q}{R^3}r = \frac{\rho r}{3\varepsilon_0}$$

21. $V = \frac{kq}{R}$

$$\Rightarrow \quad kq = VR$$

(a) $V = \frac{kq}{R^3}(1.5R^2 - 0.5r^2)$

$$= \frac{VR}{R^3}\left[\frac{3}{2}R^2 - \frac{1}{2}\left(\frac{R}{2}\right)^2\right]$$

$$= \frac{11}{8}V$$

(b) $V = \frac{kq}{r} = \frac{VR}{2R} = \frac{V}{2}$

(c) $E = \frac{kq}{R^3}\cdot r = \frac{(V_R)}{(R^3)}\left(\frac{R}{2}\right)$

$$= \frac{V}{2R} = \frac{V}{2} \quad (\text{if } R = 1\,\text{m})$$

(d) $E = \frac{kq}{r^2} = \frac{VR}{(2R)^2} = \frac{V}{4R} = \frac{V}{4}$ for $R = 1$ m

23.

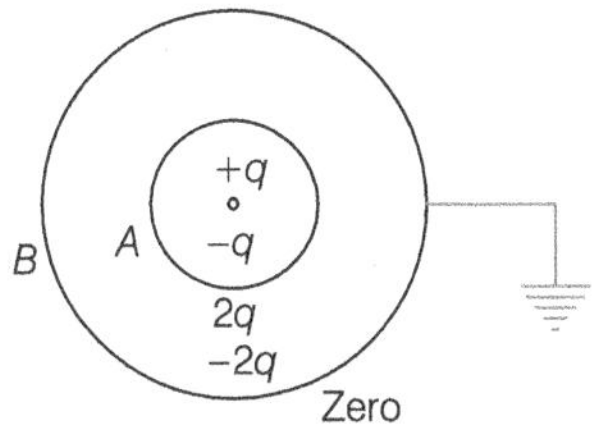

If outermost shell is earthed. Then, charge on outer surface of outermost shell in this case is always zero.

24. (a) $V_A = \frac{\sigma a}{\varepsilon_0} - \frac{\sigma b}{\varepsilon_0} + \frac{\sigma c}{\varepsilon_0}$

$$= \frac{\sigma}{\varepsilon_0}(a - b + c)$$

$$V_B = \left(\frac{\sigma a}{\varepsilon_0}\right)\left(\frac{a}{b}\right) - \frac{\sigma b}{\varepsilon_0} + \frac{\sigma c}{\varepsilon_0}$$

$$= \frac{\sigma}{\varepsilon_0}\left[\frac{a^2}{b} - b + c\right]$$

$$V_C = \left(\frac{\sigma a}{\varepsilon_0}\right)\left(\frac{a}{c}\right) - \left(\frac{\sigma b}{\varepsilon_0}\right)\left(\frac{b}{c}\right) + \frac{\sigma c}{\varepsilon_0}$$

$$= \frac{\sigma}{\varepsilon_0}\left(\frac{a^2}{c} - \frac{b^2}{c} + c\right)$$

(b) $W_{A \to C} = 0$

$\therefore \quad V_A = V_C$

or $\quad (a - b + c) = \left(\frac{a^2 - b^2}{c}\right) + c$

or $\quad a + b = c$

25. (a)

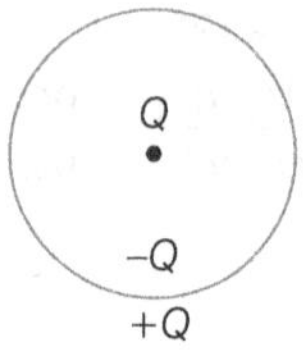

$$\sigma_i = \frac{-Q}{4\pi a^2} \text{ and } \sigma_0 = \frac{Q}{4\pi a^2}$$

(b)

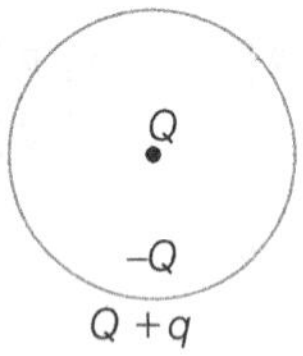

$$\sigma_i = \frac{-Q}{4\pi a^2} \text{ and } \sigma_0 = \frac{(Q+q)}{4\pi a^2}$$

(c) According to Gauss's theorem,

$$E = \frac{1}{4\pi\varepsilon_0}\frac{q_{in}}{r^2}$$

For $x \le R$, $q_{in} = Q$ and $r = x$ both cases.

$$\therefore \quad E = \frac{1}{4\pi\varepsilon_0}\frac{Q}{x^2}$$

26.

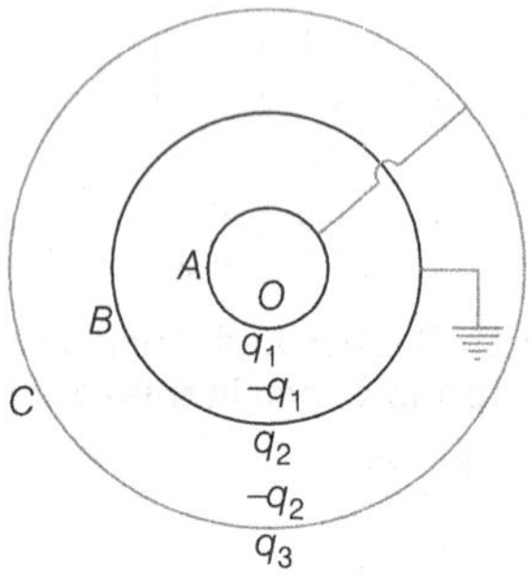

Total charge on $A + C$ is $3q$

$$\therefore \quad q_1 - q_2 + q_3 = 3q \qquad \text{...(i)}$$

$$V_B = 0$$

$$\therefore \quad \frac{kq_1}{2R} + k\left(\frac{q_2 - q_1}{2R}\right) + k\left(\frac{q_3 - q_2}{3R}\right) = 0 \qquad \text{...(ii)}$$

$$V_A = V_C$$

$$\therefore \quad \frac{kq_1}{R} + k\left(\frac{q_2 - q_1}{2R}\right) + k\left(\frac{q_3 - q_2}{3R}\right)$$

$$= \frac{kq_1}{3R} + k\left(\frac{q_2 - q_1}{3R}\right) + k\left(\frac{q_3 - q_2}{3R}\right) \qquad \text{...(iii)}$$

In the above equations, $k = \frac{1}{4\pi\varepsilon_0}$.

Solving these three equations, we can find the asked charges.

27.

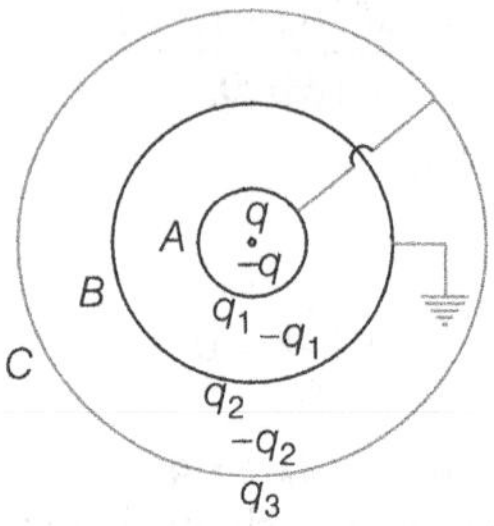

Total charge on $A + C$ is $3q$. Therefore,

$$-q + q_1 - q_2 + q_3 = 3q \qquad \text{...(i)}$$

$$V_B = 0$$

$$\therefore \quad k\left(\frac{q_2 - q_1 + q_1 - q + q}{2R}\right) + k\left(\frac{q_3 - q_2}{3R}\right) = 0 \qquad \text{...(ii)}$$

$$V_A = V_C$$

$$\therefore \quad k\left(\frac{q_1 - q + q}{R}\right) + k\left(\frac{q_2 - q_1}{2R}\right) + k\left(\frac{q_3 - q_2}{3R}\right)$$

$$= k\left(\frac{q_1 - q + q}{3R}\right) + k\left(\frac{q_2 - q_1}{3R}\right) + k\left(\frac{q_3 - q_2}{3R}\right) \qquad \text{...(iii)}$$

In the above equations,

$$k = \frac{1}{4\pi\varepsilon_0}$$

Solving above there equations, we can find q_1, q_2 and q_3.

28. (a) From Gauss's theorem

$$E = \left(\frac{1}{4\pi\varepsilon_0}\right)\left(\frac{q_{in}}{r^2}\right)$$

$$= \left(\frac{1}{4\pi\varepsilon_0}\right)\left(\frac{2Q}{r^2}\right) = \frac{Q}{2\pi\varepsilon_0 r^2}$$

(b) According to principle of generator, potential difference depends only on q_{in}.

$$\therefore \quad PD = \frac{+2Q}{4\pi\varepsilon_0}\left(\frac{1}{R} - \frac{1}{3R}\right) = \frac{Q}{3\pi\varepsilon_0 R}$$

(c) According to principle of generator, whole inner charge transfers to outer sphere.

(d) $V_{in} = 0$

$$\therefore \quad \frac{k\,q_{in}}{R} - \frac{kQ}{3R} = 0$$

$$\Rightarrow \quad q_{in} = \frac{Q}{3}$$

Unit Exercise (Cumulative)

Level 1

1. Potential, $V = \frac{Kq}{r}$

   Let the charge on each drop $= q$ and radius $= r$

   $\therefore \quad V_0 = \frac{kq}{r} = 2\text{ V}$

   Now, for combination of 512 drops

   $$\frac{4}{3}\pi R^3 = 512 \times \frac{4}{3}\pi r^3$$

   $\Rightarrow \quad R = (512 \times r^3)^{1/3} = 8r$

   Potential of bigger drop,

   $$V = \frac{K(512)q}{8r} = \frac{512}{8}\frac{Kq}{r}$$

   $$= \frac{512}{8} \times 2\text{V} = 128\text{ V}$$

2. Given, electric potential $V = 4x^2$ volts

   Therefore, electric potential is variable only along x-axis.

   We know, $E_x = \frac{-dV}{dx}$

   $\Rightarrow \quad E_x = -\frac{d(4x^2)}{dx} = -8x$

   The electric field at point (1, 0, 2) (here $x = 1$)

   $$E_x = -8\text{ V/m}$$

3. The magnitude of electric field is greatest at point B because at B the equipotential surface are closest.

4. For a particle to move in circular motion, we need a centripetal force which is not available. Force will be tangential to electric field lines.

5. When high energy X-ray beam falls,it will knock out electrons from the small metal ball due to photoelectric effect, making it positively charged and an electrostatic force in the direction of electric field acts. Therefore the ball will be deflected in the direction of electric field.

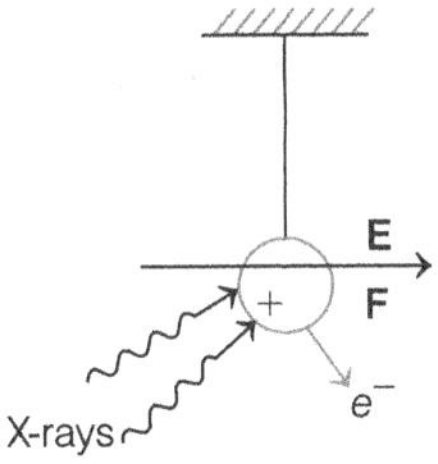

6. The earth is used as reference at zero potential in electrical circuits for practical purposes.

   For spherical capacitor, electrical potential $V = \frac{Q}{4\pi\varepsilon_0 R}$.

7.

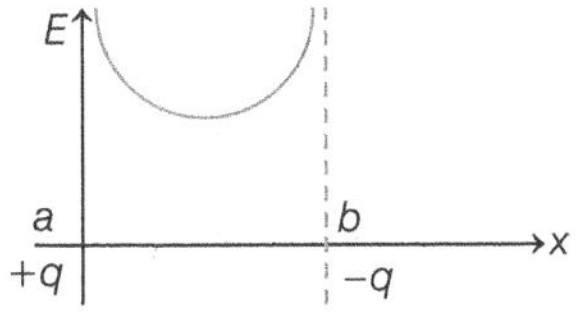

Just to the right of a, electric field is along ab ($\because$positive) and tending to infinite. Similarly, electric field just to the left of b electric field is again along ab ($\because$positive) and tending to infinite.

8.

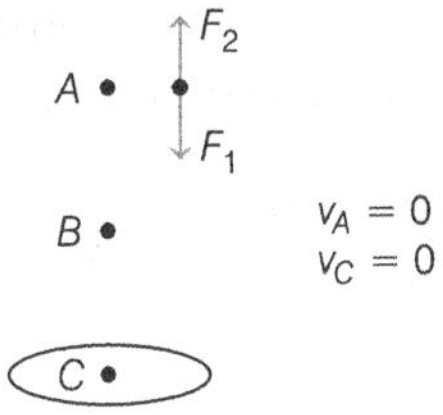

Hence, in between A and C there is a point B, where speed of the particle should be maximum.

$$F_1 = mg = \text{constant}$$

F_2 = electrostatic repulsion (which increases as the particle moves down)

From A to B kinetic energy of the particle increases and potential energy decreases. Then, from B to C kinetic energy decreases and potential energy increases.

9. Electric lines of forces of q will not penetrate the conductor.

10. $\mathbf{E}_0 = 0$

    $\Rightarrow \quad \mathbf{E}_{AB} + \mathbf{E}_{ACDB} = 0$

    $\Rightarrow \quad \mathbf{E}_{ACDB} = -\mathbf{E}_{AB}$

    $\mathbf{E}_{AB}$ will be along KO. Therefore, $\mathbf{E}_{ACDB}$ will be along OK.

11.

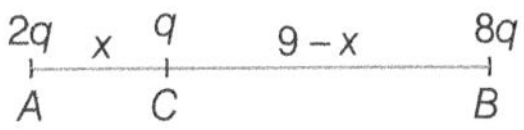

$$U = \frac{1}{4\pi\varepsilon_0} \times \frac{1}{10^{-2}}\left[\frac{(8q)(2q)}{9} + \frac{(2q)(q)}{x} + \frac{(q)(8q)}{(9-x)}\right]$$

$$= \frac{10^2 q^2}{4\pi\varepsilon_0}\left[\frac{16}{9} + \frac{2}{x} + \frac{8}{9-x}\right]$$

For potential energy to be minimum,

$$\frac{dU}{dx} = 0 \quad \Rightarrow \quad -\frac{2}{x^2} + \frac{8}{(9-x)^2} = 0$$

or $\quad \frac{9-x}{x} = 2$

or $\quad x = 3\text{ cm}$

12. $K_i + U_i = K_f + U_f$

    $$0 - q_0 V_i = K_f - q_0 V_f$$

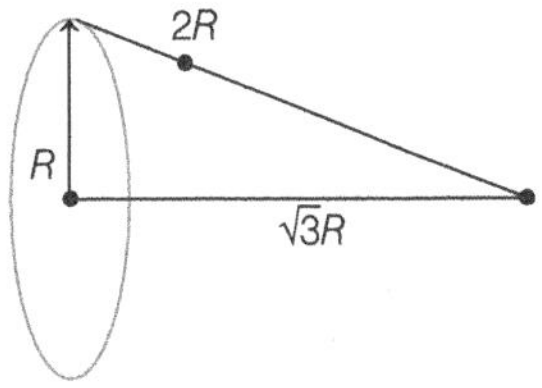

$\therefore \quad K_f = q_0 (V_f - V_i)$

$$= q_0\left[\frac{q}{4\pi\varepsilon_0}\left\{\frac{1}{R} - \frac{1}{2R}\right\}\right]$$

$$= \frac{1}{4\pi\varepsilon_0}\left(\frac{qq_0}{2R}\right)$$

13. Initially, $qE = mg$ (in opposite directions)

After changing the polarity, qE is also in the direction of mg. Therefore,

$$a = \frac{F_{net}}{m} = \frac{mg + qE}{m} = 2g \quad (\text{as } qE = mg)$$

$$= 20 \text{ m/s}^2$$

14. $v = \sqrt{\frac{2qV}{m}}$

$$v_p = \sqrt{\frac{2eV}{m}}$$

$$v_d = \sqrt{\frac{2e(2V)}{2m}}$$

$$v_\alpha = \sqrt{\frac{2(2e)(4V)}{4m}}$$

The ratio is $1:1:\sqrt{2}$.

15. $F_1 = \frac{kq_1q_2}{r^2}$ $\left(\text{as } k = \frac{1}{4\pi\varepsilon_0}\right)$

$$F_2 = \frac{kq_1'q_2'}{r^2}$$

where, $q_1' = q_2' = \frac{q_1 + q_2}{2}$

$$\therefore \quad F_2 = \frac{k(q_1 + q_2/2)^2}{r^2}$$

$$\left(\frac{q_1 + q_2}{2}\right)^2 > q_1q_2$$

This is because,

$$(q_1 + q_2)^2 = (q_1 - q_2)^2 + 4q_1q_2$$

or $(q_1 + q_2)^2 > 4q_1q_2$

or $\left(\frac{q_1 + q_2}{2}\right)^2 > q_1q_2$

$$\therefore \quad F_2 > F_1$$

16. $W = Fs\cos\theta = qEs\cos\theta$

$$\therefore \quad E = \frac{W}{qs\cos\theta} = \frac{4}{0.2 \times 2 \times \cos 60^\circ}$$

$$= 20 \text{ N/C}$$

17. At point P, for equilibrium,

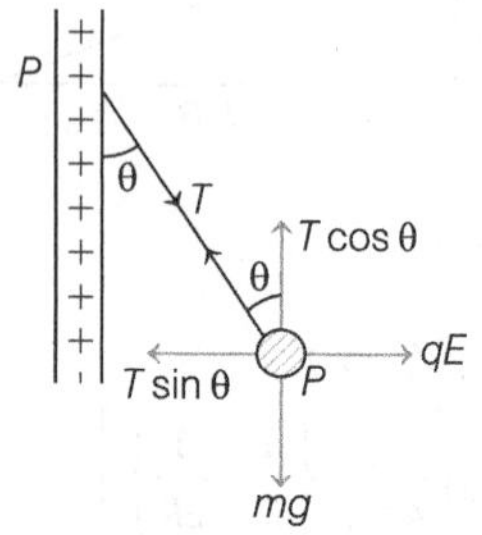

$$T\sin\theta = qE \text{ and } T\cos\theta = mg$$

Dividing, we get

$$\tan\theta = \frac{qE}{mg}$$

$$\because \quad E = \frac{\sigma}{2\varepsilon_0}$$

$$\therefore \quad \tan\theta = \frac{q}{mg}\left(\frac{\sigma}{2\varepsilon_0}\right)$$

$$\therefore \quad \sigma \propto \tan\theta$$

18. The induced charges on conducting sphere due to $+q$ charge at P are as shown in figure.

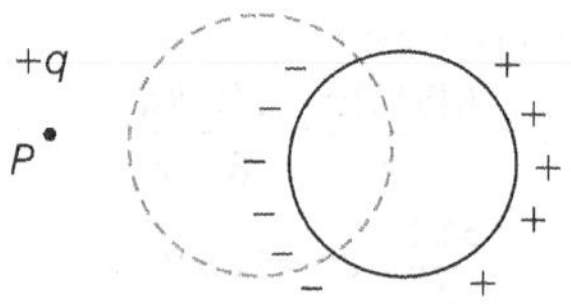

Now, net charge inside the closed dotted surface is negative. Hence, according to Gauss's theorem net flux is negative.

19. $W_{A \to B} = q(V_B - V_A)$

$$= q\left[-\int_A^B E\,dr\right] = q\int_B^A E\,dr$$

$$= q\int_{2a}^{a} \frac{\lambda}{2\pi\varepsilon_0 r}\,dr = \frac{q\lambda}{2\pi\varepsilon_0}\ln\left(\frac{1}{2}\right)$$

20. Let radius of drop is r. According to Millikan, for balance of drop

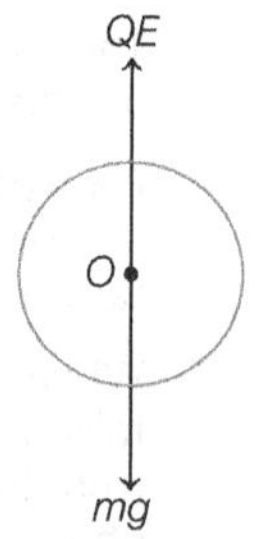

$$QE = mg$$

or $Q\frac{V}{d} = \left[\left(\frac{4}{3}\pi r^3\right)\rho\right]g$

where, V is potential difference and ρ is density of drop

$$\therefore \quad \frac{Q_1}{Q_2} = \left(\frac{r_1}{r_2}\right)^3 \times \frac{V_2}{V_1}$$

$$\therefore \quad \frac{Q}{Q_2} = \left(\frac{r}{r/2}\right)^3 \times \frac{600}{2400} = 2$$

$$\Rightarrow \quad Q_2 = Q/2$$

21. Electric field at a distance r due to infinite long wire is

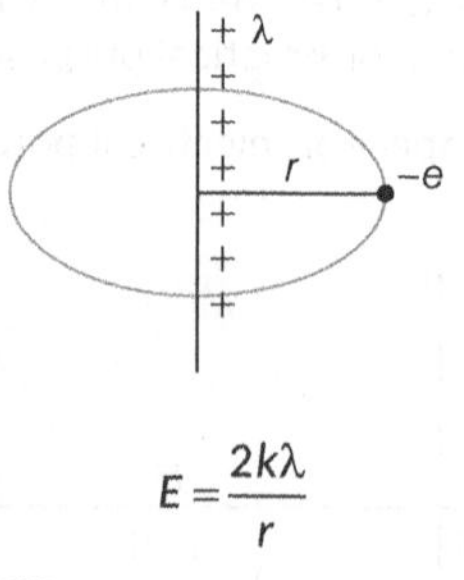

$$E = \frac{2k\lambda}{r} \quad \text{...(i)}$$

Force on electron,

$$F = qE = eE \quad \text{...(ii)}$$

or $$F = e\left(\frac{2k\lambda}{r}\right)$$

or $$F = \frac{2k\lambda e}{r} \quad \text{...(iii)}$$

This force will provide the centripetal force,

$$F = \frac{mv^2}{r} = \frac{2k\lambda e}{r} \Rightarrow v = \sqrt{\frac{2k\lambda e}{m}}$$

Kinetic energy, $KE = \frac{1}{2}mv^2$

$$= \frac{1}{2}m\left(\frac{2k\lambda e}{m}\right)$$

$= k\lambda e$ (where k is constant)

Kinetic energy will remain constant.

22. Given, $q_0 = 10\,\mu C = 10^{-5}\,C$

Let point charge of

$$q = 40\,\mu C = 4 \times 10^{-5}\,C$$

is placed at a distance x cm from origin O

According to given situation,

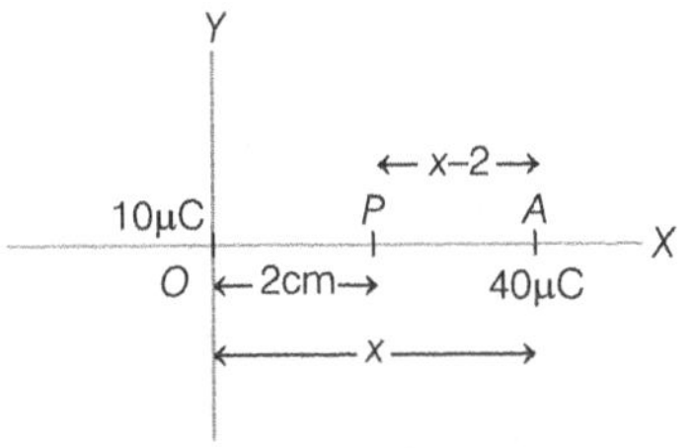

$E_{OP} = E_{AP}$

$$\Rightarrow \frac{1}{4\pi\varepsilon_0} \cdot \frac{10^{-5}}{(2 \times 10^{-2})^2} = \frac{1}{4\pi\varepsilon_0} \cdot \frac{4 \times 10^{-5}}{(x-2)^2 \times (10^{-2})^2}$$

$$\Rightarrow \frac{1}{(2 \times 10^{-2})^2} = \frac{4}{(x-2)^2 \times (10^{-2})^2}$$

$$\Rightarrow \frac{1}{2} = \frac{2}{x-2}$$

$$\Rightarrow x - 2 = 4$$

$$\Rightarrow x = 6\text{ cm}$$

23. Electric field in conducting materials is always zero. So, $E_{II} = 0$

If we make any Gaussian surface in region I and III, then it will enclose the charge Q.

Hence, by Gauss law,

$$\oint E \cdot ds = \frac{Q\text{ enclosed}}{\varepsilon_0}$$

There will be electric field.

So, $E_I \neq 0$ and $E_{III} \neq 0$

24. We can replace $-Q$ charge at origin by $+Q$ and $-2Q$.

Now, due to $+Q$ charge at every corner of cube, electric field at centre of cube is zero. So, net electric field at centre is only due to $-2Q$ charge at origin. Vector form of electric field strength,

$$E = \frac{Kqr}{r^3}$$

Here, position vector, $r = \frac{a}{2}(\hat{x} + \hat{y} + \hat{z})$

$$\Rightarrow |r| = \sqrt{\left(\frac{a}{2}\right)^2 + \left(\frac{a}{2}\right)^2 + \left(\frac{a}{2}\right)^2} = \frac{\sqrt{3}a}{2}$$

$$\Rightarrow E = \frac{1}{4\pi\varepsilon_0} \times \frac{(-2Q)a/2}{\left(\frac{a}{2}\sqrt{3}\right)} \cdot (\hat{x} + \hat{y} + \hat{z})$$

$$E = \frac{-2Q(\hat{x} + \hat{y} + \hat{z})}{3\sqrt{3}\,\pi a^2\,\varepsilon_0}$$

25. Potential at centre due to an elemental charge dq at centre of half-ring will be

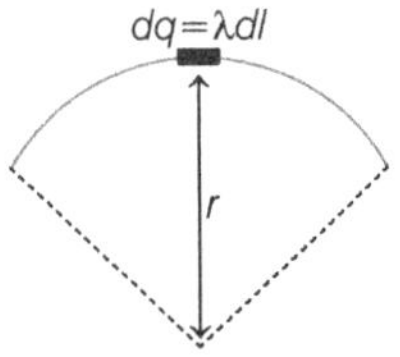

$$dV = \frac{kdq}{R} = \frac{1}{4\pi\varepsilon_0} \cdot \frac{\lambda dl}{R}$$

So, potential due to both half-rings will be

$$V = \int_{\text{ring-1}} dV + \int_{\text{ring-2}} dV$$

$$= \frac{\lambda}{4\pi\varepsilon_0 R_1}\int dl + \frac{\lambda}{4\pi\varepsilon_0 R_2}\int dl$$

$$= \frac{\lambda}{4\pi\varepsilon_0 R_1} \times \pi R_1 + \frac{\lambda}{4\pi\varepsilon_0 R_2} \times \pi R_2$$

$$= \frac{\lambda}{4\varepsilon_0} + \frac{\lambda}{4\varepsilon_0} = \frac{2\lambda}{4\varepsilon_0} = \frac{\lambda}{2\varepsilon_0}$$

26. Potential at origin will be given by

$$V = \frac{q}{4\pi\varepsilon_0}\left[\frac{1}{x_0} - \frac{1}{2x_0} + \frac{1}{3x_0} - \frac{1}{4x_0} + \dots\right]$$

$$= \frac{q}{4\pi\varepsilon_0} \cdot \frac{1}{x_0}\left[1 - \frac{1}{2} + \frac{1}{3} - \frac{1}{4} + \dots\right]$$

$$= \frac{q}{4\pi\varepsilon_0 x_0}\ln(2)$$

27. Electric potential at any point inside a hollow metallic sphere is constant. Therefore, if potential at surface is 10 V, potential at centre will also be 10 V.

28. Electric field at any point depends on both charges Q_1 and Q_2. But electric flux passing from any closed surface depends on the charged enclosed by that closed surface only.

29. $E = \frac{k\,q_{in}}{r^2}$ $\left(\text{Here, } k = \frac{1}{4\pi\varepsilon_0}\right)$

$\therefore$ $E_A = E_c = 0$

but, $E_B \neq 0$

$$V = \frac{kq}{R} \qquad (r \le R)$$

$$V = \frac{kq}{r} \qquad (r \ge R)$$

30.

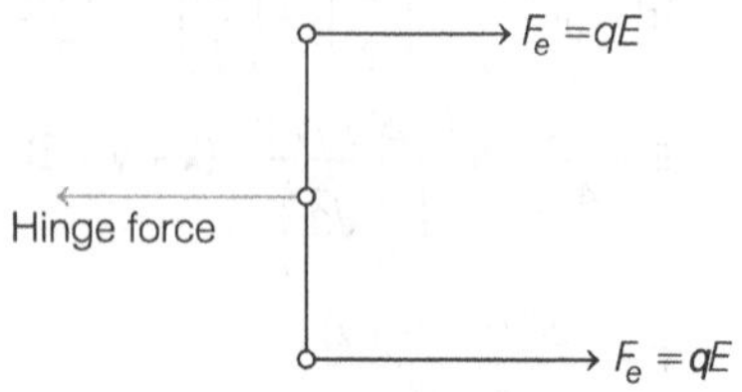

Higher force $= 2qE$ (towards left)

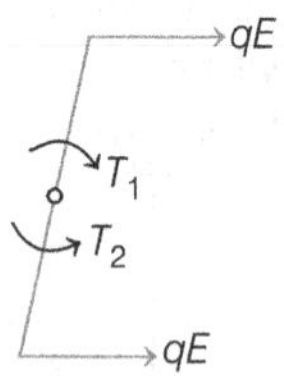

If we displace the rod, $\tau_1 = \tau_2$ or $\tau_{net} = 0$ in displaced position too. Hence, equilibrium is neutral.

31. The electric field inside the sphere

$$(r < R)\, E = \frac{1}{4\pi\varepsilon_0} \cdot \frac{Qq}{R^3} r$$

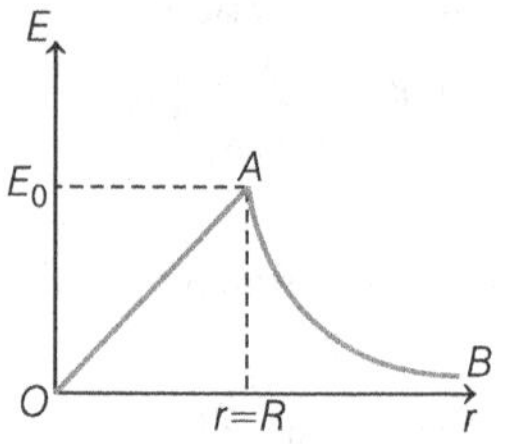

or, $E \propto r$

Outside the sphere

$$(R < r < \infty)\, E = \frac{1}{4\pi\varepsilon_0} \cdot \frac{Q}{r^2} \quad \text{or} \quad E \propto \frac{1}{r^2}$$

Hence, E increases for $r < R$ and decreases for $R < r < \infty$

32. Let Q be the charge on the ring, the negative charge $-q$ is released from point $P(0, 0, z_0)$. The electric field at P due to the charged ring will be along positive z-axis and its magnitude will be

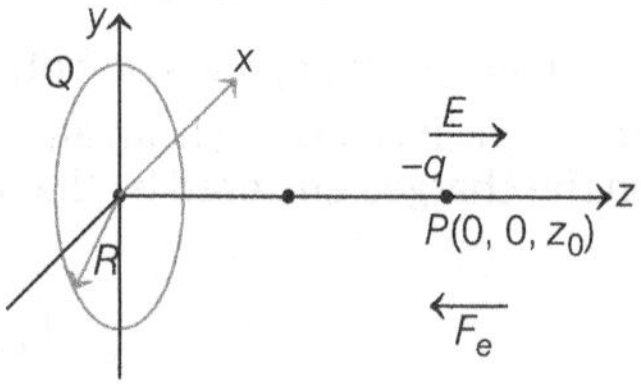

$$E = \frac{1}{4\pi\varepsilon_0} \frac{Qz_0}{(R^2 + z_0^2)^{3/2}}$$

$E = 0$, at centre of the ring because $z_0 = 0$

Force on charge at P will be towards centre as shown, and its magnitude is

$$F_e = qE = \frac{1}{4\pi\varepsilon_0} \cdot \frac{Qq}{(R^2 + z_0^2)^{3/2}} \cdot z_0 \qquad \text{...(i)}$$

Similarly, when it crosses the origin, the force is again towards centre O.

Thus, the motion of the particle is periodic for all values of z_0 lying between 0 and ∞.

Secondly, if $z_0 << R$, $(R^2 + z_0^2)^{3/2} \approx R^3$

$$F_e \approx \frac{1}{4\pi\varepsilon_0} \cdot \frac{Qq}{R^3} \cdot z_0 \qquad \text{[From Eq. (i)]}$$

i.e. the restoring force $F_3 \propto -z_0$. Hence, the motion of the particle will be simple harmonic. (Here negative sign implies that the force is towards its mean position O, where E is zero, so force on $-q$ is zero)

33.

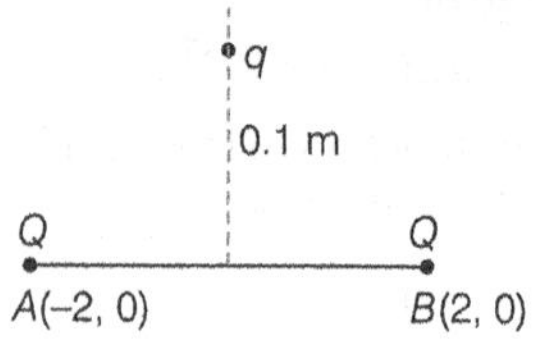

Charge q oscillates, so q is negative. Force on q is

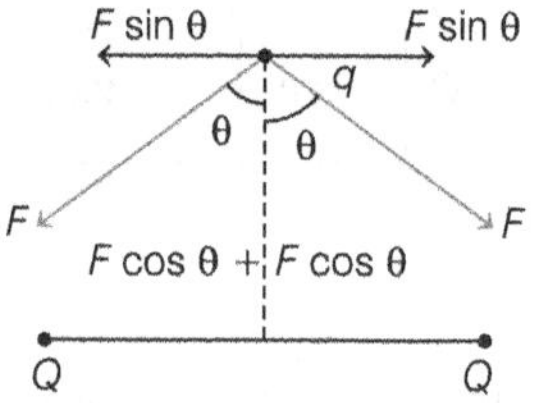

So, net force on q is

$$F_{net} = 2F\cos\theta$$

$$= 2 \times \frac{1}{4\pi\varepsilon_0} \times \frac{Qq}{y^2 + a^2} \times \cos\theta$$

$$= 2 \times \frac{1}{4\pi\varepsilon_0} \times \frac{Qq}{y^2 + a^2} \times \frac{y}{(y^2 + a^2)^{1/2}}$$

$$= \frac{1}{4\pi\varepsilon_0} \cdot \frac{2Qqy}{(y^2 + a^2)^{3/2}}$$

For $y << a$, $F_{net} = \frac{1}{4\pi\varepsilon_0} \cdot \frac{2Qqy}{a^3}$

Now, as $a = 2$ m and $y = 0.1$ m

We can use approximation

$$\therefore \quad F = \frac{1}{4\pi\varepsilon_0} \times \frac{2 \times Qqy}{a^3}$$

$$\Rightarrow 9 \times 10^{-3} = \frac{9 \times 10^9 \times 2 \times 8 \times 10^{-6} \times q \times 0.1}{2^3}$$

$$\Rightarrow \quad q = 5 \times 10^{-6}\ \text{C}$$

$$= 5\ \mu\text{C}$$

34. At $t = 0$, $y = 0.1$ m assuming an undamped SHM, amplitude of SHM is 10 cm.

35. $T = 2\pi\sqrt{\frac{m}{k}}$

Here, spring constant, $k = \frac{1}{4\pi\varepsilon_0} \cdot \frac{2Qq}{a^3}$

So, frequency, $f = \frac{1}{T}$

On substituting values, we get

$$f = 5$$

36. Charge on surface = Charge density × Area

$$= \sigma (4\pi a^2)$$

37. Net field reaching the hollow metallic inside surface is zero. Metal sphere shield effect of charge q_B. So, net force on A is zero.

38. Charge distribution on outer surface remains symmetrical even if q_A is not put at centre.

40. $F_e = \frac{1}{4\pi\varepsilon_0} \frac{q_1 q_2}{r^2} = \frac{1}{4\pi\varepsilon_0} \frac{q^2}{r^2}$

$$F_g = \frac{Gm_1 m_2}{r^2} = \frac{Gm^2}{r^2}$$

$$\therefore \quad \frac{F_e}{F_g} = \frac{(1/4\pi\varepsilon_0)}{G}\left(\frac{q}{m}\right)^2$$

Substituting the values, we can get the result.

41. Atomic weight of copper is 63.55 g.

∴ Number of atoms in 63.55 g $= 6.02 \times 10^{23}$

∴ Number of atoms in 1 g of copper,

$$N_1 = \frac{6.02 \times 10^{23}}{63.55}$$

∴ Total charge (of protons) in 1 g of protons,

$$N_2 = (29 N_1)\, e$$

According to given problem charge on each sphere,

$$q = 1\% \text{ of } N_2 = \frac{29 N_1 e}{100}$$

Substituting the values of N_1 and e, we can find q and then we can apply

$$F = \frac{1}{4\pi\varepsilon_0} \frac{q \times q}{r^2}$$

42. Distance between two given points,

$$r = \sqrt{(8-2)^2 + (-5-3)^2} = 10\,\text{m}$$

$$\therefore \quad E = \frac{1}{4\pi\varepsilon_0} \frac{q}{r^2}$$

$$= \frac{9 \times 10^9 \times 50 \times 10^{-6}}{(10)^2}$$

$$= 4.5 \times 10^3 \text{ V/m}$$

or $\quad E = 4.5$ kV/m

If charge is kept at a point whose position vector is $\mathbf{r}_1$ and we have to find $\mathbf{E}$ at a point whose position vector is $\mathbf{r}_2$. Then formula of $\mathbf{E}$ is given by,

$$\mathbf{E} = \frac{1}{4\pi\varepsilon_0} \frac{q}{r^3} (\mathbf{r}_2 - \mathbf{r}_1)$$

where, $\quad r = |\mathbf{r}_2 - \mathbf{r}_1|$

Here, $\quad \mathbf{r}_1 = \mathbf{r}_0 = (2\hat{\mathbf{i}} + 3\hat{\mathbf{j}})$ m

and $\quad \mathbf{r}_2 = \mathbf{r} = (8\hat{\mathbf{i}} - 5\hat{\mathbf{j}})$ m

Substituting the values we can get the answer.

43. There are two dipoles and their electric dipole moment **p** vectors are at right angle. We have to find electric field at their perpendicular bisectors at a distance x from their centres.

$$\therefore \quad E_{net} = \sqrt{2}(E_{\perp \text{ bisector}}) = \sqrt{2}\left[\frac{p}{4\pi\varepsilon_0 (x^2 + l^2)^{3/2}}\right]$$

$$= \frac{\sqrt{2}(q)(2l)}{4\pi\varepsilon_0 (x^2 + l^2)^{3/2}}$$

$$\therefore \quad E_{net} = \frac{ql}{\sqrt{2}\,\pi\varepsilon_0 (x^2 + l^2)^{3/2}}$$

44. Electric field at a distance x from the centre of ring (along its axis) is given by,

$$E = \frac{1}{4\pi\varepsilon_0} \frac{qx}{(R^2 + x^2)^{3/2}}$$

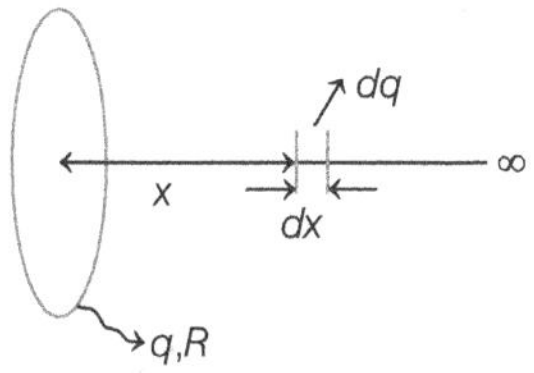

If we take a small length dx of the thread at this point, then charge on this thread will be,

$$dq = (\lambda dx)$$

Small force dF on this charge dq will be,

$$dF = (dq)E$$

If we integrate dF from $x = 0$ to $x = \infty$, then we will get the total force F. This force comes out to be,

$$F = \frac{q\lambda}{4\pi\varepsilon_0 R}$$

45. Let q_1, q_2 and q_3 be the respective charges. Then,

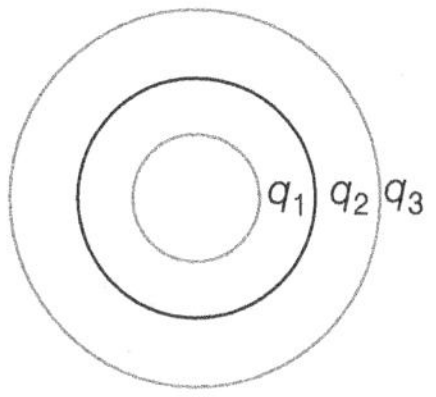

$$10 = \frac{9 \times 10^9}{10^{-2}}\left[\frac{q_1}{1} + \frac{q_2}{2} + \frac{q_3}{4}\right]$$

$$0 = \frac{9 \times 10^9}{10^{-2}}\left[\frac{q_1}{2} + \frac{q_2}{2} + \frac{q_3}{4}\right]$$

and $$40 = \frac{9 \times 10^9}{10^{-2}}\left[\frac{q_1}{4} + \frac{q_2}{4} + \frac{q_3}{4}\right]$$

Solving these equations, we get

$$q_1 = +\frac{200}{9}\times 10^{-12}\ \text{C}, q_2 = -200\times 10^{-12}\ \text{C and}$$

$$q_3 = \frac{3200}{9}\times 10^{-12}\ \text{C}$$

(a) **At $r = 1.25$ cm**

$$V = \frac{9\times 10^9}{10^{-2}}\left[\frac{(200/9)\times 10^{-12}}{1.25} - \frac{200\times 10^{-12}}{2} + \frac{(3200/9)\times 10^{-12}}{4}\right]$$

$$= 6\ \text{V}$$ **Ans.**

(b) **Potential at $r = 2.5$ cm**

$$V = \frac{9\times 10^9}{10^{-2}}\left[\frac{(200/9)\times 10^{-12}}{2.5} - \frac{200\times 10^{-12}}{2.5} + \frac{(3200/9)\times 10^{-12}}{4}\right]$$

$$= 16\ \text{V}$$ **Ans.**

(c) Electric field at $r = 1.25$ cm will be due to charge q_1 only.

$$\therefore \quad E = \frac{1}{4\pi\varepsilon_0}\cdot\frac{q_1}{r^2} = \frac{9\times 10^9\times(200/9)\times 10^{-12}}{(1.25\times 10^{-2})^2}$$

$$= 1.28\times 10^3\ \text{V/m}$$ **Ans.**

46.

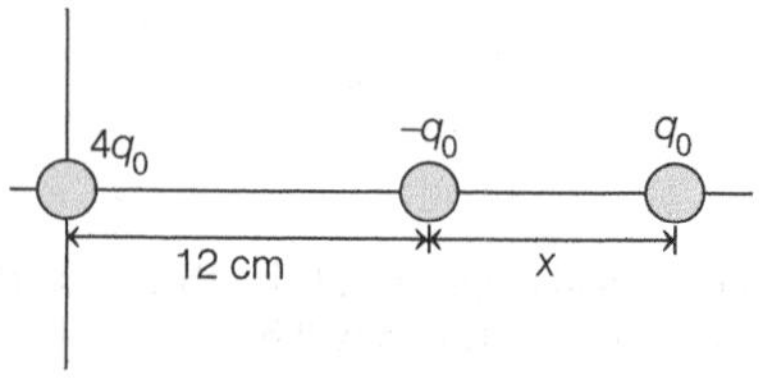

Force on proton of charge q_0 is zero in above shown situation.

As, net force on q_0 is zero,

$$\frac{k4q_0(q_0)}{(12+x)^2\times 10^{-4}} = \frac{k(q_0)(q_0)}{x^2\times 10^{-4}}$$

$$\Rightarrow \quad 4x^2 = (12+x)^2$$

$$\Rightarrow \quad x = 12\ \text{cm}$$

So, distance of proton from origin is $12 + 12 = 24$ cm.

47.

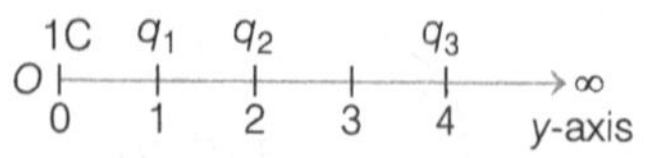

$$F_{total} = \left[\frac{kq_1q_2}{r_1^2} + \frac{kq_1q_3}{r_2^2} + \frac{kq_1q_4}{r_3^2} + \ldots\right]$$

$$\Rightarrow \quad F = (k)(10)^{-6}\left[\frac{1}{1^2} + \frac{1}{2^2} + \frac{1}{4^2} + \ldots\right]$$

$$= \frac{(k)10^{-6}}{1-\frac{1}{4}} = \frac{(9\times 10^9)\times 4\times 10^{-6}}{3}\ \text{N}$$

$$= 12\times 10^3\ \text{N}$$

Hence, the value of x is 12.

Level 2

48. Let us conserve angular momentum of $+2q$ about the point at $+Q$.

$$mv_1r_1\sin\theta_1 = mv_2r_2\sin\theta_2$$

$$(m)(v)(R)\sin 150° = m\left(\frac{v}{\sqrt{3}}\right)r_{min}\sin 90°$$

$$\therefore \quad r_{min} = \frac{\sqrt{3}}{2}R$$

49. $(v_A)_y = v \Rightarrow (v_B)_y = 2v\sin 30° = v$

Since, y-component of velocity remains unchanged. Hence electric field is along $(-\hat{\mathbf{i}})$ direction. Work done by electrostatic force in moving from A to B = change in its kinetic energy

$$\therefore \quad (eE)(2a - a) = \frac{1}{2}m(4v^2 - v^2)$$

$$E = \frac{3mv^2}{2ea} \quad \text{or} \quad \mathbf{E} = -\frac{3mv^2}{2ea}\hat{\mathbf{i}}$$

Rate of doing work done = power

$$= Fv\cos\theta$$

$$= \left(\frac{3mv^2}{2ea}\right)(e)(2v)\cos 30°$$

$$= \frac{3\sqrt{3}}{2}\frac{mv^2}{a}$$

50.

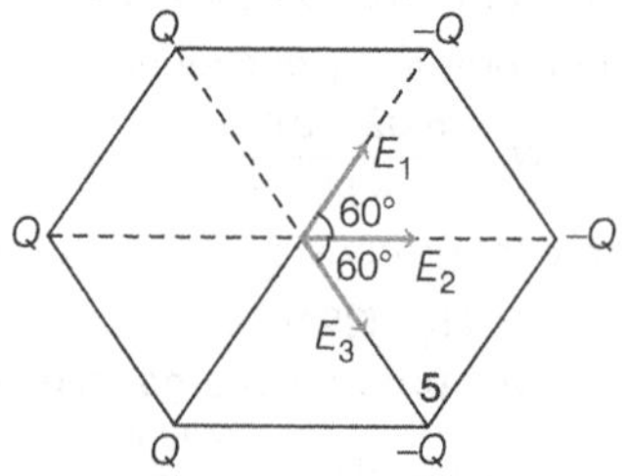

$$E = E_1 = E_2 = E_3 = \left(\frac{1}{4\pi\varepsilon_0}\right)\left(\frac{P}{x^3}\right)$$

$$= \left(\frac{1}{4\pi\varepsilon_0}\right)\frac{2Qa}{x^3}$$

Resultant of E_1 and E_3 is also equal to E along E_2

$$\therefore \quad E_{net} = 2E \quad (\text{along } E_2)$$

$$= \frac{Qa}{\pi\varepsilon_0 x^3}$$

51.

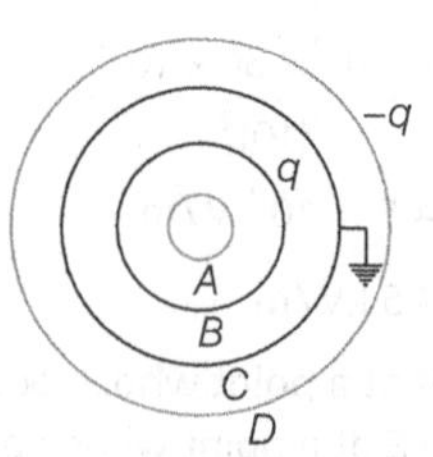

Let Q charge comes on shell-C from earth. Then,

$$V_C = 0$$

$\therefore$ $$\frac{kq}{3a} + \frac{kQ}{3a} - \frac{kq}{4a} = 0$$

Solving, we get

$$Q = -\frac{q}{4}$$

Now, $$V_A = \frac{kq}{2a} - \frac{kq/4}{3a} - \frac{kq}{4a} = \frac{kq}{6a}$$

and $$V_C = 0$$

$\therefore$ $$V_A - V_C = \frac{kq}{6a}$$

52. $\rho = \dfrac{q}{(4/3)\,\pi R^3}$

$\therefore$ $$q = \frac{4}{3}\pi\rho R^3$$

$$V_C - V_S = \frac{3}{2}\left(\frac{1}{4\pi\varepsilon_0}\cdot\frac{q}{R}\right) - \frac{1}{4\pi\varepsilon_0}\cdot\frac{q}{R}$$

$$= \frac{q}{8\pi\varepsilon_0 R}$$

Substituting the value of q, we have

$$V_C - V_S = \frac{\rho R^2}{6\,\varepsilon_0}$$

53. $W_T = 0$

$W_{F_e} = (F_e)$ (displacement in the direction of force)
= Kinetic energy of the particle.

$\therefore$ $$\frac{1}{2}mv^2 = qE\left[\frac{l}{2} - \frac{l}{2}\cos 60°\right]$$

$\therefore$ $$v = \sqrt{\frac{qEl}{m}}$$

54. Electric field, $E = -\dfrac{d\phi}{dr} = -2ar$

By Gauss' theorem, $E(4\pi r^2) = \dfrac{q}{\varepsilon_0}$

$\Rightarrow$ $$q = -8\pi\,\varepsilon_0\,ar^3$$

$$\rho = \frac{dq}{dV} = \frac{dq}{dr}\times\frac{dr}{dV}$$

$$= (-24\pi\varepsilon_0\,ar^2)\left(\frac{1}{4\pi r^2}\right)$$

$$= -6\,\varepsilon_0 a$$

55. $F_x = 0$

$\therefore$ $$a_x = 0$$

$$F_y = qE$$

$\therefore$ $$a_y = \frac{qE}{m}$$

$x = vt$ and

$$y = \frac{1}{2}a_y t^2 = \frac{1}{2}\left(\frac{qE}{m}\right)t^2$$

Substituting $t = \dfrac{x}{v}$ in expression of y, we get

$$y = \frac{1}{2}\left(\frac{qEx^2}{mv^2}\right)$$

$$KE = \frac{1}{2}m\,(v_x^2 + v_y^2)$$

where, $v_x = v$ and $v_y = a_y t = \dfrac{qE}{m}t$

56.

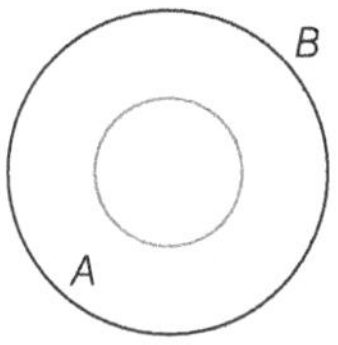

They have a common potential in the beginning. This implies that only B has the charge in the beginning.

$\therefore$ $$V = \frac{kq_B}{b}$$

or $$kq_B = Vb$$

Now, suppose q_A charge is given to A. Then,

$$V_A = \frac{kq_A}{a} + \frac{kq_B}{b} = 0$$

or $$kq_A = -a\left(\frac{kq_B}{b}\right) = -aV$$

Now, $$V_B = \frac{kq_A}{b} + \frac{kq_B}{b}$$

$$= -\frac{a}{b}V + V = V\left(1 - \frac{a}{b}\right)$$

57.

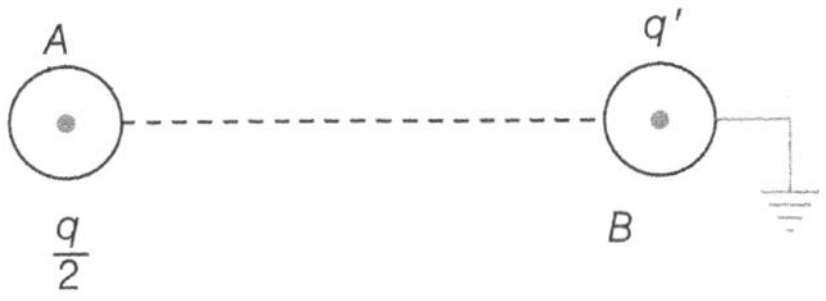

Let charge on B is q'.

$$V_B = 0$$

$\therefore$ $$\frac{(k(q/2))}{d} + \frac{kq'}{r} = 0$$

$\therefore$ $$q' = -\frac{qr}{2d}$$

58. $V = \dfrac{q}{4\pi\varepsilon_0 r} - \dfrac{q}{4\pi\varepsilon_0 3r} = \dfrac{q}{6\pi\varepsilon_0 r}$

Electric field intensity at a distance $3r$ is

$$E = \frac{q}{4\pi\varepsilon_0(3r)^2} = \frac{q}{4\pi\varepsilon_0\cdot 9r^2}$$

Thus, $$\frac{E}{V} = \frac{1}{6r}$$

$\Rightarrow$ $$E = \frac{V}{6r}$$

59. $\frac{K(2Q)}{2a}+\frac{KQ}{2a}=5\,V$ or $\frac{K(3Q)}{2a}=5$

$\therefore \quad \frac{KQ}{a}=\frac{10}{3}$

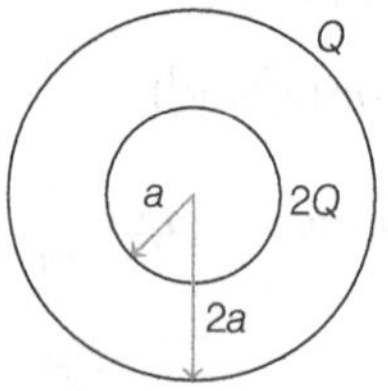

Now, potential of inner sphere

$$V_i=\frac{K2Q}{a}+\frac{KQ}{2a}=\frac{5}{2}\frac{KQ}{a}=\frac{5}{2}\left(\frac{10}{3}\right)$$

$$=\frac{25}{3}\,V$$

60. $120=\frac{Kq}{2} \Rightarrow Kq=240$ units

When it is made to touch the bigger sphere, whole charge will transfer to bigger sphere.

$\therefore \quad V=\frac{Kq}{6}=\frac{240}{6}=40$ volts

61. $q_A=\frac{Q}{2}$

When B is earthed :

$$V_B=0$$

$\therefore \quad \frac{kq_B}{a}+\frac{k(Q/2)}{d}=0$

$\therefore \quad q_B=\frac{-Qa}{2d}$

When C is earthed :

$$V_C=0$$

$\therefore \quad \frac{kq_C}{a}+\frac{k(Q/2)}{d}-\frac{k(Qa/2d)}{d}=0$

$\therefore \quad q_C=\frac{Qa^2}{2d^2}-\frac{Qa}{2d}$

62. The induced charges will be as under.

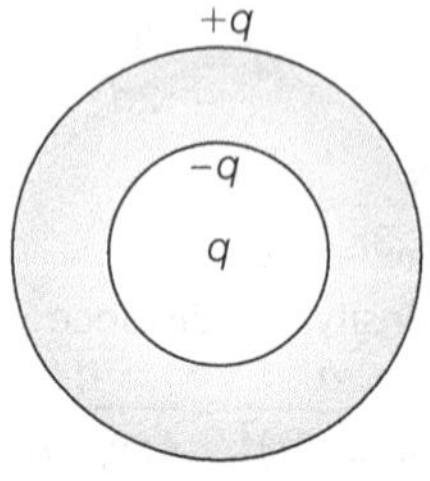

Potential at point O (the centre),

$$V_O=\frac{1}{4\pi\varepsilon_0}\left[\frac{q}{r}-\frac{q}{R}+\frac{q}{2R}\right]$$

$$=\frac{q}{4\pi\varepsilon_0}\left[\frac{1}{r}-\frac{1}{2R}\right]$$

63. Pressure inside the soap bubble is

$$p=p_0+\frac{4T}{r}$$

Due to charging, electrostatic pressure is

$$p_\sigma=\frac{\sigma^2}{2\varepsilon_0}$$

So, net pressure inside the charged bubble is

$$p_0+\frac{4T}{r}-\frac{\sigma^2}{2\varepsilon_0}$$

By given condition,

$$p_0+\frac{4T}{r}-\frac{\sigma^2}{2\varepsilon_0}=p_0$$

$$\Rightarrow \quad \frac{4T}{r}=\frac{\sigma^2}{2\varepsilon_0}=\frac{\varepsilon_0 V^2}{2r^2}$$

$$\Rightarrow \quad V=\sqrt{\frac{8Tr}{\varepsilon_0}}$$

64.

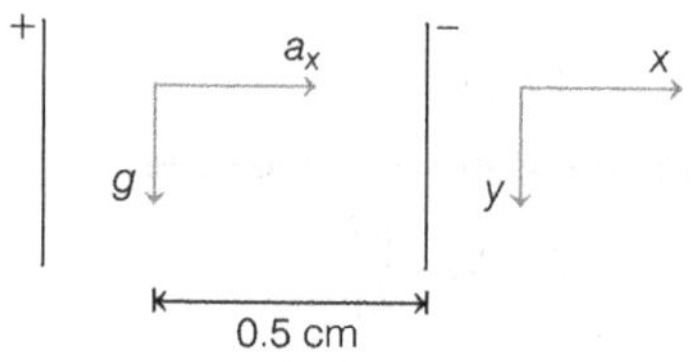

$$a_x=qE=q\left(\frac{V}{d}\right)$$

$$t=\sqrt{\frac{2S}{a_x}}=\sqrt{\frac{2(d/2)}{a_x}}=\sqrt{\frac{d}{(qV/d)}}=\frac{d}{\sqrt{qV}}$$

Just before touching the other plate,

$$v_x=\sqrt{2a_xS}=\sqrt{2\left(\frac{qV}{d}\right)\left(\frac{d}{2}\right)}=\sqrt{qV}$$

$$v_y=gt=\frac{gd}{\sqrt{qV}}$$

Velocity is at 45° with vertical. Hence,

$$\sqrt{qV}=\frac{gd}{\sqrt{qV}}$$

$$\Rightarrow \quad V=\frac{gd}{q}=\frac{9.81\times10^{-2}}{1.6\times10^{-19}}$$

$$=6.13\times10^{17}\,V$$

65.

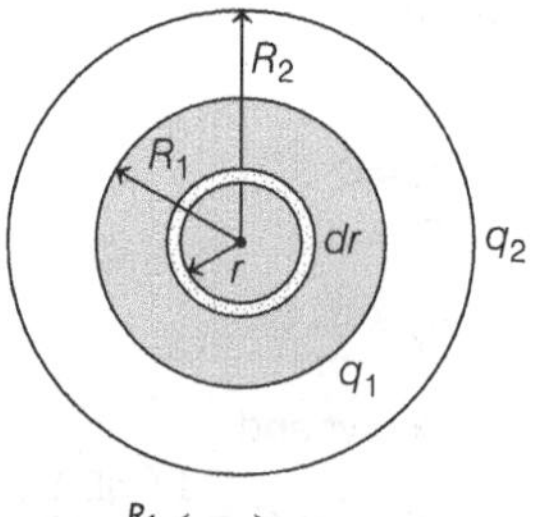

$$q_1=\int_0^{R_1}\left(\frac{\rho_0}{r}\right)(4\pi r^2 dr)$$

$$= \frac{4\pi\rho_0 R_1^2}{2} = 2\pi\rho_0 R_1^2$$

$$q_2 = (4\pi R_2^2)\,\sigma$$

Given, $\quad q_1 = q_2$

$$\therefore \quad \frac{R_2}{R_1} = \sqrt{\frac{\rho_0}{2\sigma}}$$

66. By Gauss' law,

$$\oint \mathbf{E}\cdot d\mathbf{A} = \frac{q_{net}}{\varepsilon_0} = \frac{\int_0^{r_1} \rho\, dV}{\varepsilon_0}$$

$$= \int_0^{r_1} \left(\frac{Qr}{\pi R^4}\right)(4\pi r^2)\,dr = \frac{Q r_1^4}{\varepsilon_0 R^4}$$

$$\Rightarrow \quad E\,4\pi r_1^2 = \frac{Q r_1^4}{\varepsilon_0 R^4}$$

$$\Rightarrow \quad E = \frac{Q r_1^2}{4\pi\varepsilon_0 R^4}$$

67.

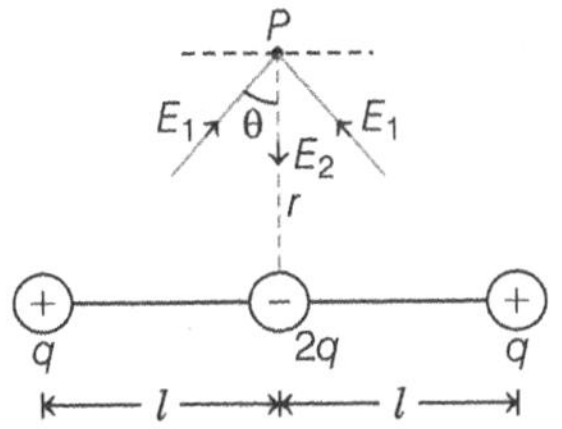

At P, vertical components of positive charges q are in same direction and horizontal components cancels out each other.

So, $\quad E_{net} = 2E_1 \cos\theta - E_2$

where, $\quad E_1$ = field of positive charge

and $\quad E_2$ = field of negative charge.

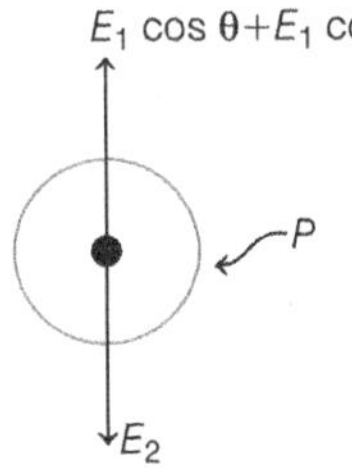

Now, $$E_1 = \frac{kq}{(\sqrt{r^2+l^2})^2}\,\frac{r}{\sqrt{r^2+l^2}}$$

and $$E_2 = \frac{k(2q)}{r^2}$$

$$\therefore \quad E_{net} = \frac{2kqr}{(r^2+l^2)^{3/2}} - \frac{2kq}{r^2}$$

$$\Rightarrow \quad E_{net} = 2kq\left\{\frac{r}{(r^2+l^2)^{3/2}} - \frac{1}{r^2}\right\}$$

For $r >> l, (r^2+l^2)^{3/2} \approx r^3$

$$\therefore \quad E_{net} = 2kq\left\{\frac{r^3-(r^2+l^2)^{3/2}}{r^2(r^2+l^2)^{3/2}}\right\}$$

$$= 2kqr^3\left\{\frac{1-(1+l^2/r^2)^{3/2}}{r^5(1+l^2/r^2)^{3/2}}\right\}$$

$$= \frac{2kqr^3}{r^5}\cdot\left\{\frac{1-\left(1+\frac{3}{2}\cdot\frac{l^2}{r^2}\right)}{1+\frac{3}{2}\cdot l^2/r^2}\right\}$$

$$= \frac{2kq\,r^3\left(-\frac{3}{2}\frac{l^2}{r^2}\right)}{r^5} = \left(-\frac{3}{2}\right)\frac{2kql^2}{r^4}$$

$$= -\frac{3}{4}\frac{k(q(2l)^2)}{r^4} = -\frac{3}{4}\frac{k\mathbf{p}}{r^4}$$

$[(\mathbf{p} = q(2l)^2 = \text{quadrupole moment})]$

$$\therefore \quad E \propto \frac{1}{r^4}$$

and it points towards negative charge.

68. Consider the given figure,

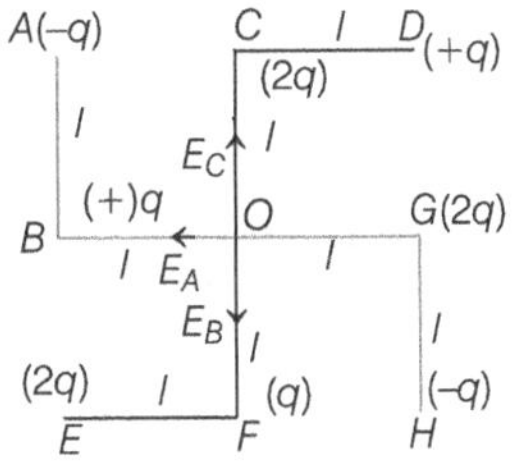

$$E_A = \frac{kq}{l^2} = E_B$$

$$E_C = \frac{kq}{(\sqrt{2}l)^2} = \frac{kq}{2l^2}$$

After resolving these electric fields and solving it, we get The magnitude of resultant electric field at O as

$$E_R = \sqrt{2}\times\frac{kq}{l^2}\left(1-\frac{1}{2\sqrt{2}}\right)$$

$$= \frac{kq}{2l^2}(2\sqrt{2}-1)$$

69. Let us consider a point M inside the cavity where electric field has to be calculated. Assume the cavity to contain similar charge distribution of positive and negative charge as the rest of sphere. Electric field at M due to uniformly distributed charge of the whole sphere of radius R_1,

$$\mathbf{E}_1 = \frac{\rho}{3\varepsilon_0}\mathbf{r}$$

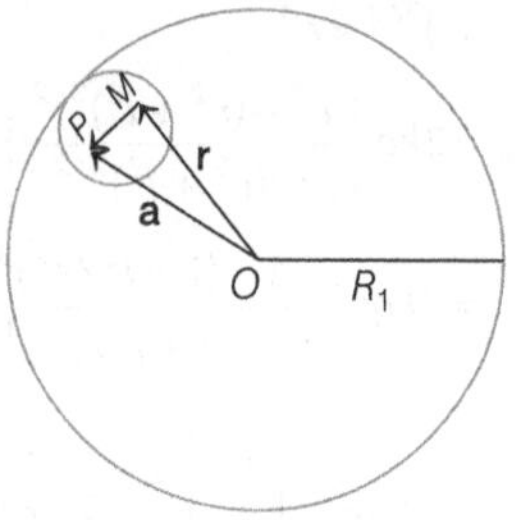

Electric field at M due to the negative charge distribution in the cavity,

$$E_2 = \frac{\rho}{3\varepsilon_0}\mathbf{MP}$$

$\therefore$ The total electric field at M inside the cavity,

$$\mathbf{E} = \mathbf{E}_1 + \mathbf{E}_2 = \frac{\rho}{3\varepsilon_0}\mathbf{r} + \frac{\rho}{3\varepsilon_0}\mathbf{MP}$$

$$\therefore \quad \mathbf{E} = \frac{\rho}{3\varepsilon_0}\mathbf{r} + \frac{\rho}{3\varepsilon_0}(\mathbf{a} - \mathbf{r}) \qquad [\because \mathbf{r} + \mathbf{MP} = \mathbf{a}]$$

$$\therefore \quad \mathbf{E} = \frac{\rho}{3\varepsilon_0}\mathbf{a}$$

Hence inside the cavity **E** is uniform and both its magnitude and direction depend on **a**.

70. $E_1 = \frac{1}{4\pi\varepsilon_0}\cdot\frac{Q}{R^2}$

$$E_2 = \frac{1}{4\pi\varepsilon_0}\cdot\frac{2Q}{R^2}$$

$$E_3 = \frac{1}{4\pi\varepsilon_0}\cdot\frac{Q/2}{R^2}$$

$$\therefore \quad E_2 > E_1 > E_3$$

71.

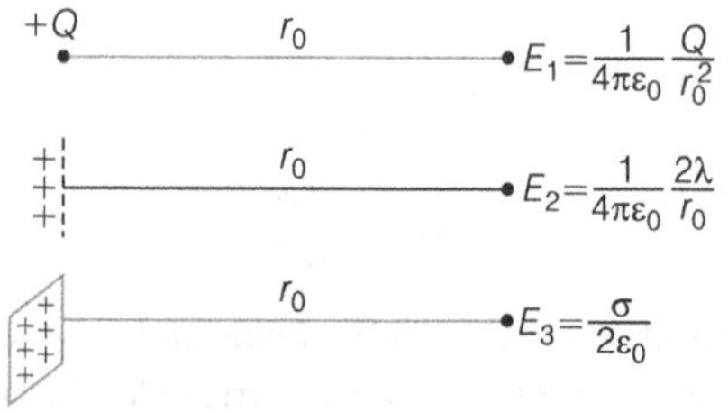

$$E_1(r_0) = E_2(r_0) = E_3(r_0) \qquad \text{(Given)}$$

(a) $E_1(r_0) = E_3(r_0)$

$$\therefore \quad \frac{1}{4\pi\varepsilon_0}\cdot\frac{Q}{r_0^2} = \frac{\sigma}{2\varepsilon_0}$$

$$\Rightarrow \quad Q = 2\pi\sigma r_0^2$$

(b) $E_2(r_0) = E_3(r_0)$

$$\frac{1}{4\pi\varepsilon_0}\frac{2\lambda}{r_0} = \frac{\sigma}{2\varepsilon_0}$$

$$\Rightarrow \quad r_0 = \frac{\lambda}{\sigma\pi}$$

(c) $E_1(r_0/2) = \frac{1}{4\pi\varepsilon_0}\frac{4Q}{r_0^2}$

$$= \frac{1}{4\pi\varepsilon_0} \times \frac{4\times 2\lambda r_0}{r_0^2} = \frac{1}{4\pi\varepsilon_0}\frac{8\lambda}{r_0}$$

$\because \quad E_1(r_0) = E_2(r_0) \therefore Q = 2\lambda r_0$

and $2E_2(r_0/2) = 2\left[\frac{1}{4\pi\varepsilon_0}\frac{4\lambda}{r_0}\right]$

$$= \frac{1}{4\pi\varepsilon_0}\frac{8\lambda}{r_0}$$

(d) $E_2(r_0/2) = \frac{1}{4\pi\varepsilon_0}\frac{2\lambda}{r_0/2}$

$$= \frac{1}{4\pi\varepsilon_0}\frac{4\lambda}{r_0} = \frac{\lambda}{\pi\varepsilon_0 r_0}$$

$$4E_3(r_0/2) = \frac{4\sigma}{2\varepsilon_0}$$

$$= \frac{2\sigma}{\varepsilon_0} = \frac{2}{\varepsilon_0}\times\frac{\lambda}{\pi r_0}$$

72. (a) $V_A = 2V = \frac{kq_A}{R} + \frac{kq_B}{2R}$

$$V_B = \frac{3}{2}V = \frac{kq_A}{2R} + \frac{kq_B}{2R}$$

Solving these two equations, we get

$$\frac{q_A}{q_B} = \frac{1}{2}$$

(b)

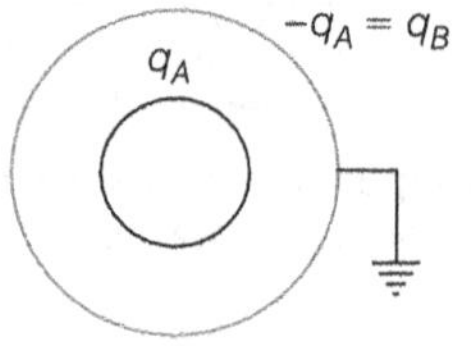

$$\frac{q_A'}{q_B'} = \frac{q_A}{-q_A} = -1$$

(c) & (d) Potential difference between A and B will remain unchanged as by earthing B, charge on A will not changed.

$$\therefore \quad V_A' - V_B' = V_A - V_B$$

$$= 2V - \frac{3}{2}V = \frac{V}{2}$$

$$\therefore \quad V_A' = \frac{V}{2} \text{ as } V_B' = 0$$

73. Using, $\oint \mathbf{E}\cdot d\mathbf{A} = \frac{q_{\text{enclosure}}}{\varepsilon_0}$

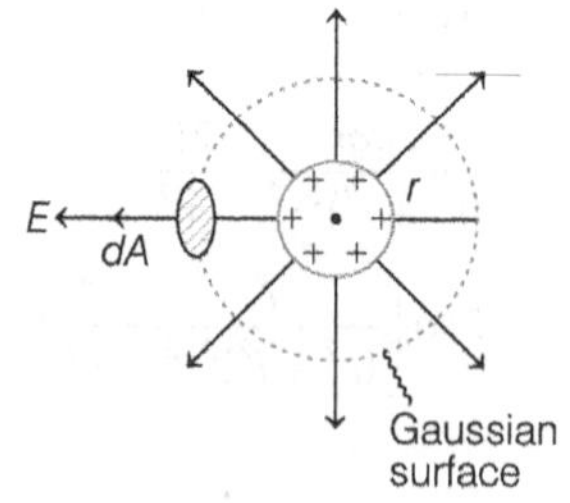

For an outside point $\mathbf{E} \parallel dA$

$\therefore \quad \mathbf{E} \cdot d\mathbf{A} = E\,dA$, as $\cos\theta = 1$

and $q_{\text{enclosure}} = \text{charge density} \times \text{volume}$

$$= \rho \times \frac{4}{3}\pi r_0^3$$

So, $$E\oint dA = \frac{\rho \times \frac{4}{3}\pi r_0^3}{\varepsilon_0}$$

$$\Rightarrow \quad E \cdot 4\pi r^2 = \frac{4}{3}\frac{\pi \rho r_0^3}{\varepsilon_0}$$

$$E = \frac{\rho r_0^3}{3\varepsilon_0 r^2} \qquad (E \text{ at } r > r_0)$$

For a point inside the sphere, $r < r_0$

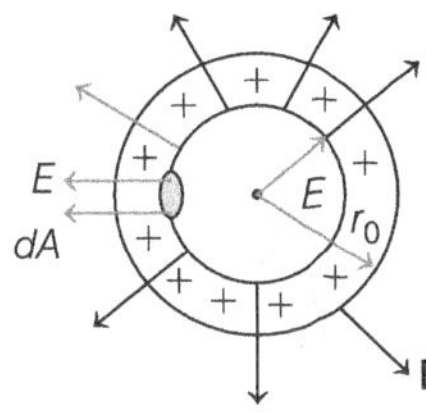

For any elemental patch dA of Gaussian sphere,

$$\mathbf{E} \parallel d\mathbf{A}$$

$$\Rightarrow \quad \theta = 0^\circ$$

$\Rightarrow \quad \mathbf{E} \cdot d\mathbf{A} = E\,dA$. Also, by symmetry E is constant.

$$\therefore \quad \oint \mathbf{E} \cdot d\mathbf{A} = \frac{q_{\text{enclosure}}}{\varepsilon_0}$$

$$\Rightarrow \quad E\oint dA = \frac{\rho \cdot \frac{4}{3}\pi r^3}{\varepsilon_0}$$

$$\Rightarrow \quad E \cdot 4\pi r^2 = \frac{\rho \frac{4}{3}\pi r^3}{\varepsilon_0}$$

$$\Rightarrow \quad E_{\text{inside}} = \frac{\rho r}{3\varepsilon_0}$$

and at $r = r_0$

$$E_{\text{surface}} = \frac{\rho r_0}{3\varepsilon_0}$$

74. $$100 = \frac{1}{4\pi\varepsilon_0} \cdot \frac{q}{(R + 0.05)} \qquad \text{...(i)}$$

$$75 = \frac{1}{4\pi\varepsilon_0} \cdot \frac{q}{(R + 0.1)} \qquad \text{...(ii)}$$

Solving these equations, we get

$$q = \frac{5}{3} \times 10^{-9}\ \text{C}$$

and $$R = 0.1\ \text{m}$$

(a) $$V = \frac{1}{4\pi\varepsilon_0} \cdot \frac{q}{R}$$

$$= \frac{(9 \times 10^9)\left(\frac{5}{3} \times 10^{-9}\right)}{0.1} = 150\ \text{V}$$

(c) $$E = \frac{1}{4\pi\varepsilon_0} \cdot \frac{q}{R^2} = \frac{V}{R} = \frac{150}{0.1}$$

$$= 1500\ \text{V/m}$$

(d) $V_{\text{centre}} = 1.5\, V_{\text{surface}}$

75. If all of the angles to the vertical (in both cases) are assumed to be small, then the sphere only have horizontal displacement and so the electric force of repulsion is always horizontal. Likewise, the small angle condition leads to $\tan\theta \approx \sin\theta \approx \theta$ for all small angles.

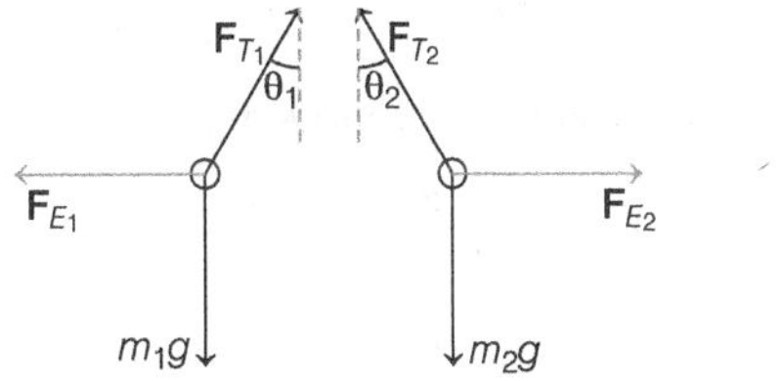

In the free body diagram, each sphere has three forces gravity, tension and the electrostatic force.

Taking to the right as the positive horizontal direction and up to the positive vertical direction. Since, the sphere 1 is in equilibrium the net force in each direction is zero.

(a) $\Sigma F_{1x} = F_{T_1} \sin\theta_1 - F_{E_1}$

$$= 0 \rightarrow F_{E_1} = F_{T_1} \sin\theta_1$$

$$\Sigma F_{1y} = F_{T_1} \cos\theta_1 - m_1 g \rightarrow F_{T_1}$$

$$= \frac{m_1 g}{\cos\theta_1} \rightarrow F_{E1} = \frac{m_1 g}{\cos\theta_1} \sin\theta_1$$

$$= m_1 g \tan\theta_1 = m_1 g \theta_1$$

A completely parallel analysis would give

$$F_{E_2} = m_2 g \theta_2$$

Since, the electric forces are a Newton's third law pair, they can be set equal to each other in magnitude.

$$F_{E_1} = F_{E_2} \rightarrow m_1 g \theta_1$$

$$= m_2 g \theta_2 \rightarrow \frac{\theta_1}{\theta_2} = \frac{m_2}{m_1} = 1$$

(b) The same analysis can be done for this case

$$F_{E_1} = F_{E_2} = m_1 g \theta_1$$

$$= m_2 g \theta_2 \rightarrow \frac{\theta_1}{\theta_2} = \frac{m_2}{m_1} = 2$$

76. When $x = q$, the situation is symmetric

Electric field at O would be zero.

Option (a) is correct.

When $x = -q$, we can think of x as $q + (-2q)$

Magnitude of electric field

$$\text{at } O = \frac{1}{4\pi\varepsilon_0} \frac{(2q)}{\left(2 \times \frac{\sqrt{3}}{2} a\right)^2}$$

$$= \frac{1}{4\pi\varepsilon_0} \frac{2q}{3a^2} = \frac{q}{6\pi\varepsilon_0 a^2}$$

Option (b) is correct.

For $x = 2q$, potential at O is

$$V_0 = 6 \times \frac{1}{4\pi\varepsilon_0} \times \frac{q}{\sqrt{3}a} + \frac{1}{4\pi\varepsilon_0} \times \frac{q}{\sqrt{3}a}$$

$$= \frac{7q}{4\sqrt{3}\pi\varepsilon_0 a}$$

Option (d) is not correct.

77. Electrostatic field at P is

$$\mathbf{E}_p = \mathbf{E}_1 + \mathbf{E}_2 = \frac{\rho}{3\varepsilon_0} \mathbf{C_1P} + \frac{(-\rho)}{3\varepsilon_0} \mathbf{C_2P}$$

where, ρ = volumetric charge density

$$= \frac{\rho}{3\varepsilon_0} (\mathbf{C_1P} - \mathbf{C_2P})$$

$$= \frac{\rho}{3\varepsilon_0} (\mathbf{C_1P} + \mathbf{PC_2})$$

$$= \frac{\rho}{3\varepsilon_0} (\mathbf{C_1C_2}) = \text{constant}$$

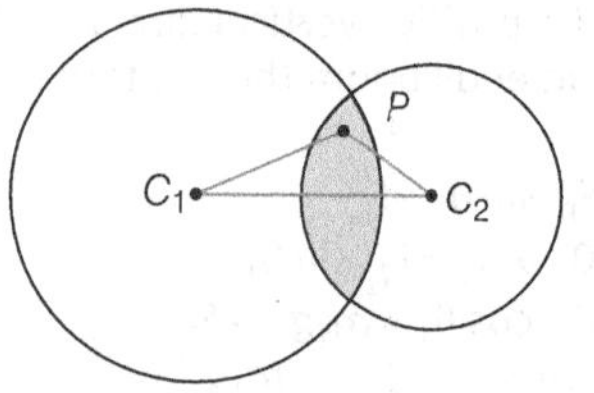

78. Potential difference in such situation depends on inner charge only. So, potential difference will remain unchanged. Hence,

$$\Delta V = V_a - V_b$$

79.

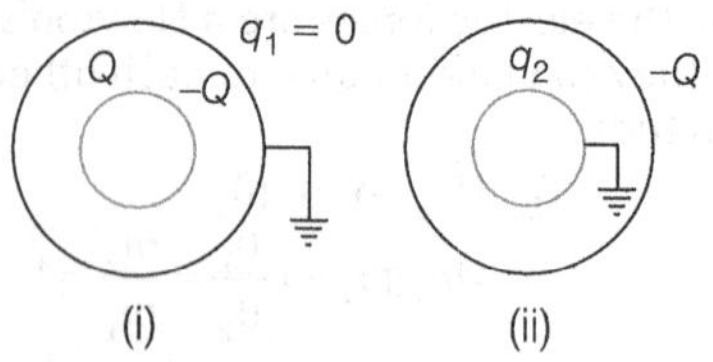

$V_{inner} = 0$ when solid sphere is earthed

$$\because \quad \frac{kq_2}{a} - \frac{kQ}{b} = 0$$

$$\because \quad q_2 = Q\left(\frac{a}{b}\right)$$

80. Whole inner charge transfers to shell.

$\therefore$ Total charge on shell $= q_2 - Q$

$$= Q\left(\frac{a}{b} - 1\right)$$

81. to 84.

Charge density $\lambda = \lambda_0 \cos\theta$, $\cos\theta$ is positive when $\theta \in \left(-\frac{\pi}{2}, \frac{\pi}{2}\right)$ and $\cos\theta$ is negative when $\theta \in \left(\frac{\pi}{2}, \frac{3\pi}{2}\right)$.

So, charge on ring is as shown

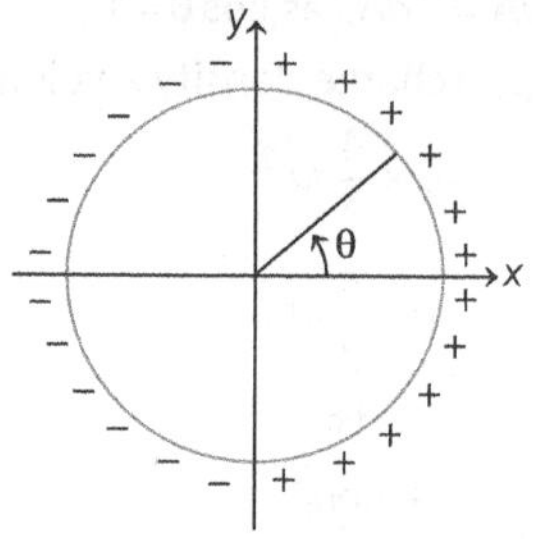

Now, for an elemental part subtending angle $d\theta$ at centre of ring, field produced at centre will be dE, given by

$$dE = \frac{kdq}{R^2}, \; dq = (Rd\theta)\lambda$$

$$= \lambda_0 \cos\theta \cdot Rd\theta$$

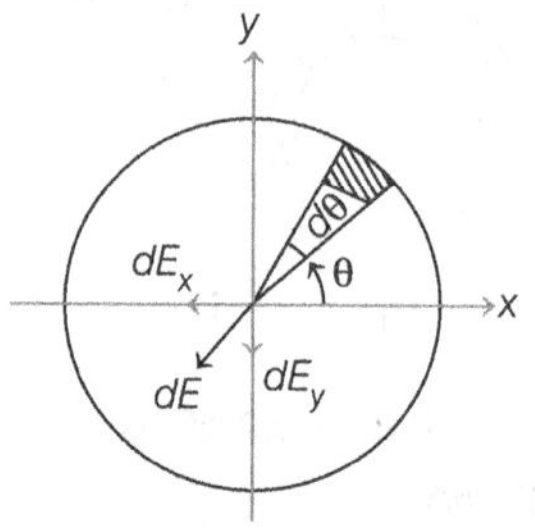

Electric field dE can be resolved into components dE_x and dE_y.

$$dE_x = dE\cos\theta,$$

$$dE_y = dE\sin\theta$$

and $$E_x = \int_0^{2\pi} dE_x = \frac{\lambda_0}{4\pi\varepsilon_0 R}\int_0^{2\pi} \cos^2\theta d\theta$$

$$= \frac{\lambda_0}{4\varepsilon_0 R}$$

Also, $$E_y = \int_0^{2\pi} dE_y = 0$$

$\Rightarrow$ E_{net} = Net field at the centre of the ring

$$= \sqrt{E_x^2 + E_y^2} = \frac{\lambda_0}{4\varepsilon_0 R}$$

Also, field on axis is

$$dE = \frac{dq}{4\pi\varepsilon_0(x^2 + R^2)} = \frac{\lambda_0 \cos\theta R d\theta}{4\pi\varepsilon_0(x^2 + R^2)}$$

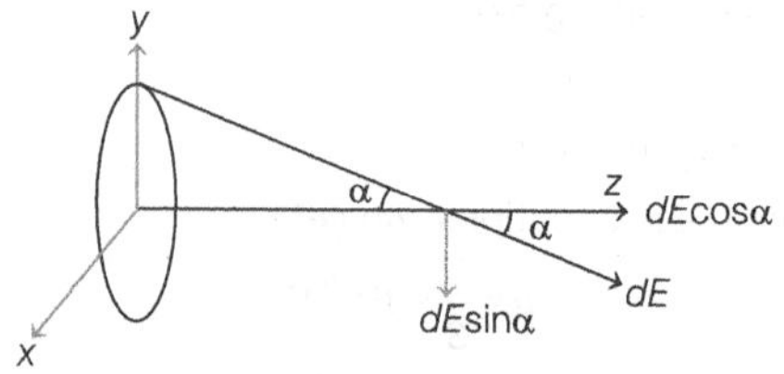

where, $\cos\alpha = \frac{x}{\sqrt{x^2 + R^2}}$

Also, $E_x = E_z = 0$

and $$E_{net} = \int_0^{2\pi} dE_x = \int_0^{2\pi} \frac{\lambda_0 R \cdot \sin\alpha \cos^2\theta}{4\pi\varepsilon_0 (x^2 + R^2)} d\theta$$

$$= \frac{\lambda_0 R \sin\alpha}{4\pi\varepsilon_0 (x^2+R^2)} \int_0^{2\pi} \cos^2\theta d\theta$$

$$= \frac{\lambda_0 R \sin\alpha}{4\varepsilon_0 (x^2+R^2)} = \frac{\lambda_0 R^2}{4\varepsilon_0 (x^2+R^2)^{3/2}}$$

and when $x >> R$, $E_{net} = \frac{\lambda_0 R^2}{4\varepsilon_0 x^3}$ along x-axis.

85. Using energy of conservation,

$$(KE + U_s + U_E)_{initial} = (k + U_s + U_E)_{final}$$

$$\Rightarrow \quad 0 + 0 + 0 = 0 + \frac{1}{2} kx_{max}^2 - QEx_{max}$$

$$\Rightarrow \quad x_{max} = \frac{2QE}{k}$$

86. At equilibrium position,

$$\Sigma F_x = -F_s + F_e = 0$$

or $$kx = QE$$

$\therefore$ $$x = \frac{QE}{k} \quad \text{(equilibrium)}$$

87. Equation of motion of block is

$$\Sigma F_x = -kx + QE = \frac{md^2x}{dt^2}$$

Let $x' = x - \frac{QE}{k}$

or $$x = x' + \frac{QE}{k}$$

So, equation of motion becomes

$$-k\left(x' + \frac{QE}{k}\right) + QE = \frac{md^2(x + QE/k)}{dt^2}$$

or $$\frac{d^2x'}{dt^2} = -\frac{k}{m}x' \quad \text{...(i)}$$

This is an equation of SHM,

$$ax' = -\omega^2 x' \quad \text{...(ii)}$$

As $$T = \frac{2\pi}{\omega} \text{ and } \omega = \sqrt{\frac{k}{m}}$$

From Eqs. (i) and (ii), we have

$$T = 2\pi\sqrt{\frac{m}{k}}$$

88. For any point $P(x, y)$ on xy-plane, electric potential at P is

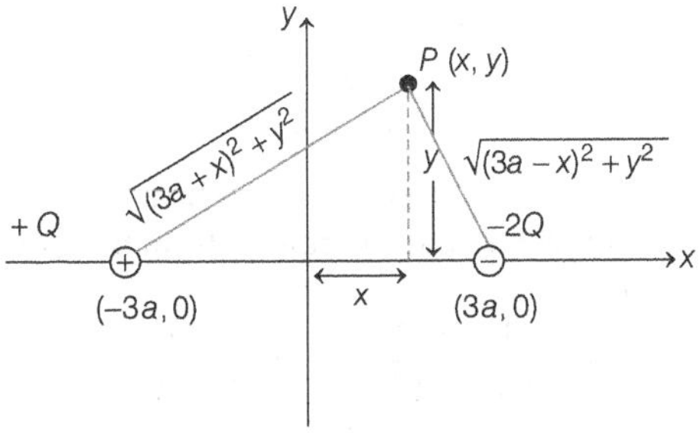

V = potential due to Q + potential due to $(-2Q)$

$$\Rightarrow V = k\left(\frac{Q}{\sqrt{(3a-x)^2+y^2}}\right) + k\left(\frac{-2Q}{\sqrt{(3a+x)^2+y^2}}\right) \quad \text{...(i)}$$

If $V_p = 0$, then

$$4[(3a-x)^2 + y^2] = (3a+x)^2 + y^2$$

$$\Rightarrow \quad (x-5a)^2 + y^2 = 16a^2$$

$\therefore$ This is a circle with centre at $(5a, 0)$ and radius $4a$.

89. Potential at point $(5a, 0) = \frac{Q}{4\pi\varepsilon_0}\left(\frac{1}{2a} - \frac{2}{8a}\right)$

$$= \frac{Q}{16\pi\varepsilon_0 a}$$

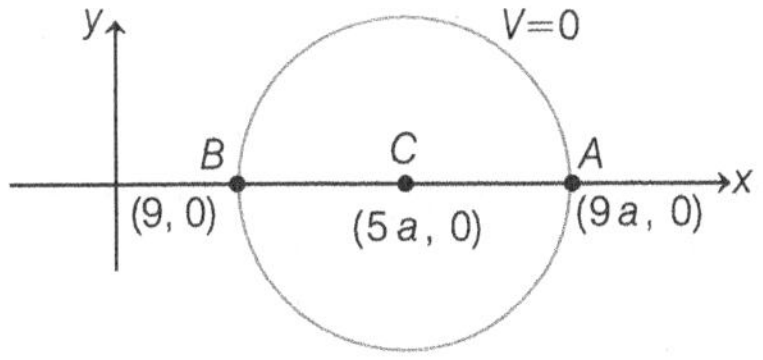

and potential at point $(9a, 0) = 0$

So, when charge is released from centre C, it moves to B i.e. from higher potential to lower potential.

90. On x-axis for $-3a < x < 3a$, we can find potential function $V(x)$ by putting $y = 0$ in Eq. (i) given in sol. 88.

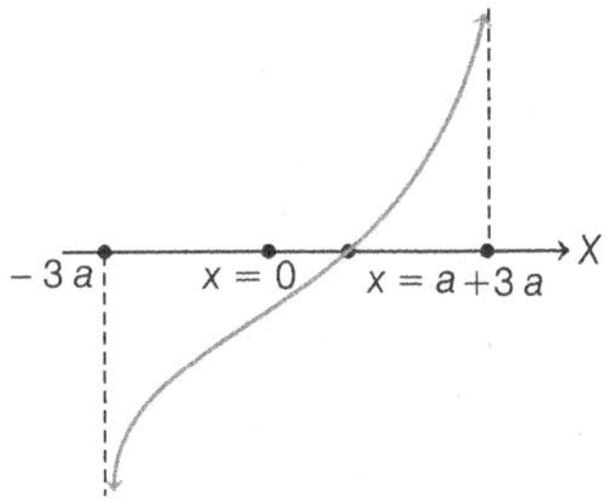

$$\therefore \quad V(x) = \frac{Q}{4\pi\varepsilon_0}\left(\frac{1}{3a-x} - \frac{2}{3a+x}\right)$$

$$\Rightarrow \quad V(x) = 0 \text{ at } x = a$$

Also, $V \to -\infty$ at $x \to -3a$

and $V \to +\infty$ at $x \to 3a$

Hence, there is no maxima or minima at $x = a$.

91 and 92. Let us consider a point P on the circle

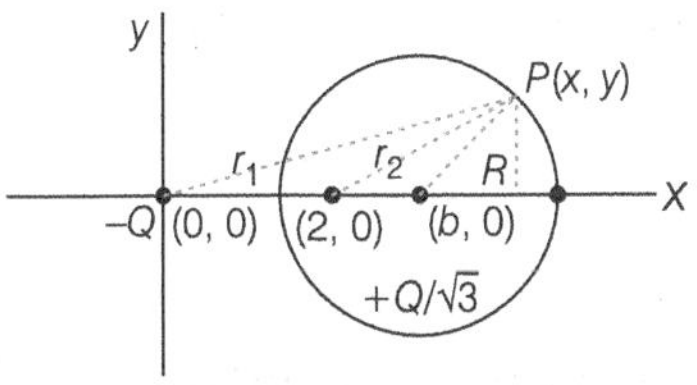

$$V_P = 0 = \frac{k(-Q)}{r_1} + \frac{kQ/\sqrt{3}}{r_2}$$

$$\Rightarrow \quad \frac{kQ}{r_1}=\frac{kQ/\sqrt{3}}{r_2} \Rightarrow \frac{1}{\sqrt{x^2+y^2}}=\frac{1}{\sqrt{3}\sqrt{(x-2)^2+y^2}}$$

$$\Rightarrow \quad 3(x-2)^2+3y^2=x^2+y^2$$

$$\Rightarrow \quad 3(x^2+4-4x)-x^2+2y^2=0$$

$$\Rightarrow \quad 2x^2+12-12x+2y^2=0$$

$$\Rightarrow \quad x^2+6-6x+y^2=0$$

$$\Rightarrow \quad (x-3)^2+y^2=(\sqrt{3})^2$$

or $\quad (x-b)^2+y^2=(\sqrt{3})^2=R^2$

$\therefore$ $R=\sqrt{3}=1.73$ and $b=3$

93. If Q_1, Q_2, Q_3 and Q_4 are all positive, then the force will be along $+y$-direction as components of forces along x-axis cancel out each other.

If Q_1, Q_2 are positive and Q_3, Q_4 are negative, the force will act along $+x$-direction as components of forces along y-axis cancel out each other.

If Q_1, Q_4 are positive and Q_2, Q_3 are negative then attractive force will dominate repulsive force and the force will be along $-y$-direction.

If Q_1, Q_3 positive and Q_2, Q_4 negative components of forces along y-axis cancel out each other. So net force on charge q along x-axis.

$P \rightarrow 3;\ Q \rightarrow 1;\ R \rightarrow 4;\ S \rightarrow 2$

94. (a) By comparing this problem with spring-block system problem suspended vertically.

Here, $\quad mg \equiv qE = 50\times10^{-6}\times5\times10^{5}=25\text{ N}$

$$X_{max}=2\,mg/K$$

Here, $X_{max}=2\,qE/K=\dfrac{2\times25}{100}=0.5\text{ m}$

or $\quad =50\text{ cm}$ **Ans.**

(b) Equilibrium position will be at $x=mg/K$. Here, it will be at $x=qE/K=\dfrac{25}{100}=0.25\text{ m}$

or $\quad 25\text{ cm}$ **Ans.**

(c) Force **QE** is constant force, which does not affect the period of oscillation of SHM.

$$\therefore \quad T=2\pi\sqrt{\frac{m}{K}}=2\pi\sqrt{\frac{4}{100}}$$

$$=\frac{2\pi}{5}\text{ s}=1.26\text{ s}$$ **Ans.**

(d) $\mu mg=0.2\times4\times10=8\text{ N}$

Therefore, here constant force will be

$$qE-\mu mg=25-8=17\text{ N}=F\text{ (say)}$$

$$X_{max}=\frac{2F}{K}=\frac{2\times17}{100}$$

$$=0.34\text{ m}$$ **Ans.**

95. Total charge on ring $=\lambda(2\pi a)=q$ (say)

Electric field at distance x from the centre of ring,

$$E=\frac{1}{4\pi\varepsilon_0}\cdot\frac{qx}{(a^2+x^2)^{3/2}}=\frac{\lambda a x}{2\varepsilon_0(a^2+x^2)^{3/2}}$$

Restoring force on $-Q$ charge in this position would be

$$F=-QE=-\left[\frac{\lambda a Q x}{2\varepsilon_0(a^2+x^2)^{3/2}}\right]$$

For $x \ll a$,

$$F=-\left[\frac{\lambda a Q}{2\varepsilon_0 a^3}\right]x=-\left[\frac{\lambda Q}{2\varepsilon_0 a^2}\right]x$$

Comparing with $\quad F=-kx$,

$$k=\frac{\lambda Q}{2\varepsilon_0 a^2}$$

$$T=2\pi\sqrt{\frac{m}{k}}=2\pi\sqrt{\frac{2\varepsilon_0 m a^2}{\lambda Q}}$$ **Ans.**

96. $qE=30\text{ N}$, vertical component of electric force $=30\sin30°=15\text{ N}$ and horizontal component of electric force $=30\cos30°=15\sqrt{3}\text{ N}$

$$a_y=\frac{mg-15}{m}=\frac{30-15}{3}=5\text{ m/s}^2 \quad \text{(downwards)}$$

$$a_x=\frac{15\sqrt{3}}{3}=5\sqrt{3}\text{ m/s}^2$$

$$T_1=\frac{2u_y}{a_y}=\frac{2\times20\sin30°}{5}=4\text{ s}$$

$$T_2=eT_1=2\text{ s}$$

Horizontal velocity after first drop

$$=(20\cos30°)+a_xT_1$$

$$=(10\sqrt{3})+(5\sqrt{3})4$$

$$=30\sqrt{3}\text{ m/s}$$

$\therefore$ Horizontal distance travelled between first drop and second drop

$$=(30\sqrt{3})T_2+\frac{1}{2}a_xT_2^2$$

$$=(30\sqrt{3})(2)+\frac{1}{2}(5\sqrt{3})(2)^2$$

$$=70\sqrt{3}\text{ m}$$ **Ans.**

97.

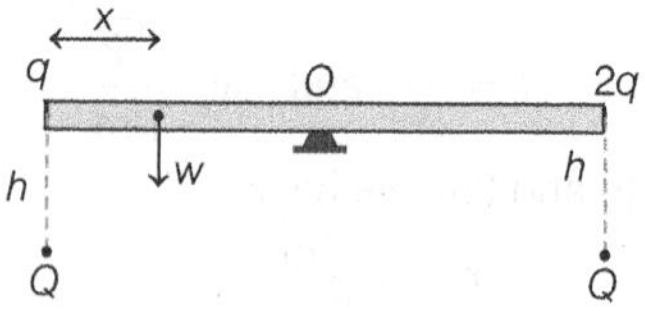

(a) Net torque on the rod about $O=0$

$$\frac{1}{4\pi\varepsilon_0}\cdot\frac{Q(2q)}{h^2}\left(\frac{L}{2}\right)+w\left(\frac{L}{2}-x\right)=\frac{1}{4\pi\varepsilon_0}\cdot\frac{Q.q}{h^2}\left(\frac{L}{2}\right)$$

$$x=\frac{L}{2}\left[1+\frac{Qq}{(4\pi\varepsilon_0)wh^2}\right]$$ **Ans.**

(b) There will be no force from the bearing if $w=$ net electrostatic repulsion from both the charges.

$$\therefore \quad w=\frac{1}{4\pi\varepsilon_0}\cdot\frac{Q(3q)}{h^2} \text{ or } h=\sqrt{\frac{3Qq}{4\pi\varepsilon_0 w}}$$ **Ans.**

98. $\mathbf{E} = -\left(\frac{\partial V}{\partial x}\hat{i} + \frac{\partial V}{\partial y}\hat{j}\right) = (-3\hat{i} - 4\hat{j})\,N/C$

$$\mathbf{a} = \frac{q\mathbf{E}}{m} = \frac{10^{-6}}{10}(-3\hat{i} - 4\hat{j})$$
$$= (-3\times 10^{-7}\,\hat{i} - 4\times 10^{-7}\,\hat{j})\,m/s^2$$

When particle crosses x-axis, $y = 0$.
Initial y-coordinate was 3.2 m.
and $a_y = -4\times 10^{-7}\,m/s^2$

$\therefore$ $y = 0$ at time $t = \sqrt{\frac{2\times 3.2}{4\times 10^{-7}}}$
$= 4000$ s

At this instant x-coordinate will be
$$x = x_i + \frac{1}{2}a_x t^2$$
$$= 2 + \frac{1}{2}(-3\times 10^{-7})(4000)^2 = -0.4\ m$$

Now, $V_i = (3\times 2) + (4\times 3.2) = 18.8$ V
$V_f = (3)(-0.4) = -1.2$ V
$\Delta V = 20$ V

$\therefore$ Speed, $v = \sqrt{\frac{2q\Delta V}{m}} = \sqrt{\frac{2\times 10^{-6}\times 20}{10}}$
$= 2.0\times 10^{-3}\,m/s$ **Ans.**

99. From work-energy theorem,

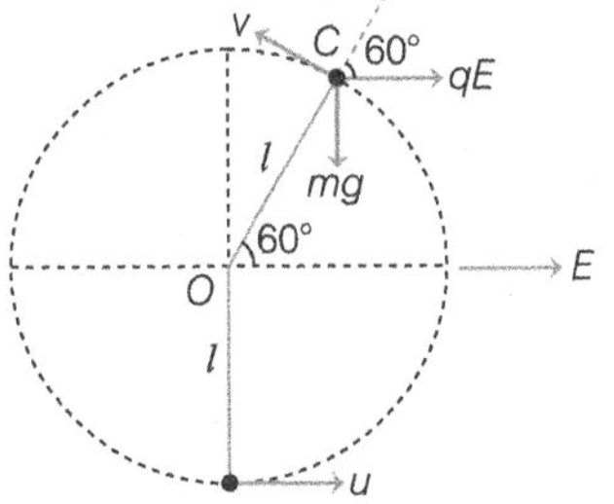

$$\frac{1}{2}m(v^2 - u^2) = -mgl(1 + \sin 60°) + qEl\cos 60°$$

Substituting the values, we get
$$u^2 - v^2 = 32.32 \quad \text{...(i)}$$

Further at C, tension in the string is zero.

Hence, $\frac{mv^2}{l} = mg\sin 60° - qE\cos 60°$

or $v^2 = 3.66$ ...(ii)

From Eqs. (i) and (ii), we get
$u = 6\,m/s$ **Ans.**

100. There are total 28 pairs of charges.
12 pairs $\rightarrow$ Q and $-$Q $\rightarrow$ distance L
12 pairs $\rightarrow$ (Q and Q) or ($-$Q and $-$Q) $\rightarrow \sqrt{2}L$
4 pairs $\rightarrow$ Q and $-$Q $\rightarrow \sqrt{3}L$

$$\therefore\ U = 12\left(\frac{1}{4\pi\varepsilon_0}\right)\left(\frac{-Q^2}{L}\right) + 12\left(\frac{1}{4\pi\varepsilon_0}\cdot\frac{Q^2}{\sqrt{2}L}\right) + 4\left(\frac{1}{4\pi\varepsilon_0}\right)\left(\frac{-Q^2}{\sqrt{3}L}\right)$$
$$= -\frac{Q^2}{\pi\varepsilon_0 L}\left(\frac{3\sqrt{6} + \sqrt{2} - 3\sqrt{3}}{\sqrt{6}}\right)$$

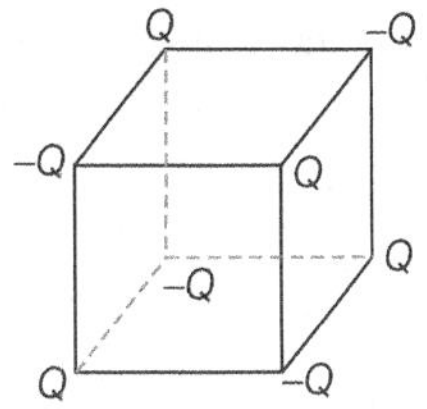

with decrease in L, potential energy will decrease. Therefore, cube should shrink as the conservative forces act in the direction of decreasing potential energy.
Increase in KE of the system = decrease in PE

or $8\left(\frac{1}{2}mv^2\right) = U_i - U_f$
$$= \frac{Q^2}{\pi\varepsilon_0}\left(\frac{3\sqrt{6} + \sqrt{2} - 3\sqrt{3}}{\sqrt{6}}\right)\left(\frac{1}{nL} - \frac{1}{L}\right)$$

or $v = \sqrt{\frac{Q^2(1-n)(3\sqrt{6} + \sqrt{2} - 3\sqrt{3})}{4nm\pi L\varepsilon_0\sqrt{6}}}$ **Ans.**

101. Let charge q_1 comes from the earth on outer shell

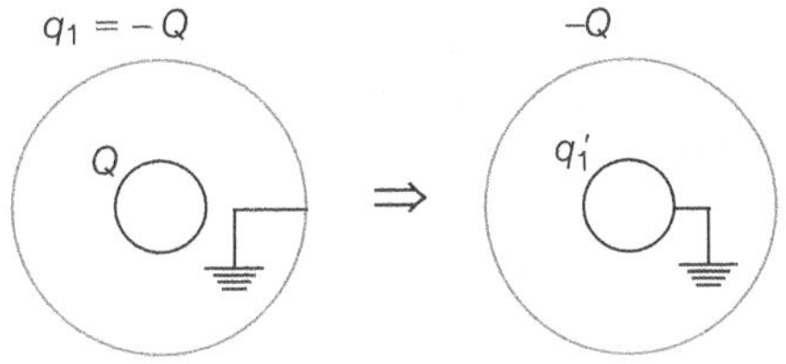

$$V_{outer} = 0$$
$$\frac{1}{4\pi\varepsilon_0}\left[\frac{Q}{2r} + \frac{q_1}{2r}\right] = 0$$

or $q_1 = -Q$

When S_2 is closed and opened,
$$V_{inner} = 0$$
$\therefore$ $\frac{1}{4\pi\varepsilon_0}\left[\frac{q_1'}{r} - \frac{Q}{2r}\right] = 0$ or $q_1' = \frac{Q}{2}$

Proceeding in the similar manner after n such operations we get,
Charge on the inner shell,
$$q_n' = \frac{Q}{(2)^n}$$

and the potential difference between the shells,
$$\Delta V = \frac{q_n'}{4\pi\varepsilon_0}\left(\frac{1}{r} - \frac{1}{2r}\right)$$
$$\Delta V = \frac{1}{(2)^{n+1}}\left[\frac{Q}{4\pi\varepsilon_0 r}\right]$$ **Ans.**

102. (a) Over charge Q_2, field intensity is infinite along negative x-axis. Therefore, Q_2 is negative. Beyond $x > (l + a)$, field intensity is positive. Therefore, Q_1 is positive.

(b) At $x = l + a$, field intensity is zero.

$$\therefore \quad \frac{kQ_1}{(l+a)^2} = \frac{kQ_2}{a^2} \text{ or } \left|\frac{Q_1}{Q_2}\right| = \left(\frac{l+a}{a}\right)^2$$

(c) Intensity at distance x from charge 2 would be

$$E = \frac{kQ_1}{(x+l)^2} - \frac{kQ_2}{x^2}$$

For E to be maximum $\frac{dE}{dx} = 0$

or $$-\frac{2kQ_1}{(x+l)^3} + \frac{2kQ_2}{x^3} = 0$$

or $$\left(1 + \frac{l}{x}\right)^3 = \frac{Q_1}{Q_2} = \left(\frac{l+a}{a}\right)^2$$

or $$1 + \frac{l}{x} = \left(\frac{l+a}{a}\right)^{2/3}$$

or $$x = \frac{l}{\left(\frac{l+a}{a}\right)^{2/3} - 1}$$

or $$b = \frac{l}{\left(\frac{l+a}{a}\right)^{2/3} - 1}$$ **Ans.**

103. Two forces will act on the tank

(a) Electrostatic force

(b) Thrust force.

Let v be the velocity at any instant. Then,

$$F_{net} = QE - mnv$$

or $$(m_0 + mnt)\frac{dv}{dt} = QE - mnv$$

or $$\int_0^v \frac{dv}{QE - mnv} = \int_0^t \frac{dt}{m_0 + mnt}$$

or $$\ln\left(\frac{QE}{QE - mnv}\right) = \ln\left(\frac{m_0 + mnt}{m_0}\right)$$

or $$\frac{QE}{QE - mnv} = \frac{m_0 + mnt}{m_0}$$

or $$v = QE\left(\frac{t}{m_0 + mnt}\right)$$ **Ans.**

104. $dF = \lambda(Rd\theta)E_0$

Perpendicular distance between two equal and opposite pairs of dF will be

$$r_\perp = 2R\sin\theta$$

$$\therefore \quad d\tau = dFr_\perp = 2\lambda R^2 E_0 \sin\theta d\theta$$

$$\therefore \quad \tau = \int_0^{\pi/2} d\tau = 2\lambda R^2 E_0 \quad \text{(clockwise)}$$

These pairs of forces will not provide net force.

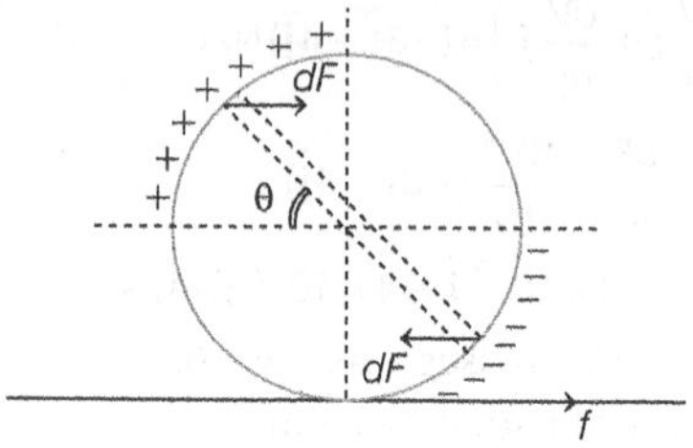

Let force of friction on ring is f in forward direction. For pure rolling to take place,

$$a = R\alpha \text{ or } \frac{f}{m} = R\left[\frac{\tau - fR}{mR^2}\right]$$

or $$f = \frac{\tau}{R} - f \text{ or } f = \frac{\tau}{2R} = \lambda RE_0$$ **Ans.**

105. Net charge between $r = a$ to $r = r$ would be

$$Q = \int_a^r \rho(4\pi r^2)\,dr = \int_a^r \frac{C}{r}(4\pi r^2)\,dr$$

$$Q = 2\pi C(r^2 - a^2)$$

$$E_r = \frac{k(Q+q)}{r^2} = \frac{k[2\pi C(r^2 - a^2) + q]}{r^2}$$

From this expression, we can see that it we put

$$C = \frac{q}{2\pi a^2} \Rightarrow E_r = \frac{kq}{a^2} = \text{constant}$$ **Ans.**

106. From conservation of energy,

$$U_i + K_i = U_f + K_f \text{ or } qV_i + K_i = qV_f + K_f$$

or $$q\left(\frac{Q}{4\pi\varepsilon_0 r}\right) + \frac{1}{2}mv^2 = q\left(\frac{Q}{4\pi\varepsilon_0 R^3}\right)\left(1.5R^2 - 0.5\frac{R^2}{4}\right) + 0$$

or $$\frac{1}{2}mv^2 = \frac{11Qq}{32\pi\varepsilon_0 R} - \frac{Qq}{4\pi\varepsilon_0 r}$$

or $$v = \sqrt{\frac{Qq}{2\pi\varepsilon_0 mR}\left(\frac{r-R}{r} + \frac{3}{8}\right)}$$ **Ans.**

107. From energy conservation principle,

$$K_i + U_i = K_f + U_f$$

or $$\frac{1}{2}mv^2 + (+q)V_i = 0 + (+q)V_f$$

or $$v = \sqrt{\frac{2q}{m}(-V_i + V_f)}$$

$$= \sqrt{\frac{2q}{m}\left[\frac{-Q}{\sqrt{10}R} + \frac{8Q}{5R} + \frac{Q}{R} - \frac{8Q}{4R}\right]\frac{1}{4\pi\varepsilon_0}}$$

$$= \sqrt{\frac{Qq}{2\pi\varepsilon_0 mR}\left(\frac{3\sqrt{10} - 5}{5\sqrt{10}}\right)}$$ **Ans.**

108. (a) $E = \frac{kQx}{(R^2 + x^2)^{3/2}}$,

$$\frac{dE}{dx} = kQ\left[\frac{(R^2 + x^2)^{3/2} - x.\frac{3}{2}(R^2 + x^2)^{1/2}(2x)}{(R^2 + x^2)^3}\right]$$

or $\frac{dE}{dx} = kQ\left[\frac{R^2 + x^2 - 3x^2}{(R^2 + x^2)^{5/2}}\right]$

or $dE = \frac{Q}{4\pi\varepsilon_0}\left[\frac{R^2 - 2x^2}{(R^2 + x^2)^{5/2}}\right]dx$

$\therefore \quad |\Delta E| = \frac{Q}{4\pi\varepsilon_0}\left[\frac{R^2 - 2x^2}{(R^2 + x^2)^{5/2}}\right]\Delta x$ **Ans.**

Here, $\Delta x = 2a$

$\therefore \quad F = |q\Delta E| = \frac{Qqa}{2\pi\varepsilon_0}\left[\frac{R^2 - 2x^2}{(R^2 + x^2)^{5/2}}\right]$

(b) $W = U_f - U_i$

$= -pE\cos 180° + pE\cos 0° = 2pE$

$= 2(q)(2a)\left[\frac{1}{4\pi\varepsilon_0}\frac{Qx}{(R^2 + x^2)^{3/2}}\right]$

$= \frac{aqQx}{\pi\varepsilon_0(R^2 + x^2)^{3/2}}$ **Ans.**

109. From conservation of mechanical energy and conservation of angular momentum about point O, we have

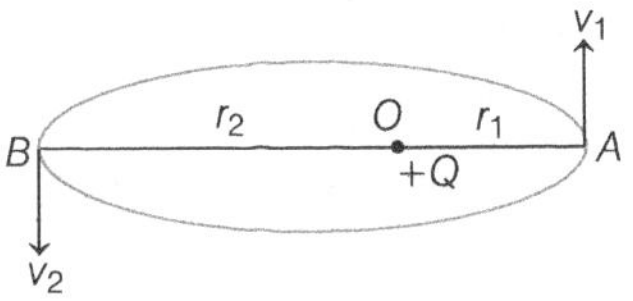

$$\frac{1}{2}mv_1^2 - \frac{1}{4\pi\varepsilon_0}\cdot\frac{Qq}{r_1} = \frac{1}{2}mv_2^2 - \frac{1}{4\pi\varepsilon_0}\cdot\frac{Q.q}{r_2} \quad \text{...(i)}$$

and $\quad mv_1r_1\sin 90° = mv_2r_2\sin 90°$

or $\quad v_1r_1 = v_2r_2$ ...(ii)

Solving these two equations, we have

$$v_1 = \sqrt{\frac{Qqr_2}{2\pi\varepsilon_0 mr_1(r_1 + r_2)}}$$

and $\quad \sqrt{\frac{Qqr_1}{2\pi\varepsilon_0 mr_2(r_1 + r_2)}} = v_2$ **Ans.**

110. (a) $F_{net} = 2F\cos\theta$

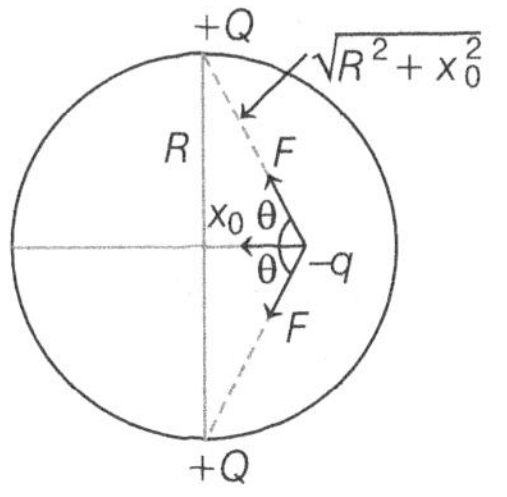

$$= \frac{2kQ.q}{(\sqrt{R^2 + x_0^2})^2}\cdot\frac{x_0}{\sqrt{R^2 + x_0^2}} \quad \left(\text{Here, } k = \frac{1}{4\pi\varepsilon_0}\right)$$

$$= \frac{2kQqx_0}{(R^2 + x_0^2)^{3/2}}$$

We can generalised the force by putting $x_0 = x$, we have

$$F = -\frac{2kQqx}{(R^2 + x^2)^{3/2}}$$ **Ans.**

(b) Motion of bead will be periodic between

$$x = \pm x_0$$ **Ans.**

(c) For $\frac{x}{R} << 1, R^2 + x^2 \approx R^2$

or $F = -\left(\frac{2kQq}{R^3}\right)x$ or $a = \frac{F}{m} = -\left(\frac{2kQq}{mR^3}\right)x$

Since $a \propto -x$, motion will be simple harmonic in nature.

Comparing with $a = -\omega^2 x$, $\omega = \sqrt{\frac{2kQq}{mR^3}}$

$x = x_0\cos\omega t$ (as the particle starts from extreme position)

$$v = \frac{dx}{dt} = -\omega x_0\sin\omega t$$ **Ans.**

(d) Velocity will become zero at $t = T/2 = \pi/\omega$

or $\quad t = \pi\sqrt{\frac{mR^3}{2kQq}}$ **Ans.**

111. Each particle will be in equilibrium under the act of three force, tension of string T, weighting, resultant electrostatic force F of the two other charges. Force F and mg are perpendicular.

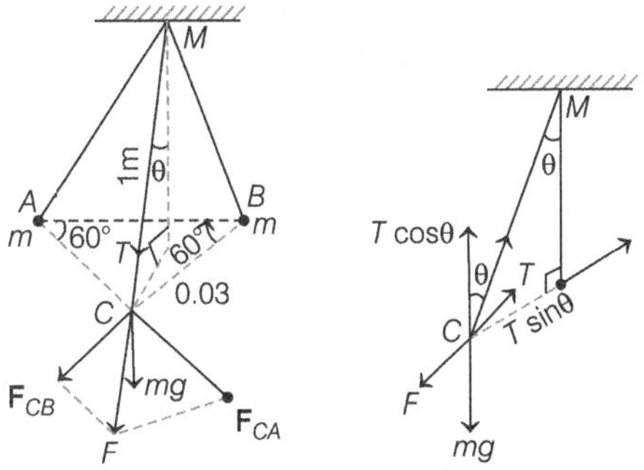

Resolving T in the direction of mg and F and applying the condition of equilibrium,

$T\cos\theta = mg$

$T\sin\theta = F$

$\therefore \quad \tan\theta = \frac{F}{mg}$ ...(i)

$$F = \sqrt{F_{CA}^2 + F_{CB}^2 + 2F_{CA}F_{CB}\cos\alpha}$$

$\therefore \quad F = \sqrt{F_{CA}^2 + F_{CA}^2 + 2F_{CA}^2 \times \frac{1}{2}}$

$$F = \sqrt{3}F_{CA} = \sqrt{3} \times \frac{kq^2}{(CA)^2} \quad \text{...(ii)}$$

$[|\mathbf{F}_{CB}| = |\mathbf{F}_{CA}|$ and $\alpha = 60°]$

Let T make an angle θ with the vertical

$$OC = \frac{2}{3}\sqrt{(0.03)^2 - (0.015)^2} = 0.0173 \text{ m}$$

$\therefore \quad OM = 0.9997$

$$\therefore \quad \tan\theta = \frac{OC}{OM} = \frac{0.0173}{0.9997} \quad \text{...(iii)}$$

From Eqs. (i), (ii) and (iii),

$$\frac{0.0173}{0.9997} = \frac{\sqrt{3} \times 9 \times 10^9 \times q^2}{(0.03)^2 \times 10^{-3} \times 9.8}$$

$$\therefore \quad q = 3.16 \times 10^{-9} \text{ C}$$

112. Given, mass of particle

$$m = 0.9 \times 10^{-3} \text{ kg}$$

Charge on particle $q = 10^{-6}$ C

$$r = 1 \text{ m}$$

$$Q = 10^{-5} \text{ C}$$

The electric field due a uniformly charged ring of radius r at a point distant x from its centre on its axis

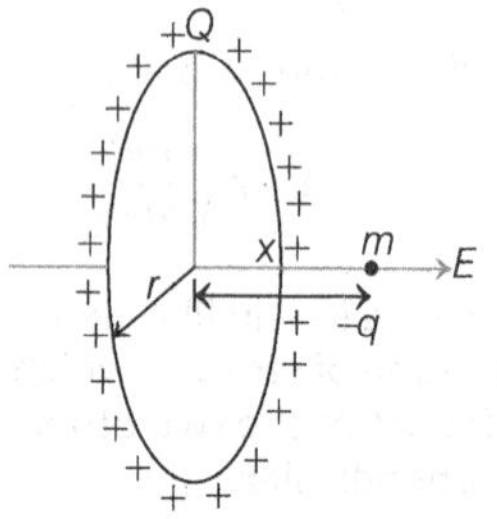

$$E = k\frac{Qx}{(r^2 + x^2)^{3/2}}$$

$\therefore$ Force on the negative charge $qE = qE$

$$\therefore \quad F = \frac{-kQq}{(r^2 + x^2)^{3/2}} \times x$$

or, $$ma = \frac{-kQq}{(r^2 + x^2)^{3/2}} \times x$$

or, $$a = -k\frac{Qq}{m(r^2 + x^2)^{3/2}} \times x$$

For $x << r$, $a = -\frac{kQq}{r^3} \times x$

Hence motion is simple harmonic in nature. Comparing the above equation with $a = -\omega^2 x$

$$\therefore \quad \omega^2 = \frac{kQq}{mr^3} \text{ or } \omega = \sqrt{\frac{kQq}{mr^3}}$$

$$\therefore \quad \frac{2\pi}{T} = \sqrt{\frac{kQq}{mr^3}}$$

$$\Rightarrow \quad T = 2\pi\sqrt{\frac{mr^3}{kQq}}$$

$$T = 2 \times 3.14\left[\frac{0.9 \times 10^{-3} \times 1^3}{9 \times 10^9 \times 10^{-5} \times 10^{-6}}\right]^{1/2}$$

$$= 6.28\,[0.01]^{1/2} = 6.28\,[0.1]$$

$$= 0.628 \text{ s}$$

113. (i) If the charges q_1 and q_2 are similar in nature, then electric force, $F = \frac{kq_1q_2}{l^2}$ will be repulsive and is opposite to tension force T'.

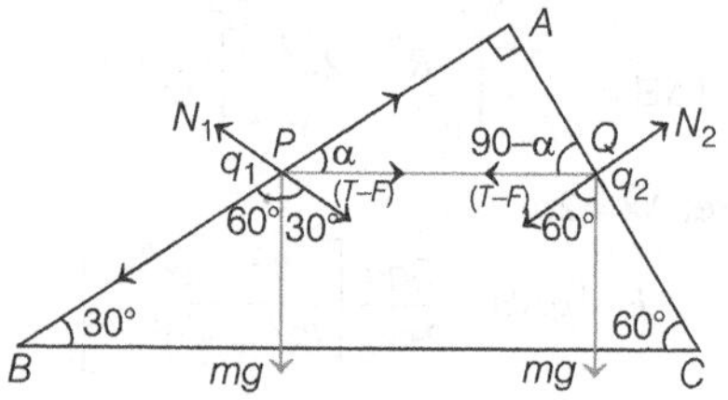

For equilibrium of bead at P

$$mg\cos 60° = (T - F)\cos\alpha \quad \text{... (i)}$$

$$N_1 = mg\cos 30° + (T - F)\sin\alpha \quad \text{...(ii)}$$

Similarly, for the equilibrium of bead at Q

$$mg\sin 60° = (T - F)\sin\alpha \quad \text{...(iii)}$$

$$N_2 = mg\cos 60° + (T - F)\cos\alpha \quad \text{...(iv)}$$

Dividing Eq. (iii) by Eq. (i), we get

$$\tan 60° = \tan\alpha$$

i.e. $$\alpha = 60° \quad \text{...(v)}$$

(ii) From Eq. (iii),

$$T - F = mg$$

or $$T = F + mg = (kq_1q_2/l^2) + mg \quad \text{...(vi)}$$

(iii) From Eq. (ii),

$$N_1 = mg\cos 30° + (T - F)\sin 60°$$

$$\therefore \quad N_1 = \sqrt{3}mg \quad \text{...(vii)}$$

From Eq. (iv),

$$N_2 = mg\cos 60° + mg\cos 60° = mg \quad \text{...(viii)}$$

Again if the chord is cut, then T becomes zero. Hence from Eq. (vi),

$$q_1q_2 = -mg\frac{l^2}{k}$$

This shows that q_1 and q_2 must have unlike charges for beads to remain stationary.

114.

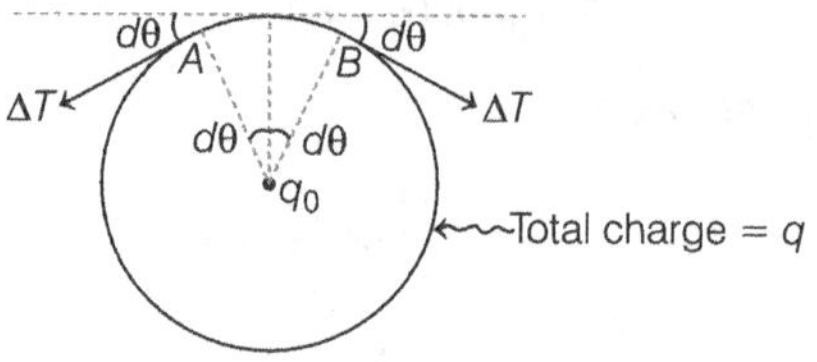

ΔT = extra tension

$2\Delta T \sin d\theta$ = repulsion on charge AB from q_0

$$= \frac{1}{4\pi\varepsilon_0}\frac{(q_{AB})\,q_0}{r^2}$$

For small value of $d\theta$,

$$\sin d\theta \approx d\theta$$

$$\therefore \quad 2\Delta T = \left(\frac{1}{4\pi\varepsilon_0}\right)\left(\frac{q}{2\pi r}\right)(2r\,d\theta)\frac{q_0}{r^2}$$

$$\therefore \quad \Delta T = \frac{qq_0}{8\pi^2\varepsilon_0 r^2}$$

115. All three particles should be in one line. Because, on any one charge two forces will act from other two charges and resultant of two forces can be zero if they are equal and opposite.

Further, q_1 and q_2 are positive and q_3 is negative. So, q_3 should lie in between q_1 and q_2 otherwise both forces on q_3 will be in same direction and their resultant can't be zero.

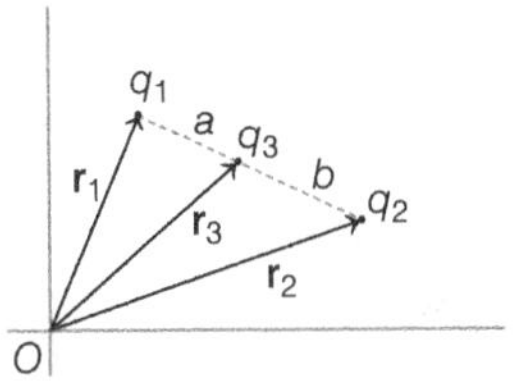

Equilibrium of q_3

$$\frac{1}{4\pi\varepsilon_0}\frac{q_1 q_3}{a^2} = \frac{1}{4\pi\varepsilon_0}\frac{q_2 q_3}{b^2}$$

or $$\frac{a}{b} = \frac{\sqrt{q_1}}{\sqrt{q_2}} \quad \text{...(i)}$$

Since position vector of q_3 divides the line joining $\mathbf{r}_1$ and $\mathbf{r}_2$ in the ratio $\frac{a}{b}$ or $\frac{\sqrt{q_1}}{\sqrt{q_2}}$. Therefore from the knowledge of vectors we can find $\mathbf{r}_3$. Hence,

$$\mathbf{r}_3 = \frac{\mathbf{r}_1\sqrt{q_2} + \mathbf{r}_2\sqrt{q_1}}{\sqrt{q_1} + \sqrt{q_2}}$$

Equilibrium of q_1

$$\frac{1}{4\pi\varepsilon_0}\frac{q_1 q_3}{a^2} = \frac{1}{4\pi\varepsilon_0}\frac{q_2 q_1}{(a+b)^2}$$

$$q_3 = \frac{q_2}{(1+b/a)^2}$$

Substituting value of $\frac{b}{a}$ from Eq. (i), we get

$$q_3 = \frac{q_1 q_2}{(\sqrt{q_1} + \sqrt{q_2})^2}$$

with sign, $$q_3 = \frac{-q_1 q_2}{(\sqrt{q_1} + \sqrt{q_2})^2}$$

Note *We will get same value of q_3 if we take equilibrium of q_2.*

116. (a)

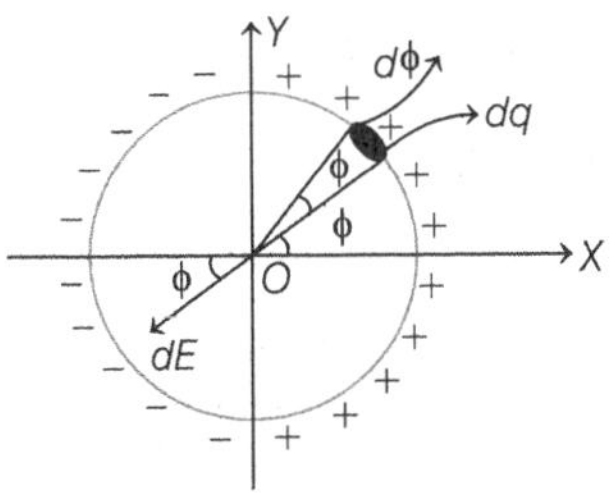

For $-\frac{\pi}{2} < \phi < \frac{\pi}{2}$, $\lambda = \text{positive}$

For $\frac{\pi}{2} < \phi < \frac{3\pi}{2}$, $\lambda = \text{negative}$

Charge dq on small element shown in figure is,

$$dq = (\lambda_\phi)(R d\phi) = (\lambda_0 \cos\phi)(R d\phi)$$

Electric field dE at centre O due to charge dq is :

$$dE = \left(\frac{1}{4\pi\varepsilon_0}\right)\frac{dq}{R^2} = \frac{(\lambda_0 \cos\phi)(R d\phi)}{4\pi\varepsilon_0 R^2}$$

$$|dE_x| = |dE\cos\phi| = \frac{(\lambda_0 \cos^2\phi)\,d\phi}{4\pi\varepsilon_0 R}$$

$$|dE_y| = |dE\sin\phi| = \frac{(\lambda_0 \sin\phi\cos\phi)\,d\phi}{4\pi\varepsilon_0 R}$$

$$|E_y| = \int_0^{2\pi} |dE_y| = 0$$

$$|E_x| = \int_0^{2\pi} |dE_x| = \frac{\lambda_0}{4\varepsilon_0 R}$$

$$\therefore \quad E = |E_x| = \frac{\lambda_0}{4\varepsilon_0 R}$$

Note *Direction of this electric field is towards negative x-axis.*

(b)

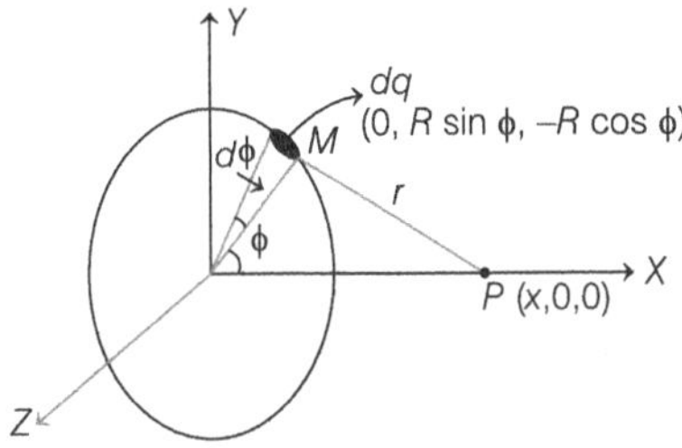

$$dq = (\lambda_\phi)(R d\phi) = (\lambda_0 \cos\phi)(R d\phi)$$

$$r = \sqrt{R^2 + x^2}$$

Electric field at P due to charge dq in vector form is,

$$d\mathbf{E} = \frac{1}{4\pi\varepsilon_0}\frac{dq}{r^3}(\mathbf{r}_P - \mathbf{r}_M)$$

$$= \left(\frac{1}{4\pi\varepsilon_0}\right)\left[\frac{(\lambda_0 \cos\phi)(R d\phi)}{(R^2 + x^2)^{3/2}}\right](x\hat{\mathbf{i}} - R\sin\phi\hat{\mathbf{j}} + R\cos\phi\hat{\mathbf{k}})$$

$$= dE_x\hat{\mathbf{i}} + dE_y\hat{\mathbf{j}} + dE_z\hat{\mathbf{k}}$$

Mathematical part for students

After finding dE_x, dE_y and dE_z, if we integrate each of them from $\phi = 0$ to $\phi = 2\pi$, we will get $|E_x|, |E_y|$ and $|E_z|$.

After doing all these calculations, we get

$$|E_x| = 0$$

$$|E_y| = 0$$

and $$|E_z| = \frac{\lambda_0 R^2}{4\varepsilon_0 (R^2 + x^2)^{3/2}}$$

$$\therefore \quad E = |E_z| = \frac{\lambda_0 R^2}{4\varepsilon_0 (R^2 + x^2)^{3/2}}$$

Note *For $x >> R$, $R^2 + x^2 \approx x^2$*

$$\therefore \quad E = \frac{R^2 \lambda_0}{4\varepsilon_0 x^3}$$

117.

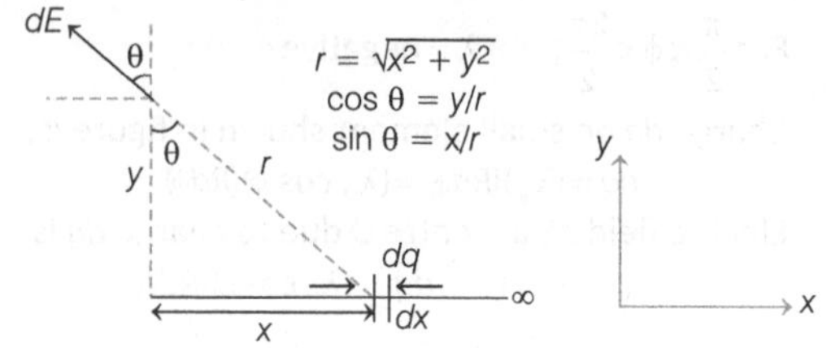

$$dq = \lambda dx$$

$$dE = \frac{1}{4\pi\varepsilon_0} \cdot \frac{dq}{r^2}$$

$$dE_x = -dE \sin\theta$$

$$dE_y = dE \cos\theta$$

If we integrate dE_x and dE_y from $x = 0$ to $x = \infty$, then we can get E_x and E_y and then finally, net electric field,

$$E = \sqrt{E_x^2 + E_y^2}$$

Final value comes out to be,

$$E = \frac{\sqrt{2}\,\lambda}{4\pi\varepsilon_0 y}$$

Note *After calculations, we can see that* $|E_x| = |E_y| = \dfrac{\lambda}{4\pi\varepsilon_0 y}$

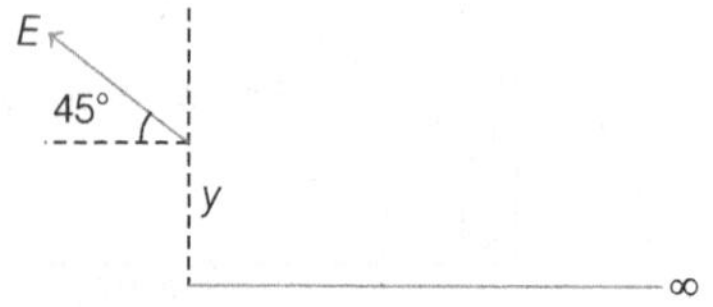

Therefore **E** *is directed at an angle 45° from the thread.*

118. (a) Using the result of Q-117,

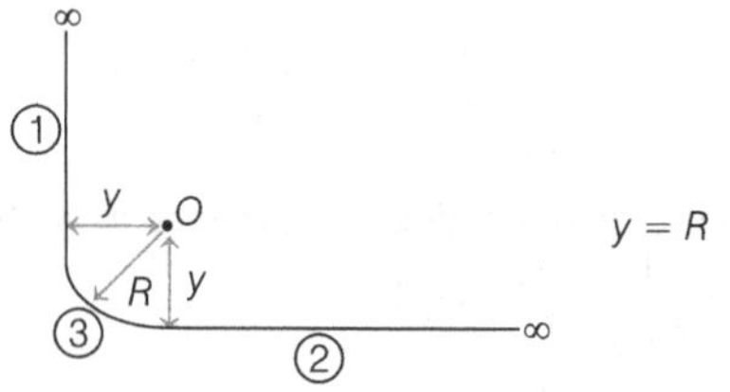

At O

E due to wire 1 $= E$ due to wire 2

$$\therefore \quad E_1 = E_2 = \frac{\sqrt{2}\,\lambda}{4\pi\varepsilon_0 y} = \frac{\sqrt{2}\,\lambda}{4\pi\varepsilon_0 R}$$

Directions of E_1 and E_2 are shown below.

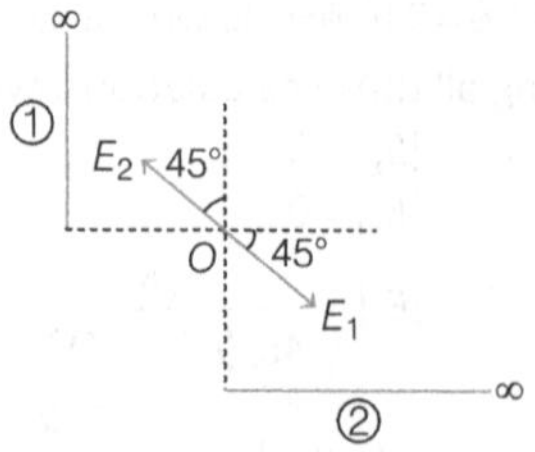

$\therefore$ Net field at O due to wires 1 and 2 will be zero. So, we have to find electric field only due to 3.

$$dq = \lambda(R\,d\phi)$$

$$dE = \frac{1}{4\pi\varepsilon_0} \cdot \frac{dq}{R^2}$$

$$dE_x = dE \cos\phi$$

$$dE_y = dE \sin\phi$$

$$\therefore \quad E_x = \int_0^{\pi/2} dE_x \quad \text{and} \quad E_y = \int_0^{\pi/2} dE_y$$

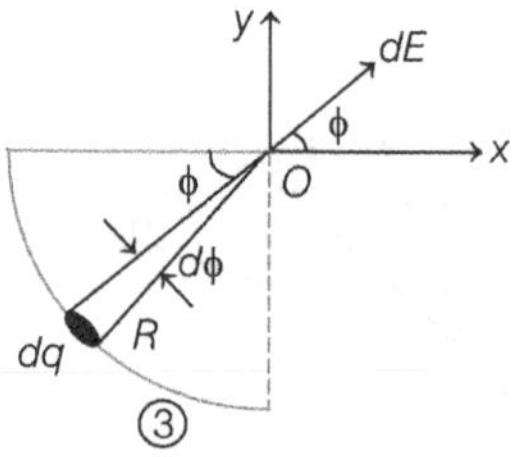

After substituting the proper values and then integrating, we get

$$E_x = E_y = \frac{\lambda}{4\pi\varepsilon_0 R}$$

$$\therefore \quad E = \sqrt{E_x^2 + E_y^2}$$

$$\text{or} \quad E = \frac{\sqrt{2}\,\lambda}{4\pi\varepsilon_0 R}$$

(b)

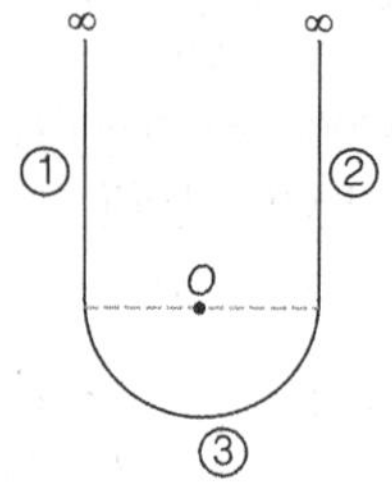

Electric field due to 1 and 2

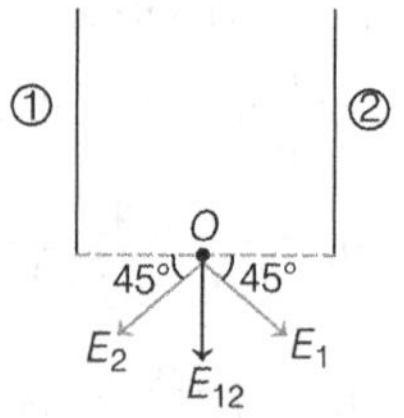

$$E_1 = E_2 = \frac{\sqrt{2}\,\lambda}{4\pi\varepsilon_0 R}$$

$$\therefore \quad E_{12} = \sqrt{E_1^2 + E_2^2} = \frac{\lambda}{2\pi\varepsilon_0 R}$$

Electric field due to 3

In part (a), we have obtained E_x and E_y due to $\frac{1}{4}$th part of ring. They both were equal.

For half ring, net x- component will cancel and net y-component will become 2-times or,

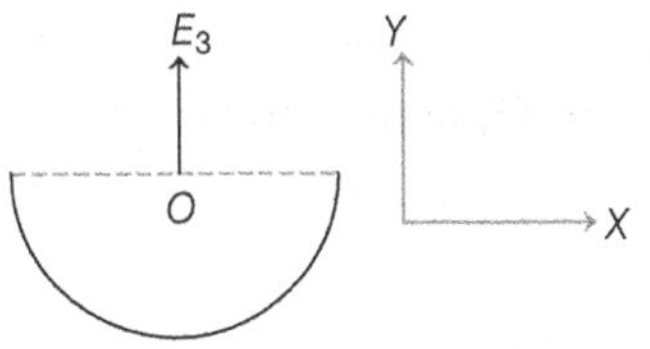

$E_3 = 2$ (E_y due to $\frac{1}{4}$th part of ring)

$$= 2\left(\frac{\lambda}{4\pi\varepsilon_0 R}\right) = \frac{\lambda}{2\pi\varepsilon_0 R} \qquad \text{(upwards)}$$

Now E_{12} and E_3 are equal and opposite.
Hence, at point O, net

$$E = 0$$

119. $\sigma = \mathbf{a} \cdot \mathbf{r} = ar\cos\theta$

Let us take **a** in the direction shown in figure.
C is the centre of differential element of ring.

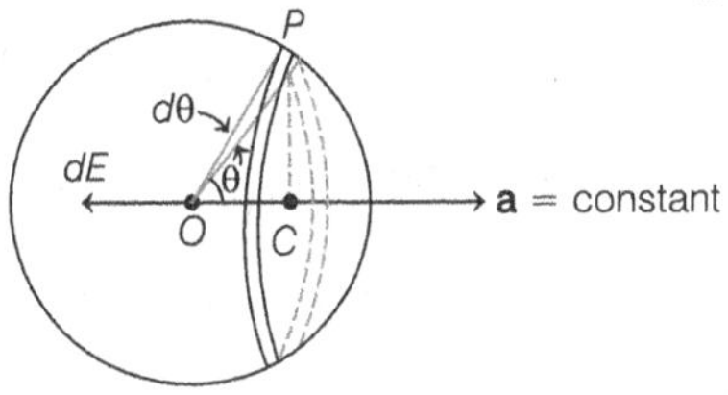

σ on ring $= ar\cos\theta$

$$CO = x\,(\text{say}) = r\cos\theta$$

$$PC = R\,(\text{say}) = r\sin\theta$$

$\Rightarrow \quad R^2 + x^2 = r^2$

Area of ring

$$dA = (2\pi R)\,r\,d\theta = (2\pi r\sin\theta)(r\,d\theta)$$

Charge on ring

$$dq = (\sigma)(dA) = (ar\cos\theta)(2\pi r\sin\theta)(r\,d\theta)$$

Using the expression,

$$E = \frac{1}{4\pi\varepsilon_0}\frac{qx}{(R^2 + x^2)^{3/2}}$$

at a point on the axis of a ring, net electric field at point O, due to differential element of ring, we have

$$dE = \frac{1}{4\pi\varepsilon_0}\left[\frac{(ar\cos\theta)(2\pi r\sin\theta)(r\,d\theta)(r\cos\theta)}{r^3}\right]$$

Integrating dE with the limits of θ from 0° to 180° we will get net electric field at O, which comes out to be,

$$E = \frac{ar}{3\varepsilon_0} \qquad \text{(towards negative } x\text{-axis)}$$

In vector form, we can write,

$$\mathbf{E} = -\frac{\mathbf{ar}}{3\varepsilon_0}$$

120. See the solution of Q -119. This time we will take a disc rather than a ring. Then we will find volume dV of that disc.

$$dq = (\rho)(dV)$$

Then we will apply the formula of electric field on the axis of a disc at some point and finally we will integrate that.

Note *In IIT-JEE we don't use the formula of electric field on the axis of a disc. So students can skip the mathematical part of this question.*

121. Small charge on element of length dx,

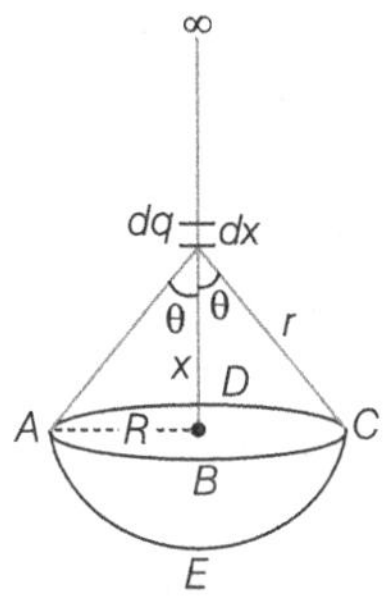

$$dq = \lambda dx$$

Flux passing through circle $ABCDA$ due to this charge 'dq',

$d\phi$ = flux passing through part of sphere AEC

$$= \left(\frac{dq}{\varepsilon_0}\right)\frac{\text{Area of }(AEC)}{\text{Area of sphere}}$$

$$= \frac{\lambda dx}{\varepsilon_0}\left[\frac{2\pi r^2(1 - \cos\theta)}{4\pi r^2}\right]$$

$$= \left[\frac{\lambda dx}{\varepsilon_0}\right]\left[\frac{1 - \dfrac{x}{\sqrt{R^2 + x^2}}}{2}\right]$$

Integrating $d\phi$ from $x = 0$ to $x = \infty$, we can find total flux which comes out to be,

$$|\phi| = \frac{\lambda R}{2\varepsilon_0}$$

Note *We can assume that charge dq is at the centre of a sphere of radius r.*

122. Electric field at some point inside a charged sphere is given by,

$$E = \frac{1}{4\pi\varepsilon_0}\frac{q}{R^3}r \qquad (r \le R)$$

$$\rho = \frac{q}{\frac{4}{3}\pi R^3}$$

$$\therefore \quad \frac{q}{4\pi R^3} = \frac{\rho}{3} \quad \therefore \quad E = \frac{\rho r}{3\varepsilon_0}$$

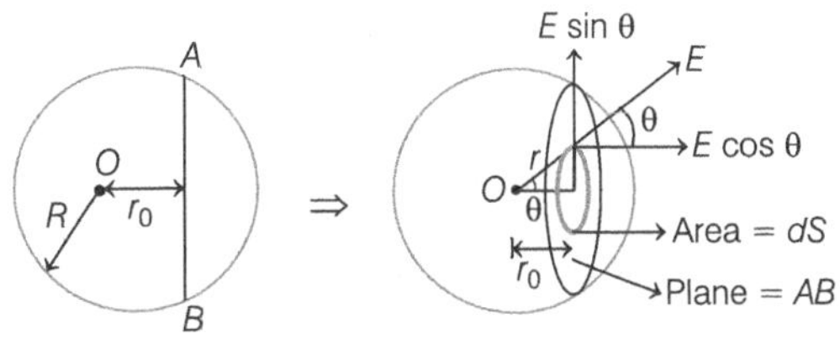

We have to find flux through AB,

$$E=\frac{\rho r}{3\varepsilon_0}$$

$E\sin\theta$ is tangential to AB. So, its flux is zero. Only $E\cos\theta$ will produce the flux. Small flux passing through small area dS (inside the plane AB) is,

$$d\phi=(E\cos\theta)\,dS$$

$$=\left(\frac{\rho r}{3\varepsilon_0}\right)\cos\theta\, dS$$

$$=\frac{\rho r_0}{3\varepsilon_0}\,dS \qquad (r\cos\theta=r_0)$$

$$\therefore \quad \phi=\frac{\rho r_0}{3\varepsilon_0}\int dS$$

$$\phi=\left(\frac{\rho r_0}{3\varepsilon_0}\right)S \qquad \text{...(i)}$$

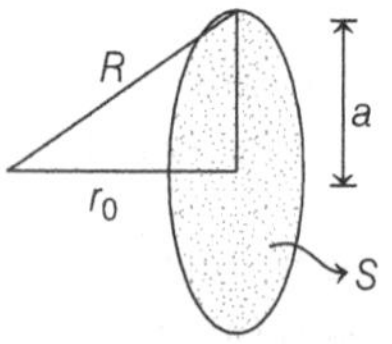

$$a=\sqrt{R^2-r_0^2}$$

$$S=\pi a^2=\pi(R^2-r_0^2)$$

Substituting value of S in Eq. (i), we get

$$\phi=\frac{\rho\pi r_0}{3\varepsilon_0}(R^2-r_0^2)$$

123.

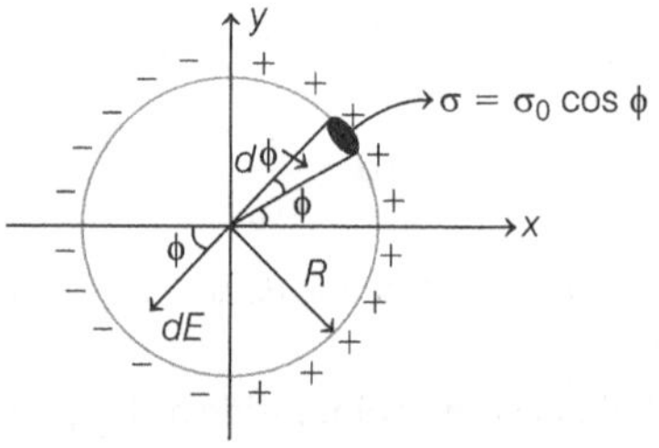

Cross-sectional view of cylinder

Let l be the length of cylinder perpendicular to plane. Area of small strip shown in figure is,

$$dA=(R\,d\phi)\,l$$

Now, $$\sigma=\frac{dq}{dA}=\frac{dq}{(Rd\phi)l}$$

$$\therefore \quad \left(\frac{dq}{l}\right)=\lambda=\sigma R d\phi$$

or $$\lambda=(\sigma_0\cos\phi)\,R\,d\phi$$

$$dE=\frac{\lambda}{2\pi\varepsilon_0 R}=\frac{(\sigma_0 R\cos\phi)\,d\phi}{2\pi\varepsilon_0 R}$$

or $$dE=\left(\frac{\sigma_0}{2\pi\varepsilon_0}\right)\cos\phi\, d\phi$$

$$dE_x=-dE\cos\phi$$

and $$dE_y=-dE\sin\phi$$

If we integrate dE_x and dE_y from 0 to 2π, we get

$$E_y=0 \quad\text{and}\quad E_x=-\frac{\sigma_0}{2\varepsilon_0}$$

Thus, $$\mathbf{E}=-\frac{\sigma}{2\varepsilon_0}\hat{\mathbf{i}}$$

124. (a)

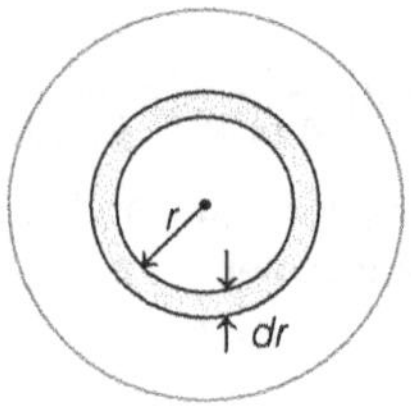

Volume of small element,

$$dV=4\pi r^2 dr$$

$$\rho=\rho_0\left(1-\frac{r}{R}\right)$$

$$\therefore \quad dq=\rho dV=4\pi\rho_0\left(r^2-\frac{r^3}{R}\right)dr$$

Integrating dq from 0 to r, we get total charge from 0 to r.

Which comes out to be,

$$q=4\pi\rho_0\left(\frac{r^3}{3}-\frac{r^4}{4R}\right) \qquad \text{...(i)}$$

Electric field for inside points ($r\le R$)

According to Gauss' theorem,

$$E=\frac{1}{4\pi\varepsilon_0}\frac{q_{\text{in}}}{r^2} \qquad \text{...(ii)}$$

$$=\frac{1}{4\pi\varepsilon_0}\left[\frac{4\pi\rho_0\left(\frac{r^3}{3}-\frac{r^4}{4R}\right)}{r^2}\right]$$

$$E=\frac{\rho_0 r}{3\varepsilon_0}\left[1-\frac{3r}{4R}\right] \qquad \text{...(iii)}$$

Electric field for outside points ($r\ge R$)

In Eq. (ii),

$$q_{\text{in}}=q_{\text{total}} \text{ or } q_T$$

Substituting $r=R$ in Eq. (i), we get

$$q_T=\frac{\pi\rho_0 R^3}{3}$$

Now, substituting $q_{\text{in}}=q_T$ in Eq. (ii), we get

$$E=\frac{\rho_0 R^3}{12\varepsilon_0 r^2}$$

(b) **Maximum value of electric field**

For outside points, $E\propto\frac{1}{r^2}$

So, E continuously decreases.

Therefore electric field is maximum inside the sphere.

For E to be maximum,

$$\frac{dE_{in}}{dr}=0 \Rightarrow \frac{d}{dr}\left[\frac{\rho_0 r}{3\varepsilon_0}\left(1-\frac{3r}{4R}\right)\right]=0$$

Solving, we get $r=\frac{2R}{3}$

Substituting this value of r in Eq. (iii), we get

$$E_{max}=\frac{\rho_0 R}{9\varepsilon_0}$$

125.

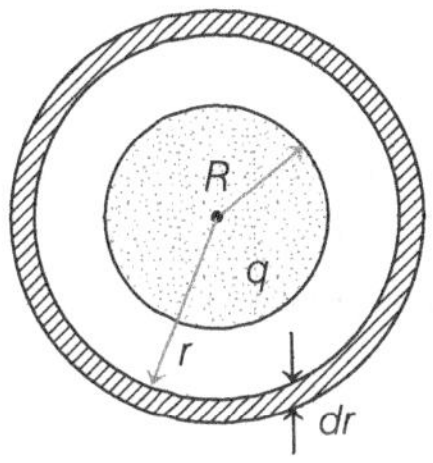

Let charge on the ball be q.

Small volume of element outside the sphere,

$dV=(4\pi r^2)dr$

$$\rho=\frac{\alpha}{r}$$

$\therefore \quad dQ=\rho\cdot dV=(4\pi\alpha r)dr$

Integrating dQ from R to r, we can find the charge between R and r, which comes out to be,

$$Q=2\pi\alpha(r^2-R^2)$$

According to Gauss' theorem, electric field at distance $r(\geq R)$,

$$E=\frac{1}{4\pi\varepsilon_0}\frac{q_{in}}{r^2}=\frac{1}{4\pi\varepsilon_0}\frac{(q+Q)}{r^2}$$

or $$E=\frac{1}{4\pi\varepsilon_0 r^2}[q+2\pi\alpha(r^2-R^2)] \quad ...(i)$$

Result is independent of r, if

$$q-2\pi\alpha R^2=0$$

$\therefore \quad q=2\pi\alpha R^2$

Substituting $q-2\pi\alpha R^2=0$ in Eq. (i), we get

$$E=\frac{\alpha}{2\varepsilon_0}$$

Note *Outside the ball charge extends upto infinity.*

126. $dV=(4\pi r^2)\,dr$

$$\rho=\rho_0 e^{-\alpha r^3}$$

$\therefore \quad dq=\rho dV=(4\pi\rho_0)r^2 e^{-\alpha r^3}dr$

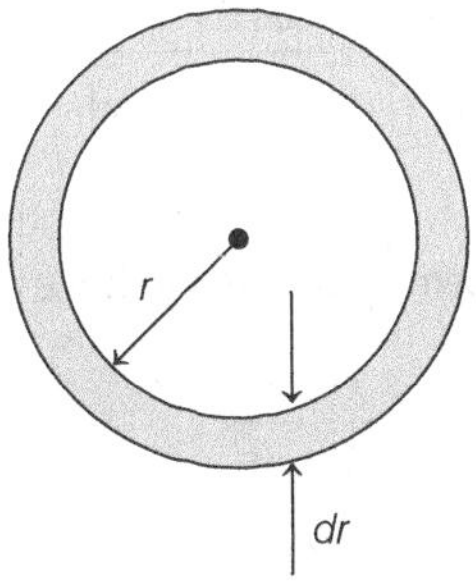

Integrating dq from 0 to r, we get total charge from 0 to r, which comes out to be,

$$q=\frac{4\pi\rho_0}{3\varepsilon_0\alpha}(1-e^{-\alpha r^3}) \quad ...(i)$$

From Gauss theorem, electric field at distance r is,

$$E=\frac{1}{4\pi\varepsilon_0}\frac{q_{in}}{r^2}$$

Substituting $q_{in}=q$ from Eq. (i), we get

$$E=\frac{\rho}{3\varepsilon_0\alpha r^2}(1-e^{-\alpha r^3})$$

For $\alpha r^3<<1$ $\quad e^{-\alpha r^3}\approx 1-\alpha r^3$

$\therefore \quad E\approx\frac{\rho_0 r}{3\varepsilon_0}$

For $\alpha r^3>>1$ $\quad e^{-\alpha r^3}\approx 0$

$\therefore \quad E=\frac{\rho_0}{3\varepsilon_0\alpha r^2}$

127.

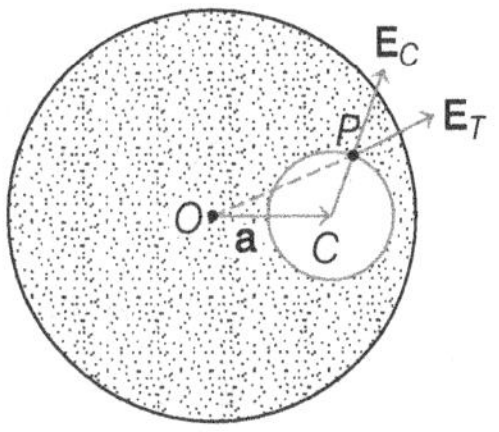

Electric field at some point inside a charged sphere is given by,

$$E=\frac{\rho r}{3\varepsilon_0} \quad \text{(See the hint of Q-122)}$$

At a point P inside the cavity,

$\mathbf{E}$ = electric field due to complete sphere $(\mathbf{E}_T)$ − electric field due to cavity $(\mathbf{E}_C)$

$$=\frac{\rho(\mathbf{OP})}{3\varepsilon_0}-\frac{\rho(\mathbf{CP})}{3\varepsilon_0}=\frac{\rho}{3\varepsilon_0}(\mathbf{OP}-\mathbf{CP})$$

$$=\frac{\rho}{3\varepsilon_0}(\mathbf{OP}+\mathbf{PC})=\frac{\rho\mathbf{a}}{3\varepsilon_0}$$

Thus, $\mathbf{E}=\frac{\rho\mathbf{a}}{3\varepsilon_0}$

128. **Note** Students are advised to see the hint of Q-127 before solving this.

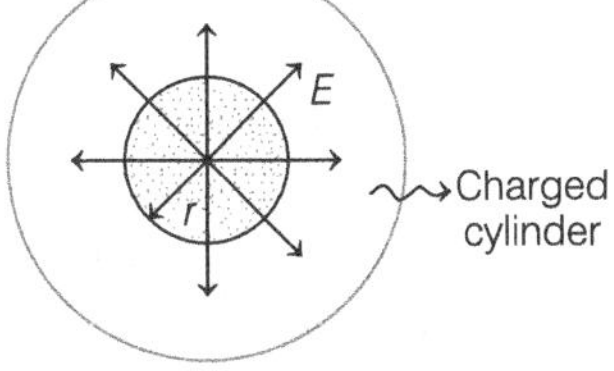

Cross-sectional view

Perpendicular to plane of paper, let the dimension of cylinder is l. Applying Gauss theorem on the closed cylindrical surface of radius r and length l gives, total electric flux $= \dfrac{q_{in}}{\varepsilon_0}$

$$\therefore \quad E(2\pi rl) = \frac{(\pi r^2 l)\rho}{\varepsilon_0}$$

$$\therefore \quad E = \frac{\rho r}{2\varepsilon_0}$$

Now we can apply the same method. What we used in earlier.

The only difference is, there it was $\dfrac{\rho r}{3\varepsilon_0}$ and here it is $\dfrac{\rho r}{2\varepsilon_0}$. So the final answer will be,

$$\mathbf{E} = \frac{\rho \mathbf{a}}{2\varepsilon_0}$$

129. Let $\mathbf{a} = a_1\hat{\mathbf{i}} + a_2\hat{\mathbf{j}} + a_3\hat{\mathbf{k}} = \text{constant}$

and $\mathbf{r} = x\hat{\mathbf{i}} + y\hat{\mathbf{j}} + z\hat{\mathbf{k}}$

$\therefore \quad V = \mathbf{a}\cdot\mathbf{r}$ (Given)

$$V = a_1x + a_2y + a_3z$$

Now, $\mathbf{E} = -\left[\dfrac{\partial V}{\partial x}\hat{\mathbf{i}} + \dfrac{\partial V}{\partial y}\hat{\mathbf{j}} + \dfrac{\partial V}{\partial z}\hat{\mathbf{k}}\right]$

$$= -[a_1\hat{\mathbf{i}} + a_2\hat{\mathbf{j}} + a_3\hat{\mathbf{k}}]$$

or $\mathbf{E} = -\mathbf{a} = \text{constant}$

Thus electric field is uniform.

130. (a) $\mathbf{E} = -\left[\dfrac{\partial V}{\partial x}\hat{\mathbf{i}} + \dfrac{\partial V}{\partial y}\hat{\mathbf{j}}\right] = -2a(x\hat{\mathbf{i}} - y\hat{\mathbf{j}})$

(b) $\mathbf{E} = -\left[\dfrac{\partial V}{\partial x}\hat{\mathbf{i}} + \dfrac{\partial V}{\partial y}\hat{\mathbf{j}}\right] = -a[y\hat{\mathbf{i}} + x\hat{\mathbf{j}}]$

Graphical representation

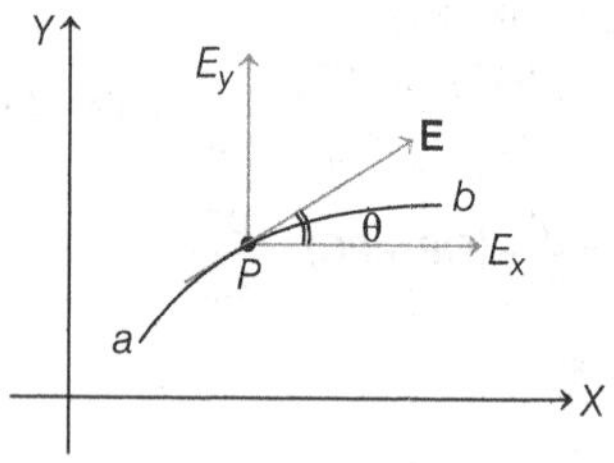

Suppose ab is an electric line of force. At some point P, value of electric field is $\mathbf{E} = E_x\hat{\mathbf{i}} + E_y\hat{\mathbf{j}}$, from the graph, we can see that

$$\tan\theta = \frac{dy}{dx} = \frac{E_y}{E_x}$$

Part (a)

$$\frac{dy}{dx} = \frac{E_y}{E_x} = \frac{2ay}{-2ax} = \frac{y}{-x} \Rightarrow \frac{dy}{y} + \frac{dx}{x} = 0$$

Integrating, we get

$$\ln y + \ln x = \text{constant}$$

or $\ln xy = \text{constant}$

or $xy = \text{constant}$

So, electric lines are rectangular hyperbolas as shown in figure.

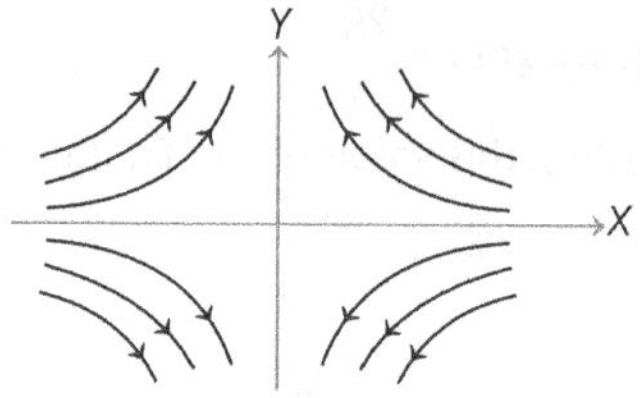

Direction of E

Let us take first quadrant, where $x > 0, y > 0$. For $a > 0$,

E_x = negative and

E_y = positive

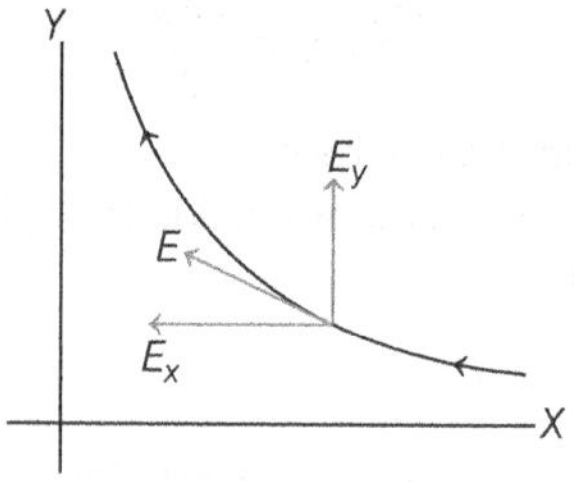

Part (b)

The same explanation can be used for this part. Here, we get equation of hyperbola and normally we do not study hyperbola equation in physics. So, students can skip this part.

131.

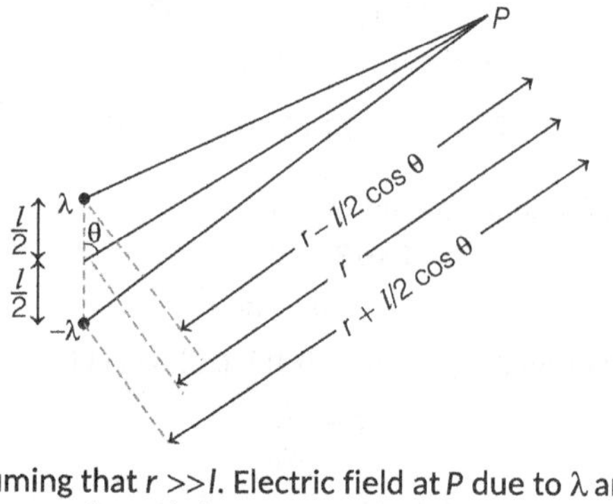

Assuming that $r >> l$. Electric field at P due to λ and $-\lambda$ are approximately in opposite directions.

Hence, electric field at P will be

$$E = \frac{\lambda}{2\pi\varepsilon_0\left[r - \dfrac{l}{2}\cos\theta\right]} - \frac{\lambda}{2\pi\varepsilon_0\left[r + \dfrac{l}{2}\cos\theta\right]}$$

$$= \frac{\lambda l\cos\theta}{2\pi\varepsilon_0\left[r^2 - \dfrac{l^2}{4}\cos^2\theta\right]}$$

Ignoring the term $\dfrac{l^2}{4}\cos^2\theta$, we have

$$E \approx \frac{\lambda l\cos\theta}{2\pi\varepsilon_0 r^2}$$

Further, $dV = -E dr$

$$\Rightarrow \quad \int_{\infty}^{r} dV = -\int_{\infty}^{r} \frac{\lambda l \cos\theta}{2\pi\varepsilon_0 r^2} dr$$

Solving this integration, we get

$$V \approx \frac{\lambda l \cos\theta}{2\pi\varepsilon_0 r}$$

132. (a)

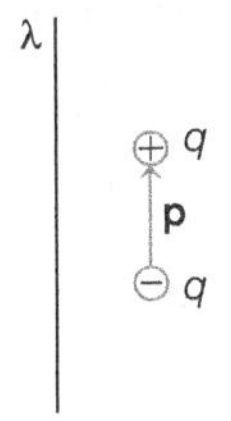

$-q$ and $+q$ are at equal distance from the charged thread. So, electric field of thread will be same. Hence, equal and opposite force will act $(= q\mathbf{E})$ on $-q$ and $+q$. So, net force will be zero.

$$\mathbf{F} = 0$$

(b)

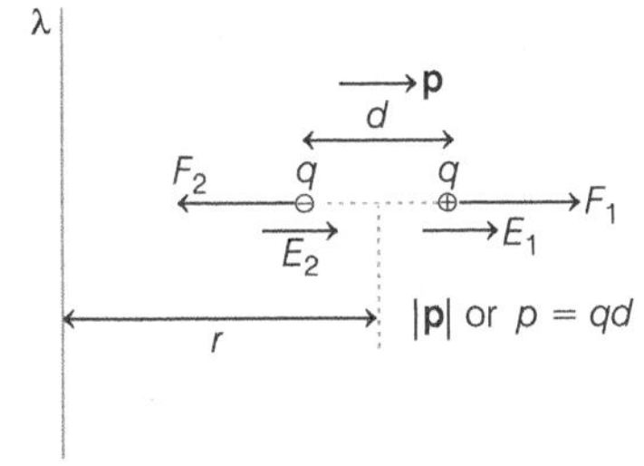

$$E_1 < E_2,\ E_1 = \frac{\lambda}{2\pi\varepsilon_0\left(r + \frac{d}{2}\right)},\ E_2 = \frac{\lambda}{2\pi\varepsilon_0\left(r - \frac{d}{2}\right)},$$

$F_1 < F_2$

∴ Net force is towards the thread, or in opposite direction of **p**,

$$F = F_2 - F_1$$

$$= \frac{q\lambda}{2\pi\varepsilon_0\left(r - \frac{d}{2}\right)} - \frac{q\lambda}{2\pi\varepsilon_0\left(r + \frac{d}{2}\right)}$$

$$= \frac{\lambda q d}{2\pi\varepsilon_0\left(r^2 - \frac{d^2}{4}\right)} \qquad (qd = p)$$

$$\approx \frac{\lambda p}{2\pi\varepsilon_0 r^2} \qquad \text{(ignoring the term } d^2/4\text{)}$$

In vector form, we can write

$$\mathbf{F} = -\frac{\lambda \mathbf{p}}{2\pi\varepsilon_0 r^2}$$

(c)

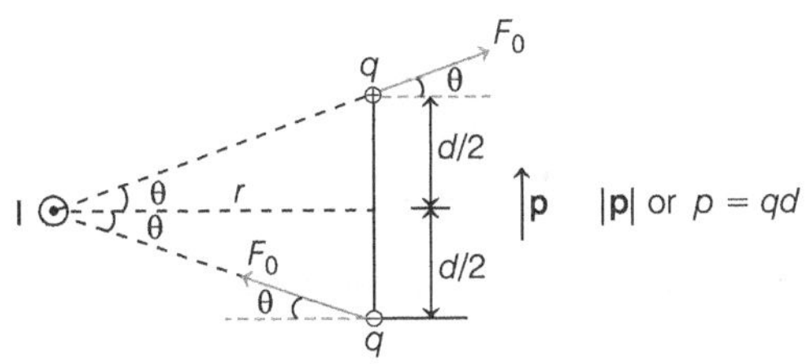

Net force $F = 2F_0 \sin\theta$ in the direction of **p**.

$$\therefore \quad F = 2F_0 \sin\theta = 2(E)(q)\sin\theta$$

$$= 2\left[\frac{\lambda}{2\pi\varepsilon_0\sqrt{r^2 + \frac{d^2}{4}}}\right](q)\frac{(d/2)}{\sqrt{r^2 + \frac{d^2}{4}}}$$

Ignoring the term $\frac{d^2}{4}$ and putting $qd = p$, we get

$$F = \frac{\lambda p}{2\pi\varepsilon_0 r^2}$$

In vector form, we can write,

$$\mathbf{F} = \frac{\lambda \mathbf{p}}{2\pi\varepsilon_0 r^2}$$

133.

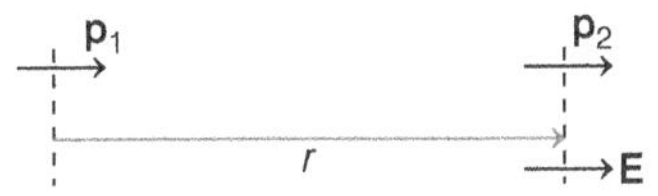

E = magnitude of electric field of $\mathbf{p}_1$ on $\mathbf{p}_2$

$$= 2\left(\frac{1}{4\pi\varepsilon_0}\right)\frac{p_1}{r^3}$$

So, potential energy of $\mathbf{p}_2$,

$$U = -\mathbf{p}_2 \cdot \mathbf{E}$$

$$= -p_2\left(\frac{2p_1}{4\pi\varepsilon_0 r^3}\right)$$

$$= -\frac{p_1 p_2}{2\pi\varepsilon_0 r^3} = -\frac{p^2}{2\pi\varepsilon_0 r^3} \qquad (p_1 = p_2 = p)$$

$$F = -\frac{dU}{dr} = \frac{3p^2}{2\pi\varepsilon_0 r^4} \qquad \text{(Given, } r = l\text{)}$$

$$\therefore \quad F = \frac{3p^2}{2\pi\varepsilon_0 l^4}$$

Substituting the values, we get

$$F \approx 2.1 \times 10^{-16}\ \text{N}$$

134. $V = -ax^3 + b$

$$\mathbf{E} = -\left[\frac{\partial V}{\partial x}\hat{\mathbf{i}} + \frac{\partial V}{\partial y}\hat{\mathbf{j}} + \frac{\partial V}{\partial z}\hat{\mathbf{k}}\right] = 3ax^2\hat{\mathbf{i}}$$

$$\therefore \quad E = E_x = 3ax^2$$

$$\Rightarrow \quad dE = (6ax)\,dx$$

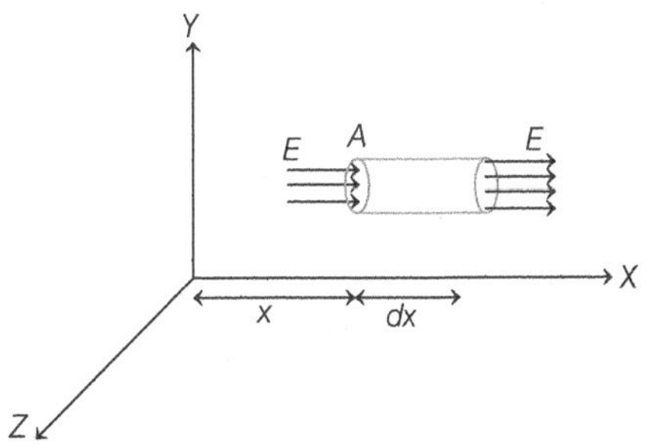

Net flux coming out from the closed surface shown in figure is,

$$(E+dE)A-EA=(dE)A$$
$$=(6ax)Adx=(6ax)dV$$

From Gauss theorem, this net flux should be equal to small charge dq inside the closed surfaces divided by ε_0. Thus,

$$(6ax)\,dV=\frac{dq}{\varepsilon_0}$$

$\Rightarrow$ $$\frac{dq}{dV}=\rho(x)=(6ax)\,\varepsilon_0$$

$\therefore$ $$\rho(x)=(6ax)\varepsilon_0$$

135.

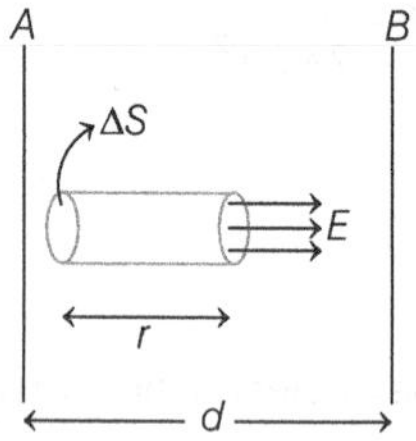

Let volume charge density between the plates is ρ.
At a distance r from plate A suppose electric field is E.
Electric field at A is given zero.
Let us take a closed surface as shown in figure.
Total flux coming out from this surface $=E\Delta S$
Let q be the charge inside this closed surface, then
$q=(\rho)$ (volume of closed surface)
or $q=\rho(r\Delta S)$
From Gauss theorem,

total flux from the closed surface $=\dfrac{q}{\varepsilon_0}$

$\therefore$ $$E\Delta S=\frac{\rho r(\Delta S)}{\varepsilon_0}$$

$\therefore$ $$E=\frac{\rho r}{\varepsilon_0}$$

Putting $r=d$ in the above expression of E, we can find electric field on other plate B. Thus,

$$E_B=\frac{\rho d}{\varepsilon_0}$$

Further, $$dV=-Edr$$

$\therefore$ $$\int_B^A dV=-\int_{r=d}^{r=0}\frac{\rho r}{\varepsilon_0}dr$$

$\therefore$ $$V_A-V_B=\Delta V=\frac{\rho d^2}{2\varepsilon_0}$$

or $$\rho=\frac{2\varepsilon_0\Delta V}{d^2}$$

136. $V=ar^2+b$

$\therefore$ $$E=-\frac{dV}{dr}=-2ar$$

Negative sign of E implies that direction of electric field is towards centre.

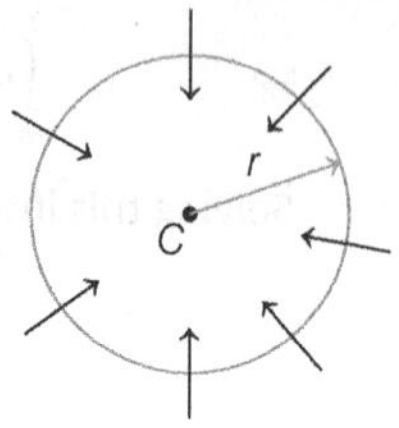

Otherwise magnitude of electric field is,

$$|E|\propto r$$

So, this is a case of charged sphere of uniform volume charge density ρ. Using the standard expression of E,

$$E=-2ar=\frac{1}{4\pi\varepsilon_0}\cdot\frac{q}{R^3}r$$

Putting $\dfrac{q}{\frac{4}{3}\pi R^3}=\rho$, we get

$$-2ar=\frac{\rho}{3\varepsilon_0}r$$

$\therefore$ $$\rho=-6\varepsilon_0 a$$

137.

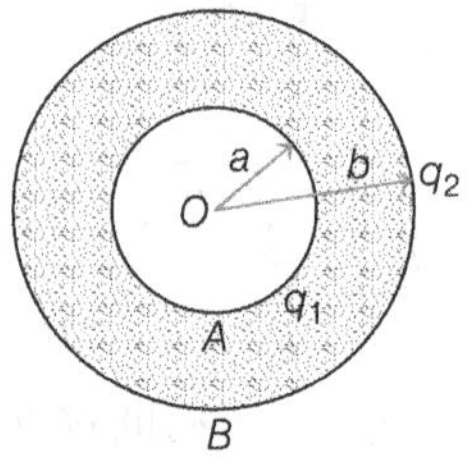

$$V_A=0$$

$\Rightarrow$ $$\frac{kq_1}{a}+\frac{kq_2}{b}=0 \qquad \left(\text{as } k=\frac{1}{4\pi\varepsilon_0}\right)$$

So, $$q_2=-\frac{b}{a}q_1$$

For $0\le r\le a$

Potential of both q_1 and q_2 is constant and this is also equal to potential of A, which is equal to zero.

$\therefore$ $$V=0$$

For $a\le r\le b$

Potential of q_2 (which is negative) is constant but potential of q_1 (which is positive) will decrease. So, net potential will increase in negative side.

$\therefore$ $$V=\frac{q_1}{4\pi\varepsilon_0 r}+\frac{q_2}{4\pi\varepsilon_0 b}$$

Substituting $\dfrac{q_2}{b}=-\dfrac{q_1}{a}$, we get

$$V=\frac{q_1}{4\pi\varepsilon_0}\left(\frac{1}{r}-\frac{1}{a}\right)$$

For $r\ge b$

As $r\to\infty$, $V\to 0$

Therefore, net potential will increase.

$$V=\frac{q_1}{4\pi\varepsilon_0 r}+\frac{q_2}{4\pi\varepsilon_0 r}$$

Substituting $q_2=-\dfrac{b}{a}q_1$, we get

$$V = \frac{q_1}{4\pi\varepsilon_0 r}\left(1 - \frac{b}{a}\right)$$

V-r graph

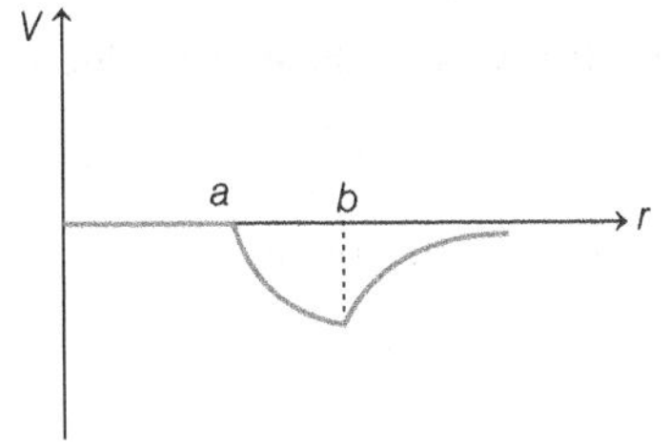

138.

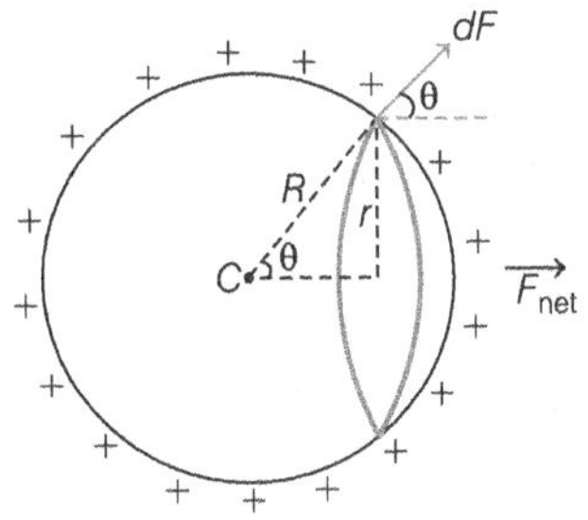

Force per unit area on the surface of charged sphere is given by,

$$\frac{1}{2}\varepsilon_0 E^2 = \frac{1}{2}\varepsilon_0\left(\frac{\sigma}{\varepsilon_0}\right)^2 = \frac{\sigma^2}{2\varepsilon_0}$$

$$r = R\sin\theta$$

Area of the element shown in figure is,

$$dA = (2\pi r)(Rd\theta)$$

$$= (2\pi R^2 \sin\theta\, d\theta)$$

$$dF = \left(\frac{\sigma^2}{2\varepsilon_0}\right)(2\pi R^2 \sin\theta d\theta)$$

Putting $\sigma = \frac{q}{4\pi R^2}$, we get

$$dF = \frac{q^2 \sin\theta \cdot d\theta}{16\,\pi\varepsilon_0 R^2}$$

$$F_{net} = \int_0^{\pi/2} dF\cos\theta$$

$$= \int_0^{\pi/2} \frac{q^2(\sin\theta\cos\theta)\,d\theta}{16\,\pi\varepsilon_0 R^2}$$

$$= \frac{q^2}{32\pi\varepsilon_0 R^2}$$

Substituting the given values, we get

$$F_{net} = 0.5 \text{ kN}$$

139. **Note** See the solution of Q-138.

For $-\frac{\pi}{2} < \theta < \frac{\pi}{2}$, $\cos\theta$ is +ve

For $\frac{\pi}{2} < \theta < \frac{3\pi}{2}$, $\cos\theta$ is −ve

Therefore, charge distribution is as shown below.

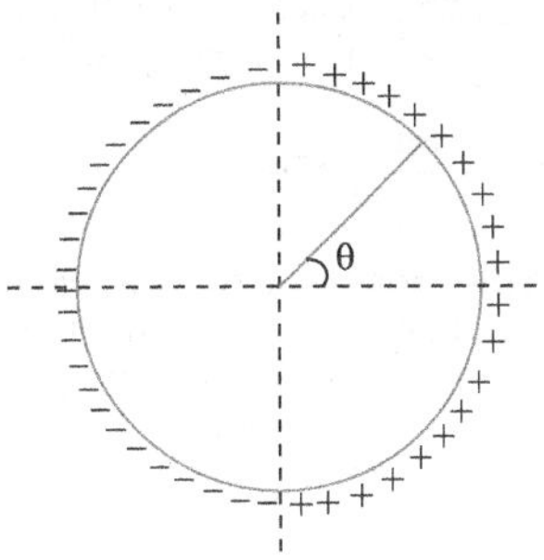

As per sol-138

$$\frac{\text{Force}}{\text{Area}} = \frac{\sigma^2}{2\varepsilon_0} = \frac{\sigma_0^2\cos^2\theta}{2\varepsilon_0}$$

$$dA = (2\pi R^2 \sin\theta)\,d\theta$$

$$\therefore \quad dF = \frac{(\sigma_0^2 \pi R^2 \sin\theta\cos^2\theta)\,d\theta}{\varepsilon_0}$$

$$\therefore \quad F_{net} = \int_0^{\pi/2} dF\cos\theta$$

After substituting the values and completing the integration, we get

$$F_{net} = \frac{\pi R^2 \sigma_0^2}{4\varepsilon_0}$$

140. (a) **Self energy**

Volume charge density,

$$\rho = \frac{q}{\frac{4}{3}\pi R^3} = \frac{3q}{4\pi R^3}$$

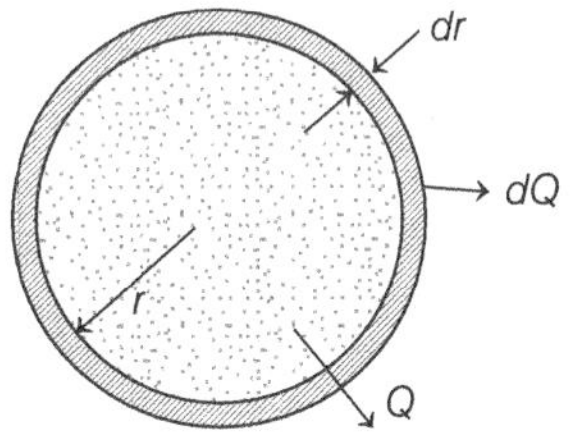

Let at some intermediate position when radius is r and charge assembled is Q.

Then, $Q = \left(\frac{4}{3}\pi r^3\right)\rho = \frac{qr^3}{R^3}$

$$dQ = \left(\frac{3qr^2}{R^3}\right)dr$$

Potential on the surface at this instant is,

$$V = \left(\frac{1}{4\pi\varepsilon_0}\right)\frac{Q}{r} = \left(\frac{1}{4\pi\varepsilon_0}\right)\left(\frac{qr^2}{R^3}\right)$$

Work done is bringing dQ charge from infinity to the surface is,

$$dW = (dQ)V$$

$$= \left[\frac{3qr^2}{R^3}\right](dr)\left(\frac{1}{4\pi\varepsilon_0}\right)\left(\frac{qr^2}{R^3}\right)$$

$$dW = \frac{3q^2}{4\pi\varepsilon_0 R^6} r^4 dr$$

Integrating dW from $r=0$ to $r=R$, we get total work done and which is also equal to the self energy of the ball. The final result comes out to be,

$$W = \frac{3q^2}{20\pi\varepsilon_0 R}$$

(b) **Energy W_2**

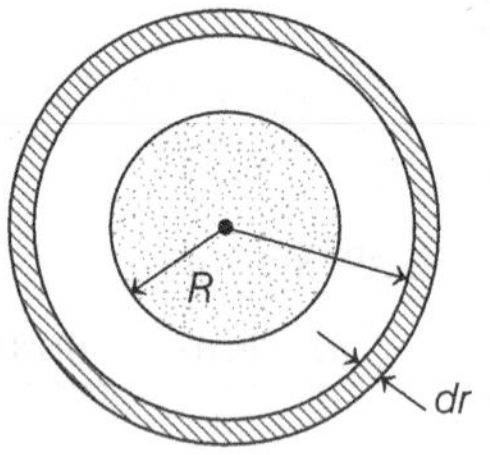

At distance r,

$$E = \frac{1}{4\pi\varepsilon_0}\frac{q}{r^2}$$

Energy density $u = \frac{1}{2}\varepsilon_0 E^2$

Small volume of the element shown in figure,

$$dV = (4\pi r^2)dr$$

Therefore small energy stored in this volume is,

$$dW_2 = (u)dV$$

When we integrate dW_2 from $r=R$ to $r=\infty$, we get total W_2. After substituting the values and completing the integration, the final results comes out to be,

$$W_2 = \frac{q^2}{8\pi\varepsilon_0 R}$$

Energy W_1

$$W_1 = W - W_2 = \frac{q^2}{40\pi\varepsilon_0 R}$$

Ratio W_1/W_2

$$\therefore \quad \frac{W_1}{W_2} = \frac{q^2/40\pi\varepsilon_0 R}{q^2/8\pi\varepsilon_0 R} = \frac{1}{5}$$

Thus, $\frac{W_1}{W_2} = \frac{1}{5}$

141. (a) Acceleration of particle,

$$a = \frac{F}{m} = \frac{qE}{m} = \frac{q}{m}(E_0 - \alpha x) \quad \text{...(i)}$$

or $v \cdot \frac{dv}{dx} = \frac{q}{m}(E_0 - \alpha x)$

$$\Rightarrow \int_0^v v\,dv = \int_0^x \frac{q}{m}(E_0 - \alpha x)\,dx$$

Solving this integration, we get

$$v = \sqrt{\frac{2q}{m}\left(E_0 x - \frac{\alpha x^2}{2}\right)}$$

Now, $v=0$ at $x=0$ at the initial point and

$$x = \frac{2E_0}{\alpha}$$

So, the distance covered is,

$$x = \frac{2E_0}{\alpha}$$

(b) From Eq. (i) acceleration at $x = \frac{2E_0}{\alpha}$ is,

$$a = -\frac{qE_0}{m}$$

So, magnitude of acceleration is,

$$|a| = \frac{qE_0}{m}$$

142. By Gauss' theorem, $E \propto \frac{q_{net}}{r^2}$

$$\Rightarrow \quad 8 = \frac{\int_0^R 4\pi r^2 kr^a dr}{\int_0^{R/2} 4\pi r^2 kr^a dr} \times \frac{(R/2)^2}{R^2}$$

$$\Rightarrow \quad 8 = \frac{\left[\frac{k4\pi r^{2+a+1}}{2+a+1}\right]_0^R}{\left[\frac{k4\pi r^{2+a+1}}{2+a+1}\right]_0^{R/2}} \times \frac{1}{4}$$

$$\Rightarrow \quad 8 \times 4 = \frac{(r^{2+a+1})_0^R}{(r^{2+a+1})_0^{R/2}} = 2^{2+a+1}$$

$$\Rightarrow \quad 2+a+1 = 5$$

$$\Rightarrow \quad a = 2$$

143. $E_{\text{Line charge}} = \frac{\lambda}{2\pi\varepsilon_0 r}$

$$\Rightarrow \quad \Delta V_{\text{Line charge}} = \int_{0.5}^{2} \frac{\lambda}{2\pi\varepsilon_0 r} dr$$

$$= \frac{\lambda}{2\pi\varepsilon_0} \ln 4 \quad \text{... (i)}$$

$$\Delta V_{\text{sphere}} = \frac{1}{4\pi\varepsilon_0}\frac{Q}{R} - \frac{1}{4\pi\varepsilon_0}\frac{Q}{2R}$$

$$= \frac{1}{4\pi\varepsilon_0}\frac{Q}{2} \quad \text{... (ii)}$$

$$\Rightarrow \quad \Delta V_{\text{net}} = \frac{\lambda}{2\pi\varepsilon_0}\ln 4 + \frac{1}{4\pi\varepsilon_0}\frac{Q}{2}$$

Substituting the given values, we get

$$\Delta V_{\text{net}} = 171\text{ V}$$

$\therefore$ The correct answer is 171.

UNIT–3

CAPACITORS

Learning Modules

1. Capacitance
2. Energy Stored in a Charged Capacitor
3. Capacitors
4. Mechanical Force on a Charged Conductor
5. Capacitors in Series and Parallel
6. Two Laws in Capacitors
7. *C-R* Circuits
8. Methods of Finding Equivalent Resistance and Capacitance

MODULE 1

Capacitance

In practice, we cannot handle free point charges or hold them fixed at desired positions. A practical way to handle a charge would be to put it on a conductor. Thus, one use of a conductor is to store electric charge (or electric potential energy). But, every conductor has its maximum limit of storing the electric charge or potential energy. Beyond that limit, the dielectric in which the conductor is placed, becomes ionized. A capacitor is a device which can store more electric charge or potential energy compared to an isolated conductor.

Capacitors have a tremendous number of applications. In the flash light used by photographers, the energy and charge stored in a capacitor are recovered quickly. In other applications, the energy is released more slowly.

Capacitance of an Isolated Conductor

When a charge q is given to a conductor, it spreads over the outer surface of the conductor. The whole conductor comes to the same potential (say V). This potential V is directly proportional to the charge q, i.e.

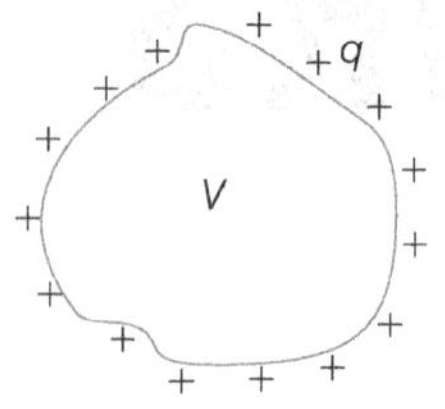

$$V \propto q$$

When the proportionality sign is removed, a constant of proportionality $\frac{1}{C}$ comes in picture.

Hence, $V = \frac{q}{C}$ or $C = \frac{q}{V}$

Here, C is called the **capacitance** of the conductor. The SI unit of capacitance is called **one farad** (1 F). One farad is equal to one coulomb per volt (1 C/V)

$\therefore$ $1\text{ F} = 1\text{ farad} = 1\text{ C/V} = 1\text{ coulomb/volt}$

Note *(i) An obvious question arises in mind that when a conductor stores electric charge and energy then why not the unit of capacitance is coulomb or joule. For example, the capacity of a storage tank is given in litres (the unit of volume) or gallons not in the name of some scientist.*

The reason is simple the capacity of tank does not depend on medium in which it is kept. While the capacity of a conductor to store charge (or energy) depends on the medium in which it is kept. It varies from medium to medium. So, it is difficult to express the capacity in terms of coulomb or joule. Because in that case we will have to mention the medium also.

For example, we will say like this, capacity of this conductor in water is 1 C in oil it is 5 C, etc. On the other hand, the C discussed above gives us an idea about the dimensions of the conductor nothing about the charge which it can store because as we said earlier also it will vary from medium to medium. By knowing the C (or the dimensions of conductor) a physics student can easily find the maximum charge which it can store, provided the medium is also given.

(ii) Farad in itself is a large unit. Microfarad (μF) is used more frequently.

Method of Finding Capacitance of a Conductor

Give a charge q to the conductor. Find potential on it due to charge q. This potential V will be a function of q and finally find q/V, which is the desired capacitance C.

Capacitance of a Spherical Conductor

When a charge q is given to a spherical conductor of radius R, the potential on it is

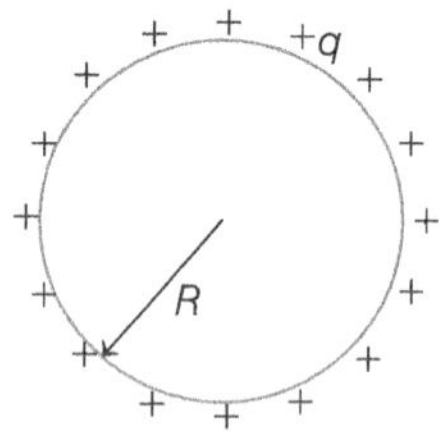

$$V = \frac{1}{4\pi\varepsilon_0} \cdot \frac{q}{R}$$

From this expression, we find that

$$\frac{q}{V} = 4\pi\varepsilon_0 R = C$$

Thus, capacitance of the spherical conductor is

$$\boxed{C = 4\pi\varepsilon_0 R}$$

From this expression, we can draw the following conclusions:

(i) $C \propto R$ or C depends on R only. Which we have already stated that C depends on the dimensions of the conductor. Moreover if two conductors have radii R_1 and R_2, then

$$\frac{C_1}{C_2} = \frac{R_1}{R_2}$$

(ii) Earth is also a spherical conductor of radius $R = 6.4 \times 10^6$ m. The capacity of earth is therefore,

$$C = \left(\frac{1}{9 \times 10^9}\right)(6.4 \times 10^6)$$

$$\approx 711 \times 10^{-6}\text{ F}$$

or $$C = 711\ \mu\text{F}$$

From here, we can see that farad is a large unit. As capacity of such a huge conductor is only 711 μF.

Important POINTS

- **Dielectric strength of an insulator** In an insulator, most of the electrons are tightly bounded with the nucleus. If an electric field is applied on this insulator, an electrostatic force acts on these electrons in the opposite direction of electric field. As electric field increases, this force also increases. After a certain value of electric field, force becomes so large that these bounded electrons are knocked out or ionized. This maximum electric field is called dielectric strength of insulator. Its unit is N/C or V/m.
- With the help of capacitance (or dimensions of conductor) and dielectric strength of an insulator we can determine the maximum charge or energy which can be stored by this conductor.

Example 1 Capacitance of a conductor is 1 μF. What charge is required to raise its potential to 100 V?

Sol. Using the equation

$$q = CV$$

We have,

$$q = (1\ \mu F)(100\ V)$$
$$= 100\ \mu C$$

Ans.

Example 2 Radius of a spherical conductor is 2m. This is kept in a dielectric medium of dielectric strength 10^6 N/C. Find

(a) the capacitance of the conductor
(b) the maximum charge which can be stored on this conductor.

Sol. (a) $C = 4\pi\varepsilon_0 R$

$$= \left(\frac{1}{9\times 10^9}\right)(2)$$
$$= 2.2 \times 10^{-10}\ F$$

Ans.

(b) Maximum electric field on the surface of spherical conductor is

$$E = \frac{1}{4\pi\varepsilon_0}\frac{q}{R^2}$$

This should not exceed 10^6 N/C.

$$\therefore \quad E_{max} = \frac{1}{4\pi\varepsilon_0}\frac{q_{max}}{R^2} = 10^6$$

$$\Rightarrow \quad q_{max} = (4\pi\varepsilon_0)R^2(10^6)$$
$$= \left(\frac{1}{9\times 10^9}\right)(2)^2(10^6)$$
$$= 4.4 \times 10^{-4}\ C$$

Ans.

Example 3 Which of the following sentence related to the relation $C = \frac{q}{V}$ is correct?

(a) If more charge q is given to a conductor, its capacitance should decrease.
(b) Ratio $\frac{q}{V}$ will remain constant for a given conductor.
(c) If more charge q is given to a conductor, its capacitance should increase.
(d) None of the above

Sol. $\frac{q}{V} = C =$ constant for a conductor which depends upon the dimensions.

Hence, option (b) is correct.

MODULE 2

Energy Stored in a Charged Capacitor

Work has to be done in charging a conductor against the force of repulsion by the already existing charges on it. This work is stored as a potential energy in the electric field of the conductor. Suppose a conductor of capacity C is charged to a potential V_0 and let q_0 be the charge on the conductor at this instant.

The potential of the conductor when (during charging) the charge on it was $q\ (< q_0)$ is

$$V = \frac{q}{C}$$

Now, work done in bringing a small charge dq at this potential is

$$dW = V dq = \left(\frac{q}{C}\right) dq$$

$\therefore$ Total work done in charging it from 0 to q_0 is

$$W = \int_0^{q_0} dW = \int_0^{q_0} \frac{q}{C} dq = \frac{1}{2}\frac{q_0^2}{C}$$

This work is stored as the potential energy,

$$\therefore \qquad U = \frac{1}{2}\frac{q_0^2}{C}$$

Further by using $q_0 = CV_0$, we can write this expression as

$$U = \frac{1}{2}CV_0^2 = \frac{1}{2}q_0 V_0$$

In general, if a conductor of capacity C is charged to a potential V by giving it a charge q, then

$$\boxed{U = \frac{1}{2}CV^2 = \frac{1}{2}\frac{q^2}{C} = \frac{1}{2}qV}$$

Redistribution of Charge

Let us take an analogous example.

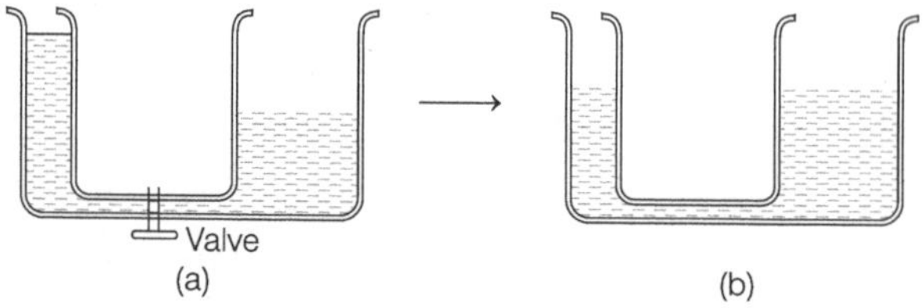

Some liquid is filled in two vessels of different sizes upto different heights. These are joined through a valve which was initially closed. When the valve is opened, the level in both the vessels becomes equal but the volume of liquid in the right side vessel is more than the liquid in the left side vessel. This is because the base area (or capacity) of this vessel is more.

Now, suppose two conductors of capacities C_1 and C_2 have charges q_1 and q_2 respectively when they are joined together by a conducting wire, charge redistributes in these conductors in the ratio of their capacities.

Charge redistributes till potential of both the conductors becomes equal. Thus, let q_1' and q_2' be the final charges on them, then

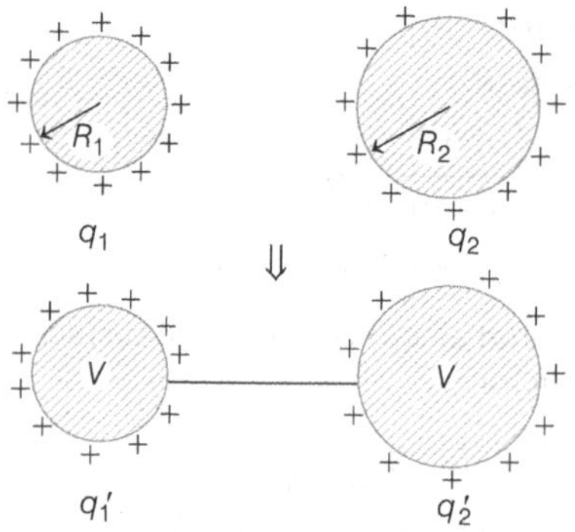

$$q' \propto C \quad \text{or} \quad \frac{q_1'}{q_2'} = \frac{C_1}{C_2}$$

and if they are spherical conductors, then

$$\frac{C_1}{C_2} = \frac{R_1}{R_2}$$

$$\therefore \qquad \frac{q_1'}{q_2'} = \frac{C_1}{C_2} = \frac{R_1}{R_2}$$

Since, the total charge is $(q_1 + q_2)$. Therefore,

$$q_1' = \left(\frac{C_1}{C_1 + C_2}\right)(q_1 + q_2) = \left(\frac{R_1}{R_1 + R_2}\right)(q_1 + q_2)$$

and $$q_2' = \left(\frac{C_2}{C_1 + C_2}\right)(q_1 + q_2) = \left(\frac{R_2}{R_1 + R_2}\right)(q_1 + q_2)$$

Note *The common potential is given by*

$$V = \frac{\text{Total charge}}{\text{Total capacity}} = \frac{q_1 + q_2}{C_1 + C_2}$$

Loss of Energy During Redistribution of Charge

We can show that, in redistribution of charges energy is always lost.

Initial potential energy, $U_i = \frac{1}{2}\frac{q_1^2}{C_1} + \frac{1}{2}\frac{q_2^2}{C_2}$

Final potential energy, $U_f = \frac{1}{2}\frac{q_{Total}}{C_{Total}} = \frac{1}{2}\frac{(q_1+q_2)^2}{C_1+C_2}$

$$\Delta U = U_i - U_f = \frac{1}{2}\left[\frac{q_1^2}{C_1} + \frac{q_2^2}{C_2} - \frac{(q_1+q_2)^2}{C_1+C_2}\right]$$

or $$\Delta U = \frac{1}{2C_1C_2(C_1+C_2)}[q_1^2C_1C_2 + q_1^2C_2^2 + q_2^2C_1^2 + q_2^2C_1C_2 - q_1^2C_1C_2 - q_2^2C_1C_2 - 2q_1q_2C_1C_2]$$

$$= \frac{C_1^2C_2^2}{2C_1C_2(C_1+C_2)}\left[\frac{q_1^2}{C_1^2} + \frac{q_2^2}{C_2^2} - \frac{2q_1q_2}{C_1C_2}\right]$$

$$= \frac{C_1C_2}{2(C_1+C_2)}[V_1^2 + V_2^2 - 2V_1V_2]$$

or $$\boxed{\Delta U = \frac{C_1C_2}{2(C_1+C_2)}(V_1-V_2)^2}$$

In the above expression C_1, C_2 and $(V_1 - V_2)^2$ are always positive. Therefore $U_i > U_f$, i.e. there is a decrease in energy. Hence, energy is always lost in redistribution of charge. Further,

$$\Delta U = 0 \quad \text{if} \quad V_1 = V_2$$

this is because no redistribution of charges takes place when both the conductors are already at same potential.

Example 1 Two isolated spherical conductors have radii 5 cm and 10 cm, respectively. They have charges of 12 μC and −3 μC. Find the charges after they are connected by a conducting wire. Also find the common potential after redistribution of charges.

Sol.

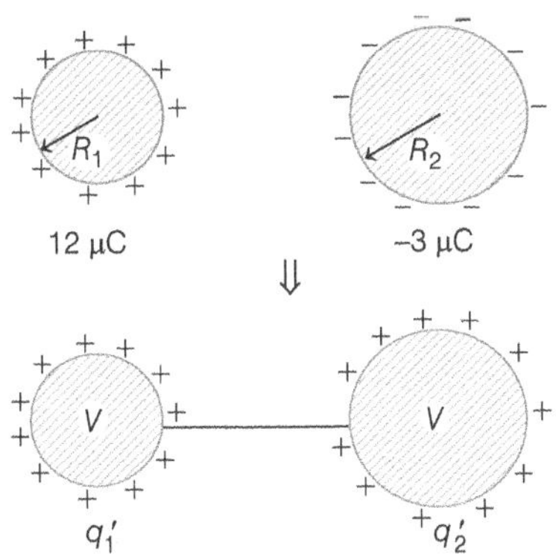

Total charge = (12 − 3) μC = 9 μC

Charge is distributed in the ratio of their capacitances (or radii in case of spherical conductors), i.e.

$$\frac{q_1'}{q_2'} = \frac{R_1}{R_2} = \frac{5}{10} = \frac{1}{2}$$

$$\therefore \quad q_1' = \left(\frac{1}{1+2}\right)(9) = 3\ \mu C$$

and $$q_2' = \left(\frac{2}{1+2}\right)(9) = 6\ \mu C$$

Common potential,

$$V = \frac{q_{Total}}{C_{Total}} = \frac{q_1+q_2}{C_1+C_2} = \frac{(9\times10^{-6})}{4\pi\varepsilon_0(R_1+R_2)}$$

$$= \frac{(9\times10^{-6})(9\times10^{9})}{(15\times10^{-2})}$$

$$= 5.4\times10^5\ V$$ **Ans.**

Example 2 An insulated conductor initially free from charge is charged by repeated contacts with a plate which after each contact is replenished to a net charge Q. If q is the charge on the conductor after first operation prove that the maximum charge which can be given to the conductor in this way is $\frac{Qq}{Q-q}$.

Sol. Let C_1 be the capacitance of plate and C_2 that of the conductor. After first contact charge on conductor is q. Therefore, charge on plate will remain $Q - q$. As the charge redistributes in the ratio of capacitances.

$$\frac{Q-q}{q} = \frac{C_1}{C_2} \quad \text{...(i)}$$

Let q_m be the maximum charge which can be given to the conductor. Then, flow of charge from the plate to the conductor will stop when

$$V_{conductor} = V_{plate}$$

$$\therefore \quad \frac{q_m}{C_2} = \frac{Q}{C_1} \Rightarrow q_m = \left(\frac{C_2}{C_1}\right)Q$$

Substituting $\frac{C_2}{C_1}$ from Eq. (i), we get

$$q_m = \frac{Qq}{Q-q}$$ **Hence Proved.**

Example 3 A conducting sphere S_1 of radius r is attached to an insulating handle. Another conducting sphere S_2 of radius R is mounted on an insulating stand. S_2 is initially uncharged. S_1 is given a charge Q, brought into contact with S_2 and removed. S_1 is recharged such that the net charge on it is again Q and it is again brought into contact with S_2 and removed. This procedure is repeated n times. **(1998 JEE)**

(a) Find the electrostatic energy of S_2 after n such contacts with S_1.

(b) What is the limiting value of this energy as $n \to \infty$?

Sol. Capacitances of conducting spheres are in the ratio of their radii. Let C_1 and C_2 be the capacitances of S_1 and S_2, then

$$\frac{C_2}{C_1} = \frac{R}{r}$$

(a) Charges are distributed in the ratio of their capacitances. Let in the first contact, charge acquired by S_2 is q_1. Therefore, charge on S_1 will be $Q - q_1$.

$$\therefore \quad \frac{q_1}{Q - q_1} = \frac{C_2}{C_1} = \frac{R}{r}$$

$$\therefore \quad q_1 = Q\left(\frac{R}{R+r}\right) \quad \text{...(i)}$$

In the second contact, S_1 again acquires the same charge Q.

Therefore, total charge in S_1 and S_2 will be

$$Q + q_1 = Q\left(1 + \frac{R}{R+r}\right)$$

This charge is again distributed in the same ratio. Therefore, charge on S_2 in second contact,

$$q_2 = Q\left(1 + \frac{R}{R+r}\right)\left(\frac{R}{R+r}\right)$$

$$= Q\left[\frac{R}{R+r} + \left(\frac{R}{R+r}\right)^2\right]$$

Similarly, $q_3 = Q\left[\frac{R}{R+r} + \left(\frac{R}{R+r}\right)^2 + \left(\frac{R}{R+r}\right)^3\right]$

and $q_n = Q\left[\frac{R}{R+r} + \left(\frac{R}{R+r}\right)^2 + \ldots + \left(\frac{R}{R+r}\right)^n\right]$

or $q_n = Q\frac{R}{r}\left[1 - \left(\frac{R}{R+r}\right)^n\right]$ ...(ii)

$$\left[S_n = \frac{a(1-r^n)}{(1-r)}\right]$$

Therefore, electrostatic energy of S_2 after n such contacts

$$= \frac{q_n^2}{2(4\pi\varepsilon_0 R)} \quad \text{or} \quad U_n = \frac{q_n^2}{8\pi\varepsilon_0 R}$$

where, q_n is given in Eq. (ii).

(b) $\frac{R}{R+r} < 1$

$$\therefore \quad \left(\frac{R}{R+r}\right)^n \to 0 \text{ as } n \to \infty$$

$$\therefore \quad q_\infty = Q\frac{R}{r}$$

$$\therefore \quad U_\infty = \frac{q_\infty^2}{2C} = \frac{Q^2R^2/r^2}{8\pi\varepsilon_0 R}$$

or $U_\infty = \frac{Q^2 R}{8\pi\varepsilon_0 r^2}$

MODULE 2 Exercise

True/False

1. No charge will flow when two conductors having the same charge are connected to each other. Is this statement true or false?

Assertion and Reason

See directions on page 25.

2. **Assertion :** Two metallic spheres are charged to the same potential. One of them is hollow and another is solid and both have the same radii. Solid sphere will have lower charge than the hollow one.

   Reason : Capacitance of metallic spheres depend on the radii of spheres. **(2023 Main)**

Single Correct Option

3. Two conducting spheres of radii R_1 and R_2 are charged with charges Q_1 and Q_2 respectively. On bringing them in contact, there is
   (a) no decrease in the energy of the system
   (b) an increase in the energy of the system if $Q_1R_2 \neq Q_2R_1$
   (c) a decrease in the energy of the system if $Q_1R_2 \neq Q_2R_1$
   (d) a decrease in the energy of the system if $Q_1R_2 = Q_2R_1$

4. Two identical metallic spheres A and B when placed at certain distance in air repel each other with a force of F. Another identical uncharged sphere C is first placed in contact with A and then in contact with B and finally placed at midpoint between spheres A and B. The force experienced by sphere C will be **(2022 Main)**
   (a) $3F/2$ (b) $3F/4$ (c) F (d) $2F$

Match the Columns

5. Let n identical drops each of radius r, capacitance c, charge q and potential v are combined to form a larger drop. Radius of big drop is R, capacitance C, charge Q and potential V. Match the following columns.

	Column I		Column II
(a)	Ratio $\frac{R}{r}$	(p)	$n^{1/3}$
(b)	Ratio $\frac{C}{c}$	(q)	$n^{2/3}$
(c)	Ratio $\frac{V}{v}$	(r)	$n^{5/3}$

Subjective Questions

6. Find the dimensions of capacitance.
7. Two conductors of capacitance 1 μF and 2 μF are charged to +10 V and −20 V. They are now connected by a conducting wire. Find
   (a) their common potential
   (b) the final charges on them
   (c) the loss of energy during redistribution of charges.
8. Two identical conducting spheres, fixed in space, attract each other with an electrostatic force of 0.108 N when separated by 50.0 cm, centre-to-centre. A thin conducting wire then connects the spheres. When the wire is removed, the spheres repel each other with an electrostatic force of 0.0360 N. What were the initial charges on the spheres?

Answers

1. False **2.** (d) **3.** (c) **4.** (b) **5.** (a-p; b-p; c-p) **6.** $[M^{-1}L^{-2}T^4A^2]$

7. (a) −10V (b) −10 μC, −20 μC (c) 3.0×10^{-4} J **8.** ± 3 μC, ∓ 1 μC

See hints on page 310.

Capacitors

Any two conductors separated by an insulator (or a vacuum) form a capacitor.

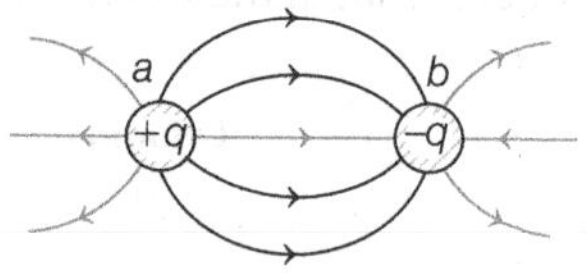

In most practical applications, each conductor initially has zero net charge, and electrons are transferred from one conductor to the other. This is called charging of the conductor. Then, the two conductors have charges with equal magnitude and opposite sign, and the net charge on the capacitor as a whole remains zero. When we say that a capacitor has charge q we mean that the conductor at higher potential has charge $+q$ and the conductor at lower potential has charge $-q$. In circuit diagram, a capacitor is represented by two parallel lines as shown in figure.

One common way to charge a capacitor is to connect the two conductors to opposite terminals of a battery. This gives a potential difference V_{ab} between the conductors, which is just equal to the PD across the terminal of the battery (= emf of the battery in most of the cases). The ratio $\frac{q}{V_{ab}}$ is called the capacitance of the capacitor. Hence,

$$C = \frac{q}{V_{ab}} \quad \text{(capacitance of a capacitor)}$$

Calculation of Capacitance

Give a charge $+q$ to one plate and $-q$ to the other plate. Then, find potential difference V between the plates. Now,

$$C = \frac{q}{V}$$

Parallel Plate Capacitor

Two metallic parallel plates of any shape but of same size and separated by a small distance constitute parallel plate capacitor. Suppose the area of each plate is A and the separation between the two plates is d. Also assume that the space between the plates contains vacuum.

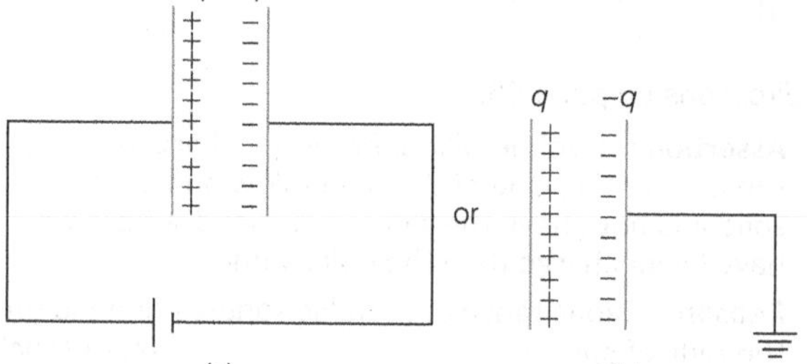

(a) (b)

We put a charge q on one plate and a charge $-q$ on the other. This can be done either by connecting one plate with the positive terminal and the other with negative terminal of a battery [as shown in Fig. (a)] or by connecting one plate to the earth and by giving a charge $+q$ to the other plate only. This charge will induce a charge $-q$ on the earthed plate. The charges will appear on the facing surfaces. The charge density on each of these surfaces has a magnitude $\sigma = q/A$.

If the plates are large as compared to the separation between them, then the electric field between the plates (at point B) is uniform and perpendicular to the plates except for a small region near the edge.

The magnitude of this uniform field E may be calculated by using the fact that both positive and negative plates produce the electric field in the same direction (from positive plate towards negative plate) of magnitude $\sigma/2\varepsilon_0$ and therefore, the net electric field between the plates will be

$$E = \frac{\sigma}{2\varepsilon_0} + \frac{\sigma}{2\varepsilon_0} = \frac{\sigma}{\varepsilon_0}$$

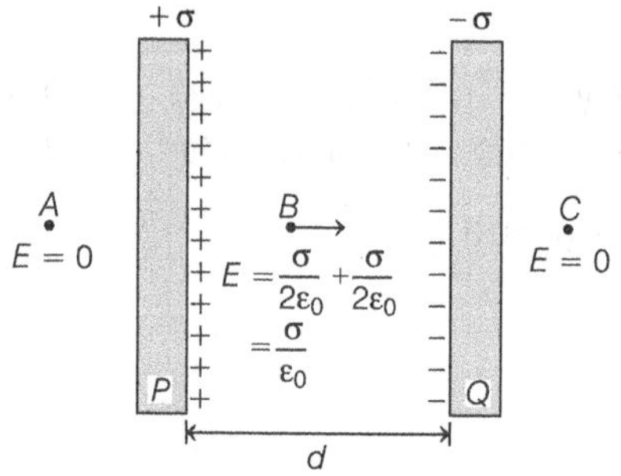

Outside the plates (at points A and C) the field due to positive sheet of charge and negative sheet of charge are in opposite directions. Therefore, net field at these points is zero.

The potential difference between the plates is

$$\therefore \quad V = E \cdot d = \left(\frac{\sigma}{\varepsilon_0}\right) d = \frac{qd}{A\varepsilon_0}$$

$\therefore$ The capacitance of the parallel plate capacitor is

$$C = \frac{q}{V} = \frac{A\varepsilon_0}{d}$$

or

$$\boxed{C = \frac{\varepsilon_0 A}{d}}$$

Note *From the above relation, it is clear that the capacitance depends only on geometrical factors (A and d).*

Effect of Dielectrics

Most capacitors have a dielectric between their conducting plates. Placing a solid dielectric between the plates of a capacitor serves the following three functions :

(i) It solves the problem of maintaining two large metal sheets at a very small separation without actual contact.

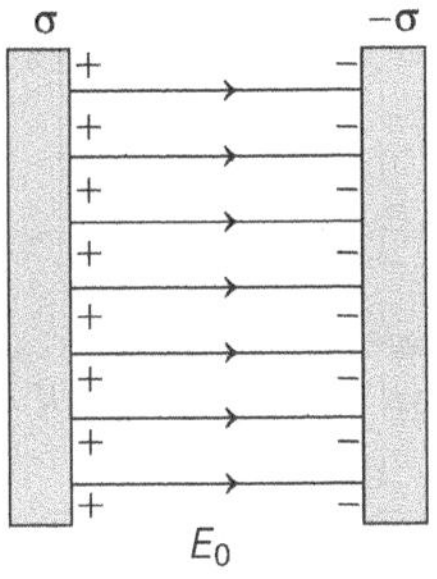

(ii) It increases the maximum possible potential difference which can be applied between the plates of the capacitor without the dielectric breakdown. Many dielectric materials can tolerate stronger electric fields without breakdown than can air.

(iii) It increases the capacitance of the capacitor.

When a dielectric material is inserted between the plates (keeping the charge to be constant) the electric field and hence the potential difference decreases by a factor K (the dielectric constant of the dielectric).

$$\therefore \quad \boxed{E = \frac{E_0}{K} \text{ and } V = \frac{V_0}{K}} \quad \text{(When } q \text{ is constant)}$$

Electric field is decreased because an induced charge of the opposite sign appears on each surface of the dielectric. This induced charge produces an electric field inside the dielectric in opposite direction and as a result net electric field is decreased. The induced charge in the dielectric can be calculated as under

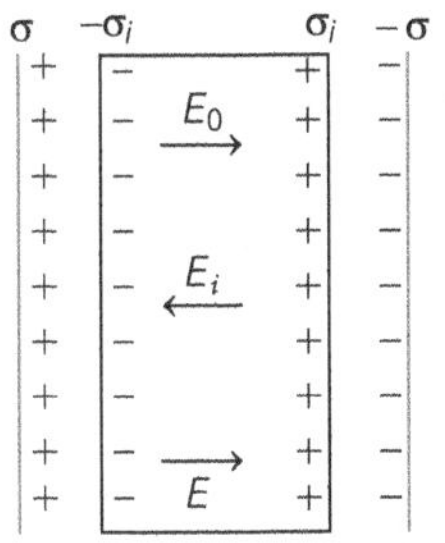

$$E = E_0 - E_i$$

or

$$\frac{E_0}{K} = E_0 - E_i$$

$$\therefore \quad E_i = E_0\left(1 - \frac{1}{K}\right)$$

Therefore,

$$\frac{\sigma_i}{\varepsilon_0} = \frac{\sigma}{\varepsilon_0}\left(1 - \frac{1}{K}\right)$$

or

$$\sigma_i = \sigma\left(1 - \frac{1}{K}\right)$$

or

$$q_i = q\left(1 - \frac{1}{K}\right)$$

For a conductor $K = \infty$. Hence,

$$q_i = q, \ \sigma_i = \sigma \quad \text{and} \quad E = 0$$

and otherwise $q_i < q$

Thus,

$$\boxed{q_i \le q, \ \sigma_i \le \sigma}$$

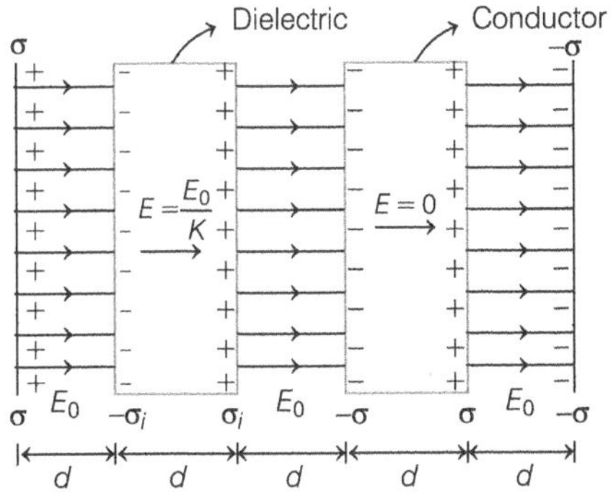

Hence, we can conclude the above discussion as under:

(i) $E_{\text{vacuum}} = E_0 = \dfrac{\sigma}{\varepsilon_0} = \dfrac{q}{A\varepsilon_0}$

(ii) $E_{\text{dielectric}} = \dfrac{E_0}{K}$ (here, K = dielectric constant)

(iii) $E_{\text{conductor}} = 0$ (as $K = \infty$)

If we plot a graph between potential and distance from positive plate, it will be as shown in figure.

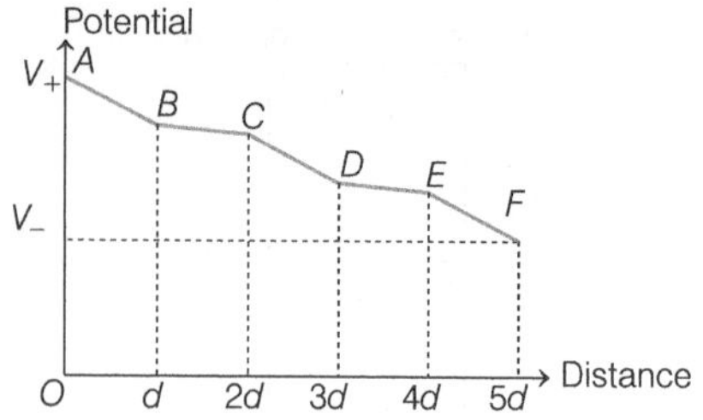

$$E = -\frac{dV}{dr} = -(\text{Slope of } V\text{-}r \text{ graph})$$

$\therefore$ Modulus of,

Slope of AB = slope of CD = slope of $EF = E_0$

Slope of $BC = \frac{E_0}{K}$ and slope of $DE = 0$

Further, the potential difference between positive and negative plate is

$$V_+ - V_- = E_0 d + \frac{E_0}{K}\cdot d + E_0 d + 0 + E_0 d$$

$$= 3E_0 d + \frac{E_0}{K}\cdot d$$

Here we have used PD $= Ed$ (in uniform electric field)

Capacitance of a Capacitor Partially Filled with Dielectric

Suppose, a capacitor is partially filled with a dielectric (dielectric constant $= K$) as shown in figure.

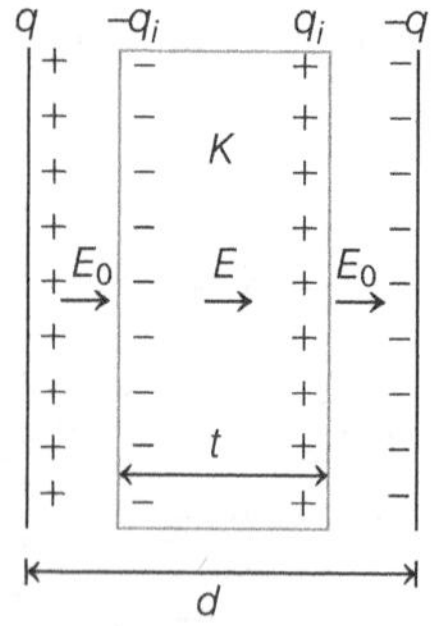

If a charge q is given to the capacitor, an induced charge q_i is developed on the dielectric.

where, $$q_i = q\left(1 - \frac{1}{K}\right)$$

Moreover, if E_0 is the electric field in the region where dielectric is absent, then electric field inside the dielectric will be $E = E_0/K$. The potential difference between the plates of the capacitor is

$$V = V_+ - V_- = Et + E_0\,(d - t)$$

$$= \frac{E_0}{K}t + E_0\,(d - t) = E_0\left(d - t + \frac{t}{K}\right)$$

$$= \frac{\sigma}{\varepsilon_0}\left(d - t + \frac{t}{K}\right) = \frac{q}{A\varepsilon_0}\left(d - t + \frac{t}{K}\right)$$

Now, as per the definition of capacitance,

$$C = \frac{q}{V} = \frac{\varepsilon_0 A}{d - t + \frac{t}{K}} \quad \text{or} \quad \boxed{C = \frac{\varepsilon_0 A}{d - t + \frac{t}{K}}}$$

Different Cases

(i) If more than one dielectric slabs are placed between the capacitor, then

$$C = \frac{\varepsilon_0 A}{(d - t_1 - t_2 - \ldots - t_n) + \left(\frac{t_1}{K_1} + \frac{t_2}{K_2} + \ldots + \frac{t_n}{K_n}\right)}$$

If $t_1 + t_2 + \ldots + t_n = d$ then first term in denominator becomes zero.

(ii) If the slab completely filles the space between the plates, then $t = d$ and therefore,

$$C = \frac{\varepsilon_0 A}{d/K} = \frac{K\varepsilon_0 A}{d}$$

(iii) If a conducting slab ($K = \infty$) is placed between the plates, then

$$C = \frac{\varepsilon_0 A}{d - t + \frac{t}{\infty}} = \frac{\varepsilon_0 A}{d - t}$$

(iv) If the space between the plates is completely filled with a conductor, then $t = d$ and $K = \infty$.

Then, $$C = \frac{\varepsilon_0 A}{d - d + \frac{d}{\infty}} = \infty$$

The significance of infinite capacitance can be explained as below :

If one of the plates of a capacitor is earthed and the second one is given a charge q, then the whole charge transfers to earth if some conducting material completely fills the space between the two plates and as the capacitance of earth is very large compared to the capacitor we can say that the capacitance has become infinite.

Alternatively, if the plates of the capacitor are connected to a battery, the current starts flowing in the circuit. Thus all the charge which enters the positive plate of the capacitor, the same charge leaves the negative plate. So, we can say, the positive plate can accept infinite amount of charge or its capacitance has become infinite.

So, we never fill the conducting material between the plates of capacitor. Otherwise the purpose of capacitor is defeated.

Important POINTS

- **Capacitance of a spherical conductor enclosed by an earthed concentric spherical shell**
  If a charge q is given to the inner spherical conductor, it spreads over the outer surface of it and a charge $-q$ appears on the inner surface of the shell. The electric field is produced only between the two. From the principle of generator, the potential difference between the two will depend upon the inner charge q only and is given by

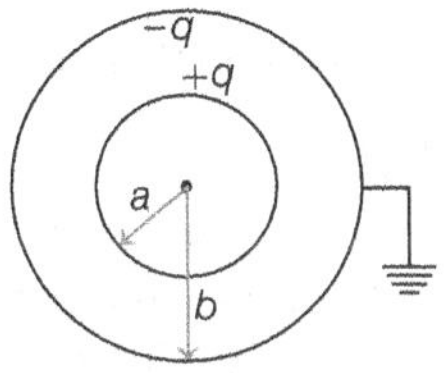

$$V = \frac{q}{4\pi\varepsilon_0}\left(\frac{1}{a} - \frac{1}{b}\right)$$

Hence, the capacitance of this system

$$C = \frac{q}{V}$$

or $$C = 4\pi\varepsilon_0\left(\frac{1}{a} - \frac{1}{b}\right) = 4\pi\varepsilon_0\left(\frac{ab}{b-a}\right)$$

From this expression we see that if $b = \infty$, $C = 4\pi\varepsilon_0 a$, which corresponds to that of an isolated sphere, i.e. the charged sphere may be regarded as a capacitor in which the outer surface has been removed to infinity.

- **Capacitance of a cylindrical capacitor** When a metallic cylinder of radius a is placed coaxially inside an earthed hollow metallic cylinder of radius $b\,(> a)$ we get cylindrical capacitor. If a charge q is given to the inner cylinder, induced charge $-q$ will reach to the inner surface of the outer cylinder. Assume that the capacitor is of very large length ($l >> b$) so that the lines of force are radial. Using Gauss's law, we can prove that

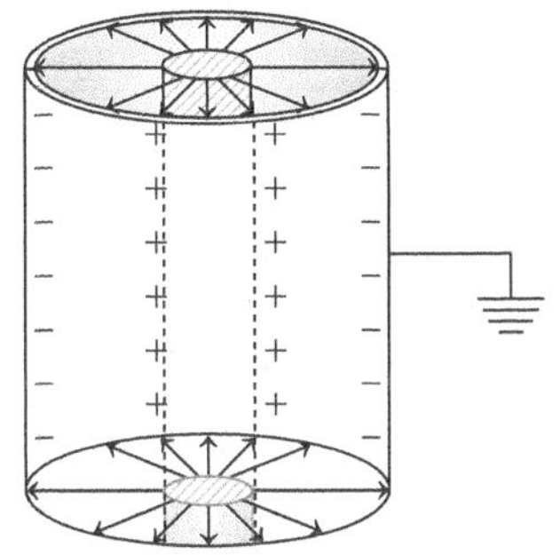

$$E(r) = \frac{\lambda}{2\pi\varepsilon_0 r} \quad \text{for} \quad a \le r \le b$$

Here, λ = charge per unit length

Therefore, the potential difference between the cylinders

$$V = -\int_b^a \mathbf{E}\cdot d\mathbf{r} = -\int_b^a \frac{\lambda}{2\pi\varepsilon_0 r}dr = \frac{\lambda}{2\pi\varepsilon_0}\ln\left(\frac{b}{a}\right)$$

$$\therefore \quad \frac{\lambda}{V} = \frac{\text{charge/length}}{\text{potential difference}} = \frac{\text{capacitance}}{\text{length}} = \frac{2\pi\varepsilon_0}{\ln(b/a)}$$

Hence, $$\text{capacitance per unit length} = \frac{2\pi\varepsilon_0}{\ln(b/a)}$$

Example 1 A parallel-plate capacitor has capacitance of 1.0 F. If the plates are 1.0 mm apart, what is the area of the plates?

Sol. $\because \quad C = \dfrac{\varepsilon_0 A}{d}$

$$\therefore \quad A = \frac{Cd}{\varepsilon_0} = \frac{(1)(10^{-3})}{8.86 \times 10^{-12}}$$
$$= 1.1 \times 10^8 \text{ m}^2$$

Example 2 Two parallel plate vacuum capacitors have areas A_1 and A_2 and equal plate spacing d. Show that when the capacitors are connected in parallel, the equivalent capacitance is the same as for a single capacitor with plate area $A_1 + A_2$ and spacing d.

Sol. $C = C_1 + C_2$ (in parallel)

$$\therefore \quad \frac{\varepsilon_0 A}{d} = \frac{\varepsilon_0 A_1}{d} + \frac{\varepsilon_0 A_2}{d}$$

or $$A = A_1 + A_2$$

Example 3 (a) Two spheres have radii a and b and their centres are at a distance d apart. Show that the capacitance of this system is

$$C = \frac{4\pi\varepsilon_0}{\frac{1}{a} + \frac{1}{b} \pm \frac{2}{d}}$$

provided that d is large compared with a and b.

(b) Show that as d approaches infinity the above result reduces to that of two isolated spheres in series.

Note : In series, $\dfrac{1}{C} = \dfrac{1}{C_1} + \dfrac{1}{C_2}$ or $C = \dfrac{C_1 C_2}{C_1 + C_2}$

Sol. (a) PD, $V = V_1 - V_2$

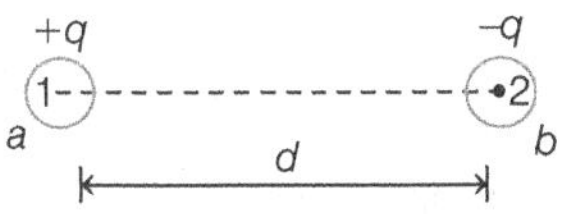

$$= \frac{1}{4\pi\varepsilon_0}\left[\left(\frac{q}{a} - \frac{q}{d}\right) - \left(\frac{-q}{b} + \frac{q}{d}\right)\right]$$

$$\therefore \quad C = \frac{q}{V} = \frac{4\pi\varepsilon_0}{\frac{1}{a} + \frac{1}{b} - \frac{2}{d}}$$

If $-q$ is given to first sphere and $+q$ to second sphere, then

$$C = \frac{4\pi\varepsilon_0}{\frac{1}{a} + \frac{1}{b} + \frac{2}{d}}$$

(b) If $d \to \infty$, then $C = \dfrac{4\pi\varepsilon_0}{\frac{1}{a} + \frac{1}{b}} = \dfrac{(4\pi\varepsilon_0)(ab)}{a+b}$

In series, $C_{net} = \dfrac{C_1 C_2}{C_1 + C_2}$

$$= \frac{(4\pi\varepsilon_0 a)(4\pi\varepsilon_0 b)}{(4\pi\varepsilon_0 a) + (4\pi\varepsilon_0 b)} = \frac{(4\pi\varepsilon_0) ab}{a+b}$$

Hence Proved.

Example 4 Three concentric conducting shells A, B and C of radii a, b and c are as shown in figure. A dielectric of dielectric constant K is filled between A and B. Find the capacitance between A and C.

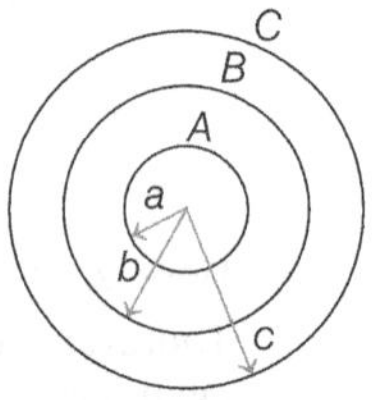

HOW TO PROCEED When the dielectric is filled between A and B, the electric field will change in this region. Therefore, the potential difference and hence the capacitance of the system will change. So, first find the electric field $E(r)$ in the region $a \le r \le c$. Then, find the PD (V) between A and C and finally the capacitance of the system will be

$$C = \frac{q}{V}$$

Here, q = charge on A

Sol. $E(r) = \dfrac{q}{4\pi\varepsilon_0 K r^2}$ for $a \le r \le b$

$= \dfrac{q}{4\pi\varepsilon_0 r^2}$ for $b \le r \le c$

Using, $dV = -\int \mathbf{E} \cdot d\mathbf{r}$

the PD between A and C is

$$\therefore \quad V = V_A - V_C = -\int_a^b \frac{q}{4\pi\varepsilon_0 K r^2} \cdot dr - \int_b^c \frac{q}{4\pi\varepsilon_0 r^2}\, dr$$

$$= \frac{q}{4\pi\varepsilon_0}\left[\frac{1}{K}\left(\frac{1}{a} - \frac{1}{b}\right) + \left(\frac{1}{b} - \frac{1}{c}\right)\right]$$

$$= \frac{q}{4\pi\varepsilon_0}\left[\frac{(b-a)}{Kab} + \frac{(c-b)}{bc}\right]$$

$$= \frac{q}{4\pi\varepsilon_0 Kabc}[c(b-a) + Ka(c-b)]$$

$\therefore$ The desired capacitance is

$$C = \frac{q}{V} = \frac{4\pi\varepsilon_0 Kabc}{Ka(c-b) + c(b-a)}$$

Ans.

Energy Stored in Charged Capacitor

A charged capacitor stores an electric potential energy in it, which is equal to the work required to charge it. This energy can be recovered if the capacitor is allowed to discharge. If the charging is done by a battery, electrical energy is stored at the expense of chemical energy of battery.

If a capacitor is given a charge q, the potential energy stored in it is

$$U = \frac{1}{2}CV^2 = \frac{1}{2}\frac{q^2}{C} = \frac{1}{2}qV$$

Here, V is the PD between the plates of capacitor.

Energy Density (u)

The potential energy of a charged conductor or a capacitor is stored in the electric field of this charged conductor (or capacitor). The energy per unit volume is called the energy density (u). Energy density in a dielectric medium is given by

$$u = \frac{1}{2}\varepsilon_0 K E^2$$

This relation shows that the energy stored per unit volume depends on E^2. If E is the electric field in a space of volume dV, then the total stored energy in an electrostatic field is given by

$$U = \frac{1}{2}\varepsilon_0 K \int E^2\, dV$$

and if E is uniform throughout the volume (electric field between the plates of a capacitor is almost uniform), then the total stored energy can be given by

$$U = u\,(\text{Total volume}) = \frac{1}{2}K\varepsilon_0 E^2 V$$

Example 5 Using the concept of energy density, find the total energy stored in a

(a) parallel plate capacitor

(b) charged spherical conductor.

Sol. (a) Electric field is uniform between the plates of the capacitor. The magnitude of this field is

$$E = \frac{\sigma}{\varepsilon_0} = \frac{q}{A\varepsilon_0}$$

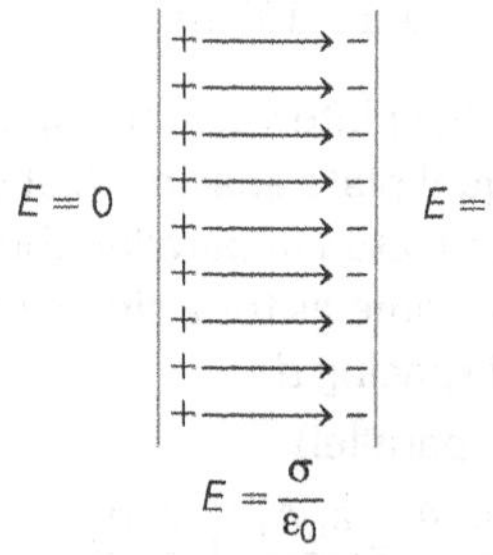

Therefore, the energy density (u) should also be constant.

$$u = \frac{1}{2}\varepsilon_0 E^2 = \frac{q^2}{2A^2\varepsilon_0}$$

$\therefore$ Total stored energy, $U = (u)$ (total volume)

$$= \left(\frac{q^2}{2A^2\varepsilon_0}\right)(A \cdot d) = \frac{q^2}{2\left(\frac{A\varepsilon_0}{d}\right)} = \frac{q^2}{2C} \quad \left(\text{as } C = \frac{A\varepsilon_0}{d}\right)$$

$$\therefore \quad U = \frac{q^2}{2C}$$ **Ans.**

(b) In case of a spherical conductor (of radius R) the excess charge resides on the outer surface of the conductor. The field inside the conductor is zero. It extends from surface to infinity. And since the potential energy is stored in the field only, it will be stored in the region extending from surface to infinity. But as the field is non-uniform, the energy density u is also non-uniform. So, the total energy will be calculated by integration. Electric field at a distance $r (\geq R)$ from the centre is

$$E = \frac{1}{4\pi\varepsilon_0} \cdot \frac{q}{r^2}$$

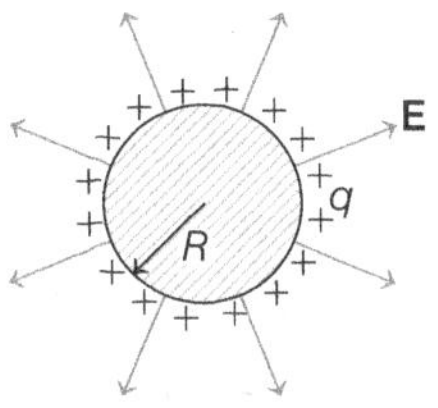

$$\therefore \quad u(r) = \frac{1}{2}\varepsilon_0 E^2$$

$$= \frac{1}{2}\varepsilon_0 \left\{\frac{1}{4\pi\varepsilon_0} \cdot \frac{q}{r^2}\right\}^2$$

Energy stored in a volume $dV = (4\pi r^2)\, dr$ is

$$dU = u\, dV$$

$\therefore$ Total energy stored,

$$U = \int_{r=R}^{r=\infty} dU$$

Substituting the values, we get

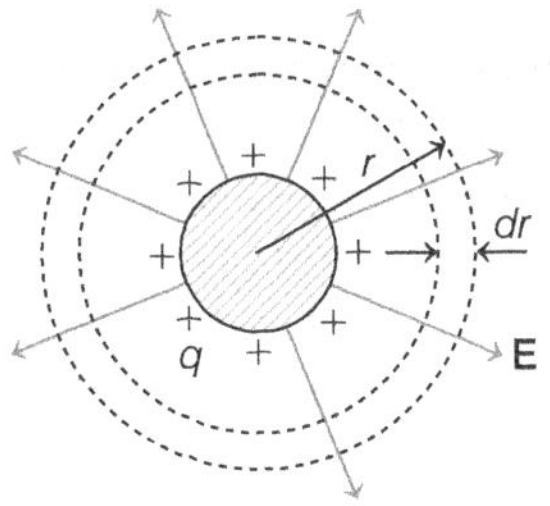

$$U = \frac{q^2}{2\,(4\pi\varepsilon_0 R)}$$

or $$U = \frac{q^2}{2C} \quad (\text{as } C = 4\pi\varepsilon_0 R)$$ **Ans.**

INTRODUCTORY Exercise

1. A capacitor has a capacitance of 7.28 μF. What amount of charge must be placed on each of its plates to make the potential difference between its plates equal to 25.0 V?
2. A parallel plate air capacitor of capacitance 245 μF has a charge of magnitude 148 mC on each plate. The plates are 3.28×10^{-8} m apart.
   (a) What is the potential difference between the plates?
   (b) What is the area of each plate?
   (c) What is the surface charge density on each plate?
3. Two parallel plates have equal and opposite charges. When the space between the plates is evacuated, the electric field is $E_0 = 3.20 \times 10^5$ V/m. When the space is filled with dielectric, the electric field is $E = 2.50 \times 10^5$ V/m.
   (a) What is the dielectric constant?
   (b) What is the charge density on each surface of the dielectric?
4. An isolated conducting sphere whose radius $R = 1$ m has a charge $q = \frac{1}{9}$ nC. The energy density at the surface of the sphere is
   (a) $\frac{\varepsilon_0}{2}$ J/m^3 (b) ε_0 J/m^3
   (c) $2\varepsilon_0$ J/m^3 (d) $\frac{\varepsilon_0}{3}$ J/m^3
5. Two protons A and B are placed in between the two plates of a parallel plate capacitor charged to a potential difference, V as shown in the figure. The forces on the two protons are identical. **(1986 JEE)**

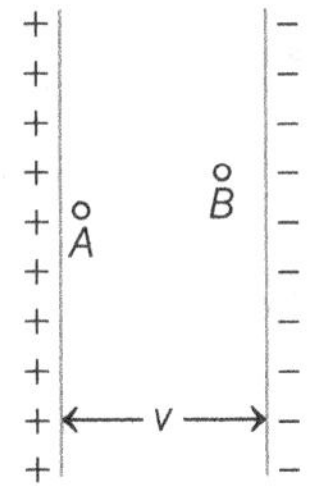

Answers

1. $\pm 182\ \mu$C
2. (a) 604 V (b) 0.9 m^2 (c) 0.164 $\frac{\text{C}}{\text{m}^2}$
3. (a) 1.28 (b) 6.2×10^{-7} $\frac{\text{C}}{\text{m}^2}$
4. (a) 5. (True)

See hints on page 311.

Typed Problems

Type 1. *Configuration of capacitor is changed and change in five quantities q, C, V, U and E is asked.*

Concept

Some problems are asked when a capacitor is charged through a battery and then the configuration of capacitor is changed :

(i) either by inserting a dielectric slab or removing the slab (if it already exists) or

(ii) by changing the distance between the plates of capacitor or

(iii) by both.

The questions will be based on the change in electric field, potential, etc.

In such problems, following two cases are possible.

Case 1 **When battery is removed after charging**

If the battery is removed after charging, then the charge stored in the capacitor remains constant.

$$q = \text{constant}$$

First of all, find the change in capacitance and according to the formula find the change in other quantities.

Ex. 1 An air capacitor is first charged through a battery. The charging battery is then removed and a dielectric slab of dielectric constant $K = 4$ is inserted between the plates. Simultaneously, the distance between the plates is reduced to half, then find change in C, E, V and U.

Sol. **Change in capacitance,** $C = \dfrac{K\varepsilon_0 A}{d} \propto \dfrac{K}{d}$

$\therefore$ Capacitance will become 8 times $\left(K = 4,\ d' = \dfrac{d}{2}\right)$

Change in electric field

$$\therefore \qquad E = E_0/K$$

$$\text{or} \qquad E \propto \frac{1}{K} \qquad \text{(if } q = \text{constant)}$$

or the electric field will become $\dfrac{1}{4}$ times its initial value.

Change in potential difference,

$$V = \frac{q}{C} \quad \text{or} \quad V \propto \frac{1}{C} \qquad \text{(if } q = \text{constant)}$$

Therefore, potential difference becomes $\dfrac{1}{8}$ times of its initial value.

Alternate method

$$V = Ed$$

Electric field has become $\dfrac{1}{4}$ times its initial value and d is reduced to half. Hence, V becomes $\dfrac{1}{8}$ times.

Change in stored potential energy,

$$U = \frac{1}{2}\frac{q^2}{C}$$

$$\text{or} \qquad U \propto \frac{1}{C} \qquad (\text{if } q = \text{constant})$$

Capacitance has become 8 times. Therefore, the stored potential energy U will become $\dfrac{1}{8}$ times.

Case 2 **When battery remains connected**

If the battery remains connected, the potential difference V becomes constant. So, in the above example, capacitance will become 8 times.

The charge stored ($q = CV$ or $q \propto C$) will also increase to 8 times. The electric field $\left(E = \dfrac{V}{d} \text{ or } E \propto \dfrac{1}{d}\right)$ becomes twice and the stored PE $\left(U = \dfrac{1}{2}CV^2 \text{ or } U \propto C\right)$ is 8 times.

Type 2. *To find self energy of a system of charges.*

Concept

The self energy of a system of charges is

$$U_s = \int_0^q V\,dq$$

This comes out to be equal to $\dfrac{q^2}{2C}$ in case of a capacitor or a conductor.

A point charge does not have any self energy.

Ex. 2 Find the electric potential energy of a uniformly charged solid sphere.

Sol. Consider a uniformly charged sphere of radius R having a total charge q_0.

The volume charge density is

$$\rho = \frac{q_0}{\frac{4}{3}\pi R^3} = \frac{3q_0}{4\pi R^3}$$

When the radius of the sphere is r, the charge contained in it is

$$q = \left(\frac{4}{3}\pi r^3\right)\rho = \left(\frac{q_0}{R^3}\right)r^3$$

The potential at the surface is

$$V = \frac{q}{4\pi\varepsilon_0 r} = \frac{q_0}{4\pi\varepsilon_0 R^3}r^2$$

The charge needed to increase the radius from r to $r + dr$ is

$$dq = \frac{3q_0}{R^3}\cdot r^2\,dr$$

$\therefore$ The self energy of the sphere is

$$U_s = \int_0^R V\,dq = \int_0^R \left(\frac{q_0}{4\pi\varepsilon_0 R^3} r^2\right)\left(\frac{3q_0}{R^3}\cdot r^2\,dr\right)$$

$$= \frac{3q_0^2}{20\pi\varepsilon_0 R}$$ **Ans.**

Ex. 3 Find the electric potential energy of a uniformly charged, thin spherical shell.

Sol. Consider a uniformly charged thin spherical shell of radius R having a total charge q_0.

Suppose at some instant a charge q is placed on the shell. The potential at the surface is

$$V = \frac{q}{4\pi\varepsilon_0 R}$$

$\therefore$ The self energy of the shell is

$$U_s = \int_0^{q_0} V\,dq$$

$$= \int_0^{q_0}\left(\frac{q}{4\pi\varepsilon_0 R}\right)dq = \frac{q_0^2}{8\pi\varepsilon_0 R}$$ **Ans.**

Type 3. *To find distribution of charges on different faces on parallel conducting plates.*

Concept

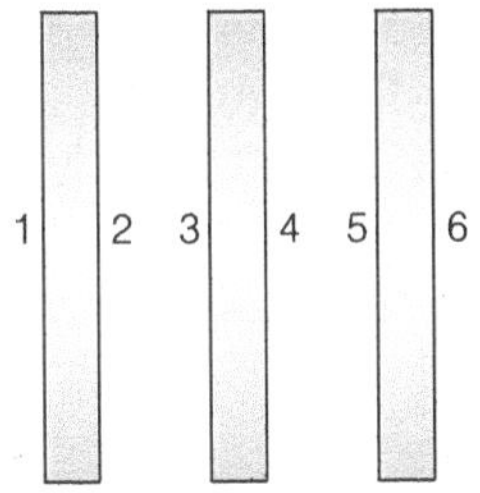

(i) $q_1 = q_6 = \frac{q_{\text{Total}}}{2}$

(ii) $q_2 = -q_3$ and $q_4 = -q_5$

Note *The above two results are proved in following example.*

(iii) Electric field between 2 and 3 is due to the charges q_2 and q_3. This electric field is given by

$$E = \frac{\sigma}{\varepsilon_0} = \frac{q/A}{\varepsilon_0} \qquad [\text{Let } |q_2| = |q_3| = q]$$

(iv) This electric field is uniform, so potential difference between any two points in this region is given by

$$V = Ed$$

(v) If two plates are connected to each other, then distance between the plates is required, otherwise there is no requirement of that.

(vi) $q_1 + q_2 =$ Charge given to first plate. Similarly, $q_3 + q_4 =$ Charge given to second plate and so on.

Ex. 4 Three parallel metallic plates each of area A are kept as shown in figure and charges q_1, q_2 and q_3 are given to them. Find the resulting charge distribution on the six surfaces, neglecting edge effects as usual.

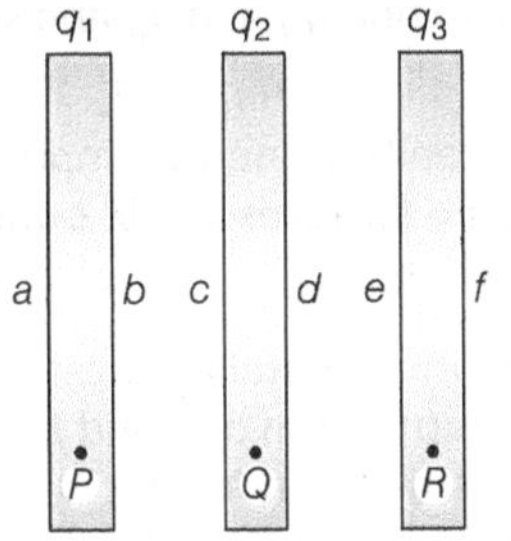

Sol. The plate separations do not affect the distribution of charge in this problem.

In the figure, $q_b = q_1 - q_a$, $q_d = q_2 - q_c$

and $q_f = q_3 - q_e$.

Electric field at point P is zero because this point is lying inside a conductor.

$$E_P = 0$$

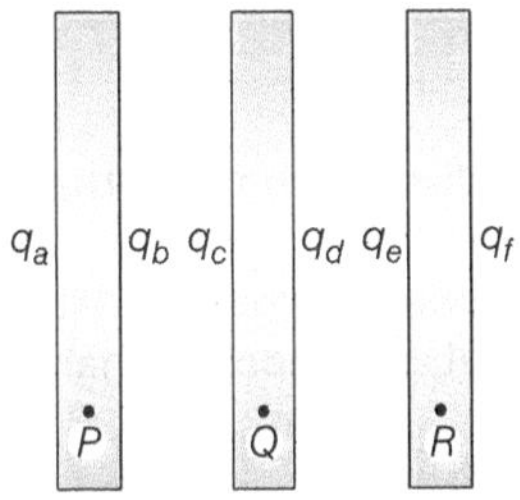

At P, charge q_a will give an electric field towards right. All other charges q_b, q_c ..., etc., will give the electric field towards left. So,

$$\frac{1}{2A\varepsilon_0}[q_a - (q_1 - q_a) - q_c - (q_2 - q_c) - q_e - (q_3 - q_e)] = 0$$

As, $\qquad E = \frac{\sigma}{2\varepsilon_0} = \frac{q}{2A\varepsilon_0}$

or $\qquad 2q_a - q_1 - q_2 - q_3 = 0$

or $\qquad q_a = \frac{q_1 + q_2 + q_3}{2} = \frac{q_{\text{Total}}}{2}$

Similarly the condition, $E_R = 0$ will give the result,

$$q_f = \frac{q_1 + q_2 + q_3}{2} = \frac{q_{\text{Total}}}{2}$$

From here we may conclude that, **half of the sum of all charges appears on each of the two outermost surfaces of the system of plates.**

Further we have a condition,

$$E_Q = 0$$

$$\frac{1}{2A\varepsilon_0}[q_a + (q_1 - q_a) + q_c - (q_2 - q_c) - q_e - (q_3 - q_e)] = 0$$

or $\qquad q_1 + 2q_c - q_2 - q_3 = 0$

$\therefore \qquad q_c = \frac{q_2 + q_3 - q_1}{2}$

$$q_b = q_1 - q_a = \frac{q_1 - q_2 - q_3}{2} = -q_c$$

Similarly, we can find q_d and q_e and show that

$$q_d = -q_e.$$

From here we can find another important result that **the pairs of opposite surfaces like b, c and d,e carry equal and opposite charges.**

Ex. 5 Three identical metallic plates are kept parallel to one another at a separation of a and b. The outer plates are connected by a thin conducting wire and a charge Q is placed on the central plate. Find final charges on all the six plate's surfaces.

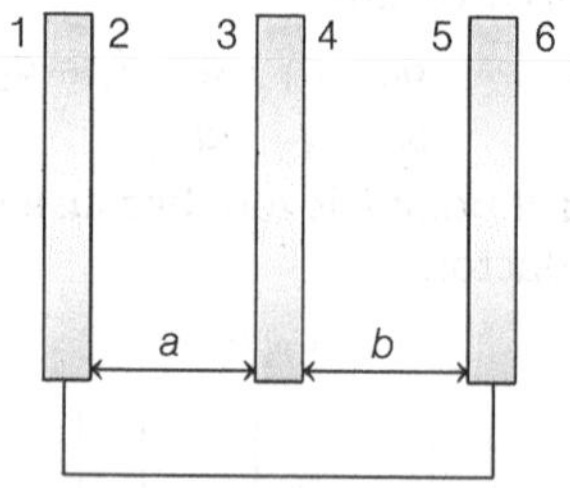

Sol. Let the charge distribution in all the six faces be as shown in figure. While distributing the charge on different faces, we have used the fact that two opposite faces have equal and opposite charges on them.

$$q_2 = q_3 = \frac{q_{Total}}{2} = \frac{Q}{2}$$

Further, total charge on surfaces 1, 2, 5 and 6 is Q.

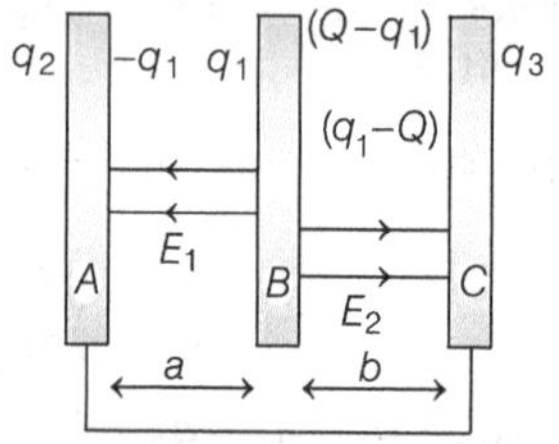

Further A and C are at same potentials. Hence,

$$V_B - V_A = V_B - V_C$$

or

$$E_1 a = E_2 b$$

$$\therefore \quad \frac{q_1}{A\varepsilon_0} \cdot a = \frac{Q - q_1}{A\varepsilon_0} \cdot b \qquad (A = \text{Area of plates})$$

$$\therefore \quad q_1 a = (Q - q_1) b$$

$$\therefore \quad q_1 = \frac{Qb}{a+b}$$

Hence, charge on different faces are as follows.

Face	Charge
1	$q_2 = \frac{Q}{2}$
2	$-q_1 = -\frac{Qb}{a+b}$
3	$q_1 = \frac{Qb}{a+b}$
4	$Q - q_1 = \frac{Qa}{a+b}$
5	$q_1 - Q = -\frac{Qa}{a+b}$
6	$q_3 = \frac{Q}{2}$

Ex. 6 Area of each plate is A. The conducting plates are connected to a battery of emf V volts. Find charges q_1 to q_6.

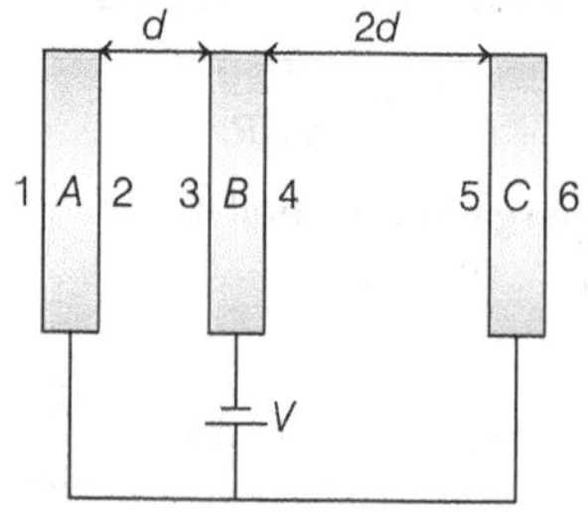

Sol. Net charge drawn from the battery is zero or

$$q_{Total} = 0$$

$$\therefore \quad q_1 = q_6 = \frac{q_{Total}}{2} = 0$$

$$V_{AB} = V \text{ with } V_A > V_B$$

$$\therefore \quad |q_2| = |q_3| = CV = \left(\frac{\varepsilon_0 A}{d}\right) V$$

$$q_2 = +\frac{\varepsilon_0 AV}{d} \text{ and } q_3 = -\frac{\varepsilon_0 AV}{d}$$

Similarly,

$$V_{CB} = V \text{ with } V_C > V_B$$

$$\therefore \quad |q_4| = |q_5| = CV = \left(\frac{\varepsilon_0 A}{2d}\right) V$$

with

$$q_5 = +\frac{\varepsilon_0 AV}{2d}$$

and

$$q_4 = -\frac{\varepsilon_0 AV}{2d}$$

Type 4. *To find total electrostatic potential energy due to spherical charged shells.*

Concept

(i) Capacity of a spherical capacitor is given by

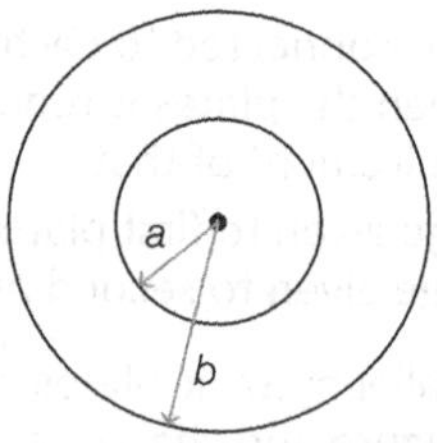

$$C = \frac{4\pi\varepsilon_0}{\frac{1}{a} - \frac{1}{b}}$$

(ii) Energy is stored only in the electric field.

(iii) $U = \frac{1}{2}\frac{q^2}{C}$

Ex. 7 In the figure shown,

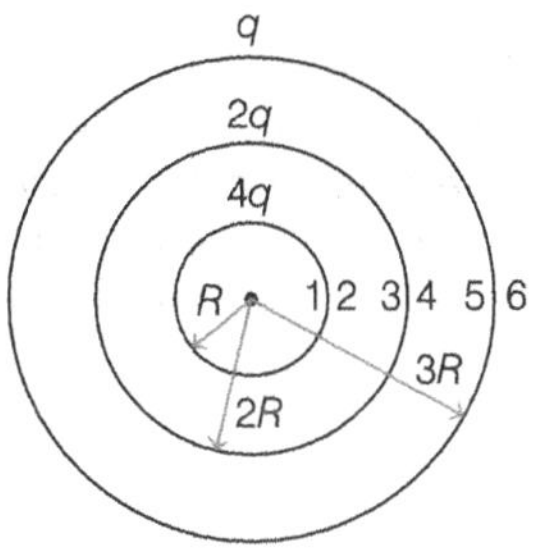

(a) Find q_1 to q_6.

(b) Total electrostatic potential energy.

Sol. (a) $q_1 = 0$

$q_2 = 4q$

$q_3 = -q_2 = -4q$

$q_4 = 2q - q_3 = 6q$

$q_5 = -q_4 = -6q$

$q_6 = q - q_5 = 7q$

(b) $U_{Total} = U_1 + U_2 + U_3$

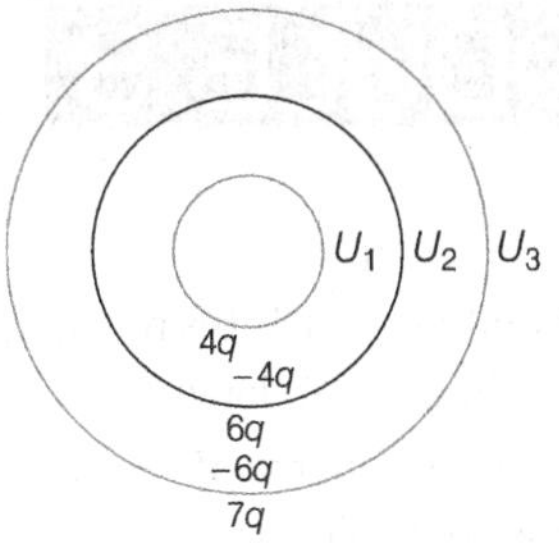

Here, $U_1 = \frac{1}{2}\frac{(4q)^2}{C_1}$

where, $C_1 = \frac{4\pi\varepsilon_0}{\frac{1}{R} - \frac{1}{2R}}$

$U_2 = \frac{1}{2}\frac{(6q)^2}{C_2}$

where, $C_2 = \frac{4\pi\varepsilon_0}{\frac{1}{2R} - \frac{1}{3R}}$

and $U_3 = \frac{1}{2}\frac{(7q)^2}{C_3}$

where, $C_3 = \frac{4\pi\varepsilon_0}{\frac{1}{3R} - \frac{1}{\infty}} = 4\pi\varepsilon_0(3R)$

MODULE 3 Exercise

Single Correct Option

1. The unit of electric permittivity is
(a) Volt/m^2 (b) Joule/C (c) Farad/m (d) Henry/m

2. Two metal spheres of radii a and b are connected by a thin wire. Their separation is very large as compared to their dimensions. The capacitance of this system is
(a) $4\pi\varepsilon_0(ab)$ (b) $2\pi\varepsilon_0(a+b)$
(c) $4\pi\varepsilon_0(a+b)$ (d) $4\pi\varepsilon_0\left(\frac{a^2+b^2}{2}\right)$

3. Two large parallel sheets charged uniformly with surface charge density σ and $-\sigma$ are located as shown in the figure. Which one of the following graphs shows the variation of electric field along a line perpendicular to the sheets as one moves from A to B ?

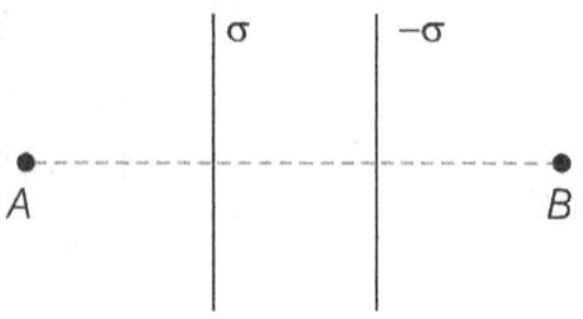

(a)
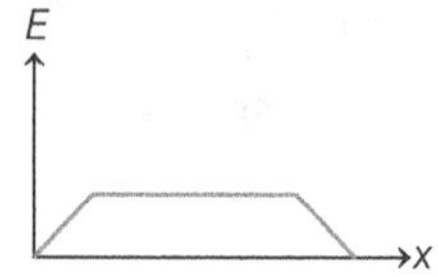

(b)
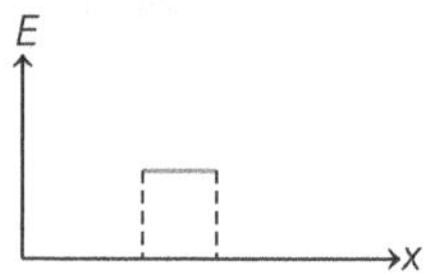

(c)
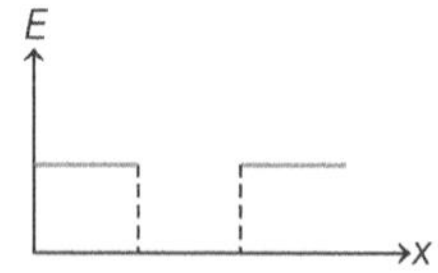

(d)
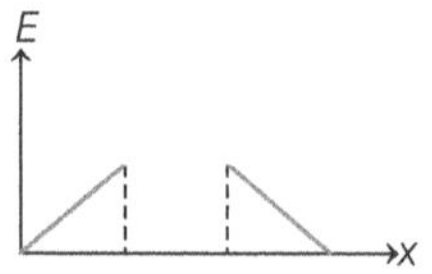

4. The capacitance and the energy stored in a charged parallel plate condenser with air between its plates are respectively C_0 and W_0. If the air is replaced by glass (dielectric constant $=5$) between the plates, the capacitance of the plates and the energy stored in it will respectively be
(a) $5C_0, 5W_0$ (b) $5C_0, \frac{W_0}{5}$ (c) $\frac{C_0}{5}, 5W_0$ (d) $\frac{C_0}{5}, \frac{W_0}{5}$

5. Two conducting plates A and B each having large surface area S (on one side) are placed parallel to each other. The plate A is given a charge q, while the plate B is neutral. Then, the electric field at a point in between the plates is

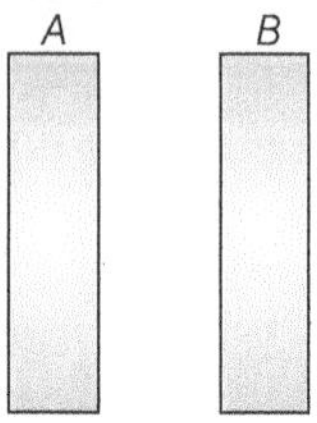

(a) $\frac{q}{2S\varepsilon_0}$ (b) $\frac{q}{S\varepsilon_0}$
(c) $\frac{2q}{S\varepsilon_0}$ (d) $\frac{3}{2}\frac{q}{S\varepsilon_0}$

6. A spherical condenser has inner and outer spheres of radii a and b, respectively. The space between the two is filled with air. The difference between the capacitances of two condensers formed when outer sphere is earthed and when inner sphere is earthed will be
(a) zero (b) $4\pi\varepsilon_0 a$
(c) $4\pi\varepsilon_0 b$ (d) $4\pi\varepsilon_0 a\left(\frac{b}{b-a}\right)$

7. Two very large thin conducting plates having same cross-sectional area are placed as shown in figure. They are carrying charges Q and $3Q$, respectively. The variation of electric field as a function at x (for $x=0$ to $x=3d$) will be best represented by

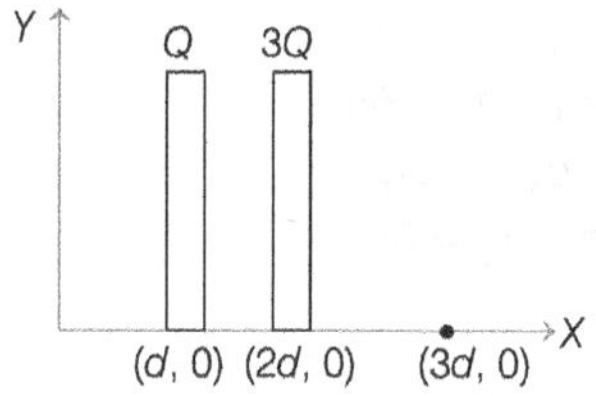

(a)
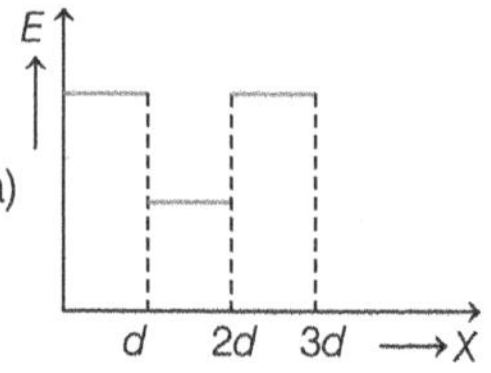

(b)
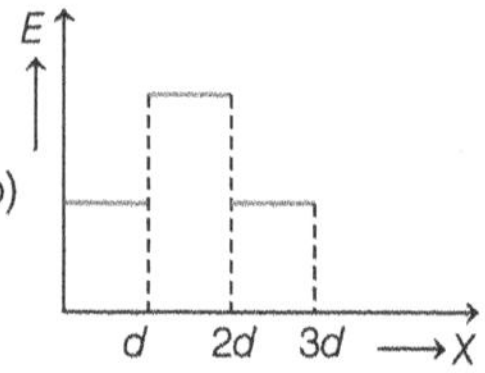

(c)
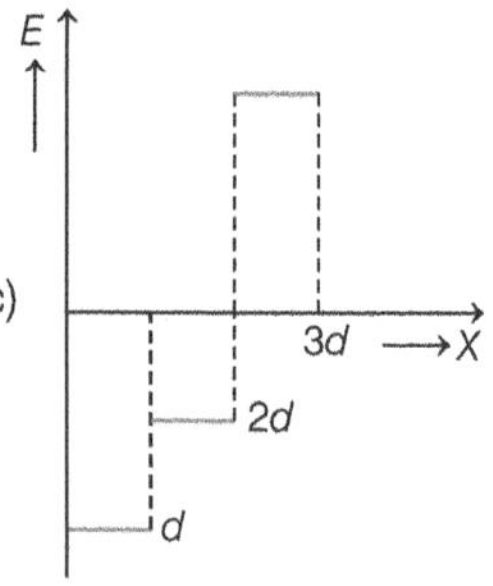

(d)
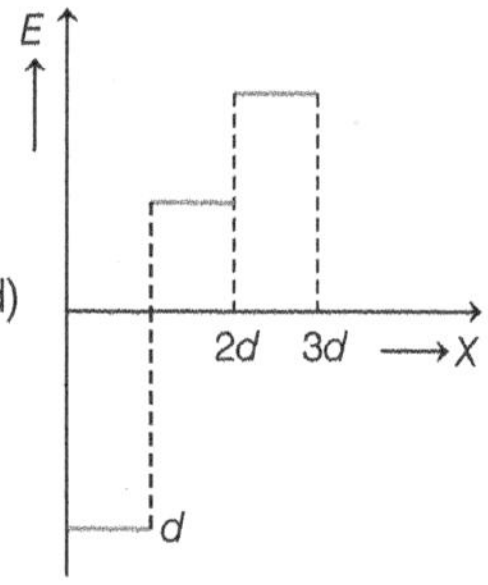

8. A capacitor of capacitance C is charged to a potential difference V from a cell and then disconnected from it. A charge $+Q$ is now given to its positive plate. The potential difference across the capacitor is now
(a) V (b) $V+\frac{Q}{C}$
(c) $V+\frac{Q}{2C}$ (d) $V-\frac{Q}{C}$, if $Q<CV$

9. The plates of a parallel plate capacitor are pulled apart with a velocity v. If at any instant their mutual distance of separation is x, then magnitude of rate of change of capacitance with respect to time varies as

(a) $\frac{1}{x}$ (b) $\frac{1}{x^2}$ (c) x^2 (d) x

10. Two identical metal plates are given positive charges Q_1 and Q_2 ($<Q_1$) respectively. If they are now brought close together to form a parallel plate capacitor with capacitance C, the potential difference between them is

(a) $(Q_1+Q_2)/2C$ (b) $(Q_1+Q_2)/C$
(c) $(Q_1-Q_2)/C$ (d) $(Q_1-Q_2)/2C$

11. A parallel plate capacitor with dielectric of dielectric constant K between the plates has a capacitance C and is charged to a potential of V volt and battery remains connected. The dielectric slab is slowly removed from the plates and then reinserted. The net work done by the system in the process is

(a) $\frac{1}{2}(k-1)CV^2$ (b) $CV^2(k-1)K$
(c) $(k-1)CV^2$ (d) zero

12. A parallel plate capacitor is made of two circular plates separated by a distance of 5 mm and with a dielectric of dielectric constant 2.2 between them. When the electric field in the dielectric is 3×10^4 V/m, the charge density of the positive plate will be close to

(a) $6\times10^{-7}\ C/m^2$ (b) $3\times10^{-7}\ C/m^2$
(c) $3\times10^4\ C/m^2$ (d) $6\times10^4\ C/m^2$

13. A parallel plate capacitor C with plates of area 1 square unit and separation d is filled with a liquid of dielectric constant $K=2$. The level of liquid is $d/3$, initially. If the liquid level decreases at a constant speed v. Then, the capacitance of capacitor after t second is

(a) $\frac{6\varepsilon_0}{5d+3vt}$ (b) $\frac{(15d+9vt)\varepsilon_0}{2d^2-3dvt-9v^2t^2}$
(c) $\frac{6\varepsilon_0}{5d-3vt}$ (d) $\frac{(15d-9vt)\varepsilon_0}{2d^2+3dvt-9v^2t^2}$

14. A solid conducting sphere of radius a having a charge q is surrounded by a concentric conducting spherical shell of inner radius $2a$ and outer radius $3a$ as shown in figure. Find the amount of heat produced when switch is closed $\left(k=\frac{1}{4\pi\varepsilon_0}\right)$

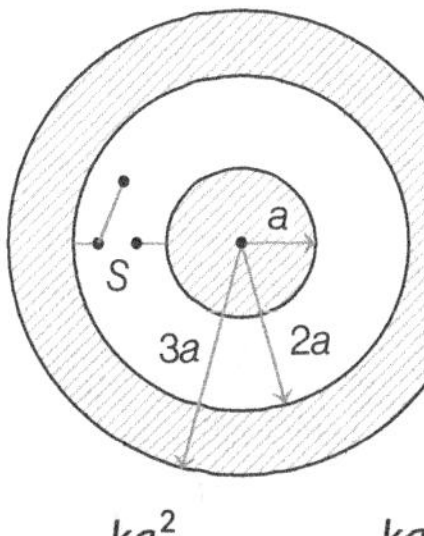

(a) $\frac{kq^2}{2a}$ (b) $\frac{kq^2}{3a}$ (c) $\frac{kq^2}{4a}$ (d) $\frac{kq^2}{6a}$

Multiple Correct Options

15. X and Y are large, parallel conducting plates close to each other. Each face has an area A. X is given a charge Q. Y is without any charge. Points A, B and C are as shown in the figure.

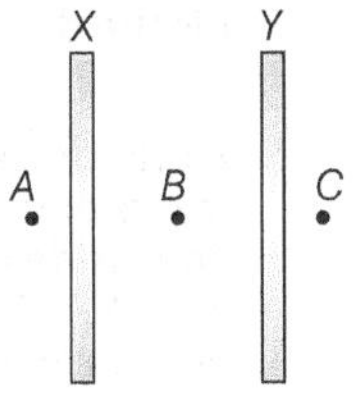

(a) The field at B is $\frac{Q}{2\varepsilon_0 A}$

(b) The field at B is $\frac{Q}{\varepsilon_0 A}$

(c) The fields at A, B and C are of the same magnitude

(d) The fields at A and C are of the same magnitude, but in opposite directions

Match the Columns

16. Four large parallel identical conducting plates are arranged as shown.

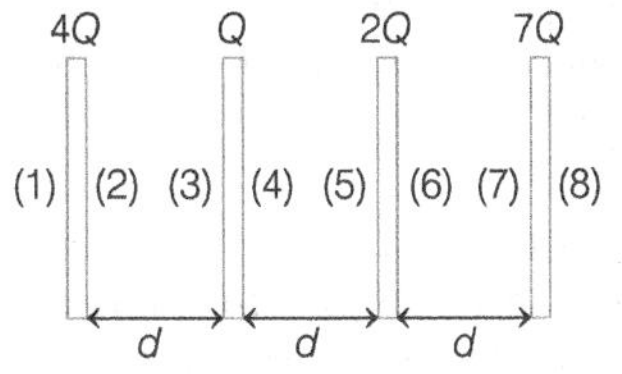

	Column I		Column II
(a)	Surfaces having charges of same magnitude and sign	(p)	1 and 8
(b)	Surfaces having positive charges	(q)	3 and 5
(c)	Uncharged surfaces	(r)	2 and 3
(d)	Charged surfaces	(s)	6 and 7

Subjective Questions

17. Charges $2q$ and $-3q$ are given to two identical metal plates of area of cross-section A. The distance between the plates is d. Find the capacitance and potential difference between the plates.

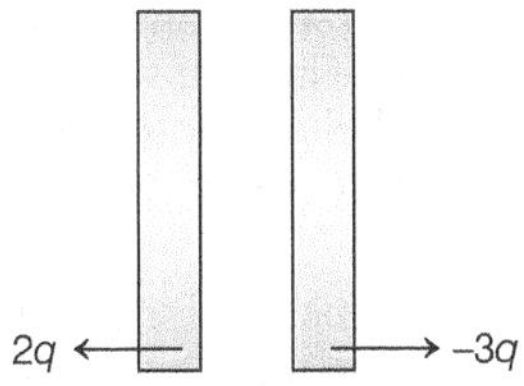

18. Find the capacitance of an isolated ball-shaped conductor of radius R_1 surrounded by an adjacent concentric layer of dielectric with dielectric constant K and outside radius R_2. **(Irodov 3.101)**

19. The gap between the plates of a parallel plate capacitor is filled with isotropic dielectric whose dielectric constant K varies linearly from K_1 to K_2 ($K_2>K_1$) in the direction perpendicular to the plates. The area of each plate equals S, the separation between the plates is equal to d. Find the capacitance of the capacitor. **(Irodov 3.104)**

20. Find the capacitance of a spherical capacitor whose electrodes have radii R_1 and R_2 ($>R_1$) and which is filled with isotropic dielectric whose dielectric

constant varies as $K = a/r$, where a is a constant and r is the distance from the centre of the capacitor. **(Irodov 3.105)**

21. A cylindrical capacitor is filled with two cylindrical layers of dielectric with dielectric constants K_1 and K_2. The inside radii of the layers are equal to R_1 and R_2 $(> R_1)$. The maximum permissible values of electric field strength are equal to E_{1m} and E_{2m} for these dielectrics. At what relationship between K, R and E_m will the voltage increase result in the field strength reaching the breakdown value for both dielectrics simultaneously? **(Irodov 3.106)**

22. There is a double-layer cylindrical capacitor whose parameters are shown in figure below. The breakdown field strength values for these dielectrics are equal to E_1 and E_2 respectively. What is the breakdown voltage of this capacitor if $K_1R_1E_1 < K_2R_2E_2$? **(Irodov 3.107)**

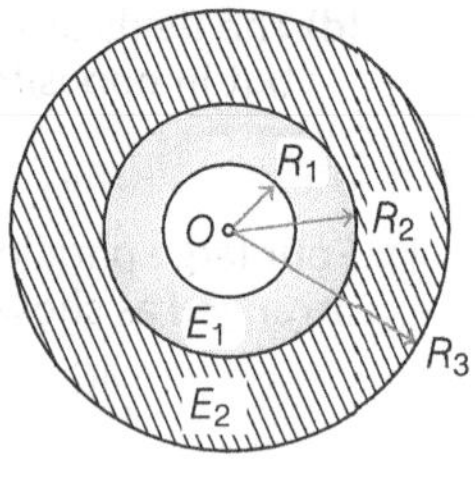

23. Two long straight wires with equal cross-sectional radii a are located parallel to each other in air. The distance between their axes equals b. Find the mutual capacitance of the wires per unit length under the condition $b >> a$. **(Irodov 3.108)**

24. Find the capacitance of a system of two identical metal balls of radius a if the distance between their centres is equal to b, with $b >> a$. The system is located in a uniform dielectric with dielectric constant K. **(Irodov 3.110)**

25. A system consists of two thin concentric metal shells of radii R_1 and R_2 with corresponding charges q_1 and q_2. Find the self energy values W_1 and W_2 of each shell, the interaction energy of the shells W_{12} and the total electric energy of the system. Given $R_2 > R_1$. **(Irodov 3.134)**

26. A spherical shell of radius R_1 with uniform charge q is expanded to a radius R_2. Find the work performed by the electric forces in this process. **(Irodov 3.137)**

27. A spherical shell of radius R_1 with a uniform charge q has a point charge q_0 at its centre. Find the work performed by the electric forces during the shell expansion from radius R_1 to radius R_2. **(Irodov 3.138)**

28. A point charge q is located at the centre O of a spherical uncharged conducting layer provided with a small orifice shown in figure. The inside and outside radii of the layer are equal to a and b respectively. What amount of work has to be performed to slowly transfer the charge q from the point O through the orifice upto infinity? **(Irodov 3.140)**

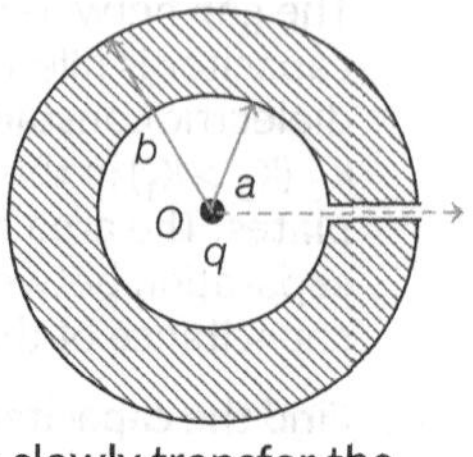

29. Three concentric conducting spherical shells of radii R, $2R$ and $3R$ carry charges Q, $-2Q$ and $3Q$, respectively.

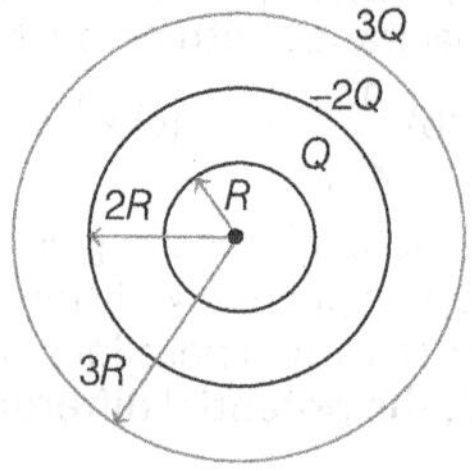

(a) Find the electric potential at $r = R$ and $r = 3R$, where r is the radial distance from the centre.

(b) Compute the electric field at $r = \frac{5}{2}R$

(c) Compute the total electrostatic energy stored in the system.
The inner shell is now connected to the external one by a conducting wire, passing through a very small hole in the middle shell.

(d) Compute the charges on the spheres of radii R and $3R$.

(e) Compute the electric field at $r = \frac{5}{2}R$.

Answers

1. (c)	2. (c)	3. (b)	4. (b)	5. (a)
6. (c)	7. (c)	8. (c)	9. (b)	10. (d)
11. (d)	12. (a)	13. (a)	14. (c)	15. (a,c,d)

16. (a-p ; b-p; c-s; d-p,q,r) **17.** $\frac{\varepsilon_0 A}{d}, \frac{5dq}{2A\varepsilon_0}$

18. $\dfrac{4\pi\varepsilon_0 K R_1}{1 + (K-1)\frac{R_1}{R_2}}$ **19.** $C = \dfrac{\varepsilon_0 S(K_2 - K_1)}{d \ln\left(\frac{K_2}{K_1}\right)}$

20. $C = \dfrac{4\pi\varepsilon_0 a}{\ln\left(\frac{R_2}{R_1}\right)}$ **21.** $K_1R_1E_{1m} = K_2R_2E_{2m}$

22. $V = E_1R_1\left[\ln\frac{R_2}{R_1} + \frac{K_1}{K_2}\ln\left(\frac{R_3}{R_2}\right)\right]$ **23.** $C = \dfrac{\pi\varepsilon_0}{\ln(b/a)}$

24. $C = 2\pi\varepsilon_0 Ka$ **25.** $W_1 = \dfrac{q_1^2}{8\pi\varepsilon_0 R_1}, W_2 = \dfrac{q_2^2}{8\pi\varepsilon_0 R_2},$

$W = \dfrac{1}{4\pi\varepsilon_0}\left[\dfrac{q_1^2}{2R_1} + \dfrac{q_2^2}{2R_2} + \dfrac{q_1q_2}{R_2}\right], W_{12} = \dfrac{1}{4\pi\varepsilon_0}\dfrac{q_1q_2}{R_2}$

26. $W = \dfrac{q^2}{8\pi\varepsilon_0}\left[\dfrac{1}{R_1} - \dfrac{1}{R_2}\right]$

27. $W = \dfrac{q\left(q_0 + \frac{q}{2}\right)}{4\pi\varepsilon_0}\left(\dfrac{1}{R_1} - \dfrac{1}{R_2}\right)$ **28.** $W = \dfrac{q^2}{8\pi\varepsilon_0}\left(\dfrac{1}{a} - \dfrac{1}{b}\right)$

29. (a) $\dfrac{Q}{4\pi\varepsilon_0 R}, \dfrac{Q}{6\pi\varepsilon_0 R}$ (b) $\dfrac{-Q}{25\pi\varepsilon_0 R^2}\hat{r}$ (c) $\dfrac{Q^2}{4\pi\varepsilon_0 R}$

(d) $Q_1 = \dfrac{Q}{2}, Q_2 = \dfrac{7Q}{2}$ (e) $\dfrac{-3Q}{50\pi\varepsilon_0 R^2}\hat{r}$

See hints on pages 311-317.

MODULE 4

Mechanical Force on a Charged Conductor

We know that similar charges repel each other, hence the charge on any part of surface of the conductor is repelled by the charge on its remaining part. The surface of the conductor thus experiences a mechanical force.

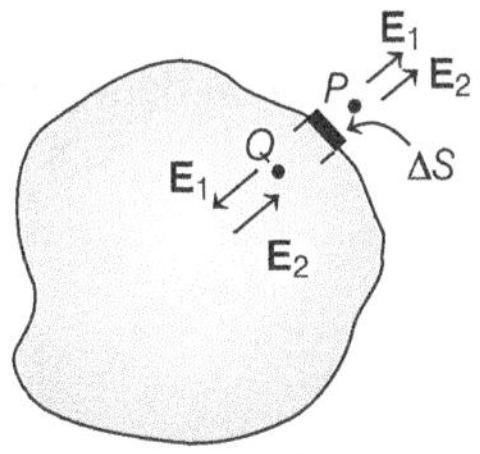

The electric field at any point P near the conductor's surface can be assumed as due to a small part of the surface of area say ΔS immediately in the neighbourhood of the point under consideration and due to the rest of the surface. Let $\mathbf{E}_1$ and $\mathbf{E}_2$ be the field intensities due to these parts respectively.

Then, total electric field, $\mathbf{E} = \mathbf{E}_1 + \mathbf{E}_2$

$\mathbf{E}$ has a magnitude σ/ε_0 at any point P just outside the conductor and is zero at point Q just inside the conductor. Thus,

$$E_1 + E_2 = \sigma/\varepsilon_0 \text{ at } P$$

and $$E_1 - E_2 = 0 \text{ at } Q$$

$$\therefore \quad E_1 = E_2 = \frac{\sigma}{2\varepsilon_0}$$

Hence, the force experienced by small surface of area ΔS due to the charge on the rest of the surface is

$$F = qE_2 = (\sigma\Delta S)(E_2) = \frac{(\sigma^2)(\Delta S)}{2\varepsilon_0}$$

$$\therefore \quad \frac{\text{Force}}{\text{Area}} = \frac{F}{\Delta S} = \frac{\sigma^2}{2\varepsilon_0} = \frac{1}{2}\varepsilon_0 E^2 \qquad \left(\text{as } E = \frac{\sigma}{\varepsilon_0}\right)$$

$$\therefore \quad \boxed{\frac{\text{Force}}{\text{Area}} = \frac{1}{2}\varepsilon_0 E^2}$$

Force between the Plates of a Capacitor

Consider a parallel plate capacitor with plate area A. Suppose a positive charge q is given to one plate and a negative charge $-q$ to the other plate. The electric field on the negative plate due to positive charge is

$$E = \frac{\sigma}{2\varepsilon_0} = \frac{q}{2A\varepsilon_0}$$

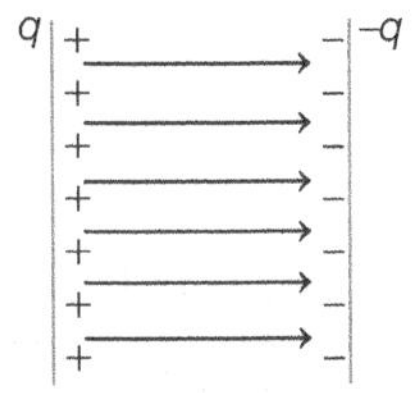

The magnitude of force on the charge in negative plate is

$$F = qE = \frac{q^2}{2A\varepsilon_0}$$

This is the force with which both the plates attract each other. Thus,

$$\boxed{F = \frac{q^2}{2A\varepsilon_0}}$$

Example 1 A capacitor is given a charge q. The distance between the plates of the capacitor is d. One of the plates is fixed and the other plate is moved away from the other till the distance between them becomes $2d$. Find the work done by the external force.

Sol. When one plate is fixed, the other is attracted towards the first with a force

$$F = \frac{q^2}{2A\varepsilon_0} = \text{constant}$$

Hence, an external force of same magnitude will have to be applied in opposite direction to increase the separation between the plates.

$$\therefore \quad W = F(2d - d) = \frac{q^2 d}{2A\varepsilon_0} \qquad \textbf{Ans.}$$

Alternate solution

$$W = \Delta U = U_f - U_i = \frac{q^2}{2C_f} - \frac{q^2}{2C_i} \qquad \text{...(i)}$$

Here, $$C_f = \frac{\varepsilon_0 A}{2d} \text{ and } C_i = \frac{\varepsilon_0 A}{d}$$

Substituting in Eq. (i), we have

$$W = \frac{q^2}{2\left(\frac{\varepsilon_0 A}{2d}\right)} - \frac{q^2}{2\left(\frac{\varepsilon_0 A}{d}\right)} = \frac{q^2 d}{2\varepsilon_0 A} \qquad \textbf{Ans.}$$

MODULE 5

Capacitors in Series and Parallel

In Series

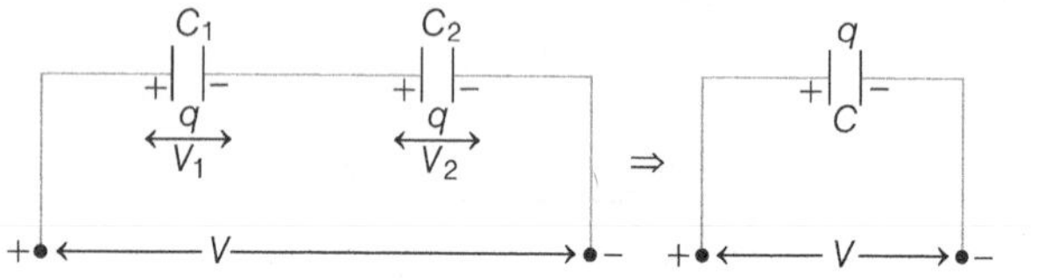

In a series connection, the magnitude of charge on all plates is same. The potential is distributed in the inverse ratio of the capacitance (as $V = q/C$ or $V \propto 1/C$). Thus, in the figure, if a potential difference V is applied across the two capacitors C_1 and C_2, then

$$\boxed{\frac{V_1}{V_2} = \frac{C_2}{C_1}}$$

or $$V_1 = \left(\frac{C_2}{C_1 + C_2}\right) V$$

and $$V_2 = \left(\frac{C_1}{C_1 + C_2}\right) V$$

Further, in the figure, $V = V_1 + V_2$

or $$\frac{q}{C} = \frac{q}{C_1} + \frac{q}{C_2}$$

or $$\frac{1}{C} = \frac{1}{C_1} + \frac{1}{C_2}$$

Here, C is the **equivalent capacitance.**

We can extend this analysis to any number of capacitors in series. We find the following result for the equivalent capacitance.

$$\boxed{\frac{1}{C} = \frac{1}{C_1} + \frac{1}{C_2} + \frac{1}{C_3} + \ldots}$$

Following points are important in case of series combination of capacitors.

(i) In a series connection, the equivalent capacitance is always less than any individual capacitance.

(ii) For the equivalent capacitance of two capacitors it is better to remember the following expression

$$C = \frac{C_1 C_2}{C_1 + C_2}$$

For example, equivalent capacitance of two capacitors $C_1 = 6\,\mu\text{F}$ and $C_2 = 3\,\mu\text{F}$ is

$$C = \frac{C_1 C_2}{C_1 + C_2} = \left(\frac{6 \times 3}{6 + 3}\right) \mu\text{F} = 2\,\mu\text{F}$$

(iii) If n capacitors of equal capacitance C are connected in series, then their equivalent capacitance is $\frac{C}{n}$.

Example 1 In the circuit shown in figure, find

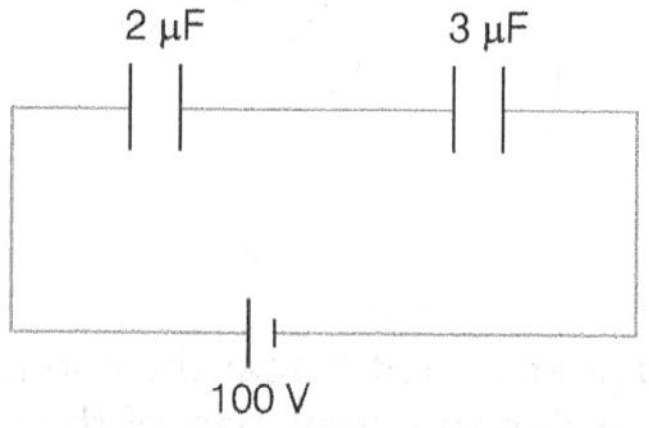

(a) the equivalent capacitance,

(b) the charge stored in each capacitor and

(c) the potential difference across each capacitor.

Sol. (a) The equivalent capacitance

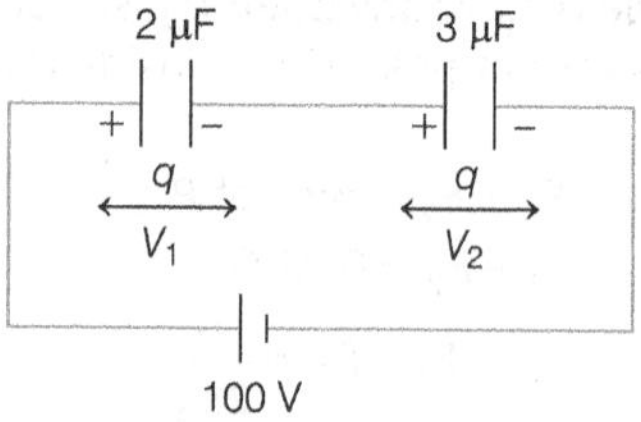

$$C = \frac{C_1 C_2}{C_1 + C_2}$$

or $$C = \frac{(2)\,(3)}{2 + 3} = 1.2\,\mu\text{F}$$ **Ans.**

(b) The charge q stored in each capacitor is

$$q = CV = (1.2 \times 10^{-6})\,(100)\ \text{C}$$
$$= 120\,\mu\text{C}$$ **Ans.**

(c) In series combination, $V \propto \frac{1}{C}$ or $\frac{V_1}{V_2} = \frac{C_2}{C_1}$

$$\therefore \quad V_1 = \left(\frac{C_2}{C_1 + C_2}\right) V = \left(\frac{3}{2 + 3}\right)(100) = 60\ \text{V}$$

and $V_2 = V - V_1 = 100 - 60 = 40\ \text{V}$ **Ans.**

In Parallel

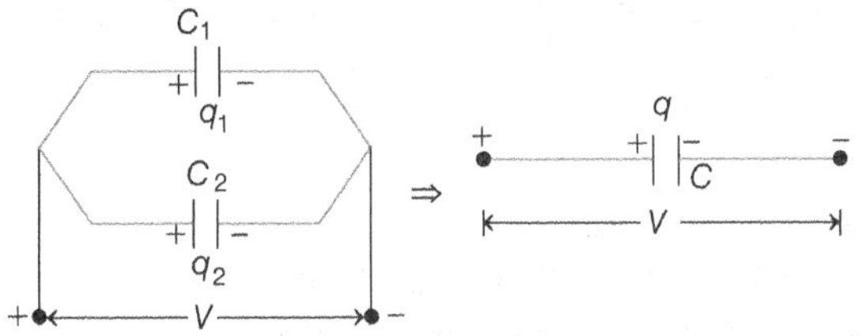

The arrangement shown in figure is called a parallel connection. In a parallel combination, the potential difference for all individual capacitors is the same and the total charge q is distributed in direct ratio of their capacitances. (as $q = CV$ or $q \propto C$ for same potential difference). Thus,

$$\boxed{\frac{q_1}{q_2} = \frac{C_1}{C_2}}$$

or $$q_1 = \left(\frac{C_1}{C_1 + C_2}\right) q$$

and $$q_2 = \left(\frac{C_2}{C_1 + C_2}\right) q$$

The parallel combination is equivalent to a single capacitor with the same total charge $q = q_1 + q_2$ and potential difference V.

Thus, $$q = q_1 + q_2$$

or $$CV = C_1 V + C_2 V$$

or $$\boxed{C = C_1 + C_2}$$

In the same way, we can show that for any number of capacitors in parallel,

$$\boxed{C = C_1 + C_2 + C_3 + \ldots}$$

In a parallel combination, the equivalent capacitance is always greater than any individual capacitance.

Example 2 In the circuit shown in figure, find

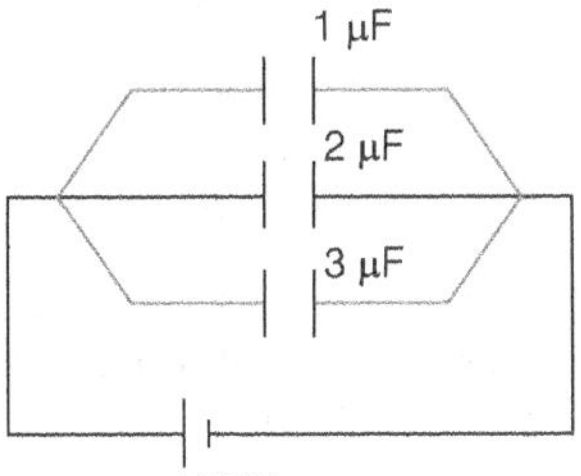

(a) the equivalent capacitance and
(b) the charge stored in each capacitor.

Sol. (a) The capacitors are in parallel. Hence, the equivalent capacitance is

$$C = C_1 + C_2 + C_3$$

or $$C = (1 + 2 + 3) = 6\ \mu F$$ **Ans.**

(b) Total charge drawn from the battery,

$$q = CV = 6 \times 100\ \mu C = 600\ \mu C$$

This charge will be distributed in direct ratio of their capacitances. Hence,

$$q_1 : q_2 : q_3 = C_1 : C_2 : C_3 = 1 : 2 : 3$$

$\therefore$ $$q_1 = \left(\frac{1}{1+2+3}\right) \times 600 = 100\ \mu C$$

$$q_2 = \left(\frac{2}{1+2+3}\right) \times 600 = 200\ \mu C$$

and $$q_3 = \left(\frac{3}{1+2+3}\right) \times 600 = 300\ \mu C.$$ **Ans.**

Alternate Solution

Since the capacitors are in parallel, the PD across each of them is 100 V. Therefore, from $q = CV$, the charge stored in 1 μF capacitor is 100 μC, in 2 μF capacitor is 200 μC and that in 3 μF capacitor is 300 μC.

INTRODUCTORY Exercise

1. In the circuit shown in figure, the ratio of charge on 5 μF and 2 μF capacitor is

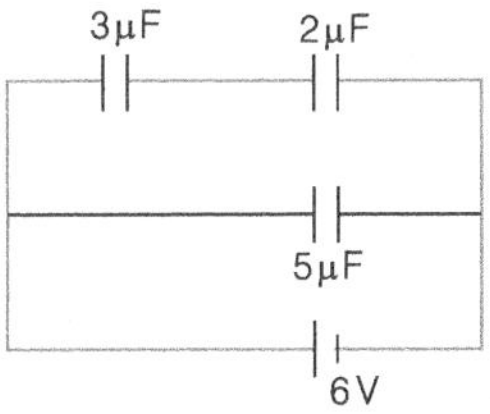

(a) 5 / 4
(b) 5 / 3
(c) 3 / 8
(d) None of the above

2. In the circuit as shown in figure if all the symbols have their usual meanings, then identify the correct statement,

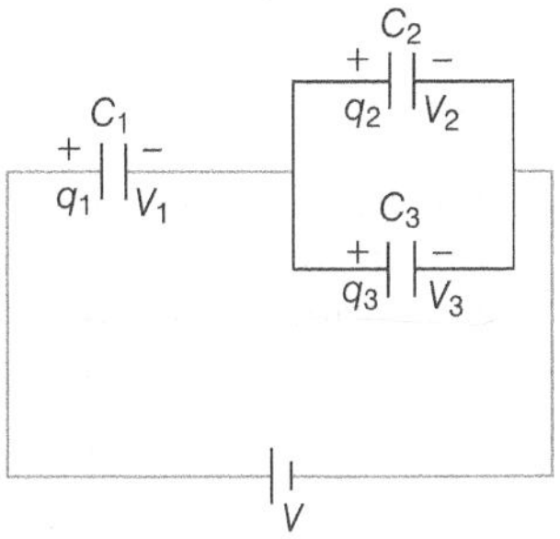

(a) $q_2 = q_3; V_2 = V_3$
(b) $q_1 = q_2 + q_3; V_2 = V_3$
(c) $q_1 = q_2 + q_3; V = V_1 + V_2 + V_3$
(d) $q_1 + q_2 + q_3 = 0; V_2 = V_3 = V - V_1$

3. Find charges on different capacitors.

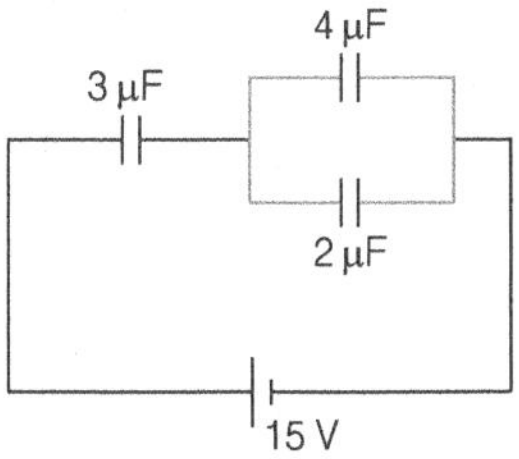

4. Find charges on different capacitors.

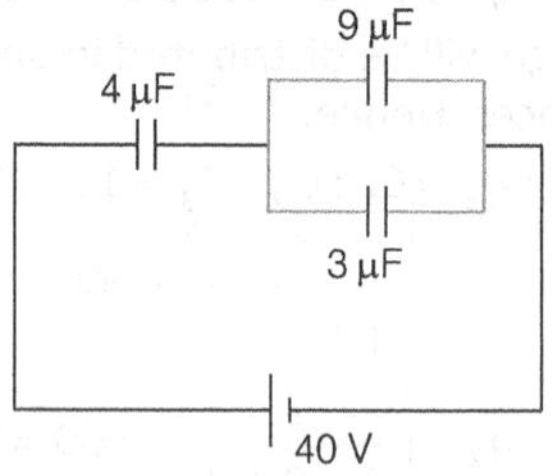

5. Find the charge stored in all capacitors.

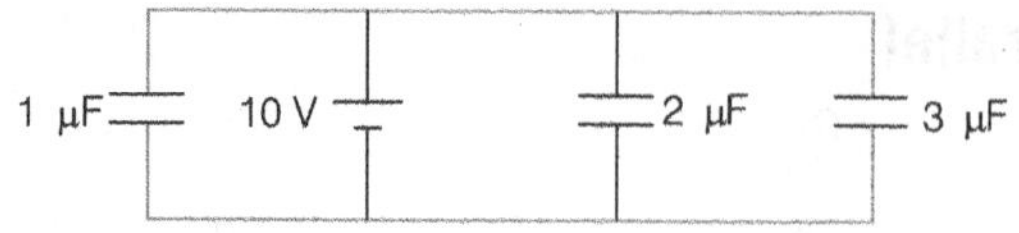

Answers

1. (d) 2. (b)
3. $q_{3\mu F} = 30\ \mu C, q_{4\mu F} = 20\ \mu C, q_{2\mu F} = 10\ \mu C$
4. $q_{4\mu F} = 120\ \mu C, q_{9\mu F} = 90\ \mu C, q_{3\mu F} = 30\ \mu C$
5. $10\ \mu C,\ 20\ \mu C,\ 30\ \mu C$

See hints on page 317.

Typed Problems

Type 1. *Method of finding values of q and V across different capacitors in a complex circuit if values across one capacitor are known.*

Concept

In series, q is same and V distributes in inverse ratio of capacity.

As, $$V = \frac{q}{C}$$

$\Rightarrow$ $$V \propto \frac{1}{C} \quad (q \text{ is same})$$

If capacitance is double, then V will be half.

In parallel, V is same and q distributes in direct ratio of capacity.

As, $$q = CV$$

$\Rightarrow$ $$q \propto C \quad (V \text{ is same})$$

If C is double, then q is also double.

Ex. 1 In the circuit shown in figure potential difference across 3 μF is 10 V. Find potential difference and charge stored in different capacitors. Also find emf of the battery E.

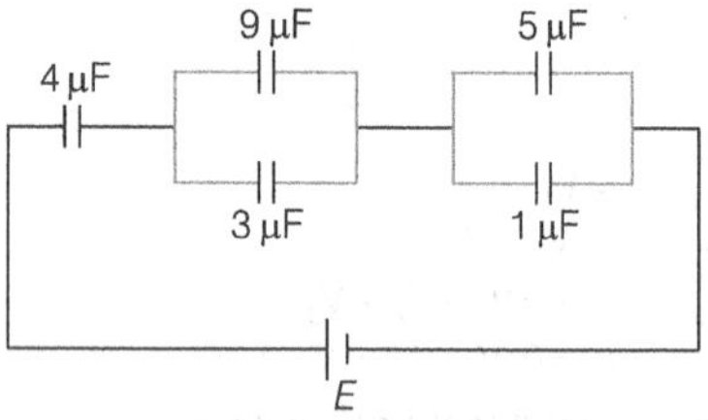

Sol. The given circuit can be simplified as below.

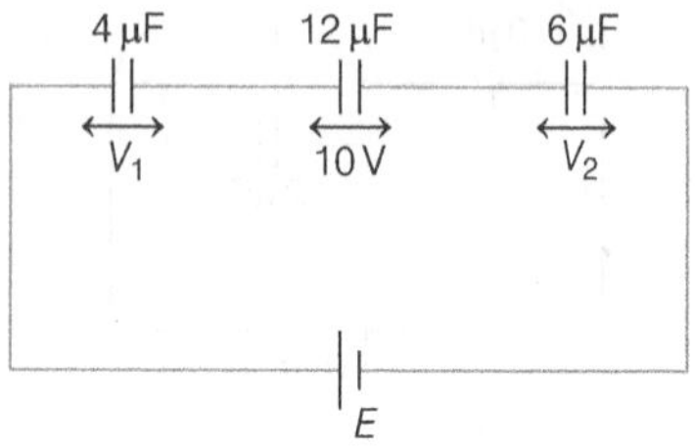

V_1 : 4 μF is $\frac{1}{3}$ rd of 12 μF. Therefore, V_1 is thrice of 10 V or 30 V.

V_2: 6 μF is half of 12 μF. Therefore, V_2 is twice of 10 V or 20 V.

$$E = V_1 + 10 + V_2$$
$$= 30 + 10 + 20$$
$$= 60\ V$$ **Ans.**

Potential difference and charge on different capacitors in tabular form are given below.

Table

Capacitance	Potential difference	Charge $q = CV$
4 μF	30 V	120 μC
9 μF	10 V	90 μC
3 μF	10 V	30 μC
5 μF	20 V	100 μC
1 μF	20 V	20 μC

Type 2. *Two or more than two capacitors are charged from different batteries and then connected in parallel.*

Concept

In parallel, they come to a common potential given by

$$V = \frac{\text{Total charge}}{\text{Total capacity}}$$

Moreover, total charge on them distributes in direct ratio of their capacitance or we can also find the final charges on the capacitors by using the equation, $q = CV$

Ex. 2 Three capacitors of capacitances 1 μF, 2 μF and 3 μF are charged by 10V, 20V and 30V, respectively. Now, positive plates of first two capacitors are connected with the negative plate of third capacitor on one side and negative plates of first two capacitors are connected with positive plate of third capacitor on the other side. Find

(a) common potential V

(b) final charges on different capacitors

Sol.

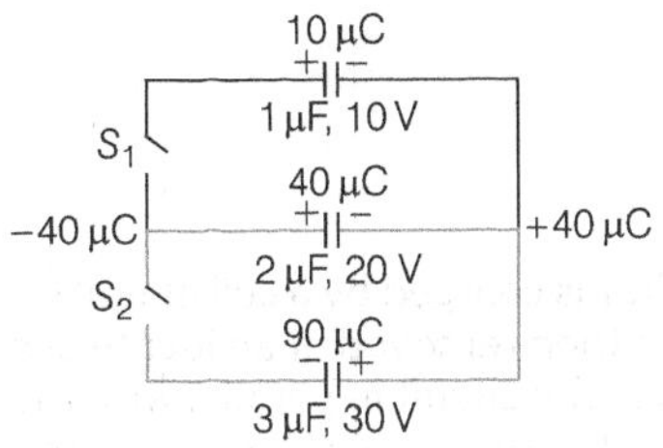

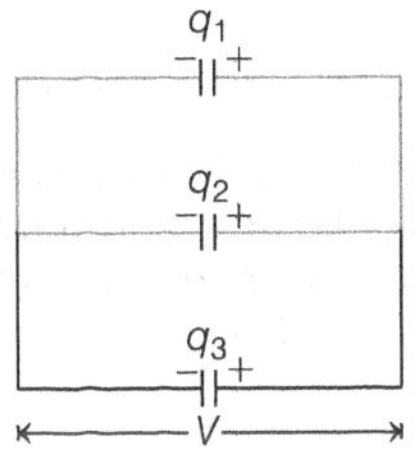

Total charge on all three capacitors =

$$(\pm 10 \pm 40 \mp 90)\ \mu C = \mp 40\ \mu C$$

Total capacitance $= (1 + 2 + 3)\ \mu F$

$$= 6\ \mu F$$

(a) Common potential,

$$V = \frac{\text{Total charge}}{\text{Total capacity}}$$

$$= \frac{40\ \mu C}{6\ \mu F} = \frac{20}{3} \text{ volt}$$

(b) $q_1 = C_1 V = (1\ \mu F)\left(\frac{20}{3}\right)$

$$= \frac{20}{3}\ \mu C$$

$$q_2 = C_2 V$$

$$= (2\ \mu F)\left(\frac{20}{3}\right)$$

$$= \frac{40}{3}\ \mu C$$

$$q_3 = C_3 V = (3\ \mu F)\left(\frac{20}{3}\right) = 20\ \mu C$$

Type 3. *To find capacitance of a capacitor filled with two or more than two dielectrics.*

Concept

PQ and *MN* are two metallic plates.

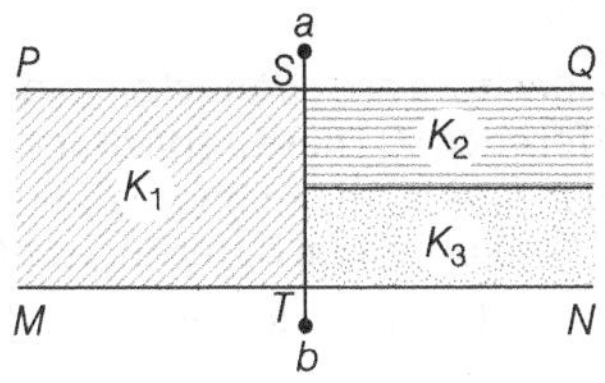

If we wish to find net capacitance between *a* and *b*, then

$$V_{PS} = V_{SQ} = V_1 \text{ (say)}$$

$$V_{MT} = V_{TN} = V_2 \text{ (say)}$$

Hence, $\quad V_{PS} - V_{MT} = V_{SQ} - V_{TN} = V_1 - V_2$

Therefore, there are two capacitors, one on right hand side and other on left hand side which are in parallel.

$\therefore \quad C = C_{RHS} + C_{LHS}$

For C_{RHS}, we can use the formula,

$$C = \frac{\varepsilon_0 A}{d - t_1 - t_2 + \frac{t_1}{K_1} + \frac{t_2}{K_2}}$$

Ex. 3 What is capacitance of the capacitor shown in figure?

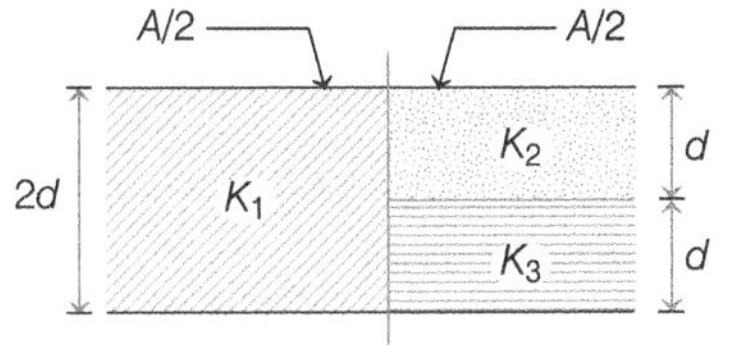

Sol. $C = C_{LHS} + C_{RHS}$

$$= \frac{K_1\ \varepsilon_0\ (A/2)}{2d} + \frac{\varepsilon_0\ (A/2)}{(2d - d - d) + (d/K_2) + (d/K_3)}$$

$$= \frac{\varepsilon_0 A}{2d}\left[\frac{K_1}{2} + \frac{K_2 K_3}{K_2 + K_3}\right]$$

MODULE 5 Exercise

Single Correct Option

1. n identical capacitors are connected in parallel to a potential difference V. These capacitors are then reconnected in series, their charges being left undisturbed. The potential difference obtained is
(a) zero (b) $(n-1)V$ (c) nV (d) n^2V

2. In the circuit shown, a potential difference of 60 V is applied across AB. The potential difference between the points M and N is

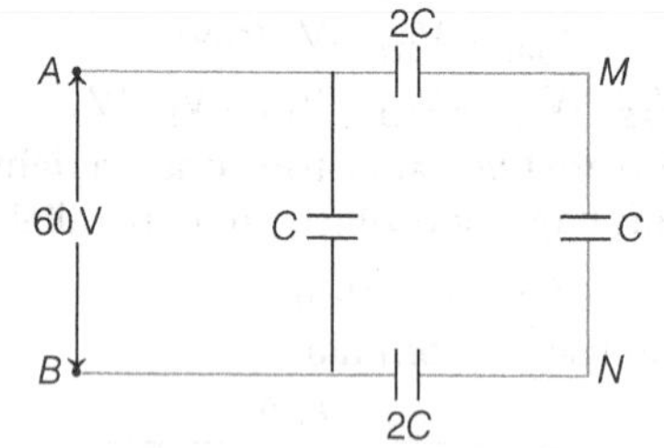

(a) 10 V (b) 15 V (c) 20 V (d) 30 V

3. In the figure a potential of +1200 V is given to point A and point B is earthed, what is the potential at the point P?

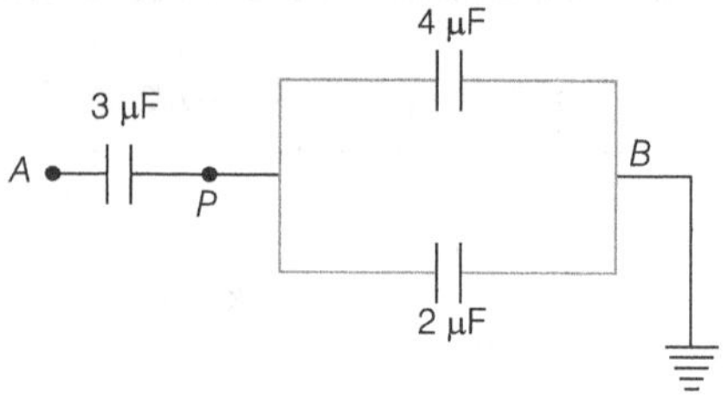

(a) 100 V (b) 200 V (c) 400 V (d) 800 V

4. In the circuit shown in figure $C = 6\,\mu F$. The charge stored in the capacitor of capacitance C is

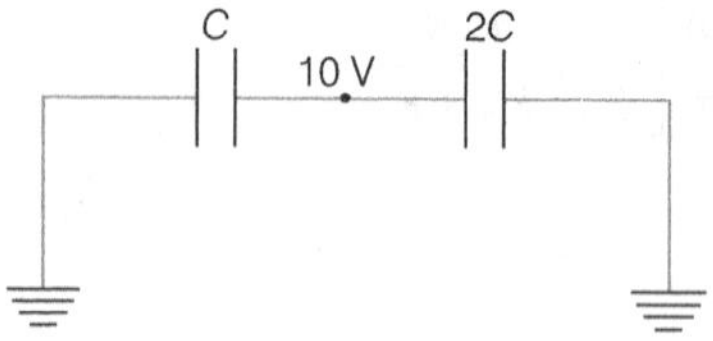

(a) zero (b) $90\,\mu C$ (c) $40\,\mu C$ (d) $60\,\mu C$

5. Four identical capacitors are connected in series with a 10 V battery as shown in the figure. The point N is earthed. The potentials of points A and B are

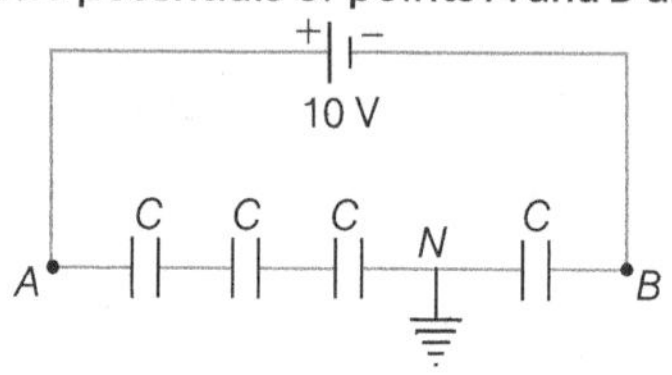

(a) 10 V, 0 V (b) 7.5 V, –2.5 V
(c) 5 V, –5 V (d) 7.5 V, 2.5 V

6. A capacitor is charged by a cell of emf E and the charging battery is then removed. If an identical capacitor is now inserted in the circuit in parallel with the previous capacitor, the potential difference across the new capacitor is
(a) $2E$ (b) E (c) $E/2$ (d) zero

7. Figure shows two capacitors connected in series and joined to a battery. The graph shows the variation in potential as one moves from left to right on the branch containing the capacitors

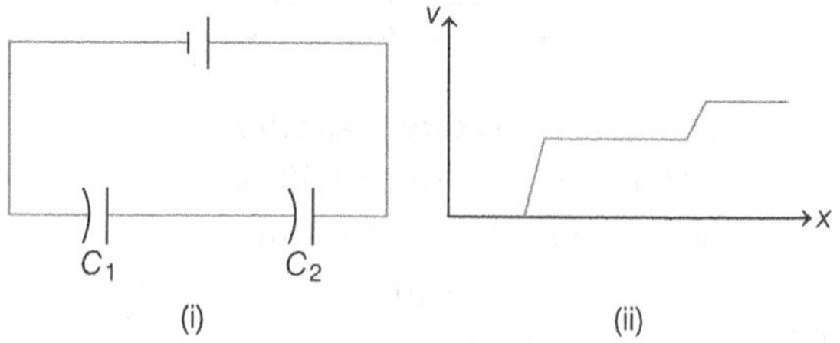

(a) $C_1 > C_2$ (b) $C_1 = C_2$ (c) $C_1 < C_2$
(d) the information is not sufficient to decide the relation between C_1 and C_2

8. Two condensers C_1 and C_2 in a circuit are joined as shown in figure. The potential of point A is V_1 and that of B is V_2. The potential of point D will be

(a) $\frac{1}{2}(V_1 + V_2)$ (b) $\frac{C_2V_1 + C_1V_2}{C_1 + C_2}$
(c) $\frac{C_1V_1 + C_2V_2}{C_1 + C_2}$ (d) $\frac{C_2V_1 - C_1V_2}{C_1 + C_2}$

9. Consider two conductors. One of them has a capacitance of 2 units and the capacitance of the other is unknown. They are charged until their potentials are 4 and 5 units, respectively. The two conductors are now connected by a wire when their common potential is found to be 4.6 units. Then, the unknown capacitance has the value (in the same units as above)
(a) 6 (b) 5 (c) 4 (d) 3

10. In the arrangement shown in figure, dielectric constant $K_1 = 2$ and $K_2 = 3$. If the capacitance across P and Q are C_1 and C_2 respectively in the two cases as shown in figure, then C_1/C_2 will be (the gaps shown are negligible)

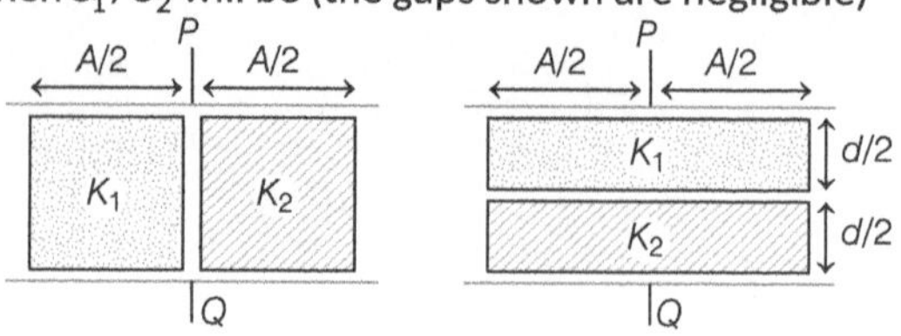

(a) 1 : 1 (b) 2 : 3 (c) 9 : 5 (d) 25 : 24

Subjective Questions

11. A 100 μF capacitor is charged to 100 V. After the charging, battery is disconnected. The capacitor is then connected in parallel to another capacitor. The final voltage is 20 V. Calculate the capacitance of second capacitor.

12. Two condensers are in parallel and the energy of the combination is 0.1 J, when the difference of potential between terminals is 2 V. With the same two condensers in series, the energy is 1.6×10^{-2} J for the same difference of potential across the series combination. What are the capacitances?

13. Several 10 pF capacitors are given, each capable of withstanding 100 V. How would you construct :
   (a) a unit possessing a capacitance of 2 pF and capable of withstanding 500 V?
   (b) a unit possessing a capacitance of 20 pF and capable of withstanding 300 V?

14. Two capacitors A and B are connected in series across a 100 V supply and it is observed that the potential difference across them are 60 V and 40 V. A capacitor of 2 μF capacitance is now connected in parallel with A and the potential difference across B rises to 90 V. Determine the capacitances of A and B.

15. A 10.0 μF parallel-plate capacitor with circular plates is connected to a 12.0 V battery.
   (a) What is the charge on each plate?
   (b) How much charge would be on the plates if their separation were doubled while the capacitor remained connected to the battery?
   (c) How much charge would be on the plates if the capacitor were connected to the 12.0 V battery after the radius of each plate was doubled without changing their separation?

16. Three capacitors having capacitances of 8.4 μF, 8.2 μF and 4.2 μF are connected in series across a 36 V battery.
   (a) What is the charge on 4.2 μF capacitor?
   (b) What is the total energy stored in all three capacitors?
   (c) The capacitors are disconnected from the battery without allowing them to discharge. They are then reconnected in parallel with each other, with the positively charged plates connected together. What is the potential difference across each capacitor in the parallel combination?
   (d) What is the total energy now stored in the capacitors?

17. Two condensers A and B each having slabs of dielectric constant $K = 2$ are connected in series. When they are connected across 230 V supply, potential difference across A is 130 V and that across B is 100 V. If the dielectric in the condenser of smaller capacitance is replaced by one for which $K = 5$, what will be the values of potential difference across them?

18. A capacitor of capacitance $C_1 = 1.0$ μF charged upto a voltage $V = 110$ V is connected in parallel to the terminals of a circuit consisting of two uncharged capacitors connected in series and possessing the capacitances $C_2 = 2.0$ μF and $C_3 = 3.0$ μF. What charge will flow through the connecting wires?

Answers

1. (c) **2.** (d) **3.** (c) **4.** (d) **5.** (b) **6.** (c) **7.** (c) **8.** (c) **9.** (d) **10.** (d)

11. 400 μF **12.** 40 mF, 10 mF **13.** (a) Five capacitors in series (b) Six rows of three capacitors in each row

14. 0.16 μF, 0.24 μF **15.** (a) 120 μC (b) 60 μC (c) 480 μC **16.** (a) 75 μC (b) 1.35 mJ (c) 10.8 V (d) 1.2 mJ

17. 78.68 V, 151.32 V **18.** 60 μC

See hints on pages 317-319.

MODULE 6

Two Laws in Capacitors

Like an electric circuit having resistances and batteries in a complex circuit containing capacitors and the batteries, charges on different capacitors can be obtained with the help of Kirchhoff 's laws.

First Law

This law is basically law of conservation of charge which is normally applied across a battery or in an isolated system.

(i) In case of a battery, both terminals of the battery supply equal amount of charge.

(ii) In an isolated system (not connected to any source of charge like terminal of a battery or earth) net charge remains constant.

For example, in the figure, the positive terminal of the battery supplies a positive charge $q_1 + q_2$. Similarly, the negative terminal supplies a negative charge of magnitude $q_3 + q_4$. Hence,

$$q_1 + q_2 = q_3 + q_4$$

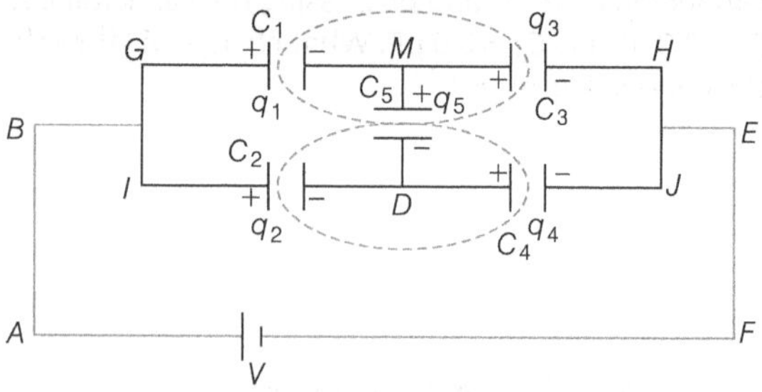

Further, the plates enclosed by the dotted lines form an isolated system, as they are neither connected to a battery terminal nor to the earth. Initially, no charge was present in these plates. Hence, after charging net charge on these plates should also be zero. Or,

$$q_3 + q_5 - q_1 = 0 \quad \text{and} \quad q_4 - q_2 - q_5 = 0$$

So, these are the three equations which can be obtained from the first law in the given circuit.

Second Law

In a capacitor, potential drops by q/C when one moves from positive plate to the negative plate and in a battery it drops by an amount equal to the emf of the battery in moving from positive terminal to negative terminal. Applying second law in loop *ABGHEFA*, we have

$$-\frac{q_1}{C_1} - \frac{q_3}{C_3} + V = 0$$

Similarly, the second law in loop *GMDIG* gives the equation,

$$-\frac{q_1}{C_1} - \frac{q_5}{C_5} + \frac{q_2}{C_2} = 0$$

Example 1 Find the charges on the three capacitors shown in figure.

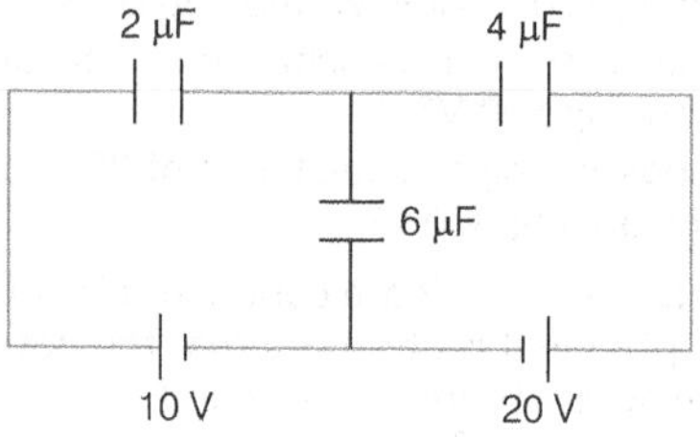

Sol. Let the charges in three capacitors are as shown in figure.

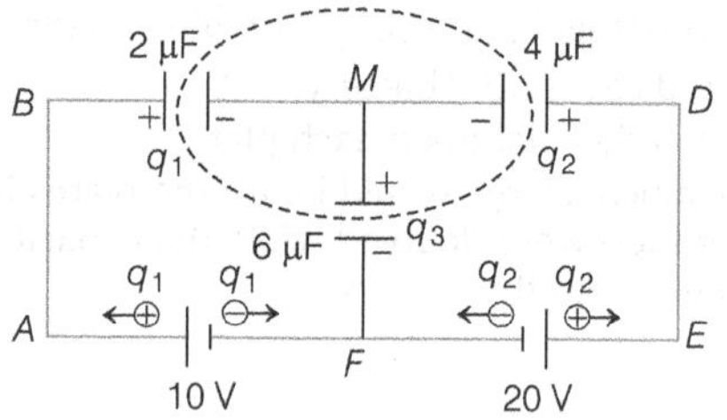

Charge supplied by 10 V battery is q_1 and that from 20 V battery is q_2. Thus,

$$q_1 + q_2 = q_3 \quad \text{...(i)}$$

This relation can also be obtained by a different method. The charges on the three plates which are in contact add to zero (plates inside the dotted line). Because these plates taken together form an isolated system which can't receive charges from the batteries. Thus,

$$q_3 - q_1 - q_2 = 0$$

or $$q_3 = q_1 + q_2$$

Applying second law in loops *BMFAB* and *MDEFM*, we have

$$-\frac{q_1}{2} - \frac{q_3}{6} + 10 = 0$$

or $$q_3 + 3q_1 = 60 \quad \text{...(ii)}$$

and $$\frac{q_2}{4} - 20 + \frac{q_3}{6} = 0$$

or $$3q_2 + 2q_3 = 240 \quad \text{...(iii)}$$

Solving the above three equations, we have

$$q_1 = \frac{10}{3}\ \mu C$$

$$q_2 = \frac{140}{3}\ \mu C \quad \text{and} \quad q_3 = 50\ \mu C$$

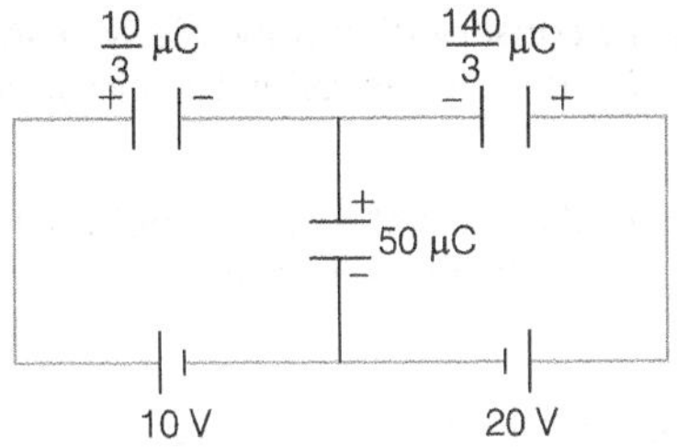

Thus, charges on different capacitors are as shown in figure.

Note *(i) In the above problem q_1, q_2 and q_3 are already in microcoulombs.*

(ii) If capacitors are in series, then charges on them are equal, provided they are initially uncharged. This can be proved by the following illustration :

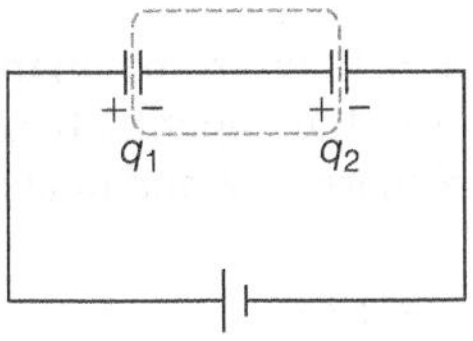

Let us suppose that charges on two capacitors are q_1 and q_2. The two plates encircled by dotted lines form an isolated system. So, net charge on them will remain constant.

$$\Sigma q_f = \Sigma q_i$$

If initially they are uncharged, then

$$\Sigma q_i = 0$$

Σq_f is also zero

or $$-q_1 + q_2 = 0$$

or $$q_1 = q_2$$

So, this proves that charges are equal if initially they are uncharged.

Typed Problems

Type 1. *Based on flow of charge when position of a switch is changed.*

Concept

From the flow of charge we mean that when a switch in a circuit is either closed or opened or it is shifted from one position to the other, then how much charge will flow through certain points of the circuit. Such problems can be solved by finding charges on different capacitors at initial and final position of the switch. The following example will illustrate the theory.

Ex. 1 What charges will flow through A, B and C in the directions shown in the figure when switch S is closed? **(Irodov 3.123)**

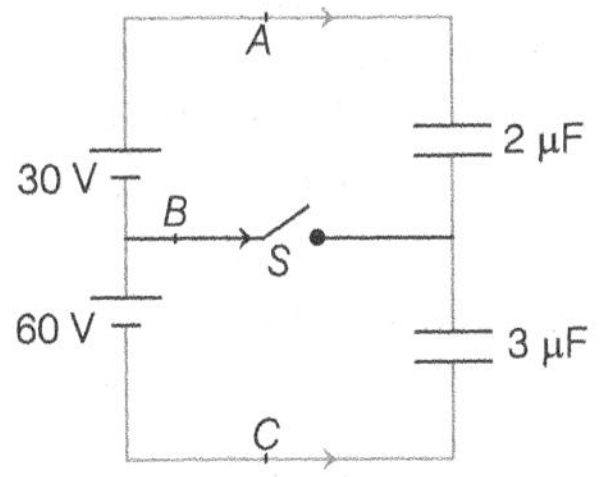

Sol. Let us draw two figures and find the charge on both the capacitors before closing the switch and after closing the switch.

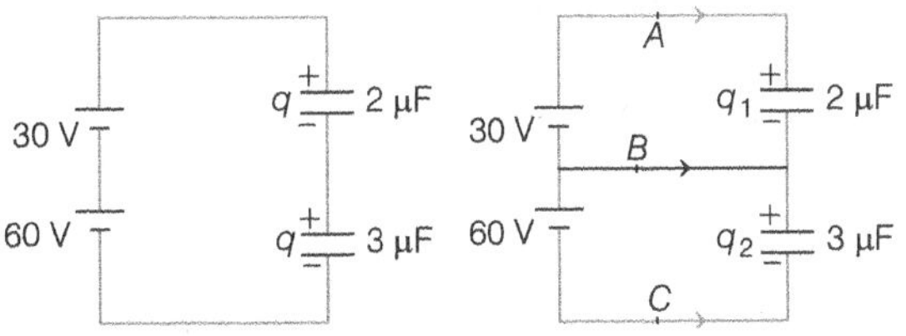

(a) (b)

Refer Fig. (a), when switch is open

Both capacitors are in series. Hence, their equivalent capacitance is

$$C_{eq} = \frac{C_1 C_2}{C_1 + C_2} = \frac{(2)(3)}{2+3} = \frac{6}{5}\ \mu F$$

Therefore, charge on both capacitors will be same. Hence, using $q = CV$, we get

$$q = (30 + 60)\left(\frac{6}{5}\right)\mu C = 108\ \mu C$$

Refer Fig. (b), when switch is closed

Let q_1 and q_2 be the charges (in μC) on two capacitors. Then, applying second law in upper and lower loops, we have

$$30 - \frac{q_1}{2} = 0 \quad \text{or} \quad q_1 = 60\ \mu C$$

$$60 - \frac{q_2}{3} = 0 \quad \text{or} \quad q_2 = 180\ \mu C$$

Charges q_1 and q_2 can be calculated alternatively by seeing that upper plate of 2 μF capacitor is connected with positive terminal of 30 V battery. Therefore, they are at the same potential. Similarly, the lower plate of this capacitor is at the same potential as that of the negative terminal of 30 V battery. So, we can say that PD across 2 μF capacitor is also 30 V.

$$\therefore \quad q_1 = (C)(PD) = (2)(30)\ \mu C = 60\ \mu C$$

Similarly, PD across 3 μF capacitor is same as that between 60 V battery. Hence,

$$q_2 = (3)(60)\ \mu C = 180\ \mu C$$

Now, let q_A charge flows from A in the direction shown. This charge goes to the upper plate of 2 μF capacitor. Initially, it had a charge $+q$ and final charge on it is $+q_1$. Hence,

$$q_1 = q + q_A$$

or $$q_A = q_1 - q = 60 - 108$$

$$= -48\ \mu C \qquad \textbf{Ans.}$$

Similarly, charge q_B goes to the upper plate of 3 μF capacitor and lower plate of 2 μF capacitor. Initially, both the plates had a charge $+q - q$ or zero. And finally they have a charge $(q_2 - q_1)$. Hence,

$$(q_2 - q_1) = q_B + 0$$

$$\therefore \quad q_B = q_2 - q_1 = 180 - 60$$

$$= 120\ \mu C \qquad \textbf{Ans.}$$

Charge q_C goes to the lower plate of 3 μF capacitor. Initially, it had a charge $-q$ and finally $-q_2$. Hence,

$$-q_2 = (-q) + q_C$$

$$\therefore \quad q_C = q - q_2 = 108 - 180$$

$$= -72\ \mu C \qquad \textbf{Ans.}$$

So, the charges will flow as shown below

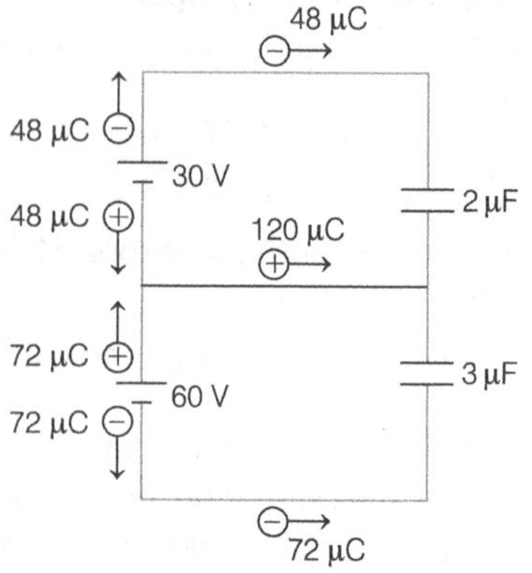

Type 2. *Based on heat generation or loss of energy during shifting of switch.*

Concept

By heat generation (or loss of energy), we mean that when a switch is shifted from one position to the other, what amount of heat will be generated (or loss will be there) in the circuit. Such problems can be solved by simple energy conservation principle. For this, remember that when a charge $+q$ flows from negative terminal to the positive terminal inside a battery of emf V, energy is supplied by the battery which is given by,

$$\boxed{E = qV}$$

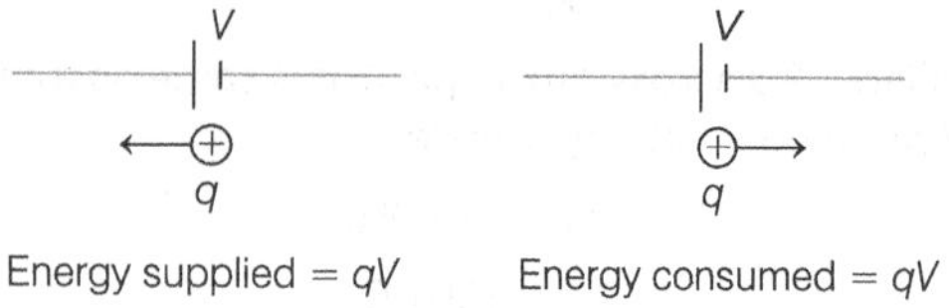

and if opposite is the case, i.e. charge $+q$ flows in opposite direction, then it consumes energy by the same amount.

Now, from energy conservation principle we can find the heat generated (or loss of energy) in the circuit in shifting the switch.

> Heat generated or loss of energy = net energy supplied by the battery/batteries − increase in stored energy in all capacitors $(= \Delta U = \Sigma U_f - \Sigma U_i)$.

Here, ΣU_i = energy stored in all the capacitors initially and

ΣU_f = energy stored in all the capacitors finally

Ex. 2 Find loss of energy in Example 1.

Sol. In Example 1, energy is supplied by 60 V battery and consumed by 30 V battery. Using $E = qV$, we have

$$\text{Energy supplied} = (72 \times 10^{-6})(60) = 4.32 \times 10^{-3}\ \text{J}$$

$$\text{Energy consumed} = (48 \times 10^{-6})(30) = 1.44 \times 10^{-3}\ \text{J}$$

Net energy supplied

$$= (4.32 - 1.44) \times 10^{-3}\ \text{J}$$

$$= 2.88 \times 10^{-3}\ \text{J}$$

$$\Sigma U_i = \frac{1}{2} C_{net} V_{net}^2 = \frac{1}{2} \times \frac{6}{5} \times 10^{-6} \times (90)^2$$

$$= 4.86 \times 10^{-3}\ \text{J}$$

and $$\Sigma U_f = \frac{1}{2} \times 2 \times 10^{-6} \times (30)^2 + \frac{1}{2} \times 3 \times 10^{-6} \times (60)^2$$

$$= 6.3 \times 10^{-3}\ \text{J}$$

$$\Delta U = \Sigma U_f - \Sigma U_i = 1.44 \times 10^{-3}\ \text{J}$$

$$\therefore \text{ Loss of energy} = (2.88 - 1.44) \times 10^{-3}\ \text{J}$$

$$= 1.44 \times 10^{-3}\ \text{J} \qquad \textbf{Ans.}$$

Ex. 3 Prove that, in charging a capacitor half of the energy supplied by the battery is stored in the capacitor and remaining half is lost during charging.

Sol. When switch S is closed, $q = CV$ charge is stored in the capacitor.

Charge transferred from the battery is also q.

Hence,

energy supplied by the battery $= qV = (CV)(V) = CV^2$.

Half of its energy, i.e. $\frac{1}{2}CV^2$ is stored in the capacitor

and the remaining 50 % or $\frac{1}{2}CV^2$ is lost.

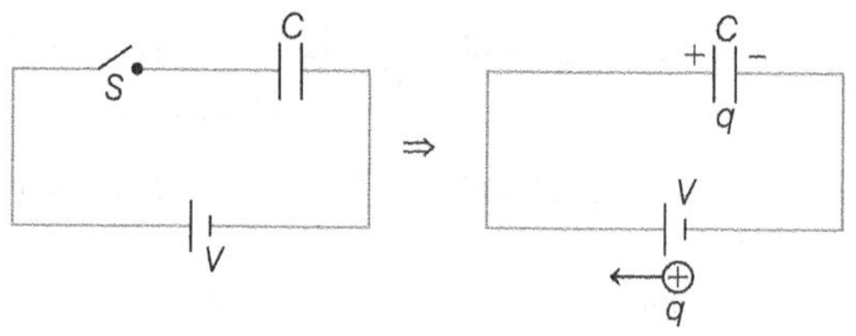

MODULE 6 Exercise

Assertion and Reason

See directions on page 25.

1. **Assertion :** In series combination, charges on two capacitors are always equal.

   Reason : If charges are same, the total potential difference applied across two capacitors will be distributed in inverse ratio of capacitances.

Single Correct Option

2. A capacitor of capacity 2 μF is charged to 100 V. What is the heat generated when this capacitor is connected in parallel to an another capacitor of same capacity?

   (a) 2.5 mJ (b) 5 mJ (c) 10 mJ (d) 4 mJ

3. Two spherical conductors each of capacity C are charged to potential V and $-V$. These are then connected by means of a fine wire. The loss of energy is

   (a) zero (b) $\frac{1}{2}CV^2$
   (c) CV^2 (d) $2\,CV^2$

4. A capacitor of capacity C is charged to a potential difference V and another capacitor of capacity $2C$ is charged to a potential difference $4V$. The charging batteries are disconnected and the two capacitors are connected with reverse polarity (i.e. positive plate of first capacitor is connected to negative plate of second capacitor). The heat produced during the redistribution of charge between the capacitors will be

   (a) $\frac{125CV^2}{3}$ (b) $\frac{50CV^2}{3}$ (c) $2CV^2$ (d) $\frac{25CV^2}{3}$

5. The flow of charge through switch S if it is closed is

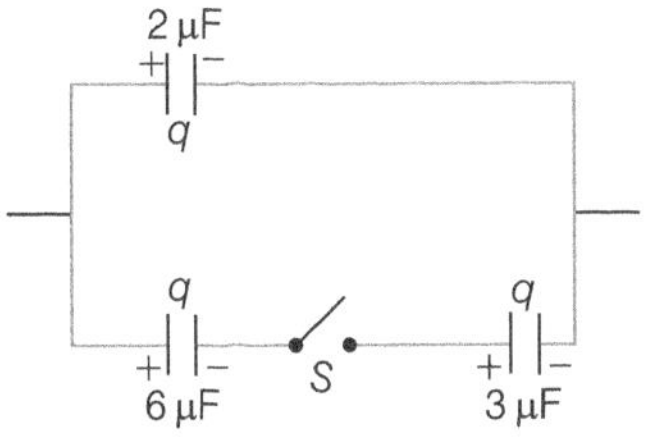

   (a) zero (b) $q/4$ (c) $2q/3$ (d) $q/3$

6. A 2 μF capacitor is charged as shown in the figure. The percentage of its stored energy dissipated after the switch S is turned to position 2 is

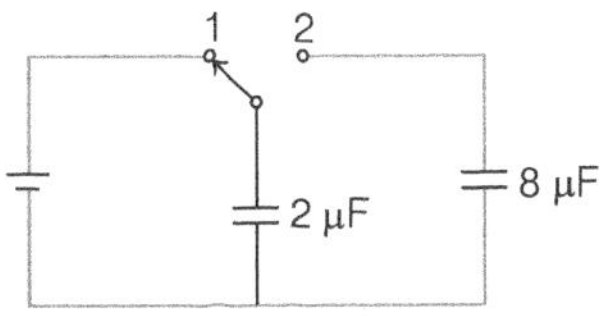

   (a) 0% (b) 20% (c) 75% (d) 80%

7. Four capacitors are connected as shown in figure to a 30 V battery. The potential difference between points a and b is

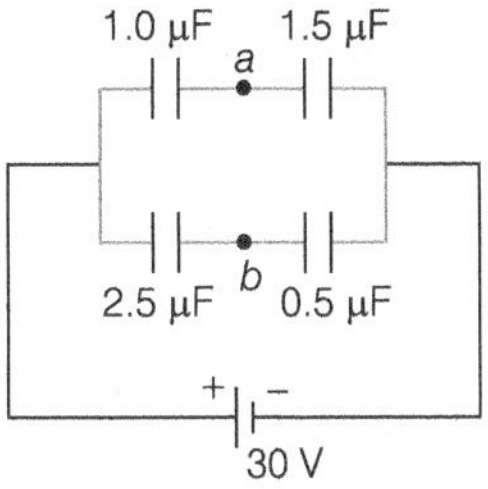

   (a) 5 V (b) 9 V (c) 10 V (d) 13 V

8. The potential difference between points a and b of circuits shown in the figure is

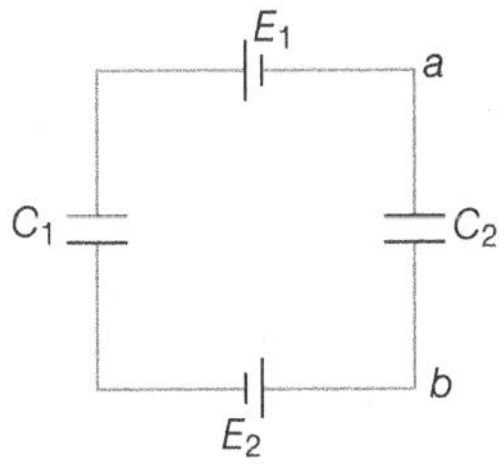

   (a) $\left(\frac{E_1+E_2}{C_1+C_2}\right)C_2$

   (b) $\left(\frac{E_1-E_2}{C_1+C_2}\right)C_2$

   (c) $\left(\frac{E_1+E_2}{C_1+C_2}\right)C_1$

   (d) $\left(\frac{E_1-E_2}{C_1+C_2}\right)C_1$

9. In the given circuit, a charge of + 80 μC is given to the upper plate of the 4 μF capacitor. Then in the steady state, the charge on the upper plate of the 3 μF capacitor is

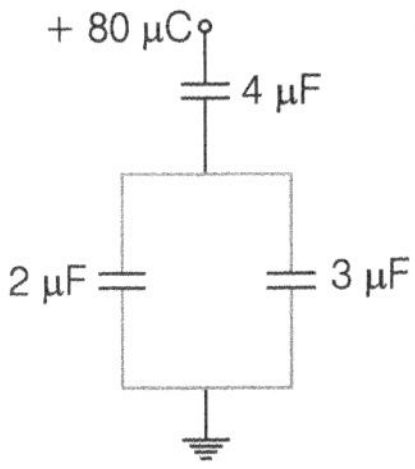

   (a) + 32 μC (b) + 40 μC
   (c) + 48 μC (d) + 80 μC

10. For the circuit shown, which of the following statements is true?

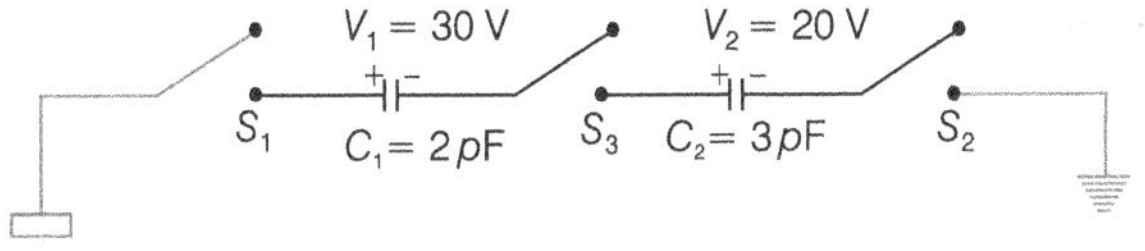

   (a) With S_1 closed, $V_1 = 15\,V, V_2 = 20\,V$
   (b) With S_3 closed, $V_1 = V_2 = 25\,V$
   (c) With S_1 and S_2 closed, $V_1 = V_2 = 0$
   (d) With S_3 closed, $V_1 = 30\,V, V_2 = 20\,V$

Subjective Questions

11. A circuit has section AB as shown in figure. The emf of the source equals $E = 10$ V, the capacitor capacitances are equal to $C_1 = 1.0$ μF and $C_2 = 2.0$ μF, and the potential

difference $V_A - V_B = 5.0$ V. Find the voltage across each capacitor. **(Irodov 3.117)**

12. Find the charges on 6 μF and 4 μF capacitors.

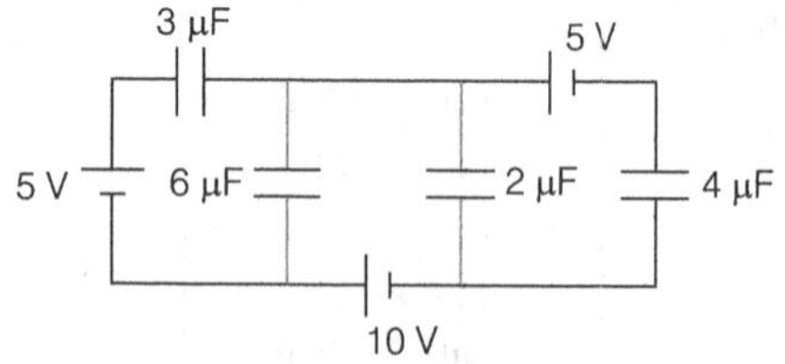

13. In figure, $C_1 = C_5 = 8.4\ \mu F$ and $C_2 = C_3 = C_4 = 4.2\ \mu F$. The applied potential difference is $V_{ab} = 220$ V.

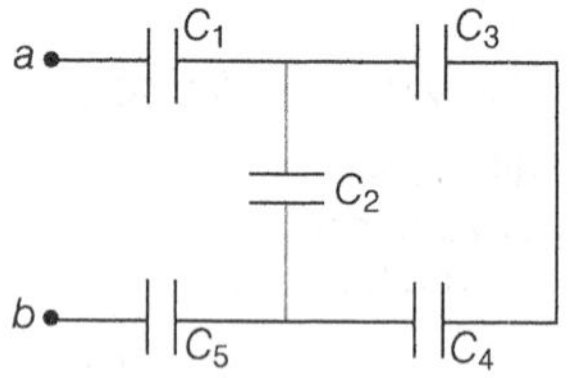

(a) What is the equivalent capacitance of the network between points *a* and *b*?

(b) Calculate the charge on each capacitor and the potential difference across each capacitor.

14. In figure, the battery has a potential difference of 20 V. Find

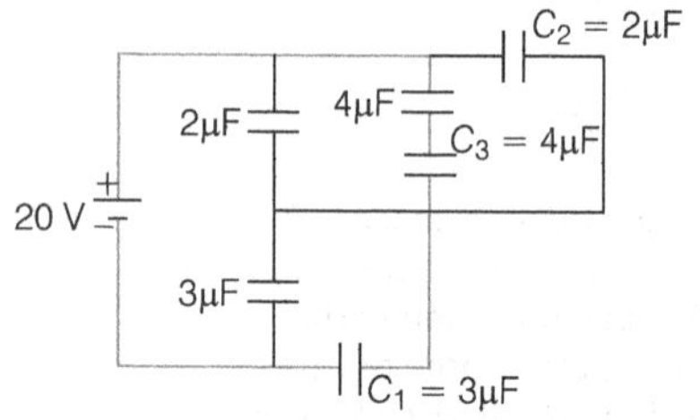

(a) the equivalent capacitance of all the capacitors across the battery and

(b) the charge stored on that equivalent capacitance. Find the charge on

(i) capacitor 1,

(ii) capacitor 2, and

(iii) capacitor 3.

15. In figure, battery *B* supplies 12V. Find the charge on each capacitor

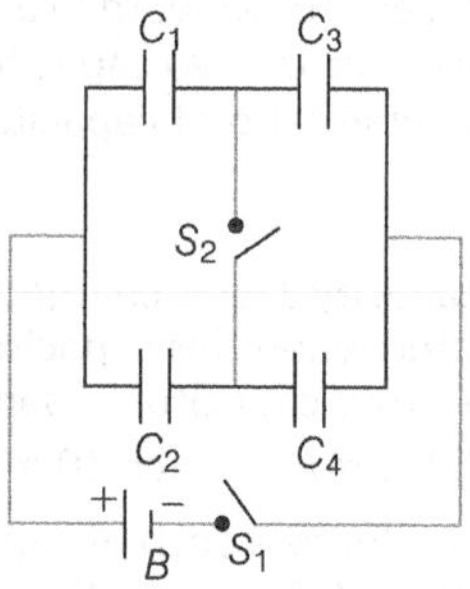

(a) first when only switch S_1 is closed and

(b) later when S_2 is also closed.

(Take $C_1 = 1.0\ \mu F, C_2 = 2.0\ \mu F, C_3 = 3.0\ \mu F$ and $C_4 = 4.0\ \mu F$)

16. When switch *S* is thrown to the left in figure, the plates of capacitor 1 acquire a potential difference V_0. Capacitors 2 and 3 are initially uncharged. The switch is now thrown to the right. What are the final charges q_1, q_2 and q_3 on the capacitors?

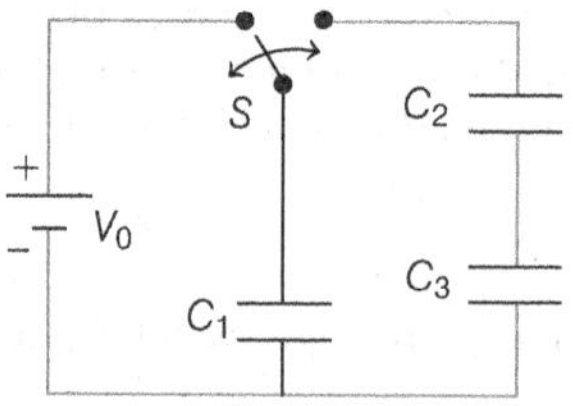

Answers

1. (d) **2.** (b) **3.** (c) **4.** (d) **5.** (a) **6.** (d) **7.** (d) **8.** (c) **9.** (c) **10.** (d)

11. 10 V, 5 V **12.** $10\ \mu C, \dfrac{40}{3}\ \mu C$

13. (a) $2.5\ \mu F$ (b) $Q_1 = 5.5 \times 10^{-4}$ C, $V_1 = 65.4$ V, $Q_2 = 3.7 \times 10^{-4}$ C, $V_2 = 88$ V, $Q_3 = Q_4 = 1.8 \times 10^{-4}$ C, $V_3 = V_4 = 44$ V, $Q_5 = Q_1$, $V_5 = V_1$

14. (a) 3 μF (b) 60 μC (c) 30 μC (d) 20 μC (e) 20 μC

15. (a) $q_1 = q_3 = 9\ \mu C$, $q_2 = q_4 = 16\ \mu C$ (b) $q_1 = 8.4\ \mu C$, $q_2 = 16.8\ \mu C$, $q_3 = 10.8\ \mu C$, $q_4 = 14.4\ \mu C$

16. $q_2 = q_3 = \dfrac{C_1 V_0}{1 + \dfrac{C_1}{C_2} + \dfrac{C_1}{C_3}}$, $q_1 = \dfrac{C_1^2 V_0}{C_1 + \left(\dfrac{C_2 C_3}{C_2 + C_3}\right)}$

See hints on pages 319-321.

MODULE 7

C-R Circuits

Charging of a Capacitor in *C-R* Circuit

To understand the charging of a capacitor in *C-R* circuit, let us first consider the charging of a capacitor without resistance.

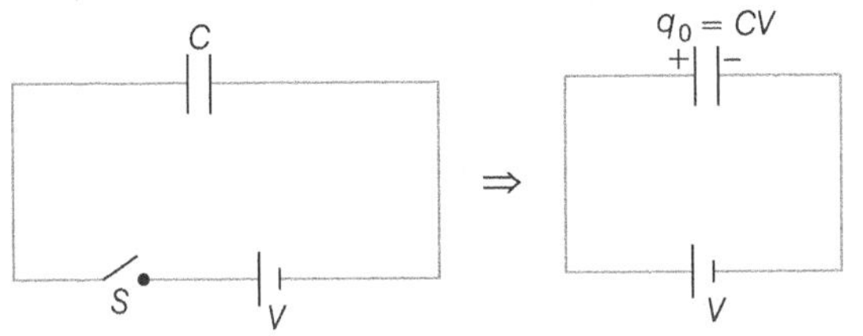

Consider a capacitor connected to a battery of emf V through a switch S. When we close the switch the capacitor gets charged immediately. Charging takes no time. A charge $q_0 = CV$ appears in the capacitor as soon as switch is closed and the *q-t* graph in this case is a straight line parallel to *t*-axis as shown in figure.

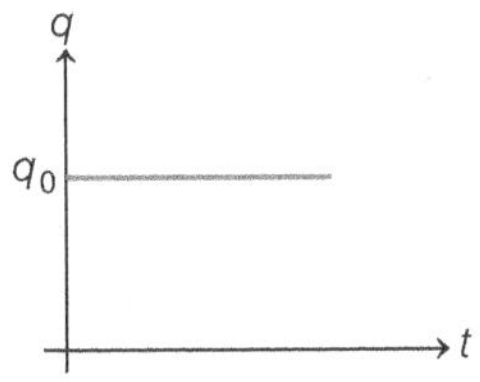

If there is some resistance in the circuit, charging takes some time. Because resistance opposes the charging (or current flow in the circuit). Final charge (called steady state charge) is still q_0 but it is acquired after a long period of time. The *q-t* equation in this case is

$$\boxed{q = q_0\,(1 - e^{-t/\tau_C})}$$

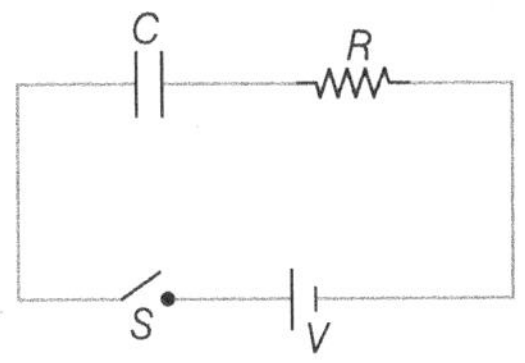

Here, $q_0 = CV$ and $\tau_C = CR =$ time constant.

q-t graph is an exponentially increasing graph. The charge q increases exponentially from 0 to q_0.

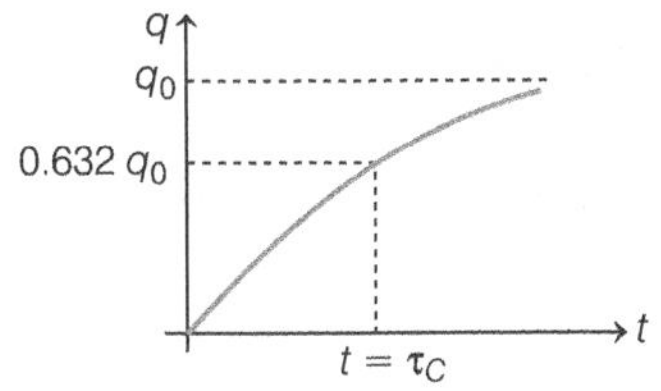

From the graph and equation, we can see that :

at $t = 0$, $q = 0$ and at $t = \infty$, $q = q_0$

Definition of τ_C

At $t = \tau_C$, $q = q_0\,(1 - e^{-1}) \approx 0.632\, q_0$

Hence, τ_C can be defined as the time in which 63.2% charging is over. Note that τ_C is the time.

Hence, $[\tau_C] = [\text{time}]$ or $[CR] = [M^0 L^0 T]$

Proof : Now, let us derive the *q-t* relation discussed above.

Suppose the switch is closed at time $t = 0$. At some instant of time, let charge in the capacitor is $q\,(< q_0)$ and it is still increasing. Therefore current is flowing in the circuit.

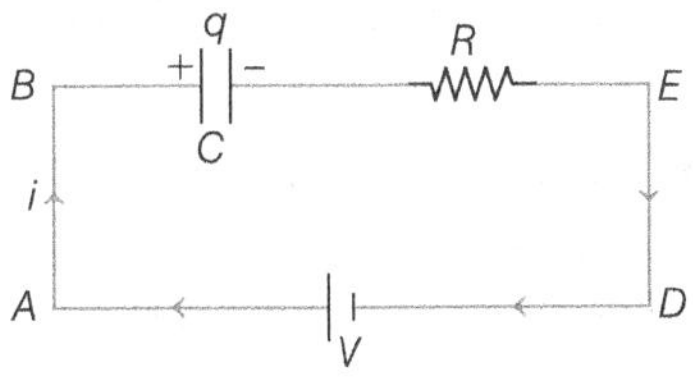

Applying loop law in *ABEDA*, we get

$$-\frac{q}{C} - iR + V = 0$$

Here, $i = \dfrac{dq}{dt}$

$$\therefore \quad -\frac{q}{C} - \left(\frac{dq}{dt}\right)R + V = 0$$

$$\therefore \quad \frac{dq}{V - \frac{q}{C}} = \frac{dt}{R} \quad \text{or} \quad \int_0^q \frac{dq}{V - \frac{q}{C}} = \int_0^t \frac{dt}{R}$$

This gives $\qquad q = CV\,(1 - e^{-\frac{t}{CR}})$

Substituting $CV = q_0$ and $CR = \tau_C$, we have

$$q = q_0\,(1 - e^{-t/\tau_C})$$

Charging Current

Current flows in a C-R circuit during charging of a capacitor. Once charging is over or the steady state condition is reached the current becomes zero. The current at any time t can be calculated by differentiating q with respect to t. Hence,

$$i = \frac{dq}{dt} = \frac{d}{dt}\{q_0 (1 - e^{-t/\tau_C})\}$$

or $$i = \frac{q_0}{\tau_C} e^{-t/\tau_C}$$

Substituting $q_0 = CV$ and $\tau_C = CR$, we have

$$i = \frac{V}{R} e^{-t/\tau_C}$$

By letting, $\frac{V}{R} = i_0$ $\boxed{i = i_0 e^{-t/\tau_C}}$

i.e. current decreases exponentially with time.

The i-t graph is as shown in figure.

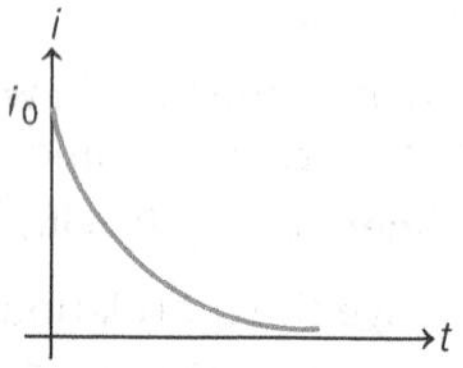

Here, $i_0 = \frac{V}{R}$ is the current at time $t = 0$. This is the current which would had been in the absence of capacitor in the circuit.

Discharging of a Capacitor in C-R Circuit

To understand discharging through a C-R circuit again we first consider the discharging without resistance.

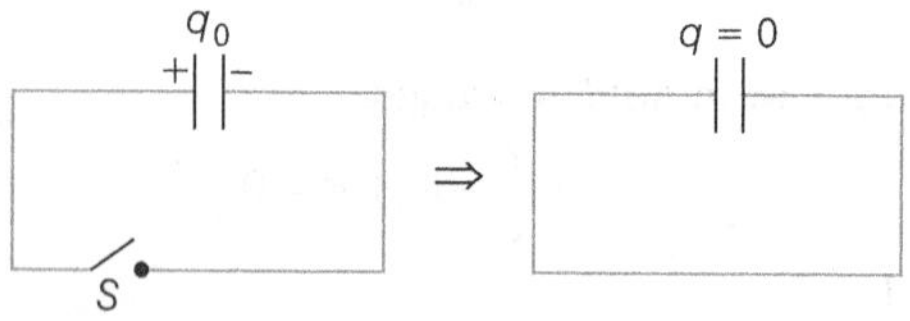

Suppose a capacitor has a charge q_0. The positive plate has a charge $+q_0$ and negative plate $-q_0$. It implies that the positive plate has deficiency of electrons and negative plate has excess of electrons. When the switch is closed, the extra electrons on negative plate immediately rush to the positive plate and net charge on both plates becomes zero. So, we can say that discharging takes place immediately.

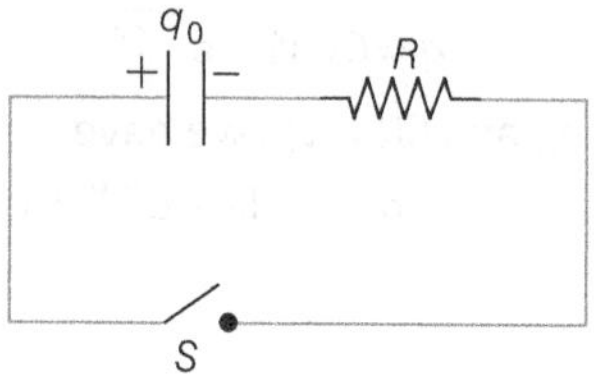

In case of a C-R circuit, discharging also takes time.

Final charge on the capacitor is still zero but after sufficiently long period of time. The q-t equation in this case is

$$\boxed{q = q_0 e^{-t/\tau_C}}$$

Thus, q decreases exponentially from q_0 to zero, as shown in figure.

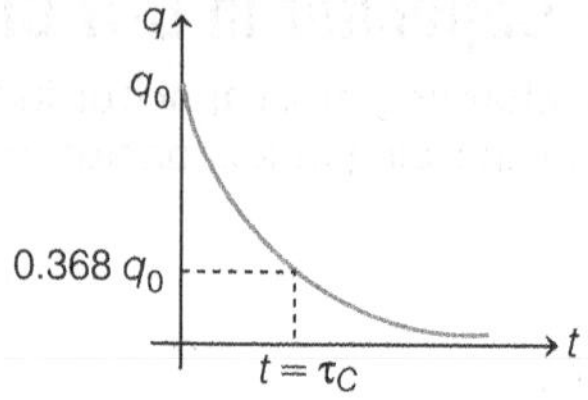

From the graph and the equation, we see that

At $t = 0$, $q = q_0$

At $t = \infty$, $q = 0$.

Definition of Time Constant (τ_C)

In case of discharging, definition of τ_C is changed.

At time $t = \tau_C$,

$$q = q_0 e^{-1} = 0.368\, q_0$$

Hence, in this case τ_C can be defined as the time when charge reduces to 36.8% of its maximum initial value q_0.

Discharging Current

During discharging, current flows in the circuit till q becomes zero. This current can be found by differentiating q with respect to t but with negative sign because charge is decreasing with time. So,

$$i = \left(-\frac{dq}{dt}\right) = -\frac{d}{dt}(q_0 e^{-t/\tau_C})$$

$$= \frac{q_0}{\tau_C} e^{-t/\tau_C}$$

By letting, $\frac{q_0}{\tau_C} = i_0$

We have, $\boxed{i = i_0 e^{-t/\tau_C}}$

This is an exponentially decreasing equation. Thus, i-t graph decreases exponentially with time from i_0 to 0. The i-t graph is as shown in figure.

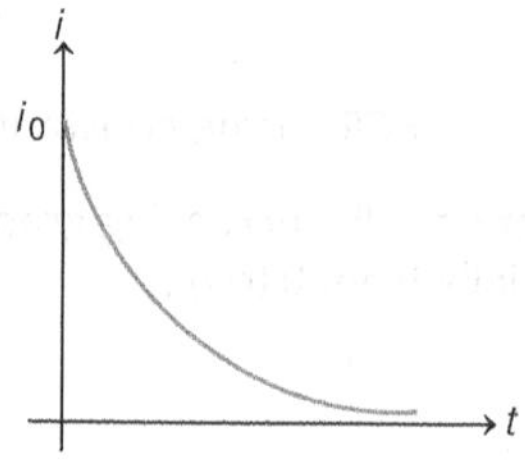

Important POINTS

1. Now, onwards we will come across the following integration very frequently. So, you can remember the direct result.

   If $\int_0^x \frac{dx}{a-bx} = \int_0^t c\,dt$, then $x = \frac{a}{b}(1-e^{-bct})$

   and if $\int_{x_0}^x \frac{dx}{a-bx} = \int_0^t c\,dt$, then $x = \frac{a}{b} - \left(\frac{a}{b} - x_0\right)e^{-bct}$

   Here, a, b and c are constants.

2. Sometimes a physical quantity x decreases from x_1 to x_2, exponentially, then x-t equation is like :

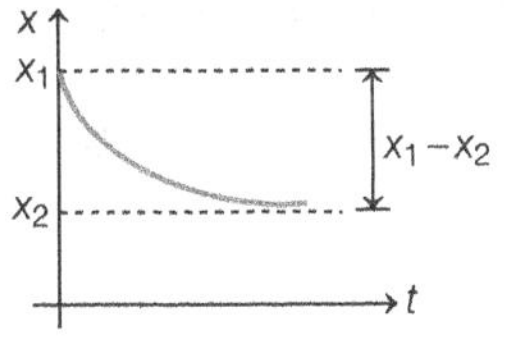

$$x = x_2 + (x_1 - x_2)\,e^{-Kt}$$

   Here, K is a constant.

   Similarly, if x increases from x_2 to x_1 exponentially, then x-t equation is

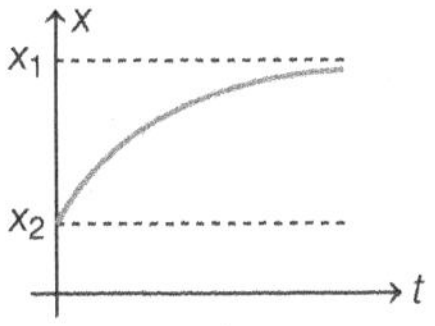

$$x = x_2 + (x_1 - x_2)(1 - e^{-Kt})$$

3. **Leakage Current Through a Capacitor**

   The space between the capacitor's plates is filled with an insulator and we assume that no current flows through it when the capacitor is connected to a battery as in figure (a) or, if the capacitor is charged and battery is removed after charging, the charge on its plates remains forever. But every insulator has some conductivity.

   On account of which some current flows through the capacitor if connected to a battery. This small current is known as the **leakage current**. Similarly, when it is charged and battery is removed after charging, the charge does not remain as it is for a long period of time. But it starts discharging. Or we can say it becomes a case of discharging of a capacitor in C-R circuit.

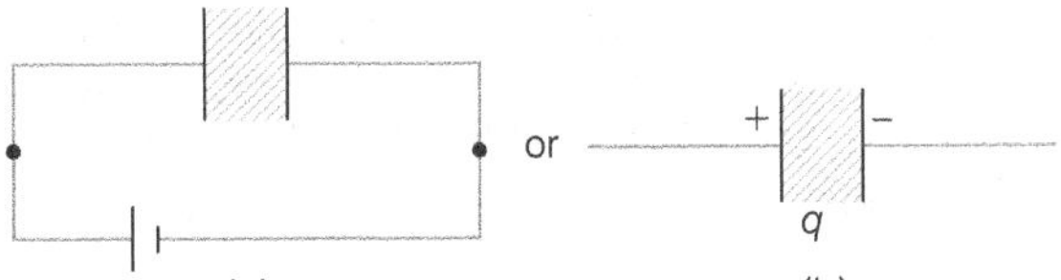

(a) (b)

   In both the cases, we will first find the resistance of the dielectric (or insulator).

$$R = \frac{l}{\sigma A} \qquad (\sigma = \text{specific conductance})$$

   Here, $l = d$

   (the distance between the plates of the capacitor)

$$\therefore \qquad \boxed{R = \frac{d}{\sigma A}}$$

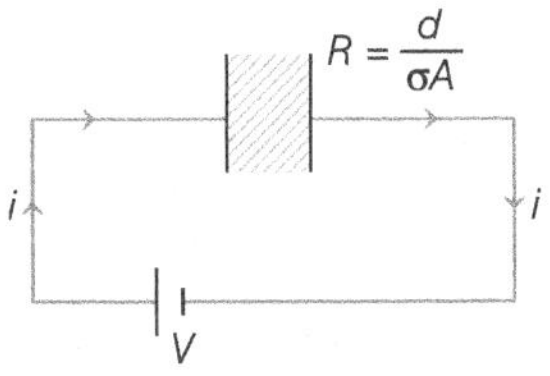

   Thus, the leakage current in the circuit shown in the figure is

$$i = \frac{V}{R}$$

   Similarly, if the capacitor is given a charge q_0 at time $t = 0$, then after time t, q charge will remain on it, where

$$\boxed{q = q_0 e^{-t/\tau_C}} \qquad \text{(Discharging of a capacitor)}$$

   Here, $\tau_C = CR = \left(\frac{K\varepsilon_0 A}{d}\right)\left(\frac{d}{\sigma A}\right)$ or $\boxed{\tau_C = \frac{K\varepsilon_0}{\sigma}}$

4. **Independent parallel circuit**

   Three circuits shown in figure are independently connected in parallel with the battery. Potential difference across each of the circuit is V. By this potential difference, capacitor C_1 is immediately charged. Capacitor C_2 is exponentially charged and current grows immediately in R_3. Thus,

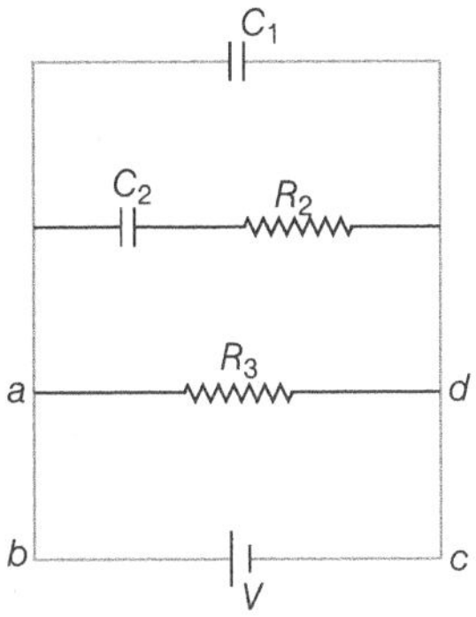

$$q_{C_1} = C_1 V \qquad \text{(immediately)}$$

$$q_{C_2} = C_2 V\left(1 - e^{-\frac{t}{C_2 R_2}}\right)$$

$$I_{R_3} = \frac{V}{R_3} \qquad \text{(immediately)}$$

Note *If any resistance or capacitance is connected between abcd, then this circuit no longer remains an independent parallel circuit.*

Typed Problems

Type 1. ***To find current and hence potential difference between two points in a wire having a capacitor.***

Concept

If charge on capacitor is constant, then current through capacitor wire is zero. If charge is variable, then current is non-zero. Magnitude of this current is

$$i = \left|\frac{dq}{dt}\right|$$

and direction of this current is towards the positive plate if charge is increasing and away from the positive plate if charge is decreasing.

Ex. 1 In the circuit shown in figure, find V_{ab} at 1s .

Sol. Charge ($q = 2t$) on capacitor is increasing. So, there is a current in the circuit from right to left. This current is given by

$$i = \frac{dq}{dt} = 2\text{ A}$$

At 1 s, $q = 2$ C.

So, at 1 s, circuit is as shown in figure.

$$V_a + \frac{2}{2} + (2)(4) + 10 = V_b$$

$\therefore$ $V_a - V_b = -19$ volt

or $V_{ab} = -19$ volt **Ans.**

Type 2. ***To find charge on different capacitors in a C-R circuit.***

Concept

In a C-R circuit, charge on different capacitors is normally asked either at $t = 0$, $t = \infty$ or $t = t$. If nothing is given in the question, then we have to find charges on capacitors at $t = \infty$ or steady state charges.

In steady state, no current flows through a wire having capacitor. But, if there is any other closed circuit then current can flow through that circuit. So, first find this current and then steady state potential difference (say V_0) across two plates of capacitor. Now,

$$q_0 = CV_0 = \text{steady state charge}$$

Ex. 2 Find potential difference across the capacitor.

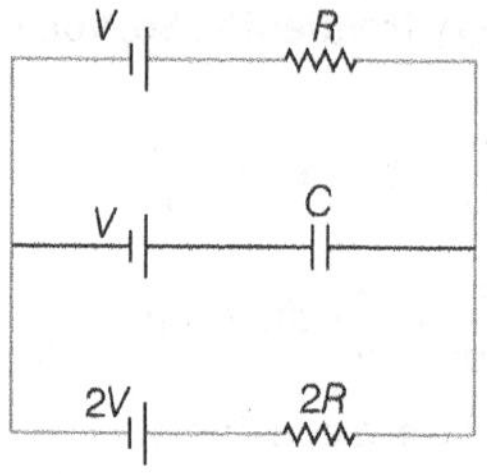

Sol. In steady state condition, no current will flow through the capacitor C. Current in the outer circuit,

$$i = \frac{2V - V}{2R + R} = \frac{V}{3R}$$

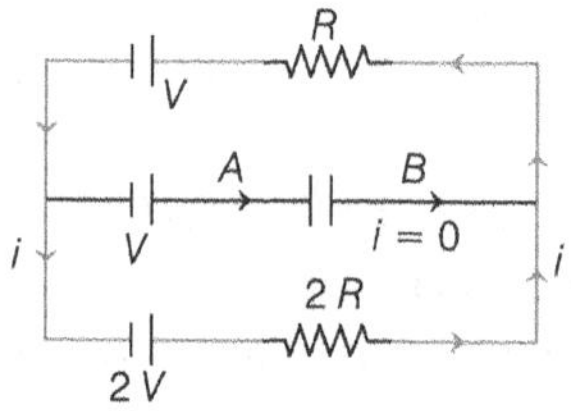

Potential difference between A and B,

$$V_A - V + V + iR = V_B$$

$$\therefore \quad V_B - V_A = V_C = iR = \left(\frac{V}{3R}\right)R = \frac{V}{3}$$ **Ans.**

Note *In this problem, charge stored in the capacitor can also be asked, which is equal to $q = C\frac{V}{3}$ with positive charge on B side and negative on A side because $V_B > V_A$.*

Ex. 3 Find the charge stored in the capacitor.

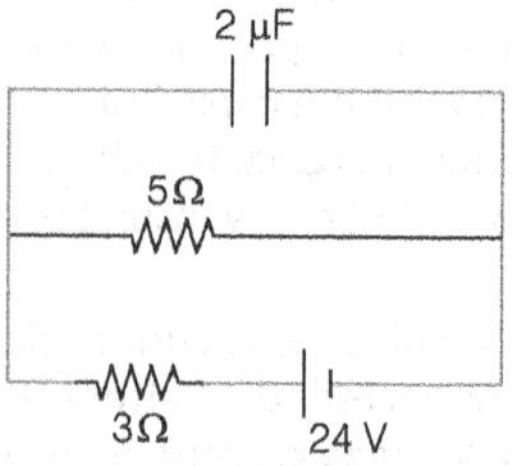

HOW TO PROCEED Some insulator is filled between the plates of the capacitor. Therefore, a capacitor does not allow current flow through it after charging is over. Hence, in the circuit current will flow through 3 Ω and 5 Ω resistances and it will not flow through the capacitor. To find the charge stored in the capacitor, we need the PD across it. So, first we will find PD across the capacitor and then apply,

$$q = CV$$

where, V = PD across the capacitor.

Sol. Current will flow in loop $ABCDA$ when charging is over and this current is

$$i = \frac{\text{Net emf}}{\text{Total resistance}} = \frac{24}{5+3} = 3\text{ A}$$

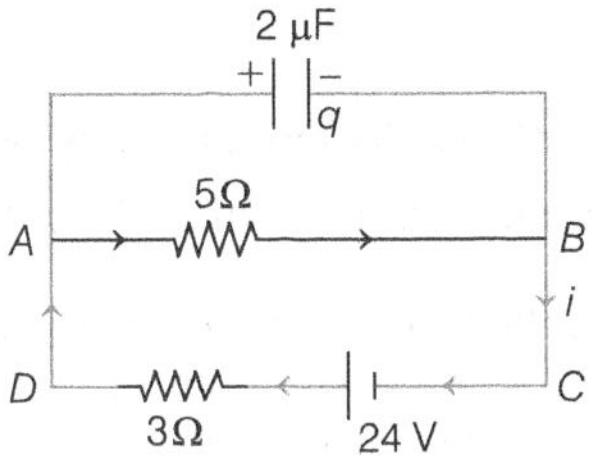

Now, PD across the capacitor is equal to the PD across the $5\,\Omega$ resistance.

Hence, $V = V_A - V_B = iR = (3)(5) = 15\text{ V}$

$\therefore$ $q = CV = (2 \times 15)\,\mu\text{C} = 30\,\mu\text{C}$ **Ans.**

Note *$V_A - V_B = 15$ V, therefore $V_A > V_B$, i.e. the positive charge will be collected on the left plate of the capacitor and negative on the right plate.*

Type 3. *Shortcut method of finding time varying functions in a C-R circuit like q or i etc.*

Concept

(i) At time $t = 0$, when capacitor is uncharged it allows maximum current to pass through it. So, it may be assumed like a conducting wire of zero resistance. With this concept, find initial values of q or i etc.

(ii) At time $t = \infty$, when capacitor is fully charged it does not allow current through it, as insulator is filled between the plates. So, its resistance may be assumed as infinite. With this concept, find steady state values of q or i at time $t = \infty$.

(iii) **Equivalent time constant** To find the equivalent time constant of a circuit, the following steps are followed :

(a) Short-circuit the battery.

(b) Find net resistance across the capacitor (suppose it is R_{net})

(c) $\tau_C = (R_{net})C$

(iv) In C-R circuit, increase or decrease of any value is always exponential. So, first make exponential graph and then write exponential equation corresponding to this graph with the time constant obtained by the method discussed above.

Ex. 4 Switch S is closed at time $t = 0$ in the circuit shown in figure.

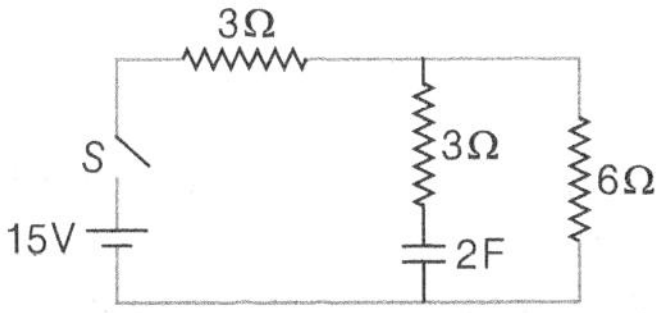

(a) Find the time varying quantities in the circuit.

(b) Find their values at time $t = 0$.

(c) Find their values at time $t = \infty$

(d) Find time constant of all time varying functions.

(e) Make their exponential graphs and write their exponential equations.

(f) Write the equations to find different time varying functions.

Sol. (a)

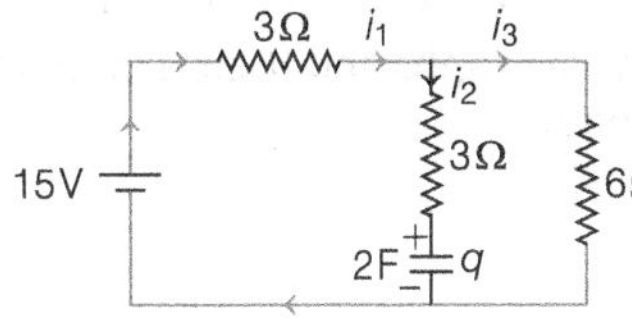

There are four times variable functions i_1, i_2, i_3 and q.

(b) At $t = 0$, equivalent resistance of capacitor is zero. So, the simple circuit is as shown below :

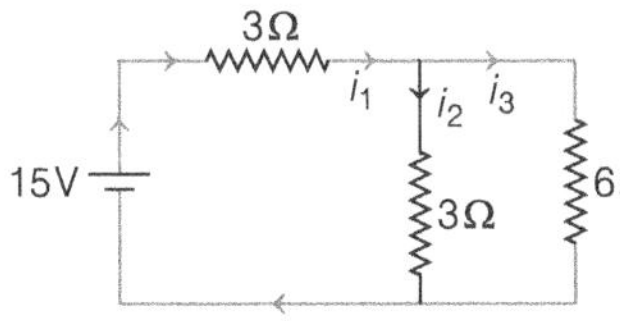

$$R_{net} = 3 + \frac{3 \times 6}{3+6} = 5\Omega$$

$$i_1 = \frac{15}{5} = 3\text{ A}$$

$$\Rightarrow \quad \frac{i_2}{i_3} = \frac{6}{3} = \frac{2}{1}$$

$$\therefore \quad i_2 = \frac{2}{2+1}(3\text{A}) = 2\text{A}$$

$$i_3 = \left(\frac{1}{2+1}\right)(3\text{A}) = 1\text{A}$$

and $q = 0$

(c) At $t = \infty$, equivalent resistance of capacitor is infinite. So, equivalent circuit is as shown below :

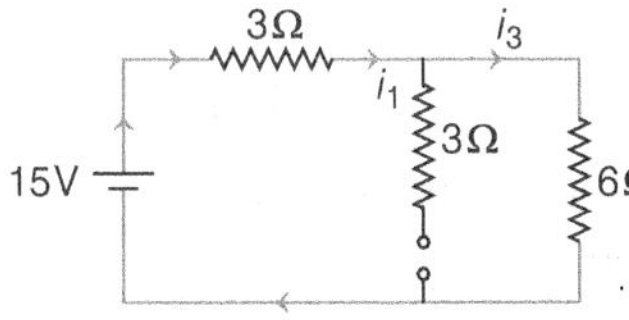

$$i_1 = i_3 = \frac{15}{3+6} = \frac{5}{3}\text{ A}$$

$$i_2 = 0$$

$$V_{2F} = V_{6\Omega} = iR = \left(\frac{5}{3}\right)(6) = 10\text{ volt}$$

$$\therefore \quad q = CV = (2)(10) = 20\text{ C}$$

(d) By short-circuiting the battery, the simplified circuit is as shown below

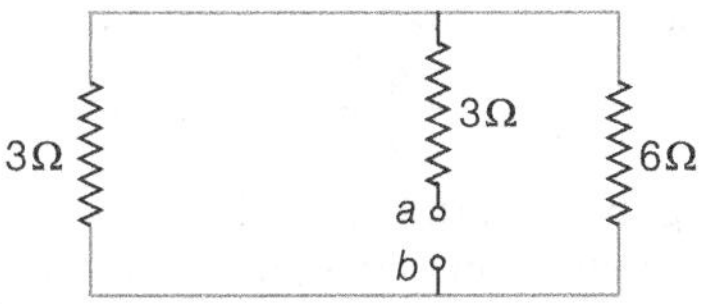

Net resistance across capacitor or ab is

$$R_{net} = 3 + \frac{3 \times 6}{3+6} = 5\Omega$$

$$\therefore \qquad \tau_C = CR_{net} = (2)(5) = 10 \text{ s}$$

(e) Exponential graphs and their exponential equations are as under.

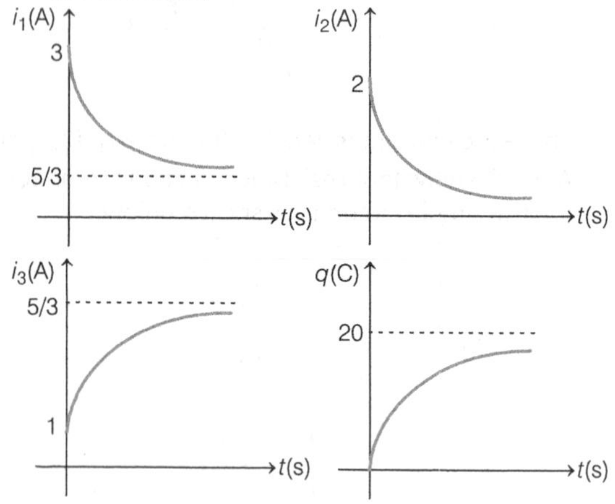

$$i_1 = \frac{5}{3} + \left(3 - \frac{5}{3}\right) e^{\frac{-t}{\tau_C}}$$

$$= \frac{5}{3} + \frac{4}{3} e^{-\frac{t}{10}}$$

$$i_2 = 2e^{-\frac{t}{\tau_C}} = 2e^{-\frac{t}{10}}$$

$$i_3 = 1 + \left(\frac{5}{3} - 1\right)(1 - e^{-\frac{t}{\tau_C}})$$

$$= 1 + \frac{2}{3}(1 - e^{-\frac{t}{10}})$$

$$q = 20(1 - e^{-\frac{t}{\tau_C}}) = 20(1 - e^{-\frac{t}{10}})$$

Note *In the above equations we can see that at any time t, $i_1 = i_2 + i_3$ and $i_2 = \frac{dq}{dt}$.*

(f) Unknowns are four: i_1, i_2, i_3 and q. So, corresponding to the figure of part (a), four equations are

$$i_1 = i_2 + i_3 \qquad \text{...(i)}$$

$$i_2 = \frac{dq}{dt} \qquad \text{...(ii)}$$

Applying loop equation in left hand side loop,

$$+15 - 3i_1 - 3i_2 - \frac{q}{2} = 0 \qquad \text{...(iii)}$$

Applying loop equation in right hand side loop,

$$+\frac{q}{2} + 3i_2 - 6i_3 = 0 \qquad \text{...(iv)}$$

Solving these equations (with some integration), we can find same time functions as we have obtained in part (e).

Type 4. ***To find final charges on different capacitors and loss of energy when position of a switch is changed (opened, closed or shifted from one position to other position).***

Concept

(i) To find current in a C-R circuit at any time t, a capacitor may be assumed a battery of emf E or $V = \frac{q}{C}$.

(ii) Difference between a normal battery and a capacitor battery is : emf of a normal battery remains constant while emf of a capacitor battery keeps on changing with q.

(iii) Before changing the position of switch, every loop of the circuit may not be balanced by Kirchhoff's second equation of potential. So, in a single loop problem rotate a charge q, either clockwise or anti-clockwise. From this charge q, some of the charges on capacitors may increase and others may decrease. To check this, always concentrate on positive plate of each capacitors. If positive charge comes towards this plate, then charge on this capacitor will increase.

(iv) With these charges, apply Kirchhoff's loop equation and find the final charges on them.

(v) In this redistribution of charges, there is some loss of energy as discussed previously.

(vi) If there are only capacitors in the circuit, then redistribution of charges is immediate and if there are resistors in the circuit, then redistribution is exponential.

(vii) If there are only capacitors, then loss is in the form of electromagnetic waves and if there are resistors in the circuit, then loss is in the form of heat.

Further, this loss is proportional to R if resistors are in series ($H = i^2Rt$ or $H \propto R$ as i is same in series) and this loss is inversely proportional to R if resistors are in parallel ($H = \frac{V^2t}{R}$ or $H \propto \frac{1}{R}$ as V is same in parallel).

Ex. 5 In the circuit shown in figure, switch S is closed at time $t = 0$. Find

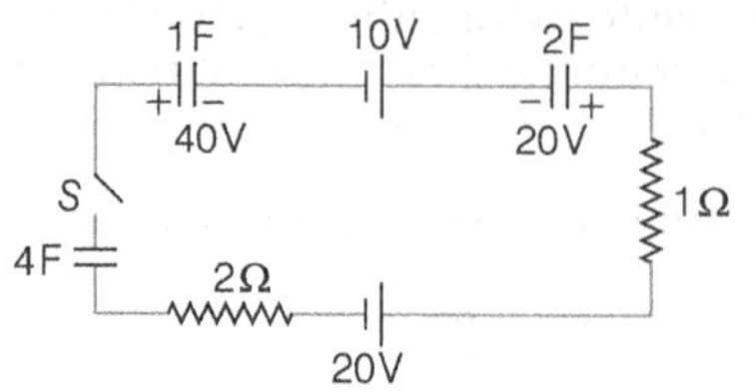

(a) Initial current at $t = 0$ and final current at $t = \infty$ in the loop.

(b) Total charge q flown from the switch.

(c) Final charges on capacitors in steady state at time $t = \infty$.

(d) Loss of energy during redistribution of charges.

(e) Individual loss across 1Ω and 2Ω resistance.

Sol. (a) **At $t = 0$**

Three capacitors may be assumed like batteries of emf 40 V, 20 V and 0 V.

$$\therefore \quad i = \frac{\text{Net emf}}{\text{Total resistance}} = \frac{40 + 20 - 10 - 20}{1 + 2}$$

$$= 10 \text{ A} \qquad \text{(anti-clockwise)}$$

At $t = \infty$

When charge redistribution is complete and loop is balanced by Kirchhoff's second equation of potential, current in the loop becomes zero, as insulator is filled between the capacitors.

(b) Redistribution current was anti-clockwise. So, we can assume that $+q$ charge rotates anti-clockwise in the loop (between time $t = 0$ and $t = \infty$). After this rotation of charges, final charges on different capacitors are as shown below.

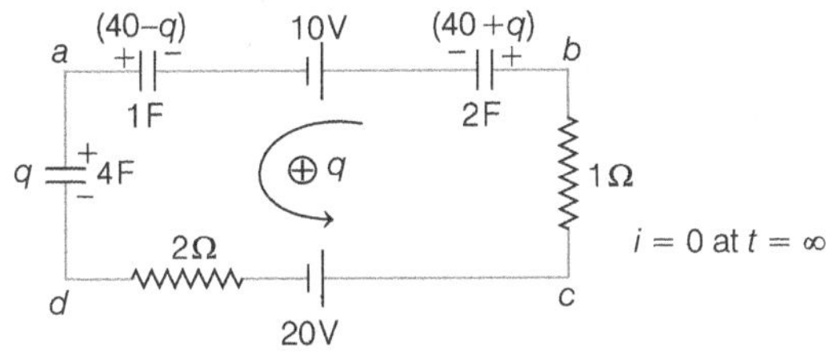

Applying loop equation in loop *abcda*,

$$-\frac{(40 - q)}{1} + 10 + \frac{(40 + q)}{2} - 20 + \frac{q}{4} = 0$$

Solving this equation, we get

$$q = \frac{120}{7} \text{ C} \qquad \textbf{Ans.}$$

(c) **Final charges**

$$q_{1F} = 40 - q = 40 - \frac{120}{7} = \frac{160}{7} \text{ C}$$

$$q_{2F} = 40 + q = 40 + \frac{120}{7} = \frac{400}{7} \text{ C}$$

$$q_{4F} = q = \frac{120}{7} \text{ C}$$

(d) **Total loss of energy during redistribution**

$$\Sigma U_i = \frac{1}{2} \times (1)(40)^2 + \frac{1}{2} \times (2)(20)^2 = 1200 \text{ J}$$

$$\Sigma U_f = \frac{1}{2} \times \frac{(160/7)^2}{1} + \frac{1}{2} \times \frac{(400/7)^2}{2} + \frac{1}{2} \times \frac{(120/7)^2}{4}$$

$$= 1114.3 \text{ J}$$

$$\Delta U = \Sigma U_f - \Sigma U_i = -85.7 \text{ J}$$

20 V battery will supply energy but 10 V battery will consume energy. So,

$$\text{Total energy supplied} = 20 \times \frac{120}{7} - 10 \times \frac{120}{7}$$

$$= 171.4 \text{ J}$$

Total heat produced = Energy supplied $- \Delta U$

$$= 171.4 - (-85.7) = 257.1 \text{ J} \qquad \textbf{Ans.}$$

(e) Resistors are in series. Hence,

$$H \propto R \quad \text{or} \quad \frac{H_1}{H_2} = \frac{R_1}{R_2} = \frac{1\Omega}{2\Omega} = \frac{1}{2}$$

$$\therefore \quad H_1 = \left(\frac{1}{1+2}\right)(257.1) = 85.7 \text{ J} \qquad \textbf{Ans.}$$

$$H_2 = \left(\frac{2}{1+2}\right)(257.1) = 171.4 \text{ J} \qquad \textbf{Ans.}$$

Note *(i) For making the calculations simple, we have taken capacities in Farad, otherwise Farad is a large unit.*

(ii) For two loop problems, we will rotate two charges q_1 and q_2.

Miscellaneous Examples

Ex. 6 In the circuit shown in figure switch *S* is closed at time $t = 0$. Find the current through different wires and charge stored on the capacitor at any time *t*.

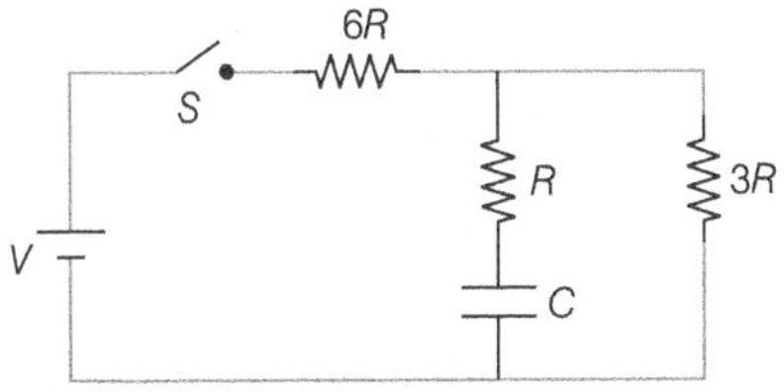

Sol. Calculation of τ_C

Equivalent resistance across capacitor after short-circuiting the battery is

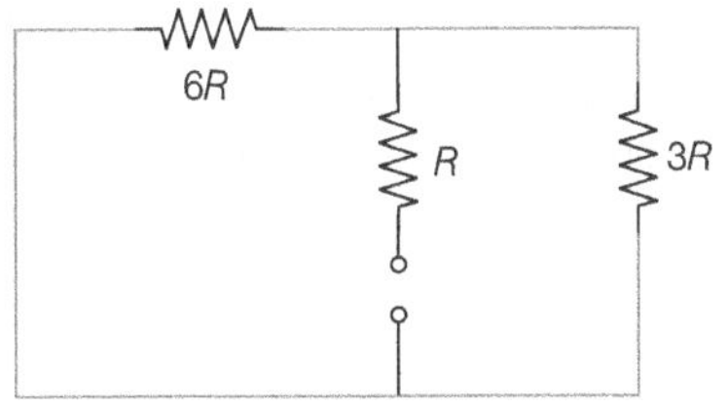

$$R_{net} = R + \frac{(6R)(3R)}{6R + 3R} = 3R$$

$$\therefore \quad \tau_C = (C)R_{net} = 3RC$$

Calculation of steady state charge q_0

At $t = \infty$, capacitor is fully charged and no current flows through it.

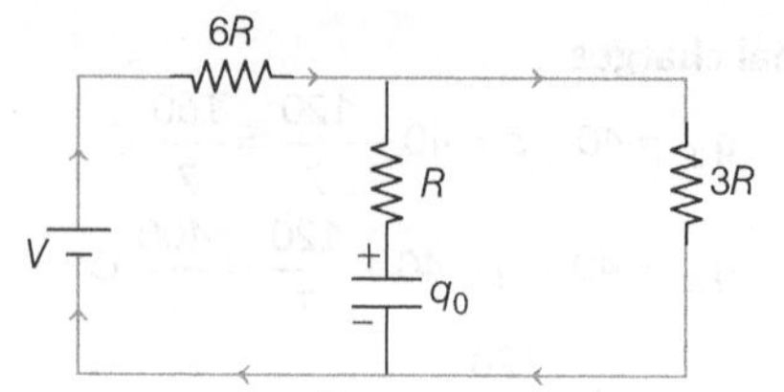

PD across capacitor = PD across 3R

$$= i\,(3R) = \left(\frac{V}{9R}\right)(3R) = \frac{V}{3}$$

$\therefore$ $$q_0 = \frac{CV}{3}$$

Now, let charge on the capacitor at any time t be q and current through it is i_1. Then,

$$q = q_0\,(1 - e^{-t/\tau_C})$$

and $$i_1 = \frac{dq}{dt} = \frac{q_0}{\tau_C}\, e^{-t/\tau_C} \qquad \text{...(i)}$$

Applying Kirchhoff's second law in loop *ACDFA*, we have

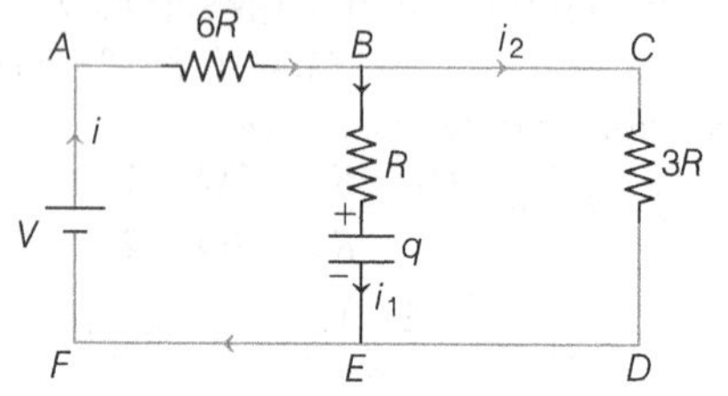

$$-6iR - 3i_2R + V = 0$$

or $$2i + i_2 = \frac{V}{3R} \qquad \text{...(ii)}$$

Applying Kirchhoff's junction law at *B*, we have

$$i = i_1 + i_2 \qquad \text{...(iii)}$$

Solving Eqs. (i), (ii) and (iii), we have

$$i_2 = \frac{V}{9R} - \frac{2}{3} i_1 = \frac{V}{9R} - \frac{2q_0}{3\tau_C}\, e^{-t/\tau_C}$$

and $$i = \frac{V}{9R} + \frac{q_0}{3\tau_C}\, e^{-t/\tau_C}$$

Ex. 7 In the circuit shown in figure, find the steady state charges on both the capacitors.

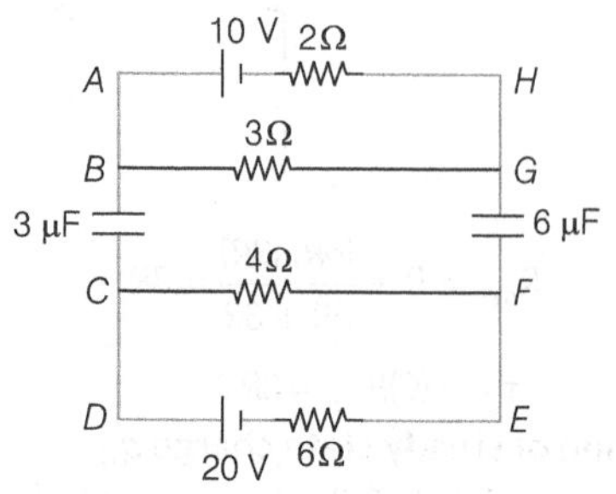

HOW TO PROCEED In steady state a capacitor offers an infinite resistance. Therefore, the two circuits *ABGHA* and *CDEFC* have no relation with each other. Hence, the battery of emf 10 V is not going to contribute any current in the lower circuit. Similarly, the battery of emf 20 V will not contribute any current in the upper circuit. So, first we will calculate the current in the two circuits, then find the potential difference V_{BG} and V_{CF} and finally we can connect two batteries of emf V_{BG} and V_{CF} across the capacitors to find the charges stored in them.

Sol. Current in the upper circuit, $i_1 = \frac{10}{3+2} = 2\text{ A}$

$$V_{BG} = V_B - V_G = 3i_1 = 3 \times 2 = 6\text{ V}$$

Current in the lower circuit, $i_2 = \frac{20}{4+6} = 2\text{ A}$

$\therefore$ $$V_{CF} = V_C - V_F = 4i_2 = 4 \times 2 = 8\text{ V}$$

Charge on both the capacitors will be same. Let it be q. Applying Kirchhoff's second law in loop *BGFCB*,

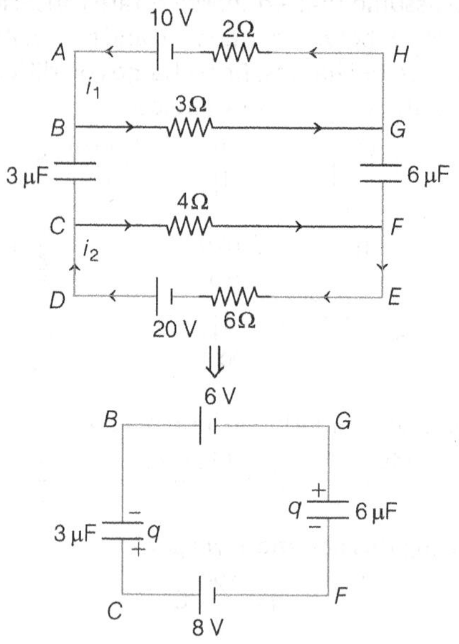

$$-6 - \frac{q}{6 \times 10^{-6}} + 8 - \frac{q}{3 \times 10^{-6}} = 0$$

or $\frac{(10^6)\,q}{2} = 2$ or $q = 4 \times 10^{-6}\text{ C}$

or $q = 4\,\mu\text{C}$ **Ans.**

Alternate solution

$$V_{net} = 8 - 6 = 2\text{V}$$

$$C_{net} = \frac{3 \times 6}{3+6} = 2\,\mu\text{F}$$

$\therefore$ $$q = C_{net} V_{net} = 4\,\mu\text{C}$$

Ex. 8 An isolated parallel plate capacitor has circular plates of radius 4.0 cm. If the gap is filled with a partially conducting material of dielectric constant *K* and conductivity $5.0 \times 10^{-14}\ \Omega^{-1}\text{m}^{-1}$. When the capacitor is charged to a surface charge density of $15\,\mu\text{C/cm}^2$, the initial current between the plates is $1.0\,\mu\text{A}$?

(a) Determine the value of dielectric constant K.
(b) If the total heat produced is 7500 J, determine the separation of the capacitor plates.

Sol. (a) This is basically a problem of discharging of a capacitor from inside the capacitor. Charge at any time t is

$$q = q_0 e^{-t/\tau_C}$$

Here, q_0 = (area of plates) (surface charge density) and discharging current,

$$i = \left(\frac{-dq}{dt}\right) = \frac{q_0}{\tau_C} \cdot e^{-t/\tau_C} = i_0 e^{-t/\tau_C}$$

Here, $i_0 = \frac{q_0}{\tau_C} = \frac{q_0}{CR}$

$$C = \frac{K\varepsilon_0 A}{d}$$

and $R = \frac{d}{\sigma A}$

$\therefore$ $CR = \frac{K\varepsilon_0}{\sigma}$

Therefore, $i_0 = \frac{q_0}{\frac{K\varepsilon_0}{\sigma}} = \frac{\sigma q_0}{K\varepsilon_0}$

$\Rightarrow$ $K = \frac{\sigma q_0}{i_0 \varepsilon_0}$

Substituting the values, we have

$$K = \frac{(5.0 \times 10^{-14})(\pi)(4.0)^2 (15 \times 10^{-6})}{(1.0 \times 10^{-6})(8.86 \times 10^{-12})}$$

$= 4.25$ **Ans.**

(b) $U = \frac{1}{2}\frac{q_0^2}{C} = \frac{1}{2}\frac{q_0^2}{\frac{K\varepsilon_0 A}{d}}$

$\therefore$ $d = \frac{2K\varepsilon_0 AU}{q_0^2}$

$$= \frac{2 \times 4.25 \times 8.86 \times 10^{-12} \times \pi \times (4.0 \times 10^{-2})^2 \times 7500}{(15 \times 10^{-6} \times \pi \times 4.0 \times 4.0)^2}$$

$= 5.0 \times 10^{-3}$ m

$= 5.0$ mm **Ans.**

MODULE 7 Exercise

Assertion and Reason

See directions on page 25.

1. **Assertion :** Discharging graphs of two C-R circuits having the same value of C is shown in figure. From the graph we can say that $\tau_{C_1} > \tau_{C_2}$.

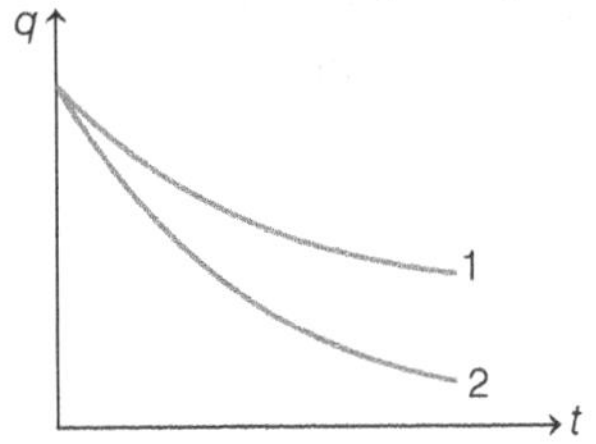

Reason : $R_1 > R_2$.

2. **Assertion :** Two capacitors are charged from the same battery and then connected as shown. A current will flow in anti-clockwise direction as soon as switch is closed.

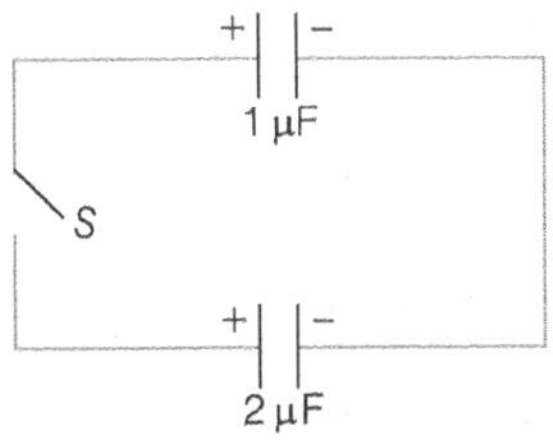

Reason : In steady state charges on two capacitors are in the ratio 1 : 2.

3. **Assertion :** In the circuit shown in figure no charge will be stored in the capacitor.

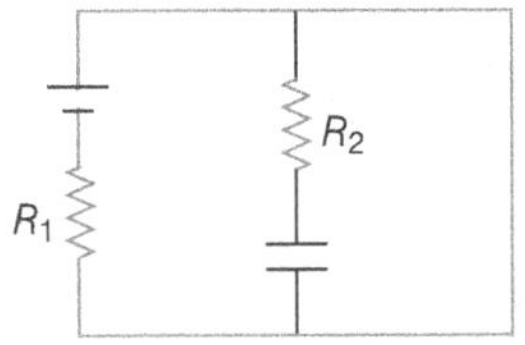

Reason : Current through R_2 will be zero.

Single Correct Option

4. When the switch is closed, the initial current through the 1 Ω resistor is

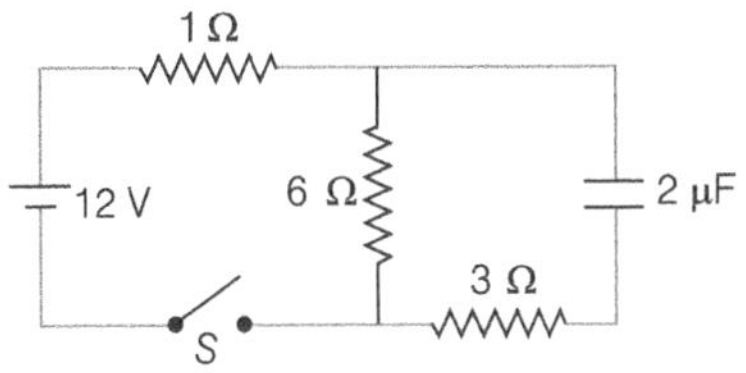

(a) 2 A
(b) 4 A
(c) 3 A
(d) 6 A

5. In the circuit, the potential difference across the capacitor is 10 V. Each resistance is of 3 Ω. The cell is ideal. The emf of the cell is

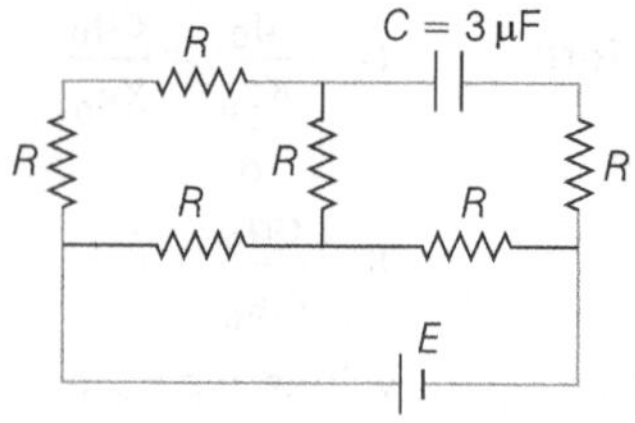

(a) 14 V (b) 16 V (c) 18 V (d) 24 V

6. The potential difference $V_A - V_B$ between points *A* and *B* for the circuit segment shown in figure at the given instant is

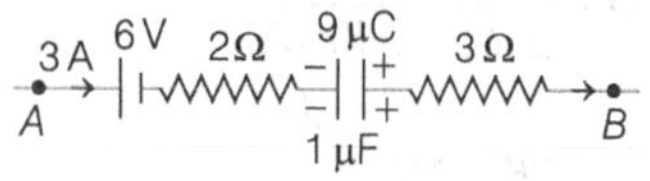

(a) 12 V (b) – 12 V (c) 6 V (d) – 6 V

7. For the circuit arrangement shown in figure, in the steady state condition charge on the capacitor is

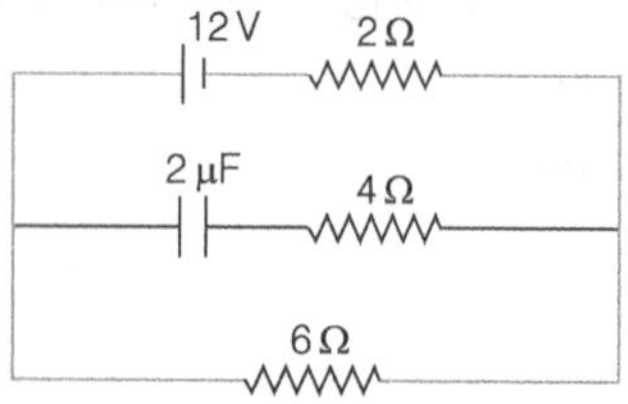

(a) 12 μC (b) 14 μC (c) 20 μC (d) 18 μC

8. Charge *Q* on a capacitor varies with voltage *V* as shown in the figure. *Q* is taken along the *X*-axis and *V* along the *Y*-axis. The area of triangle *OAB* represents

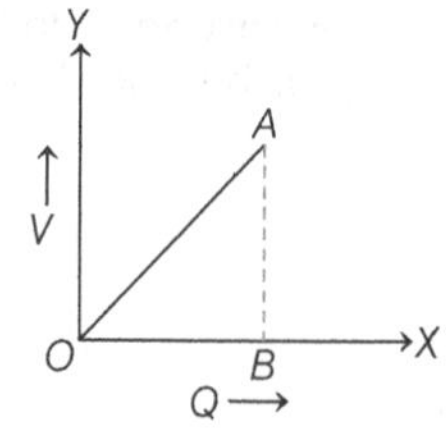

(a) capacitance
(b) time constant
(c) electric field between the plates
(d) energy stored in the capacitor

9. In the circuit shown in figure, the capacitors are initially uncharged. The current through resistor *PQ* just after closing the switch is

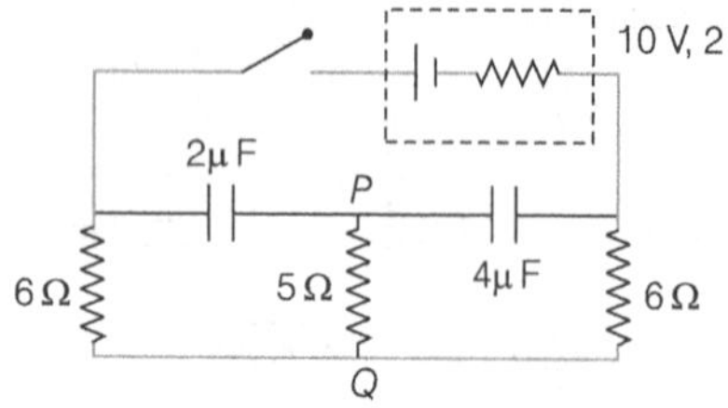

(a) 2 A from *P* to *Q* (b) 2 A from *Q* to *P*
(c) 6 A from *P* to *Q* (d) zero

10. A graph between log *I* and *t* during charging of a capacitor by a battery in series with a resistor is shown. The graphs are drawn for two circuits. R_1, R_2, C_1, C_2 and V_1, V_2 are the values of resistance, capacitance and EMF of the cell in the two circuits. If only two parameters (out of resistance, capacitance, EMF) are different in the two circuits. What may be the correct option(s)?

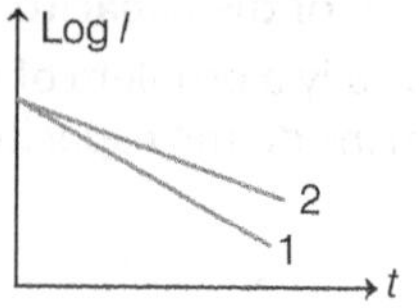

(a) $V_1 = V_2, R_1 > R_2, C_1 > C_2$ (b) $V_1 > V_2, R_1 > R_2, C_1 = C_2$
(c) $V_1 < V_2, R_1 < R_2, C_1 = C_2$ (d) $V_1 < V_2, R_1 = R_2, C_1 < C_2$

11. For the circuit shown in the figure, determine the charge on capacitor in steady state.

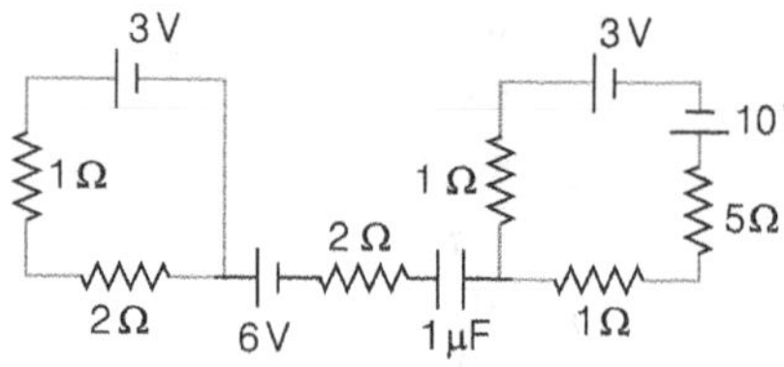

(a) 4 μC (b) 6 μC (c) 1 μC (d) Zero

12. For the circuit shown in the figure, find the charge stored on capacitor in steady state.

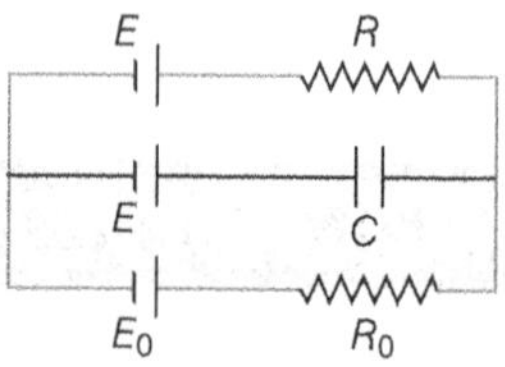

(a) $\frac{RC}{R+R_0}E$
(b) $\frac{RC}{R_0}(E-E_0)$
(c) zero
(d) $\frac{RC}{R+R_0}(E-E_0)$

13. The figure shows a graph of the current in a discharging circuit of a capacitor through a resistor of resistance 10 Ω.

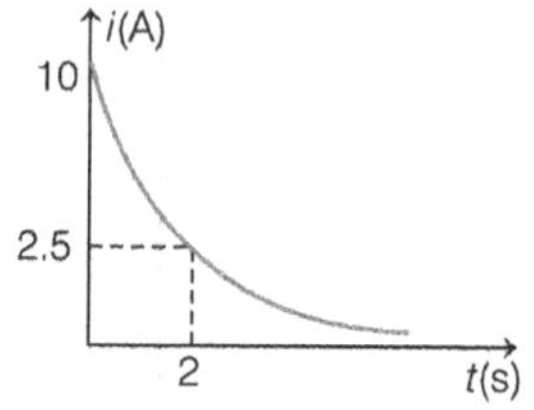

(a) The initial potential difference across the capacitor is 100 V
(b) The capacitance of the capacitor is $\frac{1}{10 \ln 2}$ F
(c) The total heat produced in the circuit will be $\frac{500}{\ln 2}$ J
(d) All of the above

14. In the circuit shown in figure switch *S* is thrown to position 1 at $t = 0$. When the current in the resistor is 1 A, it is shifted to position 2. The total heat generated in the circuit after shifting to position 2 is

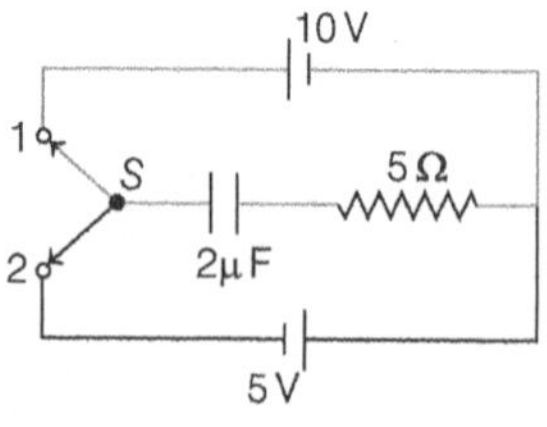

(a) zero (b) 625 μJ
(c) 100 μJ (d) None of these

15. Consider a capacitor—charging circuit. Let Q_1 be the charge given to the capacitor in time interval of 20 ms and Q_2 be the charge given in the next time interval of 20 ms. Let 10 μC charge be deposited in a time interval t_1 and the next 10 μC charge is deposited in the next time interval t_2. Then,
(a) $Q_1 > Q_2, t_1 > t_2$
(b) $Q_1 > Q_2, t_1 < t_2$
(c) $Q_1 < Q_2, t_1 > t_2$
(d) $Q_1 < Q_2, t_1 < t_2$

16. The current in 1 Ω resistance and charge stored in the capacitor are

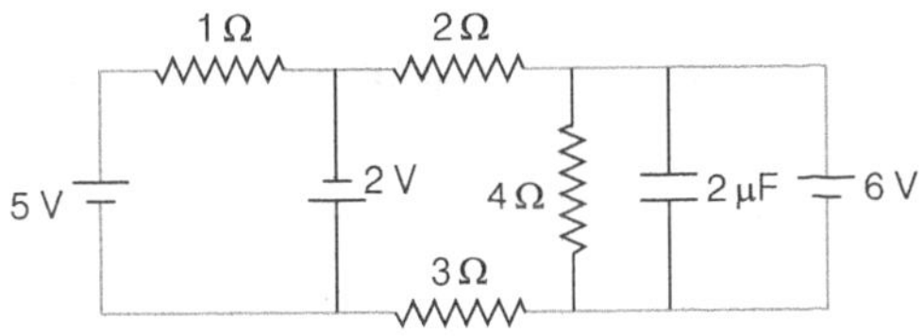

(a) 4 A, 6 μC (b) 7 A, 12 μC
(c) 4 A, 12 μC (d) 7 A, 6 μC

Subjective Questions

17. Find the charge stored in the capacitor.

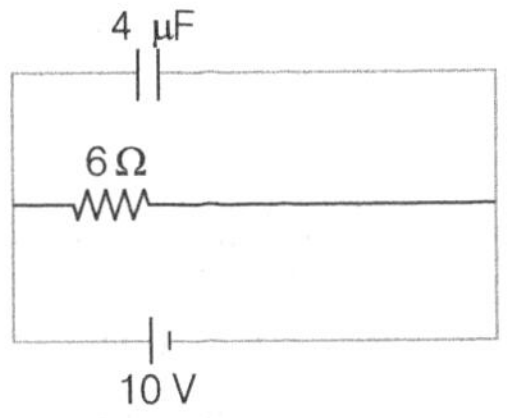

18. How many time constants will elapse before the current in a charging *R-C* circuit drops to half of its initial value?

19. A capacitor of capacitance *C* is given a charge q_0. At time $t = 0$ it is connected to an uncharged capacitor of equal capacitance through a resistance *R*. Find the charge on the first capacitor and the second capacitor as a function of time *t*. Also plot the corresponding *q-t* graphs.

20. A capacitor of capacitance *C* is given a charge q_0. At time $t = 0$, it is connected to a battery of emf *E* through a resistance *R*. Find the charge on the capacitor at time *t*. Assume that $EC > q_0$.

21. Determine the current through the battery in the circuit shown in figure.

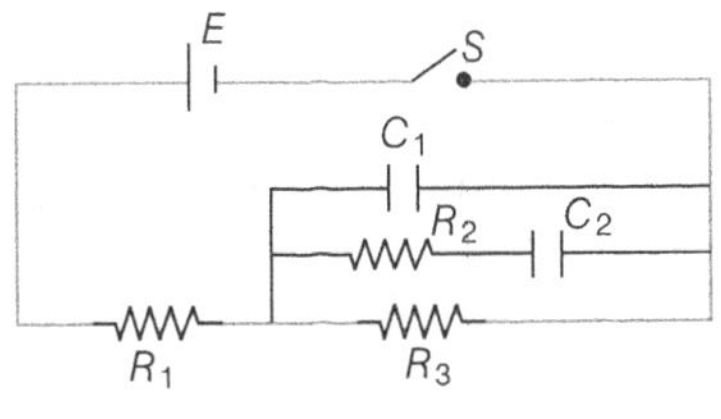

(a) immediately after the switch *S* is closed
(b) after a long time.

22. For the circuit shown in figure, find

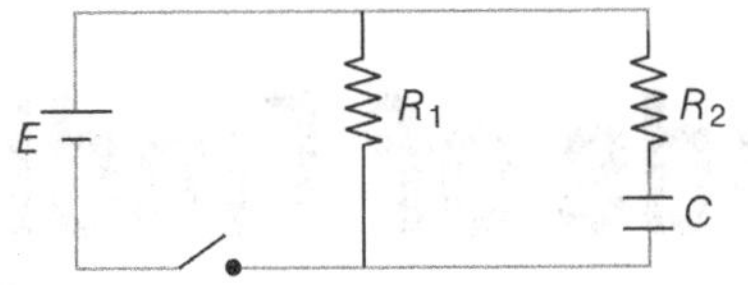

(a) the initial current through each resistor
(b) steady state current through each resistor
(c) final energy stored in the capacitor
(d) time constant of the circuit when switch is opened.

23. In the circuit shown in figure, the battery is an ideal one with emf *V*. The capacitor is initially uncharged. The switch *S* is closed at time $t = 0$.
(a) Find the charge *Q* on the capacitor at time *t*.
(b) Find the current in *AB* at time *t*. What is its limiting value as $t \to \infty$?

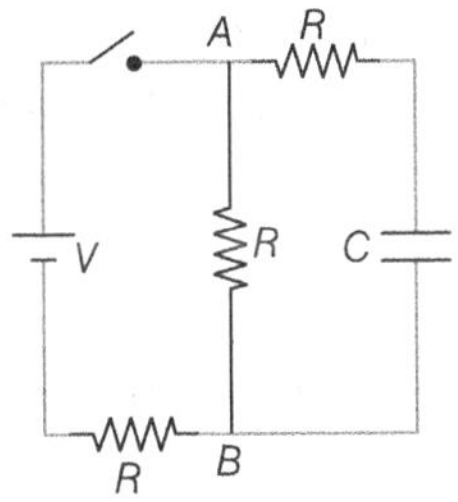

24. In the circuit shown in figure $E_1 = 2E_2 = 20$ V, $R_1 = R_2 = 10$ kΩ and $C = 1$ μF. Find the current through R_1, R_2 and *C* when
(a) *S* has been kept connected to *A* for a long time.
(b) The switch is suddenly shifted to *B*.

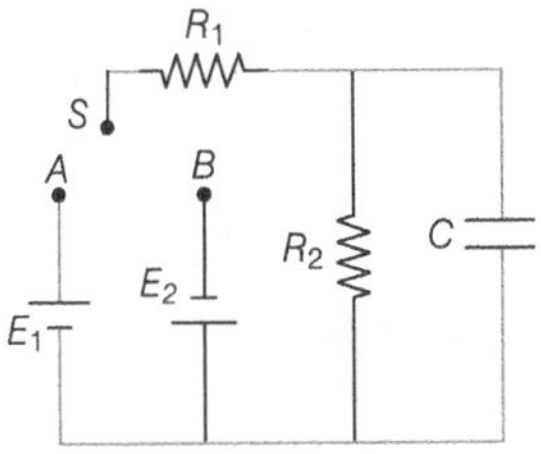

Answers

1. (a)	**2.** (d)	**3.** (b)	**4.** (b)	**5.** (a)
6. (a)	**7.** (d)	**8.** (d)	**9.** (d)	**10.** (c)
11. (d)	**12.** (d)	**13.** (d)	**14.** (c)	**15.** (b)
16. (b)	**17.** 40 μC	**18.** 0.693		

19. $q_1 = \frac{q_0}{2} + \frac{q_0}{2}e^{-2t/RC}$, $q_2 = \frac{q_0}{2}(1 - e^{-2t/RC})$

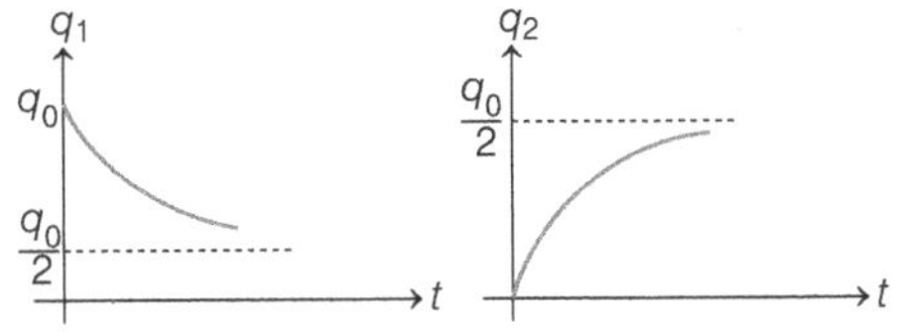

20. $CE\,(1 - e^{-t/CR}) + q_0\, e^{-t/CR}$

21. (a) E/R_1 (b) $E/(R_1 + R_3)$

22. (a) $i_1 = E/R_1, i_2 = E/R_2$ (b) $i_1 = E/R_1, i_2 = 0$ (c) $\frac{1}{2}CE^2$ (d) $C(R_1 + R_2)$

23. (a) $\frac{CV}{2}(1 - e^{-\alpha t})$ (b) $\frac{V}{2R} - \frac{V}{6R}e^{-\alpha t}, \frac{V}{2R}$ Here $\alpha = \frac{2}{3RC}$

24. (a) 1 mA, 1 mA, 0 (b) 2 mA, 1 mA, 3 mA

See hints on pages 321-323.

MODULE 8

Methods of Finding Equivalent Resistance and Capacitance

We know that, in series,

$$R_{eq} = R_1 + R_2 + \ldots + R_n$$

and $$\frac{1}{C_{eq}} = \frac{1}{C_1} + \frac{1}{C_2} + \ldots + \frac{1}{C_n}$$

and in parallel, $$\frac{1}{R_{eq}} = \frac{1}{R_1} + \frac{1}{R_2} + \ldots + \frac{1}{R_n}$$

and $$C_{eq} = C_1 + C_2 + \ldots + C_n$$

Sometimes there are circuits in which resistances/capacitors are in mixed grouping. To find R_{eq} or C_{eq} for such circuits few methods are suggested here which will help you in finding R_{eq} or C_{eq}.

1. Method of Same Potential

Give any arbitrary potentials ($V_1, V_2, \ldots$ etc.) to all terminals of capacitors/resistors. But notice that the points connected directly by a conducting wire will have at the same potential. The capacitors/resistors having the same PD are in parallel. Make a table corresponding to the figure. Now, corresponding to this table a simplified figure can be formed and from this figure C_{eq} and R_{eq} can be calculated.

Example 1 Find equivalent capacitance between points A and B as shown in figure.

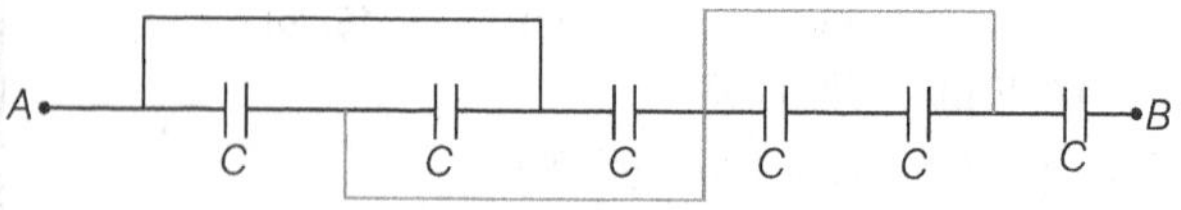

Sol. Three capacitors have PD, $V_1 - V_2$. So, they are in parallel. Their equivalent capacitance is $3C$.

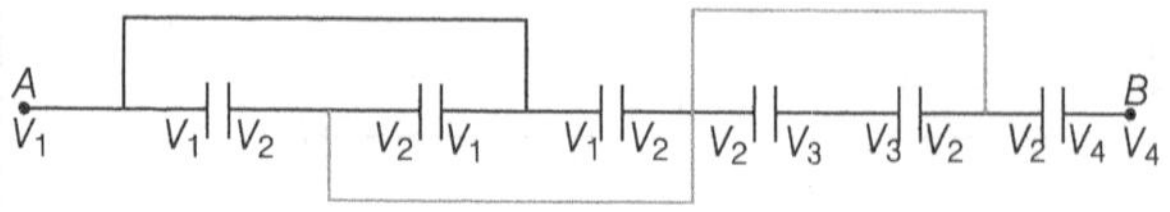

Two capacitors have PD, $V_2 - V_3$. So, their equivalent capacitance is $2C$ and lastly there is one capacitor across which PD is $V_2 - V_4$. So, let us make a table corresponding to this information.

PD	Capacitance
$V_1 - V_2$	$3C$
$V_2 - V_3$	$2C$
$V_2 - V_4$	C

Now, corresponding to this table, we make a simple figure as shown in figure.

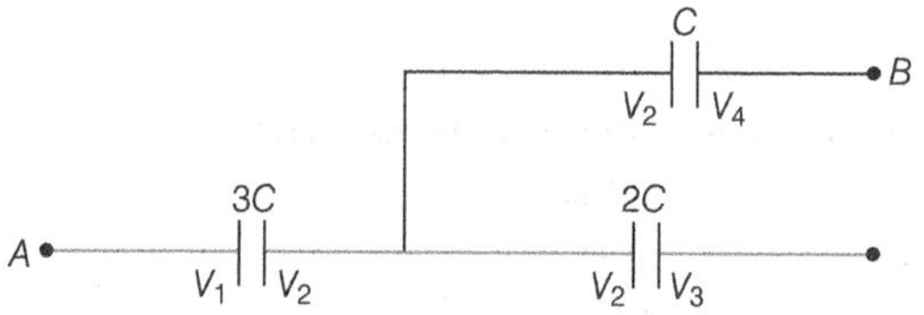

As we have to find the equivalent capacitance between points A and B, across which PD is $V_1 - V_4$. From the simplified figure, we can see that the capacitor of capacitance $2C$ is out of the circuit and points A and B are as shown. Now, $3C$ and C are in series and their equivalent capacitance is

$$C_{eq} = \frac{(3C)(C)}{3C + C} = \frac{3}{4}C$$ **Ans.**

Exercise Find equivalent capacitance between points A and B.

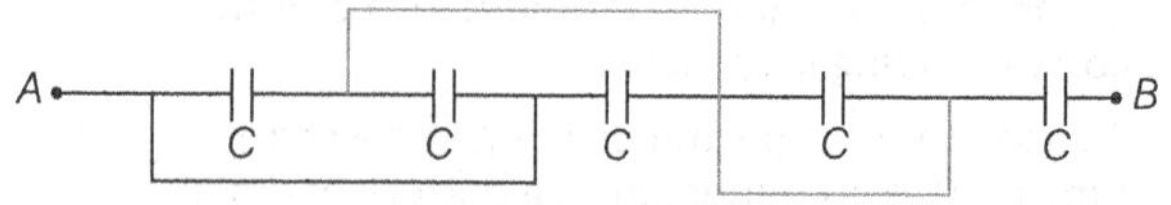

HINT In case PD across any capacitor comes out to be zero (i.e. the plates are short circuited), then this capacitor will not store any charge. So ignore this capacitor.

Ans. $\frac{3}{4}C$

Exercise Identical metal plates are located in air at equal distance d from one another as shown in figure. The area of each plate is A. Find the capacitance of the system between points P and Q if plates are interconnected as shown.

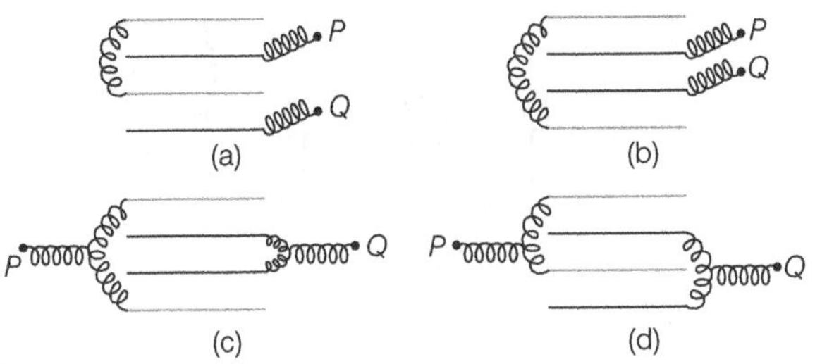

Ans. (a) $\dfrac{2}{3}\dfrac{\varepsilon_0 A}{d}$ (b) $\dfrac{3}{2}\dfrac{\varepsilon_0 A}{d}$ (c) $\dfrac{2\varepsilon_0 A}{d}$ (d) $\dfrac{3\varepsilon_0 A}{d}$

Exercise Find equivalent resistance between A and B.

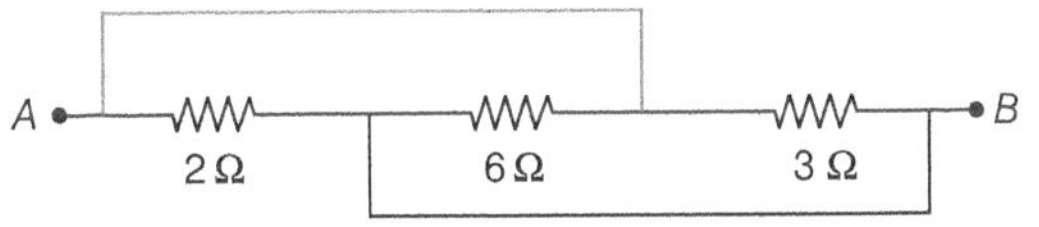

Ans. $1\,\Omega$

Exercise Find equivalent capacitance between points A and B.

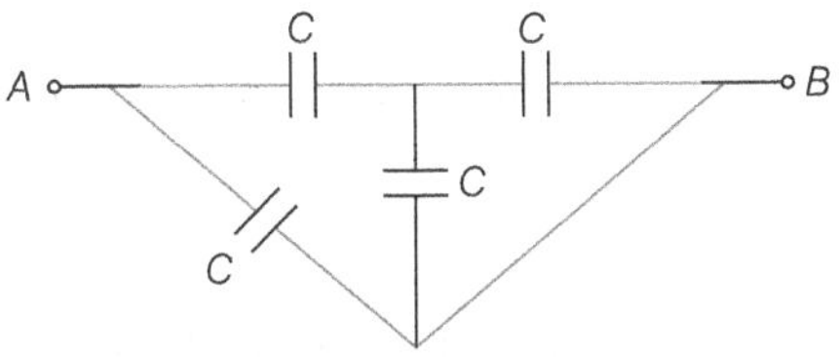

Ans. $\dfrac{5}{3}C$

2. Infinite Series Problems

This circuit consists of an infinite series of identical loops. To find C_{eq} or R_{eq} of such a series first we consider by ourself a value (say x) of C_{eq} or R_{eq}. Then, we break the chain in such a manner that only one loop is left with us and in place of the remaining portion we connect a capacitor or resistor of value x. Then, we find the C_{eq} or R_{eq} and put it equal to x. With this we get a quadratic equation in x. By solving this equation, we can find the desired value of x.

Example 2 An infinite ladder network is constructed with $1\,\Omega$ and $2\,\Omega$ resistors as shown. Find the equivalent resistance between points A and B.

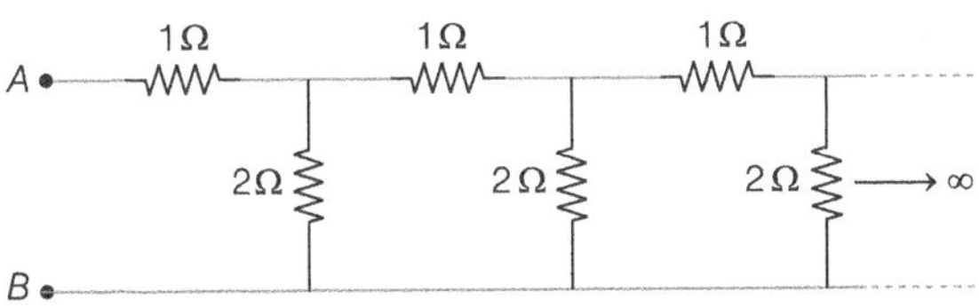

Sol. Let the equivalent resistance between A and B is x. We may consider the given circuit as shown in figure.

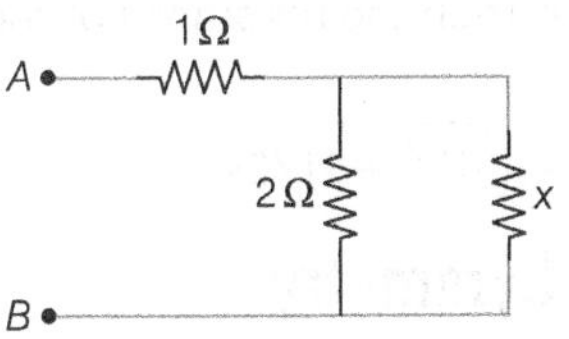

In this diagram,

$$R_{AB} = \frac{2x}{2+x} + 1$$

or $$x = \frac{2x}{2+x} + 1 \qquad (\text{as } R_{AB} = x)$$

or $$x(2+x) = 2x + 2 + x$$

or $$x^2 - x - 2 = 0$$

$$x = \frac{1 \pm \sqrt{1+8}}{2} = -1\,\Omega \text{ and } 2\,\Omega$$

Ignoring the negative value, we have $R_{AB} = x = 2\,\Omega$ **Ans.**

Note *Care should be taken while breaking the chain. It should be broken from those points from where the broken chain resembles with the original chain.*

(a)

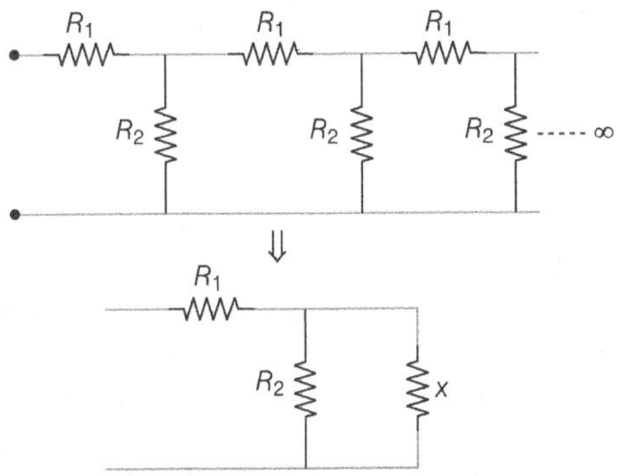

(b)

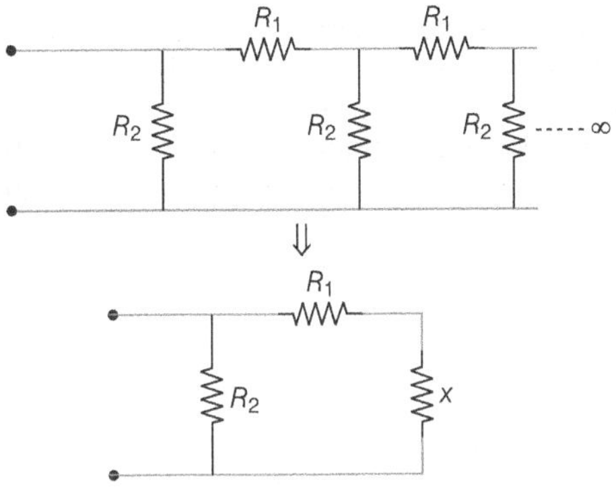

Exercise Find equivalent resistance between A and B.

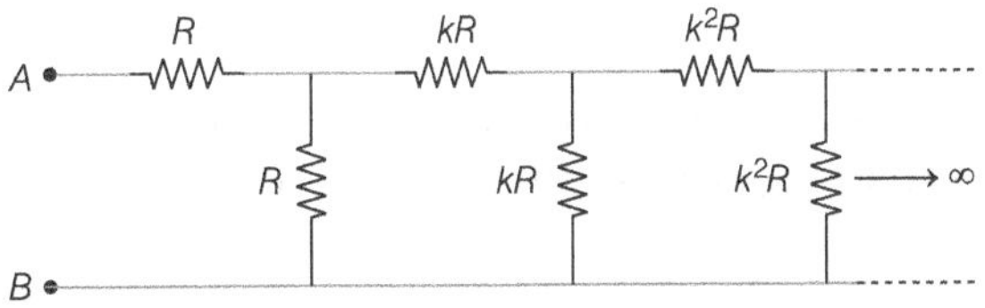

HINT Let $R_{AB} = x$, then the resistance of the broken chain will be kx.

Ans. $R\,[(2k-1)+\sqrt{4k^2+1}]\,/2k$

3. Method of Symmetry

Symmetry of a circuit can be checked in the following four manners :

First : Points which are symmetrically located about the starting and last points are at same potentials. So, the resistances/capacitors between these points can be ignored. The following example will illustrate the theory.

Example 3 Twelve resistors each of resistance r are connected together so that each lies along the edge of the cube as shown in figure. Find the equivalent resistance between

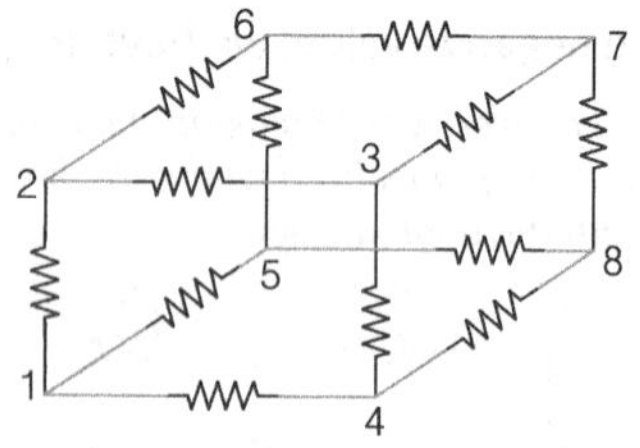

(a) 1 and 4

(b) 1 and 3

Sol. (a) **Between 1 and 4 :**

Points 2 and 5 are symmetrically located w.r.t. points 1 and 4. So, they are at same potentials.

Similarly, points 3 and 8 are also symmetrically located w.r.t. points 1 and 4. So, they are again at same potential.

Now, we have 12 resistors each of resistance r connected across 1 and 2, 2 and 3,..., etc. So, redrawing them with the assumption that 2 and 5 are at same potential and 3 and 8 are at same potential. The new figure is as shown in figure.

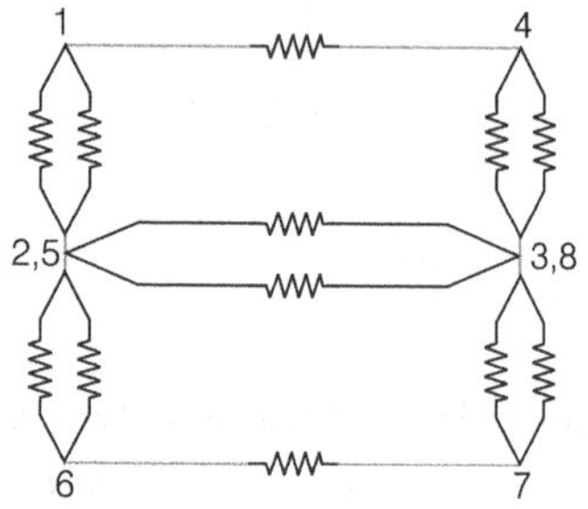

Now, we have to find the equivalent resistance between 1 and 4. So, now we can simplify the circuit as :

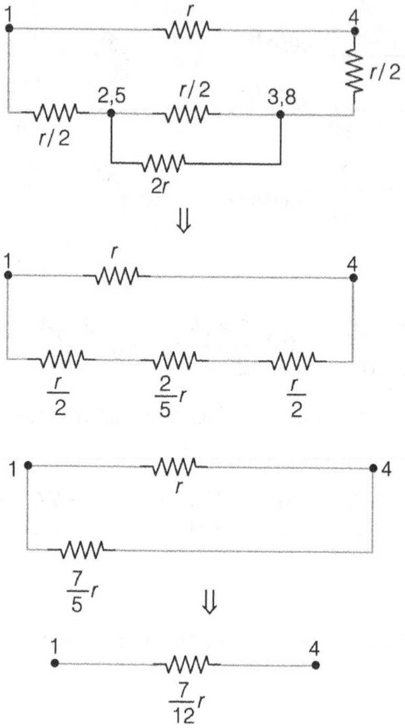

Thus, the equivalent resistance between points 1 and 4 is $\frac{7}{12}r$. **Ans.**

(b) **Between 1 and 3 :**

Points 6 and 8 are symmetrically located w.r.t. points 1 and 3. Similarly, points 2 and 4 are located symmetrically w.r.t. points 1 and 3. So, points 6 and 8 are at same potential. Similarly, 2 and 4 are at same potentials. Redrawing the simple circuit, we have the figure like :

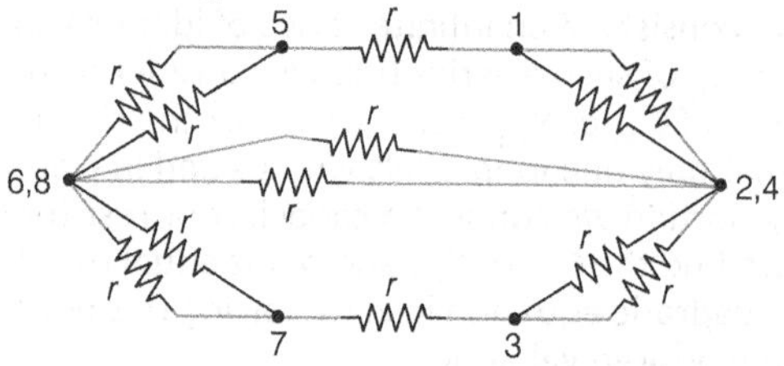

Between 1 and 3, a balanced Wheatstone bridge is formed as shown in figure.

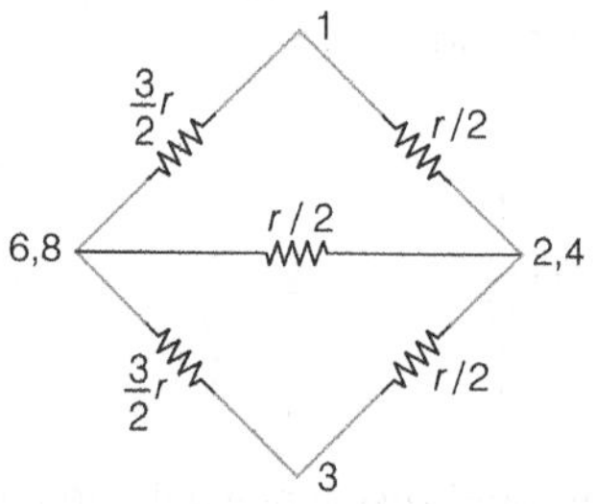

So, the resistance between 2 and 6 and between 4 and 8 can be removed.

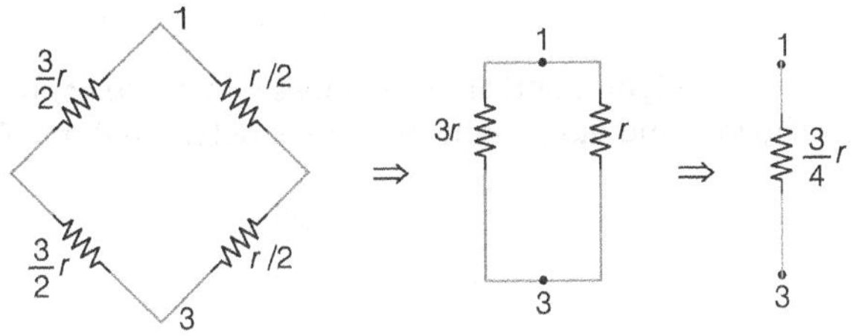

Thus, the equivalent resistance between 1 and 3 is $\frac{3}{4}r$. **Ans.**

Exercise Fourteen identical resistors each of resistance r are connected as shown. Calculate equivalent resistance between A and B.

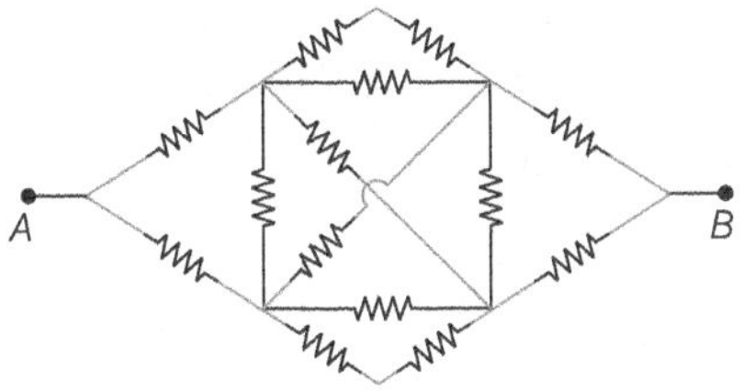

Ans. $1.2r$

Exercise Eight identical resistances r each are connected along edges of a pyramid having square base $ABCD$ as shown. Calculate equivalent resistance between A and O.

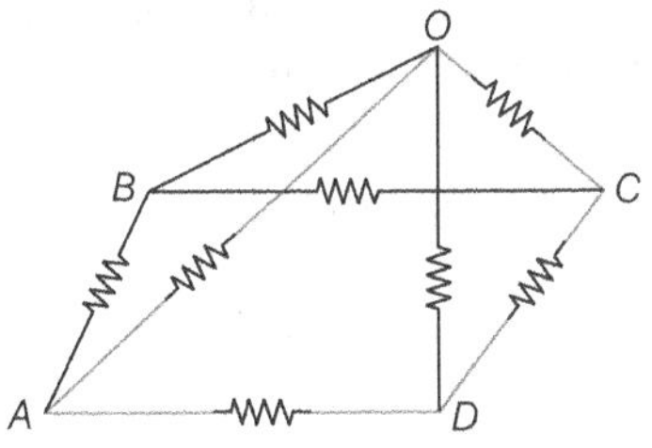

Ans. $\frac{7r}{15}$

Second : If points A and B are connected to a battery and AB is a line of symmetry, then all points lying on perpendiculars drawn to AB are at the same potential. For example,

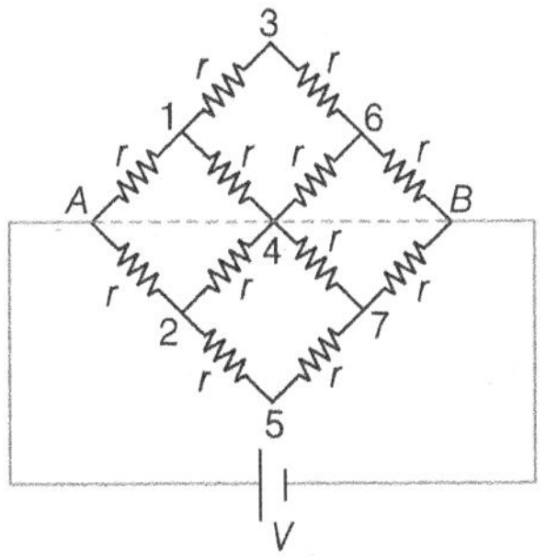

In above figure, points (1, 2), (3, 4, 5) and (6, 7) are at same potential. So, we can join these points and draw a simple circuit as shown in below figure.

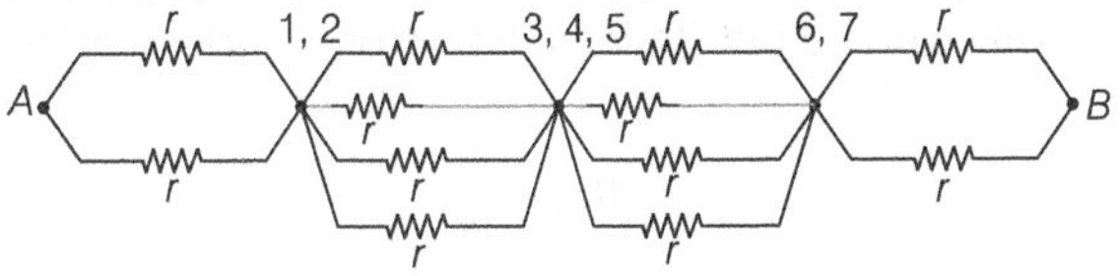

Now, the equivalent of this series combination is

$$R_{eq} = \frac{r}{2} + \frac{r}{4} + \frac{r}{4} + \frac{r}{2} = \frac{3r}{2}$$

Exercise Solve the same problem by connection removal method (will be discussed later).

Third : Even if AB is not a line of symmetry but its perpendicular bisector is, then all the points on this perpendicular bisector are at the same potential.

For example,

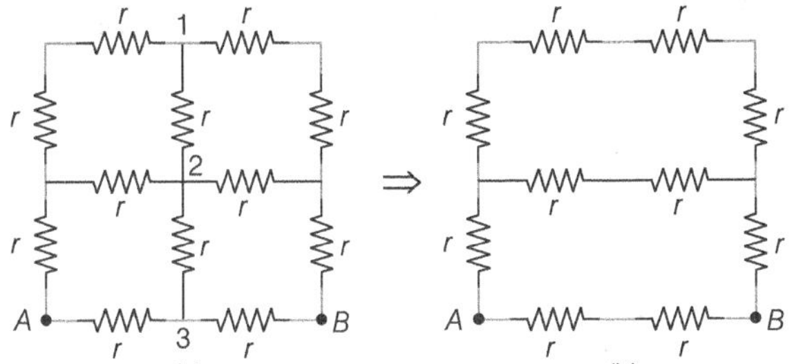

(a) (b)

In Fig. (a), AB is not a line of symmetry but, 1, 2 and 3 are line of symmetry. Hence, they are at same potential (if A and B are connected to a battery). This makes the resistors between 1 & 2 and 2 and 3 redundant because no current flows through them. So, the resistance between them can be removed [as shown in Fig. (b)]. The equivalent resistance between A and B can now be easily determined as $\frac{5r}{4}$.

Fourth : Each wire in the cube has a resistance r. We are interested in calculating the equivalent resistance between A and B.

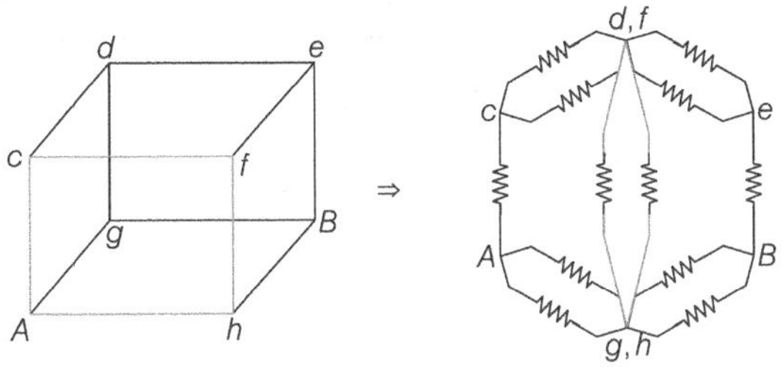

(a) (b)

This is a three-dimensional case and in place of a line of symmetry involving points A and B we locate a plane of symmetry involving A and B.

Such a plane is the plane $ABce$ and for this plane points d and f and g and h have the same potential.

The equivalent resistance between A and B can now be easily worked out (Using Wheatstone's bridge principle) as

$$R_{eq} = \frac{3r}{4}$$

4. Connection Removal Method

This method is useful when the circuit diagram is symmetric except for the fact that the input and output are reversed. That is the flow of current is a mirror image between input and output above a particular axis. In such cases, some junctions are unnecessarily made. Even if we remove that junction there is no difference in the remaining circuit or current distribution. But after removing the junction, the problem becomes very simple. The following example illustrates the theory.

Example 4 Find the equivalent resistance between points A and B.

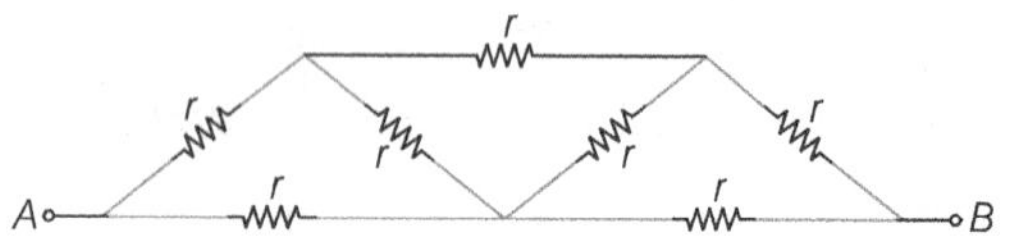

Sol.

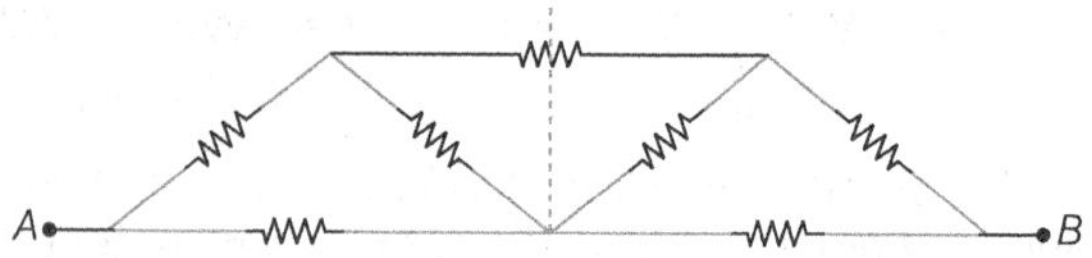

Input and output circuits are mirror images of each other about the dotted line as shown.

So, if a current i enters from A and leaves from B, it will distribute as shown below.

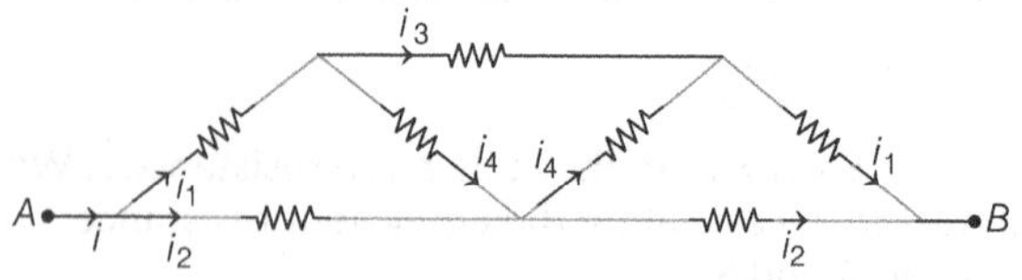

Now, we can see in figure that the junction where i_2 and i_4 are meeting can be removed easily and then the circuit becomes simple.

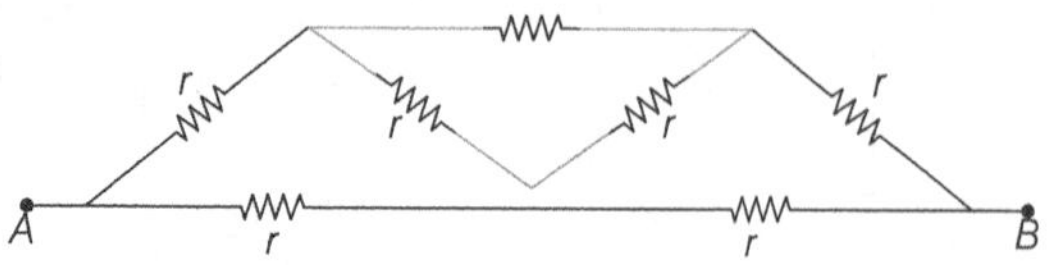

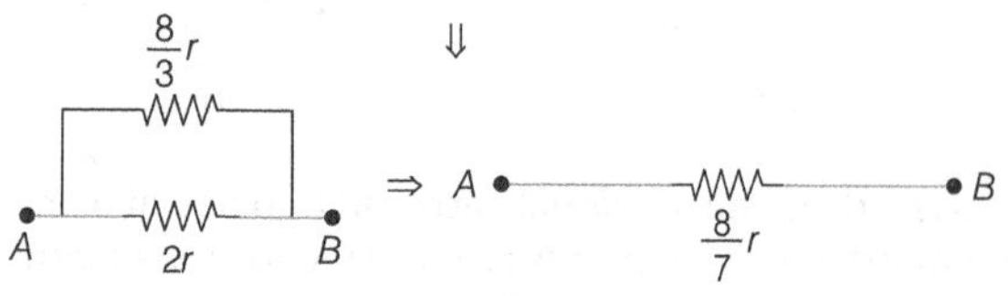

Hence, the equivalent resistance between A and B is $\frac{8}{7}r$.

Ans.

Exercise Eight identical resistances r each are connected as shown. Find equivalent resistance between A and D.

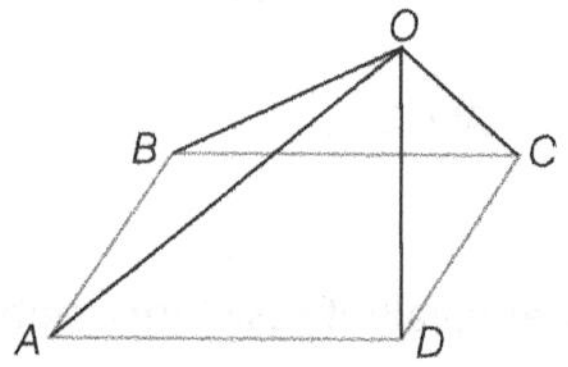

Ans. $\frac{8r}{15}$

Exercise Find equivalent resistance between A and B.

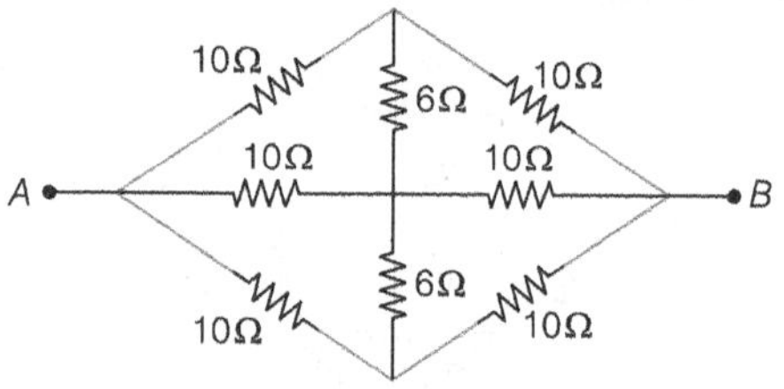

Ans. $\frac{20}{3}\Omega$.

5. Wheatstone Bridge Circuits

Wheatstone bridge in case of resistors has already been discussed in the chapter of current electricity.

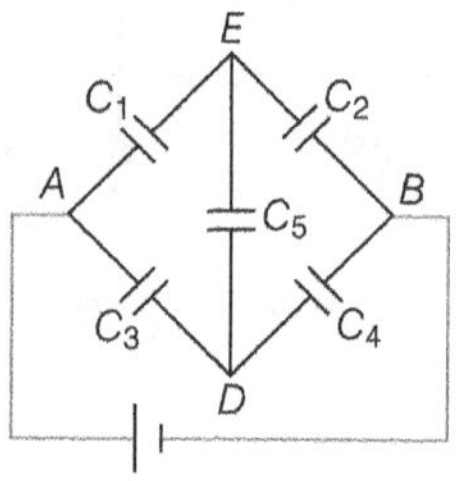

For capacitor, theory is same.

If $\frac{C_1}{C_2} = \frac{C_3}{C_4}$, bridge is said to be balanced and in that case

$$V_E = V_D$$

or $$V_E - V_D$$

or $$V_{ED} = 0$$

i.e. no charge is stored in C_5. Hence, it can be removed from the circuit.

Exercise In the circuit shown in figure, prove that $V_{AB} = 0$ if $\frac{R_1}{R_2} = \frac{C_2}{C_1}$.

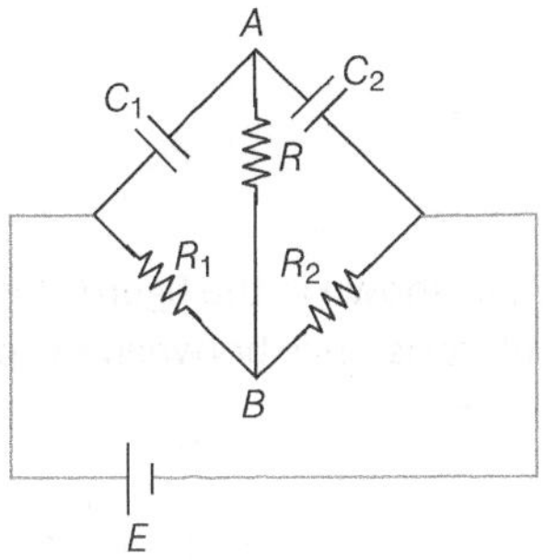

6. By Distributing Current/Charge

Sometimes none of the methods discussed above is applicable. So, this one is the last and final method which can be applied everywhere. Of course this method is a little bit lengthy but is applicable everywhere, under all conditions. In this method, we assume a main current/charge, i or q. Distribute it in different resistors/capacitors as $i_1, i_2 \ldots$ (or $q_1, q_2, \ldots$, etc.).

Using Kirchhoff's laws, we find $i_1, i_2, \ldots$ etc., (or $q_1, q_2, \ldots$, etc.) in terms of i (or q). Then, find the potential difference between starting and end points through any path and equate it with iR_{net} or q/C_{net}. By doing so, we can calculate R_{net} or C_{net}.

The following example illustrates this theory.

Example 5 Find the equivalent capacitance between A and B.

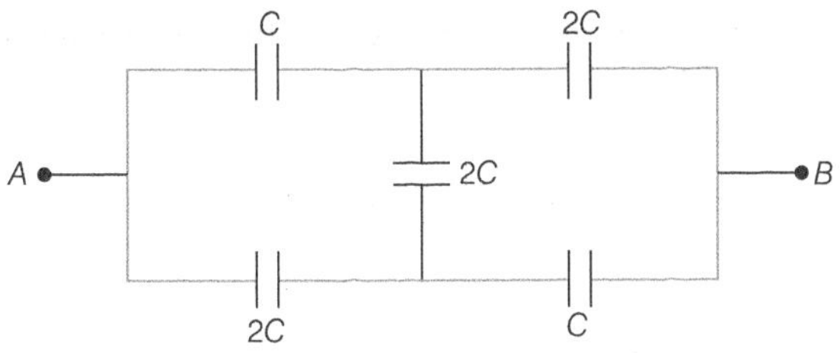

Sol. The given circuit forms a Wheatstone bridge. But the bridge is not balanced. Let us suppose that point A is connected to the positive terminal of a battery and B to the negative terminal of the same battery, so that a total charge q is stored in the capacitors.

Just by seeing input and output symmetry, we can say that charges will be distributed as shown below.

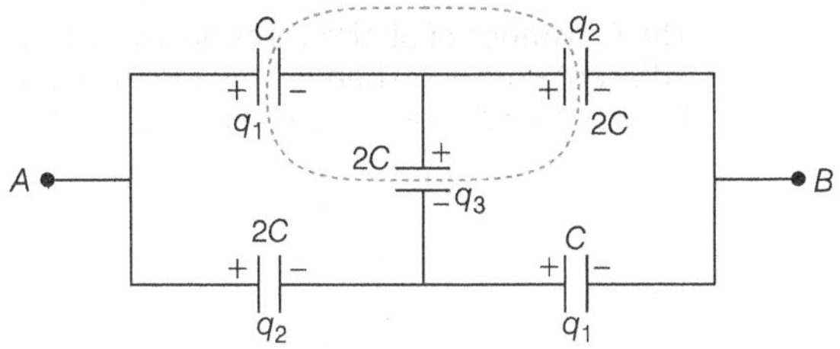

$$q_1 + q_2 = q \qquad \text{...(i)}$$

Applying second law in left loop, we have

$$-\frac{q_1}{C} - \frac{q_3}{2C} + \frac{q_2}{2C} = 0$$

or $$q_2 - q_3 - 2q_1 = 0 \qquad \text{...(ii)}$$

Plates inside the dotted line form an isolated system. Hence,

$$q_2 + q_3 - q_1 = 0 \qquad \text{...(iii)}$$

Solving these three equations, we have

$$q_1 = \frac{2}{5}q, \quad q_2 = \frac{3}{5}q$$

and $$q_3 = -\frac{q}{5}$$

Now, let C_{eq} be the equivalent capacitance between A and B. Then,

$$V_A - V_B = \frac{q}{C_{eq}} = \frac{q_1}{C} + \frac{q_2}{2C}$$

$$\therefore \quad \frac{q}{C_{eq}} = \frac{2q}{5C} + \frac{3q}{10C} = \frac{7q}{10C}$$

$$\therefore \quad C_{eq} = \frac{10}{7}C$$ **Ans.**

MODULE 8 Exercise

Single Correct Option

1. The resistance of all the wires between any two adjacent dots is R. Then, equivalent resistance between A and B as shown in the figure is

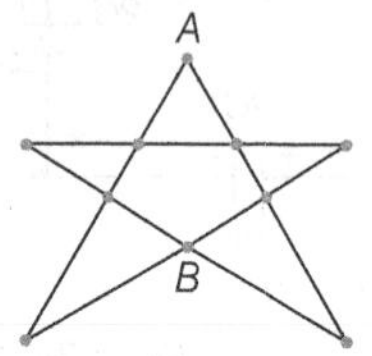

(a) $(7/3)R$ (b) $(7/6)R$
(c) $(14/8)R$ (d) None of these

2. The equivalent resistance of the hexagonal network as shown in figure between points A and B is

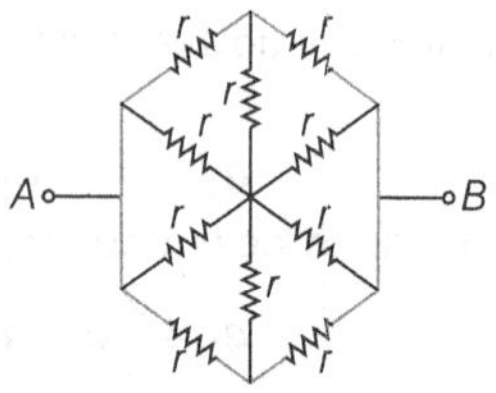

(a) r (b) $0.5\,r$
(c) $2r$ (d) $3r$

3. Each resistor shown in figure is an infinite network of resistance $1\,\Omega$. The effective resistance between points A and B is

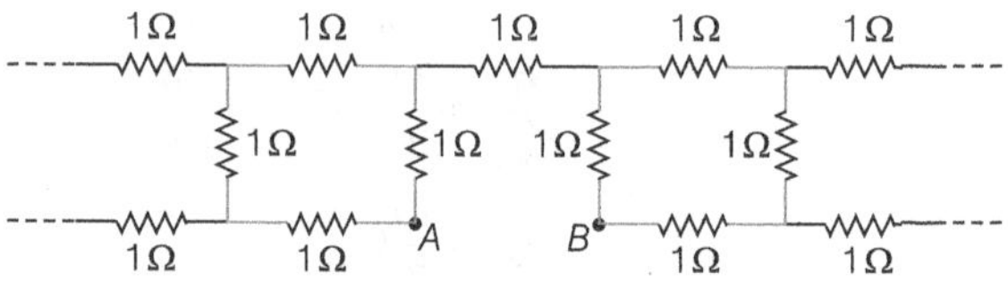

(a) less than $1\,\Omega$
(b) $1\,\Omega$
(c) more than $1\,\Omega$ but less than $3\,\Omega$
(d) $3\,\Omega$

4. In the circuit shown in figure, the total resistance between points A and B is R_0. The value of resistance R is

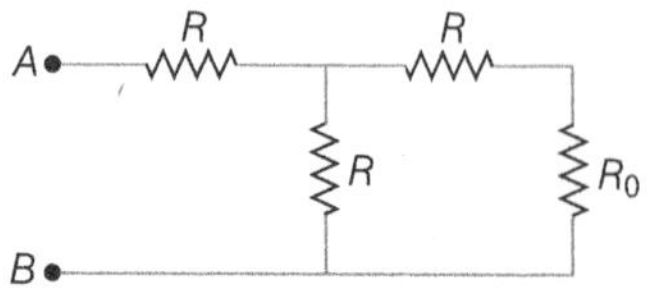

(a) R_0 (b) $\sqrt{3}\,R_0$
(c) $\frac{R_0}{2}$ (d) $\frac{R_0}{\sqrt{3}}$

5. In the circuit shown in the figure, $R = 55\,\Omega$, the equivalent resistance between the points P and Q is

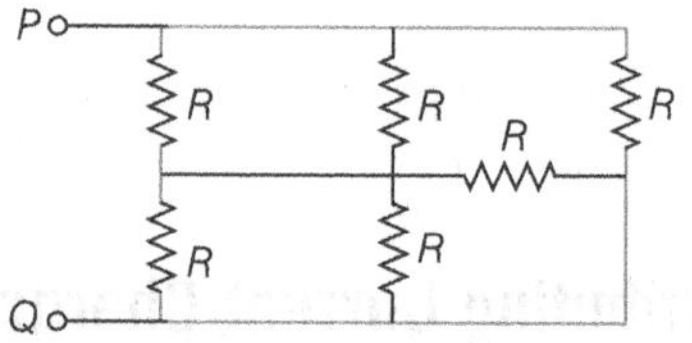

(a) $30\,\Omega$ (b) $35\,\Omega$
(c) $55\,\Omega$ (d) $25\,\Omega$

6. The equivalent resistance between the points A and B is (R is the resistance of each side of smaller square)

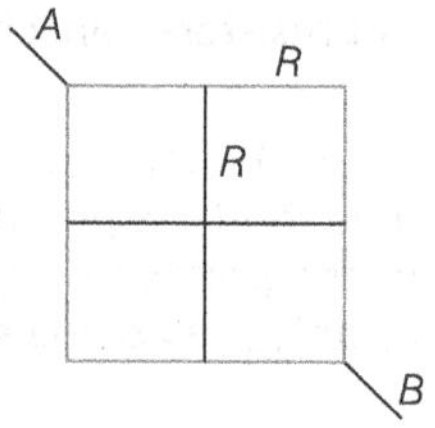

(a) R (b) $\frac{3R}{2}$ (c) $2R$ (d) $\frac{R}{2}$

7. Equivalent resistance between the points A and B is (in Ω)

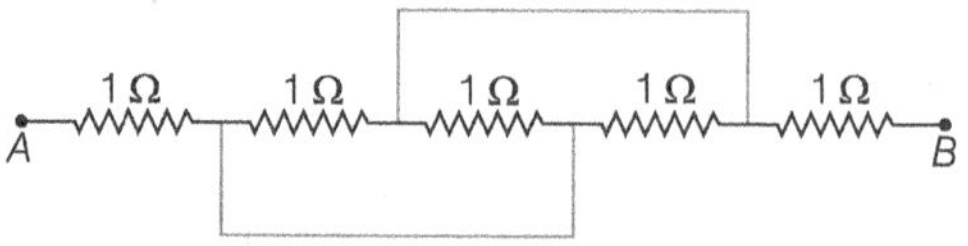

(a) $\frac{1}{5}$ (b) $\frac{5}{4}$ (c) $\frac{7}{3}$ (d) $\frac{7}{2}$

8. In the given circuit, it is observed that the current I is independent of the value of the resistance R_6. Then, the resistance values must satisfy

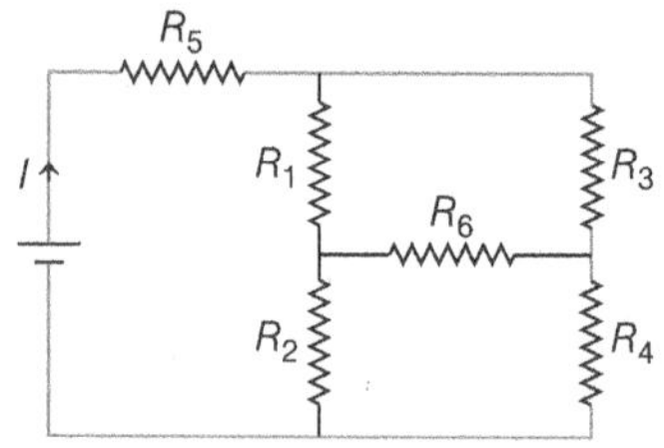

(a) $R_1R_2R_5 = R_3R_4R_6$
(b) $\frac{1}{R_5} + \frac{1}{R_6} = \frac{1}{R_1 + R_2} + \frac{1}{R_3 + R_4}$
(c) $R_1R_4 = R_2R_3$
(d) $R_1R_3 = R_2R_4$

9. A battery of internal resistance 4 Ω is connected to the network of resistances as shown in figure. In order that the maximum power can be delivered to the network, the value of R in Ω should be

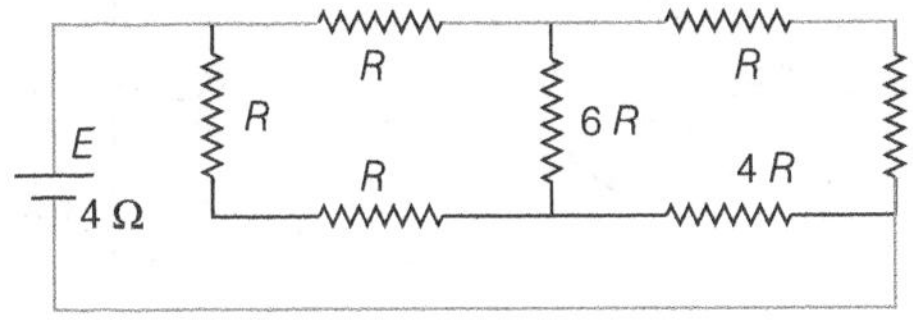

(a) $\frac{4}{9}$ (b) 2 (c) $\frac{8}{3}$ (d) 18

10. If twelve resistors each of value R are connected between points A and B, then

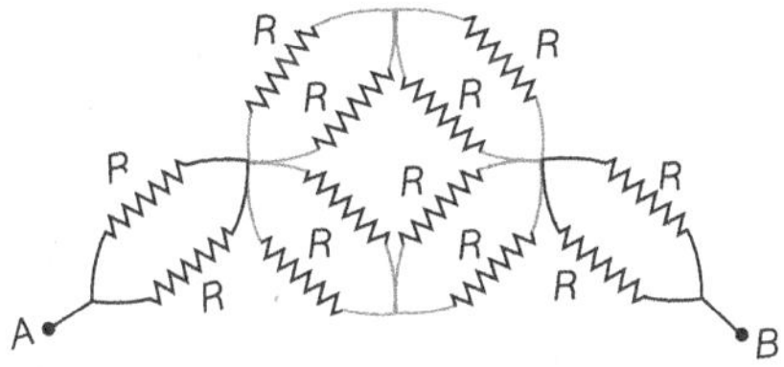

(a) $R_{AB} = R$ (b) $R_{AB} = \frac{R}{2}$
(c) $R_{AB} = \frac{3R}{2}$ (d) $R_{AB} = \frac{5R}{2}$

11. Sixteen resistances each of 5 Ω are connected as shown in the figure.

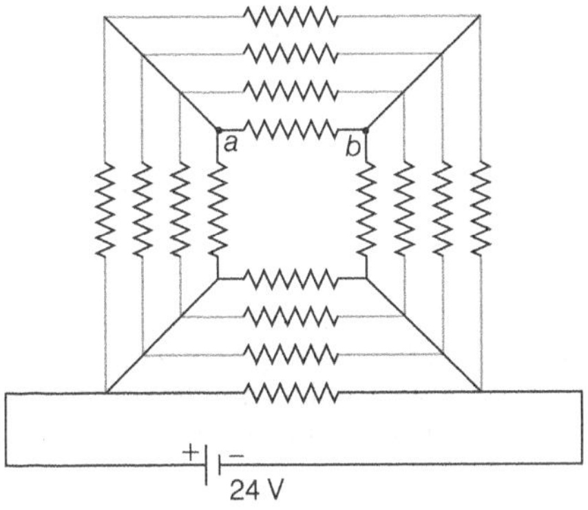

If a battery of 24 V is connected across the network, then current in branch marked ab is

(a) $\frac{8}{5}$ A (b) $\frac{48}{75}$ A (c) 5 A (d) 13 A

12. Consider three capacitors of capacitance C_1, C_2 and C_3. If C_1, C_2 and C_3 are in parallel, equivalent capacitance is 6 unit. When C_2 and C_3 are in parallel, the equivalent capacitance is 5 unit. If $C_1C_2C_3 = 6$, then the capacitances are
(a) 1, 2, 3 (b) 2, 4, 6
(c) 1, 5, 7 (d) 2, 5, 6

13. Four identical capacitors are connected as shown in diagram. When a battery of 6 V is connected between A and B, the charge stored is found to be 1.5 μC. The value of C_1 is

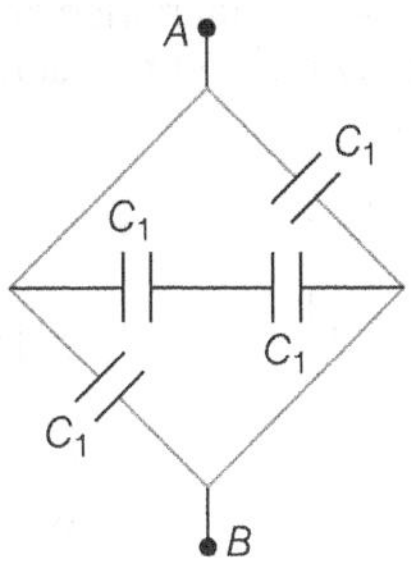

(a) 2.5 μF (b) 15 μF (c) 1.5 μF (d) 0.1 μF

14. Four ways of making a network of five capacitors of the same value are shown in four choices. Three out of four are identical. The one which is different is

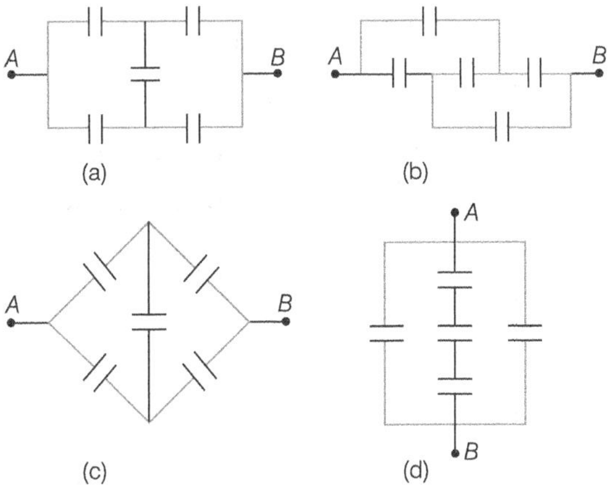

15. Find equivalent capacitance between points A and B. [Assume each conducting plate is having same dimensions and neglect the thickness of the plate, $\frac{\varepsilon_0 A}{d} = 7$ μF, where A is area of plates]

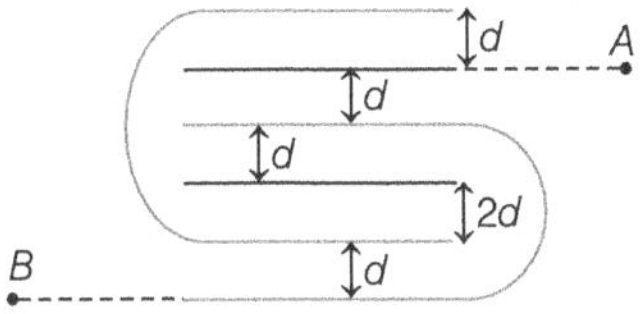

(a) 7 μF (b) 11 μF
(c) 12 μF (d) 15 μF

16. Six equal capacitors each of capacitance C are connected as shown in the figure. The equivalent capacitance between points A and B, is

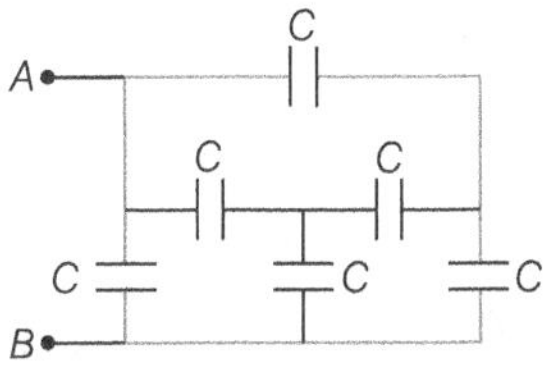

(a) 1.5 C (b) C
(c) 2C (d) 0.5 C

17. Find the equivalent capacitance across A and B for the arrangement shown in figure. All the capacitors are of capacitance C

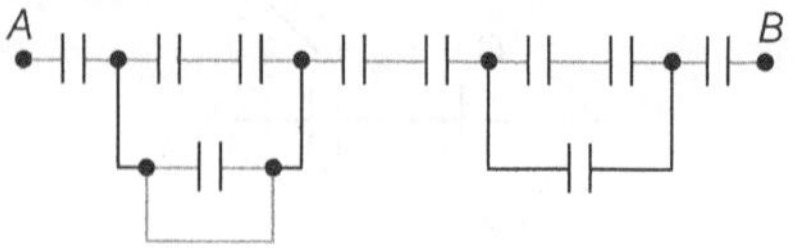

(a) $\frac{3C}{14}$ (b) $\frac{C}{8}$
(c) $\frac{3C}{16}$ (d) None of these

18. The equivalent capacitance between X and Y is

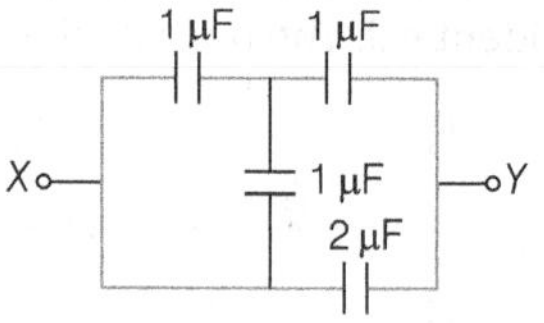

(a) $5/6\,\mu F$ (b) $7/6\,\mu F$ (c) $8/3\,\mu F$ (d) $1\,\mu F$

19. The resultant capacitance of given circuit between points P and Q is

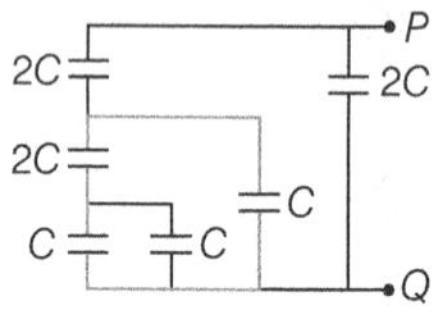

(a) $3C$ (b) $2C$ (c) C (d) $\frac{C}{3}$

20. Four condensers are joined as shown in the figure. The capacity of each is $8\,\mu F$. The equivalent capacity between the points A and B will be

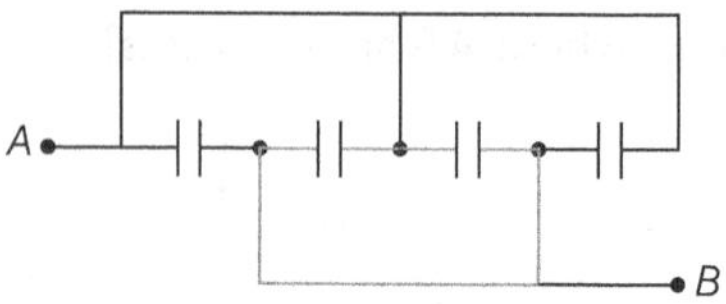

(a) $32\,\mu F$ (b) $2\,\mu F$ (c) $8\,\mu F$ (d) $16\,\mu F$

Multiple Correct Options

21. Each resistance of the network shown in figure is r. Net resistance between

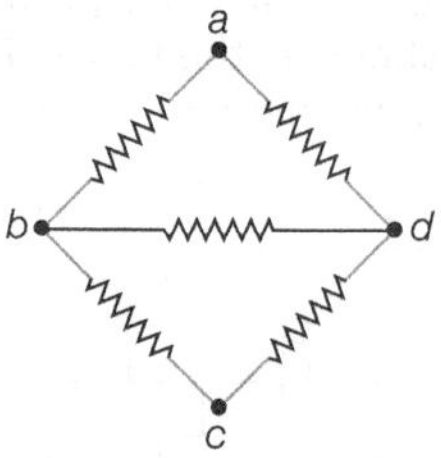

(a) a and b is $\frac{7}{3}r$ (b) a and c is r
(c) b and d is r (d) b and d is $\frac{r}{2}$

Subjective Questions

22. Find the equivalent resistance between points A and B in the following circuits :

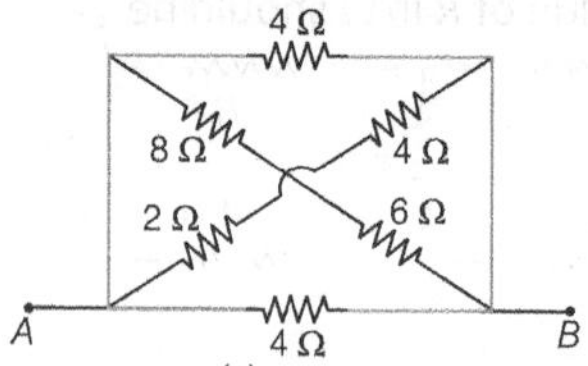

(a)

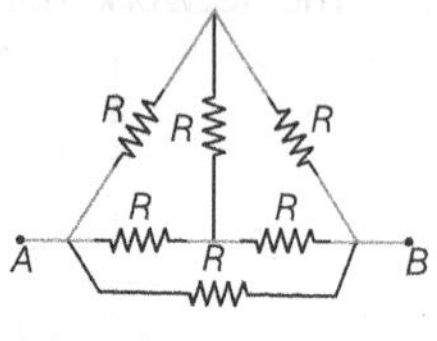

(b)

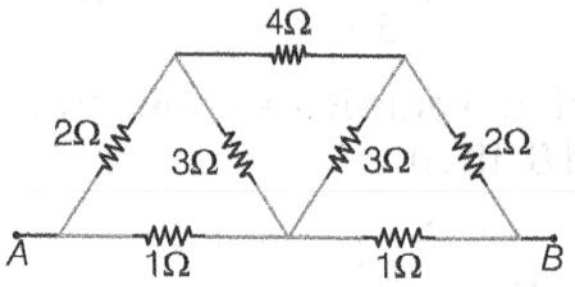

(c)

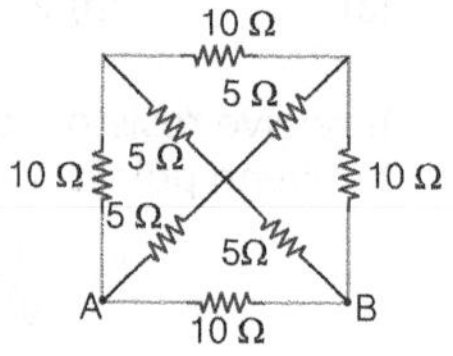

(d)

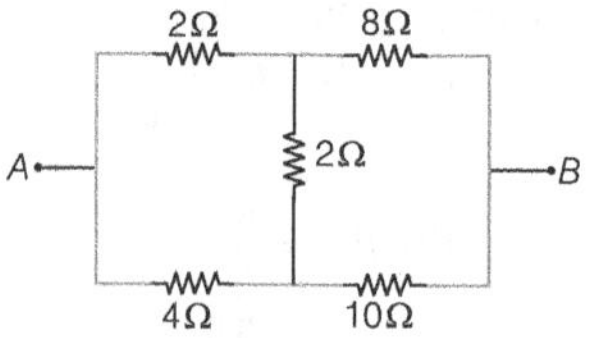

(e)

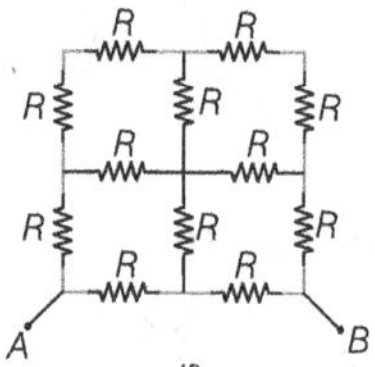

(f)

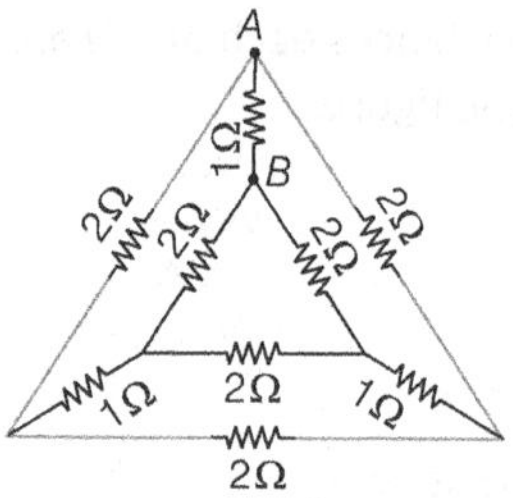

(g)

23. What will be the change in the resistance of a circuit between A and F consisting of five identical conductors, if two similar conductors are added as shown by the dashed line in figure?

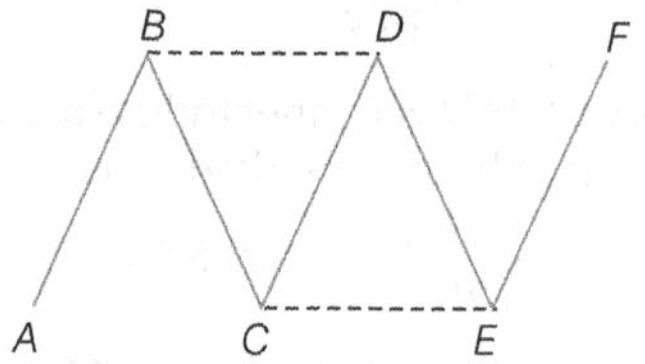

24. Find R_{AB} in the circuit, shown in figure.

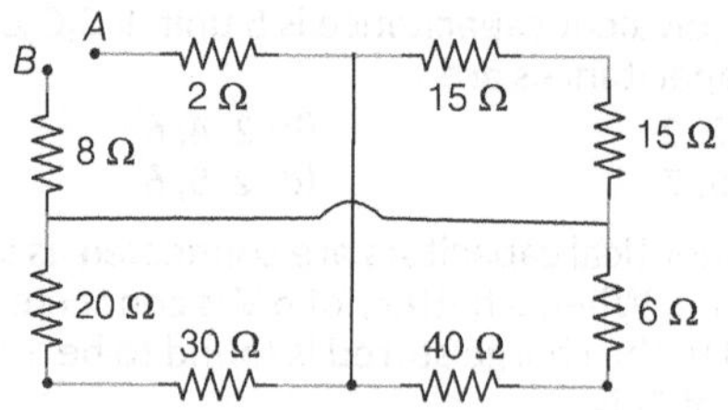

25. Find the equivalent resistance of the networks shown in figure between the points *a* and *b*.

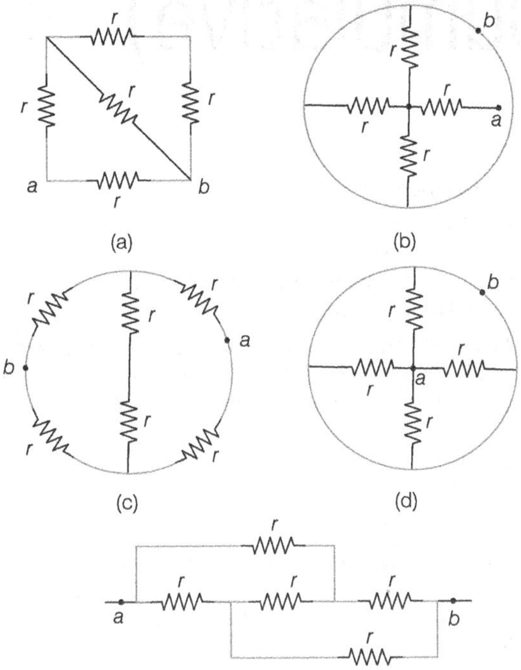

(a) (b) (c) (d) (e)

26. Find the equivalent resistance of the circuits shown in figure between the points *a* and *b*. Each resistor has a resistance *r*.

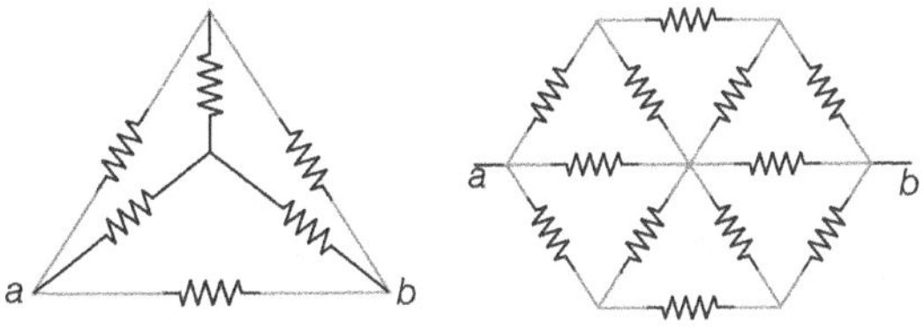

(a) (b)

27. Find the equivalent resistance of the triangular bipyramid between the points.

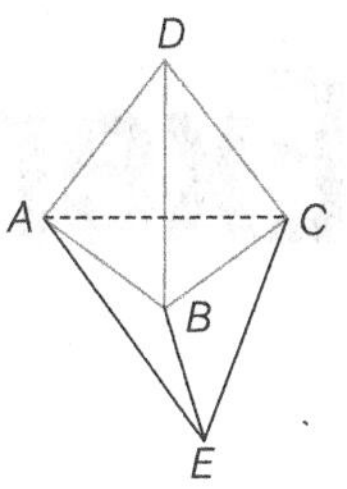

(a) *A* and *C* (b) *D* and *E*

Assume the resistance of each branch to be *R*.

28. Nine wires each of resistance *r* are connected to make a prism as shown in figure. Find the equivalent resistance of the arrangement across

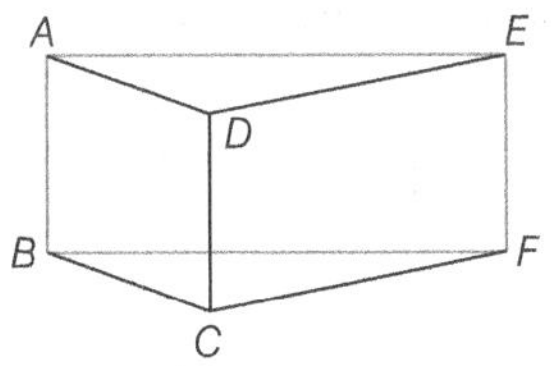

(a) *AD* (b) *AB*

29. Find equivalent capacitance between points *A* and *B*,

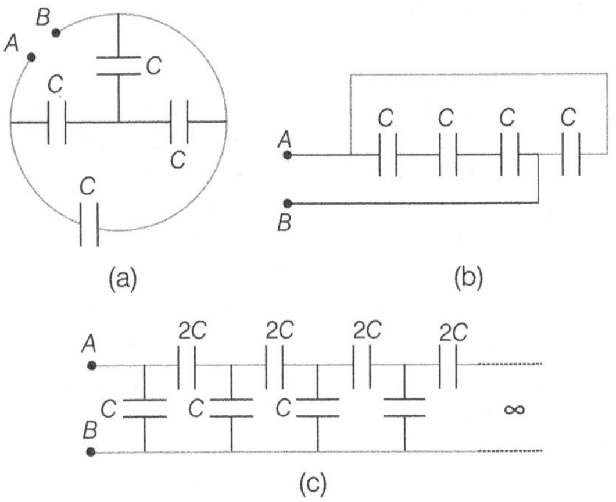

(a) (b) (c)

Answers

1. (b) 2. (b) 3. (c) 4. (d) 5. (d) 6. (b) 7. (c) 8. (c) 9. (b) 10. (c)
11. (a) 12. (a) 13. (d) 14. (d) 15. (b) 16. (c) 17. (a) 18. (c) 19. (a) 20. (a)
21. (b,d) 22. (a) $\frac{42}{31}\Omega$ (b) $\frac{R}{2}$ (c) $\frac{32}{21}\Omega$ (d) $\frac{25}{6}\Omega$ (e) 6.194 Ω (f) $\frac{5R}{4}$ (g) $\frac{5}{7}\Omega$
23. The new equivalent resistance will become 0.6 times 24. 23.32 Ω 25. (a) $\frac{5}{8}r$ (b) $\frac{4}{3}r$ (c) r (d) $\frac{r}{4}$ (e) r
26. (a) $r/2$ (b) $4r/5$ 27. (a) $\frac{2}{5}R$ (b) $\frac{2}{3}R$ 28. (a) $\frac{8}{15}r$ (b) $\frac{3}{5}r$ 29. (a) $\frac{5C}{3}$ (b) $\frac{4}{3}C$ (c) $2C$

See hints on pages 324-327.

Unit Exercise (Cumulative)

Level 1

Assertion and Reason

See directions on page 25.

1. **Assertion :** In the circuit shown in figure, when a dielectric slab is inserted in C_2, the potential difference across C_2 will decrease.

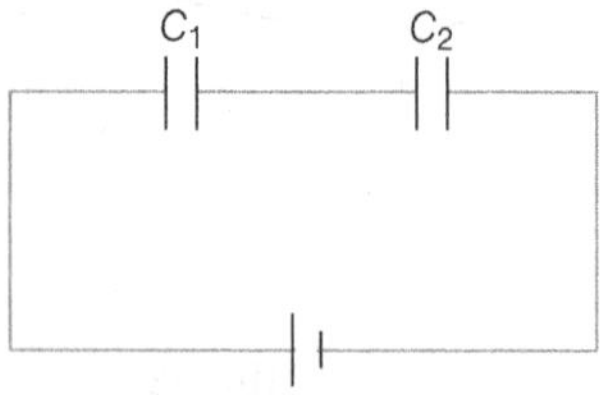

Reason : By inserting the slab a current will flow in the circuit in clockwise direction.

Single Correct Option

2. In Milikan's oil drop experiment, an oil drop of radius r and charge q is held in equilibrium between the plates of a charged parallel-plate capacitor when the potential difference is V. To keep a drop of radius $2r$ and with a charge $2q$ in equilibrium between the plates the potential difference V required is
   (a) V (b) $2V$ (c) $4V$ (d) $8V$

3. A capacitor of capacitance C carrying charge Q is connected to a source of emf E. Finally, the charge on capacitor would be
   (a) Q (b) $Q+CE$
   (c) CE (d) None of these

4. A charged capacitor is discharged through a resistance. The time constant of the circuit is η. Then, the value of time constant for the power dissipated through the resistance will be
   (a) η (b) 2η
   (c) $\eta/2$ (d) zero

5. An electron enters the region between the plates of a parallel-plate capacitor at an angle θ to the plates. The plate width is l. The plate separation is d. The electron follows the path shown, just missing the upper plate. Neglect gravity. Then,

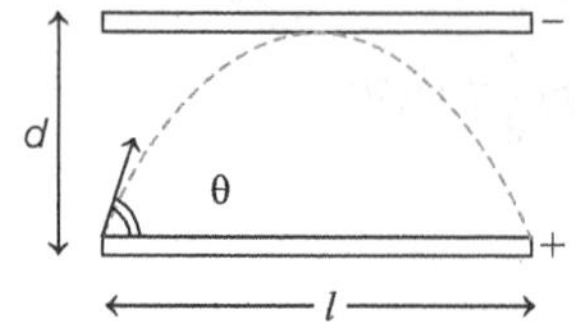

   (a) $\tan\theta = 2d/l$
   (b) $\tan\theta = 4d/l$
   (c) $\tan\theta = 8d/l$
   (d) The data given is insufficient to find a relation between d, l and θ

6. An infinite sheet of charge has a surface charge density of $10^{-7}\ C/m^2$. The separation between two equipotential surfaces whose potentials differ by 5 V is
   (a) 0.64 cm
   (b) 0.88 mm
   (c) 0.32 cm
   (d) 5×10^{-7} m

7. Consider a system composed of two metallic spheres of radii r_1 and r_2 connected by a thin wire and switch S as shown in the figure. Initially, S is in open position, and the spheres carry charges q_1 and q_2 respectively. If the switch is closed, the potential of the system is

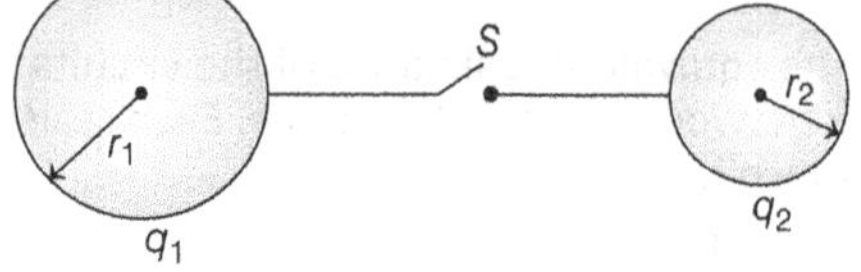

   (a) $\dfrac{1}{4\pi\varepsilon_0}\dfrac{q_1q_2}{r_1r_2}$ (b) $\dfrac{1}{4\pi\varepsilon_0}\left(\dfrac{q_1+q_2}{r_1+r_2}\right)$
   (c) $\dfrac{1}{4\pi\varepsilon_0}\left(\dfrac{q_1}{r_1}+\dfrac{q_2}{r_2}\right)$ (d) $\dfrac{1}{4\pi\varepsilon_0}\left(\dfrac{q_1+q_2}{\sqrt{r_1r_2}}\right)$

8. The distance between two plates of a capacitor is d and its capacitance is C_1, when air is the medium between the plates. If a metal sheet of thickness $\dfrac{2d}{3}$ and of the same area as plate is introduced between the plates, the capacitance of the capacitor becomes C_2. The ratio $\dfrac{C_2}{C_1}$ is **(2023 Main)**
   (a) 3 : 1 (b) 1 : 1
   (c) 4 : 1 (d) 2 : 1

9. A parallel plate capacitor has plate area 40 cm^2 and plates separation 2 mm. The space between the plates is filled with a dielectric medium of a thickness 1mm and dielectric constant 5. The capacitance of the system is **(2023 Main)**
   (a) $24\varepsilon_0$ F (b) $\dfrac{3}{10}\varepsilon_0$ F
   (c) $\dfrac{10}{3}\varepsilon_0$ F (d) $10\ \varepsilon_0$ F

10. The equivalent capacitance of the combination shown is **(2023 Main)**

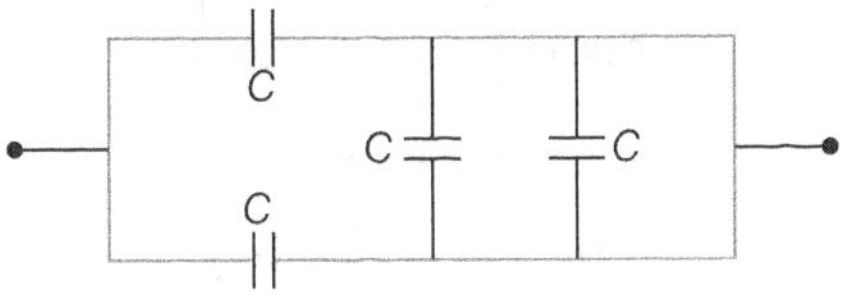

(a) $\frac{5}{3}C$ (b) $2C$
(c) $\frac{C}{2}$ (d) $4C$

11. A slab of dielectric constant K has the same cross-sectional area as the plates of a parallel plate capacitor and thickness $\frac{3}{4}d$, where d is the separation of plates. The capacitance of the capacitor when the slab is inserted between the plates will be
(Given, C_0 = capacitance of capacitor with air as medium between plates.) **(2022 Main)**

(a) $\frac{4KC_0}{3+K}$ (b) $\frac{3KC_0}{3+K}$
(c) $\frac{3+K}{4KC_0}$ (d) $\frac{K}{4+K}$

12. A parallel plate capacitor is made of two square plates of side a, separated by a distance d $(d << a)$. The lower triangular portion is filled with a dielectric of dielectric constant K, as shown in the figure.
Capacitance of this capacitor is **(2019 Main)**

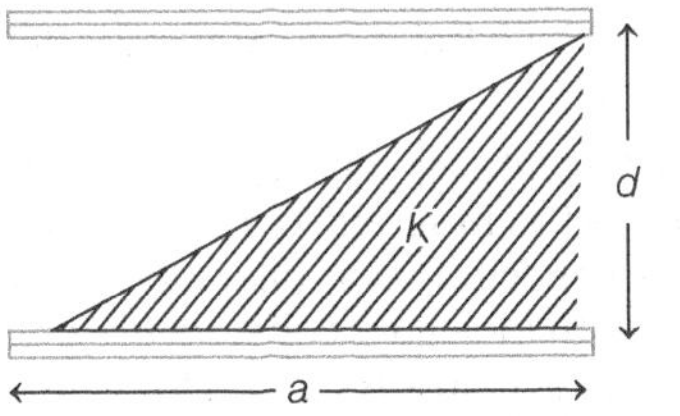

(a) $\frac{K\varepsilon_0 a^2}{2d(K+1)}$ (b) $\frac{K\varepsilon_0 a^2}{d(K-1)}\ln K$
(c) $\frac{K\varepsilon_0 a^2}{d}\ln K$ (d) $\frac{1}{2}\frac{K\varepsilon_0 a^2}{d}$

13. A parallel plate capacitor is of area 6 cm^2 and a separation 3 mm. The gap is filled with three dielectric materials of equal thickness (see figure) with dielectric constants $K_1 = 10$, $K_2 = 12$ and $K_3 = 14$. The dielectric constant of a material which give same capacitance when fully inserted in above capacitor, would be **(2019 Main)**

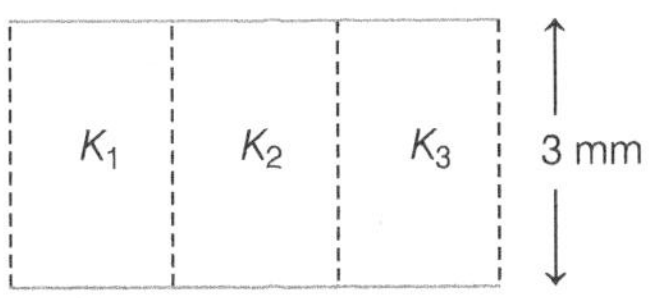

(a) 4 (b) 36
(c) 12 (d) 14

14. In the circuit shown, find C if the effective capacitance of the whole circuit is to be 0.5 μF. All values in the circuit are in μF. **(2019 Main)**

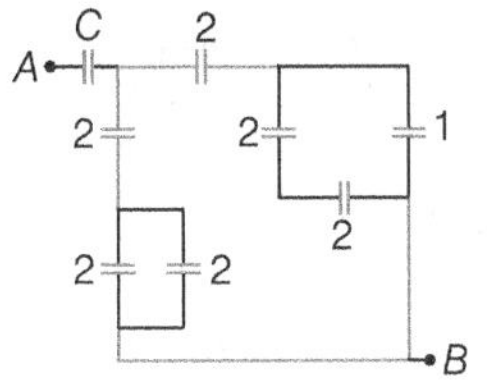

(a) $\frac{6}{5}$ μF (b) 4 μF
(c) $\frac{7}{10}$ μF (d) $\frac{7}{11}$ μF

15. A parallel plate capacitor having plates of area S and plate separation d, has capacitance C_1 in air. When two dielectrics of different relative permittivity ($\varepsilon_1 = 2$ and $\varepsilon_2 = 4$) are introduced between the two plates as shown in the figure, the capacitance becomes C_2. The ratio $\frac{C_2}{C_1}$ is **(2015 Adv.)**

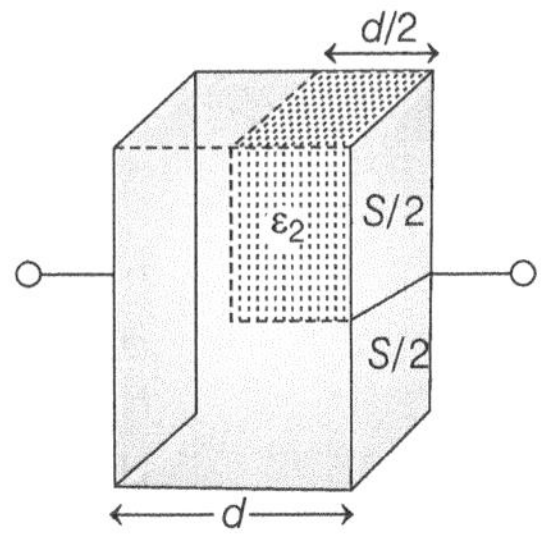

(a) 6/5 (b) 5/3
(c) 7/5 (d) 7/3

16. Three capacitors each of 4 μF are to be connected in such a way that the effective capacitance is 6 μF. This can be done by connecting them **(2016 Main)**
(a) all in series
(b) all in parallel
(c) two in parallel and one in series
(d) two in series and one in parallel

17. Seven capacitors each of capacitance 2 μF are to be connected in a configuration to obtain an effective capacitance of $\left(\frac{10}{11}\right)$ μF. Which of the combination (s) shown in figure will achieve the desired result? **(1990 JEE)**

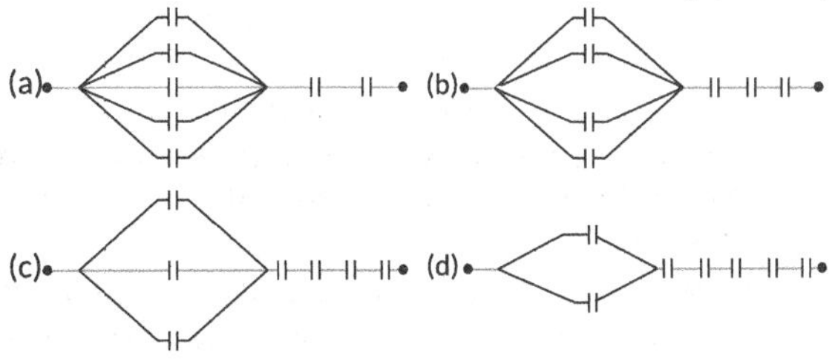

18. In the circuit shown in figure, the capacitor is charged with a cell of 5 V. If the switch is closed at $t=0$, then at $t=12$ s, charge on the capacitor is

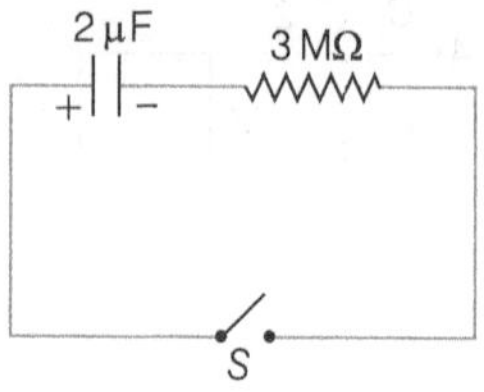

(a) $(0.37)\,10\,\mu C$ (b) $(0.37)^2\,10\,\mu C$
(c) $(0.63)\,10\,\mu C$ (d) $(0.63)^2\,10\,\mu C$

19. A capacitor C_1 is charged to a potential V and connected to another capacitor in series with a resistor R as shown. It is observed that heat H_1 is dissipated across resistance R, till the circuit reaches steady state. Same process is repeated using resistance of $2R$. If H_2 is heat dissipated in this case, then

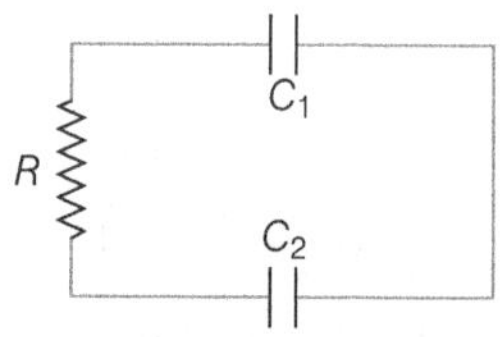

(a) $\frac{H_2}{H_1}=1$ (b) $\frac{H_2}{H_1}=4$
(c) $\frac{H_2}{H_1}=\frac{1}{4}$ (d) $\frac{H_2}{H_1}=2$

20. A leaky parallel capacitor is filled completely with a material having dielectric constant $K=5$ and electrical conductivity $\sigma=7.4\times10^{-12}\,\Omega^{-1}\text{m}^{-1}$. Charge on the plate at instant $t=0$ is $q=8.885\,\mu C$. Then, time constant of leaky capacitor is
(a) 3 s (b) 4 s
(c) 5 s (d) 6 s

21. An uncharged capacitor with a solid dielectric is connected to a similar air capacitor charged to a potential of V_0. If the common potential after sharing of charges becomes V, then the dielectric constant of the dielectric must be
(a) V_0/V (b) $\frac{V}{V_0}$ (c) $\frac{(V_0-V)}{V}$ (d) $\frac{(V_0-V)}{V_0}$

Subjective Questions

22. The dielectric to be used in a parallel-plate capacitor has a dielectric constant of 3.60 and a dielectric strength of 1.60×10^7 V/m. The capacitor is to have a capacitance of 1.25×10^{-9} F and must be able to withstand a maximum potential difference of 5500 V. What is the minimum area the plates of the capacitor may have?

23. (a) What is the steady state potential of point a with respect to point b in figure when switch S is open?

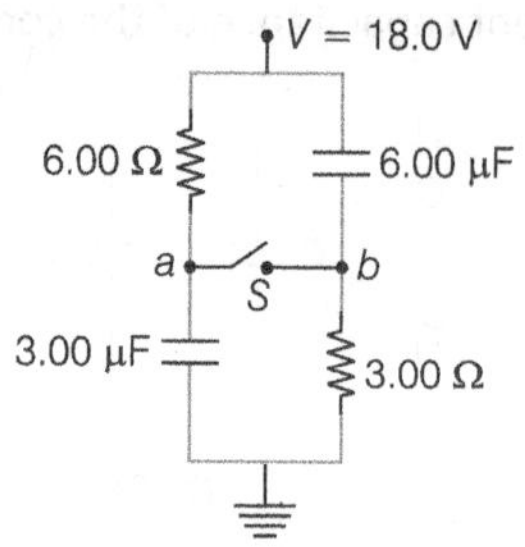

(b) Which point, a or b, is at the higher potential?
(c) What is the final potential of point b with respect to ground when switch S is closed?
(d) How much does the charge on each capacitor change when S is closed?

24. (a) What is the potential of point a with respect to point b in figure, when switch S is open?

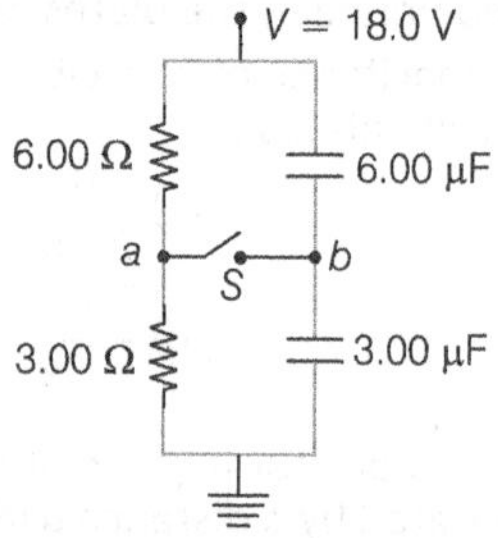

(b) Which point, a or b, is at the higher potential?
(c) What is the final potential of point b with respect to ground when switch S is closed?
(d) How much charge flows through switch S when it is closed?

25. Two parallel plate air capacitors, each of capacitance C, were connected in series to a battery with emf E. Then one of the capacitors was filled up with uniform dielectric with dielectric constant K. How many times did the electric field strength in that capacitor decrease? What amount of charge flows through the battery? **(Irodov 3.102)**

26. A capacitor of capacitance $C_1=1.0\,\mu F$ withstands the maximum voltage $V_1=6.0$ kV while another capacitor of capacitance $C_2=2.0\,\mu F$, the maximum voltage $V_2=4.0$ kV. What voltage will the system of these two capacitors withstand if they are connected in series? **(Irodov 3.114)**

27. Find the potential difference between points A and B of the system as shown in figure below, if the emf is equal to $E=110$ V and the capacitance ratio $\frac{C_2}{C_1}=2.0$. **(Irodov 3.115)**

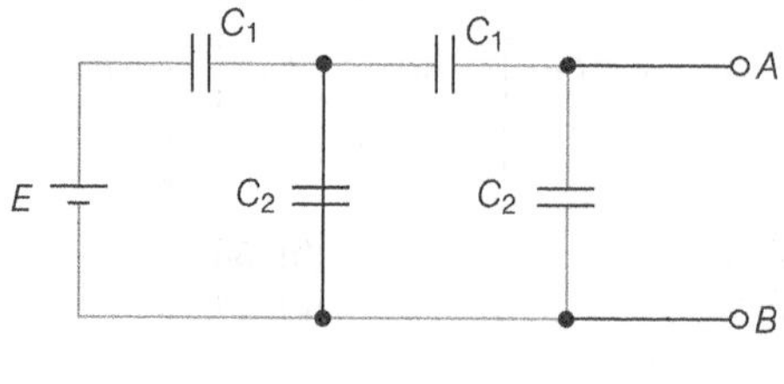

28. A capacitor of capacitance $C_1 = 1.0\ \mu F$ charged up to a voltage $V = 110$ V is connected in parallel to the terminals of a circuit consisting of two uncharged capacitors connected in series and possessing the capacitances $C_2 = 2.0\ \mu F$ and $C_3 = 3.0\ \mu F$. What charge will flow through the connecting wires? **(Irodov 3.121)**

29. What amount of heat will be generated in the circuit shown in figure after the switch S is shifted from position 1 to position 2? **(Irodov 3.133)**

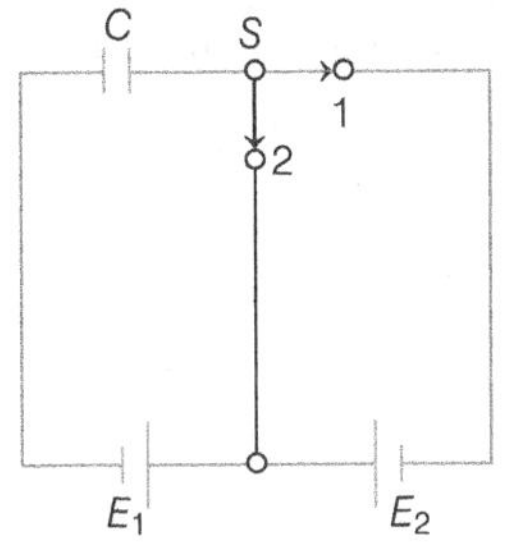

30. Each plate of a parallel plate air capacitor has an area S. What amount of work has to be performed to slowly increase the distance between the plates from x_1 to x_2 if
(a) the charge on the capacitor, which is equal to q, or
(b) the voltage across the capacitor, which is equal to V, is kept constant in the process? **(Irodov 3.141)**

31. Inside a parallel plate capacitor there is a plate parallel to the outer plates, whose thickness is equal to $\eta = 0.60$ of the gap width. When the plate is absent the capacitor capacitance equals $C = 20$ nF. First, the capacitor was connected in parallel to a constant voltage source producing $V = 200$ V, then it was disconnected from it, after which the plate was slowly removed from the gap. Find the work performed during the removal, if the plate is
(a) made of metal; (b) made of glass. **(Irodov 3.142)**

32. The space between two conducting concentric spheres of radii a and b $(a < b)$ is filled up with homogeneous poorly conducting medium. The capacitance of such a system equals C. Find the resistivity of the medium if the potential difference between the spheres, when they are disconnected from an external voltage, decreases η-times during the time interval Δt. **(Irodov 3.156)**

33. The gap between the plates of a parallel plate capacitor is filled with glass of resistivity $\rho = 100\ G\Omega \cdot m$. The capacitance of the capacitor equals $C = 4.0$ nF. Find the leakage current of the capacitor when a voltage $V = 2.0$ kV is applied to it. **(Irodov 3.160)**

34. The gap between the plates of a parallel plate capacitor is filled with an inhomogeneous poorly conducting medium whose conductivity varies linearly in the direction perpendicular to the plates from $\sigma_1 = 1.0 p\Omega^{-1}m^{-1}$ to $\sigma_2 = 2.0\ p\Omega^{-1}m^{-1}$. Each plate has an area $A = 230\ cm^2$ and the separation between the plates is $d = 2.0$ mm. Find the current flowing through the capacitor due to a voltage $V = 300$ V. **(Irodov 3.163)**

35. A capacitor with capacitance $C = 400$ pF is connected *via* a resistance $R = 650\ \Omega$ to a source of constant voltage V_0. How soon will the voltage developed across the capacitor reach a value $V = 0.90\ V_0$? **(Irodov 3.170)**

36. A capacitor filled with dielectric of dielectric constant $K = 2.1$, loses half the charge acquired during a time interval $\tau = 3.0$ min. Assuming the charge to leak only through the dielectric filler, calculate its resistivity. **(Irodov 3.171)**

Integer Type Questions

37. A parallel plate capacitor with plate area A and plate separation d is filled with a dielectric material of dielectric constant $K = 4$. The thickness of the dielectric material is x, where $x < d$.
Let C_1 and C_2 be the capacitance of the system for $x = \frac{1}{3}d$ and $x = \frac{2d}{3}$, respectively. If $C_1 = 2\ \mu F$ the value of C_2 is .... μF. **(2023 Main)**

38. A parallel plate capacitor with air between the plates has a capacitance of 15 pF. The separation between the plate becomes twice and the space between them is filled with a medium of dielectric constant 3.5. Then the capacitance becomes $\frac{x}{4}$ pF. The value of x is ...... pF. **(2023 Main)**

39. A capacitor has capacitance $5\ \mu F$ when it's parallel plates are separated by air medium of thickness d. A slab of material of dielectric constant 1.5 having area equal to that of plates but thickness $d/2$ is inserted between the plates. Capacitance of the capacitor in the presence of slab will be ...... μF. **(2023 Main)**

40. As shown in the figure, two parallel plate capacitors having equal plate area of 200 cm² are joined in such a way that $a \neq b$. The equivalent capacitance of the combination is $x\ \varepsilon_0 F$. The value of x is .... **(2023 Main)**

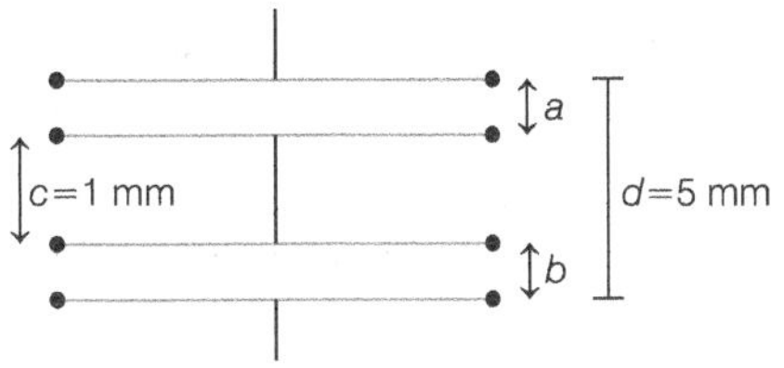

41. A parallel plate capacitor is made up of stair like structure with a plate area A of each stair and that is connected with a wire of length b, as shown in the figure. The capacitance of the arrangement is $\frac{x}{15}\frac{\varepsilon_0 A}{b}$.
The value of x is .... **(2022 Main)**

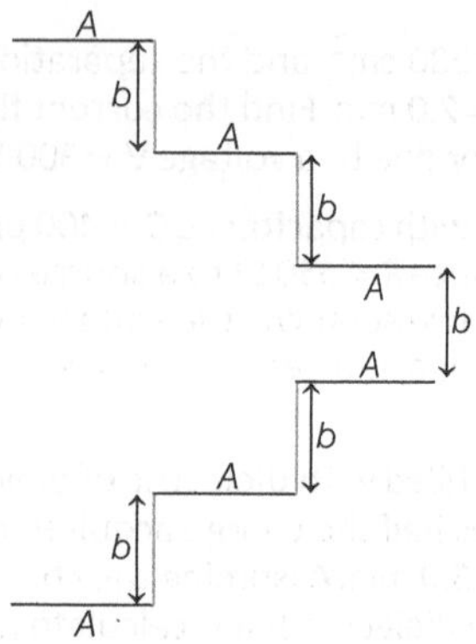

Level 2

Single Correct Option

42. The electric field on two sides of a thin sheet of charge is shown in the figure. The charge density on the sheet is

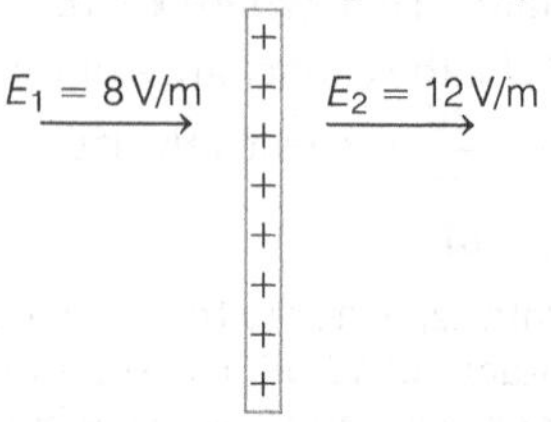

(a) $2\varepsilon_0$ (b) $4\varepsilon_0$
(c) $10\varepsilon_0$ (d) zero

43. A capacitor of capacitance C is charged by a battery of emf E and internal resistance r. A resistance $2r$ is also connected in series with the capacitor. The amount of heat liberated inside the battery by the time capacitor is 50% charged is

(a) $\frac{8}{8}E^2C$ (b) $\frac{E^2C}{6}$ (c) $\frac{E^2C}{12}$ (d) $\frac{E^2C}{24}$

44. A fully charged capacitor has a capacitance C. It is discharged through a small coil of resistance wire embedded in a thermally insulated block of specific heat capacity s and mass m. If the temperature of the block is raised by ΔT, the potential difference V across the capacitance is

(a) $\sqrt{\frac{2mC\Delta T}{s}}$ (b) $\frac{mC\Delta T}{s}$ (c) $\frac{ms\Delta T}{C}$ (d) $\sqrt{\frac{2ms\Delta T}{C}}$

45. The switch shown in the figure is closed at $t = 0$. The charge on the capacitor as a function of time is given by

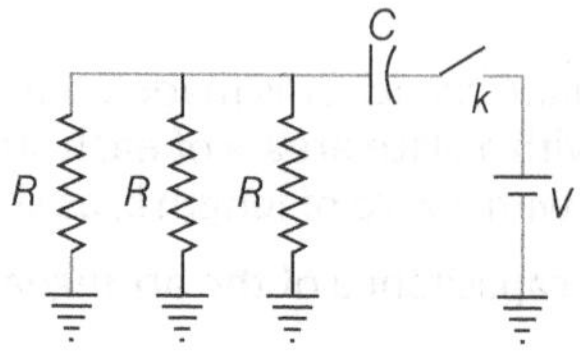

(a) $CV(1-e^{-t/RC})$ (b) $3CV(1-e^{-t/RC})$
(c) $CV(1-e^{-3t/RC})$ (d) $CV(1-e^{-t/3RC})$

46. Let C be the capacitance of a capacitor discharging through a resistor R. Suppose t_1 is the time taken for the energy stored in the capacitor to reduce to half its initial value and t_2 is the time taken for the charge to reduce to one-fourth its initial value. Then, the ratio t_1/t_2 will be

(a) 1 (b) $1/2$ (c) $1/4$ (d) 2

47. Four capacitors are connected in series with a battery of emf 10 V as shown in the figure. The point P is earthed. The potential of point A is equal in magnitude to potential of point B but opposite in sign if

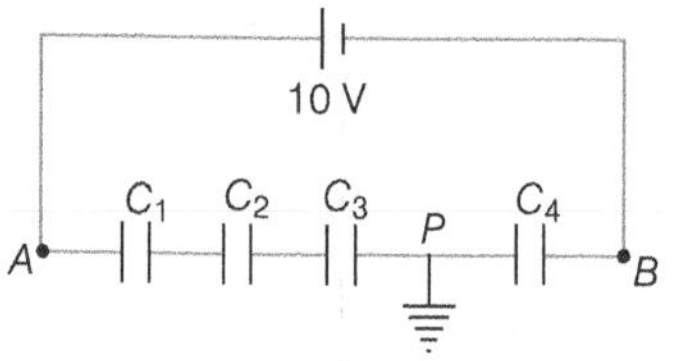

(a) $C_1 + C_2 + C_3 = C_4$ (b) $\frac{1}{C_1} + \frac{1}{C_2} + \frac{1}{C_3} = \frac{1}{C_4}$
(c) $\frac{C_1C_2C_3}{C_1^2 + C_2^2 + C_3^2} = C_4$ (d) It is never possible

48. A capacitor of capacitance 2 μF is charged to a potential difference of 5 V. Now, the charging battery is disconnected and the capacitor is connected in parallel to a resistor of 5 Ω and another unknown resistor of resistance R as shown in figure. If the total heat produced in 5 Ω resistance is 10 μJ, then the unknown resistance R is equal to

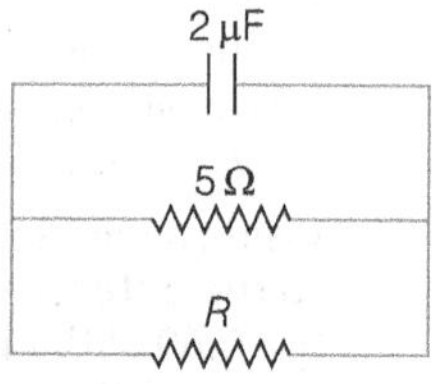

(a) 10Ω (b) 15Ω (c) $\frac{10}{3}$Ω (d) 7.5Ω

49. Consider the arrangement of three plates X, Y and Z each of the area A and separation d. The energy stored when the plates are fully charged is

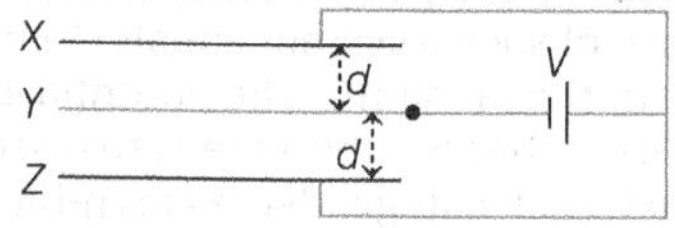

(a) $\varepsilon_0 AV^2/2d$ (b) $\varepsilon_0 AV^2/d$
(c) $2\varepsilon_0 AV^2/d$ (d) $3\varepsilon_0 AV^2/d$

50. A capacitor C is connected to two equal resistances as shown in the figure. Consider the following statements.

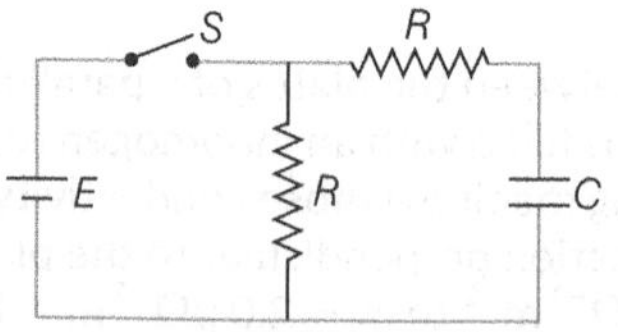

(i) At the time of charging of capacitor time constant of the circuit is $2CR$

(ii) At the time of discharging of the capacitor the time constant of the circuit is CR

(iii) At the time of discharging of the capacitor the time constant of the circuit is $2CR$

(iv) At the time of charging of capacitor the time constant of the circuit is CR

(a) Statements (i) and (ii) are correct
(b) Statements (ii) and (iii) are correct
(c) Statements (iii) and (iv) are correct
(d) Statements (i) and (iii) are correct

51. Three uncharged capacitors of capacitance C_1, C_2 and C_3 are connected to one another as shown in figure. The potential at O will be

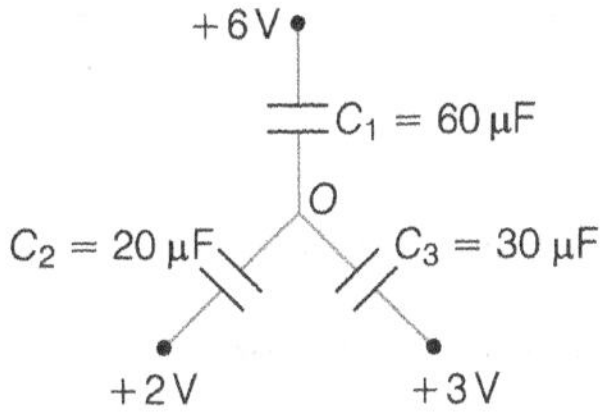

(a) 3 V (b) $\frac{49}{11}$ V
(c) 4 V (d) $\frac{3}{11}$ V

52. In the circuit shown in figure, the potential difference between the points A and B in the steady state is

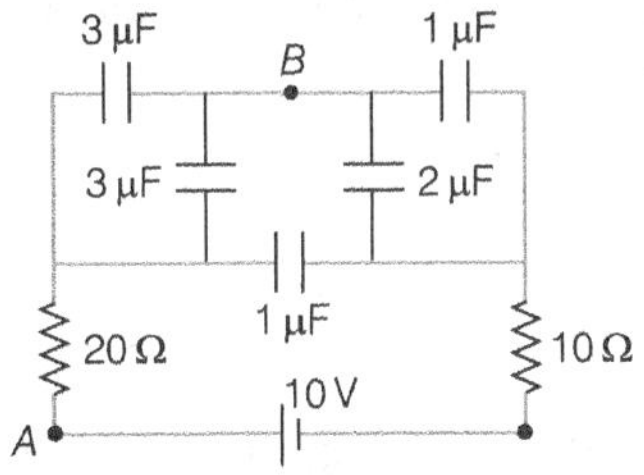

(a) zero (b) 6 V
(c) 4 V (d) $\frac{10}{3}$ V

53. Two cells, two resistors and two capacitors are connected as shown in figure. The charge on 2 μF capacitor is

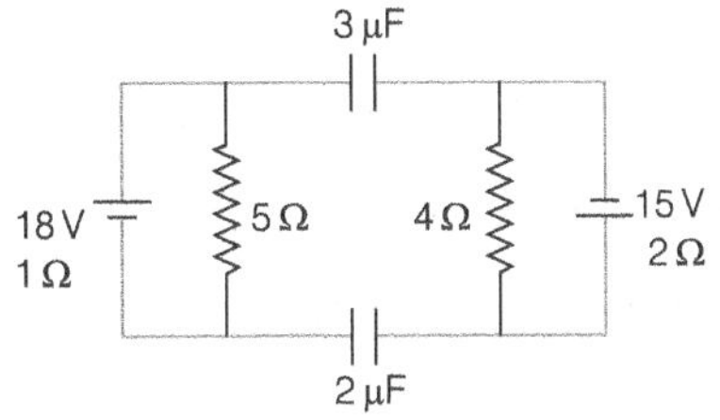

(a) 30 μC (b) 20 μC
(c) 25 μC (d) 48 μC

54. In the circuit shown, switch S is closed at $t = 0$. Let i_1 and i_2 be the current at any finite time t, then the ratio i_1/i_2

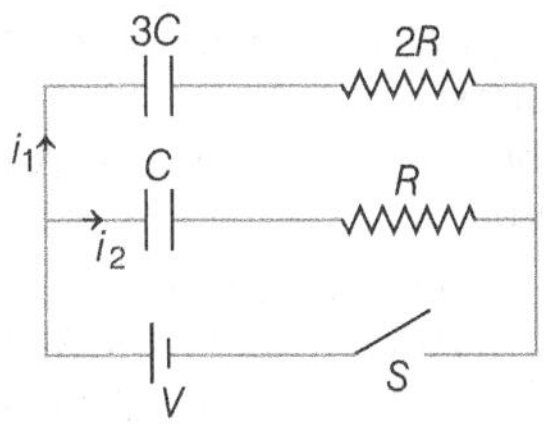

(a) is constant
(b) increases with time
(c) decreases with time
(d) first increases and then decreases

55. A charged capacitor is allowed to discharge through a resistor by closing the key at the instant $t = 0$. At the instant $t = (\ln 4)$ μs, the reading of the ammeter falls half the initial value. The resistance of the ammeter is equal to

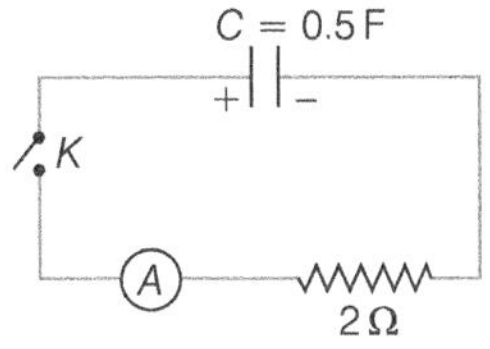

(a) 0.5 Ω (b) 1 Ω (c) 2 Ω (d) 4 Ω

56. Five identical capacitor plates are arranged such that they make four capacitors each of 2 μF. The plates are connected to a source of emf 10 V. The charge on plate C is

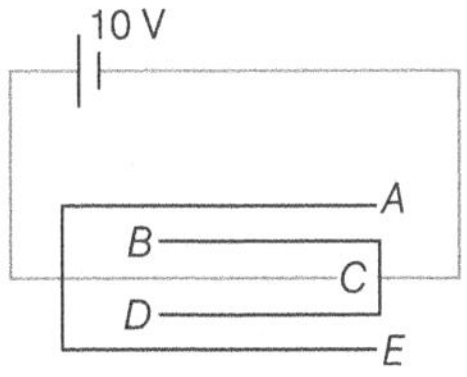

(a) +20 μC (b) +40 μC (c) +60 μC (d) +80 μC

57. The capacities and connection of five capacitors are shown in the figure. The potential difference between the points A and B is 60 V. Then, the equivalent capacity between A and B and the charge on 5 μF capacitance will be respectively

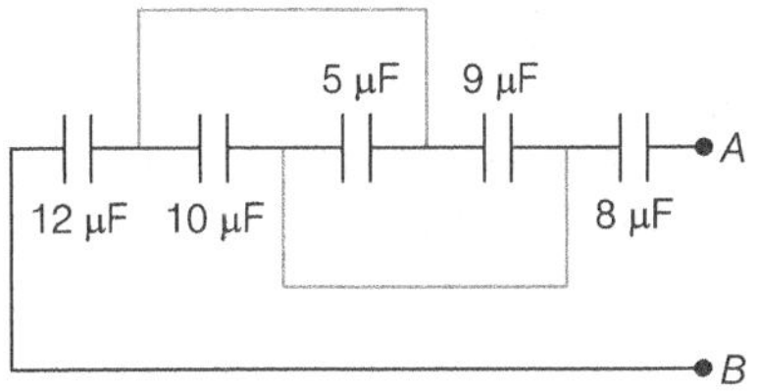

(a) 44 μF, 300 μC (b) 16 μF, 150 μC
(c) 15 μF, 200 μC (d) 4 μF, 50 μC

58. A number of condensers, each of the capacitance 1 μF and each one of which gets punctured if a potential difference just exceeding 500 V is applied, are provided. An arrangement suitable for giving

capacitance of 2 μF across which 3000 V may be applied requires

(a) 6 capacitors (b) 12 capacitors
(c) 72 capacitors (d) 2 capacitors

59. In the given circuit, charge Q_2 on the 2 μF capacitor changes as C is varied from 1 μF to 3 μF. Q_2 as a function of C is given properly by (figures are drawn schematically and are not to scale) **(2015 Main)**

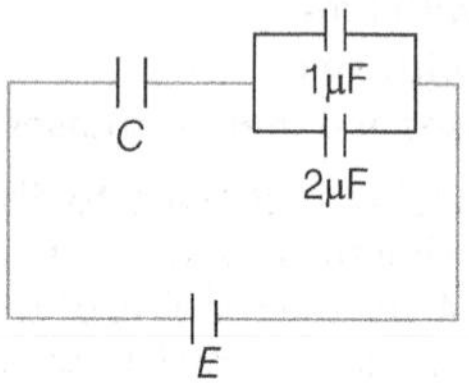

(a)
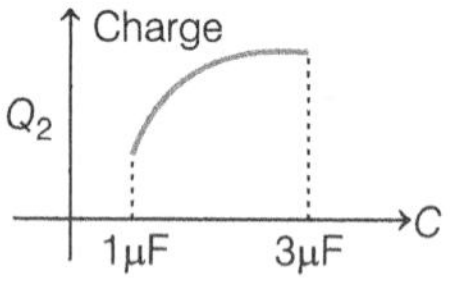

(b)
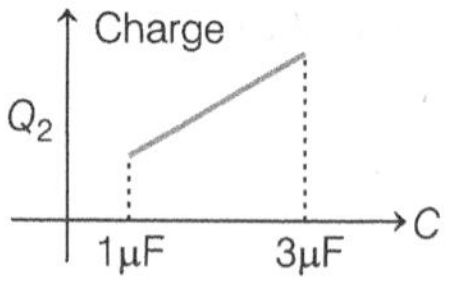

(c)
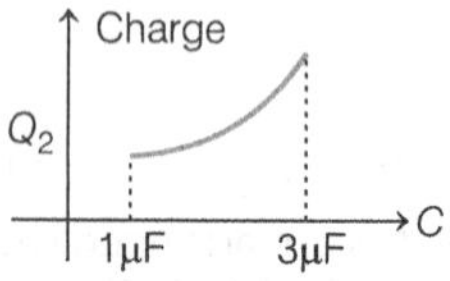

(d)
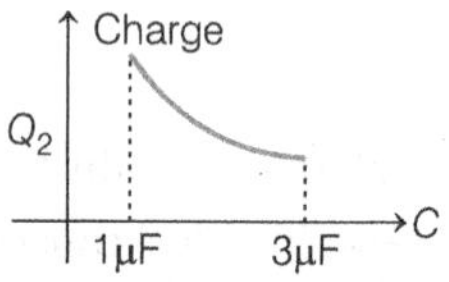

Multiple Correct Options

60. Given that potential difference across 1 μF capacitor is 10 V. Then,

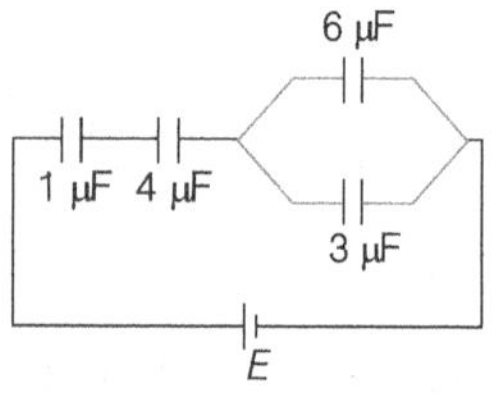

(a) potential difference across 4 μF capacitor is 40 V
(b) potential difference across 4 μF capacitor is 2.5 V
(c) potential difference across 3 μF capacitor is 5 V
(d) value of E is 50 V

61. An electrical circuit is shown in the given figure. The resistance of each voltmeter is infinite and each ammeter is 100 Ω. The charge on the capacitor of 100 μF in steady state is 4 mC. Choose correct statement(s) regarding the given circuit.

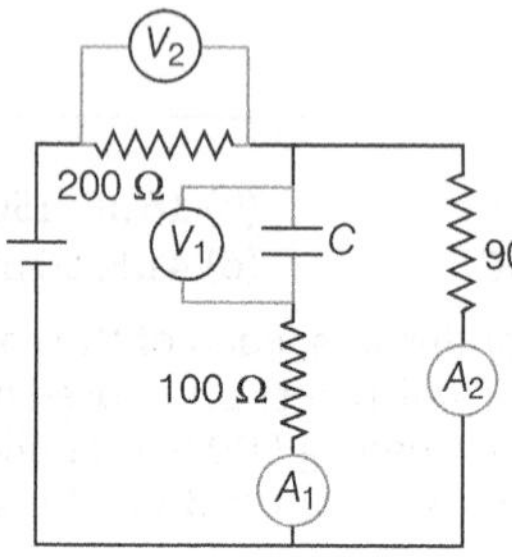

(a) Reading of voltmeter V_2 is 16 V
(b) Reading of ammeter A_1 is zero and A_2 is $\frac{1}{25}$ A
(c) Reading of voltmeter V_1 is 40 V
(d) Emf of the ideal cell is 66 V

62. In the circuit shown, A and B are equal resistances. When switch S is closed, the capacitor C charges from the cell of emf ε and reaches a steady state.

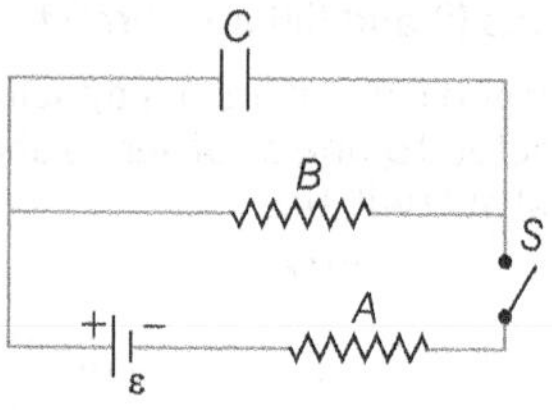

(a) During charging, more heat is produced in A than in B
(b) In steady state, heat is produced at the same rate in A and B
(c) In the steady state, energy stored in C is $\frac{1}{4}C\varepsilon^2$
(d) In the steady state energy stored in C is $\frac{1}{8}C\varepsilon^2$

63. Two large metal plates are originally oriented horizontally and separated by a distance of 3 d. A conducting wire joins them, and initially plates are uncharged.

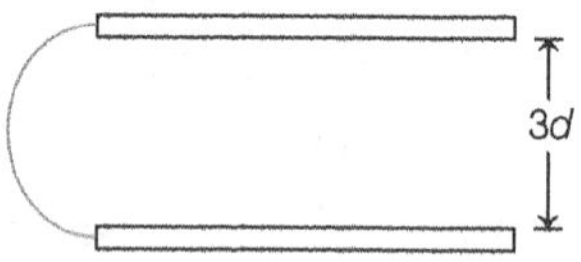

A third identical plate with charge Q is put parallel to them at a distance of d from upper plate.

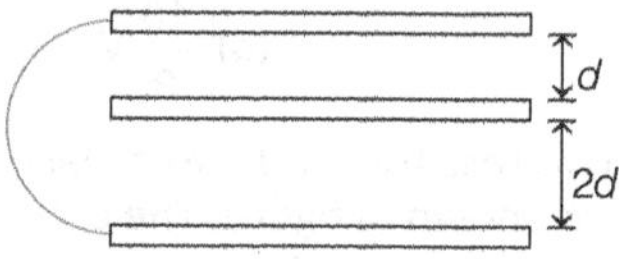

Then,

(a) charge on lower plate is $-\frac{Q}{3}$
(b) charge on upper plate is $-\frac{2Q}{3}$
(c) potential difference between upper and middle plate is $\frac{2Qd}{3\varepsilon_0 A}$
(d) potential difference between upper and lower plate is $\frac{Qd}{3\varepsilon_0 A}$

64. In the circuit shown, each capacitor has a capacitance C. The emf of the cell is E. If the switch S is closed, then

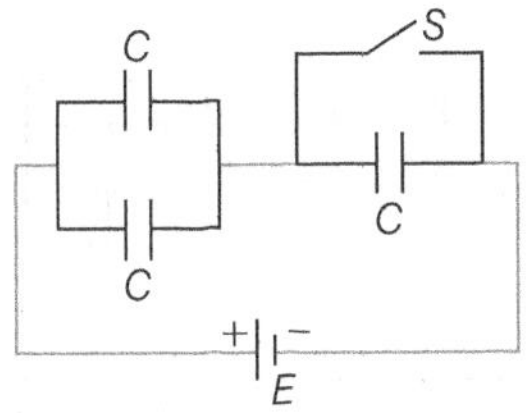

(a) positive charge will flow out of the positive terminal of the cell
(b) positive charge will enter the positive terminal of the cell
(c) the amount of the charge flowing through the cell will be $\frac{1}{3}CE$
(d) the amount of charge flowing through the cell is $\left(\frac{4}{3}\right)CE$

65. Two capacitors of 2 μF and 3 μF are charged to 150 V and 120 V, respectively. The plates of capacitor are connected as shown in the figure. An uncharged capacitor of capacity 1.5 μF falls to the free end of the wire. Then,

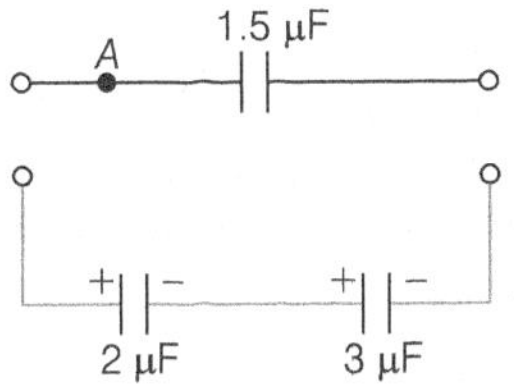

(a) charge on 1.5 μF capacitor is 180 μC
(b) charge on 2 μF capacitor is 120 μC
(c) positive charge flows through A from right to left
(d) positive charge flows through A from left to right

66. In the given circuit, the relationship between voltages of point A, B and C are

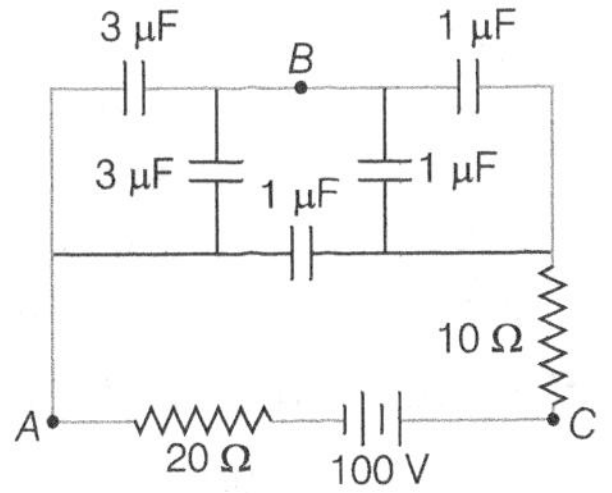

(a) $V_{AB}=25\,V$ (b) $V_{AB}=27\,V$
(c) $V_{BC}=75\,V$ (d) $V_{BC}=50\,V$

67. A capacitor of 2 F (practically not possible to have a capacity of 2 F) is charged by a battery of 6 V. The battery is removed and circuit is made as shown. Switch is closed at time $t=0$. Choose the correct options.

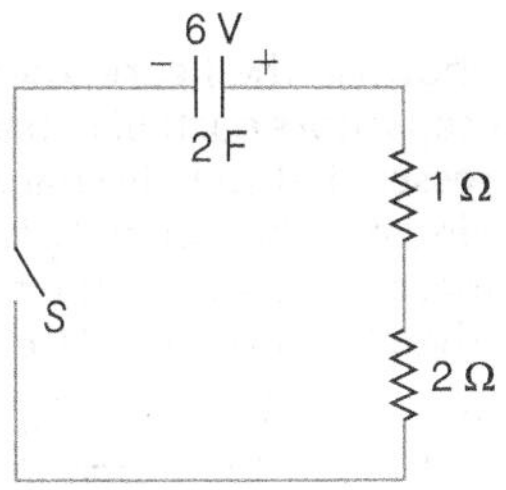

(a) At time $t=0$ current in the circuit is 2 A
(b) At time $t=(6\ln 2)$ second, potential difference across capacitor is 3 V
(c) At time $t=(6\ln 2)$ second, potential difference across 1Ω resistance is 1V
(d) At time $t=(6\ln 2)$ second, potential difference across 2Ω resistance is 2V.

68. In the circuit shown in the figure, switch S is closed at time $t=0$. Select the correct statements.

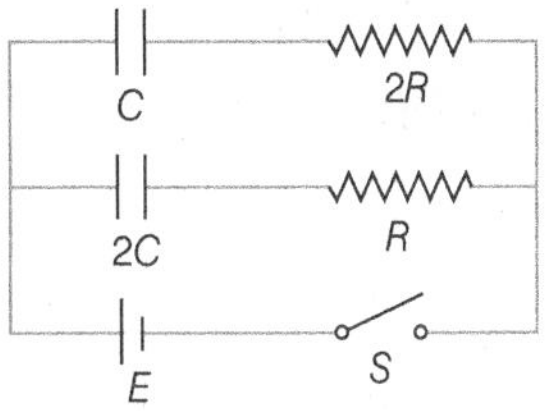

(a) Rate of increase of charge is same in both the capacitors
(b) Ratio of charge stored in capacitors C and 2C at any time t would be 1 : 2
(c) Time constants of both the capacitors are equal
(d) Steady state charges on capacitors C and 2C are in the ratio of 1 : 2

69. A parallel plate capacitor has a dielectric slab of dielectric constant K between its plates that covers 1/3 of the area of its plates, as shown in the figure. The total capacitance of the capacitor is C while that of the portion with dielectric in between is C_1. When the capacitor is charged, the plate area covered by the dielectric gets charge Q_1 and the rest of the area gets charge Q_2. The electric field in the dielectric is E_1 and that in the other portion is E_2. Choose the correct option/options, ignoring edge effects. **(2014 Adv.)**

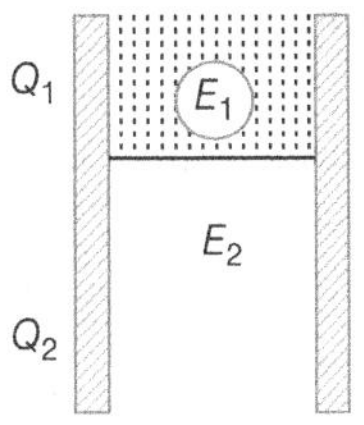

(a) $\frac{E_1}{E_2}=1$ (b) $\frac{E_1}{E_2}=\frac{1}{K}$ (c) $\frac{Q_1}{Q_2}=\frac{3}{K}$ (d) $\frac{C}{C_1}=\frac{2+K}{K}$

70. In the circuit shown in the figure, there are two parallel plate capacitors each of capacitance C. The switch S_1 is pressed first to fully charge the capacitor C_1 and then released. The switch S_2 is then pressed to charge the capacitor C_2. After some time, S_2 is released and then S_3 is pressed. After some time

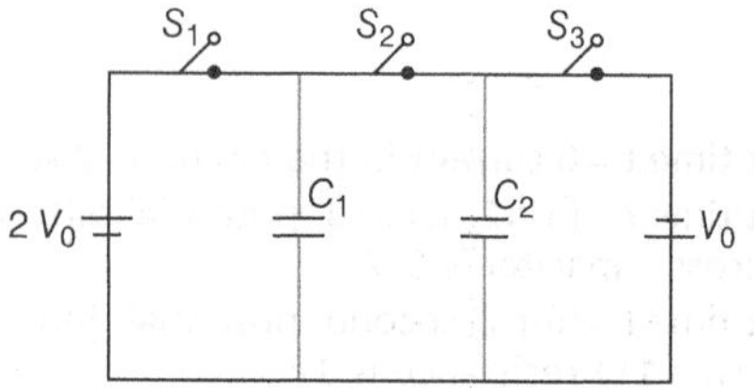

(a) the charge on the upper plate of C_1 is $2\,CV_0$
(b) the charge on the upper plate of C_1 is $+CV_0$
(c) the charge on the upper plate of C_2 is 0
(d) the charge on the upper plate of C_2 is $-CV_0$

Passage Based Questions

Passage 1 (Q.No. 71-72)

The capacitor C_1 in the figure shown initially carries a charge q_0. When the switches S_1 and S_2 are closed, capacitor C_1 is connected in series to a resistor R and a second capacitor C_2, which is initially uncharged.

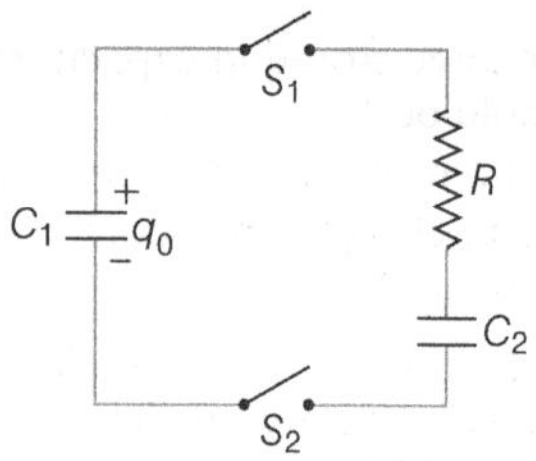

71. The charge flown through wires as a function of time t is

(a) $q_0 e^{-t/RC} + \frac{C}{C_2} q_0$ (b) $\frac{q_0 C}{C_1} \times [1 - e^{-t/RC}]$
(c) $q_0 \frac{C}{C_1} e^{-t/CR}$ (d) $q_0 e^{-t/RC}$

where, $C = \frac{C_1 C_2}{C_1 + C_2}$

72. The total heat dissipated in the circuit during the discharging process of C_1 is

(a) $\frac{q_0^2}{2C_1^2} \times C$ (b) $\frac{q_0^2}{2C}$ (c) $\frac{q_0^2 C_2}{2C_1^2}$ (d) $\frac{q_0^2}{2C_1 C_2}$

Passage 2 (Q.No. 73-74)

Figure shows a parallel plate capacitor with plate area A and plate separation d. A potential difference is being applied between the plates. The battery is then disconnected and a dielectric slab of dielectric constant K is placed in between the plates of the capacitor as shown.

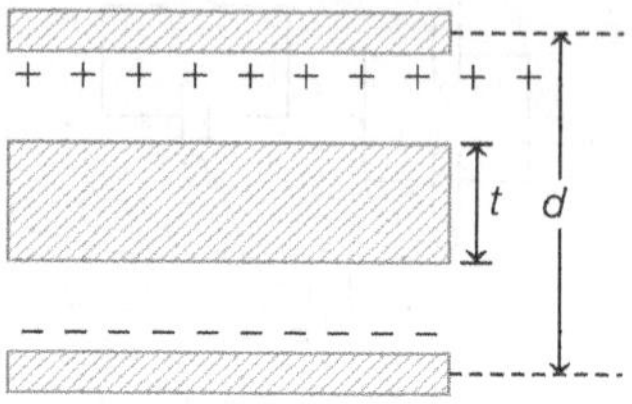

Now, answer the following questions based on above information.

73. The electric field in the gaps between the plates and the dielectric slab will be

(a) $\frac{\varepsilon_0 AV}{d}$ (b) $\frac{V}{d}$
(c) $\frac{KV}{d}$ (d) $\frac{V}{d-t}$

74. The electric field in the dielectric slab is

(a) $\frac{V}{Kd}$ (b) $\frac{KV}{d}$
(c) $\frac{V}{d}$ (d) $\frac{KV}{t}$

Passage 3 (Q.No. 75-78)

Three conducting shells A, B and C of radii a, b and c are as shown. Suppose, a source of emf is connected between shells A and C so that shell A acquires a charge q and shell C has a charge $-q$.

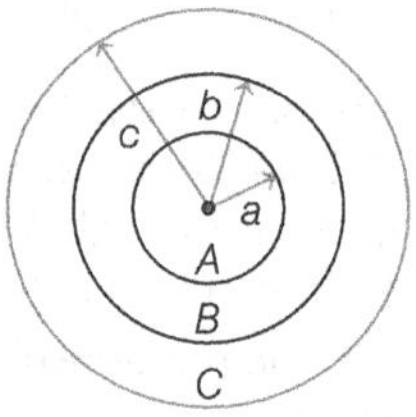

75. Potential of shell A is

(a) $\frac{q}{4\pi\varepsilon_0 a}$ (b) $\frac{q}{4\pi\varepsilon_0 c}$
(c) $\frac{q}{4\pi\varepsilon_0}\left(\frac{1}{c} - \frac{1}{a}\right)$ (d) $\frac{q}{4\pi\varepsilon_0}\left(\frac{1}{a} - \frac{1}{c}\right)$

76. Potential difference between shells A and C is

(a) $\frac{q}{4\pi\varepsilon_0 a}$ (b) $\frac{q}{4\pi\varepsilon_0 c}$
(c) $\frac{q}{4\pi\varepsilon_0}\left(\frac{1}{c} - \frac{1}{a}\right)$ (d) $\frac{q}{4\pi\varepsilon_0}\left(\frac{1}{a} - \frac{1}{c}\right)$

77. Capacity of the system is

(a) $4\pi\varepsilon_0 a$ (b) $4\pi\varepsilon_0 b$
(c) $4\pi\varepsilon_0\left(\frac{ac}{c-a}\right)$ (d) $\frac{4\pi\varepsilon_0 ac}{(a-c)}$

78. Suppose, a dielectric medium of dielectric constant K is filled between A and B. Then, potential difference between A and C is

(a) $\dfrac{q}{4\pi\varepsilon_0 Kabc}[c(b-a)+Ka(c-b)]$

(b) $\dfrac{q}{4\pi\varepsilon_0 Kabc}[Kc(b-a)+a(c-b)]$

(c) $\dfrac{q}{4\pi\varepsilon_0 abc}K[c(b-a)+a(c-b)]$

(d) $\dfrac{q}{4\pi\varepsilon_0 K}(abc)[(a-b)+(b-c)]$

Passage 4 (Q.No. 79-81)

A combination of capacitors given is charged by a cell of emf E as shown :

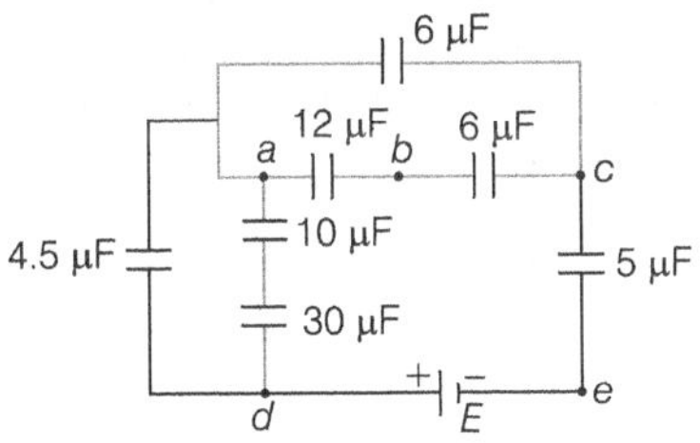

If it is given that V_{ab}, i.e. potential difference between points a and b is 4 V, then answer the given questions.

79. Potential difference between points *a* and *c* will be
(a) 4 V (b) 12 V
(c) 10 V (d) 8 V

80. Potential difference between points *d* and *a* will be
(a) 4 V (b) 12 V
(c) 10 V (d) 8 V

81. EMF *E* of the charging battery is
(a) 46 V (b) 12 V
(c) 20 V (d) 18 V

Match the Columns

82. In the figure shown, $C_1 = 4\,\mu F$ (without dielectric) and $C_2 = 4\,\mu F$ (with a dielectric slab of dielectric constant $K = 2$). Now, the same slab after removing from C_2 is filled in C_1. Then, match the following two columns.

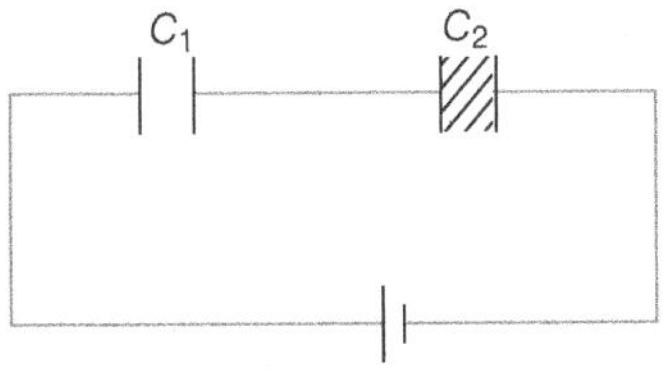

	Column I		Column II
(a)	Charge on C_2	(p)	will increase
(b)	Energy stored in C_2	(q)	will decrease
(c)	Potential difference across C_2	(r)	will remain same
(d)	Electric field between the plates of C_2	(s)	data insufficient

83. In the circuit shown in figure, match the following two columns for the flow of charge when switch is closed.

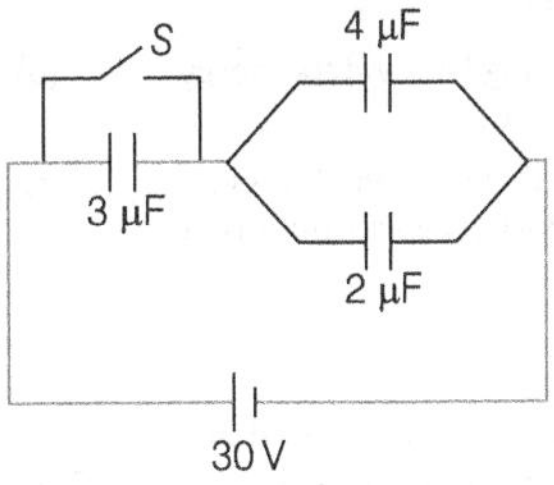

	Column I		Column II
(a)	From the battery	(p)	40 μC
(b)	From 2 μF capacitor	(q)	100 μC
(c)	From 3 μF	(r)	60 μC
(d)	From 4 μF capacitor	(s)	None of these

84. Three identical capacitors are connected in three different configurations as shown in Column II. Points *a* and *b* are connected with a battery. Match the two columns.

	Column I		Column II
(a)	Maximum charge on C_1	(p)	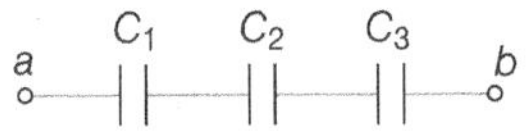
(b)	Minimum charge on C_2	(q)	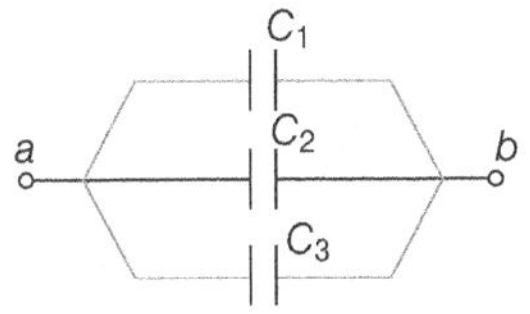
(c)	Maximum potential difference across C_1	(r)	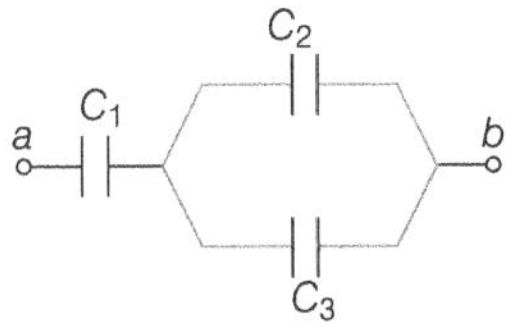
(d)	Minimum potential difference across C_1	(s)	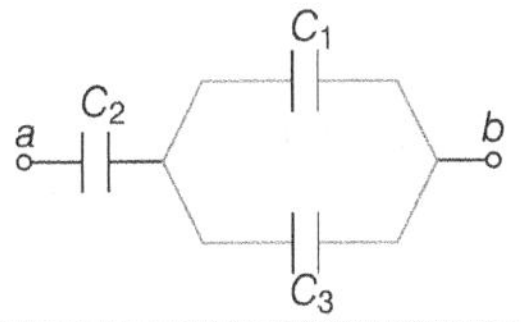

85. A capacitor *C* is charged by a battery of *V* volts. Then, it is connected to an uncharged capacitor of capacity 2*C* as shown in figure. Now, match the following two columns.

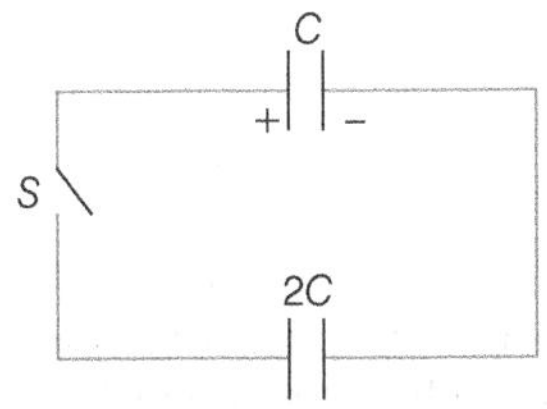

	Column I		Column II
(a)	After closing the switch energy stored in C.	(p)	$\frac{1}{9}CV^2$
(b)	After closing the switch energy stored in $2C$.	(q)	$\frac{1}{6}CV^2$
(c)	After closing the switch loss of energy during redistribution of charge.	(r)	$\frac{1}{18}CV^2$
		(s)	None of these

86. Two identical sized capacitors C_1 and C_2 are connected with a battery as shown in figure. Capacitor plates are square plates. A dielectric slab of dielectric constant $K = 2$, is filled in half the region of the two capacitors as shown :

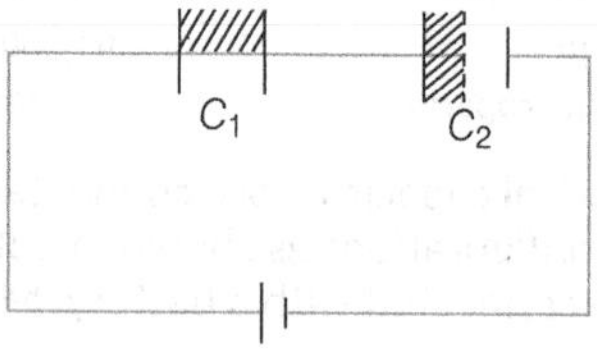

$C \rightarrow$ capacity, $q \rightarrow$ charge stored, $U \rightarrow$ energy stored. Match the following two columns.

	Column I		Column II
(a)	C_1/C_2	(p)	9 / 4
(b)	q_1/q_2	(q)	4 / 9
(c)	U_1/U_2	(r)	4 / 3
		(s)	None of these

87. Consider the situation shown in figure. The switch S is open for a long time and then closed.

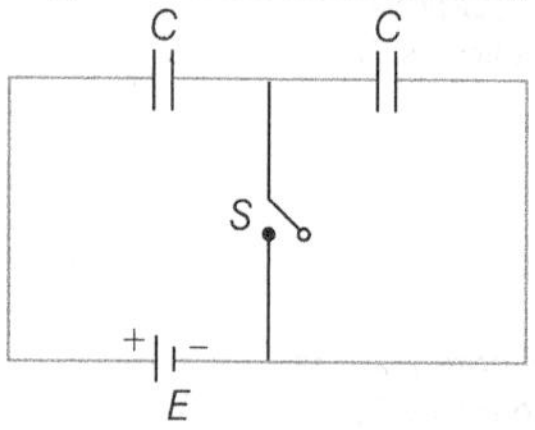

Match the Column I and Column II and mark the correct option codes.

	Column I		Column II
(a)	Charge flows through cell when S is closed.	(p)	$CE^2/2$
(b)	Work done by battery when S is closed.	(q)	$CE/2$
(c)	Change in energy stored when S is closed.	(r)	$CE^2/4$
(d)	Heat developed when S is closed.	(s)	$CE^2/8$

Subjective Questions

88. Five identical conducting plates, 1, 2, 3,4 and 5 are fixed parallel plates equidistant from each other (see figure). A conductor connects plates 2 and 5 while another conductor joins 1 and 3. The junction of 1 and 3 and the plate 4 are connected to a source of constant emf V_0. Find

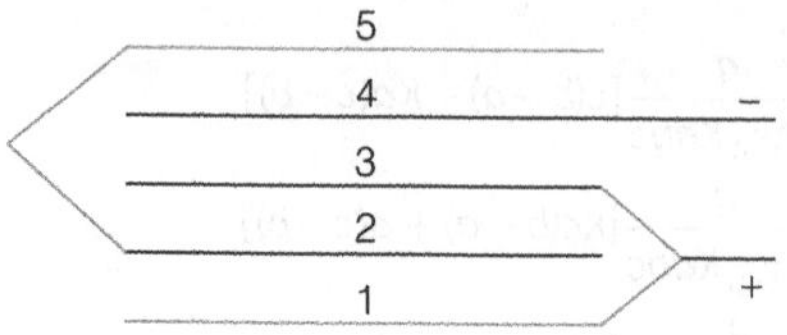

(a) the effective capacity of the system between the terminals of source.

(b) the charges on the plates 3 and 5.

Given, d = distance between any two successive plates and A = area of either face of each plate.

89. A charged capacitor C_1 is discharged through a resistance R by putting switch S in position 1 of circuit shown in figure. When discharge current reduces to I_0, the switch is suddenly shifted to position 2. Calculate the amount of heat liberated in resistance R starting from this instant.

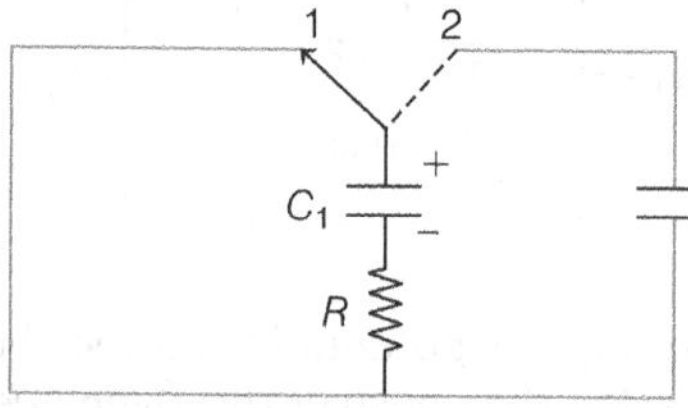

90. In the given circuit, the switch is closed in the position 1 at $t = 0$ and then moved to 2 after 250 μs. Derive an expression for current as a function of time for $t > 0$. Also plot the variation of current with time.

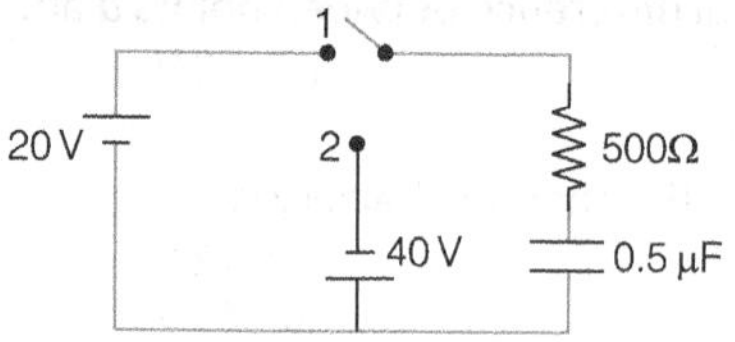

91. In the circuit shown, a time varying voltage $V = 2000t$ volt is applied where t is in second. At time $t = 5$ ms, determine the current through the resistor $R = 4\,\Omega$ and through the capacitor $C = 300\,\mu F$.

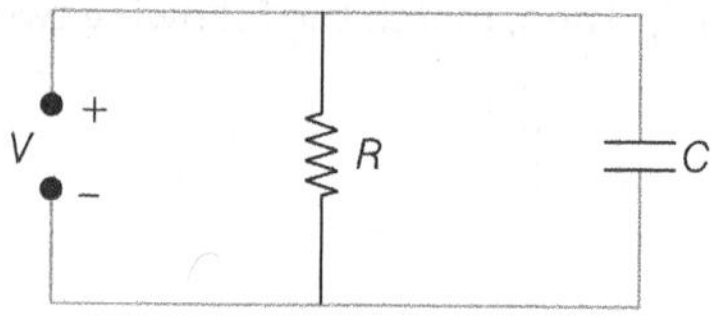

92. A capacitor of capacitance 5 μF is connected to a source of constant emf of 200 V. Then, the switch was shifted to contact 2 from contact 1. Find the amount of heat generated in the 400 Ω resistance.

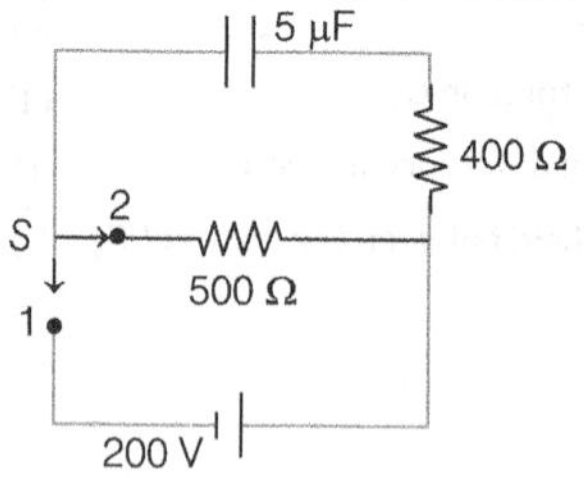

93. Analyze the given circuit in the steady state condition. Charge on the capacitor is $q_0 = 16\ \mu C$.

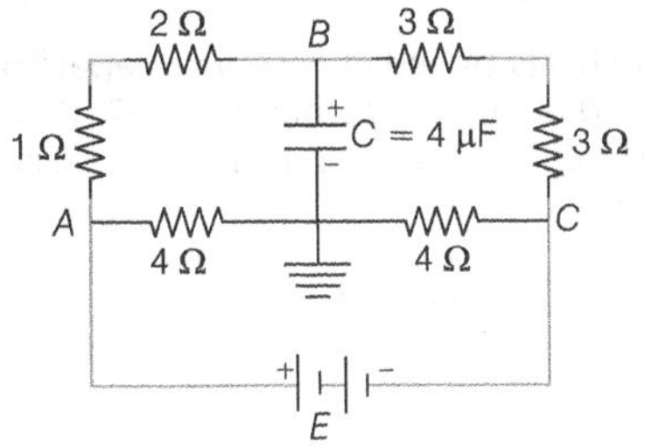

(a) Find the current in each branch
(b) Find the emf of the battery.
(c) If now the battery is removed and the points A and C are shorted. Find the time during which charge on the capacitor becomes $8\ \mu C$.

94. In the given circuit diagram, find the charges which flow through directions 1 and 2 when switch S is closed.

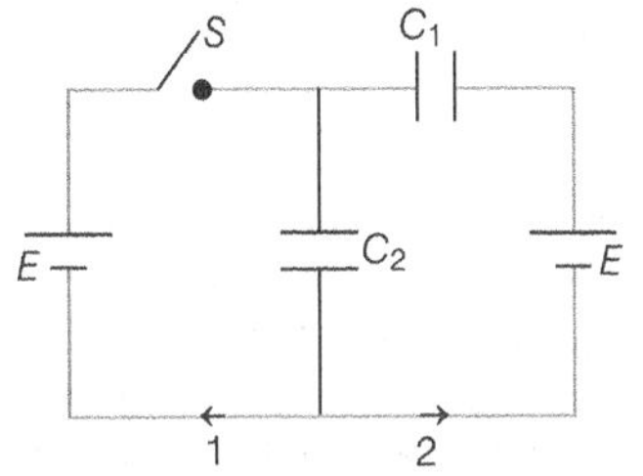

95. For the arrangement shown in the figure, the switch is closed at $t = 0$. C_2 is initially uncharged while C_1 has a charge of $2\ \mu C$.

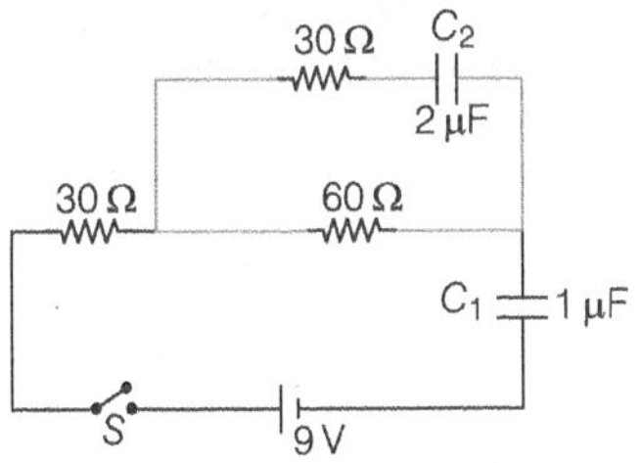

(a) Find the current coming out of the battery just after the switch is closed.
(b) Find the charge on the capacitors in the steady state condition.

96. The capacitor C_1 in the figure initially carries a charge q_0. When the switch S_1 and S_2 are closed, capacitor C_1 is connected to a resistor R and a second capacitor C_2, which initially does not carry any charge.

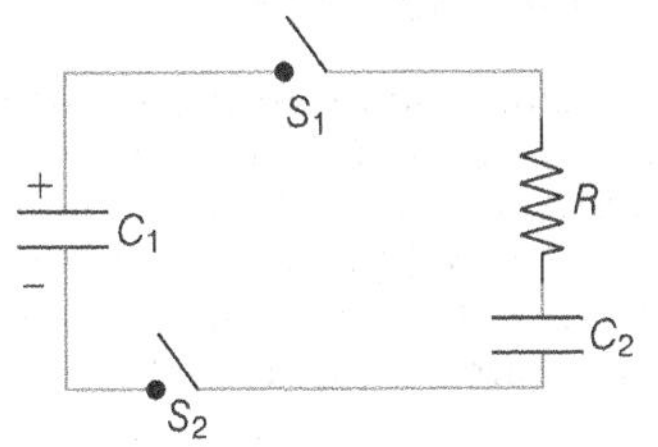

(a) Find the charges deposited on the capacitors in steady state and the current through R as a function of time.
(b) What is heat lost in the resistor after a long time of closing the switch?

97. Initially, the switch is in position 1 for a long time. At $t = 0$, the switch is moved from 1 to 2. Obtain expressions for V_C and V_R for $t > 0$.

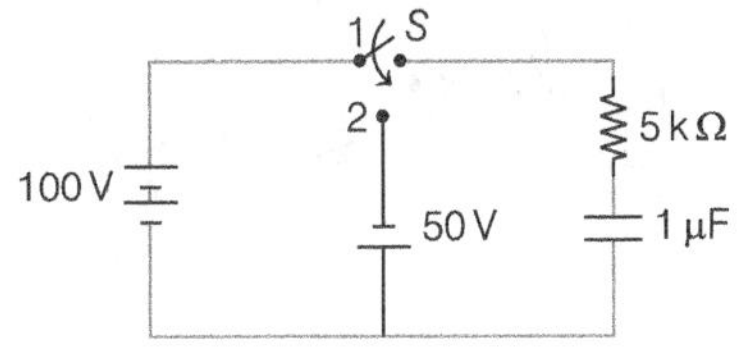

98. A parallel plate vacuum capacitor with plate area A and separation x has charges $+Q$ and $-Q$ on its plates. The capacitor is disconnected from the source of charge, so the charge on each plate remains fixed.
(a) What is the total energy stored in the capacitor?
(b) The plates are pulled apart an additional distance dx. What is the change in the stored energy?
(c) If F is the force with which the plates attract each other, then the change in the stored energy must equal the work $dW = Fdx$ done in pulling the plates apart. Find an expression for F.
(d) Explain why F is not equal to QE, where E is the electric field between the plates.

99. A spherical capacitor has the inner sphere of radius 2 cm and the outer one of 4 cm. If the inner sphere is earthed and the outer one is charged with a charge of $2\ \mu C$ and isolated. Calculate
(a) the potential to which the outer sphere is raised.
(b) the charge retained on the outer surface of the outer sphere.

100. Calculate the charge on each capacitor and the potential difference across it in the circuits shown in figure for the cases :

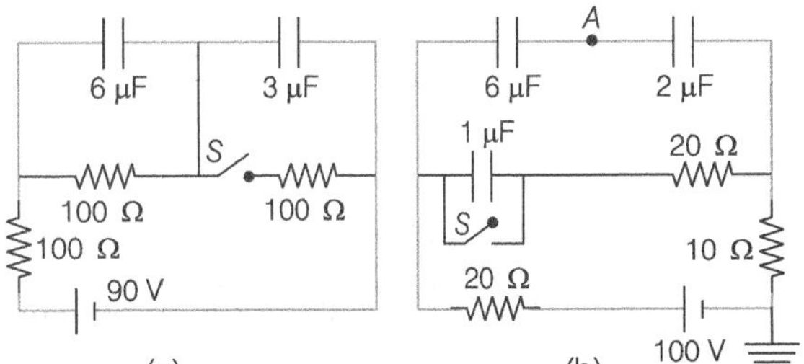

(a) (b)

(i) switch S is closed and
(ii) switch S is open.
(iii) In figure (b), what is the potential of point A when S is open?

101. In the shown network, find the charges on capacitors of capacitances $5\ \mu F$ and $3\ \mu F$, in steady state.

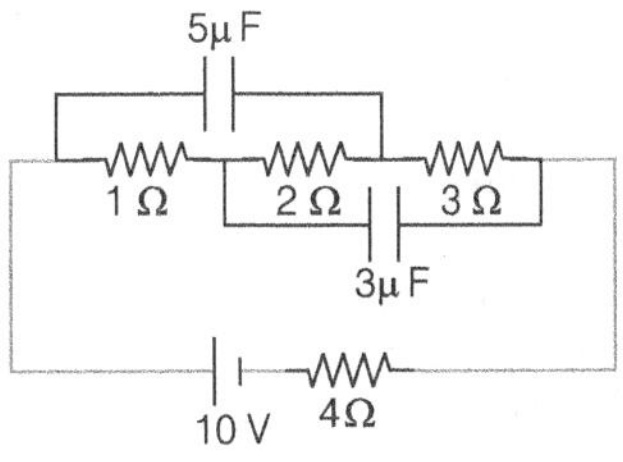

102. In the circuit shown, $E = 18$ kV, $C = 10\ \mu F$, $R_1 = 4\ M\Omega$, $R_2 = 6 M\Omega$, $R_3 = 3\ M\Omega$. With C completely uncharged, switch S is suddenly closed (at $t = 0$).

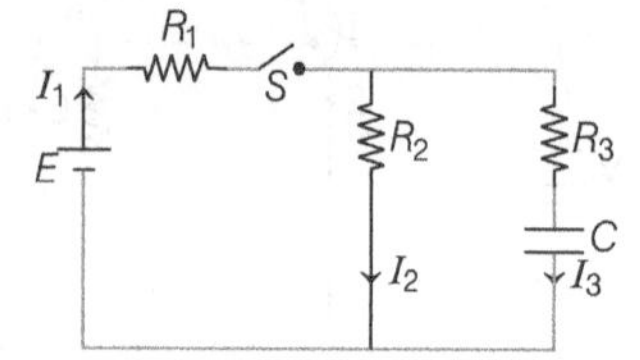

(a) Determine the current through each resistor for $t = 0$ and $t = \infty$.
(b) What are the values of V_2 (potential difference across R_2) at $t = 0$ and $t = \infty$?
(c) Plot a graph of the potential difference V_2 *versus* t and determine the instantaneous value of V_2.

103. The charge on the capacitor is initially zero. Find the charge on the capacitor as a function of time t. All resistors are of equal value R.

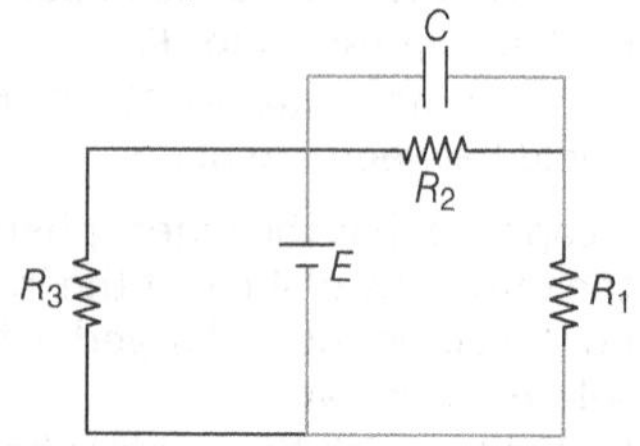

104. The capacitors are initially uncharged. In a certain time the capacitor of capacitance 2 μF gets a charge of 20 μC. In that time interval find the heat produced by each resistor individually.

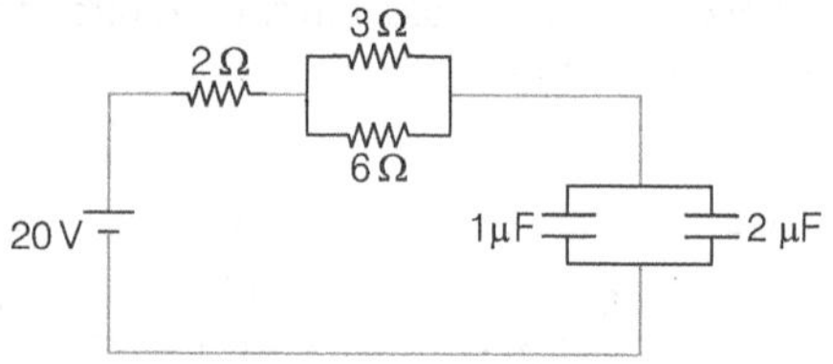

105. A capacitor of capacitance C has potential difference $E/2$ and another capacitor of capacitance $2C$ is uncharged. They are joined to form a closed circuit as shown in the figure.

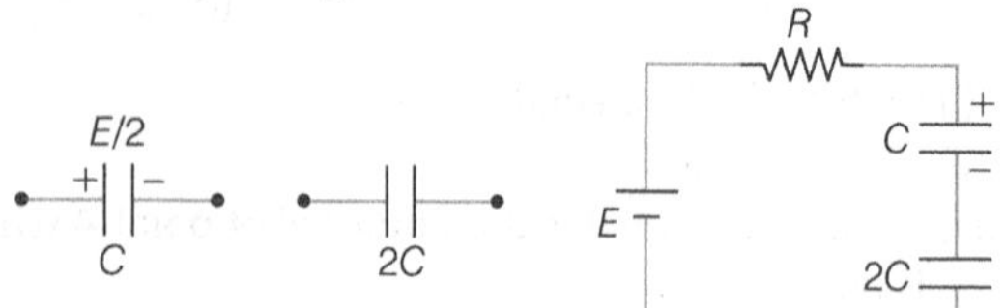

(a) Find the current in the circuit at $t = 0$.
(b) Find the charge on C as a function of time.

106. The capacitor shown in figure has been charged to a potential difference of V volt, so that it carries a charge CV with both the switches S_1 and S_2 remaining open. Switch S_1 is closed at $t = 0$. At $t = R_1C$ switch S_1 is opened and S_2 is closed. Find the charge on the capacitor at $t = 2R_1C + R_2C$.

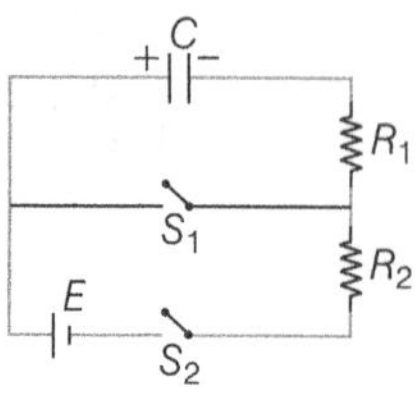

107. The switch S is closed at $t = 0$. The capacitor C is uncharged but C_0 has a charge $Q_0 = 2\ \mu C$ at $t = 0$. If $R = 100\ \Omega$, $C = 2\ \mu F$, $C_0 = 2\ \mu F$, $E = 4$ V. Calculate $i(t)$ in the circuit.

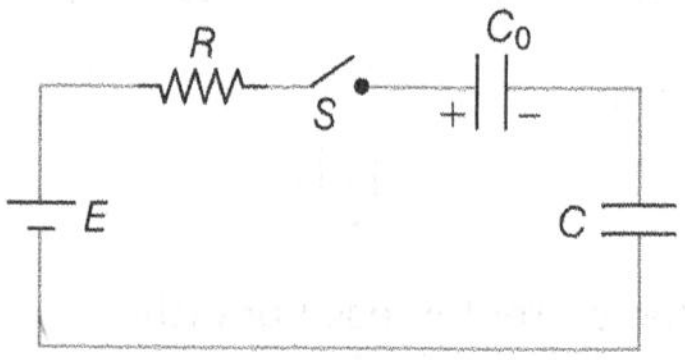

108. A time varying voltage is applied to the clamps A and B such that voltage across the capacitor plates is as shown in the figure. Plot the time dependence of voltage across the terminals of the resistance E and D.

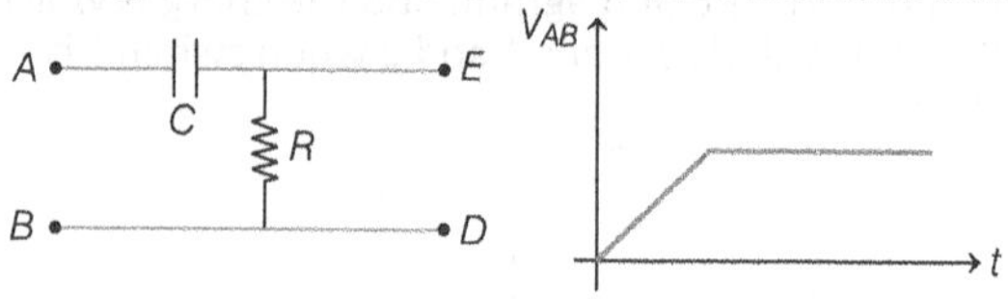

109. In the above problem if given graph is between V_{AB} and time. Then, plot graph between V_{ED} and time.

110. The capacitance of a parallel plate capacitor with plate area A and separation d is C. The space between the plates is filled with two wedges of dielectric constants K_1 and K_2, respectively. Find the capacitance of the resulting capacitor. **(1996 JEE)**

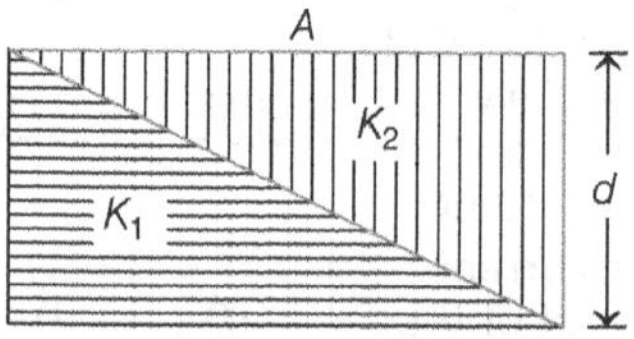

111. Two square metal plates of side 1 m are kept 0.01 m apart like a parallel plate capacitor in air in such a way that one of their edges is perpendicular to an oil surface in a tank filled with an insulating oil. The plates are connected to a battery of emf 500 V. The plates are then lowered vertically into the oil at a speed of 0.001 ms^{-1}. Calculate the current drawn from the battery during the process. (Dielectric constant of oil = 11, $\varepsilon_0 = 8.85 \times 10^{-12}\ C^2 N^{-1} m^{-1}$) **(1994 JEE)**

112. Two parallel plate capacitors A and B have the same separation $d = 8.85 \times 10^{-4}$ m between the plates. The plate area of A and B are 0.04 m^2 respectively. A slab of dielectric constant (relative permittivity) $K = 9$ has dimensions such that it can exactly fill the space between the plates of capacitor B. **(1993 JEE)**

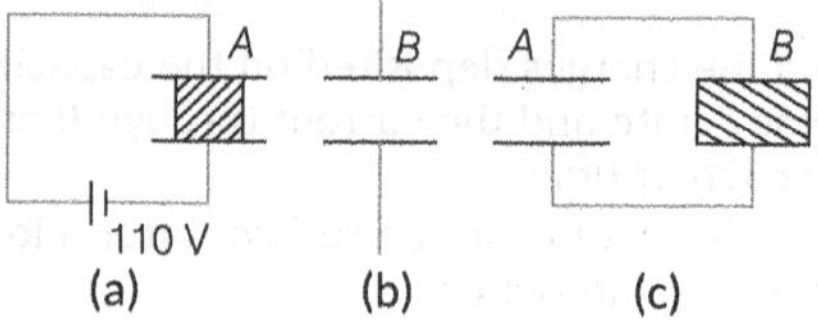

(a) The dielectric slab is placed inside A as shown in Fig. (a). A is then charged to a potential difference of 110 V. Calculate the capacitance of A and the energy stored in it.

(b) The battery is disconnected and then the dielectric slab is moved from A. Find the work done by the external agency is removing the slab from A.

(c) The same dielectric slab is now placed inside B, filling it completely. The two capacitors A and B are then connected as shown in figure (c). Calculate the energy stored in the system.

113. Find the potential difference $V_A - V_B$ between points A and B of the circuit shown in figure. **(Irodov 3.124)**

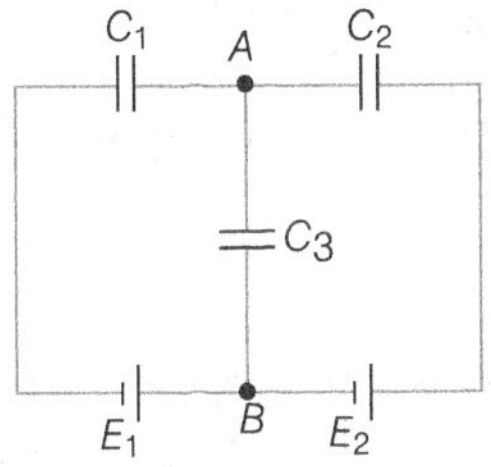

114. Determine the potential at point 1 of the circuit shown in figure, assuming the potential at the point O to be equal to zero. **(Irodov 3.125)**

Using the symmetry of the formula obtained,write the expressions for the potentials, at points 2 and 3.

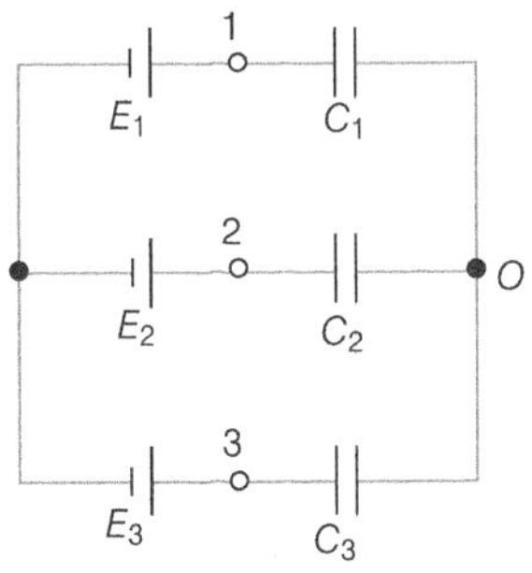

115. What amount of heat will be generated in the circuit shown in figure after the switch S is shifted from position 1 to position 2? **(Irodov 3.132)**

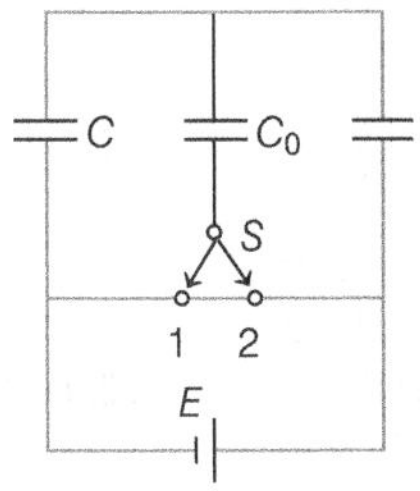

116. An air cylindrical capacitor with a DC voltage $V = 200$ V applied across it is being submerged vertically into a vessel filled with water at a velocity $v = 5.0$ m/s. The electrodes of the capacitor are separated by a distance $d = 2.0$ mm, the mean curvature radius of the electrodes is equal to $r = 50$ mm. Find the current flowing in the lead wires, if $d \ll r$. **(Irodov 3.148)**

117. A circuit consists of a source of a constant emf V, a resistance R and a capacitor with capacitance C connected in series. The internal resistance of the source is negligible. At a moment $t = 0$ the capacitance of the capacitor is abruptly decreased η-fold. Find the current flowing through the circuit as a function of time t. **(Irodov 3.172)**

118. At the moment $t = 0$ an electron leaves one plate of a parallel plate capacitor with a negligible velocity. An accelerating voltage, varying as $V = \alpha t$, where $\alpha = 100$ V/s, is applied between the plates. The separation between the plates is $l = 5.0$ cm. What is the velocity of the electron at the moment it reaches the opposite plate? **(Irodov 3.372)**

119. A proton accelerated by a potential difference V gets into the uniform electric field of a parallel plate capacitor whose plates extend over a length l in the motion direction. The field strength varies with time as $E = \alpha t$, where α is a constant. Assuming the proton to be non-relativistic, find the angle between the motion directions of the proton before and after its flight through the capacitor, the proton gets in the field at the moment $t = 0$. The edge effects are to be neglected. **(Irodov 3.373)**

Answers

Level 1

1. (b) 2. (c) 3. (c) 4. (c) 5. (b) 6. (b) 7. (b) 8. (a) 9. (c) 10. (b)
11. (a) 12. (b) 13. (c) 14. (d) 15. (d) 16. (d) 17. (a) 18. (b) 19. (a) 20. (d)
21. (c) 22. 0.0135 m^2 23. (a) 18 V (b) a is at higher potential (c) 6 V (d) – 36 μC on both the capacitors

24. (a) –6.0 V (b) b (c) 6.0 V (d) – 54.0 μC 25. $\Delta q = \left[\frac{(K-1)}{2(K+1)}\right]EC$, Electric field will also decrease by $\left(\frac{K+1}{2}\right)$ times.

26. $V_{net} = 9.0$ kV 27. $V_{AB} = 10$ volt 28. $\Delta q = 0.06$ mC 29. $Q = \frac{1}{2}CE_2^2$

30. (a) $W = \frac{q^2(x_2 - x_1)}{2\varepsilon_0 S}$ (b) $W = \frac{\varepsilon_0 SV^2(x_1 - x_2)}{2x_1 x_2}$ 31. (a) $W_1 = 1.5$ mJ (b) $W_2 = 0.8$ mJ 32. $\rho = \frac{4\pi ab(\Delta t)}{C(b-a)\ln\eta}$

33. $I = 1.5\ \mu$A 34. $I = 5$ nA 35. $t = 0.6\ \mu$s 36. $\rho = 1.4 \times 10^{13}\ \Omega$-m 37. (3) 38. (105)

39. (6) 40. (5) 41. (23)

Level 2

42. (b) 43. (d) 44. (d) 45. (c) 46. (c) 47. (b) 48. (c) 49. (b) 50. (c) 51. (b)
52. (d) 53. (a) 54. (b) 55. (c) 56. (b) 57. (d) 58. (c) 59. (a) 60. (b) 61. (b,c)
62. (a,b,d) 63. (a,b,c) 64. (a,d) 65. (a,b,d) 66. (a,c) 67. (a,b,c,d) 68. (b,c,d) 69. (a,d) 70. (b,d) 71. (b)
72. (a) 73. (b) 74. (a) 75. (d) 76. (d) 77. (c) 78. (a) 89. (b) 80. (c) 81. (a)
82. (a→q; b→p; c→p; d→p) 83. (a→s; b→p; c→r; d→s) 84. (a→q, r; b→p; c→q; d→p)
85. (a→r; b→p; c→s) 86. (a→s; b→s; c→s) 87. (a→q; b→p; c→r; d→r)

88. (a) $\frac{5}{3}\left(\frac{\varepsilon_0 A}{d}\right)$ (b) $q_3 = \frac{4}{3}\left(\frac{\varepsilon_0 AV_0}{d}\right), q_5 = \frac{2}{3}\left(\frac{\varepsilon_0 AV_0}{d}\right)$ 89. $\frac{(I_0 R)^2 C_1 C_2}{2(C_1 + C_2)}$

90. $i = (0.04e^{-4000t})$ A for $t \le 250\ \mu$s, $i = -(0.11\,e^{-4000t})$ A for $t \ge 250\ \mu$s For i - t graph, see the hints.

91. 2.5A, 0.6A 92. 44.4 mJ 93. (a) 3 A, 2.67A (b) 24 V (c) 11.1 μs

94. $q_1 = EC_2$, $q_2 = \frac{-EC_1C_2}{(C_1 + C_2)}$ 95. (a) $\frac{7}{50}$ A or $\frac{11}{50}$ A (b) $Q_1 = 9\ \mu$C, $Q_2 = 0$

96. (a) $q_1 = \left(\frac{C_1}{C_1 + C_2}\right)q_0$ and $q_2 = \left(\frac{C_2}{C_1 + C_2}\right)q_0$, $i = \frac{q_0}{RC_1}e^{-t/RC}$ (b) $\Delta H = \frac{q_0^2 C_2}{2C_1(C_1 + C_2)}$, here $C = \frac{C_1 C_2}{C_1 + C_2}$

97. $V_C = 50\,(3\,e^{-200t} - 1)$, $V_R = 150\,e^{-200t}$ 98. (a) $\frac{Q^2 x}{2\varepsilon_0 A}$ (b) $\left(\frac{Q^2}{2\varepsilon_0 A}\right)\cdot dx$ (c) $\frac{Q^2}{2\varepsilon_0 A}$

99. (a) 2.25×10^5 V (b) +1 μC

100.

		Fig. (a)		Fig. (b)		
		6 μF	3 μF	1 μF	6 μF	2 μF
(i)	PD (volts)	30	30	0	10	30
	charge (μC)	180	90	0	60	60
(ii)	PD (volts)	0	90	100	25	75
	charge (μC)	0	270	100	150	150
(iii)	$V_A = 75$ volt					

101. 15 μC, 15 μC

102. (a) At $t = 0$, $i_1 = 3$ mA, $i_2 = 1$ mA, $i_3 = 2$ mA, At $t = \infty$, $i_1 = i_2 = 1.8$ mA, $i_3 = 0$ (b) At $t = 0$, $V_2 = 6$ kV, At $t = \infty$, $V_2 = 10.8$ kV (c) $V_2 = (10.8 - 4.8\,e^{-t/54})$ kV

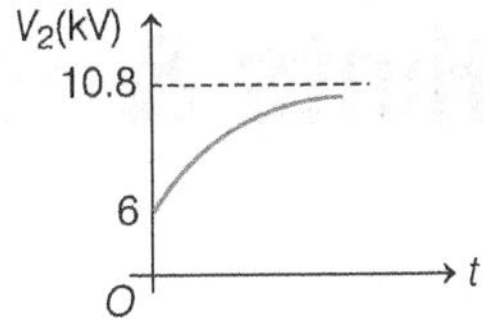

103. $q = \dfrac{CE}{2}\left(1 - e^{-\frac{2t}{CR}}\right)$

104. $H_2 = 0.225$ mJ, $H_3 = 0.15$ mJ, $H_6 = 0.075$ mJ

105. (a) $\dfrac{E}{2R}$ (b) $\dfrac{CE}{6}\,[5 - 2e^{-3t/2RC}]$

106. $EC\left(1 - \dfrac{1}{e}\right) + \dfrac{VC}{e^2}$

107. $(0.03\ e^{-10^4 t})$ A

108.

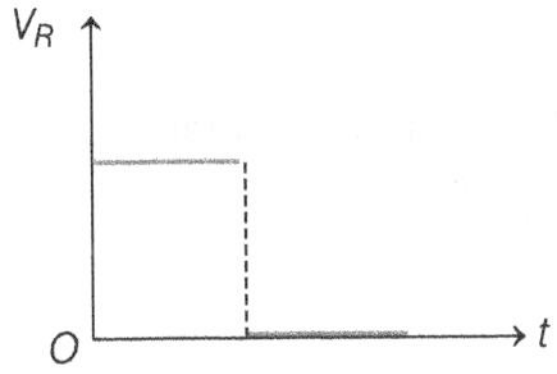

109.

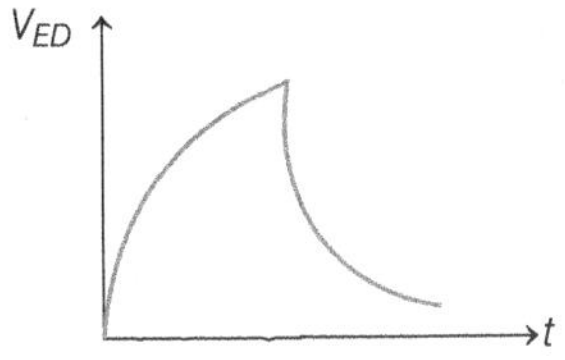

110. $\left(\dfrac{CK_1K_2}{K_2 - K_1}\ln\dfrac{K_2}{K_1}\right)$, where $C = \dfrac{\varepsilon_0 A}{d}$

111. $(4.43 \times 10^{-9}$ A)

112. (a) 2×10^{-9} F, 1.21×10^{-5} J (b) 4.84×10^{-5} J (c) 1.1×10^{-5} J

113. $V_A - V_B = \dfrac{C_2E_2 - C_1E_1}{C_1 + C_2 + C_3}$

114. $V_1 = \dfrac{E_1(C_2 + C_3) - E_2C_2 - E_3C_3}{C_1 + C_2 + C_3}$, $V_2 = \dfrac{E_2(C_1 + C_3) - E_1C_1 - E_3C_3}{C_1 + C_2 + C_3}$, $V_3 = \dfrac{E_3(C_1 + C_2) - E_1C_1 - E_2C_2}{C_1 + C_2 + C_3}$

115. $Q = E^2CC_0/(2C + C_0)$

116. $I = 0.11\ \mu$A

117. $I = (\eta - 1)\dfrac{V}{R}e^{\frac{-\eta t}{CR}}$

118. 16 km/s

119. $\tan\alpha = \dfrac{\alpha l^2}{4}\sqrt{\dfrac{m}{2eV^3}}$

See hints on pages 328-349.

Hints & SOLUTIONS

Module 2.

1. Charge does not flow if their potentials are same.
2. The capacitance of spherical conductor of radius R is given as

$$C = 4\pi\varepsilon_0 R$$

So, capacitance does not depend on its charge, it depends only on the radius of the conductor (R). Therefore, assertion is false but reason is true.

3. Potential, $V_1 = \dfrac{1}{4\pi\varepsilon_0}\dfrac{Q_1}{R_1}$

and $\quad V_2 = \dfrac{1}{4\pi\varepsilon_0}\dfrac{Q_2}{R_2}$

If $V_1 = V_2$, then $\dfrac{Q_1}{R_1} = \dfrac{Q_2}{R_2}$

or $\quad Q_1 R_2 = Q_2 R_1$

In this case, there will be no flow of charges. However, if the given spheres have unequal potentials, then the charges flows from higher potential to lower potential. Flow of charges shall be accompanied by the loss of energy.

4. Consider two metallic spheres A and B both have charge q. The force between two spheres will be

$$F = \frac{kq^2}{r^2}$$

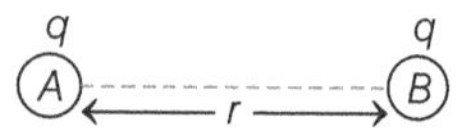

When sphere C is placed in contact with sphere A, charge on sphere A and sphere C will be $= \dfrac{q}{2}$.

Now sphere C is placed in contact with sphere B, charge on sphere B and sphere C will be $\dfrac{q + \frac{q}{2}}{2} = \dfrac{3q}{4}$

Now,

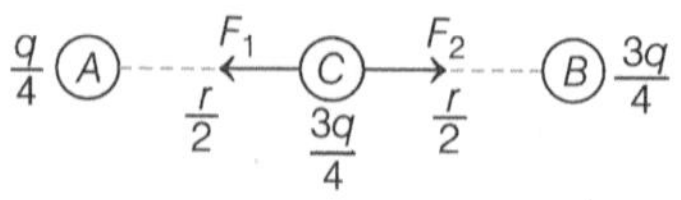

Then $\quad F_1 = \dfrac{\frac{kq}{2}\cdot\frac{3q}{4}}{\frac{r^2}{4}}$

$$F_2 = \frac{\frac{k3q}{4}\cdot\frac{3q}{4}}{\frac{r^2}{4}}$$

The force experienced by sphere C,

$$F' = F_2 - F_1 = \frac{\left(k\frac{3q}{4} - k\frac{q}{2}\right)\cdot\frac{3q}{4}}{\frac{r^2}{4}}$$

$$= \frac{3kq^2}{4r^2} = \frac{3F}{4}$$

5. Volume of big drop $= n \times$ Volume of small drop

$\therefore \quad \dfrac{4}{3}\pi R^3 = n \times \dfrac{4}{3}\pi r^3$

$\Rightarrow \quad R = n^{1/3} r$

Capacitance of big drop, $C = 4\pi\varepsilon_0 R = 4\pi\varepsilon_0 n^{1/3} r = n^{1/3} C$

Potential of big drop $V = \dfrac{1}{4\pi\varepsilon_0}\cdot\dfrac{Q}{R} = \dfrac{1}{4\pi\varepsilon_0}\cdot\dfrac{nq}{n^{1/3} r}$

$$= \frac{n}{n^{1/3}}\left(\frac{1}{4\pi\varepsilon_0}\times\frac{q}{r}\right) = n^{2/3}\, V$$

Hence, (a) $\rightarrow$ (p), (b) $\rightarrow$ (p), (c) $\rightarrow$ (q)

6. $U = \dfrac{1}{2}\dfrac{q^2}{C}$

$\therefore \quad [C] = \left[\dfrac{q^2}{U}\right] = \left[\dfrac{A^2T^2}{ML^2T^{-2}}\right]$

$$= [M^{-1}L^{-2}T^4A^2]$$

7. $q_1 = C_1V_1 = 10\,\mu C$

$$q_2 = C_2V_2 = -40\,\mu C$$

(a) $V = \dfrac{q_{Total}}{C_{Total}} = \dfrac{-30\,\mu C}{3\,\mu F} = -10$ volt

(b) $q_1' = C_1 V$ and $q_2' = C_2 V$

(c) $\Delta U = \dfrac{C_1C_2}{2(C_1 + C_2)}(V_1 - V_2)^2$

8. Let q_1 and q_2 are the initial charges. After they are connected by a conducting wire, final charge on them become.

$$q'_1 = q'_2 = \left(\frac{q_1 + q_2}{2}\right)$$

Now, given that

$$0.108 = \frac{(9\times 10^9)(q_1)(q_2)}{(0.5)^2} \quad \text{...(i)}$$

$$0.036 = \frac{(9\times 10^9)\left(\frac{q_1 + q_2}{2}\right)^2}{(0.5)^2} \quad \text{...(ii)}$$

Solving Eqs. (i) and (ii), we can find q_1 and q_2.

Module 3.

Introductory Exercise

1. $q = CV$

2. (a) $V = \frac{q}{C}$

(b) $C = \frac{\varepsilon_0 A}{d}$ $\quad \therefore \quad A = \frac{Cd}{\varepsilon_0}$

(c) $\sigma = \frac{q}{A}$

3. (a) $K = \frac{E_0}{E} = \frac{3.20 \times 10^5}{2.50 \times 10^5} = 1.28$

(b) $$\sigma_i = \sigma_0 \left(1 - \frac{1}{K}\right)$$
$$= (E_0 \varepsilon_0)\left(1 - \frac{1}{K}\right)$$
$$= (3.20 \times 10^5)(8.86 \times 10^{-12})\left(1 - \frac{1}{1.28}\right)$$
$$= 6.2 \times 10^{-7} \text{ C/m}^2$$

4. $$U = \frac{1}{2}\varepsilon_0 E^2 = \frac{1}{2}\varepsilon_0 \left(\frac{1}{4\pi\varepsilon_0}\frac{q}{R^2}\right)^2$$
$$= \frac{1}{2}\varepsilon_0 \left[\frac{9 \times 10^9 \times \frac{1}{9} \times 10^{-9}}{(1)^2}\right]^2 = \frac{\varepsilon_0}{2} \text{ J/m}^3$$

5. The electric field produced between the parallel plate capacitor is uniform. The force acting on charged particle placed in an electric field is given by $F = qE$. Hence, the forces on the two protons are identical being equal charge and uniform electric field.

Exercise

1. Capacity of parallel plate capacitor is given by
$$C = \frac{\varepsilon_0 A}{d} \Rightarrow \varepsilon_0 = \frac{Cd}{A}$$
$$\text{Unit} = \frac{\text{Farad} \times \text{metre}}{\text{metre}^2} = \frac{\text{Farad}}{\text{metre}}$$

2. $$C = C_1 + C_2$$
$$= (4\pi\varepsilon_0 a) + (4\pi\varepsilon_0 b) = 4\pi\varepsilon_0 (a + b)$$

3.

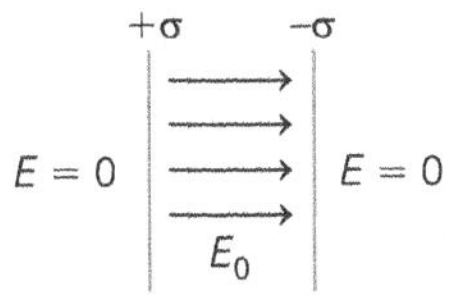

$$\Rightarrow \quad E = \frac{\sigma}{\varepsilon_0} = \text{constant}$$

4. On introducing dielectric K in a parallel plate capacitor, its capacity becomes
$$C' = KC_0$$
$$\therefore \quad C' = 5C_0$$
Also, energy stored $W_0 = \frac{q^2}{2C_0}$
$$\therefore \quad W' = \frac{q^2}{2C'} = \frac{q^2}{2 \times 5C_0}$$
$$\therefore \quad \frac{W_0}{W'} = \frac{5}{1}$$
$$\therefore \quad W' = \frac{W_0}{5}$$

5. The distribution of charge is as under.

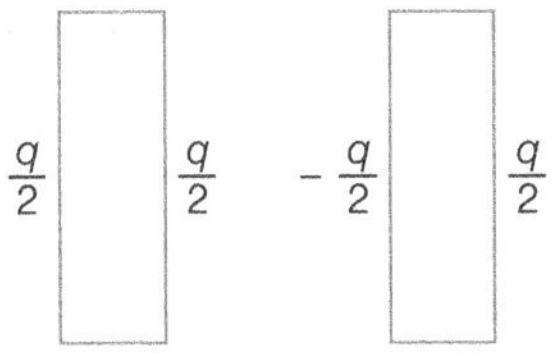

$$E = \frac{\sigma}{\varepsilon_0} = \frac{(q/2)/S}{\varepsilon_0} = \frac{q}{2S\varepsilon_0}$$

6. Capacity of spherical condenser when outer sphere is earthed.

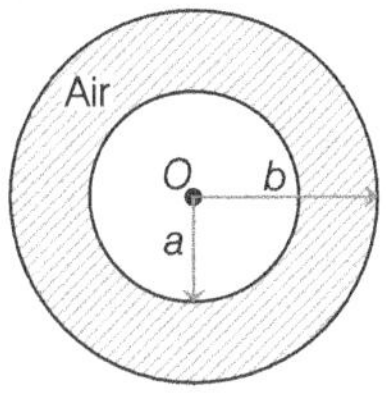

$$C_1 = 4\pi\varepsilon_0 \cdot \frac{ab}{b-a}$$

Capacity of spherical condenser when inner sphere is earthed.
$$C_2 = 4\pi\varepsilon_0 b + \frac{4\pi\varepsilon_0 ab}{b-a}$$

$\therefore$ Difference in their capacity
$$= C_2 - C_1 = 4\pi\varepsilon_0 b$$

7.

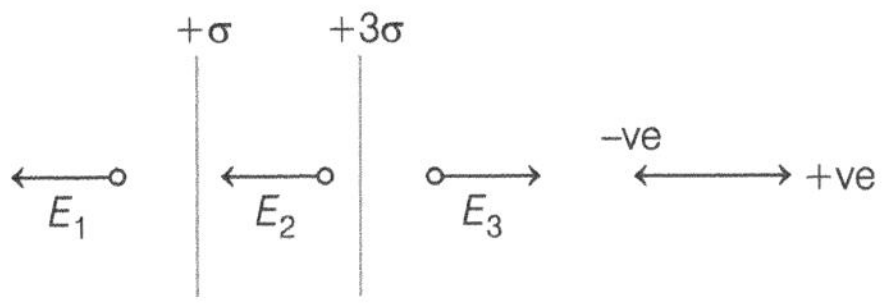

$$E_1 = \frac{\sigma}{2\varepsilon_0} + \frac{3\sigma}{2\varepsilon_0} = \frac{2\sigma}{\varepsilon_0} = E_3$$
$$E_2 = \frac{3\sigma}{2\varepsilon_0} - \frac{\sigma}{2\varepsilon_0} = \frac{\sigma}{\varepsilon_0}$$

E_1 and E_2 are in the negative direction and E_3 in positive direction.

8.

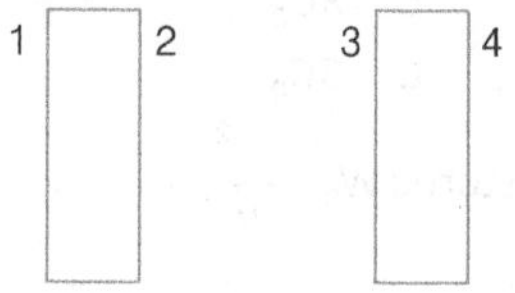

$$q_1 = q_4 = \frac{q_{Total}}{2} = \frac{CV - CV + Q}{2} = \frac{Q}{2}$$

$$q_2 = (Q + CV) - \frac{Q}{2} = \left(\frac{Q}{2} + CV\right)$$

$$q_3 = -q_2 = -\left(\frac{Q}{2} + CV\right)$$

Electric field between two plates and hence the potential difference is due to q_2 and q_3 only.

$$PD = \frac{q_2}{C} = V + \frac{Q}{2C}$$

9. $C = \frac{\varepsilon_0 A}{x}$

$$\frac{dC}{dt} = -\frac{\varepsilon_0 A}{x^2}\left(\frac{dx}{dt}\right) = -\frac{\varepsilon_0 A v}{x^2}$$

$$\left|\frac{dC}{dt}\right| \propto \frac{1}{x^2}$$

10. Electric field within the plates $\mathbf{E} = \mathbf{E}_{Q_1} + \mathbf{E}_{Q_2}$

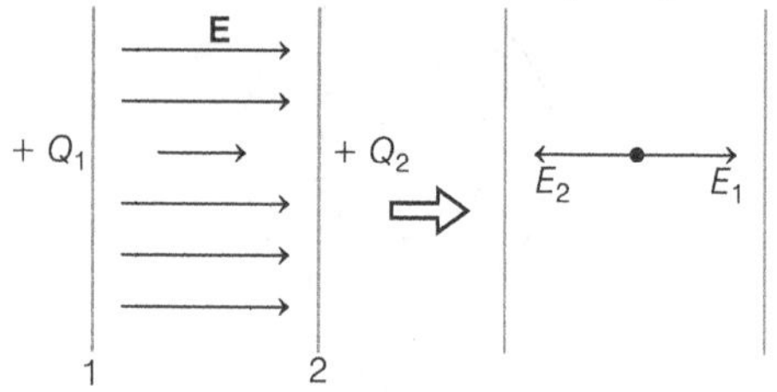

$$E = E_1 - E_2 = \frac{Q_1}{2A\varepsilon_0} - \frac{Q_2}{2A\varepsilon_0}, E = \frac{Q_1 - Q_2}{2A\varepsilon_0}$$

∴ Potential difference between the plates,

$$V_A - V_B = Ed = \left(\frac{Q_1 - Q_2}{2A\varepsilon_0}\right)d$$

$$= \frac{Q_1 - Q_2}{2\left(\frac{A\varepsilon_0}{d}\right)}$$

$$= \frac{Q_1 - Q_2}{2C}$$

11. As battery remains connected to the potential difference remains same.

$$W = U_f - U_i$$

$$= \frac{1}{2}CV^2 - \frac{1}{2}CV^2 = 0$$

12. When free space between parallel plate capacitor,

$$\varepsilon = \frac{\sigma}{\varepsilon_0 K}$$

When dielectric is introduced between parallel plate capacitor $\varepsilon' = \frac{\sigma}{\varepsilon_0 K}$.

As, electric field inside dielectric

$$\frac{\sigma}{\varepsilon_0 k} = 3 \times 10^4$$

Given, $k = 2.2$

$$\varepsilon_0 = 8.85 \times 10^{-12}$$

$$\sigma = 2.2 \times 8.85 \times 8.85 \times 10^{-12} \times 3 \times 10^4$$

$$= 6.6 \times 8.85 \times 10^{-8}$$

$$= 5.841 \times 10^{-7}$$

$$\sigma = 6 \times 10^{-7}\ C/m^2$$

13. After time t, thickness of liquid will remain $\left(\frac{d}{3} - vt\right)$.

So, capacity after t second is

$$C = \frac{\varepsilon_0(1)}{\left(d - \frac{d}{3} + vt\right) + \left(\frac{d/3 - vt}{2}\right)} \qquad \left(\text{Using } C = \frac{\varepsilon_0 A}{d - t + \frac{t}{K}}\right)$$

$$\Rightarrow C = \frac{6\varepsilon_0}{5d + 3vt}$$

14. By closing the switch whole inner charge transfers to outer shell.

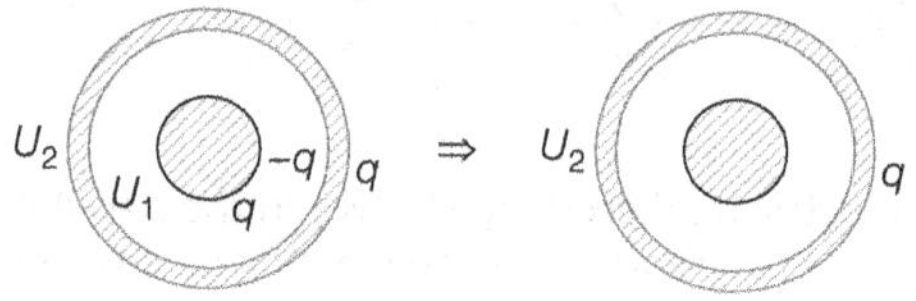

$$\text{Heat produced} = U_i - U_f$$

$$= (U_1 + U_2) - U_2$$

$$= U_1 = \frac{1}{2}\frac{q^2}{C}$$

where, $C = \frac{4\pi\varepsilon_0}{(1/a - 1/2a)} = 8\pi\varepsilon_0 a$

∴ $\text{Heat} = \frac{q^2}{16\pi\varepsilon_0 a} = \frac{kq^2}{4a}$

15.

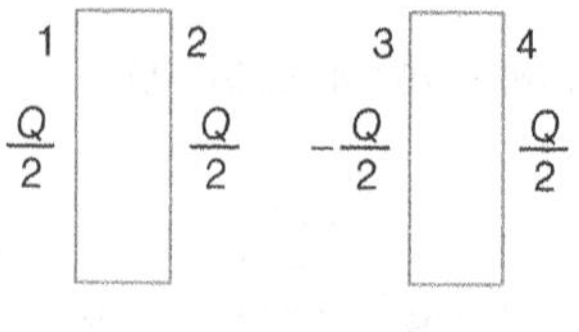

$$q_1 = q_4 = \frac{q_{Total}}{2} = \frac{Q}{2}$$

$$q_2 = Q - q_1 = \frac{Q}{2}$$

$$q_3 = -q_2 = \frac{-Q}{2}$$

$$E_A = E_1 + E_4 \quad \text{(towards left)}$$

$$= \frac{Q}{4A\varepsilon_0} + \frac{Q}{4A\varepsilon_0} = \frac{Q}{2A\varepsilon_0}$$

$$E_C = -E_A = \frac{Q}{2A\varepsilon_0} \quad \text{(towards right)}$$

$$E_B = E_2 + E_3 \quad \text{(towards right)}$$

$$= \frac{Q}{4A\varepsilon_0} + \frac{Q}{4A\varepsilon_0} = \frac{Q}{2A\varepsilon_0}$$

16. $$q_1 = q_8 = \frac{q_{\text{Total}}}{2} = 7Q$$

$$q_2 = 4Q - q_1 = -3Q$$
$$q_3 = -q_2 = +3Q$$
$$q_4 = Q - q_3 = -2Q$$
$$q_5 = -q_4 = +2Q$$
$$q_6 = 2Q - q_5 = 0$$
$$q_7 = -q_6 = 0$$

17. Charge on outermost surfaces

$$= \frac{q_{\text{Total}}}{2} = \frac{2q - 3q}{2} = -\frac{q}{2}$$

Hence, charge on different faces are as shown below.

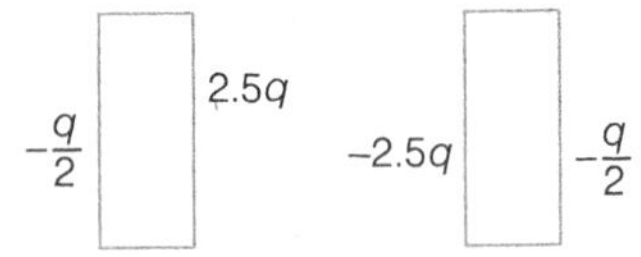

Electric field and hence potential difference between the two plates is due to $\pm 2.5\,q$.

$$\text{PD} = Ed$$

$$= \left(\frac{\sigma}{\varepsilon_0}\right) d = \left(\frac{2.5\,q\,d}{\varepsilon_0 A}\right)$$

Capacitance, $C = \dfrac{\varepsilon_0 A}{d}$

18.

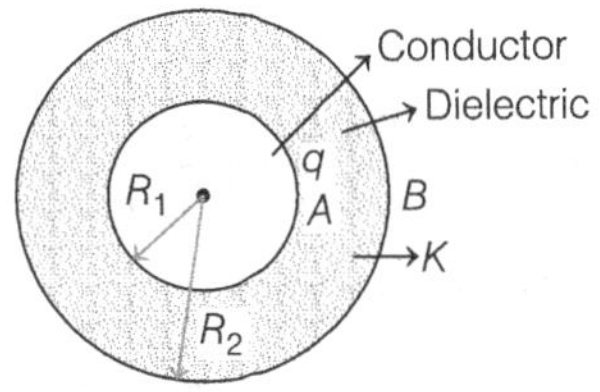

Let us assume that q charge is given to conductor A and $-q$ charge is at ∞.

Then, we find potential difference V between A and ∞.
Finally,

$$C = \frac{q}{V}$$

From A to B ($R_1 \le r \le R_2$)

$$E = \frac{1}{4\pi\varepsilon_0 K}\frac{q}{r^2}$$

$$dV = -E\,dr = -\frac{1}{4\pi\varepsilon_0 K}\frac{q}{r^2}dr$$

$$\therefore \quad \int_B^A dV = -\frac{q}{4\pi\varepsilon_0 K}\int_{R_2}^{R_1}\frac{dr}{r^2}$$

$$\therefore \quad V_A - V_B = \frac{q}{4\pi\varepsilon_0 K}\left(\frac{1}{R_1} - \frac{1}{R_2}\right) \quad \text{...(i)}$$

From B to ∞ ($r \ge R_2$)

$$E = \frac{1}{4\pi\varepsilon_0}\frac{q}{r^2}$$

$$\therefore \quad \int_\infty^B dV = -\frac{q}{4\pi\varepsilon_0}\int_\infty^{R_2}\frac{dr}{r^2}$$

$$\therefore \quad V_B - V_\infty = \frac{q}{4\pi\varepsilon_0}\left(\frac{1}{R_2}\right) \quad \text{...(ii)}$$

Adding Eqs. (i) and (ii), we get

$$V_A - V_\infty$$

or $$V = \frac{q}{4\pi\varepsilon_0 K}\left(\frac{1}{R_1} - \frac{1}{R_2}\right) + \frac{q}{4\pi\varepsilon_0}\left(\frac{1}{R_2}\right)$$

Simplifying, we get

$$C = \frac{q}{V} = \frac{4\pi\varepsilon_0 K R_1}{1 + (K-1)\frac{R_1}{R_2}}$$

19.

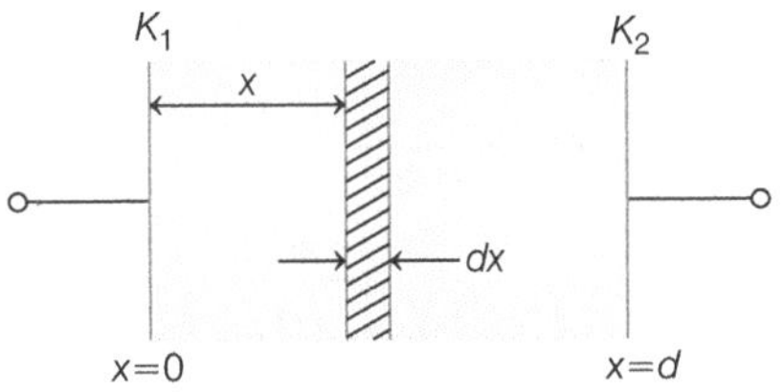

Let at x value of dielectric constant is K. Then,

$$K = K_1 + \left(\frac{K_2 - K_1}{d}\right)x$$

Now formula of C is,

$$C = \frac{\varepsilon_0 S}{(d - t_1 - t_2 - \cdots) + \left(\frac{t_1}{K_1} + \frac{t_2}{K_2} + \cdots\right)} \quad \text{...(i)}$$

If dielectric is completely filled, then

$$(d - t_1 - t_2 \ldots) = 0$$

and $$\frac{t_1}{K_1} + \frac{t_2}{K_2} + \ldots = \int_0^d \frac{dx}{K}$$

$$= \int_0^d \frac{dx}{K_1 + \left(\frac{K_2 - K_1}{d}\right)x}$$

Completing this integration and then putting the values in Eq. (i), we get

$$C = \frac{\varepsilon_0 S(K_2 - K_1)}{d \ln\left(\frac{K_2}{K_1}\right)}$$

20.

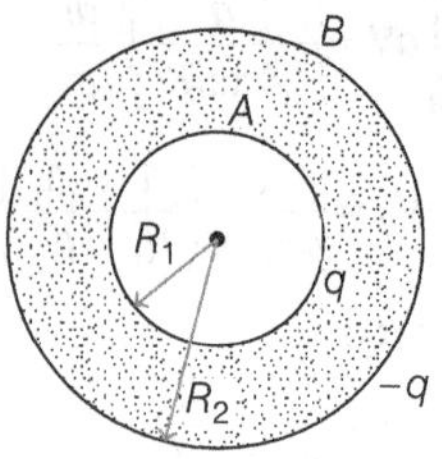

Let us take a charge q on inner conductor and $-q$ on outer conductor. Then, we find the potential difference $V = V_A - V_B$ and finally,

$$C = \frac{q}{V}$$

Electric field at a distance r ($R_1 < r < R_2$) from centre,

$$E = \frac{q}{4\pi\varepsilon_0 K r^2}$$

Substituting $K = \frac{a}{r}$, we have

$$E = \frac{q}{4\pi\varepsilon_0 ar}$$

Now, $dV = -E dr$

$$\therefore \quad \int_B^A dV = -\frac{q}{4\pi\varepsilon_0 a}\int_{R_2}^{R_1}\frac{dr}{r}$$

$$\text{or} \quad V_A - V_B \quad \text{or} \quad V = \frac{q}{4\pi\varepsilon_0 a}\ln\frac{R_2}{R_1}$$

$$\therefore \quad C = \frac{q}{V} = \frac{4\pi\varepsilon_0 a}{\ln\left(\frac{R_2}{R_1}\right)}$$

21.

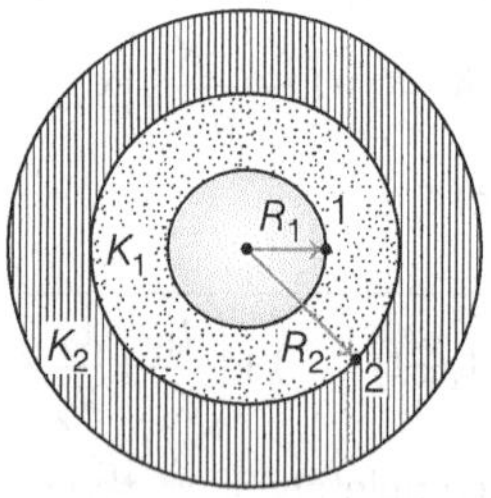

Maximum electric field in dielectric-1 will be at 1 and in dielectric-2 at 2.

$$E_{1m} = \frac{\lambda_{m_1}}{2\pi\varepsilon_0 K_1 R_1}$$

$$\Rightarrow \quad \lambda_{m_1} = 2\pi\varepsilon_0 K_1 R_1 E_{1m} \quad \text{...(i)}$$

Similarly, $$E_{2m} = \frac{\lambda_{m_2}}{2\pi\varepsilon_0 K_2 R_2}$$

$$\Rightarrow \quad \lambda_{m_2} = 2\pi\varepsilon_0 K_2 R_2 E_{2m} \quad \text{...(ii)}$$

Equating Eqs. (i) and (ii), we get

$$K_1 R_1 E_{1m} = K_2 R_2 E_{2m}$$

22. **Note** See the solution of Q-21.

$$E_1 = \frac{\lambda_{m_1}}{2\pi\varepsilon_0 K_1 R_1}$$

$$\therefore \quad \lambda_{m_1} = 2\pi\varepsilon_0 E_1 K_1 R_1$$

Similarly, $$\lambda_{m_2} = 2\pi\varepsilon_0 E_2 K_2 R_2$$

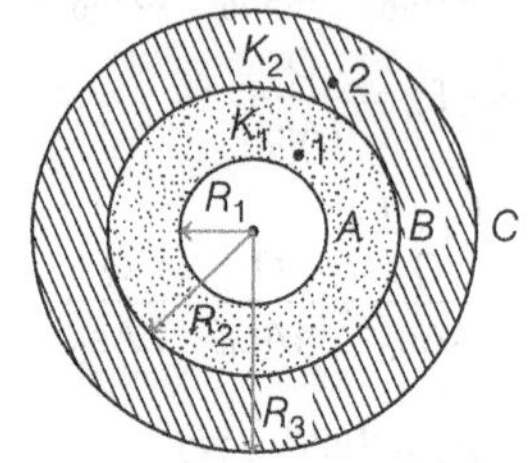

Since, $E_1 K_1 R_1 < E_2 K_2 R_2$

$$\therefore \quad \lambda_{m_1} < \lambda_{m_2}$$

So we will take safe $\lambda = \lambda_{m_1} = 2\pi\varepsilon_0 E_1 K_1 R_1$. Now with this λ, we have to find potential difference between A and C.

From A to B

$$E = \frac{\lambda}{2\pi\varepsilon_0 K_1 r} = \frac{2\pi\varepsilon_0 E_1 K_1 R_1}{2\pi\varepsilon_0 K_1 r}$$

$$= (E_1 R_1)\frac{1}{r}$$

$$dV = -E dr$$

$$\text{or} \quad \int_B^A dV = -E_1 R_1 \int_{R_2}^{R_1}\frac{dr}{r}$$

$$\text{or} \quad V_A - V_B = E_1 R_1 \ln\frac{R_2}{R_1} \quad \text{...(i)}$$

From B to C

$$E = \frac{\lambda}{2\pi\varepsilon_0 K_2 r} = \frac{2\pi\varepsilon_0 E_1 K_1 R_1}{2\pi\varepsilon_0 K_2 r}$$

$$= \left(\frac{E_1 K_1 R_1}{K_2}\right)\left(\frac{1}{r}\right)$$

$$dV = -E dr$$

$$\therefore \quad \int_C^B dV = -\left(\frac{E_1 K_1 R_1}{K_2}\right)\int_{R_3}^{R_2}\frac{dr}{r}$$

$$\therefore \quad V_B - V_C = \frac{E_1 K_1 R_1}{K_2}\ln\left(\frac{R_3}{R_2}\right) \quad \text{...(ii)}$$

Adding Eqs. (i) and (ii), we get

$$V_A - V_C \text{ or } V = E_1 R_1\left[\ln\frac{R_2}{R_1} + \frac{K_1}{K_2}\ln\left(\frac{R_3}{R_2}\right)\right]$$

23. Electric field at P due to $+\lambda$ and $-\lambda$ are additive.

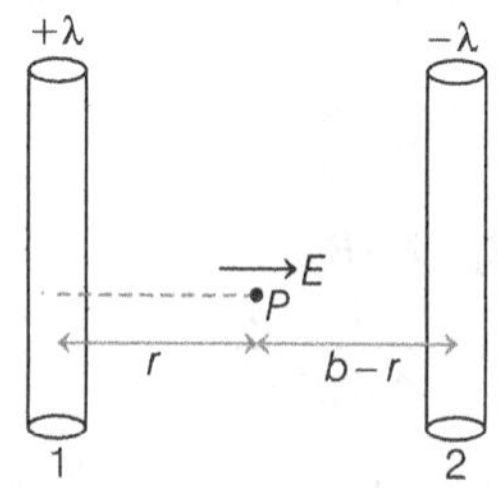

$$E = \frac{\lambda}{2\pi\varepsilon_0 r} + \frac{\lambda}{2\pi\varepsilon_0 (b-r)}$$

$$dV = -Edr = -\frac{\lambda}{2\pi\varepsilon_0}\left[\frac{1}{r}+\frac{1}{b-r}\right]$$

or $$\int_2^1 dV = -\frac{\lambda}{2\pi\varepsilon_0}\left[\int_{b-a}^{a}\left(\frac{1}{r}+\frac{1}{b-r}\right)dr\right]$$

$\therefore$ $V_1 - V_2$ or $$V = \frac{\lambda}{2\pi\varepsilon_0}\left[\ln\left(\frac{b-a}{a}\right)+\ln\left(\frac{b-a}{a}\right)\right]$$

or $$V = \frac{\lambda}{\pi\varepsilon_0}\ln\left(\frac{b-a}{a}\right)$$

For $b >> a$, $b-a \approx b$

$\therefore$ $$V = \frac{\lambda}{\pi\varepsilon_0}\ln\left(\frac{b}{a}\right)$$

So, capacitance per unit length,

$$C = \frac{\lambda}{V} = \frac{\pi\varepsilon_0}{\ln(b/a)}$$

or $$C = \frac{\pi\varepsilon_0}{\ln(b/a)}$$

24. At point P electric field due to $+q$ and $-q$ both are in the same direction. Therefore net electric field,

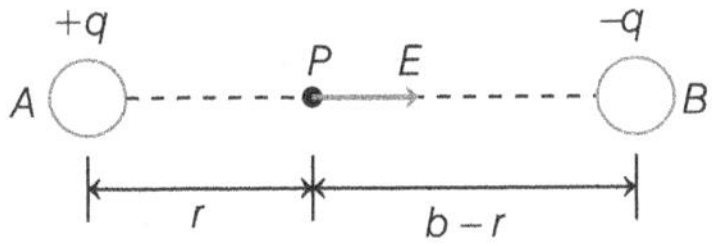

$$E = \frac{1}{4\pi\varepsilon_0 K}\frac{q}{r^2} + \frac{1}{4\pi\varepsilon_0 K}\frac{q}{(b-r)^2}$$

Now, $$dV = -Edr$$

or $$\int_B^A dV = V_A - V_B$$

or $$V = -\int_{b-a}^{a} Edr$$

Substituting the value of E we can find V, which comes out to be,

$$V = \frac{q}{2\pi\varepsilon_0 K}\left(\frac{1}{a}-\frac{1}{b}\right)$$

$$= \frac{q}{2\pi\varepsilon_0 K}\left(\frac{b-a}{ab}\right)$$

Now, $$C = \frac{q}{V} = \frac{2\pi\varepsilon_0 Kab}{b-a}$$

Since, $b >> a$

$\therefore$ $$b - a \approx b$$

or $$C = 2\pi\varepsilon_0 Ka$$

25. **Self energy of a shell-1**

Let us assume that dq charge is brought from infinity to the surface of shell when charge is q. So potential on surface is $V = \frac{1}{4\pi\varepsilon_0}\frac{q}{R_1}$.

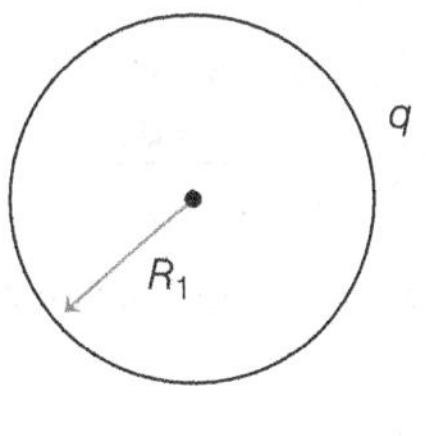

$$dW_1 = (dq)V = \frac{1}{4\pi\varepsilon_0}\frac{q}{R_1}dq$$

$\therefore$ $$W_1 = \int_0^{q_1}\frac{1}{4\pi\varepsilon_0}\frac{q}{R_1}dq \quad \text{or} \quad W_1 = \frac{q_1^2}{8\pi\varepsilon_0 R_1}$$

So self energy of this shell is,

$$W_1 = \frac{q_1^2}{8\pi\varepsilon_0 R_1}$$

Self energy of shell-2

Similarly self energy of another shell will be,

$$W_2 = \frac{q_2^2}{8\pi\varepsilon_0 R_2}$$

Total electric energy of the system

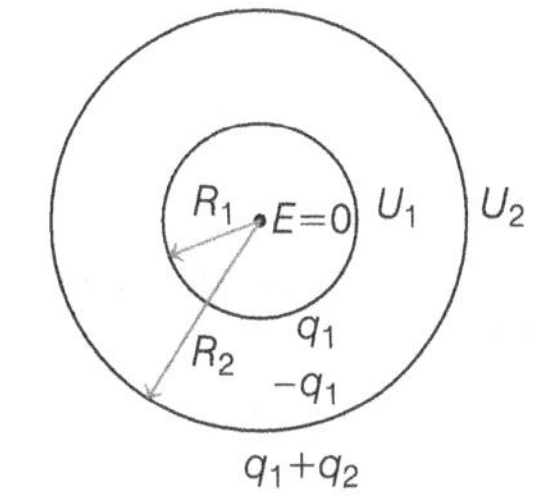

Distribution of charges is as shown in figure. Energy is stored only in the electric field. So, for $r < R_1$, energy stored is zero.

$$U_1 = \frac{1}{2}\frac{q_1^2}{C_1} \quad \text{...(i)}$$

$$U_2 = \frac{1}{2}\frac{(q_1+q_2)^2}{C_2} \quad \text{...(ii)}$$

Here, $$C_1 = \left(\frac{4\pi\varepsilon_0}{\frac{1}{R_1}-\frac{1}{R_2}}\right)$$ (spherical capacitor)

and $$C_2 = \left(\frac{4\pi\varepsilon_0}{\frac{1}{R_2}-\frac{1}{\infty}}\right)$$

(assuming that second plate is at infinity)

Substituting the values of C_1 and C_2 in Eqs. (i) and (ii), we get

Total electric energy,

$$W = U_1 + U_2$$

or $$W = \frac{1}{4\pi\varepsilon_0}\left[\frac{q_1^2}{2R_1}+\frac{q_2^2}{2R_2}+\frac{q_1 q_2}{R_2}\right]$$

Interaction energy

$$W_{12} = W - W_1 - W_2$$

or $$W_{12} = \frac{1}{4\pi\varepsilon_0}\frac{q_1 q_2}{R_2}$$

26. **Note** See the solution of Q-25. Self energy of a charged spherical shell is,

$$U \text{ or } W = \frac{q^2}{8\pi\varepsilon_0 R}$$

Here, $$U_i = \frac{q^2}{8\pi\varepsilon_0 R_1} \text{ and } U_f = \frac{q^2}{8\pi\varepsilon_0 R_2}$$

∴ Work done by electric forces,

$$W = -\Delta U = U_i - U_f$$

or $$W = \frac{q^2}{8\pi\varepsilon_0}\left[\frac{1}{R_1} - \frac{1}{R_2}\right]$$

27. **Note** See the hint of Q-25

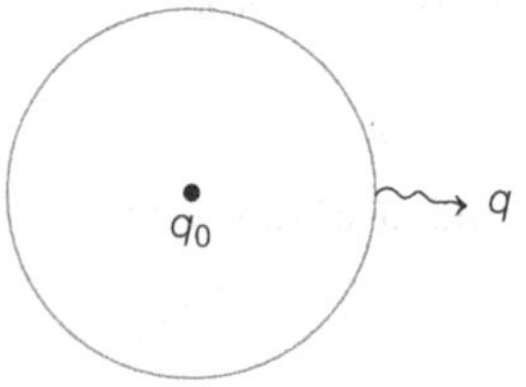

Total energy of this system,

U = self energy of shell + interaction energy of point charge and shell

or, $U = \frac{q^2}{8\pi\varepsilon_0 R}$ + (charge on shell) (potential of q_0 on shell)

So, $$U = \frac{q^2}{8\pi\varepsilon_0 R} + q\left(\frac{q_0}{4\pi\varepsilon_0 R}\right)$$

In the given problem,

$$U_i = \frac{q^2}{8\pi\varepsilon_0 R_1} + \frac{qq_0}{4\pi\varepsilon_0 R_1}$$

and $$U_f = \frac{q^2}{8\pi\varepsilon_0 R_2} + \frac{qq_0}{4\pi\varepsilon_0 R_2}$$

Work done by the electric forces,

$$W = -\Delta U = U_i - U_f$$

Substituting the values, we get

$$W = \frac{q\left(q_0 + \frac{q}{2}\right)}{4\pi\varepsilon_0}\left(\frac{1}{R_1} - \frac{1}{R_2}\right)$$

28. **Total electrostatic potential energy of the above system**

U_i = self energy of 2 + self energy of 3 + interaction energy of 1 and 2 + interaction energy of 1 and 3 + interaction energy of 2 and 3

$= \frac{q^2}{8\pi\varepsilon_0 a} + \frac{q^2}{8\pi\varepsilon_0 b}$ + $(-q)$ (potential of 1 at 2) + (q) (potential of 1 at 3) + $(-q)$ (potential of 3 at 2)

$$= \frac{q^2}{8\pi\varepsilon_0 a} + \frac{q^2}{8\pi\varepsilon_0 b} - q\left(\frac{q}{4\pi\varepsilon_0 a}\right) + q\left(\frac{q}{4\pi\varepsilon_0 b}\right) - q\left(\frac{q}{4\pi\varepsilon_0 b}\right)$$

Simplifying, we get

or $$U_i = \frac{q^2}{8\pi\varepsilon_0}\left(\frac{1}{b} - \frac{1}{a}\right)$$

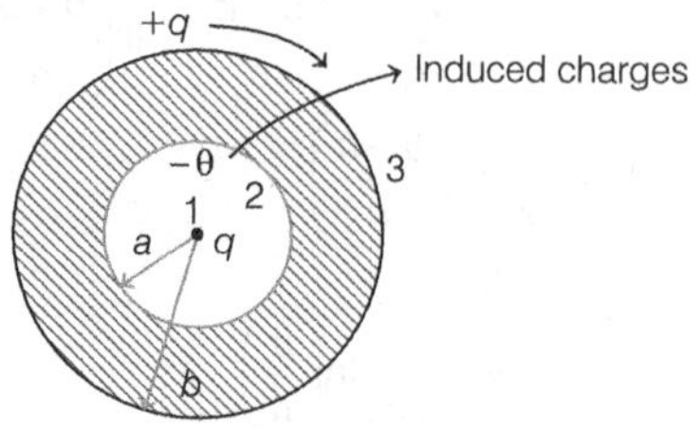

Final potential energy of system

When charge is moved to infinity induced charges also disappear. So,

$$U_f = 0$$

Now work done by external forces,

$$W = \Delta U = U_f - U_i$$

Substituting the values, we get

$$W = \frac{q^2}{8\pi\varepsilon_0}\left(\frac{1}{a} - \frac{1}{b}\right)$$

29. (a) At $r = R$,

$$V = \frac{1}{4\pi\varepsilon_0}\left[\frac{Q}{R} - \frac{2Q}{2R} + \frac{3Q}{3R}\right]$$

$$= \left(\frac{1}{4\pi\varepsilon_0}\right)\left(\frac{Q}{R}\right)$$

At $r = 3R$

$$V = \frac{1}{4\pi\varepsilon_0}\left[\frac{Q}{3R} - \frac{2Q}{3R} + \frac{3Q}{3R}\right]$$

$$= \frac{Q}{6\pi\varepsilon_0 R}$$

(b) $$E = \left(\frac{1}{4\pi\varepsilon_0}\right)\left(\frac{q_{in}}{r^2}\right)$$

$$= \left(\frac{1}{4\pi\varepsilon_0}\right)\left[\frac{Q - 2Q}{(5R/2)^2}\right] = \frac{-Q}{25\pi\varepsilon_0 R^2}$$

Negative sign implies that this electric field is radially towards centre.

(c)

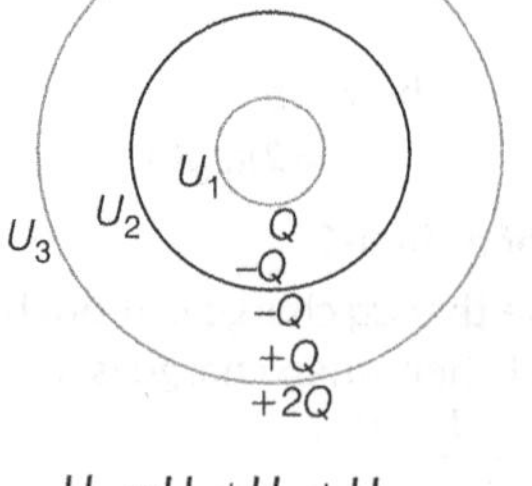

$$U_T = U_1 + U_2 + U_3$$

where, $U_1 = \frac{1}{2}\frac{Q^2}{C_1}$

where, $C_1 = 4\pi\varepsilon_0 \Big/ \left(\frac{1}{R} - \frac{1}{2R}\right)$

$U_2 = \frac{1}{2}\frac{Q^2}{C_2}$

where, $C_2 = 4\pi\varepsilon_0 \Big/ \left(\frac{1}{2R} - \frac{1}{3R}\right)$

and $U_3 = \frac{1}{2}\frac{(2Q)^2}{C_3}$

where, $C_3 = 4\pi\varepsilon_0 \Big/ \left(\frac{1}{3R} - \frac{1}{\infty}\right)$

(d)

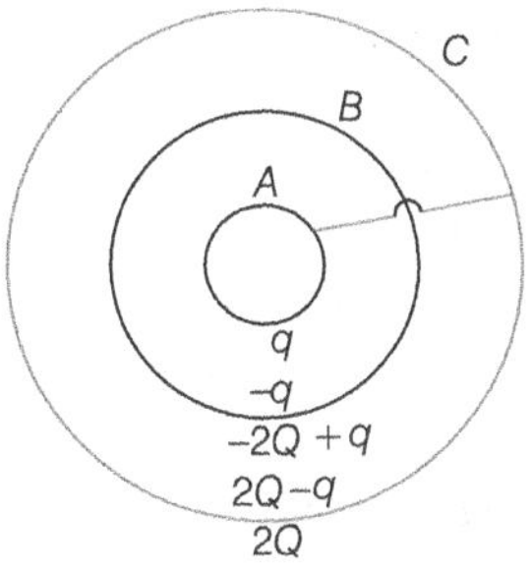

Total charge on $(A+C)$ is $3Q+Q=4Q$.

Now, $V_A = V_C$

$$\therefore \quad \frac{kq}{R} - \frac{k(2Q)}{2R} + \frac{k(4Q-q)}{3R} = \frac{k(Q-2Q+3Q)}{3R} \quad \text{...(i)}$$

Solving this equation, we get

$$q = \frac{Q}{2}$$

Now, $q_A = q = \frac{Q}{2}$

$$q_C = 4Q - q = \frac{7}{2}Q$$

(e) $E = \frac{1}{4\pi\varepsilon_0} \cdot \frac{q_{in}}{r^2}$

$$= \frac{1}{4\pi\varepsilon_0}\left[\frac{q-2Q}{(5R/2)^2}\right] = \frac{-3Q}{50\pi\varepsilon_0 R^2}$$

Minus sign indicates that electric field is radially inwards.

Module 5.

Introductory Exercise

1. $V_{32} = V_5 = 6\text{ V}$

$\therefore \quad q_5 = CV = 30\ \mu\text{C}$

$$q_{32} = \left(\frac{3\times 2}{3+2}\right) \times 6 = 7.2\ \mu\text{C}$$

$\therefore \quad \frac{q_5}{q_2} = \frac{30}{7.2}$

3. $C_{net} = 2\ \mu\text{F}, q = CV = 2\times 15 = 30\ \mu\text{C}$

Now, this q will be distributed between $4\ \mu\text{F}$ and $2\ \mu\text{F}$ in direct ratio of their capacities.

4. $C_{net} = 3\ \mu\text{F}, q = CV = 3\times 40 = 120\ \mu\text{C}$.

Now, this q will be distributed between $9\ \mu\text{F}$ and $3\ \mu\text{F}$ in the direct ratio of their capacities.

5. All three capacitors are in parallel with the battery. PD across each of them is 10 V. So, apply $q = CV$ for all of them.

Exercise

1. $V_{net} = V_1 + V_2 + \ldots$ (in series)

$= V + V + \ldots$

$= nV$

2.

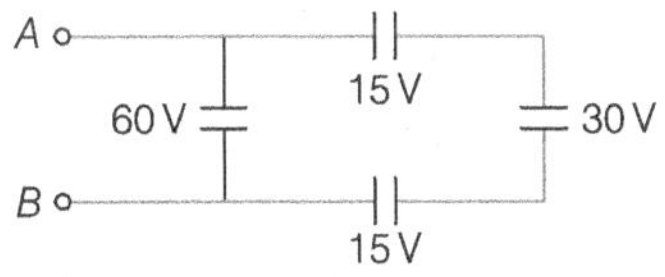

3. Given circuit can be reduced as

Let potential at P is V_P and potential at B is V_B.

As capacitors $3\ \mu\text{F}$ and $6\ \mu\text{F}$ are in series, they have same charge.

$\therefore$ Charge on $3\ \mu\text{F}$ = Charge on $6\ \mu\text{F}$

$\therefore \quad C_1V_1 = C_2V_2$

or $\quad 3(1200 - V_P) = 6(V_P - V_B)$

As B point is earthed.

So, $\quad V_B = 0$

So, $\quad 1200 - V_P = 2V_P$

$\therefore \quad V_P = 400$ volt

4. Potential difference across C is 10 V.

$\therefore \quad q = CV = 60\ \mu\text{C}$

5. All capacitors have equal capacitance. Hence, equal potential drop (=2.5 V) will take place across all capacitors.

$V_N - V_B = 2.5\text{ V}$

$0 - V_B = 2.5\text{ V}$

$\therefore \quad V_B = -2.5\text{ V}$

Further, $\quad V_A - V_N = 3(2.5)\text{ V}$

$= 7.5\text{ V}$

$\therefore \quad V_A = +7.5\text{ V}$ (as $V_N = 0$)

6. Common potential in parallel grouping

$$= \frac{\text{Total charge}}{\text{Total capacity}} = \frac{EC}{2C} = \frac{E}{2}$$

7. In series potential difference,

$$V \propto \frac{1}{C} \Rightarrow V_1 > V_2$$

$\therefore \quad C_1 < C_2$

8. As capacitors C_1 and C_2 are in series, there should be equal charge on them, i.e.

$$\text{Charge on } C_1 = \text{Charge on } C_2$$

$\therefore \quad C_1(V_A - V_D) = C_2(V_D - V_B)$

or $\quad C_1(V_1 - V_D) = C_2(V_D - V_2)$

or $\quad C_1V_1 - C_1V_D = C_2V_D - C_2V_2$

or $\quad V_D(C_1 + C_2) = C_1V_1 + C_2V_2$

$\therefore \quad V_D = \dfrac{C_1V_1 + C_2V_2}{C_1 + C_2}$

9. $\because \quad q_i = q_f$

$$C_1V_1 + C_2V_2 = (C_1 + C_2)V$$

$\therefore \quad (2)(4) + (C)5 = (2 + C)\,4.6$

$$0.4C = 1.2$$

$\therefore \quad C = 3 \text{ units}$

10. $C_1 = C_{RHS} + C_{LHS}$

$$= \frac{K_2\,\varepsilon_0(A/2)}{d} + \frac{K_1\varepsilon_0(A/2)}{d}$$

$$= \frac{\varepsilon_0 A}{2d}(K_1 + K_2) = \frac{5\varepsilon_0 A}{2d}$$

$$C_2 = \frac{\varepsilon_0 A}{d - d/2 - d/2 + \dfrac{d/2}{K_1} + \dfrac{d/2}{K_2}}$$

$$= \frac{2\varepsilon_0 A}{d}\left(\frac{K_1K_2}{K_1 + K_2}\right) = \frac{12\,\varepsilon_0 A}{5d}$$

$\therefore \quad \dfrac{C_1}{C_2} = \dfrac{25}{24}$

11. Charge, $q = CV = 10^4\ \mu C$

In parallel, common potential is given by

$$V = \frac{\text{Total charge}}{\text{Total capacity}}$$

$$20 = \frac{(10^4\ \mu C)}{(C + 100)\ \mu C}$$

Solving this equation, we get

$$C = 400\ \mu F$$

12. $0.1 = \dfrac{1}{2}(C_1 + C_2)(2)^2 \quad$ ...(i)

$$1.6 \times 10^{-2} = \frac{1}{2}\left(\frac{C_1C_2}{C_1 + C_2}\right)(2)^2 \quad \text{...(ii)}$$

Solving these two equations, we can find C_1 and C_2.

14. In series, potential difference distributes in inverse ratio of capacitance.

$\therefore \quad \dfrac{V_A}{V_B} = \dfrac{C_B}{C_A} = \dfrac{C_2}{C_1} = \dfrac{60}{40} = \dfrac{3}{2}$

$\therefore \quad C_2 = 1.5\,C_1 \quad$ ...(i)

Now, $\quad \dfrac{V_A'}{V_B'} = \dfrac{C_B'}{C_A'}$

or $\quad \dfrac{10}{90} = \dfrac{C_2}{(C_1 + 2)}$

or $\quad C_1 + 2 = 9C_2 \quad$ ...(ii)

Solving Eqs. (i) and (ii), we get

$$C_1 = 0.16\ \mu F \quad \text{and} \quad C_2 = 0.24\ \mu F$$

15. (a) $q = CV$

(b) $C = \dfrac{\varepsilon_0 A}{d}$ or $C \propto \dfrac{1}{d}$

If d is doubled, C will remain half. Hence, q will also remain half.

(c) $q = CV = \left(\dfrac{\varepsilon_0 A}{d}\right)V = \dfrac{\varepsilon_0\,(\pi R^3)V}{d}$

or $\quad q \propto R^2$

R is doubled. Hence, q will become four times or 480 μC

16. (a) $\dfrac{1}{C_{net}} = \dfrac{1}{8.4} + \dfrac{1}{8.2} + \dfrac{1}{4.2}$

$\therefore \quad C_{net} = 2.09\ \mu F$

$$q_{net} = C_{net}V = (2.09)(36)$$

$$= 75.24\ \mu C \approx 75\ \mu C$$

In series, charge remains same in all capacitors.

(b) $U_{total} = \dfrac{1}{2}C_{net}V^2$

(c) $q_{total} = (3)(75)\ \mu C = 225\ \mu C$

Now common potential in parallel,

$$V = \frac{q_{total}}{C_{total}} = \frac{225\ \mu C}{(8.4 + 8.2 + 4.2)\ \mu F}$$

$$= 10.8 \text{ volt}$$

(d) $U_{total} = \dfrac{1}{2}C_{net}V^2$

17. In series, potential difference distributes in inverse ratio of capacitance.

$\therefore \quad \dfrac{V_A}{V_B} = \dfrac{C_2}{C_1}$ or $\dfrac{130}{100} = \dfrac{C_2}{C_1}$

or $\quad C_1 = \dfrac{C_2}{1.3} = C_A$

K is made 2.5 times. Therefore, C_1 will also become 2.5 times.

$$C_1' = 2.5\,C_1 = \frac{2.5\,C_2}{1.3}$$

or $\quad \dfrac{C_1'}{C_2} = \dfrac{25}{13}$

Now, $\quad \dfrac{V_A'}{V_B'} = \dfrac{C_2}{C_1'} = \dfrac{13}{25}$

or $\quad V_A' = \left(\dfrac{13}{13 + 25}\right)(230) = 78.68\ V$

$$V_B' = 230 - V_B' = 151.32\ V$$

18. $C_{23} = \dfrac{2 \times 3}{2 + 3} = 1.2\ \mu F$

$$q_{total} = C_1V = 110\ \mu C$$

Common potential in parallel is given by

$$V = \frac{\text{Total charge}}{\text{Total capacity}} = \frac{110}{1+1.2} = 50\ V$$

$$q_{23} = (C_{23})V = 60\ \mu C$$

So, this much charge flows through the switch.

Module 6.

1. Charges are same, if initially the capacitors are uncharged.

   Further, $V = \frac{q}{C}$

   Hence, $V \propto \frac{1}{C}$

   if q is same.

2. $q = CV = 200\ \mu C$

   In parallel, the common potential is given by

$$V = \frac{\text{Total charge}}{\text{Total capacity}}$$

$$= \frac{200\ \mu C}{(2+2)\ \mu F} = 50\ V$$

   Heat loss $= U_i - U_f$

$$= \frac{1}{2}(2\times 10^{-6})(100)^2 - \frac{1}{2}(4\times 10^{-6})(50)^2$$

$$= 5\times 10^{-3}\ J$$

$$= 5\ mJ$$

3. Both the conductors carry equal and opposite charges. So, after connecting by a wire, there will be no charge in any conductor. Hence, all the stored energy will be destroyed.

$$\text{Loss of energy} = 2\left(\frac{1}{2}CV^2\right) = CV^2$$

4. Total charge $= (2C)(4V) - CV$

$$= 7CV$$

   Common potential after they are connected is

$$V_C = \frac{\text{Total charge}}{\text{Total capacitance}} = \frac{7\ CV}{2C + C} = \frac{7}{3}V$$

$$\text{Heat} = U_i - U_f = \frac{1}{2}CV^2 + \frac{1}{2}(2C)(4V)^2 - \frac{1}{2}\times 3C \times \left(\frac{7}{3}V\right)^2$$

$$= \frac{25}{3}CV^2$$

5. Equivalent capacitance of 6 μF and 3 μF is also 2 μF and charge across it is also q or circuit is balanced. Hence, there is no flow of charge.

6. $q_i = C_iV = 2V = q$

   This charge will remains constant after switch S is shifted from position 1 to position 2.

$$U_i = \frac{1}{2}\frac{q^2}{C_i} = \frac{q^2}{2\times 2} = \frac{q^2}{4}$$

$$U_f = \frac{1}{2}\frac{q^2}{C_f} = \frac{q^2}{2\times 10} = \frac{q^2}{20}$$

$$\text{Energy dissipated} = U_i - U_f = \frac{q^2}{5}$$

   The energy dissipated is 80% of the initial stored energy as % decrease in energy

$$= \frac{U_i - U_f}{U_i}\times 100\%$$

$$= \frac{q^2/5}{q^2/4}\times 100\%$$

$$= \frac{4}{5}\times 100\% = 80\%$$

7. $\frac{V_1}{V_{1.5}} = \frac{1.5}{1}$

$$\therefore \quad V_1 = \left(\frac{1.5}{1.5+1}\right)(30)$$

$$= 18\ V$$

$$\frac{V_{2.5}}{V_{0.5}} = \frac{0.5}{2.5} = \frac{1}{5}$$

$$\therefore \quad V_{2.5} = \left(\frac{1}{1+5}\right)(30)$$

$$= 5\ V$$

   Now, $|V_{ab}| = V_1 - V_{2.5}$

$$= 13\ V$$

8. $q = (E_1 + E_2)C_{net} = (E_1 + E_2)\frac{C_1C_2}{C_2 + C_2}$

$$V_{ab} = \frac{q}{C_2} = \left(\frac{E_1 + E_2}{C_1 + C_2}\right)C_1$$

9. Between 3 μF and 2 μF (in parallel), total charge of 80 μC will distribute in direct ratio of capacity.

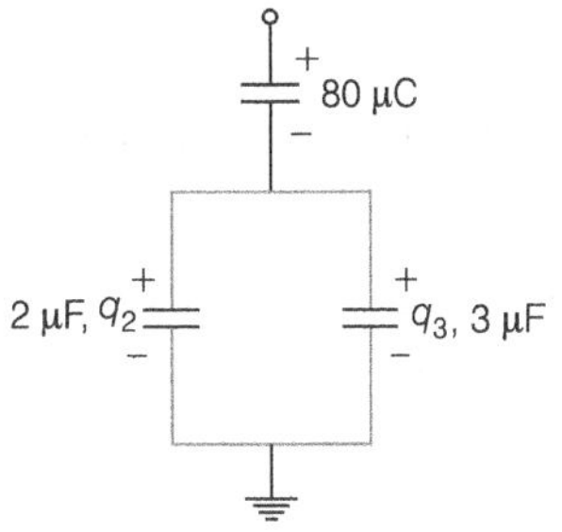

$$\frac{q_3}{q_2} = \frac{3}{2}$$

$$q_3 = \left(\frac{3}{3+2}\right)(80) = 48\ \mu C$$

10. When S_3 is closed, due to attraction with opposite charge, no flow of charge takes place through S_3. Therefore, potential difference across capacitor plates remains unchanged or $V_1 = 30$ V and $V_2 = 20$ V.

Alternate Solution

Charges on the capacitors are

$$q_1 = (30)(2) = 60\text{ pC}$$

and $$q_2 = (20)(3) = 60\text{ pC}$$

or $$q_1 = q_2 = q \text{ (say)}$$

The situation is similar as the two capacitors in series are first charged with a battery of emf 50 V and then disconnected.

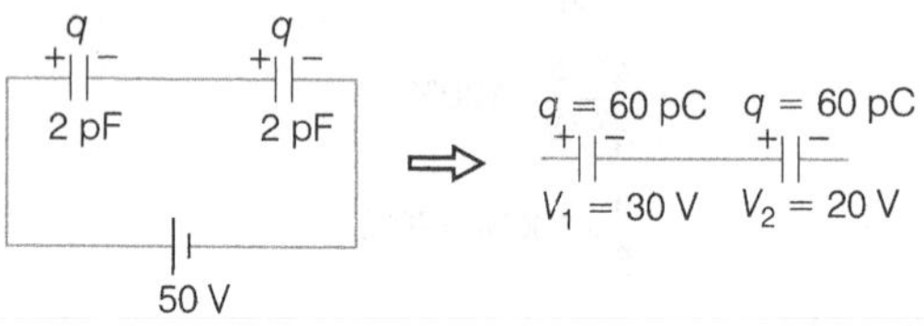

$\therefore$ When S_3 is closed,

$$V_1 = 30\text{ V} \quad \text{and} \quad V_2 = 20\text{ V}$$

11.

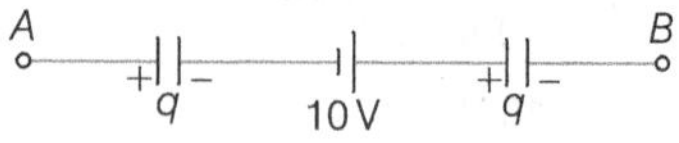

q shown in figure is in μC.

Now, $$V_A - \frac{q}{1} + 10 - \frac{q}{2} = V_B$$

or $$V_A - V_B = \frac{3q}{2} - 10 = 5$$

$\therefore$ $$q = 10\ \mu\text{C}$$

Now, $V = \frac{q}{C}$ across each capacitor.

12.

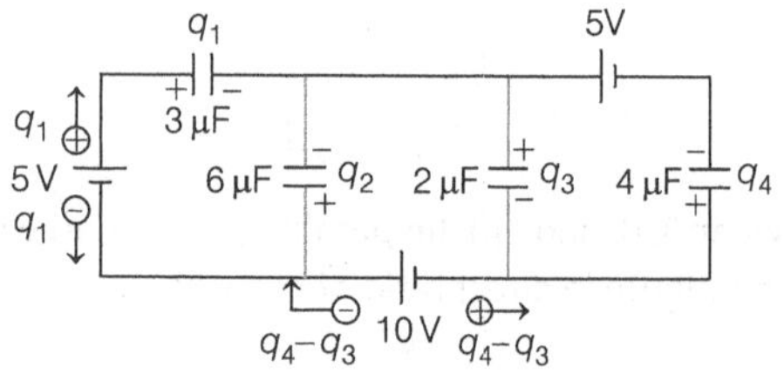

If we see the charge on positive plate of 6 μF capacitor, then

$$q_2 = -q_1 - (q_4 - q_3) \quad \text{...(i)}$$

Now, applying three loop equations, we have

$$5 - \frac{q_1}{3} + \frac{q_2}{6} = 0 \quad \text{...(ii)}$$

$$10 - \frac{q_2}{6} - \frac{q_3}{2} = 0 \quad \text{...(iii)}$$

$$5 - \frac{q_3}{2} - \frac{q_4}{4} = 0 \quad \text{...(iv)}$$

Solving these four equations, we can find q_1, q_2, q_3 and q_4.

13. (a) Simple series and parallel grouping of capacitors.

(b) $$q_{net} = C_{net}V$$
$$= (2.5 \times 10^{-6})(220)$$
$$= 5.5 \times 10^{-4}\text{ C}$$

C_1, C_5 and equivalent of other three capacitors are in series. Hence, charges across them are same.

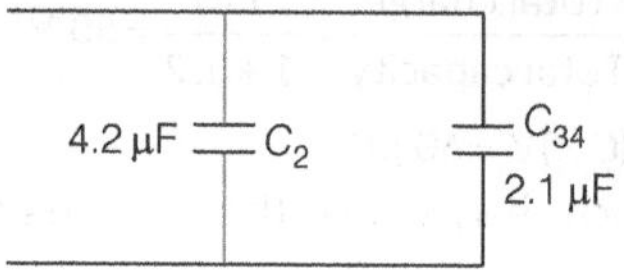

q_{total} will distribute between C_2 and C_{34} in direct ratio of capacitance.

$\therefore$ $$\frac{q_2}{q_{34}} = \frac{4.2}{2.1} = \frac{2}{1}$$

$\therefore$ $$q_2 = \frac{2}{3}(5.5 \times 10^{-4})$$
$$= 3.7 \times 10^{-4}\text{ C}$$

$$q_{34} = \frac{1}{3}(5.5 \times 10^{-4})$$
$$= 1.8 \times 10^{-4}\text{ C}$$

For finding PD across any capacitor, use the equation

$$C = \frac{q}{V}$$

14. (a) Simple circuit is as shown below.

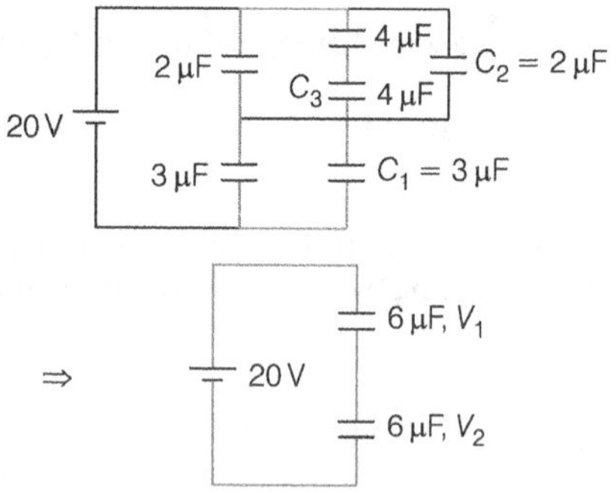

(b) $q_{net} = (C_{net})V = (34\mu\text{F})(20\text{ V}) = 60\ \mu\text{C}$

(c) Upper network and lower network both have same capacitance $= 6\ \mu\text{F}$

$$V_1 - V_2 = \frac{20}{2} = 10\text{ V}$$

$$V_{C_1} = 10\text{ V}$$

$\therefore$ $$q_{C_1} = (C_1)(V_{C_1}) = 30\ \mu\text{C}$$

(d) $V_{C_2} = 10\text{ V}$,

$\therefore$ $$q_{C_2} = (C_2)(V_{C_2}) = 20\ \mu\text{C}$$

(e) $V_{C_3} = 5\text{ V}$,

$\therefore$ $$q_{C_3} = (C_3)(V_{C_3}) = 20\ \mu\text{C}$$

15. (a)

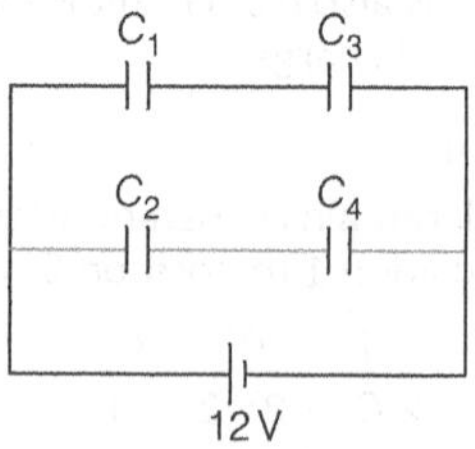

$$C_{13}=\frac{1\times 3}{1+3}=\frac{3}{4}\ \mu F$$

$$C_{24}=\frac{2\times 4}{2+4}=\frac{4}{3}\ \mu F$$

$$V_{C_1C_3}=V_{C_2C_4}=12\ V$$

$$\therefore \quad q_1=q_3=(C_{13})(V_{C_1C_3})$$

$$=\frac{3}{4}\times 12=9\ \mu C$$

$$q_2=q_4=(C_{24})(V_{C_2C_4})$$

$$=\frac{4}{3}\times 12=16\ \mu C$$

(b)

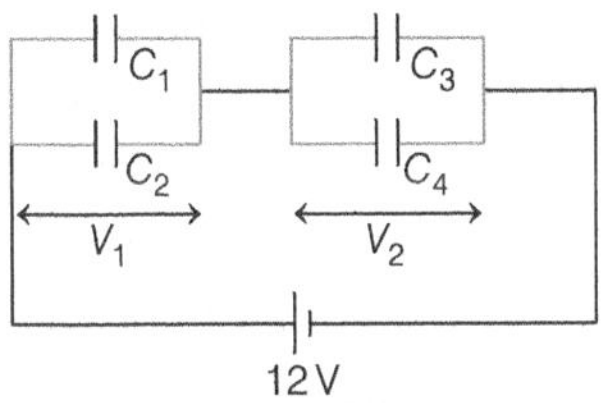

$$C_{12}=1+2=3\ \mu F$$

$$C_{34}=3+4=7\ \mu F$$

$$\frac{V_1}{V_2}=\frac{C_{34}}{C_{12}}=\frac{7}{3}$$

$$\therefore \quad V_1=\frac{7}{10}(12)=8.4\ V$$

$$V_2=12-8.4=3.6\ V$$

Now, we can apply $q=CV$ for finding charge on different capacitors.

16. $q_{total}=C_1V_0$

After switch is thrown towards right, C_{23} and C_1 are in parallel. The common potential is

$$V=\frac{\text{Total charge}}{\text{Total capacity}}=\frac{C_1V_0}{C_1+\left(\frac{C_2C_3}{C_2+C_3}\right)}$$

Now, $q_{C_1}=C_1V=\dfrac{C_1^2V_0}{C_1+\left(\frac{C_2C_3}{C_2+C_3}\right)}$

This is the same result as given in the answer.

$$q_{C_2}=q_{C_3}=C_{23}V$$

$$=\left(\frac{C_2C_3}{C_2+C_3}\right)\left[\frac{C_1V_0}{C_1+\frac{C_2C_3}{C_2+C_3}}\right]$$

$$=\frac{C_1V_0}{1+\frac{C_1}{C_2}+\frac{C_1}{C_3}}$$

Module 7.

1. In graph-1, discharging is slow.

Hence, $\tau_{C_1}>\tau_{C_2}$

Further, $\tau_C=CR$

$\therefore \quad \tau_C\propto R$ (as $C=$ constant)

2. Charge (or current) will not flow in the circuit as they have already the same potential, which is a condition of parallel grouping.

Further,

$$\frac{q_1}{q_2}=\frac{C_1V}{C_2V}=\frac{C_1}{C_2}=\frac{1}{2}$$

3. Capacitor and R_2 are short-circuited. Hence, current through R_2 is zero and capacitor is not charged.

4. At $t=0$, when capacitor is under charged, equivalent resistance of capacitor $=0$

In this case, 6Ω and 3Ω are parallel (equivalent $=2\Omega$)

$$\therefore \quad R_{net}\ (1+2)\Omega=3\Omega$$

$\therefore$ Current from battery $=\dfrac{12}{3}=4A$

$=$ Current through 1Ω resistor

5.

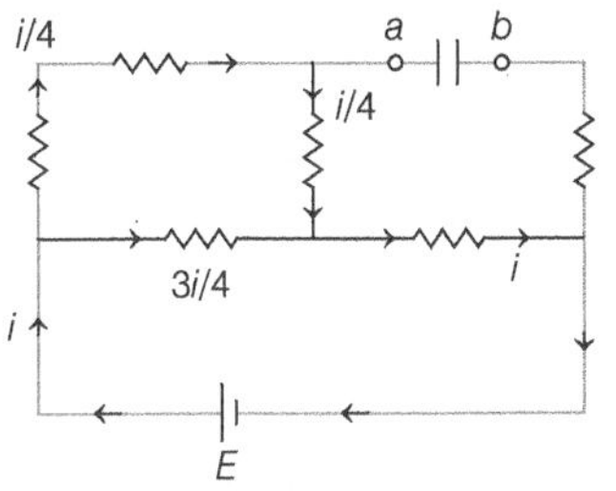

$$V_a-\frac{i}{4}R-iR=V_b$$

$$\therefore \quad V_a-V_b=\frac{5}{4}iR=10 \quad ...(i)$$

$$R_{net}=R+\frac{(3R)(R)}{(3R+R)}=\frac{7}{4}R$$

$$\therefore \quad i=\frac{E}{(7/4)R}=\left(\frac{4E}{7R}\right)$$

Substituting in Eq. (i), we have

$$\left(\frac{5}{4}\right)\left(\frac{4E}{7R}\right)\times R=10$$

$$\therefore \quad E=14\ V$$

6. $V_A-6-3\times 2+\dfrac{9}{1}-3\times 3=V_B$

$$\therefore \quad V_A-V_B=12\ V$$

7. In steady state condition, current flows from outermost loop.

$$i=\frac{12}{6+2}=1.5\ A$$

Now, $V_C=V_{6\Omega}=iR$

$$=1.5\times 6=9\ V$$

$$\therefore \quad q=CV_C=18\ \mu C$$

8. Area $=\dfrac{1}{2}QV=$ energy stored in capacitor

9. At $t=0$ when capacitors are initially uncharged, their equivalent resistance is zero. Hence, whole current passes through these capacitors.

10. Changing current is given by

$$i=i_0 e^{-t/\tau_C}$$

or $$i=\frac{V}{R}e^{-t/CR}$$

If we have take log on both sides, we have

$$\ln(i)=\ln\left(\frac{V}{R}\right)-\left(\frac{1}{CR}\right)t$$

Hence, $\ln(i)$ *versus* t graph is a straight line with slope $\left(-\frac{1}{CR}\right)$ and intercept $+\ln\left(\frac{V}{R}\right)$.

Intercepts are same, but $|\text{slope}|_1 > |\text{slope}|_2$.

11. Equal and opposite charges should transfer from two terminals of a battery. For charging of a capacitor, it should lie on a closed loop.

12.

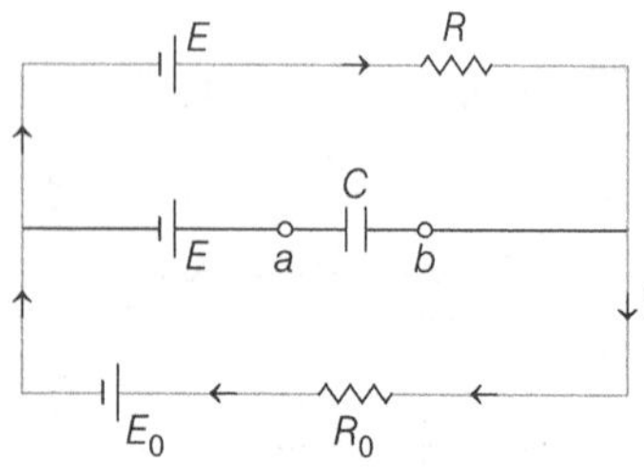

$$i=\frac{[E-E_0]}{R+R_0}$$

Now, $V_a-E+E_0+iR_0=V_b$

$\therefore$ $$V_a-V_b=(E-E_0)-iR_0$$

$$=(E-E_0)\left[1-\frac{R_0}{R+R_0}\right]$$

$$=\frac{R(E-E)}{R+R_0}$$

$\therefore$ $$q=C(V_a-V_b)=\frac{CR(E-E_0)}{R+R_0}$$

13. $V_0=i_0R=(10)(10)=100$ V

After 2 s, current becomes $\frac{1}{4}$th. Therefore, after 1 s, current will remain half also called half-life.

$$t_{1/2}=(\ln 2)\tau_C=(\ln 2)CR$$

$\therefore$ $$C=\frac{(t_{1/2})}{(\ln 2)R}=\frac{1}{10\ln 2}\text{F}$$

Total heat $$=\frac{1}{2}CV_0^2$$

$$=\frac{1}{2}\times\frac{1}{10\ln 2}(100)^2$$

$$=\frac{500}{\ln 2}\text{J}$$

14. In position-1, initial maximum current is

$$i_0=\frac{V}{R}=\frac{10}{5}=2\text{ A}$$

At the given time, given current is 1 A or half of the above value. Hence, at this is instant capacitor is also charged to half of the final value of 5 V.

Now, it is shifted to position-2 wherein steady state it is again charged to 5 V but with opposite polarity.

$$U_i=U_f=\frac{1}{2}CV^2 \qquad (\because V=5\text{ V})$$

$\therefore$ Total energy supplied by the lower battery is converted into heat. But double charge transfer (from the normal) takes place from this battery.

$\therefore$ Heat produced = Energy supplied by the battery

$$=(\Delta q)V=(2CV)(V)=2CV^2$$

$$=2\times 2\times 10^{-6}\times(5)^2$$

$$=100\times 10^{-6}\text{J}$$

$$=100\ \mu\text{J}$$

15. Initially, the rate of charging is fast.

16. $$V_{1\Omega}=5+2=7\text{ V}$$

$\therefore$ $$i_{1\Omega}=\frac{V}{R}=7\text{A}$$

$$V_{2\mu\text{F}}=6\text{ V}$$

$\therefore$ $$q_{2\mu\text{F}}=CV=12\ \mu\text{C}$$

17. Capacitor and resistor both are in parallel with the battery. PD across capacitor is 10 V. Now, apply $q=CV$.

18. $i=i_0 e^{-t/\tau_C}$

Putting $i=\frac{i_0}{2}$, we get

$$t=(\ln 2)\tau_C=(0.693)\tau_C$$

19. Both capacitors have equal capacitance. Hence, half-half charge distribute over both the capacitors.

$$q_1=q_2=\frac{q_0}{2}$$

q_1 decreases exponentially from q_0 to $\frac{q_0}{2}$ while q_2 increases exponentially from 0 to $\frac{q_0}{2}$. Corresponding graphs and equation are given in the answer.

Time constant of two exponential equations will be

$$\tau_C=(C_{net})R=\left(\frac{C}{2}\right)R=\frac{CR}{2}$$

20. $q_i=q_0$

$$q_f=EC$$

Now, charge on capacitor changes from q_i to q_f exponentially.

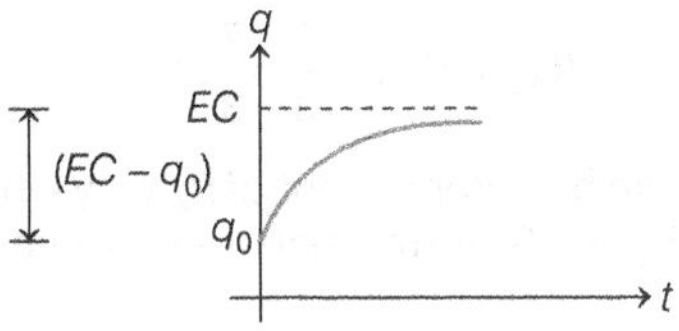

$\therefore \quad q = q_0 + (EC - q_0)(1 - e^{-t/\tau_C})$

$= EC(1 - e^{-t/\tau_C}) + q_0 e^{-t/\tau_C}$

Here, $\tau_C = CR$

21. (a) Immediately after the switch is closed whole current passes through C_1.

$\therefore \quad i = E/R_1$

(b) Long after switch is closed no current will pass through C_1 and C_2.

$\therefore \quad i = \frac{E}{R_1 + R_3}$

22. (a) At $t = 0$ equivalent resistance of capacitor is zero. R_1 and R_2 are in parallel across the battery PD across each is E.

$\therefore \quad i_{R_1} = E/R_1$

$i_{R_2} = E/R_2$

(b) In steady state, no current flow through capacitor wire. PD across R_1 is E.

$\therefore \quad i_{R1} = E/R_1$

and $\quad i_{R2} = 0$

(c) In steady state, potential difference across capacitor is E.

$\therefore \quad U = \frac{1}{2}CV^2$

$= \frac{1}{2}CE^2$

(d) When switch is opened, capacitor is discharged through resistors R_1 and R_2.

$\tau_C = CR_{net}$

$= C(R_1 + R_2)$

23. (a) In steady state,

$V_C = \frac{V}{2}$

$\therefore$ Steady state charge,

$q_0 = CV_C = \frac{CV}{2}$

For equivalent value of τ_C : We short circuit the battery and find the value of R_{net} across capacitors and then

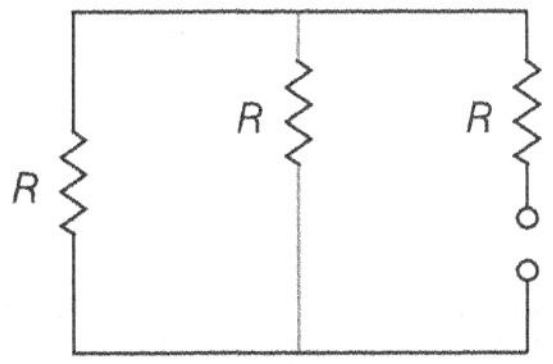

$R_{net} = \frac{3R}{2}$

$\therefore \quad \tau_C = CR_{net} = \frac{3RC}{2}$

Now, $\quad q = q_0(1 - e^{-t/\tau_C})$

(b) **At $t = 0$,** capacitor offers zero resistance.

$R_{net} = \frac{3R}{2}$

$\therefore \quad i = \frac{V}{3R/2} = \frac{2V}{3R}$

$i_C = i_{AB} = \frac{1}{2} = \frac{V}{3R}$

At t $= \infty$, capacitor offers infinite resistance. So, $i_C = 0$.

$\therefore \quad i_{battery} = i_{AB} = \frac{V}{2R}$

Now, current through AB increases exponentially from $\frac{V}{3R}$ to $\frac{V}{2R}$ with same time constant.

(i-t) graph is as shown below.

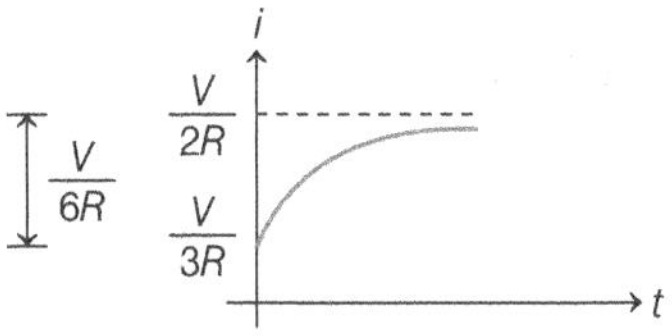

(i-t) equation corresponding to this graph is

$i = \frac{V}{3R} + \frac{V}{6R}(1 - e^{-t/\tau_C})$

24. (a) After long time, capacitor gets fully charged by E_1.

$\therefore \quad i_C = 0$ and $i_{R_1} = i_{R_2} = \frac{E_1}{R_1 + R_2}$

$= \frac{20}{20 \times 10^3} = 10^{-3}$ A $= 1$ mA

(b) In steady state (with E_1).

$V_C = V_{R_2} = V_{R_1} = \frac{E_1}{2} = 10$ V

Now, when the switch is shifted to position B, capacitor (at $t = 0$) behaves like a battery of 10 V. The circuit in that case is as shown below.

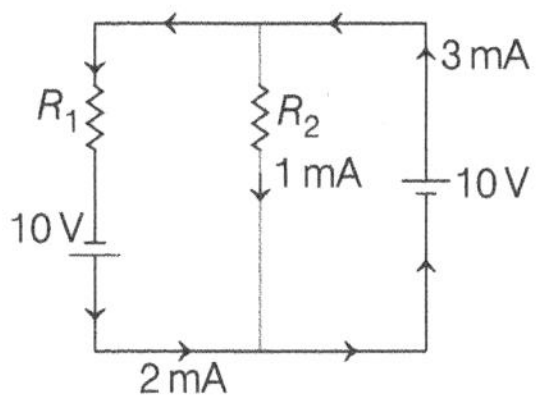

Now, with the help of Kirchhoff's laws we can find different currents. Final currents are shown in the diagram.

Module 8.

1. Wheatstone (balanced) between A and B. So, resistance between C and D can removed.

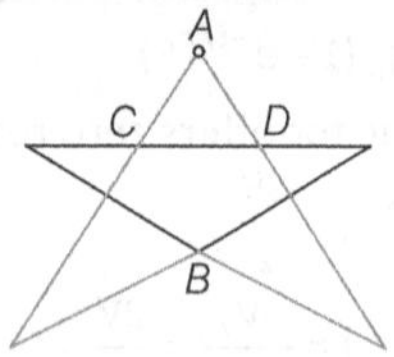

2. Two balanced Wheatstone bridges in parallel.

3. $R_{AB} = 2$ [Net resistance of infinite series] + 1

In parallel net resistance is always less than the smallest one. Hence, net resistance of infinite series is less than 1Ω.

$$\therefore \quad 1\Omega < R_{AB} < 3\Omega$$

4. $R_0 = R_{AB} = R + \frac{(R + R_0)R}{(R + R_0) + R}$

Solving this equation, we get $R = \frac{R_0}{\sqrt{3}}$

5. Simple circuit is as shown in figure,

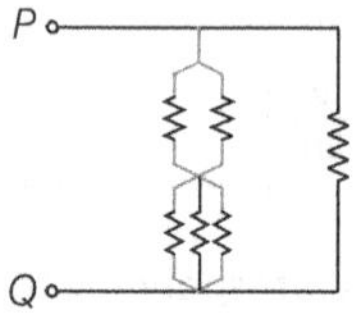

Each resistance $\rightarrow R$

6. 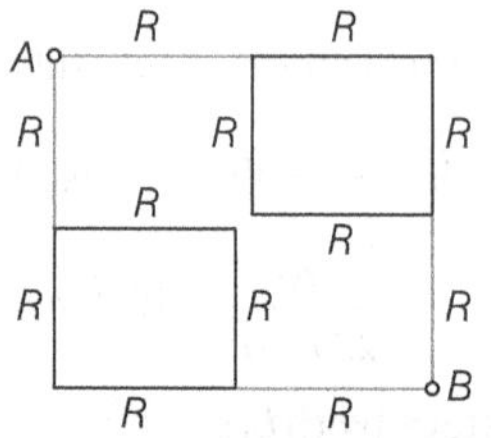

Connection can be removed from centre. $3R$ and $3R$ from two sides of AB are in parallel.

7. 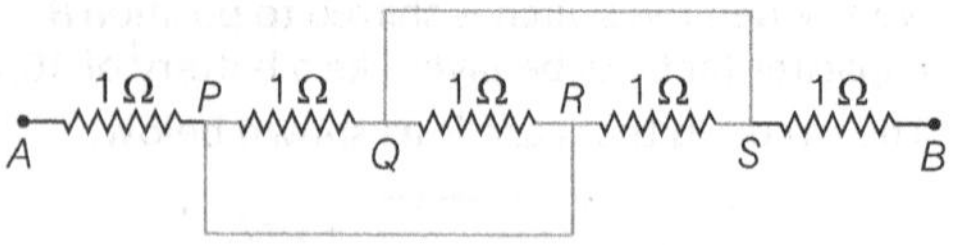

In the circuit, resistance between PQ, QR and RS are in parallel. Now, circuit reduces to

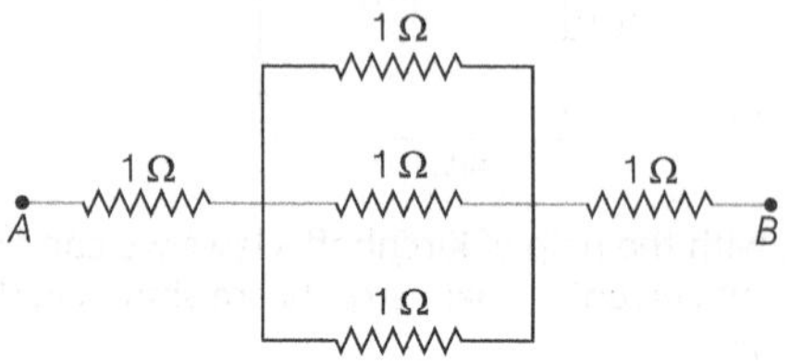

$$\therefore \quad R_{AB} = 1 + 1 + \frac{1}{3} = \frac{7}{3}\Omega$$

8. Current I can be independent of R_6 only when R_1, R_2, R_3, R_4 and R_6 form a balanced Wheatstone's bridge.

Therefore, $\frac{R_1}{R_2} = \frac{R_3}{R_4}$ or $R_1R_4 = R_2R_3$

9. The given circuit is a balanced Wheatstone's bridge.

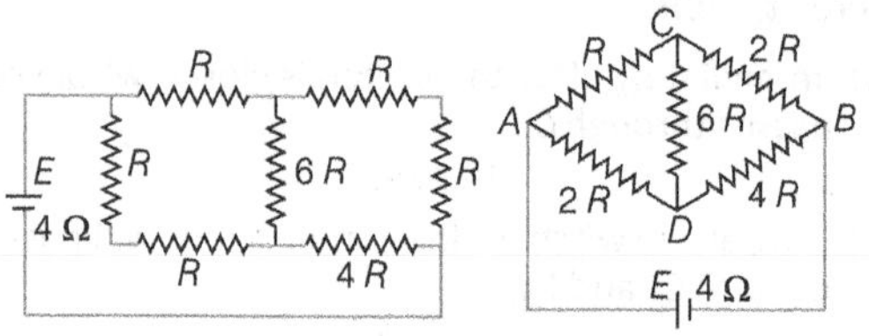

Thus, no current will flow across $6R$ of the side CD. The given circuit will now be equivalent to

For maximum power, net external resistance
= Total internal resistance

or $\quad 2R = 4$

or $\quad R = 2\Omega$

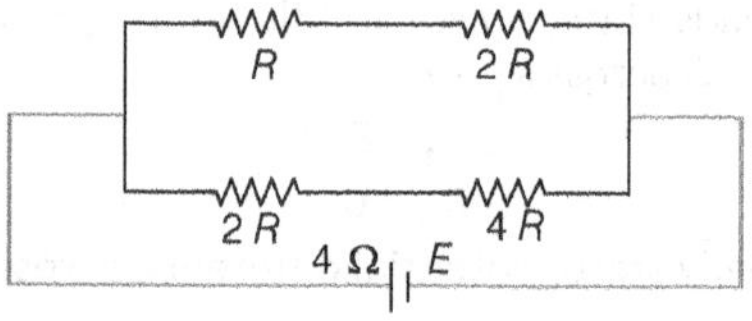

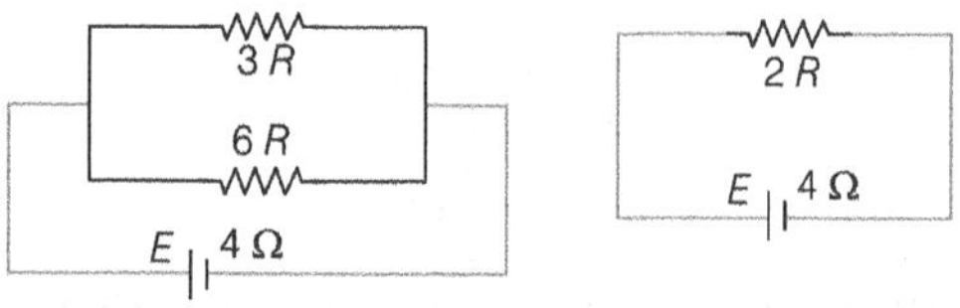

10. Given circuit can be simplified as

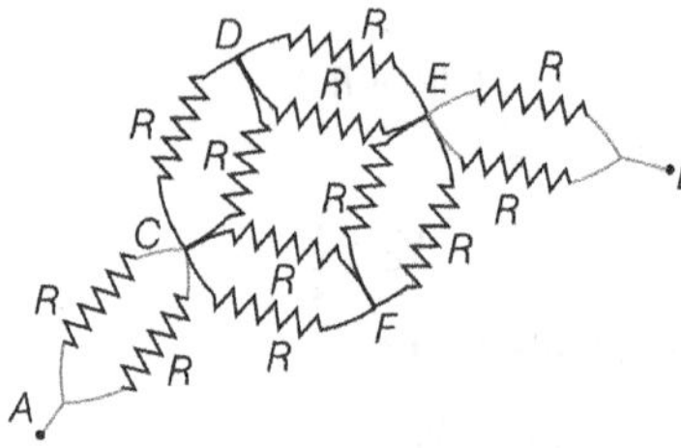

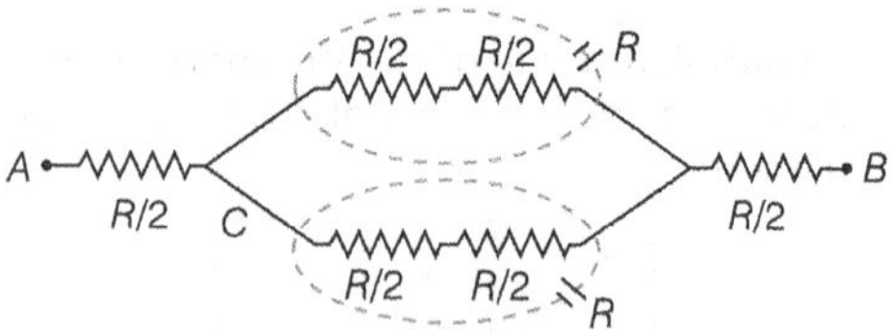

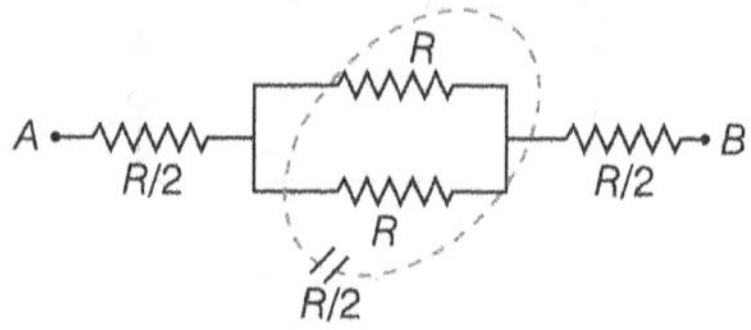

$$\Rightarrow \quad R_{AB} = \frac{3}{2}R$$

11. Given circuit is

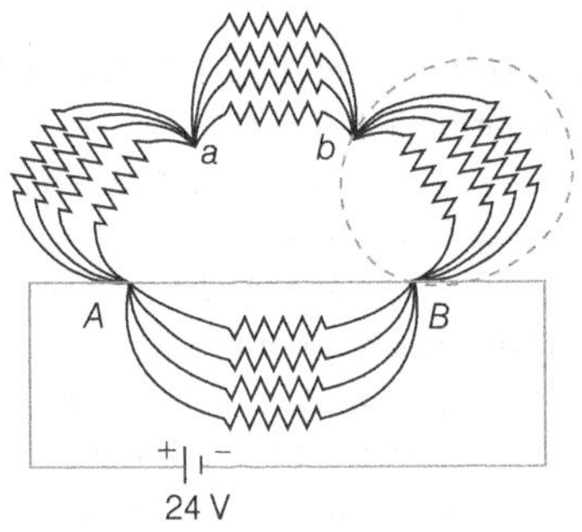

$$\frac{1}{R} = \frac{1}{5} + \frac{1}{5} + \frac{1}{5} + \frac{1}{5}$$

$$R = \frac{5}{4}\Omega$$

So, circuit can be redrawn as

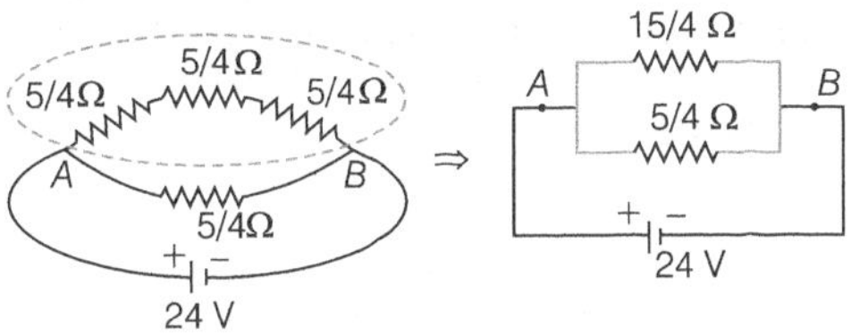

Now, as $\qquad V_{AB} = 24\text{ V}$

Current through arm of resistance $\frac{15}{4}\Omega$

$$I = \frac{V_{AB}}{R} = \frac{24}{15/4} = \frac{24 \times 4}{15}\text{ A}$$

Now, $I_{ab} = \frac{1}{4}$ of $I = \frac{1}{4} \times \frac{24 \times 4}{15} = \frac{8}{5}\text{ A}$

12. $C_1 + C_2 + C_3 = 6$ ...(i)

$C_2 + C_3 = 5$ ...(ii)

$C_1 \times C_2 \times C_3 = 6$ ...(iii)

Solving these three equations, we get

$C_1 = 1$ unit, $C_2 = 2$ unit and $C_3 = 3$ unit

13. Equivalent capacitance between A and B,

$$C_{AB} = C_1 + \frac{C_1}{2} + C_1 = \frac{5}{2}C_1$$

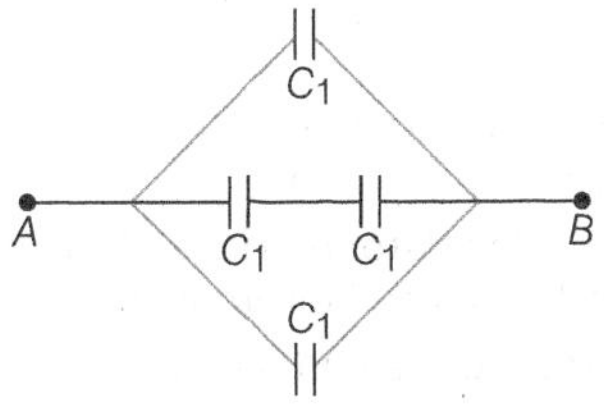

As charge, $Q = CV$

So, $\qquad 1.5\,\mu\text{C} = \frac{5}{2}C_1 \times 6$

$$\Rightarrow \quad C_1 = \frac{1.5}{15} \times 10^{-6}$$

$$= 0.1 \times 10^{-6}\text{ F} = 0.1\,\mu\text{F}$$

14. First three circuits are balanced Wheatstone bridge circuits.

15.

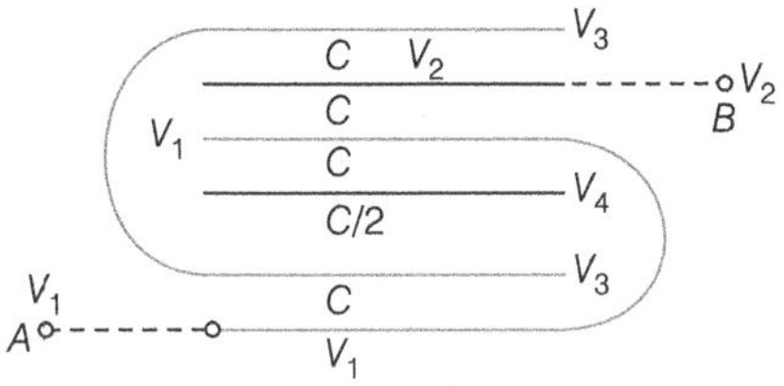

$$C = \frac{\varepsilon_0 A}{d} = 7\,\mu\text{F}$$

The equivalent circuit is as shown in figure.

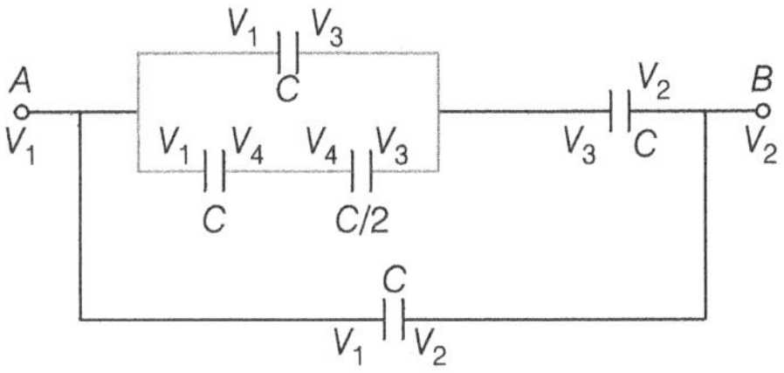

$$C_{AB} = \frac{11}{7}C = \frac{11}{7}(7\,\mu\text{F})$$

$$= 11\,\mu\text{F}$$

16. A balanced Wheatstone bridge is parallel with C.

17. Three capacitors (consisting of two loops are short-circuited).

18. The equivalent circuit is as shown below.

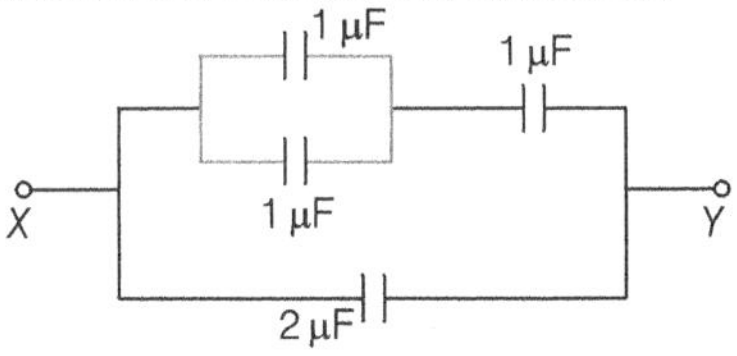

19. The given circuit can be simplified in the following way.

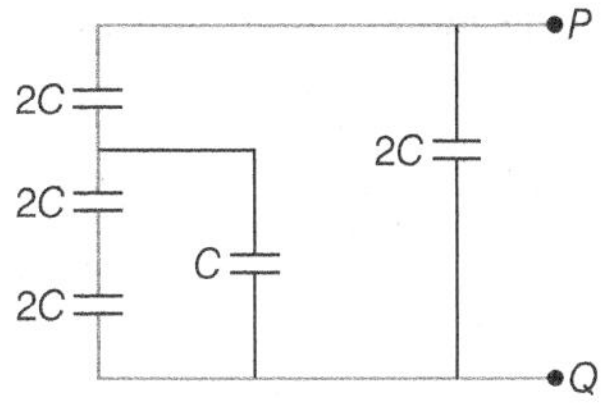

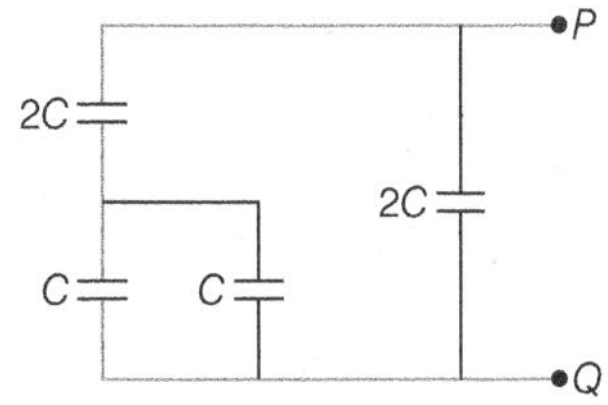

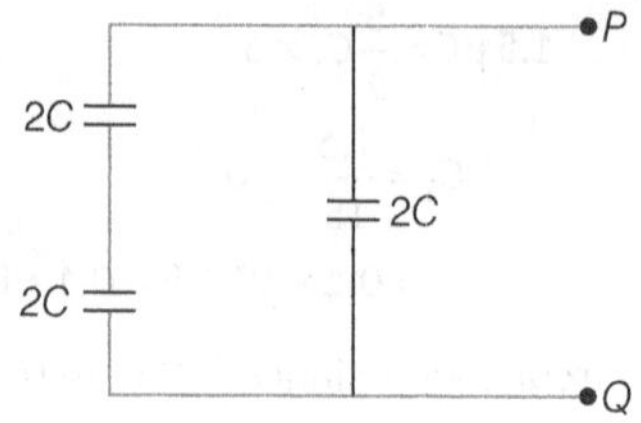

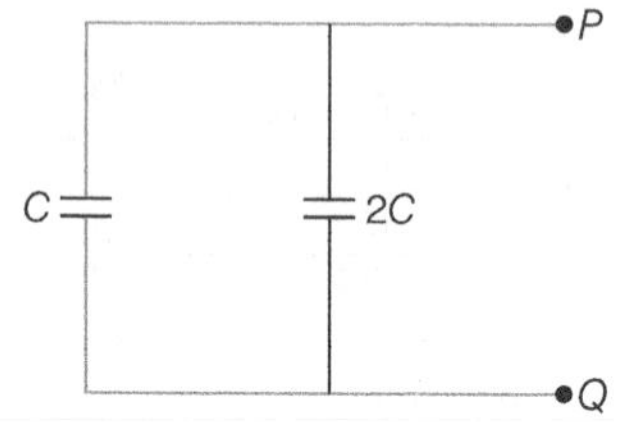

So, capacity between P and Q

$$C_{PQ} = 2C + C = 3C$$

20. Here, circuit can be redrawn as

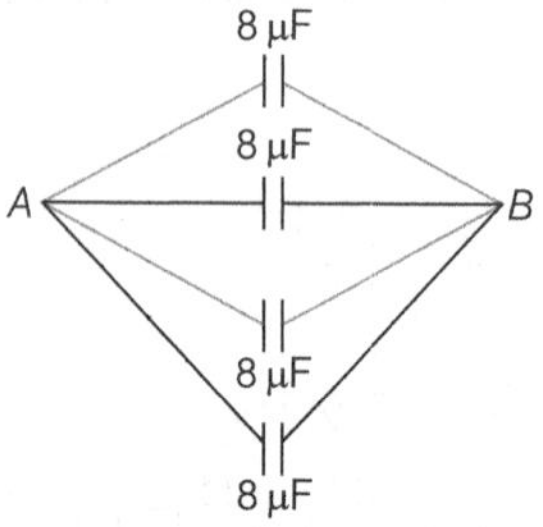

Equivalent capacitance of these capacitors,

$$C_{eq} = 8 + 8 + 8 + 8$$
$$= 32\,\mu F$$

21. Between a and c, balanced Wheatstone bridge is formed. Across all other points simple series and parallel grouping of resistors.

22. (a)

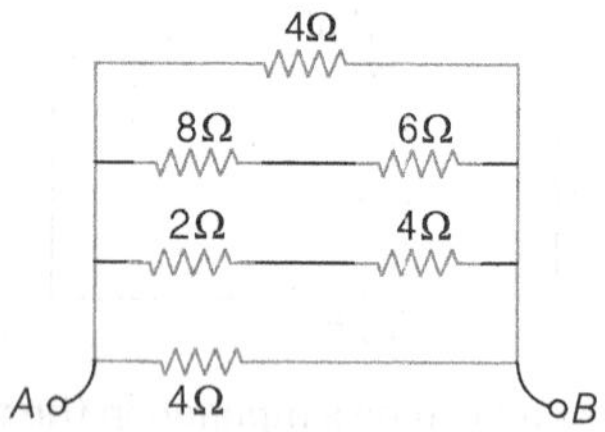

(b) A balanced Wheatstone bridge in parallel with R.

(c)

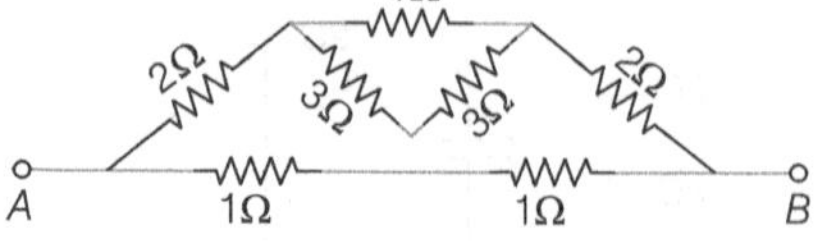

(d)

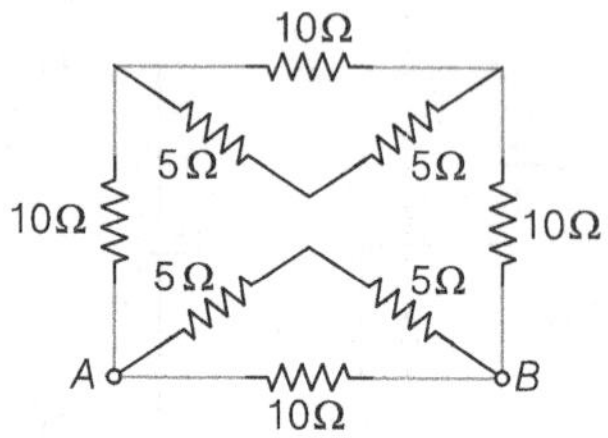

(e)

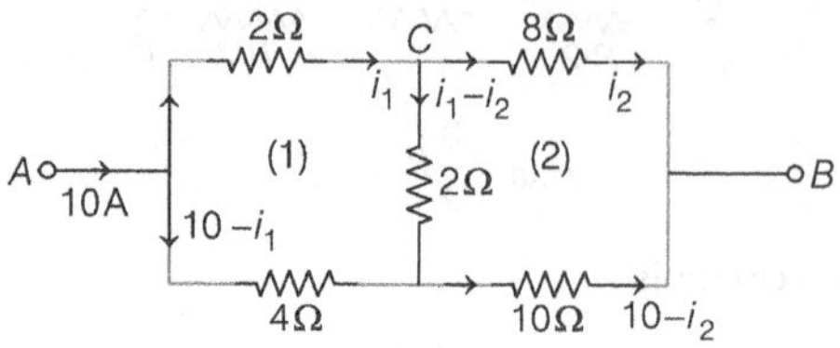

Let us take a current of 10 A between A and B

Loop equation in loop (1)

$$-2i_1 - 2(i_1 - i_2) + 4(10 - i_1) = 0 \quad \text{...(i)}$$

Loop equation in loop (2)

$$-8i_2 + 10(10 - i_2) + 2(i_1 - i_2) = 0 \quad \text{...(ii)}$$

Solving these two equations, we get

$$i_1 = 6.53\text{ A} \quad \text{and} \quad i_2 = 6.11\text{ A}$$

Now let us find V_{AB} across path ACB,

$$V_{AB} = 2i_1 + 8i_2 = 61.94\text{ V}$$

Now, $V_{AB} = i_{net} R_{net} = (10) R_{net}$

$\therefore \quad 61.94 = 10 R_{net}$

or $\quad R_{net} = 6.194\,\Omega$

(f) Two resistors in vertical middle wire can be removed.

(g) Now balanced Wheatstone bridge, in parallel with 1Ω resistance between points A and B. The encircled resistance of 2Ω can be removed from the Wheatstone bridge.

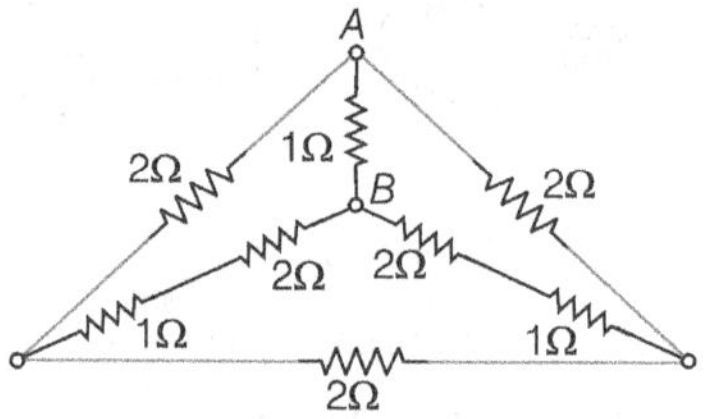

One resistor of 2Ω has already removed from the original circuit given the question. As its two ends will be the same potential (by symmetry).

23. $R_i + 5R$

If connected by two equal resistors between B and D and between C and E, the combination is a balanced Wheatstone bridge and two resistors in series.

$$\therefore \quad R_f = R + R + R = 3R = 0.6\,R_i$$

24.

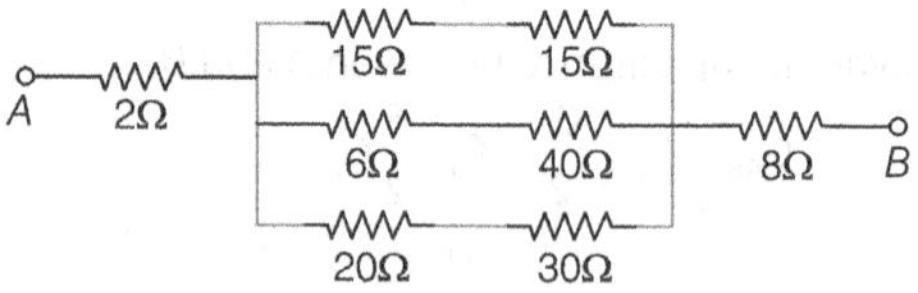

25. (b) Three resistors are in parallel. Then, one resistor in series with this combination.

(c) Balanced Wheatstone bridge. Hence, two resistors in vertical wire can be removed.

(d) All four resistors are in parallel

(e) A balanced Wheatstone bridge.

26. (a) A balanced Wheatstone bridge with one resistance in parallel.

(b)

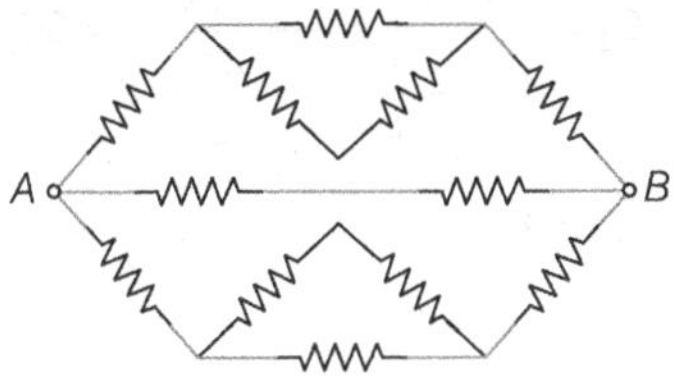

27. (a) Points D and E are symmetrically located with respect to points A and C. The circuit can be redrawn as shown in figure.

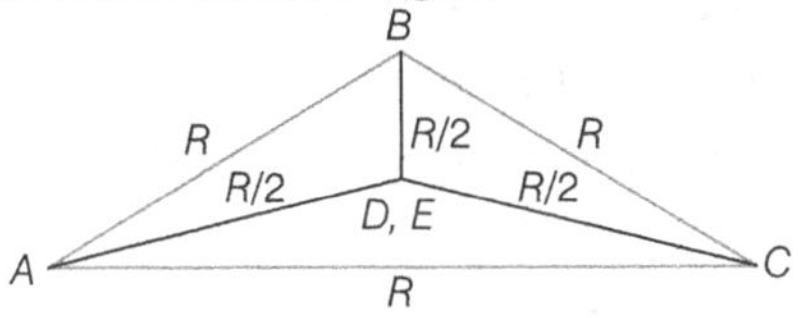

This is a combination of a balanced Wheatstone bridge in parallel with a resistance R. So, the resistance between B and D (or E) can be removed.

$$\frac{1}{R_{AC}} = \frac{1}{R} + \frac{1}{\frac{R}{2} + \frac{R}{2}} + \frac{1}{2R}$$

or $$R_{AC} = \frac{2R}{5}$$ **Ans.**

(b) With respect to D and E, points A, B and C all are symmetrically located. Hence, the simplified circuit can be drawn as shown in figure.

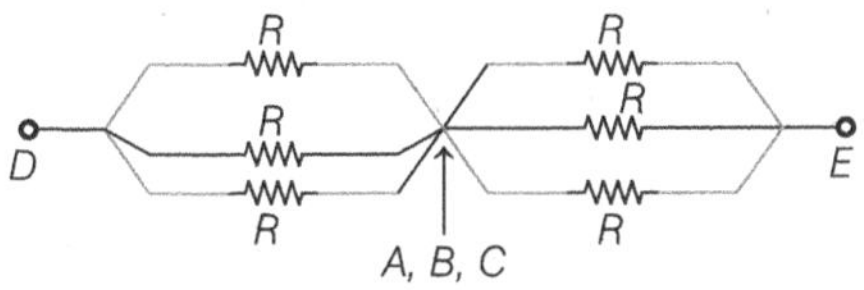

$\therefore$ $$R_{DE} = \frac{R}{3} + \frac{R}{3} = \frac{2R}{3}$$ **Ans.**

28. (a) Due to symmetry about the shaded plane, current distribution on either side of the plane will be identical and points E and F will be at same potential and no current will flow through it.

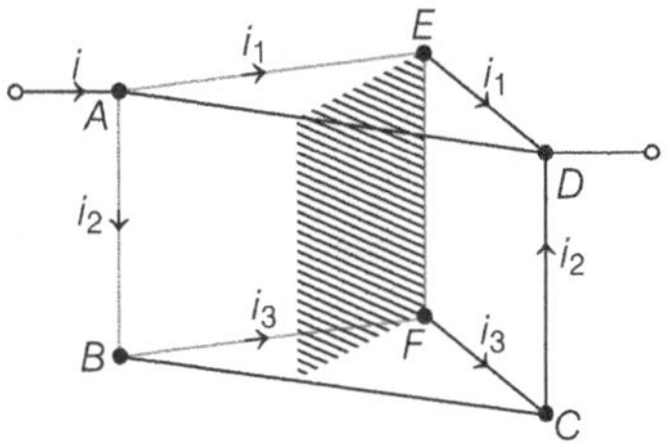

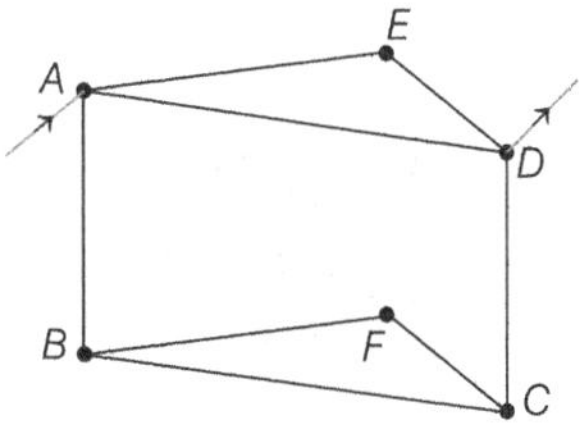

$\therefore$ $$R_{AD} = \frac{\frac{2}{3}r \times \frac{8}{3}r}{\frac{2}{3}r + \frac{8}{3}r} = \frac{8}{15}r$$ **Ans.**

(b) Redrawing the given arrangement for resistance across AB. Potentials $V_D = V_E$

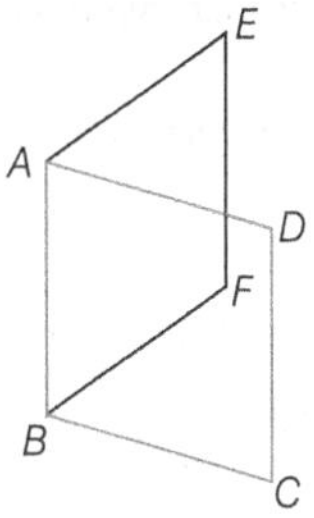

$$V_C = V_F$$

$\therefore$ No current flows through DE and CF.

$\therefore$ $$R_{AB} = \frac{\frac{3}{2}r \times r}{\frac{3}{2}r + r} = \frac{3}{5}r$$ **Ans.**

29. (a) Simple circuit is as shown below.

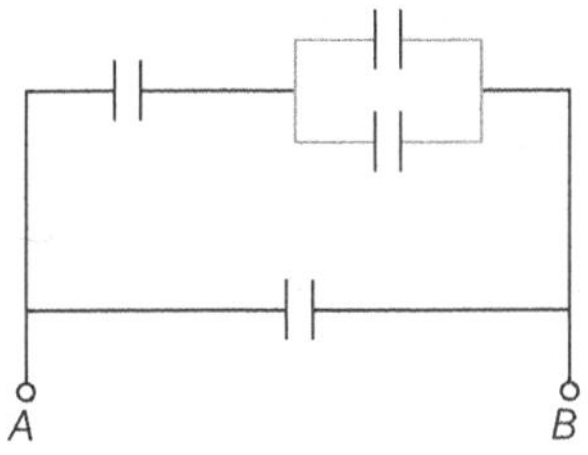

(b) The simple circuit is as shown below.

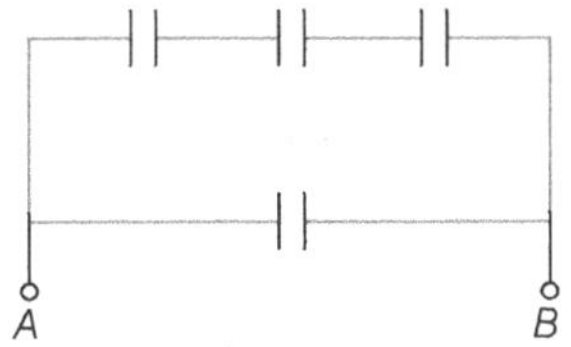

(c) Let $C_{AB} = x$. Then,

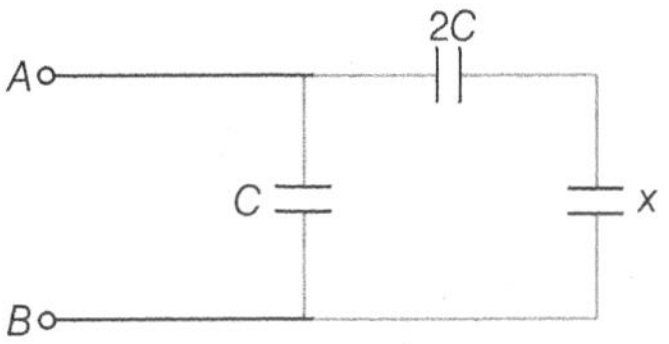

Now, $$C_{AB} = C + \frac{(2C)(x)}{2C + x}$$

or $$x = C + \frac{2Cx}{2C + x}$$

Solving this equation, we get

$$x = 2C$$

Unit Exercise (Cumulative)

Level 1

1. By inserting dielectric slab, value of C_2 will increase. In series, potential difference distributes in inverse ratio of capacitance. If capacitance C_2 is increased PD across C_2 will decrease. If C_2 is increased, charge on capacitors will also increase.

   So, positive charge or current flows in clockwise direction.

2. $\because \quad qE = mg$

$$q\left(\frac{V}{d}\right) = \left(\frac{4}{3}\pi r^3 \rho\right) g$$

$$V \propto \frac{r^3}{q}$$

3. Final potential difference $= E$

   $\therefore$ Final charge $= EC$

4. $P = i^2 R = (i_0 e^{-t/\eta})^2 R$

$$= (i_0^2 R)\, e^{-2t/\eta}$$

$$= P_0 e^{-t/(\eta/2)}$$

   Hence, the time constant is $\frac{\eta}{2}$.

5. Horizontal range,

$$R = \frac{2u_x \times u_y}{g} = l \qquad \text{...(i)}$$

   Maximum height, $H = \frac{u_y^2}{2g} = d$ ...(ii)

   Dividing Eq. (ii) by Eq. (i), we have

$$\frac{1}{4}\left(\frac{u_y}{u_x}\right) = \frac{d}{l}$$

   or

$$\frac{u_y}{u_x} = \frac{u \sin\theta}{u \cos\theta}$$

$$= \tan\theta = \frac{4d}{l}$$

6. $V = Ed = \left(\frac{\sigma}{2\varepsilon_0}\right) d$

$$\therefore \quad d = \frac{2\varepsilon_0 V}{\sigma} = 2\frac{(8.86 \times 10^{-12})(5)}{10^{-7}}$$

$$= 0.88 \times 10^{-3} \text{m}$$

$$= 0.88 \text{ mm}$$

7. $V = \frac{\text{Net charge}}{\text{Net capacity}} = \frac{q_1 + q_2}{4\pi\varepsilon_0 (r_1 + r_2)}$

8. Dielectric constant for metallic sheet, $k = \infty$

   Given, thickness of metallic plate, $t = \frac{2d}{3}$

   Distance between two plates of capacitor $= d$

   According to question, $C_1 = \frac{\varepsilon_0 A}{d}$

$$C_2 = \frac{\varepsilon_0 A}{d - t + \frac{t}{k}} = \frac{\varepsilon_0 A}{d - \frac{2d}{3} + 0} = 3C_1$$

$$\Rightarrow \quad \frac{C_2}{C_1} = 3$$

9. Given,

$$A = 40 \text{ cm}^2 = 4 \times 10^{-3} \text{ m}^2$$

$$d = 2 \text{ mm} = 2 \times 10^{-3} \text{ m}$$

   Thickness, $t = 1 \text{ mm} = 1 \times 10^{-3}$ m

   Dielectric constant, $K = 5$

   $\therefore$Capacitance (when dielectric is inserted),

$$C = \frac{\varepsilon_0 A}{(d - t) + \frac{t}{K}}$$

$$= \frac{\varepsilon_0 \times 4 \times 10^{-3}}{(2 \times 10^{-3} - 1 \times 10^{-3}) + \frac{1 \times 10^{-3}}{5}}$$

$$= \frac{4\varepsilon_0 \times 10^{-3}}{10^{-3}\left(1 + \frac{1}{5}\right)} = 4\varepsilon_0 \times \frac{5}{6} = \frac{10\,\varepsilon_0}{3} \text{ F}$$

10.

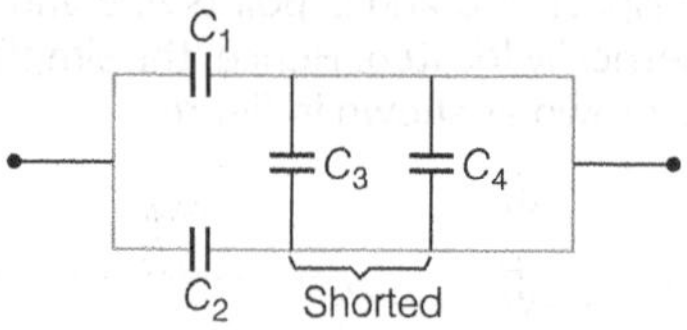

   In given circuit, C_4 and C_3 are shorted, so capacitance of circuit is,

$$C_{CQ} = C_1 + C_2 = C + C = 2C \qquad [\because C_1 \| C_2]$$

11. We have, $x + y + \frac{3d}{4} = d$

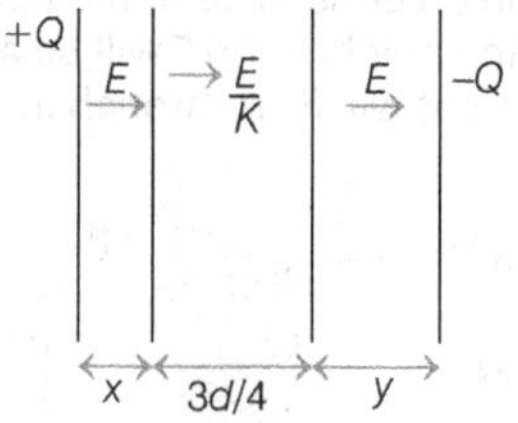

$$\Rightarrow \quad x + y = \frac{d}{4}$$

$$\Delta V = Ex + \frac{E}{K} \times \frac{3d}{4} + Ey$$

$$\Rightarrow \quad \Delta V = \frac{3Ed}{4K} + E(x + y)$$

$$\Rightarrow \quad \Delta V = E\left[\frac{3d}{4K} + \frac{d}{4}\right]$$

$$\Delta V = \frac{\sigma}{\varepsilon_0}\left[\frac{3d + dK}{4K}\right] = \frac{Qd}{A\varepsilon_0}\left[\frac{3+K}{4K}\right]$$

$$\frac{Q}{\Delta V} = C = \frac{A\varepsilon_0}{d}\left[\frac{4K}{3+K}\right]$$

$$= \frac{4KC_0}{K+3} \qquad \left[\because C_0 = \frac{A\varepsilon_0}{d}\right]$$

12. Slope, $\frac{y}{x} = \frac{d}{a}$

$$\Rightarrow \qquad y = \frac{d}{a}x \qquad \text{...(i)}$$

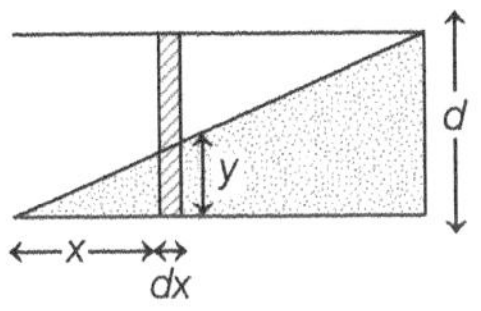

Capacitance for small separation with dielectrics is

$$\int dC = \int \frac{\varepsilon_0 a dx}{d - y + \frac{y}{K}} \qquad \text{...(ii)}$$

Substituting Eqs. (i) and (ii), we get

$$C = \varepsilon_0 a \int_0^a \frac{dx}{d + \frac{dx}{a}\left(\frac{1}{K} - 1\right)}$$

$$\Rightarrow \qquad C = \frac{\varepsilon_0 a^2 K}{d(K-1)} \ln K$$

13. In the given arrangement, capacitor can be viewed as three- different capacitors connected in parallel as shown below,

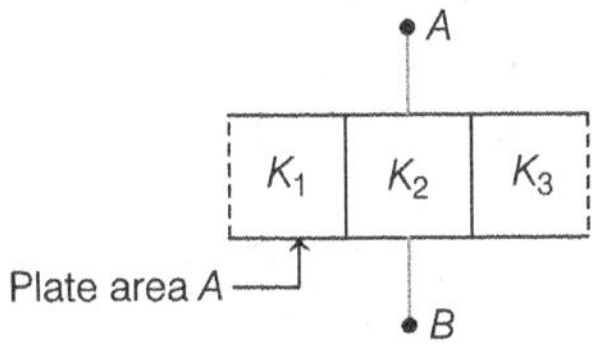

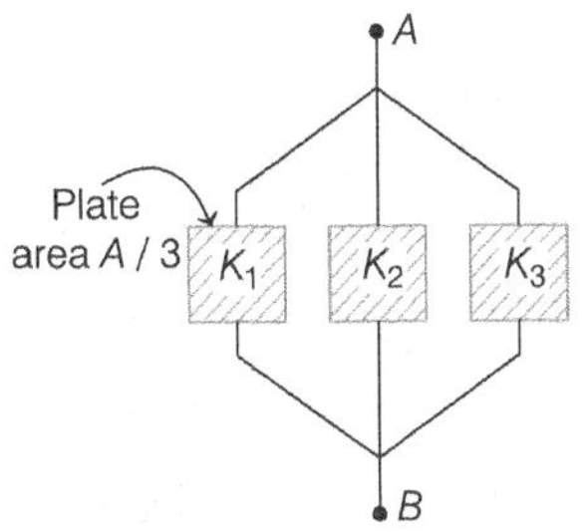

So, equivalent capacity of the system is

$$C_{eq} = C_1 + C_2 + C_3$$

$$\Rightarrow \frac{K\varepsilon_0 A}{d} = \frac{K_1\varepsilon_0 A/3}{d} + \frac{K_2\varepsilon_0 A/3}{d} + \frac{K_3\varepsilon_0 A/3}{d}$$

$$\Rightarrow \quad K = \frac{K_1}{3} + \frac{K_2}{3} + \frac{K_3}{3}$$

Here, $K_1 = 10, K_2 = 12$

and $\qquad K_3 = 14$

So, $\qquad K = \frac{10 + 12 + 14}{3}$

$\Rightarrow \qquad K = 12$

14.

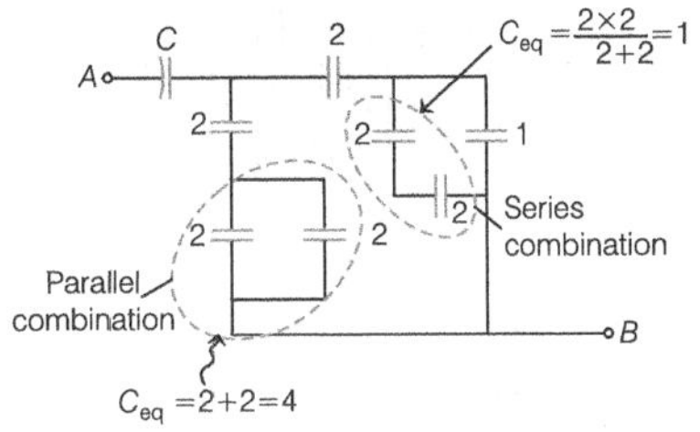

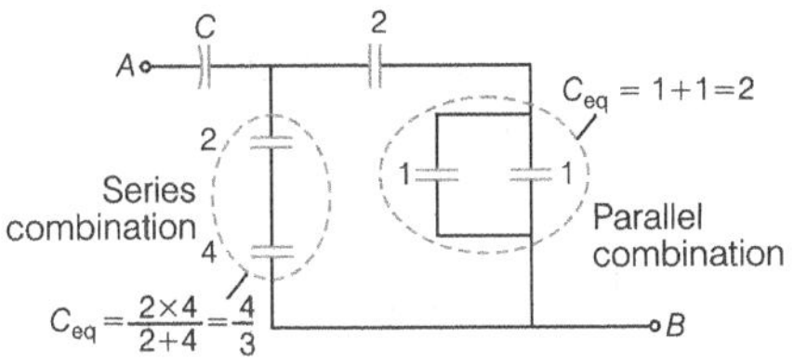

We simplify gives circuit is

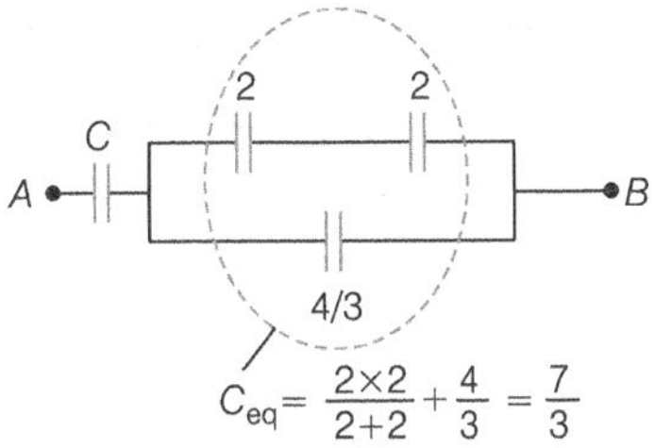

So, $\qquad C_{AB} = \frac{C \times 7/3}{C + 7/3} = \frac{1}{2} \qquad$ (given)

$$\Rightarrow \qquad \frac{7C}{3} = \frac{C}{2} + \frac{7}{6} \Rightarrow \left(\frac{14-3}{6}\right)C = \frac{7}{6}$$

$$\Rightarrow \qquad C = \frac{7}{11}\,\mu\text{F}$$

15.

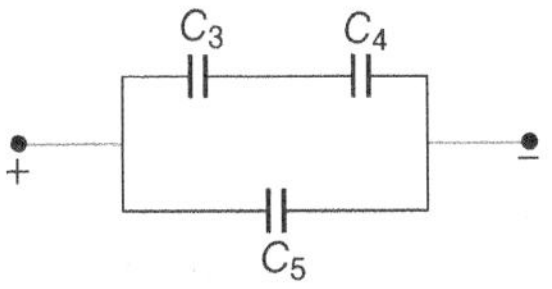

$$C_1 = \frac{\varepsilon_0 S}{d}, C_3 = \frac{2\varepsilon_0 \frac{S}{2}}{\frac{d}{2}} = \frac{2\varepsilon_0 S}{d}$$

$$\Rightarrow \qquad C_4 = \frac{4\varepsilon_0 \frac{S}{2}}{\frac{d}{2}} = \frac{4\varepsilon_0 S}{d}$$

and $\qquad C_5 = \frac{2\varepsilon_0 \frac{S}{2}}{d} = \frac{\varepsilon_0 S}{d}$

$$C_2 = \frac{C_3 C_4}{C_3 + C_4} + C_5 = \frac{4}{3}\frac{\varepsilon_0 S}{d} + \frac{\varepsilon_0 S}{d}$$

$$= \frac{7}{3}\frac{\varepsilon_0 S}{d} = \frac{7}{3}C_1 \Rightarrow \frac{C_2}{C_1} = \frac{7}{3}$$

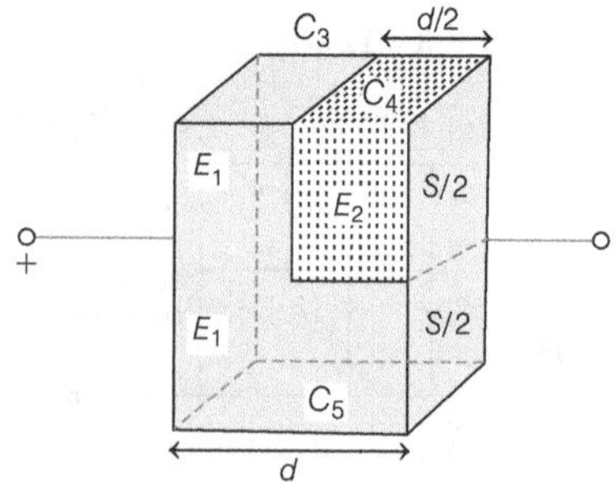

16.

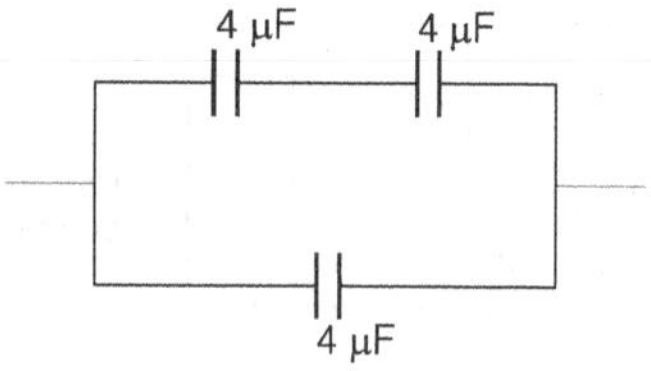

$$\text{Capacitance, } C = \frac{1}{4} + \frac{1}{4} = \frac{1}{2}$$

$$C_{eq} = 2\mu F$$

$$\Rightarrow \quad C_{eq} = 2 + 4 = 6\ \mu F$$

17. The equivalent capacitance,

$$\frac{1}{C_{eq}} = \frac{1}{2} + \frac{1}{2} + \frac{1}{2\times 5} = \frac{11}{10}$$

$$\Rightarrow \quad C_{eq} = \frac{10}{11}\mu F$$

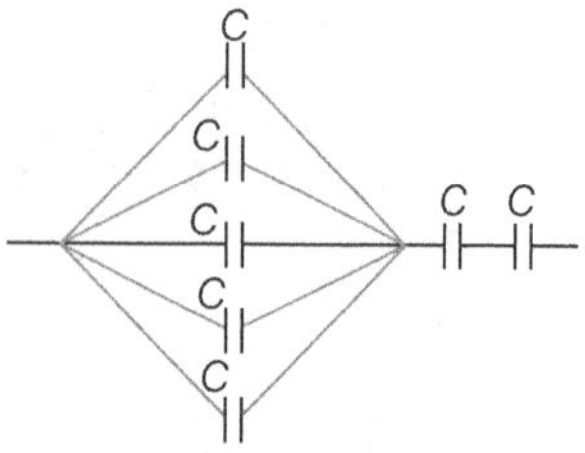

18. $\tau_C = CR = 6$ s

$$q_0 = CV = 10\ \mu C$$

Now, $q = q_0 e^{-t/\tau_C} = (10\ \mu C)\, e^{-12/6}$

$$= \left(\frac{1}{e}\right)^2 (10\ \mu C)$$

$$= (0.37)^2\,(10\ \mu C)$$

19. $H_1 = H_2 = U_i - U_f$

The only change is by increasing the resistance τ_C increase. Hence, process of redistribution of charge slows down.

20. $\tau_C = CR = \left(\frac{K\varepsilon_0 A}{d}\right)\left(\frac{d}{A\sigma}\right) \qquad \left(R = \frac{l}{\sigma A}\right)$

$$= \frac{K\varepsilon_0}{\sigma} = \frac{5\times 8.86\times 10^{-12}}{7.4\times 10^{-12}} = 6\text{ s}$$

21. Common potential, $V = \frac{\text{Total charge}}{\text{Total capacitance}}$

$$V = \frac{C_1V_1 + C_2V_2}{C_1 + C_2}$$

$$= \frac{0 + CV_0}{KC + C} = \frac{CV_0}{C(1+K)}$$

$$V = \frac{V_0}{(1+K)}$$

$$K = \frac{V_0}{V} - 1$$

$$K = \frac{V_0 - V}{V}$$

22. $d = \frac{V_{max}}{E_{max}}$

$$C = \frac{K\varepsilon_0 A}{d}$$

$$\therefore \quad A = \frac{dC}{K\,\varepsilon_0} = \frac{(V_{max})\,(C)}{(K)\,(E_{max})\,(\varepsilon_0)}$$

23. (a) $V_a = 18$ V and $V_b = 0$ as no current flow through the resistors.

$$\therefore \quad V_a - V_b = 10\text{ V}$$

(b) $V_a - V_b = +$ ve

Hence, $V_a > V_b$

(c) Current flows through two resistors,

$$i = \frac{18-0}{6+3} = 2\text{A}$$

$\therefore \quad V_b - 0 = iR = 2\times 3$

or $\quad V_b = 6$V

(d) Initially, $\quad V_{3\mu F} = V_{6\mu F} = 18$V

$\therefore \quad q_{3\mu F} = 54\ \mu C \qquad (q = CV)$

and $\quad q_{6\mu F} = 108\ \mu C$

Finally,

$$V_{6\mu F} = V_{6\Omega} = iR = 2\times 6 = 12\text{ V}$$

$\therefore \quad q_{6\mu F} = 72\ \mu C$

$V_{3\mu F} = V_{3\Omega} = 6$V

$\therefore \quad q_{3\mu F} = 18\ \mu C$

$\Delta q = q_f - q_i$

$= -36\ \mu C$ on both capacitors.

24. (a) In resistors (in series) potential drops in direct ratio of resistance and in capacitors (in series) potential drops in inverse ratio of capacitance.

$$18 - V_a = \left(\frac{6}{6+3}\right)(18)$$

$\therefore \quad V_a = 6$ V

$$18 - V_b = \left(\frac{3}{6+3}\right)(18)$$

$V_b = 12$ V

$\therefore \quad V_a - V_b = -6$V

(c) $V_{3\mu F} = V_{3\Omega} = iR$

$$=\left(\frac{18-0}{6+3}\right)(3)=6\text{ V}$$

$\therefore \quad V_b-0=6\text{ V}$

$\therefore \quad V_b=6\text{ V}$

(d) Initially,

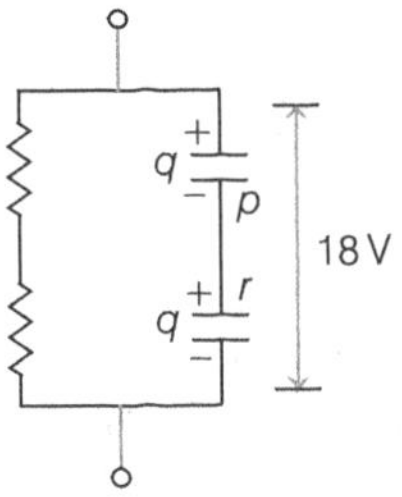

$$q=C_{net}V=(2\ \mu\text{F})(18)=36\ \mu\text{C}$$

Finally,

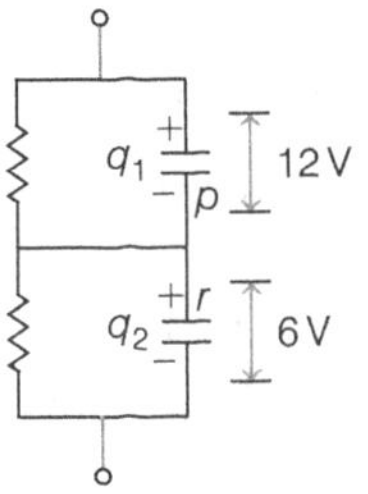

$$q_1=72\ \mu\text{C}$$

$$\Rightarrow \quad q_2=18\ \mu\text{C}$$

Charge flow from $S=$ (Final charge on plates p and r) − (Initial charges on plates p and r)

$$=(-72+18)-(-36+36)$$

$$=-54\ \mu\text{C}$$

25.

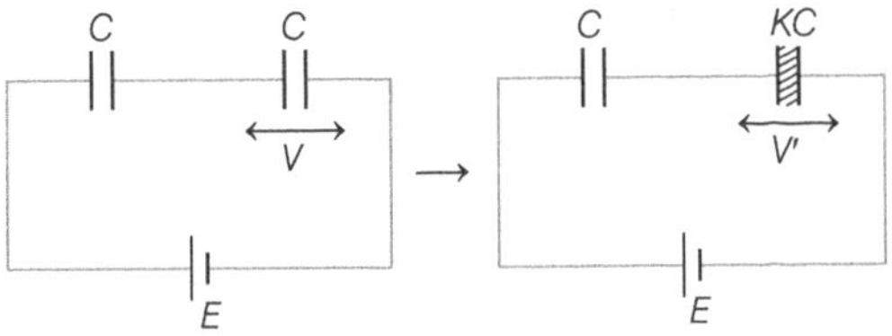

In first figure,

$$C_{net}=\frac{C}{2}$$

$$\therefore \quad q=\left(\frac{C}{2}\right)E$$

In second figure,

$$C'_{net}=\frac{(KC)C}{KC+C}=\left(\frac{K}{K+1}\right)C$$

$$\therefore \quad q'=\left(\frac{KC}{K+1}\right)E$$

Flow of charge from battery,

$$\Delta q=q'-q$$

$$=EC\left[\frac{K}{K+1}-\frac{1}{2}\right]$$

$$\Delta q=\left[\frac{(K-1)}{2(K+1)}\right]EC$$

In series potential difference distributes in inverse ratio of capacitance.

$$V=\frac{E}{2}$$

$$V'=\left(\frac{C}{C+KC}\right)E=\frac{E}{1+K}$$

$$\frac{V'}{V}=\left(\frac{2}{1+K}\right)$$

Since $K>1 \Rightarrow V'<V$

So, potential difference will decrease by $\left(\frac{K+1}{2}\right)$ times.

Electric field $=\frac{V}{d}\propto V$

Hence, electric field will also decrease by $\left(\frac{K+1}{2}\right)$ times.

26.

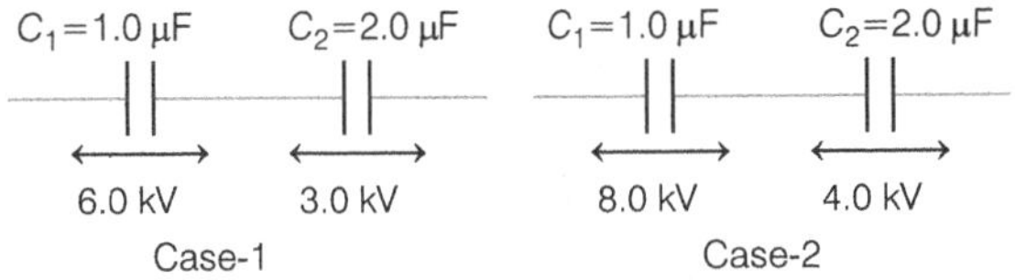

In series, potential difference is in inverse ratio of capacitance. So potential difference across C_2 will be half.

We can see that, in case-1, both capacitors are safe.

$$\therefore \quad V_{net}=9.0\text{ kV}$$

Alternate method

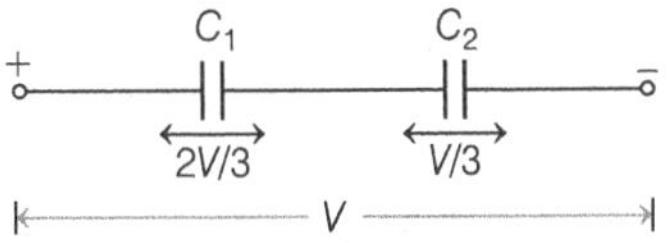

For C_1

$$\frac{2V}{3}<6.0\text{ kV} \Rightarrow V<9.0\text{ kV}$$

For C_2

$$\frac{V}{3}<4.0\text{ kV} \Rightarrow V<12\text{ kV}$$

So answer will be 9.0 kV.

27. Let $C_1=C$

Then $C_2=2C$

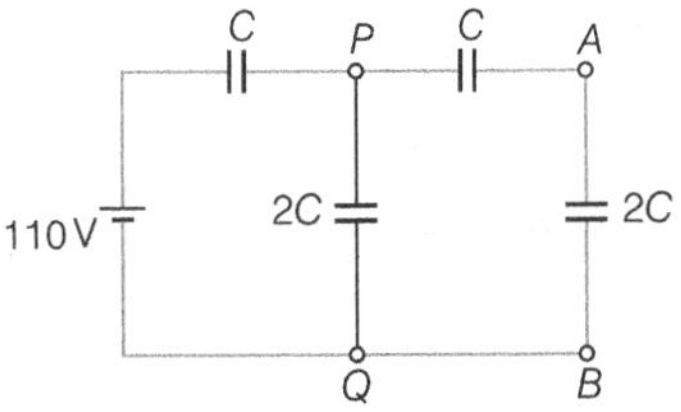

Net capacitance between P and Q is

$$2C+\frac{(2C)(C)}{(2C)+C}=\frac{8}{3}C$$

Now, 110V will distribute between C and $\frac{8}{3}C$ in inverse ratio of capacitance.

$$\therefore \quad V_{PQ}=\left[\frac{C}{C+\frac{8}{3}C}\right](110)\text{ volt}$$

$$=30\text{ volt}$$

Now, this 30 volt will further distribute between C and $2C$ in inverse ratio of capacitance. So,

$$V_{AB}=\left(\frac{C}{C+2C}\right)(30)\text{ volt}$$

$$=10\text{ volt}$$

$$\therefore \quad V_{AB}=10\text{ volt}$$

28. Initial charge on C_1,

$$q_i=C_1V_i=110\,\mu\text{C}$$

Net capacitance of C_2 and C_3 is,

$$C_{23}=\frac{C_2C_3}{C_2+C_3}$$

$$=\frac{2\times 3}{2+3}=1.2\,\mu\text{F}$$

In parallel connection between C_1 and C_{23}, final potential will be same and given by,

$$V_f=\frac{\text{Total charge}}{\text{Total capacity}}$$

$$=\frac{110\,\mu\text{C}}{(1.0+1.2)\,\mu\text{F}}=50\text{ V}$$

Final charge on C_1,

$$q_f=C_1V_f=50\,\mu\text{C}$$

$\therefore$ Flow of charge from connecting wires,

$$\Delta q=|q_i-q_f|=60\,\mu\text{C}$$

or $$\Delta q=0.06\text{ mC}$$

29. **Switch in position-1**

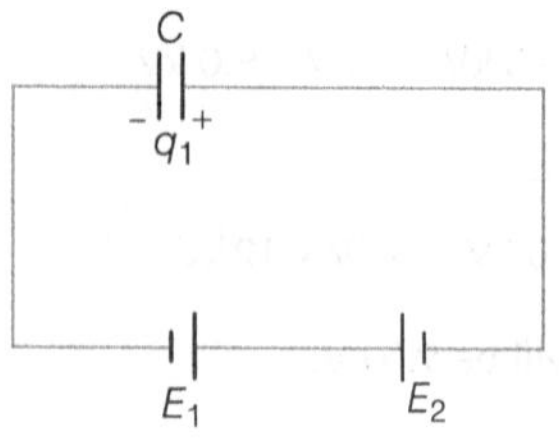

Let us suppose that $E_1>E_2$.

$$E_{net}=E_1-E_2$$

$$\therefore \quad q_1=CE_{net}=(E_1-E_2)\,C$$

Energy stored in capacitor,

$$U_1=\frac{1}{2}C(E_1-E_2)^2$$

Switch in position-2

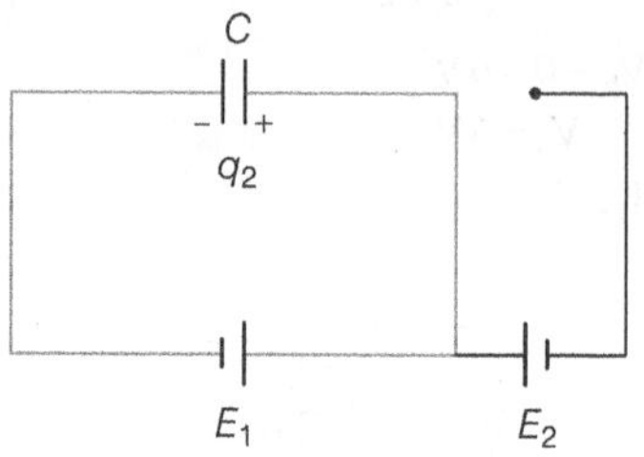

$$q_2=E_1C$$

$$q_2>q_1$$

$\therefore$ Flow of charge from battery of emf E_1 is,

$$\Delta q=(q_2-q_1)=E_2C$$

So energy supplied by the battery,

$$W=(\Delta q)E_1=E_1E_2C \quad \text{...(i)}$$

Energy stored in capacitor,

$$U_2=\frac{1}{2}CE_1^2$$

Increase in stored electrostatic potential energy of capacitor,

$$\Delta U=U_2-U_1$$

or $$\Delta U=\frac{1}{2}C[2E_1-E_2][E_2] \quad \text{...(ii)}$$

Energy loss

Q = energy supplied by the battery (W) – increase in stored potential energy of capacitor (ΔU)

Substituting the values of W and ΔU from Eqs. (i) and (ii), we get

$$Q=\frac{1}{2}CE_2^2$$

30. (a) $W=U_f-U_i$

$$=\frac{1}{2}\frac{q^2}{C_f}-\frac{1}{2}\frac{q^2}{C_i}$$

$$=\frac{q^2}{2}\left[\frac{x_2}{\varepsilon_0S}-\frac{x_1}{\varepsilon_0S}\right]$$

or $$W=\frac{q^2(x_2-x_1)}{2\varepsilon_0S}$$

(b) $$W=U_f-U_i$$

$$=\frac{1}{2}C_fV^2-\frac{1}{2}C_iV^2=\frac{V^2}{2}\left[\frac{\varepsilon_0S}{x_2}-\frac{\varepsilon_0S}{x_1}\right]$$

or $$W=\frac{\varepsilon_0SV^2(x_1-x_2)}{2x_1x_2}$$

31. (a) $\frac{\varepsilon_0A}{d}=20\text{ nF}=C_0$ (say)

With metallic plate

$$C_1=\frac{\varepsilon_0A}{d-t}=\frac{\varepsilon_0A}{0.4d}=\frac{20}{0.4}=50\text{ nF}$$

$$q_1 = C_1V = (50 \times 10^{-9})(200)\ 10^{-5}\ \text{C}$$

If charging battery is removed, then q_1 remains constant.

$$\therefore \quad W_1 = U_f - U_i = \frac{1}{2}\frac{q_1^2}{C_f} - \frac{1}{2}\frac{q_1^2}{C_i}$$

$$= \frac{1}{2}\frac{q_1^2}{C_0} - \frac{1}{2}\frac{q_1^2}{C_1}$$

$$= \frac{q_1^2}{2}\left[\frac{C_1 - C_0}{C_1C_0}\right]$$

$$= \frac{(10^{-5})^2}{2}\left[\frac{30}{1000}\right] \times \frac{1}{10^{-9}}$$

or $\quad W_1 = 1.5\ \text{mJ}$

(b) Dielectric constant of glass, $K = 6$

Capacitance with glass slab,

$$C_2 = \frac{\varepsilon_0 A}{d - 0.6d + \frac{0.6d}{6}}$$

$$= \frac{\varepsilon_0 A}{0.5\ d} = 2C_0 = 40\ \text{nF}$$

$$q_2 = C_2V = (40 \times 10^{-9})(200)$$

$$= 0.8 \times 10^{-5}\ \text{C}$$

Here, $\quad W_2 = \frac{q_2^2}{2}\left[\frac{C_2 - C_0}{C_2C_0}\right]$

$$= \frac{(0.8 \times 10^{-5})^2}{2}\left[\frac{20}{800}\right] \times \frac{1}{10^{-9}}$$

or $\quad W_2 = 0.8\ \text{mJ}$

32. $\therefore$ Resistance, $R = \frac{\rho(b-a)}{4\pi ab}$

This is a case of discharging of a capacitor. Hence charge and potential difference will decrease exponentially with time constant,

$$\tau_C = CR = \frac{\rho C(b-a)}{4\pi ab} \quad \text{...(i)}$$

Further,

$$V = V_0 e^{-t/\tau_C} \quad \text{...(ii)}$$

Here, V_0 = initial potential difference

Given, $V = \frac{V_0}{\eta}$ in $t = \Delta t$

$$\therefore \quad \frac{V_0}{\eta} = V_0\, e^{-\frac{(\Delta t)(4\pi ab)}{\rho C(b-a)}}$$

Solving this equation, we get

$$\rho = \frac{4\pi ab(\Delta t)}{C(b-a)\ln \eta}$$

33. **Note** Dielectric constant of glass $K = 6$.

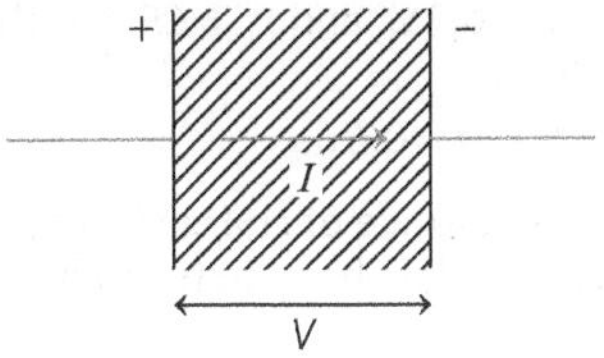

$$C = \frac{K\varepsilon_0 A}{d} = 4 \times 10^{-9}\ \text{F}$$

$$\Rightarrow \quad (A/d) = \frac{4 \times 10^{-9}}{K\varepsilon_0}$$

$$= \frac{4 \times 10^{-9}}{6 \times 8.86 \times 10^{-12}} = 75.24\ \text{m} \quad \text{...(i)}$$

$$I = \frac{V}{R} = \frac{V}{\rho\left(\frac{l}{A}\right)} = \frac{V}{\rho\left(\frac{d}{A}\right)} = \left(\frac{V}{\rho}\right)\left(\frac{A}{d}\right)$$

Substituting the given values of V_0, ρ and the value of A/d from equation (i), we get

$$I = \frac{2 \times 10^3}{100 \times 10^9} \times 75.24 = 1.5 \times 10^{-6}\ \text{A}$$

or $\quad I = 1.5\ \mu\text{A}$

34. Conductivity at distance x from one plate,

$$\sigma = \sigma_1 + \left(\frac{\sigma_2 - \sigma_1}{d}\right)x$$

Using $R = \frac{l}{\sigma A}$, small resistance of element shown in figure is,

$$dR = \frac{dx}{\sigma A}$$

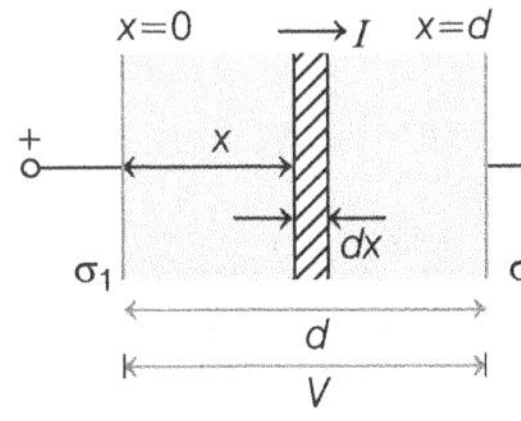

Integrating dR from $x = 0$ to $x = d$, we can find total resistance R between two plates of capacitor. Finally,

$$I = \frac{V}{R}$$

Substituting the given values and completing the integration, we get the final answer,

$$I = 5\ \text{nA}$$

35. It is a case of charging of a C-R circuit, where charge or potential difference across the capacitor will increase exponentially. Thus,

$$V = V_0(1 - e^{-t/\tau_C})$$

Substituting, $V = 0.9\,V_0$ and $\tau_C = CR$ and then substituting the given values,

we get $\quad t = 0.6\,\mu s$

36. This is a case of leakage of charge from inside. Charge will decrease to half in half life time given by,

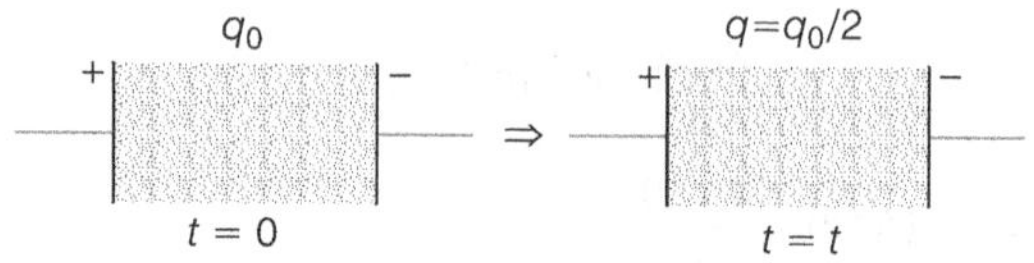

$$t = t_{1/2} = (\ln 2)\tau_C = (\ln 2)CR$$

$$= (\ln 2)\left[\frac{K\varepsilon_0 A}{d}\right]\left[\frac{\rho d}{A}\right]$$

$$t = (\ln 2)K\varepsilon_0\rho$$

$$\therefore \quad \rho = \frac{t}{(\ln 2)(K)\varepsilon_0}$$

Substituting the given values, we get

$$\rho = 1.4 \times 10^{13}\ \Omega\text{-m}$$

37. 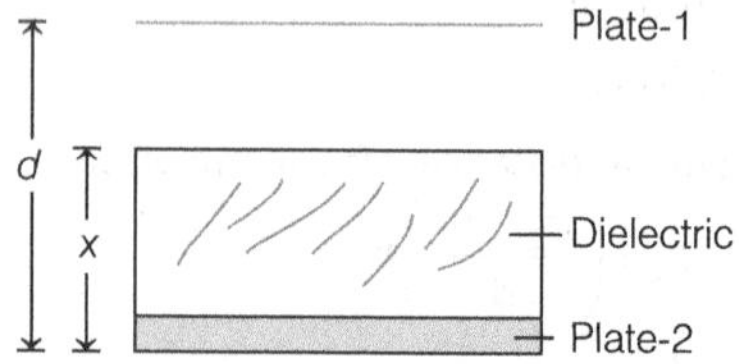

$$C_1 = \frac{\dfrac{\varepsilon_0 A}{\frac{2d}{3}} \times \dfrac{4\varepsilon_0 A}{\frac{d}{3}}}{\dfrac{\varepsilon_0 A}{\frac{2d}{3}} + \dfrac{4\varepsilon_0 A}{\frac{d}{3}}}$$

$$= \frac{18}{\frac{3}{2} + 12} \cdot \frac{\varepsilon_0 A}{d}$$

$$= 18 \times \frac{2}{27}\frac{\varepsilon_0 A}{d} = \frac{4}{3}\frac{\varepsilon_0 A}{d}$$

According to question,

$$\frac{4}{3}\frac{\varepsilon_0 A}{d} = 2 \Rightarrow \frac{\varepsilon_0 A}{d} = \frac{3}{2} \qquad \text{...(i)}$$

Now, $$C_2 = \frac{\dfrac{\varepsilon_0 A}{\frac{d}{3}} \times \dfrac{4\varepsilon_0 A}{\frac{2d}{3}}}{\dfrac{\varepsilon_0 A}{\frac{d}{3}} + \dfrac{4\varepsilon_0 A}{\frac{2d}{3}}} = \frac{18}{3+6}\frac{\varepsilon_0 A}{d}$$

$$= 2 \times \frac{\varepsilon_0 A}{d} = 2 \times \frac{3}{2} = 3\,\mu F$$

38. Initially, capacity of capacitor,

$$C = \frac{\varepsilon_0 A}{d} = 15\text{ pF} \qquad \text{[Given]}$$

When separation is doubled and space is filled with dielectric $K = 3.5$, capacity of capacitor becomes

$$C' = \frac{\varepsilon_0 K A}{2d} = \frac{K}{2}\left(\frac{\varepsilon_0 A}{d}\right)$$

$$= \frac{3.5}{2} \times 15\text{ pF} = \frac{105}{4}\text{ pF}$$

39. 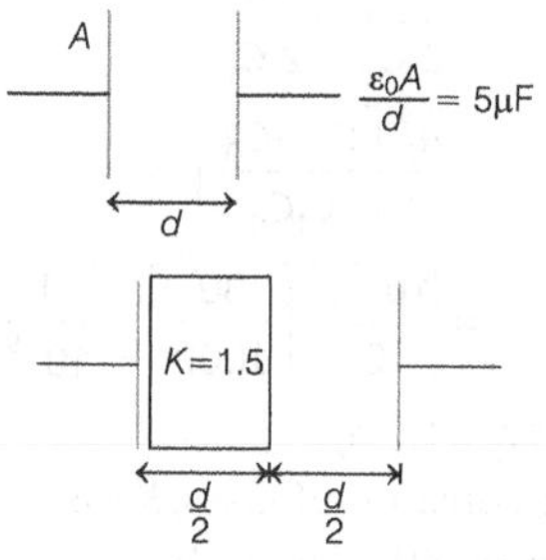

$$C_{new} = \frac{\varepsilon_0 A}{\dfrac{\left(\frac{d}{2}\right)}{1.5} + \dfrac{\left(\frac{d}{2}\right)}{1}} = \frac{\varepsilon_0 A}{\left(\frac{d}{3} + \frac{d}{2}\right)} = \frac{6\varepsilon_0 A}{5d}$$

$$= \frac{6}{5} \times 5\mu F = 6\mu F$$

40. Distance between capacitor plates

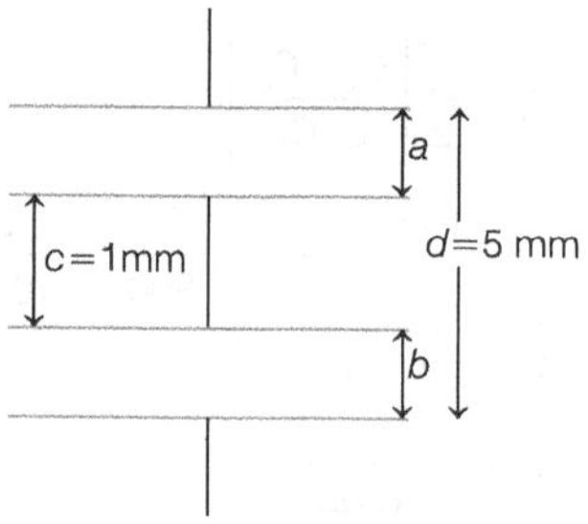

$$b = 5 - a - 1 = (4 - a)\text{ mm}$$

We know, $C = \dfrac{\varepsilon_0 A}{d}$

and in series arrangement

$$\frac{1}{C_{eq}} = \frac{1}{C_1} + \frac{1}{C_2}$$

$$\frac{1}{C_{eq}} = \frac{a}{\varepsilon_0 A} + \frac{4-a}{\varepsilon_0 A} = \frac{4(\text{mm})}{\varepsilon_0 A}$$

$$C_{eq} = \frac{\varepsilon_0 A}{4(\text{mm})}$$

Given, $\quad A = 200\text{ cm}^2$

$$C_{eq} = \frac{\varepsilon_0 \times 200 \times 10^{-4}}{4 \times 10^{-3}}$$

$$C_{eq} = 50 \times 10^{-1}\varepsilon_0 = 5\varepsilon_0\text{ F}$$

$$\therefore \quad x = 5$$

41. Given,

Plate area of each stair $= A$

Capacitance of arrangement,

$$C = \frac{x}{15}\frac{\varepsilon_0 A}{b}$$

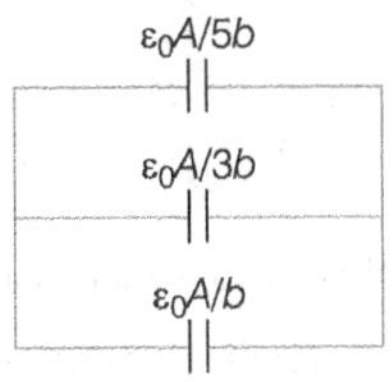

This circuit is equivalent to three capacitors in parallel. Capacitances are added to find equivalent capacity.

$$C_{eq} = \frac{\varepsilon_0 A}{b}\left(1+\frac{1}{3}+\frac{1}{5}\right) = \frac{23}{15}\frac{\varepsilon_0 A}{b}$$

Comparing with given value, we get

$\Rightarrow$ $x = 23$

42. Let E be the external field (toward right). Then,

$$E - \frac{\sigma}{2\varepsilon_0} = 8 \qquad \text{...(i)}$$

$$E + \frac{\sigma}{2\varepsilon_0} = 12 \qquad \text{...(ii)}$$

Solving these equations, we get $\sigma = 4\varepsilon_0$

43. Total energy stored $= \frac{1}{2}\frac{q^2}{C}$

$$= \frac{1}{2}\frac{(EC/2)^2}{C} = \frac{E^2C}{8}$$

Energy supplied $= qE = \left(\frac{EC}{2}\right)E = \frac{1}{2}E^2C$

$\therefore$ Energy lost $= \frac{1}{2}E^2C - \frac{1}{8}E^2C = \frac{3}{8}E^2C$

Now, this total loss is in direct ratio $r : 2r$ or $1 : 2$

$\therefore$ Energy lost in battery is $\frac{1}{3}$rd or $\frac{E^2C}{8}$.

44. $E = \left(\frac{1}{2}\right)CV^2$ ...(i)

The energy stored in capacitor is lost in form of heat energy.

$\therefore$ $H = ms\,\Delta T$ ...(ii)

From Eqs. (i) and (ii), we have

$$ms\,\Delta T = \left(\frac{1}{2}\right)CV^2$$

$\Rightarrow$ $V = \sqrt{\frac{2ms\,\Delta T}{C}}$

45. The simple circuit is as shown below.

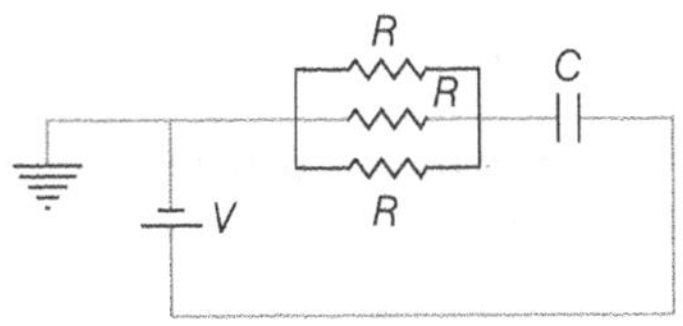

$$R_{net} = \frac{R}{3}$$

$\therefore$ $\tau_C = CR_{net} = \frac{CR}{3}$

$$q = q_0(1 - e^{-t/\tau_C}),$$

where, $q_0 = CV$

46. Energy stored in capacitor,

$$U = \frac{1}{2}\frac{q^2}{C} = \frac{1}{2C}(q_0 e^{-t/\tau})^2$$

$$= \frac{q_0^2}{2C}e^{-2t/\tau} \qquad \text{(where, } \tau = CR)$$

$$U = U_i e^{-2t/\tau}$$

$$\frac{1}{2}U_i = U_i e^{-2t_1/\tau}$$

$$\frac{1}{2} = e^{-2t_1/\tau} \Rightarrow t_1 = \frac{\tau}{2}\ln 2$$

Now, $q = q_0 e^{-t/\tau}$

$$\frac{1}{4}q_0 = q_0 e^{-t_2/\tau}$$

$$t_2 = \tau \ln 4 = 2\tau \ln 2$$

$\therefore$ $\frac{t_1}{t_2} = \frac{1}{4}$

47. Net capacitance between points A and P will be equal to the net capacitance between points P and B.

48. Total heat produced $= \frac{1}{2}CV^2$

$$= \frac{1}{2}(2\,\mu\text{F})(5)^2$$

$$= 25\,\mu\text{J}$$

Now, this should distribute in inverse ratio of resistors, as they are in parallel.

$\therefore$ $\frac{H_{5\Omega}}{H_R} = \frac{R}{5}$

or $H_{5\Omega} = \left(\frac{R}{R+5}\right)$ (Total heat)

or $10 = \left(\frac{R}{R+5}\right)(25)$

Solving this equation, we get

$$R = \left(\frac{10}{3}\right)\Omega$$

49. Two capacitors are in parallel.

$\therefore$ $U = \frac{1}{2}C_{net}V^2 = \frac{1}{2}(2C)V^2 = CV^2$

$$= \left(\frac{\varepsilon_0 A}{d}\right)V^2$$

50. During charging capacitor and resistance of its wire are independently connected with the battery. Hence,

$$\tau_C = CR$$

During discharging capacitor is discharged through both resistors (in series). Hence,

$$\tau_C = C(2R) = 2CR$$

51. In the figure,

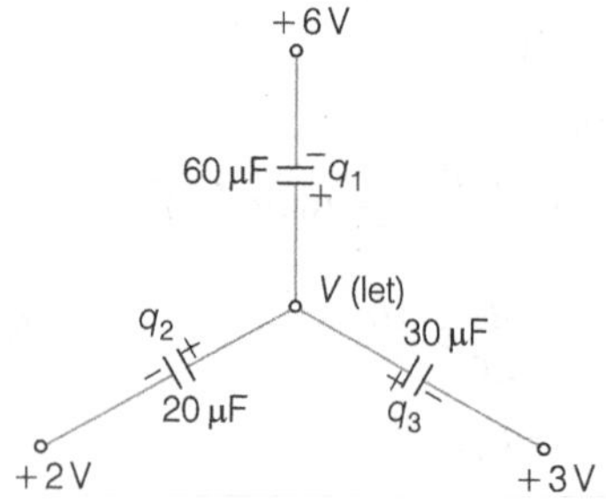

$$q_1 + q_2 + q_3 = 0$$

$$60(V-6) + 20(V-2) + 30(V-3) = 0$$

Solving this equation, we get

$$V = \frac{49}{11}\text{ V}$$

52.

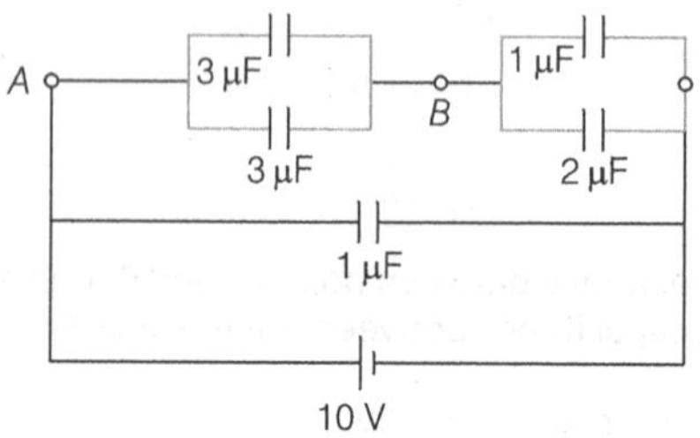

$$\frac{V_{AB}}{V_{BC}} = \frac{(C)_{BC}}{(C)_{AB}} = \frac{3}{6} = \frac{1}{2}$$

$$\therefore \quad V_{AB} = \left(\frac{1}{1+2}\right)(10)\text{ V}$$

$$= \frac{10}{3}\text{ V}$$

53. Applying Kirchhoff's loop law in outermost loop, we have

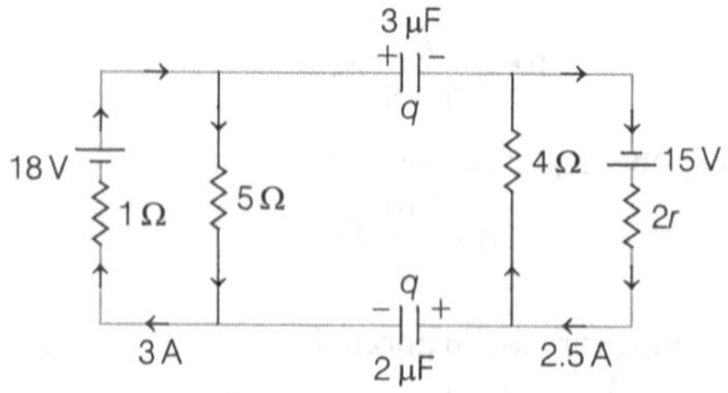

$$-\frac{q}{3} + 15 - 2\times 2.5 - \frac{q}{2} - 3\times 1 + 18 = 0$$

Solving this equation, we get

$$q = 30\ \mu\text{C}$$

54. $i_1 = \left(\frac{V}{2R}\right)e^{-\frac{t}{6CR}}$

$$\therefore \quad i_2 = \left(\frac{V}{R}\right)e^{-\frac{t}{CR}}$$

$$\therefore \quad \frac{i_1}{i_2} = \frac{e^{\frac{5t}{6CR}}}{2}$$

We can see that this ratio is increasing with time.

55. The given time is the half-life time of the exponentially decreasing equation.

$$\therefore \quad t = t_{1/2} = (\ln 2)\,\tau_C = (\ln 2)\,CR_{\text{net}}$$

$$\therefore \quad R_{\text{net}} = \frac{t}{(\ln 2)C}$$

$$= \frac{2(\ln 2)\ \mu\text{s}}{(\ln 2)(0.5\ \mu\text{F})} = 4\Omega$$

$\therefore$ Resistance of ammeter $= 2\Omega$

56. Four capacitors are in parallel charge across each is $q = CV$. Two surfaces of plate C marks two capacitors, one with B and other with D and C is connected to positive terminal of the battery. Hence,

$$q_C = 2CV = +40\ \mu\text{C}$$

57. Given circuit can be redrawn as follows.

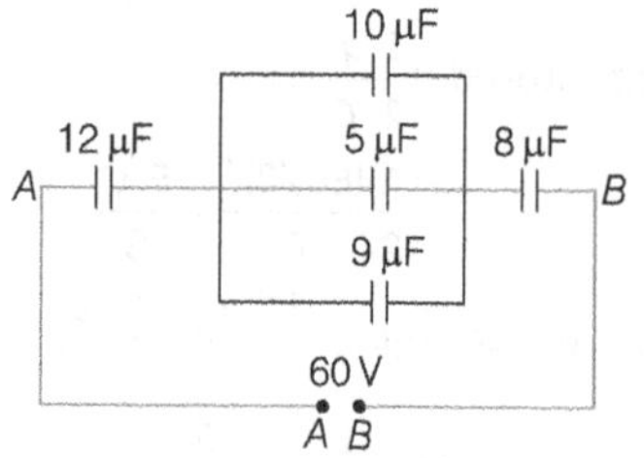

Equivalent capacitance of the circuit,

$$C_{AB} = \frac{24}{2+1+3} = 4\ \mu\text{F}$$

Total charge given by battery,

$$q = C_{AB}V = 4\times 60$$

$$= 240\ \mu\text{C}$$

Charge on 5 μF capacitor

$$q_2 = \left(\frac{5}{10+5+9}\right)\times 240$$

$$= 50\ \mu\text{C}$$

58. Minimum number of condensers in each row $= \frac{3000}{500} = 6$

If C_s is capacity of 6 condensers in a row, then

$$\frac{1}{C_s} = \frac{1}{1} + \frac{1}{1} + \frac{1}{1} + \frac{1}{1} + \frac{1}{1} + \frac{1}{1} = 6$$

$$C_s = \frac{1}{6}\ \mu\text{F}$$

Let there be m such rows in parallel,

Total capacity $= m \times C_s$

$$2 = m \times \frac{1}{6}$$

$\therefore \quad m = 12$

Total number of capacitors $= 6 \times 12 = 72$

59. Resultant of $1\,\mu F$ and $2\,\mu F$ is $3\,\mu F$. Now in series, potential difference distributes in inverse ratio of capacity.

$$\therefore \quad \frac{V_{3\mu F}}{V_C} = \frac{C}{3} \quad \text{or} \quad V_{3\mu F} = \left(\frac{C}{C+3}\right)E$$

This is also the potential difference across $2\,\mu F$.

$$\therefore \quad Q_2 = (2\,\mu F)(V_{2\mu F})$$

$$\text{or} \quad Q_2 = \left(\frac{2CE}{C+3}\right) = \left(\frac{2}{1+\frac{3}{C}}\right)E$$

From this expression of Q_2, we can see that Q_2 will increase with increase in the value of C (but not linearly). Therefore, only options (a) and (b) may be correct.

Further,

$$\frac{d}{dc}(Q_2) = 2E\left[\frac{(C+3)-C}{(C+3)^2}\right] = \frac{6E}{(C+3)^2}$$

$= $ Slope of Q_2 *versus* c graph.

i.e. slope of Q_2 *versus* C graph decreases with increase in the value of C. Hence, the correct graph is (a).

60.

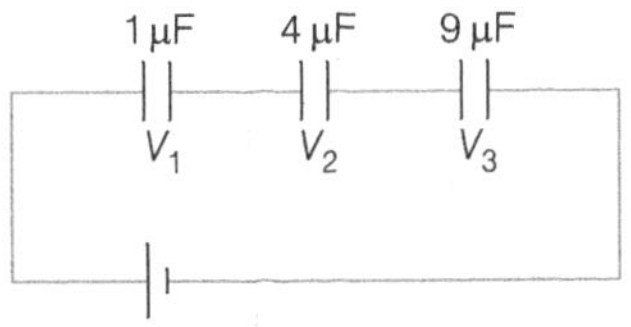

In series, $\quad V \propto \frac{1}{C} \quad$ (as $q = $ constant)

$$\therefore \quad \frac{V_2}{V_1} = \frac{1}{4}$$

$$\text{or} \quad V_2 = \frac{V_1}{4} = \frac{10}{4} = 2.5\text{ V}$$

$$\frac{V_3}{V_1} = \frac{1}{9}$$

$$\therefore \quad V_3 = \frac{V_1}{9} = \frac{10}{9}\text{ V}$$

$$\text{Now,} \quad E = \left(10 + 2.5 + \frac{10}{9}\right)\text{V}$$

61. In steady state, current through capacitor wire is zero. Current flows through $200\,\Omega$, $900\,\Omega$ and A_2.

$$V_C = \frac{q}{C} = \frac{4 \times 10^{-3}}{100 \times 10^{-6}} = 40\text{ V}$$

This is also potential drop across $900\,\Omega$ resistance and $100\,\Omega$ ammeter A_2 (Total resistance $= 1000\,\Omega$). Now, this $1000\,\Omega$ and $200\,\Omega$ are in series. Therefore,

$$V_2 = V_{200\,\Omega} = \frac{V_{1000\,\Omega}}{5}$$

$$= \frac{40}{5} = 8\text{ V}$$

$$\text{Emf} = V_{1000\,\Omega} + V_{200\,\Omega} = 48\text{ V}$$

$$i = \frac{\text{Emf}}{\text{Net resistance}}$$

$$= \frac{48}{1200} = \frac{1}{25}\text{ A}$$

62. Current through A is the main current passing through the battery. So, this current is more than the current passing through B. Hence, during charging more heat is produced in A.

In steady state,

$$i_C = 0$$

$$\text{and} \quad i_A = i_B$$

Hence, heat is produced at the same rate in A and B.

Further, in steady state

$$V_C = V_B = \frac{\varepsilon}{2}$$

$$\therefore \quad U = \frac{1}{2}CV_C^2 = \frac{1}{8}C\varepsilon^2$$

63. Centre plate makes an arrangement equivalent to system shown in figure.

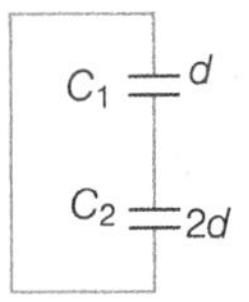

So, we have two capacitors with same potential difference V and total charge Q.

$$\text{Also,} \quad C_1 = \frac{\varepsilon_0 A}{d}, C_2 = \frac{\varepsilon_0 A}{2d}$$

$$Q_1 + Q_2 = Q$$

$$\text{and} \quad V_1 = V_2 = V$$

$$\text{Then,} \quad \frac{Q_1}{C_1} = \frac{Q_2}{C_2} \Rightarrow \frac{Q_1 d}{\varepsilon_0 A} = \frac{Q_2\,2d}{\varepsilon_0 A}$$

$$Q_1 = 2Q_2;\ 2Q_2 + Q_2 = Q$$

$$\text{So,} \quad Q_2 = \frac{Q}{3}$$

$\Rightarrow$ Charge on lower plate is $-\frac{Q}{3}$.

and $Q_1 = \frac{2Q}{3}$, charge on upper plate is $-\frac{2Q}{3}$.

$$\text{Also,} \quad \Delta V = \frac{Q_1}{C_1} = \frac{2Qd}{3\varepsilon_0 A}$$

64. $C_i = \frac{(C)(2C)}{C+2C} = \frac{2}{3}C$

$$q_i = C_i E = \frac{2}{3}EC$$

$$C_f = 2C$$

$$\therefore \quad q_f = 2EC$$

$$\Delta q = q_f - q_i$$

$$= \frac{4}{3}CE$$

65. Let $(+q)\,\mu C$ charge flows in the closed loop in clockwise direction. Then, final charges on different capacitors are as shown in figure.

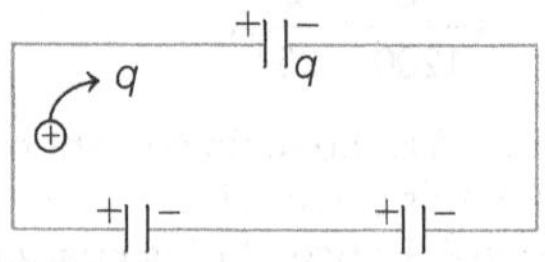

Now, applying Kirchhoff's loop law

$$\frac{360-q}{3}+\frac{300-q}{2}=\frac{q}{1.5}$$

Solving the above equation, we get

$$q=180\,\mu C$$

66. $C_{eq}=\frac{5}{2}\mu F,\ Q=CV=250\,\mu C$

Charge in $6\mu F$ branch $V_C=150\,\mu C$

$$V_{AB}=\frac{150}{6}=25\,V$$

and $\quad V_{BC}=100-25=75\,V$

67. At $t=0$, emf of the circuit = PD across the capacitor = 6 V.

$$\therefore \quad i=\frac{6}{1+2}=2\,A$$

Half-life of the circuit

$$=(\ln 2)\,\tau_C=(\ln 2)\,CR=(6\ln 2)\text{ s}.$$

In half-life time, all values get halved.

For example

$$V_C=\frac{6}{2}=3\,V$$

$$i=\frac{2}{1}=1\,A$$

$$\therefore \quad V_{1\Omega}=iR=1\,V$$

$$V_{2\Omega}=iR=2\,V$$

68. In steady state, $q_C=EC$ and $q_{2C}=2EC$

$$\tau_C=2CR \text{ of both circuits}$$

At time t, $\quad q_C=EC\,(1-e^{t/\tau_C})$

$$q_{2C}=2EC\,(1-e^{-t/\tau_C})$$

$$\therefore \quad \frac{q_C}{q_{2C}}=\frac{1}{2}$$

69. $C=C_1+C_2$

$$C_1=\frac{K\varepsilon_0 A/3}{d},\ C_2=\frac{\varepsilon_0 2A/3}{d}$$

$$\Rightarrow \quad C=\frac{(K+2)\varepsilon_0 A}{3d}$$

$$\Rightarrow \quad \frac{C}{C_1}=\frac{K+2}{K}$$

Also, $E_1=E_2=V/d$, where V is potential difference between the plates.

70. After pressing S_1 charge on upper plate of C_1 is $+2CV_0$. After pressing S_2 this charge equally distributes in two capacitors. Therefore charge an upper plates of both capacitors will be $+CV_0$.

When S_2 is released and S_3 is pressed, charge on upper plate of C_1 remains unchanged $(=+CV_0)$ but charge on upper plate of C_2 is according to new battery $(=-CV_0)$.

71. Finally, the capacitors are in parallel and total charge $(=q_0)$ distributes between them in direct ratio of capacity.

$$\therefore \quad q_{C_2}=\left(\frac{C_2}{C_1+C_2}\right)q_0 \rightarrow \text{in steady state.}$$

But this charge increases exponentially.

Hence, charge on C_2 at any time t is

$$q_{C_2}=\left(\frac{C_2\,q_0}{C_1+C_2}\right)(1-e^{-t/\tau_C})$$

Initially, C_2 is uncharged so, whatever is the charge on C_2, it is charge flown through switches.

72. Common potential in steady state when they finally come in parallel is

$$V=\frac{\text{Total charge}}{\text{Total capacity}}=\frac{q_0}{C_1+C_2}$$

Total heat dissipated $=U_i-U_f$

$$=\frac{q_0^2}{2C_1}-\frac{1}{2}(C_1+C_2)\left(\frac{q_0}{C_1+C_2}\right)^2$$

$$=\left(\frac{q_0^2}{2C_1}\right)\left(\frac{C_1C_2}{C_1+C_2}\right)$$

73. $E_{air}=E_0=\frac{V}{d}$

74. $E_{dielectric}=\frac{E_0}{K}=\frac{V}{Kd}$

75.

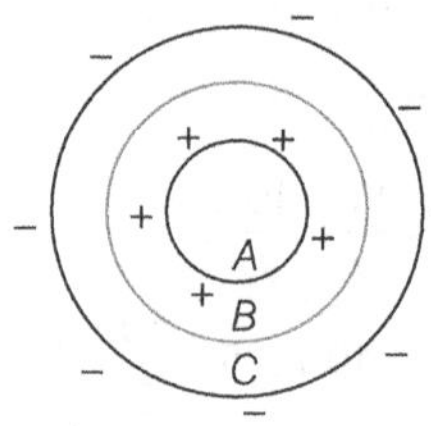

Potential on the surface of A is due to charge on A and due to charge on C.

$$V_A=\frac{q}{4\pi\varepsilon_0 a}+\frac{-q}{4\pi\varepsilon_0 c}=\frac{q}{4\pi\varepsilon_0}\left(\frac{1}{a}-\frac{1}{c}\right)$$

76. Potential of shell C is

$$V_C=\frac{q}{4\pi\varepsilon_0 c}-\frac{q}{4\pi\varepsilon_0 c}=0$$

$$\therefore \quad \Delta V=V_A-V_C=\frac{q}{4\pi\varepsilon_0}\left(\frac{1}{a}-\frac{1}{c}\right)$$

77.

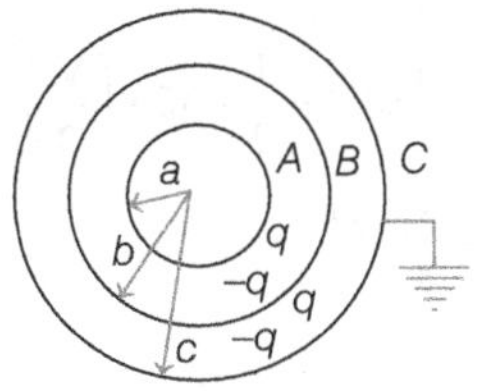

$$V_A - V_B = \frac{q}{4\pi\varepsilon_0}\left(\frac{1}{a} - \frac{1}{b}\right)$$

$$V_B - V_C = \frac{1}{4\pi\varepsilon_0}\left(\frac{1}{b} - \frac{1}{c}\right)$$

Adding these two,

$$V_A - V_C = V = \frac{q}{4\pi\varepsilon_0}\left(\frac{1}{a} - \frac{1}{c}\right)$$

Now, $c = \frac{q}{V}$

78. After inserting the dielectric, electric field from A to C is formed by using Gauss' law.

The field,

$$E(r) = \frac{q}{4\pi\varepsilon_0 K r^2}, a \le r \le b$$

and $$E(r) = \frac{q}{4\pi\varepsilon_0 r^2} \text{ for } b \le r \le c$$

So, potential difference between A and C is

$$\Delta V = V_A - V_C$$

$$= -\int_a^b \frac{q}{4\pi\varepsilon_0 K r^2} dr - \int_b^c \frac{q}{4\pi\varepsilon_0 r^2} dr$$

$$= \frac{q}{4\pi\varepsilon_0}\left\{\frac{1}{K}\left(\frac{1}{a} - \frac{1}{b}\right) + \left(\frac{1}{b} - \frac{1}{c}\right)\right\}$$

$$= \frac{q}{4\pi\varepsilon_0}\left(\frac{b-a}{kab} + \frac{c-b}{bc}\right)$$

$$= \frac{q}{4\pi\varepsilon_0 K abc}[c(b-a) + Ka(c-b)]$$

79. Considering branch ac,

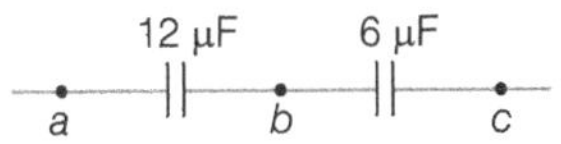

$V_{ab} = 4\text{V}$ (given)

$\Rightarrow$ $Q_{ab} = C_{ab} V_{ab} = 12\,\mu\text{F} \times 4\text{ V} = 48\,\mu\text{C}$

As, 12 μF and 6 μF capacitors are in series

$\therefore$ Charge on both must be same.

$\Rightarrow$ $Q_{bc} = 48\,\mu\text{C}$

As, $C_{bc} = 6\,\mu\text{F}$

$$V_{bc} = \frac{48}{6}\text{V} = 8\text{ V}$$

Hence $V_{ac} = V_a - V_c = V_a - V_b + V_b - V_c$

$= 4 + 8 = 12\text{ V}$

80. In given circuit, there are two portions marked I and II as shown

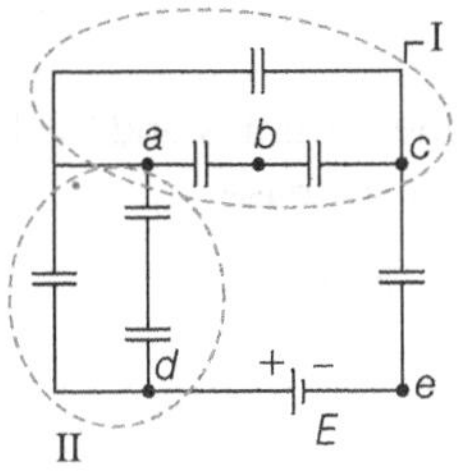

Now, considering circuit as given in figure.

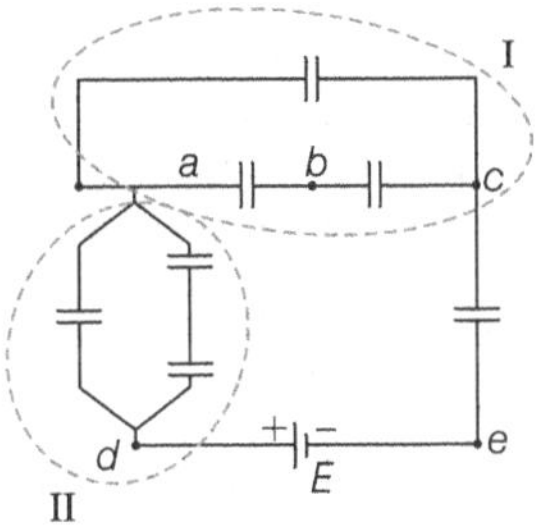

Clearly I and II parts are in series, so charge on both portions must be same.

As, $V_{ac} = 12\text{ V}$

And C_{eq} of portion marked I $= 10\,\mu\text{F}$

So charge on portion I $= V_{eq} \times C_{eq}$

$= 10 \times 12 = 120\,\mu\text{C}$

$\therefore$ Charge on portion II $= 120\,\mu\text{C}$

$\Rightarrow$ $C_{ad} \times (V_d - V_a) = 120\,\mu\text{C}$

$\Rightarrow$ $12 \times (V_d - V_a) = 120$ $\{\because C_{ad} = 12\,\mu\text{F}\}$

$\Rightarrow$ $V_d - V_a = 10\text{ V}$

81. We can consider the given circuit as

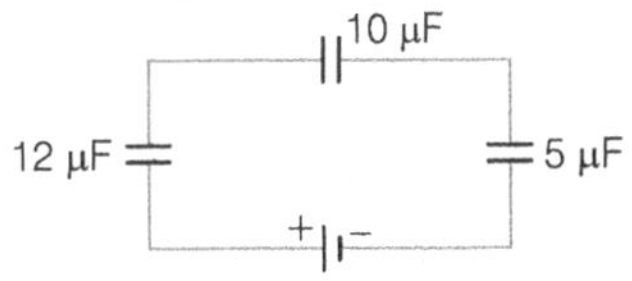

Charge on 5 μF capacitor $= 120\,\mu\text{C}$

$\therefore$ $V_c - V_e = \frac{120}{5} = 24\text{ V}$

$\therefore$ $E = 12 + 10 + 24 = 46\text{ V}$

82. (a) $C_i = \frac{4 \times 4}{4 + 4} = 2\,\mu\text{F}$

$$C_f = \frac{C_1 C_2}{C_1 + C_2} = \frac{8 \times 2}{8 + 2}$$

$= 1.6\,\mu\text{F}$

$q = CV$

Since, total capacity is decreasing. Hence, charge on both capacitors will decrease.

(b) $U_2 = \frac{1}{2}\frac{q^2}{C}$

q has become $\frac{1.6}{2}$ or 0.8 times but C is halved.

Hence, U_2 will increase.

(c) $V_2 = \frac{q}{C}$

q has become 0.8 times and C is halved.

Hence, V_2 will increase.

(d) $E_2 = \frac{V_2}{d}$ or $E_2 \propto V_2$

83. (a) $C_i = 2\ \mu F$

$\therefore \quad q_i = 60\ \mu C$

$C_f = 6\ \mu F$

$\therefore \quad q_f = 180\ \mu C$

$\therefore$ Δq from the battery $= q_f - q_i = 120\ \mu C$.

(b) Between 4 μF and 2 μF charge distributes indirect ratio of capacity. Hence, on 2 μF

$$q_i = \left(\frac{2}{2+4}\right)(60\ \mu C) = 20\ \mu C$$

$$q_f = \left(\frac{2}{2+4}\right)(180)\,60\ \mu C$$

$\therefore \quad \Delta q = \Delta f - q_i = 40\ \mu C$

(c) On 3 μF, initial charge is 60 μF and final charge is zero.

$\therefore \quad \Delta q = 60\ \mu C$

(d) On 4 μF

$$q_i = \left(\frac{4}{2+4}\right)(60\ \mu C) = 40\ \mu C$$

$$q_f = \left(\frac{4}{2+4}\right)(180\ \mu C) = 120\ \mu C$$

$\therefore \quad \Delta q = q_f - q_i = 80\ \mu C$

84. (a) In second figure, $V_{C_1} = V =$ maximum

Hence, q_{C_1} is maximum.

(b) In first figure, $V_{C_2} = \frac{V}{3}$

In second figure, $V_{C_2} = V$

In third figure, $V_{C_2} = \left(\frac{C}{2C + C}\right)V = \frac{V}{3}$

In fourth figure, $V_{C_2} = \left(\frac{2C}{C + 2C}\right)V = \frac{2V}{3}$

Now, $q = CV$

Hence, q_{C_2} is minimum in first figure.

(c) In second figure, $V_{C_1} = V =$ maximum

(d) Similar to option (b)

85. After closing the switch, the common potential is parallel.

$$V = \frac{\text{Total charge}}{\text{Total capacity}} = \frac{CV}{3C} = \frac{V}{3}$$

$$U_C = \frac{1}{2}C\left(\frac{V}{3}\right)^2 = \frac{1}{18}CV^2$$

$$U_{2C} = \frac{1}{2}(2C)\left(\frac{V}{3}\right)^2 = \frac{1}{9}CV^2$$

Loss of energy $= U_i - U_f$

$$= \frac{1}{2}CV^2 - \frac{1}{2} \times 3C \times \left(\frac{V}{3}\right)^2$$

$$= \frac{1}{3}CV^2$$

86. Let $C_0 = \frac{\varepsilon_0 A}{d}$

$$C_1 = \frac{\varepsilon_0 (A/2)}{d} + \frac{K\,\varepsilon_0 (A/2)}{d}$$

$$= \frac{C_0}{2} + C_0 = \frac{3C_0}{2}$$

$$C_2 = \frac{\varepsilon_0 A}{d - d/2 + \frac{d/2}{K}} = \frac{2\,\varepsilon_0 A}{d\left(1 + \frac{1}{K}\right)}$$

$$= \frac{4}{3}\frac{\varepsilon_0 A}{d} = \frac{4}{3}C_0$$

$\therefore \quad \frac{C_1}{C_2} = \frac{9}{8}$

Capacitors are in series.

Hence, $q_1 = q_2$ or $\frac{q_1}{q_2} = 1$

$$U = \frac{1}{2}\frac{q^2}{C}$$

or $U \propto \frac{1}{C}$ (as q is same)

$\therefore \quad \frac{U_1}{U_2} = \frac{C_2}{C_1} = \frac{8}{9}$

87. Initially capacity of system is $\frac{C}{2}$ and when switch is closed one of capacitors is short circuited and capacity finally becomes C.

Initially potential drop across each capacitor is $\frac{E}{2}$ and finally potential drop across first capacitor is E and across second is zero.

Initially,

Capacity of system $= \frac{C}{2}$

Potential drop of each capacitor $= \frac{E}{2}$

Charge stored $= Q_1 = \frac{C}{2} \times E$

Finally charge stored $= Q_2 = C \times E$

Charge flow through cell $= CE - \frac{CE}{2} = \frac{CE}{2}$

Initially energy stored, $U_i = \frac{1}{2}\left(\frac{C}{2}\right) \cdot E^2 = \frac{CE^2}{4}$

Finally energy stored, $U_f = \frac{1}{2}(C)(E^2)$

$\therefore$ Change in energy stored $= U_f - U_i = \frac{1}{4}CE^2$

$\therefore$ Work done by cell when S is closed $= \frac{1}{2}CE^2$

Heat developed = decrease in stored energy of second capacitor $= \frac{1}{4}CE^2$

88.

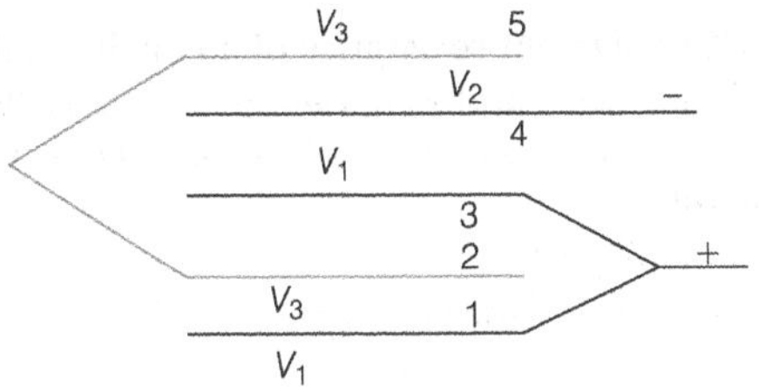

$$C = \frac{\varepsilon_0 A}{d} \quad \text{(between two successive plates)}$$

The effective capacity has to be found between V_1 and V_2.

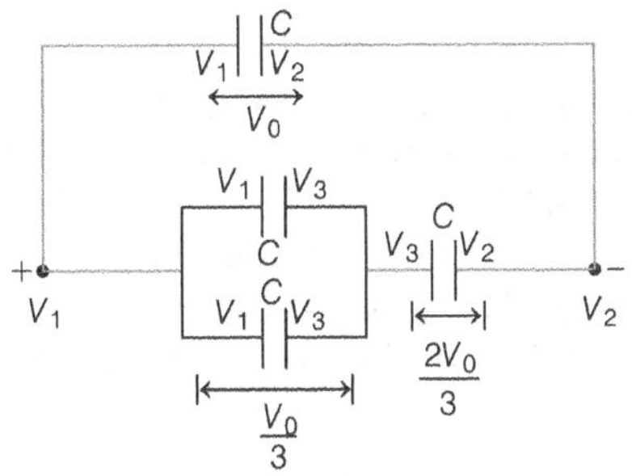

$$C_{net} = C + \frac{(2C)(C)}{2C + C} = \frac{5}{3}C = \frac{5}{3}\frac{\varepsilon_0 A}{d}$$

$$q_3 = +[C(V_1 - V_2) + C(V_1 - V_3)]$$

$$= \frac{\varepsilon_0 A}{d}\left[V_0 + \frac{V_0}{3}\right] = \frac{4\varepsilon_0 A V_0}{3d}$$

$$q_5 = C(V_3 - V_2) = \left(\frac{\varepsilon_0 A}{d}\right)\left(\frac{2V_0}{3}\right)$$

$$= \frac{2\varepsilon_0 A V_0}{3d}$$

89. Capacitor C_1 will discharge according to the equation,

$$q = q_0 e^{-t/\tau_C} \quad \text{...(i)}$$

Here, $\tau_C = C_1 R$

and discharging current :

$$i = \frac{-dq}{dt} = \frac{q_0}{\tau_C} \cdot e^{-t/\tau_C} = \frac{q_0 e^{-t/\tau_C}}{C_1 R} \quad \text{...(ii)}$$

At the given instant $i = i_0$

Therefore, from Eq. (ii)

$$q_0 e^{-t/\tau_C} = i_0 C_1 R \text{ at this instant}$$

or charge C_1 at this instant will be

$$q = (i_0 C_1 R) \quad \text{[From Eq. (ii)]}$$

Now, this charge q will later on distribute in C_1 and C_2.

$$\therefore \quad U_i = \frac{q^2}{2C_1} \quad \text{and} \quad \frac{q^2}{2(C_1 + C_2)} = U_f$$

$\therefore$ Heat generated in resistance,

$$H = U_i - U_f$$

Substituting values of q: U_i and U_f, we get

$$H = \frac{(I_0 R)^2 C_1 C_2}{2(C_1 + C_2)}$$ **Ans.**

90. Time constant of the circuit is

$$\tau_C = CR = (0.5 \times 10^{-6})(500) = 2.5 \times 10^{-4} \text{ s}$$

For $t \le 250\,\mu s$ $\quad i = i_0 e^{-t/\tau_C}$

Here, $\quad i_0 = \frac{20}{500} = 0.04$ A

$\therefore \quad i = (0.04\, e^{-4000t})$ amp

At $t = 250\,\mu s = 2.5 \times 10^{-4}$ s

$$i = 0.04\, e^{-1} = 0.015 \text{ amp}$$

At this moment PD across the capacitor,

$$V_C = 20(1 - e^{-1}) = 12.64 \text{ V}$$

So, when the switch is shifted to position 2, the current in the circuit is 0.015 A (clockwise) and PD across capacitor is 12.64 V.

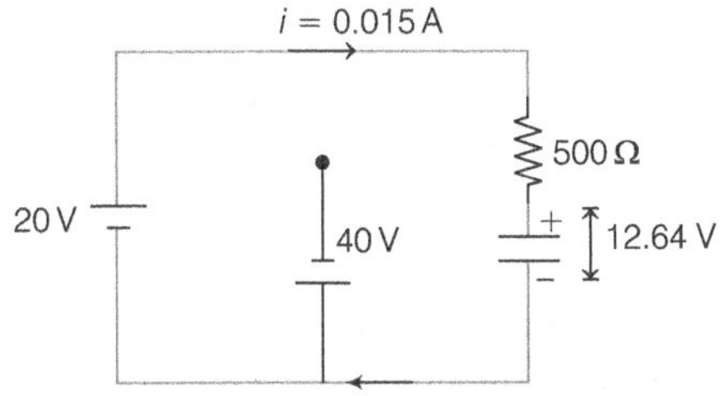

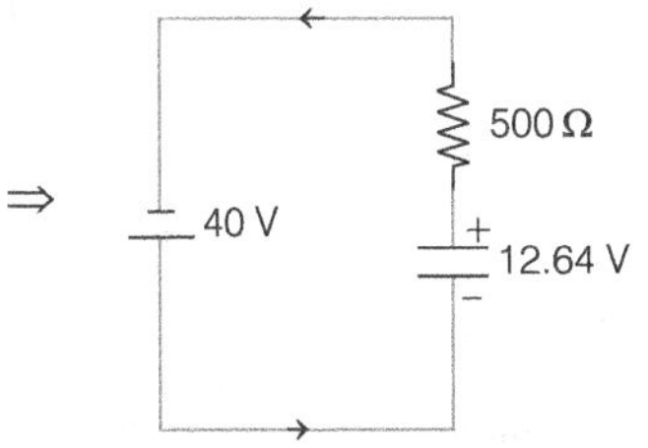

As soon as the switch is shifted to position 2, current will reverse its direction with maximum current.

$$i_0' = \frac{40 + 12.64}{500} = 0.11 \text{ A}$$

Now, it will decrease exponentially to zero.

For $t \ge 250\,\mu s$

$$i = -i_0'\, e^{-t/\tau_C} = -0.11\, e^{-4000t}$$

The (i - t) graph is as shown in figure.

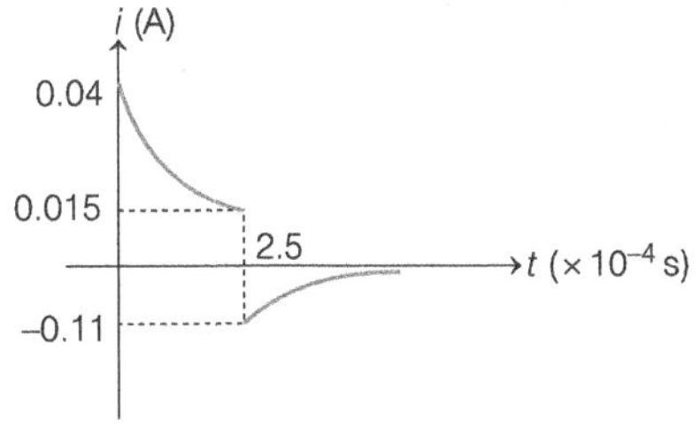

91. At $t = 5$ ms, $V = 10$ V

$$\therefore \quad i_R = \frac{V}{R} = \frac{10}{4} = 2.5 \text{ A} \qquad \textbf{Ans.}$$

Further, $q = CV = (300 \times 10^{-6})(2000t) = 0.6\,t$

$$i_C = \frac{dq}{dt} = 0.6 \text{ A} = \text{constant} \qquad \textbf{Ans.}$$

92. Potential energy stored in the capacitor,

$$U = \frac{1}{2}CV^2 = \frac{1}{2} \times 5 \times 10^{-6} \times (200)^2 = 0.1 \text{ J}$$

During discharging this 0.1 J will distribute in direct ratio of resistance,

$$\therefore \quad H_{400} = \frac{400}{400 + 500} \times 0.1$$

$$= 44.4 \times 10^{-3} \text{ J}$$

$$= 44.4 \text{ mJ} \qquad \textbf{Ans.}$$

93.

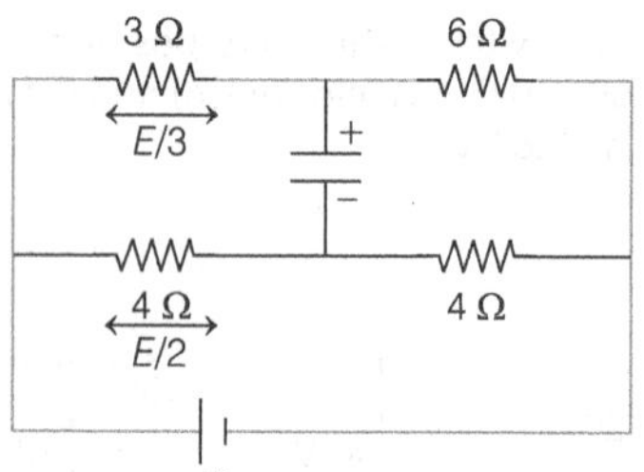

(a) Current in lower branch $= E/8 = 3$ A

Current in upper branch $= E/9 = 24/9 = 2.67$ A

(b) PD across the capacitor $= E/2 - E/3 = E/6$

From $q = CV$, we have

$$16 = (4)\frac{E}{6}$$

$$\therefore \quad E = 24 \text{ V}$$

(c) After short-circuiting the battery, we will have to find net resistance across capacitor to calculate equivalent value of τ_C in discharging.

3Ω and 6Ω are in parallel. Similarly, 4Ω and 4Ω are in parallel. They are then in series.

$$\therefore \quad R_{net} = 4\Omega,$$

$$\tau_C = CR_{net} = (4 \times 4)\ \mu\text{s} = 16\ \mu\text{s}$$

During discharging $q = q_0 e^{-t/\tau_C}$

or $\quad 8 = 16 e^{-t/16}$

Solving this equation, we get $t = 11.1\ \mu$s **Ans.**

94. Let us first find charges on both capacitors before and after closing the switch.

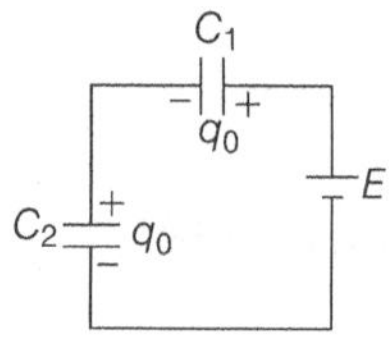

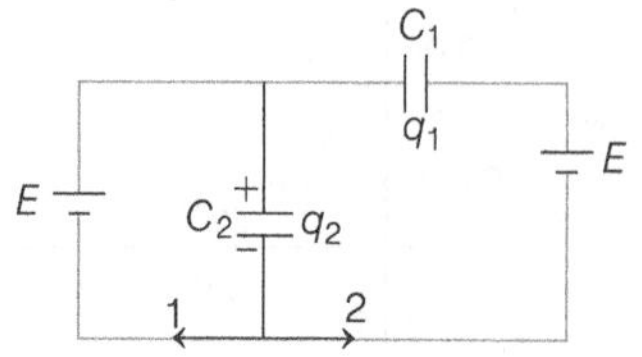

$$q_2 = EC_2 \quad \text{and} \quad q_1 = 0$$

$$q_0 = E\left(\frac{C_1 C_2}{C_1 + C_2}\right)$$

From 2, $-q_0$ charge will flow, so that charge on right hand side plate of C_1 becomes zero. From 1, q_2 charge will flow.

95. At $t = 0$, equivalent resistance of an uncharged capacitor is zero and a charged capacitor is like a battery of emf = potential difference across the capacitor.

(a) $V_{C_1} = q_1/C_1 = 2$ V

$\therefore$ Net emf of the circuit $= 9 - 2 = 7$ V

or $\quad 9 + 2 = 11$ V

Net resistance $= 30 + \dfrac{30 \times 60}{30 + 60} = 50\,\Omega$

$\therefore$ Current at $t = 0$ would be $i_0 = \dfrac{7}{50}$ A

or $\quad \dfrac{11}{50}$ A **Ans.**

(b) In steady state, no current will flow through the circuit. C_2 will therefore be short-circuited, while PD across C_1 will be 9 V.

$$\therefore \quad Q_2 = 0 \quad \text{and} \quad Q_1 = 9\ \mu\text{C} \qquad \textbf{Ans.}$$

96. (a) In steady state, capacitors will be in parallel. Charge will distribute in direct ratio of their capacity.

$$\therefore \quad q_1 = \left(\frac{C_1}{C_1 + C_2}\right) q_0$$

and

$$q_2 = \left(\frac{C_2}{C_1 + C_2}\right) q_0$$

Initial emf in the circuit is potential difference across capacitor C_1 or q_0/C_1.

Therefore, initial current would be

$$i_0 = \frac{q_0/C_1}{R} = \frac{q_0}{C_1 R}$$

Current as function of time will be $i = i_0 e^{-t/\tau_C}$

Here, $\tau_C = \left(\dfrac{C_1 C_2}{C_1 + C_2}\right) R$ **Ans.**

(b) $U_i = \dfrac{1}{2}\dfrac{q_0^2}{C_1}$ and $U_f = \dfrac{1}{2}\dfrac{q_0^2}{C_1 + C_2}$

Heat lost in the resistor

$$= U_i - U_f = \frac{q_0^2}{2}\left[\frac{C_2}{C_1(C_1 + C_2)}\right] \qquad \textbf{Ans.}$$

97. $q_i = CV_i = 100\ \mu\text{C}, q_f = CV_f = -50\ \mu\text{C}$

Therefore, charge will vary from $100\ \mu$C to $-50\ \mu$C exponentially.

$\therefore \quad q = -50 + 150\, e^{-t/\tau_C}$, Here q is in μC

$$\tau_C = CR = (10^{-6})(5 \times 10^3) = 5 \times 10^{-3}$$

$$\therefore \quad q = -50 + 150\, e^{-t/\tau_C}$$

$$V_C = \frac{q}{C} = (-50 + 150 e^{-t/\tau_C})\text{ V}$$

or $\quad V_C = 50\,(3e^{-200t} - 1)$

$$i = -\frac{dq}{dt} = \frac{150 \times 10^{-6}}{\tau_C} e^{-200t}$$

$$= \frac{150 \times 10^{-6}}{5 \times 10^{-3}} e^{-200\,t} = 30 \times 10^{-3} e^{-200\,t}$$

$V_R = iR = 150e^{-200t}$ **Ans.**

98. (a) $C = \frac{\varepsilon_0 A}{x},\ U = \frac{Q^2}{2C} = \frac{Q^2 x}{2\varepsilon_0 A}$

(b) $\frac{dU}{dx} = \frac{Q^2}{2\varepsilon_0 A}$

$\therefore \quad dU = \left(\frac{Q^2}{2\varepsilon_0 A}\right) dx$

(c) $\left(\frac{Q^2}{2\varepsilon_0 A}\right) dx = dW = Fdx$

$\therefore \quad F = \frac{Q^2}{2\varepsilon_0 A}$

(d) Because E between the plates is due to both the plates.

While $F = (Q)$ (field due to other plate) **Ans.**

99. (a) Let q be the charge on smaller sphere. Then,

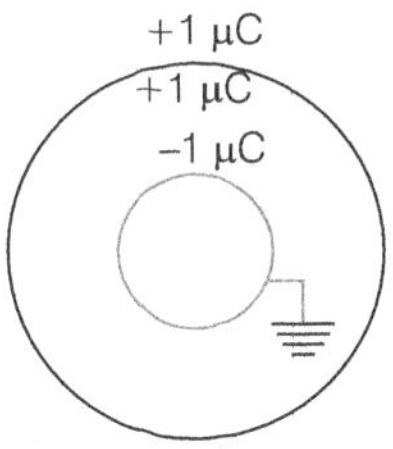

$V_{inner} = 0$

$\therefore \quad \frac{Kq}{2} + \frac{K(2)}{4} = 0$

or $\quad q = -1\,\mu C$

Now, $\quad V_{outer} = \frac{K(2-1) \times 10^{-6}}{4 \times 10^{-2}}$

$$= \frac{9 \times 10^9 \times 10^{-6}}{4 \times 10^{-2}}$$

$= 2.25 \times 10^5$ V **Ans.**

(b) Charge distribution is as shown in above figure.

100. (a) **Fig. (a)** $V_6 = \frac{90}{3} = 30\text{ V}, q_6 = 6 \times 30 = 180\,\mu C$

$V_3 = \frac{90}{3} = 30$ V

$q_3 = 30 \times 3 = 90\,\mu C$ **Ans.**

Fig. (b) Capacitor 1 μF is short-circuited. Therefore, $q_1 = 0$.

$$V_{26} = \left(\frac{20}{20 + 20 + 10}\right) \times 100 = 40 \text{ V}$$

This 40 V will distribute in inverse ratio of capacity.

$\therefore \quad V_6 = \frac{2}{8} \times 40 = 10$ V

$V_2 = \frac{6}{8} \times 40 = 30$ V

$\therefore \quad q_6 = 60\,\mu C, q_2 = 60\,\mu C$ **Ans.**

(b) **Fig. (a)** When S is open, 6 μF is short-circuited

or $\quad V_6 = 0, q_6 = 0$

and $\quad V_3 = 90\text{ V},\ q_3 = 270\,\mu C$ **Ans.**

Fig. (b) When S is open, $V_1 = 100$ V

$q_1 = 100\,\mu C$

$V_{26} = 100$ V

$\therefore \quad V_6 = \frac{2}{8} \times 100 = 25$ V

and $\quad V_2 = \frac{6}{8} \times 100 = 75$ V

$\therefore \quad q_6 = 150\,\mu C,\ q_2 = 150\,\mu C$

Further, $V_A - 0 = V_2$

$\therefore \quad V_A = V_2 = 75$ V **Ans.**

101. Current in the circuit, $i = \frac{10}{4 + 1 + 2 + 3} = 1$ A

Now, $\quad V_{5\,\mu F} = V_{1,\,2\Omega} = 3$ V

$\therefore \quad q_5 = 15\,\mu C$

Further, $\quad V_{3\mu F} = V_{2,\,3\Omega} = 5$ V

$\therefore \quad q_3 = 15\,\mu C$ **Ans.**

102. (a) At $t = 0$, when capacitor is uncharged, its equivalent resistance is zero.

$\therefore \quad R_{net} = 4 + \frac{6 \times 3}{6 + 3} = 6\,M\Omega$

or $\quad i_1 = \frac{18 \times 10^3}{6 \times 10^6} \text{ A} = 3\text{ mA}$

This will distribute in inverse ratio of resistances.

$\therefore \quad i_2 = \frac{3}{6+3} i_1 = 1$ mA

and $\quad i_3 = 2$ mA

At $t = \infty$, when capacitor is completely charged, equivalent resistance of capacitor is infinite.

$\therefore \quad i_3 = 0,$

$i_1 = i_2 = \frac{18 \times 10^3}{(4 + 6) \times 10^6} = 1.8$ mA **Ans.**

(b) At $t = 0$,

$V_2 = i_2 R_2 = (1 \times 10^{-3})(6 \times 10^6)$ V

$= 6$ kV

At $t = \infty$,

$V_2 = i_2 R_2 = (1.8 \times 10^{-3})(6 \times 10^6)$ V

$= 10.8$ kV **Ans.**

(c) To find time constant of the circuit we will have to short-circuit the battery and find resistance across capacitor. In that case, R_1 and R_2 are in parallel and they are in series with R_3.

$\therefore \quad R_{net} = 3 + \frac{4 \times 6}{4 + 6} = 5.4\,M\Omega$

$\therefore \quad \tau_C = CR_{net} = (10 \times 10^{-6})(5.4 \times 10^6)$

$= 54\,s$

$\therefore \quad V_2 = 6 + (10.8 - 6)(1 - e^{-t/\tau_C})$

$= 6 + 4.8\,(1 - e^{-t/54})$

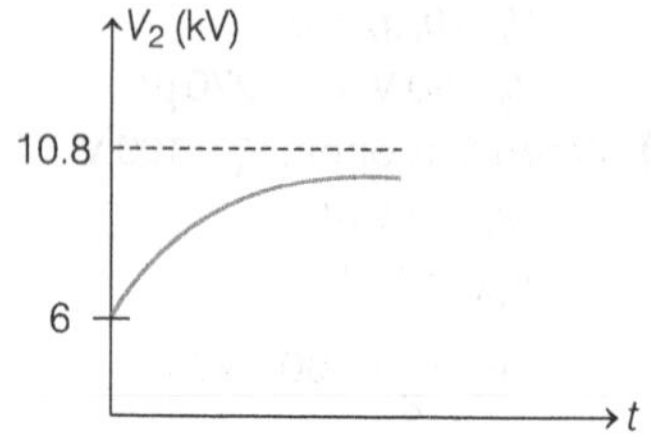

Here, V_2 is in kV and t is in second.

103. Circuit can be drawn as shown in figure.

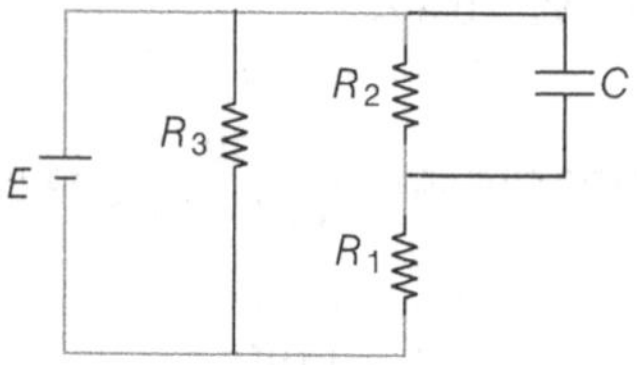

In charging of capacitor, R_3 has no role.

In steady state, potential difference across capacitor = potential difference across $R_2 = E/2$

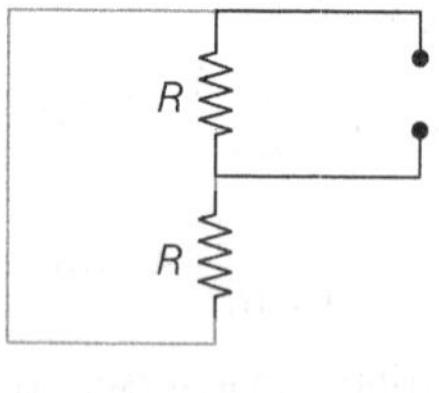

Therefore, steady state charge across capacitor

$$q_0 = \frac{CE}{2}$$

To find time constant of circuit we will have to short circuit the battery, then we will find net resistance across capacitor.

$$R_{net} = \frac{R}{2}$$

$$\Rightarrow \quad \tau_C = CR_{net} = \frac{CR}{2}$$

$\therefore$ Charge in the capacitor at time t would be

$$q = q_0(1 - e^{-t/\tau_C}) = \frac{CE}{2}\left(1 - e^{-\frac{2t}{CR}}\right)$$ **Ans.**

104. $q_2 = 20\,\mu C$

$\therefore \quad q_1 = 10\,\mu C$ (as they are in parallel)

Energy stored at this instant,

$$U = \frac{1}{2}\frac{q_1^2}{C_1} + \frac{1}{2}\frac{q_2^2}{C_2}$$

$$= \frac{1}{2} \times \frac{(10^{-5})^2}{10^{-6}} + \frac{1}{2} \times \frac{(2 \times 10^{-5})^2}{2 \times 10^{-6}}$$

$$= 1.5 \times 10^{-4}\,J$$

$$= 0.15\,mJ$$

Charge supplied by battery, $\Delta q = 30\,\mu C$

$\therefore$ Energy supplied by battery $= (\Delta q)V = 0.6\,mJ$

$\therefore$ Heat loss $= (0.6 - 0.15) = 0.45\,mJ$

Therefore, 0.45 mJ will be dissipated in the form of heat across all the resistors. In series in direct ratio of resistance ($H = i^2Rt$) and in parallel in inverse ratio of resistance.

$\therefore \quad H_2 = 0.225\,mJ, H_3 = 0.15\,mJ$

and $\quad H_6 = 0.075\,mJ$ **Ans.**

105. (a) At $t = 0$, capacitor is equivalent to a battery of emf $\frac{E}{2}$.

Net emf of the circuit $= E - E/2 = E/2$

Total resistance is R.

Therefore, current in the circuit at $t = 0$ would be

$$i = \frac{E/2}{R} = \frac{E}{2R}$$ **Ans.**

(b) Let in steady state there is total q charge on C.

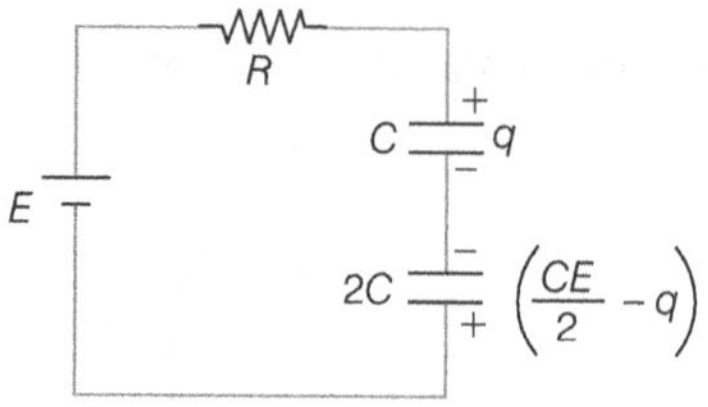

Initial charge on C was $CE/2$. Therefore, charge on $2C$ in steady state would be $\left(\frac{CE}{2} - q\right)$ with polarities as shown. This is because net charge on lower plate of C and of upper plate on $2C$ should remain constant.

Applying loop law in the circuit in steady state, we have

$$E - \frac{q}{C} + \frac{CE/2 - q}{2C} = 0$$

$$\therefore \quad q = \frac{5}{6}CE$$

Therefore, charge on C increases from $q_i = \frac{CE}{2}$ to $q_f = \frac{5CE}{6}$ exponentially.

Equivalent time constant would be

$$\tau_C = \left(\frac{C \times 2C}{C + 2C}\right)R = \frac{2}{3}CR$$

Therefore, charge as function of time would be

$$q = q_i + (q_f - q_i)(1 - e^{-t/\tau_C})$$

$$= \frac{CE}{2} + \frac{CE}{3}\left(1 - e^{-\frac{3t}{2CR}}\right)$$ **Ans.**

106. When S_1 is closed and S_2 open, capacitor will discharge. At time $t = R_1C$, one time constant, charge will remain $q_1 = \left(\frac{1}{e}\right)$ times of CV or $q_1 = \frac{CV}{e}$

When S_1 is open and S_2 closed, charge will increase (or may decrease also) from $\frac{CV}{e}$ to CE exponentially. Time constant for this would be $(R_1C + R_2C)$. Charge as function of time would be

$$q = q_i + (q_f - q_i)(1 - e^{-t/\tau_C})$$

$$q = \frac{CV}{e} + \left(CE - \frac{CV}{e}\right)(1 - e^{-t/\tau_C})$$

After total time $2R_1C + R_2C$ or $t = R_1C + R_2C$, one time constant in above equation, charge will remain

$$q = \frac{CV}{e} + \left(CE - \frac{CV}{e}\right)\left(1 - \frac{1}{e}\right)$$

$$= EC\left(1 - \frac{1}{e}\right) + \frac{VC}{e^2}$$

107. At $t = 0$, capacitor C_0 is like a battery of

$$\text{emf} = \frac{Q_0}{C_0} = 1\ \text{V}$$

Net emf of the circuit $= 4 - 1 = 3$ V

Total resistance is $R = 100\ \Omega$

$\therefore$ Initial current $= \frac{3}{100} = 0.03$ A

This current will decrease exponentially to zero.

$\therefore$ $i = 0.03\, e^{-t/\tau_C}$

Here, $\tau_C = C_{net} R = (1 \times 10^{-6})(100)$

$= 10^{-4}$ s

$\therefore$ $i = 0.03\, e^{-10^4 t}$ **Ans.**

108. From O to A

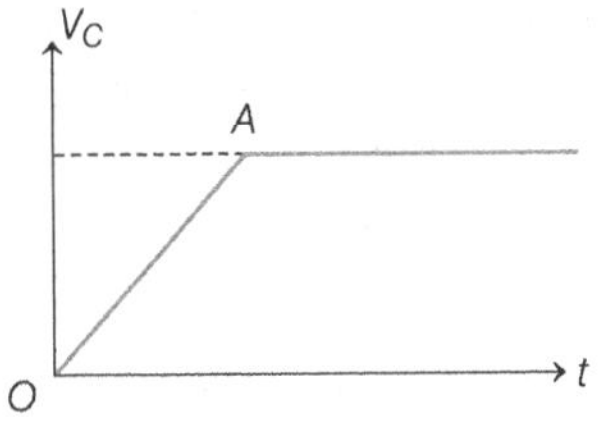

$V_C = at$ $(a = \text{constant})$

$\therefore$ $q_C = CV_C = Cat$

$\therefore$ $i = \frac{dq_C}{dt} = aC$

$V_R = iR = aCR = \text{constant}$

From *A* onwards V_C = constant

$\therefore$ $q_C = \text{constant}$

$\therefore$ $i = \frac{dq_C}{dt} = 0$ or $V_R = 0$

Therefore, V_R *versus* t graph is as shown in figure.

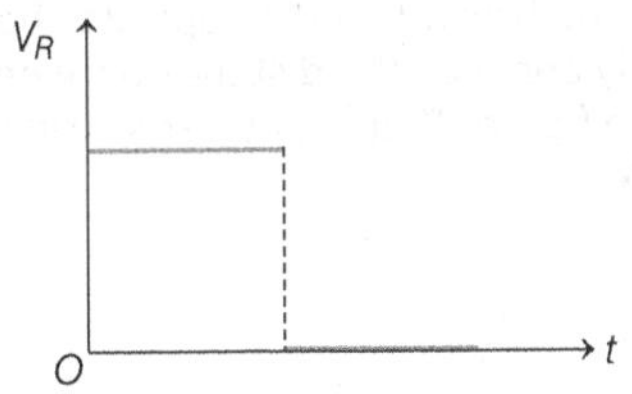

109. From **O to A** $V = at$

Here, a is a positive constant.

$\therefore$ $at = \frac{q}{C} + iR$

Differentiating w.r.t. time, we have

$$a = \frac{1}{C}\left(\frac{dq}{dt}\right) + \left(\frac{di}{dt}\right)R$$

or $\left(\frac{di}{dt}\right)R = a - \frac{i}{C}$ $\left(\text{as } i = \frac{dq}{dt}\right)$

$\therefore$ $\int_0^i \frac{di}{a - i/C} = \int_0^t \frac{dt}{R}$

$\therefore$ $i = aC(1 - e^{-t/CR})$

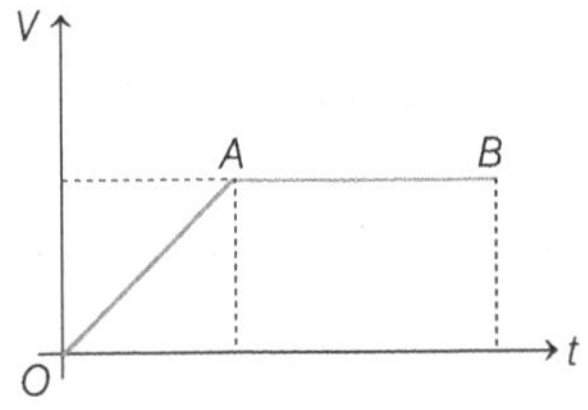

i.e. current in the circuit increases exponentially

$\therefore$ $V_{ED} = iR = aCR\,(1 - e^{-t/CR})$

or V_{ED} also increases exponentially.

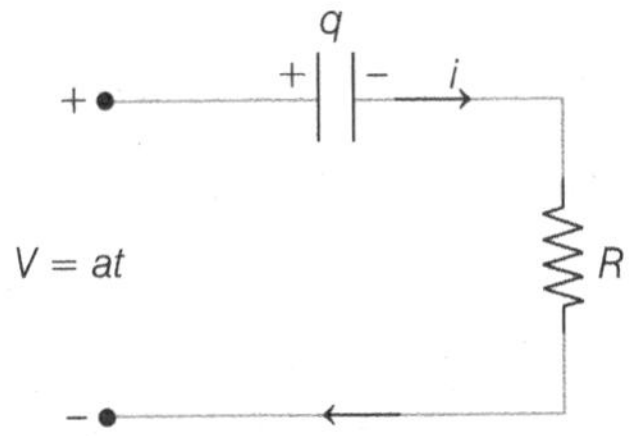

From *A* onwards When V = constant (say V_0)

$$V_0 = at \quad \text{or} \quad t = \frac{V_0}{a}$$

$\therefore$ $V_{ED} = aCR\,(1 - e^{-V_0/aCR})$

After this V_{ED} will decrease exponentially. Hence, a rough graph is as shown in figure.

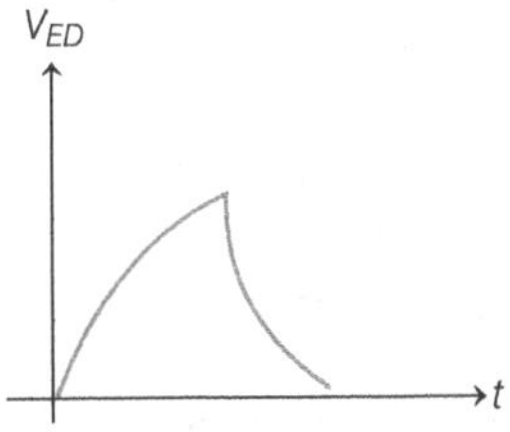

110. Let length and breadth of the capacitor be l and b respectively and d be the distance between the plates as shown in figure. Then, consider a strip at a distance x of width dx.

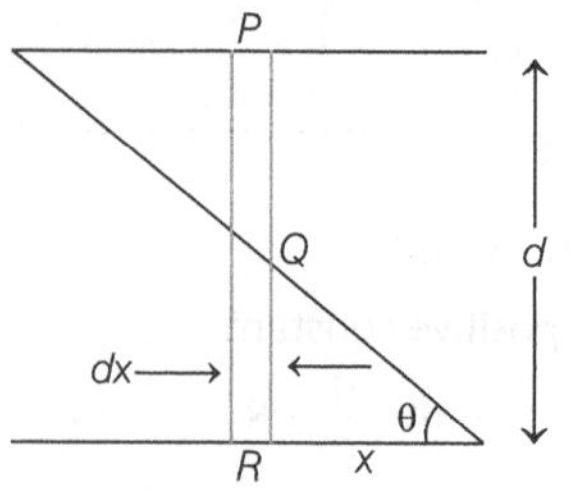

Now, $QR = x\tan\theta$

and $PQ = d - x\tan\theta$

where, $\tan\theta = d/l$

Using the formula of dielectric in capacitor,

$$C = \frac{\varepsilon_0 A}{d - t_1 - t_2 + (t_1/K_1) + (t_2/K_2)}$$

We have,

Small capacitance between P and R is

$$dC = \frac{\varepsilon_0 (bdx)}{d - PQ - PR + \frac{PQ}{K_1} + \frac{QR}{K_2}}$$

Substituting the values of PQ and QR with $\tan\theta = \frac{d}{l}$

We have,

$$dC = \frac{\varepsilon_0 A K_1 K_2}{d\{K_2 l + (K_1 - K_2)x\}}dx \qquad [bl = A]$$

Now all such small capacitors from $x = 0$ and $x = l$ will be in parallel as potential difference between them will be same.

Hence, the net capacitance of the given parallel plate capacitor is obtained by adding such infinitesimal capacitors placed parallel from $x = 0$ to $x = l$

i.e. $$C_R = \int_{x=0}^{x=l} dC$$

$$= \int_0^l \frac{\varepsilon_0 A K_1 K_2}{d\{K_2 l + (K_1 - K_2)x\}}dx$$

Finally we get $$C_R = \frac{K_1 K_2 \varepsilon_0 A}{(K_2 - K_1)d}\ln\frac{K_2}{K_1}$$

$$= \frac{CK_1K_2}{K_2 - K_1}\ln\frac{K_2}{K_1}, \text{ where } C = \frac{\varepsilon_0 A}{d}$$

111. Let a be the side of the square plate.

As shown in figure, C_1 and C_2 are in parallel. Therefore, total capacity of capacitors in the position shown is

$$C = C_1 + C_2$$

$$C = \frac{\varepsilon_0 a(a-x)}{d} + \frac{K\varepsilon_0 ax}{d}$$

$$\therefore \quad q = CV = \frac{\varepsilon_0 aV}{d}(a - x + Kx)$$

As plates are lowered in the oil, C increases or charge stored will increase.

Therefore, $$i = \frac{dq}{dt} = \frac{\varepsilon_0 aV}{d}(K-1)\cdot\frac{dx}{dt}$$

Substituting the values,

$$\varepsilon_0 = 8.85\times10^{-12}\ C^2/N\text{-}m^2$$

$$a = 1\text{ m}, V = 500\text{ volt}, d = 0.01\text{ m}, K = 11$$

and $$\frac{dx}{dt} = \text{speed of plate} = 0.001\text{ m/s}$$

We get current,

$$i = \frac{(8.85\times10^{-12})(1)(500)(11-1)(0.001)}{(0.01)}$$

$$i = 4.43\times10^{-9}\text{ A}$$

112. (a) Capacitor A is a combination of two capacitors C_K and C_O in parallel. Hence,

$$C_A = C_K + C_O = \frac{K\varepsilon_0 A}{d} + \frac{\varepsilon_0 A}{d} = (K+1)\frac{\varepsilon_0 A}{d}$$

Here, $A = 0.02\text{ m}^2$. Substituting the values, we have

$$C_A = (9+1)\frac{8.85\times10^{-12}(0.02)}{(8.85\times10^{-4})}$$

$$C_A = 2.0\times10^{-9}\text{ F}$$

Energy stored in capacitor A, when connected with a 110 V battery is

$$U_A = \frac{1}{2}C_A V^2 = \frac{1}{2}(2\times10^{-9})(110)^2$$

$$U_A = 1.21\times10^{-5}\text{ J}$$

(b) Charge stored in the capacitor

$$q_A = C_A V = (2.0\times10^{-9})(110)$$

$$\Rightarrow \quad q_A = 2.2\times10^{-7}\text{ C}$$

Now, this charge remains constant even after battery is disconnected. When the slab is removed, capacitance of A will get reduced. Let it be C_A.

$$C_A = \frac{\varepsilon_0(2A)}{d} = \frac{(8.85\times10^{-12})(0.04)}{8.85\times10^{-4}}$$

$$C_A = 0.4\times10^{-9}\text{ F}$$

Energy stored in this case would be

$$U'_A = \frac{1}{2}\frac{(q_A)^2}{C'_A} = \frac{1}{2}\frac{(2.2\times10^{-7})^2}{(0.4\times10^{-9})}$$

$$U'_A = 6.05\times10^{-5}\text{ J} > U_A$$

Therefore, work done to remove the slab would be

$$W = U'_A - U_A = (6.05 - 1.21)\times10^{-5}\text{ J}$$

or $$W = 4.84\times10^{-5}\text{ J}$$

(c) Capacity of B when filled with dielectric is

$$C_B = \frac{K\varepsilon_0 A}{d} = \frac{(9)(8.85\times10^{-12})(0.02)}{(8.85\times10^{-4})}$$

$$C_B = 1.8\times10^{-9}\text{ F}$$

These two capacitors are in parallel. Therefore, net capacitance of the system is

$$C = C'_A + C_B$$
$$= (0.4 + 1.8) \times 10^{-9} \text{ F}$$
$$C = 2.2 \times 10^{-9} \text{ F}$$

Charge stored in the system is $q = q_A$
$$= 2.2 \times 10^{-7} \text{ C}$$

Therefore, energy stored, $U = \dfrac{1}{2}\dfrac{q^2}{C}$

$$U = \frac{1}{2}\frac{(2.2 \times 10^{-7})^2}{(2.2 \times 10^{-9})}$$

or $U = 1.1 \times 10^{-5}$ J

113.

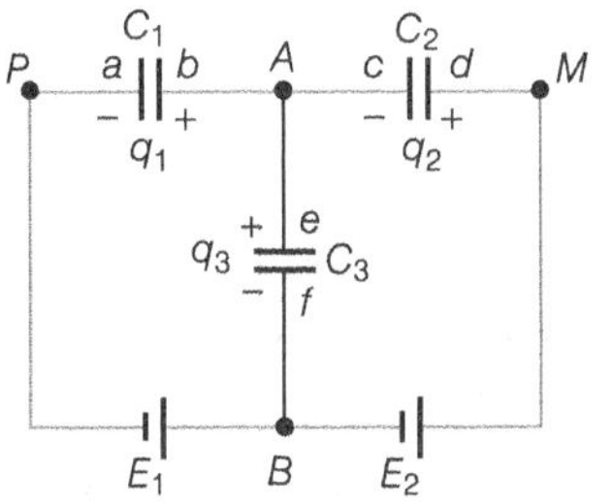

Let us assume that $V_B = 0$ and $V_A = x$. Then we have to find $V_A - V_B$ or simply x.

Since $V_B = 0$

$\therefore \quad V_P = -E_1$

and $\quad V_M = +E_2$

$$V_{C_1} = V_A - V_P = x + E_1$$

$\therefore \quad q_1 = C_1(x + E_1)$

$$V_{C_2} = V_M - V_A = E_2 - x$$

$\therefore \quad q_2 = C_2(E_2 - x)$

$$V_{C_3} = V_A - V_B = x$$

$\therefore \quad q_3 = C_3 x$

Now b, c and e plates form an isolated system. Net charge on them should be zero.

$\therefore \quad q_1 + q_3 - q_2 = 0$

or $C_1(x + E_1) + C_3 x - C_2(E_2 - x) = 0$

$$\therefore \quad x = \frac{C_2E_2 - C_1E_1}{C_1 + C_2 + C_3}$$

$$\therefore \quad V_A - V_B = \frac{C_2E_2 - C_1E_1}{C_1 + C_2 + C_3}$$

114.

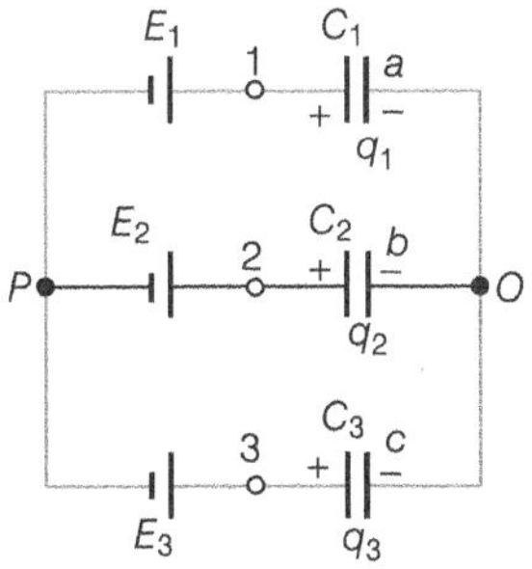

Given, $V_O = 0$

Let $\quad V_P = x$

Then, $V_1 = x + E_1, V_2 = x + E_2$

and $\quad V_3 = x + E_3$

Since $V_O = 0$

$\therefore \quad V_{C_1} = V_1 = x + E_1$

$\therefore \quad q_1 = C_1V_1 = C_1(x + E_1)$

Similarly, $\quad q_2 = C_2(x + E_2)$

and $\quad q_3 = C_3(x + E_3)$

Now, a, b and c plates make an isolated system. So, their combined net charge will be zero.

or $\quad -q_1 - q_2 - q_3 = 0$

or $\quad q_1 + q_2 + q_3 = 0$

$\Rightarrow \quad C_1(x + E_1) + C_2(x + E_2) + C_3(x + E_3) = 0$

or $$x = -\left[\frac{C_1E_1 + C_2E_2 + C_3E_3}{C_1 + C_2 + C_3}\right]$$

$\therefore \quad V_1 = x + E_1$

Putting the value of x, we get

$$V_1 = \frac{E_1(C_2 + C_3) - E_2C_2 - E_3C_3}{C_1 + C_2 + C_3}$$

Similarly, we can write the expressions of V_2 and V_3,

$$V_2 = \frac{E_2(C_1 + C_3) - E_1C_1 - E_3C_3}{C_1 + C_2 + C_3}$$

and $$V_3 = \frac{E_3(C_1 + C_2) - E_1C_1 - E_2C_2}{C_1 + C_2 + C_3}$$

115. **Switch in position-1**

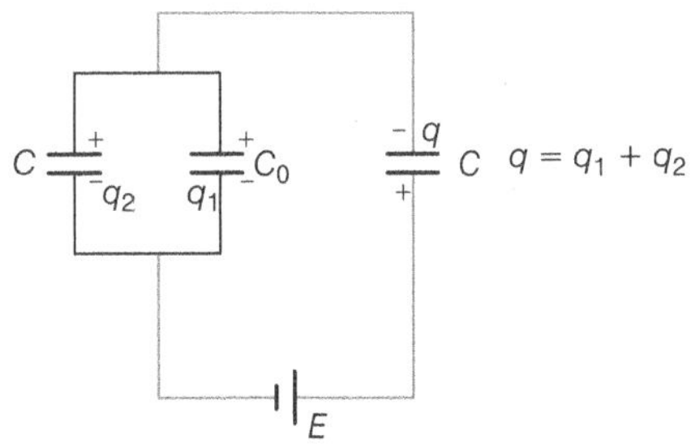

Net capacitance, $C_{net} = \dfrac{C(C + C_0)}{2C + C_0}$

Net charge, $q = C_{net}\, E = \dfrac{EC(C + C_0)}{2C + C_0}$

Switch in position-2

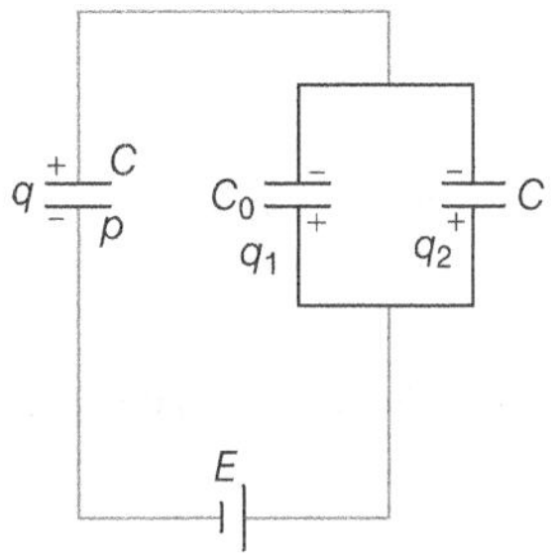

From negative terminal of the battery charge will transfer to plate p only. So, charge flow from negative terminal of the battery,

$$\Delta q = (q_p)_f - (q_p)_i$$
$$= -q - (-q_2)$$
$$= -q_1 \qquad [q_1 + q_2 = q]$$

or, we can say charge flow from the battery in normal direction is Δq or q_1. Now q will distribute among C and C_0 in direct ratio of capacitance. Hence,

$$q_1 = \left(\frac{C_0}{C + C_0}\right) q$$

Substituting the value of q, we get

$$q_1 = \frac{ECC_0}{(2C + C_0)} = \Delta q \quad \text{(also)}$$

So, energy supplied by the battery is

$$W = (\Delta q)E = \frac{E^2CC_0}{(2C + C_0)}$$

Both circuits are symmetrical. Therefore stored electrostatic potential energy in all capacitors will be same in both cases. So, energy supplied by the battery is completely lost.

$$\therefore \quad \text{Energy lost } Q = \left(\frac{E^2CC_0}{2C + C_0}\right)$$

116. Between inner and outer faces, we will have two capacitors in parallel. One is air capacitor and the other filled with water. Both capacitors are in parallel. So, the net capacity will be,

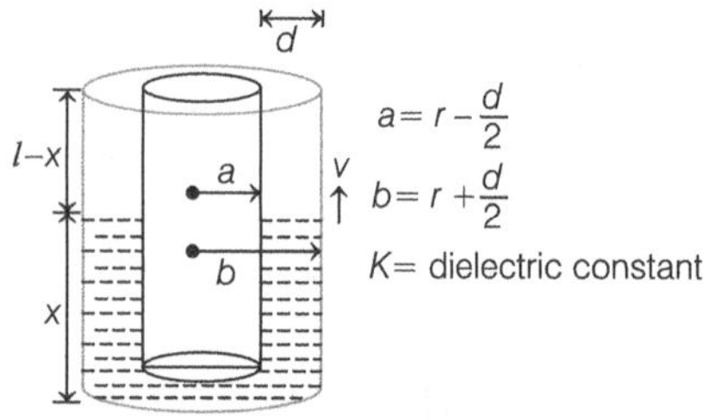

$$C = C_{air} + C_{water}$$
$$= \frac{(2\pi\varepsilon_0)(l - x)}{\ln\left[\frac{r + d/2}{r - d/2}\right]} + \frac{(2\pi K\varepsilon_0)x}{\ln\left[\frac{r + d/2}{r - d/2}\right]}$$

$$C = \frac{2\pi\varepsilon_0}{\ln\left[\frac{r + d/2}{r - d/2}\right]}[l + (K - 1)x] \qquad \text{...(i)}$$

$$\frac{r + \frac{d}{2}}{r - \frac{d}{2}} = \left(1 + \frac{d}{2r}\right)\left(1 - \frac{d}{2r}\right)^{-1}$$

Since $d << r$, so applying Binomial expansion, we have

$$\frac{r + \frac{d}{2}}{r - \frac{d}{2}} \approx \left(1 + \frac{d}{2r}\right)\left(1 - \frac{d}{2r}\right)$$

Again ignoring the expressions of $\frac{d^2}{4r^2}$, we have

$$\frac{r + \frac{d}{2}}{r - \frac{d}{2}} \approx \left(1 + \frac{d}{r}\right)$$

Further, $$\ln\left[\frac{r + \frac{d}{2}}{r - \frac{d}{2}}\right] = \ln\left(1 + \frac{d}{r}\right)$$

Since $\frac{d}{r} << 1$

$$\therefore \qquad \ln\left(1 + \frac{d}{r}\right) \approx \frac{d}{r}$$

Substituting this value in Eq. (i), we get

$$C = \frac{2\pi\varepsilon_0 r}{d}[l + (K - 1)x]$$

$$q = CV = \frac{2\pi\varepsilon_0 rV}{d}[l + (K - 1)x]$$

Here x is increasing, so q will be increasing. Hence a current will flow from the lead wires given by,

$$I = \frac{dq}{dt} = \frac{2\pi\varepsilon_0 rV(K - 1)}{d} \cdot \frac{dx}{dt}$$

Putting $\frac{dx}{dt} = v$, we get

$$I = \frac{2\pi\varepsilon_0 rV(K - 1)v}{d}$$

Taking K of water $= 81$ and substituting the other given values we finally get,

$$I = 0.11\,\mu A$$

Note (i) *Capacitance of a cylindrical capacitor per unit length (with air between the plates) is* $\frac{2\pi\varepsilon_0}{\ln(b/a)}$.

Here length of air capacitor is $(l - x)$ and the capacitor filled with water is x.

(ii) $\ln(1 + x) = x - \frac{x^2}{2} + \frac{x^3}{3} - \dots$

For $x << 1$, we can neglect the higher power terms of x. So,

$$\ln(1 + x) \approx x$$

117. At $t = 0$, $q_i = CV$

At $t = \infty$, $q_f = \left(\frac{C}{\eta}\right)V$

$$q_f < q_i$$

Since it is a C-R circuit. So charge will decrease exponentially.

q *versus* t graph is as shown in figure.

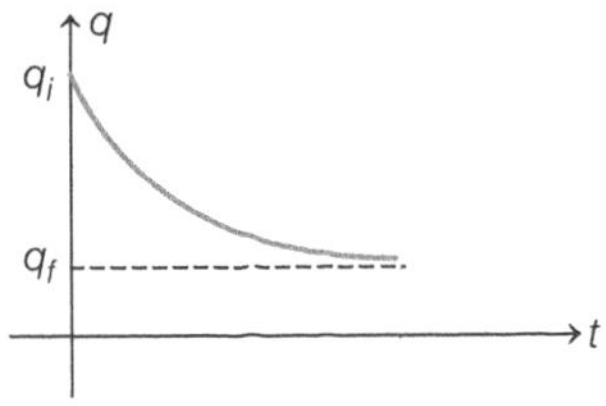

q-t equation of this graph is as under,

$$q = q_f + (q_i - q_f)e^{-t/\tau_C}$$

Now, $$I = -\frac{dq}{dt} = \left(\frac{q_i - q_f}{\tau_C}\right)e^{-t/\tau_C}$$

Here, $$\tau_C = \left(\frac{C}{\eta}\right)R$$

Substituting the values, we get

$$I = \frac{\left[CV - \frac{C}{\eta}V\right]}{\left(C\frac{R}{\eta}\right)} e^{\frac{-\eta t}{CR}}$$

or $$I = (\eta - 1)\frac{V}{R}e^{\frac{-\eta t}{CR}}$$

118. Acceleration,

$$a = \frac{qE}{m} = \frac{q(V/l)}{m} = \left(\frac{q\alpha}{ml}\right)t$$

or $$\frac{dv}{dt} = \left(\frac{q\alpha}{ml}\right)t$$

$$\Rightarrow \int_0^v dv = \left(\frac{q\alpha}{ml}\right)\int_0^t t\,dt$$

$$\therefore \quad v = \frac{q\alpha t^2}{2ml} = \frac{dx}{dt}$$

$$\therefore \quad \int_0^l dx = \frac{q\alpha}{2ml}\int_0^t t^2 dt$$

Solving this equation, we get

$$t = \left(\frac{6ml^2}{q\alpha}\right)^{1/3} \quad \text{...(i)}$$

Now, $$v = \frac{q\alpha t^2}{2ml}$$

Substituting value of t from Eq. (i), we can find v. After substituting the given values and values of q and m for an electron, we get

$$v \approx 16\,\text{km/s}$$

119.

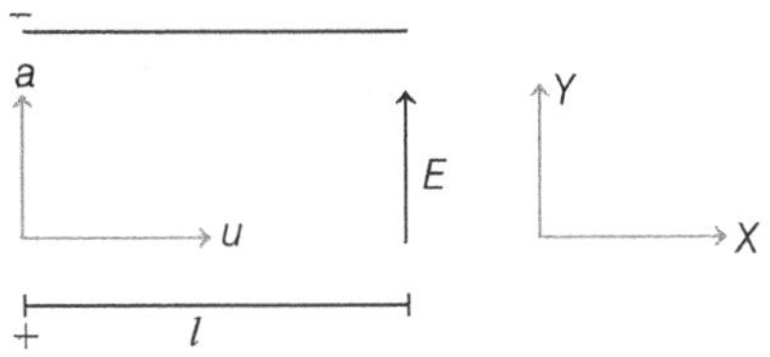

$$u_x = u, a_x = 0, u_y = 0, a_y = \frac{qE}{m} = \frac{e\alpha t}{m}$$

Time taken by the proton in reaching from one end to another end,

$$t_0 = \frac{l}{u} \quad \text{...(i)}$$

Further, $$a_y = \frac{e\alpha t}{m} = \frac{dv_y}{dt}$$

$$= \int_0^{v_y} dv_y = \int_0^{t_0} \frac{e\alpha t}{m}dt$$

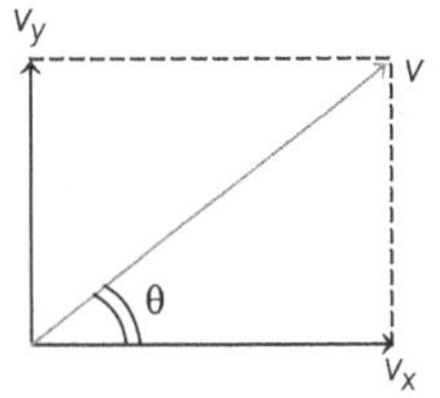

Integrating this equation and then substituting the value of t_0 from Eq. (i), we get

$$v_y = \frac{e\alpha l^2}{2mu^2}$$

So, at time t_0

$$\tan\theta = \frac{v_y}{v_x} \quad \text{...(ii)}$$

Here, $$v_y = \frac{e\alpha l^2}{2mu^2}$$

and $$v_x = u_x = u = \sqrt{\frac{2eV}{m}}$$

Substituting in Eq. (ii), we get

$$\tan\theta = \frac{\alpha l^2}{4}\sqrt{\frac{m}{2eV^3}}$$

UNIT–4

MAGNETICS

Learning Modules

1. Magnetic Force on a Moving Charge
2. Path of a Charged Particle in Uniform Magnetic Field
3. Magnetic Force on a Current Carrying Conductor
4. Magnetic Dipole
5. Biot Savart Law
6. Ampere's Circuital Law
7. Force Between Parallel Current Carrying Wires
8. Magnetic Poles and Bar Magnets
9. Earth's Magnetism
10. Magnetic Materials
11. Moving Coil Galvanometer

MODULE 1

Magnetic Force on a Moving Charge

Introduction

The fascinating attractive properties of magnets have been known since ancient times. The word magnet comes from ancient Greek place name Magnesia (the modern town Manisa in Western Turkey), where the natural magnets called lodestones were found. The fundamental nature of magnetism is the interaction of moving electric charges. Unlike electric forces which act on electric charges whether they are moving or not, magnetic forces act only on moving charges and current carrying wires.

We will describe magnetic forces using the concept of a field. A magnetic field is established by a permanent magnet, by an electric current or by other moving charges. This magnetic field, in turn, exerts forces on other moving charges and current carrying conductors. In this chapter, first we study the magnetic forces and torques exerted on moving charges and currents by magnetic fields, then we will see how to calculate the magnetic fields produced by currents and moving charges.

Magnetic Force on a Moving Charge ($\mathbf{F}_m$)

The magnetic force ($\mathbf{F}_m$) on a charge q moving with velocity $\mathbf{v}$ in a magnetic field $\mathbf{B}$ is given, both in magnitude and direction, by

$$\mathbf{F}_m = q(\mathbf{v} \times \mathbf{B}) \quad \text{...(i)}$$

Following points are worthnoting regarding the above expression.

(i) The magnitude of $\mathbf{F}_m$ is

$$F_m = Bqv \sin\theta$$

where, θ is the angle between $\mathbf{v}$ and $\mathbf{B}$.

(ii) $\mathbf{F}_m$ is zero when,

(a) $B = 0$, i.e. no magnetic field is present.

(b) $q = 0$, i.e. particle is neutral.

(c) $v = 0$, i.e. charged particle is at rest or

(d) $\theta = 0°$ or $180°$, i.e. $\mathbf{v} \uparrow\uparrow \mathbf{B}$ or $\mathbf{v} \uparrow\downarrow \mathbf{B}$

(iii) $\mathbf{F}_m$ is maximum at $\theta = 90°$ and this maximum value is Bqv.

(iv) The units of $\mathbf{B}$ must be the same as the units of F/qv. Therefore, the SI unit of B is equivalent to $\dfrac{\text{N-s}}{\text{C-m}}$. This unit is called the **tesla** (abbreviated as T), in honour of Nikola Tesla, the prominent Serbian-American scientist and inventor.

Thus,

$$1 \text{ tesla} = 1\text{T} = \frac{1\text{N-s}}{\text{C-m}} = \frac{1\text{ N}}{\text{A-m}}$$

The CGS unit of $\mathbf{B}$, the **gauss** ($1\text{G} = 10^{-4}$ T) is also in common use.

(v) In equation number (i) q is to be substituted with sign. If q is positive, magnetic force is along $\mathbf{v} \times \mathbf{B}$ and if q is negative, magnetic force is in a direction opposite to $\mathbf{v} \times \mathbf{B}$.

(vi) **Direction of $\mathbf{F}_m$** From the property of cross product we can infer that $\mathbf{F}_m$ is perpendicular to both $\mathbf{v}$ and $\mathbf{B}$ or it is perpendicular to the plane formed by $\mathbf{v}$ and $\mathbf{B}$. The exact direction of $\mathbf{F}_m$ can be given by any of the following methods:

(a) Direction of $\mathbf{F}_m$ = (sign of q) (direction of $\mathbf{v} \times \mathbf{B}$) or, as we have stated earlier also,

$\mathbf{F}_m \uparrow\uparrow \mathbf{v} \times \mathbf{B}$ if q is positive and

$\mathbf{F}_m \uparrow\downarrow \mathbf{v} \times \mathbf{B}$ if q is negative.

(b) Fleming's left hand rule According to this rule, the forefinger, the central finger and the thumb of the left hand are stretched in such a way that they are mutually perpendicular to each other. If the central finger shows the direction of velocity of positive charge ($\mathbf{v}_{+q}$) and forefinger shows the direction of magnetic field ($\mathbf{B}$), then the thumb will give the direction of magnetic force ($\mathbf{F}_m$). If instead of positive charge we have the negative charge, then $\mathbf{F}_m$ is in opposite direction.

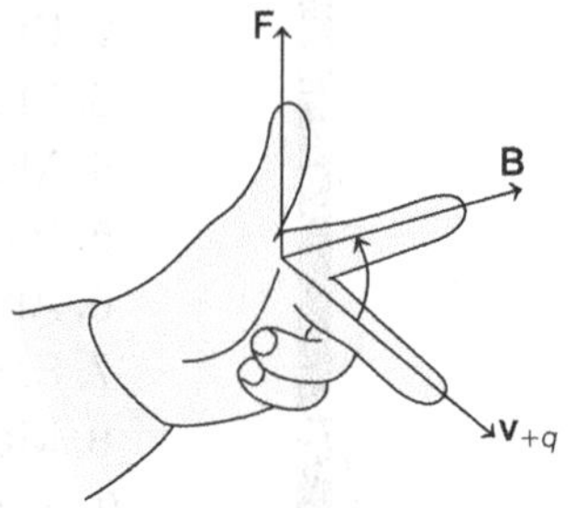

(c) Right hand rule Wrap the fingers of your right hand around the line perpendicular to the plane of $\mathbf{v}$ and $\mathbf{B}$ as shown in figure, so that they curl around with the sense of rotation from $\mathbf{v}$ to $\mathbf{B}$ through the smaller angle between them. Your thumb then points in the direction of the force $\mathbf{F}_m$ on a positive charge. (Alternatively, the direction of the force $\mathbf{F}_m$ on a positive charge is the direction in which a right hand thread screw would advance if turned the same way).

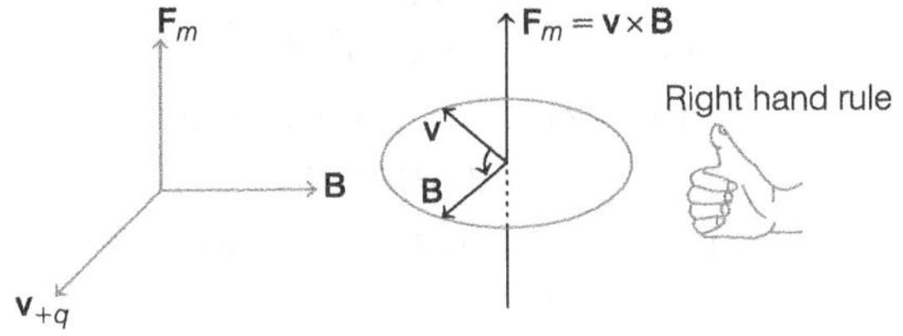

(vii) $\mathbf{F}_m \perp \mathbf{v}$ or $\mathbf{F}_m \perp \frac{d\mathbf{s}}{dt}$. Therefore, $\mathbf{F}_m \perp d\mathbf{s}$ or the work done by the magnetic force in a static magnetic field is zero.

$$W_{F_m} = 0$$

So, from work energy theorem KE and hence the speed of the charged particle remains constant in magnetic field. The magnetic force can change the direction only. It cannot increase or decrease the speed or kinetic energy of the particle.

Note *By convention the direction of magnetic field* **B** *perpendicular to the paper going inwards is shown by ⊗ and the direction perpendicular to the paper coming out is shown by ⊙.*

Direction of **B** perpendicular to paper inwards

Direction of **B** perpendicular to paper outwards

Note *Sometimes we will write only* **F** *in place of* F_m.

Example 1 A charged particle is projected in a magnetic field

$$\mathbf{B} = (3\hat{i} + 4\hat{j}) \times 10^{-2}\,\text{T}$$

The acceleration of the particle is found to be

$$\mathbf{a} = (x\hat{i} + 2\hat{j})\,\text{m/s}^2$$

Find the value of x.

Sol. As we have read, $\mathbf{F}_m \perp \mathbf{B}$

i.e. the acceleration $\mathbf{a} \perp \mathbf{B}$ or $\mathbf{a} \cdot \mathbf{B} = 0$

or $(x\hat{i} + 2\hat{j}) \cdot (3\hat{i} + 4\hat{j}) \times 10^{-2} = 0$

or $(3x + 8) \times 10^{-2} = 0$

$\therefore$ $x = -\frac{8}{3}\,\text{m/s}^2$ **Ans.**

Example 2 In a magnetic field, a proton has a velocity $\mathbf{v} = (2\hat{i} + 3\hat{j}) \times 10^6\,\text{m/s}$, it experiences a force $\mathbf{F} = -(1.28 \times 10^{-13}\,\hat{k})\,\text{N}$. When its velocity is along the z-axis, it experiences a force along the x-axis. What is the magnetic field?

HOW TO PROCEED In the second part of the question, it is given that magnetic force is along x-axis when velocity is along z-axis. Hence, magnetic field should be along negative y-direction. As in case of positive charge (here proton)

$$\boxed{\mathbf{F}_m \uparrow\uparrow \mathbf{v} \times \mathbf{B}}$$

So, let $\mathbf{B} = -B_0\hat{j}$

where, B_0 = positive constant.

Now, applying $\mathbf{F}_m = q(\mathbf{v} \times \mathbf{B})$ we can find value of B_0 from the first part of the question.

Sol. Substituting proper values in, $\mathbf{F}_m = q(\mathbf{v} \times \mathbf{B})$

We have,

$$-(1.28 \times 10^{-13}\,\hat{k}) = (1.6 \times 10^{-19})\,[(2\hat{i} + 3\hat{j}) \times (-B_0\hat{j})] \times 10^6$$

$\therefore$ $1.28 = 1.6 \times 2 \times B_0$

or $B_0 = \frac{1.28}{3.2} = 0.4$

Therefore, the magnetic field is

$\mathbf{B} = (-0.4\,\hat{j})\,\text{T}$ **Ans.**

Example 3 A magnetic field of $(4.0 \times 10^{-3}\,\hat{k})\,\text{T}$ exerts a force $(4.0\,\hat{i} + 3.0\,\hat{j}) \times 10^{-10}\,\text{N}$ on a particle having a charge $10^{-9}\,\text{C}$ and moving in the x-y plane. Find the velocity of the particle.

Sol. Given, $\mathbf{B} = (4 \times 10^{-3}\,\hat{k})\,\text{T}$, $q = 10^{-9}\,\text{C}$

and magnetic force $\mathbf{F}_m = (4.0\,\hat{i} + 3.0\,\hat{j}) \times 10^{-10}\,\text{N}$

Let velocity of the particle in x-y plane be

$$\mathbf{v} = v_x\hat{i} + v_y\hat{j}$$

Then, from the relation

$$\mathbf{F}_m = q(\mathbf{v} \times \mathbf{B})$$

We have

$$(4.0\,\hat{i} + 3.0\,\hat{j}) \times 10^{-10} = 10^{-9}[(v_x\hat{i} + v_y\hat{j}) \times (4 \times 10^{-3}\hat{k})]$$
$$= (4v_y \times 10^{-12}\,\hat{i} - 4v_x \times 10^{-12}\,\hat{j})$$

Comparing the coefficients of $\hat{i}$ and $\hat{j}$, we have

$$4 \times 10^{-10} = 4v_y \times 10^{-12}$$

$\therefore$ $v_y = 10^2\,\text{m/s} = 100\,\text{m/s}$

and $3.0 \times 10^{-10} = -4v_x \times 10^{-12}$

$\therefore$ $v_x = -75\,\text{m/s}$

$\therefore$ $\mathbf{v} = (-75\hat{i} + 100\hat{j})\,\text{m/s}$ **Ans.**

Example 4 A charged particle carrying charge $q = 1\,\mu\text{C}$ moves in uniform magnetic field with velocity $v_1 = 10^6\,\text{m/s}$ at angle 45° with x-axis in the positive xy-plane and experiences a force $F_1 = 5\sqrt{2}\,\text{mN}$ along the negative z-axis. When the same particle moves with velocity $v_2 = 10^6\,\text{m/s}$ along the z-axis, it experiences a force F_2 in y-direction. Find

(a) the magnitude and direction of the magnetic field

(b) the magnitude of the force F_2

Sol. F_2 is in y-direction when velocity is along z-axis.

Therefore, magnetic field should be along x-axis. So let,

$$\mathbf{B} = B_0\hat{i}$$

(a) Given, $\mathbf{v}_1 = \frac{10^6}{\sqrt{2}}\hat{\mathbf{i}} + \frac{10^6}{\sqrt{2}}\hat{\mathbf{j}}$

and $\mathbf{F}_1 = -5\sqrt{2} \times 10^{-3}\hat{\mathbf{k}}$

From the equation, $\mathbf{F} = q(\mathbf{v} \times \mathbf{B})$

We have

$$(-5\sqrt{2} \times 10^{-3})\hat{\mathbf{k}} = (10^{-6})\left[\left(\frac{10^6}{\sqrt{2}}\hat{\mathbf{i}} + \frac{10^6}{\sqrt{2}}\hat{\mathbf{j}}\right) \times (B_0\hat{\mathbf{i}})\right]$$

$$= -\frac{B_0}{\sqrt{2}}\hat{\mathbf{k}}$$

$$\therefore \quad \frac{B_0}{\sqrt{2}} = 5\sqrt{2} \times 10^{-3}$$

or $B_0 = 10^{-2}$ T

Therefore, the magnetic field is

$$\mathbf{B} = (10^{-2}\hat{\mathbf{i}})\text{ T}$$ **Ans.**

(b) $F_2 = B_0 q v_2 \sin 90°$

As the angle between **B** and **v** in this case is 90°.

$$\therefore \quad F_2 = (10^{-2})(10^{-6})(10^6)$$

$$= 10^{-2}\text{ N}$$ **Ans.**

MODULE 1 Exercise

Assertion and Reason

See directions on page 25.

1. **Assertion :** A charged particle enters in a magnetic field $\mathbf{B} = B_0\hat{\mathbf{i}}$ with velocity $\mathbf{v} = v_0\hat{\mathbf{i}} + v_0\hat{\mathbf{j}}$, then minimum speed of charged particle may be v_0.

   Reason : A variable acceleration will act on the charged particle.

Single Correct Option

2. If the acceleration and velocity of a charged particle moving in a constant magnetic region is given by $\mathbf{a} = a_1\hat{\mathbf{i}} + a_2\hat{\mathbf{k}}$, $\mathbf{v} = b_1\hat{\mathbf{i}} + b_2\hat{\mathbf{k}}$. [$a_1, a_2, b_1$ and b_2 are constants], then choose the **wrong** statement.
   (a) Magnetic field may be along y-axis
   (b) $a_1b_1 + a_2b_2 = 0$
   (c) Magnetic field is along x-axis
   (d) Kinetic energy of particle is always constant

3. A charge q is moving with a velocity $\mathbf{v}_1 = 1\hat{\mathbf{i}}$ m/s at a point in a magnetic field and experiences a force $\mathbf{F} = q[-\hat{\mathbf{j}} + 1\hat{\mathbf{k}}]$ N. If the charge is moving with a velocity $\mathbf{v}_2 = 1\hat{\mathbf{j}}$ m/s at the same point, it experiences a force $\mathbf{F}_2 = q(1\hat{\mathbf{i}} - 1\hat{\mathbf{k}})$ N. The magnetic induction **B** at that point is
   (a) $(\hat{\mathbf{i}} + \hat{\mathbf{j}} + \hat{\mathbf{k}})$ Wb/m^2
   (b) $(\hat{\mathbf{i}} - \hat{\mathbf{j}} + \hat{\mathbf{k}})$ Wb/m^2
   (c) $(-\hat{\mathbf{i}} + \hat{\mathbf{j}} - \hat{\mathbf{k}})$ Wb/m^2
   (d) $(\hat{\mathbf{i}} + \hat{\mathbf{j}} - \hat{\mathbf{k}})$ Wb/m^2

Subjective Questions

4. Can a charged particle be accelerated by a magnetic field. Can its speed be increased?
5. Write the dimensions of E/B. Here, E is the electric field and B the magnetic field.
6. In the relation $\mathbf{F} = q(\mathbf{v} \times \mathbf{B})$, which pairs are always perpendicular to each other.
7. If a beam of electrons travels in a straight line in a certain region. Can we say there is no magnetic field?
8. An electron beam projected along positive x-axis deflects along the positive y-axis. If this deflection is caused by a magnetic field, what is the direction of the field?
9. A charge $q = -4\ \mu$C has an instantaneous velocity $\mathbf{v} = (2\hat{\mathbf{i}} - 3\hat{\mathbf{j}} + \hat{\mathbf{k}}) \times 10^6$ m/s in a uniform magnetic field $\mathbf{B} = (2\hat{\mathbf{i}} + 5\hat{\mathbf{j}} - 3\hat{\mathbf{k}}) \times 10^{-2}$ T. What is the force on the charge?
10. A particle initially moving towards south in a vertically downward magnetic field is deflected toward the east. What is the sign of the charge on the particle?
11. An electron experiences a magnetic force of magnitude 4.60×10^{-15} N, when moving at an angle of 60° with respect to a magnetic field of magnitude 3.50×10^{-3} T. Find the speed of the electron.
12. He^{2+} ion travels at right angles to a magnetic field of 0.80 T with a velocity of 10^5 m/s. Find the magnitude of the magnetic force on the ion.

13. An electron moves through a uniform magnetic field given by $\mathbf{B} = B_x\hat{\mathbf{i}} + (3B_x)\,\hat{\mathbf{j}}$. At a particular instant, the electron has the velocity $\mathbf{v} = (2.0\hat{\mathbf{i}} + 4.0\hat{\mathbf{j}})$ m/s and the magnetic force acting on it is $(6.4 \times 10^{-19}\ \text{N})\hat{\mathbf{k}}$. Find B_x.

14. A particle with charge 7.80 μC is moving with velocity $\mathbf{v} = -(3.80 \times 10^3\ \text{m/s})\hat{\mathbf{j}}$. The magnetic force on the particle is measured to be $\mathbf{F} = +(7.60 \times 10^{-3}\ \text{N})\,\hat{\mathbf{i}} - (5.20 \times 10^{-3}\ \text{N})\,\hat{\mathbf{k}}$.
(a) Calculate the components of the magnetic field you can find from this information.
(b) Which component(s) of the magnetic field cannot be determined by the measurement of the force? Explain.
(c) Calculate the scalar product $\mathbf{B} \cdot \mathbf{F}$. What is the angle between **B** and **F**?

15. Each of the lettered points at the corners of the cube in figure represents a positive charge q moving with a velocity of magnitude v in the direction indicated. The region in the figure is in a uniform magnetic field **B**, parallel to the x-axis. Find the magnitude and direction of the force on each charge.

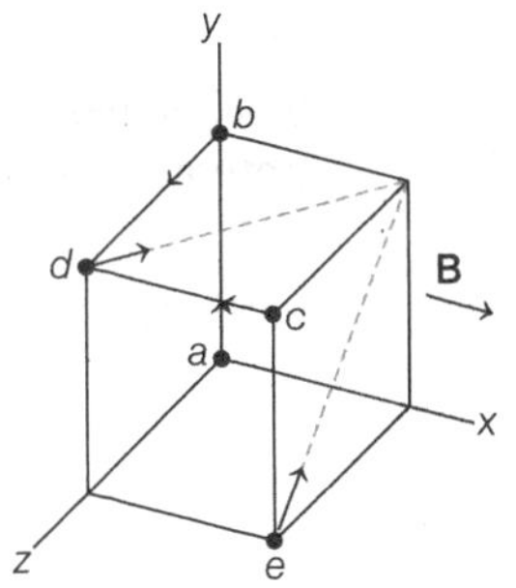

Answers

1. (d) 2. (c) 3. (a) 4. Yes, No 5. $[LT^{-1}]$ 6. (**F**, **v**), (**F**, **B**) 7. No
8. Along positive z-direction 9. $(-0.16\ \hat{\mathbf{i}} - 0.32\ \hat{\mathbf{j}} - 0.64\ \hat{\mathbf{k}})$ N 10. Positive 11. 9.47×10^6 m/s
12. 2.56×10^{-14} N 13. $B_x = -2.0$ T 14. (a) $B_x = (-0.175)$ T, $B_z = (-0.256)$ T (b) B_y (c) Zero, 90°
15. (a) $-qvB\hat{\mathbf{k}}$ (b) $+qvB\,\hat{\mathbf{j}}$ (c) Zero (d) $\dfrac{-qvB}{\sqrt{2}}\hat{\mathbf{j}}$ (e) $\left(-\dfrac{qvB}{\sqrt{2}}\right)(\hat{\mathbf{j}} + \hat{\mathbf{k}})$

See hints on page 434.

MODULE 2

Path of a Charged Particle in Uniform Magnetic Field

The path of a charged particle in uniform magnetic field depends on the angle θ (the angle between **v** and **B**). Depending on the different values of θ, the following three cases are possible.

Case 1 When θ is 0° or 180°

As we have seen previously, $\mathbf{F}_m = 0$, when θ is either 0° or 180°. Hence, path of the charged particle is a straight line (undeviated) when it enters parallel or antiparallel to magnetic field.

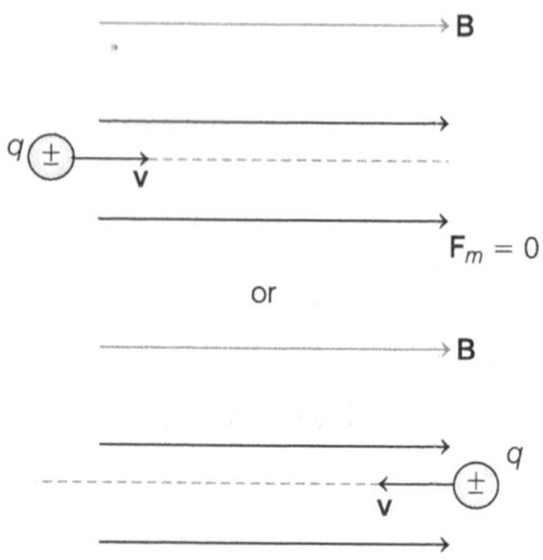

Case 2 When $\theta = 90°$

When $\theta = 90°$, the magnetic force is $F_m = Bqv \sin 90° = Bqv$. This magnetic force is perpendicular to the velocity at every instant. Hence, path is a **circle**. The necessary centripetal force is provided by the magnetic force. If r be the radius of the circle, then

$$\frac{mv^2}{r} = Bqv$$

or

$$r = \frac{mv}{Bq}$$

This expression of r can be written in the following different ways

$$\boxed{r = \frac{mv}{Bq} = \frac{p}{Bq} = \frac{\sqrt{2Km}}{Bq} = \frac{\sqrt{2qVm}}{Bq}}$$

Here, p = momentum of particle

$$K = \text{KE of particle} = \frac{p^2}{2m}$$

or

$$p = \sqrt{2Km}$$

We also know that if the charged particle is accelerated by a potential difference of V volts, it acquires a KE given by

$$K = qV$$

Further, time period of the circular path will be

$$T = \frac{2\pi r}{v} = \frac{2\pi\left(\frac{mv}{Bq}\right)}{v} = \frac{2\pi m}{Bq}$$

or

$$\boxed{T = \frac{2\pi m}{Bq}}$$

or The angular speed (ω) of the particle is $\omega = \frac{2\pi}{T} = \frac{Bq}{m}$

$$\therefore \quad \boxed{\omega = \frac{Bq}{m}}$$

Frequency of rotation is $f = \frac{1}{T}$

or

$$\boxed{f = \frac{Bq}{2\pi m}}$$

The following points are worthnoting regarding a circular path:

(i) The plane of the circle is perpendicular to magnetic field. If the magnetic field is along z-direction, the circular path is in x-y plane. The speed of the particle does not change in magnetic field.
Hence, if v_0 be the speed of the particle, then velocity of particle at any instant of time will be

$$\mathbf{v} = v_x\hat{\mathbf{i}} + v_y\hat{\mathbf{j}}$$

where, $\quad v_x^2 + v_y^2 = v_0^2$

(ii) T, f and ω are independent of v while the radius is directly proportional to v.

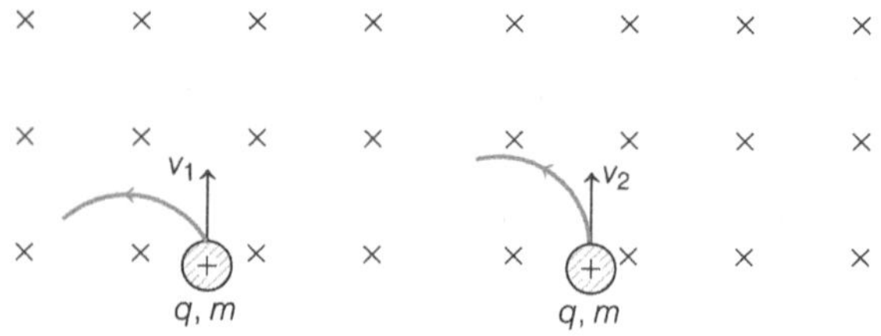

If two charged particles of equal mass and charge enter in a magnetic field **B** with different speeds v_1 and $v_2\ (> v_1)$ at right angles, then

$$T_1 = T_2$$

but $$r_2 > r_1$$

as shown in figure.

(iv) If $\theta = 90°$, between **v** and **B**, then we have seen that path is a circle. But circle is never completed if the charged particle enters perpendicularly in uniform magnetic field from outside (see examples 1, 4 and 5 in typed problems).

Circle is completed if the magnetic field is present all around (see example 2 in typed problems).

Note *Charge per unit mass $\frac{q}{m}$ is known as specific charge. It is sometimes denoted by α. So, in terms of α, the above formulae can be written as*

$$r = \frac{v}{B\alpha},\ T = \frac{2\pi}{B\alpha},\ f = \frac{B\alpha}{2\pi} \text{ and } \omega = B\alpha$$

Case 3 **When θ is other than 0°, 180° or 90°**

In this case velocity can be resolved into two components, one along **B** and another perpendicular to **B**. Let the two components be $v_{||}$ and $v_\perp$.

Then, $\quad v_{||} = v\cos\theta \quad$ and $\quad v_\perp = v\sin\theta$

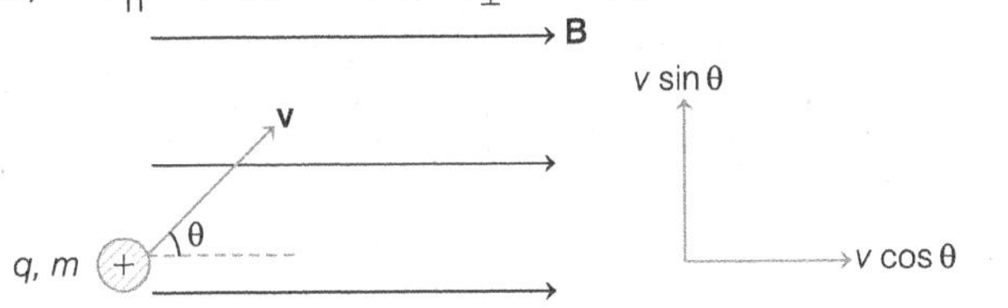

The component perpendicular to field ($v_\perp$) gives a circular path and the component parallel to field ($v_{||}$) gives a straight line path. The resultant path is a helix as shown in figure.

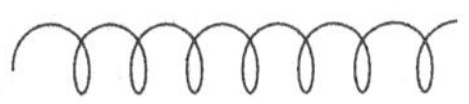

The radius of this helical path is

$$r = \frac{mv_\perp}{Bq} = \frac{mv\sin\theta}{Bq}$$

Time period and frequency do not depend on velocity and so they are given by

$$T = \frac{2\pi m}{Bq} \quad \text{and} \quad f = \frac{Bq}{2\pi m}$$

There is one more term associated with a helical path, that is **pitch (p)** of the helical path. Pitch is defined as the distance travelled along magnetic field in one complete cycle.

i.e. $$p = v_{||} T$$

or $$p = (v\cos\theta)\frac{2\pi m}{Bq}$$

$\therefore$ $$p = \frac{2\pi mv\cos\theta}{Bq}$$

Important points in helical path

(i) The plane of the circle of the helix is perpendicular to the magnetic field.

(ii) The axis of the helix is parallel to the magnetic field.

Example 1 Two particles A and B of masses m_A and m_B respectively and having the same charge are moving in a plane. A uniform magnetic field exists perpendicular to this plane. The speeds of the particles are v_A and v_B respectively and the trajectories are as shown in the figure. Then, **(2001 JEE)**

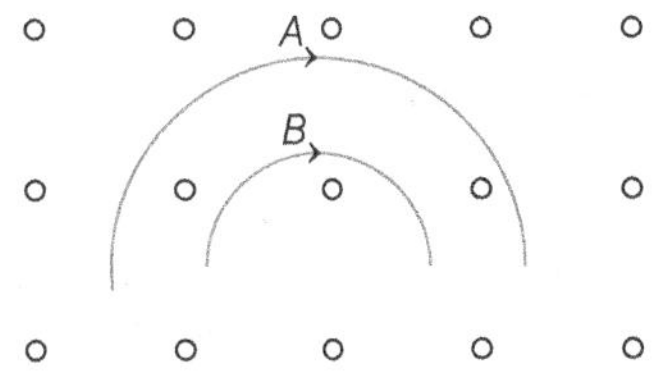

(a) $m_A v_A < m_B v_B$ (b) $m_A v_A > m_B v_B$
(c) $m_A < m_B$ and $v_A < v_B$ (d) $m_A = m_B$ and $v_A = v_B$

Sol. Radius of the circle $= \frac{mv}{Bq}$

or radius $\propto mv$ if B and q are same.

$$(\text{Radius})_A > (\text{Radius})_B$$

$\therefore$ $$m_A v_A > m_B v_B$$

$\therefore$ Correct option is (b).

Example 2 A proton, a deuteron and an α-particle having the same kinetic energy are moving in circular trajectories in a constant magnetic field. If r_p, r_d and r_α denote respectively the radii of the trajectories of these particles, then **(1997 JEE)**

(a) $r_\alpha = r_p < r_d$ (b) $r_\alpha > r_d > r_p$
(c) $r_\alpha = r_d > r_p$ (d) $r_p = r_d = r_\alpha$

Sol. Radius of the circular path is given by

$$r = \frac{mv}{Bq} = \frac{\sqrt{2Km}}{Bq}$$

Here, K is the kinetic energy of the particle.

Therefore, $r \propto \frac{\sqrt{m}}{q}$ if K and B are same.

$\therefore$ $$r_p : r_d : r_\alpha = \frac{\sqrt{1}}{1} : \frac{\sqrt{2}}{1} : \frac{\sqrt{4}}{2}$$
$$= 1 : \sqrt{2} : 1$$

Hence, $r_\alpha = r_p < r_d$

$\therefore$ Correct option is (a).

Example 3 Two particles X and Y having equal charges, after being accelerated through the same potential difference, enter a region of uniform magnetic field and describe circular paths of radii R_1 and R_2, respectively. The ratio of the mass of X to that of Y is **(1988 JEE)**

(a) $(R_1/R_2)^{1/2}$ (b) R_2/R_1

(c) $(R_1/R_2)^2$ (d) R_1/R_2

Sol. $R = \dfrac{\sqrt{2qVm}}{Bq}$

or $R \propto \sqrt{m}$

or $\dfrac{R_1}{R_2} = \sqrt{\dfrac{m_X}{m_Y}}$

or $\dfrac{m_X}{m_Y} = \left(\dfrac{R_1}{R_2}\right)^2$

∴ Correct option is (c).

INTRODUCTORY Exercise

1. A neutron, a proton, an electron and an α-particle enter a region of constant magnetic field with equal velocities. The magnetic field is along the inward normal to the plane of the paper. The tracks of the particles are labeled in figure. The electron follows track...... and the α-particle follows track .... **(1984 JEE)**

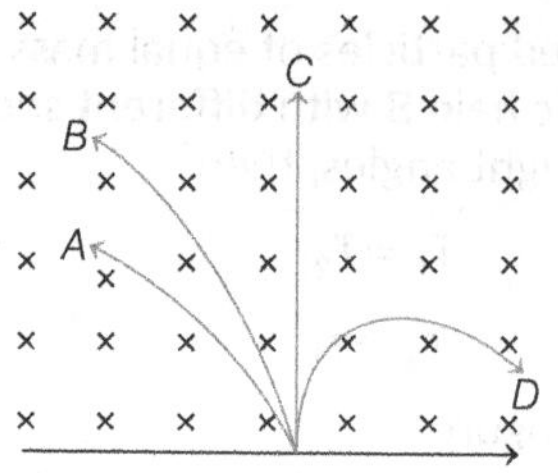

2. An electron and a proton are projected with same velocity perpendicular to a magnetic field.
(a) Which particle will describe the smaller circle?
(b) Which particle will have greater frequency?
3. An electron is accelerated through a potential difference of 100 V and then enters a region where it is moving perpendicular to a magnetic field $B = 0.2$ T. Find the radius of the circular path. Repeat this problem for a proton.

Answers

1. D, B **2.** (a) Electron (b) Electron **3.** 0.0167 cm, 0.7 cm

See hints on pages 434-435.

Important POINTS

1. **Determination of e/m of an Electron (Thomson Method)**
JJ Thomson in 1897, devised an experiment for the determination of e/m (specific charge) of the electron by using electric and magnetic fields in mutually perpendicular directions.

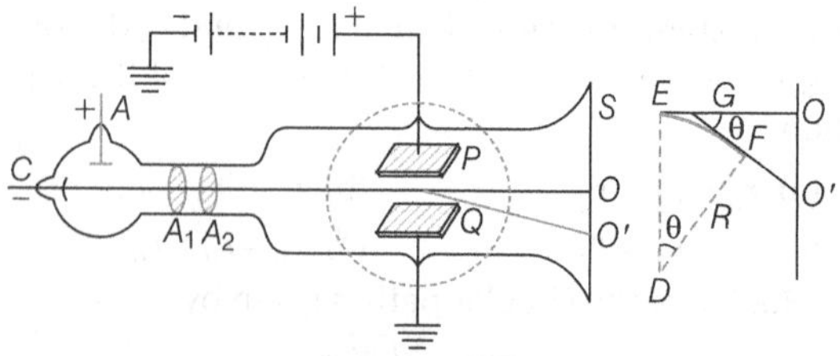

The discharge is maintained by the application of high PD between the cathode C and anode A of a discharge tube containing air at a very low pressure (~10^{-2} mm of Hg).

The electrons so produced are allowed to pass through slits A_1 and A_2 also kept at the potential of A. The beam then passes along the axis of the tube and produces a spot of light at O on the fluorescent screen S. The electric field E is applied between two horizontal plates P and Q.

The magnetic field B is applied in the direction perpendicular to the paper plane by passing the current through coils, in the region within the dotted circle. It is clear from Fleming's left hand rule, that F due to E is in upward direction, while due to B in downward direction.

Hence, fields E and B can be adjusted so that the electrons suffer no deflection and strike at point O on the screen. In this case,

$$eE = evB$$

or $$v = E/B \quad \text{...(i)}$$

Now, electric field is switched off, the electrons thus, describe the circular arc and fall at O' on the screen. In this case, the force $F = Bev$ bends the electron beam in a circular arc, such that it is balanced by the centripetal force mv^2/R.

∴ $$Bev = mv^2/R$$

or $$v = BRe/m \quad \text{...(ii)}$$

Combining Eqs. (i) and (ii), we get

$$e/m = E/RB^2 \quad \text{...(iii)}$$

As E and B are known. To find R, consider arc EF of the circular path in the magnetic field region. From the geometry, we get

$$\theta = \frac{EF}{R}$$

and $$\theta \approx \tan\theta = \frac{OO'}{GO}$$

∴ $$OO'/GO = EF/R$$

or $$R = EF \times GO/OO' \quad \text{...(iv)}$$

Practically, EF is replaced by the width of the magnetic flux region and G is taken at the middle of the region. Thomson's value for e/m was 1.7×10^{11} C/kg, which is in excellent agreement with the modern value of 1.75890×10^{11} C/kg.

2. **Cyclotron** In 1932, Lawrence developed a machine named cyclotron, for the acceleration of charged particles, such as protons or deuterons. These particles (ions) are caused to move in circular orbits by magnetic field and are accelerated by the electric field.
In its simplest form, it consists of two flat semicircular metal boxes, called dees because of their shape. These

hollow chambers have their diametric edges parallel and slightly separated from each other. An alternating potential (with frequency of the order of megacycles per second) is applied between the dees. The dees are placed between the poles of a strong electromagnet which provides a magnetic field perpendicular to the plane of the dees.

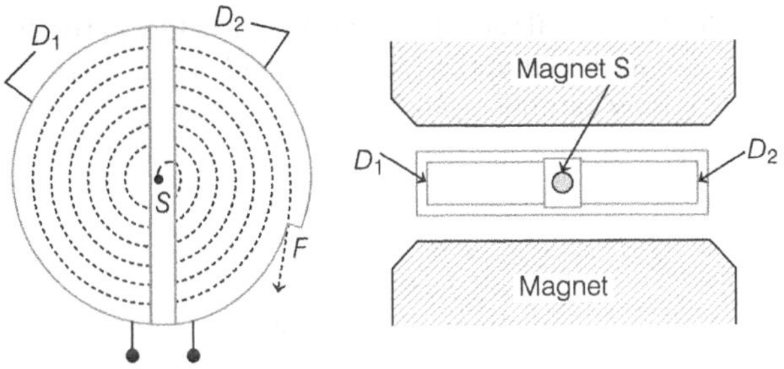

Suppose that at any particular instant the alternating potential is in the direction which makes D_1 positive and D_2 negative. A positive ion of mass m, charge q starting from the source S (of positive ion) will be attracted by the dee D_2. Let its velocity while entering in dee D_2 is v. Due to magnetic field B, it will move in a circular path of radius r inside the dee D_2, where

$$r = \frac{mv}{Bq}$$

or

$$\boxed{v = \frac{Bqr}{m}}$$

In the interior of the dee, the speed of the ion remains constant. After it has traversed half a cycle, the ion comes to the edge of D_2. If in the meantime, the potential difference between D_1 and D_2 has changed direction so that D_2 is now positive and D_1 negative, the positive ion will receive an additional acceleration, while going across the gap between the dees and speed of ion will increase. Then, it travels in a circular path of larger radius inside D_1 under the influence of magnetic field (because $r \propto v$).

After traversing a half cycle in D_1, it will reach the edge of D_1 and receive an additional acceleration between the gaps because in the meantime the direction of potential difference between the dees has changed.

The ion will continue travelling in a semicircle of increasing radii, the direction of potential difference changes every time the ion goes from D_1 to D_2 and from D_2 to D_1. The time taken by the charged particle to traverse the semicircular path in the dee is given by

$$t = \frac{T}{2} = \frac{\pi m}{Bq} \qquad \text{...(i)}$$

This relation indicates that time t is independent of the velocity of the particle and of the radius. For any given value of m/q, it is determined by the magnetic field intensity. By adjusting the magnetic field intensity the time can be made the same as that required to change the potentials.

The frequency of the oscillations required to keep the ion in phase is given by the relation

$$\boxed{f = \frac{1}{T} = \frac{1}{2t} = \frac{Bq}{2\pi m}} \qquad \text{...(ii)}$$

If the oscillation frequency is adjusted to keep the charged ion always in phase, each time the ion crosses the gap it receives an additional energy and at the same time it describes a flat spiral of increasing radius.

Eventually, the ion reaches the periphery of the dee, where it can be brought out of the chamber by means of a deflecting plate charged to a high negative potential. This attractive force draws the ion out of its spiral path and thus can be used easily. If R is the radius of the dee, kinetic energy of the ion emerging from the cyclotron is thus given by

$$K = \frac{1}{2}mv^2 = \frac{1}{2}m\,(BqR/m)^2$$

$$\boxed{K = B^2R^2q^2/2m} \qquad \text{...(iii)}$$

This relation indicates that the maximum energy attained by the ion is limited by the radius R, magnetic field B or the frequency of the alternating potential f. It is independent of the alternating voltage. It can be explained by the fact that when the voltage is low, the ion makes a large number of turns before reaching the periphery, but when the voltage is high the number of turns is small. The total energy remains same in both the cases provided B and R are unchanged.

Note *The cyclotron is used to bombard nuclei with energetic particles and study the resulting nuclear reactions. It is also used in hospitals to produce radioactive substances which can be used in diagnosis and treatment.*

Cyclotron is suitable only for accelerating heavy particles like proton, deuteron, α-particle etc. Electrons cannot be accelerated by the cyclotron because the mass of the electron is small and a small increase in energy of the electron makes the electrons move with a very high speed.

The uncharged particles (e.g. neutrons) cannot be accelerated by cyclotron.

Typed Problems

Type 1. *Based on deviation of charged particle in uniform magnetic field when $\theta = 90°$ or path is uniform circular.*

Concept

Suppose a charged particle (q, m) enters a uniform magnetic field **B** at right angles with velocity v as shown in figure. The magnetic field extends upto a length x. The path of the particle is a circle of radius r, where

$$r = \frac{mv}{Bq}$$

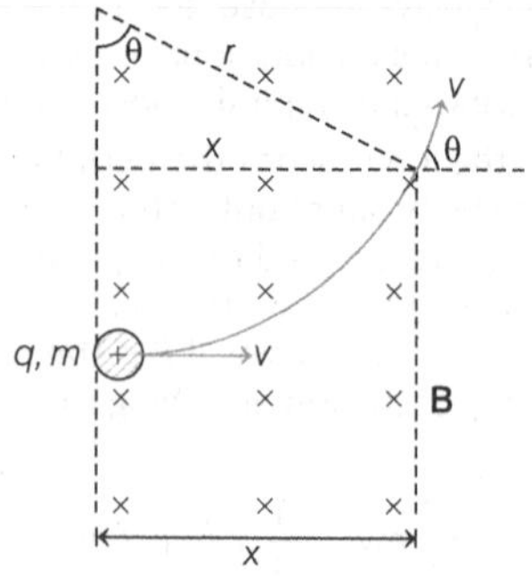

The speed of the particle in magnetic field does not change. But, it gets deviated in the magnetic field. The deviation θ can be found in two ways

(i) After time t, deviation will be

$$\boxed{\theta = \omega t = \left(\frac{Bq}{m}\right)t} \qquad \left[\text{as } \omega = \frac{Bq}{m}\right]$$

(ii) In terms of the length of the magnetic field (i.e. when the particle leaves the magnetic field) the deviation will be

$$\boxed{\theta = \sin^{-1}\left(\frac{x}{r}\right)}$$

But, since, $\sin\theta \not> 1$, this relation can be used only when $x < r$.

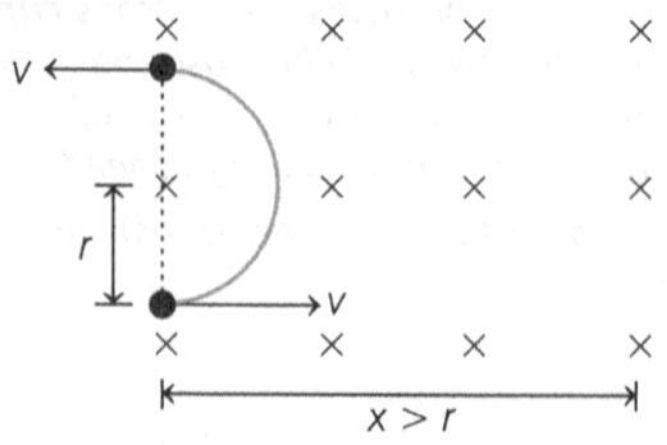

For $x \geq r$, the deviation will be 180° as shown in figure.

Ex. 1 The region between $x = 0$ and $x = L$ is filled with uniform steady magnetic field $-B_0\hat{\mathbf{k}}$ A particle of mass m, positive charge q and velocity $v_0\hat{\mathbf{i}}$ travels along x-axis and enters the region of the magnetic field. **[1999 JEE]**

Neglect the gravity throughout the question.

(a) Find the value of L if the particle emerges from the region of magnetic field with its final velocity at an angle 30° to its initial velocity.

(b) Find the final velocity of the particle and the time spent by it in the magnetic field, if the magnetic field now extends upto 2.1 L.

Sol. (a) $\theta = 30°$

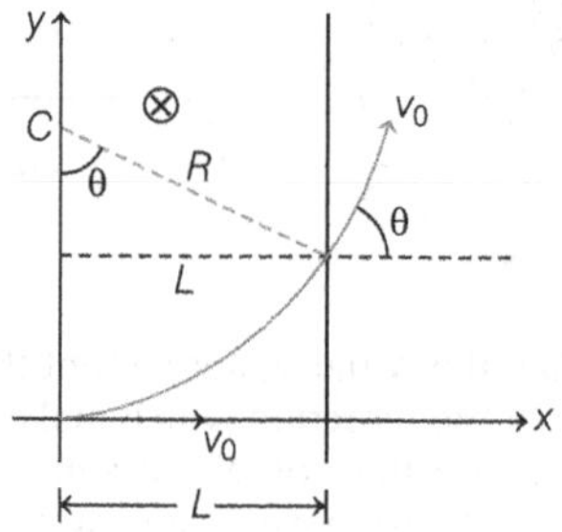

$$\sin\theta = \frac{L}{R}$$

Here, $$R = \frac{mv_0}{B_0 q}$$

$\therefore$ $$\sin 30° = \frac{L}{\frac{mv_0}{B_0 q}}$$

or $$\frac{1}{2} = \frac{B_0 qL}{mv_0}$$

$\therefore$ $$L = \frac{mv_0}{2B_0 q}$$

(b) In part (a)

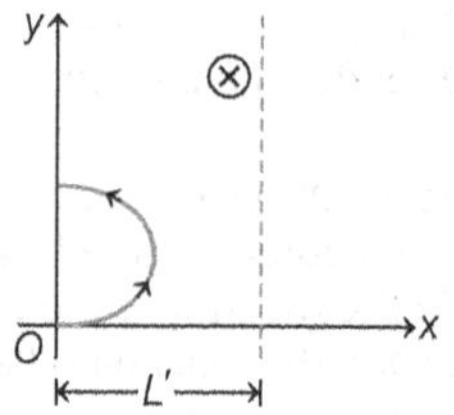

$$\sin 30° = \frac{L}{R} \quad \text{or} \quad \frac{1}{2} = \frac{L}{R}$$

or $$L = R/2$$

Now, when $L' = 2.1\,L$

$$= \frac{2.1}{2}R \Rightarrow L' > R$$

Therefore, deviation of the particle is θ = 180° as shown in figure.

$\therefore$ $$\mathbf{v}_f = -v_0\hat{\mathbf{i}}$$

and $$t_{AB} = T/2 = \frac{\pi m}{B_0 q}$$

Type 2. *To find coordinates and velocity of particle at any time t in circular path.*

Ex. 2 A particle of specific charge α enters a uniform magnetic field $\mathbf{B} = -B_0\hat{\mathbf{k}}$ with velocity $\mathbf{v} = v_0\hat{\mathbf{i}}$ from the origin. Find the time dependence of velocity and position of the particle.

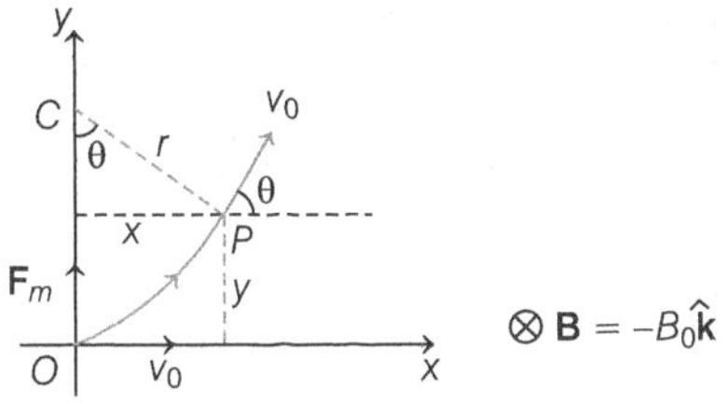

$OC = CP$ = radius of circle

HOW TO PROCEED In such type of problems first of all see the angle between **v** and **B**. Because only this angle decides the path of the particle. Here, the angle is 90°. Therefore, the path is a circle. If it is a circle, see the plane of the circle (perpendicular to the magnetic field). Here, the plane is *xy*. Then, see the sense of the rotation.

Here, it will be anti-clockwise as shown in figure, because at origin the magnetic force is along positive *y*-direction (which can be seen from Fleming's left hand rule). Find the deviation and radius of the particle.

$$\theta = \omega t = B_0\alpha t \quad \text{and} \quad r = \frac{v_0}{B_0\alpha}$$

Now, according to the figure, find $\mathbf{v}(t)$ and $\mathbf{r}(t)$.

Sol. Velocity of the particle at any time *t* is

$$\mathbf{v}(t) = v_x\hat{\mathbf{i}} + v_y\hat{\mathbf{j}} = v_0\cos\theta\hat{\mathbf{i}} + v_0\sin\theta\hat{\mathbf{j}}$$

or $$\mathbf{v}(t) = v_0\cos(B_0\alpha t)\hat{\mathbf{i}} + v_0\sin(B_0\alpha t)\hat{\mathbf{j}}$$ **Ans.**

Position of particle at time *t* is

$$\mathbf{r}(t) = x\hat{\mathbf{i}} + y\hat{\mathbf{j}}$$
$$= r\sin\theta\hat{\mathbf{i}} + (r - r\cos\theta)\hat{\mathbf{j}}$$

Substituting the values of r and θ, we have

$$\mathbf{r}(t) = \frac{v_0}{B_0\alpha}[\sin(B_0\alpha t)\hat{\mathbf{i}} + \{1 - \cos(B_0\alpha t)\}\hat{\mathbf{j}}]$$ **Ans.**

Note *In this case magnetic field **B** is present all around. So, the charged particle will complete the circle. But the question is asked at some intermediate time t.*

Type 3. *To find coordinates and velocity of particle at any time t in helical path.*

Ex. 3 A particle of specific charge α is projected from origin with velocity $\mathbf{v} = v_0\hat{\mathbf{i}} - v_0\hat{\mathbf{k}}$ in a uniform magnetic field $\mathbf{B} = -B_0\hat{\mathbf{k}}$. Find time dependence of velocity and position of the particle.

HOW TO PROCEED Here, the angle between **v** and **B** is

$$\theta = \cos^{-1}\frac{\mathbf{v}\cdot\mathbf{B}}{|\mathbf{v}||\mathbf{B}|}$$
$$= \cos^{-1}\frac{B_0v_0}{\sqrt{2}v_0\cdot B_0}$$
$$= \cos^{-1}\left(\frac{1}{\sqrt{2}}\right)$$

or $$\theta = 45°$$

Hence, the path is a helix. The axis of the helix is along *z*-axis (parallel to **B**) and plane of the circle of helix is *xy* (perpendicular to **B**). So, in *xy*-plane, the velocity components and *x* and *y*-coordinates are same as that of the above problem. The only change is along *z*-axis. Velocity component in this direction will remain unchanged while the z-coordinate of particle at time *t* would be $v_z t$.

Sol. Velocity of particle at time *t* is

$$\mathbf{v}(t) = v_x\hat{\mathbf{i}} + v_y\hat{\mathbf{j}} + v_z\hat{\mathbf{k}}$$
$$= v_0\cos(B_0\alpha t)\hat{\mathbf{i}} + v_0\sin(B_0\alpha t)\hat{\mathbf{j}} - v_0\hat{\mathbf{k}}$$ **Ans.**

v_x and v_y can be found in the similar manner as done in Example 2.

The position of the particle at time *t* would be

$$\mathbf{r}(t) = x\hat{\mathbf{i}} + y\hat{\mathbf{j}} + z\hat{\mathbf{k}}$$

Here, $$z = v_z t = -v_0 t$$

and *x* and *y* are same as in Ex. 2.

Hence, $$\mathbf{r}(t) = \frac{v_0}{B_0\alpha}[\sin(B_0\alpha t)\hat{\mathbf{i}} + \{1 - \cos(B_0\alpha t)\}\hat{\mathbf{j}}] - v_0 t\hat{\mathbf{k}}$$

Ans.

Type 4. *To find the time spent in magnetic field, deviation etc. if a charged particle enters from outside in uniform magnetic field (which extends upto large distance from point of entering).*

Ex. 4 A charged particle (q, m) enters a uniform magnetic field **B** at angle α as shown in figure with speed v_0. Find

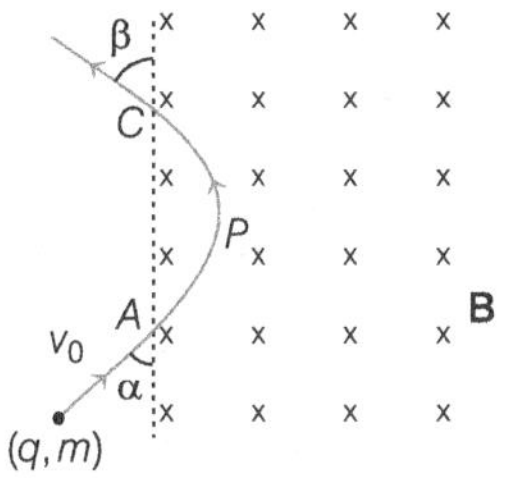

(a) the angle β at which it leaves the magnetic field.

(b) time spent by the particle in magnetic field and

(c) the distance *AC*.

Sol. (a) Here, velocity of the particle is in the plane of paper while the magnetic field is perpendicular to the paper inwards, i.e. angle between **v** and **B** is 90°. So, the path is a circle. The radius of the circle is

$$r = \frac{mv_0}{Bq}$$

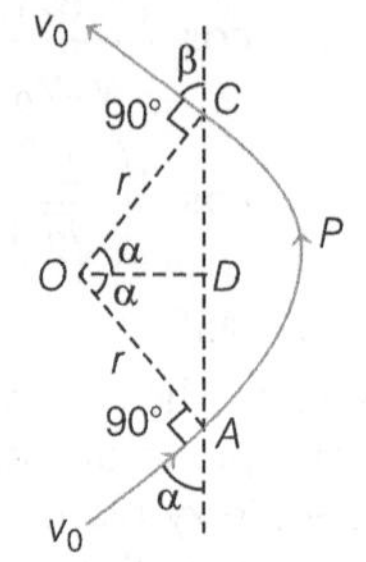

O is the centre of the circle.

In ΔAOC, $\angle OCD = \angle OAD$

or $90° - \beta = 90° - \alpha$

$\therefore$ $\beta = \alpha$ **Ans.**

(b) $\angle COD = \angle DOA = \alpha$ (as $\angle OCD = \angle OAD = 90° - \alpha$)

$\therefore$ $\angle AOC = 2\alpha$

or length $APC = r\,(2\alpha) = \dfrac{2mv_0}{Bq}\cdot\alpha$

$\therefore$ $t_{APC} = \dfrac{APC}{v_0} = \dfrac{2m\alpha}{Bq}$ **Ans.**

Alternate method

$$t_{APC} = \frac{2\alpha}{\omega} = \left(\frac{1}{\omega}\right)(2\alpha) = \left(\frac{T}{2\pi}\right)(2\alpha) = \left(\frac{\alpha}{\pi}\right)\cdot T$$

$$= \left(\frac{\alpha}{\pi}\right)\left(\frac{2\pi m}{Bq}\right) = \frac{2\alpha m}{Bq}$$ **Ans.**

(c) Distance, $AC = 2\,(AD) = 2\,(r\sin\alpha)$

$$= \frac{2mv_0}{Bq}\sin\alpha$$ **Ans.**

Ex. 5 A particle of mass $m = 1.6 \times 10^{-27}$ kg and charge $q = 1.6 \times 10^{-19}$ C enters a region of uniform magnetic field of strength 1 T along the direction shown in figure. The speed of the particle is 10^7 m/s. **(1984 JEE)**

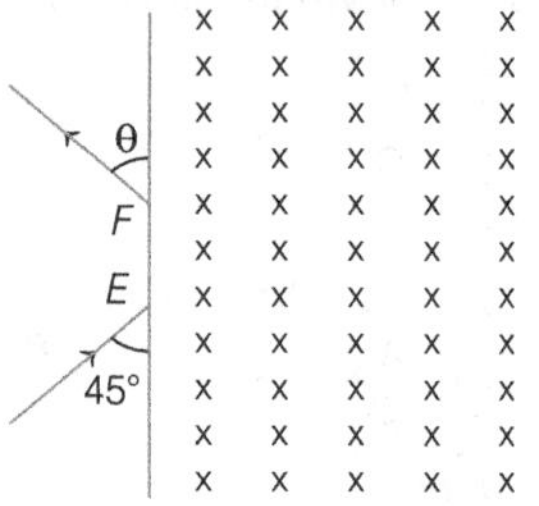

(a) The magnetic field is directed along the inward normal to the plane of the paper. The particle leaves the region of the field at the point F. Find the distance EF and the angle θ.

(b) If the direction of the field is along the outward normal to the plane of the paper, find the time spent by the particle in the region of the magnetic field after entering it at E.

Sol. Inside a magnetic field, speed of charged particle does not change. Further, velocity is perpendicular to magnetic field in both the cases hence path of the particle in the magnetic field will be circular. Centre of circle can be obtained by drawing perpendiculars to velocity (or tangent to the circular path) at E and F. Radius and angular speed of circular path would be

$$r = \frac{mv}{Bq} \quad \text{and} \quad \omega = \frac{Bq}{m}$$

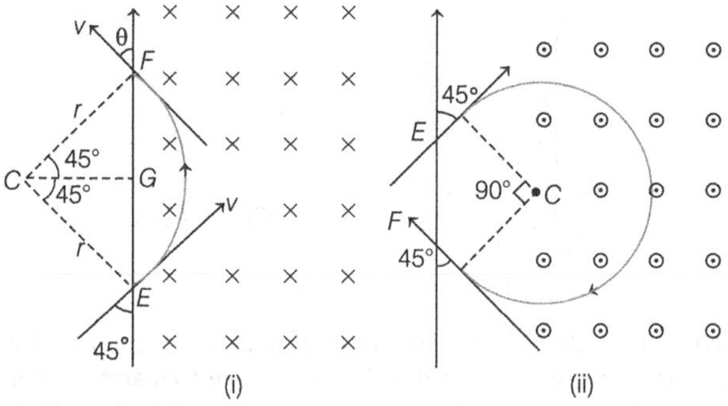

(a) **Refer figure (i)**

$\angle CFG = 90° - \theta$

and $\angle CEG = 90° - 45° = 45°$

Since, $CF = CE$

$\therefore$ $\angle CFG = \angle CEG$

or $90° - \theta = 45°$

or $\theta = 45°$

Further, $FG = GE = r\cos 45°$

$\therefore$ $EF = 2FG = 2r\cos 45° = \dfrac{2mv\cos 45°}{Bq}$

$$= \frac{2\,(1.6 \times 10^{-27})\,(10^7)\left(\frac{1}{\sqrt{2}}\right)}{(1)\,(1.6 \times 10^{-19})}$$

$= 0.14$ m

Note *That in this case particle completes 1/4th of circle in the magnetic field because the angle rotated is 90°.*

(b) **Refer figure (ii)** In this case, particle will complete $\frac{3}{4}$th of circle in the magnetic field. Hence, the time spent in the magnetic field :

$$t = \frac{3}{4}\text{ (time period of circular motion)}$$

$$= \frac{3}{4}\left(\frac{2\pi m}{Bq}\right) = \frac{3\pi m}{2Bq}$$

$$= \frac{(3\pi)\,(1.6 \times 10^{-27})}{(2)\,(1)(1.6 \times 10^{-19})}$$

$$= 4.712 \times 10^{-8}\text{ s}$$ **Ans.**

Note *From the above examples, we can see that particle never completes circular path if it enters from outside in uniform magnetic field at right angles (as in Examples 1, 4 and 5). Circle is completed if magnetic field extends all around (Example-2). Following figures explain these points more clearly. In all figures, particle is positively charged.*

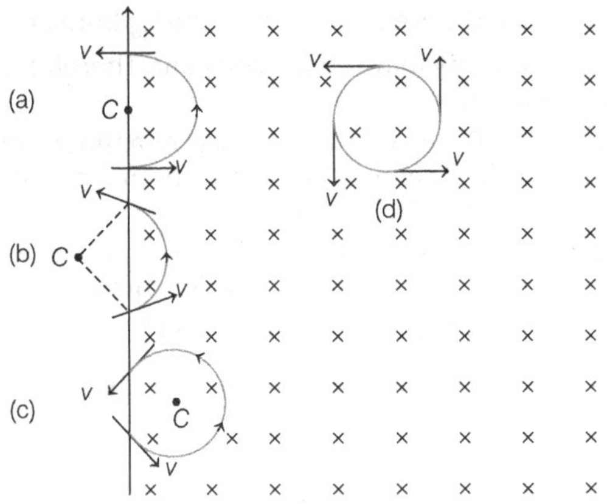

In figure (a) *Centre of circular path is lying on the boundary line of magnetic field. Deviation of the particle is 180° and time spent in magnetic field* $t = \frac{T}{2}$.

In figure (b) *Centre of circular path lies outside the magnetic field. Deviation of the particle is less than 180° and time spent in magnetic field* $t < \frac{T}{2}$.

In figure (c) *Centre of circular path lies inside the magnetic field. Deviation of the particle is more than 180° and time spent in magnetic field* $t > \frac{T}{2}$.

In figure (d) *Circle is completed because the magnetic field is present all around.*

Type 5. *Based on the concept of helical path.*

Concept

Following points are worthnoting in case of a helical path.

(i) The plane of the circle of the helix is perpendicular to the magnetic field.

(ii) The axis of the helix is parallel to the magnetic field.

(iii) The particle while moving in helical path in magnetic field touches the line passing through the starting point parallel to the magnetic field after every pitch.

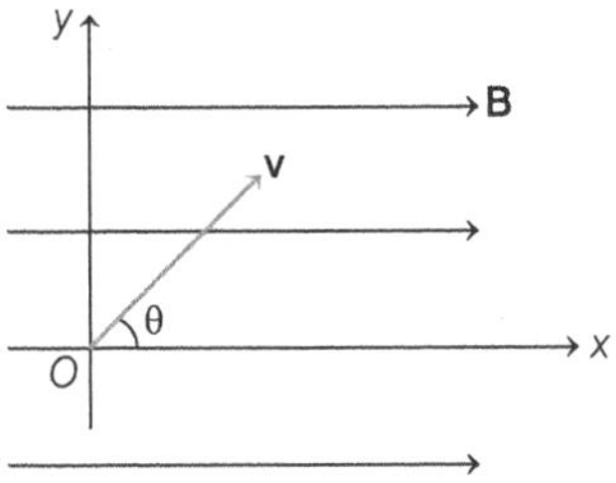

For example, a charged particle is projected from origin in a magnetic field (along *x*-direction) at angle θ from the *x*-axis as shown. As the velocity vector **v** makes an angle θ with **B**, its path is a helix. The plane of the circle of the helix is *yz* (perpendicular to magnetic field) and axis of the helix is parallel to *x*-axis. The particle while moving in helical path touches the *x*-axis after every pitch, i.e. it will touch the *x*-axis at a distance

$$\boxed{x = np} \qquad \text{where, } n = 0, 1, 2 \ldots$$

Ex. 6 An electron gun *G* emits electrons of energy 2 keV travelling in the positive *x*-direction. The electrons are required to hit the spot *S* where *GS* = 0.1 m, and the line *GS* makes an angle of 60° with the *x*-axis as shown in figure. A uniform magnetic field **B** parallel to *GS* exists in the region outside the electron gun.

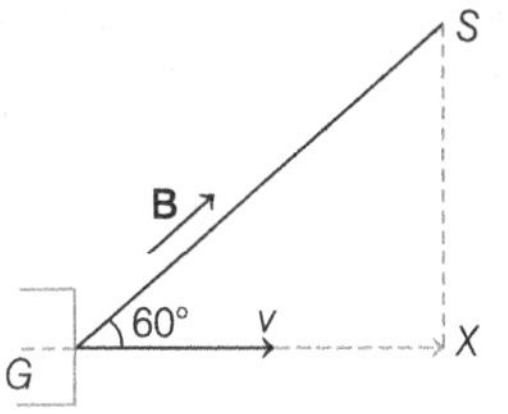

Find the minimum value of *B* needed to make the electrons hit *S*. **(1993 JEE)**

Sol. Kinetic energy of electron,

$$K = \frac{1}{2}mv^2 = 2 \text{ keV} = 2 \times 10^3 \text{ eV}$$
$$= 2 \times 1.6 \times 10^{-16} \text{ J}$$

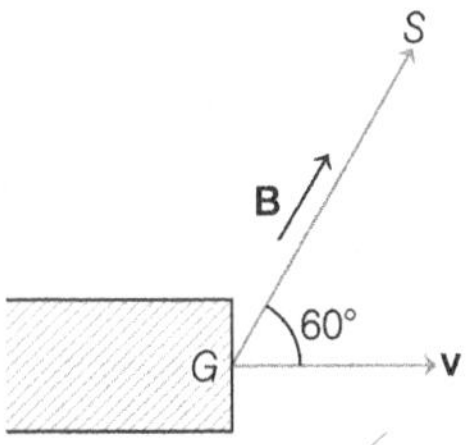

$\therefore$ Speed of electron, $v = \sqrt{\frac{2K}{m}}$

$$v = \sqrt{\frac{2 \times 2 \times 1.6 \times 10^{-16}}{9.1 \times 10^{-31}}} \text{ m/s}$$
$$= 2.65 \times 10^7 \text{ m/s}$$

Since, the velocity (**v**) of the electron makes an angle of $\theta = 60°$ with the magnetic field **B**, the path will be a helix. So, the particle will hit *S* if

$$GS = np$$

Here, $n = 1, 2, 3$ ....

$$p = \text{pitch of helix} = \frac{2\pi m}{qB} v\cos\theta$$

But for *B* to be minimum, $n = 1$

Hence, $$GS = p = \frac{2\pi m}{qB} v\cos\theta$$

$$B = B_{min} = \frac{2\pi mv\cos\theta}{q(GS)}$$

Substituting the values, we have

$$B_{min} = \frac{(2\pi)(9.1 \times 10^{-31})(2.65 \times 10^7)\left(\frac{1}{2}\right)}{(1.6 \times 10^{-19})(0.1)}$$

or $B_{min} = 4.73 \times 10^{-3}$ T **Ans.**

Type 6. *Based on the electromagnetic force on a charged particle in electric and (or) magnetic field.*

Ex. 7 A particle of charge $+q$ and mass m moving under the influence of a uniform electric field $E\hat{\mathbf{i}}$ and uniform magnetic field $B\hat{\mathbf{k}}$ follows a trajectory from P to Q as shown in figure. The velocities at P and Q are $v\hat{\mathbf{i}}$ and $-2\hat{\mathbf{j}}$. Which of the following statement(s) is/are correct ? **(1991 JEE)**

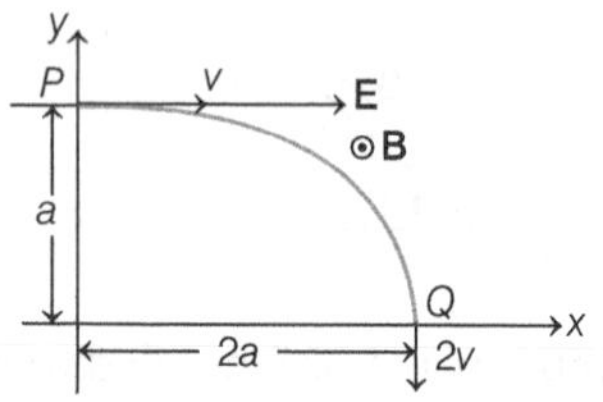

(a) $E=\frac{3}{4}\left[\frac{mv^2}{qa}\right]$

(b) Rate of work done by the electric field at P is $\frac{3}{4}\left[\frac{mv^3}{a}\right]$

(c) Rate of work done by the electric field at P is zero

(d) Rate of work done by both the fields at Q is zero

Sol. Work done by magnetic force is always zero. From work-energy theorem :

$$W_{F_e} = \Delta KE$$

or $$(qE)(2a) = \frac{1}{2}m[4v^2 - v^2]$$

or $$E = \frac{3}{4}\left(\frac{mv^2}{qa}\right)$$

$\therefore$ Option (a) is correct.

At P, rate of work done by electric field

$$= \mathbf{F}_e \cdot \mathbf{v} = (qE)(v)\cos 0°$$

$$= q\left(\frac{3}{4}\frac{mv^2}{qa}\right)v = \frac{3}{4}\left(\frac{mv^3}{a}\right)$$

Therefore, option (b) is also correct.

Rate of work done at Q :

of electric field $= \mathbf{F}_e \cdot \mathbf{v} = (qE)(2v)\cos 90° = 0$ and of magnetic field is always zero.

Therefore, option (d) is also correct.

Note that $\mathbf{F}_e = qE\hat{\mathbf{i}}$

$\therefore$ The correct options are (a), (b) and (d).

Ex. 8 A proton moving with a constant velocity passes through a region of space without any change in its velocity. If **E** and **B** represent the electric and magnetic fields, respectively. Then, this region of space may have **(1985 JEE)**

(a) $\mathbf{E} = 0, \mathbf{B} = 0$ (b) $\mathbf{E} = 0, \mathbf{B} \neq 0$

(c) $\mathbf{E} \neq 0, \mathbf{B} = 0$ (d) $\mathbf{E} \neq 0, \mathbf{B} \neq 0$

Sol. If both **E** and **B** are zero, then $\mathbf{F}_e$ and $\mathbf{F}_m$ both are zero. Hence, velocity may remain constant. Therefore, option (a) is correct.

If $\mathbf{E} = 0, \mathbf{B} \neq 0$ but velocity is parallel or antiparallel to magnetic field, then also $\mathbf{F}_e$ and $\mathbf{F}_m$ both are zero. Hence, option (b) is also correct.

If $\mathbf{E} \neq 0, \mathbf{B} \neq 0$ but $\mathbf{F}_e + \mathbf{F}_m = 0$, then again velocity may remain constant or option (d) is also correct.

$\therefore$ The correct options are (a), b) and (d).

Type 7. *Path of charged particle in both electric and magnetic fields.*

Concept

Here, normally two cases are popular. In the first case, $\mathbf{E} \uparrow\uparrow \mathbf{B}$ and particle velocity is perpendicular to both of these fields. In the second case, $\mathbf{E} \perp \mathbf{B}$ and the particle is released from rest. Let us now consider both the cases separately.

Ex. 9 When $\mathbf{E} \uparrow\uparrow \mathbf{B}$ and particle velocity is perpendicular to both of these fields.

Sol. Consider a particle of charge q and mass m released from the origin with velocity $\mathbf{v} = v_0\hat{\mathbf{i}}$ into a region of uniform electric and magnetic fields parallel to y-axis, i.e. $\mathbf{E} = E_0\hat{\mathbf{j}}$ and $\mathbf{B} = B_0\hat{\mathbf{j}}$. The electric field accelerates the particle in y-direction, i.e. y-component of velocity goes on increasing with acceleration,

$$a_y = \frac{F_y}{m} = \frac{F_e}{m} = \frac{qE_0}{m} \quad \text{...(i)}$$

The magnetic field rotates the particle in a circle in xz-plane (perpendicular to magnetic field). The resultant path of the particle is a **helix with increasing pitch**. The axis of the plane is parallel to y-axis. Velocity of the particle at time t would be

$$\mathbf{v}(t) = v_x\hat{\mathbf{i}} + v_y\hat{\mathbf{j}} + v_z\hat{\mathbf{k}}$$

Here, $$v_y = a_y t = \frac{qE_0}{m}t$$

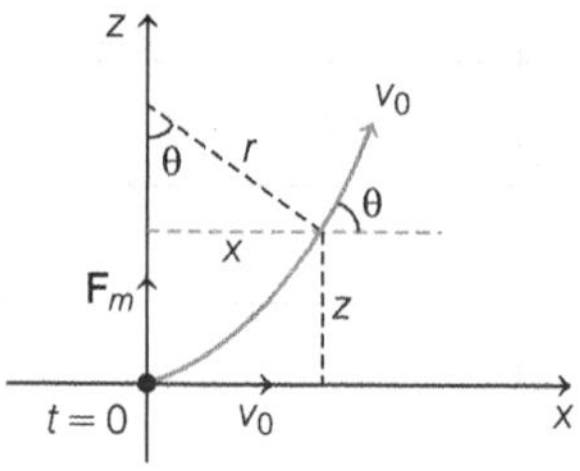

and $$v_x^2 + v_z^2 = \text{constant} = v_0^2$$

$$\theta = \omega t = \frac{Bq}{m}t$$

$$v_x = v_0\cos\theta = v_0\cos\left(\frac{Bqt}{m}\right)$$

and $$v_z = v_0\sin\theta = v_0\sin\left(\frac{Bqt}{m}\right)$$

$$\therefore \quad \boxed{\mathbf{v}(t) = v_0 \cos\left(\frac{Bqt}{m}\right)\hat{\mathbf{i}} + \left(\frac{qE_0}{m}t\right)\hat{\mathbf{j}} + v_0 \sin\left(\frac{Bqt}{m}\right)\hat{\mathbf{k}}}$$

Similarly, position vector of particle at time t can be given by

$$\mathbf{r}(t) = x\hat{\mathbf{i}} + y\hat{\mathbf{j}} + z\hat{\mathbf{k}}$$

Here, $$y = \frac{1}{2}a_y t^2 = \frac{1}{2}\left(\frac{qE_0}{m}\right)t^2$$

$$x = r\sin\theta = \left(\frac{mv_0}{Bq}\right)\sin\left(\frac{Bqt}{m}\right)$$

and $$z = r(1-\cos\theta) = \left(\frac{mv_0}{Bq}\right)\left[\left\{1-\cos\left(\frac{Bqt}{m}\right)\right\}\right]$$

$$\therefore \quad \boxed{\mathbf{r}(t) = \left(\frac{mv_0}{Bq}\right)\sin\left(\frac{Bqt}{m}\right)\hat{\mathbf{i}} + \frac{1}{2}\left(\frac{qE_0}{m}\right)t^2\hat{\mathbf{j}} + \left(\frac{mv_0}{Bq}\right)\left[\left\{1-\cos\left(\frac{Bqt}{m}\right)\right\}\right]\hat{\mathbf{k}}}$$

Note *(i) While moving in helical path the particle touches the y-axis after every T or after,*

$$t = nT, \quad \text{where } n = 0, 1, 2\ldots$$

Here, $$T = \frac{2\pi m}{Bq}$$

(ii) At $t = 0$, velocity is along positive x-axis and magnetic field is along y-axis. Therefore, magnetic force is along positive z-axis and the particle rotates in xz-plane as shown in figure.

Ex. 10 When $\mathbf{E} \perp \mathbf{B}$ and the particle is released at rest from origin.

Sol. Consider a particle of charge q and mass m emitted at origin with zero initial velocity into a region of uniform electric and magnetic fields. The field **E** is acting along x-axis and field **B** along y-axis, i.e.

$$\mathbf{E} = E_0\hat{\mathbf{i}} \quad \text{and} \quad \mathbf{B} = B_0\hat{\mathbf{j}}$$

Electric field will provide the particle an acceleration (and therefore a velocity component) in x-direction and the magnetic field will rotate the particle in xz-plane (perpendicular to **B**). Hence, at any instant of time its velocity (and hence, position) will have only x and z components. Let at time t its velocity be

$$\mathbf{v} = v_x\hat{\mathbf{i}} + v_z\hat{\mathbf{k}}$$

Net force on it at this instant is

$$\mathbf{F} = \mathbf{F}_e + \mathbf{F}_m = q\mathbf{E} + q(\mathbf{v}\times\mathbf{B})$$
$$= q[E_0\hat{\mathbf{i}} + (v_x\hat{\mathbf{i}} + v_z\hat{\mathbf{k}})\times(B_0\hat{\mathbf{j}})]$$

or $$\mathbf{F} = q(E_0 - v_zB_0)\hat{\mathbf{i}} + qv_xB_0\hat{\mathbf{k}}$$

$$\therefore \quad \mathbf{a} = \frac{\mathbf{F}}{m} = a_x\hat{\mathbf{i}} + a_z\hat{\mathbf{k}}$$

where, $$a_x = \frac{q}{m}(E_0 - v_zB_0) \quad \ldots\text{(i)}$$

and $$a_z = \frac{q}{m}v_xB_0 \quad \ldots\text{(ii)}$$

Differentiating Eq. (i) w.r.t. time, we have

$$\frac{d^2v_x}{dt^2} = -\frac{qB_0}{m}\left(\frac{dv_z}{dt}\right)$$

But, $$\frac{dv_z}{dt} = a_z = \frac{qB_0}{m}v_x$$

$$\therefore \quad \frac{d^2v_x}{dt^2} = -\left(\frac{qB_0}{m}\right)^2 v_x \quad \ldots\text{(iii)}$$

Comparing this equation with the differential equation of SHM $\left(\frac{d^2y}{dt^2} = -\omega^2 y\right)$, we get

$$\omega = \frac{qB_0}{m}$$

and the general solution of Eq. (iii) is

$$v_x = A\sin(\omega t + \phi) \quad \ldots\text{(iv)}$$

At time $t = 0$, $v_x = 0$, hence $\phi = 0$

Again, $$\frac{dv_x}{dt} = A\omega\cos\omega t \quad (\text{as } \phi = 0)$$

From Eq. (i), $a_x = \frac{qE_0}{m}$ at $t = 0$, as $v_z = 0$ at $t = 0$

$$\therefore \quad A\omega = \frac{qE_0}{m} \quad \text{or} \quad A = \frac{qE_0}{\omega m}$$

Substituting $\omega = \frac{qB_0}{m}$, we get

$$A = \frac{E_0}{B_0}$$

Therefore, Eq. (iv) becomes

$$\boxed{v_x = \frac{E_0}{B_0}\sin\omega t \quad \text{where, } \omega = \frac{qB_0}{m}}$$

Now substituting value of v_x in Eq. (ii), we get

$$\frac{dv_z}{dt} = \frac{qE_0}{m}\sin\omega t$$

$$\therefore \quad \int_0^{v_z} dv_z = \frac{qE_0}{m}\int_0^t \sin\omega t\, dt$$

or $$v_z = \frac{qE_0}{\omega m}(1-\cos\omega t)$$

Substituting $\omega = \frac{qB_0}{m}$, we get

$$\boxed{v_z = \frac{E_0}{B_0}(1-\cos\omega t)}$$

On integrating equations for v_x and v_z and knowing that at $t = 0$, $x = 0$ and $z = 0$, we get

$$x = \frac{E_0}{B_0\omega}(1 - \cos\omega t)$$

and

$$z = \frac{E_0}{B_0\omega}(\omega t - \sin\omega t)$$

These equations are the equations for a **cycloid** which is defined as the path generated by the point on the circumference of a wheel rolling on a ground.

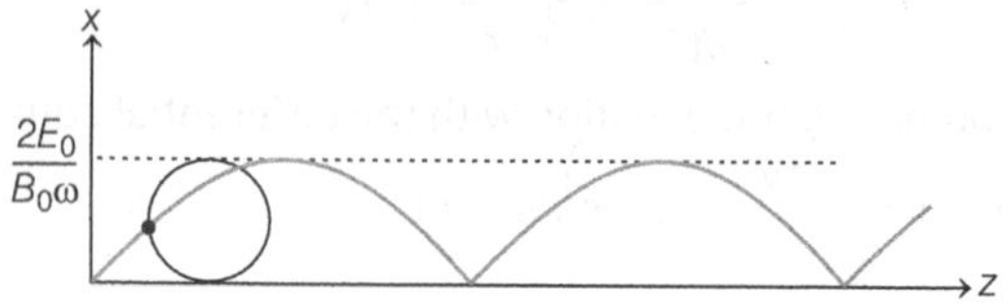

In the present case, the radius of the rolling wheel is $\frac{E_0}{B_0\omega}$, the maximum displacement along x-direction is $\frac{2E_0}{B_0\omega}$. The x-displacement becomes zero at $t = 0, 2\pi/\omega, 4\pi/\omega$, etc.

Note *Path of a charged particle in uniform electric and magnetic field will remain unchanged if*

$$\mathbf{F}_{net} = 0$$

or $$\mathbf{F}_e + \mathbf{F}_m = 0$$

or $$q\mathbf{E} + q(\mathbf{v} \times \mathbf{B}) = 0$$

or $$\mathbf{E} = -(\mathbf{v} \times \mathbf{E}) = (\mathbf{E} \times \mathbf{v})$$

Miscellaneous Examples

Ex. 11 A cyclotron's oscillator frequency is 10 MHz. What should be the operating magnetic field for accelerating protons? If the radius of its dees is 60cm, what is the kinetic energy (in MeV) of the proton beam produced by the accelerator?

($e = 1.60 \times 10^{-19}$ C, $m_p = 1.67 \times 10^{-27}$ kg, 1 MeV $= 1.6 \times 10^{-13}$ J)

Sol. Magnetic field Cyclotron's oscillator frequency should be same as the proton's revolution frequency (in circular path)

$$\therefore \quad f = \frac{Bq}{2\pi m} \quad \text{or} \quad B = \frac{2\pi mf}{q}$$

Substituting the values in SI units, we have

$$B = \frac{(2)(22/7)(1.67 \times 10^{-27})(10 \times 10^6)}{1.6 \times 10^{-19}}$$

$$= 0.67 \text{ T}$$ **Ans.**

Kinetic energy Let final velocity of proton just after leaving the cyclotron is v. Then, radius of dee should be equal to

$$R = \frac{mv}{Bq} \quad \text{or} \quad v = \frac{BqR}{m}$$

$\therefore$ Kinetic energy of proton,

$$K = \frac{1}{2}mv^2 = \frac{1}{2}m\left(\frac{BqR}{m}\right)^2 = \frac{B^2q^2R^2}{2m}$$

Substituting the values in SI units, we have

$$K = \frac{(0.67)^2(1.6 \times 10^{-19})^2(0.60)^2}{2 \times 1.67 \times 10^{-27}}$$

$$= 1.2 \times 10^{-12} \text{ J}$$

$$= \frac{1.2 \times 10^{-12}}{(1.6 \times 10^{-19})(10^6)} \text{ MeV}$$

$$= 7.5 \text{ MeV}$$ **Ans.**

Ex. 12 What is the value of B that can be set up at the equator to permit a proton of speed 10^7 m/s to circulate around the earth? [$R = 6.4 \times 10^6$ m, $m_p = 1.67 \times 10^{-27}$ kg]

Sol. From the relation

$$r = \frac{mv}{Bq}$$

We have, $$B = \frac{mv}{qr}$$

Substituting the values, we have

$$B = \frac{(1.67 \times 10^{-27})(10^7)}{(1.6 \times 10^{-19})(6.4 \times 10^6)}$$

$$= 1.6 \times 10^{-8} \text{ T}$$ **Ans.**

Ex. 13 Uniform electric and magnetic fields with strength **E** and **B** are directed along the y-axis. A particle with specific charge q/m leaves the origin in the direction of x-axis with an initial velocity v_0. Find

(a) the y-coordinate of the particle when it crosses the y-axis for nth time.

(b) the angle α between the particle's velocity vector and the y-axis at that moment. **[Irodov 3.387]**

Sol. (a) As discussed in Type-7 path of the particle is a helix of increasing pitch. The axis of the helix is parallel to y-axis (parallel to **E**) and plane of circle of the helix is xz (perpendicular to **B**). The particle will cross the y-axis nth time after,

$$t = nT = n\left(\frac{2\pi m}{Bq}\right) = \frac{2\pi mn}{Bq}$$

The y-coordinate of particle at this instant is

$$y = \frac{1}{2}a_y t^2$$

where, $$a_y = \frac{F_y}{m} = \frac{qE}{m}$$

$$\therefore \quad y = \frac{1}{2}\left(\frac{qE}{m}\right)\left(\frac{2\pi mn}{Bq}\right)^2$$

$$= \frac{2n^2 mE\pi^2}{qB^2} \quad \textbf{Ans.}$$

(b) At this moment y-component of its velocity is

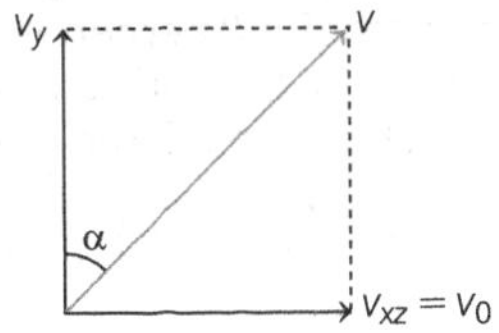

$$v_y = a_y t = \left(\frac{qE}{m}\right)\left(\frac{2\pi mn}{Bq}\right) = 2\pi n\left(\frac{E}{B}\right)$$

The angle α between particle's velocity vector and the y-axis at this moment is

$$\alpha = \tan^{-1}\left(\frac{v_{xz}}{v_y}\right)$$

Here, $\quad v_{xz} = \sqrt{v_x^2 + v_z^2} = v_0$

or $\quad \alpha = \tan^{-1}\left(\frac{Bv_0}{2\pi nE}\right)$ **Ans.**

Ex. 14 A particle of charge q and mass m is projected from the origin with velocity $\mathbf{v} = v_0\hat{\mathbf{i}}$ in a non-uniform magnetic field $\mathbf{B} = -B_0 x\hat{\mathbf{k}}$. Here, v_0 and B_0 are positive constants of proper dimensions. Find the maximum positive x-coordinate of the particle during its motion.

Sol. Magnetic field is along negative z-direction. So in the coordinate axes shown in figure, it is perpendicular to paper inwards. (⊗) Magnetic force on the particle at origin is along positive y-direction. So, it will rotate in xy-plane as shown. The path is not a perfect circle as the magnetic field is non-uniform. Speed of the particle in magnetic field remains constant. Magnetic force is always perpendicular to velocity. Let at point P (x, y), its velocity vector makes an angle θ with positive x-axis. Then, magnetic force $\mathbf{F}_m$ will be at angle θ with positive y-direction. So,

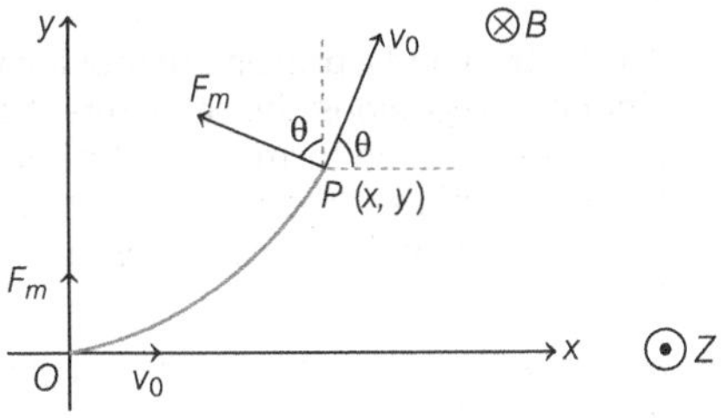

$$a_y = \left(\frac{F_m}{m}\right)\cos\theta$$

$$\therefore \quad \frac{dv_y}{dt} = \frac{(B_0 x)(qv_0\cos\theta)}{m} \qquad [F_m = Bqv_0 \sin 90°]$$

$$\therefore \quad \left(\frac{dv_y}{dx}\right)\cdot\left(\frac{dx}{dt}\right) = \left(\frac{B_0 qx}{m}\right)(v_0\cos\theta)$$

Here, $\quad \frac{dx}{dt} = v_x = v_0\cos\theta$

$$\therefore \quad \frac{dv_y}{dx} = \left(\frac{B_0 q}{m}\right)x$$

$$\therefore \quad \int_0^{v_0} dv_y = \left(\frac{B_0 q}{m}\right)\int_0^{x_{max}} x\,dx$$

$$\therefore \quad v_0 = \left(\frac{B_0 q}{m}\right)\left(\frac{x_{max}^2}{2}\right)$$

$$\therefore \quad x_{max} = \sqrt{\frac{2mv_0}{B_0 q}} \quad \textbf{Ans.}$$

Note *At maximum x-displacement velocity is along positive y-direction.*

MODULE 2 Exercise

True/False

1. An electron and a proton are moving with the same kinetic energy along the same direction. When they pass through a uniform magnetic field perpendicular to the direction of their motion, they describe circular path of the same radius. Is this statement true or false? **(1985 JEE)**

2. A charged particle enters a region of uniform magnetic field at an angle of 85° to the magnetic line of force. The path of the particle is a circle. Is this statement true or false? **(1983 JEE)**

Assertion and Reason

See directions on page 25.

3. **Assertion :** A charged particle moves along positive y-axis with constant velocity in uniform electric and magnetic fields. If magnetic field is acting along positive x-axis, then electric field should act along positive z-axis.

   Reason : To keep the charged particle undeviated the relation $\mathbf{E} = \mathbf{B} \times \mathbf{v}$ must hold good.

4. **Assertion :** If a charged particle enters from outside at right angles in uniform magnetic field. The maximum time spent in magnetic field may be $\frac{\pi m}{Bq}$.

   Reason : It cannot complete circle in the magnetic field if it enters from outside.

Single Correct Option

5. A stream of protons and α-particles of equal momenta enter a uniform magnetic field perpendicularly. The radii of their orbits are in the ratio
   (a) 1 : 1
   (b) 1 : 2
   (c) 2 : 1
   (d) 4 : 1

6. When an electron is accelerated through a potential difference V, it experiences a force F through a uniform transverse magnetic field. If the potential difference is increased to 2 V, the force experienced by the electron in the same magnetic field is
   (a) $2F$
   (b) $2\sqrt{2}F$
   (c) $\sqrt{2}F$
   (d) $4F$

7. Electric field and magnetic field in a region of space is given by $\mathbf{E} = E_0\hat{\mathbf{j}}$ and $\mathbf{B} = B_0\hat{\mathbf{j}}$. A particle of specific charge α is released from origin with velocity $\mathbf{v} = v_0\hat{\mathbf{i}}$. Then, path of the particle
   (a) is a circle
   (b) is a helix with uniform pitch
   (c) is a helix with non-uniform pitch
   (d) is cycloid

Note E_0, B_0 and v_0 are constant values.

8. An electron having kinetic energy K is moving in a circular orbit of radius R perpendicular to a uniform magnetic induction. If kinetic energy is doubled and magnetic induction tripled, the radius will become
   (a) $\frac{2R}{3}$
   (b) $\frac{\sqrt{2}}{3}R$
   (c) $\sqrt{\frac{2}{3}}R$
   (d) $\frac{2}{\sqrt{3}}R$

9. A charged particle of mass m and charge q is accelerated through a potential difference of V volts. It enters a region of uniform magnetic field B which is directed perpendicular to the direction of motion of the particle. The particle will move on a circular path of radius
   (a) $\sqrt{\frac{Vm}{2qB^2}}$
   (b) $\frac{2Vm}{qB^2}$
   (c) $\sqrt{\frac{2Vm}{q}}\left(\frac{1}{B}\right)$
   (d) $\sqrt{\frac{Vm}{q}}\left(\frac{1}{B}\right)$

10. A charged particle travels along a straight line with a speed v in a region where both electric field $\mathbf{E}$ and magnetic field $\mathbf{B}$ are present. It is given that $(\mathbf{v} \perp \mathbf{B})$.
    (a) $|\mathbf{E}| = |\mathbf{B}|$ and the two fields are perpendicular
    (b) $|\mathbf{E}| = v|\mathbf{B}|$ and the two fields are perpendicular
    (c) $|\mathbf{B}| = v|\mathbf{E}|$ and the two fields are parallel
    (d) $|\mathbf{B}| = v|\mathbf{E}|$ and the two fields are perpendicular

11. A particle of mass m and charge q is released from rest at the origin as shown in the figure. The speed of the particle when it has travelled a distance d along the z-axis is given by

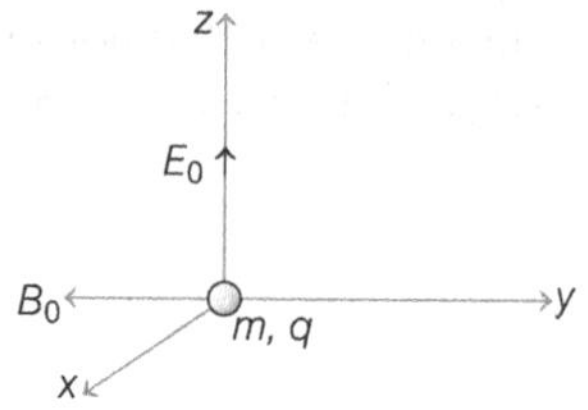

    (a) $\sqrt{\frac{2qE_0d}{m}}$
    (b) $q\sqrt{\frac{E_0d}{B_0}}$
    (c) $\sqrt{\frac{m}{qE_0d}}$
    (d) $\sqrt{\frac{2m}{qE_0d}}$

12. A charged particle begins to move from the origin in a region which has a uniform magnetic field in the x-direction and a uniform electric field in the y-direction. Its speed v when it reaches the point (x, y, z), will depend
    (a) only on x
    (b) only on y
    (c) on both x and y but not z
    (d) on x, y and z

13. An ionized gas contains both positive and negative ions initially at rest. If it is subjected simultaneously

to an electric field along the +x-direction and a magnetic field along the +z-direction, then

(a) positive ions deflect towards +y-direction and negative ions towards –y-direction
(b) all ions deflect towards +y-direction
(c) all ions deflect towards –y-direction
(d) positive ions deflect towards –y-direction and negative ions towards +y-direction

14. In the figure shown, a charge q moving with a velocity v along the x-axis enter into a region of uniform magnetic field. The minimum value of v so that the charge q is able to enter the region $x > b$ **(2002 JEE)**

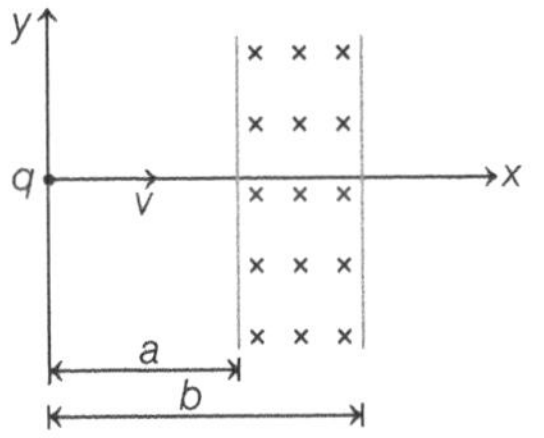

(a) $\frac{qBb}{m}$ (b) $\frac{qBa}{m}$

(c) $\frac{qB(b-a)}{m}$ (d) $\frac{qB(b+a)}{2m}$

15. A particle of mass 1×10^{-26} kg and charge $+1.6 \times 10^{-19}$ C travelling with a velocity of 1.28×10^{6} m/s along positive direction of x-axis enters a region in which a uniform electric field **E** and a uniform magnetic field **B** are present such that $E_z = -102.4$ kV/m and $B_y = 8 \times 10^{-2}$ Wb/m^2

The particle enters this region at origin at time $t = 0$. Then,

(a) net force acts on the particle is along the positive z-direction
(b) net force acts on the particle is along negative z-direction
(c) net force acting on particle is zero
(d) net force acts in xz-plane

16. A particle of mass m and having a positive charge q is projected from origin with speed v_0 along the positive x-axis in a magnetic field $\mathbf{B} = -B_0\hat{\mathbf{k}}$ where B_0 is a positive constant. If the particle passes through $(0, y, 0)$, then y is equal to

(a) $-\frac{2mv_0}{qB_0}$ (b) $\frac{mv_0}{qB}$

(c) $-\frac{mv_0}{qB}$ (d) $\frac{2mv_0}{qB_0}$

Subjective Questions

17. A deuteron (the nucleus of an isotope of hydrogen) has a mass of 3.34×10^{-27} kg and a charge of $+e$. The deuteron travels in a circular path with a radius of 6.96 mm in a magnetic field with magnitude 2.50 T.

(a) Find the speed of the deuteron.
(b) Find the time required for it to make half of a revolution.
(c) Through what potential difference would the deuteron have to be accelerated to acquire this speed?

18. A neutral particle is at rest in a uniform magnetic field **B**. At time $t = 0$, it decays into two charged particles, each of mass m.

(a) If the charge of one of the particles is $+q$, what is the charge of the other?
(b) The two particles move off in separate paths, both of them lie in the plane perpendicular to **B**. At a later time, the particles collide. Express the time from decay until collision in terms of m, B and q.

19. An electron at point A in figure has a speed $v_0 = 1.41 \times 10^{6}$ m/s. Find

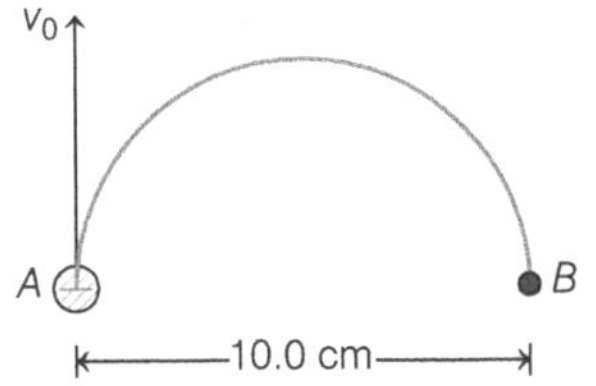

(a) the magnitude and direction of the magnetic field that will cause the electron to follow the semicircular path from A to B,
(b) the time required for the electron to move from A to B.

20. A proton moves at a constant velocity of 50 m/s along the x-axis, in uniform electric and magnetic fields. The magnetic field is $\mathbf{B} = (2.0 \text{ mT})\hat{\mathbf{j}}$. What is the electric field? **v**, **B** and **E** all are mutually perpendicular.

21. A particle having mass m and charge q is released from the origin in a region in which electric field and magnetic field are given by

$$\mathbf{B} = -B_0\hat{\mathbf{j}}$$

and

$$\mathbf{E} = E_0\hat{\mathbf{k}}$$

Find the y-component of the velocity and the speed of the particle as a function of its z-coordinate.

22. Non-relativistic protons move rectilinearly in the region of space where there are uniform mutually perpendicular electric and magnetic fields with $E = 4.0$ kV/m and $B = 50$ mT. The trajectory of the protons lies in the plane XZ and forms an angle $\phi = 30°$ with the X-axis. Find the pitch of the helical trajectory along which the protons will move after the electric field is switched OFF. **(Irodov 3.390)**

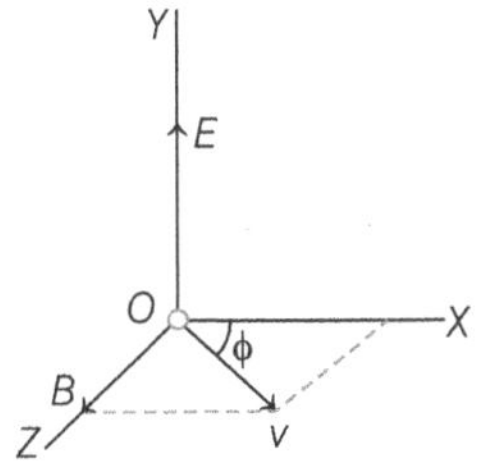

23. In a region of space, a uniform magnetic field **B** is along positive *x*-axis. Electrons are emitted from the origin with speed *v* at different angles. Show that the paraxial electrons are refocused on the *x*-axis at a distance $\frac{2\pi mv}{Be}$. Here, *m* is the mass of electron and *e* the charge on it.

24. A particle of mass *m* and charge *q* is projected into a region having a perpendicular magnetic field **B**. Find the angle of deviation of the particle as it comes out of the magnetic field, if the width of the region is

(a) $\frac{2mv}{Bq}$ (b) $\frac{mv}{Bq}$ (c) $\frac{mv}{2Bq}$

25. In a certain region, uniform electric field $\mathbf{E} = -E_0\hat{\mathbf{k}}$ and magnetic field $\mathbf{B} = B_0\hat{\mathbf{k}}$ are present. At time $t = 0$, a particle of mass *m* and charge *q* is given a velocity $\mathbf{v} = v_0\hat{\mathbf{j}} + v_0\hat{\mathbf{k}}$. Find the minimum speed of the particle and the time when it happens so.

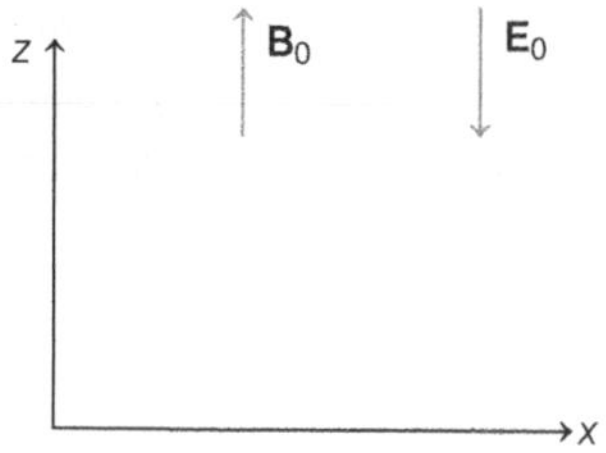

Answers

1. (False) 2. (False) 3. (a) 4. (a) 5. (c) 6. (c) 7. (c) 8. (b) 9. (c) 10. (b)
11. (a) 12. (b) 13. (c) 14. (c) 15. (c) 16. (d)
17. (a) 8.35×10^5 m/s (b) 2.62×10^{-8} s (c) 7.26 kV 18. (a) $-q$ (b) $\frac{\pi m}{Bq}$ 19. (a) 1.6×10^{-4} T into the page (b) 1.11×10^{-7}s
20. $\mathbf{E} = (-0.1 \text{ V/m})\hat{\mathbf{k}}$ 21. $v_y = 0$, $v = \sqrt{\frac{2qE_0z}{m}}$ 22. $p = 6$ cm 24. (a) π (b) π (c) $\frac{\pi}{6}$ 25. v_0, $\frac{mv_0}{qE_0}$

See hints on pages 435-436.

MODULE 3

Magnetic Force on a Current Carrying Conductor

A charged particle in motion experiences a magnetic force in a magnetic field. Similarly, a current carrying wire also experiences a force when placed in a magnetic field. This follows from the fact that the current is a collection of many charged particles in motion. Hence, the resultant force exerted by the field on the wire is the vector sum of the individual forces exerted on all the charged particles making up the current.

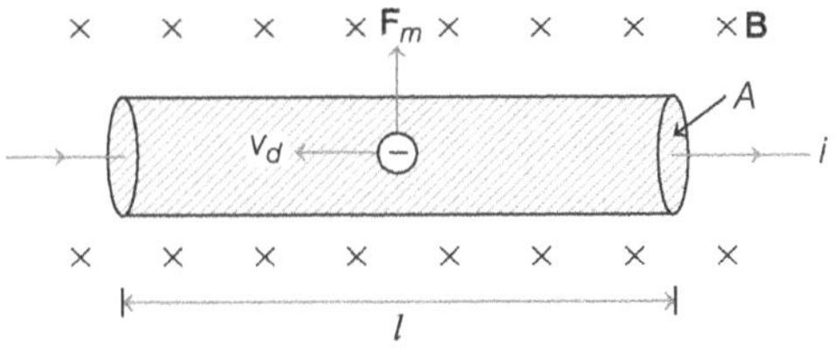

Suppose a conducting wire carrying a current i is placed in a magnetic field **B**. The length of the wire is l and area of cross-section is A. The free electrons drift with a speed v_d opposite to the direction of current. The magnetic force exerted on the electron is

$$d\mathbf{F}_m = -e\,(\mathbf{v}_d \times \mathbf{B})$$

If n be the number of free electrons per unit volume of the wire, then total number of electrons in volume Al of the wire are, nAl. Therefore, total force on the wire is

$$\mathbf{F}_m = -e\,(nAl)\,(\mathbf{v}_d \times \mathbf{B})$$

If we denote the length l along the direction of the current by **l**, then the above equation becomes

$$\boxed{\mathbf{F}_m = i\,(\mathbf{l} \times \mathbf{B})} \quad ...(i)$$

where, $neAv_d = i$

The following points are worthnoting regarding the above expression :

(i) Magnitude of $\mathbf{F}_m$ is, $F_m = ilB\sin\theta$, here θ is the angle between **l** and **B**.

F_m is zero for $\theta = 0°$ or $180°$ and maximum for $\theta = 90°$.

(ii) Here, **l** is a vector that points in the direction of the current i and has a magnitude equal to the length.

(iii) The above expression applies only to a straight segment of wire in a uniform magnetic field.

(iv) For the magnetic force on an arbitrarily shaped wire segment, let us consider the magnetic force exerted on a small segment of vector length $d\mathbf{l}$.

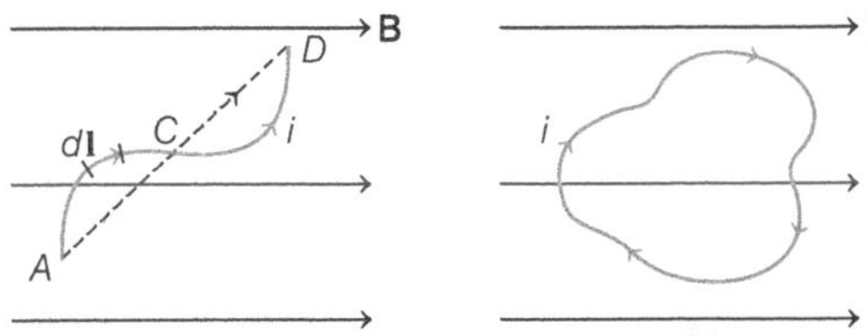

(a) (b)

$$d\mathbf{F}_m = i\,(d\mathbf{l} \times \mathbf{B}) \quad ...(ii)$$

To calculate the total force $\mathbf{F}_m$ acting on the wire shown in figure, we integrate Eq. (ii) over the length of the wire.

$$\mathbf{F}_m = i\int_A^D (d\mathbf{l} \times \mathbf{B}) \quad ...(iii)$$

Now, let us consider two special cases involving Eq. (iii). In both cases, the magnetic field is taken to be constant in magnitude and direction.

Case 1 A curved wire ACD as shown in Fig. (a) carries a current i and is located in a uniform magnetic field **B**. Because the field is uniform, we can take **B** outside the integral in Eq. (iii) and we obtain,

$$\mathbf{F}_m = i\left(\int_A^D d\mathbf{l}\right) \times \mathbf{B} \quad ...(iv)$$

But, the quantity $\int_A^D d\mathbf{l}$ represents the vector sum of all length elements from A to D. From the polygon law of vector addition, the sum equals the vector **l** directed from A to D. Thus,

$$\mathbf{F}_m = i\,(\mathbf{l} \times \mathbf{B})$$

or we can write

$$\boxed{\mathbf{F}_{ACD} = \mathbf{F}_{AD} = i\,(\mathbf{AD} \times \mathbf{B})} \text{ in uniform field.}$$

Case 2 An arbitrarily shaped closed loop carrying a current i is placed in a uniform magnetic field as shown in Fig. (b). We can again express the force acting on the loop in the form of Eq. (iv), but this time we must take the vector sum of the length elements $d\mathbf{l}$ over the entire loop,

$$\mathbf{F}_m = i\,(\oint d\mathbf{l}) \times \mathbf{B}$$

Because the set of length elements forms a closed polygon, the vector sum must be zero.

$$\therefore \quad \mathbf{F}_m = 0$$

Thus, **the net magnetic force acting on any closed current carrying loop in a uniform magnetic field is zero.**

(v) The direction of $\mathbf{F}_m$ can be given by Fleming's left hand rule as discussed in Module 1. According to this rule, the forefinger, the central finger and the thumb of the left hand are stretched in such a way that they are mutually perpendicular to each other. If the central finger shows the direction of current (or $\mathbf{l}$) and forefinger shows the direction of magnetic field($\mathbf{B}$), then the thumb will give the direction of magnetic force ($\mathbf{F}_m$).

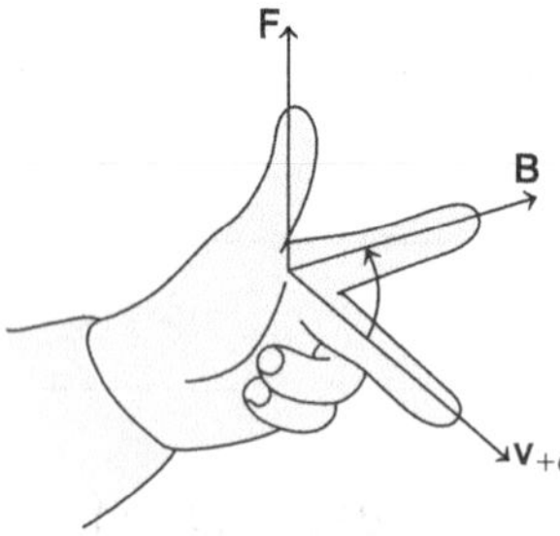

Example 1 A horizontal rod 0.2 m long is mounted on a balance and carries a current. At the location of the rod a uniform horizontal magnetic field has magnitude 0.067 T and direction perpendicular to the rod. The magnetic force on the rod is measured by the balance and is found to be 0.13 N. What is the current?

Sol. $F = ilB \sin 90°$

$$\therefore \quad i = \frac{F}{lB} = \frac{0.13}{0.2 \times 0.067}$$

$$= 9.7 \text{ A}$$ **Ans.**

Example 2 A square of side 2.0 m is placed in a uniform magnetic field $\mathbf{B} = 2.0$ T in a direction perpendicular to the plane of the square inwards. Equal current $i = 3.0$ A is flowing in the directions shown in figure. Find the magnitude of magnetic force on the loop.

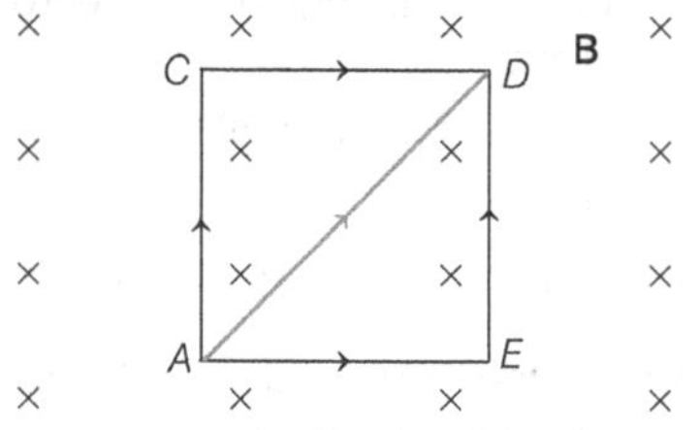

Sol. Force on wire ACD = Force on AD = Force on AED

$\therefore$ Net force on the loop $= 3\,(\mathbf{F}_{AD})$

or $\quad F_{net} = 3\,(i)\,(AD)\,(B)$

$$= (3)\,(3.0)\,(2\sqrt{2})\,(2.0) \text{ N}$$

$$= 36\sqrt{2} \text{ N}$$

Using Fleming's left hand rule we can see that the direction of this force is towards EC.

Example 3 In the figure shown a semicircular wire loop is placed in a uniform magnetic field $B = 1.0$ T. The plane of the loop is perpendicular to the magnetic field. Current $i = 2$ A flows in the loop in the directions shown. Find the magnitude of the magnetic force in both the cases (a) and (b). The radius of the loop is 1.0 m

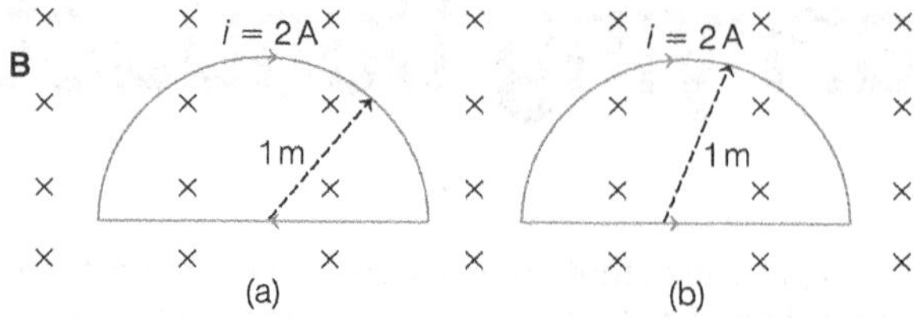

Sol. Refer figure (a) It forms a closed loop and the current completes the loop. Therefore, net force on the loop in uniform field should be zero. **Ans.**

Refer figure (b) In this case although it forms a closed loop, but current does not complete the loop. Hence, net force is not zero.

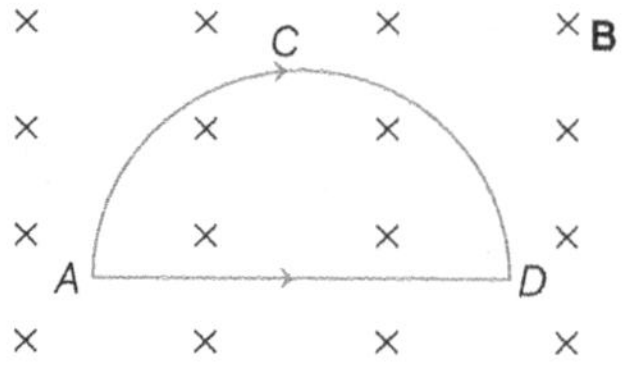

$$\mathbf{F}_{ACD} = \mathbf{F}_{AD}$$

$$\therefore \quad \mathbf{F}_{loop} = \mathbf{F}_{ACD} + \mathbf{F}_{AD} = 2\mathbf{F}_{AD}$$

$$\therefore \quad |\mathbf{F}_{loop}| = 2\,|\mathbf{F}_{AD}| = 2ilB \sin\theta \qquad [l = 2r = 2.0 \text{ m}]$$

$$= (2)\,(2)\,(2)\,(1) \sin 90° = 8 \text{ N}$$ **Ans.**

Example 4 A wire PQ of mass 10 g is at rest on two parallel metal rails. The separation between the rails is 4.9 cm. A magnetic field of 0.80 T is applied perpendicular to the plane of the rails, directed downwards. The resistance of the circuit is slowly decreased. When the resistance decreases to below 20 Ω, the wire PQ begins to slide on the rails. Calculate the coefficient of friction between the wire and the rails.

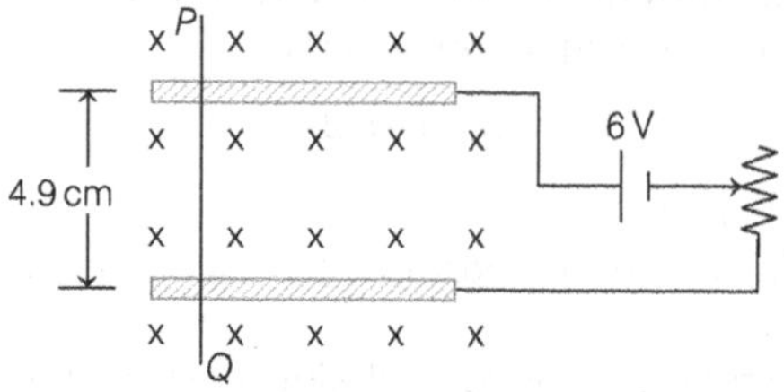

Sol. Wire PQ begins to slide when magnetic force is just equal to the force of friction, i.e.

$$\mu\, mg = ilB \sin\theta = ilB \qquad (\text{as } \theta = 90°)$$

Here, $\quad i = \frac{E}{R} = \frac{6}{20} = 0.3 \text{ A}$

$$\therefore \quad \mu = \frac{ilB}{mg} = \frac{(0.3)\,(4.9 \times 10^{-2})\,(0.8)}{(10 \times 10^{-3})\,(9.8)}$$

$$= 0.12$$ **Ans.**

MODULE 3 Exercise

Single Correct Option

1. A conductor *ab* of arbitrary shape carries current I flowing from b to a. The length vector **ab** is oriented from a to b. The force **F** experienced by this conductor in a uniform magnetic field **B** is

   (a) $\mathbf{F} = -I\,(\mathbf{ab} \times \mathbf{B})$ (b) $\mathbf{F} = I(\mathbf{B} \times \mathbf{ab})$
   (c) $\mathbf{F} = I(\mathbf{ba} \times \mathbf{B})$ (d) All of these

2. A circular loop carrying a current I is placed in the xy-plane as shown in figure. A uniform magnetic field **B** is oriented along the positive z-axis. The loop tends to

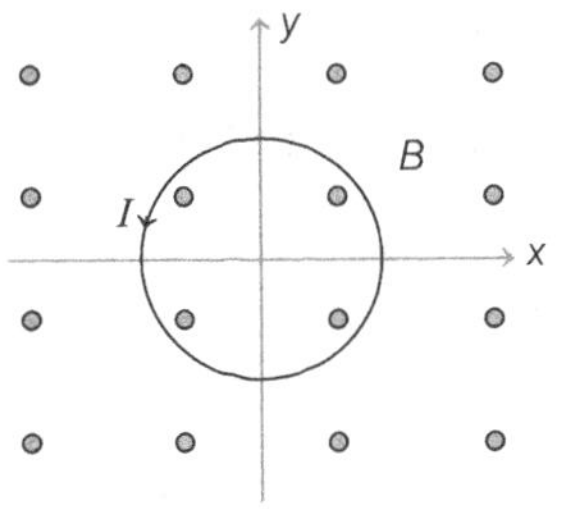

   (a) expand (b) contract
   (c) rotate about x-axis (d) rotate about y-axis

3. A current carrying wire AC is placed in a uniform transverse magnetic field, then the force on wire AC is

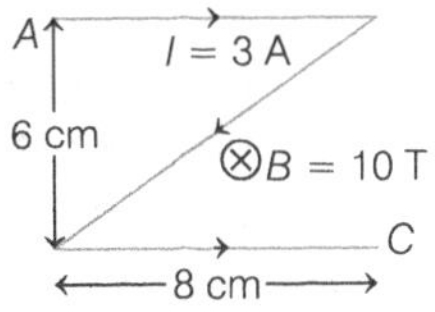

   (a) 3 N (b) 4.2 N
   (c) 6 N (d) 4 N

4. A conducting rod of length L and mass m is moving down a smooth inclined plane of inclination θ with constant speed v. A current I is flowing in the conductor perpendicular to the paper inwards. A vertically upward magnetic field **B** exists in space there. The magnitude of magnetic field B is

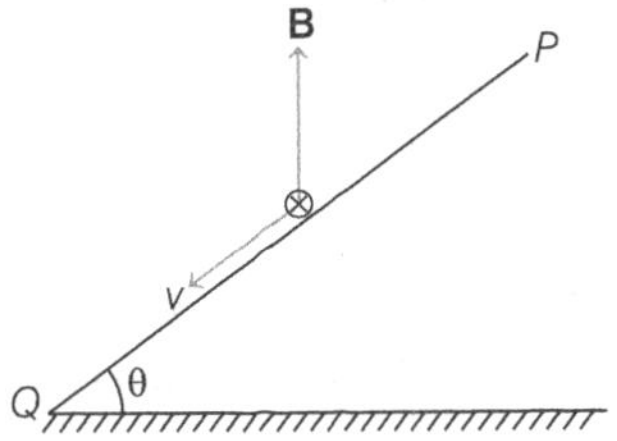

   (a) $\frac{mg}{IL}\sin\theta$ (b) $\frac{mg}{IL}\cos\theta$
   (c) $\frac{mg}{IL}\tan\theta$ (d) $\frac{mg}{IL}\sec\theta$

5. A wire $PQRS$ carrying a current I runs along three edges of a cube of side l as shown. There exists a uniform magnetic field of magnitude B along one of the sides of the cube. The magnitude of the force acting on the wire is

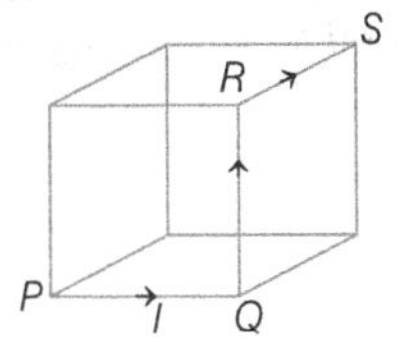

   (a) zero (b) $\sqrt{3}\,IB$ (c) $\sqrt{2}\,IlB$ (d) $2IlB$

6. A conductor of length l is placed perpendicular to a horizontal uniform magnetic field B. Suddenly, a certain amount of charge is passed through it, when it is found to jump to a height h. The amount of charge that passes through the conductor is

   (a) $\frac{m\sqrt{gh}}{Bl}$ (b) $\frac{m\sqrt{gh}}{2Bl}$
   (c) $\frac{m\sqrt{2gh}}{Bl}$ (d) None of these

7. A wire lying along y-axis from $y = 0$ to $y = 1$ m carries a current of 2 mA in the negative y-direction. The wire lies in a non-uniform magnetic field given by $\mathbf{B} = (0.3\ \text{T/m})\, y\hat{\mathbf{i}} + (0.4\ \text{T/m})\, y\hat{\mathbf{j}}$. The magnetic force on the entire wire is

   (a) $-3 \times 10^{-4}\hat{\mathbf{j}}$ N (b) $6 \times 10^{-3}\hat{\mathbf{k}}$ N
   (c) $-3 \times 10^{-4}\hat{\mathbf{k}}$ N (d) $3 \times 10^{-4}\hat{\mathbf{k}}$ N

8. In the figure, the force on the wire ABC in the given uniform magnetic field will be ($B = 2$ tesla)

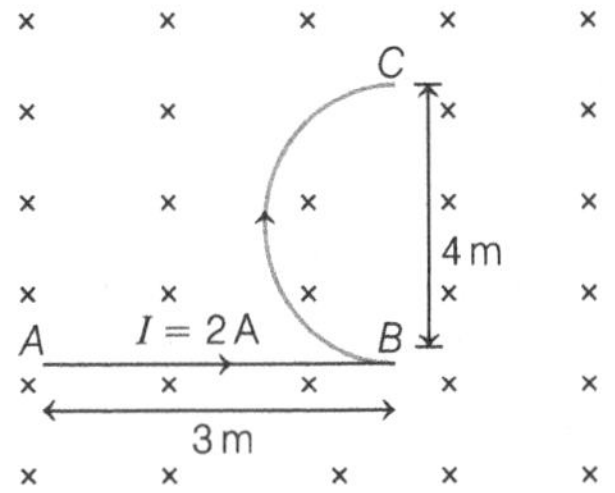

   (a) $4(3 + 2\pi)$ N (b) 20 N
   (c) 30 N (d) 40 N

9. A circular flexible loop of wire of radius r carrying a current I is placed in a uniform magnetic field B perpendicular to the plane of the circle so that wire comes under tension. If B is doubled, tension in the loop

   (a) remains unchanged (b) is doubled
   (c) is halved (d) becomes 4 times

10. The magnetic field existing in a region is given by

$$\mathbf{B} = B_0\left[1 + \frac{x}{l}\right]\hat{\mathbf{k}}$$

A square loop of edge l and carrying current I is placed with its edges parallel to the xy-axis. The magnitude of the net magnetic force experienced by the loop is

(a) $2B_0Il$ (b) zero (c) B_0Il (d) $4B_0Il$

11. A straight rod of mass m and length L is suspended from the identical springs as shown in figure. The spring is stretched a distance x_0 due to the weight of the wire.

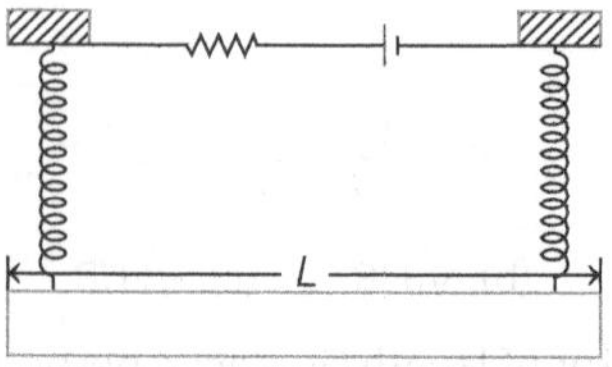

The circuit has total resistance R. When the magnetic field perpendicular to the plane of paper is switched on, springs are observed to extend further by the same distance. The magnetic field strength is

(a) $\frac{2mgR}{LE}$ (b) $\frac{mgR}{LE}$ (c) $\frac{mgR}{2LE}$ (d) $\frac{mgR}{E}$

12. A wire is bent as a parabolic curve and kept in the xy-plane. The curve can be described by the equation $x^2 = 6y$. The wire carries a current $i = 2$ A. If a uniform magnetic field $\mathbf{B} = 2 \times 10^{-3}\hat{\mathbf{k}}$ tesla is applied, force experienced by the wire is nearly

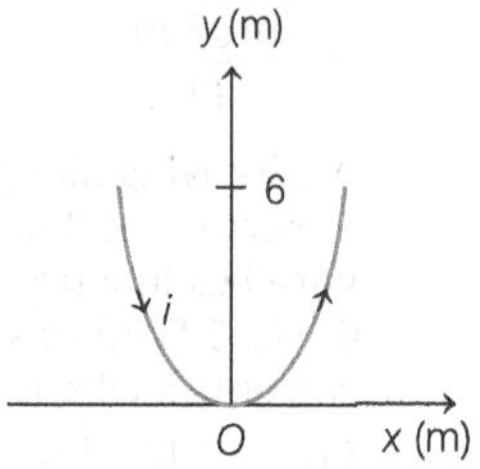

(a) $1.5\hat{\mathbf{j}}$ N (b) $-1.5\hat{\mathbf{j}}$ N (c) $-0.05\hat{\mathbf{j}}$ N (d) $2.5\hat{\mathbf{j}}$ N

Subjective Questions

13. A wire of length l carries a current i along the x-axis. A magnetic field $\mathbf{B} = B_0(\hat{\mathbf{j}} + \hat{\mathbf{k}})$ exists in the space. Find the magnitude of the magnetic force acting on the wire.

14. In the above problem will the answer change if magnetic field becomes $\mathbf{B} = B_0(\hat{\mathbf{i}} + \hat{\mathbf{j}} + \hat{\mathbf{k}})$.

15. A wire along the x-axis carries a current of 3.50 A in the negative direction. Calculate the force (expressed in terms of unit vectors) on a 1.00 cm section of the wire exerted by these magnetic fields

(a) $\mathbf{B} = -(0.65\ \text{T})\hat{\mathbf{j}}$

(b) $\mathbf{B} = +(0.56\ \text{T})\hat{\mathbf{k}}$

(c) $\mathbf{B} = -(0.31\ \text{T})\hat{\mathbf{i}}$

(d) $\mathbf{B} = +(0.33\ \text{T})\hat{\mathbf{i}} - (0.28\ \text{T})\hat{\mathbf{k}}$

(e) $\mathbf{B} = +(0.74\ \text{T})\hat{\mathbf{j}} - (0.36\ \text{T})\hat{\mathbf{k}}$

16. Find net force on the equilateral loop of side 4 m carrying a current of 2 A kept in a uniform magnetic field of 2 T as shown in figure.

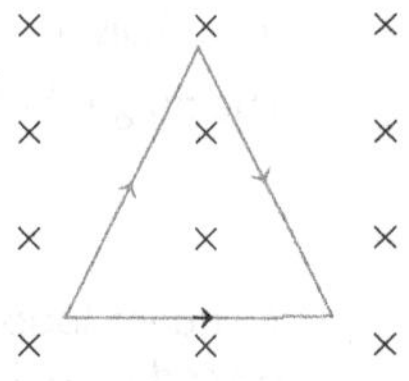

17. A wire of 62.0 cm length and 13.0 g mass is suspended by a pair of flexible leads in a uniform magnetic field of magnitude 0.440 T in figure. What are the magnitude and direction of the current required to remove the tension in the supporting leads? Take, $g = 10\ \text{m/s}^2$.

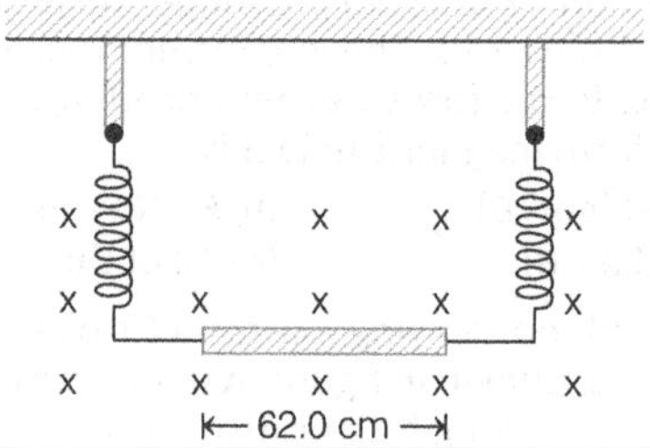

18. A thin, 50.0 cm long metal bar with mass 750 g rests on, but is not attached to, two metallic supports in a 0.450 T magnetic field as shown in figure. A battery and a resistance $R = 25.0\ \Omega$ in series are connected to the supports.

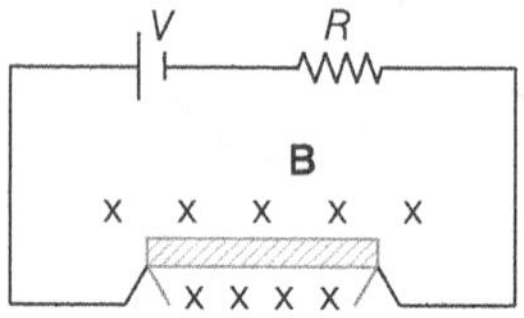

(a) What is the largest voltage the battery can have without breaking the circuit at the supports?

(b) The battery voltage has this maximum value calculated. Decreasing the resistance to 2.0Ω, find the initial acceleration of the bar.

19. In figure, the cube is 40.0 cm on each edge. Four straight segments of wire ab, bc, cd and da form a closed loop that carries a current $I = 5.00$ A, in the direction shown. A uniform magnetic field of magnitude $B = 0.020$ T is in the positive y-direction. Determine the magnitude and direction of the magnetic force on each segment.

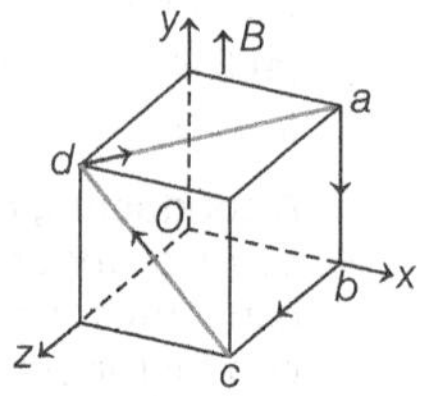

Answers

1. (d) **2.** (a) **3.** (a) **4.** (c) **5.** (c)
6. (c) **7.** (d) **8.** (b) **9.** (b) **10.** (c)
11. (b) **12.** (c) **13.** $\sqrt{2}B_0il$ **14.** No
15. (a) $(0.023\text{N})\hat{\mathbf{k}}$ (b) $(0.02\ \text{N})\hat{\mathbf{j}}$ (c) Zero (d) $(-0.0098\ \text{N})\hat{\mathbf{j}}$
(e) $(-0.013\ \text{N})\hat{\mathbf{j}} + (-0.026\ \text{N})\hat{\mathbf{k}}$
16. 32 N upwards **17.** 0.47 A from left to right
18. (a) 817 V (b) $112.8\ \text{m/s}^2$
19. $\mathbf{F}_{ab} = 0$, $\mathbf{F}_{bc} = (-0.04\text{N})\hat{\mathbf{i}}$, $\mathbf{F}_{cd} = (-0.04\ \text{N})\hat{\mathbf{k}}$,
$\mathbf{F}_{da} = (0.04\ \hat{\mathbf{i}} + 0.04\ \hat{\mathbf{k}})$ N

See hints on pages 436-438.

MODULE 4

Magnetic Dipole

Every current carrying loop is a magnetic dipole. It has two poles: south (S) and north (N). This is similar to a bar magnet. Magnetic field lines emanate from the north pole and after forming a closed path terminate on south pole. Each magnetic dipole has some magnetic moment (**M**). The magnitude of **M** is

$$|\mathbf{M}| = NiA$$

Here, N = number of turns in the loop

i = current in the loop and

A = area of cross-section of the loop.

For the direction of **M** any one of the following methods can be used:

(i) As in case of an electric dipole, the dipole moment **p** has a direction from negative charge to positive charge. In the similar manner, direction of **M** is from **south to north** pole.

The south and north poles can be identified by the sense of current. The side from where the current seems to be clockwise becomes south pole and the opposite side from where it seems anti-clockwise becomes north pole.

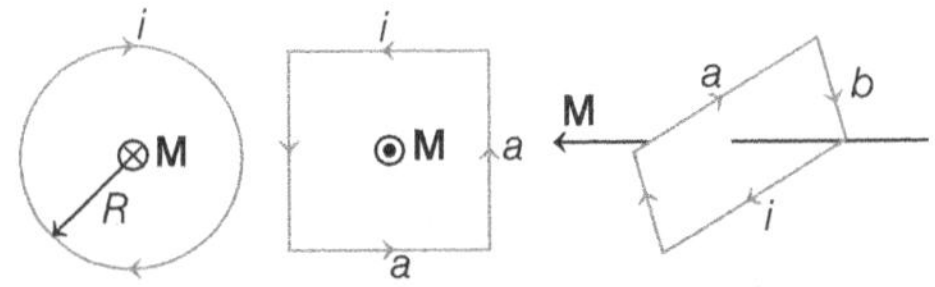

(a) (b) (c)

Now, let us find the direction and magnitude of $\mathbf{M}$ in the three loops shown in figure.

Refer figure (a) In this case, current appears to be clockwise from outside the paper, so this side becomes the south pole. From the back of the paper it seems anti-clockwise. Hence, this side becomes the north pole. As the magnetic moment is from south to north pole. It is directed perpendicular to paper inwards. Further,

$$|\mathbf{M}| = NiA = \pi R^2 i \qquad (N = 1)$$

Refer figure (b) Here, opposite is the case. South pole is into the paper and north pole is outside the paper. Therefore, magnetic moment is perpendicular to paper in outward direction. The magnitude of **M** is

$$|\mathbf{M}| = NiA = a^2 i \qquad (N = 1)$$

Refer figure (c) In this case, south pole is on the right side of the loop and north pole on the left side. Hence, **M** is directed from right to left. The magnitude of magnetic moment is

$$|\mathbf{M}| = NiA = abi \qquad (N = 1)$$

(ii) Vector **M** is along the normal to the plane of the loop. The orientation (up or down along the normal) is given by the right hand rule. Wrap your fingers of the right hand around the perimeter of the loop in the direction of current as shown in figure. Then, extend your thumb so that it is perpendicular to the plane of the loop. The thumb points in the direction of **M**.

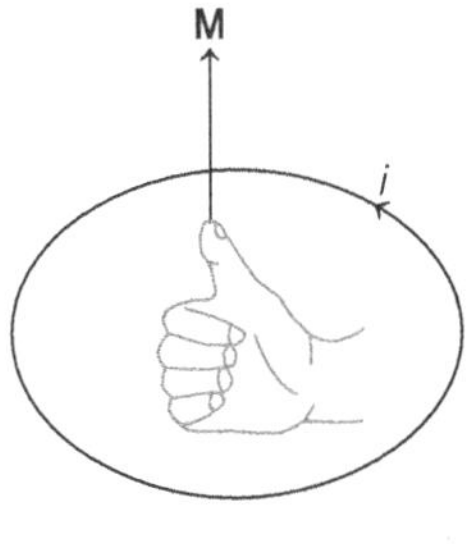

Important POINTS

- In addition to the method discussed above for finding **M** here are two more methods for calculating **M**.

Method 1 This method is useful for calculating **M** for a rectangular or square loop.

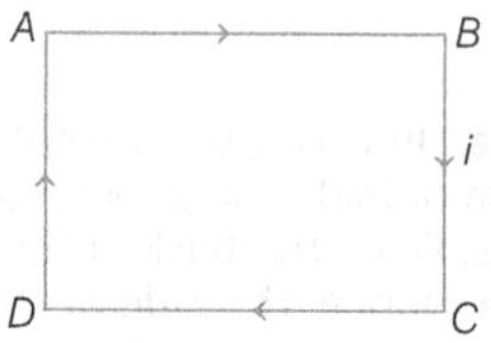

The magnetic moment (**M**) of the rectangular loop shown in figure is

$$\mathbf{M} = i(\mathbf{AB} \times \mathbf{BC}) = i(\mathbf{BC} \times \mathbf{CD}) = i(\mathbf{CD} \times \mathbf{DA}) = i(\mathbf{DA} \times \mathbf{AB})$$

Here, the cross product of any two consecutive sides (taken in order) gives the area as well as the correct direction of **M** also.

Note *If coordinates of vertices are known. Then, vector of any side can be written in terms of coordinates, e.g.*

$$\mathbf{AB} = (x_B - x_A)\hat{\mathbf{i}} + (y_B - y_A)\hat{\mathbf{j}} + (z_B - z_A)\hat{\mathbf{k}}$$

Method 2 Sometimes, a current carrying loop does not lie in a single plane. But, by assuming two equal and opposite currents in one branch (which obviously makes no change in the given circuit) two (or more) closed loops are completed in different planes. Now, the net magnetic moment of the given loop is the vector sum of individual loops.

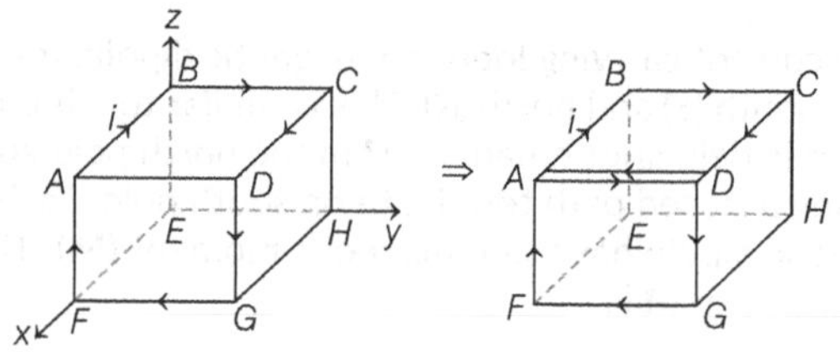

(a) (b)

For example, in Fig. (a), six sides of a cube of side l carry a current i in the directions shown. By assuming two equal and opposite currents in wire AD, two loops in two different planes (xy and yz) are completed.

$$\mathbf{M}_{ABCDA} = -il^2\hat{\mathbf{k}}$$

$$\mathbf{M}_{ADGFA} = -il^2\hat{\mathbf{i}}$$

$$\therefore \quad \mathbf{M}_{net} = -il^2(\hat{\mathbf{i}} + \hat{\mathbf{k}})$$

Example 1 A square loop $OABCO$ of side l carries a current i. It is placed as shown in figure. Find the magnetic moment of the loop.

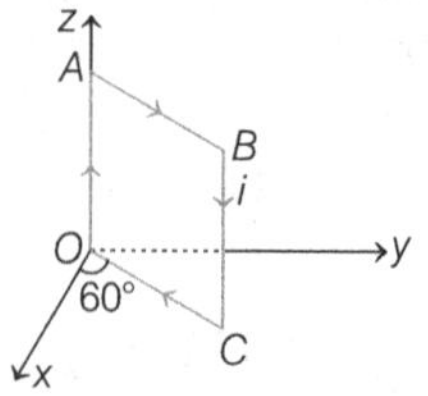

Sol. As discussed above, magnetic moment of the loop can be written as

$$\mathbf{M} = i(\mathbf{BC} \times \mathbf{CO})$$

Here, $\mathbf{BC} = -l\hat{\mathbf{k}}$,

$$\mathbf{CO} = -l\cos 60^\circ\,\hat{\mathbf{i}} - l\sin 60^\circ\,\hat{\mathbf{j}} = -\frac{l}{2}\hat{\mathbf{i}} - \frac{\sqrt{3}l}{2}\hat{\mathbf{j}}$$

$$\therefore \quad \mathbf{M} = i\left[(-l\hat{\mathbf{k}}) \times \left(-\frac{l}{2}\hat{\mathbf{i}} - \frac{\sqrt{3}l}{2}\hat{\mathbf{j}}\right)\right]$$

or $$\mathbf{M} = \frac{il^2}{2}(\hat{\mathbf{j}} - \sqrt{3}\,\hat{\mathbf{i}})$$ **Ans.**

Example 2 Find the magnitude of magnetic moment of the current carrying loop $ABCDEFA$. Each side of the loop is 10 cm long and current in the loop is $i = 2.0$ A.

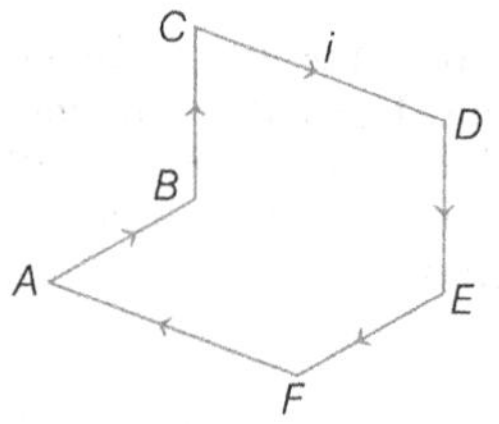

Sol. By assuming two equal and opposite currents in BE, two current carrying loops ($ABEFA$ and $BCDEB$) are formed. Their magnetic moments are equal in magnitude but perpendicular to each other. Hence,

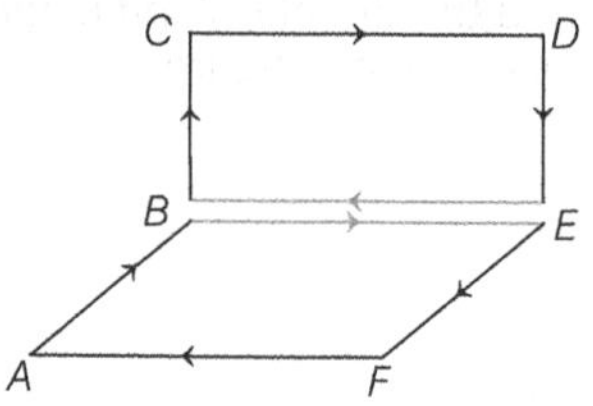

$$M_{net} = \sqrt{M^2 + M^2} = \sqrt{2}\,M$$

where, $$M = iA = (2.0)(0.1)(0.1) = 0.02 \text{ A-m}^2$$

$$\therefore \quad M_{net} = (\sqrt{2})(0.02) \text{ A-m}^2$$

$$= 0.028 \text{ A-m}^2$$ **Ans.**

Magnetic Dipole in Uniform Magnetic Field

Let us consider a rectangular ($a \times b$) current carrying loop *OACDO* placed in *xy*-plane. A uniform magnetic field

$$\mathbf{B} = B_x\hat{\mathbf{i}} + B_y\hat{\mathbf{j}} + B_z\hat{\mathbf{k}}$$

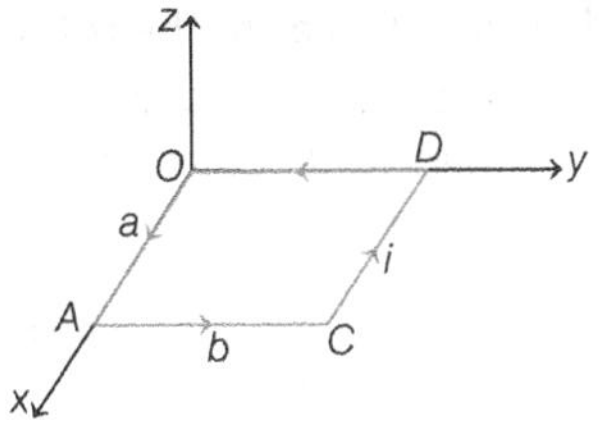

exists in space.

We are interested in finding the net force and torque in the loop.

Force : Net force on the loop is

$$\mathbf{F} = \mathbf{F}_{OA} + \mathbf{F}_{AC} + \mathbf{F}_{CD} + \mathbf{F}_{DO}$$
$$= i\,[(\mathbf{OA}\times\mathbf{B}) + (\mathbf{AC}\times\mathbf{B}) + (\mathbf{CD}\times\mathbf{B}) + (\mathbf{DO}\times\mathbf{B})]$$
$$= i\,[(\mathbf{OA} + \mathbf{AC} + \mathbf{CD} + \mathbf{DO})\times\mathbf{B}]$$
$$= \text{null vector}$$

or $|\mathbf{F}| = 0$, as $\mathbf{OA} + \mathbf{AC} + \mathbf{CD} + \mathbf{DO}$ forms a null vector.

Torque : Using $\mathbf{F} = i\,(\mathbf{l}\times\mathbf{B})$, we have

$$\mathbf{F}_{OA} = i\,(\mathbf{OA}\times\mathbf{B}) = i\,[(a\hat{\mathbf{i}})\times(B_x\hat{\mathbf{i}} + B_y\hat{\mathbf{j}} + B_z\hat{\mathbf{k}})] = ia\,[B_y\hat{\mathbf{k}} - B_z\hat{\mathbf{j}}]$$

$$\mathbf{F}_{AC} = i\,(\mathbf{AC}\times\mathbf{B}) = i\,[(b\hat{\mathbf{j}})\times(B_x\hat{\mathbf{i}} + B_y\hat{\mathbf{j}} + B_z\hat{\mathbf{k}})] = ib\,[-B_x\hat{\mathbf{k}} + B_z\hat{\mathbf{i}}]$$

$$\mathbf{F}_{CD} = i\,(\mathbf{CD}\times\mathbf{B}) = i\,[(-a\hat{\mathbf{i}})\times(B_x\hat{\mathbf{i}} + B_y\hat{\mathbf{j}} + B_z\hat{\mathbf{k}})] = ia\,[-B_y\hat{\mathbf{k}} + B_z\hat{\mathbf{j}}]$$

$$\mathbf{F}_{DO} = i\,(\mathbf{DO}\times\mathbf{B}) = i\,[(-b\hat{\mathbf{j}})\times(B_x\hat{\mathbf{i}} + B_y\hat{\mathbf{j}} + B_z\hat{\mathbf{k}})] = ib\,[B_x\hat{\mathbf{k}} - B_z\hat{\mathbf{i}}]$$

All these forces are acting at the centre of the wires. For example, $\mathbf{F}_{OA}$ will act at the centre of *OA*. When the forces are in equilibrium, net torque about any point remains the same. Let us find the torque about *O*.

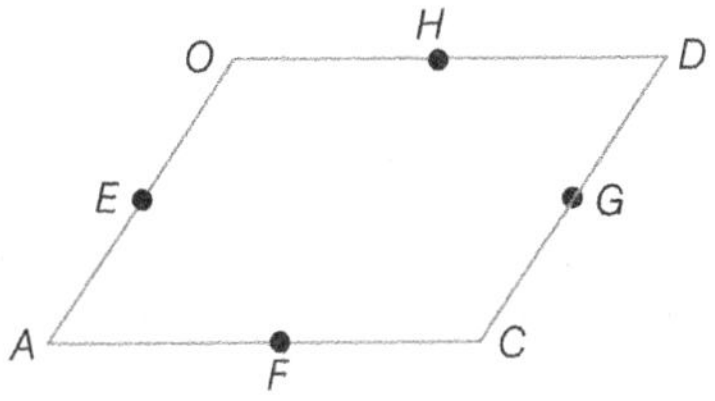

E, F, G and *H* are the mid-points of *OA, AC, CD* and *DO*, respectively.

Using $\tau = \mathbf{r}\times\mathbf{F}$, we have

$$\tau_O = (\mathbf{OE}\times\mathbf{F}_{OA}) + (\mathbf{OF}\times\mathbf{F}_{AC}) + (\mathbf{OG}\times\mathbf{F}_{CD}) + (\mathbf{OH}\times\mathbf{F}_{DO})$$

$$= \left[\left(\frac{a}{2}\hat{\mathbf{i}}\right)\times\{ia\,(B_y\hat{\mathbf{k}} - B_z\hat{\mathbf{j}})\}\right] + \left[\left(a\hat{\mathbf{i}} + \frac{b}{2}\hat{\mathbf{j}}\right)\times\{ib\,(-B_x\hat{\mathbf{k}} + B_z\hat{\mathbf{i}})\}\right]$$

$$+ \left[\left(\frac{a}{2}\hat{\mathbf{i}} + b\hat{\mathbf{j}}\right)\times\{ia\,(-B_y\hat{\mathbf{k}} + B_z\hat{\mathbf{j}})\}\right] + \left[\left(\frac{b}{2}\hat{\mathbf{j}}\right)\times ib\,(B_x\hat{\mathbf{k}} - B_z\hat{\mathbf{i}})\right]$$

$$= iab\,B_x\hat{\mathbf{j}} - iabB_y\hat{\mathbf{i}}$$

This can also be written as

$$\tau_O = (iab\hat{\mathbf{k}})\times(B_x\hat{\mathbf{i}} + B_y\hat{\mathbf{j}} + B_z\hat{\mathbf{k}})$$

Here, $iab\hat{\mathbf{k}}$ = magnetic moment of the dipole **M**

and $B_x\hat{\mathbf{i}} + B_y\hat{\mathbf{j}} + B_z\hat{\mathbf{k}} = \mathbf{B}$

$\therefore$
$$\boxed{\tau = \mathbf{M}\times\mathbf{B}}$$

Note that although this formula has been derived for a rectangular loop, it comes out to be true for any shape of loop. The following points are worthnoting regarding the torque acting on the loop in uniform magnetic field.

(i) Magnitude of τ is $MB\sin\theta$ or $NiAB\sin\theta$. Here, θ is the angle between **M** and **B**. Torque is zero when $\theta = 0°$ or $180°$ and it is maximum at $\theta = 90°$.

(ii) If the loop is free to rotate in a magnetic field, the axis of rotation becomes an axis parallel to τ passing through the centre of mass of the loop.

The above equation for the torque is very similar to that of an electric dipole in an electric field. The similarity between electric and magnetic dipoles extends even further as illustrated in the table below.

Table

S.No.	Field of similarity	Electric dipole	Magnetic dipole
1.	Magnitude	$\lvert\mathbf{p}\rvert = q\,(2d)$	$\lvert\mathbf{M}\rvert = NiA$
2.	Direction	from $-q$ to $+q$	from *S* to *N*
3.	Net force in uniform field	zero	zero
4.	Torque	$\tau = \mathbf{p}\times\mathbf{E}$	$\tau = \mathbf{M}\times\mathbf{B}$
5.	Potential energy	$U = -\mathbf{p}\cdot\mathbf{E}$	$U = -\mathbf{M}\cdot\mathbf{B}$
6.	Work done in rotating the dipole	$W_{\theta_1-\theta_2} = pE\,(\cos\theta_1 - \cos\theta_2)$	$W_{\theta_1-\theta_2} = MB\,(\cos\theta_1 - \cos\theta_2)$
7	Field along axis	$\mathbf{E} = \frac{1}{4\pi\varepsilon_0}\cdot\frac{2\mathbf{p}}{r^3}$	$\mathbf{B} = \frac{\mu_0}{4\pi}\cdot\frac{2\mathbf{M}}{r^3}$
8.	Field perpendicular to axis	$\mathbf{E} = -\frac{1}{4\pi\varepsilon_0}\cdot\frac{\mathbf{p}}{r^3}$	$\mathbf{B} = -\frac{\mu_0}{4\pi}\cdot\frac{\mathbf{M}}{r^3}$

Note *(i) In last two points r >>size of loop.*

(ii) At $\theta = 0°$ and $180°$, $\mathbf{F} = 0$ and $\tau = 0$. So the dipole is in equilibrium. At $\theta = 0°$, U is minimum. So, it is stable equilibrium position. Similarly at $\theta = 180°$, U is maximum. So, it is unstable equilibrium position.

(iii) If the loop is left from angle θ, the rotational oscillations start about an axis passing through centre of mass and parallel to τ (= $\mathbf{M}\times\mathbf{B}$) between $+\theta$ and $-\theta$.

If θ is small then these oscillations are simple harmonic in nature.

Note that the expressions for the magnetic dipole can be obtained from the expressions for the electric dipole by replacing **p** by **M** and ε_0 by $\frac{1}{\mu_0}$. Here, μ_0 is called the permeability of free space. It is related with ε_0 and speed of light *c* as

$$\boxed{c = \frac{1}{\sqrt{\varepsilon_0\mu_0}}}$$

and it has the value,

$$\mu_0 = 4\pi \times 10^{-7} \text{ T-m/A}$$

Dimensions of $\frac{1}{\sqrt{\varepsilon_0 \mu_0}}$ are same as that of speed or $[LT^{-1}]$.

Hence, $$\left[\frac{1}{\sqrt{\varepsilon_0 \mu_0}}\right] = [LT^{-1}]$$

Example 3 A circular loop of radius $R = 20$ cm is placed in a uniform magnetic field $\mathbf{B} = 2T$ in xy-plane as shown in figure. The loop carries a current $i = 1.0$ A in the direction shown in figure. Find the magnitude of torque acting on the loop.

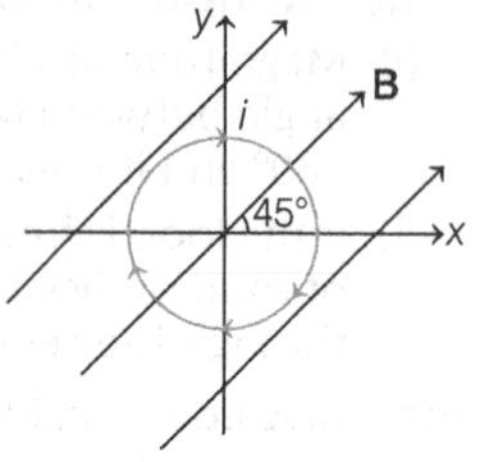

Sol. Magnitude of torque is given by

$$|\tau| = MB \sin\theta$$

Here, $M = NiA = (1)(1.0)(\pi)(0.2)^2$

$$= (0.04\,\pi) \text{ A-m}^2$$

$B = 2\text{T}$ and θ = angle between **M** and **B** = 90°

$\therefore \quad |\tau| = (0.04\,\pi)(2) \sin 90°$

$= 0.25$ N-m **Ans.**

Note *M is along negative z-direction (perpendicular to paper inwards) while* **B** *is in xy-plane. So, the angle between* **M** *and* **B** *is 90° not 45°. If we want to find torque in vector form, then we can write*

$$\mathbf{B} = 2\cos 45°\hat{i} + 2\sin 45°\hat{j} = \sqrt{2}(\hat{i} + \hat{j})\ T$$

and $\quad \mathbf{M} = -(0.04\,\pi)\,\hat{k}$ *A-m*^{2}

$\therefore \quad \tau = \mathbf{M} \times \mathbf{B} = (0.04\sqrt{2}\pi)(-\hat{j} + \hat{i})$

or $\quad \tau = 0.18(\hat{i} - \hat{j})$ **Ans.**

Example 4 In the Bohr model of the hydrogen atom, the electron circulates around the nucleus in a path of radius 5.1×10^{-11} m at a frequency of 6.8×10^{15} Hz.

What is the equivalent magnetic dipole moment?

Sol. An electron moving around the nucleus is equivalent to a current,

$$i = qf$$

The current carrying circular loop is equivalent to a magnetic dipole with magnetic dipole moment,

$$M = NiA = (Nqf\,\pi R^2)$$

Substituting the values, we have

$$M = (1)(1.6 \times 10^{-19})(6.8 \times 10^{15})(3.14)(5.1 \times 10^{-11})^2$$

$= 8.9 \times 10^{-24}$ A-m^2 **Ans.**

Direction of **M** is as shown below :

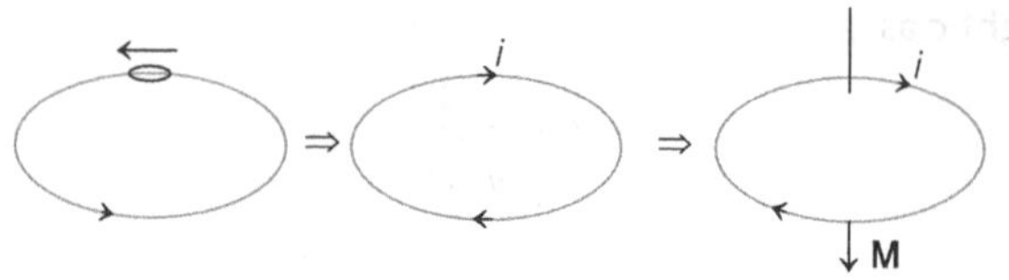

Example 5 A flat dielectric disc of radius R carries an excess charge on its surface. The surface charge density is σ. The disc rotates about an axis perpendicular to its plane passing through the centre with angular velocity ω. Find the torque on the disc if it is placed in a uniform magnetic field B directed perpendicular to the rotation axis.

Sol. Consider an annular ring of radius r and of thickness dr on this disc. Charge within this ring,

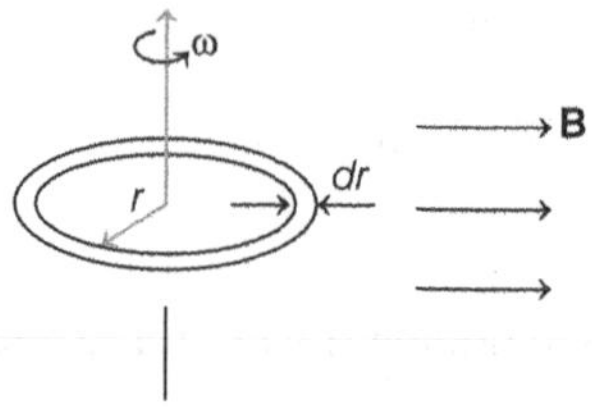

$$dq = (\sigma)(2\pi r dr)$$

As ring rotates with angular velocity ω, the equivalent current is

$$i = (dq)\text{ (frequency)}$$

$$= (\sigma)(2\pi r dr)\left(\frac{\omega}{2\pi}\right) \quad \text{or} \quad i = \sigma\omega r dr$$

Magnetic moment of this annular ring,

$M = iA = (\sigma\omega r dr)(\pi r^2)$ (along the axis of rotation)

Torque on this ring,

$$d\tau = MB \sin 90° = (\sigma\omega\pi r^3 B)\, dr$$

$\therefore$ Total torque on the disc is

$$\tau = \int_0^R d\tau = (\sigma\omega\pi B)\int_0^R r^3 dr$$

$$= \frac{\sigma\omega\pi B R^4}{4}$$

Ans.

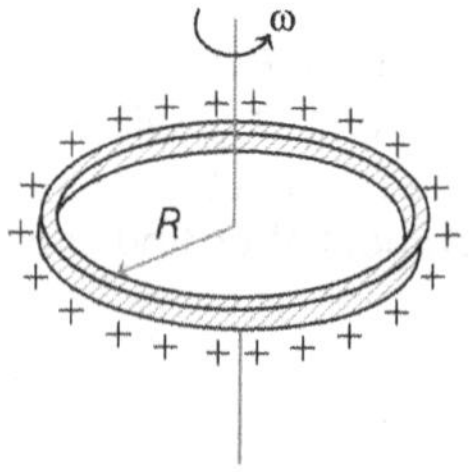

Note *Sometimes, a non-conducting charged body is rotated with some angular speed. In this case, the ratio of magnetic moment and angular momentum is constant which is equal to q/2m, where q is the charge and m the mass of the body.*

e.g. In case of a ring of mass m, radius R and charge q distributed on its circumference.

Angular momentum, $L = I\omega = (mR^2)(\omega)$ ...(i)

Magnetic moment, $M = iA = (qf)(\pi R^2)$

Here, f = *frequency* $= \frac{\omega}{2\pi}$

$\therefore \quad M = (q)\left(\frac{\omega}{2\pi}\right)(\pi R^2) = q\frac{\omega R^2}{2}$ ...(ii)

From Eqs. (i) and (ii), we get

$$\frac{M}{L} = \frac{q}{2m}$$

Although this expression is derived for simple case of a ring, it holds good for all other bodies. For example, for a disc or a sphere.

MODULE 4 Exercise

Single Correct Option

1. A non-planar closed loop of arbitrary shape carrying a current I is placed in uniform magnetic field. The force acting on the loop
   (a) is zero only for one orientation of loop in magnetic field
   (b) is zero for two symmetrically located positions of loop in magnetic field
   (c) is zero for all orientations
   (d) is never zero

2. A loop of magnetic moment **M** is placed in the orientation of unstable equilibrium position in a uniform magnetic field **B**. The external work done in rotating it through an angle θ is
   (a) $-MB(1-\cos\theta)$ (b) $-MB\cos\theta$
   (c) $MB\cos\theta$ (d) $MB(1-\cos\theta)$

3. A loop carrying current I lies in the x-y plane as shown in the figure. The unit vector $\hat{\mathbf{k}}$ is coming out of the plane of the paper. The magnetic moment of the current loop is **(2012 JEE)**

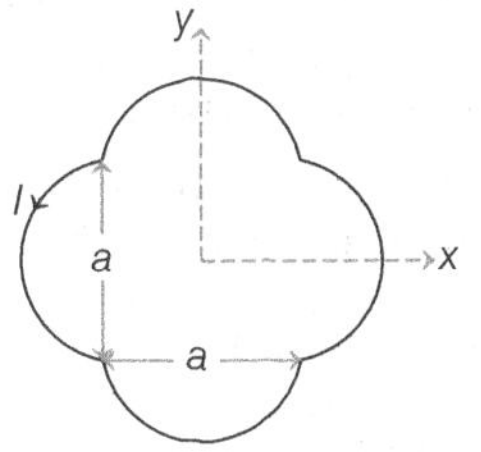

   (a) $a^2I\hat{\mathbf{k}}$ (b) $\left(\frac{\pi}{2}+1\right)a^2I\hat{\mathbf{k}}$
   (c) $-\left(\frac{\pi}{2}+1\right)a^2I\hat{\mathbf{k}}$ (d) $(2\pi+1)a^2I\hat{\mathbf{k}}$

4. A thin circular wire carrying a current I has a magnetic moment M. The shape of the wire is changed to a square and it carries the same current. It will have a magnetic moment
   (a) M (b) $\frac{4}{\pi^2}M$ (c) $\frac{4}{\pi}M$ (d) $\frac{\pi}{4}M$

5. The rectangular coil of area A is in a field B. Find the torque on the coil when it lies in the position shown and carries a current I.

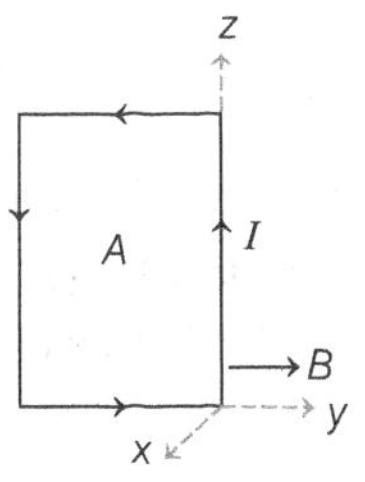

   (a) IAB in negative z-axis (b) IAB in positive z-axis
   (c) IAB in positive x-axis (d) IAB in negative x-axis

6. An insulating rod of length l carries a charge q uniformly distributed on it. The rod is pivoted at one of its ends and is rotated at a frequency f about a fixed perpendicular axis. The magnetic moment of the rod is
   (a) $\frac{\pi q f l^2}{12}$ (b) $\frac{\pi q f l^2}{2}$ (c) $\frac{\pi q f l^2}{6}$ (d) $\frac{\pi q f l^2}{3}$

7. A current carrying loop is placed in a uniform magnetic field in four different orientations, I, II, III and IV. Arrange them in the decreasing order of potential energy. **(2003 JEE)**

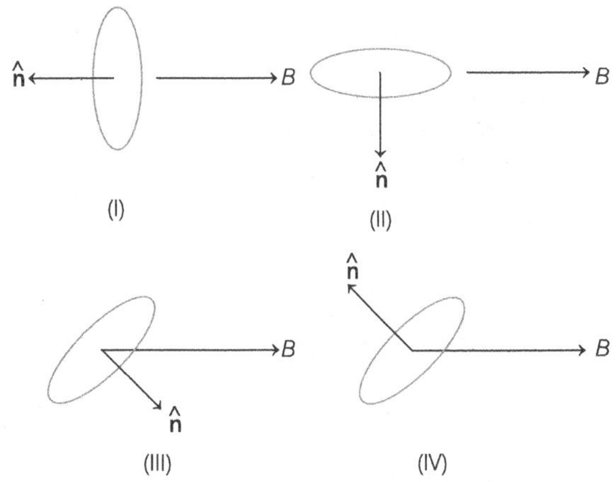

   (a) I > III > II > IV (b) I > II > III > IV
   (c) I > IV > II > III (d) III > IV > I > II

Multiple Correct Options

8. A rectangular loop ($a \times b$) carries a current i. A uniform magnetic field $\mathbf{B} = B_0\hat{\mathbf{i}}$ exists in space. Then,

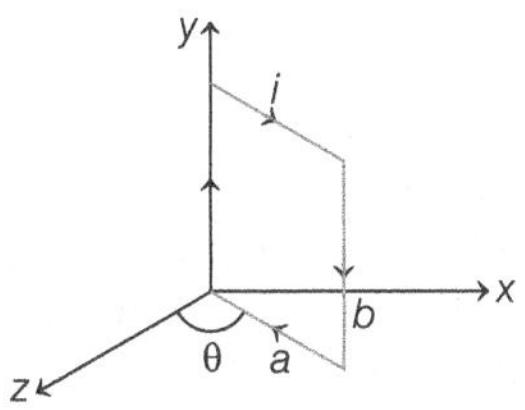

   (a) torque on the loop is $iabB_0 \sin\theta$
   (b) torque on the loop is in negative y-direction
   (c) if allowed to move the loop turn so as to increase θ
   (d) if allowed to move the loop turn so as to decrease θ

Match the Columns

9. In Column I, a current carrying loop and a uniform magnetic field are shown. Match this with Column II.

	Column I		Column II
(a)	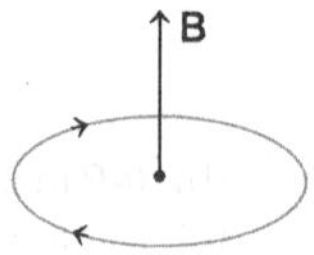	(p)	Force = 0
(b)		(q)	Maximum torque
(c)	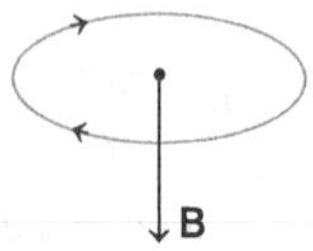	(r)	Minimum potential energy
(d)	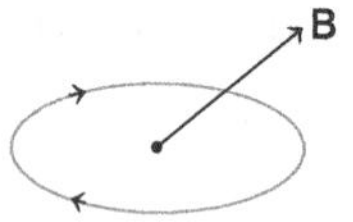	(s)	Positive potential energy

10. A circular current carrying loop of 100 turns and radius 10 cm is placed in xy-plane as shown in figure. A uniform magnetic field $\mathbf{B} = (-\hat{i} + \hat{k})$ T is present in the region. If current in the loop is 5 A, then match the Column I with Column II and select the correct option.

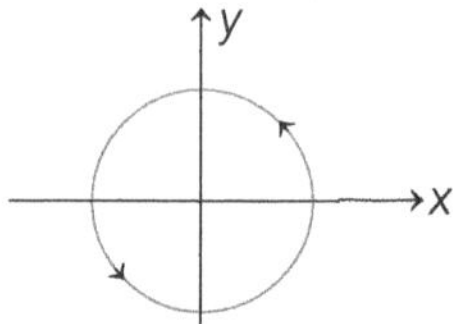

	Column I		Column II
(a)	Magnitude and direction of moment (in A - m^2) of the loop are	(p)	zero
(b)	Magnitude and direction of torque (in N-m) on the loop are	(q)	5π
(c)	Magnitude and direction of net force (in N-m) on the current loop are	(r)	along positive z-axis
		(s)	along negative y-axis

Subjective Questions

11. A charge q is uniformly distributed on a non-conducting disc of radius R. It is rotated with an angular speed ω about an axis passing through the centre of mass of the disc and perpendicular to its plane. Find the magnetic moment of the disc.

12. A circular loop of wire having a radius of 8.0 cm carries a current of 0.20 A. A vector of unit length and parallel to the dipole moment $\mathbf{M}$ of the loop is given by $0.60\,\hat{i} - 0.80\,\hat{j}$. If the loop is located in uniform magnetic field given by $\mathbf{B} = (0.25\text{ T})\hat{i} + (0.30\text{ T})\hat{k}$ find,
(a) the torque on the loop and
(b) the magnetic potential energy of the loop.

13. A length L of wire carries a current i. Show that if the wire is formed into a circular coil, then the maximum torque in a given magnetic field is developed when the coil has one turn only, and that maximum torque has the magnitude $\tau = L^2 iB / 4\pi$.

14. A coil with magnetic moment 1.45 A-m^2 is oriented initially with its magnetic moment antiparallel to a uniform 0.835 T magnetic field. What is the change in potential energy of the coil when it is rotated 180° so that its magnetic moment is parallel to the field?

15. A magnetic dipole with a dipole moment of magnitude 0.020 J/T is released from rest in a uniform magnetic field of magnitude 52 mT. When the dipole rotates through the orientations where its dipole moment is aligned with the magnetic field, its kinetic energy is 0.80 mJ.
(a) What is the initial angle between the dipole moment and the magnetic field?
(b) What is the angle when the dipole is next (momentarily) at rest?

16. In the Bohr model of the hydrogen atom, in the lowest energy state the electron revolves round the proton at a speed of 2.2×10^6 m/s in a circular orbit of radius 5.3×10^{-11} m.
(a) What is the orbital period of the electron?
(b) If the orbiting electron is considered to be a current loop, what is the current I?
(c) What is the magnetic moment of the atom due to the motion of the electron?

17. A conductor carries a constant current I along the closed path $abcdefgha$ involving 8 of the 12 edges each of length l. Find the magnetic dipole moment of the closed path.

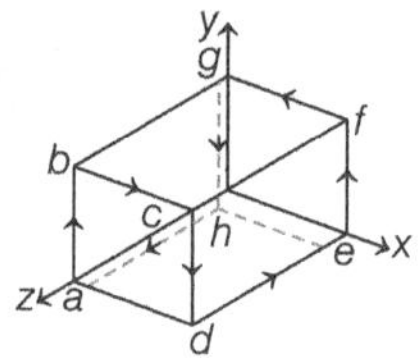

18. Given figure shows a coil bent with all edges of length 1 m and carrying a current of 1A. There exists in space a uniform magnetic field of 2 T in positive y-direction. Find the torque on the loop.

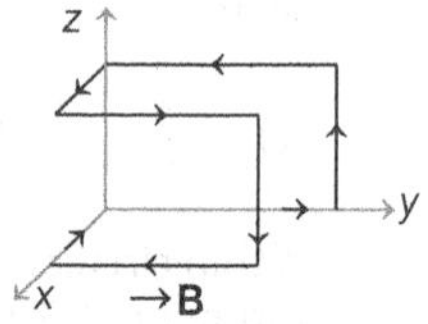

Answers

1. (c) 2. (a) 3. (b) 4. (d) 5. (b)
6. (d) 7. (c) 8. (a,b,d)
9. (a $\to$ p,s; b $\to$ p,q ; c $\to$ p,r; d $\to$ p,s)
10. (a $\to$ q,r; b $\to$ q,s; c $\to$ p) 11. $\dfrac{qR^2\omega}{4}$
12. (a) $\tau = (-9.6\hat{i} - 7.2\hat{j} + 8.0\hat{k}) \times 10^{-4}$ N-m (b) $U = -(6.0 \times 10^{-4})$ J
14. –2.42 J 15. (a) 76.7° (b) – 76.7°
16. (a) 1.5×10^{-16} s (b) 1.1 mA (c) 9.3×10^{-24} A-m^2
17. $\mathbf{M} = 2Il^2\hat{j}$ 18. Zero

See hints on pages 438-439.

MODULE 5

Biot Savart Law

In the preceding modules, we discussed the magnetic force exerted on a charged particle and current carrying conductor in a magnetic field. To complete the description of the magnetic interaction, this and the next module deals with the origin of the magnetic field. As in electrostatics, there are two methods of calculating the electric field at some point.

One is Coulomb's law which gives the electric field due to a point charge and the another is Gauss's law which is useful in calculating the electric field of a highly symmetric configuration of charge. Similarly, in magnetics, there are basically two methods of calculating magnetic field at some point.

One is **Biot Savart law** which gives the magnetic field due to an infinitesimally small current carrying wire at some point and the another is **Ampere's law**, which is useful in calculating the magnetic field of a highly symmetric configuration carrying a steady current.

We begin by showing how to use the law of Biot and Savart to calculate the magnetic field produced at some point in space by a small current element. Using this formalism and the principle of superposition, we then calculate the total magnetic field due to various current distributions.

From their experimental results, Biot and Savart arrived at a mathematical expression that gives the magnetic field at some point in space in terms of the current that produces the field. That expression is based on the following experimental observations for the magnetic field $d\mathbf{B}$ at a point P associated with a length element $d\mathbf{l}$ of a wire carrying a steady current i.

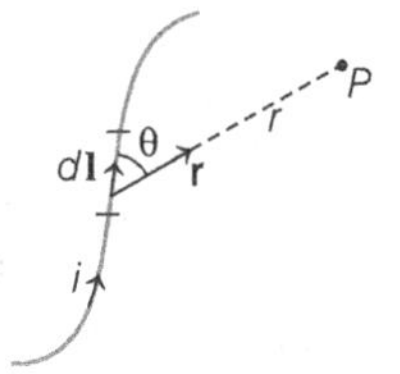

(i) The vector $d\mathbf{B}$ is perpendicular to both $d\mathbf{l}$ (which points in the direction of the current) and the unit vector $\hat{\mathbf{r}}$ directed from $d\mathbf{l}$ to P.

(ii) The magnitude of $d\mathbf{B}$ is inversely proportional to r^2, where r is the distance from $d\mathbf{l}$ to P.

(iii) The magnitude of $d\mathbf{B}$ is proportional to the current and to the magnitude dl of the length element $d\mathbf{l}$.

(iv) The magnitude of $d\mathbf{B}$ is proportional to $\sin\theta$ where θ is the angle between $d\mathbf{l}$ and $\hat{\mathbf{r}}$. These observations are summarized in mathematical formula known today as **Biot Savart law**

$$d\mathbf{B} = \frac{\mu_0}{4\pi}\frac{i\,(d\mathbf{l}\times\hat{\mathbf{r}})}{r^2} \qquad \text{...(i)}$$

Here, $\dfrac{\mu_0}{4\pi} = 10^{-7}\ \dfrac{\text{T-m}}{\text{A}}$

To find the total magnetic field $\mathbf{B}$ created at some point by a current of finite size, we must sum up contributions from all current elements that make up the current. That is, we must evaluate $\mathbf{B}$ by integrating Eq. (i).

$$\mathbf{B} = \frac{\mu_0 i}{4\pi}\int\frac{d\mathbf{l}\times\hat{\mathbf{r}}}{r^2}$$

The following points are worthnoting regarding the Biot Savart law.

(i) Magnitude of $d\mathbf{B}$ is given by

$$|d\mathbf{B}| = \frac{\mu_0}{4\pi}\frac{i\,dl\sin\theta}{r^2}$$

$|d\mathbf{B}|$ is zero at $\theta = 0°$ or $180°$ and maximum at $\theta = 90°$.

(ii) For the direction of $d\mathbf{B}$ either of the following methods can be employed.

(a) $d\mathbf{B} \uparrow\uparrow d\mathbf{l}\times\hat{\mathbf{r}}$. So, $d\mathbf{B}$ is along $d\mathbf{l}\times\mathbf{r}$.

(b) If $d\mathbf{l}$ is in the plane of paper. $d\mathbf{B} = 0$ at all points lying on the straight line passing through $d\mathbf{l}$.

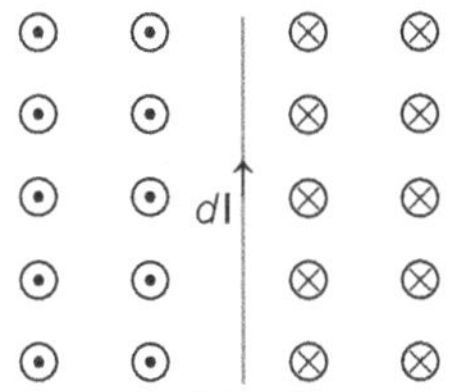

$d\mathbf{B} = 0$

The magnetic field to the right of this line is in ⊗ direction and to the left of this line is in ⊙ direction.

Applications of Biot Savart Law

Let us now consider few applications of Biot Savart law.

Magnetic Field Surrounding a Thin, Straight Conductor

According to Biot Savart law,

$$\mathbf{B} = \frac{\mu_0}{4\pi}\int\frac{i\,d\mathbf{l}\times\hat{\mathbf{r}}}{r^2} \qquad \text{...(i)}$$

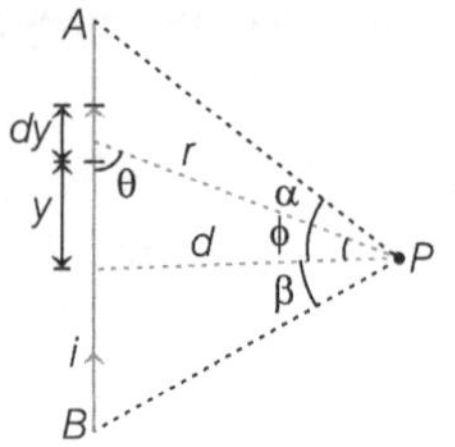

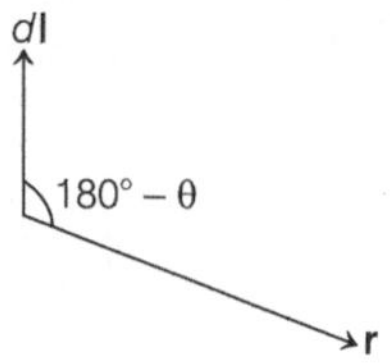

As here every element of the wire contributes to **B** in the same direction (which is here ⊗).

Eq. (i) for this case becomes,

$$B=\frac{\mu_0}{4\pi}\int\frac{idl\sin\theta}{r^2}=\frac{\mu_0 i}{4\pi}\int\frac{dy\sin(180°-\theta)}{r^2}$$

$$=\frac{\mu_0 i}{4\pi}\int\frac{dy\sin\theta}{r^2}$$

$$y=d\tan\phi \quad \text{or} \quad dy=(d\sec^2\phi)d\phi$$

$$r=d\sec\phi \quad \text{and} \quad \theta=90°-\phi$$

$$B=\frac{\mu_0 i}{4\pi}\int_{\phi=-\beta}^{\phi=\alpha}\frac{\{(d\sec^2\phi)d\phi\}\sin(90°-\phi)}{(d\sec\phi)^2}$$

or $\quad B=\frac{\mu_0 i}{4\pi d}\int_{-\beta}^{\alpha}\cos\phi\cdot d\phi \quad$ or $\quad \boxed{B=\frac{\mu_0}{4\pi}\frac{i}{d}(\sin\alpha+\sin\beta)}$

Note down the following points regarding the above equation.

(i) For an infinitely long straight wire, $\alpha=\beta=90°$

$\therefore \quad \sin\alpha+\sin\beta=2 \quad$ or $\quad \boxed{B=\frac{\mu_0}{2\pi}\frac{i}{d}}$

(ii) The direction of magnetic field at a point *P* due to a long straight wire can be found by the right hand thumb rule. If we stretch the thumb of the right hand along the current and curl our fingers to pass through *P*, the direction of the fingers at *P* gives the direction of magnetic field there.

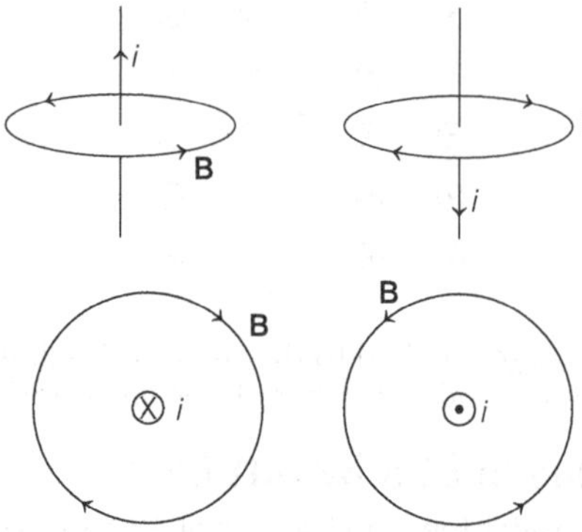

(iii) $B\propto\frac{1}{d}$, i.e. *B*-*d* graph for an infinitely long straight wire is a rectangular hyperbola as shown in the figure.

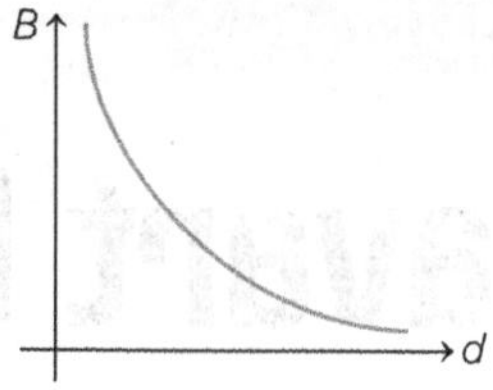

Magnetic Field on the Axis of a Circular Coil

According to Biot Savart law, magnetic field at point *P* due to small element *XY*, will be

$$dB=\left(\frac{\mu_0}{4\pi}\right)\left(\frac{idl\sin\phi}{r^2}\right) \quad \text{...(i)}$$

Since, for the circular loop ($\sin\phi=1$), we get

$$dB=\left(\frac{\mu_0}{4\pi}\right)\left(\frac{idl}{r^2}\right) \quad \text{...(ii)}$$

The direction of ***dB*** can be obtained from the property of cross product. Because ***dB*** is along ($\mathbf{dl}\times\mathbf{r}$). For the element *X*,*Y* vector ***dl*** is in ⊙ direction and for *X*'*Y*', it is in ⊗ direction.

It is perpendicular to the plane formed by (***dl***) and (**r**) and is along *PL* which is perpendicular to *PC* as shown in figure.

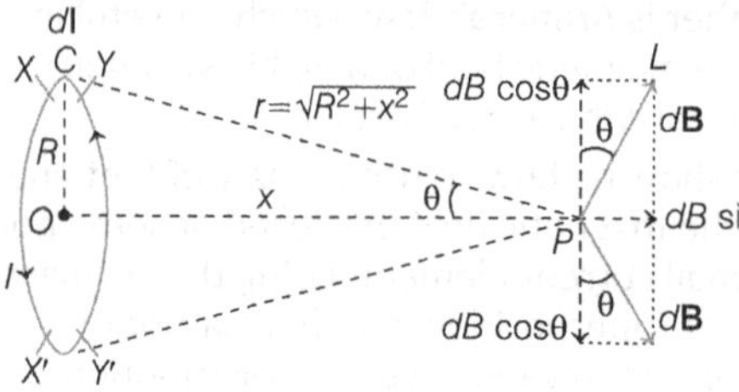

Resolving (*dB*) into rectangular components, we get Component (*dB* cos θ) is perpendicular to the axis of the loop.

Component (*dB* sin θ) is along the axis of the loop and away from the centre of loop.

As the loop is symmetrical about its axis, so every current element of length (*dl*) has another equal and opposite element on the circular loop.

Current element *XY* has equal and opposite current element *X*'*Y*'. The perpendicular components of the magnetic field due to these current elements being equal and opposite, they will cancel to each other. Hence, the total contribution of perpendicular components of the magnetic field is zero.

Also, (*dB* sin θ) component of magnetic field due to each current element of the loop is directed in the same direction. So, magnetic field at point *P* due to the whole current carrying loop is equal to the sum of (*dB* sin θ) components of magnetic field of all current elements.

Therefore, $\quad B=\Sigma(dB\sin\theta)$

$$=\int(dB\sin\theta)$$

Using Eq. (ii), we will get

$$B = \int \left(\frac{\mu_0}{4\pi}\right)\left(\frac{idl}{r^2}\right)\sin\theta$$

$$= \left(\frac{\mu_0 i \sin\theta}{4\pi r^2}\right)\int dl$$

$$= \left(\frac{\mu_0 i}{4\pi r^2}\right)\sin\theta \int dl \qquad \text{...(iii)}$$

But $\int dl = (2\pi R)$ (Circumference of the loop)

and $\sin\theta = \left(\frac{R}{r}\right)$

Putting these values in Eq. (iii), we get

$$B = \left(\frac{\mu_0 i}{4\pi r^2}\right)\left(\frac{R}{r}\right)(2\pi R)$$

$$= \left(\frac{\mu_0}{4\pi}\right)\left(\frac{2\pi i R^2}{r^3}\right)$$

Also, $r = (\sqrt{R^2 + x^2}) = [(R^2 + x^2)^{\frac{1}{2}}]$

Therefore, $B = \left(\frac{\mu_0}{4\pi}\right)\left[\frac{2\pi i R^2}{(R^2 + x^2)^{(3/2)}}\right]$

Magnetic field at a distance (x) from centre of a loop on the axis is given by the equation

$$B = \left(\frac{\mu_0}{4\pi}\right)\left[\frac{2\pi I R^2}{(R^2 + x^2)^{(3/2)}}\right]$$

Now, consider a coil having (N) number of turns.

Then magnetic field at a distance (x) from the centre of coil on its axis will be

$$B' = \left(\frac{\mu_0}{4\pi}\right)\left[\frac{2\pi N I R^2}{(R^2 + x^2)^{(3/2)}}\right]$$

Note down the following points regarding a circular current carrying loop.

(i) At the centre of the loop, $x = 0$

and $\boxed{B\,(\text{centre}) = \frac{\mu_0 N i}{2R}}$

(ii) For $x >> R$, $x^2 + R^2 \approx x^2$

$$\therefore \quad B = \frac{\mu_0 N i R^2}{2x^3} = \left(\frac{\mu_0}{4\pi}\right)\frac{(2Ni\pi R^2)}{x^3}$$

$$= \left(\frac{\mu_0}{4\pi}\right)\left(\frac{2M}{x^3}\right)$$

Here, M = magnetic moment of the loop $= NiA = Ni\pi R^2$.

This result was expected as the magnetic field on the axis of a dipole is $\frac{\mu_0}{4\pi}\frac{2M}{r^3}$.

(iii) Direction of magnetic field on the axis of a circular loop can be obtained using the right hand thumb rule. If the fingers are curled along the current, the stretched thumb will point towards the magnetic field.

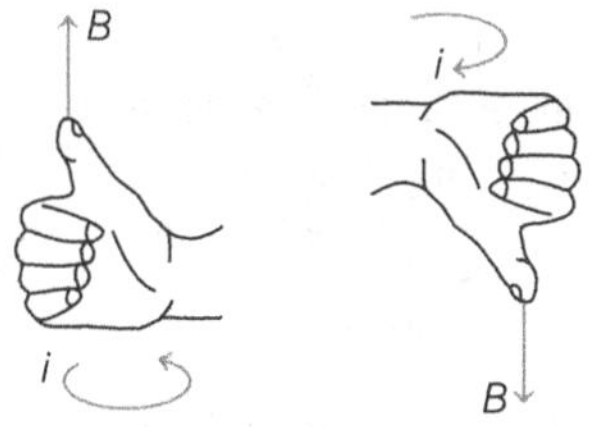

(iv) The magnetic field at a point not on the axis is mathematically difficult to calculate. We have shown qualitatively in figure the magnetic field lines due to a circular current.

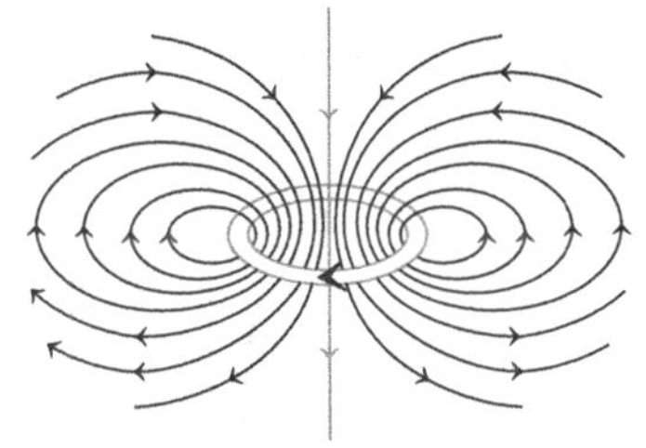

(v) Magnetic field is maximum at the centre and decreases as we move away from the centre (on the axis of the loop). The B-x graph is somewhat like shown in figure.

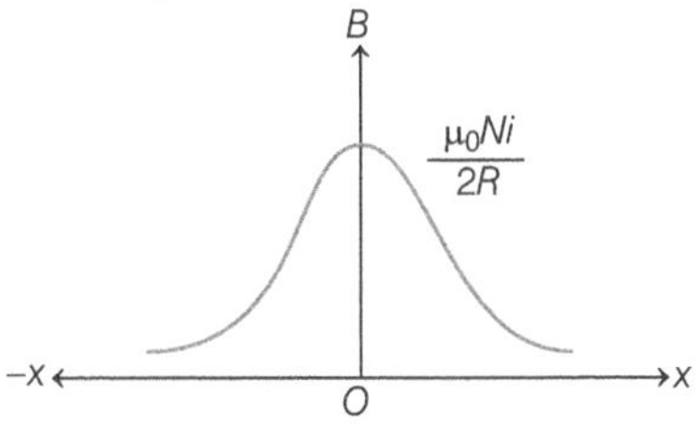

(vi) Magnetic field due to an arc of a circle at the centre is

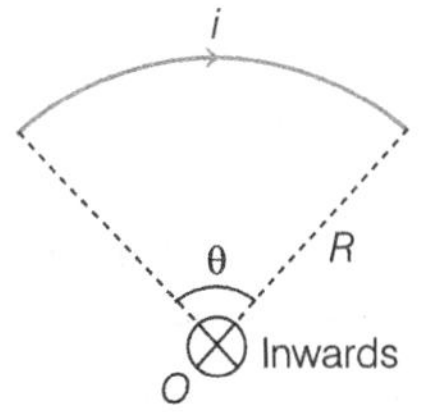

$$B = \left(\frac{\theta}{2\pi}\right)\frac{\mu_0 i}{2R} = \frac{\mu_0}{4\pi}\left(\frac{i}{R}\right)\theta$$

or $\boxed{B = \left(\frac{\mu_0}{4\pi}\right)\left(\frac{i}{R}\right)\theta}$

Here, θ is to be substituted in radians.

Field Along the Axis of a Solenoid

The name solenoid was first given by Ampere to a wire wound in a closely spaced spiral over a hollow cylindrical non-conducting core. If n is the number of turns per unit length, each carries a current i uniformly wound round a cylinder of radius R, the number of turns in length dx are ndx. Thus, the magnetic field at the axial point O due to this element dx is

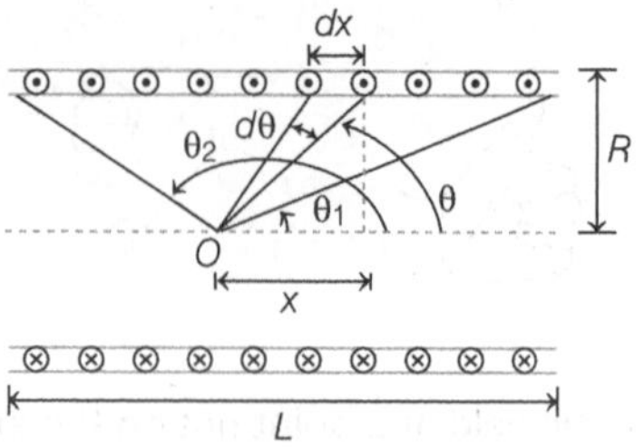

$$dB = \frac{\mu_0}{2} \frac{(indx)R^2}{(R^2 + x^2)^{3/2}} \quad \text{...(i)}$$

Its direction is along the axis of the solenoid. From the geometry, we know

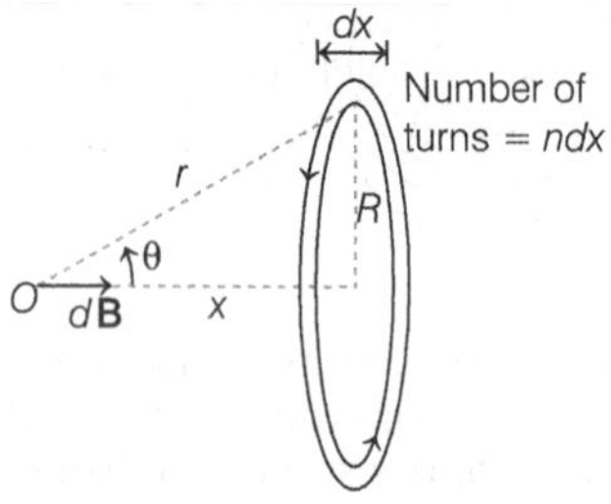

$$x = R\cot\theta$$

$$dx = -R\,\text{cosec}^2\,\theta \cdot d\theta$$

Substituting these values in Eq. (i), we get

$$dB = -\frac{1}{2}\mu_0 ni \sin\theta \cdot d\theta$$

Total field B due to the entire solenoid is

$$B = \frac{1}{2}\mu_0 ni \int_{\theta_1}^{\theta_2} (-\sin\theta)\, d\theta$$

$\therefore$

$$\boxed{B = \frac{\mu_0 ni}{2}(\cos\theta_2 - \cos\theta_1)}$$

If the solenoid is very long ($L >> R$) and the point O is chosen at the middle, i.e. if $\theta_1 = 180°$ and $\theta_2 = 0°$, then we get

$$\boxed{B\,(\text{centre}) = \mu_0 ni} \quad [\text{For } L >> R]$$

At the end of the solenoid,

$\theta_2 = 0°$, $\theta_1 = 90°$ and we get

$$\boxed{B\,(\text{end}) = \frac{1}{2}\mu_0 ni} \quad [\text{For } L >> R]$$

Thus, the field at the end of a solenoid is half the value at the centre. The field lines are as shown in figure.

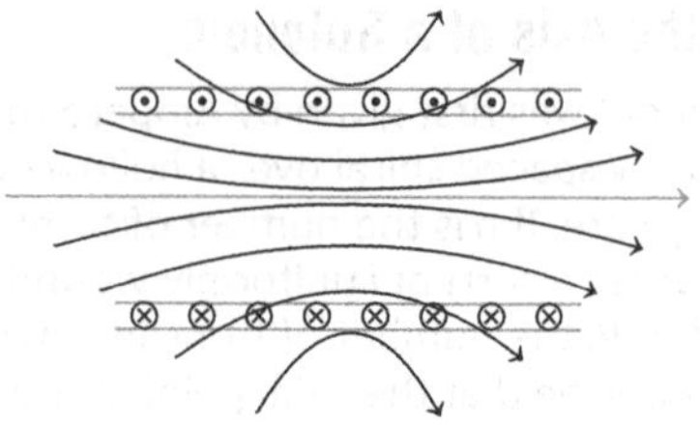

Example 1 In a high tension wire electric current runs from east to west. Find the direction of magnetic field at points above and below the wire.

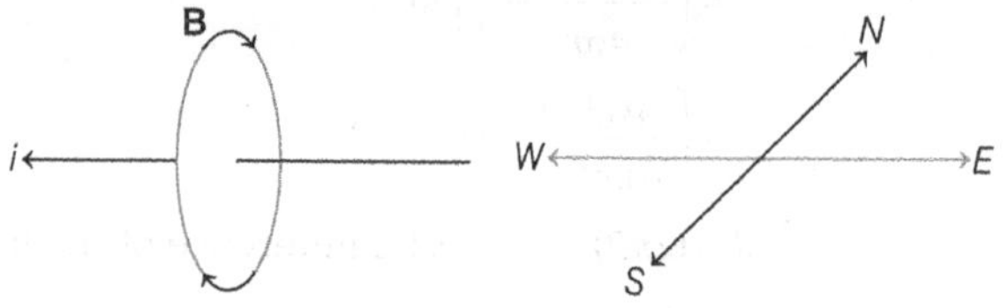

(a) (b)

Sol. When the current flows from east to west, magnetic field lines are circular round it as shown in figure (a). Therefore the magnetic field above the wire is towards north and below the wire towards south.

Example 2 A current path shaped as shown in figure produces a magnetic field at P, the centre of the arc. If the arc subtends an angle of 30° and the radius of the arc is 0.6 m, what are the magnitude and direction of the field produced at P if the current is 3.0 A.

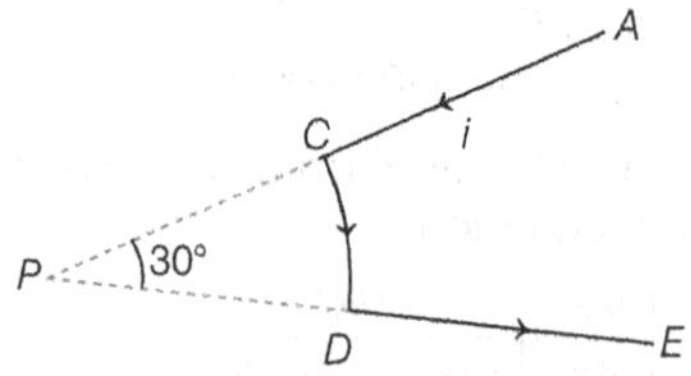

Sol. The magnetic field at P due to the straight segments AC and DE is zero.

CD is arc of circle.

$\therefore$ $$B = \left(\frac{\theta}{2\pi}\right)\left(\frac{\mu_0 i}{2R}\right) \quad (\text{as } N = 1)$$

or $$B = \left(\frac{\mu_0}{4\pi}\right)\left(\frac{i}{R}\right)\theta$$

$\therefore$ $$B = (10^{-7})\left(\frac{3.0}{0.6}\right)\left(\frac{\pi}{6}\right)$$

$$= 2.62 \times 10^{-7}\ \text{T}$$ **Ans.**

Example 3 Figure shows a current loop having two circular arcs joined by two radial lines. Find the magnetic field B at the centre O.

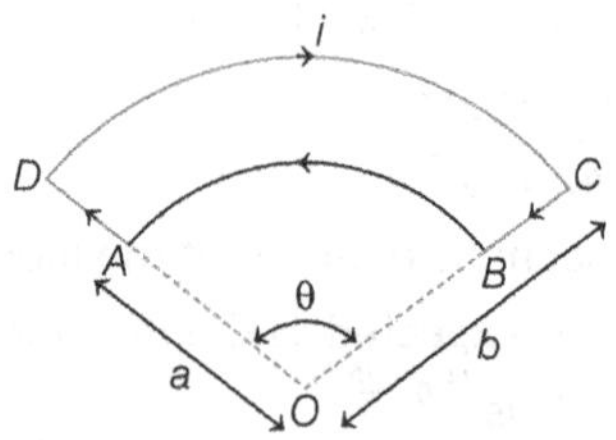

Sol. Magnetic field at point O, due to wires CB and AD will be zero.

Magnetic field due to wire BA will be

$$B_1 = \left(\frac{\theta}{2\pi}\right)\left(\frac{\mu_0 i}{2a}\right)$$

Direction of field $\mathbf{B}_1$ is perpendicular to paper outwards.

Similarly, field at O due to arc DC will be

$$B_2 = \left(\frac{\theta}{2\pi}\right)\left(\frac{\mu_0 i}{2b}\right) \qquad (B_2 < B_1)$$

Direction of field $\mathbf{B}_2$ is perpendicular to paper inwards.

The resultant field at O is

$$B = B_1 - B_2 = \frac{\mu_0 i\theta (b-a)}{4\pi ab} \qquad \textbf{Ans.}$$

Perpendicular to paper outwards.

Example 4 The magnetic field B due to a current carrying circular loop of radius 12 cm at its centre is 0.5×10^{-4} T. Find the magnetic field due to this loop at a point on the axis at a distance of 5.0 cm from the centre.

Sol. Magnetic field at the centre of a circular loop is

$$B_1 = \frac{\mu_0 i}{2R}$$

and that at an axial point, $B_2 = \dfrac{\mu_0 iR^2}{2(R^2 + x^2)^{3/2}}$

Thus, $$\frac{B_2}{B_1} = \frac{R^3}{(R^2 + x^2)^{3/2}}$$

or $$B_2 = B_1 \left[\frac{R^3}{(R^2 + x^2)^{3/2}}\right]$$

Substituting the values, we have

$$B_2 = (0.5 \times 10^{-4})\left[\frac{(12)^3}{(144 + 25)^{3/2}}\right]$$

$$= 3.9 \times 10^{-5}\ \text{T} \qquad \textbf{Ans.}$$

Example 5 A wire is shaped to a regular hexagon of side 2cm carries a current of 2A. Find the magnetic field at the centre of the hexagon.

Sol. $\because \quad \theta = 30^\circ$

$$\frac{BC}{OC} = \tan\theta \qquad (\text{as } BC = 1 \text{ cm})$$

$$\therefore \quad \frac{1}{r} = \tan 30^\circ = \frac{1}{\sqrt{3}}$$

$$\therefore \quad r = \sqrt{3} \text{ cm}$$

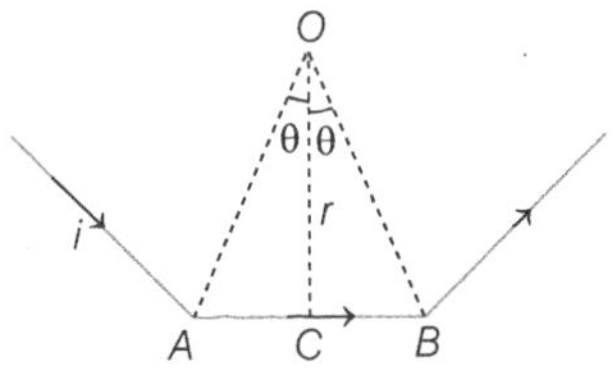

Net magnetic field at O is 6 times the magnetic field due to one side.

$$\therefore \quad B = 6\left[\frac{\mu_0}{4\pi}\frac{i}{r}(\sin\theta + \sin\theta)\right]$$

$$= \frac{6(10^{-7})(2)}{\sqrt{3} \times 10^{-2}}\left(\frac{1}{2} + \frac{1}{2}\right)$$

$$= 6.9 \times 10^{-5}\ \text{T} \qquad \textbf{Ans.}$$

Example 6 Find the magnetic field **B** at the point P in figure.

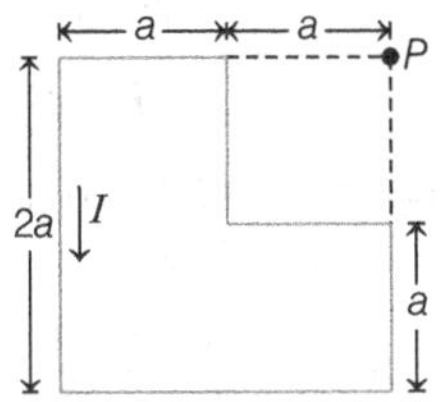

Sol. Magnetic field at P due to SM and OQ is zero. Due to QR and RS are equal and outwards. Due to MN and NO are equal and inwards.

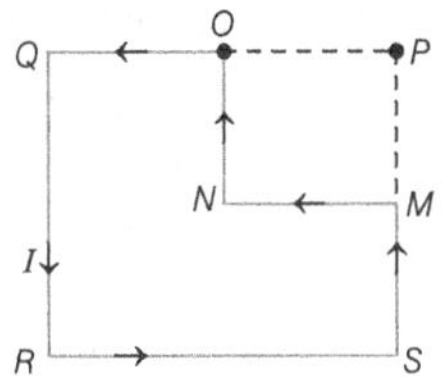

Due to QR and RS,

$$B_1 = 2\left[\frac{\mu_0}{4\pi}\frac{I}{2a}(\sin 0^\circ + \sin 45^\circ)\right]$$

$$= \frac{\mu_0 I}{4\sqrt{2}\pi a} \qquad [\text{outwards}]$$

Due to MN and NO,

$$B_2 = 2\left[\frac{\mu_0}{4\pi}\frac{I}{a}(\sin 0^\circ + \sin 45^\circ)\right]$$

$$= \frac{\mu_0 I}{2\sqrt{2}\,\pi a} \qquad [\text{inwards}]$$

$$\therefore \quad B_{net} = B_2 - B_1 = \frac{\mu_0 I}{4\sqrt{2}\,\pi a} \qquad [\text{inwards}]$$

Example 7 A long insulated copper wire is closely wound as a spiral of N turns. The spiral has inner radius a and outer radius b. The spiral lies in the xy-plane and a steady current I flows through the wire. The z-component of the magnetic field at the centre of the spiral is **(2011 JEE)**

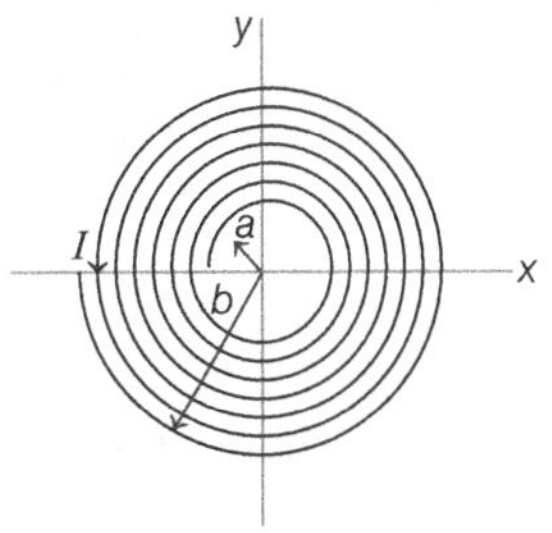

(a) $\frac{\mu_0 NI}{2(b-a)} \ln\left(\frac{b}{a}\right)$ (b) $\frac{\mu_0 NI}{2(b-a)} \ln\left(\frac{b+a}{b-a}\right)$

(c) $\frac{\mu_0 NI}{2b} \ln\left(\frac{b}{a}\right)$ (d) $\frac{\mu_0 NI}{2b} \ln\left(\frac{b+a}{b-a}\right)$

Sol. (a) If we take a small strip of dr at distance r from centre, then number of turns in this strip would be

$$dn = \left(\frac{N}{b-a}\right) dr$$

Magnetic field due to this element at the centre of the coil will be

$$dB = \frac{\mu_0 (dn) I}{2r} = \frac{\mu_0 NI}{(b-a)} \frac{dr}{2r}$$

$$\therefore \quad B = \int_{r=a}^{r=b} dB = \frac{\mu_0 NI}{2(b-a)} \ln\left(\frac{b}{a}\right)$$

$\therefore$ Correct answer is (a).

Example 8 An infinitely long conductor PQR is bent to form a right angle as shown in figure. A current I flows through PQR. The magnetic field due to this current at the point M is H_1. Now, another infinitely long straight conductor QS is connected at Q, so that current is $I/2$ in QR as well as in QS, the current in PQ remaining unchanged. The magnetic field at M is now H_2. The ratio H_1/H_2 is given by

(2000 JEE)

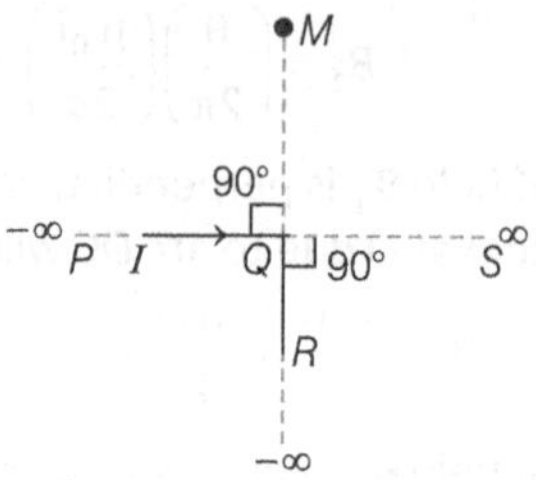

(a) 1/2 (b) 1 (c) 2/3 (d) 2

Sol. H_1 = Magnetic field at M due to PQ + Magnetic field at M due to QR

But magnetic field at M due to $QR = 0$

$\therefore$ Magnetic field at M due to PQ (or due to current I in PQ)

$$= H_1$$

Now, H_2 = Magnetic field at M due to PQ (current I) + magnetic field at M due to QS (current $I/2$) + magnetic field at M due to QR

$$= H_1 + \frac{H_1}{2} + 0 = \frac{3}{2} H_1$$

$$\therefore \quad \frac{H_1}{H_2} = \frac{2}{3}$$

$\therefore$ The correct answer is (c).

Note *Magnetic field at any point lying on the current carrying straight conductor is zero.*

MODULE 5 Exercise

Assertion and Reason

See directions on page 25.

1. **Assertion :** A charged particle is rotating in a circle. Then, magnetic field (**B**) at centre of circle and magnetic moment (**M**) produced by motion of charged particle are parallel to each other.

   Reason : **M** and **B** are always parallel to each other.

Single Correct Option

2. A current of 50 A is passed through a straight wire of length 6 cm, then the magnetic induction at a point 5 cm from both ends of the wire is (1 gauss $= 10^{-4}$ T)
   (a) 2.5 gauss
   (b) 1.25 gauss
   (c) 1.5 gauss
   (d) 3.0 gauss
3. Two long straight wires, each carrying a current I in opposite directions are separated by a distance R. The magnetic induction at a point mid-way between the wires is
   (a) zero (b) $\frac{\mu_0 I}{\pi R}$ (c) $\frac{2\mu_0 I}{\pi R}$ (d) $\frac{\mu_0 I}{4\pi R}$
4. The magnetic field at a distance x on the axis of a circular coil of radius R is $\frac{1}{8}$th of that at the centre. The value of x is
   (a) $\frac{R}{\sqrt{3}}$ (b) $\frac{2R}{\sqrt{3}}$ (c) $R\sqrt{3}$ (d) $R\sqrt{2}$
5. The straight wire AB carries a current I. The ends of the wire subtend angles θ_1 and θ_2 at the point P as shown in figure. The magnetic field at the point P is

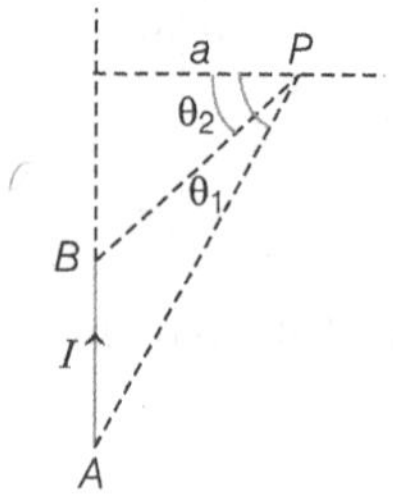

(a) $\frac{\mu_0 I}{4\pi a}(\sin\theta_1 - \sin\theta_2)$ (b) $\frac{\mu_0 I}{4\pi a}(\sin\theta_1 + \sin\theta_2)$

(c) $\frac{\mu_0 I}{4\pi a}(\cos\theta_1 - \cos\theta_2)$ (d) $\frac{\mu_0 I}{4\pi a}(\cos\theta_1 + \cos\theta_2)$

6. The figure shows three identical current carrying square loops A, B and C. Identify the correct statement related to magnetic field **B** at the centre O of the square loop. Current in each wire is I.

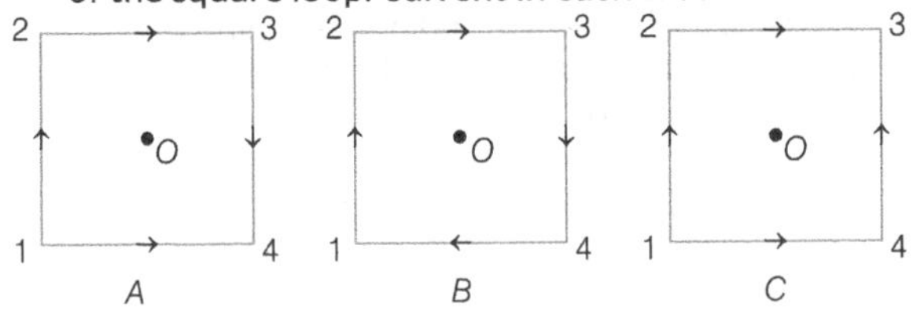

(a) **B** is zero in all cases
(b) **B** is zero only in case of C
(c) **B** is non-zero in all cases
(d) **B** is non-zero only in case of B

7. The figure shows a wire frame in xy-plane carrying a current I. The magnetic field at the point O is

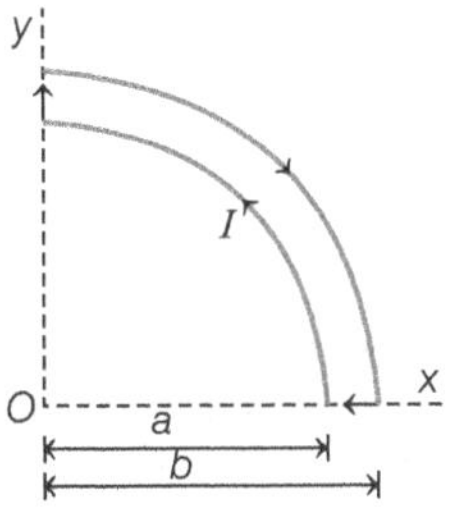

(a) $\frac{\mu_0 I}{8}\left[\frac{1}{a}-\frac{1}{b}\right]\hat{\mathbf{k}}$ (b) $\frac{\mu_0 I}{8}\left[\frac{1}{b}-\frac{1}{a}\right]\hat{\mathbf{k}}$

(c) $\frac{\mu_0 I}{4}\left[\frac{1}{a}-\frac{1}{b}\right]\hat{\mathbf{k}}$ (d) $\frac{\mu_0 I}{4}\left[\frac{1}{b}-\frac{1}{a}\right]\hat{\mathbf{k}}$

8. An electron moving in a circular orbit of radius R with frequency f. The magnetic field at the centre of the orbit is

(a) $\frac{\mu_0 ef}{2\pi R}$ (b) $\frac{\mu_0 ef}{2R}$ (c) $\frac{\mu ef^2}{2R}$ (d) zero

9. The figure shows a point P on the axis of a circular loop carrying current I. The correct direction of magnetic field vector at P due to $d\mathbf{l}$ is represented by

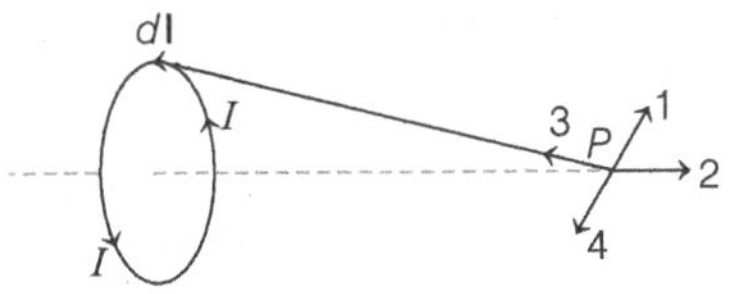

(a) 1 (b) 2
(c) 3 (d) 4

10. A battery is connected between two points A and B on the circumference of a uniform conducting ring of radius r and resistance R. One of the arcs AB of the ring subtends an angle θ at the centre. The value of the magnetic induction at the centre due to the current in the ring is

(a) proportional to $2(180° - \theta)$
(b) inversely proportional to r
(c) zero, only if $\theta = 180°$
(d) zero for all values of θ

11. A current I flows through a closed loop as shown in figure. The magnetic field at the centre O is

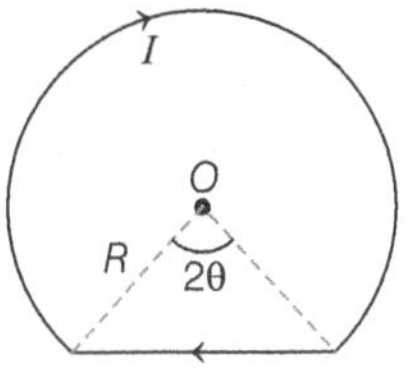

(a) $\frac{\mu_0 I}{2\pi R}(\pi - \theta + \tan\theta)$ (b) $\frac{\mu_0 I}{2\pi R}(\pi - \theta + \sin\theta)$

(c) $\frac{\mu_0 I}{2\pi R}(\theta + \sin\theta)$ (d) None of these

12. When a certain length of wire is turned into one circular loop, the magnetic induction at the centre of coil due to some current flowing is B_1. If the same wire is turned into three loops to make a circular coil, the magnetic induction at the centre of this coil for the same current will be

(a) B_1 (b) $9B_1$ (c) $3B_1$ (d) $27B_1$

13. There are two identical concentric coils of suitable radii X and Y with their planes at right angles to each other. The coil X lies in the horizontal plane while coil Y lies in the vertical plane. If the coil X carries a current of 1 A, then what value of current in coil Y be passed, so that the resultant field at the centre of the coils just balances the earth's magnetic field of 10^{-5} T inclined at 30° with the vertical?

(a) 1 A (b) $\sqrt{3}$ A
(c) 2 A (d) $(1/\sqrt{3})$ A

14. A long wire carrying a current i is bent to form a plane angle θ. The magnetic field at a point O on the bisector of this angle situated at a distance d from vertex P is

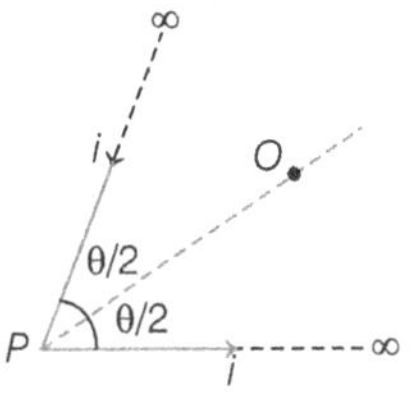

(a) $\frac{\mu_0 i}{\pi d}\cot\left(\frac{\theta}{4}\right)$, $\odot$ (b) $\frac{\mu_0 i}{2\pi d}\cot\left(\frac{\theta}{4}\right)$, $\odot$

(c) $\frac{\mu_0 i}{4\pi d}\cot\left(\frac{\theta}{2}\right)$, $\otimes$ (d) None of these

15. Two circular coils A and B of radius $\frac{5}{\sqrt{2}}$ cm and 5 cm carry currents 5 A and $5\sqrt{2}$ A, respectively. The plane

of B is perpendicular to plane of A and their centres coincide. Magnetic field at the centre is

(a) 0 (b) $4\pi\sqrt{2}\times10^{-5}$ T

(c) $4\pi\times10^{-5}$ T (d) $2\pi\sqrt{2}\times10^{-5}$ T

16. Two circular coils 1 and 2 are made from the same wire but the radius of the 1st coil is twice that of the 2nd coil. What is the ratio of potential difference applied across them, so that the magnetic field at their centres is the same?

(a) 3 (b) 4 (c) 6 (d) 2

17. An electron moves in a circular orbit with a uniform speed v. It produces a magnetic field B at the centre of the circle. The radius of the circle is proportional to

(a) $\frac{B}{v}$ (b) $\frac{v}{B}$ (c) $\sqrt{\frac{v}{B}}$ (d) $\sqrt{\frac{B}{v}}$

18. A square frame of side 1 m carrying a current I, produces a magnetic field B at its centre. The same current is passed through a circular coil having the same perimeter as the square. The magnetic field at the centre of the circular coil is B'. The ratio B/B' is

(a) $\frac{8}{\pi^2}$ (b) $\frac{8\sqrt{2}}{\pi^2}$ (c) $\frac{16}{\pi^2}$ (d) $\frac{8}{\sqrt{2}\pi^2}$

19. The magnetic field at the centre of square of side a is

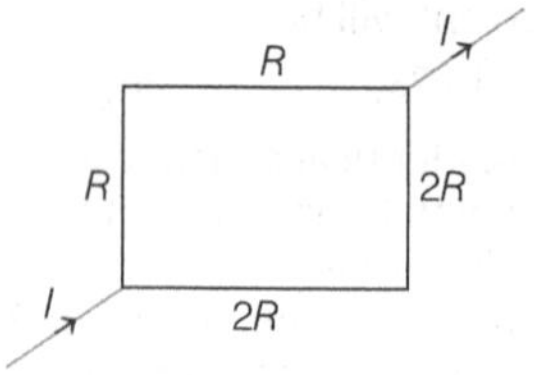

(a) $\frac{\sqrt{2}\mu_0 I}{\pi a}$ (b) $\frac{\sqrt{2}\mu_0 I}{3\pi a}$ (c) $\frac{2}{3}\frac{\mu_0 I}{a}$ (d) zero

R stands to resistance of wire.

Subjective Questions

20. (a) A conductor in the shape of a square of edge length $l = 0.4$ m carries a current $i = 10.0$ A. Calculate the magnitude and direction of magnetic field at the centre of the square.

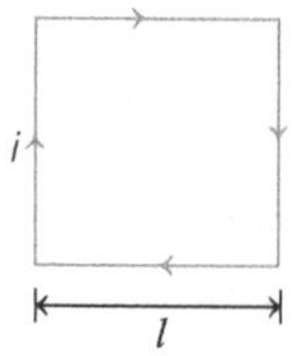

(b) If this conductor is formed into a single circular turn and carries the same current, what is the value of the magnetic field at the centre.

21. Determine the magnetic field at point P located a distance x from the corner of an infinitely long wire bent at right angle as shown in figure. The wire carries a steady current i.

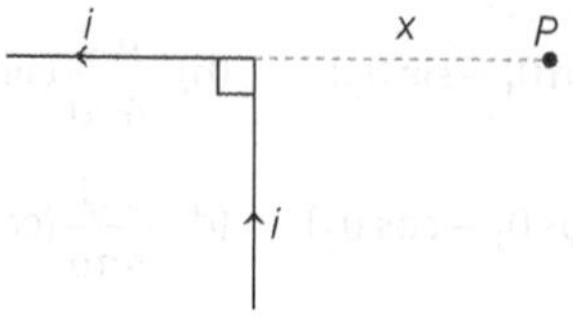

22. A conductor consists of a circular loop of radius $R = 10$ cm and two straight long sections as shown in figure. The wire lies in the plane of the paper and carries a current of $i = 7.00$ A. Determine the magnitude and direction of the magnetic field at the centre of the loop.

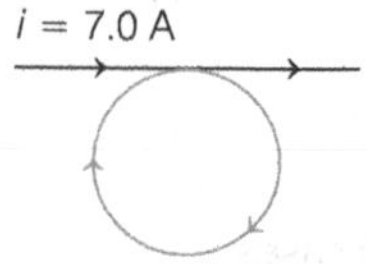

23. The segment of wire shown in figure carries a current of $i = 5.0$ A, where the radius of the circular arc is $R = 3.0$ cm. Determine the magnitude and direction of the magnetic field at the origin.

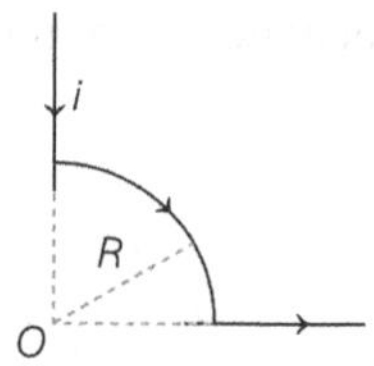

24. Consider the current carrying loop shown in figure formed of radial lines and segments of circles whose centres are at point P. Find the magnitude and direction of **B** at point P.

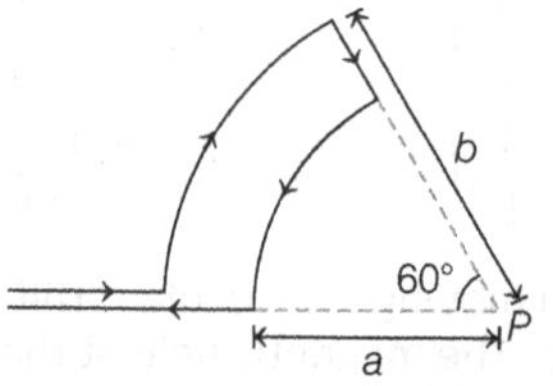

25. A very long wire carrying a current $I = 5.0$ A is bent at right angles. Find the magnetic induction at a point lying on a perpendicular normal to the plane of the wire drawn through the point of bending at a distance $l = 35$ cm from it. **(Irodov 3.227)**

26. A current $I = \sqrt{2}$ A flows in a circuit having the shape of isosceles trapezium. The ratio of the bases of the trapezium is 2. Find the magnetic induction B at symmetric point O in the plane of the trapezium. The length of the smaller base of the trapezium is 100 mm and the distance $r = 50$ mm.

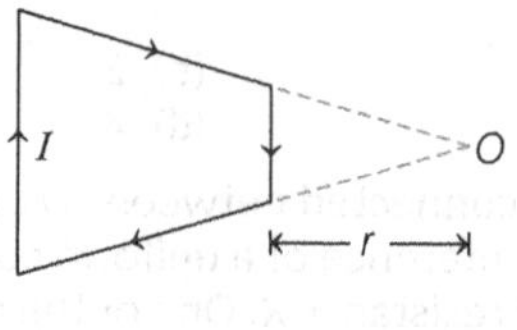

27. Two long mutually perpendicular conductors carrying currents I_1 and I_2 lie in one plane. Find the locus of points at which the magnetic induction is zero.

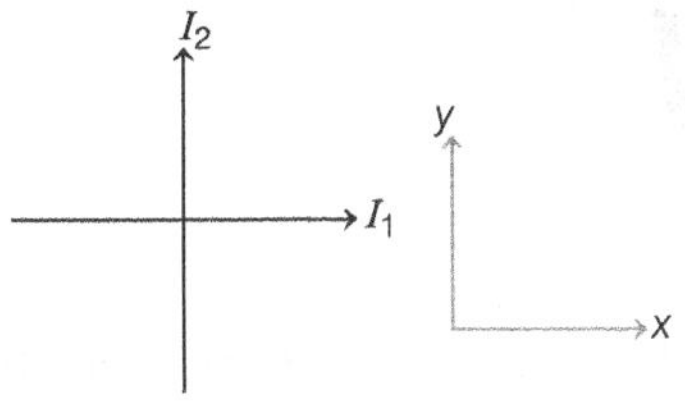

28. A wire carrying current i has the configuration as shown in figure. Two semi-infinite straight sections, both tangent to the same circle, are connected by a circular arc of central angle θ, along the circumference of the circle, with all sections lying in the same plane. What must θ be for B to be zero at the centre of the circle?

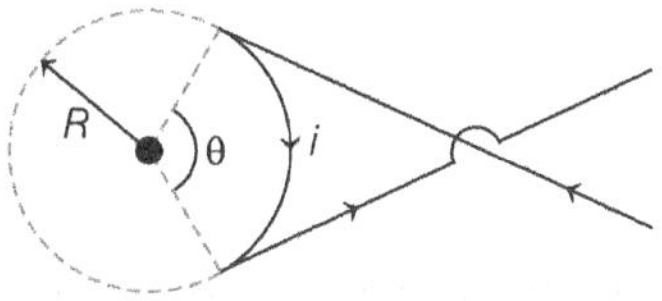

29. Two long parallel transmission lines 40.0 cm apart carry 25.0 A and 75.0 A currents. Find all locations where the net magnetic field of the two wires is zero if these currents are in

(a) the same direction

(b) the opposite direction

30. A closely wound coil has a radius of 6.0 cm and carries a current of 2.50 A. How many turns must it have if, at a point on the coil axis 6.00 cm from the centre of the coil, the magnetic field is 6.39×10^{-4} T ?

31. A circular loop of radius R carries current I_2 in a clockwise direction as shown in figure. The centre of the loop is a distance D above a long, straight wire. What are the magnitude and direction of the current I_1 in the wire if the magnetic field at the centre of loop is zero?

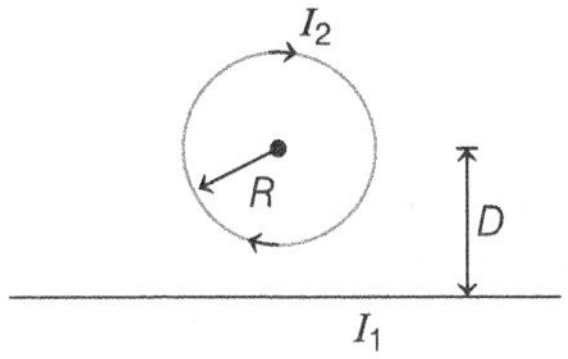

32. A closely wound, circular coil with radius 2.40 cm has 800 turns.

(a) What must the current in the coil be if the magnetic field at the centre of the coil is 0.0580 T?

(b) At what distance x from the centre of the coil, on the axis of the coil, is the magnetic field half its value at the centre?

33. Four very long, current carrying wires in the same plane intersect to form a square 40.0 cm on each side as shown in figure. Find the magnitude and direction of the current I so that the magnetic field at the centre of square is zero. Wires are insulated from each other.

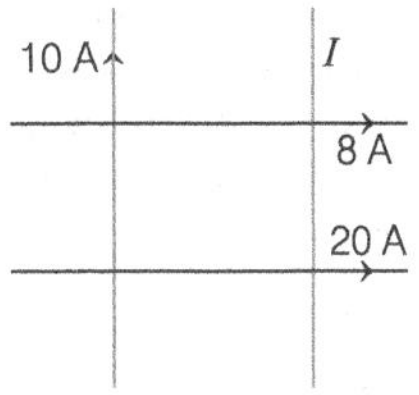

Answers

1. (c) **2.** (c) **3.** (c) **4.** (c) **5.** (a) **6.** (b) **7.** (a) **8.** (b) **9.** (a) **10.** (d)
11. (a) **12.** (b) **13.** (d) **14.** (b) **15.** (c) **16.** (b) **17.** (c) **18.** (b) **19.** (b)

20. (a) 28.3 μT into the page (b) 24.7 μT into the page **21.** $\frac{\mu_0 i}{4\pi x}$ into the page

22. 58.0 μT into the page **23.** 26.2 μT into the page **24.** $\frac{\mu_0 i}{12}\left(\frac{1}{a}-\frac{1}{b}\right)$ out of the page **25.** 2.0 μT **26.** 2×10^{-6} T

27. $y=\left(\frac{I_1}{I_2}\right)x$ **28.** 2 rad

29. (a) Between the wires, 30.0 cm from wire carrying 75.0 A (b) 20.0 cm from wire carrying 25.0 A and 60.0 cm from wire carrying 75.0 A

30. 69 **31.** $I_1=\left(\frac{\pi D}{R}\right)I_2$, towards right **32.** (a) 2.77 A (b) 0.0184 m **33.** 2.0 A toward bottom of page

See hints on pages 439-442.

MODULE 6

Ampere's Circuital Law

The electrical force on a charge is related to the electric field (caused by other charges) by the equation,

$$\mathbf{F}_e = q\mathbf{E}$$

Just like the gravitational force, the static electrical force is a conservative force. This means that the work done by the static electric force around any closed path is zero.

$$q\oint \mathbf{E}\cdot d\mathbf{l} = 0\,\text{J}$$

Hence, we have $\oint \mathbf{E}\cdot d\mathbf{l} = 0\,\text{V}$

In other words, the integral of the static (time independent) electric field around a closed path is zero.

What about the integral of the magnetic field around a closed path? That is, we want to determine the value of

$$\oint \mathbf{B}\cdot d\mathbf{l}$$

Here, we have to be careful. The quantity $\mathbf{B}\cdot d\mathbf{l}$ does not represent some physical quantity, and certainly not work. Although the static magnetic force does no work on a moving charge, we cannot conclude that the path integral of the magnetic field around a closed path is zero. We are just curious about what this analogous line integral amounts to.

The line integral $\oint \mathbf{B}\cdot d\mathbf{l}$ of the resultant magnetic field along a closed, plane curve is equal to μ_0 times the total current crossing the area bounded by the closed curve provided the electric field inside the loop remains constant. Thus,

$$\oint \mathbf{B}\cdot d\mathbf{l} = \mu_0\,(i_{\text{net}}) \quad \text{...(i)}$$

This is known as **Ampere's circuital law**.

Eq. (i) in simplified form can be written as

$$Bl = \mu_0\,(i_{\text{net}}) \quad \text{...(ii)}$$

But this equation can be used only under the following conditions.

(i) At every point of the closed path $\mathbf{B} \parallel d\mathbf{l}$.

(ii) Magnetic field has the same magnitude B at all places on the closed path.

Besides the Biot Savart law, Ampere's law gives another method to calculate the magnetic field due to a given current distribution. However, Ampere's law is more useful under certain symmetrical conditions. To illustrate the theory now let us take few applications of Ampere's circuital law.

Magnetic Field Created by a Long Thick Current Carrying Wire

A long straight wire of radius R carries a steady current i that is uniformly distributed through the cross-section of the wire.

For finding the behaviour of magnetic field due to this wire, let us divide the whole region into two parts.

One is $r \geq R$ and the another is $r < R$. Here, r is the distance from the centre of the wire.

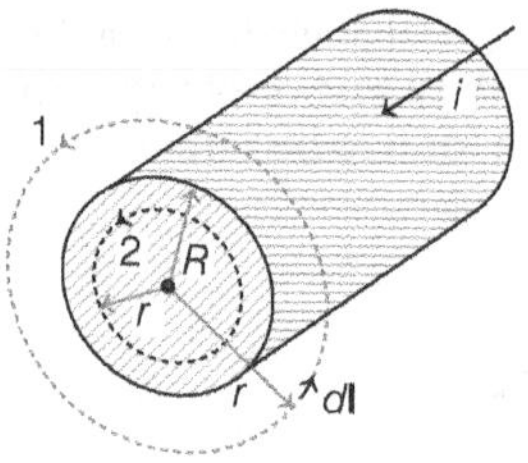

For $r \geq R$: Let us choose for our path of integration circle 1. From symmetry **B** must be constant in magnitude and parallel to $d\mathbf{l}$ at every point on this circle. Because the total current passing through the plane of the circle is i. Ampere's law gives

$$\oint \mathbf{B}\cdot d\mathbf{l} = \mu_0 i_{\text{net}}$$

or $\quad Bl = \mu_0 i_{\text{net}}$ [simplified form]

or $\quad B(2\pi r) = \mu_0 i$

$\therefore \quad B = \dfrac{\mu_0}{2\pi}\dfrac{i}{r}$ or $B \propto \dfrac{1}{r}$ [for $r \geq R$] ...(iii)

For $r < R$: Here, the current i' passing through the plane of circle 2 is less than the total current i. Because the current is uniform over the cross-section of the wire, the fraction of the current enclosed by circle 2 must equal the ratio of the area πr^2 enclosed by circle 2 to the cross-sectional area πR^2 of the wire.

$$\frac{i'}{i} = \frac{\pi r^2}{\pi R^2}$$

$$\Rightarrow \quad i' = \left(\frac{r^2}{R^2}\right) i$$

Applying Ampere's law to circle 2,

$$\oint \mathbf{B}\cdot d\mathbf{l} = \mu_0 i_{\text{net}}$$

$Bl = \mu_0\, i_{\text{net}} = \mu_0 i'$ [simplified form]

$$\therefore \quad B\,(2\pi r) = \mu_0 \left(\frac{r^2}{R^2}\right) i$$

$\therefore \quad B = \left(\dfrac{\mu_0 i}{2\pi R^2}\right) r$ or $B \propto r$ [For $r < R$] ...(iv)

The magnitude of the magnetic field *versus* r for this configuration is plotted in figure.

Note that inside the wire $B \to 0$ as $r \to 0$. Note also that Eqs. (iii) and (iv) give the same value of the magnetic field at $r = R$, demonstrating that the magnetic field is continuous at the surface of the wire.

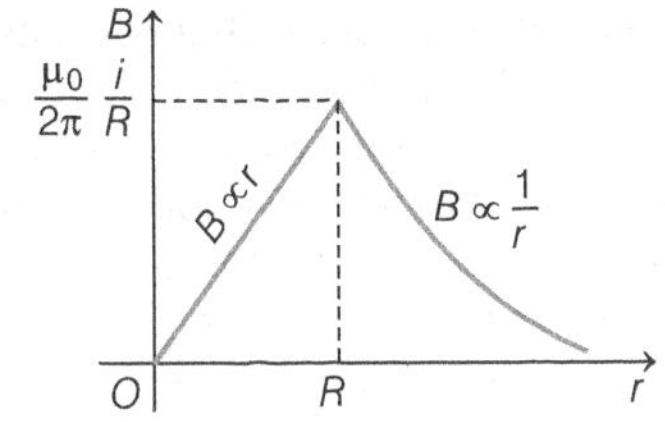

Magnetic Field of a Solenoid

A solenoid is a long wire wound in the form of a helix. With this configuration, a reasonably uniform magnetic field can be produced in the interior of the solenoid, when the solenoid carries a current. If the turns are closely spaced and the solenoid is of infinite length, the magnetic field lines are as shown in figure.

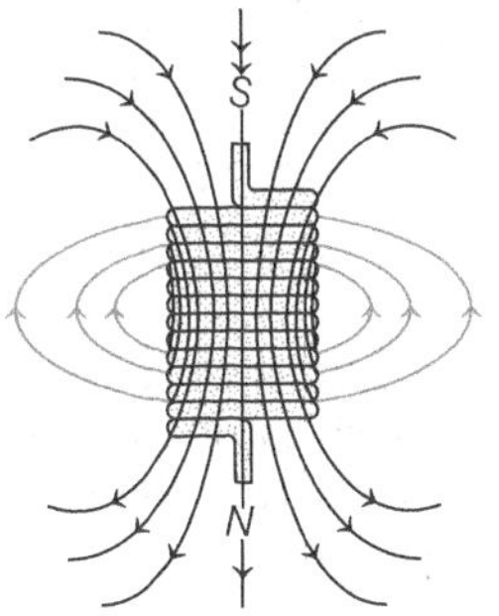

One end of the solenoid behaves like the north pole (↺) and the opposite end behaves like the south pole (↻). As the length of the solenoid increases, the interior field becomes more uniform and the exterior field becomes weaker. An ideal solenoid is approached when the turns are closely spaced and the length is much greater than the radius of the turns.

In this case, the external field is zero, and the interior field is uniform over a great volume.

We can use Ampere's law to obtain an expression for the interior magnetic field in an ideal solenoid.

Adjoining figure shows a longitudinal cross-section of part of such a solenoid carrying a current i. Because the solenoid is ideal, **B** in the interior space is uniform and parallel to the axis, and **B** in the exterior space is zero.

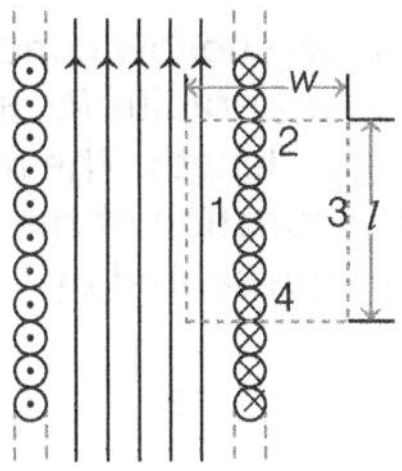

Consider the rectangular path of length l and width w as shown in figure. We can apply Ampere's law to this path by evaluating the line integral $\mathbf{B} \cdot d\mathbf{l}$ over each side of the rectangle.

$\int (\mathbf{B} \cdot d\mathbf{l})_{\text{side 3}} = 0$ as $B = 0$

$\int (\mathbf{B} \cdot d\mathbf{l})_{\text{side 2 and 4}} = 0$ as $\mathbf{B} \perp d\mathbf{l}$ or $B = 0$ along these paths

$\int (\mathbf{B} \cdot d\mathbf{l})_{\text{side 1}} = Bl$ as **B** is uniform and parallel to $d\mathbf{l}$

The integral over the closed rectangular path is therefore,

$$\oint \mathbf{B} \cdot d\mathbf{l} = Bl$$

The right side of Ampere's law involves the total current passing through the area bounded by the path of integration. In this case,

$i_{net} =$ (number of turns inside the area) (current through each turn)

$= (nl)(i)$ (n = number of turns per unit length)

Using Ampere's law, $\oint \mathbf{B} \cdot d\mathbf{l} = \mu_0 i_{net}$

or $Bl = (\mu_0)(nli)$ or $\boxed{B = \mu_0 ni}$ ...(v)

This result is same as obtained in Module 6. Eq. (v) is valid only for points near the centre (that is far from the ends) of a very long solenoid. The field near each end is half the value given by Eq. (v).

Example 1 A closed curve encircles several conductors. The line integral $\int \mathbf{B} \cdot d\mathbf{l}$ around this curve is 3.83×10^{-7} T-m.

(a) What is the net current in the conductors?

(b) If you were to integrate around the curve in the opposite directions, what would be the value of the line integral?

Sol. (a) $\int \mathbf{B} \cdot d\mathbf{l} = \mu_0 i_{net}$

$$\therefore \quad i_{net} = \frac{\int \mathbf{B} \cdot d\mathbf{l}}{\mu_0} = \frac{3.83 \times 10^{-7}}{4\pi \times 10^{-7}}$$

$= 0.3$ A

(b) In opposite direction, line integral will be negative.

Example 2 An infinitely long hollow conducting cylinder with inner radius $R/2$ and outer radius R carries a uniform current density along its length. The magnitude of the magnetic field, $|\mathbf{B}|$ as a function of the radial distance r from the axis is best represented by **(2012 JEE)**

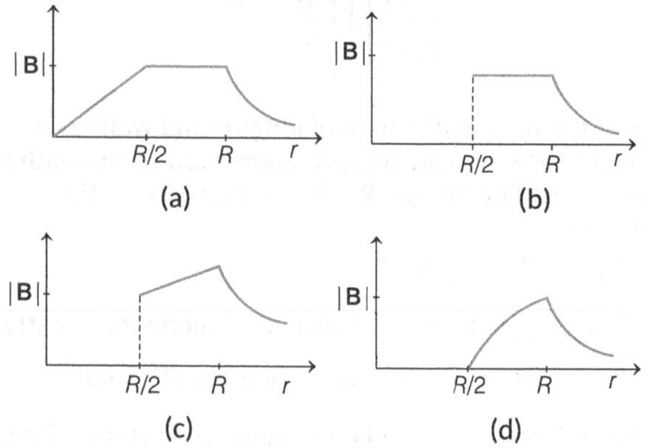

Sol. r = distance of a point from centre

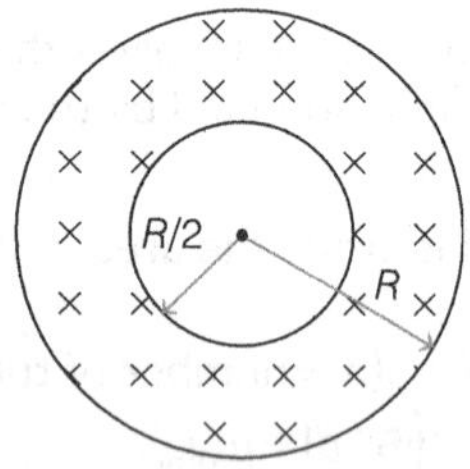

For $r \le R/2$ Using Ampere's circuital law,

$$\oint \mathbf{B} \cdot d\mathbf{l} = \mu_0 i_{net}$$

or $$Bl = \mu_0 (I_{in})$$

or $$B(2\pi r) = \mu_0 (I_{in})$$

or $$B = \frac{\mu_0}{2\pi} \frac{I_{in}}{r} \quad \text{...(i)}$$

Since, $I_{in} = 0$

$\therefore \quad B = 0$

For $\frac{R}{2} \le r \le R$ $\quad I_{in} = \left[\pi r^2 - \pi\left(\frac{R}{2}\right)^2\right]\sigma$

Here, σ = current per unit area

Substituting in Eq. (i), we have

$$B = \frac{\mu_0}{2\pi} \frac{\left[\pi r^2 - \pi \frac{R^2}{4}\right]\sigma}{r}$$

$$= \frac{\mu_0 \sigma}{2r}\left(r^2 - \frac{R^2}{4}\right)$$

At $\quad r = \frac{R}{2}, B = 0$

At $\quad r = R, B = \frac{3\mu_0 \sigma R}{8}$

For $r \ge R$ $\quad I_{in} = I_{Total} = I$ (say)

Therefore, substituting in Eq. (i), we have

$$B = \frac{\mu_0}{2\pi} \cdot \frac{I}{r} \quad \text{or} \quad B \propto \frac{1}{r}$$

$\therefore$ The correct graph is (d).

Example 3 A device called a toroid (figure) is often used to create an almost uniform magnetic field in some enclosed area. The device consists of a conducting wire wrapped around a ring (a torus) made of a non-conducting material. For a toroid having N closely spaced turns of wire, calculate the magnetic field in the region occupied by the torus, a distance r from the centre.

Sol. To calculate this field, we must evaluate $\oint \mathbf{B} \cdot d\mathbf{l}$ over the circle of radius r. By symmetry we see that the magnitude of the field is constant on this circle and tangent to it.

So, $$\oint \mathbf{B} \cdot d\mathbf{l} = Bl = B(2\pi r) \quad \text{...(i)}$$

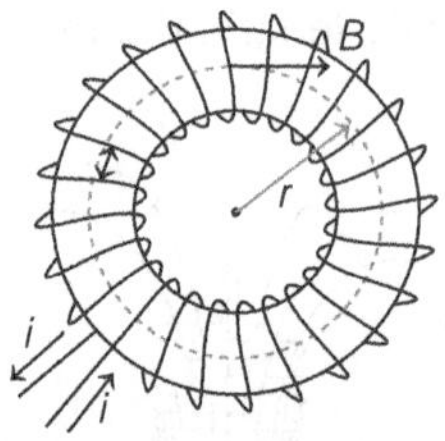

Furthermore, the circular closed path surrounds N loops of wire, each of which carries a current i. Therefore, right side of Eq. (i) is $\mu_0 Ni$ in this case.

$\therefore$ $$\oint \mathbf{B} \cdot d\mathbf{l} = \mu_0 i_{net}$$

or $$B(2\pi r) = \mu_0 Ni$$

or $$B = \frac{\mu_0 Ni}{2\pi r}$$

This result shows that $B \propto \frac{1}{r}$ and hence is non-uniform in the region occupied by torus. However, if r is very large compared with the cross-sectional radius of the torus, then the field is approximately uniform inside the torus. In that case,

$$\frac{N}{2\pi r} = n = \text{ number of turns per unit length of torus}$$

$\therefore$ $$B = \mu_0 ni$$

For an ideal toroid, in which turns are closely spaced, the external magnetic field is zero.

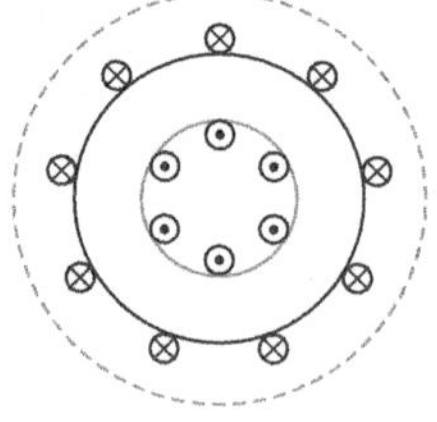

This is because the net current passing through any circular path lying outside the toroid is zero. Therefore, from Ampere's law we find that $B = 0$, in the regions exterior to the torus.

Example 4 A current is passing through a cylindrical conductor with a hole (or cavity) inside it. Show that the magnetic field inside the hole is uniform and find its magnitude and direction.

Sol. Let us find the magnetic field at point P inside the cavity at a distance r_1 from O and r_2 from C.

J = current per unit area

R = radius of cylinder

a = radius of cavity

i_1 = whole current from cylinder [say in ⊗ direction] $= J(\pi R^2)$

i_2 = current from hole [say in ⊙ direction] $= J(\pi a^2)$

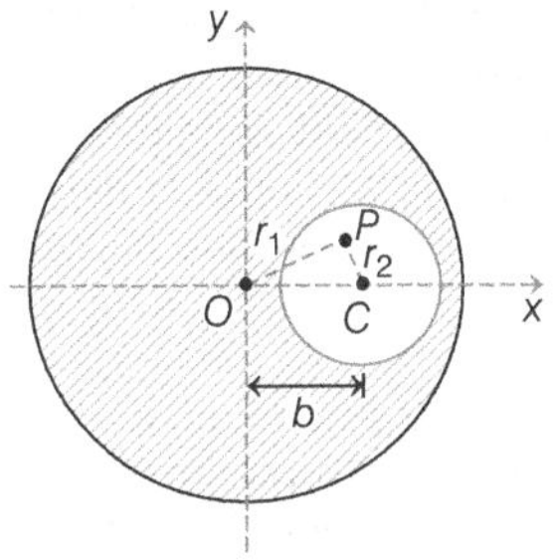

At point P magnetic field due to i_1 is B_1 (perpendicular to OP) and is $\mathbf{B}_2$ due to i_2 (perpendicular to CP) in the directions shown. Although $\mathbf{B}_1$ and $\mathbf{B}_2$ are actually at P, but for better understanding they are drawn at O and C respectively. Let B_x be the x-component of resultant of $\mathbf{B}_1$ and $\mathbf{B}_2$ and B_y its y-component. Then,

$$B_x = B_1 \sin\alpha - B_2 \sin\beta$$

$$= \left(\frac{\mu_0}{2\pi}\frac{i_1}{R^2} r_1\right)\sin\alpha - \left(\frac{\mu_0}{2\pi}\frac{i_2}{a^2}\cdot r_2\right)\sin\beta$$

$$= \left(\frac{\mu_0}{2\pi}\frac{J\pi R^2}{R^2}\cdot r_1 \sin\alpha\right) - \left(\frac{\mu_0}{2\pi}\frac{J.\pi a^2}{a^2} r_2 \sin\beta\right)$$

$$= \frac{\mu_0 J}{2}(r_1 \sin\alpha - r_2 \sin\beta) = 0$$

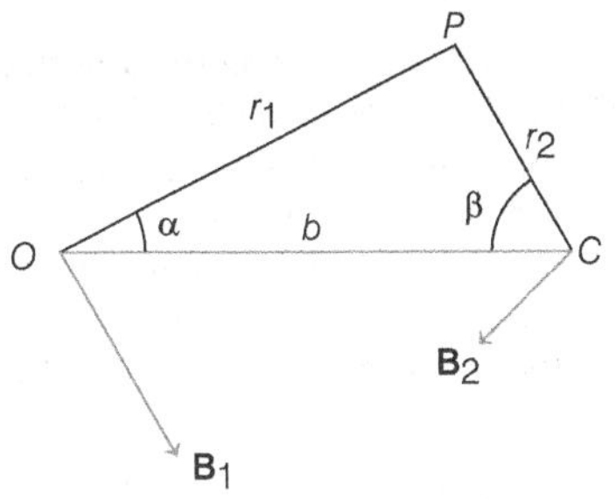

Because in ΔOPC $\quad \dfrac{r_1}{\sin\beta} = \dfrac{r_2}{\sin\alpha}$

or $\quad r_1 \sin\alpha - r_2 \sin\beta = 0$

Now, $\quad B_y = -(B_1 \cos\alpha + B_2 \cos\beta)$

$$= -\frac{\mu_0 J}{2}(r_1 \cos\alpha + r_2 \cos\beta)$$

From ΔOPC, we can see that

$$r_1 \cos\alpha + r_2 \cos\beta = b$$

or $\quad B_y = -\dfrac{\mu_0 Jb}{2}$ = constant

Thus, we can see that net magnetic field at point P is along negative y-direction and constant in magnitude.

Proved

Note *(i) That $\angle OPC$ is not necessarily 90°. At some point it may be 90°.*

(ii) At point C magnetic field due to i_2 is zero (i.e. $B_2 = 0$) while that due to i_1 is $\frac{\mu_0}{2\pi}\frac{i_1}{R^2}b$ in negative y-direction.

Substituting $i_1 = J(\pi R^2)$, we get

$$B = B_2 = \frac{\mu_0 Jb}{2} \quad \text{(along negative } y\text{-direction)}$$

This agrees with the result derived above.

MODULE 6 Exercise

Single Correct Option

1. A current I flows along the length of an infinitely long, straight, thin-walled pipe. Then, **(1993 JEE)**
   (a) the magnetic field at all points inside the pipe is the same, but not zero
   (b) the magnetic field at any point inside the pipe is zero
   (c) the magnetic field is zero only on the axis of the pipe
   (d) the magnetic field is different at different points inside the pipe

2. The figure shows the cross-section of two long coaxial tubes carrying equal currents I in opposite directions. If B_1 and B_2 are magnetic fields at points 1 and 2 as shown in figure, then

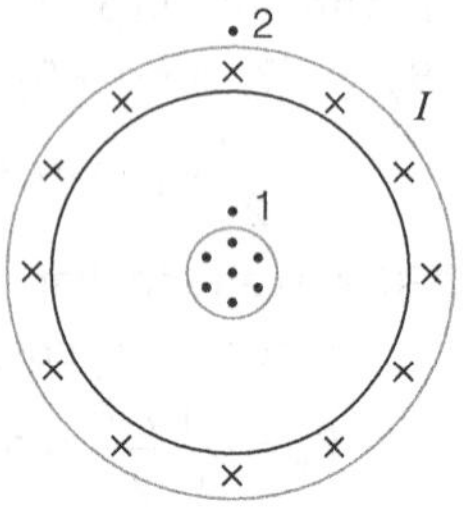

   (a) $B_1 \neq 0; B_2 = 0$ (b) $B_1 = 0; B_2 = 0$
   (c) $B_1 \neq 0; B_2 \neq 0$ (d) $B_1 = 0; B_2 \neq 0$

3. A cylindrical long wire of radius R carries a current I uniformly distributed over the cross-sectional area of the wire. The magnetic field at a distance x from the surface inside the wire is
   (a) $\frac{\mu_0 I}{2\pi(R-x)}$ (b) $\frac{\mu_0 I}{2\pi x}$
   (c) $\frac{\mu_0 I}{2\pi(R+x)}$ (d) None of these

4. Consider a coaxial cable which consists of an inner wire of radius a surrounded by an outer shell of inner and outer radii b and c, respectively. The inner wire carries a current I and outer shell carries an equal and opposite current. The magnetic field at a distance x from the axis where $b < x < c$ is
   (a) $\frac{\mu_0 I(c^2-b^2)}{2\pi x(c^2-a^2)}$ (b) $\frac{\mu_0 I(c^2-x^2)}{2\pi x(c^2-a^2)}$
   (c) $\frac{\mu_0 I(c^2-x^2)}{2\pi x(c^2-b^2)}$ (d) zero

5. Consider a coaxial cable which consists of an inner wire of radius a surrounded by an outer shell of inner and outer radii b and c, respectively. The inner wire carries a current I and the outer shell carries an equal current in opposite direction.
   (a) for $x < a, B = \frac{\mu_0 I x}{2\pi a^2}$
   (b) for $a < x < b, B = \frac{\mu_0 I}{2\pi x}$
   (c) Both are correct
   (d) Both are wrong

Multiple Correct Options

6. If a long thick cylindrical conductor carries a steady current parallel to its length, then
   (a) the electric field along the axis is zero
   (b) the magnetic field along the axis is zero
   (c) the magnetic field outside the conductor is zero
   (d) the electric field outside the conductor is zero

Subjective Questions

7. Figure given in the question is a cross-sectional view of a coaxial cable. The centre conductor is surrounded by a rubber layer, which is surrounded by an outer conductor, which is surrounded by another rubber layer. The current in the inner conductor is 1.0 A out of the page, and the current in the outer conductor is 3.0 A into the page. Determine the magnitude and direction of the magnetic field at points a and b.

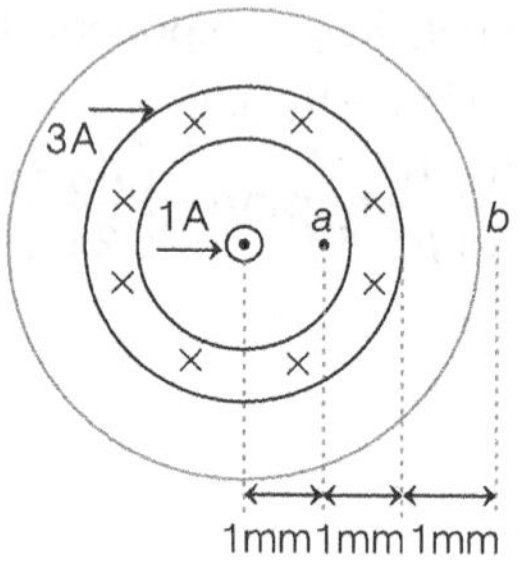

8. Figure shows, in cross-section, several conductors that carry currents through the plane of the figure. The currents have the magnitudes $I_1 = 4.0$ A, $I_2 = 6.0$ A, and $I_3 = 2.0$ A, in the directions shown. Four paths labelled a to d, are shown.

What is the line integral $\int \mathbf{B} \cdot d\mathbf{l}$ for each path? Each integral involves going around the path in the counter-clockwise direction.

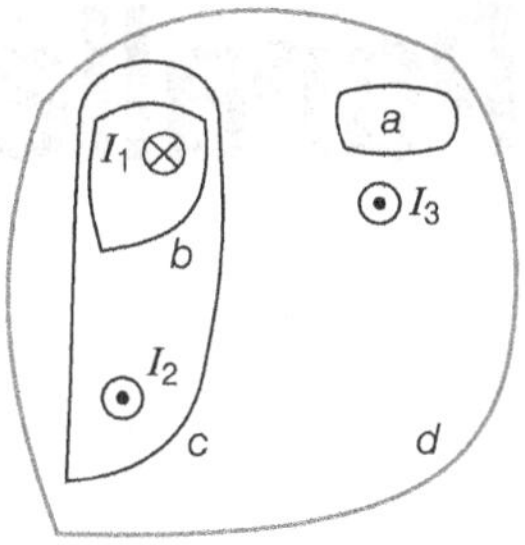

9. A long cylindrical conductor of radius R carries a current i as shown in figure. The current density J is a function of radius according to, $J = br$, where b is a constant. Find an expression for the magnetic field B

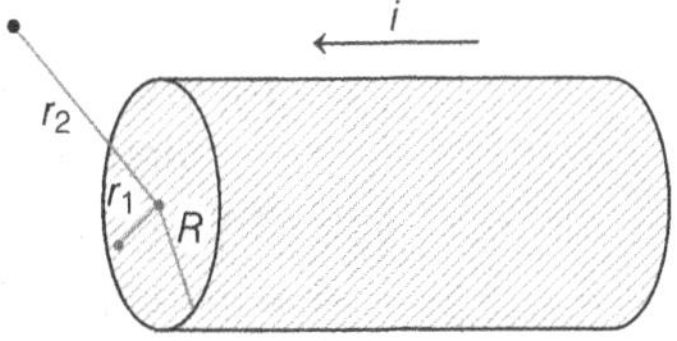

(a) at a distance $r_1 < R$ and
(b) at a distance $r_2 > R$, measured from the axis.

10. Two infinite plates shown in cross-section in figure carry λ amperes of current out of the page per unit width of plate. Find the magnetic field at points P and Q.

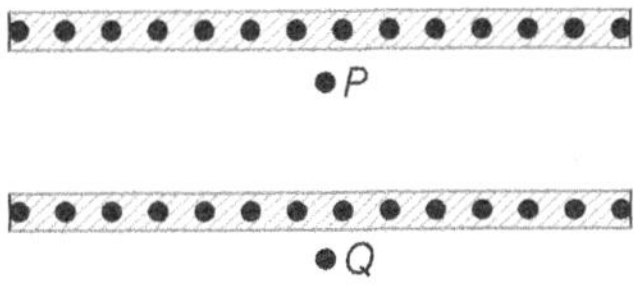

11. A current I flows along a round loop. Find the integral $\int \mathbf{B}\, d\mathbf{r}$ along the axis of the loop within the range from $-\infty$ to $+\infty$. **(Irodov 3.232)**

Answers

1. (b) **2.** (a) **3.** (d) **4.** (c) **5.** (c) **6.** (b,d)

7. 200 μT toward the top of the page, 133 μT toward the bottom of the page

8. (a) Zero (b) -5.0×10^{-6} T-m (c) 2.5×10^{-6} T-m (d) 5.0×10^{-6} T-m

9. (a) $\frac{\mu_0 b r_1^2}{3}$ (b) $\frac{\mu_0 b R^3}{3 r_2}$

10. $B_P = 0, B_Q = \mu_0 \lambda$ **11.** $\int_{-\infty}^{+\infty} \mathbf{B} \cdot d\mathbf{r} = \mu_0 I$

See hints on pages 442-443.

MODULE 7

Force Between Parallel Current Carrying Wires

Consider two long wires 1 and 2 kept parallel to each other at a distance r and carrying currents i_1 and i_2 respectively in the same direction.

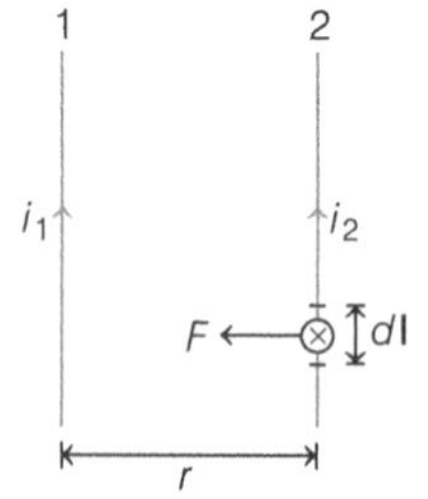

Magnetic field on wire 2 due to current in wire 1 is,

$$B = \frac{\mu_0}{2\pi} \cdot \frac{i_1}{r} \quad \text{[in } \otimes \text{ direction]}$$

Magnetic force on a small element dl of wire 2 due to this magnetic field is

$$d\mathbf{F} = i_2 (d\mathbf{l} \times \mathbf{B})$$

Magnitude of this force is $dF = i_2 [(dl)(B) \sin 90°]$

$$= i_2 (dl) \left(\frac{\mu_0}{2\pi} \frac{i_1}{r} \right) = \frac{\mu_0}{2\pi} \cdot \frac{i_1 i_2}{r} \cdot dl$$

Direction of this force is along $d\mathbf{l} \times \mathbf{B}$ or towards the wire 1.

The force per unit length of wire 2 due to wire 1 is

$$\boxed{\frac{F}{l} = \frac{\mu_0}{2\pi} \frac{i_1 i_2}{r}}$$

The same force acts on wire 1 due to wire 2. The wires attract each other if currents in the wires are flowing in the same direction and they repel each other if the currents are in opposite directions.

Example 1 Two long parallel wires are separated by a distance of 2.50 cm. The force per unit length that each wire exerts on the other is 4.00×10^{-5} N/m, and the wires repel each other. The current in one wire is 0.600 A.

(a) What is the current in the second wire?

(b) Are the two currents in the same direction or in opposite directions?

Sol. (a) $\frac{F}{l} = \left(\frac{\mu_0}{2\pi} \right) \frac{i_1 i_2}{r}$

$$\therefore \quad 4 \times 10^{-5} = \frac{(2 \times 10^{-7})(0.6)\, i_2}{2.5 \times 10^{-2}}$$

$$\therefore \quad i_2 = 8.33 \text{ A}$$

(b) Wires repel each other if currents are in opposite directions.

Example 2 Consider three long straight parallel wires as shown in figure. Find the force experienced by a 25 cm length of wire C.

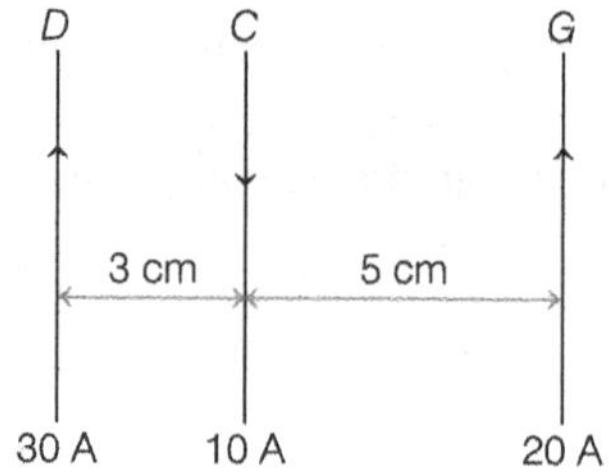

Sol. Repulsion by wire D, [towards right]

$$F_1 = \frac{\mu_0}{2\pi} \frac{i_1 i_2 l}{r}$$

$$= \frac{(2 \times 10^{-7})(30 \times 10)}{3 \times 10^{-2}} (0.25)$$

$$= 5 \times 10^{-4} \text{ N}$$

Repulsion by wire G, [towards left]

$$F_2 = \frac{(2 \times 10^{-7})(20 \times 10)}{5 \times 10^{-2}} (0.25)$$

$$= 2 \times 10^{-4} \text{ N}$$

$$\therefore \quad F_{net} = F_1 - F_2$$

$$= 3 \times 10^{-4} \text{ N} \quad \text{[towards right]}$$

Example 3 A long horizontal wire AB, which is free to move in a vertical plane and carries a steady current of 20 A, is in equilibrium at a height of 0.01 m over another parallel long wire CD which is fixed in a horizontal plane and carries a steady current of 30 A, as shown in figure. Show that when AB is slightly depressed, it executes simple harmonic motion. Find the period of oscillations. **(1994 JEE)**

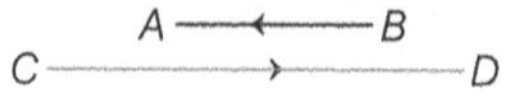

Sol. Let m be the mass per unit length of wire AB. At a height x above the wire CD, magnetic force per unit length on wire AB will be given by

$$F_m = \frac{\mu_0}{2\pi}\frac{i_1 i_2}{x} \quad \text{(upwards)} \quad ...(i)$$

Weight per unit length of wire AB is

$$F_g = mg \quad \text{(downwards)}$$

Here, m = mass per unit length of wire AB

At $x = d$, wire is in equilibrium, i.e.

$$F_m = F_g$$

or $$\frac{\mu_0}{2\pi}\frac{i_1 i_2}{d} = mg$$

or $$\frac{\mu_0}{2\pi}\frac{i_1 i_2}{d^2} = \frac{mg}{d} \quad ...(ii)$$

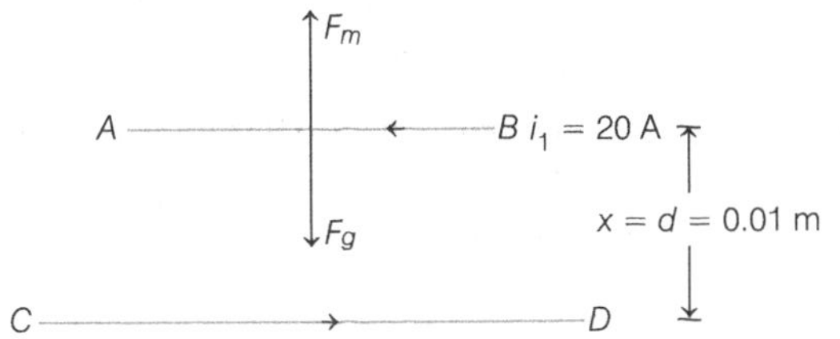

When AB is depressed, x decreases therefore, F_m will increase, while F_g remains the same. Change in magnetic force will become the net restoring force, Let AB is displaced by dx downwards.

Differentiating Eq. (i) w.r.t. x, we get

$$dF_m = -\frac{\mu_0}{2\pi}\frac{i_1 i_2}{x^2}\cdot dx \quad ...(iii)$$

i.e. restoring force, $F = dF_m \propto -dx$

Hence, the motion of wire is simple harmonic.

From Eqs. (ii) and (iii), we can write

$$dF_m = -\left(\frac{mg}{d}\right)\cdot dx \quad [\because x = d]$$

$\therefore$ Acceleration of wire, $a = -\left(\frac{g}{d}\right).dx$

Hence, period of oscillation

$$T = 2\pi\sqrt{\frac{|\text{displacement}|}{|\text{acceleration}|}} = 2\pi\sqrt{\left|\frac{dx}{a}\right|}$$

or $$T = 2\pi\sqrt{\frac{d}{g}} = 2\pi\sqrt{\frac{0.01}{9.8}}$$

or $$T = 0.2 \text{ s}$$ **Ans.**

Example 4 A straight segment OC (of length L) of a circuit carrying a current I is placed along the x-axis . Two infinitely long straight wires A and B, each extending from $z = -\infty$ to $+\infty$, are fixed at $y = -a$ and $y = +a$ respectively, as shown in the figure. If the wires A and B each carry a current I into the plane of the paper, obtain the expression for the force acting on the segment OC. What will be the force on OC if the current in the wire B is reversed? **(1992 JEE)**

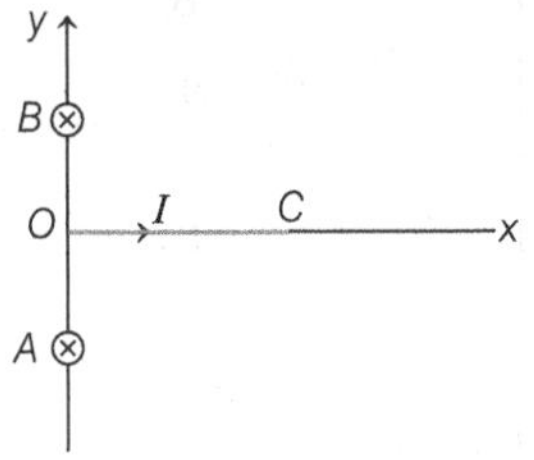

Sol. (a) Let us assume a segment of wire OC at a point P, a distance x from the centre of length dx as shown in figure.

Magnetic field at P due to current in wires A and B will be in the directions perpendicular to AP and BP respectively as shown.

$$|\mathbf{B}| = \frac{\mu_0}{2\pi}\frac{I}{AP}$$

Therefore, net magnetic force at P will be along negative y-axis as shown below

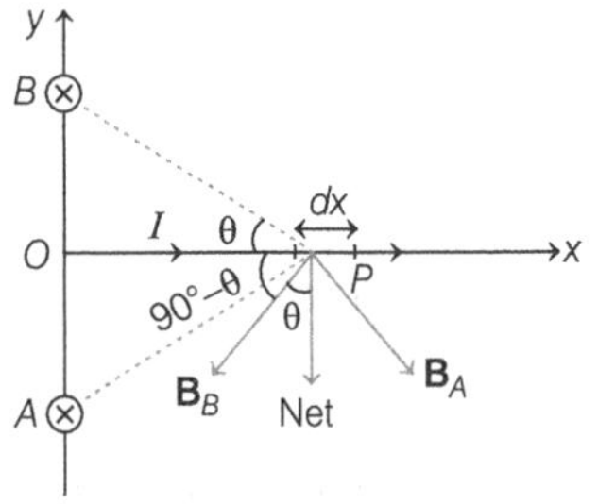

$$B_{net} = 2|\mathbf{B}|\cos\theta$$
$$= 2\left(\frac{\mu_0}{2\pi}\right)\frac{I}{AP}\left(\frac{x}{AP}\right)$$
$$B_{net} = \left(\frac{\mu_0}{\pi}\right)\frac{Ix}{(AP)^2}$$
$$B_{net} = \frac{\mu_0}{\pi}\cdot\frac{Ix}{(a^2 + x^2)}$$

Therefore, force on this element will be

$$dF = I\left\{\frac{\mu_0}{\pi}\frac{Ix}{a^2 + x^2}\right\}dx \quad \text{[in negative } z\text{-direction]}$$

$\therefore$ Total force on the wire will be

$$F = \int_{x=0}^{x=L} dF = \frac{\mu_0 I^2}{\pi}\int_0^L \frac{x\,dx}{x^2 + a^2}$$
$$= \frac{\mu_0 I^2}{2\pi}\ln\left(\frac{L^2 + a^2}{a^2}\right) \quad \text{[in negative } z\text{-axis]}$$

Hence, $$\mathbf{F} = -\frac{\mu_0 I^2}{2\pi}\ln\left(\frac{L^2 + a^2}{a^2}\right)\hat{\mathbf{k}}$$

(b) When direction of current in B is reversed net magnetic field is along the current.

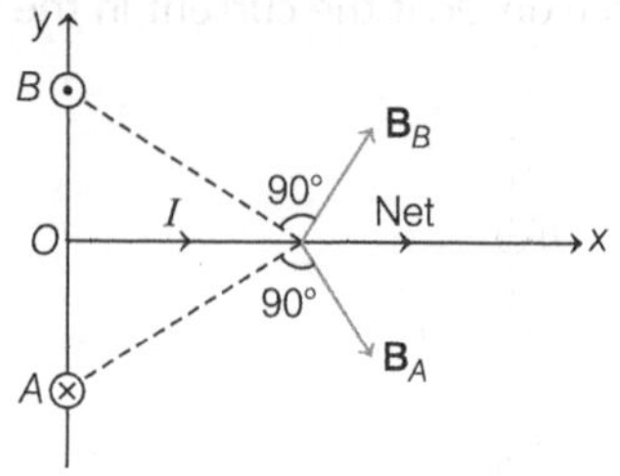

Hence, force is zero.

Example 5 Two long straight parallel wires are 2 m apart, perpendicular to the plane of the paper. The wire *A* carries a current of 9.6 A, directed into the plane of the paper. The wire *B* carries a current such that the magnetic field of induction at the point *P*, at a distance of 10 / 11 m from the wire *B*, is zero.

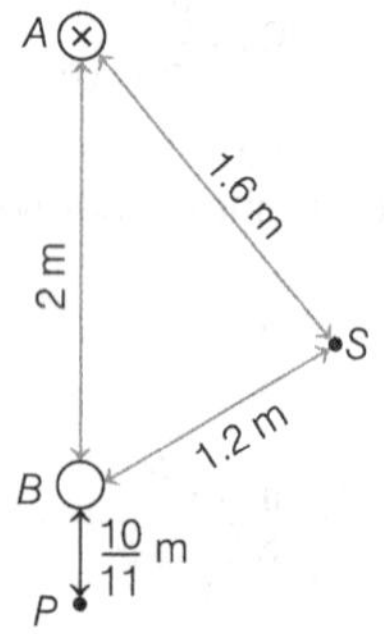

Find **(1997 JEE)**

(a) the magnitude and direction of the current in *B*.

(b) the magnitude of the magnetic field of induction at the point *S*.

(c) the force per unit length on the wire *B*.

Sol. (a) Direction of current at *B* should be perpendicular to paper outwards. Let current in this wire be i_B. Then,

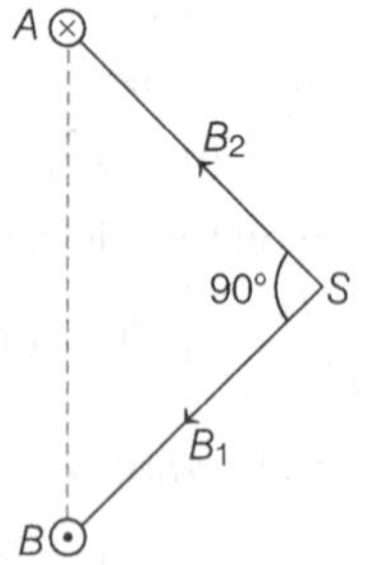

$$\frac{\mu_0}{2\pi}\frac{i_A}{\left(2+\frac{10}{11}\right)}=\frac{\mu_0}{2\pi}\frac{i_B}{(10/11)}$$

or $$i_B=\frac{10}{32}\times i_A=\frac{10}{32}\times 9.6=3\text{A}$$

(b) Since, $AS^2+BS^2=AB^2$

$\therefore \quad \angle ASB=90°$

At S : B_1 = Magnetic field due to i_A

$$=\frac{\mu_0}{2\pi}\frac{i_A}{1.6}=\frac{(2\times10^{-7})(9.6)}{1.6}$$

$$=12\times10^{-7}\text{ T}$$

B_2 = Magnetic field due to i_B

$$=\frac{\mu_0}{2\pi}\frac{i_B}{1.2}$$

$$=\frac{(2\times10^{-7})(3)}{1.2}$$

$$=5\times10^{-7}\text{ T}$$

Since, B_1 and B_2 are mutually perpendicular. Net magnetic field at *S* would be

$$B=\sqrt{B_1^2+B_2^2}$$

$$=\sqrt{(12\times10^{-7})^2+(5\times10^{-7})^2}$$

$$=13\times10^{-7}\text{ T}$$

(c) Force per unit length on wire *B* (repulsive in nature):

$$\frac{F}{l}=\frac{\mu_0}{2\pi}\frac{i_Ai_B}{r} \qquad [\because r=AB=2\text{ m}]$$

$$=\frac{(2\times10^{-7})(9.6\times3)}{2}$$

$$=2.88\times10^{-6}\text{ N/m}$$

Example 6 A current of 10 A flows around a closed path in a circuit which is in the horizontal plane as shown in the figure. The circuit consists of eight alternating arcs of radii $r_1=0.08$ m and $r_2=0.12$ m. Each subtends the same angle at the centre. **(2001 JEE)**

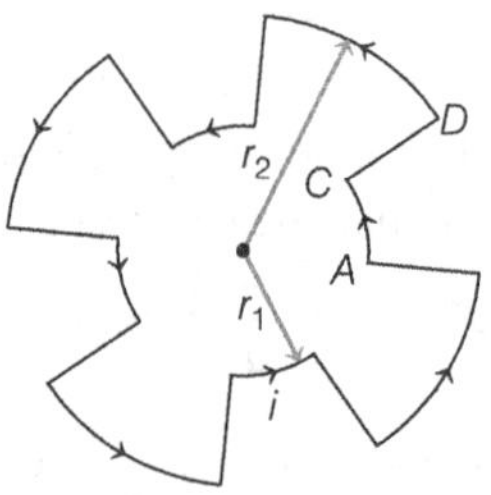

(a) Find the magnetic field produced by this circuit at the centre.

(b) An infinitely long straight wire carrying a current of 10 A is passing through the centre of the above circuit vertically with the direction of the current being into the plane of the circuit. What is the force acting on the wire at the centre due to the current in the circuit? What is the force acting on the arc *AC* and the straight segment *CD* due to the current at the centre?

Sol. (a) Given, i = 10 A, $r_1=0.08$ m and $r_2=0.12$ m. Straight portions, i.e. *CD* etc. will produce zero magnetic field at the centre. Rest eight arcs will produce the magnetic field at the centre in the same direction, i.e. perpendicular to the paper outwards or vertically upwards and its magnitude is

$$B = B_{\text{inner arcs}} + B_{\text{outer arcs}}$$
$$= \frac{1}{2}\left\{\frac{\mu_0 i}{2r_1}\right\} + \frac{1}{2}\left\{\frac{\mu_0 i}{2r_2}\right\}$$
$$= \left(\frac{\mu_0}{4\pi}\right)(\pi i)\left(\frac{r_1 + r_2}{r_1 r_2}\right)$$

Substituting the values, we have

$$B = \frac{(10^{-7})(3.14)(10)(0.08 + 0.12)}{(0.08 \times 0.12)}$$
$$B = 6.54 \times 10^{-5}\ \text{T}$$

(vertically upward or outward normal to the paper)

(b) **Force on AC**

Force on circular portions of the circuit, i.e. AC etc, due to the wire at the centre will be zero because magnetic field due to the central wire at these arcs will be tangential ($\theta = 180°$).

Force on CD

Current in central wire is also $i = 10$ A. Magnetic field at distance x due to central wire

$$B = \frac{\mu_0}{2\pi} \cdot \frac{i}{x}$$

$\therefore$ Magnetic force on element dx due to this magnetic field

$$dF = (i)\left(\frac{\mu_0}{2\pi} \cdot \frac{i}{x}\right) \cdot dx = \left(\frac{\mu_0}{2\pi}\right) i^2 \frac{dx}{x} \qquad [F = ilB \sin 90°]$$

Therefore, net force on CD is

$$F = \int_{x = r_1}^{x = r_2} dF = \frac{\mu_0 i^2}{2\pi} \int_{0.08}^{0.12} \frac{dx}{x} = \frac{\mu_0}{2\pi} i^2 \ln\left(\frac{3}{2}\right)$$

Substituting the values, $F = (2 \times 10^{-7})(10)^2 \ln(1.5)$

or $\qquad F = 8.1 \times 10^{-6}$ N $\qquad$ (inwards)

Force on wire at the centre

Net magnetic field at the centre due to the circuit is in vertical direction and current in the wire in centre is also in vertical direction. Therefore, net force on the wire at the centre will be zero. ($\theta = 180°$).

Hence,

(i) Force acting on the wire at the centre is zero.

(ii) Force on arc AC = 0.

(iii) Force on segment CD is 8.1×10^{-6} N $\qquad$ (inwards)

Example 7 Three infinitely long thin wires, each carrying current i in the same direction, are in the xy-plane of a gravity free space. The central wire is along the y-axis while the other two are along $x = \pm d$.

(a) Find the locus of the points for which the magnetic field B is zero.

(b) If the central wire is displaced along the z-direction by a small amount and released, show that it will execute simple harmonic motion. If the linear mass density of the wires is λ, find the frequency of oscillation.

Sol. (a)

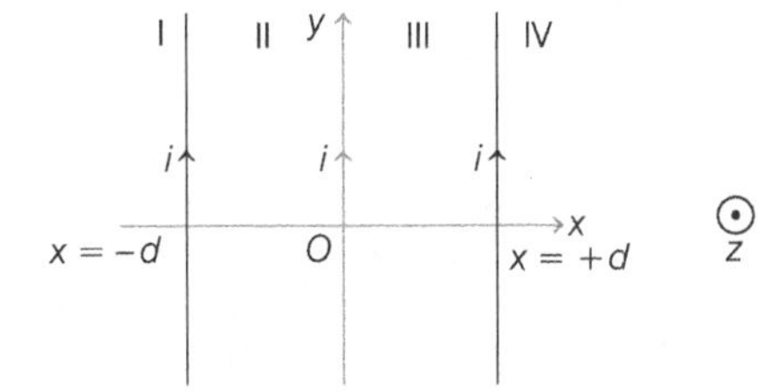

Magnetic field cannot be zero in region I and region IV because in region I magnetic field will be along positive z-direction due to all the three wires, while in region IV magnetic field will be along negative z-axis due to all the three wires. It can be zero only in region II and III.

Let magnetic field is zero on line $z = 0$ and $x = x$ (shown as dotted). The magnetic field on this line due to wires 1 and 2 will be along negative z-axis and due to wire 3 along positive z-axis. Thus,

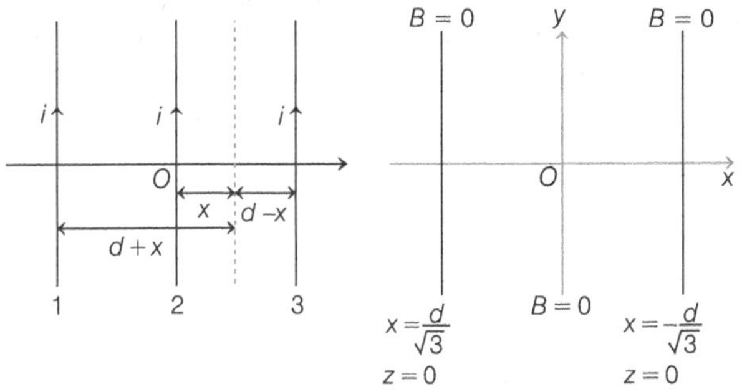

$$B_1 + B_2 = B_3$$

or $$\frac{\mu_0}{2\pi} \frac{i}{(d + x)} + \frac{\mu_0 i}{2\pi x} = \frac{\mu_0}{2\pi} \frac{i}{(d - x)}$$

or $$\frac{1}{d + x} + \frac{1}{x} = \frac{1}{d - x}$$

This equation gives $x = \pm \dfrac{d}{\sqrt{3}}$

Hence, there will be two lines $x = \dfrac{d}{\sqrt{3}}$

and $\qquad x = -\dfrac{d}{\sqrt{3}} \qquad (z = 0)$

where, magnetic field is zero **Ans.**

(b) In this part, let us change our coordinate axes system, just for better understanding.

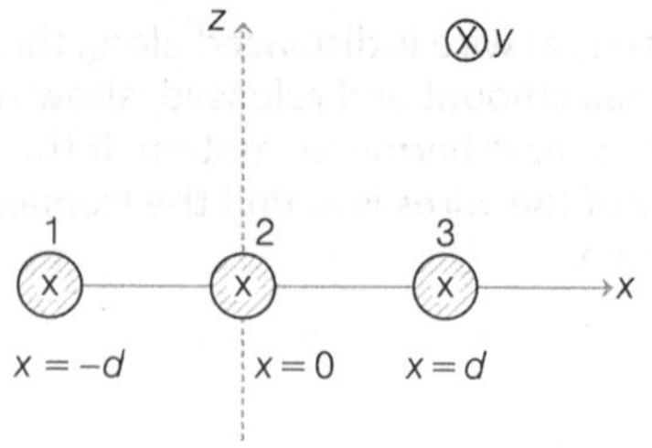

There are three wires 1, 2 and 3 as shown in figure. If we displace the wire 2 towards the z-axis, then force of attraction per unit length between wires (1 and 2) and (2 and 3) will be given by

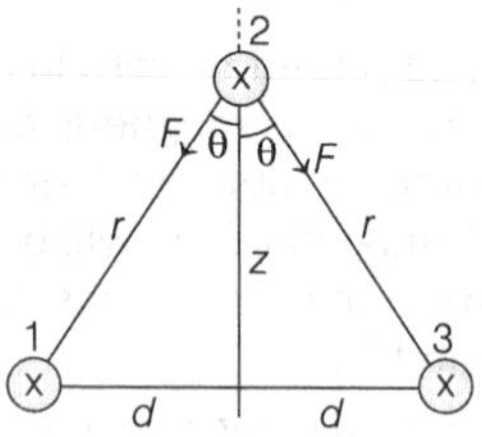

$$F = \frac{\mu_0}{2\pi}\frac{i^2}{r}$$

The components of F along x-axis will be cancelled out. Net resultant force will be towards negative z-axis (or mean position) and will be given by

$$F_{net} = 2F\cos\theta = 2\left\{\frac{\mu_0}{2\pi}\frac{i^2}{r}\right\}\frac{z}{r}$$

$$F_{net} = \frac{\mu_0}{\pi}\frac{i^2}{(z^2+d^2)}\cdot z \qquad (r^2 = z^2 + d^2)$$

If $z << d$, then

$$z^2 + d^2 \approx d^2$$

and $$F_{net} = -\left(\frac{\mu_0}{\pi}\frac{i^2}{d^2}\right)\cdot z$$

Negative sign implies that F_{net} is restoring in nature.

Therefore, $F_{net} \propto -z$

i.e. the wire will oscillate simple harmonically.

Let a be the acceleration of wire in this position and λ the mass per unit length of wire, then

$$F_{net} = \lambda a = -\left(\frac{\mu_0}{\pi}\frac{i^2}{d^2}\right)z$$

or $$a = -\left(\frac{\mu_0 i^2}{\pi\lambda d^2}\right)z$$

$\therefore$ Frequency of oscillation,

$$f = \frac{1}{2\pi}\sqrt{\left|\frac{\text{acceleration}}{\text{displacement}}\right|}$$

$$= \frac{1}{2\pi}\sqrt{\left|\frac{a}{z}\right|} = \frac{1}{2\pi}\frac{i}{d}\sqrt{\frac{\mu_0}{\pi\lambda}}$$

or $$f = \frac{i}{2\pi d}\sqrt{\frac{\mu_0}{\pi\lambda}}$$ **Ans..**

MODULE 7 Exercise

Single Correct Option

1. Two long conductors, separated by a distance d carry currents I_1 and I_2 in the same direction. They exert a force F on each other. Now, the current in one of them is increased to two times and its direction is reversed. The distance is also increased to $3d$. The new value of the force between them is
(a) $-F/3$ (b) $F/3$
(c) $2F/3$ (d) $-2F/3$

2. Figure shows three long straight wires parallel and equally spaced with identical currents. Then, the magnitude of force acting on each wire due to the other is

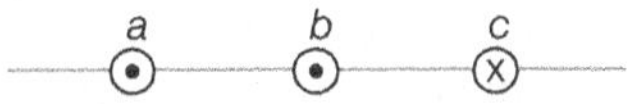

(a) $F_a > F_b > F_c$ (b) $F_b > F_c > F_a$
(c) $F_c > F_a > F_b$ (d) $F_b > F_a > F_c$

3. The segment AB of wire carrying current I_1 is placed perpendicular to a long straight wire carrying current I_2 as shown in figure. The magnitude of force experienced by the straight wire AB is

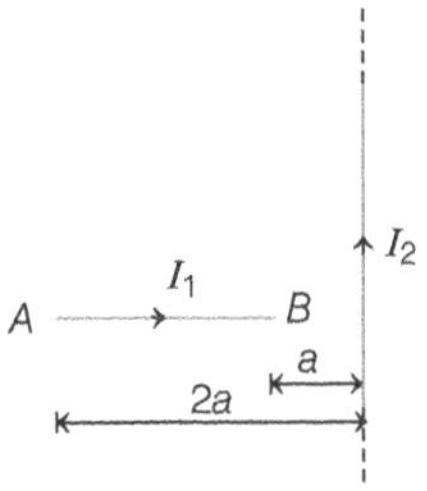

(a) $\frac{\mu_0 I_1 I_2}{2\pi} \ln 3$ (b) $\frac{\mu_0 I_1 I_2}{2\pi} \ln 2$
(c) $\frac{2\mu_0 I_1 I_2}{2\pi}$ (d) $\frac{\mu_0 I_1 I_2}{2\pi}$

4. Infinitely long wires 1 and 2 carrying currents I_1 and I_2 respectively are inclined at an angle θ to each other. What is the force on a small element dl of wire 2 at a distance r from wire 1 (as shown in figure) due to the magnetic field of wire 1?

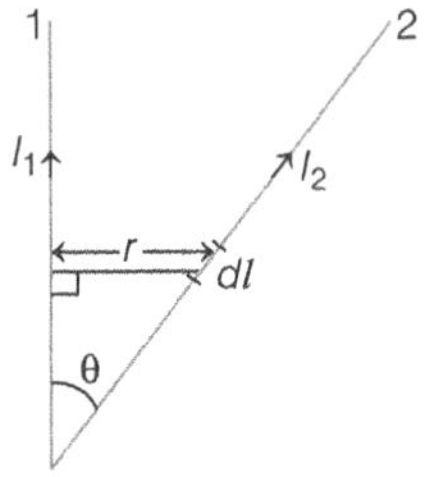

(a) $\frac{\mu_0}{2\pi r} I_1 I_2 \, dl \tan\theta$ (b) $\frac{\mu_0}{2\pi r} I_1 I_2 \, dl \sin\theta$
(c) $\frac{\mu_0}{2\pi r} I_1 I_2 \, dl \cos\theta$ (d) None of these

5. Two long straight wires P and Q carrying equal current 10 A each were kept parallel to each other at 4 cm distance. Magnitude of magnetic force experienced by 10 cm length of wire P is F_1. If distance between wires is halved and currents on them are doubled, force F_2 on 10 cm length of wire P will be **(2023 Main)**
(a) $8F_1$ (b) $10F_1$ (c) $\frac{F_1}{10}$ (d) $\frac{F_1}{8}$

6. Two parallel, long wires are kept 0.20 m apart in vacuum, each carrying current of x A in the same direction. If the force of attraction per meter of each wire is 2×10^{-6} N, then the value of x is approximately **(2022 Main)**
(a) 1 (b) 2.4 (c) 1.4 (d) 2

7. A wire X of length 50 cm carrying a current of 2 A is placed parallel to a long wire Y of length 5 m. The wire Y carries a current of 3 A. The distance between two wires is 5 cm and currents flow in the same direction. The force acting on the wire Y is **(2022 Main)**

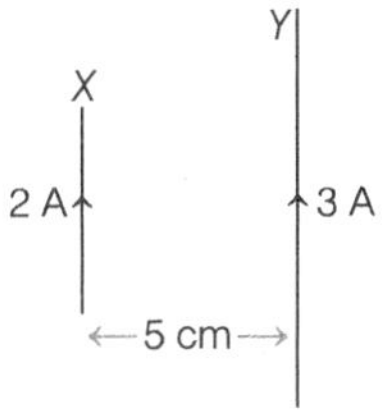

(a) 1.2×10^{-5} N directed towards wire X
(b) 1.2×10^{-4} N directed away from wire X
(c) 1.2×10^{-4} N directed towards wire X
(d) 2.4×10^{-5} N directed towards wire X

8. A rigid square loop of side a and carrying current I_2 is lying on a horizontal surface near a long current I_1, carrying wire in the same plane as shown in figure. The net force on the loop due to the wire will be **(2019 Main)**

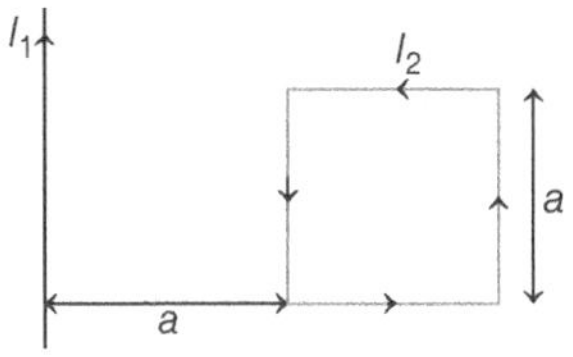

(a) repulsive and equal to $\frac{\mu_0 I_1 I_2}{2\pi}$
(b) attractive and equal to $\frac{\mu_0 I_1 I_2}{3\pi}$
(c) zero
(d) repulsive and equal to $\frac{\mu_0 I_1 I_2}{4\pi}$

9. Two thin long parallel wires separated by a distance b are carrying a current i ampere each. The magnitude

of the force per unit length exerted by one wire on the other is **(1986 JEE)**

(a) $\frac{\mu_0 i^2}{b^2}$ (b) $\frac{\mu_0 i^2}{2\pi b}$

(c) $\frac{\mu_0 i}{2\pi b}$ (d) $\frac{\mu_0 i}{2\pi b^2}$

10. A rectangular loop carrying a current i is situated near a long straight wire such that the wire is parallel to one of the sides of the loop and is in the plane of the loop. If steady current I is established in the wire as shown in the figure, the loop will **(1985 JEE)**

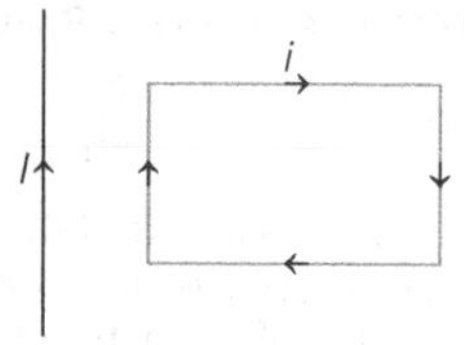

(a) rotate about an axis parallel to the wire
(b) move away from the wire
(c) move towards the wire
(d) remain stationary

Multiple Correct Options

11. There are two wires ab and cd in a vertical plane as shown in figure. Direction of current in wire ab is rightwards. Choose the correct options.

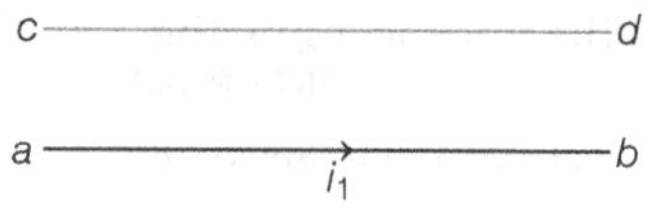

(a) If wire ab is fixed, then wire cd can be kept in equilibrium by the current in cd in leftward direction
(b) Equilibrium of wire cd will be stable equilibrium
(c) If wire cd is fixed, then wire ab can be kept in equilibrium by flowing current in cd in rightward direction
(d) Equilibrium of wire ab will be stable equilibrium

Match the Columns

12. Equal currents are flowing in four infinitely long wires. Distance between two wires is same and directions of currents are shown in figure. Match the following two columns.

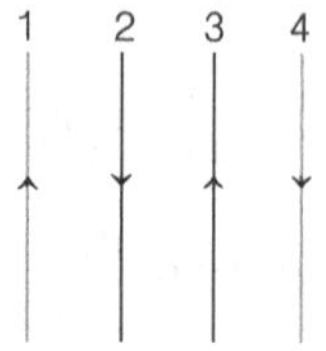

Column I		Column II	
(a)	Force on wire-1	(p)	inwards
(b)	Force on wire-2	(q)	leftwards
(c)	Force on wire-3	(r)	rightwards
(d)	Force on wire-4	(s)	zero

Subjective Questions

13. A square frame carrying a current $I = 0.90$ A is located in the same plane as a long straight wire carrying a current $I_0 = 5.0$ A. The frame side has a length $a = 8.0$ cm. The axis of the frame passing through the midpoints of opposite sides is parallel to the wire and is separated from it by the distance which is $\eta = 1.5$ times greater than the side of the frame. Find ampere force acting on the frame. **(Irodov 3.255)**

14. A direct current I flows in a long straight conductor whose cross-section has the form of a thin half-ring of radius R. The same current flows in the opposite direction along a thin conductor located on the "axis" of the first conductor. Find the magnetic force per unit length on the conductor along the axis. **(Irodov 3.257)**

15. Two long thin parallel conductors of the shape shown in figure carry direct currents I_1 and I_2. The separation between the conductors is a, the width of the right-hand conductor is equal to b. With both conductors lying in same plane, find the magnetic interaction force per unit length between them. **(Irodov 3.258)**

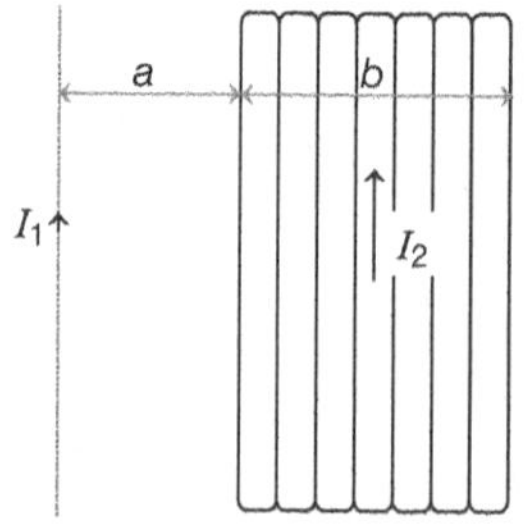

Answers

1. (d) **2.** (b) **3.** (b) **4.** (d) **5.** (a)
6. (c) **7.** (a) **8.** (d) **9.** (b) **10.** (c)
11. (a,b,c) **12.** (a → q; b → r; c → q; d → r)
13. $F = 0.4\,\mu\text{N}$ **14.** $\frac{F}{l} = \frac{\mu_0 I^2}{\pi^2 R}$
15. $\frac{F}{l} = \left(\frac{\mu_0}{4\pi}\right)\left(\frac{2I_1 I_2}{b}\right)\ln\left(1 + \frac{b}{a}\right)$

See hints on pages 443-445.

MODULE 8

Magnetic Poles and Bar Magnets

In electricity, the isolated charge q is the simplest structure that can exist. If two such charges of opposite sign are placed near each other, they form an electric dipole characterized by an electric dipole moment **p**. In magnetism isolated magnetic 'poles' which would correspond to isolated electric charges do not exist. The simplest magnetic structure is the **magnetic dipole**, characterized by a magnetic dipole moment **M**. A current loop, a bar magnet and a solenoid of finite length are examples of magnetic dipoles.

When a magnetic dipole is placed in an external magnetic field **B** a magnetic torque τ acts on it, which is given by

$$\tau = \mathbf{M} \times \mathbf{B}$$

Alternatively, we can measure **B** due to the dipole at a point along its axis a (large) distance r from its centre by the expression,

$$B = \frac{\mu_0}{4\pi} \cdot \frac{2M}{r^3}$$

A bar magnet might be viewed as two poles (North and South) separated by some distance. However, all attempts to isolate these poles fail. If a magnet is broken, the fragments prove to be dipoles and not isolated poles. If we break up a magnet into the electrons and nuclei that make up its atoms, it will be found that even these elementary particles are magnetic dipoles.

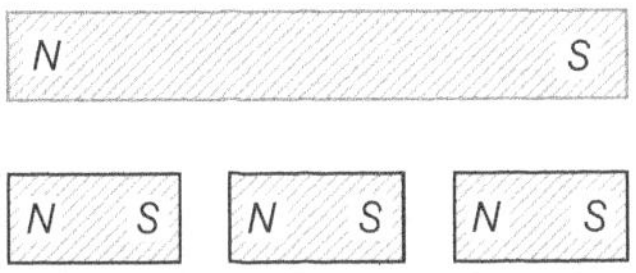

If a bar magnet is broken, each fragment becomes a small dipole.

Each current carrying loop is just like a magnetic dipole, whose magnetic dipole moment is given by

$$\mathbf{M} = Ni\mathbf{A}$$

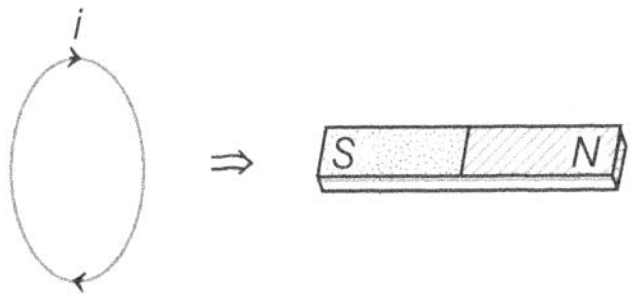

Here, N is the number of turns in the loop, i is the current and **A** represents the area vector of the current loop.

The behaviour of a current loop can be described by the following hypothetical model:

(i) There are two magnetic charges; positive magnetic charge and negative magnetic charge. We call the positive magnetic charge a north pole and the negative magnetic charge as the south pole. Every pole has a pole strength m. The unit of pole strength is A-m.

(ii) A magnetic charge placed in a magnetic field experiences a force,

$$\mathbf{F} = m\mathbf{B}$$

The force on positive magnetic charge is along the field and a force on a negative magnetic charge is opposite to the field.

(iii) A magnetic dipole is formed when a negative magnetic charge $-m$ and a positive magnetic charge $+m$ are placed at a small separation d. The magnetic dipole moment is

$$M = md$$

The direction of **M** is from $-m$ to $+m$.

Geometrical Length and Magnetic Length

In case of a bar magnet, the poles appear at points which are slightly inside the two ends. The distance between the locations of the assumed poles is called the **magnetic length** of the magnet. The actual distance between the ends is called the **geometrical length.**

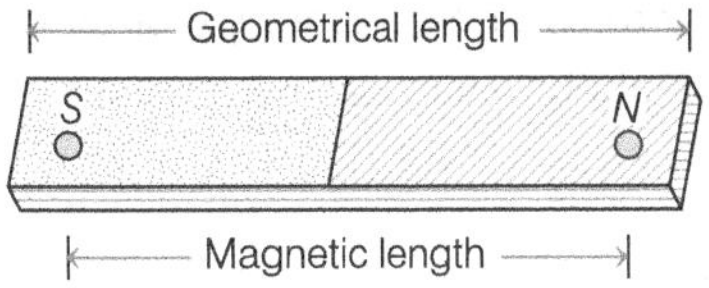

The magnetic length of a bar magnet is written as $2l$. If m be the pole strength and $2l$ the magnetic length of a bar magnet, then its magnetic moment is

$$M = 2ml$$

Important POINTS

- Current carrying loop, solenoid etc. are just like magnetic dipoles, whose dipole moment M is equal to NiA. Direction of **M** is from south pole (S) to north pole (N).
- The behaviour of a magnetic dipole (may be a bar magnet also) is similar to the behaviour of an electric dipole.
- The only difference is that the electric dipole moment **p** is replaced by magnetic dipole moment **M** and the constant $\frac{1}{4\pi\varepsilon_0}$ is replaced by $\frac{\mu_0}{4\pi}$.
- Table given below makes a comparison between an electric dipole and a magnetic dipole.

Table

S.No.	Physical quality to be compared	Electric dipole	Magnetic dipole
1.	Dipole moment	$p = q\,(2l)$	$M = m\,(2l)$
2.	Direction of dipole moment	From negative charge to the positive charge	From south to north pole
3.	Net force in uniform field	0	0
4.	Net torque in uniform field	$\tau = \mathbf{p} \times \mathbf{E}$	$\tau = \mathbf{M} \times \mathbf{B}$
5.	Field at far away point on the axis	$\frac{1}{4\pi\varepsilon_0} \cdot \frac{2p}{r^3}$ (along **p**)	$\frac{\mu_0}{4\pi} \cdot \frac{2M}{r^3}$ (along **M**)
6.	Field at far away point on perpendicular bisector	$\frac{1}{4\pi\varepsilon_0} \cdot \frac{p}{r^3}$ (opposite to **p**)	$\frac{\mu_0}{4\pi} \cdot \frac{M}{r^3}$ (opposite to **M**)
7.	Potential energy	$U_\theta = -\mathbf{p} \cdot \mathbf{E} = -pE \cos\theta$	$U_\theta = -\mathbf{M} \cdot \mathbf{B} = -MB \cos\theta$
8.	Work done in rotating the dipole	$W_{\theta_1 - \theta_2} = pE(\cos\theta_1 - \cos\theta_2)$	$W_{\theta_1 - \theta_2} = MB(\cos\theta_1 - \cos\theta_2)$

Note *In the above table, θ is the angle between field (**E** or **B**) and dipole moment (**p** or **M**).*

Example 1 Calculate the magnetic induction (or magnetic field) at a point 1 Å away from a proton, measured along its axis of spin. The magnetic moment of the proton is 1.4×10^{-26} A-m^2.

Sol. On the axis of a magnetic dipole, magnetic induction is given by

$$B = \frac{\mu_0}{4\pi} \cdot \frac{2M}{r^3}$$

Substituting the values, we get

$$B = \frac{(10^{-7})\,(2)\,(1.4 \times 10^{-26})}{(10^{-10})^3}$$

$$= 2.8 \times 10^{-3}\ \text{T} = 2.8\ \text{mT}$$ **Ans.**

Example 2 A bar magnet of magnetic moment 2.0 A-m^2 is free to rotate about a vertical axis through its centre. The magnet is released from rest from the east-west position. Find the kinetic energy of the magnet as it takes the north-south position. The horizontal component of the earth's magnetic field is $B = 25\ \mu$T. Earth's magnetic field is from south to north.

Sol. Gain in kinetic energy = loss in potential energy

Thus,

$$\text{KE} = U_i - U_f$$

As, $$U = -MB\cos\theta$$

$\therefore$ $$\text{KE} = -MB\cos\left(\frac{\pi}{2}\right) - (-MB\cos 0°)$$

$$= MB$$

Substituting the values, we have

$$\text{KE} = (2.0)\,(25 \times 10^{-6})\ \text{J}$$

$$= 50\ \mu\text{J}$$ **Ans.**

MODULE 8 Exercise

Note *Attempt this exercise after completing Module 9.*

Single Correct Option

1. A bar magnet with a magnetic moment 5.0 A-m^2 is placed in parallel position relative to a magnetic field of 0.4 T. The amount of required work done in turning the magnet from parallel to antiparallel position relative to the field direction is **(2023 Main)**

   (a) 2 J (b) zero (c) 4 J (d) 1 J

2. Two bar magnets oscillate in a horizontal plane in Earth's magnetic field with time periods of 3 s and 4 s respectively. If their moments of inertia are in the ratio of 3 : 2, then the ratio of their magnetic moments will be **(2022 Main)**

   (a) 2 : 1 (b) 8 : 3 (c) 1 : 3 (d) 27 : 16

3. A bar magnet having a magnetic moment of 2.0×10^5 JT^{-1}, is placed along the direction of uniform magnetic field of magnitude $B = 14 \times 10^{-5}$ T. The work done in rotating the magnet slowly through 60° from the direction of field is **(2022 Main)**

   (a) 14 J (b) 8.4 J (c) 4 J (d) 1.4 J

4. A small bar magnet placed with its axis at 30° with an external field of 0.06 T experiences a torque of 0.018 N-m. The minimum work required to rotate it from its stable to unstable equilibrium position is **(2020 Main)**

   (a) 6.4×10^{-2} J (b) 9.2×10^{-3} J
   (c) 7.2×10^{-2} J (d) 11.7×10^{-3} J

5. Two magnetic dipoles X and Y are placed at a separation d, with their axes perpendicular to each other. The dipole moment of Y is twice that of X. A particle of charge q is passing through their mid-point P, at angle $\theta = 45°$ with the horizontal line, as shown in figure. What would be the magnitude of force on the particle at that instant?
   (d is much larger than the dimensions of the dipole) **(2019 Main)**

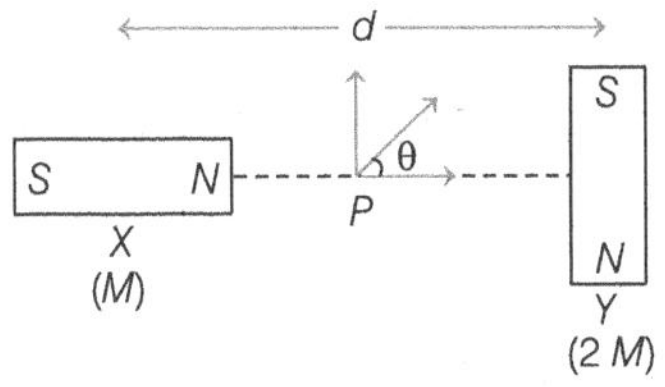

   (a) $\left(\frac{\mu_0}{4\pi}\right)\frac{M}{\left(\frac{d}{2}\right)^3} \times qv$ (b) 0

   (c) $\sqrt{2}\left(\frac{\mu_0}{4\pi}\right)\frac{M}{\left(\frac{d}{2}\right)^3} \times qv$ (d) $\left(\frac{\mu_0}{4\pi}\right)\frac{2M}{\left(\frac{d}{2}\right)^3} \times qv$

6. A short bar magnet is placed in the magnetic meridian of the Earth with north pole pointing north. Neutral points are found at a distance of 30 cm from the magnet on the East–West line, drawn through the middle point of the magnet. The magnetic moment of the magnet in A-m^2 is close to

   (Given $\frac{\mu_0}{4\pi} = 10^{-7}$ in SI units and B_H = Horizontal component of Earth's magnetic field $= 3.6 \times 10^{-5}$ T) **(2015 Main)**

   (a) 14.6 (b) 19.4 (c) 9.7 (d) 4.9

7. The magnetic field lines due to a bar magnet are correctly shown in **(2002 JEE)**

   (a)
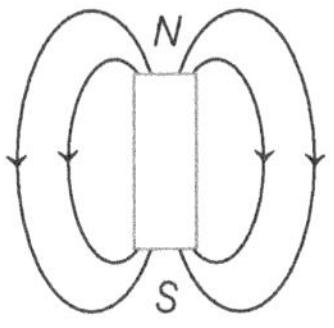

   (b)
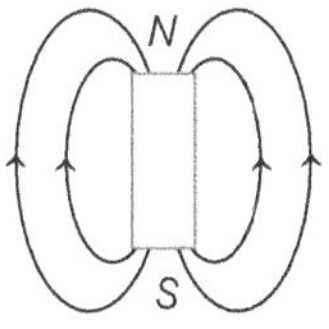

   (c)
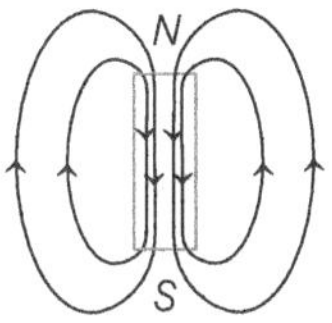

   (d)
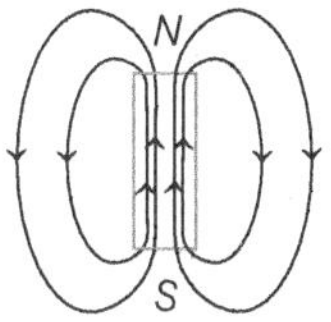

8. A magnetic needle is kept in a non-uniform magnetic field. It experiences **(1982 JEE)**

   (a) a force and a torque (b) a force but not a torque
   (c) a torque but not a force
   (d) neither a force nor a torque

Subjective Questions

9. A bar magnet is placed with its north pole pointing north and its south pole pointing south. Draw a figure to show the location of neutral points. **(1979 JEE)**

10. A bar magnet with poles 25 cm apart and of strength 14.4 amp-m rests with centre on a frictionless pivot. It is held in equilibrium at an angle of 60° with respect to a uniform magnetic field of induction 0.25 Wb/m^2, by applying a force F at right angles to its axis at a point 12 cm from pivot. Calculate F. What will happen if the force F is removed? **(1978 JEE)**

Answers

1. (c) **2.** (b) **3.** (a) **4.** (c) **5.** (b)
6. (c) **7.** (d) **8.** (a) **9.** (See the hint)
10. F = 25.98 N

See hints on pages 445-446.

MODULE 9

Earth's Magnetism

Our earth behaves as it has a powerful magnet within it. The value of magnetic field on the surface of earth is a few tenths of a gauss ($1\ G = 10^{-4}\ T$). The earth's south magnetic pole is located near the north geographic pole and the earth's north magnetic pole is located near the south geographic pole. In fact, the configuration of the earth's magnetic field is very much like the one that would be achieved by burying a gigantic bar magnet deep in the interior of the earth.

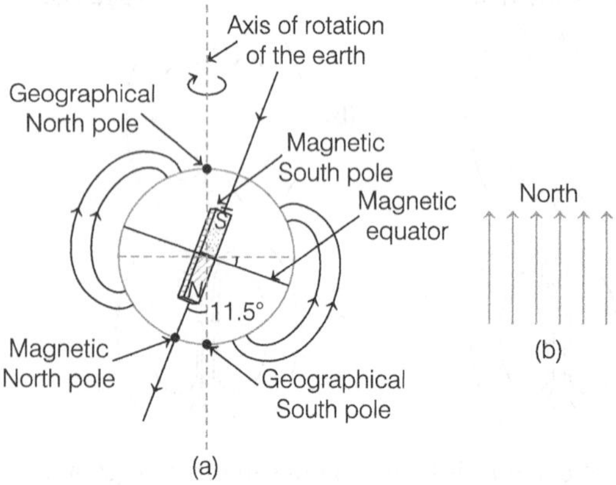

The axis of earth's magnet makes an angle of 11.5° with the earth's rotational axis.

Elements of Earth's Magnetism

There are three elements of earth's magnetism.

(i) **Angle of Declination** (α) At any point (say P) on earth's surface the vertical plane passing through the line joining the geographical north and south poles is called the **'geographical meridian'**.

At point P, there also exists the earth's magnetic field **B**. A vertical plane in the direction of **B** is called **'magnetic meridian'**.

The acute angle between the magnetic meridian and the geographical meridian is called **'angle of declination** 'α'.

(ii) **Angle of Dip** (θ) 'The angle of dip (θ) at a place is the angle between the direction of earth's magnetic field and the horizontal at that place.'

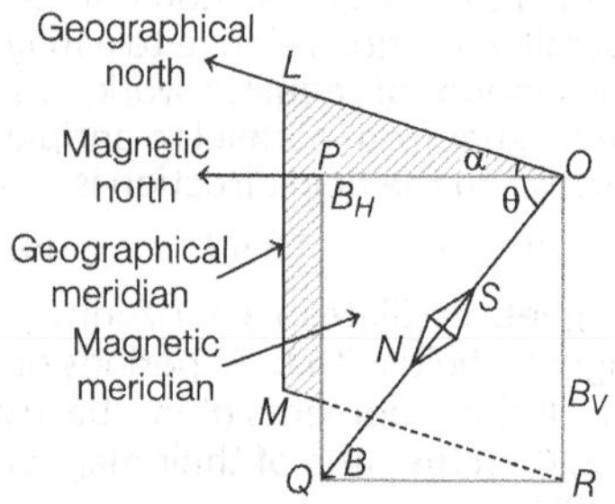

Angle of dip at some place can be measured from a magnetic needle free to rotate in a vertical plane about a horizontal axis passing through centre of gravity of the needle. At earth's magnetic poles the magnetic field of earth is vertical, i.e. angle of dip is 90°, the freely suspended magnetic needle is vertical there. At magnetic equator field is horizontal, or angle of dip is 0°. The needle is horizontal.

In northern hemisphere, the north pole of the magnetic needle inclines downwards, whereas in the southern hemisphere the south pole of the needle inclines downwards.

(iii) **Horizontal Component of Earth's Magnetic Field**

Let B be the net magnetic field at some point. B_H and B_V be the horizontal and vertical components of B. Let θ is the angle of dip at the same place, then we can see that

$$B_H = B\cos\theta \qquad \text{...(i)}$$

and

$$B_V = B\sin\theta \qquad \text{...(ii)}$$

Squaring and adding Eqs. (i) and (ii), we get

$$B = \sqrt{B_H^2 + B_V^2}$$

Further, dividing Eq. (ii) by Eq. (i), we get

$$\theta = \tan^{-1}\left(\frac{B_V}{B_H}\right)$$

By knowing B_H and θ at some place we can find B and B_V at that place.

Apparent Dip

The angle of dip at a place is determined by a dip circle. It consists of a magnetized needle capable of rotation in a vertical plane about a horizontal axis. The needle can rotate over a vertical scale graduated in degrees.

If the plane of the scale of the dip circle is not in the magnetic meridian then the needle will not indicate the correct direction of the earth's magnetic field. The angle made by the needle with the horizontal is called apparent dip.

True Dip

When the plane of scale of the dip circle is in the magnetic meridian the needle comes to rest in direction of the earth's magnetic field. The angle made by the needle with the horizontal is called true dip.

Suppose dip circle is set at an angle β to the magnetic meridian.

Horizontal component

$$B_H' = B_H \cos\beta$$

Vertical component $B_V' = B_V$ (remains unchanged)

Apparent dip is θ' where :

$$\tan\theta' = \frac{B_V'}{B_H'} = \frac{B_V}{B_H \cos\beta} = \frac{\tan\theta}{\cos\beta} \quad \left(\tan\theta = \frac{B_V}{B_H} = \text{true dip}\right)$$

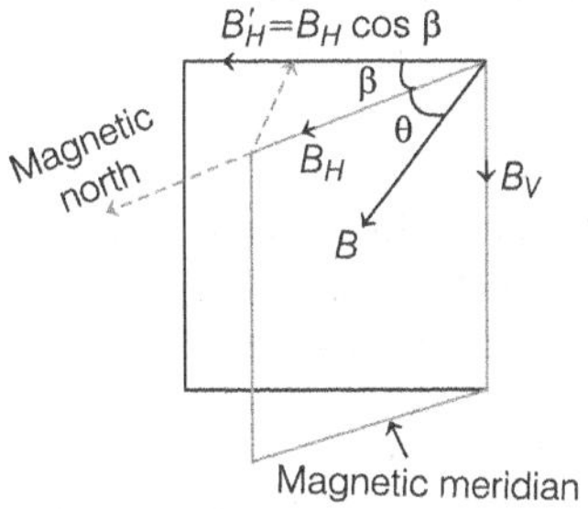

(i) For a vertical plane other than magnetic meridian $\beta > 0$ or $\cos\beta < 1$ so $\theta' > \theta$. Therefore in a vertical plane other than magnetic meridian angle of dip is more than in magnetic meridian.

(ii) For a plane perpendicular to magnetic meridian $\beta = \frac{\pi}{2}$.

Therefore $\tan\theta' = \infty$, so $\theta' = \frac{\pi}{2}$. So in a plane perpendicular to magnetic meridian dip needle will become vertical.

Magnetic Equator

(i) The angle of dip is zero.

(ii) Vertical component of earth's magnetic field becomes zero.

(iii) A freely suspended magnet will become horizontal at the magnetic equator.

(iv) At the equator earth's magnetic field is parallel to the earth's surface i.e. horizontal.

Magnetic Poles

(i) The angle of dip is 90°.

(ii) Horizontal component of earth's magnetic field becomes zero.

(iii) A freely suspended magnet will become vertical at magnetic poles.

(iv) At poles, the earth's magnetic field is perpendicular to the surface of the earth i.e. vertical.

Neutral Points

A neutral point in the magnetic field of a bar magnet is that point where the field due to the magnet is completely neutralized by the horizontal component of the earth's magnetic field. At neutral point field due to bar magnet (B) is equal and opposite to horizontal component of earth's magnetic field (B_H) or $B = B_H$. Following two possible cases arise :

(i) **The neutral point when the north pole of a magnet is towards the geographical north of the earth.**

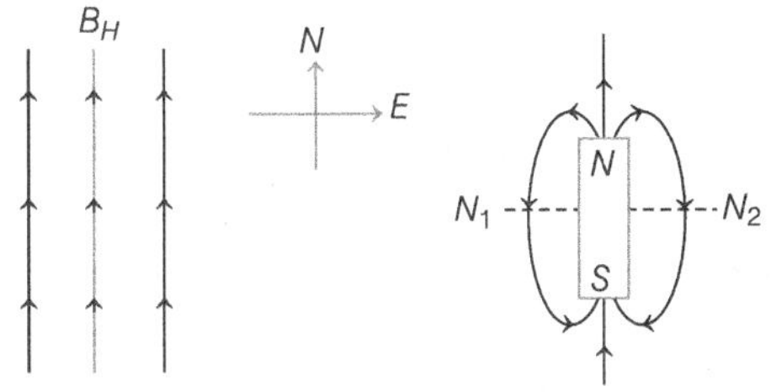

The neutral points N_1 and N_2 lie on the equatorial line where B_H and B due to bar magnet are in opposite directions. The magnetic field due to magnet at an equatorial point is

$$B = \frac{\mu_0}{4\pi} \frac{M}{(r^2 + l^2)^{\frac{3}{2}}}$$

Where M is the magnetic dipole moment of the magnet, $2l$ is its length and r is a distance of the neutral point.

At neutral point $B = B_H$.

So, $\frac{\mu_0}{4\pi} \frac{M}{(r^2 + l^2)^{\frac{3}{2}}} = B_H$

For a small bar magnet ($l^2 << r^2$)

Hence, $\frac{\mu_0}{4\pi} \frac{M}{r^2} = B_H$

From this relation we can find the distance r of neutral points N_1 and N_2 from the bar magnet.

(ii) **Neutal point when south pole of a magnet is towards geographical north of earth.**

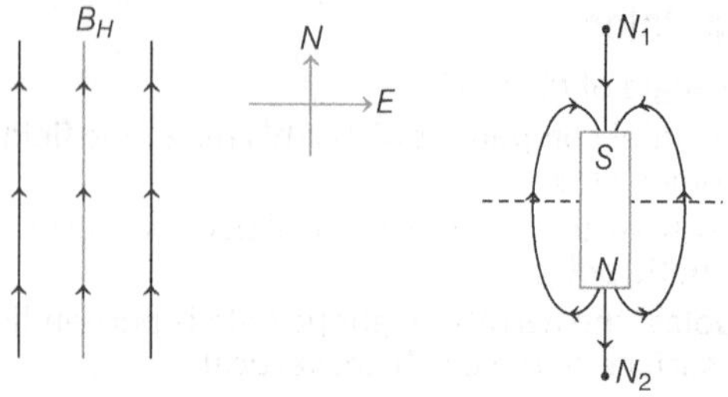

The neutral points N_1 and N_2 lie on the axial line of magnet where B_H and B due to bar magnet are in opposite directions. The magnetic field due to magnet at axial point is:

$$B = \frac{\mu_0}{4\pi} \frac{2Mr}{(r^2 - l^2)^2}$$

At neutral point $B_H = B = \frac{\mu_0}{4\pi} \frac{2Mr}{(r^2 - l^2)^2}$

For a small bar magnet ($l^2 << r^2$)

Hence, $\frac{\mu_0}{4\pi} \frac{2M}{r^2} = B_H$

From the above relation we can find the distance r of neutral points N_1 and N_2 from the magnet.

Example 1 If θ_1 and θ_2 are angles of dip in two vertical planes at right angle to each other and θ is true dip then prove that $\cot^2 \theta = \cot^2 \theta_1 + \cot^2 \theta_2$.

Sol. If the vertical plane in which dip angle is θ_1 subtends an angle β with magnetic meridian, then other vertical plane in which dip is θ_2 and is perpendicular to first will make an angle of $90° - \beta$ with magnetic meridian. If θ_1 and θ_2 are apparent dips, then

$$\tan\theta_1 = \frac{B_V}{B_H \cos\beta}$$

$$\tan\theta_2 = \frac{B_V}{B_H \cos(90° - \beta)} = \frac{B_V}{B_H \sin\beta}$$

$$\cot^2\theta_1 + \cot^2\theta_2 = \frac{1}{(\tan\theta_1)^2} + \frac{1}{(\tan\theta_2)^2}$$

$$= \frac{B_H^2 \cos^2\beta + B_H^2 \sin^2\beta}{B_V^2} = \frac{B_H^2}{B_V^2}$$

$$= \left(\frac{B\cos\theta}{B\sin\theta}\right)^2 = \cot^2\theta$$

So, $\cot^2\theta_1 + \cot^2\theta_2 = \cot^2\theta$

Example 2 In the magnetic meridian of a certain place, the horizontal component of earth's magnetic field is 0.26 G and the dip angle is 60°. Find

(a) vertical component of earth's magnetic field.

(b) the net magnetic field at this place.

Sol. Given, $B_H = 0.26$ G and $\theta = 60°$

(a) $\tan\theta = \frac{B_V}{B_H}$

$\therefore$ $B_V = B_H \tan\theta = (0.26) \tan 60°$

$= 0.45$ G **Ans.**

(b) $B_H = B\cos\theta$

$\therefore$ $B = \frac{B_H}{\cos\theta} = \frac{0.26}{\cos 60°}$

$= 0.52$ G **Ans.**

Example 3 A magnetic needle suspended in a vertical plane at 30° from the magnetic meridian makes an angle of 45° with the horizontal. Find the true angle of dip.

Sol. We have learnt that,

$$\tan\theta' = \frac{\tan\theta}{\cos\beta} = \frac{\tan\theta}{\cos 30°}$$

$\therefore$ $\theta = \tan^{-1}[\tan\theta' \cos 30°]$

$= \tan^{-1}[(\tan 45°)(\cos 30°)]$

$\approx 41°$ **Ans.**

Vibration Magnetometer

Vibration magnetometer is an instrument which is used for the following two purposes:

(i) To find magnetic moment of a bar magnet.

(ii) To compare magnetic fields of two magnets.

The construction of a vibration magnetometer is as shown in figure.The magnet shown in figure is free to rotate in a horizontal plane. The magnet stays parallel to the horizontal component of earth's magnetic field. If the magnet is now displaced through an angle θ, a restoring torque of magnitude $MH \sin\theta$ acts on it and the magnet starts oscillating.

From the theory of simple harmonic motion, we can find the time period of oscillations of the magnet.

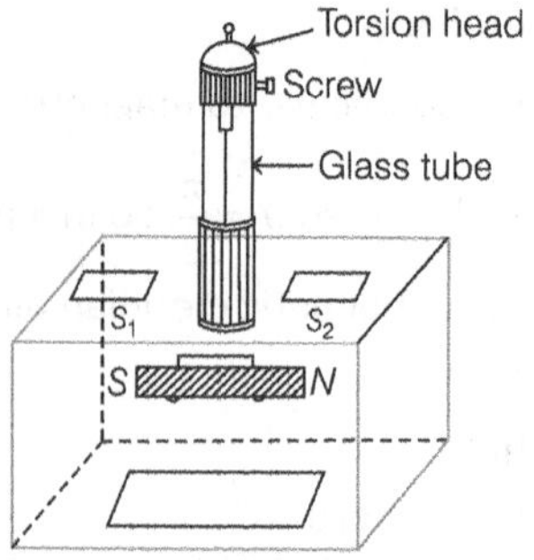

Restoring torque in displaced position is

$$\tau = -MB_H \sin\theta \quad ...(i)$$

Here, M = Magnetic moment of the magnet

and B_H = Horizontal component of earth's magnetic field. Negative sign shows the restoring nature of torque. Now since, $\tau = I\alpha$ and $\sin\theta \approx \theta$ for small angular displacement.

Thus, Eq. (i) can be written as

$$I\alpha = -MB_H\theta$$

Since, α is proportional to $-\theta$. Therefore, motion is simple harmonic in nature, time period of which will be given by

$$T = 2\pi\sqrt{\left|\frac{\theta}{\alpha}\right|} = 2\pi\sqrt{\frac{I}{MB_H}}$$

$$\therefore \quad \boxed{T = 2\pi\sqrt{\frac{I}{MB_H}}} \quad \text{...(ii)}$$

In the expression of T, I is the moment of inertia of the magnet about its axis of vibration.

(i) Measurement of Magnetic Moment : By finding time period T of vibrations of the given magnet, we can calculate magnetic moment M by the relation,

$$M = \frac{4\pi^2 I}{T^2 B_H}$$

(ii) Comparison of Two Magnetic Fields : Suppose we wish to compare the magnetic fields B_1 and B_2 at some point P due to two magnets. For this, vibration magnetometer is so placed that the centre of its magnet lies on P. Now, one of the given magnets is placed at some known distance from P in the magnetic meridian, such that point P lies on its axial line and its north pole points north. In this position, the field B_1 at P produced by the magnet will be in the direction of B_H. Hence, the magnet suspended in the magnetometer will vibrate in the resultant magnetic field $(B_H + B_1)$. Its period of vibration is noted, say it is T_1, then

$$T_1 = 2\pi\sqrt{\frac{I}{M(B_H + B_1)}}$$

Now, the first magnet is replaced by the second magnet and the second magnet is placed in the same position and again the time period is noted. If the field produced at P due to this magnet be B_2 and the new time period be T_2, then

$$T_2 = 2\pi\sqrt{\frac{I}{M(B_H + B_2)}}$$

Finally, the time period of the magnetometer under the influence of the earth's magnetic field alone is determined. Let it be T, then

$$T = 2\pi\sqrt{\frac{I}{MB_H}}$$

Solving above three equations for T, T_1 and T_2, we can show that

$$\boxed{\frac{B_1}{B_2} = \frac{(T^2 - T_1^2)T_2^2}{(T^2 - T_2^2)T_1^2}}$$

Example 4 A short bar magnet is placed with its north pole pointing north. The neutral point is 10 cm away from the centre of the magnet. If $B_H = 0.4$ G, calculate the magnetic moment of the magnet.

Sol. When north pole of the magnet points towards magnetic north, null point is obtained on perpendicular bisector of the magnet. Simultaneously, magnetic field due to the bar magnet should be equal to the horizontal component of earth's magnetic field H.

Thus, $$B_H = \frac{\mu_0}{4\pi}\cdot\frac{M}{r^3}$$

or $$M = \frac{B_H r^3}{(\mu_0/4\pi)}$$

Substituting the values, we have

$$M = \frac{(0.4\times 10^{-4})(10\times 10^{-2})^3}{10^{-7}}$$

$$= 0.4 \text{ A-m}^2$$ **Ans.**

Example 5 A magnetic needle performs 20 oscillations per minute in a horizontal plane. If the angle of dip be 30°, then how many oscillations per minute will this needle perform in vertical magnetic north-south plane and in vertical magnetic east-west plane?

Sol. In horizontal plane, the magnetic needle oscillates in horizontal component B_H.

$$\therefore \quad T = 2\pi\sqrt{\frac{I}{MB_H}}$$

In the vertical magnetic north-south plane (magnetic meridian), the needle oscillates in the total earth's magnetic field B, and in vertical magnetic east-west plane (plane perpendicular to the magnetic meridian) it oscillates only in earth's vertical component B_V. If its time period be T_1 and T_2, then

$$T_1 = 2\pi\sqrt{\frac{I}{MB}} \quad \text{and} \quad T_2 = 2\pi\sqrt{\frac{I}{MB_V}}$$

From above equations, we can find

$$\frac{T_1^2}{T^2} = \frac{B_H}{B}$$

or $$\frac{n_1^2}{n^2} = \frac{B}{B_H}$$

Similarly, $$\frac{n_2^2}{n^2} = \frac{B_V}{B_H}$$

Further, $$\frac{B}{B_H} = \sec\theta = \sec 30° = \frac{2}{\sqrt{3}}$$

and $$\frac{B_V}{B_H} = \tan\theta = \tan 30° = \frac{1}{\sqrt{3}}$$

$$\therefore \quad n_1^2 = (n)^2\left(\frac{B}{B_H}\right) = (20)^2\left(\frac{2}{\sqrt{3}}\right)$$

or $$n_1 = 21.5 \text{ oscillations/min}$$ **Ans.**

and $$n_2^2 = (n)^2\left(\frac{B_V}{B_H}\right) = (20)^2\left(\frac{1}{\sqrt{3}}\right)$$

$$\therefore \quad n_2 = 15.2 \text{ oscillations/min}$$ **Ans.**

MODULE 9 Exercise

Single Correct Option

1. At a certain place, the angle of dip is 30° and the horizontal component of earth's magnetic field is 0.5 G. The earth's total magnetic field (in G), at that certain place, is **(2022 Main)**

(a) $\frac{1}{\sqrt{3}}$ (b) $\frac{1}{2}$ (c) $\sqrt{3}$ (d) 1

2. The vertical component of the earth's magnetic field is 6×10^{-5} T at some place, where the angle of dip is 37°. The Earth's resultant magnetic field at that place will be (Given, $\tan 37° = \frac{3}{4}$) **(2022 Main)**

(a) 8×10^{-5} T (b) 6×10^{-5} T (c) 5×10^{-4} T (d) 1×10^{-4} T

3. A magnet hung at 45° with magnetic meridian makes an angle of 60° with the horizontal. The actual value of the angle of dip is **(2022 Main)**

(a) $\tan^{-1}\left(\sqrt{\frac{3}{2}}\right)$ (b) $\tan^{-1}(\sqrt{6})$ (c) $\tan^{-1}\left(\sqrt{\frac{2}{3}}\right)$ (d) $\tan^{-1}\left(\sqrt{\frac{1}{2}}\right)$

4. A magnetic compass needle oscillates 30 times per minute at a place, where the dip is 45° and 40 times per minute, where the dip is 30°. If B_1 and B_2 are respectively the total magnetic field due to the earth at the two places, then the ratio $\frac{B_1}{B_2}$ is best given by **(2019 Main)**

(a) 1.8 (b) 0.7 (c) 3.6 (d) 2.2

Subjective Question

5. A small magnetic needle performs small oscillations about an axis perpendicular to the magnetic induction vector. On changing the magnetic induction the needle's oscillation period decreased $\eta = 5.0$ times. How much and in what way was the magnetic induction changed? The oscillation damping is assumed to be negligible. **(Irodov 4.64)**

Numerical Value/Integer Type Question

6. A compass needle oscillates 20 times per minute at place where the dip is 30° and 30 times per minute where the dip is 60°. The ratio of total magnetic field due to the earth at two places respectively is $\frac{4}{\sqrt{x}}$. The value of x is .... **(2023 Main)**

Answers

1. (a) **2.** (d) **3.** (a) **4.** (b) **5.** The induction of the field increased $\eta^2 = 25$ times. **6.** 243

See hints on pages 446-447.

MODULE 10

Magnetic Materials

All substances (whether solid, liquid or gas) may be classified into three categories in terms of their magnetic properties. (i) paramagnetic, (ii) diamagnetic and (iii) ferromagnetic..

Paramagnetic Substances

Examples of such substances are platinum, aluminium, chromium, manganese, $CuSO_4$ solution etc. These substances when placed in a magnetic field, acquire a feeble magnetisation in the same direction as the applied field. Thus, the magnetic field inside the substance is slightly greater than outside to it.

Diamagnetic Substances

Examples of such substances are bismuth, antimony, gold, quartz, water, alcohol etc. These substances when placed in a magnetic field, acquire feeble magnetisation in a direction opposite to that of the applied field. Thus, the magnetic field inside the substance is smaller than that outside to it.

Ferromagnetic Substances

Examples of such substances are iron, nickel, steel, cobalt and their alloys. These substances resemble to a higher degree with paramagnetic substances as regard to their behaviour.

Some Important Terms Used in Magnetism

Relative Magnetic Permeability (μ_r)

It is the ratio of the magnetic permeability μ of the substance to the permeability of free space.

Thus, $$\mu_r = \frac{\mu}{\mu_0}$$

μ_r is a pure ratio, hence, dimensionless. For vacuum its value is 1.

μ_r can also be defined as the ratio of the magnetic field B in the substance when placed in magnetic field B_0. Thus,

$$\mu_r = \frac{B}{B_0}$$

For paramagnetic substance, $\mu_r > 1$,
For diamagnetic substance, $\mu_r < 1$ and
For ferromagnetic substance, $\mu_r >> 1$.

Magnetic Susceptibility (χ_m)

Magnetic susceptibility is a measure of how easily a substance is magnetised in a magnetic field. It is denoted by χ_m. Magnetic permeability μ_r and magnetic susceptibility χ_m are related by :

$$\mu_r = 1 + \chi_m$$

For paramagnetic substances χ_m is slightly positive. For diamagnetic substances, it is slightly negative and for ferromagnetic substances, χ_m is positive and very large.

Derivation of the Relation between μ_r and χ_m

(i) If i current passes through a long solenoid with n-number of turns per unit length, then magnetic field inside the solenoid is given by :

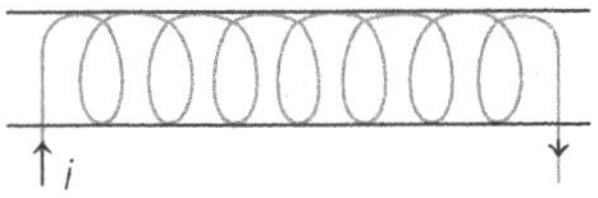

$$B_0 = \mu_0 ni$$

Here, $ni = H$ is called the magnetic intensity.

Thus, $$B_0 = \mu_0 H \quad \text{...(i)}$$

or $$B_0 \propto H$$

Note (i) *SI unit of H is* $\frac{A}{m}$.

(ii) *H is also sometimes called the magnetic field strength or simply the magnetic field (which is actually applied from outside).*

(ii) Now some magnetic material (dia, para or ferro) is placed inside the solenoid. Now the magnetic field B is not equal to B_0 (due to magnetization). Let us suppose that B is now,

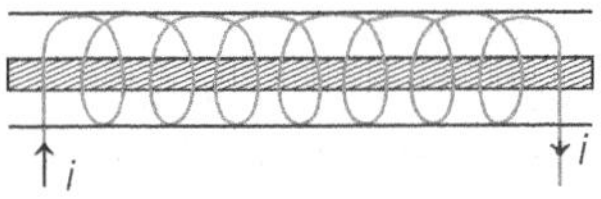

$$B = B_0 + \Delta B_0 \quad \text{...(ii)}$$

Here, $\Delta B_0 \propto M$ or I

Where, M (or I) is called intensity of magnetization, where

$$M = \frac{\text{Magnetized magnetic moment}}{\text{Volume}}$$

$\therefore \quad \Delta B_0 = \mu_0 M \quad$ (μ_0 = proportionality constant)

Further,

$$M \propto H \quad \text{or} \quad ni$$

or $\quad M = \chi_m H \quad (\chi_m = \text{proportionality constant})$

Thus, $\Delta B_0 = \mu_0 M$ or $\mu_0 \chi_m H = \chi_m B_0 \quad (\text{as } \mu_0 H = B_0)$

Substituting this value in Eq. (i), we get

$$B = B_0 + \chi_m B_0 = (1 + \chi_m) B_0$$

or $$\frac{B}{B_0} = (1 + \chi_m)$$

Thus, $\mu_r = 1 + \chi_m$ **Hence proved.**

Hysteresis Loop

The distinguishing characteristics of a ferromagnetic material is not that it can be strongly magnetised but that the intensity of magnetisation **I** is not directly proportional to the magnetic intensity **H**. If a gradually increasing magnetic intensity **H** is applied to an unmagnetised piece of iron, its magnetisation increases non-linearly until it reaches a maximum.

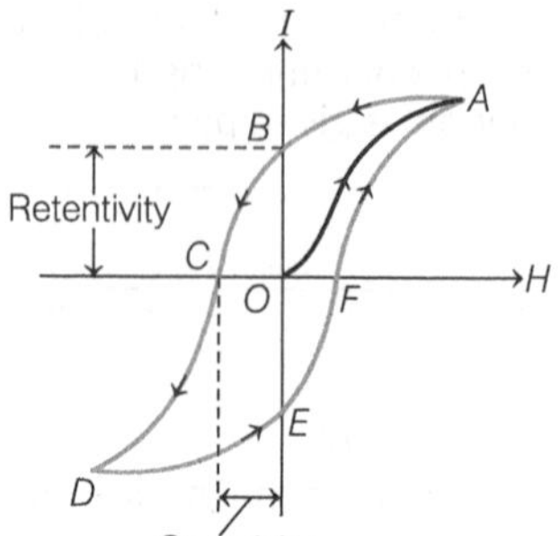

If **I** is plotted against **H**, a curve like *OA* is obtained. This curve is known as magnetisation curve. If the magnetic intensity **H** is now decreased, the **I** does not return along the path *AO* but follows path *AB*. At **H** = 0, **I** does not come to its zero value but its value is still near the saturated value. The value of **I** at this point (i.e. *OB*) is known as **remanence, remanent magnetisation or retentivity**.

On applying a reverse field the value of **I** finally becomes zero. The abscissa *OC* represents the reversed magnetic intensity needed to demagnetise the specimen. This is known as **coercivity** of the material.

If the reverse magnetic intensity *H* is further increased, a reverse magnetisation is set up which quickly reaches the saturation value. This is shown as *CD*. If **H** is now taken back from its negative saturation value to its original positive saturation value, a similar curve *DEFA* will be traced. The whole graph *ABCDEFA* thus, forms a closed loop, usually known as **hysteresis loop**.

The energy lost per unit volume of a substance in a complete cycle is equal to the area of the loop. Thus, we can conclude the following three points from the above discussion:

(i) The retentivity of a substance is a measure of the magnetisation remaining in the substance when the magnetising field is removed.

(ii) The coercivity of a substance is a measure of the reverse magnetising field required to destroy the residual magnetism of the substance.

(iii) The energy loss per unit volume of a substance in a complete cycle of magnetisation is equal to the area of the hysteresis loop.

Note *Hysteresis curve is also sometimes given between magnetised field B along y-axis and magnetic intensity H along x-axis.*

The main concept is : along x-axis you have to take H (or B) which is applied externally and along y-axis you have to take I (or B) which is magnetised in the presence of $\mathbf{B}_{ext}$.

Magnetic Properties of Soft Iron and Steel

Following three conclusions can be drawn from their hysteresis loops:

(i) Retentivity of soft iron is less but it is more than the retentivity of steel.

(ii) Coercivity of soft iron is less than the coercivity of steel.

(iii) Area of hysteresis loop (i.e. hysteresis loss) in soft iron is smaller than that in steel.

Choice of Magnetic Materials

(i) **Permanent Magnets** The materials for a permanent magnet should have

(a) High retentivity and

(b) High coercivity. As the material in this case is never put to cyclic changes of magnetisation, hence, hysteresis loss is immaterial. From the point of view of these facts **steel** is more suitable for the construction of permanent magnets than soft iron.

(ii) **Electromagnets** The materials for the construction of electromagnets should have

(a) High initial permeability

(b) Low hysteresis loss

From the view point of these facts, **soft iron** is an ideal material for this purpose.

(iii) **Transformer Cores and Telephone Diaphragms** As the magnetic material used in these cases is subjected to cyclic changes. Thus, the essential requirements for the selection of the material are

(a) High initial permeability

(b) Low hysteresis loss to prevent the breakdown

Electromagnet

As we know that a current carrying solenoid behaves like a bar magnet. If we place a soft iron rod in the solenoid, the magnetism of the solenoid increases hundreds of times and the solenoid is called an 'electromagnet'. It is a temporary magnet.

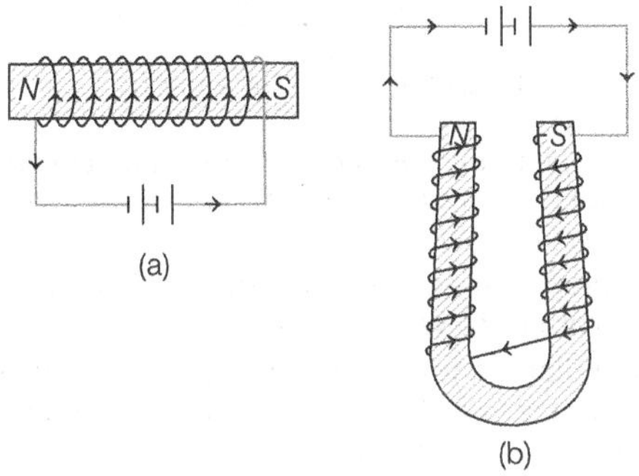

An electromagnet is made by winding closely a number of turns of insulated copper wire over a soft iron straight rod or a horse shoe rod. On passing current through this solenoid, a magnetic field is produced in the space within the solenoid.

Applications of Electromagnets

(i) Electromagnets are used in electric bell, transformer, telephone diaphragms etc.

(ii) In medical field, they are used in extracting bullets from the human body.

(iii) Large electromagnets are used in cranes for lifting and transferring big machines and parts.

Note *(i) Dia and para magnetic substances may be solid, liquid or gas. But ferro magnetic substances are only solids.*

(ii) Hysteresis loop is drawn only for ferromagnetic substances.

Explanation of Paramagnetism, Diamagnetism and Ferromagnetism

There are three properties of atoms that give rise to magnetic dipole moment.

1. The electrons moving around the nucleus in the orbits act as small current loops and contribute magnetic moments.
2. The spinning electron has an intrinsic magnetic dipole moment.
3. The nucleus contribute to magnetic moment due to the motion of charge within the nucleus.

   The magnitude of third type is about 10^{-3} times that of first two. The first two are of the same order. Still most of the magnetic moment of an atom is produced by electron spin, the net contribution of the orbital revolution is very small.

   This is because most of the electrons pair off in such a way that they produce equal and opposite orbital magnetic moment and they cancel out. Although, the electrons also try to pair up with their opposite spins but in case of spin motion of an electron it is not always possible to form equal and opposite pairs.

Paramagnetism

The property of paramagnetism is found in those substances whose atoms or molecules have an excess of electrons spinning in the same direction.

Hence, atoms of paramagnetic substances have a permanent magnetic moment and behave like tiny bar magnets. In the absence of external magnetic field, the atomic magnets are randomly oriented and net magnetic moment is thus, zero.

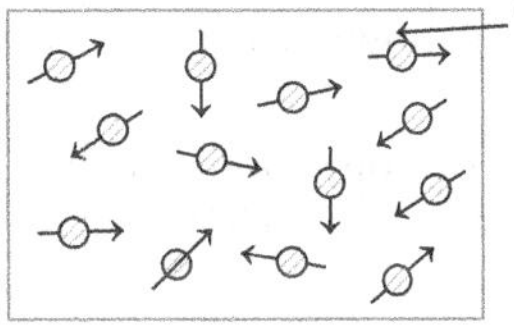

In the absence of external magnetic field

When paramagnetic substance is placed in an external magnetic field, then each atomic magnet experiences a torque which tends to turn the magnet in the direction of the field. The atomic magnets are thus, aligned in the direction of the field. Thus, the whole substance is magnetised in the direction of the external magnetic field.

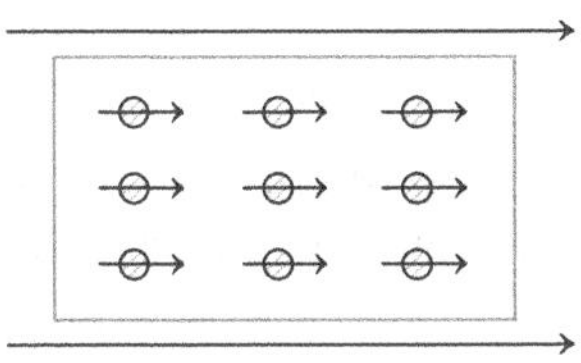

As the temperature of substance is increased, the thermal agitation disturbs the magnetic alignment of the atoms. Thus, we can say that paramagnetism is temperature dependent.

Curie's law

According to Curie's law, magnetic susceptibility of a paramagnetic substance is inversely proportional to absolute temperature T.

$$\chi_m \propto \frac{1}{T}$$

The exact law is beyond the scope of our course.

Diamagnetism

The property of diamagnetism is generally found in those substances whose atoms (or molecules) have even number of electrons which form pairs.

"The net magnetic moment of an atom of a diamagnetic substance is thus zero." When a diamagnetic substance is placed in an external magnetic field, the spin motion of electrons is so modified that the electrons which produce the magnetic moments in the direction of external field slow down while the electrons which produce magnetic moments in opposite direction get accelerated. Thus, a net magnetic moment is induced in the opposite direction of applied magnetic field. Hence, the substance is magnetised opposite to the external field.

Note *That diamagnetism is temperature independent.*

Ferromagnetism

Iron like elements and their alloys are known as ferromagnetic substances. The susceptibility of these substances is in several thousands. Like paramagnetic substances, atoms of ferromagnetic substances have a permanent magnetic moment and behave like tiny magnets. But in ferromagnetic substances the atoms form innumerable small effective regions called 'domains'.

The size of the domain vary from about 10^{-6} cm^3 to 10^{-2} cm^3. Each domain has 10^{17} to 10^{21} atoms whose magnetic moments are aligned in the same direction.

In an unmagnetised ferromagnetic specimen, the domains are oriented randomly, so that their resultant magnetic moment is zero.

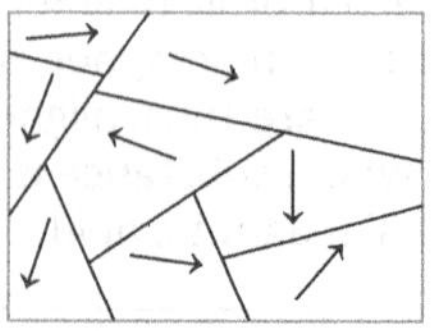

Unmagnetised

When the specimen is placed in a magnetic field, the resultant magnetisation may increase in two different ways.

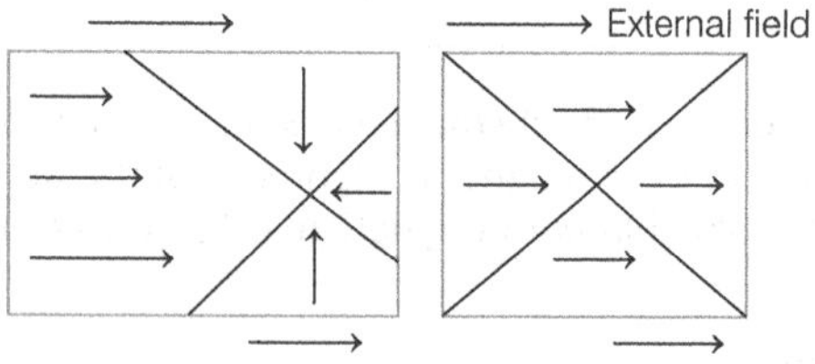

(a) (b)

(a) The domains which are oriented favourably with respect to the field increase in size. Whereas those oriented opposite to the external field are reduced.

(b) The domains rotate towards the field direction.

Note *That if the external field is weak, specimen gets magnetised by the first method and if the field is strong they get magnetised by the second method.*

Comparison of Dia, Para and Ferro Magnetic Substance

Diamagnetic Substance

(i) Diamagnetic substances are those substances which are feebly repelled by a magnet. e.g. antimony, bismuth, copper, gold, silver, quartz, mercury, alcohol, water, hydrogen, air, argon etc.

(ii) When placed in magnetic field, the lines of force tend to avoid the substance.

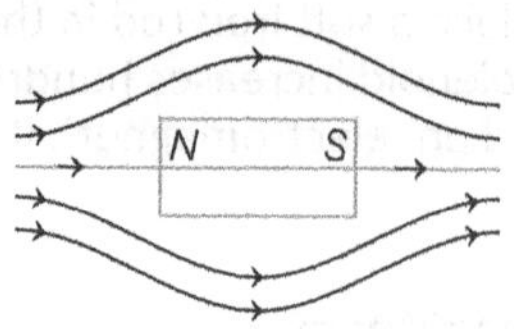

Field lines near a diamagnetic

Note *N and S shown in above figure represent north and south poles due to induced magnetic field produced in the opposite direction of external field.*

(iii) When placed in non-uniform magnetic field, it moves from stronger to weaker field.

(iv) When a diamagnetic rod is freely suspended in a uniform magnetic field, it aligns itself in a direction perpendicular to the field.

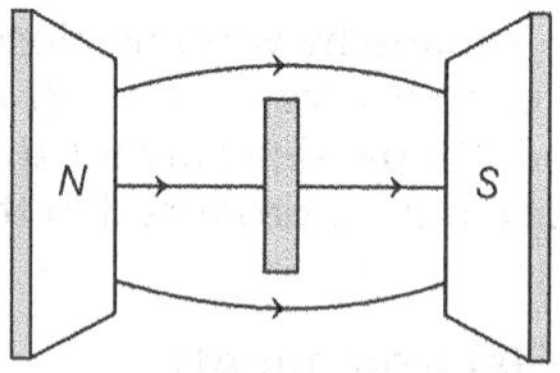

(v) When a diamagnetic substance is placed in a magnetic field, it is weakly magnetised in the direction opposite to the inducing field.

(vi) Induced dipole moment (**M**) is small negative value.

(vii) Intensity of magnetisation (**I**) has a small negative value.

(viii) Magnetic permeability μ is always less than unity.

(ix) Magnetic susceptibility is slightly negative.

(x) In non-uniform magnetic field these substances are attracted towards weaker field. Following figure illustrate this point.

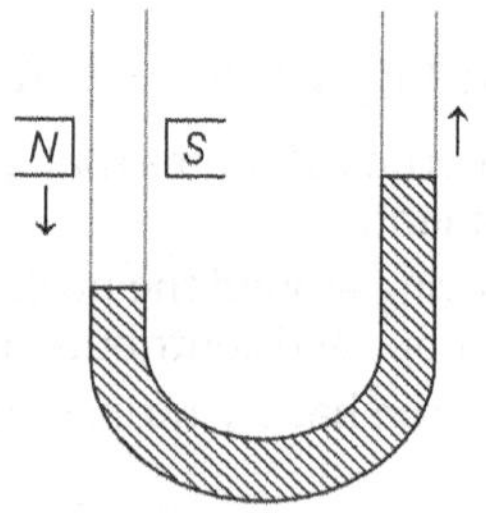

(xi) Susceptibility does not depend upon temperature.

Paramagnetic Substance

(i) Paramagnetic substances are those substances which are feebly attracted by a magnet.
e.g. aluminium, Chromium, alkali and alkaline earth metals. platinum, oxygen, etc.

(ii) The lines of force prefer to pass through the substance rather than air.

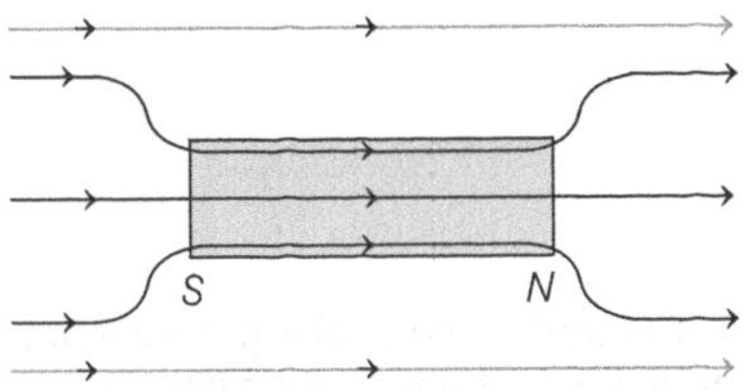

(iii) When a paramagnetic rod is freely suspended in a uniform magnetic field, it aligns itself in a direction parallel to the field.

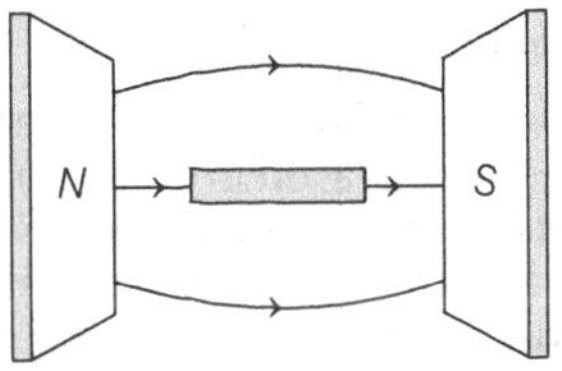

(iv) When a paramagnetic substance is placed in a magnetic field, it is weakly magnetised in the direction of the inducing field.

(v) Induced dipole moment (**M**) is a small positive value.

(vi) Intensity of magnetisation (**I**) has a small positive value.

(vii) Magnetic permeability μ is more than unity.

(viii) Magnetic susceptibility is slightly positive.

(ix) In non-uniform magnetic field these substances are attracted towards stronger fields.

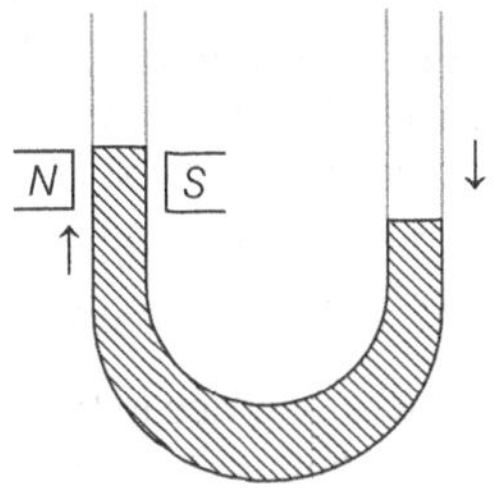

(x) Susceptibility $\chi_m \propto \frac{1}{T}$. This is curie's law.

Ferromagnetic Substance

(i) Ferromagnetic substances are those substances which are strongly attracted by a magnet.
e.g. iron, cobalt, nickel, gadolinium, dysprosium, etc.

(ii) The lines of force tend to crowd into the specimen.

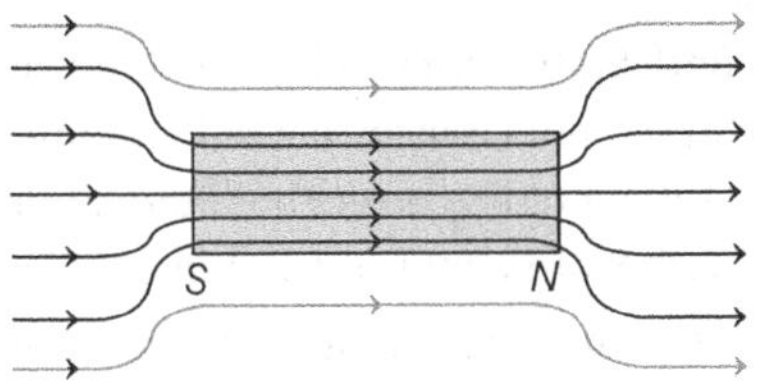

(iii) When placed in non-uniform magnetic field, it moves from weaker to stronger field (strong attraction).

(iv) When a paramagnetic rod is freely suspended in a uniform magnetic field, it aligns itself in a direction parallel to the field very quickly.

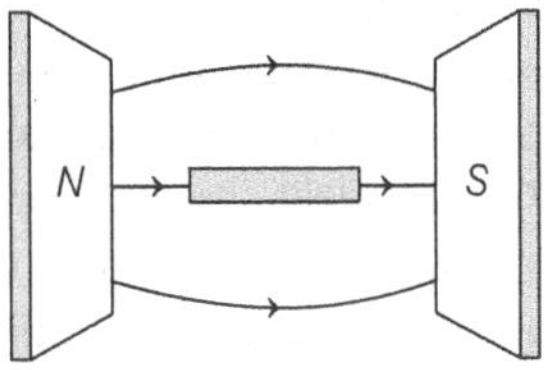

(v) When a ferromagnetic substance is placed in a magnetic field, it is strongly magnetised in the direction of the inducing field.

(vi) Induced dipole moment (**M**) is a large positive value.

(vii) Intensity of magnetisation (**I**) has a large positive value.

(viii) Magnetic permeability μ is large i.e. much more than unity.

(ix) Magnetic susceptibility is highly positive.

(x) Same properties as paramagnetic substances with strong intensity.

(xi) When temperature is increased, then after a centain temperature, domains are broken and ferromagnetic substances become paramagnetic substances.

Note *Diamagnetism is a universal property.*

MODULE 10 Exercise

Statement Type Questions

(a) Both Statement I and Statement II are true.
(b) Statement I is incorrect but Statement II is true.
(c) Statement I is correct but Statement II is false.
(d) Both Statement I and Statement II are false.

1. **Statement I :** For diamagnetic substance, $-1 \leq \chi < 0$, where χ is the magnetic susceptibility.
Statement II : Diamagnetic substances when placed in an external magnetic field, tend to move from stronger to weaker part of the field. **(2023 Main)**

2. **Statement I :** The diamagnetic property depends on temperature.
Statement II : The induced magnetic dipole moment in a diamagnetic sample is always opposite to the magnetising field. **(2023 Main)**

3. **Statement I :** Susceptibilities of paramagnetic and ferromagnetic substances increase with decrease in temperature.
Statement II : Diamagnetism is a result of motions of electrons developing magnetic moments opposite to the applied magnetic field. **(2022 Main)**

Single Correct Option

4. The space inside a straight current carrying solenoid is filled with a magnetic material having magnetic susceptibility equal to 1.2×10^{-5}. What is fractional increase in the magnetic field inside solenoid with respect to air as medium inside the solenoid? **(2022 Main)**
(a) 1.2×10^{-5} (b) 1.2×10^{-3}
(c) 1.8×10^{-3} (d) 2.4×10^{-5}

5. The susceptibility of a paramagnetic material is 99. The permeability of the material in Wb/A-m is
[Permeability of free space $\mu_0 = 4\pi \times 10^{-7}$ Wb/A-m] **(2022 Main)**
(a) $4\pi \times 10^{-7}$ (b) $4\pi \times 10^{-4}$
(c) $4\pi \times 10^{-5}$ (d) $4\pi \times 10^{-6}$

6. The soft-iron is a suitable material for making an electromagnet. This is because soft-iron has **(2022 Main)**
(a) low coercively and high retentivity
(b) low coercively and low permeability
(c) high permeability and low retentivity
(d) high permeability and high retentivity

7. A ferromagnetic material is placed in an external magnetic field. The magnetic domains **(2022 Main)**
(a) increase in size but no change in orientation
(b) have no relation with external magnetic field
(c) decrease in size and changes orientation
(d) may increase or decrease in size and change its orientation

8. A paramagnetic sample shows a net magnetisation of 6 A/m when it is placed in an external magnetic field of 0.4 T at a temperature of 4 K. When the sample is placed in an external magnetic field of 0.3 T at a temperature of 24 K, then the magnetisation will be **(2020 Main)**
(a) 1 A/m (b) 4 A/m (c) 2.25 A/m (d) 0.75 A/m

9. Magnetic materials used for making permanent magnets (*P*) and magnets in a transformer (*T*) have different properties of the following, which property best matches for the type of magnet required? **(2020 Main)**
(a) T : Large retentivity, small coercivity
(b) P : Small retentivity, large coercivity
(c) T : Large retentivity, large coercivity
(d) P : Large retentivity, large coercivity

10. A paramagnetic material has 10^{28} atoms/m^3. Its magnetic susceptibility at temperature 350 K is 2.8×10^{-4}. Its susceptibility at 300 K is **(2019 Main)**
(a) 3.726×10^{-4} (b) 3.672×10^{-4}
(c) 2.672×10^{-4} (d) 3.267×10^{-4}

11. A paramagnetic substance in the form of a cube with sides 1 cm has a magnetic dipole moment of 20×10^{-6} J/T when a magnetic intensity of 60×10^3 A/m is applied. Its magnetic susceptibility is **(2019 Main)**
(a) 3.3×10^{-4} (b) 3.3×10^{-2}
(c) 4.3×10^{-2} (d) 2.3×10^{-2}

12. A bar magnet is demagnetised by inserting it inside a solenoid of length 0.2 m, 100 turns and carrying a current of 5.2 A. The coercivity of the bar magnet is **(2019 Main)**
(a) 1200 A/m (b) 285 A/m
(c) 2600 A/m (d) 520 A/m

13. The *B-H* curve for a ferromagnet is shown in the figure. The ferromagnet is placed inside a long solenoid with 1000 turns/cm. The current that should be passed in the solenoid to demagnetise the ferromagnet completely is **(2018 Main)**

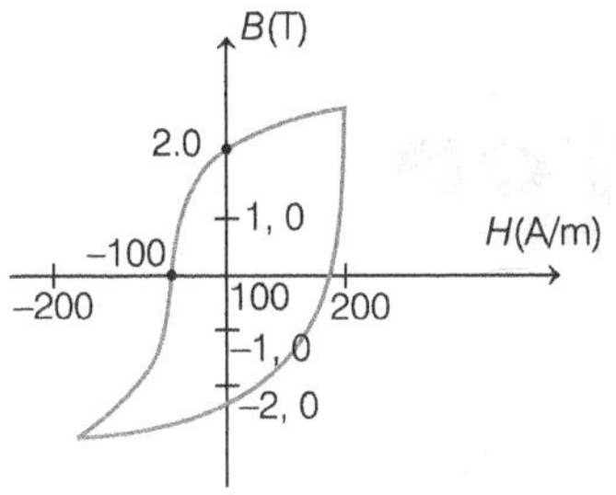

(a) 2 mA (b) 1 mA
(c) 40 μA (d) 20 μA

14. Hysteresis loops for two magnetic materials A and B are as given below : **(2016 Main)**

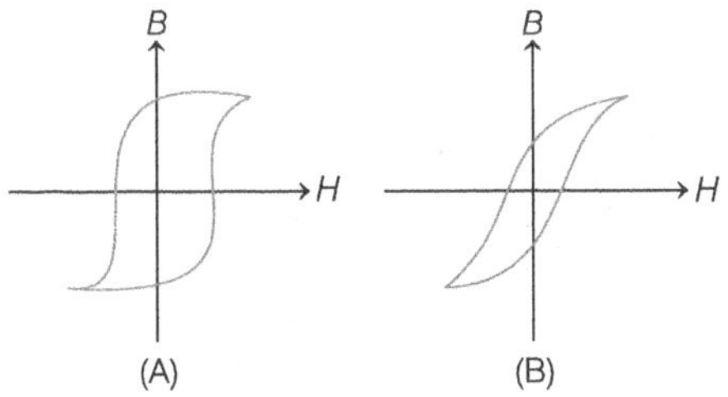

These materials are used to make magnets for electric generators, transformer core and electromagnet core. Then, it is proper to use
(a) A for electric generators and transformers
(b) A for electromagnets and B for electric generators
(c) A for transformers and B for electric generators
(d) B for electromagnets and transformers

15. A 25 cm long solenoid has radius 2 cm and 500 total number of turns. It carries a current of 15A. If it is equivalent to magnet of the same size and magnetization **M** (magnetic moment/volume), then |**M**| is **(2015 Main)**
(a) 30000π Am^{-1} (b) 3π Am^{-1}
(c) 30000 Am^{-1} (d) 300 Am^{-1}

16. The coercivity of a small magnet where the ferromagnet gets demagnetised is 3×10^3 Am^{-1}. The current required to be passed in a solenoid of length 10 cm and number of turns 100, so that the magnet gets demagnetised when inside the solenoid is **(2014 Main)**
(a) 30 mA (b) 60 mA (c) 3 A (d) 6 A

Numerical Value/Integer Type Questions

17. The value of aluminium susceptibility is 2.2×10^{-5}. The percentage increase in the magnetic field if space within a current carrying toroid is filled with aluminium is $\frac{x}{10^4}$. Then the value of x is .... **(2022, 2020 Main)**

18. A long solenoid with 1000 turns/m has a core material with relative permeability 500 and volume 10^3 cm^3. If the core material is replaced by another material having relative permeability of 750 with same volume maintaining same current of 0.75 A in the solenoid, the fractional change in the magnetic moment of the core would be approximately $\left(\frac{\chi}{499}\right)$. Find the value of χ. **(2021 Main)**

Answers

1. (a)	**2.** (b)	**3.** (a)	**4.** (a)	**5.** (c)	**6.** (c)	**7.** (d)	**8.** (d)	**9.** (d)	**10.** (d)
11. (a)	**12.** (c)	**13.** (b)	**14.** (d)	**15.** (c)	**16.** (c)	**17.** 22	**18.** 250		

See hints on pages 447-449.

MODULE 11

Moving Coil Galvanometer

The moving coil galvanometer is a device used to measure an electric current.

Principle

Action of a moving coil galvanometer is based upon the principle that when a current carrying coil is placed in a magnetic field, it experiences a torque whose magnitude depends on the magnitude of current.

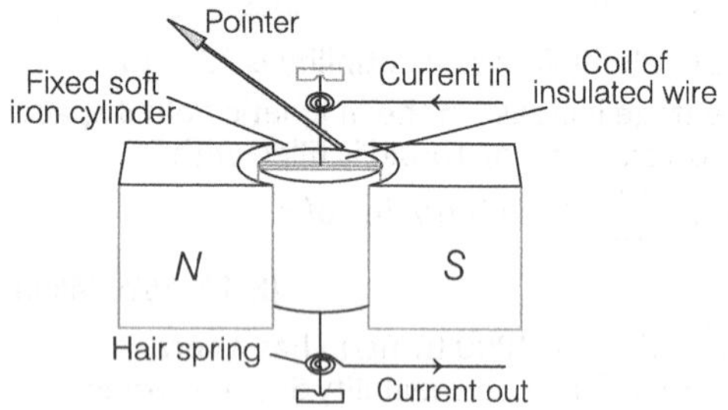

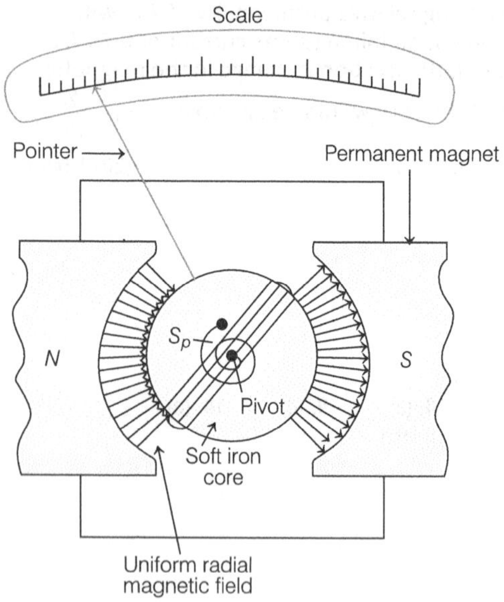

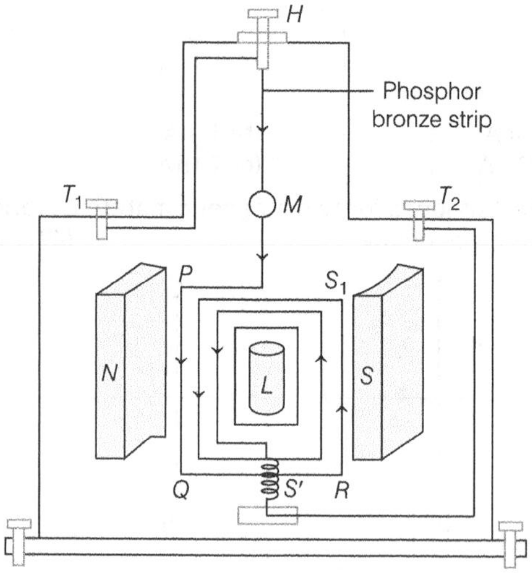

Construction

The main parts of a moving coil galvanometer are shown in figure. The galvanometer consists of a coil, with many turns free to rotate about a fixed vertical axis in a uniform radial magnetic field. There is a cylindrical soft iron core, which not only makes the field radial but also increases the strength of magnetic field.

Theory

The current to be measured is passed through the galvanometer. As the coil is in the magnetic field (of constant magnitude) it experiences a torque given by

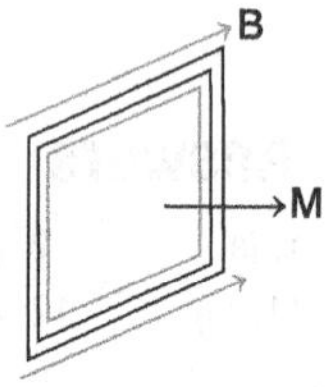

$$\tau = MB\sin\theta$$

$$= (NiA)\,B\sin\theta \qquad \text{...(i)}$$

As shown in the figure, the pole pieces are made cylindrical, the magnetic field always remains parallel to the plane of the coil. Or angle between **B** and **M** always remains 90°.

Therefore, Eq. (i) can be written as

$$\tau = NiAB \quad (\text{as } \sin\theta = \sin 90^\circ = 1)$$

Here, N = total number of turns of the coil

i = current passing through the coil

A = area of cross-section of the coil and

B = magnitude of radial magnetic field.

This torque rotates the coil. The spring S_p shown in figure provides a counter torque $k\phi$ that balances the above torque $NiAB$. In equilibrium,

$$k\phi = NiAB \quad \text{...(ii)}$$

Here, k is the torsional constant of the spring. With rotation of coil a small light mirror M (attached with phosphor bronze wire W) also rotates and equilibrium deflection ϕ can be measured by a lamp and scale arrangement.

The above Eq. (ii) can be written as

$$\boxed{i = \left(\frac{k}{NAB}\right)\phi} \quad \text{...(iii)}$$

Hence, the current i is proportional to the deflection ϕ.

Galvanometer Constant

The constant $\frac{k}{NAB}$ in Eq. (iii) is called galvanometer constant.

Hence,

$$\boxed{\text{Galvanometer constant} = \frac{k}{NAB}} \quad \text{...(iv)}$$

This constant may be found by passing a known current through the coil. Measuring the deflection ϕ and putting these values in Eq. (iii), we can find galvanometer constant.

Sensitivity of Galvanometer

Deflection per unit current (ϕ/i) is called sensitivity of galvanometer. From Eq. (iii), we can see that $\phi/i = NAB/k$. Hence,

$$\boxed{\text{Sensitivity} = \frac{\phi}{i} = \frac{NAB}{k}} \quad \text{...(v)}$$

The sensitivity of a galvanometer can be increased by

(i) increasing the number of turns in the coil N or

(ii) increasing the magnitude of magnetic field.

Example 1 A rectangular coil of area $5.0 \times 10^{-4}\,\text{m}^2$ and 60 turns is pivoted about one of its vertical sides. The coil is in a radial horizontal magnetic field of 9×10^{-3} T. What is the torsional constant of the spring connected to the coil if a current of 0.20 mA produces an angular deflection of 18°?

Sol. From the equation, $i = \left(\frac{k}{NAB}\right)\phi$

We find that torsional constant of the spring is given by

$$k = \frac{NABi}{\phi}$$

Substituting the values in SI units, we have

$$k = \frac{(60)(5.0 \times 10^{-4})(9 \times 10^{-3})(0.2 \times 10^{-3})}{18}$$

$$= 3 \times 10^{-9}\,\text{N-m/degree} \quad \textbf{Ans.}$$

MODULE 11 Exercise

Single Correct Option

1. If a shunt of $\frac{1}{10}$ th of the coil resistance is applied to a moving coil galvanometer, then its sensitivity becomes

(a) 10 fold (b) 11 fold (c) $\frac{1}{10}$ fold (d) $\frac{1}{11}$ fold

Subjective Questions

2. A coil of a moving coil galvanometer twists through 90° when a current of one microampere is passed through it. If the area of the coil is $10^{-4}\,\text{m}^2$ and it has 100 turns, calculate the magnetic field of the magnet of the galvanometer. Given, $k = 10^{-8}$ N-m / degree.

3. A galvanometer coil 5 cm × 2 cm with 200 turns is suspended vertically in a field of 5×10^{-2} T. The suspension fibre needs a torque of 0.125×10^{-7} N-m to twist it through one radian. Calculate the strength of the current required to be maintained in the coil if we require a deflection of 6°.

Answers

1. (d) **2.** 90 T **3.** 1.3×10^{-7} A

See hints on page 449.

Unit Exercise (Cumulative)

Level 1

Fill in the Blank

1. A uniform magnetic field with a slit system as shown in figure is to be used as momentum filter for high-energy charged particles. With a field B Tesla, it is found that the filter transmits α-particles each of energy 5.3 MeV. The magnetic field is increased to 2.3 B Tesla and deuterons are passed into the filter. The energy of each deuteron transmitted by the filter is ......MeV. **(1997 JEE)**

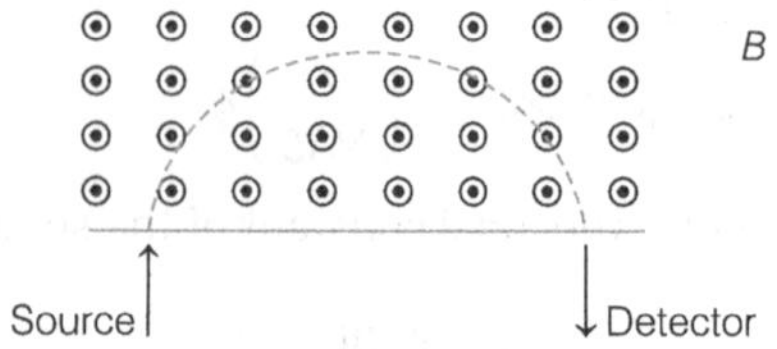

Assertion and Reason

See directions on page 25.

2. **Assertion :** A current carrying loop is placed in uniform magnetic field as shown in figure. Torque in the loop in this case is zero.

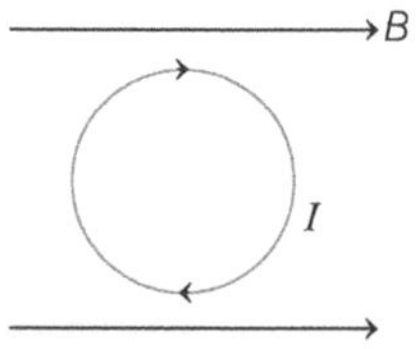

Reason : Magnetic moment vector of the loop is perpendicular to paper inwards.

3. **Assertion :** Force on current carrying loop shown in figure in magnetic field, $\mathbf{B} = (B_0 x)\hat{\mathbf{k}}$ is along positive x-axis. Here, B_0 is a positive constant.

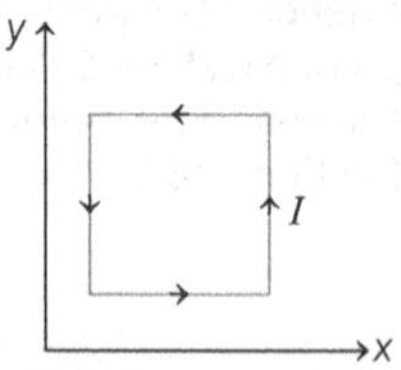

Reason : A torque will also act on the loop.

4. **Assertion :** Two infinitely long wires A and B carry unequal currents both in inward direction.

Then there is only one point (excluding the points at infinity) where net magnetic field is zero.

Reason : That point lies between points A and B.

Single Correct Option

5. The universal property among all substances is
   (a) diamagnetism (b) paramagnetism
   (c) ferromagnetism (d) non-magnetism

6. Four long straight wires are located at the corners of a square $ABCD$. All the wires carry equal currents. Current in the wires A and B are inwards and in C and D are outwards. The magnetic field at the centre O is parallel to

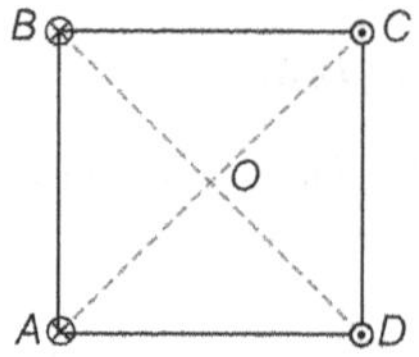

   (a) AD (b) CB
   (c) AB (d) CD

7. The figure shows a long straight wire carrying a current I_1 along the axis of a circular ring carrying a current I_2. Identify the correct statement.

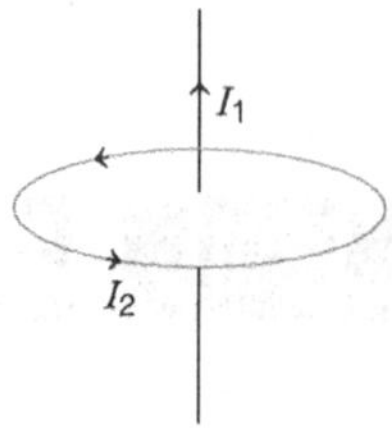

   (a) Straight wire attracts the ring
   (b) Straight wire attracts a small element of the ring
   (c) Straight wire does not attract any small element of the ring
   (d) None of the above

8. A charged particle moving along positive x-direction with a velocity v enters a region where there is a uniform magnetic field $\mathbf{B} = -B\hat{\mathbf{k}}$, from $x = 0$ to $x = d$. The particle gets deflected at an angle θ from its initial path. The specific charge of the particle is
   (a) $\dfrac{Bd}{v\cos\theta}$ (b) $\dfrac{v\tan\theta}{Bd}$
   (c) $\dfrac{B\sin\theta}{vd}$ (d) $\dfrac{v\sin\theta}{Bd}$

9. A rectangular loop of sides 10 cm and 5 cm carrying a current I of 12 A is placed in different orientations as shown in the figures below

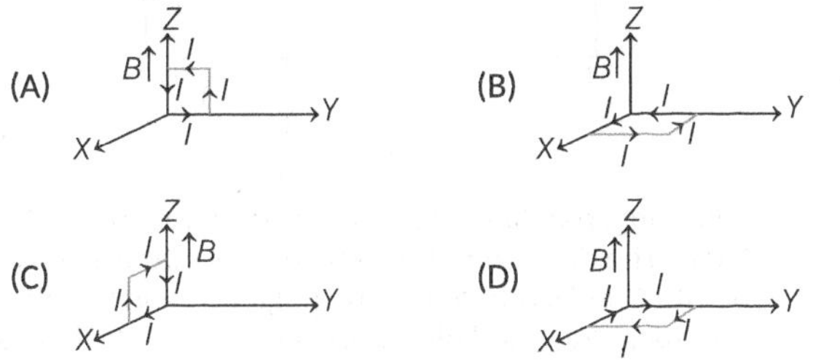

If there is a uniform magnetic field of 0.3 T in the positive z-direction, in which orientations the loop would be in (i) stable equilibrium and (ii) unstable equilibrium? **(2015 JEE)**

(a) (B) and (D), respectively
(b) (B) and (C), respectively
(c) (A) and (B), respectively
(d) (A) and (C), respectively

10. A current carrying rod AB is placed perpendicular to an infinitely long current carrying wire as shown in figure. The point at which the conductor should be hinged so that it will not rotate ($AC = CB$)

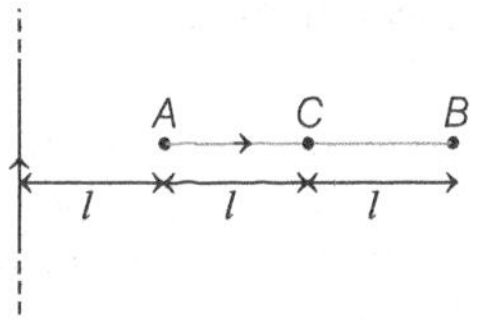

(a) A
(b) somewhere between B and C
(c) C
(d) somewhere between A and C

11. The figure shows two coaxial circular loops 1 and 2, which forms same solid angle θ at point O. If B_1 and B_2 are the magnetic fields produced at the point O due to loop 1 and 2 respectively, then

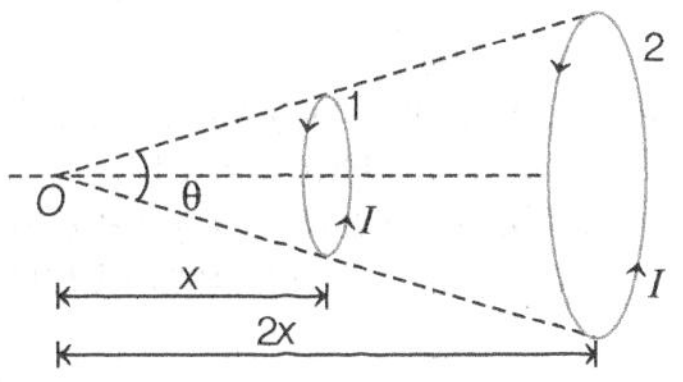

(a) $\frac{B_1}{B_2} = 1$ (b) $\frac{B_1}{B_2} = 2$
(c) $\frac{B_1}{B_2} = 8$ (d) $\frac{B_1}{B_2} = 4$

12. A tightly wound long solenoid has n turns per unit length, radius r and carries a current i. A particle having charge q and mass m is projected from a point on the axis in the direction perpendicular to the axis. The maximum speed for which particle does not strike the solenoid will be

(a) $\frac{\mu_0 qrni}{2m}$ (b) $\frac{\mu_0 qrni}{m}$
(c) $\frac{2\mu_0 qrni}{3m}$ (d) None of these

13. Two infinitely long conductors carrying equal currents are shaped as shown. The short sections are all of equal lengths. The point P is located symmetrically with respect to the two conductors. The magnetic field at P due to any one conductor is B. The total field at P is

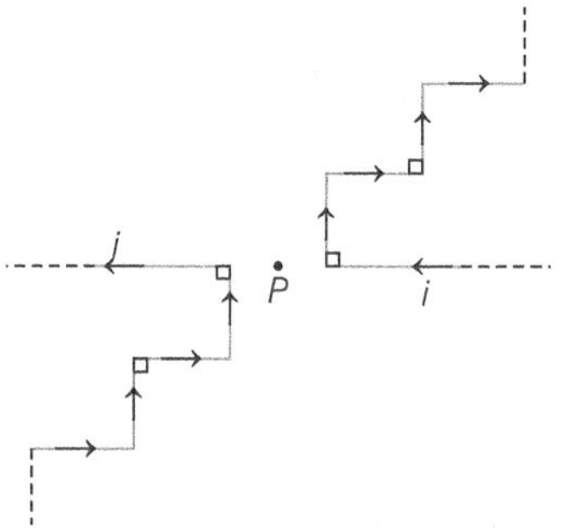

(a) zero (b) B
(c) $\sqrt{2}B$ (d) $2B$

14. A charged particle enters into a uniform magnetic field with velocity v_0 perpendicular to it. The length of magnetic field is $x = \frac{\sqrt{3}}{2}R$, where R is the radius of the circular path of the particle in the field. The magnitude of change in velocity of the particle when it comes out of the field is

(a) $2v_0$ (b) $\frac{v_0}{2}$
(c) $\frac{\sqrt{3}v_0}{2}$ (d) v_0

15. An electron moving with a speed u along the positive x-axis at $y = 0$ enters a region of uniform magnetic field $\mathbf{B} = -B_0\hat{\mathbf{k}}$, which exists to the right of y-axis. The electron exits from the region after sometime with the speed v at coordinate y, then **(2004 JEE)**

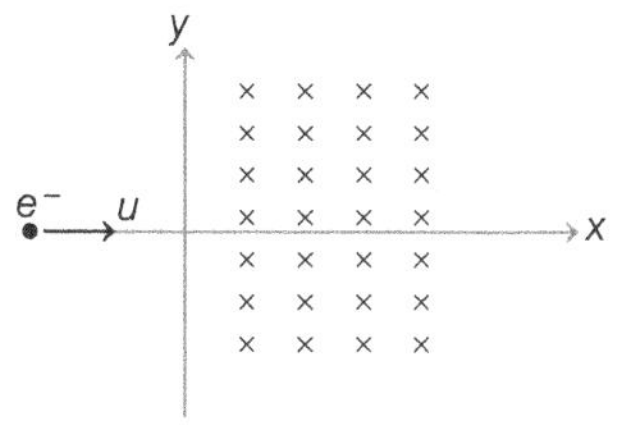

(a) $v > u, y < 0$ (b) $v = u, y > 0$
(c) $v > u, y > 0$ (d) $v = u, y < 0$

16. Two long parallel wires are at a distance $2d$ apart. They carry steady equal currents flowing out of the plane of the paper as shown. The variation of the magnetic field B along the line XX' is given by **(2000 JEE)**

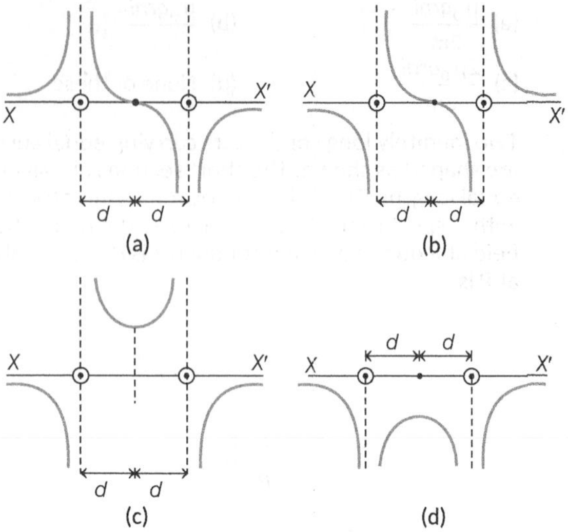

Match the Columns

17. Match List I with List II.

	List-I (Current configuration)		List-II (Magnitude of Magnetic Field at point O)
A.	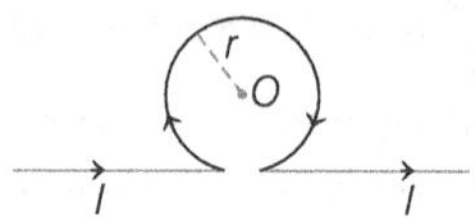	I.	$B_0 = \frac{\mu_0 I}{4\pi r}(\pi + 2)$
B.	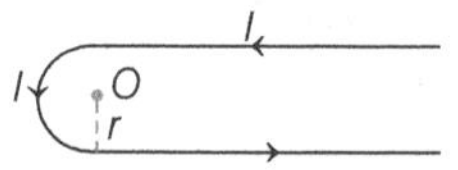	II.	$B_0 = \frac{\mu_0}{4}\frac{I}{r}$
C.	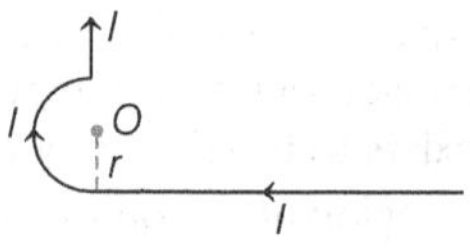	III.	$B_0 = \frac{\mu_0 I}{2\pi r}(\pi - 1)$
D.		IV.	$B_0 = \frac{\mu_0 I}{4\pi r}(\pi + 1)$

Choose the correct answer from the options given below **(2023 Main)**

(a) A-III, B-I, C-IV, D-II
(b) A-III, B-IV, C-I, D-II
(c) A-I, B-III, C-IV, D-II
(d) A-II, B-I, C-IV, D-III

Subjective Questions

18. For the situation shown in figure, find the force experienced by side MN of the rectangular loop. Also, find the torque on the loop.

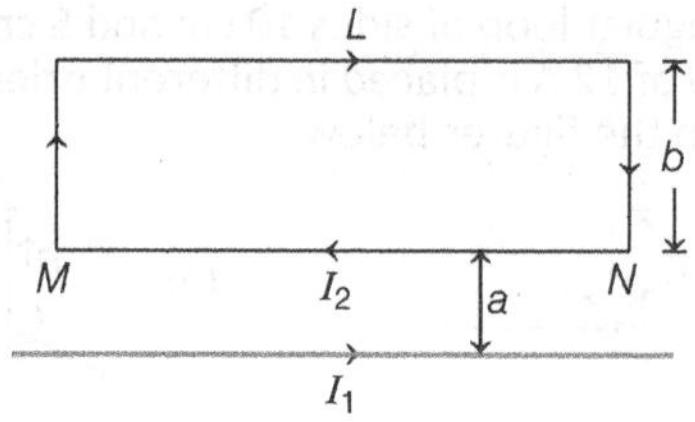

19. Four long, parallel conductors carry equal currents of 5.0 A. The direction of the currents is into the page at points A and B and out of the page at C and D. Calculate the magnitude and direction of the magnetic field at point P, located at the centre of the square.

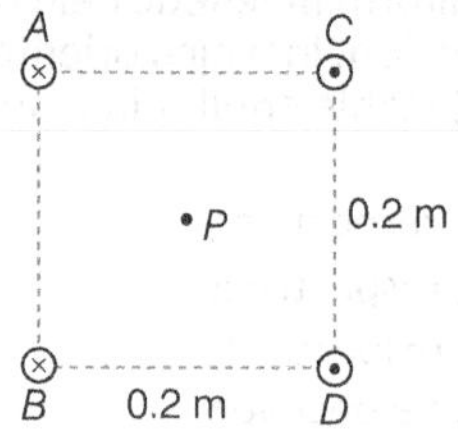

20. A current I flows along a thin wire, shaped as a regular polygon with n sides which can be inscribed into a circle of radius R. Find the magnetic induction at the centre of the polygon. Analyse the obtained expression at $n \to \infty$. **(Irodov 3.220)**

21. A current I flows along a lengthy thin-walled tube of radius R with longitudinal slit of width h. Find the induction of the magnetic field inside the tube under the condition $h \ll R$. **(Irodov 3.224)**

22. A current I flows in a long straight wire with cross-section having the form of a thin half-ring of radius R. Find the induction of the magnetic field at the point O. **(Irodov 3.225)**

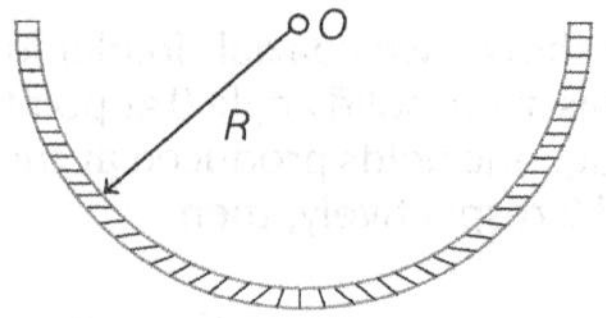

23. A single-layer coil (solenoid) has length l and cross-section radius R. Number of turns per unit length is equal to n. Find the magnetic induction at the centre of the coil when a current I flows through it. **(Irodov 3.236)**

24. Find the magnetic moment of a thin round loop with current if the radius of the loop is equal to $R = 100$ mm and the magnetic induction at its centre is equal to $B = 6.0\ \mu T$. **(Irodov 3.243)**

25. A thin insulated wire forms a plane spiral of $N = 100$ tight turns, carrying a current $I = 8$ mA. The radii of inside and outside turns are equal to $a = 50$ mm and $5 = 100$ mm. Find:
(a) the magnetic induction at the centre of the spiral
(b) the magnetic moment of the spiral with a given current. **(Irodov 3.245)**

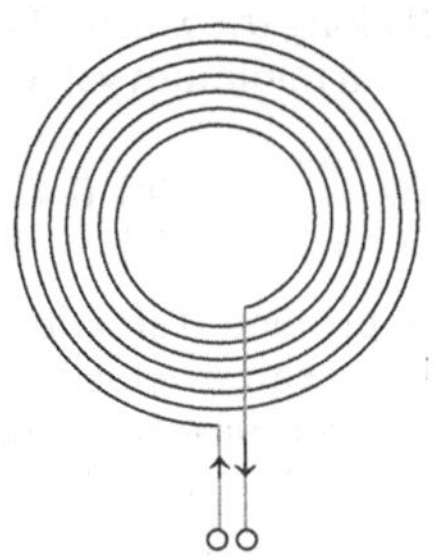

26. A non-conducting thin disc of radius R charged uniformly over one side with surface charge density σ rotates about its axis with an angular velocity ω. Find:
   (a) the magnetic induction at the centre of the disc
   (b) the magnetic moment of the disc. **(Irodov 3.246)**

27. A charge q is uniformly distributed over the volume of a uniform ball of mass m and radius R which rotates with an angular velocity ω about the axis passing through its centre. Find the respective magnetic moment and its ratio to the mechanical moment. **(Irodov 3.248)**

28. Find the magnitude and direction of a force vector acting on a unit length of a thin wire, carrying a current $I = 8.0$ A, at a point O, if the wire is bent as shown in **(Irodov 3.251)**

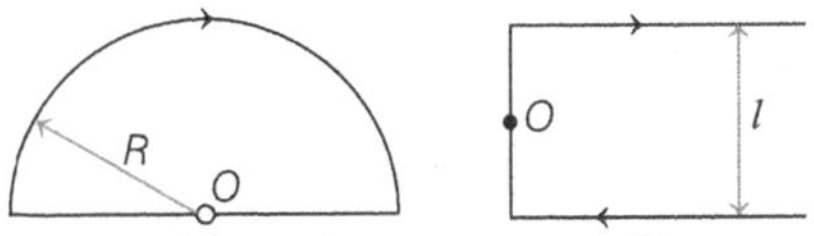

(a) (b)

   (a) Fig. (a), with curvature radius $R = 10$ cm
   (b) Fig. (b), the distance between the long parallel segments of the wire being equal to $l = 20$ cm.

29. A small coil C with $N = 200$ turns is mounted on one end of a balance beam and introduced between the poles of an electromagnet as shown in figure. The cross-sectional area of the coil is $S = 1.0\ \text{cm}^2$, the length of the arm OA of the balance beam is $l = 30$ cm. When there is no current in the coil the balance is in equilibrium. On passing a current $I = 22$ mA through the coil the equilibrium is restored by putting the additional counterweight of mass $\Delta m = 60$ mg on the balance pan. Find the magnetic induction at the spot where the coil is located. **(Irodov 3.254)**

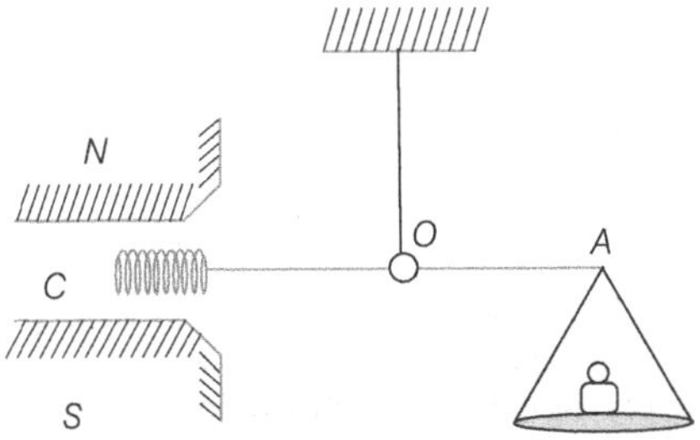

30. Find the interaction energy of two loops carrying currents I_1 and I_2 if both loops are shaped as circles of radii a and b, with $a \ll b$. The centres of the loops are located at the same point and their planes form an angle θ between them. **(Irodov 3.346)**

Level 2

Single Correct Option

31. A uniform current carrying ring of mass m and radius R is connected by a massless string as shown. A uniform magnetic field B_0 exists in the region to keep the ring in horizontal position, then the current in the ring is

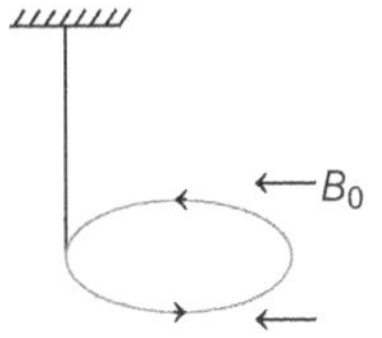

(a) $\dfrac{mg}{\pi R B_0}$ (b) $\dfrac{mg}{R B_0}$ (c) $\dfrac{mg}{3\pi R B_0}$ (d) $\dfrac{mg}{\pi R^2 B_0}$

32. A straight long conductor carries current along the positive x-axis. Identify the correct statement related to the four points $A(a, a, 0)$, $B(a, 0, a)$, $C(a, -a, 0)$ and $D(a, 0, -a)$.
   (a) The magnitude of magnetic field at all points is same
   (b) Fields at A and B are mutually perpendicular
   (c) Fields at A and C are antiparallel
   (d) All of the above

33. An equilateral triangle frame PQR of mass M and side a is kept under the influence of magnetic force due to inward perpendicular magnetic field B and gravitational field as shown in the figure. The magnitude and direction of current in the frame so that the frame remains at rest, is

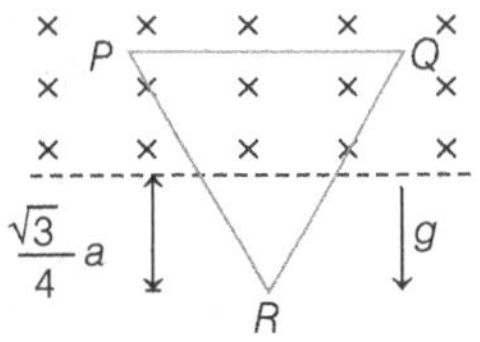

(a) $I = \dfrac{2Mg}{aB}$; anti-clockwise (b) $I = \dfrac{2Mg}{aB}$; clockwise
(c) $I = \dfrac{Mg}{aB}$; anti-clockwise (d) $I = \dfrac{Mg}{aB}$; clockwise

34. A simple pendulum with a charged bob is oscillating as shown in the figure. Time period of oscillation is T and angular amplitude is θ. If a uniform magnetic field perpendicular to the plane of oscillation is switched on, then

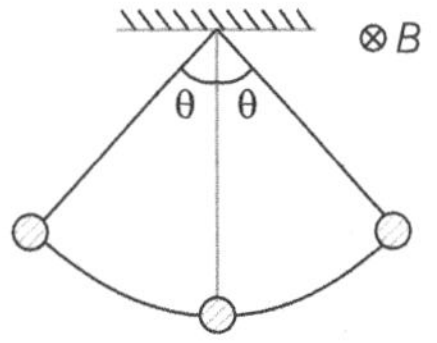

(a) T will decrease but θ will remain constant
(b) T will remain unchanged but θ will decrease
(c) Both T and θ will remain the same
(d) Both T and θ will decrease

35. Magnetic field in a region is given by $\mathbf{B} = B_0 x\hat{\mathbf{k}}$. Two loops each of side a are placed in this magnetic region in the xy-plane with one of its sides on x-axis. If F_1 is the force on loop 1 and F_2 be the force on loop 2, then

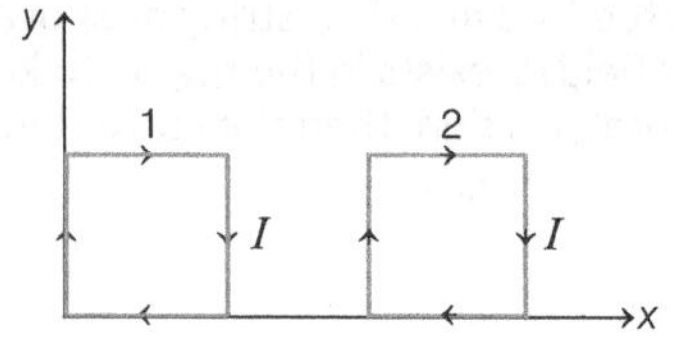

(a) $F_1 = F_2 = 0$
(b) $F_1 > F_2$
(c) $F_2 > F_1$
(d) $F_1 = F_2 \neq 0$

36. A particle having a charge of 20 μC and mass 20 μg moves along a circle of radius 5 cm under the action of a magnetic field $B = 0.1$ tesla. When the particle is at P, uniform transverse electric field is switched on and it is found that the particle continues along the tangent with a uniform velocity. Find the electric field

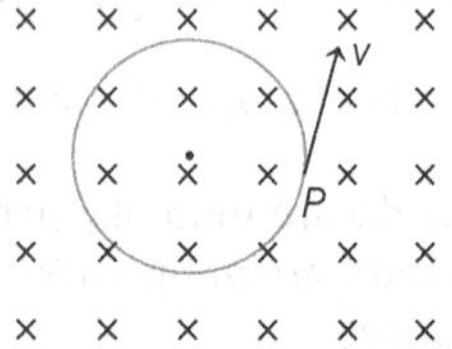

(a) 2 V/m (b) 0.5 V/m (c) 5 V/m (d) 1.5 V/m

37. Equal currents are flowing in three infinitely long wires along positive x, y and z-directions. The magnetic field at a point $(0, 0, -a)$ would be (i = current in each wire)

(a) $\frac{\mu_0 i}{2\pi a}(\hat{\mathbf{j}} - \hat{\mathbf{i}})$
(b) $\frac{\mu_0 i}{2\pi a}(\hat{\mathbf{i}} - \hat{\mathbf{j}})$
(c) $\frac{\mu_0 i}{2\pi a}(\hat{\mathbf{i}} + \hat{\mathbf{j}})$
(d) $\frac{\mu_0 i}{2\pi a}(-\hat{\mathbf{i}} - \hat{\mathbf{j}})$

38. A uniformly charged ring of radius R is rotated about its axis with constant linear speed v of each of its particles. The ratio of electric field to magnetic field at a point P on the axis of the ring distant $x = R$ from centre of ring is (c is speed of light)

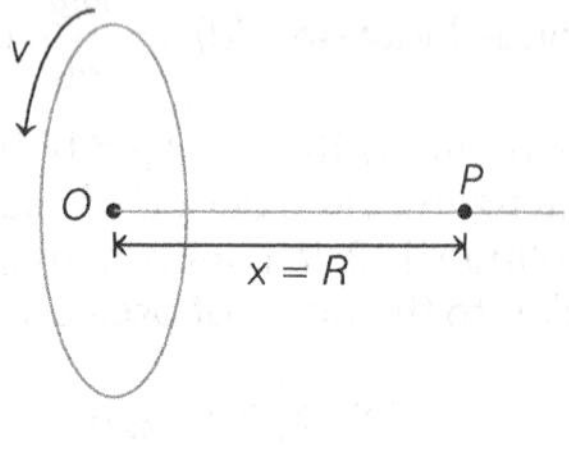

(a) $\frac{c^2}{v}$ (b) $\frac{v^2}{c}$ (c) $\frac{v}{c}$ (d) $\frac{c}{v}$

39. A charged particle P leaves the origin with speed $v = v_0$, at some inclination with the x-axis. There is a uniform magnetic field B along the x-axis. P strikes a fixed target T on the x-axis for a minimum value of $B = B_0$. P will also strike T if

(a) $B = 2B_0, v = 2v_0$
(b) $B = 2B_0, v = v_0$
(c) Both are correct
(d) Both are wrong

40. Two protons are projected simultaneously from a fixed point with the same velocity v into a region where there exists a uniform magnetic field. The magnetic field strength is B and it is perpendicular to the initial direction of v. One proton starts at time $t = 0$ and another proton at $t = \frac{\pi m}{2qB}$. The separation between them at time $t = \frac{\pi m}{qB}$ [where, m and q are the mass and charge of proton], will be approximately (neglect electrostatic force between protons)

(a) $2\frac{mv}{qB}$ (b) $\frac{\sqrt{2}\, mv}{qB}$ (c) $\frac{mv}{qB}$ (d) $\frac{mv}{2qB}$

41. A disc of radius R rotates with constant angular velocity ω about its own axis. Surface charge density of this disc varies as $\sigma = \alpha r^2$, where r is the distance from the centre of disc. Determine the magnetic field intensity at the centre of disc.

(a) $\mu_0 \alpha \omega R^3$ (b) $\frac{\mu_0 \alpha \omega R^3}{6}$ (c) $\frac{\mu_0 \alpha \omega R^3}{8}$ (d) $\frac{\mu_0 \alpha \omega R^3}{3}$

42. A rigid circular loop of radius r and mass m lies in the XY-plane on a flat table and has a current i flowing in it. At this particular place, the earth's magnetic field is $\mathbf{B} = B_x\hat{\mathbf{i}} + B_z\hat{\mathbf{k}}$. The value of i so that the loop starts tilting is

(a) $\frac{mg}{\pi r\sqrt{B_x^2 + B_z^2}}$
(b) $\frac{mg}{\pi r B_x}$
(c) $\frac{mg}{\pi r B_z}$
(d) $\frac{mg}{\pi r\sqrt{B_x B_z}}$

43. Two protons P and Q move parallel to the x-axis in opposite directions with equal speeds v. At the instant shown, the ratio of magnetic force and electric force acting on the proton P is (c = speed of light in vacuum)

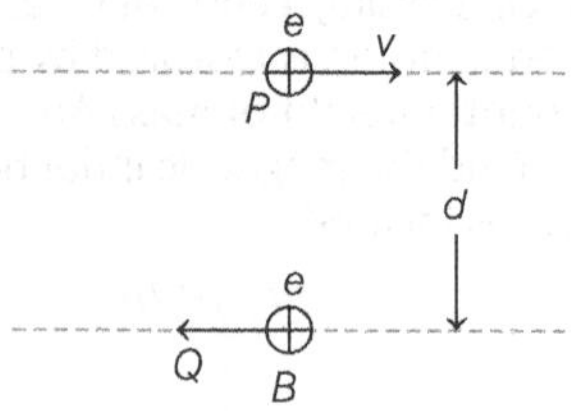

(a) $\frac{v}{c}$ (b) $\frac{v^2}{c^2}$ (c) $\frac{vd^2}{c}$ (d) $\frac{2v}{c}$

44. Two straight infinitely long current carrying wires are kept along z-axis at the coordinates $(0, a, 0)$ and $(0, -a, 0)$ respectively as shown in the figure. The current in each of the wire is equal and along negative z-axis (into the plane of the paper). The variation of magnetic field on the x-axis will be approximately

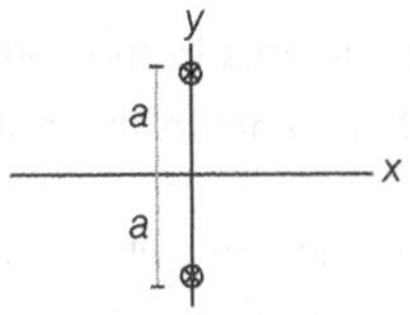

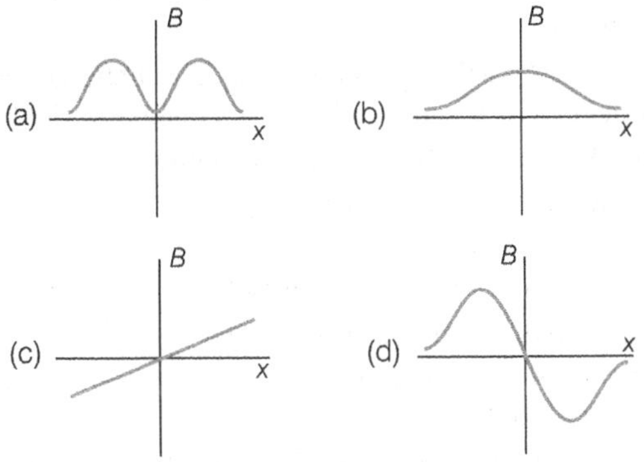

45. A particle of specific charge $q/m = (\pi)$ C/kg is projected from the origin towards positive x-axis with a velocity of 10 m/s in a uniform magnetic field $\mathbf{B} = -2\hat{\mathbf{k}}$ T. The velocity $\mathbf{v}$ of the particle after time $t = \frac{1}{6}$ s will be

(a) $(5\hat{\mathbf{i}} + 5\sqrt{3}\hat{\mathbf{j}})$ m/s (b) $10\hat{\mathbf{j}}$ m/s
(c) $(5\sqrt{3}\,\hat{\mathbf{i}} + 5\hat{\mathbf{j}})$ m/s (d) $-10\hat{\mathbf{j}}$ m/s

46. A proton goes undeflected in a crossed electric and magnetic field (the fields are perpendicular to each other) at a speed of 2.0×10^5 m/s. The velocity is perpendicular to both the fields. When the electric field is switched off, the proton moves along a circle of radius 4.0 cm. The magnitudes of the electric and the magnetic fields are : (Take mass of the proton $= 1.6 \times 10^{-27}$ kg)

(a) 10^4 N/C, 0.05 T (b) 10^5 N/C, 0.05 T
(c) 10^{-4} N/C, 0.04 T (d) 10^4 N/C, 0.04 T

47. A particle of charge q and mass m is moving along the x-axis with a velocity v and enters a region of electric field $\mathbf{E}$ and magnetic field $\mathbf{B}$ as shown in figure below. For which figure the net force on the charge may be zero?

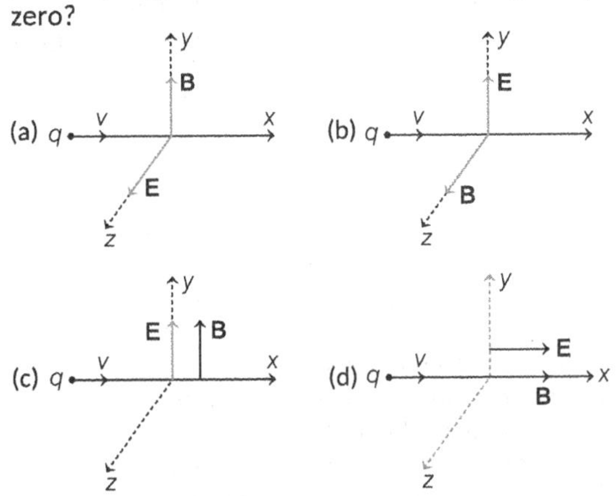

48. A magnetic field $\mathbf{B} = B_0\hat{\mathbf{j}}$ exists in the region $a < x < 2a$ and $\mathbf{B} = -B_0\hat{\mathbf{j}}$, in the region $2a < x < 3a$, where B_0 is a positive constant. A positive point charge moving with a velocity $\mathbf{v} = v_0\hat{\mathbf{i}}$, where v_0 is a positive constant, enters the magnetic field at $x = a$. The trajectory of the charge in this region can be like **(2007 JEE)**

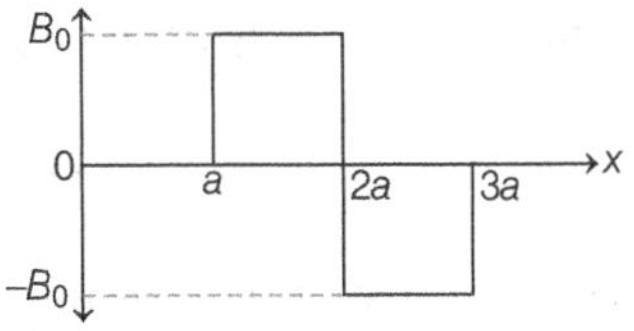

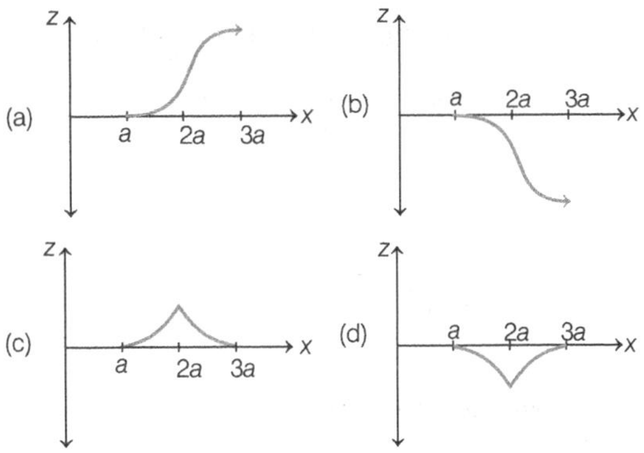

49. For a positively charged particle moving in a x-y plane initially along the x-axis, there is a sudden change in its path due to the presence of electric and/or magnetic fields beyond P. The curved path is shown in the x-y plane and is found to be non-circular. **(2003 JEE)**

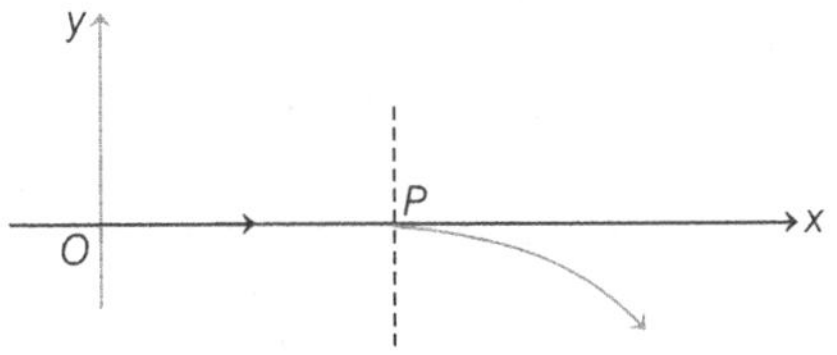

Which one of the following combinations is possible?

(a) $\mathbf{E} = 0$; $\mathbf{B} = b\hat{\mathbf{j}} + c\hat{\mathbf{k}}$ (b) $\mathbf{E} = a\hat{\mathbf{i}}$; $\mathbf{B} = c\hat{\mathbf{k}} + a\hat{\mathbf{i}}$
(c) $\mathbf{E} = 0$; $\mathbf{B} = c\hat{\mathbf{j}} + b\hat{\mathbf{k}}$ (d) $\mathbf{E} = a\hat{\mathbf{i}}$; $\mathbf{B} = c\hat{\mathbf{k}} + b\hat{\mathbf{j}}$

50. A long straight wire along the z-axis carries a current I in the negative z-direction. The magnetic vector field $\mathbf{B}$ at a point having coordinate (x, y) on the $z = 0$ plane is **(2002 JEE)**

(a) $\dfrac{\mu_0 I(y\hat{\mathbf{i}} - x\hat{\mathbf{j}})}{2\pi(x^2 + y^2)}$ (b) $\dfrac{\mu_0 I(x\hat{\mathbf{i}} + y\hat{\mathbf{j}})}{2\pi(x^2 + y^2)}$

(c) $\dfrac{\mu_0 I(x\hat{\mathbf{j}} - y\hat{\mathbf{i}})}{2\pi(x^2 + y^2)}$ (d) $\dfrac{\mu_0 I(x\hat{\mathbf{i}} - y\hat{\mathbf{j}})}{2\pi(x^2 + y^2)}$

51. A non-planar loop of conducting wire carrying a current I is placed as shown in the figure. Each of the straight sections of the loop is of length $2a$. The magnetic field due to this loop at the point P $(a, 0, a)$ points in the direction is **(2001 JEE)**

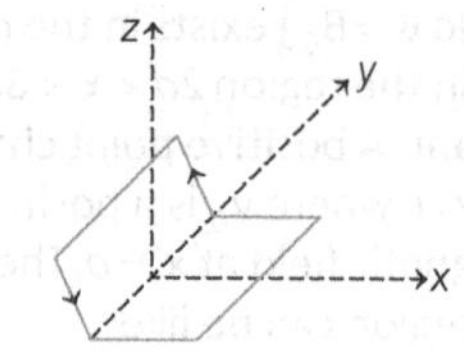

(a) $\frac{1}{\sqrt{2}}(-\hat{j}+\hat{k})$ (b) $\frac{1}{\sqrt{3}}(-\hat{j}+\hat{k}+\hat{i})$

(c) $\frac{1}{\sqrt{3}}(\hat{i}+\hat{j}+\hat{k})$ (d) $\frac{1}{\sqrt{2}}(\hat{i}+\hat{k})$

52. A circular current carrying wire having radius R is placed in XY-plane with its centre at origin O. There is a non-uniform magnetic field $\mathbf{B}=\frac{B_0 x}{2R}\hat{k}$ (here B_0 is a positive constant) is existing in the region.

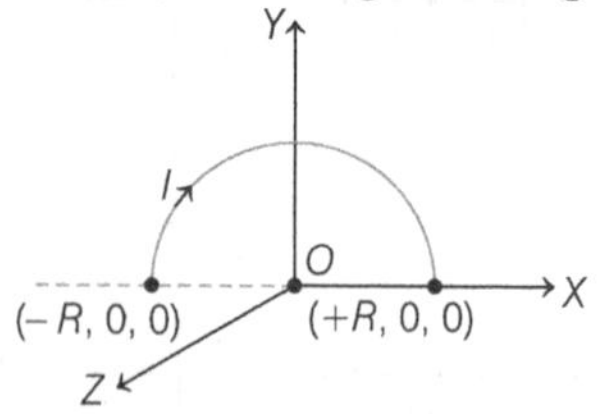

The magnetic force acting on semicircular wire will be along

(a) –X-axis (b) +Y-axis

(c) –Y-axis (d) +X-axis

53. A moving coil galvanometer has 150 equal divisions. Its current sensitivity is 10 div/mA and voltage sensitivity is 2 div/mV.In order that each division read 1 V, the resistance (in ohm) needed to be connected is series with the coil is

(a) 99995 (b) 9995 (c) 10^3 (d) 10^5

54. The resistances of three parts of a circular loop is shown in the figure. The magnetic field at the centre O is

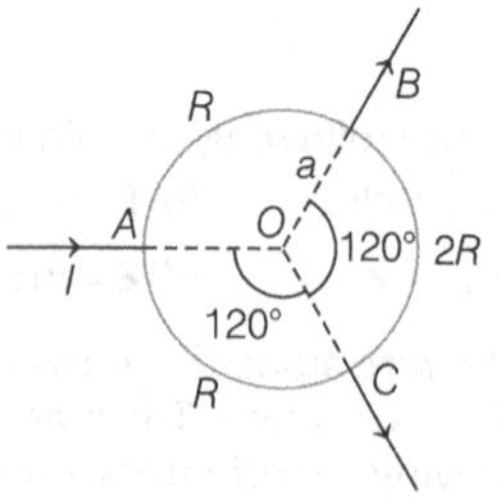

(a) $\frac{\mu_0 I}{6a}$ (b) $\frac{\mu_0 I}{3a}$

(c) $\frac{2\mu_0 I}{3a}$ (d) zero

Multiple Correct Options

55. Two circular coils of radii 5 cm and 10 cm carry currents of 2 A. The coils have 50 and 100 turns respectively and are placed in such a way that their planes as well as their centres coincide. Magnitude of magnetic field at the common centre of coils is

(a) $8\pi\times 10^{-4}$ T if currents in the coils are in same sense

(b) $4\pi\times 10^{-4}$ T if currents in the coils are in opposite sense

(c) zero if currents in the coils are in opposite sense

(d) $8\pi\times 10^{-4}$ T if currents in the coils are in opposite sense

56. A charged particle of unit mass and unit charge moves with velocity $\mathbf{v}=(8\hat{i}+6\hat{j})$ m/s in a magnetic field of $\mathbf{B}=2\hat{k}$T. Choose the correct alternative(s).

(a) The path of the particle may be $x^2+y^2-4x-21=0$

(b) The path of the particle may be $x^2+y^2=25$

(c) The path of the particle may be $y^2+z^2=25$

(d) The time period of the particle is approximately equal to 3.14 s

57. When a current carrying coil is placed in a uniform magnetic field with its magnetic moment anti-parallel to the field, then

(a) torque on it is maximum

(b) torque on it is zero

(c) potential energy is maximum

(d) dipole is in unstable equilibrium

58. A charged particle is moving along positive y-axis in uniform electric and magnetic fields

$$\mathbf{E}=E_0\hat{k} \text{ and } \mathbf{B}=B_0\hat{i}$$

Here, E_0 and B_0 are positive constants. Choose the correct options.

(a) particle may be deflected towards positive z-axis

(b) particle may be deflected towards negative z-axis

(c) particle may pass undeflected

(d) kinetic energy of particle may remain constant

59. A charged particle revolves in circular path in uniform magnetic field after accelerating by a potential difference of V volts. Choose the correct options if V is doubled.

(a) kinetic energy of particle will become two times

(b) radius in circular path will become two times

(c) radius in circular path will become $\sqrt{2}$ times

(d) angular velocity will remain unchanged

60. $abcd$ is a square. There is a current I in wire efg as shown.

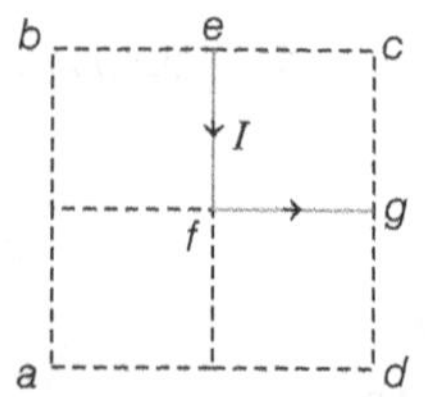

Choose the correct options.

(a) Net magnetic field at a is inwards

(b) Net magnetic field at b is zero

(c) Net magnetic field at c is outwards

(d) Net magnetic field at d is inwards

61. An electron is moving along the positive X-axis. You want to apply a magnetic field for a short time, so that the electron may reverse its direction and move parallel to the negative X-axis. This can be done by applying the magnetic field along
(a) Y-axis (b) Z-axis
(c) Y-axis only (d) Z-axis only

62. Let ε_0 denotes the dimensional formula of the permittivity of the vacuum, and μ_0 that of the permeability of the vacuum. If M = mass, L = length, T = time and I = electric current
(a) $\varepsilon_0 = [M^{-1}L^{-3}T^2 I]$
(b) $\varepsilon_0 = [M^{-1}L^{-3}T^4 I]$
(c) $\mu_0 = [MLT^{-2}I^{-2}]$
(d) $\mu_0 = [ML^2T^{-1}I]$

63. Consider three quantities $x = E/B$, $y = \sqrt{1/\mu_0\varepsilon_0}$ and $z = \frac{l}{CR}$. Here, l is the length of a wire, C is a capacitance and R is a resistance. All other symbols have standard meanings.
(a) x, y have the same dimensions
(b) y, z have the same dimensions
(c) z, x have the same dimensions
(d) None of the three pairs have the same dimensions

64. H^+, He^+ and O^{2+} all having the same kinetic energy pass through a region in which there is a uniform magnetic field perpendicular to their velocity. The masses of H^+, He^+ and O^{2+} are 1 amu, 4 amu and 16 amu respectively. Then, **(1994 JEE)**
(a) H^+ will be deflected most
(b) O^{2+} will be deflected most
(c) He^+ and O^{2+} will be deflected equally
(d) all will be deflected equally

65. From a cylinder of radius R, a cylinder of radius $R/2$ is removed as shown in the figure. Current flowing in the remaining cylinder is I. Magnetic field strength is

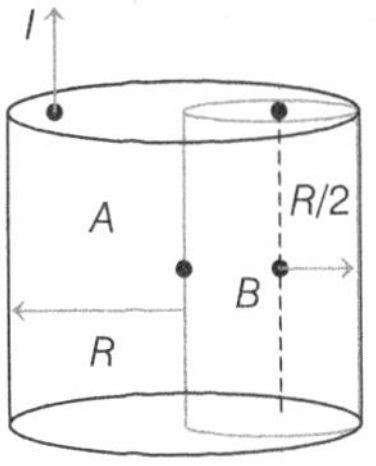

(a) zero at point A (b) zero at point B
(c) $\frac{\mu_0 I}{3\pi R}$ at point A (d) $\frac{\mu_0 I}{3\pi R}$ at point B

Passage Based Questions

Passage (Q.66 to 67)

The figure shows a circular loop of radius a with two long parallel wires (numbered 1 and 2) all in the plane of the paper. The distance of each wire from the centre of the loop is d. The loop and the wires are carrying the same current I. The current in the loop is in the counter-clockwise direction if seen from above. **(2014 Adv.)**

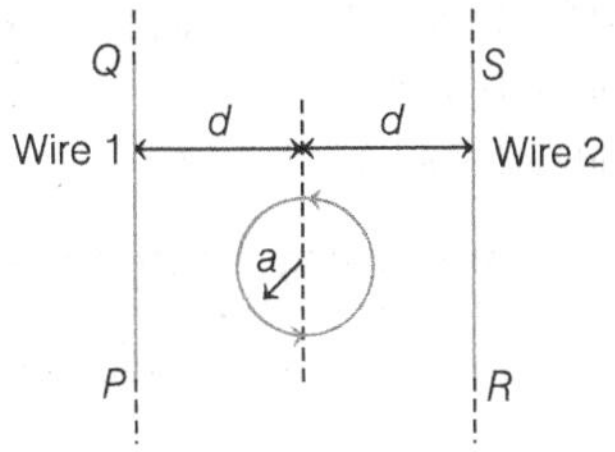

66. When $d \approx a$ but wires are not touching the loop, it is found that the net magnetic field on the axis of the loop is zero at a height h above the loop. In that case
(a) current in wire 1 and wire 2 is in the directions PQ and RS, respectively and $h \approx a$
(b) current in wire 1 and wire 2 is in the directions PQ and SR, respectively and $h \approx a$
(c) current in wire 1 and wire 2 is in the directions PQ and SR, respectively and $h \approx 1.1a$
(d) current in wire 1 and wire 2 is in the directions PQ and RS, respectively and $h \approx 1.1a$

67. Consider $d >> a$, and the loop is rotated about its diameter parallel to the wires by 30° from the position shown in the figure. If the currents in the wires are in the opposite directions, the torque on the loop at its new position will be (assume that the net field due to the wires is constant over the loop)
(a) $\frac{\mu_0 I^2 a^2}{d}$ (b) $\frac{\mu_0 I^2 a^2}{2d}$
(c) $\frac{\sqrt{3}\mu_0 I^2 a^2}{d}$ (d) $\frac{\sqrt{3}\mu_0 I^2 a^2}{2d}$

Match the Columns

68. An electron is moving towards positive x-direction. Match the following two columns for deflection of electron just after the fields are switched on. (E_0 and B_0 are positive constants)

	Column I	Column II
(a)	If magnetic field $\mathbf{B} = B_0\hat{\mathbf{j}}$ is switched on	(p) Negative y-axis
(b)	If magnetic field $\mathbf{B} = B_0\hat{\mathbf{k}}$ is switched on	(q) Positive y-axis
(c)	If magnetic field $\mathbf{B} = B_0\hat{\mathbf{i}}$ and electric field $\mathbf{E} = E_0\hat{\mathbf{j}}$ are switched on	(r) Negative z-axis
(d)	If electric field $\mathbf{E} = E_0\hat{\mathbf{k}}$ is switched on	(s) Positive z-axis

69. Four charged particles, $(-q, m)$, $(-3q, 4m)$, $(+q, m)$ and $(+2q, m)$ enter in uniform magnetic field (in inward direction) with same kinetic energy as shown in figure. Inside the magnetic field their paths are shown. Match the following two columns.

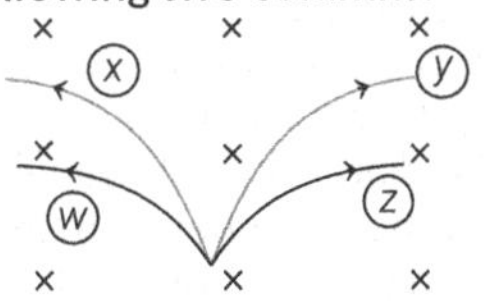

Column I		Column II	
(a)	Particle $(-q, m)$	(p)	w
(b)	Particle $(-3q, 4m)$	(q)	x
(c)	Particle $(+q, m)$	(r)	y
(d)	Particle $(+2q, m)$	(s)	z

70. Equal currents are flowing in two infinitely long wires lying along x and y-axes in the directions shown in figure. Match the following two columns.

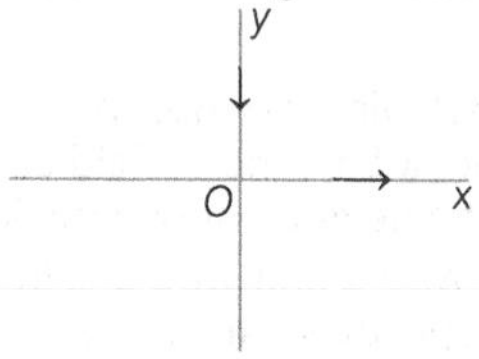

Column I		Column II	
(a)	Magnetic field at (a, a)	(p)	along positive y-axis
(b)	Magnetic field at $(-a, -a)$	(q)	along positive z-axis
(c)	Magnetic field at $(a, -a)$	(r)	along negative z-axis
(d)	Magnetic field at $(-a, a)$	(s)	zero

71. A square loop of uniform conducting wire is as shown in figure. A current I (in ampere) enters the loop from one end and exits the loop from opposite end as shown in figure.

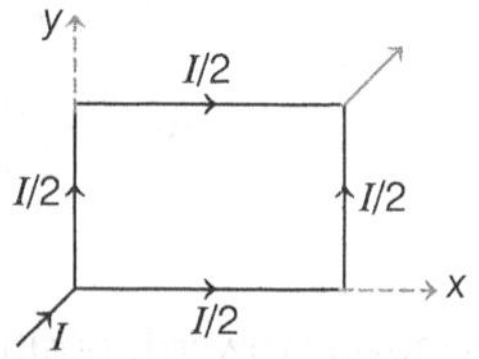

The length of one side of square loop is l metre. The wire has uniform cross-section area and uniform linear mass density.

Column I		Column II	
(a)	$\mathbf{B} = B_0\hat{\mathbf{i}}$ in tesla	(p)	magnitude of net force on loop is $\sqrt{2}B_0Il$ newton
(b)	$\mathbf{B} = B_0\hat{\mathbf{j}}$ in tesla	(q)	magnitude of net force on loop is zero
(c)	$\mathbf{B} = B_0(\hat{\mathbf{i}} + \hat{\mathbf{j}})$ in tesla	(r)	magnitude of force on loop is $2B_0Il$
(d)	$\mathbf{B} = B_0\hat{\mathbf{k}}$ in tesla	(s)	magnitude of net force on loop is B_0Il newton

72. Column II gives four situations in which three or four semi-infinite current carrying wires are placed in XY-plane as shown in the figure. The magnitude of the direction of current is shown in each figure. Column I gives statements regarding the x and y-components of magnetic field at a point P whose coordinates are $(0, 0, d)$. Match the statements in Column I with the corresponding figures in Column II.

Column I		Column II	
(a)	The x-component of magnetic field at point P is zero in	(p)	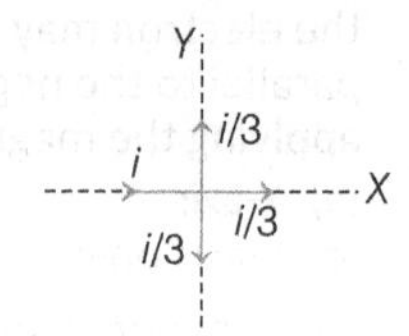
(b)	The z-component of magnetic field at point P is zero in	(q)	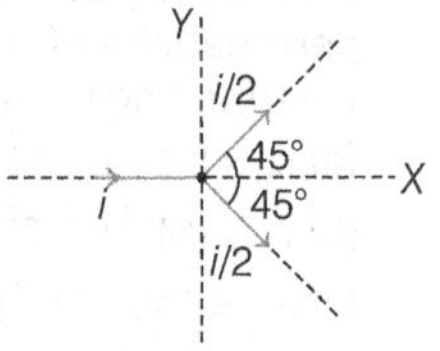
(c)	The magnitude of magnetic field at point P is $\frac{\mu_0 i}{4\pi d}$ in	(r)	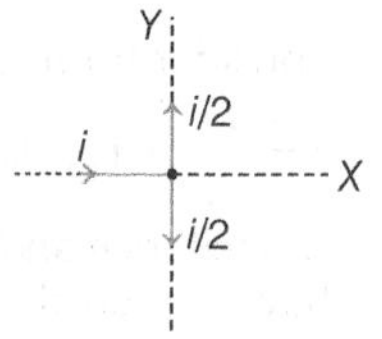
(d)	The magnitude of magnetic field at point P is less than $\frac{\mu_0 i}{2\pi d}$ in	(s)	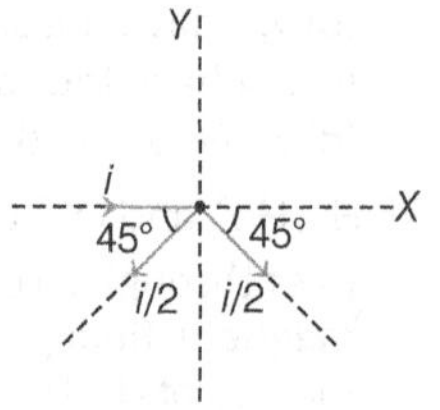

Directions (Q.Nos. 73-75) Match the information given in the three columns of the following table.

A charged particle (electron or proton) is introduced at the origin $(x = 0, y = 0, z = 0)$ with a given initial velocity $\mathbf{v}$. A uniform electric field $\mathbf{E}$ and a uniform magnetic field $\mathbf{B}$ exist everywhere. The velocity $\mathbf{v}$, electric field $\mathbf{E}$ and magnetic field $\mathbf{B}$ are given in columns 1, 2 and 3, respectively. The quantities E_0, B_0 are positive in magnitude. **(2017 Adv.)**

Column 1		Column 2		Column 3	
(I)	Electron with $\mathbf{v} = 2\frac{E_0}{B_0}\hat{\mathbf{x}}$	(i)	$\mathbf{E} = E_0\hat{\mathbf{z}}$	(P)	$\mathbf{B} = -B_0\hat{\mathbf{x}}$
(II)	Election with $\mathbf{v} = \frac{E_0}{B_0}\hat{\mathbf{y}}$	(ii)	$\mathbf{E} = -E_0\hat{\mathbf{y}}$	(Q)	$\mathbf{B} = B_0\hat{\mathbf{x}}$
(III)	Proton with $\mathbf{v} = 0$	(iii)	$\mathbf{E} = -E_0\hat{\mathbf{x}}$	(R)	$\mathbf{B} = B_0\hat{\mathbf{y}}$
(IV)	Proton with $\mathbf{v} = 2\frac{E_0}{B_0}\hat{\mathbf{x}}$	(iv)	$\mathbf{E} = E_0\hat{\mathbf{x}}$	(S)	$\mathbf{B} = B_0\hat{\mathbf{z}}$

73. In which case would the particle move in a straight line along the negative direction of Y-axis (i.e. move along $-\hat{\mathbf{y}}$)?

(a) (IV) (ii) (S)
(b) (II) (iii) (Q)
(c) (III), (ii) (R)
(d) (III) (ii) (P)

74. In which case will the particle move in a straight line with constant velocity?
(a) (II) (iii) (S) (b) (III) (iii) (P)
(c) (IV) (i) (S) (d) (III) (ii) (R)

75. In which case will the particle describe a helical path with axis along the positive z-direction?
(a) (II) (ii) (R) (b) (III) (iii) (P)
(c) (IV) (i) (S) (d) (IV) (ii) (R)

Subjective Questions

76. An equilateral triangular frame with side a carrying a current I is placed at a distance a from an infinitely long straight wire carrying a current I as shown in the figure. One side of the frame is parallel to the wire. The whole system lies in the xy-plane. Find the magnetic force **F** acting on the frame.

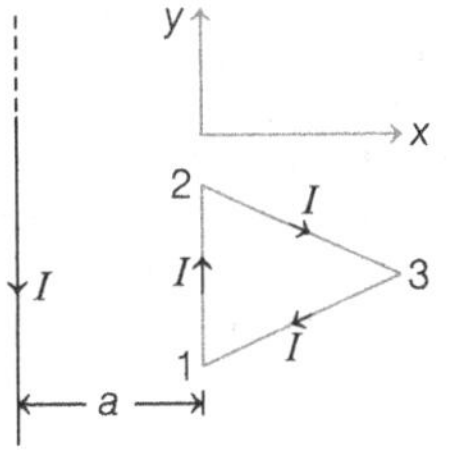

77. Find an expression for the magnetic dipole moment and magnetic field induction at the centre of a Bohr's hypothetical hydrogen atom in the nth orbit of the electron in terms of universal constants.

78. A square loop of side 6 cm carries a current of 30 A. Calculate the magnitude of magnetic field **B** at a point P lying on the axis of the loop and a distance $\sqrt{7}$ cm from centre of the loop.

79. A positively charged particle of charge 1 C and mass 40 g, is revolving along a circle of radius 40 cm with velocity 5 m/s in a uniform magnetic field with centre at origin O in xy-plane. At $t = 0$, the particle was at (0, 0.4 m, 0) and velocity was directed along positive x-direction. Another particle having charge 1 C and mass 10 g moving uniformly parallel to z-direction with velocity $\frac{40}{\pi}$ m/s collides with revolving particle at $t = 0$ and gets stuck with it. Neglecting gravitational force and colombians force, calculate x, y and z- coordinates of the combined particle at $t = \frac{\pi}{40}$ sec.

80. A positively charged particle having charge q is accelerated by a potential difference V. This particle moving along the x-axis enters a region where an electric field E exists. The direction of the electric field is along positive y-axis. The electric field exists in the region bounded by the lines $x = 0$ and $x = a$. Beyond the line $x = a$ (i.e. in the region $x \geq a$) there exists a magnetic field of strength B, directed along the positive y-axis. Find

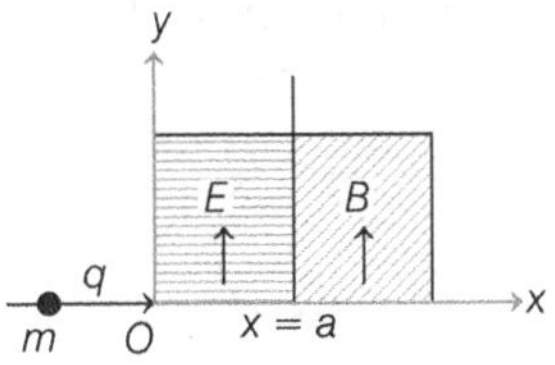

(a) at which point does the particle meet the line $x = a$
(b) the pitch of the helix formed after the particle enters the region $x \geq a$. Mass of the particle is m.

81. A charged particle having charge 10^{-6} C and mass of 10^{-10} kg is fired from the middle of the plate making an angle 30° with plane of the plate. Length of the plate is 0.17 m and it is separated by 0.1 m. Electric field $E = 10^{-3}$ N/C is present between the plates. Just outside the plates magnetic field is present. Find the velocity of projection of charged particle and magnitude of the magnetic field perpendicular to the plane of the figure, if it has to graze the plate at C and A parallel to the surface of the plate. (Neglect gravity)

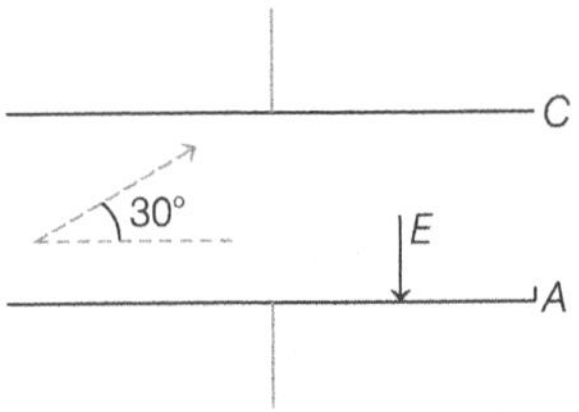

82. A circular loop of radius R is bent along a diameter and given a shape as shown in figure. One of the semicircles (KNM) lies in the xz-plane and the other one (KLM) in the yz-plane with their centres at origin. Current I is flowing through each of the semicircles as shown in figure. **(2000 JEE)**

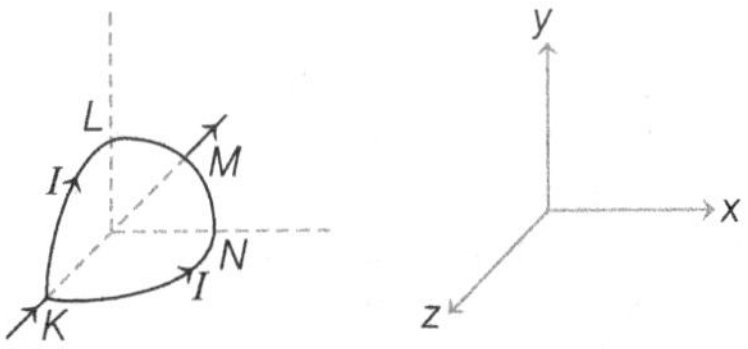

(a) A particle of charge q is released at the origin with a velocity $\mathbf{v} = -v_0\hat{\mathbf{i}}$. Find the instantaneous force **F** on the particle. Assume that space is gravity free.
(b) If an external uniform magnetic field $B_0\hat{\mathbf{j}}$ is applied, determine the force $\mathbf{F}_1$ and $\mathbf{F}_2$ on the semicircles KLM and KNM due to the field and the net force **F** on the loop.

83. A long cylindrical conductor of radius a has two cylindrical cavities of diameter a through its entire length as shown in cross-section in figure. A current I is directed out of the page and is uniform throughout the cross-section of the conductor. Find the magnitude and direction of the magnetic field in terms of μ_0, I, r and a.

(a) at point P_1 and (b) at point P_2

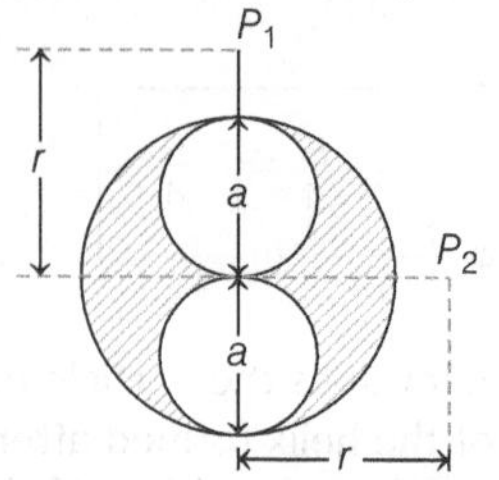

84. A uniform constant magnetic field B is directed at an angle of 45° to the x-axis in xy-plane. $PQRS$ is a rigid square wire frame carrying a steady current I_0, with its centre at the origin O. At time $t = 0$, the frame is at rest in the position shown in the figure with its sides parallel to x and y-axis. Each side of the frame has mass M and length L. **(1998 JEE)**

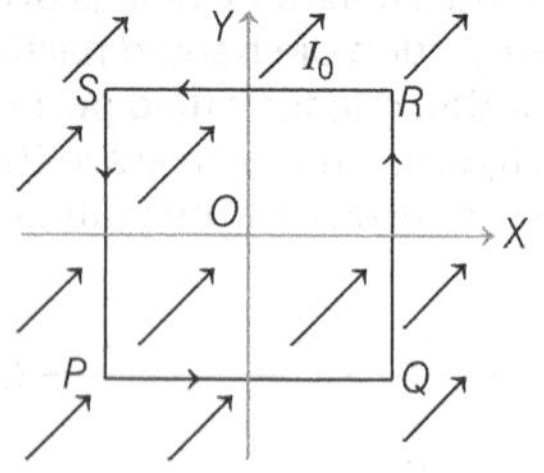

(a) What is the magnitude of torque τ acting on the frame due to the magnetic field?

(b) Find the angle by which the frame rotates under the action of this torque in a short interval of time Δt, and the axis about which the rotation occurs (Δt is so short that any variation in the torque during this interval may be neglected). Given : The moment of inertia of the frame about an axis through its centre perpendicular to its plane is $\frac{4}{3}ML^2$.

85. A ring of radius R having uniformly distributed charge Q is mounted on a rod suspended by two identical strings. The tension in strings in equilibrium is T_0. Now, a vertical magnetic field is switched on and ring is rotated at constant angular velocity ω. Find the maximum value of ω with which the ring can be rotated if the strings can withstand a maximum tension of $\frac{3T_0}{2}$. **(2003 JEE)**

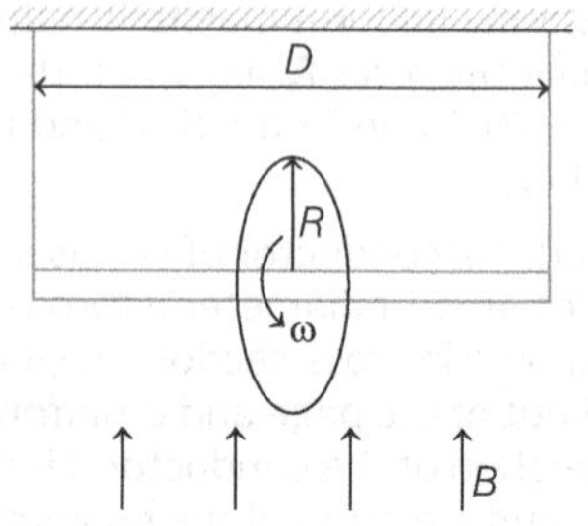

86. Figure shows a cross-section of a long ribbon of width ω that is carrying a uniformly distributed total current i into the page. Calculate the magnitude and direction of the magnetic field **B** at a point P in the plane of the ribbon at a distance d from its edge.

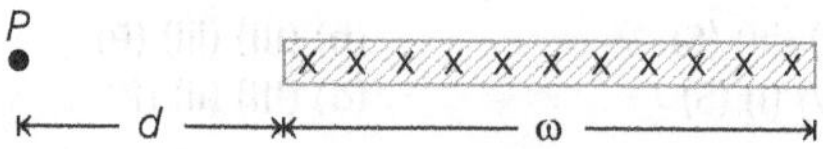

87. A particle of mass m having a charge q enters into a circular region of radius R with velocity v directed towards the centre. The strength of magnetic field is B. Find the deviation in the path of the particle.

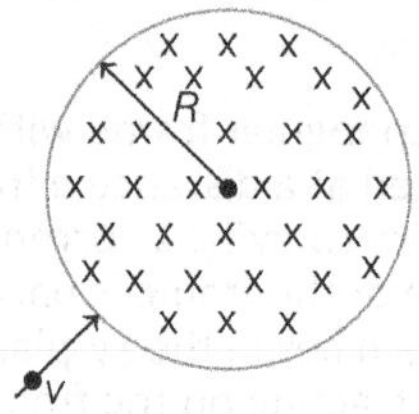

88. A thin, uniform rod with negligible mass and length 0.2 m is attached to the floor by a frictionless hinge at point P. A horizontal spring with force constant $k = 4.80$ N/m connects the other end of the rod with a vertical wall. The rod is in a uniform magnetic field $B = 0.340$ T directed into the plane of the figure. There is current $I = 6.50$ A in the rod, in the direction shown.

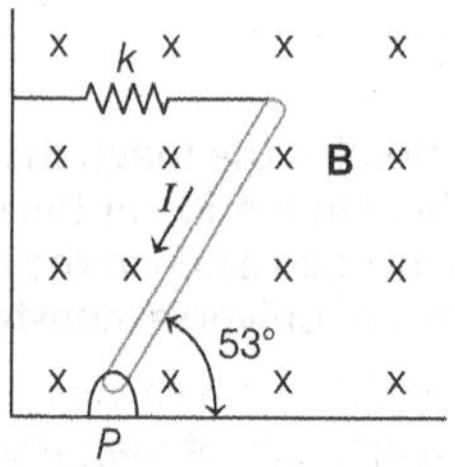

(a) Calculate the torque due to the magnetic force on the rod, for an axis at P. Is it correct to take the total magnetic force to act at the centre of gravity of the rod when calculating the torque?

(b) When the rod is in equilibrium and makes an angle of 53.0° with the floor, is the spring stretched or compressed?

(c) How much energy is stored in the spring when the rod is in equilibrium?

89. A rectangular loop $PQRS$ made from a uniform wire has length a, width b and mass m. It is free to rotate about the arm PQ, which remains hinged along a horizontal line taken as the y-axis (see figure). Take the vertically upward direction as the z-axis. A uniform magnetic field $\mathbf{B} = (3\hat{i} + 4\hat{k})B_0$ exists in the region. The loop is held in the xy-plane and a current I is passed through it. The loop is now released and is found to stay in the horizontal position in equilibrium. **(2002 JEE)**

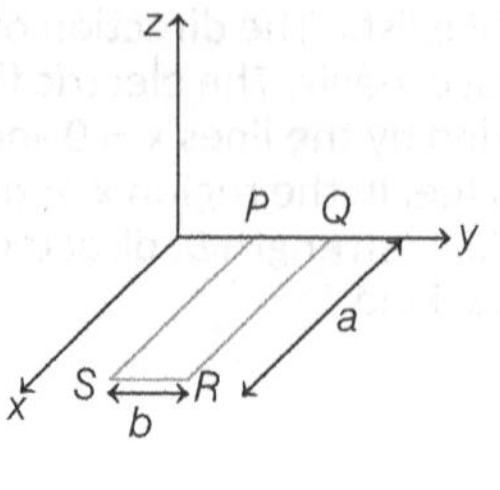

(a) What is the direction of the current I in PQ?
(b) Find the magnetic force on the arm RS.
(c) Find the expression for I in terms of B_0, a, b and m.

90. A particle of mass 1×10^{-26} kg and charge $+1.6 \times 10^{-19}$ C travelling with a velocity 1.28×10^6 m/s in the +x direction enters a region in which a uniform electric field E and a uniform magnetic field of induction B are present such that $E_x = E_y = 0$, $E_z = -102.4$ kV/m and $B_x = B_z = 0$, $B_y = 8 \times 10^{-2}$ T. The particle enters this region at the origin at time $t = 0$. Determine the location (x, y and z coordinates) of the particle at $t = 5 \times 10^{-6}$ s. If the electric field is switched off at this instant (with the magnetic field still present), what will be the position of the particle at $t = 7.45 \times 10^{-6}$ s? **(1982 JEE)**

91. A potential difference of 600 V is applied across the plates of a parallel plate condenser. The separation between the plates is 3 mm. An electron projected vertically, parallel to the plates, with a velocity of 2×10^6 m/s moves undeflected between the plates. Find the magnitude and direction of the magnetic field in the region between the condenser plates. (Neglect the edge effects). (Charge of the electron $= 1.6 \times 10^{-19}$ C) **(1981 JEE)**

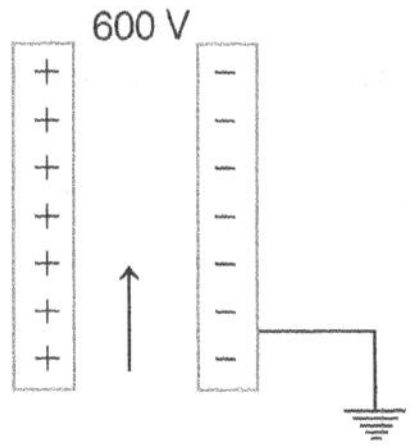

92. A coil carrying a current $I = 10$ mA is placed in a uniform magnetic field so that its axis coincides with the field direction. The single-layer winding of the coil is made of copper wire with diameter $d = 0.10$ mm, radius of turns is equal to $R = 30$ mm. At what value of the induction of the external magnetic field can the coil winding be ruptured? **(Irodov 3.252)**

93. A non-relativistic proton beam passes without deviation through the region of space where there are uniform transverse mutually perpendicular electric and magnetic fields with $E = 120$ kV/m and $B = 50$ mT. Then the beam strikes a grounded target. Find the force with which the beam acts on the target if the beam current is equal to $I = 0.80$ m. **(Irodov 3.389)**

94. Find the magnetic induction at the point O if the wire carrying a current $I = 8.0$ A has the shape shown in Fig. (a), (b), (c).
The radius of the curved part of the wire is $R = 100$ mm, the linear parts of the wire are very long. **(Irodov 3.228)**

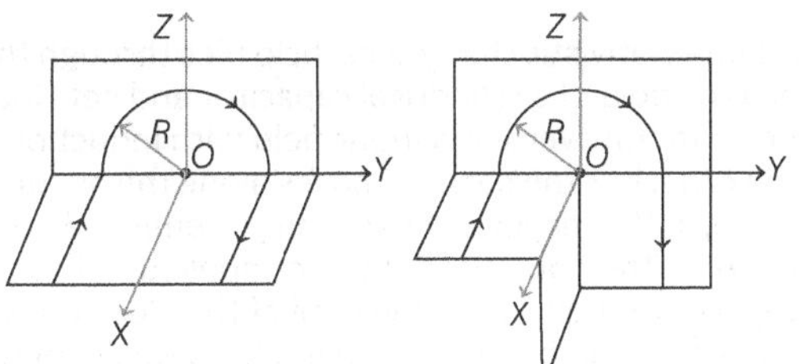

(a) (b)

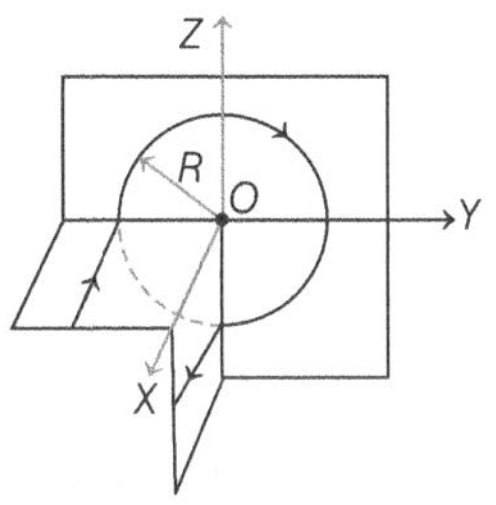

(c)

95. Inside a long straight uniform wire of round cross-section there is a long round cylindrical cavity whose axis is parallel to the axis of the wire and displaced from the latter by a distance l. A direct current of density **J** flows along the wire. Find the magnetic induction inside the cavity. Consider, in particular, the case $l = 0$. **(Irodov 3.234)**

96. A non-conducting sphere of radius $R = 50$ mm charged uniformly with surface density $\sigma = 10.0\ \mu\text{C/m}^2$ rotates with an angular velocity $\omega = 70$ rad/s about the axis passing through its centre. Find the magnetic induction at the centre of the sphere. **(Irodov 3.247)**

97. A slightly divergent beam of non-relativistic charged particles accelerated by a potential difference V propagates from a point A along the axis of a straight solenoid. The beam is brought into focus at a distance l from the point A at two successive values of magnetic induction B_1 and B_2. Find the specific charge q/m of the particles. **(Irodov 3.383)**

98. A non-relativistic electron originates at a point A lying on the axis of a straight solenoid and moves with velocity v at an angle α to the axis. The magnetic induction of the field is equal to B. Find the distance r from the axis to the point on the screen into which the electron strikes. The screen is oriented at right angles to the axis and is located at a distance l from the point A. **(Irodov 3.384)**

99. From the surface of a round wire of radius a carrying a direct current I an electron escapes with a velocity v_0 perpendicular to the surface. Find what will be the maximum distance of the electron from the axis of the wire before it turns back due to the action of the magnetic field generated by the current. **(Irodov 3.385)**

100. A non-relativistic charged particle flies through the electric field of a cylindrical capacitor and gets into a uniform transverse magnetic field with induction B. In the capacitor the particle moves along the arc of a circle, in the magnetic field, along a semi-circle of radius r. The potential difference applied to the capacitor is equal to V, the radii of the electrodes are equal to a and b, with $a < b$. Find the velocity of the particle and its specific charge q/m. **(Irodov 3.386)**

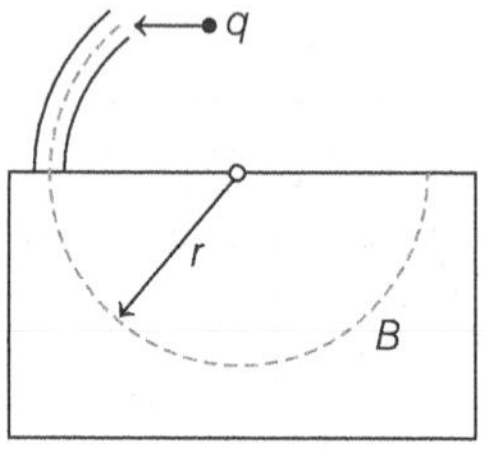

101. A beam of non-relativistic charged particles moves without deviation through the region of space A where there are transverse mutually perpendicular electric and magnetic fields with strength E and induction B. When the magnetic field is switched OFF, the trace of the beam on the screen S shifts by Δx. Knowing the distances a and b, find the specific charge q/m of the particles. **(Irodov 3.391)**

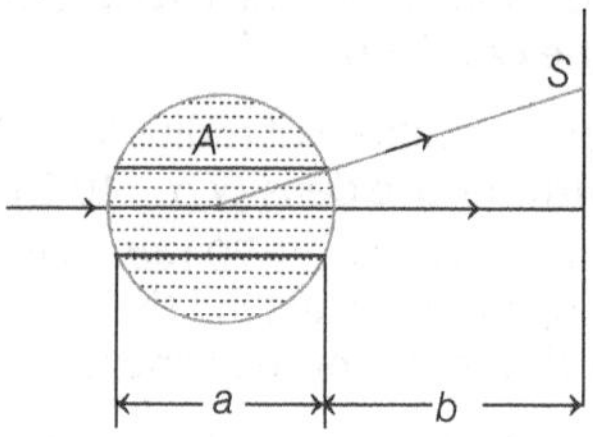

Integer Type Questions

102. As shown in the figure, three sided frame is pivoted at P and Q and hangs vertically. Its sides are of same length and have a linear density of $\sqrt{3}$ kg/m. A current of $10\sqrt{3}$ A is sent through the frame, which is in a uniform magnetic field of 2T directed upwards as shown in the figure. Then, angle through which the frame will be deflected in equilibrium is $\frac{\pi}{k}$. Find the value of k. (Given, $g = 10\ \text{m/s}^2$)

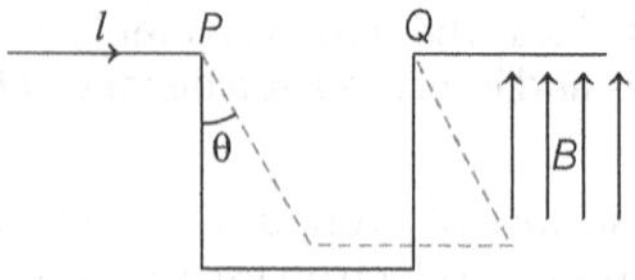

103. A non-uniform magnetic field $\mathbf{B} = B_0\left(1 + \frac{y}{d}\right)(-\hat{\mathbf{k}})$ is present in region of space in between $y = 0$ and $y = d$. The lines are shown in the figure. A particle of mass m and positive charge q is moving. Given, an initial velocity $\mathbf{v} = v_0\hat{\mathbf{i}}$. The x-component of velocity of the particle is $v_x = v_0 - \frac{kqB_0 d}{2m}$ when it leaves the field, then find the value of k.

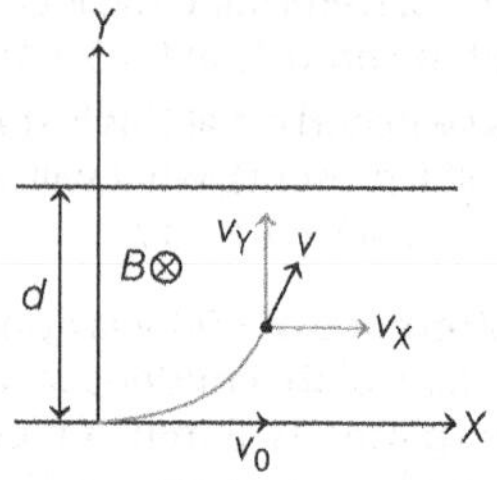

104. Two parallel wires in the plane of the paper are distance X_0 apart. A point charge is moving with speed u between the wires in the same plane at a distance X_1 from one of the wires. When the wires carry current of magnitude I in the same direction, the radius of curvature of the path of the point charge is R_1. In contrast, if the current I in the two wires have directions opposite to each other, the radius of curvature of the path is R_2. If $\frac{X_0}{X_1} = 3$, then find the value of $\frac{R_1}{R_2}$.

105. A cylindrical cavity of diameter a exists inside a cylinder of diameter $2a$ as shown in the figure. Both the cylinder and the cavity are infinitely long. A uniform current density J flows along the length. If the magnitude of the magnetic field at the point P is given by $\frac{N}{12}\mu_0 aJ$, then the value of N is **(2012 JEE)**

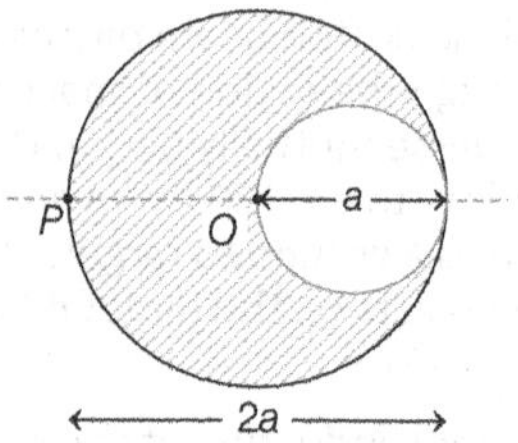

Answers

Level 1

1. 14.02 **2.** (d) **3.** (c) **4.** (b) **5.** (a) **6.** (d) **7.** (c) **8.** (d) **9.** (a) **10.** (d)
11. (b) **12.** (a) **13.** (a) **14.** (d) **15.** (d) **16.** (b) **17.** (a) **18.** $\frac{\mu_0 I_1 I_2 L}{2\pi a}$, zero

19. 20 μT toward the bottom of the square **20.** $B = \frac{n\mu_0 I \tan(\pi/n)}{2\pi R}$, $B = \frac{\mu_0 I}{2R}$, for $n \to \infty$ **21.** $B = \frac{\mu_0 hI}{4\pi^2 Rr}$

22. $B = \frac{\mu_0 I}{\pi^2 R}$ **23.** $B = \frac{\mu_0 NI}{l\sqrt{1+\left(\frac{2R}{l}\right)^2}}$ **24.** $M = 30$ mA-metre^2 **25.** (a) $B = 7\ \mu$T (b) $M = 15$ mA-metre^2

26. (a) $B = \frac{\mu_0 \sigma\omega R}{2}$ (b) $M = \frac{\pi\sigma\omega R^4}{4}$ **27.** $M = \frac{qR^2\omega}{5}$, $\frac{M}{L} = \frac{q}{2m}$ **28.** (a) $F = 0.20$ mN/metre (b) $F = 0.13$ mN/metre

29. $B = 0.4$ T **30.** $U = -\left(\frac{\mu_0 \pi a^2}{2b}\right)(I_1 I_2)\cos\theta$

Level 2

31. (a) **32.** (d) **33.** (b) **34.** (c) **35.** (d) **36.** (b) **37.** (a) **38.** (a) **39.** (c) **40.** (b)
41. (b) **42.** (b) **43.** (b) **44.** (d) **45.** (a) **46.** (a) **47.** (b) **48.** (a) **49.** (b) **50.** (a)
51. (d) **52.** (a) **53.** (b) **54.** (d) **55.** (a,c) **56.** (a,b,d) **57.** (b,c,d) **58.** (a,b,c,d) **59.** (a,c,d) **60.** (a,c,d)
61. (a,b) **62.** (b,c) **63.** (a,b,c) **64.** (a,c) **65.** (c,d) **66.** (c) **67.** (b) **68.** (a $\to$ r; b $\to$ q; c $\to$ p; d $\to$ r)
69. (a $\to$ r; b $\to$ s; c $\to$ q; d $\to$ p) **70.** (a $\to$ q; b $\to$ r; c $\to$ s; d $\to$ s) **71.** (a $\to$ s; b $\to$ s; c $\to$ q; d $\to$ p)
72. (a $\to$ p,q,r; b $\to$ p,q,r,s; c $\to$ r; d $\to$ p,q,r,s) **73.** (c) **74.** (a) **75.** (c)

76. $\mathbf{F} = \frac{\mu_0 I^2}{\pi}\left[\frac{1}{2} - \frac{1}{\sqrt{3}}\ln\left(\frac{2+\sqrt{3}}{2}\right)\right](\hat{\mathbf{i}})$ **77.** $M = \frac{neh}{4\pi m}$, $B = \frac{\mu_0 \pi m^2 e^7}{8\varepsilon_0^3 h^5 n^5}$ **78.** 2.7×10^{-4} T

79. (0.2 m, 0.2 m, 0.2 m) **80.** (a) $y = \frac{Ea^2}{4V}$ (b) $p = \frac{\pi Ea}{B}\sqrt{\frac{2m}{qV}}$ **81.** 2.0 m/s, 3.46 mT

82. (b) $\frac{-\mu_0 q v_0 I}{4R}\hat{\mathbf{k}}$ (b) $\mathbf{F}_1 = \mathbf{F}_2 = 2B_0 IR\,\hat{\mathbf{i}}$, $\mathbf{F} = 4B_0 IR\,\hat{\mathbf{i}}$ **83.** (a) $\frac{\mu_0 I}{\pi r}\left(\frac{2r^2 - a^2}{4r^2 - a^2}\right)$ to the left (b) $\frac{\mu_0 I}{\pi r}\left(\frac{2r^2 + a^2}{4r^2 + a^2}\right)$ towards the top of the page

84. (a) $I_0 L^2 B$ (b) $\frac{3}{4}\frac{I_0 B}{M}(\Delta t)^2$ **85.** $\omega_{max} = \frac{DT_0}{BQR^2}$ **86.** $B = \frac{\mu_0}{2\pi}\frac{i}{\omega}\ln\left(\frac{d+\omega}{d}\right)$ (upwards) **87.** $2\tan^{-1}\left(\frac{BqR}{mv}\right)$

88. (a) 0.0442 N-m, clockwise, yes (b) stretched (c) 7.8×10^{-3} J **89.** (a) P to Q (b) $IbB_0(3\hat{\mathbf{k}} - 4\hat{\mathbf{i}})$ (c) $\frac{mg}{6bB_0}$

90. (6.4m, 0,0), (6.4m, 0, 2m) **91.** (0.1 T, negative z-direction) **92.** $B_{max} = 8$kT **93.** $F = 20\ \mu$N

94. (a) $B = 0.3\ \mu$T (b) $B = 0.34\ \mu$T (c) $B = 0.11\ \mu$T **95.** $\mathbf{B} = \frac{\mu_0}{2}[\mathbf{J} \times \mathbf{l}]$, $\mathbf{B} = 0$ when $l = 0$ **96.** $B = 29$ pT

97. $\frac{q}{m} = \frac{8\pi^2 V}{l^2(B_2 - B_1)^2}$ **98.** $r = 2\rho\,|\sin(\phi/2)|$, where $\rho = \frac{mv}{eB}\sin\alpha$, $\phi = \frac{leB}{mv\cos\alpha}$ **99.** $x_m = ae^{v_0/b}$, where $b = \frac{\mu_0}{2\pi}\frac{e}{m}I$

100. $v = \frac{V}{rB\ln(b/a)}$, $\frac{q}{m} = \frac{V}{r^2B^2\ln(b/a)}$ **101.** $\frac{q}{m} = \frac{a(a+2b)B^2}{2E\Delta x}$ **102.** (4) **103.** (3) **104.** (3) **105.** (5)

See hints on pages 450-465.

Hints & SOLUTIONS

Module 1.

1. $|\mathbf{v}| = \sqrt{2}v_0$ = speed, which always remains constant.

2. (a) $\mathbf{B} \perp \mathbf{v}$, so it may along y-axis

(b) $\mathbf{F} \perp \mathbf{v}$,

$\therefore \quad \mathbf{a} \perp \mathbf{v} = 0$

or $\quad \mathbf{a} \cdot \mathbf{v} = 0$ or $a_1 b_1 = a_2 b_2$

(c) See the logic of option (a).

(d) Magnetic force cannot change the kinetic energy of a particle.

3. Let the magnetic field is

$$\mathbf{B} = B_1\hat{\mathbf{i}} + B_2\hat{\mathbf{j}} + B_3\hat{\mathbf{k}}$$

Applying $\mathbf{F}_m = q(\mathbf{v} \times \mathbf{B})$ two times, we have

$$q[-\hat{\mathbf{j}} + \hat{\mathbf{k}}] = q[(\hat{\mathbf{i}}) \times (B_1\hat{\mathbf{i}} + B_2\hat{\mathbf{j}} + B_3\hat{\mathbf{k}})]$$
$$= q[B_2\hat{\mathbf{k}} - B_3\hat{\mathbf{j}}]$$

Comparing two sides, we get

$$B_2 = 1 \text{ and } B_3 = 1$$

Further, $q[\hat{\mathbf{i}} - \hat{\mathbf{k}}] = q[(\hat{\mathbf{j}}) \times (B_1\hat{\mathbf{i}} + B_2\hat{\mathbf{j}} + B_3\hat{\mathbf{k}})]$

$$= q[-B_1\hat{\mathbf{k}} + B_3\hat{\mathbf{i}}]$$

Again comparing, we get

$$B_1 = 1 \quad \text{and} \quad B_3 = 1$$
$$\mathbf{B} = (\hat{\mathbf{i}} + \hat{\mathbf{j}} + \hat{\mathbf{k}}) \text{ Wb/m}^2$$

4. Magnetic force may be non-zero. Hence, acceleration due to magnetic force may be non-zero.
Magnetic force is always perpendicular to velocity. Hence, its power is always zero or work done by magnetic force is always zero. Hence, it can be change the speed of charged particle.

5. $\because qE = Bqv \sin\theta$

$\therefore \quad [E/B] = [v] = [LT^{-1}]$

6. From the property of cross product, $\mathbf{F}$ is always perpendicular to both $\mathbf{v}$ and $\mathbf{B}$.

7. May be possible that $\theta = 0°$ or $180°$ between $\mathbf{v}$ and $\mathbf{B}$, so that $\mathbf{F}_m = 0$.

8. $\mathbf{F} = q(\mathbf{v} \times \mathbf{B})$

$\mathbf{F}$ is along position y-direction. q is negative and $\mathbf{v}$ is along positive x-direction. Therefore, $\mathbf{B}$ should be along positive z-direction.

9. $\mathbf{F} = q(\mathbf{v} \times \mathbf{B})$

Here, q has to be substituted with sign.

10. Apply Fleming's left hand rule.

11. $\because \quad F = Bqv \sin\theta$

$$\therefore \quad v = \frac{F}{Bq\sin\theta}$$
$$= \frac{4.6 \times 10^{-15}}{3.5 \times 10^{-3} \times 1.6 \times 10^{-19} \times \sin 60°}$$
$$= 9.47 \times 10^6 \text{ m/s}$$

12. $\because \quad F = Bqv \sin 90°$

$$= 0.8 \times 2 \times 1.6 \times 10^{-19} \times 10^5$$
$$= 2.56 \times 10^{-14} \text{ N}$$

13. $\mathbf{F} = q(\mathbf{v} \times \mathbf{B})$

14. (a) $\mathbf{F} = q(\mathbf{v} \times \mathbf{B})$

or $[(7.6 \times 10^{-3})\hat{\mathbf{i}} - (5.2 \times 10^{-3})\hat{\mathbf{k}}]$

$$= (7.8 \times 10^{-6})[(-3.8 \times 10^3)\hat{\mathbf{j}}(B_x\hat{\mathbf{i}} + B_y\hat{\mathbf{j}} + B_z\hat{\mathbf{k}})]$$

$\therefore \quad (7.8 \times 10^{-6})(3.8 \times 10^{-3})(B_x) = (-5.2 \times 10^{-3})$

$\therefore \quad B_x = -0.175$ T

Similarly,

$$(7.8 \times 10^{-6})(-3.8 \times 10^3)(B_z) = 7.6 \times 10^{-3}$$

or $\quad B_z = -0.256$ T

(c) From the property of cross product.
$\mathbf{F}$ is always perpendicular to $\mathbf{B}$.
Hence, $\mathbf{F} \cdot \mathbf{B} = 0$

15. Apply $\mathbf{F} = q(\mathbf{v} \times \mathbf{B})$

For example, let us apply for charged particle at e.

$$\mathbf{F}_e = q\left[\left(\frac{v}{\sqrt{2}}\hat{\mathbf{j}} - \frac{v}{\sqrt{2}}\hat{\mathbf{k}}\right) \times (B\hat{\mathbf{i}})\right]$$
$$= \frac{qvB}{\sqrt{2}}(-\hat{\mathbf{j}} - \hat{\mathbf{k}})$$

Module 2.

Introductory Exercise

1. Path C is undeviated. Therefore, it is of neutron's path. From Fleming's left hand rule magnetic force on positive charge will be leftwards and on negative charge is rightwards. Therefore, track D is of electron. Among A and B one is of proton and other of α-particle.

Further, $\quad r = \dfrac{mv}{Bq} \quad$ or $\quad r \propto \dfrac{m}{q}$

Since, $\quad \left(\dfrac{m}{q}\right)_\alpha > \left(\dfrac{m}{q}\right)_P$

$\therefore \quad r_\alpha > r_P$

or track B is of α-particle.

2. (a) $r = \dfrac{mv}{Bq}$

or $r \propto m$ as other factors are same.

(b) $f = \dfrac{Bq}{2\pi m}$ or $f \propto \dfrac{1}{m}$

3. $\because$ $r = \dfrac{\sqrt{2qVm}}{Bq}$

Exercise

1. $r = \dfrac{\sqrt{2km}}{Bq}$ or $r \propto \sqrt{m}$ (k, q and B are same)

$$m_p > m_e$$

$$\Rightarrow \quad r_p > r_e$$

2. The path will be a helix. Path is circle when it enters normal to the magnetic field.

3. $\because$ $\mathbf{F}_e + \mathbf{F}_m = 0$

$\therefore$ $q\mathbf{E} + q(\mathbf{v} \times \mathbf{B}) = 0$

$\therefore$ $\mathbf{E} = -(\mathbf{v} \times \mathbf{B})$

$= (\mathbf{B} \times \mathbf{v})$

4.

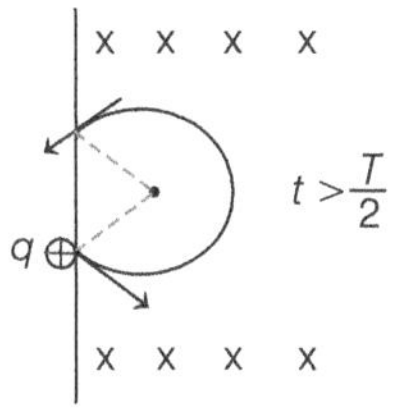

5. $r = \dfrac{P}{Bq}$

or $r \propto \dfrac{1}{q}$ (as P = constant)

6. $F = Bqv$ or $F \propto v$

Now, $v = \sqrt{\dfrac{2qV}{m}}$

$\therefore$ $F \propto \sqrt{V}$

7. Electric field (acting along $\hat{\mathbf{j}}$ direction) will change the velocity component which is parallel to $\mathbf{B}$ (which is also along $\hat{\mathbf{j}}$ direction).

$\mathbf{B} = B_0\hat{\mathbf{j}}$ and $\mathbf{v} = v_0\hat{\mathbf{j}}$ will rotate the particle in a circle. Hence, the net path is helical with variable pitch.

8. $r = \sqrt{\dfrac{2Km}{Bq}}$

$\therefore$ $r \propto \dfrac{\sqrt{K}}{B}$

9. $r = \dfrac{\sqrt{2qVm}}{Bq} = \left(\sqrt{\dfrac{2Vm}{q}}\right)\left(\dfrac{1}{B}\right)$

10. $q\mathbf{E} + q(\mathbf{v} \times \mathbf{B}) = 0$

$\Rightarrow$ $\mathbf{E} = \mathbf{B} \times \mathbf{v}$

or $\mathbf{E} \perp \mathbf{B}$

Further, $E = Bv \sin 90°$

11. Work is done only by the electrostatic force, which is equal to change in kinetic energy.

$\therefore$ $qE_0 d = \dfrac{1}{2}mv^2$

or $v = \sqrt{\dfrac{2qE_0 d}{m}}$

12. Kinetic energy or speed will increase only due to the work done by electric force which is acting along y-axis.

13. If q is positive, then due to electric field, their velocity will become towards positive x-axis.

Now, $\mathbf{F}_m = q(\mathbf{v} \times \mathbf{B})$

$\mathbf{v} \to$ towards $\hat{\mathbf{i}}$

$\mathbf{B} \to$ towards $\hat{\mathbf{k}}$

$\therefore$ $\mathbf{F}_m \to$ towards $-\hat{\mathbf{j}}$

Similarly, we can see with negative charge.

14. $(b-a)$ = radius of circular part

$= \dfrac{mv}{Bq}$

$\therefore$ $V = \dfrac{Bq(b-a)}{m}$

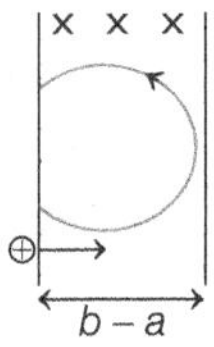

15. $\mathbf{F}_e = q\mathbf{E} = (1.6 \times 10^{-19})(-102.4 \times 10^3)\hat{\mathbf{k}}$

$= (-1.6384 \times 10^{-14}\,\hat{\mathbf{k}})$ N

$\mathbf{F}_m = q(\mathbf{v} \times \mathbf{B})$

$= (1.6 \times 10^{-19})[(1.28 \times 10^6\hat{\mathbf{i}}) \times (8 \times 10^{-2}\hat{\mathbf{j}})]$

$= (1.6384 \times 10^{-14}\hat{\mathbf{k}})$ N

Now, we can see that

$$\mathbf{F}_e + \mathbf{F}_m = 0$$

16. $y = 2$ (Radius)

17. (a) $r = \dfrac{mv}{Bq}$ $\Rightarrow$ $v = \dfrac{Bqr}{m}$

(b) $t = \dfrac{T}{2} = \dfrac{\pi m}{Bq}$

(c) $r = \dfrac{\sqrt{2qVm}}{Bq}$

$\therefore$ $V = \dfrac{r^2B^2q^2}{2qm} = \dfrac{r^2B^2q}{2m}$

18. Conservation of charge

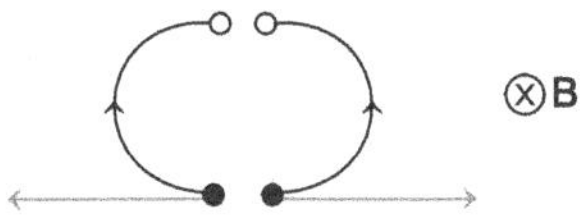

They will collide after time,

$$t = \frac{T}{2} = \frac{\pi m}{Bq}$$

19. (a) At A, magnetic force should be towards right. From Fleming's left hand rule, magnetic field should be inwards.

Further, $r = \dfrac{mv}{Bq}$

$\therefore \quad B = \dfrac{mv}{qr} \qquad (r = 5 \text{ cm})$

(b) $t_{AB} = \dfrac{T}{2} = \dfrac{\pi m}{Bq}$

20. $\mathbf{F}_e + \mathbf{F}_m = 0$

or $\quad q\mathbf{E} + q(\mathbf{v} \times \mathbf{B}) = 0$

or $\quad \mathbf{E} = -(\mathbf{v} \times \mathbf{B})$

or $\quad \mathbf{E} = (\mathbf{B} \times \mathbf{v})$

21. Work is done only by electrostatic force. Hence, from work-energy theorem

$$= \frac{1}{2}mv^2 = \text{work done by electrostatic force only}$$

$$= (qE_0)z$$

or $\quad$ Speed $v = \sqrt{\dfrac{2qE_0 z}{m}}$

Particle rotates in a plane perpendicular to **B**, i.e. in xz-plane only. Hence, $v_y = 0$

22. Angle between **v** and **B** is $90° - \phi$ or $60°$.

Initially net force is zero.

$\therefore \quad F_e = F_m$ or $qE = Bqv \sin 60°$

$\therefore \quad v = \dfrac{E}{B \sin 60°} = \dfrac{2E}{\sqrt{3}B}$

When electric field is switched OFF, path of the particle will be helical, as angle between **v** and **B** is 60°. Pitch,

$$p = (v \cos 60°)\left(\frac{2\pi m}{Bq}\right)$$

$$= \left(\frac{2E}{\sqrt{3}B}\right)\left(\frac{1}{2}\right)\left(\frac{2\pi m}{Bq}\right) = \left(\frac{2\pi mE}{\sqrt{3}B^2 q}\right)$$

Substituting the values, we get

$$p = 6 \text{ cm}$$

23.

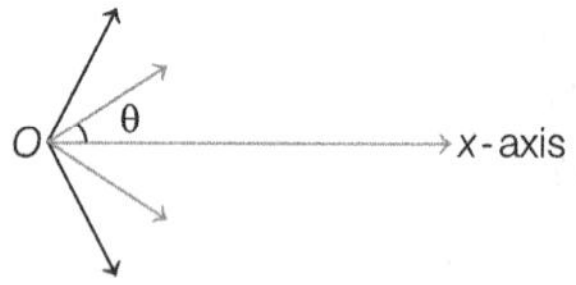

Electrons touch the x-axis again after every pitch. Therefore, the asked distance is

$$d = p = v_{11}T = (v \cos \theta)\left(\frac{2\pi m}{Bq}\right)$$

For paraxial electron $\theta \approx 0°$ and $q = e$,

$$d = \frac{2\pi mv}{Be}$$

24.

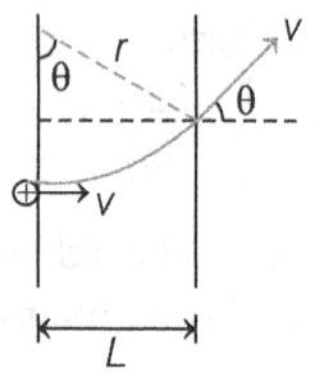

Deviation, $\theta = \sin^{-1}\left(\dfrac{L}{r}\right)$ for $L < r$

where, $r = \dfrac{mv}{Bq}$

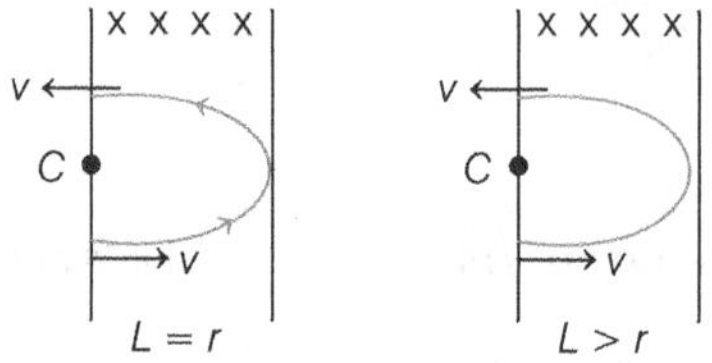

and $\qquad \theta = \pi$ if $L \le r$

25. $a_E = \dfrac{qE_0}{m}$ (along negative z-direction)

Electric field will make z-component of velocity zero. At that time speed of the particle will be minimum and that minimum speed is the other component, i.e. v_0.

This is minimum when,

$$v_z = u_z + a_z t$$

or $\quad 0 = v_0 - \dfrac{qE_0}{m}t$

or $\quad t = \dfrac{mv_0}{qE_0}$

Module 3.

1. $\mathbf{F} = I(\mathbf{l} \times \mathbf{B}) = I(\mathbf{ba} \times \mathbf{B})$

We can see that all (a), (b) and (c) options are same.

2. From Fleming's left hand rule, we can see that magnetic force is outwards on the loop.

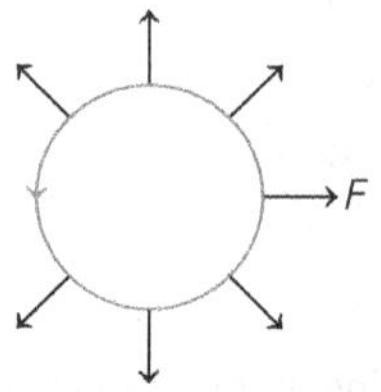

So, it tends to expand.

3. We can directly find the force on wire AC.

$$F = ilB \sin 90°$$

$$(l = AC = 10 \text{ cm} = 0.1 \text{ m})$$

$$= (3)(0.1)(10) = 3 \text{ N}$$

4. $mg \sin \theta = ILB \cos \theta$

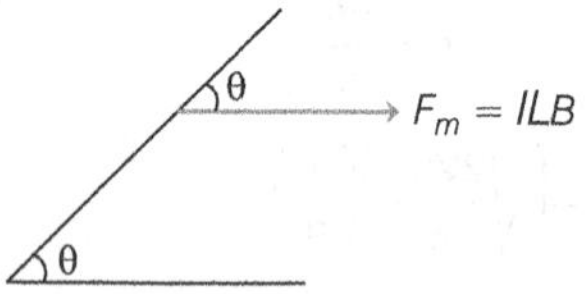

$\therefore \quad B = \dfrac{mg}{IL} \tan \theta$

5. $F_{net} = \sqrt{F^2 + F^2} = \sqrt{2}F = \sqrt{2}\, IlB$

6. Linear impulse $= mv$

or $F\Delta t = m\sqrt{2gh}$

or $(ilB)\,\Delta t = m\sqrt{2gh}$

But, $i\,\Delta t = \Delta q$

$\therefore$ $(\Delta q)\,(lB) = m\sqrt{2gh}$

Hence, $\Delta q = \dfrac{m\sqrt{2gh}}{Bl}$

7. Due to $\hat{\mathbf{j}}$ component of $\mathbf{B}$, magnetic force is zero. It is only due to $\hat{\mathbf{i}}$ component.
$\mathbf{l}$ is towards $-y$ direction, $\mathbf{B}$ is towards $\hat{\mathbf{i}}$ direction.
Hence, $\mathbf{l}\times\mathbf{B}$ (the direction of magnetic force is $+\hat{\mathbf{k}}$)

$$F = \int dF = \int_0^1 (i)(dy)(B) = \int_0^1 (2\times 10^{-3})(dy)(0.3\,y)$$
$$= 3\times 10^{-4}\ \text{N}$$

8. $|\mathbf{F}_{ABC}| + = |\mathbf{F}_{AC}| = ilB = (2)\,(5)\,(2) = 20\ \text{N}$

9.

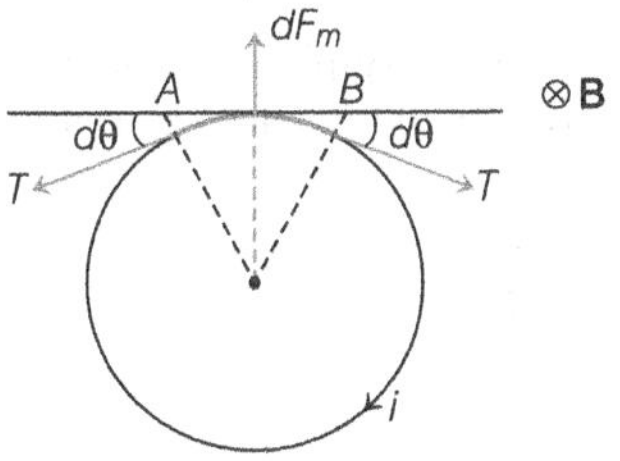

$$2T\sin d\theta = dF_m = il_{AB}B$$

or $2T\,(d\theta) = i\,(R)(2d\theta)\,B$

or $T = iRB \Rightarrow T \propto B$

10. $\mathbf{F}_{bc} + \mathbf{F}_{da} = 0$

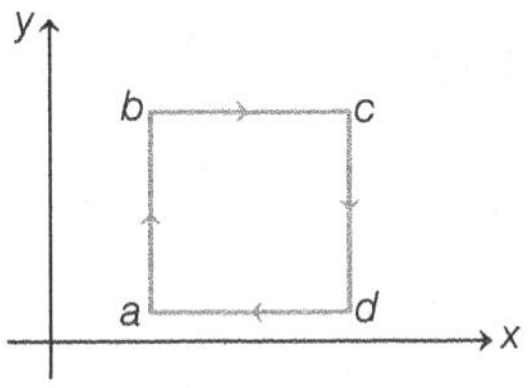

$$\mathbf{F}_{ab} = IlB_0\left(1 + \frac{x}{l}\right)$$
$$= lB_0l + lB_0x \qquad \text{(towards right)}$$
$$\mathbf{F}_{cd} = IlB_0\left[1 + \frac{x+l}{l}\right]$$
$$= 2lB_0l + lB_0x \qquad \text{(towards left)}$$

$\therefore$ $\mathbf{F}_{net} = \mathbf{F}_{cd} - \mathbf{F}_{ab} = lB_0l$ (towards left)

11. $2kx_0 = mg = ILB = \left(\dfrac{E}{R}\right)\cdot LB$

$\Rightarrow$ $B = \dfrac{mgR}{EL}$

12.

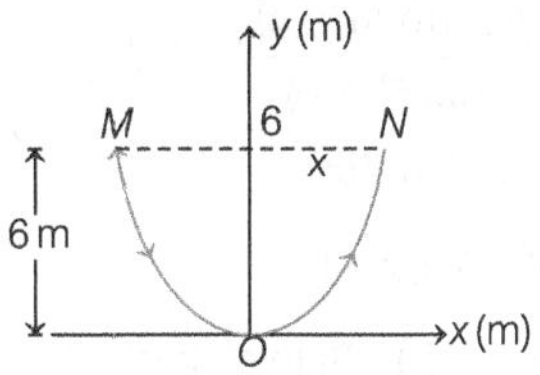

$$x = \sqrt{6y} = \sqrt{6\times 6} = 6\ \text{m}$$
$$\mathbf{F}_{MON} = \mathbf{F}_{MN} = i(\mathbf{l}\times\mathbf{B})$$
$$= 2\,[(12\,\hat{\mathbf{i}})\times(2\times 10^{-3}\,\hat{\mathbf{k}})]$$
$$= -0.048\,\hat{\mathbf{j}}$$
$$\approx (-0.05\,\hat{\mathbf{j}})\ \text{N}$$

13. $\because$ $\mathbf{l} = l\,\hat{\mathbf{i}}$

Now, $\mathbf{F} = i(\mathbf{l}\times\mathbf{B})$
$$= i\,[(l\,\hat{\mathbf{i}})\times(B_0\,\hat{\mathbf{j}} + B_0\,\hat{\mathbf{k}})]$$
$$= ilB_0(\hat{\mathbf{i}} - \hat{\mathbf{j}}) = |\mathbf{F}| = \sqrt{2}ilB_0$$

14. No, it will not change, as the new $\hat{\mathbf{i}}$ component of $\mathbf{B}$ is in the direction of $\mathbf{l}$.

15. $\because$ $i = 3.5\ \text{A}$
$$\mathbf{l} = (-10^{-2}\,\hat{\mathbf{j}})$$
Now, apply $\mathbf{F} = i\,(\mathbf{l}\times\mathbf{B})$ in all parts.

16.

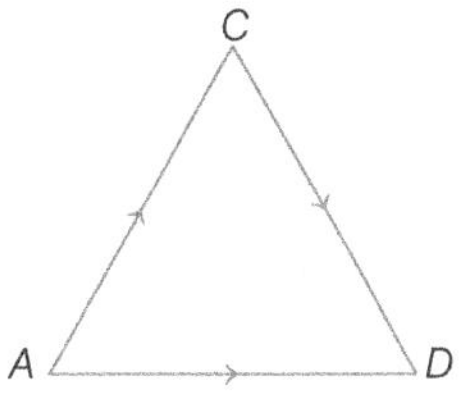

$$\mathbf{F}_{ACD} = \mathbf{F}_{AD}$$

$\therefore$ $|\mathbf{F}_{net}| = 2|\mathbf{F}_{AD}| = 2ilB$
$$= 2\times 2\times 4\times 2 = 32\ \text{N}$$

17. $W = F_m$ or $mg = ilB\sin 90°$

$\therefore$ $i = \dfrac{mg}{Bl} = \dfrac{(13\times 10^{-3})\,(10)}{0.44\times 0.62}$
$$= 0.47\ \text{A}$$

Magnetic force should be upwards to balance the weight. Hence, from Fleming's left hand rule we can see that direction of current should be from left to right.

18. (a) $ilB = mg$ or $\dfrac{V}{R}lB = mg$

or $V = \dfrac{mgR}{lB} = \dfrac{(0.75)\,(9.8)\,(25)}{0.5\times 0.45}$
$$\approx 817\ \text{V}$$

(b) $a = \dfrac{ilB - mg}{m} = \dfrac{VlB}{mR} - g \qquad \left(\text{as } i = \dfrac{V}{R}\right)$

$$=\frac{(817)(0.5)(0.45)}{(0.75)(2.0)}-9.8$$
$$\approx 112.8\ m/s^2$$

19. $i=5\ A \Rightarrow \mathbf{B}=(0.02\hat{\mathbf{j}})\ T$

Now, applying $\mathbf{F}=i(\mathbf{l}\times\mathbf{B})$ in all parts. Let us find $\mathbf{l}$ for part *cd*.

$$\mathbf{l}_{cd}=\mathbf{r}_d-\mathbf{r}_c$$
$$=(0.4\hat{\mathbf{j}}+0.4\hat{\mathbf{k}})-(0.4\hat{\mathbf{i}}+0.4\hat{\mathbf{k}})$$
$$=(0.4\hat{\mathbf{j}}-0.4\hat{\mathbf{i}})$$

Module 4.

1. In uniform **B** is force on any current carrying loop is always zero.

2. $\theta_i=180°,\ \theta_f=180°-\theta$
$$W=U_f-U_i$$
$$=-MB\cos(180°-\theta)-(-MB\cos 180°)$$
$$=MB\cos\theta-MB$$

3. **M** is outwards or along $\hat{\mathbf{k}}$ direction.
$$M=IA$$
where, $$A=a^2+\left(\frac{\pi a^2}{4}\right)2$$
$$\therefore\quad \mathbf{M}=\left(\frac{\pi}{2}+1\right)a^2 I\hat{\mathbf{k}}$$

4. $M=iA \Rightarrow M\propto A$
$$2\pi R=4L \Rightarrow \frac{R}{L}=\frac{2}{\pi}$$
$$\frac{M_2}{M_1}=\frac{A_2}{A_1}=\frac{L^2}{\pi R^2}=\left(\frac{1}{\pi}\right)\left(\frac{\pi}{2}\right)^2=\frac{\pi}{4}$$
or $$M_2=\left(\frac{\pi}{4}\right)M_1=\left(\frac{\pi}{4}\right)M$$

5. $\mathbf{M}=IA\hat{\mathbf{i}}$
$$\mathbf{B}=B\hat{\mathbf{j}}$$
$$\therefore\quad \tau=\mathbf{M}\times\mathbf{B}=IAB\hat{\mathbf{k}}$$

6. $\frac{M}{L}=\frac{q}{2m}$
$$M=\left(\frac{q}{2m}\right)(I\omega)$$
$$=\left(\frac{q}{2m}\right)\left(\frac{ml^2}{3}\right)(2\pi f)=\frac{\pi qfl^2}{3}$$

7. $U=-\mathbf{M}\mathbf{B}=-MB\cos\theta$

Here, **M** = magnetic moment of the loop
θ = angle between **M** and **B**
U is maximum when $\theta=180°$ and minimum when $\theta=0°$. So, as θ decreases from 180° to 0° its PE also decreases.

8. $\tau=\mathbf{M}\times\mathbf{B}$

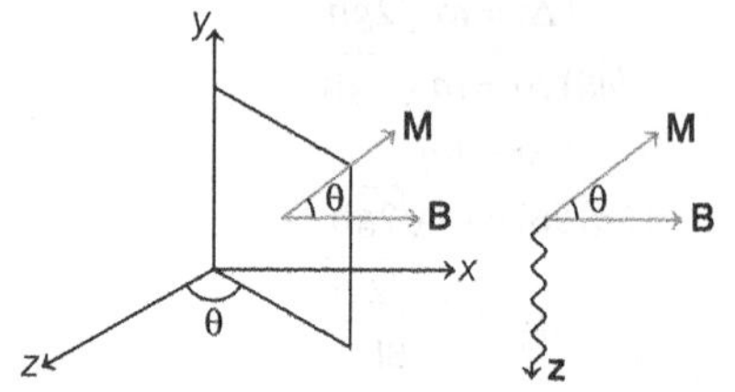

$$|\tau|=|\mathbf{M}||\mathbf{B}|\sin\theta=iAB_0\sin\theta=i(ab)B_0\sin\theta$$
τ is in the direction of $\mathbf{M}\times\mathbf{B}$, i.e. along the negative *y*-direction.
Since, τ is in negative *y*-direction, the direction of rotation of loop is such that θ decreases.

9. Force on a current carrying loop is zero for all angles. τ is maximum when Q, then angle between **M** and **B** is 90° minimum potential energy is at $\theta=0°$. Positive potential energy is for obtuse angle. Direction of **M** is obtained by screw law.

10. (a) As, the current in *xy*-plane is anti-clockwise, so moment will be along *y*-axis by right hand thumb rule.

(b) $\tau=\mathbf{M}\times\mathbf{B}$
$$\Rightarrow|\tau|=|\mathbf{M}||\mathbf{B}|\sin\theta$$
$$=\frac{5\times\pi\times(0.1)^2\times 100\sqrt{2}}{\sqrt{2}}\qquad(\text{here, }\theta=45°)$$

(c) Net force on a closed loop carrying current in a uniform magnetic field is zero.

11. $\because\quad \frac{M}{L}=\frac{q}{2m}$
$$\therefore\quad M=\left(\frac{q}{2m}\right)L=\left(\frac{q}{2m}\right)(I\omega)$$
$$=\left(\frac{q}{2m}\right)\left(\frac{1}{2}mR^2\right)(\omega)=\frac{qR^2\omega}{4}$$

12. $\because\quad M=iA=i(\pi R^2)$
$$=(0.2)(\pi)(8\times10^{-2})^2$$
$$=(4.0\times10^{-3})\ A\text{-}m^2$$
Now, $\mathbf{M}=M\hat{\mathbf{M}}=(4.0\times10^{-3})(0.6\hat{\mathbf{i}}-0.8\hat{\mathbf{j}})$

(a) $\tau=\mathbf{M}\times\mathbf{B}$
(b) $U=-\mathbf{M}\cdot\mathbf{B}$

13. $\because\ L=2\pi R \Rightarrow R=\frac{L}{2\pi}$
$$M=iA=i(\pi R^2)$$
$$=(i)(\pi)\left(\frac{L}{2\pi}\right)^2=\frac{iL^2}{4\pi}$$
$$\tau_{max}=MB\sin 90°=\frac{iBL^2}{4\pi}$$

14. $\because\ \Delta U=U_{0°}-U_{180°}$
$$=-MB\cos 0°+MB\cos 180°=-2MB$$

15. (a) From energy conservation,

$$U_\theta = K_\theta = U_{0^\circ} + K_{0^\circ}$$

or $(-MB\cos\theta) + 0 = (-MB\cos 0^\circ) + K_{0^\circ}$

Substituting the given values, we can calculate.

(b) To other side also it rotates upto the same angle.

16. (a) $T = \dfrac{2\pi r}{v}$

(b) $I = qf = (e)\left(\dfrac{v}{2\pi r}\right)$

(c) $M = IA = \left(\dfrac{ev}{2\pi r}\right)(\pi r^2) = \dfrac{evr}{2}$

17. Assume equal and opposite currents in wires *cf* and *eh*.

18. Assume equal and opposite currents in wires *PQ* and *RS*, then find **M**.

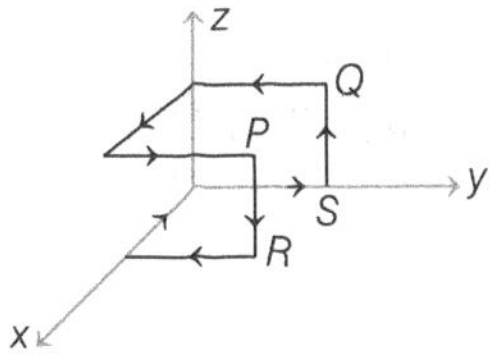

Now, $\mathbf{B}(2\hat{\mathbf{j}})$

$\Rightarrow$ $\tau = \mathbf{M} \times \mathbf{B}$

Module 5.

1. If a positive charged particle is rotating in a circle in clockwise direction, then equivalent current is also clockwise. Therefore, at centre **B** and **M** are both inwards. Hence, they are parallel. But at some other point they may not be parallel.

2.

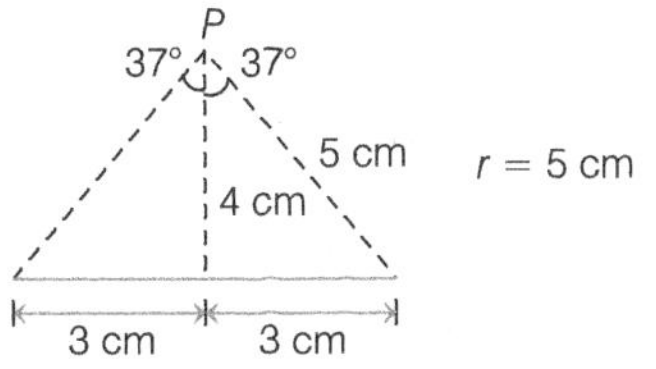

$$B = \frac{\mu_0}{4\pi}\frac{i}{r}(\sin 37^\circ + \sin 37^\circ)$$

3. Two fields are additive.

$\therefore$ $$B_{net} = 2\left[\frac{\mu_0 I}{2\pi(R/2)}\right] = \frac{2\mu_0 I}{\pi R}$$

4. $B_x = \dfrac{1}{8}B_c$

$\therefore$ $$\frac{\mu_0 NiR^2}{2(R^2 + x^2)^{3/2}} = \frac{1}{8}\left[\frac{\mu_0 Ni}{2R}\right]$$

6. In (c), two wires are producing ⊙ magnetic field and two wires are producing ⊗ magnetic field.

7. Arc of radius $a\left(\dfrac{1}{4}\text{th of circle}\right)$ produces magnetic field in $\hat{\mathbf{k}}$ direction or outwards, while arc of radius *b* produces magnetic field in $-\hat{\mathbf{k}}$ direction.

$\therefore$ $$\mathbf{B} = \frac{1}{4}\left(\frac{\mu_0 I}{2a}\right)\hat{\mathbf{k}} + \frac{1}{4}\left(\frac{\mu_0 I}{2a}\right)(-\hat{\mathbf{k}})$$

$$= \frac{\mu_0 I}{8}\left(\frac{1}{a} - \frac{1}{b}\right)\hat{\mathbf{k}}$$

8. Equivalent current,

$$i = qf = ef$$

$$B = \frac{\mu_0 i}{2R} = \frac{\mu_0 ef}{2R}$$

9. $d\mathbf{B} = \dfrac{\mu_0}{4\pi}\dfrac{i}{r^3}(d\mathbf{l} \times \mathbf{r})$

$\therefore$ $d\mathbf{B}$ is in the direction $d\mathbf{l} \times \mathbf{r}$. Hence, $d\mathbf{l}$ is outwards and **r** is from $d\mathbf{l}$ towards *P*.

10. Let l_1 and l_2 be the lengths of two parts *ARB* and *ASB* of the ring, and ρ be resistance per unit length of the conductor.

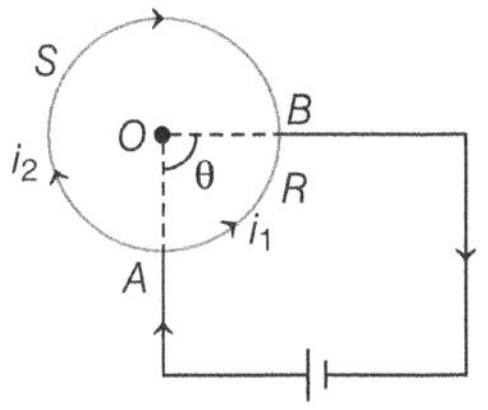

The resistance of the portion *ARB* will be $R_1 = l_1\rho$

The resistance of the portion *ASB* will be $R_2 = l_2\rho$

Let i_1 and i_2 be the currents in *ARB* and *ASB* respectively.

Since, these parts are in parallel, the potential difference across their ends will be same.

$\therefore$ $$i_1R_1 = i_2R_2$$

or $$i_1l_1\rho = i_2l_2\rho$$

or $$i_1l_1 = i_2l_2 \quad \text{...(i)}$$

According to Biot Savart's law, the magnetic fields at the centre *O* due to current through circular ring *ARB* and *ASB* will be

$$B_1 = \frac{\mu_0}{4\pi}\frac{i_1l_1\sin 90^\circ}{r^2}$$

and $$B_2 = \frac{\mu_0}{4\pi}\frac{i_2l_2\sin 90^\circ}{r^2}$$

Since, $i_1l_1 = i_2l_2$

$\therefore$ $$B_1 = B_2$$

According to right hand rule, the directions of B_1 and B_2 are opposite to each other, hence the resultant magnetic field at *O* will be zero.

11. $B_0 = B_1 + B_2$

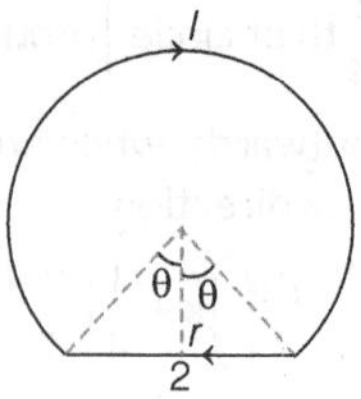

$$= \frac{\mu_0 I}{2R}\left[\frac{2\pi - 2\theta}{2\pi}\right] + \frac{\mu_0 I}{4\pi R}[\sin\theta + \sin\theta]$$

$$= \frac{\mu_0 I}{2R}\left[\frac{\pi - \theta}{\pi}\right] + \frac{\mu_0 I}{4\pi (R\cos\theta)}(2\sin\theta)$$

$$= \frac{\mu_0 I}{2\pi R}[\pi - \theta + \tan\theta]$$

12. When wire is turned into n circular loops, then magnetic induction is

$$B' = n^2 B = (3)^2 B_1 = 9B_1 \qquad \left(B \propto \frac{n}{R} \text{ and } R \propto \frac{1}{n}\right)$$

13. $\dfrac{B_y}{B_x} = \tan 30°$

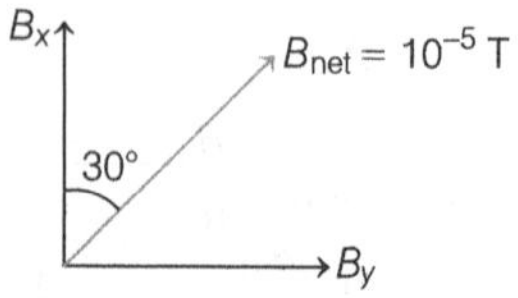

$$\therefore \quad B_y = \frac{B_x}{\sqrt{3}}$$

$$\therefore \quad i_y = \frac{i_x}{\sqrt{3}} = \frac{1}{\sqrt{3}}\text{A}$$

14. Magnetic field due to both wires is in ⊙ direction, so additive.

Due to one wire,

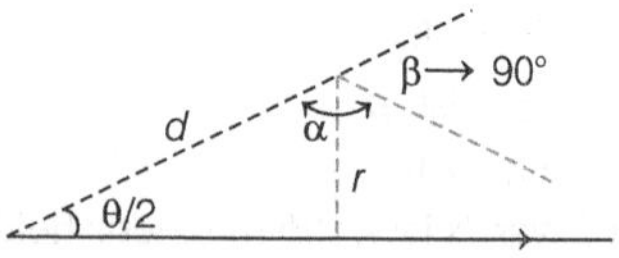

$r = d\sin\dfrac{\theta}{2}$

$\alpha = 90° - \dfrac{\theta}{2}$

$$B = \frac{\mu_0}{4\pi}\frac{i}{r}(\sin\alpha + \sin\beta)$$

$$= \frac{\mu_0}{4\pi}\frac{i}{d\sin\frac{\theta}{2}}\left[\sin\left(90° - \frac{\theta}{2}\right) + \sin 90°\right]$$

$$= \frac{\mu_0}{4\pi}\frac{i}{d\sin\frac{\theta}{2}}\left[\cos\frac{\theta}{2} + 1\right]$$

$$= \frac{\mu_0}{4\pi}\frac{i}{d\sin\frac{\theta}{2}}\left[2\cos^2\frac{\theta}{4} - 1 + 1\right]$$

$$= \frac{\mu_0}{2\pi}\frac{i}{d}\left[\frac{2\cos^2\frac{\theta}{4}}{2\sin\frac{\theta}{4}\cos\frac{\theta}{4}}\right] = \frac{\mu_0}{4\pi}\frac{i\cot\frac{\theta}{4}}{d}$$

$\therefore$ Magnetic field due to both the wires will be double of it.

15. $B_1 = \dfrac{\mu_0 i_1}{2R_1} = \dfrac{(4\pi \times 10^{-7})(5)}{(2)\left(\dfrac{5}{\sqrt{2}} \times 10^{-2}\right)}$

$$= 2\sqrt{2}\pi \times 10^{-5}\ \text{T}$$

$$B_2 = \frac{\mu_0 i_2}{2R_2} = \frac{(4\pi \times 10^{-7})(5\sqrt{2})}{(2)(5 \times 10^{-2})}$$

$$= 2\sqrt{2}\pi \times 10^{-5}\ \text{T}$$

$$B_{net} = \sqrt{B_1^2 + B_2^2}$$

16. At the centre of coil 1,

$$B_1 = \frac{\mu_0}{4\pi} \times \frac{2\pi i_1}{r_1} \qquad \text{...(i)}$$

At the centre of coil 2,

$$B_2 = \frac{\mu_0}{4\pi} \times \frac{2\pi i_2}{r_2} \qquad \text{...(ii)}$$

But, $B_1 = B_2$

$$\therefore \quad \frac{\mu_0}{4\pi}\frac{2\pi i_1}{r_1} = \frac{\mu_0}{4\pi}\frac{2\pi i_2}{r_2}$$

or $\quad \dfrac{i_1}{r_1} = \dfrac{i_2}{r_2}$

As, $r_1 = 2r_2 \Rightarrow \dfrac{i_1}{2r_2} = \dfrac{i_2}{r_2}$ or $i_1 = 2i_2$

Now, ratio of potential differences

$$\frac{V_2}{V_1} = \frac{i_2 \times r_2}{i_1 \times r_1} = \frac{i_2 \times r_2}{2i_2 \times 2r_2} = \frac{1}{4}$$

$$\therefore \quad \frac{V_1}{V_2} = \frac{4}{1}$$

17. $B = \dfrac{\mu_0 i}{2R} = \dfrac{\mu_0 (ef)}{2R} = \dfrac{(\mu_0)(e)\left(\dfrac{v}{2\pi R}\right)}{2R}$

or $\quad R \propto \sqrt{\dfrac{v}{B}}$

18. $l = 2\pi R \Rightarrow R = \dfrac{l}{2\pi}$

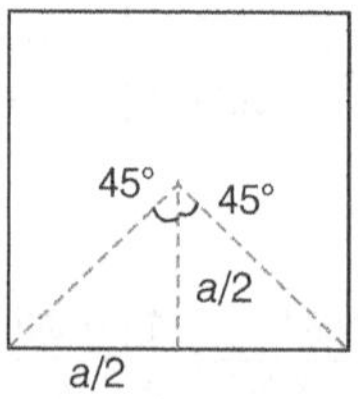

$$l = 4a \quad \Rightarrow \quad a = \frac{l}{4}$$

$$B' = \frac{\mu_0 i}{2R} = \frac{\mu_0 i}{2 \cdot \frac{l}{2\pi}} = \frac{\mu_0 \pi i}{l}$$

$$B = 4\left[\frac{\mu_0}{4\pi} \cdot \frac{i}{a/2}(\sin 45° + \sin 45°)\right]$$

$$= \frac{2\sqrt{2}\,\mu_0 i}{\pi a} = \frac{8\sqrt{2}\,\mu_0 i}{\pi l}$$

$$\therefore \quad \frac{B}{B'} = \frac{8\sqrt{2}}{\pi^2}$$

19. $$B = 2\left[\frac{\mu_0}{4\pi}\frac{2I/3}{a/2}(\sin 45° + \sin 45°) - \frac{\mu_0}{4\pi} \cdot \frac{I/3}{a/2}(\sin 45° + \sin 45°)\right]$$

$$= \frac{\sqrt{2}\,\mu_0 I}{3\pi a}$$

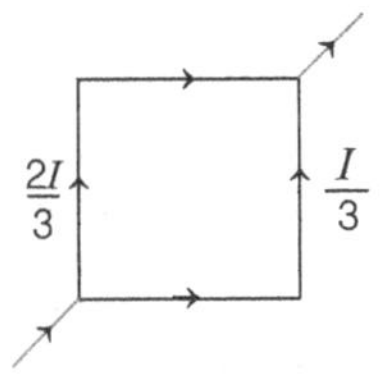

20. (a) From screw law, we can see that direction of magnetic field at centre of the square is inwards as the current is clockwise.

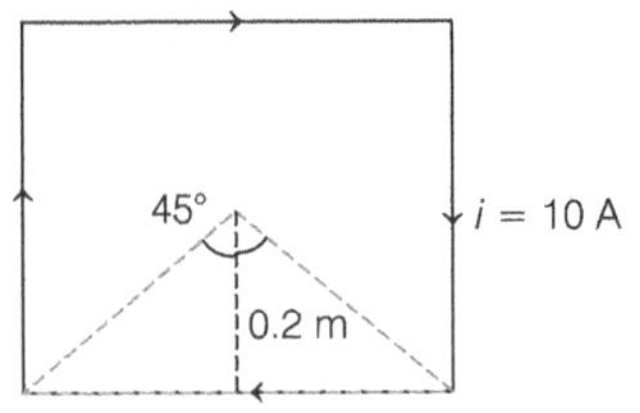

$$B = 4\left(\frac{\mu_0}{4\pi}\frac{i}{r}\right)(\sin\alpha + \sin\beta)$$

$$= \frac{4 \times 10^{-7} \times 10}{0.2}(\sin 45° + \sin 45°)$$

$$= 2.83 \times 10^{-5}\ \text{T}$$

(b) $2\pi R = 4(0.4)$

$$2R = \frac{1.6}{\pi}\ \text{m}$$

$$B = \frac{\mu_0 i}{2R} = \frac{(4\pi \times 10^{-7})(10)}{(1.6/\pi)}$$

$$= 24.7 \times 10^{-6}\ \text{T}$$

21. Magnetic field due to horizontal wire is zero. Magnetic field due to vertical is

$$B = \frac{\mu_0}{4\pi}\frac{i}{x}(\sin 90° + \sin 0°)$$

$$= \frac{\mu_0}{4\pi}\frac{i}{x} \qquad \text{(inwards)}$$

22. Both straight and circular wires will produce magnetic fields inwards.

$$\therefore \quad B = \frac{\mu_0}{2\pi}\frac{i}{R} + \frac{\mu_0 i}{2R}$$

$$= \frac{(2 \times 10^{-7})(7)}{0.1} + \frac{(4\pi \times 10^{-7})(7)}{2 \times 0.1}$$

$$= 5.8 \times 10^{-5}\ \text{T}$$

23. Magnetic field at O due to two straight wires = 0

Magnetic field due to circular wire,

$$B = \frac{1}{4}\ \text{(due to whole circle)}$$

$$= \frac{1}{4}\left(\frac{\mu_0 i}{2R}\right) \qquad \text{(inwards)}$$

$$= \frac{1}{4} \times \frac{(4\pi \times 10^{-7})(5)}{2 \times 0.03}$$

$$= 2.62 \times 10^{-5}\ \text{T}$$

24. Magnetic field at P due to straight wires = 0

Due to circular wires one is outwards (of radius a) and other is inwards. 60° means $\frac{1}{6}$ th of whole circle.

$$\therefore \quad B = \frac{1}{6}\left[\frac{\mu_0 i}{2a} - \frac{\mu_0 i}{2b}\right] \qquad \text{(outwards)}$$

25. Two magnetic fields of equal magnitude at point $P(0, 0, l)$ will be at right angles.

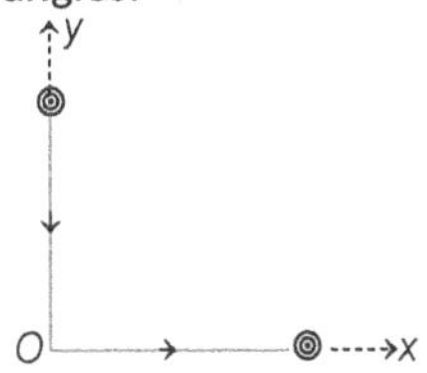

Therefore,

$$B_{net} = \sqrt{2}B = \sqrt{2}\left[\frac{\mu_0}{4\pi}\frac{I}{l}(\sin 0° + \sin 90°)\right]$$

$$= \sqrt{2}\left(\frac{\mu_0}{4\pi}\right)\left(\frac{I}{l}\right) = \sqrt{2} \times 10^{-7} \times \frac{5.0}{0.35}$$

$$= 2.0 \times 10^{-6}\ \text{T} \quad \text{or} \quad B_{net} = 2\,\mu\text{T}$$

Note To find direction of magnetic field due to a straight wire at some point we can use the basic formula of Biot-Savart law where direction of $d\mathbf{B}$ is in the direction of $d\mathbf{l} \times \mathbf{r}$. Here $d\mathbf{l}$ is in the direction of current I. We can extend the same for **B** also which is in the direction of $\mathbf{I} \times \mathbf{r}$. Here **r** of point P is along $\hat{\mathbf{k}}$. Further, **I** in the wire having current along x-axis is along $\hat{\mathbf{i}}$. Therefore, $\mathbf{I} \times \mathbf{r}$ means $-\hat{\mathbf{j}}$. So, magnetic field is along $-\hat{\mathbf{j}}$ direction. Similarly, we can find magnetic field due to other wire.

26. Magnetic fields at O due to currents in wires *ab* and *cd* are zero.

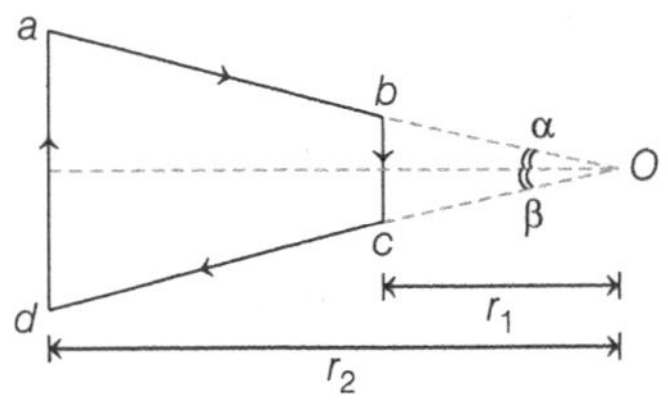

Magnetic field due to current in wire *da* (say B_2) is inwards due to current in wire *bc* (say B_1) is outwards.

$$B_1 = \frac{\mu_0}{4\pi}\frac{i}{r_1}(\sin\alpha + \sin\beta)$$

$$B_2 = \frac{\mu_0}{4\pi}\frac{i}{r_2}(\sin\alpha + \sin\beta)$$

$$B_1 > B_2 \quad \text{as} \quad r_1 < r_2$$

$$\therefore \quad B_{net} = B_1 - B_2 \qquad \text{(outwards)}$$

27. In first quadrant magnetic field due to I_1 is outwards and due to I_2 is inwards. So, net magnetic field may be zero. Similarly, in third quadrant magnetic field due to I_1 is inwards and due to I_2 magnetic field is outwards. Hence, only in first and third quadrants magnetic field may be zero.

Let magnetic field is zero at point $P(xy)$, then

$$B_{I_1} = B_{I_2}$$

$$\therefore \quad \frac{\mu_0}{2\pi}\frac{I_1}{y} = \frac{\mu_0}{2\pi}\frac{I_2}{x}$$

$$\therefore \quad y = \frac{I_1}{I_2}x$$

28. Two straight wires produces outward magnetic field by arc of circle produces inward magnetic field.

Due to straight wires,

$$B_1 = 2\left[\frac{\mu_0}{4\pi}\frac{i}{R}(\sin\theta + \sin 90°)\right]$$

$$= \frac{\mu_0 i}{2\pi R} \qquad \text{(outwards)}$$

Due to circular arc, $B_2 = \frac{\theta}{2\pi}\left(\frac{\mu_0 i}{2R}\right)$ (inwards)

For net field to be zero,

$$B_1 = B_2$$

or $$\theta = 2 \text{ rad}$$

29. (a) If currents are in the same direction, then above and below the wires magnetic fields are in the same direction. Hence, they can't produce zero magnetic field.

In between the wires, let $B = 0$ at a distance (r) cm from the wire carrying 75 A current. Then,

$$\frac{\mu_0}{2\pi}\left(\frac{75}{r}\right) = \left(\frac{\mu_0}{2\pi}\right)\left(\frac{25}{40-\pi}\right)$$

Solving, we get $r = 30$ cm.

(b) If currents are in opposite direction, then in between the wire magnetic field are in the same direction. So, they cannot produce zero magnetic field. The points should be above or below the wires, nearer to wire having smaller current. Let it is at a distance r from the wire having 25 A current. Then,

$$\frac{\mu_0}{2\pi}\left(\frac{25}{r}\right) = \left(\frac{\mu_0}{2\pi}\right)\left(\frac{75}{40+r}\right)$$

Solving this equation, we get

$$r = 20 \text{ cm}$$

30. Apply $B = \frac{\mu_0 NiR^2}{2(R^2 + x^2)^{3/2}}$

31. I_2 produces inwards magnetic field at centre. Hence, I_1 should produce outward magnetic field. Or current should be towards right. Further,

$$\frac{\mu_0 I_2}{2R} = \left(\frac{\mu_0}{2\pi}\right)\left(\frac{I_1}{D}\right)$$

$$\therefore \quad I_1 = \left(\frac{\pi D}{R}\right)I_2$$

32. (a) $B_1 = \frac{\mu_0 Ni}{2R}$

(b) $B_2 = \frac{\mu_0 NiR^2}{2(R^2 + x^2)^{3/2}}$, given $B_2 = \frac{B_1}{2}$

33. 10 A and 8 A current produce inward magnetic field. While 20 A current produces outward magnetic field. Hence, current in fourth wire should be $(20 - 10 - 8)$ A or 2 A and it should produce inward magnetic field. So, it should be downwards toward the bottom.

Module 6.

1. Using Ampere's circuital law over a circular loop of any radius less than the radius of the pipe, we can see that net current inside the loop is zero. Hence, magnetic field at every point inside the loop will be zero.

2. At distance r from centre,

$$B = \frac{\mu_0}{2\pi}\frac{(i_{in})}{r} \qquad \text{(From Ampere's circuital law)}$$

For path-1, $i_{in} \neq 0$

$$\therefore \quad B_1 \neq 0$$

For path-2, $i_{in} = 0$

$$\therefore \quad B_2 = 0$$

3. From centre, $r = (R - x)$

$$B = \frac{\mu_0}{2\pi}\frac{I}{R^2}r = \frac{\mu_0}{2\pi}\frac{I}{R^2}(R - x)$$

4. $B_x = \frac{\mu_0}{2\pi}\frac{I_{in}}{x}$

where, $$I_{in} = I - \left[\frac{I}{\pi(c^2 - b^2)}\right][\pi(x^2 - b^2)]$$

$$= \frac{I(c^2 - x^2)}{(c^2 - b^2)}$$

5. Apply Ampere's circuital law,

$$\oint \mathbf{B} \cdot d\mathbf{l} = \mu_0(i_{net})$$

$$\therefore \quad Bl = \mu_0(i_{net})$$

or $$B(2\pi x) = \mu_0 i_{net}$$

$$\therefore \quad B = \frac{\mu_0}{2\pi}\frac{i_{net}}{x}$$

Find i_{net} for different regions.

6. If current flows in a conductor, then

$E \neq 0$ (for inside points)

$E = 0$ (for outside points)

$B = \frac{\mu_0}{2\pi}\frac{i}{R^2}r$ (for inside points)

$B = 0$ at $r = 0$, i.e. at centre

$B = \frac{\mu_0}{2\pi}\frac{i}{r}$ for outside points.

7. Applying Ampere's circuital law,

$$B_A = \frac{\mu_0}{2\pi}\frac{i_{in}}{r}$$
$$= \frac{(2\times 10^{-7})(1)}{10^{-3}}$$
$$= 2\times 10^{-4}\ \text{T}$$

This is due to (·) current of 1 A. Hence, magnetic lines are circular and anti-clockwise. Hence, magnetic field is upwards.

$$B_b = \frac{\mu_0}{2\pi}\frac{i_{in}}{r}$$
$$= \frac{(2\times 10^{-7})(3-1)}{3\times 10^{-3}}$$
$$= 1.33\times 10^{-4}\ \text{T}$$

This is due to net ⊗ current. Hence, magnetic lines are clockwise.

So, magnetic field at B is downwards.

8. $\int \mathbf{B}\cdot d\mathbf{l} = \mu_0(i_{net})$

Along path (a), net current enclosed by this path is zero.

Hence, line integral $= 0$

Along path (b), i_{net} is ⊗. So, magnetic lines along this current is clockwise. But, we have to take line integral in counter clockwise direction. Hence, line integral will be negative.

9. $i = \int_0^r (2\pi r\, dr) j = \int_0^r (2\pi r\, dr)(br) = \frac{2\pi b r^3}{3}$

(a) **For $r_1 < R$**

$$B = \frac{\mu_0}{2\pi}\frac{i_{in}}{r_1} = \frac{\mu_0}{2\pi}\left(\frac{2\pi b r_1^3/3}{r_1}\right) = \frac{\mu_0 b r_1^2}{3}$$

(b) **For $r_2 > R$**

$$B = \frac{\mu_0}{2\pi}\frac{i_{in}}{r_2} = \left(\frac{\mu_0}{2\pi}\right)\left(\frac{2\pi b R^3/b}{r_2}\right) = \frac{\mu_0 b R^3}{3r_2}$$

10.

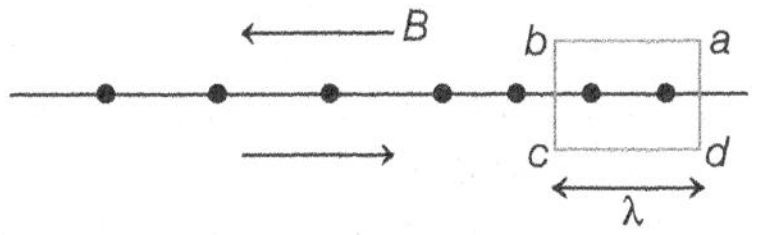

B in the above situation is given by

$$B = \frac{\mu_0\lambda}{2}$$

This result can be obtained as,

$$\oint_{abcda}\mathbf{B}\cdot d\mathbf{l} = \oint_{ab}\mathbf{B}\cdot d\mathbf{l} + \oint_{bc}\mathbf{B}\cdot d\mathbf{l} + \oint_{cd}\mathbf{B}\cdot d\mathbf{l} + \oint_{da}\mathbf{B}\cdot d\mathbf{l} = \mu_0(i_{n_0})$$

$$\Rightarrow Bl + 0 + Bl + 0 = \mu_0(\lambda l) \quad \Rightarrow \quad B = \frac{\mu_0\lambda}{2}$$

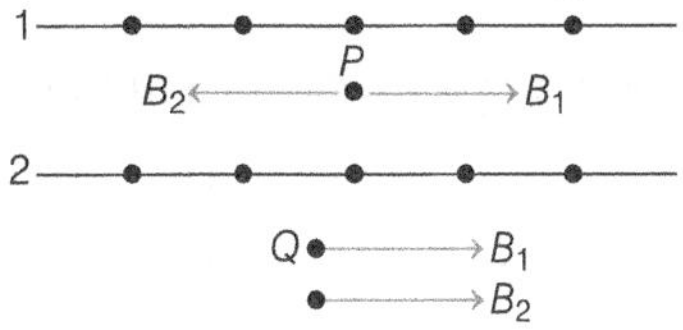

At point P, B_1 and B_2 are in opposite directions. Hence, $B_P = 0$

At point, Q, B_1 and B_2 are in same direction.

Hence, $B_Q = 2\left(\frac{\mu_0\lambda}{2}\right) = \mu_0\lambda$

11. $\int_{-\infty}^{+\infty}\mathbf{B}\cdot d\mathbf{r} = \int_{-\infty}^{+\infty}\frac{\mu_0 I R^2}{2(R^2+r^2)^{3/2}}dr$

After completing this integration, the result comes out to be,

$$\int_{-\infty}^{+\infty}\mathbf{B}\cdot d\mathbf{r} = \mu_0 I$$

Module 7.

1. Force between two conductors,

$$F \propto \frac{I_1 I_2}{d}$$

When I_1 is changed to $2I_1$ and d is changed to $3d$.

$$\therefore \quad F' \propto \frac{(2I_1)(I_2)}{3d} \propto \frac{2}{3}F$$

As direction of current is reversed so,

$$F' = -2F/3$$

2. $F_a = F_{ba} - F_{ca} = \frac{\mu_0}{2\pi}\frac{I^2}{r} - \frac{\mu_0}{2\pi}\frac{I^2}{2r} = \frac{\mu_0}{4\pi}\frac{I^2}{r}$

$$F_b = F_{ab} + F_{cb} = \frac{\mu_0}{2\pi}\frac{I^2}{r} + \frac{\mu_0}{2\pi}\frac{I^2}{r} = \frac{\mu_0}{\pi}\frac{I^2}{r}$$

$$F_c = F_{ac} + F_{bc} = \frac{\mu_0}{2\pi}\frac{I^2}{2r} + \frac{\mu_0}{2\pi}\frac{I^2}{r} = \frac{3\mu_0}{4\pi}\frac{I^2}{r}$$

$$\therefore \quad F_b > F_c > F_a$$

3. At a distance X from current I_2,

$$B = \frac{\mu_0}{2\pi}\frac{I_2}{X}$$

Magnetic force of small element dX of wire AB

$$dF = I_2(dX)B\sin 90°$$

$$\therefore \quad F = \int_{x=a}^{x=2a} dF$$

4. Magnetic field of 1 at dl

$$B = \frac{\mu_0}{4\pi}\frac{I_1}{r}[\sin 90° + \sin(90° - \theta)]$$
$$= \frac{\mu_0}{4\pi}\frac{I_1}{r}\left(2\cos^2\frac{\theta}{2}\right)$$

Now, $dF = I_2(dl) B \sin 90°$

$$= \frac{\left(\mu_0 I_1 I_2 \cos^2 \frac{\theta}{2}\right) dl}{2\pi r}$$

5. Force between two wires carrying current I_1 and I_2 is

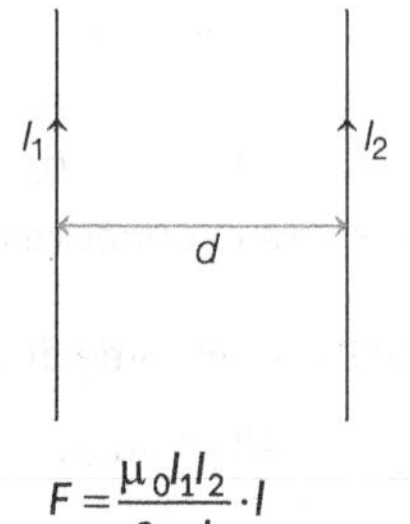

$$F = \frac{\mu_0 I_1 I_2}{2\pi d} \cdot l$$

Here, $I_1 = I_2 = 10$ A,

$d = 5 \times 10^{-2}$ m,

$l = 10 \times 10^{-2}$ m

So, $F_1 = \frac{\mu_0 I_1 I_2 l}{2\pi d}$

If distance between wires is halved and currents are doubled.

Then, $F_2 = \frac{\mu_0 I_1' I_2' l}{2\pi d'} = \frac{\mu_0 4 I_1 I_2 l}{2\pi \frac{d}{2}}$

$$= 8\left(\frac{\mu_0 I_1 I_2 l}{2\pi d}\right) = 8F_1$$

6. Given,

$\Rightarrow$ distance between wire, $d = 0.20$ m

$\Rightarrow$ $F = 2 \times 10^{-6}$ N

We know $F = \frac{\mu_0 I_1 I_2 l}{2\pi d}$

$$\frac{F}{l} = \frac{\mu_0}{2\pi} \frac{I_1 I_2}{d}$$

$\Rightarrow$ $2 \times 10^{-6} = 2 \times 10^{-7} \times \frac{x^2}{0.2}$

$\Rightarrow$ $2 = x^2 \Rightarrow x = \sqrt{2} = 1.4$

7. The given situation is shown below :

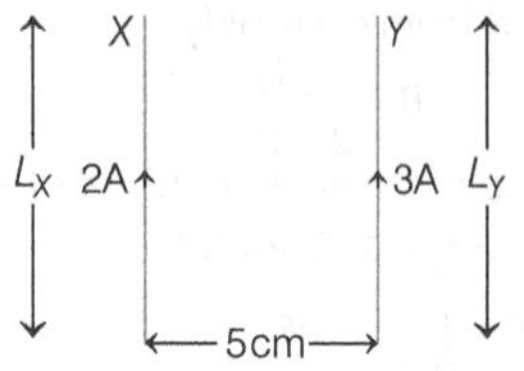

$L_X = 50$ cm $= 0.5$ m

$\Rightarrow$ $I_X = 2$ A

$L_Y = 5$ m

$\Rightarrow$ $I_Y = 3$ A

$r = 5$ cm $= 5 \times 10^{-2}$ m

Magnitude of force acting on both wires X and Y will be same and given as

$$F = \frac{\mu_0 I_X I_Y}{2\pi r} \times l_X$$

$$= 2 \times 10^{-7} \times \frac{2 \times 3 \times 0.5}{5 \times 10^{-2}}$$

$$= 1.2 \times 10^{-5} \text{ N}$$

8.

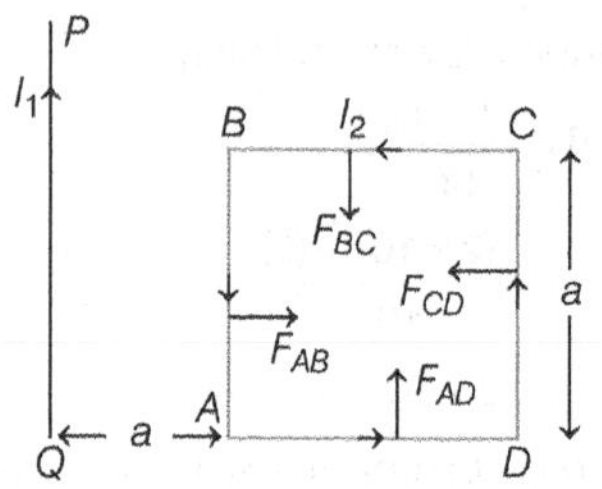

$F_{AB} = \frac{\mu_0 I_1}{2\pi a} \cdot I_2 a$ (away from wire PQ)

$F_{BC} + F_{AD} = 0$

$F_{CD} = \frac{\mu_0 I_1}{2\pi(2a)} \cdot I_2 a = \frac{\mu_0 I_1}{4\pi a} \cdot I_2 a$

(towards the wire PQ)

$\therefore$ $F_{net} = F_{AB} - F_{CD}$

$= \frac{\mu_0 I_1 I_2}{2\pi} - \frac{\mu_0 I_1 I_2}{4\pi}$ (away from wire)

$= \frac{\mu_0 I_1 I_2}{4\pi}$ (repulsive in nature)

9. Force per unit length between two wires carrying currents i_1 and i_2 at distance r is given by

$$\frac{F}{l} = \frac{\mu_0}{2\pi} \frac{i_1 i_2}{r}$$

Here, $i_1 = i_2 = i$ and $r = b$

$$\therefore \quad \frac{F}{l} = \frac{\mu_0 i^2}{2\pi b}$$

10.

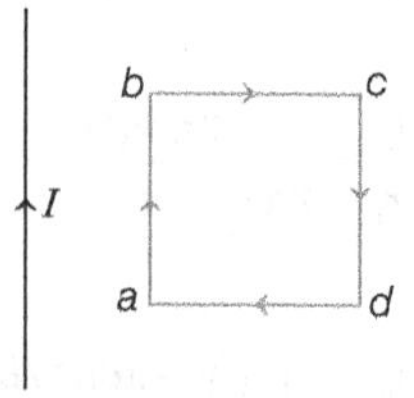

Straight wire will produce a non-uniform field to the right of it. $\mathbf{F}_{bc}$ and $\mathbf{F}_{da}$ will be calculated by integration but these two forces will cancel each other. Further force on wire *ab* will be towards the long wire and on wire *cd* will be away from the long wire.

Since the wire *ab* is nearer to the long wire, force of attraction towards the long wire will be more. Hence, the loop will move towards the wire.

11. Weight of the free wire should be balanced by the attractive or repulsive force by the fixed wire.

12. Let us take $F = \frac{\mu_0}{2\pi} \cdot \frac{i.i}{r}$

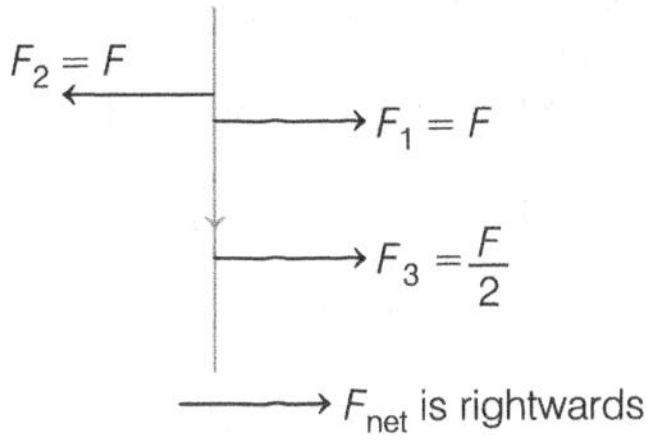

Current in same direction means attraction and current in opposite direction means repulsion. Let us find force on wire-2 by other three wires 1,3 and 4.

13. Magnetic force on BC and DA can be obtained by integration, as magnetic field of I_0 is non-uniform over these two wires. But these two forces are equal and opposite. So, net force on the frame is :

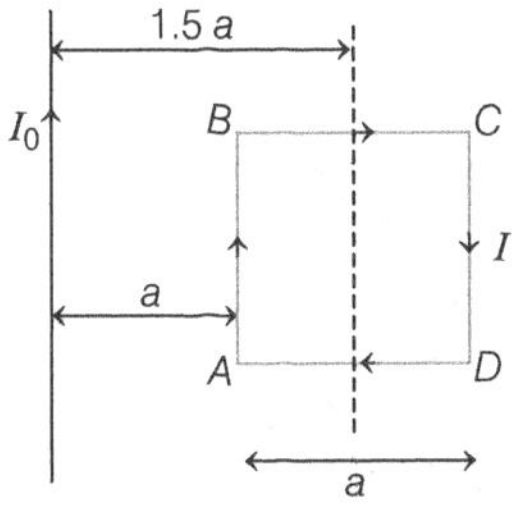

$F =$ Attractive force on AB − repulsive force on CD

Force per unit length of two parallel current carrying wires is given by, $\frac{\mu_0}{2\pi} \frac{I_1 I_2}{r}$. Hence,

$$F = \left[\frac{\mu_0}{2\pi} \frac{I I_0}{a}\right](a) - \left[\frac{\mu_0}{2\pi} \frac{I I_0}{2a}\right](a)$$

$$= \frac{\mu_0 I I_0}{4\pi}$$

Substituting the values, we get

$$F = 0.4\,\mu\text{N}$$

14. In problem 7, we have obtained the expression of B at centre due to long half ring of radius R. If current is inwards then magnetic field is towards right as shown in figure. This magnetic field is given by,

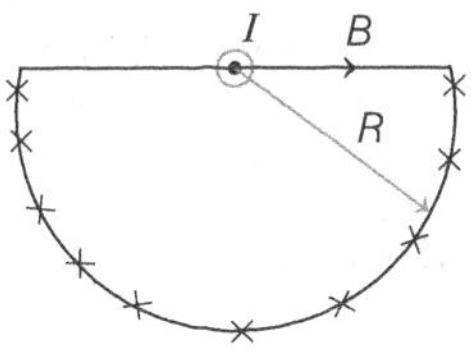

$$B = \frac{\mu_0 I}{\pi R^2}$$

Now, magnetic force per unit length on the wire at centre will be,

$$\frac{F}{l} = (I)(1)(B)\sin 90^\circ$$

$$\therefore \quad \frac{F}{l} = \frac{\mu_0 I^2}{\pi^2 R}$$

15. Force per unit length between parallel current carrying wires is given by,

$$\frac{F}{l} = \frac{\mu_0}{2\pi} \frac{I_1 I_2}{r}$$

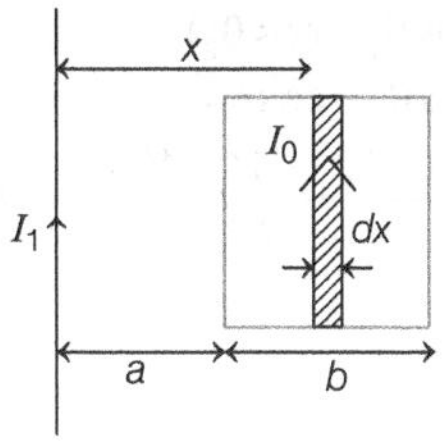

Small current I_0 in the element shown in figure is

$$I_0 = \left(\frac{I_2}{b}\right) dx$$

Force per unit length on this element will be,

$$\frac{\mu_0}{2\pi} \frac{I_1 I_0}{x}$$

Integrating this result with the limits from $x = a$ to $x = a + b$, we get total force per unit length. The final result comes out to be,

$$\frac{F}{l} = \left(\frac{\mu_0}{4\pi}\right)\left(\frac{2I_1 I_2}{b}\right)\ln\left(1 + \frac{b}{a}\right)$$

Module 8.

1. Given, magnetic moment,

$$M = 5 \text{ A-m}^2$$

Magnetic field, $B = 0.4$ T

Initial orientation, $\theta_i = 0^\circ$ (parallel to **B**)

Final orientation, $\theta_f = 180^\circ$ (anti-parallel to **B**)

Work done in rotating a dipole from θ_i position to θ_f position,

$$W = MB(\cos\theta_i - \cos\theta_f)$$
$$= 5 \times 0.4(\cos 0^\circ - \cos 180^\circ)$$
$$= 2[1 - (-1)]$$
$$\Rightarrow \quad W = 4\text{ J}$$

2. Given, $\frac{I_1}{I_2} = \frac{3}{2}, T_1 = 3\text{ s}, T_2 = 4\text{ s}$

As, $T = 2\pi\sqrt{\frac{I}{MB}} \Rightarrow T^2 = 4\pi^2 \frac{I}{MB}$

$$\Rightarrow \quad T^2 \propto \frac{I}{M} \Rightarrow M \propto \frac{I}{T^2}$$

$$\Rightarrow \quad \frac{M_1}{M_2} = \left(\frac{I_1}{I_2}\right)\left(\frac{T_2}{T_1}\right)^2 = \frac{3}{2} \times \left(\frac{4}{3}\right)^2$$
$$= \frac{3}{2} \times \frac{16}{9} = \frac{8}{3}$$

$$\Rightarrow \quad \frac{M_1}{M_2} = \frac{8}{3} \Rightarrow M_1 : M_2 = 8 : 3$$

3. Given, magnetic moment

$$M = 2.0 \times 10^5 \text{ JT}^{-1}$$
$$B = 14 \times 10^{-5}, \quad \theta_2 = 60^\circ$$

Work done is rotating a dipole for $\theta = \theta_1$ to $\theta = \theta_2$ is given by

$$W = MB(\cos\theta_1 - \cos\theta_2)$$
$$= MB(\cos 0° - \cos 60°)$$
$$= \frac{MB}{2} = \frac{2 \times 10^5 \times 14 \times 10^{-5}}{2} = 14\text{ J}$$
$$W = 14\text{ J}$$

4. Here,

$$\theta = 30°, \tau = 0.018\text{ N-m}, B = 0.6\text{ T}$$

Torque on a bar magnet

$$\tau = MB\sin\theta$$
$$\Rightarrow \quad 0.018 = M \times 0.06 \times \sin 30°$$
$$\Rightarrow \quad 0.018 = M \times 0.06 \times \frac{1}{2}$$
$$\Rightarrow \quad M = 0.6\text{ A-m}^2$$

Position of stable equilibrium ($\theta = 0°$)

Position of unstable equilibrium ($\theta = 180°$)

Minimum work required to rotate bar magnet from stable to unstable equilibrium

$$\Rightarrow \quad U_i = -MB\cos\theta_i$$
$$= -MB\cos 0° = -MB$$
$$\Rightarrow \quad U_f = -MB\cos\theta_f$$
$$= -MB\cos 180° = MB$$
$$\Rightarrow \quad \Delta U = U_f - U_i$$
$$= MB - (-MB) = 2MB$$
$$\Rightarrow \quad W = 2MB = 2 \times 0.6 \times 0.06$$
$$\Rightarrow \quad W = 7.2 \times 10^{-2}\text{ J}$$

5. $$|\mathbf{B}_X| = \frac{\mu_0}{4\pi}\frac{2M}{(d/2)^3}$$

$$|\mathbf{B}_Y| = \frac{\mu_0}{4\pi}\frac{2M}{(d/2)^3}$$

Since, $|\mathbf{B}_Y| = |\mathbf{B}_X|$

Thus, the resultant magnetic field ($\mathbf{B}_{net}$) at P will be at 45° with the horizontal.

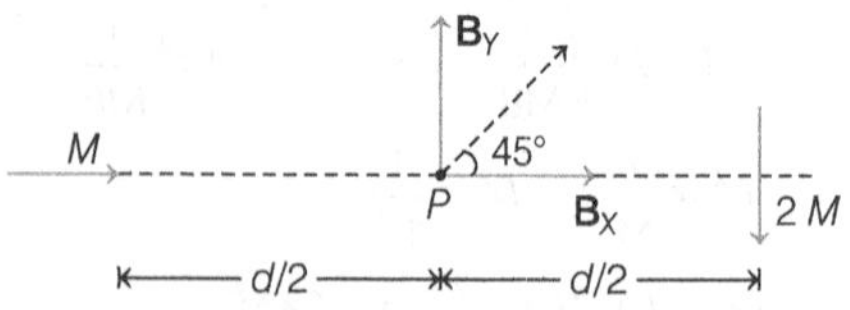

This means, direction of $\mathbf{B}_{net}$ and velocity of the charged particle is same.

∴ Force on the charged particle moving with velocity v in the presence of magnetic field which is

$$\mathbf{B} = q(\mathbf{B} \times \mathbf{B}) = q|\mathbf{v}||\mathbf{B}|\sin\theta$$

where, θ is the angle between **B** and **v**.

According to the above analysis, we get

$$\theta = 0$$
$$\therefore \quad \mathbf{F} = 0$$

Thus, magnitude of force on the particle at that instant is zero.

6. Here, $r = 30\text{ cm} = 0.3\text{ m}$

We know, $\frac{\mu_0 M}{4\pi r^3} = B_H = 3.6 \times 10^{-5}$

M = Magnetic moment

$$\Rightarrow \quad M = \frac{3.6 \times 10^{-5}}{10^{-7}}(0.3)^3$$

Hence, $M = 9.7\text{ A}^2\text{-m}$

7. Magnetic lines form closed loop. Inside magnet, these are directed from south to north-pole.

8. The force on north pole $= m\mathbf{B}_1$

The force on south pole $= m\mathbf{B}_2$

At different position, the value of magnetic field is different. So, the needle experience different forces in different direction.

Therefore, the magnetic needle experience a force as well as a torque.

9.

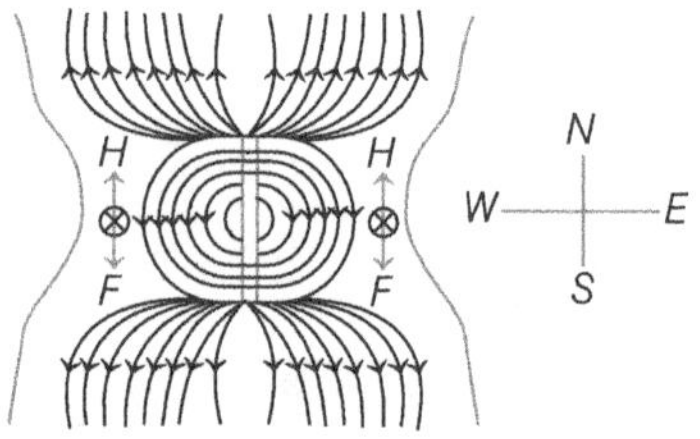

10. $M = m \cdot 2l$

$$M = 14.4 \times 0.25$$
$$\tau = \mathbf{M} \times \mathbf{B} = MB\sin\theta$$

Torque due to force $= F \times 0.12$

For equilibrium,

$$F \times 0.12 = MB\sin\theta$$
$$F = \frac{14.4 \times 0.25\sin 60}{0.12}$$
$$F = 25.98\text{ N}$$

If the force F is removed, then the torque due to magnetic field will move the bar magnet. It will start oscillating about the mean position where the angle between M and B is θ.

Module 9.

1. Given, angle of dip $(\theta) = 30°$,

$$B_H = 0.5\text{ G}$$

We have,

$$B_H = B\cos\theta$$
$$B = B_H\sec\theta = (0.5\sec 300)\text{ G}$$
$$= \left(0.5 \times \frac{2}{\sqrt{3}}\right)\text{G} = \frac{\text{G}}{\sqrt{3}}$$

Earth's total magnetic field at a certain place is $\frac{1}{\sqrt{3}}$ G.

2. Given that vertical component of the Earth's magnetic field $B_v = 6 \times 10^{-3}$ T and angle of dip, $\theta = 37°$

$$\therefore \quad B_V = B\sin\theta$$

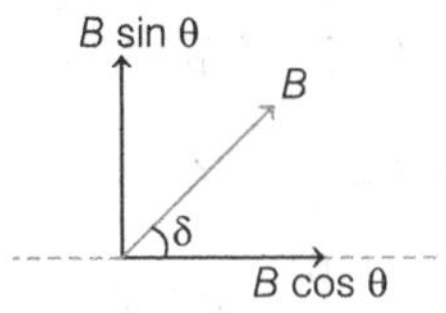

and $\sin 37° = \frac{3}{5}$

$\therefore \quad 6 \times 10^{-5} = B \times \frac{3}{5}$

$\Rightarrow \quad B = \frac{5 \times 6 \times 10^{-5}}{3} = 10^{-4}\ \text{T}$

$\Rightarrow \quad B = 1 \times 10^{-4}\ \text{T}$

3. $\tan\theta' = \frac{\tan\theta}{\cos\beta}$

$\tan\theta = \tan\theta' \cos\beta = \tan 60° \cos 45°$

$= \sqrt{3} \times \frac{1}{\sqrt{2}} = \sqrt{\frac{3}{2}}$

$\theta = \tan^{-1}\left(\sqrt{\frac{3}{2}}\right)$

4. Given, at first place, angle of dip,

$\theta_1 = 45°$

Time period, $T_1 = \frac{60}{30} = 2\ \text{s}$

At second place, angle of dip,

$\theta_2 = 30°$

Time period, $T_2 = \frac{60}{40} = \frac{3}{2}\ \text{s}$

Now, at first place,

$B_{H_1} = B_1 \cos\theta_1$

$= B_1 \cos 45° = \frac{B_1}{\sqrt{2}}$ ...(i)

and at second place,

$B_{H_2} = B_2 \cos\theta_2$

$= B_2 \cos 30° = \frac{\sqrt{3}}{2} B_2$ ...(ii)

Also, time period of a magnetic needle is given by

$T = 2\pi\sqrt{\frac{I}{MB_H}}$ ...(iii)

$\therefore \quad T \propto \sqrt{\frac{1}{B_H}}$ or $\frac{T_1}{T_2} = \sqrt{\frac{B_{H_2}}{B_{H_1}}}$ ...(iv)

By putting the values from Eqs. (i) and (ii) into Eq. (iv), we get

$\frac{2}{\frac{3}{2}} = \sqrt{\frac{\frac{\sqrt{3}B_2}{2}}{\frac{B_1}{\sqrt{2}}}}$ or $\left(\frac{4}{3}\right)^2 = \frac{\sqrt{3} \times \sqrt{2}\, B_2}{2B_1}$

$\Rightarrow \quad \frac{B_1}{B_2} = \frac{\sqrt{3} \times \sqrt{2}}{2} \times \frac{9}{16}$

$\Rightarrow \quad \frac{B_1}{B_2} = \frac{9\sqrt{3}}{16\sqrt{2}}$

$\Rightarrow \quad \frac{B_1}{B_2} = \frac{9 \times 1.732}{16 \times 1.414} = \frac{15.588}{22.624}$

$\Rightarrow \quad \frac{B_1}{B_2} = 0.689 \approx 0.7\ \text{T}$

5. In stable equilibrium position ($\theta = 0°$ between **M** and **B**), $\tau = 0$. When displaced by a smalll angle θ, net restoring torque,

$\tau = -MB \sin\theta$

For small value of θ,

$\sin\theta \approx \theta$

$\therefore \quad \tau = -(MB)\,\theta$

This is of the type $\tau = -K\theta$. So, motion is angular SHM.
Time period is given by,

$T = 2\pi\sqrt{\frac{I}{K}}$

$\therefore \quad T = 2\pi\sqrt{\frac{I}{MB}}$ (Here, $K = MB$)

$B \propto \frac{1}{T^2}$

T is decreased by 5-times. So, induction of field will increase by 25-times.

6. $T \propto \frac{1}{\sqrt{B \cos\theta}}$

$\Rightarrow \quad \frac{T_1}{T_2} = \sqrt{\frac{B_2 \cos\theta_2}{B_1 \cos\theta_1}}$

$\Rightarrow \quad \frac{60/20}{60/30} = \sqrt{\frac{B_2 \cos 60°}{B_1 \cos 30°}}$

$\Rightarrow \quad \frac{3}{2} = \sqrt{\frac{B_2}{\sqrt{3}B_1}}$

$\Rightarrow \quad \frac{9}{4} = \frac{B_2}{\sqrt{3}B_1}$

$\Rightarrow \quad \frac{B_1}{B_2} = \frac{4}{9\sqrt{3}} = \frac{4}{\sqrt{243}}$

Module 10.

1. For a diamagnetic material,

$\mu_r < 1$

$\Rightarrow \quad -1 \le \chi < 0$

Also diamagnetic material moves from stronger magnetic field to weaker magnetic fields.
Hence, statement I and II both are correct.

2. Diamagnetism is non-cooperative behaviour of orbitting electrons when exposed to an applied magnetic field.
Behaviour of the diamagnetic material is that the susceptibility is temperature independent.
Therefore, statement, I is false and statement II is true.

3. According to Curie's law, magnetic susceptibility of paramagnetic substance is inversely proportional to its temperature i.e. $\chi = C/T$, where C is Curie constant.

Same is applicable for ferromagnetic substances and the relation is given as

$$\chi = \frac{C}{T - T_C}$$

where, T_C is Curie temperature.

Thus, from above we conclude that susceptibility of paramagnetic and ferromagnetic substances increases with decrease of temperature.

Diamagnetism is a result of motions of electrons developing magnetic moments opposite to the applied magnetic field.

4. Magnetic field inside solenoid with air as core material is

$$B_1 = \mu_0 nI$$

Magnetic field with a medium inside solenoid is

$$B_2 = \mu_0 \mu_r nI$$

$$\Rightarrow \quad B_2 = \mu_0 (1 + \chi) nI$$

where, $\mu_r = 1 + \chi$

Fractional increase of field magnitude is

$$\frac{B_2 - B_1}{B_1} = \frac{\mu_0 (1+\chi) nI - \mu_0 nI}{\mu_0 nI}$$

$$\Rightarrow \quad \mu_0 \chi = 1.2 \times 10^{-5}$$

5. Susceptibility of paramagnetic material, χ is 99.

$$\mu_r = 1 + \chi = 1 + 99 = 100$$

$$\mu = \mu_0 \mu_r = \mu_0 \times 100$$

$$= 4\pi \times 10^{-7} \times 100$$

$$= 4\pi \times 10^{-5}$$

$$\mu = 4\pi \times 10^{-5} \text{ Wb/A-m}$$

6. Soft iron is used in electromagnet because of

(i) **High permeability**

As $B = \mu H$

So, higher is the value of μ, higher will be magnetic field (B) due to electromagnet.

(ii) **Low retentivity**

When current is removed then no residual magnetic field (retentivity). So retentivity after demagnetisation should be zero.

7. When external field applied, the domain in which magnetic dipole aligned opposite to magnetic field decrease in size, while when magnetic field aligned in same direction to magnetic field increase in size.

8. Given,

$$I_1 = 6 \text{ A/m}, B_1 = 0.4 \text{ T}, T_1 = 4 \text{ K}$$

$$B_2 = 0.3 \text{ T}, \quad T_2 = 24 \text{ K}, I_2 = ?$$

For paramagnetic material. According to curies law

$$\chi \propto \frac{1}{T}$$

For two temperatures T_1 and T_2,

$$\chi_1 T_1 = \chi_2 T_2$$

But $\quad \chi = \frac{I}{B} \Rightarrow \frac{I_1}{B_1} T_1 = \frac{I_2}{B_2} T_2$

$$\Rightarrow \quad \frac{6}{0.4} \times 4 = \frac{I_2}{0.3} \times 24$$

$$\Rightarrow \quad I_2 = \frac{0.3}{0.4} = 0.75 \text{ A/m}$$

$$\Rightarrow \quad I_2 = 0.75 \text{ A/m}$$

9. Permanent magnets (P) are made of materials with large retentivity and large coercivity. Transformer cores (T) are made of materials with low retentivity and low coercivity.

10. From Curie's law for paramagnetic substance, we have

Magnetic susceptiblity $\chi \propto \frac{1}{T}$

$$\therefore \quad \frac{\chi_2}{\chi_1} = \frac{T_1}{T_2} \Rightarrow \chi_2 = \frac{\chi_1 \cdot T_1}{T_2}$$

$$\Rightarrow \quad \chi_2 = \frac{2.8 \times 10^{-4} \times 350}{300} = 3.267 \times 10^{-4}$$

11. Given, side of cube $= 1 \text{ cm} = 10^{-2}$ m

$\therefore \quad$ Volume, $V = 10^{-6} \text{ m}^3$

Dipole moment, $M = 20 \times 10^{-6}$ J/T

Applied magnetic intensity,

$$H = 60 \times 10^3 \text{ A/m}$$

Intensity of magnetisation

$$I = \frac{M}{V} = \frac{20 \times 10^{-6}}{10^{-6}} = 20 \text{ A/m}$$

Now, magnetic susceptibility χ is

$$\chi = \frac{\text{Intensity of magnetisation}}{\text{Applied magnetic intensity}} = \frac{I}{H} = \frac{20}{60 \times 10^3}$$

$$\Rightarrow \chi = \frac{1}{3} \times 10^{-3} = 3.33 \times 10^{-4}$$

12. Coercivity of a bar magnet is the value of magnetic field intensity (H) that is needed to reduce magnetisation to zero.

Since, for a solenoid magnetic induction is given as,

$$B = \mu_0 nI \quad \text{...(i)}$$

where, n is the number of turns (N) per unit, length (l) and I is the current.

Also, $\quad B = \mu_0 H \quad$ ...(ii)

$\therefore$ From Eqs. (i) and (ii), we get

$$\mu_0 nI = \mu_0 H \quad \text{or} \quad H = nI = \frac{N}{l} I$$

Substituting the given values, we get

$$H = \frac{100}{0.2} \times 5.2 = 2600 \text{ A/m}$$

Thus, the value of coercivity of the bar magnet is 2600 A/m.

13. Given, number of turns,

$$n = 1000 \text{ turns/cm}$$

$$= 1000 \times 100 \text{ turns/m}$$

From figure,
Coercivity of ferromagnet,

$$H = 100 \text{ A/m}$$

Using, $H = nI$ or $100 = 10^5 \times I$

$$\therefore \quad I = \frac{100}{10^5} = 1 \text{ mA}$$

14. For electromagnets and transformers we need less hysteresis loop or small area of hysteresis loss.

15. $M = iA$

$$M = NiA \quad (\because N = \text{total number of turns})$$

$$M \text{ (mag. moment/volume)} = \frac{NiA}{Al}$$

$$= \frac{Ni}{l} = \frac{(500)15}{25 \times 10^{-2}} = 30000 \text{ Am}^{-1}$$

$$M = 30000 \text{ Am}^{-1}$$

16. For solenoid, the magnetic field needed to be demagnetised the magnet is

$$B = \mu_0 nI$$

where, $N = 100, l = 10 \text{ cm} = \frac{10}{100} \text{m} = 0.1 \text{ m}$

$$\Rightarrow 3 \times 10^3 = \frac{100}{0.1} \times I \Rightarrow I = 3\text{A}$$

17. Given value of aluminium susceptibility, $\chi = 2.2 \times 10^{-5}$

$$B = B_0(1 + \chi)$$

$$\Rightarrow \quad B - B_0 = B_0\chi \Rightarrow \frac{B - B_0}{B_0} = \chi$$

Percentage increase in magnetic field,

$$\frac{B - B_0}{B_0} \times 100 = 100\chi = 100 \times 2.2 \times 10^{-5}$$

$$= 2.2 \times 10^{-3} = \frac{22}{10^4}$$

$\therefore$ Value of $x = 22$

18. Given, number of turns,

$$n = 1000 \text{ m}^{-1}$$

Electric current, $i = 0.75$ A
Relative permeability of 1st material

$$\mu_1 = 500$$

Relative permeability of 2nd material

$$\mu_2 = 750$$

we know that $\mu_r = 1 + \chi$

$$\Rightarrow \quad \chi = \mu_r - 1$$

For 1st and 2nd material,

$$\chi_1 = \mu_1 - 1 = 500 - 1 = 499$$

and $\chi_2 = \mu_2 - 1 = 750 - 1 = 749$

Now, magnetic intensity, $M = ni$

Let M_1 and M_2 be the magnetisation and m_1 and m_2 be the magnetic moments of material 1st and 2nd.

Using, $\frac{I_1}{M} = \chi_1$ ...(i)

and $\frac{I_2}{M} = \chi_2$ ...(ii)

Dividing Eq. (i) by Eq. (ii),

$$\frac{I_1}{I_2} = \frac{\chi_1}{\chi_2} \quad \text{...(iii)}$$

We also know that,

$$\text{Magnetisation} = \frac{\text{Dipole moment in core material}}{\text{Volume}}$$

i.e. $I = \frac{m}{V}$

$$\therefore \quad \frac{I_1}{I_2} = \frac{m_1}{m_2} \quad \text{...(iv)}$$

From Eqs. (iii) and (iv), we get

$$\frac{m_1}{m_2} = \frac{\chi_1}{\chi_2} = \frac{499}{749}$$

Hence, fractional change in magnetic moment,

$$\frac{\Delta m}{m_1} = \frac{m_2 - m_1}{m_1} = \left(\frac{m_2}{m_1} - 1\right)$$

$$= \left(\frac{749}{499} - 1\right) = \frac{250}{499}$$

Thus, value of χ is 250.

Module 11.

1. Given, if a shunt of $\frac{1}{10}$th of the coil resistance is applied to a moving coil galvanometer, then

$$\frac{I_g}{I} = \frac{S}{S + G}$$

$$\frac{I_g}{I} = \frac{(G/10)}{(G/10 + G)} = \frac{1}{11}$$

Initially, sensitivity of a galvanometer,

$$\alpha_i = \theta/I_g \quad \text{...(i)}$$

Finally after shunt, it becomes

$$\alpha_f = \theta/I \quad \text{...(ii)}$$

On dividing Eq. (ii) by Eq. (i), we get

$$\therefore \quad \frac{\alpha_f}{\alpha_i} = \frac{\theta/I}{\theta/I_g} = \frac{1}{11}$$

2. $i = \left(\frac{k}{NBA}\right)\phi$

$$\therefore \quad B = \frac{k\phi}{NiA} = \frac{(10^{-8})(90)}{100 \times 10^{-6} \times 10^{-4}}$$

$$= 90 \text{ T}$$

3. $i = \left(\frac{k}{NBA}\right)\phi$

$$= \frac{(0.125 \times 10^{-7})(6)(\pi/180)}{200 \times 5 \times 10^{-2} \times 5 \times 2 \times 10^{-4}}$$

$$= 1.3 \times 10^{-7} \text{ A}$$

Unit Exercise (Cumulative)

Level 1

1. Radius of circular path is given by

$$R = \frac{mv}{Bq} = \frac{p}{Bq} = \frac{\sqrt{2Km}}{Bq}$$

Radius in both the cases are equal. Therefore,

$$\frac{\sqrt{2K_\alpha m_\alpha}}{Bq_\alpha} = \frac{\sqrt{2K_d m_d}}{2.3Bq_d}$$

$$\frac{q_d}{q_\alpha} = \frac{e}{2e} = \frac{1}{2}$$

and $\frac{m_\alpha}{m_d} = \frac{4}{2} = 2$

$$\therefore \quad K_d = \left(\frac{2.3q_d}{q_\alpha}\right)^2 \left(\frac{m_\alpha}{m_d}\right) \cdot K_\alpha$$

$$K_d = \left(2.3 \times \frac{1}{2}\right)^2 (2)(5.3) \text{ MeV},$$

$$K_d = 14.0185 \text{ MeV}$$

2. $\tau = MB \sin 90° \neq 0$

3. Magnetic field is outwards and increasing with x. So, magnetic force will also increase with x. The force on different sections are as shown in figure.

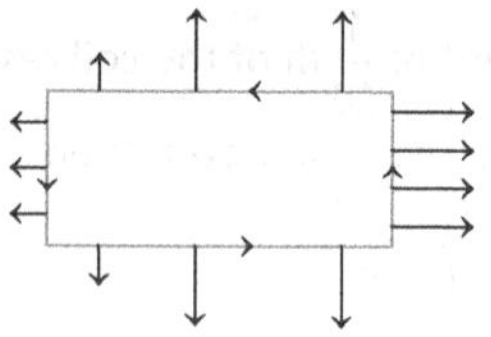

Force will act in positive x-direction. But, no torque will act.

4. To the right of B, both fields are downwards and to the left of A both fields are upwards.

6.

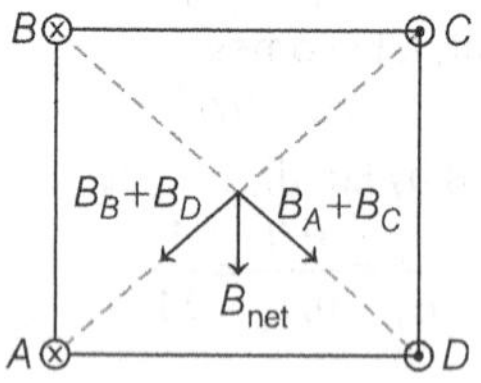

7. I_1 produces circular magnetic lines current I_2 is each small circular element is parallel to $(\theta = 0°)$ magnetic field. Hence, force is zero.

8.

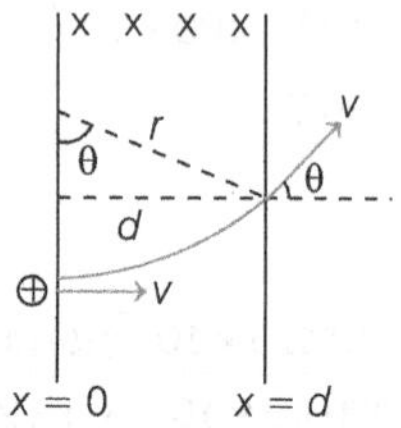

$$\sin\theta = \frac{d}{r} = \frac{d}{(mv/Bq)}$$

$$\Rightarrow \quad \frac{q}{m} = \frac{v\sin\theta}{Bd}$$

9. $\tau = M \times B = MB\sin\theta$

For stable $\tau = 0$, $\mathbf{M} \| \mathbf{B}$

For unstable $\mathbf{M} \| -\mathbf{B}$

10.

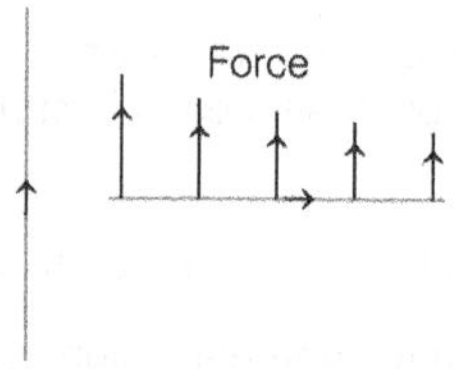

Variation of magnetic force on wire ACB is as shown in figure. Point of application of net force lies some where between A and C.

11. $B_1 = \frac{\mu_0 I R^2}{2(R^2 + x^2)^{3/2}}$

$$B_2 = \frac{\mu_0 I (2R)^2}{2[(2R)^2 + (2x)^2]^{3/2}}$$

$$\frac{B_1}{B_2} = 2$$

12. $\frac{r}{2}$ = radius of circular path

$$= \frac{mv}{Bq} = \frac{mv}{(\mu_0 ni)(q)}$$

$$\therefore \quad v = \frac{\mu_0 q r n i}{2m}$$

13. All magnetic fields are cancelling each other.

14. Deviation, $\theta = \sin^{-1}\left(\frac{L}{R}\right) = \sin^{-1}\left(\frac{\sqrt{3}}{2}\right) = 60°$

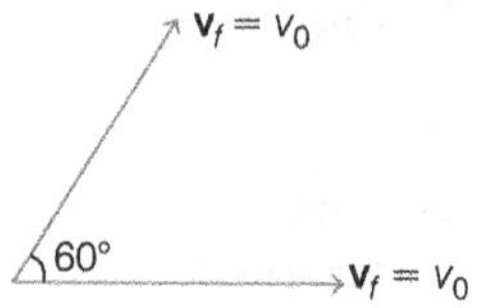

$$|\Delta \mathbf{v}| = 2v_0 \sin\frac{\theta}{2} = 2v_0 \sin 30° = v_0$$

15. Magnetic force does not change the speed of charged particle. Hence, $v = u$. Further magnetic field on the electron in the given condition is along negative y-axis in the starting. Or it describes a circular path in clockwise direction. Hence, when it exits from the field, $y < 0$.

16. If the current flows out of the paper, the magnetic field at points to the right of the wire will be upwards and to the left will be downwards as shown in figure.

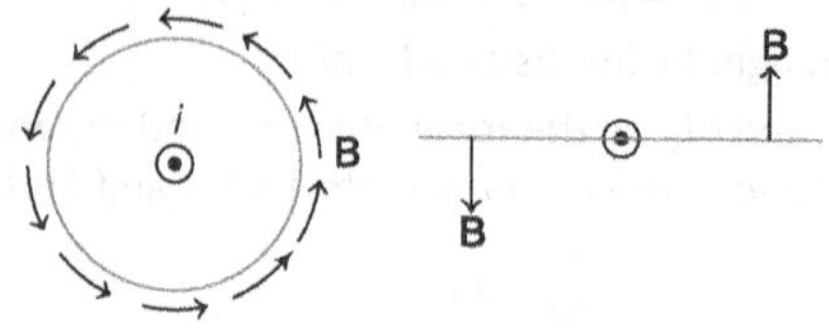

Now, let us come to the problem.

Magnetic field at $C = 0$

Magnetic field in region BX' will be upwards (+ ve) because all points lying in this region are to the right of both the wires.

Magnetic field in region AC will be upwards (+ve), because points are closer to A, compared to B. Similarly, magnetic field in region BC will be downwards (−ve).

Graph (b) satisfies all these conditions. Therefore, correct answer is (b).

17. $A \to III$

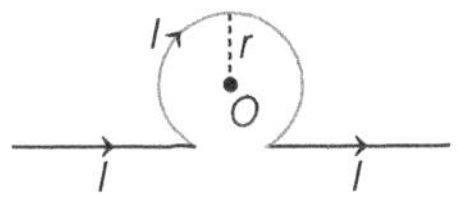

Magnetic field at centre O,

B_0 = Magnetic field due to circular wire − 2 × magnetic field due to straight wire

$$= \frac{\mu_0 I}{2r} - 2 \times \frac{\mu_0 I}{4\pi r} = \frac{\mu_0 I}{2\pi r}(\pi - 1)$$

$B \to I$

Magnetic field at point O

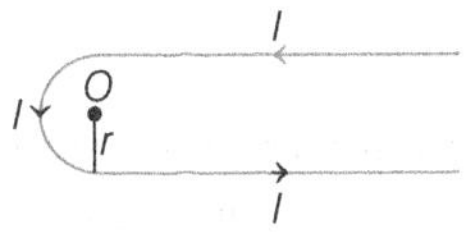

$B_0 = 2 \times$ Magnetic field due to straight wire + Magnetic field due to semi-circular wire.

$$= 2 \times \frac{\mu_0 I}{4\pi r} + \frac{1}{2} \cdot \frac{\mu_0 I}{2r} = \frac{\mu_0 I}{4\pi r}(2 + \pi)$$

$C \to IV$

Magnetic field at O

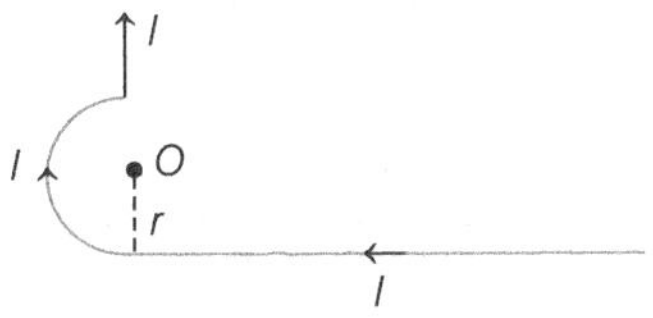

B_0 = Magnetic field due to straight wire + Magnetic field due to semi-circular wire,

$$= \frac{\mu_0 I}{4\pi r} + \frac{1}{2} \cdot \frac{\mu_0 I}{2r}$$

$$= \frac{\mu_0 I}{4\pi r}(1 + \pi)$$

$D \to II$,

Magnetic field due to straight wire will be zero.

∴ B_0 = Magnetic field due to semi-circular wire.

$$= \frac{1}{2} \times \frac{\mu_0 I}{2r} = \frac{\mu_0 I}{4r}$$

18. $\frac{F}{L} = \frac{\mu_0}{2\pi} \frac{I_1 I_2}{a}$

∴ $F = \frac{\mu_0}{2\pi} \frac{I_1 I_2}{a} L$ (Repulsion or upwards)

M of the loop is inwards and magnetic field to I_1 on the plane of loop is outwards. Hence, $\tau = 0$, as $\tau = \mathbf{M} \times \mathbf{B}$ and angle between two is 180°.

19. $B_A = B_B = B_C = B_D$

$$= B = \frac{\mu_0}{2\pi} \frac{i}{r} = \frac{(2 \times 10^{-7})(5)}{(0.2/\sqrt{2})}$$

$$= (5.0\sqrt{2}) \times 10^{-6} \text{ T}$$

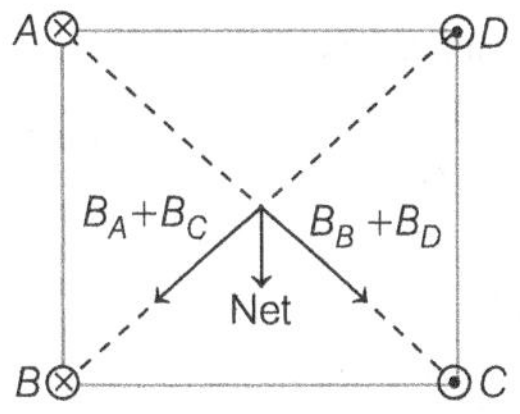

Net magnetic field

$$= \sqrt{(B_A + B_C)^2 + (B_B + B_D)^2} = 2\sqrt{2}B$$

$$= 2.0 \times 10^{-5} \text{ T}$$ (towards bottom as shown)

20. Net magnetic field at centre C,

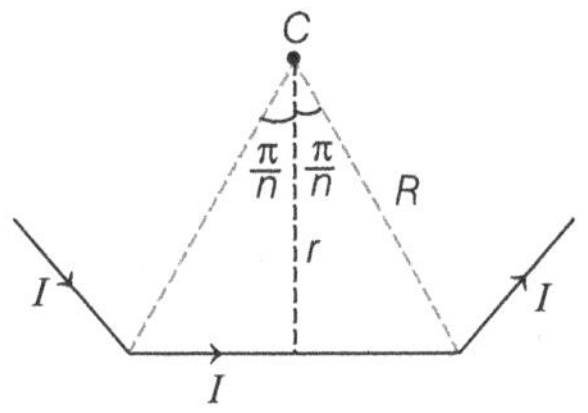

$$B = n\left[\frac{\mu_0}{4\pi} \frac{I}{r}\left(\sin\frac{\pi}{n} + \sin\frac{\pi}{n}\right)\right]$$

$$= n\left[\left(\frac{\mu_0}{4\pi}\right)\left(\frac{I}{R\cos\left(\frac{\pi}{n}\right)}\right)\left(2\sin\frac{\pi}{n}\right)\right]$$

or $$B = \frac{n\mu_0 I \tan(\pi/n)}{2\pi R}$$

This expression can also be written as

$$B = \frac{\mu_0 I}{2R}\left[\frac{\sin \pi/n}{\pi/n}\right]\left[\frac{1}{\cos \pi/n}\right]$$

As $n \to \infty, \frac{\pi}{n} \to 0$

Therefore $\frac{\sin \pi/n}{\pi/n} \to 1$

and $\cos\left(\frac{\pi}{n}\right) \to 1$

∴ $$B = \frac{\mu_0 I}{2R}$$

Note *This is the expression of magnetic field at the centre of a circular current carrying loop. The reason is obvious. As $n \to \infty$, polygon will convert into a circle.*

21. Net magnetic field at P due to figure (ii) is

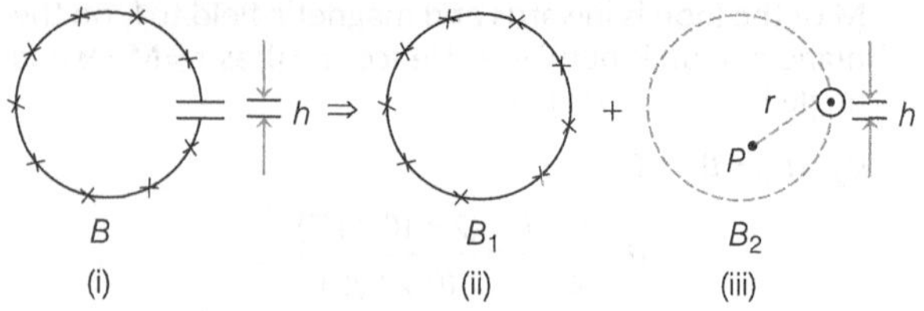

$$B_1 = 0$$

and due to figure (iii) is

$$B_2 = \frac{\mu_0}{2\pi}\frac{i}{r} \quad \text{(due to infinitely long thin wire)}$$

Here, $i \approx \left(\frac{I}{2\pi R}\right)(h)$

$$\therefore \quad B_2 = \left(\frac{\mu_0}{2\pi}\right)\left(\frac{Ih}{2\pi R}\right)\left(\frac{1}{r}\right)$$

$$B_2 = \frac{\mu_0 hI}{4\pi^2 Rr}$$

Therefore net magnetic field at P is,

$$B = B_1 + B_2 \quad \text{or} \quad B = \frac{\mu_0 hI}{4\pi^2 Rr}$$

22. $i = \left(\frac{I}{\pi}\right)d\theta$

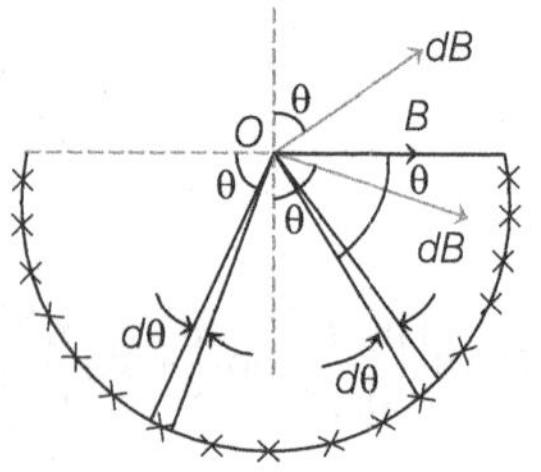

At O, $dB\cos\theta$ components are cancelled and $dB\sin\theta$ components are added (towards right).

$$\therefore \quad B = \int_0^{\pi} dB\sin\theta = \int_0^{\pi}\left(\frac{\mu_0}{2\pi}\right)\left(\frac{i}{R}\right)\sin\theta$$

$$= \int_0^{\pi}\frac{\mu_0 I}{2\pi^2 R}\sin\theta \cdot d\theta = \frac{\mu_0 I}{\pi^2 R} \qquad \left[\text{as } i = \left(\frac{I}{\pi}\right)d\theta\right]$$

$$\therefore \quad B = \frac{\mu_0 I}{\pi^2 R}$$

23. Number of turns in thickness dx,

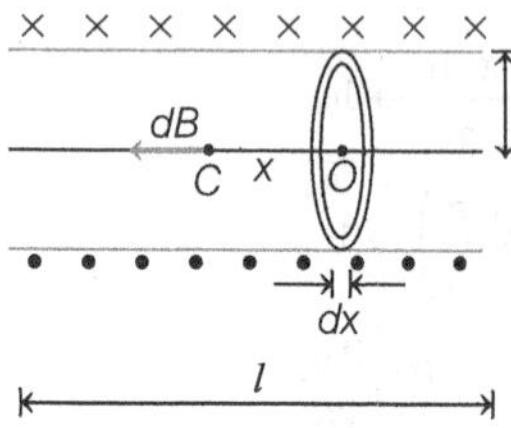

$$N = (n)\,dx$$

Point C is on the axis of circular coil of thickness dx as shown in figure. Small magnetic field dB is,

$$dB = \frac{\mu_0 NIR^2}{2(R^2 + x^2)^{3/2}} = \left[\frac{\mu_0 nIR^2}{2(R^2 + x^2)^{3/2}}\right]dx$$

Integrating dB from $x = -\frac{l}{2}$ to $x = +\frac{l}{2}$, we get total magnetic field, which comes out to be,

$$B = \frac{\mu_0 NI}{\sqrt{1 + \left(\frac{2R}{l}\right)^2}}$$

24. $B = \frac{\mu_0 I}{2R}$

$$\therefore \quad I = \frac{2BR}{\mu_0}$$

Now, magnetic moment,

$$M = IA = (I)(\pi R^2) = \left(\frac{2BR}{\mu_0}\right)(\pi R^2)$$

$$= \frac{(BR^3)}{(\mu_0/2\pi)} = \frac{(6 \times 10^{-6})(0.1)^3}{2 \times 10^{-7}}$$

$$= 3 \times 10^{-2}\ \text{A-metre}^2$$

or $\quad M = 30\ \text{mA-metre}^2$

25. In $(b - a)$ thickness total number of turns are N.

$\therefore$ Number of turns in small thickness dr as shown in figure are,

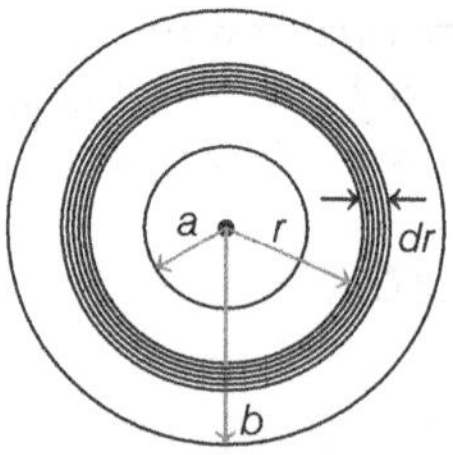

$$N_0 = \left(\frac{N}{b - a}\right)dr$$

(a) Small magnetic field at centre due to a circular coil of radius r and number of turns N_0 is,

$$dB = \frac{\mu_0 N_0 I}{2r}$$

After substituting value of N_0 and then integrating dB from $r = a$ to $r = b$, we get net magnetic field, which comes out to be,

$$B = 7\ \mu\text{T}$$

(b) Small magnetic moment due to the element shown in figure is,

$$dM = N_0 IA = N_0 I(\pi r^2)$$

After substituting the values and then integrating dM from $r = a$ to $r = b$, we get complete magnetic moment and which comes out to be,

$$M = 15\ \text{mA-metre}^2$$

26. Small area of the element shown in figure,

$$dA = (2\pi r)\,dr$$

Small charge on this element,

$$dq = (\sigma)\,dA = (2\pi\sigma r)\,dr$$

This rotating charge is equivalent to a current given by,

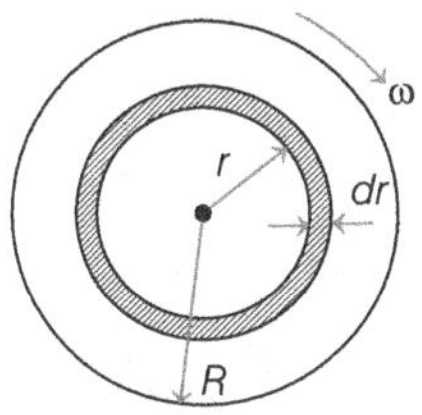

$$I = (dq)(f) \qquad [f = \text{frequency}]$$

$$= (2\pi\sigma r dr)\left(\frac{\omega}{2\pi}\right) = (\sigma\omega r dr)$$

(a) Small magnetic field at centre,

$$dB = \frac{\mu_0 I}{2r} = \frac{(\mu_0\sigma\omega)\,dr}{2}$$

Integrating dB from O to R, we get total magnetic field which comes out to be,

$$B = \frac{\mu_0\sigma\omega R}{2}$$

(b) $\frac{M}{L} = \frac{q}{2m}$

$$\therefore \quad M = \left(\frac{q}{2m}\right)(L) = \left(\frac{q}{2m}\right)(I\omega)$$

$$= \left[\frac{(\sigma)(\pi R^2)}{2m}\right]\left[\frac{1}{2}mR^2\right]\omega$$

or $$M = \frac{\pi\sigma\omega R^4}{4}$$

Alternate method of part (b)

Small magnetic moment of the element shown in figure is,

$$dM = IA = (\sigma\omega r dr)(\pi r^2)$$

$$= (\pi\sigma\omega r^3 dr)$$

Integrating dM from 0 to R, we can find total M.

27. **Magnetic moment** Volume charge density, $\rho = \frac{q}{\frac{4}{3}\pi R^3}$

This time it's a solid sphere. So, we will have to take volume of that disc. Small volume of that disc is,

$$dV = (\pi r^2)(R d\theta)$$

Small charge on this disc will be,

$$dq = \rho dV$$

Now let us use the result of Q-22 (b). There,

$$M = \frac{\pi\sigma\omega R^4}{4}$$

But $\sigma(\pi R^2) = q$

$$\therefore \quad M = \frac{q\omega R^2}{4}$$

So in this case small magnetic moment due to charge dq will be,

$$dM = \frac{(dq)\,\omega R^4}{4}$$

When we integrate dM from $\theta = 0°$ to $\theta = 180°$, we get complete magnetic moment which comes out to be,

$$M = \frac{qR^2\omega}{5}$$

Angular momentum

$$L = I\omega \quad \text{or} \quad L = \left(\frac{2}{5}mR^2\right)\omega$$

$$\therefore \quad \frac{M}{L} = \frac{q}{2m}$$

28. (a) At O magnetic field is only due to semi-circular arc,

$$B = \frac{1}{2}\left(\frac{\mu_0 I}{2R}\right) \quad \text{(perpendicular to paper inwards)}$$

$$\therefore \quad F = ILB\sin 90°$$

$$= (I)(1)\left[\frac{\mu_0 I}{4R}\right] \quad \text{or} \quad F = \frac{\mu_0 I^2}{4R}$$

Substituting the values of I and R, we get

$$F = 0.20 \text{ mN/metre}$$

(b) At O, magnetic field due to two very long straight wires,

$$B = 2\left[\frac{\mu_0}{4\pi}\frac{I}{(l/2)}(\sin 0° + \sin 90°)\right]$$

$$= \left(\frac{\mu_0 I}{\pi l}\right) \quad \text{(perpendicular to paper inwards)}$$

$$\therefore \quad F = ILB\sin 90°$$

$$= (I)(1)\left(\frac{\mu_0 I}{\pi l}\right) = \frac{\mu_0 I^2}{\pi l}$$

Substituting the given values of I and l, we get

$$F = 0.13 \text{ mN/metre}$$

29. Torque of magnetic force on coil = torque of counter weight

$$\therefore \quad NISB = (\Delta m)\,gl$$

$$\therefore \quad B = \frac{(\Delta m)gl}{NIS}$$

Substituting the values, we get

$$B = 0.4 \text{ T}$$

30. Magnetic field of I_2 at its centre,

$$B = \frac{\mu_0 I_2}{2b}$$

Magnetic moment of I_1, $M = I_1(\pi a^2)$

When planes make an angle 0°, then angle between **M** and **B** is also 0°. When planes make an angle θ, then angle between **M** and **B** is also θ. Thus,

$$U = -\mathbf{M}\cdot\mathbf{B} = -MB\cos\theta$$

or $$U = -\left(\frac{\mu_0\pi a^2}{2b}\right)(I_1 I_2)\cos\theta$$

Level 2

31. τ_{mg} about the left end (from where string is connected)

$$=|\mathbf{M}\times\mathbf{B}|=MB\sin 90°$$

or $(mgR)=(NiA)B_0=i(\pi R^2)B_0$

or $i=\dfrac{mg}{\pi RB_0}$

32. Apply screw law for finding magnetic field around a straight current carrying wire.

33. $\mathbf{F}_{MNPQ}=\mathbf{F}_{MQ}$ and this force should be upwards to balance the weight.

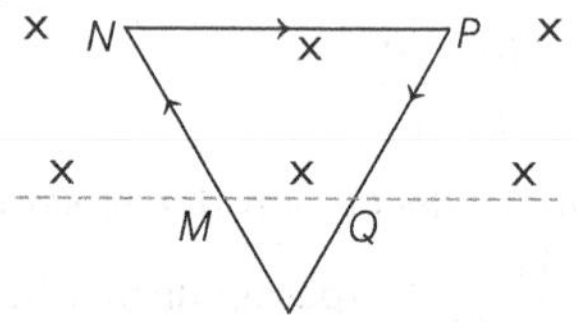

$\therefore$ $ilB=mg$, where

$$l=MQ=\frac{a}{2}$$

$\therefore$ $i(a/2)B=mg$

or $i=\dfrac{2Mg}{aB}$

Force is upwards if current is clockwise or current in MQ is towards right.

34. Magnetic force is always perpendicular to velocity. So, it will always act in radial direction which will change tension at different points. But, time period and θ will remain unchanged.

35. Force on the wires parallel to x-axis will be obtained by integration (as $B\propto x$ and x coordinates vary along these wires). But on a loop there are two such wires. Force on them will be equal and opposite.

Forces on two wires parallel to y-axis can be obtained directly (without integration) as value of B is same along these wires. But their values will be different (as x-coordinate and therefore B is different).

$F_{net}=\Delta F$ (on two wires)

$$=Ia(\Delta B)=Ia(B_0)(\Delta x)$$
$$=IaB_0(a)=IB_0a^2$$

This is indecent of x

$\therefore$ $F_1=F_2=IBla^2\neq 0$

36. $r=\dfrac{mv}{Bq}$ (during circular path)

$\therefore$ $v=\dfrac{Bqr}{m}$

Now, $qE=Bqv$

$\therefore$ $E=Bv=\dfrac{B^2qr}{m}$

$\therefore$ $E=\dfrac{(0.1)^2(20\times10^{-6})(5\times10^{-2})}{(20\times10^{-9})}$

$=0.5$ V/m

37. Magnetic field due to current in the wire along z-axis is zero. Magnetic field due to wire along x-axis is along $\hat{\mathbf{j}}$ direction and magnetic field due to wire along y-axis is along $-\hat{\mathbf{i}}$ direction. Both wires produce

$$B=\frac{\mu_0}{2\pi}\frac{i}{a}$$

38. $E=\dfrac{1}{4\pi\varepsilon_0}\dfrac{qx}{(R^2+x^2)^{3/2}}$

$$B=\frac{\mu_0 iR^2}{2(R^2+x^2)^{3/2}}$$

where, $i=qf=q\left(\dfrac{V}{2\pi R}\right)$

$$c=\frac{1}{\sqrt{\varepsilon_0\mu_0}}$$

39. For minimum value of B, it is colliding after one pitch. It will collide the target again if pitch is halved or remains same.

$$P=(v\cos\theta)\left(\frac{2\pi m}{Bq}\right) \quad \text{or} \quad P\propto\frac{v}{B}$$

40. Given, time $t=\dfrac{\pi m}{Bq}=\dfrac{T}{2}$ of proton

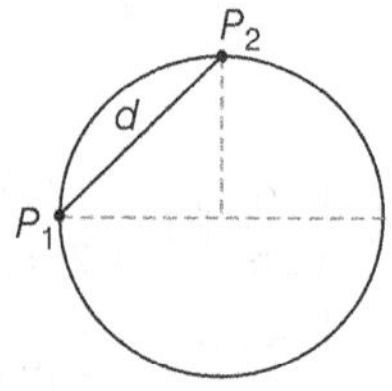

$$d=\sqrt{2}r=\frac{\sqrt{2}mv}{Bq}$$

41. $dA=(2\pi r)\,dr$

$$dq=\sigma\cdot dA=(2\pi\alpha r^3)\cdot dr$$
$$i=(dq)f=(dq)\frac{\omega}{2\pi}=(\alpha\omega r^3)\,dr$$
$$dB=\frac{\mu_0 i}{2r}=\frac{\alpha\mu_0\omega r^2 dr}{2}$$

$\therefore$ $B=\displaystyle\int_0^R dB=\frac{\mu_0\alpha\omega R^3}{6}$

42. $(mg)(R)=MB_x=I(\pi r^2)B_x \Rightarrow I=\dfrac{mg}{\pi rB_x}$

43. $F_B=\dfrac{\mu_0}{4\pi}\dfrac{q^2v^2}{d^2}$

$$F_e=\frac{1}{4\pi\varepsilon_0}\frac{qq}{d^2}$$

$\therefore$ $\dfrac{F_B}{F_e}=\varepsilon_0\mu_0v^2$

But, $\varepsilon_0\mu_0=\dfrac{1}{c^2}$

$\therefore$ $\dfrac{F_B}{F_e}=\dfrac{v^2}{c^2}$

44. At centre $B = 0$. At $x \to \infty, B \to 0$. Between zero and zero value there will be a maximum value.

Along x-axis for $x > 0, B_{net}$ is downwards.

Along x-axis for $x < 0, B_{net}$ is upwards.

45. $\theta = \omega t = \left(\frac{Bq}{m}\right) t$

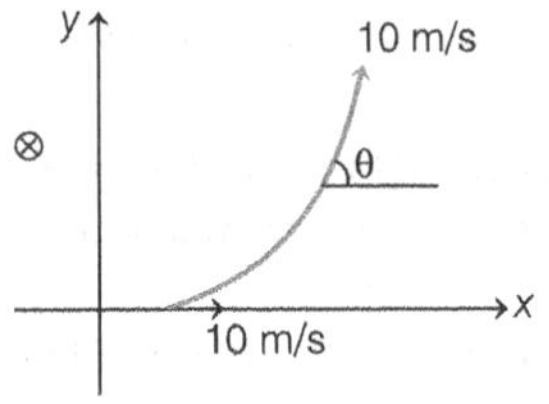

$$= 2 \times \pi \times \frac{1}{6} = \frac{\pi}{3} \text{ or } 60°$$

$$\mathbf{v} = 10 \cos 60° \hat{\mathbf{i}} + 10 \sin 60° \hat{\mathbf{j}}$$

$$= (5\hat{\mathbf{i}} + 5\sqrt{3}\hat{\mathbf{j}}) \text{ m/s}$$

46. $Bqv = qE$

$\Rightarrow \quad E = Bv$

or $\quad E = 2.0 \times 10^5 B \quad \ldots(i)$

$$r = \frac{mv}{Bq}$$

or $\quad 4 \times 10^{-2} = \frac{1.6 \times 10^{-27} \times 2.0 \times 10^5}{(B)(1.6 \times 10^{-19})} \quad \ldots(ii)$

Solving Eqs. (i) and (ii), we can find E and B.

47. $q\mathbf{E} + q(\mathbf{v} \times \mathbf{B}) = 0$

$\therefore \quad \mathbf{E} = -\mathbf{v} \times \mathbf{B}$

or $\quad \mathbf{E} = \mathbf{B} \times \mathbf{v}$

48. $\mathbf{F}_m = q(\mathbf{v} \times \mathbf{B})$

49. Electric field can deviate the path of the particle in the shown direction only when it is along negative y-direction. In the given options, **E** is either zero or along x-direction. Hence, it is the magnetic field which is really responsible for its curved path. Options (a) and (c) cannot be accepted as the path will be circular in that case. Option (d) is wrong because in that case component of net force on the particle also comes in $\hat{\mathbf{k}}$ direction which is not acceptable as the particle is moving in x-y plane. Only in option (b) the particle can move in x-y plane.

In option (d)

$$\mathbf{F}_{net} = q\mathbf{E} + q(\mathbf{v} \times \mathbf{B})$$

Initial velocity is along x-direction. So, let $\mathbf{v} = v\hat{\mathbf{i}}$

$\therefore \quad \mathbf{F}_{net} = qa\hat{\mathbf{i}} + q[(v\hat{\mathbf{i}}) \times (c\hat{\mathbf{k}} + b\hat{\mathbf{j}})]$

$$= qa\hat{\mathbf{i}} - qvc\hat{\mathbf{j}} + qvb\hat{\mathbf{k}}$$

In option (b)

$$\mathbf{F}_{net} = q(a\hat{\mathbf{i}}) + q[(v\hat{\mathbf{i}}) \times (c\hat{\mathbf{k}} + a\hat{\mathbf{i}})]$$

$$= qa\hat{\mathbf{i}} - qvc\hat{\mathbf{j}}$$

50. Magnetic field at P is **B**, perpendicular to OP in the direction shown in figure.

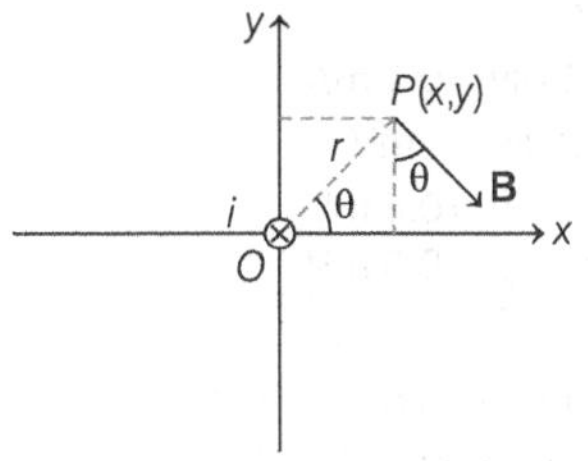

So, $\quad \mathbf{B} = B \sin\theta \hat{\mathbf{i}} - B \cos\theta \hat{\mathbf{j}}$

Here, $\quad B = \frac{\mu_0 I}{2\pi r}$

$$\sin\theta = \frac{y}{r} \quad \text{and} \quad \cos\theta = \frac{x}{r}$$

$\therefore \quad \mathbf{B} = \frac{\mu_0 I}{2\pi} \cdot \frac{1}{r^2}(y\hat{\mathbf{i}} - x\hat{\mathbf{i}})$

$$= \frac{\mu_0 I(y\hat{\mathbf{i}} - x\hat{\mathbf{j}})}{2\pi(x^2 + y^2)} \quad (\text{as } r^2 = x^2 + y^2)$$

51. The magnetic field at $P(a, 0, a)$ due to the loop is equal to the vector sum of the magnetic fields produced by loops $ABCDA$ and $AFEBA$ as shown in the figure.

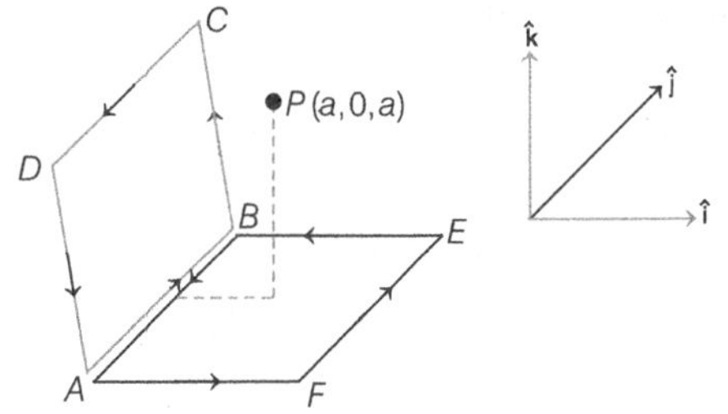

Magnetic field due to loop $ABCDA$ will be along $\hat{\mathbf{i}}$ and due to loop $AFEBA$, along $\hat{\mathbf{k}}$. Magnitude of magnetic field due to both the loops will be equal. Therefore, direction of resultant magnetic field at P will be $\frac{1}{\sqrt{2}}(\hat{\mathbf{i}} + \hat{\mathbf{k}})$.

52. $\mathbf{B} = \begin{cases} \left|\frac{B_0 x}{2R}\right| \hat{\mathbf{k}} \text{ when } & x > 0 \\ \left|\frac{B_0 x}{2R}\right| (-\hat{\mathbf{k}}) \text{ when } & x < 0 \end{cases}$

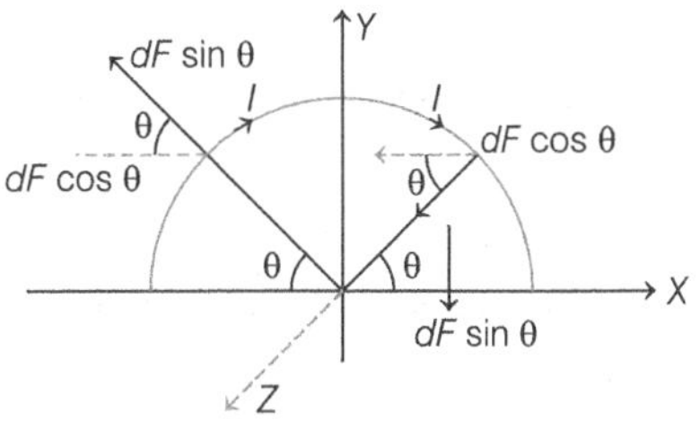

$$\mathbf{F}_{net} = F_x\hat{\mathbf{i}} + F_y\hat{\mathbf{j}}$$

$F_y = \Sigma dF_y = 0 \quad$ (from figure)

$F_x = \Sigma dF_x = -2\int_0^{\pi/2} dF \cos\theta \quad$ (from figure)

Hence, $\mathbf{F}_{net}$ is along negative X-axis.

53. 10 div = 1 mA

$\therefore$ 1 div = 0.1 mA

Similarly, 2 div = 1 mV

$\therefore$ 1 div = 0.5 mV

Now, $G = \frac{0.5 \text{ mV}}{0.1 \text{ mA}} = 5\Omega$

Number of divisions on the galvanometer scale, $n = 150$.

Current required for full scale deflection,

$$I_g = \frac{n}{\text{Current sensitivity}} = \frac{150}{10} = 15 \text{ mA}$$

$$= 15 \times 10^{-3} \text{ A}$$

Required range of voltmeter = $150 \times 1 = 150$ V

Required series resistance

$$R = \frac{V}{I_g} - G$$

$$= \frac{150}{15 \times 10^{-3}} - 5 = 9995\,\Omega$$

54. Since, the current in AB and AC will be same in magnitude and produce equal and opposite magnetic field at the centre. There will be no current in BC. Therefore, magnetic field at O will be zero.

55. $B_1 = \frac{\mu_0 N_1 i_1}{2R_1}$

$$= \frac{(4\pi \times 10^{-7})(50)(2)}{2(5 \times 10^{-2})} = 4\pi \times 10^{-4} \text{ T}$$

$$B_2 = \frac{\mu_0 N_2 i_2}{2R_2} = \frac{(4\pi \times 10^{-7})(100)(2)}{(2)(10 \times 10^{-2})}$$

$$= 4\pi \times 10^{-4} \text{ T}$$

When currents are in the same direction, then

$$B_{net} = B_1 + B_2$$

When currents are in the opposite directions, then

$$B_{net} = B_1 - B_2$$

56. $r = \frac{mV}{Bq} = \frac{(1)(10)}{(2)(1)} = 5 \text{ m}$

$$T = \frac{2\pi m}{Bq} = \frac{(2)(\pi)(1)}{(2)(1)} = \pi$$

$$= 3.14 \text{ s}$$

Plane of circle is perpendicular to $\mathbf{B}$, i.e. xy-plane.

57. $\theta = 180°$

$$\tau = MB \sin\theta = 0$$

$$U = -MB\cos\theta = +MB = \text{maximum}$$

58. $\mathbf{F}_m = q(\mathbf{v} \times \mathbf{B})$

Depending on sign of q, $\mathbf{F}_m$ may be along positive z-axis or along negative z-axis.

$$\mathbf{F}_e = q\mathbf{E}$$

Again, depending on the value of q it may be along positive z-axis or along negative z-axis.

If q is positive, $\mathbf{v} \times \mathbf{B}$ and $\mathbf{F}_m$ comes along negative z-axis also. But, $\mathbf{F}_e$ comes along positive z-axis. So, it may also pass undeflected.

59. KE $= qV$ or KE $\propto V$

$$r = \frac{\sqrt{2qVm}}{Bq} \quad \text{or} \quad r \propto \sqrt{V}$$

$$T = \frac{2\pi m}{Bq} \quad \text{or } T \text{ is independent of } V.$$

60. Point a lies to the right hand side of ef and fg. Hence, both wires produce inward magnetic field. Hence, net magnetic field is inwards.

Same logic can be applied for other points also.

61. If $\mathbf{v} \perp \mathbf{B}$, path will be a circle. Plane of that circle is perpendicular to $\mathbf{B}$. After rotating 180° in that circle particle will be moving along negative X-direction.

62. $F = \frac{1}{4\pi\varepsilon_0}\frac{q_1 q_2}{r^2}$

$$\Rightarrow \quad [\varepsilon_0] = \left[\frac{q^2}{Fr^2}\right] = \left[\frac{T^2 I^2}{MLT^{-2}L^2}\right]$$

$$= [M^{-1}L^{-3}T^4 I^2]$$

Further, $\frac{1}{\sqrt{\varepsilon_0 \mu_0}} = \text{speed of light } c$

$$\therefore \quad [\mu_0] = \frac{1}{[\varepsilon_0 c^2]} = \frac{1}{[M^{-1}L^{-3}T^4 I^2 L^2 T^{-2}]}$$

$$= [MLT^{-2}I^{-2}]$$

63. $\left[\frac{E}{B}\right] = [\text{velocity}]$

$$\left[\frac{1}{\sqrt{\mu_0 \varepsilon_0}}\right] = [\text{speed of light}]$$

$$[CR] = [\text{Time}]$$

64. $r = \frac{mv}{Bq} = \frac{p}{Bq} = \frac{\sqrt{2Km}}{Bq}$, i.e. $r \propto \frac{\sqrt{m}}{q}$

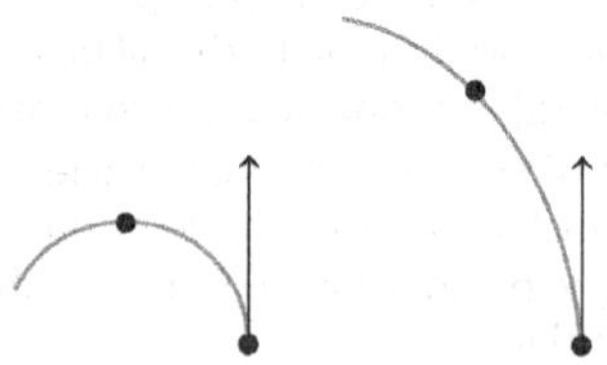

If K and B are same.

i.e. $r_{H^+} : r_{He^+} : r_{O^{2+}} = \frac{\sqrt{1}}{1} : \frac{\sqrt{4}}{1} : \frac{\sqrt{16}}{2}$

$$= 1 : 2 : 2$$

Therefore, He^+ and O^{2+} will be deflected equally but H^+ having the least radius will be deflected most.

65. Cylinder magnetic field B is

$$B = \frac{\mu_0 i r}{2\pi R^2}, \text{ for } r < a$$

$$B = \frac{\mu_0 i}{2\pi r}, \text{ for } r \geq a$$

We can consider the given cylinder as a combination of two cylinders. One of cylinder of radius R carrying current I in one direction and other of radius $\frac{R}{2}$ carrying $\frac{I}{3}$ in other direction.

At point A, $B = \frac{\mu_0(I/3)}{2\pi(R/2)} + 0 = \frac{\mu_0 I}{3\pi R}$

At point B, $B = \frac{\mu_0(4I/3)}{2(\pi R^2)} \frac{R}{2} + 0 = \frac{\mu_0 I}{3\pi R}$

66. B_R = B due to ring, B_1 = B due to wire 1

$\Rightarrow B_2 = B$ due to wire 2

In magnitudes, $B_1 = B_2 = \frac{\mu_0 I}{2\pi r}$

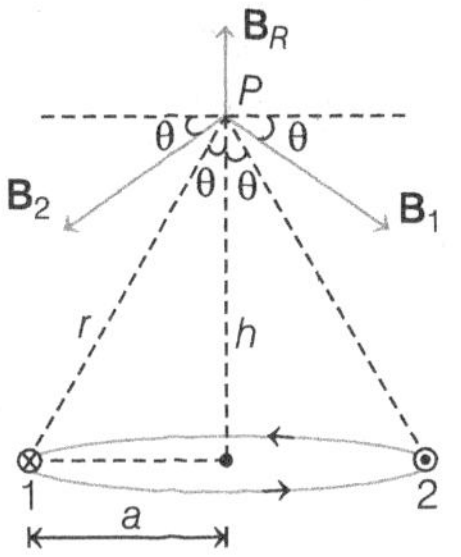

Resultant of B_1 and $B_2 = 2B_1 \cos\theta$

$$= 2\left(\frac{\mu_0 I}{2\pi r}\right)\left(\frac{h}{r}\right) = \frac{\mu_0 I h}{\pi r^2}$$

$$B_R = \frac{\mu_0 I R^2}{2(R^2 + x^2)^{3/2}} = \frac{\mu_0 I a^2}{2r^3}$$

As, $R = a, x = h$

and $a^2 + h^2 = r^2$

For zero magnetic field at P,

$$\frac{\mu_0 I h}{\pi r^2} = \frac{\mu_0 I a^2}{2r^3} \Rightarrow h \approx 1.1\, a$$

67. Magnetic field at mid-point of two wires

$= 2$ (magnetic field due to one wire)

$$= 2\left[\frac{\mu_0}{2\pi}\frac{I}{d}\right] = \frac{\mu_0 I}{\pi d} \otimes$$

Magnetic moment of loop,

$$M = IA = I\pi a^2$$

$$\text{Torque on loop} = MB \sin 30^\circ = \frac{\mu_0 I^2 a^2}{2d}$$

68. (a) $\mathbf{F} = q(\mathbf{v} \times \mathbf{B})$

q is negative, $\mathbf{v}$ is along $+\hat{\mathbf{i}}$ and $\mathbf{B}$ along $+\hat{\mathbf{j}}$.

Therefore, $\mathbf{F}$ is along negative z.

(b) Same logic is given in (a).

(c) $\mathbf{B}$ is parallel to $\mathbf{v}$. So, magnetic force is zero. Charge is negative so, electric force is opposite to $\mathbf{E}$.

(d) Charge is negative. So, electrostate force is in opposite direction of $\mathbf{E}$.

69. For direction of magnetic force apply Fleming's left hand rule. According to that w and x are positively charged particles and y and z negatively charged particle.

Secondly,

$$r = \frac{\sqrt{2Km}}{Bq}$$

$$\Rightarrow \quad r \propto \frac{\sqrt{m}}{q} \qquad (K \to \text{same})$$

70. Two currents are lying in the plane of paper. Its point is lying to the right hand side of the current carrying wire, magnetic field is inward ($-\hat{\mathbf{k}}$ direction).

If point lies to left hand side, field is outward ($\hat{\mathbf{k}}$ direction).

71. $\mathbf{F}_{ABC} = \mathbf{F}_{ADC} = \mathbf{F}_{AC}$

$$\therefore \quad \mathbf{F}_{\text{loop}} = 2\mathbf{F}_{AC}$$

$$= \left[\frac{I}{2}(\mathbf{l} \times \mathbf{B})\right]$$

$$= I(\mathbf{AC} \times \mathbf{B})$$

Hence, $\mathbf{AC} = (l\hat{\mathbf{i}} + l\hat{\mathbf{j}})$

Now, putting value of $\mathbf{B}$ we can find net force in different cases.

72. (a) The magnetic field is along negative Y-direction in case (p), (q) and (r). In case (s), there will be component of magnetic field along negative X-axis.

Thus, **(a) → (p,q,r)**

z-component of magnetic field is zero in all cases.

Thus, **(b) → (p,q,r,s)**

The magnetic field at P is $\frac{\mu_0 I}{4\pi d}$. In case (r) only.

Thus, **(c) → (r)**

The magnetic field at P is less than $\frac{\mu_0 i}{2\pi d}$ for all cases.

Thus, **(d) → (p,q,r,s)**

73. For a particle to move in negative y-direction, either its velocity must be in negative y-direction (if initial velocity $\neq 0$) and force should be parallel to velocity or it must experience a net force in negative y-direction only (if initial velocity $= 0$)

74. $\mathbf{F}_{\text{net}} = \mathbf{F}_e + \mathbf{F}_B$

$$= q\mathbf{E} + q\mathbf{v} \times \mathbf{B}$$

For particle to move in straight line with constant velocity, $\mathbf{F}_{\text{net}} = 0$

$$\therefore \quad q\mathbf{E} + q\mathbf{v} \times \mathbf{B} = 0 \quad \text{or} \quad \mathbf{E} = \mathbf{B} \times \mathbf{v}$$

75. For path to be helical with axis along positive z-direction, particle should experience a centripetal acceleration in XY-plane.

For the given set of options, only option (c) satisfy the condition. Path is helical with increasing pitch.

76. Force on wire 12 will be $\left(\frac{\mu_0}{2\pi}\frac{I}{a}\right)(Ia)$ in positive x-direction.

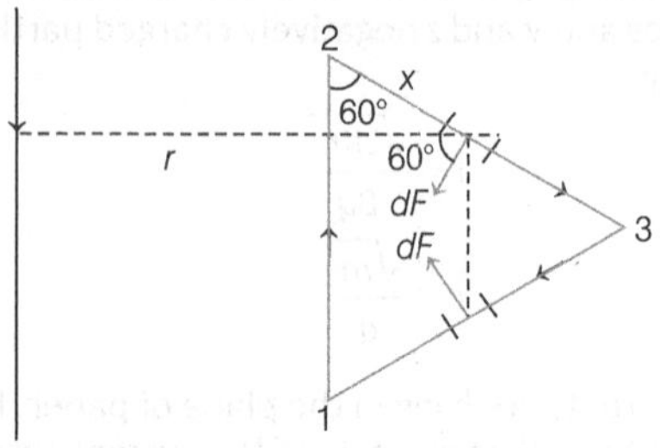

Force on 2-3 and 3-1 :

$$r = a + x\sin 60° = a + \sqrt{3}\frac{x}{2}$$

$$B_r = \frac{\mu_0}{2\pi}\frac{I}{r} = \frac{\mu_0}{2\pi}\frac{I}{a+\sqrt{3}\frac{x}{2}} = \frac{\mu_0}{\pi}\frac{I}{2a+\sqrt{3}x}$$

$$\therefore \mathbf{F}_{231} = \left[2\int_{x=0}^{x=a}(dF\cos 60°)\right](-\hat{\mathbf{i}})$$

$$= \left[2\int_0^a \frac{\mu_0}{\pi}\cdot\frac{I}{2a+\sqrt{3}x}\cdot I\cdot(dx)\frac{1}{2}\right](-\hat{\mathbf{i}})$$

$$= \left(\frac{\mu_0 I^2}{\pi}\right)\cdot\frac{1}{\sqrt{3}}[\ln(2a+\sqrt{3}x)]_0^a(-\hat{\mathbf{i}})$$

$$= \frac{\mu_0 I^2}{\sqrt{3}\pi}\ln\left(\frac{2+\sqrt{3}}{2}\right)(-\hat{\mathbf{i}})$$

and $$\mathbf{F}_{12} = \frac{\mu_0 I^2}{2\pi}(\hat{\mathbf{i}})$$

$$\therefore \mathbf{F}_{\text{Total}} = \frac{\mu_0 I^2}{\pi}\left[\frac{1}{2} - \frac{1}{\sqrt{3}}\ln\left(\frac{2+\sqrt{3}}{2}\right)\right](\hat{\mathbf{i}})$$ **Ans.**

77. **Magnetic moment** According to Bohr's hypothesis, angular momentum in nth orbit is $L = n\left(\frac{h}{2\pi}\right)$.

Further, $$\frac{M}{L} = \frac{q}{2m} \text{ or } \frac{e}{2m}$$

$\therefore$ Magnetic moment,

$$M = \left(\frac{e}{2m}\right)(L) = \left(\frac{e}{2m}\right)\left(n\frac{h}{2\pi}\right) = \left(\frac{neh}{4\pi m}\right)$$

Magnetic field induction

$$\frac{mv^2}{r} = \frac{1}{4\pi\varepsilon_0}\frac{(e)(e)}{r^2} \quad \text{...(i)}$$

From Bohr's hypothesis :

$$mvr = \frac{nh}{2\pi} \quad \text{...(ii)}$$

Solving these two equations, we find

$$v = \frac{e^2}{2\varepsilon_0 nh} \quad \text{and} \quad r = \frac{\varepsilon_0 n^2 h^2}{\pi m e^2}$$

Now, magnetic field induction at centre

$$B = \frac{\mu_0 i}{2r}$$

Here, $$i = qf = (e)\left(\frac{v}{2\pi r}\right)$$

$$\therefore \quad B = \left(\frac{\mu_0}{2r}\right)\left(\frac{ev}{2\pi r}\right) = \frac{\mu_0 e}{4\pi}\left(\frac{v}{r^2}\right)$$

$$= \left(\frac{\mu_0 e}{4\pi}\right)\left(\frac{e^2}{2\varepsilon_0 nh}\right)\left(\frac{\pi^2 m^2 e^4}{\varepsilon_0^2 n^4 h^4}\right)$$

$$= \frac{\mu_0 \pi m^2 e^7}{8\varepsilon_0^3 h^5 n^5}$$ **Ans.**

78. $r = \sqrt{9+7} = 4$ cm

$$B = \frac{\mu_0}{4\pi}\frac{i}{r}(2\sin\alpha)$$

$$= (10^{-7})\left(\frac{30}{4\times 10^{-2}}\right)\left(2\times\frac{3}{5}\right)$$

$$= 9.0\times 10^{-5}\ \text{T}$$

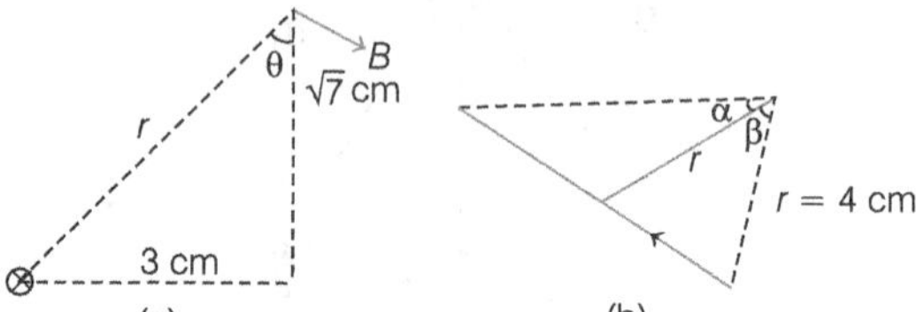

(a) (b)

$$\text{Net field} = 4B\sin\theta$$

$$= 4\times 9.0\times 10^{-5}\times\frac{3}{4}$$

$$= 2.7\times 10^{-4}\ \text{T}$$ **Ans.**

79. xy-plane $$T = \frac{2\pi r}{v} = \frac{2\pi(0.4)}{5} = \frac{8}{50}\pi$$

Since, $$T = \frac{2\pi m}{Bq} \text{ or } T \propto \frac{m}{q}$$

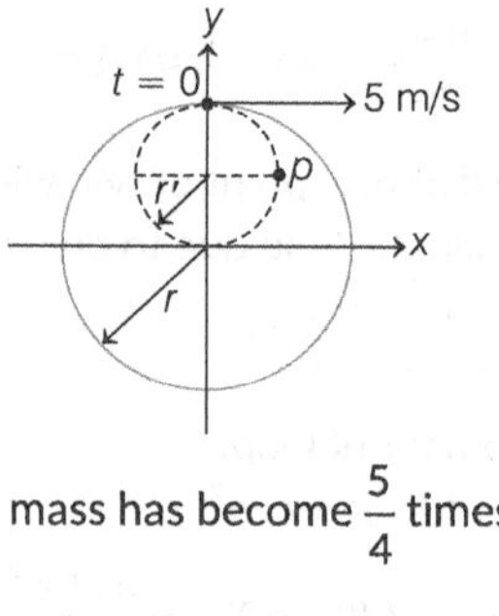

After collision mass has become $\frac{5}{4}$ times and charge two times.

$$\therefore \quad T' = \left(\frac{5}{4}\times\frac{1}{2}\right)T = \frac{5}{8}\times\frac{8}{50}\pi = \frac{\pi}{10}\ \text{s}$$

Given time $t = \frac{T'}{4}$, i.e. combined mass will complete one-quarter circle.

Further $$r = \frac{P}{Bq}$$

or $\qquad r \propto \dfrac{1}{q}$ (as $P =$ constant)

Since, charge has become two times

$\therefore \qquad r' = \dfrac{r}{2} = 0.2$ m

At $t = (\pi/40)$ second, particle will be at P in xy -plane.

$\therefore \qquad x = r' = 0.2$ m

$y = r' = 0.2$ m **Ans.**

z-coordinate Mass of combined body has become 5 times of the colliding particle. Therefore, from conservation of linear momentum, velocity component in z-direction will become $\dfrac{1}{5}$ times. Or

$$v_z = \frac{1}{5} \times \frac{40}{\pi} \text{ m/s} = \frac{8}{\pi} \text{ m/s}$$

$\therefore \qquad z = v_z t = \dfrac{8}{\pi} \times \dfrac{\pi}{40} = 0.2$ m **Ans.**

80. (a) Speed of particle at origin, $v = \sqrt{\dfrac{2qV}{m}} = v_x$

$$t = \frac{x}{v_x} = \frac{a}{\sqrt{\dfrac{2qV}{m}}} = a\sqrt{\frac{m}{2qV}}$$

$$y = \frac{1}{2} a_y t^2$$

$$= \frac{1}{2}\left[\frac{qE}{m}\right]\left[a^2 \times \frac{m}{2qV}\right]$$

$$= \frac{a^2 E}{4V}$$ **Ans.**

(b) Component parallel to **B** is

$$v_y = a_y t = \left(\frac{qE}{m}\right)\left(a\sqrt{\frac{m}{2qV}}\right) = Ea\sqrt{\frac{q}{2mV}}$$

Now, pitch = component parallel to **B** × time period

$$= v_y T = \left(Ea\sqrt{\frac{q}{2mV}}\right)\left(\frac{2\pi m}{Bq}\right)$$

$$= \frac{\pi E a}{B}\sqrt{\frac{2m}{qV}}$$ **Ans.**

81. To graze at C Using equation of trajectory of parabola,

$$y = x \tan\theta - \frac{ax^2}{2v^2 \cos^2\theta} \qquad \text{...(i)}$$

Here, $\qquad a = \dfrac{qE}{m} = \dfrac{10^{-6} \times 10^{-3}}{10^{-10}}$

$= 10 \text{ m/s}^2$

Substituting in Eq. (i), we have

$$0.05 = 0.17 \tan 30° - \frac{10 \times (0.17)^2}{2v^2 \times (\sqrt{3}/2)^2}$$

Solving this equation, we have

$v = 2$ m/s

In magnetic field, $AC = 2r$

or $\qquad 0.1 = 2r$

or $\qquad r = 0.05 \text{ m} = \dfrac{mv \cos 30°}{Bq}$

$\therefore \qquad B = \dfrac{mv \cos 30°}{(0.05)q}$

$$= \frac{(10^{-10})(2)(\sqrt{3}/2)}{(0.05)(10^{-6})}$$

$= 3.46 \times 10^{-3}$ T

$= 3.46$ mT **Ans.**

82. (a) **B** at origin $\mathbf{B}_{KLM} = \mathbf{B}_{KNM} = \dfrac{\mu_0 I}{4R}(-\hat{\mathbf{i}} + \hat{\mathbf{j}})$

Now, we can apply $\mathbf{F} = q(\mathbf{v} \times \mathbf{B})$ for finding force on it.

(b) In uniform magnetic field,

$$\mathbf{F}_{KLM} = \mathbf{F}_{KNM} = \mathbf{F}_{KM} = I(\mathbf{l} \times \mathbf{B})$$

$$= I[\{2R(-\hat{\mathbf{k}})\} \times \{B_0 \hat{\mathbf{j}}\}]$$

$$= (2B_0 IR)\,\hat{\mathbf{i}}$$

$\therefore$ Net force is two times of the above value.

83. Current per unit area,

$$\sigma = \frac{I}{\pi a^2 - \pi(a/2)^2 - (a/2)^2} = \frac{2I}{\pi a^2}$$

Total area is (πa^2). Therefore, the total current is

$$I_1 = (\sigma)(\pi a^2) = 2I$$

Cavity area is $\pi(a/2)^2$. Therefore, cavity current is

$$I_2 = (\sigma)(\pi a^2/4) = \frac{I}{2}$$

Now, the given current system can be assumed as shown below.

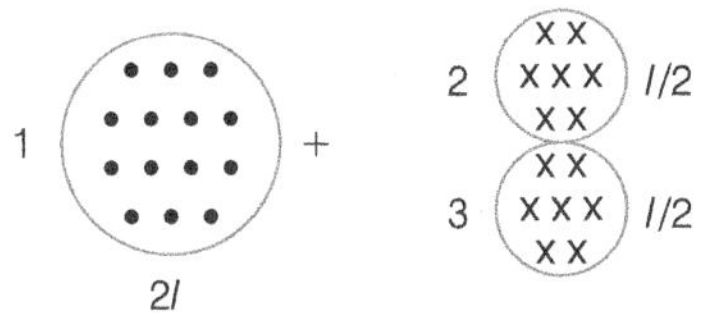

(a) **At P_1**, $\quad B_1 = \dfrac{\mu_0}{2\pi}\dfrac{2I}{5} = \dfrac{\mu_0 I}{\pi r}$ (towards left)

$B_2 = \dfrac{\mu_0}{2\pi}\dfrac{I/2}{(r - a/2)}$ (towards right)

and $\quad B_2 = \dfrac{\mu_0}{2\pi}\dfrac{I/2}{(r + a/2)}$ (towards right)

$\therefore \qquad B_{net} = B_1 - B_2 - B_2$ (towards left)

$$= \frac{\mu_0 I}{\pi}\left[\frac{1}{r} - \frac{1}{4r - 2a} - \frac{1}{4r + 2a}\right]$$

$$= \frac{\mu_0 I}{\pi}\left[\frac{16r^3 - 4a^2 - 4r^2 - 2ar - 4r^2 + 2ar}{r(16r^2 - 4a^2)}\right]$$

$$= \frac{\mu_0 I}{\pi r}\left[\frac{2r^2 - a^2}{4r^2 - a^2}\right]$$ (towards left)

(b)

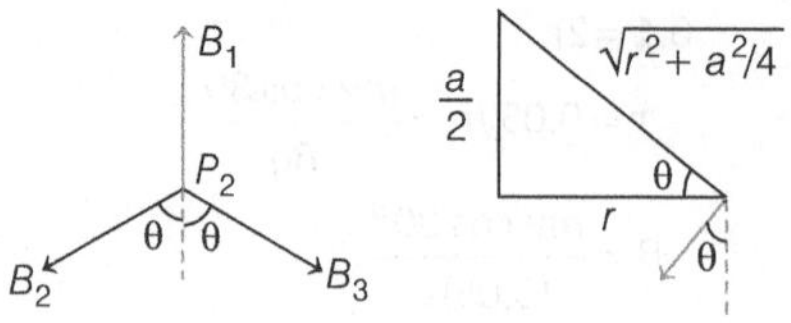

$B_{net} = B_1 - 2B_2 \cos\theta]$ (towards the top)

$$= \frac{\mu_0}{2\pi}\frac{2I}{R} - 2\left[\frac{\mu_0}{2\pi}\frac{I/2}{\sqrt{r^2+a^2/4}}\right]\frac{r}{\sqrt{r^2+a^2/4}}$$

$$= \frac{\mu_0 I}{\pi r}\left[\frac{2r^2+a^2}{4r^2+a^2}\right]$$ (towards the top)

84. Magnetic moment of the loop, $\mathbf{M} = (iA)\hat{\mathbf{j}}$

$$= (I_0 L^2)\,\hat{\mathbf{k}}$$

Magnetic field, $\mathbf{B} = (B\cos 45°)\,\hat{\mathbf{i}} + (B\sin 45°)\,\hat{\mathbf{j}}$

$$= \frac{B}{\sqrt{2}}(\hat{\mathbf{i}} + \hat{\mathbf{j}})$$

(a) Torque acting on the loop, $\tau = \mathbf{M} \times \mathbf{B}$

$$= (I_0 L^2 \hat{\mathbf{k}}) \times \left[\frac{B}{\sqrt{2}}(\hat{\mathbf{i}} + \hat{\mathbf{j}})\right]$$

$\therefore$ $$\tau = \frac{I_0 L^2 B}{\sqrt{2}}(\hat{\mathbf{j}} - \hat{\mathbf{i}})$$

or $|\tau| = I_0 L^2 B$ **Ans.**

(b) Axis of rotation coincides with the torque and since torque is along $\hat{\mathbf{j}} - \hat{\mathbf{i}}$ direction or parallel to QS. Therefore, the loop will rotate about an axis passing through Q and S as shown in the figure.

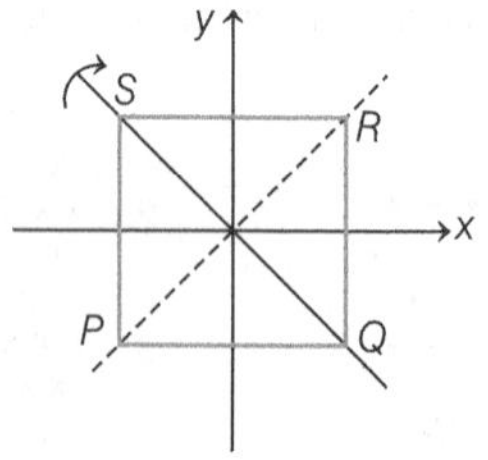

Angular acceleration, $\alpha = \frac{|\tau|}{I}$

where, I = moment of inertia of loop about QS.

$I_{QS} = I_{PR} = I_{ZZ}$

(From the theorem of perpendicular axis)

But, $I_{QS} = I_{PR}$

$\therefore$ $$2I_{QS} = I_{ZZ} = \frac{4}{3}ML^2$$

$$I_{QS} = \frac{2}{3}ML^2$$

$\therefore$ $$\alpha = \frac{|\tau|}{I} = \frac{I_0 L^2 B}{2/3 ML^2} = \frac{3}{2}\frac{I_0 B}{M}$$

$\therefore$ Angle by which the frame rotates in time Δt is

$$\theta = \frac{1}{2}\alpha(\Delta t)^2$$

or $$\theta = \frac{3}{4}\frac{I_0 B}{M}\cdot(\Delta t)^2$$ **Ans.**

85. In equilibrium,

$$2T_0 = mg \quad \text{or} \quad T_0 = \frac{mg}{2} \qquad \text{...(i)}$$

Magnetic moment, $M = iA = \left(\frac{\omega}{2\pi}Q\right)(\pi R^2)$

$$\tau = MB\sin 90° = \frac{\omega BQR^2}{2}$$

Let T_1 and T_2 be the tensions in the two strings when magnetic field is switched on $(T_1 > T_2)$.

For translational equilibrium of ring in vertical direction,

$$T_1 + T_2 = mg \qquad \text{...(ii)}$$

For rotational equilibrium,

$$(T_1 - T_2)\frac{D}{2} = \tau = \frac{\omega BQR^2}{2}$$

or $$T_1 - T_2 = \frac{\omega BQR^2}{2} \qquad \text{...(iii)}$$

Solving Eqs. (ii) and (iii), we have

$$T_1 = \frac{mg}{2} + \frac{\omega BQR^2}{2D}$$

As $T_1 > T_2$ and maximum values of T_1 can be $\frac{3T_0}{2}$,

We have

$$\frac{3T_0}{2} = T_0 + \frac{\omega_{max} BQR^2}{2D}$$

$\therefore$ $$\omega_{max} = \frac{DT_0}{BQR^2}$$ **Ans.**

86. $dB = \frac{\mu_0}{2\pi}\frac{(i/\omega).dx}{x}$

$\therefore$ $$B = \frac{\mu_0 i}{2\pi\omega}\int_d^{d+\omega}\frac{dx}{x}$$

$$B = \frac{\mu_0 i}{2\pi\omega}\ln\left(\frac{d+\omega}{d}\right)$$ (upwards) **Ans.**

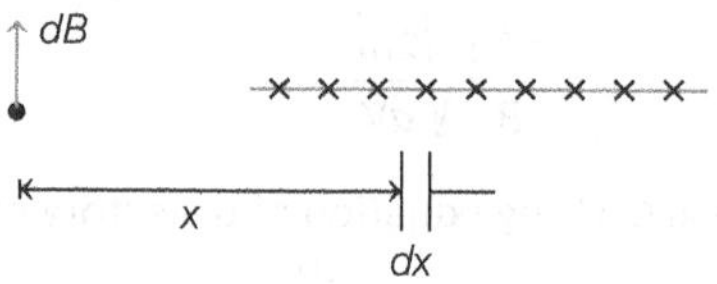

87. $\theta = \tan^{-1}\left(\frac{R}{r}\right) = \tan^{-1}\left(\frac{R}{mv/Bq}\right) = \tan^{-1}\left(\frac{BqR}{mv}\right)$

Deviation $= 2\theta = 2\tan^{-1}\left(\frac{BqR}{mv}\right)$

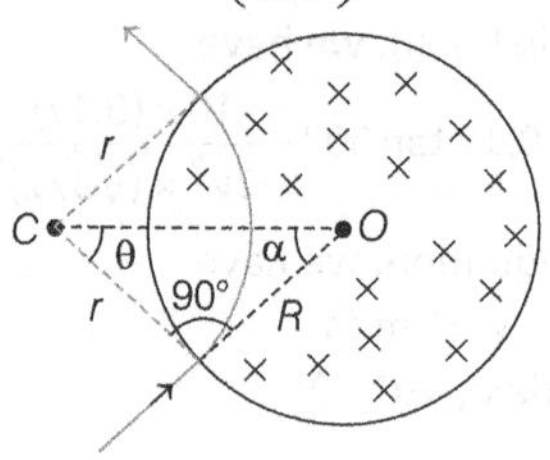

88. (a) Yes, magnetic force for calculation of torque can be assumed at centre. Since, variation of torque about P from one end of the rod to the other end comes out to be linear.

$$\therefore \quad \tau = (IlB)\left(\frac{l}{2}\right) = \frac{Il^2B}{2}$$

$$= \frac{(6.5)(0.2)^2(0.34)}{2}$$

$$= 0.0442 \text{ N-m}$$ **Ans.**

(b) Magnetic torque on rod will come out to be clockwise. Therefore, torque of spring force should be anti-clockwise or spring should be stretched.

(c) In equilibrium,

Clockwise torque of magnetic force

= anti-clockwise torque of spring force

$$\therefore 0.0442 = (kx)(l \sin 53°) = (4.8)(x)(0.2)\left(\frac{4}{5}\right)$$

or $\quad x = 0.057$ m

$$U = \frac{1}{2}kx^2 = \frac{1}{2} \times (4.8)(0.057)^2$$

$$= 7.8 \times 10^{-3} \text{ J}$$ **Ans.**

89. Let the direction of current in wire PQ is from P to Q and its magnitude be I.

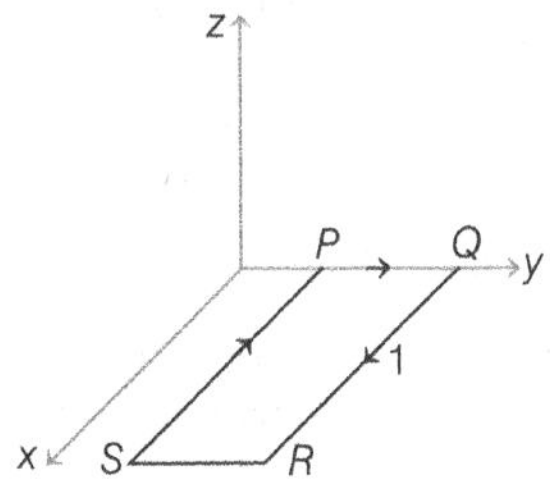

The magnetic moment of the given loop is

$$\mathbf{M} = -Iab\hat{\mathbf{k}}$$

Torque on the loop due to magnetic force is

$$\tau_1 = \mathbf{M} \times \mathbf{B}$$

$$= (-Iab\hat{\mathbf{k}}) \times (3\hat{\mathbf{i}} + 4\hat{\mathbf{k}})\,B_0\hat{\mathbf{i}}$$

$$= -3IabB_0\hat{\mathbf{j}}$$

Torque on weight of the loop about axis PQ is

$$\tau_2 = \mathbf{r} \times \mathbf{F} = \left(\frac{a}{2}\hat{\mathbf{i}}\right) \times (-mg\hat{\mathbf{k}})$$

$$= \frac{mga}{2}\hat{\mathbf{j}}$$

We see that when the current in the wire PQ is from P to Q, τ_1 and τ_2 are in opposite directions, so they can cancel each other and the loop may remain in equilibrium. So, the direction of current I in wire PQ is from P to Q. Further for equilibrium of the loop

$$|\tau_1| = |\tau_2|$$

or $$3IabB_0 = \frac{mga}{2}$$

$$I = \frac{mg}{6bB_0}$$ **Ans.**

Magnetic force on wire RS is

$$\mathbf{F} = I(\mathbf{l} \times \mathbf{B})$$

$$= I[(-b\hat{\mathbf{j}}) \times \{(3\hat{\mathbf{i}} + 4\hat{\mathbf{k}})\,B_0\}]$$

$$\mathbf{F} = IbB_0(3\hat{\mathbf{k}} - 4\hat{\mathbf{i}})$$ **Ans.**

90. $\mathbf{F}_e = q\mathbf{E} = (1.6 \times 10^{-19})(-102.4 \times 10^3)\hat{\mathbf{k}}$

$$= -(1.6384 \times 10^{-16})\hat{\mathbf{k}} \text{ N}$$

$$\mathbf{F}_m = q(\mathbf{v} \times \mathbf{B}) = 1.6 \times 10^{-19} \quad (1.28 \times 10^6\,\hat{\mathbf{i}} \times 8 \times 10^{-2}\,\hat{\mathbf{j}})$$

$$= 1.6384 \times 10^{-16}\,\hat{\mathbf{k}} \text{N}$$

$$\mathbf{F}_{net} = \mathbf{F}_e + \mathbf{F}_m = 0$$

$\therefore$ Net force on the charged particle is zero. Particle will move undeviated.

In time $t = 5 \times 10^{-6}$ s, the x-coordinate of particle will become,

$$x = v_x t = (1.28 \times 10^6)(5 \times 10^{-6}) = 6.4 \text{ m}$$

while y and z-coordinates will be zero.

At $x = 5 \times 10^{-6}$ s, electric field is switched-off. Only magnetic field is left which is perpendicular to its velocity. Hence, path of the particle will now become circular.

Plane of circle will be perpendicular to magnetic field i.e. XZ-Radius and angular velocity of circular path will become

$$r = \frac{mv}{Bq} = \frac{(10^{-26})(1.28 \times 10^6)}{(8 \times 10^{-2})(1.6 \times 10^{-19})} = 1\text{m}$$

$$\omega = \frac{Bq}{m} = \frac{(8 \times 10^{-2})(1.6 \times 10^{-19})}{(10^{-26})}$$

$$= 1.28 \times 10^6 \text{ rad/s}$$

In the remaining time i.e. $(7.45 - 5) \times 10^{-6} = 2.45 \times 10^{-6}$ s

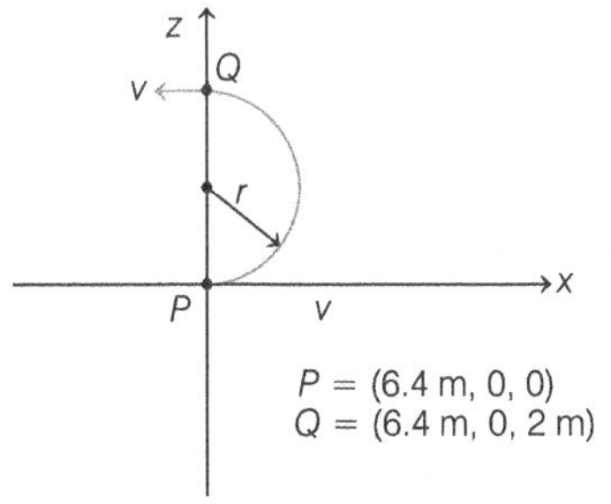

Angle rotated by particle,

$$\theta = \omega t = (1.28 \times 10^6)(2.45 \times 10^{-6})$$

$$= 3.14 \text{ rad} \approx 180°$$

$\therefore$ z-coordinate of particle will become

$$z = 2r = 2\text{m}$$

while y-coordinate will be zero.

$\therefore$ Position of particle at $t = 5 \times 10^{-6}$ s is $P \equiv (6.4 \text{ m}, 0, 0)$

and at $t = 7.45 \times 10^{-6}$ s is $Q \equiv (6.4 \text{ m}, 0, 2 \text{ m})$

91. Electron pass undeviated. Therefore,

$$|\mathbf{F}_e| = |\mathbf{F}_m| \text{ or } eE = eBv$$

or
$$B = \frac{E}{v} = \frac{V/d}{v}$$

(V = potential difference between the plates)

or
$$B = \frac{V}{dv}$$

Substituting the values, we have

$$B = \frac{600}{3 \times 10^{-3} \times 2 \times 10^{6}}$$
$$= 0.1 \text{ T}$$

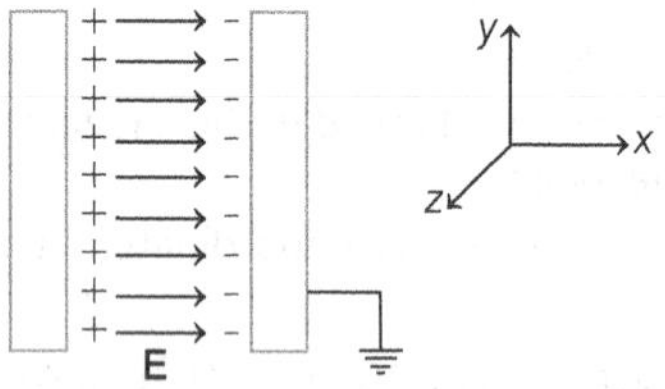

Further, direction of $\mathbf{F}_e$ should be opposite of $\mathbf{F}_m$.

or
$$e\mathbf{E} \uparrow\downarrow e(\mathbf{v} \times \mathbf{B})$$
$$\Rightarrow \quad \mathbf{E} \uparrow\downarrow \mathbf{v} \times \mathbf{B}$$

Here, **E** is in positive x-direction.

Therefore, $\mathbf{v} \times \mathbf{B}$ should be in negative x-direction or **B** should be in negative z-direction or perpendicular to paper inwards, because velocity of electron is in positive y-direction.

92. $\cos d\theta$ components of T are cancelled and $\sin d\theta$ components are added and balanced by magnetic force F on small portion AC. Thus,

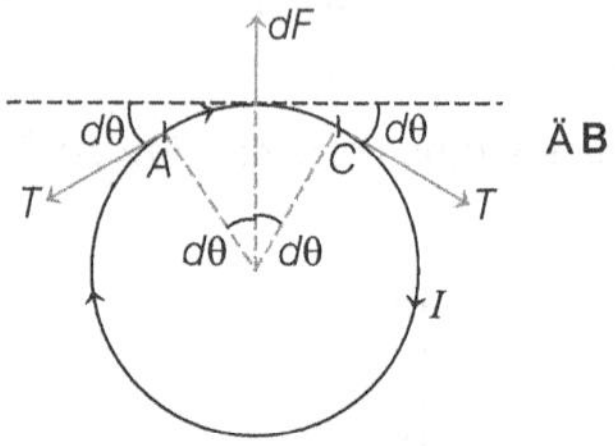

$$2T \sin d\theta = dF = I(dl)B \sin 90°$$

Here, $dl = AC = 2R\, d\theta$

For small angle, $\sin d\theta \approx d\theta$

$\therefore \quad 2T d\theta = I(2R d\theta)B$

or $\quad T = IRB$

$$\text{Stress } \sigma = \frac{T}{(\pi d^2/4)} = \frac{4T}{\pi d^2} = \frac{4IRB}{\pi d^2}$$

$$\therefore \quad B_{max} = \frac{\pi d^2 \sigma_{max}}{4IR}$$

Substituting the values, we get

$$B_{max} = 8 \text{ kT}$$

93. $F_e = F_m \Rightarrow qE = Bqv$ or $v = \frac{E}{B}$

Let n protons strike per second.

Then current,

$$I = ne \qquad (q = e)$$

or
$$n = \frac{I}{e}$$

Now force, F = rate of change of linear momentum

$$= mnv \qquad (m = \text{mass of proton})$$
$$= m\left(\frac{I}{e}\right)\left(\frac{E}{B}\right)$$

Substituting the values, we get

$$F = 20\,\mu\text{N}$$

94. (a) $$\mathbf{B} = \left(\frac{\mu_0}{4\pi}\right)\left(\frac{I}{R}\right)(\sin 0° + \sin 90°)(-\hat{\mathbf{k}}) + \frac{1}{2}\left(\frac{\mu_0 I}{2R}\right)(-\hat{\mathbf{i}}) + \left(\frac{\mu_0}{4\pi}\right)\left(\frac{I}{R}\right)(\sin 0° + \sin 90°)(-\hat{\mathbf{k}})$$

After substituting the values and then finding the magnitude of **B**,
we get

$$B = 0.3\,\mu\text{T}$$

(b) $$\mathbf{B} = \left(\frac{\mu_0}{4\pi}\right)\left(\frac{I}{R}\right)(\sin 0° + \sin 90°)(-\hat{\mathbf{k}}) + \frac{1}{2}\left(\frac{\mu_0 I}{2R}\right)(-\hat{\mathbf{i}}) + \left(\frac{\mu_0}{4\pi}\right)\left(\frac{I}{R}\right)(\sin 0° + \sin 90°)(-\hat{\mathbf{i}})$$

or
$$B = 0.34\,\mu\text{T}$$

(c) $$\mathbf{B} = \left(\frac{\mu_0}{4\pi}\right)\left(\frac{I}{R}\right)(\sin 0° + \sin 90°)(-\hat{\mathbf{k}}) + \frac{3}{4}\left(\frac{\mu_0 I}{2R}\right)(-\hat{\mathbf{i}}) + \left(\frac{\mu_0}{4\pi}\right)\left(\frac{I}{R}\right)(\sin 0° + \sin 90°)(-\hat{\mathbf{j}})$$

or
$$B = 0.11\,\mu\text{T}$$

95. Magnetic field due to a current carrying thick wire of radius R with current I at any point inside the wire is given by,

$$B = \left(\frac{\mu_0}{2\pi}\frac{I}{R^2}\right)r \qquad (r \le R)$$

If J is the currrent density then the above result can be written as,

$$B = \frac{\mu_0 J r}{2} \qquad \left(J = \frac{I}{\pi R^2}\right)$$

B, **J** and **r** are vector quantities and **J** is in the direction of current. So the above result can be written as,

$$\mathbf{B} = \frac{\mu_0}{2}(\mathbf{J} \times \mathbf{r}) \qquad \text{...(i)}$$

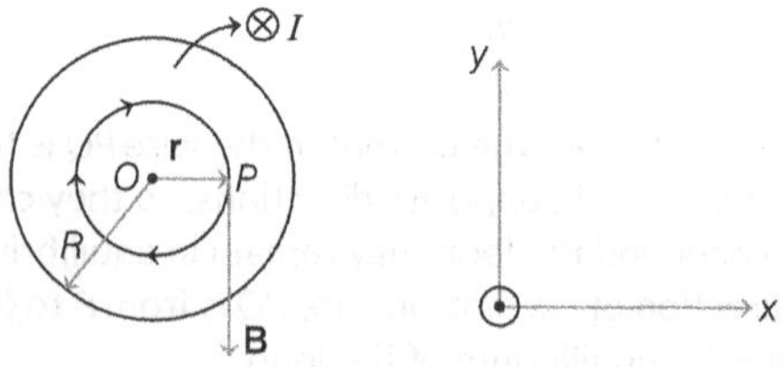

In the shown figure current I is inwards and uniformly distributed over the area πR^2. So, **J** is inwards, in the direction of $-\hat{\mathbf{k}}$. At point P magnetic field is downwards, in the direction of $-\hat{\mathbf{j}}$, as magnetic lines passing through P are circular and clockwise. Position vector of P with respect to centre O is $\mathbf{r}$, in the direction of $\hat{\mathbf{i}}$.

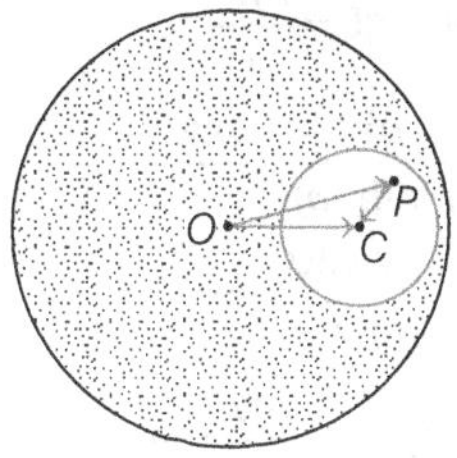

Now, we can see that $\mathbf{B}$, $\mathbf{J}$ and $\mathbf{r}$ satisfy equation (i).
Now, let us come to the given problem.
At point P, net magnetic field is,
$\mathbf{B}$ = magnetic field due to complete portion − magnetic field due to cavity

$$= \frac{\mu_0}{2}(\mathbf{J} \times \mathbf{OP}) - \frac{\mu_0}{2}(\mathbf{J} \times \mathbf{CP})$$

$$= \frac{\mu_0}{2}[\mathbf{J} \times (\mathbf{OP} + \mathbf{PC})] \qquad [\text{as } \mathbf{CP} = -\mathbf{PC}]$$

$$= \frac{\mu_0}{2}[\mathbf{J} \times \mathbf{OC}] \quad \text{or} \quad \mathbf{B} = \frac{\mu_0}{2}[\mathbf{J} \times \mathbf{l}]$$

Note *(i) The magnetic field inside cavity is constant everywhere.*
(ii) $\mathbf{B} = 0$ where $\mathbf{l} = 0$, or if C coincides with O, magnetic field will be zero.

96.

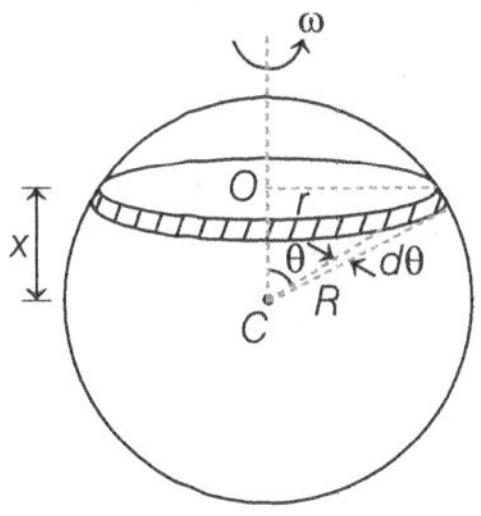

$$r = R\sin\theta \Rightarrow x = R\cos\theta$$

Small area of the element shown in figure,

$$dA = (2\pi r)R\,d\theta$$

Small charge on this element,

$$dq = (\sigma)\,dA$$

This small rotating charge is equivalent to a current carrying loop. Current on this loop is,

$$I = (dq)(f) \qquad [f = \text{frequency}]$$

Small magnetic field due to this current carrying loop at centre is,

$$dB = \frac{\mu_0 I r^2}{2(r^2 + x^2)^{3/2}}$$

Substituting the values and then integrating dB from $\theta = 0°$ to $\theta = 180°$, we get total magnetic field B at centre which comes out to be,

$$B = 29 \text{ pT}$$

97. Slightly divergent means θ between $\mathbf{v}$ and $\mathbf{B}$ is very close to 0° but not exactly zero.
So, path will be helical. But we can assume,

$$\cos\theta \approx 1$$

In this case particles are brought into focus after every pitch. So, the given distance l should be integer multiple of pitch. or,

$$l = np = n(v\cos\theta)(T)$$

Here, $v = \sqrt{\frac{2qV}{m}}$, $\cos\theta \approx 1$ and $T = \frac{2\pi m}{Bq}$

So, let $B_1 (< B_2)$, then $p_1 > p_2$ and let l is n^{th} times p_1. Then,

$$l = n\left(\sqrt{\frac{2qV}{m}}\right)\left(\frac{2\pi m}{B_1 q}\right) \qquad \text{...(i)}$$

For B_2 this length l should be $(n+1)^{\text{th}}$ times p_2.

So, $$l = (n+1)\left(\sqrt{\frac{2qV}{m}}\right)\left(\frac{2\pi m}{B_2 q}\right) \qquad \text{...(ii)}$$

We have two unknowns n and $\frac{q}{m}$.

Solving these equations, we get

$$\frac{q}{m} = \frac{8\pi^2 V}{l^2(B_2 - B_1)^2}$$

98. X-axis is also the axis of solenoid.
$v_x = v\cos\alpha$ = constant, as this component is along magnetic field.
So, time taken by the particle in reaching from A to O will be,

$$t = \frac{l}{v\cos\alpha} \qquad \text{...(i)}$$

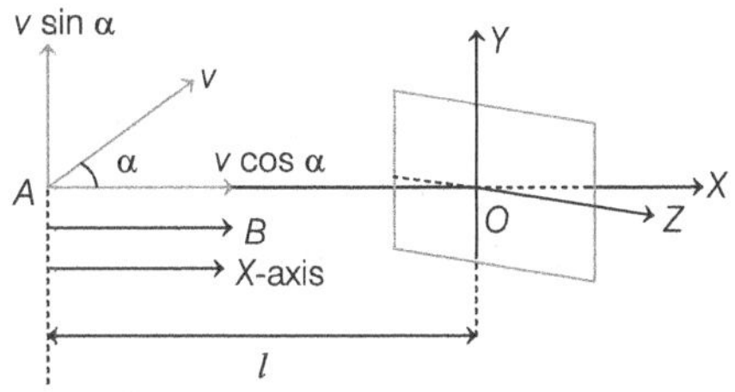

$v\sin\alpha$ and B (perpendicular to each other) will rotate the particle in circular path in YZ- plane of radius,

$$R = \frac{mv\sin\alpha}{Bq} \qquad \text{...(ii)}$$

and angular speed, $$\omega = \frac{Bq}{m} \qquad \text{...(iii)}$$

In time t particle will rotate an angle $\theta = \omega t$ from its original path.
So, in YZ-plane circular path is as shown below.

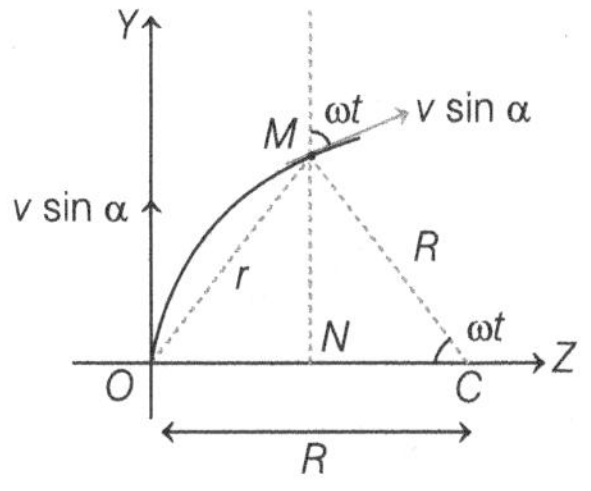

Let at time t particle is at M at a distance r from axis. Then,

$$r = \sqrt{MN^2 + ON^2} \quad \text{...(iv)}$$

Here, $MN = R \sin \omega t$ ...(v)

$ON = R - R \cos \omega t$ ...(vi)

and $q = e$ ...(vii)

Solving all these equations, we can get the answer.

99. At point P,

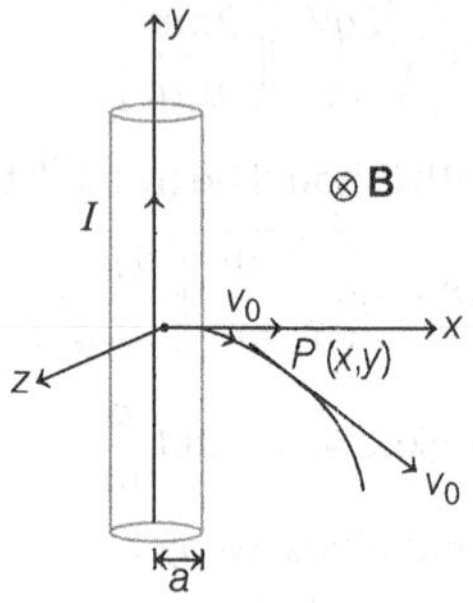

$$\mathbf{v} = v_x \hat{\mathbf{i}} + v_y \hat{\mathbf{j}}$$

$$\mathbf{B} = \frac{\mu_0}{2\pi} \frac{I}{x} (-\hat{\mathbf{k}})$$

$$\mathbf{F} = -e(\mathbf{v} \times \mathbf{B})$$

$$\mathbf{a} = \frac{\mathbf{F}}{m} = -\frac{e}{m}(\mathbf{v} \times \mathbf{B})$$

Substituting the values of $\mathbf{v}$ and $\mathbf{B}$, we get

$$a_x = \left(\frac{\mu_0 Iev_y}{2\pi mx}\right)$$

We can write, $v_y = \sqrt{v_0^2 - v_x^2}$

$$\therefore \quad a_x = -\frac{v_x dv_x}{dx} = \frac{\mu_0 Ie}{2\pi mx}\sqrt{v_0^2 - v_x^2}$$

or $$\int_{v_0}^{0} \frac{-v_x dv_x}{\sqrt{v_0^2 - v_x^2}} = \int_{a}^{x_m} \left(\frac{\mu_0 Ie}{2\pi m}\right) \frac{dx}{x}$$

Solving this integration, we get the result.

100. For cylindrical capacitor,

$$\int_B^A dV = -E dr = -\int_b^a \frac{\lambda}{2\pi\varepsilon_0 r} dr$$

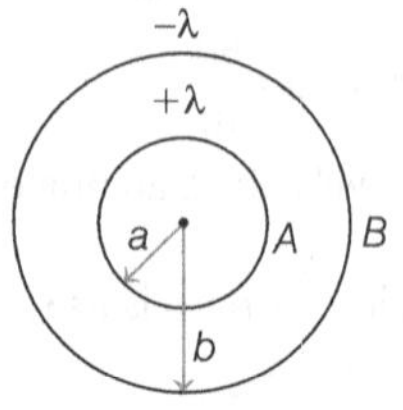

Cross-section of cylindrical capacitor

$$\therefore \quad V_A - V_B \text{ or } V = \frac{\lambda}{2\pi\varepsilon_0} \ln\left(\frac{b}{a}\right)$$

$$\therefore \quad \frac{\lambda}{2\pi\varepsilon_0} = \frac{V}{\ln\left(\frac{b}{a}\right)}$$

$$\therefore \quad E = \frac{\lambda}{2\pi\varepsilon_0 r} = \frac{V}{r\ln(b/a)}$$

In the capacitor, $qE = \frac{mv^2}{r}$

$$\therefore \quad \frac{qV}{r\ln(b/a)} = \frac{mv^2}{r}$$

$$\therefore \quad \frac{(q/m)V}{\ln(b/a)} = v^2 \quad \text{...(i)}$$

In magnetic field, $r = \frac{mv}{Bq} = \frac{v}{B(q/m)}$ ...(ii)

From these two equations, we can find v and q/m.

101. Initially, $F_e = F_m$

$$\therefore \quad qE = Bqv \text{ or } v = \frac{E}{B}$$

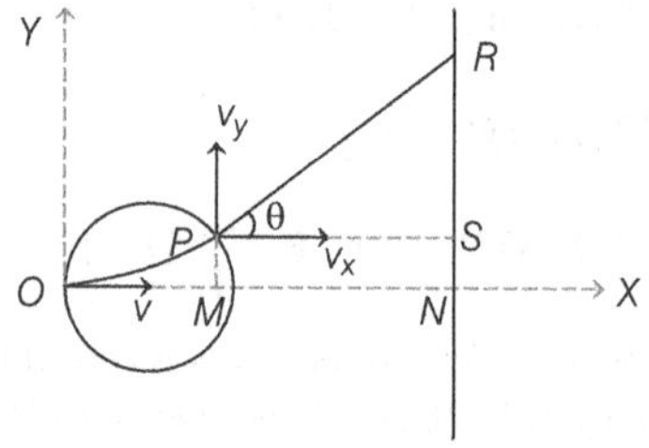

Question is mathematical. So, I am writing the steps.

(i) Write the equation of circle and parabolic path OP.

$$x = vt \quad \text{...(i)}$$

$$y = \frac{1}{2}a_y t^2 = \frac{1}{2}\left(\frac{qE}{m}\right)t^2 \quad \text{...(ii)}$$

Eliminating t from these two equations, we get

$$y = \frac{qEx^2}{2mv^2}$$

So this is the equation of parabolic path OP.

(ii) Find coordinates of P where parabola and circle intersect. Let the coordinates are (x, y).

(iii) $t = \frac{x}{v} \Rightarrow v_y = a_y t = \left(\frac{qE}{m}\right)\left(\frac{x}{v}\right)$

$$v_x = v$$

(iv) $MN = PS = (a + b) - OM = (a + b) - x$

(v) $\frac{RS}{PS} = \tan\theta = \frac{v_y}{v_x}$

$$\therefore \quad RS = (PS)\frac{v_y}{v_x}$$

(vi) $NR = \Delta x = NS + RS = PM + RS = y + RS$

102. Torque of magnetic force about PQ, ...(i)

$$\tau_m = (ILB) L \cos\theta = IL^2 B \cos\theta \quad \text{...(ii)}$$

Torque of gravitational force about PQ,

$$\tau_g = [(\lambda L) gL \sin\theta + 2(\lambda L) g (1/2) L \sin\theta]$$

$$= 2\lambda L^2 g \sin\theta$$

$$\tau_m = \tau_g$$

$$\Rightarrow \quad \tan\theta = \frac{IB}{2\lambda g} = \frac{10\sqrt{3}\times 2}{2\times\sqrt{3}\times 10} = 1$$

$$\Rightarrow \quad \theta = 45°$$

$$\therefore \quad k = 4$$

103. Let at time t, particle be at point $P(x, y)$ and its velocity be

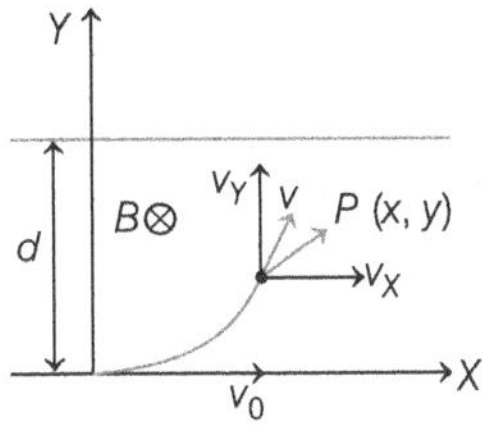

$$\mathbf{v} = (v_x\hat{\mathbf{i}} + v_y\hat{\mathbf{j}})$$

$$|\mathbf{v}| = |\mathbf{v}_0|$$

$$\Rightarrow \quad v_0^2 = v_x^2 + v_y^2$$

(work done by magnetic field is always zero, so

Then, magnetic force on the particle at point P is

$$\mathbf{F} = q(v_x\hat{\mathbf{i}} + v_y\hat{\mathbf{j}})\cdot B_0\left(1+\frac{y}{d}\right)(-\hat{\mathbf{k}})$$

$$\Rightarrow \quad -qB_0\left[1+\frac{y}{d}\right]dy = mdv_x$$

Now, when the particle will be coming out of the at that point $y = d$. Let the velocity in X-direction be v_x, then integrating we get

$$\int_{v_0}^{v_x} dv_x = -\frac{qB_0}{m}\int_0^d\left[1+\frac{y}{d}\right]dy$$

$$= -\frac{qB_0}{m}\left[d+\frac{d^2}{2d}\right]$$

$$= -\frac{3qB_0 d}{2m}$$

So, $\quad v_x = v_0 - \dfrac{3qB_0 d}{2m}$

$$\therefore \quad k = 3$$

104. Magnetic field across the wire 2 is

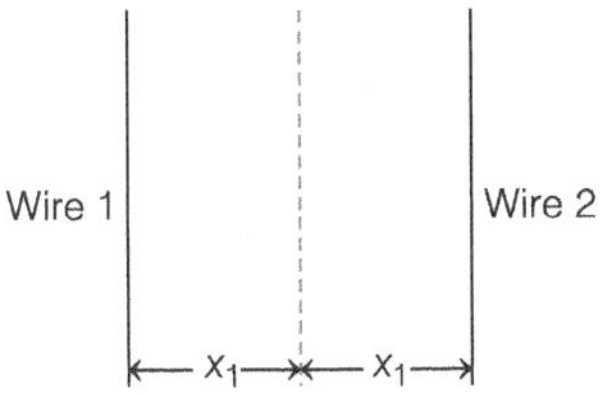

$$B_2 = \frac{\mu_0 I}{2\pi x_1} + \frac{\mu_0 I}{2\pi(x_0 - x_1)}$$

(when currents are in opposite directions)

$$B_1 = \frac{\mu_0 I}{2\pi x_1} - \frac{\mu_0 I}{2\pi(x_0 - x_1)}$$

(when currents are in same direction)

Substituting $x_1 = \dfrac{x_0}{3}\left(\text{as } \dfrac{x_0}{x_1} = 3\right)$ in above equation,

we get

Magnetic field across wire 1,

$$B_1 = \frac{3\mu_0 I}{2\pi x_0} - \frac{3\mu_0 I}{4\pi x_0} = \frac{3\mu_0 I}{4\pi x_0}$$

Radius of current carrying wire 1,

$$R_1 = \frac{mv}{qB_1} \quad \text{and} \quad B_2 = \frac{9\mu_0 I}{4\pi x_0}$$

and radius along wire 2,

$$R_2 = \frac{mv}{qB_2}$$

$\Rightarrow$ Ratio of radius of both wires,

$$\frac{R_1}{R_2} = \frac{B_2}{B_1} = \frac{9}{3} = 3$$

105. $B_R = B_T - B_C$

where, R = remaining portion, T = total portion

and $\quad C$ = cavity

$$B_R = \frac{\mu_0 I_T}{2a\pi} - \frac{\mu_0 I_C}{2(3a/2)\pi} \quad \text{...(i)}$$

$$I_T = J(\pi a^2)$$

$$I_C = J\left(\frac{\pi a^2}{4}\right)$$

Substituting the values in Eq. (i), we have

$$B_R = \frac{\mu_0}{a\pi}\left[\frac{I_T}{2} - \frac{I_C}{3}\right]$$

$$= \frac{\mu_0}{a\pi}\left[\frac{\pi a^2 J}{2} - \frac{\pi a^2 J}{12}\right]$$

$$= \frac{5\mu_0 aJ}{12}$$

$$\therefore \quad N = 5$$

UNIT-5

ELECTROMAGNETIC INDUCTION

Learning Modules

1. Faraday and Lenz Law
2. Motional Electromotive Force
3. Self Inductance and Inductor
4. Mutual Inductance
5. Growth and Decay of Current in an *L-R* Circuit
6. Oscillations in *L-C* Circuit
7. Induced Electric Field

MODULE 1

Faraday and Lenz Law

Introduction

Almost every modern device has electric circuits at its heart. We learned in the chapter of current electricity that an electromagnetic force (emf) is required for a current to flow in a circuit. But for most of the electric devices used in industry the source of emf is not a battery but an electrical generating station. In these stations other forms of energy are converted into electric energy.

For example, in a hydroelectric plant gravitational potential energy is converted into electric energy. Similarly, in a nuclear plant nuclear energy is converted into electric energy.

But how this conversion is done? Or what is the physics behind this? The branch of physics, known as **electromagnetic induction** gives the answer to all these queries.

Magnetic Field Lines and Magnetic Flux

Let us first discuss the concept of magnetic field lines and magnetic flux. We can represent any magnetic field by **magnetic field lines.** Unlike the electric lines of force, it is wrong to call them magnetic lines of force, because they do not point in the direction of the force on a charge. The force on a moving charged particle is always perpendicular to the magnetic field (or magnetic field lines) at the particle's position.

The idea of magnetic field lines is same as for the electric field lines as discussed in the chapter of electrostatics. The magnetic field at any point is tangential to the field line at that point. Where the field lines are close, the magnitude of field is large, where the field lines are far apart, the field magnitude is small. Also, because the direction of **B** at each point is unique, field lines never intersect. Unlike the electric field lines, magnetic lines form a closed loop.

Magnetic Flux

The flux associated with a magnetic field is defined in a similar manner to that used to define electric flux. Consider an element of area dS on an arbitrary shaped surface as shown in figure. If the magnetic field at this element is **B**, the magnetic flux through the element is

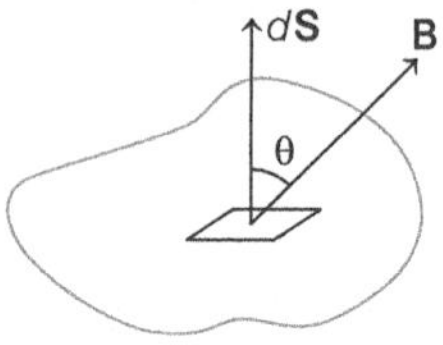

$$d\phi_B = \mathbf{B} \cdot d\mathbf{S} = BdS \cos\theta$$

Here, $d\mathbf{S}$ is a vector that is perpendicular to the surface and has a magnitude equal to the area dS and θ is the angle between **B** and $d\mathbf{S}$ at that element. In general, $d\phi_B$ varies from element to element. The total magnetic flux through the surface is the sum of the contributions from the individual area elements.

$$\therefore \qquad \phi_B = \int BdS \cos\theta = \int \mathbf{B} \cdot d\mathbf{S}$$

Note down the following points regarding the magnetic flux :

(i) Magnetic flux is a scalar quantity (dot product of two vector quantities is a scalar quantity)

(ii) The SI unit of magnetic flux is tesla-metre^2 (1 T-m^2). This unit is called **weber** (1Wb).

$$1\text{ Wb} = 1\text{ T-m}^2 = 1\text{ N-m/A}$$

Thus, unit of magnetic field is also weber/m^2 (1Wb/m^2).

or $\qquad 1\text{ T} = 1\text{ Wb/m}^2$

(iii) In the special case in which **B** is uniform over a plane surface with total area S,

$$\phi_B = BS\cos\theta$$

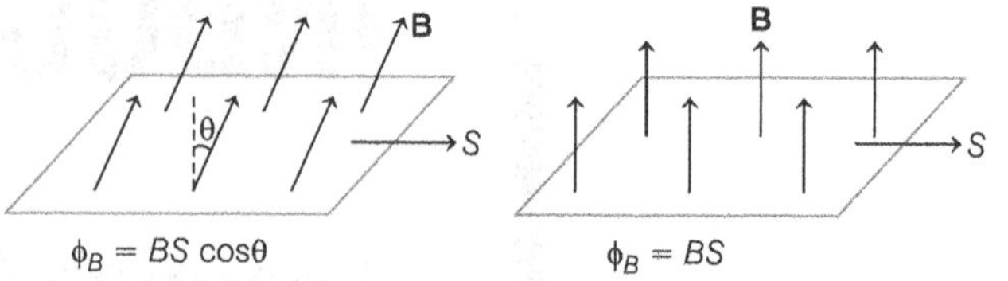

If **B** is perpendicular to the surface, then $\cos\theta = 1$ and

$$\phi_B = BS$$

(iv) If magnetic field lines are tangential at every point then magnetic flux is zero.

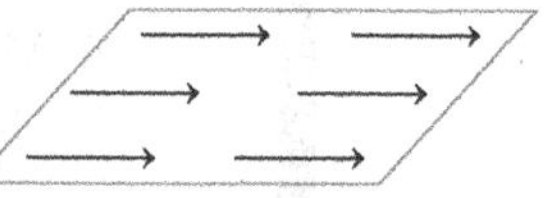

$$\phi_B = 0$$

Gauss's Law for Magnetism

In Gauss's law, the total electric flux through a closed surface is proportional to the total electric charge enclosed by the surface. For example, if a closed surface encloses an electric dipole, the total electric flux is zero because the total charge is zero.

As no magnetic monopole has ever been observed, we conclude that **the total magnetic flux through a closed surface is zero.**

$$\oint \mathbf{B} \cdot d\mathbf{S} = 0$$

Unlike electric field lines that begin and end on electric charges, magnetic field lines never have end points. Such a point would otherwise indicate the existence of a monopole. For a closed surface, the vector area element $d\mathbf{S}$ always points out of the surface. However, for an open surface we choose one of the possible sides of the surface to be the positive and use that choice consistently.

Faraday's Law

If magnetic flux passing through a closed loop is changed an induced emf is developed in the loop.

This law states that, "the induced emf in a closed loop equals the negative of the time rate of change of magnetic flux through the loop."

$$e = -\frac{d\phi_B}{dt}$$

If a circuit is a coil consisting of N loops all of the same area and if ϕ_B is the flux through one loop, an emf is induced in every loop, thus the total induced emf in the coil is given by the expression,

$$e = -N\frac{d\phi_B}{dt}$$

The negative sign in the above equations is of important physical significance, which we will discuss in **Lenz's Law.**

ϕ_B can also be represented by ϕ.

Note down the following points regarding the Faraday's law:

(i) As we have seen, induced emf is produced only when there is a change in magnetic flux passing through a loop. The flux passing through the loop is given by

$$\phi = BS\cos\theta$$

This, flux can be changed in different ways discussed below :

(a) The flux ϕ can be given as a function of time.

(b) The magnitude of **B** can change with time. In the problems if magnetic field is given a function of time, it implies that the magnetic field is changing. Thus,

$$B = B(t)$$

(c) The current producing the magnetic field can change with time. For this, the current can be given as a function of time. Hence,

$$i = i(t)$$

(d)

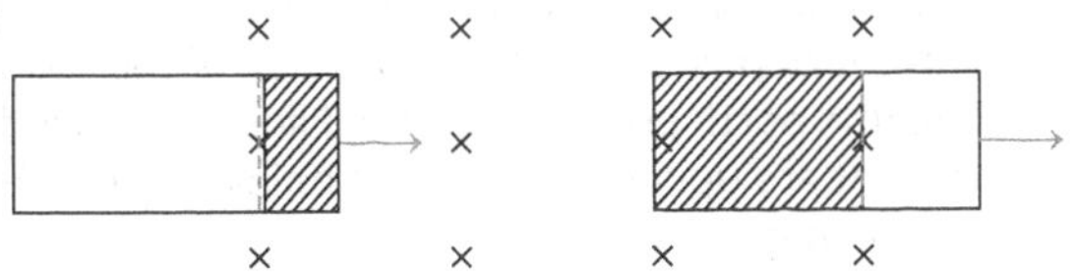

The area of the loop inside the magnetic field can change with time.This can be done by pulling a loop inside (or outside) a magnetic field.

(e) The angle θ between **B** and the normal to the loop (or **s**) can change with time.

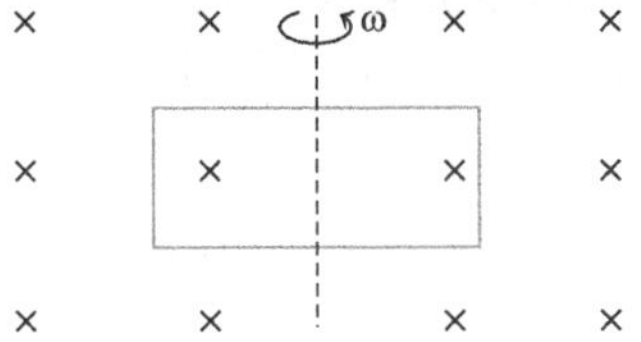

This can be done by rotating a loop in a magnetic field.

(f) Any combination of the above can occur.

(ii) When the magnetic flux passing through a loop is changed, an induced emf and hence, an induced current is produced in the circuit. If R is the resistance of the circuit, then induced current is given by

$$i = \frac{e}{R} = \frac{1}{R}\left(\frac{-d\phi_B}{dt}\right)$$

Current starts flowing in the circuit, means flow of charge takes place. Charge flown in the circuit in time dt will be given by

$$dq = i\,dt = \frac{1}{R}(-d\phi_B)$$

Thus, for a time interval Δt we can write the average values as,

$$e = -\frac{\Delta\phi_B}{\Delta t}, \quad i = \frac{1}{R}\left(\frac{-\Delta\phi_B}{\Delta t}\right) \text{ and } \Delta q = \frac{1}{R}(-\Delta\phi_B)$$

From these equations, we can see that e and i are inversely proportional to Δt while Δq is independent of Δt. It depends on the magnitude of change in flux, not the time taken in it. This can be explained by the following example.

Example 1 A square loop $ACDE$ of area 20 cm^2 and resistance $5\,\Omega$ is rotated in a magnetic field $\mathbf{B} = 2\text{T}$ through 180°, (a) in 0.01 s and (b) in 0.02 s.

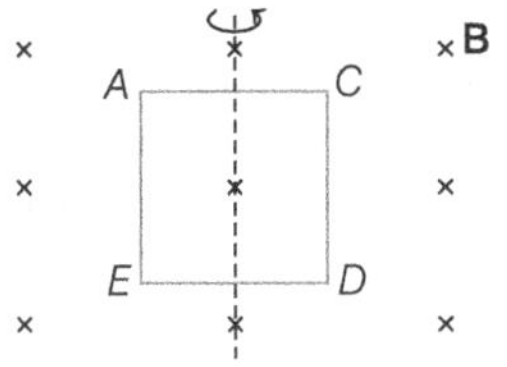

Find the magnitudes of average values of e, i and Δq in both the cases.

Sol. Let us take the area vector **S** perpendicular to plane of loop inwards. So initially, **S**↑↑**B** and when it is rotated by 180°, **S**↑↓**B**. Hence, initial flux passing through the loop,

$$\phi_i = BS\cos 0° = (2)(20 \times 10^{-4})(1) = 4.0 \times 10^{-3}\text{ Wb}$$

Flux passing through the loop when it is rotated by 180°,

$\phi_f = BS\cos 180° = (2)(20 \times 10^{-4})(-1) = -4.0 \times 10^{-3}$ Wb

Therefore, change in flux,

$$\Delta\phi_B = \phi_f - \phi_i = -8.0 \times 10^{-3} \text{ Wb}$$

(a) Given, $\Delta t = 0.01$ s, $R = 5\,\Omega$

$$\therefore \quad |e| = \left| -\frac{\Delta\phi_B}{\Delta t} \right| = \frac{8.0 \times 10^{-3}}{0.01} = 0.8 \text{ V}$$

$$i = \frac{|e|}{R} = \frac{0.8}{5} = 0.16 \text{ A}$$

and $\Delta q = i\Delta t = 0.16 \times 0.01 = 1.6 \times 10^{-3}$ C

(b) $\Delta t = 0.02$ s

$$\therefore \quad |e| = \left| -\frac{\Delta\phi_B}{\Delta t} \right| = \frac{8.0 \times 10^{-3}}{0.02} = 0.4 \text{ V}$$

$$i = \frac{|e|}{R} = \frac{0.4}{5} = 0.08 \text{ A}$$

and $\Delta q = i\Delta t = (0.08)(0.02)$
$= 1.6 \times 10^{-3}$ C

Note *Time interval Δt in part (b) is two times the time interval in part (a), so e and i are half while Δq is same.*

Example 2 A coil consists of 200 turns of wire having a total resistance of 2.0 Ω. Each turn is a square of side 18 cm, and a uniform magnetic field directed perpendicular to the plane of the coil is turned on. If the field changes linearly from 0 to 0.5 T in 0.80 s, what is the magnitude of induced emf and current in the coil while the field is changing?

Sol. From the Faraday's law,

$$\text{Induced emf, } |e| = \frac{N\Delta\phi}{\Delta t} = (NS)\frac{\Delta B}{\Delta t}$$

$$= \frac{(200)(18 \times 10^{-2})^2 (0.5 - 0)}{0.8}$$

$= 4.05$ V **Ans.**

$$\text{Induced current, } i = \frac{|e|}{R} = \frac{4.05}{2} \approx 2.0 \text{ A}$$ **Ans.**

Example 3 The magnetic flux passing through a metal ring varies with time t as : $\phi_B = 3(at^3 - bt^2)$ T-m^2 with $a = 2.00$ s^{-3} and $b = 6.00$ s^{-2}. The resistance of the ring is 3.0 Ω. Determine the current induced in the ring at $t = 2.0$ s.

Sol. Given, $\phi_B = 3(at^3 - bt^2)$

$$\therefore \quad |e| = \left| \frac{d\phi_B}{dt} \right| = 9at^2 - 6bt$$

$$\therefore \text{ Induced current, } i = \frac{|e|}{R} = \frac{9at^2 - 6bt}{3} = 3at^2 - 2bt$$

Substituting the values, we get

$$i = (3)(2)(2)^2 - (2)(3)(2)$$

$$= 12 \text{ A}$$

Lenz's Law

The negative sign in Faraday's equations of electromagnetic induction describes the direction in which the induced emf drives current around a circuit. However, that direction is most easily determined with the help of Lenz's law. This law states that:

"The direction of any magnetic induction effect is such as to oppose the cause of the effect."

For different types of problems, Lenz's law has been further subdivided into following concepts.

1. **Attraction and repulsion concept** If magnetic flux is changed by bringing a magnet and a loop (or solenoid etc.) closer to each other then direction of induced current is so produced, that the magnetic field produced by it always repels the two. Similarly, if they are moved away from each other then they are attracted towards each other. Following two examples will illustrate this.

Example 4 A bar magnet is freely falling along the axis of a circular loop as shown in figure. State whether its acceleration a is equal to, greater than or less than the acceleration due to gravity g.

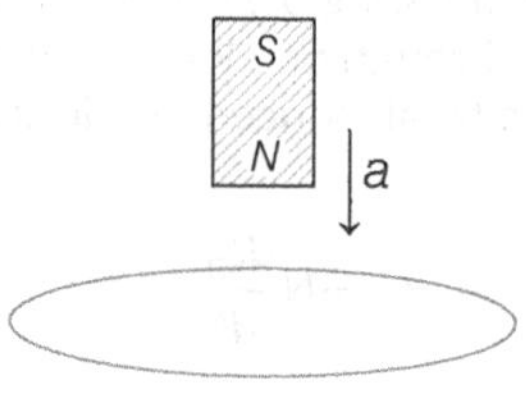

Sol. $a < g$. Because according to Lenz's law, whatever may be the direction of induced current, it will oppose the cause.

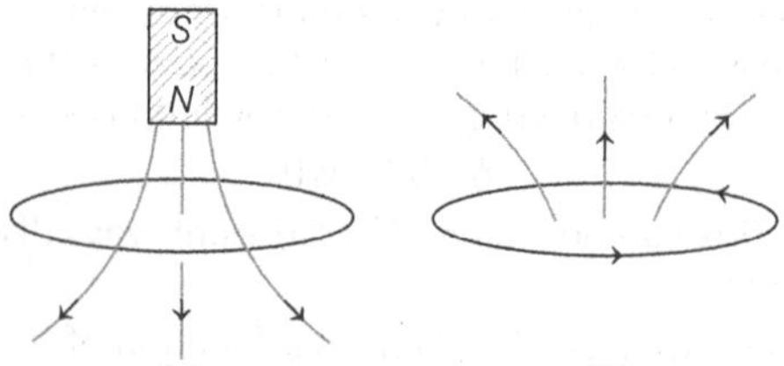

(a) (b)

Here, the cause is, the free fall of magnet and so the induced current will oppose it and the acceleration of magnet will be less than the acceleration due to gravity g. This can also be explained in a different manner. When the magnet falls downwards with its north pole downwards.

The magnetic field lines passing through the coil in the downward direction increase. Since, the induced current opposes this, the upper side of the coil will become north pole, so that field lines due to induced current are upwards. Now, like poles repel each other. Hence, $a < g$.

Example 5 A bar magnet is brought near or moved away from a solenoid as shown in figure. Will the solenoid attract or repel the magnet?

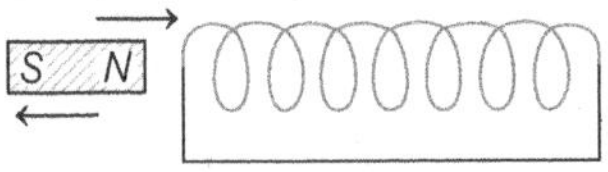

Sol. When the magnet is brought near the solenoid, according to Lenz's law, both repel each other. On the other hand, if the magnet is moved away from the solenoid, it attracts the magnet. When the magnet is brought near the solenoid, the nearer side becomes the same pole and when it is moved away it becomes the opposite pole as shown in figure.

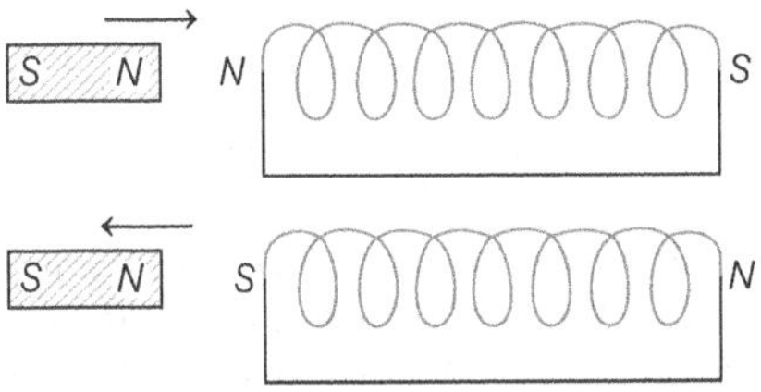

It can also be explained by increasing or decreasing field lines as discussed in example 4.

2. **Cross or dot magnetic field increasing or decreasing concept**

If cross magnetic field passing through a loop increases then induced current will produce dot magnetic field. Similarly, if dot magnetic field passing through a loop decreases then dot magnetic field is produced by the induced current. Let us take some examples in support of it.

Example 6 A circular loop is placed in magnetic field $B = 2t$. Find the direction of induced current produced in the loop.

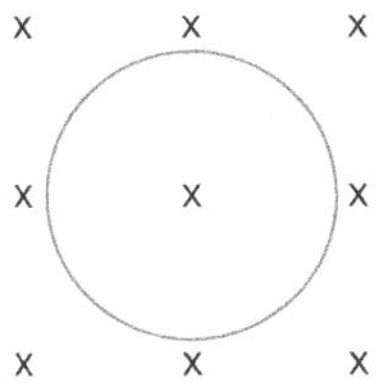

Sol. $B = 2t$, means $\otimes$ magnetic field (or we can also say cross magnetic flux) passing through the loop is increasing. So, induced current will produce dot magnetic field. To produce $\odot$ magnetic field, induced current from our side should be anti-clockwise.

Example 7 A rectangular loop is placed to the left of large current carrying straight wire as shown in figure. Current varies with time as $I = 2t$. Find direction of induced current I_{in} in the loop.

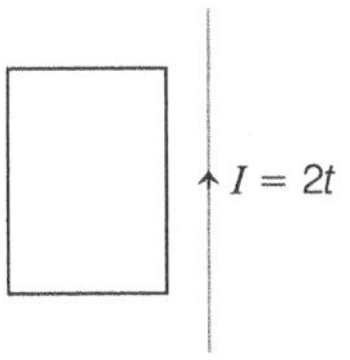

Sol. Current I will produce $\odot$ magnetic field passing through the loop. Current I is increasing, so dot magnetic field will also increase. Therefore, induced current should produce cross magnetic field. For producing cross magnetic field in the loop, induced current from our side should be clockwise.

3. **Situations where flux passing through the loop is always zero or change in flux is zero.**

Example 8 A current carrying straight wire passes inside a triangular loop as shown in figure. The current in the wire is perpendicular to paper inwards. Find the direction of the induced current in the loop if current in the wire is increased.

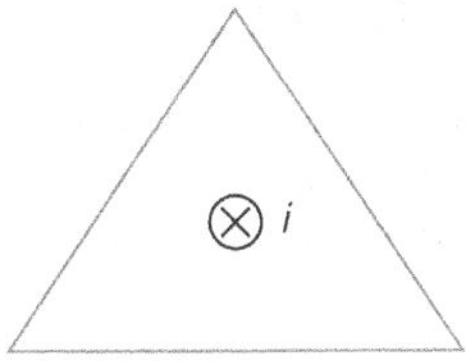

Sol. Magnetic field lines round the current carrying wire are as shown in figure.

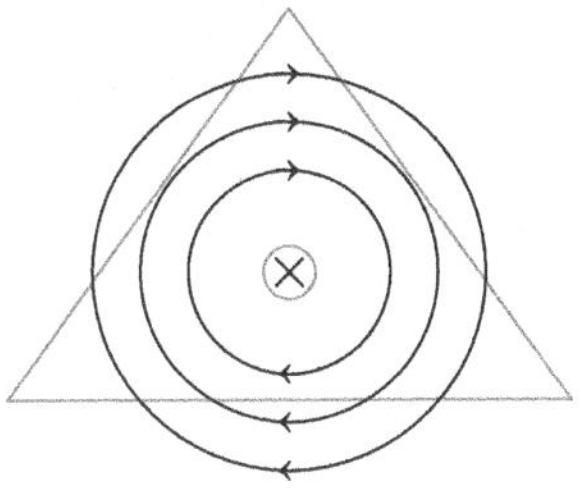

Since, the lines are tangential to the loop ($\theta = 90°$) the flux passing through the loop is always zero, whether the current is increased or decreased. Hence, change in flux is also zero. Therefore, induced current in the loop will be zero.

Example 9 A rectangular loop is placed adjacent to a current carrying straight wire as shown in figure. If the loop is rotated about an axis passing through one of its sides, find the direction of induced current in the loop.

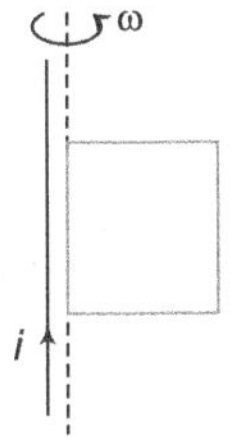

Sol. Magnetic field lines around the straight wire are circular. So, same magnetic lines will pass through loop under all conditions.

$$\Delta\phi = 0 \Rightarrow \text{emf} = 0 \Rightarrow i = 0$$

4. **Attraction or repulsion between two loops facing each other if current in one loop is changed**

Example 10 Two loops are facing each other as shown in figure. State whether the loops will attract each other or repel each other if current I_1 is increased.

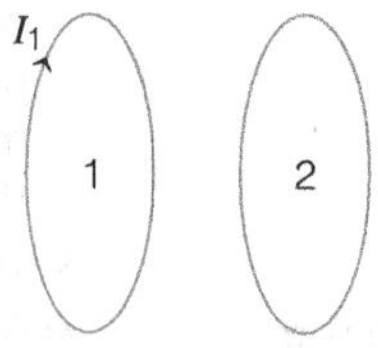

Sol. If current I_1 is increased then induced current in loop-2 (say I_2) will be in opposite direction. Now, two wires having currents in opposite directions repel each other. So, the loops will repel each other.

Important POINTS

1. **Eddy currents** Eddy currents are generated inside a conductor when it moves through a magnetic field or when the magnetic flux flowing through it varies continuously.

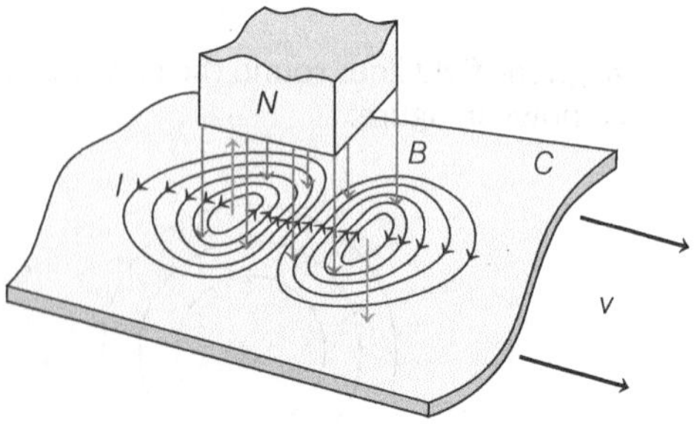

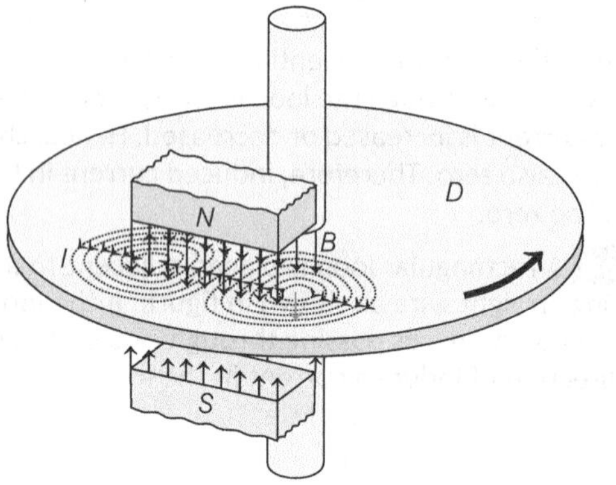

Eddy currents flow through conductors like whirling eddies in a stream and they are frequently formed in reaction to a changing magnetic field. They flow in closed loops perpendicular to the magnetic field. These are also known as **Foucault**'s Currents.

Question : Why are eddy currents undesirable?

Answer : When a conductor is moved in a magnetic field, eddy currents are generated. Eddy currents cause energy to be lost as heat.

Eddy currents can be reduced in a variety of methods, including:

(a) By laminating the metal core : Insulating materials separate the laminations of the metallic core and the plane of the laminations should be kept parallel to the magnetic field so that they cut across the routes of eddy currents.

(b) By using magnetic materials with poor electrical conductivity or high resistance.

2. **Back EMF of Motors** An electric motor converts electrical energy into mechanical energy and is based on the fact that a current carrying coil in a uniform magnetic field experiences a torque. As the coil rotates in the magnetic field, the flux linked with the rotating coil will change and hence, an emf called back emf is produced in the coil.

3. **Electric Generator or Dynamo** A dynamo converts mechanical energy (rotational kinetic energy) into electrical energy. It consists of a coil rotating in a magnetic field. Due to rotation of the coil magnetic flux linked with it changes, so an emf is induced in the coil.

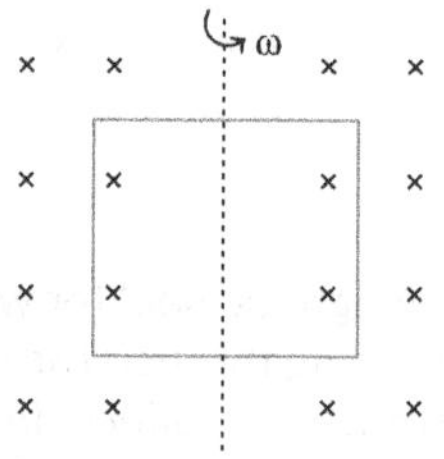

Suppose at time $t = 0$, plane of coil is perpendicular to the magnetic field.

The flux linked with it at any time t will be given by

$$\phi = NBA \cos \omega t \qquad (N = \text{number of turns in the coil})$$

$$\therefore \quad e = -\frac{d\phi}{dt} = NBA\omega \sin \omega t$$

or $\quad e = e_0 \sin \omega t$

where, $\quad e_0 = NBA\,\omega$

MODULE 1 Exercise

True/False

1. A coil of metal wire is kept stationary in a non-uniform magnetic field. An emf is induced in the coil. **(1986 JEE)**

Assertion and Reason

See directions on page 25.

2. **Assertion :** A square loop is placed in x-y plane as shown in figure. Magnetic field in the region is $\mathbf{B} = -B_0 x\hat{\mathbf{k}}$. The induced current in the loop is anti-clockwise.

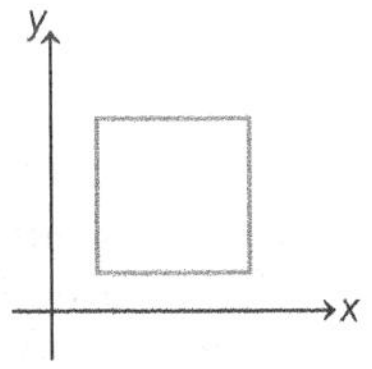

Reason : If inward magnetic field from such a loop increases, then current should be anti-clockwise.

3. **Assertion :** Magnetic field B (shown inwards) varies with time t as shown. At time t_0 induced current in the loop is clockwise.

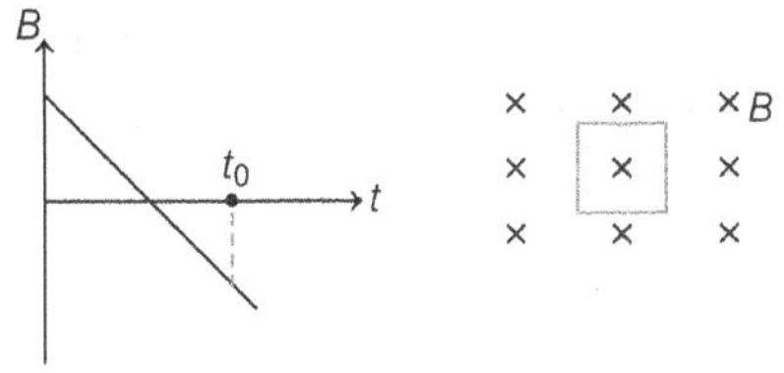

Reason : If rate of change of magnetic flux from a coil is constant, charge should flow in the coil at a constant rate.

Single Correct Option

4. Two circular loops P and Q are concentric and coplanar as shown in figure. The loop Q is smaller than P. If the current I_1 flowing in loop P is decreasing with time, then the current I_2 in the loop Q

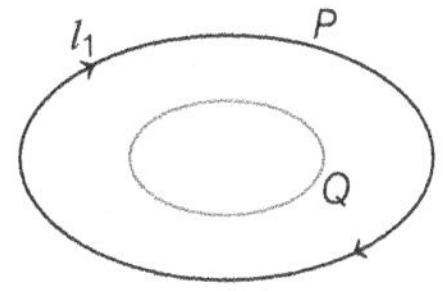

(a) flows in the same direction as that of P
(b) flows in the opposite direction as that of Q
(c) is zero
(d) None of the above

5. The variation of induced emf (e) with time (t) in a coil if a short bar magnet is moved along its axis with a constant velocity is best represented as

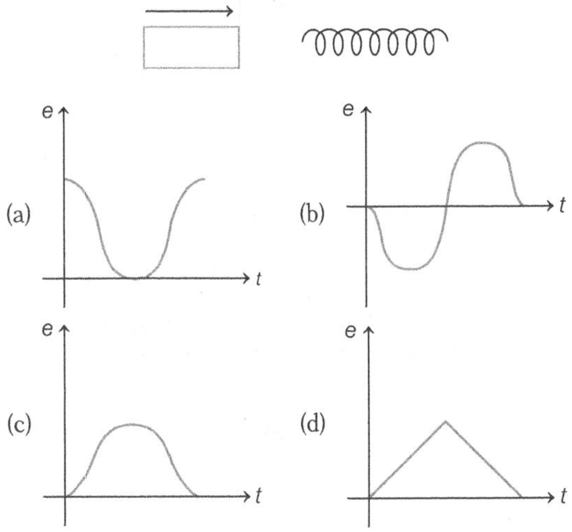

6. A circular loop of radius R, carrying current I, lies in x-y plane with its centre at origin. The total magnetic flux through x-y plane is
(a) directly proportional to I
(b) directly proportional to R
(c) inversely proportional to R
(d) zero

7. A small circular loop is suspended from an insulating thread. Another coaxial circular loop carrying a current I and having radius much larger than the first loop starts moving towards the smaller loop. The smaller loop will

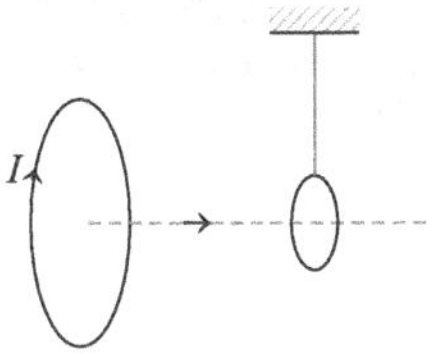

(a) be attracted towards the bigger loop
(b) be repelled by the bigger loop
(c) experience no force
(d) All of the above

8. The current i in an induction coil varies with time t according to the graph shown in the figure. Which of the following graphs shows the induced emf (ε) in the coil with time?

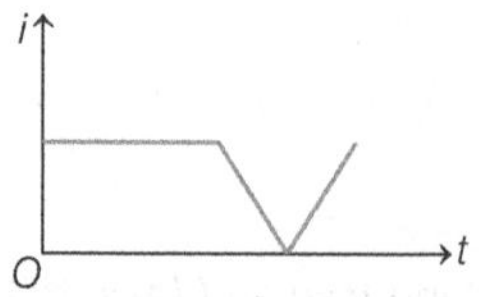

(a)
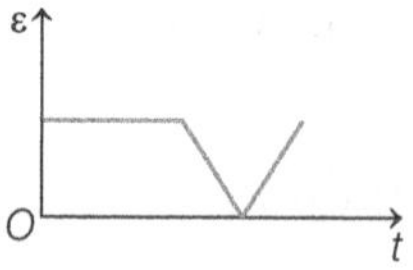

(b)
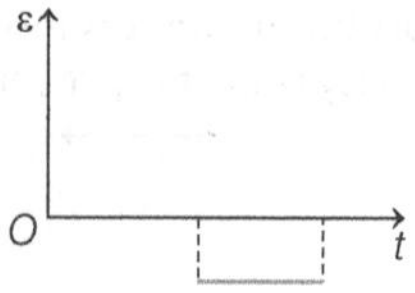

(c)
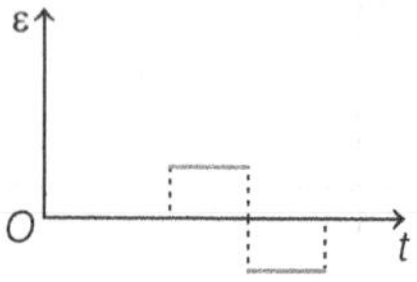

(d)
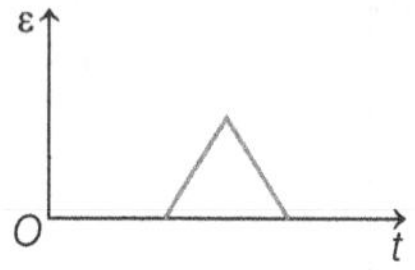

9. A magnetic flux through a stationary loop with a resistance R varies during the time interval τ as $\phi = at(\tau - t)$. Find the amount of heat generated in the loop during that time

(a) $\frac{a\tau^2}{2R}$ (b) $\frac{a^2\tau^3}{3R}$

(c) $\frac{2a^2\tau^3}{3R}$ (d) $\frac{a\tau}{3R}$

10. A magnet is taken towards a conducting ring in such a way that a constant current of 10 mA is induced in it. The total resistance of the ring is 0.5 Ω. In 5 s, the magnetic flux through the ring changes by

(a) 0.25 mWb (b) 25 mWb

(c) 50 mWb (d) 15 mWb

11. A rectangular loop of sides a and b is placed in xy-plane. A uniform but time varying magnetic field of strength $\mathbf{B} = 20t\hat{\mathbf{i}} + 10t^2\hat{\mathbf{j}} + 50\hat{\mathbf{k}}$ is present in the region. The magnitude of induced emf in the loop at time t is

(a) $20 + 20t$ (b) 20 (c) $20t$ (d) zero

12. A coil of area 10 cm^2 and 10 turns is in magnetic field directed perpendicular to the plane and changing at a rate of 10^8 gauss/s. The resistance of coil is 20 Ω. The current in the coil will be

(a) 0.5 A

(b) 5×10^{-3} A

(c) 0.05 A

(d) 5 A

13. If the instantaneous magnetic flux and induced emf produced in a coil is ϕ and E respectively, then according to Faraday's law of electromagnetic induction

(a) E must be zero if $\phi = 0$

(b) $E \neq 0$ if $\phi = 0$

(c) $E \neq 0$ but ϕ may or may not be zero

(d) $E = 0$ then ϕ must be zero

14. The figure shows a conducting ring of radius R. A uniform steady magnetic field B lies perpendicular to the plane of the ring in a circular region of radius $r (< R)$. If the resistance per unit length of the ring is λ, then the current induced in the ring when its radius gets doubled is

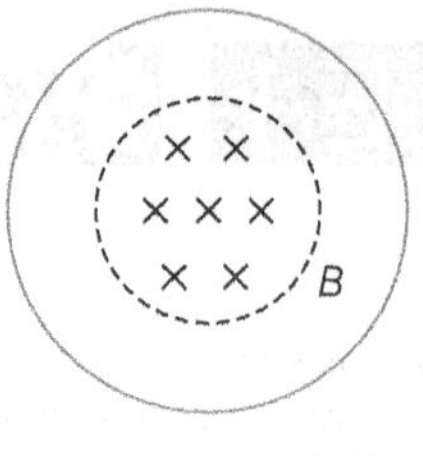

(a) $\frac{BR}{\lambda}$ (b) $\frac{2BR}{\lambda}$

(c) zero (d) $\frac{Br^2}{4R\lambda}$

15. A rectangular loop of length 2.5 m and width 2 m is placed at 60° to a magnetic field of 4 T. The loop is removed from the field in 10 s. The average emf induced in the loop during this time is **(2024 Main)**

(a) -2 V (b) -1 V

(c) $+2$ V (d) $+1$ V

16. A conducting circular loop of radius $10/\sqrt{\pi}$ cm is placed perpendicular to a uniform magnetic field of 0.5 T. The magnetic field is decreased to zero in 0.5 s at a steady rate. The induced emf in the circular loop at 0.25 s is **(2023 Main)**

(a) emf = 5 mV (b) emf = 100 mV

(c) emf = 1 mV (d) emf = 10 mV

17. The magnetic flux through a coil perpendicular to its plane is varying according to the relation $\phi = (5t^3 + 4t^2 + 2t - 5)$ weber. If the resistance of the coil is 5 ohm, then the induced current through the coil at $t = 2$ sec will be **(2022 Main)**

(a) 15.6 A (b) 16.6 A

(c) 17.6 A (d) 18.6 A

18. In a coil of resistance 100 Ω, a current is induced by changing the magnetic flux through it as shown in the figure. The magnitude of change in flux through the coil is **(2017 Main)**

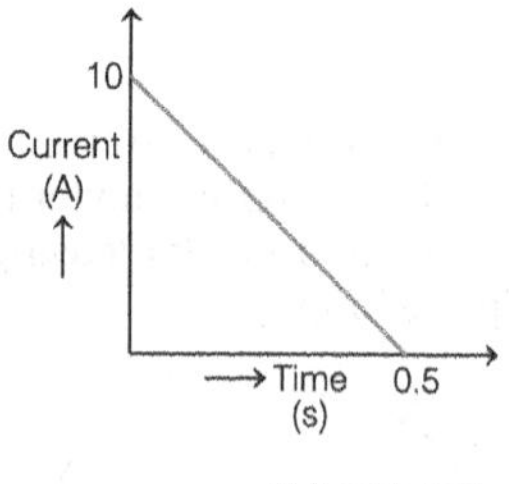

(a) 225 Wb (b) 250 Wb

(c) 275 Wb (d) 200 Wb

19. A small circular loop of wire of radius a is located at the centre of a much larger circular wire loop of radius b. The two loops are in the same plane. The outer loop of radius b carries an alternating current $I = I_0 \cos(\omega t)$. The emf induced in the smaller inner loop is nearly **(2017 Main)**

(a) $\frac{\pi\mu_0 I_0}{2} \cdot \frac{a^2}{b} \omega \sin(\omega t)$ (b) $\frac{\pi\mu_0 I_0}{2} \cdot \frac{a^2}{b} \omega \cos(\omega t)$

(c) $\pi\mu_0 I_0 \frac{a^2}{b} \omega \sin(\omega t)$ (d) $\frac{\pi\mu_0 I_0 b^2}{a} \omega \cos(\omega t)$

20. A circular loop of radius 0.3 cm lies parallel to a much bigger circular loop of radius 20 cm. The centre of the smaller loop is on the axis of the bigger loop. The distance between their centres is 15 cm. If a current of 2.0 A flows through the bigger loop, then the flux linked with smaller loop is **(2013 Main)**

(a) 9.1×10^{-11} Wb (b) 6×10^{-11} Wb

(c) 3.3×10^{-11} Wb (d) 6.6×10^{-9} Wb

Subjective Questions

21. Figure shows a conducting loop placed near a long straight wire carrying a current i as shown. If the current increases continuously, find the direction of the induced current in the loop.

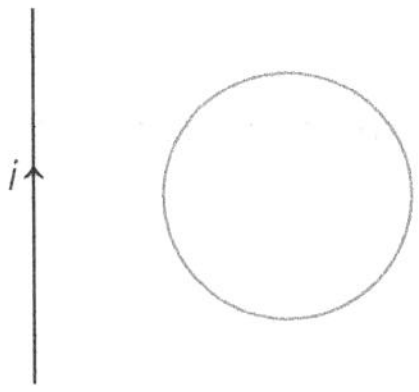

22. Write the dimensions of $\frac{d\phi_B}{dt}$.

23. A triangular loop is placed in a dot ⊙ magnetic field as shown in figure. Find the direction of induced current in the loop if magnetic field is increasing.

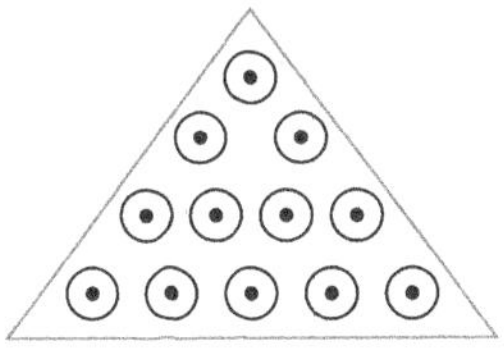

24. Two circular loops lie side by side in the same plane. One is connected to a source that supplies an increasing current, the other is a simple closed ring. Is the induced current in the ring is in the same direction as that in the loop connected to the source or opposite? What if the current in the first loop is decreasing?

25. A wire in the form of a circular loop of radius 10 cm lies in a plane normal to a magnetic field of 100 T. If this wire is pulled to take a square shape in the same plane in 0.1 s, find the average induced emf in the loop.

26. A closed coil consists of 500 turns has area 4 cm^2 and a resistance of 50 Ω. The coil is kept with its plane perpendicular to a uniform magnetic field of 0.2 Wb/m^2. Calculate the amount of charge flowing through the coil if it is rotated through 180°.

27. The magnetic field in a certain region is given by $\mathbf{B} = (4.0\,\hat{\mathbf{i}} - 1.8\,\hat{\mathbf{k}}) \times 10^{-3}$ T. How much flux passes through a 5.0 cm^2 area loop in this region if the loop lies flat on the xy-plane?

28. A coil formed by wrapping 50 turns of wire in the shape of a square is positioned in a magnetic field so that the normal to the plane of the coil makes an angle of 30°, with the direction of the field. When the magnetic field is increased uniformly from 200 μT to 600 μT in 0.4 s, an emf of magnitude 80.0 mV is induced in the coil. What is the total length of the wire?

29. A loop of wire enclosing an area S is placed in a region where the magnetic field is perpendicular to the plane. The magnetic field **B** varies with time according to the expression $B = B_0 e^{-at}$ where a is some constant. That is, at $t = 0$, the field is B_0 and for $t > 0$, the field decreases exponentially. Find the induced emf in the loop as a function of time.

30. A small coil is introduced between the poles of an electromagnet so that its axis coincides with the magnetic field direction. The cross-sectional area of the coil is equal to $S = 3.0$ mm^2, the number of turns is $N = 60$. When the coil turns through 180° about its diameter, a galvanometer connected to the coil indicates a charge $q = 4.5$ μC flowing through it. Find the magnetic induction magnitude between the poles, provided the total resistance of the electric circuit equals $R = 40\ \Omega$.

31. The magnetic field through a single loop of wire, 12 cm in radius and of 8.5 Ω resistance, changes with time as shown in figure. Calculate the emf in the loop as a function of time. Consider the time intervals

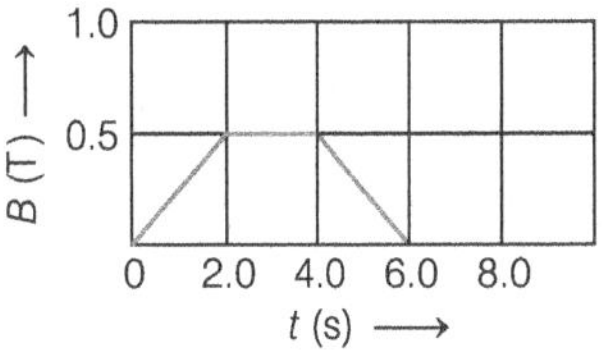

(a) $t = 0$ to $t = 2.0$ s
(b) $t = 2.0$ s to $t = 4.0$ s
(c) $t = 4.0$ s to $t = 6.0$ s.

The magnetic field is perpendicular to the plane of the loop.

Integer Type Questions

32. The magnetic field B crossing normally a square metallic plate of area 4m^2 is changing with time as shown in figure. The magnitude of induced emf in the plate during $t = 2$ s to $t = 4$ s, is .... mV. **(2023 Main)**

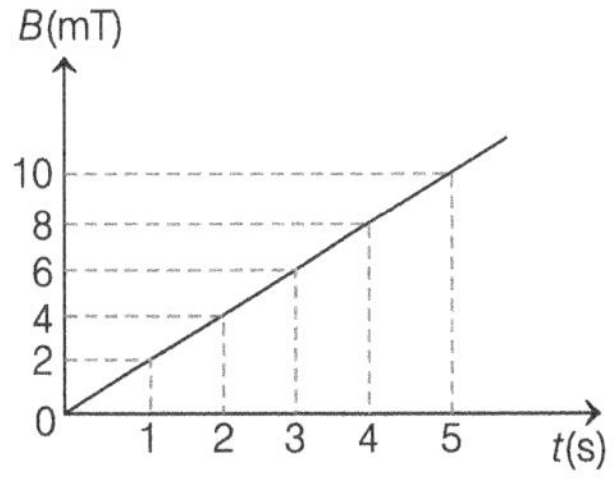

33. An insulated copper wire of 100 turns is wrapped around a wooden cylindrical core of the cross-sectional area 24 cm^2. The two ends of the wire are connected to a resistor. The total resistance in the circuit is 12 Ω. If an externally applied uniform magnetic field in the core along its axis changes from 1.5 T in one direction to 1.5 T in the opposite direction, the charge flowing through a point in the circuit during the change of magnetic field will be .... mC. **(2023 Main)**

34. A square loop of side 2.0 cm is placed inside a long solenoid that has 50 turns per centimetre and carries a sinusoidally varying current of amplitude 2.5 A and angular frequency 700 rad s^{-1}. The central axes of the loop and solenoid coincide. The amplitude of the emf induced in the loop is $x \times 10^{-4}$ V. The value of x is .... (Take, $\pi = \frac{22}{7}$) **(2023 Main)**

35. A conducting circular loop is placed in a uniform magnetic field of 0.4 T with its plane perpendicular to the field. Somehow, the radius of the loop starts expanding at a constant rate of 1 mm/s. The magnitude of induced emf in the loop at an instant when the radius of the loop is
2 cm will be .... μ V. **(2023 Main)**

36. A conducting circular loop is placed in xy-plane in presence of magnetic field $\mathbf{B} = (3t^3\,\hat{\mathbf{j}} + 3t^2\,\hat{\mathbf{k}})$ in SI unit. if the radius of the loop is 1 m, the induced emf in the loop, at time, $t = 2$ s is $n\pi$V. The value of n is .... **(2022 Main)**

37. In a coil of resistance 8 Ω, the magnetic flux due to an external magnetic field varies with time as $\phi = \frac{2}{3}(9 - t^2)$. The value of total heat produced in the coil, till the flux becomes zero, will be .... J. **(2022 Main)**

Answers

1. (False)	**2.** (d)	**3.** (b)	**4.** (a)	**5.** (b)	**6.** (d)	**7.** (b)	**8.** (c)	**9.** (b)	**10.** (b)
11. (d)	**12.** (d)	**13.** (c)	**14.** (c)	**15.** (d)	**16.** (d)	**17.** (a)	**18.** (b)	**19.** (a)	**20.** (a)

21. Anti-clockwise **22.** $[ML^2A^{-1}T^{-3}]$ **23.** Clockwise **24.** Same direction, opposite direction **25.** 6.7 V
26. 1600 μC **27.** 9.0×10^{-7} Wb **28.** 272 m **29.** $e = aB_0Se^{-at}$ **30.** 0.5 T
31. (a) 0.011V (b) Zero (c) – 0.011 V **32.** (8) **33.** (60) **34.** (44) **35.** (50) **36.** (12) **37.** (2)

See hints on pages 542-545.

MODULE 2

Motional Electromotive Force

Till now, we have considered the cases in which an emf is induced in a stationary circuit placed in a magnetic field, when the field changes with time. In this section, we describe what is called **motional emf**, which is the emf induced in a conductor moving through a constant magnetic field.

The straight conductor of length l shown in figure is moving through a uniform magnetic field directed into the page.

For simplicity we assume that the conductor is moving in a direction perpendicular to the field with constant velocity under the influence of some external agent. The electrons in the conductor experience a force

$$\mathbf{F}_m = -e\,(\mathbf{v} \times \mathbf{B})$$

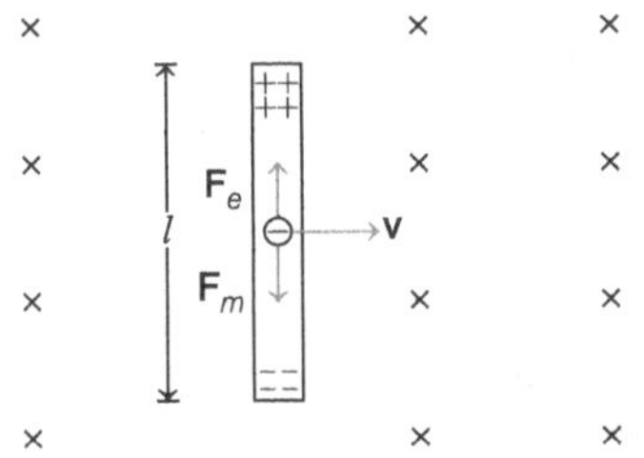

Under the influence of this force, the electrons move to the lower end of the conductor and accumulate there, leaving a net positive charge at the upper end.

As a result of this charge separation, an electric field is produced inside the conductor. The charges accumulate at both ends untill the downward magnetic force evB is balanced by the upward electric force eE. At this point, electrons stop moving. The condition for equilibrium requires that,

$$eE = evB \quad \text{or} \quad E = vB$$

The electric field produced in the conductor (once the electrons stop moving and E is constant) is related to the potential difference across the ends of the conductor according to the relationship,

$$\Delta V = El = Blv$$

$$\boxed{\Delta V = Blv}$$

where the upper end is at a higher electric potential than the lower end. Thus,

"a potential difference is maintained between the ends of a straight conductor as long as the conductor continues to move through the uniform magnetic field."

Now, suppose the moving rod slides along a stationary *U*-shaped conductor forming a complete circuit. We call this a **motional electromotive force** denoted by e, we can write

$$\boxed{e = Bvl}$$

If R is the resistance of the circuit, then current in the circuit is

$$\boxed{i = \frac{e}{R} = \frac{Bvl}{R}}$$

$V_a > V_b$. Therefore, direction of current in the loop is anti-clockwise as shown in figure.

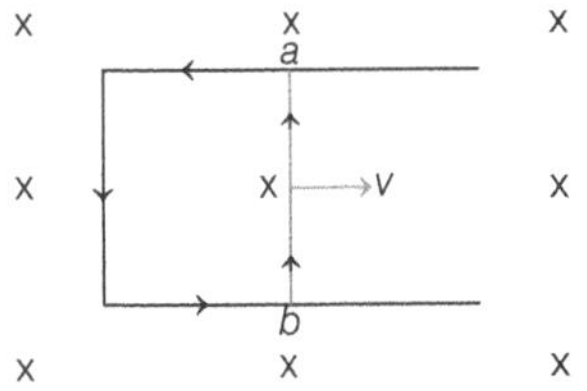

Important POINTS

- The direction of motional emf or current can be given by right hand rule.

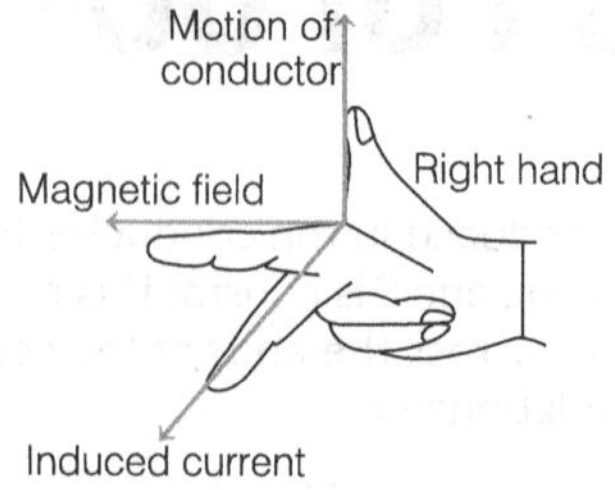

OR

Stretch your right hand.

The stretched fingers point in the direction of magnetic field. Thumb is along the velocity of conductor.

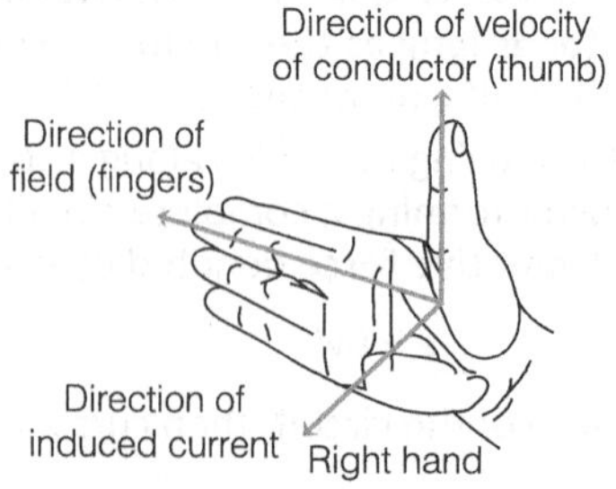

The upper side of the palm is at higher potential and lower side on lower potential. If the circuit is closed, the induced current within the conductor is along perpendicular to palm upwards.

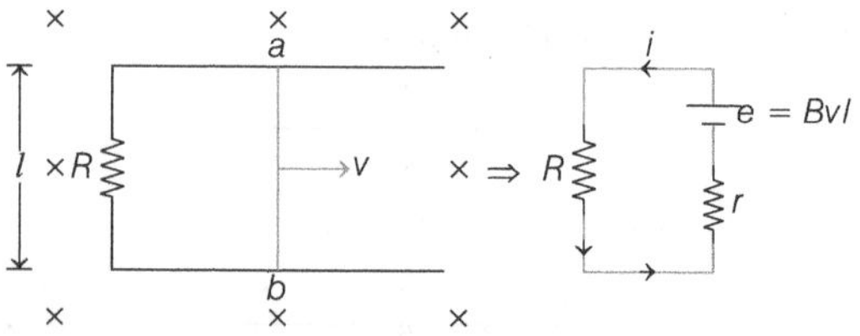

In the figure, we can replace the moving rod ab by a battery of emf Bvl with the positive terminal at a and the negative terminal at b. The resistance r of the rod ab may be treated as the internal resistance of the battery. Hence, the current in the circuit is

$$i = \frac{e}{R + r} \quad \text{or} \quad \boxed{i = \frac{Bvl}{R + r}}$$

- **Induction and energy transfers** In the above figure, if you move the conductor ab with a constant velocity v, the current in the circuit is

$$i = \frac{Bvl}{R} \qquad (r = 0)$$

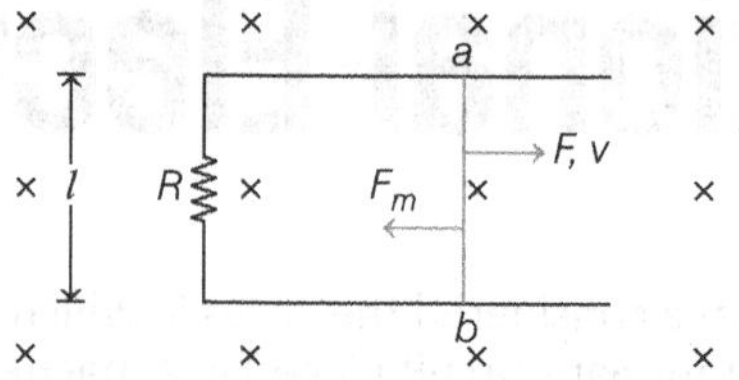

A magnetic force $F_m = ilB = \frac{B^2l^2\ v}{R}$ acts on the conductor in opposite direction of velocity. So, to move the conductor with a constant velocity v an equal and opposite force F has to be applied in the conductor. Thus,

$$F = F_m = \frac{B^2l^2v}{R}$$

The rate at which work is done by the applied force is

$$P_{\text{applied}} = Fv = \frac{B^2l^2v^2}{R}$$

and the rate at which energy is dissipated in the circuit is

$$P_{\text{dissipated}} = i^2R = \left(\frac{Bvl}{R}\right)^2 R = \frac{B^2l^2v^2}{R}$$

This is just equal to the rate at which work is done by the applied force.

- Motional emf is not a different kind of induced emf, it is exactly the induced emf described by Faraday's law, in the case in which there is a conductor moving in a magnetic field. Equation, $e = -\frac{d\phi_B}{dt}$ is best applied to problems in which there is a changing flux through a closed loop while $e = Bvl$ is applied to problems in which a conductor moves through a magnetic field. Note that, if a conductor is moving in a magnetic field but circuit is not closed, then only PD will be asked between two points of the conductor. If the circuit is closed, then current will be asked in the circuit. Now, let us see how these two are similar.

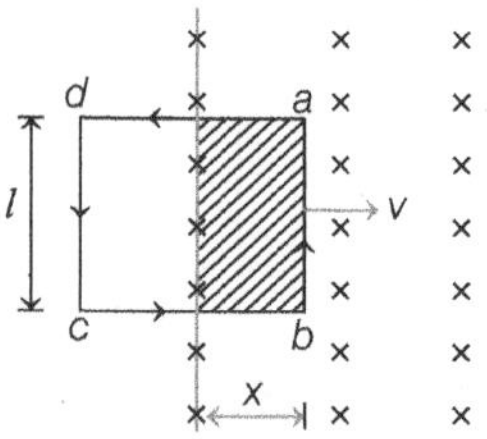

(a)

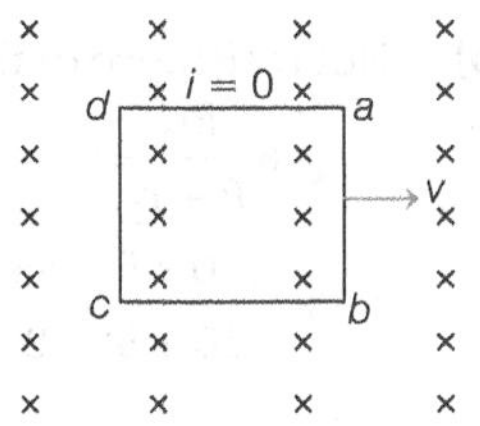

(b)

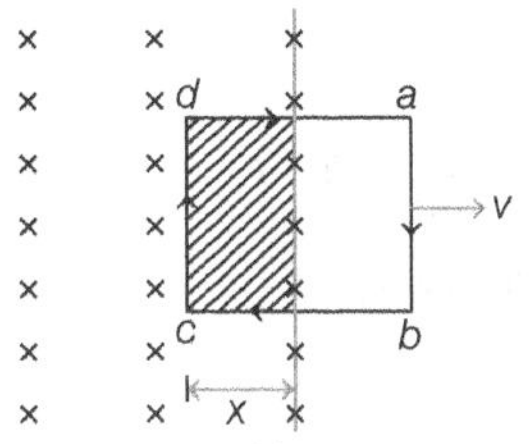

(c)

Refer figure (a) A loop *abcd* enters a uniform magnetic field B at constant speed v.

Using Faraday's equation,

$$|e| = \left| -\frac{d\phi_B}{dt} \right| = \frac{d(BS)}{dt} = \frac{d(Blx)}{dt} = Bl\frac{dx}{dt} = Blv$$

For the direction of current, we can use Lenz's law. As the loop enters the field, ⊗ magnetic field passing through the loop increases, hence, induced current should produce ⊙ magnetic field or current in the loop is anti-clockwise. From the theory of motional emf, $e = Bvl$ and using right hand rule also, current in the circuit is anti-clockwise. Thus, we see that $e = -\frac{d\phi_B}{dt}$ and $e = Bvl$ give the same result. In the similar manner, we can show that current in the loop in figure (b) is zero and in figure (c) it is clockwise.

- In general, we can say that motional emf in wire *acb* in a uniform magnetic field is the motional emf in an imaginary straight wire *ab*. Thus,

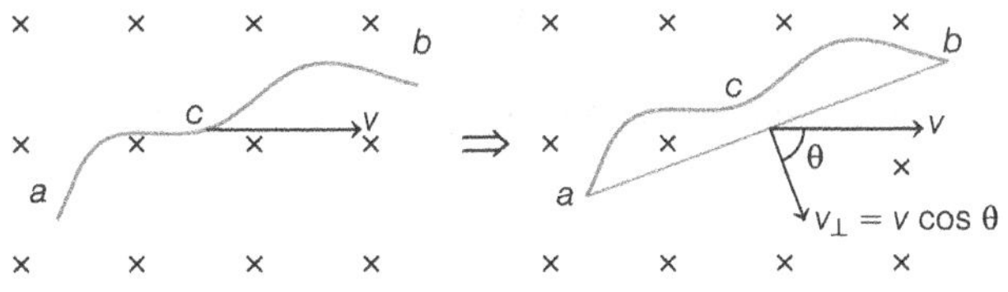

$$e_{acb} = e_{ab} = (\text{length of } ab)(v_\perp)(B)$$

Here, $v_\perp$ is the component of velocity perpendicular to both **B** and *ab*.

From right hand rule we can see that b is at higher potential and a at lower potential.

Hence, $V_{ba} = V_b - V_a = (ab)(v\cos\theta)(B)$

- $e = Bvl$ if **B**, **v** and **l** are mutually perpendicular and $e = 0$ if any two vectors are parallel to each other.
- **Motional emf induced in a rotating bar :** A conducting rod of length l rotates with a constant angular speed ω about a pivot at one end. A uniform magnetic field **B** is directed perpendicular to the plane of rotation as shown in figure. Consider a segment of rod of length dr at a distance r from O. This segment has a velocity,

$$v = r\omega$$

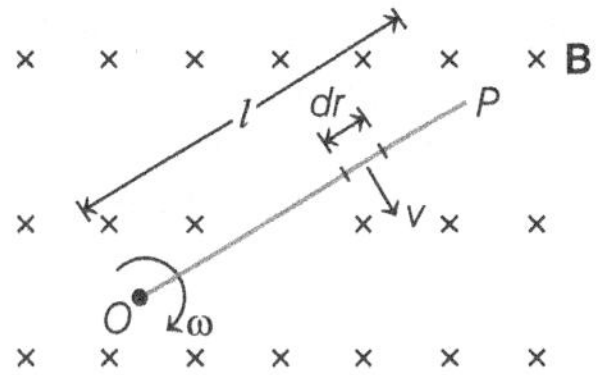

The induced emf in this segment is

$$de = Bvdr = B(r\omega)\,dr$$

Summing the emfs induced across all segments, which are in series, gives the total emf across the rod.

$$\therefore \quad e = \int_0^l de = \int_0^l Br\omega dr = \frac{B\omega l^2}{2}$$

$$\therefore \quad \boxed{e = \frac{B\omega l^2}{2}}$$

From right hand rule we can see that P is at higher potential than O. Thus,

$$V_P - V_O = \frac{B\omega l^2}{2}$$

- In the problems of induced emf (or current) if you see a conductor moving (or rotating) in a magnetic field, use the concept of $e = Bvl$ or $\frac{B\omega l^2}{2}$. It is easier than the other approach. But, if the conductor (or loop) is stationary, you have no choice. Use $e = -\frac{d\phi_B}{dt}$.
- Now onwards, the following integrations will be used very frequently.

If	$\int_0^x \frac{dx}{a - bx} = \int_0^t c\,dt$
then,	$x = \frac{a}{b}(1 - e^{-bct})$
and if	$\int_{x_0}^x \frac{dx}{a - bx} = \int_0^t c\,dt$
then,	$x = \frac{a}{b} - \left(\frac{a}{b} - x_0\right)e^{-bct}$

Here a, b and c are positive constants.

Note *In an electrical circuit, a moving or rotating wire may be assumed as a battery of emf Bvl or* $\frac{B\omega l^2}{2}$ *and then it can be solved with the help of Kirchhoff's laws. The following example will illustrate this concept.*

Example 1 Two parallel rails with negligible resistance are 10.0 cm apart. They are connected by a 5.0 Ω resistor. The circuit also contains two metal rods having resistances of 10.0 Ω and 15.0 Ω along the rails. The rods are pulled away from the resistor at constant speeds 4.00 m/s and 2.00 m/s respectively. A uniform magnetic field of magnitude 0.01 T is applied perpendicular to the, plane of the rails. Determine the current in the 5.0 Ω resistor.

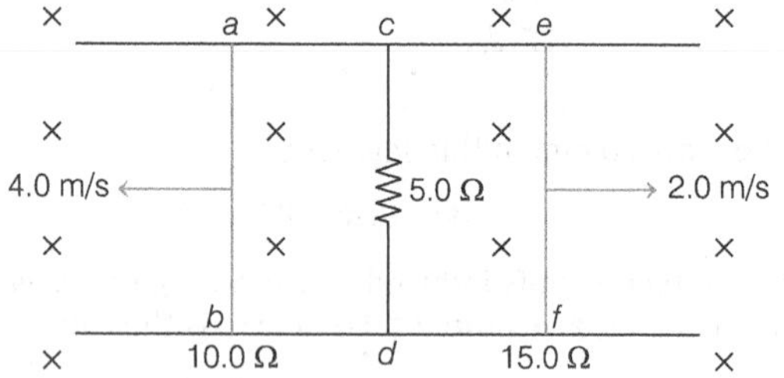

HOW TO PROCEED Here, two conductors are moving in a uniform magnetic field. So, we will use the motional approach. The rod ab will act as a source of emf,

$$e_1 = Bvl = (0.01)(4.0)(0.1) = 4 \times 10^{-3}\text{ V}$$

and internal resistance $r_1 = 10.0\ \Omega$

Similarly, rod ef will also act as a source of emf,

$$e_2 = (0.01)(2.0)(0.1) = 2.0 \times 10^{-3}\text{ V}$$

and internal resistance $r_2 = 15.0\ \Omega$.

From right hand rule we can see that, $V_b > V_a$ and $V_e > V_f$

Now, either by applying Kirchhoff's laws or applying principle of superposition (discussed in the unit of current electricity) we can find current through 5.0 Ω resistor. We will here use the superposition principle. You can solve it by using Kirchhoff's laws.

Sol. In the figures $R = 5.0\ \Omega$, $r_1 = 10\ \Omega$, $r_2 = 15\ \Omega$, $e_1 = 4 \times 10^{-3}$ V and $e_2 = 2 \times 10^{-3}$ V

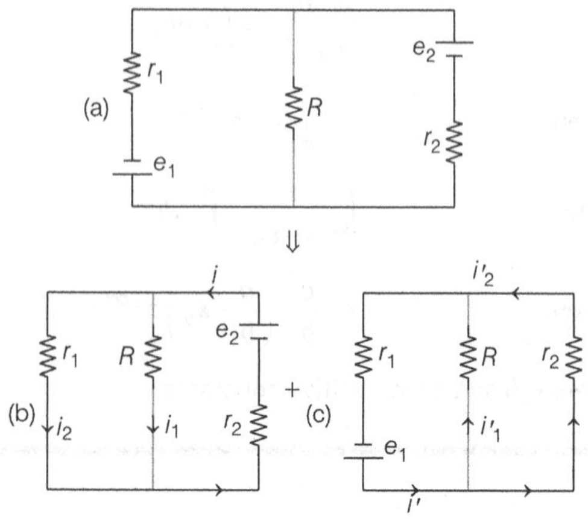

Refer figure (b) Net resistance of the circuit

$$= r_2 + \frac{Rr_1}{R + r_1}$$

$$= 15 + \frac{10 \times 5}{10 + 5} = \frac{55}{3}\ \Omega$$

$$\therefore \text{ Current, } i = \frac{e_2}{\text{Net resistance}}$$

$$= \frac{2 \times 10^{-3}}{55/3}$$

$$= \frac{6}{55} \times 10^{-3}\text{ A}$$

∴ Current through R,

$$i_1 = \left(\frac{r_1}{R + r_1}\right) i$$

$$= \left(\frac{10}{10 + 5}\right)\left(\frac{6}{55} \times 10^{-3}\right)\text{ A}$$

$$= \frac{4}{55} \times 10^{-3}\text{ A} = \frac{4}{55}\text{ mA}$$

Refer figure (c) Net resistance of the circuit

$$= r_1 + \frac{Rr_2}{R + r_2}$$

$$= 10 + \frac{5 \times 15}{5 + 15} = \frac{55}{4}\ \Omega$$

$$\therefore \quad \text{Current, } i' = \frac{e_1}{\text{Net resistance}}$$

$$= \frac{4 \times 10^{-3}}{55/4} = \frac{16}{55} \times 10^{-3}\text{ A}$$

$$\therefore \text{ Current through } R, \quad i'_1 = \left(\frac{r_2}{R + r_2}\right) i'$$

$$= \left(\frac{15}{15 + 5}\right)\left(\frac{16}{55}\right) \times 10^{-3}\text{ A}$$

$$= \frac{12}{55}\text{ mA}$$

From superposition principle net current through 5.0 Ω resistor is

$$i'_1 - i_1 = \frac{8}{55}\text{ mA from } d \text{ to } c$$ **Ans.**

Example 2 Figure shows the top view of a rod that can slide without friction. The resistor is 6.0 Ω and a 2.5 T magnetic field is directed perpendicularly downward into the paper. Let l = 1.20 m.

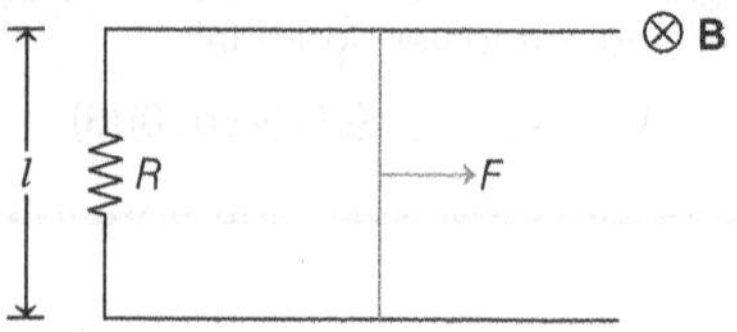

(a) Calculate the force F required to move the rod to the right at a constant speed of 2.0 m/s.
(b) At what rate is energy delivered to the resistor?
(c) Show that this rate is equal to the rate of work done by the applied force.

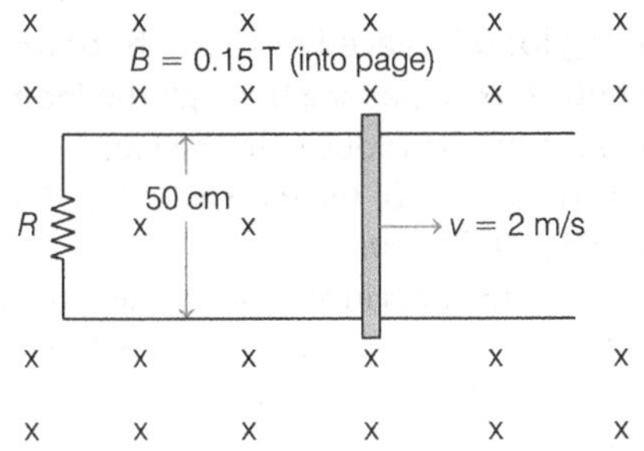

Sol. The motional emf in the rod, $e = Bvl$

or $e = (2.5)(2.0)(1.2)\text{ V} = 6.0\text{ V}$

The current in the circuit, $i = \dfrac{e}{R} = \dfrac{6.0}{6.0} = 1.0\text{ A}$

(a) The magnitude of force F required will be equal to the magnetic force acting on the rod, which opposes the motion.

$\therefore \quad F = F_m = ilB$

or $F = (1.0)(1.2)(2.5)\text{ N} = 3\text{ N}$ **Ans.**

(b) Rate by which energy is delivered to the resistor is

$P_1 = i^2R = (1)^2(6.0) = 6\text{ W}$ **Ans.**

(c) The rate by which work is done by the applied force is

$P_2 = F \cdot v = (3)(2.0) = 6\text{ W}$

$P_1 = P_2$ **Hence proved.**

INTRODUCTORY Exercise

1. A horizontal wire 0.8 m long is falling at a speed of 5 m/s perpendicular to a uniform magnetic field of 1.1 T, which is directed from east to west. Calculate the magnitude of the induced emf. Is the north or south end of the wire positive?
2. As shown in figure, a metal rod completes the circuit. The circuit area is perpendicular to a magnetic field with $B = 0.15$ T. If the resistance of the total circuit is 3 Ω, how large a force is needed to move the rod as indicated with a constant speed of 2 m/s?
3. A rod of length $3l$ is rotated with an angular velocity ω as shown in figure. The uniform magnetic field B is into the paper. Find

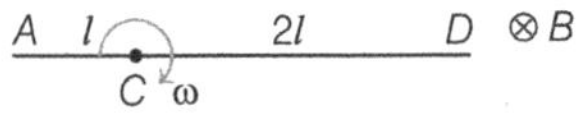

(a) $V_A - V_C$
(b) $V_A - V_D$

4. As the bar shown in figure moves in a direction perpendicular to the field, is an external force required to keep it moving with constant speed.

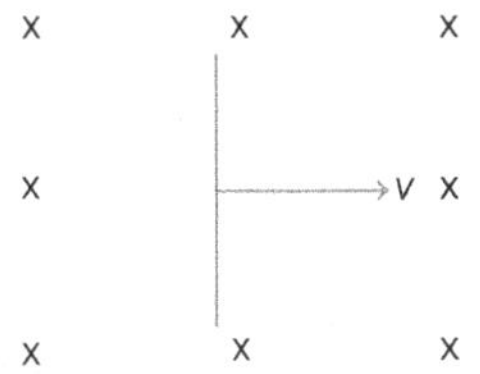

Answers

1. 4.4 V, north **2.** 0.00375 N **3.** (a) $\dfrac{B\omega l^2}{2}$ (b) $\dfrac{-3B\omega l^2}{2}$

4. No

See hints on page 545.

Typed Problems

Type 1. *Based on two methods of finding induced emf.*

Concept

Problems of induced emf or induced current can be solved by the following two methods.

Method 1 Magnitudes are given by

$$|e| = \left|N\frac{d\phi_B}{dt}\right| \quad \text{and} \quad |i| = \frac{|e|}{R}$$

Direction is given by Lenz's law.

Method 2 Magnitudes are given by

$$|e| = |Bvl| \quad \text{or} \quad |e| = \left|\frac{B\omega l^2}{2}\right|$$

and $$|i| = \frac{|e|}{R}$$

Direction is given by right hand rule.

Note *In the first method, we have to first find the magnetic flux passing through the loop and then differentiate it with respect to time. Second method is simple but it can be applied if and only if some conductor is either in translational or rotational motion.*

Ex. 1 Current in a long current carrying wire is

$$I = 2t$$

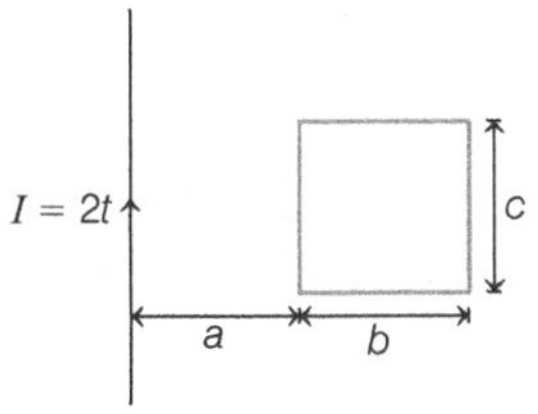

A conducting loop is placed to the right of this wire. Find

(a) magnetic flux ϕ_B passing through the loop.

(b) induced emf $|e|$ produced in the loop.

(c) if total resistance of the loop is R, then find induced current I_{in} in the loop.

Sol. Here, no conductor is in motion. So, we can apply only method-1. Further, magnetic field of straight wire is non-uniform. Therefore, magnetic flux can be obtained by integration.

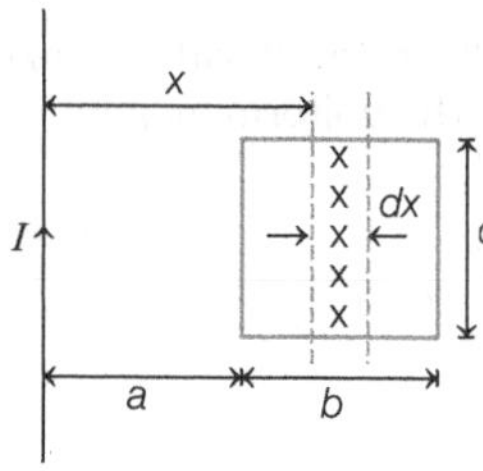

(a) At a distance x from the straight wire, magnetic field is

$$B = \frac{\mu_0}{2\pi}\frac{I}{x} \quad \text{[in } \otimes \text{ direction]}$$

Let us take a small strip of width dx.

$\therefore$ Area of this strip is

$$dS = c(dx)$$

Now, $d\mathbf{S}$ can also be assumed inwards. Or, angle between $\mathbf{B}$ and $d\mathbf{S}$ may be assumed to be 0°. Therefore, small magnetic flux passing through the strip is

$$d\phi_B = BdS \cos 0°$$
$$= \frac{\mu_0}{2\pi}\frac{I}{x} cdx$$

Total magnetic flux is

$$\phi_B = \int_{x=a}^{x=a+b} d\phi_B$$
$$= \int_a^{a+b}\left(\frac{\mu_0 Ic}{2\pi}\right)\frac{dx}{x}$$
$$= \frac{\mu_0 Ic}{2\pi}\ln\left(\frac{a+b}{a}\right)$$

Substituting the value of I, we get

$$\phi_B = \frac{\mu_0 ct}{\pi}\ln\left(\frac{a+b}{a}\right) \quad \textbf{Ans.}$$

(b) $$|e| = \left|\frac{d\phi_B}{dt}\right| = \frac{d}{dt}\left[\frac{\mu_0 ct}{\pi}\ln\left(\frac{a+b}{a}\right)\right]$$
$$= \frac{\mu_0 c}{\pi}\ln\left(\frac{a+b}{a}\right) \quad \textbf{Ans.}$$

(c) Induced current,

$$I_{in} = \frac{|e|}{R} = \frac{\mu_0 c}{\pi R}\ln\left(\frac{a+b}{a}\right)$$

Note *The main current I (= 2t) is increasing with time. Hence, ⊗ magnetic field passing through the loop will also increase. So, induced current I_{in} will produce ⊙ magnetic field. Or, induced current is anti-clockwise.*

Ex. 2 A constant current I flows through a long straight wire as shown in figure. A square loop starts moving towards right with a constant speed v.

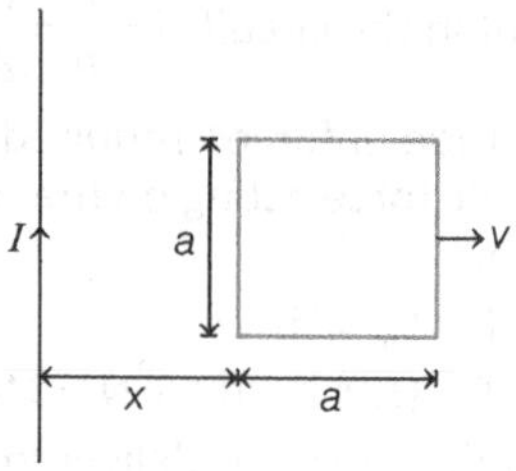

(a) Find induced emf produced in the loop as a function of x.

(b) If total resistance of the loop is R, then find induced current in the loop.

Note *In this problem, loop is in motion therefore both methods can be applied.*

Sol. Method 1

(a) Using the result of magnetic flux obtained in Ex. 1, we have

$$\phi_B = \frac{\mu_0 Ic}{2\pi}\ln\left(\frac{a+b}{a}\right)$$

Here, $a = x, b = c = a$

Substituting the values, we get

$$\phi_B = \frac{\mu_0 Ia}{2\pi}\ln\left(\frac{x+a}{x}\right)$$
$$= \frac{\mu_0 Ia}{2\pi}\ln\left(1+\frac{a}{x}\right)$$

Now, $$|e| = \left|\frac{d\phi_B}{dt}\right| = \frac{\mu_0 Ia}{2\pi}\left(\frac{x}{x+a}\right)\left(\frac{a}{x^2}\right)\frac{dx}{dt}$$

Putting $\frac{dx}{dt} = v$, we have

$$|e| = \frac{\mu_0 Ia^2}{2\pi x(x+a)}v \quad \textbf{Ans.}$$

(b) Induced current,

$$I_{in} = \frac{|e|}{R} = \frac{\mu_0 Ia^2 v}{2\pi Rx(x+a)} \quad \textbf{Ans.}$$

Note *Near the wire (towards right) value of ⊗ magnetic field is high. So, the loop is moving from higher magnetic field to lower magnetic field. or, ⊗ magnetic field passing through the loop is decreasing. Hence, induced current will produce ⊗ magnetic field or it should be clockwise..*

Method 2

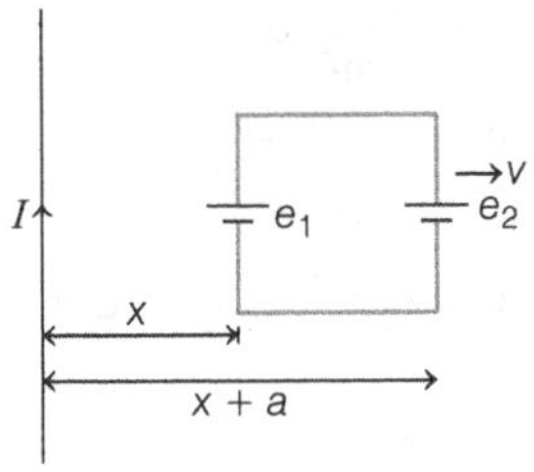

$$e_1 = B_1 vl = \frac{\mu_0}{2\pi}\frac{I}{x} va$$

$$e_2 = B_2 vl = \frac{\mu_0}{2\pi}\frac{I}{x+a} va$$

$$e_1 > e_2$$

$$\therefore \quad e_{net} = e_1 - e_2 = \frac{\mu_0 Iva}{2\pi}\left(\frac{1}{x} - \frac{1}{x+a}\right)$$

$$= \frac{\mu_0 Iva^2}{2\pi x(x+a)}$$

This is the same result as was obtained in Method 1.

Ex. 3 A conducting circular ring is rotated with angular velocity ω about point A as shown in figure. Radius of ring is a. Find

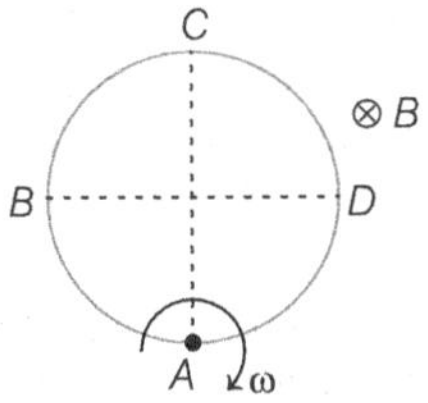

(a) potential difference between points A and C

(b) potential difference between the points A and D.

Sol. Here, the loop is rotating. So, we can applying $e = \frac{B\omega l^2}{2}$

(a)

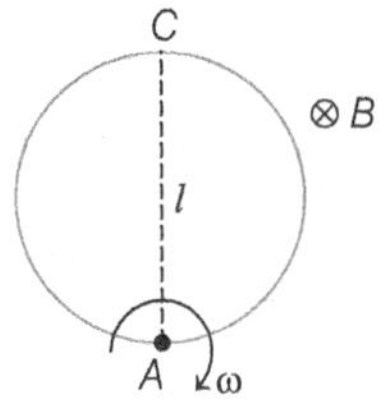

Using right hand rule, we can see that

$$V_C > V_A$$

$$\therefore \quad V_C - V_A = \frac{B\omega l^2}{2} = \frac{B\omega(2a)^2}{2} = 2B\omega a^2$$ **Ans.**

(b)

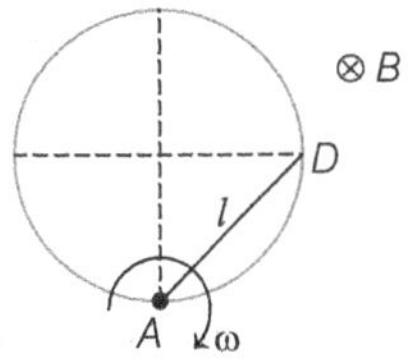

Using right hand rule, we can see that

$$V_D > V_A$$

$$\therefore \quad V_D - V_A = \frac{B\omega l^2}{2} = \frac{B\omega(\sqrt{2}\,a)^2}{2}$$

$$= B\omega a^2$$ **Ans.**

Miscellaneous Examples

Ex. 4 A conducting rod shown in figure of mass m and length l moves on two frictionless horizontal parallel rails in the presence of a uniform magnetic field directed into the page. The rod is given an initial velocity v_0 to the right and is released at $t = 0$. Find as a function of time,

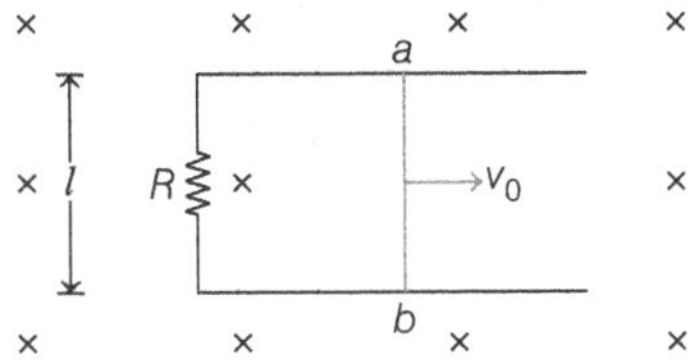

(a) the velocity of the rod

(b) the induced current and

(c) the magnitude of the induced emf.

HOW TO PROCEED The initial velocity will produce an induced emf and hence, an induced current in the circuit. The current carrying wire will now experience a magnetic force ($\mathbf{F}_m$) in opposite direction of its velocity. The force will retard the motion of the conductor. Thus,

Initial velocity $\rightarrow$ motional emf $\rightarrow$ induced current $\rightarrow$ magnetic force $\rightarrow$ retardation.

Sol. (a) Let v be the velocity of the rod at time t.

Current in the circuit at this moment is

$$i = \frac{Bvl}{R} \qquad \text{...(i)}$$

From right hand rule, we can see that this current is in counterclockwise direction.

The magnetic force is,

$$F_m = -ilB = -\frac{B^2 l^2}{R} v$$

Here, negative sign denotes that the force is to the left and retards the motion. This is the only horizontal force acting on the bar, and hence,

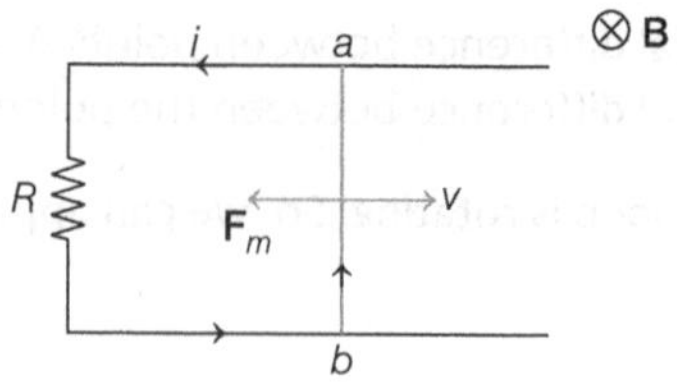

Newton's second law applied to motion in horizontal direction gives

$$m\frac{dv}{dt} = F_m = -\frac{B^2l^2}{R}v$$

$$\therefore \quad \frac{dv}{v} = -\left(\frac{B^2l^2}{mR}\right)dt$$

Integrating this equation using the initial condition that, $v = v_0$ at $t = 0$, we find that

$$\int_{v_0}^{v} \frac{dv}{v} = -\frac{B^2l^2}{mR}\int_0^t dt$$

Solving this equation, we find that

$$v = v_0 e^{-t/\tau} \quad \text{...(ii) Ans.}$$

where, $\tau = \dfrac{mR}{B^2l^2}$

This expression indicates that the velocity of the rod decreases exponentially with time under the action of the magnetic retarding force.

(b) $i = \dfrac{Bvl}{R}$

Substituting the value of v from Eq. (ii), we get

$$i = \frac{Blv_0}{R}e^{-t/\tau} \quad \textbf{Ans.}$$

(c) $$e = iR = Blv_0 e^{-t/\tau} \quad \textbf{Ans.}$$

i and e both decrease exponentially with time. v-t, i-t and e-t graphs are as shown in figure

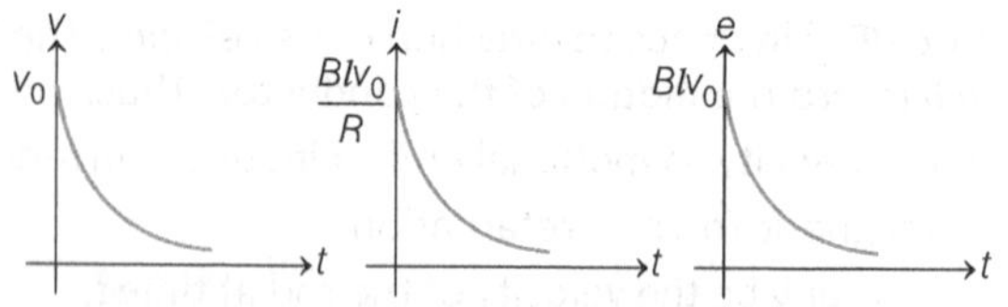

Alternate solution This problem can also be solved by energy conservation principle. Let at some instant velocity of the rod is v. As no external force is present. Energy is dissipated in the resistor at the cost of kinetic energy of the rod. Hence,

$$\left(-\frac{dK}{dt}\right) = \text{power dissipated in the resistor}$$

or $$-\frac{d}{dt}\left(\frac{1}{2}mv^2\right) = \frac{e^2}{R}$$

or $$-mv\left(\frac{dv}{dt}\right) = \frac{B^2l^2v^2}{R} \quad (\text{as } e = Bvl)$$

$$\therefore \quad \frac{dv}{v} = -\frac{B^2l^2}{mR}dt$$

$$\therefore \quad \int_{v_0}^{v} \frac{dv}{v} = -\frac{B^2l^2}{mR}\int_0^t dt$$

or $$v = v_0 e^{-t/\tau}, \text{ where } \tau = \frac{mR}{B^2l^2}$$

Ex. 5 A wire loop enclosing a semicircle of radius R is located on the boundary of a uniform magnetic field B. At the moment $t = 0$, the loop is set into rotation with a constant angular acceleration α about an axis O coinciding with a line of vector **B** on the boundary. Find the emf induced in the loop as a function of time. Draw the approximate plot of this function. The arrow in the figure shows the emf direction taken to be positive. **(Irodov 3.292)**

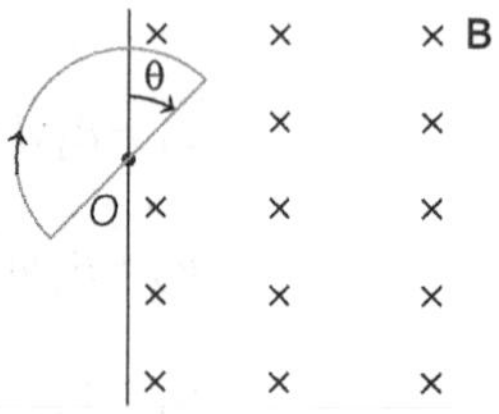

Sol. $\theta = \dfrac{1}{2}\alpha t^2$

$$\therefore \quad t = \sqrt{\frac{2\theta}{\alpha}} = \text{time taken to rotate an angle } \theta$$

when $\theta = 0$ to π, 2π to 3π, 4π to 5π etc.

⊗ magnetic field passing through the loop is increasing. Hence, current in the loop is anti-clockwise or induced emf is negative. And for, $\theta = \pi$ to 2π, 3π to 4π, 5π to 6π etc.

⊗ magnetic field passing through the loop is decreasing. Hence, current in the loop is clockwise or emf is positive. So,

$$t_1 = \text{time taken to rotate an angle } \pi = \sqrt{\frac{2\pi}{\alpha}}$$

$$t_2 = \text{time taken to rotate an angle } 2\pi = \sqrt{\frac{4\pi}{\alpha}}$$

... ...

$$t_n = \text{time taken to rotate an angle } n\pi = \sqrt{\frac{2n\pi}{\alpha}}$$

Now, from 0 to t_1 emf is negative

t_1 to t_2 emf is positive

t_2 to t_3 emf is again negative

and so on.

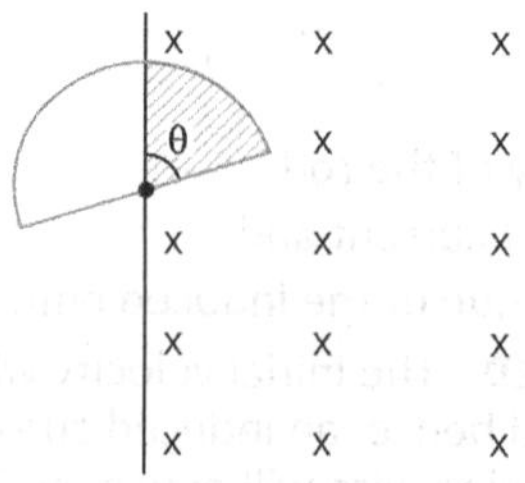

Now, at time t, angle rotated is $\theta = \frac{1}{2}\alpha t^2$

Area inside the field is

$$S = (\pi R^2)\left(\frac{\theta}{2\pi}\right) = \frac{1}{2}R^2\theta$$

or $$S = \frac{1}{4}R^2\alpha t^2$$

So, flux passing through the loop,

$$\phi = BS = \frac{1}{4}BR^2\alpha t^2$$

$$e = \left|\frac{d\phi}{dt}\right| = \frac{1}{2}BR^2\alpha t$$

$$e \propto t$$

i.e. e-t graph is a straight line passing through origin. e-t equation with sign can be written as

$$e = (-1)^n\left(\frac{1}{2}BR^2\alpha t\right)$$ **Ans.**

Here, $n = 1, 2, 3 \ldots$ is the number of half revolutions that the loop performs at the given moment t.

The e-t graph is as shown in figure.

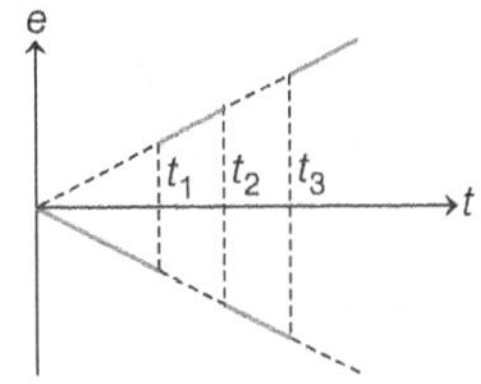

Ex. 6 A uniform wire of resistance per unit length λ is bent into a semicircle of radius a. The wire rotates with angular velocity ω in a vertical plane about a horizontal axis passing through C. A uniform magnetic field B exists in space in a direction perpendicular to paper inwards.

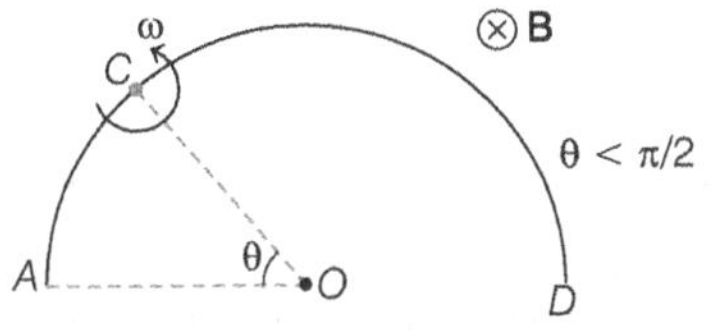

(a) Calculate potential difference between points A and D. Which point is at higher potential?

(b) If points A and D are connected by a conducting wire of zero resistance, find the potential difference between A and C.

Sol. (a) Length of straight wire AC is $l_1 = 2a\sin\left(\frac{\theta}{2}\right)$

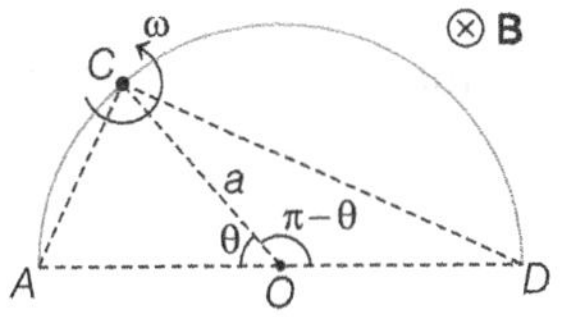

Therefore, the motional emf (or potential difference) between points C and A is

$$V_{CA} = V_C - V_A = \frac{1}{2}B\omega l_1^2 = 2a^2B\omega\sin^2\left(\frac{\theta}{2}\right) \quad \ldots\text{(i)}$$

From right hand rule, we can see that $V_C > V_A$

Similarly, length of straight wire CD is

$$l_2 = 2a\sin\left(\frac{\pi}{2} - \frac{\theta}{2}\right) = 2a\cos\left(\frac{\theta}{2}\right)$$

Therefore, the PD between points C and D is

$$V_{CD} = V_C - V_D = \frac{1}{2}B\omega l_2^2 = 2a^2B\omega\cos^2\left(\frac{\theta}{2}\right) \quad \ldots\text{(ii)}$$

with $V_C > V_D$

Eq. (ii) – Eq.(i) gives,

$$V_A - V_D = 2a^2B\omega\left(\cos^2\frac{\theta}{2} - \sin^2\frac{\theta}{2}\right)$$

$$= 2a^2B\omega\cos\theta$$ **Ans.**

A is at higher potential.

(b) When A and D are connected from a wire current starts flowing in the circuit as shown in figure :

Resistance between A and C is

$$r_1 = (\text{length of arc } AC)\ \lambda = a\theta\lambda$$

and between C and D is

$$r_2 = (\text{length of arc } CD)\ \lambda = (\pi - \theta)\ a\lambda$$

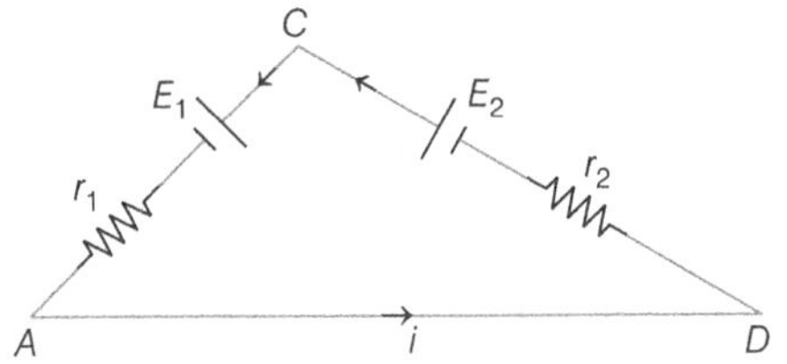

In the figure, $E_1 = 2a^2B\omega\sin^2\left(\frac{\theta}{2}\right)$

and $E_2 = 2a^2B\omega\cos^2\left(\frac{\theta}{2}\right)$

with $E_2 > E_1$

$\therefore$ Current in the circuit is

$$i = \frac{E_2 - E_1}{r_1 + r_2} = \frac{2a^2B\omega\cos\theta}{\pi a\lambda}$$

$$= \frac{2aB\omega\cos\theta}{\pi\lambda}$$

and potential difference between points C and A is

$$V'_{CA} = E_1 + ir_1 = 2a^2B\omega\sin^2\left(\frac{\theta}{2}\right) + \left(\frac{2aB\omega\cos\theta}{\pi\lambda}\right)(a\theta\lambda)$$

$$= 2a^2B\omega\left(\sin^2\frac{\theta}{2} + \frac{\theta}{\pi}\cos\theta\right)$$ **Ans.**

Note *$V_{CA} = E_1$ when no current flows through the circuit and $V'_{CA} = E_1 + ir_1$ when a current i flows in the circuit.*

MODULE 2 Exercise

True/False

1. A conducting rod AB moves parallel to the X-axis in a uniform magnetic field pointing in the positive z-direction. The end A of the rod gets positively charged. **(1987 JEE)**

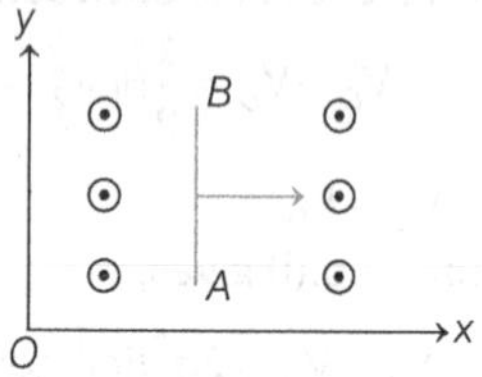

Single Correct Option

2. A conducting rod is rotated about one end in a plane perpendicular to a uniform magnetic field with constant angular velocity. The correct graph between the induced emf (e) across the rod and time (t) is

(a) 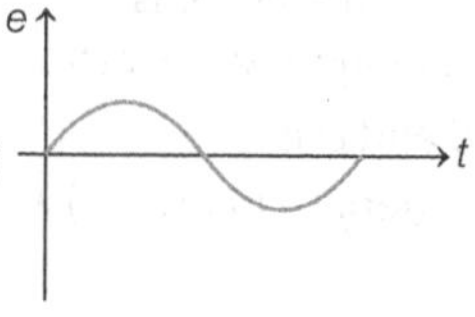

(b) 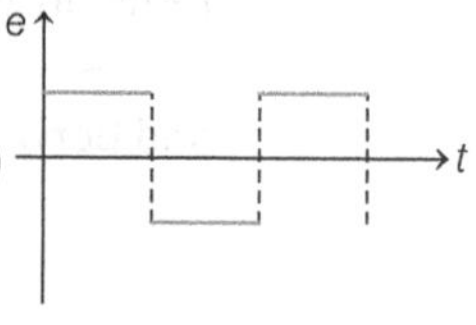

(c) 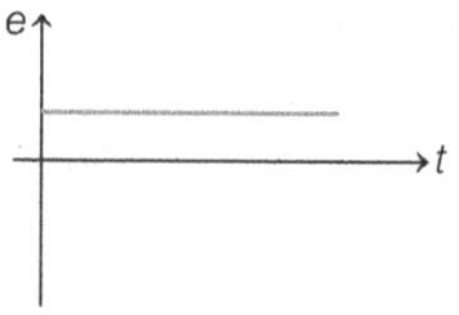

(d) 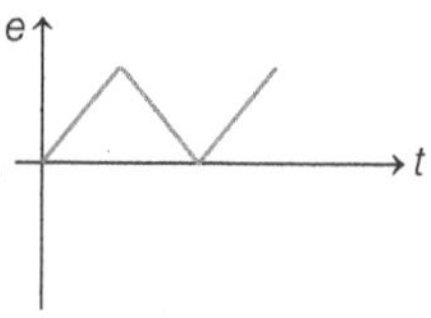

3. A semi-circular conducting ring acb of radius R moves with constant speed v in a plane perpendicular to uniform magnetic field B as shown in figure. Identify the correct statement.

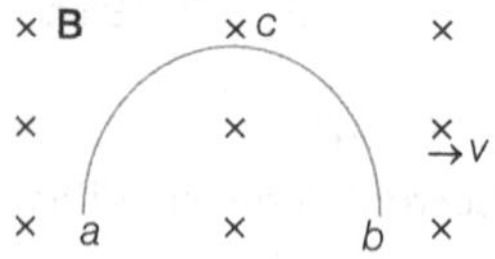

(a) $V_a - V_c = BRv$ (b) $V_b - V_c = BRv$
(c) $V_a - V_b = 0$ (d) None of these

4. A wire is bent in the form of a V shape and placed in a horizontal plane. There exists a uniform magnetic field B perpendicular to the plane of the wire. A uniform conducting rod starts sliding over the V shaped wire with a constant speed v as shown in the figure. If the wire has no resistance, the current in rod will

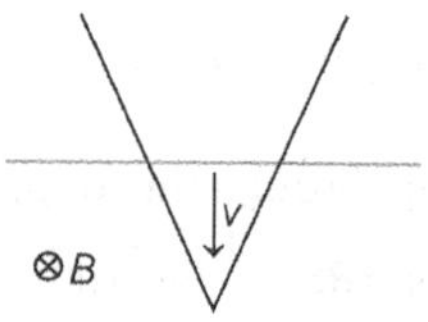

(a) increase with time (b) decrease with time
(c) remain constant (d) always be zero

5. A square loop of side b is rotated in a constant magnetic field B at angular frequency ω as shown in the figure. What is the emf induced in it?

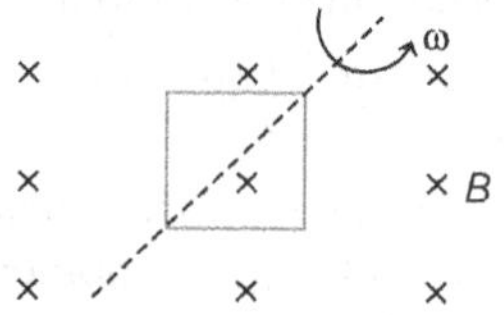

(a) $b^2 B\omega \sin \omega t$ (b) $bB\omega \sin^2 \omega t$
(c) $bB^2 \omega \cos \omega t$ (d) $b^2 B\omega$

6. In the given arrangement, the loop is moved with constant velocity v in a uniform magnetic field B in a restricted region of width a. The time for which the emf is induced in the circuit is

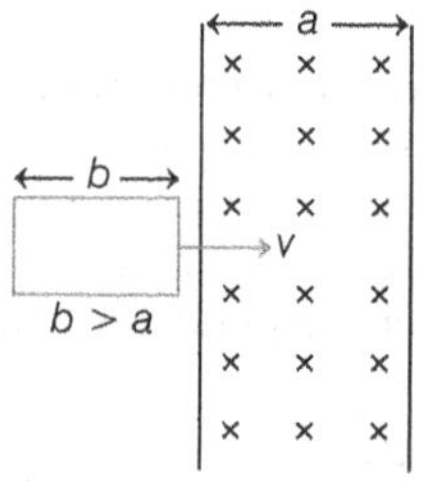

(a) $\frac{2b}{v}$ (b) $\frac{2a}{v}$ (c) $\frac{(a+b)}{v}$ (d) $\frac{2(a-b)}{v}$

7. In the figure shown, a T-shaped conductor moves with constant angular velocity ω in a plane perpendicular to uniform magnetic field **B**. The potential difference $V_A - V_B$ is

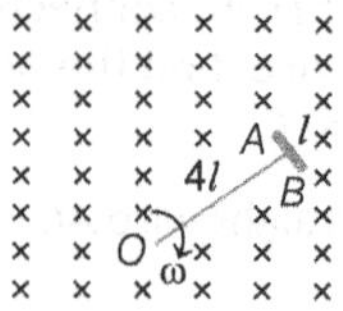

(a) zero (b) $\frac{1}{2} B\omega l^2$ (c) $2B\omega l^2$ (d) $B\omega l^2$

8. A metallic rod of length l is hinged at the point M and is rotating about an axis perpendicular to the plane of paper with a constant angular velocity ω. A uniform magnetic field of intensity B is acting in the region (as shown in the figure) parallel to the plane of paper. The potential difference between the points M and N

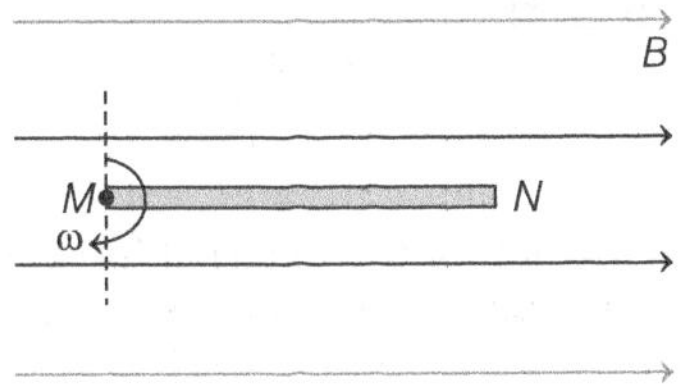

(a) is always zero
(b) varies between $\frac{1}{2}B\omega l^2$ to 0
(c) is always $\frac{1}{2}B\omega l^2$
(d) is always $B\omega l^2$

9. The number of turns in the coil of an AC generator is 5000 and the area of the coil is 0.25 m^2. The coil is rotated at the rate of 100 cycles/sec in a magnetic field of 0.2 W/m^2. The peak value of the emf generated is nearly
(a) 786 kV (b) 440 kV (c) 220 kV (d) 157.1 kV

10. A frame $CDEF$ is placed in a region where a magnetic field **B** is present. A rod of length one metre moves with constant velocity 20 m/s and strength of magnetic field is one tesla. The power spent in the process is (take $R = 0.2\,\Omega$ and all other wires and rod have zero resistance)

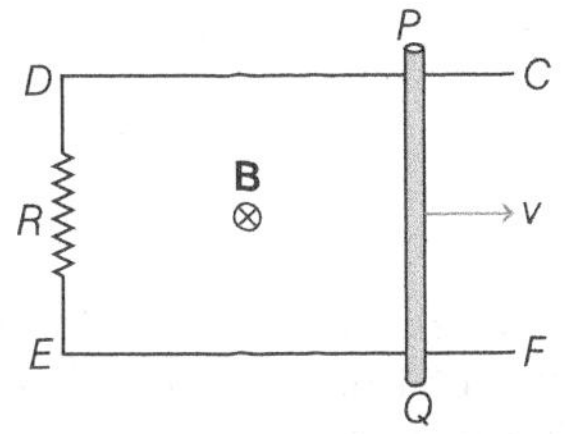

(a) 1 kW (b) 2 kW (c) 3 kW (d) 4 kW

11. An emf of 0.08 V is induced in a metal rod of length 10 cm held normal to a uniform magnetic field of 0.4 T, when moves with a velocity of **(2023 Main)**
(a) 0.5 ms^{-1} (b) 2 ms^{-1} (c) 20 ms^{-1} (d) 3.2 ms^{-1}

12. A metallic conductor of length 1 m rotates in a vertical plane parallel to east-west direction about one of its ends with an angular velocity 5 rad/s. If the horizontal component of earth's magnetic field is 0.2×10^{-4} T, then emf induced between the two ends of the conductor is **(2022 Main)**
(a) 5 μV (b) 50 μV
(c) 5 mV (d) 50 mV

13. A square loop of side 20 cm and resistance 1Ω is moved towards right with a constant speed v_0. The right arm of the loop is in a uniform magnetic field of 5 T. The field is perpendicular to the plane of the loop and is going into it. The loop is connected to a network of resistors each of value 4Ω. What should be the value of $\mathbf{v}_0$, so that a steady current of 2 mA flows in the loop ? **(2021 Main)**

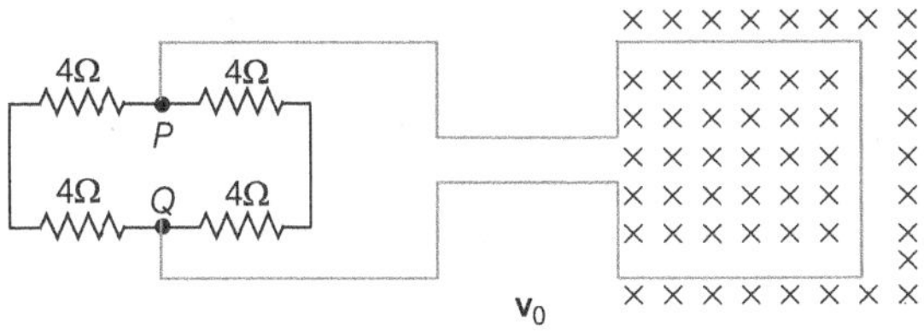

(a) 1 m/s (b) 1 cm/s
(c) 10^2 m/s (d) 10^{-2} cm/s

14. A metallic rod of length l is tied to a string of length $2l$ and made to rotate with angular speed ω on a horizontal table with one end of the string fixed. If there is a vertical magnetic field B in the region, the emf induced across the ends of the rod is **(2013 Main)**

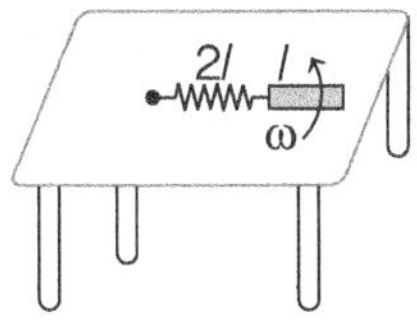

(a) $\frac{2B\,\omega l^3}{2}$ (b) $\frac{3B\,\omega l^3}{2}$ (c) $\frac{4B\omega l^2}{2}$ (d) $\frac{5B\omega l^2}{2}$

15. A metallic square loop $ABCD$ is moving in its own plane with velocity v in a uniform magnetic field perpendicular to its plane as shown in the figure. Electric field is induced **(2001 JEE)**

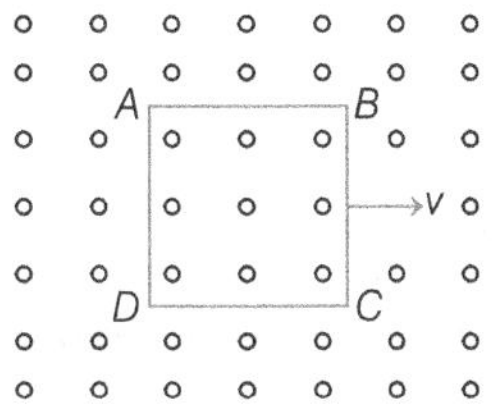

(a) in AD, but not in BC (b) in BC, but not in AD
(c) neither in AD nor in BC (d) in both AD and BC

16. A metal rod moves at a constant velocity in a direction perpendicular to its length. A constant uniform magnetic field exists in space in a direction perpendicular to the rod as well as its velocity. Select the correct statement (s) from the following. **(1998 JEE)**
(a) The entire rod is at the same electric potential.
(b) There is an electric field in the rod.
(c) The electric potential is highest at the centre of the rod and decrease towards its ends.
(d) The electric potential is lowest at the centre of the rod and increases towards its ends.

17. A thin semicircular conducting ring of radius R is falling with its plane vertical in a horizontal magnetic induction **B**. At the position MNQ the speed of the

ring is v and the potential difference developed across the ring is **(1996 JEE)**

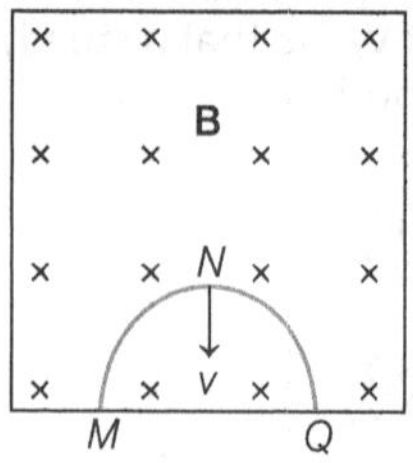

(a) zero
(b) $Bv\pi R^2/2$ and M is at higher potential
(c) πBRv and Q is at higher potential
(d) $2RBv$ and Q is at higher potential

18. A conducting square loop of side L and resistance R moves in its plane with a uniform velocity v perpendicular to one of its sides. A magnetic induction B, constant in time and space, pointing perpendicular to and into the plane of the loop exists everywhere. The current induced in the loop is **(1989 JEE)**

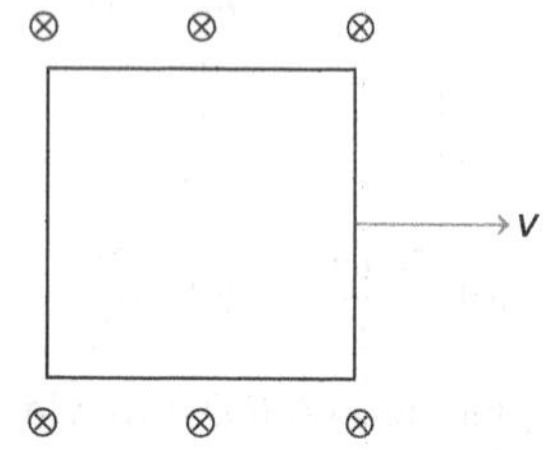

(a) BLv/R clockwise
(b) BLv/R anti-clockwise
(c) $2BLv/R$ anti-clockwise
(d) zero

Subjective Questions

19. The long straight wire in figure (a) carries a constant current i. A metal bar of length l is moving at constant velocity **v** as shown in figure. Point a is a distance d from the wire.

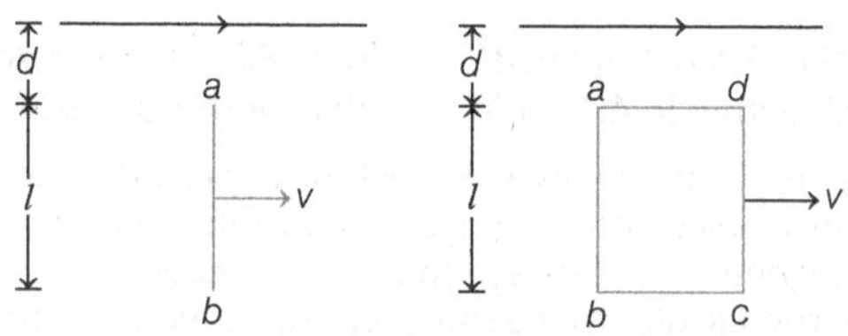

(a) (b)

(a) Calculate the emf induced in the bar.
(b) Which point a or b is at higher potential?
(c) If the bar is replaced by a rectangular wire loop of resistance R, what is the magnitude of current induced in the loop?

20. A square loop of wire with resistance R is moved at constant speed v across a uniform magnetic field confined to a square region whose sides are twice the lengths of those of the square loop.

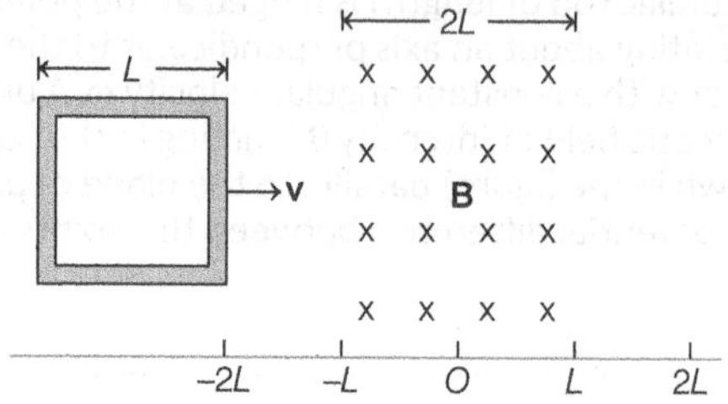

(a) Sketch a graph of the external force F needed to move the loop at constant speed, as a function of the coordinate x, from $x = -2L$ to $x = +2L$. The coordinate x is measured from the centre of the magnetic field region to the centre of the loop. It is negative when the centre of the loop is to the left of the centre of the magnetic field region. Take positive force to be to the right.
(b) Sketch a graph of the induced current in the loop as a function of x. Take counterclockwise currents to be positive.

21. A square metal wire loop of side 10 cm and resistance 1 Ω is moved with a constant velocity v_0 in a uniform magnetic field of induction $B = 2$ Wb/m^2 as shown in the figure. The magnetic field lines are perpendicular to the plane of the loop (directed into the paper). The loop is connected to a network of resistors each of value 3 Ω.

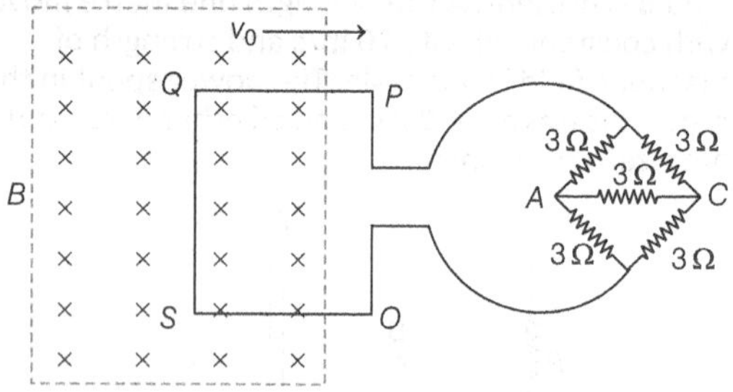

What should be the speed of the loop so as to have a steady current of 1mA in the loop? Give the direction of current in the loop. **(1983 JEE)**

Integer Type Questions

22. A metallic cube of side 15 cm is moving along y-axis at a uniform velocity of 2 ms^{-1} in a region of uniform magnetic field of magnitude 0.5T directed along z-axis. In equilibrium, the potential difference between the faces of higher and lower potential developed because of the motion through the field will be .... mV. **(2023 Main)**

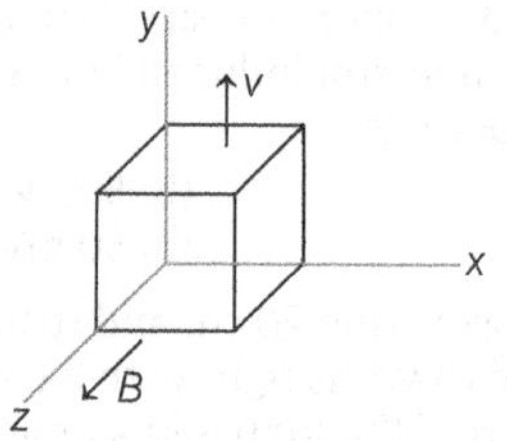

23. A 20 cm long metallic rod is rotated with 210 rpm about an axis normal to the rod passing through its one end. The other end of the rod is in contact with a circular metallic ring. A constant and uniform magnetic field 0.2 T parallel to the axis exists everywhere. The emf developed between the centre and the ring is .... mV. $\left(\text{Take, } \pi = \frac{22}{7}\right)$ **(2023 Main)**

24. A square shaped coil of area 70 cm^2 having 600 turns rotates in a magnetic field of 0.4 Wbm^{-2}, about an axis which is parallel to one of the side of the coil and perpendicular to the direction of field. If the coil completes 500 revolution in a minute, the instantaneous emf when the plane of the coil is inclined at 60° with the field, will be .... V. $\left(\text{Take, } \pi = \frac{22}{7}\right)$ **(2023 Main)**

Answers

1. (True) 2. (c) 3. (c) 4. (c) 5. (a) 6. (b) 7. (a) 8. (a) 9. (d) 10. (b)
11. (b) 12. (b) 13. (b) 14. (d) 15. (d) 16. (b) 17. (d) 18. (d)
19. (a) $\frac{\mu_0 iv}{2\pi} \ln\left(1 + \frac{l}{d}\right)$ (b) a (c) Zero

20.

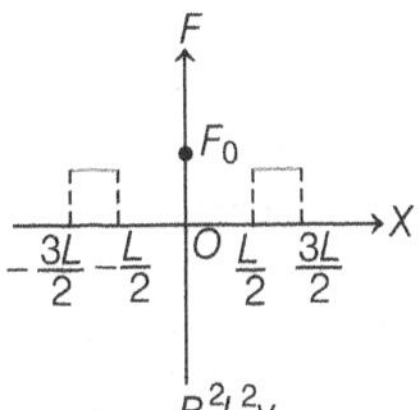

$F_0 = \frac{B^2L^2v}{R}$

(a)

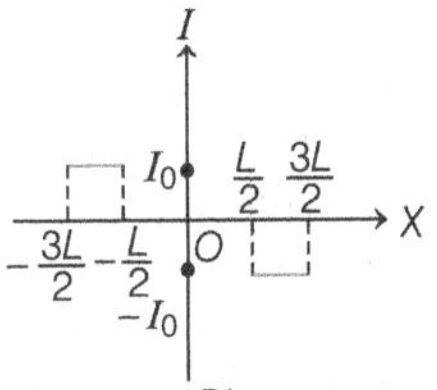

$i_0 = \frac{Blv}{R}$

(b)

21. (0.02 m/s, clockwise) 22. (150) 23. (88) 24. (44)

See hints on pages 545-547.

MODULE 3

Self Inductance and Inductor

Consider a single isolated circuit. When a current is present in the circuit, it sets up a magnetic field that causes a magnetic flux through the same circuit. This flux changes as the current in the circuit is changed. According t

o Faraday's law any change in flux in a circuit produces an induced emf in it. Such an emf is called a **self-induced emf.** The name is so called because the source of this induced emf is the change of current in the same circuit.

According to Lenz's law the self-induced emf always opposes the change in the current that caused the emf and so tends to make it more difficult for variations in current to occur. We will here like to define a term **self-inductance *L*** of a circuit which is of great importance in our proceeding discussions. It can be defined in the following two ways :

First Definition Suppose a circuit includes a coil with N turns of wire. It carries a current i. The total flux ($N\phi_B$) linked with the coil is directly proportional to the current (i) in the coil, i.e.

$$N\phi_B \propto i$$

When the proportionality sign is removed a constant L comes in picture, which depends on the dimensions and number of turns in the coil. This constant is called self-inductance. Thus,

$$N\phi_B = Li \quad \text{or} \quad \boxed{L = \frac{N\phi_B}{i}}$$

From here we can define self-inductance (L) of any circuit as, **the total flux per unit current.** The SI unit of self-inductance is henry (1H).

Second Definition If a current i is passed in a circuit and it is changed with a rate di/dt, the induced emf e produced in the circuit is directly proportional to the rate of change of current. Thus,

$$e \propto \frac{di}{dt}$$

When the proportionality constant is removed, the same constant L again comes here.

Hence, $$e = -L\frac{di}{dt}$$

The minus sign here is a reflection of Lenz's law. It says that the self-induced emf in a circuit opposes any change in the current in that circuit. From the above equation,

$$\boxed{L = \left|\frac{-e}{di/dt}\right|}$$

This equation states that, **the self-inductance of a circuit is the magnitude of self induced emf per unit rate of change of current.**

A circuit or part of a circuit, that is designed to have a particular inductance is called an **inductor**. The usual symbol for an inductor is

Thus, an inductor is a circuit element which opposes the change in current through it. It may be a circular coil, solenoid etc.

Significance of Self-inductance and Inductor

Like capacitors and resistors, inductors are among the circuit elements of modern electronics. Their purpose is to oppose any variations in the current through the circuit. An inductor plays a dormant role in a circuit so far as current is constant. It becomes active when current changes in the circuit. Every inductor has some self-inductance which depends on the size, shape and the number of turns etc. For N turns close together, it is always proportional to N^2. It also depends on the magnetic properties of the material enclosed by the circuit. When the current passing through it is changed, an emf of magnitude $L\, di/dt$ is induced across it. Later in this article, we will explore the method of finding the self-inductance of an inductor.

Potential Difference Across an Inductor

The induced emf is produced whenever there is a change in the current in the inductor. This emf always acts to oppose this change. Figure shows three cases. Assume that the inductor has negligible resistance, so the PD, $V_{ab} = V_a - V_b$ between the inductor terminals a and b is equal in magnitude to the self-induced emf.

We can find the polarities of self-induced emf across an inductor from Lenz's law.

Refer figure (a) The current is constant, and there is no self-induced emf. Hence, $V_{ab} = 0$

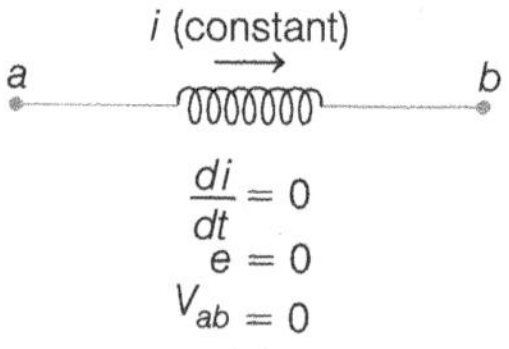

(a)

Refer figure (b) The current is increasing, so $\frac{di}{dt}$ is positive. The induced emf e must oppose the increasing current, so it must be in the sense from b to a.

a becomes the higher potential terminal and V_{ab} is positive. The direction of the emf is analogous to a battery with a as its positive terminal.

Refer figure (c) The current is decreasing and $\frac{di}{dt}$ is negative. The self-induced emf e opposes this decrease and V_{ab} is negative. This is analogous to a battery with b as its positive terminal.

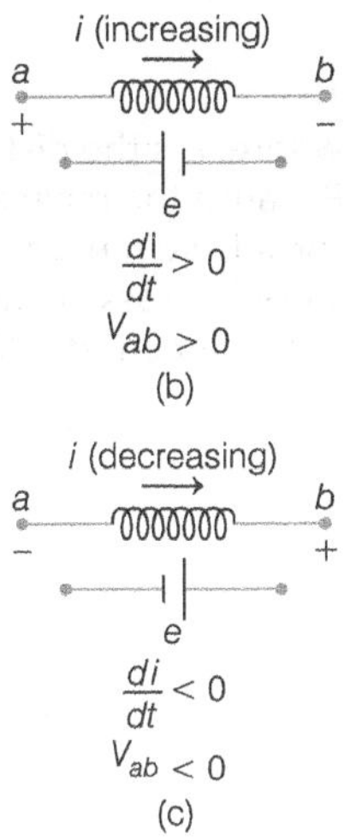

In each case, we can write the PD, V_{ab} as

$$V_{ab} = -e = L\frac{di}{dt}$$

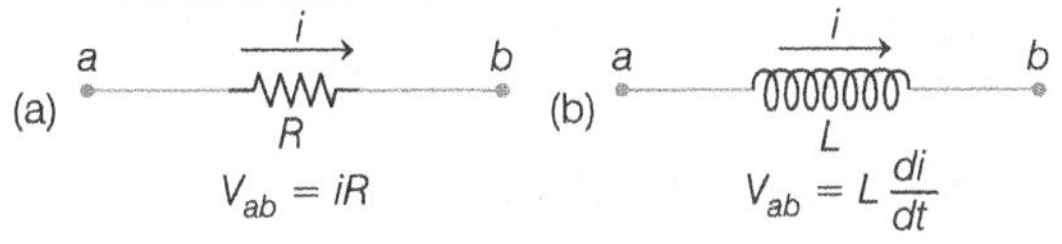

The circuit's behaviour of an inductor is quite different from that of a resistor. While a resistor opposes the current i, an inductor opposes the change (di/dt) in the current.

Kirchhoff's potential law with an inductor In Kirchhoff's potential law when we go through an inductor in the same direction as the assumed current, we encounter a voltage drop equal to $L\, di/dt$, where di/dt is to be substituted with sign.

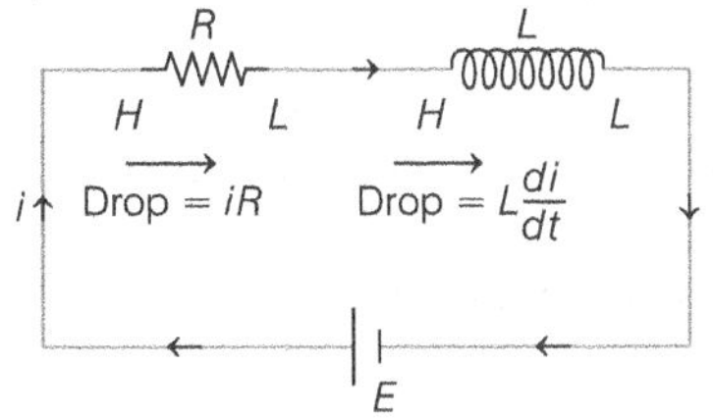

For example in the loop shown in figure, Kirchhoff's second law gives the equation.

$$E - iR - L\frac{di}{dt} = 0$$

Note *Write $-L\frac{di}{dt}$ when you cross an inductor in the direction of current and substitute (di/dt) with sign.*

Example 1 The inductor shown in the figure has inductance 0.54 H and carries a current in the direction shown that is decreasing at a uniform rate $\frac{di}{dt} = -0.03$ A/s.

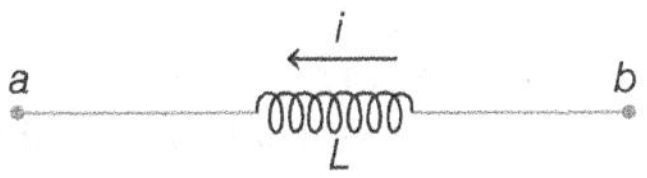

(a) Find the self-induced emf.

(b) Which end of the inductor a or b is at a higher potential?

Sol. (a) Self-induced emf,

$$|e| = \left|L\frac{di}{dt}\right| = (0.54)\,(0.03)$$

$$= 1.62 \times 10^{-2}\,\text{V}$$ **Ans.**

(b) $V_b - L\frac{di}{dt} = V_a$

$$\therefore \quad V_b - V_a = L\frac{di}{dt} = (0.54)\,(-0.03)$$

$$= -1.62 \times 10^{-2}\,\text{V}$$

Since, $V_{ba}\ (= V_b - V_a)$ is negative. It implies that $V_a > V_b$ or a is at higher potential. **Ans.**

Example 2 In the circuit diagram shown in figure, $R = 10\,\Omega$, $L = 5\text{H}$, $E = 20$ V, $i = 2$A. This current is decreasing at a rate of -1.0 A/s. Find V_{ab} at this instant.

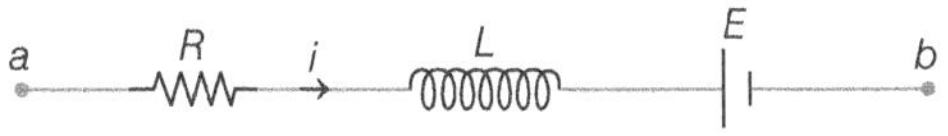

Sol. $L\frac{di}{dt} = (5)\,(-1.0) = -5$ V

Now, $\quad V_a - iR - L\frac{di}{dt} - E = V_b$

$$\therefore \quad V_{ab} = V_a - V_b = E + iR + L\frac{di}{dt}$$

$$= 20 + (2)\,(10) - 5 = 35\text{ V}$$ **Ans.**

Note *As the current is decreasing, the inductor can be replaced by a source of emf $e = \left|L\cdot\frac{di}{dt}\right| = 5$ V in such a manner that this emf supports the decreasing current, or it sends the current in the circuit in the same direction as the existing current. So, positive terminal of this source is towards b. Thus, the given circuit can be drawn as shown below,*

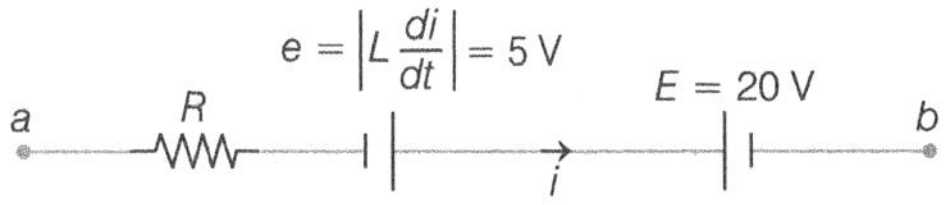

Now, we can find V_{ab}.

Method of Finding Self-inductance of a Circuit

We use the equation, $L = \frac{N\phi_B}{i}$ to calculate the inductance of given circuit.

A good approach for calculating the self-inductance of a circuit consists of the following steps:

(a) Assume that there is a current i flowing through the circuit (we can call the circuit as inductor).

(b) Determine the best known expression of magnetic field **B** produced by the current.

(c) Obtain the magnetic flux ϕ_B. Take that value of **B** which has least approximation in finding the flux ϕ_B.

(d) Now, the self-inductance can be found from $L = N\phi_B / i$.

To demonstrate this procedure, let us now take following two cases :

(i) Inductance of a Solenoid

Let us find the inductance of a uniformly wound solenoid having N turns and length l. Assume that l is much longer than the radius of the windings and that the core of the solenoid is air.

We can assume that the interior magnetic field due to a current i is uniform and given by equation,

$$B = \mu_0 ni = \mu_0 \left(\frac{N}{l}\right) i$$

where, $n = \frac{N}{l}$ is the number of turns per unit length.

The magnetic flux through each turn is,

$$\phi_B = BS = \mu_0 \frac{NS}{l} i$$

Here, S is the cross-sectional area of the solenoid. Now,

$$L = \frac{N\phi_B}{i} = \frac{N}{i}\left(\frac{\mu_0 NSi}{l}\right) = \frac{\mu_0 N^2 S}{l}$$

$\therefore$ $$\boxed{L = \frac{\mu_0 N^2 S}{l}}$$

This result shows that L depends on dimensions (S, l) and is proportional to the square of the number of turns.

$$\boxed{L \propto N^2}$$

Because $N = nl$, we can also express the result in the form,

$$L = \mu_0 \frac{(nl)^2}{l} S = \mu_0 n^2 Sl = \mu_0 n^2 V$$

or $$\boxed{L = \mu_0 n^2 V}$$

Here, $V = Sl$ is the volume of the solenoid.

(ii) Inductance of a Rectangular Toroid

A toroid with a rectangular cross-section is shown in figure. The inner and outer radii of the toroid are R_1 and R_2 and h is the height of the toroid. Applying Ampere's law for a toroid, we can show that magnetic field inside a rectangular toroid is given by

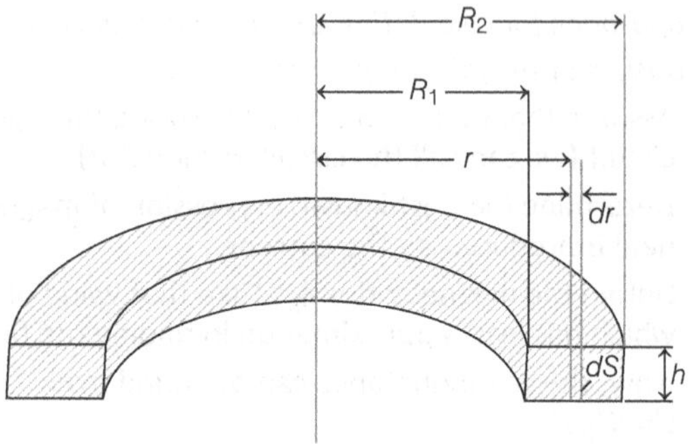

$$B = \frac{\mu_0 Ni}{2\pi r}$$

where, r is the distance from the central axis of the toroid. Because the magnitude of magnetic field changes within the toroid, we must calculate the flux by integrating over the toroid's cross-section. Using the infinitesimal cross-sectional area element $dS = hdr$ shown in the figure, we obtain

$$\phi_B = \int B\, dS = \int_{R_1}^{R_2} \left(\frac{\mu_0 Ni}{2\pi r}\right)(hdr)$$

$$= \frac{\mu_0 Nhi}{2\pi} \ln\left(\frac{R_2}{R_1}\right)$$

Now, $$L = \frac{N\phi_B}{i} = \frac{\mu_0 N^2 h}{2\pi} \ln\left(\frac{R_2}{R_1}\right)$$

or $$\boxed{L = \frac{\mu_0 N^2 h}{2\pi} \ln\left(\frac{R_2}{R_1}\right)}$$

As expected, the self-inductance is a constant determined by only the physical properties of the toroid.

Energy Stored in an Inductor

The energy of a capacitor is stored in the electric field between its plates. Similarly, an inductor has the capability of storing energy in its magnetic field.

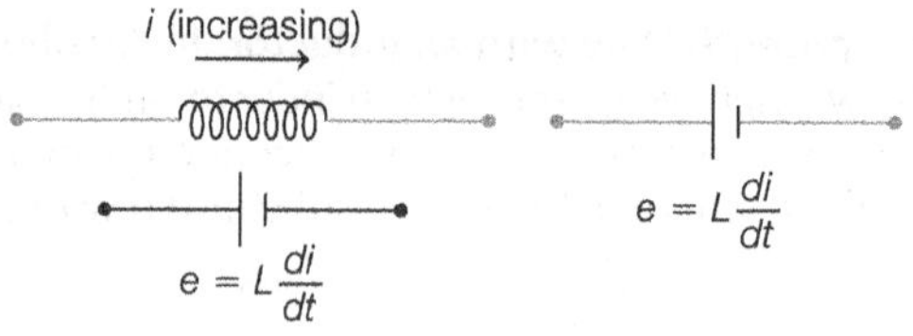

An increasing current in an inductor causes an emf between its terminals.

The work done per unit time is power.

$$P = \frac{dW}{dt} = ei = -Li\frac{di}{dt}$$

From $\quad dW = -dU$

$\therefore \quad dW = -Li\,(di)$

We have, $\quad dU = Li\, di$

The total energy U supplied while the current increases from zero to a final value i is

$$U = L\int_0^i i di = \frac{1}{2} Li^2$$

$\therefore$ $$\boxed{U = \frac{1}{2} Li^2}$$

This is the expression for the energy stored in the magnetic field of an inductor when a current i flows through it. The source of this energy is the external source of emf that supplies the current.

Note *(i) After the current has reached its final steady state value i, rate of change of current $di/dt = 0$ and no more energy is taken by the inductor.*

(ii) When the current decreases from i to zero, the inductor acts as a source that supplies a total amount of energy $\frac{1}{2}Li^2$ to the external circuit. If we interrupt the circuit suddenly by opening a switch, the current decreases very rapidly, the induced emf is very large and the energy may be dissipated as a spark across the switch.

(iii) If we compare the behaviour of a resistor and an inductor towards the current flow we can observe that energy flows into a resistor whenever a current passes through it. Whether the current is steady (constant) or varying this energy is dissipated in the form of heat. By contrast energy flows into an ideal, zero resistance inductor only when the current in the inductor increases. This energy is not dissipated, it is stored in the inductor and released when the current decreases.

(iv) As we have discussed earlier also, the energy in an inductor is actually stored in the magnetic field within the coil. We can develop relations of magnetic energy density u (energy stored per unit volume) analogous to those we obtained in electrostatics. We will concentrate on one simple case of an ideal long cylindrical solenoid. For a long solenoid its magnetic field can be assumed completely within the solenoid. The energy U stored in the solenoid when a current i is

$$U=\frac{1}{2}Li^2=\frac{1}{2}(\mu_0 n^2 V)\,i^2 \quad as \quad L=\mu_0 n^2 V$$

The energy per unit volume is $u=\frac{U}{V}$

$$\therefore \quad u=\frac{U}{V}=\frac{1}{2}\mu_0 n^2 i^2=\frac{1}{2}\frac{(\mu_0 ni)^2}{\mu_0}=\frac{1}{2}\frac{B^2}{\mu_0}$$

as $B=\mu_0 ni$

Thus, $$u=\frac{1}{2}\frac{B^2}{\mu_0}$$

This expression is similar to $u=\frac{1}{2}\varepsilon_0 E^2$ *used in electrostatics. Although, we have derived it for one special situation, it turns out to be correct for any magnetic field configuration.*

Combination of Inductances

In series If several inductances are in series so that there are no interactions through mutual inductance.

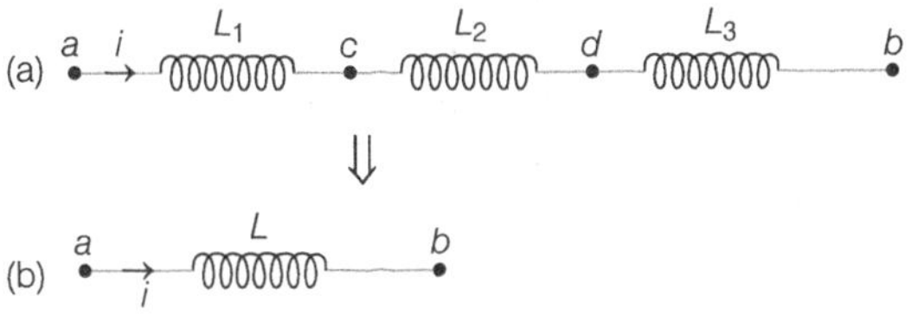

Refer figure (a)

$$V_a-V_c=L_1\frac{di}{dt}$$

$$V_c-V_d=L_2\frac{di}{dt}$$

and $$V_d-V_b=L_3\frac{di}{dt}$$

Adding all these equations, we have

$$V_a-V_b=(L_1+L_2+L_3)\frac{di}{dt} \quad \text{...(i)}$$

Refer figure (b)

$$V_a-V_b=L\frac{di}{dt} \quad \text{...(ii)}$$

Here, L = equivalent inductance.

From Eqs. (i) and (ii), we have

$$\boxed{L=L_1+L_2+L_3}$$

In parallel

Refer figure (a)

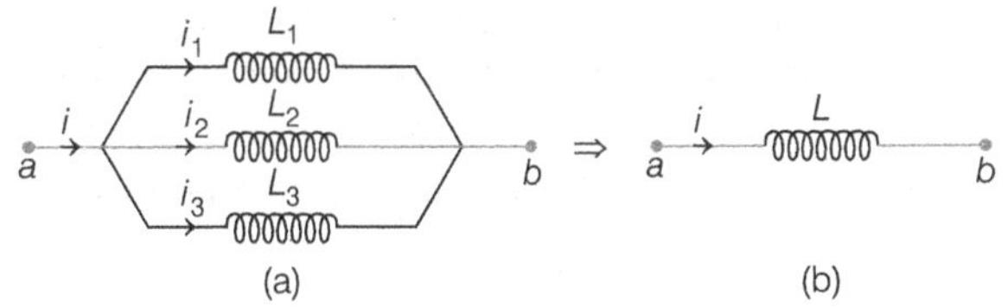

$$i=i_1+i_2+i_3$$

or $$\frac{di}{dt}=\frac{di_1}{dt}+\frac{di_2}{dt}+\frac{di_3}{dt}$$

or $$\frac{di}{dt}=\frac{V_a-V_b}{L_1}+\frac{V_a-V_b}{L_2}+\frac{V_a-V_b}{L_3} \quad \text{...(i)}$$

Refer figure (b)

$$\frac{di}{dt}=\frac{V_a-V_b}{L} \quad \text{...(ii)}$$

From Eqs. (i) and (ii),

$$\frac{1}{L}=\frac{1}{L_1}+\frac{1}{L_2}+\frac{1}{L_3}$$

Example 3 (a) Calculate the inductance of an air core solenoid containing 300 turns if the length of the solenoid is 25.0 cm and its cross-sectional area is 4.00 cm^2.

(b) Calculate the self-induced emf in the solenoid if the current through it is decreasing at the rate of 50.0 A/s.

Sol. (a) The inductance of a solenoid is given by

$$L=\frac{\mu_0 N^2 S}{l}$$

Substituting the values, we have

$$L=\frac{(4\pi\times10^{-7})(300)^2(4.00\times10^{-4})}{(25.0\times10^{-2})}\text{H}$$

$$=1.81\times10^{-4}\text{H}$$ **Ans.**

(b) $e=-L\frac{di}{dt}$

Here, $\frac{di}{dt}=-50.0$ A/s

$$\therefore \quad e=-(1.81\times10^{-4})(-50.0)$$

$$=9.05\times10^{-3}\text{ V}$$

or $e=9.05$ mV **Ans.**

Example 4 What inductance would be needed to store 1.0 kWh of energy in a coil carrying a 200 A current. (1 kWh = 3.6×10^6 J)

Sol. We have, $i = 200$ A

and $U = 1\text{ kWh} = 3.6\times10^6$ J

$$\therefore \quad L = \frac{2U}{i^2} \qquad \left(U = \frac{1}{2}Li^2\right)$$

$$= \frac{2\,(3.6\times10^6)}{(200)^2} = 180\text{ H} \qquad \textbf{Ans.}$$

Example 5 (a) What is the magnetic flux through one turn of a solenoid of self-inductance 8.0×10^{-5} H when a current of 3.0 A flows through it? Assume that the solenoid has 1000 turns and is wound from wire of diameter 1.0 mm.

(b) What is the cross-sectional area of the solenoid?

Sol. Given, $L = 8.0\times10^{-5}$ H, $i = 3.0$ A and $N = 1000$ turns

(a) From the relation, $L = \dfrac{N\phi}{i}$

The flux linked with one turn,

$$\phi = \frac{Li}{N} = \frac{(8.0\times10^{-5})\,(3.0)}{1000} = 2.4\times10^{-7}\text{ Wb}$$

(b) This $\phi = BS = (\mu_0 ni)\,(S)$

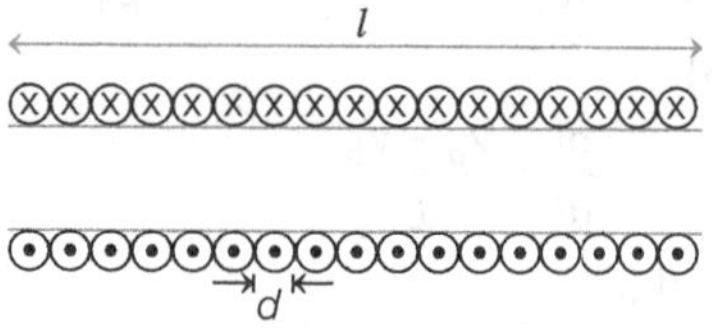

Here, n = number of turns per unit length

$$= \frac{N}{l} = \frac{N}{Nd} = \frac{1}{d}$$

$$\therefore \quad \phi = \frac{\mu_0 iS}{d}$$

$$\text{or} \quad S = \frac{\phi d}{\mu_0 i}$$

$$= \frac{(2.4\times10^{-7})\,(1.0\times10^{-3})}{(4\pi\times10^{-7})\,(3.0)}$$

$$= 6.37\times10^{-5}\text{ m}^2 \qquad \textbf{Ans.}$$

Example 6 A 10 H inductor carries a current of 20 A. How much ice at 0°C could be melted by the energy stored in the magnetic field of the inductor? Latent heat of ice is 2.26×10^3 J/kg.

Sol. Energy stored is $\frac{1}{2}Li^2$.

This energy is completely used in melting the ice.

Hence, $\frac{1}{2}Li^2 = mL_f$

Here, L_f = latent heat of fusion

Hence, mass of ice melted, $m = \dfrac{Li^2}{2L_f}$

Substituting the values, we have

$$m = \frac{(10)\,(20)^2}{2\,(2.26\times10^3)}$$

$$= 0.88\text{ kg} \qquad \textbf{Ans.}$$

Typed Problems

Type 1. *Based on potential difference across an inductor.*

Ex. 1 Two different coils have self-inductances $L_1 = 8$ mH and $L_2 = 2$ mH. The current in one coil is increased at a constant rate. The current in the second coil is also increased at the same constant rate. At a certain instant of time, the power given to the two coils is the same. At that time, the current, the induced voltage and the energy stored in the first coil are i_1, V_1 and W_1 respectively. Corresponding values for the second coil at the same instant are i_2, V_2 and W_2 respectively. Then, **(1994 JEE)**

(a) $\dfrac{i_1}{i_2} = \dfrac{1}{4}$ (b) $\dfrac{i_1}{i_2} = 4$

(c) $\dfrac{W_1}{W_2} = \dfrac{1}{4}$ (d) $\dfrac{V_1}{V_2} = 4$

Sol. Potential difference across an inductor :

$V \propto L$, if rate of change of current is constant $\left(V = -L\dfrac{di}{dt}\right)$

$$\therefore \quad \frac{V_1}{V_2} = \frac{L_1}{L_2} = \frac{8}{2} = 4$$

Power given to the two coils is same, i.e.

$$V_1 i_1 = V_2 i_2$$

$$\text{or} \quad \frac{i_1}{i_2} = \frac{V_2}{V_1} = \frac{1}{4}$$

Energy stored, $W = \frac{1}{2}Li^2$

$$\therefore \quad \frac{W_1}{W_2} = \left(\frac{L_1}{L_2}\right)\left(\frac{i_1}{i_2}\right)^2$$

$$= \left(\frac{8}{2}\right)\left(\frac{1}{4}\right)^2$$

$$= \frac{1}{4}$$

$\therefore$ The correct options are (a), (c) and (d).

Ex. 2 In the figure shown, $i_1 = 10\,e^{-2t}$ A, $i_2 = 4$ A and $V_C = 3e^{-2t}$ V. Determine **(1992 JEE)**

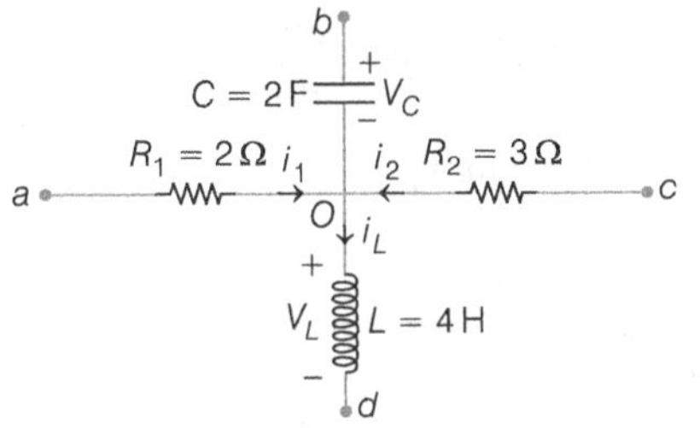

(a) i_L and V_L (b) V_{ac}, V_{ab} and V_{cd}.

Sol. (a) Charge stored in the capacitor at time t,

$$q = CV_C = (2)(3e^{-2t})$$
$$= 6e^{-2t}\text{ C}$$

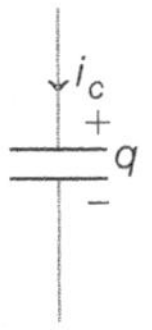

$$\therefore \quad i_c = +\frac{dq}{dt} = -12e^{-2t}\text{ A}$$

(Direction of current is from b to O)

Applying junction rule at O,

$$i_L = i_1 + i_2 + i_c$$
$$= 10e^{-2t} + 4 - 12e^{-2t}$$
$$= (4 - 2e^{-2t})\text{ A}$$
$$= [2 + 2(1 - e^{-2t})]\text{ A} \qquad \textbf{Ans.}$$

i_L *versus* time graph is as shown in figure.

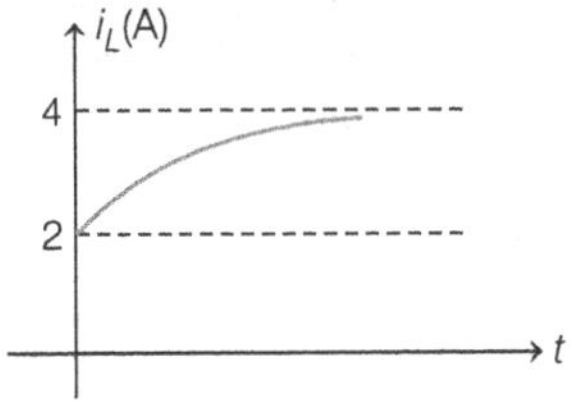

i_L increases from 2A to 4A exponentially.

$$V_L = V_{Od} = L\frac{di_L}{dt} = (4)\frac{d}{dt}(4 - 2e^{-2t})$$
$$= 16e^{-2t}\text{ V} \qquad \textbf{Ans.}$$

V_L *versus* time graph is as shown in figure.

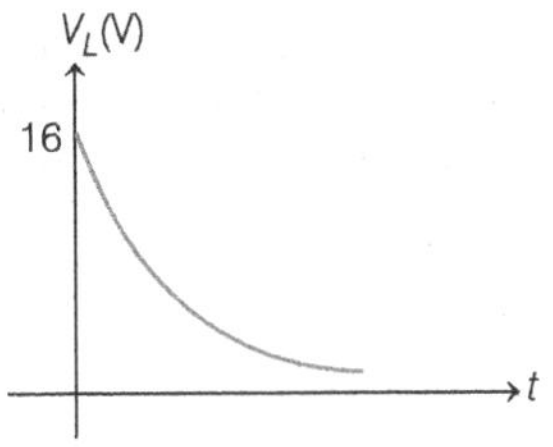

V_L decreases exponentially from 16 V to 0.

(b) $\boldsymbol{V_{ac} = V_a - V_c}$

$$V_a - i_1R_1 + i_2R_2 = V_c$$
$$\therefore \quad V_a - V_c = V_{ac} = i_1R_1 - i_2R_2$$

Substituting the values, we have

$$V_{ac} = (10e^{-2t})(2) - (4)(3)$$
$$V_{ac} = (20e^{-2t} - 12)\text{ V}$$

At $t = 0$, $V_{ac} = 8$ V and at $t = \infty$, $V_{ac} = -12$ V

Therefore, V_{ac} decreases exponentially from 8V to −12V.

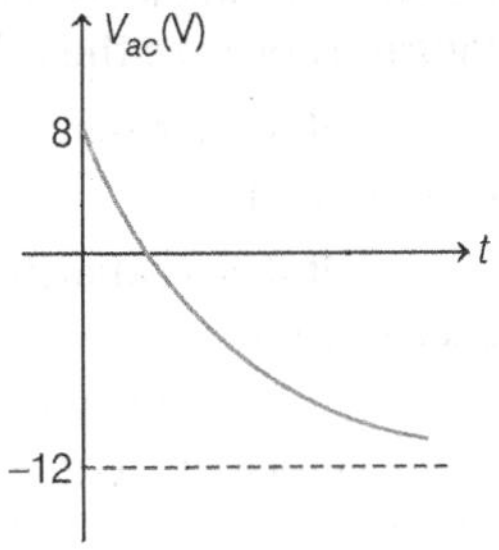

$\boldsymbol{V_{ab} = V_a - V_b}$

$$V_a - i_1R_1 + V_C = V_b$$
$$\therefore \quad V_a - V_b = V_{ab} = i_1R_1 - V_C$$

Substituting the values, we have

$$V_{ab} = (10e^{-2t})(2) - 3e^{-2t}$$

or $\quad V_{ab} = 17e^{-2t}$ V **Ans.**

Thus, V_{ab} decreases exponentially from 17 V to 0.

V_{ab} *versus* t graph is shown in figure.

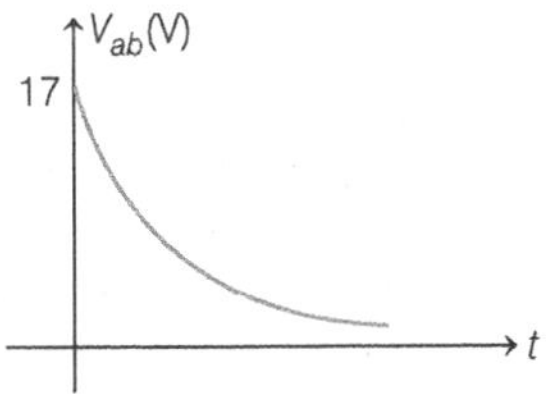

$\boldsymbol{V_{cd} = V_c - V_d}$

$$V_c - i_2R_2 - L\left(\frac{di}{dt}\right) = V_d$$
$$\therefore \quad V_c - V_d = V_{cd} = i_2R_2 + L\left(\frac{di}{dt}\right)$$

Substituting the values, we have

$$V_{cd} = (4)(3) + 16e^{-2t}$$

or $\quad V_{cd} = (12 + 16e^{-2t})$ V **Ans.**

At $t = 0$, $V_{cd} = 28$ V and at $t = \infty$, $V_{cd} = 12$ V

i.e. V_{cd} decreases exponentially from 28 V to 12 V.

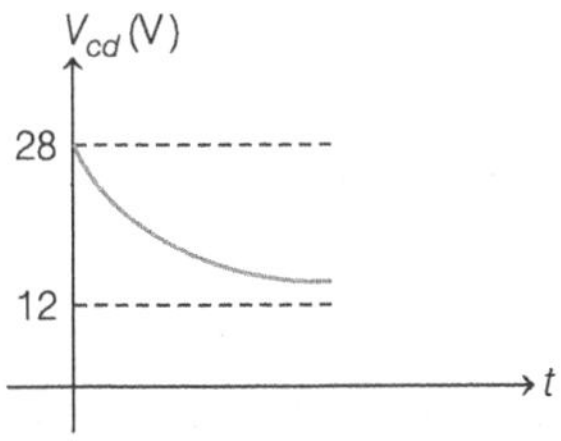

V_{cd} *versus* t graph is shown in figure.

Type 2. *Based on motion of a wire in uniform magnetic field with a circuit element like resistance, capacitor or an inductor.*

Concept

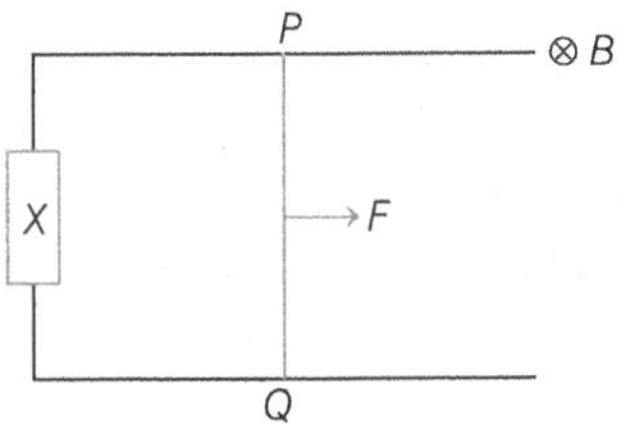

A constant force F is applied on wire PQ of length l and mass m. There is an electrical element X in the box as shown in figure. There are the following three different cases :

Case 1 If X is a resistance, then velocity of the wire increases exponentially.

Case 2 If X is a capacitor, then wire moves with a constant acceleration a $(< F/m)$.

Case 3 If X is an inductor and instead of constant force F an initial velocity v_0 is given to the wire then the wire starts simple harmonic motion with v_0 as the maximum velocity $(= \omega A)$ at mean position.

Ex. 3 In the above case if X is a resistance R, then find velocity of wire as a function of time t.

Sol. At time t suppose velocity of wire is v, then due to motional emf a current i flows in the closed circuit in anti-clockwise direction.

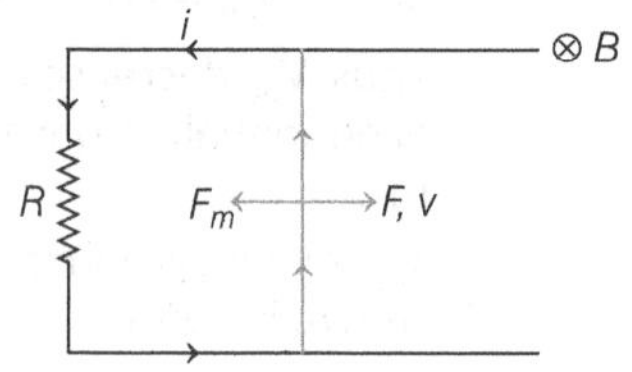

$$i = \frac{e}{R} = \frac{Bvl}{R}$$

Due to this current magnetic force will act on the wire in the direction shown in figure,

$$F_m = ilB = \left(\frac{B^2l^2}{R}\right)v$$

$$F_{net} = F - F_m = F - \left(\frac{B^2l^2}{R}\right)v = m\frac{dv}{dt}$$

$$\therefore \quad \int_0^v \frac{dv}{F - \left(\frac{B^2l^2}{R}\right)v} = \frac{1}{m}\int_0^t dt$$

Solving this equation, we get

$$v = \frac{FR}{B^2l^2}\left(1 - e^{-\frac{B^2l^2}{mR}t}\right)$$

Thus, velocity of the wire increases exponentially.

v-t graph is as shown below.

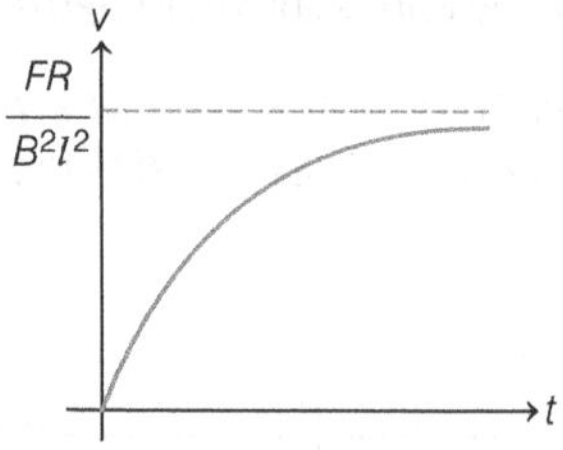

Ex. 4 If X is a capacitor C, then find the constant acceleration a of the wire.

Sol. At time t suppose velocity of wire is v. Then, due to motional emf $e = Bvl$ capacitor gets charged .

$$q = CV = C(Bvl)$$

This charge is increasing as v will be increasing. Hence, there will be a current in the circuit as shown in figure.

$$i = \frac{dq}{dt} = (BlC)\frac{dv}{dt}$$

or $\quad i = (BlC)\,a \quad \left(\text{as } \frac{dv}{dt} = a\right)$

Due to this current a magnetic force F_m will act in the direction shown in figure,

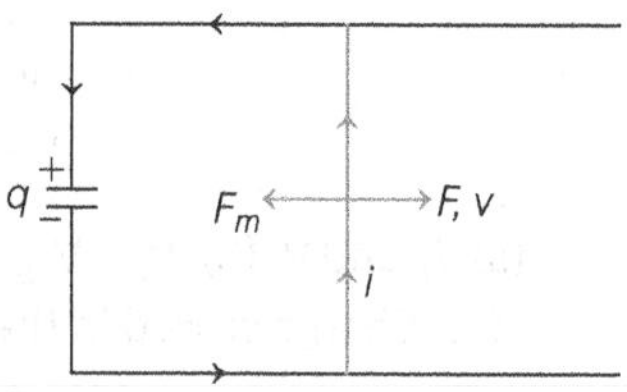

$$F_m = ilB = (B^2l^2C)\,a$$

Now, $\quad F_{net} = F - F_m$

or $\quad ma = F - (B^2l^2C)\,a$

$$\therefore \quad a = \frac{F}{m + B^2l^2C} \qquad \textbf{Ans.}$$

Now, we can see that this acceleration is constant but less than F/m.

Ex. 5 A pair of parallel horizontal conducting rails of negligible resistance shorted at one end is fixed on a table. The distance between the rails is L. A conducting massless rod of resistance R can slide on the rails frictionlessly. The rod is tied to a massless string which passes over a pulley fixed to the edge of the table. A mass m tied to the other end of the string hangs vertically. A constant magnetic field B exists perpendicular to the table. If the system is released from rest, calculate **(1997 JEE)**

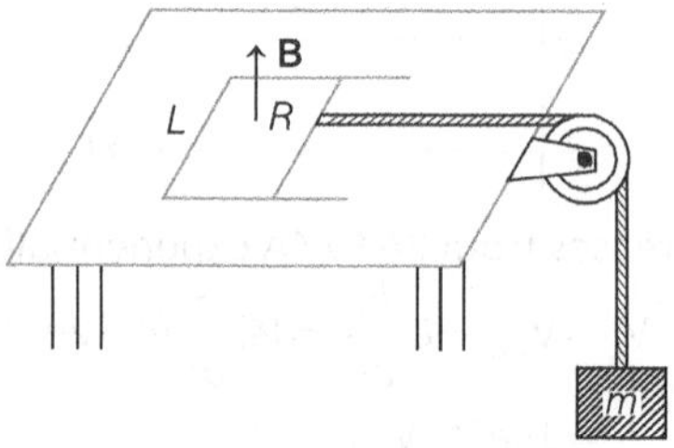

(a) the terminal velocity achieved by the rod and

(b) the acceleration of the mass at the instant when the velocity of the rod is half the terminal velocity.

Sol. (a) Let v be the velocity of the wire (as well as block) at any instant of time t.

Motional emf, $e = BvL$

Motional current, $i = \frac{e}{r} = \frac{BvL}{R}$

and magnetic force on the wire

$$F_m = iLB = \frac{vB^2L^2}{R}$$

Net force on the system at this moment will be

$$F_{net} = mg - F_m = mg - \frac{vB^2L^2}{R}$$

or $$ma = mg - \frac{vB^2L^2}{R}$$

$$a = g - \frac{vB^2L^2}{mR} \quad ...(i)$$

Velocity will acquire its terminal value, i.e. $v = v_T$ when F_{net} or acceleration a of the particle becomes zero.

Thus, $$0 = g - \frac{v_T B^2L^2}{mR}$$

or $$v_T = \frac{mgR}{B^2L^2}$$

(b) When $$v = \frac{v_T}{2} = \frac{mgR}{2B^2L^2}$$

Then from Eq. (i), acceleration of the block,

$$a = g - \left(\frac{mgR}{2B^2L^2}\right)\left(\frac{B^2L^2}{mR}\right) = g - \frac{g}{2}$$

or $$a = \frac{g}{2}$$

Ex. 6 A loop is formed by two parallel conductors connected by a solenoid with inductance L and a conducting rod of mass m which can freely (without friction) slide over the conductors. The conductors are located in a horizontal plane in a uniform vertical magnetic field B. The distance between the conductors is l.

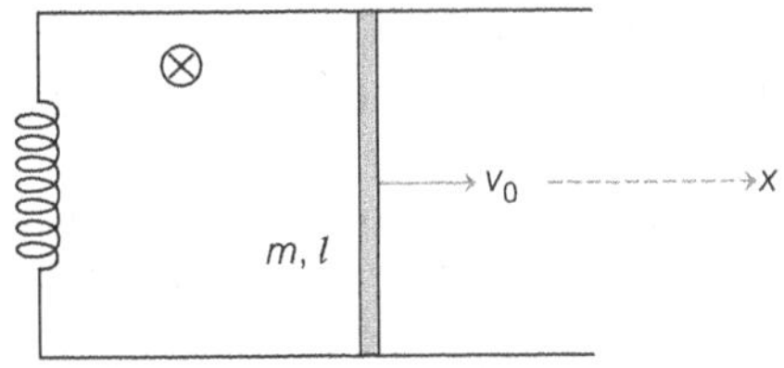

At the moment $t = 0$, the rod is imparted an initial velocity v_0 directed to the right. Find the law of its motion $x(t)$ if the electric resistance of the loop is negligible.

Sol. Let at any instant of time, suppose velocity of the rod is v towards right. The current in the circuit is i. In the figure,

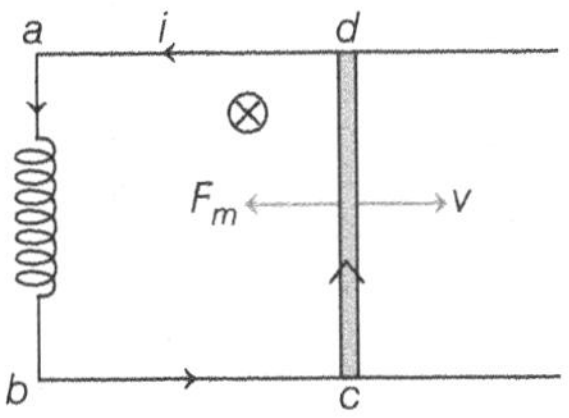

$$V_a - V_b = V_d - V_c$$

or $$L\frac{di}{dt} = Bvl = Bl\frac{dx}{dt} \quad \left(\text{as } v = \frac{dx}{dt}\right)$$

i.e. $$Ldi = Bldx$$

Integrating on both sides, we get

$$Li = Blx$$

or $$i = \frac{Bl}{L}x \quad ...(i)$$

Magnetic force on the rod at this instant is

$$F_m = ilB = \frac{B^2l^2}{L}x \quad ...(ii)$$

Since, this force is in opposite direction of $\mathbf{v}$, so from Newton's second law we can write

$$m\left(\frac{d^2x}{dt^2}\right) = -\frac{B^2l^2}{L}x$$

or $$\left(\frac{d^2x}{dt^2}\right) = -\frac{B^2l^2}{mL}x$$

Comparing this with equation of SHM,

$$\frac{d^2x}{dt^2} = -\omega^2 x$$

We have, $$\omega = \frac{Bl}{\sqrt{mL}}$$

Therefore, the rod will oscillate simple harmonically with angular frequency $\omega = \frac{Bl}{\sqrt{mL}}$. At time $t = 0$, rod was at $x = 0$ and it was moving towards positive x-axis. Hence, x-t equation of the rod is

$$x = A\sin\omega t \quad ...(iii)$$

To find A, we use the fact that at $t = 0$, v or $\frac{dx}{dt}$ has a value v_0. Hence,

$$\frac{dx}{dt} = v = A\omega\cos\omega t$$

or $$A\omega = v_0 \quad (\text{at } t = 0)$$

or $$A = \frac{v_0}{\omega}$$

Substituting in Eq. (iii), we have

$$x = \frac{v_0}{\omega}\sin\omega t, \text{ where } \omega = \frac{Bl}{\sqrt{mL}}$$ **Ans.**

Alternate method of finding *A*

At $x = A$, $v = 0$, i.e. whole of its kinetic energy is converted into magnetic energy. Thus,

$$\frac{1}{2}Li^2 = \frac{1}{2}mv_0^2$$

Substituting value of i from Eq. (i), with $x = A$, we have

$$L\left(\frac{Bl}{L}A\right)^2 = mv_0^2$$

or $$A = \frac{\sqrt{mL}}{Bl}v_0 = \frac{v_0}{\omega}$$

as $$\omega = \frac{Bl}{\sqrt{mL}}$$ **Ans.**

MODULE 3 Exercise

Single Correct Option

1. In figure, if the current i decreases at a rate α, then $V_A - V_B$ is

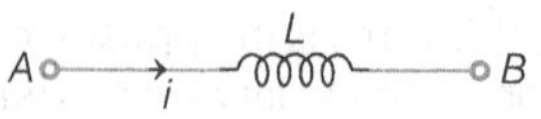

(a) zero (b) $-\alpha L$
(c) αL (d) No relation exists

2. In the figure shown, V_{ab} at $t = 1$ s is

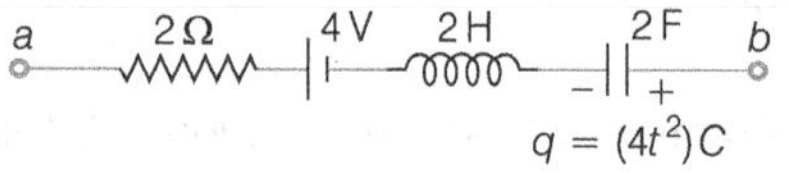

(a) 30 V (b) -30 V
(c) 20 V (d) -20 V

3. In the given branch AB of a circuit, a current, $I = (10t + 5)$ A is flowing, where t is time in second. At $t = 0$, the potential difference $(V_A - V_B)$ between points A and B is

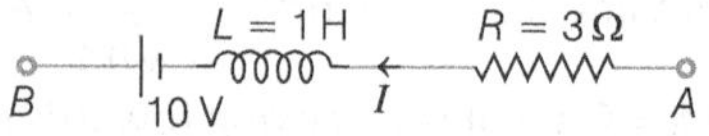

(a) 15 V (b) -5 V (c) -15 V (d) 5 V

4. A long solenoid has self-inductance L. If its length is doubled keeping total number of turns constant, then its new self-inductance will be

(a) $\frac{L}{2}$ (b) $2L$
(c) L (d) $\frac{L}{4}$

5. The self-induced emf of a coil is 25 V. When the current in it is changed at uniform rate from 10 A to 25 A in 1s, the change in the energy of the inductance is **(2019 Main)**

(a) 437.5 J (b) 740 J
(c) 637.5 J (d) 540 J

6. When the current in the portion of the circuit shown in the figure is 2A and increasing at the rate of 1 A/s, the measured potential difference $V_{ab} = 8$ V. However, when the current is 2 A and decreasing at the rate of 1 A/s, the measured potential difference $V_{ab} = 4$ V. The values of R and L are

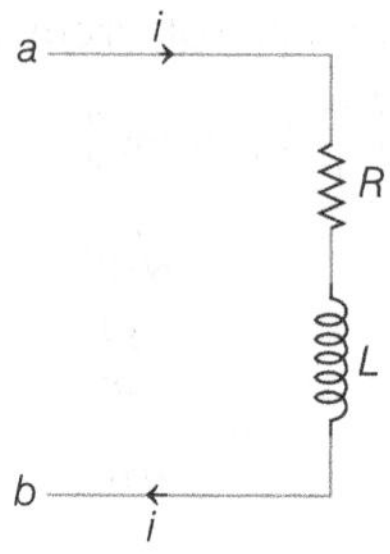

(a) $3\,\Omega$ and 2 H respectively
(b) $3\,\Omega$ and 3 H respectively
(c) $2\,\Omega$ and 1 H respectively
(d) $3\,\Omega$ and 1 H respectively

Multiple Correct Options

7. Choose the correct options.
(a) SI unit of magnetic flux is henry-ampere
(b) SI unit of coefficient of self-inductance is J/A
(c) SI unit of coefficient of self-inductance is $\frac{\text{volt - second}}{\text{ampere}}$
(d) SI unit of magnetic induction is weber

Subjective Questions

8. The current through an inductor of 1H is given by $i = 3t \sin t$. Find the voltage across the inductor.

9. In the figure shown $i = 10e^{-4t}$ A. Find V_L and V_{ab}.

10. The current (in Ampere) in an inductor is given by $I = 5 + 16t$, where t is in seconds. The self-induced emf in it is 10 mV. Find
(a) the self-inductance, and
(b) the energy stored in the inductor and the power supplied to it at $t = 1$s

11. (a) Calculate the self-inductance of a solenoid that is tightly wound with wire of diameter 0.10 cm, has a cross-sectional area 0.90 cm^2 and is 40 cm long.
(b) If the current through the solenoid decreases uniformly from 10 A to 0 A in 0.10 s, what is the emf induced between the ends of the solenoid?

12. The potential difference across a 150 mH inductor as a function of time is shown in figure. Assume that the initial value of the current in the inductor is zero. What is the current when $t = 2.0$ ms ? and $t = 4.0$ ms ?

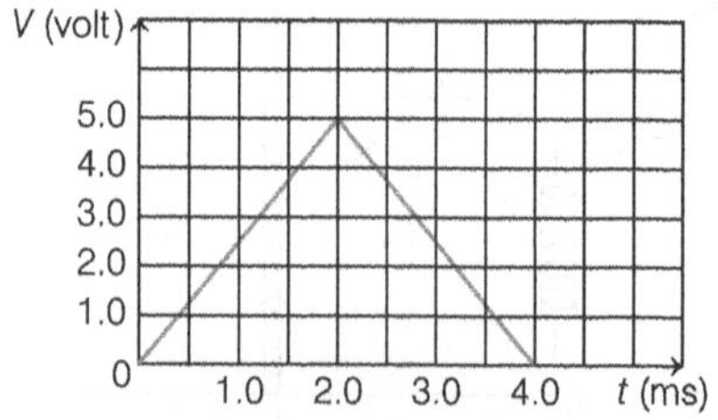

13. At the instant when the current in an inductor is increasing at a rate of 0.0640 A/s, the magnitude of the self-induced emf is 0.0160 V.
(a) What is the inductance of the inductor?
(b) If the inductor is a solenoid with 400 turns, what is the average magnetic flux through each turn when the current is 0.720 A?

14. How many metres of a thin wire are required to manufacture a solenoid of length $l_0 = 100$ cm and inductance $L = 1.0$ mH if the solenoid's cross-sectional diameter is considerably less than its length? **(Irodov 3.315)**

15. Find the inductance of a solenoid of length l_0 whose winding is made of copper wire of mass m. The winding resistance is equal to R. The solenoid diameter is considerably less than its length. **(Irodov 3.316)**

Integer Type Question

16.

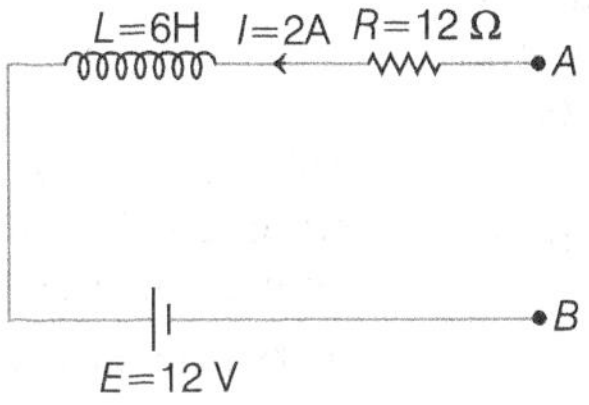

As per the given figure, if $dI/dt = -1$ A/s, then the value of V_{AB} at this instant will be .... V. **(2023 Main)**

Answers

1. (b) **2.** (b) **3.** (a) **4.** (a) **5.** (a) **6.** (a) **7.** (a,c) **8.** $3(t\cos t + \sin t)$
9. $-80\,e^{-4t}$, $-40\,e^{-4t}$ **10.** (a) 0.625 mH (b) 0.137 J, 0.21 J/s **11.** (a) 4.5×10^{-5} H (b) 4.5×10^{-3} V
12. 3.33×10^{-2} A, 6.67×10^{-2} A **13.** (a) 0.25 H (b) 4.5×10^{-4} Wb **14.** 100 m **15.** $L = \left(\frac{\mu_0}{4\pi}\right)\left(\frac{mR}{l_0 \rho \rho_0}\right)$ **16.** (30)

See hints on pages 548-549.

MODULE 4

Mutual Inductance

We have already discussed in the previous unit, the magnetic interaction between two wires carrying steady currents. The current in one wire causes a magnetic field, which exerts a force on the current in the second wire.

An additional interaction arises between two circuits when there is a changing current in one of the circuits.

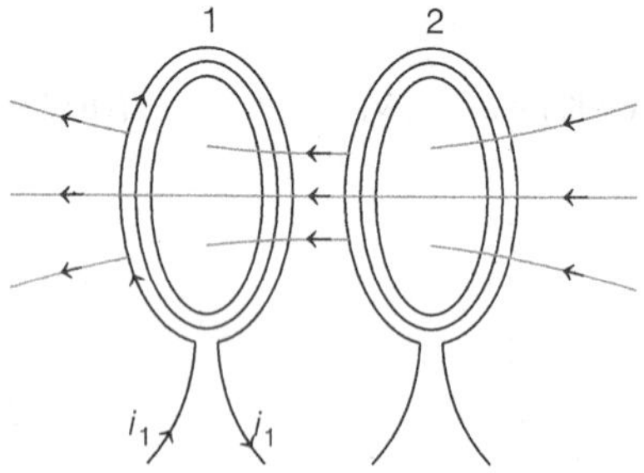

Consider two neighbouring coils of wire as shown in figure. A current flowing in coil 1 produces magnetic field and hence, a magnetic flux through coil 2. If the current in coil 1 changes, the flux through coil 2 changes as well. According to Faraday's law this induces an emf in coil 2. In this way, a change in the current in one circuit can induce a current in the second circuit. This phenomenon is known as **mutual induction**. Like the self-inductance (L), two circuits have mutual inductance (M). It also have two definitions as under:

First Definition Suppose the circuit 1 has a current i_1 flowing in it. Then, total flux $N_2\phi_{B_2}$ linked with circuit 2 is proportional to the current in 1. Thus,

$$N_2\,\phi_{B_2} \propto i_1$$

or
$$N_2\phi_{B_2} = Mi_1$$

Here, the proportionality constant M is known as the mutual inductance M of the two circuits.

Thus,
$$\boxed{M = \frac{N_2\phi_{B_2}}{i_1}}$$

From this expression, M can be defined as the **total flux $N_2\phi_{B_2}$ linked with circuit 2 per unit current in circuit 1.**

Second Definition If we change the current in circuit 1 at a rate di_1/dt, an induced emf e_2 is developed in circuit 2, which is proportional to the rate di_1/dt. Thus,

$$e_2 \propto di_1/dt$$

or
$$e_2 = -Mdi_1/dt$$

Here, the proportionality constant is again M. Minus sign indicates that e_2 is in such a direction that it opposes any change in the current in circuit 1. From the above equation,

$$\boxed{M = \left|\frac{-e_2}{di_1/dt}\right|}$$

This equation states that, **the mutual inductance of two circuits is the magnitude of induced emf e_2 per unit rate of change of current di_1/dt.**

Note down the following points regarding the mutual inductance:

1. The SI unit of mutual inductance is **henry** (1H).
2. M depends upon closeness of the two circuits, their orientations and sizes and the number of turns etc.
3. **Reciprocity theorem :**

$$M_{21} = M_{12} = M$$
$$e_2 = -M(di_1/dt)$$
and
$$e_1 = -M(di_2/dt)$$
$$M_{12} = \frac{N_2\phi_{B_2}}{i_1}$$
and
$$M_{21} = \frac{N_1\phi_{B_1}}{i_2}$$

4. A good approach for calculating the mutual inductance of two circuits consists of the following steps:
   (a) Assume any one of the circuits as primary (first) and the other as secondary (second).
   (b) Pass a current i_1 through the primary circuit.
   (c) Determine the magnetic field **B** produced by the current i_1.
   (d) Obtain the magnetic flux ϕ_{B_2}.
   (e) With this flux, the mutual inductance can be found from the relation,

$$M = \frac{N_2\,\phi_{B_2}}{i_1}$$

Mutual Inductance of a Solenoid Surrounded by a Coil

Figure shows a coil of N_2 turns and radius R_2 surrounding a long solenoid of length l_1, radius R_1 and number of turns N_1.

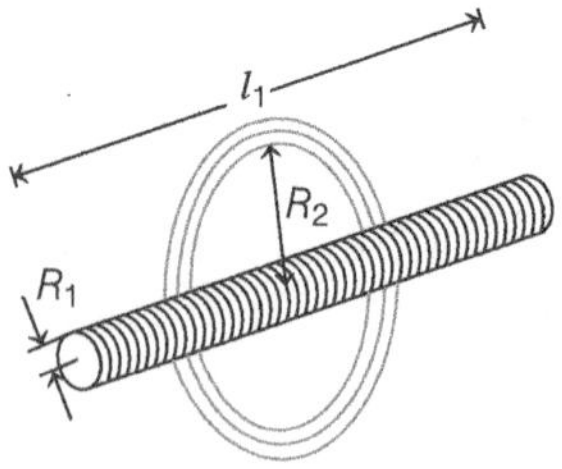

To calculate M between them, let us assume a current i_1 in the solenoid.

There is no magnetic field outside the solenoid and the field inside has magnitude,

$$B = \mu_0 n_1 i_1 = \mu_0 \left(\frac{N_1}{l_1}\right) i_1$$

and is directed parallel to the solenoid's axis. The magnetic flux ϕ_{B_2} through the surrounding coil is, therefore,

$$\phi_{B_2} = B(\pi R_1^2) = \frac{\mu_0 N_1 i_1}{l_1} \pi R_1^2$$

Now, $$M = \frac{N_2 \phi_{B_2}}{i_1} = \left(\frac{N_2}{i_1}\right)\left(\frac{\mu_0 N_1 i_1}{l_1}\right)\pi R_1^2 = \frac{\mu_0 N_1 N_2 \pi R_1^2}{l_1}$$

$\therefore$ $$\boxed{M = \frac{\mu_0 N_1 N_2 \pi R_1^2}{l_1}} \quad \text{or} \quad \boxed{M \propto N_1 N_2}$$

Note that M is independent of the radius R_2 of the surrounding coil. This is because solenoid's magnetic field is confined to its interior.

In principle, we can also calculate M by finding the magnetic flux through the solenoid produced by the current in the surrounding coil. This approach is much more difficult, because ϕ_{B_1} is so complicated. However, since $M_{12} = M_{21}$, we do know the result of this calculation.

Exercise : Two closely wound concentric circular coils have number of turns N_1 and N_2. Their radius is R. Find their mutual inductance and prove that $M \propto N_1 N_2$.

Ans. $M = \dfrac{\mu_0 N_1 N_2 \pi R}{2}$

Example 1 A straight solenoid has 50 turns per cm in primary and total 200 turns in the secondary. The area of cross-section of the solenoids is 4 cm^2. Calculate the mutual inductance. Primary is tightly kept inside the secondary.

Sol. The magnetic field at any point inside the straight solenoid of primary with n_1 turns per unit length carrying a current i_1 is given by the relation,

$$B = \mu_0 n_1 i_1$$

The magnetic flux through the secondary of N_2 turns each of area S is given as

$$N_2 \phi_2 = N_2 (BS) = \mu_0 n_1 N_2 i_1 S$$

$\therefore$ $$M = \frac{N_2 \phi_2}{i_1} = \mu_0 n_1 N_2 S$$

Substituting the values, we get

$$M = (4\pi \times 10^{-7})\left(\frac{50}{10^{-2}}\right)(200)\,(4 \times 10^{-4})$$

$$= 5.0 \times 10^{-4}\ \text{H}$$ **Ans.**

Example 2 Two solenoids A and B spaced close to each other and sharing the same cylindrical axis have 400 and 700 turns, respectively. A current of 3.50 A in coil A produced an average flux of 300 μT-m^2 through each turn of A and a flux of 90.0 μT-m^2 through each turn of B.

(a) Calculate the mutual inductance of the two solenoids.

(b) What is the self-inductance of A?

(c) What emf is induced in B when the current in A increases at the rate of 0.5 A/s?

Sol. (a) $$M = \frac{N_B \phi_B}{i_A} = \frac{(700)(90 \times 10^{-6})}{3.5}$$

$$= 1.8 \times 10^{-2}\ \text{H}$$ **Ans.**

(b) $$L_A = \frac{N_A \phi_A}{i_A} = \frac{(400)(300 \times 10^{-6})}{3.5}$$

$$= 3.43 \times 10^{-2}\ \text{H}$$ **Ans.**

(c) $$|e_B| = M\left(\frac{di_A}{dt}\right) = (1.8 \times 10^{-2})(0.5)$$

$$= 9.0 \times 10^{-3}\ \text{V}$$ **Ans.**

Example 3 A rectangular loop is kept near an infinitely long wire as shown in figure.

(a) Find value of one sided mutual inductance between them.

(b) If current in straight wire is changed at a rate of α(A/s). Then find induced emf in the loop.

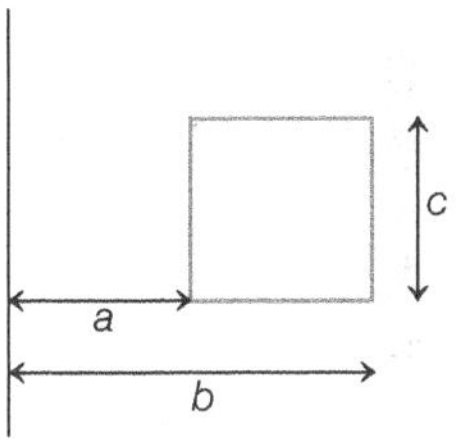

Note *(i) If number of turns N or N_1 or N_2 are one then L or M may be defined as, flux per unit current*

$$\boxed{L \text{ or } M = \frac{\phi}{i}}$$

In case of L, current i is in its own wire and in case of M this is the current of neighbouring coil.

(ii) The word one sided is used because current in straight wire will produce flux in the loop. Or change in current in the wire will produce induced emf (or current) in the loop. But the reverse is not possible.

Sol. (a) $$M = \frac{\text{Flux through the loop}}{\text{Current in wire}} = \frac{\phi}{i}$$

$$= \frac{\int d\phi}{i} = \frac{\int BdS \cos 0^\circ}{i} = \frac{\int_a^b \left(\frac{\mu_0}{2\pi}\frac{i}{x}\right) cdx}{i}$$

$$= \frac{\mu_0 c}{2\pi} \ln\left(\frac{b}{a}\right)$$ **Ans.**

(b) $$|e| = M\left|\frac{di}{dt}\right|_{\text{wire}}$$

$$= \frac{\mu_0 \alpha c}{2\pi} \ln\left(\frac{b}{a}\right)$$ **Ans.**

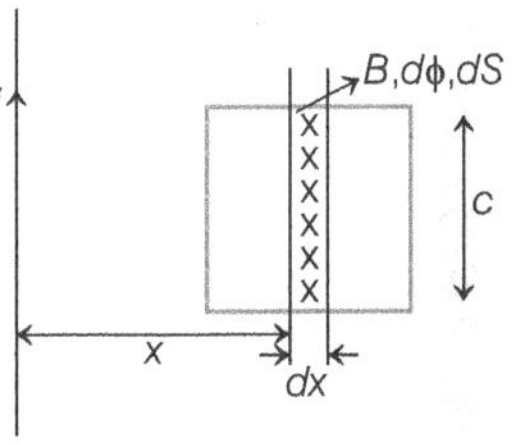

Note *For detailed solution you can also see, Ex. 1 of Type 1 in Module 2.*

MODULE 4 Exercise

Single Correct Option

1. When the number of turns in the two circular coils closely wound are doubled (in both), their mutual inductance becomes
   (a) four times (b) two times
   (c) remains same (d) sixteen times

2. A pair of coils of turns n_1 and n_2 are kept close together. Current passing through the first is reduced at a rate of r, and emf 3 mV is developed across the other coil. If the second coil carries current which is then reduced at the rate $2r$, the emf produced across the first coil will be
   (a) $\frac{6n_1}{n_2}$ mV (b) $\frac{6n_2}{n_1}$ mV
   (c) 6 mV (d) 3/2 mV

3. Two coils have a mutual inductance of 0.005 H. The current changes in the first coil according to the equation $I = I_0 \sin \omega t$, where $I_0 = 10$ A and $\omega = 100\pi$ rad/s. The maximum value of emf (in volt) in the second coil is
   (a) 2π (b) 5π
   (c) π (d) 4π

4. A circuit consists of a circular loop of radius R kept in the plane of paper and an infinitely long current carrying wire kept perpendicular to the plane of paper and passing through the centre of loop. The mutual inductance of wire and loop will be

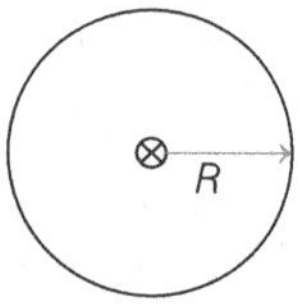

   (a) $\frac{\mu_0 \pi R}{2}$ (b) 0 (c) $\mu_0 \pi R^2$ (d) $\frac{\mu_0 R^2}{2}$

5. A small square loop of wire of side l is placed inside a large circular loop of radius r. The loops are coplanar and their centres coincide. The mutual inductance of the system is proportional to
   (a) $\frac{l^2}{r}$ (b) $\frac{l^2}{r^2}$ (c) $\frac{r}{l^2}$ (d) $\frac{r^2}{l}$

6. What is the mutual inductance of a two loop system as shown with centre separation l?

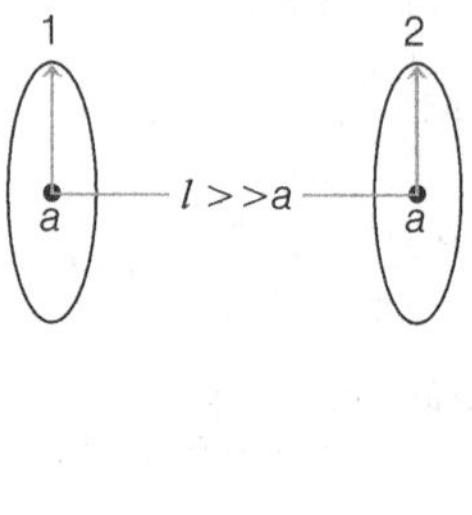

   (a) $\frac{\mu_0 \pi a^4}{8l^3}$ (b) $\frac{\mu_0 \pi a^4}{4l^3}$ (c) $\frac{\mu_0 \pi a^4}{6l^3}$ (d) $\frac{\mu_0 a^4 \pi}{2l^3}$

7. Find the mutual inductance in the arrangement, when a small circular loop of wire of radius R is placed inside a large square loop of wire of side L ($L >> R$). The loops are coplanar and their centres coincide **(2023 Main)**

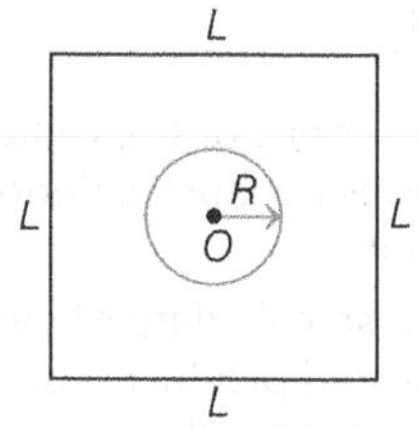

   (a) $M = \frac{2\sqrt{2}\mu_0 R^2}{L}$ (b) $M = \frac{2\sqrt{2}\mu_0 R}{L^2}$
   (c) $M = \frac{\sqrt{2}\mu_0 R}{L^2}$ (d) $M = \frac{\sqrt{2}\mu_0 R^2}{L}$

8. Two coils P and Q are separated by some distance. When a current of 3 A flows through coil P, a magnetic flux of 10^{-3} Wb passes through Q. No current is passed through Q. When no current passes through P and a current of 2 A passes through Q, the flux through P is **(2019 Main)**
   (a) 6.67×10^{-3} Wb (b) 6.67×10^{-4} Wb
   (c) 3.67×10^{-3} Wb (d) 3.67×10^{-4} Wb

9. Two circular coils can be arranged in any of the three situations shown in the figure. Their mutual inductance will be **(2001 JEE)**

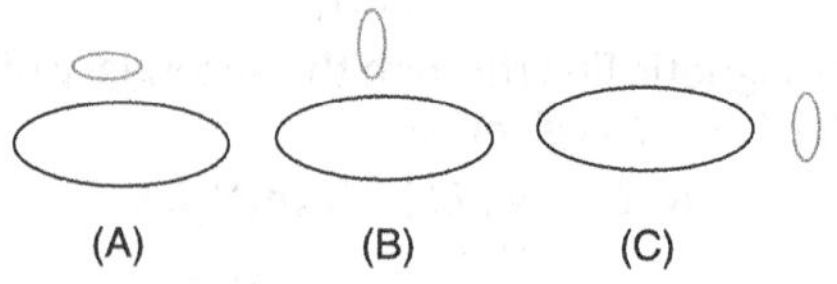

   (a) maximum in situation (A).
   (b) maximum in situation (B).
   (c) maximum in situation (C).
   (d) the same in all situations.

Subjective Questions

10. Calculate the mutual inductance between two coils when a current of 4 A changes to 12 A in 0.5 s in primary and induces an emf of 50 mV in the secondary. Also calculate the induced emf in the secondary if current in the primary changes from 3 A to 9 A is 0.02 s.

11. A primary coil has 600 turns which produces 5×10^{-3} Wb / turn of flux when 3 A current flows in this coil. This produced 6×10^{-3} Wb/turn in 1000 turns secondary coil.

When the switch is opened, the current drops to zero in 0.2 s in primary. Find
(a) mutual inductance,
(b) the induced emf in the secondary,
(c) the self-inductance of the primary coil.

12. Two coils have mutual inductance $M = 3.25 \times 10^{-4}$ H. The current i_1 in the first coil increases at a uniform rate of 830 A/s.
(a) What is the magnitude of the induced emf in the second coil? Is it constant?
(b) Suppose that the current described is in the second coil rather than the first. What is the induced emf in the first coil?

Integer Type Question

13. Two concentric circular coils with radii 1 cm and 1000 cm, and number of turns 10 and 200 respectively are placed coaxially with centers coinciding. The mutual inductance of this arrangement will be .... $\times 10^{-8}$ H. (Take, $\pi^2 = 10$)
(2023 Main)

Answers

1. (a) **2.** (c) **3.** (b) **4.** (b) **5.** (a) **6.** (d) **7.** (a) **8.** (b) **9.** (a)
10. 3.125 mH, 0.9375 V **11.** (a) 2 H (b) 30 V (c) 1 H **12.** (a) 0.27 V, Yes (b) 0.27 V **13.** (4)

See hints on pages 549-550.

MODULE 5

Growth and Decay of Current in an *L*-*R* Circuit

Growth of Current

If a battery of emf E is connected with a resistance R then current $i_0 = \frac{E}{R}$ is developed in the circuit immediately. If there is an inductor in the above circuit then the growth of current is exponential.

Let us consider a circuit consisting of a battery of emf E, a coil of self-inductance L and a resistor R.

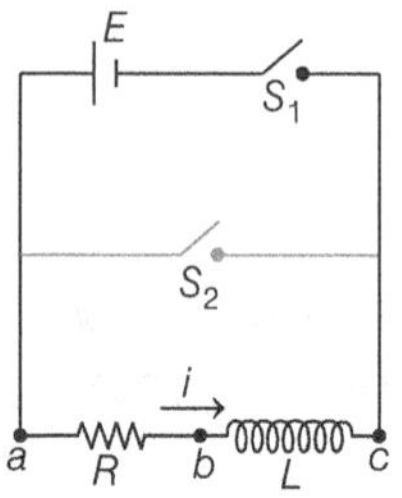

The resistor R may be a separate circuit element, or it may be the resistance of the inductor windings. By closing switch S_1, we connect R and L in series with constant emf E. Let i be the current at some time t after switch S_1 is closed and di/dt be its rate of increase at that time. Applying Kirchhoff's loop rule starting at the negative terminal and proceeding counterclockwise around the loop

$$E - V_{ab} - V_{bc} = 0 \quad \text{or} \quad E - iR - L\frac{di}{dt} = 0$$

$\therefore$
$$\int_0^i \frac{di}{E - iR} = \int_0^t \frac{dt}{L}$$

or
$$i = \frac{E}{R}\left(1 - e^{-\frac{Rt}{L}}\right)$$

By letting $E/R = i_0$ and $L/R = \tau_L$, the above expression reduces to

$$\boxed{i = i_0\,(1 - e^{-t/\tau_L})}$$

Here, $i_0 = E/R$ is the current at $t = \infty$. It is also called the steady state current or the maximum current in the circuit.

And $\tau_L = \frac{L}{R}$ is called **time constant** of the *L*-*R* circuit. At a time equal to one time constant the current has risen to $(1 - 1/e)$ or about 63% of its final value i_0.

The *i*-*t* graph is as shown in figure.

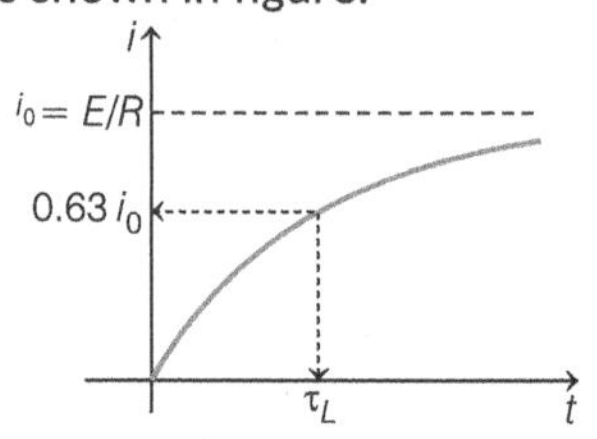

Note that the final current i_0 does not depend on the inductance L, it is the same as it would be if the resistance R alone were connected to the source with emf E.

Let us have an insight into the behaviour of an L-R circuit from energy considerations.

The instantaneous rate at which the source delivers energy to the circuit $(P = Ei)$ is equal to the instantaneous rate at which energy is dissipated in the resistor $(= i^2R)$ plus the rate at which energy is stored in the inductor $\left(= iV_{bc} = Li\frac{di}{dt}\right)$ or

$$\frac{d}{dt}\left(\frac{1}{2}Li^2\right) = Li\frac{di}{dt}.$$

Thus, $$\boxed{Ei = i^2R + Li\frac{di}{dt}}$$

Decay of Current

Now suppose switch S_1 in the circuit shown in figure has been closed for a long time and that the current has reached its steady state value i_0. Resetting our stopwatch to redefine the initial time we close switch S_2 at time $t = 0$ and at the same time we should open the switch S_1 to by pass the battery.

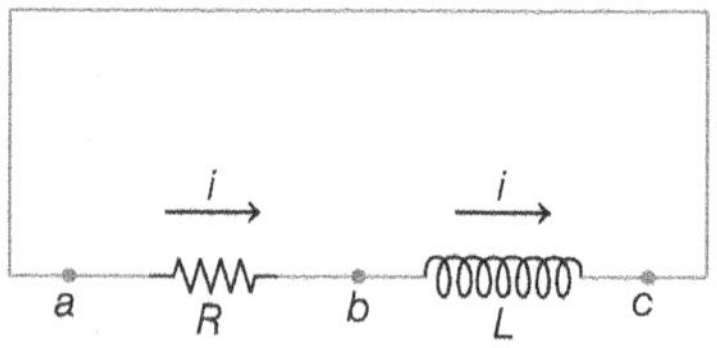

The current through L and R does not instantaneously go to zero but decays exponentially. To apply Kirchhoff's loop rule and to find current in the circuit at time t, let us draw the circuit once more.

Applying loop rule, we have

$$(V_a - V_b) + (V_b - V_c) = 0 \qquad (\text{as } V_a = V_c)$$

or
$$-iR - L\frac{di}{dt} = 0$$

$\therefore$
$$\frac{di}{i} = -\frac{R}{L}dt$$

$\therefore$
$$\int_{i_0}^{i} \frac{di}{i} = -\frac{R}{L}\int_0^t dt$$

$\therefore$
$$\boxed{i = i_0 e^{-t/\tau_L}}$$

where, τ_L = time constant $= \frac{L}{R}$ is the time for current to decrease to $1/e$ times or about 37% of its original value.

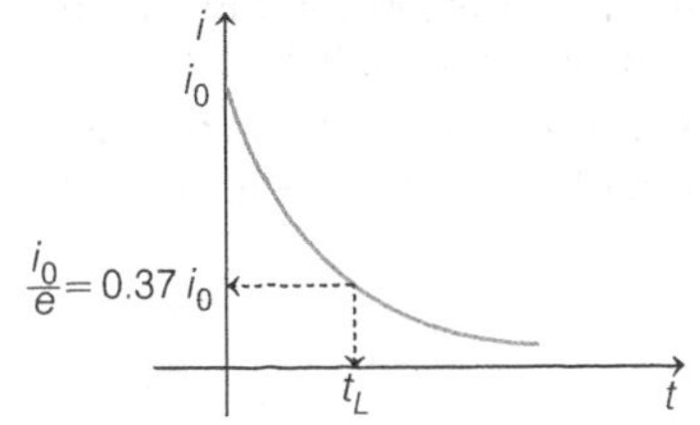

The i-t graph is as shown in above figure.

The energy that is needed to maintain the current during this decay is provided by energy stored in the magnetic field. Thus, the rate at which energy is dissipated in the resistor = rate at which the stored energy decreases in magnetic field of inductor

or $$i^2R = -\frac{dU}{dt} = -\frac{d}{dt}\left(\frac{1}{2}Li^2\right) = Li\left(-\frac{di}{dt}\right)$$

or $$i^2R = Li\left(\frac{-di}{dt}\right)$$

Example 1 A coil of resistance 20 Ω and inductance 0.5 H is connected to a DC 200 V supply. Calculate the rate of increase of current

(a) at the instant of closing the switch and
(b) after one time constant.
(c) Find the steady state current in the circuit.

Sol. (a) This is the case of growth of current in an L-R circuit. Hence, current at time t is given by

$$i = i_0\,(1 - e^{-t/\tau_L})$$

Rate of increase of current,

$$\frac{di}{dt} = \frac{i_0}{\tau_L}e^{-t/\tau_L}$$

At $t = 0$, $$\frac{di}{dt} = \frac{i_0}{\tau_L} = \frac{E/R}{L/R} = \frac{E}{L}$$

Substituting the value, we have

$$\frac{di}{dt} = \frac{200}{0.5} = 400 \text{ A/s}$$ **Ans.**

(b) At $t = \tau_L$,

$$\frac{di}{dt} = (400)\,e^{-1} = (0.37)\,(400)$$
$$= 148 \text{ A/s}$$ **Ans.**

(c) The steady state current in the circuit,

$$i_0 = \frac{E}{R} = \frac{200}{20} = 10 \text{ A}$$ **Ans.**

Example 2 A 5 H inductor is placed in series with a 10 Ω resistor. An emf of 5 V being suddenly applied to the combination. Using these values prove the principle of conservation of energy, for time equal to the time constant.

Sol. This is growth of current in L-R circuit.

$$\therefore \quad i = i_0\,(1 - e^{-t/\tau_L})$$

Here, $$i_0 = \frac{E}{R}$$

and $$\tau_L = \frac{L}{R}$$

After one time constant ($t = \tau_L$), current in the circuit is

$$i = \frac{E}{R}\left(1 - \frac{1}{e}\right)$$
$$= \frac{5}{10}\left(1 - \frac{1}{e}\right) = 0.316 \text{ A}$$

The rate at which the energy is delivered by the battery is

$$P_1 = Ei = (5)\,(0.316) = 1.58 \text{ W} \quad \text{...(i)}$$

At this time rate by which energy is dissipated in the resistor is

$$P_2 = i^2R = (0.316)^2\,(10) = 0.998 \text{ W} \quad \text{...(ii)}$$

The rate at which energy is stored in the inductor is

$$P_3 = \frac{d}{dt}\left(\frac{1}{2}Li^2\right) = Li\left(\frac{di}{dt}\right)$$

Here, $$\frac{di}{dt} = \frac{i_0}{\tau_L}e^{-1} = \frac{E}{eL} \quad \text{(after one time constant)}$$

Substituting the values, we get

$$P_3 = (L)\,(i)\left(\frac{E}{eL}\right) = \frac{Ei}{e}$$
$$= \frac{5 \times 0.316}{2.718} = 0.582 \text{ W} \quad \text{...(iii)}$$

From Eqs. (i), (ii) and (iii), we can see that,

$$P_1 = P_2 + P_3$$

It is the same as required by the principle of conservation of energy.

INTRODUCTORY Exercise

1. Show that $\frac{L}{R}$ has units of time.
2. A coil of inductance 2 H and resistance 10 Ω are in a series circuit with an open key and a cell of constant 100 V with negligible resistance. At time $t = 0$, the key is closed. Find
   (a) the time constant of the circuit.
   (b) the maximum steady current in the circuit.
   (c) the current in the circuit at $t = 1$ s.
3. In the simple L-R circuit, can the emf induced across the inductor ever be greater than the emf of the battery used to produce the current?

Answers

2. (a) 0.2 s (b) 10A (c) 9.93A **3.** No

See hints on pages 550-551.

Typed Problems

Type 1. *Problems of L-R circuit in more than one loop.*

Concept

At time $t = 0$, when there is zero current in the circuit, an inductor offers infinite resistance and at $t = \infty$, when steady state is reached an ideal inductor (of zero resistance) offers zero resistance.

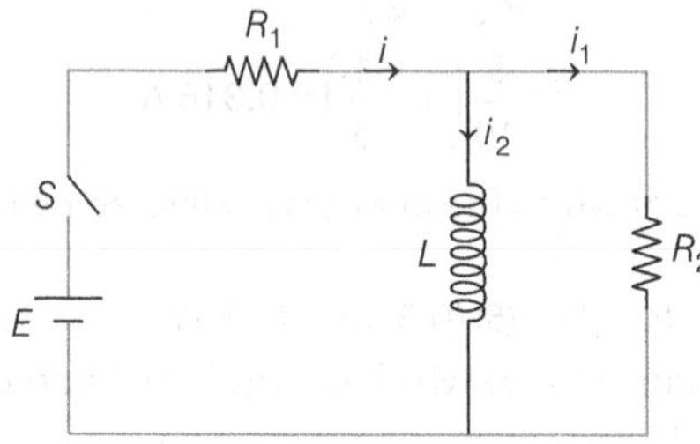

Thus, in the circuit shown, if switch S is closed at time $t = 0$, then

$$i_2 = 0$$

and $$i = i_1 = \frac{E}{R_1 + R_2}$$

as initially the inductor offers infinite resistance

and at $t = \infty$,

$$i_1 = 0, \quad \text{while} \quad i = i_2 = \frac{E}{R_1}$$

as in steady state the inductor offers zero resistance.

Ex. 1 For the circuit shown in figure, $E = 50$ V, $R_1 = 10\,\Omega$, $R_2 = 20\,\Omega$, $R_3 = 30\,\Omega$ and $L = 2.0$ mH. Find the current through R_1 and R_2.

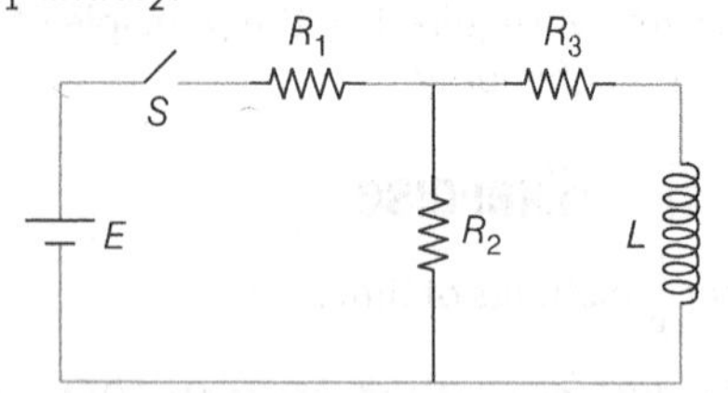

(a) Immediately after switch S is closed.
(b) A long time after S is closed.
(c) Immediately after S is reopened.
(d) A long time after S is reopened.

Sol. (a) Resistance offered by inductor immediately after switch is closed will be infinite. Therefore, current through R_3 will be zero and

$$\text{current through } R_1 = \text{current through } R_2 = \frac{E}{R_1 + R_2}$$

$$= \frac{50}{10 + 20} = \frac{5}{3}\,\text{A} \qquad \textbf{Ans.}$$

(b) After long time of closing the switch, resistance offered by inductor will be zero.

In that case R_2 and R_3 are in parallel, and the resultant of these two is then in series with R_1. Hence,

$$R_{net} = R_1 + \frac{R_2 R_3}{R_2 + R_3}$$

$$= 10 + \frac{(20)\,(30)}{20 + 30} = 22\,\Omega$$

Current through the battery (or through R_1)

$$= \frac{E}{R_{net}} = \frac{50}{22}\,\text{A} \qquad \textbf{Ans.}$$

This current will distribute in R_2 and R_3 in inverse ratio of resistance. Hence,

$$\text{Current through } R_2 = \left(\frac{50}{22}\right)\left(\frac{R_3}{R_2 + R_3}\right)$$

$$= \left(\frac{50}{22}\right)\left(\frac{30}{30 + 20}\right) = \frac{15}{11}\,\text{A} \qquad \textbf{Ans.}$$

(c) Immediately after switch is reopened, the current through R_1 will become zero.

But current through R_2 will be equal to the steady state current through R_3, which is equal to,

$$\left(\frac{50}{22} - \frac{15}{11}\right)\text{A} = 0.91\,\text{A} \qquad \textbf{Ans.}$$

(d) A long after S is reopened, current through all resistors will be zero.

Ex. 2 An inductor of inductance $L = 400$ mH and resistors of resistances $R_1 = 2\Omega$ and $R_2 = 2\Omega$ are connected to a battery of emf $E = 12$ V as shown in the figure. The internal resistance of the battery is negligible. The switch S is closed at time $t = 0$.

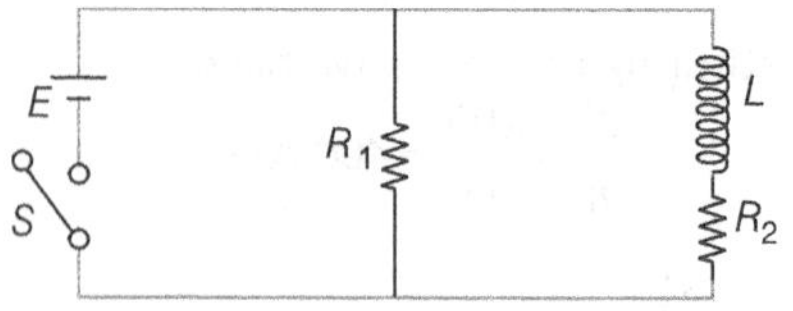

What is the potential drop across L as a function of time? After the steady state is reached, the switch is opened. What is the direction and the magnitude of current through R_1 as a function of time? **(2001 JEE)**

Sol. (a) Given, $R_1 = R_2 = 2\,\Omega$, $E = 12$ V and $L = 400$ mH $= 0.4$ H.

Two parts of the circuit are in parallel with the applied battery.

So, the given circuit can be broken as :

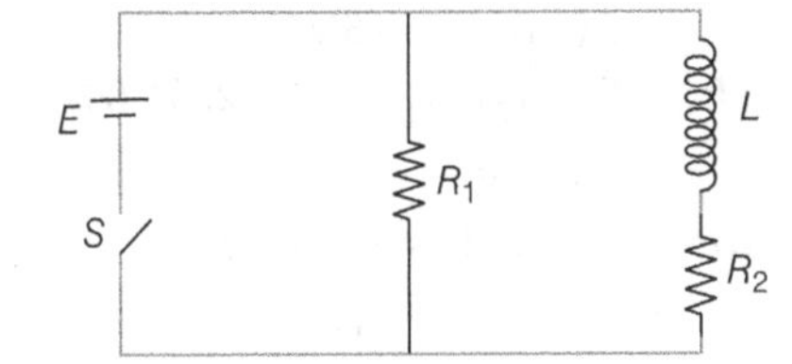

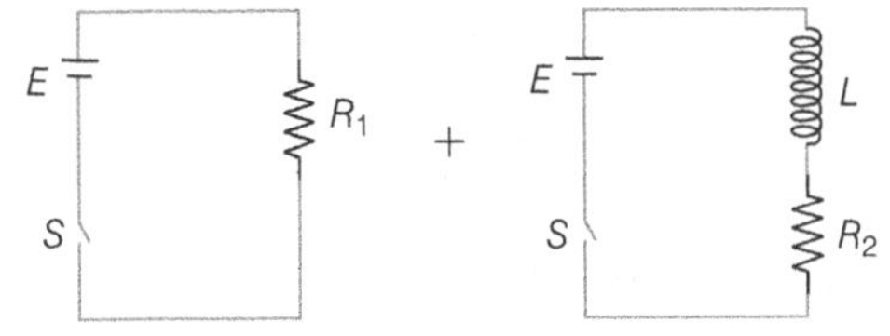

(a) (b)

Now refer Fig. (b)

This is a simple *L-R* circuit, whose time constant

$$\tau_L = L/R_2 = \frac{0.4}{2} = 0.2\text{ s}$$

and steady state current

$$i_0 = \frac{E}{R_2} = \frac{12}{2} = 6\text{ A}$$

Therefore, if switch *S* is closed at time $t = 0$, then current in the circuit at any time *t* will be given by

$$i(t) = i_0\,(1 - e^{-t/\tau_L})$$

$$i\,(t) = 6(1 - e^{-t/0.2})$$

$$= 6(1 - e^{-5t}) = i \qquad \text{(say)}$$

Therefore, potential drop across *L* at any time *t* is

$$V = \left|L\frac{di}{dt}\right| = L(30e^{-5t}) = (0.4)(30)e^{-5t}$$

or $V = 12e^{-5t}$ volt **Ans.**

(b) The steady state current in *L* or R_2 is

$$i_0 = 6\text{ A}$$

Now, as soon as the switch is opened, current in R_1 is reduced to zero immediately. But in *L* and R_2 it decreases exponentially. The situation is as follows :

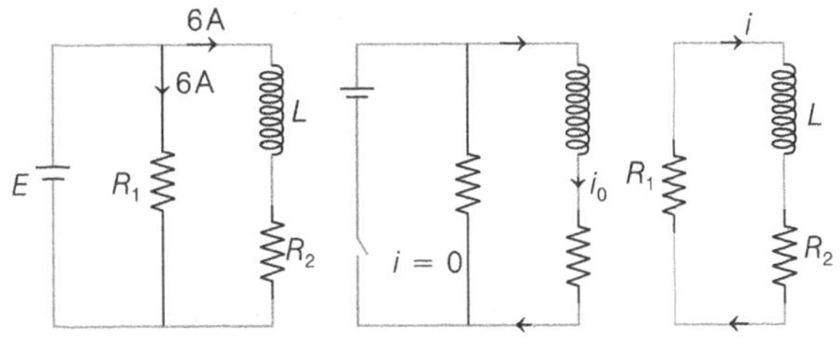

Steady state condition (c) | $t = 0$ *S* is open (d) | $t = t$ (e)

Refer figure (e)

Time constant of this circuit would be

$$\tau_L' = \frac{L}{R_1 + R_2} = \frac{0.4}{(2+2)} = 0.1\text{ s}$$

$\therefore$ Current through R_1 at any time *t* is

$$i = i_0 e^{-t/\tau_L'} = 6e^{-t/0.1}$$

or $i = 6e^{-10t}$ A

Direction of current in R_1 is as shown in figure or clockwise.

Ex. 3 A solenoid has an inductance of 10 H and a resistance of 2 Ω. It is connected to a 10 V battery. How long will it take for the magnetic energy to reach 1/4 of its maximum value? **(1996 JEE)**

Sol. $U = \frac{1}{2}Li^2$, i.e. $U \propto i^2$

U will reach $\frac{1}{4}$th of its maximum value when current is reached half of its maximum value. In *L-R* circuit, equation of current growth is written as

$$i = i_0(1 - e^{-t/\tau_L})$$

Here, i_0 = Maximum value of current

τ_L = Time constant = L/R

$$\tau_L = \frac{10\text{ H}}{2\Omega} = 5\text{ s}$$

Therefore, $i = \frac{i_0}{2} = i_0(1 - e^{-t/5})$

or $\frac{1}{2} = 1 - e^{-t/5}$ or $e^{-t/5} = \frac{1}{2}$

or $-t/5 = \ln\left(\frac{1}{2}\right)$ or $t/5 = \ln(2) = 0.693$

$\therefore$ $t = (5)(0.693)$ or $t = 3.465$ s **Ans.**

Ex. 4 A circuit containing a two position switch *S* is shown in figure. **(1991 JEE)**

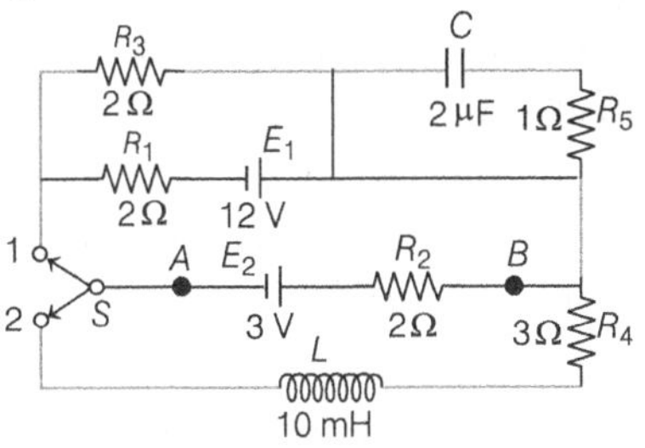

(a) The switch *S* is in position 1. Find the potential difference $V_A - V_B$ and the rate of production of joule heat in R_1.

(b) If now the switch *S* is put in position 2 at $t = 0$. Find (i) steady current in R_4 and (ii) the time when current in R_4 is half the steady value. Also calculate the energy stored in the inductor *L* at that time.

Sol. (a) In steady state, no current will flow through capacitor.

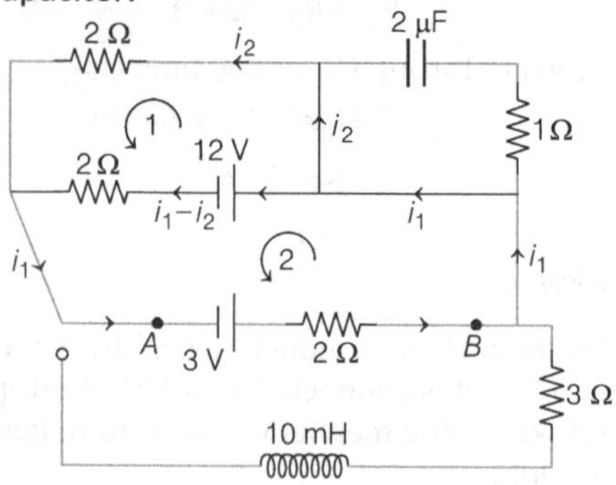

Applying Kirchhoff's second law in loop 1,

$$-2i_2 + 2(i_1 - i_2) + 12 = 0$$

∴ $2i_1 - 4i_2 = -12$

or $i_1 - 2i_2 = -6$ ...(i)

Applying Kirchhoff's second law in loop 2,

$$-12 - 2(i_1 - i_2) + 3 - 2i_1 = 0$$

∴ $4i_1 - 2i_2 = -9$ ...(ii)

Solving Eqs. (i) and (ii), we get

$$i_2 = 2.5 \text{ A} \quad \text{and} \quad i_1 = -1 \text{ A}$$

Now, $V_A + 3 - 2i_1 = V_B$

or $V_A - V_B = 2i_1 - 3$

$$= 2(-1) - 3 = -5 \text{ V}$$ **Ans.**

$$P_{R_1} = (i_1 - i_2)^2 R_1 = (-1 - 2.5)^2 (2)$$

$$= 24.5 \text{ W}$$ **Ans.**

(b) **In position 2** Circuit is as below

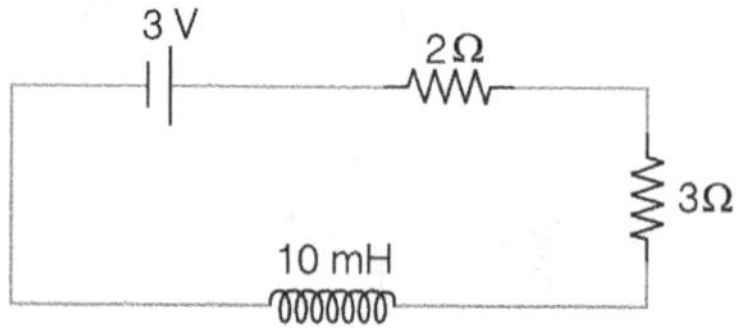

Steady current in R_4,

$$i_0 = \frac{3}{3+2} = 0.6 \text{ A}$$

Time when current in R_4 is half the steady value,

$$i = i_0(1 - e^{-t/\tau_L})$$

$i = i_0/2$ at $t = t_{1/2}$, where

$$t_{1/2} = \tau_L (\ln 2) = \frac{L}{R} \ln(2) = \frac{(10 \times 10^{-3})}{5} \ln(2)$$

$$= 1.386 \times 10^{-3} \text{ s}$$ **Ans.**

$$U = \frac{1}{2} Li^2 = \frac{1}{2}(10 \times 10^{-3})(0.3)^2$$

$$= 4.5 \times 10^{-4} \text{ J}$$ **Ans.**

Miscellaneous Examples

Ex. 5 A sensitive electronic device of resistance 175 Ω is to be connected to a source of emf by a switch. The device is designed to operate with a current of 36 mA, but to avoid damage to the device, the current can rise to no more than 4.9 mA in the first 58 μs after the switch is closed. To protect the device it is connected in series with an inductor.

(a) What emf must the source have?

(b) What inductance is required?

(c) What is the time constant?

Sol. (a) Given, $R = 175\ \Omega$ and peak value current

$$i_0 = 36 \times 10^{-3} \text{ A}$$

Applied voltage,

$$V = i_0 R = (175)(36 \times 10^{-3}) \text{ volt}$$

$$= 6.3 \text{ V}$$ **Ans.**

(b) From the relation,

$$i = i_0 (1 - e^{-t/\tau_L})$$

We have, $(4.9) = (36)[1 - e^{-t/\tau_L}]$

or $e^{-t/\tau_L} = 1 - \frac{4.9}{36} = 0.864$

∴ $\frac{t}{\tau_L} = -\ln(0.864) = 0.146$

or $\frac{t}{L/R} = 0.146$

∴ $\frac{Rt}{L} = 0.146$

or $L = \frac{Rt}{0.146} = \frac{(175)(58 \times 10^{-6})}{0.146}$

$$= 7.0 \times 10^{-2} \text{ H}$$ **Ans.**

(c) Time constant of the circuit,

$$\tau_L = \frac{L}{R} = \frac{7.0 \times 10^{-2}}{175}$$

$$= 4.0 \times 10^{-4} \text{ s}$$ **Ans.**

Ex. 6 A battery of emf E and of negligible internal resistance is connected in an L-R circuit as shown in figure. The inductor has a piece of soft iron inside it. When steady state is reached the piece of soft iron is abruptly pulled out suddenly so that the inductance of the inductor decreases to nL with $n < 1$ with battery remaining connected. Calculate

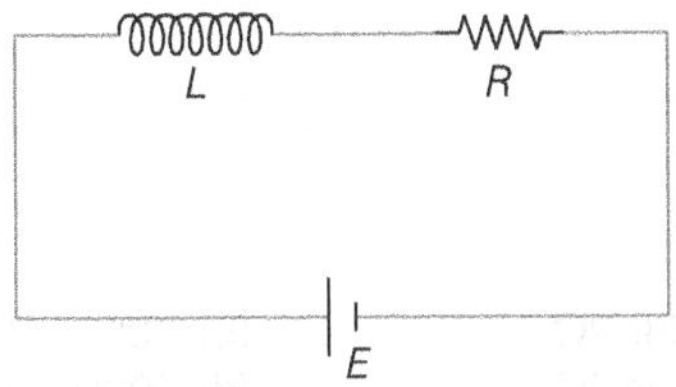

(a) current as a function of time assuming $t = 0$ at the instant when piece is pulled.

(b) the work done to pull out the piece.

(c) thermal power generated in the circuit as a function of time.

(d) power supplied by the battery as a function of time.

HOW TO PROCEED When the inductance of an inductor is abruptly changed, the flux passing through it remains constant.

$$\phi = \text{constant}$$

$$\therefore \quad Li = \text{constant} \qquad \left(L = \frac{\phi}{i}\right)$$

Sol. (a) At time $t = 0$, steady state current in the circuit is $i_0 = E/R$. Suddenly, L reduces to nL $(n < 1)$, so current in the circuit at time $t = 0$ will increase to $\frac{i_0}{n} = \frac{E}{nR}$.

Let i be the current at time t.

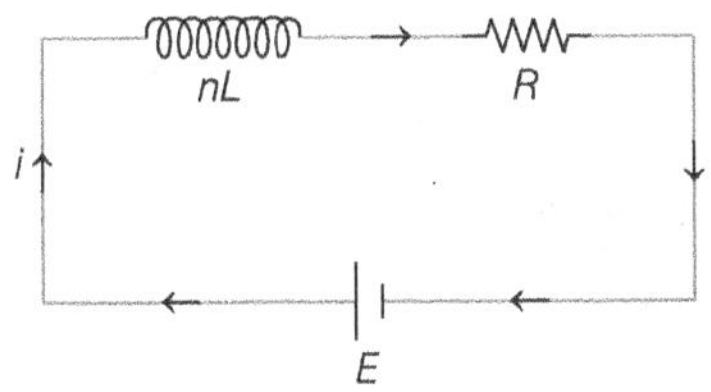

Applying Kirchhoff's loop rule, we have

$$E - nL\left(\frac{di}{dt}\right) - iR = 0$$

$$\therefore \quad \frac{di}{E - iR} = \frac{1}{nL}dt$$

$$\therefore \quad \int_{i_0/n}^{i} \frac{di}{E - iR} = \frac{1}{nL}\int_0^t dt$$

Solving this equation, we get

$$i = i_0 + \left(\frac{i_0}{n} - i_0\right)e^{-t/\tau_L} \qquad \textbf{Ans.}$$

Here, $\quad i_0 = \frac{E}{R}$

and $\quad \tau_L = \frac{nL}{R}$

From the i-t equation, we get

$$i = \frac{i_0}{n} \text{ at } t = 0$$

and $i = i_0$ at $t = \infty$

The i-t graph is as shown in figure.

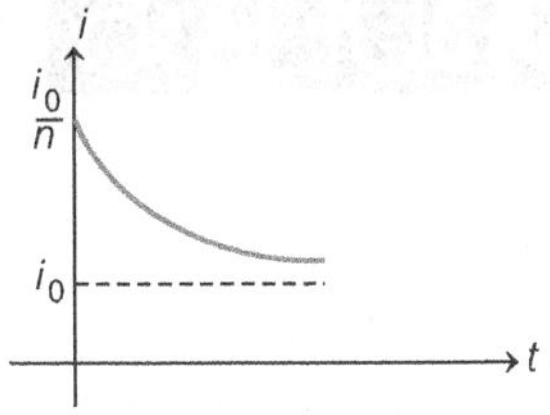

Note *At $t = 0$, current in the circuit is $\frac{i_0}{n}$. Current in the circuit in steady state will be again i_0. So, it will decrease exponentially from $\frac{i_0}{n}$ to i_0. From the i-t graph, the equation can be formed without doing any calculation.*

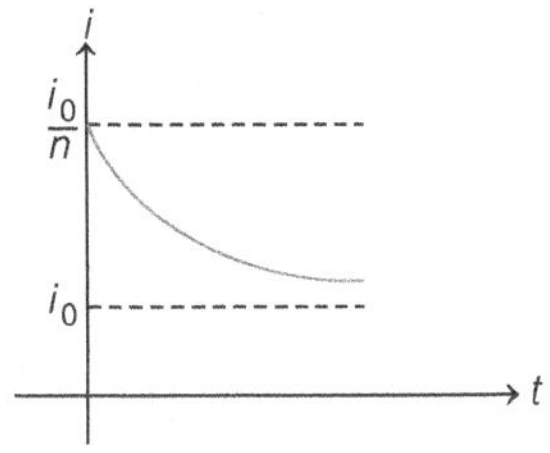

⇓

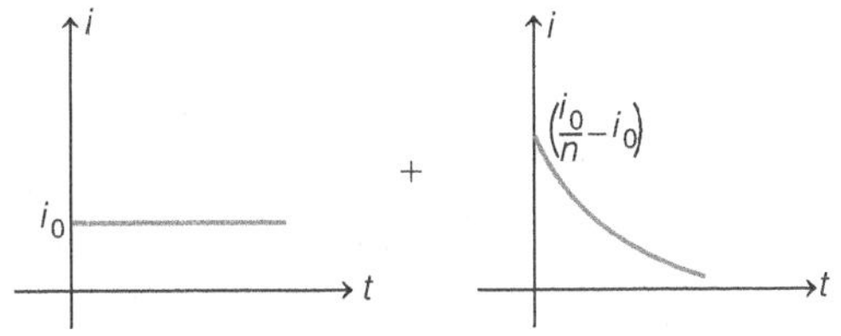

$$\therefore \quad i = i_0 + \left(\frac{i_0}{n} - i_0\right)e^{-t/\tau_L}$$

(b) Work done to pull out the piece,

$$W = U_f - U_i = \frac{1}{2}L_f i_f^2 - \frac{1}{2}L_i i_i^2$$

$$= \frac{1}{2}(nL)\left(\frac{E}{nR}\right)^2 - \frac{1}{2}(L)\left(\frac{E}{R}\right)^2$$

$$= \frac{1}{2}L\left(\frac{E}{R}\right)^2\left(\frac{1}{n} - 1\right)$$

$$= \frac{1}{2}L\left(\frac{E}{R}\right)^2\left(\frac{1 - n}{n}\right) \qquad \textbf{Ans.}$$

(c) Thermal power generated in the circuit as a function of time is

$$P_1 = i^2R \qquad \textbf{Ans.}$$

Here, i is the current calculated in part (a).

(d) Power supplied by the battery as a function of time is

$$P_2 = Ei \qquad \textbf{Ans.}$$

MODULE 5 Exercise

Assertion and Reason

See directions on page 25.

1. In the given circuit switch is closed at $t = 0$.

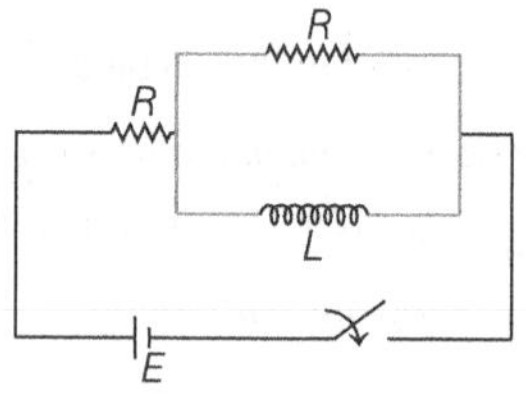

Assertion : At $t = 0$, current through battery is $\frac{E}{2R}$ and at $t = \infty$, current through battery will be $\frac{E}{R}$.

Reason : At $t = 0$, inductor will behave like open circuit and at $t = \infty$, inductor will behave like short circuit.

Single Correct Option

2. When a certain circuit consisting of a constant emf E, an inductance L and a resistance R is closed, the current in it increases with time according to curve 1. Only one parameter (E, L or R) is changed and the increase in current follows curve 2, when the circuit is closed second time. Which parameter was changed and in what direction?

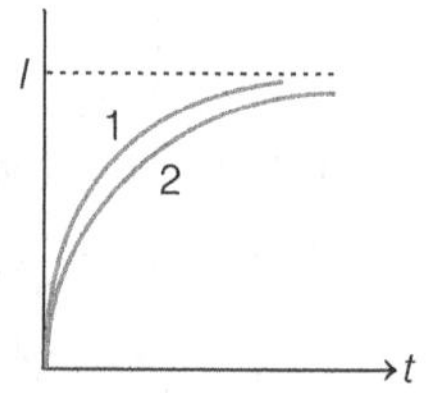

(a) L is increased (b) L is decreased
(c) R is increased (d) E is decreased

3. An emf of 15 volts is applied in a circuit containing 5 H inductance and 10 Ω resistance. The ratio of the currents at time $t = \infty$ and $t = 1$ s is

(a) $\frac{e^{1/2}}{e^{1/2} - 1}$ (b) $\frac{e^2}{e^2 - 1}$ (c) $1 - e^{-1}$ (d) e^{-1}

4. A coil has an inductance of 50 mH and a resistance of 0.3 Ω. If a 12 V emf is applied across the coil, the energy stored in the magnetic field after the current has built up to its steady state value is

(a) 40 J (b) 40 mJ (c) 20 J (d) 20 mJ

5. A constant voltage is applied to a series R-L circuit by closing the switch. The voltage across inductor ($L = 2$ H) is 20 V at $t = 0$ and drops to 5 V at 20 ms. The value of R in Ω is

(a) $100 \ln 2\,\Omega$ (b) $100(1 - \ln 2)\,\Omega$
(c) $100 \ln 4\,\Omega$ (d) $100(1 - \ln 4)\,\Omega$

6. In figure, final value of current in 10 Ω resistor, when plug of key K is inserted is

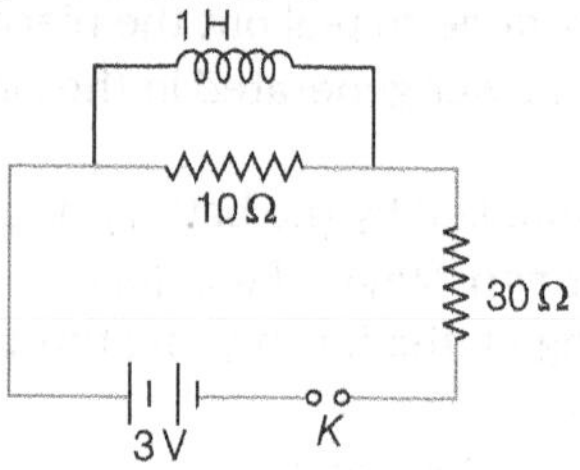

(a) $\frac{3}{10}$ A (b) $\frac{3}{20}$ A (c) $\frac{3}{11}$ A (d) zero

7. In the circuit shown in figure, $L = 10$ H, $R = 5\,\Omega$, $E = 15$ V. The switch S is closed at $t = 0$. At $t = 2$ s, the current in the circuit is

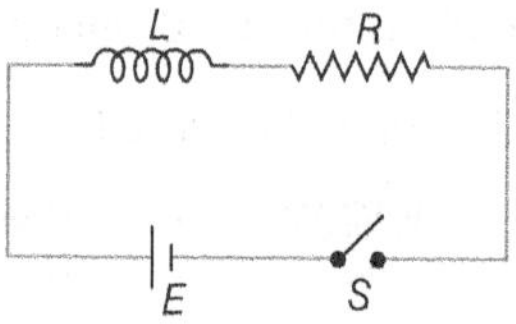

(a) $3\left(1 - \frac{1}{e}\right)$ A (b) $3\left(1 - \frac{1}{e^2}\right)$ A
(c) $3\left(\frac{1}{e}\right)$ A (d) $3\left(\frac{1}{e^2}\right)$ A

8. In the circuit shown in the figure, what is the value of I_1 just after pressing the key K ?

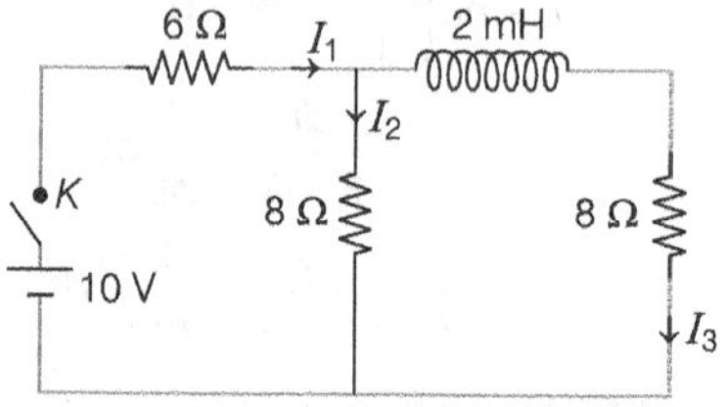

(a) 5/7 A (b) 5/11 A
(c) 1 A (d) None of these

9. In the circuit shown, what is the energy stored in the coil at steady state?

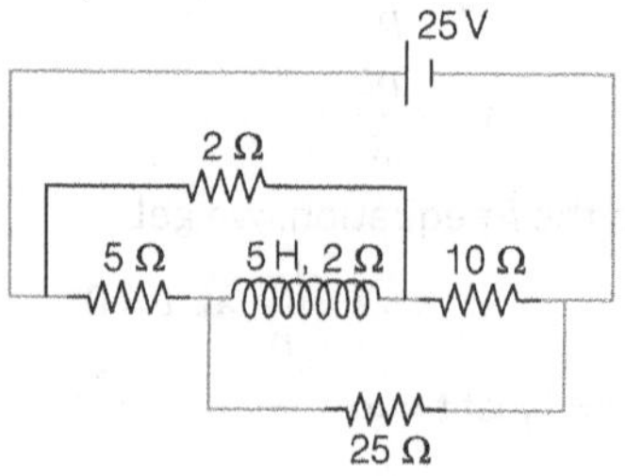

(a) 21.3 J (b) 42.6 J (c) Zero (d) 213 J

10. Figure shows an L-R circuit. When the switch S is closed, the currents through resistors R_1, R_2 and R_3 are I_1, I_2 and I_3 respectively. The values of I_1, I_2 and I_3 at $t = 0$ s are

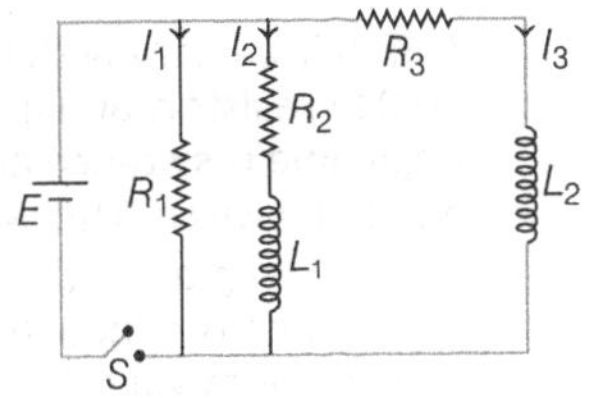

(a) $I_1 = I_2 = I_3 = 0$

(b) $I_1 = \frac{E}{R_1}, I_2 = I_3 = 0$

(c) $I_1 = 0, I_2 = \frac{E}{R_2}, I_3 = \frac{E}{R_3}$

(d) $I_1 = \frac{E}{R_1}, I_2 = \frac{E}{R_2 + L_1}, I_3 = \frac{E}{R_3 + L_3}$

11. The figure shows an L-R circuit, the time constant for the circuit is

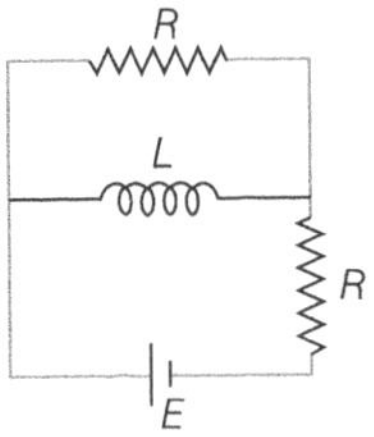

(a) $\frac{L}{2R}$ (b) $\frac{2L}{R}$

(c) $\frac{2R}{L}$ (d) $\frac{R}{2L}$

12. Switch S is closed at $t = 0$, in the circuit shown. The change in flux in the inductor ($L = 500$ mH) from $t = 0$ to an instant when it reaches steady state is

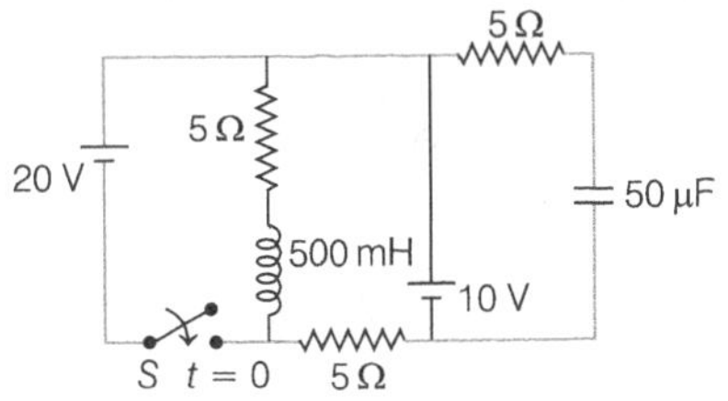

(a) 2 Wb (b) 1.5 Wb
(c) 0 Wb (d) None of these

13. When a choke coil carrying a steady current is short-circuited, the current in it decreases to $\beta (< 1)$ times its initial value in a time T. The time constant of the choke coil is

(a) $\frac{T}{\beta}$ (b) $\frac{T}{\ln\left(\frac{1}{\beta}\right)}$ (c) $\frac{T}{\ln \beta}$ (d) $T \ln \beta$

14. In the steady state condition, the rate of heat produced in a choke coil is P. The time constant of the choke coil is τ. If now the choke coil is short-circuited, then the total heat dissipated in the coil is

(a) $P\tau$ (b) $\frac{1}{2} P\tau$

(c) $\frac{P\tau}{\ln 2}$ (d) $P\tau \ln 2$

15. In figure, the switch is in the position 1 for a long time, then the switch is shifted to position 2 at $t = 0$. At this instant the value of i_1 and i_2 are

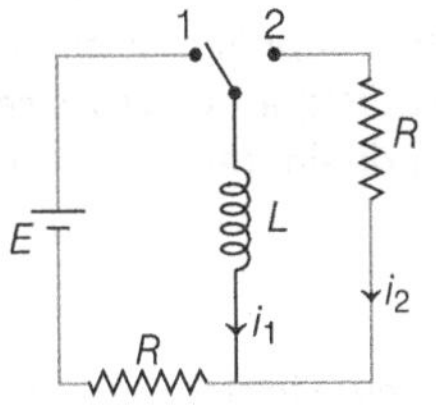

(a) $\frac{E}{R}, 0$ (b) $\frac{E}{R}, \frac{-E}{R}$

(c) $\frac{E}{2R}, \frac{-E}{2R}$ (d) None of these

16. In the circuit shown, the key (K) is closed at $t = 0$, the current through the key at the instant $t = 10^{-3} \ln 2$, is

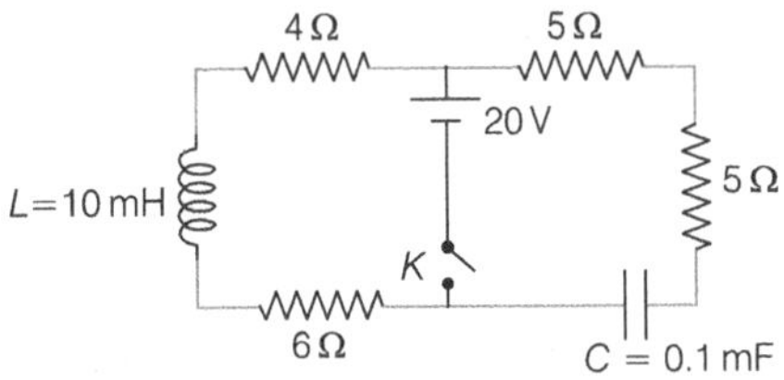

(a) 2 A (b) 8 A (c) 4 A (d) zero

17. Switch S of the circuit shown in figure is closed at $t = 0$. If e denotes the induced emf in inductor and i, the current flowing through the circuit at time t, which of the following graphs is correct?

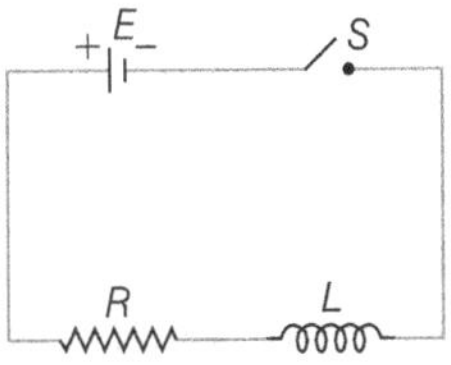

(a)

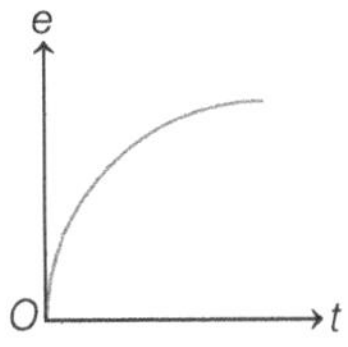

(b)

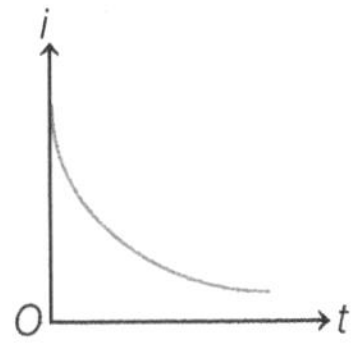

(c)

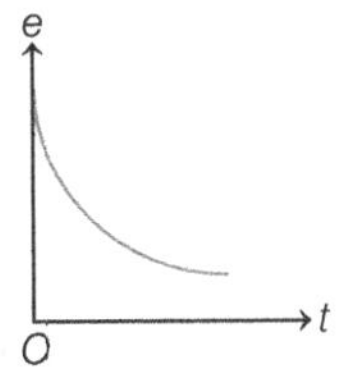

(d)

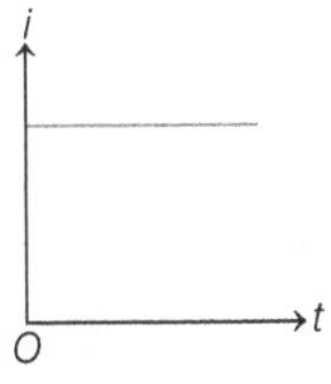

Multiple Correct Options

18. In the circuit shown in figure, circuit is closed at time $t = 0$. At time $t = \ln(2)$ second

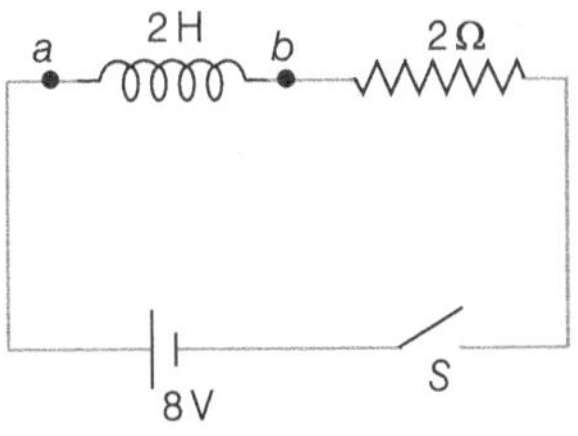

(a) rate of energy supplied by the battery is 16 J/s
(b) rate of heat dissipated across resistance is 8 J/s
(c) rate of heat dissipated across resistance is 16 J/s
(d) $V_a - V_b = 4$ V

Match the Columns

19. In the circuit shown in figure, switch is closed at time $t = 0$. Match the following two columns.

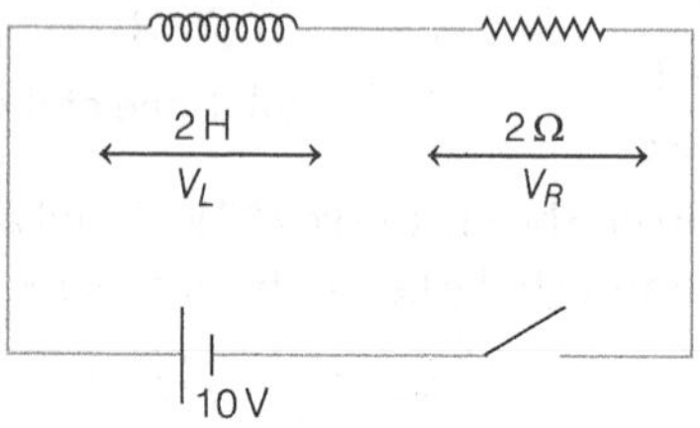

	Column I		Column II
(a)	V_L at $t = 0$	(p)	zero
(b)	V_R at $t = 0$	(q)	10 V
(c)	V_L at $t = 1$ s	(r)	$\frac{10}{e}$ V
(d)	V_R at $t = 1$ s	(s)	$\left(1 - \frac{1}{e}\right)$ 10 V

Subjective Questions

20. The switch in figure is closed at time $t = 0$. Find the current in the inductor and the current through the switch as functions of time thereafter.

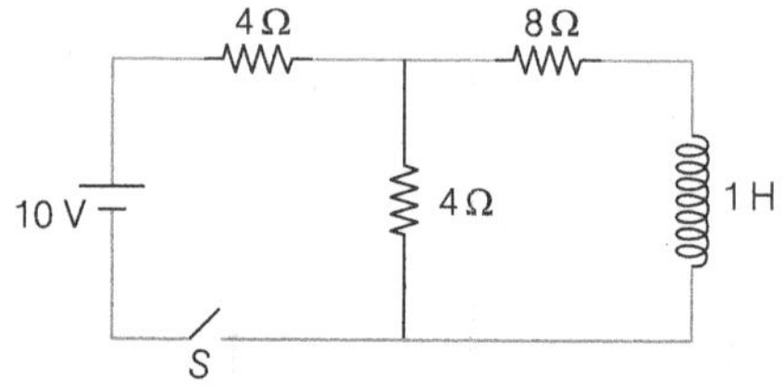

21. A coil of inductance 1 H and resistance 10 Ω is connected to a resistanceless battery of emf 50 V at time $t = 0$. Calculate the ratio of the rate at which magnetic energy is stored in the coil to the rate at which energy is supplied by the battery at $t = 0.1$ s.

22. A 3.56 H inductor is placed in series with a 12.8 Ω resistor. An emf of 3.24 V is then suddenly applied across the *R-L* combination.
(a) At 0.278s after the emf is applied what is the rate at which energy is being delivered by the battery?
(b) At 0.278 s, at what rate is energy appearing as thermal energy in the resistor?
(c) At 0.278 s, at what rate is energy being stored in the magnetic field?

23. A 35.0 V battery with negligible internal resistance, a 50.0 Ω resistor, and a 1.25 mH inductor with negligible resistance are all connected in series with an open switch. The switch is suddenly closed.
(a) How long after closing the switch will the current through the inductor reach one-half of its maximum value?
(b) How long after closing the switch will the energy stored in the inductor reach one-half of its maximum value?

24. A solenoid of inductance *L* with resistance *r* is connected in parallel to a resistance *R*. A battery of emf *E* and of negligible internal resistance is connected across the parallel combination as shown in the figure. At time $t = 0$, switch *S* is opened, calculate

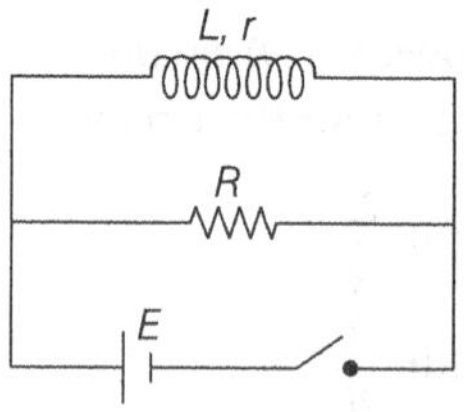

(a) current through the solenoid after the switch is opened.
(b) amount of heat generated in the solenoid

25. In the given circuit, find the current through the 5 mH inductor in steady state.

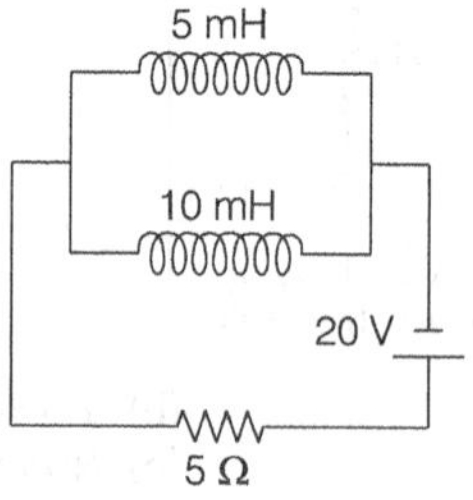

Answers

1. (a)	**2.** (a)	**3.** (b)	**4.** (a)	**5.** (c)
6. (d)	**7.** (a)	**8.** (a)	**9.** (c)	**10.** (b)
11. (b)	**12.** (b)	**13.** (b)	**14.** (b)	**15.** (b)
16. (a)	**17.** (c)	**18.** (a,b,d)		

19. (a → q ; b → p ; c → r ; d → s)
20. (0.5) $(1 - e^{-10t})$A and $[1.5 - 0.25\, e^{-10t}]$ A
21. 0.37 **22.** (a) 518 mW (b) 328 mW (c) 191 mW
23. (a) 17.3 μs (b) 30.7 μs
24. (a) $\frac{E}{r} e^{-\left(\frac{R+r}{L}\right)t}$ (b) $\frac{E^2 L}{2r(R+r)}$ **25.** $\frac{8}{3}$ A

See hints on pages 551-553.

MODULE 6

Oscillations in *L*-*C* Circuit

If a charged capacitor C is short-circuited through an inductor L, the charge and current in the circuit start oscillating simple harmonically. If the resistance of the circuit is zero, no energy is dissipated as heat.

We also assume an idealized situation in which energy is not radiated away from the circuit. With these idealizations-zero resistance and no radiation, the oscillations in the circuit persist indefinitely and the energy is transferred from the capacitor's electric field to the inductor's magnetic field and back.

The total energy associated with the circuit is constant. This is analogous to the transfer of energy in an oscillating mechanical system from potential energy to kinetic energy and back, with constant total energy.

Let us now derive an equation for the oscillations in an L-C circuit.

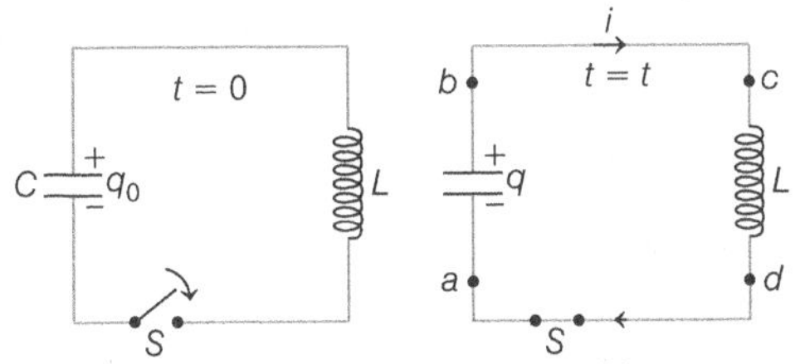

(a) (b)

Refer figure (a) A capacitor is charged to a PD, $V_0 = \dfrac{q_0}{C}$

Here, q_0 is the maximum charge on the capacitor. At time $t = 0$, it is connected to an inductor through a switch S. At time $t = 0$, switch S is closed.

Refer figure (b) When the switch is closed, the capacitor starts discharging. Let at time t charge on the capacitor is $q\,(< q_0)$ and since, it is further decreasing there is a current i in the circuit in the direction shown in figure.

Later we will see that, as the charge is oscillating there may be a situation when q will be increasing, but in that case direction of current is also reversed and the equation remains unchanged.

Applying Kirchhoff's loop law in the loop *abcda* :

$$+\frac{q}{C} - L\left(\frac{di}{dt}\right) = 0$$

or $$\frac{q}{C} = L\left(\frac{di}{dt}\right) \quad \text{...(i)}$$

Now, as the charge is decreasing,

$$i = \left(\frac{-dq}{dt}\right) \quad \text{or} \quad \frac{di}{dt} = -\frac{d^2q}{dt^2}$$

Substituting in Eq. (i), we get

$$\frac{q}{C} = -L\left(\frac{d^2q}{dt^2}\right)$$

or $$\boxed{\frac{d^2q}{dt^2} = -\left(\frac{1}{LC}\right)q} \quad \text{...(ii)}$$

This is the standard equation of simple harmonic motion $\left(\dfrac{d^2x}{dt^2} = -\omega^2 x\right)$.

Here, $$\boxed{\omega = \frac{1}{\sqrt{LC}}} \quad \text{...(iii)}$$

The general solution of Eq. (ii), is $q = q_0 \cos(\omega t \pm \phi)$

For example in our case $\phi = 0$ as $q = q_0$ at $t = 0$.

Hence, $$q = q_0 \cos \omega t \quad \text{...(iv)}$$

Thus, we can say that charge in the circuit oscillates simple harmonically with angular frequency given by Eq. (iii).

$$\therefore \quad \boxed{\omega = \frac{1}{\sqrt{LC}},\ f = \frac{\omega}{2\pi} = \frac{1}{2\pi\sqrt{LC}} \text{ and } T = \frac{1}{f} = 2\pi\sqrt{LC}}$$

The oscillations of the L-C circuit are an electromagnetic analog to the mechanical oscillations of a block-spring system.

Table below shows a comparison of oscillations of a mass-spring system and an L-C circuit.

Table

S.No.	Mass spring system	Inductor-capacitor circuit
1.	Displacement (x)	Charge (q)
2.	Velocity $(v) = \frac{dx}{dt}$	Current $i = \frac{dq}{dt}$
3.	Acceleration $(a) = \frac{dv}{dt}$	Rate of change of current $\left(= \frac{di}{dt}\right)$
4.	$\frac{d^2x}{dt^2} = -\omega^2 x$, where $\omega = \sqrt{\frac{k}{m}}$	$\frac{d^2q}{dt^2} = -\omega^2 q$, where $\omega = \frac{1}{\sqrt{LC}}$
5.	$x = A\sin(\omega t \pm \phi)$ or $x = A\cos(\omega t \pm \phi)$	$q = q_0 \sin(\omega t \pm \phi)$ or $q = q_0 \cos(\omega t \pm \phi)$
6.	$v = \frac{dx}{dt} = \pm\,\omega\sqrt{A^2 - x^2}$	$i = \frac{dq}{dt} = \pm\,\omega\sqrt{q_0^2 - q^2}$
7.	$a = \frac{dv}{dt} = -\omega^2 x$	Rate of change of current $= \frac{di}{dt} = -\omega^2 q$
8.	Kinetic energy $= \frac{1}{2}mv^2$	Magnetic energy $= \frac{1}{2}Li^2$
9.	Potential energy $= \frac{1}{2}kx^2$	Potential energy $= \frac{1}{2}\frac{q^2}{C}$
10.	$\frac{1}{2}mv^2 + \frac{1}{2}kx^2 = \text{constant} = \frac{1}{2}kA^2 = \frac{1}{2}mv_{max}^2$	$\frac{1}{2}Li^2 + \frac{1}{2}\frac{q^2}{C} = \text{constant} = \frac{1}{2}\frac{q_0^2}{C} = \frac{1}{2}Li_{max}^2$
11.	$\lvert v_{max}\rvert = A\omega$	$i_{max} = q_0\omega$
12	$\lvert a_{max}\rvert = \omega^2 A$	$\left\lvert\left(\frac{di}{dt}\right)_{max}\right\rvert = \omega^2 q_0$
13.	$\frac{1}{k}$	C
14.	m	L
15.	At mean position $x = 0$, $a = 0$, v = maximum	At mean position $q = 0$, $\frac{di}{dt} = 0$, i = maximum
16.	At extreme positions $x = \pm A$, a = maximum, $v = 0$	At extreme positions $q = \pm q_0$, $\frac{di}{dt}$ = maximum, $i = 0$

A graphical description of the energy transfer between the inductor and the capacitor in an *L-C* circuit is shown in the figure. The right side of the figure shows the analogous energy transfer in the oscillating block-spring system.

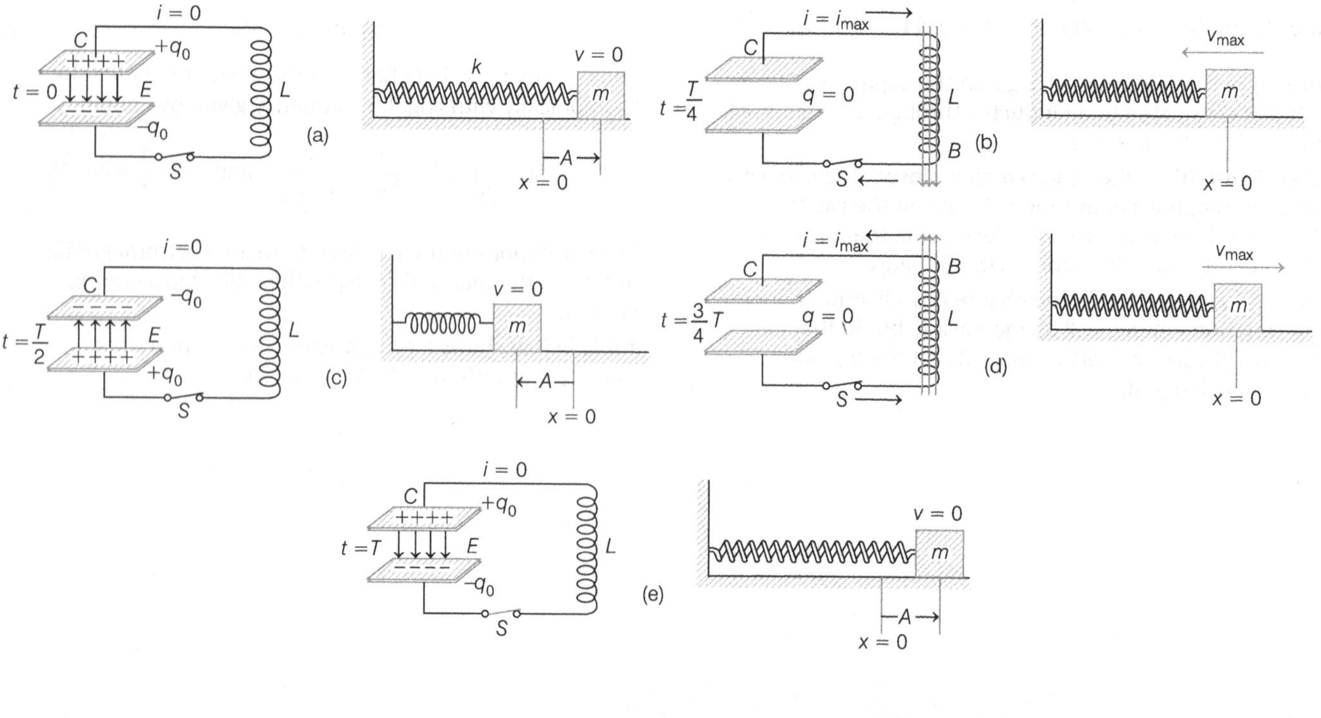

Note *In L-C oscillations, q, i and $\frac{di}{dt}$ all oscillate simple harmonically with same angular frequency ω. But the phase difference between q and i or between i and $\frac{di}{dt}$ is $\frac{\pi}{2}$, while that between i and $\frac{di}{dt}$ is π. Their amplitudes are q_0, $q_0\omega$ and $\omega^2 q_0$ respectively. Let*

$$q = q_0 \cos \omega t, \text{ then}$$

$$i = \frac{dq}{dt} = -q_0\omega \sin \omega t \text{ and}$$

$$\frac{di}{dt} = -q_0\omega^2 \cos \omega t$$

Potential energy across capacitor (U_C) and across inductor (U_L) also oscillate with double the frequency. The different graphs are as shown in figure given below.

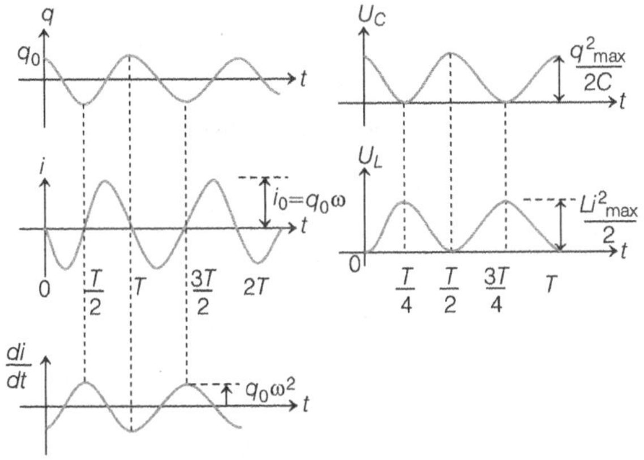

Example 1 A capacitor of capacitance 25 μF is charged to 300 V. It is then connected across a 10 mH inductor. The resistance in the circuit is negligible.

(a) Find the frequency of oscillation of the circuit.

(b) Find the potential difference across capacitor and magnitude of circuit current 1.2 ms after the inductor and capacitor are connected.

(c) Find the magnetic energy and electric energy at t = 0 and t = 1.2 ms.

Sol. (a) The frequency of oscillation of the circuit is

$$f = \frac{1}{2\pi\sqrt{LC}}$$

Substituting the given values, we have

$$f = \frac{1}{2\pi\sqrt{(10\times10^{-3})(25\times10^{-6})}} = 318.3 \text{ Hz} \quad \textbf{Ans.}$$

(b) Charge across the capacitor at time t will be

$$q = q_0 \cos \omega t \quad \text{and} \quad i = \frac{dq}{dt} = -q_0\omega \sin \omega t$$

Here, $q_0 = CV_0 = (25\times10^{-6})(300) = 7.5\times10^{-3}$ C

Now, charge in the capacitor after $t = 1.2\times10^{-3}$ s is

$$q = (7.5\times10^{-3}) \cos(2\pi\times318.3)(1.2\times10^{-3}) \text{ C}$$

$$= -5.53\times10^{-3} \text{ C}$$

∴ PD across capacitor,

$$V = \frac{|q|}{C} = \frac{5.53\times10^{-3}}{25\times10^{-6}} = 221.2 \text{ volt} \quad \textbf{Ans.}$$

The magnitude of current in the circuit at $t = 1.2\times10^{-3}$ s is

$$|i| = q_0\omega \sin \omega t$$

$$= (7.5\times10^{-3})(2\pi)(318.3) \sin(2\pi\times318.3)(1.2\times10^{-3}) \text{ A}$$

$$= 10.13 \text{ A} \quad \textbf{Ans.}$$

(c) **At t = 0** Current in the circuit is zero.

Hence, $U_L = 0$

Charge in the capacitor is maximum.

Hence, $$U_C = \frac{1}{2}\frac{q_0^2}{C}$$

or $$U_C = \frac{1}{2}\times\frac{(7.5\times10^{-3})^2}{(25\times10^{-6})}$$

$$= 1.125 \text{ J} \quad \textbf{Ans.}$$

∴ Total energy, $E = U_L + U_C = 1.125$ J

At t = 1.2 ms

$$U_L = \frac{1}{2}Li^2$$

$$= \frac{1}{2}(10\times10^{-3})(10.13)^2$$

$$= 0.513 \text{ J}$$

$$\therefore \quad U_C = E - U_L = 1.125 - 0.513$$

$$= 0.612 \text{ J} \quad \textbf{Ans.}$$

Otherwise U_C can be calculated as

$$U_C = \frac{1}{2}\frac{q^2}{C}$$

$$= \frac{1}{2}\times\frac{(5.53\times10^{-3})^2}{(25\times10^{-6})}$$

$$= 0.612 \text{ J}$$

Example 2 In an L-C circuit, L = 3.3 H and C = 840 pF. At t = 0, charge on the capacitor is 105 μC and maximum. Compute the following quantities at t = 2.0 ms:

(a) The energy stored in the capacitor

(b) The total energy in the circuit

(c) The energy stored in the inductor.

Sol. Given, $L = 3.3$ H, $C = 840\times10^{-12}$ F and

$q_0 = 105\times10^{-6}$ C

The angular frequency of L-C oscillations is

$$\omega = \frac{1}{\sqrt{LC}} = \frac{1}{\sqrt{3.3\times840\times10^{-12}}}$$

$$= 1.9\times10^4 \text{ rad/s}$$

Charge stored in the capacitor at time t would be

$$q = q_0 \cos \omega t$$

(a) At $t = 2 \times 10^{-3}$ s,

$$q = (105 \times 10^{-6}) \cos [1.9 \times 10^4][2 \times 10^{-3}]$$
$$= 100.3 \times 10^{-6} \text{ C}$$

∴ Energy stored in the capacitor,

$$U_C = \frac{1}{2}\frac{q^2}{C} = \frac{(100.3 \times 10^{-6})^2}{2 \times 840 \times 10^{-12}}$$
$$= 6.0 \text{ J}$$ **Ans.**

(b) Total energy in the circuit,

$$U = \frac{1}{2}\frac{q_0^2}{C} = \frac{(105 \times 10^{-6})^2}{2 \times 840 \times 10^{-12}}$$
$$= 6.56 \text{ J}$$ **Ans.**

(c) Energy stored in the inductor in the given time
= total energy in circuit − energy stored in capacitor

$$= (6.56 - 6.0) \text{ J}$$
$$= 0.56 \text{ J}$$ **Ans.**

Example 3 An inductor of inductance 2.0 mH is connected across a charged capacitor of capacitance 5.0 μF and the resulting *L-C* circuit is set oscillating at its natural frequency. Let *Q* denotes the instantaneous charge on the capacitor and *I* the current in the circuit. It is found that the maximum value of *Q* is 200 μC. **(1998 JEE)**

(a) When $Q = 100$ μC, what is the value of $|dI/dt|$?
(b) When $Q = 200$ μC, what is the value of *I*?
(c) Find the maximum value of *I*.
(d) When *I* is equal to one-half of its maximum value, what is the value of $|Q|$?

Sol. Here, Q_0 = maximum value of $Q = 200\,\mu\text{C}$

$$= 2 \times 10^{-4} \text{ C}$$

$$\omega = \frac{1}{\sqrt{LC}} = \frac{1}{\sqrt{(2 \times 10^{-3})(5.0 \times 10^{-6})}}$$
$$= 10^4 \text{s}^{-1}$$

(a) $$\left|\frac{dI}{dt}\right| = \omega^2 Q = (10^4)^2(100 \times 10^{-6})$$
$$= 10^4 \text{ A/s}$$

(b) When $Q = Q_0 = Q_{max}$
$$I = 0$$

(c) $$I_{max} = Q_0\,\omega = (2.0 \times 10^{-4})(10^4)$$
$$I_{max} = 2.0 \text{ A}$$

(d) From energy conservation,

$$\frac{1}{2}LI_{max}^2 = \frac{1}{2}LI^2 + \frac{1}{2}\frac{Q^2}{C}$$

or $$Q = \sqrt{LC(I_{max}^2 - I^2)}$$

$$I = \frac{I_{max}}{2} = 1.0 \text{ A}$$

∴ $$Q = \sqrt{(2.0 \times 10^{-3})(5.0 \times 10^{-6})(2^2 - 1^2)}$$
$$Q = \sqrt{3} \times 10^{-4} \text{ C}$$

or $$Q = 1.732 \times 10^{-4} \text{ C}$$

MODULE 6 Exercise

Single Correct Option

1. An *L-C* circuit contains a 0.60 H inductor and a 25 μF capacitor. What is the rate of change of the current (in amp/second) when the charge on the capacitor is 3.0×10^{-5} C?
(a) 2.0 (b) 4.0
(c) 3.0 (d) 6.0

2. A capacitor of capacity 2 μF is charged to a potential difference of 12 V. It is then connected across an inductor of inductance 0.6 mH. The current in the circuit at a time when the PD across the capacitor is 6.0 V is
(a) 0.6 A (b) 1.2 A
(c) 2.4 A (d) 3.6 A

3. Comparing the *L-C* oscillations with the oscillations of a spring-block system (force constant of spring = *k* and mass of block = *m*), the physical quantity *mk* is similar to
(a) *CL* (b) $\frac{1}{CL}$
(c) $\frac{C}{L}$ (d) $\frac{L}{C}$

4. In an *LC* circuit, the capacitor has maximum charge q_0. The value of $\left(\frac{dI}{dt}\right)_{max}$ is

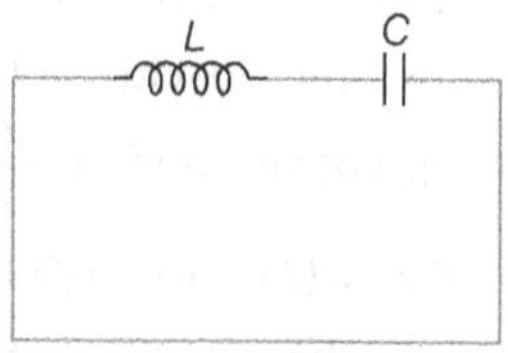

(a) $\frac{q_0}{LC}$ (b) $\frac{q_0}{\sqrt{LC}}$
(c) $\frac{q_0}{LC} - 1$ (d) $\frac{q_0}{LC} + 1$

5. The capacitor of an oscillatory circuit is enclosed in a container. When the container is evacuated, the frequency is 10 kHz. When the container is filled with a gas, the frequency changes by 50 Hz. The dielectric constant of the gas is
(a) 2.01 (b) 1.01
(c) 2.001 (d) 1.001

6. In an oscillation of L-C circuit, the maximum charge on the capacitor is Q. The charge on the capacitor, when the energy is stored equally between the electric and magnetic field is
(a) $\frac{Q}{2}$ (b) $\frac{Q}{\sqrt{2}}$
(c) $\frac{Q}{\sqrt{3}}$ (d) $\frac{Q}{3}$

Subjective Questions

7. Show that $\sqrt{LC}$ has units of time.

8. While comparing the L-C oscillations with the oscillations of spring-block system, with whom the magnetic energy can be compared and why?

9. In an L-C circuit, $L = 0.75$ H and $C = 18\ \mu$F,
(a) At the instant when the current in the inductor is changing at a rate of 3.40 A/s, what is the charge on the capacitor?
(b) When the charge on the capacitor is 4.2×10^{-4} C, what is the induced emf in the inductor?

10. An L-C circuit consists of a 20.0 mH inductor and a 0.5 μF capacitor. If the maximum instantaneous current is 0.1 A, what is the greatest potential difference across the capacitor?

11. In the L-C circuit shown, $C = 1\ \mu$F. With capacitor charged to 100 V, switch S is suddenly closed at time $t = 0$. The circuit then oscillates at 10^3 Hz.

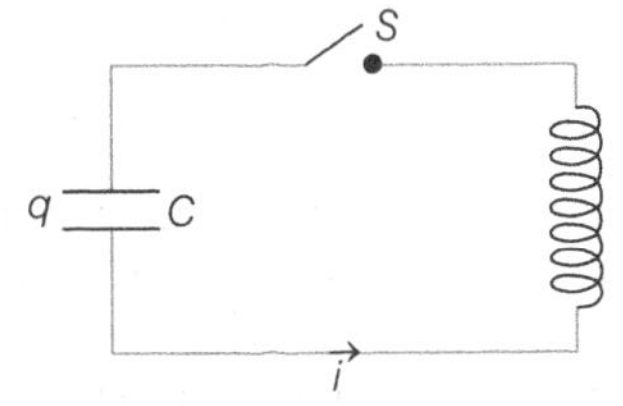

(a) Calculate ω and T
(b) Express q as a function of time
(c) Calculate L
(d) Calculate the average current during the first quarter-cycle.

12. An L-C circuit consists of an inductor with $L = 0.0900$ H and a capacitor of $C = 4 \times 10^{-4}$ F. The initial charge on the capacitor is 5.00 μC, and the initial current in the inductor is zero.
(a) What is the maximum voltage across the capacitor?
(b) What is the maximum current in the inductor?
(c) What is the maximum energy stored in the inductor?
(d) When the current in the inductor has half its maximum value, what is the charge on the capacitor and what is the energy stored in the inductor?

13. In an oscillating L-C circuit in which $C = 4.00\ \mu$F, the maximum potential difference across the capacitor during the oscillations is 1.50 V and the maximum current through the inductor is 50.0 mA.
(a) What is the inductance L?
(b) What is the frequency of the oscillations?
(c) How much time does the charge on the capacitor take to rise from zero to its maximum value?

Answers

1. (a) **2.** (a) **3.** (d) **4.** (a) **5.** (b)
6. (b)

8. With KE as $v \Leftrightarrow i$ and $m \Leftrightarrow L$. Therefore, $\frac{1}{2}mv^2 = \frac{1}{2}Li^2$

9. (a) 45.9 μC (b) 23.3 V **10.** 20.0 V

11. (a) 6.25×10^3 rad/s, 10^{-3} s (b) $10^{-4} \cos(6.25 \times 10^3 t)$
(c) 0.0253 H (d) 0.4 A

12. (a) 1.25×10^{-2} V (b) 8.33×10^{-4} A
(c) 3.125×10^{-8} J (d) 4.33×10^{-6} C, 7.8×10^{-9} J

13. (a) 3.6 mH (b) 1.33 kHz (c) 0.188 ms

See hints on pages 553-554.

MODULE 7

Induced Electric Field

When a conductor moves in a magnetic field, we can understand the induced emf on the basis of magnetic forces on charges in the conductor as described in Module 2. But an induced emf also occurs when there is a changing flux through a stationary conductor. What is it that pushes the charges around the circuit in this type of situation?

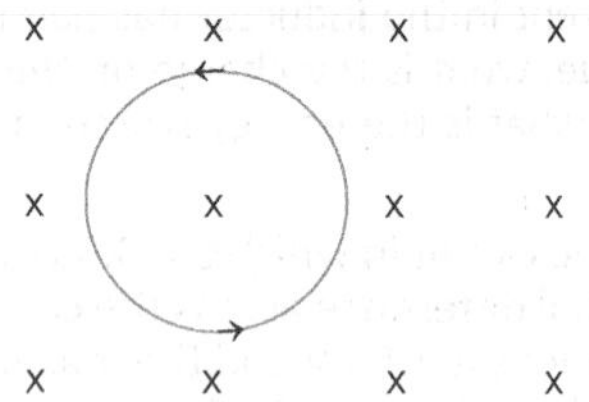

As an example, let's consider the situation shown in figure. A conducting circular loop is placed in a magnetic field which is directed perpendicular to the paper inwards. When the magnetic field changes with time (suppose it increases with time) the magnetic flux ϕ_B also changes and according to Faraday's law the induced emf $e = -\frac{d\phi_B}{dt}$ is produced in the loop. If the total resistance of the loop is R, the induced current in the loop is given by

$$i = \frac{e}{R}$$

But what force makes the charges move around the loop? It can't be the magnetic force, because the charges are not moving in the magnetic field.

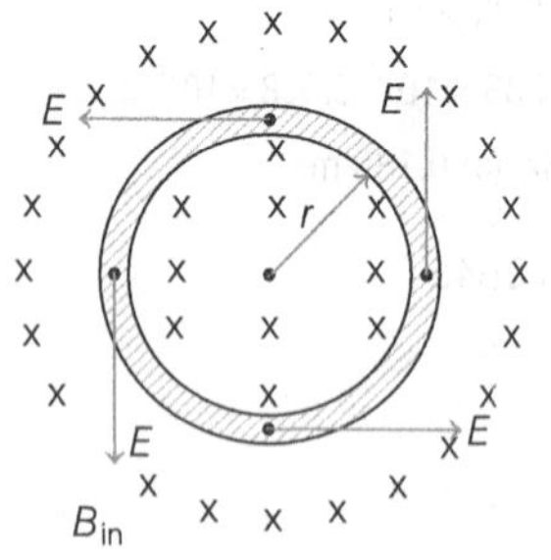

Actually, there is an **induced electric field** in the conductor **caused by the changing magnetic flux.**

This electric field has the following important properties:

1. It is non-conservative in nature. The line integral of **E** around a closed path is not zero. This line integral is given by

$$\oint \mathbf{E} \cdot d\mathbf{l} = -\frac{d\phi_B}{dt} \qquad \text{...(i)}$$

2. Because of symmetry, the electric field **E** has the same magnitude at every point on the circle and is tangent to it at each point. The directions of **E** at several points on the loop are shown in figure.
3. Being a non-conservative field, the concept of potential has no meaning for such a field.
4. This field is different from the electrostatic field produced by stationary charges (which is conservative in nature).
5. The relation $\mathbf{F} = q\mathbf{E}$ is still valid for this field.
6. This field can vary with time.

So, **a changing magnetic field also acts as a source of electric field.**

Note 1. *For symmetrical situations (as shown in figure) Eq. (i), in simplified form can be written as*

$$El = \left|\frac{d\phi_B}{dt}\right| = S\left|\frac{dB}{dt}\right| \qquad \text{...(ii)}$$

Here, l is the length of closed loop in which electric field is to be calculated and S is the area in which magnetic field is changing.

2. *Direction of electric field is the same as the direction of induced current. For example, in the above figure ⊗ magnetic field was increasing. So, the induced current in the loop will produce magnetic field in ⊙ direction. Or, induced current should be anticlockwise. We can see that electric field is also anticlockwise.*

In most of our cases magnetic field will change in a cylindrical (or circular) region. So electric field lines will be circular and value of E can be obtained by using Eq. (ii).

Example 1 The magnetic field at all points within the cylindrical region whose cross-section is indicated in the accompanying figure start increasing at a constant rate α. Find the magnitude of electric field as a function of r, the distance from the geometric centre of the region.

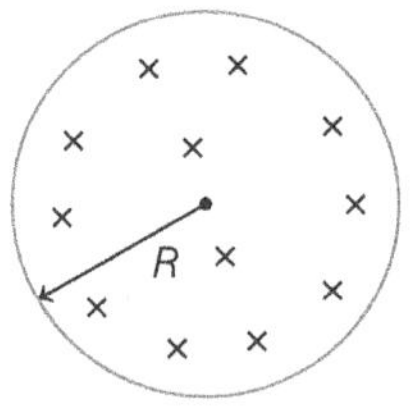

Sol. **For $r \le R$**

Using $El = S\left|\frac{dB}{dt}\right|$ or $E(2\pi r) = (\pi r^2)\,\alpha$

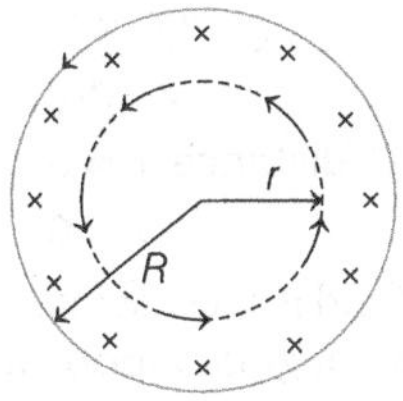

$\therefore \quad E = \dfrac{r\alpha}{2}$

$\therefore \quad E \propto r$, i.e. E-r graph is a straight line passing through origin.

At $r = R$, $E = \dfrac{R\alpha}{2}$

For $r \geq R$

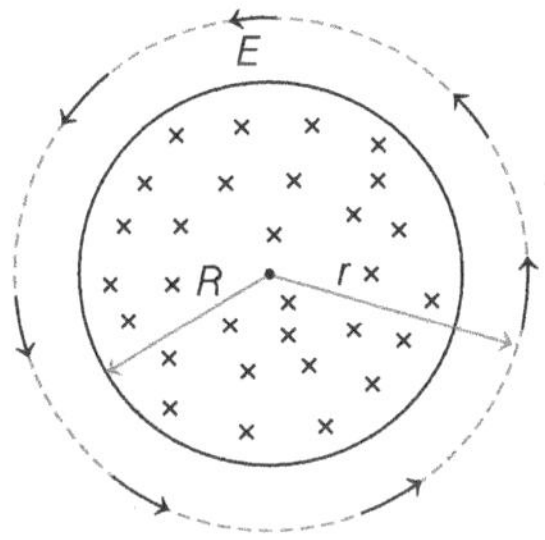

Using $El = S\left|\dfrac{dB}{dt}\right|$,

$$\therefore \quad E\,(2\pi r) = (\pi R^2)\,(\alpha)$$

$$\therefore \quad E = \frac{\alpha R^2}{2r}$$

$\therefore \quad E \propto \dfrac{1}{r}$, i.e. E-r graph is a rectangular hyperbola.

The E-r graph is as shown in figure.

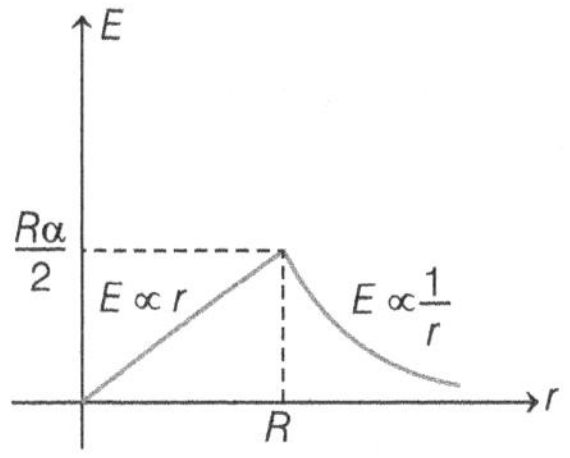

The direction of electric field is shown in above figure.

Example 2 A long thin solenoid has 900 turns/metre and radius 2.50 cm. The current in the solenoid is increasing at a uniform rate of 60 A/s. What is the magnitude of the induced electric field at a point?

(a) 0.5 cm from the axis of the solenoid.

(b) 1.0 cm from the axis of the solenoid.

Sol. $\because \quad B = \mu_0 ni$

$$\therefore \quad \frac{dB}{dt} = \mu_0 n \frac{di}{dt}$$

$$= (4\pi \times 10^{-7})\,(900)\,(60)$$

$$= 0.068 \text{ T/s}$$

Using the result of electric field derived in above problem (as both points lie inside the solenoid).

$$E = \frac{r}{2}\left(\frac{dB}{dt}\right)$$

$$\text{(a) } E = \left(\frac{0.5 \times 10^{-2}}{2}\right)(0.068)$$

$$= 1.7 \times 10^{-4} \text{ V/m}$$

$$\text{(b) } E = \frac{(1.0 \times 10^{-2})}{2}(0.068)$$

$$= 3.4 \times 10^{-4} \text{ V/m}$$

Example 3 A uniform but time-varying magnetic field $B(t)$ exists in a circular region of radius a and is directed into the plane of the paper as shown. The magnitude of the induced electric field at point P at a distance r from the centre of the circular region **(2000 JEE)**

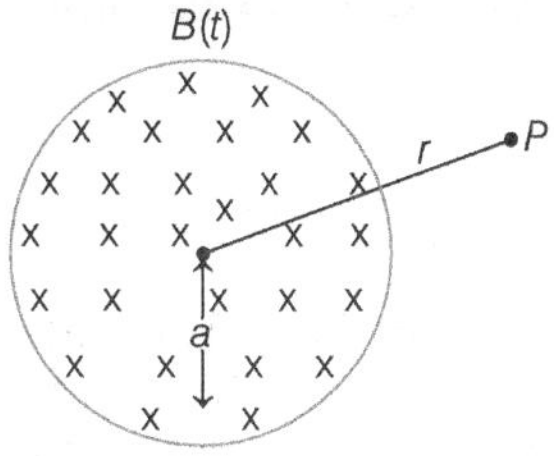

(a) is zero
(b) decreases as $1/r$
(c) increases as r
(d) decreases as $1/r^2$

Sol. $\displaystyle\oint \mathbf{E} \cdot d\mathbf{l} = \left|\frac{d\phi}{dt}\right| = S\left|\frac{dB}{dt}\right|$

or $\quad E(2\pi r) = \pi a^2 \left|\dfrac{dB}{dt}\right|$

$$\therefore \quad E = \frac{a^2}{2r}\left|\frac{dB}{dt}\right|$$

$\therefore$ The correct option is (b).

Example 4 The magnetic field **B** at all points within a circular region of radius R is uniform in space and directed into the plane of the page in figure. If the magnetic field is increasing at a rate dB/dt, what are the magnitude and direction of the force on a stationary positive point charge q located at points a, b and c? (Point a is a distance r above the centre of the region, point b is a distance r to the right of the centre, and point c is at the centre of the region.)

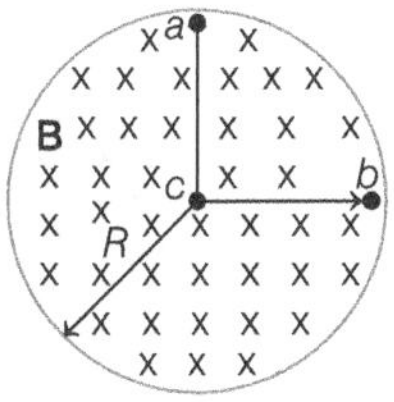

Sol. Inside the circular region at distance r,

$$El = \frac{d\phi}{dt} = S\left(\frac{dB}{dt}\right)$$

$$\therefore \quad E(2\pi r) = (\pi r^2)\cdot\frac{dB}{dt}$$

$$\therefore \quad E = \frac{r}{2}\frac{dB}{dt}$$

$$F = qE = \frac{qr}{2}\frac{dB}{dt}$$

At points a and b, distance from centre is r.

$$\therefore \quad F = \frac{qr}{2}\frac{dB}{dt}$$

At point C, distance $r = 0$

$$\therefore \quad F = 0$$

$\otimes$ magnetic field is increasing. Hence, induced current in an imaginary loop passing through a and b should produce $\odot$ magnetic field. Hence, induced current through an imaginary circular loop passing through a and b should be anti-clockwise. Force on positive charge is in the direction of induced current. Hence, force at a is towards left and force at b is upwards.

MODULE 7 Exercise

Single Correct Option

1. Identify the **incorrect** statement. Induced electric field
   (a) is produced by varying magnetic field
   (b) is non-conservative in nature
   (c) cannot exist in a region not occupied by magnetic field
   (d) None of the above

2. A uniform but increasing with time magnetic field exists in a cylindrical region. The direction of force on an electron at P is

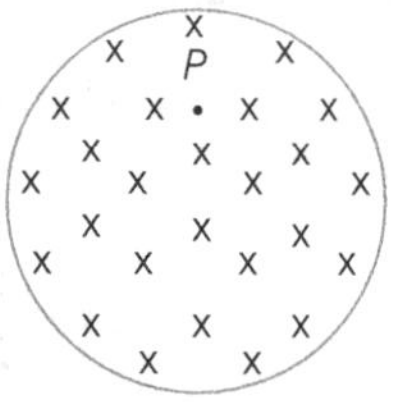

   (a) towards right
   (b) towards left
   (c) into the plane of paper
   (d) out of the plane of paper

3. The figure shows a non-conducting ring of radius R carrying a charge q. In a circular region of radius r, a uniform magnetic field **B** perpendicular to the plane of the ring varies at a constant rate $\frac{dB}{dt} = \beta$. The torque acting on the ring is

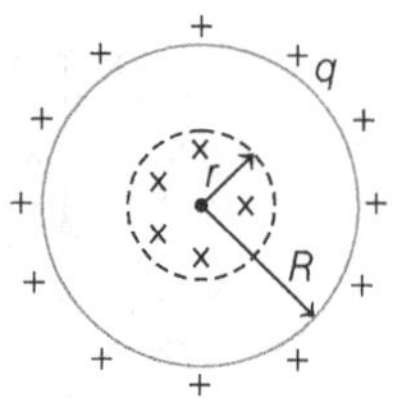

   (a) $\frac{1}{2}qr^2\beta$ (b) $\frac{1}{2}qR^2\beta$
   (c) $qr^2\beta$ (d) zero

4. A non-conducting ring having charge q uniformly distributed over its circumference is placed on a rough horizontal surface. A vertical time varying magnetic field $B = 4t^2$ is switched on at time $t = 0$. Mass of the ring is m and radius is R. The ring starts rotating after 2 s, the coefficient of friction between the ring and the table is
   (a) $\frac{4qmR}{g}$ (b) $\frac{2qmR}{g}$ (c) $\frac{8qR}{mg}$ (d) $\frac{qR}{2mg}$

Multiple Correct Options

5. Magnetic field in a cylindrical region of radius R in inward direction is as shown in figure.

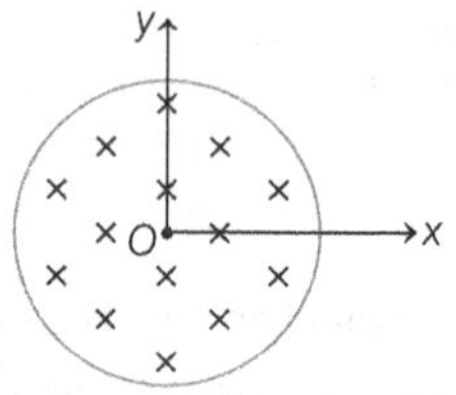

   (a) an electron will experience no force kept at $(2R, 0, 0)$ if magnetic field increases with time
   (b) in the above situation, electron will experience the force in negative y-axis
   (c) If a proton is kept at $\left(0, \frac{R}{2}, 0\right)$ and magnetic field is decreasing, then it will experience the force in positive x-direction
   (d) if a proton is kept at $(-R, 0, 0)$ and magnetic field is increasing, then it will experience force in negative y-axis

Passage Based Questions

Passage (Q.No 6 to Q.No 8)

A uniform but time varying magnetic field $B = (2t^3 + 24t)$ T is present in a cylindrical region of radius $R = 2.5$ cm as shown in figure.

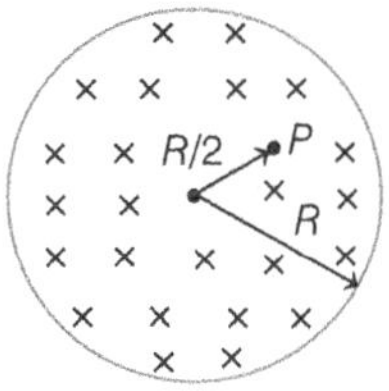

6. The force on an electron at P at $t = 2.0$ s is
   (a) 96×10^{-21} N (b) 48×10^{-21} N
   (c) 24×10^{-21} N (d) zero

7. The variation of electric field at any instant as a function of distance measured from the centre of cylinder in first problem is

(a)
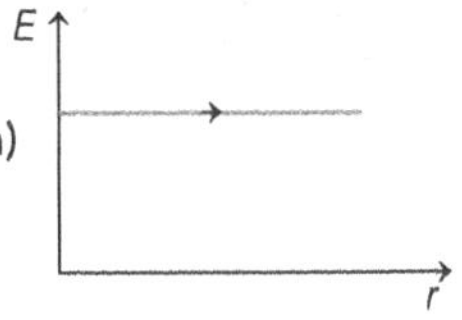

(b)
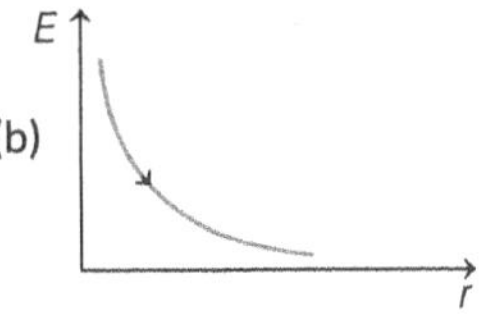

(c)
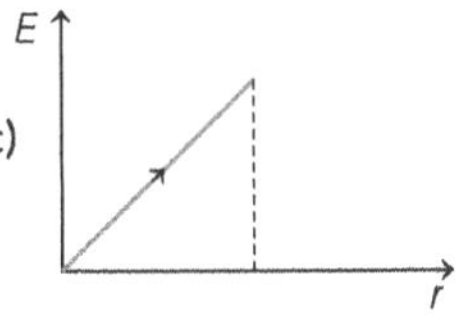

(d)
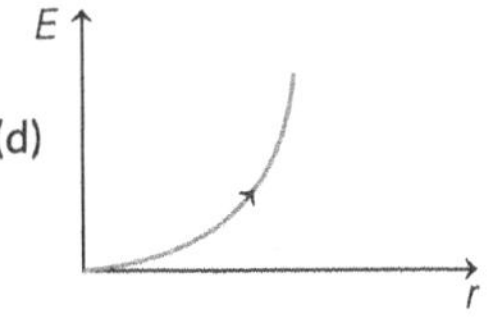

8. In the previous problem, the direction of circular electric lines at $t = 1$ s is
   (a) clockwise
   (b) anti-clockwise
   (c) no current is induced
   (d) cannot be predicted

Subjective Questions

9. A long solenoid of cross-sectional area 5.0 cm^2 is wound with 25 turns of wire per centimetre. It is placed in the middle of a closely wrapped coil of 10 turns and radius 25 cm as shown.

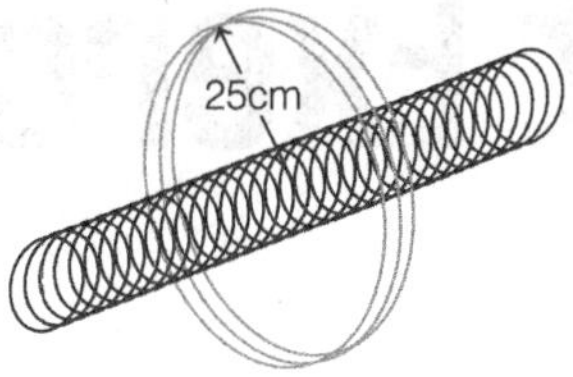

   (a) What is the emf induced in the coil when the current through the solenoid is decreasing at a rate -0.20 A/s?
   (b) What is the electric field induced in the coil?

10. For the situation described in figure, the magnetic field in cylindrical region changes with time according to

$$B = (2.00t^3 - 4.00t^2 + 0.8)\text{ T}$$

and $r_2 = 2R = 5.0$ cm

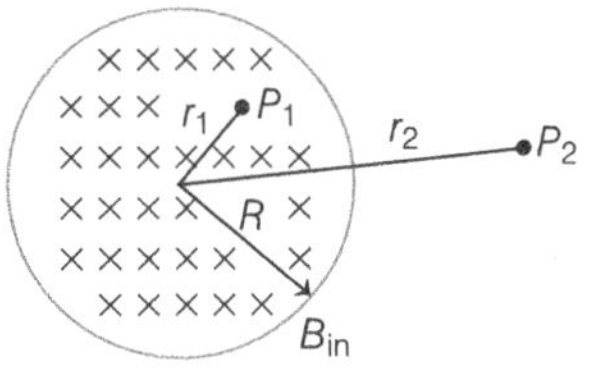

   (a) Calculate the force on an electron located at P_2 at $t = 2.00$ s
   (b) What are the magnitude and direction of the electric field at P_1 when $t = 3.00$ s and $r_1 = 0.02$ m.

Answers

1. (c)	**2.** (a)	**3.** (a)	**4.** (c)
5. (b,c,d)	**6.** (b)	**7.** (c)	**8.** (b)

9. (a) 3.14×10^{-6} V (b) 2.0×10^{-6} V/m

10. (a) 8.0×10^{-21} N (downward and to the right perpendicular to r_2)
(b) 0.3 V/m (upwards and to the left perpendicular to r_1)

See hints on pages 554-555.

Unit Exercise (Cumulative)

Level 1

Fill in the Blank

1. A uniformly wound solenoidal coil of self-inductance 1.8×10^{-4} H and resistance 6 Ω is broken up into two identical coils. These identical coils are then connected in parallel across a 15 V battery of negligible resistance. The time constant for the current in the circuit is ...... s and the steady state current through the battery is ... A. **(1989 JEE)**

True/False

2. An inductor is connected to a battery through a switch. The emf induced in the inductor is much larger when the switch is opened as compared to the emf induced when the switch is closed. Is this statement true or false?

Assertion and Reason

See directions on page 25.

3. **Assertion :** In parallel, current distributes in inverse ratio of inductance

$$i \propto \frac{1}{L}$$

Reason : In electrical circuits, an inductor can be treated as a resistor.

4. **Assertion :** A conducting loop shown in figure is rotated in a uniform magnetic field with constant angular velocity ω as shown in figure. At time $t = 0$, plane of the loop is perpendicular to the magnetic field. Induced emf produced in the loop is maximum when plane of loop is parallel to magnetic field.

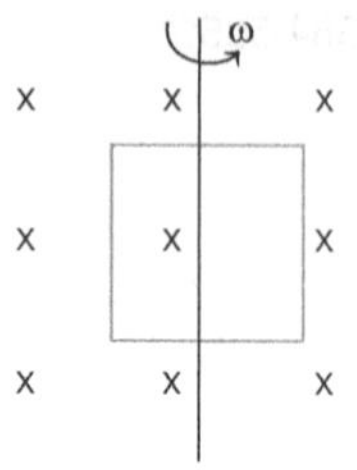

Reason : When plane of loop is parallel to magnetic field, magnetic flux passing through the loop is zero.

Single Correct Option

5. A conducting wire frame is placed in a magnetic field which is directed into the paper. The magnetic field is increasing at a constant rate.

The directions of induced current in wires *AB* and *CD* are

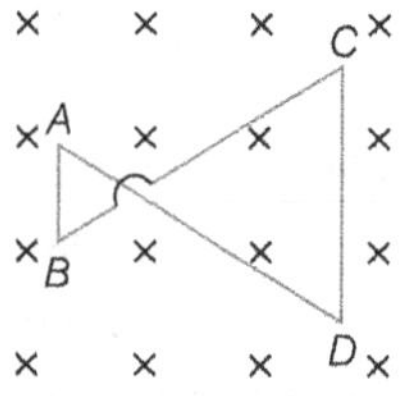

(a) *B* to *A* and *D* to *C* (b) *A* to *B* and *C* to *D*
(c) *A* to *B* and *D* to *C* (d) *B* to *A* and *C* to *D*

6. The ring *B* is coaxial with a solenoid *A* as shown in figure. As the switch *S* is closed at $t = 0$, the ring *B*

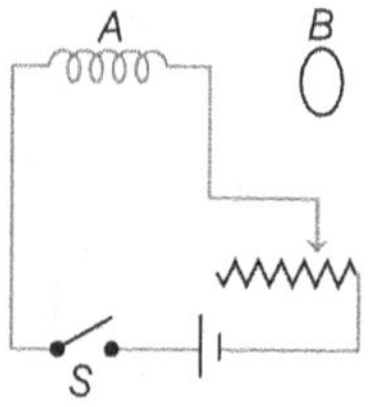

(a) is attracted towards *A*
(b) is repelled by *A*
(c) is initially repelled and then attracted
(d) is initially attracted and then repelled

7. In the circuit shown in figure the jockey *J* is being pulled towards right so that the resistance in the circuit is increasing. Its value at some instant is 5 Ω. The current in the circuit at this instant will be

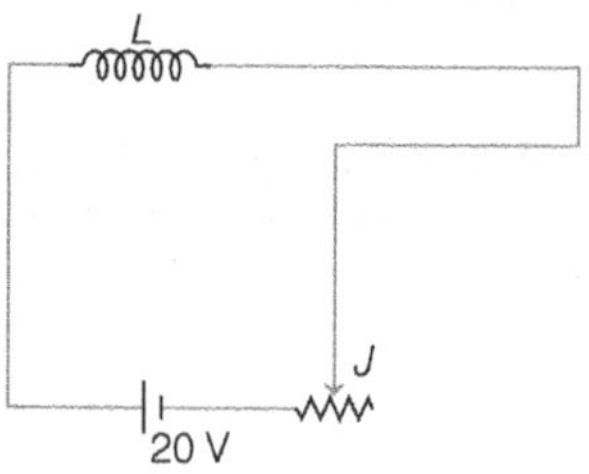

(a) 4 A
(b) less than 4 A
(c) more than 4 A
(d) may be less than or more than 4 A depending on the value of *L*

8. A uniformly wound long solenoid of inductance *L* and resistance *R* is cut into two parts in the ratio $\eta : 1$, which are then connected in parallel. The combination is then connected to a cell of emf *E*. The time constant of the circuit is

(a) $\dfrac{L}{R}$ (b) $\dfrac{L}{(\eta+1)R}$ (c) $\left(\dfrac{\eta}{\eta+1}\right)\dfrac{L}{R}$ (d) $\left(\dfrac{\eta+1}{\eta}\right)\dfrac{L}{R}$

9. The armature of a DC motor has 20 Ω resistance. It draws a current of 1.5 A when run by 200 V DC supply. The value of back emf induced in it will be

(a) 150 V (b) 170 V (c) 180 V (d) 190 V

10. Two ends of an inductor of inductance L are connected to two parallel conducting wires. A rod of length l and mass m is given velocity v_0 as shown. The whole system is placed in perpendicular magnetic field B. Find the maximum current in the inductor. (Neglect gravity and friction)

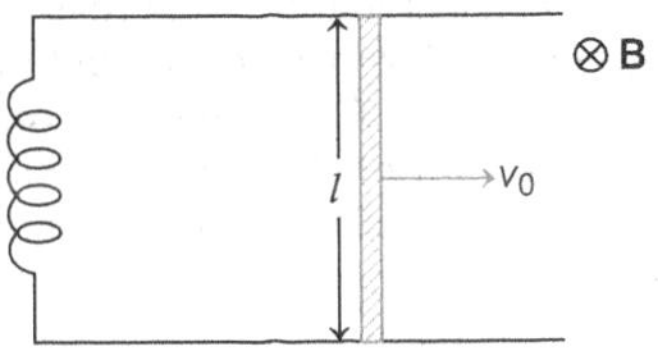

(a) $\dfrac{mv_0}{L}$ (b) $\sqrt{\dfrac{m}{L}}v_0$

(c) $\dfrac{mv_0^2}{L}$ (d) None of these

11. A flat circular coil of n turns, area A and resistance R is placed in a uniform magnetic field B. The plane of coil is initially perpendicular to B. When the coil is rotated through an angle of 180° about one of its diameter, a charge Q_1 flows through the coil. When the same coil after being brought to its initial position, is rotated through an angle of 360° about the same axis a net charge Q_2 flows through it. Then, Q_2/Q_1 is

(a) 1 (b) 2

(c) 1/2 (d) 0

12. An inductor L, a resistance R and two identical bulbs, B_1 and B_2 are connected to a battery through a switch S as shown in the figure. Which of the following statements gives the correct description of the happenings when the switch S is closed?

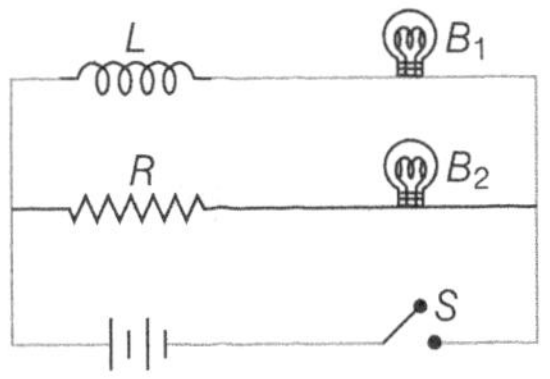

(a) The bulb B_2 lights up earlier than B_1 and finally both the bulbs shine equally bright

(b) B_1 lights up earlier and finally both the bulbs acquire equal brightness

(c) B_2 lights up earlier and finally B_1 shines brighter than B_2

(d) B_1 and B_2 light up together with equal brightness all the time

13. Figure shows two bulbs B_1 and B_2 resistor R and inductor L, when the switch S is turned off

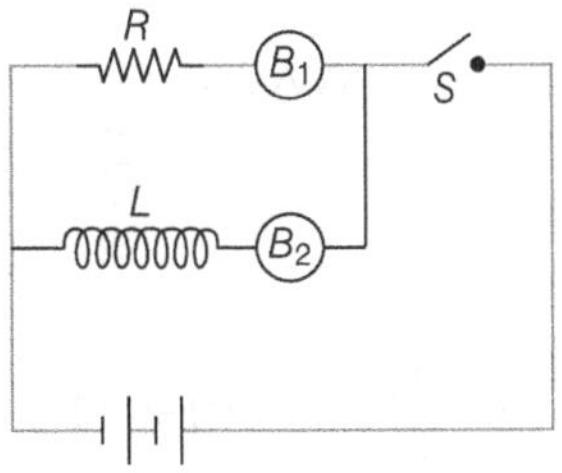

(a) both B_1 and B_2 die out promptly

(b) both B_1 and B_2 die out with some delay

(c) B_2 dies out promptly, but B_1 with some delay

(d) B_1 dies out promptly but B_2 with some delay

14. In the circuit shown in figure initially the switch is in position 1 for a long time, then suddenly at $t = 0$, the switch is shifted to position 2. It is required that a constant current should flow in the circuit, the value of resistance R in the circuit

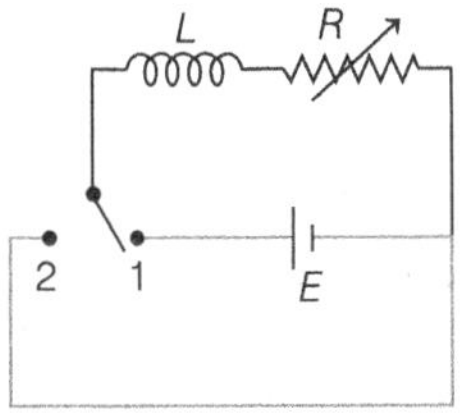

(a) should be decreased at a constant rate

(b) should be increased at a constant rate

(c) should be maintained constant

(d) Not possible

15. A conducting straight wire PQ of length l is fixed along a diameter of a non-conducting ring as shown in the figure. The ring is given a pure rolling motion on a horizontal surface such that its centre of mass has a velocity v. There exists a uniform horizontal magnetic field B in horizontal direction perpendicular to the plane of ring. The magnitude of induced emf in the wire PQ at the position shown in the figure will be

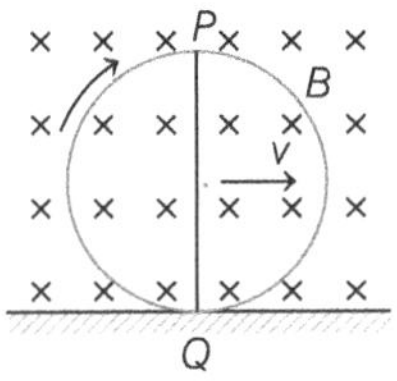

(a) Bvl (b) $2Bvl$ (c) $3Bvl/2$ (d) zero

16. A coil of inductance 1 H and resistance 100 Ω is connected to a battery of 6V. Determine approximately. **(2022 Main)**

(I) The time elapsed before the current acquires half of its steady - state value

(II) The energy stored in the magnetic field associated with the coil at an instant 15 ms after the circuit is switched on (Given $\ln 2 = 0.693, e^{-3/2} = 0.25$)

(a) $t = 10$ ms; $U = 2$ mJ (b) $t = 10$ ms; $U = 1$ mJ

(c) $t = 7$ ms; $U = 1$ mJ (d) $t = 7$ ms; $U = 2$ mJ

17. A 20 H inductor coil is connected to a 10 ohm resistance in series as shown in figure. The time at which rate of dissipation of energy (Joule's heat) across resistance is equal to the rate at which magnetic energy is stored in the inductor, is **(2019 Main)**

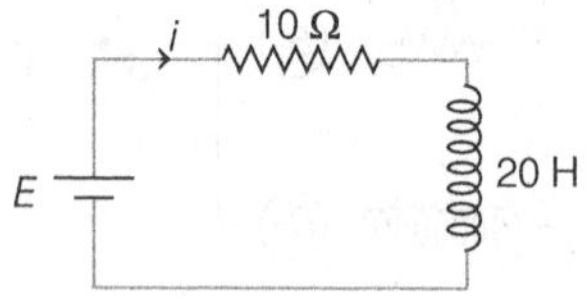

(a) $\frac{2}{\ln 2}$ (b) $\frac{1}{2}\ln 2$ (c) $2\ln 2$ (d) $\ln 2$

18. Consider the *L-R* circuit shown in the figure. If the switch *S* is closed at $t = 0$, then the amount of charge that passes through the battery between $t = 0$ and $t = \frac{L}{R}$ is **(2019 Main)**

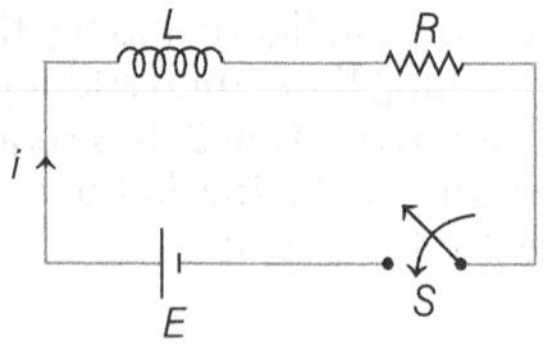

(a) $\frac{2.7EL}{R^2}$ (b) $\frac{EL}{2.7R^2}$ (c) $\frac{7.3EL}{R^2}$ (d) $\frac{EL}{7.3R^2}$

19. An emf of 20 V is applied at time $t = 0$ to a circuit containing in series 10 mH inductor and 5Ω resistor. The ratio of the currents at time $t = \infty$ and at $t = 40$ s is close to (Take $e^2 = 7.389$) **(2019 Main)**

(a) 1.06 (b) 1.15 (c) 0.84 (d) 1.46

20. An *L-C-R* circuit is equivalent to a damped pendulum. In an *L-C-R* circuit, the capacitor is charged to Q_0 and then connected to the *L* and *R* as shown.

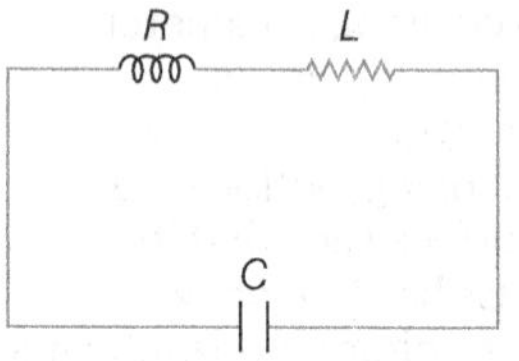

If a student plots graphs of the square of maximum charge (Q^2_{max}) on the capacitor with time (t) for two different values L_1 and L_2 ($L_1 > L_2$) of L, then which of the following represents this graph correctly? (plots are schematic and not drawn to scale) **(2015 Main)**

(a)
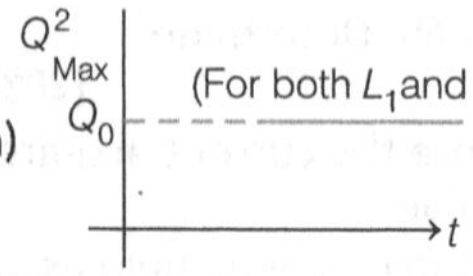

(b)
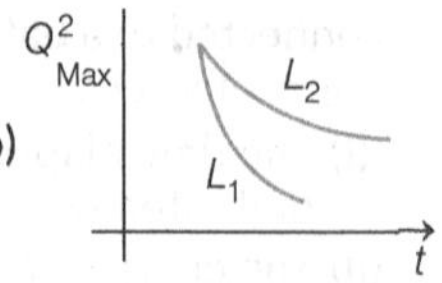

(c)
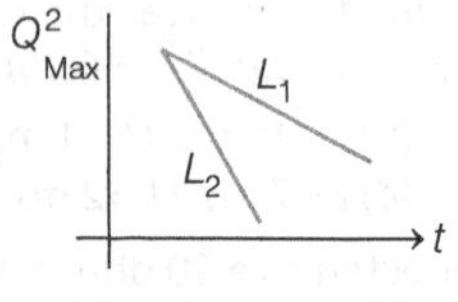

(d)
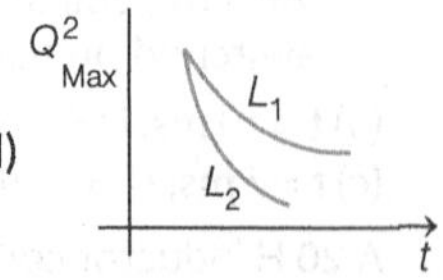

21. An inductor ($L = 0.03$ H) and a resistor ($R = 0.15$ kΩ) are connected in series to a battery of 15V EMF in a circuit shown below. The key K_1 has been kept closed for a long time. Then at $t = 0$, K_1 is opened and key K_2 is closed simultaneously. At $t = 1$ms, the current in the circuit will be ($e^5 \cong 150$) **(2015 Main)**

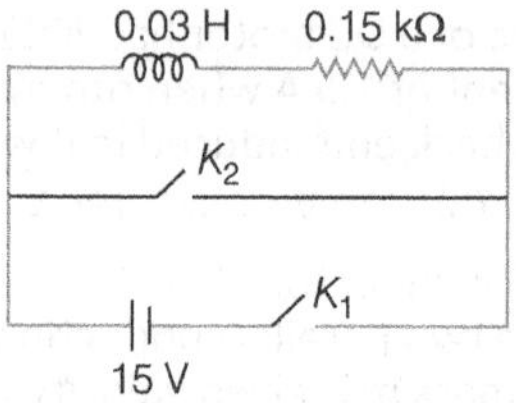

(a) 100 mA (b) 67 mA
(c) 0.67mA (d) 6.7 mA

22. A conducting rod of length $L = 0.1$ m is moving with a uniform speed $v = 0.2$ m/s on conducting rails in a magnetic field $B = 0.5$ T as shown. On one side, the end of the rails is connected to a capacitor of capacitance $C = 20\ \mu F$. Then, the charges on the capacitor's plates are

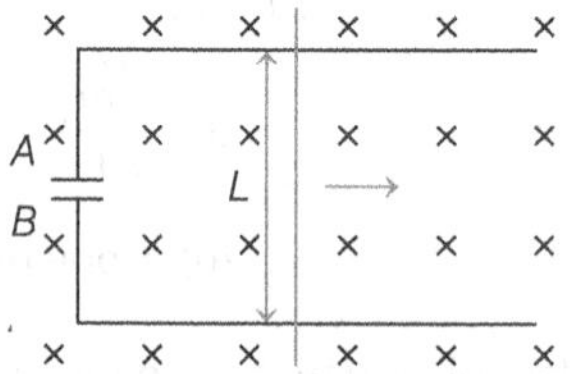

(a) $q_A = 0 = q_B$
(b) $q_A = +20\ \mu C$ and $q_B = -20\ \mu C$
(c) $q_A = +0.2\ \mu C$ and $q_B = -0.2\ \mu C$
(d) $q_A = -0.2$ C and $q_B = -0.2\ \mu C$

23. Initially, the switch is in position 1 for a long time and then shifted to position 2 at $t = 0$ as shown in figure. Just after closing the switch, the magnitude of current through the capacitor is

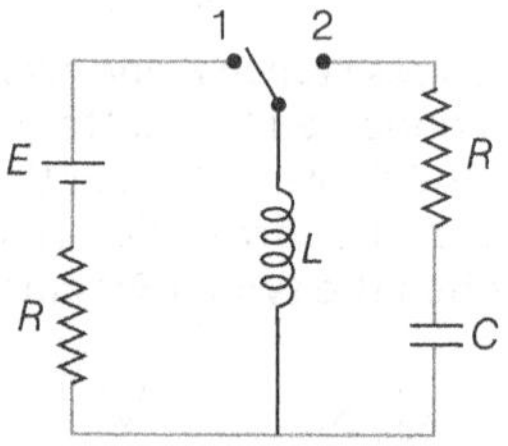

(a) zero (b) $\frac{E}{2R}$
(c) $\frac{E}{R}$ (d) None of these

24. When the switch *S* is closed at $t = 0$, identify the correct statement just after closing the switch as shown in figure

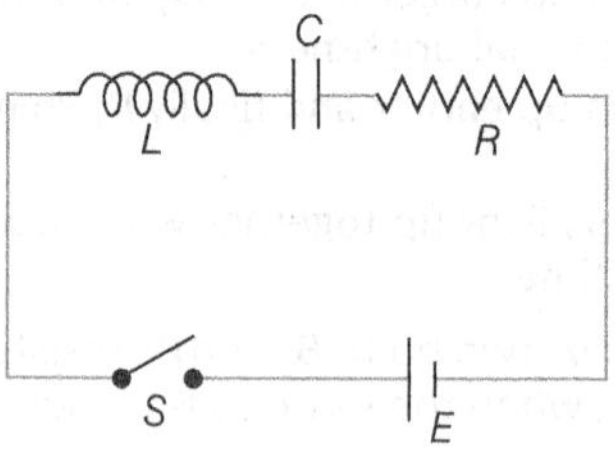

(a) The current in the circuit is maximum
(b) Equal and opposite voltages are dropped across inductor and resistor
(c) The entire voltage is dropped across inductor
(d) All of the above

25. A square loop of side l and resistance R is placed in a uniform magnetic field B acting normally to the plane of the loop. If we attempt to pull it out of the field with a constant velocity v, the power needed is

(a) $BRlv$ (b) $\frac{B^2l^2v^2}{R}$ (c) $\frac{Bl^2v^2}{R}$ (d) $\frac{Bvl}{R}$

26. In the L-C circuit shown in figure, the current is in the direction shown in the figure and charges on the capacitor plates have sign shown in the figure. At this time,

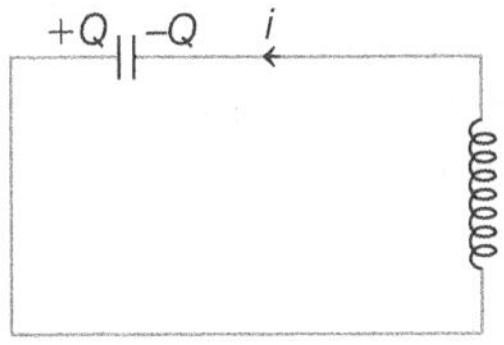

(a) i as well as Q are increasing
(b) i as well as Q are decreasing
(c) i is increasing but Q is decreasing
(d) i is decreasing but Q is increasing

27. The switch in the circuit is in position a for a long time. At $t = 0$, the switch is moved from a to b. The current through the inductor will reach its first maximum after moving the switch in a time

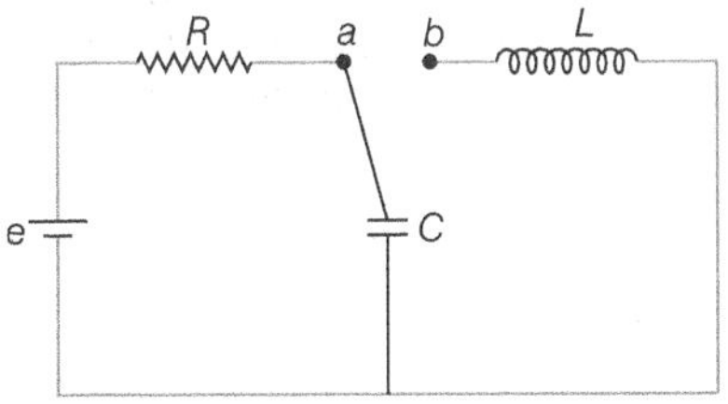

(a) $2\pi\sqrt{LC}$ (b) $\frac{1}{4}\sqrt{LC}$ (c) $\frac{\pi}{2}\sqrt{LC}$ (d) $\pi\sqrt{LC}$

28. An inductor ($L = 100$ mH), a resistor ($R = 100\ \Omega$) and a battery ($E = 100$ V) are initially connected in series as shown in the figure. After a long time the battery is disconnected by short circuiting the points A and B. The current in the circuit 1 ms after the short circuit is (in A)

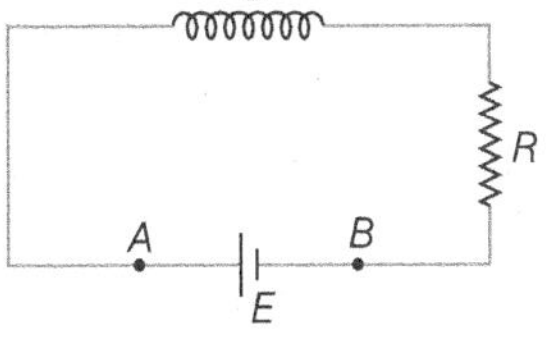

(a) e (b) 0.1
(c) 1 (d) $1/e$

29. The figure shows certain wire segments joined together to form a coplanar loop. The loop is placed in a perpendicular magnetic field in the direction going into the plane of the figure. The magnitude of the field increases with time. I_1 and I_2 are the currents in the segments ab and cd. Then, **(2009 JEE)**

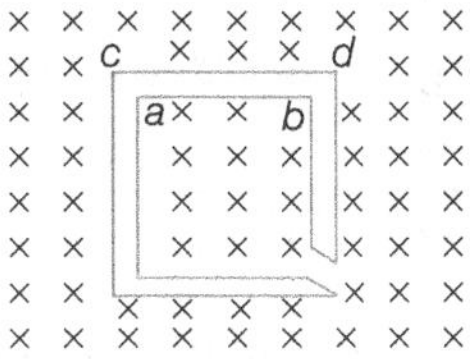

(a) $I_1 > I_2$
(b) $I_1 < I_2$
(c) I_1 is in the direction ba and I_2 is in the direction cd
(d) I_1 is in the direction ab and I_2 is in the direction dc

30. In the circuit shown here, the point C is kept connected to point A till the current flowing through the circuit becomes constant. Afterward, suddenly point C is disconnected from point A and connected to point B at time $t = 0$. Ratio of the voltage across resistance and the inductor at $t = L/R$ will be equal to **(2014 JEE)**

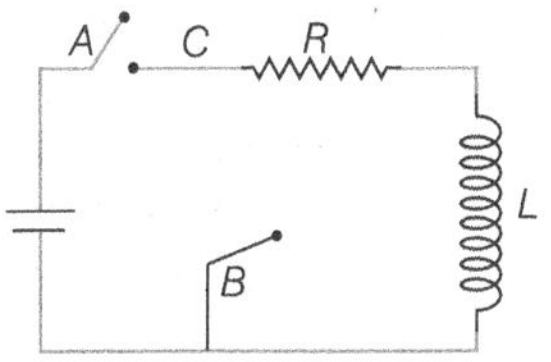

(a) $\frac{e}{1-e}$ (b) 1 (c) −1 (d) $\frac{1-e}{e}$

31. A circular loop of radius 0.3 cm lies parallel to a much bigger circular loop of radius 20 cm. The centre of the smaller loop is on the axis of the bigger loop. The distance between their centres is 15 cm. If a current of 2.0 A flows through the bigger loop, then the flux linked with smaller loop is

(a) 9.1×10^{-11} Wb (b) 6×10^{-11} Wb
(c) 3.3×10^{-11} Wb (d) 6.6×10^{-9} Wb

Subjective Questions

32. In the circuit shown in the figure, $E = 50.0$ V, $R = 250\ \Omega$ and $C = 0.500\ \mu$F. The switch S is closed for a long time, and no voltage is measured across the capacitor. After the switch is opened, the voltage across the capacitor reaches a maximum value of 150 V. What is the inductance L?

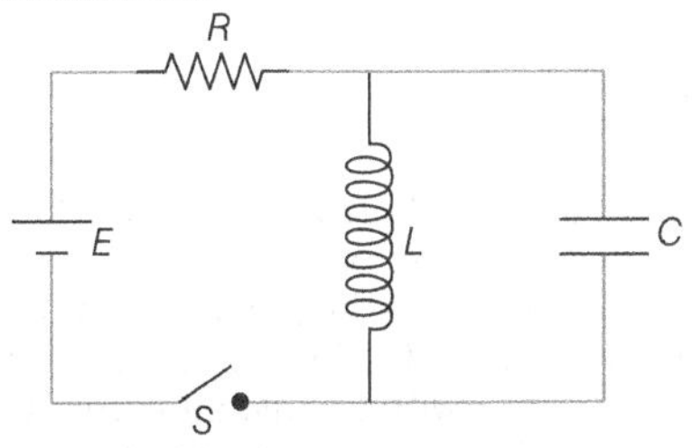

33. The conducting rod ab shown in figure makes contact with metal rails ca and db. The apparatus is in a

uniform magnetic field 0.800 T, perpendicular to the plane of the figure.

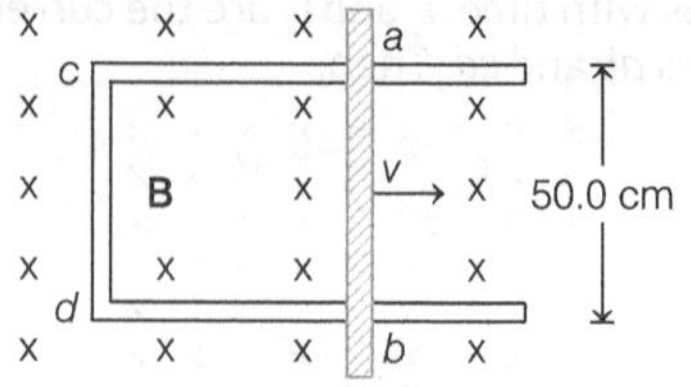

(a) Find the magnitude of the emf induced in the rod when it is moving toward the right with a speed 7.50 m/s.

(b) In what direction does the current flow in the rod?

(c) If the resistance of the circuit *abdc* is 1.50 Ω (assumed to be constant), find the force (magnitude and direction) required to keep the rod moving to the right with a constant speed of 7.50 m/s. You can ignore friction.

(d) Compare the rate at which mechanical work is done by the force (Fv) with the rate at which thermal energy is developed in the circuit (I^2R).

34. A rod of mass m and resistance R slides on frictionless and resistanceless rails a distance l apart that include a source of emf E_0. (see figure). The rod is initially at rest. Find the expression for the

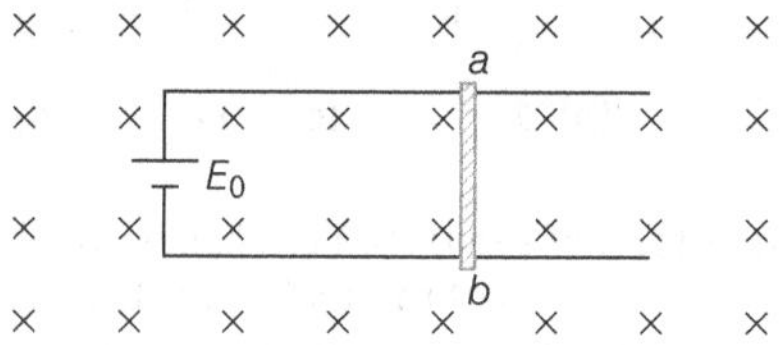

(a) velocity of the rod $v(t)$

(b) current in the loop $i(t)$

35. In the circuit shown A and B are two cells of same emf E but different internal resistances r_1 and $r_2 (r_1 > r_2)$ respectively. Find the value of R such that the potential difference across the terminals of cell A is zero, a long time after the key K is closed. **(2004 JEE)**

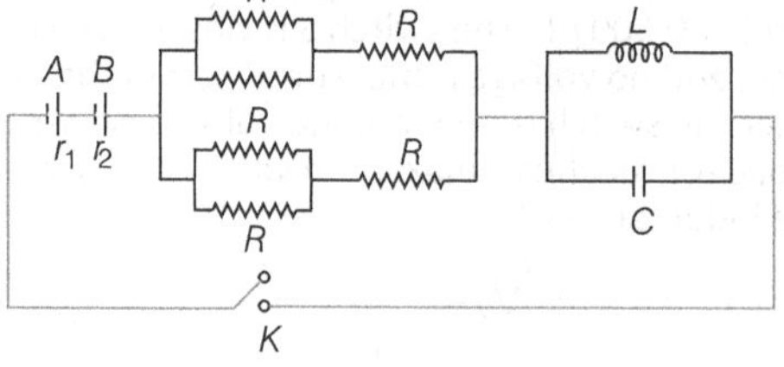

36. A wire bent as a parabola $y = \alpha x^2$ is located in a uniform magnetic field of induction B, the vector B being perpendicular to the plane x-y. At the moment $t = 0$ a connector starts sliding translationwise from the parabola apex with a constant acceleration a. Find the emf of electromagnetic induction in the loop thus formed as a function of y. **(Irodov 3.288)**

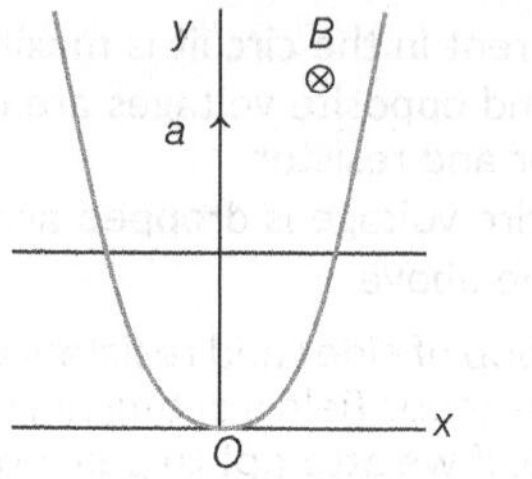

37. A long straight wire carries a current I_0. At distances a and b from it there are two other wires, parallel to the former one, which are interconnected by a resistance R. A connector slides without friction along the wires with a constant velocity v. Assuming the resistances of the wires, the connector, the sliding contacts, and the self-inductance of the frame to be negligible, find:

(a) the magnitude and the direction of the current induced in the connector

(b) the force required to maintain the connector's velocity constant. **(Irodov 3.301)**

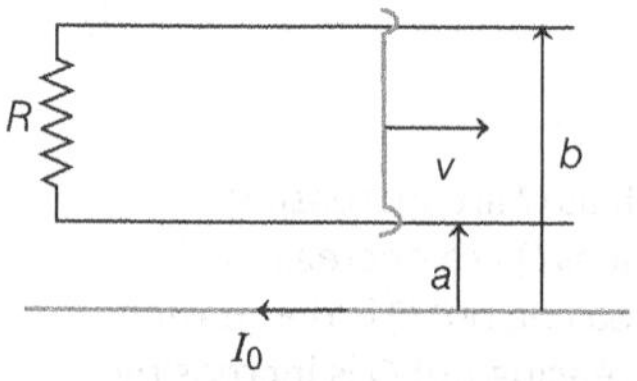

38. A rectangular loop with a sliding connector of length l is located in a uniform magnetic field perpendicular to the loop plane. The magnetic induction is equal to B. The connector has an electric resistance R, the sides PQ and MN have resistances R_1 and R_2 respectively. Neglecting the self-inductance of the loop, find the current flowing in the connector during its motion with a constant velocity v. **(Irodov 3.289)**

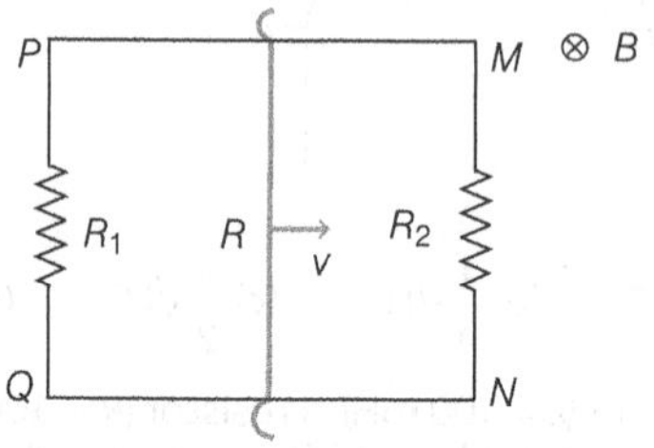

39. A thin wire AC shaped as a semi-circle of diameter $d = 20$ cm rotates with a constant angular velocity $\omega = 100$ rad/s in a uniform magnetic field of induction $B = 5.0$ mT with $\omega \uparrow\uparrow$ **B**. The rotation axis passes through the end A of the wire and is perpendicular to the diameter AC. Find the value of a line integral $\int \mathbf{E}\, d\mathbf{r}$ along the wire from point A to point C. Generalize the obtained result. **(Irodov 3.291)**

40. A copper connector of mass m slides down two smooth copper bars, set at an angle α to the horizontal. At the top, the bars are interconnected through a resistance R. The separation between the bars is equal to l. The system is located in a uniform magnetic field of induction B, perpendicular to the

plane in which the connector slides. The resistances of the bars, the connector and the sliding contacts, as well as the self-inductance of the loop, are assumed to be negligible. Find the steady-state velocity of the connector. **(Irodov 3.296)**

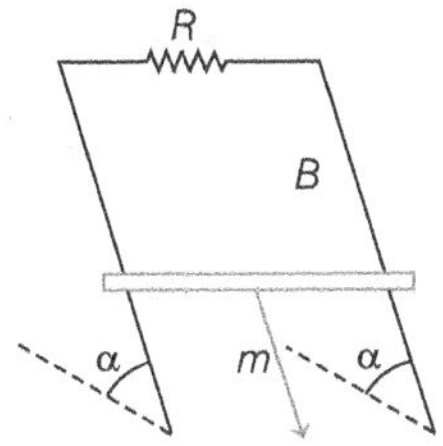

41. A conducting rod AB of mass m slides without friction over two long conducting rails separated by a distance l. At the left end, the rails are interconnected by a resistance R. The system is located in a uniform magnetic field perpendicular to the plane of the loop. At the moment $t = 0$, the rod AB starts moving to the right with an initial velocity v_0. Neglecting the resistances of the rails and the rod AB, as well as the self-inductance, find:
(a) the distance covered by the rod until it comes to a standstill
(b) the amount of heat generated in the resistance R during this process. **(Irodov 3.302)**

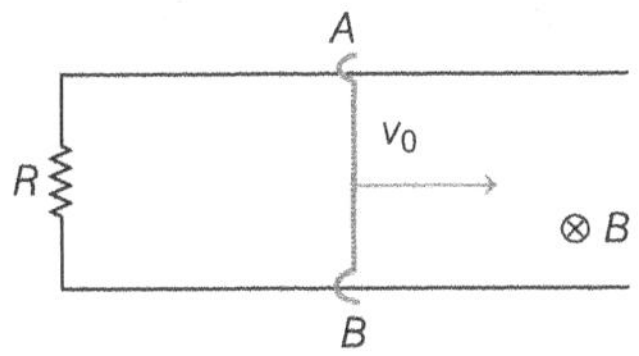

42. Figure illustrates plane figures made of thin conductors which are located in a uniform magnetic field directed away from a reader beyond the plane of the drawing. The magnetic induction starts diminishing. Find how the currents induced in these loops are directed. **(Irodov 3.304)**

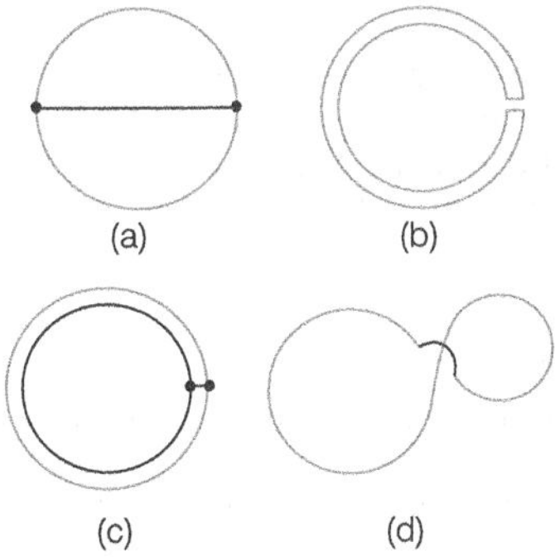

43. A plane loop shown in figure is shaped as two squares with sides $a = 20$ cm and $b = 10$ cm and is introduced into a uniform magnetic field at right angles to the loop's plane. The magnetic induction varies with time as $B = B_0 \sin \omega t$, where $B_0 = 10$ mT and $\omega = 100\ s^{-1}$. Find the amplitude of the current induced in the loop if its resistance per unit length is equal to $\rho = 50\ m\Omega/m$. The inductance of the loop is to be neglected. **(Irodov 3.305)**

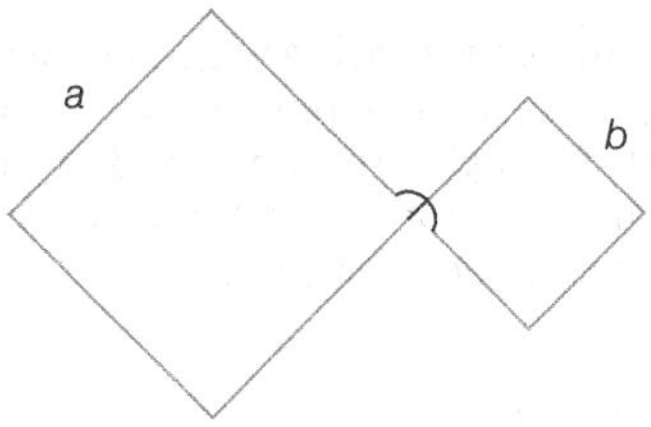

44. A plane spiral with a great number N of turns wound tightly to one another is located in a uniform magnetic field perpendicular to the spiral's plane. The outside radius of the spiral's turns is equal to a. The magnetic induction varies with time as $B = B_0 \sin \omega t$, where B_0 and ω are constants. Find the amplitude of emf induced in the spiral.

45. There are two stationary loops with mutual inductance M. The current in one of the loops starts to be varied as $I_1 = \alpha t$, where α is a constant, t is time. Find the time dependence $I_2(t)$ of the current in the other loop whose inductance is L_2 and resistance R. **(Irodov 3.334)**

46. In an oscillating circuit consisting of a parallel plate capacitor and an inductance coil with negligible active resistance the oscillations with energy W are sustained. The capacitor plates were slowly drawn apart to increase the oscillation frequency η-fold. What work was done in the process? **(Irodov 4.97)**

47. In an oscillating circuit shown in figure the coil inductance is equal to $L = 2.5$ mH and the capacitors have capacitances $C_1 = 2.0\ \mu F$ and $C_2 = 3.0\ \mu F$. The capacitors were charged to a voltage $V_m = 180$ V, and then the switch S was closed. Find: **(Irodov 4.98)**

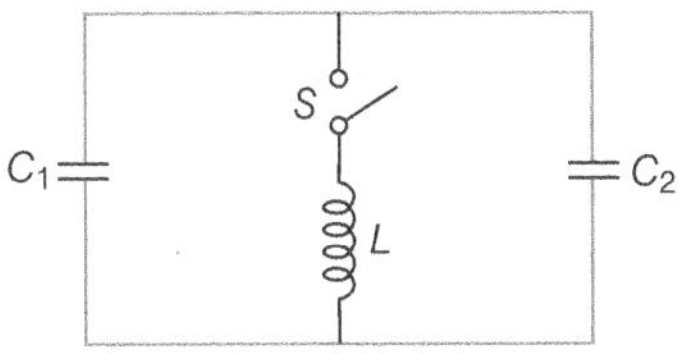

(a) the time period of oscillations
(b) the peak value of the current flowing through the coil.

48. An oscillating circuit consists of an inductance coil L and a capacitor with capacitance C. The resistance of the coil and the lead wires is negligible. The coil is placed in a permanent magnetic field so that the total flux passing through all the turns of the coil is equal to ϕ. At the moment $t = 0$ the magnetic field was switched OFF. Assuming the switching OFF time to be negligible compared to the natural oscillation period of the circuit, find the circuit current as a function of time t. **(Irodov 4.100)**

Level 2

Single Correct Option

49. A conducting rod is moving with a constant velocity v over the parallel conducting rails which are connected at the ends through a resistor R and capacitor C as shown in the figure. Magnetic field B is into the plane. Consider the following statements.

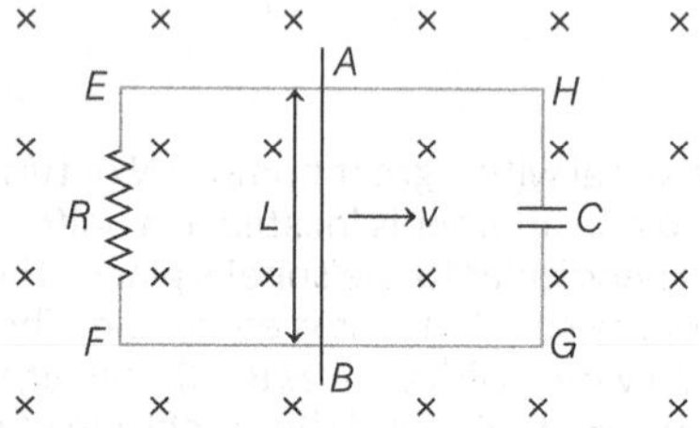

(i) Current in loop $AEFBA$ is anti-clockwise
(ii) Current in loop $AEFBA$ is clockwise
(iii) Current through the capacitor is zero
(iv) Energy stored in the capacitor is $\frac{1}{2}CB^2L^2v^2$

Which of the following options is correct?
(a) Statements (i) and (iii) are correct
(b) Statements (ii) and (iv) are correct
(c) Statements (i), (iii) and (iv) are correct
(d) None of the above

50. A rod is rotating with a constant angular velocity ω about point O (its centre) in a magnetic field B as shown. Which of the following figure correctly shows the distribution of charge inside the rod?

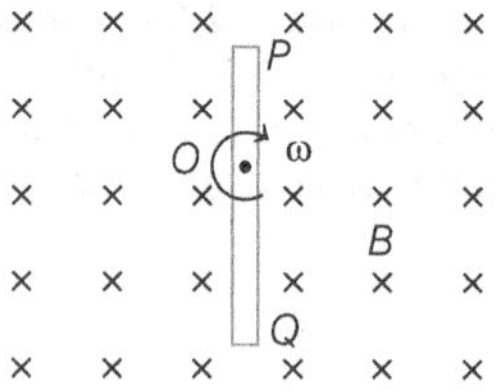

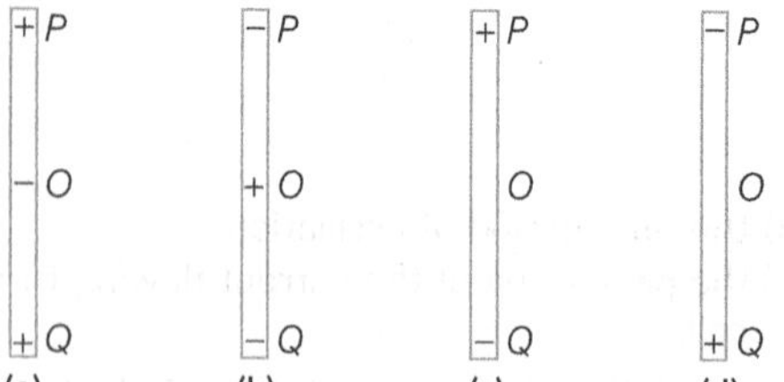

(a) (b) (c) (d)

51. A straight conducting rod PQ is executing SHM in xy-plane from $x = -d$ to $x = +d$. Its mean position is $x = 0$ and its length is along y-axis. There exists a uniform magnetic field B from $x = -d$ to $x = 0$ pointing inward normal to the paper and from $x = 0$ to $= +d$ there exists another uniform magnetic field of same magnitude B but pointing outward normal to the plane of the paper. At the instant $t = 0$, the rod is at $x = 0$ and moving to the right. The induced emf (ε) across the rod PQ vs time (t) graph will be

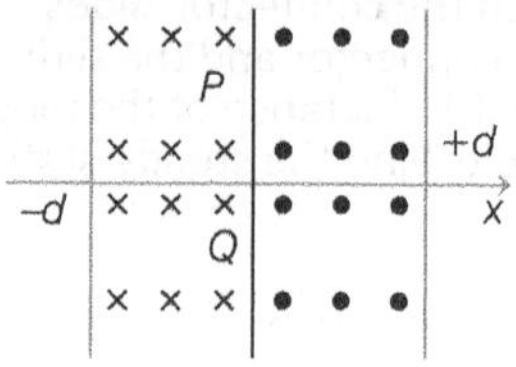

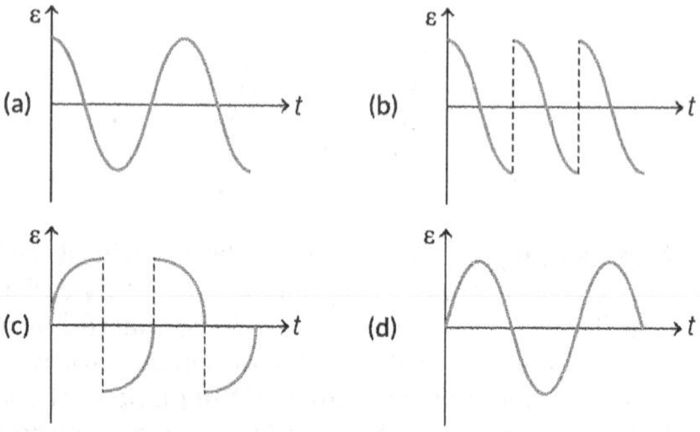

52. Two parallel long straight conductors lie on a smooth plane surface. Two other parallel conductors rest on them at right angles so as to form a square of side a. A uniform magnetic field B exists at right angles to the plane containing the conductors. Now, conductors start moving outward with a constant velocity v_0 at $t = 0$. Then, induced current in the loop at any time t is (λ is resistance per unit length of the conductors)

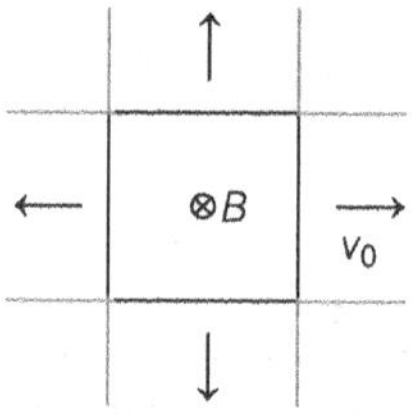

(a) $\frac{aBv_0}{\lambda(a + v_0t)}$
(b) $\frac{aBv_0}{2\lambda}$
(c) $\frac{Bv_0}{\lambda}$
(d) $\frac{Bv_0}{2\lambda}$

53. A conducting square loop is placed in a magnetic field B with its plane perpendicular to the field. Now the sides of the loop start shrinking at a constant rate α. The induced emf in the loop at an instant when its side is a, is
(a) $2a\alpha B$
(b) $a^2\alpha B$
(c) $2a^2\alpha B$
(d) $a\alpha B$

54. A uniform but time varying magnetic field exists in a cylindrical region as shown in the figure. The direction of magnetic field is into the plane of the paper and its magnitude is decreasing at a constant rate of 2×10^{-3} T/s. A particle of charge 1 μC is moved slowly along a circle of radius 1 m by an external force. The plane of the circle lies in the plane of the paper and it is concentric with the cylindrical region. The work done by the external force in moving this charge along the circle will be

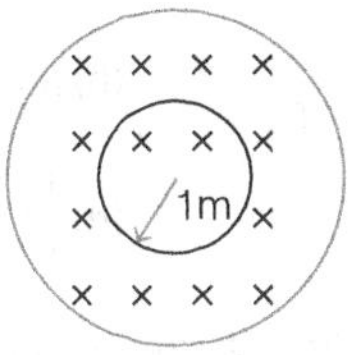

(a) zero (b) $2\pi \times 10^{-9}$ J
(c) $\pi \times 10^{-9}$ J (d) $4\pi \times 10^{-6}$ J

55. *PQ* is an infinite current carrying conductor. *AB* and *CD* are smooth conducting rods on which a conductor *EF* moves with constant velocity *v* as shown. The force needed to maintain constant speed of *EF* is

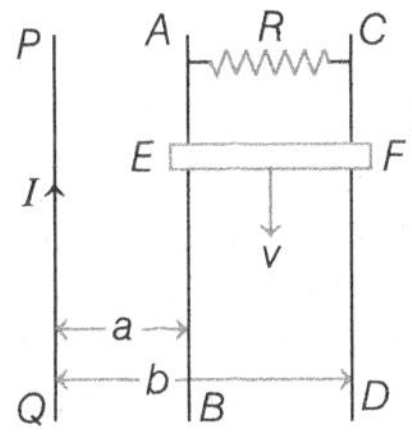

(a) $\frac{1}{vR}\left[\frac{\mu_0 Iv}{2\pi}\ln\frac{(b)}{(a)}\right]^2$ (b) $\frac{v}{R}\left[\frac{\mu_0 Iv}{2\pi}\ln\frac{(a)}{(b)}\right]^2$

(c) $\frac{v}{R}\left[\frac{\mu_0 Iv}{2\pi}\ln\frac{(b)}{(a)}\right]^2$ (d) None of these

56. Two metallic rings of radius *R* are rolling on a metallic rod. A magnetic field of magnitude *B* is applied in the region. The magnitude of potential difference between points *A* and *C* on the two rings (as shown), will be

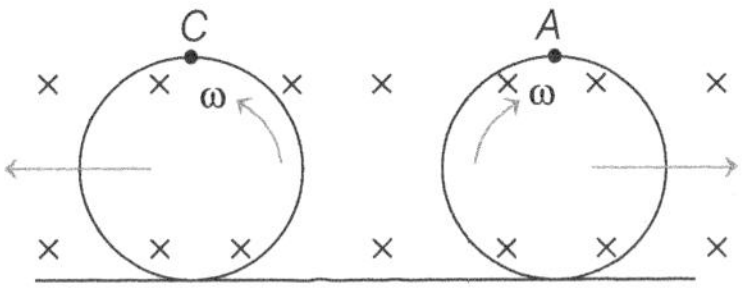

(a) 0 (b) $4B\omega R^2$
(c) $8B\omega R^2$ (d) $2B\omega R^2$

57. In the figure, magnetic field points into the plane of paper and the conducting rod of length *l* is moving in this field such that the lowest point has a velocity v_1 and the topmost point has the velocity $v_2 (v_2 > v_1)$. The emf induced is given by

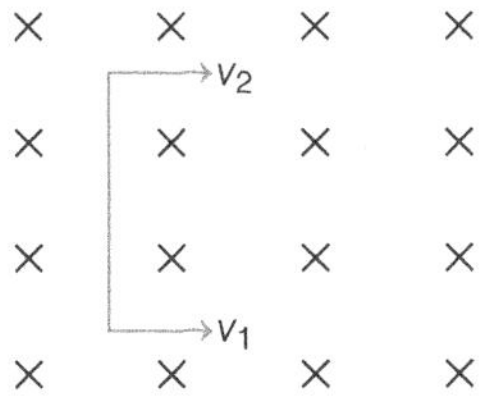

(a) $Bv_1 l$ (b) $Bv_2 l$
(c) $\frac{1}{2}B(v_2 + v_1)l$ (d) $\frac{1}{2}B(v_2 - v_1)l$

58. Find the current passing through battery immediately after key (*K*) is closed. It is given that initially all the capacitors are uncharged. (Given that $R = 6\Omega$ and $C = 4\mu F$)

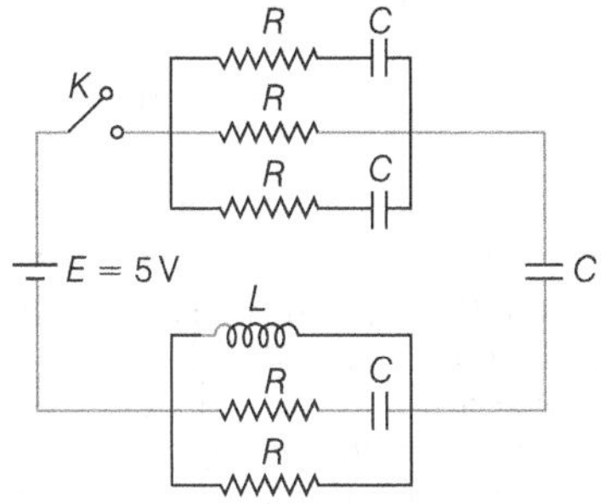

(a) 1 A (b) 5 A
(c) 3 A (d) 2 A

59. A loop shown in the figure is immersed in the varying magnetic field $B = B_0 t$, directed into the page. If the total resistance of the loop is *R*, then the direction and magnitude of induced current in the inner circle is

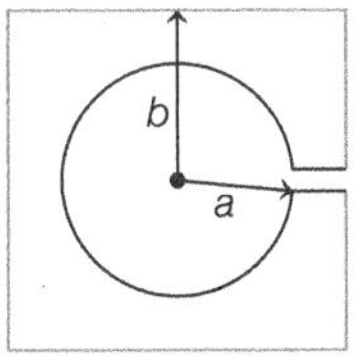

(a) clockwise $\frac{B_0(\pi a^2 - b^2)}{R}$

(b) anti-clockwise $\frac{B_0\pi(a^2 + b^2)}{R}$

(c) clockwise $\frac{B_0(\pi a^2 + 4b^2)}{R}$

(d) clockwise $\frac{B_0(4b^2 - \pi a^2)}{R}$

60. A square loop of side *a* and a straight long wire are placed in the same plane as shown in figure. The loop has a resistance *R* and inductance *L*. The frame is turned through 180° about the axis *OO′*. What is the electric charge that flows through the loop?

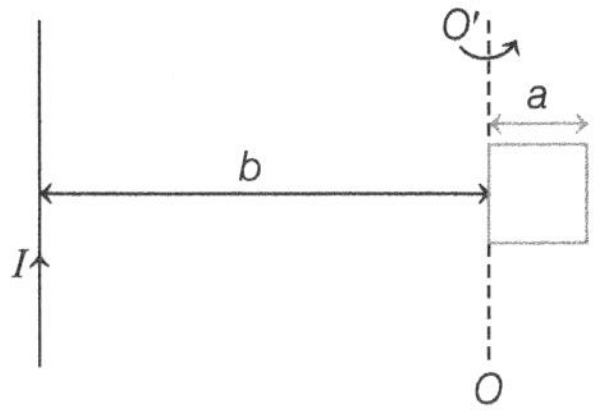

(a) $\frac{\mu_0 Ia}{2\pi R}\ln\left(\frac{2a+b}{b}\right)$ (b) $\frac{\mu_0 Ia}{2\pi R}\ln\left(\frac{b}{b^2 - a^2}\right)$

(c) $\frac{\mu_0 Ia}{2\pi R}\ln\left(\frac{a+2b}{b}\right)$ (d) None of these

61. A circuit element is placed in a closed box. At time $t = 0$, a constant current generator supplying a current of 1 A is connected across the box. Potential difference across the box varies according to graph as shown in the figure. The element in the box is

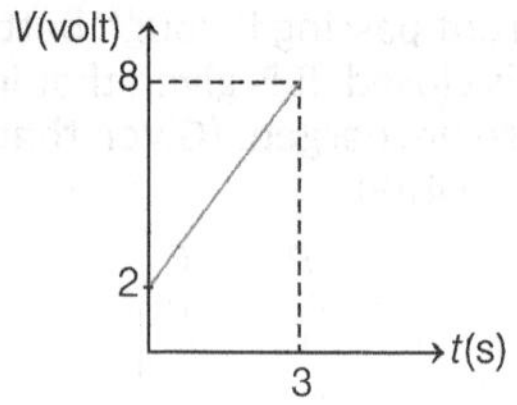

(a) a resistance of 2Ω (b) a battery of emf 6 V
(c) an inductance of 2 H (d) a capacitance of 0.5 F

62. A uniform but time varying magnetic field is present in a circular region of radius R. The magnetic field is perpendicular and into the plane of the loop and the magnitude of field is increasing at a constant rate α. There is a straight conducting rod of length $2R$ placed as shown in figure.

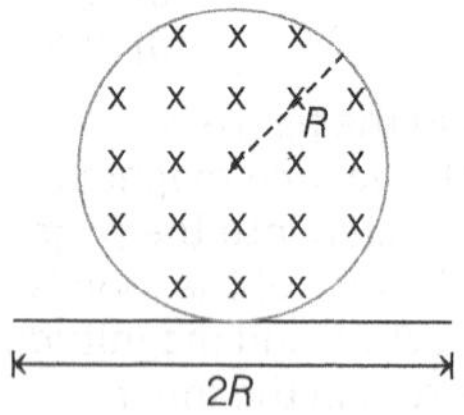

The magnitude of induced emf across the rod is

(a) $\pi R^2\alpha$ (b) $\frac{\pi R^2\alpha}{2}$ (c) $\frac{R^2\alpha}{\sqrt{2}}$ (d) $\frac{\pi R^2\alpha}{4}$

63. A uniform magnetic field B exists in a cylindrical region of radius 10 cm as shown in figure. A uniform wire of length 80 cm and resistance 4.0 Ω is bent into a square frame and is placed with one side along a diameter of the cylindrical region. If the magnetic field increases at a constant rate of 0.010 T/s, the current induced in the frame is

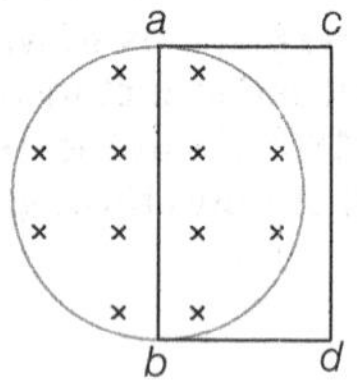

(a) 3.9×10^{-5} A (b) 2.1×10^{-2} A
(c) 2.6×10^{-5} A (d) 1.5×10^{-4} A

64. In the circuit given below,

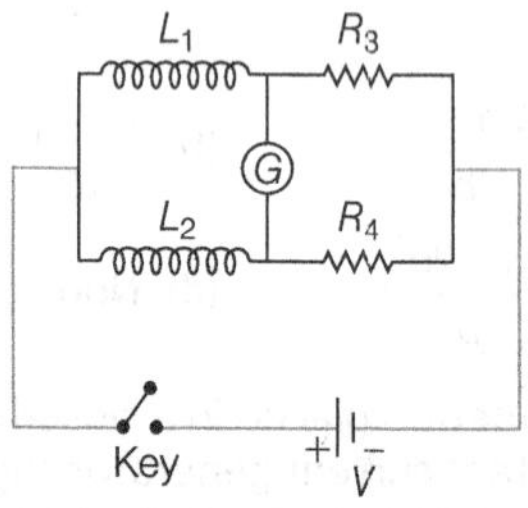

Inductors L_1 and L_2 having resistances of R_1 and R_2 are connected with R_3 and R_4. At $t=0$ key is closed, the galvanometer always shows a zero deflection when

(a) $L_1=L_2$ and $\frac{R_1}{R_2}=\frac{R_3}{R_4}$ (b) $R_1=R_2$ and $\frac{R_3}{R_4}=\frac{L_1}{L_2}$

(c) $\frac{L_1}{L_2}=\frac{R_1}{R_2}=\frac{R_3}{R_4}$ (d) $\frac{L_1}{L_2}=\frac{R_2}{R_1}=\frac{R_3}{R_4}$

65. The magnetic field in a region is given by $\mathbf{B}=B_0\left(1+\frac{x}{a}\right)\hat{\mathbf{k}}$. A square loop of edge length d is placed with its edge along the X- axis and Y-axis. The loop is moved with a constant velocity $\mathbf{v}=v_0\hat{\mathbf{i}}$. The emf induced in the loop at $t=0$ is

(a) $\frac{v_0B_0d^2}{a}$ (b) $\frac{v_0B_0d^3}{a^2}$

(c) v_0B_0d (d) zero

66. A semicircular loop of radius R is rotated about its straight edge which divides the space into two region one having a uniform magnetic field **B** and the other having no field. If initially the plane of loop is perpendicular to **B** (see figure) and if current flowing from O to A be taken as positive, the correct plot of induced current *versus* time for one time period is

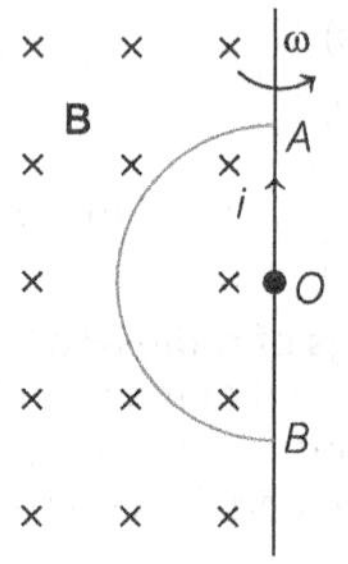

(a)
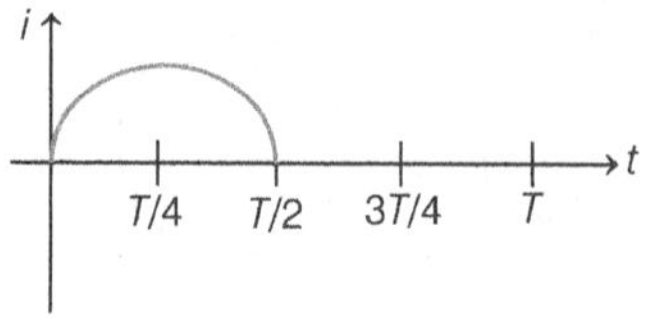

(b)
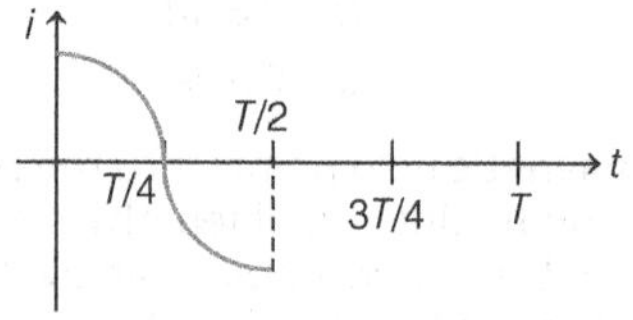

(c)
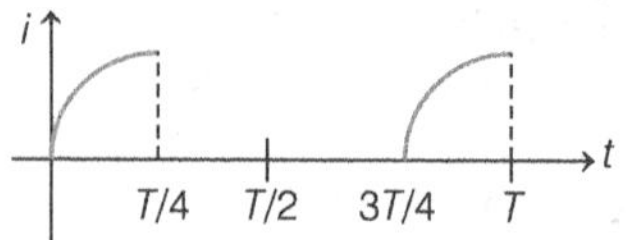

(d)
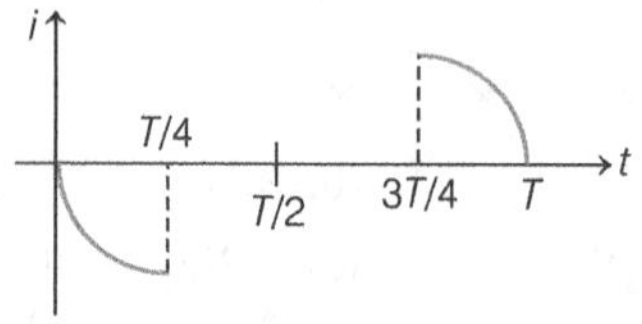

67. A rod of length 10 cm made up of conducting and non-conducting material (black part is non-conducting). The rod is rotated with constant angular velocity 10 rad/s about point O, in a constant magnetic field 2T as shown in the figure. The induced emf between the point A and B of rod will be

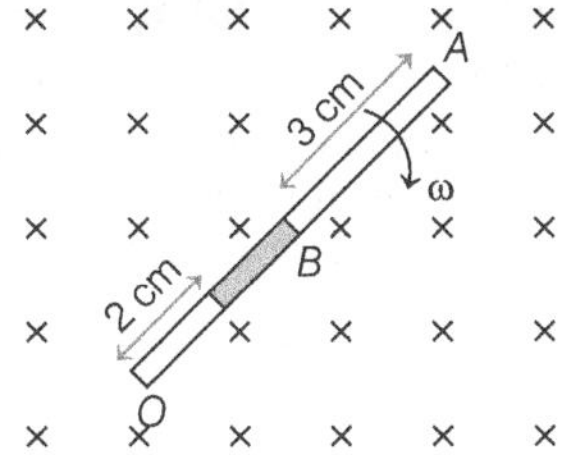

(a) 0.029 V (b) 0.1 V
(c) 0.051 V (d) 0.064 V

68. In the given arrangement, slider, PQ has given an initial speed of v_0.

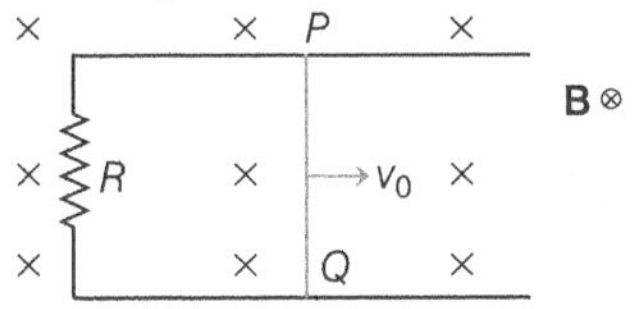

Then, graph of log v *versus* time t is given by

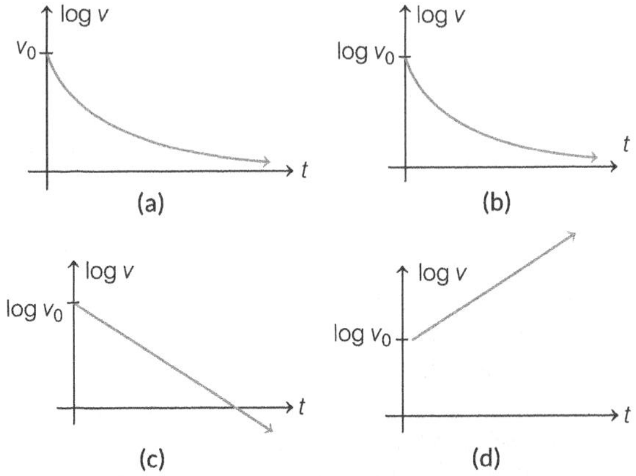

69. A triangular wire frame (each side = 2 m) is placed in a region of time varying magnetic field having $\frac{dB}{dt} = \sqrt{3}$ T/s. The magnetic field is perpendicular to the plane of the triangle. The base of the triangle AB has a resistance 1Ω while the other two sides have resistance 2Ω each. The magnitude of potential difference between the points A and B will be

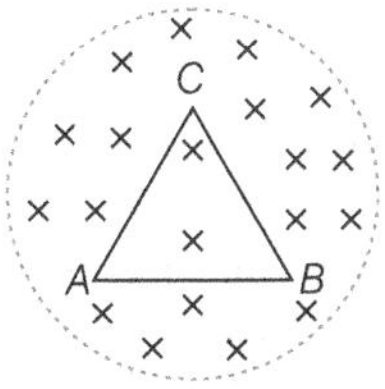

(a) 0.4 V (b) 0.6 V
(c) 1.2 V (d) None of these

70. In a cylindrical region, uniform magnetic field which is perpendicular to the plane of the figure is increasing with time and a conducting rod PQ is placed in the region. Then,

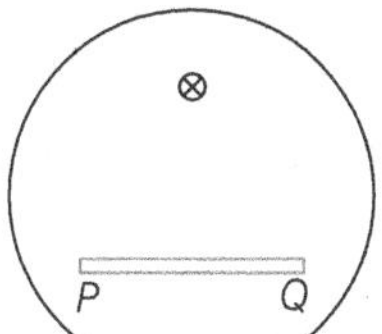

(a) P will be at higher potential than Q
(b) Q will be at higher potential than P
(c) both P and Q will be equipotential
(d) no emf will be developed

71. A cylindrical region of uniform magnetic field exists perpendicular to the plane of paper which is increasing at a constant rate $\frac{dB}{dt} = \alpha$. A rod is placed along the diameter of the cylinder. Induced emf between the ends of the rod is

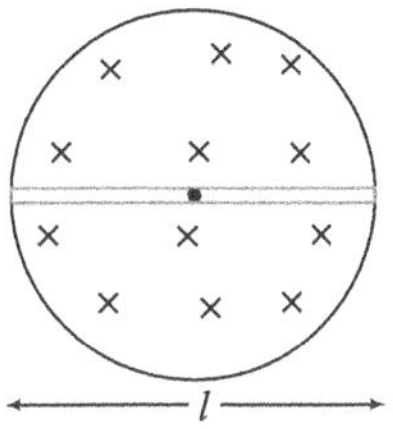

(a) $\frac{\pi \alpha l^2}{2}$ (b) $\pi l^2 \alpha$
(c) $\frac{\pi \alpha l^2}{4}$ (d) zero

Multiple Correct Options

72. The loop shown moves with a velocity v in a uniform magnetic field of magnitude B, directed into the paper. The potential difference between points P and Q is e. Then,

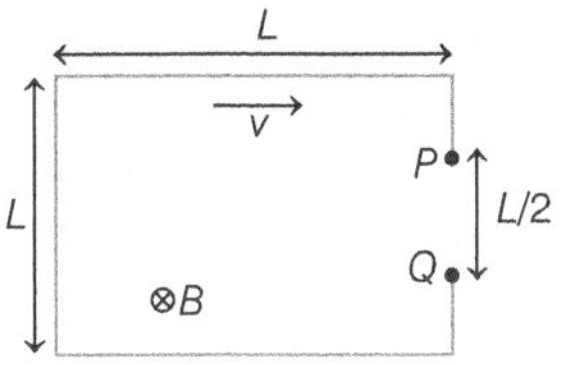

(a) $e = \frac{1}{2} BLv$
(b) $e = BLv$
(c) P is positive with respect to Q
(d) Q is positive with respect to P

73. An infinitely long wire is placed near a square loop as shown in figure. Choose the correct options.

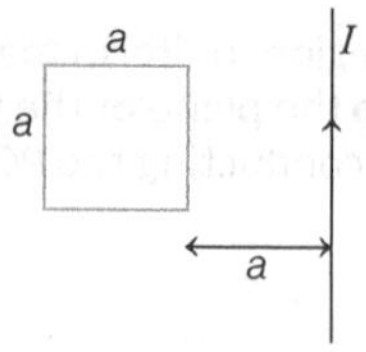

(a) The mutual inductance between the two is $\frac{\mu_0 a}{2\pi} \ln(2)$

(b) The mutual inductance between the two is $\frac{\mu_0 a^2}{2\pi} \ln(2)$

(c) If a constant current is passed in the straight wire in upward direction and loop is brought close to the wire, then induced current in the loop is clockwise

(d) In the above condition, induced current in the loop is anti-clockwise

74. Two circular coils are placed adjacent to each other. Their planes are parallel and currents through them i_1 and i_2 are in same direction. Choose the correct options.

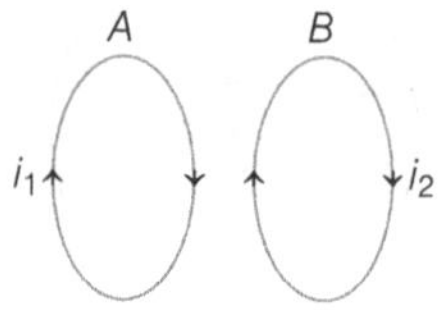

(a) When A is brought near B, current i_2 will decrease

(b) In the above process, current i_2 will increase

(c) When current i_1 is increased, current i_2 will decrease

(d) In the above process, current i_2 will increase

75. A coil of area 2 m^2 and resistance 4Ω is placed perpendicular to a uniform magnetic field of 4 T. The loop is rotated by 90° in 0.1 second. Choose the correct options.

(a) Average induced emf in the coil is 8 V

(b) Average induced current in the circuit is 20 A

(c) 2 C charge will flow in the coil in above period

(d) Heat produced in the coil in the above period can't be determined from the given data

76. In L-C oscillations,

(a) time period of oscillation is $\frac{2\pi}{\sqrt{LC}}$

(b) maximum current in circuit is $\frac{q_0}{\sqrt{LC}}$

(c) maximum rate of change of current in circuit is $\frac{q_0}{LC}$

(d) maximum potential difference across the inductor is $\frac{q_0}{2C}$. Here, q_0 is maximum charge on capacitor

77. In the figure shown, q is in coulomb and t in second. At time $t = 1$ s

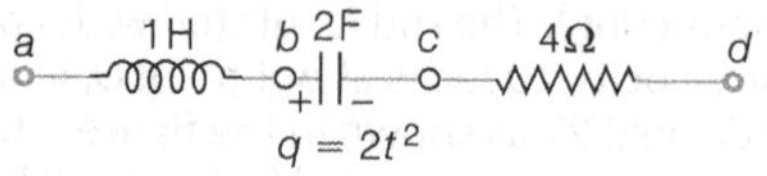

(a) $V_a - V_b = 4$ V
(b) $V_b - V_c = 1$ V
(c) $V_c - V_d = 16$ V
(d) $V_a - V_d = 20$ V

78. An equilateral triangular conducting frame is rotated with angular velocity ω in a uniform magnetic field B as shown. Side of triangle is l. Choose the correct options.

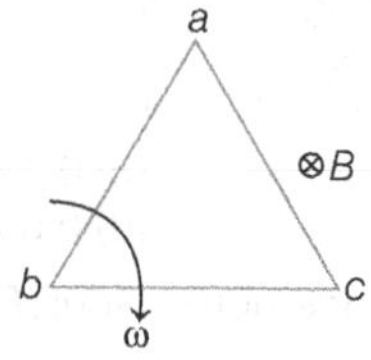

(a) $V_a - V_c = 0$
(b) $V_a - V_c = \frac{B\omega l^2}{2}$
(c) $V_a - V_b = \frac{B\omega l^2}{2}$
(d) $V_c - V_b = -\frac{B\omega l^2}{2}$

79. In the figure shown, the magnetic field starts diminishing. Then, the induced current

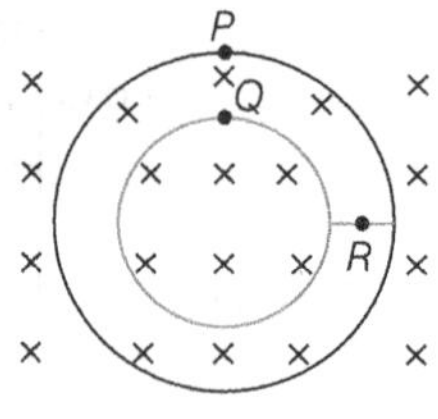

(a) at point P is clockwise

(b) at point Q is anti-clockwise

(c) at point Q is clockwise

(d) at point R is zero

80. A resistance R and a capacitor C is connected with a metal wire to form a loop and this is placed in a magnetic field that points vertically into the loop and is decreasing at a constant rate $\frac{dB}{dt} = -k$.

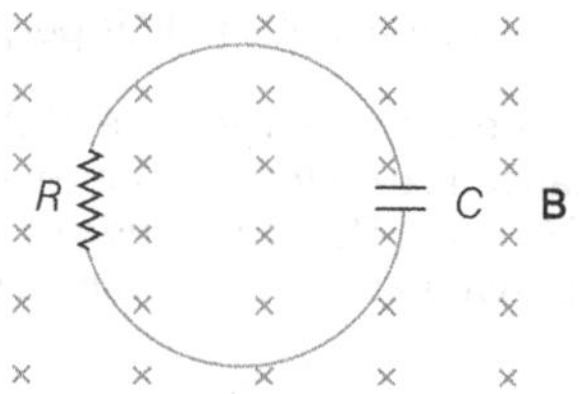

Choose the correct statements.

(a) Induced emf is $\pi a^2 k$, where a = radius of loop

(b) Charge stored in capacitor is $C\pi a^2 k$ coulombs

(c) Upper plate of capacitor is negatively charged

(d) Changing magnetic field induces an electric field that pushes charge

Passage Based Questions

Passage 1 (Q. No. 81 to 84)

A thin non-conducting ring of mass m, radius a carrying a charge q can rotate freely about its own axis which is vertical. At the initial moment, the ring was at rest in horizontal position and no magnetic field was present. At instant $t = 0$, a uniform magnetic field is switched on which is vertically downward and increases with time according to the law $B = B_0 t$. Neglecting magnetism induced due to rotational motion of ring.

81. The magnitude of induced emf on the closed surface of ring will be

(a) $\pi a^2 B_0$ (b) $2a^2 B_0$ (c) zero (d) $\frac{1}{2}\pi a^2 B_0$

82. The magnitude of an electric field on the circumference of the ring is

(a) aB_0 (b) $2aB_0$ (c) $\frac{1}{2}aB_0$ (d) zero

83. Angular acceleration of ring is

(a) $\frac{qB_0}{2m}$ (b) $\frac{qB_0}{4m}$ (c) $\frac{qB_0}{m}$ (d) $\frac{2qB_0}{m}$

84. Find instantaneous power developed by electric force acting on the ring at $t = 1$ s.

(a) $\frac{2q^2B_0^2a^2}{14m}$ (b) $\frac{q^2B_0^2a^2}{8m}$ (c) $\frac{3q^2B_0^2a^2}{m}$ (d) $\frac{q^2B_0^2a^2}{4m}$

Passage 2 (Q. No. 85 to 87)

Figure shows a conducting rod of negligible resistance that can slide on smooth U-shaped rail made of wire of resistance $1\ \Omega/m$. Position of the conducting rod at $t = 0$ is shown. A time dependent magnetic field $B = 2t$ tesla is switched on at $t = 0$.

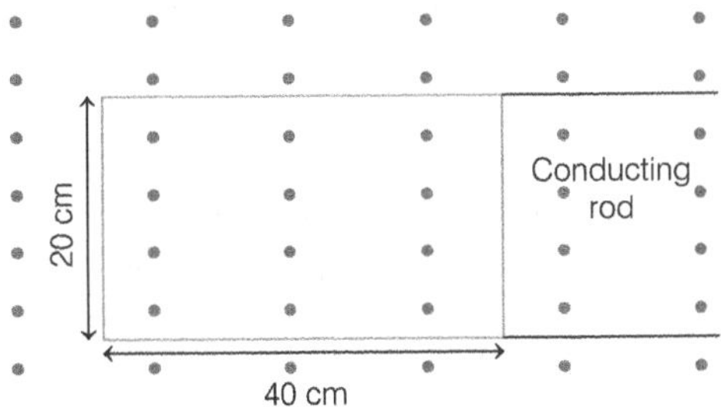

85. The current in the loop at $t = 0$ due to induced emf is

(a) 0.16 A, clockwise (b) 0.08 A, clockwise
(c) 0.16 A, anti-clockwise (d) zero

86. At $t = 0$, when the magnetic field is switched on, the conducting rod is moved to the left at constant speed 5 cm/s by some external means. At $t = 2$ s, net induced emf has magnitude

(a) 0.12 V (b) 0.08 V (c) 0.04 V (d) 0.02 V

87. The magnitude of the force required to move the conducting rod at constant speed 5 cm/s at the same instant $t = 2$ s, is equal to

(a) 0.096 N (b) 0.12 N (c) 0.08 N (d) 0.064 N

Passage 3 (Q. No. 88 to 90)

Two parallel vertical metallic rails AB and CD are separated by 1 m. They are connected at the two ends by resistances R_1 and R_2 as shown in the figure. A horizontal metallic bar L of mass 0.2 kg slides without friction, vertically down the rails under the action of gravity. There is a uniform horizontal magnetic field of 0.6 T perpendicular to the plane of the rails. It is observed that when the terminal velocity is attained, the powers dissipated in R_1 and R_2 are 0.76 W and 1.2 W, respectively ($g = 9.8\ m/s^2$)

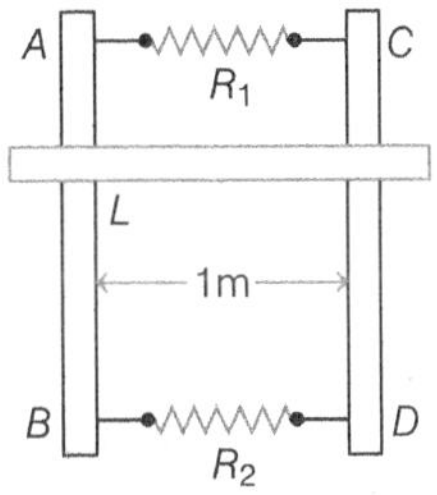

88. The terminal velocity of the bar L will be

(a) 2 m/s (b) 3 m/s
(c) 1 m/s (d) None of these

89. The value of R_1 is

(a) $0.47\ \Omega$ (b) $0.82\ \Omega$
(c) $0.12\ \Omega$ (d) None of these

90. The value of R_2 is

(a) $0.6\ \Omega$ (b) $0.5\ \Omega$
(c) $0.4\ \Omega$ (d) $0.3\ \Omega$

Passage 4 (Q. No 91 to 93)

Initially the capacitor is charged to a potential of 5 V and then connected to position 1 with the shown polarity for 1s. After 1s, it is connected to the inductor at position 2.

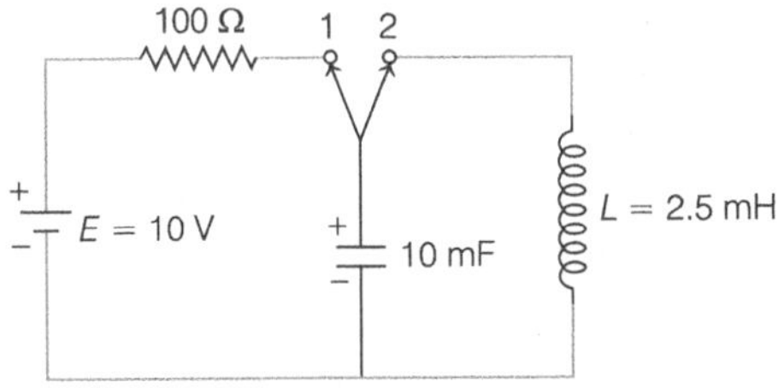

91. The potential difference across the capacitor after 1s of its connection to position 1 is

(a) $5\left[2 + \frac{1}{e}\right]$ V (b) $5\left[2 - \frac{1}{e}\right]$ V
(c) $5\left[1 + \frac{2}{e}\right]$ V (d) None of these

92. The maximum current flowing in the L-C circuit when the capacitor is connected across the inductor is

(a) $\left[2 - \frac{1}{e}\right] \times 10^4$ A (b) $\left[1 + \frac{2}{e}\right] \times 10^4$ A
(c) $\left[1 - \frac{2}{e}\right] \times 10^4$ A (d) None of these

93. The frequency of LC oscillation is

(a) $\left[\frac{20}{\pi}\right]$ Hz (b) $\left[\frac{2}{\pi}\right]$ Hz (c) $\left[\frac{40}{\pi}\right]$ Hz (d) $\left[\frac{100}{\pi}\right]$ Hz

Passage 5 (Q. No. 94 to 96)

The capacitor of capacitance C can be charged (with the help of a resistance R) by a voltage source V, by closing switch S_1 while keeping switch S_2 open. The capacitor can be connected in series with an inductor L by closing switch S_2 and opening S_1. **(2006 JEE)**

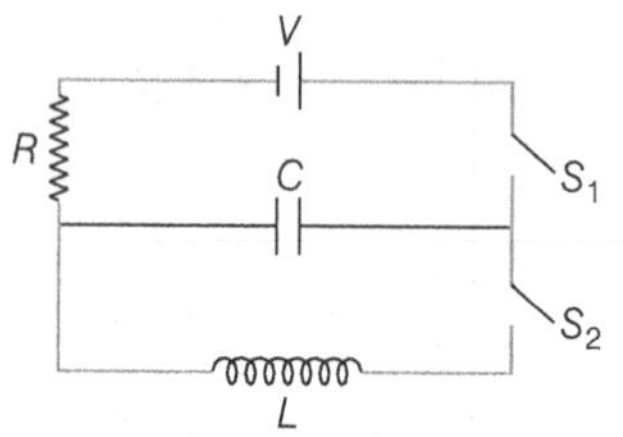

94. Initially, the capacitor was uncharged. Now S_1 is closed and S_2 is kept open. If time constant of this circuit is τ, then
(a) after time interval τ, charge on the capacitor is $CV/2$
(b) after time interval 2τ, charge on the capacitor is $CV(1-e^{-2})$
(c) the work done by the voltage source will be half of the heat dissipated when the capacitor is fully charged
(d) after time interval 2τ, charge on the capacitor is $CV(1-e^{-1})$

95. After the capacitor gets fully charged, S_1 is opened and S_2 is closed so that the inductor is connected in series with the capacitor. Then,
(a) at $t=0$, energy stored in the circuit is purely in the form of magnetic energy.
(b) at any time $t>0$, current in the circuit is in the same direction.
(c) at $t>0$, there is no exchange of energy between the inductor and capacitor.
(d) at any time $t>0$, maximum instantaneous current in the circuit may be $V\sqrt{\frac{C}{L}}$.

96. If the total charge stored in the LC circuit is Q_0, then for $t \geq 0$,

(a) the charge on the capacitor is $Q=Q_0 \cos\left(\frac{\pi}{2}+\frac{t}{\sqrt{LC}}\right)$

(b) the charge on the capacitor is $Q=Q_0 \cos\left(\frac{\pi}{2}-\frac{t}{\sqrt{LC}}\right)$

(c) the charge on the capacitor is $Q=-LC\frac{d^2Q}{dt^2}$

(d) the charge on the capacitor is $Q=-\frac{1}{\sqrt{LC}}\frac{d^2Q}{dt^2}$

Match the Columns

97. Match the following two columns.

	Column I		Column II
(a)	Magnetic induction	(p)	$[MT^{-2}A^{-1}]$
(b)	Coefficient of self-induction	(q)	$[L^2T^{-2}]$
(c)	LC	(r)	$[ML^2T^{-2}A^{-2}]$
(d)	Magnetic flux	(s)	None of these

98. In an L-C oscillation circuit, $L=1$ H, $C=\frac{1}{4}$ F and maximum charge in the capacitor is 4 C. Match the following two columns. Note that in Column II, all values are in SI units.

	Column I		Column II
(a)	Maximum current in the circuit	(p)	16
(b)	Maximum rate of change of current in the circuit	(q)	4
(c)	Potential difference across inductor when $q=2C$	(r)	2
(d)	Potential difference across capacitor when rate of change of current is half its maximum value	(s)	8

99. In the circuit shown in figure, switch remains closed for long time. It is opened at time $t=0$. Match the following two columns at $t=(\ln 2)$ second.

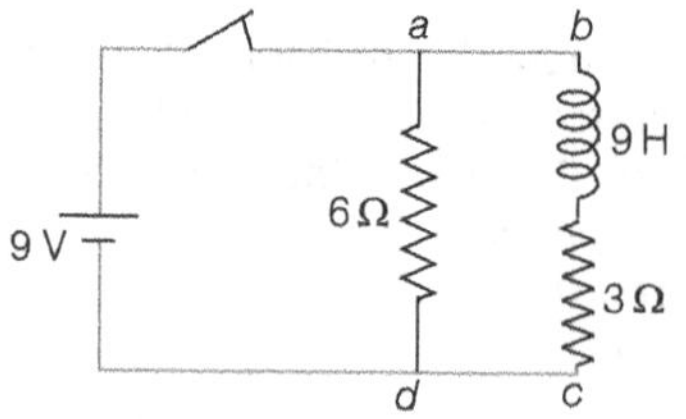

	Column I		Column II
(a)	Potential differences across inductor	(p)	9 V
(b)	Potential difference across 3Ω resistance	(q)	4.5 V
(c)	Potential difference across 6Ω resistance	(r)	6 V
(d)	Potential difference between points b and c	(s)	None of these

100. Magnetic flux passing through a coil of resistance 2 Ω is as shown in figure. Match the following two columns. In Column II all physical quantities are in SI units.

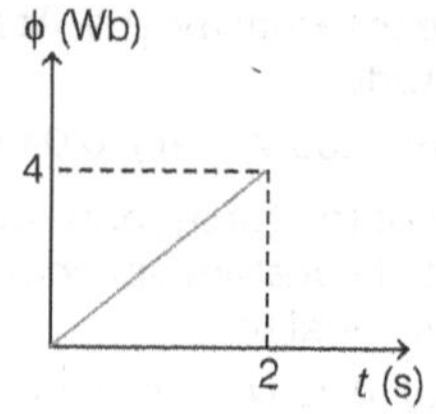

	Column I		Column II
(a)	Induced emf produced	(p)	4
(b)	Induced current	(q)	1
(c)	Charge flow in 2 s	(r)	8
(d)	Heat generation in 2 s	(s)	2

101. A square loop is placed near a long straight current carrying wire as shown. Match the following two columns.

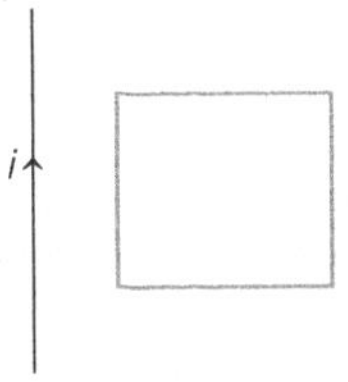

	Column I		Column II
(a)	If current is increased	(p)	induced current in loop is clockwise
(b)	If current is decreased	(q)	induced current in loop is anti-clockwise
(c)	If loop is moved away from the wire	(r)	wire will attract the loop
(d)	If loop is moved towards the wire	(s)	wire will repel the loop

102. Magnetic flux in a circular coil of resistance 10Ω changes with time as shown in the figure.

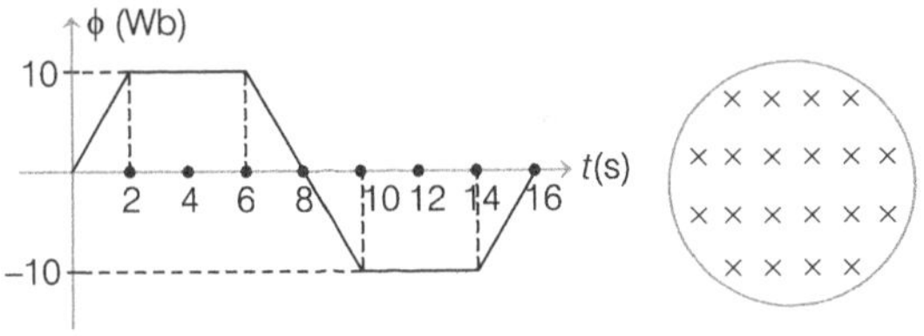

Now, match the following columns and select the correct option.

	Column I		Column II
(a)	At 1 s induced current is	(p)	Clockwise
(b)	At 5 s induced current is	(q)	Anti-clockwise
(c)	At 9 s induced current is	(r)	Zero
(d)	At 15 s induced current is	(s)	2A

Subjective Questions

103. In the circuit diagram shown, initially there is no energy in the inductor and the capacitor. The switch is closed at $t = 0$. Find the current I as a function of time if $R = \sqrt{L/C}$.

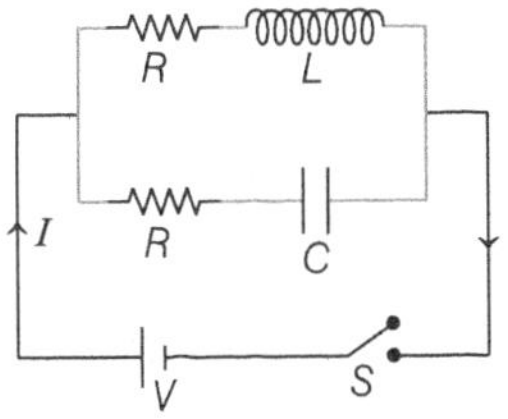

104. A rod of length $2a$ is free to rotate in a vertical plane, about a horizontal axis O passing through its mid-point. A long straight, horizontal wire is in the same plane and is carrying a constant current i as shown in figure. At initial moment of time, the rod is horizontal and starts to rotate with constant angular velocity ω, calculate emf induced in the rod as a function of tim

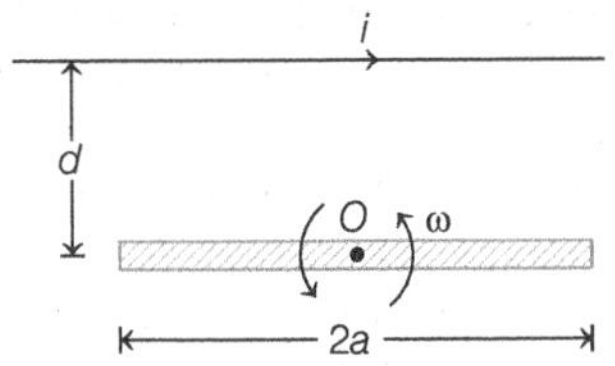

105. In the circuit shown, switch S is closed at time $t = 0$. Find the current through the inductor as a function of time t.

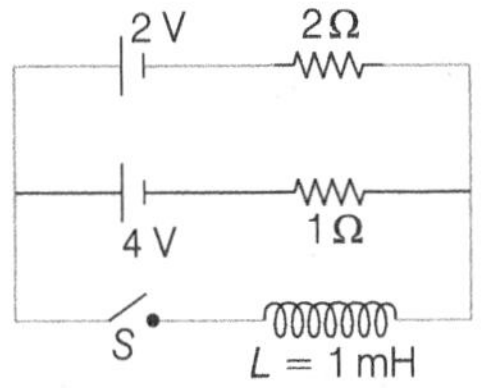

106. In the circuit shown in figure, $E = 120$ V, $R_1 = 30.0\,\Omega$, $R_2 = 50.0\,\Omega$ and $L = 0.200$ H. Switch S is closed at $t = 0$. Just after the switch is closed.

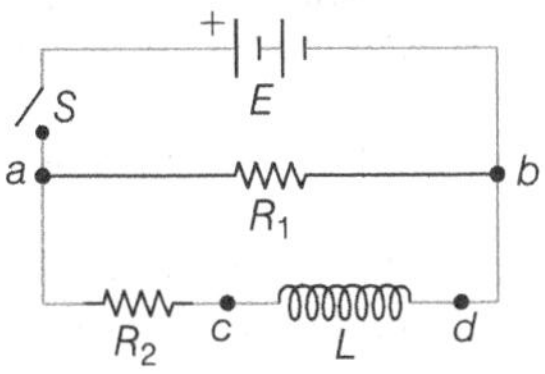

(a) What is the potential difference V_{ab} across the resistance R_1?
(b) Which point, a or b, is at higher potential?
(c) What is the potential difference V_{cd} across the inductor L?
(d) Which point, c or d, is at a higher potential?
The switch is left closed for a long time and then is opened. Just after the switch is opened
(e) What is the potential difference V_{ab} across the resistor R_1?
(f) Which point a or b, is at a higher potential?
(g) What is the potential difference V_{cd} across the inductor L?
(h) Which point, c or d, is at a higher potential?

107. Two capacitors of capacitances $2C$ and C are connected in series with an inductor of inductance L. Initially, capacitors have charge such that $V_B - V_A = 4V_0$ and $V_C - V_D = V_0$. Initial current in the circuit is zero. Find

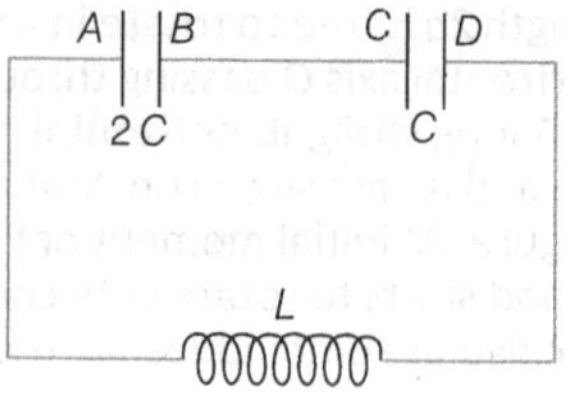

(a) maximum current that will flow in the circuit,
(b) potential difference across each capacitor at that instant,
(c) equation of current flowing towards left in the inductor.

108. A 1.00 mH inductor and a 1.00 μF capacitor are connected in series. The current in the circuit is described by $i = 20\,t$, where t is in second and i is in ampere. The capacitor initially has no charge. Determine
(a) the voltage across the inductor as a function of time,
(b) the voltage across the capacitor as a function of time,
(c) the time when the energy stored in the capacitor first exceeds that in the inductor.

109. A non-conducting ring of mass m and radius R has a charge Q uniformly distributed over its circumference. The ring is placed on a rough horizontal surface such that plane of the ring is parallel to the surface. A vertical magnetic field $B = B_0 t^2$ tesla is switched on. After 2 s from switching on the magnetic field the ring is just about to rotate about vertical axis through its centre.
(a) Find friction coefficient μ between the ring and the surface.
(b) If magnetic field is switched off after 4 s, then find the angle rotated by the ring before coming to stop after switching off the magnetic field.

110. Two parallel long smooth conducting rails separated by a distance l are connected by a movable conducting connector of mass m. Terminals of the rails are connected by the resistor R and the capacitor C as shown in figure. A uniform magnetic field B perpendicular to the plane of the rail is switched on. The connector is dragged by a constant force F. Find the speed of the connector as a function of time if the force F is applied at $t = 0$. Also, find the terminal velocity of the connector.

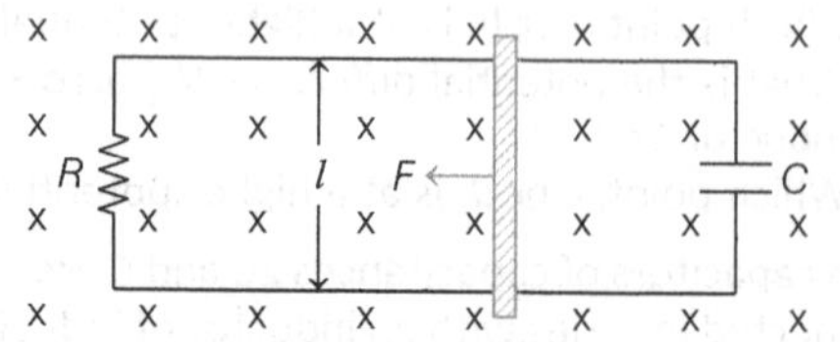

111. A circuit containing capacitors C_1 and C_2, shown in the figure is in the steady state with key K_1 closed and K_2 opened. At the instant $t = 0$, K_1 is opened and K_2 is closed.

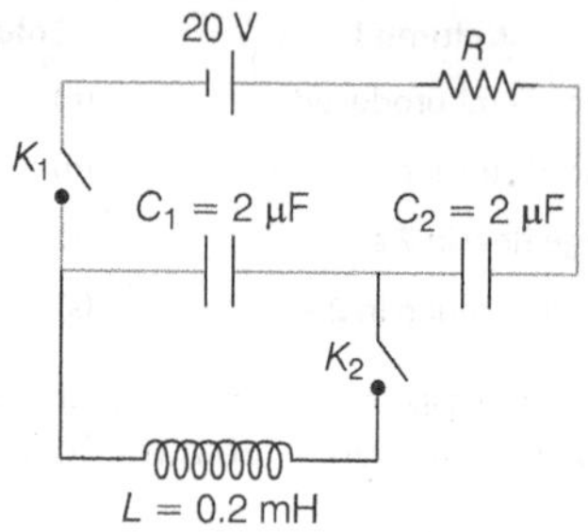

(a) Find the angular frequency of oscillations of L-C circuit.
(b) Determine the first instant t, when energy in the inductor becomes one third of that in the capacitor.
(c) Calculate the charge on the plates of the capacitor at that instant.

112. Two metal bars are fixed vertically and are connected on the top by a capacitor C. A sliding conductor of length l and mass m slides with its ends in contact with the bars. The arrangement is placed in a uniform horizontal magnetic field directed normal to the plane of the figure. The conductor is released from rest. Find the displacement $x(t)$ of the conductor as a function of time t.

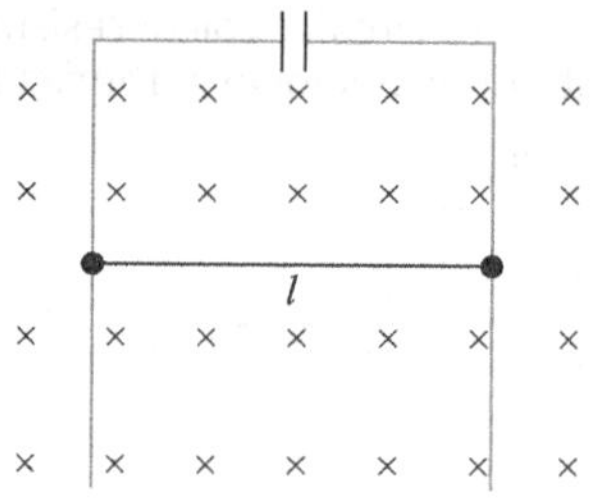

113. A conducting light string is wound on the rim of a metal ring of radius r and mass m. The free end of the string is fixed to the ceiling. A vertical infinite smooth conducting plane is always tangent to the ring as shown in the figure. A uniform magnetic field B is applied perpendicular to the plane of the ring. The ring is always inside the magnetic field. The plane and the string are connected by a resistance R. When the ring is released, find

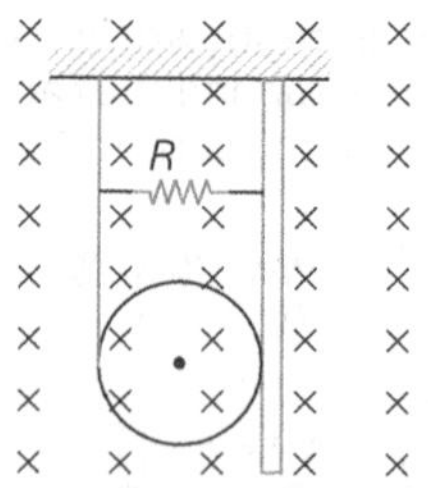

(a) the current in the resistance R as a function of time.
(b) the terminal velocity of the ring.

114. A conducting frame *abcd* is kept in a vertical plane. A conducting rod *ef* of mass *m* and length *l* can slide smoothly on it remaining always horizontal. The resistance of the loop is negligible and inductance is constant having value *L*. The rod is left from rest and allowed to fall under gravity and inductor has no initial current. A magnetic field of constant magnitude *B* is present throughout the loop pointing inwards. Determine

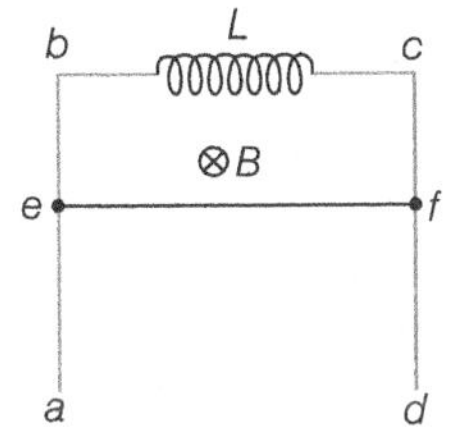

(a) position of the rod as a function of time assuming initial position of the rod to be at $x = 0$ and vertically downward as the positive *x*-axis.
(b) the maximum current in the circuit.
(c) maximum velocity of the rod

115. A rectangular loop with a sliding conductor of length *l* is located in a uniform magnetic field perpendicular to the plane of loop. The magnetic induction perpendicular to the plane of loop is equal to *B*. The part *ad* and *bc* has electric resistance R_1 and R_2, respectively. The conductor starts moving with constant acceleration a_0 at time $t = 0$. Neglecting the self-inductance of the loop and resistance of conductor. Find

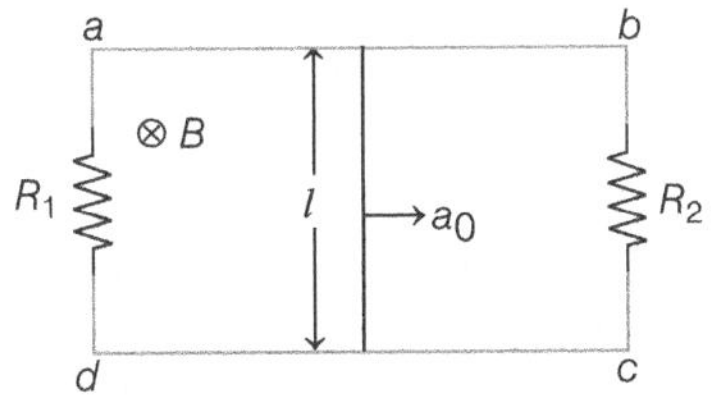

(a) the current through the conductor during its motion.
(b) the polarity of *abcd* terminal.
(c) external force required to move the conductor with the given acceleration.

116. A conducting circular loop of radius *a* and resistance per unit length *R* is moving with a constant velocity v_0, parallel to an infinite conducting wire carrying current i_0. A conducting rod of length $2a$ is approaching the centre of the loop with a constant velocity $\frac{v_0}{2}$ along the direction of the current. At the instant $t = 0$, the rod comes in contact with the loop at *A* and starts sliding on the loop with the constant velocity. Neglecting the resistance of the rod and the self-inductance of the circuit, find the following when the rod slides on the loop.

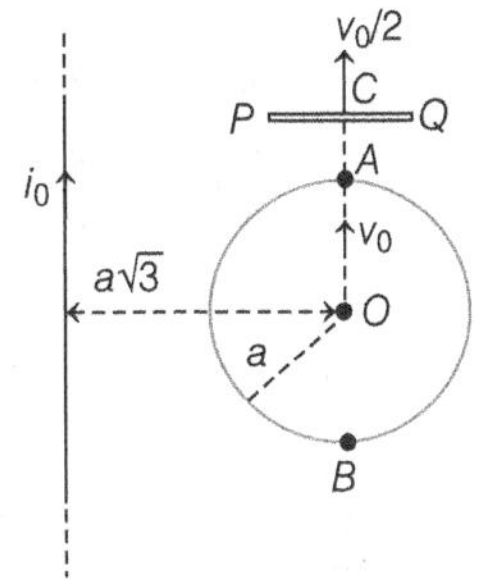

(a) The current through the rod when it is at a distance of $\left(\frac{a}{2}\right)$ from the point *A* of the loop.
(b) Force required to maintain the velocity of the rod at that instant.

117. U-frame *ADCB* and a sliding rod *PQ* of resistance *R*, starts moving with velocities *v* and $2v$ respectively, parallel to a long wire carrying current i_0. When the distance $AP = l$ at $t = 0$, determine the current through the inductor of inductance *L* just before connecting rod *PQ* loses contact with the U-frame.

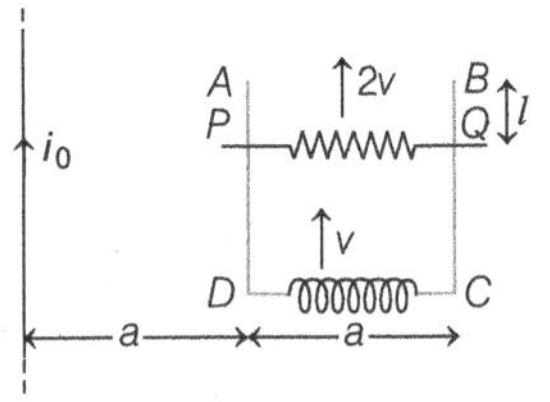

118. A uniform magnetic field of induction *B* fills a cylindrical volume of radius *R*. A rod *AB* of length $2l$ is placed as shown in the figure. If *B* is changing at the rate $\frac{dB}{dt}$, find the emf that is produced by the change of magnetic field and that acts between the ends of the rod.

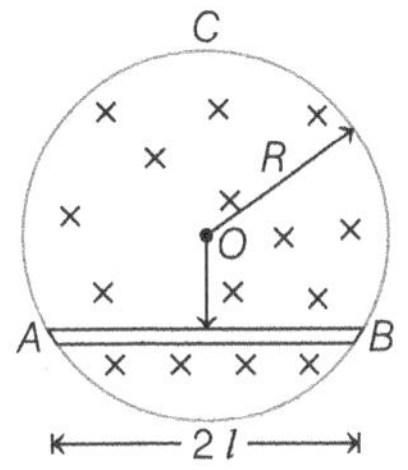

119. A rectangular frame *ABCD*, made of a uniform metal wire has a straight connection between *E* and *F* made of the same wire as shown in the figure. *AEFD* is a square of side 1m and $EB = FC = 0.5$ m. The entire circuit is placed in a steadily increasing uniform magnetic field directed into the plane of the paper and normal to it. The rate of change of the magnetic field is 1 T/s. The resistance per unit length is 1 Ω/m. Find the magnitude and directions of the current in segments *AE*, *BE* and *EF*.

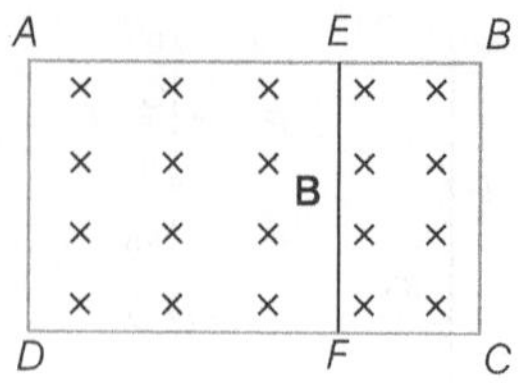

120. A metal rod OA of mass m and length r is kept rotating with a constant angular speed ω, with a uniform magnetic field B applied in perpendicular downward direction. An inductor L and resistance R are connected through a switch S across points O and C on the ring over which rods free end is rotating in vertical plane.

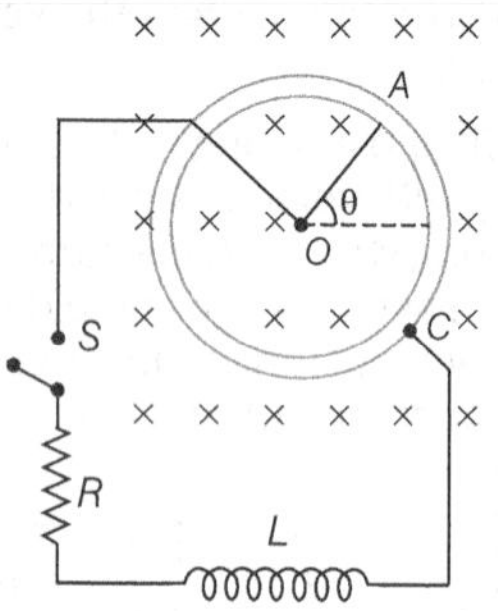

Neglect the resistance of the rod, ring and connecting wires. Obtain the value of torque required to maintain the constant speed of the rotating rod when current in the circuit reaches its steady state. At $t = 0$, switch is closed. **(1995 JEE)**

121. A magnetic field $B = (B_0 y / a)\hat{\mathbf{k}}$ is acting into the paper in the $+z$ direction. B_0 and a are positive constants. A square loop $EFGH$ of side a, mass m and resistance R in XY- plane starts falling under the influence of gravity. Note the directions of x and y in the figure. Find **(1999 JEE)**

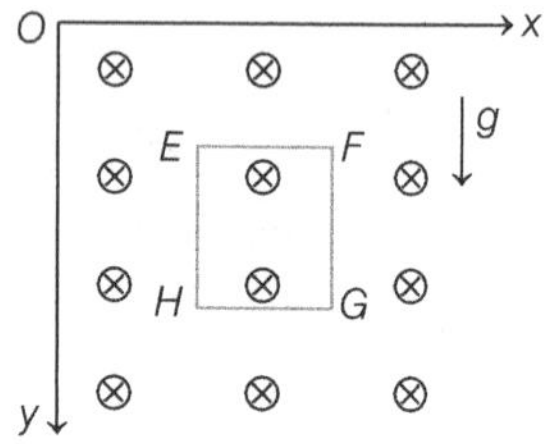

(a) the induced current in the loop and indicate its direction.
(b) the total Lorentz force acting on the loop and indicate its direction.
(c) an expression for the speed of the loop $v(t)$ and its terminal velocity.

122. Two long parallel horizontal rails, a distance d apart and each having a resistance λ per unit length, are joined at one end by a resistance R. A perfectly conducting rod MN of mass m is free to slide along the rails without friction (see figure). There is a uniform magnetic field of induction B normal to the plane of the paper and directed into the paper. A variable force F is applied to the rod MN such that, as the rod moves, a constant current i flows through R. **(1988 JEE)**

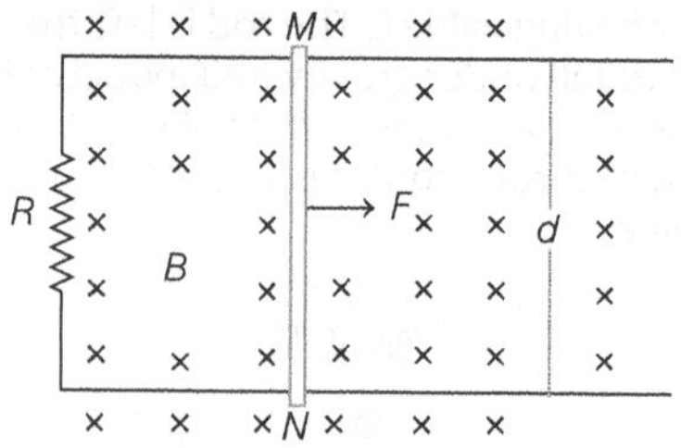

(a) Find the velocity of the rod and the applied force F as functions of the distance x of the rod.
(b) What fraction of the work done per second is converted into heat.

123. Two infinitely long parallel wires carrying currents $I = I_0 \sin \omega t$ in opposite directions are placed a distance $3a$ apart. A square loop of side a of negligible resistance with a capacitor of capacitance C is placed in the plane of wires as shown. Find the maximum current in the square loop. Also sketch the graph showing the variation of charge on the upper plate of the capacitor as a function of time for one complete cycle taking anti-clockwise direction for the current in the loop as positive. **(2003 JEE)**

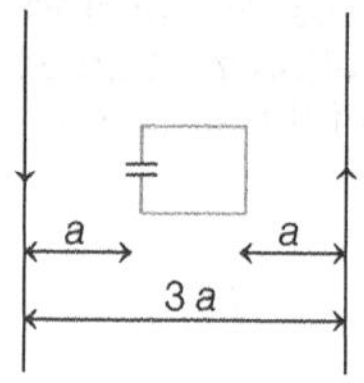

124. A metal bar AB can slide on two parallel thick metallic rails separated by a distance l. A resistance R and an inductance L are connected to the rails as shown in the figure. A long straight wire, carrying a constant current I_0 is placed in the plane of the rails as shown. The bar AB is held at rest at a distance x_0 from the long wire. At $t = 0$, it made to slide on the rails away from the wire. Answer the following questions. **(2002 JEE)**

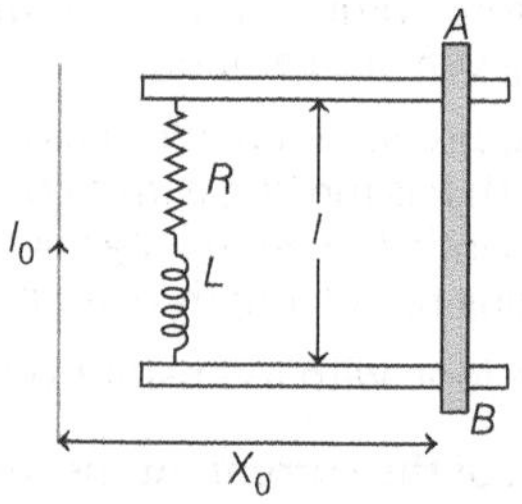

(a) Find a relation among i, $\frac{di}{dt}$ and $\frac{d\phi}{dt}$, where i is the current in the circuit and ϕ is the flux of the magnetic field due to the long wire through the circuit.

(b) It is observed that at time $t=T$, the metal bar AB is at a distance of $2x_0$ from the long wire and the resistance R carries a current i_1. Obtain an expression for the net charge that has flown through resistance R from $t=0$ to $t=T$.

(c) The bar is suddenly stopped at time T. The current through resistance R is found to be $i_1/4$ at time $2T$. Find the value of L/R in terms of the other given quantities.

125. A metal disc of radius $a = 25$ cm rotates with a constant angular velocity $\omega = 130$ rad/s about its axis. Find the potential difference between the centre and the rim of the disc if **(Irodov 3.290)**

(a) the external magnetic field is absent

(b) the external uniform magnetic field of induction $B = 5.0$ mT is directed perpendicular to the disc.

126. A metal rod of mass m can rotate about a horizontal axis O, sliding along a circular conductor of radius a. The arrangement is located in a uniform magnetic field of induction B directed perpendicular to the ring plane. The axis and the ring are connected to an emf source to form a circuit of resistance R. Neglecting the friction, circuit inductance, and ring resistance, find the law according to which the source emf must vary to make the rod rotate with a constant angular velocity ω. **(Irodov 3.295)**

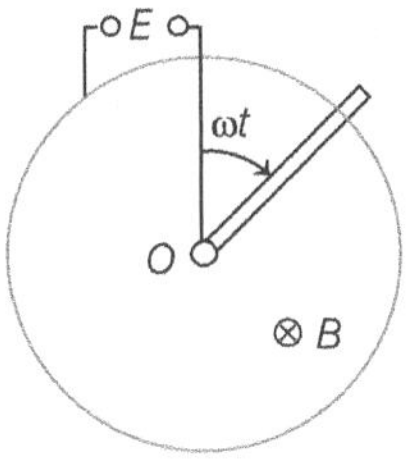

127. A Π-shaped conductor is located in a uniform magnetic field perpendicular to the plane of the conductor and varying with time at the rate $\hat{\mathbf{B}} = 0.10$ T/s. A conducting connector starts moving with an acceleration $a = 10$ cm/s^2 along the parallel bars of the conductor. The length of the connector is equal to $l = 20$ cm. Find the emf induced in the loop $t = 2.0$ s after the beginning of the motion, if at the moment $t = 0$ the loop area and the magnetic induction are equal to zero. The inductance of the loop is to be neglected. **(Irodov 3.307)**

128. A long solenoid of cross-sectional radius a has a thin insulated wire ring tightly put on its winding; one half of the ring has the resistance η times that of the other half. The magnetic induction produced by the solenoid varies with time as $B = bt$, where b is a constant. Find the magnitude of the electric field strength in the ring. **(Irodov 3.310)**

129. A thin non-conducting ring of mass m carrying a charge q can freely rotate about its axis. At the initial moment the ring was at rest and no magnetic field was present. Then a practically uniform magnetic field was switched ON, which was perpendicular to the plane of the ring and increased with time according to a certain law $\mathbf{B}(t)$. Find the angular velocity ω of the ring as a function of the induction $\mathbf{B}(t)$. **(Irodov 3.311)**

130. A thin wire ring of radius a and resistance r is located inside a long solenoid so that their axes coincide. The length of the solenoid is equal to l, its cross-sectional radius is b. At a certain moment the solenoid was connected to a source of a constant voltage V. The total resistance of the circuit is equal to R. Assuming the inductance of the ring to be negligible, find the maximum value of the radial force acting per unit length of the ring. **(Irodov 3.312)**

131. In the middle of a long solenoid there is a coaxial ring of square cross-section, made of conducting material with resistivity ρ. The thickness of the ring is equal to h, its inside and outside radii are equal to a and b respectively. Find the current induced in the ring if the magnetic induction produced by the solenoid varies with time as $B = \beta t$, where β is a constant. The inductance of the ring is to be neglected. **(Irodov 3.314)**

132. An oscillating circuit consisting of a capacitor with capacitance C and a coil of inductance L maintains free undamped oscillations with voltage amplitude across the capacitor equal to V_m. For an arbitrary moment of time find the relation between the current I in this circuit and the voltage V across the capacitor. Solve this problem using Ohm's law and then the energy conservation law. **(Irodov 4.95)**

133. An oscillating circuit consists of a capacitor with capacitance C, a coil of inductance L with negligible resistance, and a switch. Initially the capacitor was charged to a voltage V_m and then at the moment $t = 0$ the switch was closed. Find:

(a) the current $I(t)$ in the circuit as a function of time

(b) the potential difference across the coil at the moment when the electric energy of the capacitor is equal to that of the energy in the coil. **(Irodov 4.96)**

134. An electric circuit shown in figure has a negligibly small active resistance. The left hand capacitor was charged to a voltage V_0 and then at the moment $t = 0$, the switch S was closed. Find the time dependence of the voltages in left and right capacitors. **(Irodov 4.99)**

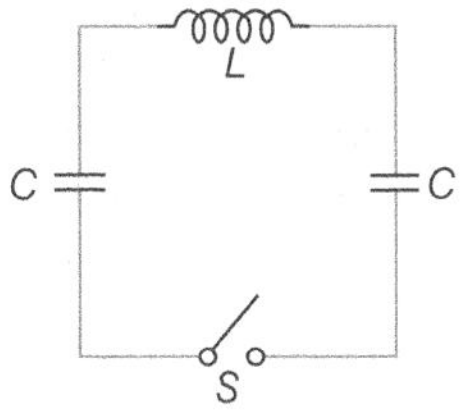

Integer Type Questions

135. A square loop of side 2 cm with a capacitor of capacity 1 μF is located centrally between two current carrying wires separated by a distance of 6 cm as shown in the figure:

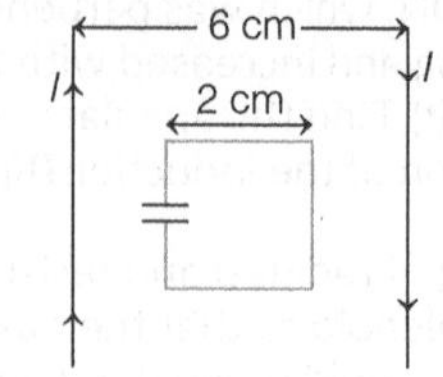

When current in wires is $I = 5\sin 50\pi t$ amperes, then maximum value of the current in square loop is $10^{-k}\pi^2 \log 2$ amperes. Find the value of $k-2$.

136. Three identical resistors with resistance $R = 12\Omega$ and two identical inductors with self inductance $L = 5$ mH are connected to an ideal battery with emf of 12 V as shown in figure. The current through the battery long after the switch has been closed will be .... A. **(2023 Main)**

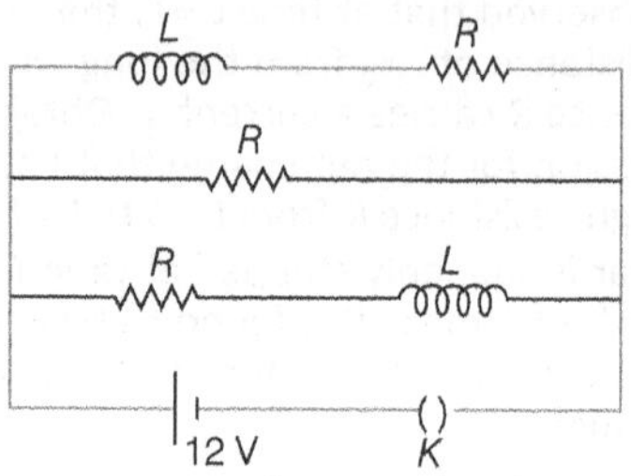

137. Two inductors L_1 (inductance 1 mH, internal resistance 3Ω) and L_2 (inductance 2 mH, internal resistance 4Ω), and a resistor R (resistance 12Ω) are all connected in parallel across a 5 V battery. The circuit is switched on at time $t = 0$. The ratio of the maximum to the minimum current (I_{max}/I_{min}) drawn from the battery is .... **(2016 Adv.)**

Answers

Level 1

1. $(3\times10^{-5}$ s, 10 A) **2.** (True) **3.** (c) **4.** (b) **5.** (a) **6.** (b) **7.** (c) **8.** (a) **9.** (b)

10. (b) **11.** (d) **12.** (c) **13.** (b) **14.** (d) **15.** (a) **16.** (c) **17.** (c) **18.** (b) **19.** (a)

20. (d) **21.** (c) **22.** (c) **23.** (c) **24.** (c) **25.** (b) **26.** (c) **27.** (c) **28.** (d) **29.** (d)

30. (c) **31.** (a) **32.** 0.28 H **33.** (a) 3 V (b) *b* to *a* (c) 0.8 N towards right (d) both are 6 W

34. (a) $v = \frac{E_0}{Bl}\left(1 - e^{-\frac{B^2l^2}{mR}t}\right)$ (b) $i = \frac{E_0 - Blv}{R}$ **35.** $\frac{4}{3}(r_1 - r_2)$ **36.** $e = By\sqrt{\frac{8a}{\alpha}}$

37. (a) $I = \frac{\mu_0 I_0 v}{2\pi R}\ln\left(\frac{b}{a}\right)$ anticlockwise (b) $F = \left[\frac{\mu_0 I_0}{2\pi}\ln\left(\frac{b}{a}\right)\right]^2\frac{v}{R}$ **38.** $I = \frac{Bvl}{R + \frac{R_1R_2}{R_1+R_2}}$ **39.** $\int_A^C E\,dr = -10$ mV

40. $v = \frac{mgR\sin\alpha}{B^2l^2}$ **41.** (a) $s = \frac{mRv_0}{B^2l^2}$ (b) $Q = \frac{1}{2}mv_0^2$ **42.** See the hints **43.** $I_0 = 0.5$ A

44. $e_0 = \frac{\pi B_0\omega Na^2}{3}$ **45.** $I_2 = \frac{\alpha M}{R}\left(1 - e^{-\frac{Rt}{L_2}}\right)$ **46.** $W_{ext} = (\eta^2 - 1)W$ **47.** (a) 0.7 ms (b) 8A **48.** $I = \frac{\phi}{L}\cos\frac{t}{\sqrt{LC}}$

Level 2

49. (c) **50.** (a) **51.** (b) **52.** (c) **53.** (a) **54.** (b) **55.** (a) **56.** (b) **57.** (c) **58.** (a)

59. (d) **60.** (d) **61.** (d) **62.** (d) **63.** (a) **64.** (c) **65.** (a) **66.** (d) **67.** (c) **68.** (c)

69. (a) **70.** (b) **71.** (d) **72.** (a,c) **73.** (a,c) **74.** (a,c) **75.** (b,c,d) **76.** (b,c) **77.** (a,b,c) **78.** (a,c)

79. (a,c,d) **80.** (a,b,d) **81.** (a) **82.** (c) **83.** (a) **84.** (d) **85.** (a) **86.** (b) **87.** (c) **88.** (c)

89. (a) **90.** (d) **91.** (b) **92.** (a) **93.** (d) **94.** (b) **95.** (d) **96.** (c)

97. ($a \to p$; $b \to r$; $c \to s$; $d \to s$) **98.** ($a \to s$; $b \to p$; $c \to s$; $d \to s$) **99.** ($a \to s$; $b \to q$; $c \to p$; $d \to p$)

100. ($a \to s$; $b \to q$; $c \to s$; $d \to p$) **101.** ($a \to q,s$; $b \to p,r$; $c \to p,r$; $d \to q,s$) **102.** ($a \to q$; $b \to r$; $c \to p$; $d \to q$)

103. $I = \frac{V}{R}$ **104.** $e = \frac{\mu_0 i\omega}{2\pi\sin\omega t}\left[\frac{d}{\sin\omega t}\ln\left(\frac{d - a\sin\omega t}{d + a\sin\omega t}\right) + 2a\right]$ **105.** $i = 5(1 - e^{-2000t/3})$ A

106. (a) 120 V (b) *a* (c) 120 V (d) *c* (e) –72V (f) *b* (g) –192V (h) *d*

107. (a) $q_0\omega$ (b) $3V_0 : 3V_0$ (c) $i = q_0\sin\omega t$. Here, $q_0 = 2CV_0$ and $\omega = \sqrt{\frac{3}{2LC}}$ **108.** (a) 20 mV (b) 10^7t^2 V (c) 63.2 μs

109. (a) $\frac{2B_0QR}{mg}$ (b) $\frac{B_0Q}{m}$ **110.** $V = \frac{FR}{B^2l^2}(1 - e^{-\alpha t})$ Here, $\alpha = \frac{B^2l^2}{mR + RB^2l^2C}$, $V_T = \frac{FR}{B^2l^2}$

111. (a) 5×10^4 rad/s (b) 1.05×10^{-5} s (c) $10\sqrt{3}\,\mu C$ **112.** $x = \dfrac{mgt^2}{2(m + CB^2l^2)}$ **113.** (a) $i = \dfrac{mg}{2Br}\left(1 - e^{-\frac{2B^2r^2}{mR}t}\right)$ (b) $v_T = \dfrac{mgR}{4B^2r^2}$

114. (a) $x = \dfrac{v_0}{\omega}(1 - \cos \omega t)$, $\omega = \dfrac{Bl}{\sqrt{mL}}$ (b) $i_{max} = \dfrac{2mg}{Bl}$ (c) $v_{max} = v_0 = \dfrac{g\sqrt{mL}}{Bl}$

115. (a) $i = \dfrac{Bla_0t}{R_1R_2}(R_1 + R_2)$ (b) Polarity of a and b is positive. Polarity of c and d is negative (c) $F_{ext} = a_0m + \dfrac{B^2l^2ta_0}{R_1R_2}(R_1 + R_2)$

116. (a) $i = \dfrac{9v_0\mu_0i_0}{16aR\pi^2}\ln(3)$ (b) $\dfrac{9\mu_0^2i_0^2v_0}{32aR\pi^3}(\ln 3)^2$ **117.** $i = \left(\dfrac{e}{R}\right)[1 - e^{-lR/Lv}]$, where $e = \dfrac{\mu_0 i_0 v}{2\pi}\ln(2)$ **118.** $\varepsilon = \sqrt{R^2 - l^2}\left(\dfrac{dB}{dt}\right)l$

119. Current in segment AE is $\left(\dfrac{7}{22}\right)$ A from E to A, current in segment BE is $\left(\dfrac{6}{22}\right)$ A from B to E and current in segment EF is $\left(\dfrac{1}{22}\right)$ A from F to E.

120. $\dfrac{1}{2}mgr\cos\theta + \dfrac{1}{4}\cdot\dfrac{B^2r^4\omega}{R}$ **121.** (a) $\dfrac{B_0av}{R}$, counter clockwise (b) $-\dfrac{B_0^2a^2v}{R}\hat{j}$, upwards (c) $\dfrac{g}{K}(1 - e^{-Kt})$, $\dfrac{mgR}{B_0^2a^2}$

122. (a) $\dfrac{(R + 2\lambda x)i}{Bd}$, $\dfrac{2\lambda i^2m}{B^2d^2}(R + 2\lambda x) + idB$ (b) $\left[1 + \dfrac{2\lambda mi}{B^3d^3}(R + 2\lambda x)\right]^{-1}$ **123.** $i_{max} = \dfrac{\mu_0 a l_0 \omega^2 C\ln(2)}{\pi}$

124. (a) $\dfrac{d\phi}{dt} = iR + L\dfrac{di}{dt}$ (b) $\Delta q = \dfrac{1}{R}\left(\dfrac{\mu_0 I_0 l}{2\pi}\ln(2) - Li_1\right)$ (c) $\dfrac{T}{\ln 4}$ **125.** (a) $\Delta V = 3.0$ nV (b) $\Delta V = 20$ mV

126. $e_{applied} = \dfrac{1}{2aB}[2mgR\sin\omega t - \omega a^2B^2]$ **127.** $e = 12$ mV **128.** $E = \dfrac{ab(\eta - 1)}{2(\eta + 1)}$ **129.** $\omega = \dfrac{q}{2m}B(t)$

130. $F_{max} = \dfrac{\mu_0 a^2 V^2}{4rRlb^2}$ **131.** $I = \dfrac{\beta h}{4\rho}(b^2 - a^2)$ **132.** $V^2 + \dfrac{I^2L}{C} = V_m^2$

133. (a) $I = \pm I_m \sin\omega t$, where $I_m = V_m\sqrt{C/L}$ and $\omega = 1/\sqrt{LC}$ (b) $V_L = V_C = V_m/\sqrt{2}$

134. LHS capacitor $V = \dfrac{V_0}{2} + \dfrac{V_0}{2}\cos\omega t$, RHS capacitor $V = \dfrac{V_0}{2} - \dfrac{V_0}{2}\cos\omega t$ **135.** (9) **136.** (3) **137.** (8)

See hints on pages 555-578.

Hints & SOLUTIONS

Module 1.

1. If field is non-uniform from position point of view, no emf will be induced. If it is non-uniform from time point of view, emf will be induced.

2. Due to non-uniform magnetic field (a function of x) magnetic flux passing through the loop obtained by integration. But that remains constant with time.

   Hence, $\frac{d\phi}{dt}=0$ or $e=0$

   Magnetic field is along $-\hat{\mathbf{k}}$ direction or in ⊗ magnetic is increasing. Hence, induced current should produce ⊙ magnetic field. Or induced current should be anti-clockwise.

3. At time t_0, magnetic field is negative or and increasing.

   Hence, induced current will produce ⊗ magnetic field. Or induced current should be clockwise.

   If $\frac{d\phi}{dt}=$ constant. Then, $e=$ constant

   ∴ i or rate of flow of charge is constant.

4. Magnetic field through Q (by I_2) is downwards. By decreasing I_1, downward magnetic field through Q will decrease. Hence, induced current in Q should produce magnetic field in same direction.

5. Polarity of emf will be opposite in the two cases while entering and while leaving the coil. Only in option (b) polarity is changing. Hence, the correct option is (b).

6. Total magnetic flux passing through whole of the x-y plane will be zero, because magnetic lines form a closed loop. So, as many lines will move in $-z$ direction same will return to $+z$ direction from the x-y plane.

7. From Lenz's law, induced effects always oppose the cause due to which they are produced. So, when the first loop is moved towards the smaller loop, it will face repulsion.

8. $e=L\left|\frac{di}{dt}\right|=L$ (Slope of i-t graph)

   Initially, slope $=0 \Rightarrow e=0$

   Then in remaining two regions slopes are constants but of opposite signs. Hence, induced emfs are constants but of opposite signs.

9. $e=\left|\frac{d\phi}{dt}\right|=(a\tau-2at)$

   $$i=\frac{e}{R}=\frac{a\tau-2at}{R}$$

   $$H=\int_0^{\tau} i^2Rdt$$

10. $\Delta q=\frac{\Delta\phi}{R}$ or $i\Delta t=\frac{\Delta\phi}{R}$

    ∴ $\Delta\phi=i\,(\Delta t)R=(10\times10^{-3})(5)(0.5)$

    $=25\times10^{-3}$ Wb

11. $\mathbf{S}=(ab)\hat{\mathbf{k}}\rightarrow$ perpendicular to xy-plane

    $\phi=\mathbf{B}\cdot\mathbf{S}=(50)(ab)=$ constant

    $\frac{d\phi}{dt}=0$

    ∴ $e=0$

12. ∵ $i=\frac{e}{R}=\frac{N(\Delta\phi/\Delta t)}{R}$

    $=\frac{NS(\Delta B/\Delta t)}{R}$

    $=\frac{10\,(10\times10^{-4})(10^4)}{20}$

    $=5$ A

13. For $E\neq0$, ϕ must change

    or $\frac{d\phi}{dt}\neq0$

14. Even if radius is doubled, flux is not going to change.

15. $E_{avg}=\frac{\Delta\phi}{\Delta t}=\frac{BA\cos\theta}{10}$

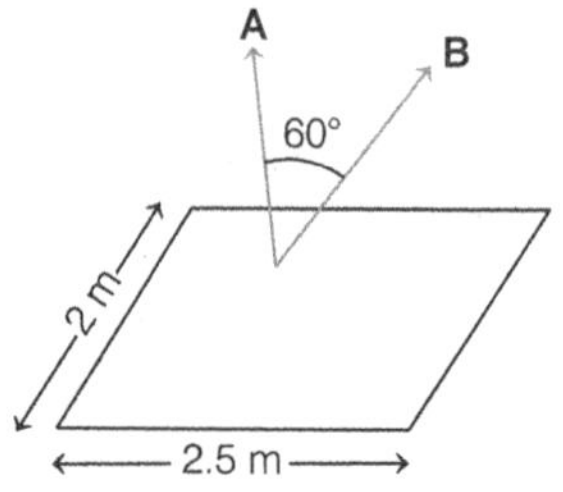

    $=4\times2\times\frac{5}{2}\times\frac{\cos60^\circ}{10}$

    $=4\times5\times\frac{1}{10\times2}=1$ V

16. It is given,

    In 0.5 s, magnetic field of 0.5 T becomes zero.

    So, rate of decay of magnetic field,

    $$\frac{\Delta B}{\Delta t}=\frac{0-0.5}{0.5}=-1\ \text{T/s}$$

    (Negative sign shows field is reducing)

    Hence, magnetic field at 0.25 s will be

    $$\frac{\Delta B}{0.25}=-1\ \text{T/s}$$

$$\Rightarrow \quad \frac{B_2 - B_1}{0.25} = -1$$

or $\quad B_2 = 0.5 - 0.25$

$= 0.25 \text{ T}$

So, magnitude of induced emf in coil at $t = 0.25$ s will be

$$\varepsilon = \frac{d}{dt}\phi = \left(\frac{dB}{dt}\right)\cdot A = \frac{\Delta B}{\Delta t}\cdot \pi r^2$$

$$= \frac{0.25}{0.25} \times \pi \times \left(\frac{10}{\sqrt{\pi}} \times 10^{-2}\right)^2$$

$$= 10^{-2} \text{ V} = 10 \times 10^{-3} \text{ V}$$

$$= 10 \text{ mV}$$

17. $\phi = 5t^3 + 4t^2 + 2t - 5$

$$e = -\frac{d\phi}{dt}$$

$$|e| = \frac{d\phi}{dt}$$

$$|e| = 15t^2 + 8t + 2$$

$R = 5\Omega$ (Given)

$$\text{So, } i = \frac{|e|}{R} = \frac{15t^2 + 8t + 2}{5}$$

$$= 3t^2 + 1.6t + 0.4$$

$= 3.2^2 + 1.6 \times 2 + 0.4$ $\quad (\because t = 2 \text{ s})$

$$= 15.6 \text{ A}$$

18. Induced current, $I = \frac{e}{R}$

Here, e = induced emf $= \frac{d\phi}{dt}$

$$I = \frac{e}{R} = \left(\frac{d\phi}{dt}\right)\cdot\frac{1}{R}$$

$$d\phi = IRdt \Rightarrow \phi = \int IRdt$$

Here, R is constant.

$$\therefore \quad \phi = R\int Idt$$

$\int I \cdot dt$ = Area under I-t graph

$$= \frac{1}{2} \times 10 \times 0.5 = 2.5$$

$$\therefore \quad \phi = R \times 2.5 = 100 \times 2.5$$

$$= 250 \text{ Wb}$$

19. Mutual inductance, $M = \frac{\mu_0 \pi N_1 N_2 a^2}{2b}$

Hence, $\quad M = \frac{\mu_0 \pi a^2}{2b}$ ...(i) $\quad [\because N_1 = N_2 = 1]$

and given, $I = I_0 \cos \omega t$ ... (ii)

Now according to Faraday's second law, induced emf,

$$e = -M\frac{dI}{dt}$$

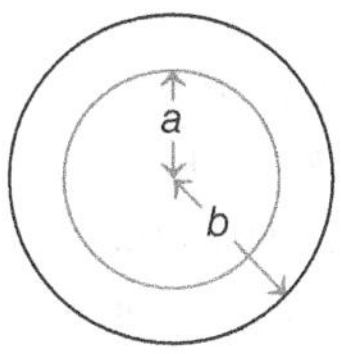

From Eq. (ii),

$$e = \frac{-\mu_0 \pi a^2}{2b}\frac{d}{dt}(I_0 \cos \omega t)$$

$$e = \frac{\pi \mu_0 I_0}{2}\cdot\frac{a^2}{b}\omega \sin \omega t$$

20. Magnetic field at the centre of smaller loop

$$B = \frac{\mu_0 i R_2^2}{2(R_2^2 + x^2)^{3/2}}$$

Area of smaller loop $S = \pi R_1^2$

$\therefore$ Flux through smaller loop $\phi = BS$

Substituting the values, we get,

$$\phi \approx 9.1 \times 10^{-11} \text{ Wb}$$

21. $\otimes$ magnetic field passing through loop is increasing. Hence, induced current will produce magnetic field. So, induced current should be anti-clockwise.

22. $\left|\frac{d\phi_B}{dt}\right|$ = [Potential or EMF]

$$= [ML^2A^{-1}T^{-3}]$$

23. $\odot$ is increasing. Hence, $\otimes$ is produced by the induced current. So, it is clockwise.

24. By increasing the current in loop-1, magnetic field in ring-2 in downward direction will increase. Hence, induced current in ring-2 should produce upward magnetic field. Or current in ring should be in the same direction.

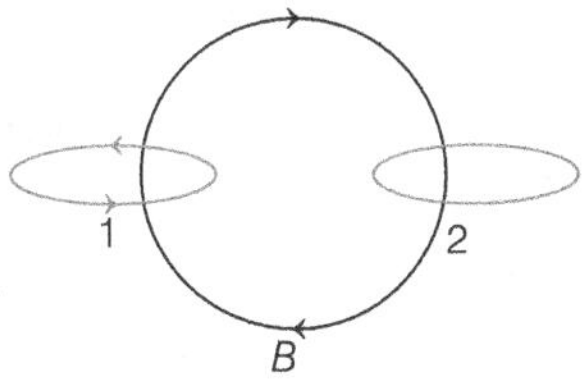

25. $2\pi R = 4L$

$$\therefore \quad L = \frac{\pi R}{2} = \frac{(\pi)(10)}{2} = (5\pi) \text{ cm}$$

$$\Delta S = S_i - S_f = (\pi R^2) - L^2$$

$$= \pi(0.1)^2 - (5\pi)^2 \times 10^{-4}$$

$$= 0.0067 \text{ m}^2$$

$$e = \frac{\Delta\phi}{\Delta t} = B\left(\frac{\Delta S}{\Delta t}\right)$$

$$= \frac{100 \times 0.0067}{0.1}$$

$$= 6.7 \text{ V}$$

26. $\Delta\phi = 2(NBS)$

$$\Delta q = \frac{\Delta\phi}{R} = \frac{2NBS}{R}$$

$$= \frac{2 \times 500 \times 0.2 \times 4 \times 10^{-4}}{50}$$

$$= 1.6 \times 10^{-3} \text{C}$$

27. $\mathbf{S} = [(5 \times 10^{-4})\hat{\mathbf{k}}] \text{m}^2$

$\phi = |\mathbf{B} \cdot \mathbf{S}| = 9 \times 10^{-7} \text{Wb}$

28. $e = \frac{N\Delta\phi}{\Delta t} = \left(\frac{NS\Delta B}{\Delta t}\right) \cos 30°$

$\therefore \quad S = \left(\frac{e\Delta t}{N\Delta B}\right) \sec 30°$

$$= \left[\frac{(80 \times 10^{-3})(0.4)}{(50)(400 \times 10^{-6})}\right]\left[\frac{2}{\sqrt{3}}\right]$$

$$= 1.85 \text{ m}$$

$\therefore$ Side of square $= 1.36$ m

Total length of wire $= 50(4 \times 1.36)$

$= 272$ m

29. $\phi = BS = B_0 S e^{-at}$

Induced emf $= \left|\frac{d\phi}{dt}\right| = aB_0 S e^{-at}$

30. Similar to above problem

$$\Delta q = \frac{2NBS}{R}$$

$\therefore \quad B = \frac{(\Delta q)R}{2NS}$

$$= \frac{(4.5 \times 10^{-6})(40)}{(2)(60)(3 \times 10^{-6})} = 0.5 \text{ T}$$

31. $e = \frac{d\phi}{dt} = S\left(\frac{dB}{dt}\right) = \pi R^2$ (Slope of B -t graph)

(a) $e = (\pi)(0.12)^2 \left(\frac{0.5}{2}\right) = 0.011$ V/m

(b) Slope of B-t graph is zero. Hence,

$e = 0$

(c) Slope is just opposite to the slope of part (a).

32. We know that induced emf,

$$e = \frac{d\phi}{dt}$$

$$e = \frac{dBA}{dt} = \frac{AdB}{dt}$$

$e = A \times$ slope of B-t curve

$$= A\left[\frac{8-4}{4-2}\right] = 4 \times \frac{4}{2}$$

$e = 8$ mV

33. $\Delta Q = \frac{\Delta\phi}{R} = \frac{2NBA}{R}$

$$= \frac{2 \times 100 \times 1.5 \times 24 \times 10^{-4}}{12}$$

$$= 6 \times 10^{-2} \text{C} = 60 \text{ mC}$$

34. Given, side of square loop,

$a = 2.0 \text{ cm} = 2 \times 10^{-2}$ m

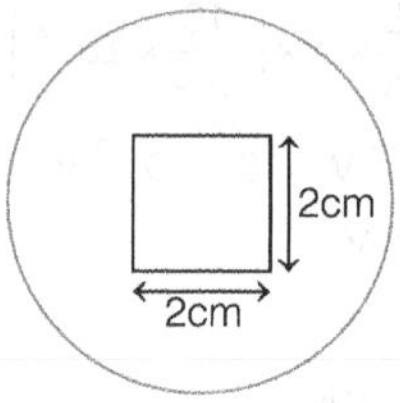

Area of loop

$= (2 \times 10^{-2})^2 = 4 \times 10^{-4} \text{ m}^2$

Amplitude of current $= 2.5$ A

Angular frequency $= 700$ rad/s

Number of turns in solenoid

$= 50$ turns per cm

$= \frac{50}{10^{-2}}$ turns per m

$B_{\text{due to solenoid}} = \mu_0 nI$

$\phi_{\text{through square}} = \mu_0 nI \times A$

Emf induced, $\varepsilon = \mu_0 nA \times \frac{dI}{dt}$

$= \mu_0 nA I_0 \omega \cos \omega t$ $\quad [\because I = I_0 \sin \omega t]$

$|\varepsilon| = \mu_0 nAI_0\omega$

$= 4\pi \times 10^{-7} \times \frac{50}{10^{-2}} \times 4 \times 10^{-4} \times 2.5 \times 700$

$= 44 \times 10^{-4}$ V

35. Magnetic field $(B) = 0.4$ T

Expansion in the radius of loop

$\left(\frac{dr}{dt}\right) = 1 \text{ mm/s} = 10^{-3} \text{m/s}$

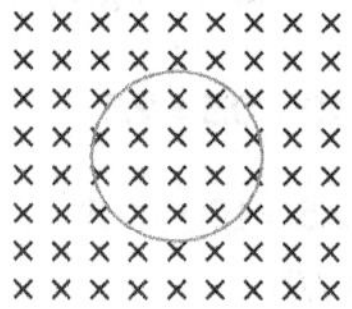

Thus, $\frac{dA}{dt} = 2\pi r \frac{dr}{dt}$

emf induced, $\varepsilon = \left(-\frac{d\phi}{dt}\right) = \frac{BdA}{dt}$

$= 0.4\, 2\pi r \frac{dr}{dt}$

$|\varepsilon|_{r = 2\text{cm}} = 0.4 \times 2 \times \pi \times 2 \times 10^{-2} \times 10^{-3}$ V

$= 16\pi\mu\text{V} = 50.24\,\mu\text{V}$

$\simeq 50\,\mu\text{V}$

36. Magnetic flux, $\phi = \mathbf{B} \cdot \mathbf{A}$

$$= (3t^3\hat{\mathbf{j}} + 3t^2\hat{\mathbf{k}}) \cdot (\pi(1)^2\hat{\mathbf{k}}) = 3t^2\pi$$

Induced emf, $\varepsilon = \left|\frac{d\phi}{dt}\right|$

$$= \frac{d(3t^2\pi)}{dt} = 6t\pi$$

$$\varepsilon = 6 \times 2 \times \pi = 12\pi \qquad (\because t = 2\pi)$$

$$n = 12$$

37. The magnetic flux is given by

$$\phi = \frac{2}{3}(9 - t^2)$$

When the flux becomes zero then

$$\phi = \frac{2}{3}(9 - t^2) = 0$$

$$9 - t^2 = 0 \quad \Rightarrow \quad t = 3 \text{ sec}$$

The emf will be

$$e = \frac{-d\phi}{dt} = -\frac{2}{3}(0 - 2t) = \frac{4t}{3}$$

$$\text{Heat produced in 3 sec} = \int_0^t \frac{e^2}{r} dt$$

$$= \int_0^3 \frac{\left(\frac{4t}{3}\right)^2 dt}{r} \qquad \left(\because e = \frac{4t}{3}\right)$$

$$= \frac{16}{9 \times 8}\int_0^3 t^2 dt$$

$$= \frac{2}{9}\left[\frac{t^3}{3}\right]_0^3 = 2 \text{ J}$$

Module 2.

Introductory Exercise

1. $\because \; e = Bvl = 1.1 \times 5 \times 0.8$

$$= 4.4 \text{ V}$$

Apply right hand rule for polarity of this emf.

2. $\because$

$$e = Bvl$$

$$i = \frac{e}{R} = \frac{Bvl}{R}$$

$$F = ilB = \frac{B^2l^2v}{R}$$

$$= \frac{(0.15)^2 (0.5)^2 (2)}{3}$$

$$= 0.00375 \text{ N}$$

3. $V_A - V_C = \frac{B\omega l^2}{2}$

$$V_D - V_C = \frac{B\omega(2l)^2}{2}$$

From these two equations, we find

$$V_A - V_D = -3B\omega l^2/2$$

4. Circuit is not closed. So, current is zero or magnetic force is zero.

Exercise

1. Magnetic force on free electrons will be towards B. Therefore, at B, there is excess of electrons (means negative charge) and at A, there is defficiency of electrons (means positive charge).

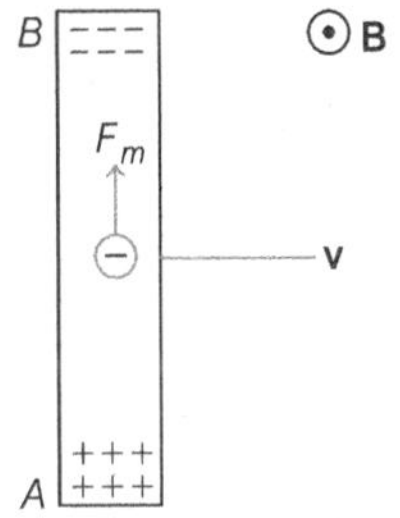

2. $e = \frac{B\omega l^2}{2} = \text{constant}$

3. For wire ab, velocity vector is parallel to $\mathbf{l}$.

4. $i = \frac{Bvl}{R}$

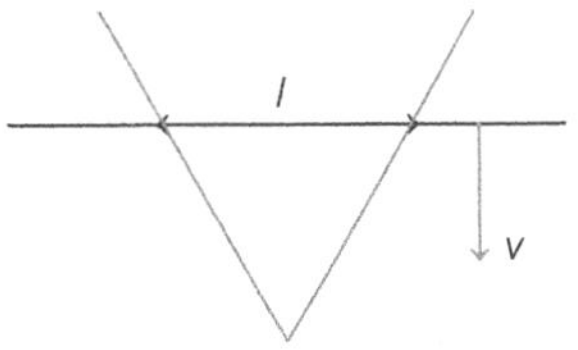

Let λ = resistance per unit length of conducting rod, then

$$i = \frac{Bvl}{\lambda l} = \frac{Bv}{\lambda} = \text{constant}$$

5. At time t, angle rotated by loop is $\theta = \omega t$. This is also the angle between $\mathbf{B}$ and $\mathbf{S}$. Then,

$$\phi = BS\cos\theta$$

$$= Bb^2 \cos \omega t$$

$$e = \left|\frac{d\phi}{dt}\right| = b^2 B\omega \sin \omega t$$

6. The emf will induce when flux changes i.e. when the area changes. In the given problem, flux changes from situations (i) to (ii) (see figure).

But no flux will change from situations (ii) to (iii) (as area inside the magnetic field remains unchanged). Again from situations (iii) to (iv), flux changes.

The total time take from situations (i) to (iv) is

$$T_0 = \frac{(a + b)}{v}$$

But during this time, flux does not change from situations (ii) to (iii) which takes a time

$$t = \frac{(b-a)}{v}$$

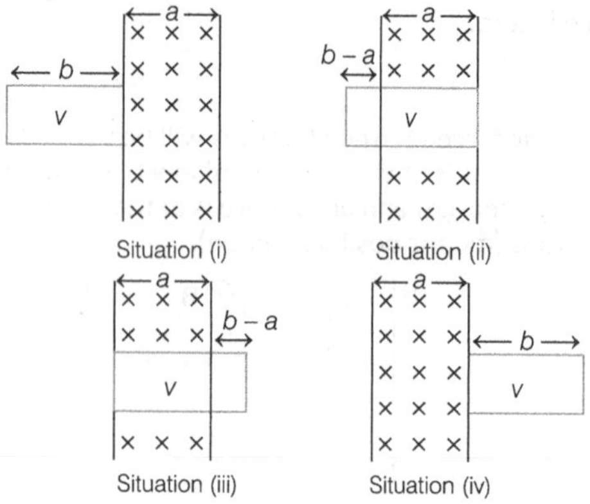

Hence, the time during which flux changes is

$$T = T_0 - t = \frac{(a+b)}{v} - \frac{(b-a)}{v}$$

$$T = \frac{2a}{v}$$

7. In AB, **l** is parallel to its **v**. Hence, PD $= 0$

8.

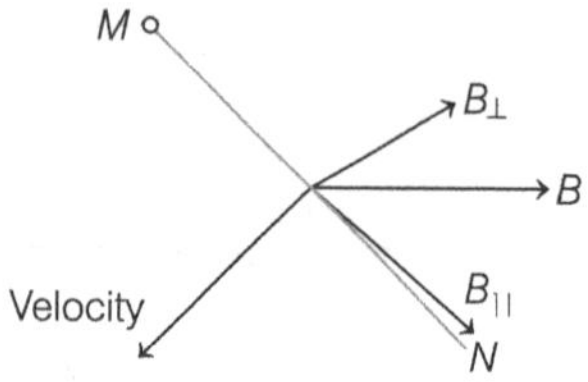

$B_{||}$ is parallel to MN (or **l**) and $B_{\perp}$ is parallel or antiparallel to velocity.

9. Peak value of emf generated in generator is given by

$$e_0 = \omega NBA$$
$$= (2\pi f) NBA$$
$$= 2 \times 3.14 \times 100 \times 5000 \times 0.2 \times 0.25$$
$$= 157.1 \text{ kV}$$

10. $\because$ $e = Bvl$

$$i = \frac{e}{R} = \frac{Bvl}{R}$$

$$F = ilB = \frac{B^2 l^2 v}{R}$$

$$P = F \cdot v = \frac{B^2 l^2 v^2}{R}$$

$$= \frac{(1 \times 1 \times 20)^2}{0.2} = 2000 \text{ W}$$

$$= 2 \text{ kW}$$

11. When a rod is moving perpendicular to a uniform magnetic field, magnitude of induced emf is;

$$E = Bl\, V$$

Here $E = 0.08$ V

$l = 10$ cm $= 10 \times 10^{-2}$ m

$v = ?, B = 0.4$ T

So, velocity v is

$$v = \frac{E}{Bl} = \frac{0.08}{0.4 \times 10 \times 10^{-2}}$$
$$= 2 \text{ m/s}$$

12. Given,

$$\omega = 5 \text{ rad/s}$$
$$B_H = 0.2 \times 10^{-4} \text{ T}$$

emf induced,

$$e = \frac{1}{2} B_H \omega l^2$$

$\Rightarrow$ $$e = \frac{1}{2} \times 0.2 \times 10^{-4} \times 5 \times 1^2$$

$$= 50 \times 10^{-6} \text{ V}$$

$\Rightarrow$ $e = 50\,\mu\text{V}$

13. According to given circuit diagram, equivalent resistance between point P and Q.

$$R_{PQ} = (4+4) \,||\, (4+4)$$
$$= \frac{8 \times 8}{8+8} = 4\Omega$$

The equivalent circuit can be drawn as,

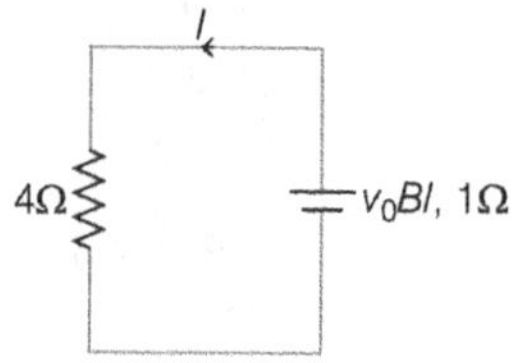

Equivalent resistance,

$$R_{eq} = 4 + 1 = 5\Omega$$

Induced emf, $e = Bv_0 l$

Induced current, $I = e / R_{eq}$

$\Rightarrow$ $$I = \frac{Bv_0 l}{R_{eq}}$$

$\Rightarrow$ $$2 \times 10^{-3} = \frac{5 \times v_0 \times 0.2}{5}$$

$\Rightarrow$ $$v_0 = 10^{-2} \text{ m/s}$$

$$= 1 \text{ cm/s}$$

14. $e = \int_{2l}^{3l} (\omega x) B dx$

$$= \frac{B\omega}{2} [x^2]_{2l}^{3l} = B\omega \frac{[(3l)^2 - (2l)^2]}{2} = \frac{5Bl^2\omega}{2}$$

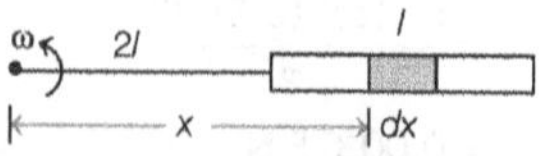

15. A motional emf or potential difference is produced in both AD and BC, thus an electric field is induced in both AD and BC.

16. A motional emf, $e = Blv$ is induced in the rod. Or we can say a potential difference is induced between the two ends of the rod AB, with A at higher potential and B at

lower potential. Due to this potential difference, there is an electric field in the rod.

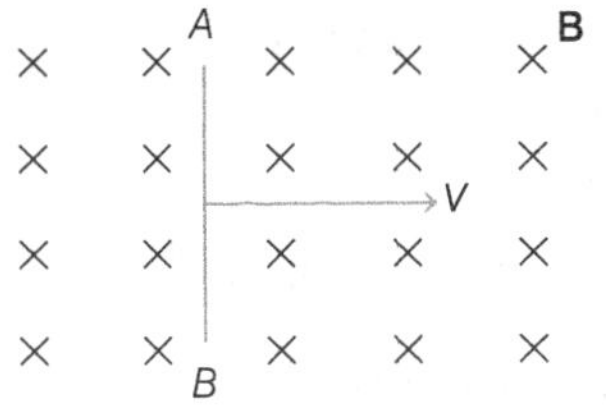

17. Induced motional emf in MNQ is equal to the motional emf in an imaginary wire MQ i.e.

$$e_{MNQ} = e_{MQ} = Bvl = Bv(2R) \qquad [l = MQ = 2R]$$

Therefore, the potential difference developed across the ring is $2RBv$ with Q at higher potential.

18. Net change in magnetic flux passing through the coil is zero.

$\therefore$ Current (or emf) induced in the loop is zero.

19. (a) At a distance x from the wire, magnetic field over the wire ab is

$$B = \frac{\mu_0}{2\pi}\frac{i}{x}$$

$$dV = Bvdx = \left(\frac{\mu_0 i}{2\pi x}\right) vdx$$

$$\therefore \text{ Total emf} = \int_{x=d}^{x=d+l} dV$$

(b) Magnetic field due to current i over the wire ab is inwards. Velocity of wire ab is towards right. Applying right hand rule, we can see that a point is at higher potential.

(c) Net change in flux through the loop $abcd$ is zero. Hence, induced emf is zero. So, induced current is zero.

20. Induced emf ($e = BvL$) and therefore induced current is developed only during entering and during existing from the magnetic field.

$$i = \frac{e}{R} = \frac{BvL}{R}$$

$$F = iLB = \frac{B^2L^2v}{R}$$

Further, magnetic force always opposes the change. Hence, external force is always positive.

During entering into the field, ⊗ magnetic field increases. Hence, induced current should produce ⊙ magnetic field. Or it should be anti-clockwise.

During existing from the magnetic field case is just opposite.

21. Given network forms a balanced Wheatstone's bridge. The net resistance of the circuit is therefore $3\Omega + 1\Omega = 4\Omega$. Emf of the circuit is Bv_0l. Therefore, current in the circuit would be

$$i = \frac{Bv_0 l}{R} \text{ or } v_0 = \frac{iR}{Bl}$$

$$= \frac{(1\times 10^{-3})(4)}{2\times 0.1} = 0.02 \text{ m/s}$$

Cross magnetic field passing through the loop is decreasing. Therefore, induced current will produce magnetic field in cross direction. Or direction of induced current is clockwise.

22. Given,

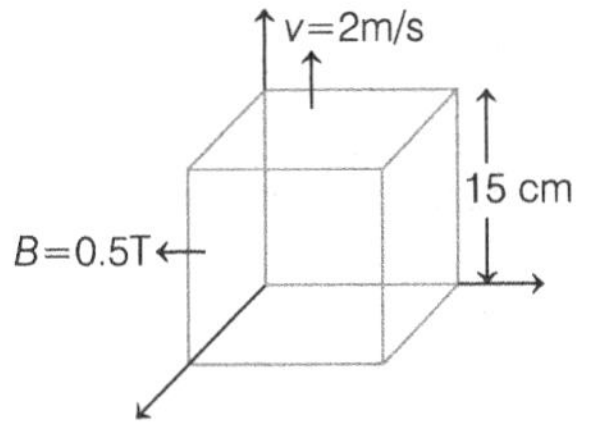

$$l = 15 \text{ cm} = 0.15\text{m}$$

$$\text{Emf} = Bvl$$

$$= 0.5 \times 2 \times 0.15 = 0.15\text{V}$$

$$= 150 \text{ mV}$$

23. Given, angular speed of metallic rod,

$$\omega = 2\pi \times \frac{210}{60} \text{ rad/s}$$

$$= 7\pi \text{ rad/s}$$

Length of rod, $l = 20$ cm $= 0.2$ m

Magnetic field, $B = 0.2$ T

The emf induced between the centre and the ring,

$$e = \frac{1}{2}B\omega l^2$$

$$= \frac{1}{2} \times 0.2 \times 7\pi \times (0.2)^2$$

$$= 0.088 \text{ V} = 88 \text{ mV}$$

24. Given, area,

$$A = 70 \text{ cm}^2$$

$$= 70 \times 10^{-4} \text{m}^2$$

Number of turns, $N = 600$

Magnetic field, $B = 0.4$ Wbm^{-2}

Angular velocity,

$$\omega = 500 \text{ rpm} = 500 \times \frac{\pi}{30} \text{ rad/s}$$

$$\Rightarrow \quad \omega = \frac{50\pi}{3} \text{ rad/s}$$

The angle between the normal to coil and the direction of magnetic field,

$$\theta = 90° - 60° = 30°$$

Induced emf in rotating coil is given as

$$e = N\omega BA\sin\theta$$

$$= 600 \times \frac{50\pi}{3} \times 0.4 \times 70 \times 10^{-4} \times \sin 30°$$

$$\Rightarrow \quad e = 44 \text{ V}$$

Module 3.

1. $V_A - V_B = L\frac{di}{dt} = L(-\alpha) = -\alpha L$

2. $i = \frac{dq}{dt} = (8t)\text{A}, \quad \frac{di}{dt} = 8 \text{ A/s}$

At $\quad t = 1\text{ s}, q = 4\text{C}, i = 8\text{ A}$

and $\quad \frac{di}{dt} = 8\text{ A/s}$

Charge on capacitor is increasing. So, charge on positive plate is also increasing. Hence, direction of current is towards left.

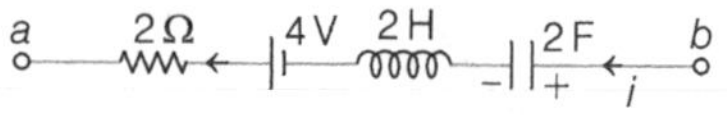

Now, $\quad V_a + 2\times 8 - 4 + 2\times 8 + \frac{4}{2} = V_b$

$\therefore \quad V_a - V_b = -30\text{ V}$

3. $I = (10t + 5)\text{ A}$

$$\frac{dI}{dt} = 10\text{ A/s} = \text{constant}$$

At, $\quad t = 0, I = 5\text{ A}$

Now, $\quad V_A - 3\times 5 - 1\times 10 + 10 = V_B$

$\therefore \quad V_A - V_B = 15\text{ V}$

4. $L = \frac{\mu_0 N^2 R^2 \pi}{l}$

$\Rightarrow \quad L \propto \frac{1}{l}$

By doubling the length, L will remain half.

5. Energy stored in an inductor of inductance L and current I is given by

$$E = \frac{1}{2}LI^2$$

When current is being changed from I_1 to I_2, change in energy will be

$$\Delta E = E_2 - E_1 = \frac{1}{2}LI_2^2 - \frac{1}{2}LI_1^2 \quad \text{...(i)}$$

As I_1 and I_2 are given, we need to find value of L. Now, induced emf in a coil is

$$\varepsilon = L\frac{dI}{dt}$$

Here, $\quad \varepsilon = 25\text{ V}, \; dI = I_2 - I_1 = (25 - 10) = 15\text{ A}$

and $\quad dt = 1\text{ s}$

$\Rightarrow \quad 25 = L \times \frac{15}{1}$

or $\quad L = \frac{25}{15} = 5/3\text{H}$

Putting values of L, I_1 and I_2 in Eq. (i), we get

$$\Delta E = \frac{1}{2}\times\frac{5}{3}\times[25^2 - 10^2]$$

$$= \frac{1}{2}\times\frac{5}{3}\times 525$$

$$\Delta E = 437.5\text{ J}$$

6. $V_a - iR - L\frac{di}{dt} = V_b$

$\therefore \quad V_a - V_b = iR + L\frac{di}{dt}$

According to given conditions,

$$8 = 2R + L \quad \text{...(i)}$$

$$4 = 2R - L \quad \text{...(ii)}$$

Solving these two equations, we get

$$R = 3\Omega \text{ and } L = 2\text{H}$$

7. (a) $L = \frac{N\phi}{i} \Rightarrow \phi = \frac{Li}{N}$

So, SI unit of flux is Henry-ampere.

(c) $L = \frac{-e}{\Delta i/\Delta t} = -\frac{-e\Delta t}{\Delta i}$

Hence, SI unit of L is $\frac{\text{V-s}}{\text{ampere}}$.

8. $|e| = \left|L\frac{\Delta i}{\Delta t}\right|$ or $\left|L\frac{di}{dt}\right|$

Here, $\quad L = 1\text{H}$

and $\quad \frac{di}{dt} = 3[\sin t + t\cos t]$

$\therefore \quad |e| = 3(t\cos t + \sin t)$

9. $V_L = +L\frac{di}{dt} = (2)\frac{d}{dt}(10e^{-4t})$

$$= -80\,e^{-4t}$$

Further, $\quad V_a - iR - L\frac{di}{dt} = V_b$

$\therefore \quad V_a - V_b = iR + L\frac{di}{dt}$

or $\quad V_{ab} = (10e^{-4t})(4) - 80e^{-4t}$

$$= -40e^{-4t}$$

10. (a) $dI/dt = 16\text{ A/s}$

$\therefore \quad L = \left|\frac{e}{dI/dt}\right| = \frac{10\times 10^{-3}}{16}$

$$= 0.625\times 10^{-3}\text{ H} = 0.625\text{ mH}$$

(b) At $t = 1$ s, $I = 21$ A

$$U = \frac{1}{2}LI^2 = \frac{1}{2}\times(0.625\times 10^{-3})(21)^2$$

$$= 0.137\text{ J}$$

$$P = Ei = (10\times 10^{-3})(21)$$

$$= 0.21\text{ J/s}$$

11. (a)

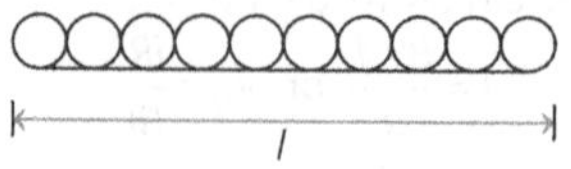

$$L=\frac{\mu_0 N^2 S}{l}$$

$$N=\frac{l}{d}$$

$$\therefore \quad L=\frac{\mu_0 l S}{d^2}=\frac{(4\pi\times 10^{-7})(0.4)(0.9\times 10^{-4})}{(0.1\times 10^{-2})^2}$$

$$=4.5\times 10^{-5}\text{ H}$$

(b) $e=\left|L\frac{\Delta i}{\Delta t}\right|=\frac{(4.5\times 10^{-5})(10)}{0.1}$

$$=4.5\times 10^{-3}\text{ V}$$

12. $V_L=L\frac{di}{dt}$

$\therefore \quad di=\frac{1}{L}(V_L dt)$

$\therefore \quad \int di=i=\frac{1}{L}\int V_L dt$

or $\quad i=\frac{1}{L}$ (area under V_L *versus* t graph)

(a) **At $t=2$ ms**

$$i=(150\times 10^{-3})^{-1}\left(\frac{1}{2}\times 2\times 10^{-3}\times 5\right)$$

$$=3.33\times 10^{-2}\text{ A}$$

(b) **At $t=4$ ms**

Area is just double. Hence, current is also double.

13. (a) $L=\left|\frac{e}{di/dt}\right|=\frac{0.016}{0.064}=0.25$ H

(b) $L=\frac{N\phi}{i}$

$$\therefore \quad \phi=\frac{Li}{N}=\frac{(0.25)(0.72)}{400}$$

$$=4.5\times 10^{-4}\text{ Wb}$$

14. Let total number of turns in the solenoid are N and cross-sectional radius of solenoid is r. Then, length of wire is,

$$l=(2\pi r)N \quad \text{...(i)}$$

Number of turns per unit length,

$$n=\frac{N}{l_0} \quad \text{...(ii)}$$

If I current is passed through the solenoid, then

$$N\phi=NBS=N(\mu_0 nI)(\pi r^2)$$

$$=\frac{\mu_0 N^2\pi I r^2}{l_0} \quad \left(n=\frac{N}{l_0}\right)$$

But $N\phi=LI$

$$\therefore \quad LI=\frac{\mu_0 N^2\pi I r^2}{l_0}$$

$$\therefore \quad N=\frac{1}{r}\sqrt{\frac{Ll_0}{\mu_0\pi}}$$

Now $l=(2\pi r)N$

$$\therefore \quad l=\sqrt{\frac{4\pi L l_0}{\mu_0}}$$

Substituting the given values, we get

$$l=100\text{ m}$$

15. From the result of Q-14,

Length of solenoid wire,

$$l=\sqrt{\frac{4\pi L l_0}{\mu_0}} \quad \text{...(i)}$$

Let resistivity of copper is ρ and density is ρ_0. Then,

$$R=\frac{\rho l}{A}$$

$$\therefore \quad l=\frac{RA}{\rho}=\frac{R}{\rho}\left[\frac{m}{\rho_0 l}\right]$$

$$\therefore \quad l=\sqrt{\frac{mR}{\rho\rho_0}} \quad \text{...(ii)}$$

Equating Eqs. (i) and (ii), we get

$$L=\left(\frac{\mu_0}{4\pi}\right)\left(\frac{mR}{l_0\rho\rho_0}\right)$$

16. $V_A-IR-\frac{LdI}{dt}-12=V_B$

$$\Rightarrow \quad V_A-V_B=2\times 12+6(-1)+12$$

$$\Rightarrow \quad V_A-V_B=24-6+12$$

$$\Rightarrow \quad V_A-V_B=30\text{ V}$$

$$\Rightarrow \quad \Delta V=30\text{ V}$$

Module 4.

1. $M\propto N_1 N_2$

2. $M_{12}=M_{21}$, when rate of charge of current is doubled, induced emf also becomes two times.

3. $\frac{dI}{dt}=I_0\,\omega\cos\omega t$

$$e=M\frac{dI}{dt}=MI_0\,\omega\cos\omega t$$

$$\therefore \quad e_{max}=MI_0\,\omega$$

$$=0.005\times 10\times 100\pi$$

$$=(5\pi)\text{ V}$$

4. If current is passed through the straight wire, magnetic lines are circular and tangential to the loop. So, no flux is linked with the loop.

5. Let us pass a current i through circular loop. Magnetic field at its centre is,

$$B=\frac{\mu_0 i}{2r}$$

Area of square loop, $S=l^2$

$$\phi=BS=\frac{\mu_0 i l^2}{\pi r}$$

$$M=\frac{\phi}{i}=\frac{\mu_0 l^2}{\pi r} \quad \text{or} \quad M\propto\frac{l^2}{r}$$

6. Let i_1 current is passed through coil 1, then magnetic field at the centre of coil 2 is,

$$B_2 = \frac{\mu_0 i_1 a^2}{2(l^2 + a^2)^{3/2}} \qquad \text{(as } l >> a\text{)}$$

$$B_2 = \frac{\mu_0 i_1 a^2}{2l^3}$$

Flux induced in coil 2 due to coil 1,

$$\phi_{21} = Mi_1$$

Also,

$$\phi_{21} = B_1 A_2$$

$$\phi_{21} = \frac{\mu_0 i_1 a^2}{2l^3}(\pi a^2)$$

$$\frac{\mu_0 i_1 \pi a^4}{2l^3} = Mi_1$$

Mutual inductance of two loop system,

$$M = \frac{\mu_0 a^4 \pi}{2l^3}$$

7.

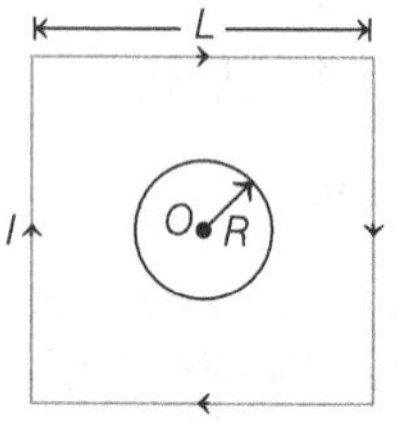

Mutual inductance,

$$M = \frac{\text{Number of flux linkages}}{\text{Current}}$$

$$= \frac{B_1 A_2}{I_1}$$

B_1 = Field of square due to current I in it

$$= \left(\frac{\mu_0}{4\pi} \times \frac{2I \sin 45°}{L/2}\right) \times 4 = \frac{\mu_0 2\sqrt{2} I}{\pi L}$$

So, $$M = \frac{\frac{\mu_0 2\sqrt{2} I}{\pi L} \times \pi R^2}{I} = \frac{\mu_0 2\sqrt{2} R^2}{L}$$

8. $M_{PQ} = M_{QP}$

$$\Rightarrow \qquad \frac{N_P \phi_{PQ}}{I_Q} = \frac{N_Q \phi_{QP}}{I_P} \qquad \text{...(i)}$$

Here, $N_P = N_Q = 1$,

$$\phi_{QP} = 10^{-3} \text{ Wb}$$

$$I_Q = 2\text{A}, I_P = 3\text{A}$$

Substituting values in Eq (i), we get

$$\phi_{PQ} = \frac{I_Q \cdot \phi_{QP}}{I_P} = \frac{2}{3} \times 10^{-3}$$

$$= 0.667 \times 10^{-3}$$

$$= 6.67 \times 10^{-4} \text{ Wb}$$

9. When current flows in any of the coils, the flux linked with the other coil will be maximum in the first case.

Therefore, mutual inductance will be maximum in case (A).

10. $\because \quad M = \frac{e_2}{di_1/dt} = \frac{(50 \times 10^{-3})}{(8/0.5)}$

$$= 3.125 \times 10^{-3} \text{ H} = 3.125 \text{ mH}$$

$$e_1 = M\left(\frac{\Delta i_2}{\Delta t}\right) = \frac{(3.125 \times 10^{-3})(6)}{0.02}$$

$$= 0.9375 \text{ V}$$

11. (a) $M = \frac{N_2 \phi_2}{i_1} = \frac{(1000)(6 \times 10^{-23})}{3} = 2 \text{ H}$

(b) $|e_2| = M\left(\frac{\Delta i_1}{\Delta t}\right) = \frac{(2)(3)}{0.2} = 30 \text{ V}$

(c) $L_1 = \frac{N_1 \phi_1}{i_1} = \frac{(600)(5 \times 10^{-3})}{3} = 1\text{H}$

12. (a) $|e_2| = M\left(\frac{di_1}{dt}\right)$

$$= (3.24 \times 10^{-4})(830)$$

$$= 0.27 \text{ V}$$

(b) Result will remain same.

13. Given, $a = 1000$ cm, $b = 1$ cm

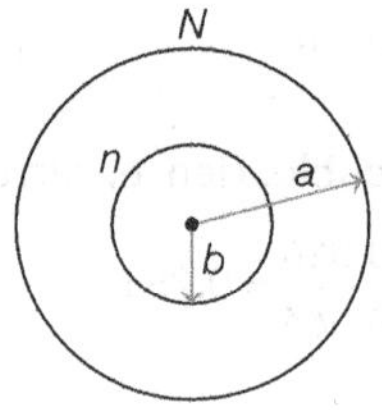

$b << a$

We will take larger coil as primary

$$B = \frac{\mu_0 i_p N}{2a}$$

Flux passing through secondary coil,

$$\phi_s = BA = \frac{\mu_0 i_p N}{2a} \times \pi b^2 \times n$$

Mutual inductance,

$$M = \frac{\phi_s}{i_p} \Rightarrow M = \frac{\mu_0 N n \pi b^2}{2a}$$

$$M = \frac{4\pi \times 10^{-7} \times 200 \times 10 \times \pi \times 1 \times 10^{-4}}{2 \times 1000 \times 10^{-2}}$$

$$= 4\pi^2 \times 10^{-9}$$

or $\quad M = 4 \times 10^{-8}$ H $\qquad$ (using $\pi^2 = 10$)

Module 5.

Introductory Exercise

1. $\because \quad E = \frac{1}{2}Li^2 = i^2 Rt$

$\therefore \quad \frac{L}{R}$ has the units of time.

2. (a) $\tau_L = \frac{L}{R} = \frac{2}{10} = 0.2\text{ s}$

(b) $i_0 = \frac{E}{R} = \frac{100}{10} = 10\text{ A}$

(c) $i = i_0(1 - e^{-t/\tau_L})$

$$= 10(1 - e^{-1/0.2})$$

$$= 9.93\text{ A}$$

3. $E = V_R + V_L$

Exercise

1. When current through inductor becomes constant at $t = \infty$, it behaves like a conducting wire.

2. Time constant $\left(= \frac{L}{R}\right)$ has increased as charging slows down. Further steady state current $\left(= \frac{E}{R}\right)$ is same. So, the correct option L is increased.

3. $\tau_L = \frac{L}{R} = \frac{1}{2}\text{ s}$

$$i = i_0(1 - e^{-t/\tau_L})$$

$$\therefore \quad \frac{i_0}{i} = \frac{1}{1 - e^{-t/\tau_L}}$$

Substituting, $t = 1\text{ s}$ and $\tau_L = \frac{1}{2}\text{ s}$

We get, $\frac{i_0}{i} = \frac{e^2}{e^2 - 1}$

4. $i_0 = \frac{E}{R} = \frac{12}{0.3} = 40\text{ A}$

$$U = \frac{1}{2} L i_0^2$$

$$= \frac{1}{2} \times 50 \times 10^{-3} (40)^2$$

$$= 40\text{ J}$$

5. Value remains $\frac{1}{4}$th in 20 ms times. Hence, two half-lives are equal to 20 ms. So, one half-life is 10 ms.

$$t_{1/2} = (\ln 2)\,\tau_C = (\ln 2)\frac{L}{R}$$

$$\therefore \quad R = \frac{(\ln 2) L}{t_{1/2}} = \frac{(\ln 2)(2)}{10 \times 10^{-3}}$$

$$= (100 \ln 4)\,\Omega$$

6. In steady state, whole current passes through the inductor.

7. $\tau_L = \frac{L}{R} = 2\text{ s}$

$$i_0 = \frac{E}{R} = 3\text{A}, t = 2\text{ s}$$

$$i = i_0(1 - e^{-t/\tau_L})$$

Substituting the given values, we can find i.

8. Just after pressing the key K, the inductor offers infinite resistance.

9. Four resistances form a balanced Wheatstone bridge.

10. At $t = 0, L_1$ and L_2 will produce infinite resistance. So, whole of the current will pass through R_1.

11. By short-circuiting the battery, net resistance across inductor is $\frac{R}{2}$ (R and R in parallel).

$$\therefore \quad \tau_{net} = \frac{L}{R_{net}} = \frac{2L}{R}$$

12. Initial current $= \frac{10}{10} = 1\text{ A}$

$$\therefore \quad \phi_i = L(I_i) = 500\text{ mWb}$$

$$= 0.5\text{ Wb}$$

Final current $= \frac{20}{5} = 4\text{A}$

$$\phi_f = L(I_f) = (0.5) \times 4$$

$$= 2\text{ Wb}$$

$$\therefore \quad \Delta\phi = 1.5\text{ Wb}$$

13. $i = i_0 e^{-t/\tau_L}$

$$\beta i_0 = i_0 e^{-T/\tau_L}$$

$$\therefore \quad \tau_L = \frac{T}{\ln(1/\beta)}$$

14. $P = i_0^2 R \;\Rightarrow\; i_0^2 = \frac{P}{R} \;\Rightarrow\; \tau = \frac{L}{R}$

$$\therefore \quad L = \tau R$$

Heat dissipated $= \frac{1}{2} L i_0^2 = \frac{1}{2}(\tau R)\left(\frac{P}{R}\right) = \frac{1}{2} P\tau$

15. At $t = 0, i = E/R$

Now, this current will decay in closed loop in anti-clockwise direction. So, $|i_2| = i_2 = E/R$ in upward or opposite direction.

Hence, $i_2 = -\frac{E}{R}$

16. $\tau_L = \frac{L}{R} = \frac{0.01}{10} = 10^{-3}\text{ s}$

$$\tau_C = CR = (0.1 \times 10^{-3})(10)$$

$$= 10^{-3}\text{ s}$$

$$(i_0)_L = \frac{20}{10} = 2\text{ A}$$

$$(i_0)_C = \frac{20}{10} = 2\text{ A}$$

The given time is the half-life time of both the circuits.

$$\therefore \quad i_L = i_C = \frac{2}{2} = 1\text{ A}$$

or total current is 2 A.

17. At $t = 0$, induced emf e is maximum and is equal to applied emf, but current i is zero.

After lapse of time, current through the circuit increases but induced emf decreases.

18. $\tau_L = \frac{L}{R} = \frac{2}{2} = 1\text{ s}$

$$t_1 = (\ln 2)\,\tau_L = (\ln 2)\text{ s}$$

Hence, the given time is half-life time.

$$\therefore \quad i = \frac{i_0}{2} = \frac{8/2}{2} = 2\text{ A}$$

Rate of energy supplied by battery

$$= Ei = 8 \times 2 = 16\text{ J/s}$$

$$P_R = i^2R = (2)^2(2) = 8\text{ J/s}$$

$$V_a - V_b = E - iR = 8 - 2 \times 2 = 4\text{V}$$

19. $V_L = Ee^{-t/\tau_L} = 10e^{-t/\tau_L}$

$$\tau_L = \frac{L}{R} = 1\text{ s}$$

$$\therefore \quad V_L = 10\,e^{-t}$$

$$V_R = E - V_L = 10\,(1 - e^{-t})$$

Now, we can put $t = 0$ and $t = 1$ second.

20. At $t = 0$, inductor offers infinite resistance. Hence, current through inductor wire is zero. Whole current passes through two resistors of 4Ω each.

$$i_1 = \frac{10}{4+4} = 1.25\text{ A}$$

At $t = \infty$, inductor offers zero resistance.

$$R_{net} = 4 + \frac{8 \times 4}{8+4}$$

$$= 6.67\ \Omega$$

So, main current,

$$i_2 = \frac{10}{R_{net}} = 1.5\text{ A}$$

This distributes in 4Ω and 8Ω in inverse ratio of resistance. Hence, current through 4Ω is 1A and through 8Ω is 0.5A.

For equivalent τ_L of the circuit R_{net} across inductor after short-circuiting, the battery is 10Ω.

$$\therefore \quad \tau_L = \frac{1}{R_{net}} = \frac{1}{10} = 0.1\text{ s}$$

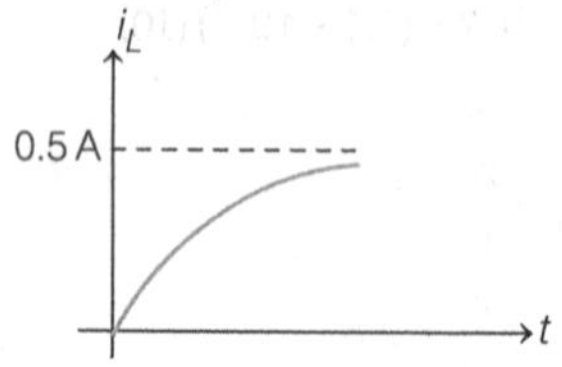

$$i_L = 0.5\,(1 - e^{-\frac{t}{0.1}})$$

$$= 0.5\,(1 - e^{-10t})$$

$$i = 1.25 + 0.25\,(1 - e^{-t/0.1})$$

$$= 1.5 - 0.25\,e^{-10t}$$

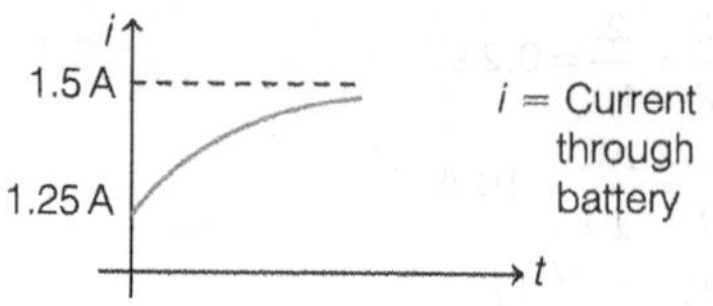

21. $\tau_L = \frac{L}{R} = 0.1\text{ s}$

The given time $t = 0.1$ s is one time constant.

The desired ratio is $\frac{iV_L}{iE} = \frac{V_L}{E}$ $\quad (\because P = Vi)$

After on time constant $V_L = \frac{E}{e}$ as

$$V_L = Ee^{-t/\tau_L}$$

Hence, the desired ratio is $\frac{1}{e} \approx 0.37$.

22. $i_0 = \frac{V}{R} = \frac{3.24}{12.8} = 0.253\text{ A}$

$$\tau_L = \frac{L}{R} = \frac{3.56}{12.8} = 0.278\text{ s}$$

(a) After one time constant ($t = 0.278\text{ s} = \tau_C$)

$$i = \left(1 - \frac{1}{e}\right) i_0 \approx 0.63\, i_0$$

$$= 0.16\text{ A}$$

Power supplied by battery $= Ei$

$$P = (3.24)\,(0.16)$$

$$= 0.518\text{ W}$$

(b) $P_R = i^2R = (0.16)^2\,(12.8) = 0.328\text{ W}$

(c) $P_L = P - P_R = 0.191\text{ W}$

23. (a) After one half-life,

$$t = t_{1/2} = (\ln 2)\,\tau_L$$

$$= 0.693\frac{L}{R}$$

$$= \frac{(0.693)\,(1.25 \times 10^{-3})}{50}$$

$$= 1.73 \times 10^{-5}\text{ s}$$

(b) $\left(\frac{1}{2}Li^2\right) = \left(\frac{1}{2}Li_0^2\right)/2$

$$\therefore \quad i = \frac{i_0}{\sqrt{2}}$$

Now, apply

$$i = i_0\,(1 - e^{-t/\tau_L})$$

where, $\tau_L = \frac{L}{R}$

24. Steady state current developed in the inductor

$$= \frac{E}{r} = i_0\ \text{(say)}$$

(a) Now this current decreases to zero exponentially through r and R.

$$\therefore \quad i = i_0\,e^{-t/\tau_L}$$

where, $\tau_L = \dfrac{L}{R+r}$

(b) Energy stored in inductor,

$$U_0 = \frac{1}{2}Li_0^2 = \left(\frac{1}{2}L\right)\left(\frac{E}{r}\right)^2$$

Now, this energy dissipates in r and R in direct ratio of resistances.

$$\therefore \quad H_r = \left(\frac{r}{R+r}\right)U_0$$

$$= \frac{E^2L}{2r\,(R+r)}$$

25. In steady state, main current from the battery is

$$i_0 = \frac{E}{R} = \frac{20}{5} = 4\text{ A}$$

Now, this current distributes in inverse ratio of inductor.

$$\therefore \quad i_5 = \left(\frac{10}{10+5}\right)(4\text{A})$$

$$= \frac{8}{3}\text{ A}$$

Module 6.

1. $\dfrac{di}{dt} = \omega^2 q = \dfrac{q}{LC} = \dfrac{3.0\times 10^{-5}}{0.6\times 25\times 10^{-6}}$

$$= 2\text{ A/s}$$

2. $q = (2\ \mu\text{F})(6.0\text{ V}) = 12\ \mu\text{C}$,

$$q_0 = (2\ \mu\text{F})(12\text{ V}) = 24\ \mu\text{C}$$

$$\omega = \frac{1}{\sqrt{LC}} = \frac{1}{\sqrt{0.6\times 10^{-3}\times 2\times 10^{-6}}}$$

$$= 2.9\times 10^4\text{ rad/s}$$

$$i = \omega\sqrt{q_0^2 - q^2}$$

$$= (2.9\times 10^4)(10^{-6})\sqrt{(24)^2 - (12)^2}$$

$$= 0.6\text{ A}$$

3. m is similar to L and k is similar to $\dfrac{1}{C}$.

$\therefore$ mk is similar to $\dfrac{L}{C}$.

4. $(V_C)_{max} = (V_L)_{max}$

$$\frac{q_0}{C} = L\left(\frac{dI}{dt}\right)_{max}$$

$$\therefore \quad \left(\frac{dI}{dt}\right)_{max} = \frac{q_0}{LC}$$

5. $\because \quad \omega = \dfrac{1}{\sqrt{LC}}$

or $\quad f \propto \dfrac{1}{\sqrt{C}}$

$$\therefore \quad \frac{f_2}{f_1} = \sqrt{\frac{C_1}{C_2}} = \frac{10^4 - 50}{10^4} = \sqrt{\frac{1}{K}}$$

(where, K = dielectric constant)

$\therefore \quad K = 1.01$

6. Total energy $= \dfrac{1}{2}\dfrac{Q^2}{C}$

At the given instant half of the energy is in electric field and rest half in magnetic field.

$$\therefore \quad \frac{1}{2}\frac{q^2}{C} = \frac{1}{2}\left[\frac{1}{2}\frac{Q^2}{C}\right]$$

$$\Rightarrow \quad q = \frac{Q}{\sqrt{2}}$$

7. $U = \dfrac{1}{2}Li^2 = \dfrac{1}{2}\dfrac{q^2}{C}$

$$\therefore \quad \sqrt{LC} = \frac{q}{i} = \frac{it}{i} = t$$

9. (a) $V_L = V_C$

$$L\frac{di}{dt} = \frac{q}{C}$$

$$q = (LC)\frac{di}{dt} = (0.75\times 18\times 10^{-6})(3.4)$$

$$= 45.9\times 10^{-6}\text{ C}$$

$$= 45.9\ \mu\text{C}$$

(b) $\quad V_L = V_C = \dfrac{q}{C}$

$$= \frac{4.2\times 10^{-4}}{18\times 10^{-6}}$$

$$= 23.3\text{ V}$$

10. $\dfrac{1}{2}Li_{max}^2 = \dfrac{1}{2}CV_{max}^2$

$$\therefore \quad V_{max} = \left(\sqrt{\frac{L}{C}}\right)i_{max}$$

$$= \left(\sqrt{\frac{20\times 10^{-3}}{0.5\times 10^{-6}}}\right)(0.1)$$

$$= 20\text{ V}$$

11. (a) $\omega = \dfrac{2\pi}{f}, T = \dfrac{1}{f}$

(b) At $t = 0$, $q = q_0 = CV_0 = (100\ \mu\text{C})$

Now, $q = q_0\cos\omega t$

(c) $\omega = \dfrac{1}{\sqrt{LC}} \Rightarrow L = \dfrac{1}{\omega^2 C}$

(d) $|i| = \left|\dfrac{dq}{dt}\right| = q_0\,\omega\sin\omega t$

Average value of current in first quarter cycle

$$= \frac{\int_0^{T/4} i\,dt}{T/4}$$

12. (a) $\frac{1}{2}\frac{q_i^2}{C}+\frac{1}{2}LI_i^2=\frac{1}{2}CV_0^2$

$\therefore \quad V_0=\frac{q_i}{C}$ (as $I_i=0$)

(b) $\frac{1}{2}CV_0^2=\frac{1}{2}LI_0^2$

$\therefore \quad I_0=\sqrt{\frac{C}{L}}V_0$

(c) $U_{max}=\frac{1}{2}LI_0^2$

(d) $U_L=\frac{1}{2}L\left(\frac{I_0}{2}\right)^2$

$$U_C=\frac{1}{2}\frac{q^2}{C}=U_{max}-U_L$$

13. (a) $\frac{1}{2}Li_0^2=\frac{1}{2}CV_0^2$

$\therefore \quad L=\frac{CV_0^2}{i_0^2}=\frac{(4\times 10^{-6})(1.5)}{(50\times 10^{-3})^2}$

$=3.6\times 10^{-3}$ H

(b) $f=\frac{1}{2\pi\sqrt{LC}}$

$$=\frac{1}{2\pi\sqrt{(3.6\times 10^{-3})(4\times 10^{-6})}}$$

$=0.133\times 10^4$ Hz

$=1.33$ kHz

(c) $t=\frac{T}{4}=\frac{1}{4f}$

$$=\frac{1}{4\times 1.33\times 10^3}\text{ s}$$

$=0.188\times 10^{-3}$ s

$=0.188$ ms

Module 7.

1.

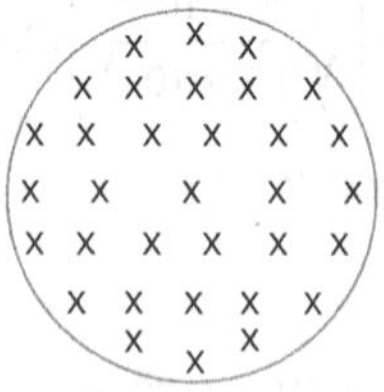

If magnetic field in the shown cylindrical region is changing, then induced electric field exists even outside the cylindrical regions also where magnetic field does not exist.

2. ⊗ magnetic field is increasing. Therefore, induced electric lines are circular and anti-clockwise. Force on negative charge is opposite to electric field.

3. $El=\frac{d\phi}{dt}=S\frac{dB}{dt}$

$$E(2\pi R)=(\pi r^2)(\beta)$$

$\therefore \quad E=\frac{r^2}{2R}\beta \Rightarrow F=qE$

and $\quad \tau=FR=qER=\frac{1}{2}qr^2\beta$

4. $El=\frac{d\phi}{dt}$

or $\quad E(2\pi R)=\pi R^2\cdot\frac{dB}{dt}=\pi R^2(8t)$

$\therefore \quad E=4Rt$

$F=qE=4qRt$ (tangential)

$\tau_F=4qR^2t$

$\tau_f=(\mu mgR)$

When $\tau_F>\tau_f$, ring will start rotating.

At 2 s,

$$8qR^2=\mu mgR$$

$\Rightarrow \quad \mu=\frac{8qR}{mg}$

5. If ⊗ magnetic field increases, then induced electric lines are anti-clockwise. If ⊗ magnetic field decreases, then induced electric lines are clockwise (both inside and outside the cylindrical region).
On positive charge, force is in the direction of **E**. On negative charge, force is in the opposite direction of **E**.

6. $\frac{dB}{dt}=(6t^2+24)$ T/s

At $t=2$ s, $\frac{dB}{dt}=48$ T/s

$$El=\frac{d\phi}{dt}=S\left(\frac{dB}{dt}\right)$$

or $\quad E(2\pi r)=\pi r^2\left(\frac{dB}{dt}\right)$

$\therefore \quad E=\frac{r}{2}\cdot\frac{dB}{dt}$ ...(i)

$$F=qE=\frac{qr}{2}\frac{dB}{dt}$$

$$=\frac{(1.6\times 10^{-19})(1.25\times 10^{-2})}{2}(48)$$

$$=48\times 10^{-21}\text{ N}$$

7. From Eq. (i) of previous problem, we can see that

$$E\propto r$$

i.e. E - r graph is a straight line passing through origin.

8. ⊗ Magnetic field is increasing. Hence, ⊙ magnetic field is produced by a conducting circular loop placed there. For producing, magnetic field induced current should be anti-clockwise. Direction of induced circular electric lines are also anti-clockwise.

9. (a) In the theory we have already derived mutual inductance between solenoid and coil,

$$M = \frac{\mu_0 N_1 N_2 (\pi R_1^2)}{l_1} = \frac{\mu_0 N_1 N_2 S_1}{l_1}$$

$$|e_2| = \left| M \frac{di}{dt} \right| = \left| \frac{\mu_0 N_1 N_2 S_1}{l_1} \frac{di_1}{dt} \right|$$

$$= \frac{(4\pi \times 10^{-7})(25)(10)(5\times 10^{-4})(0.2)}{10^{-2}}$$

$$= 3.14 \times 10^{-6}\ \text{V}$$

(b) $$El = \left| \frac{d\phi}{dt} \right| = e$$

$\therefore$ $$E = \frac{e}{l} = \frac{e}{2\pi R_2} = \frac{3.14\times 10^{-6}}{(2\pi)(0.25)}$$

$$= 2\times 10^{-6}\ \text{V/m}$$

10. (a) **At P_2**

$$El = \frac{d\phi}{dt}$$

$$E(2\pi r_2) = \pi R^2 \cdot \frac{dB}{dt}$$

$\therefore$ $$E = \frac{R^2}{2r_2}\left(\frac{dB}{dt}\right)$$

$$F = qE = \frac{eR^2}{2r_2}(6t^2 - 8t)$$

Substituting the values, we have

$$F = \frac{(1.6\times 10^{-19})(2.5\times 10^{-2})}{2\times 5\times 10^{-2}}[6(2)^2 - 8(2)]$$

$$= 8.0\times 10^{-21}\ \text{N}$$

⊗ magnetic field at the given instant is increasing. Hence, induced current in an imaginary circular loop passing through P_2 should produced ⊙ magnetic field. Or, current in this should be anti-clockwise. Hence, electrons should move in clockwise direction. Or electron at P_2 should experience force in downward direction (perpendicular to r_2).

(b) **At P_1**

$$El = \frac{d\phi}{dt}$$

$\therefore$ $$E(2\pi r_1) = s\frac{dB}{dt} = (\pi r_1^2)\left(\frac{dB}{dt}\right)$$

$\therefore$ $$E = \frac{r_1}{2}\left(\frac{dB}{dt}\right) = \frac{r_1}{2}[6t^2 - 8t]$$

$$= \frac{0.02}{2}[6(3)^2 - 8(3)]$$

$$= 0.3\ \text{V/m}$$

As discussed in the above part, direction of electric field is in the direction of induced current (anti-clockwise) in an imaginary circular conducting loop passing through P_1.

Unit Exercise (Cumulative)

Level 1

1. Inductance of the circuit

$$L = \frac{0.9\times 10^{-4}}{2} = 0.45\times 10^{-4}\ \text{H} \qquad \text{(in parallel)}$$

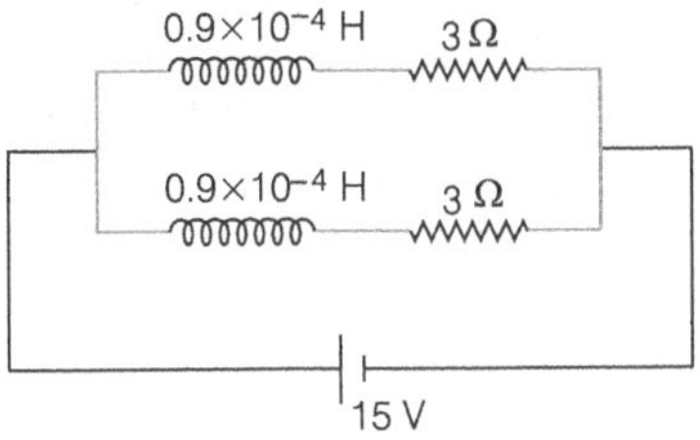

Resistance of the circuit

$$R = 3/2 = 1.5\,\Omega \qquad \text{(in parallel)}$$

$$\therefore \tau_L \text{ (time constant)} = \frac{L}{R} = 3.0\times 10^{-5}\ \text{s}$$

Steady state current in the circuit through the battery

$$i_0 = \frac{V}{R} = \frac{15}{1.5} = 10\ \text{A}$$

2. When switch is opened, current suddenly decreasing from steady state value to zero. When switch is closed, it takes time to increase from 0 to steady state value.

$$e = \left| L\frac{\Delta i}{\Delta t} \right|$$

Δt in second case is large. Hence, induced emf is less.

3. $V_{L_1} = V_{L_2}$

$\therefore$ $$L\frac{di_1}{dt} = L_2\frac{di_2}{dt}$$

or $$L_1 di_1 = L_2 di_2$$

or $$L_1 i_1 = L_2 i_2$$

or $$i \propto \frac{1}{L}$$

4. $\phi = BA\cos\omega t$

$$|e| = \left|\frac{d\phi}{dt}\right| = B\omega A\sin\omega t$$

ϕ and $|e|$ have a phase difference of 90°. When ϕ is zero, $|e|$ will be maximum.

5. Inward magnetic field is increasing. Therefore, induced current in both the loops should be anti-clockwise. But as the area of loop on right side is more, induced emf in this will be more compared to the left side loop

$\left(e = -\frac{d\phi}{dt} = -A\frac{dB}{dt}\right)$. So, net current in the complete loop will be as given

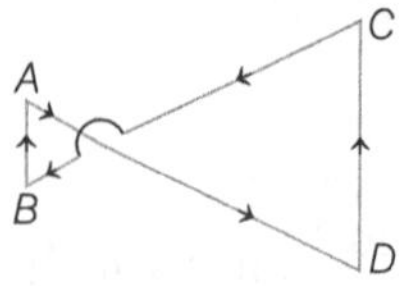

6. Current increases with time. So, flux passing through B will increase with time. From Lenz's law, it should have a tendency to move away from the coil to decrease flux.

7. Since, you are decreasing current in the circuit by increasing resistance of the circuit.
Induced emf across inductor will support 20 V battery. Hence, net emf of the circuit is greater than 20 V. Or current in the circuit is more than 4 A.

8. $L_1 = \left(\frac{\eta}{\eta+1}\right)L, R_1 = \left(\frac{\eta}{\eta+1}\right)R$

$$L_2 = \left(\frac{1}{\eta+1}\right)L, R_2 = \left(\frac{1}{\eta+1}\right)R$$

$$L_{net} = \frac{L_1 L_2}{L_1 + L_2}$$

Similarly, $R_{net} = \frac{R_1 R_2}{R_1 + R_2}$

$$\tau_L = \frac{L_{net}}{R_{net}} = \frac{L}{R}$$

9. Back emf = Applied voltage potential drop across armature coil

$$= 200 - iR = 200 - 1.5 \times 20 = 170 \text{ V}$$

10. $\frac{1}{2}mv_0^2 = \frac{1}{2}Li_{max}^2$

$\therefore$ $i_{max} = \sqrt{\frac{m}{L}}\, v_0$

11. In second position, $\Delta\phi = 0$

$\therefore$ $|Q_2| = \frac{\Delta\phi}{R} = 0$

12. In L-R circuit current grows exponentially.

13. Current through B_2 will decay exponentially but this current will pass through B_1 also.

14. In decay of current through L-R circuit, current can not remain constant.

15.

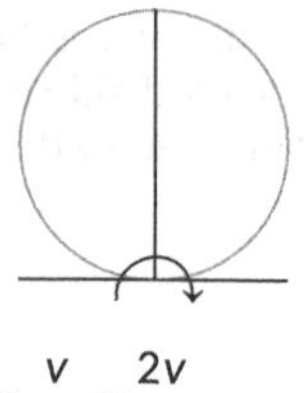

$$\omega = \frac{v}{R} = \frac{v}{l/2} = \frac{2v}{l}$$

$$e = \frac{B\omega l^2}{2} = \frac{B\left(\frac{2v}{l}\right)l^2}{2} = Bvl$$

16. For R-L circuit,

$$i = \frac{E}{R}(1 - e^{-t/\tau}) = i_0(1 - e^{-t/\tau})$$

when $i = \frac{i_0}{2}$

$\Rightarrow$ $\frac{E}{2R} = \frac{E}{R}(1 - r^{-t/\tau})$

Solving we get, $t = \tau \ln 2$

$\Rightarrow$ $t = \frac{L}{R}\ln 2 = \frac{1}{100}\, 0.693$

$$= 0.00693 = 7 \text{ ms}$$

Now, i (15 ms)

$$= \frac{E}{R}\left(1 - e^{\frac{15}{10}}\right) = \frac{E}{R}(1 - e^{-3/2})$$

$\Rightarrow$ $i = \frac{6}{100}\left(1 - \frac{1}{4}\right) = \frac{3}{4} \times \frac{6}{100}$

$$= \frac{18}{400} = \frac{9}{200} \text{ A}$$

As, $U = \frac{1}{2}Li^2 = \frac{1}{2} \times 1 \times \left(\frac{9}{200}\right)^2$

$$\simeq 10^{-3} \text{ J} = 1 \text{ mJ}$$

17. Given, circuit is a series L-R circuit
In an L-R circuit, current increases as

$$i = \frac{E}{R}\left(1 - e^{\frac{-R}{L}t}\right)$$

Now, energy stored in inductor is

$$U_L = \frac{1}{2}Li^2$$

where, L = self inductance of the coil
and energy dissipated by resistor is

$$U_R = i^2 R$$

Given, rate of energy stored in inductor is equal to the rate of energy dissipation in resistor. So, after differentiating, we get

$$iL\frac{di}{dt} = i^2 R$$

$\Rightarrow$ $\frac{di}{dt} = \frac{R}{L}i$

$\Rightarrow$ $\frac{E}{R}\cdot\frac{R}{L}e^{-\frac{R}{L}t} = \frac{R}{L}\cdot\frac{E}{R}\left(1 - e^{-\frac{R}{L}t}\right)$

$\Rightarrow$ $2e^{-\frac{R}{L}t} = 1$

$\Rightarrow$ $e^{-\frac{R}{L}t} = \frac{1}{2}$

Taking log on both sides, we have

$$\Rightarrow \quad \frac{-R}{L}t = \ln\left(\frac{1}{2}\right)$$

$$\Rightarrow \quad \frac{R}{L}t = \ln 2$$

$$\Rightarrow \quad t = \frac{L}{R}\ln 2 = \frac{20}{10}\ln 2$$

$$\Rightarrow \quad t = 2\ln 2$$

Given, phase difference is $\frac{\pi}{4}$.

18. In an L-R circuit, current during charging is given by

$$I = I_0\left(1 - e^{-\frac{R}{L}t}\right)$$

where, $I_0 = \frac{E}{R}$ = saturation current

So, we have $\frac{dq}{dt} = I = I_0\left(1 - e^{-\frac{R}{L}t}\right)$

$$\Rightarrow \quad dq = I_0\left(1 - e^{-\frac{R}{L}t}\right)dt$$

$$q = \int_{t=0}^{t=\frac{L}{R}} I_0\left(1 - e^{-\frac{R}{L}t}\right)dt$$

or $$q = \frac{EL}{2.7R^2} \qquad [\because e \approx 2.72]$$

19. The current in L-R series circuit is given as

$$I = \frac{V}{R}\left(1 - e^{-\frac{tR}{L}}\right); \text{At } t = \infty$$

$$I_\infty = \frac{20}{5}\left(1 - e^{\frac{-\infty}{L/R}}\right) = 4 \qquad \text{...(i)}$$

At $t = 40$ s,

$$I_{40} = 4\left(1 - e^{\frac{-40\times 5}{10\times 10^{-3}}}\right) = 4(1 - e^{-20,000}) \qquad \text{...(ii)}$$

Dividing Eq. (i) by Eq. (ii), we get

$$\frac{I_\infty}{I_{40}} = \frac{1}{1 - e^{-20,000}} \simeq 1.06$$

20.

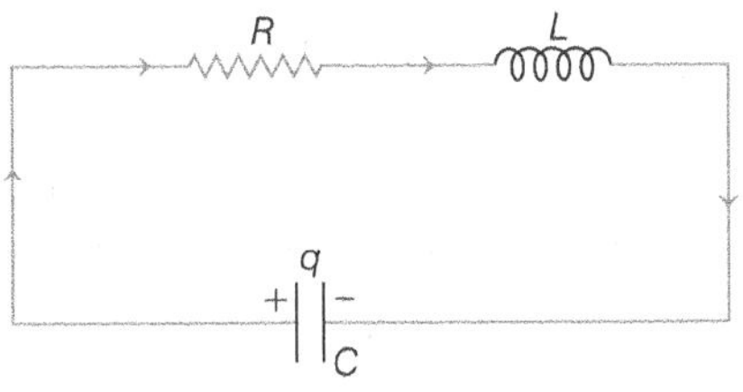

At a time t, suppose charge on capacitor is q and current in the circuit is i, then applying Kirchhoff's loop law, we have

$$\frac{q}{C} - iR - L\frac{di}{dt} = 0$$

Putting $\quad i = -\frac{dq}{dt}$ and $\frac{di}{dt} = -\frac{d^2q}{dt^2}$

In the above equation, we have

$$\frac{d^2q}{dt^2} + \frac{R}{L}\left(\frac{dq}{dt}\right) + \frac{q}{LC} = 0 \qquad \text{...(i)}$$

Comparing this equation with the standard differential equation of damped oscillation, $\frac{d^2X}{dt^2} + \frac{b}{m}\frac{dX}{dt} + \frac{K}{m}X = 0$

Which has a general solution of amplitude, $A = A_0e^{\frac{-bt}{2m}}$

The general solution of Eq. (i) will be

$$Q_{max} = Q_0e^{\frac{-Rt}{2L}} \text{ or } Q_{max}^2 = Q_0^2e^{\frac{-Rt}{L}}$$

Hence, Q_{max}^2 *versus* time graph is exponentially decreasing graph. Lesser the value of self inductance, faster will be the damping.

21. Steady state current i_0 was already flowing in the L-R circuit when K_1 was closed for a long time. Here,

$$i_0 = \frac{V}{R} = \frac{15\text{ V}}{150\,\Omega}$$

$$= 0.1\text{ A}$$

Now, K_1 is opened and K_2 is closed. Therefore, this i_0 will decrease exponentially in the L-R circuit. Current i at time t will be given by

$$i = i_0e^{\frac{-t}{\tau_L}}$$

where, $\tau_L = \frac{L}{R}$

$$\therefore \quad i = i_0e^{\frac{-Rt}{L}}$$

Substituting the values, we have

$$i = (0.1)e^{\frac{-(0.15\times 10^3)(10^{-3})}{(0.03)}} = (0.1)(e^{-5})$$

$$= \frac{0.1}{150} = 6.67\times 10^{-4}\text{ A}$$

$$= 0.67\text{ mA}$$

22. From right hand rule, we can see that

$$V_A > V_B$$

$\therefore$ q_A is positive and q_B is negative.

$$q = CV = C\,(Bvl)$$

$$= (20\times 10^{-6})(0.5)(0.2)(0.1)$$

$$= 0.2\times 10^{-6}\text{ C}$$

$$= 0.2\,\mu\text{C}$$

23. Steady state current through inductor in $\frac{E}{R}$.

So, at $t = 0$, current in closed loop (confiding of capacitor) will remain same.

24. At $t = 0, V_L = -E$

25. $P = F\cdot v = Fm\cdot v$

$$=ilB\cdot v=\left(\frac{B\,v\,l}{R}\right)\cdot lB\cdot v$$

$$=\frac{B^2l^2v^2}{R}$$

26. Current or positive charge is flowing through negatively charged plate. Hence, Q is decreasing. Total energy, $\frac{1}{2}\frac{Q^2}{C}+\frac{1}{2}Li^2$ remains the same. Hence, current i should increase.

27. At $t=0$, charge is maximum, current will be maximum when charge is zero and it will take $\frac{T}{4}$ time for the charge to decrease from maximum to zero value. Hence,

$$t=\frac{T}{4}=\frac{2\pi/\omega}{4}=\frac{\pi}{2\omega}=\frac{\pi}{2}\sqrt{LC}$$

28. Steady state current,

$$i_0=\frac{E}{R}=1\text{ A}$$

Time constant, $t=\frac{L}{R}=1\text{ ms}$

After one time constant or 1 ms, current will be $\frac{1}{e}$ times i_0. Or $\frac{1}{e}$ A.

29. Cross ⊗ magnetic field passing from the closed loop is increasing. Therefore, from Lenz's law induced current will produce dot ⊙ magnetic field. Hence, induced current is anti-clockwise in both loops. Greater the area larger will be the induced emf.

30. After connecting C to B hanging the switch, the circuit will act like an L-R discharging circuit.

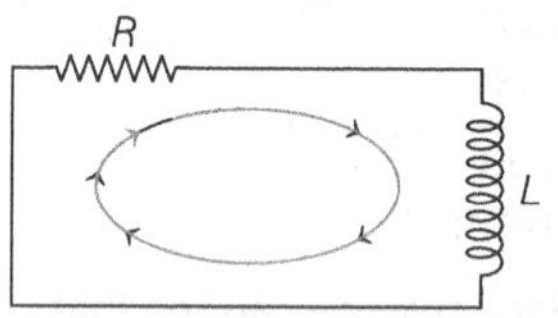

Applying Kirchhoff's loop equation,

$$V_R+V_L=0$$

$$\Rightarrow\quad V_R=-V_L$$

$$\Rightarrow\quad \frac{V_R}{V_L}=-1$$

31. Magnetic field at the centre of a small loop

$$B=\frac{\mu_0 IR_2^2}{2(R_2^2+x^2)^{3/2}}$$

Area of smaller loop $S=\pi R_1^2$

Flux through smaller loop $=BS$

$$=\frac{\pi\mu_0 IR_2^2R_1^2}{2(R_2^2+x^2)^{3/2}}$$

$$\Rightarrow\quad \phi=9.1\times10^{-11}\text{ Wb}$$

32. In steady state when switch was closed,

$$i_0=E/R=(1/5)\text{A}=0.2\text{ A}$$

After switch is opened, it becomes L-C circuit in which peak value current is 0.2 A.

$$\therefore\quad \frac{1}{2}Li_0^2=\frac{1}{2}CV_0^2 \quad\text{or}\quad L=\frac{V_0^2}{i_0^2}.C$$

$$=\frac{(150)^2}{(0.2)^2}\times0.5\times10^{-6}$$

$$=0.28\text{ H}$$ **Ans.**

33. (a) $e=Bvl=0.8\times7.5\times0.5=3$ V

(b) Current will flow in anti-clockwise direction, as magnetic field in ⊗ direction passing through the closed loop is increasing. Therefore, induced current will produce magnetic field in ⊙ direction.

(c) $F=F_m=ilB=\frac{e}{R}lB=\left(\frac{Bvl}{R}\right)Bl=\frac{B^2l^2}{R}v$

$$=\frac{(0.8)^2(0.5)^2}{1.5}\times7.5=0.8\text{ N}$$ **Ans.**

(d) $Fv=0.8\times7.5=6$ W **Ans.**

$$i^2R=\left(\frac{Bvl}{R}\right)^2R=\frac{B^2l^2}{R}\cdot v^2$$

$$=\frac{(0.8)^2\,(0.5)^2}{1.5}\times(7.5)^2=6\text{ W}$$ **Ans.**

So, we can see that both rates are equal.

34. (a) Let at time t velocity of rod be v (towards right) and current in the circuit is i (from a to b). The magnetic force on it is ilB (towards right). Writing the equation of motion of the rod,

$$m\cdot\frac{dv}{dt}=ilB=\left(\frac{E_0-Blv}{R}\right)lB$$

$$\int_0^v\frac{dv}{\frac{E_0Bl}{mR}-\frac{B^2l^2}{mR}v}=\int_0^t dt$$

$$\therefore\quad v=\frac{E_0}{Bl}\left(1-e^{-\frac{B^2l^2}{mR}t}\right)$$ **Ans.**

(b) $i=\frac{E_0-Blv}{R}$

35. After a long time, resistance across an inductor becomes zero while resistance across capacitor becomes infinite. Hence, net external resistance,

$$R_{\text{net}}=\frac{\frac{R}{2}+R}{2}=\frac{3R}{4}$$

Current through the batteries,

$$i=\frac{2E}{\frac{3R}{4}+r_1+r_2}$$

Given that potential across the terminals of cell A is zero.

$$\therefore \quad E - ir_1 = 0$$

or
$$E - \left(\frac{2E}{\frac{3R}{4} + r_1 + r_2} \right) r_1 = 0$$

Solving this equation, we get,

$$R = \frac{4}{3}(r_1 - r_2)$$

36. $v = \sqrt{2as} = \sqrt{2ay}$ ...(i)

$$y = \alpha x^2$$

$$\therefore \quad x = \sqrt{\frac{y}{\alpha}}$$

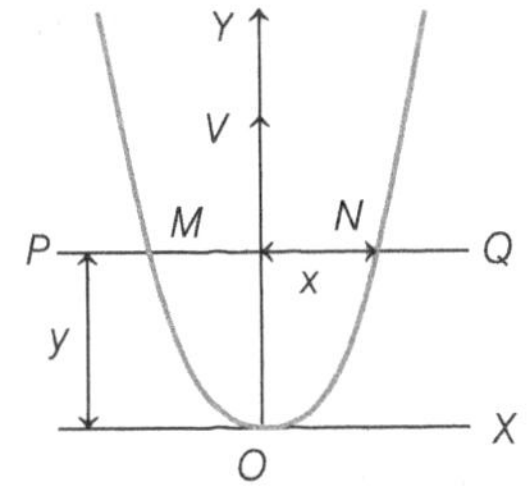

Length of wire $MN = 2x = 2\sqrt{\frac{y}{\alpha}}$

$\therefore$ Motional emf in wire MN will be,

$$e = Bvl = B(\sqrt{2ay})\left(2\sqrt{\frac{y}{\alpha}}\right)$$

or
$$e = By\sqrt{\frac{8a}{\alpha}}$$

Note *MP and NQ are outside the closed loop. Hence, we have not taken motional emf along them.*

37. (a) At a distance x from the wire, magnetic field,

$$B = \frac{\mu_0}{2\pi}\frac{I_0}{x} = f(x)$$

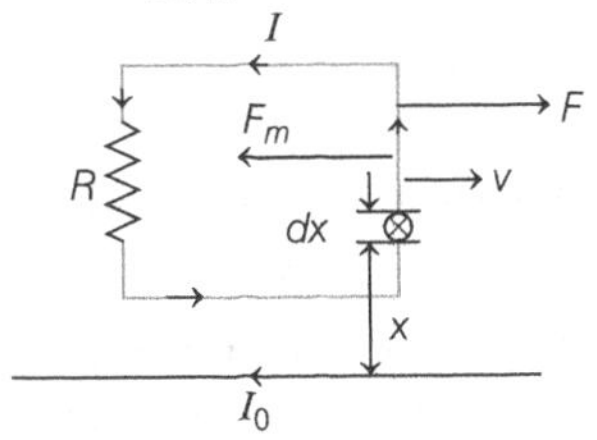

Small motional emf induced in length dx,

$$de = Bvl = \left(\frac{\mu_0 v}{2\pi}\right)\left(\frac{I_0}{x}\right)dx$$

$$\therefore \quad e = \int_{x=a}^{x=b} de = \frac{\mu_0 v I_0}{2\pi}\ln\left(\frac{b}{a}\right)$$

$\therefore$ Induced current,

$$I = \frac{e}{R}$$

or
$$I = \frac{\mu_0 I_0 v}{2\pi R}\ln\left(\frac{b}{a}\right)$$

Direction of I

⊗ magnetic field in the closed loop is increasing. So, according to Lenz law, induced current I should produce ⊙ magnetic field. Hence the current I should be anticlockwise as shown in figure.

(b) Small magnetic force on current carrying wire of length dx.

$$dF_m = IlB = (I)\left(\frac{\mu_0}{2\pi}\right)\left(\frac{I_0}{x}\right)dx$$

$$\therefore \quad F_m = \int_{x=a}^{x=b} dF_m = \frac{\mu_0 I I_0}{2\pi}\ln\left(\frac{b}{a}\right) \quad ...(i)$$

From Fleming's left hand rule we can see that this force is leftwards as shown.

To maintain connector's velocity constant same magnitude of force F will have to be applied rightwards. Substituting value of I from part (a) in equation (i), we get

$$F = \left[\frac{\mu_0 I_0}{2\pi}\ln\left(\frac{b}{a}\right)\right]^2 \frac{v}{R}$$

38. The moving connector will behave like a battery of motional emf $E = Bvl$.

The equivalent electrical circuit is as shown below

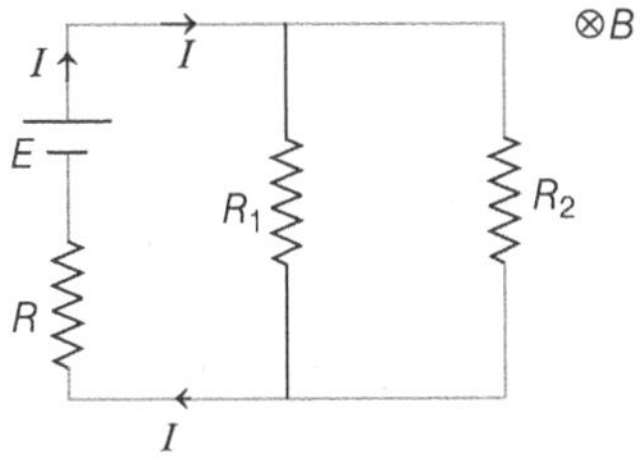

Net resistance of the above circuit,

$$R_{net} = R + \frac{R_1 R_2}{R_1 + R_2}$$

$$\therefore \quad I = \frac{\text{Net emf}}{\text{Net resistance}}$$

or
$$I = \frac{Bvl}{R + \frac{R_1 R_2}{R_1 + R_2}}$$

39.

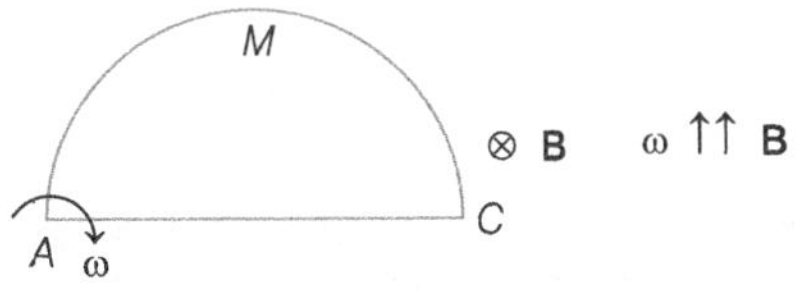

Between A and C, there are two conducting wires AMC and AC and both are in parallel. So, potential difference across both of them will be same. So, we can find potential difference across A and C, which is nothing but rotational emf. Thus,

$$V_C - V_A = \frac{B\omega l^2}{2} \qquad (l = d)$$

$$= \frac{B\omega d^2}{2}$$

$$dV = -Edr$$

$$\therefore \quad \int_A^C Edr = -\int_A^C dV = V_A - V_C = -\frac{B\omega d^2}{2}$$

Substituting the values, we get

$$\int_A^C Edr = -10 \text{ mV}$$

40. When $mg \sin\alpha$ (down the plane) is equal to resistive magnetic force (up the plane), net force becomes zero. So, net acceleration is zero, or velocity is constant. Thus,

$$IlB \sin 90° = mg \sin\alpha$$

$$\Rightarrow \quad \left(\frac{Bvl}{R}\right)(lB) = mg \sin\alpha$$

or $$v = \frac{mgR \sin\alpha}{B^2 l^2}$$

41. (a) When velocity of rod AB is v.

Current, $I = \dfrac{\text{Motional emf}}{\text{Resistance}} = \dfrac{Bvl}{R}$ (anticlockwise)

Magnetic retarding force towards left,

$$F = IlB = \left(\frac{B^2 l^2}{R}\right) v$$

$$\therefore \text{ Retardation } a = -\frac{F}{m} = -\left(\frac{B^2 l^2}{mR}\right) v$$

or $$v \cdot \frac{dv}{ds} = -\left(\frac{B^2 l^2}{mR}\right) v$$

$$\Rightarrow \quad \int_0^s ds = -\frac{mR}{B^2 l^2} \int_{v_0}^0 dv$$

$$\therefore \quad s = \frac{mRv_0}{B^2 l^2}$$

(b) Heat generated in resistance R,

Q = Initial mechanical energy − final mechanical energy

$$= \frac{1}{2} m v_0^2 - 0$$

or $$Q = \frac{1}{2} m v_0^2$$

42. (a) ⊗ magnetic field is decreasing. So, induced current will produce ⊗ magnetic field. Hence induced current should be clockwise in both halves.

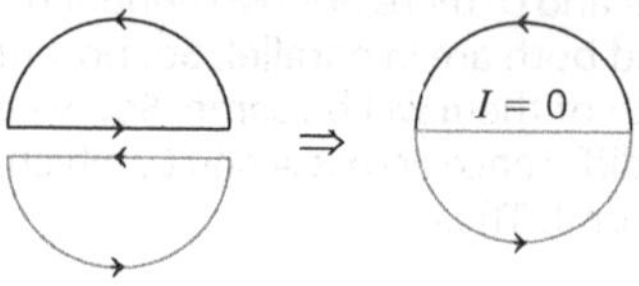

(b)

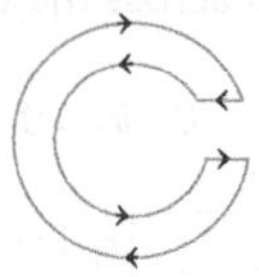

In bigger loop, flux linked will be more. Hence bigger loop will determine the current direction.

(c)

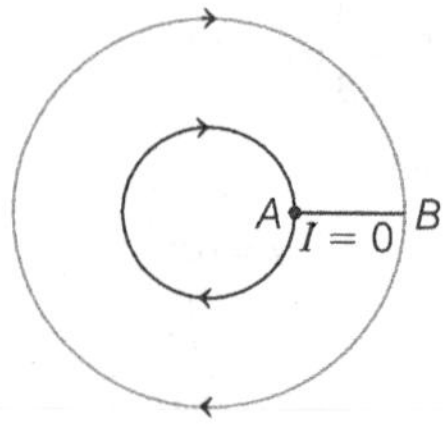

If we apply Kirchhoff's junction law at A or B, current in AB will be zero.

(d)

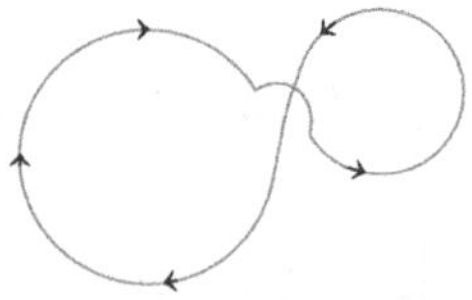

Current in bigger loop will dominate.

43. Note Refer the solution of Q-42 (d).

$$|e_{net}| = |e_1 - e_2| = \left|\frac{d\phi_1}{dt} - \frac{d\phi_2}{dt}\right| \qquad \left(e = -\frac{d\phi}{dt}\right)$$

$$= S_1 \frac{dB}{dt} - S_2 \frac{dB}{dt} = (S_1 - S_2)\frac{dB}{dt}$$

$$= (a^2 - b^2) B_0 \,\omega \cos \omega t$$

$$I = \frac{e_{net}}{R_{net}} = \frac{(a^2 - b^2) B_0 \,\omega \cos \omega t}{\rho(4a + 4b)}$$

$$= \frac{(a - b) B_0 \omega \cos \omega t}{4\rho}$$

$\therefore$ Amplitude of current,

$$I_0 = \frac{(a - b) B_0 \omega}{4\rho}$$

Substituting the given values, we get

$$I_0 = 0.5 \text{ A}$$

44. Let us consider an element of radius r with thickness dr. Number of turns in this element (say N_0),

$$N_0 = \left(\frac{N}{a}\right) dr$$

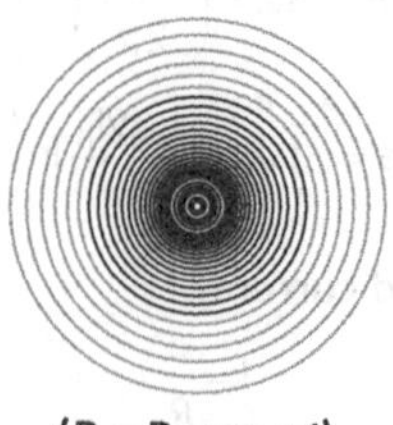

($B = B_0 \cos \omega t$)

Small induced emf in this element,

$$de = -N_0 \frac{d\phi}{dt} = -N_0 \frac{d}{dt}(\pi r^2 B)$$

$$= -\frac{(\pi N r^2) dr}{a} \frac{dB}{dt}$$

$$= -\frac{(\pi B_0 \omega N r^2)\, dr}{a} \cos \omega t$$

Integrating de from $r = 0$ to $r = a$, we get total emf, which comes out to be,

$$e = -\frac{\pi B_0 \omega N a^2}{3} \cos \omega t$$

So, amplitude of emf,

$$e_0 = \frac{\pi B_0 \omega N a^2}{3}$$

45. EMF in secondary coil,

$$|e_2| = M \frac{dI_1}{dt} = \alpha M$$

Now, the simple circuit of secondary coil is

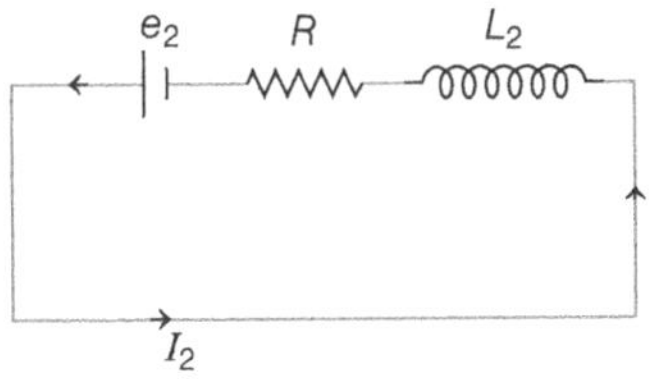

So, I_2 will increase exponentially.

$$I_2 = \frac{e_2}{R}(1 - e^{-t/\tau_L})$$

where, $\tau_L = \frac{L_2}{R}$

Substituting the values of e_2, we get

$$I_2 = \frac{\alpha M}{R}\left(1 - e^{-\frac{Rt}{L_2}}\right)$$

46. $\omega' = \eta\omega$ or $\frac{1}{\sqrt{LC'}} = (\eta)\left(\frac{1}{\sqrt{LC}}\right)$

$$\Rightarrow \quad C' = \frac{C}{\eta^2}$$

Let q_0 be the maximum charge. Then,

$$W = \frac{1}{2}\frac{q_0^2}{C} \quad \text{and} \quad W' = \frac{1}{2}\frac{q_0^2}{C'}$$

Work done by the external agent,

$$W_{ext} = W' - W$$

Putting the values, we get

$$W_{ext} = (\eta^2 - 1) W$$

Note *Maximum charge q_0 will remain unchanged.*

47. (a) Capacitors are in parallel. Therefore,

$$C = C_1 + C_2$$

$$\therefore \quad \omega = \frac{2\pi}{T} = \frac{1}{\sqrt{LC}} = \frac{1}{\sqrt{L(C_1 + C_2)}}$$

$$\therefore \quad T = 2\pi\sqrt{L(C_1 + C_2)}$$

Substituting the given values, we get

$$T = 0.7 \text{ ms}$$

(b) From energy conservation,

$$\frac{1}{2} C V_m^2 = \frac{1}{2} L I_m^2$$

$$\Rightarrow \quad I_m = \sqrt{\frac{C}{L}} V_m = \left(\sqrt{\frac{C_1 + C_2}{L}}\right) V_m$$

Here, $V_m = 180$ V

Substituting the values, we get

$$I_m = 8 \text{A}$$

48. L = Total flux per unit current

$$= \frac{\phi}{I_0} \quad \therefore \quad I_0 = \frac{\phi}{L}$$

At $t = 0$, current is I_0. Therefore I-t equation will be a cos equation.

Thus, $\quad I = I_0 \cos \omega t$

Here, $I_0 = \frac{\phi}{L}$ and $\omega = \frac{1}{\sqrt{LC}}$

$$\therefore \quad I = \frac{\phi}{L} \cos \frac{t}{\sqrt{LC}}$$

Level 2

49. $V_C = Bvl$

$$\therefore \quad q = CV_C = BvlC = \text{constant}$$

$$\therefore \quad I_C = \frac{dq}{dt} = 0$$

$$U_C = \frac{1}{2} CV^2 = \frac{1}{2} CB^2 l^2 v$$

50. From right hand rule, we can see that P and Q points are at higher potential than O.

51. At mean position, velocity is maximum. Hence, motional emf Bvl is also maximum. v oscillates simple harmonically. Hence, motional emf will also move simple harmonically. Further, polarity of induced emf will keep on changing.

52. At $t = t$ side of square,

$$l = (a + 2v_0 t)$$

Area, $\quad S = l^2 = (a + 2v_0 t)^2$

$$\phi = BS = B\,(a + 2v_0 t)^2$$

$$e = \frac{d\phi}{dt} = 4Bv_0\,(a + 2v_0 t)$$

$$R = \lambda[4l] = 4\lambda(a + 2v_0 t)$$

$$\therefore \quad i = \frac{e}{R} = \frac{Bv_0}{\lambda}$$

53. **At time t**

Side of square $l = l_i - dt$

$$S = l^2 = (l_i - \alpha t)^2$$

At given time

$$I = I_i - \alpha t = a$$

$$\phi = BS = B\,(I_i - \alpha t)^2$$

$$e = \left|\frac{d\phi}{dt}\right| = 2B\alpha\,(I_i - \alpha t)$$

But, $(I_i - \alpha t) = a$

$$\therefore \quad e = 2a\,\alpha\,B$$

54. $El = \frac{d\phi}{dt} = S\frac{dB}{dt}$

$$\therefore \quad E\,(2\pi r) = (\pi r^2)\frac{dB}{dt}$$

or

$$E = \frac{r}{2}\frac{dB}{dt}$$

$$F = qE = \frac{qr}{2}\frac{dB}{dt}$$

$$W = Fd = (2\pi r)$$

$$= \pi r^2 q\left(\frac{dB}{dt}\right)$$

$$= \left(\frac{22}{7}\right)(1)^2\,(10^{-6})\,(2\times 10^{-3})$$

$$= 2\pi \times 10^{-9}\,\text{J}$$

55. $B_x = \frac{\mu_0}{2\pi}\frac{I}{x}$

$$de = B_x\,v dx = \frac{\mu_0}{2\pi}\frac{I}{x}v\,dx$$

$$e = \int_a^b de = \frac{\mu_0 I v}{2\pi}\ln\left(\frac{b}{a}\right)$$

$$i = \frac{e}{R} = \frac{\mu_0 I v}{2\pi R}\ln\left(\frac{b}{a}\right) = \text{Induced current}$$

$$dF = (i)\,(dx)\,B_x = \left[\frac{\mu_0 I v}{2\pi R}\ln\left(\frac{b}{a}\right)\right]\left[\frac{\mu_0}{2\pi}\frac{I}{x}\right]dx$$

$$F = \int_a^b dF$$

56. $V_A - V_O = \frac{B\omega(2R)^2}{2} = 2B\omega R^2$ ...(i)

$V_O - V_C = \frac{B\omega(2R)^2}{2} = 2B\omega R^2$ ...(ii)

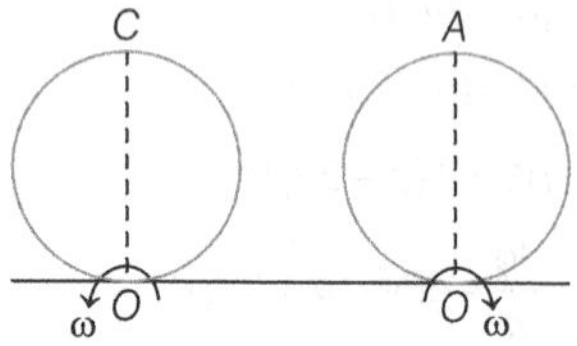

Adding these two equations, we get

$$V_A - V_C = 4B\omega R^2$$

57.

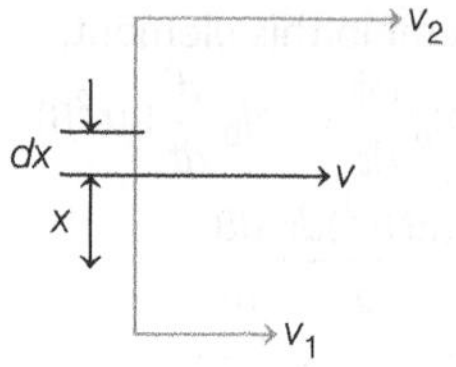

$$v = v_1 + \left(\frac{v_2 - v_1}{l}\right)x$$

Small potential difference $= Bv\,(dx)$

$\therefore$ Total potential difference $= \int_0^l Bv dx$

$$= \frac{1}{2}B\,(v_1 + v_2)\,l$$

58. At time $t = 0$, resistance offered by a capacitor $= 0$ and resistance offered by an inductor $= \alpha$.

$$R_{net} = \frac{R}{2} + \frac{R}{3} = \frac{5R}{6} = 5\,\Omega$$

$\therefore$ Current from the battery,

$$i = \frac{E}{R_{net}} = \frac{5}{5} = 1\,\text{A}$$

59. $|e| = \frac{d\phi}{dt} = S\frac{dB}{dt}$

$$= (4b^2 - \pi a^2)\,B_0$$

$$i = \frac{|e|}{R} = \frac{(4b^2 - \pi a^2)\,B_0}{R}$$

⊗ magnetic field is increasing. So,

⊙ magnetic field is produced.

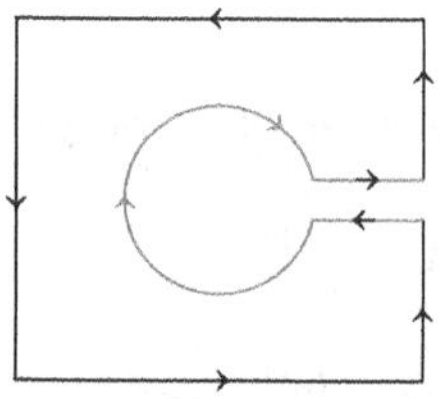

60. $\phi_i = \int_b^{b+a}\frac{\mu_0}{2\pi}\frac{I}{x}(adx)$

$$= \frac{\mu_0 Ia}{2\pi}\ln\left(\frac{b+a}{b}\right)$$

Similarly,

$$\phi_f = \frac{\mu_0 Ia}{2\pi}\ln\left(\frac{b-a}{b}\right)$$

$$\Delta\phi = |Q_i - Q_f| = \frac{\mu_0 Ia}{2\pi}\ln\left(\frac{b+a}{b-a}\right)$$

$$\Delta q = \frac{\Delta\phi}{R} = \frac{\mu_0 la}{2\pi R}\ln\left(\frac{b+a}{b-a}\right)$$

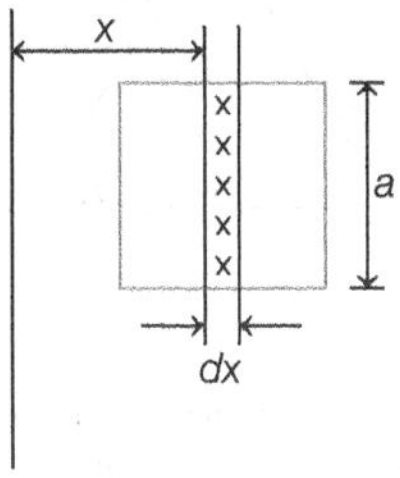

61. For a capacitor,

$$i = \left(\frac{dq}{dt}\right) = C\cdot\left(\frac{dV}{dt}\right),$$

i.e. $i = \text{constant if } \frac{dV}{dt} = \text{constant}$

$$\therefore \quad C = \frac{i}{\left(\frac{dV}{dt}\right)} = \frac{1}{2} = 0.5\ \text{F}$$

62. $V_{ba} = \int \mathbf{E}\cdot d\mathbf{l} = \frac{1}{4}\left(\frac{d\phi}{dt}\right)$

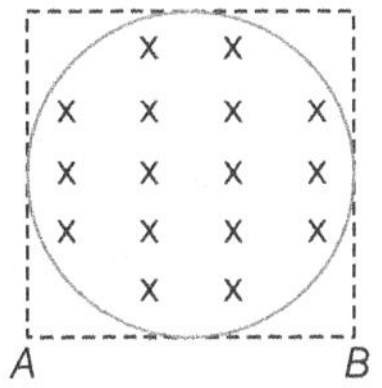

$$= \frac{1}{4}(\pi R^2)\cdot\frac{dB}{dt} = \frac{\pi R^2\alpha}{4}$$

63. emf, $|e| = \left|\frac{d\phi}{dt}\right| = S(dB/dt) = \frac{\pi r^2}{2}\left(\frac{dB}{dt}\right)$

$$i = \frac{|e|}{R} = \frac{\pi r^2}{2R}\left(\frac{dB}{dt}\right)$$

$$= \frac{\pi(0.1)^2}{2\times 4}\times 0.01$$

$$= 3.9\times 10^{-5}\ \text{A}$$

64. Given circuit is,

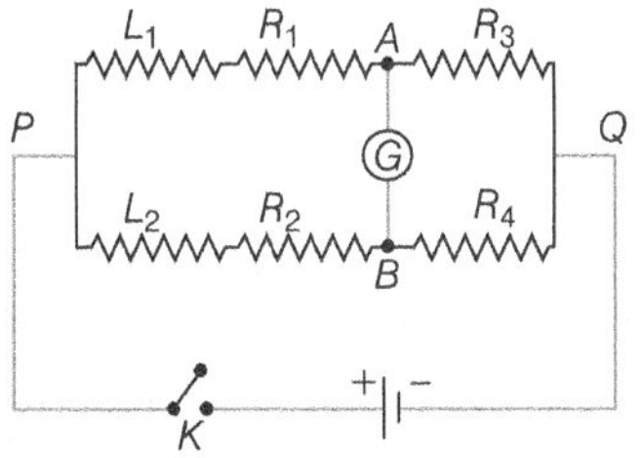

As galvanometer shows no deflection, $V_A = V_B$.

So, $V_{PA} = V_{PB}$

$$L_1\frac{di}{dt} + i_1R_1 = L_2\frac{di_2}{dt} + i_2R_2 \quad \text{...(i)}$$

and $V_{AQ} = V_{BQ}$

$$i_1R_3 = i_2R_4 \quad \text{...(ii)}$$

Substituting for i_1 in Eq. (i) from Eq. (ii), we get

$$L_1\frac{R_4}{R_3}\cdot\frac{di_2}{dt} + i_2\frac{R_1}{R_3} = L_2\frac{di_2}{dt} + i_2R_2$$

At $t = 0$, $i_2 = 0$.

So, $L_1\frac{R_4}{R_3} - L_2 = 0$

$$\Rightarrow \quad \frac{L_1}{L_2} = \frac{R_3}{R_4}$$

At $t = \infty$,

$$\frac{di_2}{dt} = 0 \Rightarrow i_2 = \frac{V}{R_2 + R_4}$$

$$\Rightarrow \quad R_2 - \frac{R_4R_1}{R_3} = 0 \Rightarrow \frac{R_1}{R_2} = \frac{R_3}{R_4}$$

So, combining both, we have

$$\frac{L_1}{L_2} = \frac{R_1}{R_2} = \frac{R_3}{R_4}$$

65. Magnetic field at side (1),

$$B_1 = B_0$$

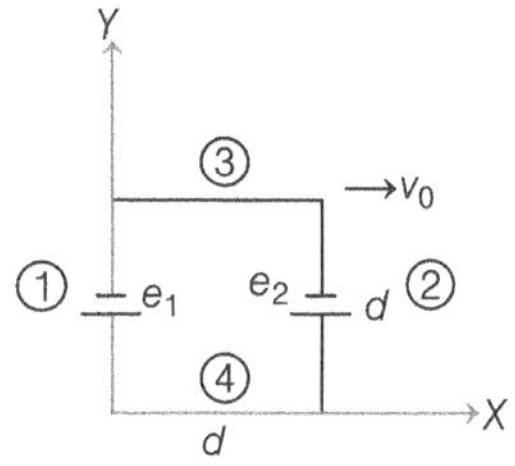

Induced emf in (1),

$$e_1 = B_0 v_0 d$$

Magnetic field at side (2),

$$B_2 = B_0\left(1 + \frac{d}{a}\right)$$

Induced emf in (2),

$$e_2 = B_0\left(1 + \frac{d}{a}\right)v_0 d$$

Induced emf in (3) and (4) will be zero.

Net emf, $e = e_2 - e_1$

$$= \frac{B_0 v_0 d^2}{a}$$

66. As, emf induced when a loop rotates in a magnetic field is ($e = NAB\,\omega\sin\omega t$), which is sinusoidal function. Therefore, induced current will also be sinusoidal as shown in the figure.

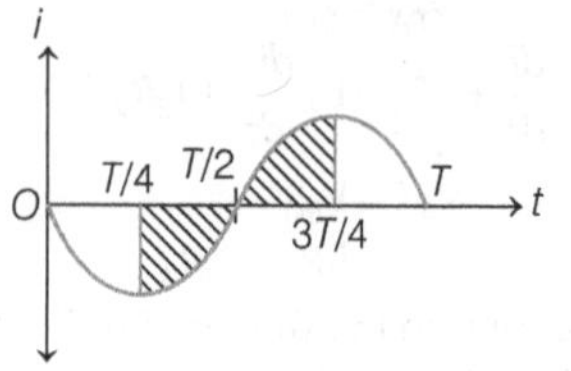

The curve will be negative, because initially the flux decreases from $t=0$ to $T/4$ and then increases from $t=\frac{3T}{4}$ to $\frac{T}{4}$.

But from $t=\frac{T}{4}$ to $\frac{3T}{4}$, there will be no induced current as there is no field on the right side.
Hence, the current produced will be as shown in the figure.

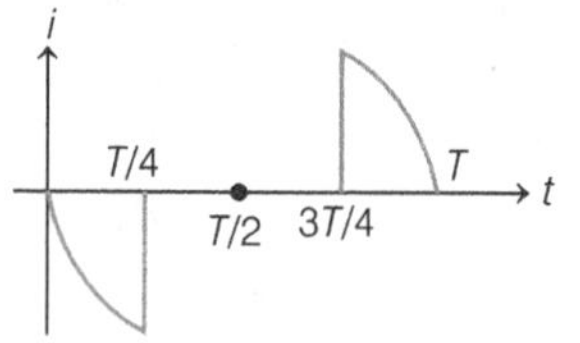

67. $x_2=0.10$ m

The emf induced between the point A and B of rod,

$$e=\int_{x_1=0.07}^{x_2=0.10} B(\omega x)\,dx=\omega B\int_{0.07}^{0.1} x\,dx=\omega B\left[\frac{x^2}{2}\right]_{0.07}^{0.1}$$

$$=2\times 10\times\left[\frac{(0.1)^2}{2}-\frac{(0.07)^2}{2}\right]$$

$$=0.05092\approx 0.051\text{ V}$$

68. Kinetic energy of slider is dissipated in form of heat in resistor.

$$\therefore \qquad -\frac{d}{dt}(\text{KE})=\frac{\varepsilon^2}{R}$$

$$\Rightarrow \qquad -\frac{d}{dt}\left(\frac{1}{2}mv^2\right)=\frac{B^2l^2v^2}{R}$$

$$\Rightarrow \qquad \frac{dv}{v}=\frac{-B^2l^2}{mR}\cdot dt$$

$$\text{or} \qquad \int_{v_0}^{v}\frac{dv}{v}=\frac{-B^2l^2}{mR}\int_0^t dt$$

$$\Rightarrow \qquad \log v=\frac{-B^2l^2}{mR}\cdot t+\log v_0$$

which is best shown by graph (c).

69. Induced emf,

$$\varepsilon=\left|\frac{d\phi}{dt}\right|=\frac{d}{dt}(B\cdot A)=\frac{A\,dB}{dt}$$

$$=\sqrt{3}\times\sqrt{3}=3\text{ V}$$

Induced current $=\frac{3}{5}=0.6$ A

Potential difference between the point A and B,

$$\Delta V_{AB}=1-(0.6)(1)=0.4\text{ V}$$

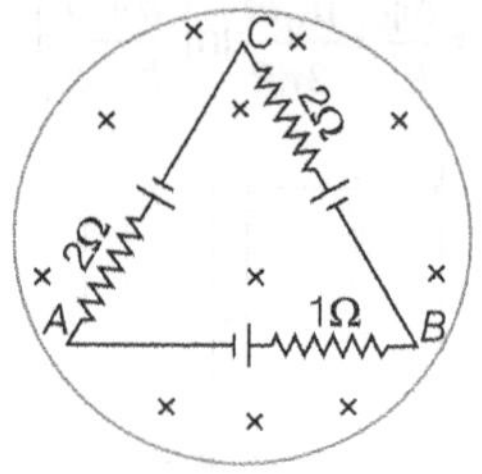

70. Consider a cylindrical region of uniform magnetic field which is perpendicular to the plane.

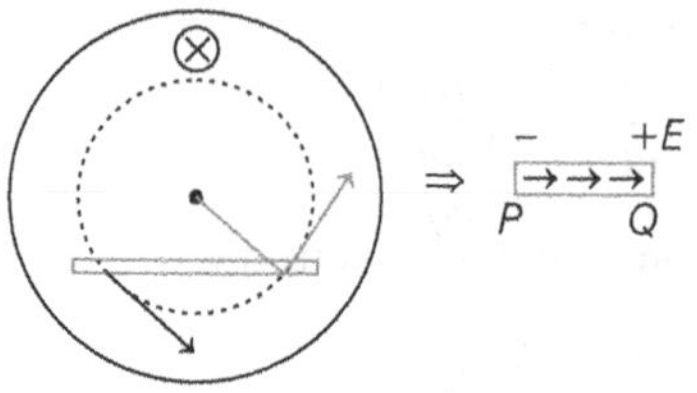

So, change in magnetic flux induced emf at high potential, i.e. positive across Q while low potential, i.e. negative across P in the conducting rod PQ.

71. Consider a cylindrical region of uniform magnetic field exist perpendicular to the plane of paper.

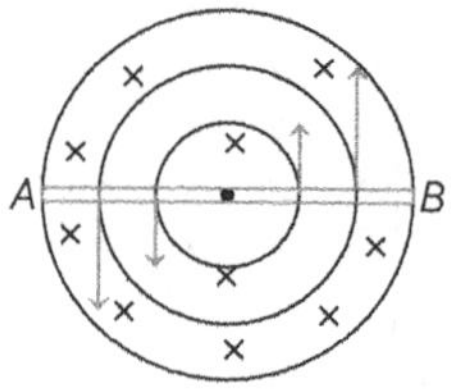

$\mathbf{E}\perp d\mathbf{l}$ at every point to the plane of paper.
So, $\Delta V_{AB}=E\,dl\cos 90^\circ=\text{zero}$
Thus, induced emf between the ends of the rod is zero.

72. $e=Bvl$, where $l=\frac{L}{2}$

For polarity of this motional emf, we can use right hand rule.

73. (a)

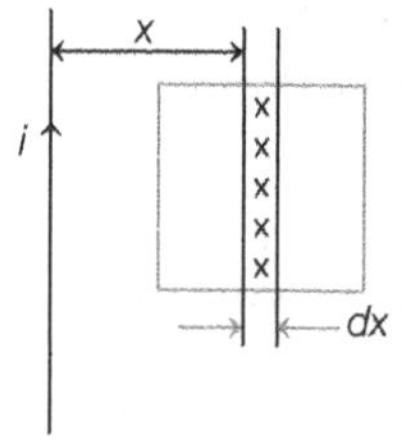

$$B_x=\frac{\mu_0}{2\pi}\frac{i}{x}$$

$$d\phi=(B_x)\,dS=\left(\frac{\mu_0}{2\pi}\frac{i}{x}\right)(a\,dx)$$

$$\phi=\int_a^{2a} d\phi=\frac{\mu_0 ia}{2\pi}\ln 2$$

$$M=\frac{\phi}{i}=\frac{\mu_0 a}{2\pi}\ln 2$$

(c)

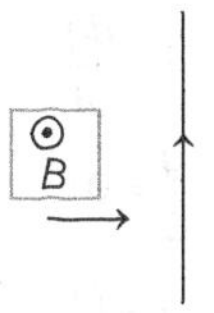

Wire produces ⊙ magnetic field over the loop. If the loop is brought closer to the wire, ⊙ magnetic field passing through the loop increases. Hence, induced current produces ⊗ magnetic field so, induced current is clockwise.

74. According to Lenz's law, induced effects always oppose the change i_1 and i_2 both are in same direction. Hence, magnetic lines from B due to both currents are from right to left. By bringing A closer to B or increasing i_1 right to left magnetic field from B will increase. So, i_2 should decrease.

75. $\phi_i = BS\cos 0° = (4)(2) = 8$ Wb

$$\phi_f = BS\cos 90° = 0$$
$$\Delta\phi = 8 \text{ Wb}$$
$$|e| = \frac{\Delta\phi}{\Delta t} = \frac{8}{0.1} = 80 \text{ V}$$
$$i = \frac{|e|}{R} = 20 \text{ A}$$
$$\Delta q = \frac{\Delta\phi}{R} = 2 \text{ C}$$

This current is not constant. So, we cannot find the heat generated unless current function with time is not known.

76. $i_{max} = \omega q_0 = \left(\frac{1}{\sqrt{LC}}\right) q_0$

$$\left(\frac{di}{dt}\right)_{max} = \omega^2 q_0 = \left(\frac{1}{LC}\right) q_0$$

77. $q = 2t^2$

$$i = \frac{dq}{dt} = 4t$$
$$\frac{di}{dt} = 4 \text{ A/s}$$

At $t = 1$ s, $q = 2$ C, $i = 4$ A

and $\frac{di}{dt} = 4$ A/s

$$V_a - V_b = L\frac{di}{dt}$$
$$= 1 \times 4 = 4 \text{ V}$$
$$V_b - V_c = \frac{q}{C} = \frac{2}{2} = 1 \text{ V}$$
$$V_c - V_d = iR$$
$$= 4 \times 4 = 16 \text{ V}$$

$V_a - V_d$ is summation of above three, i.e. 21 V.

78. $V_a - V_c = 0$ as **l** is parallel to **v**.

$$V_a - V_b = V_c - V_b = \frac{B\omega l^2}{2}$$

79. According to Lenz's law, induced current at point P and Q is clockwise.

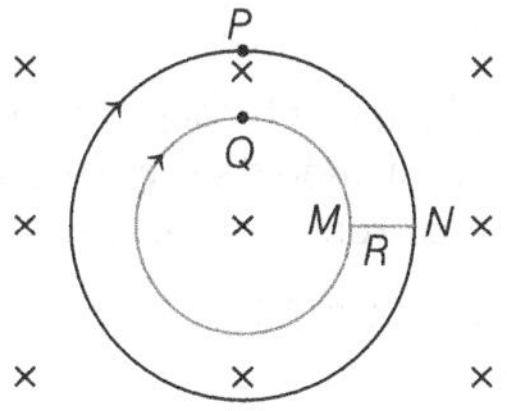

If we apply Kirchhoff's first law at M or N, we can see that

$$I_R = 0$$

80. (a) Induced emf $= -\frac{d}{dt}NBA$

$$= -\pi a^2\left(\frac{d}{dt}B\right) = \pi a^2 k$$

(b) Charge stored

$$Q = CE = C\pi a^2 k$$

(c) B points into plane of paper and is decreasing so induced emf also produces a field that points into the paper.

∴ Current flows clockwise and so upper plate is positive.

(d) Changing B induces E and this pushes on charges in the wire

81. $|e| = \left|\frac{d\phi}{dt}\right| = S\frac{dB}{dt} = (\pi a^2)B_0$

82. $El = \frac{d\phi}{dt}$

∴ $E(2\pi a) = (\pi a^2)B_0$

or $E = \frac{1}{2}aB_0$

83. $F = qE = \frac{1}{2}qaB_0$

$$\tau = Fa = \frac{1}{2}qa^2B_0$$
$$\alpha = \frac{\tau}{I} = \frac{\left(\frac{1}{2}qa^2B_0\right)}{ma^2} = \frac{qB_0}{2m}$$

84. $\omega = \alpha t = \left(\frac{qB_0}{2m}\right)t$

$$P = \tau\omega = \left(\frac{1}{2}qa^2B_0\right)\left(\frac{qB_0}{2m}\right)$$
$$= \frac{q^2B_0^2a^2}{4m}$$

85. $|e| = \frac{d\phi}{dt} = S\frac{dB}{dt}$

$$= (0.2 \times 0.4)(2)$$
$$= 0.16 \text{ V}$$

$$i = \frac{|e|}{R} = \frac{0.16}{(1)(40+40+20)\times 10^{-2}}$$

$$= 0.16 \text{ A}$$

⊙ Magnetic field passing through the loop is increasing. So, induced current should produce ⊗ magnetic field. Hence, induced current is clockwise.

86. At $t = 2$ s, rod will move 10 cm. Hence, 40 cm side will become 30 cm.

$$|e| = e_1 \text{ (say)} = S\left(\frac{dB}{dt}\right)$$

$$(0.2\times 0.3)(2) = 0.12 \text{ V}$$

At $t = 2$ s, $B = 4$ T

$$\therefore \quad e_2 = Bvl = (4)(5\times 10^{-2})(0.2)$$

$$= 0.04 \text{ V}$$

$$\therefore \quad e_{net} = e_1 - e_2 = 0.08 \text{ V}$$

87. $i = \dfrac{e_{net}}{R} = \dfrac{0.08}{(1)(30+30+20)\times 10^{-2}}$

$$= 0.1 \text{ A}$$

$$F = ilB$$

$$= (0.1)(0.2)(4) = 0.08 \text{ N}$$

88. to 90.

At terminal velocity,

$$iLB = mg$$

$$\therefore \quad i = \frac{mg}{LB} = \frac{0.2\times 9.8}{1\times 0.6}$$

$$i = 3.27 \text{ A} \quad \text{...(i)}$$

$$e = BvL \quad (v = \text{terminal velocity})$$

$$= (0.6)(v)(1)$$

$$e = 0.6 \text{ v}$$

$$P_{R_1} = \frac{e^2}{R_1}$$

$$\therefore \quad 0.76 = \frac{0.36\, v^2}{R_1} \quad \text{...(ii)}$$

$$P_{R_2} = \frac{e^2}{R_2}$$

$$\therefore \quad 1.2 = \frac{0.36\, v^2}{R_2} \quad \text{...(iii)}$$

R_1 and R_2 are in parallel.

$$\therefore \quad R_{net} = \frac{R_1R_2}{R_1+R_2} \quad \text{...(iv)}$$

$$i = \frac{e}{R_{net}} \quad \text{...(v)}$$

Solving these five equations, we can get the results.

91. Initial charge $= CV_0 = Q_0 = 10\times 10^{-3}\times 5 = 50$ mC

When capacitor is connected to position 1,

$$E - IR - \frac{q}{C} = 0$$

Here, $\quad I = \dfrac{dq}{dt}$

$$\therefore \quad E - \frac{dq}{dt}R - \frac{q}{C} = 0$$

or $\quad \displaystyle\int_0^t \frac{1}{RC}\,dt = \int_{50}^{q} \frac{dq}{EC-q}$

$$\Rightarrow \quad q = 50\,[2 - e^{-t}]\text{ mC}$$

Voltage across the capacitor at that time $t = 1$ s, we get

$$V = \frac{q}{C} = \frac{50\,[2-1/e]}{10}$$

$$= 5\,[2-(1/e)] \text{ V}$$

92. The maximum current flowing in the L-C circuit.

$$\frac{1}{2}LI^2 = \frac{1}{2}CV^2 \Rightarrow I = \left[2 - \frac{1}{e}\right]\times 10^4 \text{ A}$$

93. Frequency of L-C oscillation

i.e. $\quad \dfrac{1}{2\pi\sqrt{LC}} = \dfrac{1}{2\pi\sqrt{10\times 10^{-3}\times 2.5\times 10^{-3}}}$

$$= \frac{100}{\pi}\text{ Hz}$$

94. Charge on capacitor at time t is

$$q = q_0\,(1 - e^{-t/\tau})$$

Here, $q_0 = CV$ and $t = 2\tau$

$$\therefore \quad q = CV\,(1 - e^{-2\tau/\tau})$$

$$= CV\,(1 - e^{-2})$$

95. From conservation of energy,

$$\frac{1}{2}LI_{max}^2 = \frac{1}{2}CV^2$$

$$\therefore \quad I_{max} = V\sqrt{\frac{C}{L}}$$

96. Comparing the L-C oscillations with normal SHM, we get

$$\frac{d^2Q}{dt^2} = -\omega^2 Q$$

Here, $\quad \omega^2 = \dfrac{1}{LC}$

$$\therefore \quad Q = -LC\frac{d^2Q}{dt^2}$$

97. (a) $B = \dfrac{F}{il}$

$$[B] = \left[\frac{MLT^{-2}}{AL}\right] = [MT^{-2}A^{-1}]$$

(b) $U = \dfrac{1}{2}Li^2$

$$\therefore \quad [L] = \left[\frac{U}{i^2}\right] = \left[\frac{ML^2T^{-2}}{A^2}\right]$$

$$= [ML^2T^{-2}A^{-2}]$$

(c) $\omega = \frac{1}{\sqrt{LC}}$

$\Rightarrow \quad [LC] = \left[\frac{L}{\omega^2}\right] = [T^2]$

(d) $[\phi] = [BS] = [MT^{-2}A^{-1}L^2]$

98. $\omega = \frac{1}{\sqrt{LC}} = 2 \text{ rad/s}$

(a) $i_{max} = \omega q_0 = 8 \text{ A}$

(b) $\left|\frac{di}{dt}\right|_{max} = \omega^2 q_0 = 16 \text{ A/s}$

(c) $V_L = V_C = \frac{q}{C} = \frac{2}{1/4} = 8 \text{ V}$

(d) $V_C = V_L = L\frac{di}{dt} = (1)\left(\frac{16}{2}\right) = 8 \text{ V}$

99. Steady state current through inductor,

$$i_0 = \frac{9}{3} = 3 \text{ A}$$

Now, this current decays exponentially across inductor and two resistors.

$$\tau_L = \frac{L}{R} = \frac{9}{6+3} = 1 \text{ s}$$

$$t_{1/2} = (\ln 2)\,\tau_L = (\ln 2) \text{ s}$$

Given time is half-life time. Hence, current will remain 1.5 A.

$$i = i_0 e^{-t/\tau_L} = 3e^{-t}$$

$$\left(-\frac{di}{dt}\right) = 3e^{-t}$$

In the beginning $\left(\frac{-di}{dt}\right) = 3 \text{ A/s}$

After one half-life time $\left(\frac{-di}{dt}\right) = 1.5 \text{ A/s}$

(a) $V_L = L\left(\frac{-di}{dt}\right) = 9 \times 1.5 = 13.5 \text{ V}$

(b) $V_{3\Omega} = iR = 1.5 \times 3 = 4.5 \text{ V}$

(c) $V_{6\Omega} = iR = 1.5 \times 6 = 9 \text{ V}$

(d) $V_{bc} = V_L - V_{3\Omega} = 9 \text{ V}$

100. $\phi = 2t$

(a) $e = \frac{d\phi}{dt} = 2 \text{ V}$

(b) $i = \frac{e}{R} = 1 \text{ A} = \text{constant}$

(c) $\Delta q = i\Delta t = 1 \times 2 = 2 \text{ C}$

(d) $H = i^2 R\Delta t = (1)^2(2)(2) = 4\text{J}$

101. (a) If current is increased, ⊗ magnetic field passing through loop will increase. So, induced current will produce ⊙ magnetic field. Hence, induced current is anti-clockwise.

Now, i and I currents in PQ are in opposite directions. Hence, they will repel each other.

Same logic can be applied for (b) part.

(c) situation is similar to (b) situation and (d) situation is similar to (a) situation.

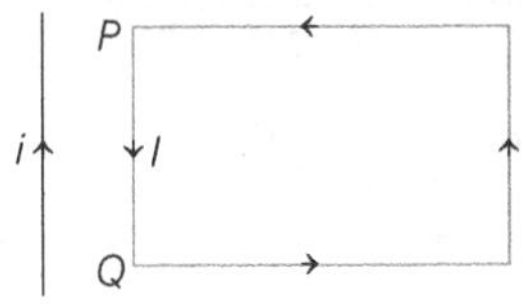

102. (a) At 1s, ⊗ flux is increasing. So, induced current will produce ⊙ magnetic field or induced current is anticlockwise.

$$i = \frac{(d\phi/dt)}{R} = \frac{5}{10} = \frac{1}{2} \text{ A}$$

103. $\tau_L = \frac{L}{R}$

and $\quad \tau_C = CR, \quad \frac{\tau_C}{\tau_L} = \frac{CR^2}{L} = \frac{C}{L}\cdot\frac{L}{C} = 1$

$\therefore \quad \tau_L = \tau_C$

$\therefore$ For the given condition $\tau_L = \tau_C = \tau$ (say)

Now, in L-R circuit

$$I_1 = \frac{V}{R}(1 - e^{-t/\tau})$$

In C-R circuit, $\quad I_2 = \frac{V}{R}e^{-t/\tau}$

$\therefore \quad I = I_1 + I_2 = \frac{V}{R} = \text{constant}$ **Ans.**

104. $\theta = \omega t$

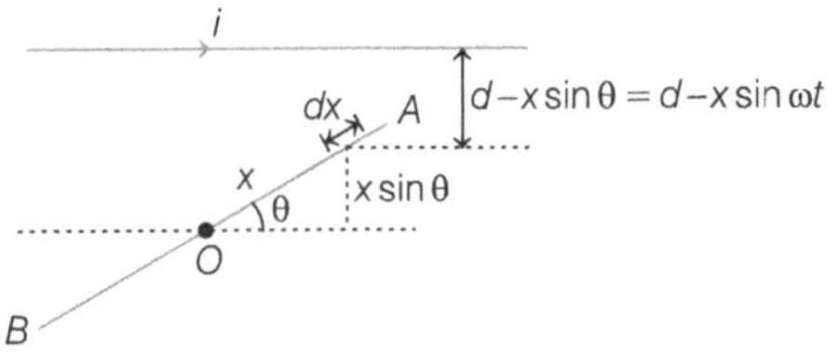

$$de = B(\omega x)\,dx$$

Here, $\quad B = \frac{\mu_0}{2\pi}\frac{i}{d - x\sin\omega t}$

$\therefore \quad de = \frac{\mu_0 i\omega}{2\pi}\frac{x}{d - x\sin\omega t}\cdot dx$

$$V_{OA} = V_O - V_A = \int_0^a de = \frac{\mu_0 i\omega}{2\pi}\int_0^a \frac{x}{d - x\sin\omega t}dx$$

$$= = -\frac{\mu_0 i\omega}{2\pi\sin\omega t}\left[\frac{d}{\sin\omega t}\ln\left(\frac{d - a\sin\omega t}{d}\right) + a\right]$$

Similarly, $\quad V_{OB} = V_O - V_B = \frac{\mu_0 i\omega}{2\pi}\int_0^a \frac{x}{d + x\sin\omega t}dx$

$$= \frac{\mu_0 i\omega}{2\pi\sin\omega t}\left[a - \frac{d}{\sin\omega t}\ln\left(\frac{d + a\sin\omega t}{d}\right)\right]$$

$\therefore \quad V_{AB} = V_{OB} - V_{OA}$

$$= \frac{\mu_0 i\omega}{2\pi\sin\omega t}\left[2a + \frac{d}{\sin\omega t}\ln\left(\frac{d - a\sin\omega t}{d + a\sin\omega t}\right)\right]$$

Note *This function is discontinuous at $\omega t = n\pi$.*

105. At $t = 0$, Current through inductor will be zero.

At $t = \infty$, net emf $= \dfrac{2/2 + 4/1}{1/2 + 1/1} = \dfrac{10}{3}$ V

Net resistance $= \dfrac{2 \times 1}{2 + 1} = \dfrac{2}{3}\,\Omega$

$\therefore \quad i = \dfrac{10/3}{2/3} = 5$ A

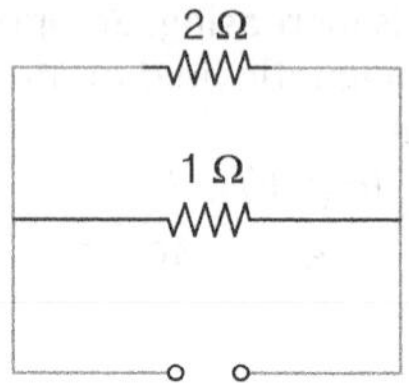

To find equivalent time constant short circuit, both the batteries and find net resistance across inductor.

$$R_{net} = \frac{2 \times 1}{2 + 1} = \frac{2}{3}\,\Omega$$

$$\therefore \quad \tau_L = \frac{L}{R_{net}} = \frac{1 \times 10^{-3}}{2/3} = \frac{3}{2000}\text{ s}$$

Current through inductor will increase exponentially from 0 to 5 A.

$$\therefore \quad i = 5\left(1 - e^{-\frac{2000\,t}{3}}\right)$$

106. These are two independent parallel circuits across the battery.

(a) $V_{ab} = E = 120$ volt (at all instants)

(b) a is at higher potential.

(c) V_{cd} will decrease exponentially from 120 V to zero.
$\therefore V_{cd} = 120$ volt, just after the switch is closed.

(d) c will be at higher potential.

(e) When switch is opened, current through R_1 will immediately become zero. While through R_2, will decrease to zero from the value $\dfrac{E}{R_2} = 2.4\text{ A} = i_0$ (say), exponentially. Path of this decay of current will be *cdbac*.
$\therefore$ Just after the switch is opened,
$$V_{ab} = -i_0R_1 = -2.4 \times 30 = -72\text{ volt}$$

(f) Point b is at higher potential.

(g) $V_{cd} = -i_0(R_1 + R_2) = -2.4\,(80) = -192$ volt

(h) This time point d will be at higher potential.

107. $q_1 = 8CV_0$ and $q_2 = CV_0$

$$q_1 + q_2 = 9CV_0$$

In the absence of inductor, this $9C_0V$ will distribute as $6\,CV_0$ in $2C$ and $3CV_0$ in C. Thus, mean position of q_1 is $6\,CV_0$ and mean position of q_2 is $3CV_0$.

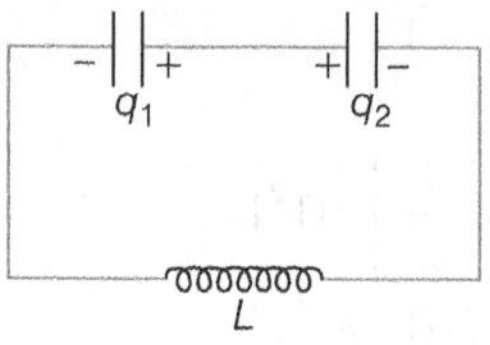

At $t = 0$, q_1 is $2CV_0$ more than its mean position and q_2 is $2CV_0$ less.

Thus, $\quad q_0 = 2CV_0$

$$C_{net} = \frac{2C}{3}$$

$$\therefore \quad \omega = \frac{1}{\sqrt{LC_{net}}} = \sqrt{\frac{3}{2LC}}$$

(a) $I_{max} = q_0\,\omega$

(b) $V_1 = \dfrac{6CV_0}{2C} = 3V_0$ and $V_2 = \dfrac{3CV_0}{C} = 3V_0$

(c) $i = q_0 \sin \omega t$

108.

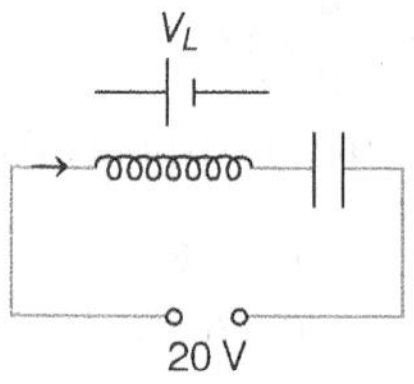

(a) $V_L = L\dfrac{di}{dt} = (1 \times 10^{-3})\dfrac{d}{dt}(20t) = 0.02$ V
$$= 20\text{ mV}$$

(b) $q = \int_0^t i\,dt = \int_0^t (20t)\,dt = 10t^2$

$$V = \frac{q}{C} = \frac{10t^2}{10^{-6}} = (10^{+7}t^2)\text{ V}$$

(c) $\dfrac{q^2}{2C} > \dfrac{1}{2}Li^2$

or $\quad \dfrac{(10t^2)^2}{2 \times 10^{-6}} > \dfrac{1}{2} \times 10^{-3} \times (20t)^2$

or $\quad t > 63.2 \times 10^{-6}$ s

or $\quad t > 63.2\ \mu$s **Ans.**

109. (a) Magnitude of induced electric field due to change in magnetic flux is given by

$$\oint \mathbf{E} \cdot d\mathbf{l} = \frac{d\phi}{dt} = S \cdot \frac{dB}{dt}$$

or $\quad El = \pi R^2(2B_0t) \quad \left(\dfrac{dB}{dt} = 2B_0t\right)$

Here, E = induced electric field due to change in magnetic flux

or $\quad E(2\pi R) = 2\pi R^2B_0t \quad$ or $\quad E = B_0Rt$

Hence, $\quad F = QE = B_0QRt$

This force is tangential to ring. Ring starts rotating when torque of this force is greater than the torque due to maximum friction ($f_{max} = \mu mg$) or when

$$\tau_F \geq \tau_{f_{max}}$$

Taking the limiting case

$$\tau_F = \tau_{f_{max}} \quad \text{or} \quad FR = (\mu mg)R$$

or $\quad F = \mu mg \quad$ or $\quad B_0 QRt = \mu mg$

It is given that ring starts rotating after 2.

So, putting $t = 2$, we get

$$\mu = \frac{2B_0 RQ}{mg}$$ **Ans.**

(b) After 2

$$\tau_F > \tau_{f_{max}}$$

Therefore, net torque is

$$\tau = \tau_F - \tan_{f_{max}} = B_0 QR^2 t - \mu mgR$$

Substituting $\mu = \frac{2B_0 QR}{mg}$, we get

$$\tau = B_0 QR^2(t-2)$$

or $$I\left(\frac{d\omega}{dt}\right) = B_0 QR^2(t-2)$$

or $$mR^2\left(\frac{d\omega}{dt}\right) = B_0 QR^2(t-2)$$

or $$\int_0^{\omega} d\omega = \frac{B_0 Q}{m}\int_2^4 (t-2)\,dt$$

or $$\omega = \frac{2B_0 Q}{m} \qquad \text{...(i)}$$

Now, magnetic field is switched off, i.e only retarding torque is present due to friction. So, angular retardation will be

$$\alpha = \frac{\tau_{f_{max}}}{I} = \frac{\mu mgR}{mR^2} = \frac{\mu g}{R}$$

Therefore, applying

$$\omega^2 = \omega_0^2 - 2\alpha\theta$$

or $$0 = \left(\frac{2B_0 Q}{m}\right)^2 - 2\left(\frac{\mu g}{R}\right)\theta$$

or $$\theta = \frac{2B_0^2 Q^2 R}{\mu m^2 g}$$

Substituting $\mu = \frac{2B_0 RQ}{mg}$

We get $$\theta = \frac{B_0 Q}{m}$$ **Ans.**

110. Let v be the velocity of connector at some instant of time. Then,

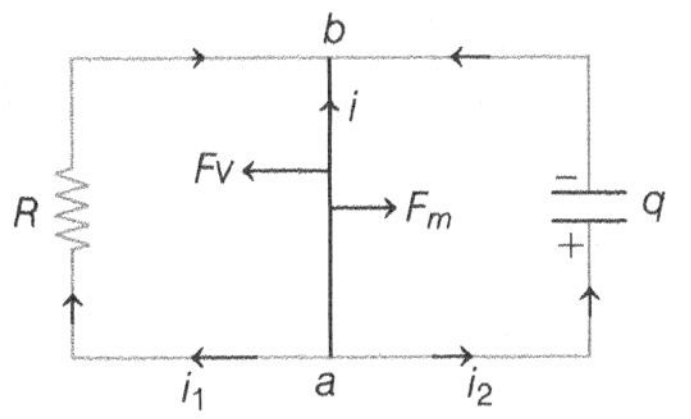

$$V_{ab} = Bvl,\ i_1 = \frac{Bvl}{R},\ q = C(Bvl)$$

$$\therefore \quad i_2 = \frac{dq}{dt} = CBl\frac{dv}{dt}$$

Now, $$i = i_1 + i_2 = \frac{Bvl}{R} + CBl\frac{dv}{dt}$$

Magnetic force, $$F_m = ilB = \frac{B^2 l^2}{R}\cdot v + B^2 l^2 C\cdot\frac{dv}{dt}$$

Further, $$F_{net} = F - F_m$$

or $$m\frac{dv}{dt} = F - \frac{B^2 l^2}{R}v - B^2 l^2 C\frac{dv}{dt}$$

$$\therefore \quad \int_0^v \frac{dv}{F - \frac{B^2 l^2}{R}v} = \int_0^t \frac{dt}{m + B^2 l^2 C}$$

Integrating we get,

$$v = \frac{FR}{B^2 l^2}\left[1 - e^{-\left(\frac{B^2 l^2}{mR + RB^2 l^2 C}\right)t}\right]$$

Terminal velocity in this case is : $v_T = \frac{FR}{B^2 l^2}$ **Ans.**

111. With key K_1 closed, C_1 and C_2 are in series with the battery in steady state.

$\therefore \quad C_{net} = 1\,\mu F$ or $q_0 = C_{net}\,V = 20\,\mu C$

(a) With K_1 opened and K_2 closed, charge on C_2 will remain as it is, while charge on C_1 will oscillate in L-C_1 circuit.

$$\omega = \frac{1}{\sqrt{LC_1}} = \frac{1}{\sqrt{0.2\times10^{-3}\times2\times10^{-6}}}$$

$$= 5\times10^4 \text{ rad/s}$$ **Ans.**

(b) Since, at $t = 0$, charge is maximum $(= q_0)$. Therefore, current will be zero.

$$\frac{1}{2}Li^2 = \frac{1}{3}\left(\frac{1}{2}\frac{q^2}{C}\right) \quad \text{or} \quad i = \frac{q}{\sqrt{3LC}} = \frac{q\omega}{\sqrt{3}}$$

From the expression,

$$i = \omega\sqrt{q_0^2 - q^2}$$

We have, $$\frac{q\omega}{\sqrt{3}} = \omega\sqrt{q_0^2 - q^2} \quad \text{or} \quad q = \frac{\sqrt{3}}{2}q_0$$

Since at $t = 0$, charge is maximum or q_0, so we can write

$$q = q_0 \cos\omega t \quad \text{or} \quad \frac{\sqrt{3}q_0}{2} = q_0\cos\omega t$$

or $$\omega t = \frac{\pi}{6} \quad \text{or} \quad t = \frac{\pi}{6\omega} = \frac{\pi}{6\times5\times10^4}$$

$$= 1.05\times10^{-5}\text{ s}$$

(c) $$q = \frac{\sqrt{3}}{2}q_0 = \frac{\sqrt{3}}{2}\times20 = 10\sqrt{3}\,\mu C$$ **Ans**

112. Let v be the velocity at some instant. Then, motional emf, $V = Bvl$

Charge stored in capacitor $q = CV = (CBl)v$

Current in the wire $= \frac{dq}{dt} = (CBl)\frac{dv}{dt}$

Magnetic force, $F_m = ilB = (CB^2l^2)\dfrac{dv}{dt}$ (upwards)

$\therefore$ Net force, $F_{net} = mg - F_m$

or $m\dfrac{dv}{dt} = mg - (CB^2l^2)\dfrac{dv}{dt}$

$\therefore$ $\dfrac{dv}{dt}$ = acceleration, $a = \dfrac{mg}{m + CB^2l^2}$

Since, a = constant

$\therefore$ $x = \dfrac{1}{2}at^2 = \dfrac{mgt^2}{2(m + CB^2l^2)}$

113. Let at time t velocity of ring be v (downwards)

$$e = Bv\,(2r) = 2B\,v\,r$$

(Two batteries of emf $2Bvr$ are connected in parallel)

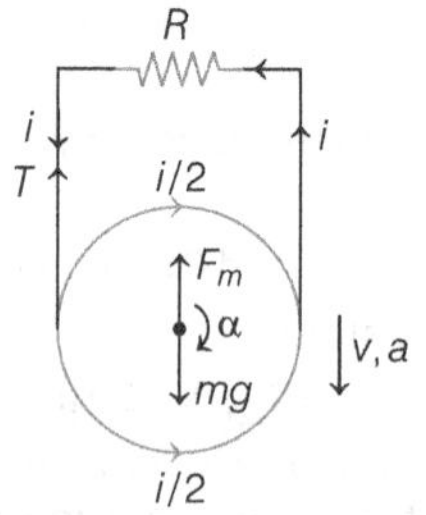

$\therefore$ $i = \dfrac{e}{R} = \dfrac{2Bvr}{R}$

Now, $a = \dfrac{mg - F_m - T}{m}$

Here, $F_m = 2\left[\left(\dfrac{i}{2}\right)(2r)B\right] = 2irB = \dfrac{4B^2r^2v}{R}$

$\therefore$ $a = g - \dfrac{4B^2r^2v}{mR} - \dfrac{T}{m}$ ...(i)

$\alpha = \dfrac{Tr}{mr^2} = \dfrac{T}{mr}$ ...(ii)

$a = r\,\alpha = \dfrac{T}{m}$ ...(iii)

From Eqs. (i), (ii) and (iii), we get

$$a = \frac{g}{2} - \frac{2B^2r^2v}{mR}$$

or $\displaystyle\int_0^v \frac{dv}{\dfrac{g}{2} - \dfrac{2B^2r^2v}{mR}} = \int_0^t dt$

or $v = \dfrac{mgR}{4B^2r^2}\left(1 - e^{-\frac{2B^2r^2}{mR}t}\right)$ **Ans.**

$$i = \frac{2Bvr}{R} = \frac{mg}{2Br}\left(1 - e^{-\frac{2B^2r^2}{mR}t}\right)$$

and $v_T = \dfrac{mgR}{4B^2r^2}$ **Ans.**

114. (a) Suppose v be the velocity of rod ef when it has fallen a distance, x. Then,

$$V_{fe} = V_{cb} \text{ or } Bvl = L(di/dt)$$

or $B(dx/dt)l = L(di/dt)$ or $Bl(dx) = L(di)$

Integrating, we get $Li = Blx$

or $i = \left(\dfrac{Bl}{L}\right)x$ ...(i)

Now, magnetic force opposite to displacement x will be

$$F = F_m = ilB = \left(\frac{B^2l^2}{L}\right)x$$

A constant downward force is mg.

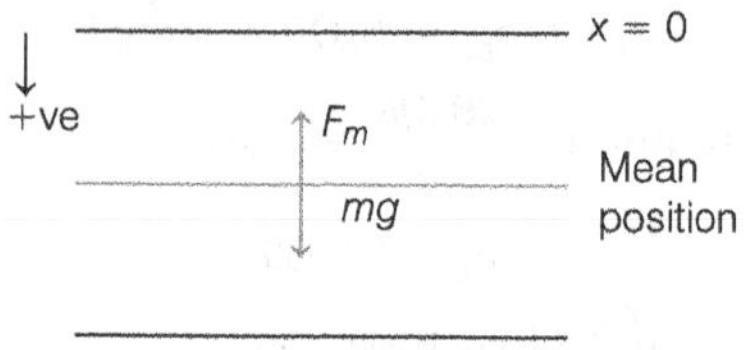

So, this is similar situation like spring-block system in vertical position. In which a force $F = kx$ acts upwards and a constant force mg acts downwards. Hence, the wire will execute SHM, where

$$k = \frac{B^2l^2}{L}$$

Amplitude will be at $F_m = mg$

or $\left(\dfrac{B^2l^2}{L}\right)A = mg \Rightarrow A = \dfrac{mgL}{B^2l^2}$

At $t = 0$, rod is in its extreme position. Therefore, if we write the equation from mean position we will write,

$$X = -A\cos\omega t$$

But, $x = X + A = A - A\cos\omega t = A(1 - \cos\omega t)$

where, $\omega = \sqrt{\dfrac{k}{m}} = \sqrt{\dfrac{B^2l^2}{mL}}$ **Ans.**

(b) From Eq. (i),

$$i_{max} = \left(\frac{Bl}{L}\right)x_{max}$$

Here, $x_{max} = 2A = \dfrac{2mgL}{B^2l^2}$

$\therefore$ $i_{max} = \left(\dfrac{Bl}{L}\right)\left(\dfrac{2mgL}{B^2l^2}\right) = \dfrac{2mg}{Bl}$ **Ans.**

(c) Maximum velocity,

$$v_0 = \omega A = \left(\sqrt{\frac{B^2l^2}{mL}}\right)\left(\frac{mgL}{B^2l^2}\right) = g\sqrt{\frac{mL}{B^2l^2}}$$

$= \dfrac{g\sqrt{mL}}{Bl}$ **Ans.**

115. (a) At time t, $v = a_0t$

Motional emf, $V = Bvl = Ba_0lt$

Total resistance $= \dfrac{R_1R_2}{R_1 + R_2}$

$\therefore$ $i = \dfrac{(Ba_0lt)(R_1 + R_2)}{R_1R_2}$ **Ans.**

(b) From right hand rule, we can see that points a and b will be at higher potential and c and d at lower potentials.

(c) $$F_m = ilB = \frac{B^2 l^2 a_0 t}{R_1 R_2}(R_1 + R_2)$$

Let F be the external force applied, then,

$$F - F_m = ma_0$$

$$\therefore \quad F = F_m + ma_0 = \frac{B^2 l^2 a_0 t}{R_1 R_2}(R_1 + R_2) + ma_0$$

116. (a) At the given instant,

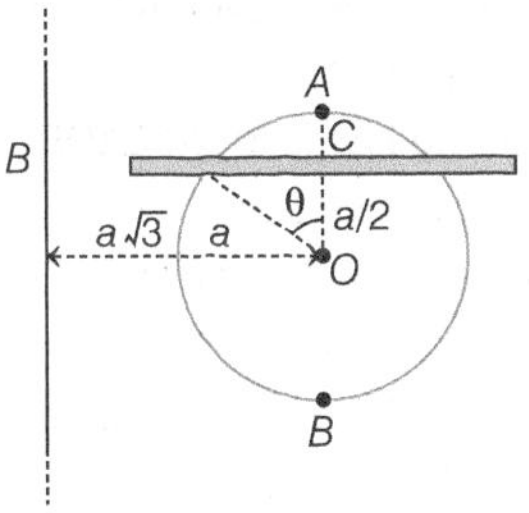

$$AC = \frac{a}{2}, \quad OC = \frac{a}{2}$$

and $$\cos\theta = \frac{a/2}{a} = \frac{1}{2}, \quad \theta = \frac{\pi}{3}$$

$\therefore$ Velocity of rod

$$= \left(+\frac{v_0}{2}\right) \text{ along the direction of current.}$$

Emf induced across the ends M and N

$$E_{rod} = \int_{a\sqrt{3} - \frac{a\sqrt{3}}{2}}^{a\sqrt{3} + \frac{a\sqrt{3}}{2}} v_{rod} B\, dx$$

$$= v_{rod} \int_{a\frac{\sqrt{3}}{2}}^{3a\frac{\sqrt{3}}{2}} \frac{\mu_0 i_0}{2\pi x}\, dx$$

$$= \left(\frac{v_0}{2}\right)\frac{\mu_0 i_0}{2\pi} \ln\left(\frac{3}{1}\right)$$

with end M at higher potential.

Since, the effective length of both the arcs MAN and MBN is MN.

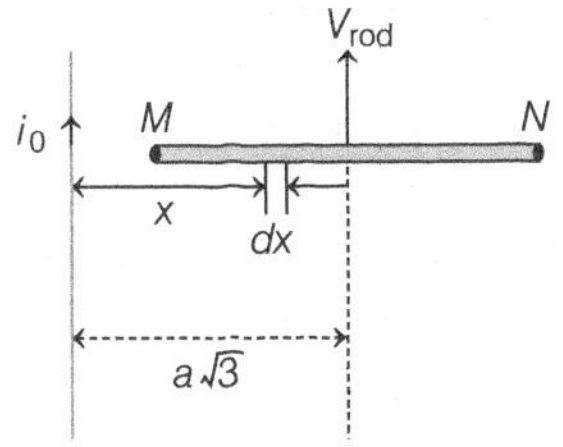

$$\therefore \quad E_{MAN} = E_{MBN} = v_{loop}\frac{\mu_0 i_0}{2\pi}\ln 3 = v_0 \frac{\mu_0 i_0}{2\pi}\ln 3$$

with point M at higher potential.

Resistance of arc MAN

$$\Rightarrow \quad R_1 = (R)2(a\theta) = 2aR\frac{\pi}{3}$$

$\Rightarrow$ Resistance of arc MBN

$$\Rightarrow \quad R_2 = (R)a(2\pi - 2\theta) = 4aR\frac{\pi}{3}$$

Equivalent circuit at the given instant is shown in the figure.

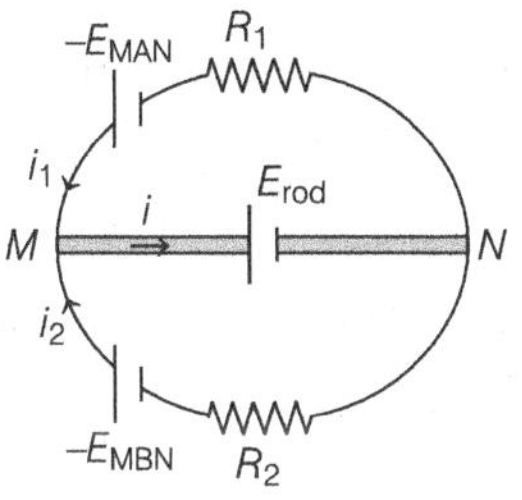

Current through the rod MN,

$$i = (i_1 + i_2) = \left(\frac{E_{MAN} - E_{rod}}{R_1}\right) + \left(\frac{E_{MBN} - E_{rod}}{R_2}\right)$$

$$i = (E_{MAN} - E_{rod})\left[\frac{1}{R_1} + \frac{1}{R_2}\right]$$

$$= \frac{v_0 \mu_0 i_0 (\ln 3)}{4\pi}\left[\left(\frac{1}{2} + \frac{1}{4}\right)\frac{3}{aR\pi}\right]$$

$$= \frac{9 v_0 i_0 \mu_0}{16 aR\pi^2}\ln(3)$$ **Ans.**

(b) Force on the rod

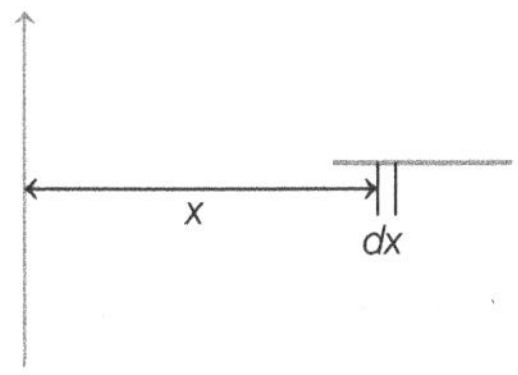

$$F_{rod} = \int_{\frac{a\sqrt{3}}{2}}^{\frac{3a\sqrt{3}}{2}} i\, dxB = \frac{i\mu_0 i_0}{2\pi}\ln 3 = \frac{9\mu_0^2 i_0^2 v_0}{32 aR\pi^3}(\ln 3)^2$$

117. Since, PQ and DC both cut the lines of field.

$\therefore$ Motional emf will be induced across both of them.

Integrating, potential difference across dx

$$\Rightarrow \quad \int de = \int_a^{2a} v\left(\frac{\mu_0 i_0}{2\pi x}\right)dx$$

$$e_{DC} = \frac{v\mu_0 i_0}{2\pi}\ln 2 \text{ with } D \text{ at higher potential}$$

$$e_{PQ} = \frac{2v\mu_0 i_0}{2\pi}\ln 2 \text{ with } P \text{ at higher potential}$$

The relative velocity of the rod PQ w.r.t. U frame

$$v_{rel} = 2v - v = v$$

Now, time taken by it to loose the contact $t = \frac{l}{v}$

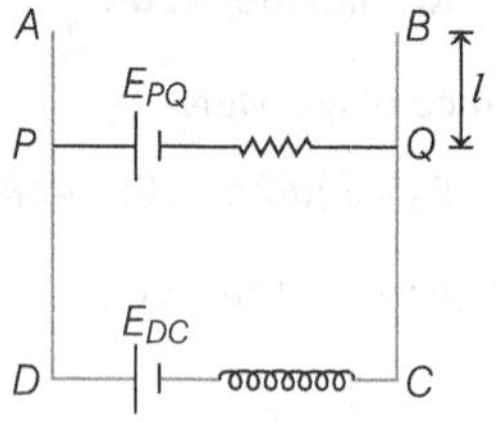

From equivalent electrical network

Net emf in the closed loop $QPDC$.

$$e = e_{PQ} - e_{DC} = \frac{v\mu_0 i_0}{2\pi} \ln 2$$

Growth of current in the L-R circuit is given by

$$i = i_0(1 - e^{-tR/L}) = \left(\frac{e}{R}\right)(1 - e^{-tR/L})$$

At time $t = \frac{l}{v}$

$$\Rightarrow \quad i = \left(\frac{e}{R}\right)(1 - e^{-\frac{Rl}{vL}})$$

118. Consider a point on the circumference of a circle of radius $r\,(r < R)$. Let E be the induced electric field along the tangents to the circle. Then, $E = \frac{r}{2} \cdot \frac{dB}{dt}$

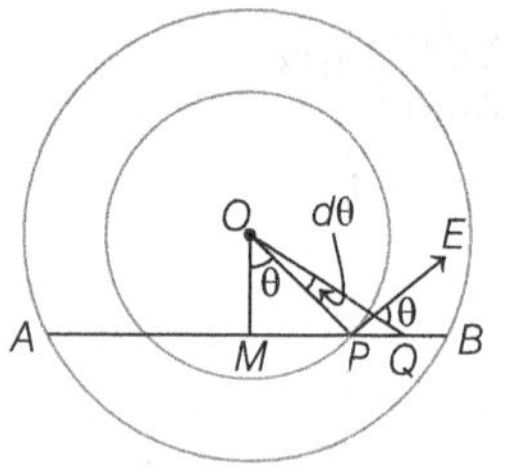

Now, consider a point P in the rod and a small distance dl along AB.

So, $$\varepsilon = \int \mathbf{E} \cdot d\mathbf{l} = \int \frac{l}{2} \cdot \frac{dB}{dt}\, dl \cos\theta$$

$$= \frac{(OM)}{2} \cdot \frac{dB}{dt} \int dl = \frac{\sqrt{R^2 - l^2}}{2} \left(\frac{dB}{dt}\right) 2l$$

$$\Rightarrow \quad \varepsilon = \sqrt{R^2 - l^2} \left(\frac{dB}{dt}\right) l \qquad \left(\because \int dl = 2l\right)$$

119. Induced emf in loops $AEFD$ and $EBCF$ would be

$$e_1 = \left|\frac{d\phi_1}{dt}\right| = S_1 \left|\frac{dB}{dt}\right|$$

$$= (1 \times 1) = 1\,\text{V}$$

Similarly, $$e_2 = \left|\frac{d\phi_2}{dt}\right| = S_2 \left|\frac{dB}{dt}\right|$$

$$= (0.5 \times 1) \times 1\,\text{V}$$

$$= 0.5\,\text{V}$$

Now, current is increasing, induced current will produce the magnetic field in direction, hence e_1 and e_2 will be applied as shown.

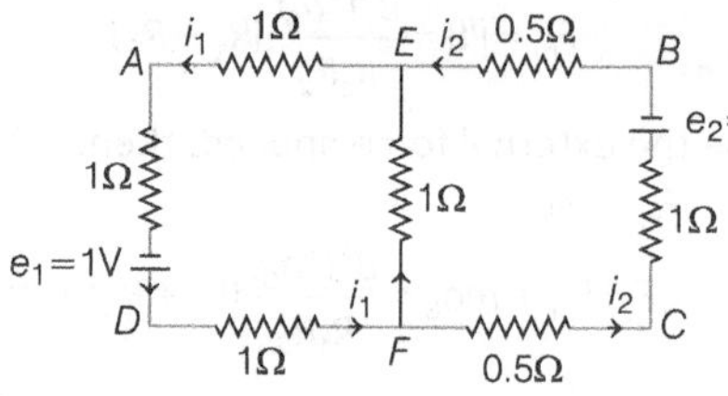

By Kirchhoff's first law at junction F, we get

$$i_1 = i + i_2 \quad \text{...(i)}$$

Kirchhoff's second law is loop $FEADF$

$$3i_1 + i = 1 \quad \text{...(ii)}$$

Kirchhoff's second law is loop $FEBCF$

$$2i_2 - i = 0.5 \quad \text{...(iii)}$$

On solving Eqs. (i), (ii) and (iii), we get

$$i_1 = \left(\frac{7}{22}\right)\text{A}$$

$$i_2 = \left(\frac{6}{22}\right)\text{A}$$

$$i = \left(\frac{1}{22}\right)\text{A}$$

Therefore, current in segment AE is $\left(\frac{7}{22}\right)$A from E to A, current in segment BE is $\left(\frac{6}{22}\right)$A from B to E and current in segment EF is $\left(\frac{1}{22}\right)$A from F to E.

120. Induced emf, $\varepsilon = \frac{1}{2} Br^2\omega$

Under steady state

$$i = \frac{\varepsilon}{R} = \frac{1}{2}\left(\frac{Br^2\omega}{R}\right)$$

Force on rod due to this induced current,

$$F = BIL = B \times \frac{1}{2}\frac{Br^2\omega}{R} \times r$$

$$= \frac{1}{2}\frac{B^2 r^3 \omega}{R}$$

Torque due to this force $= F \times$ distance

$$= F \times \frac{r}{2} = \frac{1}{4} \cdot \frac{B^2 r^4 \omega}{R}$$

Torque due to weight of rod at θ angular position

$$= mg\cos\theta \times \frac{r}{2}$$

So, total torque required $= \frac{mgr\cos\theta}{2} + \frac{1}{4}\frac{B^2 r^4 \omega}{R}$

121. When the side EF is at a distance y from the X-axis, magnetic flux passing through the loop is

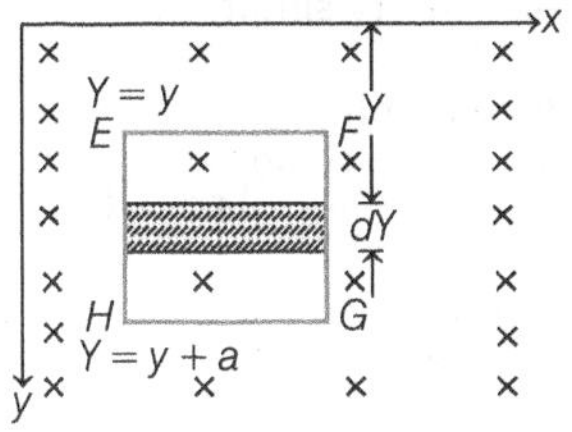

⊗ z-direction

$$\mathbf{B} = \left(\frac{B_0 y}{a}\right)\hat{\mathbf{k}}$$

$$\phi = \int d\phi = \int_{Y=y}^{Y=y+a} \frac{B_0 Y}{a}(a\,dY)$$

$$\phi = \frac{B_0}{2}[(y+a)^2 - y^2]$$

(a) Induced emf is

$$e = \left|\frac{-d\phi}{dt}\right| = \left|-\frac{B_0}{2}[2(y+a)-2y]\frac{dy}{dt}\right|$$

$$e = B_0 a\frac{dy}{dt} \Rightarrow e = B_0 va$$

where, $v = \frac{dy}{dt}$ = speed of loop

$\therefore$ Induced current, $i = \frac{e}{R} = \frac{B_0 av}{R}$

Direction $|\mathbf{B}| \propto y$ i.e. as the loop comes down ⊗ magnetic field passing through the loop increases, therefore the induced current will produce ⊙, magnetic field or the induced current in the loop will be **counter-clockwise.**

Alternate solution (of part a)

Motional emf in EH and FG = 0 as $\mathbf{v} \| \mathbf{l}$

Motional emf in EF is

$$e_1 = \left(\frac{B_0 y}{a}\right)(a)v = B_0 yv \qquad (\because e = Blv)$$

Similarly, motional emf in GH will be

$$e_2 = \left\{\frac{B_0(y+a)}{a}\right\}(a)(v) = B_0(a+y)v$$

Polarities of e_1 and e_2 are shown below.

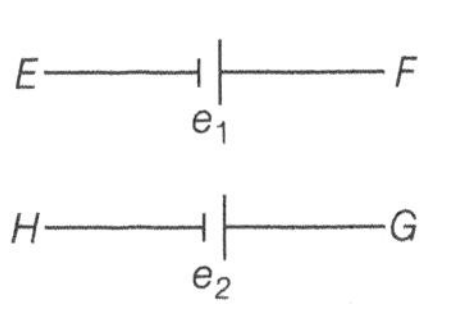

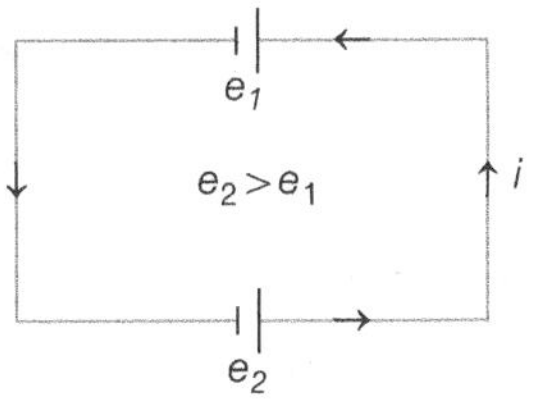

Net emf, $e = e_2 - e_1$, $e = B_0 av$

$$\therefore \quad i = \frac{e}{R} = \frac{B_0 av}{R}$$

and direction of current will be counter-clockwise.

(b) **Total Lorentz force on the loop :**

We have seen in part (a) that induced current passing through the loop (when its speed is v) is

$$i = \frac{B_0 av}{R}$$

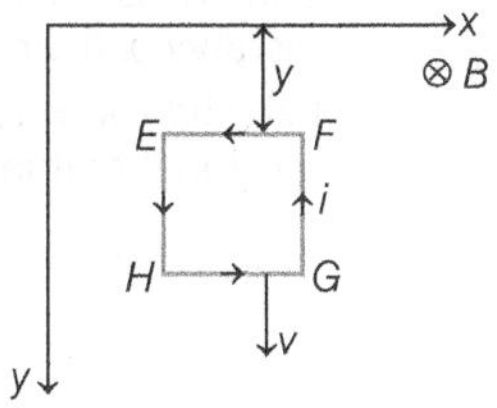

Now, magnetic force on EH and FG are equal in magnitude and in opposite directions, hence they cancel each other and produce no force on the loop.

$$F_{EF} = ilB\sin 90° = \left(\frac{B_0 av}{R}\right)(a)\left(\frac{B_0 y}{a}\right) \quad \text{(downwards)}$$

$$= \frac{B_0^2 avy}{R}$$

and $$F_{GH} = \left(\frac{B_0 av}{R}\right)(a)\left(\frac{B_0(y+a)}{a}\right) \quad \text{(upwards)}$$

$$= \left(\frac{B_0^2 av}{R}\right)(y+a)$$

$$F_{GH} > F_{EF}$$

$\therefore$ Net Lorentz force on the loop

$$\mathbf{F} = F_{GH} - F_{EF} = \frac{B_0^2 a^2 v}{R} \quad \text{(upwards)}$$

$$\mathbf{F} = -\frac{B_0^2 a^2 v}{R}\hat{\mathbf{j}}$$

(c) Net force on the loop will be

$$F = \text{weight} - \text{Lorentz force} \quad \text{(downwards)}$$

or $$F = mg - \frac{B_0^2 a^2 v}{R}$$

or $$m\left(\frac{dv}{dt}\right) = mg - \left(\frac{B_0^2 a^2}{R}\right)v$$

$$\therefore \quad \frac{dv}{dt} = g - \left(\frac{B_0^2 a^2}{mR}\right)v = g - Kv$$

where, $K = \frac{B_0^2 a^2}{mR}$ = constant

or $\frac{dv}{g - Kv} = dt$ or $\int_0^v \frac{dv}{g - Kv} = \int_0^t dt$

This equation gives $v = \frac{g}{K}(1 - e^{-Kt})$

Here, $K = \left(\frac{B_0^2 a^2}{mR}\right)$

i.e. speed of the loop is increasing exponentially with time t. Its terminal velocity will be

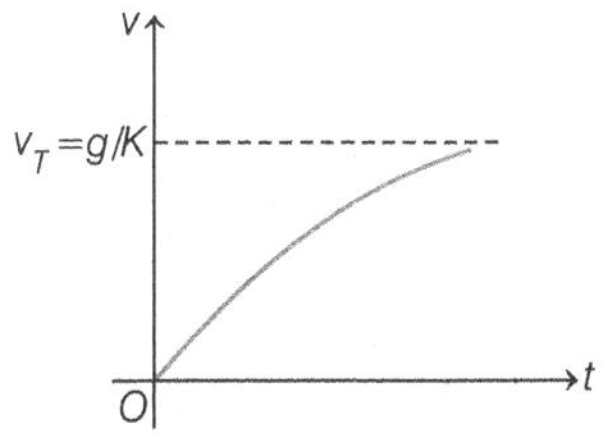

$$v_T = \frac{g}{K} = \left(\frac{mgR}{B_0^2 a^2}\right)$$

at $t \to \infty$

122. (a) Total resistance of the circuit as a function of distance x from resistance R is $R_{net} = R + 2\lambda x$

Let v be the velocity of rod at this instant, then motional emf induced across the rod, $e = Bvd$

$\therefore$ Current $i = \frac{e}{R_{net}} = \frac{Bvd}{R + 2\lambda x}$

$\Rightarrow$ $v = \frac{(R + 2\lambda x)i}{Bd}$

Net force on the rod,

$$F_{net} = m\frac{dv}{dt} = \frac{2\lambda im}{Bd} \cdot \frac{dx}{dt}$$

but $\frac{dx}{dt} = v = \frac{(R + 2\lambda x)i}{Bd}$

$$F_{net} = \frac{2\lambda i^2 m}{B^2 d^2}(R + 2\lambda x)$$

This net force is equal to $F - F_m$

where $F_m = idB$

$$\therefore \quad F = F_{net} + F_m = \frac{2\lambda i^2 m}{B^2 d^2}(R + 2\lambda x) + idB$$

(b) Work done by F per second

= power of this force $= Fv$

or $P_1 = \left[\frac{2\lambda i^2 m}{B^2 d^2}(R + 2\lambda x) + idB\right]\left[\frac{(R + 2\lambda x)i}{Bd}\right]$

Heat produced per second in the circuit

$$P_2 = i^2 R_{net} = i^2 (R + 2\lambda x)$$

The desired fraction is,

$$f = \frac{P_2}{P_1} = \left[1 + \frac{2\lambda mi}{B^3 d^3}(R + 2\lambda x)\right]^{-1}$$

123. (a) For an elemental strip of thickness dx at a distance x from left wire, net magnetic field (due to both wires)

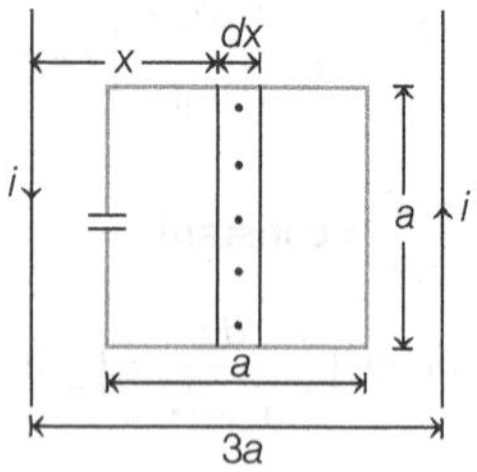

$$B = \frac{\mu_0}{2\pi}\frac{I}{x} + \frac{\mu_0}{2\pi}\frac{I}{3a - x} \quad \text{(outwards)}$$

$$= \frac{\mu_0 I}{2\pi}\left(\frac{1}{x} + \frac{1}{3a - x}\right)$$

Magnetic flux in this strip,

$$d\phi = BdS = \frac{\mu_0 I}{2\pi}\left(\frac{1}{x} + \frac{1}{3a - x}\right) a\,dx$$

$\therefore$ Total flux,

$$\phi = \int_a^{2a} d\phi = \frac{\mu_0 Ia}{2\pi}\int_a^{2a}\left(\frac{1}{x} + \frac{1}{3a - x}\right)dx$$

or $\phi = \frac{\mu_0 Ia}{\pi}\ln(2)$

$$\phi = \frac{\mu_0 a \ln(2)}{\pi}(I_0 \sin \omega t) \quad \text{...(i)}$$

Magnitude of induced emf,

$$e = \left|-\frac{d\phi}{dt}\right| = \frac{\mu_0 a I_0 \omega \ln(2)}{\pi}\cos \omega t$$

$$= e_0 \cos \omega t$$

where, $e_0 = \frac{\mu_0 a I_0 \omega \ln(2)}{\pi}$

Charge stored in the capacitor,

$$q = Ce = Ce_0 \cos \omega t \quad \text{...(ii)}$$

and current in the loop

$$i = \frac{dq}{dt} = C\omega e_0 \sin \omega t \quad \text{...(iii)}$$

$$i_{max} = C\omega e_0 = \frac{\mu_0 a I_0 \omega^2 C \ln(2)}{\pi}$$

(b) Magnetic flux passing through the square loop

$\phi \propto \sin \omega t$ [From Eq. (i)]

i.e. ⊙ magnetic field passing through the loop is increasing at $t = 0$. Hence, the induced current will produce ⊗ magnetic field (from Lenz's law). Or the current in the circuit at $t = 0$ will be clockwise (or negative as per the given convention). Therefore, charge on upper plate could be written as,

$q = +q_0 \cos \omega t$ [From Eq. (ii)]

Here, $q_0 = Ce_0 = \frac{\mu_0 a C I_0 \omega \ln(2)}{\pi}$

The corresponding q-t graph is shown in figures.

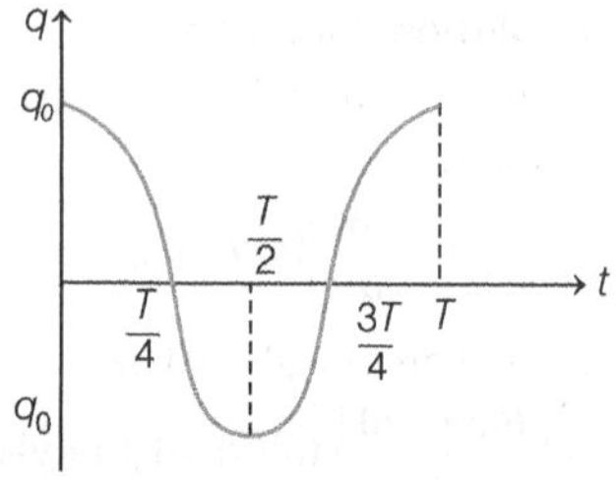

124. (a) Applying Kirchhoff's second law, we get

$$\frac{d\phi}{dt} - iR - L\frac{di}{dt} = 0$$

or

$$\frac{d\phi}{dt} = iR + L\frac{di}{dt} \quad \text{...(i)}$$

This is the desired relation between i, $\frac{di}{dt}$ and $\frac{d\phi}{dt}$.

(b) Eq. (i) can be written as

$$d\phi = iRdt + Ldi$$

Integrating, we get

$$\Delta\phi = R\Delta q + Li_1$$

$$\Rightarrow \quad \Delta q = \frac{\Delta\phi}{R} - \frac{Li_1}{R} \quad \text{...(ii)}$$

Here, $\Delta\phi = \phi_f - \phi_i = \int_{x = 2x_0}^{x = x_0} \frac{\mu_0}{2\pi}\frac{I_0}{x} l\,dx$

$$=\frac{\mu_0 i_0 l}{2\pi}\ln(2)$$

So, from Eq. (ii) charge flown through the resistance upto time $t = T$, when current i_1, is

$$\Delta q = \frac{1}{R}\left[\frac{\mu_0 i_0 l}{2\pi}\ln(2) - Li_1\right]$$

(c) This is the case of current decay in an L-R circuit. Thus,

$$i = i_0 e^{-t/\tau_L} \quad \text{...(iii)}$$

Here, $i = \frac{i_1}{4}, i_0 = i_1$

$$t = (2T - T) = T \text{ and } \tau_L = \frac{L}{R}$$

Substituting these values in Eq. (iii), we get

$$\tau_L = \frac{L}{R} = \frac{T}{\ln 4}$$

125. (a) Let E be the electric field radially outwards at distance r from centre.

Force eE on free electron towards centre (opposite to electric field) will provide the necessary centripetal force to electron. Thus,

$$eE = mr\omega^2 \quad (m = \text{mass of electron})$$

$$\therefore \quad E = \left(\frac{m\omega^2}{e}\right)r$$

$$dV = -E\,dr$$

or $$\int_{\text{rim}}^{\text{centre}} dV = -\int_a^0 \left(\frac{m\omega^2}{e}\right) r\,dr \text{ or } V_{\text{centre}} - V_{\text{rim}}$$

or $$\Delta V = \frac{m\omega^2 a^2}{2e}$$

Substituting the values, we get

$$\Delta V = 3.0 \text{ nV}$$

Note *Electric lines are radially outwards and electric lines flow from higher potential to lower potential. Hence,*

$$V_{\text{centre}} > V_{\text{rim}}$$

(b) Disc may be assumed as made up of infinite conducting wires connected in parallel between O to C. In parallel potential difference is same. Hence we can find potential difference across any one of them.

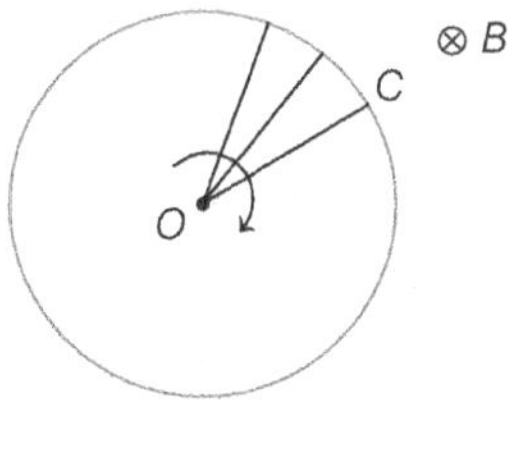

Potential difference across any one of them will be,

$$\Delta V = \text{rotational emf}$$

$$= \frac{B\omega l^2}{2} \quad (l = a)$$

$$= \frac{B\omega a^2}{2}$$

Substituting the values, we get

$$\Delta V = 20 \text{ mV}$$

Note *Potential difference in first part is negligible as compared to potential difference in second part.*

126. For constant ω, net torque should be zero, or torque of mg about O should be equal to torque of magnetic force about O.

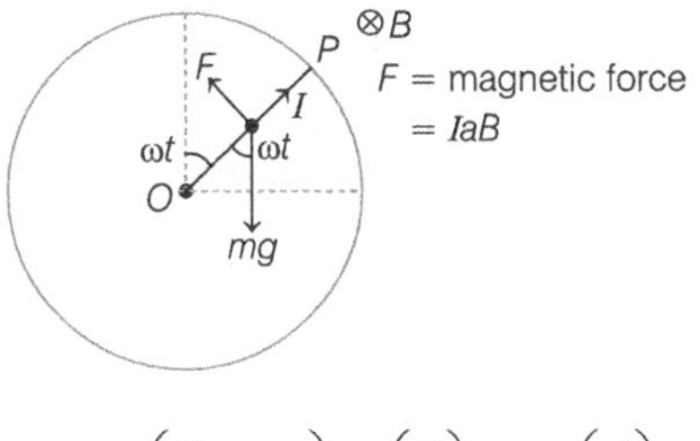

$$\Rightarrow \quad mg\left(\frac{a}{2}\sin\omega t\right) = F\left(\frac{a}{2}\right) = (IaB)\left(\frac{a}{2}\right)$$

$$\therefore \quad I = \frac{mg\sin\omega t}{aB}$$

Net emf required for this current,

$$e_{\text{net}} = IR = \frac{mgR\sin\omega t}{aB}$$

For magnetic force F in the direction shown in figure, current in the wire should be in the direction from O to P. Rotational emf $\frac{B\omega a^2}{2}$ will provide current in the same direction. Hence,

$$e_{\text{net}} = \frac{mgR\sin\omega t}{aB} = \frac{B\omega a^2}{2} + e_{\text{applied}}$$

$$\therefore \quad e_{\text{applied}} = \frac{1}{2aB}[2mgR\sin\omega t - \omega a^2 B^2]$$

127.

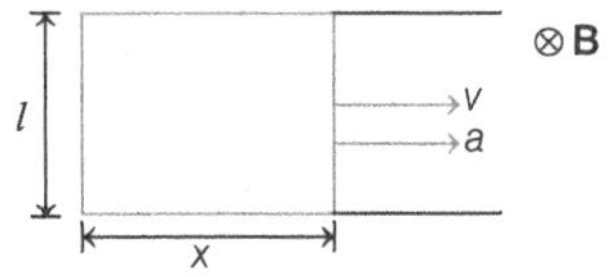

At time t,

$$\phi = BS = Blx$$

Here, B and x are variables.

$$|e| = \left|-\frac{d\phi}{dt}\right| = lx\left(\frac{dB}{dt}\right) + Bl\left(\frac{dx}{dt}\right) \quad \text{...(i)}$$

Here, $x = \frac{1}{2}at^2$

$$\frac{dx}{dt} = v = at$$

$$\frac{dB}{dt} = 0.1 \text{ T/s}$$

and $$B = \left(\frac{dB}{dt}\right)t$$

Substituting all these values in equation (i), we get

$$e=\left(\frac{3\omega l t^2}{2}\right)\frac{dB}{dt}$$

For the given values, emf comes out to be,

$$e=12\,\text{mV}$$

128. $\phi=BS=(bt)(\pi a^2)$

$$|e|=\left|-\frac{d\phi}{dt}\right|=\pi a^2 b=e \quad \text{(say)}$$

$\therefore$
$$I=\frac{e}{R_{net}}=\frac{\pi a^2 b}{r+\eta r} \quad \text{...(i)}$$

The equivalent circuit is as shown below.

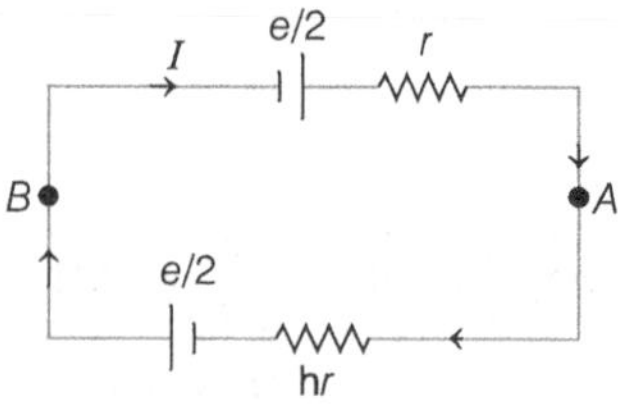

Writing potential equation two times,

$$V_A-V_B=\frac{e}{2}-Ir \quad \text{...(ii)}$$

$$V_A-V_B=-\frac{e}{2}+\eta I r \quad \text{...(iii)}$$

Adding equations (ii) and (iii), we get

$$V_A-V_B=\frac{Ir}{2}(\eta-1) \quad \text{...(iv)}$$

Now, $$V_A-V_B=\int_{\text{half ring}}\mathbf{E}\cdot d\mathbf{l}=El=E(\pi a) \quad \text{...(v)}$$

Equating equations (iv) and (v), we get

$$E=\frac{Ir}{2\pi a}(\eta-1)$$

Substituting value of I from equation (i), we get

$$E=\frac{ab(\eta-1)}{2(\eta+1)}$$

129. Induced electric field,

$$\left|\oint \mathbf{E}\cdot d\mathbf{l}\right|=\left|\frac{d\phi}{dt}\right|$$

$\Rightarrow$
$$E(2\pi r)=(\pi r^2)\frac{d}{dt}B(t)$$

Here r is radius of ring,

$\therefore$
$$E=\frac{r}{2}\frac{d}{dt}B(t)$$

Torque, $$\tau=Fr=qEr$$

$$=\frac{qr^2}{2}\frac{d}{dt}B(t)$$

$$\alpha=\frac{d\omega}{dt}=\frac{\tau}{I}$$

$$=\frac{\tau}{mr^2}=\frac{q}{2m}\frac{d}{dt}B(t)$$

or
$$d\omega=\frac{q}{2m}dB(t)$$

$\therefore$
$$\int d\omega=\frac{q}{2m}\int_0^{B(t)}dB(t)$$

$\therefore$
$$\omega=\frac{q}{2m}B(t)$$

130. Solenoid is an L-R circuit. Current in solenoid will be given by,

$$I_0=\frac{V}{R}(1-e^{-t/\tau_L}) \quad \text{...(i)}$$

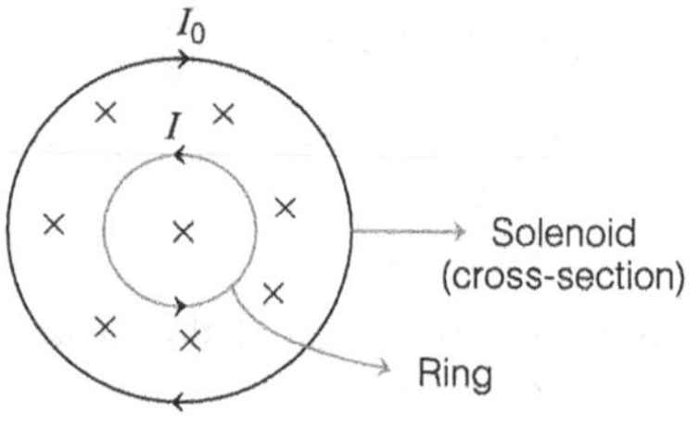

Suppose this current is clockwise as shown. Then its magnetic field inside the solenoid is in ⊗ direction and given by,

$$B=\mu_0 n I_0$$

or
$$B=\frac{\mu_0 n V(1-e^{-t/\tau_L})}{R} \quad \text{...(ii)}$$

Flux passing through the ring,

$$\phi=BS=B(\pi a^2)$$

or
$$\phi=\frac{\mu_0 n V\pi a^2(1-e^{-t/\tau_L})}{R}$$

Induced emf in the ring

$$|e| \text{ or } e=\left|-\frac{d\phi}{dt}\right|=\frac{\mu_0 n V\pi a^2}{\tau_L R}e^{-t/\tau_L}$$

$\therefore$ Induced current I in the loop,

$$I=\frac{e}{r}=\frac{\mu_0 n V\pi a^2}{\tau_L R r}e^{-t/\tau_L} \quad \text{...(iii)}$$

Note *I_0 is increasing, so ⊗ B will increase. Hence induced current I will produce ⊙ magnetic field. So, I will be anticlockwise.*

Now I current in the ring is lying in ⊗ magnetic field. So, radial force will act on each part of the ring in inward direction (Fleming's left hand rule). This force per unit length (say F) will be,

$$F=I(1)B$$

or
$$F=IB$$

Now, substituting the values of B and I from equations (ii) and (iii), we can get F and then by putting $\frac{dF}{dt}=0$, we can get the maximum value of F which comes out to be,

$$F_{max}=\frac{\mu_0 a^2 V^2}{4rRlb^2}$$

Note *(i) For students concept is more important rather than the calculations.*

(ii) Self-inductance of the solenoid, comes out to be,

$$L = (\mu_0 n^2) l(\pi b^2)$$

$$\therefore \quad \tau_L = \frac{L}{R}$$

(iii) F is product of B and I.

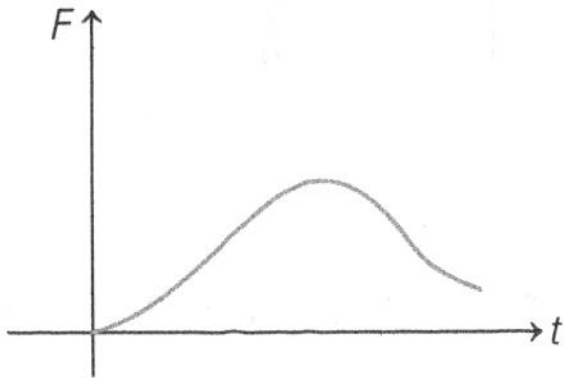

Here B is exponentially increasing and I is exponentially decreasing.

At $t = 0, B = 0 \Rightarrow F = 0$

At $t = \infty, I = 0 \Rightarrow F = 0$

So, value of F will first increase and then decrease and in between we will get a maximum value. F versus t graph is as shown in figure.

131. Magnetic flux passing through small element shown in figure is,

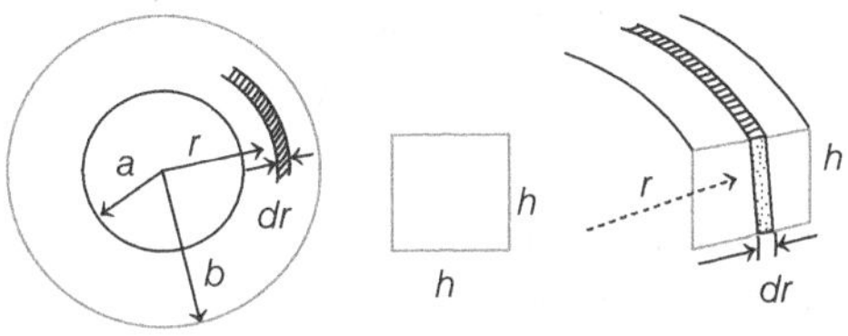

Ring Cross-section of ring

$$d\phi = B(\pi r^2) = (\beta t)(\pi r^2)$$

$$e \text{ or } |e| = \left|-\frac{d\phi}{dt}\right| = \beta\pi r^2$$

Small current, $dI = \frac{e}{dR}$

Here, $dR = \frac{\rho l}{A} = \frac{\rho(2\pi r)}{(hdr)}$

$$\therefore \quad dI = \frac{(\beta\pi r^2)(hdr)}{(2\pi\rho r)} = \left(\frac{Bh}{2\rho}\right) rdr$$

Integrating dI from $r = a$ to $r = b$, we get the total current which comes out to be,

$$I = \frac{\beta h}{4\rho}(b^2 - a^2)$$

Note *Inside the long solenoid a ring of inner radius a and outer radius b of cross-section as a square $(h \times h)$ is kept. Current in solenoid is variable. Therefore flux passing through the ring is variable.*

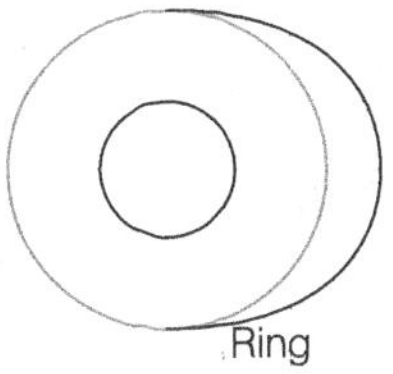

132. Method-1 Let potential difference across capacitor at time t is V. At $t = 0$, it was V_m. Therefore,

$$V = V_m \cos \omega t \quad \text{...(i)}$$

Charge on capacitor,

$$q = CV = CV_m \cos \omega t$$

$$I = \frac{dq}{dt} = -\omega CV_m \sin \omega t$$

Here, $\omega = \frac{1}{\sqrt{LC}}$

$$\therefore \quad I = -\sqrt{\frac{C}{L}} V_m \sin \omega t \quad \text{...(ii)}$$

From Eq. (i), $\cos \omega t = \frac{V}{V_m}$ ...(iii)

and from Eq. (iv), $\sin \omega t = -\frac{I}{V_m}\sqrt{\frac{L}{C}}$ ...(iv)

Squaring and adding Eqs. (iii) and (iv), we get

$$V^2 + \frac{I^2 L}{C} = V_m^2$$

Method-2 $\frac{1}{2}CV^2 + \frac{1}{2}LI^2 = \text{constant}$

Initially, $I = 0$ and $V = V_m$

$$\therefore \quad \frac{1}{2}CV_m^2 = \frac{1}{2}CV^2 + \frac{1}{2}LI^2$$

Simplifying this equation, we get the same result.

133. (a) At $t = 0$, potential difference across capacitor is V_m. Hence potential difference across capacitor at time t will be,

$$V = V_m \cos \omega t$$

$\therefore$ Charge on capacitor,

$$q = CV = CV_m \cos \omega t$$

$$I = \frac{dq}{dt} = -C\omega V_m \sin \omega t$$

Putting $\omega = \frac{1}{\sqrt{LC}}$, we get

$$I = -V_m\sqrt{\frac{C}{L}} \sin \omega t$$

More appropriate answer is,

$$I = \pm I_m \sin \omega t$$

where, $I_m = V_m\sqrt{\frac{C}{L}}$ and $\omega = \frac{1}{\sqrt{LC}}$

(b) Half of the initial energy $\frac{1}{2}CV_m^2$ is across capacitor and half in the magnetic field.

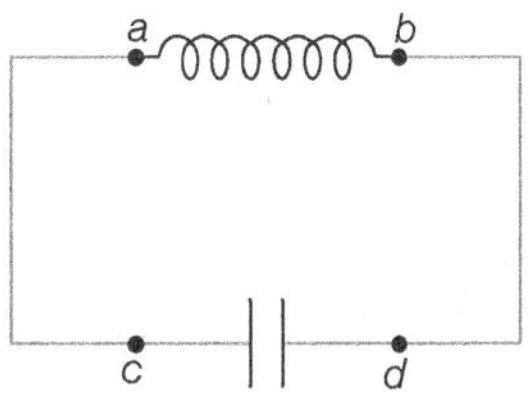

$V_{cd} = V_{ab} \Rightarrow V_C = V_L$

$\therefore \quad \frac{1}{2}CV^2 = \frac{1}{2}\left[\frac{1}{2}CV_m^2\right]$ or $V = \frac{V_m}{\sqrt{2}}$

This is potential difference across capacitor. So,

$$V_L = V_C = \frac{V_m}{\sqrt{2}}$$

134. Capacitors are in series during L-C oscillations.

$$\therefore \quad C_{net} = \frac{C}{2}$$

Angular frequency of L-C oscillations will be,

$$\omega = \frac{1}{\sqrt{LC_{net}}} = \sqrt{\frac{2}{LC}}$$

In the absence of inductor charge will equally distribute in both capacitors.

$\therefore$ Initially given potential difference will also distribute equally in both capacitor.

So, $V = \frac{V_0}{2}$ (in steady state across both capacitors)

Left hand side capacitor

At $t = 0, V = V_0$

So, V-t graph is as shown below.

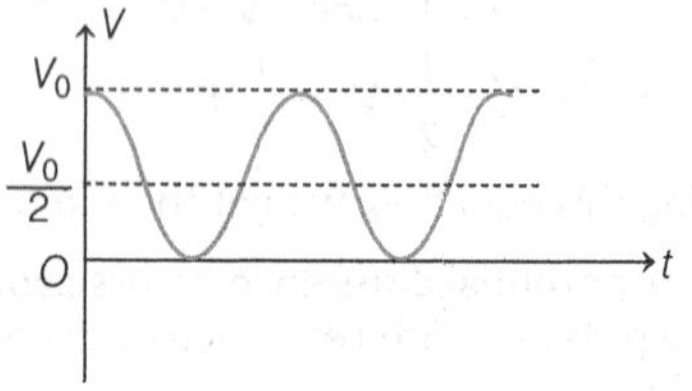

This is a cos graph with mean position at $\frac{V_0}{2}$.

Thus, $V = \frac{V_0}{2} + \frac{V_0}{2}\cos\omega t$

Right hand side capacitor

At $t = 0, V = 0$

So, V-t graph is as shown below.

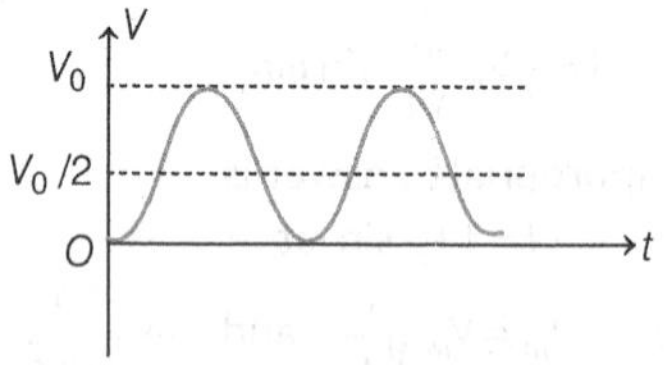

This is $-$cos graph with mean position at $V = V_0/2$. Thus,

$$V = \frac{V_0}{2} - \frac{V_0}{2}\cos\omega t$$

135. Flux through loop is

$$\phi = \int_0^{2a} \frac{\mu_0}{4\pi} 2Ia\left(\frac{1}{x} + \frac{1}{3a - x}\right)dx$$

So, induced emf is

$$\varepsilon = -\frac{d\phi}{dt} = \frac{-\mu_0}{\pi} \cdot aI_0\omega \cdot \log 2 \cdot \cos\omega t$$

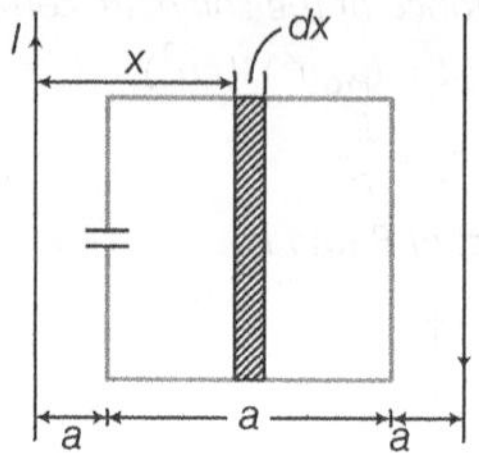

Charge on the capacitor,

$$Q = C\varepsilon = \frac{C\mu_0}{\pi} \cdot aI_0\omega \cdot \log 2 \cdot \cos\omega t$$

So, current in the loop,

$$= \frac{dQ}{dt} = \frac{d}{dt}\frac{C\mu_0}{\pi} \cdot I_0\omega \cdot a\log 2 \cdot \cos\omega t$$

$$= \frac{\mu_0}{\pi} CI_0\omega^2 a\log 2 \cdot \sin\omega t$$

So, maximum current in loop is

$$I_{max} = \frac{\mu_0}{\pi} \cdot CI_0\omega^2 \cdot a\log 2$$

Substituting the values, we have

$$I_{max} = 4 \times 10^{-7} \times 1 \times 10^{-6} \times 5 \times 2500 \times \pi^2 \times 2 \times 10^{-2} \times \log 2$$

$$= 10^{-11} \times \pi^2 \times \log 2$$

So, $\quad k - 2 = 9$

136. After a long time, induced emf in the coil will be zero. So, we have equivalent circuit,

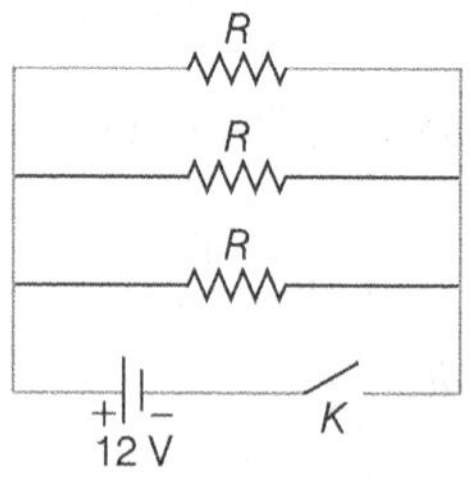

Equivalent resistance R_{eq} of circuit is given by,

$$\frac{1}{R_{eq}} = \frac{1}{R_1} + \frac{1}{R_2} + \frac{1}{R_3} = \frac{3}{12} = \frac{1}{4}$$

or $\quad R_{eq} = 4\Omega$

Circuit current,

$$I = \frac{V}{R_{eq}} = \frac{12}{4} = 3\text{ A}$$

137. At $t = 0, I_{min} = \frac{5}{12}$

At $\quad t = \infty$

$$I_{max} = \frac{5}{R_{eq}} = \frac{5}{3/2} = \frac{10}{3} \quad \left[\because \frac{1}{R_{eq}} = \frac{1}{3} + \frac{1}{4} + \frac{1}{12} = \frac{8}{12}\right]$$

$$\therefore \quad \frac{I_{max}}{I_{min}} = \frac{10}{3} \times \frac{12}{5} = 8$$

UNIT-6

ALTERNATING CURRENT

Learning Modules

1. Alternating Currents and Phasors
2. Current and Potential Relations
3. Power in an AC Circuit

MODULE 1

Alternating Currents and Phasors

Introduction

A century ago, one of the great technological debates was whether the electrical distribution system should be AC or DC. Thomas Edison favoured direct current (DC), that is, steady current that does not vary with time. George Westinghouse favoured alternating current (AC), with sinusoidally varying voltages and currents.

He argued that transformers can be used to step the voltage up or down with AC but not with DC. Low voltages are safer for consumer use, but high voltages and correspondingly low currents are best for long distances power transmission to minimize i^2R losses in the cables. Eventually, Westinghouse prevailed, and most present day household and industrial power distribution systems operate with alternating current.

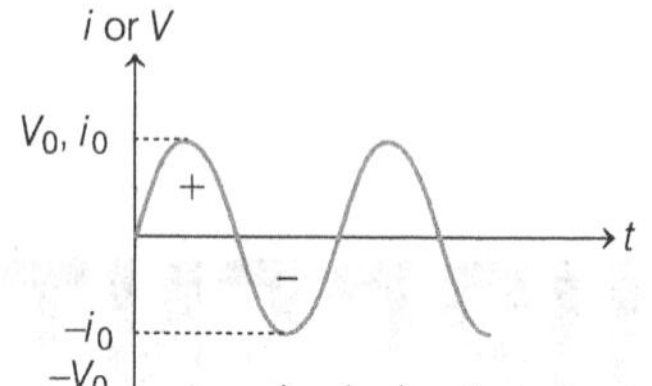

$i = i_0 \sin \omega t$
or
$V = V_0 \sin \omega t$

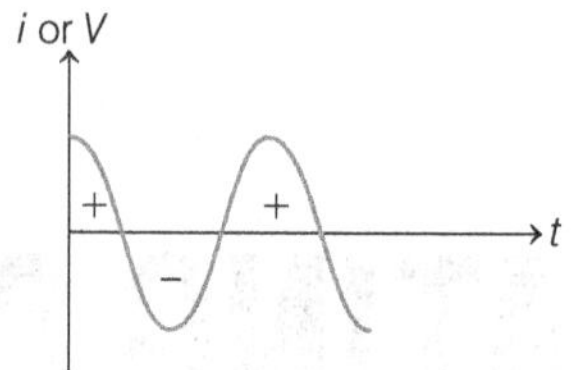

$i = i_0 \cos \omega t$
or
$V = V_0 \cos \omega t$

A time varying current or voltage may be periodic and non-periodic. If the current or voltage varies periodically as sine or cosine function of time, the current or voltage is said to be sinusoidal and is what we usually mean by it.

Alternating Currents

The basic principle of the AC generator is a direct consequence of Faraday's law of induction. When a conducting loop is rotated in a magnetic field at constant angular frequency ω a sinusoidal voltage (emf) is induced in the loop. This instantaneous voltage is of type :

$$V = V_0 \sin \omega t \quad \text{...(i)}$$

The usual circuit diagram symbol for an AC source is shown in figure.

In Eq. (i), V_0 is the maximum output voltage of the AC generator or the **voltage amplitude** and ω is the **angular frequency** equal to 2π times the frequency f.

$$\omega = 2\pi f$$

The frequency of AC in India is 50 Hz, i.e.

$$f = 50 \text{ Hz}$$

So, $$\omega = 2\pi f \approx 314 \text{ rad/s}$$

The time of one cycle is known as **time period *T***, the number of cycles per second the frequency f.

$$T = \frac{1}{f} \quad \text{or} \quad T = \frac{2\pi}{\omega}$$

A sinusoidal current might be described as

$$i = i_0 \sin \omega t$$

If an alternating current is passed through an ordinary ammeter or voltmeter, it will record the mean value for the complete cycle, as the quantity to be measured varies with time. The **average** value of current for one cycle is

$$\langle i \rangle_{\text{One cycle}} = (i_{av})_T = \frac{\int_0^T i\,dt}{\int_0^T dt}$$

$$= \frac{\int_0^{2\pi/\omega} (i_0 \sin \omega t)\, dt}{\int_0^{2\pi/\omega} dt} = 0$$

Thus, $$\langle i \rangle_{\text{One cycle}} = 0$$

Similarly, the average value of the voltage (or emf) for one cycle is zero.

$$\langle V \rangle_{\text{One cycle}} = 0$$

Since, these averages for the whole cycle are zero, the DC instrument will indicate zero deflection. In AC, the average value of current is defined as its average taken over positive half of the cycle. Hence,

$$\langle i \rangle_{\text{Half cycle}} = (i_{av})_{T/2} = \frac{\int_0^{T/2} i\,dt}{\int_0^{T/2} dt} = \frac{\int_0^{\pi/\omega} (i_0 \sin \omega t)\, dt}{\int_0^{\pi/\omega} dt} = \frac{2}{\pi} i_0$$

This is sometimes simply written as i_{av}. Hence,

$$i_{av} = \langle i \rangle_{\text{Half cycle}} = \frac{2}{\pi} i_0 \approx 0.637\, i_0$$

Similarly, $$V_{av} = \frac{2}{\pi} V_0 \approx 0.637\, V_0$$

A more useful way to describe a quantity is the **root mean square (rms)** value. We square the instantaneous current, take the average (mean) value of i^2 and finally take the square root of that average. This procedure defines the root-mean-square current denoted as i_{rms}. Even when i is negative, i^2 is always positive so i_{rms} is never zero (unless i is zero at every instant). Hence,

$$\langle i^2 \rangle_{\text{One cycle}} = \frac{\int_0^T i^2 dt}{\int_0^T dt} = \frac{\int_0^{2\pi/\omega} (i_0^2 \sin^2 \omega t)\, dt}{\int_0^{2\pi/\omega} dt} = \frac{i_0^2}{2}$$

$$\therefore \quad i_{rms} = \sqrt{\langle i^2 \rangle_{\text{One cycle}}} = \frac{i_0}{\sqrt{2}} \approx 0.707\, i_0$$

Thus, $$i_{rms} = \frac{i_0}{\sqrt{2}} \approx 0.707\, i_0$$

Similarly, we get

$$V_{rms} = \frac{V_0}{\sqrt{2}} \approx 0.707\, V_0$$

RMS value is the main value which is used for heat generation across a resistance.

Thus, when we speak of our house hold power supply as 220 V AC, this means that the rms voltage is 220 V and its voltage amplitude is

$$V_0 = \sqrt{2}\, V_{rms} = 311 \text{ V}$$

Form Factor

The ratio, $= \dfrac{\text{rms value}}{\text{average value}} = \dfrac{V_0/\sqrt{2}}{2V_0/\pi} = \dfrac{\pi}{2\sqrt{2}} = 1.11$

is known as form factor.

The different values i_0, i_{av} and i_{rms} are shown in figure.

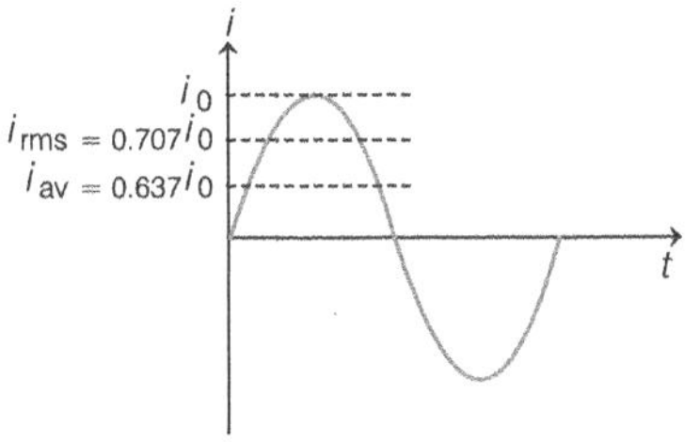

Note (i) *There are four values of i or V in an AC. These are :*

(a) Instantaneous value. This value will be a function of time.

(b) Peak value i_0 or V_0.

(c) Average value and

(d) RMS value.

(ii) *The average value of $\sin \omega t$, $\cos \omega t$, $\sin 2\omega t$, $\cos 2\omega t$, etc., in one complete cycle is zero because it is positive in half of the time and negative in rest half of the time. Thus, in one cycle :*

$$\langle \sin \omega t \rangle = \langle \cos \omega t \rangle = \langle \sin 2\omega t \rangle = \langle \cos 2\omega t \rangle = 0$$

If $\quad i = i_0 \sin \omega t$

then $\quad \langle i \rangle = \langle i_0 \sin \omega t \rangle = i_0 \langle \sin \omega t \rangle = 0$

(iii) *The average value of $\sin^2 \omega t$ and $\cos^2 \omega t$ in one complete cycle is $\frac{1}{2}$.*

or $\quad \langle \sin^2 \omega t \rangle_T = \langle \cos^2 \omega t \rangle_T = \frac{1}{2}$

If $\quad i^2 = i_0^2 \sin^2 \omega t$

then $\quad \langle i^2 \rangle_T = \langle i_0^2 \sin^2 \omega t \rangle_T = i_0^2 \langle \sin^2 \omega t \rangle_T = \frac{i_0^2}{2}$

(iv) *Like SHM, general expressions of time functions of current/voltage in an sinusoidal AC are*

$$i = i_0 \sin (\omega t \pm \phi)$$
$$V = V_0 \sin (\omega t \pm \phi)$$

or $\quad i = i_0 \cos (\omega t \pm \phi)$

and $\quad V = V_0 \cos (\omega t \pm \phi)$

(v) *Average value of current or voltage over a half cycle can be zero also. This depends on the time interval (of course T/2) over which average value is to be found. Think why?*

(vi) *If $y = f(t)$, then*

$\langle y \rangle_{t_1 \text{ to } t_2}$ = *average value of y between t_1 and t_2*

$$= \frac{\sum_{t_1}^{t_2} y}{(t_2 - t_1)} = \frac{\int_{t_1}^{t_2} y\, dt}{t_2 - t_1} = \frac{\text{area under } y\text{-}t \text{ graph from } t_1 \text{ to } t_2}{t_2 - t_1}$$

→ *If y = constant = C, then*

$$\langle y \rangle_{\text{any interval}} = C$$

→ *If y is a linear function of t or y-t graph is a straight line (passing through origin or not passing through origin), then*

$$\langle y \rangle_{t_1 \text{ to } t_2} = \frac{y_f + y_i}{2} = \frac{y_{t_2} + y_{t_1}}{2}$$

Phasors

If an AC generator is connected to a series circuit containing resistors, inductors and capacitors and we want to know the amplitude and time characteristics of the alternating current. To simplify our analysis of circuits containing two or more elements, we use graphical constructions called **phasor diagrams.** In these constructions, alternating (sinusoidal) quantities, such as current and voltage are rotating vectors called **phasors.**

In these diagrams, the instantaneous value of a quantity that varies sinusoidally with time is represented by the projection onto a vertical axis (if it is a sine function) or onto a horizontal axis (if it is a cosine function) of a vector with a length equal to the amplitude (i_0) of the quantity. The vector rotates counterclockwise with constant angular velocity ω

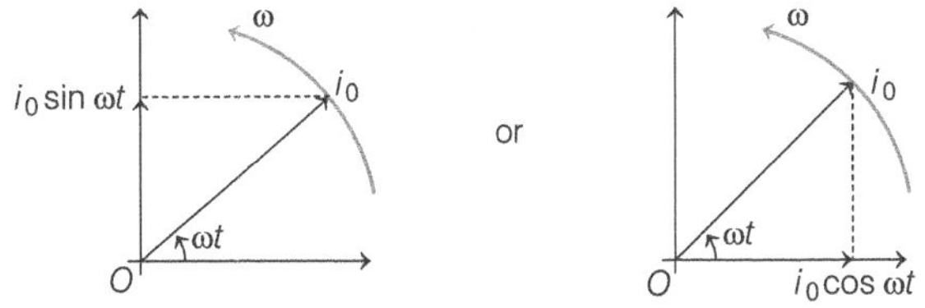

A phasor is not a real physical quantity with a direction in space, such as velocity, momentum or electric field. Rather, it is a geometric entity that helps us to describe and analyze physical quantities that vary sinusoidally with time.

Example 1 Show that average heat produced during a cycle of AC is same as produced by DC with $i = i_{rms}$.

Sol. For an AC, $i = i_0 \sin \omega t$

Therefore, instantaneous value of heat produced in time dt across a resistance R is

$$dH = i^2 R dt = i_0^2 R \sin^2 \omega t \, dt$$

$\therefore$ Average value of heat produced during a cycle,

$$H_{av} = \frac{\int_0^T dH}{\int_0^T dt} = \frac{\int_0^{2\pi/\omega} (i_0^2 R \sin^2 \omega t) \, dt}{\int_0^{2\pi/\omega} dt}$$

$$= \frac{i_0^2}{2} R \left(\frac{2\pi}{\omega}\right) = i_{rms}^2 RT$$

i.e. AC produces same heating effects as DC of value $i = i_{rms}$.

Example 2 If the current in an AC circuit is represented by the equation,

$$i = 5 \sin (300t - \pi/4)$$

Here, t is in second and i in ampere. Calculate

(a) peak and rms value of current

(b) frequency of AC

(c) average current

Sol. (a) As in case of AC,

$$i = i_0 \sin (\omega t \pm \phi)$$

$\therefore$ The peak value, $i_0 = 5$ A **Ans.**

and $$i_{rms} = \frac{i_0}{\sqrt{2}} = \frac{5}{\sqrt{2}}$$

$= 3.535$ A **Ans.**

(b) Angular frequency, $\omega = 300$ rad/s

$$\therefore \quad f = \frac{\omega}{2\pi} = \frac{300}{2\pi}$$

≈ 47.75 Hz **Ans.**

(c) $$i_{av} = \left(\frac{2}{\pi}\right) i_0 = \left(\frac{2}{\pi}\right)(5)$$

$= 3.18$ A **Ans.**

Example 3 Sinusoidal i - t graph is as shown in figure.

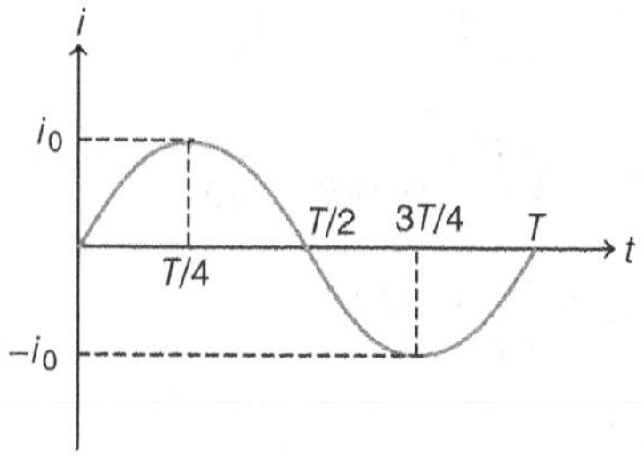

Find average value of i in time interval :

(a) $\frac{T}{4}$ to $\frac{3T}{4}$

(b) $\frac{T}{2}$ to T

where $T = \frac{2\pi}{\omega}$

Sol. (a) $$\langle i \rangle_{\frac{T}{4}-\frac{3T}{4}} = \frac{\text{Area of graph between } \frac{T}{4} \text{ and } \frac{3T}{4}}{\frac{3T}{4} - \frac{T}{4}}$$

$$= \frac{0}{T/2}$$

$= 0$ **Ans.**

(b) $\langle i \rangle_{\frac{T}{2}-T} = -\langle i \rangle_{0-\frac{T}{2}} = -\frac{2}{\pi} i_0$ **Ans.**

MODULE 1 Exercise

Assertion and Reason

See directions on page 25.

1. **Assertion :** Average value of current in the given graph in 6s is 3 A.

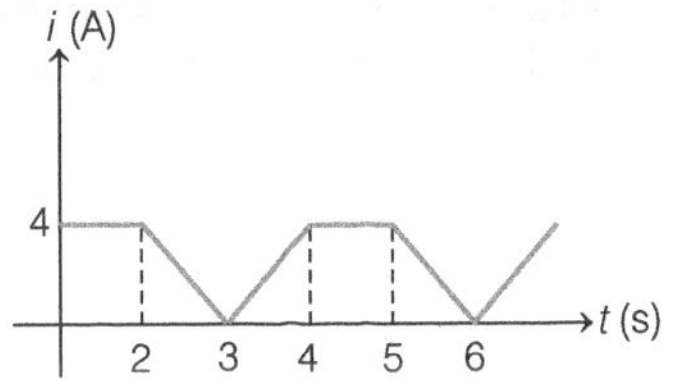

Reason : Average value can't be greater than the peak value of any function.

2. **Assertion :** An AC can be transmitted over long distances without much power loss.

Reason : An AC can be stepped up or down with the help of a transformer.

Single Correct Option

3. A complex current wave is given by $i = (5 + 5 \sin 100\,\omega t)$ A. Its average value over one time period is given as
(a) 10 A
(b) 5 A
(c) $\sqrt{50}$ A
(d) 0

4. In a certain circuit, current changes with time according to $i = 2\sqrt{t}$. Root mean square value of current between $t = 2$ to $t = 4$ s will be
(a) 3 A
(b) $3\sqrt{3}$ A
(c) $2\sqrt{3}$ A
(d) $\sqrt{3}$ A

5. A steady current of magnitude I and an AC current of peak value I are allowed to pass through identical resistors for the same time. The ratio of heat produced in the two resistors will be
(a) 2 : 1
(b) 1 : 2
(c) 1 : 1
(d) None of these

6. The current flowing through an AC circuit is given by $I = 5\sin(120\pi t)$A
How long will the current take to reach the peak value starting from zero? **(2022 Main)**
(a) $\frac{1}{60}$ s
(b) 60 s
(c) $\frac{1}{120}$ s
(d) $\frac{1}{240}$ s

7. A resistance of 40 Ω is connected to a source of alternating current rated 220 V, 50 Hz. Find the time taken by the current to change from its maximum value of rms value. **(2022 Main)**
(a) 2.5 ms
(b) 1.25 ms
(c) 2.5 s
(d) 0.25 s

Subjective Question

8. The current in a certain circuit varies with time as shown in figure. Find the average current and the rms current in terms of I_0.

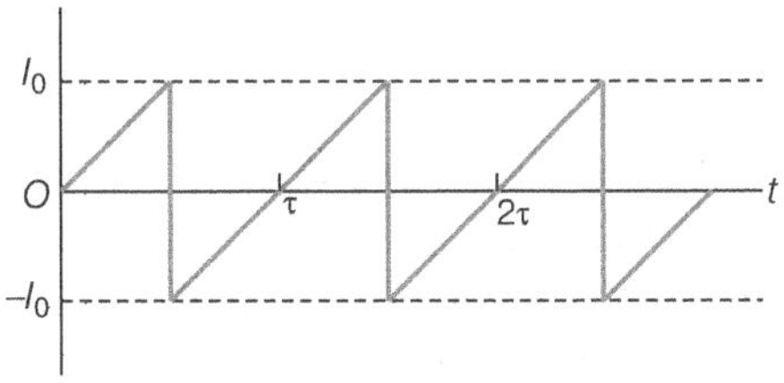

Integer Type Question

9. An alternating current at any instant is given by $i = \left[6 + \sqrt{56}\sin\left(100\pi t + \frac{\pi}{3}\right)\right]$ A. The rms value of the current is......A. **(2024 Main)**

Answers

1. (b) 2. (a) 3. (b) 4. (c) 5. (a) 6. (d) 7. (a) 8. Zero, $\frac{I_0}{\sqrt{3}}$ 9. (8)

See hints on page 614.

MODULE 2

Current and Potential Relations

In this module, we will derive voltage current relations for individual circuit elements carrying a sinusoidal current. We will consider resistors, inductors and capacitors.

Resistor in an AC Circuit

Consider a resistor with resistance R through which there is a sinusoidal current given by

$$i = i_0 \sin\omega t \quad \text{...(i)}$$

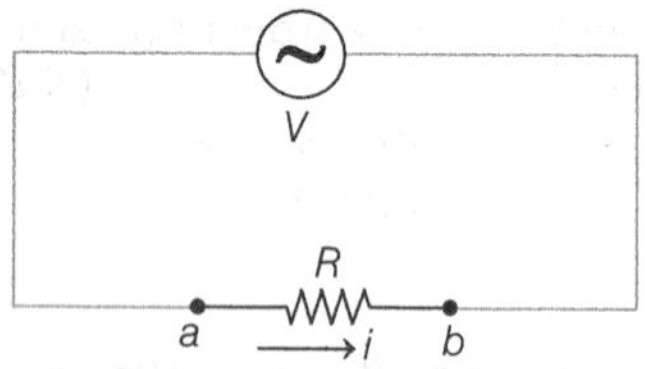

Here, i_0 is the current amplitude (maximum current). From Ohm's law, the instantaneous PD between points a and b is

$$V_R = iR = (i_0 R) \sin\omega t$$

We can write as

$$i_0 R = V_0 \text{, the voltage amplitude}$$

$$\therefore \quad V_R = V = V_0 \sin\omega t \quad \text{...(ii)}$$

From Eqs. (i) and (ii), we can see that current and voltage are **in phase** if only resistance is in the circuit. Figure (a) shows graphs of i and V_R as functions of time.

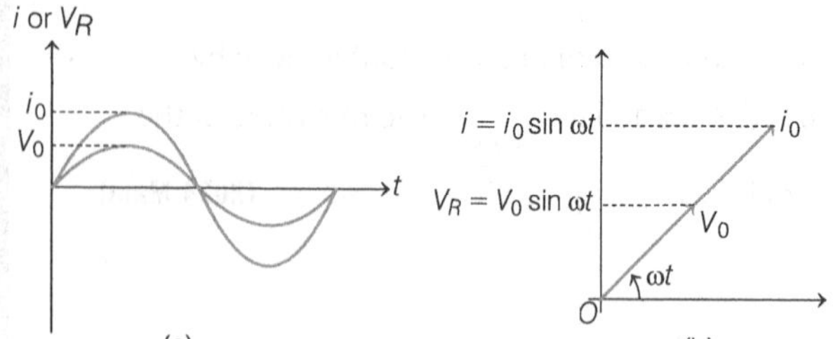

(a) (b)

The corresponding phasor diagram is shown in figure (b).

Because i and V_R are in phase and have the same frequency, the current and voltage phasors rotate together, they are parallel at each instant. Their projection on vertical axis represents the instantaneous current and voltage respectively.

Note *Direction of an alternating current is not shown in a circuit, as it keeps on changing. In the figure, the direction of instantaneous current is only shown.*

Capacitor in an AC Circuit

If a capacitor of capacitance C is connected across the alternating source, the instantaneous charge on the capacitor is

$$q = CV_C = CV_0 \sin\omega t$$

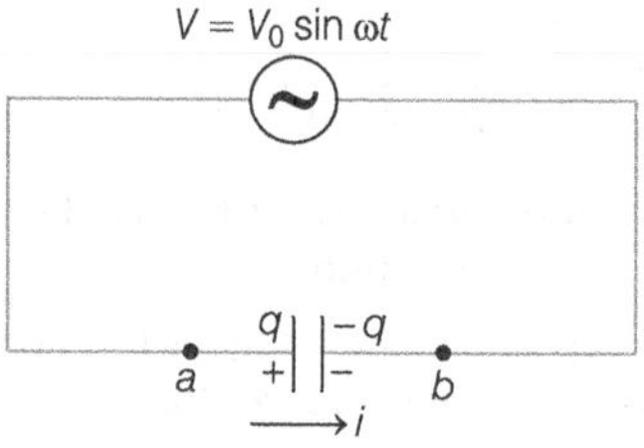

and the instantaneous current i passing through it is given by

$$i = \frac{dq}{dt} = CV_0\omega \cos \omega t$$

$$= \frac{V_0}{1/\omega C} \sin (\omega t + \pi/2)$$

or

$$i = i_0 \sin (\omega t + \pi/2)$$

Here, $$\boxed{V_0 = (i_0)\left(\frac{1}{\omega C}\right)}$$

This relation shows that the quantity $\frac{1}{\omega C}$ is the effective AC resistance or the **capacitive reactance** of the capacitor and is represented as X_C. It has unit as ohm. Thus,

$$\boxed{X_C = \frac{1}{\omega C}}$$

It is clear that the **current leads the voltage by 90°** or the potential drop across the capacitor lags the current passing it by 90°.

Figure (a) shows V and i as functions of time t.

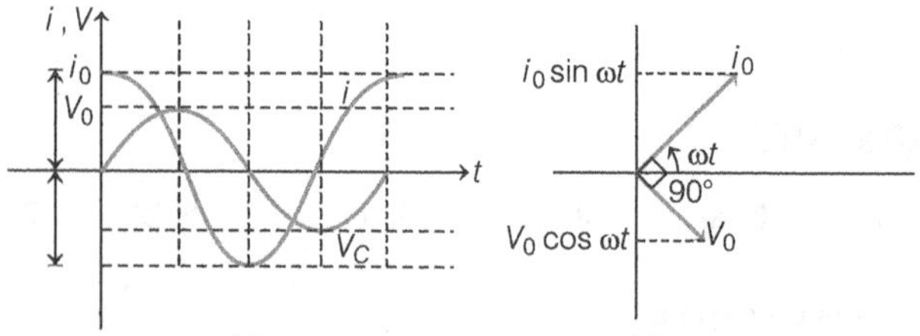

(a) (b)

The phasor diagram in figure (b) shows that voltage phasor is behind the current phasor by a quarter cycle or 90°.

Inductor in an AC Circuit

Consider a pure inductor of self-inductance L and zero resistance connected to an alternating source. Again we assume that an instantaneous current $i = i_0 \sin\omega t$ flows through the inductor. Although, there is no resistance, there is a potential difference V_L between the inductor terminals a and b because the current varies with time giving rise to a self-induced emf.

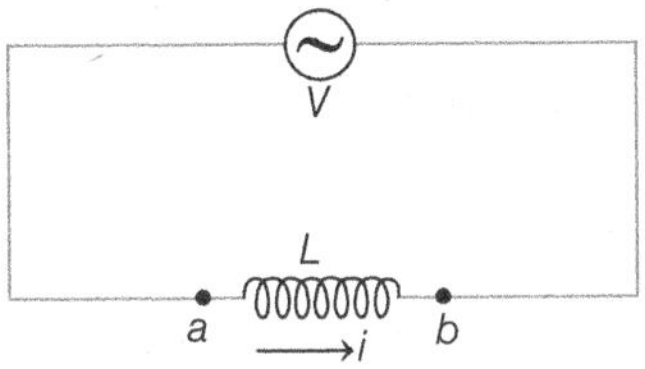

$$V_a - L\frac{di}{dt} = V_b \quad \text{or} \quad V_a - V_b = V_{ab} = L\left(\frac{di}{dt}\right)$$

or $$V_L = L\frac{di}{dt} = Li_0\omega\cos\omega t$$

or $$V_L = V = V_0 \sin\left(\omega t + \frac{\pi}{2}\right) \quad \text{...(i)}$$

Here, $$\boxed{V_0 = i_0(\omega L)} \quad \text{...(ii)}$$

or $$i_0 = \frac{V_0}{\omega L}$$

$\therefore$ $$\boxed{i = \frac{V_0}{\omega L}\sin\omega t} \quad \text{...(iii)}$$

Eq. (iii) shows that effective AC resistance, i.e. **inductive reactance of inductor** is

$$\boxed{X_L = \omega L}$$

and the maximum current,

$$\boxed{i_0 = \frac{V_0}{X_L}}$$

The unit of X_L is also ohm.

From Eqs. (i) and (iii), we see that **the voltage across the inductor leads the current passing through it by 90°.**

Figure (a) shows V and i as functions of time.

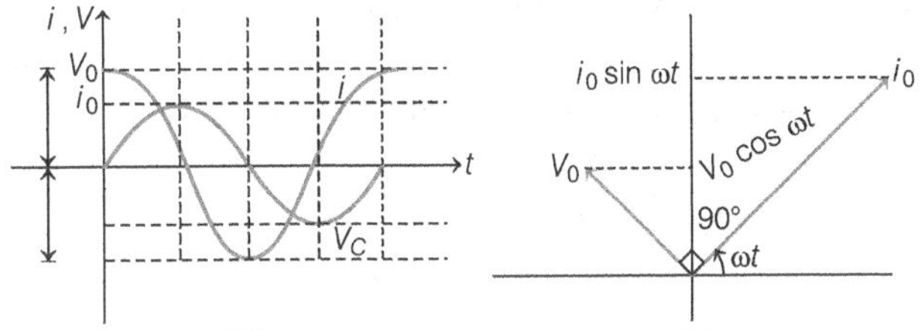

(a) (b)

Phasor diagram in figure (b) shows that V_L or V leads the current i by 90°.

Important POINTS

- **Circuit elements with AC**

Circuit elements	Amplitude relation	Resistance or Reactance	Phase of V
Resistor	$V_0 = i_0 R$	R	in phase with i
Capacitor	$V_0 = i_0 X_C$	$X_C = \frac{1}{\omega C}$	lags i by 90°
Inductor	$V_0 = i_0 X_L$	$X_L = \omega L$	leads i by 90°

- In DC, $\omega = 0$, therefore, $X_L = 0$ and $X_C = \infty$
- Figure shows the graphs of R, X_L and X_C as functions of angular frequency ω.

  R does not depend upon ω.

$$X_L = \omega L \Rightarrow X_L \propto \omega$$

$$X_C = \frac{1}{\omega C} \Rightarrow X_C \propto \frac{1}{\omega}$$

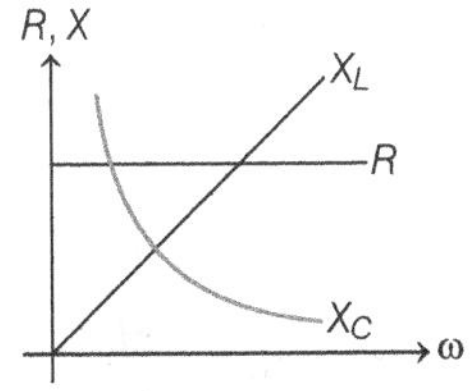

- If an oscillating voltage of a given amplitude V_0 is applied across an inductor, the resulting current will have a smaller amplitude i_0 for larger value of ω. Since, X_L is proportional to ω, a high frequency voltage applied to the inductor gives only a small current while a lower frequency voltage of the same amplitude gives rise to a larger current. Inductors are used in some circuit applications, such as power supplies and radio interference filters to block high frequencies while permitting lower frequencies to pass through. A circuit device that uses an inductor for this purpose is called a **low pass filter**.
- The capacitive reactance of a capacitor is inversely proportional to the capacitance C and the angular frequency ω. The greater the capacitance and the higher the frequency, the smaller is the capacitive reactance X_C. Capacitors tend to pass high frequency current and to block low frequency current, just the opposite of inductors. A device that passes signals of high frequency is called a **high pass filter**.
- Remember that we can write, $V_R = iR$, $(V_0)_R = i_0 R$, $(V_0)_L = i_0 X_L$ and $(V_0)_C = i_0 X_C$ but can't write (for instantaneous values).

$$V_L = iX_L$$

or $$V_C = iX_C$$

This is because there is a phase difference between the voltage and current functions in both an inductor and a capacitor.

Example 1 A 100 Ω resistance is connected in series with a 4H inductor. The voltage across the resistor is $V_R = (2.0\ \text{V})\sin(10^3 \text{rad/s})\,t$:

(a) Find the expression of circuit current
(b) Find the inductive reactance
(c) Derive an expression for the voltage across the inductor.

Sol. (a) $i = \dfrac{V_R}{R} = \dfrac{(2.0\ \text{V})\sin(10^3\ \text{rad/s})t}{100}$

$= (2.0 \times 10^{-2}\ \text{A})\sin(10^3\ \text{rad/s})t$ **Ans.**

(b) $X_L = \omega L = (10^3\ \text{rad/s})(4\ \text{H})$

$= 4.0 \times 10^3\ \Omega$ **Ans.**

(c) The amplitude of voltage across inductor,

$V_0 = i_0 X_L = (2.0 \times 10^{-2}\ \text{A})(4.0 \times 10^3\ \Omega)$

$= 80\ \text{V}$ **Ans.**

In an AC, voltage across the inductor leads the current by 90° or $\pi/2$ rad. Hence,

$V_L = V_0 \sin(\omega t + \pi/2)$

$= (80\ \text{V})\sin\left\{(10^3\ \text{rad/s})\,t + \dfrac{\pi}{2}\ \text{rad}\right\}$

Note *That the amplitude of voltage across the resistor (= 2.0 V) is not same as the amplitude of the voltage across the inductor (= 80 V), even though the amplitude of the current through both devices is the same.*

Phasor Algebra

The complex quantities normally employed in AC circuit analysis, can be added and subtracted like coplanar vectors. Such coplanar vectors which represent sinusoidally time varying quantities are known as **phasors**.

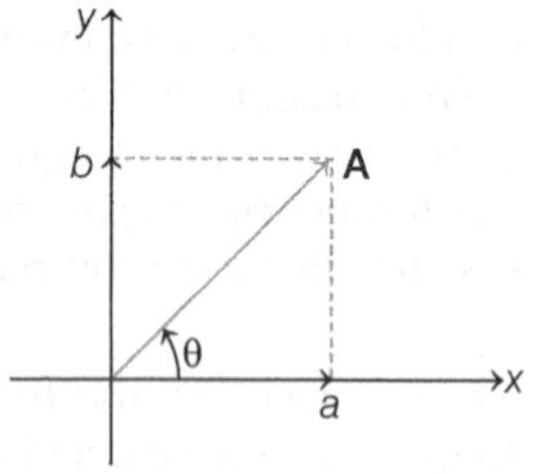

In Cartesian form, a phasor **A** can be written as

$$\mathbf{A} = a + jb$$

where, a is the x-component and b is the y-component of phasor **A**.

The magnitude of **A** is $|\mathbf{A}| = \sqrt{a^2 + b^2}$

and the angle between the direction of phasor **A** and the positive x-axis is

$$\theta = \tan^{-1}\left(\frac{b}{a}\right)$$

When a given phasor **A**, the direction of which is along the x-axis is multiplied by the operator j, a new phasor $j\mathbf{A}$ is obtained which will be 90° anti-clockwise from **A**, i.e. along y-axis. If the operator j is multiplied now to the phasor $j\mathbf{A}$, a new phasor $j^2\mathbf{A}$ is obtained which is along negative x-axis and having same magnitude as of **A**. Thus,

$$j^2\mathbf{A} = -\mathbf{A}$$

$\therefore$ $$j^2 = -1 \quad \text{or} \quad j = \sqrt{-1}$$

Now, using the j operator, let us discuss different circuits of an AC.

Series *L-R* Circuit

As we know, potential difference across a resistance in AC is in phase with current and it leads in phase by 90° with current across the inductor.

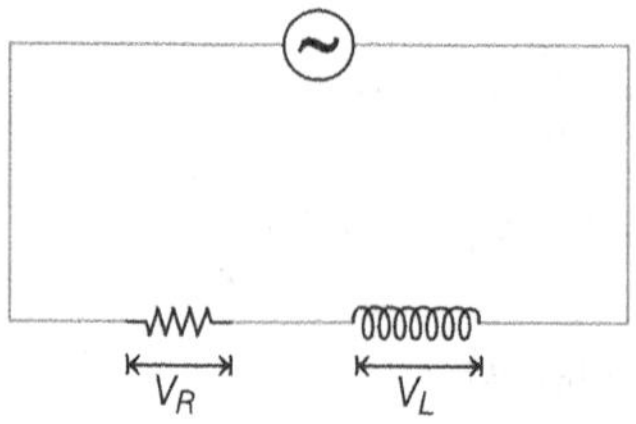

Suppose in phasor diagram current is taken along positive x-direction. Then, V_R is also along positive x-direction and V_L along positive y-direction, so, we can write

$$\mathbf{V} = \mathbf{V}_R + j\mathbf{V}_L = iR + j(iX_L)$$
$$= iR + j(i\omega L) \qquad (\text{as } X_L = \omega L)$$
$$= i\mathbf{Z}$$

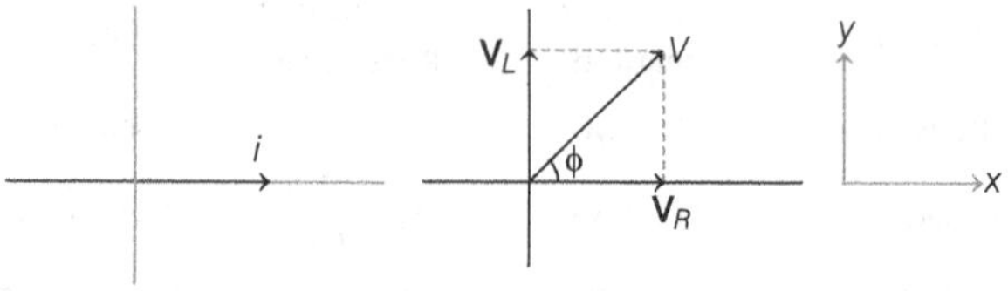

Here, $\mathbf{Z} = R + jX_L = R + j(\omega L)$ is called as **impedance** of the circuit. Impedance plays the same role in AC circuits as the ohmic resistance does in DC circuits. The modulus of impedance is

$$|\mathbf{Z}| = \sqrt{R^2 + (\omega L)^2}$$

The potential difference leads the current by an angle,

$$\phi = \tan^{-1}\frac{|\mathbf{V}_L|}{|\mathbf{V}_R|}$$
$$= \tan^{-1}\left(\frac{X_L}{R}\right)$$

or $$\phi = \tan^{-1}\left(\frac{\omega L}{R}\right)$$

Series *C-R* Circuit

Potential difference across a capacitor in AC lags in phase by 90° with the current in the circuit.

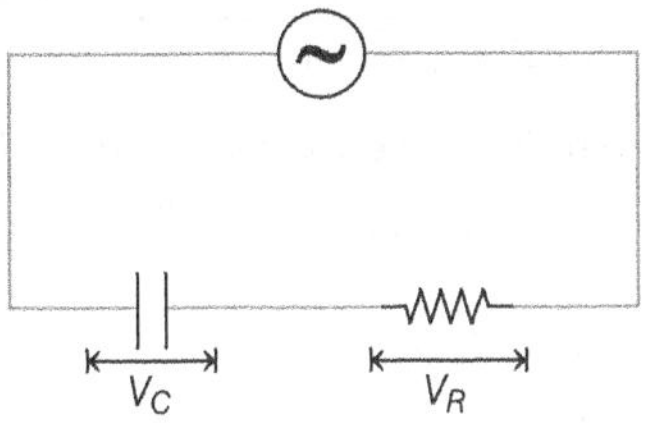

Suppose in phasor diagram current is taken along positive x-direction. Then, V_R is also along positive x-direction but V_C is along negative y-direction. So, we can write

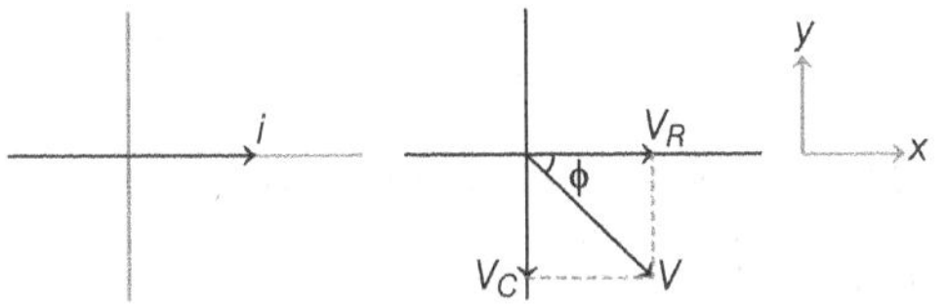

$$\mathbf{V} = \mathbf{V}_R - j\mathbf{V}_C = iR - j(iX_C)$$

$$= iR - j\left(\frac{i}{\omega C}\right) = i\mathbf{Z}$$

Here, impedance is $\mathbf{Z} = R - j\left(\frac{1}{\omega C}\right)$

The modulus of impedance is

$$|\mathbf{Z}| = \sqrt{R^2 + \left(\frac{1}{\omega C}\right)^2}$$

and the potential difference lags the current by an angle,

$$\phi = \tan^{-1}\left|\frac{V_C}{V_R}\right| = \tan^{-1}\frac{X_C}{R}$$

$$= \tan^{-1}\left(\frac{1/\omega C}{R}\right)$$

or

$$\phi = \tan^{-1}\left(\frac{1}{\omega RC}\right)$$

Series *L-C-R* Circuit

Potential difference across an inductor leads the current by 90° in phase while that across a capacitor, it lags in phase by 90°.

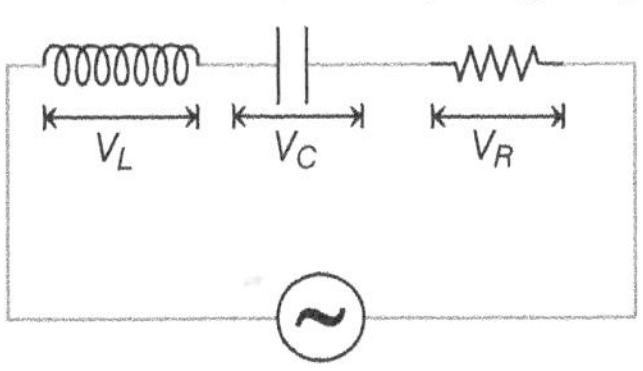

Suppose in a phasor diagram current is taken along positive x-direction. Then, V_R is along positive x-direction, V_L along positive y-direction and V_C along negative y-direction.

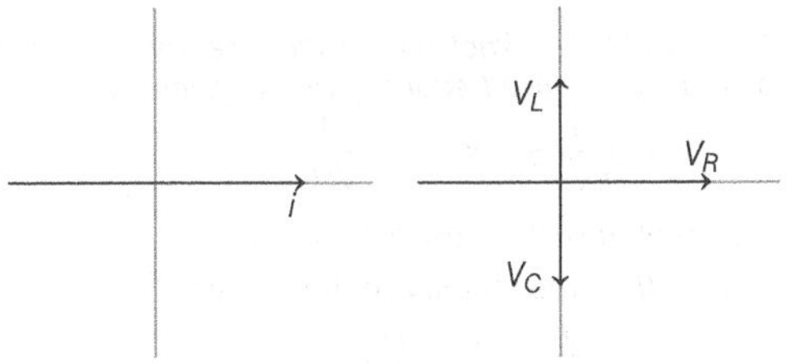

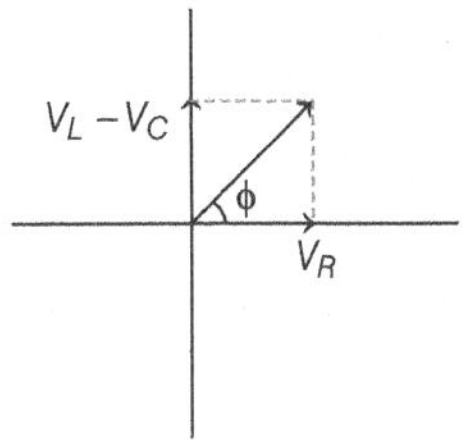

Let us assume that $X_L > X_C$

or $\quad V_L > V_C$

So, we can write $\quad V = V_R + jV_L - jV_C$

$$= iR + j(iX_L) - j(iX_C)$$

$$= iR + j[i(X_L - X_C)] = iZ$$

Here, impedance is $Z = R + j(X_L - X_C)$

$$= R + j\left(\omega L - \frac{1}{\omega C}\right)$$

The modulus of impedance is $|\mathbf{Z}| = \sqrt{R^2 + \left(\omega L - \frac{1}{\omega C}\right)^2}$

and the potential difference leads the current by an angle,

$$\phi = \tan^{-1}\left|\frac{V_L - V_C}{V_R}\right|$$

$$= \tan^{-1}\left(\frac{X_L - X_C}{R}\right)$$

or

$$\phi = \tan^{-1}\left(\frac{\omega L - \dfrac{1}{\omega C}}{R}\right)$$

Note *Let us take the most general case of a series L-C-R circuit in an AC.*

$$|\mathbf{Z}| = \sqrt{R^2 + (X_L \sim X_C)^2}$$

If $X_L = X_C$ *or* $\omega L = \frac{1}{\omega C}$ *or* $\omega = \frac{1}{\sqrt{LC}}$

or $\quad f = \frac{1}{2\pi\sqrt{LC}}$

the modulus of impedance

$$|\mathbf{Z}| = R$$

and the current is in phase with voltage, i.e. if $V = V_0 \sin \omega t$, *then*

$$i = i_0 \sin \omega t$$

where, $\quad i_0 = \frac{V_0}{|\mathbf{Z}|} = \frac{V_0}{R}$

Such a condition is known as resonance and frequency known as resonance frequency and is given by

$$f = \frac{1}{2\pi\sqrt{LC}} \quad \text{or} \quad \omega = \frac{1}{\sqrt{LC}}$$

The current in such a case is maximum.

If $X_L > X_C$, then the modulus of the impedance

$$|Z| = \sqrt{R^2 + (X_L - X_C)^2}$$

and the voltage leads the current by an angle given by

$$\phi = \tan^{-1}\left(\frac{X_L - X_C}{R}\right)$$

i.e. if $V = V_0 \sin \omega t$, then $i = i_0 \sin(\omega t - \phi)$

where, $i_0 = \frac{V_0}{|Z|}$

If $X_C > X_L$, then the modulus of the impedance is

$$|Z| = \sqrt{R^2 + (X_C - X_L)^2}$$

and the current leads the voltage by an angle given by

$$\phi = \tan^{-1}\left(\frac{X_C - X_L}{R}\right)$$

i.e. if $V = V_0 \sin \omega t$, then $i = i_0 \sin(\omega t + \phi)$

where, $i_0 = \frac{V_0}{|Z|}$

Important POINTS

- $i_0 = \frac{V_0}{|Z|}$, $i_{rms} = \frac{V_{rms}}{|Z|}$.

  But in general $i \neq \frac{V}{|Z|}$.
- In *L-C-R* circuit, whenever voltage across various elements is asked, find only rms values unless stated in the question for the peak or instantaneous value.

  The rms values are

  $$V_R = i_{rms} R, \; V_L = i_{rms} X_L \text{ and } V_C = i_{rms} X_C$$
- **Voltage magnification in series resonance circuit**

  At resonance $\left(f = \frac{1}{2\pi\sqrt{LC}} \text{ or } \omega = \frac{1}{\sqrt{LC}}\right)$, the PD across the inductor and the capacitor are equal and 180° out of phase and therefore, cancel out. Hence, the applied emf is merely to overcome the resistance opposition only. If an inductance or capacitance of very large reactance (X_L or X_C) is connected with $X_L = X_C$ (at resonance) then PD across them increases to a very high value. The ratio is known as **voltage magnification** and is given by,

  $$\frac{\text{PD across inductance (or capacitance)}}{\text{Applied emf}} = \frac{i_{rms}(\omega L)}{i_{rms}(R)} = \frac{\omega L}{R}$$

  or

  $$\frac{i_{rms}\left(\frac{1}{\omega C}\right)}{i_{rms}(R)} = \frac{1}{\omega CR}$$

  This ratio is greater than unity.
- **Response curves of series circuit** The impedance of an *L-C-R* circuit depends on the frequency. The dependence is shown in figure. The frequency is taken on logarithmic scale because of its wide range.

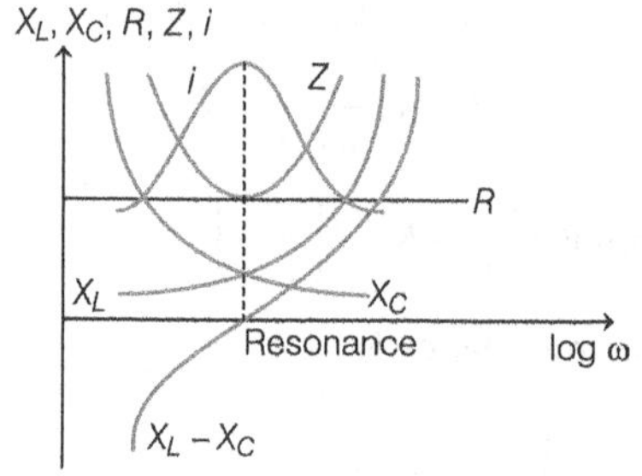

  From the figure, we can see that at resonance,

  (i) $X_L = X_C$ or $\omega = \frac{1}{\sqrt{LC}}$

  (ii) $Z = Z_{min} = R$ and

  (iii) i is maximum.

Note *Here, by Z we mean the modulus of* ***Z*** *and i means i_{rms}.*

- **Acceptor circuit** If the frequency of the AC supply can be varied (e.g. in radio or television signal), then in series L-C-R circuit, at a frequency $f = 1/2\pi\sqrt{LC}$ maximum current flows in the circuit and have a maximum PD across its inductance (or capacitance). This is the method by which a radio or television set is tuned at a particular frequency. The circuit is known as **acceptor circuit**.

Example 2 An alternating emf 200 virtual (means rms) volts at 50 Hz is connected to a circuit of resistance 1 Ω and inductance 0.01 H. What is the phase difference between the current and the emf in the circuit? Also, find the virtual current in the circuit.

Sol. In case of an L-R AC circuit, the voltage leads the current in phase by an angle,

$$\phi = \tan^{-1}\left(\frac{X_L}{R}\right)$$

Here, $X_L = \omega L = (2\pi f L)$

$= (2\pi)(50)(0.01) = \pi\,\Omega$

and $R = 1\,\Omega$

$\therefore \quad \phi = \tan^{-1}(\pi) \approx 72.3°$ **Ans.**

Further, $i_{rms} = \dfrac{V_{rms}}{|Z|} = \dfrac{V_{rms}}{\sqrt{R^2 + X_L^2}}$

Substituting the values, we have

$$i_{rms} = \frac{200}{\sqrt{(1)^2 + (\pi)^2}} = 60.67 \text{ A}$$ **Ans.**

Example 3 A resistance and inductance are connected in series across a voltage,

$$V = 283 \sin 314\, t$$

The current is found to be $4 \sin(314t - \pi/4)$. Find the values of the inductance and resistance.

Sol. In L-R series circuit, current lags the voltage by an angle,

$$\phi = \tan^{-1}\left(\frac{X_L}{R}\right)$$

Here, $\phi = \dfrac{\pi}{4}$

$\therefore \quad X_L = R$ or $\omega L = R$ $(\omega = 314 \text{ rad/s})$

$\therefore \quad 314\, L = R$ ...(i)

Further, $V_0 = i_0 |Z|$

$\therefore \quad 283 = 4\sqrt{R^2 + X_L^2}$

or $R^2 + (\omega L)^2 = \left(\dfrac{283}{4}\right)^2 = 5005.56$

or $2R^2 = 5005.56$ (as $\omega L = R$)

$\therefore \quad R \approx 50\,\Omega$ **Ans.**

and from Eq. (i),

$L = 0.16$ H **Ans.**

Example 4 Find the voltage across the various elements, i.e. resistance, capacitance and inductance which are in series and having values 1000 Ω, 1 μF and 2.0 H, respectively. Given emf is $V = 100\sqrt{2} \sin 1000\, t$ volt.

Sol. The rms value of voltage across the source,

$$V_{rms} = \frac{100\sqrt{2}}{\sqrt{2}} = 100 \text{ V}$$

$\therefore \quad \omega = 1000$ rad/s

$$\therefore \quad i_{rms} = \frac{V_{rms}}{|Z|} = \frac{V_{rms}}{\sqrt{R^2 + (X_L \sim X_C)^2}}$$

$$= \frac{V_{rms}}{\sqrt{R^2 + \left(\omega L - \dfrac{1}{\omega C}\right)^2}}$$

$$= \frac{100}{\sqrt{(1000)^2 + \left(1000 \times 2 - \dfrac{1}{1000 \times 1 \times 10^{-6}}\right)^2}}$$

$= 0.0707$ A

The current will be same in the circuit. Therefore,

PD across resistor, $V_R = i_{rms} R = 0.0707 \times 1000 = 70.7$ V

PD across inductor, $V_L = i_{rms} X_L = 0.0707 \times 1000 \times 2$

$= 141.4$ V and

PD across capacitor,

$$V_C = i_{rms} X_C = 0.0707 \times \frac{1}{1 \times 1000 \times 10^{-6}}$$

$= 70.7$ V **Ans.**

Note *The rms voltages do not add directly as $V_R + V_L + V_C = 282.8$ V which is not the source voltage 100 V. The reason is that these voltages are not in phase and can be added by vector or by phasor algebra.*

$$V = \sqrt{V_R^2 + (V_L \sim V_C)^2}$$

Important POINTS

1. Frequency of AC in India is 50 Hz.
2. The AC is converted into DC with the help of rectifier while DC is converted into AC with the help of inverter.
3. An AC cannot produce electroplating or electrolysis.
4. The AC is measured by hot wire ammeter.
5. An AC can be stepped up or down with the help of a transformer.
6. An AC can be transmitted over long distances without much power loss.
7. An AC can be regulated by using choke coil without any significant waste of energy.
8. In an AC (sinusoidal), current or voltage can have the following four values
   (i) instantaneous value
   (ii) peak value (i_0 or V_0)
   (iii) rms value (i_{rms} or V_{rms})
   (iv) **average value :** In full cycle, average value is zero while in half cycle it may or may not be zero.
9. $X_C = \dfrac{1}{\omega C}$

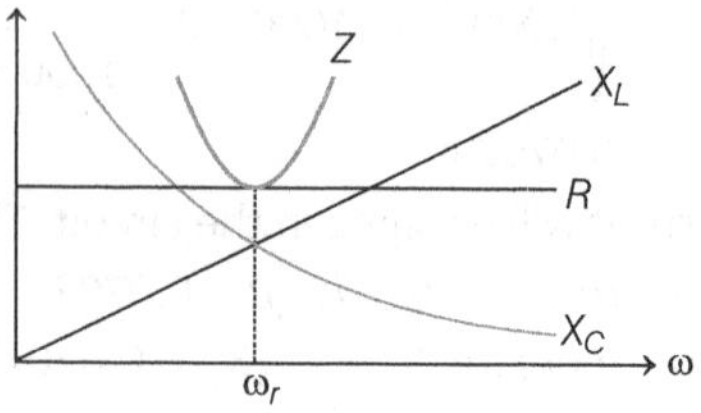

$\Rightarrow \quad X_C \propto \dfrac{1}{\omega} \Rightarrow X_L = \omega L \Rightarrow X_L \propto \omega$

R does not depend on ω. It is a constant.

At $\omega = \omega_r : X_C = X_L$ and $Z = Z_{min} = R$

10. For $\omega > \omega_r, X_L > X_C$.
    Hence, voltage will lead the current or circuit is inductive.
    For $\omega < \omega_r, X_C > X_L$.
    Hence, current will lead the voltage function or circuit is capacitive.
    At $\omega = \omega_r, X_C = X_L$.
    Hence, current function and voltage function are in same phase. Circuit is neither capacitive or inductive.
11.

Conditions	Phase angle
$R = 0$	90°
$X_C = X_L \neq 0 \;\; R \neq 0$	0°
$X_C = X_L = 0 \;\; R \neq 0$	0°
$\omega = \omega_r$	0°

In all other cases, phase difference between current function and voltage function is

$$0° < \phi < 90°$$

If $X_C > X_L, \phi = \tan^{-1}\left(\dfrac{X_C - X_L}{R}\right)$ or $\cos^{-1}\left(\dfrac{R}{Z}\right)$

If $X_L > X_C, \phi = \tan^{-1}\left(\dfrac{X_L - X_C}{R}\right)$ or $\cos^{-1}\left(\dfrac{R}{Z}\right)$

12. **Quality Factor (Q-factor)**
    The Q, or quality factor of a resonant circuit is a measure of 'goodness' or quality of a resonant circuit.
    A higher value corresponds to a more narrow bandwidth which is desirable in many applications.

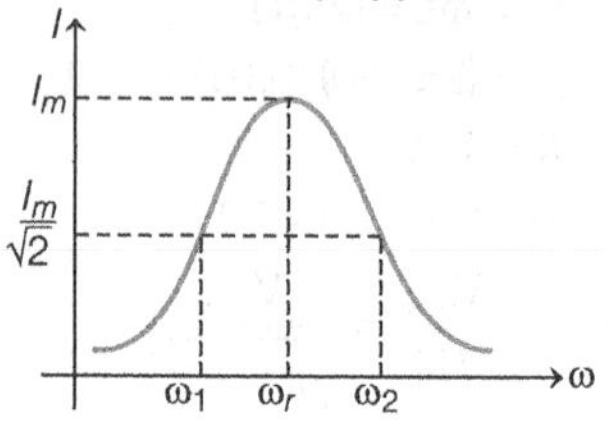

In the figure :

At ω_r current is maximum (say I_m).

At ω_1 and ω_2 current is $\dfrac{I_m}{\sqrt{2}}$.

$$Q\text{-factor} = \frac{\omega_r}{\Delta\omega} = \frac{\omega_r}{\omega_r - \omega_1}$$

Now, $\quad \omega_r = \dfrac{1}{\sqrt{LC}}$

At ω_1 and ω_2 current is $\dfrac{1}{\sqrt{2}}$ times of I_m.

So Z will be $\sqrt{2}$ times of $Z_{min} (= R)$.

Hence Z^2 will be 2-times.

$\therefore \quad R^2 + (X_L \sim X_C)^2 = 2R^2$

$\therefore \quad X_L - X_C = R$

$\therefore \quad \omega_2 L - \dfrac{1}{\omega_2 C} = R$

or $\quad \omega_2^2 LC - \omega_2 CR - 1 = 0$

or $\quad \omega_2 = \dfrac{CR \pm \sqrt{C^2R^2 + 4LC}}{2LC}$

or $\quad \omega_2 = \dfrac{CR + \sqrt{C^2R^2 + 4LC}}{2LC}$

Further, $\quad X_C - X_L = R$

or $\quad -\omega_1^2 LC - \omega_1 CR + 1 = 0$ which gives

$$\omega_1 = \frac{\sqrt{C^2R^2 + 4LC} - CR}{2LC}$$

$\Rightarrow \quad \omega_2 - \omega_1 = \dfrac{2CR}{2LC} = \dfrac{R}{L}$

Hence, Q-factor $= \dfrac{\omega_r}{\omega_2 - \omega_1} = \dfrac{1/\sqrt{LC}}{R/L}$

$$= \frac{1}{R}\sqrt{\frac{L}{C}}$$

MODULE 2 Exercise

Assertion and Reason

See directions on page 25.

1. **Assertion :** In an AC circuit, potential difference across the capacitor may be greater than the applied voltage.

   Reason : $V_C = IX_C$, whereas $V = IZ$ and X_C can be greater than Z also.

2. **Assertion :** In series L-C-R circuit, voltage will lead the current function for frequency greater than the resonance frequency.

   Reason : At resonance frequency, phase difference between current function and voltage function is zero.

3. **Assertion :** Resonance frequency will decrease in L-C-R series circuit if a dielectric slab is inserted in between the plates of the capacitor.

   Reason : By doing so, capacity of capacitor will increase.

4. **Assertion :** In series L-C-R circuit, if a ferromagnetic rod is inserted inside an inductor, current in the circuit may increase or decrease.

   Reason : By doing so X_L will increase.

5. **Assertion :** In an L-R series circuit in AC, current in the circuit will decrease with increase in frequency.

   Reason : Phase difference between current function and voltage function will increase with increase in frequency.

Single Correct Option

6. As the frequency of an alternating current increases, the impedance of the circuit
   (a) increases continuously
   (b) decreases continuously
   (c) remains constant
   (d) None of the above

7. Phasor diagram of a series AC circuit is shown in figure. Then,

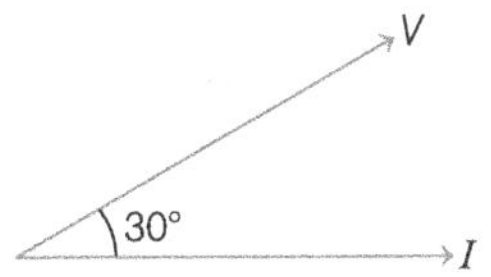

   (a) the circuit must be containing resistor and capacitor only
   (b) the circuit must be containing resistor and inductor only
   (c) the circuit must be containing all three elements L, C and R
   (d) the circuit cannot have only capacitor and inductor

8. A capacitor becomes a perfect insulator for
   (a) alternating current
   (b) direct current
   (c) both (a) and (b)
   (d) None of the above

9. In a purely resistive AC circuit,
   (a) voltage leads current
   (b) voltage lags current
   (c) voltage and current are in same phase
   (d) nothing can be said

10. In an AC circuit, the impedance is $\sqrt{3}$ times the reactance, then the phase angle is
    (a) 60° (b) 30°
    (c) zero (d) None of these

11. An L-C-R series circuit has a maximum current of 5 A. If $L = 0.5$ H and $C = 8\ \mu$F, then the angular frequency of AC voltage is
    (a) 500 rad/s (b) 5000 rad/s
    (c) 400 rad/s (d) 250 rad/s

12. In the series L-C-R circuit, the voltmeter and ammeter readings are

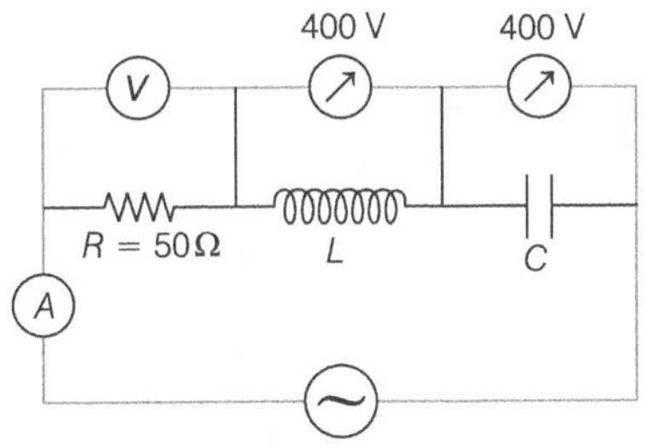

    (a) $V = 100$ V, $I = 2$ A
    (b) $V = 100$ V, $I = 5$ A
    (c) $V = 50$ V, $I = 2$ A
    (d) $V = 50$ V, $I = 1$ A

13. An AC voltage is applied across a series combination of L and R. If the voltage drop across the resistor and inductor are 20 V and 15 V respectively, then applied peak voltage is
    (a) 25 V (b) 35 V
    (c) $25\sqrt{2}$ V (d) $5\sqrt{7}$ V

14. When an AC signal of frequency 1 kHz is applied across a coil of resistance 100 Ω, then the applied voltage leads the current by 45°. The inductance of the coil is
    (a) 16 mH (b) 12 mH
    (c) 8 mH (d) 4 mH

15. A 50 Hz AC source of 20 V is connected across R and C as shown in figure. The voltage across R is 12 V. The voltage across C is

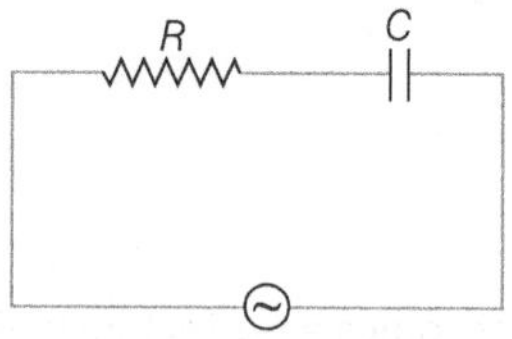

(a) 8 V
(b) 16 V
(c) 10 V
(d) Not possible to determine unless value of R and C are given

16. In a series resonant R-L-C circuit, if L is increased by 25% and C is decreased by 20%, then the resonant frequency will
(a) increase by 10% (b) decrease by 10%
(c) remain unchanged (d) increase by 2.5%

17. Which of the shown graphs may represent the reactance of a series L-C combination?

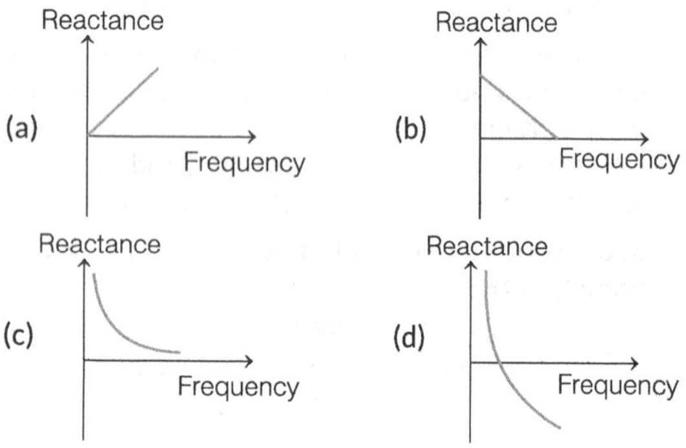

18. An alternating voltage $V = 200\sqrt{2}\sin(100t)$ is connected to a 1 μF capacitor through an AC ammeter. The reading of the ammeter shall be
(a) 10 mA (b) 20 mA
(c) 40 mA (d) 80 mA

Subjective Questions

19. (a) What is the reactance of a 2.00 H inductor at a frequency of 50.0 Hz?
(b) What is the inductance of an inductor whose reactance is 2.00 Ω at 50.0 Hz?
(c) What is the reactance of a 2.00 μF capacitor at a frequency of 50.0 Hz?
(d) What is the capacitance of a capacitor whose reactance is 2.00 Ω at 50.0 Hz?

20. An electric lamp which runs at 100 V DC and consumes 10 A current is connected to AC mains at 150 V, 50 Hz cycles with a choke coil in series. Calculate the inductance and drop of voltage across the choke. Neglect the resistance of choke.

21. A circuit operating at $\frac{360}{2\pi}$ Hz contains a 1 μF capacitor and a 20 Ω resistor. How large an inductor must be added in series to make the phase angle for the circuit zero? Calculate the current in the circuit if the applied voltage is 120 V.

22. A 300 Ω resistor, a 0.250 H inductor, and a 8.00 μF capacitor are in series with an AC source with voltage amplitude 120 V and angular frequency 400 rad/s.
(a) What is the current amplitude?
(b) What is the phase angle of the source voltage with respect to the current? Does the source voltage lag or lead the current?
(c) What are the voltage amplitudes across the resistor, inductor, and capacitor?

23. A 5.00 H inductor with negligible resistance is connected across an AC source. Voltage amplitude is kept constant at 60.0 V but whose frequency can be varied. Find the current amplitude when the angular frequency is
(a) 100 rad/s (b) 1000 rad/s
(c) 10000 rad/s

Answers

1. (a) **2.** (b) **3.** (a) **4.** (a) **5.** (b) **6.** (d) **7.** (d) **8.** (b) **9.** (c) **10.** (d)
11. (a) **12.** (a) **13.** (c) **14.** (a) **15.** (b) **16.** (c) **17.** (d) **18.** (b)
19. (a) 628 Ω (b) 6.374 mH (c) 1.59 kΩ (d) 1.59 mF **20.** 0.036 H, 111.8 V **21.** 7.7 H, 6 A
22. (a) 0.326 A (b) 35.3°, lagging (c) 97.8 V, 32.6 V, 102 V **23.** (a) 0.12 A (b) 1.2×10^{-2} A (c) 1.2×10^{-3} A

See hints on pages 615-616.

MODULE 3

Power in an AC Circuit

In case of a steady current, the rate of doing work is given by

$$P = Vi$$

In an alternating circuit, current and voltage both vary with time. Suppose in an AC, the voltage is leading the current by an angle ϕ. Then, we can write

$$V = V_0 \sin\omega t$$

and $$i = i_0 \sin(\omega t - \phi)$$

The instantaneous value of power in that case is

$$P = Vi = V_0 i_0 \sin\omega t \sin(\omega t - \phi)$$

or $$P = V_0 i_0 \left[\sin^2 \omega t \cos\phi - \frac{1}{2}\sin 2\omega t \sin\phi\right] \quad \text{...(i)}$$

Now, the average rate of doing work (power) in one cycle will be

$$\langle P \rangle_{\text{One cycle}} = \frac{\int_0^{T = 2\pi/\omega} P\,dt}{\int_0^{T = 2\pi/\omega} dt} \quad \text{...(ii)}$$

Substituting the value of P from Eq. (i) in Eq. (ii) and then integrating it with proper limits, we get

$$\langle P \rangle_{\text{One cycle}} = \frac{1}{2} V_0 i_0 \cos\phi = \frac{V_0}{\sqrt{2}} \cdot \frac{i_0}{\sqrt{2}} \cos\phi$$

or $$\boxed{\langle P \rangle_{\text{One cycle}} = V_{rms} i_{rms} \cos\phi}$$

Here, the term $\cos\phi$ is known as **power factor**.

It is said to be leading if current leads voltage, lagging if current lags voltage. Thus, a power factor of 0.5 lagging means current lags the voltage by 60° (as $\cos^{-1} 0.5 = 60°$).

The product of V_{rms} and i_{rms} gives the **apparent power**. While the **true power** is obtained by multiplying the apparent power by the power factor $\cos\phi$. Thus,

Apparent power $= V_{rms} \times i_{rms}$

and True power = apparent power × power factor

For $\phi = 0°$, the current and voltage are in phase. The power is thus, maximum ($= V_{rms} \times i_{rms}$). For $\phi = 90°$, the power is zero. The current is then stated **wattless**. Such a case will arise when resistance in the circuit is zero. The circuit is purely inductive or capacitive.

Conditions	Phase angle	Power factor
$R = 0$	90°	0
$X_C = X_L \neq 0 \;\; R \neq 0$	0°	1
$X_C = X_L = 0 \;\; R \neq 0$	0°	1
$\omega = \omega_r$	0°	1

Important POINTS

Let us consider a choke coil (used in tube lights) of large inductance L and low resistance R. The power factor for such a coil is given by

$$\cos\phi = \frac{R}{Z} = \frac{R}{\sqrt{R^2 + \omega^2 L^2}} \approx \frac{R}{\omega L} \quad (\text{as } R << \omega L)$$

As $R << \omega L$, $\cos\phi$ is very small. Thus, the power absorbed by the coil $V_{rms}\, i_{rms} \cos\phi$ is very small. On account of its large impedance $Z = \sqrt{R^2 + \omega^2 L^2}$, the current passing through the coil is very small. Such a coil is used in AC circuits for the purpose of adjusting current to any required value without waste of energy. The only loss of energy is due to hysteresis in the iron core, which is much less than the loss of energy in the resistance that can also reduce the current instead of a choke coil.

Example 1 A 750 Hz, 20V source is connected to a resistance of 100 Ω, an inductance of 0.1803 H and a capacitance of 10 μF all in series. Calculate the time in which the resistance (thermal capacity 2J/°C) will get heated by 10°C.

Sol. The impedance of the circuit,

$$Z = \sqrt{R^2 + (X_L - X_C)^2}$$

$$= \sqrt{R^2 + \left\{(2\pi f L) - \frac{1}{(2\pi f C)}\right\}^2}$$

$$= \sqrt{(100)^2 + \left\{(2 \times 3.14 \times 750 \times 0.1803) - \frac{1}{(2 \times 3.14 \times 750 \times 10^{-5})}\right\}^2} = 834\,\Omega$$

In case of an AC,

$$\langle P \rangle = V_{rms} i_{rms} \cos\phi = (V_{rms})\left(\frac{V_{rms}}{Z}\right)\left(\frac{R}{Z}\right) = \left(\frac{V_{rms}}{Z}\right)^2 R$$

$$=\left(\frac{20}{834}\right)^2 \times 100 = 0.0575\ \text{J/s}$$

Now, $\langle P\rangle \times t = S\Delta\theta$

Here, S = thermal capacity

$$\therefore \quad t = \frac{S\Delta\theta}{\langle P\rangle} = \frac{2\times 10}{0.0575} = 348\ \text{s} \qquad \textbf{Ans.}$$

Note *In AC, the whole energy or power is consumed by resistance.*

Example 2 In an *L-C-R* series circuit, $R = 150\,\Omega$, $L = 0.750$ H and $C = 0.0180\,\mu$F. The source has voltage amplitude $V = 150$ V and a frequency equal to the resonance frequency of the circuit.

(a) What is the power factor?

(b) What is the average power delivered by the source?

(c) The capacitor is replaced by one with $C = 0.0360\,\mu$F and the source frequency is adjusted to the new resonance value. Then, what is the average power delivered by the source?

Sol. (a) At resonance frequency, $X_L = X_C$, $Z = R$ and power factor $\cos\phi = \frac{R}{Z} = 1.0$

(b) $P = V_{rms} I_{rms} \cos\phi$

$$= V_{rms}\left(\frac{V_{rms}}{2}\right)(1) = \frac{V_{rms}^2}{R} = \frac{(150/\sqrt{2})^2}{150}$$

$$= 75\ \text{W}$$

(c) Again, $P = \frac{V_{rms}^2}{R} = 75$ W

INTRODUCTORY Exercise

1. If a 0.03 H inductor, a $10\,\Omega$ resistor and a $2\,\mu$F capacitor are connected in series. At what frequency will they resonate? What will be the phase angle at resonance?
2. An arc lamp consumes 10 A at 40 V. Calculate the power factor when it is connected with a suitable value of choke coil required to run the arc lamp on AC mains of 200 V (rms) and 50 Hz.

Answers

1. 650 Hz, 0° **2.** 0.2

See hints on page 616.

Important POINTS

1. In an series *L-C-R* circuit,

(i) Capacitive reactance, $X_C = \frac{1}{\omega C}$

(ii) Inductive reactance, $X_L = \omega L$

(iii) Impedance, $Z = \sqrt{R^2 + (X_C - X_L)^2}$

(iv) If $X_C > X_L$, current leads and if $X_L > X_C$, voltage leads by an angle ϕ given by

$$\cos\phi = \frac{R}{Z} \quad \text{and} \quad \tan\phi = \frac{X_C \sim X_L}{R}$$

(v) Instantaneous power = instantaneous current × instantaneous voltage

(vi) Average power $= V_{rms}\, i_{rms} \cos\phi$, where

$$\cos\phi = \frac{R}{Z} = \text{power factor.}$$

Note *Power is also equal to $P = i_{rms}^2 R$, but this is not equal to $P \neq \frac{V_{rms}^2}{R}$, this is because $V_{rms} = i_{rms} Z$ and $\cos\phi = \frac{R}{Z}$. If we substitute in $P = V_{rms}\, i_{rms} \cos\phi$, then we get the first relation but not the second one. This implies that power is consumed only across resistance. If $Z = R$ then we can also write $P = \frac{V_{rms}^2}{R}$.*

(vii) $i_0 = \frac{V_0}{Z}$ or $i_{rms} = \frac{V_{rms}}{Z}$

(viii) $(V_C)_{rms} = (i_{rms}) X_C$, $(V_L)_{rms} = (i_{rms}) X_L$ and $(V_R)_{rms} = (i_{rms}) R$

(ix) $V = \sqrt{V_R^2 + (V_C \sim V_L)^2}$

Here, V is the rms value of applied voltage

V_R is the rms value of voltage across resistance.

V_C across capacitor and V_L across inductor etc.

2. $\omega = \omega_r = \frac{1}{\sqrt{LC}}$ is called resonance frequency.

3. At $\omega = \omega_r$,

(i) $X_L = X_C$

(ii) Z = minimum value $= R$

(iii) i_{rms} is maximum which is equal to $\frac{V_{rms}}{Z_{min}}$ or $\frac{V_{rms}}{R}$

(iv) i_0 is also maximum which is equal to $\frac{V_0}{Z_{min}}$ or $\frac{V_0}{R}$

(v) Power factor $\cos\phi = 1$

4. In one complete cycle, power is consumed only by resistance. No power is consumed by a capacitor or an inductor.

5. Wattless current : Power in AC is given by :

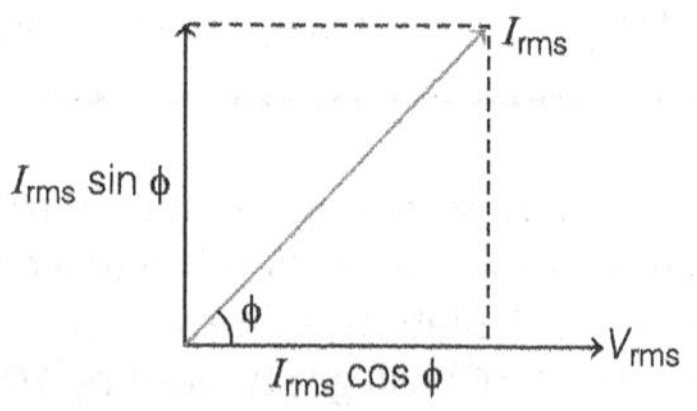

$$P = V_{rms} I_{rms} \cos\phi$$

Phase angle, ϕ is the angle between V_{rms} and I_{rms}.

So I_{rms} can be resolved to two components, $I_{rms}\cos\phi$ and $I_{rms}\sin\phi$ as shown.

$I_{rms}\cos\phi$ is along the direction of V_{rms}.

But, $I_{rms} \sin \phi$ is perpendicular to V_{rms} and so does not produce any power in the circuit.

Therefore $I_{rms} \sin \phi$ is the wattless component of current.

6. **Transformer** In different situations we need different potential difference. Transformer is a device which is either used to increase or decrease the voltage in AC circuits through mutual induction. A transformer consists of two coils wound on the same core.

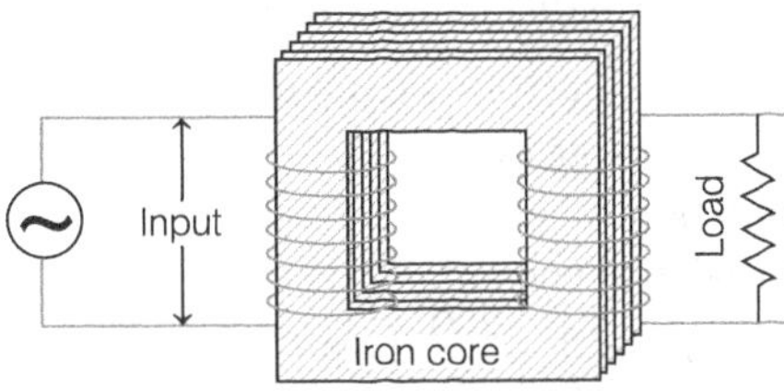

The coil connected to input is called primary while the other connected to output is called secondary coil. An alternating current passing through the primary creates a continuously changing flux through the core. This changing flux induces an alternating emf in the secondary.

As magnetic lines of force are closed curves, so rate of change of magnetic flux $\frac{d\phi}{dt}$ may be assumed same on both sides. Therefore,

$$e_s = -N_s \frac{d\phi}{dt} \quad \text{...(i)}$$

$$e_p = -N_p \frac{d\phi}{dt} \quad \text{...(ii)}$$

Dividing Eq. (i) by Eq. (ii), we get

$$\frac{e_s}{e_p} = \frac{N_s}{N_p}$$

In an ideal transformer, there is no loss of power. Hence,

$$ei = \text{constant}$$

$$\therefore \quad \frac{e_S}{e_P} = \frac{N_S}{N_P} = \frac{i_P}{i_S}$$

The following points are noteworthy in transformers :

(i) In **step-up** transformer, $N_S > N_P$. It increases voltage and reduces current

(ii) In **step-down** transformer, $N_P > N_S$. It increases current and reduces voltage

(iii) It works only on AC

(iv) A transformer cannot increase (or decrease) voltage and current simultaneously.

(v) Some power is always lost due to eddy currents, hysteresis, etc.

(vi) If efficiency of a transformer is 100% then,

$$\text{Power output} = \text{Power input}$$

$$\Rightarrow \quad e_s i_s = e_p i_p$$

$$\Rightarrow \quad \frac{e_s}{e_p} = \frac{i_p}{i_s} = \frac{N_s}{N_p}$$

If efficiency is 90%. Then,

Power output = 90% of power input

$$\Rightarrow \quad e_s i_s = 0.9\, e_p i_p$$

$$\Rightarrow \quad \frac{e_s}{e_p} = \frac{0.9 i_p}{i_s} = \frac{N_s}{N_p}$$

Typed Problems

Type 1. *Based on a real inductor.*

Concept

An ideal inductor wire has zero resistance ($R = 0$). So, in an AC, only X_L is the impedance. But, a real inductor has some resistance also. In DC, total resistance is only R. In AC, total resistance called impedance is $Z = \sqrt{R^2 + X_L^2}$

Therefore,

$$I_{DC} = \frac{V_{DC}}{R} \quad \text{and} \quad I_{AC} = \frac{V_{AC}}{Z} = \frac{V_{AC}}{\sqrt{R^2 + X_L^2}}$$

If $\quad V_{AC} = V_{DC}, \quad$ then $\quad I_{AC} < I_{DC}$

Ex. 1 A current of 4 A flows in a coil when connected to a 12V DC source. If the same coil is connected to a 12 V, 50 rad/s AC source, a current of 2.4 A flows in the circuit. Determine the inductance of the coil. Also, find the power developed in the circuit if a 2500 μF capacitor is connected in series with the coil.

Sol. (i) A coil consists of an inductance (L) and a resistance (R).

In DC, only resistance is effective. Hence,

$$R = \frac{V}{i} = \frac{12}{4} = 3\,\Omega$$

In AC, $\quad i_{rms} = \frac{V_{rms}}{Z} = \frac{V_{rms}}{\sqrt{R^2 + \omega^2 L^2}}$

$$\therefore \quad L^2 = \frac{1}{\omega^2}\left[\left(\frac{V_{rms}}{i_{rms}}\right)^2 - R^2\right]$$

$$\therefore \quad L = \frac{1}{\omega}\sqrt{\left(\frac{V_{rms}}{i_{rms}}\right)^2 - R^2}$$

Substituting the values, we have

$$L = \frac{1}{50}\sqrt{\left(\frac{12}{2.4}\right)^2 - (3)^2} = 0.08 \text{ H}$$ **Ans.**

(ii) When capacitor is connected to the circuit, the impedance is

$$Z = \sqrt{R^2 + (X_L - X_C)^2}$$

Here, $R = 3\,\Omega$

$$X_L = \omega L = (50)(0.08) = 4\,\Omega$$

and $X_C = \dfrac{1}{\omega C} = \dfrac{1}{(50)(2500 \times 10^{-6})} = 8\,\Omega$

$\therefore$ $Z = \sqrt{(3)^2 + (4-8)^2} = 5\,\Omega$

Now, $\langle P \rangle = V_{rms} i_{rms} \cos\phi$

$$= V_{rms} \times \frac{V_{rms}}{Z} \times \frac{R}{Z}$$

$$= \left(\frac{V_{rms}}{Z}\right)^2 \times R$$

Substituting the values, we have

$$\langle P \rangle = \left(\frac{12}{5}\right)^2 \times 3 = 17.28\text{ W}$$ **Ans.**

Type 2. *Different time functions in AC.*

Concept

In an *L-C-R* series circuit, there are total five functions of time, V, I, V_R, V_C and V_L. Now, the following points are important in these functions.

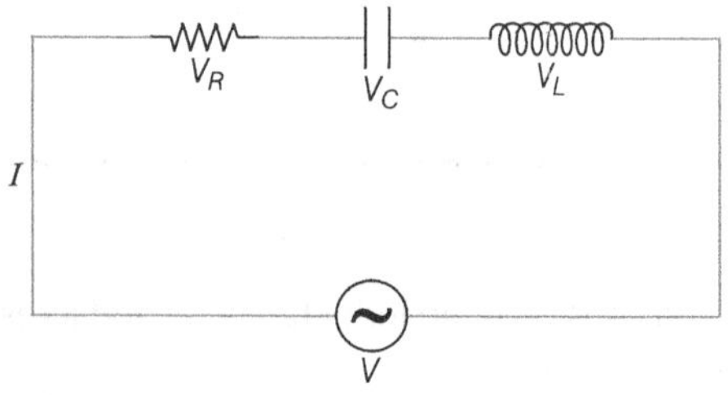

(i) V and I have a phase difference of ϕ where
$$0° \le \phi \le 90°$$
(ii) V_R and I are in same phase
(iii) V_C lags behind I by 90°
(iv) V_L leads I by 90°
(v) The functions, $V = V_R + V_C + V_L$ (all the time)

Ex. 2 In the diagram shown in figure, V function is given.

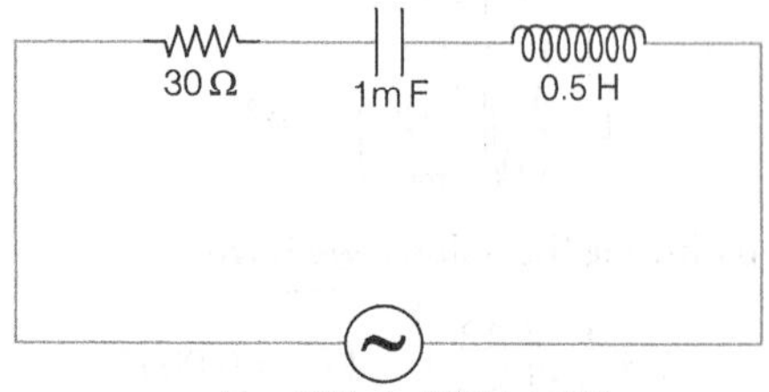

$V = 200 \sin (100\,t + 30°)$

Find other four functions of time I, V_C, V_R and V_L. Also, find power consumed in the circuit, V is given in volts and ω in rad/s.

Sol. Given, $\omega = 100$ rad/s

$$X_L = \omega L = 50\,\Omega$$

$$X_C = \frac{1}{\omega C} = \frac{1}{100 \times 10^{-3}} = 10\,\Omega$$

$$Z = \sqrt{R^2 + (X_L - X_C)^2}$$

$$= \sqrt{(30)^2 + (50-10)^2}$$

$$= 50\,\Omega$$

Current function

Maximum value of current,

$$I_0 = \frac{V_0}{Z} = \frac{200}{50} = 4\text{ A}$$

$X_L > X_C$, therefore voltage leads the current by a phase difference ϕ, where

$$\cos\phi = \frac{R}{Z} = \frac{30}{50} = \frac{3}{5}$$

or $\phi = 53°$

$\therefore$ $I = 4 \sin(100t + 30° - 53°)$

or $I = 4 \sin(100\,t - 23°)$ **Ans.**

V_R, V_C and V_L functions

Maximum value of $V_R = I_0 R = 4 \times 30 = 120$ volt, V_R and I are in same phase.

Therefore,

$$V_R = 120 \sin(100\,t - 23°)$$ **Ans.**

Maximum value of $V_C = I_0 X_C$

$$= 4 \times 10 = 40\text{ volt}$$

Now, V_C function lags the current function by 90°.

Therefore,

$$V_C = 40 \sin(100\,t - 23° - 90°)$$

or $V_C = 40 \sin(100\,t - 113°)$ **Ans.**

Maximum value of $V_L = I_0 X_L = 4 \times 50 = 200$ volt.

V_L function leads the current function by 90°. Therefore,

$$V_L = 200 \sin(100\,t - 23° + 90°)$$

or $V_L = 200 \sin(100\,t + 67°)$ **Ans.**

Note *We can check at any time that,*

$$V = V_R + V_L + V_C$$ *(functions)*

Power

Power is consumed in an AC circuit only across a resistance and this power is given by

$$P = V_{rms} I_{rms} \cos\phi$$

$$= I_{rms}^2 R$$

Let us use the first formula,

$$P = \left(\frac{200}{\sqrt{2}}\right)\left(\frac{4}{\sqrt{2}}\right)\left(\frac{3}{5}\right)$$

$$= 240\text{ watt}$$ **Ans.**

Type 3. *Parallel circuits.*

Concept

Two or more than two sine or cosine functions of same ω can be added by vector method. Actually, their amplitudes are added by vectors method.

Ex. 3 In the circuit shown in figure,

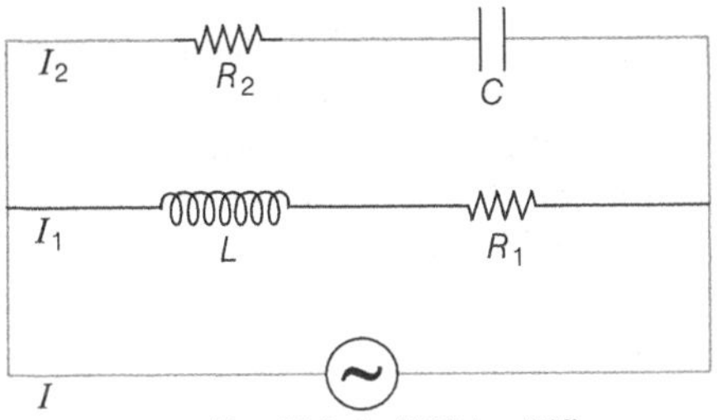

$V = 200 \sin(100t + 30°)$

$R_1 = 30\,\Omega, R_2 = 40\,\Omega, L = 0.4$ H and $C = \frac{1}{3}$ mF.

Find seven functions of time $I, I_1, I_2, V_{R_1}, V_L, V_{R_2}$ and V_C. Also, find total power consumed in the circuit. In the given potential function, V is in volts and ω in rad/s.

Sol. Circuit 1 (containing L and R_1)

I_1 : $X_L = \omega L = 100 \times 0.4 = 40\,\Omega$

$R_1 = 30\,\Omega$

$\therefore \quad Z_1 = \sqrt{R_1^2 + X_L^2}$

$= \sqrt{(30)^2 + (40)^2}$

$= 50\,\Omega$

Maximum value of current, $I_1 = (I_1)_0 = \frac{V_0}{Z_1} = \frac{200}{50} = 4$ A

Since, there is only X_L, so voltage function will lead the current function by an angle ϕ_1, where

$$\cos\phi_1 = \frac{R_1}{Z_1} = \frac{30}{50} = \frac{3}{5}$$

$\therefore \quad \phi_1 = 53°$

$\therefore \quad I_1 = 4\sin(100t + 30° - 53°)$

or $\quad I_1 = 4\sin(100t - 23°)$ **Ans.**

V_{R_1} : V_{R_1} function is in phase with I_1 function.

Maximum value of V_{R_1} = (maximum value of I_1) (R_1)

$= (4)(30) = 120$ volt

$\therefore \quad V_{R_1} = 120\sin(100t - 23°)$ **Ans.**

V_L : V_L function is 90° ahead of I_1 function.

Maximum value of V_L = (maximum value of I_1)(X_L)

$= (4)(40) = 160$ volt

$\therefore \quad V_L = 160\sin(100t - 23° + 90°)$

or $\quad V_L = 160\sin(100t + 67°)$ **Ans.**

Power In this circuit, power will be consumed only across R_1. This power is given by

$$P_{R_1} = (\text{rms value of } I_1)^2 R_1$$

$$= \left(\frac{4}{\sqrt{2}}\right)^2 (30)$$

$= 240$ watt

Circuit 2 (containing C and R_2)

$$I_2 : X_C = \frac{1}{\omega C} = \frac{1}{100 \times \frac{1}{3} \times 10^{-3}} = 30\,\Omega$$

$R_2 = 40\,\Omega$

$\therefore \quad Z_2 = \sqrt{R_2^2 + X_C^2}$

$= \sqrt{(40)^2 + (30)^2} = 50\,\Omega$

Maximum value of $I_2 = (I_2)_0 = \frac{V_0}{Z_2} = \frac{200}{50} = 4$ A

Since, there is only X_C. So I_2 function will lead the V function by an angle ϕ_2, where

$$\cos\phi_2 = \frac{R_2}{Z_2} = \frac{40}{50} = \frac{4}{5}$$

$\therefore \quad \phi_2 = 37°$

$\therefore \quad I_2 = 4\sin(100t + 30° + 37°)$

$= 4\sin(100t + 67°)$ **Ans.**

V_{R_2} : V_{R_2} function is in phase with I_2 function.

Maximum value of V_{R_2} = (maximum value of I_2) (R_2)

$= 4 \times 40 = 160$ volt

$\therefore \quad V_{R_2} = 160\sin(100t + 67°)$ **Ans.**

V_C : V_C function lags I_2 function by 90°

Maximum value of V_C = (Maximum value of I_2)(X_C)

$= 4 \times 30 = 120$ volt

$\therefore \quad V_C = 120\sin(100t + 67° - 90°)$

or $\quad V_C = 120\sin(100t - 23°)$ **Ans.**

Power In this circuit, power will be consumed only across R_2 and this power is given by

$$P_{R_2} = (\text{rms value of } I_2)^2 R_2$$

$$= \left(\frac{4}{\sqrt{2}}\right)^2 (40) = 320 \text{ W}$$

$\therefore$ Total power consumed in the circuit,

$P = P_{R_1} + P_{R_2}$

$= (240 + 320)$ W

$= 560$ W **Ans.**

I : $\quad I = I_1 + I_2$

$I = 4\sin(100t - 23°) + 4\sin(100t + 67°)$

Now, the amplitudes can be added by vector method.

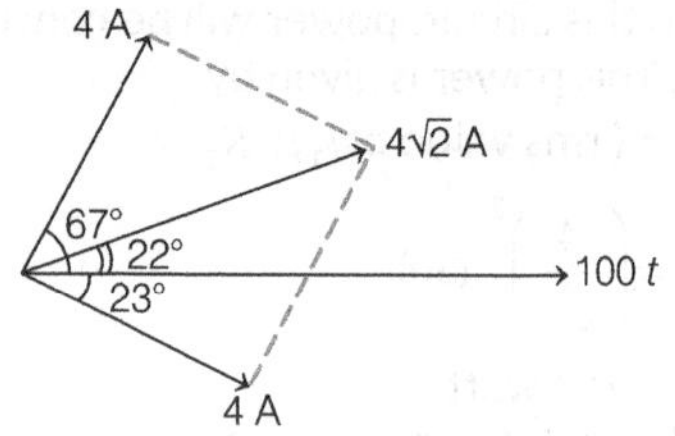

Resultant of 4A and 4A at 90° is $4\sqrt{2}$ A at 45° from both currents or at 22° from 100t line.

$\therefore \quad I = 4\sqrt{2}\sin(100\,t + 22°)$ **Ans.**

Type 4. *Based on transformers and wattless current.*

Ex. 4 A transformer consisting of 300 turns in the primary and 150 turns in the secondary gives output power of 2.2 kW. If the current in the secondary coil is 10 A, then the input voltage and current in the primary coil are **(2019 Main)**

(a) 440 V and 5 A (b) 220 V and 20 A
(c) 220 V and 10 A (d) 440 V and 20 A

Sol. Given, Number of turns in primary, $N_1 = 300$

Number of turns in secondary, $N_2 = 150$

Output power, $P_2 = 2.2\text{ kW} = 2.2 \times 10^3$ W

Current in secondary coil, $I_2 = 10$ A

Output power, $P_2 = I_2 V_2$

$$\Rightarrow \quad V_2 = \frac{P_2}{I_2} = \frac{2.2 \times 10^3}{10} = 220\text{V} \quad \text{... (i)}$$

We know that, $\dfrac{N_1}{N_2} = \dfrac{\text{Input voltage}}{\text{Output voltage}} = \dfrac{V_1}{V_2}$

$$\Rightarrow \quad V_1 = \left(\frac{N_1}{N_2}\right) V_2$$

$$\Rightarrow \quad V_1 = \left(\frac{300}{150}\right) \times (220\text{ V}) \quad \text{[using Eq. (i)]}$$

$$V_1 = 440\text{ V} \quad \text{... (ii)}$$

Again, $\dfrac{V_1}{V_2} = \dfrac{I_2}{I_1} \Rightarrow I_1 = \left(\dfrac{V_2}{V_1}\right) I_2$

$$= \frac{220}{440} \times 10 \quad \text{[using Eqs. (i) and (ii)]}$$

$$\Rightarrow \quad I_1 = 5\text{ A}$$

Hence, the correct option is (a).

Note *Here efficiency of transformer is assumed* 100 %.

Ex. 5 A power transmission line feeds input power at 2300V to a step-down transformer with its primary windings having 4000 turns. The output power is delivered at 230 V by the transformer. If the current in the primary of the transformer is 5A and its efficiency is 90%, the output current would be **(2019 Main)**

(a) 45 A (b) 50 A (c) 25 A (d) 35 A

Sol. The efficiency of the transformer is

$$\eta = \frac{\text{Output (secondary) power}}{\text{Input (primary) power}}$$

$$\Rightarrow \quad \eta = \frac{V_S \cdot I_S}{V_P \cdot I_P} \times 100$$

$$\Rightarrow \quad \eta = \frac{(230)(I_S)}{(2300)(5)} \times 100$$

$$90 = \frac{230\,I_S}{(2300) \times 5} \times 100$$

$$\Rightarrow \quad I_S = 45\text{ A}$$

Hence, the correct option is (a).

Ex. 6 In an AC, phase difference between current and voltage is 45° wattless current is 4A and the applied voltage is 220 V. Find average power consumed in the circuit in one cycle.

Sol. Wattless current is given by $I_{rms}\sin\phi$

Since, $\phi = 45°$

$\therefore \quad I_{rms}\cos\phi = I_{rms}\sin\phi = 4\text{A}$ (given)

Now, $\quad p_{av} = V_{rms}(I_{rms}\cos\phi)$

Substituting the values, we get

$p_{av} = 220 \times 4 = 880$ watt **Ans.**

Miscellaneous Examples

Ex. 7 An AC circuit consists of a 220 Ω resistance and a 0.7H choke. Find the power absorbed from 220 V and 50 Hz source connected in this circuit if the resistance and choke are joined

(a) in series (b) in parallel.

Sol. (a) In series, the impedance of the circuit is

$$Z=\sqrt{R^2+\omega^2L^2}=\sqrt{R^2+(2\pi fL)^2}$$
$$=\sqrt{(220)^2+(2\times3.14\times50\times0.7)^2}$$
$$=311\,\Omega$$

$\therefore \quad i_{rms}=\frac{V_{rms}}{Z}=\frac{220}{311}=0.707$ A

and $\quad \cos\phi=\frac{R}{Z}=\frac{220}{311}=0.707$

$\therefore$ The power absorbed in the circuit,

$$P=V_{rms}i_{rms}\cos\phi=(220)(0.707)(0.707)$$
$$=110.08 \text{ W}$$ **Ans.**

(b) When the resistance and choke are in parallel, the power is absorbed in resistance circuit only.

$\therefore \quad P=\frac{V_{rms}^2}{R}=\frac{(220)^2}{220}=220$ W **Ans.**

Ex. 8 A sinusoidal voltage of frequency 60 Hz and peak value 150 V is applied to a series *L-R* circuit, where $R=20\,\Omega$ and $L=40$ mH.

(a) Compute T, ω, X_L, Z and ϕ

(b) Compute the amplitudes of current, V_R and V_L

Sol. (a) $T=\frac{1}{f}=\frac{1}{60}$ s **Ans.**

$\omega=2\pi f=(2\pi)(60)=377$ rad/s **Ans.**

$X_L=\omega L=(377)(0.040)$

$=15.08\,\Omega$ **Ans.**

$Z=\sqrt{X_L^2+R^2}=\sqrt{(15.08)^2+(20)^2}$

$=25.05\,\Omega$ **Ans.**

$\phi=\tan^{-1}\left(\frac{X_L}{R}\right)=\tan^{-1}\left(\frac{15.08}{20}\right)$

$=\tan^{-1}(0.754)$

$=37°$ **Ans.**

(b) Amplitudes (maximum value) are

$i_0=\frac{V_0}{Z}=\frac{150}{25.05}\approx 6$ A **Ans.**

$(V_0)_R=i_0R=(6)(20)=120$ V **Ans.**

$(V_0)_L=i_0X_L$

$=(6)(15.08)=90.5$ V **Ans.**

Note $V_0=\sqrt{(V_0)_R^2+(V_0)_L^2}$

Ex. 9 For the circuit shown in figure, find the current functions through each element.

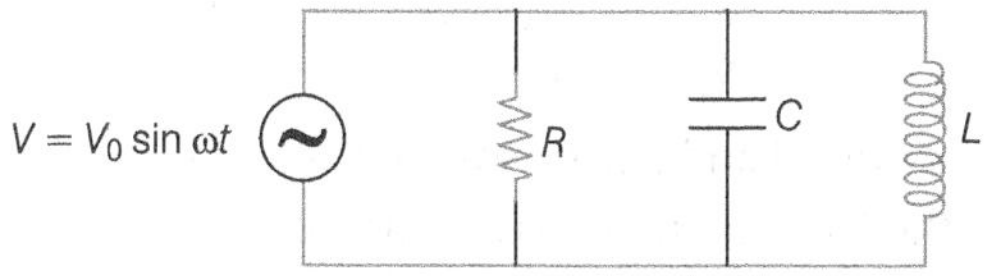

Sol. $\quad i_R=\frac{V_0}{R}\sin\omega t\equiv(i_0)_R\sin\omega t$

$i_L=\frac{V_0}{\omega L}\sin(\omega t-90°)=-\frac{V_0}{\omega L}\cos\omega t$

$=-\frac{V_0}{X_L}\cos\omega t=-(i_0)_L\cos\omega t$

and $\quad i_C=V_0\omega C\sin(\omega t+90°)=V_0\omega C\cos\omega t$

$=\frac{V_0}{X_C}\cos\omega t=(i_0)_C\cos\omega t$

where, the reactances X_L and X_C are as defined.

Ex. 10 In the above problem find the total instantaneous current through the source, and find expressions for phase angle of this current and the impedance of the circuit.

Sol. For the total current, we have

$$i=i_R+i_L+i_C=V_0\left[\frac{1}{R}\sin\omega t+\left(\frac{1}{X_C}-\frac{1}{X_L}\right)\cos\omega t\right]$$

Using the trigonometric identity,

$$A\sin\theta+B\cos\theta=\sqrt{A^2+B^2}\sin(\theta+\phi)$$

where, $\phi=\tan^{-1}(B/A)$

We can write, $i=i_0\sin(\omega t+\phi)$

Here, $\quad i_0=\frac{V_0}{Z}$

where, $\quad \frac{1}{Z}=\sqrt{\left(\frac{1}{R}\right)^2+\left(\frac{1}{X_C}-\frac{1}{X_L}\right)^2}$

and $\quad \tan\phi=\frac{\left(\frac{1}{X_C}-\frac{1}{X_L}\right)}{(1/R)}$

Ex. 11 An *L-C-R* series circuit with 100 Ω resistance is connected to an AC source of 200 V and angular frequency 300 rad/s. When only the capacitance is removed, the current lags behind the voltage by 60°. When only the inductance is removed, the current leads the voltage by 60°. Calculate the current and the power dissipated in the *L-C-R* circuit.

Sol. When capacitance is removed, then

$\tan\phi=\frac{X_L}{R}$ or $\tan 60°=\frac{X_L}{R}$

$\therefore \quad X_L=\sqrt{3}R$ ...(i)

When inductance is removed, then

$$\tan\phi = \frac{X_C}{R} \quad \text{or} \quad \tan 60° = \frac{X_C}{R}$$

$$\therefore \quad X_C = \sqrt{3}\,R \qquad \text{...(ii)}$$

From Eqs. (i) and (ii), we see that $X_C = X_L$

So, the *L-C-R* circuit is in resonance.

Hence, $Z = R$

$$\therefore \quad i_{rms} = \frac{V_{rms}}{Z} = \frac{200}{100} = 2\text{ A}$$ **Ans.**

$$\langle P \rangle = V_{rms}\, i_{rms} \cos\phi$$

At resonance current and voltage are in phase,

or $\phi = 0°$

$$\therefore \quad \langle P \rangle = (200)(2)(1) = 400\text{ W}$$ **Ans.**

Ex. 12 A series *L-C-R* circuit containing a resistance of $120\,\Omega$ has resonance frequency 4×10^5 rad/s. At resonance the voltages across resistance and inductance are 60 V and 40 V, respectively. Find the values of *L* and *C*. At what angular frequency the current in the circuit lags the voltage by $\pi/4$?

Sol. At resonance, $X_L - X_C = 0$ and $Z = R = 120\,\Omega$

$$\therefore \quad i_{rms} = \frac{(V_R)_{rms}}{R} = \frac{60}{120} = \frac{1}{2}\text{ A}$$

Also, $$i_{rms} = \frac{(V_L)_{rms}}{\omega L}$$

$$\therefore \quad L = \frac{(V_L)_{rms}}{\omega i_{rms}} = \frac{40}{(4 \times 10^5)\left(\frac{1}{2}\right)}$$

$$= 2.0 \times 10^{-4}\text{ H}$$

$$= 0.2\text{ mH}$$ **Ans.**

The resonance frequency is given by

$$\omega = \frac{1}{\sqrt{LC}} \quad \text{or} \quad C = \frac{1}{\omega^2 L}$$

Substituting the values, we have

$$C = \frac{1}{(4 \times 10^5)^2 (2.0 \times 10^{-4})}$$

$$= 3.125 \times 10^{-8}\text{ F}$$ **Ans.**

Current lags the voltage by 45°, when

$$\tan 45° = \frac{\omega L - \frac{1}{\omega C}}{R}$$

Substituting the values of *L*, *C*, *R* and tan 45°, we get

$$\omega = 8 \times 10^5\text{ rad/s}$$ **Ans.**

Ex. 13 A choke coil is needed to operate an arc lamp at 160V (rms) and 50 Hz. The lamp has an effective resistance of $5\,\Omega$ when running at 10A (rms). Calculate the inductance of the choke coil. If the same arc lamp is to be operated on 160V (DC), what additional resistance is required for the same current.

Sol. For lamp,

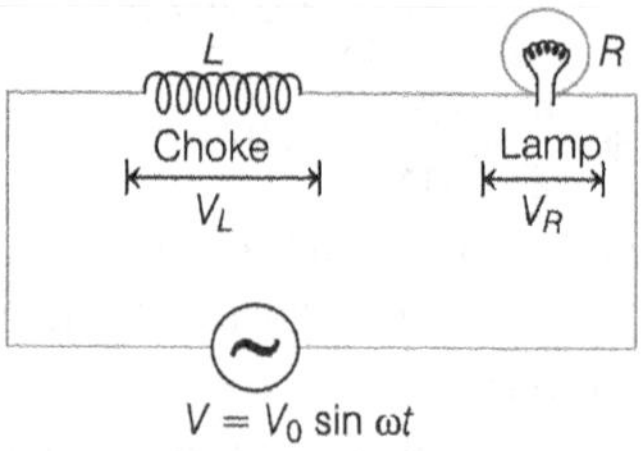

$$(V_{rms})_R = (i_{rms})(R) = 10 \times 5 = 50\text{ V}$$

In series, $$(V_{rms})^2 = (V_{rms})_R^2 + (V_{rms})_L^2$$

$$\therefore \quad (V_{rms})_L = \sqrt{(V_{rms})^2 - (V_{rms})_R^2}$$

$$= \sqrt{(160)^2 - (50)^2} = 152\text{ V}$$

As, $$(V_{rms})_L = (i_{rms})X_L = (i_{rms})(2\pi f L)$$

$$\therefore \quad L = \frac{(V_{rms})_L}{(2\pi f)(i_{rms})}$$

Substituting the values, we get

$$L = \frac{152}{(2\pi)(50)(10)} = 4.84 \times 10^{-2}\text{ H}$$ **Ans.**

Now, when the lamp is operated at 160 V, DC and instead of choke let an additional resistance R' is put in series with it, then

$$V = i(R + R')$$

or $$160 = 10(5 + R')$$

$$\therefore \quad R' = 11\,\Omega$$ **Ans.**

MODULE 3 Exercise

Assertion and Reason

See directions on page 25.

1. **Assertion :** At some given instant I_1 and I_2 both are 2 A each. Then, I at this instant should be zero.

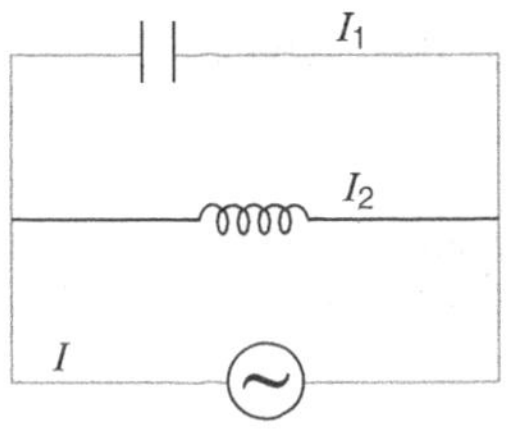

Reason : There is a phase difference of π between I_1 and I_2 functions.

Single Correct Option

2. If V_0 and I_0 are the peak voltage and current across the resistor in a series L-C-R circuit, then the power dissipated in the circuit is (Power factor $= \cos\theta$)
   (a) $\frac{V_0 I_0}{2}$ (b) $\frac{V_0 I_0}{\sqrt{2}}$
   (c) $V_0 I_0 \cos\theta$ (d) $\frac{V_0 I_0}{2}\cos\theta$

3. Voltage applied to an AC circuit and current flowing in it is given by

$$V = 200\sqrt{2}\sin\left(\omega t + \frac{\pi}{4}\right)$$

and $$i = -\sqrt{2}\cos\left(\omega t + \frac{\pi}{4}\right)$$

Then, power consumed in the circuit will be
   (a) 200 W (b) 400 W
   (c) $200\sqrt{2}$ W (d) None of these

4. In a transformer, the output current and voltage are respectively 4 A and 20 V. If the ratio of number of turns in the primary to secondary is 2 : 1, what is the input current and voltage?
   (a) 2 A and 40 V (b) 8 A and 10 V
   (c) 4 A and 10 V (d) 8 A and 40 V

Multiple Correct Options

5. In an AC circuit, the power factor
   (a) is unity when the circuit contains an ideal resistance only
   (b) is unity when the circuit contains an ideal inductance only
   (c) is zero when the circuit contains an ideal resistance only
   (d) is zero when the circuit contains an ideal inductance only

6. In an AC series circuit, $R = 10\,\Omega$, $X_L = 20\,\Omega$ and $X_C = 10\,\Omega$. Then, choose the correct options
   (a) Voltage function will lead the current function
   (b) Total impedance of the circuit is $10\sqrt{2}\,\Omega$
   (c) Phase angle between voltage function and current function is 45°
   (d) Power factor of circuit is $\frac{1}{\sqrt{2}}$

7. In the previous problem further choose the correct options.
   (a) The given values are at frequency less than the resonance frequency
   (b) The given values are at frequency more than the resonance frequency
   (c) If frequency is increased from the given value, impedance of the circuit will increase
   (d) If frequency is decreased from the given value, current in the circuit may increase or decrease

Match the Columns

8. Match the following two columns for a series AC circuit.

	Column I		Column II
(a)	Only C in the circuit	(p)	current will lead
(b)	Only L in the circuit	(q)	voltage will lead
(c)	Only R in the circuit	(r)	$\phi = 90°$
(d)	R and C in the circuit	(s)	$\phi = 0°$

9. For an L-C-R series AC circuit, match the following two columns.

	Column I		Column II
(a)	If resistance is increased	(p)	current will increase
(b)	If capacitance is increased	(q)	current will decrease
(c)	If inductance is increased	(r)	current may increase or decrease
(d)	If frequency is increased	(s)	power may decrease or increase

Subjective Questions

10. Voltage and current for a circuit with two elements in series are expressed as

$$V(t) = 170\sin(6280t + \pi/3)\ \text{volt}$$
$$i(t) = 8.5\sin(6280\,t + \pi/2)\ \text{amp}$$

   (a) Plot the two waveforms.
   (b) Determine the frequency in Hz.
   (c) Determine the power factor stating its nature.
   (d) What are the values of the elements?

11. An L-C-R series circuit with $L = 0.120$ H, $R = 240\ \Omega$, and $C = 7.30\ \mu F$ carries an rms current of 0.450 A with a frequency of 400 Hz.
(a) What are the phase angle and power factor for this circuit?
(b) What is the impedance of the circuit?
(c) What is the rms voltage of the source?
(d) What average power is delivered by the source?
(e) What is the average rate at which electrical energy is converted to thermal energy in the resistor?
(f) What is the average rate at which electrical energy is dissipated (converted to other forms) in the capacitor?
(g) In the inductor?

12. A coil is in series with a 20 μF capacitor across a 230 V, 50 Hz supply. The current taken by the circuit is 8 A and the power consumed is 200 W. Calculate the inductance of the coil if the current in the circuit is
(a) leading
(b) lagging

13. Two impedances Z_1 and Z_2 when connected separately across a 230V, 50 Hz supply consume 100 W and 60 W at power factor of 0.5 lagging and 0.6 leading respectively. If these impedances are now connected in series across the same supply, find
(a) total power absorbed and overall power factor
(b) the value of reactance to be added in series so as to raise the overall power factor to unity.

14. A wire, shaped as a semi-circle, of radius a rotates about an axis OO' with an angular velocity ω in a uniform magnetic field of induction B. The rotation axis is perpendicular to the field direction. The total resistance of the circuit is equal to R. Neglecting the magnetic field of the induced current, find the mean amount of thermal power being generated in the loop during a rotation period. **(Irodov 3.298)**

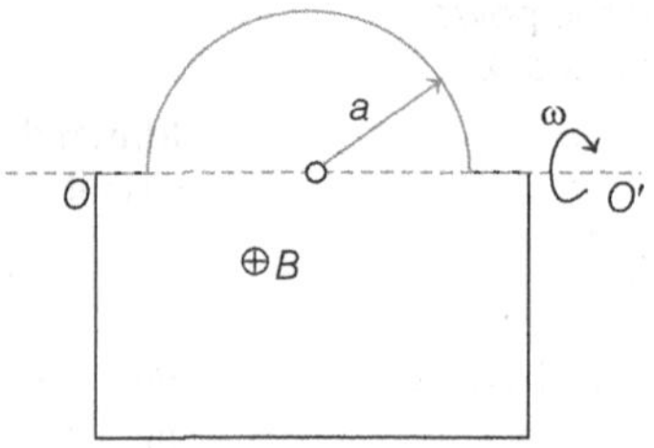

Answers

1. (a) 2. (d) 3. (d) 4. (a) 5. (a,d) 6. (a,b,c,d) 7. (b,c,d)
8. (a → p,r; b → q,r; c → s; d → p) 9. (a → q, s ; b → r,s ; c → r, s ; d → r, s)
10. (b) 1000 Hz (c) $\frac{\sqrt{3}}{2}$, leading (d) $R = 17.32\Omega$, $C = 15.92\mu F$
11. (a) 45.8°, voltage leads the current 0.697 (b) 343 Ω (c) 155 V (d) 48.6 W (e) 48.6 W (f) 0 (g) 0
12. (a) 0.416 H (b) 0.597 H 13. (a) 99 W, 0.92 leading (b) 194.2 Ω
14. $P_{mean} = \frac{(\pi\omega a^2 B)^2}{8R}$

See hints on pages 616-618.

Unit Exercise (Cumulative)

Level 1

Single Correct Option

1. If L and R be the inductance and resistance of the choke coil, then identify the correct statement.
   (a) L is very high as compared to R
   (b) R is very high as compared to L
   (c) Both L and R are high
   (d) Both L and R are low

2. In the circuit shown in figure, if both the bulbs B_1 and B_2 are identical, then

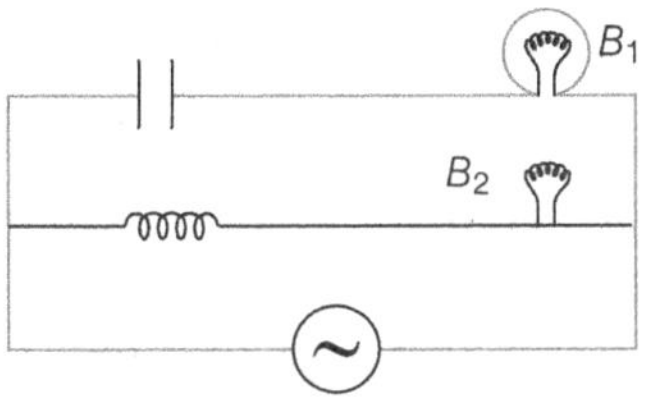

   (a) Both bulbs have equal brightness
   (b) B_2 will be brighter than B_1
   (c) As the frequency is increased, brightness of B_1 will increase and that of B_2 will decrease
   (d) As the frequency is decreased, the brightness of B_1 will increase and that of B_2 will decrease

3. The figure represents the voltage applied across a pure inductor. The diagram which correctly represents the variation of current i with time t is given by

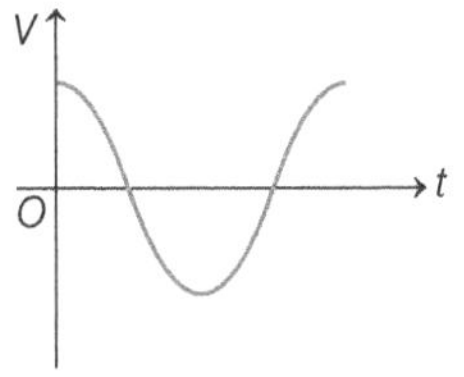

(a)
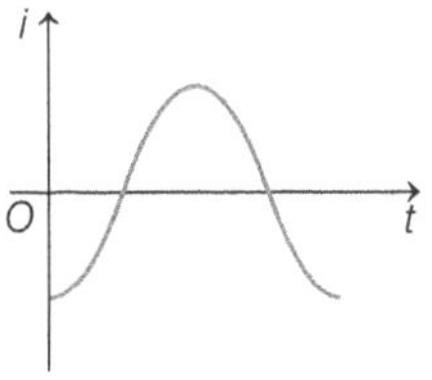

(b)
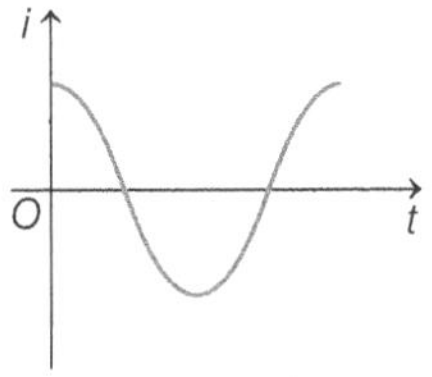

(c)
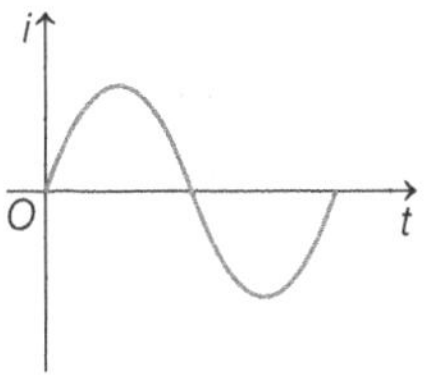

(d)
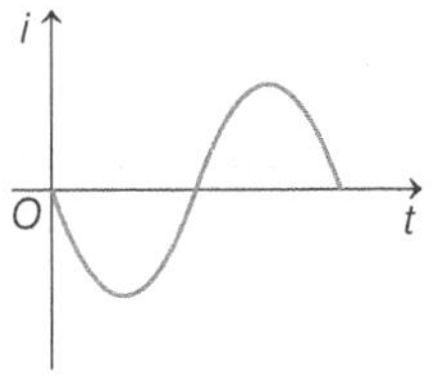

4. In the given circuit, the current drawn from the source is

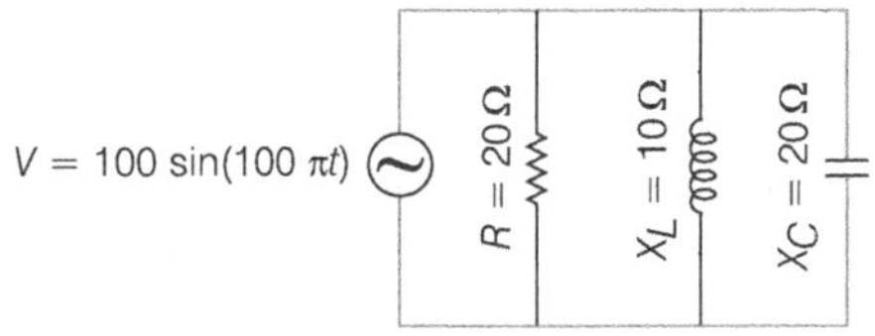

   (a) 20 A (b) 10 A (c) 5 A (d) $5\sqrt{2}$ A

5. An alternating emf is applied across a parallel combination of a resistance R, capacitance C and an inductance L. If I_R, I_L and I_C are the currents through R, L and C respectively, then the diagram which correctly represents, the phase relationship among I_R, I_L, I_C and source emf E, is given by

(a)
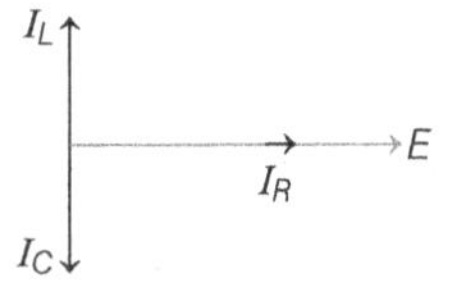

(b)
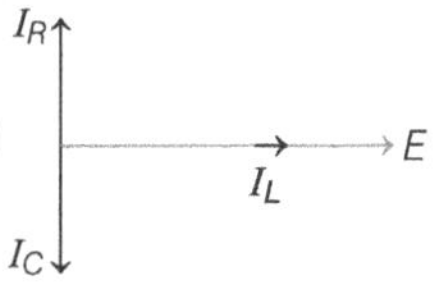

(c)
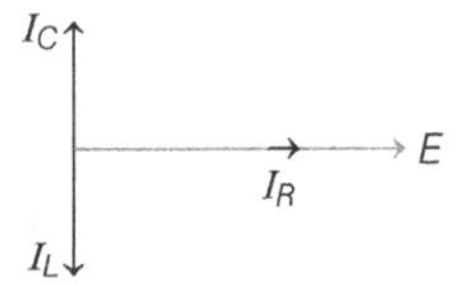

(d)
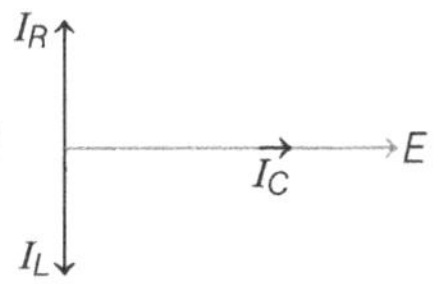

6. In a series L-C-R circuit, the voltage across resistance, capacitance and inductance is 10 V each. If the capacitance is short circuited, the voltage across the inductance will be
   (a) $\frac{10}{\sqrt{2}}$ V (b) 10 V
   (c) $10\sqrt{2}$ V (d) 20 V

7. In the circuit shown, rms current is 11 A. The potential difference across the inductor is

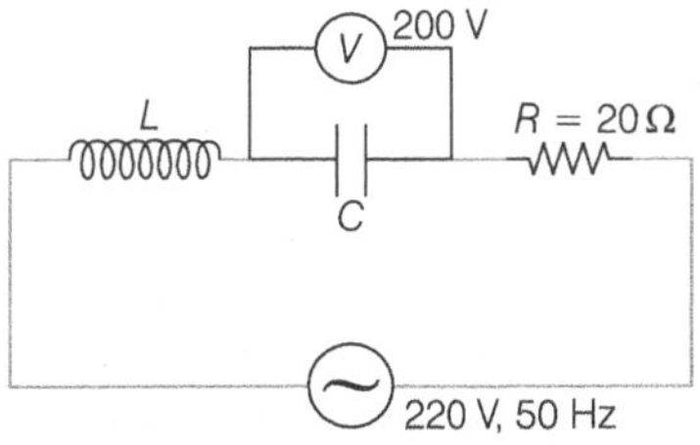

   (a) 220 V (b) 0 V
   (c) 300 V (d) 200 V

8. In figure, which voltmeter reads zero, when ω is equal to the resonant frequency of series L-C-R circuit?

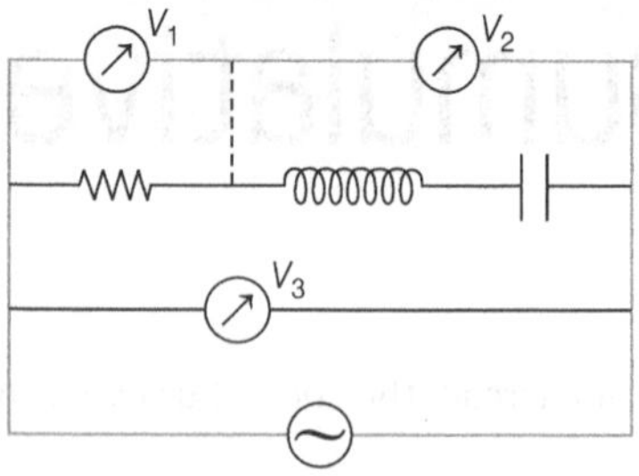

(a) V_1 (b) V_2
(c) V_3 (d) None of these

9. As per the given graph, choose the correct representation for curve A and curve B.

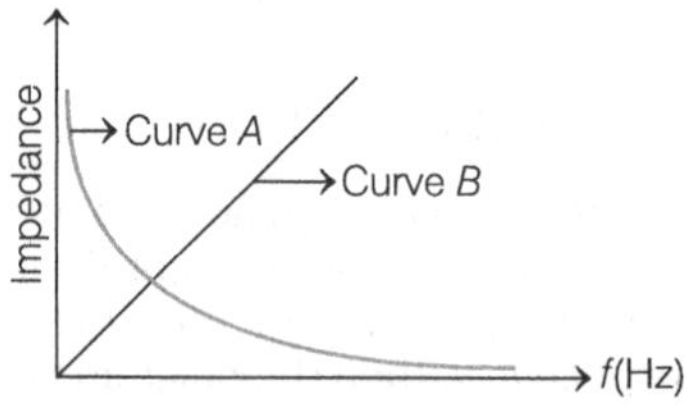

{Where, X_C = reactance of pure capacitive circuit connected with AC source
X_L = Reactance of pure inductive circuit connected with AC source
R = Impedance of pure resistive circuit connected with AC source.
Z = Impedance of the L-C-R series circuit} **(2023 Main)**

(a) $A = X_L, B = R$ (b) $A = X_L, B = Z$
(c) $A = X_C, B = X_L$ (d) $A = X_C, B = R$

10. In an AC circuit, an inductor, a capacitor and a resistor are connected in series with $X_L = R = X_C$. Impedance of this circuit is **(2021 Main)**

(a) $2R^2$ (b) zero (c) R (d) $R\sqrt{2}$

11. For an R-L-C circuit driven with voltage of amplitude v_m and frequency $\omega_0 = \frac{1}{\sqrt{LC}}$, the current exhibits resonance. The quality factor, Q is given by **(2018 Main)**

(a) $\frac{CR}{\omega_0}$ (b) $\frac{\omega_0 L}{R}$ (c) $\frac{\omega_0 R}{L}$ (d) $\frac{R}{\omega_0 C}$

12. A circuit connected to an AC source of emf $e = e_0 \sin(100t)$ with t in seconds, gives a phase difference of $\frac{\pi}{4}$ between the emf e and current i.
Which of the following circuits will exhibit this? **(2019 Main)**

(a) RC circuit with $R = 1\,k\Omega$ and $C = 1\,\mu F$
(b) RL circuit with $R = 1\,k\Omega$ and $L = 1$ mH
(c) RC circuit with $R = 1\,k\Omega$ and $C = 10\,\mu F$
(d) RL circuit with $R = 1\,k\Omega$ and $L = 10$ mH

13. For the given figures, choose the correct options **(2023 Main)**

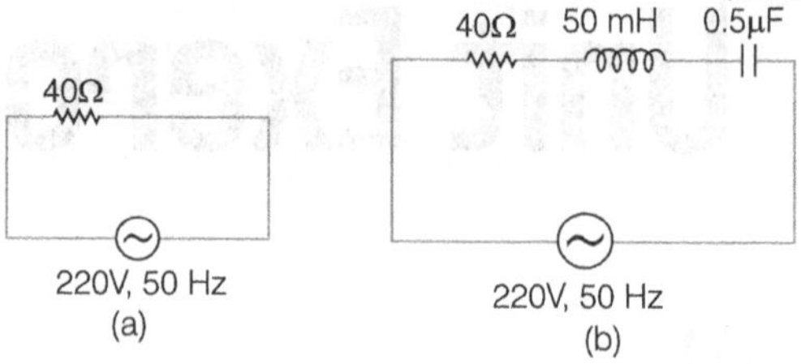

(a) The rms current in circuit (b) can be larger than that in (a)
(b) The rms current in circuit (a) is always equal to that in (b)
(c) At resonance, current in circuit (b) is less than that in (a)
(d) The rms current in circuit (b) can never be larger than that in (a)

14. A transformer operating at primary voltage 8 kV and secondary voltage 160 V serves a load of 80 kW. Assuming the transformer to be ideal with purely resistive load and working on unity power factor, the loads in the primary and secondary circuit would be **(2022 Main)**

(a) $800\,\Omega$ and $1.06\,\Omega$ (b) $10\,\Omega$ and $500\,\Omega$
(c) $800\,\Omega$ and $0.32\,\Omega$ (d) $1.06\,\Omega$ and $500\,\Omega$

15. A transformer consisting of 300 turns in the primary and 150 turns in the secondary gives output power of 2.2 kW. If the current in the secondary coil is 10 A, then the input voltage and current in the primary coil are **(2019 Main)**

(a) 440 V and 5 A
(b) 220 V and 20 A
(c) 220 V and 10 A
(d) 440 V and 20 A

16. In the circuit shown in figure, the power consumed is

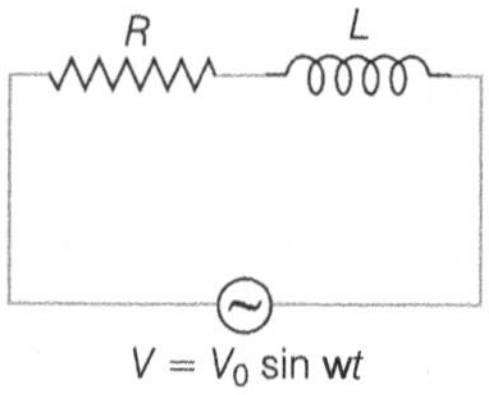

(a) zero (b) $\frac{V_0^2}{2R}$
(c) $\frac{V_0^2 R}{2(R^2 + \omega^2 L^2)}$ (d) None of these

17. In a series C-R circuit shown in figure, the applied voltage is 10 V and the voltage across capacitor is found to be 8 V. The voltage across R, and the phase difference between current and the applied voltage will respectively be

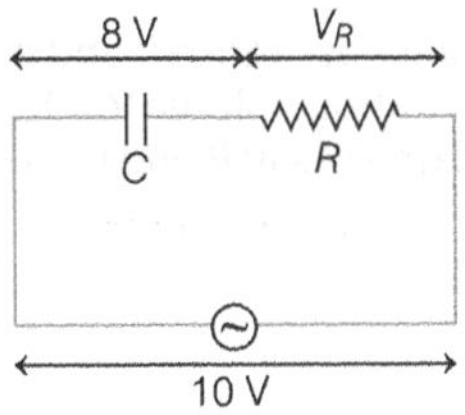

(a) 6 V, $\tan^{-1}\left(\frac{4}{3}\right)$
(b) 3 V, $\tan^{-1}\left(\frac{3}{4}\right)$
(c) 6V, $\tan^{-1}\left(\frac{3}{4}\right)$
(d) None of these

18. An AC voltage source described by $V = 10\cos(\pi/2)t$ is connected to a $1\,\mu F$ capacitor as shown in figure. The key K is closed at $t = 0$. The time ($t > 0$) after which the magnitude of current I reaches its maximum value for the first time is

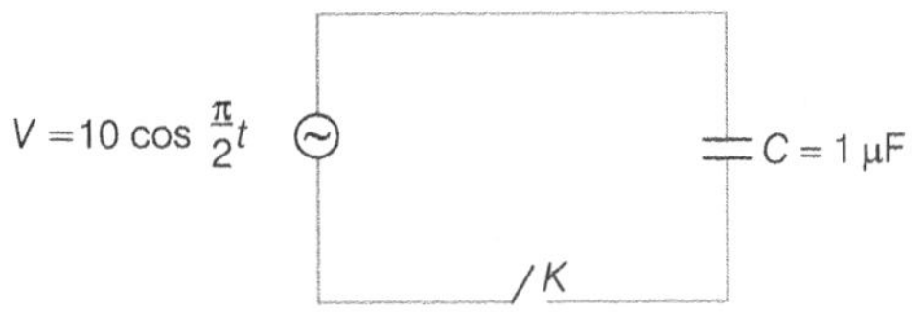

(a) 1 s
(b) 2 s
(c) 3 s
(d) 4 s

Subjective Questions

19. A series circuit has an impedance of 60.0 Ω and a power factor of 0.720 at 50.0 Hz. The source voltage lags the current.
(a) What circuit element, an inductor or a capacitor, should be placed in series with the circuit to raise its power factor?
(b) What size element will raise the power factor to unity?

20. A 300 Ω resistor is connected in series with a 0.800 H inductor. The voltage across the resistor as a function of time is $V_R = (2.50\text{ V})\cos[(950\text{ rad/s})t]$.
(a) Derive an expression for the circuit current.
(b) Determine the inductive reactance of the inductor.
(c) Derive an expression for the voltage V_L across the inductor.

21. Draw the approximate voltage vector diagrams in the electric circuits shown in figures (a) and (b). The external voltage V is assumed to be alternating of frequency ω **(Irodov 4.123)**

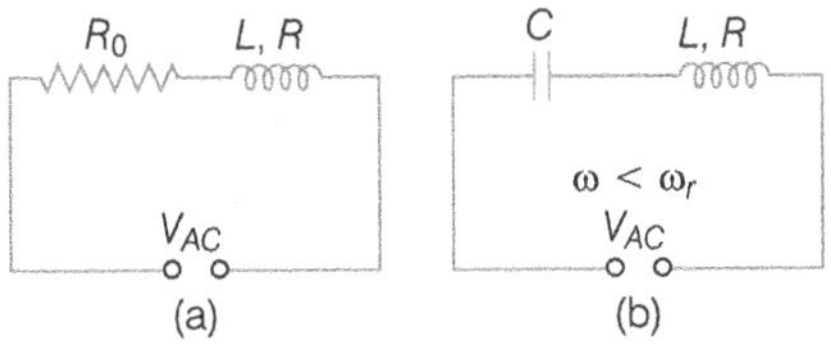

22. A coil with inductive resistance $X_L = 30\,\Omega$ and impedance $Z = 50\,\Omega$ is connected to the mains with effective voltage value $V = 100$ V. Find the phase difference between the current and the voltage, as well as the heat power generated in the coil. **(Irodov 4.137)**

23. A series circuit consisting of an inductance-free resistance $R = 0.16\text{ k}\Omega$ and a coil with active resistance is connected to the mains with effective voltage $V = 220$ V. Find the heat power generated in the coil if the effective voltage values across the resistance R and the coil are equal to $V_1 = 80$ V and $V_2 = 180$ V respectively. **(Irodov 4.141)**

24. A coil and an inductance-free resistance $R = 25\,\Omega$ are connected in parallel to the AC mains. Find the heat power generated in the coil provided a current $I = 0.90$ A is drawn from the mains. The coil and the resistance R carry currents $I_1 = 0.50$ A and $I_2 = 0.60$ A respectively. **(Irodov 4.142)**

25. An alternating current of frequency $\omega = 314\text{ s}^{-1}$ is fed to a circuit consisting of a capacitor of capacitance $C = 73\,\mu F$ and an active resistance $R = 100\,\Omega$ connected in parallel. Find the impedance of the circuit. **(Irodov 4.143)**

Integer Type Questions

26. An inductor of inductance $2\,\mu H$ is connected in series with a resistance, a variable capacitor and an AC source of frequency 7 kHz. The value of capacitance for which maximum current is drawn into the circuit is $\frac{1}{x}$ F, where the value of x is ....
$\left(\text{Take, } \pi = \frac{22}{7}\right)$ **(2023 Main)**

27. An inductor of 0.5 mH, a capacitor of 20 μF and resistance of 20 Ω are connected in series with a 220 V AC source. If the current is in phase with the emf, the amplitude of current of the circuit is $\sqrt{x}$ A. The value of x is .... **(2023 Main)**

28. A series L-C-R circuit consists of $R = 80\Omega$, $X_L = 100\Omega$, and $X_C = 40\Omega$. The input voltage is $2500\cos(100\,\pi t)$ V. The amplitude of current, in the circuit is .... A. **(2023 Main)**

29. As shown in the figure an inductor of inductance 200 mH is connected to an AC source of emf 220 V and frequency 50 Hz. The instantaneous voltage of the source is 0 V when the peak value of current is $\frac{\sqrt{a}}{\pi}$ A. The value of a is .... **(2022 Main)**

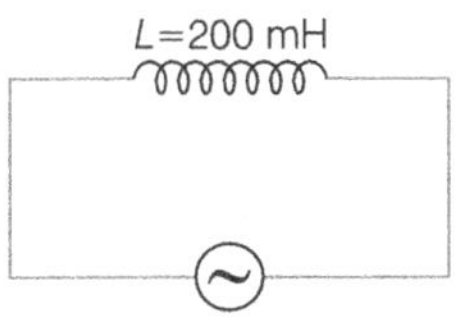

30. A series L-C-R circuit with $R = \frac{250}{11}\,\Omega$ and $X_L = \frac{70}{11}\,\Omega$ is connected across a 220 V, 50 Hz supply. The value of

capacitance needed to maximize the average power of the circuit will be .... μF. $\left(\text{Take, } \pi = \frac{22}{7}\right)$ **(2022 Main)**

31. In a series *L-R* circuit, power of 400 W is dissipated from a source of 250 V, 50 Hz. The power factor of the circuit is 0.8. In order to bring the power factor to unity, a capacitor of value *C* is added in series to the *L* and *R*. Taking the value *C* as $\left(\frac{n}{3\pi}\right)$ μF, then value of *n* is .... **(2022 Main)**

32. An AC circuit has an inductor and a resistor of resistance *R* in series, such that $X_L = 3R$. Now, a capacitor is added in series such that $X_C = 2R$. The ratio of new power factor with the old power factor of the circuit is $\sqrt{5} : x$. The value of *x* is .... **(2021 Main)**

33. An ideal transformer with purely resistive load operates at 12 kV on the primary side. It supplies electrical energy to a number of nearby houses at 120 V. The average rate of energy consumption in the houses served by the transformer is 60 kW. The value of resistive load (R_S) required in the secondary circuit will be .... mΩ. **(2023 Main)**

34. In an AC generator, a rectangular coil of 100 turns each having area $14 \times 10^{-2}\ m^2$ is rotated at 360 rev/min about an axis perpendicular to a uniform magnetic field of magnitude 3.0 T. The maximum value of the emf produced will be .... V. $\left(\text{Take, } \pi = \frac{22}{7}\right)$ **(2023 Main)**

35. A common transistor radio set requires 12 V (DC) for its operation. The DC source is constructed by using a transformer and a rectifier circuit, which are operated at 220 V (AC) on standard domestic AC supply. The number of turns of secondary coil are 24, then the number of turns of primary are .... **(2021 Main)**

Level 2

Single Correct Option

36. A capacitor and resistor are connected with an AC source as shown in figure. Reactance of capacitor is $X_C = 3\,\Omega$ and resistance of resistor is 4Ω. Phase difference between current *I* and I_1 is $\left[\tan^{-1}\left(\frac{3}{4}\right) = 37°\right]$

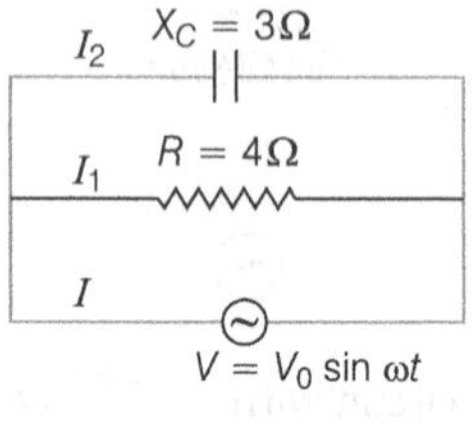

(a) 90° (b) zero
(c) 53° (d) 37°

37. A circuit contains resistance *R* and an inductance *L* in series. An alternating voltage $V = V_0 \sin \omega t$ is applied across it. The currents in *R* and *L* respectively will be

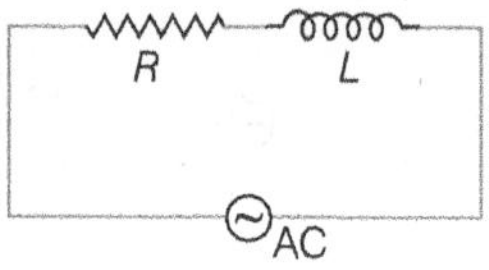

(a) $I_R = I_0 \cos \omega t, I_L = I_0 \cos \omega t$
(b) $I_R = -I_0 \sin \omega t, I_L = I_0 \cos \omega t$
(c) $I_R = I_0 \sin \omega t, I_L = -I_0 \cos \omega t$
(d) None of the above

38. In the circuit shown in figure, the AC source gives a voltage $V = 20 \cos (2000\,t)$. Neglecting source resistance, the voltmeter and ammeter readings will be

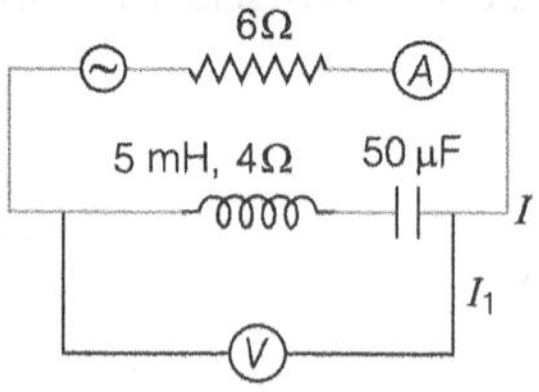

(a) 0 V, 2.0 A (b) 0 V, 1.4 A
(c) 5.6 V, 1.4 A (d) 8 V, 2.0 A

39. A signal generator supplies a sine wave of 200 V, 5 kHz to the circuit shown in the figure. Then, choose the **wrong** statement.

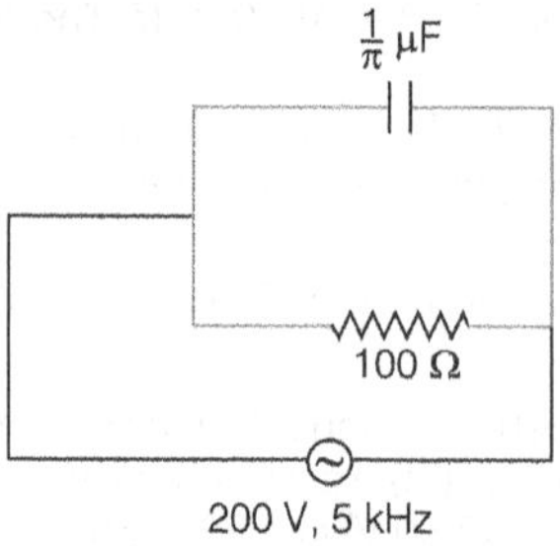

(a) The current in the resistive branch is 2.0 A
(b) The current in the capacitive branch is 0.126 A
(c) Total line current is ≈2.83 A
(d) Current in both the branches is same

40. For the circuit as shown in figure, if the value of rms current is 2.2 A, the power factor of the box is

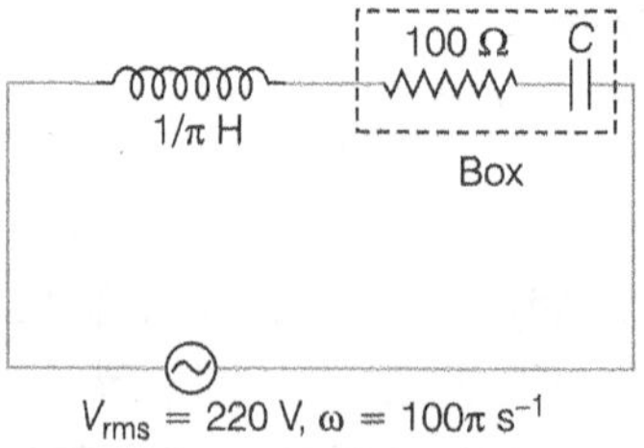

(a) $\frac{1}{\sqrt{2}}$ (b) 1 (c) $\frac{\sqrt{3}}{2}$ (d) $\frac{1}{2}$

41. An AC voltage $V = V_0 \sin 100\, t$ is applied to the circuit, the phase difference between current and voltage is found to be $\pi/4$, then

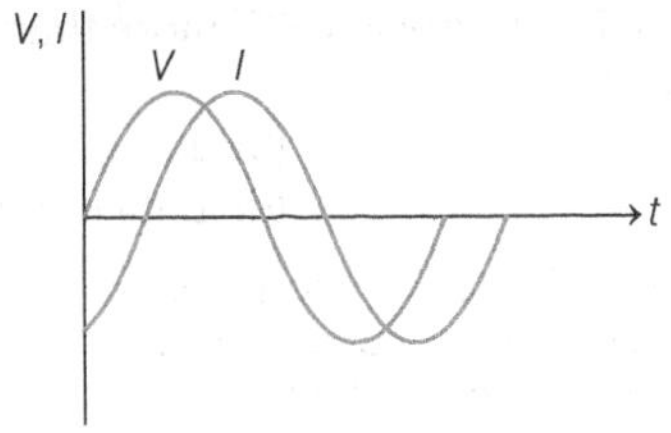

(a) $R = 100\,\text{k}\Omega, C = 1\,\mu\text{F}$
(b) $R = 1\,\text{k}\Omega, C = 10\,\mu\text{F}$
(c) $R = 10\,\text{k}\Omega, L = 1\,\text{H}$
(d) $R = 1\,\text{k}\Omega, L = 10\,\text{H}$

42. In series *L-C-R* circuit, voltage drop across resistance is 8 V, across inductor is 6 V and across capacitor is 12 V. Then,

(a) voltage of the source will be leading in the circuit
(b) voltage drop across each element will be less than the applied voltage
(c) power factor of the circuit will be 3 /4
(d) None of the above

43. Consider an *L-C-R* circuit as shown in figure with an AC source of peak value V_0 and angular frequency ω. Then, the peak value of current through the AC source is

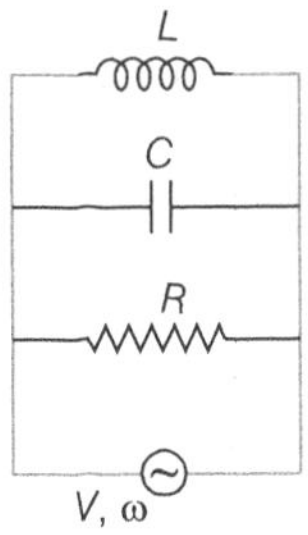

(a) $\dfrac{V_0}{\sqrt{\dfrac{1}{R^2} + \left(\omega L - \dfrac{1}{\omega C}\right)^2}}$
(b) $V_0 \sqrt{\dfrac{1}{R^2} + \left(\omega C - \dfrac{1}{\omega L}\right)^2}$
(c) $\dfrac{V_0}{\sqrt{R^2 + \left(\omega L - \dfrac{1}{\omega C}\right)^2}}$
(d) None of these

44. The adjoining figure shows an AC circuit with resistance R, inductance L and source voltage V_s. Then,

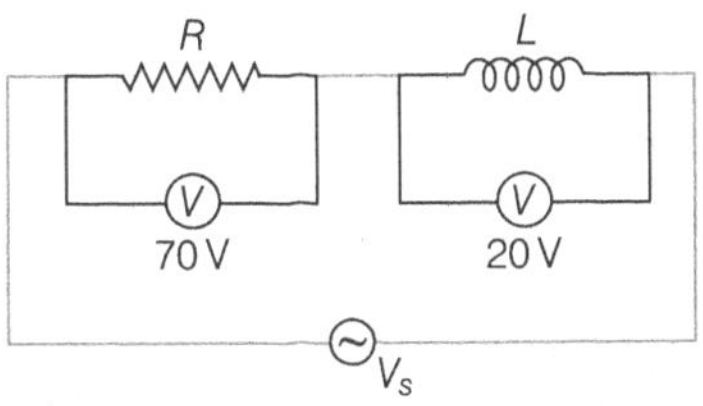

(a) the source voltage $V_s = 72.8$ V
(b) the phase angle between current and source voltage is $\tan^{-1}(7/2)$
(c) Both (a) and (b) are correct
(d) Both (a) and (b) are wrong

45. When an alternating voltage of 220 V is applied across a device P, a current of 0.25 A flows through the circuit and it leads the applied voltage by an angle $\pi/2$ radian. When the same voltage source is connected across another device Q, the same current is observed in the circuit but in phase with the applied voltage. What is the current when the same source is connected across a series combination of P and Q?

(a) $\dfrac{1}{4\sqrt{2}}$ A lagging in phase by $\pi/4$ with voltage
(b) $\dfrac{1}{4\sqrt{2}}$ A leading in phase by $\pi/4$ with voltage
(c) $\dfrac{1}{\sqrt{2}}$ A leading in phase by $\pi/4$ with voltage
(d) $\dfrac{1}{4\sqrt{2}}$ A leading in phase by $\pi/2$ with voltage

46. In a parallel *L-C-R* circuit as shown in figure if I_R, I_L, I_C and I represent the rms values of current flowing through resistor, inductor, capacitor and the source, then choose the appropriate correct answer.

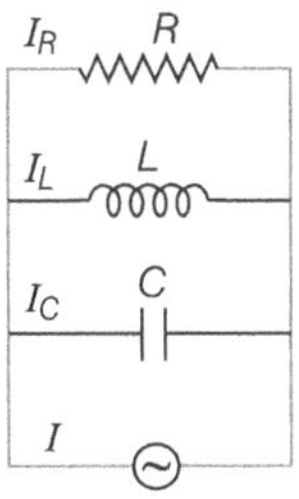

(a) $I = I_R + I_L + I_C$
(b) $I = I_R + I_L - I_C$
(c) I_L or I_C may be greater than I
(d) None of these

47. In a series *L-C-R* circuit, current in the circuit is 11 A when the applied voltage is 220 V. Voltage across the capacitor is 200 V. If the value of resistor is 20 Ω, then the voltage across the unknown inductor is

(a) zero
(b) 200 V
(c) 20 V
(d) None of these

48. In a series *L-C* circuit, the applied voltage is V_0. If ω is very low, then the voltage drop across the inductor V_L and capacitor V_C are

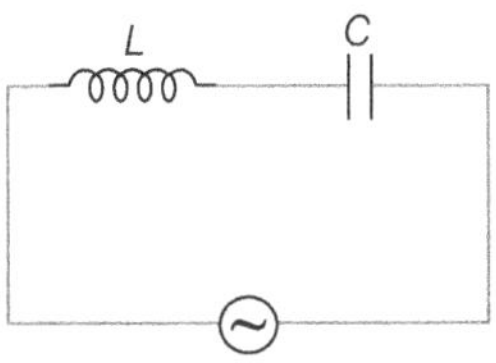

(a) $V_L = \dfrac{V_0}{2}; V_C = \dfrac{V_0}{2}$
(b) $V_L = 0; V_C = V_0$
(c) $V_L = V_0; V_C = 0$
(d) $V_L = -V_C = \dfrac{V_0}{2}$

49. A coil, a capacitor and an AC source of rms voltage 24 V are connected in series. By varying the frequency of the source, a maximum rms current of 6 A is observed. If coil is connected to a DC battery of emf 12 volt and internal resistance 4 Ω, then current through it in steady state is

(a) 2.4 A (b) 1.8 A (c) 1.5 A (d) 1.2 A

50. An AC voltage source $V = V_0 \sin \omega t$ is connected across resistance R and capacitance C as shown in figure. It is given that $R = 1/\omega C$. The peak current is I_0. If the angular frequency of the voltage source is changed to $\omega/\sqrt{3}$, then the new peak current in the circuit is

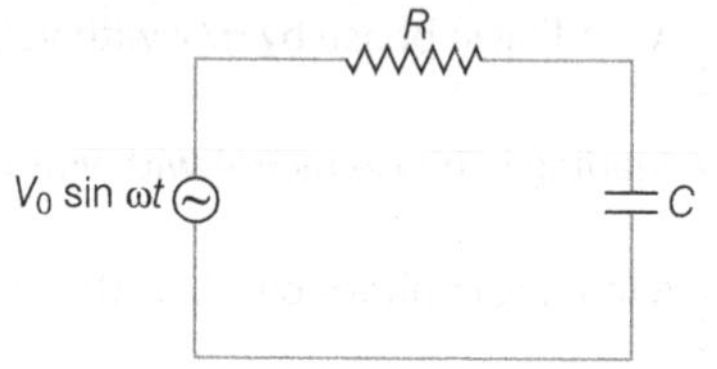

(a) $\frac{I_0}{2}$ (b) $\frac{I_0}{\sqrt{2}}$ (c) $\frac{I_0}{\sqrt{3}}$ (d) $\frac{I_0}{3}$

51. Which of the following option is correct?

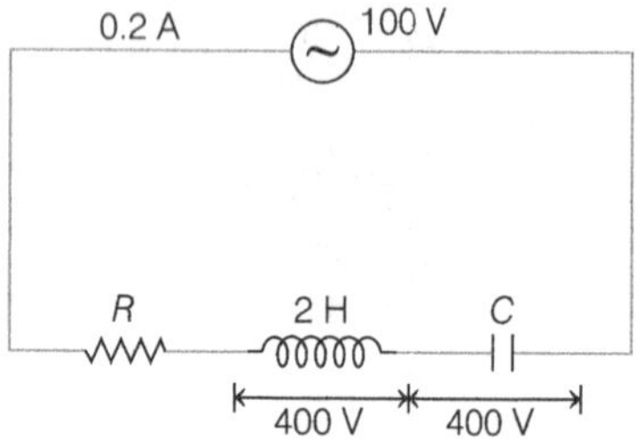

(a) $R = 400\,\Omega, C = 0.5$ F (b) $R = 500\,\Omega, C = 0.5\,\mu$F
(c) $R = 500\,\Omega, C = 1\,\mu$F (d) $R = 400\,\Omega, C = 0.1\,\mu$F

52. When an AC voltage of variable frequency is applied to series L-C-R circuit, the current in the circuit is the same at 4 kHz and 9 kHz. The current in the circuit is maximum at

(a) 5 kHz (b) 6.5 kHz
(c) 4.2 kHz (d) 6 kHz

53. An ideal choke takes a current of 8 A when connected to an AC source of 100 V and 50 Hz. A pure resistor under the same conditions takes a current of 10 A. If two are connected in series to an AC supply of 100 V and 40 Hz, then the current in the series combination of above resistor and inductor is

(a) 10 A (b) 5 A
(c) $5\sqrt{2}$ A (d) $10\sqrt{2}$ A

54. An AC source is connected with a resistance (R) and an uncharged capacitance (C) in series. The potential difference across the resistor is in phase with the initial potential difference across the capacitor for the first time at the instant (assume that at $t = 0$, emf is zero).

(a) $\frac{\pi}{4\omega}$ (b) $\frac{2\pi}{\omega}$ (c) $\frac{\pi}{2\omega}$ (d) $\frac{3\pi}{2\omega}$

where, ω is the angular frequency.

55. Current through an AC series L-C-R circuit is 2 A if operated at resonant frequency, and 1 A if operated at 50% less than resonant frequency. The current (in A) if the frequency is 100% more than the resonant frequency, is

(a) $\sqrt{2}$ (b) 1
(c) $\sqrt{3}$ (d) Data insufficient

56. In the given AC circuit, when switch S is at position 1, the source emf leads current by $\pi/6$. Now, if the switch is at position 2, then

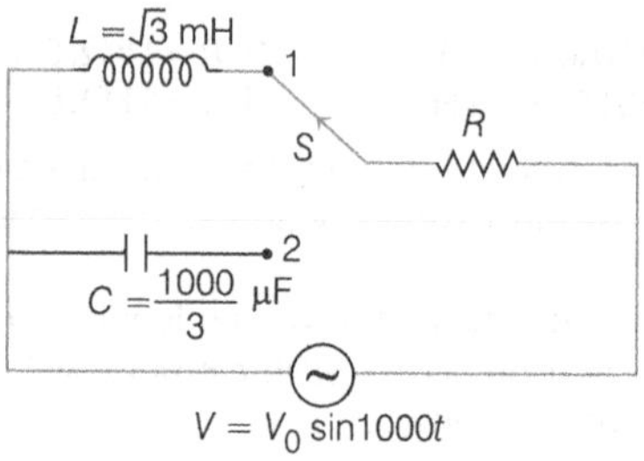

(a) current leads source emf by $\frac{\pi}{4}$

(b) current leads source emf by $\frac{\pi}{3}$

(c) source emf leads current by $\frac{\pi}{4}$

(d) source emf leads current by $\frac{\pi}{3}$

57. In the given AC circuit,

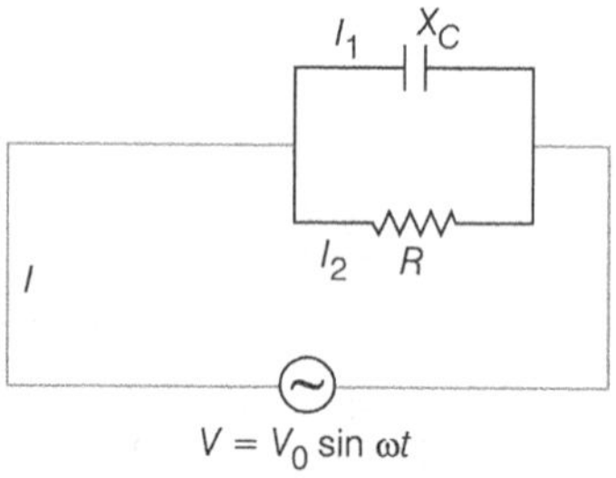

(a) current I_2 and V are in same phase
(b) current I_2 leads I_1 by 90°
(c) current I leads I_2 by $\theta < 90°$
(d) current I leads I_1 by $\theta < 90°$

58. If a DC of value a ampere is superimposed on an alternating currrent $I = b \sin \omega t$ flowing through a wire, what is the effective value of the resulting current in the circuit?

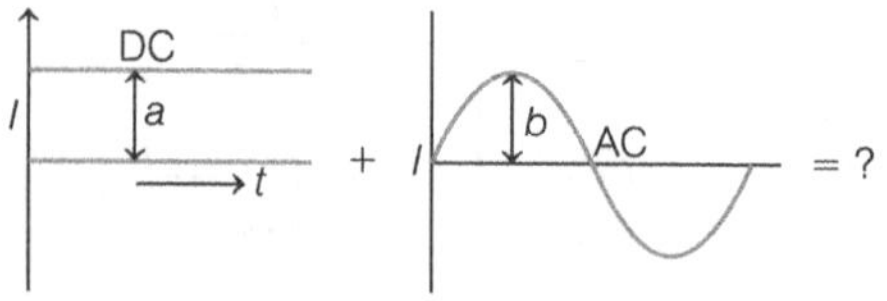

(a) $\left(a^2 - \frac{1}{2}b^2\right)^{1/2}$ (b) $(a^2 + b^2)^{1/2}$

(c) $\left(\frac{a^2}{2} + b^2\right)^{1/2}$ (d) $\left(a^2 + \frac{b^2}{2}\right)^{1/2}$

59. An AC source of angular frequency ω is fed across a resistor R and a capacitor C in series. The current registered is I. If now the frequency of source is changed to $\frac{\omega}{3}$ (but maintaining the same voltage), the current in the circuit is found to be halved. Calculate the ratio of reactance to resistance at the original frequency ω

(a) $\sqrt{\frac{3}{5}}$ (b) $\sqrt{\frac{2}{5}}$ (c) $\sqrt{\frac{1}{5}}$ (d) $\sqrt{\frac{4}{5}}$

60. In the given circuit, R is a pure resistor, L is a pure inductor, S is a 50 V, 50 Hz AC source and A is an AC ammeter. With either K_1 or K_2 alone closed, the ammeter reading is I. If the source is changed to 100 V, 100 Hz, then ammeter reading with K_1 alone closed and with K_2 alone closed will be respectively

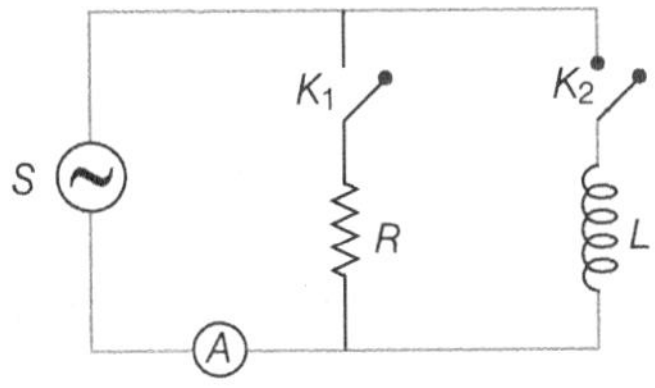

(a) $I, \frac{I}{2}$ (b) $I, 2I$

(c) $2I, I$ (d) $2I, \frac{I}{2}$

61. Time constant of the given circuit is τ. If the battery is replaced by an AC source having voltage $V = V_0 \cos \omega t$, power factor of the circuit will be

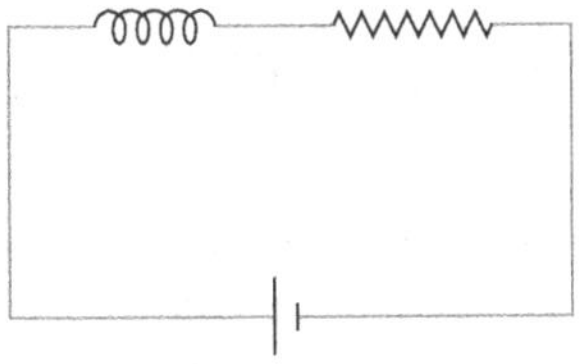

(a) $\omega\tau$ (b) $\frac{1}{\sqrt{1+(\omega\tau)^2}}$

(c) $\sqrt{1+(\omega\tau)^2}$ (d) None of these

62. Circuit element X when connected to an AC supply of peak voltage 100 V gives a peak current of 5 A which is in phase with the voltage. A second element Y when connected to the same AC supply also gives the same value of peak current which lags behind the voltage by $\frac{\pi}{2}$. If X and Y are connected in series to the same supply, what will be the rms value of the current in ampere? **(2022 Main)**

(a) $\frac{10}{\sqrt{2}}$ (b) $\frac{5}{\sqrt{2}}$ (c) $5\sqrt{2}$ (d) $\frac{5}{2}$

Multiple Correct Options

63. In a R-L-C series circuit shown, the readings of voltmeters V_1 and V_2 are 100 V and 120 V. Choose the correct statement(s).

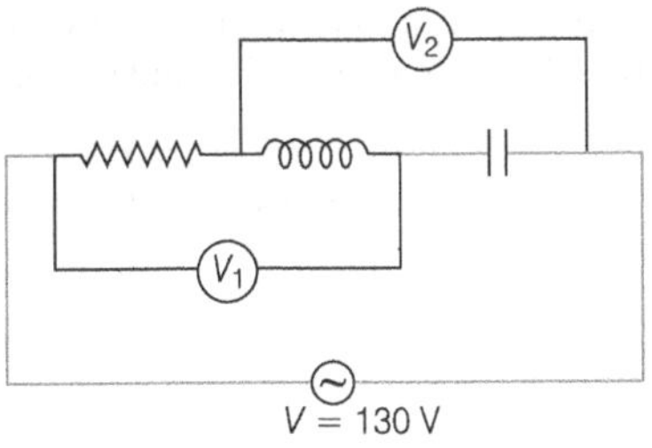

(a) Voltage across resistor, inductor and capacitor are 50 V, 86.6 V and 206.6 V respectively

(b) Voltage across resistor, inductor and capacitor are 10 V, 90 V and 30 V respectively

(c) Power factor of the circuit is $\frac{5}{13}$

(d) Circuit is capacitive in nature

64. Current in an AC circuit is given by $i = 3 \sin \omega t + 4 \cos \omega t$, then

(a) rms value of current is 5 A

(b) mean value of this current in positive one-half cycle will be $\frac{6}{\pi}$

(c) if voltage applied is $V = V_m \sin \omega t$, then the circuit may contain resistance and capacitance

(d) if voltage applied is $V = V_m \cos \omega t$, then the circuit may contain resistance and inductance only

65. A tube light of 60 V, 60 W rating is connected across an AC source of 100 V and 50 Hz frequency. Then,

(a) an inductance of $\frac{2}{5\pi}$ H may be connected in series

(b) a capacitor of $\frac{250}{\pi}$ μF may be connected in series to it

(c) an inductor of $\frac{4}{5\pi}$ H may be connected in series

(d) a resistance of 40 Ω may be connected in series

66. In the circuit shown in figure,

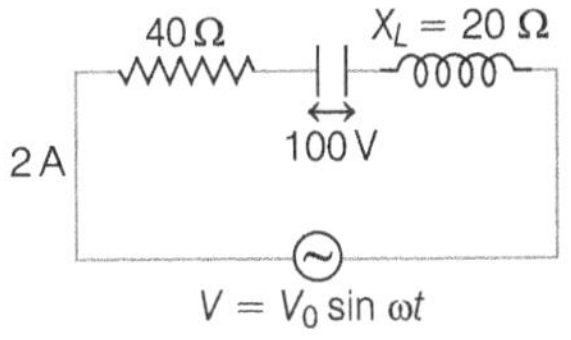

(a) $V_R = 80$ V (b) $X_C = 50\ \Omega$ (c) $V_L = 40$ V (d) $V_0 = 100$ V

67. A series R-C circuit is connected to AC voltage source. Consider two cases; (A) when C is without a dielectric medium and (B) when C is filled with dielectric of constant 4. The current I_R through the resistor and voltage V_C across the capacitor are compared in the two cases. Which of the following is/are true? **(2011 JEE)**

(a) $I_R^A > I_R^B$ (b) $I_R^A < I_R^B$ (c) $V_C^A > V_C^B$ (d) $V_C^A < V_C^B$

68. For the figure shown, $R = 100\ \Omega, L = \frac{2}{\pi}$ H and $C = \frac{8}{\pi}$ μF are connected in series with an AC source of 200 V

and frequency f. V_1 and V_2 are two hot wire voltmeters. If the readings of V_1 and V_2 are same, then

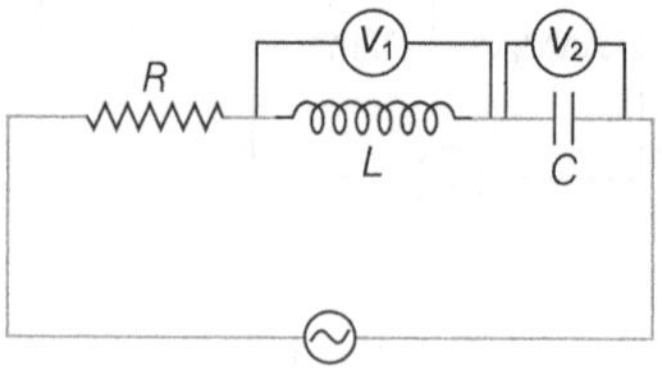

(a) $f = 125$ Hz (b) $f = 250\pi$ Hz
(c) current through R is 2 A (d) $V_1 = V_2 = 1000$ V

69. In the given circuit, the AC source has $\omega = 100$ rad/s. Considering the inductor and capacitor to be ideal, the correct choice(s) is(are) **(2012 JEE)**

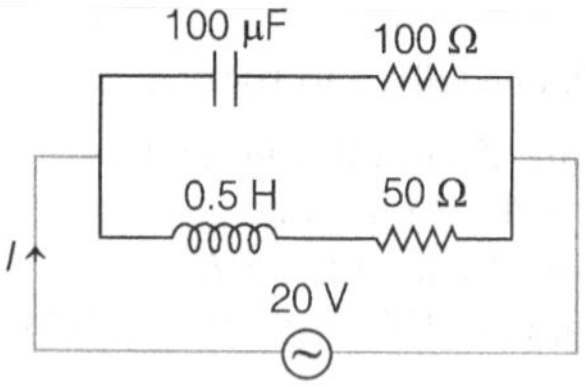

(a) The current through the circuit, I is 0.34 A
(b) The current through the circuit, I is $0.3\sqrt{2}$ A
(c) The voltage across 100 Ω resistor $= 10\sqrt{2}$ V
(d) The voltage across 50 Ω resistor = 10 V

70. The instantaneous voltages at three terminals marked X, Y and Z are given by $V_X = V_0 \sin \omega t$,

$$V_Y = V_0 \sin\left(\omega t + \frac{2\pi}{3}\right) \text{and } V_Z = V_0 \sin\left(\omega t + \frac{4\pi}{3}\right).$$

An ideal voltmeter is configured to read rms value of the potential difference between its terminals. It is connected between points X and Y and then between Y and Z. The reading(s) of the voltmeter will be **(2017 Adv.)**

(a) $(V_{YZ})_{rms} = V_0\sqrt{\frac{1}{2}}$
(b) $(V_{XY})_{rms} = V_0\sqrt{\frac{3}{2}}$
(c) independent of the choice of the two terminals
(d) $(V_{XY})_{rms} = V_0$

71. In the circuit shown, $L = 1\,\mu$H, $C = 1\,\mu$F and $R = 1\,k\Omega$. They are connected in series with an AC source $V = V_0 \sin \omega t$ as shown. Which of the following options is/are correct? **(2017 Adv.)**

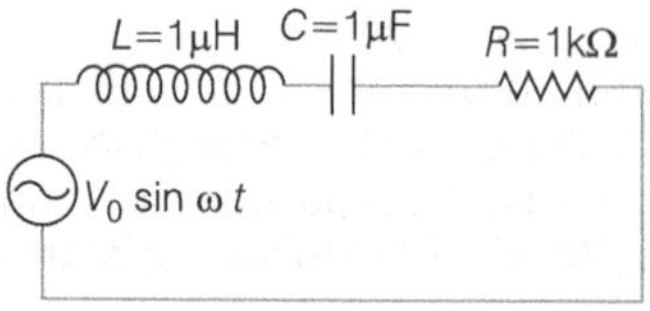

(a) At $\omega \approx 0$, the current flowing through the circuit becomes nearly zero
(b) The frequency at which the current will be in phase with the voltage is independent of R
(c) The current will be in phase with the voltage if $\omega = 10^4$ rads^{-1}
(d) At $\omega >> 10^6$ rads^{-1}, the circuit behaves like a capacitor

72. At time $t = 0$, terminal A in the circuit shown in the figure is connected to B by a key and an alternating current $I(t) = I_0 \cos(\omega t)$, with $I_0 = 1$ A and $\omega = 500$ rad s^{-1} starts flowing in it with the initial direction shown in the figure. At $t = 7\pi/6\omega$, the key is switched from B to D. Now onwards only A and D are connected. A total charge Q flows from the battery to charge the capacitor fully. If $C = 20\,\mu$F, $R = 10\,\Omega$ and the battery is ideal with emf of 50 V, identify the correct statement(s). **(2014 Adv.)**

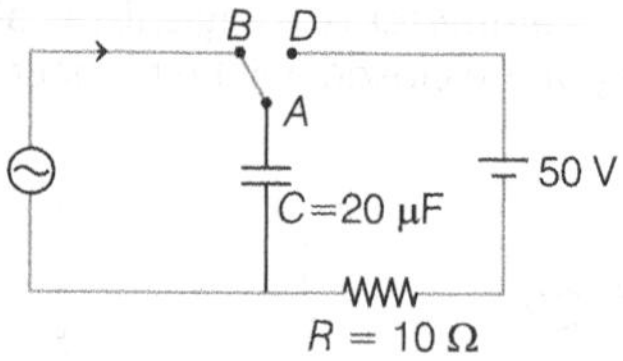

(a) Magnitude of the maximum charge on the capacitor before $t = \frac{7\pi}{6\omega}$ is 1×10^{-3} C
(b) The current in the left part of the circuit just before $t = \frac{7\pi}{6\omega}$ is clockwise
(c) Immediately after A is connected to D, the current in R is 10 A
(d) $Q = 2 \times 10^{-3}$ C

Passage Based Questions

Passage (Q.No. 73 to 75)

It is known to all of you that the impedance of a circuit is dependent on the frequency of source. In order to study the effect of frequency on the impedance, a student in a lab took 2 impedance boxes P and Q and connected them in series with an AC source of variable frequency. The emf of the source is constant at 10 V. Box P contains a capacitance of 1 μF in series with a resistance of 32 Ω. And the box Q has a coil of self-inductance 4.9 mH and a resistance of 68 Ω in series. He adjusted the frequency so that the maximum current flows in P and Q. Based on his experimental set up and the reading by him at various moment, answer the following questions.

73. The angular frequency for which he detects maximum current in the circuit is
(a) $10^5/7$ rad/s (b) 10^4 rad/s
(c) 10^5 rad/s (d) $10^4/7$ rad/s

74. Impedance of box P at the above frequency is
(a) 70Ω (b) 77Ω
(c) 90Ω (d) 100Ω

75. Power factor of the circuit at maximum current is
(a) 1/2 (b) 1 (c) 0 (d) $1/\sqrt{2}$

Match the Columns

76. Applied AC voltage is given as

$$V = V_0 \sin \omega t$$

Corresponding to this voltage, match the following two columns.

	Column I		Column II
(a)	$I = I_0 \sin \omega t$	(p)	only R circuit
(b)	$I = -I_0 \cos \omega t$	(q)	only L circuit
(c)	$I = I_0 \sin (\omega t + \pi/6)$	(r)	may be C-R circuit
(d)	$I = I_0 \sin (\omega t - \pi/6)$	(s)	may be L-C-R circuit

77. In the circuit shown in figure, match the following two columns. In Column II, quantities are given in SI units.

$X_C = 30\ \Omega$, $X_L = 15\ \Omega$

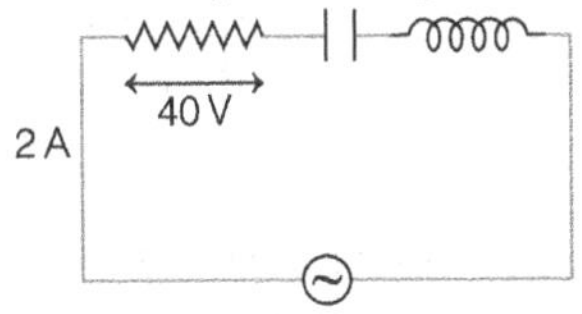

	Column I		Column II
(a)	Value of resistance R	(p)	60
(b)	Potential difference across capacitor	(q)	20
(c)	Potential difference across inductor	(r)	30
(d)	Applied potential difference	(s)	None of the above

78. You are given many resistances, capacitors and inductors. These are connected to a variable DC voltage source (*the first two circuits*) or an AC voltage source of 50 Hz frequency (*the next three circuits*) in different ways as shown in **Column II**. When a current I (*steady state for DC or rms for AC*) flows through the circuit, the corresponding voltage V_1 and V_2 (*indicated in circuits*) are related as shown in **Column I**. **(2010 JEE)**

	Column I		Column II
(A)	$I \neq 0$, V_1 is proportional to I	(p)	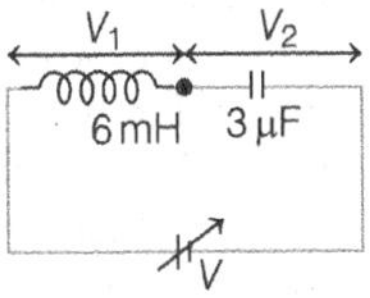
(B)	$I \neq 0$, $V_2 > V_1$	(q)	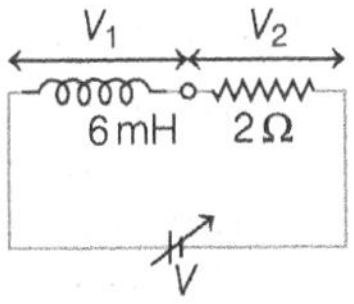
(C)	$V_1 = 0$, $V_2 = V$	(r)	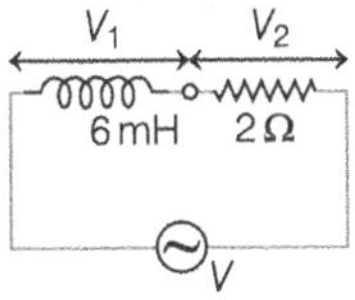
(D)	$I \neq 0$, V_2 is proportional to I	(s)	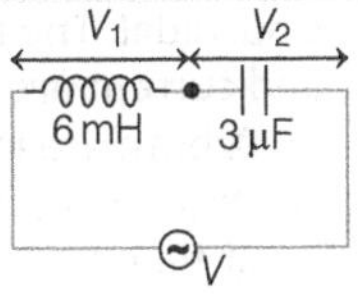
		(t)	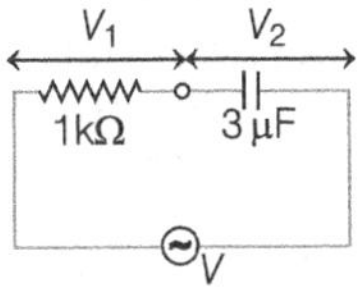

Subjective Questions

79. In the figure shown, the reading of voltmeters are $V_1 = 40$ V, $V_2 = 40$ V and $V_3 = 10$ V. Find

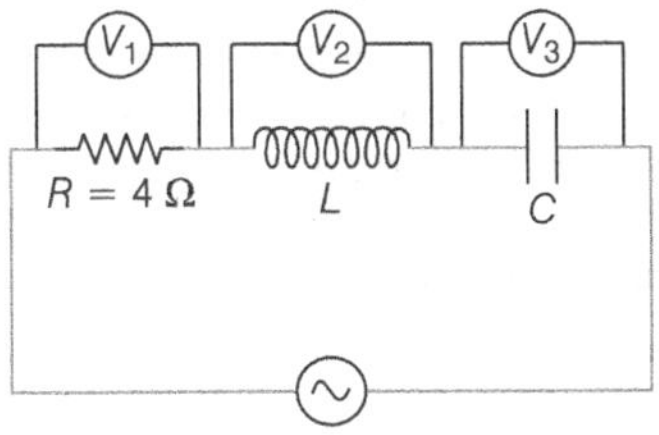

$$E = E_0 \sin\left(100\ \pi t + \frac{\pi}{6}\right)$$

(a) the peak value of current
(b) the peak value of emf
(c) the value of L and C

80. In the circuit shown in figure power factor of box is 0.5 and power factor of circuit is $\sqrt{3}/2$. Current leading the voltage. Find the effective resistance of the box.

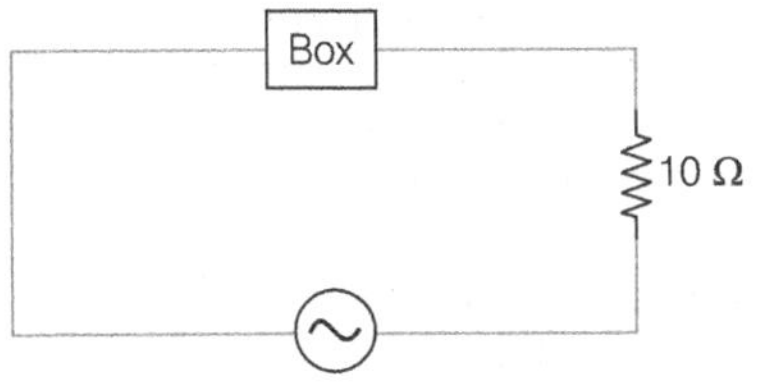

81. A circuit element shown in the figure as a box is having either a capacitor or an inductor. The power factor of the circuit is 0.8, while current lags behind the voltage. Find

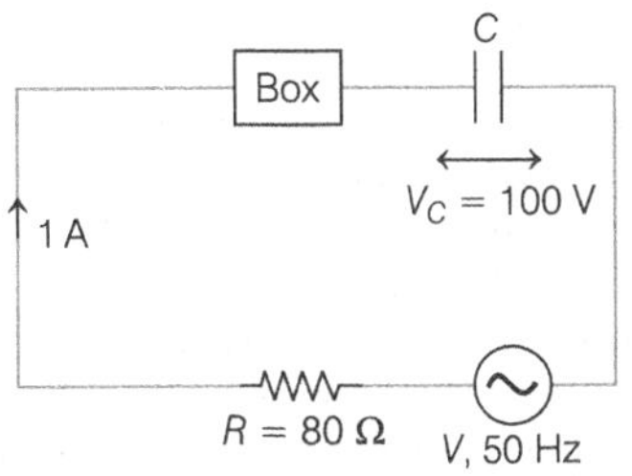

(a) the source voltage V,
(b) the nature of the element in box and find its value.

82. The maximum values of the alternating voltages and current are 400 V and 20 A respectively in a circuit

connected to 50 Hz supply and these quantities are sinusoidal. The instantaneous values of the voltage and current are $200\sqrt{2}$ V and 10 A, respectively. At $t = 0$, both are increasing positively.

(a) Write down the expression for voltage and current at time t.

(b) Determine the power consumed in the circuit.

83. An L-C circuit consists of an inductor coil with $L = 5.00$ mH and a 20.0 μF capacitor. There is negligible resistance in the circuit. The circuit is driven by a voltage source with $V = V_0 \cos \omega t$. If $V_0 = 5.00$ mV and the frequency is twice the resonance frequency, determine

(a) the maximum charge on the capacitor

(b) the maximum current in the circuit

(c) the phase relationship between the voltages across the inductor, the capacitor and the source.

84. A coil having a resistance of 5 Ω and an inductance of 0.02 H is arranged in parallel with another coil having a resistance of 1 Ω and an inductance of 0.08 H. Calculate the power absorbed when a voltage of 100 V at 50 Hz is applied.

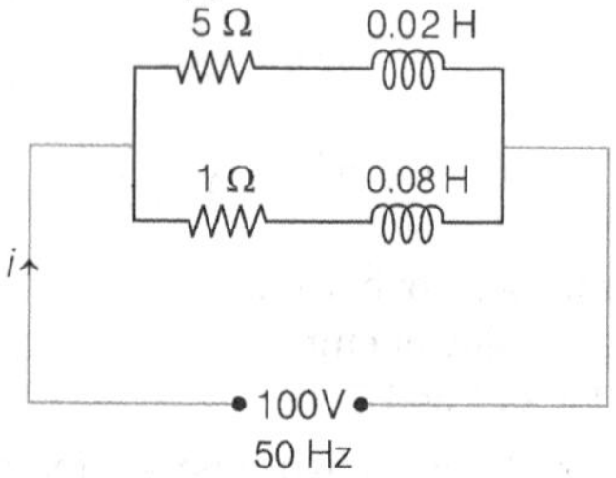

85. A circuit takes a current of 3 A at a power factor of 0.6 lagging when connected to a 115 V-50 Hz supply. Another circuit takes a current of 5 A at a power factor of 0.707 leading when connected to the same supply. If the two circuits are connected in series across a 230 V, 50 Hz supply, then calculate

(a) the current

(b) the power consumed and

(c) the power factor

86. A coil with active resistance R and inductance L was connected at the moment $t = 0$ to a source of voltage $V = V_m \cos \omega t$. Find the current in the coil as a function of time t. **(Irodov 4.118)**

87. A long one-layer solenoid tightly wound of wire with resistivity ρ has n turns per unit length. The thickness of the wire insulation is negligible. The cross-sectional radius of the solenoid is equal to a. Find the phase difference between current and alternating voltage fed to the solenoid with frequency f. **(Irodov 4.120)**

88. A series circuit consisting of a capacitor with capacitance C, a resistance R, and a coil with inductance L and negligible active resistance is connected to an oscillator whose frequency can be varied without changing the voltage amplitude. Find the frequency at which the voltage amplitude is maximum **(Irodov 4.125)**

(a) across the capacitor

(b) across the coil.

89. An alternating voltage with frequency $\omega = 314$ s^{-1} and amplitude $V_m = 180$ V is fed to a series circuit consisting of a capacitor and a coil with active resistance $R = 40\ \Omega$ and inductance $L = 0.36$ H. At what value of the capacitor's capacitance will the voltage amplitude across the coil be maximum? What is this amplitude equal to? What is the corresponding voltage amplitude across the condenser? **(Irodov 4.126)**

90. A capacitor with capacitance C whose interelectrode space is filled up with poorly conducting medium with active resistance R is connected to a source of alternating voltage $V = V_m \cos \omega t$. Find the time dependence of the steady-state current flowing in lead wires. The resistance of the wires is to be neglected. **(Irodov 4.127)**

91. Draw the approximate vector diagrams of currents in the circuits shown in figure (a), (b) and (c). The voltage applied across the points A and B is assumed to be sinusoidal; the parameters of each circuit are so chosen that the total current I lags in phase behind the external voltage by an angle ϕ. **(Irodov 4.144)**

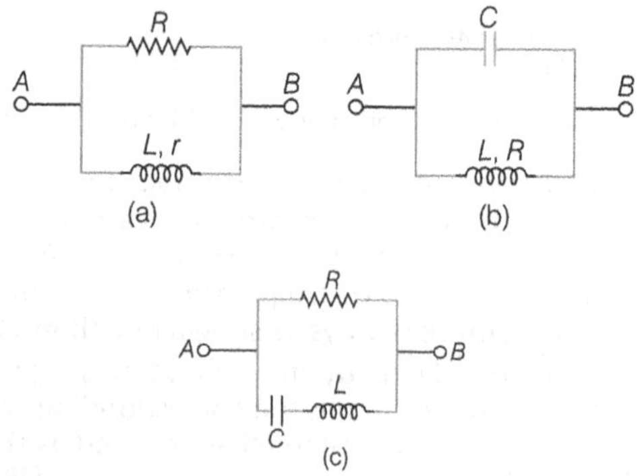

92. A coil with inductance $L = 0.70$ H and active resistance $r = 20\ \Omega$ is connected in series with an inductance-free resistance R. An alternating voltage with effective value $V = 220$ V and frequency $\omega = 314$ s^{-1} is applied across the terminals of this circuit. At what value of the resistance R will the maximum heat power be generated in the circuit? What is it equal to? **(Irodov 4.138)**

93. A circuit consisting of a capacitor and a coil in series is connected to the mains. Varying the capacitance of the capacitor, the heat power generated in the coil was increased $n = 1.7$ times. How much (in per cent) was the value of $\cos \phi$ changed in the process? **(Irodov 4.139)**

Answers

Level 1

1. (a) **2.** (c) **3.** (c) **4.** (d) **5.** (c) **6.** (a) **7.** (d) **8.** (b) **9.** (c) **10.** (c)
11. (b) **12.** (c) **13.** (d) **14.** (c) **15.** (a) **16.** (c) **17.** (a) **18.** (a)
19. (a) Inductor (b) 0.133 H **20.** (a) (8.33 mA) cos (950 rad/s) t (b) 760 Ω (c) –(6.33 V) sin (950 rad/s)t
21. See the hints **22.** (i) 37° (ii) 0.16 kW **23.** 30 W **24.** 2.5 W **25.** 40 Ω **26.** (3872) **27.** (242) **28.** (25)
29. (242) **30.** (500) **31.** (400) **32.** (1) **33.** (240) **34.** (1584) **35.** (440)

Level 2

36. (c) **37.** (d) **38.** (c) **39.** (b) **40.** (a) **41.** (b) **42.** (d) **43.** (b) **44.** (a) **46.** (b)
46. (c) **47.** (b) **48.** (b) **49.** (c) **50.** (b) **51.** (b) **52.** (d) **53.** (c) **54.** (d) **55.** (b)
56. (a) **57.** (a) **58.** (d) **59.** (a) **60.** (c) **61.** (b) **62.** (d) **63.** (a,c,d) **64.** (c,d) **65.** (c,d)
66. (a,b,c) **67.** (b,c) **68.** (a,c,d) **69.** (a,c) **70.** (b,c) **71.** (a,b) **72.** (c,d) **73.** (a) **74.** (b) **75.** (b)
76. (a → p,s ; b → q ; c → r,s; d → s) **77.** (a → q ; b → p ; c → r ; d → s) **78.** (a → r,s,t ; b → q,r,s,t ; c → p,q ; d → q,r,s,t)
79. (a) $10\sqrt{2}$ A (b) $50\sqrt{2}$ V (c) $\frac{1}{25\pi}$ H (d) $\frac{1}{100\pi}$ F **80.** 5 Ω **81.** (a) 100 V (b) Inductor, $L = \frac{1.6}{\pi}$ H
82. (a) $V = 400 \sin(100\pi t + \pi/4)$, $i = 20 \sin(100\pi t + \pi/6)$ (b) $P = 3864$ W **83.** (a) 33.4 nC (b) 0.211 mA
(c) Source and inductor voltages in phase. Phase difference between V_L and V_C is 180°. **84.** 797 W
85. (a) 5.5 A (b) 1.188 kW (c) 0.939 lag **86.** See the hints **87.** $\tan\phi = \frac{X_L}{R} = \frac{\mu_0 \pi^2 fa}{4n\rho}$
88. (a) $\omega = \sqrt{\omega_r^2 - 2\beta^2}$ (b) $\omega = \omega_r^2/\sqrt{\omega_r^2 - 2\beta^2}$, where $\omega_r^2 = 1/LC$, $\beta = R/2L$
89. (i) 28 μF (ii) 0.54 kV (iii) 0.51 kV **90.** $I = I_m \cos(\omega t + \phi)$, where $I_m = \frac{V_m}{R}\sqrt{1 + (\omega RC)^2}$ and $\phi = \tan^{-1}(\omega CR)$
91. See the hints **92.** 0.20 kΩ, 0.11 kW **93.** % change = 30%

See hints on pages 618-630.

Hints & SOLUTIONS

Module 1.

1. Average value $= \dfrac{\text{total area under } i\text{-}t \text{ graph}}{\text{total time interval}}$

$$= \frac{8+2+2+4+2}{6}$$

$$= \frac{18}{6} = 3\text{ A}$$

3. Average value of 5 sin 100 ωt is zero. But average value of 5A (= constant current) is 5A. Hence, average value of total given function is 5A.

4. $$i_{rms}^2 = \frac{\int i^2\, dt}{\int dt} = \frac{\int_2^4 (4t)\, dt}{\int_2^4 dt} = \frac{4\int_2^4 t\, dt}{2} = 2\left[\frac{t^2}{2}\right]_2^4$$

$$= 12\text{ A}^2$$

$$i_{rms} = 2\sqrt{3}\text{ A}$$

5. $H_{DC} = I^2Rt$

$$H_{AC} = I_{rms}^2 Rt = \left(\frac{I}{\sqrt{2}}\right)^2 Rt = \frac{I^2 Rt}{2}$$

$$\therefore \quad \frac{H_{DC}}{H_{AC}} = \frac{2}{1}$$

6. Given, $\omega = 120\pi$

$$\therefore \quad T = \frac{2\pi}{\omega} = \frac{2\pi}{120\pi} = \frac{1}{60}\text{ s}$$

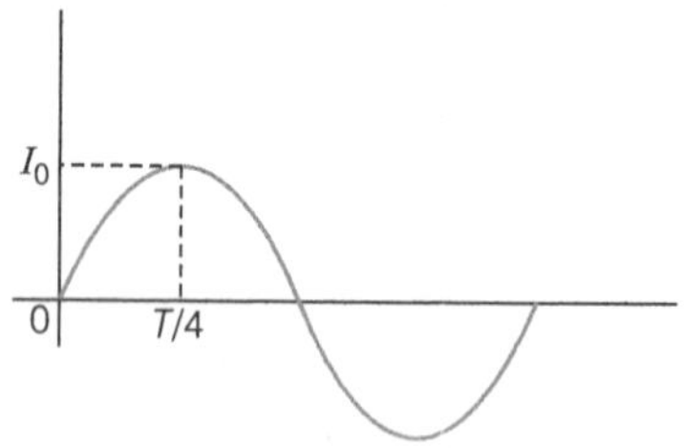

So, required time $= \dfrac{T}{4} = \dfrac{1}{240}$ s

7. When $I = I_0$

$$I_0 \sin(\omega t_1 + \phi) = I_0$$

$$\Rightarrow \quad \sin(\omega t_1 + \phi) = 1$$

$$\sin(\omega t_1 + \phi) = \sin\frac{\pi}{2}$$

$$\omega t_1 + \phi = \frac{\pi}{2} \quad \text{... (i)}$$

When $I = \dfrac{I_0}{\sqrt{2}}$

$$\Rightarrow \quad I_0 \sin(\omega t_2 + \phi) = \frac{I_0}{\sqrt{2}}$$

$$\Rightarrow \quad \sin(\omega t_2 + \phi) = \frac{1}{\sqrt{2}}$$

$$\Rightarrow \quad \omega t_2 + \phi = \frac{\pi}{4} \quad \text{...(ii)}$$

Subtracting Eq. (ii) from Eq. (i), we get

$$\omega(t_1 - t_2) = \frac{\pi}{4}$$

$$\Rightarrow \quad 2\pi f(t_1 - t_2) = \frac{\pi}{4}$$

$$\Rightarrow \quad t_1 - t_2 = \frac{1}{8f} = \frac{1}{400}\text{ s}$$

$$= 2.5\text{ ms}$$

8. Average current will be zero as positive and negative half cycles are symmetrical RMS current can also be obtained from 0 to $\tau/2$.

$$I = \left(\frac{I_0}{\tau/2}\right)t = \left(\frac{2I_0}{\tau}\right)t$$

$$\Rightarrow \quad I^2 = \left(\frac{4I_0^2}{\tau^2}\right)t^2$$

$$\Rightarrow \quad \langle I^2\rangle_{0-\tau/2} = \frac{\int_0^{\tau/2} \frac{4I_0^2}{\tau^2} t^2 dt}{\tau/2}$$

$$\text{or} \quad \langle I^2\rangle_{0-\tau/2} = \frac{I_0^2}{3}$$

$$\Rightarrow \quad I_{rms} = \sqrt{\frac{I_0^2}{3}} = \frac{I_0}{\sqrt{3}}$$

9. Given,

$$i = \left[6 + \sqrt{56}\sin\left(100\pi t + \frac{\pi}{3}\right)\right]$$

$$\langle i^2 \rangle = \langle 6^2 \rangle + \left\langle 56\sin^2\left(100\,\pi t + \frac{\pi}{3}\right)\right\rangle$$

$$+ \left\langle 2 \times 6 \times \sqrt{56}\sin\left(100\,\pi t + \frac{\pi}{3}\right)\right\rangle$$

$$= 36 + \left(56 \times \frac{1}{2}\right) + 0 = 64$$

$$\therefore \quad i_{rms} = \sqrt{\langle i^2 \rangle} = 8\text{ A}$$

Note $\langle \sin(\omega t + \phi) \rangle_T = 0$

$\langle \sin^2(\omega t + \phi) \rangle_T = \dfrac{1}{2}$

Module 2.

1. $Z=\sqrt{R^2+(X_C-X_L)^2}$

From this expression, we can see that X_C may be greater than Z also.

2. At resonance frequency f_r,

$$X_C=X_L$$

Now, $X_L=2\pi fL$ or $X_L\propto f$

For $f>f_r$, $X_L>X_C$

At resonance, $X_L=X_C \Rightarrow Z=R$

$\therefore \quad \cos\phi=\frac{R}{Z}=1$ or $\phi=0°$

3. $\omega=\frac{1}{\sqrt{LC}}$

By inserting a slab, C will increase. So, ω will decrease.

4. $Z=\sqrt{R^2+(X_C\sim X_L)^2}$

X_L will increase. So, Z may increase or decrease, depending on the value of X_C. Therefore, current may decrease or increase.

5. $I=\frac{V}{Z}=\frac{V}{\sqrt{R^2+X_L^2}}=\frac{V}{\sqrt{R^2+(2\pi fL)^2}}$

with increase in frequency, I will decrease.

$$\tan\phi=\frac{X_L}{R}=\frac{2\pi fL}{R}$$

with increase in frequency $\tan\phi$ and therefore ϕ will increase.

6.

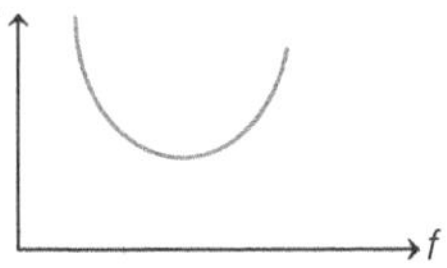

Impedance first decreases, then increases. At resonance frequency Z is minimum.

7. In case of only capacitor and inductor phase difference between current and voltage should be 90°.

8. $X_C=\frac{1}{\omega C}$

$\omega=0$ for DC

$\therefore \quad X_C=\infty$

or it becomes a perfect insulator.

10. $\sin\phi=\frac{X}{Z}=\frac{1}{\sqrt{3}}$

$\therefore \quad \phi=\sin^{-1}\left(\frac{1}{\sqrt{3}}\right)$

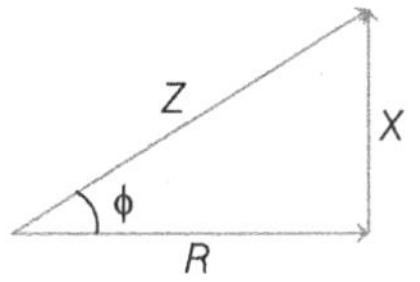

11. $\omega=\frac{1}{\sqrt{LC}}=\frac{1}{\sqrt{0.5\times 8\times 10^{-6}}}=500$ rad/s

12. The situation shown is resonance condition.

$\therefore \quad V=V_R=100$ volt

$Z=R=50\,\Omega$

$\therefore \quad i=\frac{V}{R}=\frac{100}{50}=2$ A

13. $V=\sqrt{V_R^2+V_L^2}$

$=\sqrt{(20)^2+(15)^2}=25$ V

But this is the rms value.

$\therefore \quad$ Peak value $=\sqrt{2}\,V_{rms}=25\sqrt{2}$ V

14. 45° phase angle means,

$X_L=R$

$\therefore \quad (2\pi fL)=R$

$\therefore \quad L=\frac{R}{2\pi f}=\frac{100}{(2\pi)(10^3)}$

$=0.0159$ H

≈ 16 mH

15. $V_C=\sqrt{V^2-V_R^2}=\sqrt{(20)^2-(12)^2}=16$ V

16. $\omega=\frac{1}{\sqrt{LC}}$

$$\omega'=\frac{1}{\sqrt{(1.25L)(0.8\,C)}}=\frac{1}{\sqrt{LC}}=\omega$$

17. $Z=X_C\sim X_L$

At resonance $Z=0$ as $X_C=X_L$

18. Reading of ammeter $=i_{rms}$

$$i_{rms}=\frac{V_{rms}}{X_C}=\frac{V_0}{\sqrt{2}}\times\omega C$$

$$=\frac{200\sqrt{2}\times 100\times(1\times 10^{-6})}{\sqrt{2}}$$

$=2\times 10^{-2}$ A

$=20$ mA

19. (a) $X_L=2\pi fL$

(b) $L=\frac{X_L}{2\pi f}$

(c) $X_C=\frac{1}{2\pi fC}$

(d) $C=\frac{1}{2\pi fX_C}$

20. $V_L=\sqrt{V^2-V_R^2}$

$=\sqrt{(150)^2-(100)^2}=111.8$ V

$V_L=IX_L=I(2\pi fL)$

$V_L = IX_L = I(2\pi fL)$

$\therefore \quad L = \frac{V_L}{2\pi fI} = \frac{111.8}{2\pi \times 50 \times 10} = 0.036\text{ H}$

21. $\phi = 0$, if

$X_L = X_C$ or $2\pi fL = \frac{1}{2\pi fC}$

$\therefore \quad L = \frac{1}{(2\pi f)^2 C} = \frac{1}{(360)^2 \times 10^{-6}} = 7.7\text{ H}$

$Z = R$ (When $X_L = X_C$)

$\therefore \quad I = \frac{V}{Z} = \frac{V}{R} = \frac{120}{20}$

$= 6\text{ A}$

22. $X_L = \omega L = 100\,\Omega$

$X_C = \frac{1}{\omega C} = 312.5\,\Omega$

$Z = \sqrt{R^2 + (X_C - X_L)^2}$

$= \sqrt{(300)^2 + (312.5 - 100)^2}$

$= 368\,\Omega$

(a) $I_0 = \frac{V_0}{Z} = \frac{120}{368} = 0.326\text{ A}$

(b) Since, $X_C > X_L$, voltage lags the current by an angle given by

$$\phi = \cos^{-1}\left(\frac{R}{Z}\right) = \cos^{-1}\left(\frac{300}{368}\right) = 35.3°$$

(c) $(V_0)_R = I_0 R = (0.326)\,300 = 97.8\text{ V}$

$(V_0)_L = I_0 X_L = (0.326)\,(100) = 32.6\text{ V}$

$(V_0)_C = I_0 X_C = (0.326)\,(312.5) = 102\text{ V}$

23. $I_0 = \frac{V_0}{X_L} = \frac{V_0}{\omega L}$

Module 3.

Introductory Exercise

1. $f = \frac{1}{2\pi\sqrt{LC}}$ = resonance frequency

$= \frac{1}{2\pi\sqrt{0.03 \times 2 \times 10^{-6}}}$

$= 650\text{ Hz}$

At resonance, $X_L = X_C$

and $Z = R$

$\therefore \quad \cos\phi = \frac{R}{Z} = 1$

or $\phi = 0°$

2. $R = \frac{V}{I} = \frac{40}{10} = 4\Omega$

$Z = \frac{V}{I} = \frac{200}{10} = 20\,\Omega$

Power factor, $\cos\phi = \frac{R}{Z} = \frac{4}{20}$

$= 0.2$

Exercise

2. $\because \quad P = V_{rms} I_{rms} \cos\theta$

$= \left(\frac{V_0}{\sqrt{2}}\right)\left(\frac{I_0}{\sqrt{2}}\right)\cos\theta$

$= \frac{V_0 I_0}{2}\cos\theta$

3. $i = \sqrt{2}\sin\left[\omega t + \frac{\pi}{4} + \frac{\pi}{2}\right]$

Hence, phase difference between V and i is $\frac{\pi}{2}$. So, power consumed = 0

4. $\frac{N_S}{N_P} = \frac{I_P}{I_S} = \frac{V_S}{V_P}$

$V_P = V_i = \left(\frac{N_P}{N_S}\right)V_0$

$= \frac{2}{1} \times 20 = 40\text{ V}$

$I_P = \left(\frac{N_S}{N_P}\right)I_S = \left(\frac{1}{2}\right)(4) = 2\text{ A}$

5. Power factor, $\cos\phi = \frac{R}{Z}$

When circuit contains only resistance, then

$Z = R \Rightarrow \cos\phi = 1$

When circuit contains only inductance, then

$R = 0$

$\Rightarrow \quad \cos\phi = 0$

6. (a) $X_L > X_C$, hence voltage function will lead the current function.

(b) $Z = \sqrt{R^2 + (X_L - X_C)^2}$

$= \sqrt{(10)^2 + (20 - 10)^2}$

$= 10\sqrt{2}\,\Omega$

(c) $\cos\phi = \frac{R}{Z} = \frac{1}{\sqrt{2}}$

Hence, $\phi = 45°$

(d) Power factor $= \cos\phi = \frac{R}{Z} = \frac{1}{\sqrt{2}}$

7. (b) At resonance frequency (ω_r)

$X_L > X_C$

In the given values, $X_L > X_C$. Hence,

$\omega > \omega_r$

As, $X_L = \omega L \Rightarrow X_L \propto \omega$

and $X_C = \frac{1}{\omega C} \Rightarrow X_C \propto \frac{1}{\omega}$

(c) If frequency is increased from the given value, X_L will further increase. So, $X_L - X_C$ will increase. Hence, net impedance will increase.

(d) If frequency is decreased from the given value, then X_C will increase and X_L will decrease.
So, $X_L - X_C$ may be less than the previous value or $X_C - X_L$ may be greater than the previous. So, Z may either increase or decrease. Hence, current may decrease or increase.

9. $P = I_{rms}^2 R$

By increasing R, current i_{rms} will decrease but the power, $P = I_{rms}^2 R$ may increase or decrease.

10. (a)

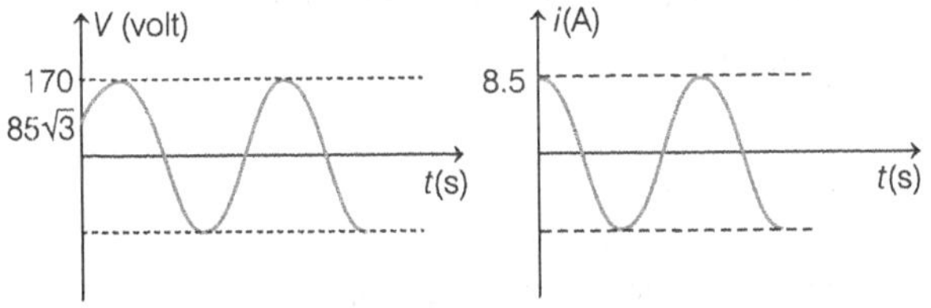

(b) $f = \frac{\omega}{2\pi} = \frac{6280}{2 \times 3.14} = 1000\,\text{Hz}$

(c) $\phi = \frac{\pi}{2} - \frac{\pi}{3} = \frac{\pi}{6}$ or $30°$

Power factor $= \cos\phi = \cos 30°$
$= \frac{\sqrt{3}}{2}$

From the given functions of V and i, we can see that current function leads the voltage function.

(d) $Z = \frac{170}{8.5} = 20\,\Omega$ ...(i)

$\cos\phi = \frac{\sqrt{3}}{2} = \frac{R}{Z} = \frac{R}{20}$

$\therefore \quad R = 10\sqrt{3}\,\Omega = 17.32\,\Omega$

$X_C = \sqrt{Z^2 - R^2} = 10\,\Omega$

$C = \frac{1}{\omega X_C} = \frac{1}{6280 \times 10}\,\text{F}$

$= 15.92 \times 10^{-6}\,\text{F}$

11. $X_L = 2\pi f L = 301\,\Omega$

and $X_C = \frac{1}{2\pi f C} = 55\,\Omega$

$Z = \sqrt{R^2 + (X_L - X_C)^2}$

$= \sqrt{(240)^2 + (301 - 55)^2}$

$= 343\,\Omega$

(a) $\phi = \cos^{-1}\left(\frac{R}{Z}\right) = \cos^{-1}\left(\frac{240}{343}\right) = 45.8°$

$\cos\phi = \frac{R}{Z} = \frac{240}{343} = 0.697$

Since, $X_L > X_C$, voltage leads the current.

(b) Impedance $= Z = 343\,\Omega$

(c) $V_{rms} = I_{rms} Z = 0.45 \times 343 = 155\,\text{V}$

(d) $P = I_{rms}^2 R = (0.45)^2 (240) = 48.6\,\text{W}$

(e) $P = P_R = 48.6\,\text{W}$

(f) $P_C = 0$

(g) $P_L = 0$

12. $P = V_{rms} i_{rms} \cos\phi$

or $\quad 200 = 230 \times 8 \times \cos\phi$

$\therefore \quad \cos\phi = 0.108$

or $\quad \phi = 83.8°$

Further, $\quad P = i_{rms}^2 R$

$\therefore \quad R = \frac{P}{i_{rms}^2} = \frac{200}{(8)^2} = 3.125\,\Omega$

(a) $\tan\phi = \frac{X_C - X_L}{R}$

$\therefore \quad \frac{1}{2\pi f C} - (2\pi f L) = R\tan\phi$

$\therefore \quad L = \frac{1}{(2\pi f)^2 C} - \frac{R\tan\phi}{2\pi f}$

$= \frac{1}{(2\pi \times 50)^2 \times 20 \times 10^{-6}} - \frac{3.125\tan 83.8°}{2\pi \times 50}$

$= 0.416\,\text{H}$

(b) $\tan\phi = \frac{X_L - X_C}{R}$

or $\quad 2\pi f L - \frac{1}{(2\pi f C)} = R\tan\phi$ **Ans.**

$L = \frac{1}{(2\pi f)^2 C} + \frac{R\tan\phi}{(2\pi f)}$

$= \frac{1}{(2 \times \pi \times 50)^2 \times 20 \times 10^{-6}} + \frac{3.125\tan 83.8°}{(2\pi \times 50)}$

$= 0.597\,\text{H}$ **Ans.**

13. (a) $0.5 = \frac{R_1}{Z_1}$

Further, $\quad P = V_{rms} i_{rms} \cos\phi$

or $\quad 100 = 230 \times \frac{230}{Z_1} \times 0.5$

$\therefore \quad Z_1 = 264.5\,\Omega$

and $\quad R_1 = 132.25\,\Omega$

Further, $\quad X_L = \sqrt{Z_1^2 - R_1^2} = \frac{\sqrt{3}}{2} Z_1$

$= 229\,\Omega$

In second case, $0.6 = \frac{R_2}{Z_2}$

and $\quad 60 = \frac{230 \times 230}{Z_2} \times 0.6$

$\therefore \quad Z_2 = 529\,\Omega$

and $\quad R_2 = 317.4\,\Omega$

Further, $\quad X_C = \sqrt{Z_2^2 - R_2^2}$

$$= 423.2\,\Omega$$

When connected in series,

$$R = R_1 + R_2 = 449.65\,\Omega$$

$$X_C - X_L = 194.2$$

$$\therefore \quad Z = \sqrt{(449.65)^2 + (194.2)^2}$$

$$= 489.79\,\Omega$$

Power factor, $\cos\phi = \frac{R}{Z} = 0.92$ (leading)

$$P = V_{rms} i_{rms} \cos\phi$$

$$= (230)\left(\frac{230}{489.79}\right)(0.92)$$

$$= 99\text{ W}$$ **Ans.**

(b) Since, $X_C - X_L = 194.2\,\Omega$

Therefore, if $194.2\,\Omega$ inductive reactance is to be added in series, then it will become only R circuit and power factor will become unity.

14. Area of loop $S = \frac{\pi a^2}{2}$

At time $t = 0$, let us assume that area vector and magnetic field vector are parallel. At time t loop and therefore area vector will rotate an angle, $\theta = \omega t$. So, angle between **S** and **B** will become $\theta = \omega t$.

Magnetic flux passing through the loop,

$$\phi = BS\cos\theta = \frac{B\pi a^2 \cos\omega t}{2}$$

$$e = -\frac{d\phi}{dt} = \frac{B\omega\pi a^2 \sin\omega t}{2}$$

Current $I = \frac{e}{R} = \frac{B\omega\pi a^2 \sin\omega t}{2R}$

This is an alternating current having only resistance in the circuit. Thus power factor will be 1. So,

$$P_{mean} = I_{rms}^2 R \quad \text{...(i)}$$

Here, $I_0 = \frac{B\omega\pi a^2}{2R}$

$$\therefore \quad I_{rms} = \frac{I_0}{\sqrt{2}} = \frac{B\omega\pi a^2}{2\sqrt{2}R}$$

Substituting in equation (i), we get

$$P_{mean} = \frac{(\pi\omega a^2 B)^2}{8R}$$

Unit Exercise (Cumulative)

Level 1

1. An ideal choke coil should have almost zero internal resistance. Otherwise, it will consume some power.

2. As frequency is increased, impedance of upper circuit will decrease as $\left(X_C \propto \frac{1}{\omega}\right)$. So, current of upper circuit will increase and brightness of upper circuit will increase. Opposite is the case with lower circuit.

3. V_L function is cos function, which is 90° ahead of the current function. Hence, current function should be sin function.

4.

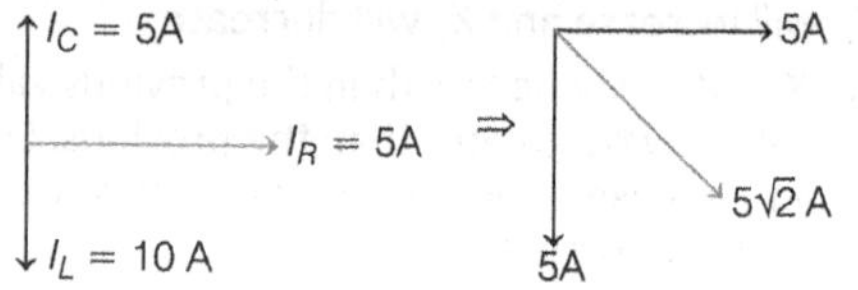

$$I_R = \frac{100}{20} = 5\text{ A},\ I_L = \frac{100}{10} = 10\text{ A} \text{ and } I_C = \frac{100}{20} = 5\text{ A}$$

5. Current in inductance I_L lags behind, current in resistance I_R is in phase with voltage, while current in capacitance I_C leads by a phase difference of $\pi/2$.

6. $X_C = X_L = R$

$$V = \sqrt{V_R^2 + (V_L - V_C)^2}$$

But, $\quad V_L = V_C$

$\therefore \quad V = V_R = 10$

When capacitor is short circuited, $V_C = 0$

and $\quad V_R = V_L \quad$ (as $R = X_L$)

$\therefore \quad V = \sqrt{V_R^2 + V_L^2}$

or $\quad 10 = \sqrt{2}\,V_L$

$\therefore \quad V_L = \frac{10}{\sqrt{2}}$ volt

7. $V_C = 200$ V (given)

$V_R = (11)(20) = 220$ V

$V = 220$ V (applied)

$\therefore \quad V_L = 200$ V

Because $\quad V = \sqrt{V_R^2 + (V_L - V_C)^2}$

8. $X_L = X_C$ at resonance V_L and V_C are out of phase. Hence,

$$V_{LC} \text{ or } V_2 = 0$$

9.

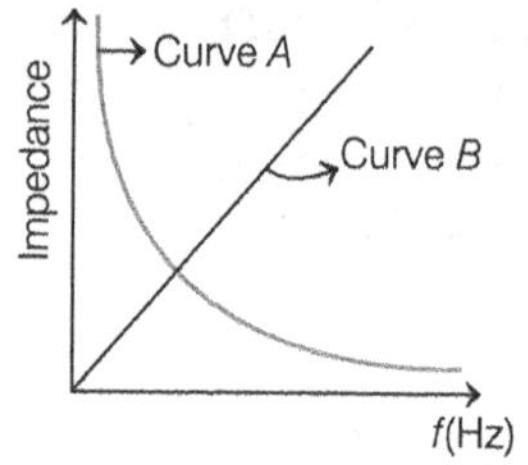

$$X_L = \omega L = 2\pi f L$$

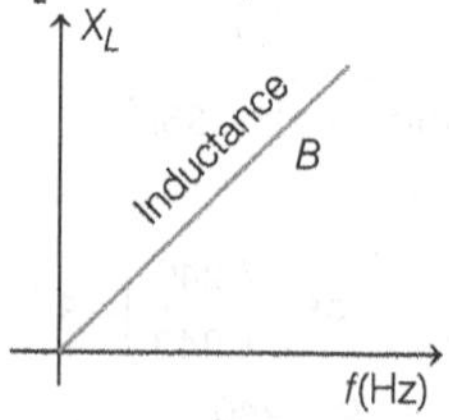

$$X_C = \frac{1}{\omega C} = \frac{1}{2\pi f C}$$

$$R = \text{constant}$$

$$A \to X_C \qquad X_C = A$$

$$B \to X_L \qquad X_L = B$$

Hence, option (c) is correct.

10. Impedance in *L-C-R* circuit is given as,

$$Z = \sqrt{(X_L - X_C)^2 + R^2}$$

$$\because \quad X_L = X_C = R \qquad (\because Z = R)$$

11. $\dfrac{\omega_0}{\omega_1 - \omega_2}$

Here, $\omega_1 - \omega_2 = \text{bandwidth} = \dfrac{R}{L}$

Substituting the values, we get

$$Q = \frac{\omega_0 L}{R}$$

12. Since, $\tan\dfrac{\pi}{4} = 1 = \dfrac{X_C \text{ or } X_L}{R}$

$\therefore$ For *R-C* circuit, we have

$$1 = \frac{1}{C\omega R} \Rightarrow \omega = \frac{1}{CR} \qquad \text{...(i)}$$

Similarly, for *R-L* circuit, we have

$$1 = \frac{\omega L}{R} \Rightarrow \omega = \frac{R}{L} \qquad \text{...(ii)}$$

Given, $\omega = 100$ rad/s

Thus, again by substituting the given values of *R*, *C* or *L* option wise in the respective Eqs. (i) and (ii), we get that only for option (c),

$$\omega = \frac{1}{CR} = \frac{1}{10 \times 10^{-6} \times 10^3}$$

or $\qquad \omega = 100$ rad/s

13. In circuit (b),

Inductive reactance

$$X_L = 2\pi f L$$

$$= 2\pi \times 50 \times 50 \times 10^{-3} \Omega$$

$$= 15.7 \Omega$$

and capacitive reactance,

$$X_C = \frac{1}{2\pi f C} = \frac{1}{2\pi \times 50 \times 0.5 \times 10^{-6}}$$

$$= 6366 \Omega$$

As $X_L \neq X_C$, there is no resonance in circuit (b). So, impedance of circuit (b) is always greater than (a).

$\Rightarrow I_{rms}$ is always smaller in circuit (b).

14. Power at primary coil,

$$P_B = \frac{V_P^2}{R_P} = \frac{(8 \times 10^3)^2}{R_P} = 80 \times 10^3$$

$$\Rightarrow \qquad R_P = 800 \Omega$$

Similarly, for secondary coil,

$$\frac{V_S^2}{R_S} = P \Rightarrow \frac{(160)^2}{R_S} = 80 \times 10^3$$

$$\Rightarrow \qquad R_S = 0.32 \Omega$$

15. Given, number of turns in primary,

$$N_1 = 300$$

Number of turns in secondary,

$$N_2 = 150$$

Output power,

$$P_2 = 2.2 \text{ kW}$$

$$= 2.2 \times 10^3 \text{ W}$$

Current in secondary coil,

$$I_2 = 10 \text{ A}$$

Output power, $P_2 = I_2 V_2$

$$\Rightarrow \qquad V_2 = \frac{P_2}{I_2} = \frac{2.2 \times 10^3}{10} = 220 \text{ V} \qquad \text{...(i)}$$

We know that,

$$\frac{N_1}{N_2} = \frac{\text{Input voltage}}{\text{Output voltage}} = \frac{V_1}{V_2}$$

$$\Rightarrow \qquad V_1 = \left(\frac{N_1}{N_2}\right) V_2$$

$$\Rightarrow \qquad V_1 = \left(\frac{300}{150}\right) \times (220 \text{ V}) \qquad \text{[using Eq. (i)]}$$

$$V_1 = 440 \text{ V} \qquad \text{...(ii)}$$

Again, $\dfrac{V_1}{V_2} = \dfrac{I_2}{I_1}$

$$\Rightarrow \qquad I_1 = \left(\frac{V_2}{V_1}\right) I_2 = \frac{220}{440} \times 10 \qquad \text{[using Eqs. (i) and (ii)]}$$

$$\Rightarrow \qquad I_1 = 5 \text{ A}$$

16. $P = I_{rms}^2 R = \left(\dfrac{V_{rms}}{Z}\right)^2 R$

$$= \left[\frac{(V_0/\sqrt{2})^2}{R^2 + \omega^2 L^2}\right] R = \frac{V_0^2 R}{2(R^2 + \omega^2 L^2)}$$

17. $V_R = \sqrt{V^2 - V_C^2}$

$$= \sqrt{(10)^2 - (8)^2} = 6 \text{ V}$$

$$\tan\phi = \frac{X_C}{X_R} = \frac{V_C}{V_R} = \frac{8}{6} = \frac{4}{3}$$

18. Current will lead the voltage function by 90° voltage function is cos function. Therefore, current function will be – sin function.

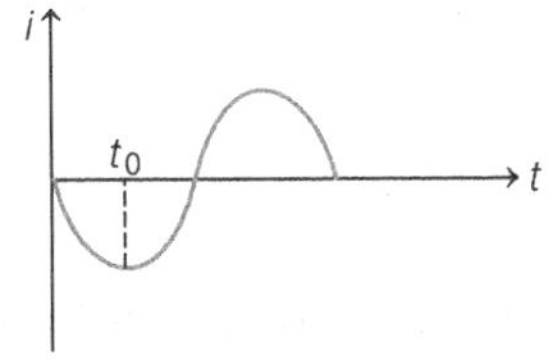

$$t_0 = \frac{T}{4} = \frac{(2\pi/\omega)}{4} = \frac{\pi}{2\omega}$$

$$= \frac{\pi}{2(\pi/2)} = 1\text{ s}$$

19. (a) Voltage lags

$\therefore$ $X_C > X_L$

Power factor, $\cos\phi = \frac{R}{Z}$

$$= \frac{R}{\sqrt{R^2 + (X_C - X_L)^2}}$$

To increase the power factor denominator should decrease. Hence, X_L should increase. Therefore, an inductor is required to be connected.

(b) $\cos\phi = \frac{R}{Z} = 0.72$

$\therefore$ $R = 0.72\,Z = 0.72 \times 60 = 43.2\,\Omega$

$$(X_C - X_L) = \sqrt{(60)^2 - (43.2)^2}$$

$$= 41.64\,\Omega$$

New inductor of inductance 41.64 Ω should be added in the circuit.

$$L = \frac{X_L}{2\pi f} = \frac{41.64}{2\pi(50)} = 0.133\text{ H}$$

20. (a) $I_0 = \frac{(V_0)_R}{R} = \frac{2.5}{300}$

$$= 8.33 \times 10^{-3}\text{ A}$$

$$= 8.33\text{ mA}$$

Current function and V_R function are in phase. Hence,

$$I = (8.33\text{ mA})\cos[(950\text{ rad/s})t]$$

(b) $X_L = \omega L = 950 \times 0.8 = 760\,\Omega$

(c) $(V_0)_L = I_0 X_L$

$$= (8.33 \times 10^{-3})(760)$$

$$= 6.33\text{ V}$$

Now, V_L function leads the current (or V_R) function by 90°.

$\therefore$ $V_L = 6.33\cos(950t + 90°)$

$$= -6.33\sin(950t)$$

21. (a) V_{R_0} will be in phase with main current. Further V_{LR} will lead the main current by some angle $\theta < 90°$. So, an approximate graph is as shown below.

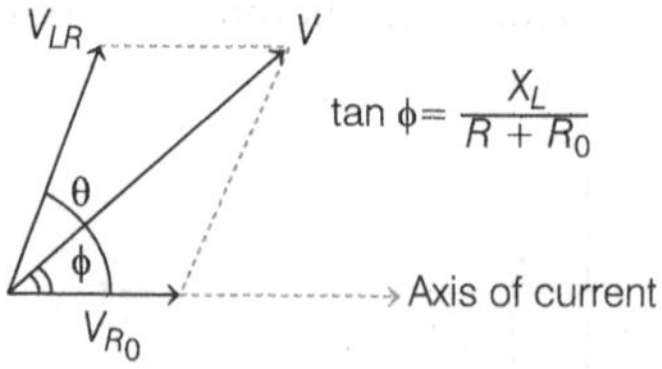

(b) V_C will lag the main current by 90° and V_{LR} will lead the main current by an angle $\theta < 90°$.
So, an approximate graph is as shown below.

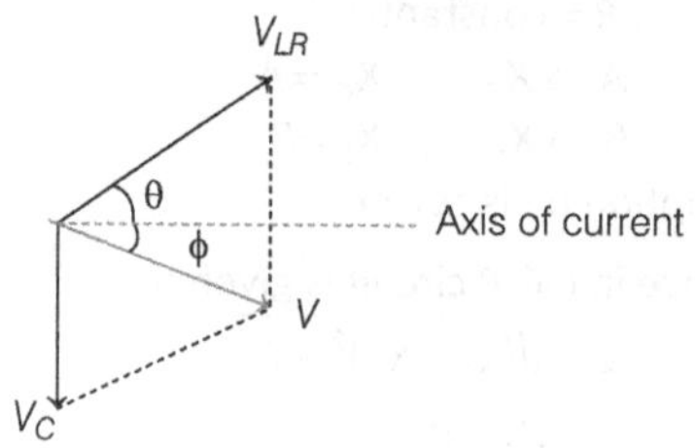

$$\tan\phi = \frac{X_C - X_L}{R}$$

Note *$\omega < \omega_r$, where ω_r is resonance frequency. So, $X_C > X_L$ or current will lead the applied voltage V.*

22. (i) $R = \sqrt{Z^2 - X_L^2} = 40\,\Omega$

$$\cos\phi = \frac{R}{Z} = \frac{40}{50} = \frac{4}{5}$$

$\therefore$ $\phi = 37°$

Note *It is an inductive circuit. Hence, current lags the voltage.*

(ii) Power, $P = V_{rms} I_{rms} \cos\phi$

$$= (V_{rms})\left(\frac{V_{rms}}{Z}\right)\left(\frac{R}{Z}\right)$$

$$= \frac{V_{rms}^2 R}{Z^2} = \frac{(100)^2(40)}{(50)^2}$$

$$= 160\text{ W} \quad\text{or}\quad P = 0.16\text{ kW}$$

23. Let r be the resistance of coil.

$$\frac{V_R}{V_{LR}} = \frac{I_{rms}R}{I_{rms}Z_{Lr}} = \frac{80}{180}$$

$\therefore$ $Z_{Lr} = \frac{180}{80}R = \frac{180}{80} \times 0.16 = 0.36\text{ k}\Omega$

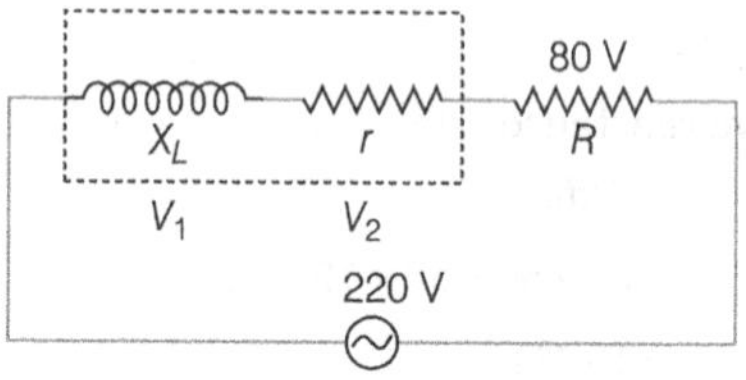

$$V_1^2 + V_2^2 = (180)^2 \quad \text{...(i)}$$

$$V_1^2 + (80 + V_2)^2 = (220)^2 \quad \text{...(ii)}$$

Solving these two equations, we get

$$V_1 \approx 170\text{ V}$$

and $V_2 = 60\text{ V}$

Now, V_1 and V_2 will be in the ratio of X_L and r.

$\therefore$ $\frac{X_L}{r} = \frac{170}{60} \approx 2.83$ ...(iii)

$$Z_{Lr} = \sqrt{X_L^2 + r^2} = 0.36\text{ k}\Omega \quad \text{...(iv)}$$

Solving Eqs. (iii) and (iv), we get

$$r \approx 0.12\text{ k}\Omega = 120\,\Omega$$

Now power across coil,

$$P = \frac{V_2^2}{r} = \frac{(60)^2}{120} = 30\text{ W}$$

24. Let r be the resistance of coil and ϕ the phase angle between applied voltage V and current I_1. I_2 will be in phase with the applied voltage. Hence angle between I_1 and I_2 is also ϕ.

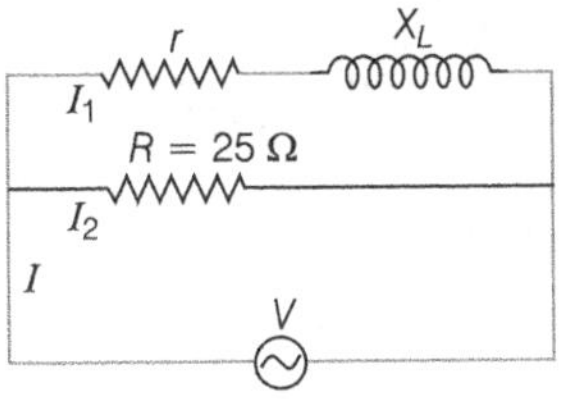

Functions, $I = I_1 + I_2$

So, their peak values or rms (or effective or given) values can be added by vector method.

Thus, $(0.9)^2 = (0.5)^2 + (0.6)^2 + 2(0.5)(0.6)\cos\phi$

$$\Rightarrow \quad \cos\phi = \frac{1}{3}$$

Now, $(I_2)_{rms} = \frac{V_{rms}}{R} \Rightarrow 0.6 = \frac{V_{rms}}{25}$

$$\Rightarrow \quad V_{rms} = 15\text{ V}$$

Similarly, $(I_1)_{rms} = \frac{(V_{rms})}{Z_{Lr}}$

$$\Rightarrow \quad 0.5 = \frac{15}{Z_{Lr}} \Rightarrow Z_{Lr} = 30\,\Omega$$

$$\cos\phi = \frac{r}{Z_{Lr}} \quad \text{or} \quad \frac{1}{3} = \frac{r}{30}$$

$$\Rightarrow \quad r = 10\,\Omega$$

So, power consumed by coil,

$$P = (I_1)^2_{rms}\, r = (0.5)^2(10) \quad \text{or} \quad P = 2.5\text{ W}$$

25. Functions I_1 and V are in same phase. I_2 leads V by 90°. So, I_2 also leads I_1 by 90°, as I_1 and V are in same phase.

Functions $I = I_1 + I_2$

So, their rms values can be added by vector method which have a phase difference of 90°. Thus,

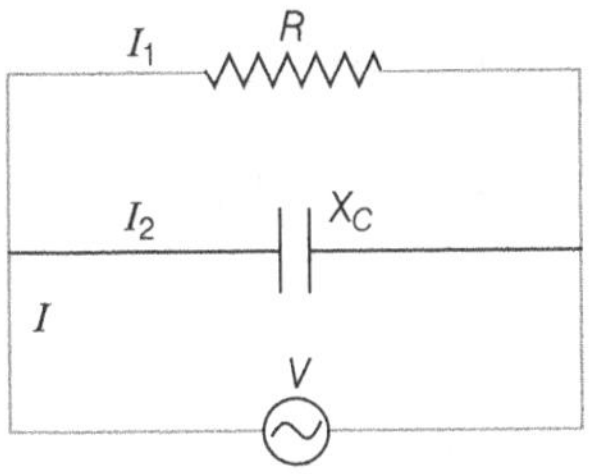

$$I_{rms} = \sqrt{(I_1)^2_{rms} + (I_2)^2_{rms}}$$

or $$\frac{V_{rms}}{Z} = \sqrt{\left(\frac{V_{rms}}{R}\right)^2 + \left(\frac{V_{rms}}{X_C}\right)^2}$$

$$\Rightarrow \quad \frac{1}{Z} = \sqrt{\frac{1}{R^2} + (\omega C)^2} = \frac{\sqrt{1 + (\omega CR)^2}}{R}$$

or $$Z = \frac{R}{\sqrt{1 + (\omega CR)^2}}$$

Substituting the values, we get

$$Z = 40\,\Omega$$

26. Current drawn in L-C-R series circuit is maximum at resonance.

And at resonance, $\omega = \frac{1}{\sqrt{LC}}$

Substituting given values, $2\pi(7000) = \frac{1}{\sqrt{2 \times 10^{-6} \times C}}$

$$\Rightarrow \quad C = \frac{1}{3872}\text{ F}$$

27. Given, current in phase with the emf, then electrical circuit is in resonance.

Current, $I_{rms} = \frac{V_{rms}}{R} = \frac{220}{20} = 11\text{A}$

Amplitude of current,

$$I_0 = \sqrt{2} \times I_{rms}$$
$$= \sqrt{2} \times 11 = \sqrt{2} \times \sqrt{121}$$
$$\Rightarrow \quad I_0 = \sqrt{242}\text{ A}$$

28. Given,

$$R = 80\,\Omega,\ X_L = 100\,\Omega,$$
$$X_C = 40\,\Omega,\ V_0 = 2500\text{ V}$$

Amplitude of current,

$$I_0 = \frac{V_0}{Z} = \frac{V_0}{\sqrt{R^2 + (X_L - X_C)^2}}$$
$$= \frac{2500}{\sqrt{(80)^2 + (100 - 40)^2}}$$
$$= \frac{2500}{\sqrt{(80)^2 + (60)^2}} = \frac{2500}{100}$$
$$\Rightarrow \quad I_0 = 25\text{ A}$$

29. Given, $L = 200\text{ mH} = 0.2\text{ H}$,

$$V_{rms} = 220\text{ V},$$
$$f = 50\text{ Hz}$$

$\therefore$ Inductive reactance,

$$X_L = 2\pi fL = 2\pi \times 50 \times 0.2$$
$$= 20\pi\,\Omega$$

Peak current,

$$I_0 = \frac{V_0}{X_L} = \frac{\sqrt{2}\,V_{rms}}{20\,\pi} = \frac{\sqrt{2} \times 220}{20\,\pi}$$

$$\Rightarrow \quad I_0 = \frac{11\sqrt{2}}{\pi} \Rightarrow I_0 = \frac{\sqrt{242}}{\pi}\text{ A} \qquad \text{...(i)}$$

Since, in inductive circuit, there is a phase difference of $\frac{\pi}{2}$ between alternating current and voltage. Hence, when instantaneous voltage is zero, then current reaches at its peak value, which is given by Eq. (i),

$$I_0 = \frac{\sqrt{242}}{\pi} = \frac{\sqrt{a}}{\pi}\text{ A} \qquad \text{(given)}$$

$$\therefore \quad a = 242$$

30. For maximum power in *L-C-R* circuit, power factor $= \cos\theta = 1$

$$\therefore \quad \frac{R}{Z} = 1 \Rightarrow R^2 = Z^2$$

$$\Rightarrow \quad R^2 = (X_L - X_C)^2 + R^2$$

$$X_L = X_C$$

$$\frac{70}{11} = \frac{1}{100\pi \times C}$$

$$\Rightarrow \quad C = \frac{11}{7000\pi} = 500 \times 10^{-6}\,\text{F}$$

$$= 500\,\mu\text{F}$$

31. For *L-R* circuit $P = V_{rms} I_{rms} \cos\phi$

$$\Rightarrow \quad 400 = \frac{V_{rms}^2}{Z} \times 0.8$$

$$\Rightarrow \quad Z = \frac{(250)^2}{400} \times 0.8 = 125\,\Omega$$

$$\cos\phi = \frac{R}{Z}$$

$$R = Z\cos\phi = 125 \times 0.8 = 100\,\Omega$$

Now, $\quad Z^2 = X_L^2 + R^2$

$$\Rightarrow \quad 125^2 = X_L^2 + 100^2$$

$$\Rightarrow \quad X_L = 75\,\Omega$$

At resonance $X_L = X_C$

$$\Rightarrow \quad 75 = \frac{1}{\omega C} \quad \Rightarrow \quad 75 = \frac{1}{2\pi \times 50 \times C}$$

$$\Rightarrow \quad C = \frac{400}{3\pi}\,\mu\text{F} \quad \Rightarrow \quad n = 400$$

32. Power factor for *R-L* circuit is given by

$$\cos\phi = \frac{R}{\sqrt{R^2 + X_L^2}} = \frac{R}{\sqrt{R^2 + (3R^2)}} = \frac{1}{\sqrt{10}}$$

Power factor for *L-C-R* circuit

$$\cos\phi' = \frac{R}{\sqrt{R^2 + (X_L^2 - X_C^2)}}$$

$$= \frac{R}{\sqrt{R^2 + R^2}} = \frac{1}{\sqrt{2}}$$

$$\therefore \quad \frac{\cos\phi'}{\cos\phi} = \frac{\sqrt{10}}{\sqrt{2}} = \frac{\sqrt{5}}{1}$$

$$\therefore \quad x = 1$$

33. Given, $\dfrac{V_S}{V_P} = \dfrac{N_S}{N_P}$

$$\frac{120}{12000} = \frac{N_S}{N_P}$$

$$\Rightarrow \quad \frac{N_S}{N_P} = \frac{1}{100} \qquad \text{...(i)}$$

For an ideal transformer, input power = Output power

and power is given by

$$P = VI$$

$$I_P V_P = I_S V_S = 60000\,\text{W}$$

$$12000 \times I_P = 60000$$

$$I_P = 5\,\text{A}$$

Now, $\quad R_P = \dfrac{V_P}{I_P} = \dfrac{12000}{5} = 2400\,\Omega$

For secondary coil,

$$P_S = V_S \cdot I_S$$

$$\Rightarrow \quad I_S = \frac{P_S}{V_S} = \frac{60000}{120} = 500\text{A}$$

$$\therefore \quad \frac{V_P}{V_S} = \frac{I_P R_S}{I_S R_S}$$

$$\Rightarrow \quad R_S = \frac{I_P}{I_S} \times \frac{V_S}{V_P} \times R_P$$

$$= \frac{I_P}{I_S} \times \frac{N_S}{N_P} \times R_P = \frac{5}{500} \times \frac{1}{100} \times 2400$$

$$R_S = 240 \times 10^{-3}\,\Omega = 240\,\text{m}\Omega$$

34. Number of turns in rectangular coil,

$$N = 100$$

Area, $A = 14 \times 10^{-2}\,\text{m}^2$

Rotation frequency,

$$f = 360\ \text{rev/min} = \frac{360}{60}\ \text{rev/s} = 6\ \text{rev/s}$$

Magnetic field, $B = 3.0\,\text{T}$

∴ Maximum value of emf produced,

$$e = NBA\omega = NBA(2\pi f)$$

$$= 100 \times 3 \times 14 \times 10^{-2} \times 2 \times \frac{22}{7} \times 6$$

$$= 1584\,\text{V}$$

35. We know that, $\dfrac{N_P}{N_S} = \dfrac{V_P}{V_S}$

Since, $\quad N_S = 24,\ V_P = 220\,\text{V}$

and $\quad V_S = 12\,\text{V} \quad \Rightarrow \quad \dfrac{N_P}{24} = \dfrac{220}{12}$

$$\Rightarrow \quad N_P = \frac{220 \times 24}{12} = 440$$

36. $I_2 = \dfrac{V}{X_C} = \dfrac{V}{3}$ (here V = rms value)

$$I_1 = \frac{V}{R} = \frac{V}{4}$$

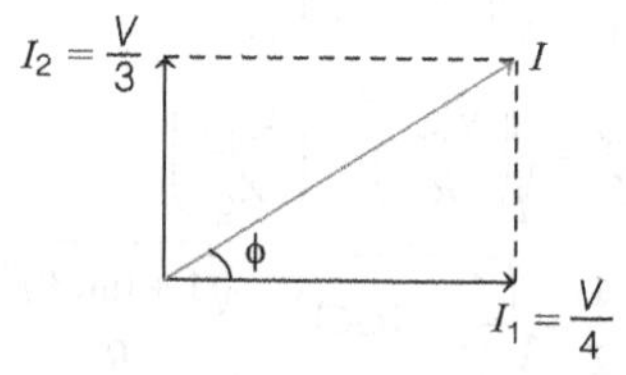

I_2 is 90° ahead of applied voltage function and I_1 is in phase with it.

$$\tan\phi = \frac{V/3}{V/4} = \frac{4}{3}$$

$$\therefore \qquad \phi = 53°$$

37. I_R and I_L are in same phase and phase difference between them and applied voltage lies between 0° and 90°.

38. $X_L = \omega L = (5\times10^{-3})(2000) = 10\,\Omega$

$$X_C = \frac{1}{\omega C} = \frac{1}{(2000)(50\times10^{-6})} = 10\,\Omega$$

Since, $X_L = X_C$ circuit is in resonance.

$$Z = R = (6+4) = 10\,\Omega$$

$$I_{rms} = \frac{V_{rms}}{Z}$$

$$= \frac{(20/\sqrt{2})}{10} = 1.414\text{ A}$$

This is also the reading of ammeter.

$$V = 4I_{rms}$$

$$\approx 5.6\text{ volt}$$

39. $I_R = \frac{V_{rms}}{R} = \frac{200}{100} = 2.0\text{ A}$

$$X_C = \frac{1}{2\pi fC} = \frac{1}{(2\pi)(5\times10^3)\left(\frac{1}{\pi}\times10^{-6}\right)}$$

$$= 100\,\Omega$$

$$\therefore \qquad I_C = \frac{V_{rms}}{X_C} = \frac{200}{100} = 2\text{ A}$$

I_C is 90° ahead of the applied voltage and I_R is in phase with the applied voltage. Hence, there is a phase difference of 90° between I_R and I_C too.

$$\therefore \qquad I = \sqrt{I_R^2 + I_C^2} = \sqrt{(2)^2 + (2)^2}$$

$$= 2.83\text{ A}$$

40. As, rms value of current,

$$I_{rms} = \sqrt{2}\text{ A} \quad \text{and} \quad V_{rms} = 220\text{ V}$$

So, net impedance across *L-C-R* circuit,

$$Z = \sqrt{R^2 + (X_L - X_C)^2}$$

$$= \sqrt{(100)^2 + [\omega L - X_C]^2}$$

$$= \sqrt{(100)^2 + \left[\left(100\times\pi\times\frac{1}{\pi}\right) - (X_C)\right]}$$

$$\left(\frac{220}{2.2}\right)^2 = (100)^2 + ((100) - (X_C))^2$$

$$\Rightarrow \qquad X_C = 100\,\Omega$$

$$\text{As,} \qquad \tan\phi = \frac{X_C}{R} = \frac{100}{100} = 1$$

$$\Rightarrow \qquad \phi = \tan^{-1}(1)$$

$$\phi = 45°$$

Power factor, $\cos\phi = \frac{1}{\sqrt{2}}$

41. *V* function is sin function. *I* function is ahead of *V* function. Hence, the circuit should be capacitive in nature.

Further, $\phi = 45°$

$$\therefore \qquad X_C = R$$

$$\text{or} \qquad \omega C = R$$

$$\text{or} \qquad C = \frac{R}{\omega} = \frac{R}{100} = 0.01\,R$$

In option (b), this condition is satisfied.

42. $V = \sqrt{V_R^2 + (V_C - V_L)^2} = 10\text{ V}$

$V_C > V_L$, hence current leads the voltage.

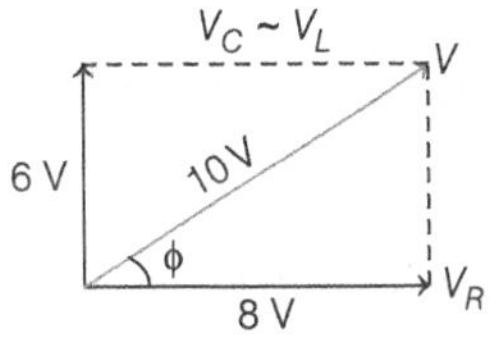

$$\text{Power factor} = \cos\phi = \frac{8}{10} = 0.8$$

43. See the hint of miscellaneous example numbers 9 and 10 of module 3.

44. $V_S = \sqrt{V_R^2 + V_L^2} = \sqrt{(70)^2 + (20)^2} = 72.8\text{ V}$

$$\tan\phi = \frac{X_L}{R} = \frac{V_L}{V_R} = \frac{20}{70} = \frac{2}{7}$$

45. In first case, $X_C = \frac{V}{I} = \frac{220}{0.25} = 880\,\Omega$

In the second case, $R = \frac{V}{I} = \frac{220}{0.25} = 880\,\Omega$

In the combination of *P* and *Q*,

$$\tan\phi = \frac{X_C}{R} = 1$$

$$\therefore \qquad \phi = 45°$$

Since the circuit is capacitive, current leads the voltage. Further,

$$Z = \sqrt{R^2 + X_C^2} = 880\sqrt{2}\,\Omega$$

$$I = \frac{V}{Z} = \frac{220}{880\sqrt{2}} = \frac{1}{4\sqrt{2}}\text{ A}$$

46. See the hints of miscellaneous example numbers 9 and 10 of module 3.

47. $i = \frac{V}{R}$, i.e. circuit is in resonance because $V_R = IR = 220$ volt $= V_{applied}$.

Hence, $\qquad V_C = V_L = 200\text{ V}$

48. $X_L = \omega L$

If ω is very low, then $X_L \approx 0$

$$\therefore \qquad V_L \approx 0 \quad \text{or} \quad V = V_C = V_0$$

49. $I_{max} = \frac{V}{R}$ (at resonance)

$$6 = \frac{24}{R}$$

$\therefore$ $R = 4\Omega$

$$I_{DC} = \frac{V}{R+r} = \frac{12}{4+4} = 1.5\text{ A}$$

50. $R = \frac{1}{\omega C} = X_C$

$\therefore$ $Z = \sqrt{R^2 + X_C^2}$

$= \sqrt{2}R$ (as $X_C = R$)

$$I_0 = \frac{V_0}{Z} = \frac{V_0}{\sqrt{2}R} \quad \text{...(i)}$$

When ω becomes $\frac{1}{\sqrt{3}}$ times, X_C will become $\sqrt{3}$ times or $\sqrt{3}R$.

$$Z' = \sqrt{(R^2) + (\sqrt{3}R)^2} = 2R$$

$$I_0' = \frac{V_0}{Z'} = \frac{V_0}{2R} = \frac{I_0}{\sqrt{2}}$$

51. Circuit is in resonance,

$\therefore$ $V = V_R = IR$

or $R = \frac{V}{I} = \frac{100}{0.2} = 500\Omega$

$$X_L = \frac{V_L}{I} = \frac{400}{0.2}$$

$= 2000\Omega = \omega L$

$\therefore$ $\omega = 1000$ rad/s

Further, $X_L = X_C = 2000\Omega = \frac{1}{\omega C}$

$\therefore$ $C = \frac{1}{2000\omega} = 0.5\,\mu\text{F}$

52. $I_1 = I_2$

$\therefore$ $\omega_1 L - \frac{1}{\omega_1 C} = \frac{1}{\omega_2 C} - \omega_2 L$

Solving this, we get

$$\frac{1}{\sqrt{LC}} = \text{resonance frequency} = \sqrt{\omega_1 \omega_2}$$

At resonance frequency current will be maximum.

53. $X_L = \frac{100}{8}\Omega = 12.5\Omega$ with 50 Hz frequency. With 40 Hz frequency

$$X_L' = 12.5 \times \frac{40}{50} \quad (\text{as } X_L \propto \omega)$$

$= 10\Omega$

$$R = \frac{100}{10} = 10\Omega$$

$\therefore$ $Z = \sqrt{R^2 + (X_L')^2} = 10\sqrt{2}\,\Omega$

$$I = \frac{100}{10\sqrt{2}} = 5\sqrt{2}\text{ A}$$

54. Let $V = V_0 \sin \omega t$ (as $V = 0$ at $t = 0$)

Then, $V_R = V_0 \sin \omega t$ and $V_C = V_0 \sin(\omega t - \pi/2)$

V and V_R are in same phase. While V_C lags V (or V_R) by 90°.

Now V_R is in same phase with initial potential difference across the capacitor for the first time when,

$$\omega t = -\frac{\pi}{2} + 2\pi = \frac{3\pi}{2}$$

$\therefore$ $t = \frac{3\pi}{2\omega}$

55. At resonance, $X_C = X_L$

If ω is reduced to half, then

$$X_C' = 2X_C \quad \left(\text{as } X_C \propto \frac{1}{\omega}\right)$$

and $X_L' = \frac{X_L}{2}$ (as $X_L \propto \omega$)

If ω is made two times, then

$$X_C' = \frac{X_C}{2} \text{ and } X_L' = 2X_L$$

i.e. value of Z will remain unchanged. Hence, current should be 1 A.

56. In Position-1,

$$\tan\frac{\pi}{6} = \frac{\omega L}{R} = \frac{(1000)(\sqrt{3} \times 10^{-3})}{R}$$

or $\frac{1}{\sqrt{3}} = \frac{\sqrt{3}}{R}$

$\therefore$ $R = 3\Omega$

In Position-2,

$$\tan\phi = \frac{X_C}{R} = \frac{(1/\omega C)}{R}$$

$$= \frac{1\Big/\left(1000 \times \frac{1000}{3} \times 10^{-6}\right)}{3} = 1$$

$\therefore$ $\phi = \frac{\pi}{4}$

57. The given AC circuit is the combination of two pure parallel circuits with the applied voltage. In which I_2 is in phase with V and I_1 leads V by 90°.

58. As, current at any instant in the circuit will be

$$I = I_{DC} + I_{AC} = a + b\sin\omega t$$

So, effective current,

$$I_{eff} = \left[\frac{\int_0^T I^2 dt}{\int_0^T dt}\right]^{1/2} = \left[\frac{1}{T}\int_0^T (a + b\sin\omega t)^2\, dt\right]^{1/2}$$

i.e. $I_{eff} = \left[\frac{1}{T}\int_0^T (a^2 + 2ab\sin\omega t + b^2\sin^2\omega t)\,dt\right]^{1/2}$

But $\frac{1}{T}\int_0^T \sin\omega t\, dt = 0$

and $\quad \frac{1}{T}\int_0^T \sin^2 \omega t\, dt = \frac{1}{2}$

So, $\quad I_{eff} = \left[a^2 + \frac{b^2}{2}\right]^{1/2}$

59. Current is halved, means impedance is doubled.

$$X_C \propto \frac{1}{\omega}$$

$\therefore$ When ω is made $\frac{1}{3}$rd, X_C will become 3 times.

$$Z' = 2Z$$

$\therefore \quad \sqrt{R^2 + (3X_C)^2} = 2\sqrt{R^2 + X_C^2}$

$\therefore \quad \frac{X_C}{R} = \sqrt{\frac{3}{5}}$

60. By doubling the frequency, R remains unchanged but X_L doubles ($X_L \propto \omega$). At the same time V is also doubled. So, current through R will become two times and current through L will remain the same.

61. $\frac{L}{R} = \tau \Rightarrow L = \tau R$

$$\cos\phi = \frac{R}{Z} = \frac{R}{\sqrt{R^2 + (\omega L)^2}}$$

$$= \frac{R}{\sqrt{R^2 + (\omega \tau R)^2}} = \frac{1}{\sqrt{1 + (\omega\tau)^2}}$$

62. For element X,

$$R = \frac{E}{I} = \frac{100}{5} = 20\,\Omega$$

For element Y,

$$X_L = 20\,\Omega$$

When X and Y are connected in series,

$$I_0 = \frac{E_0}{Z} = \frac{E_0}{\sqrt{X_L^2 + R^2}}$$

$$= \frac{100}{20\sqrt{2}} = \frac{5}{\sqrt{2}}\,\text{A}$$

The rms value of the current will be

$$I_{rms} = \frac{I_0}{\sqrt{2}} = \frac{5}{2}\,\text{A}$$

63. $\sqrt{V_R^2 + V_L^2} = 100 \quad$ ...(i)

$V_L \sim V_C = 120 \quad$ ...(ii)

$\sqrt{V_R^2 + (V_L \sim V_C)^2} = 130 \quad$ ...(iii)

Solving these three equations, we get

$$V_R = 50\text{ V}, V_L = 86.6\text{ V}$$

and $\quad V_C = 206.6\text{ V}$

Power factor $= \cos\phi = \frac{R}{Z} = \frac{V_R}{V} = \frac{50}{130} = \frac{5}{13}$

Since $V_C > V_L$, circuit is capacitive in nature.

64. $i = 5\sin(\omega t + 53°)$

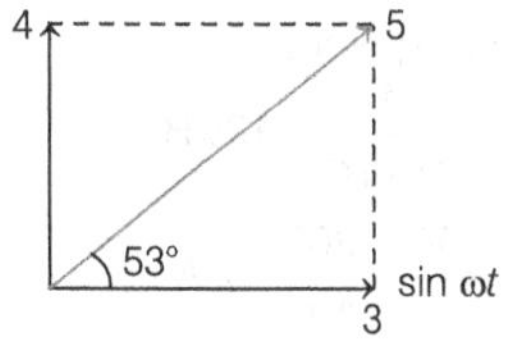

$$i_0 = 5\,\text{A}$$

$\therefore \quad i_{rms} = \frac{i_0}{\sqrt{2}} = \frac{5}{\sqrt{2}}\,\text{A}$

Mean value of current in positive half cycle is

$$\frac{2}{\pi} i_0 = \left(\frac{2}{\pi}\right)(5) = \left(\frac{10}{\pi}\right)\text{A}$$

In $V = V_m \sin\omega t$, current $i = 5\sin(\omega t + 53°)$ leads the voltage function. Hence, circuit is capacitive in nature. Same is the case with part (d).

65. $P_R = V_R\, i$

$\therefore \quad i = \frac{P_R}{V_R} = \frac{60}{60} = 1\text{ A}$

Now, $\quad V_L = \sqrt{V^2 - V_R^2}$

$$= \sqrt{(100)^2 - (60)^2}$$

$$= 80\text{ V} = i X_L = i(2\pi f L)$$

$$L = \frac{80}{2\pi f i} = \frac{80}{(2\pi)(50)(1)} = \frac{4}{5\pi}\text{H}$$

If we connect another resistance R in series, then it should consume 40 V, so that remaining 60 V is used by the tube light.

$$R = \frac{V}{i} = \frac{40}{1} = 40\,\Omega$$

66. (a) $V_R = IR = 80\text{ V}$

(b) $X_C = \frac{V_C}{I} = \frac{100}{2} = 50\,\Omega$

(c) $V_L = IX_L = 40\text{ V}$

(d) $V = V_{rms} = \sqrt{V_R^2 + (V_C - V_L)^2}$

$$= \sqrt{(80)^2 + (100-40)^2} = 60\text{ V}$$

$\therefore \quad V_0 = \sqrt{2}\, V_{rms} = 60\sqrt{2}\text{ V}$

67. $Z = \sqrt{R^2 + X_C^2} = \sqrt{R^2 + \left(\frac{1}{\omega C}\right)^2}$

In case (b) capacitance C will be more. Therefore, impedance Z will be less. Hence, current will be more.

$\therefore$ Option (b) is correct.

Further, $V_C = \sqrt{V^2 - V_R^2} = \sqrt{V^2 - (IR)^2}$

In case (b), since current I is more.

Therefore, V_C will be less.

$\therefore$ Option (c) is correct.

$\therefore$ Correct options are (b) and (c).

68. At resonating condition, we have

$$V_1 = V_2 \Rightarrow X_L = X_C$$

So, resonating frequency,

$$f = \frac{1}{2\pi\sqrt{LC}} = 125\,\text{Hz}$$

$$I_0 = \frac{V_0}{R} = \frac{200}{100} = 2\,\text{A}$$

$$V_1 = V_2 = IX_L = I(\omega L)$$

$$= IL\frac{1}{\sqrt{LC}} = 1000\,\text{V}$$

69.

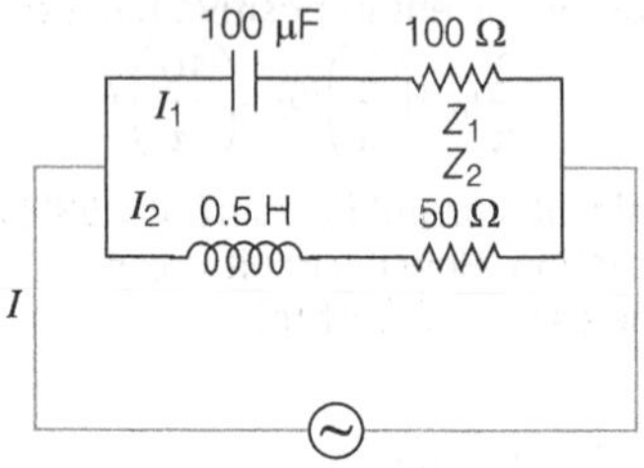

Circuit 1

$$X_C = \frac{1}{\omega C} = 100\,\Omega$$

$$\therefore \quad Z_1 = \sqrt{(100)^2 + (100)^2}$$

$$= 100\sqrt{2}\,\Omega$$

$$\phi_1 = \cos^{-1}\left(\frac{R_1}{Z_1}\right) = 45°$$

In this circuit current leads the voltage.

$$I_1 = \frac{V}{Z_1} = \frac{20}{100\sqrt{2}} = \frac{1}{5\sqrt{2}}\,\text{A}$$

$$V_{100\,\Omega} = (100)\,I_1 = (100)\frac{1}{5\sqrt{2}}\,\text{V}$$

$$= 10\sqrt{2}\,\text{V}$$

Circuit 2

$$X_L = \omega L = (100)(0.5) = 50\,\Omega$$

$$Z_2 = \sqrt{(50)^2 + (50)^2} = 50\sqrt{2}\,\Omega$$

$$\phi_2 = \cos^{-1}\left(\frac{R_2}{Z_2}\right) = 45°$$

In this circuit, voltage leads the current.

$$I_2 = \frac{V}{Z_2} = \frac{20}{50\sqrt{2}} = \frac{\sqrt{2}}{5}\,\text{A}$$

$$V_{50\,\Omega} = (50)\,I_2 = 50\left(\frac{\sqrt{2}}{5}\right)$$

$$= 10\sqrt{2}\,\text{V}$$

Further, I_1 and I_2 have a mutual phase difference of 90°.

$$\therefore \quad I = \sqrt{I_1^2 + I_2^2} = 0.34$$

70. $V_{XY} = V_X - V_Y$

$$= V_0 \sin\left(\omega t + \frac{2\pi}{3}\right) - V_0 \sin \omega t$$

$$= V_0 \sin\left(\omega t + \frac{2\pi}{3}\right) + V_0 \sin(\omega t + \pi)$$

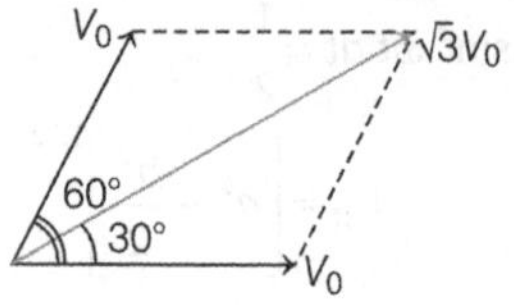

$$\Rightarrow \quad \phi = \pi - \frac{2\pi}{3} = \frac{\pi}{3}$$

$$\Rightarrow \quad V_{XY} = \sqrt{3}V_0 \sin(\omega t + 30°)$$

$$(V_{XY})_{rms} = \frac{\sqrt{3}V_0}{\sqrt{2}}.$$

Similarly we can find $(V_{YZ})_{rms}$

Therefore, the reading of the voltmeter will be independent of choice of two terminals.

71. At $\omega \approx 0$, $X_C = \frac{1}{\omega C} = \infty$.

Therefore, current is nearly zero.

Further at resonance frequency, current and voltage are in phase. This resonance frequency is given by

$$\omega_r = \frac{1}{\sqrt{LC}} = \frac{1}{\sqrt{10^{-6} \times 10^{-6}}}$$

$$= 10^6\,\text{rad/s}$$

We can see that this frequency is independent of R.

Further, $X_L = \omega L$, $X_C = \frac{1}{\omega C}$

At, $\omega = \omega_r = 10^6\,\text{rad/s}$, $X_L = X_C$.

For $\omega > \omega_r$, $X_L > X_C$. So, circuit is inductive.

72. $\frac{dQ}{dt} = I$

$$\Rightarrow \quad Q = \int I\,dt = \int (I_0 \cos \omega t)\,dt$$

$$\therefore \quad Q_{max} = \frac{I_0}{\omega} = \frac{1}{500} = 2 \times 10^{-3}\,\text{C}$$

$$\text{At} \quad t = \frac{7\pi}{6\omega} \quad \text{or} \quad \omega t = \frac{7\pi}{6}$$

Current comes out to be negative from the given expression. So, current is anti-clockwise. Charge supplied by source from $t = 0$ to $t = \frac{7\pi}{6\omega}$

$$Q = \int_0^{\frac{7\pi}{6\omega}} \cos(500t)\,dt$$

$$= \left[\frac{\sin 500t}{500}\right]_0^{\frac{7\pi}{6\omega}} = \frac{\sin\frac{7\pi}{6}}{500} = -1\,\text{mC}$$

Just after switching

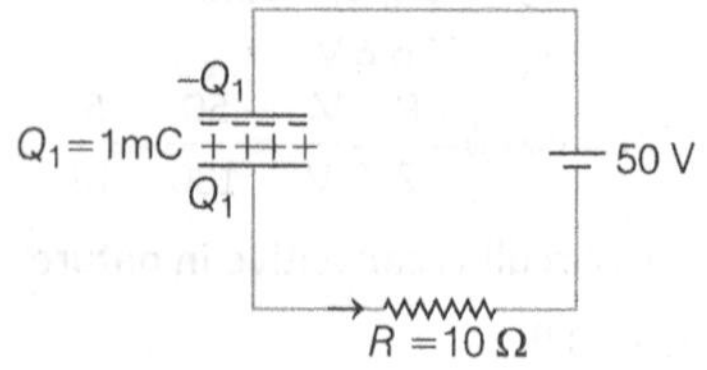

In steady state

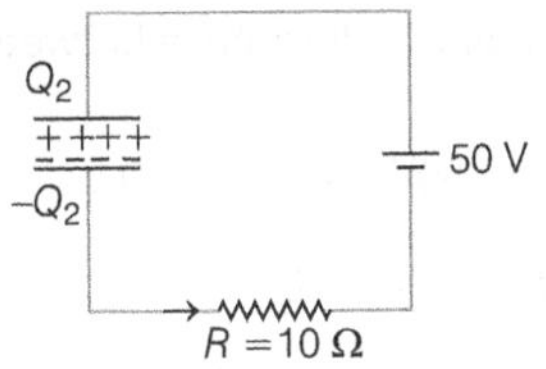

Apply Kirchhoff's loop law, just after changing the switch to position D

$$50+\frac{Q_1}{C}-IR=0$$

Substituting the values of Q_1, C and R, we get

$$I=10\text{ A}$$

In steady state $Q_2=CV=1\text{ mC}$

$\therefore$ Net charge flown from battery $=2\text{ mC}$

$$=2\times10^{-3}\text{ C}$$

73. $\omega=\frac{1}{\sqrt{LC}}$

$$=\frac{1}{\sqrt{4.9\times10^{-3}\times10^{-6}}}$$

$$=\frac{10^5}{7}\text{ rad/s}$$

74. $X_C=\frac{1}{\omega C}=\frac{1}{\left(\frac{10^5}{7}\right)(10^{-6})}=70\,\Omega$

$$Z_P=\sqrt{R_P^2+X_C^2}$$

$$=\sqrt{(32)^2+(70)^2}$$

$$\approx 77\,\Omega$$

75. At maximum current means at resonance,

$$X_L=X_C, Z=R$$

$\therefore$ Power factor $=\cos\phi=\frac{R}{Z}=1$

76. (a) $\phi=0°$ between voltage function and current function.

(b) $I=I_0\sin(\omega t-90°)$
i.e. $\phi=90°$ and voltage function leads the current function.

(c) Current function leads the voltage function. So,

$$X_C>X_L$$

(d) Voltage function leads the current function. So,

$$X_L>X_C$$

77. (a) $R=\frac{V_R}{I}=\frac{40}{2}=20\,\Omega$

(b) $V_C=IX_C=2\times30=60\text{ V}$

(c) $V_L=IX_L=2\times15=30\text{ V}$

(d) $V=\sqrt{V_R^2+(V_C-V_L)^2}=50\text{ V}$

78. In circuit (p) I can't be non zero in steady state.
In circuit (q)
$V_1=0$ and $V_2=2I=V$ (also)
In circuit (r) $V_1=X_LI=(2\pi fL)I$

$$=(2\pi\times50\times6\times10^{-3})I=1.88I$$

$$V_2=2I$$

In circuit (s) $V_1=X_LI=1.88\,I$

$$V_2=X_CI=\left(\frac{1}{2\pi fC}\right)I$$

$$=\left(\frac{1}{2\pi\times50\times3\times10^{-6}}\right)I$$

$$=(1061)I$$

In circuit (t)

$$V_1=IR=(1000)I;\ V_2=X_CI=(1061)I$$

79. (a) $i_{rms}=\frac{V_1}{R}=\frac{40}{4}=10\text{ A}$

$$i_0=\sqrt{2}i_{rms}=10\sqrt{2}\text{ A}$$ **Ans.**

(b) $\frac{E_0}{\sqrt{2}}=\sqrt{V_1^2+(V_2-V_3)^2}=50\text{ V}$

$\therefore$ $E_0=50\sqrt{2}\text{ V}$

(c) $X_L=(\omega L)=\frac{V_2}{i_{rms}}=\frac{40}{10}=4\,\Omega$

$\therefore$ $L=\frac{4}{\omega}=\frac{4}{100\pi}\text{H}=\frac{1}{25\pi}\text{H}$ **Ans.**

$$X_C=\frac{1}{\omega C}=\frac{V_1}{i_{rms}}=\frac{10}{10}=1\,\Omega$$

$\therefore$ $C=\frac{1}{X_C\,\omega}=\frac{1}{100\pi}\text{F}$ **Ans.**

80. $\cos\phi_1=0.5$

$\therefore$ $\phi_1=60°$

$$\cos\phi_2=\frac{\sqrt{3}}{2}$$

$\therefore$ $\phi_2=30°$

Let R be the effective resistance of the box. Then,

$$\tan\phi_1=\frac{X_C}{R}\quad\text{or}\quad\sqrt{3}=\frac{X_C}{R}\qquad\text{...(i)}$$

$$\tan\phi_2=\frac{X_C}{R+10}\quad\text{or}\quad\frac{1}{\sqrt{3}}=\frac{X_C}{R+10}\qquad\text{...(ii)}$$

From these two equations, we get $R=5\,\Omega$

81. (a) $V_R=IR=80\text{ V}, V_C=100\text{ V}$

and $V_L=IX_L=160\text{ V}$

$\therefore$ $V=\sqrt{V_R^2+(V_L-V_C)^2}=100\text{ V}$ **Ans.**

Note *Value of X_L have been taken from part (b).*

(b) Since the current is lagging behind, there should be an inductor in the box.

$$X_C=\frac{V_C}{I}=100\,\Omega$$

Now, $0.8 = \frac{R}{Z} = \frac{80}{\sqrt{(80)^2 + (X_L - 100)^2}}$

Solving, we get

$$X_L = 160\,\Omega \quad \text{or} \quad \omega L = 160$$

$\therefore \quad (2\pi f L) = 160$

$\therefore \quad L = \frac{160}{2\pi f} = \frac{160}{(2\pi)\times 50} = \frac{1.6}{\pi}\,\text{H}$

82. (a)

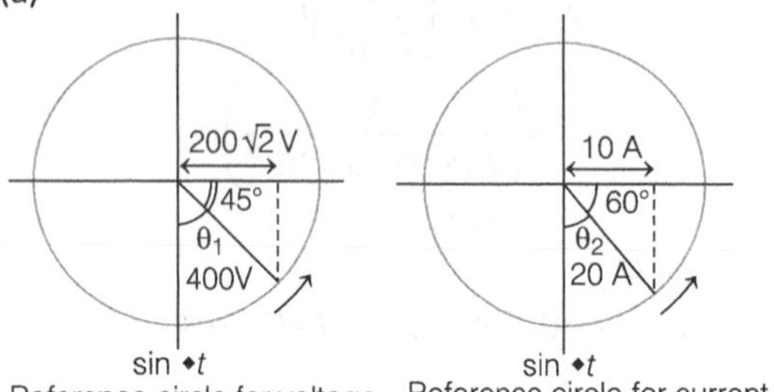

Reference circle for voltage Reference circle for current

$$\omega = 2\pi f = (100\pi)\,\text{rad/s}$$

From the above two figures, we can write

$V = 400\sin(\omega t + \theta_1) = 400\sin\left[100\pi t + \frac{\pi}{4}\right]$ **Ans.**

$i = 20\sin(\omega t + \theta_2) = 20\sin\left[100\pi t + \frac{\pi}{6}\right]$ **Ans.**

(b) Phase difference between V and i

$$\phi = (\pi/4 - \pi/6) = \frac{\pi}{12} \text{ or } 15^\circ$$

$$P = V_{rms} i_{rms} \cos\phi = \left(\frac{400}{\sqrt{2}}\right)\left(\frac{20}{\sqrt{2}}\right)\cos 15^\circ$$

$= 3864$ W **Ans.**

83. $\omega = \frac{2}{\sqrt{LC}} = \frac{2}{\sqrt{5\times 10^{-3} \times 20\times 10^{-6}}}$

$= 6324.5$ rad/s

$X_L = \omega L = (6324.5)(5\times 10^{-3}) = 31.62\,\Omega$

$X_C = \frac{1}{\omega C} = \frac{1}{6324.5 \times 20\times 10^{-6}} = 7.9\,\Omega$

$\therefore \quad Z = X_L - X_C = 23.72\,\Omega$

(a) Maximum voltage across capacitor

$= i_0 X_C = (0.211)(7.9) = 1.67$ mV

$\therefore$ Maximum charge

$q_0 = (20\times 10^{-6})(1.67\times 10^{-3})$

$= 33.4$ nC

(b) $i_0 = \frac{V_0}{Z} = \frac{5}{23.72}$ mA $= 0.211$ mA

(c) Since $X_L > X_C$, current in the circuit will lag behind the applied voltage by $\pi/2$.

Further voltage across the inductor will lead this current by $\pi/2$.

Therefore, applied voltage and voltage across inductor are in phase.

Voltage across the capacitor will lag the circuit current by $\pi/2$.

Therefore, phase difference between V_L and V_C will be 180°.

84. $X_{L_1} = \omega L_1 = (2\pi\times 50)(0.02) = 6.28\,\Omega$

$\therefore \quad Z_1 = \sqrt{R_1^2 + X_{L_1}^2}$

$= \sqrt{(5)^2 + (6.28)^2} = 8.0\,\Omega$

$P_1 = (I_{rms})_1 V_{rms}\cos\phi_1 = \left(\frac{100}{8}\right)(100)\left(\frac{5}{8}\right)$

$= 781.25$ W

$X_{L_2} = \omega L_2 = (2\pi\times 50)(0.08) = 25.13\,\Omega$

$\therefore \quad Z_2 = \sqrt{R_2^2 + X_{L_2}^2} = 25.15\,\Omega$

$\therefore \quad P_2 = (i_{rms})_2 V_{rms}\cos\phi_2$

$= \left(\frac{100}{25.15}\right)(100)\left(\frac{1}{25.15}\right)$

$= 15.8$ W

$\therefore \quad P_{Total} = P_1 + P_2 = 797$ W

85. $Z_1 = \frac{115}{3} = 38.33\,\Omega$

$\cos\phi_1 = \frac{R_1}{Z_1} \Rightarrow R_1 = Z_1\cos\phi_1$

$= (38.33)(0.6) = 23\,\Omega$

$X_L = \sqrt{Z_1^2 - R_1^2} = 30.67\,\Omega$

$Z_2 = \frac{115}{5} = 23\,\Omega$

$R_2 = Z_2\cos\phi_2 = (23)(0.707)$

$= 16.26$

$X_C = \sqrt{Z_2^2 - R_2^2} = \sqrt{(23)^2 - (16.26)^2}$

$= 16.26\,\Omega$

When connected in series,

$R = R_1 + R_2 = 39.26\,\Omega$

$X_L - X_C = 14.41\,\Omega$

$\therefore \quad Z = \sqrt{R^2 + (X_L - X_C)^2} = 41.82\,\Omega$

(a) $i = \frac{V}{Z} = \frac{230}{41.82} = 5.5$ A

(b) $P = i^2 R = (5.5)^2(39.26) = 1187.6\text{ W} \approx 1.188$ kW

(c) Power factor $= \cos\phi = \frac{R}{Z} = \frac{39.26}{41.92} = 0.939$

Since $X_L > X_C$, this power factor is lagging.

86. This question seems to be a simple L-R circuit of an AC in which applied voltage should lead the current by an angle,

$$\phi = \tan^{-1}\left(\frac{X_L}{R}\right) = \tan^{-1}\left(\frac{\omega L}{R}\right)$$

So, current function should be like,

$$I = I_m\cos(\omega t - \phi) \quad \text{...(i)}$$

where, $I_m = \dfrac{V_m}{Z}$

and $Z = \sqrt{R^2 + X_L^2}$, where $X_L = \omega L$

But this approach is wrong, as switch is closed at $t = 0$. So current in the circuit should be zero at $t = 0$. But according to equation (i), current is not zero at $t = 0$. Thus we will have to solve it from the basics. Let at time t circuit is as shown below.

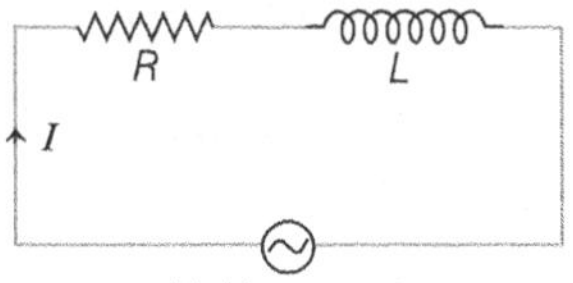

$V = V_m \cos \omega t$

$$V = V_R + V_L$$

$$\therefore \quad V_m \cos \omega t = IR + L\frac{dI}{dt}$$

Solution of this differential equation is out of syllabus from JEE point of view. So, students can leave its mathematical part.

87. **Inductance of solenoid**

Let l be the length of solenoid. Then total number of turns,

$$N = nl$$

If I current is passed through the solenoid, then

$$B = \mu_0 nI$$

Area of one turn, $S = \pi a^2$

Flux from one turn,

$$\phi = BS = (\mu_0 nI \pi a^2)$$

Now, $L = \dfrac{N\phi}{I} = (\mu_0 n^2 \pi a^2 l)$

Resistance of solenoid

In unit length there are n - turns.

Therefore if d is the diameter of wire of solenoid, then

$$nd = 1 \quad \text{or} \quad d = \frac{1}{n}$$

Now, $R = \dfrac{\rho l_0}{A}$ (l_0 = length of wire of solenoid)

$$= \frac{\rho (2\pi a) nl}{(\pi d^2 / 4)} \qquad \left(d = \frac{1}{n}\right)$$

$$= 8\rho a n^3 l$$

$$X_L = \omega L = (2\pi f L) = (2\mu_0 n^2 \pi^2 a^2 l f)$$

Now, $\tan\phi = \dfrac{X_L}{R} = \dfrac{\mu_0 \pi^2 f a}{4n\rho}$

Note *This is an inductive circuit. So, current lags.*

88. (a) $(V_C)_m = I_m X_C = \left(\dfrac{V_m}{Z}\right) X_C$

$$= \left[\frac{V_m}{\sqrt{R^2 + \left(\omega L - \dfrac{1}{\omega C}\right)^2}}\right] \frac{1}{\omega C}$$

$$= \frac{V_m}{\sqrt{(R\omega C)^2 + (\omega^2 LC - 1)^2}}$$

For $(V_C)_m$ to be maximum, denominator should be minimum, or,

$$\frac{d}{d\omega}[(R\omega C)^2 - (\omega^2 LC - 1)^2] = 0$$

From here, we can find the value of ω.

(b) Same method can be used for maximum value of $(V_L)_m$, where $(V_L)_m = I_m X_L = \left(\dfrac{V_m}{Z}\right) X_L$

$$= \left[\frac{V_m}{\sqrt{R^2 + \left(\omega L - \dfrac{1}{\omega C}\right)^2}}\right] (\omega L)$$

89. (i) $(V_{LR})_m = I_m \sqrt{R^2 + (\omega L)^2} = \left(\dfrac{V_m}{Z}\right)\sqrt{R^2 + (\omega L)^2}$

$$\left[\frac{V_m}{\sqrt{R^2 + \left(\omega L - \dfrac{1}{\omega C}\right)^2}}\right] \sqrt{R^2 + (\omega L)^2}$$

C is variable and we have to maximise $(V_{LR})_m$. Therefore, square term in denominator should be zero. Or, $X_L = X_C$

or $\quad C = \dfrac{1}{\omega^2 L} = 28\,\mu\text{F}$

(ii) Then maximum value of $(V_{LR})_m$ will be

$$\frac{V_m}{R}\sqrt{R^2 + (\omega L)^2} \quad \text{or} \quad V_m \sqrt{1 + \left(\frac{\omega L}{R}\right)^2}$$

Substituting the values, we get

$$(V_{LR})_m = 0.54 \text{ kV}$$

(iii) $(V_C)_m = I_m X_C = \left(\dfrac{V_m}{R}\right) X_C$ or $\left(\dfrac{V_m}{R}\right) X_L$

Substituting the values, we get

$$(V_C)_m = 0.51 \text{ kV}$$

90. R and C in this case can be assumed in parallel across the applied voltage as shown below.

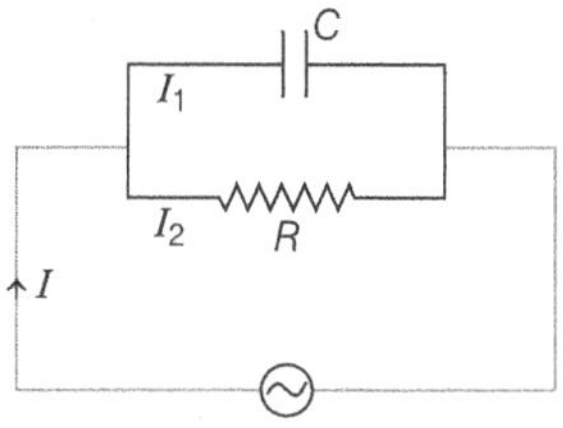

$V = V_m \cos \omega t$

$$I_1 = \frac{V_m}{X_C}\cos\left(\omega t + \frac{\pi}{2}\right)$$

$$= (\omega C V_m)\cos\left(\omega t + \frac{\pi}{2}\right)$$

$$I_2 = \frac{V_m}{R}\cos \omega t$$

$$I = I_1 + I_2 = (\omega C V_m)\cos\left(\omega t + \frac{\pi}{2}\right) + \frac{V_m}{R}\cos\omega t$$

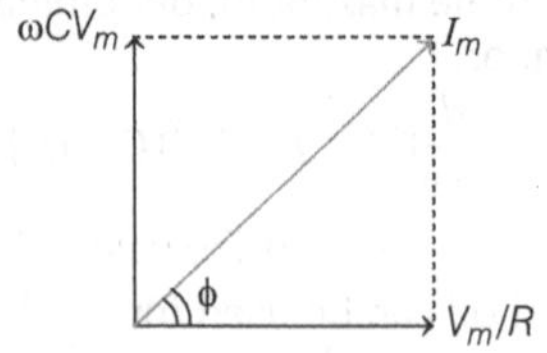

or $\quad I = I_m \cos(\omega t + \phi)$

Here, $\quad I_m = \sqrt{\left(\frac{V_m}{R}\right)^2 + (\omega CV_m)^2}$

$$= \frac{V_m}{R}\sqrt{1 + (\omega CR)^2}$$

and $\quad \tan\phi = \frac{\omega CV_m}{V_m/R} = \omega CR$

or $\quad \phi = \tan^{-1}(\omega CR)$

91. (a) V and I_R are in same phase.

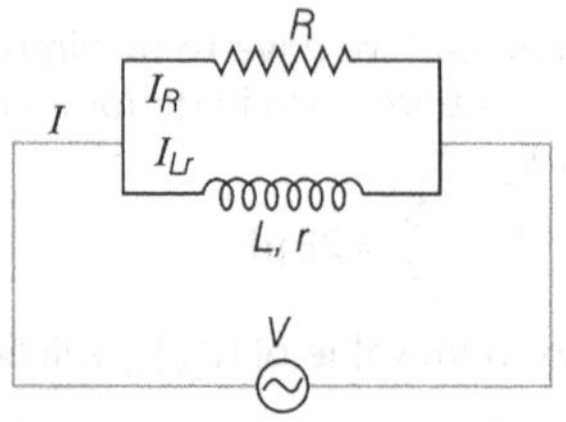

I lags with V by a phase angle ϕ (given in the question).

Let I_{Lr} lags with V by a phase angle θ. Then I_{Lr} will also lag with I_R by the same angle θ, as I_R and V are in same phase.

Functions, $\quad I = I_R + I_{Lr}$

and they are sinusoidal. So, they can be added by vector method as shown below.

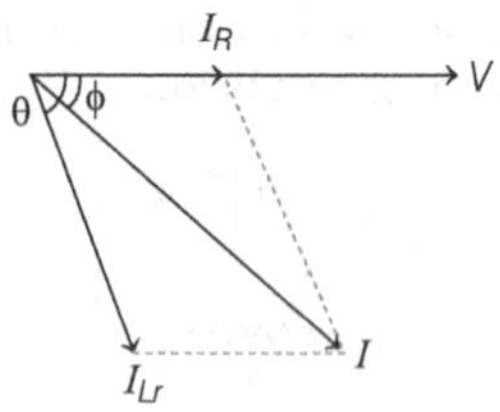

(b)

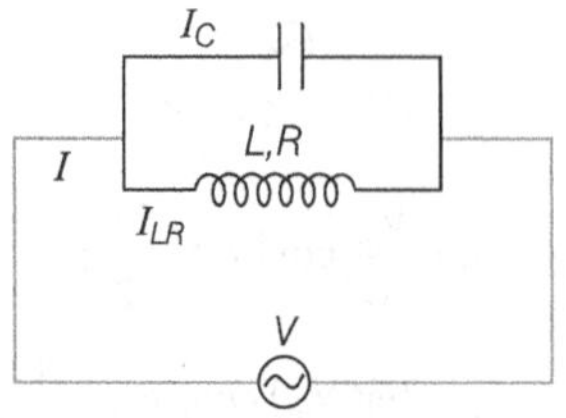

Same logics can be applied as was discussed in part (a).

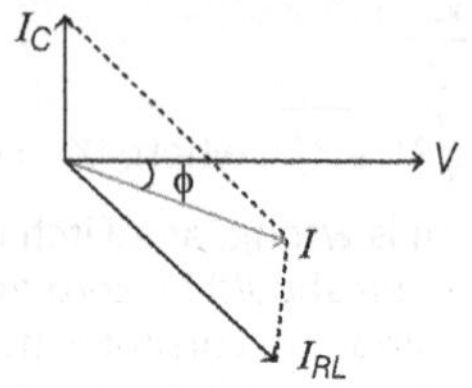

(c)

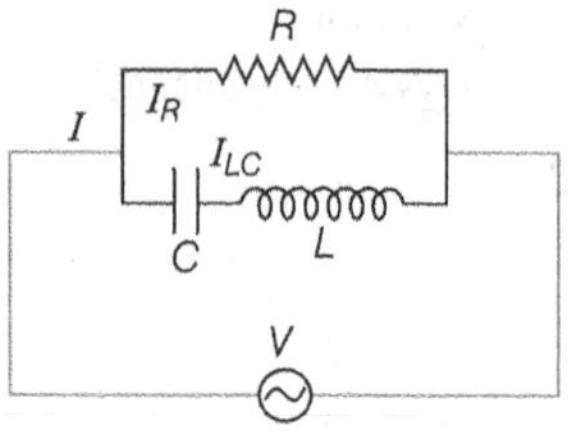

As per the given conditions, $X_L > X_C$.

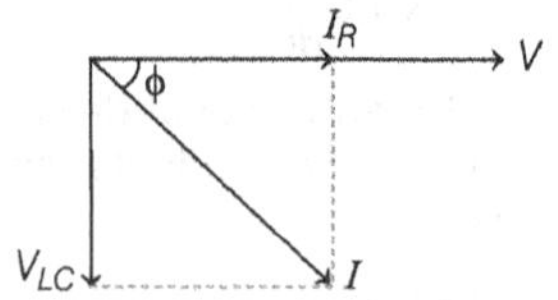

92. Power, $P = V_{rms} I_{rms} \cos\phi$

$$= (V_{rms})\left(\frac{V_{rms}}{Z}\right)\left(\frac{R}{Z}\right) = \frac{v_{rms}^2 R}{Z^2}$$

$\Rightarrow \quad P = \frac{V_{rms}^2 R_{net}}{Z^2}$

Here it is $\quad P = \frac{V^2(R+r)}{(R+r)^2 + (\omega L)^2}$

For P to be maximum, put

$$\frac{dP}{dR} = 0$$

93. In the above problem, we have proved that

$$P = \frac{V_{rms}^2 R}{Z^2} \qquad \text{...(i)}$$

Now $\cos\phi = \frac{R}{Z}, \quad \therefore \quad Z = \frac{R}{\cos\phi}$

Substituting in equation (i), we get

$$P = \frac{V_{rms}^2 \cos^2\phi}{R} \quad \text{or} \quad \cos\phi = \frac{\sqrt{PR}}{V_{rms}}$$

Here, R and V_{rms} are unchanged. So, we can write

$$\cos\phi = K\sqrt{P} \qquad \left[K = \frac{\sqrt{R}}{V_{rms}}\right]$$

$\therefore$ Percentage change in $\cos\phi$

$$= \left(\frac{\text{final value} - \text{initial value}}{\text{initial value}}\right) \times 100$$

$$= \left(\frac{K\sqrt{1.7P} - K\sqrt{P}}{K\sqrt{P}}\right) \times 100$$

$$= (\sqrt{1.7} - 1) \times 100$$

or % change $= 30\%$